Manual on Uniform Traffic Control Devices

for Streets and Highways

11th Edition

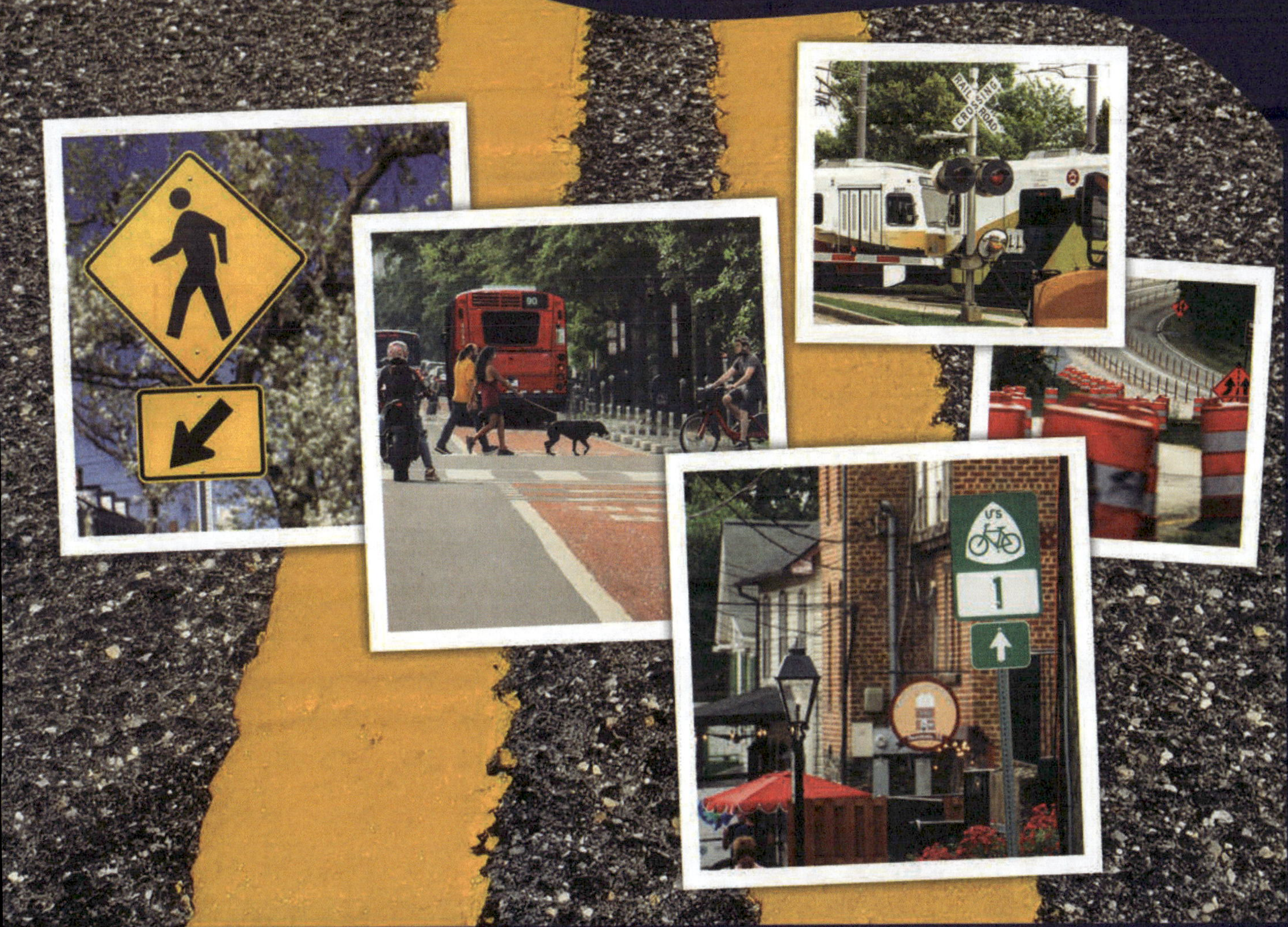

December 2023

U.S. Department of Transportation

Federal Highway Administration

MANUAL ON UNIFORM TRAFFIC CONTROL DEVICES
TABLE OF CONTENTS

Page

CHAPTER 2B.   REGULATORY SIGNS, BARRICADES, AND GATES

GENERAL

SIGNING FOR RIGHT-OF-WAY AT INTERSECTIONS

SPEED LIMIT SIGNS AND PLAQUES

CHAPTER 2E. GUIDE SIGNS—FREEWAYS AND EXPRESSWAYS

GUIDE SIGNING FOR INTERCHANGES

OTHER GUIDE SIGNS

SIGNS FOR ROUTE DIVERSION BY VEHICLE CLASS

INTERFACE WITH CONVENTIONAL ROADWAYS

CHAPTER 2F.   TOLL ROAD SIGNS

GENERAL

REGULATORY SIGNS

CHAPTER 2G. PREFERENTIAL AND MANAGED LANE SIGNS

CHAPTER 2H.   GENERAL INFORMATION SIGNS

CHAPTER 2I.   GENERAL SERVICE SIGNS

CHAPTER 2J.   SPECIFIC SERVICE SIGNS

CHAPTER 2K.   TOURIST-ORIENTED DIRECTIONAL SIGNS

PART 3   MARKINGS

CHAPTER 3C. CROSSWALK MARKINGS

CHAPTER 3D. CIRCULAR INTERSECTION MARKINGS

CHAPTER 3E. PREFERENTIAL LANE MARKINGS FOR MOTOR VEHICLES

CHAPTER 3F. MARKINGS FOR TOLL PLAZAS

CHAPTER 4G. FLASHING OPERATION OF TRAFFIC CONTROL SIGNALS

CHAPTER 4H. BICYCLE SIGNALS

CHAPTER 4I. PEDESTRIAN CONTROL FEATURES

CHAPTER 4J. PEDESTRIAN HYBRID BEACONS

CHAPTER 4K. ACCESSIBLE PEDESTRIAN SIGNALS AND DETECTORS

CHAPTER 4L. RECTANGULAR RAPID FLASHING BEACONS

CHAPTER 4M. TRAFFIC CONTROL SIGNALS FOR EMERGENCY-VEHICLE ACCESS

CHAPTER 4N. HYBRID BEACONS FOR EMERGENCY-VEHICLE ACCESS

CHAPTER 4O. TRAFFIC CONTROL SIGNALS FOR ONE-LANE, TWO-WAY FACILITIES

CHAPTER 4P. TRAFFIC CONTROL SIGNALS FOR FREEWAY ENTRANCE RAMPS

CHAPTER 4Q. TRAFFIC CONTROL FOR MOVABLE BRIDGES

CHAPTER 4R. HIGHWAY TRAFFIC SIGNALS AT TOLL PLAZAS

CHAPTER 4S. FLASHING BEACONS

CHAPTER 4T. LANE-USE CONTROL SIGNALS

CHAPTER 4U. IN-ROADWAY WARNING LIGHTS

PART 5   TRAFFIC CONTROL DEVICE CONSIDERATIONS FOR AUTOMATED VEHICLES

CHAPTER 5A.   GENERAL

CHAPTER 5B.   PROVISIONS FOR TRAFFIC CONTROL DEVICES

PART 6   TEMPORARY TRAFFIC CONTROL

CHAPTER 6A.   GENERAL

CHAPTER 6B.   TEMPORARY TRAFFIC CONTROL ELEMENTS

CHAPTER 6C.   PEDESTRIAN AND WORKER SAFETY

CHAPTER 6D.   FLAGGER CONTROL

CHAPTER 6E. ONE-LANE, TWO-WAY TRAFFIC CONTROL

CHAPTER 6F. TEMPORARY TRAFFIC CONTROL ZONE SIGNS – GENERAL

CHAPTER 6G. TTC ZONE REGULATORY SIGNS

CHAPTER 6H. TTC ZONE WARNING SIGNS

CHAPTER 6N. TYPES OF TEMPORARY TRAFFIC CONTROL ZONE ACTIVITIES

CHAPTER 6O. CONTROL OF TRAFFIC THROUGH TRAFFIC INCIDENT MANAGEMENT AREAS

CHAPTER 6P. TYPICAL APPLICATIONS

PART 7   TRAFFIC CONTROL FOR SCHOOL AREAS

CHAPTER 7A. GENERAL

CHAPTER 7B. SIGNS

CHAPTER 7C. MARKINGS

CHAPTER 7D. CROSSING SUPERVISION

PART 8   TRAFFIC CONTROL FOR RAILROAD AND LIGHT RAIL TRANSIT GRADE CROSSINGS

CHAPTER 8D. FLASHING-LIGHT SIGNALS, AUTOMATIC GATES, AND TRAFFIC CONTROL SIGNALS

CHAPTER 8E. PATHWAY AND SIDEWALK GRADE CROSSINGS

PART 9   TRAFFIC CONTROL FOR BICYCLE FACILITIES

CHAPTER 9A. GENERAL

CHAPTER 9B. REGULATORY SIGNS

CHAPTER 9C.   WARNING SIGNS AND OBJECT MARKERS

CHAPTER 9D.   GUIDE AND SERVICE SIGNS

CHAPTER 9E.   MARKINGS

FIGURES   **Page**

TABLES   Page

PART 1
GENERAL

CHAPTER 1A. GENERAL

Section 1A.01 Purpose of the MUTCD

Support:

01   The purpose of the MUTCD is to establish uniform national criteria for the use of traffic control devices that meet the needs and expectancy of road users on all streets, highways, pedestrian and bicycle facilities, and site roadways open to public travel.

02   This purpose is achieved through the following objectives:

    A. Promote safety, inclusion, and mobility for all users of the road network;
    B. Promote efficiency through creating national uniformity in the meaning and appearance of traffic control devices;
    C. Promote national consistency in the use, installation, and operation of traffic control devices; and
    D. Provide basic principles for traffic engineers to use in making decisions regarding the use, installation, operation, maintenance, and removal of traffic control devices.

03   Uniformity of the meaning of traffic control devices is vital to their effectiveness. Uniformity means treating similar situations in a similar way. Uniformity of devices simplifies the task of the road user because it aids in recognition and understanding, thereby reducing perception/reaction time. Uniformity assists road users, law enforcement officers, and traffic courts by giving everyone the same interpretation. Uniformity assists public highway officials through efficiency in manufacture, installation, maintenance, and administration.

04   The use of uniform traffic control devices also requires uniform and appropriate application.

05   The applicability of the MUTCD to facilities open to public travel is independent of the type of ownership or jurisdiction (public or private) and the source of funding (Federal, State, local, or private).

06   This Manual presumes the user of the MUTCD has sufficient working knowledge, professional training and experience, and education in the principles of traffic engineering. Other resources can be consulted to understand the basis for decisions that are made in which engineering study or judgment will be applied.

Section 1A.02 Traffic Control Devices – General Description

Support:

01   As defined in Section 1C.02 of this Manual, traffic control devices include all signs, signals, markings, channelizing devices, or other devices that use colors, shapes, symbols, words, sounds, and/or tactile information for the primary purpose of communicating a regulatory, warning, or guidance message to road users on a street, highway, pedestrian facility, bikeway, pathway, or site roadway open to public travel.

02   Infrastructure elements that restrict the road user's travel paths or vehicle speeds, such as islands, curbs, speed humps, and other raised roadway surfaces, are not traffic control devices. Transverse or longitudinal rumble strips are also not traffic control devices. Operational devices associated with the application of traffic control strategies such as fencing, roadway lighting, barriers, and attenuators are shown in this Manual for context, but their design, application, and usage are not specified since they are not traffic control devices.

03   Certain types of signs and other devices that do not have any traffic control purpose are sometimes placed within the highway right-of-way by or with the permission of the public agency or the official having jurisdiction over the street or highway. These signs and other devices are not considered to be traffic control devices and provisions regarding their design and use are not included in this Manual. Among these signs and other devices are the following:

    A. Devices whose purpose is to assist highway maintenance personnel, such as markers to guide snowplow operators, devices that identify culvert and drop inlet locations, and devices that precisely identify highway locations for maintenance or mowing purposes;
    B. Devices whose purpose is to assist fire or law enforcement personnel, such as markers that identify fire hydrant locations, signs that identify fire or water district boundaries, speed measurement pavement markings, small indicator lights to assist in enforcement of red light violations, and photo enforcement systems;
    C. Devices whose purpose is to assist utility company personnel and highway contractors, such as markers that identify underground utility locations;
    D. Signs posting local non-traffic ordinances; and
    E. Signs giving civic organization meeting information.

Section 1A.03 <u>Target Road Users</u>

Support:

01   Traffic control devices can be targeted at operators of motor vehicles, including driving automation systems, and at vulnerable road users.

02   Targeted operators of motor vehicles include motorists, public transportation operators, truck drivers, and motorcyclists. Targeted users also include vulnerable road users, who have little to no protection from crash forces. These users are defined in Title 23, U.S.C. 148(a). They include bicyclists and pedestrians, including persons with disabilities. Pedestrians with disabilities might be blind or vision-impaired, have mobility limitations, or other impairments. Protection of vulnerable users is a priority in this Manual as directed in Section 11135 of the Infrastructure Investment and Jobs Act.

03   Operators of motor vehicles and vulnerable road users are both likely to be present on roadways where adjacent land use suggests that trips could be served by varied modes. Application of traffic control devices on these roadways requires careful consideration of measures to set and design for appropriate speeds; separation of various users in time and space; improvement of connectivity and access for pedestrians, bicyclists, and transit riders, including for people with disabilities; and implementation of safety countermeasures.

Section 1A.04 <u>Use of the MUTCD</u>

Support:

01   Traffic control device principles in the MUTCD are developed for and used by individuals who are duly authorized and qualified to conduct traffic control device activities.

Standard:

02   **Where the content of this Manual requires a decision for implementation, such decisions shall be made by an engineer, or an individual under the supervision of an engineer, who has the appropriate levels of experience and expertise to make the traffic control device decision. Those decisions shall be made using engineering judgment or engineering study, as required by the MUTCD provision.**

Support:

03   Section 1C.02 contains definitions of "engineering study" and "engineering judgment."

Guidance:

04   *In making traffic control device decisions, individuals should consider the impacts of the decision on the following: safety and operational efficiency (mobility) of all road users at that location, the effective use of agency resources, cost-effectiveness, and enforcement and education aspects of traffic control devices.*

Support:

05   Throughout this Manual the headings Standard, Guidance, Option, and Support, the meanings of which are defined in Section 1C.01, are used to classify the nature of the text that follows. Figures and tables, including the notes contained therein, supplement the text and might constitute a Standard, Guidance, Option, or Support. The user needs to refer to the appropriate text to classify the nature of the figure, table, or note contained therein.

Guidance:

06   *Except when a specific numeral is required or recommended by the text of a Section of this Manual, numerals displayed on the images of devices in the figures that specify quantities such as times, distances, speed limits, and weights should be regarded as examples only. When installing any of these devices, the numerals should be appropriately altered to fit the specific situation.*

07   *Similarly, destination names, route numbers, and State route shields that are displayed on the images of devices in the figures should be regarded as examples only. When installing any of these devices, the destination names, route numbers, and State route shields should be appropriately altered to fit the specific situation.*

Support:

08   The information contained in Paragraphs 9 and 10 of this Section will be useful when reference is being made to a specific portion of text in this Manual.

09   There are nine Parts in this Manual and each Part includes one or more Chapters. Each Chapter includes one or more Sections. Parts are identified by a single-digit numerical identification, such as "Part 2 – Signs." Chapters are identified by the Part number and a letter, such as "Chapter 2B – Regulatory Signs." Sections are identified by the Chapter number and letter followed by a decimal point and a 2-digit number, such as "Section 2B.03 – Size of Regulatory Signs." In some Chapters, the Sections are grouped together by subject into unnumbered sub-chapters with a heading, such as "Signing for Right-of-Way at Intersections" (for Sections 2B.06 through 2B.20).

10   Each Section includes one or more paragraphs. The paragraphs are indented and are identified by a number. Paragraphs are counted from the beginning of each Section without regard to the intervening text headings (Standard, Guidance, Option, or Support) or any intervening text in embedded Figures or Tables. Some paragraphs have lettered or numbered items. As an example of how to cite this Manual, the phrase "[n]ot less than 40 feet beyond the stop line" that appears in Section 4D.08 of this Manual would be referenced in writing as "Section 4D.08, Par.1, A.1," and would be verbally referenced as "Item A.1 of Paragraph 1 of Section 4D.08."

Section 1A.05 <u>Relation to Other Publications</u>

Standard:

01   **To the extent that they are incorporated by specific reference, the latest editions of the following publications, shall be a part of this Manual: "Standard Highway Signs" publication (FHWA), and "Color Specifications for Retroreflective Sign and Pavement Marking Materials" (appendix to Subpart F of Part 655 of Title 23 of the Code of Federal Regulations).**

Support:

02   The "Standard Highway Signs" publication includes standard alphabets and symbols and arrows for signs and pavement markings.

03   The MUTCD is not a roadway design manual, and engineers seeking guidance on design should refer to appropriate roadway design guides recognized by the Federal Highway Administration as needed for the design application.

04   Other publications are referenced in this Manual as useful resources, but they are not regulatory in nature and are not independently legally enforceable.

Section 1A.06 <u>Uniform Vehicle Code – Rules of the Road</u>

Support:

01   The "Uniform Vehicle Code" (UVC) is one of the publications referenced in the MUTCD. The UVC contains a model set of motor vehicle codes and traffic laws for use throughout the United States, the intent of which is to promote national uniformity in these laws. The Rules of the Road contained in the UVC are intended to be recommendations for States to adopt in their State statutes and are not independently legally enforceable.

Guidance:

02   *The actions required of road users to obey regulatory devices should be specified by State statute, or in cases not covered by State statute, in local ordinances or resolutions. Such statutes, ordinances, and resolutions should be consistent with the "Uniform Vehicle Code."*

CHAPTER 1B. LEGAL REQUIREMENTS FOR TRAFFIC CONTROL DEVICES

Section 1B.01 National Standard

Standard:

01   **The Manual on Uniform Traffic Control Devices for Streets and Highways (MUTCD) is incorporated by reference in 23 Code of Federal Regulations (CFR), Part 655, Subpart F and shall be recognized as the national standard for all traffic control devices installed on any street, highway, bikeway, or site roadway open to public travel (see definition in Section 1C.02) in accordance with 23 U.S.C. 109(d) and 402(a).**

02   **In accordance with 23 CFR 655.603(a), the MUTCD shall apply to all of the following types of facilities:**

    A. Any street, roadway, or bikeway open to public travel, either publicly or privately owned;

    B. Streets and roadways on sites that are off the public right-of-way that are open to public travel without full-time access restrictions. Examples include roadways within shopping centers, office parks, airports, sports arenas, other similar business and/or recreation facilities, governmental office complexes, schools, universities, recreational parks, and other similar publicly-owned complexes and/or recreation facilities. The above-described examples of streets and roadways are referred to in this Manual as site roadways open to public travel;

    C. Publicly-owned toll roads, including those under the jurisdiction of a public agency, public authority, or public-private partnership;

    D. Privately-owned toll roads where the public is allowed to travel without access restriction. This includes gated toll roads or roadways where the general public is able to pay to access the facility; and

    E. Grade crossings of publicly-owned roadways with railroads or light rail transit.

03   **The MUTCD shall not apply to the following types of facilities:**

    A. Roadways within private gated properties where access to the general public is restricted at all times;

    B. Grade crossings of privately-owned roadways with railroads; and

    C. Parking areas, including the driving aisles within those parking areas, that are either publicly or privately owned.

Support:

04   The policies and procedures of the Federal Highway Administration (FHWA) to obtain basic uniformity of traffic control devices are as described in 23 CFR 655, Subpart F.

05   Section 15-116 of the UVC (see Section 1A.06) states, "No person shall install or maintain in any area of private property used by the public any sign, signal, marking, or other device intended to regulate, warn, or guide traffic unless it conforms with the State manual and specifications adopted under Section 15-104." Adoption by agencies of such a provision through statute or ordinance can help maintain the integrity of official traffic control devices and provide continuity of uniformity at locations that are not subject to the provisions of this Manual.

Section 1B.02 State Adoption and Conformance

Support:

01   All States have officially adopted the National MUTCD either in its entirety, with supplemental provisions, or as a separate published document. The National MUTCD has also been adopted by the National Park Service, the U.S. Forest Service, the U.S. Military Command, the Bureau of Indian Affairs, the Bureau of Land Management, and the U.S. Fish and Wildlife Service.

Standard:

02   **States or other Federal agencies that have their own MUTCDs or Supplements shall revise these MUTCDs or Supplements to be in substantial conformance with changes to the National MUTCD within 2 years of the effective date of the Final Rule for the changes [23 CFR 655.603(b)(3)]. Substantial conformance of such State or other Federal agency MUTCDs or Supplements shall be as defined in 23 CFR 655.603(b)(1).**

03   **For the purposes of Paragraph 2 of this Section, policies, directives, specifications, standard drawings, or similar documents that are issued by an agency and that change or modify Standard, Guidance, or Option provisions in this Manual shall be considered as supplements to the MUTCD and shall also be revised to be in substantial conformance with the National MUTCD.**

Section 1B.03 Compliance of Devices

Standard:

01   **The U.S. Secretary of Transportation, under authority granted by the Highway Safety Act of 1966, decreed that traffic control devices on all streets and highways open to public travel in accordance with 23 U.S.C. 109(d) and 402(a) in each State shall be in substantial conformance with the Standards issued or endorsed by the FHWA.**

Support:

02   23 CFR 655.603 also requires traffic control devices on all streets, highways, bikeways, and site roadways open to public travel in each State be in substantial conformance with standards issued or endorsed by the Federal Highway Administrator.

Standard:

03   **After the effective date of a new edition of the MUTCD or a revision thereto, or after the adoption thereof by the State, whichever occurs later, new or reconstructed devices installed shall comply with the new edition or revision, as required by 23 CFR 655.603.**

04   **In cases involving Federal-aid projects for new construction, reconstruction, resurfacing, restoration, or rehabilitation of a facility to which this Manual applies, the traffic control devices installed (temporary or permanent) shall comply with the most recent edition of the National MUTCD before that highway is opened or re-opened to the public for unrestricted travel [23 CFR 655.603(d)(2) and (d)(3)].**

05   **Unless a particular device is no longer serviceable (see definition in Section 1C.02), non-compliant devices on existing highways and bikeways shall be brought into compliance with the current edition of the National MUTCD as part of the systematic upgrading of substandard traffic control devices (and installation of new required traffic control devices) required pursuant to the Highway Safety Program, 23 U.S.C. §402(a).**

Support:

06   The FHWA has the authority to establish other target compliance dates for implementation of particular changes to the MUTCD [23 CFR 655.603(d)(1)].

Standard:

07   **The target compliance dates established by the FHWA shall be as shown in Table 1B-1.**

08   **Design, application, and placement of traffic control devices other than those adopted in this Manual shall be prohibited unless the provisions of Sections 1B.04 through 1B.08 are followed regarding official interpretations, experiments, changes to the MUTCD, and interim approvals granted by the FHWA.**

Support:

09   Many of the provisions in this Manual that are explicitly prohibitive have been included to address practices that have been shown to be ineffective, unsafe, or inconsistent with uniformity. A provision of mandatory or recommended practice represents the accepted and established practice that promotes uniformity and consistency. The absence of a provision in this Manual that explicitly prohibits a particular practice, use, design, application, operation, or other aspect of a traffic control device does not, in itself, constitute acceptability or permission to use the device in a manner not provided for in this Manual.

Table 1B-1. Target Compliance Dates Established by the FHWA

MUTCD Section(s)	Subject Area	Specific Provision	Compliance Date
2B.64	Weight Limit Signs	Paragraph 14 - requirement for additional Weight Limit sign with the advisory distance or directional legend in advance of applicable section of highway or structure	5 years from the effective date of this edition of the MUTCD
2C.25	Low Clearance Signs (W12-2)	Paragraph 1 - Required posting of the Low Clearance Advance (W12-2) sign in advance of the structure	5 years from the effective date of this edition of the MUTCD
2C.25	Low Clearance Signs (W12-2a, W12-2b)	Paragraph 8 - Recommended posting of Low Clearance Overhead (W12-2a or 12-2b) signs on an arch or other structure under which the clearance varies greatly	5 years from the effective date of this edition of the MUTCD
3A.05	Maintaining Minimum Retroreflectivity	Implementation and continued use of a method that is designed to maintain retroreflectivity of longitudinal pavement markings (see Paragraph 1 of Section 3A.05)	September 6, 2026
8B.16	High-Profile Grade Crossings	Paragraphs 3 and 7 - Recommended installation of Low Ground Clearance and/or Vehicle Exclusion signs and detour signs for vehicles with low ground clearances that might hang up on high-profile grade crossings at locations with a known history	5 years from the effective date of this edition of the MUTCD
8D.09 through 8D.12	Highway Traffic Signals at or Near Grade Crossings	Assessment and determination of appropriate treatment to achieve compliance (preemption, movement prohibition, pre-signals, queue cutter signals)	10 years from the effective date of this edition of the MUTCD

Guidance:

10   *Agencies should contact the FHWA when considering employing a practice or application that is not explicitly addressed in this Manual to ensure continued compliance with the provisions in this Manual.*

Support:

11   The FHWA reviews and interprets the provisions in this Manual for agencies on an as-needed basis, which can lead to the issuance of official interpretations (see Section 1B.04), or interim approvals (see Section 1B.07).

Standard:

12   **A non-compliant traffic control device that is being replaced or refurbished because it is damaged, missing, or no longer serviceable (see definition in Section 1C.02) for any reason shall be replaced with a compliant device, except as provided for in Paragraph 13 of this Section.**

Option:

13   A non-compliant traffic control device may be replaced in kind when engineering judgment indicates it is more appropriate because:

    A. One compliant device in the midst of a series of adjacent non-compliant devices would be confusing to road users, and/or

    B. The schedule for replacement of the whole series of non-compliant devices will result in achieving timely compliance with the MUTCD.

Section 1B.04 Interpretations

Support:

01   The FHWA issues authoritative interpretations of this Manual when necessary to provide clarity in response to unique situations for device application or general requests for clarification of a provision.

02   An interpretation includes a consideration of the application and operation of standard traffic control devices, the official meanings of standard traffic control devices, or the variations from standard device designs and design requirements.

Guidance:

03   *Requests for an interpretation of this Manual should contain the following information:*

    A. A concise statement of the interpretation being sought;

    B. A description of the condition that provoked the need for an interpretation;

    C. Any illustration that would be helpful to understand the request; and

    D. Any supporting research data that is pertinent to the item to be interpreted.

Support:

04   Section 1B.08 contains information on submitting a request for interpretation.

Section 1B.05 Experimentation

Support:

01   Requests for experimentation (see Section 1B.08) include consideration of field deployment for the purpose of testing or evaluating a new traffic control device, its application or manner of use, or a provision not specifically described in this Manual.

Standard:

02   **A traffic control device or application that does not comply with the provisions of this Manual shall not be used on any street, highway, bikeway, or site roadway open to public travel (see definition in Section 1C.02) without first receiving official approval to experiment from the FHWA's Office of Transportation Operations.**

Support:

03   A request for permission to experiment (see Section 1B.08) will be considered only when submitted by the public agency or toll facility authority responsible for the operation of the road or street on which the experiment is to take place. For a site roadway open to public travel, the request will be considered only if it is submitted by the private owner or official having jurisdiction.

04   A request for experimentation with a novel device or application across multiple jurisdictions as a single experiment with a common hypothesis, evaluation plan, and evaluation team will be considered when submitted jointly by all the authorities responsible for operation of the roads or streets on which the experiment is to take place. Similarly, a request to add experimental sites to an experimentation approved for another jurisdiction will be considered when submitted jointly by the all the authorities for operation of the roads or streets on which the experiment is then to take place.

05   Manufacturers or inventors of novel devices are encouraged to engage the services of a qualified traffic engineer or other professional who is versed in traffic control devices. Early engagement during the concept and development processes will help ensure the efficacy of the device with regard to human factors, operational, safety, and other considerations prior to an agency requesting experimentation.

06   In some cases, an off-roadway closed-course or laboratory study might be required before a request for experimentation can be considered. The purpose of such a study is to determine whether testing the experimental device or application in an open-road setting could result in an undue safety risk.

Guidance:

07   *Before requesting permission to experiment with a new device or application, an owner of a site roadway open to public travel should first check for any laws, regulations, and/or directives covering the application of the MUTCD that might apply.*

Option:

08   An agency may request a preliminary assessment of the viability of a potential request for experimentation by submitting an abstract that briefly describes the experimental concept.

Support:

09   A diagram indicating the process for requesting and conducting experimentations with traffic control devices is shown in Figure 1B-1.

Figure 1B-1. Process for Requesting and Conducting Experimentations for New Traffic Control Devices

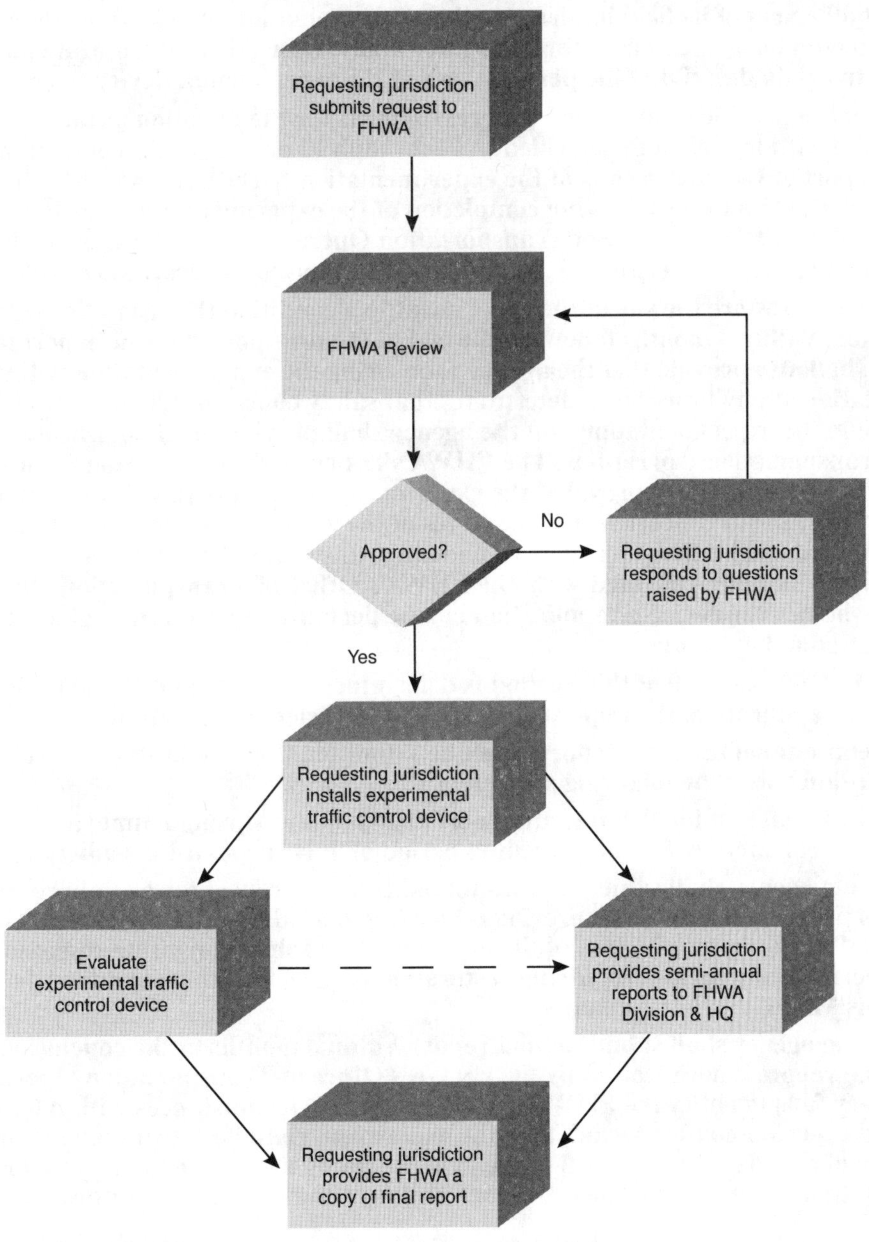

Standard:

10   The request for permission to experiment shall contain the following:

A. A statement indicating the nature of the problem and a hypothesis establishing the premise of the experiment.

B. A description of the proposed change to the traffic control device or application of the traffic control device, including the manner in which it deviates from the provisions of this Manual, and how it is expected to be an improvement over existing provisions.

C. Illustrations that would help to explain the traffic control device or use of the traffic control device.

D. Any supporting data explaining how the traffic control device was developed, including if it has been tested, in what ways it was found to be adequate or inadequate, and how this choice of device or application was derived.

E. Comparison of the proposed device to other compliant devices or treatments, either individually or in combination, that address the same condition, if applicable.

F. A legally-binding statement that the experimental device or application is in the public domain, in accordance with Paragraph 16 of this Section.

G. The time period and location(s) of the experiment.

H. Control sites for comparison purposes or justification for not using control sites.

I. A detailed research and evaluation plan that provides for close monitoring of the experimentation, throughout all stages of its field implementation. The evaluation plan shall include an appropriate evaluation methodology, such as before and after analysis, or other appropriate methodology as well as quantitative data describing the performance of the experimental device.

J. An agreement to provide semi-annual progress reports for the duration of the experimentation, in accordance with the schedule provided in Paragraph 12 of this Section, and an agreement to provide a report of the final results of the experimentation to the FHWA's Office of Transportation Operations within 3 months following completion of the experimentation (see Paragraph 14 of this Section). The FHWA's Office of Transportation Operations shall have the right to terminate approval of an agency's experiment if reports are not received in accordance with this schedule.

K. An agreement to restore the site of the experiment to a condition that complies with the provisions of this Manual within 3 months following the end of the time period of the experiment. This agreement shall also provide that the agency sponsoring the experimentation will terminate the experimentation at any time that it determines that safety concerns are directly or indirectly attributable to the experimentation and the agency shall provide timely notification to the FHWA's Office of Transportation Operations. The FHWA's Office of Transportation Operations shall have the right to terminate approval of the experimentation at any time if there is an indication of safety or operational concerns, or if the terms of the approval are not being adhered to. If, as a result of the experimentation, a request is made that this Manual be changed to include the device or application being experimented with, the FHWA's Office of Transportation Operations will determine whether the device or application can be permitted to remain in place until an official rulemaking action has occurred.

11   Where an item in Paragraph 10 of this Section is determined to not be applicable to the type of experiment, device, or application, the request shall provide sufficient explanation.

12   The required semi-annual progress reports shall be submitted throughout the course of an approved experiment in accordance with the following schedule:

A. No later than August 1st for the preceding period of January through June; and

B. No later than February 1st for the preceding period of July through December.

13   The experimenting agency shall submit a semi-annual progress report for any approved experiment even if no work was performed during the previous reporting period. Failure to submit two consecutive progress reports shall result in termination of the experiment and shall constitute rescission of the FHWA's approval to the experimenting agency, requiring restoration of the site(s) to a condition that complies with the provisions of this Manual within 3 months.

14   The experimenting agency shall submit a final report within 3 months of the conclusion of an approved experiment. If a final report is not received by the FHWA's Office of Transportation Operations, and the experimenting agency fails to notify the FHWA of any mitigating circumstances within 6 months of the end of the approved experimentation period, then the experiment shall be considered terminated and shall constitute rescission of the FHWA's approval to the experimenting agency, requiring restoration of the site(s) to a condition that complies with the provisions of this Manual within 3 months.

Support:

15   Under certain circumstances the FHWA Office of Transportation Operations might allow an experimental device or device application that has been shown to be effective and without safety concerns to remain in use after the experiment has ended. This typically would occur if the device or application is actively being considered for interim approval under the provisions of Section 1B.07.

Standard:

16   **A request for experimentation that involves a new traffic control device or a new application of an existing traffic control device shall include from the agency conducting the experiment, the manufacturer and/or developer of the device, and the supplier of the device, a legally-binding statement certifying that the traffic control device is not protected by a patent, trademark, or copyright in accordance with Section 1D.06, and that the traffic control device is in the public domain and can be used freely in traffic control device design and application without infringement or claim of trade secret misappropriation. The legally-binding statement shall also state that the agency conducting the experiment, the manufacturer and/or developer of the device, and the supplier of the device are aware that if patent, trademark, or copyright protection is established in the future for the device or application, such action will result in its removal from the MUTCD, cancellation of its interim approval, or cancellation of the authorization for experimentation.**

Support:

17   For the purpose of the Standard in Paragraph 16 of this Section, traffic control device refers to those aspects of a sign, signal, marking or other device which regulates, warns, or guides traffic. The limitation on patent, trademark, or copyright protection does not include the legal protection of individual elements of such devices. For example, manufacturing methods, assembly methods, or individual components of such devices can be protected, whereas the traffic control device cannot be subject to protection so long as it remains in this Manual. As a further example, an internal circuit board for an electronic traffic control device can be legally protected, but the electronic traffic control device itself or its operational function cannot be legally protected by any of the above forms of intellectual property rights.

Section 1B.06 Changes to the MUTCD

Support:

01   Continuing advances in technology and approaches to traffic safety will produce changes in the highway, vehicle, and road-user proficiency; therefore, portions of the system of traffic control devices in this Manual will require updating. It is important to have a procedure for recognizing these developments and for introducing new ideas and modifications into the system.

02   A change includes consideration of a new device to replace a present standard device, an additional device to be added to the list of standard devices, or a revision to a traffic control device application or placement criteria.

Guidance:

03   *Requests for a change to this Manual (see Section 1B.08) should contain the following information:*

    A. A statement indicating what change is proposed;
    B. Any illustration that would be helpful to understand the request; and
    C. Any supporting research data that is pertinent to the item to be reviewed.

Support:

04   Requests for a change to this Manual will be evaluated to consider the potential safety and operational benefits of the requested change and be considered for inclusion in a futurefor consideration in the next rulemaking to issue a new edition or revision of the Manual. A diagram indicating the process for incorporating new traffic control devices into this Manual is shown in Figure 1B-2.

Section 1B.07 Interim Approvals

Support:

01   Interim approval allows for provisional use, pending official rulemaking, of a new traffic control device, a revision to the application or manner of use of an existing traffic control device, or a provision not specifically described in this Manual.

02   The FHWA issues an interim approval by official memorandum signed by the Associate Administrator for Operations and posts this memorandum on the MUTCD Web site.

03   Interim approval allows for the optional use of a traffic control device or application and does not create a new mandate or recommendation for its use. Interim approval includes conditions that jurisdictions, toll facility operators, or owners of site roadways open to public travel agree to comply with in order to use the traffic control device or application until an official rulemaking action has occurred.

Figure 1B-2. Process for Incorporating
New Traffic Control Devices into the MUTCD

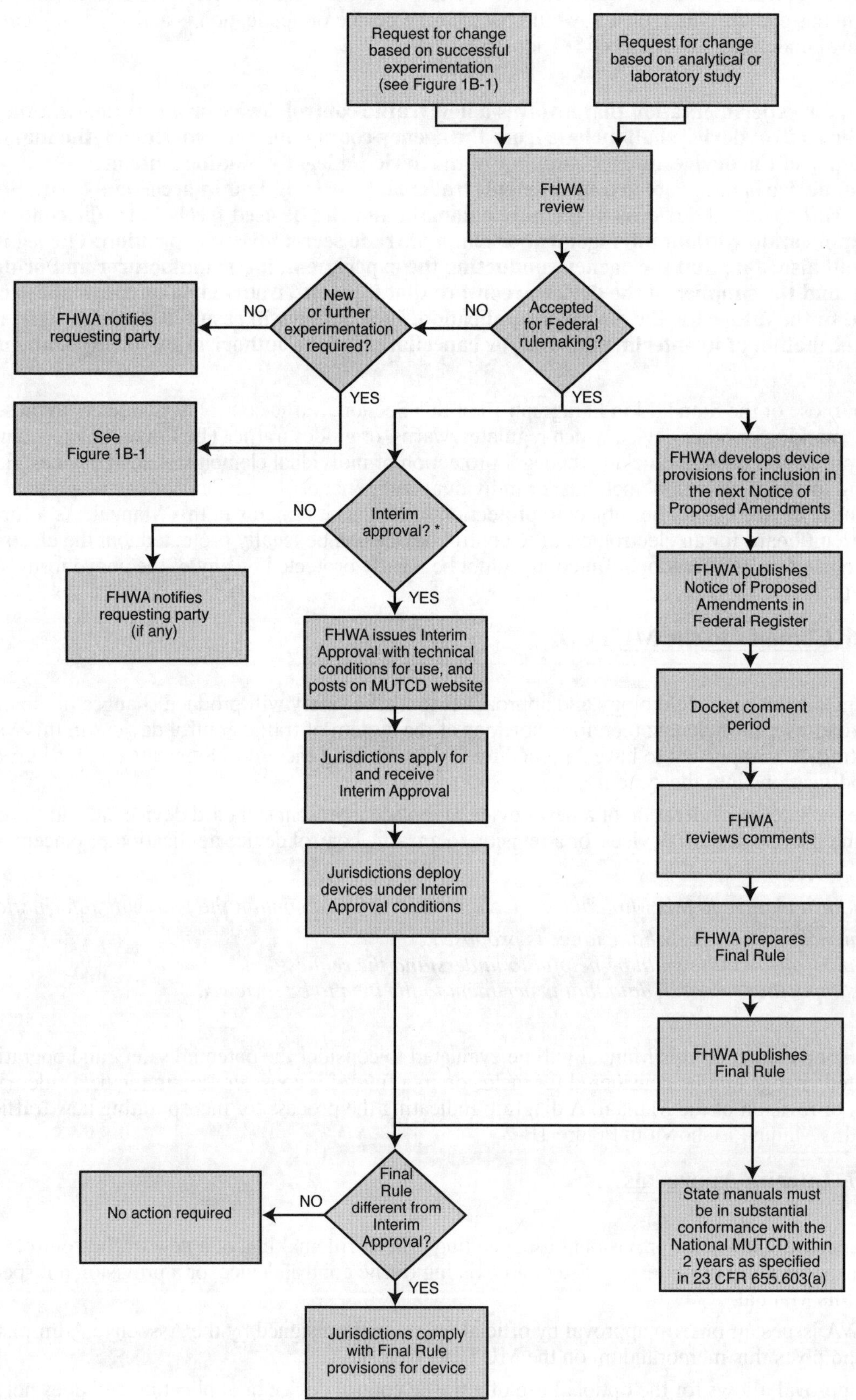

*A change request to allow the use of a new traffic control device is considered for Interim Approval when FHWA determines that the change will have a substantial national safety or operational benefit and that it would not be in the public interest to wait the intervening time period before the device could otherwise be officially incorporated into the MUTCD through rulemaking.

04   The issuance by FHWA of an interim approval might result in the traffic control device or application being proposed for adoption in the next scheduled rulemaking process to issue a new edition or revision of this Manual. If the device or application under interim approval is not proposed in the next rulemaking for a new edition or revision, then a statement of the status of the interim approval, whether it is to be rescinded or remain in effect, will be included in the Federal Register notice for the rulemaking.

05   Interim approval is considered based on the results of experimentation, and/or results of analytical or laboratory studies with a traffic control device or application that analytically demonstrates a device effectively communicates its intended meaning. Interim approval considerations include an assessment of relative risks, benefits, costs, impacts, and other factors.

06   Section 1B.08 contains information on submitting a request for interim approval.

07   Interim approval is ordinarily considered only after published authoritative research and experimentation sufficiently demonstrate that the device or application provides a significant safety or operational improvement. Individual experiments by various jurisdictions, without a research report on the overall findings of the experimental device or application, will not ordinarily qualify for issuance of an interim approval.

08   Interim approval ordinarily is not considered based solely on non-U.S. experience with a new traffic control device or application. Differences in regulations, enforcement and penalties, and driver licensing requirements, among other factors, can result in dissimilar road-user behavior. Additionally, due to variations in conventions for traffic control device design, a non-U.S. traffic control device concept might need to be adapted to U.S. criteria to ensure consistency with the provisions and principles of this Manual. However, documented non-U.S. experience can be considered in the development of requests for experimentation (see Section 1B.05) and within the evaluation plan for traffic control device research.

Standard:

09   **A jurisdiction, toll facility operator, or owner of a site roadway open to public travel that desires to use a traffic control device or application for which FHWA has issued an interim approval shall request and receive permission from FHWA in writing prior to applying the device or application.**

10   **The request to place a traffic control device or application under an existing interim approval shall contain the following:**

    A. A description of where the device or application will be used, such as a list of specific locations or highway segments or types of situations, or a statement of the intent to use the device or application jurisdiction-wide;

    B. An agreement to abide by the specific conditions for use of the device or application as contained in the FHWA's interim approval memorandum;

    C. An agreement to maintain and continually update a list of locations where the device or application has been installed; and

    D. An agreement to:

        1. Restore the site(s) of the interim approval to a condition that complies with the provisions in this Manual within 3 months following the issuance of a final rule on this traffic control device or application; and

        2. Terminate use of the device or application installed under the interim approval at any time that it determines that safety concerns are directly or indirectly attributable to the device or application. The FHWA's Office of Transportation Operations shall have the right to terminate the interim approval at any time if there is an indication of safety, operational, or other concerns.

Option:

11   A State may submit a request for permission to use a device or application under an existing interim approval for all jurisdictions in that State, as long as the request contains the information required in Paragraph 9 of this Section.

Standard:

12   **A jurisdiction, toll facility operator, or owner of a site roadway open to public travel that elects to use a device or application under a statewide interim approval shall inform the State of its use of the device or application.**

13   **Under a statewide interim approval, the respective jurisdictions, toll facility operators, and owners of site roadways open to public travel shall maintain and continually update a record of all locations on their roads where the device or application is implemented (see Item C of Paragraph 9 of this Section) and shall furnish this information to the State.**

Section 1B.08 <u>Requesting Official Interpretations, Experiments, Changes to the MUTCD, or Interim Approvals</u>

Guidance:

01   *A local jurisdiction, toll facility operator, or owner of a site roadway open to public travel that is requesting permission to experiment or permission to use a device or application under an existing interim approval should first check for any State laws, regulations, and/or directives covering the application of the MUTCD provisions that might apply.*

Standard:

02   **Except as provided in Paragraph 3 of this Section, requests for an interpretation, permission to experiment, a change to the MUTCD, granting of an interim approval, or permission to use an existing interim approval shall be submitted electronically to the Federal Highway Administration (FHWA), Office of Transportation Operations, MUTCD team, at the following e-mail address: MUTCDofficialrequest@dot.gov.**

Option:

03   If electronic submittal is not possible, requests for an interpretation, permission to experiment, a change to the MUTCD, granting of an interim approval, or permission to use an existing interim approval may instead be mailed to the Office of Transportation Operations, HOTO-1, Federal Highway Administration, 1200 New Jersey Avenue, SE, Washington, DC 20590.

Support:

04   Communications regarding other MUTCD matters that are not related to official requests will receive quicker attention if they are submitted electronically to the MUTCD Team Leader or to the appropriate individual MUTCD technical lead team member. Their e-mail addresses are available through the links contained on the "MUTCD Team" page on the MUTCD Web site at http://mutcd.fhwa.dot.gov/team.htm.

05   For additional information concerning interpretations, experimentation, changes, or interim approvals, visit the MUTCD Web site at http://mutcd.fhwa.dot.gov.

CHAPTER 1C. DEFINITIONS, ACRONYMS, AND ABBREVIATIONS USED IN THIS MANUAL

Section 1C.01 Definitions of Headings Used in this Manual

Standard:

01   When used in this Manual, the text headings of Standard, Guidance, Option, and Support shall be defined as follows:

    A. Standard—a statement of required, mandatory, or specifically prohibitive practice regarding a traffic control device. In limited, location-specific cases, the results of a documented engineering study (see Section 1D.03) might indicate a deviation from one or more requirements of a Standard provision to be appropriate. All Standard statements are labeled, and the text appears in bold type. The verb "shall" is typically used. The verbs "should" and "may" are not used in Standard statements. Standard statements are sometimes modified by Option statements.

    B. Guidance—a statement of recommended practice in typical situations, with deviations allowed if engineering judgment or engineering study (see Section 1D.03) indicates the deviation to be appropriate. All Guidance statements are labeled, and the text appears in unbold italic type. The verb "should" is typically used. The verbs "shall" and "may" are not used in Guidance statements. Guidance statements are sometimes modified by Option statements.

    C. Option—a statement of practice that is a permissive condition and carries no requirement or recommendation. Option statements sometimes contain allowable modifications to a Standard or Guidance statement. All Option statements are labeled, and the text appears in unbold type. The verb "may" is typically used. The verbs "shall" and "should" are not used in Option statements.

    D. Support—an informational statement that does not convey any degree of mandate, recommendation, authorization, prohibition, or enforceable condition. Support statements are labeled, and the text appears in unbold type. The verbs "shall," "should," and "may" are not used in Support statements.

Section 1C.02 Definitions of Words and Phrases Used in this Manual

Standard:

01   Unless otherwise defined in this Section, or in other Parts of this Manual, words or phrases shall have the meaning(s) as defined in the "Uniform Vehicle Code," "AASHTO Transportation Glossary (Highway Definitions)," or other appropriate publications.

02   Where a term that is defined in this Section or elsewhere in this Manual has a different definition in another resource or in common use, the definition herein shall govern for purposes of the applicability of the provisions of this Manual.

03   The following words and phrases, when used in this Manual, shall have the following meanings:

    1. Accessible Pedestrian Signal—a device that communicates information about pedestrian signal timing in a non-visual format such as audible tones and/or speech messages and vibrating surfaces.

    2. Accessible Pedestrian Signal Detector—a device designated to assist the pedestrian who has vision or physical disabilities in activating the pedestrian phase.

    3. Active Grade Crossing—a grade crossing equipped with automatic traffic control devices, such as flashing-light signals, gates, and/or traffic control signals, that are activated upon the detection of approaching rail traffic.

    4. Actuated—a type of traffic control signal operation in which some or all signal phases are operated on the basis of actuation.

    5. Actuation—initiation of, a change in, or an extension of a traffic signal phase or a sign legend through the operation of any type of detector.

    6. Advance Preemption—the notification of approaching rail traffic that is forwarded to the highway traffic signal controller unit or assembly by the railroad or light rail transit equipment in advance of the activation of the railroad or light rail transit warning devices.

    7. Advance Preemption Time—the period of time that is the difference between the required maximum highway traffic signal preemption time and the activation of the railroad or light rail transit warning devices.

    8. Advisory Speed—a recommended speed for all vehicles operating on a section of highway and based on the highway design, operating characteristics, and conditions.

    9. Agency—an organization with the responsibility for providing, maintaining, and/or operating a public or private road system.

    10. Alley—a street or highway intended to provide access to the rear or side of lots or buildings in urban areas and not intended for the purpose of through vehicular traffic.

11. **Annual Average Daily Traffic (AADT)**—the total volume of traffic passing a point or segment of a highway facility in both directions for one year divided by the number of days in the year. Normally, periodic daily traffic volumes are adjusted for hours of the day counted, days of the week, and seasons of the year to arrive at annual average daily traffic.

12. **Application**—in regard to a traffic control device, the act of deciding to use a device, generally or at a particular location for a particular condition.

13. **Approach**—all lanes of traffic moving toward an intersection or a midblock location from one direction, including any adjacent parking lane(s).

14. **Arterial Highway (Street)**—a general term denoting a highway primarily used by through traffic, usually on a continuous route or a highway designated as part of an arterial system.

15. **Automated Vehicle**—see Driving Automation System.

16. **Automatic Lane**—see Exact Change Lane within the definition of Toll Collection.

17. **Average Daily Traffic (ADT)**—the average 24 hour volume, being the total volume during a stated period divided by the number of days in that period. Normally, this would be periodic daily traffic volumes over several days, not adjusted for days of the week or seasons of the year.

18. **Average Day**—a day representing traffic volumes normally and repeatedly found at a location, typically a weekday when volumes are influenced by employment or a weekend day when volumes are influenced by entertainment or recreation.

19. **Backplate**—see Signal Backplate.

20. **Barrier-Separated Lane**—a preferential lane or other special purpose lane that is separated from the adjacent general-purpose lane(s) by a physical barrier.

21. **Beacon**—a highway traffic signal with one or more signal indications that operates in a flashing mode. Types of beacons include:

   (a) **Emergency-Vehicle Hybrid Beacon**—a special type of beacon (see Hybrid Beacon).

   (b) **Intersection Control Beacon**—a beacon used only at an intersection to control two or more directions of travel.

   (c) **Pedestrian Hybrid Beacon**—a special type of beacon (see Hybrid Beacon).

   (d) **Rectangular Rapid-Flashing Beacon (RRFB)**—a pedestrian-activated and/or bicycle-activated device comprising two horizontally arranged, rapidly flashed, rectangular-shaped yellow indications that is used to provide supplemental emphasis for a pedestrian, school, or trail crossing warning sign at a marked crosswalk across an uncontrolled approach.

   (e) **Speed Limit Sign Beacon**—a beacon used only to supplement a SPEED LIMIT sign.

   (f) **Stop Beacon**—a beacon used only to supplement a STOP sign, a DO NOT ENTER sign, or a WRONG WAY sign.

   (g) **Warning Beacon**—a beacon used only to supplement an appropriate warning or regulatory sign or marker.

22. **Bicycle**—a pedal-powered vehicle upon which the human operator sits.

23. **Bicycle Box**—a designated area on the approach to a signalized intersection, between an advance motorist stop line and the crosswalk or intersection, intended to provide bicyclists a visible place to wait in front of stopped motorists during the red signal phase.

24. **Bicycle Facilities**—a general term denoting improvements and provisions that accommodate or encourage bicycling, including parking and storage facilities, and shared roadways not specifically defined for bicycle use.

25. **Bicycle Lane**—a portion of a roadway that has been designated for preferential or exclusive use by bicyclists. A typical bicycle lane is delineated from the adjacent general-purpose lane(s) by longitudinal pavement markings and bicycle lane symbol or word markings and, if used, signs. Other types of bicycle lanes include:

   (a) **Buffer-Separated Bicycle Lane**—a bicycle lane that is separated from the adjacent general-purpose lane(s) by a pattern of standard longitudinal pavement markings that is wider than a normal or wide lane line marking.

   (b) **Counter-Flow Bicycle Lane**—a one-directional bicycle lane that provides a lawful path of travel for bicycles in the opposite direction from general traffic on a roadway that allows general traffic to travel in only one direction. Counter-flow bicycle lanes are designated by the traffic control devices used for other bicycle lanes.

   (c) **Separated Bicycle Lane**—an exclusive facility for bicyclists that is located within or directly adjacent to the roadway and that is physically separated from motor vehicle traffic with a vertical element. Separated bicycle lanes are differentiated from other bicycle lanes by a vertical element.

26. **Bicycle Signal Face**—a signal face that displays only bicycle symbol signal indications, that exclusively controls a bicycle movement from a designated bicycle lane or from a separate facility such as a shared-use path, and that displays signal indications that are applicable only to the bicycle movement.

27. **Bicycle Symbol Signal Indication**—a red, yellow, or green signal indication that displays a bicycle symbol rather than a circular or arrow indication.

28. **Bikeway**—a generic term for any road, street, path, or way that in some manner is specifically designated for bicycle travel, regardless of whether such facilities are designated for the exclusive use of bicycles or are to be shared with other transportation modes.

29. **Blank-Out Sign**—a sign that displays a single predetermined message only when activated. When not activated, the sign legend is not visible.

30. **Buffer-Separated Lane**—a preferential lane or other special purpose lane that is separated from the adjacent general-purpose lane(s) by a pattern of standard longitudinal pavement markings that is wider than a normal or wide lane line marking. The buffer area might include rumble strips, textured pavement, or channelizing devices such as tubular markers or traversable curbs, but does not include a physical barrier.

31. **Business Identification Sign Panel**—a panel containing a word legend or logo used to identify a business on a Specific Service sign.

32. **Busway**—a traveled way that is used exclusively by buses.

33. **Cantilevered Signal Structure**—a structure, also referred to as a mast arm, that is rigidly attached to a vertical pole and is used to provide overhead support of highway traffic signal faces or grade crossing signal units.

34. **Center Line Markings**—the yellow pavement marking line(s) that delineates the separation of traffic lanes that have opposite directions of travel on a roadway. These markings need not be at the geometrical center of the pavement.

35. **Changeable Message Sign**—a sign that is capable of displaying more than one message (one of which might be a "blank" display), changeable manually, by remote control, or by automatic control. Electronic-display changeable message signs are referred to as Dynamic Message Signs in the National Intelligent Transportation Systems (ITS) Architecture and are referred to as Variable Message Signs in the National Electrical Manufacturers Association (NEMA) standards publication.

36. **Channelizing Line**—a solid wide or double white line marking used to form islands where traffic in the same direction of travel is permitted on both sides of the island.

37. **Circular Intersection**—an intersection that has an island, generally circular in design, located in the center of the intersection where traffic passes to the right of the island. Circular intersections include roundabouts, rotaries, and traffic circles.

38. **Circulatory Roadway**—the roadway within a circular intersection on which traffic travels in a counterclockwise direction around an island in the center of the circular intersection.

39. **Clear Storage Distance**—when used in Part 8, the distance available for vehicle storage measured between 6 feet from the rail nearest the intersection to the intersection stop line or the normal stopping point on the highway. At skewed grade crossings and intersections, the 6-foot distance shall be measured perpendicular to the nearest rail either along the center line or edge line of the highway, as appropriate, to obtain the shorter distance. Where exit gates are used, the distance available for vehicle storage is measured from the point where the rear of the vehicle would be clear of the exit gate arm. In cases where the exit gate arm is parallel to the track(s) and is not perpendicular to the highway, the distance is measured either along the center line or edge line of the highway, as appropriate, to obtain the shorter distance.

40. **Clear Zone**—the total roadside border area, starting at the edge of the traveled way, that is available for an errant driver to stop or regain control of a vehicle. This area might consist of a shoulder, a recoverable slope, and/or a non-recoverable, traversable slope with a clear run-out area at its toe.

41. **Collector Highway**—a term denoting a highway that in rural areas connects small towns and local highways to arterial highways, and in urban areas provides land access and traffic circulation within residential, commercial, and business areas and connects local highways to the arterial highways.

42. **Conflict Monitor**—a device used to detect and respond to improper or conflicting signal indications and improper operating voltages in a traffic controller assembly.

43. **Constant Warning Time Detection**—a means of detecting rail traffic that provides relatively uniform warning time for the approach of through rail traffic that is not accelerating or decelerating after being detected.

44. **Contiguous Lane**—a lane, preferential or otherwise, that is separated from the adjacent lane(s) only by a normal or wide lane line marking.

45. **Controller Assembly**—a complete electrical device mounted in a cabinet for controlling the operation of a highway traffic signal.

46. **Controller Unit**—that part of a controller assembly that is devoted to the selection and timing of the display of signal indications.

47. **Conventional Road**—a street or highway other than an expressway or freeway.

48. **Counter-Flow Lane**—a lane operating in a direction opposite to the normal flow of traffic designated for peak direction of travel during at least a portion of the day. Counter-flow lanes are usually separated from the off-peak direction lanes by tubular markers or other flexible channelizing devices, temporary lane separators, or movable or permanent barrier.

49. **Crashworthy**—the ability of a roadside safety hardware device or appurtenance to minimize risks to vehicle occupants by allowing a vehicle impacting the appurtenance to be slowed before stopping, redirected, or to continue without significant resistance. Section 1D.11 contains additional information about crashworthiness.

50. **Crosswalk**—(a) that part of a roadway at an intersection included within the connections of the lateral lines of the sidewalks on opposite sides of the highway measured from the curbs or in the absence of curbs, from the edges of the traversable roadway, and in the absence of a sidewalk on one side of the roadway, the part of a roadway included within the extension of the lateral lines of the sidewalk at right angles to the center line; (b) any portion of a roadway at an intersection or elsewhere distinctly indicated as a pedestrian crossing by pavement marking lines on the surface, which might be supplemented by contrasting pavement texture, style, or color.

51. **Crosswalk Lines**—white pavement marking lines that identify a crosswalk.

52. **Cycle Length**—the time required for one complete sequence of signal indications.

53. **Dark Mode**—the lack of all signal indications at a signalized location. The dark mode is most commonly associated with power failures, ramp meters, hybrid beacons, beacons, and some movable bridge signals.

54. **Dedicated Lane**—A lane on a freeway or expressway that provides access to: (a) either an exit lane or the mainline, but not both, at a freeway or expressway exit, or (b) only one roadway at a freeway or expressway split.

55. **Delineator**—a retroreflective device mounted at the side of the roadway in a series to indicate the alignment of the roadway, especially at night or in adverse weather.

56. **Design Vehicle**—the longest vehicle permitted by statute of the road authority (State or other) on that roadway.

57. **Designated Bicycle Route**—a system of bikeways designated by the jurisdiction having authority with appropriate directional and informational route signs, with or without specific bicycle route numbers.

58. **Detectable**—having a continuous edge within 6 inches of the surface so that pedestrians with vision disabilities can sense its presence and receive usable guidance information.

59. **Detector**—a device used for determining the presence or passage of motor vehicles, bicycles, or pedestrians.

60. **Detection Plate**—a smooth continuous plate used on pedestrian channelizing devices to facilitate the use of low-vision canes for pedestrians with vision disabilities. The bottom edge of the detection plate shall be no more than 2 inches above the walkway and the top edge of the detection plate shall be at least 8 inches above the walkway. The detection plate shall share the same vertical plane as the hand trailing edge of the pedestrian channelizing device.

61. **Diagnostic Team**—a group of knowledgeable representatives of the parties of interest in a grade crossing or group of grade crossings (see 23 CFR Part 646.204).

62. **Downstream**—a term that refers to a location that is encountered by traffic subsequent to an upstream location as it flows in an "upstream to downstream" direction. For example, "the downstream end of a lane line separating the turn lane from a through lane on the approach to an intersection" is the end of the lane line that is closest to the intersection.

63. **Driveway**—an access from a roadway to a building, site, or abutting property.

64. **Driving Aisle**—circulation area for motor vehicles within a parking area, typically between rows of parking spaces. Driving aisles provide one-way or two-way travel. Driving aisles are exempted from compliance with MUTCD provisions.

65. **Driving Automation System**—technology that automates some or all aspects of the driving task to assist or replace the human vehicle operator. Section 5A.03 contains descriptions of the automation levels.

66. **Dropped Lane**—see Lane Drop.

67. **Dual-Arrow Signal Section**—a type of signal section designed to include both a yellow arrow and a green arrow.
68. **Dynamic Envelope**—the clearance required for light rail transit traffic or a train and its cargo overhang due to any combination of loading, lateral motion, or suspension failure (see Figure 8C-3).
69. **Dynamic Exit Gate Operating Mode**—a mode of operation where the exit gate operation is based on the presence of vehicles within the minimum track clearance distance.
70. **Dynamic Message Sign**—see Changeable Message Sign.
71. **Edge Line Markings**—white or yellow pavement marking lines that delineate the right or left edge(s) of a traveled way.
72. **Electronic Toll Collection (ETC) Account Only Lane**—a non-attended toll lane that is restricted to use only by vehicles with a registered toll payment account.
73. **Emergency-Vehicle Hybrid Beacon**—see Hybrid Beacon.
74. **Emergency-Vehicle Traffic Control Signal**—see Highway Traffic Signal.
75. **Engineer**—see Professional Engineer.
76. **Engineering Judgment**—the evaluation of available pertinent information including, but not limited to, the safety and operational efficiency of all road users, and the application of appropriate principles, provisions, and practices as contained in this Manual and other sources, for the purpose of deciding upon the design (see Section 1D.03), use, installation, or operation of a traffic control device. Engineering judgment shall be exercised by a professional engineer (see definition in this Section) with appropriate traffic engineering expertise, or by an individual working under the supervision of such an engineer, through the application of procedures and criteria established by the engineer. Documentation of engineering judgment is not required.
77. **Engineering Study**—the analysis and evaluation of available pertinent information including, but not limited to, the safety and operational efficiency of all road users, and the application of appropriate principles, provisions, and practices as contained in this Manual and other sources, for the purpose of deciding upon the design (see Section 1D.03), use, installation, or operation of a traffic control device. An engineering study shall be performed by a professional engineer (see definition in this Section) with appropriate traffic engineering expertise, or by an individual working under the supervision of such an engineer, through the application of procedures and criteria established by the engineer. An engineering study shall be documented in writing.
78. **Entrance Gate**—an automatic gate that can be lowered across the lanes approaching a grade crossing to block road users from entering the grade crossing.
79. **Exclusive Alignment**—a light rail transit track(s) or a bus rapid transit busway that is grade-separated or protected by a fence or traffic barrier. No grade crossings exist along the track(s) or busway. Motor vehicles, bicycles, and pedestrians are prohibited within the right-of-way. Subways and elevated structures are included within this definition.
80. **Exit Gate**—an automatic gate that can be lowered across the lanes departing a grade crossing to block road users from entering the grade crossing by driving in the opposing traffic lanes.
81. **Exit Gate Clearance Time**—for Four-Quadrant Gate systems at grade crossings, the amount of time provided to delay the descent of the exit gate arm(s) after entrance gate arm(s) begin to descend.
82. **Exit Gate Operating Mode**—for Four-Quadrant Gate systems at grade crossings, the mode of control used to govern the operation of the exit gate arms.
83. **Expressway**—a divided highway with partial control of access.
84. **Fail-Safe**—when used in Part 8, a railroad signal design philosophy applied to a system or device such that the result of a hardware failure or the effect of a software error shall either prohibit the system or device from assuming or maintaining an unsafe state or shall cause the system or device to assume a state that is known to be safe.
85. **Flagger**—a person who actively controls the flow of vehicular traffic into and/or through a temporary traffic control zone using hand-signaling devices or an Automated Flagger Assistance Device (AFAD).
86. **Flasher**—a device used to turn highway traffic signal indications on and off at a repetitive rate of approximately once per second.
87. **Flashing**—an operation in which a light source, such as a traffic signal indication or LEDs in a sign, is turned on and off repetitively.
88. **Flashing-Light Signals**—a warning device consisting of two red signal indications arranged horizontally that are activated to flash alternately when rail traffic is approaching or present at a grade crossing.
89. **Flashing Mode**—a mode of operation in which at least one traffic signal indication in each vehicular signal face of a highway traffic signal is turned on and off repetitively.

90. **Four-Quadrant Gate System**—an exit gate system that includes entrance and exit gates that control and block road users on all lanes entering and exiting the grade crossing.
91. **Freeway**—a divided highway with full control of access.
92. **Full-Actuated**—a type of traffic control signal operation in which all signal phases function on the basis of actuation.
93. **Gate**—an automatically-operated or manually-operated traffic control device that is used to physically obstruct road users such that they are discouraged from proceeding past a particular point on a roadway or pathway, or such that they are discouraged from entering a particular grade crossing, ramp, lane, roadway, or facility.
94. **General-Purpose Lane**—a highway lane or set of lanes, other than a Managed Lane (see definition in this Section) or a Preferential Lane (see definition in this Section), that all or most of the traffic that is allowed on that highway is also allowed to use. Certain classes of vehicles, such as commercial vehicles or vehicles exceeding a certain weight or size, might be prohibited from using one or more of the general-purpose lanes. A general-purpose lane might also be restricted to certain uses, such as passing or turning or as an auxiliary lane.
95. **Gore Area**—see Physical Gore and Theoretical Gore.
96. **Grade Crossing**—the general area where a highway and a railroad and/or light rail transit route cross at the same level, within which are included the tracks, highway, and traffic control devices for traffic traversing that area.
97. **Grade Crossing Warning System**—the flashing-light signals, with or without automatic gates, together with the necessary control equipment used to inform road users of the approach or presence of rail traffic at a grade crossing.
98. **Guide Sign**—a sign that shows route designations, highway names, destinations, directions, distances, services, points of interest, or other geographical, recreational, or cultural information.
99. **High-Occupancy Vehicle (HOV)**—a motor vehicle carrying at least two (or more than two if the signs for a specific roadway indicate a higher minimum occupancy requirement) persons, including carpools, vanpools, and buses.
100. **Highway**—a general term for denoting a public way for purposes of travel by vehicles and vulnerable road users, including the entire area within the right-of-way.
101. **Highway-Light Rail Transit Grade Crossing**—the general area where a highway and a light rail transit route cross at the same level, within which are included the light rail transit tracks, highway, and traffic control devices for traffic traversing that area.
102. **Highway-Rail Grade Crossing**—the general area where a highway and a railroad cross at the same level, within which are included the railroad tracks, highway, and traffic control devices for highway traffic traversing that area.
103. **Highway Traffic Signal**—a power-operated traffic control device by which traffic is warned or directed to take some specific action. These devices do not include power-operated signs (except as provided in Chapters 4S and 4T), steadily illuminated raised pavement markers, gates, flashing-light signals (see Section 8D.02), warning lights (see Section 6L.07), or steady-burning electric lamps. Highway traffic signals include:

    (a) **Flashing Beacon**—see Beacon.
    (b) **In-Roadway Warning Lights**—a special type of highway traffic signal installed in the roadway surface to warn road users that they are approaching a condition on or adjacent to the roadway that might not be readily apparent and might require the road users to reduce speed and/or come to a stop.
    (c) **Lane-Use Control Signal**—a signal face or comparable display on a full-matrix Changeable Message Sign (see Chapters 2L and 4T) displaying indications to permit or prohibit the use of specific lanes of a roadway or a shoulder where driving is sometimes permitted, or to indicate the impending prohibition of such use.
    (d) **Traffic Control Signal (Traffic Signal)**—a highway traffic signal placed at intersections, movable bridges, fire stations, midblock crosswalks, alternating one-way sections of a single lane road, private driveways, or other locations that require conflicting traffic to be directed to stop and permitted to proceed in an orderly manner. These devices do not include pedestrian hybrid beacons (see Chapter 4J) or emergency-vehicle hybrid beacons (see Chapter 4N). Traffic control signals include vehicular signal indications, pedestrian signal indications, and bicycle symbol signal indications. Special traffic control signals include:

        (1) **Emergency-Vehicle Traffic Control Signal**—a traffic control signal that directs all conflicting traffic to stop in order to permit the driver of an authorized emergency vehicle to proceed into the roadway or intersection.

    (2) **Movable Bridge Traffic Control Signal**—a traffic control signal installed at a movable bridge to notify traffic to stop during periods when the roadway is closed to allow the bridge to open.

    (3) **Portable Traffic Control Signal**—a temporary component of a traffic control signal on a mobile support with one or more signal faces that is designed so that it can be easily transported, deployed, or relocated as part of a temporary traffic control signal, or during construction and maintenance as a temporary part of a permanent traffic control signal installation.

    (4) **Pre-Signal**—traffic control signal faces that are located upstream from a signalized intersection and are operated in conjunction with the traffic control signal faces at the downstream signalized intersection in a manner that is designed to keep the area between the stop line for the upstream traffic control signal faces and the stop line for the downstream signalized intersection clear of queued vehicles. When used in conjunction with a grade crossing, the pre-signal is operated for the purpose of preventing vehicles from queuing within the minimum track clearance distance. Supplemental near-side traffic control signal faces for the downstream signalized intersection are not considered to be pre-signals.

    (5) **Queue Cutter Signal**—an independently-controlled traffic control signal (not operated in conjunction with the traffic control signal faces at a downstream signalized intersection) located at a grade crossing that controls traffic in one direction only on the roadway for the purpose of keeping the minimum track clearance distance clear of vehicles. The display of red signal indications is activated from a downstream queue detection system, by time of day, by approaching rail traffic, by an approaching bus on a busway, or by a combination of any of these methods.

    (6) **Ramp Control Signal**—a traffic control signal installed to control the merging flow of traffic onto a freeway at an entrance ramp or at a freeway-to-freeway ramp connection.

    (7) **Temporary Traffic Control Signal**—a traffic control signal that is installed for a limited time period using fixed or portable traffic control signal units.

104. **HOV Lane**—any preferential lane designated for exclusive use by high-occupancy vehicles for all or part of a day—including a designated lane on a freeway, other highway, street, or independent roadway on a separate right-of-way.

105. **Hybrid Beacon**—a special type of beacon that is intentionally placed in a dark mode (no indications displayed) between periods of operation and, when operated, displays both steady and flashing traffic control signal indications. Hybrid beacons include:

    (a) **Emergency-Vehicle Hybrid Beacon**—used to warn and control traffic at an unsignalized location to assist authorized emergency vehicles in entering or crossing a street or highway.

    (b) **Pedestrian Hybrid Beacon**—used to warn and control traffic at an unsignalized location to assist pedestrians in crossing a street or highway at a marked crosswalk.

106. **Identification Marker**—a shape, color, and/or pictograph that is used as a visual identifier for a destination guide signing system of a community wayfinding system or a shared-use path system for an area.

107. **Inherently Low Emission Vehicle (ILEV)**—any kind of vehicle that, because of inherent properties of the fuel system design, will not have significant evaporative emissions, even if its evaporative emission control system has failed.

108. **In-Roadway Warning Lights**—see Highway Traffic Signal.

109. **Interchange**—a system of interconnecting roadways providing for traffic movement between two or more highways that do not intersect at grade.

110. **Interchange Lane Drop**—see Lane Drop.

111. **Preemption Interconnection**—the electrical connection between the railroad or light rail transit active warning system and the highway traffic signal controller assembly for the purpose of preemption.

112. **Intermediate Interchange**—an interchange with an urban or rural route that is not a major or minor interchange as defined in this Section.

113. **Intersection**—intersection is defined as follows:

    (a) The area embraced within the prolongation or connection of the lateral curb lines, or if none, the lateral boundary lines of the roadways of two highways that join one another at, or approximately at, right angles, or the area within which vehicles traveling on different highways that join at any other angle might come into conflict.

    (b) The junction of an alley, driveway, or site roadway with a public roadway or highway shall not constitute an intersection, unless the public roadway or highway at said junction is controlled by a traffic control device.

    (c) If a highway includes two roadways separated by a median, then every crossing of each roadway of such divided highway by an intersecting highway shall be a separate intersection if the opposing left-turn paths cross and there is sufficient interior storage for the design vehicle (see Figure 2A-5).

    (d) At a location controlled by a traffic control signal, regardless of the distance between the separate intersections as defined in (c) above:

        (1) If a stop line, yield line, or crosswalk has not been designated on the roadway (within the median) between the separate intersections, the two intersections and the roadway (median) between them shall be considered as one intersection;

        (2) Where a stop line, yield line, or crosswalk is designated on the roadway on the intersection approach, the area within the crosswalk and/or beyond the designated stop line or yield line shall be part of the intersection; and

        (3) Where a crosswalk is designated on a roadway on the departure from the intersection, the intersection shall include the area extending to the far side of such crosswalk.

114. **Intersection Control Beacon**—see Beacon.

115. **Interval**—the part of a signal cycle during which signal indications do not change.

116. **Island**—a defined area between traffic lanes for control of vehicular movements, for toll collection, or for pedestrian or bicyclist refuge. It includes all end protection and approach treatments. Within an intersection area, a median or an outer separation is considered to be an island.

117. **Jughandle Turn**—a left-turn or U-turn that, in conjunction with special geometry, is made by initially making a right-turn or diverging to the right. With other special geometry, a right-turn or U-turn makes a jughandle turn by initially making a left-turn or diverging to the left.

118. **Lane Drop**—a through lane that becomes a mandatory turn lane on a conventional roadway, or a through lane that becomes a mandatory exit lane on a freeway or expressway. The end of an acceleration lane and reductions in the number of through lanes that do not involve a mandatory turn or exit are not considered lane drops.

119. **Lane Line Markings**—white pavement marking lines that delineate the separation of traffic lanes that have the same direction of travel on a roadway.

120. **Lane Reduction**—elimination of a through lane by a gradual narrowing of the travel pavement (taper) through physical construction or pavement markings at which traffic in the lane being eliminated must merge into the adjacent through lane and continue in the same direction of travel. A lane reduction can occur outside the influence of an intersection or interchange, or within an interchange a short distance downstream of the gore of an exit ramp. Through lanes that become a mandatory turn or exit are considered lane drops rather than lane reductions.

121. **Lane-Use Control Signal**—see Highway Traffic Signal.

122. **Legend**—see Sign Legend.

123. **Lens**—see Signal Lens.

124. **Light Rail Transit Traffic (Light Rail Transit Equipment)**—every device in, upon, or by which any person or property can be transported on light rail transit tracks, including single-unit light rail transit cars (such as streetcars and trolleys) and assemblies of multiple light rail transit cars coupled together.

125. **Loading Zone**—a specially marked, signed or designated area for the loading or unloading of vehicles (passenger or freight).

126. **Locomotive Horn**—an air horn, steam whistle, or similar audible warning device (see 49 CFR Part 229.129) mounted on a locomotive or control cab car. The terms "locomotive horn," "train whistle," "locomotive whistle," and "train horn" are used interchangeably in the railroad industry.

127. **Logo**—a distinctive emblem or trademark that identifies a commercial or non-commercial business, program, or organization.

128. **Longitudinal Markings**—pavement markings that are generally placed parallel and adjacent to the flow of traffic such as lane lines, center lines, edge lines, channelizing lines, and others.

129. **Louver**—see Signal Louver.

130. **Low-Volume Rural Road**—A category of paved or unpaved conventional or special-purpose roadways having an AADT of less than 400 vehicles and lying outside of built-up or urbanized areas of cities, towns, and communities.

131. **Major Interchange**—an interchange with another freeway or expressway, or an interchange with a high-volume multi-lane highway, principal urban arterial, or major rural route where the interchanging traffic is heavy or includes many road users unfamiliar with the area.

132. **Major Street**—the street normally carrying the higher volume of vehicular traffic.

133. **Malfunction Management Unit**—see Conflict Monitor.

134. **Managed Lane**—a highway lane or set of lanes, or a highway facility, for which variable operational strategies such as direction of travel, tolling, pricing, and/or vehicle type or occupancy requirements are implemented and managed in real-time in response to changing conditions. Managed lanes are typically buffer-separated or barrier-separated lanes parallel to the general-purpose lanes of a highway in which access is restricted to designated locations. There are also some highways on which all lanes are managed.

135. **Manual Lane**—see Attended Lane within the definition of Toll Collection.

136. **Maximum Highway Traffic Signal Preemption Time**—the maximum amount of time needed following initiation of the preemption sequence for the highway traffic signals to complete the timing of the right-of-way transfer time, queue clearance time, and separation time.

137. **Median**—the portion of a highway separating opposing directions of the traveled way or the area between two roadways of a divided highway measured from edge of traveled way to edge of traveled way. The median excludes turn lanes. The median width might be different between intersections, interchanges, and at opposite approaches of the same intersection.

138. **Minimum Track Clearance Distance**—the length along a highway over the track(s) where a vehicle could be struck by rail traffic. The minimum track clearance distance is measured from a point upstream from the track(s) on the approach to the grade crossing to a point downstream from the track(s) on the departure from the grade crossing. The length along the highway between the two points is the minimum track clearance distance.

139. **Minor Interchange**—an interchange where traffic is local and very light, such as interchanges with land service access roads. Where the sum of the exit volumes is estimated to be lower than 100 vehicles per day in the design year, the interchange is classified as local.

140. **Minor Street**—the street normally carrying the lower volume of vehicular traffic.

141. **Mixed-Use Alignment**—a light rail transit track(s), a busway, or a bus only lane(s) where the light rail transit (LRT) or bus rapid transit (BRT) vehicles operate in mixed traffic with all types of road users. This includes streets, transit malls, and pedestrian malls where the right-of-way is shared. In a mixed-use alignment, the light rail transit or the bus rapid transit traffic does not have the right-of-way over other road users at grade crossings and intersections. If the LRT traffic or buses are controlled by traffic control signals or LRT signal faces at an intersection with a roadway, the alignment is considered to be mixed-use even if some of the approaches to the intersection are used exclusively by LRT traffic or buses.

142. **Movable Bridge Resistance Gate**—a type of traffic gate, which is located downstream of the movable bridge warning gate, that provides a physical deterrent to vehicle and/or pedestrian traffic when placed in the appropriate position.

143. **Movable Bridge Signal**—see Highway Traffic Signal.

144. **Movable Bridge Warning Gate**—a type of traffic gate designed to warn, but not primarily to block, vehicle and/or pedestrian traffic when placed in the appropriate position.

145. **Multi-Lane**—more than one lane moving in the same direction. A multi-lane street, highway, or roadway has a basic cross-section comprised of two or more through lanes in one or both directions. A multi-lane approach has two or more lanes moving toward the intersection, including turning lanes.

146. **Neutral Area**—the paved area between the channelizing lines separating an entrance or exit ramp or a channelized turn lane or channelized entering lane from the adjacent through lane(s).

147. **Object Marker**—a device used to mark obstructions within or adjacent to the roadway.

148. **Occupancy Requirement**—any restriction that regulates the use of a facility or one or more lanes of a facility for any period of the day based on a specified minimum number of persons in a vehicle.

149. **Occupant**—a person driving or riding in a car, truck, bus, or other vehicle.

150. **On-Street Parking**—parking within or along, and accessed directly from, a public roadway or a site roadway open to public travel.

151. **Open-Road ETC Lane**—a non-attended lane that is designed to allow toll payments to be electronically collected from vehicles traveling at normal highway speeds. Open-Road ETC lanes are typically physically separated from the toll plaza, often following the alignment of the mainline lanes, with toll plaza lanes for cash toll payments being on a different alignment after diverging from the mainline lanes or a subset thereof.

152. **Open-Road Tolling Point**—the location along an Open-Road ETC lane at which roadside or overhead detection and receiving equipment are placed and vehicles are electronically assessed a toll.

153. **Opposing Traffic**—vehicles that are traveling in the opposite direction. At an intersection, vehicles entering from an approach that is approximately straight ahead would be considered to be opposing traffic, but vehicles entering from approaches on the left or right would be considered to be conflicting traffic rather than opposing traffic.

154. **Option Lane**—A lane on a freeway, expressway, or conventional road multi-lane exit or multi-lane split that widens on the approach to allow access, without changing lanes, to:

   (a) Both an exit lane and the mainline at a freeway or expressway exit; or
   (b) Both diverging roadways at a freeway, expressway, or conventional road split.

155. **Overhead Sign**—a sign that is placed such that a portion or the entirety of the sign or its support is directly above the roadway or shoulder such that vehicles travel below it. Typical installations include signs placed on cantilever arms that extend over the roadway or shoulder, signs placed on sign support structures that span the entire width of the pavement, signs placed on mast arms or span wires either independently or that also support traffic control signals, and signs placed on highway bridges that cross over the roadway.

156. **Parking Area**—a parking lot or parking garage that is separated from a roadway. Parallel, perpendicular, or angle parking spaces along a roadway are not considered a parking area.

157. **Parking Space**—an area marked or designated for storage of a vehicle while the driver is not present.

158. **Preemption Clearance Interval**—the part of a traffic signal sequence displayed as a result of a preemption request when vehicles are provided the opportunity to clear the railroad or light rail transit tracks, or a movable bridge, prior to the arrival of the train or boat for which the traffic signal is being preempted.

159. **Preemption Time Variability**—the result that occurs when the traffic signal controller enters the Preemption Clearance Interval with less than the maximum design Right-of-Way Transfer Time or the speed of a train approaching the grade crossing varies.

160. **Passive Grade Crossing**—a grade crossing where none of the automatic traffic control devices associated with an Active Grade Crossing Warning System are present and at which the traffic control devices consist entirely of signs and/or markings.

161. **Pathway**—a general term denoting a public way for purposes of travel by authorized users outside the traveled way and physically separated from the roadway by an open space or barrier and either within the highway right-of-way or within an independent alignment. Pathways include shared-use paths, but do not include sidewalks.

162. **Pathway Grade Crossing**—the general area where a pathway and railroad and/or light rail transit tracks cross at the same level, within which are included the tracks, pathway, and traffic control devices for pathway traffic traversing that area.

163. **Paved**—having a roadway surface that has both a structural (weight bearing) and a sealing purpose for the roadway, such as a bituminous surface treatment, mixed bituminous concrete, or Portland cement concrete.

164. **Pedestrian**—a person on foot, in a wheelchair, on other devices determined by local law to be equivalent, which might include skates or a skateboard.

165. **Pedestrian Change Interval**—an interval during which the flashing UPRAISED HAND (symbolizing DONT WALK) signal indication is displayed.

166. **Pedestrian Clearance Time**—the time provided for a pedestrian crossing in a crosswalk, after leaving the curb or edge of pavement, to travel to the far side of the traveled way or to a median.

167. **Pedestrian Facility**—a general term denoting a location where improvements and provisions have been made to accommodate or encourage pedestrian activity.

168. **Pedestrian Hybrid Beacon**—see Hybrid Beacon.

169. **Pedestrian Signal Head**—a signal head, which contains the symbols WALKING PERSON (symbolizing WALK) and UPRAISED HAND (symbolizing DONT WALK), that is installed to direct pedestrians at a traffic control signal.

170. **Permissive Mode**—a mode of traffic control signal operation in which left or right turns are permitted to be made after yielding to pedestrians, if any, and/or opposing traffic, if any. When a CIRCULAR GREEN signal indication is displayed, both left and right turns are permitted unless otherwise prohibited by another traffic control device. When a flashing YELLOW ARROW or flashing RED ARROW signal indication is displayed, the turn indicated by the arrow is permitted.

171. **Physical Gore**—a longitudinal point where a physical barrier or the lack of a paved surface inhibits road users from crossing from a ramp or channelized turn lane or channelized entering lane to the adjacent through lane(s) or vice versa.
172. **Pictograph**—a pictorial representation used to identify a governmental jurisdiction, an area of jurisdiction, a governmental or other public transportation agency or provider, a military base or branch of service, a governmental-approved university or college, a governmental-approved institution, or a toll payment system.
173. **Plaque**—a traffic control device intended to communicate specific information to road users through a word, symbol, or arrow legend that is placed immediately adjacent to a sign to supplement the message on the sign. The difference between a plaque and a sign is that a plaque cannot be used alone. The designation for a plaque includes a "P" suffix.
174. **Platoon**—a group of vehicles or pedestrians traveling together as a group, either voluntarily or involuntarily, because of traffic signal controls, geometrics, or other factors.
175. **Portable Traffic Control Signal**—see Highway Traffic Signal.
176. **Post-Exit Ramp Lane Reduction**—see Lane Reduction.
177. **Post-Mounted Sign**—a sign that is placed to the side of the roadway such that no portion of the sign or its support is directly above the roadway or shoulder.
178. **Posted Speed Limit**—a speed limit determined by law or regulation and displayed on Speed Limit signs.
179. **Preemption**—the transfer of normal operation of a traffic control signal or a hybrid beacon to a special control mode of operation.
180. **Preferential Lane**—a highway lane or set of lanes, or a highway facility, reserved for the exclusive use of one or more specific types of vehicles or of vehicles with a specific minimum number of occupants.
181. **Pre-Signal**—see Highway Traffic Signal.
182. **Pretimed Operation**—a type of traffic control signal operation in which none of the signal phases function on the basis of actuation.
183. **Primary Signal Face**—one of the required or recommended minimum number of signal faces for a given approach or separate turning movement, but not including near-side signal faces required as a result of the far-side signal faces exceeding the maximum distance from the stop line.
184. **Principal Legend**—place names, street names, and route numbers displayed on guide signs.
185. **Priority Control**—a means by which the assignment of right-of-way is obtained or modified.
186. **Private Road**—see Site Roadways Open to Public Travel.
187. **Professional Engineer (P.E.)**—An individual who has fulfilled education and experience requirements and passed examinations that, under State licensure laws, permit the individual to offer engineering services within areas of expertise directly to the public.
188. **Protected Mode**—a mode of traffic control signal operation in which left or right turns are permitted to be made only when a left or right GREEN ARROW signal indication is displayed.
189. **Public Road**—any road, street, or similar facility under the jurisdiction of and maintained by a public agency and open to public travel.
190. **Push Button**—a button to activate a device or signal timing for pedestrians, bicyclists, or other road users.
191. **Push Button Information Message**—a recorded message that can be actuated by pressing a push button when the walk interval is not timing and that provides the name of the street that the crosswalk associated with that particular push button crosses and can also provide other information about the intersection signalization or geometry.
192. **Push Button Locator Tone**—a repeating sound that informs approaching pedestrians that a push button exists to actuate pedestrian timing or receive additional information and that enables pedestrians with vision disabilities to locate the push button.
193. **Queue Clearance Time**—when used in Part 8, the time required for the design vehicle of maximum length stopped just inside the minimum track clearance distance to start up and move through and clear the entire minimum track clearance distance.
194. **Queue Cutter Signal**—see Highway Traffic Signal.
195. **Quiet Zone**—a segment of a rail line, within which is situated one or a number of consecutive public highway-rail grade crossings at which locomotive horns are not routinely sounded per 49 CFR Part 222.

196. **Rail Traffic**—every device in, upon, or by which any person or property can be transported on rails or tracks and to which all other traffic must yield the right-of-way by law at grade crossings, including trains, one or more locomotives coupled (with or without cars), other railroad equipment, and light rail transit operating in exclusive or semi-exclusive alignments. Light rail transit operating in a mixed-use alignment, to which other traffic is not required to yield the right-of-way by law, is a vehicle and is not considered to be rail traffic.

197. **Raised Pavement Marker**—a device mounted on or in a road surface that has a height generally not exceeding approximately 1 inch above the road surface for a permanent marker, or not exceeding approximately 2 inches above the road surface for a temporary flexible marker, and that is intended to be used as a positioning guide and/or to supplement or substitute for pavement markings. Raised pavement markers might also be recessed into or flush with the pavement surface.

198. **Ramp Control Signal**—see Highway Traffic Signal.

199. **Red Clearance Interval**—an interval that follows a yellow change interval and precedes the next conflicting green interval.

200. **Regulatory Sign**—a sign that gives notice to road users of traffic laws or regulations.

201. **Retroreflectivity**—a property of a surface that allows a large portion of the light coming from a point source to be returned directly back to a point near its origin.

202. **Road**—see Roadway.

203. **Road User**—a vehicle operator, bicyclist, or pedestrian, including persons with disabilities, within the highway or on a site roadway open to public travel.

204. **Roadway**—that portion of a highway improved, designed, or ordinarily used for vehicular travel and parking lanes, but exclusive of the sidewalk, berm, or shoulder even though such sidewalk, berm, or shoulder is used by persons riding bicycles or other human-powered vehicles. In the event a highway includes two or more separate roadways, the term roadway as used in this Manual shall refer to any such roadway separately, but not to all such roadways collectively.

205. **Roadway Network**—a geographical arrangement of intersecting roadways.

206. **Roundabout**—a circular intersection with yield control at entry, which permits a vehicle on the circulatory roadway to proceed, and with deflection of the approaching vehicle counter-clockwise around a central island.

207. **Rumble Strip**—a series of intermittent, narrow, transverse areas of rough-textured, slightly raised, or depressed road surface that extend across the travel lane to alert vehicle operators to unusual traffic conditions or are located along the shoulder, along the roadway center line, or within islands formed by pavement markings to alert road users that they are leaving the travel lanes.

208. **Rural Highway**—a type of roadway normally characterized by lower volumes, higher speeds, fewer turning conflicts, and less conflict with pedestrians.

209. **Scanning Graphic**—a graphic designed for scanning by machine, and includes bar codes, quick-response (QR) codes or other matrix bar code formats, or similar graphics.

210. **School**—a public or private educational institution recognized by the state education authority for one or more grades K through 12 or as otherwise defined by the State.

211. **School Zone**—a designated roadway segment approaching, adjacent to, and beyond school buildings or grounds, or along which school related activities occur.

212. **Semi-Actuated**—a type of traffic control signal operation in which at least one, but not all, signal phases function on the basis of actuation.

213. **Semi-Exclusive Alignment**—a light rail transit track(s) or a bus rapid transit busway that is in a separate right-of-way or that is along a street or railroad right-of-way where motor vehicles, bicycles, and pedestrians have limited access and cross only at designated locations, such as at grade crossings where road users must yield the right-of-way to the light rail transit or the bus rapid transit traffic.

214. **Separate Turn Signal Face**—a signal face that exclusively controls a turn movement and that displays signal indications that are applicable only to the turn movement.

215. **Separation Time**—the component of maximum highway traffic signal preemption time during which the minimum track clearance distance is clear of vehicular traffic prior to the arrival of rail traffic.

216. **Serviceable**—a condition in which a traffic control device appears (day and night) and operates as intended, beyond which it requires replacement due to damage or wear. Whether a device is serviceable will depend on the type of device under consideration. In general, if the device is capable of being serviced with minimal effort or replacement parts so that it continues to appear and operate as intended, and the device is otherwise substantially intact, then it can be considered to be in serviceable condition. If the device is damaged or not operational beyond reasonable repair, then it is likely no longer serviceable.

217. **Shared Roadway**—a roadway that is officially designated and marked as a bicycle route, but which is open to motor vehicle travel and upon which no bicycle lane is designated.

218. **Shared Turn Signal Face**—a signal face, for controlling both a turn movement and the adjacent through movement, that always displays the same color of circular signal indication that the adjacent through signal face or faces display.

219. **Shared-Use Path**—a bikeway outside the traveled way and physically separated from motorized vehicular traffic by an open space or barrier and either within the highway right-of-way or within an independent alignment. Shared-use paths are also used by pedestrians (including skaters, users of manual and motorized wheelchairs, and joggers) and other authorized motorized and non-motorized users.

220. **Shoulder**—a longitudinal area contiguous with the traveled way that is used for accommodation of stopped vehicles for emergency use and for lateral support of base and surface courses, and that is graded for emergency stopping. A shoulder might be paved or unpaved. A paved shoulder might be opened to part-time travel by some or all vehicles and might also be available for use by pedestrians and/or bicycles in the absence of other pedestrian or bicycle facilities.

221. **Sidewalk**—that portion of a street between the curb line, or the lateral line of a roadway, and the adjacent property line or on easements of private property that is paved or improved and intended for use by pedestrians.

222. **Sidewalk Extension**—a pedestrian facility at an intersection or midblock crosswalk which extends the sidewalk by physically and visually narrowing the roadway.

223. **Sidewalk Grade Crossing**—the portion of a highway-rail grade crossing or of a highway-light rail transit grade crossing where a sidewalk and railroad tracks or a sidewalk and light rail transit tracks cross at the same level, within which are included the tracks, sidewalk, and traffic control devices for sidewalk users traversing that area.

224. **Sign**—with regard to controlling traffic, any traffic control device that is intended to communicate specific information to road users through a word, symbol, and/or arrow legend. Signs do not include highway traffic signals, pavement markings, delineators, or channelization devices. Signs whose purpose is unrelated to traffic control are addressed in Section 1A.02.

225. **Sign Assembly**—a group of signs, located on the same support(s), that supplement one another in conveying information to road users.

226. **Sign Illumination**—either internal or external lighting that shows similar color by day or night. Street or highway lighting shall not be considered as meeting this definition.

227. **Sign Legend**—all word messages, logos, pictographs, and symbol and arrow designs that are intended to convey specific meanings. The border, if any, on a sign is not considered to be a part of the legend.

228. **Sign Panel**—a separate panel or piece of material containing a word, logo, pictograph, symbol, and/or arrow legend that is affixed to the face of a sign.

229. **Signal**—See Highway Traffic Signal.

230. **Signal Backplate**—a thin strip of material that extends outward from and parallel to a signal face on all sides of a signal housing to provide a background for improved visibility of the signal indications.

231. **Signal Coordination**—the establishment of timed relationships between adjacent traffic control signals.

232. **Signal Dimming**—a reduction of the light output from a signal indication, hybrid beacon, or rectangular rapid-flashing beacon indication, typically for nighttime conditions, to a value that is below the minimum specified intensity for daytime conditions. If a variety of intensity levels are used during daytime conditions and all of the various levels (including the lowest of the intensities) are above the minimum specified intensity for daytime conditions, this would not be considered to be signal dimming.

233. **Signal Face**—an assembly of one or more signal sections that is provided for controlling one or more traffic movements on a single approach.

234. **Signal Head**—an assembly of one or more signal faces that is provided for controlling traffic movements on one or more approaches.

235. **Signal Housing**—that part of a signal section that protects the light source and other required components.

236. **Signal Indication**—the illumination of a signal lens or equivalent device.

237. **Signal Lens**—that part of the signal section that redirects the light coming directly from the light source and its reflector, if any.

238. **Signal Louver**—a device that can be mounted inside a signal visor to restrict visibility of a signal indication from the side or to limit the visibility of the signal indication to a certain lane or lanes, or to a certain distance from the stop line.

239. **Signal Phase**—the right-of-way, yellow change, and red clearance intervals in a cycle that are assigned to an independent traffic movement or combination of movements.

240. **Signal Section**—the assembly of a signal housing, signal lens, if any, and light source with necessary components to be used for displaying one signal indication.

241. **Signal Sequence (Sequence of Indications)**—the order of appearance of signal indications during successive intervals of a signal cycle.

242. **Signal System**—two or more traffic control signals operating in signal coordination.

243. **Signal Timing**—the amount of time allocated for the display of a signal indication.

244. **Signal Visor**—that part of a signal section that directs the signal indication specifically to approaching traffic and reduces the effect of direct external light entering the signal lens.

245. **Signing**—individual signs or a group of signs, not necessarily on the same support(s), that supplement one another in conveying information to road users.

246. **Simultaneous Preemption**—notification of approaching rail traffic is forwarded to the highway traffic signal controller unit or assembly and railroad or light rail transit active warning devices at the same time.

247. **Site Roadways Open to Public Travel**—Roadways and bikeways on sites of shopping centers, office parks, airports, schools, universities, sports arenas, recreational parks, and other similar business, governmental, and/or recreation facilities that are publicly or privately owned but where the public is allowed to travel without full-time access restrictions. Two types of roadways are not included in this definition: (1) roadways where access is restricted at all times by gates and/or guards to residents, employees, or other specifically-authorized persons; and (2) private highway-rail grade crossings. Site roadways open to public travel do not include parking areas (see definition in this Section), including the driving aisles (see definition in this Section) within those parking areas.

248. **Special-Purpose Road**—a low-volume, low-speed road that serves recreational areas or resource development activities.

249. **Speed**—speed is defined based on the following classifications:

   (a) **Average Speed**—the summation of the instantaneous or spot-measured speeds at a specific location of vehicles divided by the number of vehicles observed.

   (b) **Design Speed**—a selected speed used to determine the various geometric design features of a roadway.

   (c) **85th-Percentile Speed**—the speed at or below which 85 percent of the motor vehicles travel.

   (d) **Operating Speed**—a speed at which a typical vehicle or the overall traffic operates. Operating speed might be defined with speed values such as the average, pace, or 85th-percentile speeds.

   (e) **Pace**—the 10 mph speed range representing the speeds of the largest percentage of vehicles in the traffic stream.

250. **Speed Limit**—the maximum (or minimum) speed applicable to a section of highway as established by law or regulation.

251. **Speed Zone**—a section of highway with a speed limit that is established by law or regulation, but which might be different from a legislatively-specified statutory speed limit.

252. **Splitter Island**—a median island used to separate opposing directions of traffic entering and exiting a roundabout.

253. **Station Crossing**—a pathway grade crossing that is associated with a station platform.

254. **Statutory Speed Limit**—a speed limit established by legislative action (such as Federal or State law) that typically is applicable for a particular class of highways with specified design, functional, jurisdictional, and/or location characteristics and that is not necessarily displayed on Speed Limit signs.

255. **Steady (Steady Mode)**—the continuous display of a signal indication for the duration of an interval, signal phase, or consecutive signal phases.

256. **Stop Line**—a solid white pavement marking line extending across approach lanes to indicate the point at which a stop is intended or required to be made.

257. **Street**—see Highway.

258. **Supplemental Signal Face**—a signal face that is not a primary signal face but which is provided for a given approach or separate turning movement to enhance visibility or conspicuity.

259. **Swing Gate**—a self-closing fence-type gate designated to swing open away from the track area and return to the closed position upon release.

260. **Symbol**—the approved design of a pictorial or graphical representation of a specific traffic control message for signs, pavement markings, traffic control signals, or other traffic control devices, as shown in the MUTCD.

261. **Temporary Traffic Control Signal**—see Highway Traffic Signal.

262. **Temporary Traffic Control Zone**—an area of a highway, pedestrian or bicycle facility where road user conditions are changed because of a work zone or incident by the use of temporary traffic control devices, flaggers, uniformed law enforcement officers, or other authorized personnel.

263. **Theoretical Gore**—a longitudinal point at the upstream end of a neutral area at an exit ramp or channelized turn lane where the channelizing lines that separate the ramp or channelized turn lane from the adjacent through lane(s) begin to diverge, or a longitudinal point at the downstream end of a neutral area at an entrance ramp or channelized entering lane where the channelizing lines that separate the ramp or channelized entering lane from the adjacent through lane(s) intersect each other.

264. **Through Train**—a train movement that continues without stopping or reversing direction throughout the entire length of the rail traffic detection circuit length approaching a highway-rail grade crossing.

265. **Timed Exit Gate Operating Mode**—a mode of operation where the exit gate descent at a grade crossing is based on a predetermined time interval.

266. **Toll Booth**—a shelter where a toll attendant is stationed to collect tolls or issue toll tickets. A toll booth is located adjacent to a toll lane and is typically set on a toll island.

267. **Toll Collection**—manual or electronic methods and elements used to collect a fee for use of a toll facility. Toll collection methods include:

    (a) **Electronic Toll Collection (ETC)**—a cashless system for automated collection of tolls from moving or stopped vehicles through wireless technologies such as radio-frequency communication or optical scanning. ETC systems are classified as one of the following:

        (1) Systems that require users to have registered toll accounts, with the use of equipment inside or on the exterior of vehicles, such as a transponder or barcode decal, that communicates with or is detected by roadside or overhead receiving equipment, or with the use of license plate optical scanning, to automatically deduct the toll from the registered user account,

        (2) Systems that do not require users to have registered toll accounts because vehicle license plates are optically scanned and invoices for the toll amount are typically sent through postal mail to the address of the vehicle owner, or

        (3) Systems that allow electronic toll collection for both registered and non-registered toll accounts.

    (b) **Open-Road Tolling (ORT)**—a system designed to allow electronic toll collection (ETC) from vehicles traveling at posted speeds. Open-road tolling might be used on toll roads or toll facilities in conjunction with toll plazas. Open-road tolling is also typically used on managed lanes and on toll facilities that only accept payment by ETC.

    (c) **Manual Toll Collection**—a system of toll collection from stopped vehicles through acceptance of cash, toll tickets, tokens, or credit cards, and may involve issuance of receipts. Toll collection may be by a machine or toll booth attendant.

        (1) **Toll-Ticket System**—a toll system in which the user of a toll road must stop to receive a ticket from a machine or toll booth attendant upon entering the toll facility. The ticket denotes the user's point of entry and, upon exiting the toll system, the user surrenders the ticket and is charged a toll based on the distance traveled between the points of entry and exit.

        (2) **Attended Lane (Manual Lane)**—a toll lane adjacent to a toll booth occupied by a human toll collector who makes change, issues receipts, and performs other toll-related functions. Attended lanes at toll plazas typically require vehicles to stop to pay the toll.

        (3) **Exact Change Lane (Automatic Lane)**—a non-attended toll lane that has a receptacle into which road users deposit coins totaling the exact amount of the toll. Exact Change lanes at toll plazas typically require vehicles to stop to pay the toll.

268. **Toll Island**—a raised island on which a toll booth or other toll collection and related equipment are located.

269. **Toll Lane**—an individual lane located within a toll plaza in which a toll payment is collected or, for toll-ticket systems, a toll ticket is issued.

270. **Toll Plaza**—the location at which tolls are collected consisting of a grouping of toll booths, toll islands, toll lanes, and, typically, a canopy. Toll plazas might be located on highway mainlines or on interchange ramps. A mainline toll plaza is sometimes referred to as a barrier toll plaza because it interrupts the traffic flow.

271. **Toll Road (Facility)**—a road or facility that is open to traffic only by payment of a user toll or fee.

272. **Traffic**—pedestrians, bicyclists, ridden or herded animals, vehicles, streetcars, and other conveyances either singularly or together while using for purposes of travel any highway or site roadway open to public travel.

273. **Traffic Control Device**—all signs, signals, markings, channelization devices, or other devices that use colors, shapes, symbols, words, sounds, and/or tactile information for the primary purpose of communicating a regulatory, warning, or guidance message to road users on a street, highway, pedestrian facility, bikeway, pathway, or site roadway open to public travel. Section 1A.02 contains information regarding items that are not traffic control devices.

274. **Traffic Control Signal (Traffic Signal)**—see Highway Traffic Signal.

275. **Train**—one or more locomotives coupled, with or without cars, that operates on rails or tracks and to which all other traffic must yield the right-of-way by law at highway-rail grade crossings.

276. **Transverse Markings**—pavement markings that are generally placed perpendicular and across the flow of traffic such as shoulder markings; word, symbol, and arrow markings; stop lines; crosswalk lines; parking space markings; and others.

277. **Traveled Way**—the portion of the roadway for the movement of vehicles, exclusive of the shoulders, berms, sidewalks, and parking lanes.

278. **Turn Bay**—a lane for the exclusive use of turning vehicles that is formed on the approach to the location where the turn is to be made. In most cases where turn bays are provided, drivers who desire to turn must move out of a through lane into the newly-formed turn bay in order to turn. A through lane that becomes a turn lane is considered to be a lane drop rather than a turn bay.

279. **Two-Stage Bicycle Turn Box**—a designated area at an intersection intended to provide bicyclists a place to wait for traffic to clear before proceeding in a different direction of travel.

280. **Uncontrolled Approach**—an approach on which vehicles are not controlled by a traffic control signal, hybrid beacon, STOP sign, or YIELD sign.

281. **Upstream**—a term that refers to a location that is encountered by traffic prior to a downstream location as it flows in an "upstream to downstream" direction. For example, "the upstream end of a lane line separating the turn lane from a through lane on the approach to an intersection" is the end of the line that is furthest from the intersection.

282. **Urban Street**—a type of street normally characterized by relatively low speeds, wide ranges of traffic volumes, narrower lanes, frequent intersections and driveways, significant pedestrian traffic, and more businesses and houses.

283. **Variable Message Sign**—see Changeable Message Sign.

284. **Vehicle**—every device in, upon, or by which any person or property can be transported or drawn upon a highway, except trains and light rail transit operating in exclusive or semi-exclusive alignments. Light rail transit equipment operating in a mixed-use alignment, to which other traffic is not required to yield the right-of-way by law, is a vehicle.

285. **Vibrotactile Pedestrian Device**—an accessible pedestrian signal feature that communicates, by touch, information about pedestrian timing using a vibrating surface.

286. **Visibility-Limited Signal Face or Visibility-Limited Signal Section**—a type of signal face or signal section designed (or shielded, hooded, or louvered) to restrict the visibility of a signal indication from the side, to a certain lane or lanes, or to a certain distance from the stop line.

287. **Walk Interval**—an interval during which the WALKING PERSON (symbolizing WALK) signal indication is displayed.

288. **Warning Light**—a portable, powered, yellow, lens-directed, enclosed light that is used in a temporary traffic control zone in either a steady burn or a flashing mode.

289. **Warning Sign**—a sign that gives notice to road users of a situation that might not be readily apparent.

290. **Warrant**—a warrant describes a threshold condition based upon average or normal conditions that, if found to be satisfied as part of an engineering study, shall result in analysis of other traffic conditions or factors to determine whether a traffic control device or other improvement is justified. Warrants are not a substitute for engineering judgment. The fact that a warrant for a particular traffic control device is met is not conclusive justification for the installation of the device.

291. **Wayside Horn System**—a stationary horn (or a series of horns) located at a grade crossing that is used in conjunction with train-activated or light rail transit-activated warning systems to provide audible warning of approaching rail traffic to road users on the highway or pathway approaches to a grade crossing, either as a supplement or alternative to the sounding of a locomotive horn.

292. **Worker**—a person on foot whose duties place him or her within the right-of-way of a street, highway, or pathway, such as: construction and maintenance forces; survey crews; utility crews;

responders to incidents within the right-of-way; and law enforcement personnel when directing traffic, investigating crashes, and handling lane closures, obstructed roadways, and disasters within the right-of-way.

293. Wrong-Way Arrow—a slender, elongated, white pavement marking arrow placed upstream from the ramp terminus to indicate the correct direction of traffic flow. Wrong-way arrows are intended primarily to warn wrong-way road users that they are going in the wrong direction.

294. Yellow Change Interval—the first interval following the green or flashing arrow interval during which the steady yellow signal indication is displayed.

295. Yield Line—a row of solid white isosceles triangles pointing toward approaching vehicles extending across approach lanes to indicate the point at which the yield is intended or required to be made.

Section 1C.03 <u>Meanings of Acronyms and Abbreviations Used in this Manual</u>

Standard:

01 The following acronyms and abbreviations, when used in this Manual, shall have the following meanings:

1. AADT—annual average daily traffic
2. AASHTO—American Association of State Highway and Transportation Officials
3. AC—alternating current
4. ADA—Americans with Disabilities Act
5. ADAS—Advanced Driver Assistance Systems
6. ADS—Automated Driving System
7. ADT—average daily traffic
8. AFAD—Automated Flagger Assistance Device
9. ANSI—American National Standards Institute
10. AREMA—American Railway Engineering and Maintenance-of-Way Association
11. AV—automated vehicle
12. $cd/lx/m^2$—candelas per lux per square meter
13. CFR—Code of Federal Regulations
14. CMS—changeable message sign
15. dBA—A-weighted decibels
16. DC—direct current
17. DDT—Dynamic Driving Task
18. EPA—Environmental Protection Agency
19. ETC—electronic toll collection
20. EV—electric vehicle
21. FHWA—Federal Highway Administration
22. FRA—Federal Railroad Administration
23. ft—foot or feet
24. FTA—Federal Transit Administration
25. HOV—high-occupancy vehicle
26. IEEE—Institute of Electrical and Electronics Engineers
27. IES—Illuminating Engineering Society
28. ILEV—inherently low emission vehicle
29. in—inch(es)
30. ISEA—International Safety Equipment Association
31. ITE—Institute of Transportation Engineers
32. ITS—intelligent transportation systems
33. L—taper length
34. LED—light-emitting diode
35. LP—liquified petroleum
36. LRT—light rail transit
37. mi—mile(s)
38. MPH or mph—miles per hour
39. MUTCD—Manual on Uniform Traffic Control Devices for Streets and Highways
40. N—length of one line segment plus one gap of a broken line
41. NCEES—National Council of Examiners for Engineering and Surveying
42. NCHRP—National Cooperative Highway Research Program
43. ODD—Operational Design Domain
44. OPM—U.S. Office of Personnel Management

45. **ORT—open-road tolling**
46. **PCMS—portable changeable message sign**
47. **PRT—perception-response time**
48. **RRFB—rectangular rapid-flashing beacon**
49. **RV—recreational vehicle**
50. **SAE—Society of Automotive Engineers**
51. **SHV—Specialized Hauling Vehicle**
52. **SPF—safety performance function**
53. **TA—Typical Application**
54. **TDD—telecommunication device for the deaf**
55. **TRB—Transportation Research Board**
56. **TTC—temporary traffic control**
57. **U.S.—United States**
58. **U.S.C.—United States Code**
59. **USDOT—United States Department of Transportation**
60. **UVC—Uniform Vehicle Code**
61. **VPH or vph—vehicles per hour**
62. **V2I—vehicle-to-infrastructure**

CHAPTER 1D. PROVISIONS APPLICABLE TO TRAFFIC CONTROL DEVICES IN GENERAL

Section 1D.01 Purpose and Principles of Traffic Control Devices

Support:

01   The purpose of traffic control devices, as well as the principles for their use, is to promote highway safety, inclusion and mobility of all road users, and efficiency by providing for the orderly movement of road users on streets, highways, bikeways, and site roadways open to public travel throughout the Nation. Section 1A.03 contains additional information on target road users.

02   This Manual contains the basic principles that govern the design and use of traffic control devices for all streets, highways, bikeways, and site roadways open to public travel (see definition in Section 1C.02) regardless of type or class or the public agency, official, or owner having jurisdiction. The text of this Manual specifies the restriction on the use of a device if it is intended for limited application or for a specific system. It is important that these principles be given primary consideration in the selection and application of each device.

Guidance:

03   *To be effective, a traffic control device should:*

    A. *Fulfill a need;*
    B. *Command attention;*
    C. *Convey a clear, simple meaning;*
    D. *Command respect from road users; and*
    E. *Give adequate time for proper response.*

04   *Design, placement, operation, maintenance, and uniformity are aspects that should be carefully considered in order to maximize the ability of a traffic control device to be consistent with the five principles listed in Paragraph 3 of this Section. Vehicle speed and road-user types should be carefully considered as an element that governs the design, operation, placement, and location of various traffic control devices.*

05   *The proper use of traffic control devices should provide the road user with the information necessary to safely, efficiently, and lawfully use the streets, highways, pedestrian facilities, and bikeways.*

Standard:

06   **Traffic control devices used on site roadways open to public travel shall have the same shape, color, and meaning as those required by the MUTCD for use on public highways, except as provided otherwise elsewhere in this Manual. Sign size exceptions are noted in each Part as applicable.**

Section 1D.02 Responsibility and Authority for Traffic Control Devices

Standard:

01   **The responsibility for the design, placement, operation, maintenance, and uniformity of traffic control devices in compliance with the provisions of this Manual shall rest with the public agency or the official having jurisdiction, or, in the case of site roadways open to public travel, with the private owner or private official having jurisdiction.**

02   **All regulatory traffic control devices shall be supported by laws, ordinances, or regulations.**

03   **Traffic control devices, public announcements or notices, and other signs or messages within the highway right-of-way shall be placed only as authorized by a public authority or the official having jurisdiction, or, in the case of site roadways or private toll roads open to public travel, by the private owner or private official having jurisdiction, for the purpose of regulating, warning, or guiding traffic.**

04   **When the public agency or the official having jurisdiction over a street or highway or, in the case of site roadways open to public travel, the private owner or private official having jurisdiction, has granted proper authority, others such as contractors and public utility companies shall be allowed to install temporary traffic control devices in temporary traffic control zones. Such traffic control devices shall comply with the provisions of this Manual.**

05   **Signs and other devices that do not have any traffic control purpose that are placed within the highway right-of-way shall not be located where they will interfere with, or detract from, traffic control devices.**

Support:

06   States are encouraged to adopt, through policy or legislation, the provisions of 23 CFR 750.108 that restrict outdoor advertising from resembling traffic control devices.

Section 1D.03 Engineering Study and Engineering Judgment

Support:

01   Definitions of professional engineer, engineering study, and engineering judgment are provided in Section 1C.02.

02   The application of engineering study and engineering judgment is a fundamental principle of the use of traffic control devices. It is for this reason that, in most cases, the selection of a particular device is not required by a Standard provision, but is determined by engineering study or engineering judgment. Many Standard provisions in this Manual specifically require, by explicit language in the individual provisions or by implication, the application of engineering study or engineering judgment in applying those Standards. Site-specific conditions might result in the determination that it is impossible or impracticable to comply with a Standard at that location. In such a case, a deviation from the requirement of a particular Standard at that location might be the only possibility. In such limited, specific cases, the deviation is allowed, provided that the agency or official having jurisdiction fully documents, through an engineering study, the engineering basis for the deviation.

Standard:

03   **This Manual describes the application of traffic control devices, but shall not be a legal requirement for their installation.**

Support:

04   The MUTCD does not mandate, and is not intending to imply, that an engineer must make the final decision whether to implement or execute the determination or advice of an engineer by installing or constructing the traffic control device to the engineer's specification in the field. Rather, the engineer, individual under supervision of an engineer, or other individual as duly authorized by State law to engage in the practice of engineering, develops an engineering-based solution that includes the specifications for selection and placement of traffic control devices, but the responsibility for a final decision to implement that solution rests with the agency having jurisdiction over the roadway, after consultation with and based on advice from the engineer.

Guidance:

05   *The decision to use a particular device at a particular location should be made on the basis of either an engineering study or the application of engineering judgment by an engineer, someone under the direct supervision of an engineer, or other individual as duly authorized by State law to engage in the practice of engineering. Thus, while this Manual provides Standards, Guidance, and Options for design and application of traffic control devices, this Manual should not be considered a substitute for engineering judgment. Engineering judgment should be exercised in the selection and application of traffic control devices, as well as in the location and design of roads and streets that the devices complement.*

06   *Early in the processes of location and design of roads and streets, engineers should coordinate such location and design with the design and placement of the traffic control devices to be used with such roads and streets.*

07   *Jurisdictions, or owners of site roadways or private toll roads open to public travel, with responsibility for traffic control that do not have an engineer on their staff who is trained and/or experienced in traffic control devices should seek engineering assistance from others, such as the State transportation agency, their county, a nearby large city, or a traffic engineering consultant.*

Support:

08   The provisions of this Manual are intended to be interpreted and applied by engineers or those under the supervision of an engineer. The construction of the provisions of this Manual, therefore, are informed by bases referenced in Paragraphs 9 and 10 of this Section.

09   The National Council of Examiners for Engineering and Surveying (NCEES) has defined the practice of engineering as "any service or creative work requiring engineering education, training, and experience in the application of engineering principles and the interpretation of engineering data to engineering activities that potentially impact the health, safety, and welfare of the public." The practice of engineering is, therefore, subject to regulation in the public interest and is regulated by the State licensing boards in order to safeguard the health, safety, and welfare of the public. The NCEES has defined an engineer as "an individual who is qualified to practice engineering by reason of engineering education, training, and experience in the application of engineering principles and the interpretation of engineering data."

10   The U.S. Office of Personnel Management (OPM) has defined the professional knowledge of engineering as "the comprehensive, in-depth knowledge of mathematical, physical, and engineering sciences applicable to a specialty field of engineering that characterizes a full 4-year engineering program leading to a bachelor's degree, or the equivalent." The OPM has defined professional ability to apply engineering knowledge as "the ability to (a) apply fundamental and diversified professional engineering concepts, theories, and practices to achieve engineering objectives with versatility, judgment, and perception; (b) adapt and apply methods and techniques of related scientific disciplines; and (c) organize, analyze, interpret, and evaluate scientific data in the solution of engineering problems."

11   Requisite technical training in the application of the principles of the MUTCD might be available from the State's Local Technical Assistance Program (LTAP) for needed engineering guidance and assistance.

Section 1D.04 Design of Traffic Control Devices

Guidance:

01   *Devices should be designed so that features such as size, shape, color, composition, lighting or retroreflection, and contrast are combined to draw attention to the devices; so that size, shape, color, and simplicity of message combine to produce a clear meaning; so that legibility and size combine with placement to provide adequate time for response; and so that uniformity, size, legibility, and reasonableness of the message combine to command respect.*

Option:

02   Except for symbols and colors, minor modifications in the specific design elements of a device may be made based on an engineering study or engineering judgment, in accordance with Paragraph 3 of this Section, provided the essential appearance characteristics are preserved.

Guidance:

03   *Aspects of the standard design of a traffic control device should not be modified unless there is a demonstrated need in unusual circumstances, based on an engineering study or engineering judgment.*

Support:

04   An example of acceptably modifying the design of a device would be to modify the Combination Horizontal Alignment/Intersection (W1-10) sign to show intersecting side roads on both sides rather than on just one side of the major road within the curve.

Section 1D.05 Color Code

Support:

01   The following color code establishes general meanings for 11 colors of a total of 13 colors that have been identified as being appropriate for use in conveying traffic control information.

Standard:

02   **The general meaning of the 13 colors shall be as follows:**

    A. **Black—regulation**
    B. **Blue—road-user services guidance, tourist information, and evacuation route**
    C. **Brown—recreational and cultural interest area guidance**
    D. **Coral—reserved for future designation (see Paragraph 4 of this Section)**
    E. **Fluorescent Pink—incident management**
    F. **Fluorescent Yellow-Green—pedestrian warning, bicycle warning, playground warning, school bus warning, and school warning**
    G. **Green—indicated movements or actions permitted and direction guidance**
    H. **Light Blue—reserved for future designation (see Paragraph 4 of this Section)**
    I. **Orange—temporary traffic control**
    J. **Purple—restricted to use only by vehicles with registered electronic toll collection (ETC) accounts**
    K. **Red—stop or prohibition**
    L. **White—regulation**
    M. **Yellow—warning**

03   **These colors shall be used only as prescribed for the specific devices or applications throughout this Manual.**

Support:

04   The two colors for which general meanings have not yet been assigned are being reserved for future applications that will be determined only by the FHWA after consultation with the States, the engineering community, and the general public. The meanings described in this Section are of a general nature. More specific assignments of colors are given in the individual Parts of this Manual relating to each class of devices.

05   Tolerance limits for each color are contained in 23 CFR Part 655, Appendix to Subpart F and are available at the Federal Highway Administration's MUTCD Web site at http://mutcd.fhwa.dot.gov.

Section 1D.06 Public Domain, Copyrights, and Patents

Standard:

01   **Traffic control device design or application provisions contained in this Manual shall be in the public domain. Traffic control devices contained in this Manual shall not be protected by a patent, trademark, or copyright, except for the Interstate Shield, 511 Travel Information pictograph, National Scenic Byway graphic, and any items under the stewardship of or owned by FHWA.**

02   **A traffic control device design or application shall not be eligible for official experimentation (see Section 1B.05) or interim approval (see Section 1B.07) unless it is in the public domain. Express abandonment of any and all forms of proprietary protection, such as patents, trademarks, or copyrights, related to the design and application of the traffic control device shall satisfy the requirement for the traffic control device to be in the public domain.**

03   **The requirement for the traffic control device to be in the public domain shall not apply to individual components used in the assembly or manufacture of the traffic control device.**

Support:

04   The limitation on patented, trademarked, or copyrighted traffic control devices, applies to the message that the device conveys to the road user. If a patent or other protection covers the device's communication to the road user by virtue of its appearance, audible message, or other aspects of the message conveyed (such as the order in which traffic control signal indications change from green to yellow and red), then the device is considered to be protected and not in the public domain. Such a device is precluded from inclusion in this Manual. The purpose of this limitation is to ensure uniformity of the messaging of individually approved traffic control devices. This limitation does not apply to other aspects of a device (such as internal controls, circuitry, electronics, mechanics, or housing) so long as the appearance, audible message, or other aspects of the message conveyed, including the manner of conveyance, remain freely reproducible by all without infringing on any proprietary rights or interests. This Manual does not prohibit such other aspects of a traffic control device that meet the legal requirements from being protected through patent, trademark, or copyright; and does not restrict components, parts, manufacturing processes, or similar aspects of traffic control devices from being patented or otherwise protected. Examples of acceptable protected traffic control device components or parts might include: sign sheeting or retroreflectivity technology, internal electronic components of traffic signal controllers, and breakaway sign support mechanisms.

05   Pictographs, as defined in Section 1C.02, are embedded in traffic control devices, but the pictographs themselves are not considered traffic control devices for the purposes of Paragraph 4 of this Section.

06   Business identification logos, as defined in Section 1C.02, are embedded in traffic control devices, but the logos themselves are not considered traffic control devices for the purposes of Paragraph 4 of this Section.

Section 1D.07 Advertising

Standard:

01   **Traffic control devices or their supports shall not bear any advertising message or any other message that is not related to traffic control.**

Support:

02   Acknowledgment signs (see Section 2H.13), Specific Service signs (see Chapter 2J), and Tourist-Oriented Directional signs (see Chapter 2K) are not considered advertising.

Section 1D.08 Abbreviations Used on Traffic Control Devices

Standard:

01   **When the word messages shown in Table 1D-1 need to be abbreviated in connection with traffic control devices, the abbreviations shown in Table 1D-1 shall be used.**

02   **When the word messages shown in Table 1D-2 need to be abbreviated on a portable changeable message sign, the abbreviations shown in Table 1D-2 shall be used. Unless indicated by an asterisk, these abbreviations shall only be used on portable changeable message signs.**

Guidance:

03   *The abbreviations for the words listed in Table 1D-2 that also show a prompt word should not be used on a portable changeable message sign (or on a static sign if indicated in Table 1D-2 by an asterisk) unless the prompt word shown in Table 1D-2 either precedes or follows the abbreviation, as applicable.*

Standard:

04   **The abbreviations shown in Table 1D-3 shall not be used in connection with traffic control devices because of their potential to be misinterpreted by road users.**

Guidance:

05   *If Table 1D-1 or 1D-2 indicates that more than one abbreviation is allowed for a given word or phrase, the same abbreviation should be used throughout a single jurisdiction.*

06   *Except as otherwise provided in Table 1D-1 or 1D-2 or unless necessary to avoid confusion, periods, commas, apostrophes, question marks, ampersands, and other punctuation marks or characters that are not letters or numerals should not be used in any abbreviation.*

Table 1D-1. Acceptable Abbreviations
General Abbreviations

Word Message	Standard Abbreviation		Word Message	Standard Abbreviation
Afternoon / Evening	PM		Mile(s)	MI
Alternate	ALT		Miles per Hour	MPH
AM Radio	AM		Minimum	MIN
Avenue	Ave, Av*		Minute(s)	MIN, MINS
Bicycle(s)	BIKE, BIKES		Morning / Late Night	AM
Boulevard	Blvd*		Mount	Mt**
Bridge	(See Table 1D-2)		Mountain	Mtn**
CB Radio	CB		National	Natl**
Center	Ctr**		North	N
Circle	Cir*		Northeast	NE
Civil Defense	CD		Northwest	NW
Compressed Natural Gas	CNG		Parkway	Pkwy*
Court	Ct*		Pedestrian(s)	PED, PEDS
Crossing (other than highway-rail)	X-ING		Place	Pl*
Drive	Dr*		Pounds	LBS
East	E		Road	Rd*
Electric Vehicle	EV		Saint	St**
Expressway	Expwy*		South	S
Feet	FT		Southeast	SE
FM Radio	FM		Southwest	SW
Freeway	Fwy*		State, county, or other non-US or non-Interstate numbered route	(See Table 1D-2)
Hazardous Material(s)	HAZMAT, HAZMATS		Street	St*
High Occupancy Vehicle(s)	HOV		Telephone	PHONE
Highway	Hwy*		Temporary	TEMP
Hospital	HOSP		Terrace	Ter*
Hour(s)	HR, HRS		Thruway	Thwy*
Information	INFO		Ton(s)	T
Inherently Low Emission Vehicle	ILEV		Trail	Tr*
International	Intl		Turnpike	Tpk*
Interstate	(See Table 1D-2)		Two-Way Intersection, Two-Way Traffic	2-WAY
Junction / Intersection	JCT		US Numbered Route	(See Table 1D-2)
Lane	(See Table 1D-2)		West	W
Liquified Petroleum Gas	LP-GAS			
Maximum	MAX			

Days of the Week

Day	Standard Abbreviation		Day	Standard Abbreviation
Sunday	SUN		Thursday	THURS***
Monday	MON		Friday	FRI
Tuesday	TUES***		Saturday	SAT
Wednesday	WED			

* Abbreviation shall not be used for any application other than the name of a roadway. See Table 2D-3 for complete list of street name descriptors. Examples include: Bayshore Fwy, Cross County Hwy, Mid-County Pkwy

** Abbreviation shall not be used for any application other than as a descriptor or title within a proper name. Examples include: Vestal Ctr, Mt Hope, Pocono Mtn, Eldorado Natl Forest, St Louis

*** Tuesday and Thursday may be abbreviated on a Changeable Message Sign (CMS) to TUE and THU, respectively, when the number of Characters in a message to be displayed cannot be practically reduced through rewording to fit the number of characters supported by the CMS, such as might occur at times on a portable CMS.

Note: Abbreviations shown in upper- and lower-case lettering may be in all upper-case lettering when displayed on a changeable message sign with lower resolution that will not accommodate lower-case letter forms. See Chapter 2L of this Manual.

Table 1D-2. Abbreviations that Shall be Used Only for Temporary Messages on Portable Changeable Message Signs (Sheet 1 of 2)

Word Message	Standard Abbreviation	Prompt Word Preceding the Abbreviation	Prompt Word Following the Abbreviation	Example
Access	ACCS	—	Road	ACCS ROAD
Ahead	AHD	Fog	—	FOG AHD
Blocked	BLKD	Lane	—	2 LANES BLKD
Bridge	BR*	[Name]	—	BAY BR
Cannot	CANT	—	—	—
Center	CNTR	—	Lane	CNTR LANE, CNTR LN
Chemical	CHEM	—	Spill	CHEM SPILL
Condition	COND	Traffic	—	TRAFFIC COND
Congested	CONG	Traffic	—	TRAFFIC CONG AHD
Construction	CONST	—	Ahead	CONST AHEAD
Crossing	XING	—	—	PED XING
Do Not	DONT	—	—	—
Downtown	DWNTN	—	Traffic	DWNTN TRAFFIC
Eastbound	EAST	Route Number, Road Name	—	I-4 EAST
	E-BND	—	Lane, Traffic	E-BND LANE
Emergency	EMER	—	—	EMER VEHICLES
Entrance, Enter	ENT	—	—	ENT TO I-90
Exit	EX	Next	—	NEXT EX
Express	EXP	—	Lane	EXP LANE OPEN
Frontage	FRNTG	—	Road	FRNTG RD
Hazardous	HAZ	—	Driving	HAZ DRIVING
Highway-Rail Grade Crossing	RR XING	—	—	RR XING
Interstate	I-*	—	[Number]	I-80
It Is	ITS	—	—	—
Lane(s) (travel lanes of a highway)	LN, LNS	Right, Left, Center	—	LEFT LN ONLY 2 RIGHT LNS
Left	LFT	Keep, Next	—	NEXT LFT
	LFT	—	Lane	LFT LANE
Local	LOC	—	Traffic	LOC TRAFFIC ONLY
Lower	LWR	—	Level	LWR LEVEL
Maintenance	MAINT	—	—	ROAD MAINT
Major	MAJ	—	Crash	MAJ CRASH
Minor	MNR	—	Crash	MNR CRASH
Normal	NORM	—	—	—
Northbound	NORTH	Route Number, Road Name	—	US 1 NORTH
	N-BND	—	Lane, Traffic	N-BND TRAFFIC
Oversized	OVRSZ	—	Load	OVRSZ LOAD
Parking	PKING	—	—	—
Pavement	PVMT	Icy	—	ICY PVMT
Prepare	PREP	—	To Stop	PREP TO STOP
Quality	QLTY	Air	—	AIR QLTY
Right	RT	Keep, Next	—	KEEP RT
	RT	—	Lane	RT LANE
Road Work	RD WK	—	Ahead, [Distance]	RD WK 1 MILE
Route	RTE	Best	—	BEST RTE
Service	SERV	—	—	SERV AREA OPEN
Shoulder	SHLDR	—	—	SHLDR CLOSED

Table 1D-2. Abbreviations that Shall be Used Only for Temporary Messages on Portable Changeable Message Signs (Sheet 2 of 2)

Word Message	Standard Abbreviation	Prompt Word Preceding the Abbreviation	Prompt Word Following the Abbreviation	Example
Slippery	SLIP	—	—	—
Southbound	SOUTH	Route Number, Road Name	—	CA 1 SOUTH
	S-BND	—	Lane, Traffic	S-BND TRAFFIC
Speed	SPD	—	—	SPD LIMIT
State, County, or other non-U.S. or non-Interstate numbered route	[Route Abbreviation determined by highway agency]*	—	[Number]**	NY 7, CR 43
Tires With Lugs	LUGS	—	—	—
Traffic	TRAF	—	—	—
Travelers	TRVLRS	—	—	—
Two-Wheeled Vehicles	CYCLES	—	—	—
Upper	UPR	—	Level	UPR LEVEL
U.S. Numbered Route	US*	—	[Number]**	US 202
Vehicle(s)	VEH, VEHS	—	—	—
Warning	WARN	—	—	—
Westbound	WEST	Route Number, Road Name	—	IL 53 WEST
	W-BND	—	Lane, Traffic	W-BND LANES
Will Not	WONT	—	—	—

* Abbreviation, when accompanied by the prompt word, may be used on traffic control devices other than portable message signs. See Table 1D-1 for uses and format.

** A space and no hyphen shall be placed between the abbreviation and the number of the route.

Note: See Chapter 2L of this Manual for additional information on changeable message signs.

Table 1D-3. Unacceptable Abbreviations

Abbreviation	Intended Word	Common Misinterpretation
ACC	Accident	Access (Road)
CLRS	Clears	Colors
DLY	Delay	Daily
FDR	Feeder	Federal
L	Left	Lane (Merge)
LT	Light (Traffic)	Left
PARK	Parking	Park
POLL	Pollution (Index)	Poll
RED	Reduce	Red
STAD	Stadium	Standard
WRNG	Warning	Wrong

Section 1D.09 Placement and Operation of Traffic Control Devices

Standard:

01 **Before any highway, site roadway open to public travel (see definition in Section 1C.02), detour, or temporary route is opened to public travel, all traffic control devices necessary for safe operation shall be in place.**

Option:

02 Temporary traffic control devices, as provided for in Part 6 of this Manual, may be used in place of permanent devices that have yet to be installed for safe operation.

Guidance:

03 *Placement of a traffic control device should be within the road user's view so that adequate visibility is provided. To aid in conveying the proper meaning, the traffic control device should be appropriately positioned with respect to the location, object, or situation to which it applies. The location and legibility of the traffic control device should be such that a road user has adequate time to make the proper response in both day and night conditions.*

04 *Traffic control devices should be placed and operated in a uniform and consistent manner as part of maintaining uniformity in traffic control.*

Support:

05 Inconsistent placement or use of a device can result in disrespect for the device at locations where the device is needed and appropriate.

Guidance:

06 *Unnecessary traffic control devices should be removed. The fact that a device is in good physical condition should not be a basis for deferring needed removal or change.*

Support:

07 Section 2A.02 contains information on excessive use of signs and other considerations that can reduce their effectiveness and the effectiveness of other traffic control devices.

Section 1D.10 Maintenance of Traffic Control Devices

Guidance:

01 *Functional maintenance of traffic control devices should be used to determine if certain devices need to be changed to meet current traffic conditions.*

02 *Physical maintenance of traffic control devices should be performed to retain the legibility and visibility of the device, and to retain the proper functioning of the device.*

Support:

03 Clean, legible, properly-mounted devices in good working condition command the respect of road users.

Section 1D.11 Crashworthiness of Traffic Control Devices and Other Roadside Appurtenances

Standard:

01 **In accordance with various Sections of this Manual, certain traffic control devices and their supports, and/or related appurtenances shall be crashworthy (see definition in Section 1C.02). Crashworthiness provisions in this Manual shall apply to all streets, highways, and site roadways open to public travel.**

Support:

02 Roadside appurtenances include permanent and portable sign supports, other permanent or temporary traffic control devices, and other roadside fixtures that are not traffic control devices, such as longitudinal barriers, bridge railings, and crash cushions, within the clear zone. Crashworthiness of a device or appurtenance is determined by nationally established standards such as the "Manual for Assessing Safety Hardware" (MASH), 2016, AASHTO. Information on the FHWA's policy on crashworthiness of devices on the National Highway System and other roadways is available at the FHWA Office of Safety Web site at https://safety.fhwa.dot.gov/roadway_dept/countermeasures/reduce_crash_severity/policy_memo_guidance.cfm.

CHAPTER 2A. GENERAL

Section 2A.01 Function and Purpose of Signs

Support:

01   This Manual contains Standards, Guidance, and Options for the signing of all types of highways, and site roadways open to public travel. The functions of signs are to provide regulations, warnings, and guidance information for road users. Words, symbols, and arrows are used to convey the messages. Signs are not typically used to confirm rules of the road (see Paragraph 4 of this Section).

02   Detailed sign requirements are located in the following Chapters of Part 2:

Chapter 2B—Regulatory Signs, Barricades, and Gates

Chapter 2C—Warning Signs and Object Markers

Chapter 2D—Guide Signs for Conventional Roads

Chapter 2E—Guide Signs for Freeways and Expressways

Chapter 2F—Toll Road Signs

Chapter 2G—Preferential and Managed Lane Signs

Chapter 2H—General Information Signs

Chapter 2I—General Service Signs

Chapter 2J—Specific Service Signs

Chapter 2K—Tourist-Oriented Directional Signs

Chapter 2L—Changeable Message Signs

Chapter 2M—Recreational and Cultural Interest Area Signs

Chapter 2N—Emergency Management Signs

03   Definitions and acronyms that are applicable to signs are provided in Chapter 1C.

Guidance:

04   *Permanent signs should not be used on a frequent basis to confirm rules of the road or statutes. Instead, when determined necessary to advise of new regulations as part of an educational campaign, temporary signs or messages should be used instead of permanent signs. These temporary signs or messages should be used sparingly and only at strategic locations, and should be considered only as a supporting element of a larger educational campaign rather than as the primary source of notification. If engineering judgment determines a need for a permanent sign to distinguish between differing requirements of similar statutes in different jurisdictions, then a sign should be located in the vicinity of the jurisdictional boundary, and should be located away from warning, directional, and higher-priority regulatory signs, so as not to contribute to sign clutter (see Section 2A.20).*

Section 2A.02 Standardization of Application

Support:

01   It is recognized that urban traffic conditions differ from those in rural environments, and in many instances signs are applied and located differently. Where pertinent and practical, this Manual sets forth separate recommendations for urban and rural conditions.

02   Low-volume rural roads typically include access to rural residences, agricultural, recreational, resource management and development (such as mining, logging, and grazing), and local roads in rural areas. On low-volume rural roads, the use of traffic control devices is limited to essential information regarding regulation, warning, and guidance. On low-volume rural roads, it is important to consider the needs of unfamiliar road users for occasional, recreational, and commercial transportation purposes.

Guidance:

03   *Signs should be used only where justified by engineering judgment or studies, as provided in Section 1D.03.*

04   *Results from traffic engineering studies of physical and traffic safety or operational factors should indicate the locations where signs are deemed necessary or desirable.*

05   *Roadway geometric design and sign application should be coordinated so that signing can be effectively placed to give the road user any necessary regulatory, warning, guidance, and other information.*

Standard:

06   **Each standard sign (see Paragraph 1 of Section 2A.04) shall be displayed only for the specific purpose as prescribed in this Manual. Before any new highway, site roadway open to public travel (see definition in Section 1C.02), detour, or temporary route is opened to public travel, all necessary signs shall be in place. Signs required by road conditions or restrictions shall be removed when those conditions cease to exist or the restrictions are withdrawn.**

Section 2A.03 Classification of Signs

Standard:

01   **Signs shall be defined by their function as follows:**
   A. Regulatory signs give notice of traffic laws or regulations.
   B. Warning signs give notice of a situation that might not be readily apparent.
   C. Guide signs show route designations, destinations, directions, distances, services, points of interest, and other geographical, recreational, or cultural information.

Support:

02   Barricades are described in Sections 2B.75 and 6K.07.

03   Gates are described in Section 2B.76.

04   Object markers are described in Section 2C.70.

Section 2A.04 Design of Signs

Support:

01   This Manual shows many standard signs and object markers approved for use on streets, highways, bikeways, and pedestrian crossings. Standard signs and object markers have a standardized design, shape, background, and legend as shown in this Manual.

02   In the provisions for individual standard signs and object markers, the general appearance of the legend, color, and size are shown in the accompanying tables and illustrations, and are not always detailed in the text.

03   Detailed drawings of standard signs, object markers, alphabets, symbols, and arrows (see Figure 2D-3) are contained in the "Standard Highway Signs" publication. Section 1A.05 contains information regarding this publication.

04   The basic requirements of a sign are that it be legible to those for whom it is intended and that it be understandable in time to allow for a proper response. Desirable attributes include:
   A. High visibility by day and night; and
   B. High legibility (adequately-sized letters, symbols, or arrows, and a short legend for quick comprehension by a road user approaching a sign).

05   Standardized colors and shapes are specified so that the several classes of traffic signs can be promptly recognized. Simplicity and uniformity in design, position, and application are essential for a sign to be effective.

Standard:

06   **The term legend shall include all word messages and symbol and arrow designs that are intended to convey specific meanings.**

07   **Uniformity in design shall include shape, color, dimensions, legends, letter style, borders, and illumination or retroreflectivity.**

08   **Standardization of these designs does not preclude further improvement by minor modifications to the orientation of symbols (see Section 2A.09), width of borders, or layout of word messages, but all shapes and colors shall be as indicated.**

09   **All symbols (see Section 2A.09) shall be unmistakably similar to, or mirror images of, the adopted symbol signs, all of which are shown in the "Standard Highway Signs" publication (see Section 1A.05). Symbols and colors shall not be modified unless otherwise provided in this Manual. All symbols, colors, or other design features for signs not shown in the "Standard Highway Signs" publication (see Section 1A.05) shall follow the procedures for experimentation and change described in Chapter 1B.**

10   **Where a standard word message is applicable, the wording shall be as provided in this Manual.**

11   **In situations where word messages are necessary other than those provided in this Manual (see Paragraph 15 of this Section), the signs shall be of the same shape and color as standard signs of the same functional type.**

12   **Where the legend of a standard sign is a symbol or a combination of a symbol and words, an alternative word legend shall not be allowed in place of the symbol, except as otherwise provided in this Manual.**

13   **Where a standard sign provided in this Manual or the "Standard Highway Signs" publication (see Section 1A.05) is applicable, an alternative legend sign or alternative sign design shall not be allowed in place of the standardized legend or design except as provided in this Manual.**

14   **Where a standard sign provided in this Manual or the "Standard Highway Signs" publication (see Section 1A.05) is applicable, but the legend is variable, such as for destination names, an alternative sign design or dimensions shall not be allowed in place of the standardized design for the non-variable elements except as provided in this Manual.**

Option:

15   State and local highway agencies and owners of site roadways open to public travel may develop special word legend signs in situations where engineering judgment determines roadway conditions make it necessary to provide road users with additional regulatory, warning, or guidance information, such as when road users need to be notified of special regulations or warned about a situation that might not be readily apparent. Unlike colors that have not been assigned or symbols that have not been approved for signs, new word legend signs may be used without the need for experimentation.

Support:

16   The message conveyed by some special word legend signs might be unclear to the road user. Although experimentation is not required for such word legends, they might still warrant an evaluation to determine comprehension or possible misinterpretation of the intended message by the road user.

17   Scanning graphics are graphics designed for scanning by machine, and include bar codes, quick-response (QR) codes or other matrix bar-code formats, or similar graphics.

Standard:

18   **Unless otherwise provided in this Manual for a specific sign or as provided in Paragraph 19 of this Section, telephone numbers, Internet addresses, e-mail addresses, domain names, uniform resource locators (URL), metadata tags ("hash-tags"), and scanning graphics (see Paragraph 17 of this Section) for the purpose of obtaining information (other than those for maintenance or inventory purposes per the provisions of Paragraphs 21 through 23 of this Section) shall not be displayed on any sign, plaque, sign panel, or changeable message sign.**

Option:

19   Internet addresses, e-mail addresses, telephone numbers, scanning graphics, or other graphics for the purpose of conveying information may be displayed on the face of signs, plaques, sign panels, and changeable message signs that are oriented away from or otherwise not readily visible to operators of motor vehicles but rather are intended for viewing only by pedestrians, occupants of parked vehicles, and driving automation systems.

Standard:

20   **Pictographs (see definition in Section 1C.02) shall not be displayed on signs except as specifically provided in this Manual for a particular type of sign. Pictographs shall be simple, dignified, and devoid of any advertising and shall not contain any scanning graphics (see Paragraph 17 of this Section) for the purpose of conveying information. When used to represent a political jurisdiction (a State, county, or municipal corporation) the pictograph shall be the official designation adopted by the jurisdiction, except as provided otherwise in this Manual. When used to represent any other type of jurisdiction, the pictograph shall be the official designation adopted by the jurisdiction. When used to represent a college or university, the pictograph shall be the official seal adopted by the institution. College or university pictographs shall not include pictorial representations of university or college programs, or athletic mascots.**

21   **No items other than official traffic control signs, inventory stickers or decals, sign installation dates, manufacturer name, sign sizes, sign designations, anti-vandalism stickers, inventory or maintenance codes, and maintenance-related scanning graphics shall be mounted on the back of a sign.**

Option:

22   The date of fabrication, sign designation, sign size, and/or manufacturer name may be displayed on the front of a sign face in accordance with the provisions of Paragraph 23 of this Section.

Standard:

23   **If displayed on the sign face, the date of fabrication, sign designation, sign size, manufacturer name, or similar maintenance and inventory information shall be completely within the border or inset along the bottom edge of the sign. The letter height or scanning graphic shall not exceed ¾ of the width of the border or inset or, if no border is used, shall not exceed 1.75 inches and shall be within 2 inches of the edge of the sign. The color of the lettering within the border shall be the same as the color of the sign background. The color of the lettering or scanning graphic within the inset shall be the same as the color of the sign border. For changeable message signs or blank-out signs, such information, if displayed, shall be embossed in a non-contrasting color in the housing of the sign.**

Section 2A.05 Shapes

Standard:

01 **Particular shapes, as shown in Table 2A-1, shall be used exclusively for specific signs or series of signs, unless otherwise provided in this Manual for a particular sign or class of signs.**

02 **The Crossbuck is a shape exclusive to the Grade Crossing (R15-1) sign and shall not be obscured by mounting a different shape sign on the back of the Crossbuck (see Section 8B.03).**

Guidance:

03 *Shapes that are exclusive to a particular sign (such as an octagon for STOP, a pennant for NO PASSING ZONE, or a circle for Railroad Advance) should not be obscured by another sign mounted on the back of the same assembly protruding or extending beyond the edge of the sign with the exclusive shape. The following methods should be considered in lieu of mounting a sign on the back of another sign that would obscure the exclusive shape of the sign:*

    A. Install the signs on separate mountings to maintain the exclusive shape.

    B. Increase the size of the sign with the exclusive shape so the sign installed on the back does not obscure its shape.

    C. Increase the mounting height of the sign with the exclusive shape to allow the installation of a back-mounted sign below the bottom edge while still ensuring the minimum required mounting height for the lower sign.

04 *Where the lateral space available in which to install a standard sign is constrained, such as mounting on a narrow median barrier or adjacent to a retaining wall, the following methods should be considered to maintain the shape of the sign:*

    A. Angle the sign up to 45 degrees toward the roadway while still maintaining adequate legibility.

    B. Install the sign at a different location that still provides adequate advance warning, supplementing the sign with a Distance plaque (see Section 2C.61), if appropriate.

    C. Reduce the size of the sign, but supplement it with a duplicate sign on the opposite side of the roadway (see Section 2A.11).

    D. In addition to either angling or reducing the size of the sign, supplement it with a duplicate warning sign and Distance plaque at an upstream location.

    E. Mount the sign asymmetrically on the sign support, such as when the support is mounted on a bridge parapet or railing, such that the edge of the sign does not overhang the roadway, shoulder, or other areas used by bicyclists or pedestrians.

Option:

05 Where the shape of the sign cannot be maintained due to lateral constraints, the following methods may be considered:

    A. For warning signs or other types of signs displayed in a horizontally-oriented rectangle, the legend may be displayed in a vertically-oriented rectangle.

    B. When mounted overhead, the word legend for a standard warning sign may be displayed in a horizontally-oriented rectangle.

Table 2A-1. Use of Sign Shapes

Shape	Signs
Octagon*	Stop (R1-1)**
Equilateral Triangle (downward-pointing)*	Yield (R1-2)**
Circle*	Grade Crossing Advance Warning (W10-1)**
Pennant (Isosceles Triangle with longer axis horizontal, pointed right)*	No Passing Zone (W14-3)**
Pentagon (upward-pointing)*	School (S1-1) (squared bottom corners)** County Route (M1-6) (tapered lower sides)**
Crossbuck (two rectangles in a perpendicular "X" configuration)*	Grade Crossing (R15-1)**
Diamond	Warning Series
Rectangle (including square)	Regulatory Series Guide Series*** Warning Series
Trapezoid*	Recreational and Cultural Interest Area Guide Series (isosceles or right-angled) National Forest Route Sign (M1-1) (isosceles)**

* This shape shall be limited exclusively to the sign(s) indicated.

** This sign shall be exclusively the shape shown.

*** Guide series includes general service, specific service, tourist-oriented directional, general information, recreational and cultural interest area, and emergency management signs.

Note: Signs with standardized designs shall not be modified to accommodate a different shape except as provided in this Manual.

Support:

06   Provisions for mounting height of signs that overhang any portion of the traveled way are contained in Section 2A.15.

07   Provisions for lateral offset are contained in Section 2A.16.

Standard:

08   **Modifications to sign shapes, such as cutting off the left and right points of a diamond, shall not be allowed.**

Option:

09   Where the methods described in Paragraph 3 of this Section are impracticable, the legend of the warning sign may be displayed in a vertically-oriented rectangle.

Section 2A.06 Colors

Standard:

01   **The colors to be used on signs and their specific uses on signs shall be as provided in the applicable Sections of this Manual. The color coordinates and values shall be as described in 23 CFR, Part 655, Subpart F, Appendix.**

02   **Colors (see Section 1D.05) shall be consistent across the face of a sign or a sign panel. Color gradients (smooth or defined gradual transitions either within a color or transition to another color) shall not be allowed, except as specifically provided in Section 2J.03 for business identification sign panels.**

Support:

03   Common uses of sign colors are shown in Table 2A-2. Color schemes on specific signs are shown in the illustrations located in each applicable Chapter.

04   Whenever white is specified in this Manual or in the "Standard Highway Signs" publication (see Section 1A.05) as a color, it is understood to include silver-colored retroreflective coatings or elements that reflect white light.

05   The colors coral and light blue are being reserved for uses that will be determined in the future by the Federal Highway Administration.

06   Information regarding color coding of destinations on guide signs, including community wayfinding signs, is contained in Chapter 2D.

Option:

07   The approved fluorescent version of the standard red, yellow, green, or orange color may be used as an alternative to the corresponding standard color.

Section 2A.07 Dimensions

Support:

01   The "Standard Highway Signs" publication (see Section 1A.05) prescribes design details for different sizes of each sign or plaque depending on the type of traffic facility, including bikeways. Smaller sizes are designed to be used on bikeways and some other off-road applications. Larger sizes are designed for use on freeways and expressways, and can also be used in oversized applications to enhance road user safety and convenience on other facilities, especially on multi-lane divided highways and on undivided highways having five or more lanes of traffic and/or high speeds. The intermediate sizes are designed to be used on other highway types. Minimum sizes of signs and plaques for specific applications are prescribed in the various sign size tables in each Chapter of this Manual.

Standard:

02   **The sign dimensions prescribed in the sign size tables in the various Parts and Chapters in this Manual and in the "Standard Highway Signs" publication (see Section 1A.05) shall be used unless engineering judgment determines that other sizes are appropriate in accordance with the following. Except as provided in Paragraph 3 of this Section, where engineering judgment determines that sizes smaller than the prescribed dimensions are appropriate for use, the sign dimensions shall not be less than the minimum dimensions specified in this Manual. The sizes shown in the Minimum columns that are smaller than the sizes shown in the Conventional Road columns in the various sign size tables in this Manual shall only be used on low-speed roadways, alleys, site roadways open to public travel, and on low-volume rural roads with operating speeds of 30 mph or less; and only where the reduced legend size would be adequate for the regulation or warning or where physical conditions preclude the use of larger sizes.**

Table 2A-2. Common Uses of Sign Colors

Type of Sign	Legend								Background										
	Black	Green	Red	White	Yellow	Orange	Fluorescent Yellow-Green	Fluorescent Pink	Black	Blue	Brown	Green	Orange[1]	Red[1]	White	Yellow[1]	Purple	Fluorescent Yellow-Green	Fluorescent Pink
Regulatory	X		X	X					X					X	X				
Prohibitive			X	X[2]										X[2]	X				
Permissive		X													X				
Warning	X															X			
Pedestrian	X															X		X	
Bicycle	X															X		X	
Guide				X								X							
Interstate Route				X						X					X				
State Route	X														X				
U.S. Route	X														X				
County Route					X					X									
Forest Route				X							X								
Street Name				X								X							
Destination				X								X							
Reference Location				X								X							
Information				X								X			X				
Evacuation Route				X								X							
Road User Service				X								X							
Recreational				X							X	X							
Temporary Traffic Control	X												X						
Incident Management	X												X						X
School	X																	X	
ETC-Account Only	X																X[5]		

Changeable Message Signs																				
Regulatory			X[4]	X					X										
Warning				X					X										
Temporary Traffic Control				X	X				X										
Guide				X					X			X[3]							
Motorist Services				X					X	X[3]									
Incident Management				X		X			X										
School, Pedestrian, Bicycle				X	X				X										

[1] Fluorescent versions of these background colors may also be used.

[2] Legend and background color combination for use only as identified for specific signs in this Manual or Standard Highway Signs.

[3] These alternative background colors would be provided by blue or green lighted pixels such that the entire CMS would be lighted, not just the legend.

[4] Red is used only for the circle and diagonal or other red elements of a similar static regulatory sign.

[5] The use of purple on signs is restricted per the provisions of Chapter 2F.

Notes:

1. The purpose of the information in this table is to provide a general overview of common color combinations. The color combinations and orientations for signs with standardized designs shall not be modified. For signs with unique legends, the shape and color shall be the same as standard signs of the same functional type.

2. The colors shown for changeable message signs are for those with electronic displays.

Option:

03　For alleys with restrictive physical conditions and vehicle use that limits installation of the Minimum size sign (or the Conventional Road size sign if no Minimum size is shown), both the sign height and the sign width may be decreased by up to 6 inches.

Guidance:

04　*The sizes shown in the Freeway and Expressway columns in the various sign size tables in this Manual should also be used for other higher-speed applications on conventional roads, based upon engineering judgment, to provide larger signs for increased visibility and recognition.*

05　*The sizes shown in the Oversized columns in the various sign size tables in this Manual should be used for those special applications where speed, volume, or other factors result in conditions where increased emphasis, improved recognition, or increased legibility is needed, as determined by engineering judgment or study.*

06　*Except as provided in Paragraph 7 of this Section, and where specifically prohibited in this Manual, increases above the minimum prescribed sizes should be used where greater legibility or emphasis is needed. If signs larger than the prescribed sizes are used, the overall sign dimensions should be increased in 6-inch increments.*

Standard:

07　**Where a maximum allowable sign size is prescribed, increases in sign size above the maximum size shall not be allowed.**

08　**Where engineering judgment determines that sizes that are different from the minimum prescribed dimensions are appropriate for use, standard shapes and colors shall be used. Standard proportions shall be retained as much as practicable.**

Guidance:

09　*Except where specifically prohibited in this Manual, when supplemental plaques are installed with larger-sized signs, a corresponding increase in the size of the plaque and its legend should also be made. The resulting plaque size should be approximately in the same relative proportion to the larger-sized sign as the conventional-sized plaque is to the conventional-sized sign.*

Section 2A.08　Word Messages

Standard:

01　**Except as otherwise provided in this Manual, all word messages shall be aligned horizontally across a sign, reading left to right.**

02　**Except as provided in Section 2A.04, all word messages shall use standard wording as shown in this Manual and in the "Standard Highway Signs" publication (see Section 1A.05).**

03　**All sign lettering, numerals, and other characters shall be of the Standard Alphabets as provided in the "Standard Highway Signs" publication (see Section 1A.05), unless otherwise provided in this Manual.**

04　**The sign lettering for names of places, streets, and highways shall be composed of a combination of lower-case letters with initial upper-case letters. The sign lettering for other legends shall be composed of upper-case letters, unless otherwise provided in this Manual for a particular sign or type of message.**

05　**Except as provided in Chapter 2E of this Manual, when a mixed-case legend is used, the nominal loop height of the lower-case letters shall be ¾ of the height of the initial upper-case letter.**

06　**The unique letter forms for each of the Standard Alphabet series shall not be stretched, compressed, warped, or otherwise manipulated.**

Support:

07　Section 2D.03 contains information regarding the acceptable methods of modifying the length of a word for a given letter height and series.

Guidance:

08　*Word messages should be as brief as practical to convey a clear, simple meaning, and the lettering should be large enough to provide the necessary legibility distance. A minimum specific ratio of 1 inch of letter height per 30 feet of legibility distance should be used.*

09　*Abbreviations (see Section 1D.08) should be kept to a minimum, except as otherwise prescribed in this Manual.*

10　*Word messages should not contain periods, apostrophes, question marks, ampersands, or other punctuation or characters that are not letters, numerals, or hyphens unless necessary to avoid confusion.*

Support:

11   Diacritical marks on words or names that are adapted to English are not normally needed on signs for comprehension or navigational purposes.

Option:

12   A legend in a secondary language, in addition to English, may be displayed on the face of signs, plaques, sign panels, and changeable message signs that are oriented away from or otherwise not readily visible to operators of motor vehicles, but rather are intended for viewing only by pedestrians and occupants of parked vehicles.

Guidance:

13   *The solidus (slanted line or forward slash) is intended to be used for fractions only and should not be used to separate words on the same line of legend. Instead, a hyphen should be used for this purpose, such as "TRUCKS - BUSES."*

Standard:

14   **Fractions shall be displayed with the numerator and denominator diagonally arranged about the solidus. The overall height of the fraction is measured from the top of the numerator to the bottom of the denominator, each of which is vertically aligned with the upper and lower ends of the solidus. The overall height of the fraction shall be determined by the height of the numerals within the fraction, and shall be 1.5 times the height of an individual numeral within the fraction.**

15   **Except as otherwise provided in this Manual, distances shall be displayed on signs using fractions of a mile rather than decimals.**

Support:

16   The "Standard Highway Signs" publication (see Section 1A.05) contains details regarding the layouts of fractions on signs.

Guidance:

17   *When initials are used to represent an abbreviation for separate words (such as "U S" for a United States route), the initials should be separated by a space of between ½ and ¾ of the letter height of the initials.*

18   *When an Interstate route is displayed in text form instead of using the route shield, a hyphen should be used for clarity, such as "I-50."*

Support:

19   Letter height is expressed in terms of the height of an upper-case letter. For mixed-case legends (those composed of an initial upper-case letter followed by lower-case letters), the height of the lower-case letters is derived from the specified height of the initial upper-case letter based on a prescribed ratio. Letter heights for mixed-case legends might be expressed in terms of both the upper- and lower-case letters, or in terms of the initial upper-case letter alone. When the height of a lower-case letter is specified or determined from the prescribed ratio, the reference is to the nominal loop height of the letter. The term loop height refers to the portion of a lower-case letter that excludes any ascending or descending stems or tails of the letter, such as with the letters "d" or "q." The nominal loop height is equal to the actual height of a non-rounded lower-case letter whose form does not include ascending or descending stems or tails, such as the letter "x." The rounded portions of a lower-case letter extend slightly above and below the baselines projected from the top and bottom of such a non-rounded letter so that the appearance of a uniform letter height within a word is achieved. The actual loop height of a rounded lower-case letter is slightly greater than the nominal loop height and this additional height is excluded from the expression of the lower-case letter height.

Section 2A.09 <u>Symbols</u>

Standard:

01   **Symbol designs shall in all cases be unmistakably similar to those shown in this Manual and in the "Standard Highway Signs" publication (see Section 1A.05).**

Option:

02   Although most standard symbols are oriented facing left, mirror images of these symbols may be used where the reverse orientation might better convey to road users a direction of movement.

Support:

03   New symbol designs are adopted by the Federal Highway Administration based on research evaluations to determine road user comprehension, sign conspicuity, and sign legibility.

Option:

04   State and/or local highway agencies may conduct research studies to determine road user comprehension, sign conspicuity, and sign legibility in compliance with the provisions for official experimentation (see Section 1B.05) when a new symbol design is under consideration.

Support:

05   Sometimes a change from word messages to symbols requires significant time for public education and transition. Therefore, this Manual sometimes includes the practice of using educational plaques to accompany new symbol signs.

Guidance:

06   *New standard warning or regulatory symbol signs should be accompanied by an educational plaque where engineering judgment determines that the plaque will improve road user comprehension during the transition from word message to symbol signs.*

Option:

07   Educational plaques may be left in place as long as they are in serviceable condition.

Standard:

08   **A symbol used for a given category of signs (regulatory, warning, or guide) shall not be used for a different category of signs, except as specifically authorized in this Manual.**

09   **A recreational and cultural interest area symbol (see Chapter 2M) shall not be used on streets or highways outside of recreational and cultural interest areas.**

10   **A recreational and cultural interest area symbol (see Chapter 2M) shall not be used on any regulatory or warning sign on any street, road, or highway.**

Support:

11   Section 2M.07 contains provisions for the use of recreational and cultural interest area symbols to indicate prohibited activities or items in non-road applications.

Section 2A.10 Sign Borders

Standard:

01   **Unless otherwise provided, signs shall have a border of the same color as the legend in order to outline their distinctive shape and thereby give them easy recognition and a finished appearance.**

02   **The corners of all sign borders shall be rounded, except for STOP signs.**

Guidance:

03   *A dark border on a light background should be set in from the edge, while a light border on a dark background should extend to the edge of the sign. A border for 30-inch signs with a light background should be from ½ to ¾ inch in width, ½ inch from the edge. For similar signs with a light border, a width of 1 inch should be used. For other sizes, the border width should be of similar proportions, but should not exceed the stroke-width of the major lettering of the sign. On signs exceeding 72 x 120 inches in size, the border should be 2 inches wide. On unusually large signs with oversized letter heights, route shields, or other legend elements, the border should be 2.5 inches wide and should not exceed 3 inches in width. Except for STOP signs and as otherwise provided in Section 2E.14, the corners of the sign should be rounded to a radius that is concentric with that of the border.*

Support:

04   Section 2A.12 contains information regarding the use of light-emitting diode (LED) units within the border of a sign.

Section 2A.11 Enhanced Conspicuity for Standard Signs

Option:

01   Based upon engineering judgment, where the improvement of the conspicuity of a standard regulatory, warning, or guide sign is desired, any of the following methods may be used, as appropriate, to enhance the sign's conspicuity (see Figure 2A-1):

    A. Increasing the size of a standard regulatory, warning, or guide sign.

    B. Dual signing of a standard regulatory, warning, or guide sign by adding a second identical sign on the left-hand side of the roadway at the same location.

    C. Adding a solid yellow or fluorescent yellow rectangular header panel above a standard regulatory sign, with the width of the panel corresponding to the width of the standard regulatory sign. A legend of "NOTICE," "STATE LAW," or other appropriate text may be added in black letters within the header panel for a period of time determined by engineering judgment.

    D. Adding a NEW plaque (see Section 2C.60) above a new standard regulatory or warning sign, for a period of time in accordance with Paragraph 3 of this Section, to call attention to the new sign.

Figure 2A-1. Examples of Enhanced Conspicuity for Signs

A – W16-15P plaque above a regulatory or warning sign if the regulation or condition is new

B – Red or orange flags above a regulatory, warning, or guide sign

C – W16-18P plaque above a regulatory sign

D – Solid yellow, solid fluorescent yellow, or diagonally striped black and yellow (or black and fluorescent yellow) strip of retroreflective sheeting around a warning sign

E – Vertical retroreflective strip on sign support

F – Supplemental beacon

G – LEDs in border

    E. Adding one or more red or orange flags (cloth or retroreflective sheeting) above a standard regulatory or warning sign, with the flags oriented at 45 degrees to the vertical.

    F. Adding a solid yellow, a solid fluorescent yellow, or a diagonally-striped black and yellow (or black and fluorescent yellow) strip of retroreflective sheeting at least 3 inches wide around the perimeter of a standard warning sign. This may be accomplished by affixing the standard warning sign on a background that is 6 inches larger than the size of the standard warning sign.

    G. Adding a Warning Beacon (see Section 4S.03) to a standard regulatory (other than a STOP, DO NOT ENTER, WRONG WAY, or a Speed Limit sign), warning, or guide sign.

    H. Adding a Speed Limit Sign Beacon (see Section 4S.04) to a standard Speed Limit sign.

    I. Adding a Stop Beacon (see Section 4S.05) to a STOP, DO NOT ENTER, or WRONG WAY sign.

    J. Adding a rectangular rapid-flashing beacon (see Chapter 4L) to a Pedestrian, School, or Trail warning sign at an uncontrolled marked crosswalk.

    K. Adding light-emitting diode (LED) units within the symbol, legend, or border of a standard regulatory, warning, or guide sign, as provided in Section 2A.12.

    L. Adding a strip of retroreflective material to the sign support in accordance with the provisions of Paragraph 5 of this Section.

    M. Using other methods that are specifically allowed for certain signs as described elsewhere in this Manual.

Support:

02   Sign conspicuity improvements can also be achieved by removing non-essential and illegal signs from the right-of-way (see Section 1D.02), and by relocating signs to provide better spacing. Section 2A.20 contains information on excessive use of signs.

Guidance:

03   *If a NEW plaque is used, it should remain in place for a period of time determined by engineering judgment, but not more than 12 months.*

Standard:

04   **Strobe lights shall not be used to enhance the conspicuity of highway signs.**

05   **If a strip of retroreflective material is used on the sign support, it shall be at least 2 inches in width, it shall be placed for the full length of the support from the sign to within 2 feet above the near edge of the roadway, and its color shall match the background color of the sign, except that the color of the strip for the YIELD and DO NOT ENTER signs shall be red. The retroreflective strip shall not display any legend or other information.**

06   **For a post-mounted sign installation, placing a duplicate sign in the same assembly facing the same direction of traffic shall not be permitted as a method of enhancing conspicuity.**

Section 2A.12 LEDs Used for Conspicuity Enhancement on Standard Signs

Support:

01   This Section regarding light-emitting diode (LED) units applies to the use of illuminated elements that supplement a sign legend to enhance the conspicuity of the sign.

02   LED units that are used to illuminate the full sign display, background, or legend are changeable message signs (CMS), which are covered in Chapters 2B, 2C, and 2L, and Part 7.

03   The application of LED units in compliance with Paragraph 8 of this Section does not create a changeable message sign because the legend of the sign is always displayed when the LED units are not illuminated. Changeable message or blank-out signs whose legends change or extinguish by means of illuminated elements are addressed elsewhere in this Manual.

Option:

04   Light-emitting diode (LED) units may be used individually within the symbol, legend, or border of a sign to enhance the sign conspicuity and legibility (see Section 2A.11).

05   Except as provided in Paragraph 11 of this Section, LED units may either operate continuously or be actuated.

Standard:

06   **Where LED units are used to enhance the conspicuity of a sign, the sign shall otherwise comply with the requirements for retroreflection and illumination for nighttime viewing (see Section 2A.21).**

07   **Except as provided in Paragraphs 16 and 17 of this Section, and for changeable message signs, neither individual LEDs nor groups of LEDs shall be placed within the background area of a sign.**

08   **The application of LEDs to display sign legends or symbols shall use a maximum pitch of 20 millimeters to cover the stroke width of the letter or symbol.**

09   The LEDs shall not protrude outside the sign border or legend when used in such applications, shall have a maximum diameter of ¼ inch, and shall be the following colors based on the type of sign:

    A. White or red, with STOP, YIELD, DO NOT ENTER, or WRONG WAY signs.
    B. White, with other regulatory signs.
    C. White or yellow, with warning signs.
    D. White or green, with guide signs.
    E. White, yellow, or orange, with temporary traffic control signs.
    F. White, yellow, or fluorescent yellow-green, with school area or pedestrian or bicycle warning signs.

10   If flashed, all LED units shall flash simultaneously at a steady rate between 50 and 60 times per minute. All the LED units in a sign legend or border shall be illuminated simultaneously with no sequential (chasing) or variable flash rates (dancing), except as otherwise allowed in this Manual. A cluster of LEDs shall not be used within the border of a sign.

11   Where used in STOP or YIELD signs, flashing LED units shall operate continuously. Actuation of the LED units shall not be allowed.

12   Flashing LED units shall not be used within the legend or border of a Speed Limit sign to indicate that the displayed speed limit is in effect.

13   LED units shall not be used within the legend or border of a sign in conjunction with the phrase WHEN FLASHING in its legend or on a supplemental WHEN FLASHING plaque (see Item E in Paragraph 1 of Section 4S.03 for the use of Warning Beacons to indicate when a regulatory or warning message is in effect).

14   Where LED units are used along the edge of a sign, at least one LED unit shall be placed along each edge of the sign, in addition to one LED unit at each corner of the sign, so that the distinct outline of the sign shape is recognized under nighttime viewing conditions. The LED units along each side of the sign shall be spaced approximately equidistantly. For a circular sign shape, the number of LED units shall clearly form the appearance of a circle and not be perceived as some other shape.

15   The uniformity of the sign design shall be maintained without any decrease in visibility, legibility, or driver comprehension during either daytime or nighttime conditions. The LED units shall have the capability to be dimmed automatically by a timing mechanism or a device sensitive to ambient light (photoelectric cell) such that the LEDs do not reduce the visibility of the sign legend.

Option:

16   For STOP, YIELD, DO NOT ENTER, and WRONG WAY signs, LEDs may be placed within the border or within one border width within the background of the sign.

Support:

17   Section 6D.02 contains information about STOP/SLOW paddles used by flaggers. Section 7D.02 contains information about STOP paddles used by adult crossing guards.

18   Other methods of enhancing the conspicuity of standard signs are described in Section 2A.11.

Section 2A.13 Standardization of Location

Support:

01   Standardization of position cannot always be attained in practice. Examples of heights and lateral locations of signs for typical installations are illustrated in Figure 2A-2, and examples of locations for some typical signs at intersections are illustrated in Figure 2A-3 and all four sheets in Figure 2A-4.

02   Examples of advance signing on intersection approaches are illustrated in all four sheets in Figure 2A-4. Chapters 2B, 2C, and 2D contain provisions regarding the application of regulatory, warning, and guide signs, respectively.

Standard:

03   Signs requiring separate decisions by the road user shall be spaced sufficiently far apart for the appropriate decisions to be made.

Guidance:

04   *One of the factors considered when determining the appropriate spacing of signs should be the posted or 85th-percentile speed.*

05   *Except as provided in Paragraph 8 of this Section, signs should be located on the right-hand side of the roadway where they are easily recognized and understood by road users. Signs in other locations should be considered only as supplementary to signs in the normal locations, except as otherwise provided in this Manual.*

06   *Signs should be individually installed on separate posts or mountings except where:*

    A. One sign supplements another;

Figure 2A-2. Examples of Heights and Lateral Locations of Sign Installations

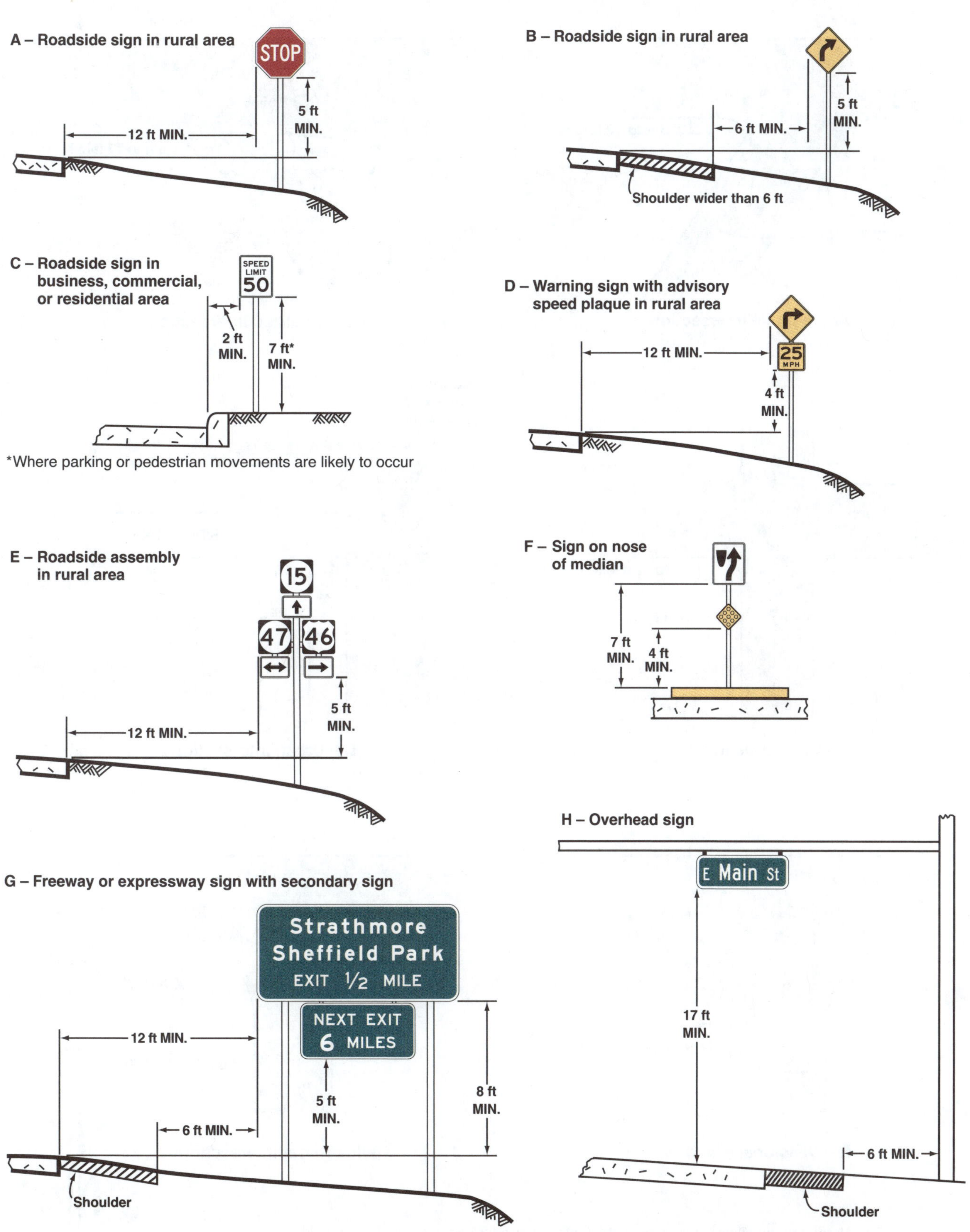

Note: See Section 2A.16 for reduced lateral offset distances that may be used in areas where lateral offsets are limited, and in business, commercial, or residential areas where sidewalk width is limited or where existing poles are close to the curb.

Figure 2A-3. Examples of Locations for Some Typical Signs at Intersections

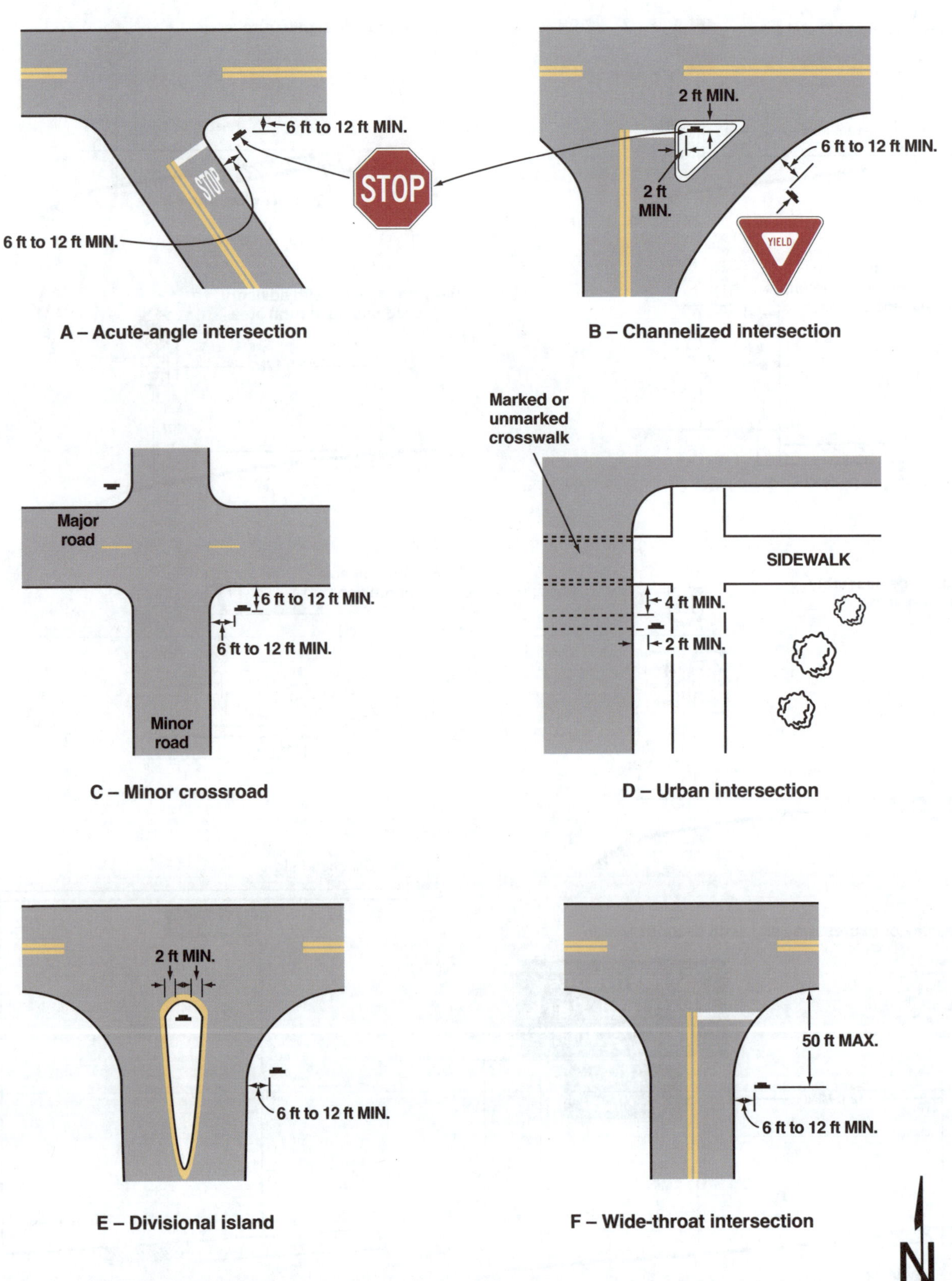

Note: Lateral offset is a minimum of 6 feet measured from the edge of the shoulder, or 12 feet measured from the edge of the traveled way. See Section 2A.16 for lower minimums that may be used in urban areas, or where lateral offset space is limited.

Figure 2A-4. Relative Locations of Regulatory, Warning, and Guide Signs on an Intersection Approach (Sheet 1 of 4)

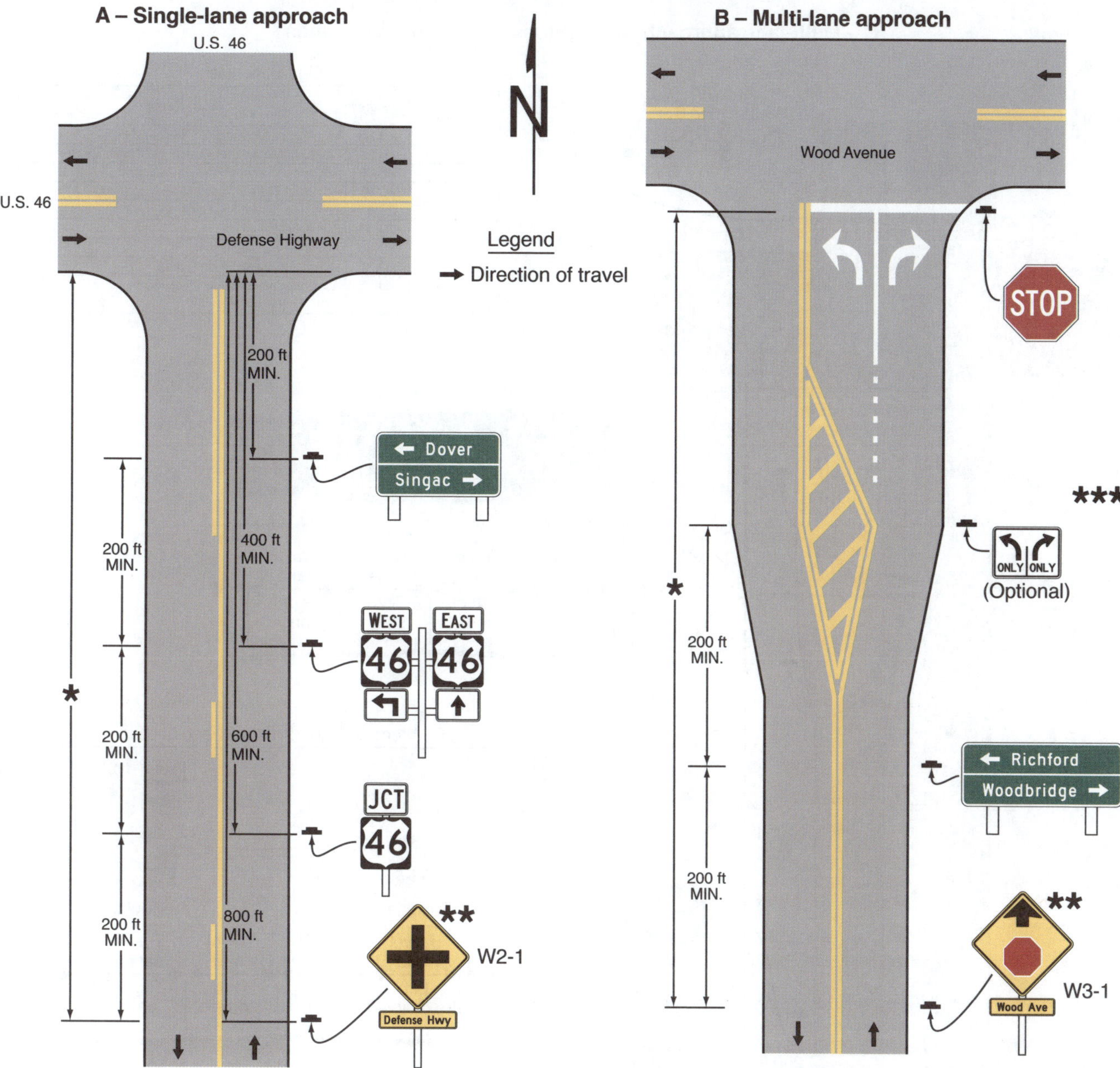

*   See Table 2C-3 for the recommended minimum distance

**   See Section 2C.41 for the application of the W2-1 sign and Section 2C.35 for the application of the W3-1 sign

***   See Section 2B.27 for the application of Intersection Lane Control signs

Note: See Chapter 2D for information on guide signs and Part 3 for information on pavement markings

Figure 2A-4. Relative Locations of Regulatory, Warning, and Guide Signs on an Intersection Approach (Sheet 2 of 4)

C – Multi-lane approach with optional movement turn lanes

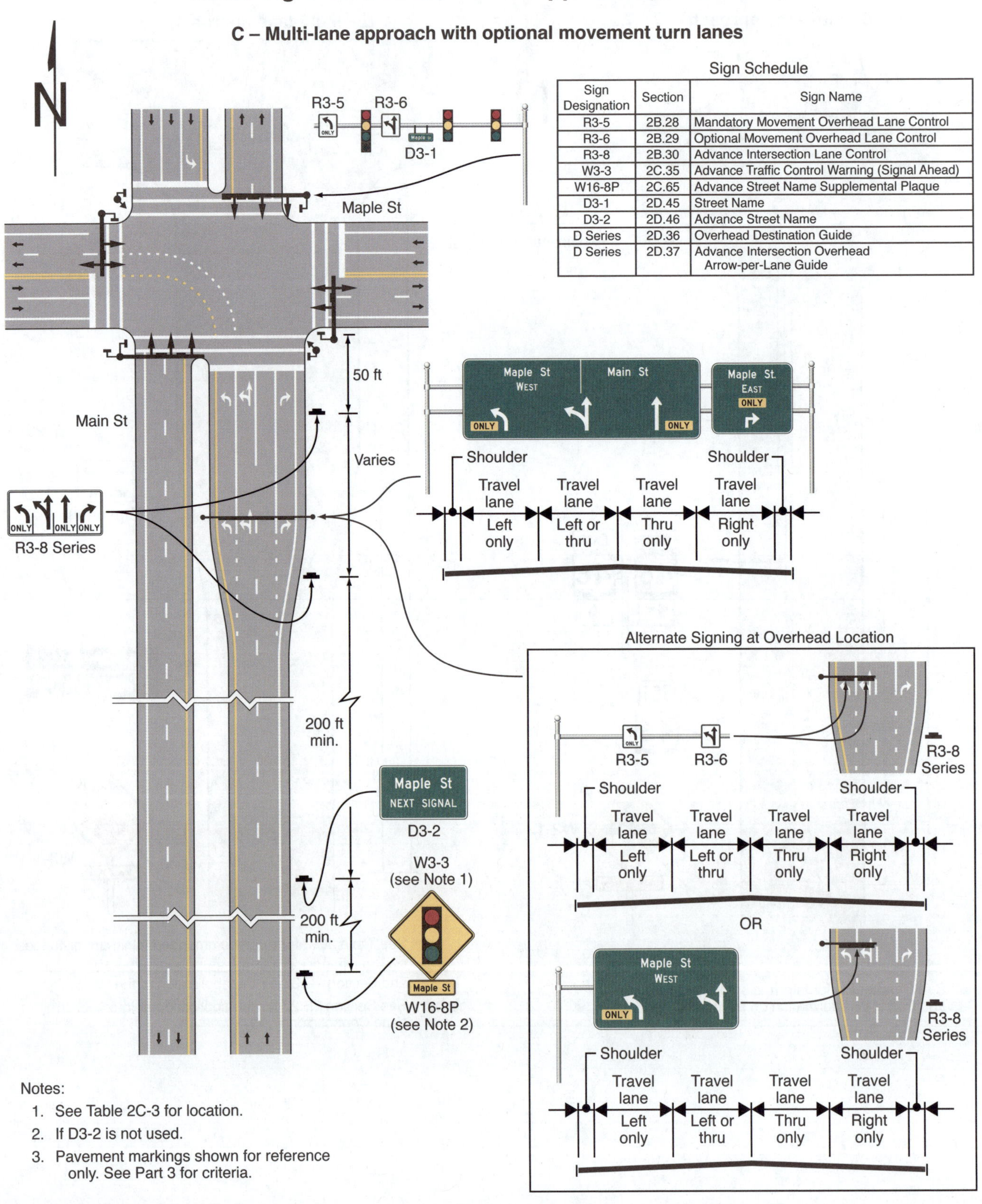

Sign Designation	Section	Sign Name
R3-5	2B.28	Mandatory Movement Overhead Lane Control
R3-6	2B.29	Optional Movement Overhead Lane Control
R3-8	2B.30	Advance Intersection Lane Control
W3-3	2C.35	Advance Traffic Control Warning (Signal Ahead)
W16-8P	2C.65	Advance Street Name Supplemental Plaque
D3-1	2D.45	Street Name
D3-2	2D.46	Advance Street Name
D Series	2D.36	Overhead Destination Guide
D Series	2D.37	Advance Intersection Overhead Arrow-per-Lane Guide

Notes:

1. See Table 2C-3 for location.

2. If D3-2 is not used.

3. Pavement markings shown for reference only. See Part 3 for criteria.

Legend

➡ Direction of travel

Figure 2A-4. Relative Locations of Regulatory, Warning, and Guide Signs on an Intersection Approach (Sheet 3 of 4)

D – Multi-lane approach with mandatory movement turn lanes

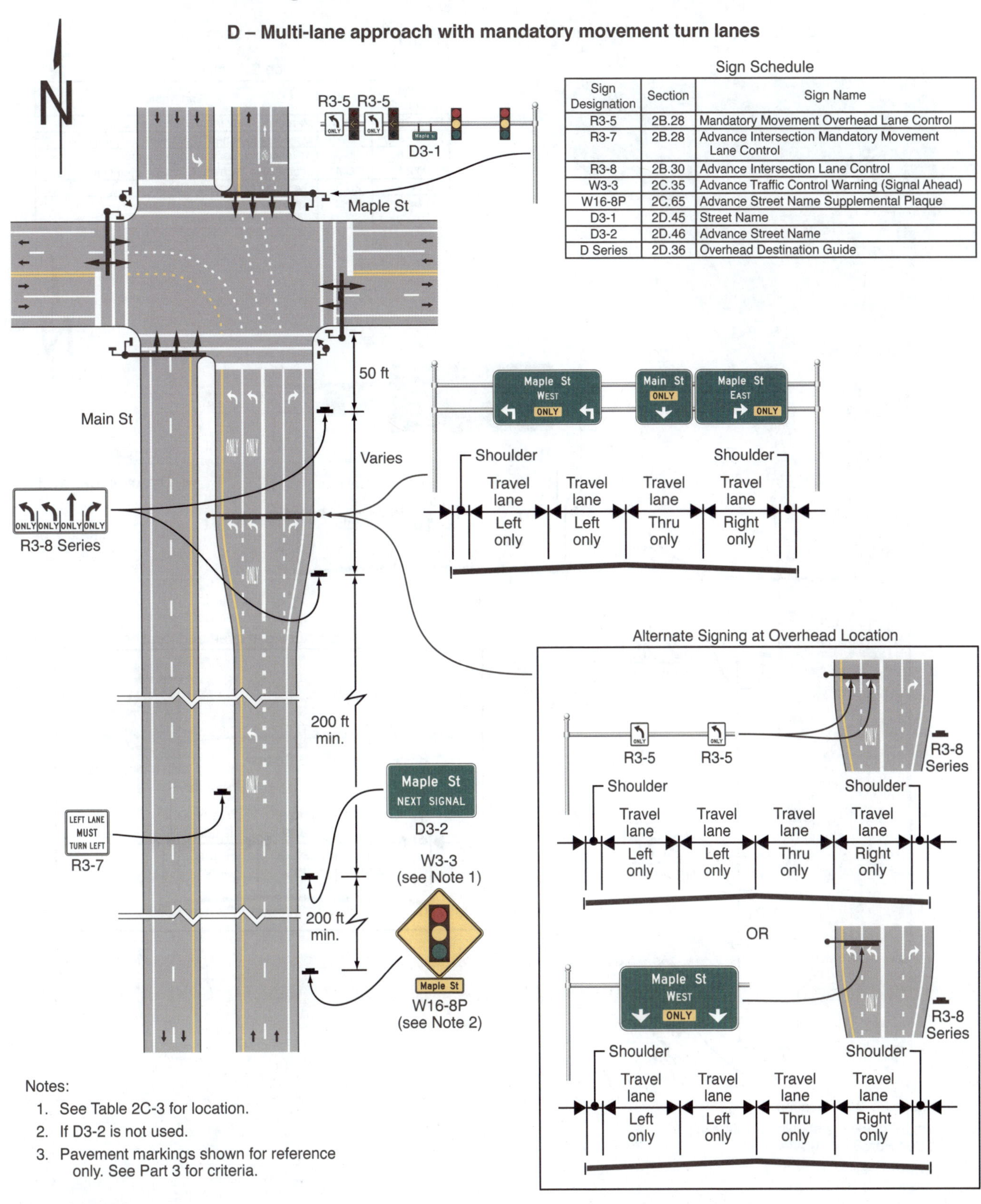

Sign Schedule

Sign Designation	Section	Sign Name
R3-5	2B.28	Mandatory Movement Overhead Lane Control
R3-7	2B.28	Advance Intersection Mandatory Movement Lane Control
R3-8	2B.30	Advance Intersection Lane Control
W3-3	2C.35	Advance Traffic Control Warning (Signal Ahead)
W16-8P	2C.65	Advance Street Name Supplemental Plaque
D3-1	2D.45	Street Name
D3-2	2D.46	Advance Street Name
D Series	2D.36	Overhead Destination Guide

Notes:

1. See Table 2C-3 for location.
2. If D3-2 is not used.
3. Pavement markings shown for reference only. See Part 3 for criteria.

Legend

➤ Direction of travel

Figure 2A-4. Relative Locations of Regulatory, Warning, and Guide Signs on an Intersection Approach (Sheet 4 of 4)

E – Multi-lane approach to a circular intersection with optional movement turn lanes

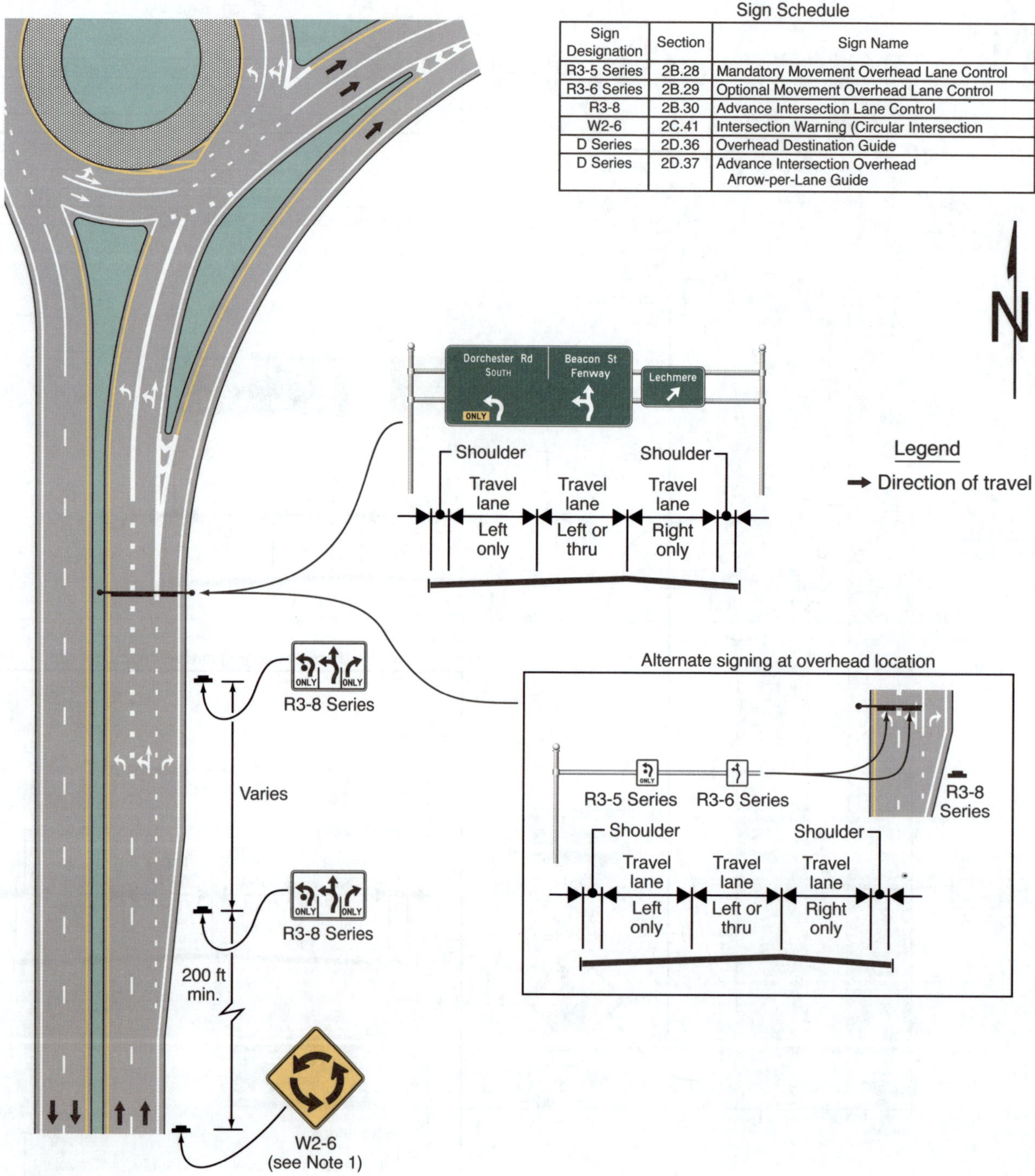

Sign Designation	Section	Sign Name
R3-5 Series	2B.28	Mandatory Movement Overhead Lane Control
R3-6 Series	2B.29	Optional Movement Overhead Lane Control
R3-8	2B.30	Advance Intersection Lane Control
W2-6	2C.41	Intersection Warning (Circular Intersection
D Series	2D.36	Overhead Destination Guide
D Series	2D.37	Advance Intersection Overhead Arrow-per-Lane Guide

Notes:

1. See Table 2C-3 for location.
2. Pavement markings shown for reference only. See Part 3 for criteria.

    B.   *Route or directional signs are grouped to clarify information to motorists;*
    C.   *Regulatory signs that do not conflict with each other are grouped, such as Turn Prohibition signs posted with ONE WAY signs or a parking regulation sign posted with a Speed Limit sign; or*
    D.   *Street Name signs are posted with a STOP or YIELD sign.*

07   *Signs should be located so that they:*

    A.   *Are outside the clear zone unless placed on a crashworthy (see definition in Section 1C.02) support,*
    B.   *Optimize nighttime visibility,*
    C.   *Minimize the effects of mud splatter and debris,*
    D.   *Do not obscure each other,*
    E.   *Do not obscure the sight distance to approaching vehicles on the major street for drivers who are stopped on minor-street approaches, and*
    F.   *Are not hidden from view.*

08   *Except for STOP, YIELD, DO NOT ENTER, and WRONG WAY signs, or as otherwise provided in this Manual, where a sign on a one-way roadway indicates an action intended exclusively or primarily for a road user in the left-hand lane or at the left-hand side of that roadway, such as LEFT LANE MUST TURN LEFT (R3-7) or LEFT LANE ENDS (W9-1), the sign should be located on the left-hand side of the roadway. In the case of a divided road, the sign should be located in the median.*

Option:

09   Signs located on the left-hand side of a one-way roadway or in the median of a divided road, in accordance with Paragraph 8 of this Section, may be supplemented by an identical sign located on the right-hand side of the roadway.

Support:

10   The clear zone (see definition in Section 1C.02) is the total roadside border area, starting at the edge of the traveled way, available for an errant driver to stop or regain control of a vehicle. The width of the clear zone is dependent upon traffic volumes, speeds, and roadside geometry. Additional information can be found in the "Roadside Design Guide," 4th Edition, 2011, AASHTO.

Guidance:

11   *With the increase in traffic volumes and the need to provide road users regulatory, warning, and guidance information, an order of priority for sign installation should be established.*

Support:

12   An order of priority is especially critical where space is limited for sign installation and there is a demand for several different types of signs. Overloading road users with too much information is not desirable. Priority according to type of sign will depend on the specific situation and conditions of the site at which the signs are to be installed. For example, in the vicinity of an exit ramp, guide signs and warning signs for the exit ramp might take precedence over regulatory signs that confirm rules of the road, such as a STATE LAW-NO HANDHELD PHONE USE BY DRIVER sign, or a mainline Speed Limit sign where there is no change in the speed zone.

Guidance:

13   *Because regulatory and warning information is typically more critical to the road user than guidance information, regulatory and warning signing whose locations are critical should be displayed rather than guide signing in cases where conflicts occur. In such cases, the guide sign should be relocated to another appropriate location where it will still be effective. In other cases, such as at a decision point, the guide sign should take precedence over other signs whose locations are not as critical to an immediate decision or action necessary by the road user. In all cases, careful attention should be given to minimizing sign clutter (see Section 2A.20). Community wayfinding and acknowledgment guide signs should have a lower priority as to placement than other guide signs. Signs conveying information of a less-critical nature should be moved to less-critical locations or omitted.*

Option:

14   Under some circumstances, such as on curves to the right, signs may be placed on median islands or on the left-hand side of the road. A supplementary sign located on the left-hand side of the roadway may be used on a multi-lane road where traffic in a lane to the right might obstruct the view to the right.

Guidance:

15   *In urban areas where crosswalks exist, signs should not be placed within 4 feet in advance of the crosswalk (see Drawing D in Figure 2A-3).*

Section 2A.14 <u>Overhead Sign Installations</u>

Guidance:

01   *Overhead signs should be used on freeways and expressways, at locations where some degree of lane-use control is desirable, and at locations where space is not available at the roadside.*

Support:

02   The operational requirements of the present highway system are such that overhead signs have value at many locations. The factors to be considered for the installation of overhead sign displays are not definable in specific numerical terms. In some cases, overhead mounting of a sign might be required by other provisions of this Manual.

Option:

03   The following conditions (not in priority order) may be considered in an engineering study to determine if overhead signs would be beneficial:

    A. Traffic volume at or near capacity,
    B. Complex interchange design,
    C. Three or more lanes in each direction,
    D. Restricted sight distance,
    E. Closely-spaced interchanges,
    F. Multi-lane exits,
    G. Large percentage of trucks,
    H. Street lighting background,
    I. High-speed traffic,
    J. Consistency of sign message location through a series of interchanges,
    K. Insufficient space for post-mounted signs,
    L. Junction of two freeways, and
    M. Left-side exit ramps.

04   Over-crossing structures may be used to support overhead signs.

Support:

05   Under some circumstances, the use of over-crossing structures as sign supports might be the only practical solution that will provide adequate viewing distance. The use of such structures as sign supports might eliminate the need for the foundations and sign supports along the roadside.

Section 2A.15 <u>Mounting Height</u>

Standard:

01   **The provisions of this Section shall apply unless specifically stated otherwise for a particular sign or object marker elsewhere in this Manual.**

Support:

02   It might be necessary to use larger minimum mounting heights than those prescribed in this Manual to ensure appropriate crash performance of sign installations that are required to be crashworthy (see Section 1D.11).

03   In addition to the provisions of this Section, information affecting the minimum mounting height of signs as a function of crash performance can be found in the "Roadside Design Guide," 4th Edition, 2011, AASHTO.

Standard:

04   **In rural areas, the minimum height, measured vertically from the bottom of the sign to the elevation of the near edge of the pavement, of signs installed at the side of the road shall be 5 feet (see Figure 2A-2).**

05   **In business, commercial, or residential areas where parking, bicyclist, or pedestrian movements are likely to occur, or where the view of the sign might be obstructed, the minimum height, measured vertically from the bottom of the sign to the top of the curb, or in the absence of curb, measured vertically from the bottom of the sign to the elevation of the near edge of the traveled way, of signs installed at the side of the road shall be 7 feet (see Figure 2A-2).**

Option:

06   The height to the bottom of a secondary sign mounted below another sign may be 1 foot less than the height specified in Paragraphs 4 and 5 of this Section.

Standard:

07   **The minimum height of signs, measured vertically from the bottom of the sign to the sidewalk shall be 7 feet.**

08   **If the bottom of a secondary sign that is mounted below another sign is mounted lower than 7 feet above a pedestrian sidewalk or pathway (see Section 6C.02), the secondary sign shall not project more than 4 inches into the pedestrian facility.**

Support:

09   Section 9A.02 contains provisions for the minimum mounting height of signs on shared-use paths.

Option:

10   Signs that are placed 30 feet or more from the edge of the traveled way may be installed with a minimum height of 5 feet, measured vertically from the bottom of the sign to the elevation of the near edge of the pavement.

Standard:

11   **Directional signs on freeways and expressways shall be installed with a minimum height of 7 feet, measured vertically from the bottom of the sign to the elevation of the near edge of the pavement. All route signs, warning signs, and regulatory signs on freeways and expressways shall be installed with a minimum height of 7 feet, measured vertically from the bottom of the sign to the elevation of the near edge of the pavement. If a secondary sign is mounted below another sign on a freeway or expressway, the major sign shall be installed with a minimum height of 8 feet and the secondary sign shall be installed with a minimum height of 5 feet, measured vertically from the bottom of the sign to the elevation of the near edge of the pavement.**

12   **Where large signs having an area exceeding 50 square feet are installed on multiple breakaway posts, the clearance from the ground to the bottom of the sign shall be at least 7 feet.**

Option:

13   A route sign assembly (see Section 2D.29) consisting of a route sign and auxiliary signs may be treated as a single sign for the purposes of this Section.

14   The mounting height may be adjusted when supports are located near the edge of the right-of-way on a steep backslope in order to avoid the sometimes less desirable alternative of placing the sign closer to the roadway.

Standard:

15   **Signs that are post-mounted on a median barrier that overhang any portion of the traveled way shall be mounted with a vertical clearance that complies with that of overhead signs.**

16   **Overhead signs shall provide a vertical clearance of not less than 17 feet to the sign, light fixture, or sign bridge over the entire width of the pavement and shoulders, except where the structure on which the overhead signs are to be mounted or other structures along the roadway near the sign structure have a lesser vertical clearance.**

Option:

17   If the vertical clearance of other structures along the roadway near the sign structure is less than 16 feet, the vertical clearance to an overhead sign structure or support may be as low as 1 foot higher than the vertical clearance of the other structures in order to improve the visibility of the overhead signs.

18   In special cases the clearance to overhead signs may be reduced if necessary because of substandard dimensions in tunnels and other major structures such as double-deck bridges.

Guidance:

19   *While a maximum mounting height for signs is generally not prescribed in this Manual, agencies should ensure that signs are not mounted at such a height as to be out of the road user's normal field of vision (see Paragraph 3 of Section 1D.09), especially in urban settings where signs are mounted on traffic signal or light poles.*

Support:

20   Figure 2A-2 illustrates some examples of the mounting height requirements contained in this Section.

Section 2A.16 Lateral Offset

Standard:

01   **For overhead sign supports, the minimum lateral offset from the edge of the shoulder (or if no shoulder exists, from the edge of the pavement) to the near edge of overhead sign supports (cantilever or sign bridges) shall be 6 feet. Overhead sign supports shall have a barrier or crash cushion to shield them if they are within the clear zone.**

02   **Post-mounted sign and object marker supports shall be crashworthy (see Section 1D.11) if within the clear zone.**

Guidance:

03   *For post-mounted signs, the minimum lateral offset should be 12 feet from the edge of the traveled way. If a shoulder wider than 6 feet exists, the minimum lateral offset for post-mounted signs should be 6 feet from the edge of the shoulder.*

04   *Supports for signs mounted laterally behind a longitudinal barrier should be placed so that the near edge of the support is located beyond the deflection distance of the longitudinal barrier.*

Support:

05   The minimum lateral offset requirements for object markers are provided in Chapter 2C.

06   The minimum lateral offset is intended to keep trucks and cars that use the shoulders from striking the signs or supports. The minimum lateral offset requirements do not supersede the requirement for crashworthiness (see Paragraph 2 of this Section) if the sign is located within the clear zone.

Guidance:

07   *All supports should be located as far as practical from the edge of the shoulder. Advantage should be taken to place signs behind existing roadside barriers, on over-crossing structures, or other locations that minimize the exposure of the traffic to sign supports.*

Option:

08   Lesser lateral offsets may be used on connecting roadways or ramps at interchanges, but not less than 6 feet from the edge of the traveled way.

09   On conventional, low-volume rural, and special-purpose roads in areas where it is impractical to locate a sign with the lateral offset prescribed by this Section because of roadside features such as terrain or vegetation, a lateral offset of at least 2 feet may be used.

10   A lateral offset of at least 1 foot from the face of the curb may be used in business, commercial, or residential areas where sidewalk width is limited or where existing poles are close to the curb.

Guidance:

11   *Overhead sign supports and post-mounted sign and object marker supports should not intrude into the usable width of a sidewalk or other pedestrian facility.*

Support:

12   Guidance for maintaining sign shape in laterally-constrained conditions is described in Section 2A.05.

13   Figures 2A-2 and 2A-3 illustrate some examples of the lateral offset requirements contained in this Section.

Section 2A.17 Orientation

Guidance:

01   *Unless otherwise provided in this Manual, signs should be vertically mounted at right angles to the direction of, and facing, the traffic that they are intended to serve.*

02   *Where mirror reflection from the sign face is encountered to such a degree as to reduce legibility, the sign should be turned slightly away from the road. Signs that are placed 30 feet or more from the pavement edge should be turned toward the road. On curved alignments, the angle of placement should be determined by the direction of approaching traffic rather than by the roadway edge at the point where the sign is located.*

Option:

03   On grades, sign faces may be tilted forward or back from the vertical position to improve the viewing angle.

Section 2A.18 Posts and Mountings

Standard:

01   **Sign posts, foundations, and mountings shall be so constructed as to hold signs in a proper and permanent position, and to resist swaying in the wind or displacement by vandalism.**

Support:

02   The latest edition of AASHTO's "Specifications for Structural Supports for Highway Signs, Luminaires, and Traffic Signals" contains additional information regarding posts and mounting.

Option:

03   Where permitted, signs may be placed on existing supports used for other purposes, such as highway traffic signal supports, highway lighting supports, and utility poles.

Support:

04   Section 2A.11 contains criteria for enhanced conspicuity of standard signs.

05   Sections 2A.15 and 2A.16 contain lateral and height placement criteria for signs placed on existing supports.

Standard:

06   **If mounted to the sign support, equipment for powering electronic components of a sign, including solar panels, shall be mounted so as not to compromise the crashworthy performance of the sign installation (see Section 1D.11). Such equipment shall be mounted so as not to obscure the shape of the sign.**

Section 2A.19 <u>Maintenance</u>

Guidance:

01   *Maintenance activities should consider proper position, cleanliness, legibility, and daytime and nighttime visibility (see Sections 2A.21 and 2A.22). Damaged or deteriorated signs, gates, or object markers should be replaced.*

02   *To assure adequate maintenance, a schedule for inspecting (both day and night), cleaning, and replacing signs, gates, and object markers should be established. Employees of highway, law enforcement, and other public agencies whose duties require that they travel on the roadways should be encouraged to report any damaged, deteriorated, or obscured signs, gates, or object markers at the first opportunity.*

03   *Steps should be taken to see that weeds, trees, shrubbery, and construction, maintenance, and utility materials and equipment do not obscure the face of any sign or object marker.*

04   *A regular schedule of replacement of lighting elements for illuminated signs should be maintained.*

Section 2A.20 <u>Excessive Use of Signs</u>

Guidance:

01   *Signs should be used and located judiciously, minimizing their proliferation in order to maintain their effectiveness. Regulatory and warning signs should be used conservatively because these signs, if used to excess, tend to lose their effectiveness. Route signs and directional guide signs for primary routes and destinations should be used frequently at strategic locations because their use promotes efficient operations by keeping road users informed of their location. In all cases, however, sign clutter (see Paragraph 2 of this Section) should be avoided and minimized as much as practicable.*

Support:

02   Sign clutter is the proliferation of sign installations or assemblies along the roadway or roadside, either separately or grouped, to such an extent that adequate spacing between installations necessary for orderly processing of the sign messages by the driver cannot be achieved. Sign clutter can reduce the effectiveness of one or more signs in a sequence of signs.

03   The basic role of traffic control devices is to provide only as much information to the road user as necessary to promote the safe and efficient operation of streets and highways. Sign clutter can result from the overuse of MUTCD-compliant signs and or signs that display information unrelated to traffic operation, navigation, or transportation information. Examples of such signs would include, but are not limited to, those displaying the birthplace or home of a noted person, local sports team accomplishments, population information, and self-described qualities of a community such as "friendly" or "open for business."

Guidance:

04   *Signs and other traffic control devices should be installed and maintained from a systematic standpoint rather than individually. When a new sign is installed, the existing signs in the vicinity should be considered for replacement, relocation, or removal as a result of the new sign that is installed. Existing systems of signs should be reviewed periodically for evidence of sign clutter and adjustments should be made accordingly.*

Support:

05   Section 2A.13 contains information regarding an order of priority for signs where available spacing along the roadway is limited.

Section 2A.21 <u>Retroreflection and Illumination</u>

Support:

01   There are many materials currently available for retroreflection and various methods currently available for the illumination of signs and object markers. New materials and methods continue to emerge. New materials and methods can be used as long as the signs and object markers meet the standard requirements for color, both by day and by night.

02   This Section applies to visibility of signs at night or in low-light or adverse weather conditions, whose legends are otherwise visible under typical daytime viewing conditions.

Standard:

03　Regulatory, warning, and guide signs (see Section 2A.03), and object markers, shall be retroreflective or illuminated to show the same shape and similar color by both day and night, unless otherwise provided in this Manual for a particular sign or group of signs.

04　Where the color black is specified for the legend or background of a sign, an opaque and non-retroreflective material shall be used.

05　The requirements for sign illumination shall not be considered to be satisfied by street or highway lighting.

Option:

06　Sign elements may be illuminated by the means shown in Table 2A-3.

07　Retroreflection of sign elements may be accomplished by the means shown in Table 2A-4.

Support:

08　Information regarding the use of retroreflective material on the sign support is contained in Section 2A.11.

Section 2A.22 Maintaining Minimum Retroreflectivity

Support:

01　Retroreflectivity is one of several factors associated with maintaining nighttime sign visibility (see Section 2A.21).

Standard:

02　Public agencies or officials having jurisdiction shall use an assessment or management method that is designed to maintain sign retroreflectivity at or above the minimum levels in Table 2A-5.

Support:

Table 2A-3. Illumination of Sign Elements

Means of Illumination	Sign Element to be Illuminated
Light behind the sign face	• Symbol or word message • Background • Symbol, word message, and background (through a translucent material)
Attached or independently-mounted light source designed to direct essentially uniform illumination onto the sign face	• Entire sign face
Light-emitting diodes (LEDs)	• Symbol or word message • Entire sign border
Other devices or treatments that highlight the sign shape, color, or message: 　Luminous tubing 　Fiber optics 　Incandescent light bulbs 　Luminescent panels	• Symbol or word message • Entire sign face

Table 2A-4. Retroreflection of Sign Elements

Means of Retroreflection	Sign Element
Prismatic reflector "buttons" or similar units	Symbol Word message Border
A material that has a smooth, sealed outer surface over a microstructure that reflects light	Symbol Word message Border Background

03　Compliance with the Standard in Paragraph 2 of this Section is achieved by having a method in place and using the method to maintain the minimum levels established in Table 2A-5. Provided that an assessment or management method is being used, an agency or official having jurisdiction would be in compliance with the Standard in Paragraph 2 of this Section even if there are some individual signs that do not meet the minimum retroreflectivity levels at a particular point in time.

Guidance:

04　*Except for those signs specifically identified in Paragraph 5 of this Section, one or more of the methods described in "Maintaining Traffic Sign Retroreflectivity," (FHWA-SA-07-020, Revised 2013), FHWA, or a method developed based on an engineering study, should be used to maintain sign retroreflectivity at or above the minimum levels in Table 2A-5. Signs that are identified through the agency's method as being below the minimum levels should be replaced.*

Option:

05　Highway agencies may exclude the following signs from the retroreflectivity maintenance guidelines described in this Section:

　A. Parking, Standing, and Stopping (R7 and R8 series) signs;
　B. Walking/Hitchhiking/Crossing (R9 series, R10-1 through R10-4b) signs;
　C. Acknowledgment signs; and
　D. Bikeway signs that are intended for exclusive use by bicyclists or pedestrians.

Table 2A-5. Minimum Maintained Retroreflectivity Levels[1]

Sign Color	Beaded Sheeting Type (ASTM D4956)			Prismatic Sheeting	Additional Criteria
	I	II	III		
White on Green	W*; G ≥ 7	W*; G ≥ 15	W*; G ≥ 25	W ≥ 250; G ≥ 25	Overhead
	W*; G ≥ 7	W ≥ 120; G ≥ 15			Post-mounted
White on Blue	W*; B ≥ 3	W*; B ≥ 5	W*; B ≥ 12	W ≥ 250; B ≥ 12	Overhead
	W*; B ≥ 3	W ≥ 120; B ≥ 7			Post-mounted
White on Brown	W*; Br ≥ 1	W*; Br ≥ 5	W*; Br ≥ 10	W ≥ 350; Br ≥ 10	Overhead
	W*; Br ≥ 1	W ≥ 150; Br ≥ 5			Post-mounted
Black on Yellow or Black on Orange	Y*; O*	Y ≥ 50; O ≥ 50			[2]
	Y*; O*	Y ≥ 75; O ≥ 75			[3]
White on Red	W ≥ 35; R ≥ 7				[4]
Black on White	W ≥ 50				–

[1] The minimum maintained retroreflectivity levels shown in this table are in units of cd/lx/m^2 measured at an observation angle of 0.2° and an entrance angle of -4.0°.
[2] For word legend and fine symbol signs measuring at least 48 inches and for all sizes of bold symbol signs
[3] For word legend and fine symbol signs measuring less than 48 inches
[4] Minimum sign contrast ratio ≥ 3:1 (white retroreflectivity ÷ red retroreflectivity)
* This sheeting type shall not be used for this color for this application

Bold Symbol Signs

- W1-1,2 – Turn and Curve
- W1-3,4 – Reverse Turn and Curve
- W1-5 – Winding Road
- W1-6,7 – Large Arrow
- W1-8 – Chevron
- W1-10 – Intersection in Curve
- W1-11 – Hairpin Curve
- W1-15 – 270 Degree Loop
- W2-1 – Cross Road
- W2-2,3 – Side Road
- W2-4,5 – T and Y Intersection
- W2-6 – Circular Intersection
- W2-7,8 – Double Side Roads

- W3-1 – Stop Ahead
- W3-2 – Yield Ahead
- W3-3 – Signal Ahead
- W4-1 – Merge
- W4-2 – Lane Ends
- W4-3 – Added Lane
- W4-5 – Entering Roadway Merge
- W4-6 – Entering Roadway Added Lane
- W6-1,2 – Divided Highway Begins and Ends
- W6-3 – Two-Way Traffic
- W10-1,2,3,4,11,12 – Grade Crossing Advance Warning

- W11-2 – Pedestrian Crossing
- W11-3,4,16-22 – Large Animals
- W11-5 – Farm Equipment
- W11-6 – Snowmobile Crossing
- W11-7 – Equestrian Crossing
- W11-8 – Fire Station
- W11-10 – Truck Crossing
- W12-1 – Double Arrow
- W16-5P,6P,7P – Pointing Arrow Plaques
- W20-7 – Flagger
- W21-1 – Worker

Fine Symbol Signs (symbol signs not listed as bold symbol signs)

Special Cases

- W3-1 – Stop Ahead: Red retroreflectivity ≥ 7
- W3-2 – Yield Ahead: Red retroreflectivity ≥ 7; White retroreflectivity ≥ 35
- W3-3 – Signal Ahead: Red retroreflectivity ≥ 7; Green retroreflectivity ≥ 7
- W3-5 – Speed Reduction: White retroreflectivity ≥ 50
- For non-diamond shaped signs, such as W14-3 (No Passing Zone), W4-4P (Cross Traffic Does Not Stop), or W13-1P,2,3,6,7 (Speed Advisory Signs), use the largest sign dimension to determine the proper minimum retroreflectivity level.

Section 2A.23 Median Opening Treatments for Divided Highways

Guidance:

01   *A divided highway crossing should be signed and marked as separate intersections when both of the following conditions are present:*

    A. *The paths of opposing left turns from the divided highway cross each other (see Figure 2A-5), and*

    B. *There is adequate storage in the interior approaches for the design vehicles expected to cross the divided highway.*

02   *If either one or both of the conditions in Paragraph 1 of this Section do not exist, the divided highway crossing should be signed and marked as a single intersection.*

03   *At the crossing of two divided highways, engineering judgment should be used to determine the number of separate intersections.*

Figure 2A-5. Intersection Configuration at a Divided Highway Crossing

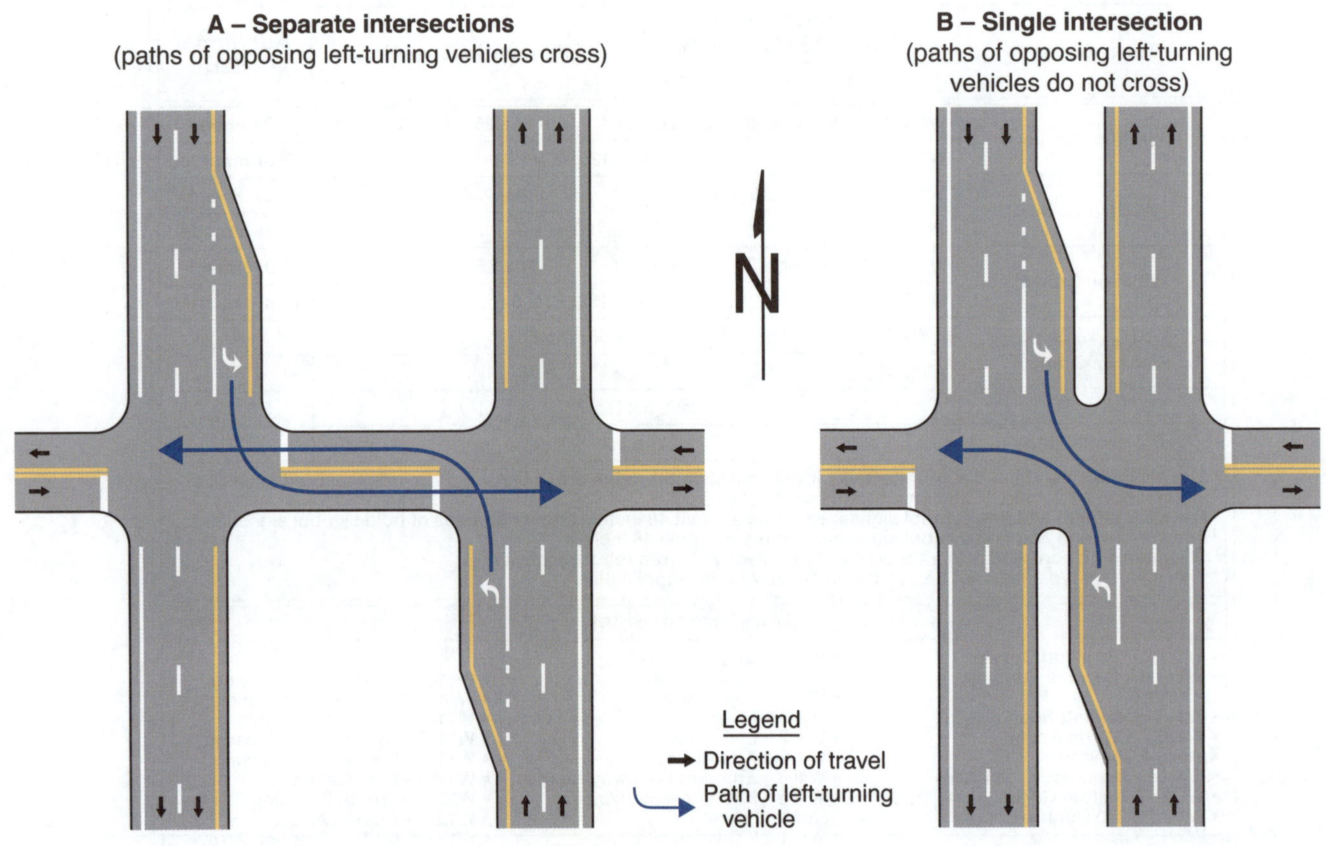

Support:

04   Divided highway crossings with median widths between 30 feet and 85 feet might function as either one or two intersections depending upon the interaction of the opposing left-turn vehicle paths and the available interior storage in the median for a crossing vehicle. Other factors that could determine whether a divided highway crossing is operating as one or two intersections include:

A. The geometric design of the divided highway crossing,
B. The use of positive offset mainline left-turn lanes,
C. The length of the median opening (as measured parallel to the center line of the divided highway),
D. The geometric design of the median noses,
E. Other roadway geometric considerations such as a skewed side street approach or a variable median width,
F. Intersection sight distance,
G. The physical characteristics of the design vehicle, and
H. The observed prevailing driver behavior with regard to opposing left-turn path interaction.

CHAPTER 2B. REGULATORY SIGNS, BARRICADES, AND GATES

Chapter 2B Subchapter and Section Organization

GENERAL

Section 2B.01 Application of Regulatory Signs
Standard:

01 **Regulatory signs shall be used to inform road users of selected traffic laws or regulations and to indicate the applicability of the legal requirements.**

02 **Regulatory signs shall be installed at or near where the regulations apply. The signs shall clearly indicate the requirements imposed by the regulations and shall be designed and installed to provide adequate visibility and legibility in order to obtain compliance.**

03 **Regulatory signs shall be retroreflective or illuminated (see Section 2A.21).**

Section 2B.02 Design of Regulatory Signs
Standard:

01 **Regulatory signs shall be rectangular unless specifically designated otherwise in this Manual. Regulatory signs shall be designed in accordance with the sizes, shapes, colors, and legends contained in the "Standard Highway Signs" publication (see Section 1A.05).**

Support:

02 The use of educational plaques to supplement symbol signs is described in Section 2A.09.

03 The use of LEDs in the border or legend of regulatory signs is described in Section 2A.12.

Standard:

04 **LED signs displaying a part-time prohibitory message incorporating a red circle and diagonal of a static sign shall display a red symbol that approximates the same red circle and diagonal as closely as possible. The symbol of the action to be prohibited shall be displayed in white LEDs on a black background.**

05 **A regulatory sign displayed entirely with LEDs and incorporated within the border of a larger full-matrix changeable message sign shall display the regulatory sign legend in the size, shape, color, and legend of the standard regulatory sign.**

Section 2B.03 Size of Regulatory Signs
Standard:

01 **Except as provided in Section 2A.07, the minimum sizes for regulatory signs shall be as shown in Table 2B-1.**

Support:

02 Section 2A.07 contains information regarding the applicability of the various columns in Table 2B-1.

Standard:

03 **Except as provided in Paragraphs 5 and 6 of this Section, the minimum sizes for regulatory signs facing traffic on multi-lane conventional roads shall be as shown in the Multi-Lane column of Table 2B-1.**

04 **The minimum size of regulatory signs applied on low-volume rural roads with operating speeds of 30 mph or less shall be as shown in the Minimum column of Table 2B-1.**

Option:

05 Where the posted speed limit is 35 mph or less on a multi-lane highway or street, other than for a STOP sign, the minimum size shown in the Single Lane column in Table 2B-1 may be used.

06 Where a regulatory sign, other than a STOP sign, is placed on the left-hand side of a multi-lane roadway in addition to the installation of the same regulatory sign on the right-hand side or the roadway, the minimum size shown in the Single Lane column in Table 2B-1 may be used for both the sign on the right-hand side and the sign on the left-hand side of the roadway.

Guidance:

07 *The minimum sizes for regulatory signs facing traffic on exit and entrance ramps at major interchanges connecting an Expressway or Freeway with an Expressway or Freeway (see Section 2E.11) should be as shown in the column of Table 2B-1 that corresponds to the mainline roadway classification (Expressway or Freeway). If a minimum size is not provided in the Freeway column, the minimum size in the Expressway column should be used. If a minimum size is not provided in the Freeway or Expressway Column, the size in the Oversized column should be used.*

08 *The minimum sizes for all regulatory signs facing traffic on exit and entrance ramps at all other classifications of interchanges (see Section 2E.11) should be the sizes shown in Table 2B-1 in the Conventional Road Single Lane column for single-lane ramps and in the Multi-Lane column for multi-lane ramps.*

Table 2B-1. Regulatory Sign and Plaque Sizes (Sheet 1 of 6)

Sign or Plaque	Sign Designation	Section	Conventional Road		Expressway	Freeway	Minimum	Oversized
			Single Lane	Multi-Lane				
Stop	R1-1	2B.04	30 x 30	36 x 36	36 x 36	—	30 x 30*	48 x 48
Yield	R1-2	2B.05	36 x 36 x 36	48 x 48 x 48	48 x 48 x 48	60 x 60 x 60	30 x 30 x 30*	—
To Oncoming Traffic (plaque)	R1-2aP	2B.18	24 x 18	24 x 18	36 x 30	48 x 36	24 x 18	—
To Traffic in Circle (plaque)	R1-2bP	2B.18	24 x 15	24 x 15	—	—	24 x 15	36 x 24
To All Lanes (plaque)	R1-2cP	2B.18	24 x 15	24 x 15	—	—	24 x 15	36 x 24
All Way (plaque)	R1-3P	2B.04	18 x 6	18 x 6	—	—	—	30 x 12
Yield Here to Pedestrians	R1-5	2B.19	—	36 x 36	—	—	—	36 x 36
Stop Here for Pedestrians	R1-5b	2B.19	—	36 x 36	—	—	—	36 x 36
Yield Here to (Stop Here for) Trail Crossing	R1-5d,5e	2B.19	—	36 x 42	—	—	—	—
In-Street Pedestrian Crossing - Yield (Stop)	R1-6,6a	2B.20	12 x 36	12 x 36	—	—	—	—
In-Street Trail Crossing - Yield (Stop)	R1-6d,6e	2B.20	12 x 36	12 x 36	—	—	—	—
Overhead Pedestrian Crossing - Yield (Stop)	R1-9,9a	2B.20	90 x 24	90 x 24	—	—	—	—
Overhead Trail Crossing	R1-9d,9e	2B.20	72 x 24	72 x 24	—	—	—	—
Except Right Turn (plaque)	R1-10P	2B.04	24 x 18	24 x 18	—	—	—	—
Speed Limit	R2-1	2B.21	24 x 30	30 x 36	36 x 48	48 x 60	18 x 24	30 x 36
Truck Speed Limit (plaque)	R2-2P	2B.22	24 x 24	24 x 24	36 x 36	48 x 48	—	36 x 36
Bus Speed Limit (plaque)	R2-2aP	2B.22	24 x 24	24 x 24	36 x 36	48 x 48	—	36 x 36
Truck-Bus Speed Limit (plaque)	R2-2bP	2B.22	24 x 30	24 x 30	36 x 42	48 x 54	—	36 x 42
Vehicles Over X Tons Speed Limit (plaque)	R2-2cP	2B.22	24 x 30	24 x 30	36 x 42	48 x 54	—	36 x 42
Night Speed Limit (plaque)	R2-3P	2B.23	24 x 24	24 x 24	36 x 36	48 x 48	—	36 x 36
Minimum Speed Limit (plaque)	R2-4P	2B.24	24 x 24	24 x 24	36 x 36	48 x 48	—	36 x 36
Combined Maximum and Minimum Speed Limits	R2-4a	2B.24	24 x 48	24 x 48	36 x 72	48 x 96	—	36 x 72
Unless Otherwise Posted (plaque)	R2-5P	2B.21	24 x 18	24 x 18	36 x 24	36 x 24	—	36 x 24
Citywide (plaque)	R2-5aP	2B.21	24 x 6	24 x 6	—	—	—	30 x 9
Neighborhood (plaque)	R2-5bP	2B.21	24 x 6	24 x 6	—	—	—	30 x 9
Residential (plaque)	R2-5cP	2B.21	24 x 6	24 x 6	—	—	—	30 x 9
Fines Higher (plaque)	R2-6P	2B.25	24 x 18	24 x 18	36 x 24	48 x 36	—	36 x 24
Fines Double (plaque)	R2-6aP	2B.25	24 x 18	24 x 18	36 x 24	48 x 36	—	36 x 24
$XX Fine (plaque)	R2-6bP	2B.25	24 x 18	24 x 18	36 x 24	48 x 36	—	36 x 24
Begin Higher Fines Zone	R2-10	2B.25	24 x 30	24 x 30	36 x 48	48 x 60	—	36 x 48
End Higher Fines Zone	R2-11	2B.25	24 x 30	24 x 30	36 x 48	48 x 60	—	36 x 48
End Variable Speed Limit	R2-13	2B.21	24 x 30	24 x 30	36 x 48	48 x 60	—	36 x 48
End Truck Speed Limit	R2-14	2B.21	24 x 30	24 x 30	36 x 48	48 x 60	—	36 x 48
Movement Prohibition	R3-1,2,3, 4,18,27	2B.26	24 x 24	36 x 36	36 x 36	—	—	48 x 48
Movement Prohibition - Trucks	R3-1b	2B.26	24 x 36	24 x 36	36 x 54	36 x 54	—	—
Movement Prohibition - Trucks Buses	R3-1c	2B.26	24 x 42	24 x 42	36 x 60	36 x 60	—	—
Movement Prohibition - Trucks Over X Tons	R3-1d	2B.26	24 x 48	24 x 48	36 x 66	36 x 66	—	—
Movement Prohibition - Except Buses	R3-1e	2B.26	24 x 36	24 x 36	36 x 54	36 x 54	—	—
Movement Prohibition - Except Buses Taxis	R3-1f	2B.26	24 x 42	24 x 42	36 x 66	36 x 66	—	—
Movement Prohibition - Time and Day	R3-1g	2B.26	24 x 36	24 x 36	36 x 54	36 x 54	—	—

Table 2B-1. Regulatory Sign and Plaque Sizes (Sheet 2 of 6)

Sign or Plaque	Sign Designation	Section	Conventional Road		Expressway	Freeway	Minimum	Oversized
			Single Lane	Multi-Lane				
Movement Prohibition - Multiple Times and Day	R3-1h	2B.26	24 x 42	24 x 42	36 x 66	36 x 66	—	—
Mandatory Movement Lane Control	R3-5,5a	2B.28	30 x 36	30 x 36	—	—	—	—
Left Lane (plaque)	R3-5bP	2B.28	30 x 12	30 x 12	—	—	—	—
HOV 2+ (plaque)	R3-5cP	2B.28	24 x 12	24 x 12	—	—	—	—
Taxi Lane (plaque)	R3-5dP	2B.28	30 x 12	30 x 12	—	—	—	—
Right Lane (plaque)	R3-5fP	2B.28	30 x 12	30 x 12	—	—	—	—
Bus Lane (plaque)	R3-5gP	2B.28	30 x 12	30 x 12	—	—	—	—
Optional Movement Lane Control Thru and Turn	R3-6	2B.29	30 x 36	30 x 36	—	—	—	—
Optional Movement U and Left Turn	R3-6a	2B.29	30 x 36	30 x 36	—	—	—	—
Optional Movement Left Turns	R3-6b	2B.29	30 x 36	30 x 36	—	—	—	—
Right (Left) Lane Must Turn Right (Left)	R3-7	2B.28	30 x 30	36 x 36	48 x 48	—	—	48 x 48
Except Buses (plaque)	R3-7aP	2B.28	24 x 12	24 x 12	—	—	—	—
Except Bicycles (plaque)	R3-7bP	2B.28	24 x 12	24 x 12	—	—	—	—
Advance Intersection Lane Control	R3-8,8a,8b, 8xa, 8xb,8xc	2B.30	Varies x 30	Varies x 30	—	—	—	Varies x 36
Two-Way Left Turn Only (overhead)	R3-9a	2B.32	30 x 36	30 x 36	—	—	—	—
Two-Way Left Turn Only (post-mounted)	R3-9b	2B.32	24 x 36	24 x 36	—	—	—	36 x 48
Begin (plaque)	R3-9cP	2B.33	24 x 12	24 x 12	—	—	—	36 x 18
End (plaque)	R3-9dP	2B.33	24 x 12	24 x 12	—	—	—	36 x 18
Reversible Lane Control (overhead)	R3-9e	2B.34	108 x 48	108 x 48	—	—	—	—
Reversible Lane Control (post-mounted)	R3-9f	2B.34	30 x 42	36 x 54	—	—	—	—
Advance Reversible Lane Control Transition	R3-9g,9h	2B.34	108 x 36	108 x 36	—	—	—	—
End Reverse Lane	R3-9i	2B.34	108 x 48	108 x 48	—	—	—	—
Lane For Left Turn Only	R3-19	2B.28	30 x 24	30 x 24	—	—	—	—
Lane for U Turn Only	R3-19a	2B.28	30 x 24	30 x 24	—	—	—	—
Lane For U and Left Turns Only	R3-19b	2B.28	30 x 30	30 x 30	—	—	—	—
Begin Right (Left) Turn Lane	R3-20	2B.28	24 x 36	24 x 36	—	—	—	—
All Turns (U-Turn) from Right Lane	R3-23,23a	2B.35	60 x 36	60 x 36	—	—	—	—
All Turns (U-Turn) Directional	R3-24,24b, 25,25b,26a	2B.35	72 x 18	72 x 18	—	—	—	—
U-Turns and Left Turns Directional	R3-24a, 25a,26	2B.35	60 x 24	60 x 24	—	—	—	—
Right (Left) Lane Must Exit	R3-33	2B.31	—	—	78 x 36	78 x 36	—	—
Right (Left) Lane Must Exit	R3-33a	2B.31	—	—	42 x 60	42 x 60	—	—
Do Not Pass	R4-1	2B.36	24 x 30	24 x 30	36 x 48	48 x 60	18 x 24	36 x 48
Pass With Care	R4-2	2B.37	24 x 30	24 x 30	36 x 48	48 x 60	18 x 24	36 x 48
Slower Traffic Keep Right	R4-3	2B.38	24 x 30	24 x 30	36 x 48	48 x 60	18 x 24	36 x 48
Trucks Use Right Lane	R4-5	2B.38	24 x 30	24 x 30	36 x 48	48 x 60	—	36 x 48
Keep Right	R4-7,7a,7b	2B.39	24 x 30	24 x 30	36 x 48	48 x 60	18 x 24	36 x 48
Narrow Keep Right	R4-7c	2B.39	18 x 30	18 x 30	—	—	—	—
Keep Left	R4-8,8a,8b	2B.39	24 x 30	24 x 30	36 x 48	48 x 60	18 x 24	36 x 48
Narrow Keep Left	R4-8c	2B.39	18 x 30	18 x 30	—	—	—	—
Stay in Lane	R4-9	2B.40	24 x 30	24 x 30	36 x 48	48 x 60	18 x 24	36 x 48
Runaway Vehicles Only	R4-10	2B.41	48 x 48	48 x 48	—	—	—	—

Table 2B-1. Regulatory Sign and Plaque Sizes (Sheet 3 of 6)

Sign or Plaque	Sign Designation	Section	Conventional Road		Expressway	Freeway	Minimum	Oversized
			Single Lane	Multi-Lane				
Slow Vehicles with XX or More Following Vehicles Must Use Turn-Out	R4-12	2B.42	42 x 24	42 x 24	—	—	—	72 x 42
Slow Vehicles Must Use Turn-Out Ahead	R4-13	2B.42	42 x 24	42 x 24	—	—	—	—
Slow Vehicles Must Turn Out	R4-14	2B.42	30 x 42	30 x 42	—	—	—	—
Keep Right Except to Pass	R4-16	2B.38	24 x 30	24 x 30	36 x 48	48 x 60	18 x 24	36 x 48
Do Not Drive on Shoulder	R4-17	2B.43	24 x 30	24 x 30	36 x 48	48 x 60	18 x 24	36 x 48
Do Not Pass on Shoulder	R4-18	2B.43	24 x 30	24 x 30	36 x 48	48 x 60	18 x 24	36 x 48
All Traffic	R4-20	2B.44	24 x 30	24 x 30	36 x 48	48 x 60	—	36 x 48
Right (Left) Turn Only	R4-21	2B.44	24 x 30	24 x 30	—	—	—	—
Do Not Enter	R5-1	2B.46	30 x 30	36 x 36	36 x 36	48 x 48	—	36 x 36
Wrong Way	R5-1a	2B.47	36 x 24	42 x 30	36 x 24	42 x 30	30 x 18	42 x 30
No Trucks	R5-2	2B.45	24 x 24	24 x 24	30 x 30	36 x 36	—	36 x 36
Except Local Deliveries (plaque)	R5-2aP	2B.45	24 x 12	24 x 12	30 x 15	36 x 18	—	36 x 18
No Thru Trucks	R5-2b	2B.45	24 x 30	24 x 30	30 x 36	36 x 48	—	36 x 48
No Motor Vehicles	R5-3	2B.45	24 x 24	24 x 24	—	—	24 x 24	—
No Commercial Vehicles	R5-4	2B.45	24 x 30	24 x 30	36 x 48	36 x 48	—	—
No Vehicles with Lugs	R5-5	2B.45	24 x 30	24 x 30	36 x 48	48 x 60	—	—
No Bicycles	R5-6	2B.45	24 x 24	24 x 24	30 x 30	36 x 36	24 x 24*	48 x 48
No Non-Motorized Traffic	R5-7	2B.45	30 x 24	30 x 24	42 x 24	48 x 30	—	42 x 24
No Motor-Driven Cycles	R5-8	2B.45	30 x 24	30 x 24	42 x 24	48 x 30	—	42 x 24
No Pedestrians, Bicycles, Motor-Driven Cycles	R5-10	2B.45	30 x 36	30 x 36	—	—	—	—
No Pedestrians, Bicycles, Motor-Driven Cycles On Freeway	R5-10a	2B.45	30 x 36	30 x 36	—	—	—	—
No Pedestrians or Bicycles	R5-10b	2B.45	30 x 18	30 x 18	—	—	—	—
No Pedestrians	R5-10c	2B.45	24 x 12	24 x 12	—	—	—	—
Authorized Vehicles Only	R5-11	2B.45	30 x 24	30 x 24	—	—	—	—
No Thru Traffic	R5-12	2B.45	24 x 30	24 x 30	—	—	—	30 x 36
One Way	R6-1	2B.49	36 x 12	48 x 18	48 x 18	48 x 18	—	72 x 24
One Way	R6-2	2B.49	24 x 30	30 x 36	36 x 48	48 x 60	18 x 24	36 x 48
Divided Highway Crossing	R6-3,3a	2B.50	30 x 24	30 x 24	36 x 30	—	—	36 x 30
Roundabout Circulation (plaque)	R6-5P	2B.51	30 x 30	30 x 30	—	—	—	—
Begin One Way	R6-6	2B.49	24 x 30	30 x 36	—	—	—	—
End One Way	R6-7	2B.49	24 x 30	30 x 36	—	—	—	—
Parking Restrictions	R7-1,2,2a,3, 4,4a, 5,6,8,10, 107,108	2B.52, 2B.53	12 x 18	12 x 18	—	—	—	—
Van Accessible (plaque)	R7-8aP	2B.52, 2B.53	12 x 6	12 x 6	—	—	—	—
Parking Fee Station, Multispace Meter	R7-20	2B.52, 2B.53	24 x 18	24 x 18	—	—	—	—
Metered Parking	R7-21	2B.52, 2B.53	12 x 18	12 x 18	—	—	—	—
Mobile Parking Payment (plaque)	R7-21aP	2B.52, 2B.53	12 x 6	12 x 6				
Metered Parking	R7-22	2B.52, 2B.53	12 x 18	12 x 18	—	—	—	—
No Parking Bus Stop	R7-107a	2B.52, 2B.53	12 x 24	12 x 24	—	—	—	—
No Parking Bus Stop (with transit pictograph)	R7-107b	2B.52, 2B.53	12 x 30	12 x 30	—	—	—	—

Table 2B-1. Regulatory Sign and Plaque Sizes (Sheet 4 of 6)

Sign or Plaque	Sign Designation	Section	Conventional Road		Expressway	Freeway	Minimum	Oversized
			Single Lane	Multi-Lane				
No Parking Except Electric Vehicles	R7-111	2B.52, 2B.53	12 x 18	12 x 18	—	—	—	—
No Parking Except Electric Vehicles (part-time)	R7-111a	2B.52, 2B.53	12 x 18	12 x 18	—	—	—	—
Electric Vehicle Parking (time limit)	R7-112	2B.52, 2B.53	12 x 18	12 x 18	—	—	—	—
Electric Vehicle Parking (time limit part-time)	R7-112a	2B.52, 2B.53	12 x 18	12 x 18	—	—	—	—
Electric Vehicle Parking (time limit part-time)	R7-112b	2B.52, 2B.53	12 x 21	12 x 21	—	—	—	—
No Parking Except While Charging	R7-113	2B.52, 2B.53	12 x 18	12 x 18	—	—	—	—
Vehicle Must Be Plugged In (plaque)	R7-113aP	2B.52, 2B.53	12 x 6	12 x 6	—	—	—	—
Vacate Stall When Charging Completed (plaque)	R7-113bP	2B.52, 2B.53	12 x 6	12 x 6	—	—	—	—
Vehicle Charging Only (time limit)	R7-114	2B.52, 2B.53	12 x 18	12 x 18	—	—	—	—
Vehicle Charging Only (time limit, part-time)	R7-114a	2B.52, 2B.53	12 x 18	12 x 18	—	—	—	—
Vehicle Charging Only (time limit part-time)	R7-114b	2B.52, 2B.53	12 x 21	12 x 21	—	—	—	—
No Parking/Restricted Parking (combined sign)	R7-200	2B.52, 2B.53	24 x 18	24 x 18	—	—	—	—
No Parking/Restricted Parking (combined sign)	R7-200a	2B.52, 2B.53	12 x 36	12 x 36	—	—	—	—
Tow Away Zone (plaque)	R7-201P, 201aP	2B.52, 2B.53	12 x 6	12 x 6	—	—	—	—
This Side of Sign (plaque)	R7-202P	2B.52, 2B.53	12 x 6	12 x 6	—	—	—	—
Snow Emergency Route	R7-203	2B.52, 2B.53	18 x 24	18 x 24	—	—	—	24 x 30
No Parking on Pavement	R8-1	2B.52, 2B.53	24 x 30	24 x 30	36 x 48	48 x 60	—	36 x 48
No Parking Except on Shoulder	R8-2	2B.52, 2B.53	24 x 30	24 x 30	36 x 48	48 x 60	—	36 x 48
No Parking (symbol)	R8-3	2B.52, 2B.53	24 x 24	30 x 30	36 x 36	48 x 48	12 x 12	36 x 36
No Parking	R8-3a	2B.52, 2B.53	24 x 30	24 x 30	36 x 36	48 x 48	18 x 24	36 x 36
On Pavement	R8-3c	2B.52, 2B.53	24 x 36	24 x 36	—	—	18 x 30	36 x 54
On Bridge	R8-3d	2B.52, 2B.53	24 x 36	24 x 36	—	—	18 x 30	36 x 54
On Tracks	R8-3e	2B.52, 2B.53	24 x 36	24 x 36	—	—	18 x 30	36 x 54
Except on Shoulder	R8-3f	2B.52, 2B.53	24 x 36	24 x 36	—	—	18 x 30	36 x 54
Emergency Parking Only	R8-4	2B.55	30 x 24	30 x 24	30 x 24	48 x 36	—	48 x 36
No Stopping on Pavement	R8-5	2B.53, 2B.54	24 x 30	24 x 30	36 x 48	48 x 60	—	36 x 48
No Stopping Except on Shoulder	R8-6	2B.53, 2B.54	24 x 30	24 x 30	36 x 48	48 x 60	—	36 x 48
Emergency Stopping Only	R8-7	2B.55	30 x 24	30 x 24	48 x 36	48 x 36	—	48 x 36
Walk on Left Facing Traffic	R9-1	2B.56	18 x 24	18 x 24	—	—	—	—
Cross Only at Crosswalks	R9-2	2B.57	12 x 18	12 x 18	—	—	—	—
No Pedestrian Crossing (symbol)	R9-3	2B.57	18 x 18	18 x 18	24 x 24	30 x 30	—	30 x 30
No Pedestrian Crossing	R9-3a	2B.57	12 x 18	12 x 18	—	—	—	—

Table 2B-1. Regulatory Sign and Plaque Sizes (Sheet 5 of 6)

Sign or Plaque	Sign Designation	Section	Conventional Road		Expressway	Freeway	Minimum	Oversized
			Single Lane	Multi-Lane				
Use Crosswalk (plaque)	R9-3bP	2B.57	18 x 12	18 x 12	—	—	—	—
No Hitchhiking (symbol)	R9-4	2B.56	18 x 18	18 x 18	—	—	—	24 x 24
No Hitchhiking	R9-4a	2B.56	18 x 24	18 x 24	—	—	12 x 18	—
No Skaters	R9-13	2B.45	18 x 18	18 x 18	24 x 24	30 x 30	—	30 x 30
No Equestrians	R9-14	2B.45	18 x 18	18 x 18	24 x 24	30 x 30	—	30 x 30
No Snowmobiles	R9-15	2B.45	18 x 18	18 x 18	24 x 24	30 x 30	—	30 x 30
No All-Terrian Vehicles	R9-16	2B.45	18 x 18	18 x 18	24 x 24	30 x 30	—	30 x 30
Except on Shoulder (plaque)	R9-19P	2B.45	18 x 12	18 x 12	24 x 18	30 x 24	—	30 x 24
Cross Only On Green	R10-1	2B.58	12 x 18	12 x 18	—	—	—	—
Pedestrian Signs	R10-2,3, 3b,3c,3d,4	2B.58	9 x 12	9 x 12	—	—	—	—
Pedestrian Signs	R10-3a,3e,3f, 3g,3h,3i,4a	2B.58	9 x 15	9 x 15	—	—	—	—
Left on Green Arrow Only	R10-5	2B.59	30 x 36	30 x 36	30 x 36	—	24 x 30	48 x 60
Stop Here on Red	R10-6	2B.59	24 x 36	24 x 36	—	—	—	36 x 48
Stop Here on Red	R10-6a	2B.59	24 x 30	24 x 30	—	—	—	36 x 42
Do Not Block Intersection	R10-7	2B.59	24 x 30	24 x 30	—	—	—	—
Use Lane with Green Arrow	R10-8	2B.59	30 x 36	30 x 36	36 x 42	—	—	60 x 72
Left (Right) Turn Signal	R10-10	2B.59	24 x 30	24 x 30	—	—	—	30 x 36
U- Turn Signal	R10-10a	2B.59	24 x 30	24 x 30	—	—	—	30 x 36
No Turn on Red	R10-11	2B.60	24 x 30	24 x 30	—	—	—	36 x 48
No Turn on Circular Red	R10-11a	2B.60	24 x 30	24 x 30	—	—	—	36 x 48
No Turn on Red	R10-11b	2B.60	24 x 24	24 x 24	—	—	—	36 x 36
No Turn on Red Except From Right Lane	R10-11c	2B.60	30 x 36	30 x 36	—	—	—	—
No Turn on Red From This Lane	R10-11d	2B.60	30 x 42	30 x 42	—	—	—	—
Left Turn Yield on Green	R10-12	2B.59	30 x 36	30 x 36	—	—	—	—
Left Turn Yield on Flashing Yellow Arrow	R10-12a	2B.59	30 x 36	30 x 36	—	—	—	—
Left Turn Yield to Bicycle	R10-12b	2B.59	30 x 36	30 x 36	—	—	—	—
Emergency Signal	R10-13	2B.59	36 x 24	36 x 24	—	—	—	42 x 30
Emergency Signal - Stop on Flashing Red	R10-14	2B.59	36 x 42	36 x 42	—	—	—	—
Emergency Signal - Stop on Flashing Red (overhead)	R10-14a	2B.59	60 x 24	60 x 24	—	—	—	—
Stop Here on Flashing Red	R10-14b	2B.59	24 x 36	24 x 36	—	—	—	36 x 48
Turning Vehicles Yield to Pedestrians	R10-15	2B.59	30 x 30	30 x 30	—	—	—	—
Turning Vehicles Stop for Pedestrians	R10-15a	2B.59	30 x 30	30 x 30	—	—	—	—
U-Turn Yield to Right Turn	R10-16	2B.59	30 x 36	30 x 36	—	—	—	—
Right on Red Arrow After Stop	R10-17a	2B.60	30 x 36	30 x 36	—	—	—	36 x 48
Traffic Laws Photo Enforced	R10-18	2B.69	36 x 24	36 x 24	48 x 30	54 x 36	—	54 x 36
Traffic Signal Photo Enforced	R10-18a	2B.69	30 x 42	30 x 42	30 x 42	—	—	36 x 54
Photo Enforced (symbol plaque)	R10-19P	2B.69	24 x 12	24 x 12	36 x 18	48 x 24	—	48 x 24
Photo Enforced (plaque)	R10-19aP	2B.69	24 x 18	24 x 18	36 x 24	48 x 36	—	48 x 36
MON—FRI (and times) (3 lines) (plaque)	R10-20aP	2B.60	24 x 24	24 x 24	—	—	—	—
SUNDAY (and times) (2 lines) (plaque)	R10-20aP	2B.60	24 x 18	24 x 18	—	—	—	—
Crosswalk - Stop on Red	R10-23	2B.59	24 x 30	24 x 30	—	—	—	—

Table 2B-1. Regulatory Sign and Plaque Sizes (Sheet 6 of 6)

Sign or Plaque	Sign Designation	Section	Conventional Road		Expressway	Freeway	Minimum	Oversized
			Single Lane	Multi-Lane				
Stop on Red - Yield on Flashing Red After Stop	R10-23a	2B.59	24 x 30	24 x 30	—	—	—	—
Push Button For Warning Lights - Wait for Gap in Traffic	R10-25	2B.58	9 x 12	9 x 12	—	—	—	—
Left Turn Yield on Flashing Red Arrow After Stop	R10-27	2B.59	30 x 36	30 x 36	—	—	—	—
XX Vehicles per Green	R10-28	2B.61	24 x 30	24 x 30	—	—	—	—
XX Vehicles per Green Each Lane	R10-29	2B.61	36 x 24	36 x 24	—	—	—	—
Right Turn on Red Must Yield to U-Turn	R10-30	2B.60	30 x 36	30 x 36	—	—	—	—
At Signal (plaque)	R10-31P	2B.59	24 x 9	24 x 9	—	—	—	—
Push Button for 2 Seconds for Extra Crossing Time	R10-32P	2B.58	9 x 12	9 x 12	—	—	—	—
Keep Off Median	R11-1	2B.62	24 x 30	24 x 30	—	—	—	—
Road Closed	R11-2,2a, 2b,2c	2B.63	48 x 30	48 x 30	—	—	—	—
Road Closed - Local Traffic Only	R11-3,3a, 3b,4	2B.63	60 x 30	60 x 30	—	—	—	—
Weight Limit	R12-1, 2	2B.64	24 x 30	24 x 30	36 x 48	—	—	36 x 48
Weight Limit - Axle, Gross	R12-4	2B.64	36 x 24	36 x 24	—	—	—	—
Weight Limit	R12-5	2B.64	24 x 36	24 x 36	36 x 48	48 x 60	—	—
Weight Limit - Specialized Hauling Vehicles	R12-6	2B.64	30 x 42	36 x 48	36 x 48	48 x 60	—	48 x 60
Weight Limit - Emergency Vehicles	R12-7	2B.64	30 x 36	30 x 36	48 x 60	48 x 60	—	48 x 60
Weight Limit - Emergency Vehicles (plaque)	R12-7aP	2B.64	30 x 30	30 x 30	48 x 48	48 x 48	—	48 x 48
Weigh Station	R13-1	2B.65	72 x 54	72 x 54	96 x 72	132 x 90	—	—
Truck Route	R14-1	2B.66	24 x 18	24 x 18	—	—	—	—
Hazardous Material	R14-2,3	2B.67	24 x 24	24 x 24	30 x 30	36 x 36	—	42 x 42
National Network	R14-4,5	2B.68	30 x 30	30 x 30	36 x 36	36 x 36	—	42 x 42
Move Over or Reduce Speed	R16-3	2B.71	—	60 x 48	84 x 60	102 x 72	—	84 x 60
Minor Crashes Move Vehicles from Travel Lanes	R16-4	2B.70	—	60 x 42	84 x 54	96 x 60	—	84 x 54
Lights On When Using Wipers or Raining	R16-5,6	2B.73	24 x 30	24 x 30	36 x 48	48 x 60	—	36 x 48
Turn On Headlights Next XX Miles	R16-7	2B.73	60 x 18	60 x 18	96 x 30	132 x 36	—	96 x 30
Turn On, Check Headlights	R16-8,9	2B.73	42 x 18	42 x 18	60 x 30	78 x 36	—	60 x30
Begin, End Daytime Headlight Section	R16-10,11	2B.73	60 x 18	60 x 18	96 x 30	120 x 36	—	96 x 30
No Hand-Held Phone Use By Driver	R16-15	2B.72	—	—	72 x 48	72 x 48	—	—
No Hand-Held Phone Use By Driver	R16-15a	2B.72	30 x 42	30 x 42	—	—	—	—

* See Table 9A-1 for minimum size required for signs on bicycle facilities

Notes: 1. Larger signs may be used when appropriate

      2. Dimensions in inches are shown as width x height

Section 2B.04 STOP Sign (R1-1) and ALL-WAY Plaque (R1-3P)

Standard:

01   **When it is determined that a full stop is always required on an approach to an intersection, a STOP (R1-1) sign (see Figure 2B-1) shall be used.**

02   **Secondary legends shall not be used on STOP sign faces.**

03   **The STOP sign shall not be displayed using a changeable message sign.**

04   **At intersections where all approaches are controlled by STOP signs (see Section 2B.12), an ALL-WAY (R1-3P) supplemental plaque (see Figure 2B-1) shall be mounted below each STOP sign. The ALL-WAY plaque shall have a white legend and border on a red background.**

05   **Supplemental plaques with legends such as 2-WAY, 3-WAY, 4-WAY, or other numbers of ways shall not be used with STOP signs.**

Support:

06   The use of the CROSS TRAFFIC DOES NOT STOP (W4-4P series) and other plaques with variations of this legend is described in Section 2C.66.

Guidance:

07   *The TRAFFIC FROM LEFT (RIGHT) DOES NOT STOP (W4-4aP) plaque or ONCOMING TRAFFIC DOES NOT STOP (W4-4bP) plaque should be used at intersections where STOP signs control all but one approach to the intersection, unless the only non-stopped approach is from a one-way street.*

Option:

08   The EXCEPT RIGHT TURN (R1-10P) plaque (see Figure 2B-1) may be mounted below the STOP sign if an engineering study determines that a special combination of geometry and traffic volumes is present that makes it possible for right-turning traffic on the approach to be allowed to enter the intersection without stopping.

Support:

09   The design and application of Stop Beacons are described in Section 4S.05.

Section 2B.05 YIELD Sign (R1-2)

Support:

01   The YIELD sign requires road users to yield the right-of-way to other traffic on certain approaches to an intersection or on a two way approach to a one way section of roadway, such as a narrow bridge or underpass. Vehicles controlled by a YIELD sign need to slow down to a speed that is reasonable for the existing conditions or stop when necessary to avoid interfering with conflicting traffic.

Standard:

02   **The YIELD (R1-2) sign (see Figure 2B-1) shall not be displayed using a changeable message sign.**

Figure 2B-1. STOP and YIELD Signs and Plaques

SIGNING FOR RIGHT-OF-WAY AT INTERSECTIONS

Section 2B.06 General Considerations

Support:

01   Unsignalized intersections represent the most common form of intersection right-of-way control. Selection of control type might be impacted by specific requirements of State law or local ordinances.

02   Roundabouts and traffic circles are circular intersection designs and are not traffic control devices. The decision to convert an intersection from a conventional intersection to a circular intersection is an engineering design decision and not a traffic control device decision. As such, criteria for conversion from a conventional intersection to a circular intersection are not included in the MUTCD.

Guidance:

03   *The type of traffic control used at an unsignalized intersection should be the least restrictive that provides appropriate levels of safety and efficiency for all road users.*

Support:

04   Some types of right-of-way control that can exist at an unsignalized intersection in order from the least restrictive to the most restrictive are the following:

    A. No intersection control (see Section 2B.09): There are no right-of-way traffic control devices on any of the approaches to the intersection.

    B. Yield control (see Section 2B.10): YIELD signs are placed on all approaches (for a circular intersection), on opposing approaches for a four-leg intersection, on a single approach for a three-leg intersection, or in the median of a divided highway. The YIELD signs are placed on the minor road.

    C. Minor road stop control (see Section 2B.11): STOP signs are typically placed on opposing approaches (for a four-leg intersection) or on a single approach (for a three-leg intersection). The STOP signs are normally placed on the minor road. Section 2B.07 contains guidance on selecting the minor road.

    D. All-way stop control (see Section 2B.12): STOP signs are placed on all approaches to the intersection.

Guidance:

05   *When selecting a form of intersection control, the following factors should be considered:*

    A. Motor vehicle, bicycle, and pedestrian traffic volumes on all approaches; where the term units/day or units/hour is indicated, it should be the total of motor vehicle, bicycle, and pedestrian volume;

    B. Driver yielding behavior with regard to all modes of conflicting traffic, including bicyclists and pedestrians;

    C. Number and angle of approaches;

    D. Approach speeds;

    E. Sight distance available on each approach;

    F. Reported crash experience; and

    G. The presence of a grade crossing near the intersection.

Standard:

06   **YIELD or STOP signs shall not be used for speed control.**

Support:

07   Appropriate traffic calming or other speed control measures are available to control vehicle speeds, such as those that do not have the potential to diminish the effectiveness of traffic control devices when used for their specified purpose.

Standard:

08   **Because the potential for conflicting commands could create driver confusion, YIELD or STOP signs shall not be used in conjunction with any traffic control signal operation, except in the following cases:**

    A. If the signal indication for an approach is a flashing red at all times;

    B. If a minor street or driveway is located within or adjacent to the area controlled by the traffic control signal, but does not require separate traffic signal control because an extremely low potential for conflict exists; or

    C. If a channelized turn lane is separated from the adjacent travel lanes by an island and the channelized turn lane is not controlled by a traffic control signal.

09   **STOP signs and YIELD signs shall not be installed on different approaches to the same unsignalized intersection if those approaches conflict with or oppose each other, except as provided for in Items A and B in Paragraph 3 of Section 2B.10.**

10   **Portable or part-time STOP or YIELD signs shall not be used except for emergency and temporary traffic control zone purposes.**

11   **A portable or part-time (folding) STOP sign that is manually placed into view and manually removed from view shall not be used during a power outage to control a signalized approach unless the maintaining agency establishes that the signal indication that will first be displayed to that approach upon restoration of power is a flashing red signal indication and that the portable STOP sign will be manually removed from view prior to resuming stop-and-go operation of the traffic control signal.**

Option:

12   A portable or part-time (folding) STOP sign that is electrically or mechanically operated such that it only displays the stop message during a power outage and ceases to display the stop message upon restoration of power may be used during a power outage to control a signalized approach.

Support:

13   The use of STOP signs at grade crossings is described in Sections 8B.04 and 8B.05.

14   Section 9B.01 contains provisions regarding the assignment of priority where a shared-use path crosses a roadway.

Section 2B.07 Determining the Minor Road for Unsignalized Intersections

Guidance:

01   *The selection of the minor road to be controlled by YIELD or STOP signs should be based on one or more of the following criteria:*

   A. *A roadway intersecting a designated through or numbered highway,*
   B. *A roadway with the lower functional classification,*
   C. *A roadway with the lower traffic volume,*
   D. *A roadway with the lower speed limit, and/or*
   E. *A roadway that intersects with a roadway that has a higher priority for one or more modes of travel.*

02   *When two roadways that have relatively equal volumes, speeds, and/or other characteristics intersect, the following factors should be considered in selecting the minor road for installation of YIELD or STOP signs:*

   A. *Controlling the direction that conflicts the most with established pedestrian crossing activity or school walking routes;*
   B. *Controlling the direction that has obscured vision, dips, or bumps that already require drivers to use lower operating speeds; and*
   C. *Controlling the direction that has the best sight distance from a controlled position to observe conflicting traffic.*

Section 2B.08 Right-of-Way Intersection Control Considerations

Guidance:

01   *Before converting to a more restrictive form of right-of-way control at an unsignalized intersection, the following alternative treatments to address safety, operational, or other concerns should be among those to be considered:*

   A. *Where yield or stop controlled, installing Yield Ahead or Stop Ahead signs on the appropriate approaches to the intersection;*
   B. *Removing parking on one or more approaches;*
   C. *Removing sight distance obstructions;*
   D. *Installing signs along the major street to warn road users approaching the intersection;*
   E. *Relocating the stop line(s) and making other changes to improve the sight distance at the intersection;*
   F. *Installing measures designed to reduce speeds on the approaches;*
   G. *Installing an Intersection Control Beacon (see Section 4S.02) or Stop Beacon (see Section 4S.05) at the intersection to supplement STOP sign control;*
   H. *Installing a Warning Beacon (see Section 4S.03) on warning signs in advance of a stop-controlled intersection on major-street and/or minor-street approaches;*
   I. *Adding one or more lanes on a minor-street approach to reduce the number of vehicles per lane on the approach;*
   J. *Revising the geometrics at the intersection to channelize vehicular movements and reduce the time required for a vehicle to complete a movement, which could also assist pedestrians;*
   K. *Revising the geometrics at the intersection to add pedestrian median refuge islands and/or curb extensions;*
   L. *Installing roadway lighting if a disproportionate number of crashes occur at night;*
   M. *Restricting one or more turning movements on a full-time or part-time basis if alternate routes are available;*

   N. *Installing on the major street a pedestrian-actuated device: Warning Beacon (see Section 4S.03), rectangular rapid-flashing beacon (see Section 4L.01), or In-Roadway Warning Lights (see Chapter 4U), if pedestrian safety is the major concern;*
   O. *If the warrant is satisfied, installing all-way stop control;*
   P. *Installing a pedestrian hybrid beacon (see Chapter 4J) on the major street to address pedestrian safety;*
   Q. *Installing a circular intersection; and*
   R. *Employing other alternatives, depending on conditions at the intersection.*

Section 2B.09 <u>No Intersection Control</u>

Guidance:

01   *The decision not to use intersection control should be based on engineering judgment.*

Option:

02   The following factors may be considered:
   A. Intersection sight distance is adequate on all approaches.
   B. All approaches to the intersection are a single lane and there are no separate turn lanes.
   C. The combined motor vehicle, bicycle, and pedestrian volume (existing or projected) entering the intersection from all approaches averages less than 1,000 units per day or 80 units in the peak hour.
   D. There are no marked crosswalks or bicycle lanes on any approach.
   E. None of the approaches to the intersection are for a through highway, main road, or higher functional classification.
   F. The angle of intersection is between 90 and 75 degrees.
   G. The functional classification of the intersecting streets is either the intersection of two local streets or the intersection of a local street with a collector street.

Section 2B.10 <u>Yield Control</u>

Guidance:

01   *At intersections where a full stop is not necessary at all times, consideration should first be given to using less restrictive measures such as YIELD signs.*

02   *Yield control should be considered when engineering judgment indicates that all of the following conditions exist:*
   A. *Intersection sight distance is adequate on the approaches to be controlled by YIELD signs.*
   B. *All approaches to the intersection are a single lane and there are no separate turn lanes.*
   C. *One of the following crash-related criteria applies:*
   D. *For changing from no intersection control to yield control, there have been two or more reported crashes in the previous 12 months that are susceptible to correction by the installation of a YIELD sign.*
   E. *For changing from minor road stop control to yield control, there have been two or fewer reported crashes in the previous 12 months.*
   F. *The combined motor vehicle, bicycle, and pedestrian volume entering the intersection averages less than 1,800 units per day or 140 units in the peak hour.*
   G. *The angle of intersection is between 90 and 75 degrees.*
   H. *The functional classification of the intersecting streets is either the intersection of two local streets or the intersection of a local street with a collector street.*

Option:

03   YIELD signs may be installed at an intersection when any of the following conditions apply:
   A. At the second intersection of a divided highway crossing or median break functioning as two separate intersections (see Figure 2B-19). In this case, a YIELD sign may be installed at the entrance to the second intersection.
   B. For a channelized turn lane that is separated from the adjacent travel lanes by an island, even if the adjacent lanes at the intersection are controlled by a highway traffic control signal or by a STOP sign.
   C. At an intersection where a special problem exists and where engineering judgment indicates the problem to be susceptible to correction by the use of the YIELD sign.
   D. Facing the entering roadway for a merge-type movement if engineering judgment indicates that control is needed because acceleration geometry and/or sight distance is not adequate for merging traffic operation.
   E. On low-volume rural roads if engineering judgment indicates that a YIELD sign would provide adequate control.
   F. On an approach to an intersection where the only permissible movement is a right-turn movement with an intersection geometry similar to a channelized right-turn lane or an approach to a roundabout.

Guidance:

04   *The YIELD signs should be installed on opposing minor-street approaches (for a four-leg intersection) or on the minor-street approach (for a three-leg intersection). When two intersecting roadways have relatively equal volumes, speeds, and other characteristics, yield control should be installed on the approach that conflicts the most with established pedestrian crossing activity, school walking routes, or bicycle crossing activity.*

Standard:

05   **A YIELD sign shall be used to require road users to yield the right-of-way to other traffic at the entrance to a roundabout. YIELD signs at roundabouts shall be used to control the approach roadways and shall not be used to control the circulatory roadway.**

06   **YIELD signs shall not be placed on all of the approaches to an intersection, except at roundabouts.**

Section 2B.11 Minor Road Stop Control

Guidance:

01   *Stop control on the minor-road approach or approaches to an intersection should be considered when engineering judgment indicates that one or more of the following conditions exist:*

    A. *A restricted view exists that requires road users to stop in order to adequately observe conflicting traffic on the through street or highway.*
    B. *Crash records indicate that:*
        1. *For a four-leg intersection, there are three or more reported crashes in a 12-month period or six or more reported crashes in a 36-month period. The crashes should be susceptible to correction by installation of minor-road stop control.*
        2. *For a three-leg intersection, there are three or more reported crashes in a 12-month period or five or more reported crashes in a 36-month period. The crashes should be susceptible to correction by installation of minor-road stop control.*
    C. *The intersection is of a lower functional classification road with a higher functional classification road.*
    D. *Conditions that previously supported the installation of all-way stop control no longer exist.*

02   *On low-volume rural roads, a STOP sign should be considered at an intersection where engineering judgment indicates that Item C in Paragraph 1 of this Section is applicable or where the intersection has inadequate sight distance for the operating vehicle speeds.*

Section 2B.12 All-Way Stop Control

Support:

01   The provisions in the following sections describe warrants for the recommended engineering study to determine all-way stop control. Warrants are not a substitute for engineering judgment. The fact that a warrant for a particular traffic control device is met is not conclusive justification to install or not install all-way stop control. Because each intersection will have unique characteristics that affect its operational performance or safety, it is the engineering study for a given intersection that is ultimately the basis for a decision to install or not install all-way stop control.

02   All-way stop controls at intersections with substantially differing approach volumes can reduce the effectiveness of these devices for all roadway users.

Guidance:

03   *The decision to establish all-way stop control at an unsignalized intersection should be based on an engineering study. The engineering study for all-way stop control should include an analysis of factors related to the existing operation and safety at the intersection, the potential to improve these conditions, and the applicable factors contained in the following all-way stop control warrants:*

    A. *All-Way Stop Control Warrant A: Crash Experience (see Section 2B.13)*
    B. *All-Way Stop Control Warrant B: Sight Distance (see Section 2B.14)*
    C. *All-Way Stop Control Warrant C: Transition to Signal Control or Transition to Yield Control at a Circular Intersection (see Section 2B.15)*
    D. *All-Way Stop Control Warrant D: 8-Hour Volume (Vehicles, Pedestrians, Bicycles) (see Section 2B.16)*
    E. *All-Way Stop Control Warrant E: Other Factors (see Section 2B.17)*

Option:

04   The decision to install all-way stop control on site roadways open to public travel may be based on engineering judgment.

Standard:

05   **The satisfaction of an all-way stop control warrant or warrants shall not in itself require the installation of all-way stop control at an unsignalized intersection.**

Section 2B.13 All-Way Stop Control Warrant A: Crash Experience

Option:

01   All-way stop control may be installed at an intersection where an engineering study indicates that:

A. For a four-leg intersection, there are five or more reported crashes in a 12-month period or six or more reported crashes in a 36-month period that were of a type susceptible to correction by the installation of all-way stop control.

B. For a three-leg intersection, there are four or more reported crashes in a 12-month period or five or more reported crashes in a 36-month period that were of a type susceptible to correction by the installation of all-way stop control.

Section 2B.14 All-Way Stop Control Warrant B: Sight Distance

Option:

01   All-way stop control may be installed at an intersection where an engineering study indicates that sight distance on the minor-road approaches controlled by a STOP sign is not adequate for a vehicle to turn onto or cross the major (uncontrolled) road.

Support:

02   At such a location, a road user, after stopping, cannot see conflicting traffic and is not able to negotiate the intersection unless conflicting cross traffic is also required to stop.

Section 2B.15 All-Way Stop Control Warrant C: Transition to Signal Control or Transition to Yield Control at a Circular Intersection

Option:

01   All-way stop control may be installed at locations where all-way stop control is an interim measure that can be installed to control traffic while arrangements are being made for the installation of a traffic control signal (see Chapter 4C) at the intersection or for the installation of yield control at a circular intersection.

Section 2B.16 All-Way Stop Control Warrant D: 8-Hour Volume (Vehicles, Pedestrians, Bicycles)

Option:

01   All-way stop control may be installed at an intersection where an engineering study indicates:

A. The combined motor vehicle, bicycle, and pedestrian volume entering the intersection from the major-street approaches is at least 300 units per hour for each of any 8 hours of a typical day; and

B. The combined motor vehicle, bicycle, and pedestrian volume entering the intersection from the minor-street approaches is at least 200 units per hour for each of any of the same 8 hours.

02   If the 85th-percentile approach speed of the major-street traffic exceeds 40 mph, the minimum vehicular volume warrants may be reduced to 70 percent of the values given in Items A and B in Paragraph 1 of this Section.

Section 2B.17 All-Way Stop Control Warrant E: Other Factors

Option:

01   All-way stop control may be installed at an intersection where an engineering study indicates that all-way stop control is needed due to other factors not addressed in the other all-way stop control warrants. Such other factors may include, but are not limited to, the following:

A. The need to control left-turn conflicts,

B. An intersection of two residential neighborhood collector (through) streets of similar design and operating characteristics where all-way stop control would improve traffic operational characteristics of the intersection, or

C. Where pedestrian and/or bicyclist movements support the installation of all-way stop control.

Section 2B.18 STOP Sign or YIELD Sign Placement

Standard:

01   **The STOP or YIELD sign shall be installed on the near side of the intersection on the right-hand side of the approach to which it applies. When the STOP or YIELD sign is installed at this required location and the sign visibility is restricted, a Stop Ahead sign (see Section 2C.35) shall be installed in advance of the STOP sign or a Yield Ahead sign (see Section 2C.35) shall be installed in advance of the YIELD sign.**

02   **The STOP or YIELD sign shall be located as close as practicable to the intersection it regulates, while optimizing its visibility to the road user it is intended to regulate.**

03   **STOP signs and YIELD signs shall not be mounted on the same post.**

Support:

04   Section 2A.05 contains information about mounting signs back-to-back with a STOP or YIELD sign.

Guidance:

05   *STOP or YIELD signs should not be placed farther than 50 feet from the edge of the pavement of the intersected roadway (see Drawing F in Figure 2A-3).*

06   *Supplemental plaques used in conjunction with a STOP or YIELD sign should be limited to those specified for such use in this Manual.*

Option:

07   Where drivers proceeding straight ahead must yield to traffic approaching from the opposite direction, such as at a one-lane bridge, a TO ONCOMING TRAFFIC (R1-2aP) plaque (see Figure 2B-1) may be mounted below the YIELD sign.

08   Where drivers must yield to traffic in a multi-lane roundabout, a TO TRAFFIC IN CIRCLE (R1-2bP) or TO ALL LANES (R1-2cP) plaque (see Figure 2B-1) may be mounted below the YIELD sign.

Support:

09   Figure 2A-3 shows examples of some typical placements of STOP signs and YIELD signs.

10   Section 2A.13 contains additional information about separate and combined mounting of other signs with STOP or YIELD signs.

Guidance:

11   *Stop lines that are used to supplement a STOP sign should be located as described in Section 3B.19. Yield lines that are used to supplement a YIELD sign should be located as described in Section 3B.19.*

12   *Where there is a marked crosswalk at the intersection, the STOP sign should be installed in advance of the edge of the crosswalk that is nearest to the approaching traffic.*

13   *Except at roundabouts and channelized right-turn lanes, where there is a marked crosswalk at the intersection, the YIELD sign should be installed in advance of the edge of the crosswalk that is nearest to the approaching traffic.*

14   *Where two roads intersect at an acute angle, the STOP or YIELD sign should be positioned at an angle, or shielded, so that the legend is out of view of traffic to which it does not apply.*

15   *If a raised splitter island is available on the left-hand side of a multi-lane roundabout approach, an additional YIELD sign should be placed on the left-hand side of the approach.*

Option:

16   If a raised splitter island is available on the left-hand side of a single-lane roundabout approach, an additional YIELD sign may be placed on the left-hand side of the approach.

17   At wide-throat intersections or where two or more approach lanes of traffic exist on the signed approach, an additional STOP or YIELD sign may be installed on the left-hand side of the road and/or a stop or yield line may be used to improve observance of the right-of-way control. At channelized intersections or at divided roadways separated by a median or divisional island, the additional STOP or YIELD sign may be placed on a channelizing island, or in the median or on the divisional island. An additional STOP or YIELD sign may also be placed overhead facing the approach at the intersection to improve observance of the right-of-way control.

Standard:

18   **More than one STOP sign or more than one YIELD sign shall not be placed on the same support facing in the same direction.**

Option:

19   For a yield-controlled channelized right-turn movement onto a roadway without an acceleration lane and for an entrance ramp onto a freeway or expressway without an acceleration lane, a NO MERGE AREA (W4-5aP) supplemental plaque (see Section 2C.45) may be mounted below a Yield Ahead (W3-2) sign and/or below a YIELD (R1-2) sign when engineering judgment indicates that road users would expect an acceleration lane to be present.

Section 2B.19 <u>Yield Here To Pedestrians Signs and Stop Here For Pedestrians Signs (R1-5 Series)</u>

Support:

01   The R1-5 series signs are intended to mitigate the scenario that can place pedestrians at risk by blocking other drivers' view of pedestrians and by blocking the pedestrians' view of the vehicles approaching in the adjacent lanes.

Standard:

02   **Yield Here to (Stop Here for) Pedestrians (R1-5, R1-5a, R1-5b, R1-5c, R1-5d, and R1-5e) signs (see Figure 2B-2) shall be used if yield (stop) lines are used in advance of a marked crosswalk only where it crosses an uncontrolled multi-lane approach. The Stop Here for Pedestrians signs shall only be used where the law specifically requires that a driver must stop for a pedestrian in a crosswalk. The legend STATE LAW shall not be displayed on the R1-5 series signs.**

Guidance:

03   *If yield (stop) lines and Yield Here to (Stop Here for) Pedestrians signs are used in advance of a crosswalk that crosses an uncontrolled multi-lane approach, the signs should be placed 20 to 50 feet in advance of the nearest edge of the crosswalk (see Section 3B.19 and Figure 3B-16).*

Standard:

04   **When used with a School Crossing assembly within school zones (see Part 7), the R1-5a and R1-5c signs shall be used in place of the R1-5 and R1-5b signs in accordance with Paragraph 2 of this Section.**

05   **When used with a Trail Crossing assembly (see Section 2C.54), the R1-5d and R1-5e signs shall be used in place of the R1-5 and R1-5b signs in accordance with Paragraph 2 of this Section.**

Guidance:

06   *When Yield Here to (Stop Here for) Pedestrians signs are provided in advance of a crosswalk across an multi-lane approach, parking should be prohibited in the area between the yield (stop) line and the crosswalk.*

07   *Yield (stop) lines and Yield Here to (Stop Here for) Pedestrians signs should not be used in advance of crosswalks that cross an approach to or departure from a roundabout.*

Option:

08   Yield Here to (Stop Here for) Pedestrians signs may be used in accordance with Paragraphs 2 through 4 of this Section even if yield (stop) lines are not used.

09   A Pedestrian Crossing (W11-2) warning sign may be placed overhead or may be post-mounted with a diagonal downward-pointing arrow (W16-7P) plaque at the crosswalk location where Yield Here to (Stop Here for) Pedestrians signs have been installed in advance of the crosswalk.

Standard:

10   **If a W11-2 sign is post-mounted at the crosswalk location where a Yield Here to (Stop Here for) Pedestrians sign is used on the approach, the Yield Here to (Stop Here for) Pedestrians sign shall not be placed on the same post as the W11-2 sign.**

Option:

11   An advance Pedestrian Crossing (W11-2) warning sign with an AHEAD or a distance supplemental plaque may be used in conjunction with a Yield Here to (Stop Here for) Pedestrians sign on the approach to the same crosswalk.

12   In-Street Pedestrian Crossing signs and Yield Here to (Stop Here for) Pedestrians signs may be used together at the same crosswalk.

Section 2B.20 <u>In-Street and Overhead Pedestrian and Trail Crossing Signs (R1-6 and R1-9 Series)</u>

Option:

01   The In-Street Pedestrian Crossing (R1-6 or R1-6a) sign (see Figure 2B-2), In-Street Trail Crossing (R1-6d or R1-6e) sign (see Figure 2B-2), the Overhead Pedestrian Crossing (R1-9 or R1-9a) sign (see Figure 2B-2), or the Overhead Trail Crossing (R1-9d or R1-9e) sign (see Figure 2B-2) may be used to remind road users of laws regarding right-of-way at an unsignalized crosswalk. The legend STATE LAW may be displayed at the top of the R1-6 series and R1-9 series signs if applicable. On the R1-6 series signs, the legends STOP or YIELD may be used instead of the appropriate STOP sign or YIELD sign symbol.

02   Highway agencies may develop and apply criteria for determining the applicability of In-Street Pedestrian Crossing signs.

Figure 2B-2. Unsignalized Pedestrian Crosswalk Signs

★ The legend STATE LAW is optional. A fluorescent yellow-green background color may be used instead of yellow for this sign.

Signs are not shown in proportion to their designated sizes.

Standard:

03   **The STOP FOR legend shall only be used in States where the State law specifically requires that a driver must stop for a pedestrian or a bicyclist in a crosswalk.**

04   **If used, In-Street Pedestrian or Trail Crossing signs shall only be placed in the roadway at the crosswalk location on the center line, on a median island, on a lane line, or on an edge line.**

05   **The In-Street Pedestrian or Trail Crossing sign shall not be post-mounted on the left-hand or right-hand side of the roadway.**

Support:

06   Section 3I.02 contains information about the use of tubular markers to provide additional emphasis for a pedestrian crossing.

Standard:

07   **If used, the Overhead Pedestrian or Trail Crossing sign shall be placed over the roadway at the crosswalk location.**

08   **When used at an uncontrolled crossing, the In-Street or Overhead Pedestrian Crossing sign shall be used only as a supplement to a Pedestrian Crossing (W11-2) warning sign with a diagonal downward-pointing arrow (W16-7P) plaque at the crosswalk location.**

09   **When used at an uncontrolled crossing, the In-Street or Overhead Trail Crossing sign shall be used only as a supplement to a Trail Crossing (W11-15) warning sign with a diagonal downward-pointing arrow (W16-7P) plaque at the crosswalk location.**

10   **An In-Street or Overhead Pedestrian or Trail Crossing sign shall not be placed in advance of the crosswalk to educate road users about the State law prior to reaching the crosswalk, nor shall it be installed as an educational display that is not near any crosswalk.**

Guidance:

11   *If an island (see Chapter 3J) is available, the In-Street Pedestrian or Trail Crossing sign, if used, should be placed on the island.*

Option:

12   In-Street Pedestrian or Trail Crossing signs may be mounted back-to-back in the median or on the center line of an undivided roadway.

Standard:

13   **The In-Street Pedestrian or Trail Crossing sign and the Overhead Pedestrian Crossing or Trail sign shall not be used at crosswalks on approaches controlled by a traffic control signal, pedestrian hybrid beacon, or an emergency-vehicle hybrid beacon.**

14   **Except where the In-Street Crossing sign is placed on a physical island, the sign support shall be designed to bend over and then bounce back to its normal vertical position when struck by a vehicle.**

Option:

15   The In-Street and Overhead Pedestrian and Trail Crossing sign may be used at intersections or midblock pedestrian crossings with flashing beacons.

Support:

16   The provisions of Section 2A.15 concerning mounting height are not applicable for the In-Street Pedestrian Crossing sign. Section 2A.18 contains information about sign mounting methods.

Standard:

17   **The top of an In-Street Pedestrian or Trail Crossing sign shall be a maximum of 4 feet above the pavement surface. The top of an In-Street Pedestrian or Trail Crossing sign placed in an island shall be a maximum of 4 feet above the island surface.**

Option:

18   The In-Street Pedestrian Crossing or Trail Crossing signs may be used seasonally to prevent damage in winter because of plowing operations, and may be removed at night if the pedestrian activity at night is minimal.

19   Both sign mounting types, In-Street Crossing (R1-6 series) signs and Overhead Crossing (R1-9 series) signs, may be used together at the same crosswalk.

SPEED LIMIT SIGNS AND PLAQUES

Section 2B.21 Speed Limit Sign (R2-1)

Support:

01   In general, the maximum speed limits applicable to rural and urban roads are established:

   A. Statutorily – a maximum speed limit applicable to a particular class of road, such as freeways or city streets, that is established by State law; or
   B. As speed zones – based on engineering studies.

02   State statutory limits might restrict the maximum speed limit that can be established on a particular road, notwithstanding what an engineering study might indicate.

03   Agencies with designated authorities to set speed limits, which include States, and sometimes local jurisdictions, can establish non-statutory speed limits or designate reduced speed zones using an engineering study. Setting appropriate speed limits is especially important to ensure safety for all road users in varying types of contexts, particularly on roadways where adjacent land use suggests that trips could be served by varied modes. These situations include urban and suburban non-freeway arterials or rural arterials that serve as main streets in smaller communities, consistent with the context classifications of urban core, urban, suburban, and rural towns found in "A Policy on Geometric Design of Highways and Streets," 2018 Edition, AASHTO. When setting a speed limit, a range of factors such as land-use context, pedestrian and bicyclist activity, crash history, intersection spacing, driveway density, roadway geometry, roadside conditions, roadway functional classification, traffic volume, and observed speeds can influence the speed limit determined in the engineering study. The engineering study will determine which of the recommended factors will prevail in setting the speed limit.

04   Jurisdictions can use speed limit setting tools and methods such as expert systems and those consistent with the safe system approach as part of the required engineering study for a non-statutory speed limit. As speed limit setting tools vary, jurisdictions need to be aware of their limitations and advantages, possible variation between the tools and the need to explore gaps or weaknesses of tools, and weigh the output accordingly in consideration of setting speed limits.

05   To achieve desired operating speeds, agencies often implement other speed management strategies concurrently with setting speed limits, such as traffic calming measures, geometric design features, speed safety cameras, and increased enforcement.

Standard:

06   **Speed zones (other than statutory speed limits) shall only be established on the basis of an engineering study that has been performed in accordance with traffic engineering practices. The engineering study shall consider the roadway context.**

Guidance:

07   *Among the factors that should be considered when conducting an engineering study for establishing or reevaluating speed limits within speed zones are the following:*

   A. *Roadway environment (such as roadside development, number and frequency of driveways and access points, and land use), functional classification, public transit volume and location or frequency of stops, parking practices, and pedestrian and bicycle facilities and activity;*
   B. *Roadway characteristics (such as lane widths, shoulder condition, grade, alignment, median type, and sight distance);*
   C. *Geographic context (such as an urban district, rural town center, non-urbanized rural area, or suburban area), and multi-modal trip generation;*
   D. *Reported crash experience for at least a 12-month period;*
   E. *Speed distribution of free-flowing vehicles including the pace, median (50th-percentile), and 85th-precentile speeds; and*
   F. *A review of past speed studies to identify any trends in operating speeds.*

08   *When the 85th-percentile speed is appreciably greater than the posted speed limit, and the roadway context does not support setting a higher speed limit, the engineering study should consider whether changes to geometric features, enforcement, and/or other speed-reduction countermeasures might improve compliance with the posted speed limit. A similar approach should be used if the results of past speed studies indicate that the 85th-percentile speed has consistently increased.*

09   *On urban and suburban arterials, and on rural arterials that serve as main streets through developed areas of communities, the 85th-percentile speed should not be used to set speed limits without consideration of all factors described in Paragraph 7 of this Section.*

10   *On a freeway, expressway, or rural highway (outside urbanized locations or conditions), the speed limit that is posted within a speed zone should be within 5 mph of the 85th-percentile speed of free-flowing motor-vehicle traffic under the following conditions:*

   A.   *All factors described in Paragraph 7 of this Section have been considered and determined to be non-mitigating, and*

   B.   *The measures described in Paragraph 8 of this Section have been considered to the extent practicable.*

11   *State and local agencies should conduct engineering studies to reevaluate non-statutory speed limits on segments of their roadways that have undergone significant changes since the last review (such as changes to roadway context, the addition or elimination of parking or driveways, changes in the number of travel lanes, changes in the configuration of bicycle lanes, changes to road geometrics, changes in traffic control signal coordination, or significant changes in traffic volumes).*

12   *Speed studies for signalized intersection approaches should be taken outside the influence area of the traffic control signal, which is generally considered to be approximately 1/2 mile, to avoid obtaining skewed results for the speed distribution. If the signal spacing is less than 1 mile, the speed study should be at approximately the middle of the segment.*

Standard:

13   **The Speed Limit (R2-1) sign (see Figure 2B-3) shall display the limit established by law, ordinance, regulation, or as adopted by the authorized agency based on an engineering study. The speed limits displayed shall be in multiples of 5 mph.**

14   **Speed Limit (R2-1) signs, indicating speed limits for which posting is required by law, shall be located at the points of change from one speed limit to another.**

15   **At the downstream end of the section to which a particular speed limit applies, a Speed Limit sign showing the next speed limit shall be installed.**

16   **Speed Limit signs indicating the statutory speed limits shall be installed at entrances to the State and, where appropriate, at jurisdictional boundaries in urban areas.**

Guidance:

17   *Additional Speed Limit signs should be installed beyond interchanges and major intersections and at other locations where it is necessary to remind road users of the speed limit that is applicable.*

Support:

18   The "Traffic Control Devices Handbook" contains suggested criteria on the spacing of speed limit signs.

Option:

19   If a jurisdiction has a policy of installing Speed Limit signs in accordance with statutory requirements only on the streets that enter a city, neighborhood, or residential area to indicate the speed limit that is applicable to the entire city, neighborhood, or residential area unless otherwise posted, a CITYWIDE (R2-5aP), NEIGHBORHOOD (R2-5bP), or RESIDENTIAL (R2-5cP) plaque may be mounted above the Speed Limit sign and an UNLESS OTHERWISE POSTED (R2-5P) plaque may be mounted below the Speed Limit sign (see Figure 2B-3).

Support:

20   Section 2C.40 contains information about the use of speed zone signs to inform road users of a reduced or variable speed zone to provide advance notice to comply with the posted speed limit ahead.

Option:

21   If a W3-5b sign is posted to provide notice of a variable speed zone, an END VARIABLE SPEED LIMIT (R2-13) sign (see Figure 2B-3) may be installed at the downstream end of the zone to provide notice to road users of the termination of the speed zone.

Standard:

22   **If a W3-5c sign is posted to provide notice of a truck speed zone, an END TRUCK SPEED LIMIT (R2-14) sign (see Figure 2B-3) shall be installed at the downstream end of the zone to provide notice to road users of the termination of the speed zone.**

Guidance:

23   *An advisory speed plaque (see Section 2C.59) mounted below a warning sign should be used to warn road users of an advisory speed for a roadway condition. A Speed Limit sign should not be used for this purpose.*

24   *Advance traffic control warning signs (see Section 2C.35), intersection warning signs (see Section 2C.41), and/or other traffic control devices are appropriate warning prior to a signalized intersection. A Speed Limit sign should not be used for this purpose.*

Figure 2B-3. Speed Limit Signs and Plaques

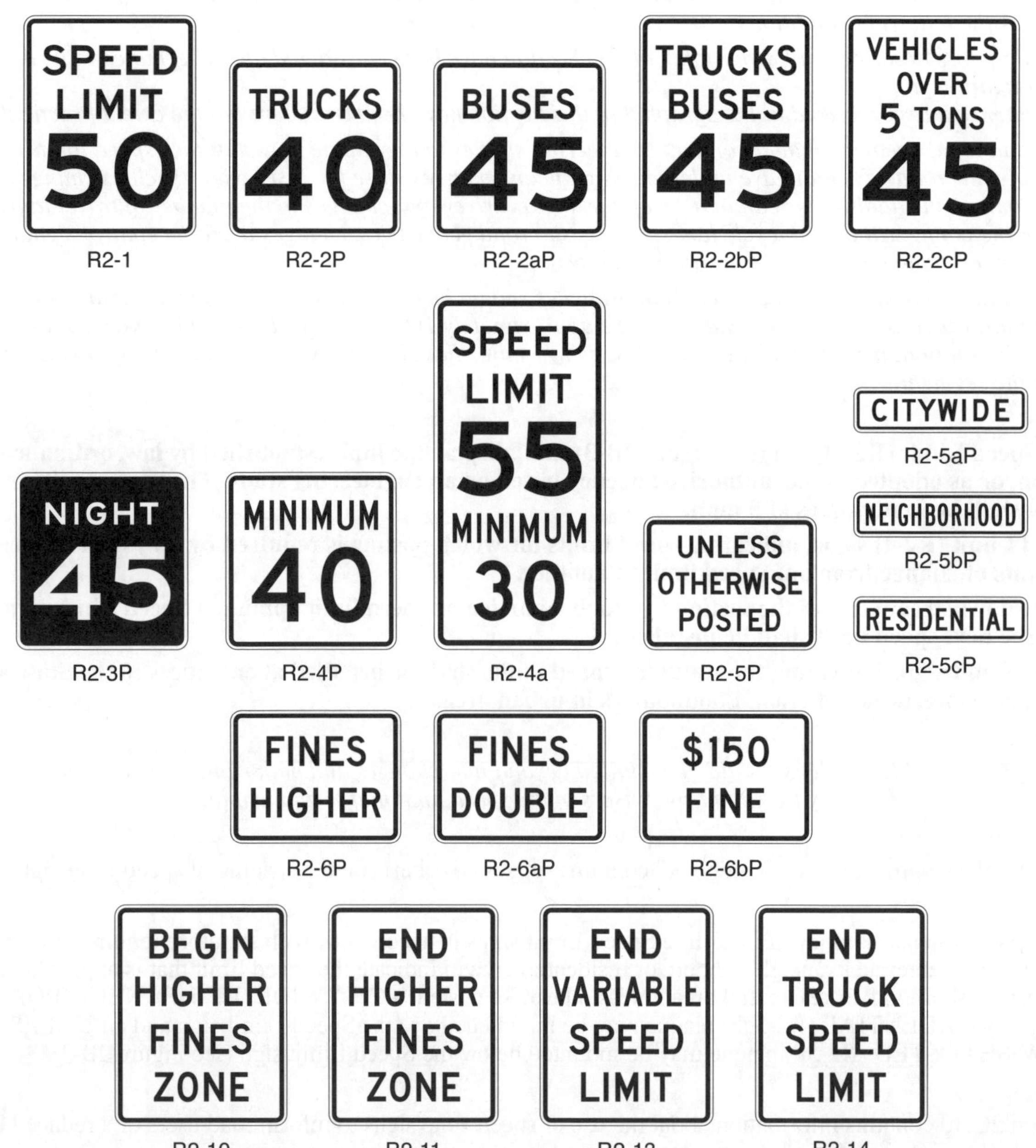

Option:

25   Two types of Speed Limit signs may be used: one to designate passenger car speeds, including any nighttime information or maximum or minimum speed limit that might apply; and the other to show any special speed limits for trucks and other vehicles.

Guidance:

26   *No more than three speed limits should be displayed on any one Speed Limit sign or assembly.*

Option:

27   A variable speed limit sign that changes the speed limit for traffic and ambient conditions may be installed provided that the appropriate speed limit is displayed at the proper times and locations in accordance with Paragraphs 9 and 10 of this Section.

Standard:

28   **The variable speed limit sign legend "SPEED LIMIT" shall be a black legend on a white retroreflective background. The variable speed limit legend shall be displayed in white LEDs on an opaque black background.**

Support:

29   Section 2C.13 contains information about the use of a Vehicle Speed Feedback plaque mounted below a Speed Limit sign that displays to approaching drivers the speed at which they are traveling.

30   Advisory speed signs and plaques are discussed in Sections 2C.12 and 2C.59. Temporary traffic control zone speed signs are discussed in Part 6. The WORK ZONE (G20-5aP) plaque intended for installation above a Speed Limit sign is discussed in Section 6G.08. School Speed Limit signs are discussed in Section 7B.05.

Section 2B.22 Vehicle Speed Limit Plaques (R2-2P Series)

Standard:

01   **Where a special speed limit applies to certain classes of vehicles, the Truck Speed Limit (R2-2P) plaque, Bus Speed Limit (R2-2aP) plaque, Truck-Bus Speed Limit (R2-2bP) plaque, or Vehicles over X Tons Speed Limit (R2-2cP) plaque (see Figure 2B-3) shall be displayed below the Speed Limit (R2-1) sign, except as provided in Paragraph 2 of this Section.**

Option:

02   The legend of a Vehicle Speed Limit (R2-2P series) plaque may be combined in a single sign and displayed below the SPEED LIMIT XX legend, similar to the Combined Maximum and Minimum Speed Limits (R2-4a) sign (see Section 2B.24).

03   A different vehicle class legend may be substituted on the R2-2P series plaque for other classes of vehicles not included in Paragraph 1 of this Section.

Section 2B.23 Night Speed Limit Plaque (R2-3P)

Standard:

01   **Where different speed limits are prescribed for day and night, both limits shall be posted.**

Guidance:

02   *A Night Speed Limit (R2-3P) plaque (see Figure 2B-3) should be reversed using a white retroreflective legend and border on a black background.*

Option:

03   A Night Speed Limit plaque may be combined with or installed below the standard Speed Limit (R2-1) sign.

Section 2B.24 Minimum Speed Limit Plaque (R2-4P) and Combined Maximum and Minimum Speed Limits Sign (R2-4a)

Standard:

01   **A Minimum Speed Limit (R2-4P) plaque (see Figure 2B-3) shall be displayed only in combination with a Speed Limit sign. Where used, the R2-4P plaque shall be mounted below a Speed Limit (R2-1) sign.**

Option:

02   Where engineering judgment determines that slow speeds on a highway might impede the normal and reasonable movement of traffic, the Minimum Speed Limit (R2-4P) plaque may be installed below a Speed Limit (R2-1) sign to indicate the minimum legal speed. In lieu of a sign assembly with the R2-1 sign and R2-4P plaque, the Combined Maximum and Minimum Speed Limits (R2-4a) sign may be used.

Section 2B.25 Higher Fines Signs and Plaque (R2-6P, R2-10, and R2-11)

Standard:

01   **Except as provided in Paragraph 3 of this Section, if increased fines are imposed for traffic violations within a designated zone of a roadway, a BEGIN HIGHER FINES ZONE (R2-10) sign (see Figure 2B-3) or a FINES HIGHER (R2-6P) plaque (see Figure 2B-3) shall be used to provide notice to road users.**

02   **If an R2-10 sign or an R2-6P plaque is posted to provide notice of increased fines for traffic violations, an END HIGHER FINES ZONE (R2-11) sign (see Figure 2B-3) shall be installed at the downstream end of the zone to provide notice to road users of the termination of the increased fines zone.**

Option:

03   The BEGIN HIGHER FINES ZONE (R2-10) sign or FINES HIGHER (R2-6P) plaque may be omitted where the higher fines zone is established by statute.

Guidance:

04   *The BEGIN HIGHER FINES ZONE sign or FINES HIGHER plaque should be located at the beginning of the temporary traffic control zone, school zone, or other applicable designated zone and just beyond any interchanges, major intersections, or other major traffic generators.*

05   *Agencies should limit the use of the Higher Fines signs and plaque to locations where work is actually underway, or to locations where the roadway, shoulder, or other conditions, including the presence of a school zone and/or a reduced school speed limit zone, require a speed reduction or extra caution on the part of the road user.*

Standard:

06   **The Higher Fines signs and plaque shall have a black legend and border on a white rectangular background. All supplemental plaques mounted below the Higher Fines signs and plaque shall have a black legend and border on a white rectangular background.**

07   **The FINES HIGHER plaque shall be mounted below an applicable regulatory or warning sign in a temporary traffic control zone (see Section 6G.08), a school zone (see Section 7B.06), or other applicable designated zone.**

Option:

08   Alternate legends such as BEGIN (or END) DOUBLE FINES ZONE may also be used for the R2-10 and R2-11 signs.

09   The legend FINES HIGHER on the R2-6P plaque may be replaced by FINES DOUBLE (R2-6aP), $XX FINE (R2-6bP), or another legend appropriate to the specific regulation (see Figure 2B-3).

10   The following may be mounted below an R2-10 sign or R2-6P plaque:

   A. A supplemental plaque specifying the times that the higher fines are in effect (similar to the S4-1P plaque shown in Figure 7B-1),
   B. A supplemental plaque WHEN CHILDREN (WORKERS) ARE PRESENT, or
   C. A supplemental plaque WHEN FLASHING (similar to the S4-4P plaque shown in Figure 7B-1) if used in conjunction with a Speed Limit Sign Beacon (see Section 4S.04).

MOVEMENT AND LANE CONTROL SIGNS AND PLAQUES

Section 2B.26 Movement Prohibition Signs (R3-1 through R3-4, R3-18, and R3-27)

Standard:

01 **Movement Prohibition signs (see Figure 2B-4) shall be installed where specific movements are prohibited at an intersection approach except as provided in Paragraphs 13 and 17 of this Section.**

Guidance:

02 *Movement Prohibition signs should only be used to prohibit a turn or through movement from an entire approach and should not be used to designate movements that are required or permitted from a specific lane or lanes on a multi-lane approach.*

03 *Movement Prohibition signs should be placed where they will be most easily seen by road users who might be intending to make the movement.*

04 *If a No Right Turn (R3-1) sign (see Figure 2B-4) is used, at least one should be placed either over the roadway or at a right-hand corner of the intersection.*

05 *If a No Left Turn (R3-2) sign (see Figure 2B-4) is used, at least one should be placed over the roadway, at the far left corner of the intersection, on a median, or in conjunction with the STOP sign or YIELD sign located on the near right corner.*

06 *Except as provided in Item C in Paragraph 11 of this Section for signalized locations, if a NO TURNS (R3-3) sign (see Figure 2B-4) is used, two signs should be used, one at a location specified for a No Right Turn sign and one at a location specified for a No Left Turn sign.*

07 *If a No U-Turn (R3-4) sign (see Figure 2B-4) or a combination No U or Left Turn (R3-18) sign (see Figure 2B-4) is used, at least one should be used at a location specified for a No Left Turn sign.*

08 *If both left turns and U-turns are prohibited, the combination No U or Left Turn (R3-18) sign should be used instead of separate R3-2 and R3-4 signs.*

Support:

09 Sections 2B.27 through 2B.30 contain information regarding lane control signs that indicate the required or permitted movements from individual lanes.

Guidance:

10 *If a No Straight Through (R3-27) sign (see Figure 2B-4) is used, at least one should be placed either over the roadway or at a location where it can be seen by road users who might be intending to travel straight through the intersection.*

11 *If turn prohibition signs are installed in conjunction with traffic control signals:*

   A. The No Right Turn sign should be installed adjacent to a signal face viewed by road users in the right-hand lane.

   B. The No Left Turn (or No U-Turn or combination No U or Left Turn) sign should be installed adjacent to a signal face viewed by road users in the left-hand lane.

   C. A NO TURNS sign should be placed adjacent to a signal face viewed by all road users on that approach, or two signs should be used.

Option:

12 If turn prohibition signs are installed in conjunction with traffic control signals, an additional turn prohibition sign may be post-mounted to supplement the sign mounted overhead.

13 Where ONE WAY signs are used (see Section 2B.49), No Left Turn and No Right Turn signs may be omitted.

14 Where the movement restriction applies to certain vehicle classes, signs incorporating a supplementary legend, modified as appropriate, may be used to indicate the specific vehicle class restriction (R3-1b through R3-1d) or exception (R3-1e and R3-1f) (see Figure 2B-4). When the movement restriction applies during certain time periods only, the following Movement Prohibition signing alternatives may be used and are listed in order of preference:

   A. A blank-out or changeable message sign (see Chapter 2L) that displays the prohibited movement only during the time that the movement prohibition is applicable, especially at signalized intersections.

   B. Permanently-mounted signs incorporating a supplementary legend showing the hours and days during which the prohibition is applicable (R3-1g and R3-1h) (see Figure 2B-4).

   C. Portable signs, installed by proper authority, located off the roadway at each corner of the intersection. The portable signs are only to be used during the time that the movement prohibition is applicable.

Figure 2B-4. Movement Prohibition and Lane Control Signs and Plaques (Sheet 1 of 2)

✱ The diamond symbol may be used instead of the "HOV" word message. The minimum vehicle occupancy level may vary, such as 2+, 3+, 4+. The words "LANE" or "ONLY" may be used with this sign when appropriate.

Figure 2B-4. Movement Prohibition and Lane Control Signs and Plaques (Sheet 2 of 2)

R3-18

R3-19

R3-19a

R3-19b

R3-20L

R3-20R

R3-27

RIGHT LANE MUST EXIT

R3-33

R3-33a

Standard:

15   **The blank-out part-time electronic-display Movement Prohibition sign shall consist of a red circle and diagonal with a white prohibited movement on an opaque black background.**

Option:

16   Movement Prohibition signs may be omitted at a ramp entrance to an expressway or a channelized intersection where the design is such as to indicate clearly the one-way traffic movement on the ramp or turning lane.

Standard:

17   **The No Left Turn (R3-2) sign, the No U-Turn (R3-4) sign, and the combination No U or Left Turn (R3-18) sign shall not be used at approaches to roundabouts to prohibit drivers from turning left onto the circulatory roadway of a roundabout.**

Support:

18   At roundabouts, the use of R3-2, R3-4, or R3-18 signs to prohibit left turns onto the circulatory roadway might confuse drivers about the possible legal turning movements around the roundabout. Roundabout Circulation (R6-5P) plaques (see Section 2B.51) and/or ONE WAY (R6-1 or R6-2) signs are appropriate to indicate the travel direction within a roundabout.

Section 2B.27 Intersection Lane Control Signs (R3-5 through R3-8)

Standard:

01   **Intersection Lane Control signs (see Figure 2B-4), if used, shall require road users in certain lanes to turn, shall permit turns from a lane where such turns would otherwise not be permitted, shall require a road user to stay in the same lane and proceed straight through an intersection, or shall indicate permitted movements from a lane.**

Support:

02   Intersection Lane Control signs have three applications:

   A. Mandatory Movement Lane Control (R3-5 series and R3-7 series) signs,
   B. Optional Movement Lane Control (R3-6 series) signs, and
   C. Advance Intersection Lane Control (R3-8 series) signs.

Guidance:

03   *When Intersection Lane Control signs are mounted overhead, each sign used should be placed over the lane or a projection of the lane to which it applies.*

04   *On signalized approaches where through lanes that become mandatory turn lanes, multiple-lane turns that include shared lanes for through and turning movements, or other lane-use regulations are present that would be unexpected by unfamiliar road users, overhead Intersection Lane Control signs should be installed at the signalized location over the appropriate lanes or projections thereof and in advance of the intersection over the appropriate lanes.*

05   *Where overhead mounting on the approach is impracticable for the Advance and/or Intersection lane Control signs, one of the following alternatives should be employed:*

    A. *At locations where through lanes become mandatory turn lanes, a Mandatory Movement Lane Control (R3-7) sign should be post-mounted on the left-hand side of the roadway where a through lane is becoming a mandatory left-turn lane on a one-way street or where a median of sufficient width for the signs is available, or on the right-hand side of the roadway where a through lane is becoming a mandatory right-turn lane.*

    B. *At locations where a through lane is becoming a mandatory left-turn lane on a two-way street where a median of sufficient width for the signs is not available, and at locations where multiple-lane turns that include shared lanes for through and turning movements are present, an Advance Intersection Lane Control (R3-8 series) sign should be post-mounted in a prominent location in advance of the intersection, and consideration should be given to the use of an oversized version in accordance with Table 2B-1.*

06   *Use of an overhead sign for one approach lane should not require installation of overhead signs for the other lanes of that approach.*

Option:

07   Intersection Lane Control signs may be omitted where:

    A. A turn bay has been provided by physical construction or pavement markings, and
    B. Only the road users using such turn bays are permitted to make a turn in that direction.

08   At roundabouts, Intersection Lane Control (R3-5, R3-6, and R3-8 series) signs may display any of the arrow symbol options shown in Figure 2B-5.

Section 2B.28 Mandatory Movement Lane Control Signs (R3-5, R3-5a, R3-7, R3-19 Series, and R3-20) and Plaques

Standard:

01   **Mandatory Movement Lane Control (R3-5, R3-5a, and R3-7) signs (see Figure 2B-4), if used, shall indicate only the single vehicle movement that is required from the lane.**

02   **The Mandatory Movement Lane Control (R3-5 and R3-5a) symbol signs shall include the legend ONLY and shall be mounted overhead over the specific lanes to which they apply (see Section 2B.27). The R3-7 sign shall be for post-mounting only. The R3-7 sign shall not be mounted at the far side of the intersection.**

03   **When the mandatory movement applies to lanes exclusively designated for HOV traffic, the HOV 2+ (R3-5cP) supplemental plaque shall be used. When the mandatory movement applies to lanes that are not HOV facilities, but are lanes exclusively designated for buses and/or taxis, the TAXI LANE (R3-5dP) and/ or BUS LANE (R3-5gP) supplemental plaques shall be used.**

Figure 2B-5. Intersection Lane Control Sign Arrow Options for Roundabouts

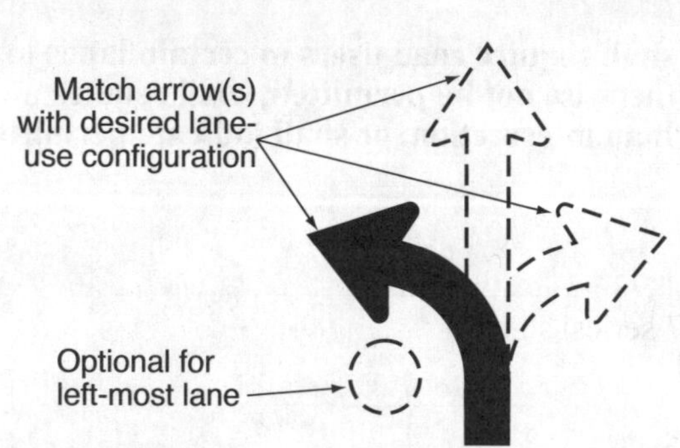

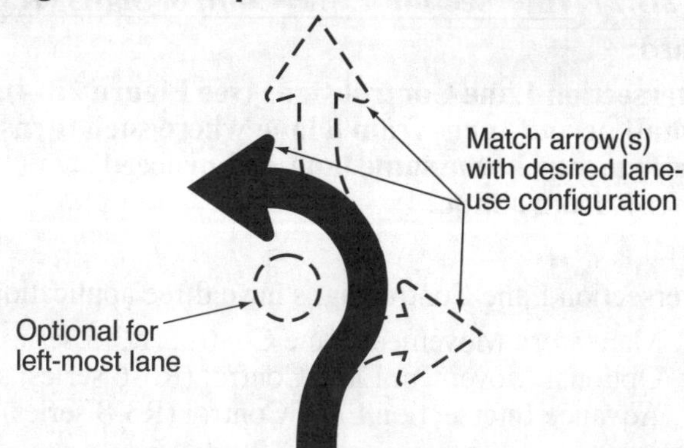

04 **If used, the Mandatory Movement Lane Control (R3-7) sign shall be located in advance of the intersection, such as near the upstream end of the mandatory movement lane, and/or at the near side of the intersection where the regulation applies.**

Guidance:

05 *The use of the Mandatory Movement Lane Control (R3-7) word message sign should be limited to only locations that are adjacent to the full-width portion of a mandatory turn lane. The R3-7 sign should not be installed adjacent to a through lane in advance of a turn bay taper or adjacent to a turn bay taper.*

06 *Mandatory Movement Lane Control signs should be accompanied by lane-use arrow markings, especially where traffic volumes are high, where there is a high percentage of commercial vehicles, or where other distractions exist.*

07 *Where the restriction does not apply to buses or bicycles an EXCEPT BUSES (R3-7aP) or EXCEPT BICYCLES (R3-7bP) plaque should be used.*

Option:

08 The Through Only (R3-5a) sign may be used to require a road user in a particular lane to proceed straight through an intersection.

09 The diamond symbol may be used instead of the word message HOV on the R3-5cP supplemental plaque.

10 Where a mandatory left or U-turn lane is added at a median location, a LANE FOR LEFT TURN ONLY (R3-19) or LANE FOR U TURN ONLY (R3-19a) sign may be post-mounted on the median at the beginning of the taper. Where a U turn and a left turn are both allowed, a LANE FOR U AND LEFT TURNS ONLY (R3-19b) sign may be used. Where a R3-19 series sign is used, Mandatory Movement Lane Control signs along the turn lane in the median may be omitted.

11 The R3-19 series signs may be used where the added median turn lane is separated from the through lanes by a channelizing or divisional island.

12 On an approach to a mandatory turn lane where traffic regularly enters the shoulder to access the turn lane inappropriately, creating safety or operational issues, a DO NOT DRIVE ON SHOULDER (R4-17) sign (see Section 2B.43) may be used to supplement the standard Mandatory Movement Lane Control (R3-5 and/or R3-7 series) signs.

Section 2B.29 <u>Optional Movement Lane Control Signs (R3-6 Series)</u>

Standard:

01 **Optional Movement Lane Control (R3-6, R3-6a and R3-6b) signs (see Figure 2B-4), if used, shall be used for two or more movements from a specific lane or to emphasize permitted movements. The Optional Movement Lane Control sign shall be mounted overhead over the specific lane to which it applies.**

02 **If used, the Optional Movement Lane Control signs shall indicate all permissible movements from specific lanes.**

03 **Because more than one movement is permitted from the lane, the word message ONLY shall not be used on an Optional Movement Lane Control sign.**

04 **Optional Movement Lane Control signs shall be used for two or more movements from a specific lane where a movement, not allowed by State statute or local ordinance, is permitted.**

05 **The Optional Movement Lane Control signs shall not be used alone to effect a turn prohibition.**

Guidance:

06 *If used, the Optional Movement Lane Control sign should be located overhead in advance of the intersection, such as near the upstream end of an adjacent mandatory movement lane, and/or overhead at the intersection where the regulation applies.*

Section 2B.30 <u>Advance Intersection Lane Control Signs (R3-8 Series)</u>

Option:

01 Advance Intersection Lane Control (R3-8, R3-8a, and R3-8b) signs (see Figure 2B-4) may be used to indicate the configuration of all lanes ahead.

02 The word messages ONLY, THRU, HOV 2+, TAXI, BUS, or BIKE, or the bicycle symbol, may be used within the border in combination with the arrow symbols of the R3-8 sign series. The R3-5cP, R3-5dP, and R3-5gP supplemental plaques may be installed at the top outside border of the R3-8 sign over the applicable lane designation on the sign. The diamond symbol may be used instead of the word message HOV. The minimum allowable vehicle occupancy requirement may vary based on the level established for a particular facility.

03   Where a bicycle lane is between two general-purpose lanes the R3-8 series signs may be modified to show the bicycle lane with a white legend on a black background in accordance with designs of the R3-8x series signs (see Figure 2B-4).

Guidance:

04   *When used, an Advance Intersection Lane Control sign should be placed at an adequate distance in advance of the intersection, either along the lane tapers or at the beginning of the turn lane so that road users can select the appropriate lane (see Figure 2A-4).*

Option:

05   An Advance Intersection Lane Control sign may be repeated closer to the intersection along the approach for additional emphasis.

Standard:

06   **An Advance Intersection Lane Control (R3-8 series) sign shall not be mounted at the far side of an intersection to which it applies.**

07   **Where three or more approach lanes are available to traffic, Advance Intersection Lane Control (R3-8 series) signs, if used, shall be post-mounted in advance of the intersection and shall not be mounted overhead.**

08   **When only the two outermost lanes of the roadway are shown on a R3-8 sign, the R3-5bP or R3-5fP plaque shall be mounted above the R3-8 sign.**

Section 2B.31 RIGHT (LEFT) LANE MUST EXIT Signs (R3-33, R3-33a)

Option:

01   A RIGHT (LEFT) LANE MUST EXIT (R3-33) sign (see Figure 2B-4) may be used to supplement an overhead EXIT ONLY guide sign to inform road users that traffic in the right-hand (left-hand) lane of a roadway that is approaching a grade-separated interchange is required to depart the roadway on the exit ramp at the next interchange.

02   The R3-33a sign (see Figure 2B-4) may be used in place of the R3-33 sign where the roadside width is limited and will not accommodate the R3-33 sign.

Support:

03   Section 2C.50 contains information regarding a warning sign that can be used in advance of lane drops at grade-separated interchanges.

Section 2B.32 Two-Way Left-Turn-Only Signs (R3-9a and R3-9b) and Plaques

Guidance:

01   *A Two-Way Left-Turn-Only (R3-9a or R3-9b) sign (see Figure 2B-6) should be used in conjunction with the required pavement markings where a non-reversible lane is reserved for the exclusive use of left-turning vehicles in either direction and is not used for passing, overtaking, or through travel.*

Option:

02   The post-mounted R3-9b sign may be used as an alternate to or a supplement to the overhead R3-9a sign. The legend BEGIN or END may be used within the border of the main sign itself, or on an R3-9cP or R3-9dP plaque (see Figure 2B-6) mounted immediately above it.

Support:

03   Signing is especially helpful to drivers in areas where the two-way left-turn-only maneuver is new, in areas subject to environmental conditions that frequently obscure the pavement markings, and on peripheral streets with two-way left-turn-only lanes leading to an extensive system of routes with two-way left-turn-only lanes.

Section 2B.33 BEGIN and END Plaques (R3-9cP and R3-9dP)

Option:

01   The BEGIN (R3-9cP) or END (R3-9dP) plaque (see Figure 2B-6), mounted directly above a regulatory sign, may be used to inform road users of the location where a regulatory condition begins or ends.

Section 2B.34 Reversible Lane Control Signs (R3-9e through R3-9i)

Option:

01   A reversible lane may be used for through traffic (with left turns either permitted or prohibited) in alternating directions during different periods of the day, and the lane may be used for exclusive left turns in one or both directions during other periods of the day as well. Reversible Lane Control (R3-9e through R3-9i) signs (see Figure 2B-6) may be either static type or changeable message type. These signs may be either post-mounted or overhead. Lane-use control signals (see Chapter 4T) may also be used for reversible lanes.

Figure 2B-6. Center and Reversible Lane Control Signs and Plaques

R3-9a

R3-9b

R3-9cP

R3-9dP

R3-9e

R3-9f

END REVERSE LANE AT Colorado Blvd	BEGIN REVERSE LANE AT Colorado Blvd
OR	OR
END REVERSE LANE 400 FEET	BEGIN REVERSE LANE 400 FEET
R3-9g	R3-9h

END REVERSE LANE — ONLY

R3-9i

Standard:

02 **Where it is determined by an engineering study that lane-use control signals or physical barriers are not necessary, the lane shall be controlled by overhead Reversible Lane Control signs (see Figure 2B-7).**

03 **Post-mounted Reversible Lane Control signs shall be used only as a supplement to overhead signs or signals. Post-mounted signs shall be identical in design to the overhead signs and an additional legend such as CENTER LANE shall be added to the top of the sign (R3-9f) to indicate which lane is controlled.**

Option:

04 Reversing traffic flow may be controlled with pavement markings and Reversible Lane Control signs (without the use of lane-use control signals), when all of the following conditions are met:

  A. Only one lane is being reversed,
  B. An engineering study indicates that the use of Reversible Lane Control signs alone would result in an acceptable level of safety and efficiency, and
  C. There are no unusual or complex operations in the reversible lane pattern.

Standard:

05 **Reversible Lane Control signs shall contain the legend or symbols designating the allowable uses of the lane and the time periods such uses are allowed. Where symbols and legends are used, their meanings shall be as shown in Table 2B-2.**

06 **Reversible Lane Control signs shall consist of a white background with a black legend and border, except for the R3-9e sign, where the color red is used for the X symbol.**

07 **Symbol signs, such as the R3-9e sign, shall consist of the appropriate symbol in the upper portion of the sign with the appropriate times of the day and days of the week below it. All times of the day and days of the week shall be accounted for on the sign to eliminate confusion to the road user.**

Table 2B-2. Meanings of Symbols and Legends on Reversible Lane Control Signs

Symbol / Word Message	Meaning
Red X on white background	Lane closed
Upward-pointing black arrow on white background (if left turns are permitted, the arrow shall be modified to show left / through arrow)	Lane open for through travel and any turns not otherwise prohibited
Black two-way left-turn arrows on white background and legend ONLY	Lane may be used only for left turns in either direction (i.e., as a two-way left-turn lane)
Black single left-turn arrow on white background and legend ONLY	Lane may be used only for left turns in one direction (without opposing left turns in the same lane)

Figure 2B-7. Location of Reversible Two-Way Left-Turn Signs

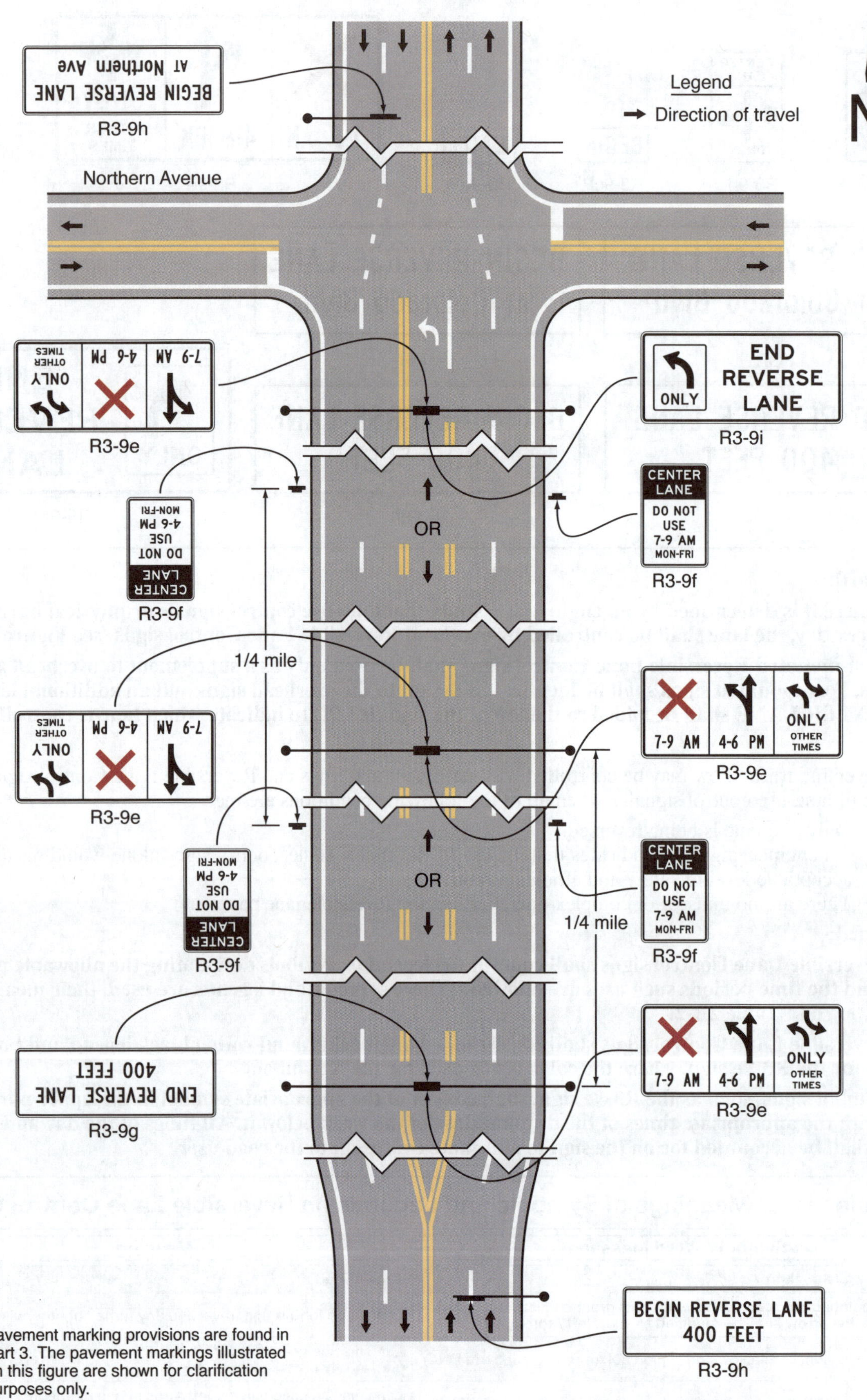

Note: Pavement marking provisions are found in Part 3. The pavement markings illustrated on this figure are shown for clarification purposes only.

08   **In situations where more than one message is conveyed to the road user, such as on the R3-9e sign, the sign legend shall be arranged as follows:**

    A. The prohibition or restriction message is the primary legend and shall be on the top for word message signs and to the far left for symbol signs,

    B. The permissive use message shall be displayed as the second legend, and

    C. The OTHER TIMES message shall be displayed at the bottom for word message signs and to the far right for symbol signs.

Option:

09   The symbol signs may also include a downward-pointing arrow with the legend THIS LANE. The term OTHER TIMES may be used for either the symbol or word message sign.

Standard:

10   **A Reversible Lane Control sign shall be mounted over the approximate center of the lane that is being reversed.**

11   **If the vertical or horizontal alignment is curved to the degree that a driver would be unable to see at least one sign, and preferably two signs, then additional overhead signs shall be installed. The placement of the signs shall be such that the driver will have a definite indication of the lanes specifically reserved for use at any given time. Special consideration shall be given to major generators introducing traffic between the normal sign placement.**

12   **Transitions at the entry to and exit from a section of roadway with reversible lanes shall include advance signs to notify or warn drivers of the boundaries of the reversible lane controls. The R3-9g or R3-9h signs (see Figure 2B-6) shall be used for this purpose.**

Option:

13   More than one End Reverse Lane (R3-9i) sign (see Figure 2B-6) may be used at the termination of the reversible lane to emphasize the importance of the message.

14   Where longitudinal barriers separate opposing directions of traffic, the R3-9g or R3-9h signs may be omitted.

Standard:

15   **Flashing beacons, if used to supplement the overhead Reversible Lane Control signs, shall comply with the applicable requirements for flashing beacons in Chapter 4S.**

16   **When used in conjunction with Reversible Lane Control signs, the Turn Prohibition (R3-1 through R3-4 and R3-18) signs shall be mounted overhead and separate from the Reversible Lane Control signs. The Turn Prohibition signs shall be designed and installed in accordance with Section 2B.26.**

Guidance:

17   *For additional emphasis, a supplemental plaque stating the distance of the prohibition, such as NEXT 1 MILE, should be added to the Turn Prohibition signs that are used in conjunction with Reversible Lane Control signs.*

18   *If used, overhead signs should be located at intervals not greater than ¼ mile. The bottom of the overhead Reversible Lane Control signs should not be more than 19 feet above the pavement grade.*

19   *Where more than one sign is used at the termination of a reversible lane, they should be at least 250 feet apart. Longer distances between signs are appropriate for streets with speeds over 35 mph, but the separation should not exceed 1,000 feet.*

20   *Because left-turning vehicles have a significant impact on the safety and efficiency of a reversible lane operation, if a mandatory left-turn lane or two-way left-turn lane cannot be incorporated into the lane-use pattern for a particular peak or off-peak period, consideration should be given to prohibiting left turns and U-turns during that time period.*

21   *Reversible Lane Control signs and parking signs should be consistent in message during the same operational periods.*

Section 2B.35 Jughandle Signs (R3-23, R3-24, R3-25, and R3-26 Series)

Support:

01   A jughandle turn is a left turn or U-turn that because of special geometry is made by initially making a right turn. This type of turn can increase the operational efficiency of a roadway by eliminating the need for mandatory left-turn lanes and can increase the operational efficiency of a traffic control signal by eliminating the need for protected left-turn phases. A jughandle turn can also provide an opportunity for trucks and commercial vehicles to make a U-turn where the median and roadway are not of sufficient width to accommodate a traditional U-turn by these vehicles.

02   Figure 2B-8 shows the various signs that can be used for signing jughandle turns. Figure 2B-9 shows examples of regulatory and destination guide signing for various types of jughandle turns.

Standard:

03   **On multi-lane roadways, since road users generally anticipate that they need to be in the left-hand lane when approaching a location where they desire to turn left or make a U-turn, an ALL TURNS FROM RIGHT LANE (R3-23) or a U TURN FROM RIGHT LANE (R3-23a) sign (see Figure 2B-8) shall be installed in advance of the location to inform drivers that left turns and/or U-turns will be made from the right-hand lane.**

Option:

04   Where a median of sufficient width is available, supplemental regulatory or guide signs may also be placed on the left-hand side of the roadway.

Standard:

05   **The R3-24 series sign with an upward diagonal arrow pointing to the right if the jughandle entrance is designed as an exit ramp (see Drawings A and B in Figure 2B-9), or the R3-25 series sign with a horizontal arrow pointing to the right if the jughandle entrance is designed as an intersection, shall be installed on the right-hand side of the roadway at the entrance to the jughandle. The legend on the sign shall be ALL TURNS, U TURN, or U AND LEFT TURNS, as appropriate.**

06   **If the jughandle is designed such that the jughandle entrance is downstream of the location where the turn would normally have been made (see Drawing C in Figure 2B-9), the R3-26 series sign with an arrow pointing straight upward shall be installed on the right-hand side of the roadway at the intersection to inform road users that they need to proceed straight through the intersection in order to make a left turn or U-turn. The legend on the sign shall be U TURN or U AND LEFT TURNS, as appropriate.**

Support:

07   The R3-24, R3-25, and R3-26 series of signs are designed to be mounted below conventional guide signs.

08   Section 2C.12 contains information regarding the use of advisory exit and ramp speed signs for exit ramps.

09   Section 2D.40 contains information regarding the use of guide signs for jughandles.

Figure 2B-8. Jughandle Regulatory Signs

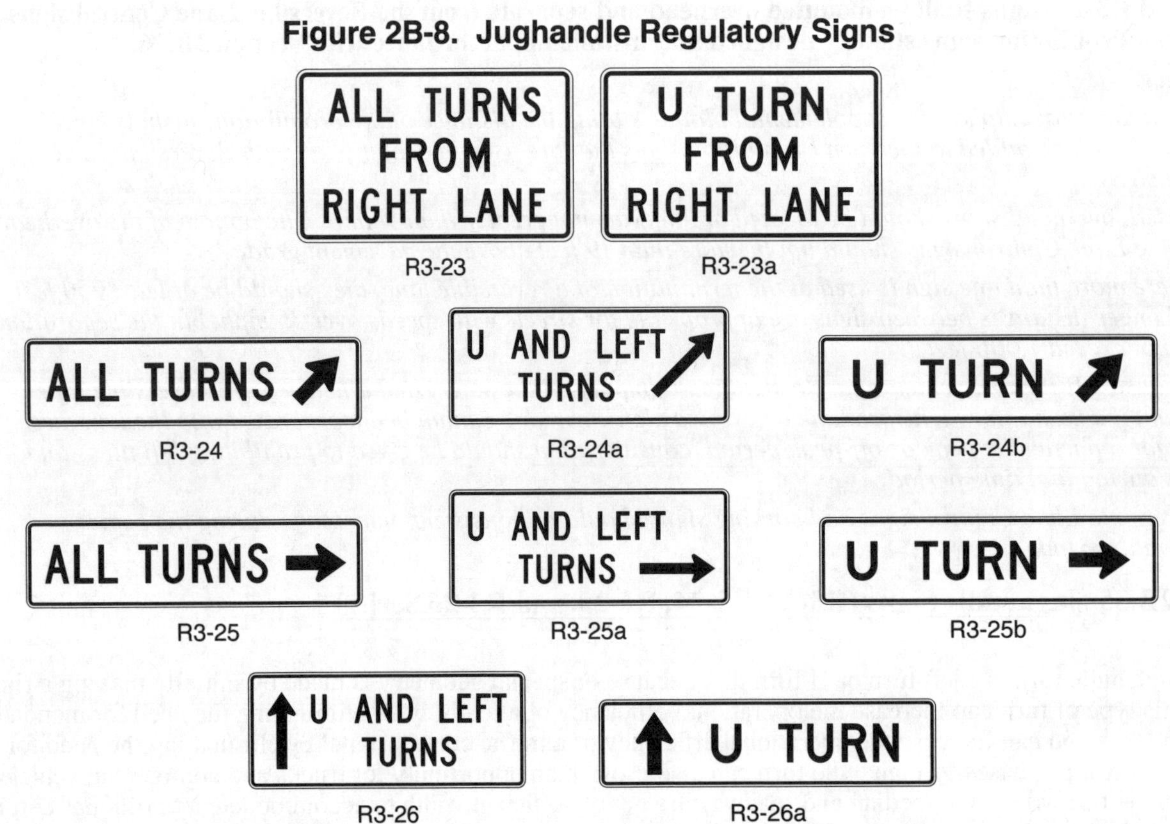

Figure 2B-9. Examples of Applications of Jughandle Regulatory and Guide Signing
(Sheet 1 of 3)

A – Turns made prior to the intersection

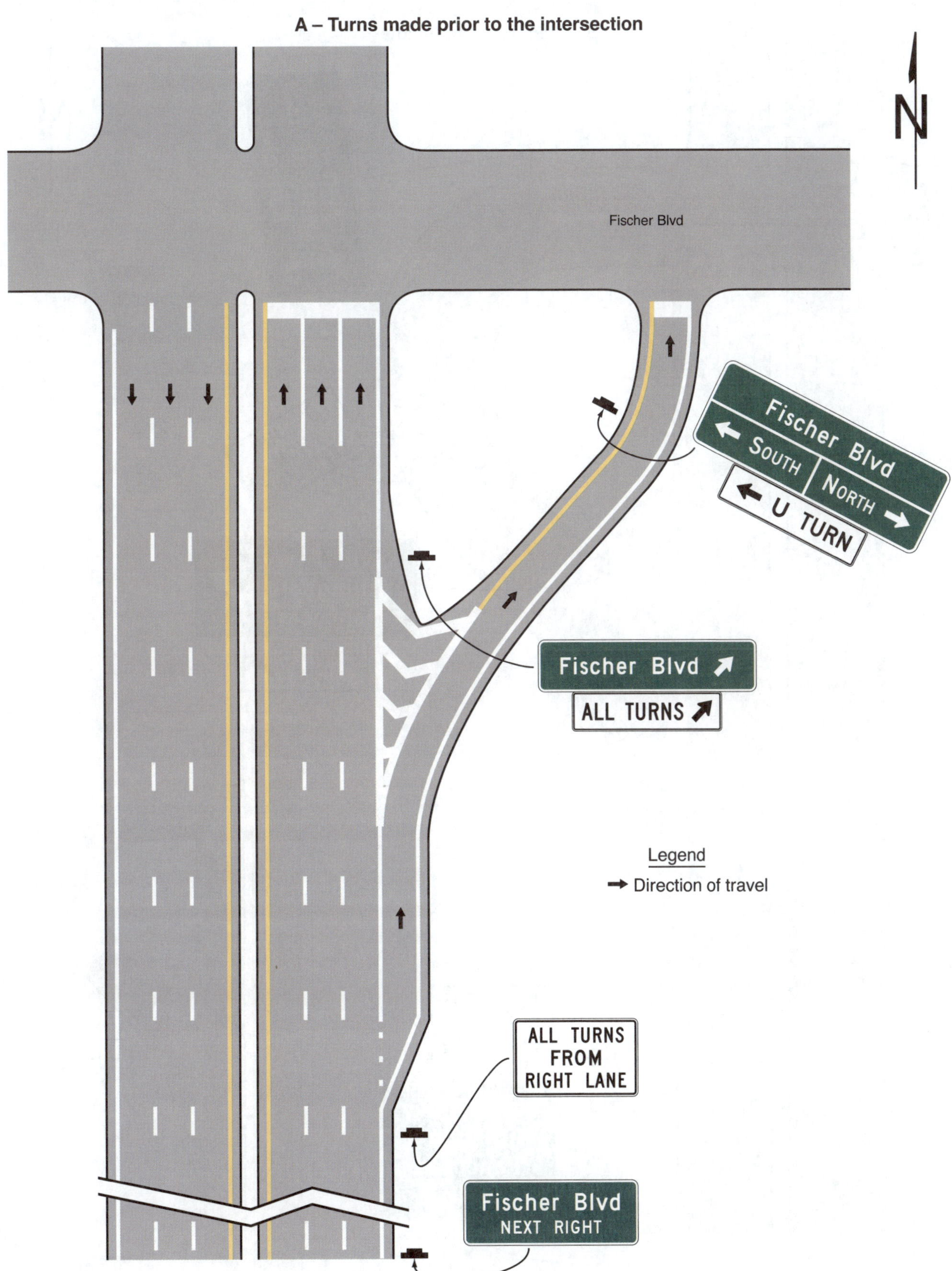

Figure 2B-9. Examples of Applications of Jughandle Regulatory and Guide Signing
(Sheet 2 of 3)

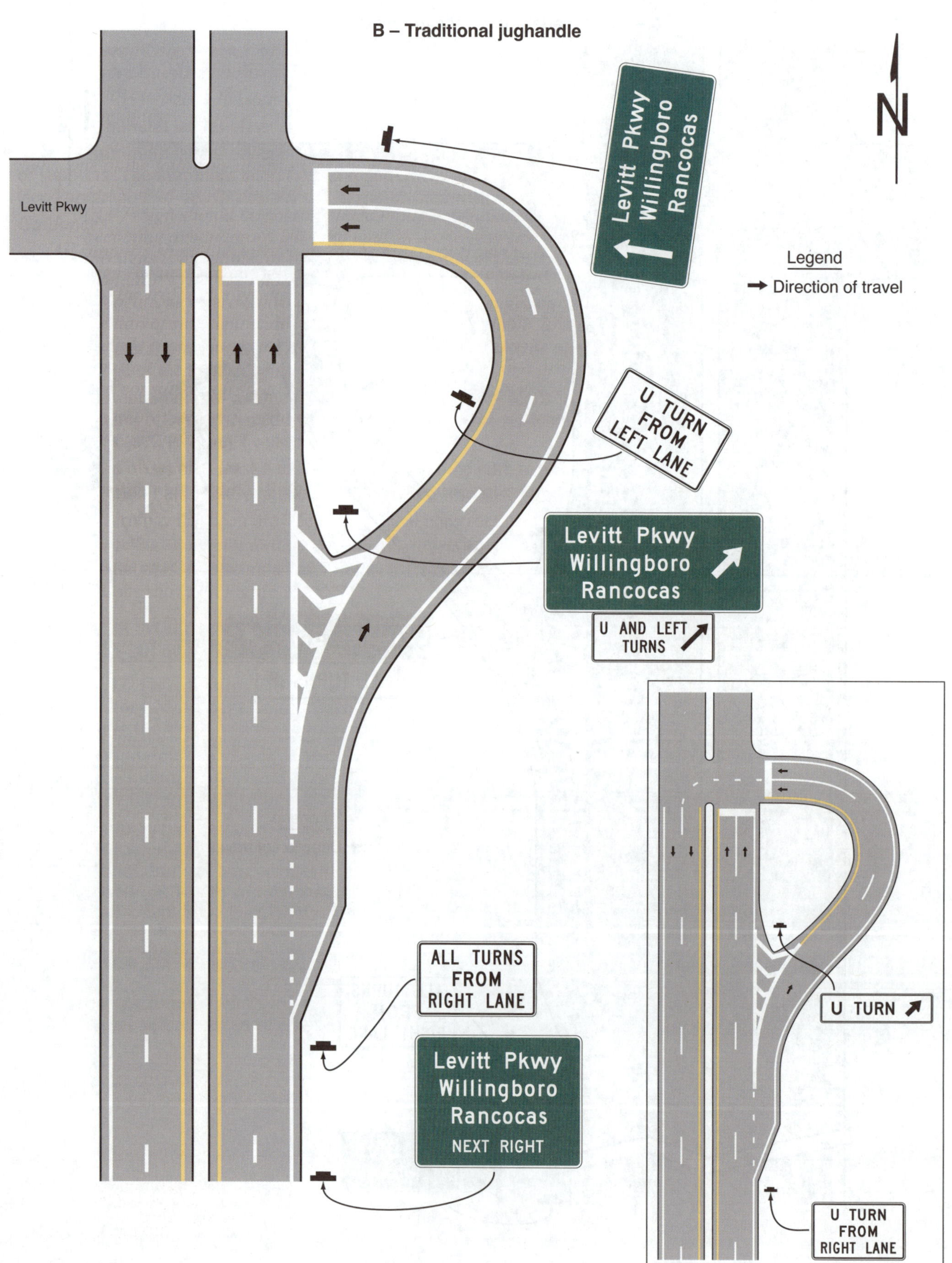

Figure 2B-9. Examples of Applications of Jughandle Regulatory and Guide Signing
(Sheet 3 of 3)

C – Turns made beyond the intersection

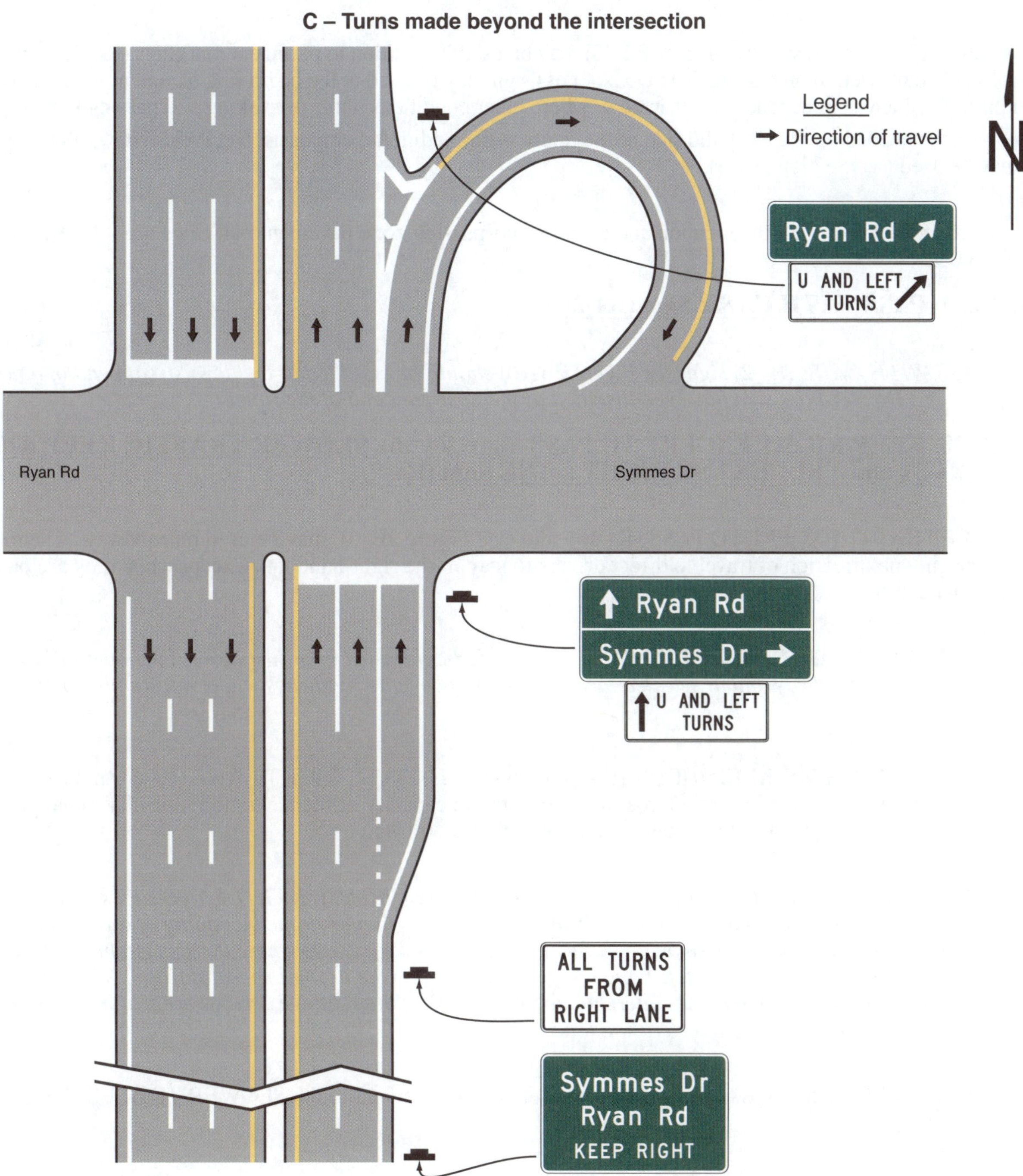

PASSING, KEEP RIGHT, AND SLOW TRAFFIC SIGNS

Section 2B.36 DO NOT PASS Sign (R4-1)

Option:

01   The Do Not Pass (R4-1) sign (see Figure 2B-10) may be used in addition to pavement markings (see Section 3B.03) to emphasize the restriction on passing. The Do Not Pass sign may be used at the beginning of, and at intervals within, a zone through which sight distance is restricted or where other conditions make overtaking and passing inappropriate.

02   If signing is needed on the left-hand side of the roadway for additional emphasis, NO PASSING ZONE (W14-3) signs may be used (see Section 2C.53).

Support:

03   Standards for determining the location and extent of no-passing zone pavement markings are set forth in Section 3B.03.

Section 2B.37 PASS WITH CARE Sign (R4-2)

Guidance:

01   *The PASS WITH CARE (R4-2) sign (see Figure 2B-10) should be installed at the downstream end of a no-passing zone if a Do Not Pass sign has been installed at the upstream end of the zone.*

Section 2B.38 KEEP RIGHT EXCEPT TO PASS Sign (R4-16), SLOWER TRAFFIC KEEP RIGHT Sign (R4-3), and TRUCKS USE RIGHT LANE Sign (R4-5)

Option:

01   The KEEP RIGHT EXCEPT TO PASS (R4-16) sign (see Figure 2B-10) may be used on roadways where there are two lanes in one direction of travel to direct drivers to stay in the right-hand lane except when they are passing another vehicle.

Guidance:

02   *If used, the KEEP RIGHT EXCEPT TO PASS sign should be installed at or just beyond the beginning of a two-lane section of roadway and at selected locations along two-lane roadways where additional emphasis is needed.*

Option:

03   The SLOWER TRAFFIC KEEP RIGHT (R4-3) or the TRUCKS USE RIGHT LANE (R4-5) sign (see Figure 2B-10) may be used on multi-lane through roadways to improve capacity or reduce unnecessary lane changing due to the presence of slower vehicles that impede the normal flow of traffic.

Guidance:

04   *If used, the SLOWER TRAFFIC KEEP RIGHT sign or the TRUCKS USE RIGHT LANE sign should be installed at or just beyond the beginning of a multi-lane roadway section or at the beginning of an extra lane provided for trucks and/or other slow-moving traffic, and at selected locations where there is a tendency on the part of some road users to drive in the left-hand lane(or lanes) below the normal speed of traffic. These signs should not be used on the approach to an interchange or through an interchange area where traffic is entering or exiting, or along deceleration or acceleration lanes.*

Option:

05   The TRUCKS USE RIGHT LANE sign may be used as a supplement to the SLOWER TRAFFIC KEEP RIGHT sign.

Guidance:

06   *If an extra lane has been provided for trucks and other slow-moving traffic, a Lane Ends sign (see Section 2C.47) should be installed in advance of the point where the extra lane ends. Appropriate pavement markings should be installed at both the upstream and downstream ends of the extra lane (see Section 3B.12 and Figure 3B-14).*

Section 2B.39 Keep Right and Keep Left Signs (R4-7 Series and R4-8 Series)

Option:

01   The Keep Right (R4-7) sign (see Figure 2B-10) may be used at locations where it is necessary for traffic to pass only to the right-hand side of a roadway feature or obstruction. The Keep Left (R4-8) sign (see Figure 2B-10) may be used at locations where it is necessary for traffic to pass only to the left-hand side of a roadway feature or obstruction.

Figure 2B-10. Passing, Keep Right, and Slow Traffic Signs

R4-1

R4-2

R4-3

R4-5

R4-7

R4-7a

R4-7b

R4-7c

R4-8

R4-8a

R4-8b

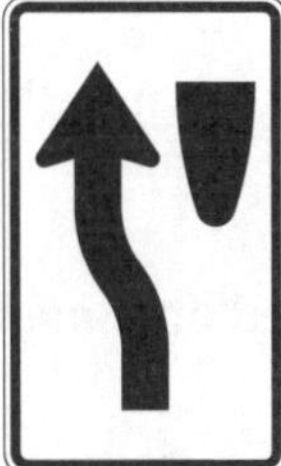

R4-8c

R4-9

R4-10

R4-12

R4-13

R4-14

R4-16

R4-17

R4-18

R4-20

R4-21

Guidance:

02   *At locations where it is not readily apparent that traffic is required to keep to the right, a Keep Right sign should be used.*

Standard:

03   **If Keep Right signs are installed at the start of a median or at a median opening, they shall be placed as close as practicable to the approach ends of the medians, and shall be visible to traffic on the divided highway and angled toward the applicable crossroad approach as shown in Figure 2B-20.**

Guidance:

04   *If used, the Keep Right sign should be mounted on the face of or just in front of a pier or other obstruction separating opposite directions of traffic in the center of the highway such that traffic will have to pass to the right-hand side of the sign.*

05   *Where the approach end of the island channelizes traffic away from the approach direction, the word legend (R4-7a, R4-7b, R4-8a, or R4-8b) signs (see Figure 2B-10) should be used instead of the symbol (R4-7 or R4-8) signs to emphasize the degree of curvature away from the approach direction (see Figure 2B-11).*

06   *Where a regulatory sign is used within the central island of a neighborhood traffic circle to direct traffic counter-clockwise around the central island, the Keep Right with diagonal arrow (R4-7b) sign should be used (see Figure 2B-24). The mounting height of the sign should be at least 4 feet, measured vertically from the bottom of the sign to the elevation of the near edge of the traveled way.*

Standard:

07   **The Keep Right (Left) sign shall not be installed on the right-hand (left-hand) side of the roadway in a position where traffic must pass to the left-hand (right-hand) side of the sign.**

Option:

08   The Keep Right sign may be omitted at intermediate ends of divisional islands and medians.

09   Word message KEEP RIGHT (LEFT) with an arrow (R4-7a or R4-7b) signs (see Figure 2B-10) may be used instead of the R4-7 or R4-8 symbol signs.

10   A narrow Keep Right (R4-7c) sign (see Figure 2B-10) may be installed on the approach end of a median island that is less than 4 feet wide at the point where the sign is to be located.

Standard:

11   **A narrow Keep Right (R4-7c) sign shall not be installed on a median island that has a width of 4 feet or more at the point where the sign is to be located.**

Option:

12   The Keep Right sign may be installed in the median of a divided highway crossing that functions as a single intersection such that it is visible to traffic on the divided highway and angled as needed toward the applicable crossroad approach as shown in Figure 2B-20.

Support:

13   Section 2B.49 provides more information about the use of the Keep Right sign in combination with or in lieu of ONE-WAY signs at divided highway crossings.

Section 2B.40 STAY IN LANE Sign (R4-9)

Option:

01   A STAY IN LANE (R4-9) sign (see Figure 2B-10) may be used on multi-lane highways to direct road users to stay in their lane until conditions permit shifting to another lane.

Guidance:

02   *If a STAY IN LANE sign is used, it should be accompanied by a solid double white lane line(s) to prohibit lane changing.*

Section 2B.41 RUNAWAY VEHICLES ONLY Sign (R4-10)

Guidance:

01   *A RUNAWAY VEHICLES ONLY (R4-10) sign (see Figure 2B-10) should be installed near a truck escape (or runaway truck) ramp entrance to discourage other road users from entering the ramp.*

Figure 2B-11. Examples of Keep Right and Keep Left Sign Placement

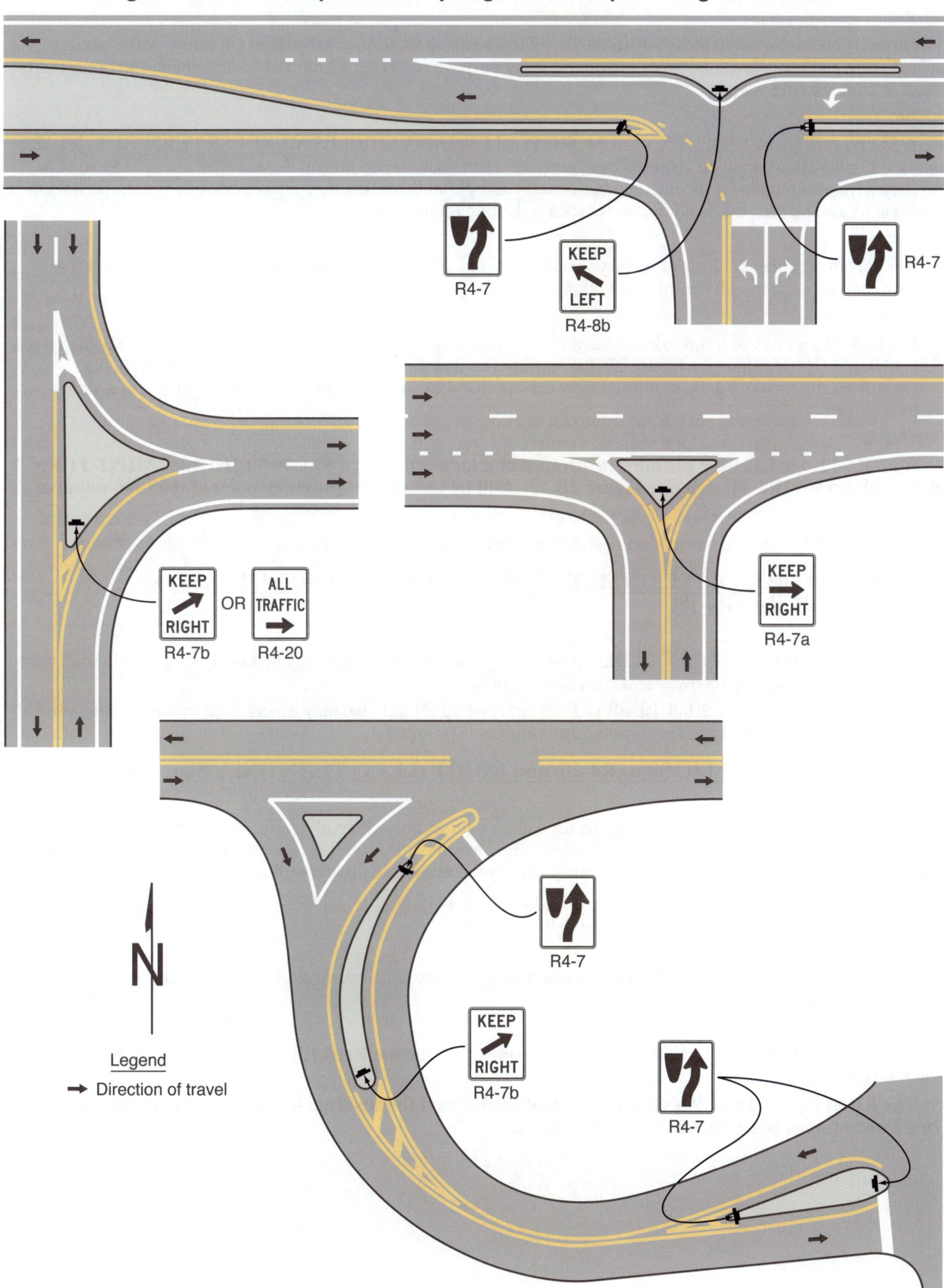

Section 2B.42 Slow Vehicle Turn-Out Signs (R4-12, R4-13, and R4-14)

Support:

01   On two-lane highways in areas where traffic volumes and/or vertical or horizontal curvature make passing difficult, turn-out areas are sometimes provided for the purpose of giving a group of faster vehicles an opportunity to pass a slow-moving vehicle.

Option:

02   A SLOW VEHICLES WITH XX OR MORE FOLLOWING VEHICLES MUST USE TURN-OUT (R4-12) sign (see Figure 2B-10) may be installed in advance of a turn-out area to inform drivers who are driving so slow that they have accumulated a specific number of vehicles behind them that they are required by the traffic laws of that State to use the turn-out to allow the vehicles following them to pass.

Support:

03   The specific number of vehicles displayed on the R4-12 sign provides law enforcement personnel with the information they need to enforce this regulation.

Option:

04   If an R4-12 sign has been installed in advance of a turn-out area, a SLOW VEHICLES MUST USE TURN-OUT AHEAD (R4-13) sign (see Figure 2B-10) may also be installed downstream from the R4-12 sign, but upstream from the turn-out area, to remind slow drivers that they are required to use a turn-out that is a short distance ahead.

Standard:

05   **If an R4-12 sign has been installed in advance of a turn-out area, a SLOW VEHICLES MUST TURN OUT (with arrow) (R4-14) sign (see Figure 2B-10) shall be installed at the entry point of the turn-out area.**

Support:

06   Section 2D.54 contains information regarding advance information signs for slow vehicle turn-out areas.

Section 2B.43 DO NOT DRIVE ON SHOULDER Sign (R4-17) and DO NOT PASS ON SHOULDER Sign (R4-18)

Option:

01   The DO NOT DRIVE ON SHOULDER (R4-17) sign (see Figure 2B-10) may be installed to inform road users that using the shoulder of a roadway as a travel lane is prohibited.

02   The DO NOT PASS ON SHOULDER (R4-18) sign (see Figure 2B-10) may be installed to inform road users that using the shoulder of a roadway to pass other vehicles is prohibited.

Section 2B.44 ALL TRAFFIC Sign (R4-20) and RIGHT (LEFT) TURN ONLY Sign (R4-21)

Option:

01   The ALL TRAFFIC (R4-20) sign may be used at an intersection where all traffic on the approach to the intersection must turn in the direction indicated and the Movement Prohibition (see Section 2B.26) and/or ONE WAY (see Section 2B.49) signs do not adequately convey the allowable direction of travel.

02   The RIGHT (LEFT) TURN ONLY (R4-21) sign may be used at or on an approach to an intersection where all traffic on that approach must turn in the direction indicated.

Guidance:

03   *The RIGHT (LEFT) TURN ONLY sign should not be used for a channelized turn lane separated from the adjacent travel lanes by an island.*

Standard:

04   **The ALL TRAFFIC sign shall not be used to substitute for the Keep Right (R4-7 series) or Keep Left (R4-8 series) signs.**

05   **The RIGHT (LEFT) TURN ONLY sign shall not be used to substitute for the Mandatory Movement Lane Control signs (see Sections 2B.27 and 2B.28).**

SELECTIVE EXCLUSION SIGNS AND PLAQUES

Section 2B.45 Selective Exclusion Signs and Plaques

Option:

01   Selective Exclusion signs (see Figure 2B-12) may be used to provide notice to road users that State or local statutes or ordinances exclude designated types of traffic from using particular roadways or facilities.

Standard:

02   **Selective Exclusion signs shall clearly indicate the type of traffic that is excluded.**

Support:

03   Typical exclusion messages include:

A. No Trucks (R5-2),
B. NO MOTOR VEHICLES (R5-3),
C. NO COMMERCIAL VEHICLES (R5-4),
D. NO VEHICLES WITH LUGS (R5-5),
E. No Bicycles (R5-6),
F. NO NON-MOTORIZED TRAFFIC (R5-7),

Figure 2B-12. Selective Exclusion Signs and Plaques

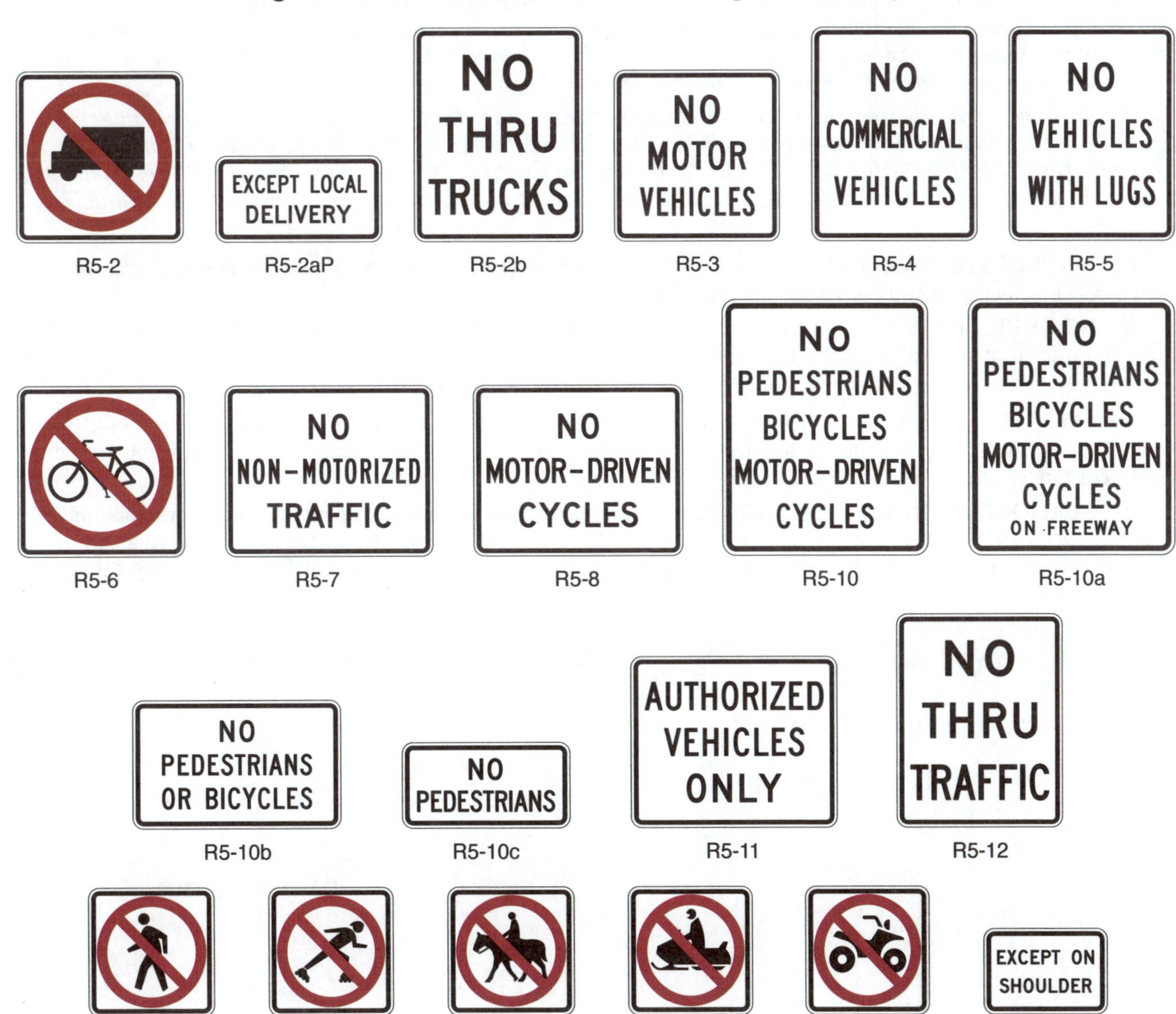

G. NO MOTOR-DRIVEN CYCLES (R5-8),
H. No Pedestrian Crossing (R9-3),
I. No Skaters (R9-13),
J. No Equestrians (R9-14),
K. No Snowmobiles (R9-15),
L. No All-Terrain Vehicles (R9-16),
M. Hazardous Material (R14-3) (see Section 2B.67),
N. NO THRU TRAFFIC (R5-12),
O. NO THRU TRUCKS (R5-2b),
P. EXCEPT ON SHOULDER (R9-19P) plaque, and
Q. EXCEPT LOCAL DELIVERY (R5-2aP) plaque.

Option:

04   Appropriate combinations or groupings of these legends into a single sign, such as NO PEDESTRIANS BICYCLES MOTOR-DRIVEN CYCLES (R5-10 and R5-10a) or NO PEDESTRIANS OR BICYCLES (R5-10b), may be used.

Guidance:

05   *If an exclusion is governed by vehicle weight, a Weight Limit sign (see Section 2B.64) should be used instead of a Selective Exclusion sign.*

06   *If used on a ramp to a freeway or expressway where pedestrian and bicyclist travel are prohibited by law or regulation, the NO PEDESTRIANS OR BICYCLES (R5-10b) sign should be installed in a location where it is clearly visible to any pedestrian or bicyclist attempting to enter the limited access facility from a street intersecting the ramp. In locations where a freeway or expressway is accessed from a ramp from a roadway parallel to the freeway or expressway, the sign should be placed in a location that clearly indicates the prohibition applies only to the freeway or expressway or to the ramp.*

07   *The Selective Exclusion sign should be placed on the right-hand side of the roadway at an appropriate distance from the intersection so as to be clearly visible to all road users turning into the roadway that has the exclusion. The NO PEDESTRIANS (R5-10c) or No Pedestrian Crossing (R9-3) sign (see Section 2B.57) should be installed so as to be clearly visible to pedestrians who are at a location where an alternative route is available.*

Option:

08   The NO PEDESTRIANS (R5-10c) or No Pedestrian Crossing (R9-3) sign may also be used at underpasses or elsewhere where pedestrian facilities are not provided.

09   The NO THRU TRAFFIC (R5-12) or NO THRU TRUCKS (R5-2b) signs may be used at locations to prohibit through traffic from using a particular roadway or facility.

10   The EXCEPT LOCAL DELIVERY (R5-2aP) plaque may be mounted below the R5-2 or R5-2b sign.

11   The EXCEPT ON SHOULDER (R9-19P) plaque may be used where such modes are allowed on a shoulder but not on the traveled way and placed at intersections with other roads and established paths or trails, where such vehicles or modes are expected to enter the highway.

12   The AUTHORIZED VEHICLES ONLY (R5-11) sign may be used at median openings and other locations to prohibit vehicles from using the median opening or facility unless they have special permission (such as law enforcement vehicles or emergency vehicles) or are performing official business (such as highway agency vehicles).

DO NOT ENTER, WRONG WAY, ONE WAY, AND RELATED SIGNS AND PLAQUES

Section 2B.46 DO NOT ENTER Sign (R5-1)

Standard:

01 The DO NOT ENTER (R5-1) sign (see Figure 2B-13) shall be used at the following locations:

   A. Where a two-way roadway becomes a one-way roadway (see Figure 2B-18);
   B. The intersection of an interchange exit ramp with a crossroad as specified in Section 2B.48 (see Figure 2B-15);
   C. The intersection of a channelized or turning roadway with a two-way undivided crossroad; and
   D. Except as provided in Paragraph 4 of this Section, an intersection with a divided highway where the crossing functions as two separate intersections (see Figure 2B-14).

Guidance:

02 *A DO NOT ENTER sign should be installed at other locations where additional emphasis is needed where wrong-way movements are prominent or where the intersecting angle of roadways is such that the visibility of ONE WAY signs alone does not sufficiently convey the restriction.*

Option:

03 A DO NOT ENTER sign may be installed at an intersection with a divided highway where the crossing functions as a single intersection as shown in Figure 2B-20.

04 A DO NOT ENTER sign may be omitted on a low-speed urban street that is a divided highway at a crossing that functions as two separate intersections.

05 An EXCEPT BICYCLES (R3-7bP) plaque (see Figure 2B-4) may be used with a DO NOT ENTER sign when counter-flow bicycle traffic is allowed.

Guidance:

06 *The DO NOT ENTER sign, if used, should be placed directly in view of a road user at the point where a road user could wrongly enter a divided highway, one-way roadway, or ramp. The sign should be mounted facing traffic that might enter the roadway or ramp in the wrong direction.*

07 *At a crossing with a divided highway that functions as a single intersection; the sign, if used, should be placed on the outside edge side of the roadway facing traffic that might enter the roadway in the wrong direction.*

08 *If the DO NOT ENTER sign would be visible to traffic to which it does not apply, the sign should be turned away from, or shielded from, the view of that traffic.*

Option:

09 A second DO NOT ENTER sign may be used, particularly where traffic approaches from an intersecting roadway (see Figure 2B-14).

Figure 2B-13. DO NOT ENTER, WRONG WAY, ONE WAY, and Related Signs and Plaques

Figure 2B-14. Locations of DO NOT ENTER and WRONG WAY Signing for Divided Highway Crossings that Function as Two Separate Intersections

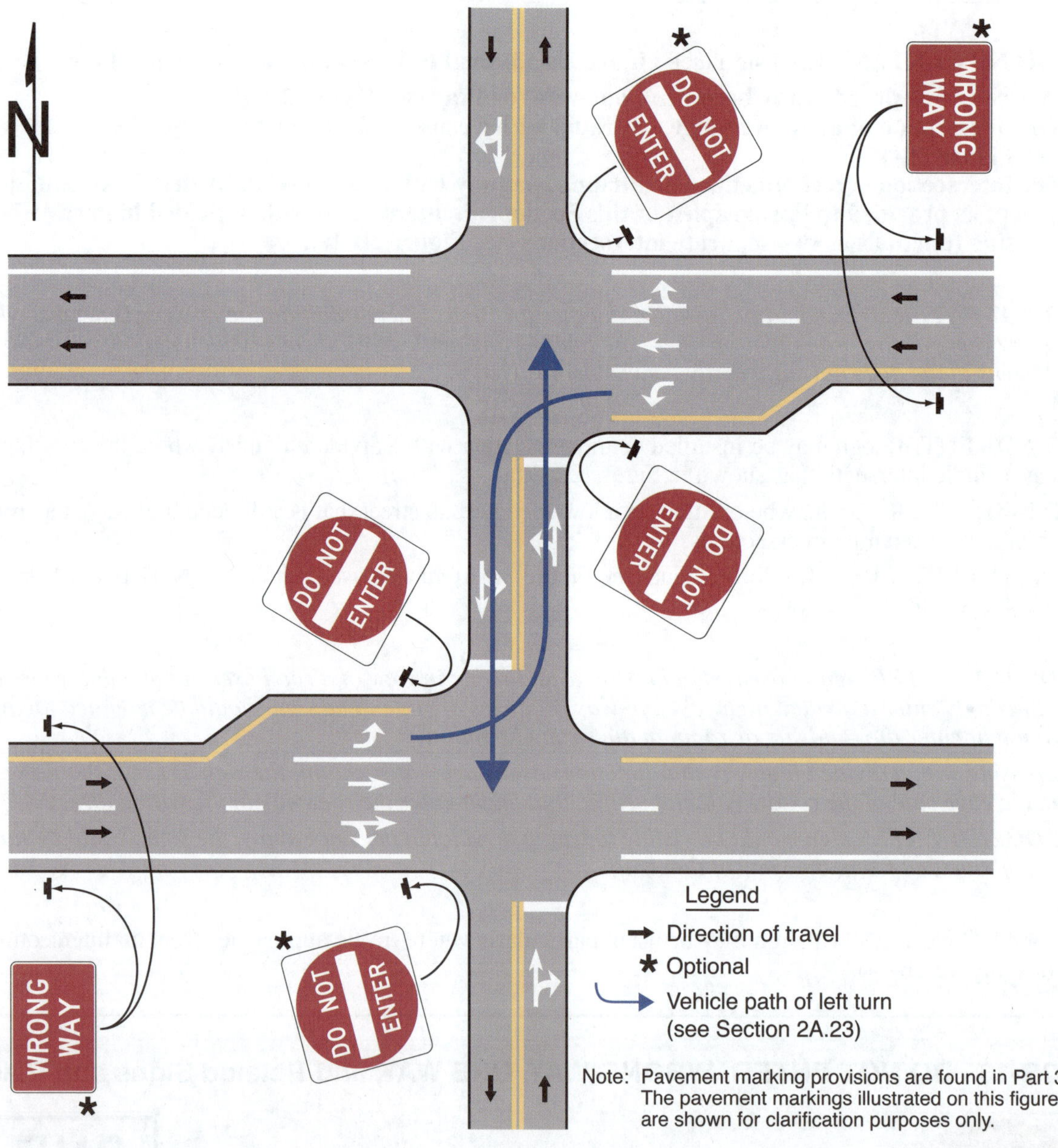

Support:

10   Section 2B.48 contains information regarding an optional lower mounting height for DO NOT ENTER signs that are located along an exit ramp facing a road user who is traveling in the wrong direction.

11   Section 2A.12 contains the provisions for the use of continuously-operated or actuated LEDs to enhance the conspicuity of signs.

Section 2B.47 WRONG WAY Sign (R5-1a)

Option:

01   The WRONG WAY (R5-1a) sign (see Figure 2B-13) may be used as a supplement to the DO NOT ENTER sign where a crossroad intersects a one-way roadway in a manner that does not physically discourage or prevent wrong-way entry (see Figures 2B-14 and 2B-20).

Guidance:

02   *If used, the WRONG WAY sign should be placed at a location along the one-way roadway farther from the crossroad than the DO NOT ENTER sign (see Section 2B.46).*

03   *The WRONG WAY sign should be placed on the same side of the road as the DO NOT ENTER sign.*

Support:

04   Section 2B.48 contains information regarding an optional lower mounting height for WRONG WAY signs that are located along an exit ramp facing a road user who is traveling in the wrong direction.

05   Section 2A.12 contains the provisions for the use of continuously-operated or actuated LEDs to enhance the conspicuity of signs.

Section 2B.48 Wrong-Way Traffic Control at Interchange Ramps

Standard:

01   **At interchange exit ramp terminals where the ramp intersects a crossroad in such a manner that wrong-way entry could inadvertently be made, the following signs shall be used (see Figure 2B-15):**

    A. At least one ONE WAY sign for each direction of travel on the crossroad shall be placed where the exit ramp intersects the crossroad.

    B. At least one DO NOT ENTER sign shall be conspicuously placed near the downstream end of the exit ramp in positions appropriate for full view of a road user starting to enter wrongly from the crossroad.

    C. At least one WRONG WAY sign shall be placed on the exit ramp facing a road user traveling in the wrong direction.

Guidance:

02   *In addition, the following pavement markings should be used (see Figure 2B-15):*

    A. On two-lane paved crossroads at interchanges, solid double yellow lines should be used as a center line for an adequate distance on both sides approaching the ramp intersections.

    B. Where crossroad channelization or ramp geometrics do not make wrong-way movements difficult, a lane-use arrow should be placed in each lane of an exit ramp near the crossroad terminal where it will be clearly visible to a potential wrong-way road user.

Option:

03   The following traffic control devices may be used to supplement the signs and pavement markings described in Paragraphs 1 and 2 of this Section:

    A. Additional ONE WAY signs may be placed, especially on two-lane rural crossroads, appropriately in advance of the ramp intersection to supplement the required ONE WAY sign(s).

    B. Additional WRONG WAY signs may be used.

    C. Slender, elongated wrong-way arrow pavement markings (see Figure 3B-21) intended primarily to warn wrong-way road users that they are traveling in the wrong direction may be placed upstream from the ramp terminus (see Figure 2B-15) to indicate the correct direction of traffic flow. Wrong-way arrow pavement markings may also be placed on the exit ramp at appropriate locations near the crossroad junction to indicate wrong-way movement. The wrong-way arrow markings may consist of pavement markings or bidirectional red-and-white raised pavement markers or other units that show red to wrong-way road users and white to other road users.

    D. Lane-use arrow pavement markings may be placed on the exit ramp and crossroad near their intersection to indicate the permissive direction of flow.

    E. Freeway entrance signs (see Section 2D.50) may be used.

    F. Lane control signs or movement prohibition signs may be used on the approaches to the exit ramp.

    G. A Keep Right (R4-7 or R4-7c) may be used on a ramp median nose for wrong-way traffic control.

Guidance:

04   *On interchange entrance ramps where the ramp merges with the through roadway and the design of the interchange does not clearly make evident the direction of traffic on the separate roadways or ramps, a ONE WAY sign visible to traffic on the entrance ramp and through roadway should be placed on each side of the through roadway near the entrance ramp merging point as illustrated in Figure 2B-16.*

Option:

05   On interchange entrance ramps where the ramp merges with the through roadway and the design of the interchange does not clearly make evident the direction of traffic on the separate roadways or ramps a No Left Turn (R3-2) sign may be located on the left-hand side of the entrance ramp at the gore (see Figure 2B-16). If a No Left Turn (R3-2) sign is located on the left-hand side a supplemental R3-2 sign may be installed on the right-hand side of the entrance ramp.

06   On interchange entrance ramps where the ramp merges with the through roadway and the design clearly indicates the direction of flow, a ONE WAY sign may be placed visible to traffic on the entrance ramp and/or a NO TURNS (R3-3) sign may be placed visible to traffic on the entrance ramp and through roadway at the gore area as illustrated in Figure 2B-16.

Figure 2B-15. Example of Application of Regulatory Signing and Pavement Markings at an Exit Ramp Termination to Deter Wrong-Way Entry

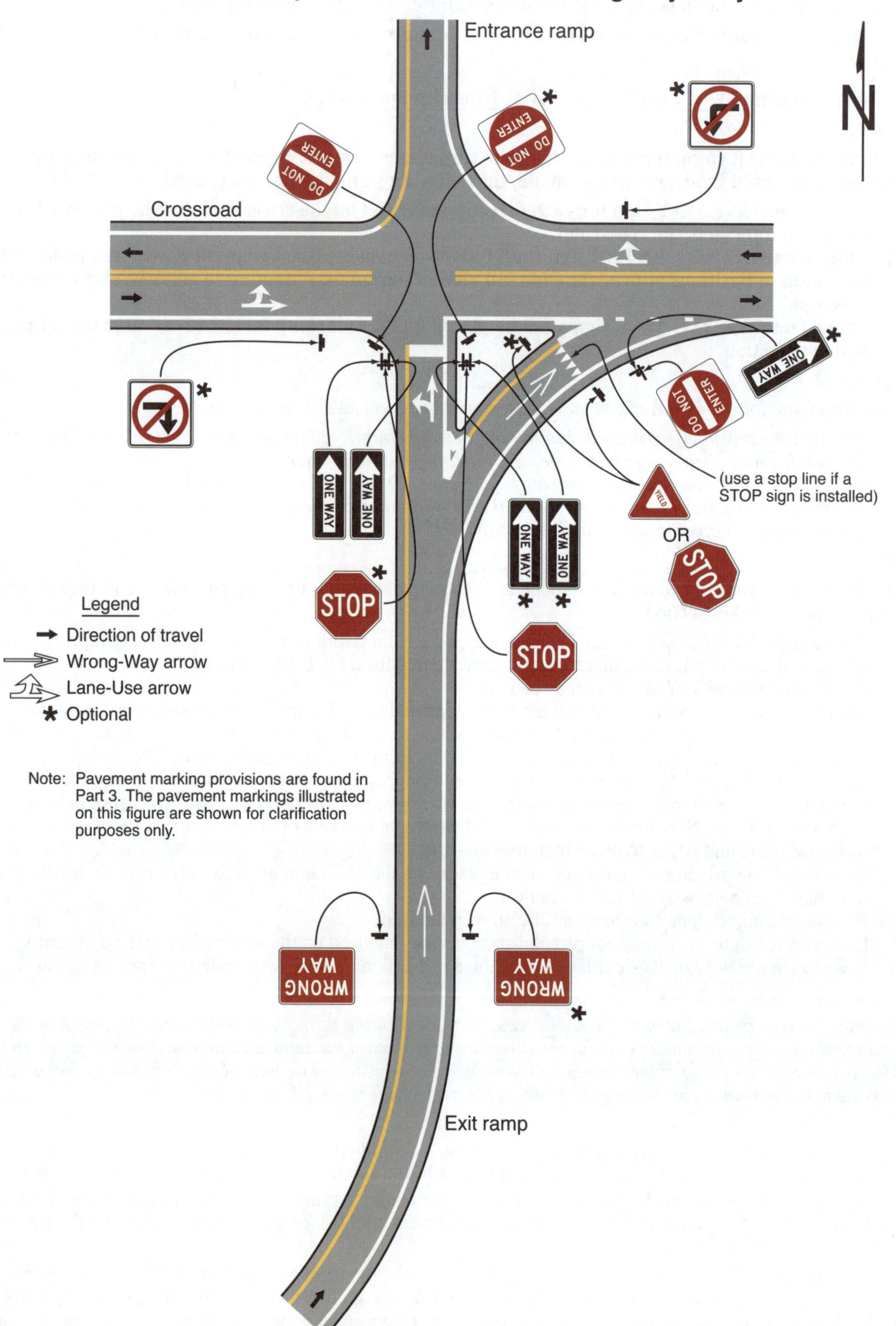

Figure 2B-16. Example of Application of Regulatory Signing and Pavement Markings at an Entrance Ramp Terminal

A – Design does not clearly indicate the direction of flow

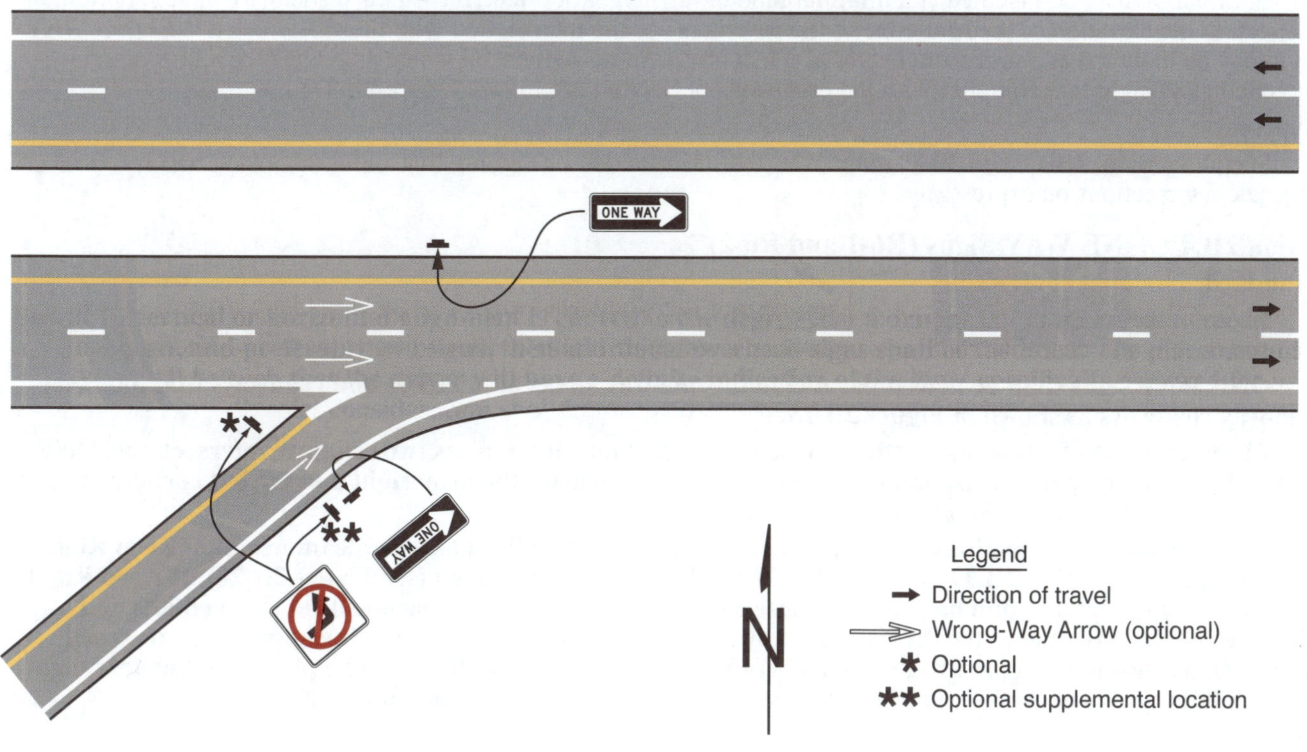

B – Design clearly indicates the direction of flow

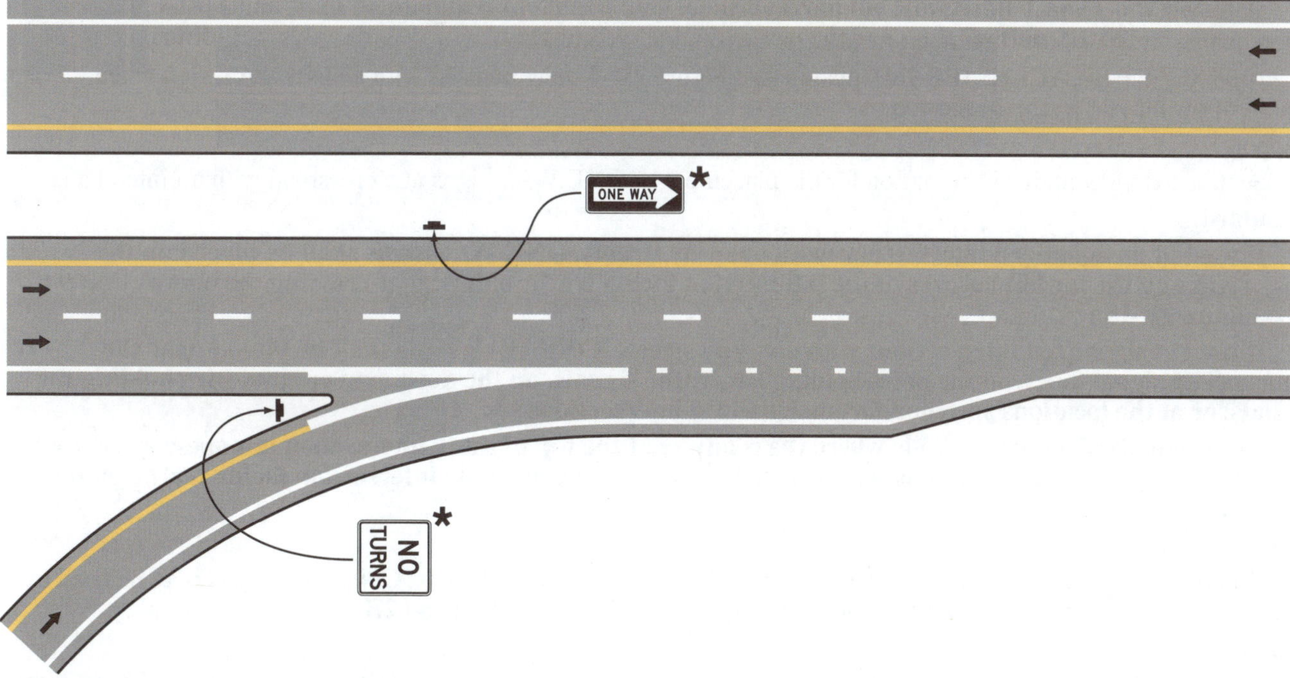

07   Where there are no parked cars, pedestrian activity, or other obstructions such as snow or vegetation, and if an engineering study indicates that a lower mounting height would address wrong-way movements on freeway or expressway exit ramps, a DO NOT ENTER sign(s) and/or a WRONG WAY sign(s) that is located along the exit ramp at a location downstream from the intersection with the crossroad facing a road user who is traveling in the wrong direction may be installed at a minimum mounting height of 3 feet, measured vertically from the bottom of the sign to the elevation of the near edge of the pavement. At the intersection with the crossroad, a WRONG WAY sign may be mounted at a minimum height of 3 feet on the same support on which a DO NOT ENTER sign is mounted at a height that complies with the provisions of Section 2A.15 (see Figure 2B-17).

Support:

08   Sections 2B.46, 2B.47, and 2B.49 contain further information on signing to avoid wrong-way movements at at-grade intersections on expressways.

Section 2B.49 ONE WAY Signs (R6-1 and R6-2)

Standard:

01   **Except as provided in Paragraph 6 of this Section, the ONE WAY (R6-1 or R6-2) sign (see Figure 2B-13) shall be used to indicate streets or roadways upon which vehicular traffic is allowed to travel in one direction only.**

02   **ONE WAY signs shall be placed parallel to the one-way street at all alleys and roadways that intersect one-way roadways as shown in Figure 2B-18.**

03   **At the crossing of a roadway with a divided highway that functions as two separate intersections, ONE WAY signs shall be placed, visible to each crossroad approach, on the near right and far left corners of each intersection with the directional roadways (see Figure 2B-19).**

04   **At the crossing of a roadway with a divided highway that functions as a single intersection Keep Right (R4-7) signs (see Section 2B.39) and/or ONE WAY signs shall be installed (see Figure 2B-20). If Keep Right signs are installed, they shall be placed as close as practicable to the approach ends of the medians and shall be visible to traffic on the divided highway and angled (as needed) toward the applicable crossroad approach as shown in Figure 2B-20. If ONE WAY signs are installed, they shall be placed on the near right and far left corners of the intersection and shall be visible to each crossroad approach.**

Option:

05   At the crossing of a roadway with a divided highway, regardless of function as a single or separate intersections, ONE WAY signs may also be placed on the far right corner of the intersection as shown in Figures 2B-19 and 2B-20.

06   ONE WAY signs may be omitted on the one-way roadways of divided highways, where the design of interchanges indicates the direction of traffic on the separate roadways.

07   An EXCEPT BICYCLES (R3-7bP) plaque (see Figure 2B-4) may be used with a ONE WAY sign when counter-flow bicycle traffic is allowed.

Support:

08   Section 2B.48 contains information for the placement of ONE WAY signs at a crossroad with an interchange.

Standard:

09   **If used at unsignalized intersections with one-way streets, ONE WAY signs shall be placed on the near right and the far left corners of the intersection facing traffic entering or crossing the one-way street (see Figure 2B-18).**

10   **If used at signalized intersections with one-way streets, ONE WAY signs shall be placed near the appropriate signal faces, on the poles holding the traffic signals, on the mast arm or span wire holding the signals, or at the locations specified for unsignalized intersections.**

11   **At unsignalized T-intersections where the roadway at the top of the T-intersection is a one-way roadway, ONE WAY signs shall be placed on the near-right and the far side of the intersection facing traffic on the stem approach (see Figure 2B-18).**

Option:

12   Where the central island of a roundabout allows for the installation of signs, ONE WAY signs may be used to direct traffic counter-clockwise around the central island (see Figures 2B-22 and 2B-23).

Guidance:

13   *Where used on the central island of a roundabout, the mounting height of a ONE WAY sign should be at least 4 feet, measured vertically from the bottom of the sign to the elevation of the near edge of the traveled way.*

Option:

14   The BEGIN ONE WAY (R6-6) sign (see Figure 2B-13) may be used to notify road users of the beginning point of a one direction of travel restriction on the street or roadway. The END ONE WAY (R6-7) sign (see Figure 2B-13) may be used to notify road users of the ending point of a one direction of travel restriction on the street or roadway.

Figure 2B-17. Examples of Low-Mounted WRONG WAY Signs with DO NOT ENTER Signs for Wrong-Way Traffic Control

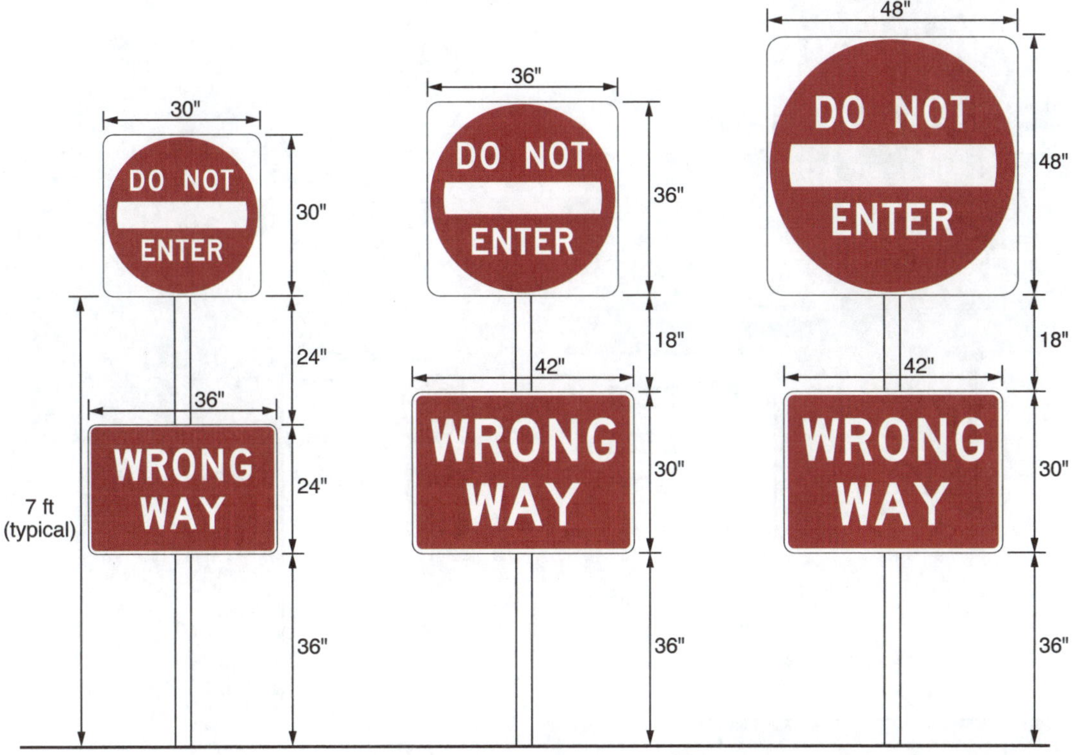

For 7-ft mounting height of primary sign

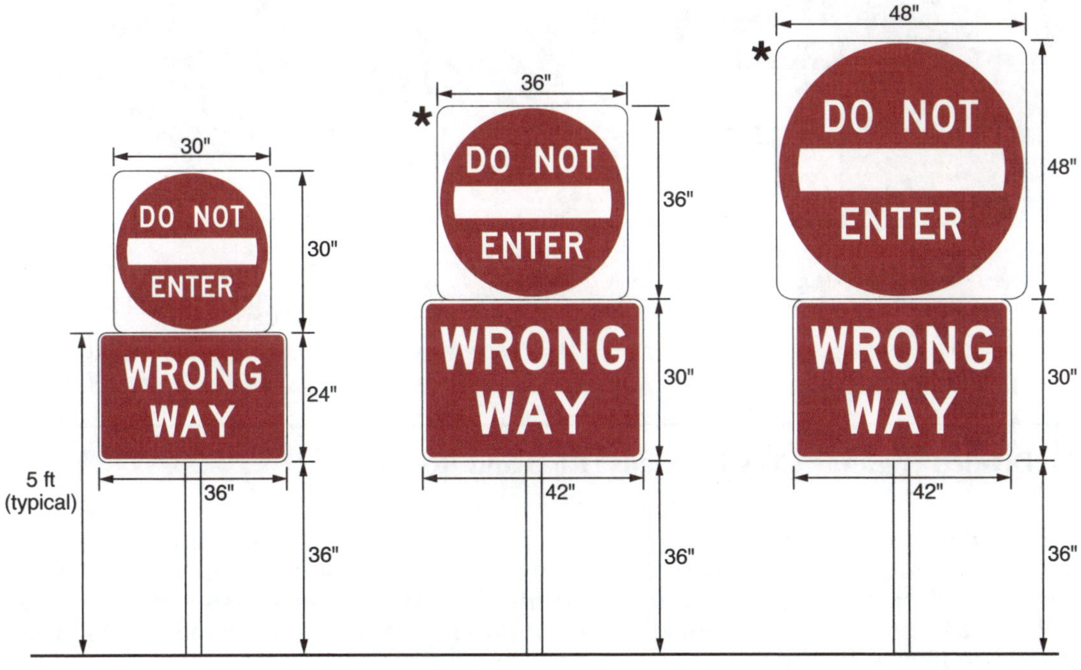

For 5-ft mounting height of primary sign

* Sign is higher than the 5-ft mounting height as the size of the 42" x 30" WRONG WAY sign is controlling.

Figure 2B-18. Locations of ONE WAY Signs

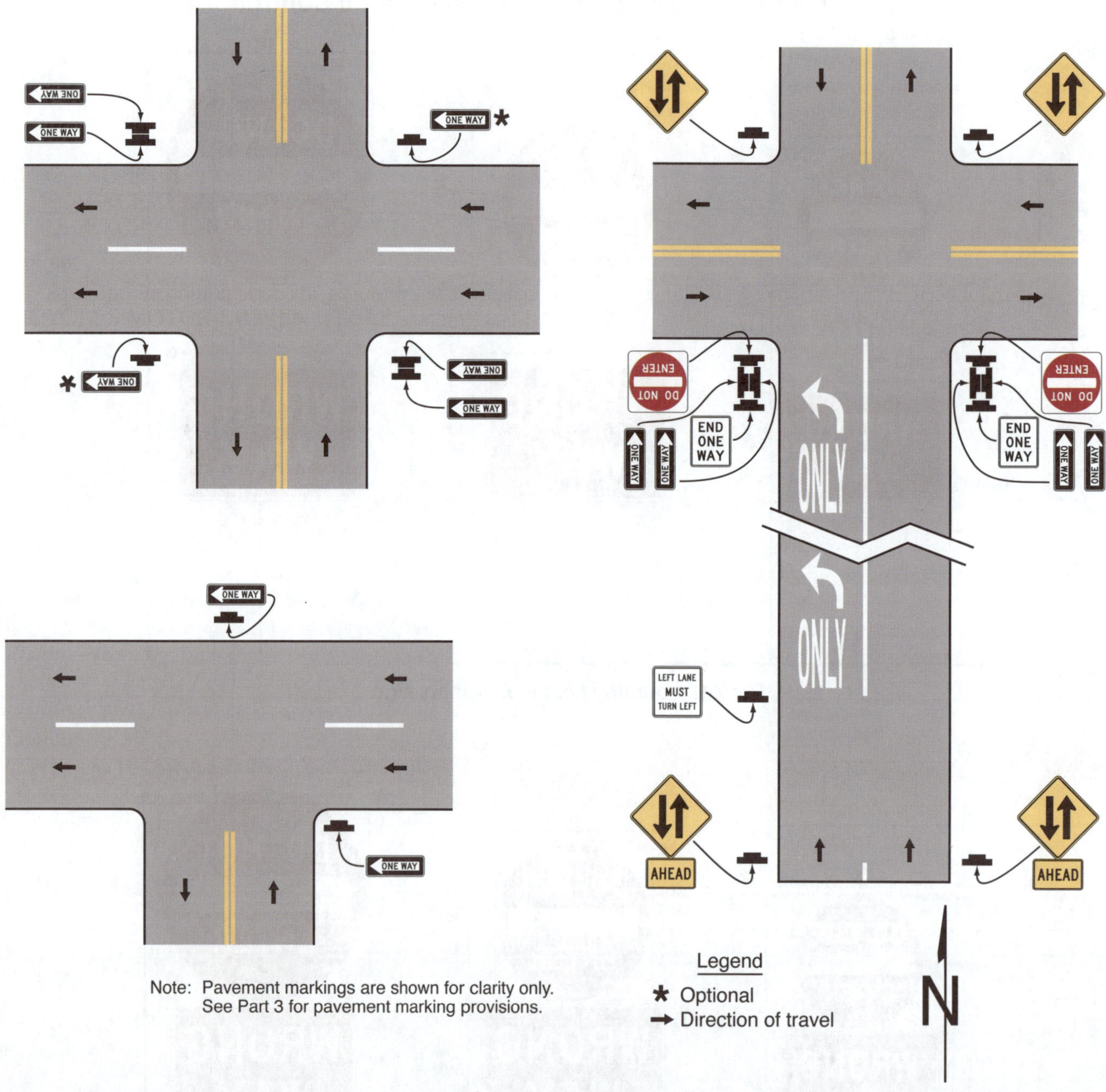

Note: Pavement markings are shown for clarity only.
See Part 3 for pavement marking provisions.

Section 2B.50 Divided Highway Crossing Signs (R6-3 and R6-3a)

Standard:

01   **On unsignalized minor-street approaches from which both left turns and right turns are permitted onto a divided highway at a crossing that functions as two separate intersections (see Section 2A.23), except as provided in Paragraph 2 of this Section, a Divided Highway Crossing (R6-3 or R6-3a) sign (see Figure 2B-13) shall be used to advise road users that they are approaching an intersection with a divided highway (see Figure 2B-19).**

Option:

02   If the divided highway has a traffic volume of less than 400 AADT and a speed limit of 25 mph or less, at a crossing that functions as two separate intersections, the Divided Highway Crossing signs facing the unsignalized minor-street approaches may be omitted.

03   A Divided Highway Crossing sign may be used on signalized minor-street approaches from which both left turns and right turns are permitted onto a divided highway to advise road users that they are approaching an intersection with a divided highway.

Figure 2B-19. ONE WAY Signing for Divided Highway Crossings that Function as Two Separate Intersections

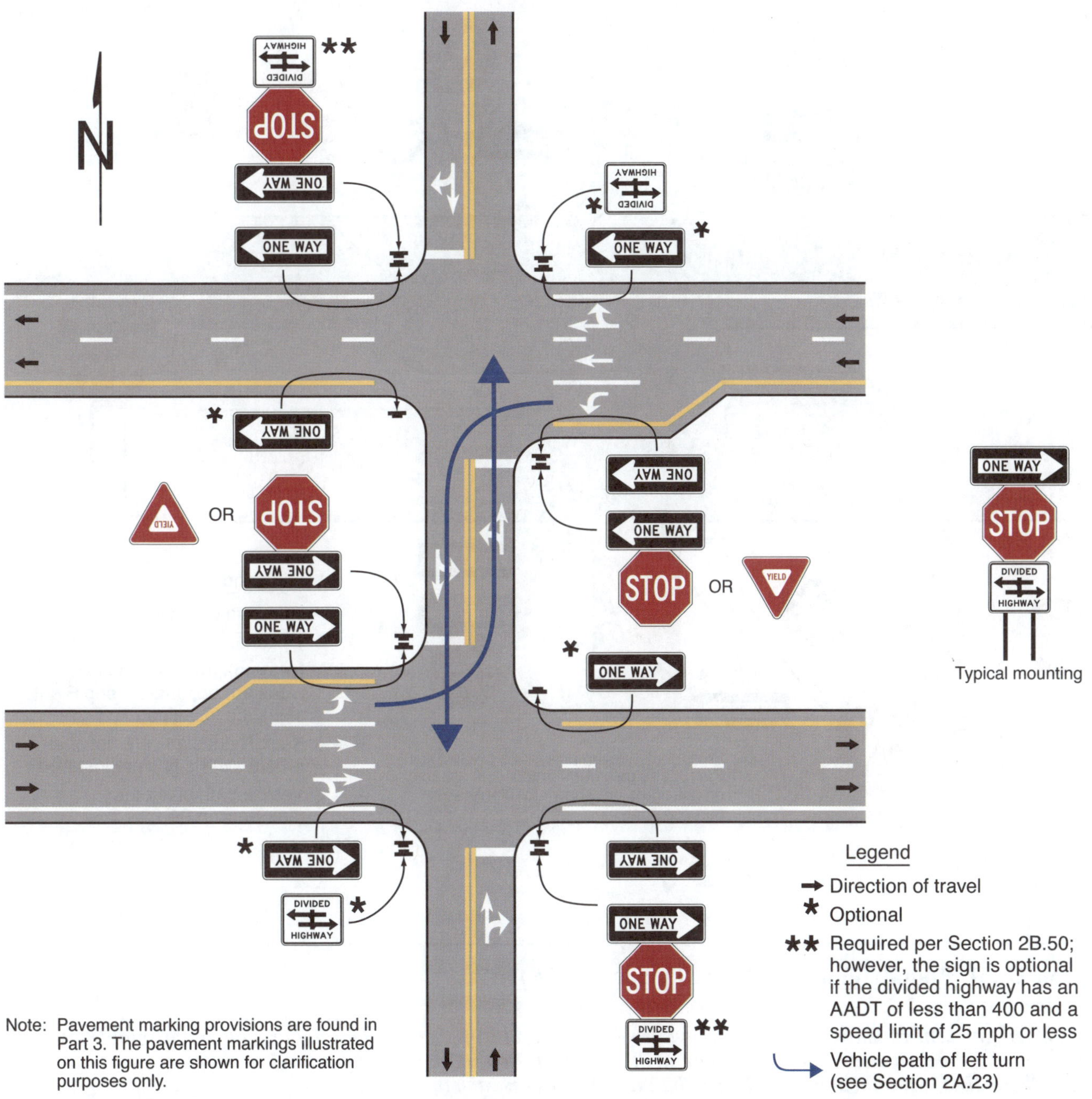

Note: Pavement marking provisions are found in Part 3. The pavement markings illustrated on this figure are shown for clarification purposes only.

Standard:

04   **If a Divided Highway Crossing sign is used at a four-leg intersection, the R6-3 sign shall be used. If used at a T-intersection, the R6-3a sign shall be used.**

05   **The Divided Highway Crossing sign shall be located on the near right corner of the intersection, mounted beneath a STOP or YIELD sign or on a separate support.**

Option:

06   An additional Divided Highway Crossing sign may be installed on the left-hand side of the approach to supplement the Divided Highway Crossing sign on the near right corner of the intersection.

Figure 2B-20. ONE WAY, DO NOT ENTER, and WRONG WAY Signing for Divided Highway Crossings that Function as a Single Intersection

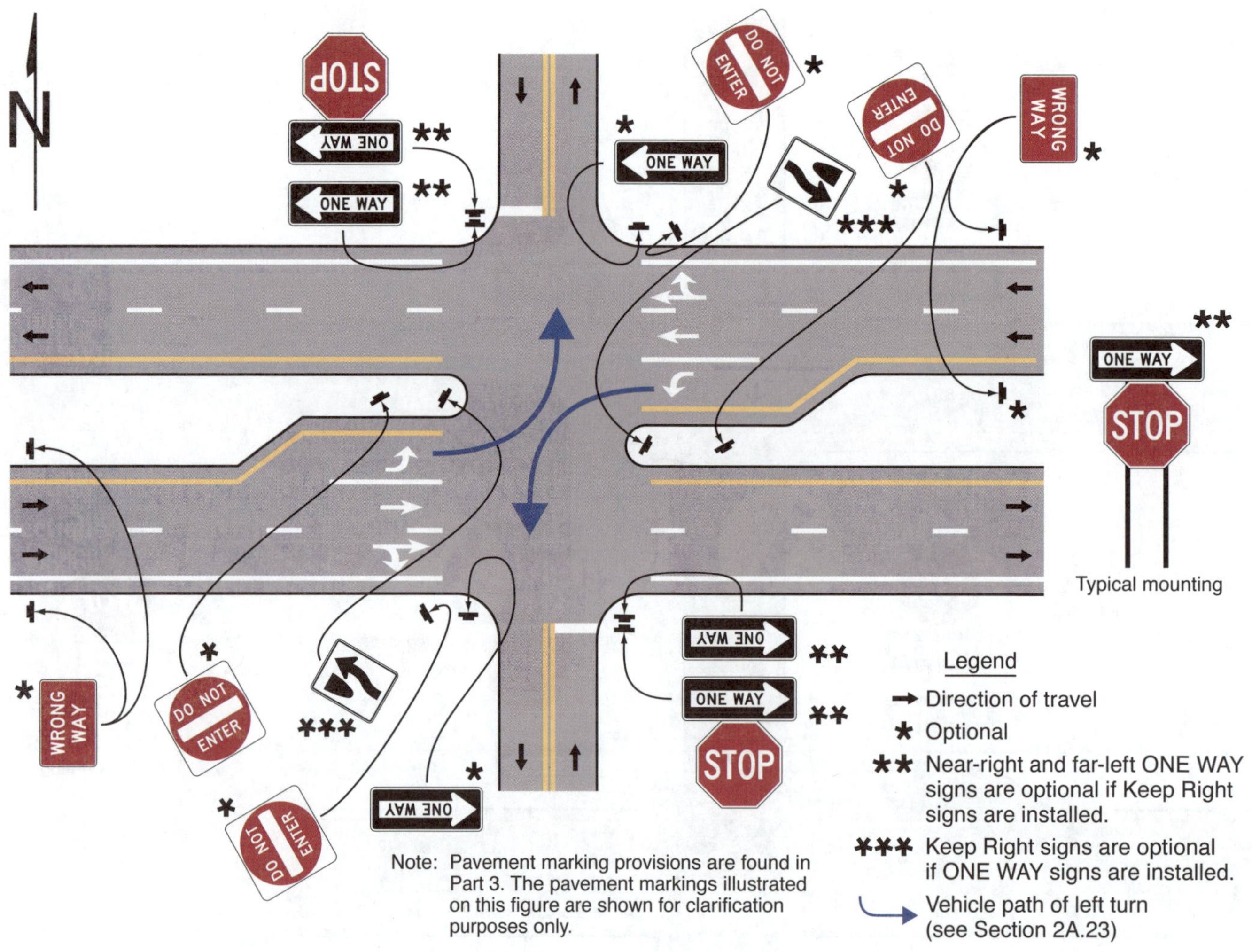

Section 2B.51 <u>Roundabout Circulation Plaque (R6-5P)</u>

Guidance:

01   *Where the central island of a roundabout or neighborhood traffic circle does not provide a reasonable place to install a sign as provided elsewhere in this Chapter, Roundabout Circulation (R6-5P) plaques (see Figure 2B-13) should be placed below the YIELD signs on each approach.*

Support:

02   Paragraph 6 of Section 2B.39 contains information about the use of a Keep Right (R4-7b) sign in the central island of a neighborhood traffic circle.

03   Paragraph 12 of Section 2B.49 contains information about the use of a ONE WAY (R6-1 or R6-2) sign in the central island of a roundabout.

Option:

04   At roundabouts where ONE WAY signs have been installed in the central island, Roundabout Circulation plaques may be placed below the YIELD signs on approaches to roundabouts to supplement the central island signs.

05   The Roundabout Circulation plaque may be used at any type of circular intersection.

Support:

06   Examples of regulatory and warning signs for roundabouts and neighborhood traffic circles are shown in Figures 2B-21 through 2B-24.

Figure 2B-21. Example of Regulatory and Warning Signs for a Mini-Roundabout

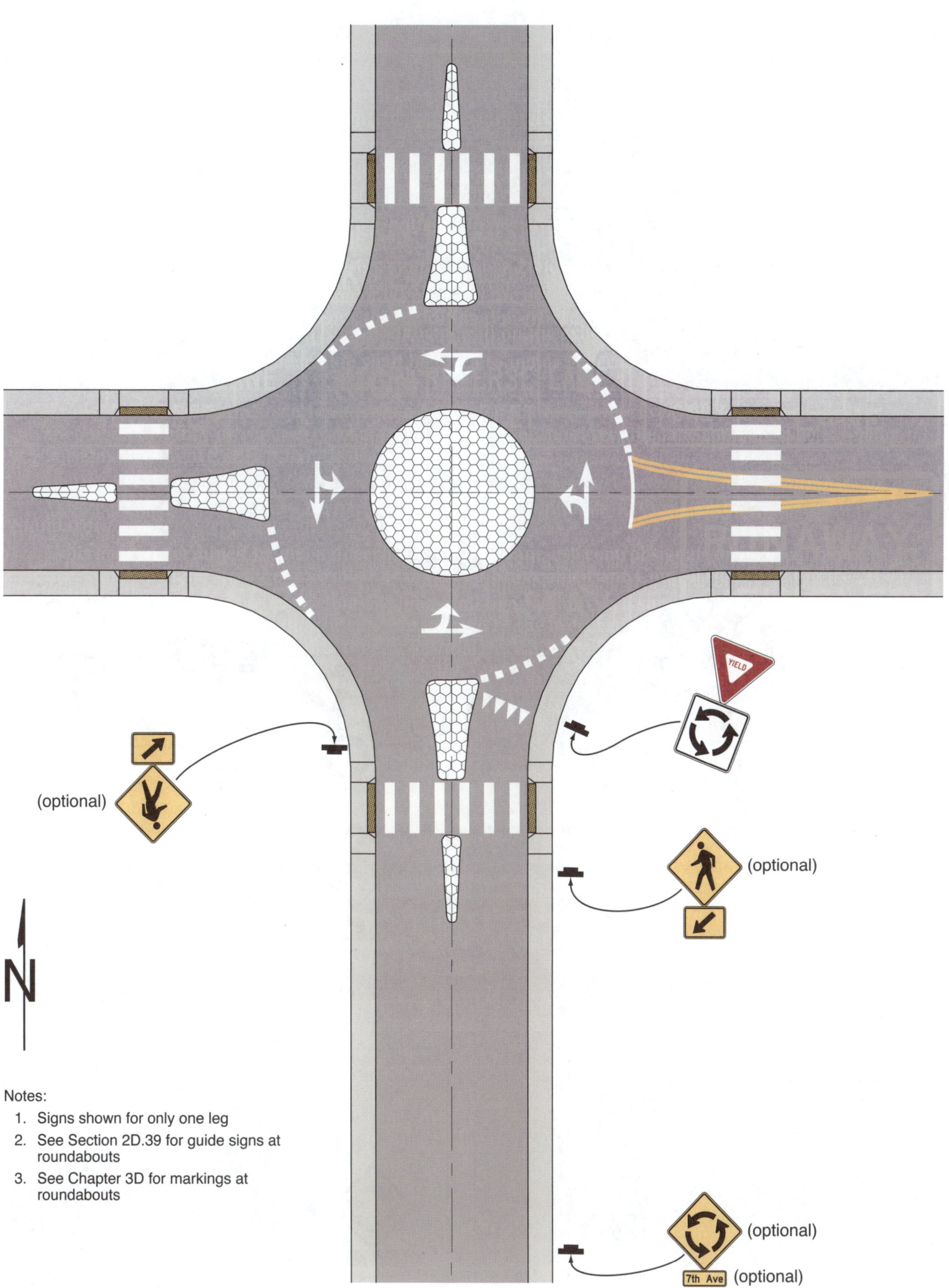

Notes:

1. Signs shown for only one leg
2. See Section 2D.39 for guide signs at roundabouts
3. See Chapter 3D for markings at roundabouts

Figure 2B-22. Example of Regulatory and Warning Signs for a One-Lane Roundabout

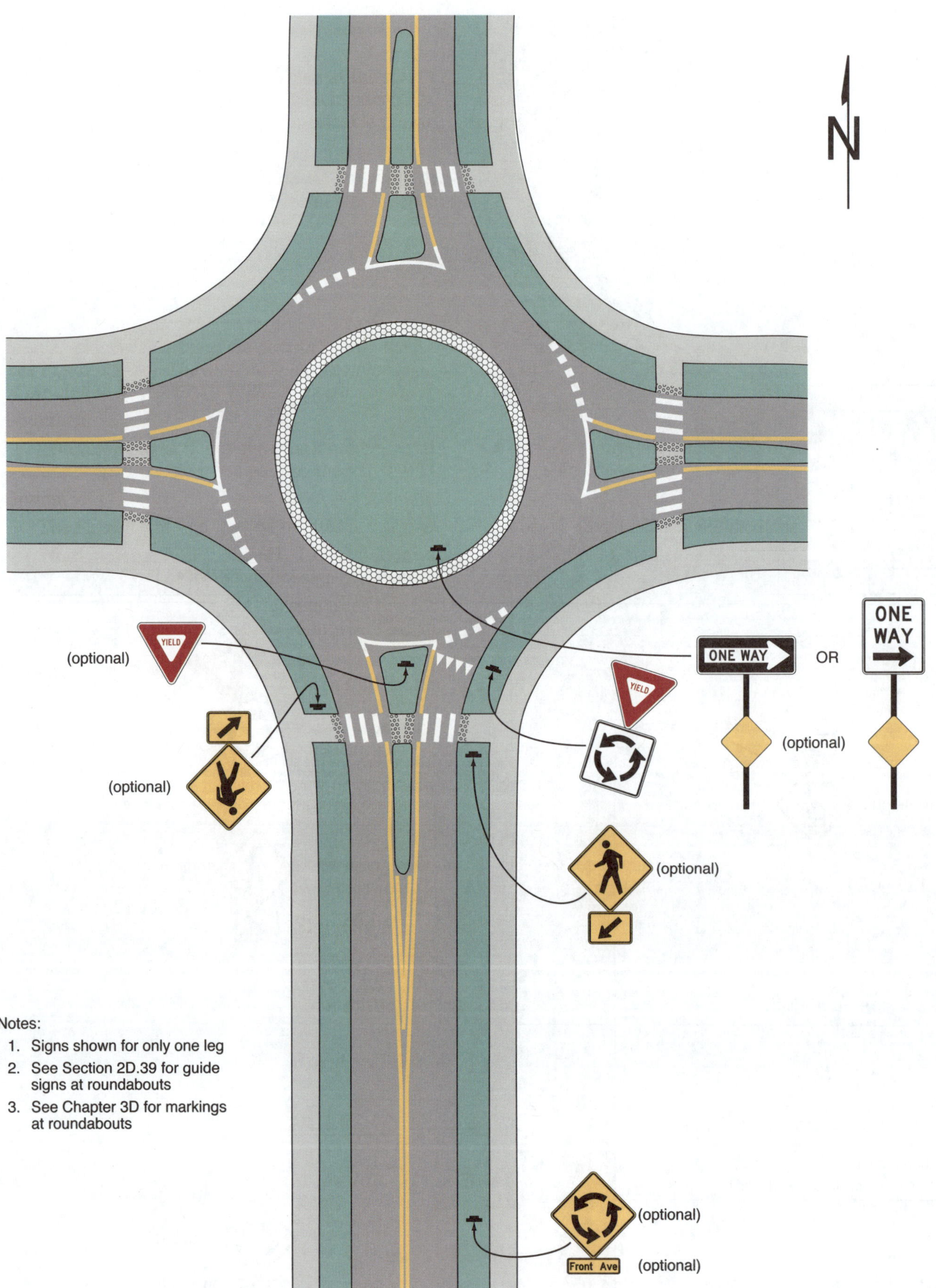

Figure 2B-23. Example of Regulatory and Warning Signs for a Two-Lane Roundabout with Consecutive Double Lefts

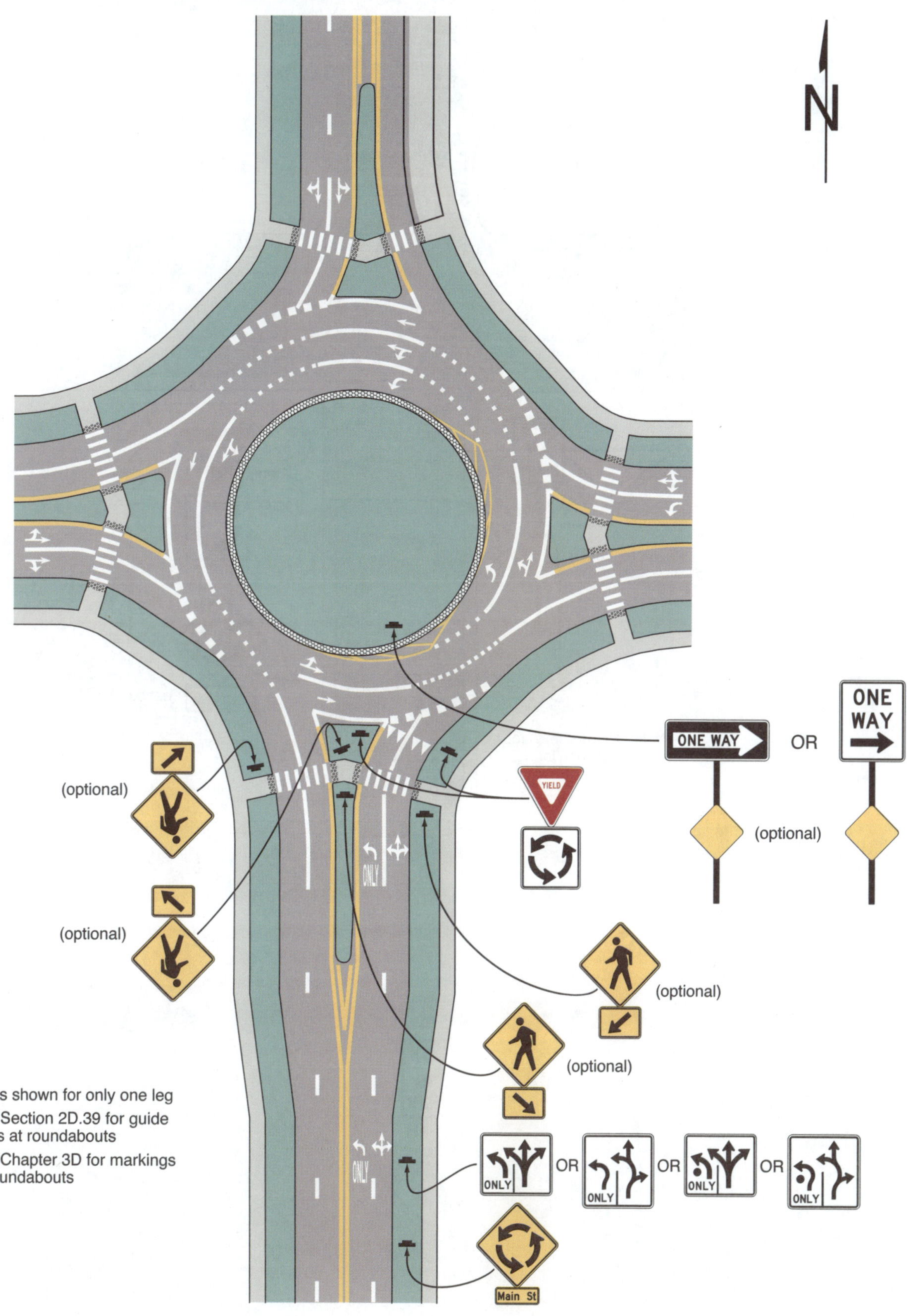

Notes:

1. Signs shown for only one leg
2. See Section 2D.39 for guide signs at roundabouts
3. See Chapter 3D for markings at roundabouts

Figure 2B-24. Example of Regulatory and Warning Signs for a Neighborhood Traffic Circle

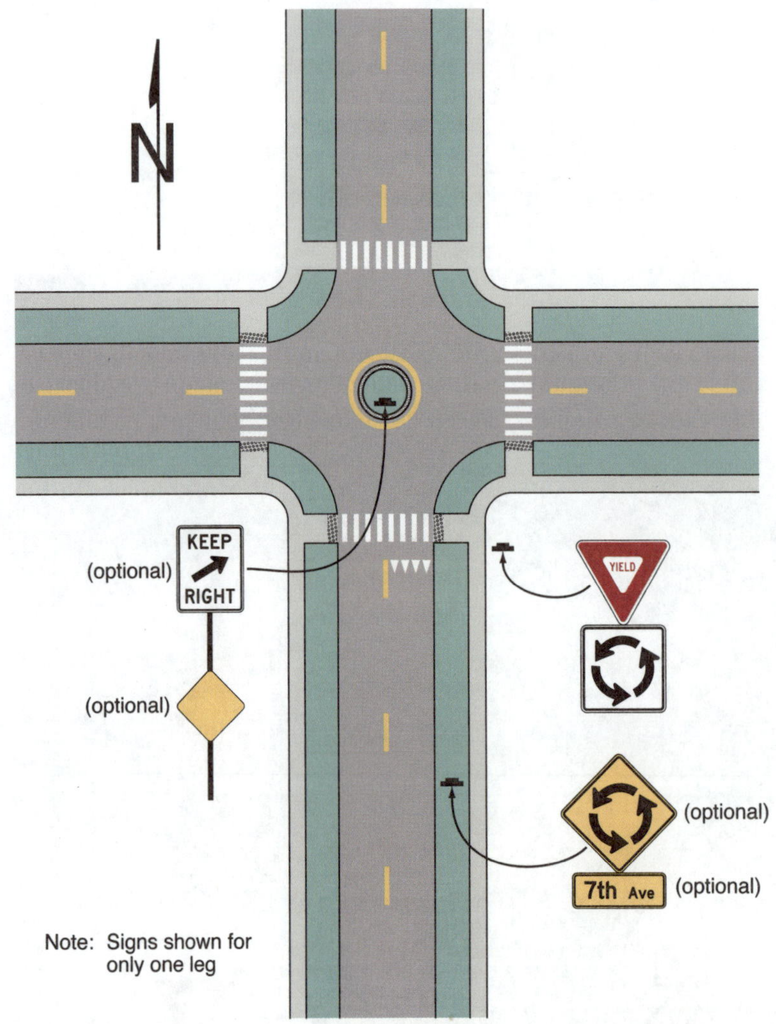

PARKING, STANDING, STOPPING, AND EMERGENCY RESTRICTION SIGNS

Section 2B.52 Parking, Standing, and Stopping Signs (R7 and R8 Series)

Support:

01   Parking signs pertain to the parking, standing, and stopping of vehicles along the roadway and in designated parking areas. They cover a wide variety of regulations, and only general guidance can be provided here. The word "standing" when used on the R7 and R8 series of signs refers to the practice of a driver keeping the vehicle in a stationary position while continuing to occupy the vehicle. The word "stopping" when used on the R7 and R8 series signs refers to any vehicle, occupied by a driver or not, that stops other than to avoid conflict with other traffic or to comply with official direction. Other types of activities such as active loading, active passenger loading, and/or waiting might be established in State or local codes for use on R7 and R8 series signs.

02   Parking signs are categorized as either (1) prohibiting parking or (2) permitting parking with restrictions on how parking is allowed.

03   The types of parking, standing, or stopping prohibitions that might be encountered include, but are not limited to:

   A. Prohibited at all times;
   B. Prohibited only at certain times of the day and/or days of the week;
   C. Prohibited with exceptions, such as for bus stops, loading/unloading zones, persons with disabilities, or electric vehicle charging stations; or
   D. Prohibited under certain conditions, such as Snow Emergency Routes.

04   Permissive parking signs allowing parking with restrictions include, but are not limited to:

   A. Parking only allowed for limited time duration (such as 30 minutes or for 1 hour);
   B. Metered parking requiring payment at an individual or a multi-space parking meter, or through electronic means such as by telephone or mobile application.;
   C. Parking only for specific persons (such as those with disabilities or patrons or employees of a business) or specific vehicle types (such as electric vehicles, police/government vehicles, motorcycles, bicycles, or taxis);
   D. Angled or back-in angled parking when it is not commonly used in the area;
   E. Parking programs such as neighborhood/residential permits, school areas, or special events; and
   F. Emergency parking or stopping only.

Section 2B.53 Design of Parking, Standing, and Stopping Signs

Standard:

01   **Parking, standing, or stopping signs (see Figure 2B-25) shall be rectangular or square.**

02   **Public agencies shall follow established law (State law, local ordinance, or regulation) as adopted by the authorized agency regarding what messages are allowed on parking signs.**

03   **The legend on parking signs shall state applicable regulations. Parking signs shall comply with the standards of shape, color, and location.**

04   **Prohibitive parking signs (see Drawing A in Figure 2B-25 for some commonly used examples) shall be used where parking is prohibited at all times or at specific times. Except as otherwise provided in this Section, parking signs shall have a red legend and border on a white background and, when the parking prohibition symbol is used, the symbol "P" shall be black.**

05   **Permissive parking signs (see Drawing B in Figure 2B-25) shall be used where only time-limited parking or parking in a particular manner is allowed. Permissive parking signs shall have a green legend and border on a white background.**

Guidance:

06   *Parking information, should be displayed from top to bottom of the sign, as applicable, in the following order:*

   A. *The restriction or prohibition;*
   B. *The times of the day that it is applicable, if not all hours;*
   C. *The days of the week that it is applicable, if not every day;*
   D. *Qualifying or supplementary information;*
   E. *Exemptions to the restriction of prohibition; and*
   F. *Any tow-away message or symbol.*

07   *If the parking regulation applies to a limited area or zone, the limits of the regulation should be shown by arrows or supplemental plaques. If arrows are used and if the sign is at the end of a parking zone, there should be a single-headed arrow pointing in the direction that the regulation is in effect. If the sign is at an intermediate point in a zone, there should be a double-headed arrow pointing both ways. When a single sign is used at the transition point between two parking zones, it should display a right arrow and a left arrow pointing in the direction that the respective regulations apply.*

Figure 2B-25. Parking, Standing, and Stopping Signs and Plaques
(R7 and R8 Series) (Sheet 1 of 2)

A – Prohibited parking, standing, and stopping signs and plaques

R7-1

R7-2

R7-2a

R7-3

R7-4

R7-4a

R7-6

R7-107

R7-107a

R7-107a

R7-107b

R7-201P

R7-201aP

R7-202P

R7-203

R8-1

R8-2

R8-3

R8-3a

R8-3c

R8-3d

R8-3e

R8-3f

R8-5

R8-6

B – Permissive parking signs and plaques

R7-5

R7-10

R7-20

R7-21

R7-21aP

R7-22

R7-108

Figure 2B-25. Parking, Standing, and Stopping Signs and Plaques
(R7 and R8 Series) (Sheet 2 of 2)

C – Combination parking signs

R7-200

R7-200a

D – Accessible parking signs and plaques

R7-8   R7-8aP

E – Electric vehicle parking and charging signs and plaques

R7-111   R7-111a   R7-112   R7-112a   R7-112b

R7-113   R7-113aP / R7-113bP   R7-114   R7-114a   R7-114b

Standard:

08   **The times and days for which the parking regulations are in effect shall be posted if they are not in effect at all times of day or all days of the week.**

Option:

09   As an alternate to the use of arrows to show designated restriction zones, the following word messages may be used: BEGIN, END, HERE TO CORNER, HERE TO ALLEY, and THIS SIDE OF SIGN.

10   The R8 series signs (see Drawing A in Figure 2B-25) may be used where sufficient notice of a parking prohibition is satisfied by the use of single signs and are not needed to designate the beginning and end of a zone in which parking is prohibited or restricted. In rural and certain other areas the legends NO PARKING ON PAVEMENT (R8-1) or NO STOPPING ON PAVEMENT (R8-5) are generally suitable and may be used where parking or stopping is allowed on an unpaved shoulder or border adjacent to the paved portion of the road. If a roadway has an adjacent paved shoulder on which parking or stopping is allowed, the legend NO PARKING EXCEPT ON SHOULDER (R8-2) or NO STOPPING EXCEPT ON SHOULDER (R8-6) may be used. The R8-3 symbol sign or the word message NO PARKING may be used to prohibit any parking along a roadway. Word legend supplemental plaques may be mounted below the NO PARKING signs or the word legend may be incorporated within signs whose sizes are increased accordingly. The R8-3 series signs may include word legends such as ON PAVEMENT (R8-3c), ON BRIDGE (R8-3d), ON TRACKS (R8-3e), and EXCEPT ON SHOULDERS (R8-3f).

Guidance:

11   *Where special parking restrictions are imposed during heavy snowfall or a declared snow emergency, a Snow Emergency Route (R7-203) sign (see Drawing A in Figure 2B-25) should be installed. The legend should be modified to display the specific regulations. The upper section of the sign should display the designation as a snow emergency route in a white legend and border on a red background.*

12   *If a fee is charged for on-street parking standing payments are made at a multi-space parking meter, instead of individual parking meters for each parking space, Metered Parking (R7-21 and R7-22) signs (see Drawing B in Figure 2B-25) should be used to define the area where the multi-space parking meter applies. The Multi-Space Parking Meter (R7-20) sign (see Drawing B in Figure 2B-25) should be used at the meter location to direct road users to the meter.*

Option:

13 Where payments can be made electronically, such as by telephone or mobile application, the Mobile Parking Payment (R7-21aP) plaque (see Drawing B in Figure 2B-25) may be installed below or as part of the legend of a Metered Parking sign.

Standard:

14 **If the metered parking is subject to a maximum time limit, the appropriate time limit (number of hours or minutes) shall be displayed on the Metered Parking (R7-21 and R7-22) signs and, except as provided in Paragraph 15 of this Section, on the Multi-space Parking Meter (R7-20) signs.**

Option:

15 Where the maximum time limit varies by the time of the day or by the day of the week, the display of the time limits may be omitted from the R7-20 sign and, instead, be displayed on the multi-space parking meter so that they are visible to pedestrians as they make payments.

Standard:

16 **Where parking spaces are reserved for persons with disabilities, the Accessible Parking (R7-8) sign (see Drawing D in Figure 2B-25) shall be used to designate the space and shall display the official International Symbol of Accessibility.**

17 **Where parking spaces that are reserved for persons with disabilities are designed to accommodate wheelchair vans, a VAN ACCESSIBLE (R7-8aP) plaque (see Drawing D in Figure 2B-25) shall be mounted below the R7-8 sign.**

Guidance:

18 *Where parking spaces are designated for parking of electric vehicles, an Electric Vehicle Parking (R7-111 series, R7-112 series, and R7-113) sign (see Drawing E of Figure 2B-25) should be installed adjacent to the designated spaces. Where there is no time limit, the R7-111 series sign should be used. Where parking is subject to a time limit, the R7-112 series sign should be used.*

19 *Where parking spaces are only designated for charging of electric vehicles, an R7-113 sign or R7-114 series sign (see Drawing E in Figure 2B-25) should be installed adjacent to the designated spaces.*

20 *Where additional restrictions apply while a vehicle occupies the designated space, the R7-113P series plaques should be installed below the R7-113 sign or the R7-114 series signs.*

Option:

21 Where parking is prohibited during certain hours and time-limited parking or parking in a particular manner is allowed during certain other time periods, the red Parking Prohibition and green Permissive Parking signs may be designed as follows (see Drawing C in Figure 2B-25):

A. Two 12 x 18-inch parking signs may be used with the red Parking Prohibition (R7-1) sign installed above or to the left of the green Permissive Parking (R7-108) sign; or

B. A single sign (R7-200 or R7-200a) may be used.

22 At the transition point between two parking zones, a single sign (R7-200 or R7-200a) or two signs mounted side-by-side may be used.

23 The words NO PARKING may be used as an alternative to the No Parking symbol (see the R7-2a sign in Drawing A in Figure 2B-25).

24 Alternate designs for the R7-107 sign may be developed such as the R7-107a sign (see Drawing A in Figure 2B-25). Alternate designs may include, on a single sign, a transit logo, an approved bus symbol, a parking prohibition, the words BUS STOP, and an arrow. The reverse side of the R7-107 series signs may display bus routing information for pedestrians.

25 A Tow-Away Zone (R7-201P or R7-201aP) plaque (see Drawing A in Figure 2B-25) may be mounted below any parking prohibition sign. The word legend TOW-AWAY ZONE may be incorporated into the parking prohibition sign in lieu of using a separate plaque.

26 The R7-201P plaque may have a black or red symbol and border on a white background.

Guidance:

27 *When a legend other than that on the standard parking signs is necessary, letter height, symbol size, and basic sign layout should be consistent with the those shown on the standard parking signs as detailed in the "Standard Highway Signs" publication (see Section 1A.05.)*

28 *In general, the letter height of the principal legend on parking signs sized for urbanized applications should be at least 2 inches.*

Section 2B.54 Placement of Parking, Standing, and Stopping Signs

Support:

01   The efficacy of parking, standing, and stopping signs, when used on conventional roads in urbanized or developed environments, depends on their visibility and consistent placement along a street or within a particular block. It is often impracticable for the entire legend to be legible from similar distances as for other types of signs. Therefore, it is important that their conventional form be recognizable from an adequate distance such that the road user can obtain the information upon closer inspection.

Guidance:

02   *When signs with arrows are used to indicate the extent of the restricted zones, the signs should be set at an angle of not less than 30 degrees or more than 45 degrees with the line of traffic flow in order to be visible to approaching traffic.*

03   *When signs are placed at the head of perpendicular parking stalls, the signs should be parallel to the roadway facing the parking stall.*

04   *Spacing of signs should be based on legibility, conspicuity, and sign orientation.*

05   *If the zone is long, signs should be used at intermediate points within the zone.*

06   *If the signs are mounted at an angle of 90 degrees to the curb line, two signs should be mounted back to back at the transition point between two parking zones, each with an appended THIS SIDE OF SIGN (R7-202P) supplemental plaque (see Drawing A in Figure 2B-25).*

07   *If the signs are mounted at an angle of 90 degrees to the curb line, signs without any arrows or appended plaques should be used at intermediate points within a parking zone, facing in the direction of approaching traffic. Otherwise, the standards of placement should be the same as for signs using directional arrows.*

Option:

08   Blanket parking regulations that apply to an entire jurisdiction may, if legal, be posted in the vicinity of the jurisdictional boundary lines. Blanket parking regulations that apply to a posted zone or district may, if legal, be posted at the entry points to the zone or district.

Section 2B.55 Emergency Restriction Signs (R8-4 and R8-7)

Standard:

01   **Emergency Restriction signs (see Figure 2B-26) shall be rectangular and shall have a black legend and border on a white background.**

Option:

02   The EMERGENCY PARKING ONLY (R8-4) sign or the EMERGENCY STOPPING ONLY (R8-7) sign may be used to discourage or prohibit shoulder parking, particularly where scenic or other attractions create a tendency for road users to desire to stop temporarily.

Support:

03   Section 8B.07 contains information for the use of the DO NOT STOP ON TRACKS (R8-8) sign (see Figure 8B-1) to discourage or prohibit parking or stopping on railroad or light rail transit tracks.

Figure 2B-26. Emergency Restriction Signs

R8-4   R8-7

PEDESTRIAN SIGNS

Section 2B.56 <u>WALK ON LEFT FACING TRAFFIC and No Hitchhiking Signs (R9-1, R9-4, and R9-4a)</u>

Option:

01   The WALK ON LEFT FACING TRAFFIC (R9-1) sign (see Figure 2B-27) may be used on highways where no sidewalks are provided.

Guidance:

02   *If used, the WALK ON LEFT FACING TRAFFIC sign should be installed on the right-hand side of the road where pedestrians walk on the pavement or shoulder in the absence of pedestrian pathways or sidewalks.*

Option:

03   The No Hitchhiking (R9-4) sign (see Figure 2B-27) may be used to prohibit standing in or adjacent to the roadway for the purpose of soliciting a ride. The R9-4a word message sign (see Figure 2B-27) may be used as an alternate to the R9-4 symbol sign.

Section 2B.57 <u>Pedestrian Crossing Signs (R9-2 and R9-3)</u>

Option:

01   Pedestrian Crossing signs (see Figure 2B-27) may be used to limit pedestrian crossing to specific locations.

Standard:

02   **If used, Pedestrian Crossing signs shall be installed to face pedestrian approaches.**

Option:

03   Where crosswalks are clearly defined, the CROSS ONLY AT CROSSWALKS (R9-2) sign may be used to prohibit pedestrians from crossing at locations away from crosswalks.

04   The No Pedestrian Crossing (R9-3) sign may be used to prohibit pedestrians from crossing a roadway at an undesirable location or in front of a school or other public building where a crossing is not designated.

05   The NO PEDESTRIAN CROSSING (R9-3a) word message sign may be used as an alternate to the R9-3 symbol sign. The USE CROSSWALK (R9-3bP) supplemental plaque, along with an arrow, may be installed below either sign to designate the direction of the crossing.

Support:

06   Pedestrians with vision disabilities might need features other than traffic control devices to provide effective communication of the prohibition of pedestrian crossing.

Guidance:

07   *The R9-3bP plaque should not be installed in combination with educational plaques.*

Section 2B.58 <u>Traffic Signal Pedestrian and Bicyclist Actuation Signs (R10-1 through R10-4 and R10-24 through R10-26)</u>

Standard:

01   **Where manual actuation of a traffic signal is required for pedestrians or bicyclists to call a signal phase to cross the roadway, traffic signal signs applicable to pedestrian actuation (see Figure 2B-27) or bicyclist actuation (see Figure 9B-1) shall be mounted immediately above or incorporated into the push button detector units (see Section 4I.05).**

Support:

02   Traffic signal signs applicable to pedestrians include:
    A. CROSS ONLY ON GREEN (symbolic circular green) (R10-1),
    B. CROSS ONLY ON (symbolic walk indication) SIGNAL (R10-2),
    C. Push Button for Walk Signal (R10-3 series), and
    D. Push Button for Green Signal (R10-4 series).

Option:

03   The following signs may be used as an alternate for the R10-3 and R10-4 signs:
    A. Push Button to Cross Street Wait for Walk Signal (R10-3a); or
    B. Push Button to Cross Street Wait for Green Signal (R10-4a).

04   The name of the street to be crossed may be substituted for the word STREET in the legends on the R10-3a and R10-4a signs.

Guidance:

05   *The finger in the push button symbol on the R10-3, R10-3a, R10-4, and R10-4a signs should point in the same direction as the arrow on the sign.*

Figure 2B-27. Pedestrian Signs and Plaques (Sheet 1 of 2)

R9-1

R9-2

R9-3

R9-3a

R9-3bP

R9-4

R9-4a

R10-1

R10-2

R10-3

R10-3a

R10-3b

R10-3c

R10-3d

R10-3e

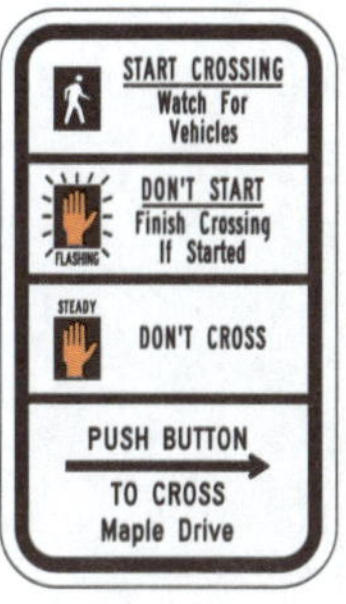

R10-3f

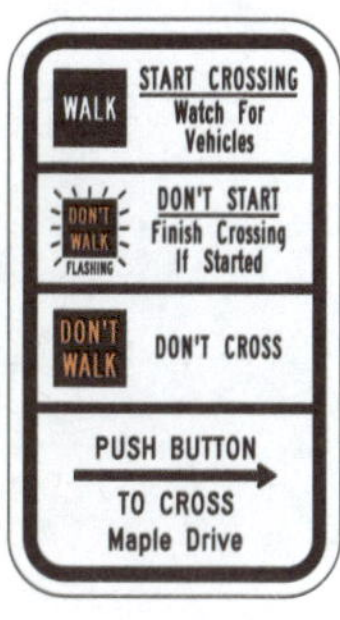

R10-3g

R10-3h

Figure 2B-27. Pedestrian Signs and Plaques (Sheet 2 of 2)

R10-3i

R10-4

R10-4a

R10-25

R10-32P

Option:

06   Where symbolic pedestrian signal indications are used, an educational sign (R10-3b) may be used instead of the R10-3 sign to improve pedestrian understanding of pedestrian indications at signalized intersections. Where word-legend pedestrian signal indications are being retained for the remainder of their useful service life, the legends WALK/DONT WALK may be substituted for the symbols on the educational sign R10-3b, thus creating educational sign R10-3c. The R10-3d educational sign may be used to inform pedestrians that the pedestrian clearance time is sufficient only for the pedestrian to cross to the median at locations where pedestrians cross in two stages using a median refuge island. The R10-3e educational sign may be used where countdown pedestrian signals have been provided. In order to assist the pedestrian in understanding which push button to push, the R10-3f through R10-3i educational signs that provide the name of the street to be crossed may be used instead of the R10-3b through R10-3e educational signs.

07   The R10-24 or R10-26 sign (see Section 9B.20) may be used where a push button detector has been installed exclusively to actuate a green phase for bicyclists.

08   The R10-25 sign (see Figure 2B-27) may be used where a push button detector has been installed for pedestrians to activate In-Roadway Warning Lights (see Chapter 4U) or flashing beacons that have been added to the pedestrian warning signs.

Support:

09   Section 4I.05 contains information regarding the application of the R10-32P plaque.

TRAFFIC SIGNAL SIGNS AND PLAQUES

Section 2B.59 Traffic Signal Signs and Plaques (R10-5 through R10-30)

Option:

01   To supplement traffic signal control, traffic signal (R10-5 through R10-30) signs (see Figure 2B-28) may be used to regulate road users.

02   Traffic signal signs may be installed at certain locations to clarify signal control. Among the legends that may be used for this purpose are:

    A. LEFT (RIGHT) ON GREEN ARROW ONLY (R10-5),
    B. STOP HERE ON RED (R10-6 or R10-6a) for observance of stop lines,
    C. DO NOT BLOCK INTERSECTION (R10-7) for avoidance of traffic obstructions,
    D. USE LANE(S) WITH GREEN ARROW (R10-8) for obedience to lane-use control signals (see Chapter 4T),
    E. LEFT (RIGHT) TURN SIGNAL (R10-10),
    F. U TURN SIGNAL (R10-10a) for exclusive control of a U-turn movement,
    G. U TURN YIELD TO RIGHT TURN (R10-16),
    H. LEFT (RIGHT) TURN YIELD ON GREEN (symbolic circular green) (R10-12),
    I. LEFT (RIGHT) TURN YIELD ON FLASHING YELLOW ARROW (R10-12a), and
    J. LEFT (RIGHT) TURN YIELD ON FLASHING RED ARROW AFTER STOP (R10-27).

Guidance:

03   *If used, the LEFT ON GREEN ARROW ONLY sign, the LEFT TURN SIGNAL sign, the LEFT TURN YIELD ON GREEN (symbolic circular green)) sign, the LEFT TURN YIELD ON FLASHING YELLOW ARROW sign, or the LEFT TURN YIELD ON FLASHING RED ARROW AFTER STOP) sign should be located adjacent to the left-turn signal face.*

04   *If used, the RIGHT ON GREEN ARROW ONLY sign, the RIGHT TURN SIGNAL sign, the RIGHT TURN YIELD ON FLASHING YELLOW ARROW sign, or the RIGHT TURN YIELD ON FLASHING RED ARROW AFTER STOP sign should be located adjacent to the right-turn signal face.*

05   *A U TURN YIELD TO RIGHT TURN (R10-16) sign should be installed near the left-turn signal face if U-turns are allowed on a protected left-turn movement on an approach from which a right-turn GREEN ARROW signal indication is simultaneously being displayed to drivers making a right turn from the conflicting approach to their left.*

Option:

06   If used, a U TURN SIGNAL (R10-10a) sign may be installed adjacent to the signal face that exclusively controls a U-turn movement.

07   If needed for additional emphasis, an additional LEFT TURN YIELD ON GREEN (symbolic circular green) (R10-12) sign with an AT SIGNAL (R10-31P) supplemental plaque (see Figure 2B-28) may be installed in advance of the intersection.

08   In situations where traffic control signals are coordinated for progressive timing, the Traffic Signal Speed (I1-1) sign may be used (see Section 2H.04).

Standard:

09   **The CROSSWALK—STOP ON RED (symbolic circular red) (R10-23) and STOP ON STEADY RED-YIELD ON FLASHING RED AFTER STOP (R10-23a) signs (see Figure 2B-28) shall only be used in conjunction with pedestrian hybrid beacons (see Section 4J.02).**

10   **The EMERGENCY SIGNAL (R10-13) sign (see Figure 2B-28) shall be used in conjunction with emergency-vehicle traffic control signals (see Section 4M.02).**

11   **The EMERGENCY SIGNAL—STOP ON FLASHING RED (R10-14 or R10-14a) sign (see Figure 2B-28) shall be used in conjunction with emergency-vehicle hybrid beacons (see Section 4N.02).**

Option:

12   If needed for extra emphasis, a STOP HERE ON FLASHING RED (R10-14b) sign may be installed with an emergency-vehicle hybrid beacon.

Standard:

13   **The Left Turn Yield to Bicycles (R10-12b) sign shall be limited to applications where the conflicting bicyclist movement would be unexpected in direction, location, or similar condition that would tend to violate the expectation of a turning motorist.**

Figure 2B-28. Traffic Signal Signs and Plaques (Sheet 1 of 2)

R10-5

R10-6

R10-6a

R10-7

R10-8

R10-10

R10-10a

R10-11

R10-11a

R10-11b

R10-11c

R10-11d

R10-12

R10-12a

R10-12b

R10-13

R10-14

R10-14a

R10-14b

Figure 2B-28. Traffic Signal Signs and Plaques (Sheet 2 of 2)

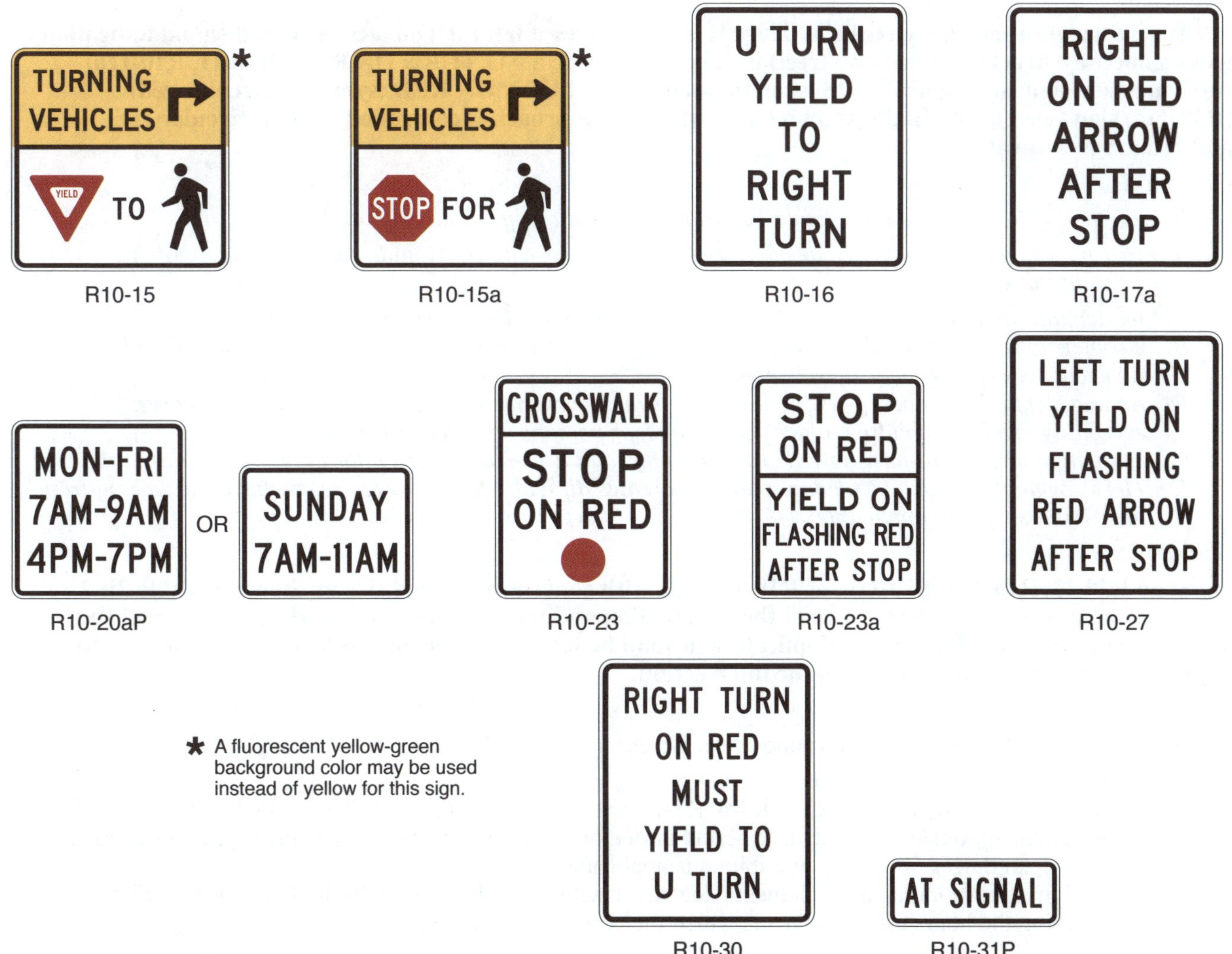

Guidance:

14   *The Left Turn Yield to Bicycles sign should be located adjacent to the left-turn signal face.*

Option:

15

If needed for additional emphasis, an additional Left Turn Yield to Bicycles sign with an AT SIGNAL (R10-31P) supplemental plaque (see Figure 2B-28) may be installed in advance of the intersection for motor vehicles.

16   Where conditions might warrant additional emphasis to drivers turning at a signalized intersection where potential pedestrian conflicts might not be readily apparent, a Turning Vehicles Yield to (Stop for) Pedestrians (R10-15, R10-15a) sign (see Figure 2B-28) may be used.

Standard:

17   **The Turning Vehicles Stop for Pedestrians (R10-15a) sign shall only be used in jurisdictions where laws, ordinances or resolutions specifically require that a driver must stop for a pedestrian.**

Guidance:

18   *The R10-15 series signs, where used, should be placed as follows:*

    A. *On the near right corner of the signalized intersection for right-turning vehicles.*
    B. *On the far left corner of the signalized intersection for the left-turning vehicles onto a two-way street.*
    C. *On the near left corner of the signalized intersection for left-turning vehicles from a one-way street onto a one-way street.*

Section 2B.60 No Turn on Red Signs (R10-11 Series, R10-17a, and R10-30)

Standard:

01 **Where a right turn on a circular red signal indication (or a left turn on a circular red signal indication from a one-way street to a one-way street) is to be prohibited, a NO TURN ON RED (R10-11, R10-11b) word message sign (see Figure 2B-28) shall be used. A NO TURN ON RED (symbolic circular red) (R10-11a) sign (see Figure 2B-28) shall be used when the approach is controlled by both circular red and red arrow indications.**

Guidance:

02 *If used, the No Turn on Red sign should be installed near the appropriate signal head.*

03 *A No Turn on Red sign should be considered when an engineering study finds that one or more of the following conditions exists:*

A. *Inadequate sight distance to vehicles approaching from the left (or right, if applicable);*
B. *Geometrics or operational characteristics of the intersection that might result in unexpected conflicts;*
C. *An exclusive pedestrian or bicycle phase;*
D. *An unacceptable number of conflicting pedestrian movements with right-turn-on-red maneuvers, especially involving children, older pedestrians, or persons with disabilities;*
E. *More than three right-turn-on-red crashes reported in a 12-month period for the particular approach; or*
F. *The skew angle of the intersecting roadways creates difficulty for drivers to see traffic approaching from their left (or right, if applicable).*

Standard:

04 **If an R10-11, R10-11a, R10-11b, or R10-17a sign with conventional road size as shown in Table 2B-1 is used on an approach on the far side of the intersection and the distance between the stop line and the sign is greater than 120 feet, then a duplicate sign shall be located on the near side of the intersection to supplement the sign on the far side of the intersection.**

Option:

05 When a no-turn-on-red restriction applies during certain time periods only, the following alternatives may be used:

A. Movement Prohibition (R3-1, R3-2, R3-4, R3-18, and R3-27) signs or NO TURN ON RED signs displayed by using a blank-out sign for the time period or one or more portion(s) of a particular cycle of the traffic control signal during which the prohibition is applicable; or
B. Static signs incorporating a supplemental legend or with a supplemental R10-20aP plaque (see Figure 2B-28) showing the hours and days during which the prohibition is applicable.

06 White LEDs may be used in the border and activated during periods of turn prohibition to enhance the sign conspicuity.

07 On signalized approaches with more than one right-turn lane, a NO TURN ON RED EXCEPT FROM RIGHT LANE (R10-11c) sign (see Figure 2B-28) may be post-mounted at the intersection or a NO TURN ON RED FROM THIS LANE (with down arrow) (R10-11d) sign (see Figure 2B-28) may be mounted over the approximate center of the lane from which turns on red are prohibited.

Guidance:

08 *Where turns on red are permitted and the signal indication is a steady RED ARROW, the RIGHT (LEFT) ON RED ARROW AFTER STOP (R10-17a) sign (see Figure 2B-28) should be installed adjacent to the RED ARROW signal indication.*

Option:

09 A RIGHT TURN ON RED MUST YIELD TO U-TURN (R10-30) sign (see Figure 2B-28) may be installed to remind road users that they must yield to conflicting U-turn traffic on the street or highway onto which they are turning right on a red signal after stopping.

Section 2B.61 <u>Ramp Metering Signs (R10-28 and R10-29)</u>

Option:

01   When ramp control signals (see Chapter 4P) are used to meter traffic on a freeway or expressway entrance ramp, regulatory signs with legends appropriate to the control may be installed adjacent to the ramp control signal faces.

02   For entrance ramps with only one controlled lane, an XX VEHICLE(S) PER GREEN (R10-28) sign (see Figure 2B-29) may be used to inform road users of the number of vehicles that are permitted to proceed during each short display of the green signal indication. For entrance ramps with more than one controlled lane, an XX VEHICLE(S) PER GREEN EACH LANE (R10-29) (see Figure 2B-29) sign may be used to inform road users of the number of vehicles that are permitted to proceed from each lane during each short display of the green signal indication.

Support:

03   Chapter 2L contains provisions for the use of blank-out or changeable message signs when the metering is limited by time, day, or condition.

Figure 2B-29. Ramp Metering Signs

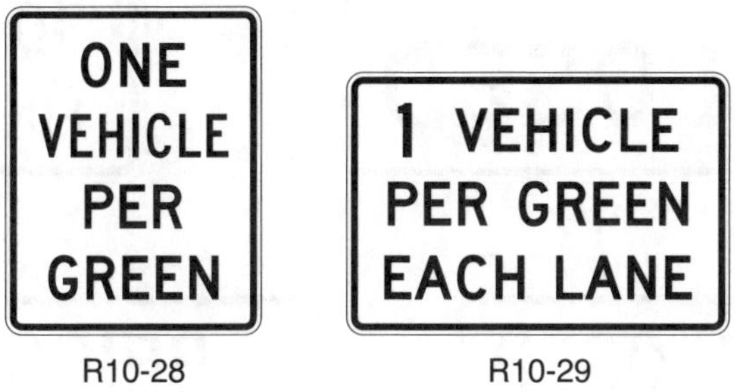

R10-28

R10-29

ROAD CLOSED AND WEIGHT LIMIT SIGNS

Section 2B.62 KEEP OFF MEDIAN Sign (R11-1)

Option:

01   The KEEP OFF MEDIAN (R11-1) sign (see Figure 2B-30) may be used to prohibit driving into or parking on the median.

Guidance:

02   *The KEEP OFF MEDIAN sign should be installed on the left-hand side of the roadway within the median at random intervals as needed wherever there is a tendency for encroachment.*

Figure 2B-30. Road Closed and Weight Limit Signs

R11-1

R11-2

R11-3

R11-3a

R11-3b

R11-4

R12-1

R12-2

R12-4

R12-5

R12-6

R12-7

R12-7aP

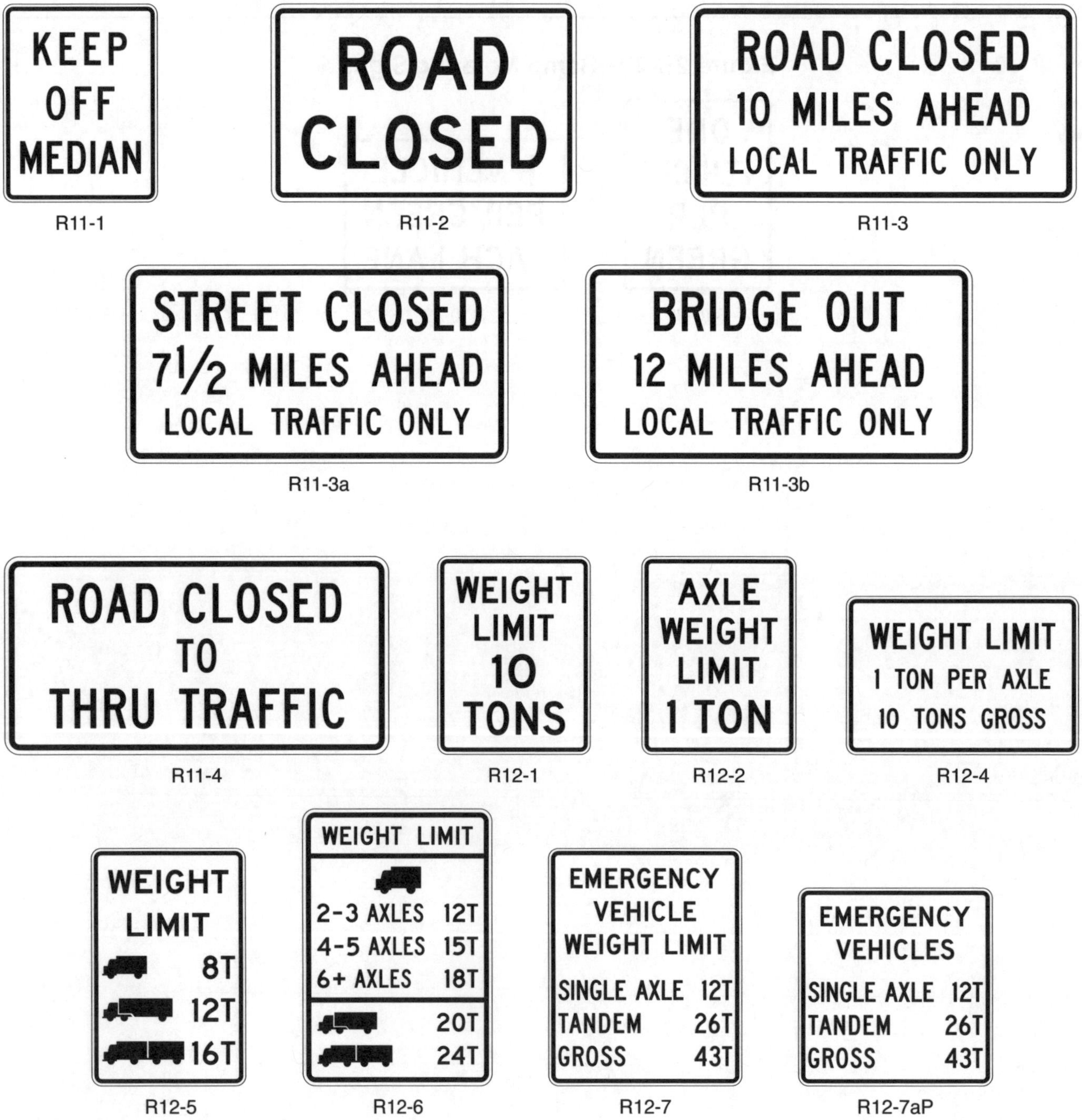

Section 2B.63 ROAD CLOSED Sign (R11-2) and LOCAL TRAFFIC ONLY Signs (R11-3 Series, R11-4)

Guidance:

01   *The ROAD CLOSED (R11-2) sign should be installed where roads have been closed to all traffic (except authorized vehicles).*

02   *ROAD CLOSED—LOCAL TRAFFIC ONLY (R11-3) or ROAD CLOSED TO THRU TRAFFIC (R11-4) signs should be used where through traffic is not permitted, or for a closure some distance beyond the sign, but where the highway is open for local traffic up to the point of closure.*

Standard:

03   **The Road Closed (R11-2, R11-3 series, and R11-4) signs (see Figure 2B-30) shall be designed as horizontal rectangles. These signs shall be preceded by the applicable Advance Road Closed warning sign with the secondary legend AHEAD and, if applicable, an Advance Detour warning sign (see Section 6H.04).**

Option:

04   An intersecting street name or a well-known destination may be substituted for the XX MILES AHEAD legend in urban areas.

05   The word message BRIDGE OUT may be substituted for the ROAD CLOSED legend where applicable.

06   Where conditions allow for bicycle travel on the road beyond the point of closure to motor vehicles, an EXCEPT BICYCLES (R3-7bP) plaque (see Figure 2B-4) may be used with the ROAD CLOSED sign.

Section 2B.64 Weight Limit Signs (R12-1 through R12-7)

Standard:

01   **Weight limit signs (see Figure 2B-30) shall be used to indicate a section of highway or structure that has a vehicle weight restriction.**

Guidance:

02   *The units shown on any weight limit sign should be consistent within a State or region with respect to pounds or tons.*

Option:

03   Where the restriction applies to axle weight rather than gross load, the legend AXLE WEIGHT LIMIT XX TONS or AXLE WEIGHT LIMIT XX LBS (R12-2) may be used.

04   In areas where multiple regulations are applicable, such as limiting both axle weight and gross vehicle weight, a WEIGHT LIMIT XX TONS PER AXLE, XX TONS GROSS (R12-4) sign combining the necessary messages on a single sign may be used.

05   Posting of specific load limits may be accomplished by use of the Weight Limit (R12-5) symbol sign. A sign containing the legend WEIGHT LIMIT on the top two lines, and showing up to three different truck symbols and their respective weight limits for which restrictions apply may be used, with the weight limits displayed to the right of each symbol as XX T. A bottom line of legend stating GROSS WT may be included if needed for enforcement purposes.

Support:

06   A specialized hauling vehicle is a single unit truck with multiple closely-spaced axles. Examples include dump trucks, construction vehicles, solid waste trucks and other hauling trucks. Specialized hauling vehicles typically have 4 to 7 axles.

Option:

07   The Weight Limit (R12-6) sign may be used to indicate vehicle weight restrictions for specialized hauling vehicles.

Standard:

08   **The symbols shown on the R12-5 and R12-6 Weight Limit sign shall apply to all trucks of that configuration (single-unit, single-trailer or multi-trailer) regardless of the shape of the vehicle. Symbolic representations of other vehicle shapes or modifications of standard symbols shall not be used.**

Option:

09   The facility type (such as "BRIDGE") may be added to the legend of the sign to clarify the specific applicability of the weight limit.

Standard:

10   **If the R12-5 sign depicts only one single-unit vehicle symbol, the weight limit associated with that single-unit vehicle symbol shall apply to all single-unit vehicles, regardless of number of axles.**

11   **The weight limit associated with the single-trailer vehicle symbol shall apply to all single-trailer vehicles, regardless of number of axles or vehicle shape.**

12   **The weight limit associated with the multi-trailer vehicle symbol shall apply to all multi-trailer vehicles with two or more trailers, regardless of number of axles or vehicle shape.**

13   **A weight limit sign (see Figure 2B-30) shall be located at the applicable section of highway or structure.**

14   **An additional weight limit sign, with an advisory distance or directional legend, shall be located in advance of the applicable section of highway or structure so that prohibited vehicles can detour or turn around prior to the limit zone.**

Support:

15   An emergency vehicle is designed to be used under emergency conditions to transport personnel and equipment to support the suppression of fires and mitigation of other hazardous situations. Emergency vehicles are typically operated by fire departments and are primarily equipped for firefighting, but are also used to respond to and mitigate other hazardous situations in an emergency. They can create higher load effects compared to non-emergency vehicles of similar weight.

Option:

16   The Emergency Vehicle Weight Limit (R12-7) sign carrying the legend EMERGENCY VEHICLE WEIGHT LIMIT SINGLE AXLE XX TONS, TANDEM XX TONS, and GROSS XX TONS may be used to indicate vehicle weight restrictions for emergency vehicles.

Standard:

17   **When the emergency-vehicle weight limit is displayed in the same assembly as the primary weight limit sign, the Emergency Vehicle Weight Limit (R12-7aP) plaque shall be mounted below.**

Section 2B.65 Weigh Station Sign (R13-1)

Guidance:

01   *An R13-1 sign with the legend TRUCKS OVER XX TONS MUST ENTER WEIGH STATION NEXT RIGHT (see Figure 2B-31) should be used to direct appropriate traffic into an inspection station.*

02   *The R13-1 sign should be supplemented by the D8 series of guide signs (see Section 2D.51).*

Section 2B.66 TRUCK ROUTE Sign (R14-1)

Guidance:

01   *The TRUCK ROUTE (R14-1) sign (see Figure 2B-31) should be used to mark a route that has been designated to allow truck traffic.*

Support:

02   Section 2D.20 contains information regarding the use of the TRUCK (M4-4P) auxiliary plaque (see Figure 2D-5) on a designated numbered alternative route.

Section 2B.67 Hazardous Material Signs (R14-2 and R14-3)

Option:

01   The Hazardous Material Route (R14-2) sign (see Figure 2B-31) may be used to identify routes that have been designated by proper authority for vehicles transporting hazardous material.

02   On routes where the transporting of hazardous material is prohibited, the Hazardous Material Prohibition (R14-3) sign (see Figure 2B-31) may be used.

Guidance:

03   *If used, the Hazardous Material Prohibition sign should be installed on a street or roadway at a point where vehicles transporting hazardous material have the opportunity to take an alternate route.*

Section 2B.68 National Network Signs (R14-4 and R14-5)

Support:

01 The signing of the National Network routes for trucking is optional.

Standard:

02 **When a National Network route is signed, the National Network (R14-4) sign (see Figure 2B-31) shall be used.**

Option:

03 The National Network Prohibition (R14-5) sign (see Figure 2B-31) may be used to identify routes, portions of routes, and ramps where trucks are prohibited. The R14-5 sign may also be used to mark the ends of designated routes.

Figure 2B-31. Truck Signs

OTHER REGULATORY SIGNS AND PLAQUES

Section 2B.69 Photo Enforced Signs and Plaques (R10-18, R10-18a, R10-19P, R10-19aP)

Option:

01   A Traffic Laws Photo Enforced (R10-18) sign (see Figure 2B-32) may be installed at a jurisdictional boundary to advise road users that some of the traffic regulations within that jurisdiction are being enforced by photographic equipment.

02   A Traffic Signal Photo Enforced (R10-18a) sign (see Figure 2B-32) may be installed in advance of or at a traffic signal to advise road users that compliance with the signal is enforced by photographic equipment. A Signal Ahead (W3-3) sign and a Traffic Signal Photo Enforced (R10-18a) sign may be used on the same approach provided that they are on separate supports.

03   A Photo Enforced (R10-19P) plaque or a PHOTO ENFORCED (R10-19aP) word message plaque (see Figure 2B-32) may be mounted below a regulatory sign to advise road users that the regulation is being enforced by photographic equipment.

Figure 2B-32. Photo Enforcement Signs and Plaques

Standard:

04   **The Traffic Signal Photo Enforced (R10-18a) sign shall not be installed on approaches to signalized locations where red-light cameras are not present on any of the approaches to the signalized location.**

05   **A Traffic Signal Photo Enforced (R10-18a) sign shall not be installed on the same support in combination with a Signal Ahead (W3-3) sign.**

06   **If used below a regulatory sign, the Photo Enforced (R10-19P or R10-19aP) plaque shall be a rectangle with a black legend and border on a white background.**

Section 2B.70 Move Vehicles from Travel Lanes Sign (R16-4)

Option:

01   A STATE LAW MINOR CRASHES MOVE VEHICLES FROM TRAVEL LANES (R16-4) sign (see Figure 2B-33) may be installed in accordance with the provisions of Section 2A.01 to require motorists to move their vehicle out of the travel lanes if they have been involved in a crash.

02   If the specific requirements of a State law vary, the word legend of the R16-4 sign may be modified to reflect the appropriate law.

Figure 2B-33. Other Regulatory Signs

Section 2B.71 Move Over or Reduce Speed Sign (R16-3)

Option:

01   A STATE LAW MOVE OVER OR REDUCE SPEED FOR VEHICLES STOPPED ON SHOULDER (R16-3) sign (see Figure 2B-33) may be installed in accordance with the provisions of Section 2A.01 to require motorists to change lanes and/or reduce speed when passing stopped emergency vehicles on the shoulder.

02   If the specific requirements of a State law vary, the word legend of the R16-3 sign may be modified to reflect the appropriate law.

Section 2B.72 No Hand-Held Phone Use by Driver Signs (R16-15 and R16-15a)

Option:

01   A STATE LAW NO HAND-HELD PHONE USE BY DRIVER (R16-15 or R16-15a) sign (see Figure 2B-33) may be installed in accordance with the provisions of Section 2A.01 to notify drivers that they are prohibited from using hand-held telephones while driving.

02   If the specific requirements of a State law vary, the word legend of the R16-15 series signs may be modified to reflect the appropriate law.

Section 2B.73 Headlight Use Signs (R16-5 through R16-11)

Support:

01   Some States require road users to turn on their vehicle headlights under certain weather conditions, as a safety improvement measure on roadways experiencing high crash rates, or in special situations such as when driving through a tunnel.

02   Figure 2B-34 shows the various signs that can be used for informing motorists of these requirements.

Option:

03   A LIGHTS ON WHEN USING WIPERS (R16-5) sign or a LIGHTS ON WHEN RAINING (R16-6) sign may be installed in accordance with the provisions of Section 2A.01 to inform road users of State laws regarding headlight use. Although these signs are typically installed facing traffic entering the State just inside the State border, they also may be installed at other locations within the State.

Guidance:

04   *If a particular section of roadway has been designated as a safety improvement zone within which headlight use is required, a TURN ON HEADLIGHTS NEXT XX MILES (R16-7) sign or a BEGIN DAYTIME HEADLIGHT SECTION (R16-10) sign should be installed at the upstream end of the section, and an END DAYTIME HEADLIGHT SECTION (R16-11) sign should be installed at the downstream end of the section.*

Option:

05   A TURN ON HEADLIGHTS (R16-8) sign may be installed to require road users to turn on their headlights in special situations such as when driving through a tunnel. A CHECK HEADLIGHTS (R16-9) sign may be installed downstream from the special situation to inform drivers that using their headlights is no longer required.

Figure 2B-34. Headlight Use Signs

Section 2B.74 Seat Belt Symbol

Guidance:

01   *The seat belt symbol should not be used alone. If used, the seat belt symbol should be incorporated into regulatory sign messages for mandatory seat belt use.*

Support:

02   The seat belt symbol is illustrated in the "Standard Highway Signs" publication (see Section 1A.05).

BARRICADES AND GATES

Section 2B.75 Barricades

Option:

01   Barricades may be used to mark any of the following conditions:

    A. The end of a roadway,

    B. A ramp or lane that is closed for operational purposes, or

    C. The permanent or semi-permanent closure or termination of a roadway.

Standard:

02   **When used to warn and alert road users of the terminus of a roadway, other than in temporary traffic control zones, barricades shall meet the design criteria of Section 6K.07 for a Type 3 Barricade, except that the colors of the stripes shall be retroreflective white and retroreflective red.**

Option:

03   An end-of-roadway marker or markers may be used as described in Section 2C.73.

Guidance:

04   *Appropriate advance warning signs (see Chapter 2C) should be used.*

Section 2B.76 Gates

Support:

01   Gates described in this section used for weather or other emergency conditions are typically permanently installed to enable the gate to be immediately deployed as needed to prohibit the entry of traffic to the highway segment(s).

02   A gate typically features a gate arm that is moved from a vertical to a horizontal position or is rotated in a horizontal plane from parallel to traffic to perpendicular to traffic. Traffic is obstructed and required to stop when the gate arm is placed in a horizontal position perpendicular to traffic. Another type of gate consists of a segment of fence (usually on rollers) that swings open and closed, or that is retracted to open and then extended to close.

03   Gates are sometimes used to enforce a required stop. Some examples of such uses are the following:

    A. Parking facility entrances and exits,

    B. Private community entrances and exits,

    C. Military base entrances and exits,

    D. Toll plaza lanes,

    E. Movable bridges (see Chapter 4Q),

    F. Automated Flagger Assistance Devices (see Chapter 6L), and

    G. Grade crossings (see Part 8).

04   Gates are sometimes used to periodically close a roadway or a ramp. Some examples of such uses are the following:

    A. Closing ramps to implement counter-flow operations for evacuations,

    B. Closing ramps that lead to reversible lanes, and

    C. Closing roadways for weather events such as snow, ice, or flooding, or for other emergencies.

Standard:

05   **Except as provided in Paragraph 6 of this Section, gate arms, if used, shall be fully retroreflective on both sides, have vertical stripes alternately red and white at 16-inch intervals measured horizontally as shown in Figure 8D-1. The width (which becomes the height of the retroreflective sheeting when the gate is in the down position) of the retroreflective sheeting on the front of the gate arm shall be at least 4 inches.**

Option:

06   If used on a one-way roadway or ramp, the retroreflective sheeting may be omitted on the side of the gate (or rolling fence) facing away from approaching traffic.

07   Where gate arms are used to block off ramps into reversible lanes or to redirect approaching traffic, the red and white striping may be angled such that the stripes slope downward at an angle of 45 degrees toward the side of the gate arm on which traffic is to pass.

Standard:

08   **The gate arm shall extend across the approaching lane or lanes of traffic to effectively block motor vehicle, bicycle, and/or pedestrian travel as appropriate.**

Guidance:

09   *When a gate that is rotated in a horizontal plane is in the position where it is parallel to traffic (indicating that the roadway is open), the outer end of the gate arm should be rotated to the downstream direction (from the perspective of traffic in the lane adjacent to the gate support) to prevent spearing if the gate is struck by an errant vehicle.*

Standard:

10   **If red lights are attached to a traffic gate, the red lights shall be steadily illuminated or flashed only during the period when the gate is in the horizontal or closed position and when the gate is in the process of being opened or closed.**

11   **Except as provided in Paragraph 6 of this Section, rolling sections of fence, if used, shall include either a horizontal strip of retroreflective sheeting on both sides of the fence with vertical stripes alternately red and white at 16-inch intervals measured horizontally to simulate the appearance of a gate arm in the horizontal position, or one or more Type 4 object markers (see Section 2C.73), or both. If a horizontal strip of retroreflective sheeting is used, the bottom of the sheeting shall be located 3.5 to 4.5 feet above the roadway surface.**

CHAPTER 2C. WARNING SIGNS AND OBJECT MARKERS

Chapter 2C Subchapter and Section Organization

GENERAL

Section 2C.01 Application of Warning Signs
Standard:

01 **The use of warning signs shall be based on an engineering study or on engineering judgment.**

02 **Warning signs shall be retroreflective or illuminated (see Section 2A.21).**

Guidance:

03 *The use of warning signs should be kept to a minimum as the unnecessary use of warning signs tends to breed disrespect for all signs. In situations where the condition or activity is seasonal or temporary, the warning sign should be removed or covered when the condition or activity does not exist.*

Section 2C.02 Design of Warning Signs
Standard:

01 **Except as provided in Paragraph 2 of this Section or unless specifically designated otherwise, all warning signs shall be diamond-shaped (square with one diagonal vertical) with a black legend and border on a yellow background. Warning signs shall be designed in accordance with the sizes, shapes, colors, and legends contained in the "Standard Highway Signs" publication (see Section 1A.05).**

Option:

02 A warning sign that is larger than the size shown in the Oversized column in Table 2C-1 for that particular sign may be diamond-shaped or may be rectangular or square in shape.

Support:

03 The use of a shape other than diamond-shaped is typical for overhead installations.

04 Section 2A.05 contains information on allowable methods to accommodate a diamond-shaped warning sign where the lateral space available in which to install a diamond-shaped warning sign is constrained, such as in urban locations, when mounting on a narrow median barrier or adjacent to a retaining wall, including the display of the standard legend in a vertically oriented rectangle.

05 The use of LEDs in the border and legend of warning signs is described in Section 2A.12.

Option:

06 Except for symbols on warning signs, minor modifications may be made to the design provided that the essential appearance characteristics are met. Modifications may be made to the symbols shown on combined horizontal alignment/intersection signs (see Section 2C.09) and intersection warning signs (see Section 2C.41) in order to approximate the geometric configuration of the intersecting roadway(s).

07 Word message warning signs other than those provided in this Manual may be developed and installed by State and local highway agencies for conditions otherwise not addressed by standard signs (see Section 2A.04).

08 Warning signs regarding conditions associated with pedestrians, bicyclists, and playgrounds and their related plaques may have a black legend and border on a yellow or fluorescent yellow-green background.

Standard:

09 **Warning signs regarding conditions associated with school buses and schools and their related supplemental plaques shall have a black legend and border on a fluorescent yellow-green background (see Section 7B.01).**

Option:

10 Consistent with the provisions of Section 4S.03, a Warning Beacon may be used in combination with a standard warning sign.

Section 2C.03 Size of Warning Signs and Plaques
Standard:

01 **Except as provided in Section 2A.07, the sizes for warning signs shall be as shown in Table 2C-1.**

Support:

02 Section 2A.07 contains information regarding the applicability of the various columns in Table 2C-1.

Standard:

03 **Except as provided in Paragraph 5 of this Section, the minimum size for all diamond-shaped warning signs facing traffic on a multi-lane conventional road where the posted speed limit is higher than 35 mph shall be 36 x 36 inches.**

Table 2C-1. Warning Sign and Plaque Sizes (Sheet 1 of 4)

Sign or Plaque	Sign Designation	Section	Conventional Road		Expressway	Freeway	Minimum	Oversized
			Single Lane	Multi-Lane				
Horizontal Alignment	W1-1,2,3,4,5	2C.07	30 x 30*	36 x 36	36 x 36	36 x 36	—	48 x 48
One-Direction Large Arrow	W1-6	2C.10	48 x 24	48 x 24	60 x 30	60 x 30	—	60 x 30
Two-Direction Large Arrow	W1-7	2C.43	48 x 24	48 x 24	—	—	—	60 x 30
Chevron Alignment	W1-8	2C.08	18 x 24	18 x 24	30 x 36	36 x 48	—	24 x 30
Combination Horizontal Alignment/Intersection	W1-10,10a,10b, 10c,10d,10e	2C.09	36 x 36	36 x 36	36 x 36	48 x 48	—	—
Hairpin Curve	W1-11	2C.07	30 x 30	30 x 30	36 x 36	48 x 48	—	48 x 48
Truck Rollover	W1-13	2C.11	36 x 36	36 x 36	36 x 36	48 x 48	—	48 x 48
270-degree Curve	W1-15	2C.07	30 x 30	30 x 30	36 x 36	48 x 48	—	48 x 48
Intersection Warning	W2-1,2,3,3a, 4,5,6,7,8	2C.41	30 x 30	30 x 30	36 x 36	—	24 x 24	48 x 48
Traffic Entering When Flashing	W2-10	2C.42	36 x 36	36 x 36	48 x 48	—	—	—
Traffic Approaching When Flashing	W2-11	2C.42	36 x 36	36 x 36	48 x 48	—	—	—
Stop, Yield, Signal Ahead	W3-1,2,3	2C.35	30 x 30	30 x 30	48 x 48	48 x 48	30 x 30	—
Be Prepared to Stop	W3-4	2C.35	36 x 36	36 x 36	48 x 48	48 x 48	30 x 30	—
Reduced Speed Limit Ahead	W3-5	2C.40	36 x 36	36 x 36	48 x 48	48 x 48	—	—
XX MPH Speed Zone Ahead	W3-5a	2C.40	36 x 36	36 x 36	48 x 48	48 x 48	—	—
Variable Speed Zone Ahead	W3-5b	2C.40	36 x 36	36 x 36	48 x 48	48 x 48	—	—
XX MPH Truck Speed Zone Ahead	W3-5c	2C.40	36 x 36	36 x 36	48 x 48	48 x 48	—	—
Draw Bridge	W3-6	2C.36	36 x 36	36 x 36	48 x 48	—	—	60 x 60
Ramp Meter Ahead	W3-7	2C.37	36 x 36	36 x 36	—	—	—	—
Ramp Metered When Flashing	W3-8	2C.37	36 x 36	36 x 36	—	—	—	—
Merge	W4-1	2C.45	36 x 36	36 x 36	48 x 48	48 x 48	30 x 30*	—
Lane Ends	W4-2	2C.47	36 x 36	36 x 36	48 x 48	48 x 48	30 x 30*	—
Added Lane	W4-3	2C.46	36 x 36	36 x 36	48 x 48	48 x 48	30 x 30*	—
Cross Traffic Does Not Stop (plaque)	W4-4P	2C.66	24 x 12	24 x 12	36 x 18	—	—	48 x 24
Traffic From Left (Right) Does Not Stop (plaque)	W4-4aP	2C.66	24 x 12	24 x 12	36 x 18	—	—	48 x 24
Oncoming Traffic Does Not Stop (plaque)	W4-4bP	2C.66	24 x 12	24 x 12	36 x 18	—	—	48 x 24
Entering Roadway Merge	W4-5	2C.45	36 x 36	36 x 36	48 x 48	48 x 48	—	—
No Merge Area (plaque)	W4-5aP	2C.45	18 x 24	18 x 24	24 x 30	24 x 30	—	—
Entering Roadway Added Lane	W4-6	2C.46	36 x 36	36 x 36	48 x 48	48 x 48	—	—
Heavy Merge from Right (Left)	W4-7	2C.49	36 x 36	36 x 36	48 x 48	48 x 48	—	—
Single Lane Transition	W4-8	2C.48	36 x 36	36 x 36	36 x 36	48 x 48	—	48 x 48
Road Narrows	W5-1	2C.17	36 x 36	36 x 36	48 x 48	48 x 48	30 x 30*	—
Narrow Bridge	W5-2	2C.18	36 x 36	36 x 36	48 x 48	48 x 48	30 x 30*	—
Narrow Underpass	W5-2a	2C.18	36 x 36	36 x 36	48 x 48	48 x 48	30 x 30*	—
One Lane Bridge	W5-3	2C.19	36 x 36	36 x 36	48 x 48	48 x 48	30 x 30*	—
One Lane Underpass	W5-3a	2C.19	36 x 36	36 x 36	48 x 48	48 x 48	30 x 30*	—
Divided Highway	W6-1	2C.20	36 x 36	36 x 36	48 x 48	48 x 48	—	—
Divided Highway Ends	W6-2	2C.21	36 x 36	36 x 36	48 x 48	48 x 48	—	—
Two-Way Traffic	W6-3	2C.51	36 x 36	36 x 36	48 x 48	48 x 48	—	—
Two-Way Traffic (3-Lane)	W6-5	2C.52	36 x 36	36 x 36	48 x 48	—	—	—
Two-Way Traffic (3-Lane)	W6-5a	2C.52	36 x 36	36 x 36	48 x 48	—	—	—
Hill	W7-1	2C.14	30 x 30*	36 x 36	36 x 36	36 x 36	24 x 24*	48 x 48
Hill with Grade	W7-1a	2C.14	30 x 30*	36 x 36	36 x 36	36 x 36	24 x 24*	48 x 48

Table 2C-1. Warning Sign and Plaque Sizes (Sheet 2 of 4)

Sign or Plaque	Sign Designation	Section	Conventional Road		Expressway	Freeway	Minimum	Oversized
			Single Lane	Multi-Lane				
Use Low Gear (plaque)	W7-2P	2C.64	24 x 18	24 x 18	—	—	—	—
Trucks Use Lower Gear (plaque)	W7-2bP	2C.64	24 x 18	24 x 18	—	—	—	—
XX% Grade (plaque)	W7-3P	2C.64	24 x 18	24 x 18	—	—	—	—
Next XX Miles (plaque)	W7-3aP	2C.61	24 x 18	24 x 18	—	—	—	—
XX% Grade, XX Miles (plaque)	W7-3bP	2C.64	24 x 18	24 x 18	—	—	—	—
Runaway Truck Ramp XX Miles	W7-4	2C.15	84 x 48	84 x 48	120 x 72	120 x 72	—	—
Runaway Truck Ramp Entrance Direction	W7-4b	2C.15	84 x 54	84 x 54	120 x 78	120 x 78	—	—
Truck Escape Ramp	W7-4c	2C.15	78 x 60	78 x 60	78 x 60	78 x 60	—	—
Sand, Gravel, Paved (plaques)	W7-4dP, 4eP,4fP	2C.15	24 x 12	24 x 12	24 x 12	24 x 12	—	—
Hill Blocks View	W7-6	2C.16	30 x 30*	36 x 36	36 x 36	—	—	48 x 48
Bump or Dip	W8-1,2	2C.26	30 x 30*	36 x 36	36 x 36	48 x 48	24 x 24*	48 x 48
Pavement Ends	W8-3	2C.28	36 x 36	36 x 36	48 x 48	—	30 x 30*	—
Soft Shoulder	W8-4	2C.29	36 x 36	36 x 36	48 x 48	48 x 48	24 x 24*	48 x 48
Slippery When Wet	W8-5	2C.30	30 x 30*	36 x 36	36 x 36	48 x 48	24 x 24*	48 x 48
Road Condition (plaques)	W8-5P,5bP,5cP	2C.30	24 x 18	24 x 18	30 x 24	36 x 30	—	36 x 30
Ice (plaque)	W8-5aP	2C.30	24 x 12	24 x 12	30 x 18	30 x 18	—	—
Truck Crossing	W8-6	2C.54	36 x 36	36 x 36	36 x 36	48 x 48	24 x 24*	48 x 48
Loose Gravel	W8-7	2C.30	36 x 36	36 x 36	36 x 36	—	24 x 24*	48 x 48
Rough Road	W8-8	2C.30	36 x 36	36 x 36	36 x 36	48 x 48	24 x 24*	48 x 48
Low Shoulder	W8-9	2C.29	36 x 36	36 x 36	36 x 36	48 x 48	24 x 24*	48 x 48
Uneven Lanes	W8-11	2C.30	36 x 36	36 x 36	36 x 36	48 x 48	—	48 x 48
No Center Line	W8-12	2C.32	36 x 36	36 x 36	36 x 36	48 x 48	—	—
Bridge Ices Before Road	W8-13	2C.30	36 x 36	36 x 36	36 x 36	48 x 48	24 x 24*	48 x 48
Fallen Rocks	W8-14	2C.30	30 x 30*	36 x 36	36 x 36	48 x 48	24 x 24*	48 x 48
Grooved Pavement	W8-15	2C.31	30 x 30*	36 x 36	36 x 36	48 x 48	24 x 24*	48 x 48
Motorcycle (plaque)	W8-15aP	2C.31	24 x 18	24 x 18	30 x 24	36 x 30	—	36 x 30
Metal Bridge Deck	W8-16	2C.31	30 x 30*	36 x 36	36 x 36	48 x 48	24 x 24*	48 x 48
Shoulder Drop-Off	W8-17	2C.29	30 x 30*	36 x 36	36 x 36	48 x 48	24 x 24*	48 x 48
Shoulder Drop-Off (plaque)	W8-17P	2C.29	24 x 18	24 x 18	30 x 24	36 x 30	—	36 x 30
Road May Flood	W8-18	2C.34	36 x 36	36 x 36	36 x 36	48 x 48	24 x 24*	48 x 48
Depth Gauge	W8-19	2C.34	12 x 72	12 x 72	—	—	—	—
Gusty Winds Area	W8-21	2C.34	36 x 36	36 x 36	36 x 36	48 x 48	24 x 24*	48 x 48
Fog Area	W8-22	2C.34	36 x 36	36 x 36	36 x 36	48 x 48	24 x 24*	48 x 48
No Shoulder	W8-23	2C.29	36 x 36	36 x 36	36 x 36	48 x 48	24 x 24*	48 x 48
Shoulder Ends	W8-25	2C.29	30 x 30*	36 x 36	36 x 36	48 x 48	24 x 24*	48 x 48
Road Ends	W8-26	2C.24	30 x 30	36 x 36	—	—	—	—
Street Ends	W8-26a	2C.24	30 x 30	36 x 36	—	—	—	—
Right (Left) Lane Ends	W9-1	2C.47	36 x 36	36 x 36	36 x 36	48 x 48	30 x 30*	48 x 48
Lanes Merge	W9-4	2C.48	36 x 36	36 x 36	36 x 36	48 x 48	30 x 30*	48 x 48
Right (Left) Lane for Exit Only	W9-7	2C.50	132 x 72	132 x 72	132 x 72	132 x 72	—	—
Bicycle	W11-1	2C.54	30 x 30	30 x 30	36 x 36	—	—	48 x 48
Pedestrian	W11-2	2C.55	30 x 30*	36 x 36	36 x 36	—	—	48 x 48
Large Animals	W11-3,4,16,17, 18,19,20,21,22	2C.55	30 x 30*	36 x 36	36 x 36	—	24 x 24*	48 x 48
Farm Vehicle	W11-5	2C.54	30 x 30*	36 x 36	36 x 36	—	24 x 24*	48 x 48
Snowmobile	W11-6	2C.55	30 x 30*	36 x 36	36 x 36	—	24 x 24*	48 x 48
Equestrian	W11-7	2C.55	30 x 30*	36 x 36	36 x 36	—	24 x 24*	48 x 48

Table 2C-1. Warning Sign and Plaque Sizes (Sheet 3 of 4)

Sign or Plaque	Sign Designation	Section	Conventional Road		Expressway	Freeway	Minimum	Oversized
			Single Lane	Multi-Lane				
Emergency Vehicle	W11-8	2C.54	30 x 30*	36 x 36	36 x 36	—	24 x 24*	48 x 48
Handicapped	W11-9	2C.55	30 x 30*	36 x 36	36 x 36	—		48 x 48
Truck	W11-10	2C.54	30 x 30*	36 x 36	36 x 36	—	24 x 24*	48 x 48
Golf Cart	W11-11	2C.54	30 x 30*	36 x 36	36 x 36	—	24 x 24*	48 x 48
Emergency Signal Ahead (plaque)	W11-12P	2C.54	36 x 30	36 x 30	36 x 30	—	—	—
Horse-Drawn Vehicle	W11-14	2C.54	30 x 30*	36 x 36	36 x 36	—	24 x 24*	48 x 48
Trail Crossing	W11-15	2C.54	30 x 30*	36 x 36	36 x 36	—	24 x 24*	48 x 48
Trail Crossing	W11-15a	2C.54	30 x 30*	36 x 36	36 x 36	—	24 x 24*	48 x 48
Trail Crossing (plaque)	W11-15P	2C.54	24 x 18	24 x 18	30 x 24	—	—	36 x 30
Double Arrow	W12-1	2C.23	30 x 30*	36 x 36	36 x 36	—	—	—
Low Clearance Advance	W12-2	2C.25	36 x 36	36 x 36	48 x 48	48 x 48	30 x 30*	—
Low Clearance Overhead	W12-2a	2C.25	84 x 24	84 x 24	84 x 24	84 x 24	—	—
Low Clearance - Lane Overhead	W12-2b	2C.25	102 x 24	102 x 24	102 x 24	102 x 24	—	—
Advisory Speed (plaque)	W13-1P	2C.59	18 x 18	18 x 18	24 x 24	30 x 30	—	30 x 30
Advisory Speed Confirmation (plaque)	W13-1aP	2C.59	48 x 15	48 x 15	60 x 18	60 x 18	48 x 15	72 x 24
Advisory Exit or Ramp Speed	W13-2,3	2C.12	24 x 30	24 x 30	36 x 48	36 x 48	—	48 x 60
Combination Horizontal Alignment/Advisory Exit or Ramp Speed Loop	W13-6,7,8,9	2C.12	24 x 42	24 x 42	36 x 66	36 x 66	—	48 x 84
Combination Horizontal Alignment/Advisory Exit or Ramp Speed Turn	W13-10,11	2C.12	24 x 36	24 x 36	36 x 54	36 x 54	—	48 x 72
Combination Horizontal Alignment/Advisory Exit or Ramp Speed - Truck Rollover	W13-12,13	2C.12	24 x 42	24 x 42	36 x 66	36 x 66	—	48 x 84
Vehicle Speed Feedback Sign	W13-20	2C.13	24 x 30	30 x 36	36 x 48	48 x 60	—	—
Vehicle Speed Feedback (plaque)	W13-20aP	2C.13	24 x 18	30 x 24	36 x 30	48 x 36	—	—
Dead End, No Outlet	W14-1,2	2C.24	30 x 30*	36 x 36	36 x 36	—	24 x 24*	48 x 48
Dead End, No Outlet (with arrow)	W14-1a,2a	2C.24	36 x 9	36 x 9	—	—	—	—
No Passing Zone	W14-3	2C.53	48 x 48 x 36	48 x 48 x 36	64 x 64 x 48	64 x 64 x 48	40 x 40 x 30	64 x 64 x 48
Playground	W15-1	2C.56	30 x 30*	36 x 36	36 x 36	—	24 x 24*	48 x 48
In Road (plaque)	W16-1P	2C.67	18 x 12	18 x 12	24 x 18	—	—	24 x 18
In Street (plaque)	W16-1aP	2C.67	18 x 12	18 x 12	24 x 18	—	—	24 x 18
XX Feet (2-line plaque)	W16-2P	2C.61	24 x 18	24 x 18	30 x 24	30 x 24	—	30 x 24
XX Ft (1-line plaque)	W16-2aP	2C.61	24 x 12	24 x 12	—	—	—	30 x 18
XX Miles (2-line plaque)	W16-3P	2C.61	30 x 24	30 x 24	—	—	—	—
XX Miles (1-line plaque)	W16-3aP	2C.61	30 x 12	30 x 12	—	—	—	—
Next XX Feet (plaque)	W16-4P	2C.61	30 x 24	30 x 24	—	—	—	—
Supplemental Arrow (plaque)	W16-5P,6P	2C.62	21 x 15	21 x 15	—	—	—	30 x 21
Diagonal Downward Arrow (plaque)	W16-7P	2C.63	21 x 15	21 x 15	—	—	—	30 x 21
Dual Downward Diagonal Arrow (plaque)	W16-7aP	2C.63	21 x 15	21 x 15	—	—	—	30 x 21
Advance Street Name (1-line plaque)	W16-8P	2C.65	Varies x 8	Varies x 8	—	—	—	—
Advance Street Name (2-line plaque)	W16-8aP	2C.65	Varies x 15	Varies x 15	—	—	—	—
Ahead (plaque)	W16-9P	2C.55	24 x 12	24 x 12	30 x 18	—	—	30 x 18
Photo Enforced (symbol plaque)	W16-10P	2C.69	24 x 12	24 x 12	36 x 18	—	—	48 x 24
Photo Enforced (plaque)	W16-10aP	2C.69	24 x 18	24 x 18	36 x 30	—	—	48 x 36

Table 2C-1. Warning Sign and Plaque Sizes (Sheet 4 of 4)

Sign or Plaque	Sign Designation	Section	Conventional Road		Expressway	Freeway	Minimum	Oversized
			Single Lane	Multi-Lane				
Traffic Circle (plaque)	W16-12P	2C.41	24 x 18	24 x 18	—	—	—	—
Roundabout (plaque)	W16-12aP	2C.41	24 x 12	24 x 12	—	—	—	—
When Flashing (plaque)	W16-13P	2C.55	24 x 18	24 x 18	—	—	—	—
New (plaque)	W16-15P	2C.60	24 x 12	24 x 12	—	—	—	—
Notice (plaque)	W16-18P	2A.11	24 x 12	24 x 12	—	—	—	—
Except Bicycles (plaque)	W16-20P	2C.68	24 x 12	24 x 12	—	—	—	—
Speed Hump	W17-1	2C.27	30 x 30*	36 x 36	—	—	24 x 24*	48 x 48
No Traffic Signs	W18-1	2C.33	30 x 30*	36 x 36	—	—	24 x 24*	36 x 36
Freeway Ends XX Miles	W19-1	2C.22	—	—	—	144 x 48	—	—
Expressway Ends XX Miles	W19-2	2C.22	—	—	144 x 48	—	—	—
Freeway Ends	W19-3	2C.22	—	—	—	48 x 48	—	—
Expressway Ends	W19-4	2C.22	—	—	48 x 48	—	—	—
All Traffic Must Exit	W19-5	2C.22	—	—	90 x 48	90 x 48	—	—
New Traffic Pattern Ahead	W23-2	2C.38	36 x 36	36 x 36	—	—	—	—
New Signal Operation Ahead	W23-2a	2C.38	36 x 36	36 x 36	48 x 48	48 x 48	—	—
Oncoming Traffic Has (May Have) Extended Green	W25-1,2	2C.44	24 x 30	24 x 30	—	—	—	—
Watch for Stopped Traffic	W26-1	2C.39	36 x 36	36 x 36	48 x 48	48 x 48	—	—

* The minimum size required for diamond-shaped warning signs facing traffic on multi-lane conventional roads shall be 36 x 36 per Section 2C.03

Notes:

1. Larger signs may be used when appropriate
2. Dimensions are shown as width x height, in inches

04 **The minimum size for supplemental warning plaques that are not included in Table 2C-1 shall be as shown in Table 2C-2.**

Option:

05 If a diamond-shaped warning sign is placed on the left-hand side of a multi-lane roadway to supplement the installation of the same warning sign on the right-hand side of the roadway, the minimum size identified in the Single Lane column in Table 2C-1 may be used.

06 Signs and plaques larger than those shown in Tables 2C-1 and 2C-2 may be used (see Section 2A.11).

Guidance:

Table 2C-2. Minimum Size of Supplemental Warning Plaques

Size of Warning Sign	Size of Supplemental Plaque			
	Rectangular			Square
	1 Line	2 Lines	Arrow	
24 x 24	24 x 12	24 x 18	21 x 15	18 x 18
30 x 30	24 x 12	24 x 18	21 x 15	18 x 18
36 x 36	30 x 18	30 x 24	30 x 21	24 x 24
48 x 48	30 x 18	30 x 24	30 x 21	24 x 24

Notes:

1. Larger supplemental plaques may be used when appropriate
2. Dimensions are shown as width x height, in inches

07 *The minimum size for all diamond-shaped warning signs facing traffic on exit and entrance ramps at major interchanges connecting an Expressway or Freeway with an Expressway or Freeway (see Section 2E.11) should be the size identified in Table 2C-1 for the mainline roadway classification (Expressway or Freeway). If a minimum size is not provided in the Freeway column, the Expressway size should be used. If a minimum size is not provided in the Freeway or the Expressway column, the Oversized size should be used.*

08 *The minimum size for all diamond-shaped warning signs facing traffic on exit and entrance ramps at all other interchanges (see Section 2E.11) should be 36 x 36 inches.*

09 *The typical size of warning signs used on low-volume rural roads with operating speeds of 30 mph or less should be in accordance with the minimum column of Table 2C-1.*

Section 2C.04 Placement of Warning Signs

Support:

01   Information on the placement of warning signs is contained in Sections 2A.13 through 2A.18.

02   The time needed for detection, recognition, decision, and reaction is called the Perception-Response Time (PRT). Table 2C-3 is provided as an aid for determining warning sign location. The distances shown in Table 2C-3 can be adjusted for roadway features, other signing, and to improve visibility.

Guidance:

03   *Warning signs should be placed so that they provide an adequate PRT. The distances contained in Table 2C-3 should be applied with engineering judgment.*

04   *Minimum spacing between warning signs with different messages should be based on the estimated PRT for driver comprehension of and reaction to the second sign.*

05   *The effectiveness of the placement of warning signs should be periodically evaluated under both day and night conditions.*

Table 2C-3. Guidelines for Advance Placement of Warning Signs

Posted or 85th-Percentile Speed	Condition A: Speed reduction and lane changing in heavy traffic[2]	Condition B: Deceleration to the listed advisory speed (mph) for the condition								
		0[3]	10[4]	20[4]	30[4]	40[4]	50[4]	60[4]	70[4]	80[4]
20 mph	225 ft	115 ft	N/A[5]	—	—	—	—	—	—	—
25 mph	325 ft	155 ft	N/A[5]	N/A[5]	—	—	—	—	—	—
30 mph	460 ft	200 ft	N/A[5]	N/A[5]	—	—	—	—	—	—
35 mph	565 ft	250 ft	N/A[5]	N/A[5]	N/A[5]	—	—	—	—	—
40 mph	670 ft	305 ft	100 ft[6]	100 ft[6]	N/A[5]	—	—	—	—	—
45 mph	775 ft	360 ft	125 ft	100 ft[6]	100 ft[6]	N/A[5]	—	—	—	—
50 mph	885 ft	425 ft	200 ft	175 ft	125 ft	100 ft[6]	—	—	—	—
55 mph	990 ft	495 ft	275 ft	225 ft	200 ft	125 ft	N/A5	—	—	—
60 mph	1,100 ft	570 ft	350 ft	325 ft	275 ft	200 ft	100 ft[6]	—	—	—
65 mph	1,200 ft	645 ft	450 ft	400 ft	350 ft	275 ft	200 ft	100 ft[6]	—	—
70 mph	1,250 ft	730 ft	525 ft	500 ft	450 ft	375 ft	275 ft	150 ft	—	—
75 mph	1,350 ft	820 ft	625 ft	600 ft	550 ft	475 ft	375 ft	250 ft	100 ft[6]	—
80 mph	1,475 ft	910 ft	725 ft	700 ft	625 ft	550 ft	450 ft	350 ft	200 ft	—
85 mph	1,600 ft	1,010 ft	825 ft	800 ft	750 ft	675 ft	575 ft	450 ft	300 ft	150 ft

[1] The distances are adjusted for a sign legibility distance of 180 feet for Condition A. The distances for Condition B (with the exception of the potential stop condition) have been adjusted for a sign legibility distance of 250 feet, which is appropriate for an alignment warning symbol sign. For Conditions A and B, warning signs with less than 6-inch legend or more than four words, a minimum of 100 feet should be added to the advance placement distance to provide adequate legibility of the warning sign.

[2] Typical conditions are locations where the road user must use extra time to adjust speed and change lanes in heavy traffic because of a complex driving situation. Typical signs are Merge and Right Lane Ends. The distances are determined by providing the driver a PRT of 14.0 to 14.5 seconds for vehicle maneuvers (2018 AASHTO Policy, Table 3-3, Decision Sight Distance, Avoidance Maneuver E) and adjusted for a legibility distance of 180 feet for the appropriate sign.

[3] Typical condition is the warning of a potential stop situation. Typical signs are Stop Ahead, Yield Ahead, Signal Ahead, and Intersection Warning signs. The distances are based on the 2018 AASHTO Policy, Table 3-1, Stopping Sight Distance, providing a PRT of 2.5 seconds, a deceleration rate of 11.2 feet/second[2].

[4] Typical conditions are locations where the road user must decrease speed to maneuver through the warned condition. Typical signs are Turn, Curve, Reverse Turn, or Reverse Curve. The distance is determined by providing a 2.5 second PRT, a vehicle deceleration rate of 10 feet/second[2], and adjusted for a sign legibility distance of 250 feet.

[5] No suggested distances are provided for these speeds, as the placement location is dependent on site conditions and other signing. An alignment warning sign may be placed anywhere from the point of curvature up to 100 feet in advance of the curve. However, the alignment warning sign should be installed in advance of the curve and at least 100 feet from any other signs.

[6] The minimum advance placement distance is listed as 100 feet to provide adequate spacing between signs.

Note: Warning signs that advise road users about conditions that are not related to a specific location, such as Deer Crossing or SOFT SHOULDER, can be installed in an appropriate location, based on engineering judgment.

HORIZONTAL ALIGNMENT WARNING SIGNS AND PLAQUES

Section 2C.05 Horizontal Alignment Warning Signs – General

Support:

01   A variety of horizontal alignment warning signs (see Figure 2C-1), pavement markings (see Chapter 3B), and delineation (see Chapter 3G) can be used to advise motorists of a change in the roadway alignment. Uniform application of these traffic control devices with respect to the amount of change in the roadway alignment conveys a consistent message establishing driver expectancy and promoting effective roadway operations. The design and application of horizontal alignment warning signs to meet those requirements are addressed in Sections 2C.05 through 2C.12.

Figure 2C-1. Horizontal Alignment Signs and Plaques

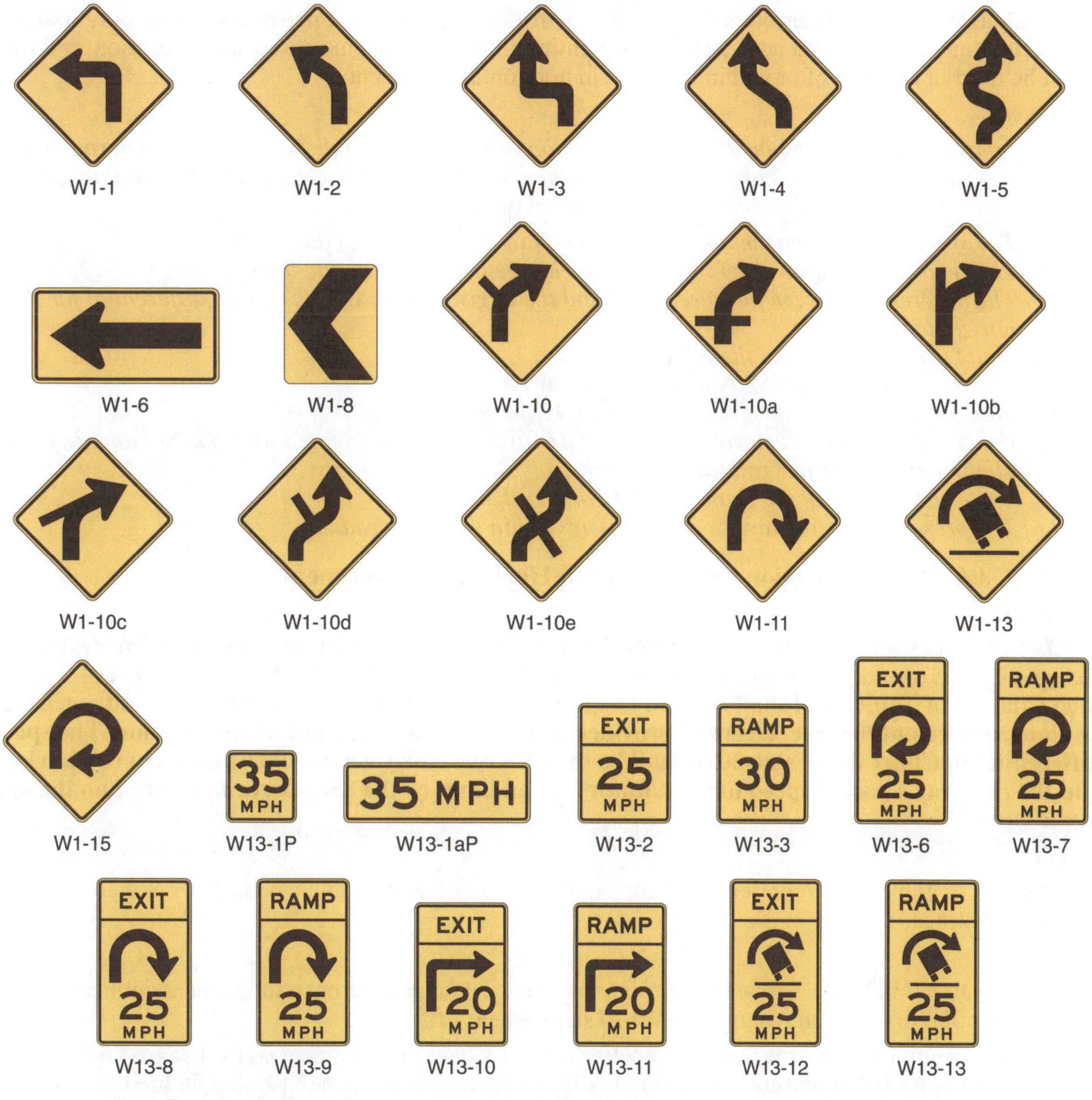

W1-1   W1-2   W1-3   W1-4   W1-5

W1-6   W1-8   W1-10   W1-10a   W1-10b

W1-10c   W1-10d   W1-10e   W1-11   W1-13

W1-15   W13-1P   W13-1aP   W13-2   W13-3   W13-6   W13-7

W13-8   W13-9   W13-10   W13-11   W13-12   W13-13

Note: Turn arrows and reverse turn arrows may be substituted for the curve arrows and reverse curve arrows on the W1-10 series signs where appropriate.

02   The following list identifies treatments that might be used in advance of or within a change in horizontal alignment:

    A. Horizontal alignment (Turn (W1-1), Curve (W1-2, W1-10 series, W1-11, W1-13, W1-15), Reverse Turn (W1-3), Reverse Curve (W1-4), Winding Road (W1-5), Exit Speed (W13-2), Ramp Speed (W13-3), and Combination Horizontal Alignment (Advisory Exit or Ramp Speed W13-6 through W13-11)) signs (see Sections 2C.07, 2C.09, and 2C.12)

    B. Advisory Speed (W13-1P) plaque (see Section 2C.59)

    C. Chevron Alignment (W1-8) signs (see Section 2C.08)

    D. Delineators (see Chapter 3G)

    E. One-Direction Large Arrow (W1-6) sign (see Section 2C.10)

    F. Raised Retroreflective Pavement Markers (see Sections 3B.15 through 3B.17)

    G. Sign or marking conspicuity enhancements (see Section 2A.11)

    H. Wide edge lines (see Section 3A.04)

    I. Pavement word, symbol and arrow markings (symbol or words) (see Sections 3B.20 through 3B.22)

    J. Rumble strips (see Chapter 3K)

    K. Vehicle Speed Feedback Sign (see Section 2C.13)

    L. Speed reduction markings (see Section 3B.28)

03   In addition, considerations other than traffic control devices, such as improved surface friction (high friction surface treatments), pavement edge treatments, lighting improvements, increased superelevation, and rumble strips, might be used in advance of or within a change in horizontal alignment.

Guidance:

04   *Except as provided in Section 2C.06, the selection of traffic control devices used to warn road users of a change in horizontal alignment or to provide guidance in navigating the change in horizontal alignment should be based on consideration of one or more of the following factors:*

    A. *The speed of traffic on the approach to the change in horizontal alignment*

    B. *The recommended advisory speed for the change in horizontal alignment*

    C. *The difference between the speed limit and the advisory speed, or the speed differential for the change in horizontal alignment*

    D. *Daily traffic volumes on the roadway*

    E. *The typical mix of vehicle types on the roadway*

    F. *Sight distance throughout the change in horizontal alignment*

    G. *Other types of traffic control devices that are used in advance of and within the change in horizontal alignment on the same roadway segment*

    H. *The crash history of the change in horizontal alignment*

    I. *The presence of driveways or intersections within the curve radius*

Section 2C.06 Device Selection for Changes in Horizontal Alignment

Standard:

01   **The criteria shown in Chart A of Table 2C-4 shall be used to determine the need for devices for changes in horizontal alignment. If the use of a device or devices is indicated by Chart A of Table 2C-4, then Chart B of Table 2C-4 shall be used to specify the type(s) of devices to be used in advance of, and/ or along, a horizontal curve, except as provided in Paragraphs 3, 5, and 6 of this Section. The speed differential in Chart B of Table 2C-4 shall be the difference between the horizontal curve's advisory speed and the roadway's posted speed limit, statutory speed limit, or the 85th-percentile speed on the approach to the curve.**

Support:

02   Chart A of Table 2C-4 represents existing AADT, type of roadway, and whether or not there are existing markings.

Option:

03   A One-Direction Large Arrow (W1-6) sign may be used in place of or to supplement delineators (see Chapter 3G) or Chevron Alignment (W1-8) signs when:

    A. Site conditions limit the number of delineators or Chevron Alignment signs that are visible; or

    B. The number of delineators or Chevron Alignment signs that can be installed within the change in horizontal alignment is less than the number determined by the spacing specified in Sections 2C.08 or 3G.04.

04   Additional or supplemental devices may be used for a change in horizontal alignment on the basis of engineering judgment.

Table 2C-4. Application of Traffic Control Devices for Changes in Horizontal Alignment

A - Determination of the Need for Devices for Changes in Horizontal Alignment[1]

Roadway Type	AADT			
	Less than 1,000	1,000-2,999	3,000-3,999	Greater than 3,999
Freeways and Expressways	Required	Required	Required	Required
Arterial or Collector without Pavement Markings	Optional	Recommended	Required	Required
Arterial or Collector with Pavement Markings[2]	Optional	Recommended	Recommended	Required
All other roadways	Optional	Optional	Optional	Optional

[1] If devices are determined to be needed, the selection of the device(s) is based on Chart B below.

[2] An arterial or collector is considered to have pavement markings when either a center line, edge lines, or both are present.

B - Selection of Devices for Changes in Horizontal Alignment

Speed Differential[3]	Devices for Change in Horizontal Alignment[3]
5 mph	Pavement markings or advance horizontal alignment warning sign on paved roadways. Advance horizontal alignment warning sign on unpaved roadways.[4]
10 mph	Advance horizontal alignment warning sign
15 mph	Delineators[5] and advance horizontal alignment warning sign
20 mph or more	Chevrons[5] and advance horizontal alignment warning sign

[3] The provisions for the use of Horizontal Alignment warning signs and devices are contained in Section 2C.06. The need for devices is determined by Chart A above.

[4] A roadway is considered to have pavement markings when either a center line, edge lines, or both are present.

[5] Section 2C.06 contains information about the use of a One Direction Large Arrow (W1-6) sign in place of or to supplement delineators and chevrons.

---

05   Devices for changes in horizontal alignment may be omitted when the speed limit on the approach to an alignment change is 20 mph or less.

06   Devices for changes in horizontal alignment may be omitted on urban streets with an AADT of 1,000 vehicles per day or less.

Support:

07   For purposes of selecting traffic control devices for changes in horizontal alignment, an arterial or collector is considered to have pavement markings when either a center line, edge lines, or both are present. Warrants for center lines and edge lines are provided in Sections 3B.02 and 3B.10, respectively.

Section 2C.07 Horizontal Alignment Signs (W1-1 through W1-5, W1-11, and W1-15)

Standard:

01   **If Table 2C-4 indicates that a horizontal alignment sign (see Figure 2C-1) is required, recommended, or allowed, the sign installed in advance of the curve shall be a Curve (W1-2) sign unless a different sign is recommended or allowed by the provisions of this Section.**

Guidance:

02   *A Turn (W1-1) sign should be used instead of a Curve (W1-2) sign in advance of a horizontal curve that has an advisory speed of 30 mph or less.*

03   *Where there are two changes in roadway alignment in opposite directions that are separated by a tangent distance of less than 600 feet, the Reverse Turn (W1-3) sign should be used instead of multiple Turn (W1-1) signs or the Reverse Curve (W1-4) sign should be used instead of multiple Curve (W1-2) signs.*

Support:

04   Figure 2C-2 provides examples of warning signs used for turns and curves.

Option:

05   A Winding Road (W1-5) sign may be used instead of multiple Turn (W1-1) or Curve (W1-2) signs where there are three or more changes in roadway alignment each separated by a tangent distance of less than 600 feet.

06   A NEXT XX MILES (W7-3aP) supplemental distance plaque (see Section 2C.61) may be installed below the Winding Road sign where continuous roadway curves exist for a specific distance.

Figure 2C-2. Examples of Warning Signs for Changes in Horizontal Alignment (Sheet 1 of 2)

A – Curve

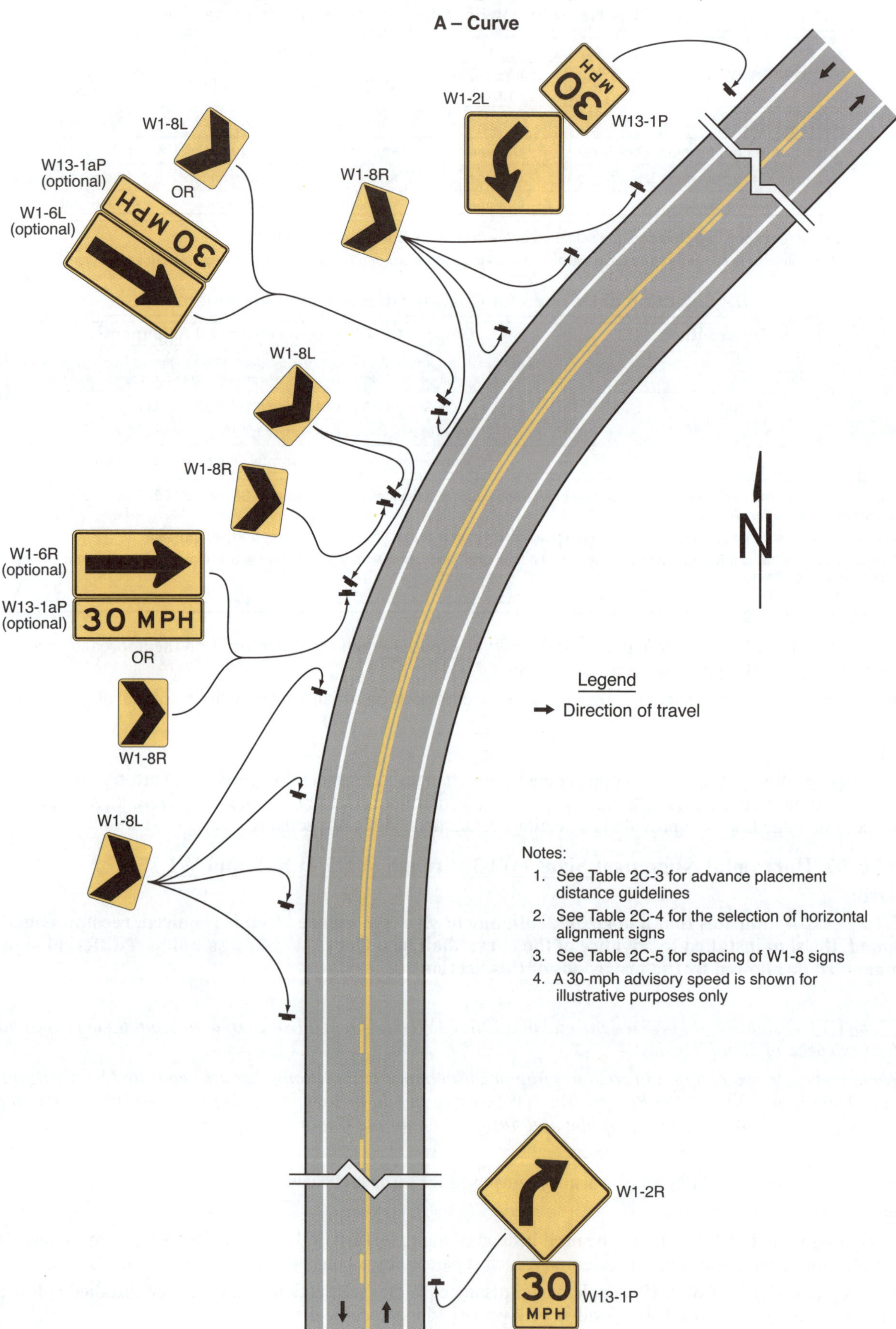

Figure 2C-2. Examples of Warning Signs for Changes in Horizontal Alignment (Sheet 2 of 2)

B – Turn

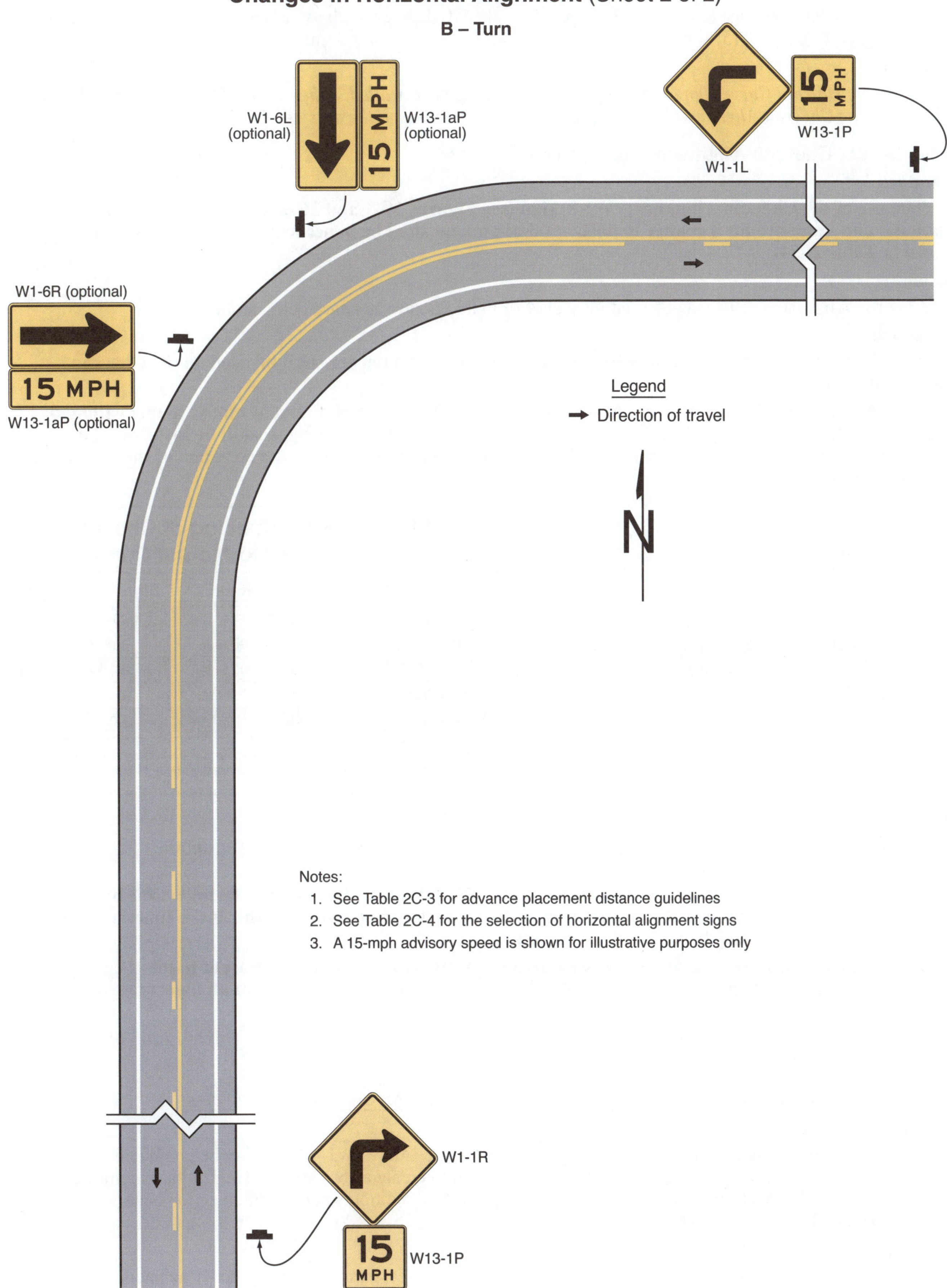

07   If the curve has a change in horizontal alignment of 135 degrees or more, the Hairpin Curve (W1-11) sign may be used instead of a Turn or Curve sign.

08   If the curve has a change of direction of approximately 270 degrees, such as on a cloverleaf interchange ramp, the 270-degree Loop (W1-15) sign may be used instead of a Turn or Curve sign.

Guidance:

09   *When the Hairpin Curve sign or the 270-degree Loop sign is installed, either a One-Direction Large Arrow (W1-6) sign or Chevron Alignment (W1-8) signs should be installed on the outside of the turn or curve.*

Section 2C.08 Chevron Alignment Sign (W1-8)

Standard:

01   **The use of the Chevron Alignment (W1-8) sign (see Figures 2C-1 and 2C-2) to provide additional emphasis and guidance for a change in horizontal alignment shall be in accordance with the information shown in Table 2C-4.**

Option:

02   Chevron Alignment signs may be used instead of or in addition to standard delineators.

Standard:

03   **The Chevron Alignment sign shall be a vertical rectangle. No border shall be used on the Chevron Alignment sign.**

04   **If used, Chevron Alignment signs shall be installed on the outside of a turn or curve, in line with and at approximately a right angle to approaching traffic. Chevron Alignment signs shall be installed at a minimum height of 4 feet, measured vertically from the bottom of the sign to the elevation of the near edge of the traveled way.**

Guidance:

05   *The approximate spacing of Chevron Alignment signs on the turn or curve measured from the point of curvature (PC) should be as shown in Table 2C-5.*

06   *The Chevron Alignment signs should be visible for a sufficient distance to provide the road user with adequate time to react to the change in alignment.*

Option:

07   LEDs may be used to enhance the conspicuity of Chevron Alignment signs (see Section 2A.12).

Table 2C-5. Typical Spacing of Chevron Alignment Signs on Horizontal Curves

Advisory Speed	Curve Radius	Sign Spacing
15 mph or less	Less than 200 feet	40 feet
20 to 30 mph	200 to 400 feet	80 feet
35 to 45 mph	401 to 700 feet	120 feet
50 to 60 mph	701 to 1,250 feet	160 feet
More than 60 mph	More than 1,250 feet	200 feet

Note: The relationship between the curve radius and the advisory speed shown in this table should not be used to determine the advisory speed.

Standard:

08   **The LEDs used in the Chevron Alignment sign shall consist of yellow LEDs outlining the chevron symbol.**

09   **Chevron Alignment signs shall not be placed on the far side of a T-intersection facing traffic on the stem approach to warn drivers that a through movement is not physically possible, as this is the function of a Two-Direction (or One-Direction) Large Arrow sign.**

10   **Chevron Alignment signs shall not be used to mark obstructions within or adjacent to the roadway, including the beginning of guardrails or barriers, as this is the function of an object marker (see Section 2C.70).**

11   **Chevron Alignment signs directing traffic to the right shall not be used in the central island of a roundabout or a neighborhood traffic circle.**

Section 2C.09 Combination Horizontal Alignment/Intersection Signs (W1-10 Series)

Option:

01   The Turn (W1-1) sign, the Curve (W1-2) sign, and the Reverse Curve (W1-4) sign may be combined with the Cross Road (W2-1) sign or the Side Road (W2-2 or W2-3) sign to create a combination Horizontal Alignment/Intersection (W1-10 series) sign (see Figure 2C-1) that depicts the condition where an intersection occurs within or immediately adjacent to a turn or curve.

Support:

02   Section 2C.65 contains information about the use of an advance street name plaque to identify an intersecting road.

Guidance:

03   *Elements of the combination Horizontal Alignment/Intersection sign related to horizontal alignment should comply with the provisions of Section 2C.07, and elements related to intersection configuration should comply with the provisions of Section 2C.41. The symbol design should approximate the configuration of the intersecting roadway(s). No more than one Cross Road or two Side Road symbols should be displayed on any one combination Horizontal Alignment/Intersection sign.*

Standard:

04   **The use of the combination Horizontal Alignment/Intersection sign shall be in accordance with the provisions of Section 2C.07 for the appropriate Turn or Curve sign .**

Section 2C.10 <u>One-Direction Large Arrow Sign (W1-6)</u>

Option:

01   A One-Direction Large Arrow (W1-6) sign (see Figure 2C-1) may be used either as a supplement or alternative to Chevron Alignment signs or delineators in order to delineate a change in horizontal alignment (see Figure 2C-2).

02   A One-Direction Large Arrow (W1-6) sign may be used to supplement a Turn (W1-1) or Reverse Turn (W1-3) sign (see Figure 2C-2) to emphasize the abrupt curvature.

Standard:

03   **The One-Direction Large Arrow sign shall be a horizontal rectangle with an arrow pointing to the left or right.**

04   **If used, the One-Direction Large Arrow sign shall be installed on the outside of a turn or curve in line with and at approximately a right angle to approaching traffic.**

05   **The One-Direction Large Arrow sign shall not be used where there is no alignment change in the direction of travel, such as at the beginnings and ends of medians or at center piers.**

06   **The One-Direction Large Arrow sign directing traffic to the right shall not be used in the central island of a roundabout or a neighborhood traffic circle.**

Guidance:

07   *The One-Direction Large Arrow sign should be visible for a sufficient distance to provide the road user with adequate time to react to the change in alignment.*

Section 2C.11 <u>Truck Rollover Sign (W1-13)</u>

Option:

01   A Truck Rollover (W1-13) sign (see Figure 2C-1) may be used as a supplement to a horizontal alignment warning sign to warn drivers of vehicles with a high center of gravity, such as trucks, tankers, and recreational vehicles, of a curve or turn where there are:

   A. Past incidents of truck rollovers at the specific location,
   B. High volumes of trucks, or
   C. A speed differential (see Section 2C.06) that might pose a greater risk for vehicles with high centers of gravity.

Guidance:

02   *Where engineering judgment determines the need for the installation of a Truck Rollover (W1-13) sign, it should be located downstream of the horizontal alignment warning sign in advance of the curve.*

Standard:

03   **If a Truck Rollover (W1-13) sign is used, it shall be accompanied by an Advisory Speed (W13-1P) plaque indicating the recommended speed for vehicles with a higher center of gravity.**

Option:

04   The Truck Rollover sign may include conspicuity enhancements, or may be a blank-out sign, activated by the detection of an approaching vehicle with a high center of gravity that is traveling in excess of the recommended speed for the condition.

Support:

05   The curved arrow on the Truck Rollover sign shows the direction of roadway curvature. The truck tips in the opposite direction.

Section 2C.12 Advisory Exit and Ramp Speed Signs (W13-2 and W13-3) and Combination Horizontal Alignment/Advisory Exit and Ramp Speed Signs (W13-6 through W13-13)

Standard:

01   **Where an advisory speed is posted in advance of a freeway or expressway exit, the Advisory Exit Speed (W13-2) sign (see Figure 2C-1) shall be used.**

02   **Where an advisory speed is posted in advance of a conventional road ramp or to another roadway or roadside facility, the Advisory Ramp Speed (W13-3) sign (see Figure 2C-1) shall be used.**

03   **An Advisory Exit Speed or Advisory Ramp Speed sign shall be used when the difference between the mainline roadway speed limit and the exit or ramp advisory speed in the vicinity of the departure is 20 mph or greater.**

Guidance:

04   *An Advisory Exit Speed or Advisory Ramp Speed sign should be used when the difference between the mainline roadway speed limit and the exit or ramp advisory speed in the vicinity of the departure is 15 mph.*

Option:

05   An Advisory Exit Speed or Advisory Ramp Speed sign may be used based on engineering judgment when the difference between the mainline roadway speed limit and the exit or ramp advisory speed in the vicinity of the departure is 10 mph or less.

06   The Combination Horizontal Alignment/Advisory Exit Speed (W13-6, W13-8, and W13-10) signs (see Figure 2C-1) may be used in lieu of the Advisory Exit Speed (W13-2) sign, and the combination Horizontal Alignment/ Advisory Ramp Speed (W13-7, W13-9, and W13-11) signs (see Figure 2C-1) may be used in lieu of the Advisory Ramp Speed (W13-3) sign.

07   The Combination Truck Rollover/Advisory Exit Speed and Truck Rollover/Advisory Ramp Speed (W13-12 and W13-13) signs (see Figure 2C-1) may be used in lieu of the W13-2 and W13-3 signs, respectively, if the tip over condition is in the vicinity of the gore.

Standard:

08   **Roadway geometrics represented on the Combination Horizontal Alignment/Advisory Exit and Combination Horizontal Alignment/Advisory Ramp Speed signs (see Figure 2C-1) shall be limited to the standard signs shown in this Manual.**

Guidance:

09   *If used, the Advisory Exit Speed sign or the Combination Horizontal Alignment/Advisory Exit Speed sign should be installed along the deceleration lane. The Advisory Exit Speed or the Combination Horizontal Alignment/Advisory Exit signs should be visible in time for the road user to decelerate and make an exiting maneuver.*

10   *Regulatory Speed Limit signs (see Section 2B.21) should not be located in the vicinity of exit ramps or deceleration lanes, particularly where they will conflict with the advisory speed displayed on the Advisory Exit or Ramp Speed signs.*

Support:

11   Section 2C.06 contains provisions for the determination of the displayed advisory speed.

12   Table 2C-3 lists recommended advance sign placement distances for deceleration to various advisory speeds.

Option:

13   Where there is a need to remind road users of the recommended advisory speed, a horizontal alignment warning sign with an advisory speed plaque displaying the same advisory speed may be installed at a downstream location along the ramp.

Guidance:

14   *If the ramp curvature changes to the extent that it warrants a lower advisory speed, a horizontal alignment warning sign with the new advisory speed should be displayed in advance of the change in curvature.*

Option:

15   The One-Direction Large Arrow (W1-6) sign may be installed beyond the exit gore on the outside of the curve to provide additional warning of an immediate change in curvature. When used in conjunction with the exit speed, the One-Direction Large Arrow (W1-6) sign may be supplemented with a Confirmation Advisory Speed (W13-1aP) plaque (see Figure 2C-1) when the plaque is not used with the Exit Gore (E5-1 series) sign.

Guidance:

16   *The horizontal alignment symbol displayed on the Combination Horizontal Alignment/Advisory Exit and Ramp Speed signs should be consistent with the horizontal geometry of the ramp.*

Support:

17   *Examples of advisory speed signing for exit ramps are shown in Figure 2C-3.*

Figure 2C-3. Examples of Exit Ramp Advisory Speed and Other Warning Signs (Sheet 1 of 5)

A – Loop ramp with constant controlling curvature along ramp proper

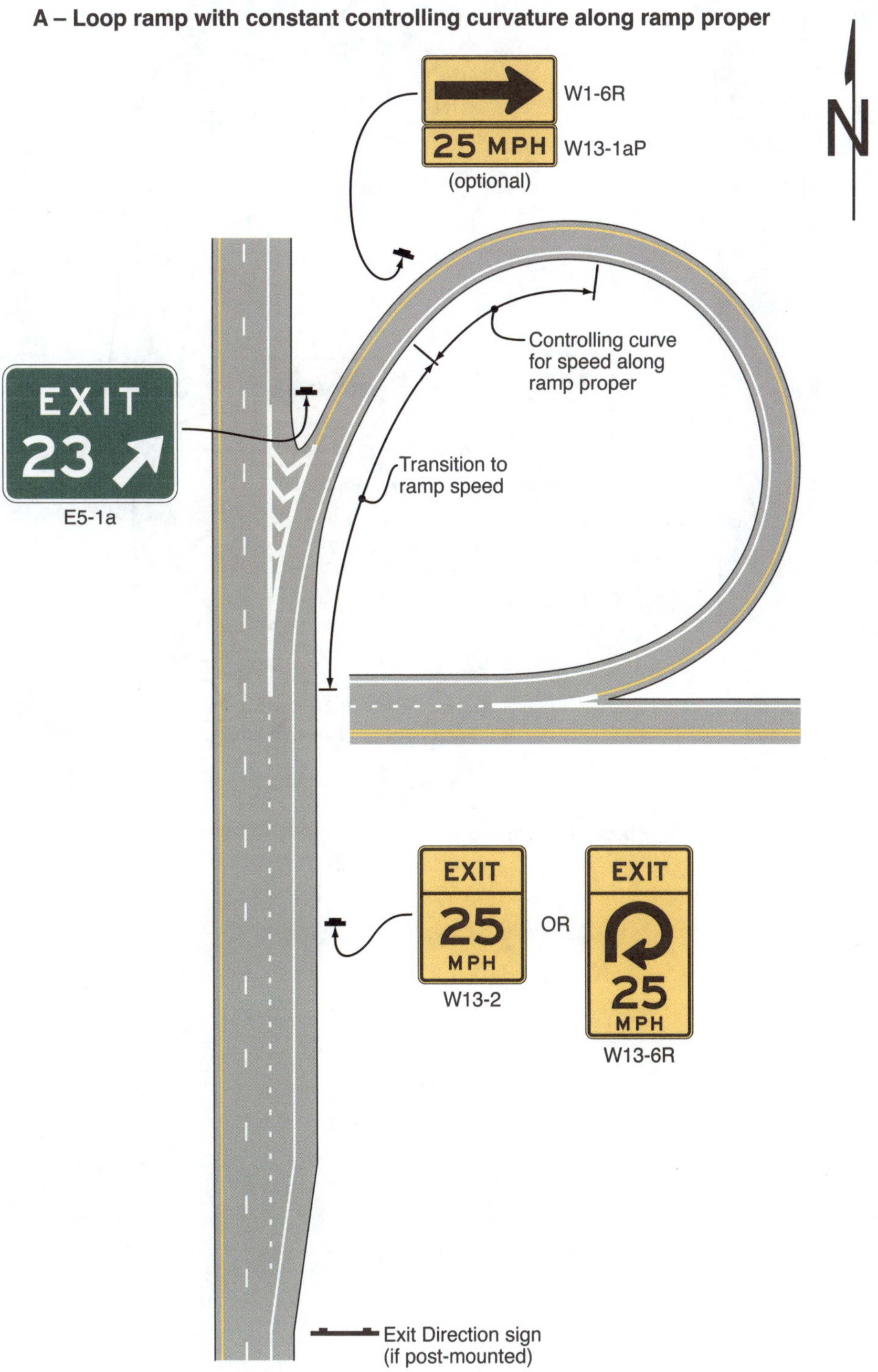

Figure 2C-3. Examples of Exit Ramp Advisory Speed and Other Warning Signs (Sheet 2 of 5)

B – Loop ramp with downstream limiting curvature

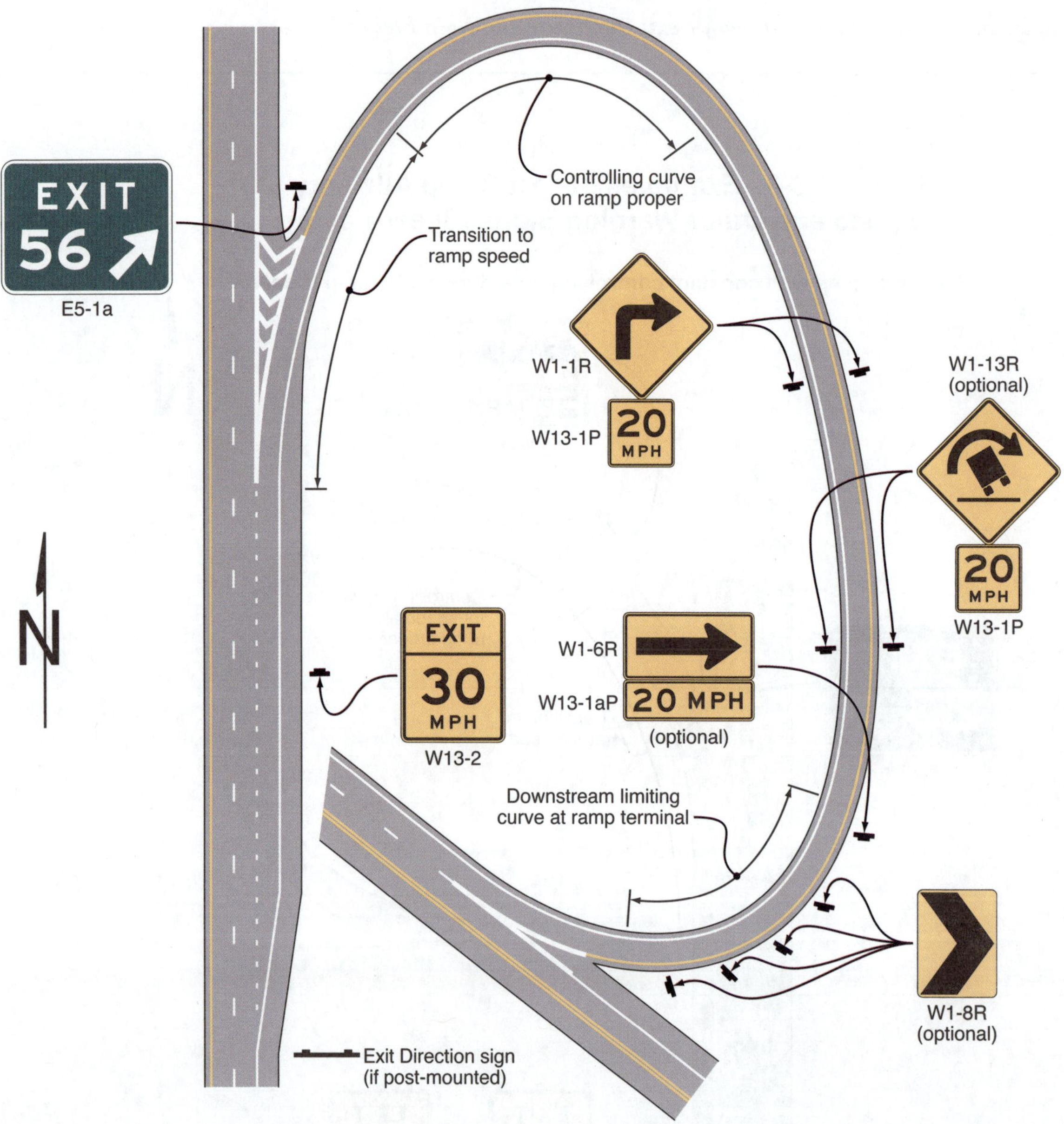

Figure 2C-3. Examples of Exit Ramp Advisory Speed and Other Warning Signs (Sheet 3 of 5)

C – Directional ramp

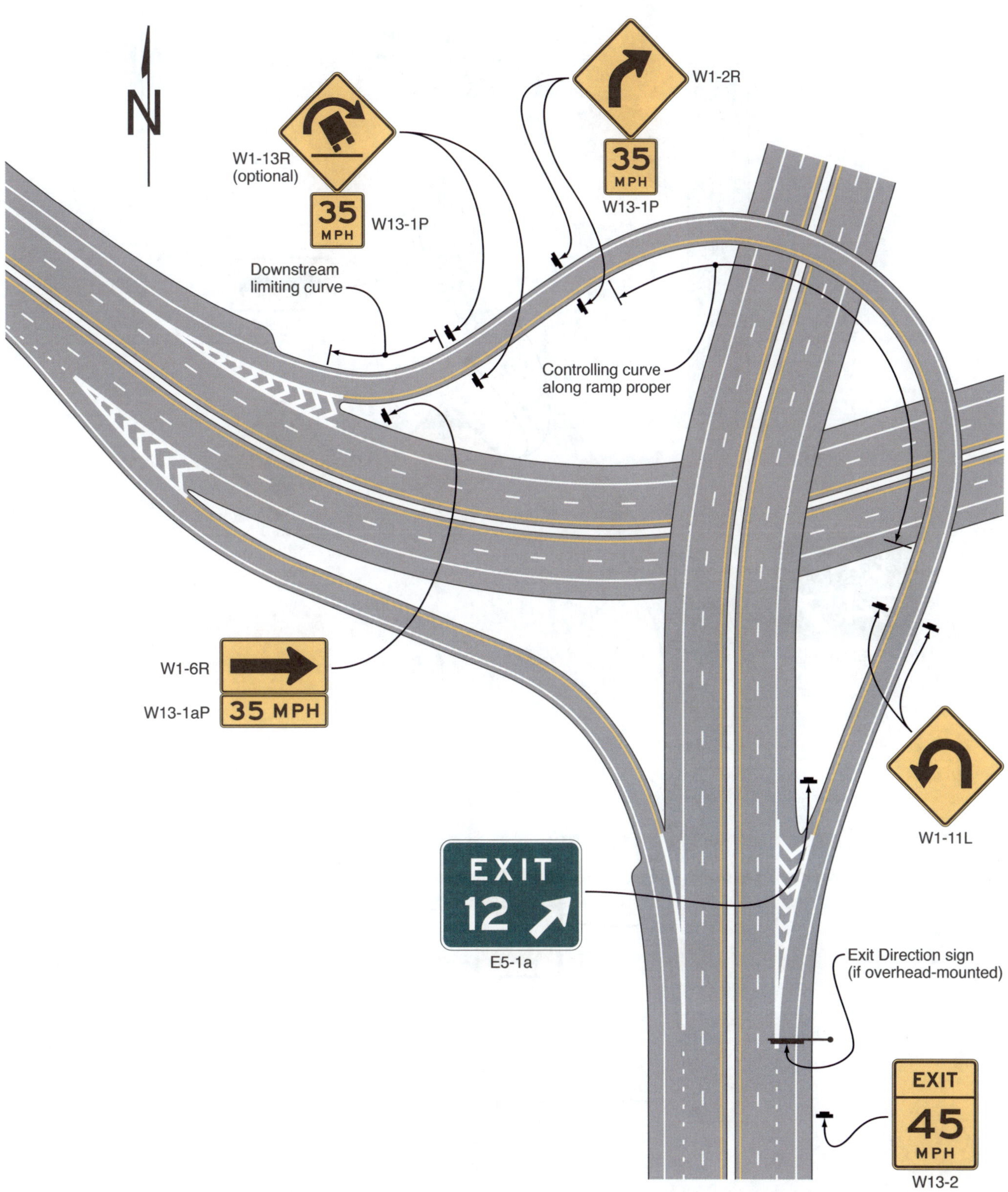

Figure 2C-3. Examples of Exit Ramp Advisory Speed and Other Warning Signs (Sheet 4 of 5)

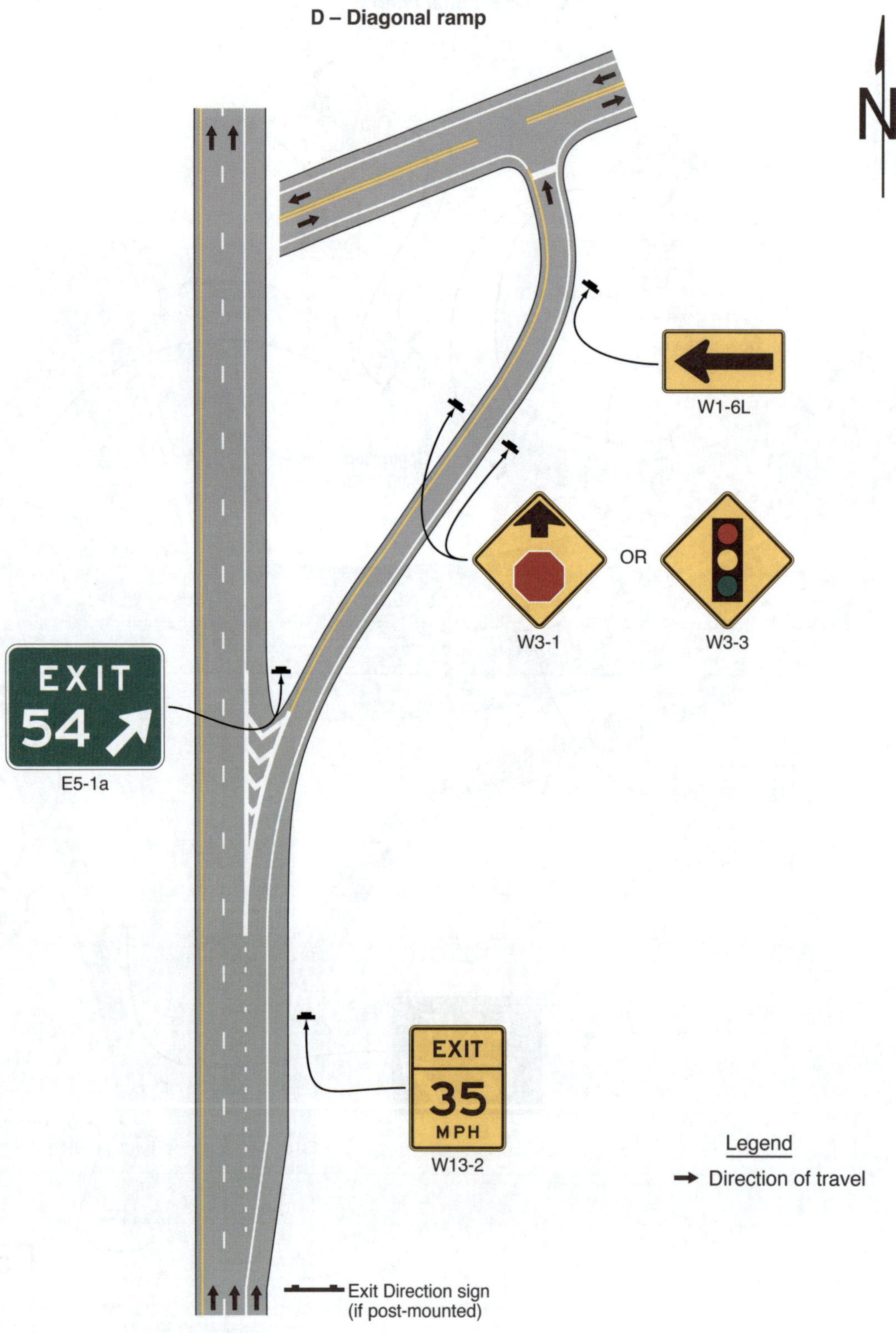

Figure 2C-3. Examples of Exit Ramp Advisory Speed and Other Warning Signs (Sheet 5 of 5)

E – Short ramp length with limiting curve near ramp terminal

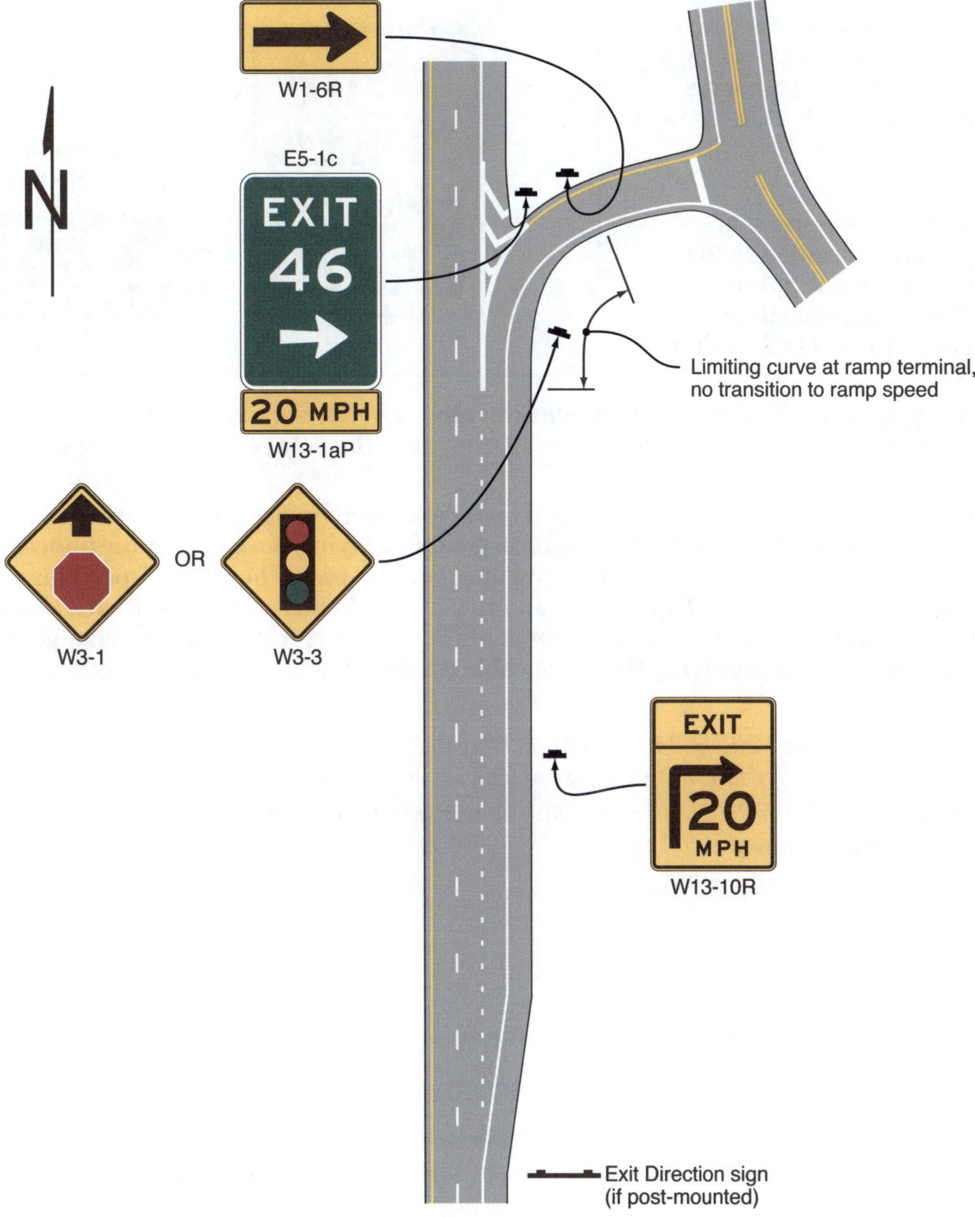

Section 2C.13 <u>Vehicle Speed Feedback Sign and Plaque (W13-20 and W13-20aP)</u>

Option:

01   A Vehicle Speed Feedback (W13-20) sign or (W13-20aP) plaque (see Figure 2C-4) that displays the speed of an approaching vehicle to the vehicle operator may be used to provide warning to drivers of their speed in relation to either a speed limit (R2-1) sign or a horizontal alignment warning sign assembly with a posted advisory speed.

Standard:

02   **When used to display the speed of an approaching vehicle in relation to the posted speed limit, the Vehicle Speed Feedback (W13-20aP) plaque shall be mounted below a Speed Limit (R2-1) sign (see Section 2B.21).**

03   **When used to supplement a horizontal alignment warning sign advisory speed, the Vehicle Speed Feedback (W13-20) sign shall be an independent installation near the point of curvature of a horizontal curve (see Section 2C.06).**

04   **The legend YOUR SPEED shall be a black legend on a yellow retroreflective background, except as provided in Sections 6H.01 and 7B.01. The changeable legend displaying the speed of the approaching vehicle shall be a yellow luminous legend on a black opaque background. The vehicle speed displayed on the changeable portion of the sign shall be displayed as an integer. The Vehicle Speed Feedback sign and plaque shall not flash, strobe, change color, or use other animated elements integrated into the changeable legend display. When no vehicles are approaching, the changeable display shall not display a legend.**

Guidance:

05   *The changeable portion of the Vehicle Speed Feedback legend should be approximately the same height, width, and stroke of those on the Speed Limit sign it supplements or is mounted below.*

06   *When a W13-20aP plaque is used with a Speed Limit sign it should be approximately the same width as the Speed Limit sign it is mounted below.*

VERTICAL GRADE WARNING SIGNS AND PLAQUES

Section 2C.14 Hill Signs (W7-1 and W7-1a)

Guidance:

01   *The Hill (W7-1) sign (see Figure 2C-5) should be used in advance of a downgrade where the length, percent of grade, horizontal curvature, and/or other physical features require special precautions on the part of road users.*

02   *The Hill sign and supplemental grade (W7-3P) plaque (see Figure 2C-5 and Section 2C.64) used in combination, or the W7-1a sign used alone, should be installed in advance of downgrades for the following conditions:*

    A. *5% grade that is more than 3,000 feet in length,*
    B. *6% grade that is more than 2,000 feet in length,*
    C. *7% grade that is more than 1,000 feet in length,*
    D. *8% grade that is more than 750 feet in length, or*
    E. *9% grade that is more than 500 feet in length.*

03   *These signs should also be installed for steeper grades or where crash experience and field observations indicate a need.*

04   *Supplemental plaques (see Sections 2C.57 and 2C.64) and larger signs should be used for emphasis or where special hill characteristics exist. On longer grades, the use of the Hill sign with a distance (W7-3aP) plaque or the combination distance/grade (W7-3bP) plaque (see Figure 2C-5) at periodic intervals of approximately 1-mile spacing should be considered.*

Option:

05   A USE LOW GEAR (W7-2P) or TRUCKS USE LOWER GEAR (W7-2bP) supplemental plaque (see Figure 2C-5) may be used to indicate a situation where downshifting as well as braking might be advisable.

Figure 2C-5. Vertical Grade Signs and Plaques

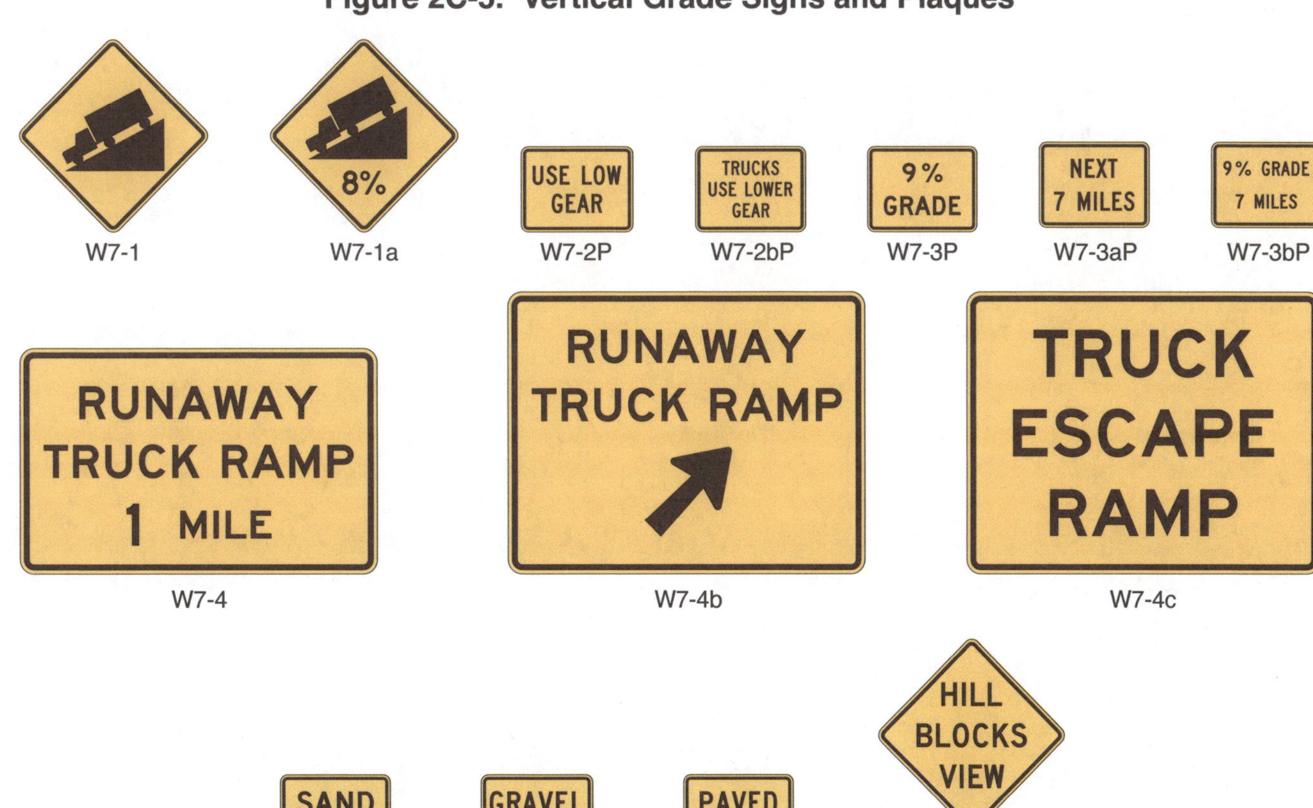

Section 2C.15 Truck Escape Ramp Signs (W7-4 Series)

Guidance:

01   *Where applicable, truck escape (or runaway truck) ramp advance warning signs (see Figure 2C-5) should be located approximately 1 mile and approximately ½ mile in advance of the grade, and of the escape ramp. An additional W7-4b or W7-4c sign should be placed at the gore.*

02   *A RUNAWAY VEHICLES ONLY (R4-10) sign (see Section 2B.41) should be installed near the escape ramp entrance to discourage other road users from entering the ramp. No Parking (R8-3) signs should be placed near the ramp entrance.*

Standard:

03   **When truck escape ramps are installed, at least one of the W7-4 series signs shall be used.**

Option:

04   A SAND (W7-4dP), GRAVEL (W7-4eP), or PAVED (W7-4fP) supplemental plaque (see Figure 2C-5) may be used to describe the ramp surface. State and local highway agencies may develop appropriate word message signs for the specific situation.

Section 2C.16 HILL BLOCKS VIEW Sign (W7-6)

Option:

01   A HILL BLOCKS VIEW (W7-6) sign (see Figure 2C-5) may be used on the approach to a crest vertical curve where the vertical curvature provides inadequate stopping sight distance at the posted speed limit.

Guidance:

02   *When a vertical curve results in a sight distance obstruction to a specific condition beyond the crest of the vertical curve, the warning sign for the specific condition beyond the vertical crest should be used rather than the HILL BLOCKS VIEW sign.*

03   *When a HILL BLOCKS VIEW sign is used, it should be supplemented by an Advisory Speed (W13-1P) plaque (see Figure 2C-1) indicating the recommended speed for traveling over the hillcrest based on available stopping sight distance.*

ROADWAY GEOMETRY WARNING SIGNS

Section 2C.17 ROAD NARROWS Sign (W5-1)

Guidance:

01   *Except as provided in Paragraph 2 of this Section, a ROAD NARROWS (W5-1) sign (see Figure 2C-6) should be used in advance of a transition on two-lane roads where the pavement width is reduced abruptly to a width such that vehicles traveling in opposite directions cannot simultaneously travel through the narrow portion of the roadway without reducing speed.*

Figure 2C-6. Miscellaneous Warning Signs

Option:

02   The ROAD NARROWS (W5-1) sign may be omitted on low-volume local streets that have speed limits of 30 mph or less.

03   Additional emphasis may be provided by the use of object markers and delineators (see Sections 2C.70 through 2C.73 and Chapter 3G). The Advisory Speed (W13-1P) plaque (see Figure 2C-1 and Section 2C.59) may be used to indicate the recommended speed.

Section 2C.18 NARROW BRIDGE and NARROW UNDERPASS Signs (W5-2 and W5-2a)

Guidance:

01   *A NARROW BRIDGE (W5-2) sign (see Figure 2C-6) should be used in advance of any bridge or culvert having a two-way roadway horizontal clearance of 16 to 18 feet, or any bridge or culvert having a roadway horizontal clearance less than the width of the approach travel lanes. Where these conditions exist for an underpass, a NARROW UNDERPASS (W5-2a) sign (see Figure 2C-6) should be used.*

02   *Additional emphasis should be provided by the use of object markers, delineators, and/or pavement markings.*

Option:

03   A NARROW BRIDGE sign may be used in advance of a bridge or culvert on which the approach shoulders are narrowed or eliminated. Where these conditions exist for an underpass, a NARROW UNDERPASS sign may be used.

04   The NARROW BRIDGE or NARROW UNDERPASS sign may be omitted on low-volume rural roads where there is adequate sight distance to the bridge, culvert, or underpass on both approaches.

Section 2C.19 ONE LANE BRIDGE and ONE LANE UNDERPASS Signs (W5-3 and W5-3a)

Guidance:

01   *A ONE LANE BRIDGE (W5-3) sign (see Figure 2C-6) should be used on two-way roadways in advance of any bridge or culvert:*

    A. *Having a roadway horizontal clearance of less than 16 feet, or*
    B. *Having a roadway horizontal clearance of less than 18 feet when commercial vehicles constitute a high proportion of the traffic, or*
    C. *Having a roadway horizontal clearance of 18 feet or less where the sight distance on the approach is less than that shown in Condition A of Table 2C-3.*

02   *Where these conditions exist for an underpass, a ONE LANE UNDERPASS (W5-3a) sign (see Figure 2C-6) should be used.*

03   *Additional emphasis should be provided by the use of object markers, delineators, and/or pavement markings.*

Option:

04   The ONE LANE BRIDGE or ONE LANE UNDERPASS sign may be omitted on low-volume rural roads where there is adequate sight distance to the bridge, culvert, or underpass on both approaches.

05   STOP (R1-1) or YIELD (R1-2) signs (see Sections 2B.04 and 2B.05) and related pavement markings (see Sections 3B.21 and 3B.22) may be used when conditions A, B, or C in Paragraph 1 of this Section apply.

Section 2C.20 Divided Highway Sign (W6-1)

Guidance:

01   *A Divided Highway (W6-1) sign (see Figure 2C-6) should be used on the approaches to a section of highway (not an intersection or junction) where the opposing flows of traffic are separated by a median or other physical barrier.*

Standard:

02   **The Divided Highway (W6-1) sign shall not be used instead of a Keep Right (R4-7 series) sign on the approach end of a median island.**

Section 2C.21 Divided Highway Ends Sign (W6-2)

Guidance:

01   *A Divided Highway Ends (W6-2) sign (see Figure 2C-6) should be used in advance of the end of a section of physically divided highway (not an intersection or junction) as a warning of two-way traffic ahead.*

02   *The Two-Way Traffic (W6-3) sign (see Section 2C.51) should be used to give warning and notice of the transition to a two-lane, two-way section.*

Section 2C.22 <u>Freeway or Expressway Ends Signs (W19 Series)</u>

Option:

01   A FREEWAY ENDS XX MILES (W19-1) sign or a FREEWAY ENDS (W19-3) sign (see Figure 2C-6) may be used in advance of the end of a freeway.

02   An EXPRESSWAY ENDS XX MILES (W19-2) sign or an EXPRESSWAY ENDS (W19-4) sign (see Figure 2C-6) may be used in advance of the end of an expressway.

03   The rectangular W19-1 and W19-2 signs may be post-mounted or may be mounted overhead for increased emphasis.

Guidance:

04   *If the reason that the freeway is ending is that the next portion of the freeway is not yet constructed and as a result all traffic must use an exit ramp to leave the freeway, an ALL TRAFFIC MUST EXIT (W19-5) sign (see Figure 2C-6) should be used in addition to the Freeway Ends signs in advance of the downstream end of the freeway.*

Section 2C.23 <u>Double Arrow Sign (W12-1)</u>

Option:

01   The Double Arrow (W12-1) sign (see Figure 2C-6) may be used to advise road users that traffic is permitted to pass on either side of an island, obstruction, or gore in the roadway. Traffic separated by this sign may either rejoin or change directions.

Guidance:

02   *If used on an island, the Double Arrow sign should be mounted near the approach end.*

03   *If used in front of a pier or obstruction, the Double Arrow sign should be mounted on the face of, or just in front of, the pier or obstruction. Where stripe markings are used on the pier or obstruction, they should be discontinued to leave a 3-inch space around the outside of the Double Arrow sign.*

Section 2C.24 <u>DEAD END, NO OUTLET, and ROAD ENDS Signs (W14-1, W14-1a, W14-2, W14-2a, W8-26, and W8-26a)</u>

Option:

01   The DEAD END (W14-1) sign (see Figure 2C-6) may be used at the entrance to a single road or street that terminates without intersecting another street. The NO OUTLET (W14-2) sign (see Figure 2C-6) may be used at the entrance to a road or road network from which there is no other exit.

02   DEAD END (W14-1a) or NO OUTLET (W14-2a) signs (see Figure 2C-6) may be used in combination with Street Name (D3-1) signs (see Section 2D.45) to warn turning traffic that the cross street ends in the direction indicated by the arrow.

03   At locations where the cross street does not have a name, a W14-1a or W14-2a sign may be used alone in place of a street name sign.

Guidance:

04   *When the W14-1 or W14-2 sign is used, the sign should be posted as near as practicable to the entry point or at a sufficient advance distance to permit the road user to avoid the dead end or no outlet condition by turning at the nearest intersecting street.*

Standard:

05   **The DEAD END (W14-1a) or NO OUTLET (W14-2a) sign shall not be used instead of the W14-1 or W14-2 signs where traffic can proceed straight through the intersection into the dead end street or no outlet area.**

Option:

06   The ROAD ENDS XX FT (W8-26) or STREET ENDS XX FT (W8-26a) sign (see Figure 2C-11) may be used on the approach to the end of a conventional road or street where the terminus is not apparent.

Support:

07   Information about the use of Type 4 object markers to mark the end of the road or street is contained in Section 2C.73.

Standard:

08   **The W8-26 and W8-26a signs shall not be used in place of a W14-1 or W14-2 sign at the entrance to such a road or street.**

Support:

09   Section 2C.22 contains information on signs for use on the approach to the end of a freeway or expressway.

Section 2C.25 Low Clearance Signs (W12-2, W12-2a, and W12-2b)

Standard:

01   **The Low Clearance Advance (W12-2) sign (see Figure 2C-6) shall be used to warn road users of vertical clearances less than 14 feet 6 inches, or vertical clearances less than 12 inches above the statutory maximum vehicle height, whichever is greater.**

Guidance:

02   *The actual clearance should be displayed on the Low Clearance (W12-2, W12-2a, and W12-2b) sign to the nearest 1 inch not exceeding the actual clearance. However, in areas that experience changes in temperature causing frost action, a reduction, not exceeding 3 inches, should be used for this condition.*

03   *Clearances should be evaluated periodically to determine if additional low clearance signing is necessary, particularly when resurfacing operations have occurred, on routes onto which over-height vehicles are normally directed under the permit process, and structures that are susceptible to catastrophic failure when struck by over-height vehicles.*

04   *The W12-2 sign with a supplemental distance plaque should also be placed at the nearest intersecting road or wide point in the road at which a vehicle can detour or turn around.*

05   *Where there is a need to warn of a low clearance on an intersecting road or off a freeway or expressway exit, a rectangular warning sign with an appropriate word legend should be used rather than a W12-2 sign.*

Option:

06   The Low Clearance Overhead (W12-2a or W12-2b) sign (see Figure 2C-6) may be installed on the structure to supplement the advance warning sign.

07   In cases where physical conditions on a structure limit the width such that the W12-2a or W12-2b signs are physically unable to fit, a W12-2 sign may be installed overhead on the structure or post-mounted in front of the structure, in addition to the required W12-2 sign at the advance location.

Guidance:

08   *In the case of an arch, or other structure under which the clearance varies greatly, two or more Low Clearance Overhead (W12-2a or 12-2b) signs should be installed on the structure itself to give information as to the clearances over the low clearance portions of the roadway.*

Standard:

09   **If used, the Low Clearance Overhead (W12-2b) sign shall be placed over a lane or shoulder to indicate the portion of the structure with low clearance if the posted clearance does not apply to the entire structure.**

Guidance:

10   *The clearance shown on the Low Clearance Advance sign should match the clearance on the W12-2a or W12-2b sign or, if there are multiple W12-2b signs, should match the lowest clearance.*

ROADWAY AND WEATHER-CONDITION WARNING SIGNS AND PLAQUES

Section 2C.26 BUMP and DIP Signs (W8-1 and W8-2)

Guidance:

01   *BUMP (W8-1) and DIP (W8-2) signs (see Figure 2C-7) should be used in advance of a sharp rise or depression in the profile of the road.*

Option:

02   These signs may be supplemented with an Advisory Speed plaque (see Figure 2C-1 and Section 2C.59).

Guidance:

03   *The DIP sign should not be used in advance of a short stretch of depressed alignment that might momentarily hide a vehicle.*

04   *A short stretch of depressed alignment that might momentarily hide a vehicle should be treated as a no-passing zone when center line striping is provided on a two-lane or three-lane road (see Section 3B.03).*

Section 2C.27 SPEED HUMP Sign (W17-1)

Guidance:

01   *The SPEED HUMP (W17-1) sign (see Figure 2C-7) should be used in advance of a vertical deflection in the roadway that is designed to limit the speed of traffic.*

02   *If used, the SPEED HUMP sign should be supplemented by an Advisory Speed plaque (see Figure 2C-1 and Section 2C.59).*

Option:

03   If a series of speed humps exists in close proximity, an Advisory Speed plaque may be eliminated on all but the first SPEED HUMP sign in the series.

04   The legend SPEED BUMP may be used instead of the legend SPEED HUMP on the W17-1 sign.

Support:

05   Speed humps generally provide more gradual vertical deflection than speed bumps. Speed bumps limit the speed of traffic more severely than speed humps. Other forms of speed humps include speed tables and raised crosswalks or intersections. However, these differences in engineering terminology are not well known by the public, so for signing purposes these terms are interchangeable.

06   Sections 3B.29 and 3B.30 contain information about the use of markings at and in advance of speed humps.

Section 2C.28 PAVEMENT ENDS Sign (W8-3)

Guidance:

01   *A PAVEMENT ENDS (W8-3) sign (see Figure 2C-7) should be used where a paved surface changes to either a gravel treated surface or an earth road surface.*

Option:

02   An Advisory Speed plaque (see Figure 2C-1 and Section 2C.59) may be used when the change in roadway condition requires a reduced speed.

Section 2C.29 Shoulder Signs (W8-4, W8-9, W8-17, W8-23, and W8-25)

Option:

01   The SOFT SHOULDER (W8-4) sign (see Figure 2C-7) may be used to warn of a soft shoulder condition.

02   The LOW SHOULDER (W8-9) sign (see Figure 2C-7) may be used to warn of a shoulder condition where there is an elevation difference of 3 inches or less between the shoulder and the travel lane.

Guidance:

03   *The Shoulder Drop Off (W8-17) sign (see Figure 2C-7) should be used where an unprotected shoulder drop-off, adjacent to the travel lane, exceeds 3 inches in depth for a significant continuous length along the roadway, based on engineering judgment.*

Option:

04   A SHOULDER DROP-OFF (W8-17P) supplemental plaque (see Figure 2C-7) may be mounted below the W8-17 sign.

05   The NO SHOULDER (W8-23) sign (see Figure 2C-7) may be used to warn road users that a shoulder does not exist along a portion of the roadway.

Figure 2C-7. Roadway and Weather Condition Signs and Plaques

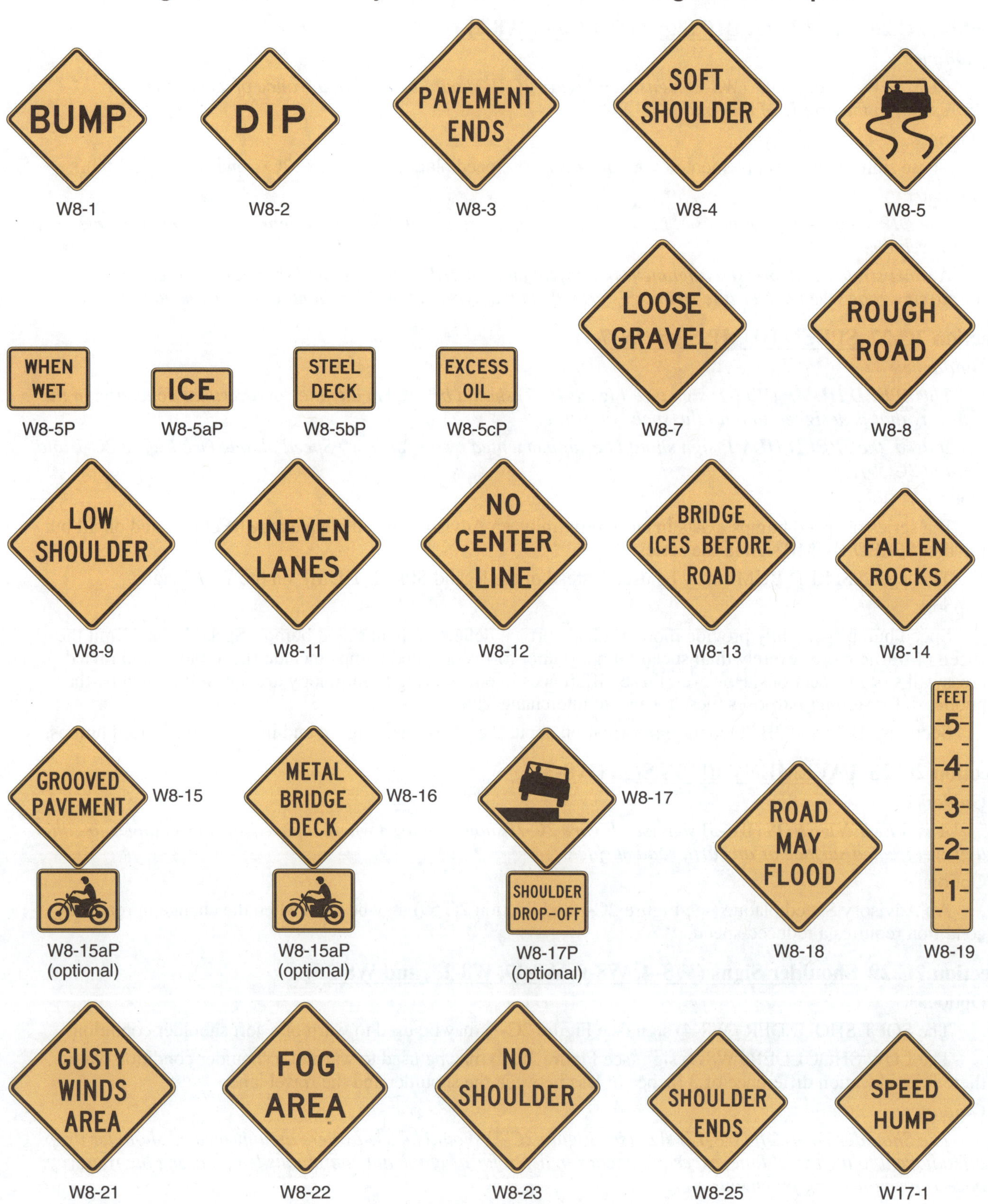

06   The SHOULDER ENDS (W8-25) sign (see Figure 2C-7) may be used to warn road users that a shoulder is ending.

Guidance:

07   *Additional shoulder signs should be placed at appropriate intervals along the road where the condition continually exists.*

Section 2C.30 Surface Condition Signs (W8-5, W8-7, W8-8, W8-11, W8-13, and W8-14)

Option:

01   The Slippery When Wet (W8-5) sign (see Figure 2C-7) may be used to warn of unexpected slippery conditions. Supplemental plaques (see Figure 2C-7) with legends such as ICE, WHEN WET, STEEL DECK, or EXCESS OIL may be used with the W8-5 sign to indicate the reason that the slippery conditions might be present.

02   The LOOSE GRAVEL (W8-7) sign (see Figure 2C-7) may be used to warn of loose gravel on the roadway surface.

03   The ROUGH ROAD (W8-8) sign (see Figure 2C-7) may be used to warn of a rough roadway surface.

04   An UNEVEN LANES (W8-11) sign (see Figure 2C-7) may be used to warn of a difference in elevation between travel lanes.

05   The BRIDGE ICES BEFORE ROAD (W8-13) sign (see Figure 2C-7) may be used in advance of bridges to advise bridge users of winter weather conditions. The BRIDGE ICES BEFORE ROAD sign may be removed or covered during seasons of the year when its message is not relevant.

06   The FALLEN ROCKS (W8-14) sign (see Figure 2C-7) may be used in advance of an area that is adjacent to a hillside, mountain, or cliff where rocks frequently fall onto the roadway.

Guidance:

07   *When used, Surface Condition signs should be placed in advance of the beginning of the affected section (see Table 2C-3), and additional signs should be placed at appropriate intervals along the road where the condition exists.*

Section 2C.31 Warning Signs and Plaque for Motorcyclists (W8-15, W8-15aP, and W8-16)

Support:

01   The signs and plaques described in this Section are intended to give motorcyclists advance notice of surface conditions that might adversely affect their ability to maintain control of their motorcycle under wet or dry conditions. The use of some of the advance surface condition warning signs described in Section 2C.30, such as Slippery When Wet, LOOSE GRAVEL, or ROUGH ROAD, can also be helpful to motorcyclists if those conditions exist.

Option:

02   If a portion of a street or highway features a roadway pavement surface that is grooved or textured instead of smooth, such as a grooved skid resistance treatment for a horizontal curve or a brick pavement surface, a GROOVED PAVEMENT (W8-15) sign (see Figure 2C-7) may be used to provide advance warning of this condition to motorcyclists, bicyclists, and other road users. Alternate legends such as TEXTURED PAVEMENT or BRICK PAVEMENT may also be used on the W8-15 sign.

03   If a bridge or a portion of a bridge includes a metal or grated surface, a METAL BRIDGE DECK (W8-16) sign (see Figure 2C-7) may be used to provide advance warning of this condition to motorcyclists, bicyclists, and other road users.

04   A Motorcycle (W8-15aP) plaque (see Figure 2C-7) may be mounted below or above a W8-15 or W8-16 sign if the warning is intended to be directed primarily to motorcyclists.

Section 2C.32 NO CENTER LINE Sign (W8-12)

Option:

01   The NO CENTER LINE (W8-12) sign (see Figure 2C-7) may be used to warn of a roadway without center line pavement markings.

Section 2C.33 NO TRAFFIC SIGNS Sign (W18-1)

Option:

01   The NO TRAFFIC SIGNS (W18-1) sign (see Figure 2C-6) may be used only on low-volume rural roads to advise road users that no signs are installed along the distance of the road. The sign may be installed at the point where road users would enter the low volume road or where, based on engineering judgment, the road user might need this information.

02   A W7-3aP (see Figure 2C-5), W16-2P (see Figure 2C-16), or W16-9P (see Figure 2C-16) supplemental plaque with the legend NEXT XX MILES, XX FEET, or AHEAD may be installed below the W18-1 sign when appropriate.

Section 2C.34 <u>Weather Condition Signs (W8-18, W8-19, W8-21, and W8-22)</u>

Option:

01   The ROAD MAY FLOOD (W8-18) sign (see Figure 2C-7) may be used to warn road users that a section of roadway is subject to frequent flooding. A Depth Gauge (W8-19) sign (see Figure 2C-7) may also be installed within a roadway section that frequently floods.

Guidance:

02   *If used, the Depth Gauge sign should be in addition to the ROAD MAY FLOOD sign and should be mounted at the appropriate height to indicate the depth of the water at the deepest point on the roadway.*

Option:

03   The GUSTY WINDS AREA (W8-21) sign (see Figure 2C-7) may be used to warn road users that wind gusts frequently occur along a section of highway that are strong enough to impact the stability of trucks, recreational vehicles, and other vehicles with high centers of gravity. A NEXT XX MILES (W7-3aP) supplemental plaque (see Figure 2C-5) may be mounted below the W8-21 sign to inform road users of the length of roadway that frequently experiences strong wind gusts.

04   The FOG AREA (W8-22) sign (see Figure 2C-7) may be used to warn road users that foggy conditions frequently reduce visibility along a section of highway. A NEXT XX MILES (W7-3aP) supplemental plaque (see Figure 2C-5) may be mounted below the W8-22 sign to inform road users of the length of roadway that frequently experiences foggy conditions.

Support:

05   Chapter 2L contains provisions for the use of blank-out or changeable message signs that can be activated by detection of the applicable condition.

TRAFFIC CONTROL AND INTERSECTION WARNING SIGNS AND PLAQUES

Section 2C.35 Advance Traffic Control Signs (W3-1, W3-2, W3-3, and W3-4)

Standard:

01   **The Stop Ahead (W3-1), Yield Ahead (W3-2), and Signal Ahead (W3-3) Advance Traffic Control signs (see Figure 2C-8) shall be installed on an approach to a primary traffic control device that is not visible for a sufficient distance to permit the road user to respond to the device (see Table 2C-3). The visibility criteria for a traffic control signal shall be based on having a continuous view of at least two signal faces for the distance specified in Table 4D-2.**

Guidance:

02   *Where intermittent obstructions occur, engineering judgment should determine the treatment to be implemented.*

Support:

03   Figure 2A-4 shows examples of the typical placement of an Advance Traffic Control sign.

04   Permanent obstructions causing the limited visibility might include roadway alignment or structures. Intermittent obstructions might include foliage or parked vehicles.

Option:

05   An Advance Traffic Control sign may be used for additional emphasis of the primary traffic control device, even when the visibility distance to the device is satisfactory.

Support:

06   Section 2C.65 contains information about the use of an advance street name plaque to identify an intersecting road.

Option:

07   A BE PREPARED TO STOP (W3-4) sign (see Figure 2C-8) may be used to warn of stopped traffic caused by a traffic control signal.

08   A Warning Beacon (see Section 4S.03) or yellow LEDs within the border of the sign may be used with an Advance Traffic Control or BE PREPARED TO STOP sign.

Standard:

09   **When a BE PREPARED TO STOP sign is used in advance of a traffic control signal, it shall be used in addition to a Signal Ahead sign and shall be placed downstream from the Signal Ahead sign.**

Guidance:

10   *When a Warning Beacon is interconnected with a traffic control signal or queue detection system, the BE PREPARED TO STOP sign should be supplemented with a WHEN FLASHING (W16-13P) plaque (see Figure 2C-16).*

Figure 2C-8. Advance Traffic Control Signs

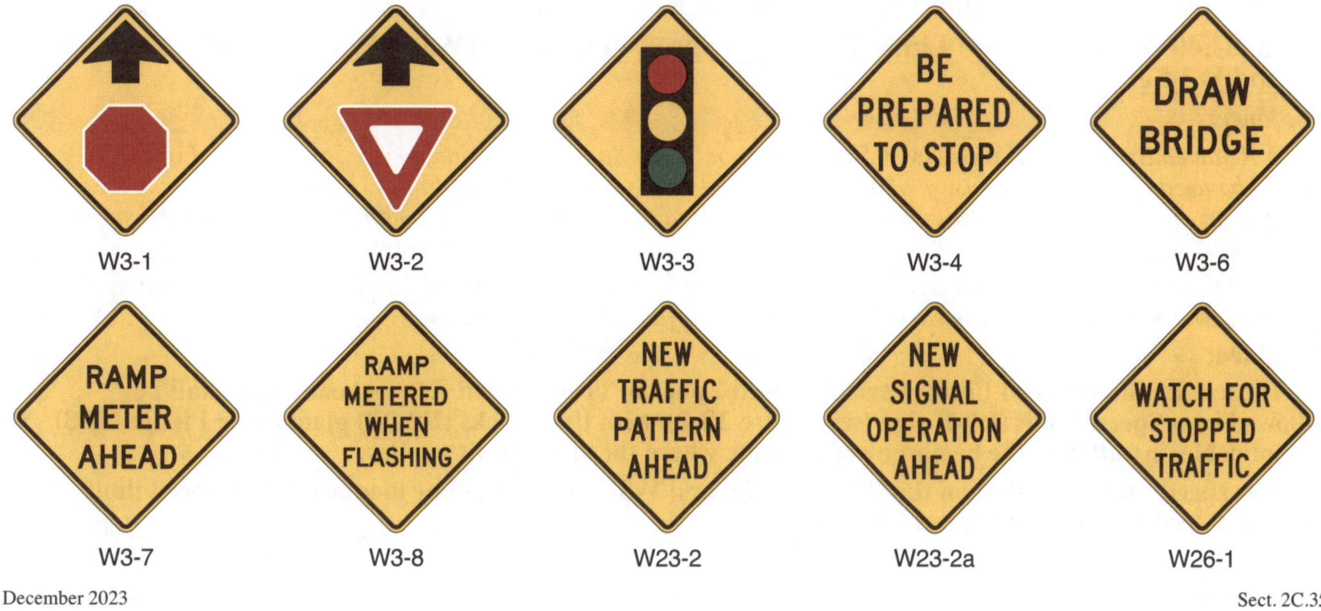

Support:

11   Section 2C.45 contains information regarding the use of a NO MERGE AREA (W4-5aP) supplemental plaque
in conjunction with a Yield Ahead sign.

Section 2C.36 DRAW BRIDGE Sign (W3-6)

Standard:

01   **A DRAW BRIDGE (W3-6) sign (see Figure 2C-8) shall be used in advance of movable bridge signals
and gates (see Section 4Q.02) to give warning to road users.**

Section 2C.37 Advance Ramp Control Signal Signs (W3-7 and W3-8)

Option:

01   A RAMP METER AHEAD (W3-7) sign (see Figure 2C-8) may be used to warn road users that a freeway
entrance ramp is metered and that they will encounter a ramp control signal (see Chapter 4P).

Guidance:

02   *When the ramp control signals are operated only during certain periods of the day, a RAMP METERED
WHEN FLASHING (W3-8) sign (see Figure 2C-8) should be installed in advance of the ramp control signal near
the entrance to the ramp, or on the arterial on the approach to the ramp, to alert road users to the presence and
operation of ramp meters.*

Standard:

03   **The RAMP METERED WHEN FLASHING sign shall be supplemented with a Warning Beacon
(see Section 4S.03) that flashes when the ramp control signal is in operation.**

Section 2C.38 NEW TRAFFIC PATTERN and NEW SIGNAL OPERATION AHEAD Signs (W23-2 and W23-2a)

Option:

01   A NEW TRAFFIC PATTERN AHEAD (W23-2) sign (see Figure 2C-8) may be used on the approach to
an intersection or along a section of roadway to provide advance warning of a change in traffic patterns, such as
revised lane usage or roadway geometry.

02   A NEW SIGNAL OPERATION AHEAD (W23-2a) sign (see Figure 2C-8) may be used on the approach to
a signalized intersection to provide advance warning of a change in signal phasing.

Guidance:

03   *The NEW TRAFFIC PATTERN or NEW SIGNAL OPERATION AHEAD sign should be removed when
the traffic pattern returns to normal, when the changed pattern is no longer considered to be new, or within
12 months.*

Section 2C.39 WATCH FOR STOPPED TRAFFIC Sign (W26-1)

Option:

01   The WATCH FOR STOPPED TRAFFIC (W26-1) sign (see Figure 2C-8) may be used to warn road users
of the possibility of vehicles stopping abruptly in the travel lane due to recurring congested conditions.

Section 2C.40 Reduced Speed Limit Ahead and Speed Zone Signs (W3-5, W3-5a, W3-5b, and W3-5c)

Guidance:

01   *A Reduced Speed Limit Ahead (W3-5 or W3-5a) or Truck Speed Zone Ahead (W3-5c) sign (see Figure 2C-9)
should be used to inform road users of a reduced speed zone where the speed limit is being reduced by more than
10 mph, or where engineering judgment indicates the need for advance notice to comply with the posted speed
limit ahead.*

02   *A VARIABLE SPEED ZONE AHEAD (W3-5b) sign (see Figure 2C-9) should be used to inform road users
of a zone where the speed limit is varied by time of day or as conditions change.*

Standard:

03   **If used, Reduced Speed Limit, Variable Speed Zone, or Truck Speed Zone Ahead signs shall be
followed by a Speed Limit (R2-1) sign (see Figure 2B-3), with the Trucks (R2-2P) plaque (see Figure 2B-3)
if applicable, installed at the beginning of the zone where the speed limit applies.**

04   **The speed limit displayed on the W3-5, W3-5a, and W3-5c signs shall be identical to the speed limit
displayed on the subsequent Speed Limit sign.**

Figure 2C-9. Reduced Speed Limit Ahead and Speed Zone Signs

Section 2C.41 Intersection Warning Signs (W2-1 through W2-8)

Option:

01   A Cross Road (W2-1), Side Road (W2-2, W2-3, or W2-3a), T-Intersection (W2-4), or Y-Intersection (W2-5) sign (see Figure 2C-10) may be used in advance of an intersection to indicate the presence of an intersection and the possibility of turning or entering traffic.

Figure 2C-10. Intersection Warning Signs and Plaques

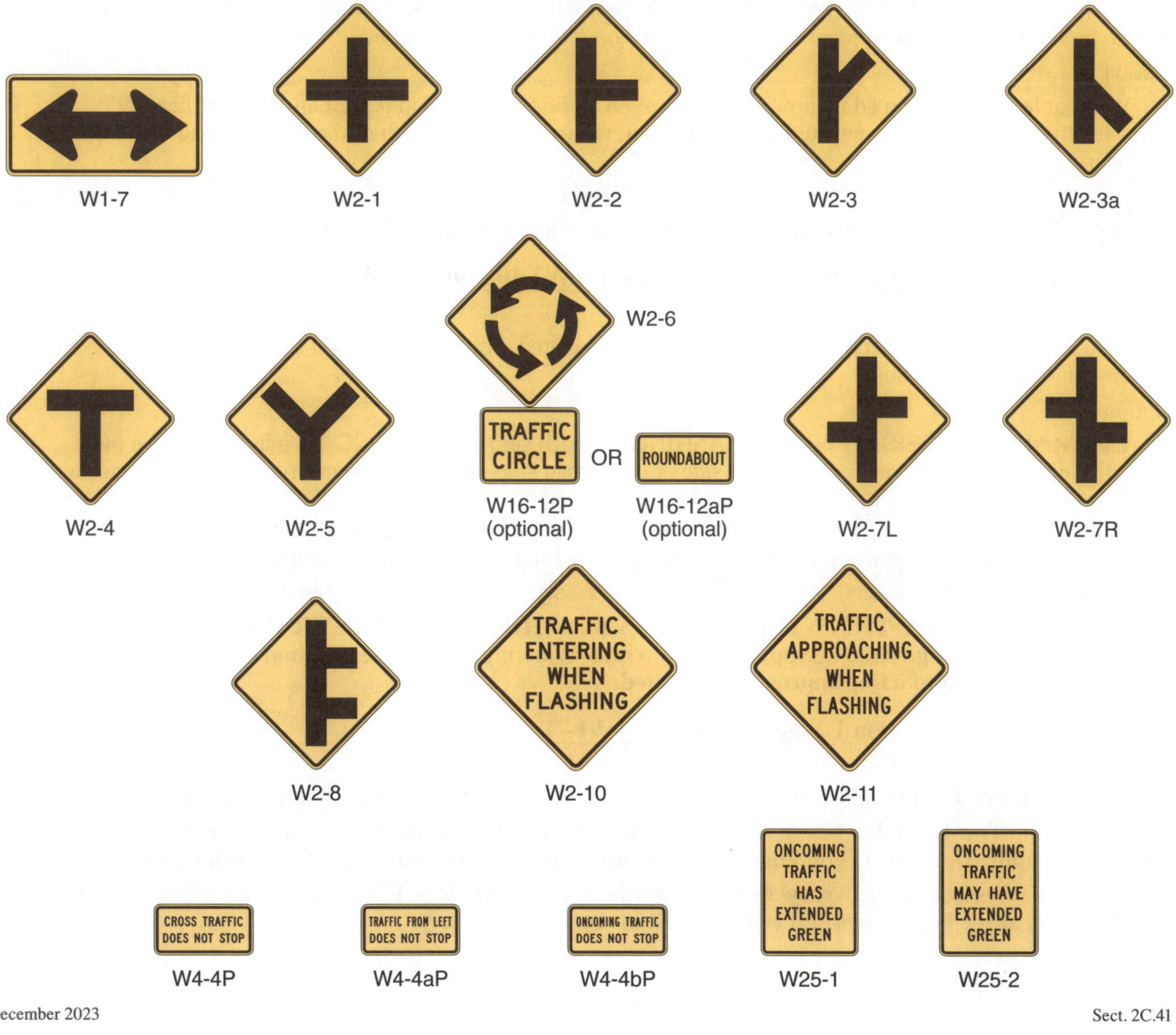

W1-7   W2-1   W2-2   W2-3   W2-3a

W2-6

W2-4   W2-5   TRAFFIC CIRCLE W16-12P (optional)   OR   ROUNDABOUT W16-12aP (optional)   W2-7L   W2-7R

W2-8   TRAFFIC ENTERING WHEN FLASHING W2-10   TRAFFIC APPROACHING WHEN FLASHING W2-11

CROSS TRAFFIC DOES NOT STOP W4-4P   TRAFFIC FROM LEFT DOES NOT STOP W4-4aP   ONCOMING TRAFFIC DOES NOT STOP W4-4bP   ONCOMING TRAFFIC HAS EXTENDED GREEN W25-1   ONCOMING TRAFFIC MAY HAVE EXTENDED GREEN W25-2

02   The Circular Intersection (W2-6) sign (see Figure 2C-10) may be installed in advance of a circular intersection (see Figures 2B-21 through 2B-23).

Guidance:

03   *If an approach to a circular intersection has a statutory or posted speed limit of 40 mph or higher, the Circular Intersection (W2-6) sign should be installed in advance of the circular intersection.*

Option:

04   An educational plaque (see Figure 2C-10) with a legend such as TRAFFIC CIRCLE (W16-12P) or ROUNDABOUT (W16-12aP) may be mounted below a Circular Intersection sign.

Support:

05   Section 2C.65 contains information about the use of an advance street name plaque to identify an intersecting road.

Guidance:

06   *The Intersection Warning sign should illustrate and depict the general configuration of the intersecting roadway, such as a cross road, side road, T-intersection, or Y-intersection.*

07   *Intersection Warning signs, other than the Circular Intersection (W2-6) sign, the T-intersection (W2-4) sign, and the Grade Crossing and Intersection Advance Warning (W10-2, W10-3, W10-4, W10-11, and W10-12) signs (see Figure 8B-4) should not be used on approaches controlled by STOP signs, YIELD signs, or signals.*

08   *If an Intersection Warning sign is used where the side roads are not opposite of each other, the Offset Side Roads (W2-7) sign (see Figure 2C-10) should be used instead of the Cross Road sign.*

09   *If an Intersection Warning sign is used where two closely-spaced side roads are on the same side of the highway, the Double Side Roads (W2-8) sign (see Figure 2C-10) should be used instead of the Side Road sign.*

10   *No more than two side roads should be depicted on the same side of the highway on a W2-7 or W2-8 sign, and no more than three side roads should be depicted on a W2-7 or W2-8 sign.*

Option:

11   When at least one side road is shown, the stem of an additional side road representing a significantly lower relative volume may be depicted using a line that is two-thirds the width of the through road based on engineering judgment.

Support:

12   Figure 2A-4 shows examples of the typical placement of an Intersection Warning sign.

Section 2C.42 <u>Actuated Advance Intersection Signs (W2-10 and W2-11)</u>

Support:

01   Actuated Advance Intersection signs are typically associated with restricted sight distance and gap selection at stop-controlled intersections.

Option:

02   The TRAFFIC ENTERING WHEN FLASHING (W2-10) sign (see Figure 2C-10) may be used on the uncontrolled through roadway approach to a side or cross road stop-controlled intersection to warn of entering traffic from the side or cross road.

03   The TRAFFIC APPROACHING WHEN FLASHING (W2-11) sign (see Figure 2C-10) may be used on the side road stop-controlled approach to warn of traffic approaching on the uncontrolled through road.

Standard:

04   **When used, the TRAFFIC ENTERING WHEN FLASHING sign and the TRAFFIC APPROACHING WHEN FLASHING sign shall be supplemented with a Warning Beacon (see Section 4S.03) that activates when a vehicle on a conflicting approach is detected.**

Section 2C.43 <u>Two-Direction Large Arrow Sign (W1-7)</u>

Standard:

01   **The Two-Direction Large Arrow (W1-7) sign (see Figure 2C-10) shall be a horizontal rectangle.**

02   **If used, the Two-Direction Large Arrow sign shall be installed on the far side of a T-intersection in line with, and at approximately a right angle to, traffic approaching from the stem of the T-intersection.**

03   **The Two-Direction Large Arrow sign shall not be used where there is no change in the direction of travel such as at the beginnings and ends of medians or at center piers.**

Guidance:

04   *The Two-Direction Large Arrow sign should be visible for a sufficient distance to provide the road user with adequate time to react to the intersection configuration.*

Section 2C.44 <u>Traffic Signal Oncoming Extended Green Signs (W25-1 and W25-2)</u>

Standard:

01   **At locations where either a W25-1 or a W25-2 sign is required based on the provisions in Section 4F.01, the W25-1 or W25-2 sign (see Figure 2C-10) shall be installed near the left-most signal face for the approach.**

MERGING, TWO-WAY TRAFFIC, AND NO PASSING WARNING SIGNS AND PLAQUES

Section 2C.45 Merge Signs and Plaque (W4-1, W4-5, and W4-5aP)

Option:

01   A Merge (W4-1) sign (see Figure 2C-11) may be used to warn road users on the major roadway that merging movements might be encountered in advance of a point where lanes from two separate roadways converge as a single traffic lane and no turning conflict occurs.

02   A Merge sign may also be installed on the side of the entering roadway to warn road users on the entering roadway of the merge condition.

Guidance:

03   *The Merge sign should be installed on the side of the major roadway where merging traffic will be encountered and in such a position as to not obstruct the road user's view of entering traffic.*

Figure 2C-11. Merging and Passing Signs and Plaques

04   *When a Merge sign is installed on a major roadway, the symbol should be oriented right or left as appropriate to depict the side from which the merge occurs, with the arrow representing the major roadway and the curved stem representing the entering roadway (see Figure 2C-11)*

05   *When a Merge sign is installed on an entering roadway that curves before merging with the major roadway, such as a ramp with a curving horizontal alignment as it approaches the major roadway, the Entering Roadway Merge (W4-5) sign (see Figure 2C-11) should be used to better portray the actual geometric conditions to road users on the entering roadway.*

06   *Where two roadways of approximately equal importance converge and merging movements are required, a Merge sign should be placed on each roadway.*

07   *The Merge sign should not be used where two roadways converge and merging movements are not required.*

Standard:

08   **The Merge sign shall not be used in place of a Lane Ends (W4-2) sign (see Section 2C.47) where lanes of traffic moving on a single roadway must merge because of a reduction in the actual or usable pavement width.**

Option:

09   An Entering Roadway Merge (W4-5) sign with a NO MERGE AREA (W4-5aP) supplemental plaque (see Figure 2C-11) mounted below it may be used to warn road users on an entering roadway that they will encounter an abrupt merging situation without an acceleration lane at the downstream end of the ramp.

10   A Merge (W4-1) sign with a NO MERGE AREA (W4-5aP) supplemental plaque mounted below it may be used to warn road users on the major roadway that traffic on an entering roadway will encounter an abrupt merging situation without an acceleration lane at the downstream end of the ramp.

11   For a yield-controlled channelized right-turn movement onto a roadway without an acceleration lane, a NO MERGE AREA (W4-5aP) supplemental plaque may be mounted below a Yield Ahead (W3-2) sign and/or below a YIELD (R1-2) sign when engineering judgment indicates that road users would expect an acceleration lane to be present.

Support:

12   Examples of the use of Merge (W4-1) signs are shown in Drawing A in Figure 2C-12.

Section 2C.46 Added Lane Signs (W4-3 and W4-6)

Guidance:

01   *The Added Lane (W4-3) sign (see Figure 2C-11) should be installed in advance of a point where two roadways converge and merging movements are not required. When possible, the Added Lane sign should be placed such that it is visible from both roadways; if this is not possible, an Added Lane sign should be placed on the side of each roadway.*

02   *When an Added Lane (W4-3) sign is installed on a major roadway, the symbol should be oriented right or left as appropriate to depict the side from which the entering roadway converges, with the straight arrow representing the major roadway and the curved arrow representing the entering roadway. The sign should be located on the side of the major roadway from which the entering roadway converges.*

03   *When an Added Lane sign is to be installed on a roadway that curves before converging with another roadway that has a tangent alignment at the point of convergence, the Entering Roadway Added Lane (W4-6) sign (see Figure 2C-11) should be used to better portray the actual geometric conditions to road users on the curving roadway.*

Support:

04   Examples of the use of Added Lane (W4-3) and Entering Roadway Added Lane (W4-6) signs are shown in Drawing B in Figure 2C-12.

Section 2C.47 Lane Ends Signs (W4-2 and W9-1)

Support:

01   The Lane Ends (W4-2) and RIGHT (LEFT) LANE ENDS (W9-1) signs (see Figure 2C-11) are used to warn of the reduction in the number of traffic lanes in the direction of travel.

02   The sequence of the W4-2 and W9-1 signs is illustrated in Figure 2C-13.

Guidance:

03   *The Lane Ends (W4-2) sign should be installed at the advance placement distance in accordance with Table 2C-3.*

Figure 2C-12. Examples of Merge and Added Lane Sign Placement for Entering and Converging Roadways

A – Example for converging and entering roadways

B – Examples for an added lane

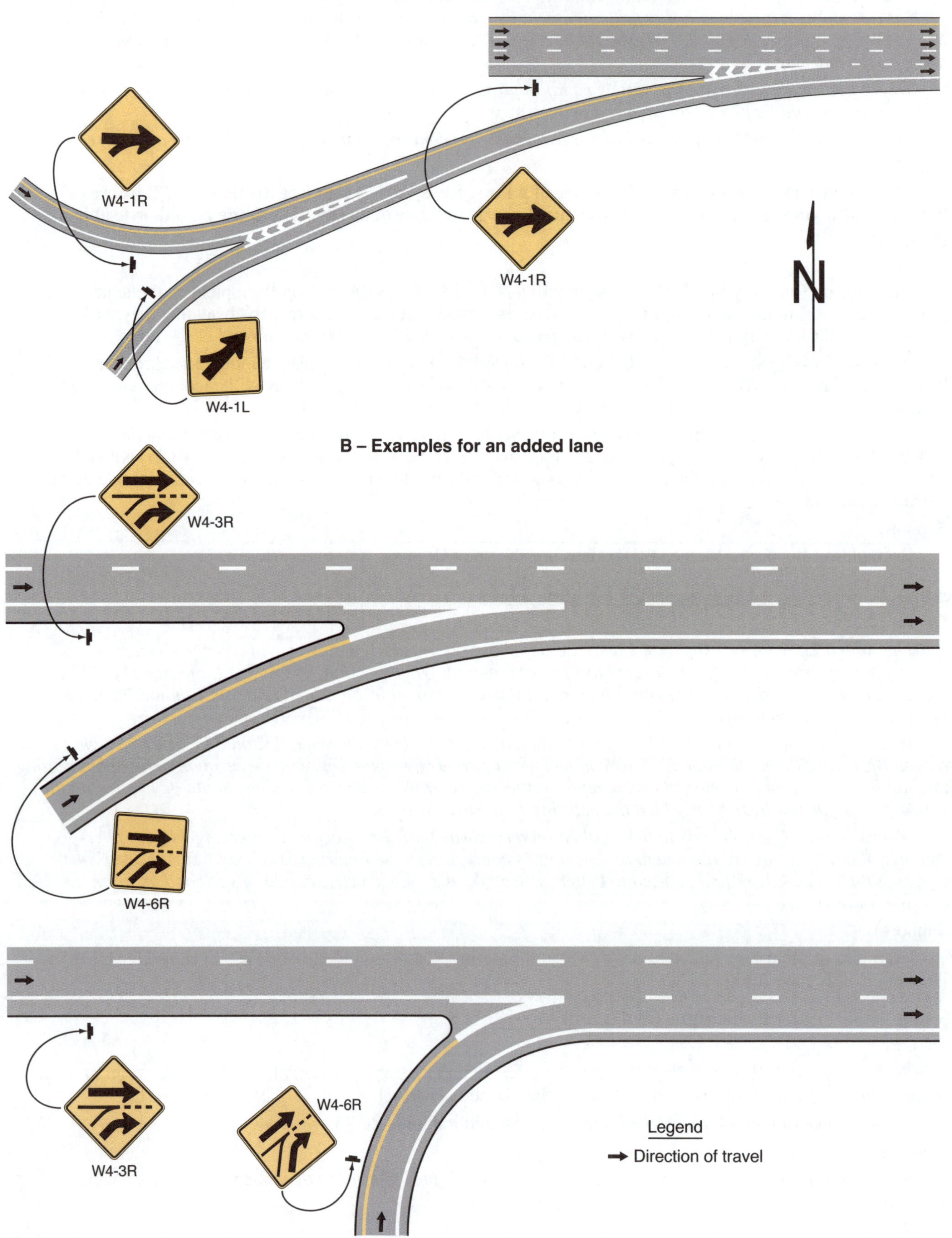

Figure 2C-13. Example Sequences for Lane Ends and Lane Merge Signs (Sheet 1 of 5)

A – Freeway or expressway - lane ends

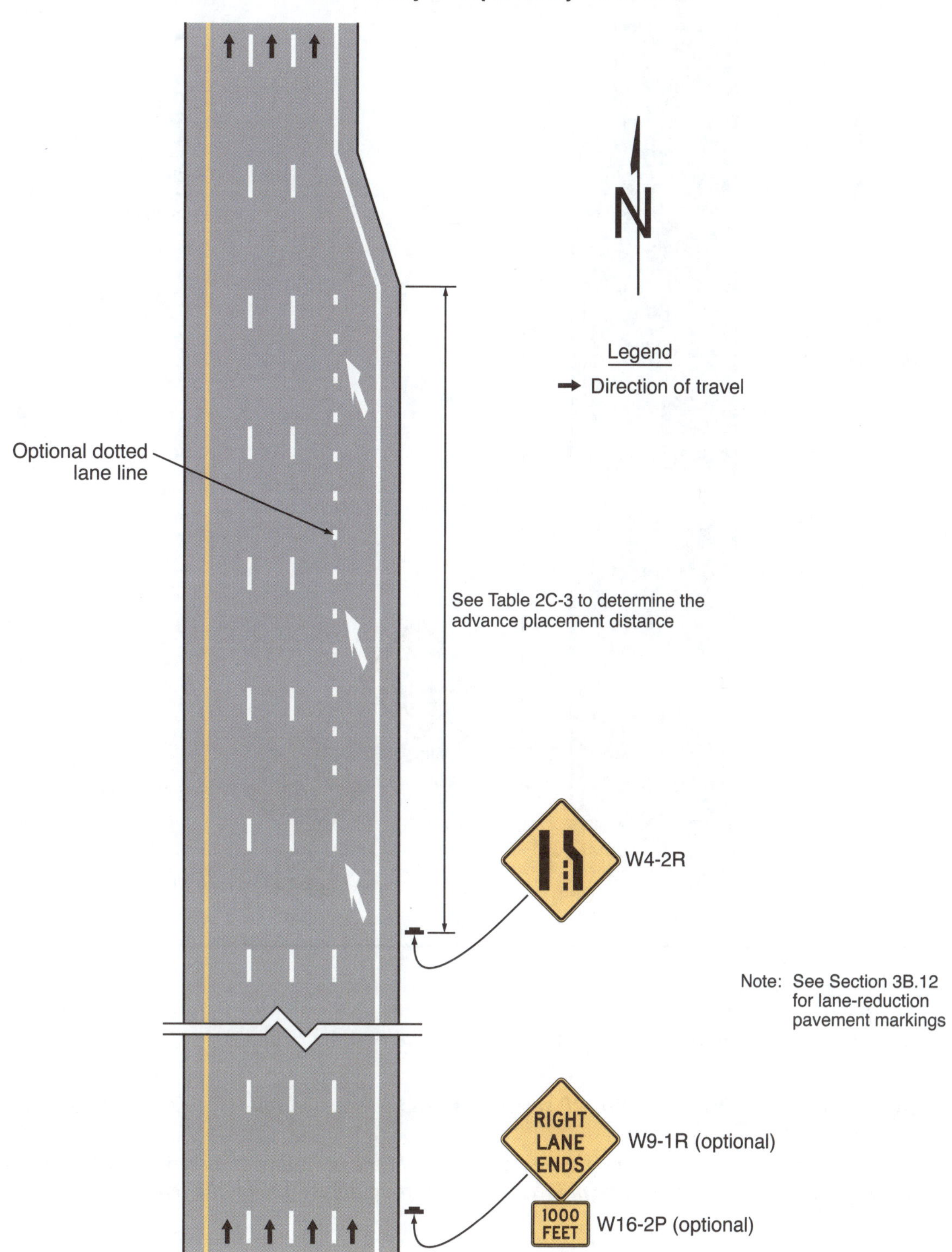

Figure 2C-13. Example Sequences for Lane Ends and Lane Merge Signs (Sheet 2 of 5)

B – Right lane on conventional road - lane ends

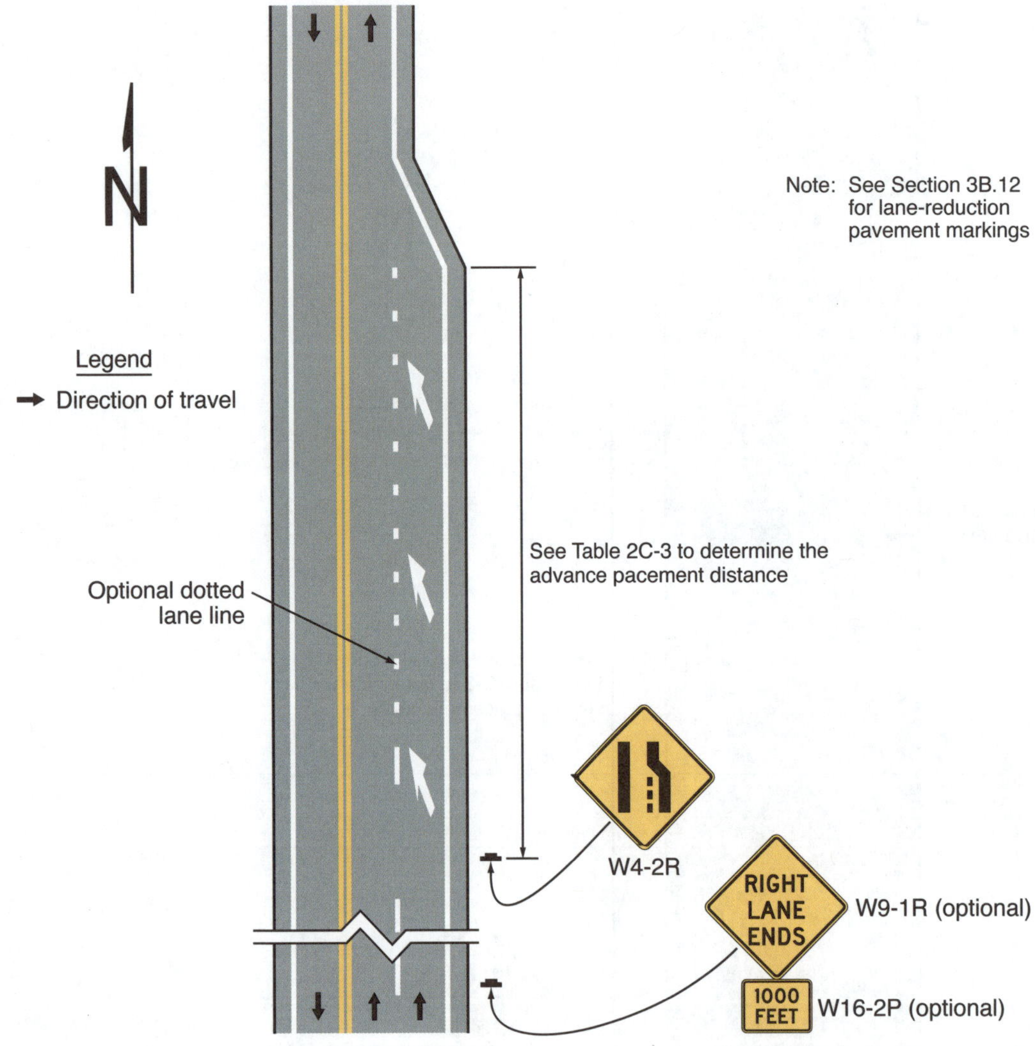

Option:

04   A RIGHT (LEFT) LANE ENDS (W9-1) sign may be installed in advance of the Lane Ends sign to provide additional warning that a lane is ending and that a merging maneuver will be required.

Guidance:

05   *If a W9-1 sign is installed, a Distance (W16-2P series or W16-3P series) plaque (see Figure 2C-16) should be installed below the W9-1 sign.*

06   *On one-way streets or on divided highways where the left-hand lane is ending and the width of the median will permit, the W9-1 and W4-2 signs should be placed facing approaching traffic on the left-hand side or median.*

Option:

07   Where a lane ends a distance beyond the intersection that is less than the advance placement distance indicated in Table 2C-3, the W4-2 sign may be located at the far side of the intersection (see Sheet 4 of Figure 2C-13).

Guidance:

08   *When the W4-2 sign is located at the far side of the intersection in accordance with Paragraph 7 of this Section, the W9-1 sign should be placed upstream of the intersection with the appropriate distance plaque.*

Support:

09   Section 3B.12 contains information regarding the use of pavement markings in conjunction with a lane reduction.

Figure 2C-13. Example Sequences for Lane Ends and Lane Merge Signs (Sheet 3 of 5)

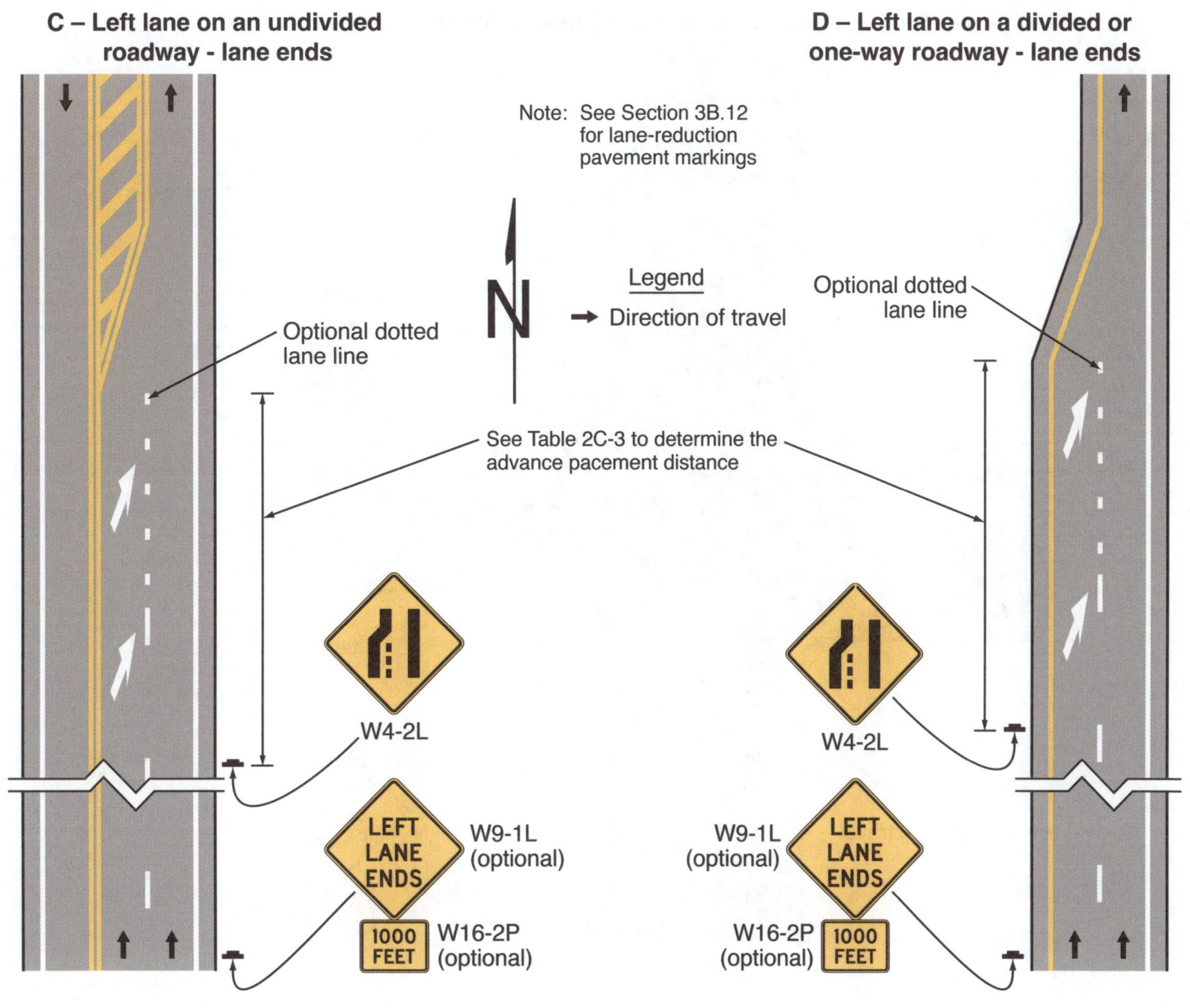

Guidance:

10   *Lane Ends signs should not be installed in advance of the downstream end of an acceleration lane.*

Standard:

11   **The W4-2 and W9-1 signs shall not be used in dropped lane situations. In dropped lane situations on conventional roads at intersections, regulatory signs (see Section 2B.28) shall be used to inform road users that a through lane becomes a mandatory turn lane.**

Section 2C.48 Lanes Merge Signs (W9-4 and W4-8)

Support:

01   The LANES MERGE (W9-4) and Single-Lane Transition (W4-8) signs (see Figure 2C-11) are used to warn of a merge of two lanes to one in the same direction of travel with a merging maneuver required for each lane (see Sheet 5 of Figure 2C-13). This type of merge is for a geometric condition where both approach lanes merge into a single lane, not where one lane merges into the other. Section 6H.08 contains information about the use of the late merge sign.

Guidance:

02   *The Single-Lane Transition (W4-8) sign should be located at the advance placement distance in accordance with Table 2C-3.*

Option:

03   The Lanes Merge (W9-4) sign may be used in advance of the W4-8 sign to provide additional warning that both lanes form a single lane and that a merging maneuver is needed for the traffic in each lane.

Figure 2C-13. Example Sequences for Lane Ends and Lane Merge Signs (Sheet 4 of 5)

E – Lane ends beyond major intersection or side road

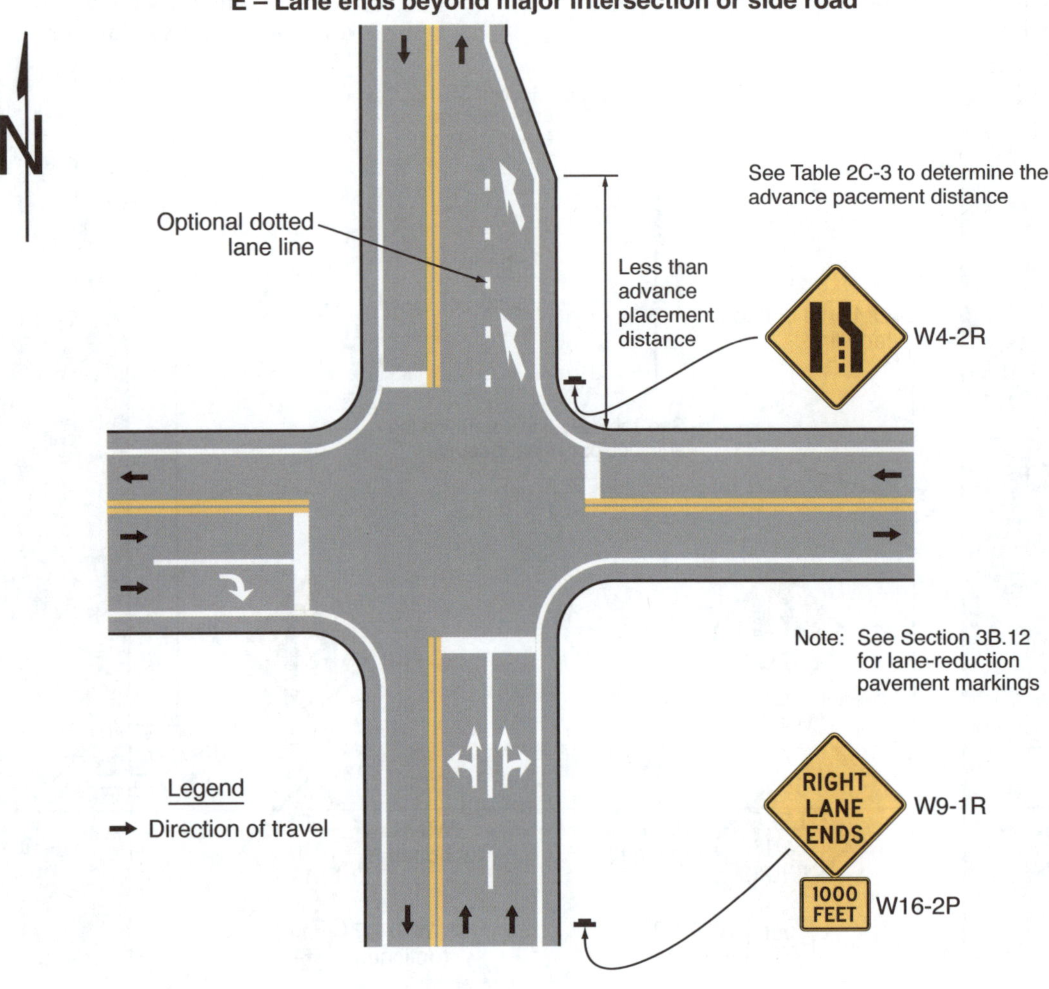

Section 2C.49 HEAVY MERGE FROM LEFT (RIGHT) Sign (W4-7)

Option:

01   The HEAVY MERGE FROM LEFT (RIGHT) (W4-7) sign (see Figure 2C-11) may be used to supplement a W4-1 sign at multilane approaches to congested areas to inform road users that it is desirable for through traffic to move out of a lane that will be occupied by a high volume of entering traffic. If used, the W4-7 sign may be supplemented with a W16-2P series or W16-3P series plaque (see Section 2C.61).

Standard:

02   **If used, the W4-7 sign shall be installed at a location upstream from the location of the W4-1 sign.**

Section 2C.50 RIGHT (LEFT) LANE FOR EXIT ONLY Sign (W9-7)

Option:

01   The RIGHT (LEFT) LANE FOR EXIT ONLY (W9-7) sign (see Figure 2C-11) may be used to provide advance warning to road users that traffic in the right-hand (left-hand) lane of a roadway will be required to depart the roadway at the next exit.

Guidance:

02   *If used, the W9-7 sign should be installed upstream from the first overhead guide sign that contains an EXIT ONLY sign panel or upstream from the first RIGHT (LEFT) LANE MUST EXIT (R3-33) regulatory sign, if used, whichever is farther upstream from the exit.*

Option:

03   A legend or plaque displaying the distance may be added to the W9-7 sign where the distance along the dropped lane between the sign and the exit ramp is greater than 1 mile.

Figure 2C-13. Example Sequences for Lane Ends and Lane Merge Signs (Sheet 5 of 5)

**F – Reduction of two lanes to one lane in the same direction
of travel with merging maneuvers for each lane**

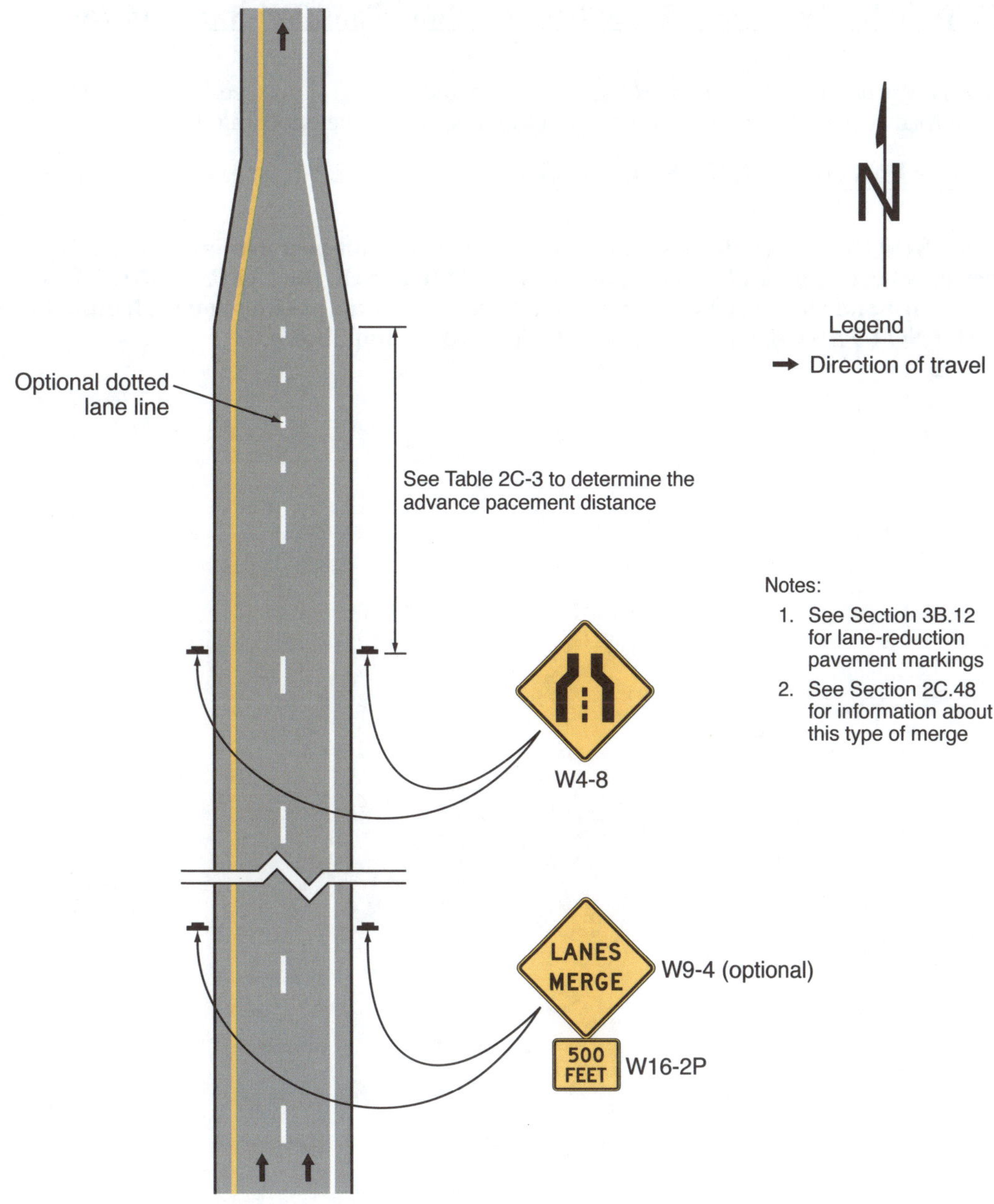

Support:

04   Section 2B.31 contains information regarding a regulatory sign that can be used for lane drops at grade-separated interchanges.

Section 2C.51 Two-Way Traffic Sign (W6-3)

Guidance:

01   *A Two-Way Traffic (W6-3) sign (see Figure 2C-11) should be used to warn road users of a transition from a multi-lane divided section of roadway to a two-lane, two-way section of roadway.*

02   *A Two-Way Traffic (W6-3) sign with an AHEAD (W16-9P) plaque (see Figure 2C-16) should be used to warn road users of a transition from a one-way street to a two-lane, two-way section of roadway (see Figure 2B-18).*

Option:

03   The Two-Way Traffic sign may be used at intervals along a two-lane, two-way roadway and may be used to supplement the Divided Highway (Road) Ends (W6-2) sign discussed in Section 2C.21.

Support:

04   Section 6H.17 contains information on a Narrow Two-Way Traffic (W6-4) sign for use in temporary traffic control situations.

Section 2C.52 Two-Way Traffic on a Three-Lane Roadway Signs (W6-5 and W6-5a)

Option:

01   The Two-Way Traffic on a Three-Lane Roadway (W6-5 and W6-5a) signs (see Figure 2C-11) may be installed along three-lane roadways with two lanes in one direction and one in the opposing direction.

Section 2C.53 NO PASSING ZONE Sign (W14-3)

Standard:

01   **The NO PASSING ZONE (W14-3) sign (see Figure 2C-11) shall be a pennant-shaped isosceles triangle with its longer axis horizontal and pointing to the right. When used, the NO PASSING ZONE sign shall be installed on the left-hand side of the roadway at the beginning of no-passing zones identified by pavement markings or DO NOT PASS signs or both (see Sections 2B.36 and 3B.03).**

MISCELLANEOUS WARNING SIGNS AND PLAQUES

Section 2C.54 Vehicular Traffic Warning Signs (W8-6, W11-1, W11-5, W11-8, W11-10, W11-11, W11-12P, W11-14, W11-15, and W11-15a)

Option:

01   Vehicular Traffic Warning (W8-6, W11-1, W11-5, W11-8, W11-10, W11-11, W11-12P, W11-14, W11-15, and W11-15a) signs (see Figure 2C-14) may be used to alert road users to locations where unexpected entries into the roadway by trucks, bicycles, farm vehicles, emergency vehicles, golf carts, horse-drawn vehicles, or other vehicles might occur. The TRUCK CROSSING (W8-6) word message sign may be used as an alternate to the Truck (W11-10) symbol sign.

Support:

02   These locations might be relatively confined or might occur randomly over a segment of roadway.

Guidance:

03   *Vehicular Traffic Warning signs should be used only at locations where the road user's sight distance is restricted, or the condition, activity, or entering traffic would be unexpected.*

04   *If the condition or activity is seasonal or temporary, the Vehicular Traffic Warning sign should be removed or covered when the condition or activity does not exist.*

Option:

05   The Trail Crossing (W11-15) sign may be used where both bicyclists and pedestrians might be crossing the roadway, such as at an intersection with a shared-use path. A TRAIL X-ING (W11-15P) supplemental plaque (see Figure 2C-14) may be mounted below the W11-15 sign. The TRAIL CROSSING (W11-15a) sign may be used to warn of shared-use path crossings where pedestrians, bicyclists, and other user groups might be crossing the roadway.

06   The W11-1, W11-15, and W11-15a signs and their related supplemental plaques may have a fluorescent yellow-green background with a black legend and border.

07   Supplemental plaques (see Figure 2C-16 and Section 2C.57) with legends such as AHEAD, XX FEET, NEXT XX MILES, IN STREET, or IN ROAD may be mounted below Vehicular Traffic Warning signs to provide advance notice to road users of unexpected entries.

Guidance:

08   *If used in advance of a trail crossing, a W11-15 or W11-15a sign should be supplemented with an AHEAD or XX FEET plaque to inform road users that they are approaching a point where crossing activity might occur.*

Figure 2C-14. Vehicular Traffic Warning Signs and Plaques

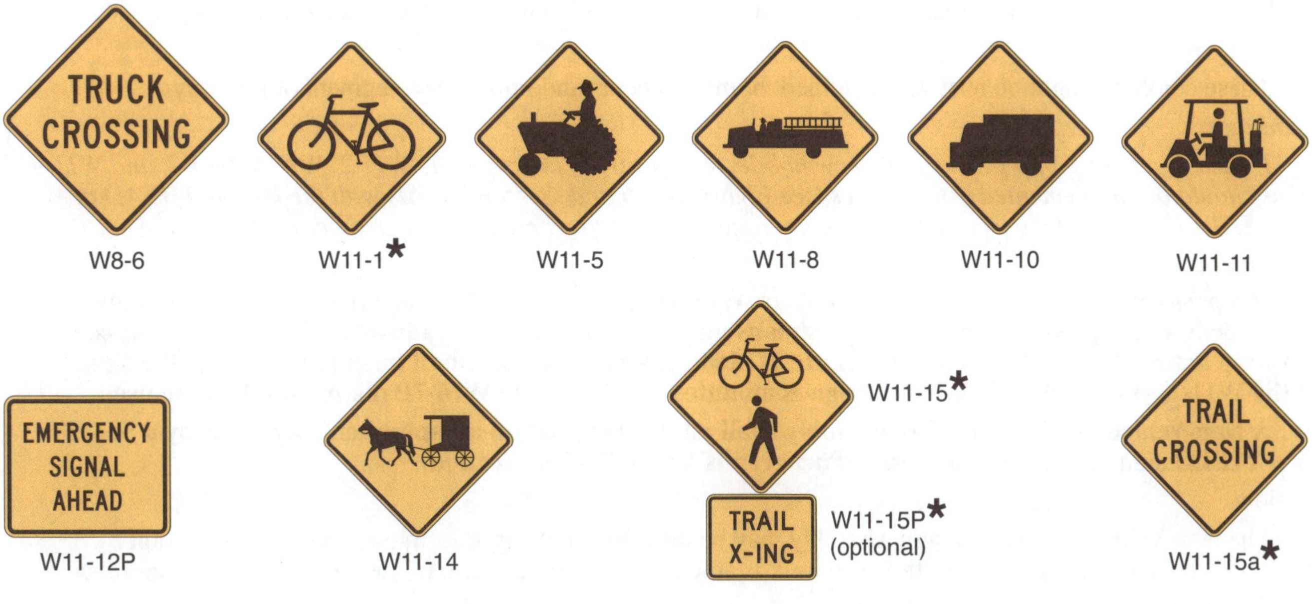

* A fluorescent yellow-green background color may be used for this sign or plaque.

Standard:

09   **If a post-mounted W11-1, W11-11, W11-15, or W11-15a sign is placed at the location of the crossing point where golf carts, pedestrians, bicyclists, or other shared-use path users might be crossing the roadway, a diagonal downward-pointing arrow (W16-7P) plaque (see Figure 2C-16 and Section 2C.63) shall be mounted below the sign. If the W11-1, W11-11, W11-15, or W11-15a sign is mounted overhead, the W16-7P supplemental plaque shall not be used.**

10   **A Vehicular Traffic Warning sign assembly shall not be installed on an approach controlled by a STOP or a YIELD sign, except as provided in Paragraphs 11 and 12 of this Section.**

Option:

11   The Vehicular Traffic Warning sign assembly may be installed on an approach to a circular intersection controlled by a YIELD sign where the crosswalk is at least 20 feet in advance of the yield point at the entrance to the circulatory roadway.

12   At a signalized or stop-controlled intersection the Vehicular Traffic Warning sign assembly may be installed on an approach to a channelized right turn lane controlled by a YIELD sign where the crosswalk is at least 20 feet in advance of the yield point.

13   The crossing location identified by a W11-1, W11-11, W11-15, or W11-15a sign may be defined with crosswalk markings (see Chapter 3C).

Standard:

14   **The Emergency Vehicle (W11-8) sign (see Figure 2C-14) with the EMERGENCY SIGNAL AHEAD (W11-12P) supplemental plaque (see Figure 2C-14) shall be placed in advance of all emergency-vehicle traffic control signals (see Chapter 4M).**

Option:

15   The Emergency Vehicle (W11-8) sign, or a word message sign indicating the type of emergency vehicle (such as rescue squad), may be used in advance of the emergency-vehicle station when no emergency-vehicle traffic control signal is present.

16   A Warning Beacon (see Section 4S.03) may be used with any Vehicular Traffic Warning sign to indicate specific periods when the condition or activity is present or is likely to be present, or to provide enhanced sign conspicuity.

17   A supplemental WHEN FLASHING (W16-13P) plaque (see Figure 2C-16) may be used with any Vehicular Traffic Warning sign that is supplemented with a Warning Beacon to indicate specific periods when the condition or activity is present or is likely to be present.

Section 2C.55 Non-Vehicular Warning Signs (W11-2, W11-3, W11-4, W11-6, W11-7, W11-9, and W11-16 through W11-22)

Option:

01   Non-Vehicular Warning (W11-2, W11-3, W11-4, W11-6, W11-7, W11-9, and W11-16 through W11-22) signs (see Figure 2C-15) may be used to alert road users in advance of locations where unexpected entries into the roadway might occur or where shared use of the roadway by pedestrians, animals, or equestrians might occur.

Support:

02   These conflicts might be relatively confined, or might occur randomly over a segment of roadway.

Guidance:

03   *If used in advance of a pedestrian, snowmobile, or equestrian crossing, the W11-2, W11-6, W11-7, and W11-9 signs should be supplemented with plaques (see Figure 2C-16 and Section 2C.61) with the legend AHEAD or XX FEET to inform road users that they are approaching a point where crossing activity might occur.*

Standard:

04   **If a post-mounted W11-2, W11-6, W11-7, or W11-9 sign is placed at the location of the crossing point where pedestrians, snowmobilers, or equestrians might be crossing the roadway, a diagonal downward-pointing arrow (W16-7P) plaque (see Figure 2C-16 and Section 2C.63) shall be mounted below the sign. If the W11-2, W11-6, W11-7, or W11-9 sign is mounted overhead, the W16-7P plaque shall not be used.**

05   **A Non-Vehicular Warning sign assembly shall not be installed on an approach controlled by a STOP or a YIELD sign, except as provided in Paragraphs 6 and 7 of this Section.**

Option:

06   The Non-Vehicular Warning sign assembly may be installed on an approach to a circular intersection controlled by a YIELD sign where the crosswalk is at least 20 feet in advance of the yield point at the entrance to a circulatory roadway.

Figure 2C-15. Non-Vehicular Warning Signs

** A fluorescent yellow-green background color may be used for this sign or plaque.*

07   At a signalized or stop-controlled intersection the Non-Vehicular Warning sign assembly may be installed on an approach to a channelized right turn lane controlled by a YIELD sign where the crosswalk is at least 20 feet in advance of the yield point.

08   A Pedestrian Crossing (W11-2) sign may be placed overhead or may be post-mounted with a diagonal downward-pointing arrow (W16-7P) plaque at the crosswalk location where Yield Here To (Stop Here For) Pedestrians signs (see Section 2B.19) have been installed in advance of the crosswalk.

Standard:

09   **If a W11-2 sign has been post-mounted at the crosswalk location where a Yield Here To (Stop Here For) Pedestrians sign is used on the approach, the Yield Here To (Stop Here For) Pedestrians sign shall not be placed on the same post as the W11-2 sign.**

Option:

10   An advance Pedestrian Crossing (W11-2) sign with an AHEAD or a distance supplemental plaque may be used in conjunction with a Yield Here To (Stop Here For) Pedestrians sign on the approach to the same crosswalk.

11   The crossing location identified by a W11-2, W11-6, W11-7, or W11-9 sign may be defined with crosswalk markings (see Chapter 3C).

12   The W11-2 and W11-9 signs and their related supplemental plaques may have a fluorescent yellow-green background with a black legend and border.

Guidance:

13   *When a fluorescent yellow-green background is used, a systematic approach featuring one background color within a zone or area should be used. The mixing of standard yellow and fluorescent yellow-green backgrounds within a selected site area should be avoided.*

Option:

14   A Warning Beacon (see Section 4S.03) may be used with any Non-Vehicular Warning sign to indicate specific periods when the condition or activity is present or is likely to be present, or to provide enhanced sign conspicuity.

15   A supplemental WHEN FLASHING (W16-13P) plaque (see Figure 2C-16) may be used with any Non-Vehicular Warning sign that is supplemented with a Warning Beacon to indicate specific periods when the condition or activity is present or is likely to be present.

Section 2C.56 <u>Playground Sign (W15-1)</u>

Option:

01   The Playground (W15-1) sign (see Figure 2C-15) may be used to give advance warning of a designated children's playground that is located adjacent to the road.

02   The Playground sign may have a fluorescent yellow-green background with a black legend and border.

Guidance:

03   *If the access to the playground area requires a roadway crossing, the application of crosswalk pavement markings (see Chapter 3C) and a Non-Vehicular Warning sign (see Section 2C.55) should be considered.*

SUPPLEMENTAL WARNING PLAQUES

Section 2C.57 Use of Supplemental Warning Plaques

Option:

01   A supplemental warning plaque (see Figure 2C-16) may be displayed with a warning or regulatory sign when engineering judgment indicates that road users require additional warning information beyond that contained in the main message of the warning or regulatory sign.

Standard:

02   **Supplemental warning plaques shall be used only in combination with and installed on the same post(s) as warning or regulatory signs. They shall not be mounted alone or displayed alone.**

03   **Unless otherwise provided in this Manual for a particular plaque, supplemental warning plaques shall be mounted below the sign they supplement.**

Section 2C.58 Design of Supplemental Warning Plaques

Standard:

01   **A supplemental warning plaque used with a warning sign shall have the same legend, border, and background color as the warning sign with which it is displayed. A supplemental warning plaque used with a regulatory sign shall have a black legend and border on a yellow background.**

02   **Supplemental warning plaques shall be square or rectangular.**

Figure 2C-16. Supplemental Warning Plaques

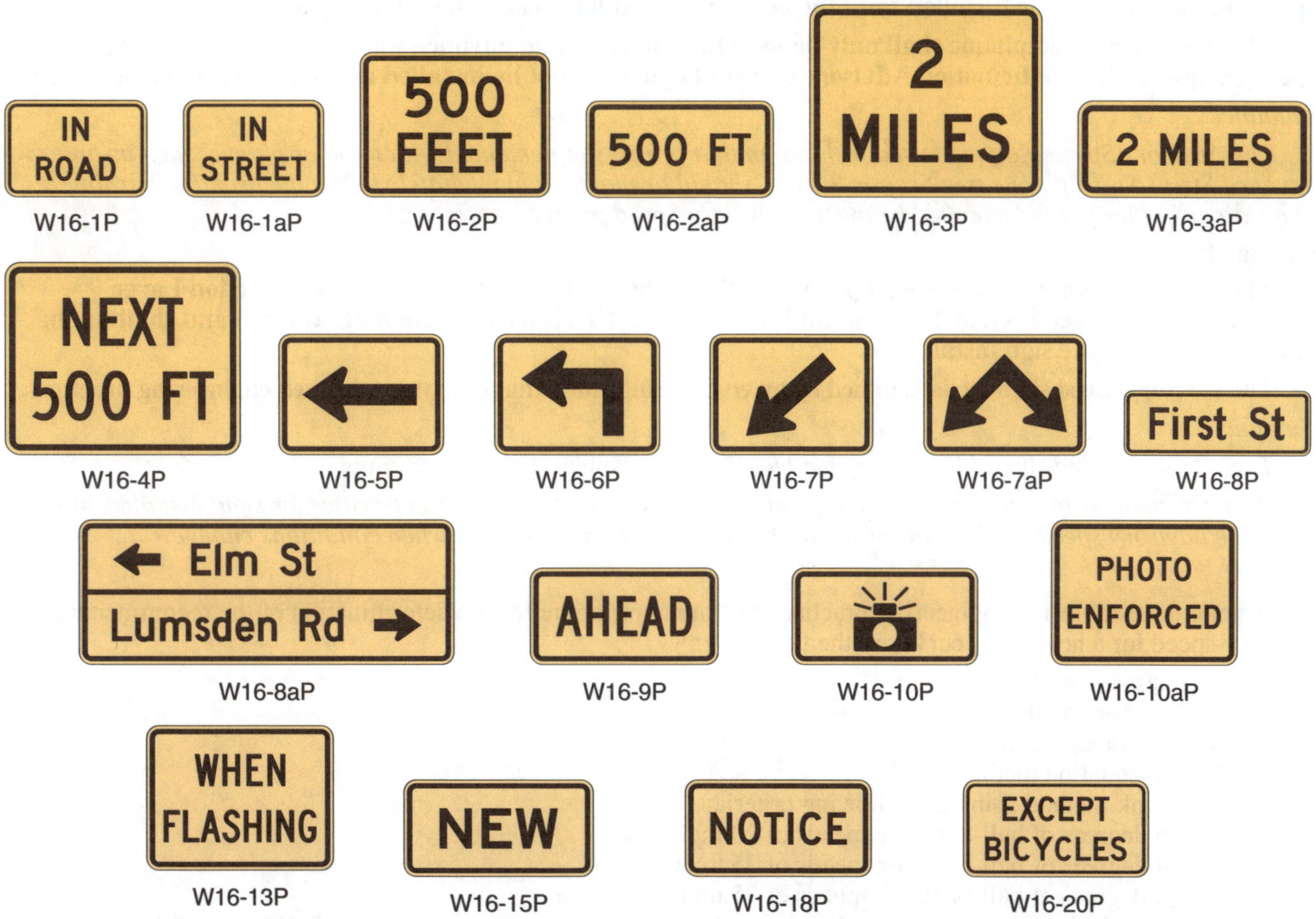

Note: The background color (yellow or fluorescent yellow-green) shall match the color of the warning sign that it supplements.

Section 2C.59 Advisory Speed Plaque (W13-1P) and Confirmation Advisory Speed Plaque (W13-1aP)

Option:

01   The Advisory Speed (W13-1P) plaque (see Figure 2C-1) may be used to supplement an advance warning sign to indicate the advisory speed for a condition.

02   The Confirmation Advisory Speed (W13-1aP) plaque (see Figure 2C-1) may be used to supplement a One-Direction Large Arrow (W1-6) sign on the outside of a turn or curve in line with and at approximately a right angle to approaching traffic.

Standard:

03   **The use of the Advisory Speed and Confirmation Advisory Speed plaques for horizontal curves shall be in accordance with Section 2C.06 and Table 2C-6. The speed differential in Table 2C-6 shall be the difference between the advisory speed for the horizontal curve and the posted speed limit, statutory speed limit, or the 85th percentile speed on the approach to the curve. The Advisory Speed plaque shall also be used where an engineering study indicates a need to advise road users of the advisory speed for other roadway conditions.**

Table 2C-6. Use of Advisory Speed Plaque for Horizontal Alignment Changes

Speed Differential	Use of Advisory Speed Plaque (W13-1P)[1]
5 mph	Optional
10 mph	Recommended
15 mph or more	Required

[1] See Section 2C.59

04   **The speed displayed on the Advisory Speed and Confirmation Advisory Speed plaques shall be a multiple of 5 mph.**

05   **Except in emergencies or when the condition is temporary, an Advisory Speed or Confirmation Advisory Speed plaque shall not be installed until the advisory speed has been determined by an engineering study.**

06   **The Advisory Speed plaque shall only be used to supplement an advance warning sign. The Advisory Speed plaque or the Confirmation Advisory Speed plaque shall not be installed as a separate sign installation.**

Guidance:

07   *The Advisory Speed plaque, if used with a sign that is also supplemented with another plaque, such as an Advance Street Name plaque (see Section 2C.65), should be mounted immediately below the primary warning sign with any other plaque mounted below the Advisory Speed plaque.*

Standard:

08   **The Confirmation Advisory Speed plaque shall only be used to supplement a One-Direction Large Arrow (W1-6) sign (see Section 2C.10) or an Exit Gore (E5-1 series) (see Section 2E.26) sign and shall not be installed as a separate sign installation.**

09   **The advisory speed shall be determined by an engineering study that follows established engineering practices.**

Guidance:

10   *The advisory speed should be determined based on free-flowing traffic conditions.*

11   *Because changes in conditions, such as roadway geometrics, surface characteristics, or sight distance, might affect the advisory speed, each location should be evaluated periodically or when conditions change.*

Support:

12   Among the established engineering practices that are appropriate for the determination of the recommended advisory speed for a horizontal curve are the following:

    A. Compass method
    B. Safety-based method
    C. Accelerometer method
    D. Design equation method
    E. Ball-bank method using the following criteria:
        1.   16 degrees of ball-bank for speeds of 20 mph or less
        2.   14 degrees of ball-bank for speeds of 25 to 30 mph
        3.   12 degrees of ball-bank for speeds of 35 mph and higher

13   The 16, 14, and 12 degrees of ball-bank criteria are comparable to the current AASHTO horizontal curve design guidance. Research has shown that drivers often exceed existing posted advisory curve speeds by 7 to 10 mph.

Section 2C.60 NEW Plaque (W16-15P)

Option:

01   A NEW (W16-15P) plaque (see Figure 2C-16) may be mounted above a regulatory sign when a new regulation takes effect in order to alert road users to the new traffic regulation. A NEW plaque may also be mounted above an advance warning sign (such as a Signal Ahead sign for a newly-installed traffic control signal) for a warning of a new traffic condition.

Guidance:

02   *The NEW plaque should be removed no later than 6 months after it was installed.*

Section 2C.61 Distance Plaques (W16-2 Series, W16-3 Series, W16-4P, and W7-3aP)

Option:

01   The Distance Ahead (W16-2 series and W16-3 series) plaques (see Figure 2C-16) may be used to inform the road user of the distance to the condition indicated by the warning sign.

02   The Next Distance (W7-3aP and W16-4P) plaques (see Figures 2C-5 and 2C-16) may be used to inform road users of the length of roadway over which the condition indicated by the warning sign exists.

Section 2C.62 Supplemental Arrow Plaques (W16-5P and W16-6P)

Guidance:

01   *If the condition indicated by a warning sign is located on an intersecting road and the distance between the intersection and condition is not sufficient to provide adequate advance placement of the warning sign, a Supplemental Arrow (W16-5P or W16-6P) plaque (see Figure 2C-16) should be used below the warning sign.*

Standard:

02   **Supplemental Arrow plaques shall have the same legend design as the Advance Turn Arrow and Directional Arrow auxiliary signs (see Sections 2D.26 and 2D.28) except that they shall have a black legend and border on a yellow or fluorescent yellow-green background, as appropriate.**

Section 2C.63 Diagonal Downward-Pointing Arrow Plaques (W16-7P and W16-7aP)

Support:

01   Diagonal downward-pointing arrow (W16-7P and W16-7aP) plaques (see Figure 2C-16) are used with certain Vehicular Traffic Warning signs (see Section 2C.54) and certain Non-Vehicular Warning signs (see Section 2C.55), and School Crossing signs (see Section 7B.03) to indicate the specific location of a crossing point.

02   The W16-7P plaque contains a single arrow pointing diagonally down to the right or left, toward the roadway, depending on which side of the roadway it is located.

Option:

03   A W16-7aP plaque may be used with a single crossing sign located on a narrow median separating two roadways with traffic in the same direction where the crossing traverses both roadways.

Section 2C.64 Hill-Related Plaques (W7-2 Series and W7-3 Series)

Guidance:

01   *Hill-Related (W7-2 series and W7-3 series) plaques (see Figure 2C-5) or other appropriate legends and larger signs should be used for emphasis or where special hill characteristics exist.*

02   *On longer grades, the use of a distance (W7-3aP or W7-3bP) plaque (see Figure 2C-5) at periodic intervals of approximately 1-mile spacing should be considered.*

Section 2C.65 Advance Street Name Plaques (W16-8P and W16-8aP)

Option:

01   An Advance Street Name (W16-8P or W16-8aP) plaque (see Figure 2C-16) may be used with any Intersection (W1-10 series, W2 series, W10-2, W10-3, or W10-4) or Advance Traffic Control (W3 series) sign to identify the name of the intersecting street.

Standard:

02   **The lettering on Advance Street Name plaques shall be composed of a combination of lower-case letters with initial upper-case letters.**

03   **If two street names are used on the Advance Street Name plaque, a directional arrow pointing in the direction of the street shall be placed next to each street name. Arrows pointing to the left shall be placed to the left of the street name, and arrows pointing to the right shall be placed to the right of the street name.**

Guidance:

04   *If two street names are used on the Advance Street Name plaque, the street names and associated arrows should be displayed in the following order:*

    A. *For a single intersection, the name of the street to the left should be displayed above the name of the street to the right; or*

    B. *For two sequential intersections, such as where the plaque is used with an Offset Side Roads (W2-7) or a Double Side Road (W2-8) sign, the name of the first street encountered should be displayed above the name of the second street encountered, and the arrow associated with the second street encountered should be an advance arrow, such as the arrow shown on the W16-6P arrow plaque (see Figure 2C-16).*

Section 2C.66 Traffic Does Not Stop Plaques (W4-4P Series)

Option:

01   The CROSS TRAFFIC DOES NOT STOP (W4-4P) plaque (see Figure 2C-10) may be used in combination with a STOP sign when engineering judgment indicates that conditions are present that are causing or could cause road users to misinterpret the intersection as an all-way stop.

02   The TRAFFIC FROM LEFT (RIGHT) DOES NOT STOP (W4-4aP) or ONCOMING TRAFFIC DOES NOT STOP (W4-4bP) plaque may be used when such messages more accurately describe the traffic controls established at the intersection.

Guidance:

03   *The W4-4aP and W4-4bP plaques should be used at intersections where STOP signs control all but one approach to the intersection, unless the only non-stopped approach is from a one-way street.*

Standard:

04   **If a W4-4P series plaque is used, it shall be mounted below the STOP sign.**

Support:

05   Section 9C.06 contains information for Bicycle Cross Traffic warning plaques that can be used below STOP signs on crossroads or driveways that intersect with bicycle facilities.

Section 2C.67 IN ROAD and IN STREET Plaques (W16-1P and W16-1aP)

Option:

01   In situations where there is a need to warn drivers to watch for other slower forms of transportation traveling along the highway, such as bicycles, pedestrians, golf carts, horse-drawn vehicles, or farm machinery, an IN ROAD (W16-1P) plaque or IN STREET (W16-1aP) plaque (see Figure 2C-16) may be used.

Standard:

02   **The background color of the W16-1P or W16-1aP plaque shall match the background color of the warning sign with which it is displayed. If a W16-1P or W16-1aP plaque is used, it shall be mounted below either a Vehicular Traffic Warning sign (see Section 2C.54) or a Non-Vehicular Warning sign (see Section 2C.55), and shall not be mounted alone.**

Support:

03   Section 9B.14 contains information about the use of a Bicycles Allowed Use of Full Lane (R9-20) sign to inform drivers of the presence of bicycles in the roadway or where bicyclists are expected or preferred to use the full lane.

Section 2C.68 EXCEPT BICYCLES Plaque (W16-20P)

Option:

01   Where it is desired to notify bicyclists that the conditions depicted by a warning sign are not applicable to bicycles, the EXCEPT BICYCLES (W16-20P) supplemental warning plaque (see Figure 2C-16) may be mounted below the warning sign.

Support:

02   Examples of warning signs with which an EXCEPT BICYCLES (W16-20P) plaque can be mounted include DEAD END (W14-1) or NO OUTLET (W14-2) signs.

Section 2C.69 Photo Enforced Plaques (W16-10P and W16-10aP)

Option:

01   A Photo Enforced (W16-10P) plaque or a PHOTO ENFORCED (W16-10aP) word message plaque (see Figure 2C-16) may be mounted below a warning sign to advise road users that the regulations associated with the condition being warned about (such as a traffic control signal or a toll plaza) are being enforced by photographic equipment.

OBJECT MARKERS

Section 2C.70 Object Marker Design and Placement Height

Support:

01 Types 1, 2, and 3 object markers are used to mark obstructions within or adjacent to the roadway. Type 4 object markers are used to mark the end of a roadway.

Standard:

02 **When used, object markers (see Figure 2C-17) shall not have a border and shall consist of an arrangement of one or more of the following types:**

Type 1—a diamond-shaped sign, at least 18 inches on a side, consisting of either a yellow (OM1-1) or black (OM1-2) sign with nine yellow retroreflective devices, each with a minimum diameter of 3 inches, mounted symmetrically on the sign, or an all-yellow retroreflective sign (OM1-3).

Type 2—either a marker (OM2-1V or OM2-1H) consisting of three yellow retroreflective devices, each with a minimum diameter of 3 inches, arranged either horizontally or vertically on a white sign measuring at least 6 x 12 inches; or an all-yellow horizontal or vertical retroreflective sign (OM2-2V or OM2-2H), measuring at least 6 x 12 inches.

Type 3—a striped marker, 12 x 36 inches, consisting of a vertical rectangle with alternating black and retroreflective yellow stripes sloping downward at an angle of 45 degrees toward the side of the obstruction on which traffic is to pass. The minimum width of the yellow and black stripes shall be 3 inches.

Type 4—a diamond-shaped sign, at least 18 inches on a side, consisting of either a red (OM4-1) or black (OM4-2) sign with nine red retroreflective devices, each with a minimum diameter of 3 inches, mounted symmetrically on the sign, or an all-red retroreflective sign (OM4-3).

Support:

03 Type 3 object markers with stripes that begin at the upper right side and slope downward to the lower left side are designated as right object markers (OM3-R). Object markers with stripes that begin at the upper left side and slope downward to the lower right side are designated as left object markers (OM3-L). Object markers with chevron stripes that slope downward to both the lower left and lower right sides are designated as center object markers (OM3-C).

Guidance:

04 *When used for marking obstructions within the roadway or obstructions that are 8 feet or less from the shoulder or curb, the minimum mounting height, measured from the bottom of the object marker to the elevation of the near edge of the traveled way, should be 4 feet.*

05 *When used to mark obstructions more than 8 feet from the shoulder or curb, the clearance from the ground to the bottom of the object marker should be at least 4 feet.*

06 *Object markers should not present a vertical or horizontal clearance obstacle for pedestrians.*

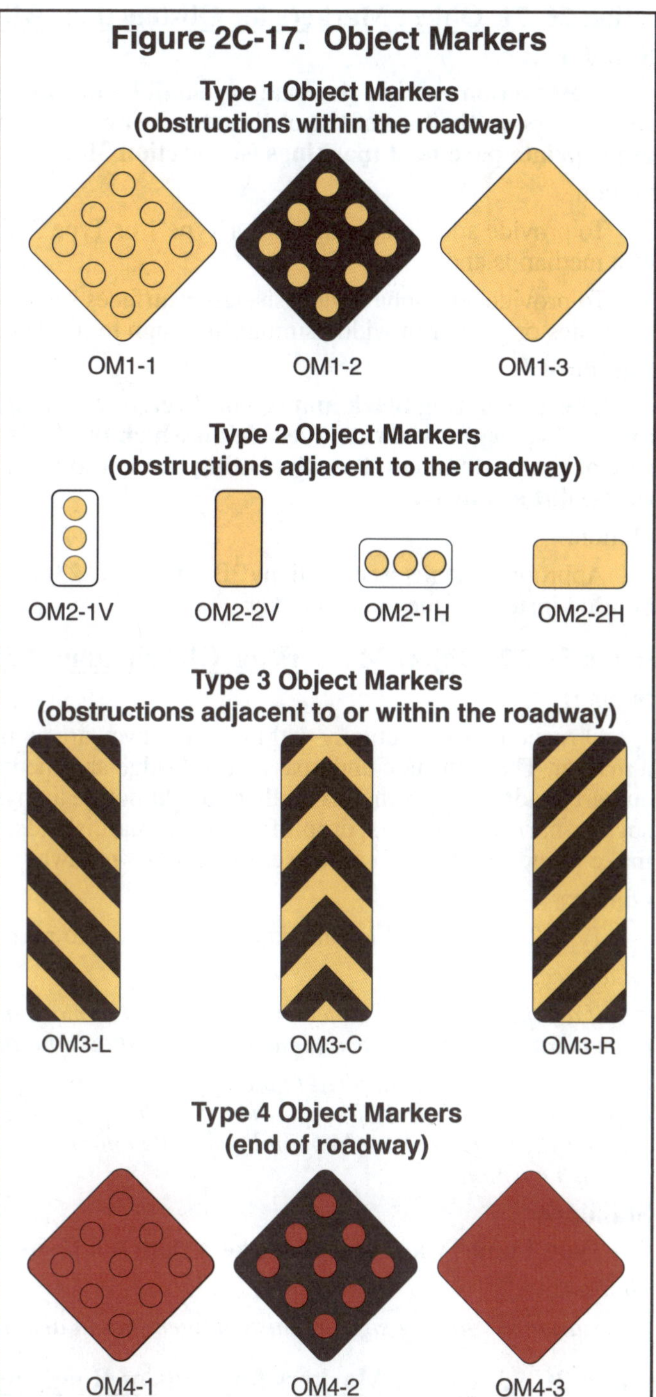

Figure 2C-17. Object Markers

Option:

07   When object markers or markings are applied to an obstruction that by its nature requires a lower or higher mounting, the vertical mounting height may vary according to need.

Support:

08   Section 9C.09 contains information regarding the use of object markers on shared-use paths.

Section 2C.71 Object Markers for Obstructions within the Roadway

Standard:

01   **Obstructions within the roadway shall be marked with a Type 1 or Type 3 object marker. In addition to markers on the face of the obstruction, warning of approach to the obstruction shall be given by appropriate pavement markings (see Section 3B.13).**

Option:

02   To provide additional emphasis, a Type 1 or Type 3 object marker may be installed at or near the approach end of a median island.

03   To provide additional emphasis, large surfaces such as bridge piers may be painted with diagonal stripes, 12 inches or greater in width, similar in design to the Type 3 object marker.

Standard:

04   **The alternating black and retroreflective yellow stripes (OM3-L, OM3-R) shall be sloped down at an angle of 45 degrees toward the side on which traffic is to pass the obstruction. If traffic can pass to either side of the obstruction, the alternating black and retroreflective yellow stripes (OM3-C) shall form chevrons that point upwards.**

Option:

05   Appropriate signs (see Sections 2B.40 and 2C.23) directing traffic to one or both sides of the obstruction may be used instead of the object marker.

Section 2C.72 Object Markers for Obstructions Adjacent to the Roadway

Support:

01   Obstructions not actually within the roadway are sometimes so close to the edge of the road that they need a marker. These include underpass piers, bridge abutments, handrails, ends of traffic barriers, utility poles, and culvert headwalls. In other cases there might not be a physical object involved, but other roadside conditions exist, such as narrow shoulders, drop-offs, gores, small islands, and abrupt changes in the roadway alignment, that might make it undesirable for a road user to leave the roadway, and therefore would create a need for a marker.

Option:

02   Type 2 or Type 3 object markers may be used to mark an obstruction adjacent to the roadway.

Guidance:

03   *If a Type 2 or Type 3 object marker is used to mark an obstruction adjacent to the roadway, the edge of the object marker that is closest to the road user should be installed in line with the closest edge of the obstruction.*

04   *When a marker is applied to the approach ends of guardrail or crash cushion terminals it should have the appearance of a Type 3 object marker and should be directly affixed, without a substrate, to the approach end of the guardrail or crash cushion and generally conform to the size and shape of the approach end of the guardrail or crash cushion.*

Standard:

05   **Type 1 and Type 4 object markers shall not be used to mark obstructions adjacent to the roadway.**

Guidance:

06   *Standard warning signs in this Chapter should also be used where applicable.*

Section 2C.73 Object Markers for Ends of Roadways

Support:

01   The Type 4 object marker is used to warn and alert road users of the end of a roadway in other than construction or maintenance areas.

Standard:

02   **If an object marker is used to mark the end of a roadway, a Type 4 object marker shall be used.**

Option:

03   The Type 4 object marker may be used in instances where there are no alternate vehicular paths.

04   Where conditions warrant, more than one marker, or a larger marker with or without a Type 3 Barricade (see Section 2B.75), may be used at the end of the roadway.

Standard:

05   **The minimum mounting height, measured vertically from the bottom of a Type 4 object marker to the elevation of the near edge of the traveled way, shall be 4 feet.**

Guidance:

06   *Appropriate advance warning signs in this Chapter should be used.*

(This page intentionally left blank)

CHAPTER 2D. GUIDE SIGNS—CONVENTIONAL ROADS

Chapter 2D Subchapter and Section Organization

GENERAL DESIGN

Section 2D.01 Scope of Conventional Road Guide Sign Standards and Application
Standard:

01   **The provisions of this Chapter shall apply to any road or street other than expressways and freeways, except as otherwise provided in this Manual.**

Support:

02   Guide signs direct road users along streets and highways; inform them of intersecting routes; direct them to cities, towns, villages, or other important destinations; identify nearby rivers and streams, parks, forests, and historical sites; and provide information that will help them along their way in the most simple and direct manner possible.

Guidance:

03   *The selection of primary or control destinations (those displayed consistently over longer distances along a route) displayed on guide signs should be meaningful to road users in navigation and orientation. The destinations selected should be identifiable on official maps.*

04   *The familiarity of the road users with the road should be considered in determining the need for guide signs on low-volume roads.*

Support:

05   Low-volume roads generally do not require guide signs to the extent that they are needed on higher classes of roads. Because guide signs are typically only beneficial as a navigational aid for road users who are unfamiliar with a low-volume road, guide signs might not be needed on low-volume roads that serve only local traffic.

06   Guide signs, other than Street Name signs, generally are not used on low-volume rural roads except as needed to guide road users back to the major roadways.

Guidance:

07   *If used on low-volume roads, destination names should be as specific and descriptive as possible. Destinations such as campgrounds, ranger stations, recreational areas, and the like should be clearly indicated so that they are not interpreted to be communities or locations with road user services.*

Option:

08   Guide signs may be used on low-volume roads at intersections to provide information for road users returning to a higher class of roads.

Support:

09   Chapter 2A addresses placement, location, and other general criteria for signs.

Section 2D.02 Color, Retroreflection, and Illumination
Support:

01   Requirements for illumination, retroreflection, and color are stated under the specific headings for individual guide signs or groups of signs. General provisions are given in 2A.06, 2A.21, and 2A.22.

Standard:

02   **Except as otherwise provided in this Manual for individual signs or groups of signs, guide signs on streets and highways shall have a white message and border on a green background. All messages, borders, and legends shall be retroreflective and all backgrounds shall be retroreflective or illuminated.**

Support:

03   Color coding is sometimes used to help road users distinguish between multiple potentially confusing destinations. Examples of valuable uses of color coding include guide signs for roadways approaching or inside an airport property with multiple terminals serving multiple airlines, and community wayfinding guide signs for various traffic generator destinations within a community or area.

Standard:

04   **Except as otherwise provided in this Manual, different color sign backgrounds shall not be used to provide color coding of destinations. The color coding shall be accomplished by the use of different colored square or rectangular sign panels on the face of the guide signs (see Figure 2D-1).**

Option:

05   The different colored sign panels on the face of a sign may include a black or white (whichever provides the better contrast with the panel color) letter, numeral, or other appropriate designation to identify an airport terminal or other destination.

Support:

06   Section 2D.55 contains specific provisions regarding Community Wayfinding guide signs.

Figure 2D-1. Examples of Color-Coded Destination Guide Signs

Section 2D.03 <u>Size of Signs</u>

Standard:

01　**Except as provided in Section 2A.07, the minimum sizes of conventional road guide signs that have standardized designs shall be as shown in Table 2D-1.**

Support:

02　Section 2A.07 contains information regarding the applicability of the various columns in Table 2D-1.

Option:

03　Signs larger than those shown in Table 2D-1 may be used (see Section 2A.07).

Support:

04　For other guide signs, the legends are so variable that a standardized design or size is not appropriate. The sign size is determined primarily by the length of the message, and the size of lettering and spacing necessary for proper legibility.

Option:

05　Reduced letter height, reduced interline spacing, and reduced edge spacing may be used on guide signs if sign size must be limited by factors such as lane width or vertical or lateral clearance.

Guidance:

06　Reduced spacing between the letters or words on a line of legend should not be used as a means of reducing the overall size of a guide sign, except where determined necessary by engineering judgment to meet unusual lateral-space constraints. In such cases, the legibility distance of the sign legend should be the primary consideration in determining whether to reduce the spacing between the letters or the words or between the words and the sign border, or to reduce the letter height.

07　When a reduction in the prescribed size is necessary, the design used should be as similar as possible to the design for the standard size.

Section 2D.04 <u>Lettering Style</u>

Standard:

01　**The design of upper-case letters, lower-case letters, numerals, route shields, and spacing shall be as provided in the "Standard Highway Signs" publication (see Section 1A.05).**

02　**The lettering for names of places, streets, and highways on conventional road guide signs shall be a combination of lower-case letters with initial upper-case letters (see Section 2A.08). The nominal loop height of the lower-case letters shall be ¾ the height of the initial upper-case letter. When a mixed-case legend letter height is specified referring only to the initial upper-case letter, the height of the lower-case letters that follow shall be determined by this proportion. When the height of a lower-case letter is referenced, the reference is made to the nominal loop height. The height of the initial upper-case letter shall also be determined by this proportion.**

03　**All other word legends on conventional road guide signs shall be in upper-case letters.**

04　**The unique letter forms for each of the Standard Alphabet series shall not be stretched, compressed, warped, or otherwise manipulated. Modifications to the length of a word for a given letter height and series shall be accomplished only by the methods described in Section 2D.03.**

Table 2D-1. Conventional Road Guide Sign and Plaque Sizes (Sheet 1 of 2)

Sign or Plaque	Designation	Section	Conventional Road	Minimum	Oversized
Interstate Route (1 or 2 digits)	M1-1,1a	2D.11	24 x 24	—	36 x 36
Interstate Route (3 digits)	M1-1,1a	2D.11	30 x 24	—	45 x 36
Off-Interstate Route (1 or 2 digits)	M1-2,3	2D.11	24 x 24	—	36 x 36
Off-Interstate Route (3 digits)	M1-2,3	2D.11	30 x 24	—	45 x 36
U.S. Route (1 or 2 digits)	M1-4	2D.11	24 x 24	—	36 x 36
U.S. Route (3 digits)	M1-4	2D.11	30 x 24	—	45 x 36
State Route (1 or 2 digits)	M1-5	2D.11	24 x 24	—	36 x 36
State Route (3 digits)	M1-5	2D.11	30 x 24	—	45 x 36
County Route	M1-6	2D.11	24 x 24	—	36 x 36
Forest Route	M1-7	2D.11	24 x 24	18 x 18	36 x 36
Junction (plaque)	M2-1P	2D.13	21 x 15	—	30 x 21
Combination Junction (2 route signs)	M2-2	2D.14	60 x 48*	—	—
Cardinal Direction (plaque)	M3-1P,2P,3P,4P	2D.15	24 x 12	—	36 x 18
Alternate (plaque)	M4-1P,1aP	2D.17	24 x 12	—	36 x 18
By-Pass (plaque)	M4-2P	2D.18	24 x 12	—	36 x 18
Business (plaque)	M4-3P	2D.19	24 x 12	—	36 x 18
Truck (plaque)	M4-4P	2D.20	24 x 12	—	36 x 18
To (plaque)	M4-5P	2D.21	24 x 12	—	36 x 18
End (plaque)	M4-6P	2D.22	24 x 12	—	36 x 18
Temporary (plaque)	M4-7P,7aP	2D.24	24 x 12	—	36 x 18
Emergency Route	M4-11	2D.59	30 x 30	—	—
Emergency Route	M4-11a	2D.59	30 x 30	—	—
Emergency Route To (plaque)	M4-11bP, 11cP	2D.59	30 x 18	—	—
End Emergency Route	M4-12	2D.59	24 x 18	—	—
Begin (plaque)	M4-14P	2D.23	24 x 12	—	36 x 18
Advance Turn Arrow (plaque)	M5-1P,2P,3P	2D.26	21 x 15	—	30 x 21
Lane Designation (plaque)	M5-4P,5P,6P	2D.27	24 x 18	—	36 x 24
Directional Arrow (plaque)	M6-1P,2P,2aP,3P,4P,5P,6P,7P	2D.28	21 x 15	—	30 x 21
National Scenic Byway	M10-1	2D.57	24 x 24	—	—
National Scenic Byway (plaque)	M10-1aP	2D.57	24 x 12	—	—
Byway Identification	M10-2	2D.58	24 x 24	—	—
Byway Identification (plaque)	M10-2aP	2D.58	24 x 12	—	—
State Scenic Byway System	M10-3	2D.58	24 x 24	—	—
State Scenic Byway - Simple Graphic and Byway Identification	M10-3a	2D.58	24 x 24	—	—
Scenic Byway (plaque)	M10-3bP	2D.58	24 x 12	—	—
National Historic Trail - Identification	M11-1	2D.58	24 x 24	—	—
National Historic Trail - Historic Route (plaque)	M11-1aP	2D.58	24 x 12	—	—
National Historic Trail - Crossing (plaque)	M11-1bP	2D.58	24 x 12	—	—
National Historic Trail - Auto Tour Route (plaque)	M11-1cP	2D.58	24 x 12	—	—
National Historic Trail - Distance (plaque)	M11-1dP	2D.58	24 x 12	—	—
Destination (1 line)	D1-1	2D.36	Varies x 18	—	—
Destination and Distance (1 line)	D1-1a	2D.36	Varies x 18	—	—
Circular Intersection Destination (1 line)	D1-1d	2D.39	Varies x 18	—	—
Circular Intersection Departure Guide	D1-1e	2D.39	Varies x 42*	—	—
Destination (2 lines)	D1-2	2D.36	Varies x 30	—	—
Destination and Distance (2 lines)	D1-2a	2D.36	Varies x 30	—	—
Circular Intersection Destination (2 lines)	D1-2d	2D.39	Varies x 30	—	—

Table 2D-1. Conventional Road Guide Sign and Plaque Sizes (Sheet 2 of 2)

Sign or Plaque	Designation	Section	Conventional Road	Minimum	Oversized
Destination (3 lines)	D1-3	2D.36	Varies x 42	—	—
Destination and Distance (3 lines)	D1-3a	2D.36	Varies x 42	—	—
Circular Intersection Destination (3 lines)	D1-3d	2D.39	Varies x 42	—	—
Circular Intersection Diagrammatic Destination	D1-5	2D.39	Varies x 72*	—	—
Circular Intersection Diagrammatic Destination, Right-Turn Bypass	D1-5a	2D.39	Varies x 78*	—	—
Distance (1 line)	D2-1	2D.43	Varies x 18	—	—
Distance (2 lines)	D2-2	2D.43	Varies x 30	—	—
Distance (3 lines)	D2-3	2D.43	Varies x 42	—	—
Street Name (1 line)	D3-1,1a	2D.45	Varies x 12	Varies x 8	Varies x 18
Overhead Street Name (1 line)	D3-1,1a	2D.45	Varies x 24	—	—
Street Name (2 lines)	D3-1,1a	2D.45	Varies x 24	Varies x 15	Varies x 33
Overhead Street Name (2 lines)	D3-1,1a	2D.45	Varies x 48	—	—
Advance Street Name (2 lines)	D3-2	2D.46	Varies x 30	—	—
Advance Street Name (3 lines)	D3-2	2D.46	Varies x 42	—	—
Advance Street Name (4 lines)	D3-2	2D.46	Varies x 54	—	—
Parking Area Directional	D4-1	2D.47	30 x 24	18 x 15	—
Park - Ride	D4-2	2D.48	30 x 36	24 x 30	36 x 48
Advance Weigh Station Distance	D8-1	2D.51	78 x 60	60 x 48	96 x 72
Weigh Station Ahead	D8-1a	2D.51	66 x 48	48 x 36	—
Weigh Station Advance Direction	D8-2	2D.51	84 x 72	66 x 54	108 x 90
Weigh Station Entrance Direction	D8-3	2D.51	66 x 60	48 x 42	84 x 78
Crossover	D13-1,2	2D.52	60 x 30	—	78 x 42
Freeway Entrance	D13-3	2D.50	48 x 30	—	—
Freeway Entrance (Directional)	D13-3a	2D.50	48 x 42	—	—
Combination Lane Use / Destination	D15-1	2D.38	Varies x 96	—	—
Next Truck Lane	D17-1	2D.53	42 x 48	—	60 x 66
Advance Truck Lane	D17-2	2D.53	42 x 42	—	60 x 54
Next Passing Lane	D17-3	2D.53	42 x 48	—	60 x 66
Advance Passing Lane	D17-4	2D.53	42 x 42	—	60 x 54
Advance Emergency Turn-Out	D17-5	2D.54	60 x 36	—	78 x 54
Emergency Turn-Out (Directional)	D17-6	2D.54	60 x 36	—	78 x 60
Advance Slow Vehicle Turn-Out	D17-7	2D.54	72 x 36	—	96 x 54

*The size shown is for a typical sign. The size should be determined based on the amount of legend required for the sign.

Notes: 1. Larger signs may be used when appropriate

2. Dimensions in inches are shown as width x height

Section 2D.05 Size of Lettering

Support:

01   Sign legibility is a direct function of letter size and spacing. Legibility distance has to be sufficient to give road users enough time to read and comprehend the sign. Under optimum conditions, a guide sign message can be read and understood in a brief glance. The legibility distance takes into account factors such as inattention, blocking of view by other vehicles, unfavorable weather, inferior eyesight, or other causes for delayed or slow reading. Where conditions permit, repetition of guide information on successive signs gives the road user more than one opportunity to obtain the information needed.

Standard:

02   **Design layouts for conventional road guide signs showing interline spacing, edge spacing, and other specification details shall be as shown in the "Standard Highway Signs" publication (see Section 1A.05).**

03   **Except as otherwise provided in this Manual, the principal legend on post-mounted guide signs shall be in letters and numerals at least 6 inches in height for all upper-case letters, or a combination of 6 inches in height for upper-case letters and 4.5 inches in nominal loop height (see Section 2D.04) for lower-case letters. On low-volume roads with speeds of 25 mph or less, and on urban streets with speeds of 25 mph or less, the principal legend on post-mounted guide signs shall be in letters at least 4 inches in height for all upper-case letters, or a combination of 4 inches in height for upper-case letters and 3 inches in nominal loop height for lower-case letters.**

04   **Except as otherwise provided in this Manual, the principal legend on overhead guide signs shall be in letters and numerals at least 6 inches in height for all upper-case letters, or a combination of 6 inches in height for upper-case letters and 4.5 inches in nominal loop height (see Section 2D.04) for lower-case letters.**

Guidance:

05   *Lettering sizes should be consistent on any particular class of highway.*

06   *The minimum lettering and numeral sizes provided in this Manual (see Table 2D-2) should be exceeded where conditions indicate a need for greater legibility.*

Section 2D.06 **Amount of Legend**

Support:

01   The longer the legend on a guide sign, the longer it will take road users to recognize and comprehend it, regardless of letter size.

Guidance:

02   *Except where otherwise provided in this Manual, guide signs should be limited to no more than three lines of destinations, which include place names, route numbers, street names, and cardinal directions. Where two or more signs are included in the same overhead display, the amount of legend should be further minimized. Where appropriate, a distance message or action information, such as an exit number, NEXT RIGHT, or directional arrows, should be provided on guide signs in addition to the destinations.*

Section 2D.07 **Abbreviations**

Support:

01   The use of commonly recognized abbreviations for certain words can be useful in reducing the reading time and improve quicker comprehension of a sign message. Descriptors and directional or quadrant orientations for street names and destinations, such as Boulevard (Blvd), North (N), and Southwest (SW), are some examples of commonly recognized abbreviations. Examples of the use of some guide sign abbreviations are shown in Figure 2D-2.

Standard:

02   **The words NORTH, SOUTH, EAST, and WEST shall not be abbreviated when used to indicate cardinal directions of numbered or named highways on guide signs.**

Guidance:

03   *Abbreviations should be kept to a minimum; however, they are useful when complete destination messages produce excessively long signs. If used, abbreviations should be unmistakably recognized by road users (see Section 1D.08). Longer commonly used words that are not part of a proper name and are readily recognizable, such as street name descriptors (such as Street, Boulevard, or Avenue), should be abbreviated as provided in Table 2D-3 to expedite recognition of the sign legend by reducing the amount and complexity of the legend. Shorter street name descriptors, such as those shown in Table 2D-4, should not be abbreviated.*

04   *Periods, apostrophes, question marks, ampersands, or other punctuation or characters that are not letters, numerals, or hyphens should not be used in abbreviations, unless necessary to avoid confusion.*

05   *The solidus is intended to be used for fractions only and should not be used to separate words on the same line of legend. Instead, a hyphen should be used for this purpose, such as "TRUCKS – BUSES."*

Table 2D-2. Recommended Minimum Letter and Numeral Sizes for Conventional Road Guide Signs According to Speed* (Sheet 1 of 2)
A - Post-Mounted Signs

Type of Sign	Single-Lane			Multi-Lane		
	Less than 30 mph	30-40 mph	Greater than 40 mph	Less than 30 mph	30-40 mph	Greater than 40 mph
A. Intersection or Interchange Advance Guide Signs and Entrance Direction Guide Signs						
Interstate or Off-Interstate Business Route Signs						
Numerals**	6	9	14	9	9	14
1- or 2-Digit Shields	18 x 18	24 x 24	36 x 36	24 x 24	24 x 24	36 x 36
3-Digit Shields	22.5 x 18	30 x 24	45 x 36	30 x 24	30 x 24	45 x 36
U.S. or State Route Signs						
Numerals	9	12	18	12	12	18
1- or 2-Digit Shields	18 x 18	24 x 24	36 x 36	24 x 24	24 x 24	36 x 36
3-Digit Shields	22.5 x 18	30 x 24	45 x 36	30 x 24	30 x 24	45 x 36
County Route Signs						
Numerals	6	8	10	8	8	10
1-, 2-, or 3-Digit Shields	18 x 18	24 x 24	36 x 36	24 x 24	24 x 24	36 x 36
U.S. or State Route Text Identification (Examples: U S 56, Md 2)						
Numerals & Letters	8	12	15	10	12	15
Cardinal Directions (NORTH, SOUTH, EAST, WEST)						
First Letter - Upper-Case	6	8	10	8	8	10
Rest of Word - Upper-Case	5	6	8	6	6	8
Auxiliary and Alternative Route Legends (Examples: JCT, TO, ALT, BUSINESS)						
Words - Upper-Case	5	6	8	6	6	8
Names of Destinations or Roads (Examples: Springfield, Main St, 2nd Ave)						
Leading Upper-Case Letter or Numerals	6	8	10.67	8	10.67	13.33
Following Lower-Case Letters or Ordinals**	4.5	6	8	6	8	10
Distance or Action Messages (Examples: 2 MILES, 1/2 MILE, KEEP RIGHT)						
Distance Numerals	6	6	8	6	8	10
Distance Fraction Numerals	4.5	4.5	6	4.5	6	8
Distance Words - Upper-Case	4.5	4.5	6	4.5	6	8
Action Message Words - Upper-Case	6	6	8	6	8	10
B. Destination and Other Guide Signs						
Names of Destinations or Roads (Examples: Springfield, Main St, 2nd Ave)						
Leading Upper-Case Letter or Numerals	4	6	8	6	8	10.67
Following Lower-Case Letters or Ordinals***	3	4.5	6	4.5	6	8
Distance or Action Messages (Examples: 2 MILES, 1/2 MILE, KEEP RIGHT)						
Distance Numerals	5	6	8	5	6	8
Distance Fraction Numerals	4	4.5	6	4	4.5	6
Distance Words - Upper-Case	4	4.5	6	4	4.5	6
Action Message Words - Upper-Case	5	6	8	6	6	8

Table 2D-2. Recommended Minimum Letter and Numeral Sizes for Conventional Road Guide Signs According to Speed* (Sheet 2 of 2)
B - Overhead-Mounted Signs

Type of Sign	Less than 35 mph	35-55 mph	Greater than 55 mph
A. Intersection or Interchange Advance Guide Signs and Entrance Direction Guide Signs			
Interstate, U.S., State, or Off-Interstate Business Route Signs			
Numerals**	6	9	14
1- or 2-Digit Shields	18 x 18	24 x 24	36 x 36
3-Digit Shields	22.5 x 18	30 x 24	45 x 36
U.S. or State Route Signs			
Numerals	9	12	18
1- or 2-Digit Shields	18 x 18	24 x 24	36 x 36
3-Digit Shields	22.5 x 18	30 x 24	45 x 36
County Route Signs			
Numerals	6	8	10
1-, 2-, or 3-Digit Shields	18 x 18	24 x 24	36 x 36
U.S. or State Route Text Identification (Examples: U S 56, Md 2)			
Numerals & Letters	8	12	15
Cardinal Directions (NORTH, SOUTH, EAST, WEST)			
First Letter - Upper-Case	6	8	12
Rest of Word - Upper-Case	5	6	10
Auxiliary and Alternative Route Legends (Examples: JCT, TO, ALT, BUSINESS)			
Words - Upper-Case	5	6	10
Names of Destinations or Roads (Examples: Springfield, Main St, 2nd Ave)			
Leading Upper-Case Letter or Numerals	6	8 (min.) 10.67 (des.)	13.33 (min.) 16 (des.)
Following Lower-Case Letters or Ordinals**	4.5	6 (min.) 8 (des.)	10 (min.) 12 (des.)
Distance or Action Messages (Examples: 2 MILES, 1/2 MILE, KEEP RIGHT)			
Distance Numerals	6	6 (min.) 8 (des.)	12 (min.) 15 (des.)
Distance Fraction Numerals	4.5	4.5 (min.) 6 (des.)	10 (min.) 12 (des.)
Distance Words - Upper-Case	4.5	4.5 (min.) 6 (des.)	10 (min.) 12 (des.)
Action Message Words - Upper-Case	6	6 (min.) 8 (des.)	12 (min.) 15 (des.)
B. Destination and Other Guide Signs			
Names of Destinations or Roads (Examples: Springfield, Main St, 2nd Ave)			
Leading Upper-Case Letter or Numerals	6	8 (min.) 10.67 (des.)	13.33 (min.) 16 (des.)
Following Lower-Case Letters or Ordinals**	4.5	6 (min.) 8 (des.)	10 (min.) 12 (des.)
Distance or Action Messages (Examples: 2 MILES, 1/2 MILE, KEEP RIGHT)			
Distance Numerals	6	6 (min.) 8 (des.)	12 (min.) 15 (des.)
Distance Fraction Numerals	4.5	4.5 (min.) 6 (des.)	10 (min.) 12 (des.)
Distance Words - Upper-Case	4.5	4.5 (min.) 6 (des.)	10 (min.) 12 (des.)
Action Message Words - Upper-Case	6	6 (min.) 8 (des.)	12 (min.) 15 (des.)

* Except as provided otherwise in this Manual

** Minimum size listed for 3-digit shields. Larger numeral sizes used for 1-digit, some 2-digit, and some 3-digit shields. See the Standard Highways Signs publication for more information on Route Sign numeral heights and Standard Alphabet series.

*** Lower-case letter height (loop height) is determined by the initial upper-case letter height (see Sec. 2A.08)

Notes: 1. Sizes are shown in inches and where applicable are shown as width x height

2. For Street Name (D3-1 Series) signs, see Table 2D-6

3. The 18-inch route shield size is not for independent use, such as in Directional or Confirmation Assemblies.

Figure 2D-2. Examples of Uses of Abbreviations on Guide Signs

A – Cardinal directions and orientations

"South" is the street name and shall not be abbreviated.

"EAST" is the cardinal direction of travel and shall not be abbreviated.

"Avenue" is the street name descriptor with a recognizable abbreviation* and should be abbreviated.

"South" is a cardinal direction and may be abbreviated.

"EAST" is the cardinal direction of travel and shall not be abbreviated.

"Road" is the street name descriptor with a recognizable abbreviation* and should be abbreviated.

B – Quadrant and cardinal directions

"Southeast" is a quadrant cardinal orientation and should be abbreviated.

"Boulevard" is the street name and shall not be abbreviated.

"Southeast" is the street name and shall not be abbreviated.

"Boulevard" is the street name descriptor with a recognizable abbreviation* and should be abbreviated.

C – Other descriptors within proper names

OR

"Peak" is a geographical feature as a descriptor within a proper name. When a geographical feature descriptor does not have a recognizable abbreviation, it shall not be abbreviated.

"Mount" is a geographical feature as a descriptor within a proper name. When a geographical feature descriptor has a recognizable abbreviation, it may be abbreviated.

* See Tables 1D-1 and 2D-3 for the abbreviations that are acceptable for use on signs

Section 2D.08 Arrows

Support:

01   Arrows are used for lane assignment and to indicate the direction toward designated routes or destinations. Figure 2D-3 shows the various standard arrow designs that have been approved for use on guide signs. Detailed drawings are shown for these arrows in the "Standard Highway Signs" publication (see Section 1A.05).

Standard:

02   **Except for Overhead Arrow-per-Lane signs (see Section 2D.37), on overhead signs where it is desirable to indicate a lane to be followed, a down arrow shall be positioned over the approximate center of the lane and shall point vertically downward toward the approximate center of that lane. Down arrows shall be used only on overhead guide signs that restrict the use of specific lanes to traffic bound for the destination(s) and/or route(s) indicated by these arrows. Down arrows shall not be used unless an arrow can be located over and pointed to the approximate center of each lane that can be used to reach the destination displayed on the sign.**

03   **If down arrows are used, having more than one down arrow pointing to the same lane on a single overhead sign (or on multiple signs on the same overhead sign structure) shall not be permitted.**

04   **Where a roadway is leaving the through lanes, a directional arrow shall point upward at an angle that approximates the alignment of the exit roadway in the vicinity of the point of departure.**

05   **The Type E directional arrow for circular intersections shall not be used on any sign that is not associated with a circular intersection**

Guidance:

06   *The Type A directional arrow should be used on guide signs on freeways, expressways, and conventional roads to indicate the direction to a specific destination or group of destinations, except as otherwise provided in this Section and in Section 2E.18.*

07   *When a directional arrow in a vertical, upward-pointing orientation is placed to the side of a group of destinations to indicate a through movement, the Type A directional arrow should be used. When a directional arrow in a vertical, upward-pointing orientation is placed to the side of a single destination or under a destination or group of destinations, the Type B directional arrow should be used.*

08   *The Type B directional arrow should be used on guide signs on conventional roads when placed at any angle to the side of a single destination or when placed in a horizontal orientation to the side of a group of destinations.*

09   *The Type C advance turn directional arrow should be used on conventional road guide signs placed in advance of an intersection where a turn must be made to reach a posted destination or group of destinations.*

Table 2D-3. Acceptable Abbreviations for Street Name Descriptors

Descriptor	Standard Abbreviation	Descriptor	Standard Abbreviation
Avenue	Ave	Northwest	NW*
Boulevard	Blvd	Parkway	Pkwy
Bypass	Byp	Place	Pl
Causeway	Cswy	Plaza	Plz
Circle	Cir	Road	Rd
Corner	Cor	Route	Rte
Court	Ct	South	S*
Crescent	Cres	Southeast	SE*
Drive	Dr	Southwest	SW*
East	E*	Square	Sq
Expressway	Expwy	Street	St
Extension	Ext	Terrace	Ter
Freeway	Fwy	Thruway	Thwy
Highway	Hwy	Trafficway	Trfwy
Lane	La, Ln	Trail	Tr
Landing	Lndg	Turnpike	Tpk
North	N*	West	W*
Northeast	NE*		

* For pre-directional or post-directional designations or cardinal orientations, such as E Main St or 3rd St SW

Table 2D-4. Street Name Descriptors Not Acceptable for Abbreviation

Descriptor	Descriptor
Alley	Oval
Belt	Pass
Beltway	Passage
Close	Path
Cove	Ridge
Edge	Row
Gate	Run
Green	Trace
Grove	Turn
Hill	View
Loop	Vista
Mews	Walk

Figure 2D-3. Arrows for Use on Guide Signs

Directional arrows

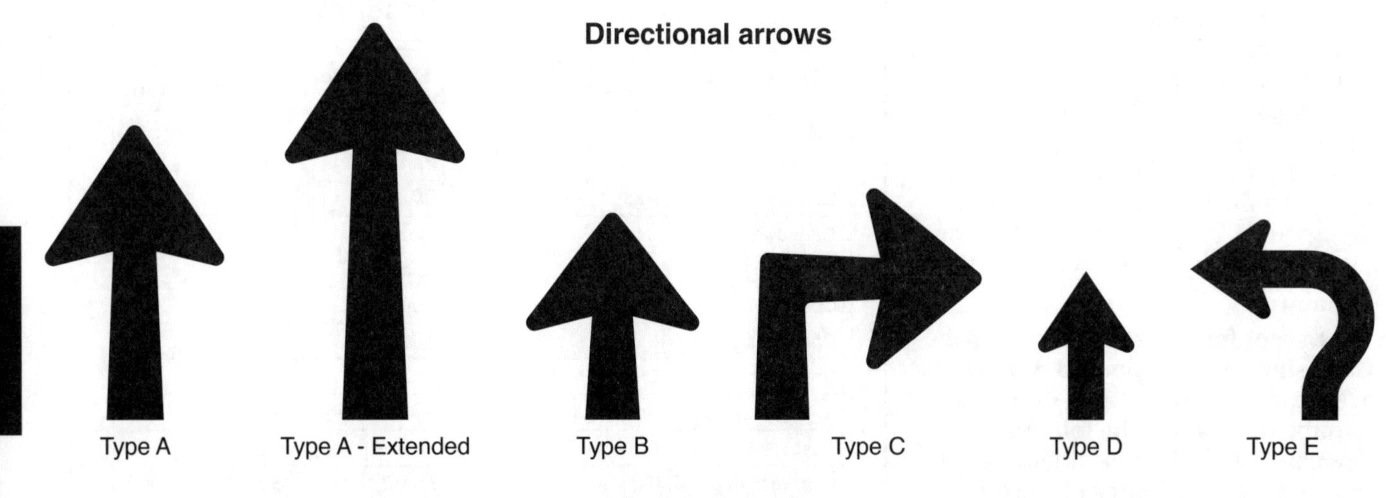

Down arrow

Note: The "Standard Highway Signs"
       publication contains the details
       of these arrow designs.

10   *The Type D directional arrow should be used primarily for sign applications other than guide signs, except as provided in Paragraph 15 of this Section.*

11   *If the Type E directional arrow is used, the principles set forth in Sections 2D.26 through 2D.29 should be followed.*

Option:

12   The Type A-Extended directional arrow may be used on guide signs where additional emphasis regarding the direction is needed relative to the amount of legend on the sign.

13   The Type C directional arrow may be used to the side of the legend of an overhead guide sign to accentuate a sharp turn exit maneuver from a mainline roadway (see Section 2E.25 for additional information regarding Exit Direction signs for low advisory ramp speeds).

14   On conventional roads on the approach to an intersection where the Combination Lane-Use/Destination overhead guide sign (see Section 2D.38) is not used, the Type C advance turn directional arrow may be used beneath the legend of an overhead guide sign to indicate the fact that a turn must be made from a mandatory movement lane over which the sign is placed to reach the destination or destinations displayed on the sign.

15   The Type D directional arrow may be used on post-mounted guide signs on conventional roads with lower operating speeds if the height of the text on the sign is 8 inches or less. Type D arrows may be used on a Street Name (D3-1 only) sign displaying two street names to indicate the different direction of travel for each street.

16   The Type E directional arrow may be used on guide signs on approaches to circular intersections to represent the intended driver paths to destinations involving left-turn movements around the circulatory island.

17   The directional and down arrows shown in Figure 2D-3 may be used on signs other than guide signs for the purposes of providing directional guidance and lane assignment.

Guidance:

18   *Arrows used on guide signs to indicate the directions toward designated routes or destinations should be pointed at the appropriate angle to clearly convey the direction to be taken. A horizontally-oriented directional arrow design should be used at right-angle intersections.*

19   *On a post-mounted guide sign, a directional arrow for a straight-through movement should point upward. Except as provided in Section 2D.50, for a turn, the arrow on a guide sign should point horizontally or at an upward angle that approximates the sharpness of the turn.*

20   *At an exit, an arrow should be placed at the side of the sign that will reinforce the movement of exiting traffic. The directional arrow design should be used.*

Standard:

21   **If used, the Type C advance turn directional arrow shall display a right or left arrow, the shaft of which is bent at a 90-degree or oblique angle.**

Option:

22   Arrows may be placed below the principal sign legend or on the appropriate side of the legend that is consistent with the direction of the movement.

23   On a post-mounted sign at an exit where placement of the arrow to the side of the legend farthest from the roadway would create an unusually wide sign that limits the road user's view of the arrow, the directional arrow may be placed at the bottom portion of the sign, centered under the legend.

Guidance:

24   *The width across the arrowhead for the Types A, B, and C directional arrows should be between 1.5 and 1.75 times the height of the upper-case letters of the principal legend on the sign. The width across the arrowhead for the Type D directional arrow should be at least equal to the height of the upper-case letters of the principal legend on the sign. For down arrows used on overhead signs, the width across the arrowhead should be approximately 2 times the height of the upper-case letters of the principal legend on the sign.*

Support:

25   Section 2D.37 contains the provisions for arrows used in Overhead Arrow-per-Lane signs on approaches to conventional road intersections. Section 2D.41 contains the provisions for arrows used in Diagrammatic Advance guide signing on approaches to conventional road intersections other than circular intersections. Section 2D.39 contains the provisions for diagrammatic arrows used in Destination signs on the approaches to circular intersections (see Figure 2D-11).

26   The "Standard Highway Signs" publication (see Section 1A.05) contains design details and standardized sizes of the various arrows based on ranges of letter heights of principal legends.

ROUTE SIGNS AND AUXILIARY PLAQUES

Section 2D.09 Numbered Highway Systems

Support:

01   The purpose of numbering and signing highway systems is to identify routes and facilitate travel.

02   The Interstate and United States (U.S.) highway systems are numbered by the American Association of State Highway and Transportation Officials (AASHTO) upon recommendations of the State highway organizations because the respective States own these systems. State and county road systems are numbered by the appropriate authorities.

03   The basic policy for numbering the Interstate and U.S. highway systems is contained in the following Purpose and Policy statements published by AASHTO:

    A. "Establishment and Development of United States Numbered Highways," and
    B. "Establishment of a Marking System of the Routes Comprising the National System of Interstate and Defense Highways."

Guidance:

04   *The principles of these policies should be followed in establishing the highway systems described in Paragraph 3 of this Section and any other systems, with effective coordination between adjacent jurisdictions. Care should be taken to avoid the use of numbers or other designations that have been assigned to Interstate, U.S., or State routes in the same geographic area. Overlapping numbered routes should be kept to a minimum.*

Standard:

05   **Route systems shall be given preference in this order: Interstate, United States, State, and county. The preference shall be given by installing the highest-priority route number on the top or the left of the sign, except as provided in Paragraph 6 of this Section.**

06   **Interstate route numbering shall be approved by the FHWA.**

Option:

07   The prioritization of route systems may be modified when a different prioritization would better accommodate the expectancy of the road user and provide more effective direction, such as for separate decision points for routes that are encountered in a particular order.

Support:

08   Section 2D.56 contains information regarding the signing of unnumbered highways to enhance route guidance and facilitate travel.

Section 2D.10 Route Signs and Auxiliary Plaques

Standard:

01   **Except as provided in Paragraph 9 of Section 2D.29, all numbered highway routes shall be identified by route signs and auxiliary plaques.**

02   **The signs for each system of numbered highways, which are distinctive in shape and color, shall be used only on that system and the approaches thereto.**

Option:

03   Route signs and auxiliary plaques may be proportionally enlarged where greater conspicuity or legibility is needed.

Support:

04   Route signs are typically mounted in assemblies with auxiliary plaques.

05   Section 2D.57 contains information regarding the signing for National Scenic Byways.

06   Section 2D.58 contains information regarding the signing for State-designated scenic byways, historic trails, and auto tour routes.

Section 2D.11 Design of Route Signs

Standard:

01   **The design of standard route signs shall conform to the designs provided in the "Standard Highway Signs" publication (see Section 1A.05). The design of other route signs shall be established by the authority having jurisdiction and shall be in general conformance with the designs provided in the "Standard Highway Signs" publication.**

02   **Interstate Route (M1-1 and M1-1a) signs (see Figure 2D-4) shall be used on all Interstate routes and in connection with Route Sign assemblies on intersecting highways.**

Figure 2D-4. Route Signs

Interstate Route Sign
M1-1

Interstate Route Sign
M1-1a

Off-Interstate Business Route Sign
M1-2 (Loop)

Off-Interstate Business Route Sign
M1-3 (Spur)

U.S. Route Sign
M1-4

State Route Sign
M1-5

County Route Sign
M1-6

Forest Route Sign
M1-7

03   **Except as otherwise provided in this Manual, a 24 x 24-inch minimum sign size shall be used for Interstate route numbers with one or two digits, and a 30 x 24-inch minimum sign size shall be used for Interstate route numbers having three digits.**

Option:

04   When the Interstate Route sign is used in a Route Sign assembly (see Section 2D.29), the M1-1a sign, containing the State name in white upper-case letters on a blue background as detailed in the "Standard Highway Signs" publication (see Section 1A.05), may be used in place of the M1-1 sign.

Standard:

05   **Use of the M1-1a sign shall be limited to Route Sign assemblies.**

06   **Off-Interstate Business Route (M1-2 and M1-3) signs (see Figure 2D-4) shall consist of a cutout shield displaying the number of the connecting Interstate route and the words BUSINESS and either LOOP (when the route rejoins the same Interstate route) or SPUR (when the route leaves the corresponding Interstate route and does not rejoin) in upper-case letters. The legend and border shall be white on a green background, and the shield shall be the same shape and dimensions as the Interstate Route sign. In no instance shall the word INTERSTATE appear on the Off-Interstate Business Route sign.**

Option:

07   The Off-Interstate Business Route sign may be used on a major highway that is not a part of the Interstate system, but one that serves the business area of a city from an interchange on the system.

Standard:

08   **U.S. Route signs (see Figure 2D-4) shall consist of black numerals on a white shield surrounded by a rectangular black background without a border. This sign shall be used on all U.S. routes and in connection with Route Sign assemblies on intersecting highways.**

09   **A 24 x 24-inch minimum sign size shall be used for U.S. route numbers with one or two digits, and a 30 x 24-inch minimum sign size shall be used for U.S. route numbers having three digits.**

10   **State Route signs shall be designed by the individual State highway agencies.**

11   **The legend on State Route signs shall conform to the Standard Alphabets contained in the "Standard Highway Signs" publication (see Section 1A.05).**

Guidance:

12   *State Route signs (see Figure 2D-4) should be rectangular and should be approximately the same size as the U.S. Route sign. State Route signs should also be similar to the U.S. Route sign by containing approximately the same size black numerals on a white area surrounded by a rectangular black background without a border, and should be devoid of complex graphics. The shape of the white area should be circular in the absence of any determination to the contrary by the individual State concerned.*

13   *Where U.S. or State Route signs are used as components of guide signs, only the distinctive shape of the shield itself and the route numerals within should be used. The rectangular background upon which the distinctive shape of the shield is mounted, such as the black area around the outside of the shields on the M1-4 and standard M1-5 signs, should not be included on the guide sign. Where U.S. or State Route signs are used as components of other signs of non-contrasting background colors, the rectangular background should be used so that recognition of the distinctive shape of the shield can be maintained.*

Standard:

14   **If county road authorities elect to establish and identify a special system of important county roads, a statewide policy for such signing shall be established that includes a uniform numbering system to uniquely identify each route. The County Route (M1-6) sign (see Figure 2D-4) shall consist of a pentagon shape with a yellow county name and route number and border on a blue background. County Route signs shall be a minimum size of 24 x 24 inches.**

15   **If a jurisdiction uses letters instead of numbers to identify routes, all references to numbered routes in this Chapter shall be interpreted to also include lettered routes.**

Guidance:

16   *If used with other route signs in common assemblies, the County Route sign should be of a size compatible with that of the other route signs.*

Standard:

17   **The design of the National Forest Route (M1-7) sign (see Figure 2D-4) shall be as detailed in the "Standard Highway Signs" publication (see Section 1A.05). Route signs for other park and forest roads shall be designed with an appropriate level of distinctiveness and adequate legibility, but in general compliance with the design principles for route signs and of a size compatible with other route signs used in common assemblies.**

Section 2D.12 Design of Route Sign Auxiliary Plaques

Standard:

01   **Route sign auxiliary plaques displaying word legends, except the Junction (M2-1P) auxiliary plaque, shall have a minimum standard size of 24 x 12 inches. The Junction auxiliary plaque and those auxiliary plaques displaying arrows shall have a minimum standard size of 21 x 15 inches. All route sign auxiliary plaques shall match the color combination of the route sign that they supplement.**

Guidance:

02   *The background, legend, and border of a route sign auxiliary plaque should have the same colors as those of the route sign with which the auxiliary plaque is mounted in a Route Sign assembly (see Section 2D.29). For a route sign design that uses multiple background colors, such as the Interstate Route sign, the background color of the corresponding auxiliary plaque should be that of the background area on which the route number is placed on the route sign.*

Option:

03   A route sign and any auxiliary plaques used with it may be combined on a single sign as a guide sign.

Standard:

04   **If a route sign and its auxiliary plaques are combined to form a single guide sign, the background color of the sign shall be green and the design shall comply with the basic principles for the design of guide signs. The auxiliary messages shall be white legends placed directly on the green background. Auxiliary plaques shall not be mounted directly to a guide sign or other type of sign.**

Support:

05   Chapter 2F contains information regarding auxiliary plaques for toll highways.

Section 2D.13 Junction Auxiliary Plaque (M2-1P)

Standard:

01   **The Junction (M2-1P) auxiliary plaque (see Figure 2D-5) shall display the abbreviated legend JCT and shall be mounted at the top of an assembly (see Section 2D.30) directly above the route sign, the sign for an alternative route (see Section 2D.17) that is part of the route designation, or the Cardinal Direction auxiliary plaque where access is available only to one direction of the intersected route. The minimum size of the Junction auxiliary plaque shall be 21 x 15 inches for compatibility with auxiliary plaques displaying arrow symbols.**

Section 2D.14 Combination Junction Sign (M2-2)

Option:

01   As an alternative to the standard Junction assembly where more than one route is to be intersected or joined, a rectangular guide sign may be used displaying the word JUNCTION above the route numbers.

Standard:

02   **The Combination Junction (M2-2) sign (see Figure 2D-5) shall have a green background with white border and lettering for the word JUNCTION.**

Guidance:

03   *The Combination Junction sign should comply with the specific provisions of Section 2D.11 regarding the incorporation of the route signs as components of guide signs.*

04   *Although the size of the Combination Junction sign will depend on the number of routes involved, the numerals should be large enough for clear legibility and should be of a size comparable with those in the individual route signs.*

Section 2D.15 Cardinal Direction Auxiliary Plaques (M3-1P through M3-4P)

Guidance:

01   *Cardinal Direction auxiliary plaques (see Figure 2D-5) displaying the legend NORTH, EAST, SOUTH, or WEST should be used to indicate the general direction of the entire route.*

Standard:

02   **To improve the readability and recognition of the cardinal directions, the first letter of the cardinal direction words shall be ten percent larger, rounded up to the nearest whole number size.**

03   **If used, the Cardinal Direction auxiliary plaque shall be mounted directly above a route sign or, if used, an auxiliary plaque for an alternative route.**

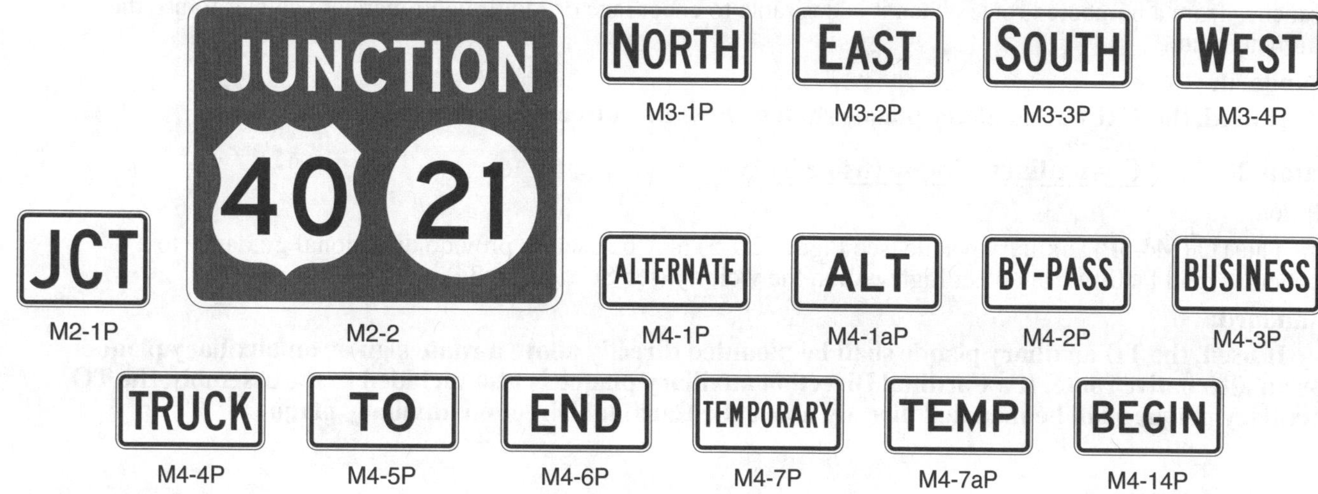

Figure 2D-5. Route Sign Auxiliary Plaques and Combination Junction Sign

Section 2D.16 <u>Alternative Route Auxiliary Plaques (M4-1P through M4-4P)</u>

Option:

01   Alternative Route auxiliary plaques (see Figure 2D-5) displaying legends such as ALTERNATE, BY-PASS, BUSINESS, or TRUCK, may be used to indicate an alternate route of the same number between two points on that route.

Standard:

02   **If used, the Alternative Route auxiliary plaques shall be mounted directly above a route sign.**

Section 2D.17 <u>ALTERNATE Auxiliary Plaques (M4-1P and M4-1aP)</u>

Option:

01   The ALTERNATE (M4-1P) or the ALT (M4-1aP) auxiliary plaque (see Figure 2D-5) may be used to indicate an officially designated alternate routing of a numbered route between two points on that route.

Standard:

02   **If used, the ALTERNATE or ALT auxiliary plaque shall be mounted directly above a route sign.**

03   **The M4-1P series plaques shall not be used to sign an alternative routing that is not officially designated and incorporated into the numbered highway system, such as alternative routings for incident management or emergency detours.**

Guidance:

04   *The shorter (time or distance) or better-constructed route should retain the regular route number, and the longer or worse-constructed route should be designated as the alternate route.*

Section 2D.18 <u>BY-PASS Auxiliary Plaque (M4-2P)</u>

Option:

01   The BY-PASS (M4-2P) auxiliary plaque (see Figure 2D-5) may be used to designate a route that branches from the numbered route through a city, bypasses a part of the city or congested area, and rejoins the numbered route beyond the city.

Standard:

02   **If used, the BY-PASS auxiliary plaque shall be mounted directly above a route sign.**

Section 2D.19 <u>BUSINESS Auxiliary Plaque (M4-3P)</u>

Option:

01   The BUSINESS (M4-3P) auxiliary plaque (see Figure 2D-5) may be used to designate an alternate route that branches from a numbered route, passes through the business portion of a city, and rejoins the numbered route beyond that area.

Standard:

02   **If used, the BUSINESS auxiliary plaque shall be mounted directly above a route sign.**

Section 2D.20 <u>TRUCK Auxiliary Plaque (M4-4P)</u>

Option:

01   The TRUCK (M4-4P) auxiliary plaque (see Figure 2D-5) may be used to designate an alternate route that branches from a numbered route, when it is desirable to encourage or require commercial vehicles to use the alternate route.

Standard:

02   **If used, the TRUCK auxiliary plaque shall be mounted directly above a route sign.**

Section 2D.21 <u>TO Auxiliary Plaque (M4-5P)</u>

Option:

01   The TO (M4-5P) auxiliary plaque (see Figure 2D-5) may be used to provide directional guidance to a particular road facility from other highways in the vicinity (see Section 2D.34).

Standard:

02   **If used, the TO auxiliary plaque shall be mounted directly above a route sign or an auxiliary plaque for an alternative route. If a Cardinal Direction auxiliary plaque is also included in the assembly, the TO auxiliary plaque shall be mounted directly above the Cardinal Direction auxiliary plaque.**

Section 2D.22 END Auxiliary Plaque (M4-6P)

Guidance:

01   *The END (M4-6P) auxiliary plaque (see Figure 2D-5) should be used where the route being traveled ends, usually at a junction with another route.*

Standard:

02   **If used, the END auxiliary plaque shall be mounted either directly above a route sign or above a sign for an alternative route that is part of the designation of the route being terminated.**

Section 2D.23 BEGIN Auxiliary Plaque (M4-14P)

Option:

01   The BEGIN (M4-14P) auxiliary plaque (see Figure 2D-5) may be used where a route begins, usually at a junction with another route.

Standard:

02   **If used, the BEGIN auxiliary plaque shall be mounted at the top of the first Confirming assembly (see Section 2D.33) for the route that is beginning.**

Guidance:

03   *If a BEGIN auxiliary plaque is included in the first Confirming assembly, a Cardinal Direction auxiliary plaque should also be included in the assembly.*

Standard:

04   **If a Cardinal Direction auxiliary plaque is also included in the assembly, the BEGIN auxiliary plaque shall be mounted directly above the Cardinal Direction auxiliary plaque.**

Section 2D.24 TEMPORARY Auxiliary Plaques (M4-7P and M4-7aP)

Option:

01   The TEMPORARY (M4-7P) or the TEMP (M4-7aP) auxiliary plaque (see Figure 2D-5) may be used for an interim period to designate a section of highway that is not planned as a permanent part of a numbered route, but that connects completed portions of that route.

Standard:

02   **If used, the TEMPORARY or TEMP auxiliary plaque shall be mounted directly above the route sign, above a Cardinal Direction auxiliary plaque, or above an auxiliary plaque for an alternate route that is a part of the route designation.**

03   **TEMPORARY or TEMP auxiliary plaques shall be promptly removed when the temporary route is abandoned.**

Section 2D.25 Temporary Detour Signs and Auxiliary Plaques

Support:

01   Chapter 6F contains information regarding Temporary Detour signs and auxiliary plaques.

Section 2D.26 Advance Turn Arrow Auxiliary Plaques (M5-1P, M5-2P, and M5-3P)

Standard:

01   **If used, the Advance Turn Arrow auxiliary plaque (see Figure 2D-6) shall be mounted directly below the route sign in Advance Route Turn assemblies, and shall display a right or left arrow, the shaft of which is bent at a 90-degree angle (M5-1P) or at an oblique angle (M5-2P).**

02   **If used, the Circular Intersection Advance Turn Arrow (M5-3P) auxiliary plaque (see Figure 2D-6) shall be used only on the approach to a circular intersection to depict a movement along the circulatory roadway around the central island and to the left, relative to the approach roadway and entry into the intersection.**

Guidance:

03   *If the M5-3P plaque is used, then this arrow type should also be used consistently on any regulatory lane-use signs (see Chapter 2B), Destination signs (see Section 2D.36), and pavement markings (see Part 3) for a particular destination or movement.*

Figure 2D-6. Advance Turn and Directional Arrow Auxiliary Plaques

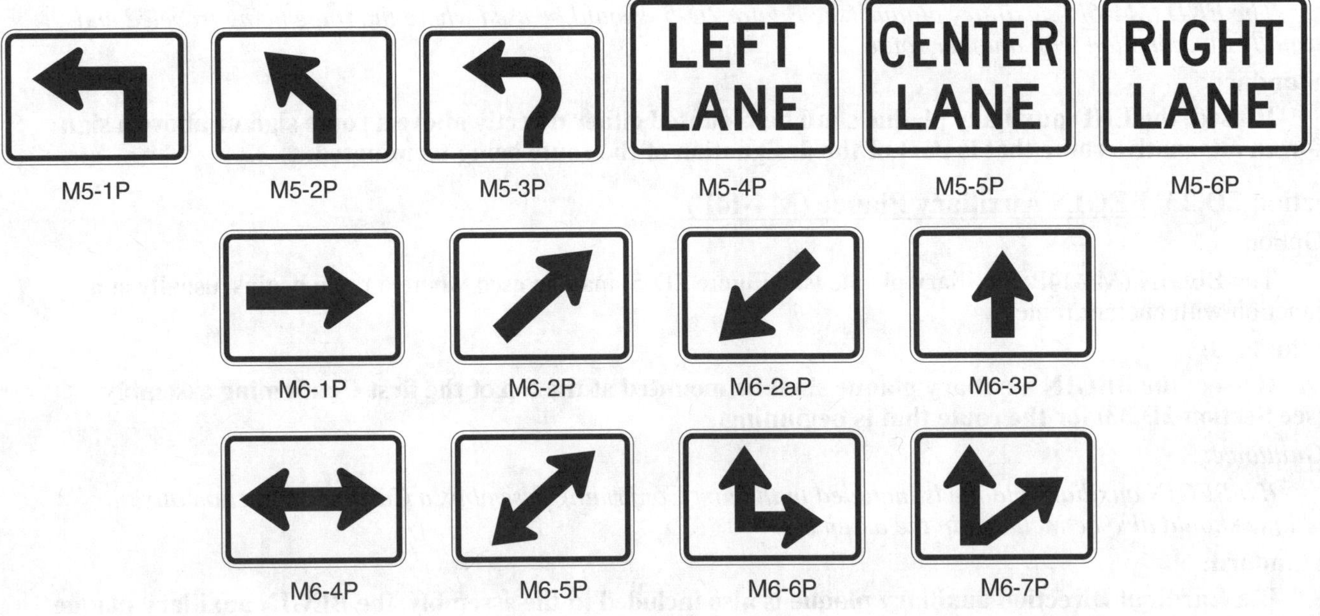

Section 2D.27 Lane Designation Auxiliary Plaques (M5-4P, M5-5P, and M5-6P)

Option:

01   A Lane Designation (M5-4P, M5-5P, or M5-6P) auxiliary plaque (see Figure 2D-6) may be mounted directly below the route sign in an Advance Route Turn assembly on multi-lane roadways to allow road users to move into the appropriate lane prior to reaching the intersection or interchange.

Standard:

02   **If used, the Lane Designation auxiliary plaques shall be used only where the designated lane is a mandatory movement lane and shall be located adjacent to the full-width portion of the mandatory movement lane. The Lane Designation auxiliary plaques shall not be installed adjacent to a through lane in advance of a lane that is being added or along the taper for a lane that is being added.**

Section 2D.28 Directional Arrow Auxiliary Plaques (M6 Series)

Standard:

01   **If used, the Directional Arrow auxiliary plaque (see Figure 2D-6) shall be mounted below the route sign and any other auxiliary plaques in Directional assemblies (see Section 2D.32), and shall display a single-headed or double-headed arrow pointing in the general direction that the route follows.**

02   **A Directional Arrow auxiliary plaque that displays a double-headed arrow shall not be mounted in any Directional assembly in advance of or at a circular intersection.**

Option:

03   The diagonal downward-pointing arrow auxiliary (M6-2aP) plaque may be used in a Directional assembly at the far corner of an intersection to indicate the immediate entry point to a freeway or expressway entrance ramp (see Section 2D.50).

Standard:

04   **The M6-2aP plaque shall not be used on the approach to or on the near side of an intersection, such as to designate an approach lane.**

SIGN ASSEMBLIES

Section 2D.29 Route Sign Assemblies

Standard:

01   **A Route Sign assembly shall consist of a route sign and auxiliary plaques that further identify the route and indicate the direction. Except as provided in Paragraph 9 of this Section, Route Sign assemblies shall be installed on all approaches to numbered routes that intersect with other numbered routes.**

02   **Where two or more routes follow the same section of highway, the route signs for Interstate, U.S., State, and county routes shall be mounted in that order from the left in horizontal arrangements and from the top in vertical arrangements. Subject to this order of precedence, route signs for lower-numbered routes shall be placed at the left or top.**

03   **Within groups of assemblies, information for routes intersecting from the left shall be mounted at the left in horizontal arrangements and at the top or center of vertical arrangements. Similarly, information for routes intersecting from the right shall be at the right or bottom, and for straight-through routes at the center in horizontal arrangements or top in vertical arrangements.**

04   **Route Sign assemblies shall be mounted in accordance with the general specifications for signs (Chapter 2A), with the lowest sign in the assembly at the height prescribed for single signs.**

Guidance:

05   *Assemblies for two or more routes, or for different directions on the same route, should be mounted in groups on a common support.*

06   *Where more than four route signs would be needed in a single Advance Route Turn or Directional assembly, the route signs should instead be mounted in a guide sign to minimize the need for repetition of the same information on multiple Cardinal Direction and Directional Arrow auxiliary plaques (see Figure 2D-7).*

Option:

07   Route Sign assemblies may be installed on the approaches to numbered routes on unnumbered roads and streets that carry an appreciable amount of traffic destined for the numbered route.

08   If engineering judgment indicates that groups of assemblies that include overlapping routes or multiple turns might be confusing, route signs or auxiliary signs may be omitted or combined, provided that clear directions are given to road users.

09   Route Sign assemblies may be omitted for routes that are part of an agency's internal numbering system, such as for maintenance or other purposes, and are not publicly mapped or intended to be used for navigational purposes by the general public. Similarly, numbered routes that are not maintained during certain times of year, such as not being plowed during winter months, may be omitted from Route Sign assemblies.

Support:

10   Figure 2D-8 shows typical placements of route signs.

Section 2D.30 Junction Assembly

Standard:

01   **A Junction assembly shall consist of a Junction auxiliary plaque (see Section 2D.13) and a route sign. The route sign shall display the number of the intersected or joined route.**

02   **The Junction assembly shall be installed in advance of every intersection where a numbered route is intersected or joined by another numbered route.**

Guidance:

03   *In urban areas, the Junction assembly should be installed in the block preceding the intersection. In urban areas where speeds are low, the Junction assembly should not be installed more than 300 feet in advance of the intersection.*

04   *In rural areas, the Junction assembly should be installed at least 400 feet in advance of the intersection. In rural areas, the minimum distance between a Junction assembly and either a Destination sign or an Advance Route Turn assembly should be 200 feet.*

05   *Where speeds are high, greater spacings should be used.*

Option:

06   Where two or more routes are to be indicated, a single Junction auxiliary plaque may be used for the assembly and all route signs grouped in a single mounting, or a Combination Junction (M2-2) sign (see Section 2D.14) may be used.

Figure 2D-7. Examples of Consolidation of Route Sign Assemblies into Guide Signs (Sheet 1 of 2)

A – Minor roadway approach to a stop-controlled intersection

B – Major roadway approach or approach to a signalized intersection

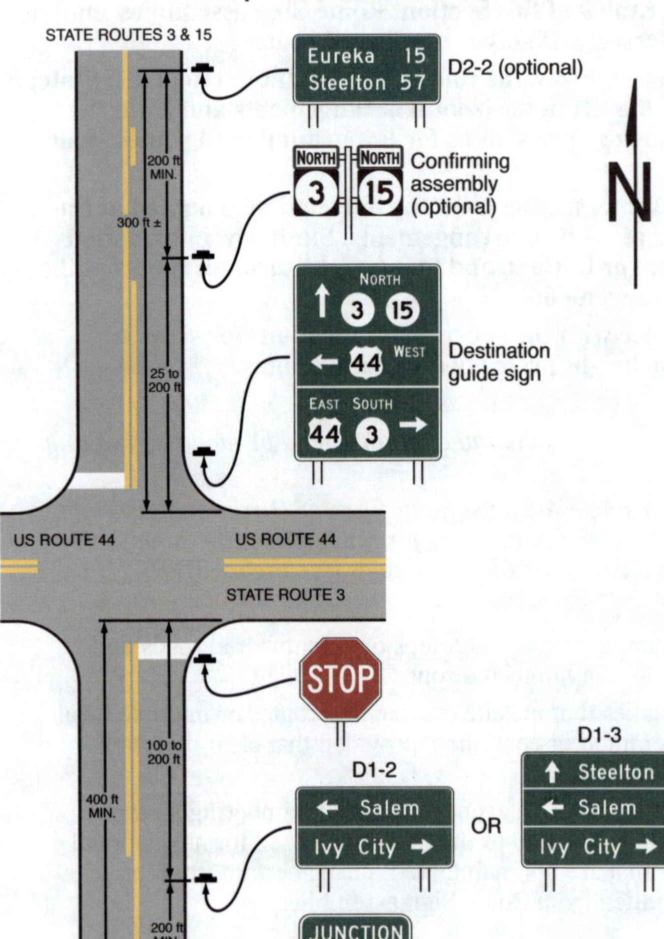

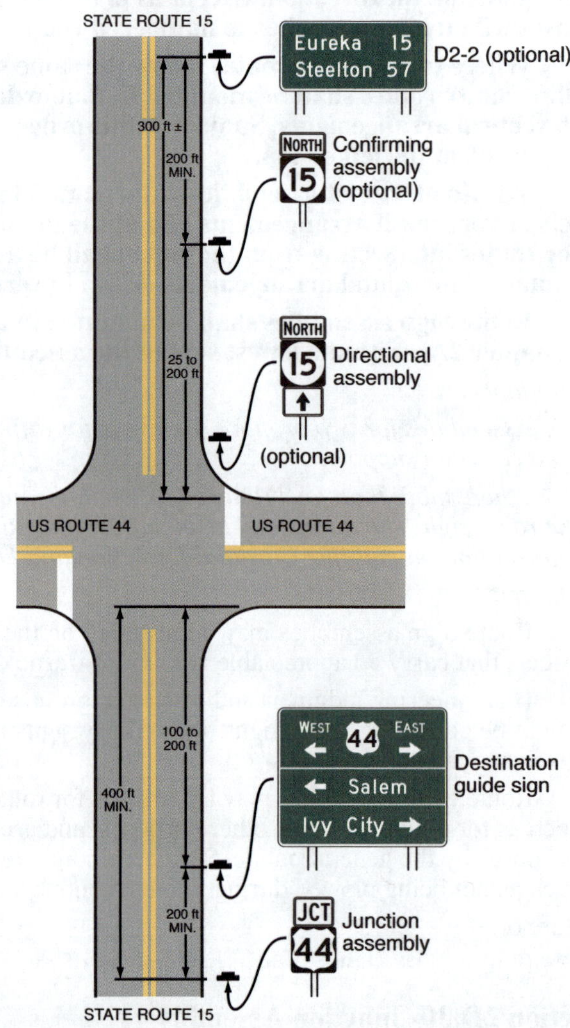

Notes:

1. Only one direction of travel and associated signs shown.

2. The spacings shown on this figure are for rural intersections. See Sections 2D.29, 2D.30, 2D.31, 2D.32, 2D.33, 2D.42, and 2D.44 for low-speed and/or urban conditions.

Figure 2D-7. Examples of Consolidation of Route Sign Assemblies into Guide Signs (Sheet 2 of 2)

C – Major roadway approach or approach to a signalized intersection where overlapping routes join or separate

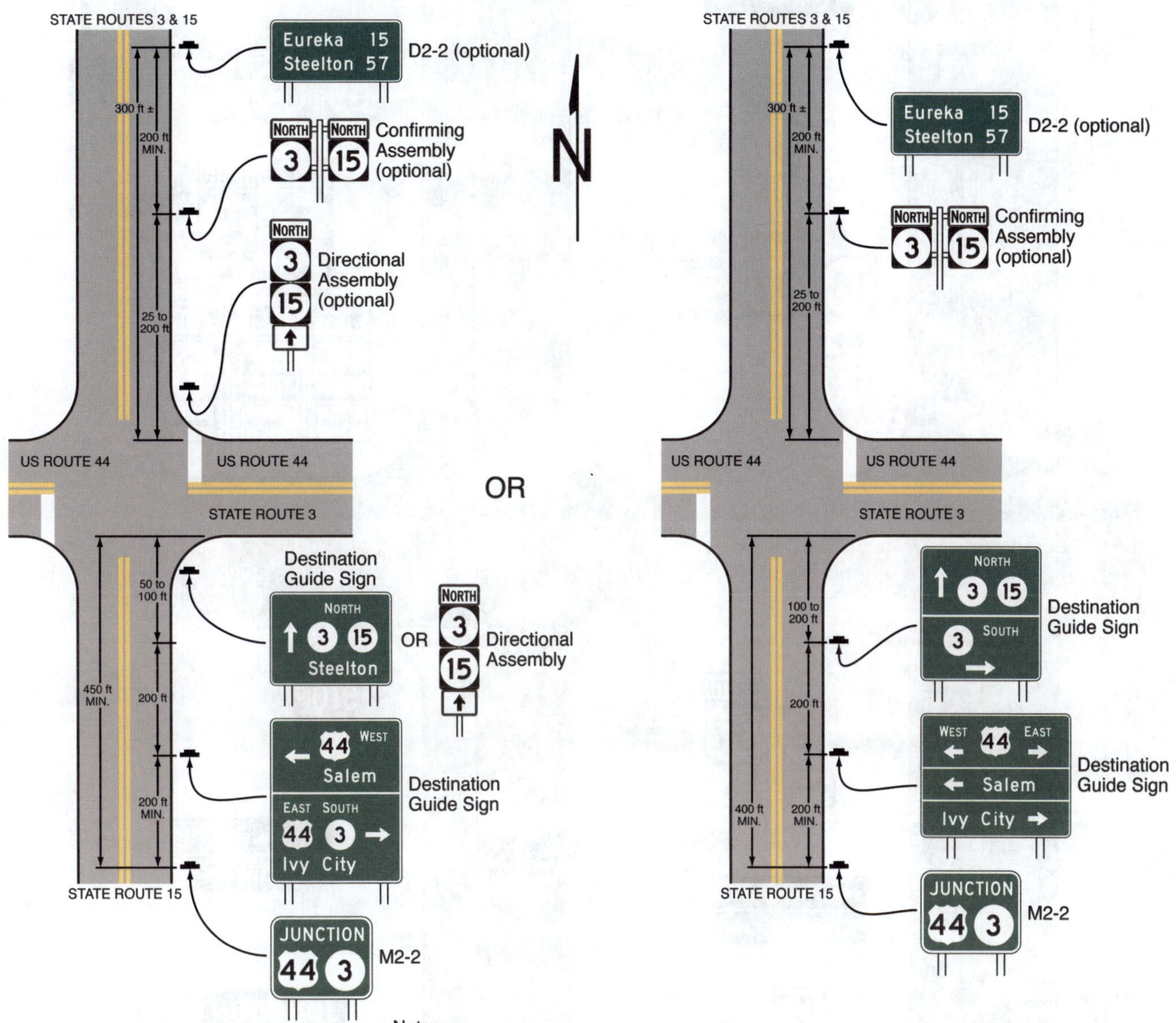

Notes:

1. Only one direction of travel and associated signs shown.
2. The spacings shown on this figure are for rural intersections. See Sections 2D.29, 2D.30, 2D.31, 2D.32, 2D.33, 2D.42, and 2D.44 for low-speed and/or urban conditions.

Figure 2D-8. Illustration of Directional Assemblies and Other Route Signs (Sheet 1 of 4)

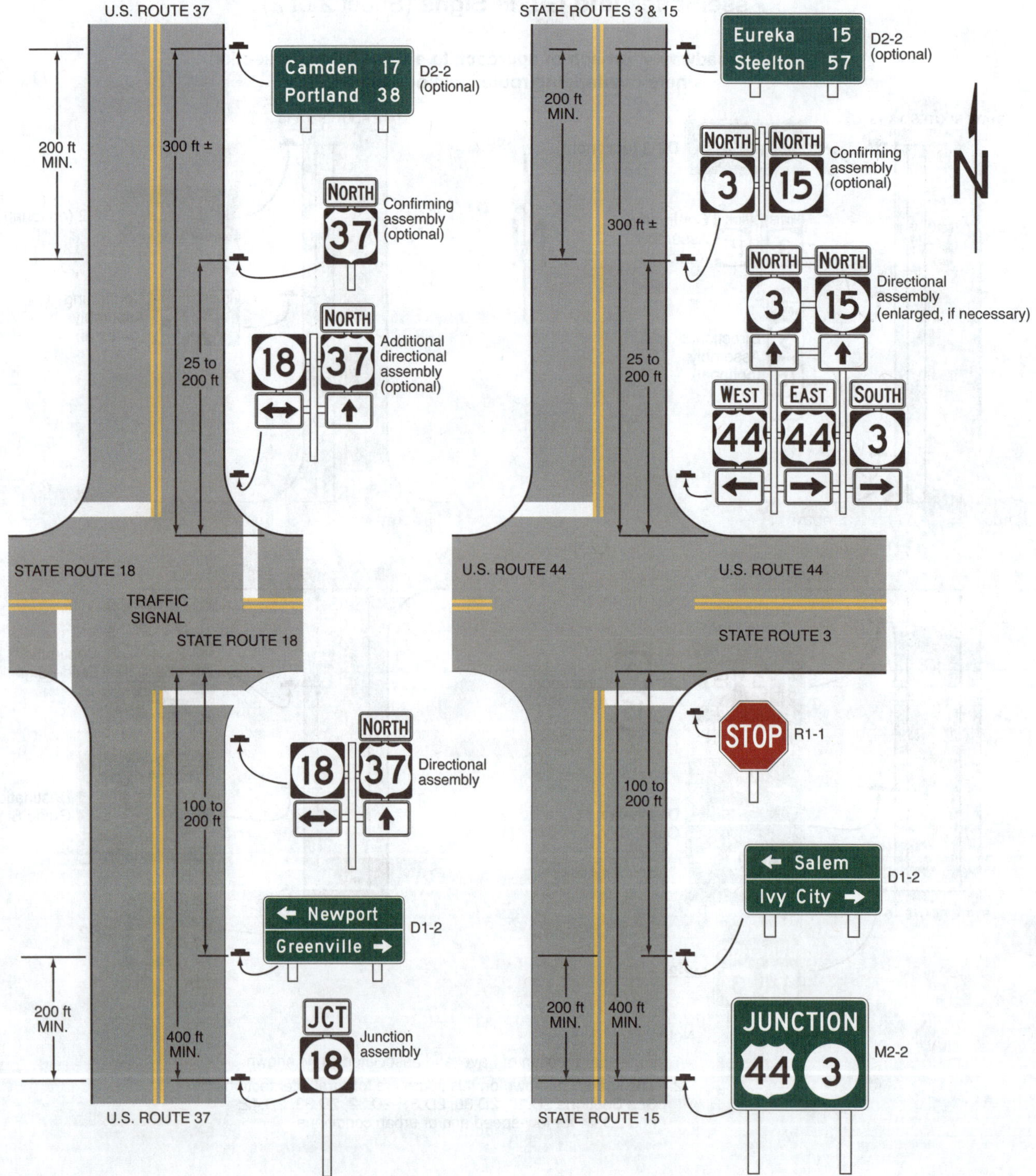

Notes:

1. Only the signs associated with the northbound direction of travel are shown on this figure.

2. The spacings shown on this figure are for rural intersections. See Sections 2D.29, 2D.30, 2D.31, 2D.32, 2D.33, 2D.42, and 2D.44 for low-speed and/or urban conditions.

Figure 2D-8. Illustration of Directional Assemblies and Other Route Signs (Sheet 2 of 4)

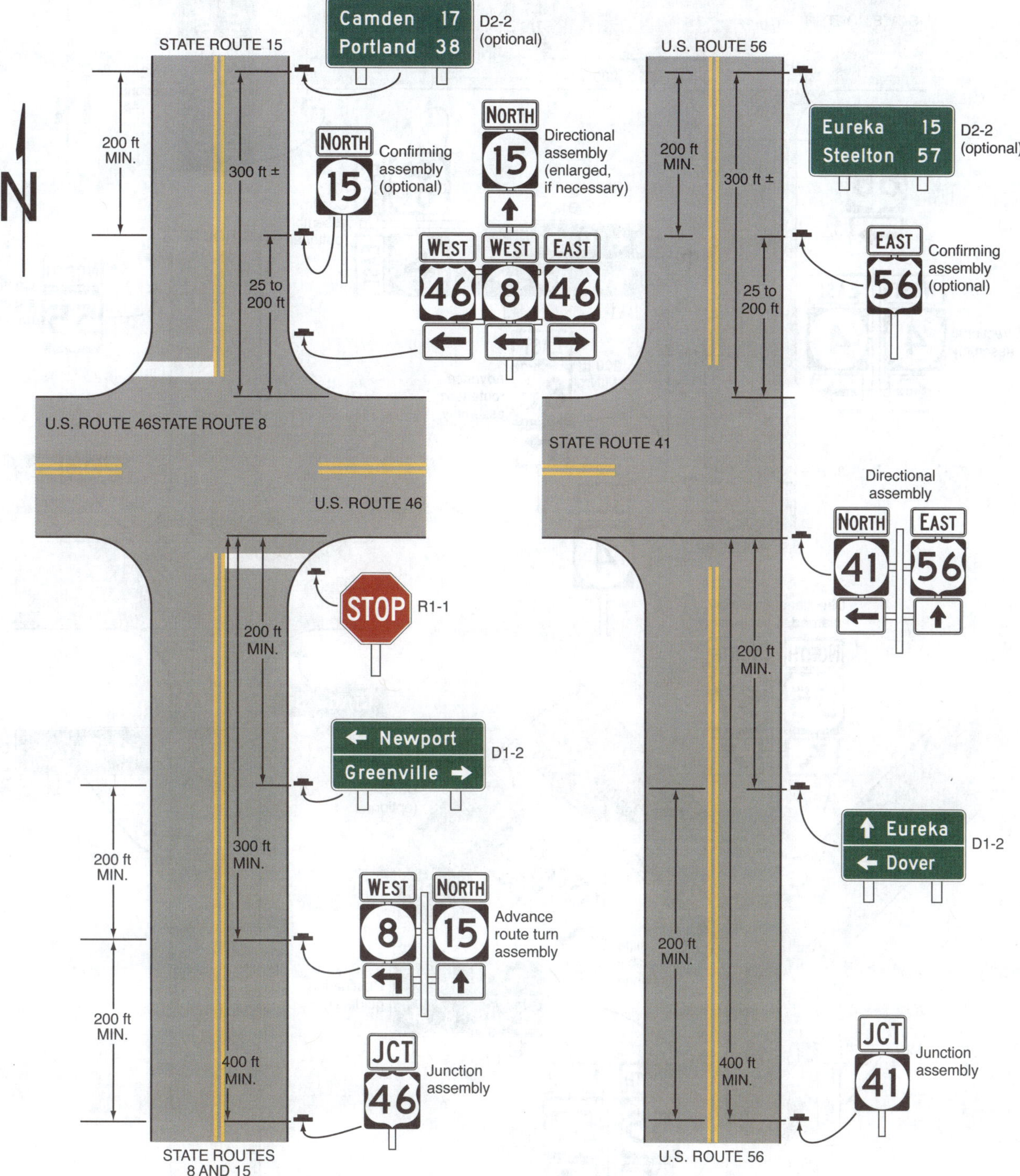

Notes:

1. Only the signs associated with the northbound direction of travel are shown on this figure.

2. The spacings shown on this figure are for rural intersections. See Sections 2D.29, 2D.30, 2D.31, 2D.32, 2D.33, 2D.42, and 2D.44 for low-speed and/or urban conditions.

Figure 2D-8. Illustration of Directional Assemblies and Other Route Signs (Sheet 3 of 4)

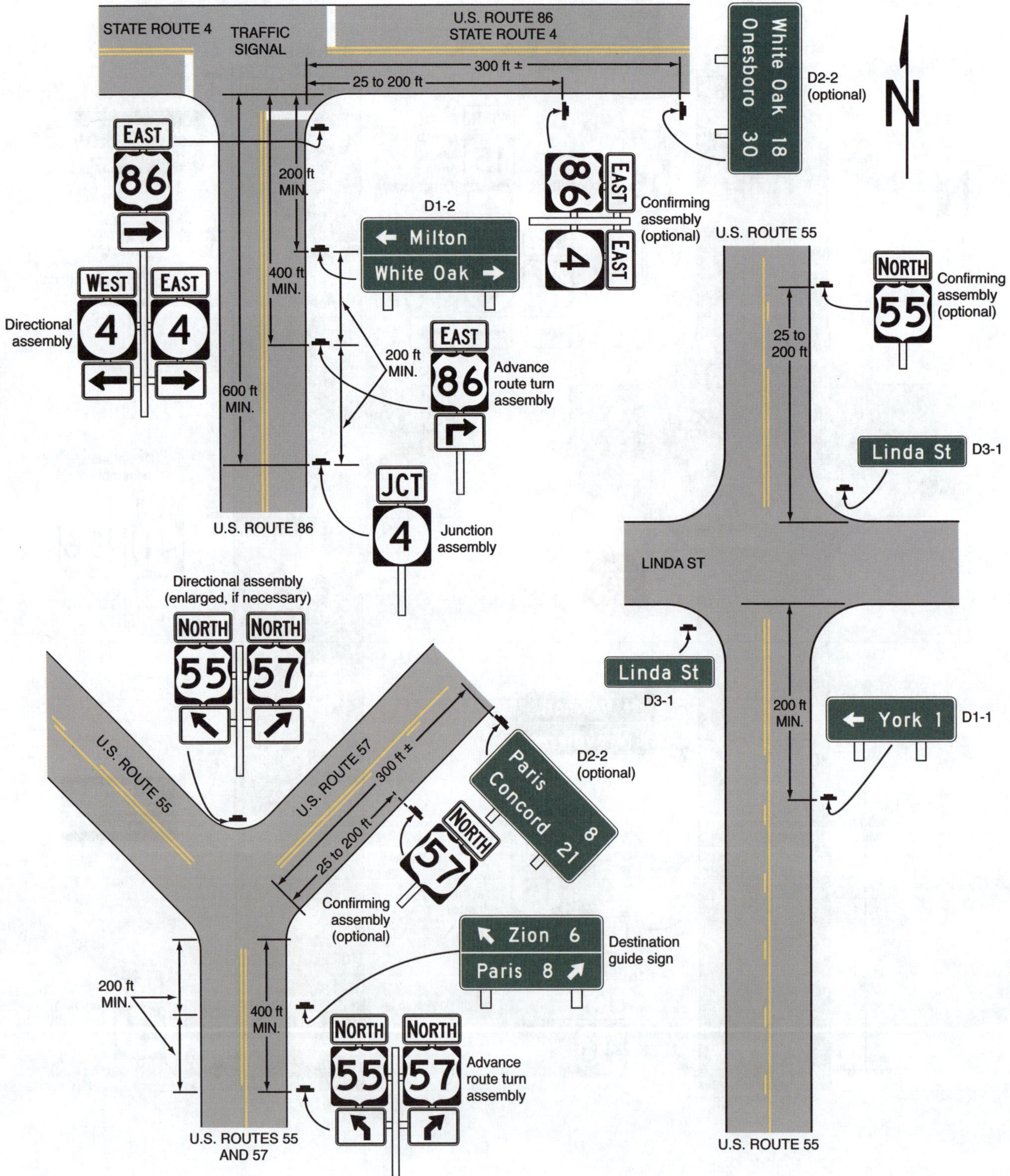

Notes:

1. Only the signs associated with the northbound direction of travel are shown on this figure.

2. The spacings shown on this figure are for rural intersections. See Sections 2D.29, 2D.30, 2D.31, 2D.32, 2D.33, 2D.42, and 2D.44 for low-speed and/or urban conditions.

Figure 2D-8. Illustration of Directional Assemblies and Other Route Signs (Sheet 4 of 4)

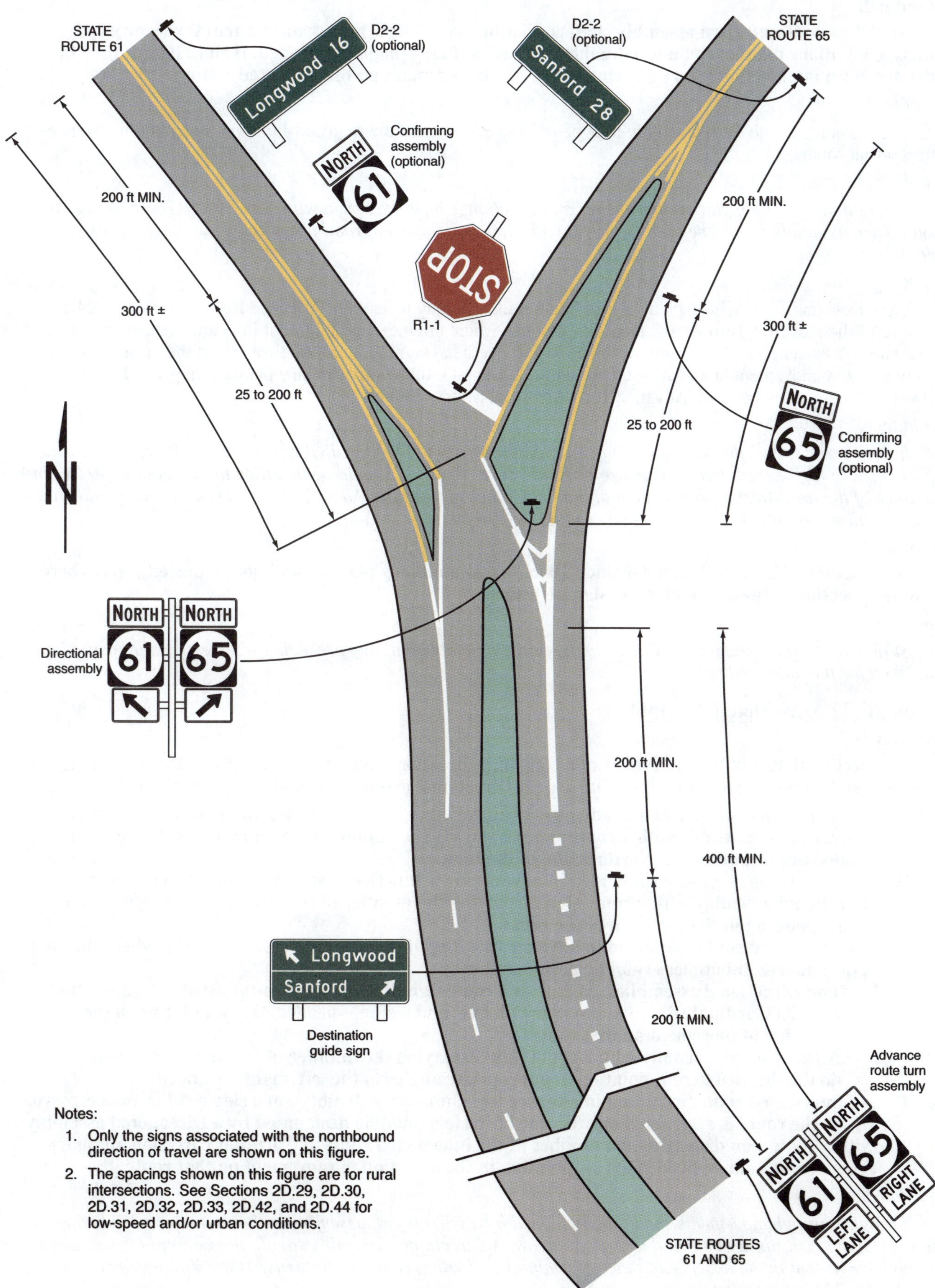

Notes:

1. Only the signs associated with the northbound direction of travel are shown on this figure.

2. The spacings shown on this figure are for rural intersections. See Sections 2D.29, 2D.30, 2D.31, 2D.32, 2D.33, 2D.42, and 2D.44 for low-speed and/or urban conditions.

Section 2D.31 Advance Route Turn Assembly

Standard:

01   **An Advance Route Turn assembly shall consist of a route sign, an Advance Turn Arrow or word message auxiliary plaque, and a Cardinal Direction auxiliary plaque, if needed. It shall be installed in advance of an intersection where a turn must be made to remain on the indicated route.**

Option:

02   The Advance Route Turn assembly may be used to supplement the required Junction assembly in advance of intersecting routes.

Guidance:

03   *Where a multi-lane highway approaches an interchange or intersection with a numbered route, the Advance Route Turn assembly should be used to provide advance notice so that road users know the correct lane(s) from which to make their turn.*

Option:

04   Lane Designation auxiliary plaques (see Section 2D.27) may be used in Advance Route Turn Assemblies in place of the Advance Turn Arrow auxiliary plaques where engineering judgment indicates that specific lane information associated with each route is needed and overhead signing is impracticable and the designated lane is a mandatory movement lane. An assembly with the Lane Designation auxiliary plaques may supplement or substitute for an assembly with Advance Turn Arrow auxiliary plaques.

Guidance:

05   *In low-speed areas, the Advance Route Turn assembly should be installed not less than 200 feet in advance of the turn. In high-speed areas, the Advance Route Turn assembly should be installed not less than 300 feet in advance of the turn. In rural areas, the minimum distance between an Advance Route Turn assembly and either a Destination sign or a Junction assembly should be 200 feet.*

Standard:

06   **An assembly that includes an Advance Turn Arrow auxiliary plaque shall not be placed where there is an intersection between it and the designated turn.**

Guidance:

07   *Sufficient distance should be allowed between the assembly and any preceding intersection that could be mistaken for the indicated turn.*

Section 2D.32 Directional Assembly

Standard:

01   **A Directional assembly shall consist of a Cardinal Direction auxiliary plaque, if needed; a route sign; and a Directional Arrow auxiliary plaque. The uses of Directional assemblies shall comply with the following:**

   A. Turn movements (indicated in advance by an Advance Route Turn assembly) shall be marked by a Directional assembly with a route sign displaying the number of the turning route and a single-headed arrow pointing in the direction of the turn.

   B. The beginning of a route (indicated in advance by a Junction assembly) shall be marked by a Directional assembly with a route sign displaying the number of that route and a single-headed arrow pointing in the direction of the route.

   C. An intersected route (indicated in advance by a Junction assembly) on a crossroad where the route is designated on both legs shall be designated by:

   1. Two Directional assemblies, each with a route sign displaying the number of the intersected route, a Cardinal Direction auxiliary plaque, and a single-headed arrow pointing in the direction of movement on that route; or

   2. A Directional assembly with a route sign displaying the number of the intersected route and a double-headed arrow, pointing at appropriate angles to the left, right, or ahead.

   D. An intersected route (indicated in advance by a Junction assembly) on a side road or on a crossroad where the route is designated only on one of the legs shall be designated by a Directional assembly with a route sign displaying the number of the intersected route, a Cardinal Direction auxiliary plaque, and a single-headed arrow pointing in the direction of movement on that route.

Guidance:

02   *Straight-through movements should be indicated by a Directional assembly with a route sign displaying the number of the continuing route and a vertical arrow. A Directional assembly should not be used for a straight-through movement in the absence of other assemblies indicating right or left turns, as the Confirming assembly sign beyond the intersection normally provides adequate guidance.*

03   *Directional assemblies should be located on the near right corner of the intersection. At major intersections and at Y or offset intersections, additional Directional assemblies should be installed on the far right or left corner to confirm the near-side assemblies. When the near-corner position is impractical for Directional assemblies, the far right corner should be the preferred alternative, with oversized signs, if necessary, for legibility. Where unusual conditions exist, the location of a Directional assembly should be determined by engineering judgment with the goal being to provide the best possible combination of view and safety.*

Support:

04   It is more important that guide signs be readable, and that the information and direction displayed thereon be readily understood, at the appropriate time and place than to be located with absolute uniformity.

05   Figure 2D-8 shows typical placements of Directional assemblies.

Section 2D.33 Confirming or Reassurance Assemblies

Standard:

01   **If used, Confirming or Reassurance assemblies shall consist of a Cardinal Direction auxiliary plaque and a route sign. Where the Confirming or Reassurance assembly is for an alternative route, the appropriate auxiliary plaque for an alternative route (see Section 2D.16) shall also be included in the assembly.**

Guidance:

02   *A Confirming assembly should be installed just beyond intersections of numbered routes. It should be placed 25 to 200 feet beyond the far shoulder or curb line of the intersected highway.*

03   *If used, Reassurance assemblies should be installed between intersections in urban areas as needed, and beyond the built-up area of any incorporated city or town.*

04   *Route signs for either confirming or reassurance purposes should be spaced at such intervals as necessary to keep road users informed of their routes.*

Section 2D.34 Trailblazer Assembly

Support:

01   Trailblazer assemblies provide directional guidance to a particular road facility from other highways in the vicinity. This guidance is accomplished by installing Trailblazer assemblies at strategic locations to indicate the direction to the nearest or most convenient point of access. The use of the word TO indicates that the road or street where the sign is posted is not a part of the indicated route, and that a road user is merely being directed progressively to the route.

Standard:

02   **A Trailblazer assembly shall consist of a TO auxiliary plaque (see Section 2D.21), a route sign for a numbered or named highway (see Section 2D.56) or an identification sign for a byway, historic trail, or auto tour route sign (see Sections 2D.57 and 2D.58), and a single-headed Directional Arrow auxiliary plaque pointing in the direction leading to the route. Where the Trailblazer assembly is for an alternative route, the appropriate auxiliary plaque for an alternative route (see Section 2D.16) shall also be included in the assembly.**

Option:

03   A Cardinal Direction auxiliary plaque (see Section 2D.15) may be used in a Trailblazer assembly where the direction leading to the route provides access only to one direction of travel for that route.

Guidance:

04   *The TO auxiliary plaque, Cardinal Direction auxiliary plaque, and Directional Arrow auxiliary plaque should be of the standard size provided for auxiliary plaques of their respective type. The route sign should be the size provided in Section 2D.11.*

Option:

05   Trailblazer assemblies may be installed with other Route Sign assemblies, or alone, in the immediate vicinity of the designated facilities.

DESTINATION AND DISTANCE SIGNS

Section 2D.35 Destination and Distance Signs

Support:

01   In addition to guidance by route numbers, it is desirable to supply the road user information concerning the destinations that can be reached by way of numbered or unnumbered routes. This is done by means of Destination signs and Distance signs.

Option:

02   Route shields and cardinal directions may be included on the Destination sign with the destinations and arrows.

Guidance:

03   *If Route shields and cardinal directions are included on a Destination sign, the height of the route shields should be at least two times the height of the upper-case letters of the principal legend and not less than 18 inches, and the letter height of cardinal directions should be at least the minimum letter height specified for these signs.*

04   *If used, destination names on low-volume rural roads should be as specific and descriptive as possible. Destinations such as campgrounds, ranger stations, and recreational areas should be clearly indicated so that they are not interpreted to be communities or locations with road user services.*

Section 2D.36 Destination Signs (D1 Series)

Standard:

01   **Except on approaches to interchanges (see Section 2D.49), the Destination (D1-1 through D1-3) signs (see Figure 2D-9), if used, shall be a horizontal rectangle displaying the name of a city, town, village, or other traffic generator, and a directional arrow.**

Option:

02   The distance (see Section 2D.43) to the place named may also be displayed on the Destination (D1-1a through D1-3a) signs (see Figure 2D-9). If several destinations are to be displayed at a single point, the several names may be placed on a single sign with an arrow (and the distance, if desired) for each name. If more than one destination lies in the same direction, a single arrow may be used for such a group of destinations.

Guidance:

03   *Adequate separation should be made between any destinations or group of destinations in one direction and those in other directions by suitable design of the arrow, spacing of lines of legend, heavy lines entirely across the sign, or separate signs.*

Support:

04   Separation of destinations by direction by the use of a horizontal separator line can enhance the readability of a Destination sign by relating an arrow and its corresponding destination(s) and by eliminating the need for multiple arrows that point in the same direction and excessive space between lines of legend.

Standard:

05   **Except as otherwise provided in this Manual, an arrow pointing to the right shall be at the extreme right of the sign, and an arrow pointing left or up shall be at the extreme left. The distance numerals, if used, shall be placed to the right of the destination names.**

Option:

06   An arrow pointing up may be placed at the extreme right of the sign when the sign is mounted to the left of the traffic to which it applies.

Guidance:

07   *Unless a sloping arrow will convey a clearer indication of the direction to be followed, the directional arrows should be horizontal or vertical.*

08   *If several individual name signs are assembled into a group, all signs in the assembly should be of the same horizontal width.*

09   *Destination signs should be used:*

    A. *At the intersections of U.S. or State numbered routes with Interstate, U.S., or State numbered routes; and*

    B. *At points where they serve to direct traffic from U.S. or State numbered routes to the business section of towns, or to other destinations reached by unnumbered routes.*

Figure 2D-9. Destination and Distance Signs

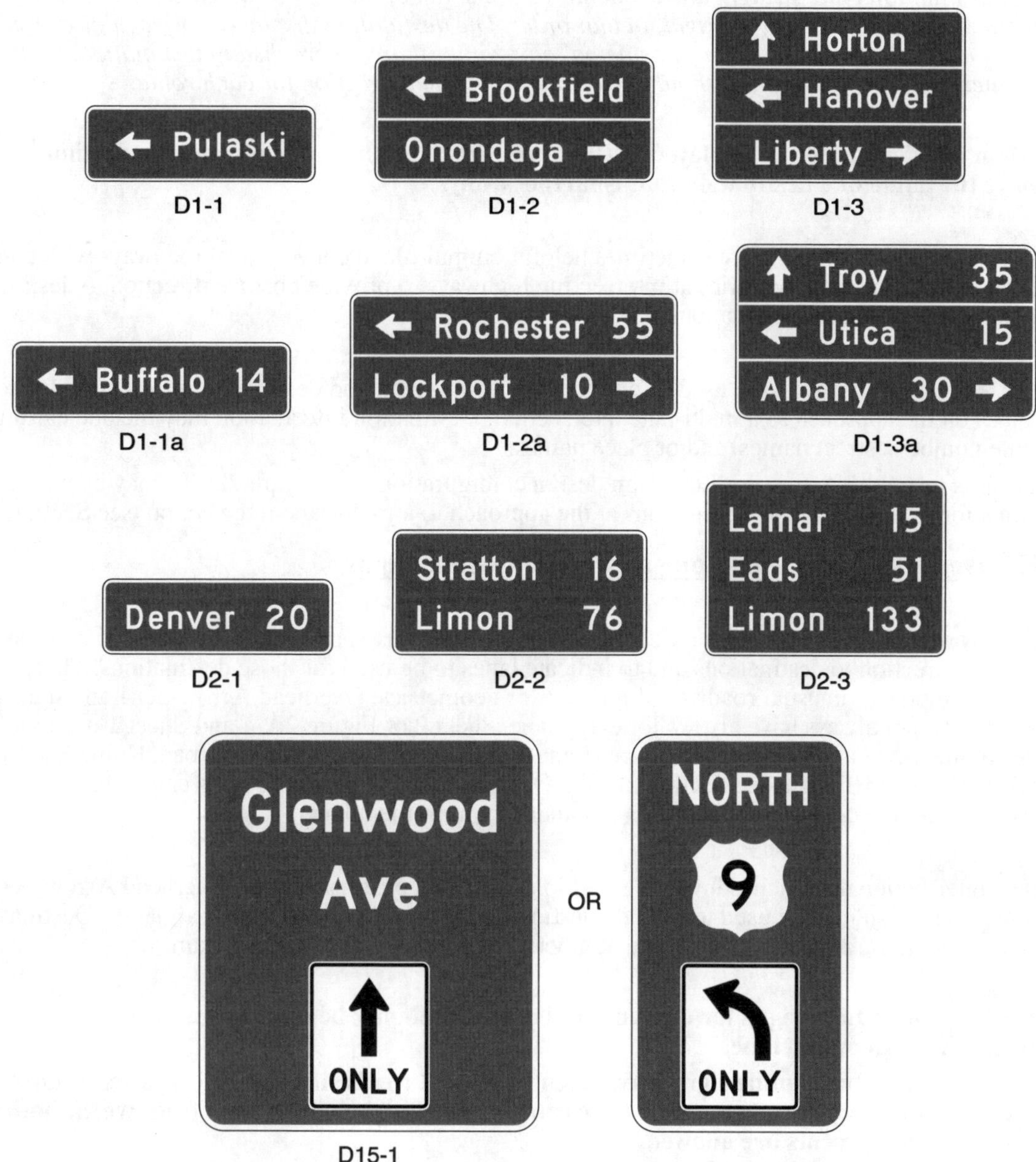

Standard:

10   **Where a total of three or fewer destinations are displayed on the Advance guide (see Section 2E.23) and Supplemental guide (see Section 2E.51) signs, no more than three destination names shall be displayed on a Destination sign. Where four destinations are displayed on the Advance guide and Supplemental guide signs, no more than four destination names shall be displayed on a Destination sign.**

Guidance:

11   *If space permits, four destinations should be displayed on two separate signs at two separate locations.*

Option:

12   *Where space does not permit, or where all four destinations are in one direction, a single sign may be used. Where a single sign is used and all destinations are in the same direction, the arrow may be placed below the destinations for the purpose of enhancing the conspicuity of the arrow.*

Standard:

13   **Where a single four-name sign assembly is used, a heavy line approximating the width of the sign border entirely across the sign or separate signs shall be used to separate destinations by direction.**

Guidance:

14   *The closest destination lying straight ahead should be at the top of the sign or assembly, and below it the closest destinations to the left and to the right, in that order. The destination displayed for each direction should ordinarily be the next county seat or the next principal city, rather than a more distant destination. In the case of overlapping routes, only one destination should be displayed in each direction for each route.*

Standard:

15   **If more than one destination is displayed in the same direction, the name of a nearer destination shall be displayed above the name of a destination that is farther away.**

Support:

16   Overhead destination guide signs are sometimes helpful on multi-lane conventional roadways with complex or unusual roadway alignments or geometrics at intersecting highways to provide positive direction to destinations and to assign lanes to be used for destinations.

Option:

17   Overhead Destination signs may be used to provide lane assignment and destination information for some or all of the lanes on the approach to a multi-lane intersection. Destination information may include cardinal directions, route numbers, street names, and/or place names.

18   Overhead signs using the Arrow-per-Lane sign design configuration (see Figure 2D-10) may be used to provide lane assignments for some or all lane destinations at the approach to a multi-lane intersection (see Section 2D.37).

Section 2D.37 Overhead Arrow-per-Lane Destination Guide Signs

Support:

01   Overhead Arrow-per-Lane destination guide signs are sometimes used on multi-lane conventional roadways to provide positive direction to destinations and to indicate lanes to be used for those destinations. These locations typically include complex or unusual roadway alignments or geometrics. Overhead Arrow-per-Lane signs on conventional roads do not always have arrows for every lane. Sheet 2 of Figure 2A-4 and Sheet 1 of Figure 2D-10 show examples of the use of an Overhead Arrow-per-Lane Guide sign on a conventional road. Unlike the Combined Lane-Use/Destination (D15-1) sign (see Section 2D.38), Overhead Arrow-per-Lane signs can be used to provide lane assignments where the designated lane is not a mandatory movement lane.

Option:

02   At complex intersection approaches involving multiple lanes and destinations, an Overhead Arrow-per-Lane destination guide sign may be used to provide destination information for some or all lanes. Destination information may include cardinal direction, route numbers, street names, and/or place names.

Standard:

03   **Overhead Arrow-per-Lane signs for conventional roads shall only be used for multi-lane approaches to intersections that have an option lane.**

04   **Overhead Arrow-per-Lane guide signs used on conventional roads shall include as a minimum one arrow above each mandatory turn lane and a bifurcated arrow for the option lane from which both the through and turning movements are allowed.**

Guidance:

05   *Displaying an arrow over each through movement lane that does not allow turning should be considered for providing additional positive guidance.*

Standard:

06   **Overhead Arrow-per-Lane signs for conventional roads shall be designed in accordance with the following criteria:**

    A. The shaft of each arrow shall be located over the approximate center of the lane to which it applies.

    B. Arrows for continuing through lanes shall be vertically upward-pointing (see Figure 2D-10).

    C. The arrow for a lane that must turn shall be curved in the direction of the turn and shall be accompanied by a black-on-yellow ONLY (E11-1b) sign panel (see Figure 2E-17) adjacent to the lower end of the arrow shaft.

    D. The arrow for an optional exit lane that also carries the through route shall have a single shaft that bifurcates into a vertically upward-pointing arrow and a curving arrow corresponding to the configuration of the through and turn lanes.

    E. A vertical white line shall be used to separate the route shields and destinations for the two diverging movements from each other.

    F. The number of lanes displayed on a sign shall correspond to the number of lanes being signed for at the location of that sign. An advance sign shall not depict lanes that are added downstream of a sign location.

Figure 2D-10. Examples of Overhead Signs at an Intersection (Sheet 1 of 3)

A – Combination of mandatory and optional movement lanes

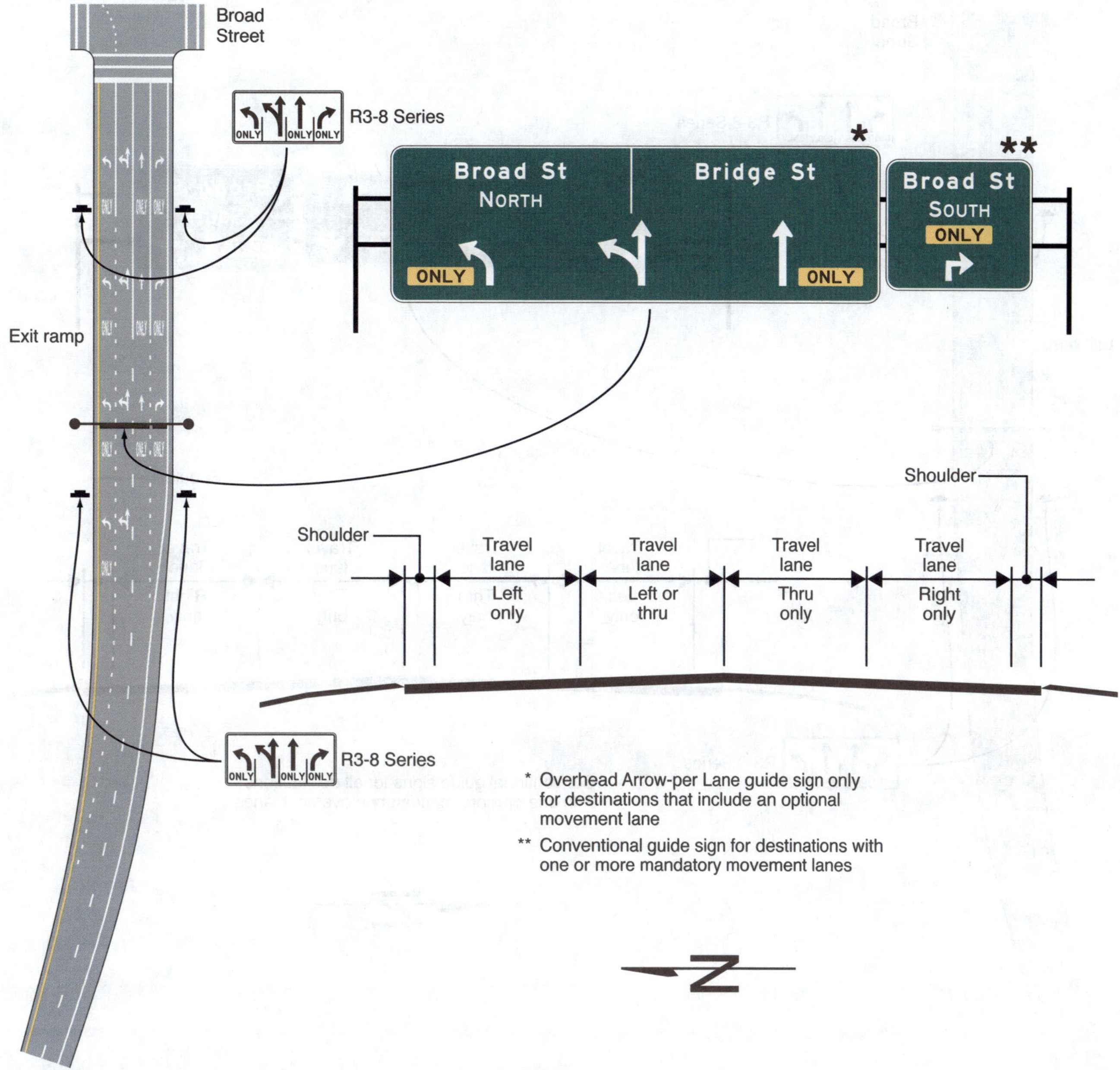

Guidance:

07   *Overhead Arrow-per-Lane guide signs used on conventional roads should be designed in accordance with the following additional criteria:*

   A. *No more than one destination should be displayed for each movement, and no more than three destinations should be displayed per sign.*

   B. *The arrowhead(s) for the diverging movement should be positioned lower on the sign than the arrowhead(s) for the movement that continues straight ahead.*

   C. *Route shields, cardinal directions, and destinations should be positioned on the sign such that they are clearly related to the arrowhead(s) for the movement to which they apply.*

   D. *The vertical white line that is used to separate the route shields and destinations for the two diverging movements from each other should not descend below the top of the arrowheads for the through lanes, and should be positioned approximately halfway between the diverging arrowheads for the optional movement lane.*

08   *Destination information should be kept to a minimum necessary to provide positive guidance without overloading the road user.*

Figure 2D-10. Examples of Overhead Signs at an Intersection (Sheet 2 of 3)

B – Mandatory movement lanes

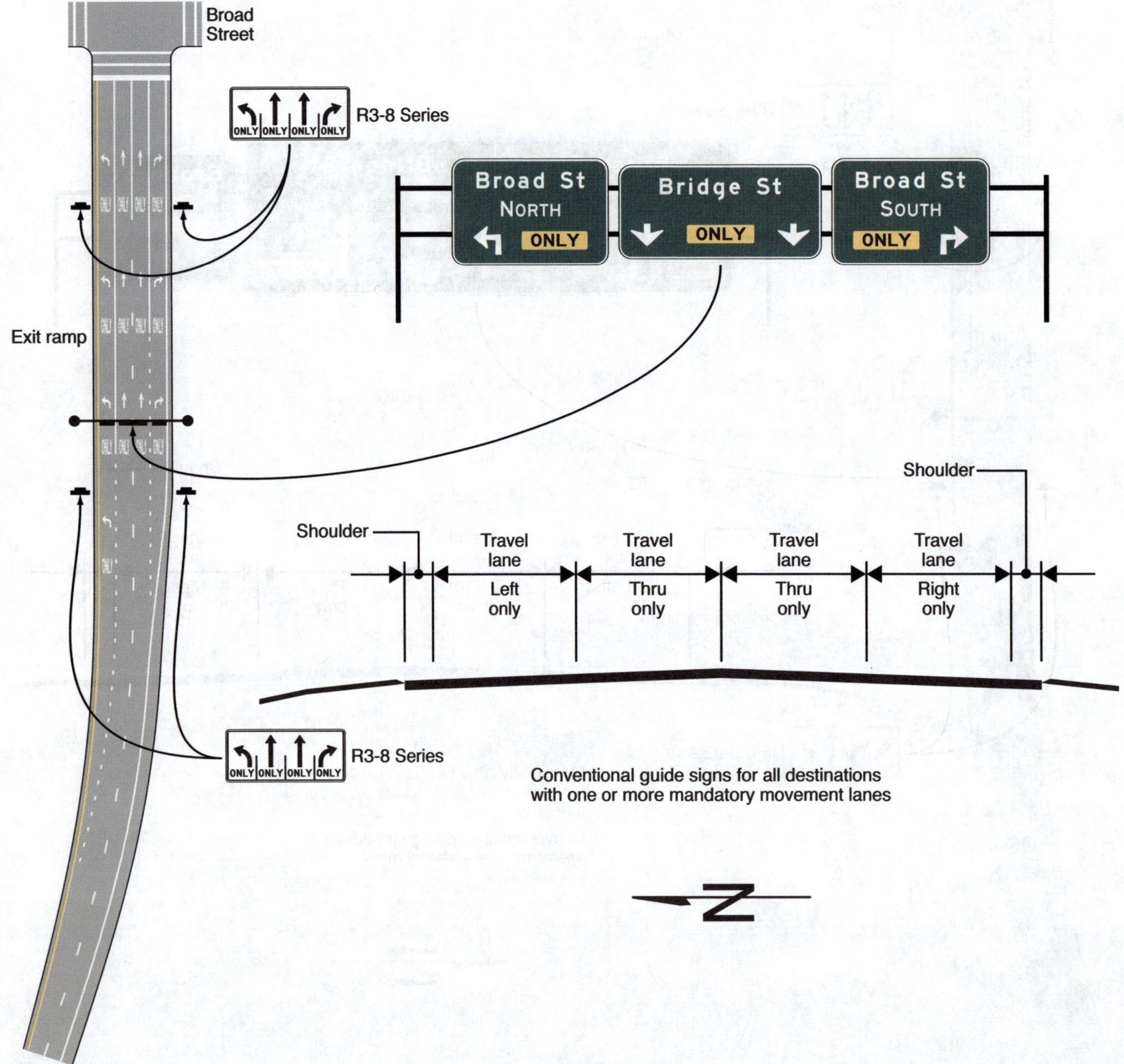

Standard:

09   **The minimum height of arrows on an Overhead Arrow-per-Lane sign used on a conventional road shall be as shown in Table 2D-5.**

Guidance:

10   *When letter heights and other sign legend elements are enlarged there should be an corresponding increase in the arrow size used.*

Option:

11   Curved-stem arrows may be substituted on Overhead Arrow-per-Lane signs on multi-lane approaches to a circular intersection with an option lane (see Section 2D.39).

Table 2D-5. Overhead Arrow-per-Lane Arrow Height Based on Principal Legend Letter Height

Principal Legend Letter Height	Straight Arrow	Turn Arrow
10.67 - 13.33	32	24
8	24	18

Note: Letter and arrow heights are shown in inches.

Figure 2D-10. Examples of Overhead Signs at an Intersection (Sheet 3 of 3)

C – Mandatory movement lanes with dual turn lanes

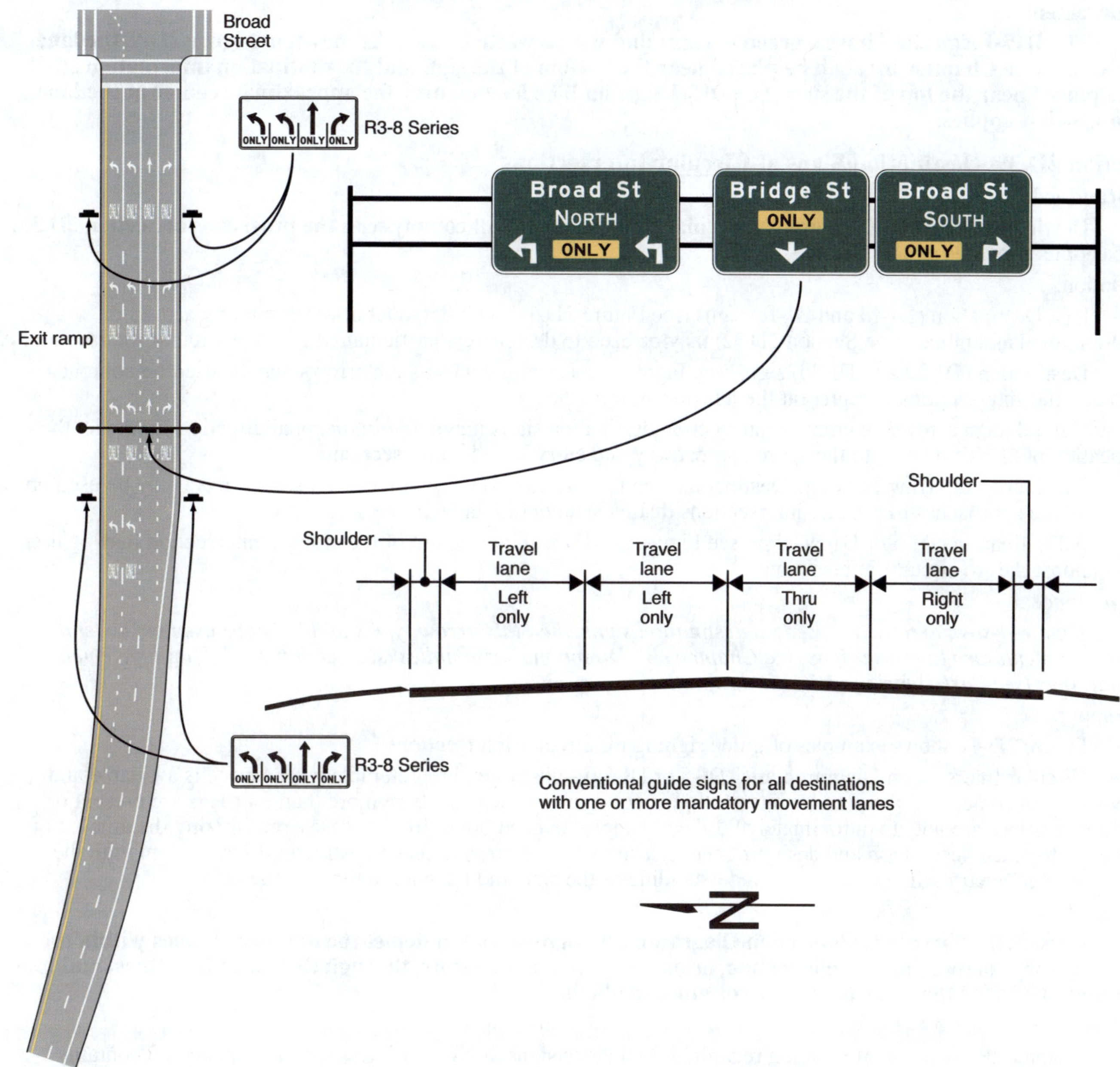

Section 2D.38 Combination Lane-Use/Destination Overhead Guide Sign (D15-1)

Option:

01   At intersection approaches involving multiple turn lanes and destinations, a Combination Lane-Use/ Destination (D15-1) overhead guide sign (see Figure 2D-9) that combines a lane-use regulatory sign with destination information such as a cardinal direction, a route number, a street name, and/or a place name may be used.

Support:

02   At such locations, the combined information on the D15-1 signs can be even more effective than separate lane-use and guide signs for conveying to unfamiliar drivers which lane or lanes to use for a particular destination.

03   Figure 2D-9 shows an example of a D15-1 sign that combines lane-use and street name information and an example of a D15-1 sign that combines lane-use, cardinal direction, and route number information.

Standard:

04   **The Combination Lane-Use/Destination (D15-1) overhead guide sign shall be used only where the designated lane is a mandatory movement lane. The D15-1 sign shall not be used for lanes with optional movements.**

05   **The D15-1 sign shall have a green background with a white border. As shown in Figure 2D-9, the lane-use sign (see Chapter 2B) shall be placed near the bottom of the sign and the destination information shall be placed near the top of the sign. The D15-1 sign shall be located over the approximate center of the lane to which it applies.**

Section 2D.39 <u>Destination Signs at Circular Intersections</u>

Standard:

01   **Destination signs that are used at circular intersections shall comply with the provisions of Section 2D.36, except as provided in this Section.**

Option:

02   Exit Destination (D1-1d and D1-1e) signs (see Figure 2D-11) with diagonal upward-pointing arrows or Directional assemblies (see Section 2D.32) may be used to designate a particular exit from a circular intersection.

03   Destination (D1-2d and D1-3d) signs (see Figure 2D-11) with curved-stem arrows may be used on approaches to circular intersections to represent the left-turn movements.

04   Curved-stem arrows on circular intersection destination signs may point in diagonal directions to depict the location of an exit relative to the approach roadway and entry into the intersection.

05   An Overhead Arrow-per-Lane Destination sign (see Section 2D.37) with curved-stem arrows may be used on multi-lane approaches to circular intersections that have an option lane.

06   A Destination (D1-5 or D1-5a) sign (see Figure 2D-11) with a diagram of the circular intersection may be used on approaches to circular intersections.

Guidance:

07   *If curved-stem arrows are used on destination signs, then this arrow type should also be used consistently on any regulatory lane-use signs (see Chapter 2B), Directional assemblies (see Section 2D.32), and pavement markings (see Part 3) for a particular destination or movement.*

Support:

08   Figure 2D-12 shows examples of guide signing for circular intersections.

09   Circular Intersection Diagrammatic (D1-5 or D1-5a) signs might be preferable where space is available and where the geometry of the circular intersection is non-typical, such as where more than four legs are present or where the legs are not at approximately 90-degree angles to each other. In such cases, minimizing the amount of legend for each destination and designing the sign so that the arrows for each destination clearly align with the roadway geometry will aid road user understanding of the sign and navigation through the area.

Standard:

10   **If used, the Circular Intersection Diagrammatic signs shall not depict the number of lanes within the circulatory roadway of the intersection, or on its approaches or exits, through the use of lane lines, multiple arrow shafts for the same movement, or other methods.**

Support:

11   Chapter 2B contains information regarding regulatory signs at circular intersections, Chapter 2C contains information regarding warning signs at circular intersections, and Chapter 3D contains information regarding pavement markings at circular intersections.

Section 2D.40 <u>Destination Signs at Jughandles</u>

Standard:

01   **Destination signs that are used at jughandles shall comply with the provisions of Section 2D.36.**

Support:

02   Section 2B.35 contains information regarding regulatory signs for jughandle turns. Figure 2B-9 shows examples of regulatory and destination guide signing for various types of jughandle turns.

Section 2D.41 <u>Destination Signs at Intersections with Indirect Turning Movements</u>

Guidance:

01   *A system of guide signs along with associated lane markings should be used to direct traffic through intersections with indirect turning movements.*

Support:

02   Figure 2D-13 shows examples of destination guide signing for intersections with indirect turning movements.

Figure 2D-11. Destination Signs for Circular Intersections

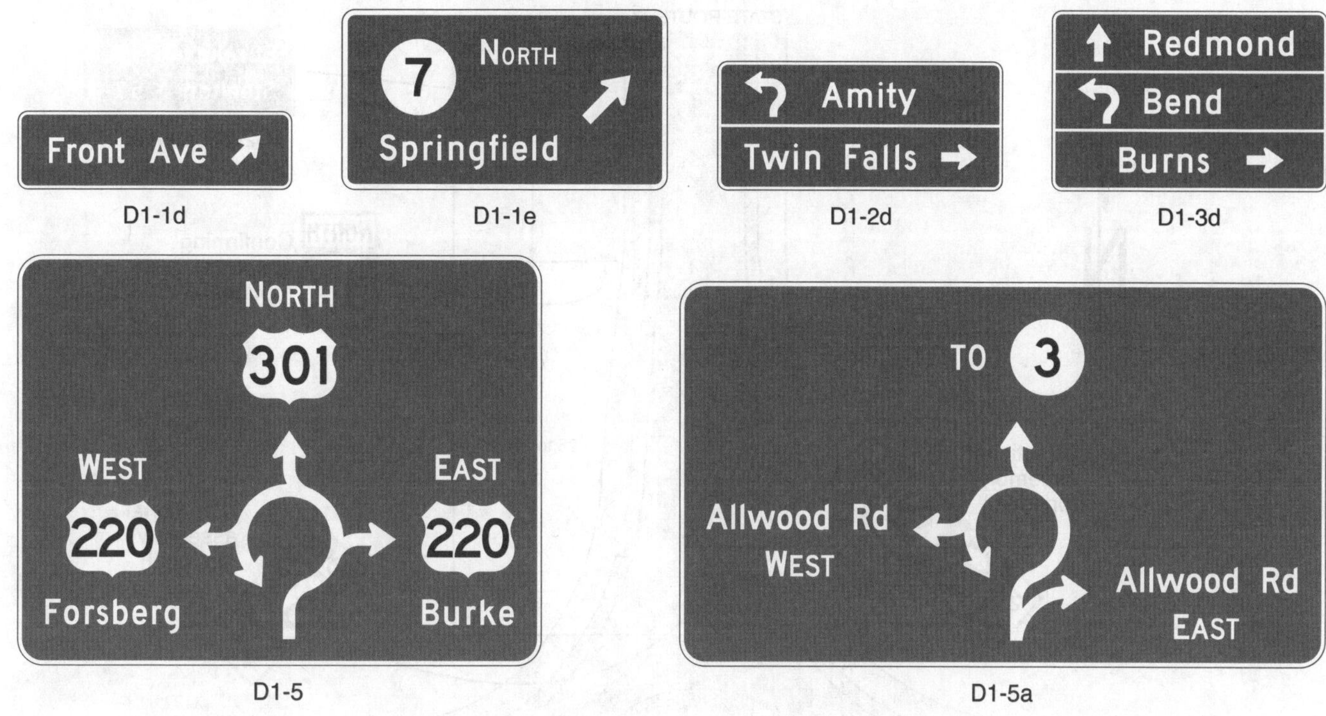

Section 2D.42 Location of Destination Signs

Guidance:

01   *When used in high-speed areas, Destination signs should be located 200 feet or more in advance of the intersection, and following any Junction or Advance Route Turn assemblies that might be required. In rural areas, the minimum distance between a Destination sign and either an Advance Route Turn assembly or a Junction assembly should be 200 feet.*

Option:

02   In urban areas, advance distances shorter than those specified in Paragraph 1 of this Section may be used.

03   Because the Destination sign is of lesser importance than the Junction, Advance Route Turn, or Directional assemblies, the Destination sign may be eliminated where the distance in which to provide adequate sign spacing is limited.

Support:

04   Figure 2D-8 shows typical placements of Destination signs.

Section 2D.43 Distance Signs (D2 Series)

Standard:

01   **If used, the Distance (D2-1 through D2-3) signs (see Figure 2D-9) shall be a horizontal rectangle of a size appropriate for the required legend, displaying the names of no more than three cities, towns, junctions, or other traffic generators, and the distance (to the nearest mile) to those places.**

02   **The distance numerals shall be placed to the right of the destination names as shown in Figure 2D-9.**

Guidance:

03   *The distance displayed should be selected on a case-by-case basis by the jurisdiction that owns the road or by statewide policy. A well-defined central area or central business district should be used where one exists. In other cases, the layout of the community should be considered in relation to the highway being signed and the decision based on where it appears that most drivers would feel that they are in the center of the community in question.*

04   *The top name on the Distance sign should be that of the next place on the route having a post office or a railroad station, a route number or name of an intersected highway, or any other significant geographical identity. The bottom name on the sign should be that of the next major destination or control city. If three destinations are displayed, the middle line should be used to indicate communities of general interest along the route or important route junctions.*

Option:

05   The choice of names for the middle line may be varied on successive Distance signs to give road users additional information concerning communities served by the route.

Figure 2D-12. Examples of Guide Signs for Circular Intersections (Sheet 1 of 4)

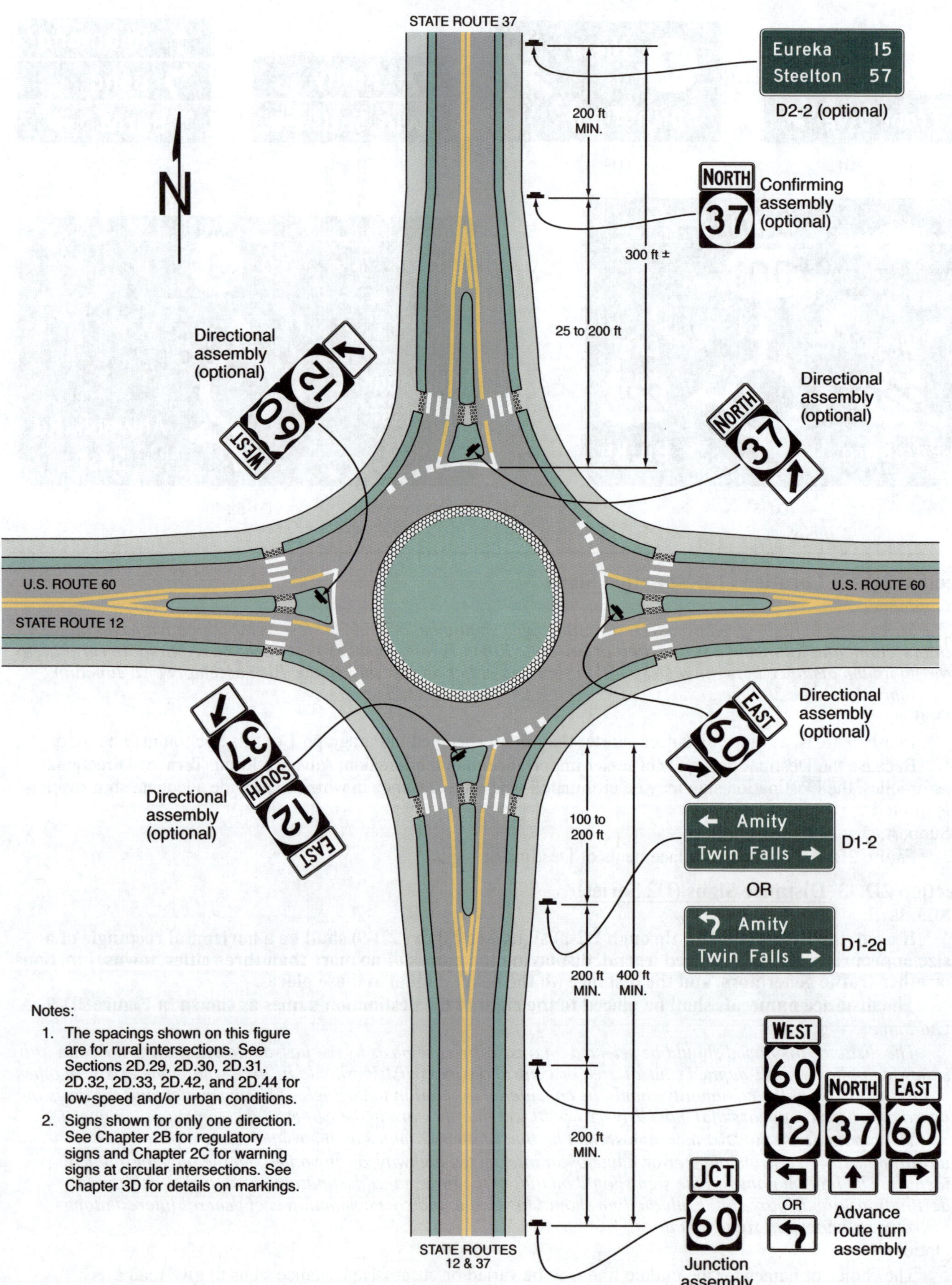

Notes:

1. The spacings shown on this figure are for rural intersections. See Sections 2D.29, 2D.30, 2D.31, 2D.32, 2D.33, 2D.42, and 2D.44 for low-speed and/or urban conditions.

2. Signs shown for only one direction. See Chapter 2B for regulatory signs and Chapter 2C for warning signs at circular intersections. See Chapter 3D for details on markings.

Figure 2D-12. Examples of Guide Signs for Circular Intersections (Sheet 2 of 4)

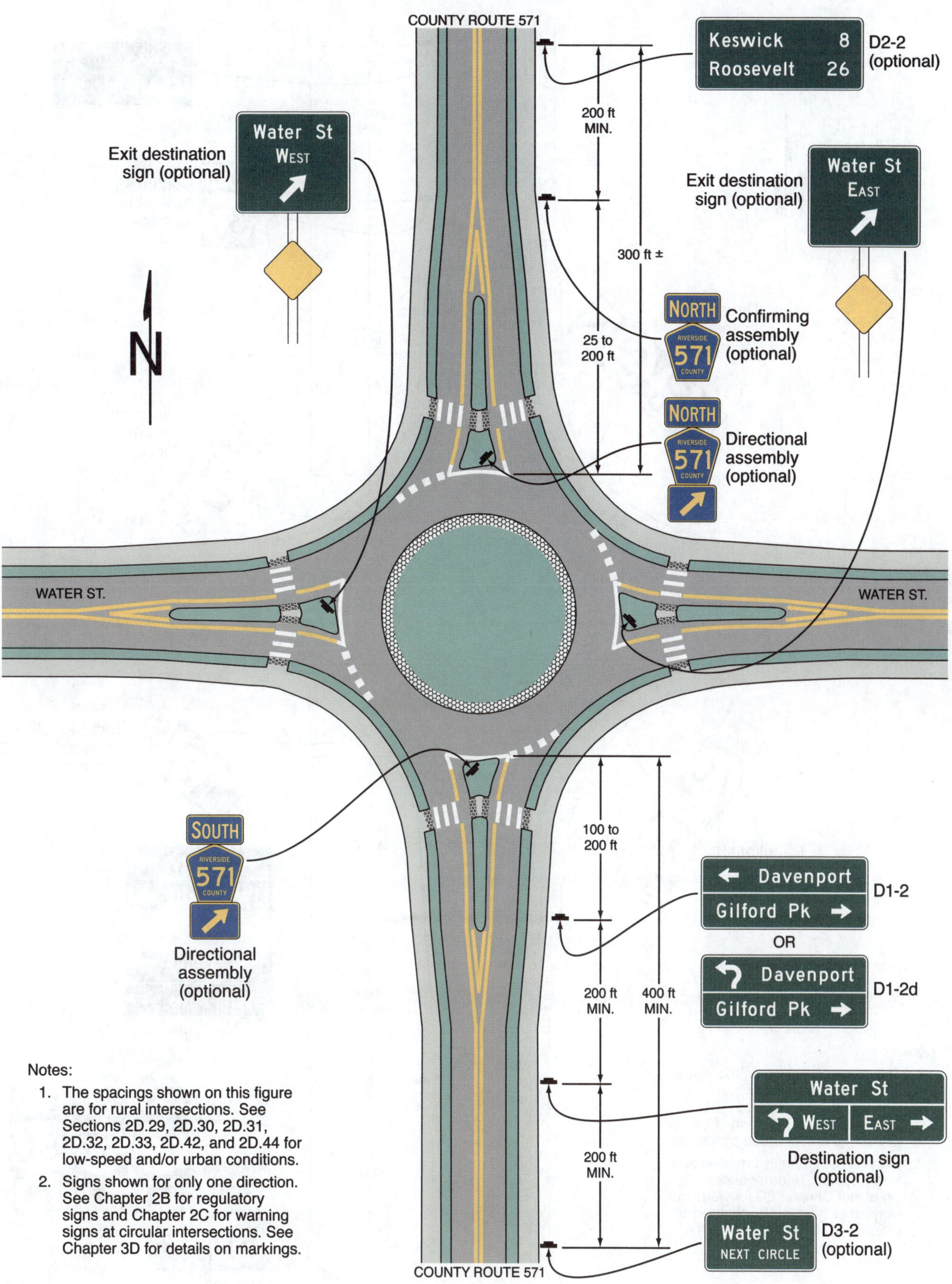

Notes:

1. The spacings shown on this figure are for rural intersections. See Sections 2D.29, 2D.30, 2D.31, 2D.32, 2D.33, 2D.42, and 2D.44 for low-speed and/or urban conditions.

2. Signs shown for only one direction. See Chapter 2B for regulatory signs and Chapter 2C for warning signs at circular intersections. See Chapter 3D for details on markings.

Figure 2D-12. Examples of Guide Signs for Circular Intersections (Sheet 3 of 4)

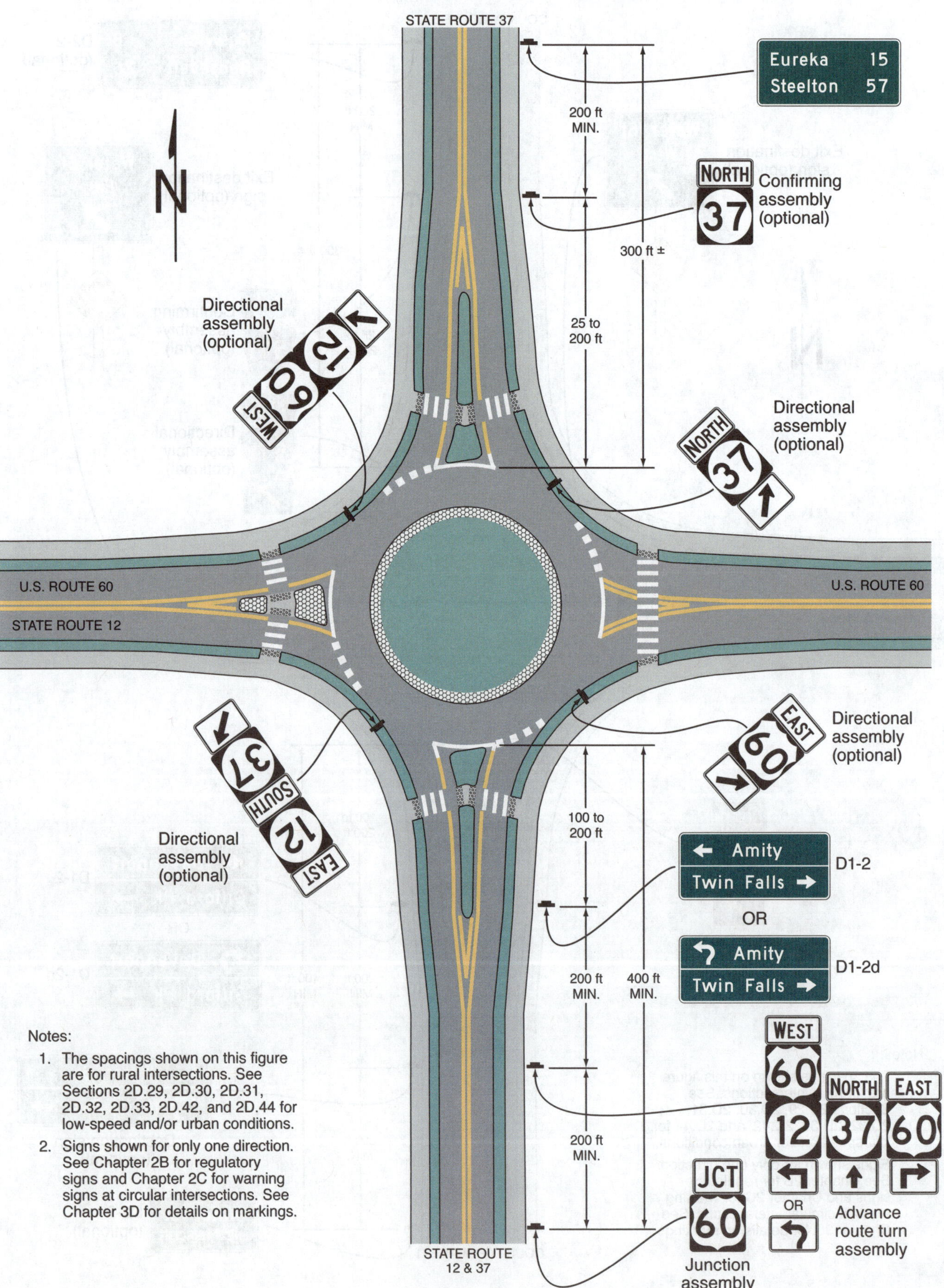

Notes:

1. The spacings shown on this figure are for rural intersections. See Sections 2D.29, 2D.30, 2D.31, 2D.32, 2D.33, 2D.42, and 2D.44 for low-speed and/or urban conditions.

2. Signs shown for only one direction. See Chapter 2B for regulatory signs and Chapter 2C for warning signs at circular intersections. See Chapter 3D for details on markings.

Figure 2D-12. Examples of Guide Signs for Circular Intersections (Sheet 4 of 4)

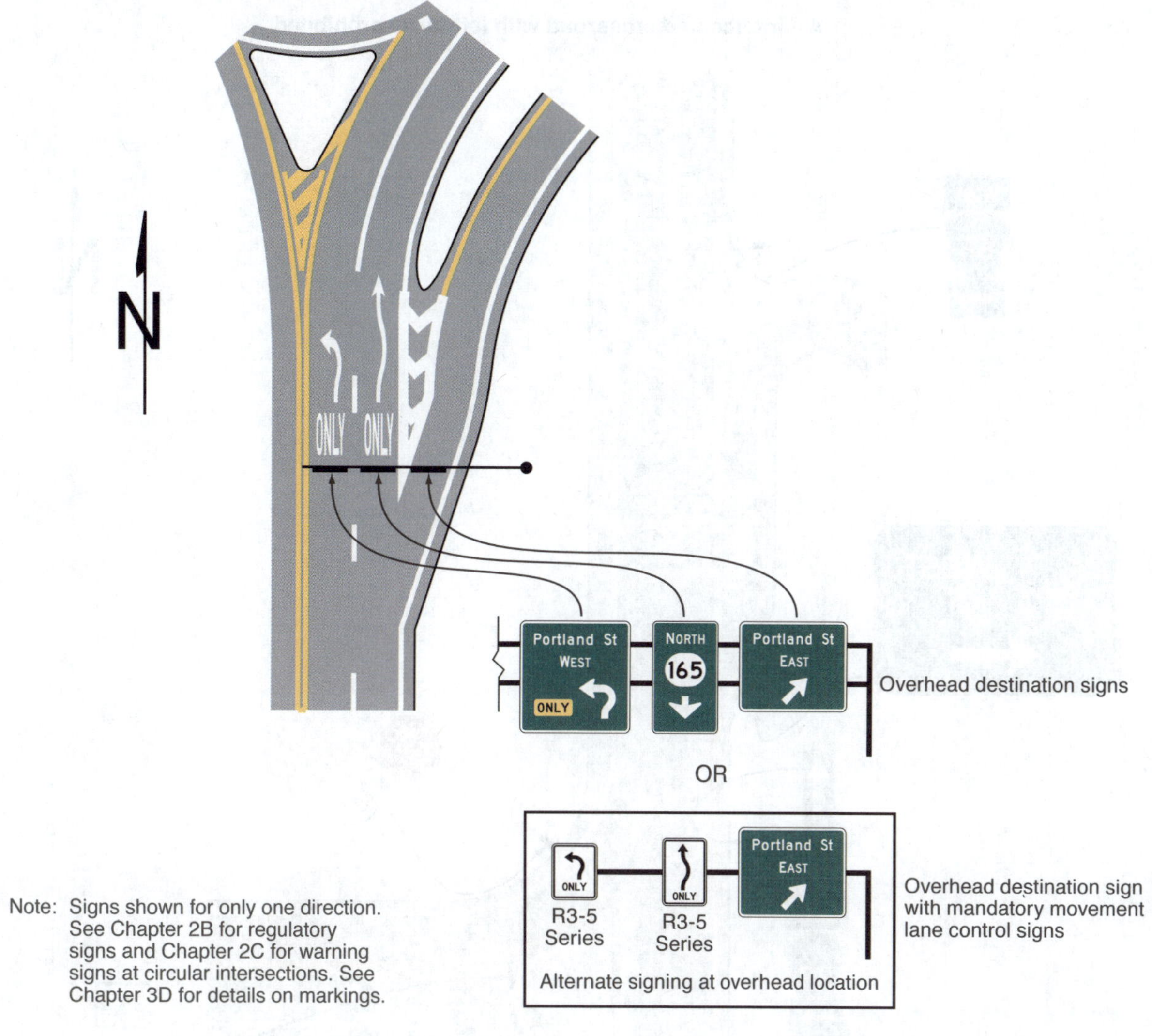

Guidance:

06   *The control city should remain the same on all successive Distance signs throughout the length of the route until that city is reached.*

Option:

07   If more than one distant point may properly be designated, such as where the route divides at some distance ahead to serve two destinations of similar importance, and if these two destinations cannot appear on the same sign, the two names may be alternated on successive signs.

08   On a route continuing into another State, destinations in the adjacent State may be displayed.

Section 2D.44 Location of Distance Signs

Guidance:

01   *If used, Distance signs should be installed on important routes leaving municipalities and just beyond intersections of numbered routes in rural areas. If used, they should be placed just outside the municipal limits or at the edge of the built-up area if it extends beyond the limits.*

02   *Where overlapping routes separate a short distance from the municipal limits, the Distance sign at the municipal limits should be omitted. The Distance sign should be installed approximately 300 feet beyond the separation of the two routes.*

03   *Where, just outside of an incorporated municipality, two routes are concurrent and continue concurrently to the next incorporated municipality, the top name on the Distance sign should be that of the place where the routes separate; the bottom name should be that of the city to which the greater part of the through traffic is destined.*

Support:

04   Figures 2D-7 and 2D-8 show typical placements of Distance signs.

Figure 2D-13. Examples of Signing for Intersections with Indirect Left Turns (Sheet 1 of 3)

A – Intercepted crossroad with left turns prohibited

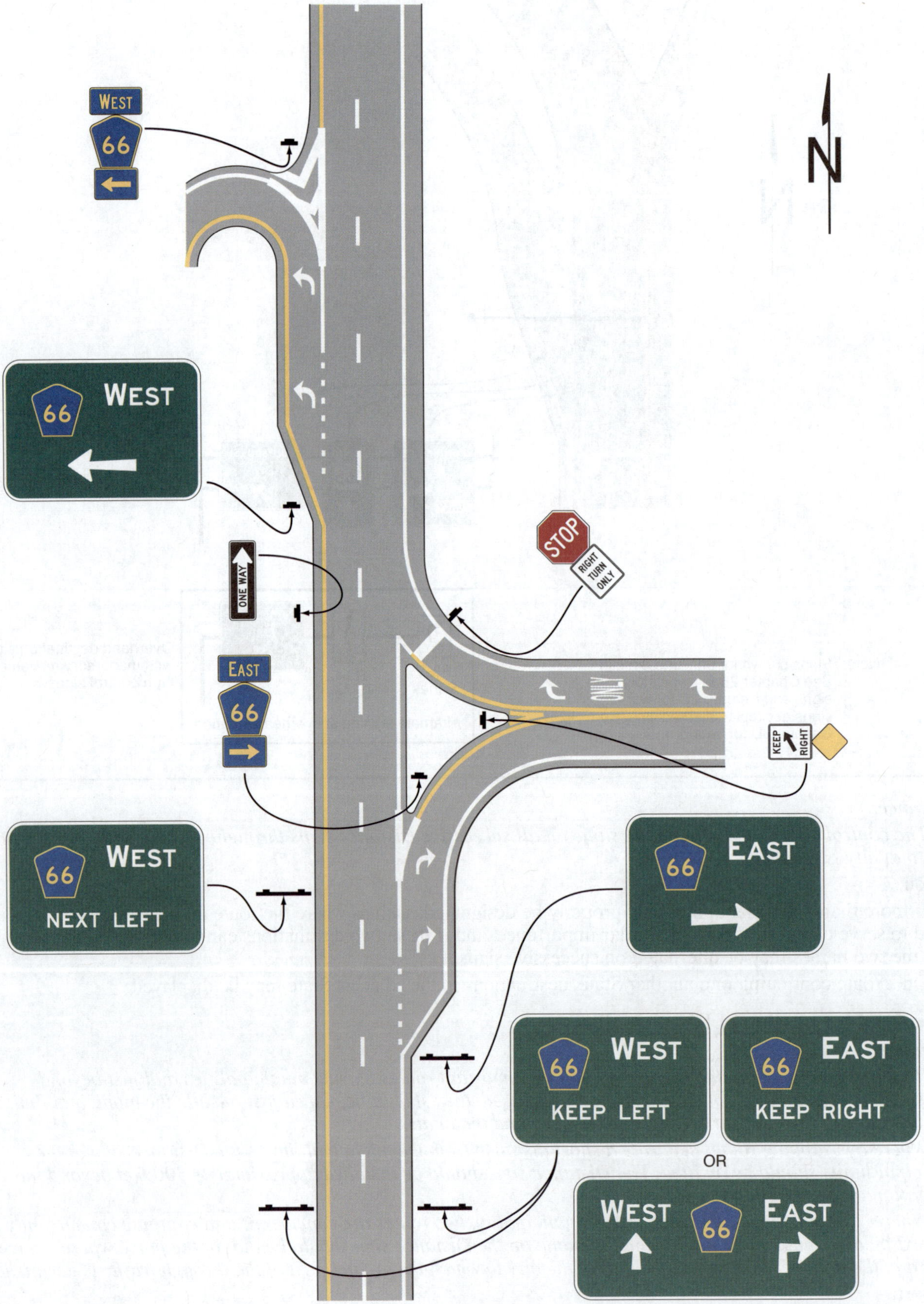

Figure 2D-13. Examples of Signing for Intersections with Indirect Left Turns (Sheet 2 of 3)

B – Intercepted crossroad with left turns prohibited

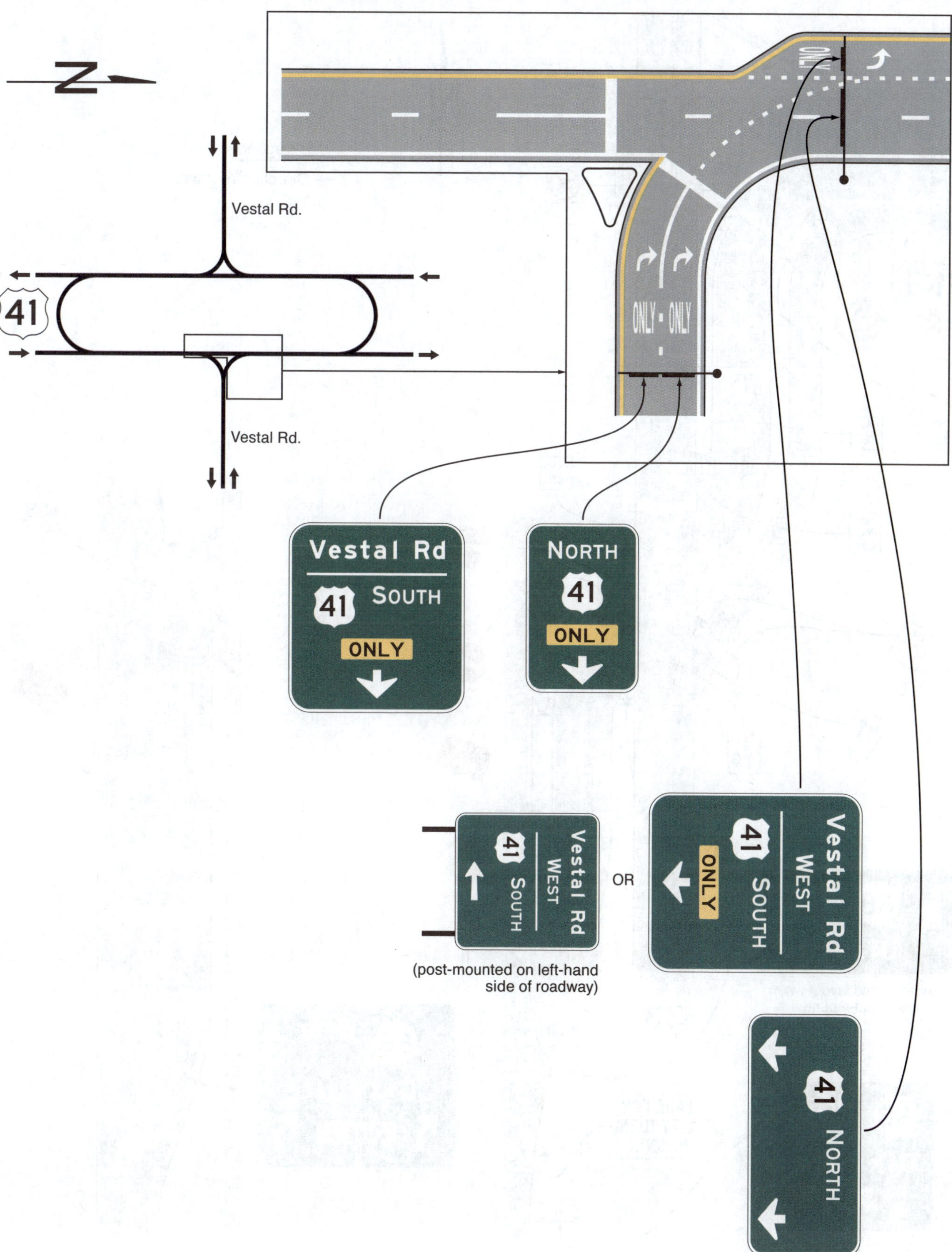

Figure 2D-13. Examples of Signing for Intersections with Indirect Left Turns (Sheet 3 of 3)

C – Continuous flow intersection

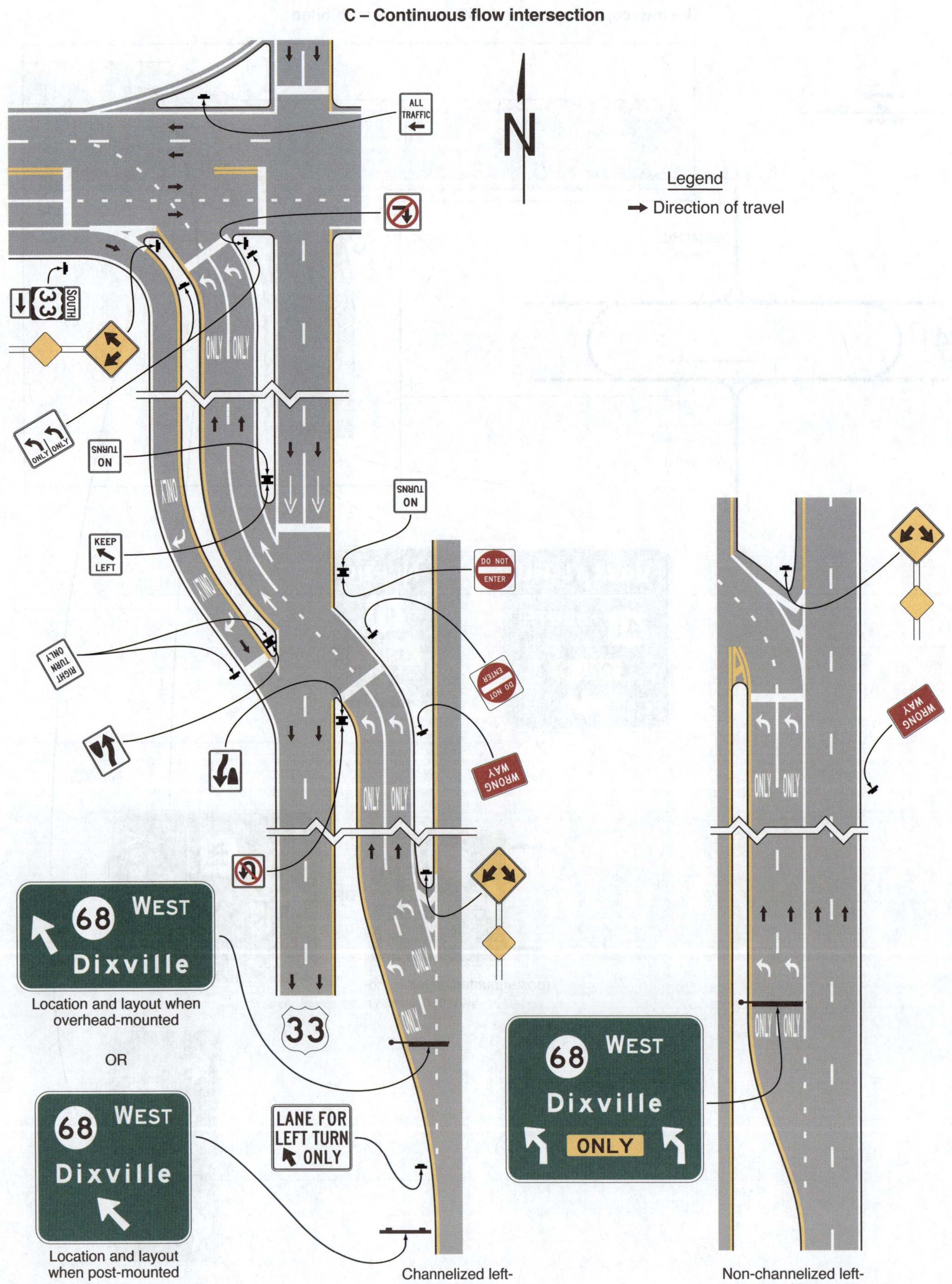

STREET NAME AND PARKING SIGNS

Section 2D.45 Street Name Signs (D3-1 and D3-1a)

Support:

01   Street Name signs at intersections and along roadways provide road users with important navigation information. Section 2H.10 contains information about signs used to identify the names of grade-separated streets, railways, bikeways, or other transportation facilities.

Guidance:

02   *Street Name (D3-1 or D3-1a) signs (see Figure 2D-14) should be installed in urban areas at all street intersections regardless of other route signs that might be present and should be installed in rural areas to identify important roads that are not otherwise signed.*

03   *To minimize wrong-way movements onto freeway or expressway exit ramps, Street Name signs should not be used at the intersection of a freeway or expressway exit ramp with the crossroad to display the name of the freeway or expressway to traffic on the crossroad.*

Option:

04   For streets that are part of a U.S., State, or county numbered route, a D3-1a Street Name sign (see Figure 2D-14) that incorporates a route shield may be used to assist road users who might not otherwise be able to associate the name of the street with the route number.

Standard:

05   **The lettering for names of streets and highways on Street Name signs shall be composed of a combination of lower-case letters with initial upper-case letters (see Section 2A.08).**

Guidance:

06   *The determination of letter heights to be used on Street Name signs should be based on, but not limited to, the following considerations:*

    A. Use of Advance Street Name signs (see Section 2D.46);
    B. Number of lanes on the intersection approach;
    C. Length of turn lanes;
    D. Distance the Street Name sign is located across the intersection (if a sign is not provided on the near side of the intersection).

07   *Letter heights on street name signs should be as shown in Table 2D-6.*

Figure 2D-14. Street Name and Parking Signs

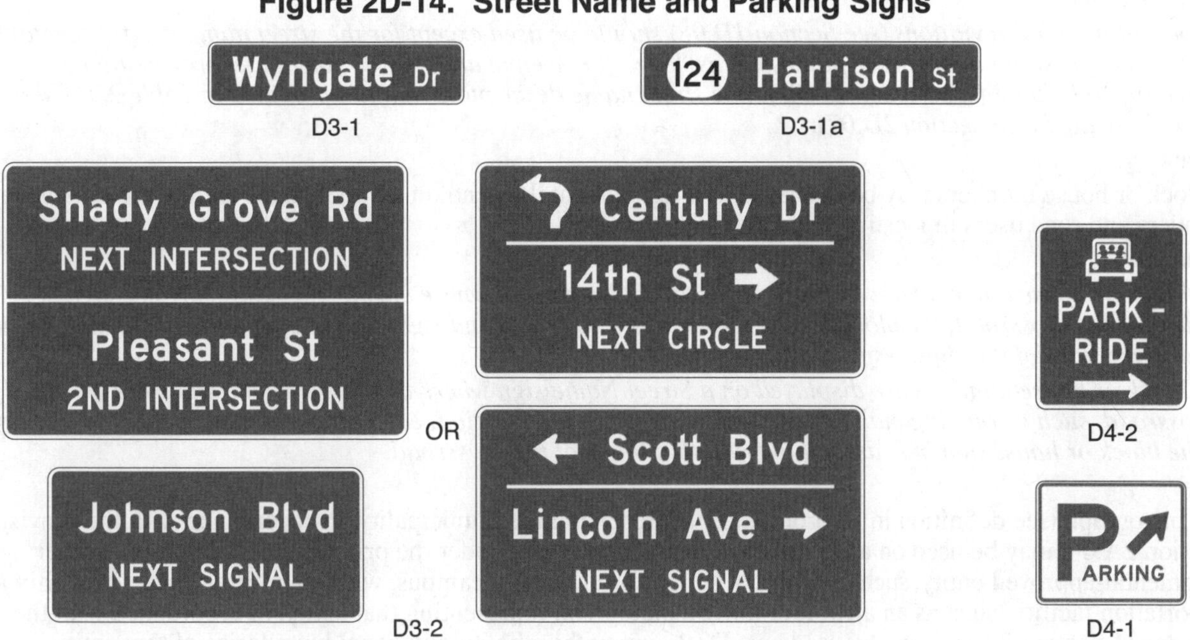

Table 2D-6. Minimum Letter Heights on Street Name Signs

Type of Mounting	Type of Street or Highway	Speed Limit	Recommended Minimum Letter Height *	
			Initial Upper-Case	Lower-Case
Overhead	All types	All speed limits	12 inches	9 inches
Post-mounted	Multi-lane	More than 40 mph	8 inches	6 inches
Post-mounted	Multi-lane	40 mph or less	6 inches	4.5 inches
Post-mounted	2-lane	All speed limits	6 inches**	4.5 inches**

* Letter heights shown are for the street name. Descriptors or other supplementary legend may be displayed in smaller lettering of at least 3 inches.

** On two-lane local streets with speed limits of 25 mph or less, 4-inch initial upper-case letters with 3-inch lower-case letters may be used.

Option:

08   For two-lane local roadways with speed limits of 25 mph or less, the lettering on post-mounted Street Name signs may be composed of initial upper-case letters at least 4 inches in height and lower-case letters at least 3 inches in nominal loop height.

Support:

09   The recommended minimum letter heights for Street Name signs are summarized in Table 2D-6. The speed limits specified and the recommended minimum letter heights provided in this Section apply to the roadway that each Street Name sign faces rather than to the street that has its name displayed on the Street Name sign. The letter heights specified in Table 2D-6 are the initial upper-case letter of a mixed-case legend.

10   A minimum upper-case letter height of 12 inches with a lower-case nominal loop height of 9 inches is recommended for all overhead Street Name signs regardless of posted speed limit as Street Name signs generally require greater legibility distances for road users to properly react.

Option:

11   Each Street Name sign in a sign assembly may use different letter heights determined by the speed limit of the street that each sign faces.

12   The letter height of the street name descriptor (such as St, Ave, or Rd), the directional legend (such as NW), or any other supplemental legend (such as block or house numbers) on the D3-1 and D3-1a signs may be smaller than that of the street name itself.

Guidance:

13   *The letter height of the street name descriptor, the directional legend, or any other supplemental legend on the D3-1 and D3-1a signs should be at least two-thirds of the letter height of the street name itself, but not less than 3 inches for the initial upper-case letters and not less than 2.25 inches for the nominal loop height of the lower-case letters.*

14   *Conventional abbreviations (see Section 1D.08) should be used except for the street name itself. Acceptable abbreviations for street name descriptors such as "Ave" for Avenue and "Blvd" for Boulevard should be as provided in Table 2D-3 (see Section 2D.07). The street name descriptors that are provided in Table 2D-4 should not be abbreviated (see Section 2D.07).*

Option:

15   Block or house numbers may be displayed as a supplemental legend on a Street Name sign to aid emergency responders and road users in locating addresses.

Guidance:

16   *If block or house numbers are displayed on a Street Name sign where only a single Street Name sign is provided for the crossroad, the block or house numbers for the left and right blocks should be positioned at the left and right sides of the sign, respectively.*

17   *If block or house numbers are displayed on a Street Name sign where two Street Name signs are provided for the crossroad, such as on diagonally opposite corners of an intersection, each Street Name sign should display only the block or house numbers associated with that block of the crossroad.*

Option:

18   A pictograph (see definition in Section 1C.02) representing the municipality, in accordance with the provisions of Section 2A.04, may be used on a D3-1 sign. For street networks under the primary jurisdiction of another governmental-approved entity, such as within a college or university campus, within a military base, or within a transportation facility (such as an airport or port), a pictograph representing that entity, in accordance with the provisions of Section 2A.04, may be used on a D3-1 sign within the jurisdictional boundaries of that entity.

Standard:

19   **Pictographs shall not be displayed on D3-1a or Advance Street Name (D3-2) signs (see Section 2D.46).**

20   **If a pictograph is used on a D3-1 sign, the height and width of the pictograph shall not exceed the upper-case letter height of the principal legend of the sign.**

Guidance:

21   *The pictograph should be positioned to the left of the street name.*

22   *Pictographs should not be used on a D3-1 sign that contains directional arrows.*

Standard:

23   **The Street Name sign shall be retroreflective or illuminated in accordance with the provisions of Section 2A.21.**

Option:

24   The border may be omitted from a post-mounted Street Name sign.

Guidance:

25   *The decision to omit the border from a post-mounted Street Name sign should be based on such factors as the visual complexity of the environment and the degree of conspicuity needed to provide for adequate recognition of the sign by the road user.*

Option:

26   An alternative background color (see Paragraph 28 of this Section) other than the standard guide sign color of green may be used for Street Name (D3-1 or D3-1a) signs where the highway agency determines this is necessary to assist road users in determining jurisdictional authority for roads.

Standard:

27   **Alternative background colors shall not be used for Advance Street Name (D3-2) signs (see Section 2D.46).**

28   **The only acceptable alternative background colors for Street Name (D3-1 or D3-1a) signs shall be blue, brown, or white. Regardless of whether green, blue, or brown is used as the background color for Street Name (D3-1 or D3-1a) signs, the legend (and border, if used) shall be white. For Street Name signs that use a white background, the legend (and border, if used) shall be black.**

Guidance:

29   *An alternative background color for Street Name signs, if used, should be applied to the Street Name (D3-1 or D3-1a) signs on all roadways under the jurisdiction of a particular highway agency.*

30   *In business or commercial areas and on principal arterials, Street Name signs should be placed at least on diagonally opposite corners. In residential areas, at least one Street Name sign should be mounted at each intersection. Signs naming both streets should be installed at each intersection. They should be mounted with their faces parallel to the streets they name.*

31   *Where used, Street Name signs should display their legends on both the front and back sides of the sign to facilitate navigation for pedestrians.*

Option:

32   To optimize visibility, Street Name signs may be mounted overhead. Street Name signs may also be placed above a regulatory or STOP or YIELD sign with no required vertical separation.

Guidance:

33   *In urban or suburban areas, especially where Advance Street Name signs for signalized and other major intersections are not used, the use of overhead Street Name signs should be strongly considered.*

Option:

34   At intersection crossroads where the same road has two different street names for each direction of travel, both street names may be displayed on the same Street Name (D3-1) sign along with Type D directional arrows, except where one arrow would point in a direction opposing the flow of traffic on a one-way street or where a turn in the direction of the arrow is not allowed.

35   On lower-speed roadways, historic street name signs within locally identified historic districts that are consistent with the criteria contained in 36 CFR 60.4 for such structures and districts may remain in service without complying with the provisions of Paragraphs 3, 4, 6, 9, 12 through 14, and 18 through 20 of this Section.

Guidance:

36   *Streets or segments of a street that have been memorialized or dedicated should not use a second Street Name sign to display the memorial or dedication name (see Section 2D.56). When signed, the Memorial or Dedication sign should be located to minimize its conspicuity to and potential for confusion.*

Support:

37   Information regarding the use of street names on supplemental plaques for use with intersection-related warning signs is contained in Section 2C.65.

38   Information regarding the identification of overcrossing and undercrossing roadways at grade separations is contained in Section 2H.10.

Section 2D.46 Advance Street Name Signs (D3-2 Series)

Support:

01   Advance Street Name (D3-2) signs (see Figure 2D-14) identify a downstream intersection. Although this is often the next intersection, it could also be several intersections away in cases where the next signalized intersection is referenced.

Standard:

02   **Advance Street Name (D3-2) signs, if used, shall supplement rather than be used instead of the Street Name (D3-1 or D3-1a) signs at the intersection.**

Option:

03   Advance Street Name (D3-2) signs may be installed in advance of signalized or unsignalized intersections to provide road users with advance information to identify the name(s) of the next intersecting street to prepare for crossing traffic and to facilitate timely deceleration and/or lane changing in preparation for a turn.

Guidance:

04   *On arterial highways in rural areas, Advance Street Name signs should be used in advance of all signalized intersections and in advance of all intersections with mandatory turn lanes.*

05   *In urban areas, Advance Street Name signs should be used in advance of all signalized intersections on major arterial streets, except where signalized intersections are so closely spaced that advance placement of the signs is impracticable.*

06   *The heights of the letters on Advance Street Name signs should comply with the provisions of Section 2D.05.*

Standard:

07   **If used, Advance Street Name signs shall have a white legend and border on a green background. Alternative background colors shall not be used on Advance Street Name signs.**

08   **If used, Advance Street Name signs shall provide the name(s) of the intersecting street(s) on the top line(s) of the legend and the distance to the intersecting streets or messages such as NEXT SIGNAL, NEXT INTERSECTION, NEXT CIRCLE, or directional arrow(s) on the bottom line of the legend.**

09   **Pictographs shall not be displayed on Advance Street Name signs.**

Option:

10   Directional arrow(s) may be placed to the right or left of the street name or message such as NEXT SIGNAL, as appropriate, rather than on the bottom line of the legend. Curved-stem arrows may be used on Advance Street Name signs on approaches to circular intersections.

11   For intersecting crossroads where the same road has a different street name for each direction of travel, the different street names may be displayed on the same Advance Street Name sign along with directional arrows.

12   In advance of two closely-spaced intersections where it is impracticable to install separate Advance Street Name signs, the Advance Street Name sign may include the street names for both intersections along with appropriate supplemental legends for both street names, such as NEXT INTERSECTION, 2ND INTERSECTION, or NEXT LEFT and NEXT RIGHT, or directional arrows.

Guidance:

13   *If two street names are used on the Advance Street Name sign, the street names should be displayed in the following order:*

     A. *For a single intersection where the same road has a different street name for each direction of travel, the name of the street to the left should be displayed above the name of the street to the right; or*

     B. *For two closely-spaced intersections, the name of the first street encountered should be displayed above the name of the second street encountered, and the arrow associated with the second street encountered should be an advance arrow, such as the arrow shown on the W16-6P arrow plaque (see Figure 2C-16).*

Option:

14   An Advance Street Name (W16-8P or W16-8aP) plaque (see Section 2C.65) with black legend on a yellow background, installed to supplement an Intersection (W2 series) or Advance Traffic Control (W3 series) warning sign may be used instead of an Advance Street Name guide sign.

Section 2D.47 <u>Parking Area Guide Sign (D4-1)</u>

Option:

01   The Parking Area (D4-1) guide sign (see Figure 2D-14) may be used to show the direction to a nearby public parking area or parking facility.

Standard:

02   **The smaller size of 18 x 15 inches for the Parking Area guide sign shall be limited to minor, low-speed streets.**

Guidance:

03   *If used, the Parking Area guide sign should be installed on major thoroughfares at the nearest point of access to the parking facility and where it can advise drivers of a place to park. The sign should not be used more than four blocks from the parking area.*

Section 2D.48 <u>PARK - RIDE Sign (D4-2)</u>

Option:

01   A PARK - RIDE (D4-2) sign (see Figure 2D-14) may be used to direct road users to park-and-ride facilities.

Standard:

02   **The signs shall display the word message PARK - RIDE and direction information (arrow or word message).**

Option:

03   PARK - RIDE signs may display the local transit pictograph and/or carpool symbol.

Standard:

04   **If used, the local transit pictograph and/or carpool symbol shall be located in the top part of the sign above the message PARK - RIDE. In no case shall the vertical dimension of the local transit pictograph and/or carpool symbol exceed 18 inches.**

Guidance:

05   *If the function of the parking facility is to provide parking for persons using public transportation, the local transit pictograph should be used on the guide sign. If the function of the parking facility is to serve carpool riders, the carpool symbol should be used on the guide sign. If the parking facility serves both functions, both the pictograph and carpool symbol should be used.*

Standard:

06   **These signs shall have a white legend and border on a rectangular green background. The carpool symbol shall be as shown for the D4-2 sign. The color of the local transit pictograph shall be selected by the local transit authority.**

Option:

07   To increase the target value and contrast of the local transit pictograph, and to allow the local transit pictograph to retain its distinctive color and shape, the pictograph may be included within a white border or placed on a white background.

FREEWAY INTERCHANGE APPROACH SIGNS

Section 2D.49 Signing on Conventional Roads on Approaches to Interchanges

Support:

01   Because there are a number of different ramp configurations that are commonly used at interchanges with conventional roads, drivers on the conventional road cannot reliably predict whether they will be required to turn left or right in order to enter the correct ramp to access the freeway or expressway in the desired direction of travel. Consistently applied signing for conventional road approaches to freeway or expressway interchanges is highly desirable.

Standard:

02   **On multi-lane conventional roads approaching an interchange, guide signs shall be provided to identify which direction of turn is to be made and/or which specific lane to use for ramp access to each direction of the freeway or expressway.**

Guidance:

03   *The signing of conventional roads with one lane of traffic approaching an interchange should consist of a sequence containing the following signs (see Figure 2D-15):*
    A. Junction Assembly
    B. Destination sign
    C. Directional Assembly or Entrance Direction sign for the first ramp
    D. Advance Route Turn Assembly or Advance Entrance Direction sign with an advance turn arrow
    E. Directional Assembly or Entrance Direction sign for the second ramp

Standard:

04   **If used, the Entrance Direction sign shall consist of a white legend and border on a green background. It shall contain the freeway or expressway route shield(s), cardinal direction, and directional arrow(s).**

Option:

05   The Entrance Direction sign may contain a destination(s) and/or an action message such as NEXT RIGHT.

06   At minor interchanges (see Section 2E.30), the following sequence of signs may be used (see Figure 2D-16):
    A. Junction Assembly
    B. Directional Assembly for the first ramp
    C. Directional Assembly for the second ramp

Guidance:

07   *On multi-lane conventional roads approaching an interchange, the sign sequence should contain the following signs (see Figures 2D-17 through 2D-19):*
    A. Junction Assembly
    B. Advance Entrance Direction sign(s) for both directions (if applicable) of travel on the freeway or expressway
    C. Entrance Direction sign for first ramp
    D. Advance Turn Assembly
    E. Entrance Direction sign for the second ramp

Support:

08   Advance Entrance Direction signs are used to direct road users to the appropriate lane(s).

Standard:

09   **The Advance Entrance Direction sign shall consist of a white legend and border on a green background. It shall contain the freeway or expressway route shield(s) and cardinal direction(s).**

Option:

10   The Advance Entrance Direction sign may have destinations, directional arrows, and/or an action message such as KEEP LEFT, NEXT LEFT, or SECOND RIGHT. Signs in this sequence may be mounted overhead to improve visibility as shown in Figures 2D-17 through 2D-19.

Support:

11   A post-mounted Advance Entrance Direction diagrammatic sign (see Figure 2D-20), within the sequence of approach guide signing described in Paragraphs 3, 6, and 7 of this Section, might be helpful in depicting the location of a freeway or expressway entrance ramp that is in close proximity to an intervening intersection on the same side of the approach roadway and where signing for only the ramp might cause confusion to road users.

Figure 2D-15. Example of Interchange Crossroad Guide Signing for a One-Lane Approach

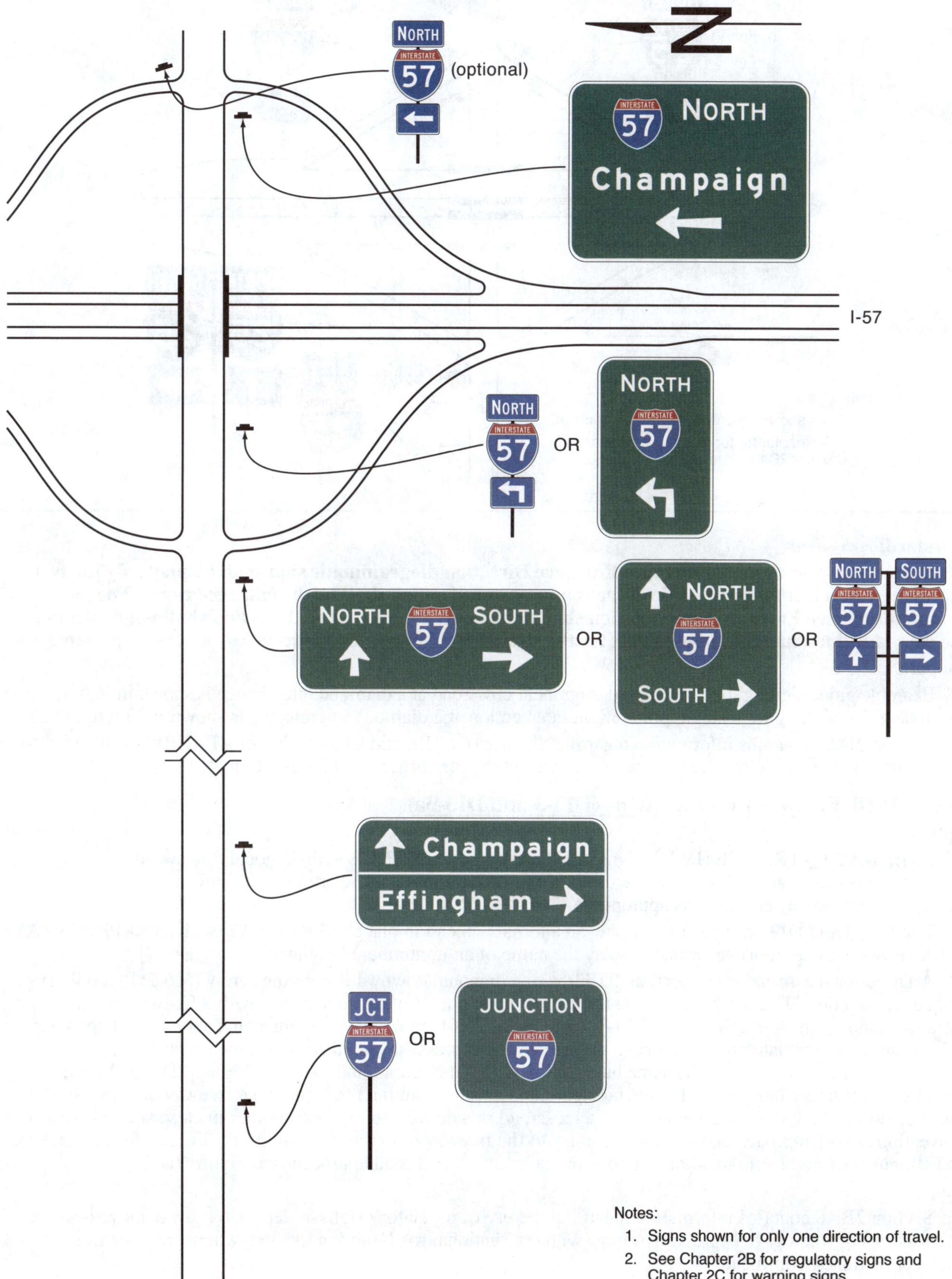

Notes:
1. Signs shown for only one direction of travel.
2. See Chapter 2B for regulatory signs and Chapter 2C for warning signs.

Figure 2D-16. Example of Minor Interchange Crossroad Guide Signing

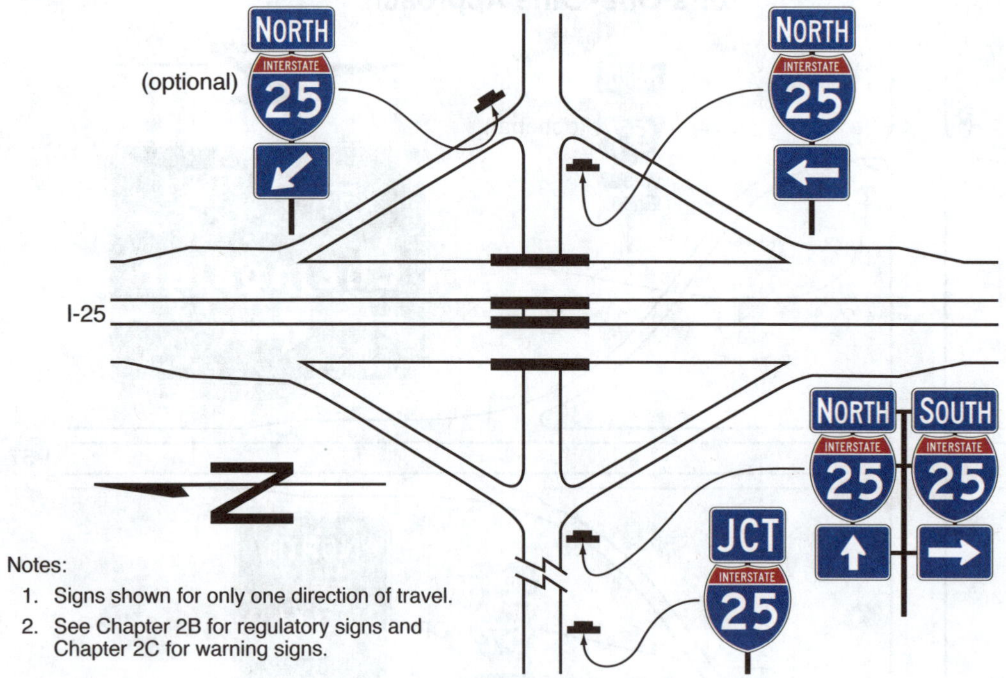

Standard:

12   **If used, the post-mounted Advance Entrance Direction diagrammatic sign shall display only the two successive turns from the same side of the roadway, one of which shall be the entrance ramp. The post-mounted Advance Entrance Direction sign shall depict only the successive turns and shall not depict lane use with lane lines, multiple arrow shafts for the approach roadway, action messages, or other representations.**

Support:

13   Example guide signing for a transposed-alignment crossroad at a diamond interchange is shown in Figure 2D-21. Example guide signing for a single-point urban intersection at a diamond interchange is shown in Figure 2D-22.

14   Section 2D.50 contains information regarding the use of a Directional assembly or a FREEWAY ENTRANCE sign to mark the entrance to a freeway or expressway at the far corner of an intersection.

Section 2D.50 Freeway Entrance Signs (D13-3 and D13-3a)

Option:

01   FREEWAY ENTRANCE (D13-3) signs or FREEWAY ENTRANCE with diagonal downward-pointing arrow (D13-3a) signs (see Figure 2D-18) may be used on entrance ramps near the crossroad to inform road users of the freeway or expressway entrance, as appropriate.

02   The D13-3 and D13-3a signs may display an alternate legend in place of FREEWAY, such as EXPRESSWAY or PARKWAY, as appropriate, or may display the name of an unnumbered highway.

03   A Directional assembly (see Section 2D.32) with a diagonal downward-pointing arrow (M6-2aP) auxiliary plaque (see Section 2D.28) may be used at the far left-hand corner of an intersection with a freeway or expressway entrance ramp as an alternative to the D13-3a sign, facing left-turning traffic on the conventional road approach to indicate the immediate point of entry to the freeway or expressway and distinguish the entrance ramp from an adjoining exit ramp terminal at the same intersection with the conventional road (see Figure 2D-18). A similar Directional assembly may be used at the far right-hand corner of an intersection with a freeway or expressway entrance ramp where the entrance ramp and a crossroad or side road follow one another in close succession on the conventional road approach and the point of entry to the freeway or expressway might be difficult for the road user to distinguish from the crossroad or side road on the conventional road approach (see Figure 2D-20).

Support:

04   Section 2B.48 contains information regarding the use of regulatory signs to deter wrong-way movements at intersections of freeway or expressway ramps with conventional roads, and in the area where entrance ramps intersect with the mainline lanes.

Figure 2D-17. Examples of Multi-Lane Crossroad Guide Signing for a Diamond Interchange

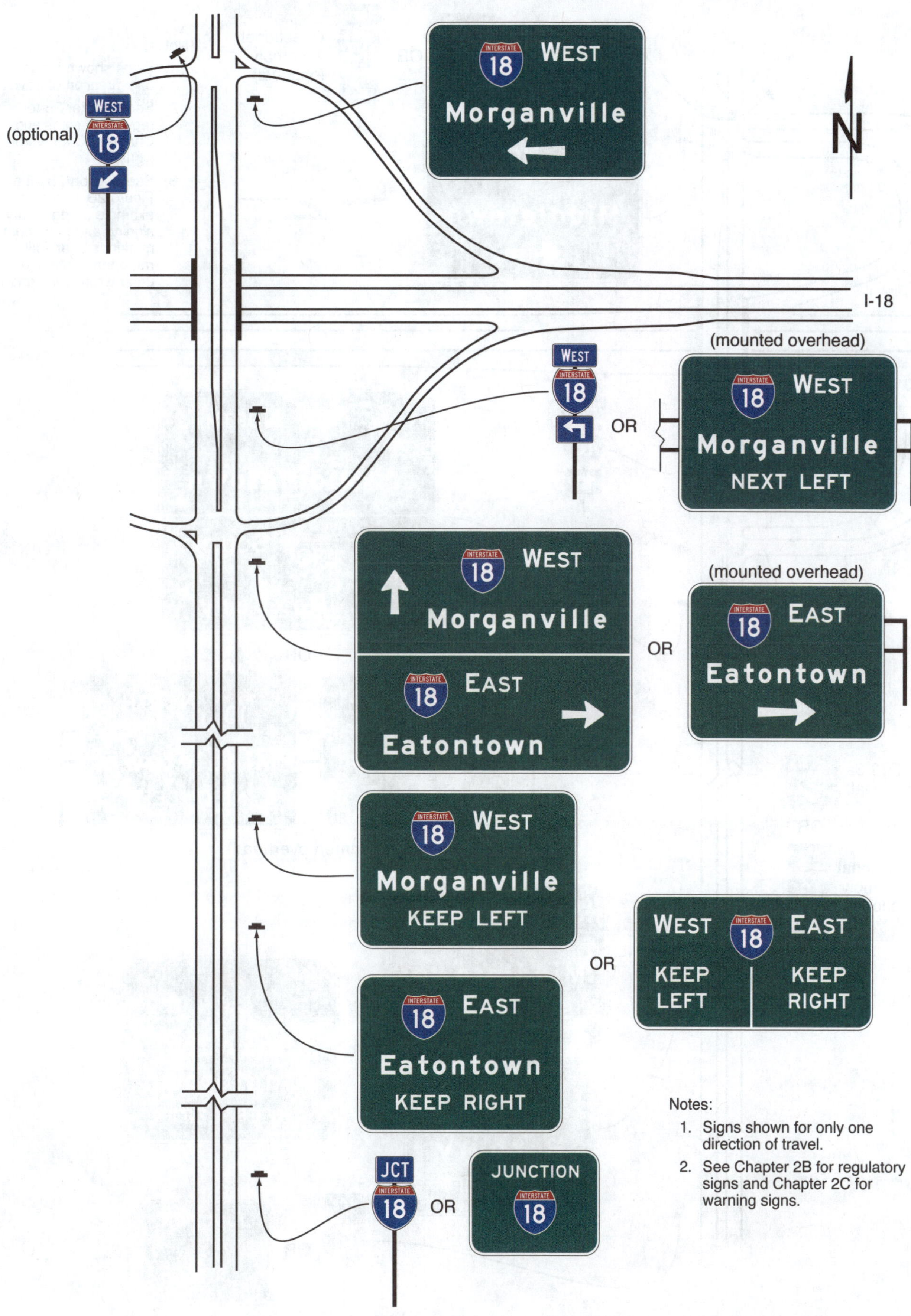

Notes:

1. Signs shown for only one direction of travel.

2. See Chapter 2B for regulatory signs and Chapter 2C for warning signs.

Figure 2D-18. Examples of Multi-Lane Crossroad Guide Signing for a Partial Cloverleaf Interchange

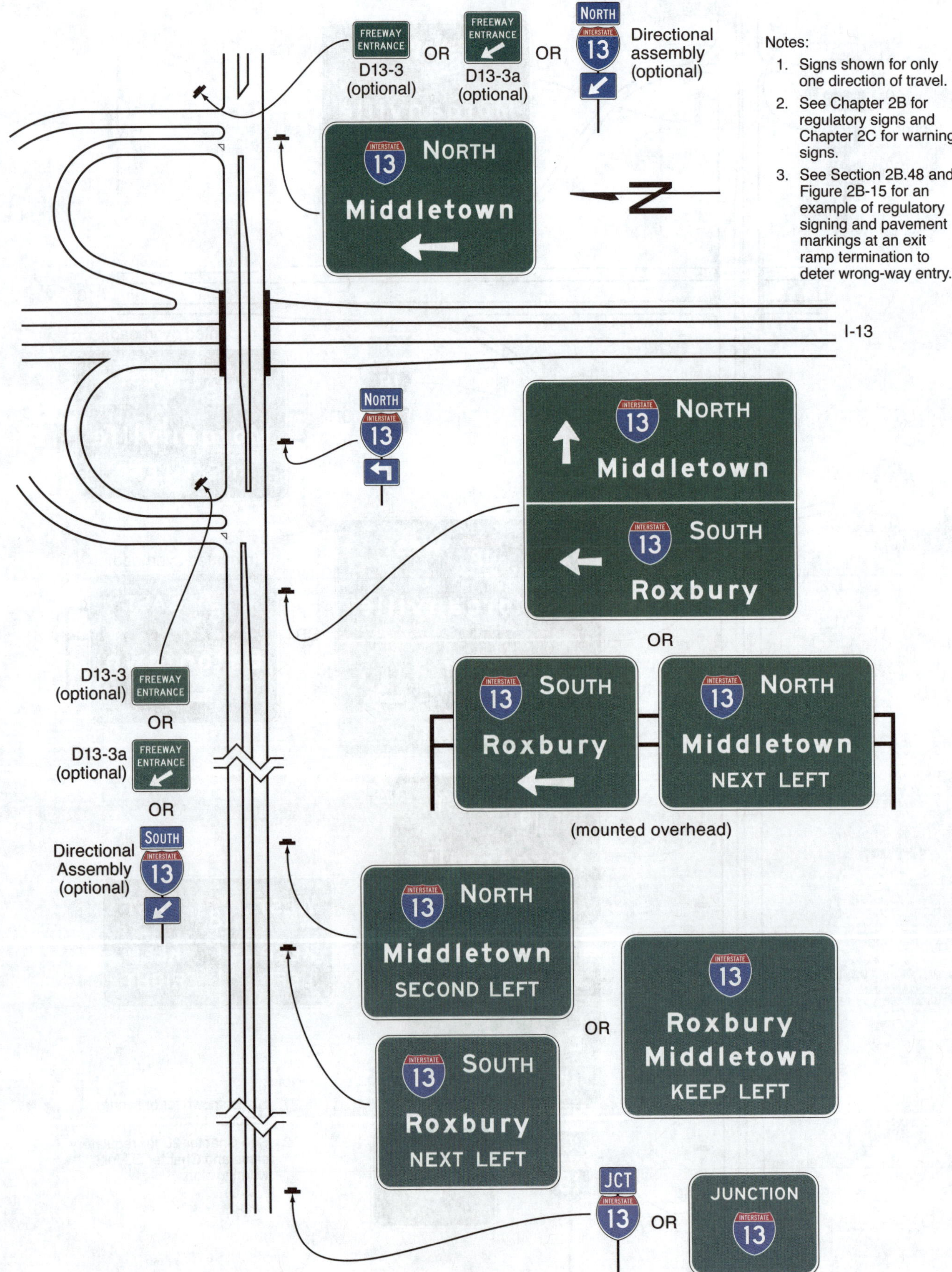

Figure 2D-19. Examples of Multi-Lane Crossroad Signing for a Cloverleaf Interchange

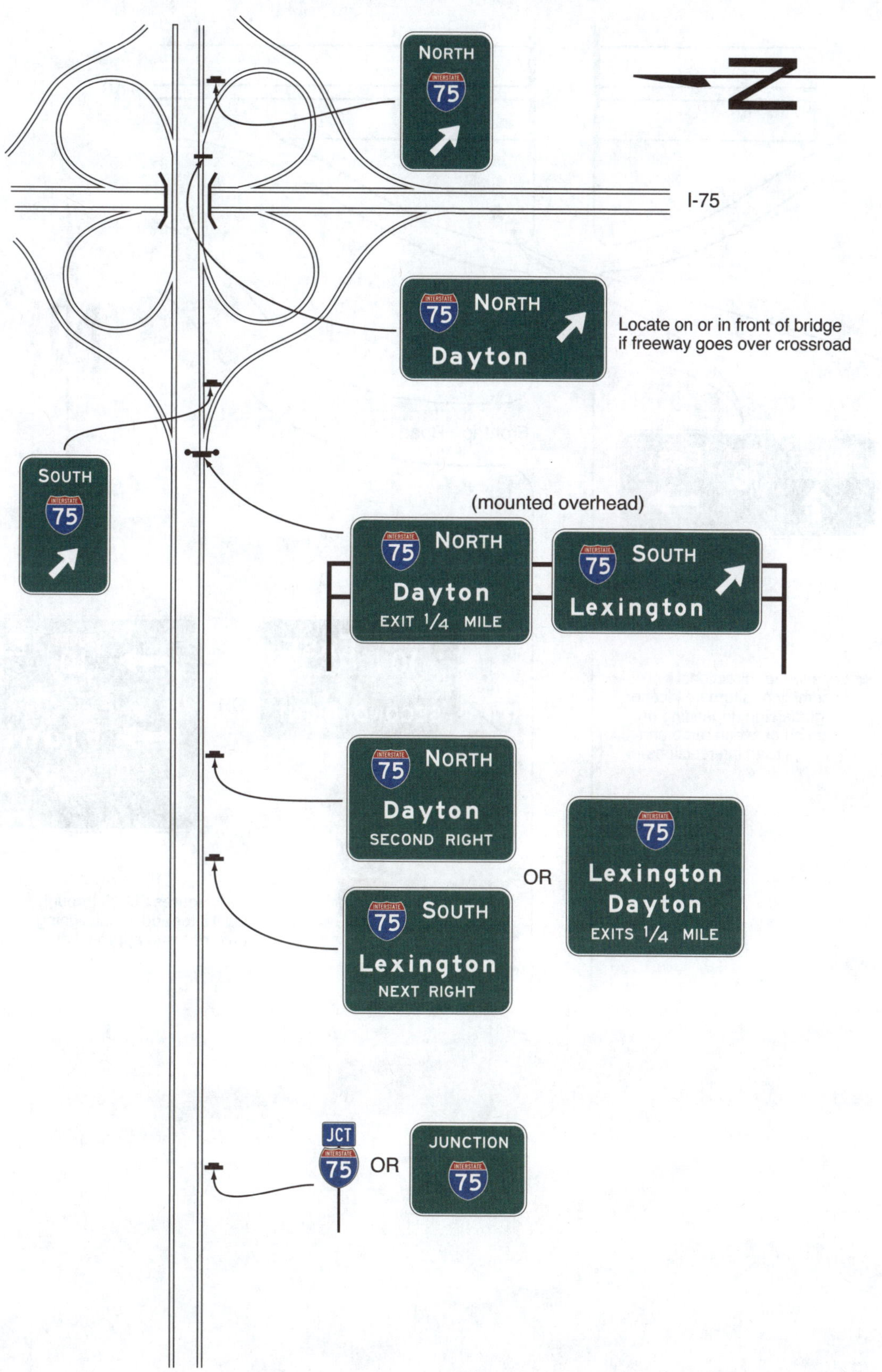

Figure 2D-20. Example of Crossroad Guide Signing for an Entrance Ramp with a Nearby Frontage Road

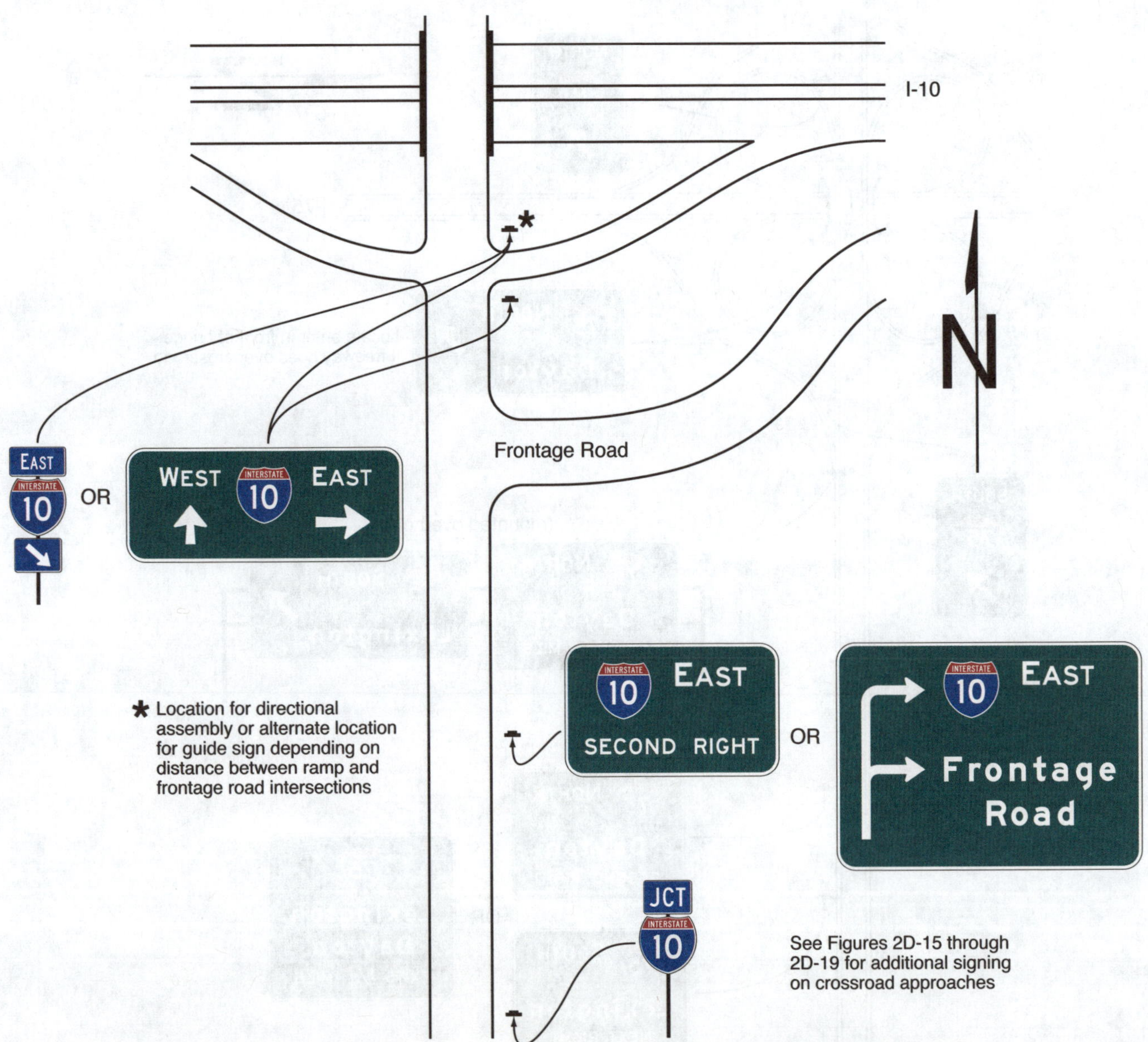

Figure 2D-21. Example of Transposed Alignment Crossroad Guide Signing at a Diamond Interchange

Figure 2D-22. Example of Crossroad Intersection Guide Signs for a Single-Point Urban Interchange

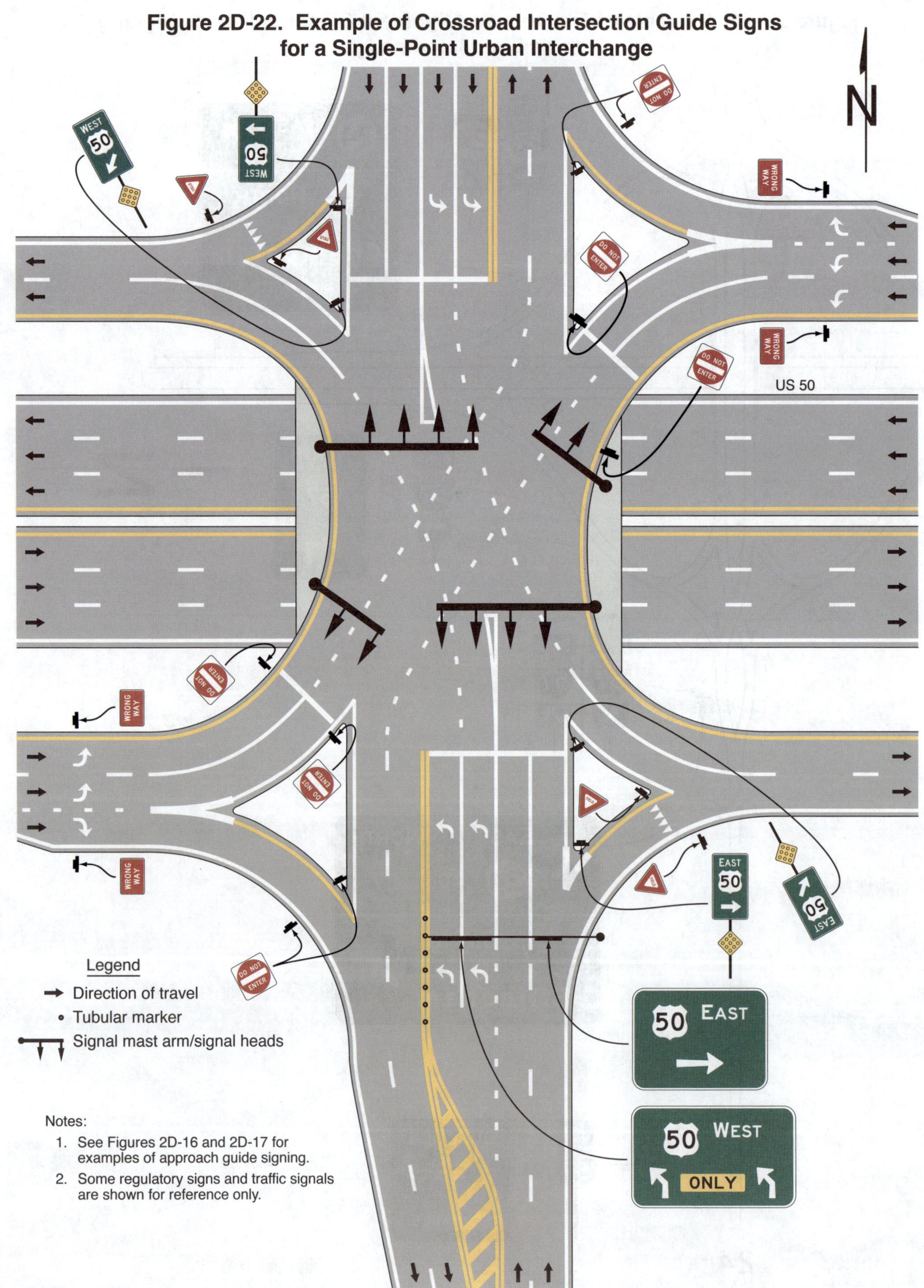

WEIGH STATION, CROSSOVER, TRUCK AND PASSING LANE, AND EMERGENCY AND SLOW VEHICLE TURN-OUT SIGNS

Section 2D.51 WEIGH STATION Signing (D8 Series)

Support:

01   Independent facilities or areas have been added along many highways where certain commercial vehicles are directed to stop to be weighed and/or inspected. These areas are sometimes permanent, such as in a roadside area, or temporary mobile facilities deployed along the roadway.

02   The general concept for signing permanent Weigh Stations is similar to signing Rest Areas (see Section 2I.05) because in both cases traffic using either area remains within the right-of-way.

Standard:

03   **The standard sequence of signs for a Weigh Station on a conventional highway shall include three basic signs (see Figure 2D-23):**

    A. Weigh Station Advance (D8-1) sign,
    B. Weigh Station Advance Direction (D8-2) sign, and
    C. Weigh Station Entrance Direction (D8-3) sign.

Guidance:

04   *A Gore sign with the same basic legend as the Weigh Station Entrance Direction (D8-3) sign should also be used to emphasize the entrance to the weigh station.*

Option:

05   Where State law requires trucks of a certain weight to enter the Weigh Station, a Weigh Station (R13-1) regulatory sign (see Section 2B.65) may be located following the Advance Weigh Station Ahead sign (see Figure 2D-23).

06   Where only commercial vehicle inspections are conducted in the inspection area, the WEIGH STATION legend of the D8 series signs may be replaced with the alternate legend, COMMERCIAL VEHICLE INSPECTION.

Guidance:

07   *The Weigh Station Advance Direction (D8-2) Sign or the Weigh Station Advance (D8-1) sign should display, either on the sign or on a supplemental plaque or sign panel, the changeable legend OPEN or CLOSED.*

Standard:

08   **When the WEIGH STATION legend of the D8 series signs is replaced with the COMMERCIAL VEHICLE INSPECTION legend, as provided in Paragraph 6 of this Section, the WEIGH STATION legend of the R13-1 sign shall likewise be replaced with the alternate legend.**

Section 2D.52 Crossover Signs (D13-1 and D13-2)

Option:

01   Crossover signs may be installed on divided highways to identify median openings not otherwise identified by warning or other guide signs.

Standard:

02   **A Crossover (D13-1) sign (see Figure 2D-24) shall not be used to identify a median opening that is permitted to be used only by official or authorized vehicles.**

Guidance:

03   *If used, the Crossover sign should be installed immediately beyond the median opening, either on the right-hand side of the roadway or in the median.*

Option:

04   The Advance Crossover (D13-2) sign (see Figure 2D-24) may be installed in advance of the Crossover sign to provide advance notice of the crossover.

Guidance:

05   *The distance displayed on the Advance Crossover sign should be 1 MILE, ½ MILE, or ¼ MILE, unless unusual conditions require some other distance. If used, the sign should be installed either on the right-hand side of the roadway or in the median at approximately the distance displayed on the sign.*

Figure 2D-23. Example of Weigh Station Signing – Conventional Road

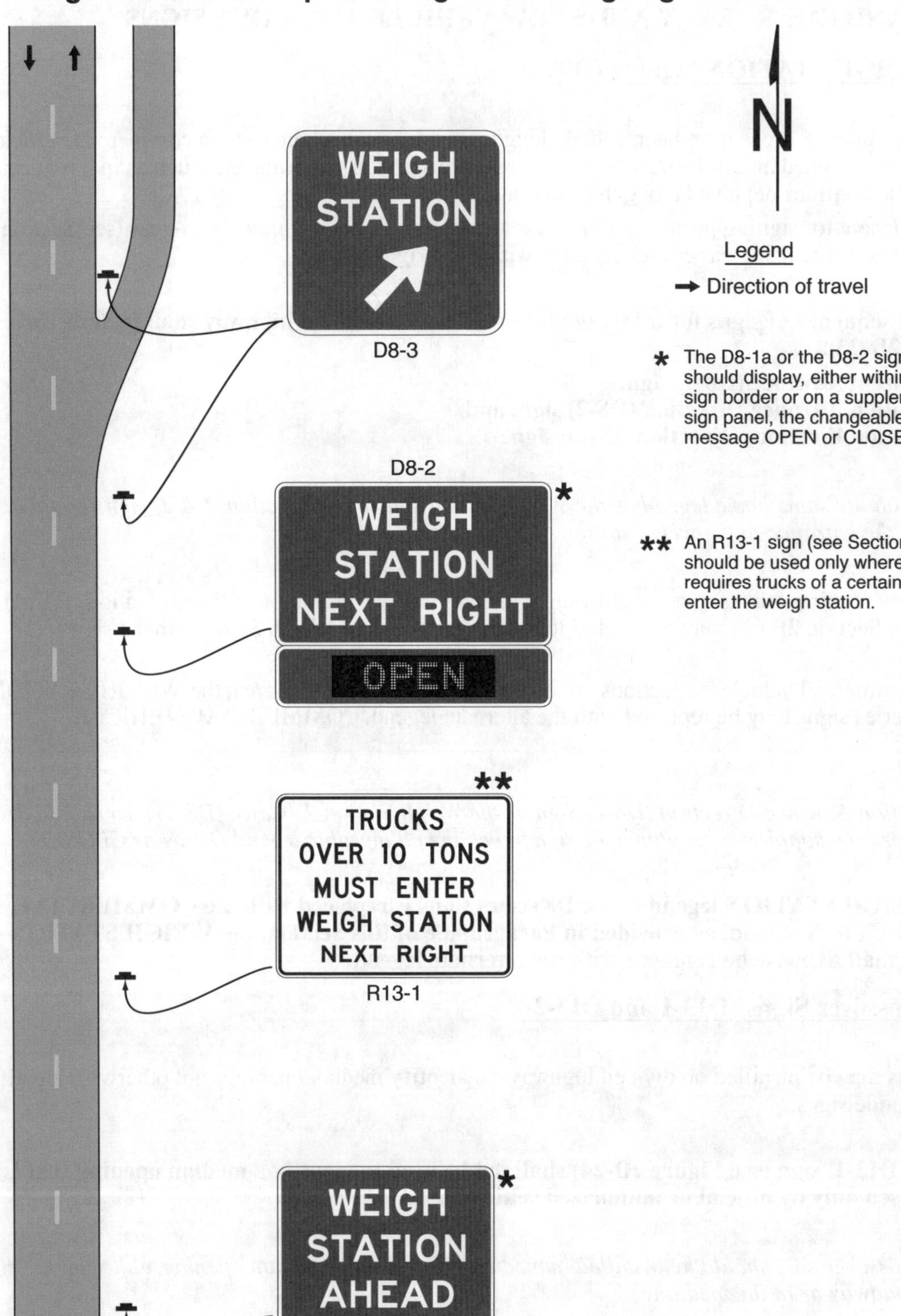

Figure 2D-24. Crossover Signs

D13-1

D13-2

Section 2D.53 Truck and Passing Lane Signs (D17-1 through D17-4)

Guidance:

01   *If an extra lane has been provided to the right-hand side of the travel lane for use by trucks and other slow-moving traffic, an Advance Truck Lane (D17-2) sign (see Figure 2D-25) should be installed in advance of the lane.*

02   *If a series of truck lanes is provided along a highway, a Next Truck Lane (D17-1) sign (see Figure 2D-25) should be installed after each truck lane segment.*

03   *If an extra lane has been provided to the left-hand side of the travel lane for passing slower moving vehicles in the travel lane, an Advance Passing Lane Advance (D17-4) sign (see Figure 2D-25) should be installed in advance of the lane.*

04   *If a series of passing lanes are provided along a highway, a Next Passing Lane (D17-3) sign (see Figure 2D-25) should be installed after each passing lane segment.*

Support:

05   An example of signing for a truck lane is shown in Figure 2D-26. An example of signing for an intermittent passing lane is shown in Figure 2D-27.

06   Section 2B.38 contains information regarding regulatory signs for these types of lanes.

Figure 2D-25. Truck and Passing Lane Signs

D17-1

D17-2

D17-3

D17-4

Figure 2D-26. Example of Signing for a Truck Lane

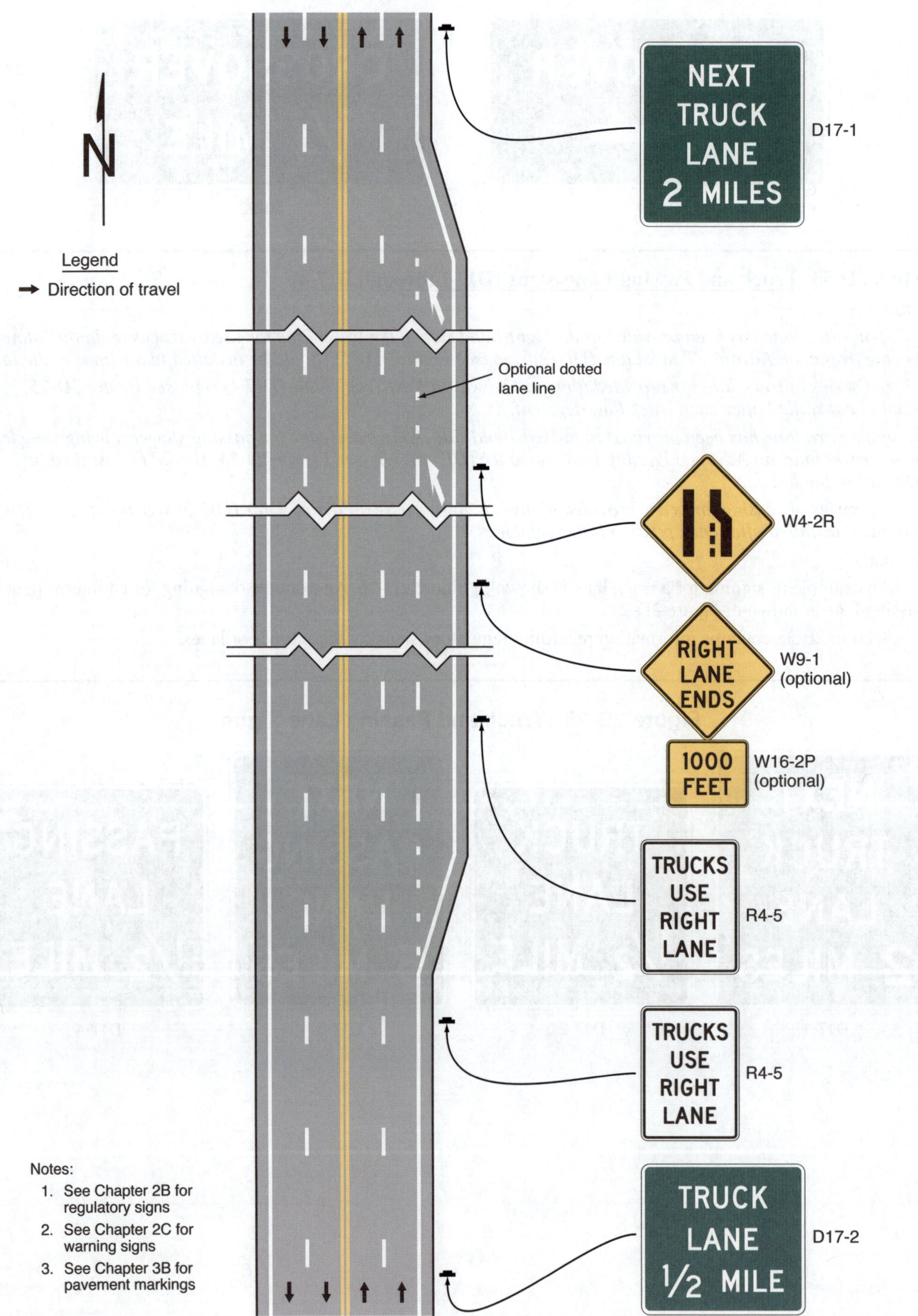

Figure 2D-27. Example of Signing for an Intermittent Passing Lane

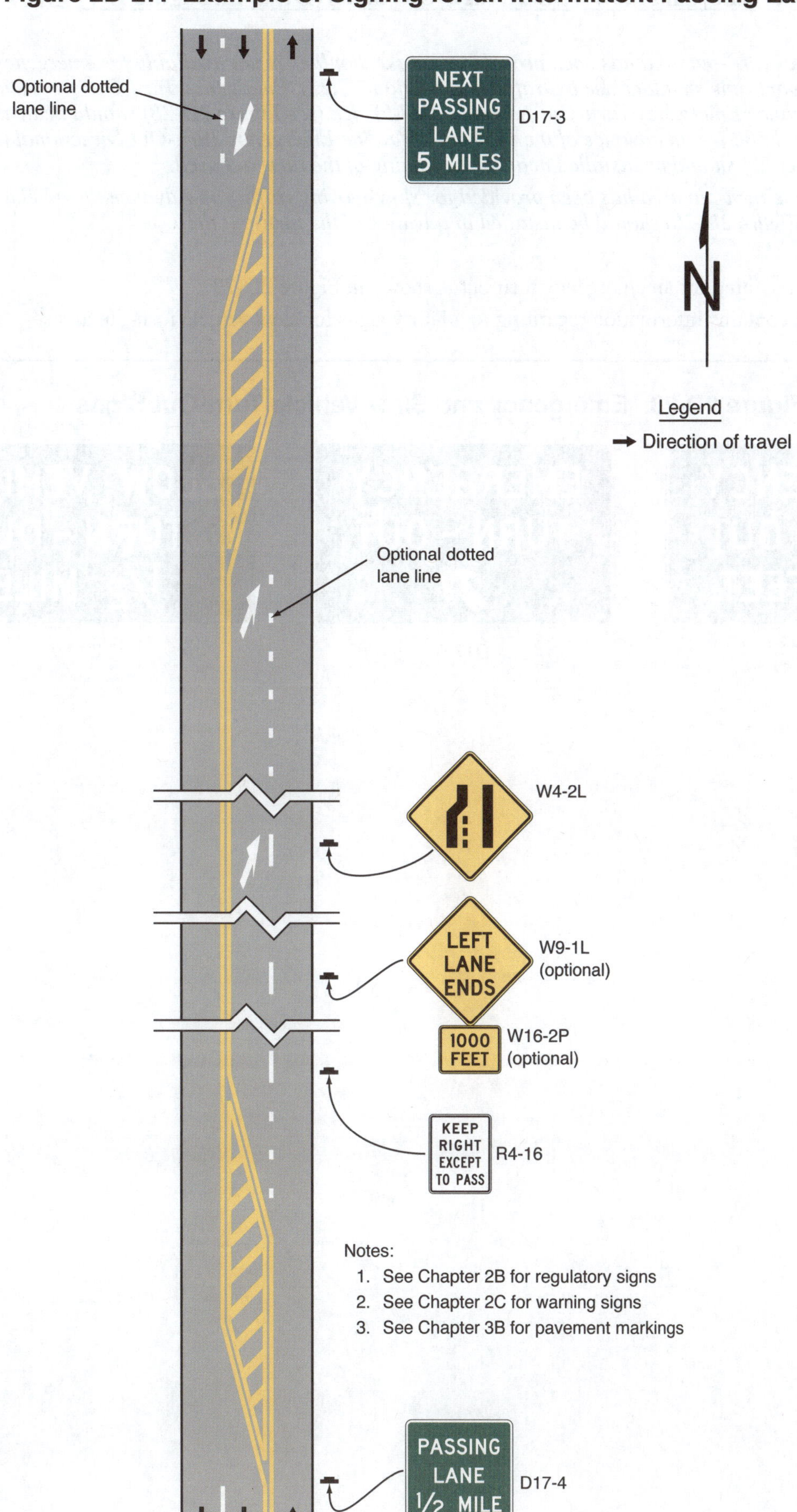

Notes:

1. See Chapter 2B for regulatory signs
2. See Chapter 2C for warning signs
3. See Chapter 3B for pavement markings

Section 2D.54 Emergency and Slow Vehicle Turn-Out Signs (D17-5 through D17-7)

Guidance:

01   *If an emergency turn-out area has been provided where a shoulder is not available for emergency stopping or where there is part-time shoulder use by traffic (see Section 2G.23), Emergency Turn-Out signs should be installed. The Advance Emergency Turn-Out Advance (D17-5) sign (see Figure 2D-28) should be installed between ¼ mile and 500 feet in advance of the turn-out area. The Emergency Turn-Out Directional (D17-6) sign (see Figure 2D-28) should be installed near the beginning of the turn-out area.*

02   *If a slow vehicle turn-out area has been provided for slow-moving traffic, an Advance Slow Vehicle Turn-Out (D17-7) sign (see Figure 2D-28) should be installed in advance of the turn-out area.*

Support:

03   An example of signing for an emergency turn-out is shown in Figure 2D-29.

04   Section 2B.42 contains information regarding regulatory signs for slow vehicle turn-out areas.

Figure 2D-28. Emergency and Slow Vehicle Turn-Out Signs

D17-5

D17-6

D17-7

Figure 2D-29. Example of Signing for an Emergency Turn-Out

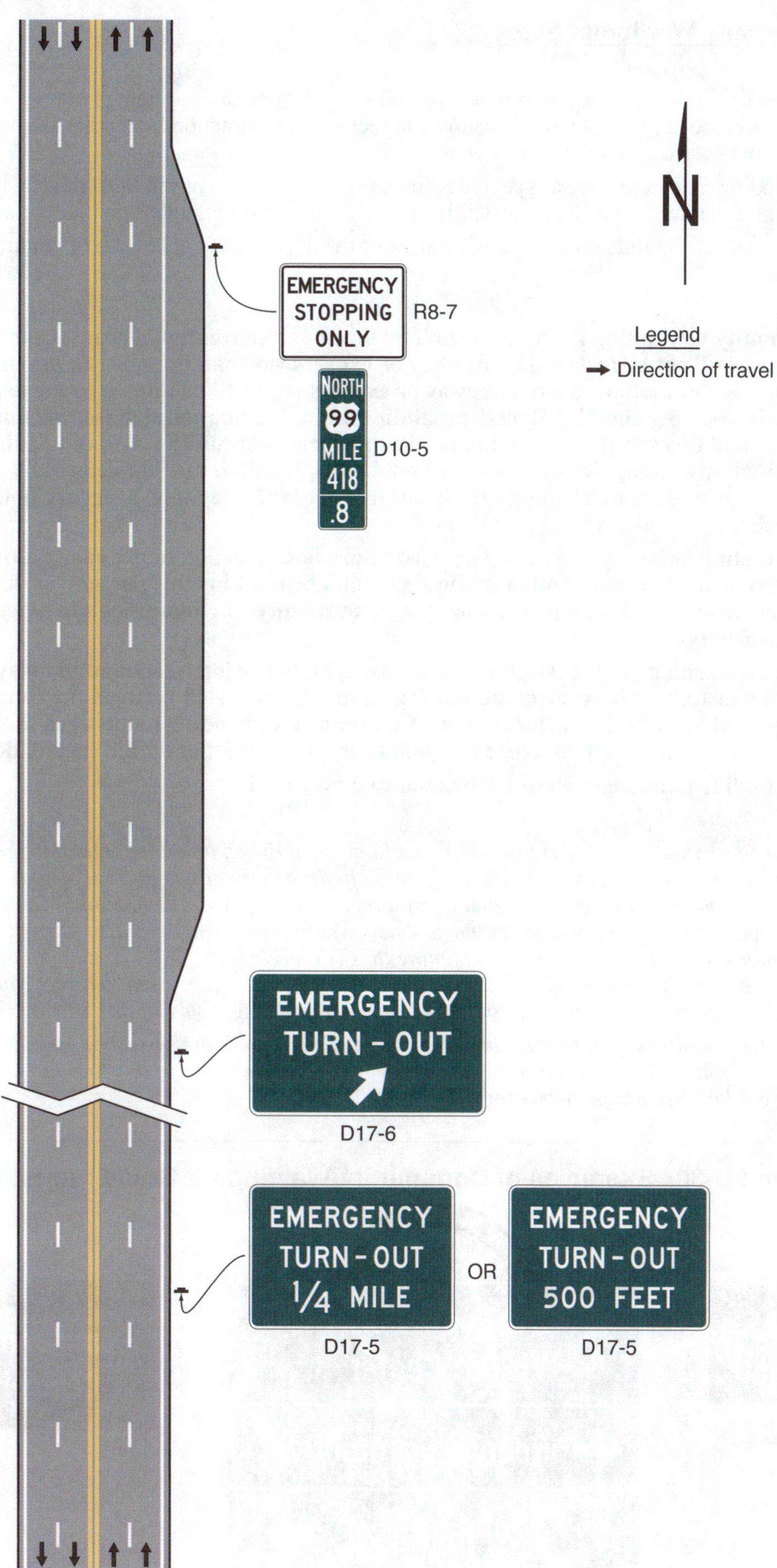

OTHER GUIDE SIGNS

Section 2D.55 Community Wayfinding Signs

Support:

01   Community wayfinding guide signs are part of a coordinated and continuous system of signs that direct tourists and other road users to key civic, cultural, visitor, and recreational attractions and other similar secondary destinations within a city or a local urbanized or downtown area.

02   Community wayfinding guide signs are a type of destination guide sign for conventional roads with a common color and/or identification marker for destinations within an overall wayfinding guide sign plan for an area.

03   Figures 2D-30 through 2D-32 illustrate various examples of the design and application of community wayfinding guide signs.

Standard:

04   **The use of community wayfinding guide signs shall be limited to conventional roads. Community wayfinding guide signs shall not be installed on freeway or expressway mainlines or ramps. Direction to community wayfinding destinations from a freeway or expressway shall be limited to the use of a Supplemental guide sign (see Section 2E.51) on the mainline and a Destination sign (see Section 2D.36) on the ramp to direct road users to the area or areas within which community wayfinding guide signs are used. The individual wayfinding destinations shall not be displayed on the Supplemental guide and Destination signs except where the destinations are in accordance with the State or agency policy on Supplemental guide signs.**

05   **Community wayfinding guide signs shall not be used to provide direction to primary destinations or highway routes or streets. Destination or other guide signs shall be used for this purpose as described elsewhere in this Chapter and shall have priority over any community wayfinding sign in placement, prominence, and conspicuity.**

06   **Because regulatory, warning, and other guide signs have a higher priority, community wayfinding guide signs shall not be installed where adequate spacing cannot be provided between the community wayfinding guide sign and other higher-priority signs. Community wayfinding guide signs shall not be installed in a position where they would obscure the road users' view of other traffic control devices.**

07   **Community wayfinding guide signs shall not be mounted overhead.**

Guidance:

08   *If used, a community wayfinding guide sign system should be established on a local municipal or equivalent jurisdictional level or for an urbanized area of adjoining municipalities or equivalent that form an identifiable geographic entity that is conducive to a cohesive and continuous system of signs. Community wayfinding guide signs should not be used on a regional or statewide basis where infrequent or sparse placement does not contribute to a continuous or coordinated system of signing that is readily identifiable as such to the road user. In such cases, Destination or other guide signs detailed in this Chapter should be used to direct road users to an identifiable area in which the type of eligible destination described in Paragraph 1 of this Section is located.*

09   *When a system of community wayfinding guide signs is being considered, the system of existing guide signs should be evaluated for applicability and general compliance with the provisions of this Manual to ensure road user directional guidance is adequately being addressed.*

Figure 2D-30. Examples of Community Wayfinding Guide Signs

Figure 2D-31. Example of a Community Wayfinding Guide Sign System Showing Direction from a Freeway or Expressway

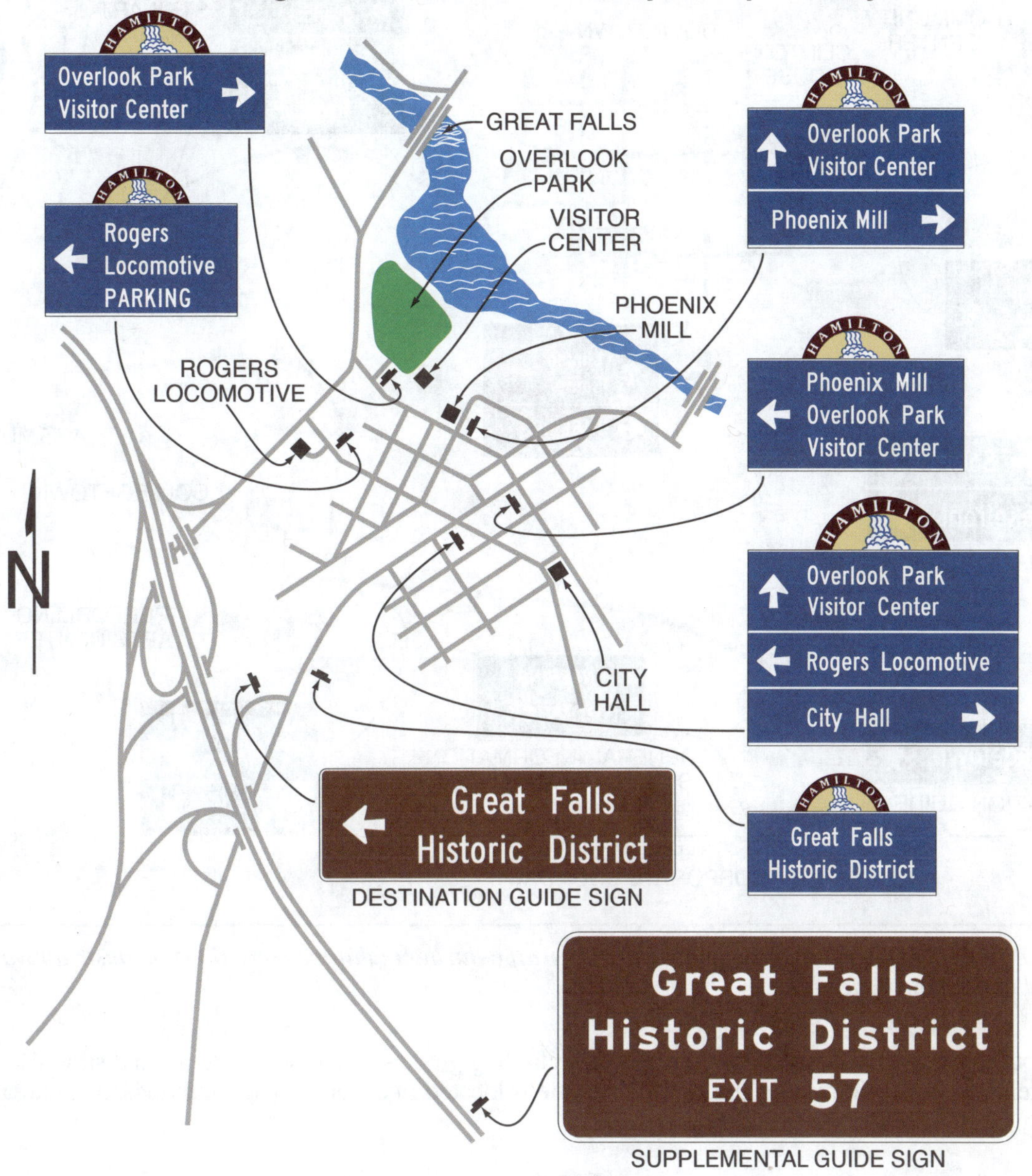

Support:

10   The specific provisions of this Section regarding the design of community wayfinding sign legends apply to vehicular community wayfinding signs and do not apply to those signs that are intended only to provide information or direction to pedestrians or other users of a sidewalk or roadside area.

Guidance:

11   *Because pedestrian wayfinding signs typically use smaller legends that are inadequately sized for viewing by vehicular traffic and because they can provide direction to pedestrians that might conflict with that appropriate for vehicular traffic, wayfinding signs designed for and intended to provide direction to pedestrians or other users of a sidewalk or other roadside area should be located to minimize their conspicuity to vehicular traffic. Such signs should be located as far as practicable from the street, such as at the far edge of the sidewalk. Where locating such signs farther from the roadway is impracticable, the pedestrian wayfinding signs should have their conspicuity to vehicular traffic minimized by employing one or a combination of the following methods:*

A. Locating signs away from intersections where high-priority traffic control devices are present.

B. Facing the pedestrian message toward the sidewalk and away from the street.

C. Cantilevering the sign over the sidewalk if the pedestrian wayfinding sign is mounted at a height consistent with vehicular traffic signs, removing the pedestrian wayfinding signs from the line of sight in a sequence of vehicular signs.

Figure 2D-32. Example of a Color-Coded Community Wayfinding Guide Sign System

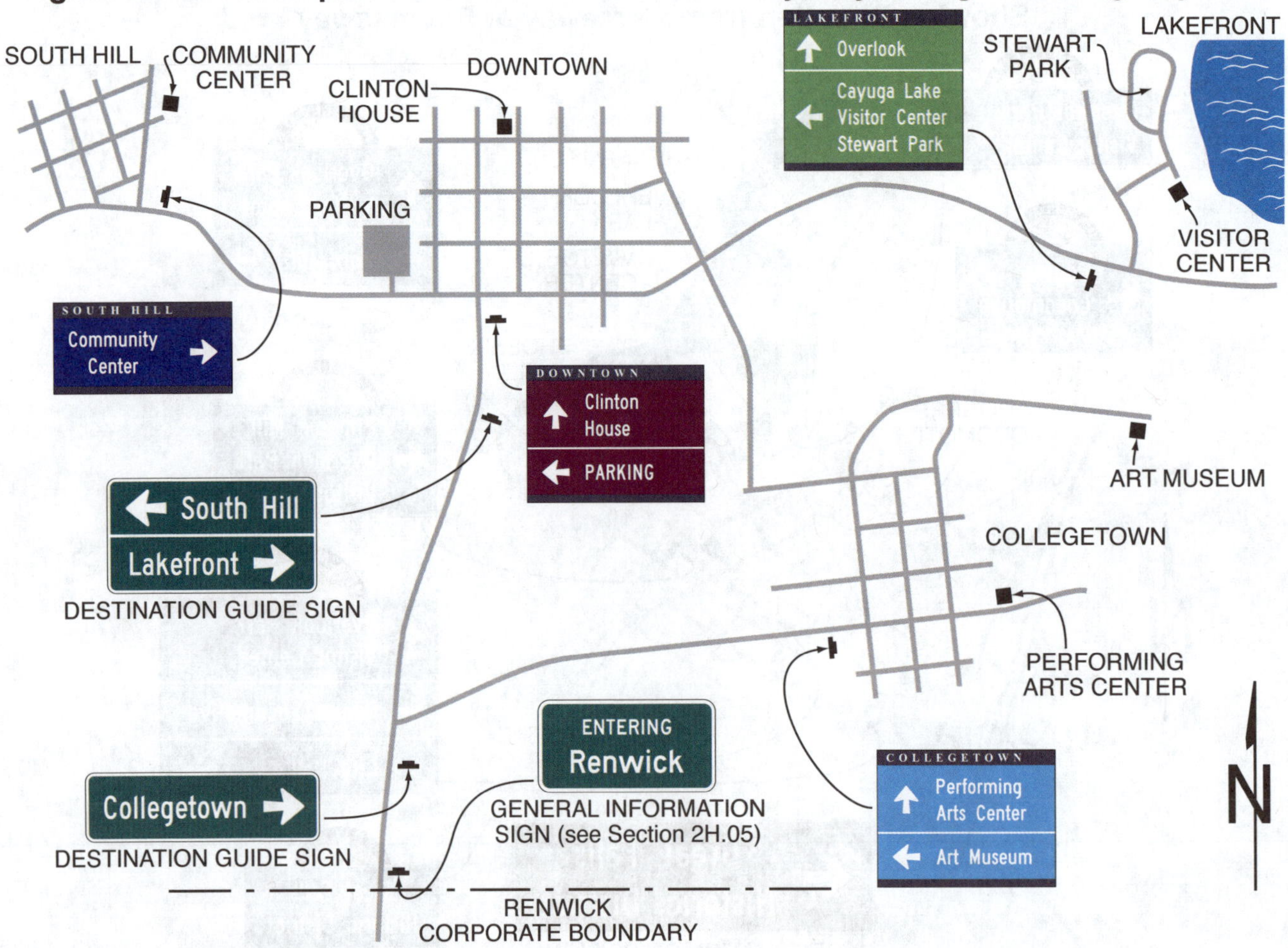

12   *To further minimize their conspicuity to vehicular traffic during nighttime conditions, pedestrian wayfinding signs should not be retroreflective.*

Support:

13   Color coding is sometimes used on community wayfinding guide signs to help road users distinguish between multiple potentially confusing traffic generator destinations located in different neighborhoods or subareas within a community or area.

Option:

14   At the boundaries of the geographical area within which community wayfinding guide signing is used, an informational guide sign may be posted to inform road users about the presence of wayfinding signing and to identify the meanings of the various color codes or pictographs that are being used.

Standard:

15   **These informational guide signs shall have a white legend and border on a green background and shall have a design similar to that illustrated in Figure 2D-1 and shall be consistent with the basic design principles for guide signs. These informational guide signs shall not be installed on freeway or expressway mainlines or ramps.**

16   **The color coding or a pictograph of the identification markers of the community wayfinding guide signing system shall be included on the informational guide sign posted at the boundary of the community wayfinding guide signing area. The color coding or pictographs shall apply to a specific, identifiable neighborhood or geographical subarea within the overall area covered by the community wayfinding guide signing. Color coding or pictographs shall not be used to distinguish between different types of destinations that are within the same designated neighborhood or subarea. The color coding shall be accomplished by the use of different colored square or rectangular panels on the face of the informational guide sign, each positioned to the left of the neighborhood or named geographic area to which the color-coding panel applies. The height of the colored square or rectangular panels shall not exceed 2 times the height of the upper-case letters of the principal legend on the sign.**

Option:

17   The different colored square or rectangular panels may include either a black or a white (whichever provides the better contrast with the color of the panel) letter, numeral, or other appropriate designation to identify the destination.

18   Except for the informational guide sign posted at the boundary of the wayfinding guide sign area, community wayfinding guide signs may use background colors other than green in order to provide a color identification for the wayfinding destinations by geographical area within the overall wayfinding guide signing system. Color-coded community wayfinding guide signs may be used with or without the boundary informational guide sign displaying corresponding color-coding panels described in Paragraphs 13 through 16 of this Section. Except as provided in Paragraph 19 of this Section, in addition to the colors that are approved in this Manual for use on official traffic control signs (see Section 2A.06), other background colors may also be used for the color coding of community wayfinding guide signs.

Standard:

19   **The standard colors of red, orange, yellow, purple, or the fluorescent versions thereof, fluorescent yellow-green, and fluorescent pink shall not be used as background colors for community wayfinding guide signs, in order to minimize possible confusion with critical, higher-priority regulatory and warning sign color meanings readily understood by road users.**

20   **The minimum contrast value of legend color to background color for community wayfinding guide signs shall be at least 0.70 (or 70%).**

21   **All messages, borders, legends, and backgrounds of community wayfinding guide signs and any identification markers shall be retroreflective (see Sections 2A.21 and 2A.22).**

22   **Community wayfinding guide signs, exclusive of any identification marker used, shall be rectangular in shape.**

Guidance:

23   *Simplicity and uniformity in design, position, and application as described in Section 2A.04 are important and should be incorporated into the community wayfinding guide sign design and location plans for the area.*

24   *Community wayfinding guide signs should be limited to three destinations per sign (see Section 2D.06).*

25   *Abbreviations (see Section 1D.08) should be kept to a minimum, and should include only those that are commonly recognized and understood.*

26   *Horizontal lines of a color that contrasts with the sign background color should be used to separate groups of destinations by direction from each other.*

Support:

27   The basic requirement for all highway signs, including community wayfinding signs, is that they be legible to those for whom they are intended and that they be understandable in time to permit a proper response. Section 2A.04 contains additional information on the design of signs, including desirable attributes of effective designs.

Guidance:

28   *Word messages should be as brief as practical and the lettering should be large enough to provide the necessary legibility distance.*

Standard:

29   **The minimum specific ratio of letter height to legibility distance shall comply with the provisions of Section 2A.08. The size of lettering used for destination and directional legends on community wayfinding signs shall comply with the provisions of minimum letter heights as provided in Section 2D.05.**

30   **Interline and edge spacing shall comply with the provisions of Section 2D.05.**

31   **Except as provided in Paragraph 34 of this Section, the lettering style used for destination and directional legends on community wayfinding guide signs shall comply with the provisions of Section 2D.04.**

32   **The lettering for destinations on community wayfinding guide signs shall be a combination of lower-case letters with initial upper-case letters (see Section 2D.04). All other word messages on community wayfinding guide signs shall be in all upper-case letters.**

Guidance:

33   *Except as provided in Paragraphs 34 and 35 of this Section, letters, numerals, and other characters should be composed of the Standard Alphabets as detailed in the "Standard Highway Signs" publication (see Section 1C.05).*

Option:

34   A lettering style other than the Standard Alphabets provided in the "Standard Highway Signs" publication (see Section 1C.05) may be used on community wayfinding guide signs if an engineering study determines that the legibility and recognition values for the chosen lettering style meet or exceed the values for the Standard Alphabets for the same legend height and stroke width.

Standard:

35   **If a lettering style other than the Standard Alphabets is used, the alternative lettering style shall be conventional in form. The letters, numerals, and other characters shall not be italic, oblique, script, highly decorative, or of other unusual forms.**

36   **In accordance with Section 2A.04, except for signs that are designed and located with the intent to be viewed only by pedestrians, bicyclists stopped out of the flow of traffic, or occupants of parked vehicles, Internet and e-mail addresses, including domain names and uniform resource locators (URL), and scanning graphics for the purpose of obtaining information (see Section 2A.04), shall not be displayed on any community wayfinding guide sign or sign assembly.**

37   **The arrow location and priority order of destinations shall follow the provisions described in Sections 2D.08 and 2D.36. Arrows shall be of the designs provided in Section 2D.08.**

Option:

38   Pictographs (see definition in Section 1C.02) may be used on community wayfinding guide signs.

Standard:

39   **If a pictograph is used, its height shall not exceed 2 times the height of the upper-case letters of the principal legend on the sign.**

40   **Except for pictographs, symbols that are not approved in this Manual for use on guide signs shall not be used on community wayfinding guide signs.**

41   **Business logos, commercial graphics, or other forms of advertising (see Section 1D.07) shall not be used on community wayfinding guide signs or sign assemblies.**

Option:

42   Other graphics that specifically identify the wayfinding system, including identification markers, may be used on the overall sign assembly and sign supports.

Support:

43   An identification marker consists of a shape, color, and/or pictograph that is used as a visual identifier for the community wayfinding guide signing system for an area. Figure 2D-30 shows examples of identification marker designs that can be used with community wayfinding guide signs.

Option:

44   An identification marker may be used in a community wayfinding guide sign assembly, or may be incorporated into the overall design of a community wayfinding guide sign, as a means of visually identifying the sign as part of an overall system of community wayfinding signs and destinations.

Standard:

45   **The sizes and shapes of identification markers shall be smaller than the community wayfinding guide signs themselves. Identification markers shall not be designed to have an appearance that could be mistaken by road users as being a traffic control device.**

Guidance:

46   *The area of the identification marker should not exceed ⅕ of the area of the community wayfinding guide sign with which it is mounted in the same sign assembly.*

Section 2D.56 <u>Signing of Named Highways for Mapping and Address Purposes</u>

Support:

01   A highway name is the officially designated name of a freeway, expressway, or conventional road for navigational, official mapping, and address purposes. Some highways are named in addition to or in lieu of being assigned a highway route number. Memorial, honorary, ceremonial, or other secondary names, such as touring route and byway names, are not considered to be highway names.

Option:

02   Guide signs may contain street or highway names if the purpose is to enhance driver communication and guidance; however, they are to be considered as supplemental information to route numbers.

Standard:

03   **Highway names shall not replace official numeral designations.**

04   **Memorial, honorary, or other secondary names (see Section 2M.10) shall not appear on supplemental signs or on any other information sign on or along the highway or its intersecting routes.**

05   **The use of route signs shall be restricted to signs officially used for guidance of traffic in accordance with this Manual and the "Purpose and Policy" statement of the American Association of State Highway and Transportation Officials that applies to Interstate and U.S. numbered routes.**

Option:

06   Unnumbered routes having major importance to proper guidance of traffic may be signed if carried out in accordance with the aforementioned policies. For unnumbered highways, a name to enhance route guidance may be used where the name is applied consistently throughout its length.

Guidance:

07   *Only one name should be used to identify any highway, whether numbered or unnumbered.*

Section 2D.57 National Scenic Byways Sign and Plaque (M10-1 and M10-1aP)

Support:

01   Certain roads have been designated by the U.S. Secretary of Transportation as National Scenic Byways or All-American Roads based on their archeological, cultural, historic, natural, recreational, or scenic qualities.

Option:

02   State and local highway agencies may install the National Scenic Byways (M10-1) sign or (M10-1aP) plaque (see Figure 2D-33) at entrance points to a route that has been recognized by the U.S. Secretary of Transportation as a National Scenic Byway or an All-American Road. The M10-1 sign may be installed as independent Directional (see Section 2D.32) or Confirming (see Section 2D.33) assemblies at periodic intervals along the designated route and near intersections where the designated route turns or follows a different numbered highway. The M10-1aP plaque may be installed below a route sign in a Confirming assembly. At locations where roadside features have been developed to enhance the traveler's experience such as rest areas, historic sites, interpretive facilities, or scenic overlooks, the National Scenic Byways sign or plaque may be placed on the associated sign assembly to inform travelers that the site contributes to the byway travel experience.

Guidance:

03   *Where the byway is identified only by the National Scenic Byways sign, the Directional assembly should consist of the M10-1 sign and an M5 series or M6 series auxiliary plaque when indication of a turn is necessary to remain on the byway route.*

Figure 2D-33. National Scenic Byways Sign and Plaque, and Examples of Use

A – National Scenic Byways sign and plaque

B – Examples of use of the National Scenic Byways sign and plaque

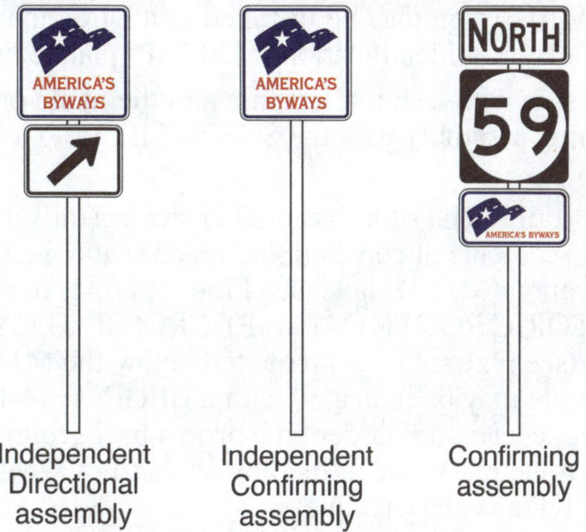

04   *Where the name of the byway is to be displayed on identification signs or plaques along the byway route, the name should be displayed in a Directional or Confirming assembly as follows (see Figure 2D-34):*

   A.   *On a Byway Identification (M10-2aP) plaque (see Section 2D.58) mounted below the M10-1 sign; or*
   B.   *On a Byway Identification (M10-2) sign (see Section 2D.58) with the M10-1aP plaque mounted below the sign.*

05   *In either case, the size of the National Scenic Byways (M10-1) sign, (M10-1aP) plaque, Byway Identification (M10-2) sign, and Byway Identification (M10-2aP) plaque should be consistent with that specified for route signs (see Section 2D.10) for the roadway classification.*

06   *Where the name of the byway is to be displayed along the byway route as provided in Paragraph 4 of this Section, the byway Directional or Confirming assemblies should be located separately from any Route Sign assemblies or Destination guide signs.*

Standard:

07   **When a National Scenic Byways sign is installed on a National Scenic Byway or an All-American Road, the design shown for the M10-1 sign or M10-1aP plaque in Figure 2D-33 shall be used. Use of this design shall be limited to routes that have been designated as a National Scenic Byway or All-American Road by the U.S. Secretary of Transportation.**

08   **If used, the M10-1 sign or M10-1aP plaque shall be placed such that the highway route signs have primary visibility for the road user.**

09   **The M10-1 sign or the M10-1aP plaque shall not be installed as sign panels on a guide sign or as part of a guide sign assembly.**

Section 2D.58 State-Designated Scenic Byway, Historic Trail, and Auto Tour Route Signs

Support:

01   Signing for State-designated scenic byways, historic trails, and auto tour routes, is similar in concept to that for National Scenic Byways as provided in Section 2D.57.

02   Named highways are officially designated and shown on official maps and serve the purpose of providing route guidance, primarily on unnumbered highways, and property addresses. A highway designated as a trail, auto tour route, or byway is not considered to be a highway name for the purposes of highway signing or road user navigation and orientation. Section 2D.56 contains provisions for the signing of named highways.

03   Section 1D.09 provides information on the authority for placement of traffic control devices within the highway right-of-way.

Guidance:

04   *Route Sign assemblies and Destination guide signs should have priority in visibility and location over signing related to historic trails, auto tour routes, and byways.*

Option:

05   Identification signs for a State scenic byway may be installed along conventional roads that have been designated as part of a State scenic byway system. A Byway Identification (M10-2) sign (see Figure 2D-34) with the name of the byway displayed may be installed in a Directional or Confirming assembly with the SCENIC BYWAY (M10-3bP) plaque (see Figure 2D-34) mounted below the M10-2 sign.

06   Where a National Scenic Byway is part of a State scenic byway system, the National Scenic Byways (M10-1aP) plaque (see Section 2D.57) may be installed in a Directional or Confirming assembly below the Byway Identification (M10-2) sign or State Scenic Byway (M10-3 or M10-3a) sign (see Figure 2D-34) for the State scenic byway.

07   A State Scenic Byway System (M10-3) sign may be installed in a Directional or Confirming assembly with the name of the byway displayed on a Byway Identification (M10-2aP) plaque below the sign (see Figure 2D-34).

08   A State Scenic Byway (M10-3a) sign with a simple graphic and the name of the byway displayed may be installed in a Directional or Confirming assembly with the SCENIC BYWAY (M10-3bP) plaque mounted below the M10-3a sign.

09   Identification signs for a historic trail, such as the National Historic Trails administered by the National Park Service, may be installed along segments of conventional roads that coincide with the original route of the trail. National Historic Trail Identification (M11-1) signs (see Figure 2D-34) may be installed in a Directional or Confirming assembly with a HISTORIC ROUTE (M11-1aP), CROSSING (M11-1bP), or AUTO TOUR ROUTE (M11-1cP) auxiliary plaque (see Figure 2D-34) mounted below the M11-1 sign. The beginning and end of a historic trail route or auto tour route may be indicated with a BEGIN (M4-14P) or END (M4-6P) auxiliary plaque (see Figure 2D-5) with a white legend and border on a brown background mounted above the historic trail identification sign. The length of the route may be identified by a NEXT XX MILES (M11-1dP) auxiliary plaque mounted below the M11-1aP or M11-1cP auxiliary plaque.

Figure 2D-34. Byway Identification, State Scenic Byway, and National Historic Trail Signs and Plaques, and Examples of Use

A – Byway Identification, State Scenic Byway, and National Historic Trails signs and plaques

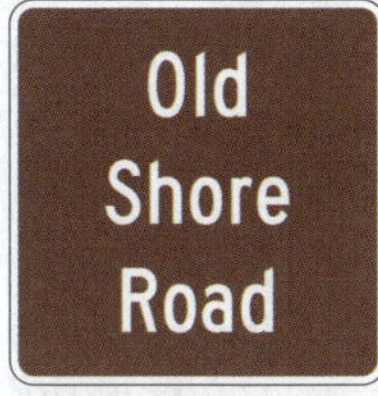

M10-2

Old Shore
Road

M10-2aP

M10-3

M10-3a

SCENIC
BYWAY

M10-3bP

M11-1

HISTORIC
ROUTE

M11-1aP

CROSSING

M11-1bP

AUTO TOUR
ROUTE

M11-1cP

NEXT
25 MILES

M11-1dP

B – Examples of Directional assemblies for National and State Scenic Byways

Byway identification emphasized

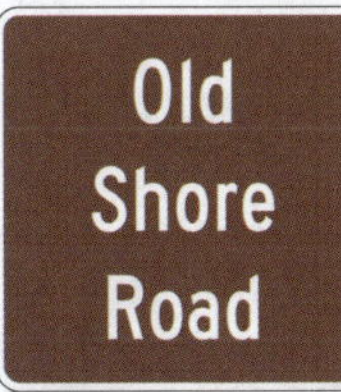

State Scenic
Byway

State Scenic Byway
that is also designated as
a National Scenic Byway

State Scenic
Byway

Byway system emphasized

National Scenic
Byway system

State Scenic
Byway system

C – Examples of Directional and Confirming assemblies for National Historic Trails

Independent
Directional
assembly

Independent Confirming assemblies

Guidance:

10   *The design and size of historic trail and State scenic byway identification or system signs should comply with the general provisions and principles for route signs (see Section 2D.10). Designs should be simple, dignified, and devoid of complex graphics. The size of the signs should not exceed the size of the route signs used along a particular route.*

Standard:

11   **Scenic byway, historic trail, and auto tour route sign designs shall not have a similar design to or resemble a highway route sign**

Guidance:

12   *Where used, historic trail and State scenic byway identification signs should be installed as Directional (see Section 2D.32) or Confirming (see Section 2D.33) assemblies at independent locations, separate from other Route Sign assemblies and Destination guide signs. Where used, Confirming assemblies for the trail or byway should be installed at less frequent intervals than Confirming assemblies for the numbered route.*

Support:

13   Where all or part of the original route of a historic trail does not follow a roadway, an auto tour route is sometimes established along a conventional road in the general vicinity of the historical route of the trail. Examples include auto tour routes following other routes that parallel the original routes of the Lewis and Clark National Historic Trail, the Oregon National Historic Trail, and the Santa Fe National Historic Trail. The auto tour route is shown on touring maps along State or other highways and provides access to sites on the trail from those highways.

14   A system of signing providing direction along conventional roads for a historic trail with an auto tour route is shown in Figure 2D-35. Examples of Destination and Supplemental guide signs providing direction to historic trail sites from a freeway or expressway interchange are shown in Figure 2D-36.

Guidance:

15   *Signing for historic trails should be limited to Destination signs for the sites related to the trail and to Directional and Confirming assemblies for the original portions of the trail itself. If an auto tour route has been designated along other highways to provide access to sites along the original trail as described in Paragraph 13 of this Section, then the signing should be limited to Destination signs for those sites and directional signing to access the original route of the trail. Identification signs for the auto tour route should not be installed. Instead, direction along the auto tour route should rely on the touring map and other directional signs for the highways that the auto tour route follows.*

Standard:

16   **Identification signs for historic trails, auto tour routes, and scenic byways shall not be installed on freeways or expressways, except as necessary to provide continuity between discontinuous segments of conventional roadways that are designated as a trail, auto tour route, or byway, for which the freeway or expressway provides the only connection between the segments. If installed on freeways or expressways, the identification signs shall be installed as independent trailblazer assemblies (see Sections 2D.34 and 2E.55) and shall not be installed with other route signs or Confirming assemblies or on guide signs. If installed on freeways or expressways, the trailblazer assemblies for the trail, auto tour route, or byway shall be installed at less frequent intervals than Confirming assemblies for the highway route.**

17   **Identification signs for historic trails, auto tour routes, and scenic byways shall not be installed as sign panels on a guide sign or as part of a guide sign assembly.**

Figure 2D-35. Example of Guide and Directional Signing for a National Historic Trail

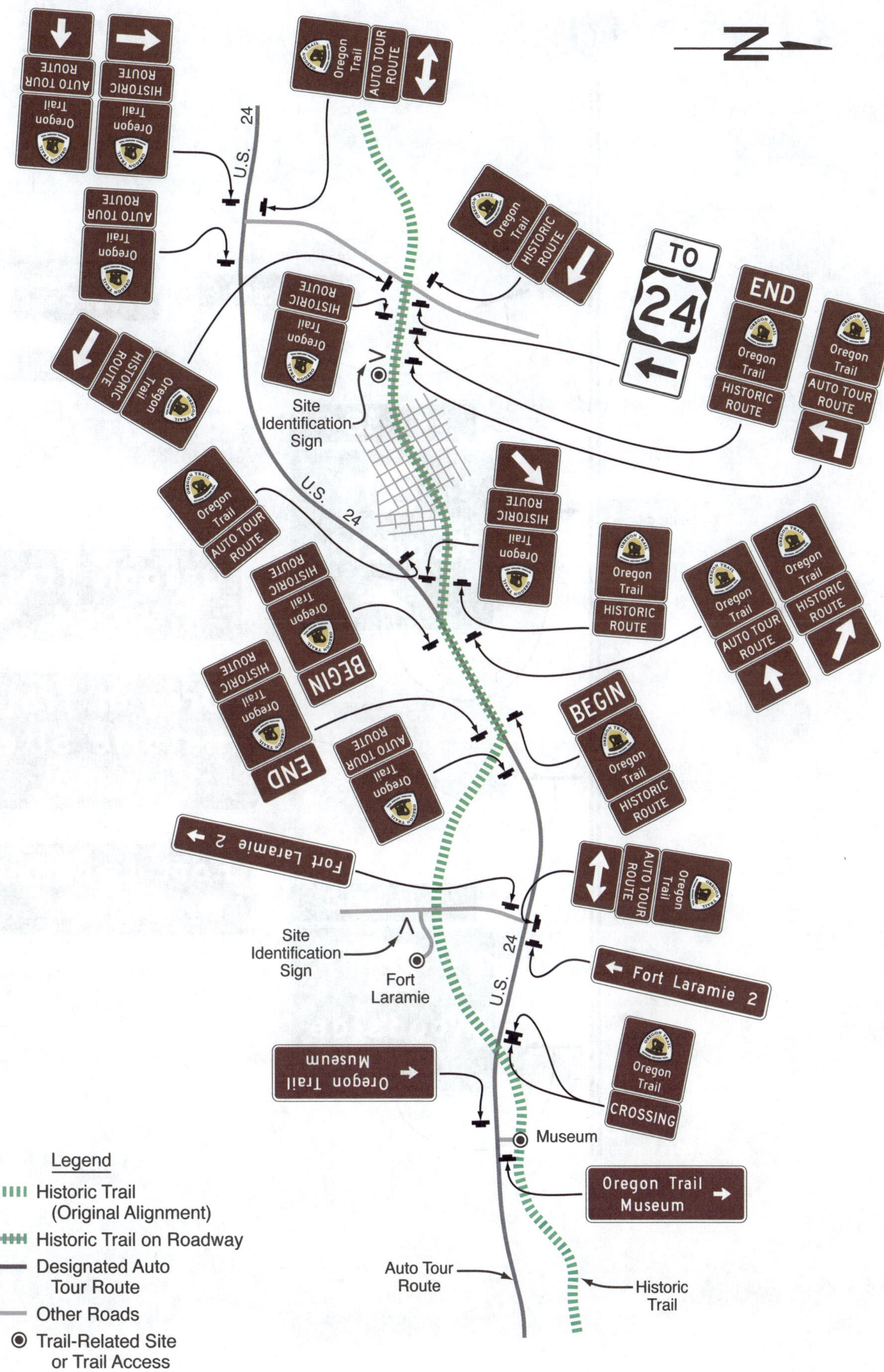

Figure 2D-36. Example of Historic Trails Signing from an Expressway or Freeway

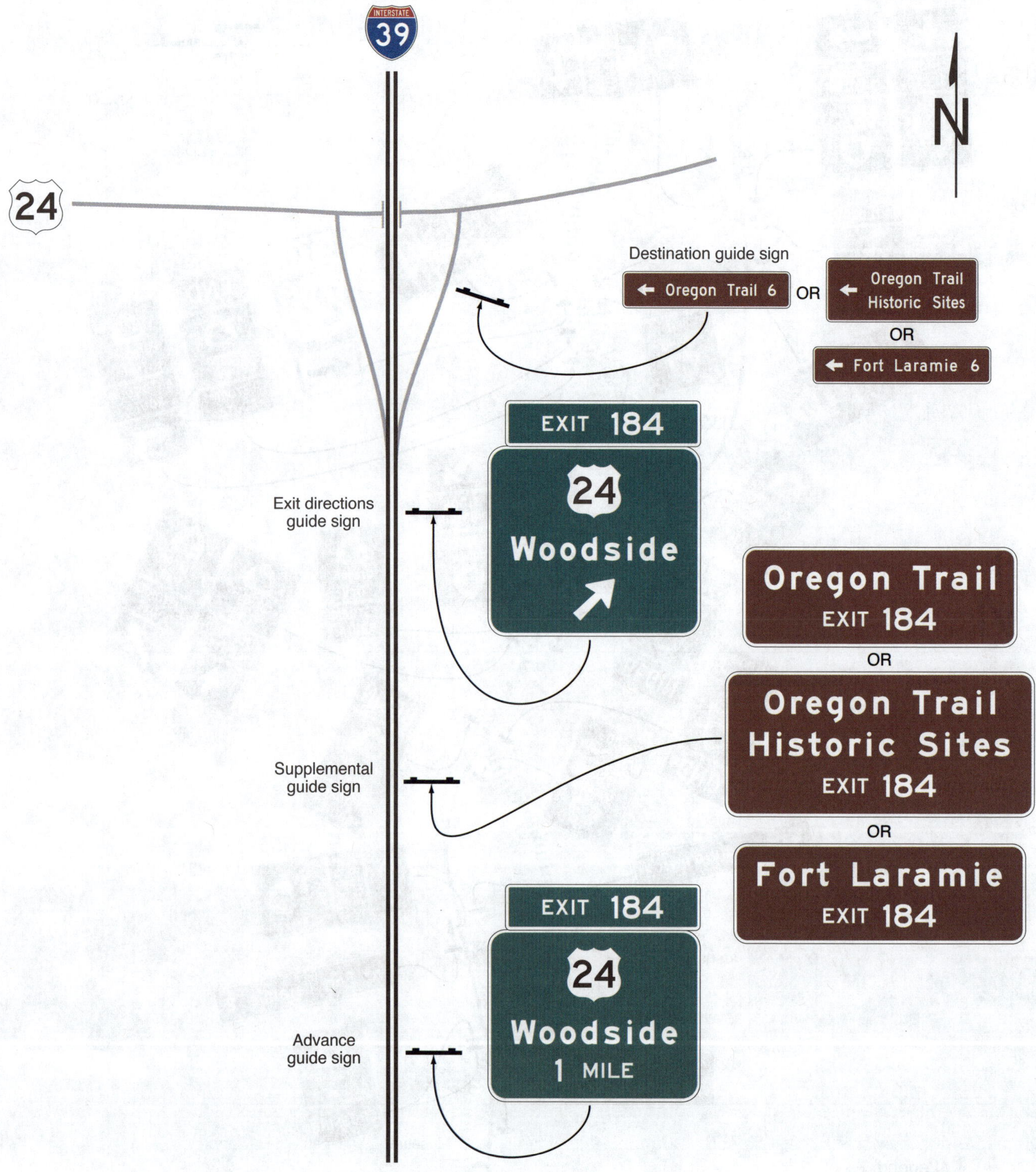

Section 2D.59   Emergency Routing Signs and Plaques (M4-11 and M4-12 Series)

Support:

01   As part of an agency's transportation incident management plan it is sometimes desirable to permanently sign routes that provide rerouting of traffic around highway segments susceptible to traffic incidents. Permanently-installed Emergency Routing signs and plaques (see Figure 2D-37) provide direction on conventional roads from an exit located upstream of an area that can be susceptible to traffic incidents back to the original route at a point downstream of the incident-susceptible area.

Option:

02   EMERGENCY ROUTE (M4-11 and M4-11a) signs and EMERGENCY ROUTE TO (M4-11bP and M4-11cP) plaques mounted on a directional assembly may be permanently installed on conventional roadways to provide trailblazing along a designated diversionary route to bypass a traffic incident.

Support:

03   The purpose of Emergency Routing signs and plaques is for corridor management along routes that have recurring incidents and have reasonable rerouting paths available. These signs are intended to be permanently installed to provide instant rerouting guidance to road users when traffic congestion or backups first begin even before emergency responders could provide temporary traffic control for rerouting traffic. These signs can be used as a stand-alone system or be a part of a larger system which might also incorporate other devices such as changeable message signs. These signs provide road users assurance that a diversionary route will lead them back to their original route of travel.

Standard:

04   **Emergency Routing signs and plaques shall only be installed at departure points and along diversion routes for directing road users around highway segments in areas that are more susceptible to traffic incidents (see Figure 2D-38). EMERGENCY ROUTE and EMERGENCY ROUTE TO signs shall be placed at each turning decision point along the designated route until it rejoins the original route or until other directional signs leading back to the original route are provided.**

05   **Emergency Routing signs shall have a white legend and border on a green background.**

Option:

06   For additional emphasis the legend EMERGENCY ROUTE or EMERGENCY ROUTE TO may be displayed in a yellow panel with black letters near the top of the sign (see Figure 2D-37).

Standard:

07   **Orange or pink shall not be used as alternate colors on permanently-installed signs or plaques for rerouting traffic during an incident or other event. If a route shield is displayed as part of the message, the wording of the sign or plaque shall be EMERGENCY ROUTE TO as shown in Figure 2D-37.**

Option:

08   An EMERGENCY ROUTE TO plaque with either a white legend and border on a green background (M4-11bP) or black legend and border on a yellow background (M4-11cP) may be added to the top of a conventional Route assembly on a diversion route to provide direction back to the original route downstream of the incident (see Figures 2D-37 and 2D-38).

09   A combination warning and regulatory message sign with flashing beacons mounted above and the legend, WHEN FLASHING [ROUTE] CLOSED AHEAD/ USE EMERGENCY ROUTE [NEXT RIGHT or EXIT XX] may be used in advance of the emergency route entrance to inform road users of the closure and require exiting the primary route and use of the emergency route (see Figures 2D-37 and 2D-38).

10   The End Emergency Route (M4-12) sign may be use at a point along the emergency route just prior to the junction where traffic is to reenter the primary route past the closed section of roadway (see Figure 2D-37 and 2D-38).

Figure 2D-37. Signs and Plaques for Rerouting Because of Traffic Incidents

M4-11

M4-11bP

M4-12

M4-11a

M4-11cP

Examples of guide signs for rerouting

Example of advance sign for rerouting

Figure 2D-38. Example of Permanent Guide Signing for Rerouting Because of Traffic Incidents

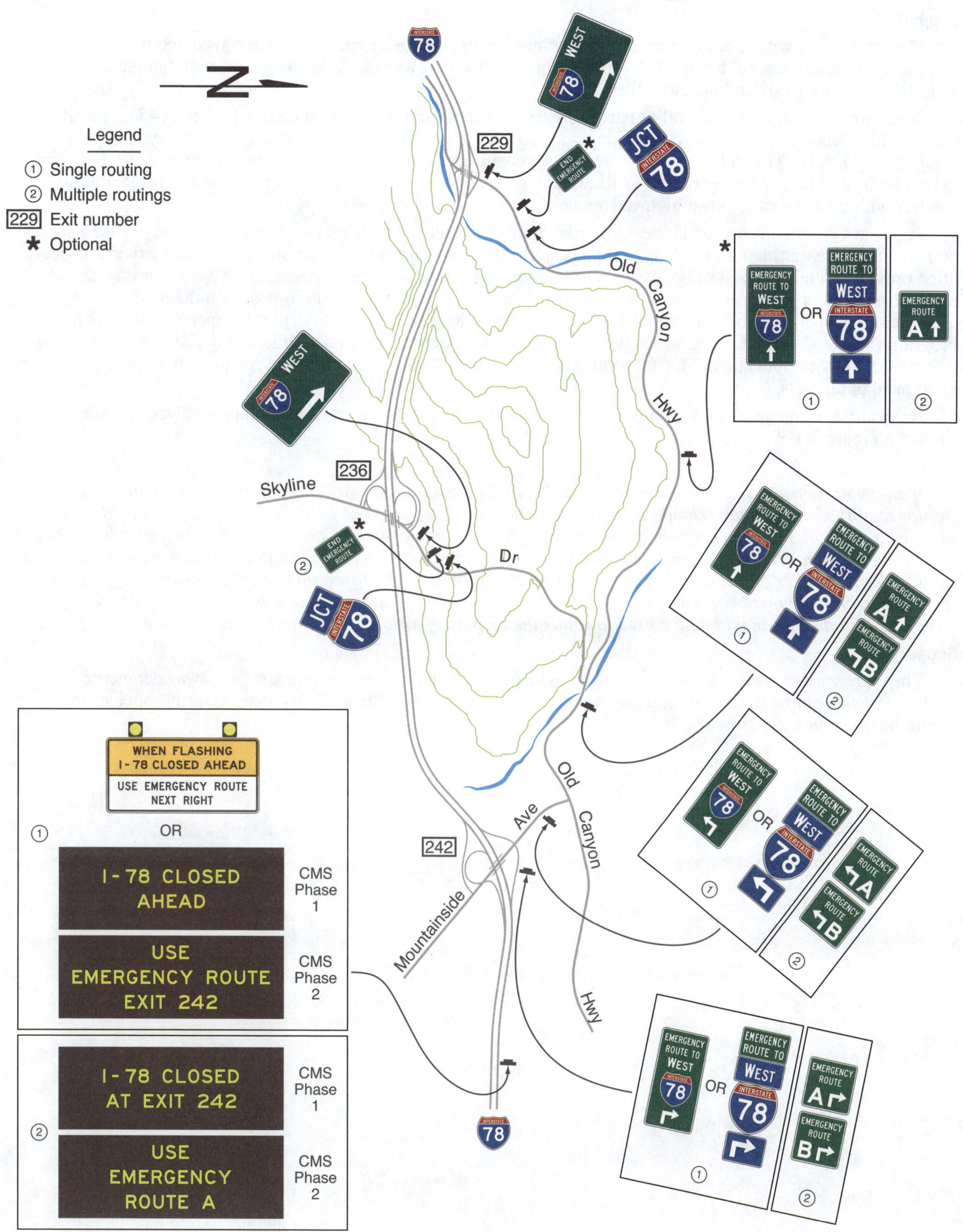

SIGNING AT AIRPORTS

Section 2D.60 Signing at Airports

Support:

01   Many roadways within airport facilities (including terminal curbside roadways) are considered to be conventional roads because they typically have frequent driveways and at-grade intersections and might have pedestrian activity along and/or across them.

02   Some airport roadways have full or partial control of access and operating speeds higher than 45 mph and thus would be classified as freeways or expressways for signing purposes (see Chapter 2E). Freeway or expressway conditions typically exist on the approaches to the airport from other highways; on the approaches to access points to terminals, parking, and other patron facilities; and on roadways that provide exits from the airport facility to connect with the local or regional highway network.

03   Roadways within airports and other similarly-contained roadway networks with multiple closely-spaced access points to multiple destinations (such as terminals, parking facilities, rental car facilities, and other airport services) often present challenges for the application of guide signing. Closely-spaced signs, excessive sign messaging either co-located or in succession, and the resulting excessive informational load imposed on the road user are of particular concern for such roadways. The Transportation Research Board's Airport Cooperative Research Program Report 52, "Wayfinding and Signing Guidelines for Airport Terminal and Landside," contains additional information on the application of traffic control devices to the unique geometrics and roadway environment that are typical of airports.

04   An example of major guide signing on the approaches to and within an airport facility roadway network is shown in Figure 2D-39.

Guidance:

05   *If adequate sign spacing cannot be provided due to the site and roadway characteristics of an airport or similar facility, then measures should be taken to reduce the speeds of vehicles on the roadway to provide road users with adequate time to comprehend and respond to the sign messages. Consideration should also be given to increasing the sign letter heights to provide greater viewing distances and decision times. Where a single terminal serves a large number of airlines, the airline information should be displayed on separate signs that appear in sequence to limit the number of airlines displayed on a single sign or at a single location. Changeable message signs (see Chapter 2L) should not be used to rotate the display of airlines to an approaching road user.*

Support:

06   There are various methods that can be used to help reduce vehicle speeds, including roadway geometric changes, implementing traffic calming measures, and increased enforcement. Provisions on setting speed limits are found in Section 2B.21.

Figure 2D-39. Example of a System of Major Guide Signs for an Airport Roadway Network (Sheet 1 of 3)

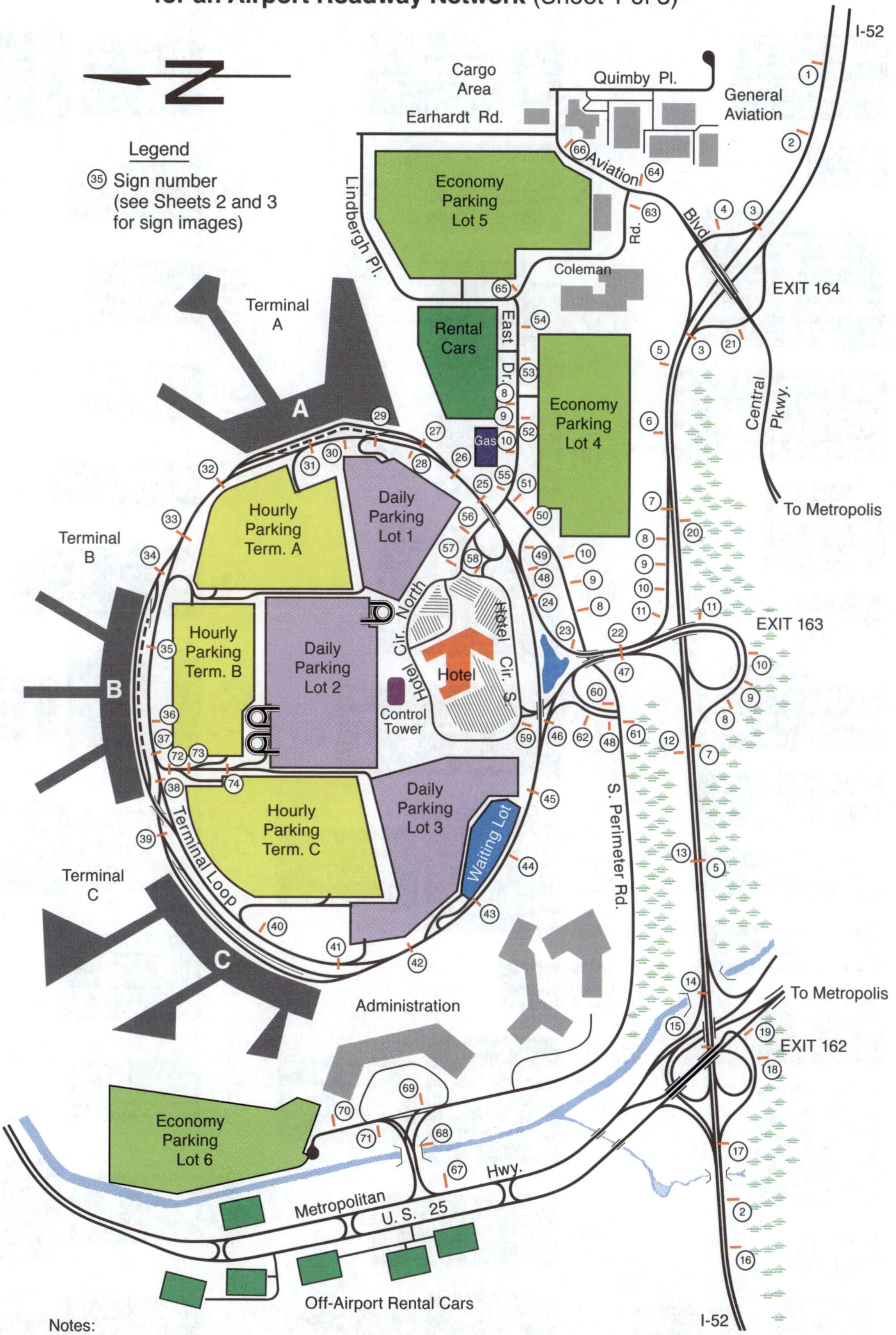

Notes:

1. Major guide signs are shown in this figure for illustrative purposes. Other guide signs (such as advance signs), as well as regulatory and warning signs, have been omitted for clarity.

2. Terminal loop is a one-way clockwise roadway.

Figure 2D-39. Example of a System of Major Guide Signs for an Airport Roadway Network (Sheet 2 of 3)

Figure 2D-39. Example of a System of Major Guide Signs for an Airport Roadway Network (Sheet 3 of 3)

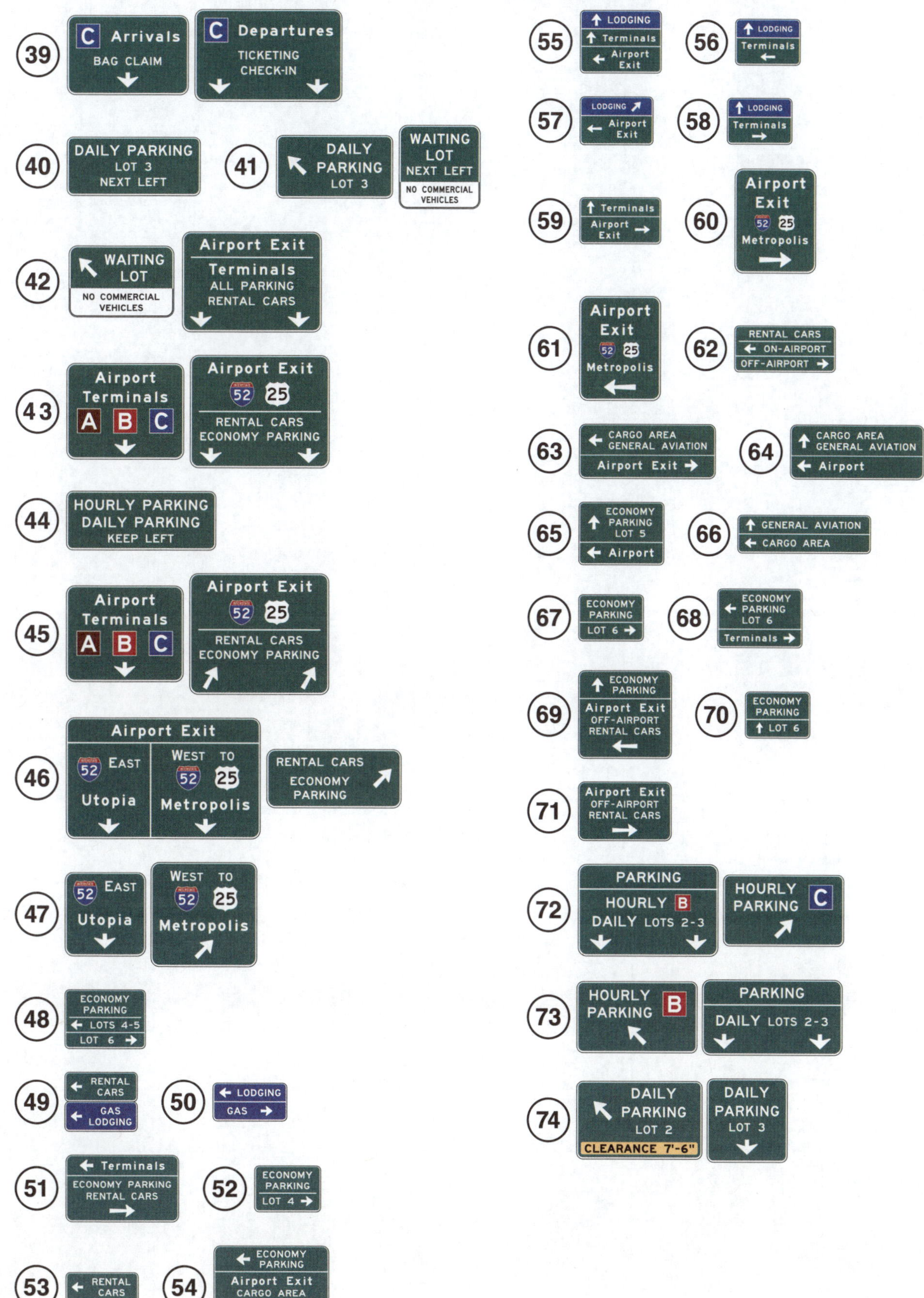

(This page intentionally left blank)

CHAPTER 2E. GUIDE SIGNS—FREEWAYS AND EXPRESSWAYS

Chapter 2E Subchapter and Section Organization

GENERAL

Section 2E.01 Scope of Freeway and Expressway Guide Sign Standards

Support:

01 The provisions of this Chapter provide a uniform and effective system of signing for high-volume, high-speed motor vehicle traffic on freeways and expressways. The requirements and specifications for expressway signing exceed those for conventional roads (see Chapter 2D), but are less than those for freeway signing. Since there are many geometric design variables to be found in existing roads, a signing concept commensurate with prevailing conditions is the primary consideration. Section 1C.02 includes definitions of freeway and expressway.

02 Guide signs for freeways and expressways are primarily identified by the name of the sign rather than by an assigned sign designation. Guidelines for the design of guide signs for freeways and expressways are provided in the "Standard Highway Signs" publication (see Section 1A.05).

Standard:

03 **The provisions of this Chapter shall apply to any highway that meets the definition of freeway or expressway facilities.**

Section 2E.02 Freeway and Expressway Signing Principles

Support:

01 The development of a signing system for freeways and expressways is approached on the premise that the signing is primarily for the benefit and direction of road users who are unfamiliar with the route or area. The signing furnishes road users with clear instructions for orderly progress to their destinations. Sign installations are an integral part of the facility and, as such, are best planned concurrently with the development of highway location and geometric design. For optimal results, plans for signing are analyzed during the earliest stages of preliminary design, and details are correlated as final design is developed. The excessive signing found on many major highways usually is the result of using a multitude of signs that are too small and that are poorly designed and placed to accomplish the intended purpose.

02 Freeway and expressway signing is to be considered and developed as a planned system of installations. An engineering study is sometimes necessary for proper solution of the problems of many individual locations, but, in addition, consideration of an entire route is necessary.

Guidance:

03 *Road users should be guided with consistent signing on the approaches to interchanges, when they drive from one State to another, and when driving through rural or urban areas. Because geographical, geometric, and operating factors regularly create significant differences between urban and rural conditions, the signing should take these conditions into account.*

04 *Guide signs on freeways and expressways should serve distinct functions as follows:*

A. *Give directions to destinations, or to streets or highway routes, at intersections or interchanges;*
B. *Furnish advance notice of the approach to intersections or interchanges;*
C. *Direct road users into appropriate lanes in advance of diverging or merging movements;*
D. *Identify routes and directions on those routes;*
E. *Show distances to destinations;*
F. *Indicate access to general motorist services, rest, scenic, and recreational areas; and*
G. *Provide other information of navigational value to the road user.*

Section 2E.03 Guide Sign Classification

Support:

01 Freeway and expressway guide signs are classified and addressed as follows:

A. Interchange signs (see Sections 2E.21 through 2E.23 and 2E.25 through 2E.44);
B. Interchange Sequence signs (see Section 2E.24);
C. Post-Interchange signs (see Sections 2E.47 through 2E.49);
D. Community Interchanges Identification signs (see Section 2E.52);
E. Next Exits signs (see Section 2E.53);
F. Weigh Station signs (see Section 2E.54);
G. Route signs and Trailblazer Assemblies (see Section 2E.55);
H. At-Grade Intersection signs (see Section 2E.58);
I. General Information signs (see Chapter 2H);
J. Reference Location signs (see Sections 2H.11 and 2H.12);
K. General Service signs (see Chapter 2I);

L.   Rest and Scenic Area signs (see Section 2I.05);
M. Tourist Information and Welcome Center signs (see Section 2I.08);
N.   Radio Information, Travel Information, and Roadside Assistance signs (see Sections 2I.09 through 2I.13);
O.   Carpool and Ridesharing signs (see Section 2I.14);
P.   Specific Service signs (see Chapter 2J); and
Q.   Recreational and Cultural Interest Area signs (see Chapter 2M).

Section 2E.04 Characteristics of Urban Signing

Support:

01   Urban conditions are characterized not so much by city limits or other arbitrary boundaries as by the following features:

A.   Mainline roadways with more than two lanes in each direction;
B.   High traffic volumes on the through roadways;
C.   High volumes of traffic entering and leaving interchanges;
D.   Interchanges that are closely spaced;
E.   Roadway and interchange lighting;
F.   Three or more interchanges serving the major city;
G.   A loop, circumferential, or spur route serving a sizable portion of the urban population; and
H.   Visual clutter from roadside development.

02   Operating conditions and road geometrics on urban freeways and expressways usually make special sign treatments desirable, including:

A.   Use of Interchange Sequence signs (see Section 2E.24);
B.   Use of sign spreading to the maximum extent possible (see Section 2E.43);
C.   Elimination of General Service or Specific Service signing (see Chapters 2I and 2J);
D.   Reduction to a minimum of post-interchange signs (see Section 2E.47);
E.   Display of advance signs at distances closer to the interchange, with appropriate adjustments in the legend (see Section 2E.23);
F.   Use of overhead signs on roadway structures and independent sign supports (see Section 2E.19);
G.   Use of Overhead Arrow-per-Lane guide signs in advance of interchanges with option lanes (see Section 2E.40), or Diagrammatic Advance guide signs in advance of interchanges with complex geometric configurations of ramp departures (see Section 2E.41); and
H.   Frequent use of street names as the principal message in guide signs.

03   Lower speeds, which are often characteristic of urban operations, do not justify lower signing standards. Typical traffic patterns are more complex for the road user to negotiate, and large, easy-to-read legends are, therefore, just as necessary as on rural highways.

Section 2E.05 Characteristics of Rural Signing

Support:

01   Rural areas ordinarily have greater distances between interchanges, which permits adequate spacing for the sequences of signs on the approach to and departure from each interchange. However, the absence of traffic in adjoining lanes and on entering or exiting ramps often adds monotony or inattention to rural driving. This increases the importance of signs that call for decisions or actions.

Guidance:

02   *Where there are long distances between interchanges and the alignment is relatively unchanging, signs should be positioned for their best effect on road users. The tendency to group all signing in the immediate vicinity of rural interchanges should be avoided by considering the entire route in the development of signing plans. Extra effort should be given to the placement of signs at natural target locations to command the attention of the road user, particularly when the message requires an action by the road user.*

Section 2E.06 Signing of Named Highways

Guidance:

01   *Signing of named highways on freeways and expressways should comply with the provisions of Section 2D.56.*

Support:

02   Section 2M.10 contains information regarding memorial or dedication signing of routes, bridges, or highway components.

Section 2E.07 <u>Designation of Destinations</u>

Standard:

01   **The direction of a freeway and the major destinations or control cities along it shall be clearly identified through the use of appropriate destination legends (see Section 2D.35). Successive freeway guide signs shall provide continuity in destination names and consistency with available map information. At any decision point, a given destination shall be indicated by way of only one route (see Figure 2E-1).**

Guidance:

02   *Control city legends should be used in the following situations along a freeway:*

    A. At interchanges between freeways;
    B. At separation points of overlapping freeway routes;
    C. On directional signs on intersecting routes, to guide traffic entering the freeway;
    D. On Pull-Through signs; and
    E. On the bottom line of post-interchange distance signs.

Support:

03   Continuity of destination names is also useful on expressways serving long-distance or intrastate travel.

04   The determination of major destinations or control cities is important to the quality of service provided by the freeway. Control cities on freeway guide signs are selected by the States and are contained in the "Guidelines for the Selection of Supplemental Guide Signs for Traffic Generators Adjacent to Freeways, 4th Edition/Guide Signs, Part II: Guidelines for Airport Guide Signing/Guide Signs, Part III: List of Control Cities for Use in Guide Signs on Interstate Highways," published by and available from the American Association of State and Highway Transportation Officials.

Figure 2E-1. Designation of Destination for Interchanges in Opposing Directions of Travel

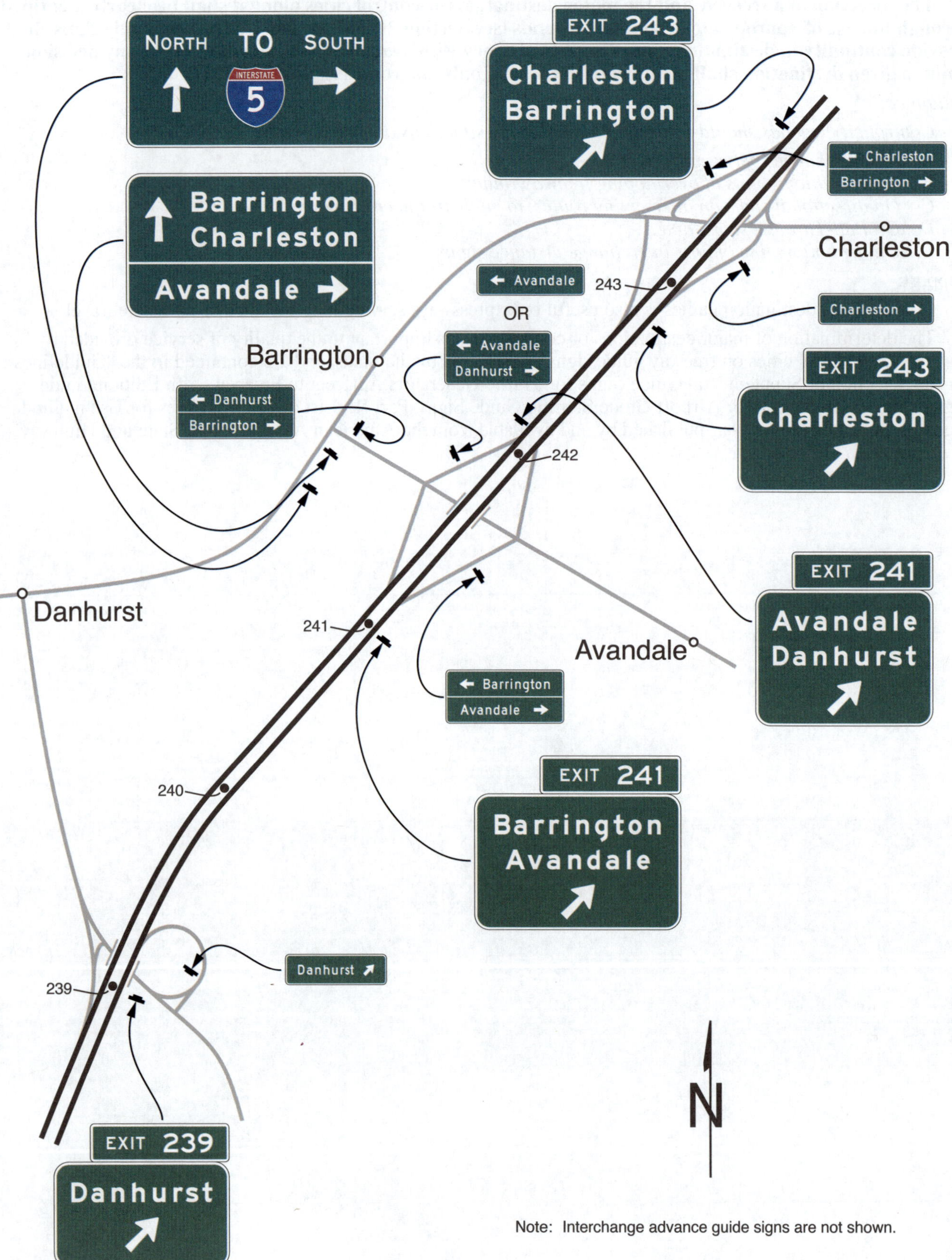

SIGN DESIGN

Section 2E.08 General

Support:

01   Effective signs are legible to road users approaching them, and are readable and comprehensible in the viewing time provided to permit proper responses. Desired design characteristics include: (a) long visibility distances; (b) large lettering, symbols, and arrows; and (c) short legends.

Section 2E.09 Color of Guide Signs

Standard:

01   **Guide signs on freeways and expressways, except as otherwise provided in this Manual, shall have white letters, symbols, arrows, and borders on a green background.**

Support:

02   Color requirements for route signs and trailblazers; for signs with blank-out or changeable messages; for signs for services, rest areas, park and recreational areas; and for certain miscellaneous signs are provided in the individual Sections dealing with the particular sign or sign group.

Section 2E.10 Retroreflection or Illumination

Standard:

01   **Letters, numerals, symbols, arrows, and borders of all guide signs shall be retroreflective. The background of all guide signs that are not independently illuminated shall be retroreflective.**

Support:

02   Where there is no serious interference from extraneous light sources, retroreflective post-mounted signs usually provide adequate nighttime visibility.

03   On freeways and expressways where much driving at night is done with low-beam headlights, the amount of headlight illumination incident to an overhead sign display is relatively small.

Guidance:

04   *Overhead sign installations should be illuminated (see Section 2A.21) unless an engineering study shows that retroreflection alone will perform effectively. The type of illumination chosen should provide effective and reasonably uniform illumination of the sign face and message.*

Section 2E.11 Interchange Classification

Support:

01   For signing purposes, interchanges are classified as major, intermediate, and minor. Minimum letter and numeral sizes based on interchange classification are contained in Tables 2E-2 and 2E-4. Descriptions of these classifications are as follows:

    A. Major interchanges are subdivided into two categories: (a) interchanges with other expressways or freeways, or (b) interchanges with high-volume multi-lane highways, principal urban arterials, or major rural routes where the volume of interchanging traffic is heavy or includes many road users unfamiliar with the area.

    B. Intermediate interchanges are those with urban and rural routes not in the category of major or minor interchanges.

    C. Minor interchanges include those where traffic is local and very light, such as interchanges with land service access roads. Where the sum of exit volumes is estimated to be lower than 100 vehicles per day in the design year, the interchange is classified as minor.

Section 2E.12 Size of Signs and Letters

Standard:

01   **Except as provided in Section 2A.07, the sizes of freeway and expressway guide signs that have standardized designs shall be as shown in Table 2E-1.**

Support:

02   Section 2A.07 contains information regarding the applicability of the various columns in Table 2E-1.

Option:

03   Signs larger than those shown in Table 2E-1 may be used (see Section 2A.07).

Table 2E-1. Freeway or Expressway Guide Sign and Plaque Sizes (Sheet 1 of 3)

Sign or Plaque	Sign Designation	Section	Minimum Size
Interchange Advance Guide (1 destination)	E1-1	2E.23	Varies
Interchange Advance Guide (2 destinations)	E1-2	2E.23	Varies
Interchange Advance Guide (3 destinations)	E1-3	2E.23	Varies
Exit Number (plaque)			
1-, 2-Digit Exit Number	E1-5P	2E.23	114 x 30
3-Digit Exit Number	E1-5aP	2E.23	132 x 30
1-, 2-Digit Exit Number (with single-letter suffix)	E1-5bP	2E.23	138 x 30
3-Digit Exit Number (with single-letter suffix)	E1-5cP	2E.23	156 x 30
1-, 2-Digit Exit Number (with dual-letter suffix)	E1-5dP	2E.23	168 x 30
3-Digit Exit Number (with dual-letter suffix)	E1-5eP	2E.23	186 x 30
Left Exit Number (plaque)			
1-, 2-Digit Exit Number	E1-5fP	2E.23	114 x 54
3-Digit Exit Number	E1-5gP	2E.23	132 x 54
1-, 2-Digit Exit Number (with single-letter suffix)	E1-5hP	2E.23	138 x 54
3-Digit Exit Number (with single-letter suffix)	E1-5iP	2E.23	156 x 54
1-, 2-Digit Exit Number (with dual-letter suffix)	E1-5jP	2E.23	168 x 54
3-Digit Exit Number (with dual-letter suffix)	E1-5kP	2E.23	186 x 54
Left (plaque)	E1-5mP	2E.23	72 x 30
Next Exit (1 line) (plaque)	E2-1P	2E.46	Varies x 24
Next Exit (2 lines) (plaque)	E2-1aP	2E.46	Varies x 36
Supplemental (1 destination)	E3-1	2E.51	Varies
Supplemental (2 destinations)	E3-2	2E.51	Varies
Exit Direction (1 destination)	E4-1	2E.25	Varies
Exit Direction (2 destinations)	E4-2	2E.25	Varies
Exit Direction (3 destinations)	E4-3	2E.25	Varies
Exit Gore	E5-1	2E.26	72 x 60
Exit Gore (with exit number)			
1-, 2-Digit Exit Number	E5-1a	2E.26	78 x 60
3-Digit Exit Number	E5-1a	2E.26	96 x 60
1-Digit Exit Number (with single-letter suffix)	E5-1a	2E.26	90 x 60
2-Digit Exit Number (with single-letter suffix)	E5-1a	2E.26	108 x 60
3-Digit Exit Number (with single-letter suffix)	E5-1a	2E.26	126 x 60
1-Digit Exit Number (with dual-letter suffix)	E5-1a	2E.26	120 x 60
2-Digit Exit Number (with dual-letter suffix)	E5-1a	2E.26	138 x 60
3-Digit Exit Number (with dual-letter suffix)	E5-1a	2E.26	156 x 60
Exit Number (plaque)			
1-, 2-Digit Exit Number	E5-1bP	2E.26	42 x 30
3-Digit Exit Number	E5-1bP	2E.26	60 x 30
1-Digit Exit Number (with single-letter suffix)	E5-1bP	2E.26	54 x 30
2-Digit Exit Number (with single-letter suffix)	E5-1bP	2E.26	72 x 30
3-Digit Exit Number (with single-letter suffix)	E5-1bP	2E.26	90 x 30
1-Digit Exit Number (with dual-letter suffix)	E5-1bP	2E.26	84 x 30
2-Digit Exit Number (with dual-letter suffix)	E5-1bP	2E.26	102 x 30
3-Digit Exit Number (with dual-letter suffix)	E5-1bP	2E.26	120 x 30
Narrow Exit Gore	E5-1c	2E.26	60 x 90*
Pull-Through	E6-1	2E.27	Varies
Pull-Through (Destination)	E6-1a	2E.27	Varies
Pull-Through (Down Arrows)	E6-2	2E.27	Varies

Table 2E-1. Freeway or Expressway Guide Sign and Plaque Sizes (Sheet 2 of 3)

Sign or Plaque	Sign Designation	Section	Minimum Size
Pull-Through (Destination, Down Arrows)	E6-2a	2E.27	Varies
Post-Interchange Distance	E7-1	2E.48	Varies
Post-Interchange Distance	E7-2	2E.48	Varies
Post-Interchange Distance	E7-3	2E.48	Varies
Post-Interchange Travel Time	E7-4	2E.49	Varies
Distance and Travel Time	E7-5	2E.50	Varies
Comparative Travel Time	E7-6	2E.50	Varies
Interchange Sequence (2 interchanges)	E9-1	2E.24	Varies
Interchange Sequence (3 interchanges)	E9-2	2E.24	Varies
Next Exits (1 destination)	E9-3	2E.53	Varies
Next Exits (2 destinations)	E9-3a	2E.53	Varies
Community Interchanges (2 interchanges)	E9-4	2E.52	Varies
Community Interchanges (3 interchanges)	E9-5	2E.52	Varies
Exit Only (with arrow)	E11-1,1d	2E.28	174** x 36
Exit	E11-1a	2E.28	66 x 18
Only	E11-1b	2E.28	66 x 18
Exit Only	E11-1c	2E.28	120 x 18
Exit Only (with two arrows)	E11-1e,1f	2E.28	222** x 36
Left (panel)	E11-2	2E.24	60 x 18
Exit Direction Advisory Speed (panel)	E13-2	2E.25	162 x 24
Interstate Route (1, 2 digits)	M1-1	2E.55	36 x 36
Interstate Route (3 digits)	M1-1	2E.55	45 x 36
Off-Interstate Route (1, 2 digits)	M1-2,3	2E.55	36 x 36
Off-Interstate Route (3 digits)	M1-2,3	2E.55	45 x 36
U.S. Route (1, 2 digits)	M1-4	2E.55	36 x 36
U.S. Route (3 digits)	M1-4	2E.55	45 x 36
State Route (1, 2 digits)	M1-5	2D.11	36 x 36
State Route (3 digits)	M1-5	2D.11	45 x 36
County Route	M1-6	2D.11	36 x 36
Forest Route	M1-7	2D.11	36 x 36
Eisenhower Interstate System	M1-10,10a	2E.56	36 x 36
Junction (plaque)	M2-1P	2D.13	30 x 21
Combination Junction (2 route signs)	M2-2	2D.14	60 x 48*
Cardinal Direction (plaque)	M3-1P,2P,3P,4P	2D.15	36 x 18
Alternate (plaque)	M4-1P,1aP	2D.17	36 x 18
By-Pass (plaque)	M4-2P	2D.18	36 x 18
Business (plaque)	M4-3P	2D.19	36 x 18
Truck (plaque)	M4-4P	2D.20	36 x 18
To (plaque)	M4-5P	2D.21	36 x 18
End (plaque)	M4-6P	2D.22	36 x 18
Temporary (plaque)	M4-7P,7aP	2D.24	36 x 18
Begin (plaque)	M4-14P	2D.23	36 x 18
Advance Turn Arrow (plaque)	M5-1P,2P,3P	2D.26	30 x 21
Lane Designation (plaque)	M5-4P,5P,6P	2D.27	36 x 24
Directional Arrow (plaque)	M6-1P,2P, 2aP,3P,4P,5P,6P,7P	2D.28	30 x 21
National Scenic Byway	M10-1	2D.57	24 x 24
National Scenic Byway (plaque)	M10-1aP	2D.57	24 x 12
Destination (1 line)	D1-1	2D.36	Varies x 24

Table 2E-1. Freeway or Expressway Guide Sign and Plaque Sizes (Sheet 3 of 3)

Sign or Plaque	Sign Designation	Section	Minimum Size
Destination and Distance (1 line)	D1-1a	2D.36	Varies x 24
Destination (2 lines)	D1-2	2D.36	Varies x 42
Destination and Distance (2 lines)	D1-2a	2D.36	Varies x 42
Destination (3 lines)	D1-3	2D.36	Varies x 60
Destination and Distance (3 lines)	D1-3a	2D.36	Varies x 60
Distance (1 line)	D2-1	2D.43	Varies x 24
Distance (2 lines)	D2-2	2D.43	Varies x 36
Distance (3 lines)	D2-3	2D.43	Varies x 48
Street Name (1 line)	D3-1,1a	2D.45	Varies x 18
Overhead Street Name (1 line)	D3-1,1a	2D.45	Varies x 24
Street Name (2 lines)	D3-1,1a	2D.45	Varies x 33
Overhead Street Name (2 lines)	D3-1,1a	2D.45	Varies x 48
Advance Street Name (2 lines)	D3-2	2D.46	Varies x 36
Advance Street Name (3 lines)	D3-2	2D.46	Varies x 48
Advance Street Name (4 lines)	D3-2	2D.46	Varies x 66
Park - Ride	D4-2	2D.48	36 x 48
Advance Weigh Station Distance	D8-1	2E.54	96 x 72 (F) 78 x 60 (E)
Weigh Station Advance Direction	D8-2	2E.54	108 x 90 (F) 84 x 72 (E)
Weigh Station Entrance Direction	D8-3	2E.54	84 x 78 (F) 66 x 60 (E)
Crossover	D13-1,2	2D.52	78 x 42
Freeway Entrance	D13-3	2D.50	48 x 30
Freeway Entrance (Directional)	D13-3a	2D.50	48 x 42
Combination Lane Use / Destination	D15-1	2D.38	Varies x 96
Next Truck Lane	D17-1	2D.53	60 x 66
Advance Truck Lane	D17-2	2D.53	60 x 54
Next Passing Lane	D17-3	2D.53	60 x 66
Advance Passing Lane	D17-4	2D.53	60 x 54
Advance Emergency Turn-Out	D17-5	2D.54	78 x 54
Emergency Turn-Out (Directional)	D17-6	2D.54	78 x 60
Advance Slow Vehicle Turn-Out	D17-7	2D.54	96 x 54

* The size shown is for a typical sign as illustrated in the figures in Chapters 2D and 2E. The size should be determined based on the amount of legend required for the sign.

** The width shown represents the minimum dimension. The width shall be increased as appropriate to match the width of the guide sign.

Notes: 1. Larger signs may be used when appropriate

2. Dimensions in inches are shown as width x height

3. Where two sizes are shown, the larger size is for freeways (F) and the smaller size is for expressways (E)

Standard:

04   The nominal loop height of the lower-case letters shall be ¾ of the height of the initial upper-case letter (see Paragraph 3 of Section 2D.05 for additional information on the specification of letter heights). Other word legends such as cardinal directions, action messages, and special characters shall be composed of all upper-case letters with a minimum letter height of 8 inches. Interline and edge spacing shall be as provided in Section 2E.13.

05   For all freeway and expressway signs that do not have a standardized design, the message dimensions shall be determined first, and the outside sign dimensions secondarily. Minimum numeral and letter sizes for expressway guide signs according to interchange classification, type of sign, and component of sign

legend shall be as shown in Tables 2E-2 and 2E-3. Minimum numeral and letter sizes for freeway guide signs according to interchange classification, type of sign, and component of sign legend shall be as shown in Tables 2E-4 and 2E-5. The minimum numeral and letter sizes for overhead-mounted expressway and freeway guide signs shall be those shown in the "Overhead" columns of Tables 2E-2 and 2E-4, respectively, except where a larger minimum numeral or letter height is provided in the columns for the applicable type of interchange (major, intermediate, or minor).

06   All names of places, streets, and highways on freeway and expressway guide signs shall be composed of lower-case letters with initial upper-case letters. The letters and the numerals used shall be FHWA Standard Alphabet Series E (modified) as provided in the "Standard Highway Signs" publication (see Section 1A.05).

07   Lettering size on freeway and expressway signs shall be the same for both rural and urban conditions.

Table 2E-2. Minimum Letter and Numeral Sizes for Expressway Guide Signs According to Interchange Classification

Type of Sign	Type of Interchange (see Section 2E.11)				Overhead*
	Major		Intermediate	Minor	
	Category a	Category b			
A. Advance Guide, Exit Direction, and Overhead Guide Signs					
Exit Number Plaques					
Words	10	10	10	8	10
Numerals & Letters	15	15	15	12	15
Interstate Route Signs					
Numerals	14**	—	—	—	14**
1- or 2-Digit Shields	36 x 36	—	—	—	36 x 36
3-Digit Shields	45 x 36	—	—	—	45 x 36
U.S. or State Route Signs					
Numerals	18	18	18	12	18
1- or 2-Digit Shields	36 x 36	36 x 36	36 x 36	24 x 24	36 x 36
3-Digit Shields	45 x 36	45 x 36	45 x 36	30 x 24	45 x 36
U.S. or State Route Text Identification (Example: US 56)					
Numerals & Letters	18	15	15	12	15
Cardinal Directions					
First Letters	18	15	12	10	15
Rest of Word	15	12	10	8	12
Auxiliary and Alternative Route Legends (Examples: JCT, TO, ALT, BUSINESS)					
Words	15	12	10	8	12
Names of Destinations					
Upper-Case Letters	20	16	13.33	10.67	16
Lower-Case Letters	15	12	10	8	12
Distance Numbers	18	15	12	10	15
Distance Fraction Numerals	12	10	10	8	10
Distance Words	12	10	10	8	10
Action Message Words	10	10	10	8	10
B. Gore Signs					
Words	10	10	10	8	—
Numerals & Letters	12	12	12	10	—

* Where a larger size is shown for the interchange classification of the interchange, that larger size is used for overhead-mounted guide signs for that interchange.

** Minimum size listed for 3-digit shields. Larger numeral sizes used for 1-digit, some 2-digit, and some 3-digit shields. See the Standard Highways Signs publication for more information on Route Sign numeral heights and Standard Alphabet series.

Note: Sizes are shown in inches and where applicable are shown as width x height

Table 2E-3. Minimum Letter and Numeral Sizes for Expressway Guide Signs According to Sign Type

Type of Sign	Minimum Size
A. Pull-Through Signs	
Destinations — Upper-Case Letters	13.33
Destinations — Lower-Case Letters	10
Route Signs	
Numerals	14*
1- or 2-Digit Shields	36 x 36
3-Digit Shields	45 x 36
Cardinal Directions — First Letters	12
Cardinal Directions — Rest of Word	10
B. Supplemental Guide Signs	
Exit Number — Words	8
Exit Number — Numerals and Letters	12
Place Names — Upper-Case Letters	10.67
Place Names — Lower-Case Letters	8
Action Messages	8
Route Signs	
Numerals	9*
1- or 2-Digit Shield	24 x 24
3-Digit Shield	30 x 24
C. Interchange Sequence or Community Interchanges Identification Signs	
Words — Upper-Case Letters	10.67
Words — Lower-Case Letters	8
Numerals	10.67
Fraction Numerals	8
Route Signs	
Numerals	9*
1- or 2-Digit Shield	24 x 24
3-Digit Shield	30 x 24
D. Next XX Exits Sign	
Place Names — Upper-Case Letters	10.67
Place Names — Lower-Case Letters	8
NEXT XX EXITS — Words	8
NEXT XX EXITS — Number	12

Type of Sign	Minimum Size
E. Distance Signs	
Words — Upper-Case Letters	8
Words — Lower-Case Letters	6
Numerals	8
Route Signs	
Numerals	6*
1- or 2-Digit Shield	18 x 18
3-Digit Shield	22.5 x 18
F. General Service Signs (see Chapter 2I)	
Exit Number — Words	8
Exit Number — Numerals and Letters	12
Services	8
G. Rest Area, Scenic Area, and Roadside Area Signs (see Chapter 2I)	
Words	10
Distance Numerals	12
Distance Fraction Numerals	8
Distance Words	8
Action Message Words	10
H. Reference Location Signs (see Chapter 2H)	
Words	4
Numerals	10
I. Boundary and Orientation Signs (see Chapter 2H)	
Words — Upper-Case Letters	8
Words — Lower-Case Letters	6
J. Next Exit and Next Services Signs	
Words and Numerals	8
K. Exit Only Signs	
Words	12
L. Overhead Arrow-per-Lane and Diagrammatic Signs	
See Table 2E-5	

* Minimum size listed for 3-digit shields. Larger numeral sizes used for 1-digit, some 2-digit, and some 3-digit shields. See the Standard Highways Signs publication for more information on Route Sign numeral heights and Standard Alphabet series.

Note: Sizes are shown in inches and where applicable are shown as width x height

Support:

08   Sign size is determined primarily in terms of the length of the message and the size of the lettering necessary for proper legibility. Letter style and height, and arrow design have been standardized for freeway and expressway signs to assure uniform and effective application.

09   Designs for upper-case and lower-case FHWA Standard Alphabets, together with tables of recommended letter spacing, are shown in the "Standard Highway Signs" publication (see Section 1A.05).

Guidance:

10   *Freeway lettering sizes (see Tables 2E-4 and 2E-5) should be used when expressway geometric design is comparable to freeway standards.*

11   *Other sign letter size requirements not specifically identified elsewhere in this Manual should be guided by these specifications. Abbreviations should be kept to a minimum, except as provided in Section 2E.16.*

Table 2E-4. Minimum Letter and Numeral Sizes for Freeway Guide Signs According to Interchange Classification

Type of Sign	Type of Interchange (see Section 2E.11)				
	Major		Intermediate	Minor	Overhead
	Category a	Category b			
A. Advance Guide, Exit Direction, and Overhead Guide Signs					
Exit Number Plaques					
Words	10	10	10	10	10
Numerals & Letters	15	15	15	15	15
Interstate Route Signs					
Numerals	18/14*	—	—	—	14*
1- or 2-Digit Shields	48 x 48/ 36 x 36	—	—	—	36 x 36
3-Digit Shields	60 x 48/ 45 x 36	—	—	—	45 x 36
U.S. or State Route Signs					
Numerals	24/18	18	18	12	18
1- or 2-Digit Shields	48 x 48/ 36 x 36	36 x 36	36 x 36	24 x 24	36 x 36
3-Digit Shields	60 x 48/ 45 x 36	45 x 36	45 x 36	30 x 24	45 x 36
U.S. or State Route Text Identification (Example: US 56)					
Numerals & Letters	18	18/15	15	12	15
Cardinal Directions					
First Letters	18	15	15	10	15
Rest of Words	15	12	12	8	12
Auxiliary and Alternative Route Legends (Examples: JCT, TO, ALT, BUSINESS)					
Words	15	12	12	8	12
Names of Destinations					
Upper-Case Letters	20	20	16	13.33	16
Lower-Case Letters	15	15	12	10	12
Distance Numbers	18	18/15	15	12	15
Distance Fraction Numerals	12	12/10	10	8	10
Distance Words	12	12/10	10	8	10
Action Message Words	12	12/10	10	8	10
B. Gore Signs					
Words	12	12	12	8	—
Numeral & Letters	18	18	18	12	—

* Minimum size listed for 3-digit shields. Larger numeral sizes used for 1-digit, some 2-digit, and some 3-digit shields. See the Standard Highways Signs publication for more information on Route Sign numeral heights and Standard Alphabet series.

Notes: 1. Sizes are shown in inches and where applicable are shown as width x height

2. Slanted line (/) signifies separation of desirable and minimum sizes

Support:

12   A sign mounted over a particular roadway lane to which it applies might have to be limited in horizontal dimension to the width of the lane, so that another sign can be placed over an adjacent lane. The necessity to maintain proper vertical clearance might also place a further limitation on the size of the overhead sign and the legend that can be accommodated.

Table 2E-5. Minimum Letter and Numeral Sizes for Freeway Guide Signs According to Sign Type

Type of Sign	Minimum Size
A. Pull-Through Signs	
Destinations — Upper-Case Letters	16
Destinations — Lower-Case Letters	12
Route Signs	
Numerals	14*
1- or 2-Digit Shields	36 x 36
3-Digit Shields	45 x 36
Cardinal Directions — First Letter	15
Cardinal Directions — Rest of Word	12
B. Supplemental Guide Signs	
Exit Number Words	10
Exit Number Numerals and Letters	15
Place Names — Upper-Case Letters	13.33
Place Names — Lower-Case Letters	10
Action Messages	8
Route Signs	
Numerals	9*
1- or 2-Digit Shield	24 x 24
3-Digit Shield	30 x 24
C. Interchange Sequence or Community Interchanges Identification Signs	
Words — Upper-Case Letters	13.33
Words — Lower-Case Letters	10
Numerals	13.33
Fraction Numerals	10
Route Signs	
Numerals	9*
1- or 2-Digit Shield	24 x 24
3-Digit Shield	30 x 24
D. Next X Exits Sign	
Place Names — Upper-Case Letters	13.33
Place Names — Lower-Case Letters	10
NEXT X EXITS — Words	10
NEXT X EXITS — Number	15
E. Distance Signs	
Words — Upper-Case Letters	8
Words — Lower-Case Letters	6
Numerals	8
Route Signs	
Numerals	6*
1- or 2-Digit Shield	18 x 18
3-Digit Shield	22.5 x 18

Type of Sign	Minimum Size
F. General Service Signs (see Chapter 2I)	
Exit Number Words	10
Exit Number Numerals and Letters	15
Services	10
G. Rest Area, Scenic Area, and Roadside Area Signs (see Chapter 2I)	
Words	12
Distance Numerals	15
Distance Fraction Numerals	10
Distance Words	10
Action Message Words	12
H. Reference Location Signs (see Chapter 2H)	
Words	4
Numerals	10
I. Boundary and Orientation Signs (see Chapter 2H)	
Words — Upper-Case Letters	8
Words — Lower-Case Letters	6
J. Next Exit and Next Services Signs	
Words and Numerals	8
K. Exit Only Signs	
Words	12
L. Overhead Arrow-per-Lane Signs*	
Arrowhead (Type D Directional Arrow)	21
Arrow Shaft Width	7.75
Arrow Height	
Through	48/40
Left Only	36/30
Right Only	36/30
Optional-Diverge (Through with Left or Right)	48/40
Optional-Split (Left and Right)	44/33.33
Vertical Separator Width	2
Vertical Space between Vertical Separator and Top of Nearest Arrow	6.5/5.0
Horizontal Space between Vertical Separator and Top of Nearest Through Arrow	12/9
Horizontal Space between Arrow Shaft and EXIT and ONLY Panels	12
EXIT and ONLY Panels	54 x 18
M. Diagrammatic Signs	
Arrowhead (Type D Directional Arrow)	13.5
Stem Height to Upper Point of Departure	30
Horizontal Space between Arrowhead and Route Shield or Destination	12

* Minimum size listed for 3-digit shields. Larger numeral sizes used for 1-digit, some 2-digit, and some 3-digit shields. See the Standard Highways Signs publication for more information on Route Sign numeral heights and Standard Alphabet series.

** Overhead Arrow-per-Lane sign example layouts and design elements sizing are provided in the Standard Highway Sign publication. Sizes shown as XX/XX correspond to 20-inch/16-inch destination letter legend sizes respecitvely.

Note: Sizes are shown in inches and where applicable are shown as width x height.

Section 2E.13 Interline and Edge Spacing

Guidance:

01   *Interline spacing of upper-case letters should be approximately ¾ of the average of upper-case letter heights in adjacent lines of letters.*

02   *The spacings to the top and bottom borders should be equal to the average of the letter height of the adjacent line of letters. The lateral spacing to the vertical borders should be essentially the same as the height of the largest letter.*

Section 2E.14 Sign Borders

Guidance:

01   *For guide signs larger than 120 x 72 inches, the border should have a width of 2 inches. For smaller guide signs, a border width of 1.25 inches should be used. On unusually large signs with oversized letter heights, route shields, or other legend elements, the border should be 2.5 inches wide and should not exceed 3 inches in width. In all cases, the width of the border should not exceed the stroke width of the lettering of the principal legend on the sign.*

02   *Corner radii of sign borders should be approximately ⅛ of the minimum sign dimension on guide signs, except that the radii should not exceed 12 inches on any sign.*

Support:

03   The "Standard Highway Signs" publication (see Section 1A.05) contains detailed information on border widths and corner radii for ranges of sign sizes.

Option:

04   The sign material in the area outside of the corner radius may be trimmed.

Section 2E.15 Amount of Legend on Guide Signs

Guidance:

01   *No more than two destination names or street names should be displayed on any Interchange Advance Guide sign or Exit Direction sign. A city name and street name on the same sign should be avoided. Where two or three signs are placed on the same supports, destinations or street names should be limited to one per sign, or to a total of three in the display. Sign legends should not exceed three lines of copy, exclusive of the exit number and action or distance information.*

Support:

02   Where only one interchange serves a community, the intersecting street name is generally superfluous to the city name on the Interchange Advance guide and Exit Direction signs. Where a community is served by multiple interchanges, the city name is typically displayed on either a Community Interchanges Identification sign (see Section 2E.52) or a Next Exits sign (see Section 2E.53) . Each interchange is then identified by its intersecting roadway name on the Interchange Advance guide and Exit Direction signs rather than by the city name.

Section 2E.16 Abbreviations

Standard:

01   **The use of abbreviations on freeway and expressway guide signs shall comply with the provisions of Section 2D.07 of this Manual.**

Section 2E.17 Symbols

Support:

01   Symbols are not normally displayed on freeway and expressway guide signs. One exception is the PARK – RIDE Supplemental guide sign (see Section 2E.51), which displays the Carpool symbol. In some cases, General Information symbols (see Chapter 2H) might be included in the legend of a guide sign to shorten an unusually lengthy legend on the sign.

Guidance:

02   *When a General Information symbol is incorporated into the legend of a guide sign, all components of the legend should be balanced in size and arrangement for maximum legibility. The General Information (I series) sign, rather than the symbol alone, should be placed as a sign panel within the guide sign so that adequate recognition of the symbol is provided by the border. The General Information sign panel should be positioned to the left of the legend to which it applies. The size of the General Information sign panel should be similar in size to that specified for a route shield for the type of guide sign on which it is displayed.*

Section 2E.18 Arrows for Interchange Guide Signs

Standard:

01   **Arrows used on interchange guide signs shall be of the types shown in Figure 2D-3 and shall comply with the provisions of this Section and Section 2D.08.**

02   **Except on Overhead Arrow-per-Lane guide signs (see Section 2E.40) and on Exit Direction signs for lane drops (see Section 2E.28), and except as provided in Paragraph 5 of this Section, directional arrows on all overhead and post-mounted Exit Direction signs shall point diagonally upward. Directional arrows on overhead Exit Direction signs shall be located on the side of the sign consistent with the direction of the exiting movement. Directional arrows on post-mounted Exit Direction signs shall be located at the bottom portion of the sign and centered under the legend.**

Option:

03   On overhead Exit Direction signs that are located fully over the tapered portion of the exit ramp at the theoretical gore, and where a directional arrow to the side of the legend farthest from the roadway might create an unusually wide sign that limits the road user's view of the arrow, the directional arrow may be placed at the bottom portion of the sign, centered under the legend.

Standard:

04   **Directional arrows on guide signs for multi-lane exits shall be positioned below the legend over the approximate center of each lane to which the arrow applies (see Figure 2E-38).**

05   **Down arrows shall only be used on overhead signs to indicate a lane to be followed and shall be positioned over the approximate center of each lane pointing vertically downward toward the approximate center of that lane. Down arrows shall be used only on overhead guide signs that restrict the use of specific lanes to traffic bound for the destination(s) and/or route(s) indicated by these arrows. Down arrows shall not be used unless an arrow can be located over and pointed to the approximate center of each lane that can be used to reach the destination displayed on the sign.**

06   **If down arrows are used, having more than one down arrow pointing to the same lane on a single overhead sign (or on multiple signs on the same overhead sign structure) shall not be permitted.**

Support:

07   Directional and down arrows for use on guide signs are shown in Figure 2D-3. Detailed drawings and standardized sizes based on ranges of letter heights for these arrows are provided in the "Standard Highway Signs" publication (see Section 1A.05). Information on the dimensions for arrows used in Overhead Arrow-per-Lane and Diagrammatic Advance guide signing is also provided in the "Standard Highway Signs" publication (see Section 1A.05).

INSTALLATION

Section 2E.19 Overhead Sign Installations

Support:

01   Specifications for the design and construction of structural supports for signs have been standardized by the American Association of State Highway and Transportation Officials (AASHTO). Overcrossing structures can often serve for the support of overhead signs, and might in some cases be the only practical location that will provide adequate viewing distance. Use of these structures as sign supports will eliminate the need for additional sign supports along the roadside. Conditions that might warrant the installation of overhead signs are given in Section 2A.14 and throughout this Chapter. Vertical clearance of overhead signs is discussed in Section 2A.15.

Section 2E.20 Lateral Offset

Standard:

01   **Except where shielded by a rigid traffic barrier, the minimum lateral offset outside the usable roadway shoulder for post-mounted freeway and expressway signs or for overhead sign supports, either to the right-hand or left-hand side of the roadway, shall be 6 feet. This minimum clearance shall also apply outside of a curb. If located within the clear zone, the signs shall be mounted on crashworthy (see definition in Section 1C.02) supports or shielded by appropriate crashworthy barriers.**

Guidance:

02   *Where practicable, a sign should not be less than 10 feet from the edge of the nearest traffic lane. Large guide signs especially should be farther removed, preferably 30 feet or more from the nearest traffic lane.*

03   *Where an expressway median is 12 feet or less in width, consideration should be given to spanning both roadways without a center support.*

04   *Where an overhead sign support cannot be placed sufficiently far away from the line of traffic, it should either be designed to minimize the impact forces, or be adequately shielded by a traffic barrier of suitable design.*

Standard:

05   **Butterfly-type sign supports and other overhead non-crashworthy sign supports shall not be installed in gores or other unshielded locations within the clear zone.**

Option:

06   Lesser clearances, but not generally less than 6 feet, may be used on connecting roadways or ramps at interchanges.

GUIDE SIGNING FOR INTERCHANGES

Section 2E.21 Interchange Guide Signs

Support:

01   For some applications, guide signing for interchanges depends upon the interchange classifications that are described in Section 2E.11. Provisions on guide signing for interchanges that are based on interchange classifications are found in Sections 2E.23 through 2E.26, 2E.46 through 2E.48, and 2E.51 through 2E.53.

Standard:

02   **The signs at interchanges and on their approaches shall include Advance Interchange guide signs and Exit Direction signs. Consistent destination messages shall be displayed on these signs.**

Guidance:

03   *New destination information should not be introduced into the major sign sequence for one interchange, nor should destination information be dropped.*

04   *Guide signs placed in advance of an interchange deceleration lane should be spaced at least 800 feet apart.*

05   *Use of Supplemental guide signing should be minimized as provided in Section 2E.51.*

Support:

06   Figure 2E-2 shows a typical sequence of interchange guide signs.

07   In some instances the interchange that provides the most direct or preferred access to a destination might be different in opposing directions of travel due to circumstances such as the configuration of the crossroads, or the fact that an interchange is a partial interchange.

Guidance:

08   *For each direction of travel, guide signing to a destination should be via the exit with the most direct or preferred access, even when this results in a destination being served by different interchanges for opposing directions of travel (see Figure 2E-1).*

Section 2E.22 Interchange Exit Numbering

Standard:

01   **Interchange exit numbering shall use the reference location sign exit numbering method. The consecutive exit numbering method shall not be used. The exit numbers shall correspond to the posted Reference Location or Enhanced Reference Location signs.**

Support:

02   Reference location sign exit numbering assists road users in determining their destination distances and travel mileage, assists road users in reporting their location in the event of an incident or breakdown, assists responders in responding to incidents, and assists highway agencies because the exit numbering sequence does not have to be changed if new interchanges are added to a route.

03   Interchange exit numbering provides valuable orientation for the road user on a freeway or expressway. The feasibility of numbering interchanges or exits on an expressway will depend largely on the extent to which grade separations are provided. Where there is appreciable continuity of interchange facilities, interrupted only by an occasional intersection at grade, the numbering will be helpful to the expressway user.

Standard:

04   **Interchange exit numbering shall be used in signing each freeway interchange exit. Interchange exit numbers shall be displayed with each Interchange Advance Guide sign, Exit Direction sign, and Exit Gore sign. The exit number shall be displayed on a separate plaque on top of the Interchange Advance Guide or Exit Direction sign. The Exit Number (E1-5P series) plaque (see Figure 2E-9) shall include the word EXIT(S) and the appropriate exit number(s) in a single-line format.**

05   **Suffix letters shall only be used to supplement exit numbers where there is more than one exit associated with the reference mile points of the freeway. Suffix letters shall not be used for an exit ramp for the purpose of identifying a downstream ramp split providing access to multiple highways or different directions on the same highway. The suffix letter shall also be included on the Exit Number plaque and shall be separated from the exit number by a space having a width of between ½ and ¾ of the height of the suffix letter. The suffix letters assigned shall be in ascending alphabetical order starting with the letter A for ramps in the direction of travel with increasing exit numbers, and in descending alphabetical order ending in the letter A in the opposite direction of travel. Exit numbers shall not include the cardinal direction initials corresponding to the directions of the cross route. The minimum numeral and letter sizes shall be as given in Tables 2E-2 through 2E-5. If used, the exit numbering system for expressways shall comply with the provisions prescribed for freeways.**

Figure 2E-2. Typical Sequence of Interchange Guide Signs

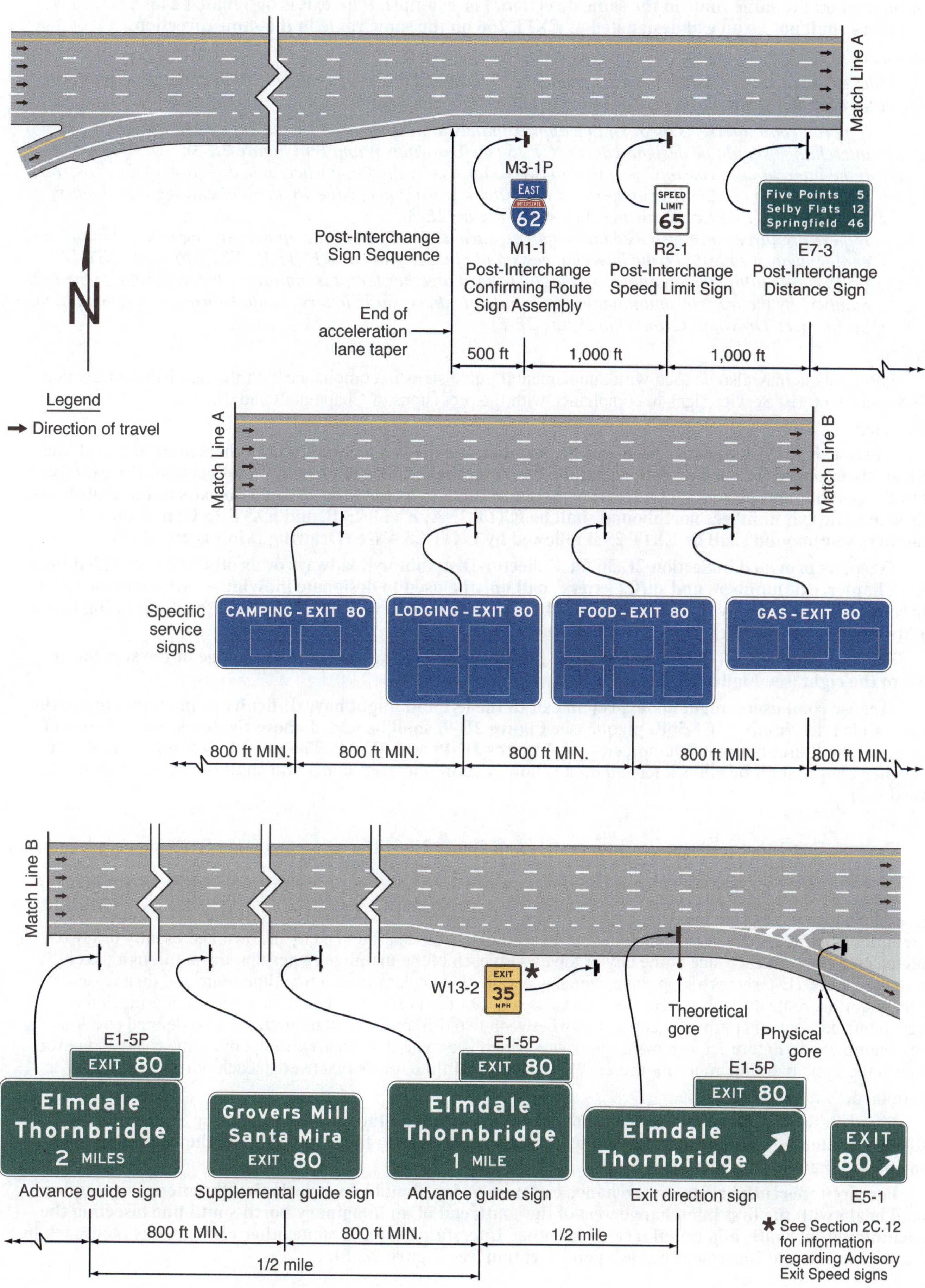

06   **Where suffix letters are used for exit numbering, an exit of the same number without a suffix letter shall not be used on the same route in the same direction. For example, if an exit is designated as EXIT 256 A, then there shall not be an exit designated as EXIT 256 on the same route in the same direction.**

Guidance:

07   *To the extent practical, exit numbering should be determined based upon the location of the crossroad with respect to reference location signs as given in the following examples:*

    A. *If a crossroad intersects the mainline approximately at or after Mile 15 and before Mile 16, the interchange should be designated as EXIT 15 (see Drawings A and B in Figure 2E-3).*

    B. *If the interchange crossroad is split into two roadways by direction where one direction of the crossroad is downstream of Mile 18 and the other direction is upstream of Mile 18, the interchange exit number should be EXIT 18 (see Drawings A and B in Figure 2E-3).*

    C. *If there are three closely-spaced interchanges, such as less than 1 mile apart, starting before Mile 16 and ending near or at Mile 17, the interchanges should be designated as EXIT 15, EXIT 16, and EXIT 17.*

    D. *If there are multiple interchanges so closely spaced together that it is impracticable to designate the exit numbers by the freeway mainline reference mile numbers, suffix letters should be used as provided in this Section (see Drawings C and D in Figure 2E-3).*

Option:

08   Exit numbers may also be used with Supplemental guide signs in compliance with the provisions of Section 2E.51, and Motorist Service signs in compliance with the provisions of Chapters 2I and 2J.

Standard:

09   **Where exit suffix letters are used and the number of exits is not equal in both directions of travel, the exit suffix lettering for each direction shall be based on the number of exits in that direction. For example, if in the northbound direction of a freeway there are three exits for Mile 25 and two exits in the southbound direction, the exit numbers northbound shall be EXIT 25 A, EXIT 25 B, and EXIT 25 C; and the exit numbers southbound shall be EXIT 25 B followed by EXIT 25 A (see Drawing D in Figure 2E-3).**

10   **Except as provided in Section 2E.36 for Collector-Distributor Roadways or as otherwise provided for in this Chapter, exit numbers and suffix letters shall only be used to designate individual exit departure points directly from the freeway mainline. Exit numbers and suffix letters shall not be used for designating ramp splits into two ramps after leaving the mainline.**

11   **The Exit Number (E1-5P) plaque shall be positioned above the top right-hand edge of the sign for an exit to the right (see Figure 2E-9).**

12   **Because road users might not expect an exit to the left and might have difficulty in maneuvering to the left, a Left Exit Number (E1-5bP) plaque (see Figure 2E-9) shall be added above the top left-hand edge of the sign for all numbered left-hand exits (see Figures 2E-18 and 2E-34). The word LEFT on the Left Exit Number plaque shall be a black legend on a yellow rectangular sign panel and shall be centered above the word EXIT.**

Support:

13   Example Exit Number plaque designs are shown in Figure 2E-9. The incorporation of Exit Number plaques on guide signs is illustrated in Figures 2E-9, 2E-12, 2E-14, 2E-35, and 2E-41.

14   Figure 2E-4 provides an example of Interstate route loops and spurs around major metropolitan areas. The general plan for numbering interchange exits is shown in Figures 2E-5 through 2E-8. Figure 2E-5 shows a circumferential route, which is a route that makes a complete circle around a city or town and usually has two interchanges (one on each side of the city or town) with each of the mainline routes that travel through the city or town. Figure 2E-6 shows a loop route, which is a route that departs from a mainline route and then rejoins the same mainline route at a subsequent point downstream. For the purpose of Interstate route numbering, a three-digit Interstate route that provides connectivity between two different Interstate routes is also defined as a loop (see Figure 2E-4). Figure 2E-7 shows a spur route, which is a route that departs from a mainline route and never rejoins the same mainline route. Figure 2E-8 shows two mainline routes that overlap each other.

Standard:

15   **Regardless of whether a mainline route originates within a State or crosses into the State from an adjacent State, the southernmost or westernmost terminus within that State shall be the beginning point for interchange exit numbering.**

16   **For circumferential routes, interchange exit numbering shall be in a clockwise direction. The numbering shall begin with the first interchange west of the south end of an imaginary north-south line bisecting the circumferential route, at a radial freeway or other Interstate route, or some other conspicuous landmark in the circumferential route near a south polar location (see Figure 2E-5).**

Figure 2E-3. Examples of Interchange Exit Numbering

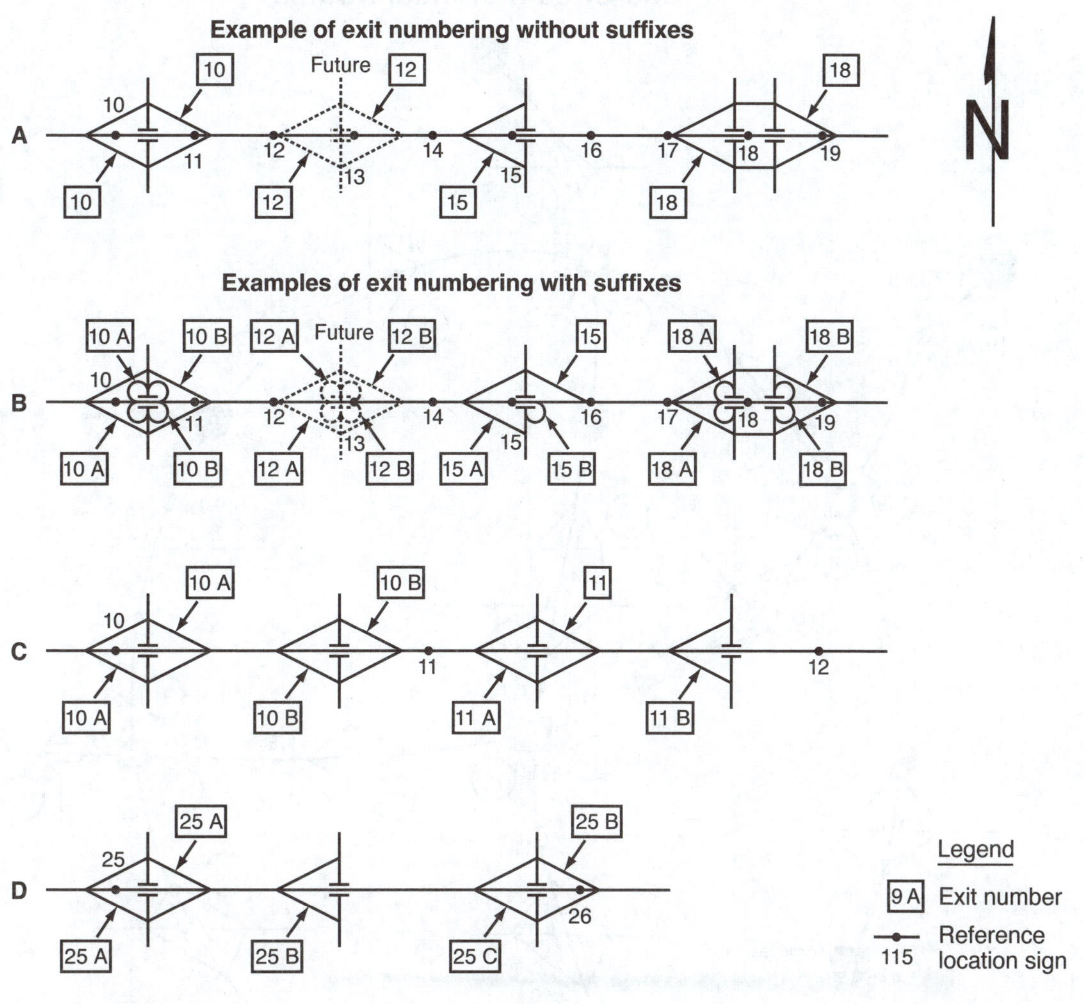

Figure 2E-4. Example of Interstate Loops and Spurs

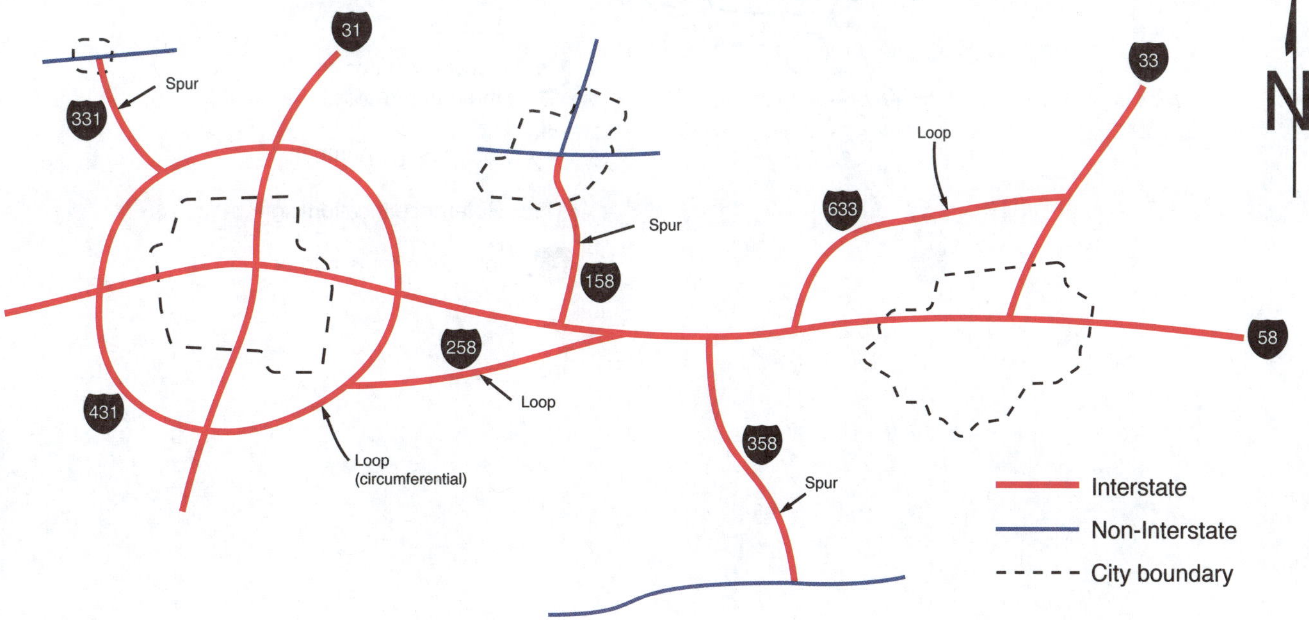

Figure 2E-5. Example of Interchange Numbering for Mainline and Circumferential Routes

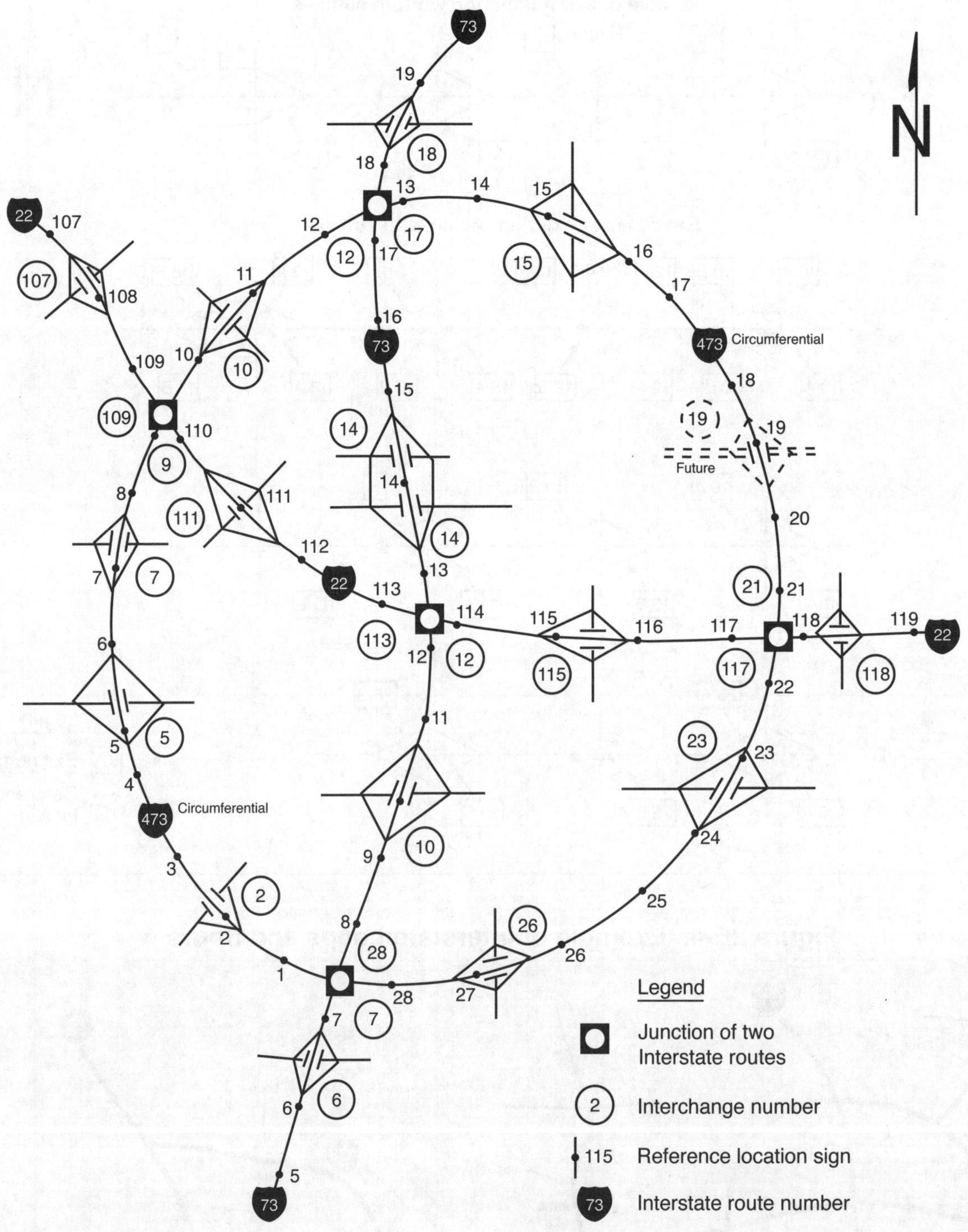

Figure 2E-6. Example of Interchange Numbering for Mainline and Loop Routes

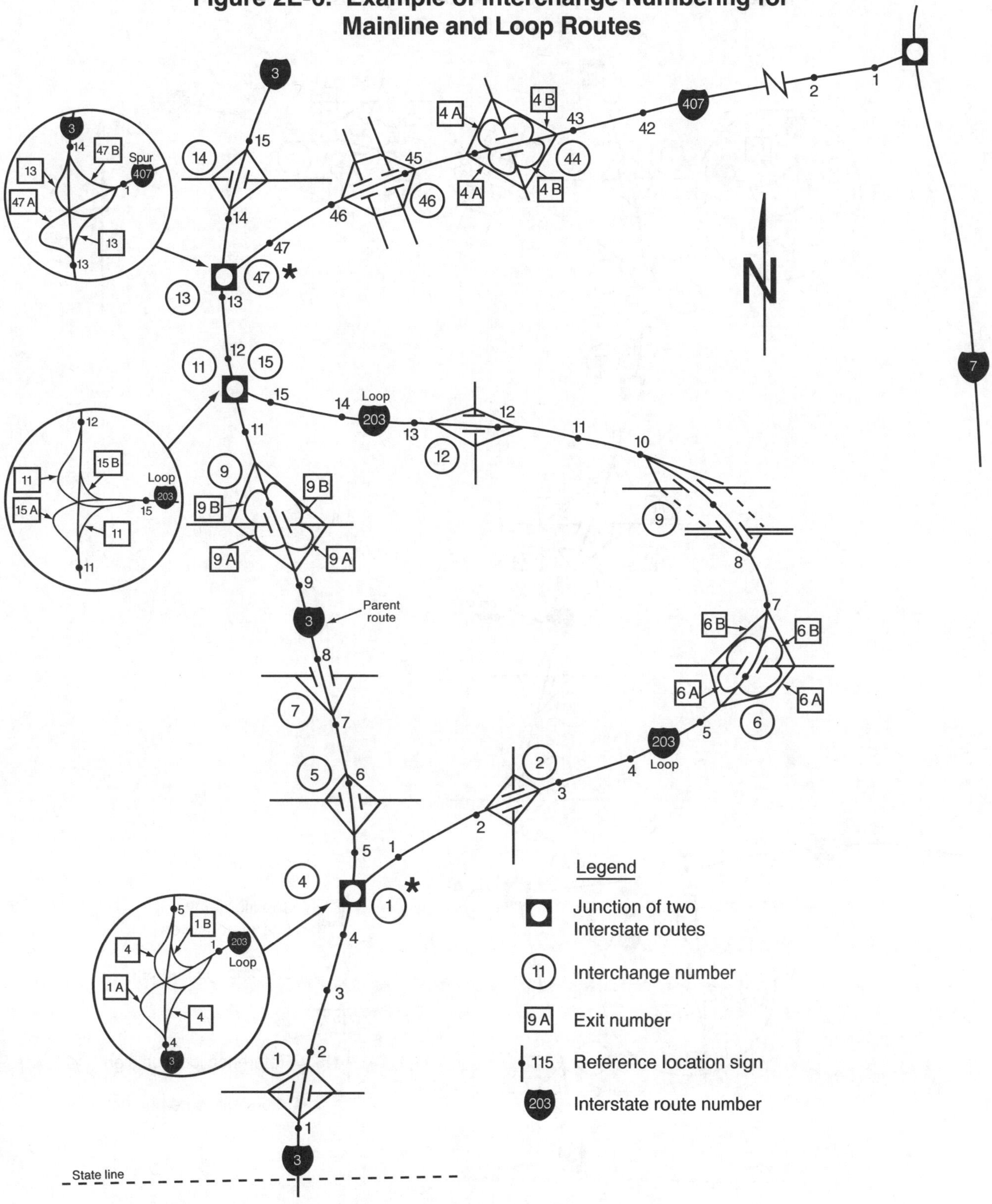

***** The freeway/freeway interchange where the beginning of the loop or spur route intersects with the mainline route may be called either Exit 1 or Exit 0 on the loop or spur route.

Figure 2E-7. Example of Interchange Numbering for Mainline and Spur Routes

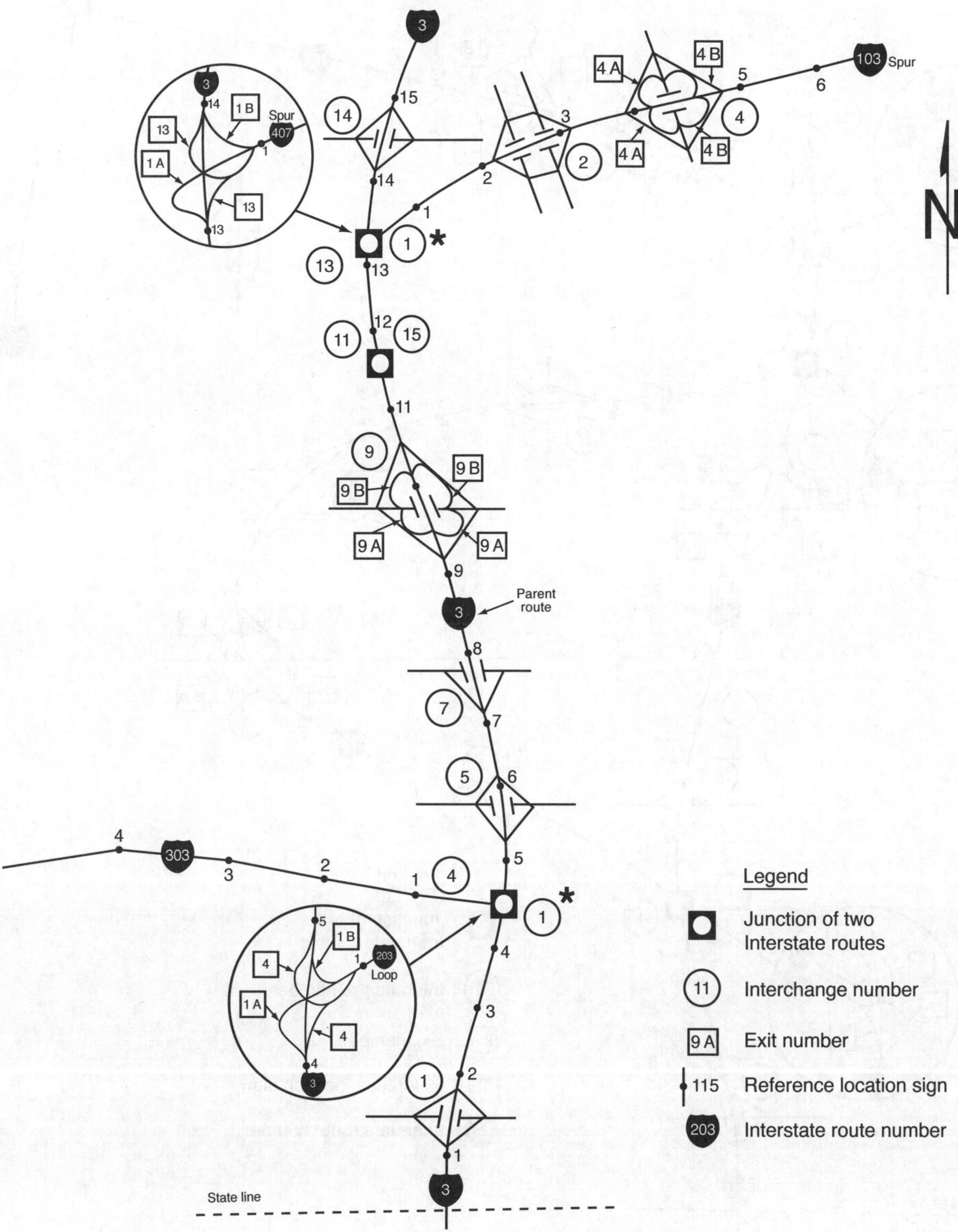

* The freeway/freeway interchange where the beginning of the loop or spur route intersects with the mainline route may be called either Exit 1 or Exit 0 on the loop or spur route.

Figure 2E-8. Example of Interchange Numbering for Overlapping Routes

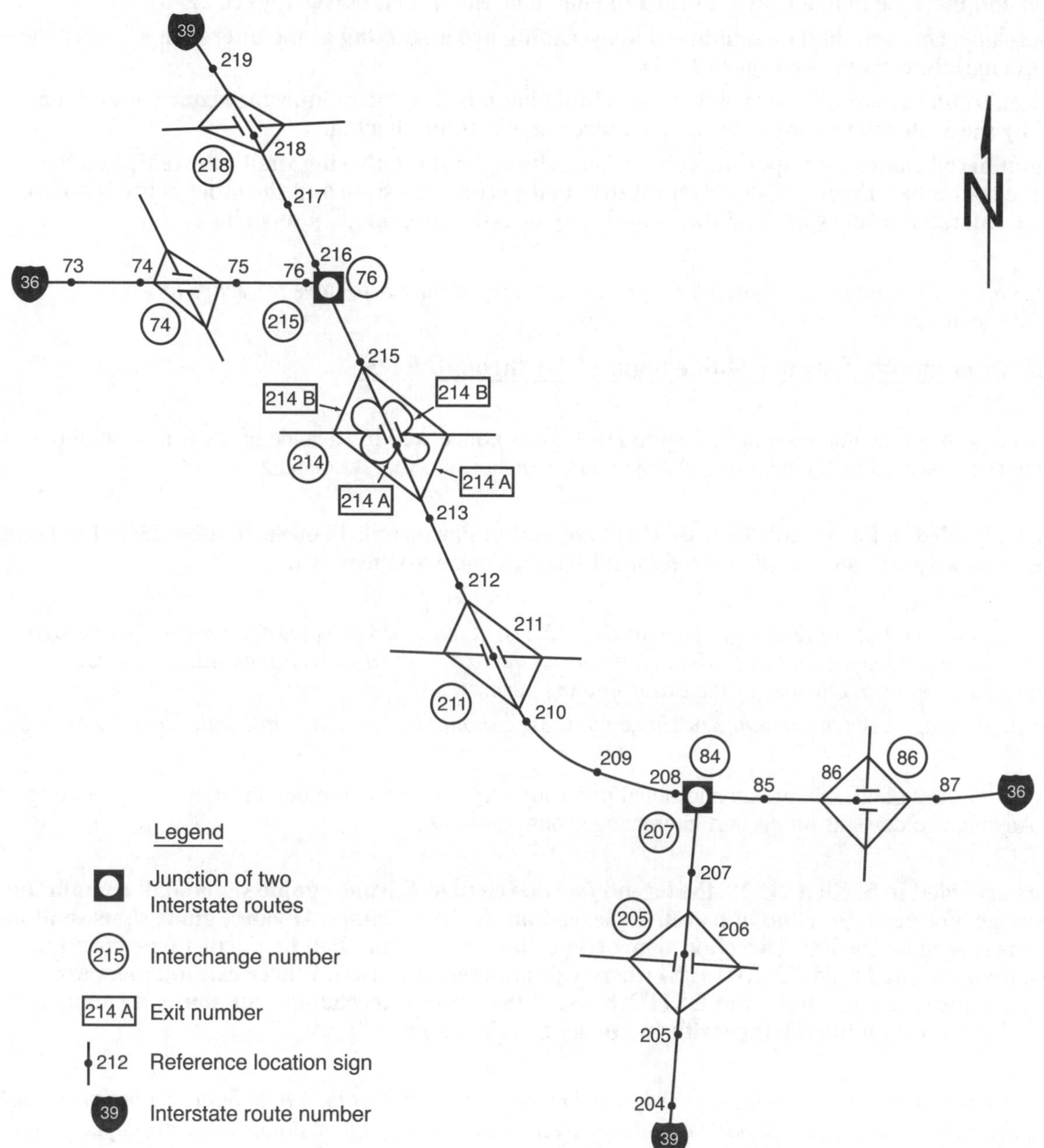

17   The interchange exit numbers on loop routes shall begin at the loop interchange nearest the south or west junction and increase in magnitude toward the north or east junction (see Figure 2E-6).

18   Spur route interchanges shall be numbered in ascending order starting at the interchange where the spur leaves the mainline route (see Figure 2E-7).

19   If a circumferential, loop, or spur route crosses State boundaries, the numbering sequence shall be coordinated by the States to provide continuous interchange exit numbering.

20   Where numbered routes overlap, continuity of interchange exit numbering shall be established for only one of the routes (see Figure 2E-8). If one of the routes is an Interstate and the other route is not an Interstate, the Interstate route shall maintain continuity of exit interchange numbering.

Guidance:

21   *The route chosen for continuity of interchange exit numbering should also have reference location sign continuity (see Figure 2E-8).*

Section 2E.23 <u>Interchange Advance Guide Signs (E1-1 through E1-3)</u>

Support:

01   An Interchange Advance guide sign (see Figure 2E-9) gives notice well in advance of the exit point of the principal destinations served by the next interchange and the distance to that interchange.

Standard:

02   **Except as provided in Paragraph 16 of this Section, and in Paragraph 18 of Section 2E.25, at least one Interchange Advance guide sign shall be used for all interchange classifications.**

Guidance:

03   *At major and intermediate interchanges (see Section 2E.11), at least two Interchange Advance guide signs should be used, placed at ½ mile and at 1 mile in advance of the exit. A third Interchange Advance guide sign should be placed at 2 miles in advance of the exit if spacing permits.*

04   *At minor interchanges, the Interchange Advance guide sign should be located ½ to 1 mile from the exit gore.*

Support:

05   Sections 2E.29 through 2E.44 contain additional provisions regarding the number, location, and mounting of Interchange Advance guide signs for certain interchange configurations.

Standard:

06   **Except as provided in Section 2E.28, the legend on Interchange Advance guide signs shall contain the distance message. For each direction of travel, the legend on the Interchange Advance guide signs shall be the same as the legend on the Exit Direction sign, except that the last line shall be the distance message. The distance message shall read XX MILE(S) where exit numbers are used. Where exit numbers are not used, the distance message shall read EXIT XX MILE(S) for an interchange with one exit ramp, and EXITS XX MILE(S) for an interchange with two or more exit ramps.**

Guidance:

07   *Where an Interchange Advance guide sign is located more than 1,000 feet to ½ mile but not more than 1 mile from the exit, the distance displayed should be to the nearest ¼ mile. Where the distance to be displayed on an Interchange Advance guide sign is 1,000 feet or less, the distance should be displayed in feet, rather than miles, to the nearest 100 feet.*

Standard:

08   **When a distance is displayed in miles, fractions of a mile, rather than decimals, shall be displayed in all cases.**

09   **For numbered exits, the exit number used with the Interchange Advance guides signs shall be displayed using an Exit Number plaque above and abutting the Interchange Advance guide sign.**

10   **For numbered exits to the right, an Exit Number (E1-5P through E1-5eP) plaque (see Figure 2E-9) shall be added to the top right-hand edge of the sign.**

11   **For numbered exits to the left, a Left Exit Number (E1-5fP through E1-5kP) plaque (see Figure 2E-9) shall be added above the top left-hand edge of the sign (see Figures 2E-18 and 2E-34).**

12   **For unnumbered exits to the left, a LEFT (E1-5mP) plaque (see Figure 2E-9) shall be added to the top left-hand edge of the sign, abutting the sign.**

Figure 2E-9. Examples of Interchange Advance Guide Signs, Exit Number Plaques, and LEFT Plaque

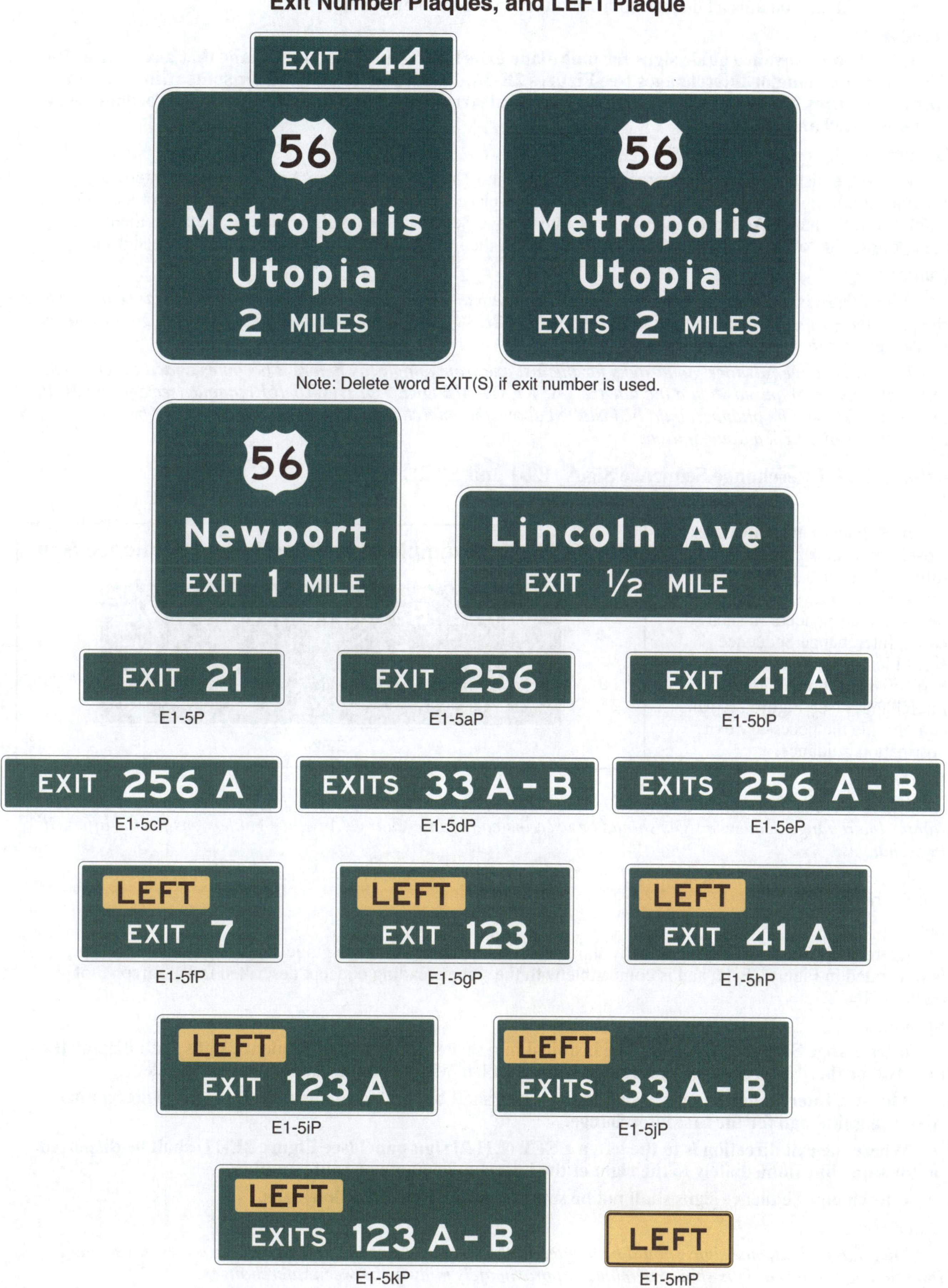

Support:

13   Section 2E.22 contains additional information regarding exit numbering.

Standard:

14   **Interchange Advance guide signs for multi-lane exits having an optional exit lane that also carries the through route at major interchanges (see Figures 2E-36, 2E-37, and 2E-42) and for splits with an option lane (see Figures 2E-38 and 2E-39) shall be Overhead Arrow-per-Lane signs designed in accordance with Sections 2E.39 and 2E.40.**

Option:

15   Where the distance between interchanges is more than 1 mile, but less than 2 miles, the first Interchange Advance guide sign may be closer than 2 miles, but not placed so as to overlap the signing for the preceding exit. Duplicate Interchange Advance guide signs or Interchange Sequence Series signs may be placed in the median on the opposite side of the roadway and are not included in the minimum requirements of interchange signing.

Guidance:

16   *Where there is less than 800 feet between the theoretical gores of successive interchange entrance or exit ramps, Interchange Sequence Series signs (see Section 2E.24) should be used instead of Interchange Advance guide signs for the affected interchanges.*

17   *The Interchange Advance guide signs for the last exit from a highway before it becomes a facility on which toll payments are required should include the LAST EXIT BEFORE TOLL (W16-16P) plaque (see Section 2F.10 and Figure 2F-4). The plaque should be installed above the Interchange Advance guide signs, but below the Exit Number or LEFT plaque, if used.*

Section 2E.24 Interchange Sequence Signs (E9-1 and E9-2)

Support:

01   Interchanges are sometimes closely spaced, particularly through large urban areas, so that typical guide signs cannot be adequately spaced. In such cases, Interchange Sequence signs identifying the next two (E9-1) or three (E9-2) interchanges (see Figure 2E-10) can provide the necessary exit destination guidance.

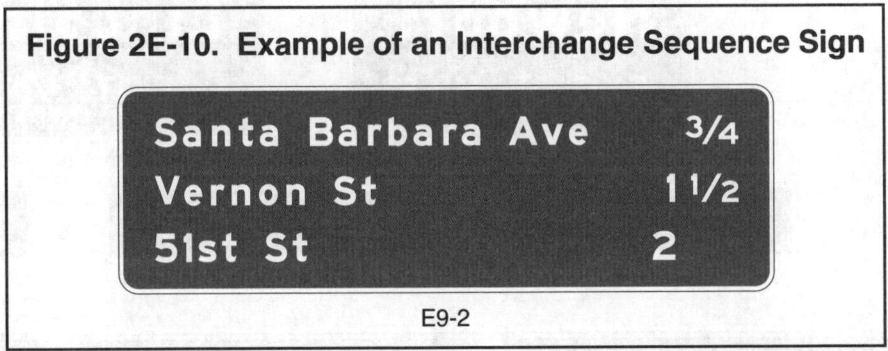

Figure 2E-10. Example of an Interchange Sequence Sign

Guidance:

02   *Where there is less than 800 feet between the theoretical gores of successive interchange entrance or exit ramps, Interchange Sequence signs should be used instead of Interchange Advance guide signs for the affected interchanges.*

03   *If used, Interchange Sequence (E9-1 or E9-2) signs should be used over the entire length of a route in an urban area.*

Support:

04   Interchange Sequence signs generally supplement Interchange Advance guide signs. Signing of this type is illustrated in Figure 2E-11, and is compatible with the sign spreading concept described in Paragraph 3 of Section 2E.43.

Standard:

05   **Interchange Sequence signs shall be installed in a series. Interchange Sequence signs shall display the next two or three interchanges by name or route number with distances to the nearest ¼ mile.**

06   **The first Interchange Sequence sign in the series shall be located in advance of the first Interchange Advance guide sign for the first interchange.**

07   **Where the exit direction is to the left, a LEFT (E11-2) sign panel (see Figure 2E-17) shall be displayed on the same line immediately to the right of the interchange name or route number.**

08   **Interchange Sequence signs shall not be substituted for Exit Direction signs.**

Guidance:

09   *Interchange Sequence signs should be located in the median. After the first sign of the series, subsequent Interchange Sequence signs should be placed approximately midway between interchanges.*

Figure 2E-11. Example of Using a Series of Interchange Sequence Signs for Closely-Spaced Interchanges

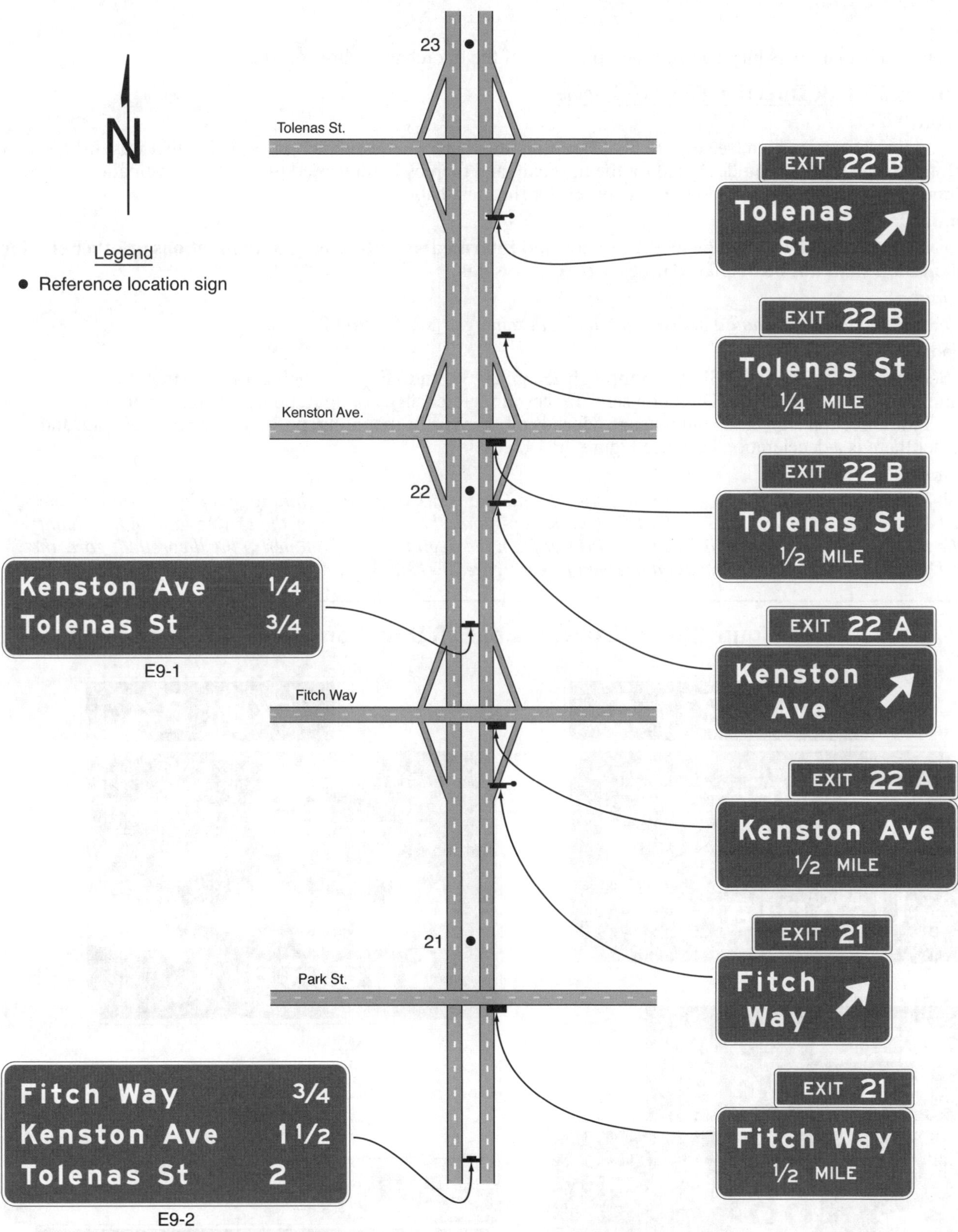

Standard:

10   **Interchange Sequence signs located in the median shall be installed at overhead sign height (see Section 2A.14).**

Option:

11   Interchange numbers may be displayed to the left of the interchange name or route number.

Section 2E.25 Exit Direction Signs (E4 Series)

Support:

01   The Exit Direction sign (see Figure 2E-12) repeats the route and destination information that was displayed on the Interchange Advance guide sign(s) for the next exit, and thereby assures road users of the destination served and indicates whether they exit to the right or left for that destination.

Standard:

02   **Exit Direction signs shall be used at major and intermediate interchanges. Populations or other similar information shall not be displayed on Exit Direction signs.**

Guidance:

03   *Exit Direction signs should be used at minor interchanges (see Section 2E.30).*

Support:

04   Sections 2E.28, 2E.30, 2E.31, 2E.33 through 2E.35, 2E.38, and 2E.40 through 2E.42 illustrate the use, location, and mounting of Exit Direction signs for certain interchange configurations. The placement location of the Exit Direction sign at the interchange depends on the type of mounting, post-mounted or overhead, and whether there is a deceleration lane (see Figure 2E-13).

Guidance:

05   *When post-mounted, the Exit Direction sign should be installed at the beginning of the deceleration lane taper. When mounted overhead, the Exit Direction sign should be installed over the exiting lane in the vicinity of the theoretical gore. If there is less than 300 feet from the beginning of the taper to the theoretical gore, the Exit Direction sign should be installed overhead (see Figure 2E-13).*

Figure 2E-12. Examples of Exit Direction Signs

Figure 2E-13. Exit Direction Sign Placement

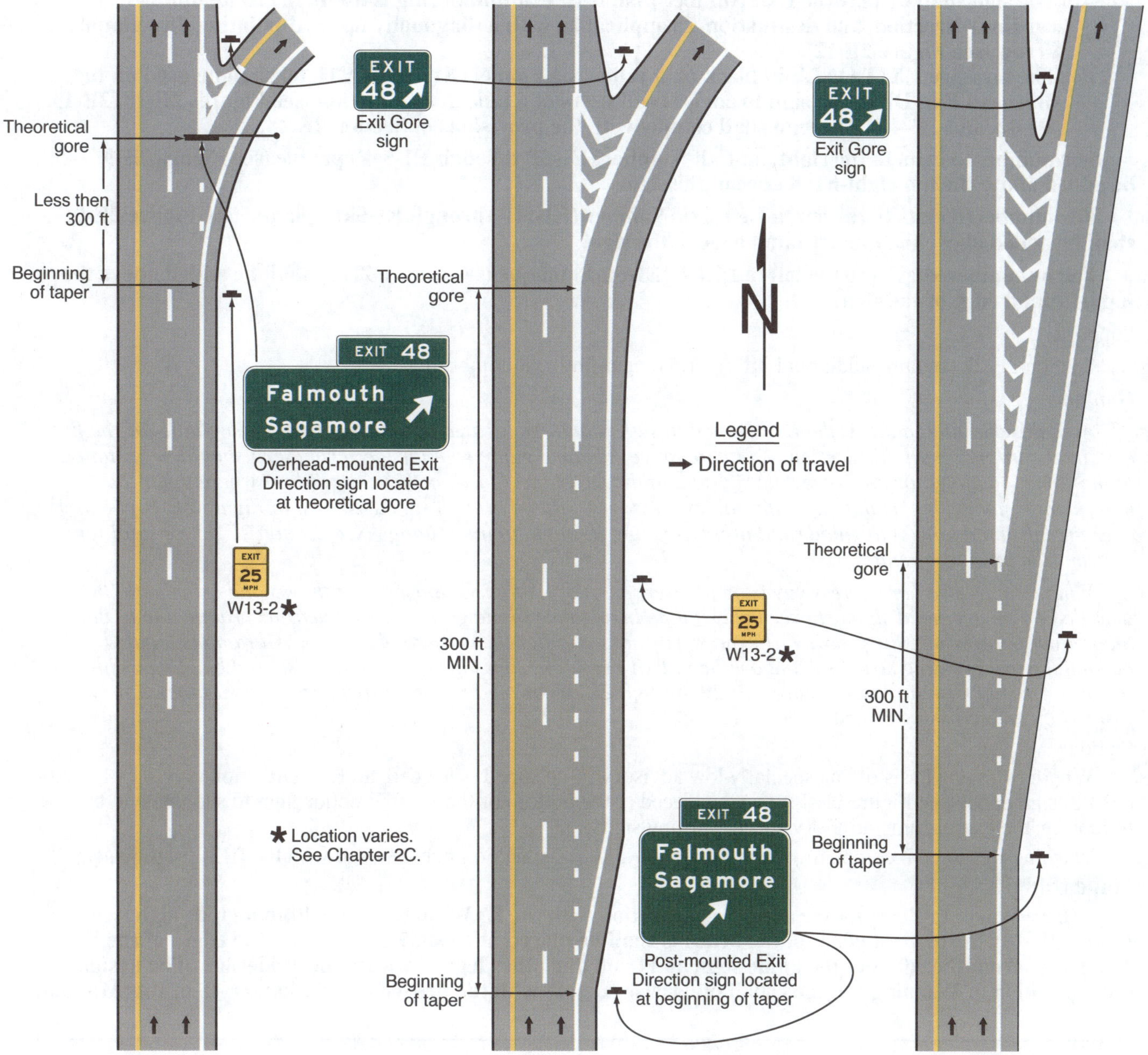

Standard:

06 **Except where Overhead Arrow-per-Lane guide signs are used (see Sections 2E.40 and 2E.42, and Paragraph 7 of this Section), where a through lane is being terminated (dropped) at an exit, the Exit Direction sign shall be placed overhead at the theoretical gore (see Figures 2E-18, 2E-19, 2E-33, 2E-42, and 2E-46).**

07 **Except as provided in Paragraph 4 of Section 2E.40, where Overhead Arrow-per-Lane guide signs are used for the Interchange Advance guide sign(s) for a multi-lane exit having an optional exit lane that also carries the through route or for a split with an option lane (see Section 2E.40), an Overhead Arrow-per-Lane guide sign shall also be used instead of the Exit Direction sign and located near, but not downstream from, the point where the outside edge of the dropped lane begins to diverge from the main roadway (see Figures 2E-36 through 2E-38). The Overhead Arrow-per-Lane guide sign shall be designed in accordance with the provisions of Section 2E.40.**

08   **The following provisions shall govern the design and application of overhead Exit Direction signs:**

   A. **The sign shall display the Exit Number plaque (if exit numbering is used), the route number, cardinal direction, and destination, as applicable, with a diagonally upward-pointing directional arrow (see Figure 2E-12).**

   B. **The message EXIT ONLY in black on a yellow sign panel (E11-1d or E11-1e) shall be used on the overhead Exit Direction sign to advise road users of a lane drop situation (see Figures 2E-18, 2E-19, 2E-42, and 2E-44). The sign shall comply with the provisions of Section 2E.28.**

09   **For numbered exits to the right, an Exit Number (E1-5P through E1-5eP) plaque (see Figure 2E-9) shall be added above the top right-hand edge of the sign.**

10   **For numbered exits to the left, a Left Exit Number (E1-5fP through E1-5kP) plaque (see Figure 2E-9) shall be added above the top left-hand edge of the sign.**

11   **For unnumbered exits to the left, a LEFT (E1-5mP) plaque (see Figure 2E-9) shall be added above the top left-hand edge of the sign.**

Support:

12   Section 2E.22 contains additional information regarding exit numbering.

Guidance:

13   *At multi-exit interchanges, the Exit Direction sign should be located directly over the exiting lane for the first exit, in accordance with this Section. An Interchange Advance guide sign for the second exit should be installed at the same location, normally over the right-hand through lane. Only for those conditions where the through movement is not evident should a confirmatory message (a Pull-Through sign as shown in Figure 2E-16) be used over the left-hand lane(s) to guide road users traveling through an interchange (see Section 2E.43 for additional information on sign spreading).*

14   *Where the freeway or expressway is on an overpass, the Exit Direction sign for the second exit should be installed on an overhead support over the exit lane in advance of the gore point, as near as practicable to the theoretical gore. Where the freeway or expressway passes under the crossroad and the exit ramp is located beyond the overcrossing structure, the overhead Exit Direction sign for the second exit should be placed either on the overcrossing structure (see Figures 2E-29 through 2E-31) or on a separate structure located immediately in front of the overcrossing structure.*

Option:

15   Where extra emphasis of an especially low advisory ramp speed is needed, an Exit Direction Advisory Speed (E13-2) sign panel (see Figure 2E-14) may be placed at the bottom of the Exit Direction sign to supplement, but not to replace, the exit or ramp advisory speed warning signs.

16   Warning Beacons in compliance with Paragraph 17 of this Section may be used with the E13-2 sign panel.

Standard:

17   **Where Warning Beacons are used in conjunction with the E13-2 sign panel within a guide sign (see Figure 2E-14), the nearest edges of the beacons shall be placed at least 12 inches from the edges of the E13-2 sign panel, from the edges of the guide sign, and from any other legend within the guide sign. The design and operation of Warning Beacons shall otherwise comply with the provisions of Chapter 4S of this Manual.**

Figure 2E-14. Examples of Exit Direction Signs with Advisory Speed Panels and Flashing Yellow Beacons

Exit Direction sign with E13-2 sign panel

OR

Exit Direction sign with E13-2 sign panel and flashing yellow beacons

Option:

18   In cases, where sight distance is restricted because of structures or unusual alignment, principally in urban areas, making it impossible to locate the Exit Direction sign without violating the required minimum spacing between major guide signs (see Section 2E.23), Interchange Sequence signs (see Section 2E.24) may be substituted for an Interchange Advance guide sign.

Guidance:

19   *At the last exit from a highway before it becomes a facility on which toll payments are required, the LAST EXIT BEFORE TOLL (W16-16P) plaque (see Section 2F.10 and Figure 2F-4) should be installed above the Exit Direction sign, but below the Exit Number or LEFT plaque, if used.*

Section 2E.26 Exit Gore Signs and Plaque (E5-1 Series)

Support:

01   The Exit Gore sign (see Figure 2E-15) in the gore indicates the exiting point or the place of departure from the main roadway. Consistent application of this sign at each exit is important to provide adequate visibility of the departure of the exit roadway from the main roadway.

Standard:

02   **The gore shall be defined as the area located between the main roadway and the ramp just beyond where the ramp branches from the main roadway. An Exit Gore sign shall be located in the gore for each ramp that departs from the main roadway of a freeway or expressway, or departs from a collector-distributor roadway, and shall display the word EXIT (E5-1) if interchange exit numbering is not used or EXIT XX (E5-1a or E5-1c) if interchange exit numbering is used, and an appropriate diagonally upward-pointing arrow. If suffix letters are used for exit numbering at a multi-exit interchange, the suffix letter shall also be included on the Exit Gore (E5-1a or E5-1c) sign or Exit Gore Number (E5-1bP) plaque and shall be separated from the exit number by a space having a width of between ½ and ¾ of the height of the suffix letter. Breakaway or yielding supports shall be used.**

Guidance:

03   *The arrow should be aligned to approximate the angle of departure. Each gore should be treated similarly, whether the interchange has one exit roadway or multiple exits.*

Option:

04   The Narrow Exit Gore (E5-1c) sign (see Figure 2E-15) may be used in gore areas of limited width where the width of the Exit Gore (E5-1a) sign would not permit sufficient lateral offset (see Section 2A.16), such as for ramp departures that are nearly parallel to the main roadway where the Exit Gore sign would be mounted on a narrow island or barrier. Where the E5-1c sign is mounted at a height of 14 feet or more from the roadway, the directional arrow may point diagonally downward.

Guidance:

05   *The E5-1c sign should not be used in gore areas where an E5-1a sign could be installed with sufficient lateral offset.*

Figure 2E-15. Exit Gore Signs and Plaques

Option:

06   Where extra emphasis of an especially low advisory ramp speed is needed, the Confirmation Advisory Speed (W13-1aP) plaque (see Section 2C.59) indicating the advisory speed may be mounted below the Exit Gore sign (see Figure 2E-15) to supplement, but not to replace, the exit or ramp advisory speed warning signs.

07   To improve the visibility of the gore for exiting drivers, a Type 1 object marker (see Chapter 2C) may be installed 4 feet above the ground line on each sign support below the Exit Gore sign.

08   An Exit Gore Number (E5-1bP) plaque (see Figure 2E-15) may be installed above an existing Exit Gore (E5-1) sign when an unnumbered exit is converted to a numbered exit until such time as an E5-1 sign is being replaced for other reasons (see Paragraph 9 of this Section).

Standard:

09   **An Exit Gore (E5-1a) sign shall be used when the replacement of an existing assembly of an E5-1 sign and E5-1bP plaque becomes necessary.**

Section 2E.27 Pull-Through Signs (E6-1 Series and E6-2 Series)

Support:

01   Pull-Through (E6-1 series and E6-2 series) signs (see Figure 2E-16) are overhead guide signs intended for through traffic.

Guidance:

02   *Pull-Through signs should be used where the geometrics of a given interchange are such that it is not clear to the road user as to which is the through roadway, or where additional route guidance is desired. Pull-Through signs with down arrows should be used where the alignment of the through lanes is curved and the exit direction is straight ahead, where the number of through lanes is not readily evident, and at multi-lane exits where there is a reduction in the number of through lanes. Pull-Through signs should not be used at exits with option lanes where full-width Overhead Arrow-per-Lane signs are being used.*

Standard:

03   **When used, Pull-Through signs shall display the route shield and the cardinal direction for the through route.**

Option:

04   Pull-Through signs may display the control city and down arrows (see Figure 2E-16 and Section 2E.18).

Support:

05   Sections 2E.28, 2E.39, and 2E.40 contain information regarding the use of Overhead Arrow-per-Lane guide signs at multi-lane exits where there is a reduction in the number of through lanes and a through lane becomes an interior option lane for through or exiting traffic.

Figure 2E-16. Examples of Pull-Through Signs

E6-1

E6-1a

Note: The E6-2 and E6-2a sign designs correspond to the E6-1 and E6-1a sign designs, respectively, but with the addition of a down arrow on each.

Section 2E.28 <u>Signing for Interchange Lane Drops without an Optional Exit Lane</u>

Standard:

01 **The provisions of this Section shall only apply to lane drops at exits that do not have an optional exit lane. At exits that have an optional exit lane in addition to the dropped lane, the provisions of Sections 2E.39 through 2E.42 shall apply.**

02 **Except as provided in Paragraph 15 of this Section, major guide signs for all lane drops at interchanges shall be mounted overhead. An EXIT ONLY sign panel shall be used for all interchange lane drops at which the through route is carried on the main roadway.**

03 **Except on Overhead Arrow-per-Lane and Diagrammatic Advance guide signs (see Sections 2E.39 through 2E.41), the EXIT ONLY (down arrow) (E11-1 or E11-1f) sign panel (see Figure 2E-17) shall be used on all overhead Advance guide signs of lane drops (see Figures 2E-18, 2E-19, and 2E-34). The number of arrows on each sign shall correspond to the number of dropped lanes at the location of each sign. Placement of the down arrow shall comply with the provisions of Section 2E.18.**

04 **For lane drops, the bottom portion of the overhead Exit Direction sign shall be yellow with a black border and shall include a diagonally upward-pointing black directional arrow (left or right, as appropriate) for each lane dropped at the exit (see Figures 2E-18 and 2E-19). The sign shall be designed and placed so that each arrow is located over the approximate center of each lane being dropped. Except as provided in Paragraph 5 of this Section, the words EXIT and ONLY shall be positioned to the left and right, respectively, of the arrow on the E11-1d sign panel (see Figure 2E-17) for a single-lane drop. For a two-lane drop, the words EXIT ONLY shall be located between the two arrows on the E11-1e sign panel (see Figure 2E-17). The number of arrows on the sign shall correspond to the number of dropped lanes at the location of the sign.**

Option:

05 Where an existing sign structure length or adjacent signs constrain the width or placement of the Interchange Advance guide sign on that structure, the down arrow may be positioned to the right or left of the words EXIT ONLY, instead of between the words, to allow for the positioning of the arrow over the approximate center of the lane. Where the width of the Exit Direction sign extends over the adjacent lane, the directional arrow may be placed to the right of the words EXIT ONLY for an exit to the right, or to the left of the words EXIT ONLY for an exit to the left, to allow for the positioning of the arrow over the dropped lane.

06 EXIT ONLY messages of either the combination of E11-1a and E11-1b, or the E11-1c sign panels (see Figure 2E-17) may be used to retrofit existing signing to warn of a lane drop situation ahead.

Standard:

07 **If used to retrofit an existing guide sign, the E11-1a and E11-1b sign panels (see Figure 2E-17) shall be placed on either side of a white down arrow on an Interchange Advance guide sign and on either side of a white directional arrow on an Exit Direction sign. The E11-1c sign panel (see Figure 2E-17), if used to retrofit an existing Interchange Advance guide sign, shall be placed between the lower destination message and the white down arrow.**

Figure 2E-17. EXIT ONLY and LEFT Sign Panels

Figure 2E-18. Guide Signs for a Single-Lane Exit to the Left with a Dropped Lane

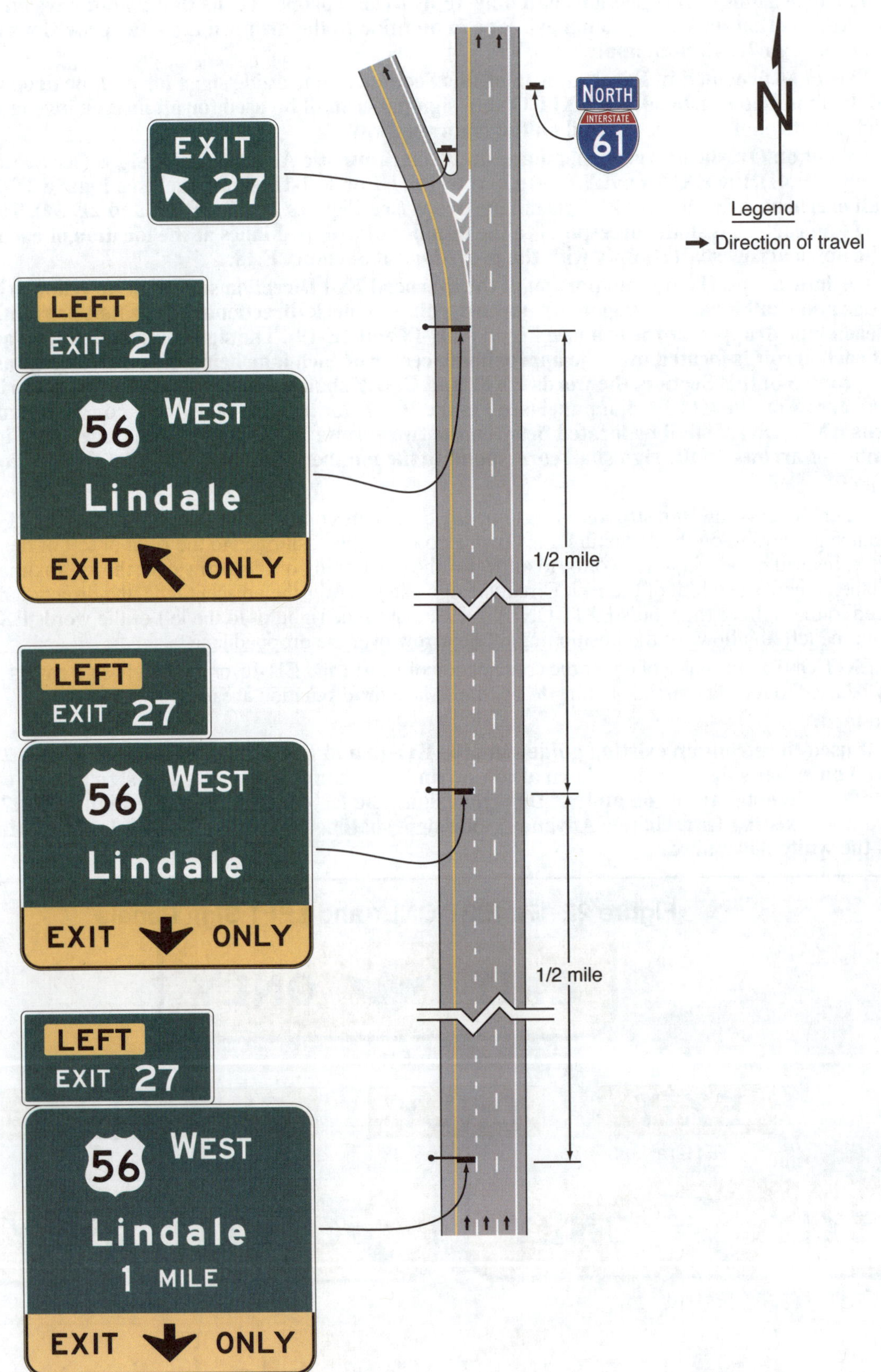

Figure 2E-19. Guide Signs for a Single-Lane Exit to the Right with a Dropped Lane

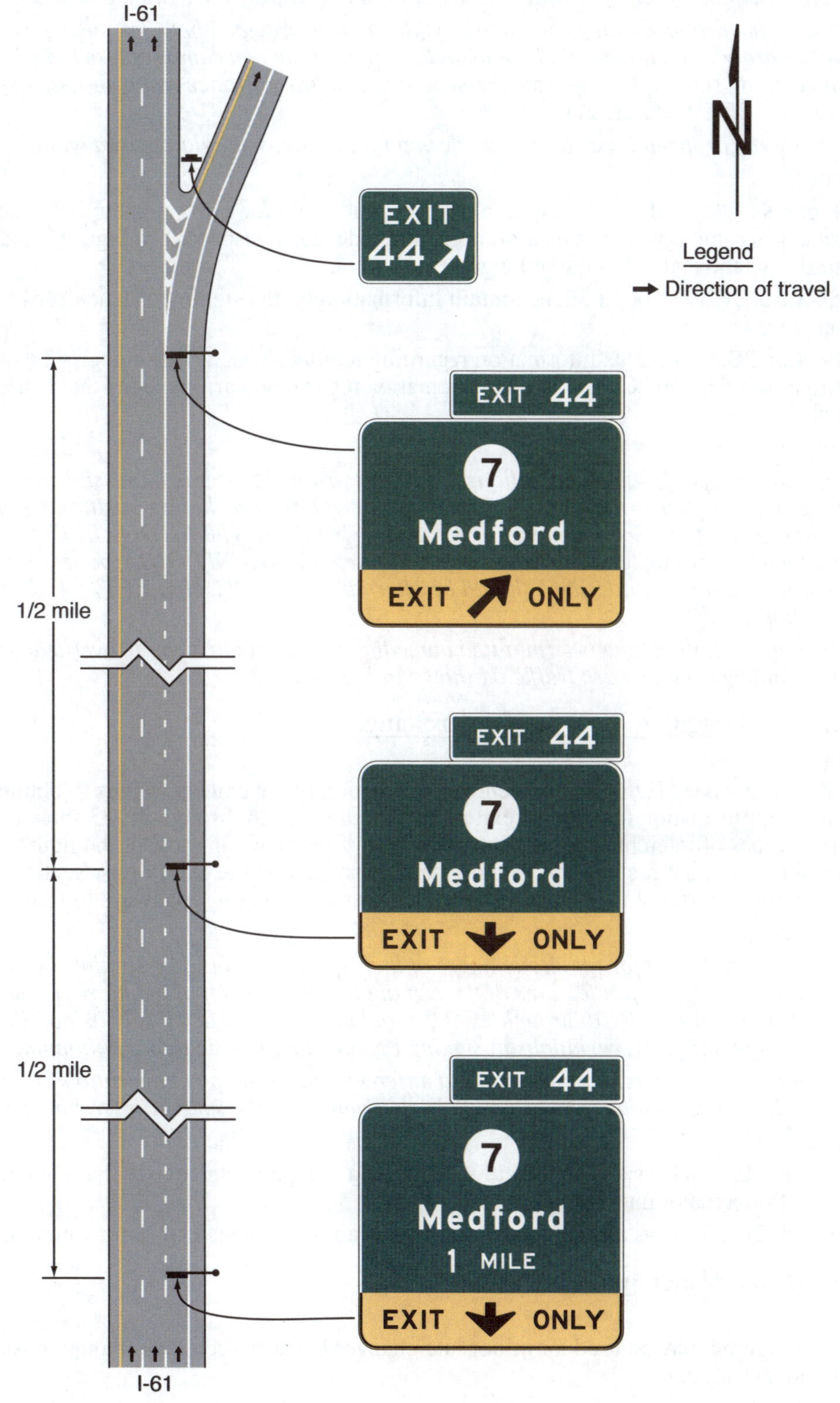

Guidance:

08   *Except as provided in Paragraph 9 of this Section for an auxiliary lane, Interchange Advance guide signs for lane drops within 1 mile of the interchange should not display the distance message.*

09   *Where the dropped lane is an auxiliary lane that is provided between successive entrance and exit ramps of two separate interchanges and the distance between the two ramps is less than 1 mile, the first Interchange Advance guide sign in the sequence downstream from the entrance ramp should display the distance message (see Figures 2E-20 and 2E-21).*

10   *Where the dropped lane carries the through route, signs should be used without the EXIT ONLY sign panel.*

Support:

11   Figures 2E-20 and 2E-21 show examples of guide sign for a dropped auxiliary lane between separate interchanges using post-mounted and overhead guide signs, respectively. Figure 2E-22 shows guides signs used for an auxiliary lane that is ½ mile or longer.

12   Sections 2E.39 through 2E.42 contain information on the signing of lane drops at exits that also have an option lane.

13   Section 2B.31 contains information regarding regulatory signs that can also be used for freeway lane drop situations and Section 2C.50 contains information regarding warning signs that can also be used for freeway lane drop situations.

Guidance:

14   *In limited cases in which conditions are so constrained that it is impossible to locate an Interchange Advance guide sign either overhead or partly over the dropped lane, precluding positioning of the down arrow as provided in Paragraph 3 of this Section, a sign panel displaying the legend RIGHT (LEFT) LANE ONLY in a black legend on a yellow background should be substituted for the EXIT ONLY panel on that sign. In such cases, the Interchange Advance guide signs should be alternated with RIGHT (LEFT) LANE FOR EXIT ONLY (W9-7) signs (see Section 2C.50).*

15   *Where a mainline lane is terminated immediately after an exit ramp, overhead and/or post mounted warning signs should be used to warn traffic as shown in Figure 2E-23.*

Section 2E.29 Signing by Type of Interchange

Support:

01   Road users need signs to help identify the location of the exit, as well as to obtain route, direction, and destination information for specific exit ramps. Figures 2E-26 through 2E-33 show examples of guide signs for common types of interchanges. The interchange layouts shown in most of the figures illustrate only the major guide signs for one direction of traffic on the freeway and on the exit ramps. Section 2D.49 contains information regarding the signing of the crossroad approaches and connecting roadways to freeways and expressways.

Guidance:

02   *The signing layout for all interchanges of the same type should be similar. For the purpose of uniform application, the significant features of the signing layout for each of the more frequent types of interchanges (illustrated in Figures 2E-26 through 2E-33) should be followed as closely as possible. Even when unusual geometric features exist, variations in signing layout should be held to a minimum.*

03   *Where a single interchange combines a different type of ramp configuration for each direction of travel, the main roadway major guide signing should be determined by the specific interchange type for that direction of travel.*

Support:

04   Figure 2E-24 shows an example of signing for a complex interchange that combines intermediate interchange ramps within a major interchange.

05   Figure 2E-25 shows an example of signing for an interchange exit ramp with a downstream split.

Section 2E.30 Minor Interchange

Option:

01   Less signing may be used for minor interchanges because such interchanges customarily serve low volumes of mostly local traffic.

Support:

02   An example of guide signs for a minor interchanges is shown in Figure 2E-26.

Standard:

03   **In accordance with the provisions of Sections 2E.23 and 2E.26, at least one Interchange Advance guide sign and an Exit Gore sign shall be used at a minor interchange.**

Guidance:

04   *An Exit Direction sign in compliance with Section 2E.25 should also be used.*

Figure 2E-20. Example of Overhead Guide Signs for a Dropped Auxiliary Lane between Separate Interchange Ramps

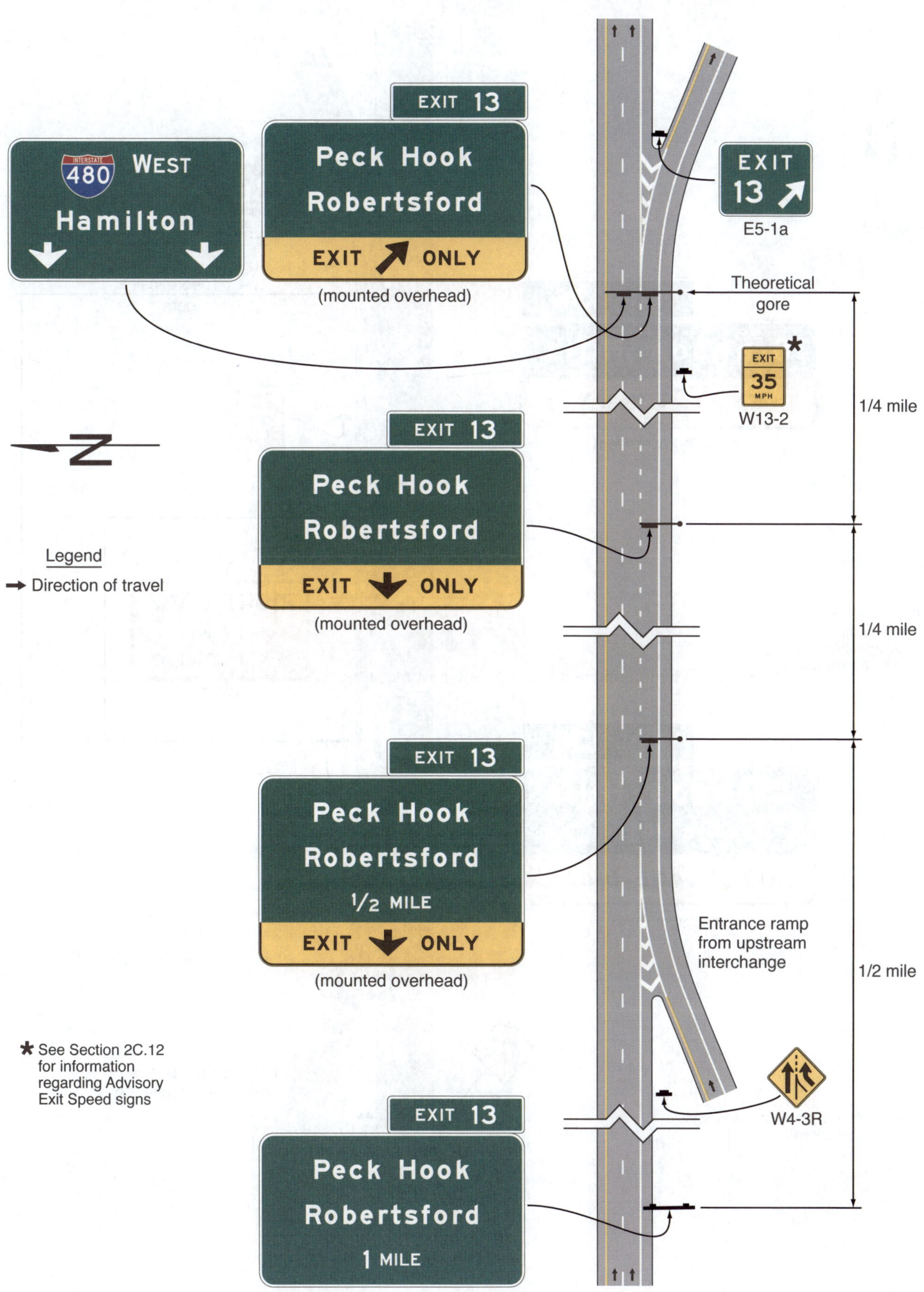

Figure 2E-21. Example of Post-Mounted Advance Guide and Supplemental Warning Signs for a Dropped Auxiliary Lane between Separate Interchange Ramps

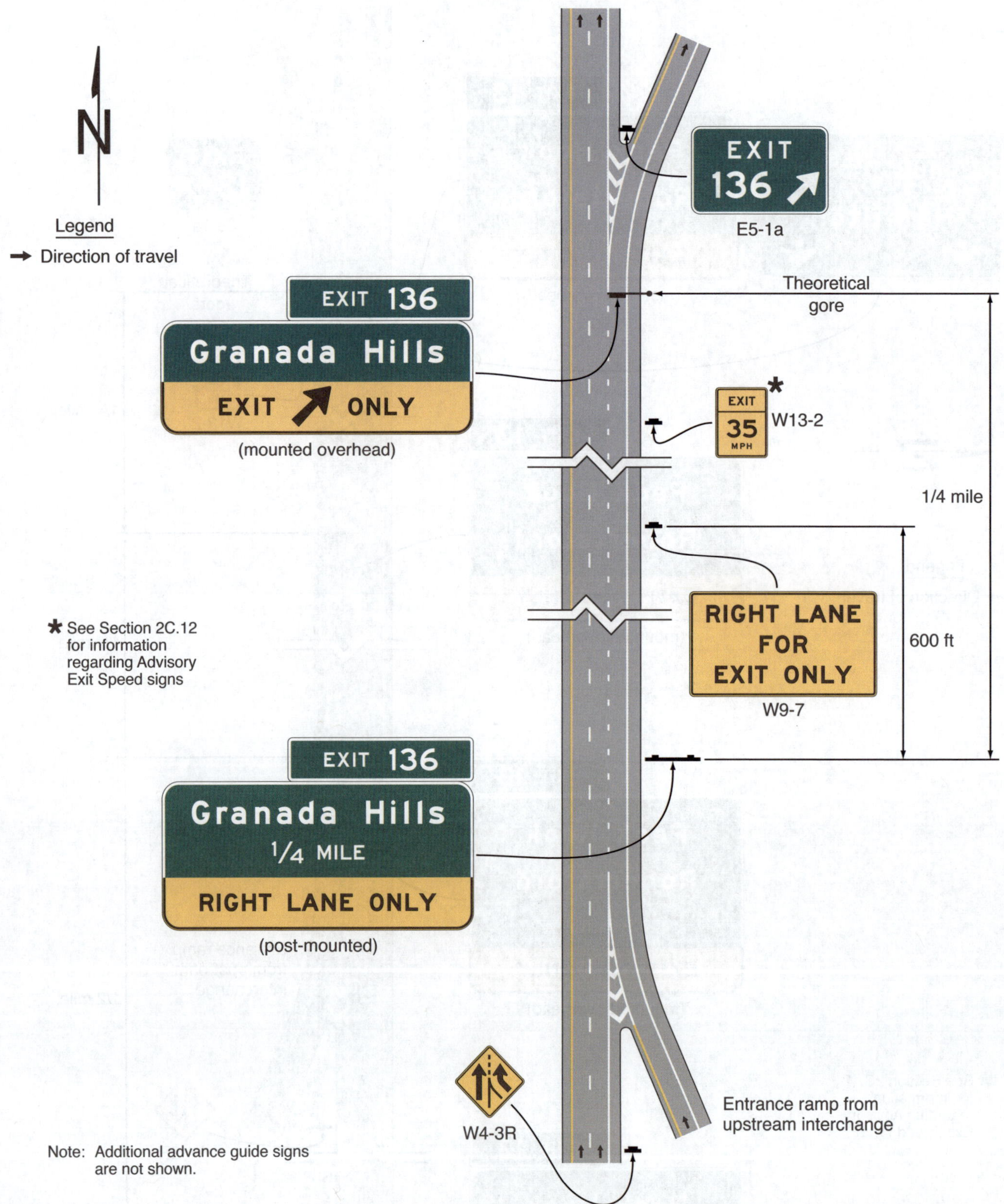

Figure 2E-22. Example of Guide Signs for an Auxiliary Lane of at Least One-Half Mile in Length

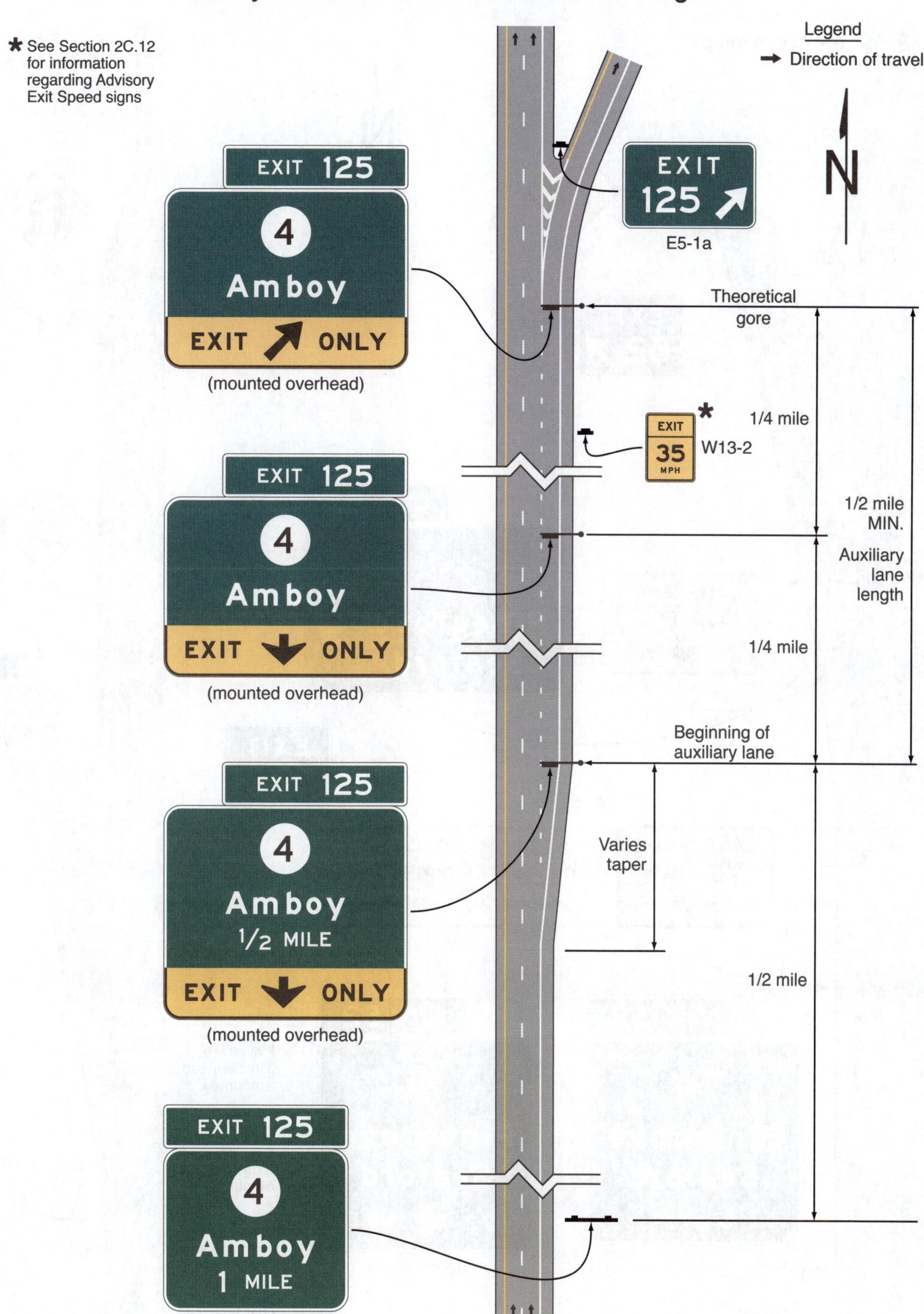

Figure 2E-23. Examples of Signing for Mainline Terminations within an Interchange

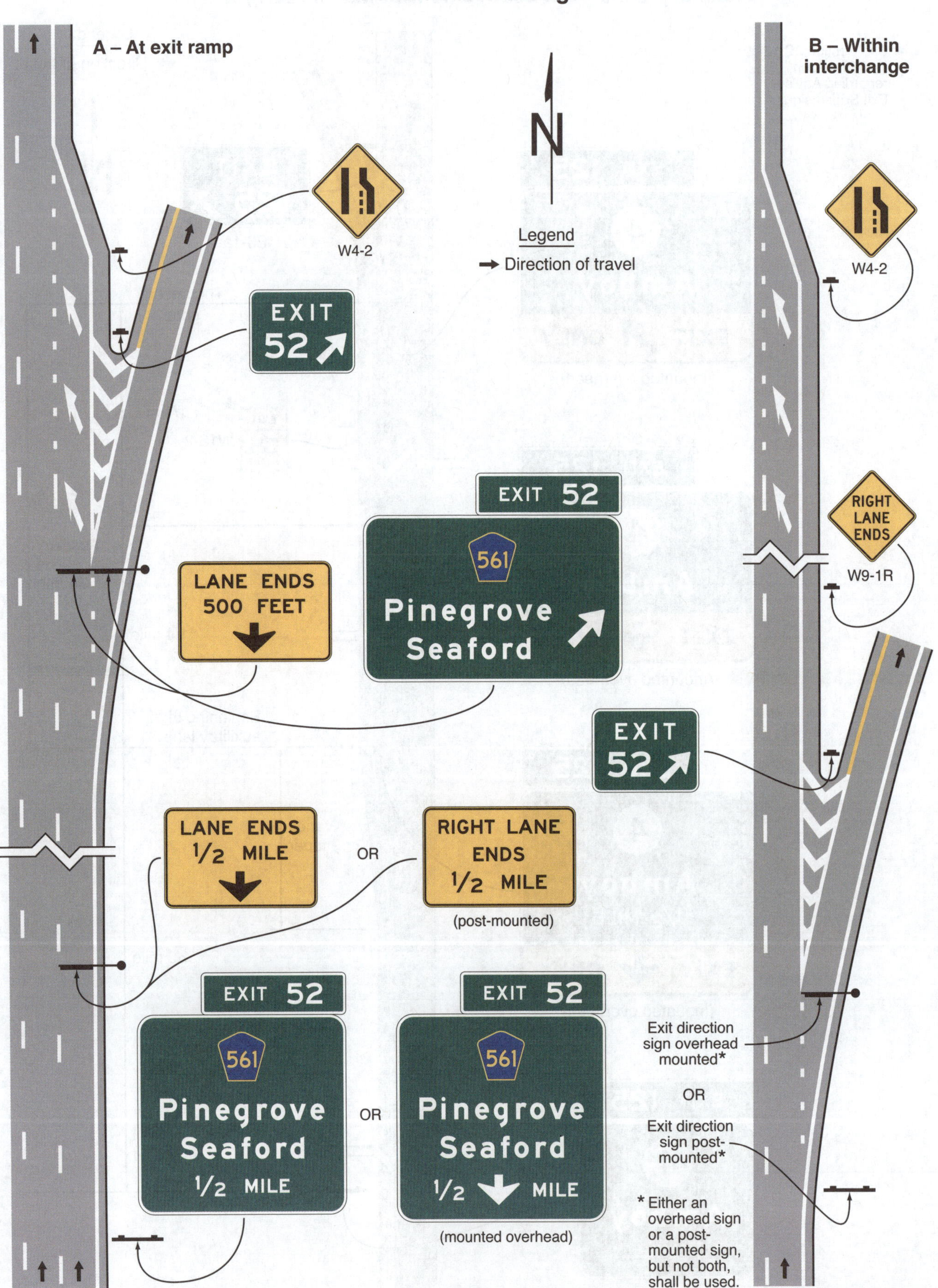

Figure 2E-24. Example of Signing for an Intermediate Interchange within a Major Interchange

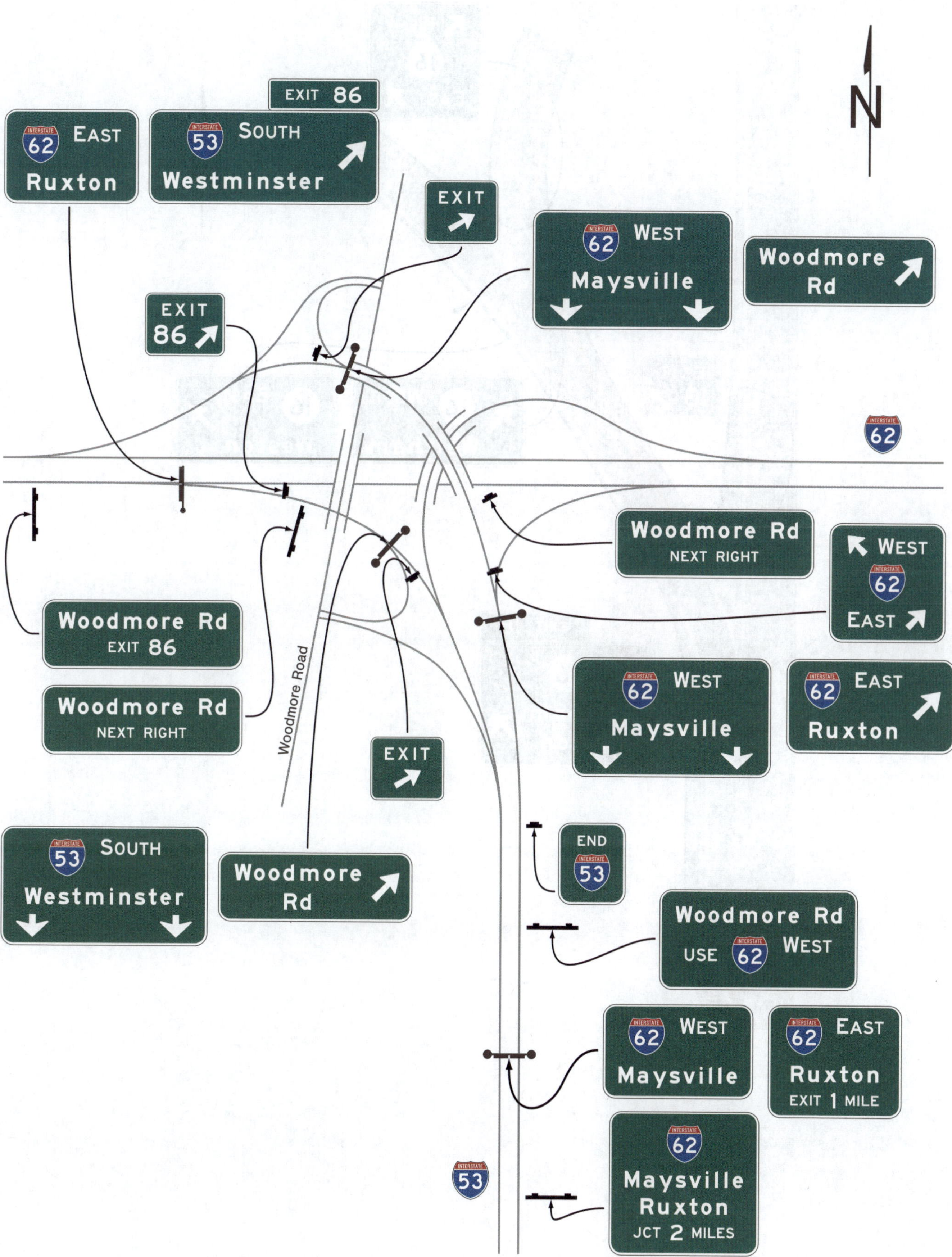

Figure 2E-25. Examples of Signing for an Interchange Exit Ramp with a Downstream Split (Sheet 1 of 2)

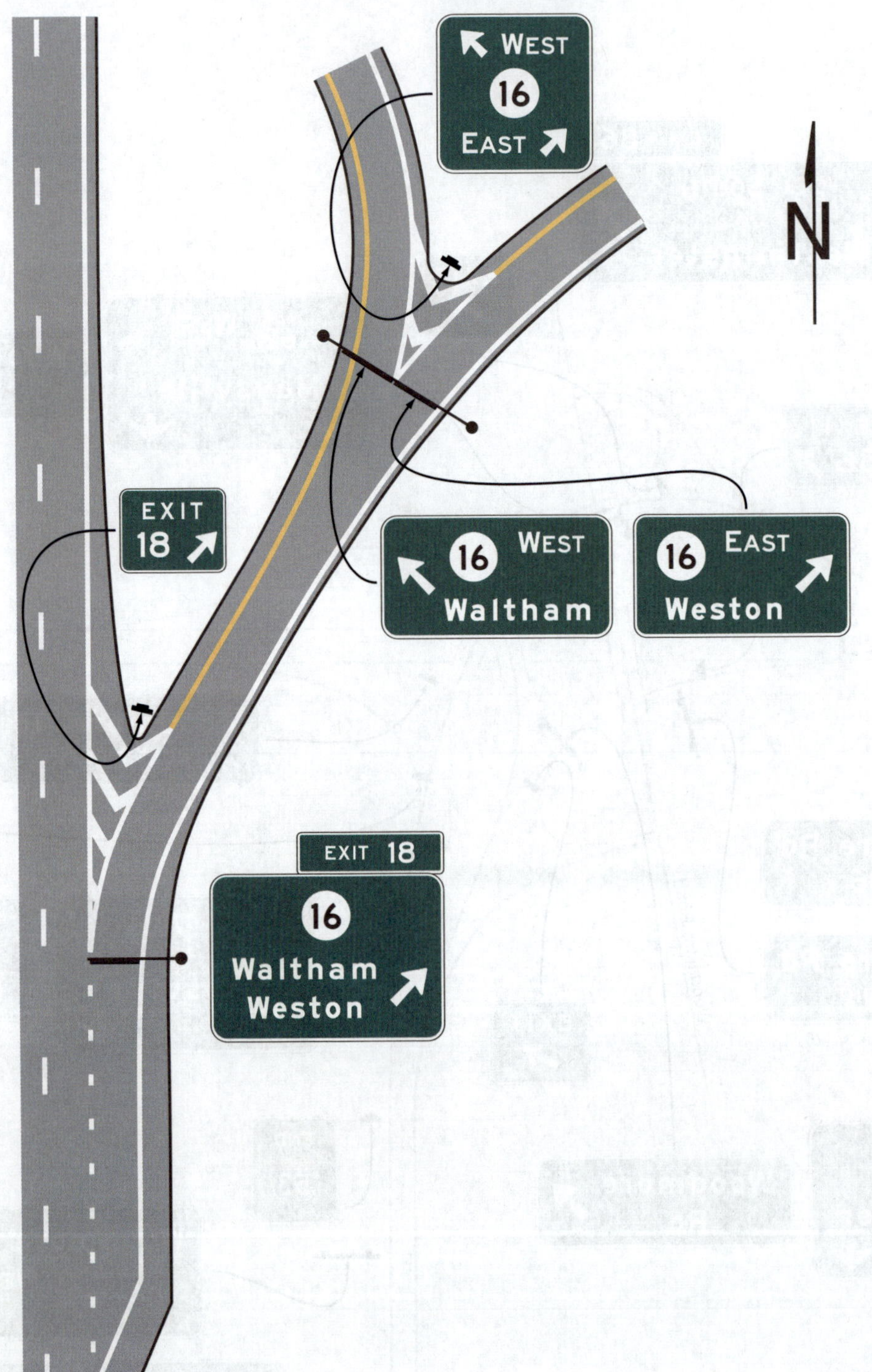

Figure 2E-25. Examples of Signing for an Interchange Exit Ramp with a Downstream Split (Sheet 2 of 2)

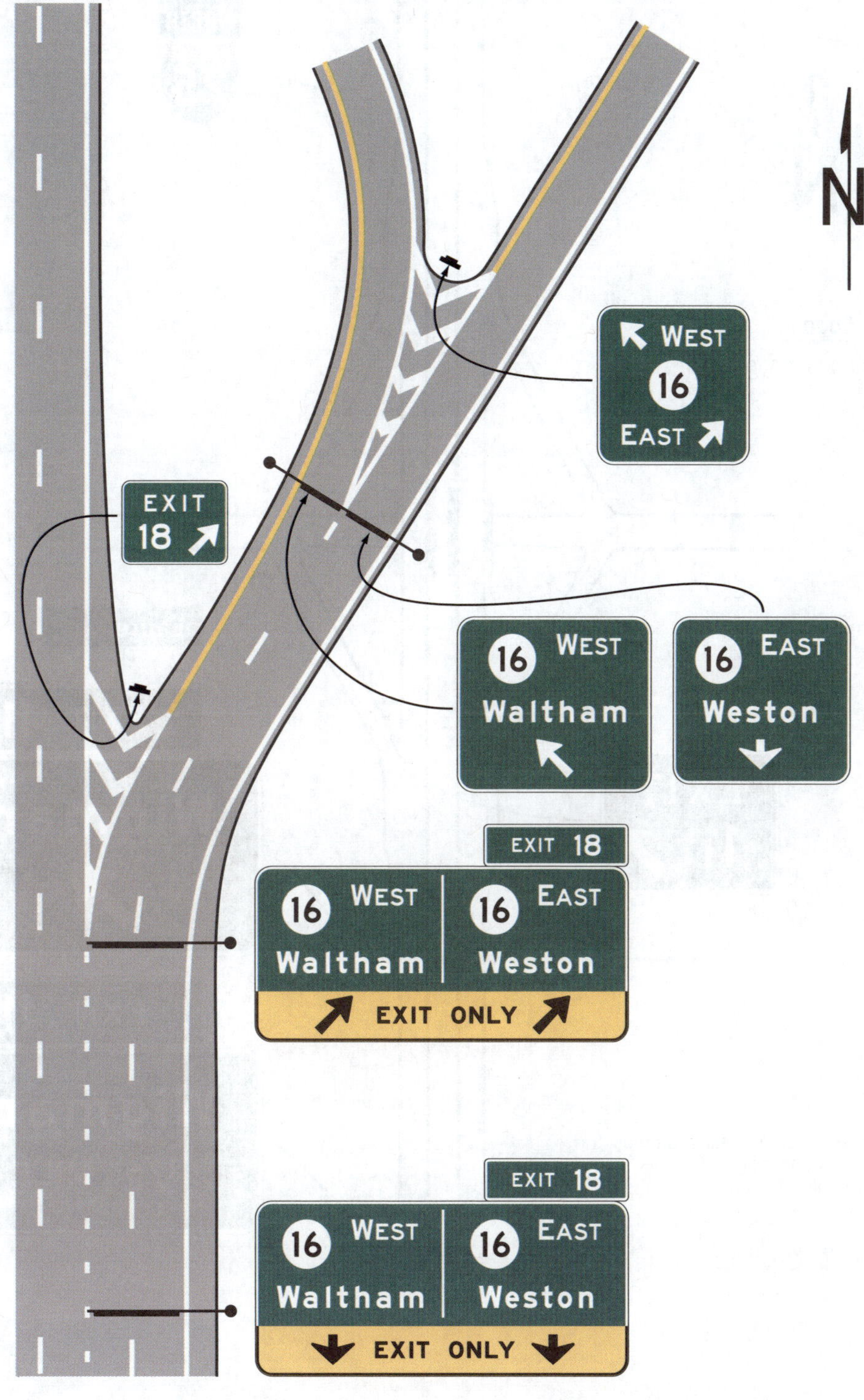

Figure 2E-26. Examples of Guide Signs for a Minor Interchange

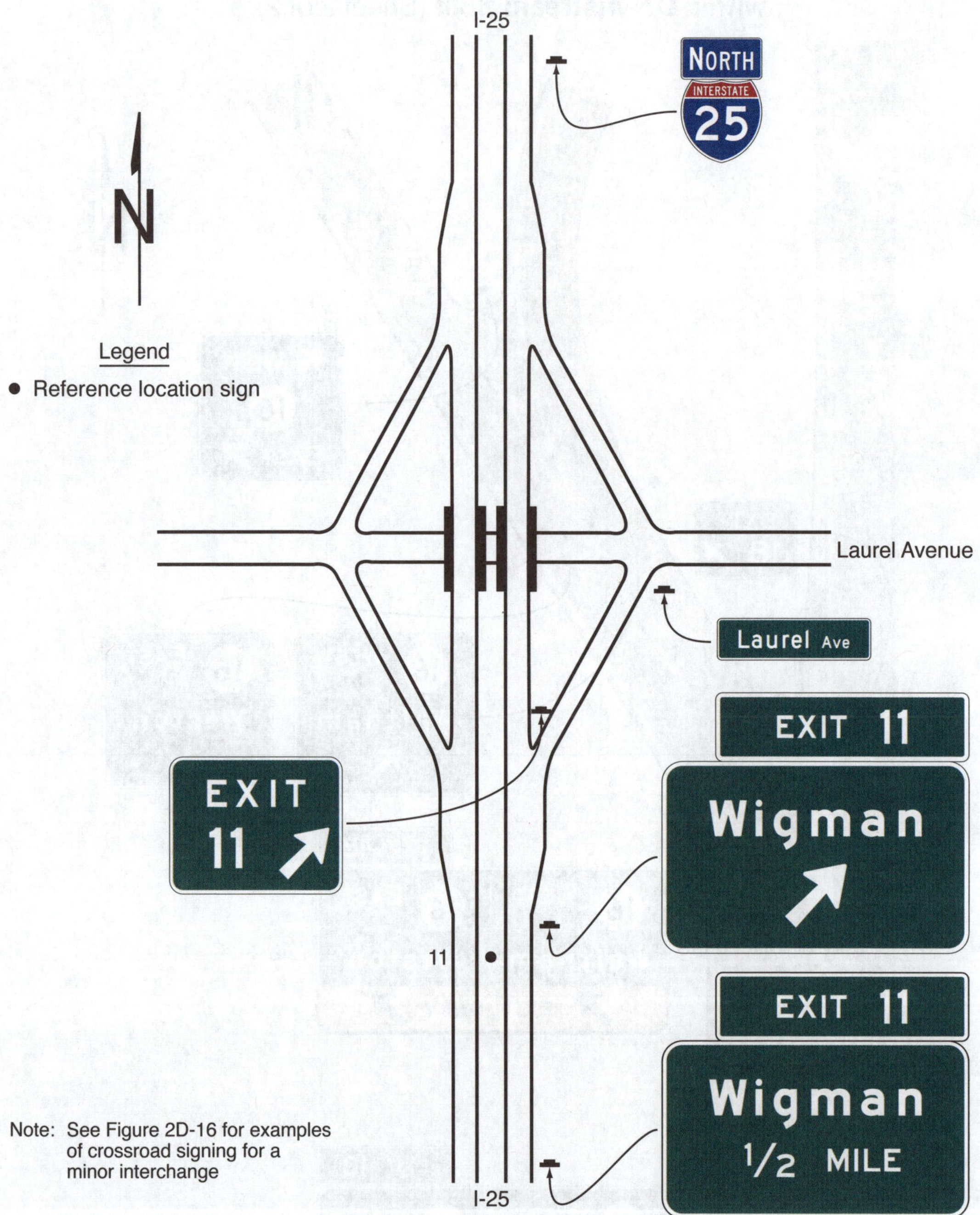

Section 2E.31 Diamond Interchange

Support:

01   An example of guide signs for a diamond interchange is shown in Figure 2E-27.

02   The typical diamond interchange ramp departs from the main roadway such that a speed reduction generally is not necessary in order for a driver to negotiate an exit maneuver from the main roadway onto the ramp roadway. Section 2C.12 contains provisions for the use of an Advisory Exit Speed (W13-2) sign for situations where a speed reduction is necessary.

Guidance:

03   *When a speed reduction is not necessary, an Advisory Exit Speed sign should not be used.*

04   *The Advisory Exit Speed sign, if used, should be located along the deceleration lane or along the ramp such that it is visible to the driver far enough in advance to allow the driver to decelerate before reaching the curve associated with the exiting maneuver. Use and placement of the Advisory Exit Speed sign should otherwise comply with Section 2C.12 of this Manual.*

Option:

05   A Stop Ahead (W3-1) or Signal Ahead (W3-3) warning sign (see Section 2C.35) may be placed, where engineering judgment indicates a need, along the ramp in advance of the crossroad, to give notice to the driver.

Guidance:

06   *When used on two-lane ramps, Stop Ahead or Signal Ahead signs should be used in pairs with one sign on each side of the ramp.*

07   *Where the exit ramp allows traffic to turn in either direction onto the crossroad, a Destination (D1 series) sign (see Section 2D.36) that includes each destination displayed on the Advance, Exit Direction, and Supplemental guide signs along the main roadway for that exit should be placed along the ramp.*

Section 2E.32 Diamond Interchange in Urban Area

Support:

01   An example of guide signs for a diamond interchange in an urban area is shown in Figure 2E-28. This example includes the use of the Community Interchanges Identification sign (see Section 2E.52), which might be useful if two or more interchanges serve the same community.

02   In urban areas, street names are often displayed as the principal message in destination signs.

Option:

03   If interchanges are too closely spaced to locate the Interchange Advance guide signs at the distances specified in Section 2E.23, they may be placed closer to the exit with the distances displayed adjusted accordingly.

Section 2E.33 Cloverleaf Interchange

Support:

01   A cloverleaf interchange has two exits for each direction of travel. The exits are closely spaced and have common Advance guide signs. An example of guide signs for a cloverleaf interchange is shown in Figure 2E-29.

Guidance:

02   *The Advance guide signs should include two place names, one corresponding to each exit ramp, with the name of the place served by the first exit on the upper line.*

Standard:

03   **An overhead guide sign assembly shall be placed at the theoretical gore of the first exit ramp, with an Exit Direction sign for the first exit and an Interchange Advance guide sign for the second exit, as shown in Figure 2E-29. The second exit shall be indicated by an overhead Exit Direction sign over the auxiliary lane.**

04   **Interchanges with more than one exit from the main roadway shall be numbered as described in Section 2E.22 with an appropriate suffix.**

05   **Diagrammatic Advance signs shall not be used for cloverleaf interchanges except as otherwise provided in Section 2E.41.**

Guidance:

06   *Where the main roadway passes under the crossroad and the exit roadway is located beyond the overcrossing structure, the placement of the overhead Exit Direction sign for the second exit should comply with Section 2E.25 (see Figure 2E-29).*

Figure 2E-27. Example of Guide Signs for a Diamond Interchange

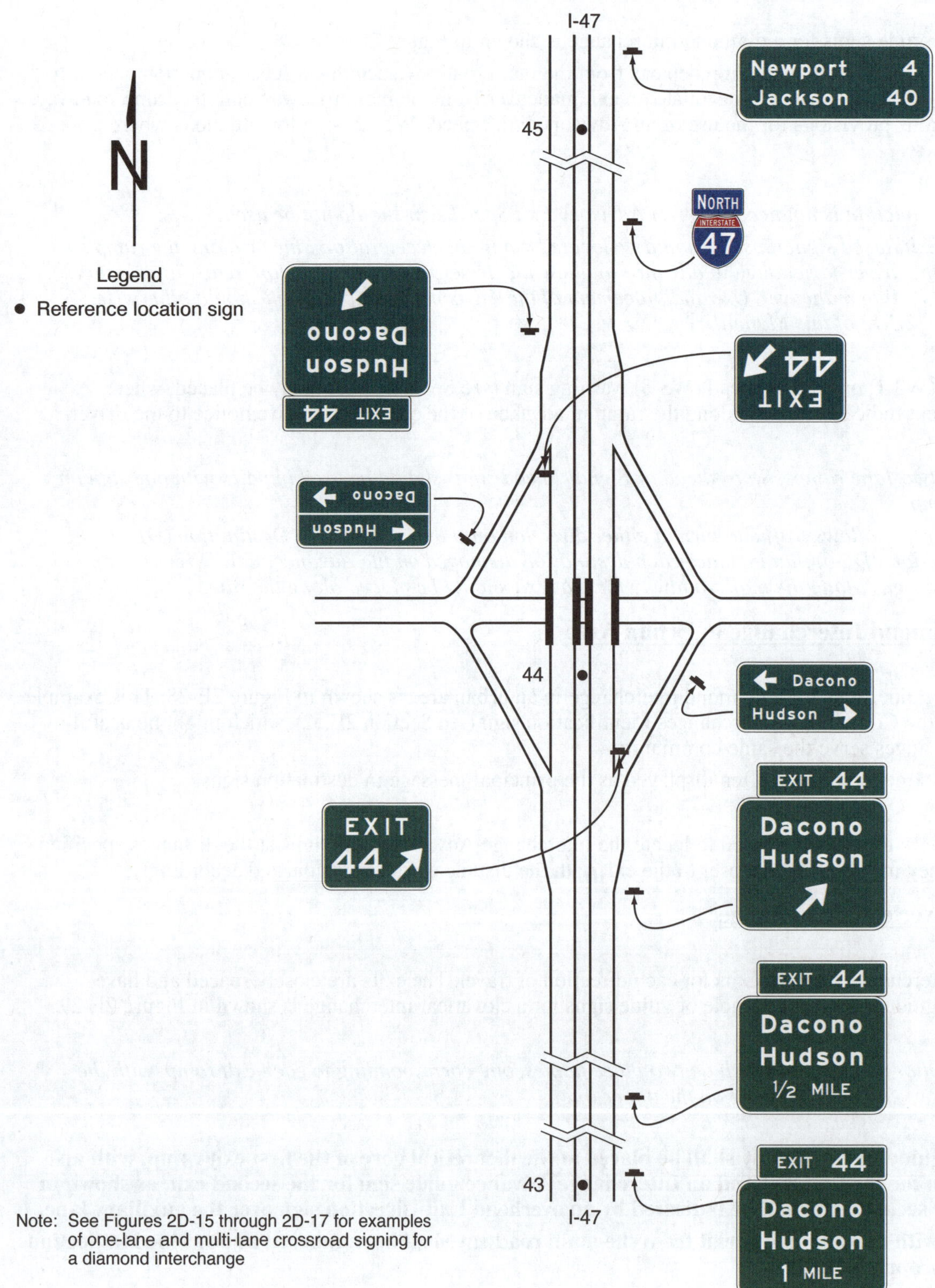

Note: See Figures 2D-15 through 2D-17 for examples
of one-lane and multi-lane crossroad signing for
a diamond interchange

Figure 2E-28. Example of Guide Signs for a Diamond Interchange in an Urban Area

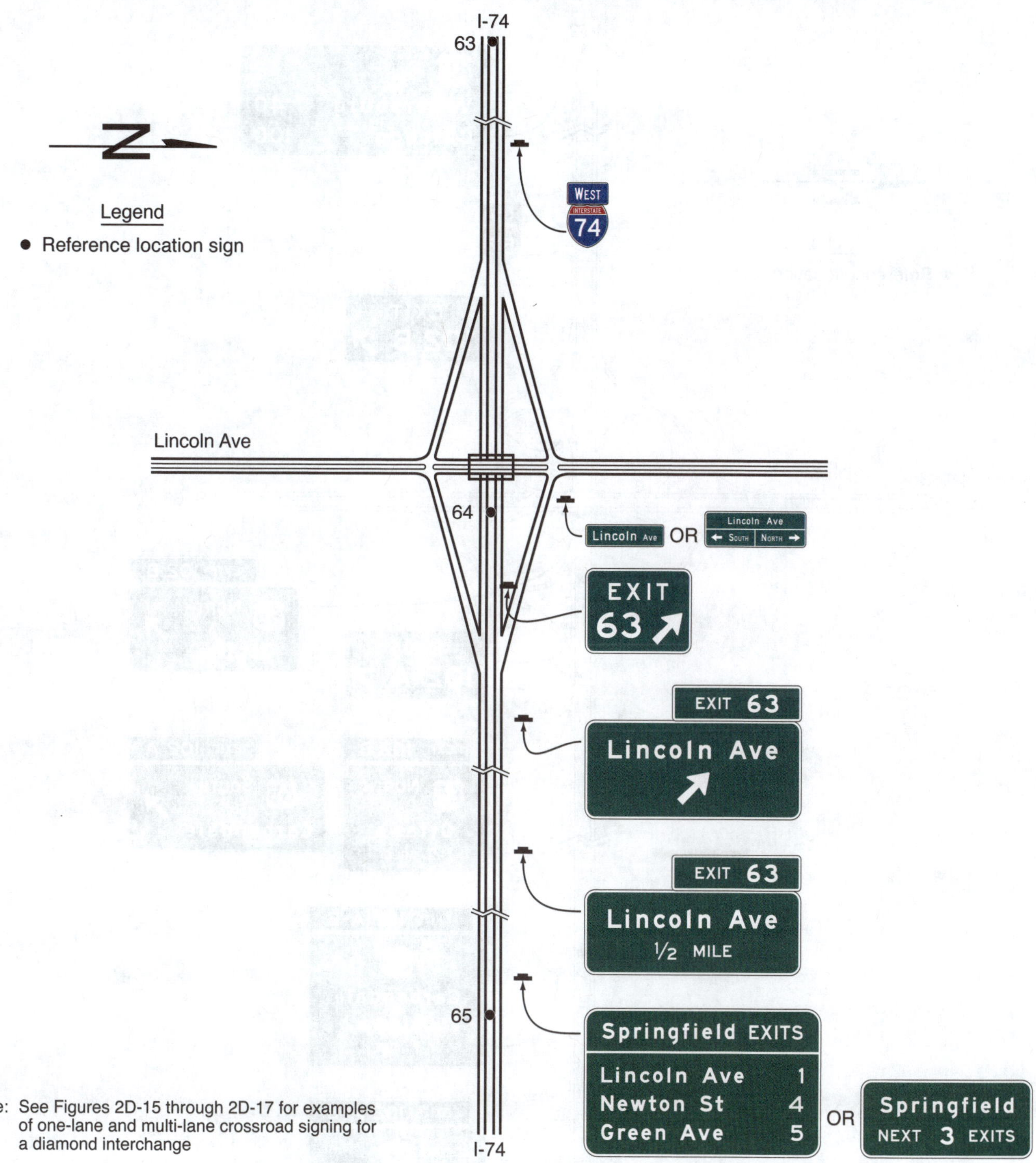

Note: See Figures 2D-15 through 2D-17 for examples
of one-lane and multi-lane crossroad signing for
a diamond interchange

Figure 2E-29. Example of Guide Signs for a Full Cloverleaf Interchange

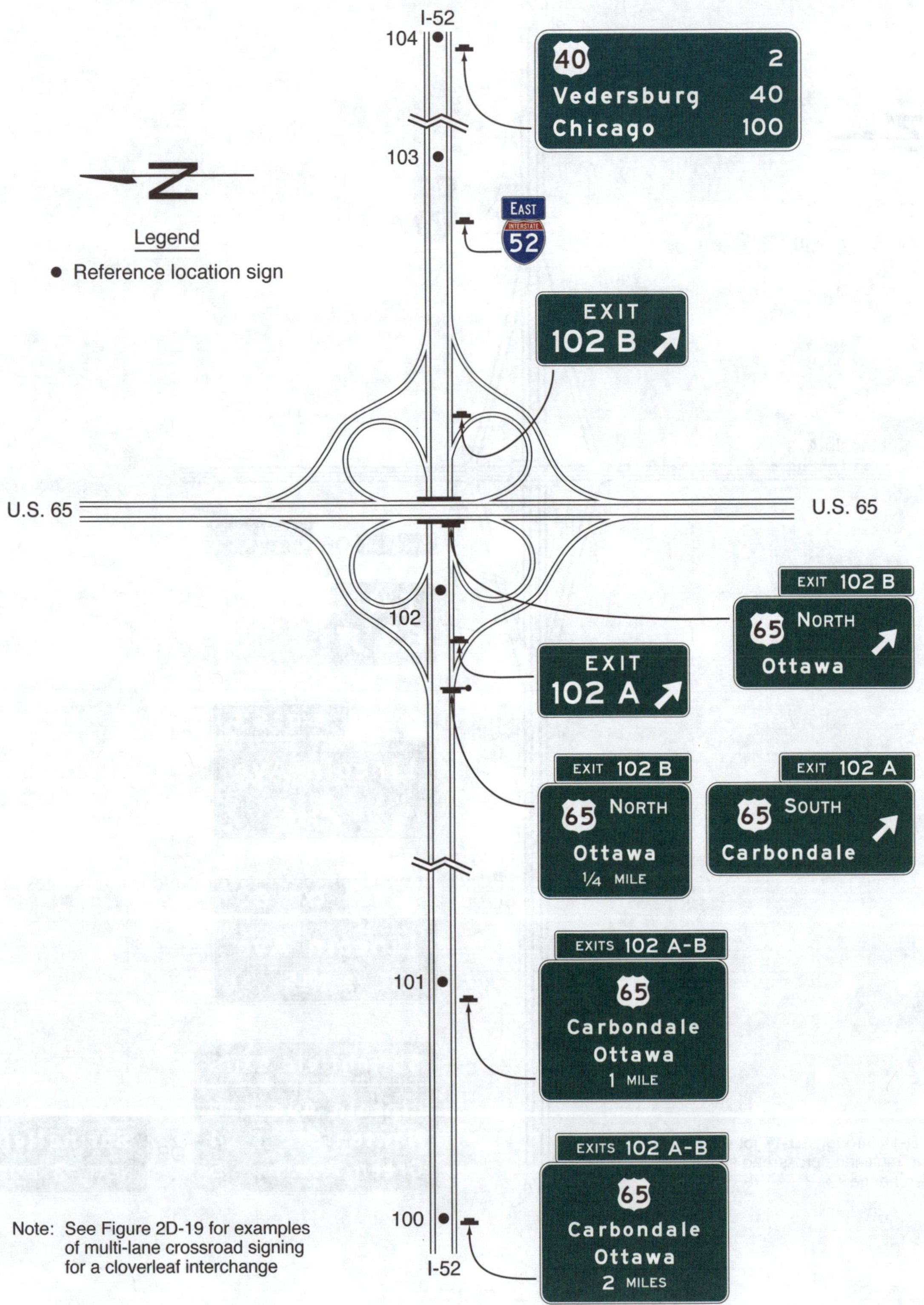

Section 2E.34 Cloverleaf Interchange with Collector-Distributor Roadways

Support:

01   An example of guide signs for a full cloverleaf interchange with collector-distributor roadways is shown in Figure 2E-30.

Guidance:

02   *Destination names and route numbers shown on the collector-distributor roadway signing should be the same as those used on the upstream Interchange Advance guide signs on the main roadway.*

Standard:

03   **Exit Direction signs at exits from the collector-distributor roadways shall be overhead and located at the theoretical gore of the collector-distributor roadway and the exit ramp.**

Guidance:

04   *Exits from the collector-distributor roadways should be numbered with an appropriate suffix. If the exits from a collector-distributor roadway are numbered, the Interchange Advance guide and Exit Direction signs on the main roadway should include, in addition to two place names, their corresponding exit number and suffixes with the plural EXITS in the Exit Number (E1-5P series) plaque. If only the exit from the main roadway is numbered, the Interchange Advance guide and Exit Direction signs on the main roadway should use the singular EXIT in the Exit Number plaque. If interchange exit numbering is not used, the Interchange Advance guide signs on the main roadway should use the singular EXIT in the distance messages.*

Section 2E.35 Partial Cloverleaf Interchange

Support:

01   An example of guide signs for a partial cloverleaf interchange is shown in Figure 2E-31.

Guidance:

02   *For a partial cloverleaf with only one exit roadway in a direction of travel, where the main roadway passes under the crossroad and the exit roadway is located beyond the overcrossing structure, the overhead Exit Direction sign should be placed either on the overcrossing structure (see Figure 2E-31) or on a separate structure located immediately in front of the overcrossing structure.*

Support:

03   Partial cloverleaf interchanges with successive exit ramps from the same direction of travel are signed the same as cloverleaf interchanges for that direction of travel (see Section 2E.33).

Section 2E.36 Collector-Distributor Roadways for Successive Interchanges

Support:

01   Examples of guide signs for a collector-distributor roadway that provides access to multiple interchanges are shown in Figure 2E-32. Section 2J.09 contains provisions for General Service and Specific Service signs.

Guidance:

02   *Where access to successive interchanges is provided from a single collector-distributor roadway, the number of lines of destination information displayed on the major guide signs on the main roadway approach to the collector-distributor roadway should comply with the provisions of Section 2E.15. Where additional destinations are displayed on the main roadway, those destinations should be displayed on Supplemental guide signs (see Section 2E.51) on the approach to the collector-distributor roadway.*

03   *Where exit numbering is used, the exit numbers for exits accessed from the collector-distributor roadway should be displayed on the main roadway guide signs.*

04   *An Exit Gore sign (see Section 2E.26) should be placed in the gore where the collector-distributor roadway departs from the main roadway.*

05   *Interchange guide signing along the collector-distributor roadway should comply with the provisions for interchange signing in this Chapter.*

Section 2E.37 Freeway-to-Freeway Interchanges

Support:

01   Freeway-to-freeway interchanges are major decision points where the effect of taking a wrong ramp cannot be easily corrected. Reversing direction on the connecting freeway or reentering to continue on the intended course is usually not possible. Examples of guide signs for freeway-to-freeway interchanges are shown in Figure 2E-33.

Guidance:

02   *The sign messages should contain only the route shield, cardinal direction, and the name of the next control city on the route. Arrows should point as indicated in Section 2D.08, except where Overhead Arrow-per-Lane or Diagrammatic Advance signs are used in accordance with the provisions of Sections 2E.39 through 2E.41.*

Figure 2E-30. Example of Guide Signs for a Full Cloverleaf Interchange with Collector-Distributor Roadways

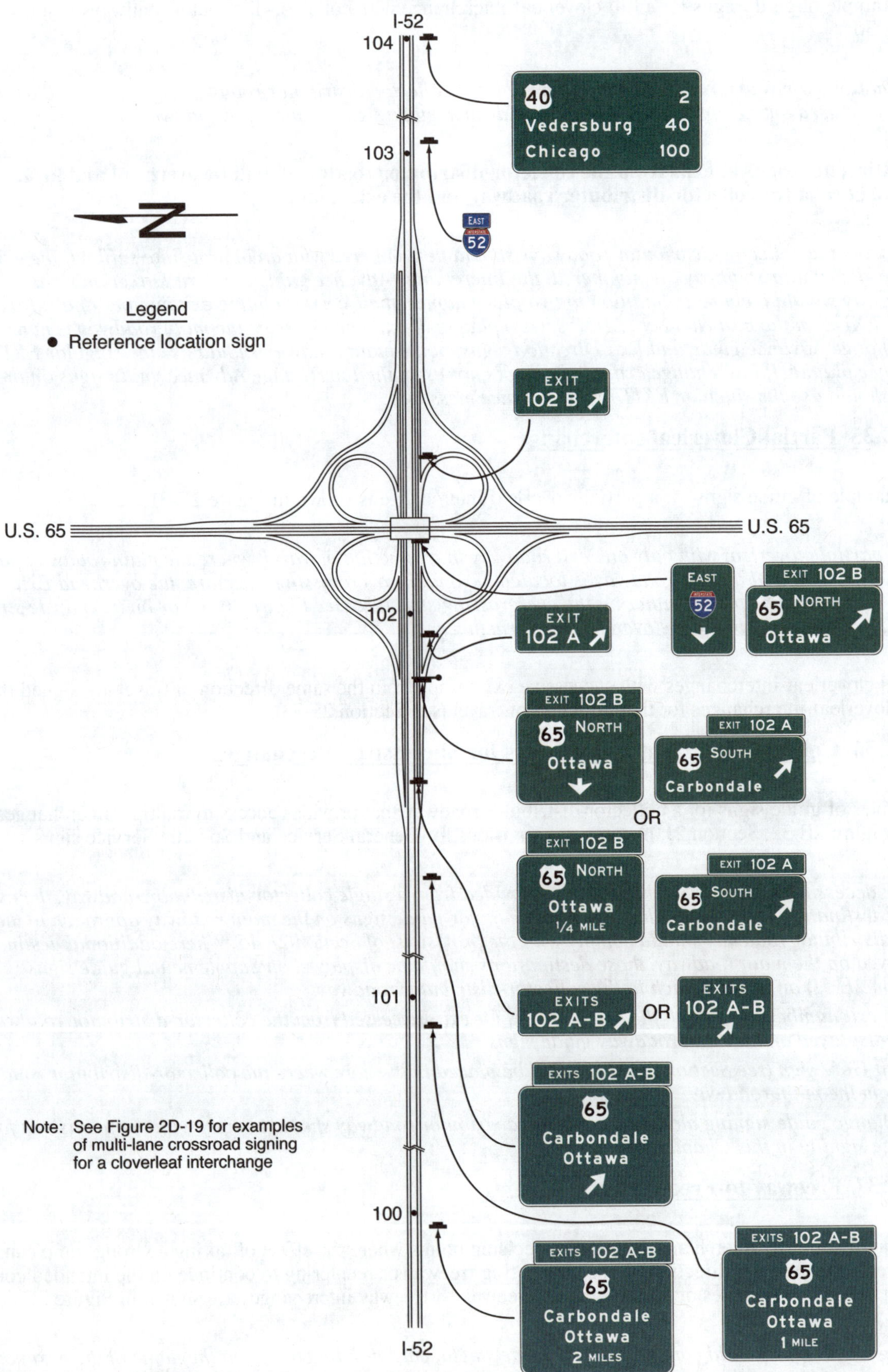

Figure 2E-31. Example of Guide Signs for a Partial Cloverleaf Interchange

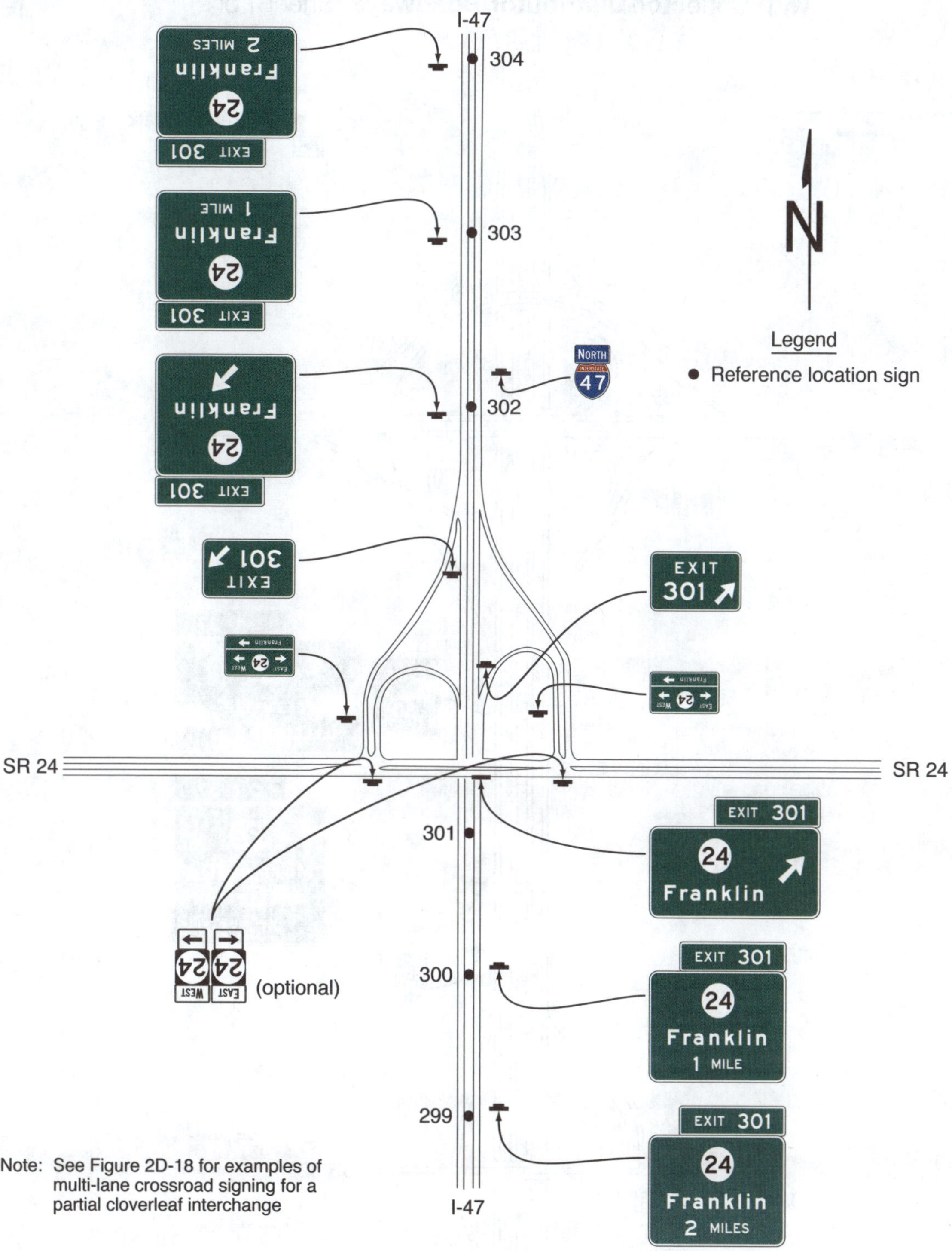

Figure 2E-32. Examples of Guide Signs for Successive Interchanges with Collector-Distributor Roadways (Sheet 1 of 2)

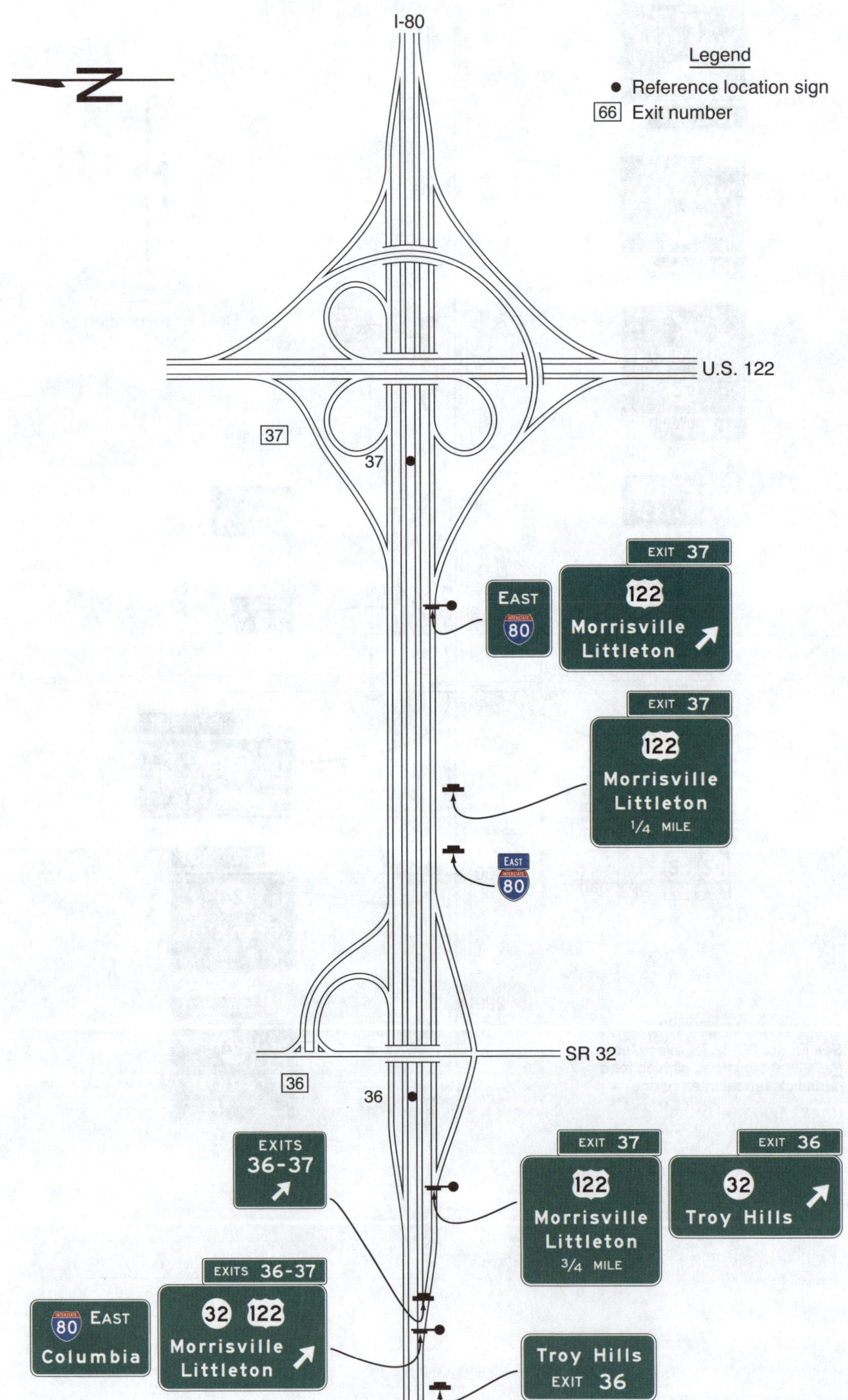

Figure 2E-32. Examples of Guide Signs for Successive Interchanges with Collector-Distributor Roadways (Sheet 2 of 2)

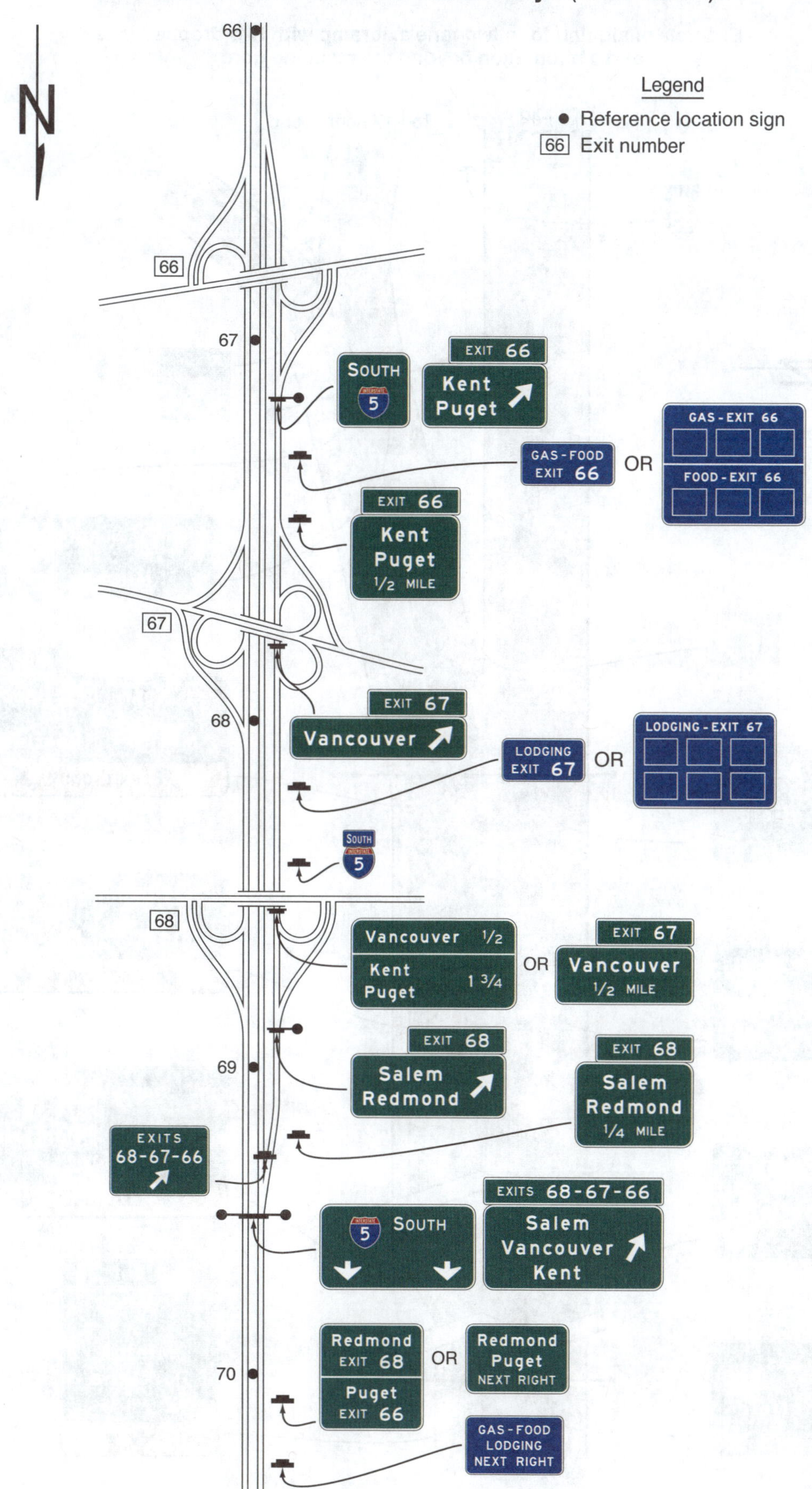

Figure 2E-33. Examples of Guide Signs for a Freeway-to-Freeway Interchange
(Sheet 1 of 2)

A – Example of signing for a two-lane exit ramp with two dropped lanes and a bifurcation beyond the mainline gore

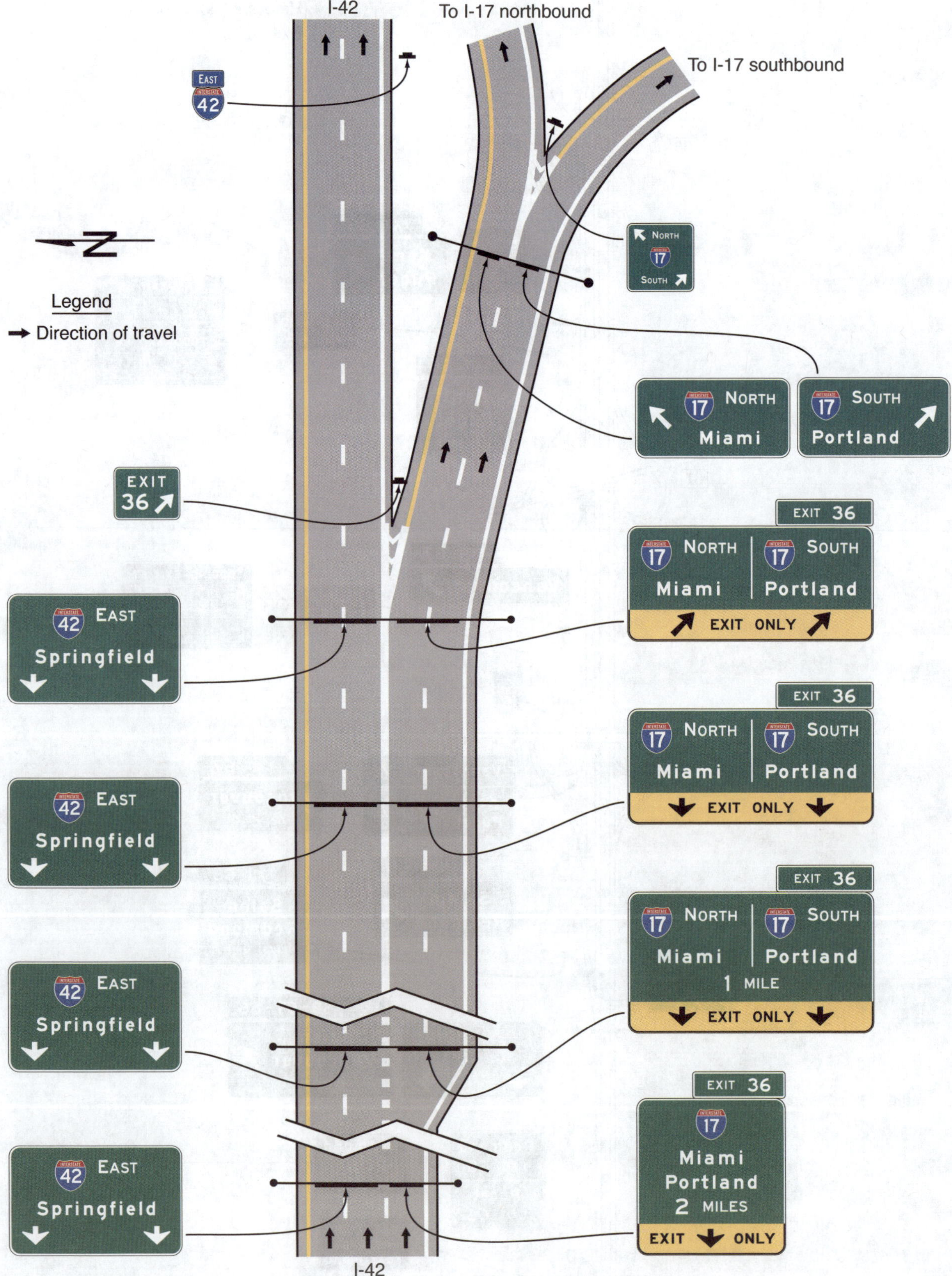

Figure 2E-33. Examples of Guide Signs for a Freeway-to-Freeway Interchange
(Sheet 2 of 2)

B – Example of signing for successive exit ramps with a dropped lane at the second exit

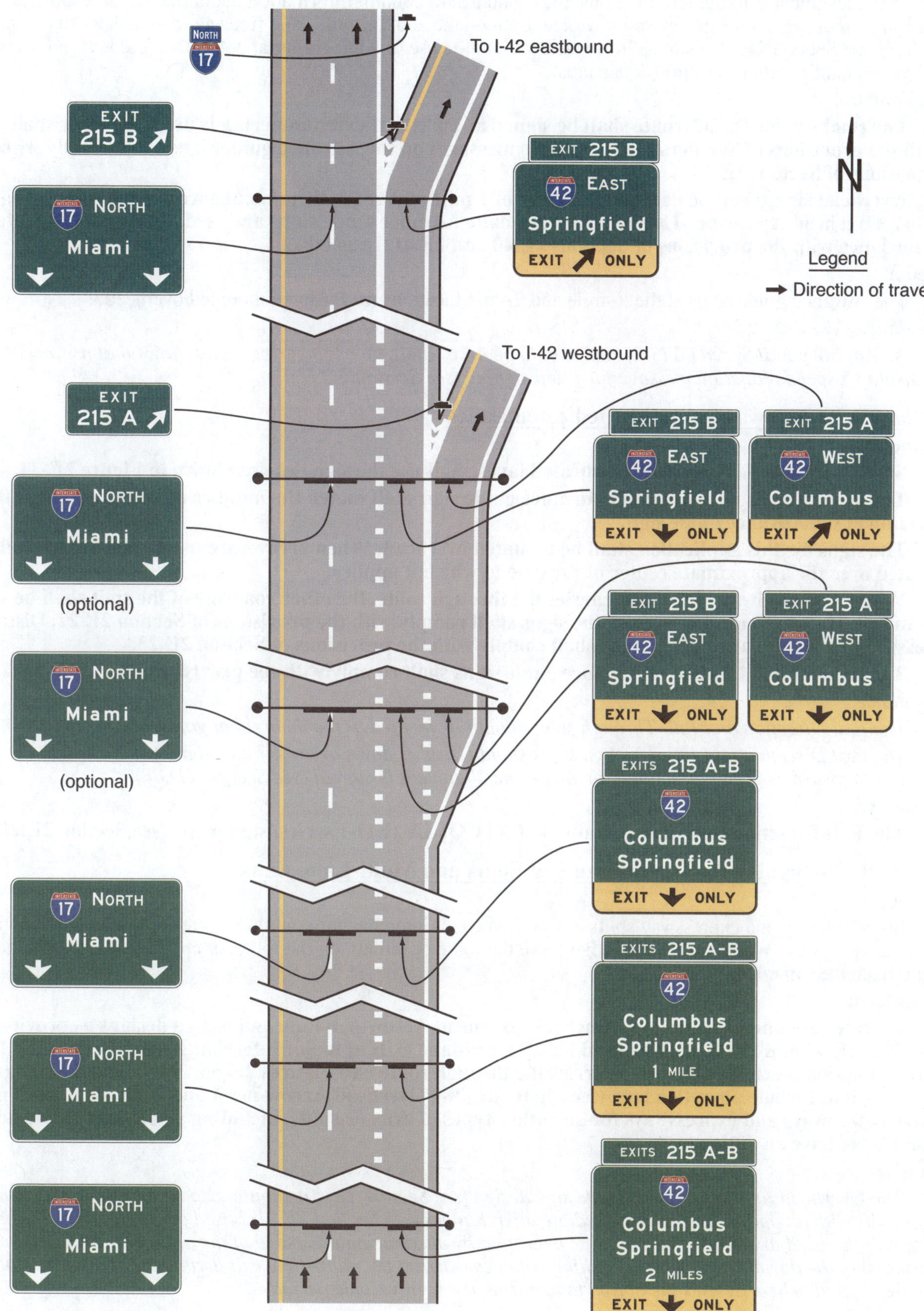

Support:

03   An off-route movement is the movement that does not follow the through route. Drivers might not expect the off-route movement to be to the left or an optional lane at a split (see Figures 2E-38 and 2E-39). Section 2E.22 contains information about the use of the Left Exit Number (E1-5fP through E1-5kP) plaque at splits where the off-route movement is to the left. Sections 2E.39 and 2E.40 contain information about the use of Overhead Arrow-per-Lane guide signs for freeway splits with an option lane and for multi-lane freeway-to-freeway exits having an option lane. Section 2E.41 contains information about the use of a Diagrammatic Advance guide sign for complex geometric configurations at ramp departures.

Standard:

04   **The roadway for the off-route shall be signed as an exit. If exit numbering is used, the signs shall comply with the provisions of Section 2E.22. Distance messages on the Advance guide signs shall comply with the provisions of Section 2E.23.**

05   **Overhead signs shall be used at a distance of 1 mile and at the theoretical gore of each connecting ramp. When Overhead Arrow-per-Lane or Diagrammatic Advance guide signs are used, they shall be located in accordance with the provisions of Sections 2E.40 and 2E.41, respectively.**

Option:

06   The Advance guide signs at the ½-mile and 2-mile locations may also be mounted overhead.

Guidance:

07   *An Advisory Exit Speed (W13-2) sign should be used where an engineering study shows that it is necessary to display a speed reduction message for ramp signing (see Section 2C.12).*

Section 2E.38 Freeway Split with Dedicated Lanes

Standard:

01   **Signing for freeway splits with dedicated lanes shall use the sign designs shown in Figure 2E-34.**

02   **The arrows on each Interchange Advance guide sign shall match the number of lanes present at the location of the Advance guide sign.**

03   **The signs for this application shall be mounted overhead. When arrows are used, each arrow shall be located over the approximate center of the lane to which it applies.**

04   **Where one roadway of the split carries the through route, the other roadway of the split shall be signed as an exit. If exit numbering is used, the signs shall comply with the provisions of Section 2E.22. Distance messages on the Advance guide signs shall comply with the provisions of Section 2E.23.**

05   **The number and location of Advance guide signs shall comply with the provisions of Section 2E.23.**

Guidance:

06   *The Exit Direction and Pull-Through signs should be located at the theoretical gore.*

07   *The Exit Direction and Pull Through signs should display down arrows if the alignment is straight or diagonal upward-pointing directional arrows if the alignment is curved (see Section 2D.08).*

Standard:

08   **The Exit Direction sign shall contain the EXIT ONLY (E11-1 series) sign panel (see Section 2E.28).**

Section 2E.39 Signing for Option Lanes at Splits and Multi-Lane Exits

Support:

01   Some freeway and expressway splits or multi-lane exit interchanges contain an interior option lane serving both movements in which traffic can either leave the route or remain on the route, or choose either destination at a split, from the same lane.

Standard:

02   **On freeways and expressways, either the Overhead Arrow-per-Lane guide sign designs as provided in Sections 2E.40 and 2E.41 shall be used for all multi-lane exits at major interchanges (see Section 2E.11) that have an optional exit lane that also carries the through route (see Figures 2E-36, 2E-37, and 2E-42) and for all splits that include an option lane (see Figure 2E-38). Overhead Arrow-per-Lane guide signs shall not be used on freeways and expressways for any other types of exits or splits, including single-lane exits and splits that do not have an option lane.**

Guidance:

03   *The Overhead Arrow-per-Lane guide sign design (see Section 2E.40) should also be considered for multi-lane exits with an option lane at intermediate interchanges (see Section 2E.11) based on such factors as the extent of the need to optimize the mainline operation by maximizing the usage of the option lane, the extent of the period(s) of the day during which the exiting volumes warrant the multi-lane exit arrangement, and the nature of the traffic that primarily uses the option lane during the high-volume periods.*

Figure 2E-34. Example of Guide Signs for a Split with Dedicated Lanes

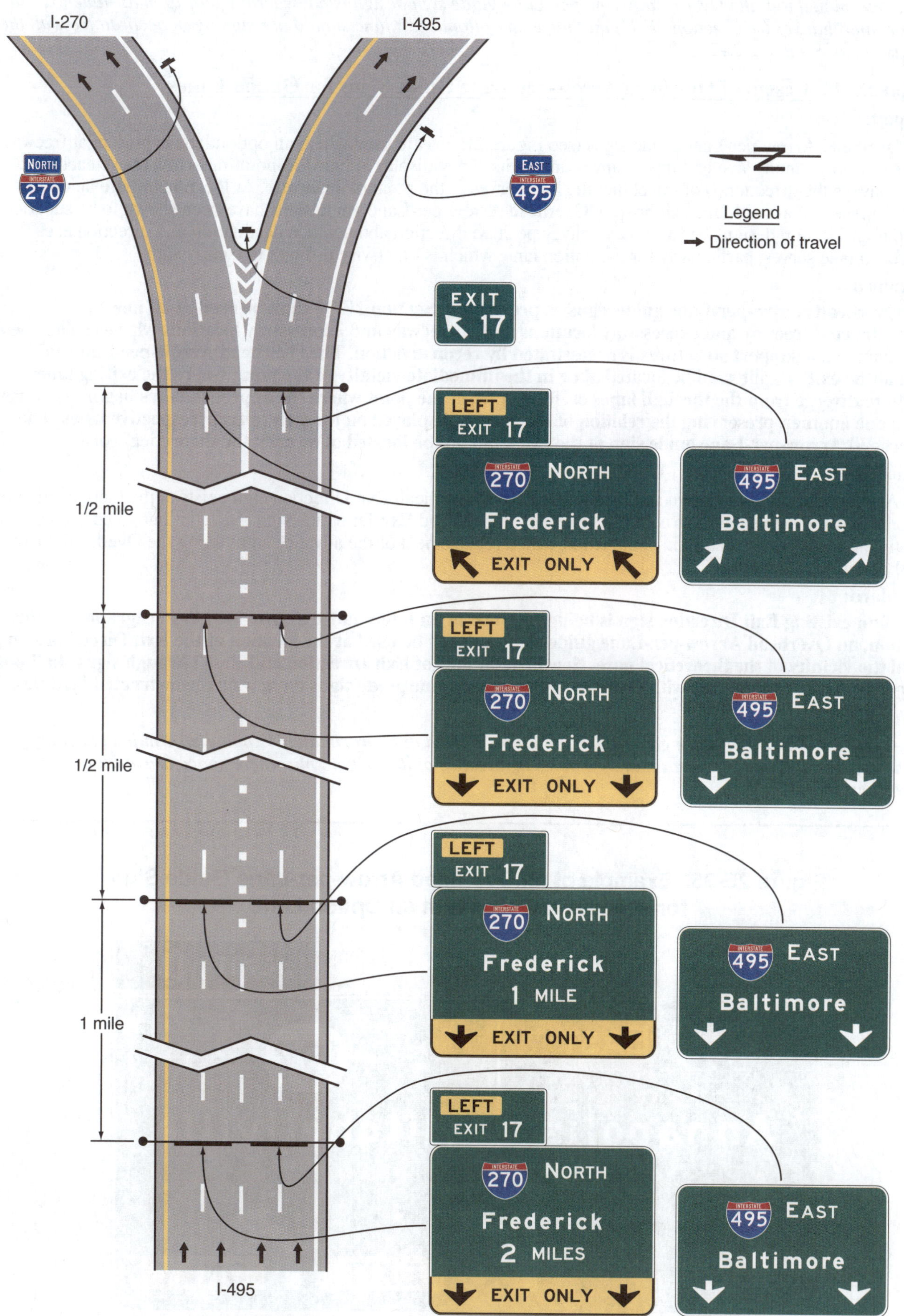

04   *Signing at intermediate interchanges (see Section 2E.11) that have an optional exit lane at which it has been determined that the Overhead Arrow-per-Lane guide sign design is not warranted or at multi-lane exits at minor interchanges (see Section 2E.11) that have an optional exit lane should use signing in accordance with the provisions of Section 2E.42.*

Section 2E.40 Design of Overhead Arrow-per-Lane Guide Signs for Option Lanes

Support:

01   Overhead Arrow-per-Lane guide signs (see Figure 2E-35) are used where an option lane is present at freeway and expressway multi-lane exit interchanges and splits. They display an upward-pointing arrow above each lane that conveys the direction(s) of travel that the lane serves at the point of departure. At locations where an option lane is present at a multi-lane exit or split, Overhead Arrow-per-Lane guide signs have been shown to be superior to other guide sign designs because they convey positive direction about which destination and direction each approach lane serves, particularly for the option lane, which is otherwise difficult to clearly sign.

Standard:

02   **Overhead Arrow-per-Lane guide signs as provided in Section 2E.39 shall be used at all new or reconstructed freeway and expressway locations and at freeway and expressway locations where replacement of existing sign support structures is necessitated by reconstruction. The Overhead Arrow-per-Lane guide sign at the exit or split shall be located at or in the immediate vicinity of the point where the exiting lanes begin to diverge from the through lanes or, for a split, at the point where the approach lanes begin to diverge from one another, preserving the relation of the arrows displayed on the sign to their respective lanes. The Overhead Arrow-per-Lane guide sign at the exit shall not be located at or near the theoretical gore.**

Option:

03   At existing or non-reconstructed locations where an overhead Exit Direction sign exists at the theoretical gore, and the existing sign support structure is retained, an overhead Exit Direction sign may continue to be used on the existing sign support structure in conjunction with a replacement of the advance signs using the Overhead Arrow-per-Lane guide sign design.

Standard:

04   **If an existing Exit Direction sign is being retained at an interchange as provided in Paragraph 3 of this Section, an Overhead Arrow-per-Lane guide sign shall not be used at the location of the Exit Direction sign at or in the vicinity of the theoretical gore. New installations of Exit Direction and Pull-Through signs shall not be permitted in conjunction with Overhead Arrow-per-Lane guide signs on new or reconstructed facilities.**

Guidance:

05   *Overhead Arrow-per-Lane guide signs should be located at approximately ½ mile and 1 mile in advance of the exit or split, and at approximately 2 miles in advance of the exit or split where space is available and conditions allow.*

Figure 2E-35. Example of an Overhead Arrow-per-Lane Guide Sign for a Multi-Lane Exit with an Option Lane

Standard:

06   **Overhead Arrow-per-Lane guide signs used on freeways and expressways shall be designed in accordance with the following criteria:**

    A. **Except as provided in Section 2E.42 for partial width Overhead Arrow-per-Lane signs, the sign shall include an upward-pointing (vertical, curved, or bifurcated) arrow for each lane of the approach to the split or exit.**

    B. **The shaft of each arrow shall be located over the approximate center of the lane to which it applies.**

    C. **Arrows for continuing through lanes shall be vertically upward-pointing (see Figure 2E-36) unless the continuing through lanes are on a significantly curved alignment beyond the theoretical gore (see Figure 2E-37).**

    D. **The arrow for a lane that must exit shall be curved in the direction of the exit and shall be accompanied by black-on-yellow EXIT (E11-1a) and ONLY (E11-1b) sign panels adjacent to the lower end of the arrow shaft. The E11-1a and E11-1b sign panels shall not be used for a split of two overlapping routes where neither of the diverging routes is designated as an exit. Where the through lanes curve and the exit continues on a straight alignment, upward-pointing vertical arrows shall be used for the exiting movement and curved arrows for the through movement (see Figure 2E-37).**

    E. **The arrow for an optional exit lane that also carries the through route shall have a single shaft that bifurcates into a vertically upward-pointing arrow and a curving arrow corresponding to the configuration of the through and exit lanes.**

    F. **For splits with an option lane, the arrow for the lane from which either direction of the split can be accessed shall have a single shaft that bifurcates into two upward-pointing curving arrows (see Figure 2E-38).**

    G. **A vertical white line shall be used to separate the route shields and destinations for the two diverging movements from each other.**

    H. **The distance to the exit or split shall be displayed below the off-movement destination on the advance signs at the 1-mile and 2-mile locations.**

    I. **The number of lanes displayed on a sign shall correspond to the number of lanes at the location of that sign. An advance sign shall not depict lanes that are added downstream of the sign location.**

    J. **For numbered exits, the Exit Number (E1-5P) or Left Exit Number (E1-5bP) plaque shall be used at the top of the sign in accordance with Section 2E.23. For unnumbered exits to the left, a LEFT (E1-5mP) plaque shall be added on the top left-hand edge of and adjacent to the sign.**

Guidance:

07   *Overhead Arrow-per-Lane guide signs used on freeways and expressways should be designed in accordance with the following additional criteria:*

    A. No more than one destination should be displayed for each movement, and no more than two destinations should be displayed per sign.

    B. The arrowhead(s) for the diverging movement should be positioned lower on the sign than the arrowhead(s) for the movement that continues straight ahead, independent of which movement carries the through route. Where the movements are freeway or expressway splits rather than exits, the arrowheads should be positioned at approximately the same height on the sign.

    C. Route shields, cardinal directions, and destinations should be positioned on the sign such that they are clearly related to the arrowhead(s) for the movement to which they apply.

    D. The cardinal direction should be placed adjacent to the route shield for exits or splits leading in a single cardinal direction.

    E. The vertical white line that is used to separate the route shields and destinations for the two diverging movements from each other should not descend below the top of the arrowheads for the through lanes, and should be positioned approximately halfway between the diverging arrowheads for the optional movement lane (see Figure 2E-35).

Standard:

08   **Overhead Arrow-per-Lane guide signs shall not be used to depict a downstream split of an exit ramp on a sign located on the mainline.**

Support:

09   Specific guidelines for more detailed design of Overhead Arrow-per-Lane guide signs are contained in the "Standard Highway Signs" publication (see Section 1A.05).

Figure 2E-36. Example of Overhead Arrow-per-Lane Guide Signs for a Two-Lane Exit to the Right with an Option Lane

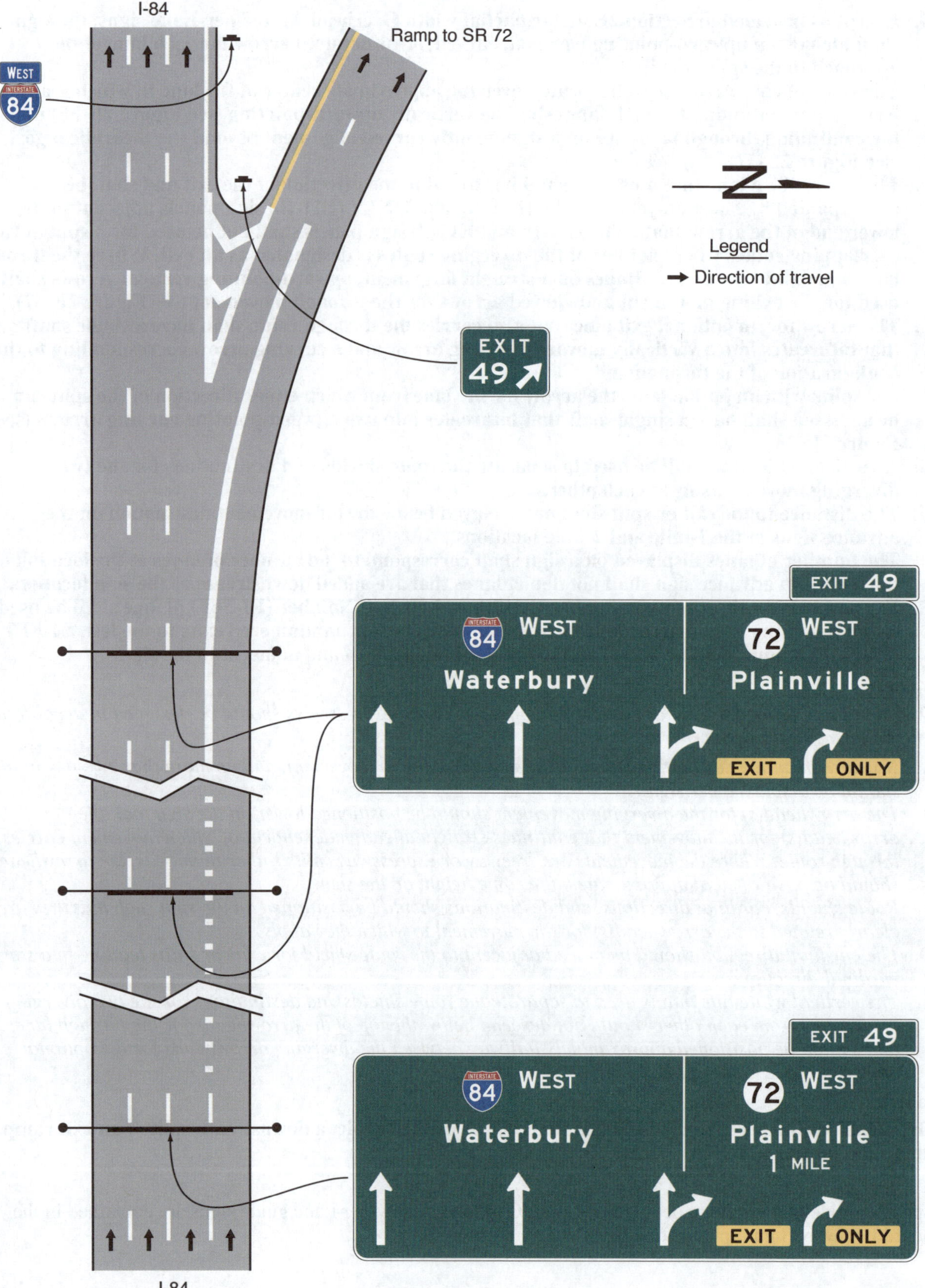

Figure 2E-37. Example of Overhead Arrow-per-Lane Guide Signs for a Two-Lane Exit to the Right with an Option Lane (Through Lanes Curve to the Left)

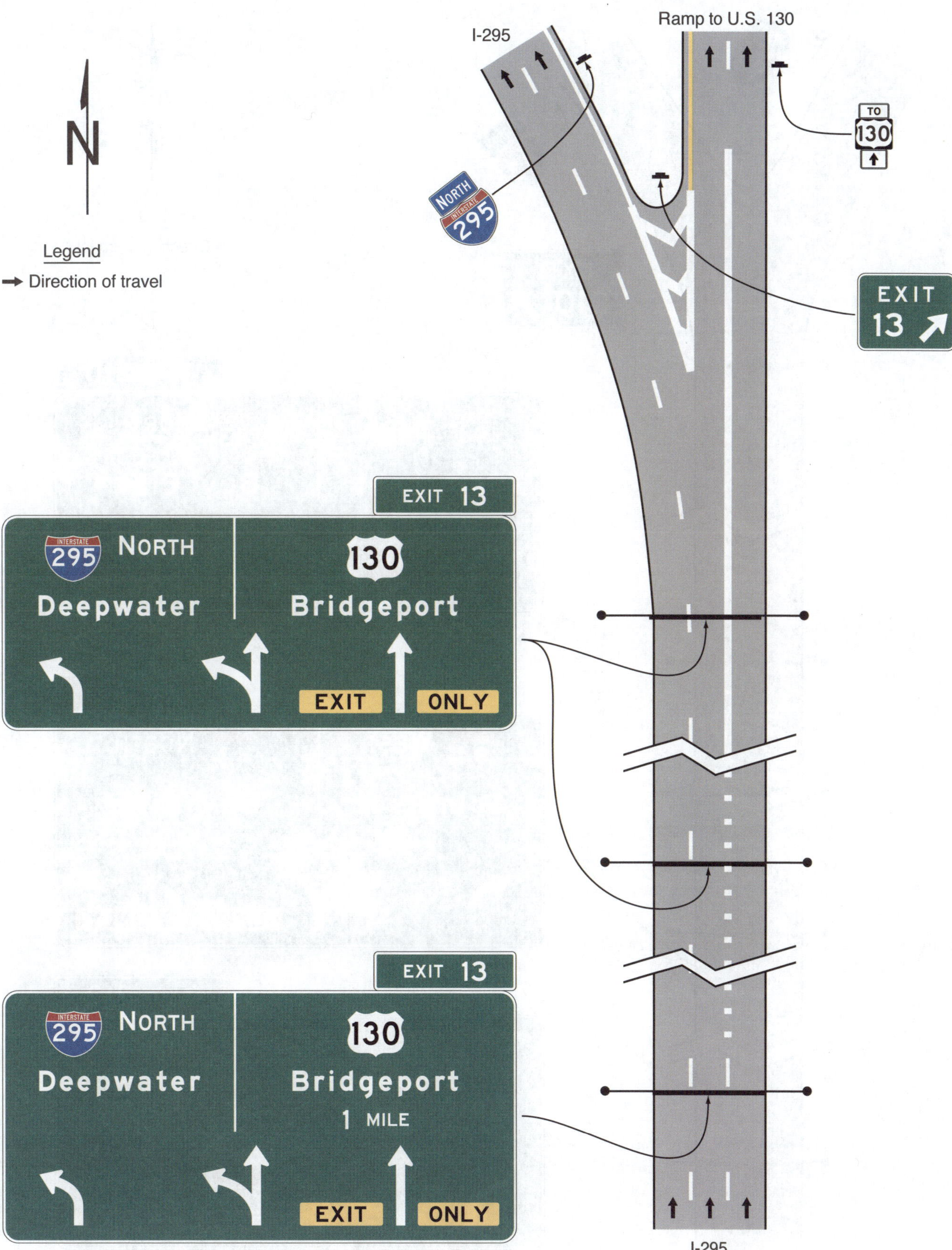

Figure 2E-38. Example of Overhead Arrow-per-Lane Guide Signs for a Split with an Option Lane

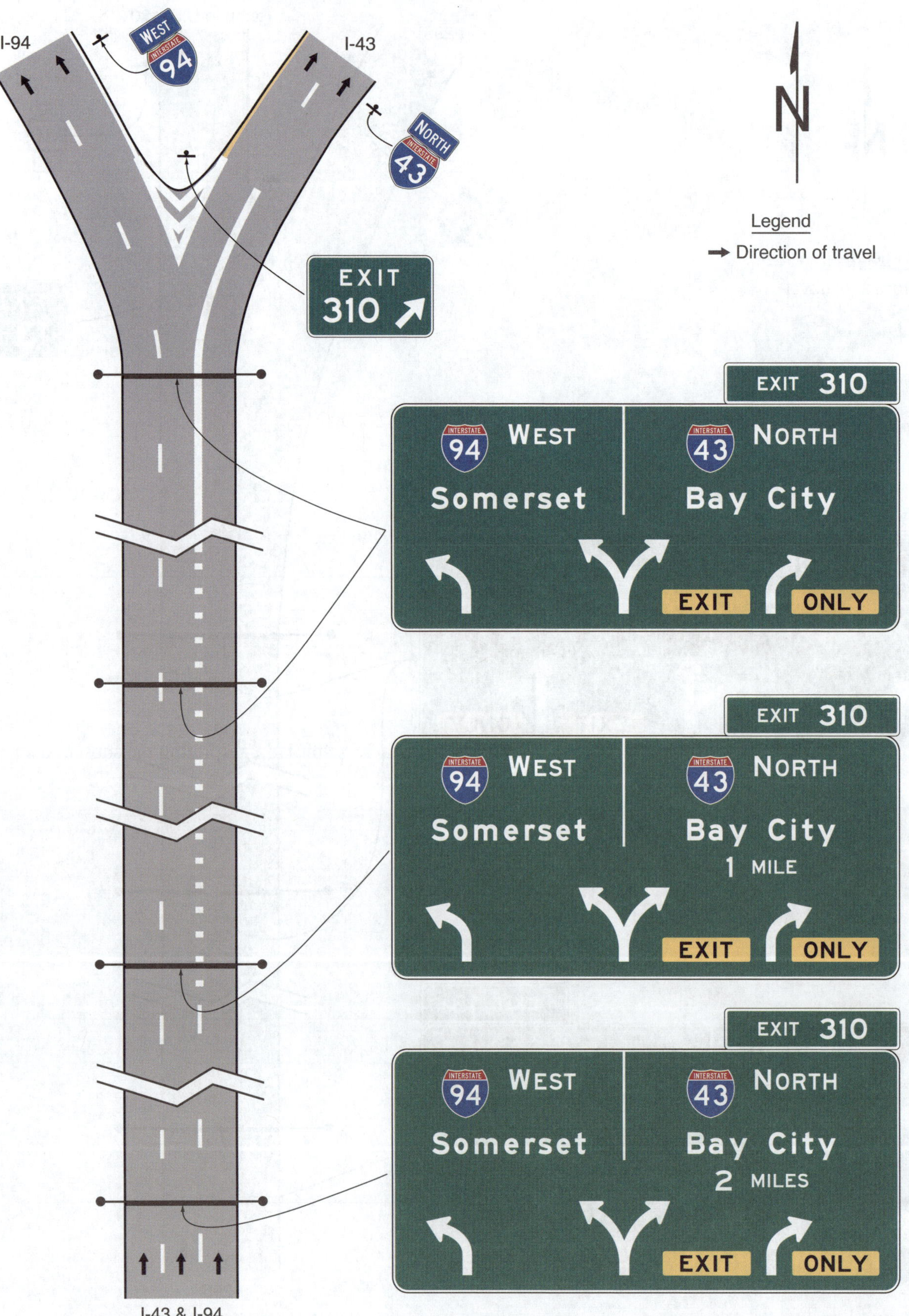

Standard:

10   **The arrow heights for Overhead Arrow-per-Lane guide signs on freeways and expressways shall be as shown in Table 2E-6.**

Table 2E-6. Overhead Arrow-per-Lane Arrow Height Based on Principal Legend Letter Height

Principle Legend Letter Height	Through Arrow	Turn Arrow	Through with Turn Arrow	Split Arrow
20	48	36	48	44
16 or less	40	30	40	33.33

Note: Letter and arrow heights are shown in inches.

Option:

11   Where extra emphasis of an especially low advisory ramp speed is needed, an EXIT XX MPH (E13-2) sign panel (see Figure 2E-14) may be placed below the applicable destination legend to supplement, but not to replace, the exit or ramp advisory speed warning signs.

12   Warning Beacons in compliance with the provisions of Section 2E.25 may be used with the E13-2 sign panel.

Support:

13   An example of guide signing for a narrow gore at a split with an option lane is shown in Figure 2E-39, and an example of guide signing for a narrow gore at a two-lane exit with an option lane is shown in Figure 2E-40.

Option:

14   Where there is 800 feet or more between the beginning of the lane diverge and the theoretical gore, signs indicating the destinations allowed by each lane may be added in the vicinity of the theoretical gore to reinforce positive guidance (see Figures 2E-39 and 2E-40).

Section 2E.41 Design of Freeway and Expressway Diagrammatic Advance Guide Signs

Support:

01   The Diagrammatic Advance guide sign (see Figure 2E-41) is a guide sign that shows a simplified graphic view of the exit departure arrangement in relationship to the main highway at an interchange. Its purpose is to provide advance notice of complex or unexpected road geometry or ramp departures at an interchange and/or depict successive decision points where additional context might be helpful to interpreting the subsequent primary Interchange Advance guide signs. Unlike Diagrammatic signs that were included in previous editions of this Manual, the Diagrammatic Advance guide sign does not depict which or the number of specific lanes that serve a particular destination or depict lanes added or reduced.

Option:

02   A Diagrammatic Advance guide sign may be used in advance of the interchange guide sign sequence, or in lieu of an Interchange Advance guide sign located 2 miles in advance of the exit, to supplement conventional or Overhead Arrow-per-Lane guide signs used for a downstream interchange.

Standard:

03   **Diagrammatic Advance guide signs shall be designed in accordance with the following criteria:**

    A. The graphic legend shall be of a plan view showing a simplified schematic graphic of the relative through and off-ramp movements.

    B. No symbols or route shields shall be used as a substitute for arrowheads.

    C. They shall not be installed at the Exit Direction sign location (see Section 2E.25).

    D. The EXIT ONLY sign panel shall not be used on Diagrammatic Advance guide signs in advance of the interchange.

    E. For numbered exits, the Exit Number (E1-5P) or Left Exit Number (E1-5bP) plaque shall be used at the top of the sign in accordance with Section 2E.22. For unnumbered left exits, the LEFT (E1-5aP) plaque shall be used at the top left edge of the sign.

    F. The graphic shall not depict deceleration or auxiliary lanes.

    G. Arrow shafts shall not contain lane lines.

    H. Destination legends for off-movements shall be positioned to the side of the arrow from which the ramp departs.

Figure 2E-39. Example of Guide Signing for a Narrow Gore at a Split with an Option Lane

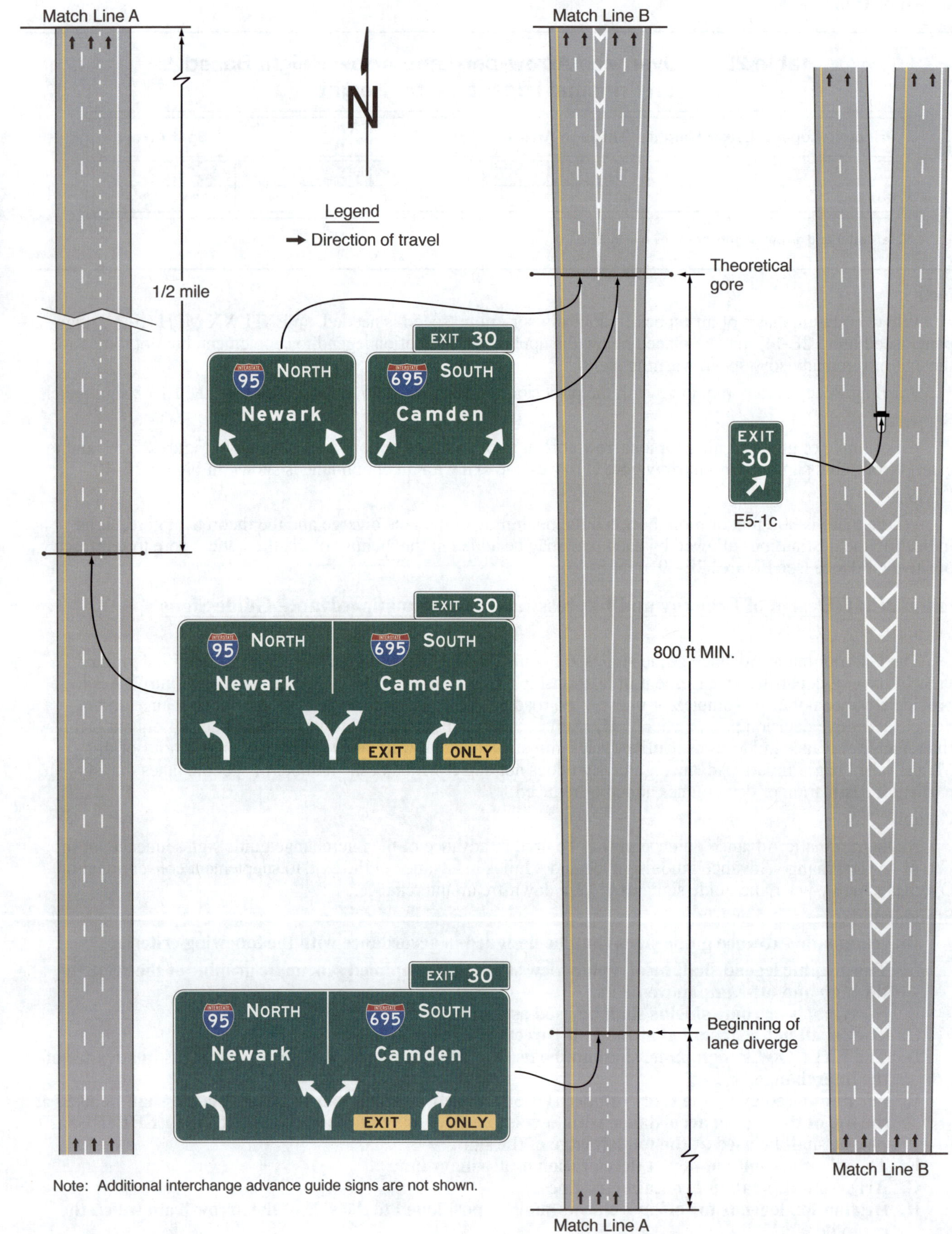

Note: Additional interchange advance guide signs are not shown.

Figure 2E-40. Example of Guide Signing for a Narrow Gore at a Two-Lane Exit with an Option Lane

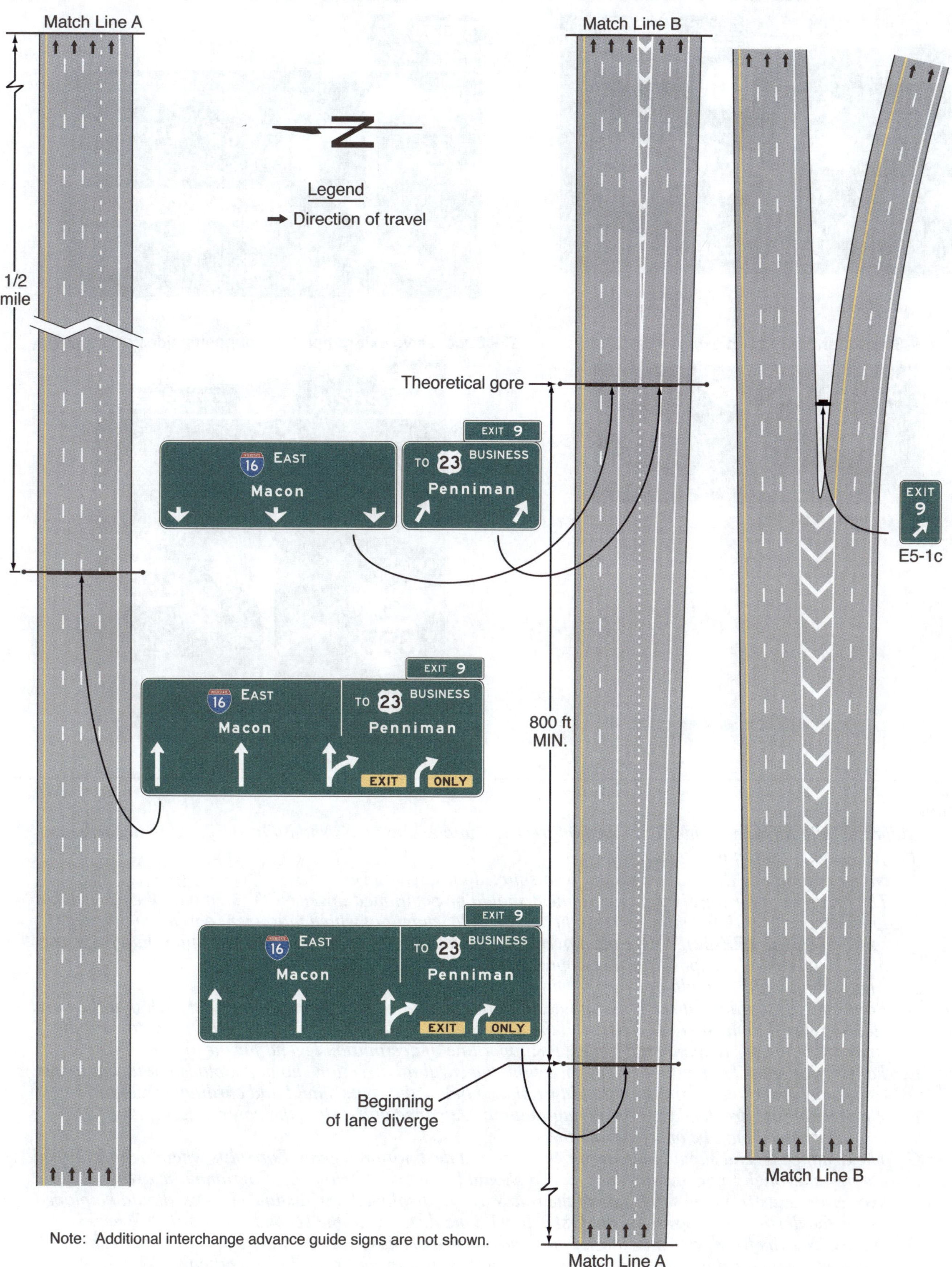

Note: Additional interchange advance guide signs are not shown.

Figure 2E-41. Examples of Diagrammatic Advance Guide Signs

A – Through route on an off-movement

B – Through route with complex geometry and reduced speed on an off-movement

C – Split at terminus of an expressway route

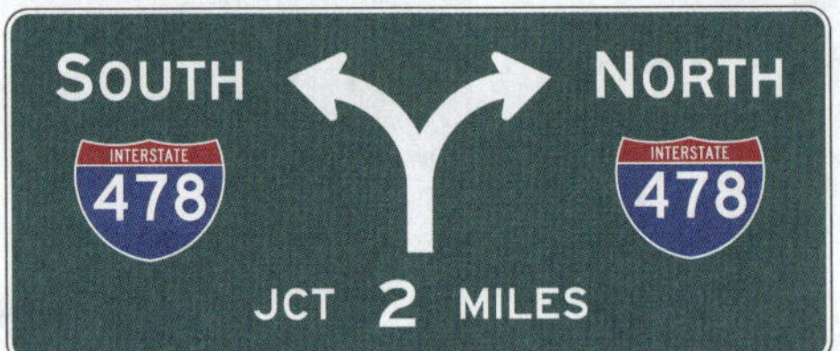

D – Successive exit ramps from opposite sides of a roadway

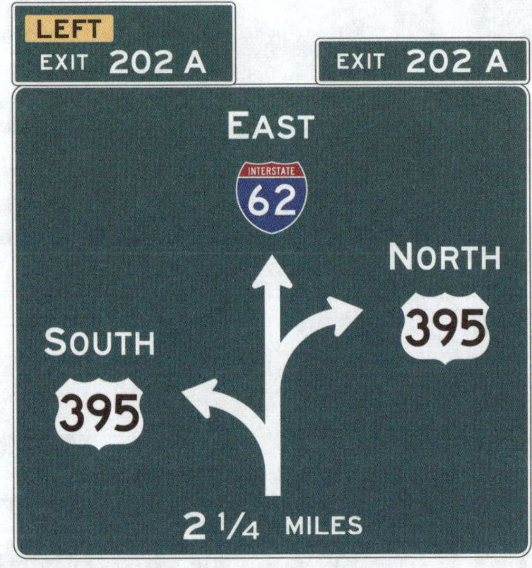

Guidance:

04 *Diagrammatic Advance guide signs used on freeways and expressways should be designed in accordance with the following additional criteria:*

A. No more than one destination should be displayed for each movement.

B. The arrowhead for the diverging movement should be positioned lower on the sign than the arrowhead for the movement that continues straight ahead, independent of which movement carries the through route (see Figure 2E-42). Where the movements are freeway or expressway splits rather than exits, the arrowheads should be positioned at approximately the same height on the sign.

C. Arrow shaft widths should not vary for different movements.

D. Route shields, cardinal directions, and destinations should be positioned on the sign such that they are clearly related to the arrowhead(s), and the arrowhead for the off movement should point toward the route shield or, for unnumbered routes, the upper line of destination legend for the off movement.

E. For exits or splits leading in a single direction, the cardinal direction should be placed adjacent to the route shield, and the destination should be placed below the route shield and cardinal direction.

F. Where two exits are displayed on a Diagrammatic Advance guide sign, the control destination for the through route should be omitted from the sign.

G. The distance legend should be placed below the exit destination legend. For splits where neither direction carries a through route, the distance legend should be centered below the diagrammatic arrow. Where successive exits from the same side of the roadway are displayed, the distance legend should be placed below the destination legend for the first exit, with the distance to the second exit omitted. Where successive exits from opposite sides of the roadway are displayed, the distance to the first exit should be centered below the diagrammatic arrow, with the distance to the second exit omitted.

Figure 2E-42. Example of Diagrammatic Advance Guide Sign Use

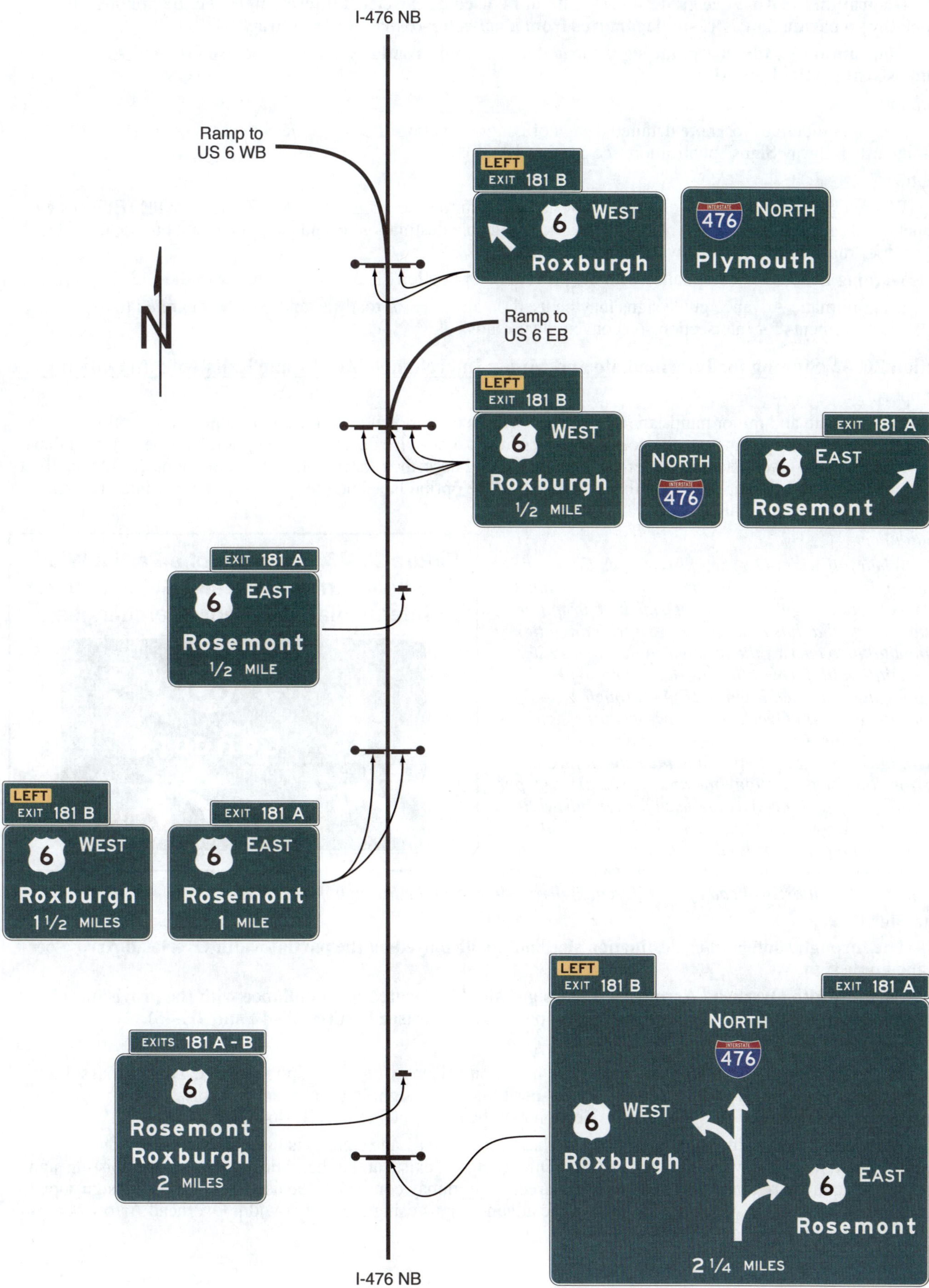

Standard:

05   **Diagrammatic Advance guide signs shall not be used at cloverleaf interchanges for the purpose of depicting separate downstream departures from a collector-distributor roadway.**

06   **Diagrammatic Advance guide signs located on the main roadway shall not be used to depict a downstream split of an exit ramp.**

Support:

07   Specific guidelines for more detailed design of Diagrammatic Advance guide signs are contained in the "Standard Highway Signs" publication (see Section 1A.05).

Option:

08   Where extra emphasis of an especially low advisory ramp speed is needed, an EXIT XX MPH (E13-2) sign panel (see Figure 2E-14) may be placed below the applicable destination legend to supplement, but not to replace, the exit or ramp advisory speed warning signs.

09   Warning Beacons in compliance with the provisions of Section 2E.25 may be used with the E13-2 sign panel.

10   Diagrammatic Advance guide signs may be used on any class of roadway and may be modified to depict relative movements for intersections on conventional roads.

Section 2E.42 <u>Signing for Intermediate and Minor Interchange Multi-Lane Exits with an Option Lane</u>

Support:

01   Intermediate and minor multi-lane exits might have an operational need for the presence of an option lane for only the peak period during which excessive queues might otherwise develop if the option lane were not available. In such cases, the Overhead Arrow-per-Lane guide signing described for option lanes in Sections 2E.39 and 2E.40 might not be practical, depending on the level of use of the option lane and the spacing of nearby interchanges, particularly in non-rural areas.

Guidance:

02   *When full-width Overhead Arrow-per-Lane guide signing is not practical, as described in Paragraph 1 of this Section, signing for an intermediate or minor interchange that has a multi-lane exit with an option lane that also carries the through route should use a partial-width form of the Overhead Arrow-per-Lane guide sign (see Figures 2E-43 through 2E-45). The partial-width Overhead Arrow-per-Lane sign should display arrows only for the option lane and the mandatory exit lane(s) using the same bifurcated arrow type for the option lane and curved arrow type for the exit only lane(s) as are used for the full-width Overhead Arrow-per-Lane sign. The legend displayed for the exit movement should be clearly aligned with the arrows pointing in the direction of the exit and not with the vertical arrow head of the bifurcated arrow depicting the through movement.*

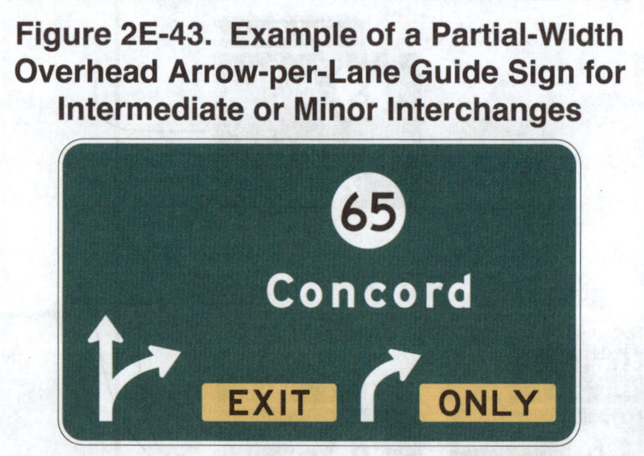

Figure 2E-43. Example of a Partial-Width Overhead Arrow-per-Lane Guide Sign for Intermediate or Minor Interchanges

Standard:

03   **The through route and/or destination shall not be displayed on the partial-width Overhead Arrow-per-Lane guide sign.**

04   **Partial-width Overhead Arrow-per-Lane signs shall be located in compliance with the provisions of Section 2E.40 for full-width Overhead Arrow-per-Lane signs (see Figures 2E-44 and 2E-45).**

Option:

05   At an intermediate or minor interchange that has a multi-lane exit with an option lane that also carries the through route, where full-width Overhead Arrow-per-Lane guide signing is not practical, conventional signing as provided in Paragraphs 7 through 9 of this Section may be used (see Figures 2E-46 and 2E-47).

06   When either full-width or partial-width Overhead Arrow-per-Lane signing is used at existing or non-reconstructed locations where an overhead Exit Direction sign exists at the theoretical gore, and the existing sign support structure is retained, an overhead Exit Direction sign may continue to be used on the existing sign support structure in conjunction with a replacement of the advance signs using the partial-width Overhead Arrow-per-Lane guide sign design (see Figure 2E-44).

Figure 2E-44. Example of Signing for a Two-Lane Intermediate or Minor Interchange Exit with an Option Lane and a Dropped Lane using Partial-Width Overhead Arrow-per-Lane Signs

Figure 2E-45. Example of Signing for a Two-Lane Intermediate or Minor Interchange Exit with Option and Auxiliary Lanes using Partial-Width Overhead Arrow-per-Lane Signs

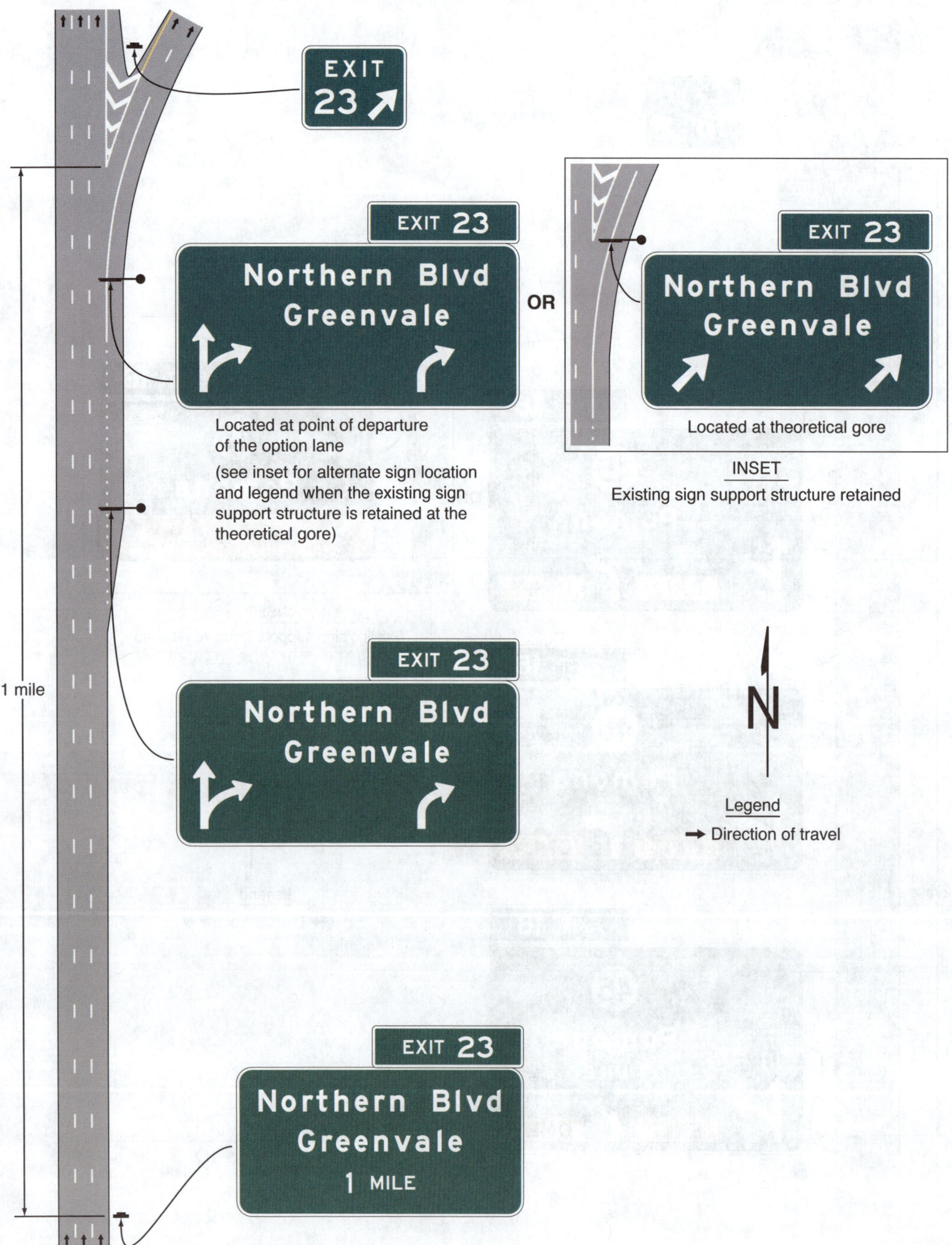

Figure 2E-46. Example of Signing for a Two-Lane Intermediate or Minor Interchange Exit with an Option Lane and a Dropped Lane

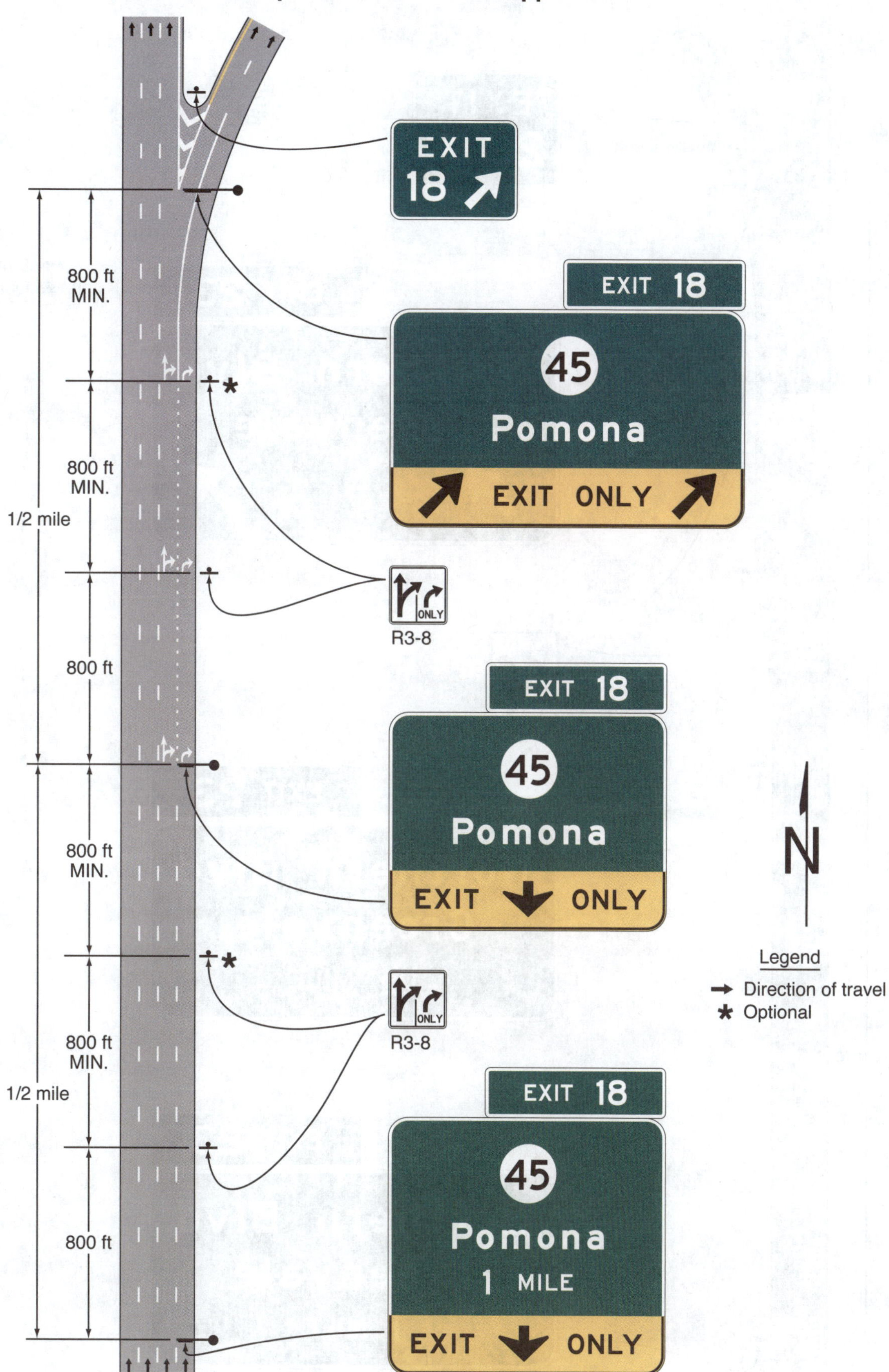

Figure 2E-47. Example of Signing for a Two-Lane Intermediate or Minor Interchange Exit with Option and Auxiliary Lanes

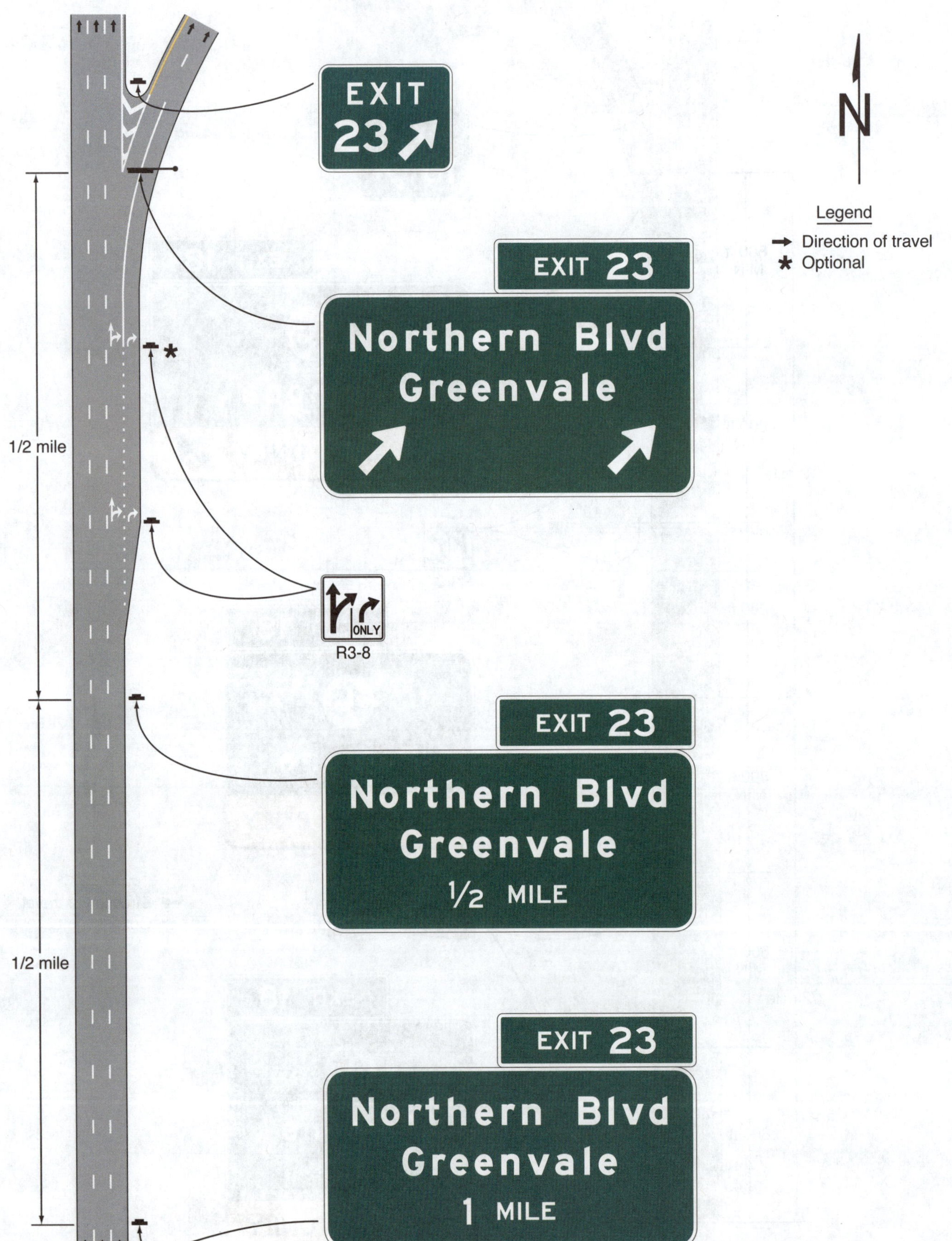

Guidance:

07 *When conventional signing is used, the option lane should not be signed on the Interchange Advance guide signs. For such exits that involve the addition of an auxiliary lane that is not present at the Interchange Advance guide sign locations, but do not involve a lane drop (see Figure 2E-47), a sequence of post-mounted or overhead-mounted Interchange Advance guide signs should be used, located in accordance with the interchange classification (see Section 2E.11). The Exit Direction sign should be located at the theoretical gore and should display a diagonally upward-pointing directional arrow above each lane that departs from the mainline alignment. The Exit Direction sign should not contain the EXIT ONLY legend.*

08 *For such interchanges that also have a lane drop (see Figure 2E-46), the Interchange Advance guide and Exit Direction signs should follow the provisions of Section 2E.28. The Exit Direction sign should be located at the theoretical gore and should contain the EXIT ONLY (E11-1e) sign panel.*

09 *Where the modified Overhead Arrow-per-Lane guide signs are not used, the presence of the option lane should be conveyed by the use of post-mounted lane-use (R3-8 series) signs (see Section 2B.30). When used, the R3-8 signs should be of an appropriate size for their application to optimize their conspicuity. The signs should be located in succession with the Interchange Advance guide signs, where the option and exit lanes have developed (see Figure 2E-46). In cases where the exiting lane or lanes have not developed and the option lane is created by the addition of an auxiliary lane that exits, the R3-8 signs should be located only adjacent to where the lanes have been fully developed and not in advance of the lane or along its transition (see Figure 2E-47).*

Support:

10 The use of a down arrow on overhead freeway or expressway guide signs has been shown to be misinterpreted by road users as an indication of a dedicated lane.

Standard:

11 **Interchange Advance guide signs that are mounted overhead shall not display a down arrow over an option lane.**

Section 2E.43 <u>Number of Signs at an Overhead Installation and Sign Spreading</u>

Guidance:

01 *If overhead signs are warranted, as set forth in Section 2A.13, the number of signs at these locations should be limited to only those essential in communicating pertinent destination information to the road user. Exit Direction signs for a single exit and the Interchange Advance Guide signs should have only one sign with one or two destinations. Regulatory signs, such as speed limits, should not be used in conjunction with overhead guide sign installations. Because road users have limited time to read and comprehend sign messages, there should not be more than three guide signs displayed at any one location either on the overhead structure or its support.*

Option:

02 At overhead locations, more than one sign may be installed to advise of a multiple exit condition at an interchange. If the roadway ramp or crossroad has complex or unusual geometrics, additional signs with confirming messages may be provided to properly guide the road user.

Support:

03 Sign spreading is a concept where major overhead signs are spaced so that road users are not overloaded with a group of signs at a single location. Figure 2E-48 illustrates an example of sign spreading.

Guidance:

04 *Where overhead signing is used, sign spreading should be used at all single-exit interchanges and to the extent possible at multi-exit interchanges. Sign spreading should be accomplished by use of the following:*

 A. The Exit Direction sign should be the only guide sign used in the vicinity of the gore (other than the Exit Gore sign). It should be located overhead near the theoretical gore and generally on an overhead sign support structure.

 B. The Interchange Advance guide sign to indicate the next interchange exit should be placed near the crossroad location. If the crossroad goes over the mainline, the Interchange Advance guide sign should be placed on the overcrossing structure or on a separate structure immediately in front of the overcrossing structure.

Figure 2E-48. Example of Sign Spreading

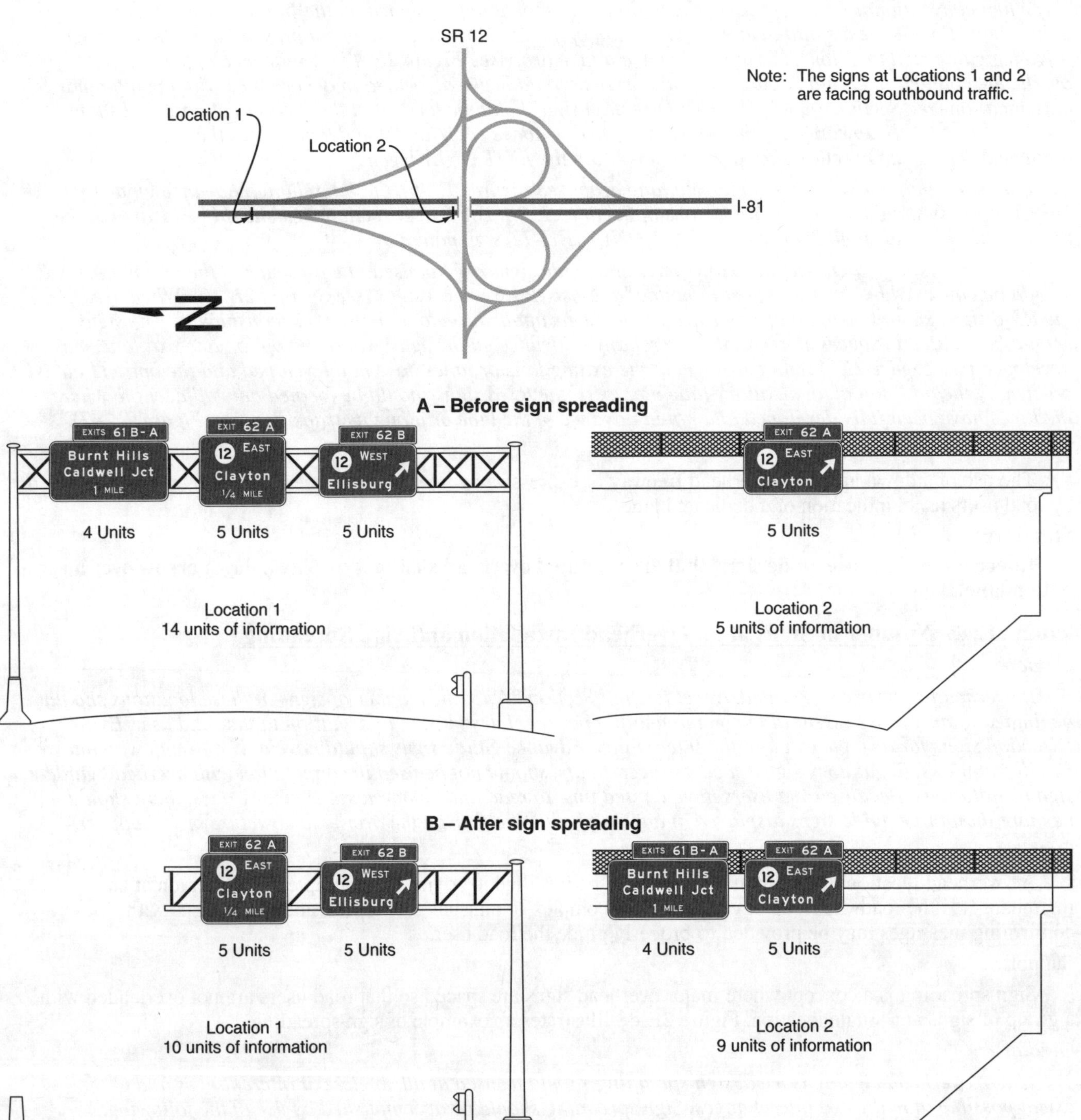

Section 2E.44 Closely-Spaced Interchanges

Support:

01   Section 2E.43 contains information regarding sign spreading where the Exit Direction sign and the Interchange Advance guide sign for the next interchange are mounted overhead. Sign spreading is particularly beneficial where interchanges are closely spaced and overhead signing is used in conjunction with Interchange Sequence signs as provided in Paragraph 2 of this Section.

Guidance:

02   *Interchange Sequence signs (see Section 2E.24) should be used at closely-spaced interchanges. When used, they should identify and show street names and distances for the next two or three exits as shown in Figure 2E-11.*

Standard:

03   **Interchange Advance guide signs for closely-spaced interchanges shall show information for only one interchange.**

Section 2E.45 Guide Signing in Tunnels and Similar Structures

Support:

01   The application of the provisions for freeway and expressway guide signs in tunnels and other similar structures can present unique challenges not encountered elsewhere due to the extended and continuous distances of constrained vertical and horizontal clearances in which to place signs. The effect of these constraints is particularly evident when there are interchange exit ramps inside the tunnel that require guide signing. As a result, it might not always be possible to use the typical layouts for guide signs inside a tunnel. In addition, interchange guide signs might need to be limited to one destination only, with other destinations displayed separately on Supplemental guide signs (see Section 2E.51). Acceptable methods to modify the layout of a sign to fit the space available in a tunnel are provided in Paragraph 2 of this Section.

Option:

02   Overhead-mounted guide signs in tunnels, or in other similar structures with extended constrained vertical and horizontal clearances, may be modified in accordance with the following when needed to accommodate limited vertical clearance available for signs:

    A. Some sign legend elements may be arranged side by side, such as by placing route shields to the left of the destination instead of above.

    B. The Exit Number plaque (see Section 2E.22) may be placed at the right-hand edge of the sign for right exits or at the left-hand edge of the sign for left exits instead of at the top edge of the sign. The legend of the Exit Number plaque may use a reduced letter height of not less than 6 inches for the word EXIT(S) and not less than 12 inches for numerals and suffixes.

    C. Destination and roadway names may be displayed in reduced letter heights of not less than 10.67 inches, when determined acceptable based on consideration of reduced speed and other relevant factors, while maintaining adequate space between the legend and edges of the sign to ensure legibility and quick recognition.

    D. Unusually long destination and roadway names that cannot be adequately shortened or otherwise acceptably abbreviated may be displayed using Series D letters in lieu of Series E(modified).

Standard:

03   **Applicability of the provisions of Paragraph 2 of this Section shall be limited to those signs within the limits of the tunnel or other similar structure and shall not be extended to the approaches to or departures from the tunnel.**

Support:

04   Unlike typical guide signs that are exposed to rain, guide signs in tunnels accumulate grime and residue quickly. This accumulation can reduce visual contrast between legend and background and reduce the retroreflectivity of the sign sheeting. Therefore, guide signs in tunnels generally need more maintenance.

Guidance:

05   *Overhead signs in tunnels should have external or internal sign illumination to ensure adequate visibility between scheduled maintenance and cleanings.*

06   *One or more Interchange Sequence signs (see Section 2E.24) should be used on the approach to the tunnel entrance to display the distances to the next interchanges that have exit ramps inside the tunnel or immediately following the end of the tunnel.*

07   *Supplementary pavement markings, such as word, arrow, and/or route shield markings, should be considered inside the tunnel in addition to the basic lane and edge line markings.*

OTHER GUIDE SIGNS

Section 2E.46 Next Exit Plaques (E2-1P and E2-1aP)

Option:

01   Where the distance to the next interchange is unusually long, a Next Exit (E2-1P or E2-1aP) plaque (see Figure 2E-49) may be installed to inform road users of the distance to the next interchange.

Figure 2E-49. Next Exit Plaques

Guidance:

02   *The Next Exit plaque should not be used unless the distance between successive interchanges is more than 5 miles.*

03   *Where the Next Exit plaque is used, the E2-1P plaque should be used where the width of the Interchange Advance guide sign is equal to or greater than the width of the E2-1P plaque. The E2-1aP plaque should be used where the width of the E2-1P plaque exceeds the width of the Interchange Advance guide sign.*

Standard:

04   **The Next Exit plaque shall display the legend NEXT EXIT XX MILES. If the Next Exit plaque is used, it shall be placed below the Interchange Advance guide sign nearest the interchange. It shall be mounted so as to not adversely affect the breakaway feature of the sign support structure.**

Section 2E.47 Post-Interchange Signs

Guidance:

01   *If space between interchanges permits, as in rural areas, and where undue repetition of messages will not occur, a fixed sequence of signs should be displayed beginning 500 feet beyond the downstream end of the acceleration lane. At this point a Route Sign assembly should be installed followed by a Speed Limit sign and a Distance sign, each at a spacing of 1,000 feet (see Figure 2E-2).*

02   *If space between interchanges does not permit placement of these three post-interchange signs without encroaching on or overlapping the Advance guide signs necessary for the next interchange, or in rural areas where the interchanging traffic is primarily local, one or more of the post-interchange signs should be omitted.*

Option:

03   Usually the Distance sign will be of less importance than the other two signs and may be omitted, especially if Interchange Sequence signs are used. If the sign for through traffic on an overhead assembly already contains the route sign, the post-interchange route sign assembly may also be omitted.

Section 2E.48 Post-Interchange Distance Signs (E7-1 through E7-3)

Standard:

01   **If used, the Post-Interchange Distance sign (see Figure 2E-50) shall consist of a one-line, two-line, or three-line sign displaying the names of significant destination points and the distances to those points. The top line of the sign shall identify the next meaningful interchange with the name of the community near or through which the route passes, or if there is no community, the route number or name of the intersected highway.**

Support:

02   The minimum sizes of the route shields identifying a significant destination point are prescribed in Tables 2E-3 and 2E-5.

Option:

03   The text identification of a route may be displayed instead of a route shield, such as "U S XX," "[State abbreviation] XX" (such as "Del XX"), or "County XX."

Guidance:

04   *If a second line is used, it should be reserved for communities of general interest that are located on or immediately adjacent to the route or for major traffic generators along the route.*

Option:

05   The choice of names for the second line, if it is used, may be varied on successive Distance signs to give road users maximum information concerning communities served by the route.

Standard:

06   **The third, or bottom line, shall contain the name and distance to a control city (if any) that has national significance for travelers using the route.**

Guidance:

07   *Distances to the same destinations should not be shown more frequently than at 5-mile intervals. The distances displayed on these signs should be the actual distance to the destination points and not to the exit from the freeway or expressway. The distance displayed for each community should comply with the provisions of Section 2D.43.*

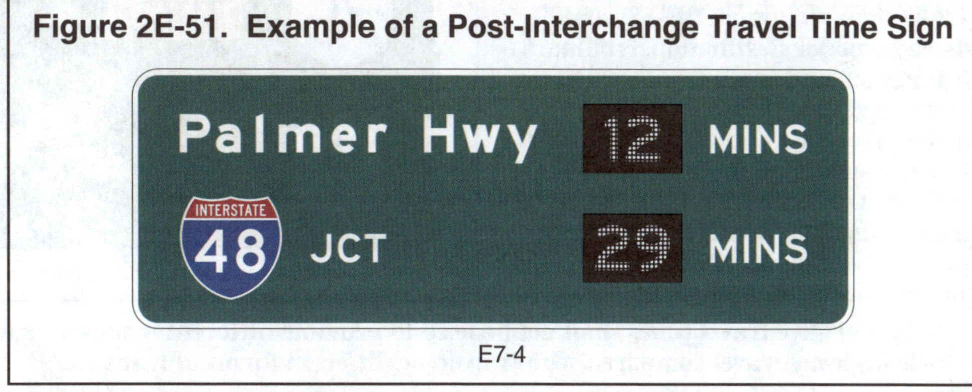

Figure 2E-50. Example of a Post-Interchange Distance Sign

Note: The one-line version (E7-1) and two-line version (E7-2) of the Post-Interchange Distance signs are not shown.

Section 2E.49 Post-Interchange Travel Time Sign (E7-4)

Support:

01   At certain locations, it might be more meaningful to recurrent road users to display the travel time rather than the distance to a destination. Such instances might be areas of adverse roadway conditions due to weather, such as in mountain passes or high elevations, congestion that occurs during peak travel seasons, or recurring congestion.

02   Section 2E.50 contains information on Distance and Travel Time and Comparative Travel Time signs.

Standard:

03   **If used, the Post-Interchange Travel Time (E7-4) sign (see Figure 2E-51) shall replace of the Post-Interchange Distance sign in the series of post-interchange signs (see Section 2E.47).**

04   **The Post-Interchange Travel Time sign shall comply with the provisions of Paragraph 1 of Section 2E.47 with the following exceptions:**

Figure 2E-51. Example of a Post-Interchange Travel Time Sign

   A. **The distance shall be replaced with a changeable message element to display the current travel time to the applicable destination; and**

   B. **The abbreviation MINS shall follow the changeable message element.**

05   **Travel times shall not be used on Interchange guide signs (see Section 2E.21).**

Section 2E.50 Distance and Travel Time Sign (E7-5) and Comparative Travel Time Sign (E7-6)

Support:

01   Some locations might benefit from a travel time message displayed with the distance, or comparative travel times displayed for alternative routes to a common destination. These locations are often in advance of an urbanized area where interchanges become more closely spaced and/or in advance of a circumferential or other alternative route(s) where the road user can decide to divert depending on the destination. Nonetheless, these signs are typically located in advance of a decision point where the road user can divert to an alternate route to avoid recurring congestion.

02   Section 2E.49 contains information on Post-Interchange Travel Time signs.

03   Section 2G.19 contains information on Comparative Travel Time signs for parallel lanes within the same highway route, such as for general-purpose lanes and managed lanes.

Standard:

04   **The Distance and Travel Time (E7-5) sign (see Figure 2E-52) shall display a major destination, landmark, or junction; a distance message; and a travel time message, each on a separate line. The distance units shall be displayed in the distance message. The travel time shall be displayed in a changeable message element and the abbreviation MINS shall follow the changeable message element. The Distance and Travel Time sign shall not display distance and time to more than one destination or junction.**

05   **The Comparative Travel Time (E7-6) sign (see Figure 2E-52) shall display a major destination, landmark, or junction, and two alternative routes with travel time messages. Each alternative route and associated travel time message shall be on a separate line. The travel time shall be displayed in a changeable message element and the abbreviation MINS shall follow the changeable message element.**

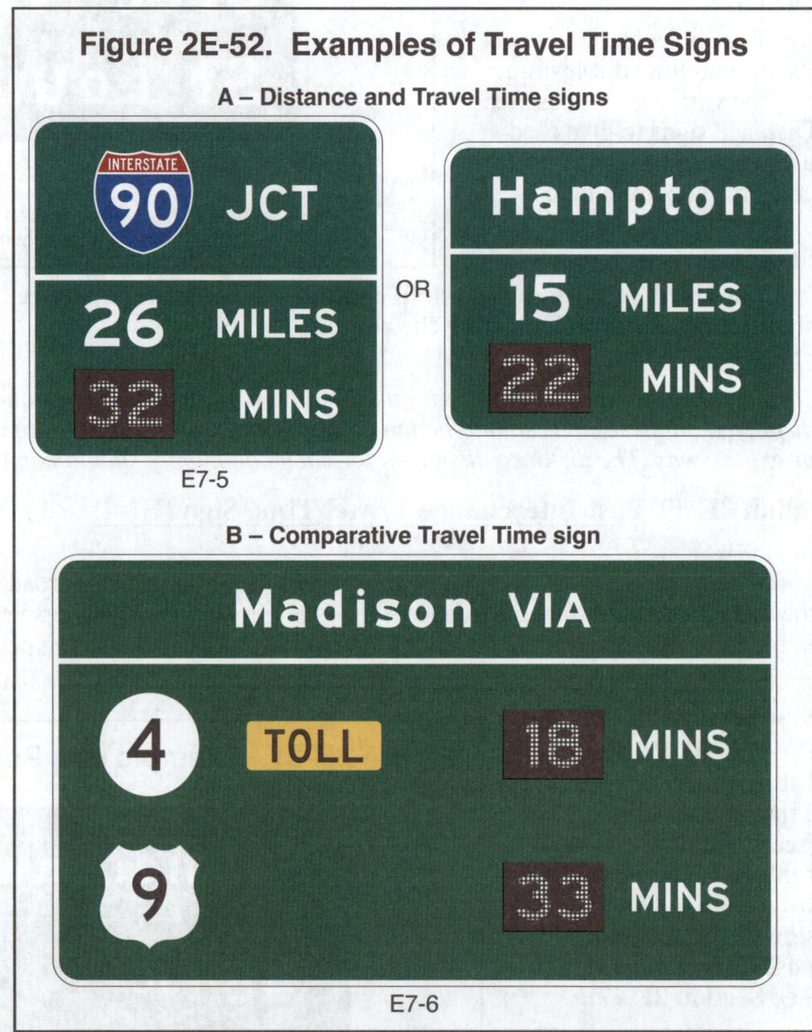

06   **Comparative travel times shall not be used to promote different modes of travel, such as personal vehicle highway travel compared with transit, or different forms of transit.**

Guidance:

07   *Where used, the Distance and Travel Time sign should be located between interchanges and away from the sequence of interchange guide signs or other major signs.*

08   *Where used, the Comparative Travel Time sign should be located in advance of the sequence of interchange guide signs to provide adequate time for the road user to decide whether to reroute.*

Section 2E.51 Supplemental Guide Signs (E3 Series)

Support:

01   Supplemental guide signs (see Figure 2E-53) can be used to provide information regarding destinations accessible from an interchange, other than places displayed on the standard interchange signing. However, such Supplemental guide signing can reduce the effectiveness of other more important guide signing because of the possibility of overloading the road user's capacity to receive visual messages and make appropriate decisions. "The AASHTO Guidelines for the Selection of Supplemental Guide Signs for Traffic Generators Adjacent to Freeways" is incorporated by reference in this Section.

Guidance:

02   *Use of Supplemental guide signs should be limited to situations where there is a demonstrated need to sign for more destinations from an interchange than those that are displayed on the Interchange Advance guide and Exit Direction signs.*

03   *A Supplemental guide sign should not be installed unless a destination meets the criteria established by the State or agency policy. States and other agencies should adopt an appropriate policy for installing Supplemental guide signs using the "AASHTO Guidelines for the Selection of Supplemental Guide Signs for Traffic Generators Adjacent to Freeways." In developing policies for such signing, such items as population, amount of traffic generated, distance from the route, and the significance of the destination, should be taken into account.*

04   *No more than one Supplemental guide sign should be used on each interchange approach.*

05   *A Supplemental guide sign should display no more than two destinations and no more than three lines of destination names. Destination names should be followed by the interchange number (and suffix), or if interchanges are not numbered, by the legend NEXT RIGHT or SECOND RIGHT or both, as appropriate. The Supplemental guide sign should be installed as an independent guide sign assembly.*

06   *Where two or more Interchange Advance guide signs are used, the Supplemental guide sign should be installed approximately midway between two of the Interchange Advance guide signs. If only one Interchange Advance guide sign is used, the Supplemental guide sign should follow it by at least 800 feet. If the interchanges are numbered, the interchange number should be used for the action message.*

07   *A Supplemental guide sign should not be installed in the same location with or where it will detract from guide signs for a different interchange.*

Standard:

08   **No more than two supplemental traffic generator destinations shall be signed from a single interchange approach and four from a single interchange along the main roadway (see Paragraphs 4 and 5 of this Section regarding the number of Supplemental guide signs at an interchange and the number of destinations displayed on a Supplemental guide sign).**

09   **Supplemental guide signs shall not be placed at the same location as Interchange Advance guide, Exit Direction, or other signs related to an exit or interchange.**

10   **Guide signs for park-and-ride facilities shall be considered as Supplemental guide signs (see Figure 2E-54).**

11   **Guide signs for recreational or cultural interest destinations (see Chapter 2M) shall be considered as Supplemental guide signs, except where the interchange provides direct access to such a destination and the destination is instead displayed on the Interchange Advance guide and Exit Direction signs.**

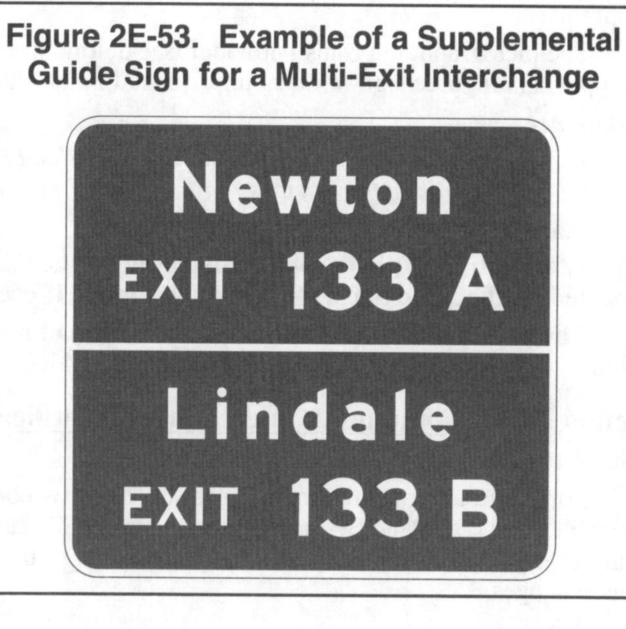

Figure 2E-53. Example of a Supplemental Guide Sign for a Multi-Exit Interchange

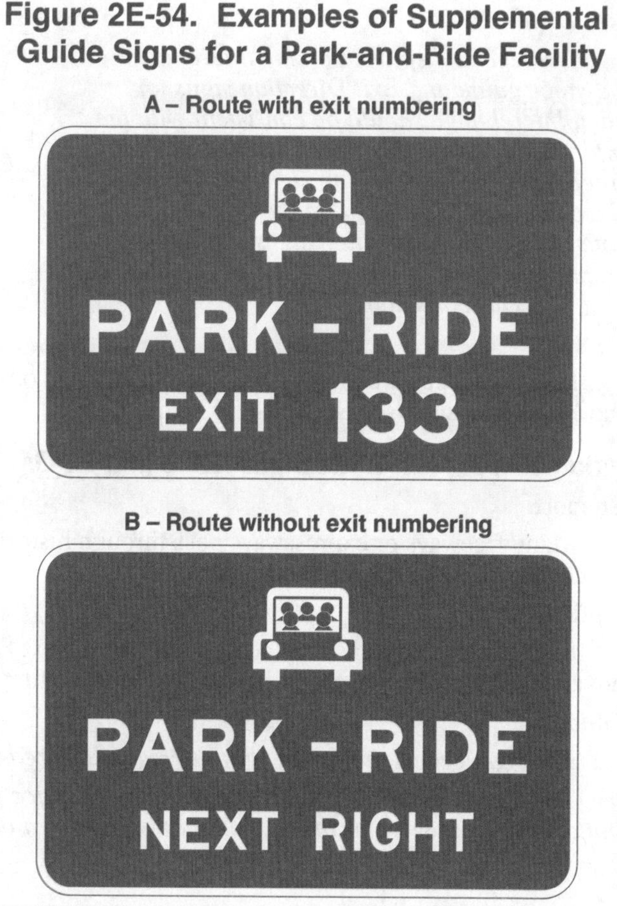

Figure 2E-54. Examples of Supplemental Guide Signs for a Park-and-Ride Facility

Option:

12   The pictograph of a transit provider (see definition in Section 1C.02) may be displayed on the Park – Ride Supplemental guide sign or on a Supplemental guide sign for a transit facility.

Guidance:

13   *The use of a transit pictograph and/or the carpool symbol on the PARK – RIDE Supplemental guide sign should comply with the provisions of Paragraph 5 of Section 2D.48.*

Standard:

14   **When a transit pictograph is displayed on the PARK – RIDE Supplemental guide sign, it shall be located on the same line as the carpool symbol, if used, above the word legend.**

15   **The maximum dimension (height or width) of a pictograph on a sign shall not exceed two times the upper-case letter height of the destination or PARK – RIDE legend.**

Section 2E.52 Community Interchanges Identification Signs (E9-4 and E9-5)

Support:

01   For suburban or rural communities served by two or three interchanges, Community Interchanges Identification (E9-4 and E9-5) signs (see Figure 2E-55) reduce the amount of information displayed on the Interchange Advance guide and Exit Direction signs by eliminating repetition of the same destinations for separate interchanges.

Guidance:

02   *In these cases, the name of the community followed by the word Exits should be displayed on the top line; the lines below should display the destination, road name or route number, and the corresponding distances to the nearest ¼ mile.*

03   *The sign should be located in advance of the first Interchange Advance guide sign for the first interchange within the community (see Figure 2E-56).*

04   *The legend displayed on the Interchange Advance guide and Exit Direction signs for each interchange should be consistent with the interchange names displayed on the Community Interchanges Identification sign. The name of the community displayed on the Community Interchanges Identification signs should be omitted from the legends of the Interchange Advance guide and Exit Direction signs.*

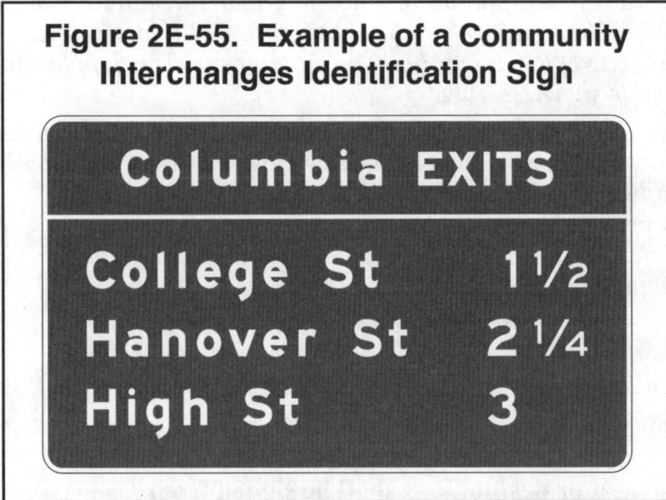

Figure 2E-55. Example of a Community Interchanges Identification Sign

Option:

05   If interchanges are not conveniently identifiable or if there are more than three interchanges to be identified, the Next Exits sign (see Section 2E.53) may be used.

Section 2E.53 Next Exits Signs (E9-3 and E9-3a)

Support:

01   Many freeways or expressways pass through historical or recreational regions, or urban areas served by a succession of several interchanges.

Option:

02   Such regions or areas may be indicated by a Next Exits (E9-3 or E9-3a) sign (see Figure 2E-57) located in advance of the Advance guide sign or signs for the first interchange.

Guidance:

03   *The sign legend should identify the region or area followed by the words NEXT XX EXITS.*

04   *The legend displayed on the Interchange Advance guide and Exit Direction signs for each interchange should not display the region or area name that is displayed on the Next Exits sign (see Figure 2E-58).*

Figure 2E-56. Example of the Use of a Community Interchanges Identification Sign

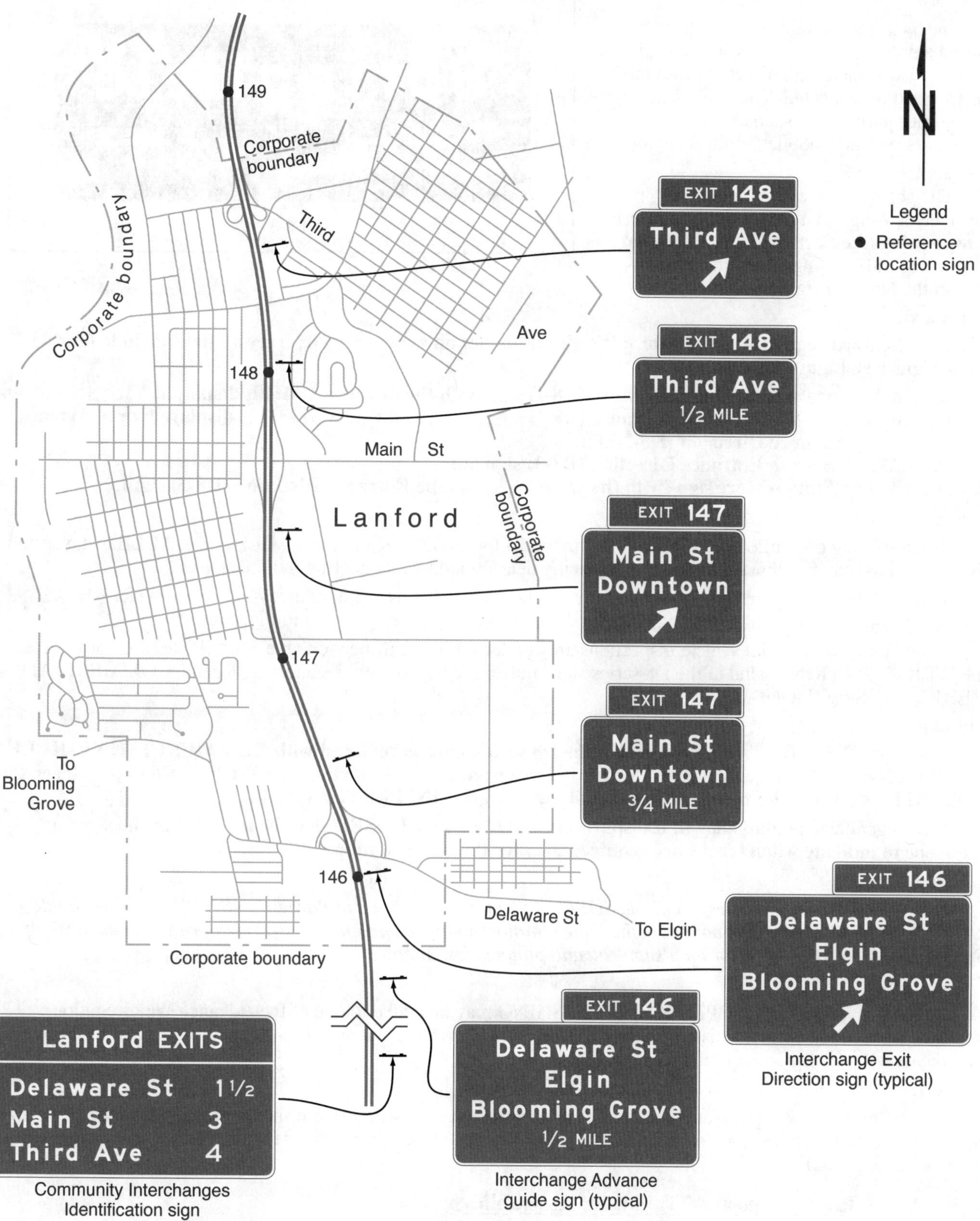

Section 2E.54 Weigh Station Signing

Support:

01   Independent facilities or areas have been added along many highways where certain commercial vehicles are directed to stop to be weighed or inspected. These areas are sometimes permanent, such as in a roadside area, or temporary mobile facilities deployed along the roadway.

02   The general concept for signing permanent Weigh Stations is similar to Rest Area signing (see Section 2I.05) because in both cases traffic using either area remains within the highway right-of-way.

Figure 2E-57. Example of a Next Exits Sign

E9-3

Standard:

03   **The standard sequence of signs for a Weigh Station on an expressway or freeway shall include four basic signs (see Figure 2E-59):**

    A. An Advance Weigh Station Distance (D8-1) sign with the distance 1 MILE displayed,

    B. An Advance Weigh Station Distance (D8-1) sign with the distance ½ MILE displayed, or a Weigh Station Advance Direction (D8-2) sign,

    C. A Weigh Station Entrance Direction (D8-3) sign, and

    D. A Weigh Station Gore sign (with the same legend as the Entrance Direction (D8-3) sign).

Option:

04   When spacing of 1 mile and ½ mile are not practical for the D8-1 signs, the 1 MILE and ½ MILE distances on the D8-1 signs may be adjusted to match the spacing determined by engineering judgment.

05   Where State law requires trucks of a certain weight to enter the weigh station, a Weigh Station (R13-1) regulatory sign (see Section 2B.65) may be added to the sign sequence as shown in Figure 2E-59.

06   Where only commercial vehicle inspections are conducted in the inspection area and vehicles are not weighed, the WEIGH STATION legend of the D8 series signs may be replaced with the alternate legend, COMMERCIAL VEHICLE INSPECTION AREA.

Standard:

07   **When the WEIGH STATION legend of the D8 series signs is replaced with COMMERCIAL VEHICLE INSPECTION AREA legend as provided for in Paragraph 6 of this Section, the WEIGH STATION legend of the R13-1 sign shall be replaced with the alternate legend INSPECTION AREA.**

08   **A changeable legend display that displays either OPEN or CLOSED shall be included in the signing sequence to indicate when trucks are required to enter the weigh station.**

Guidance:

09   *The required changeable legend display OPEN or CLOSED describe in Paragraph 8 of this Section should be displayed within and at the bottom of the Weigh Station Advance Direction (D8-2) sign or the Advance Weigh Station Distance (D8-1) sign, or on a supplemental plaque or sign panel.*

Option:

10   A plaque with the legend OPEN WHEN FLASHING may be added to one of the Advance Weigh Station Distance signs along with associated flashing beacons, in place of the changeable legend OPEN or CLOSED sign, to indicate when commercial vehicles are required to enter the weigh station.

Support:

11   Weigh Station Area sign layouts for freeway and expressway applications are shown in the "Standard Highway Signs" publication (see Section 1A.05). An example of weigh station signing for use on freeways and expressways is shown in Figure 2E-59.

Section 2E.55 Route Signs and Trailblazer Assemblies

Guidance:

01   *Route signs (see Figure 2E-60) should be incorporated as cut-out shields or other distinctive shapes on large directional guide signs. Where the Interstate shield is displayed in an assembly or on the face of a guide sign with U.S. or State Route signs, the Interstate numeral should be at least equal in size to the numerals on the other Route signs. The use of independent Route signs should be limited primarily to route confirmation assemblies.*

Figure 2E-58. Example of the Use of a Next Exits Sign

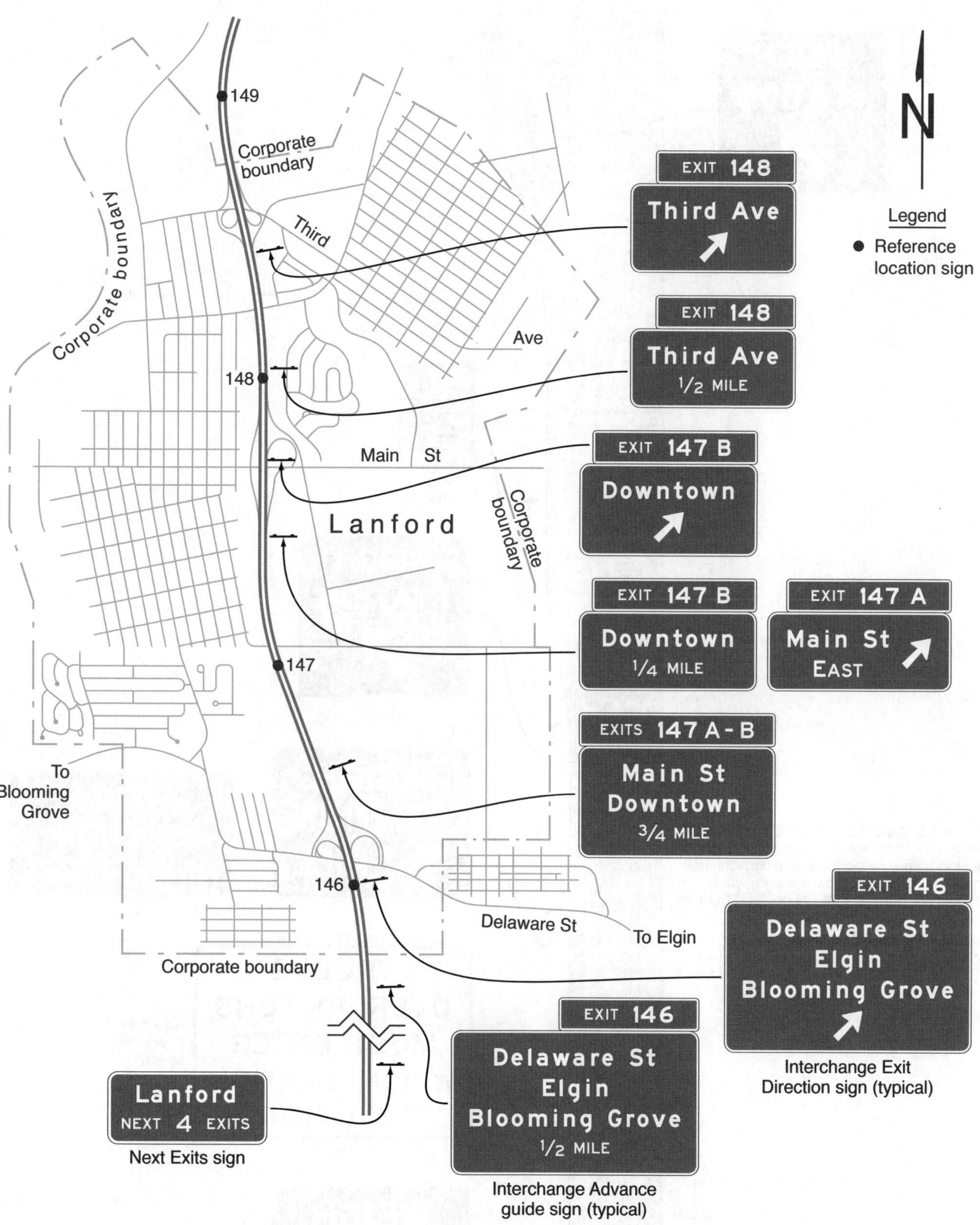

Interchange Exit
Direction sign (typical)

Next Exits sign

Interchange Advance
guide sign (typical)

Figure 2E-59. Example of Weigh Station Signing on Freeways

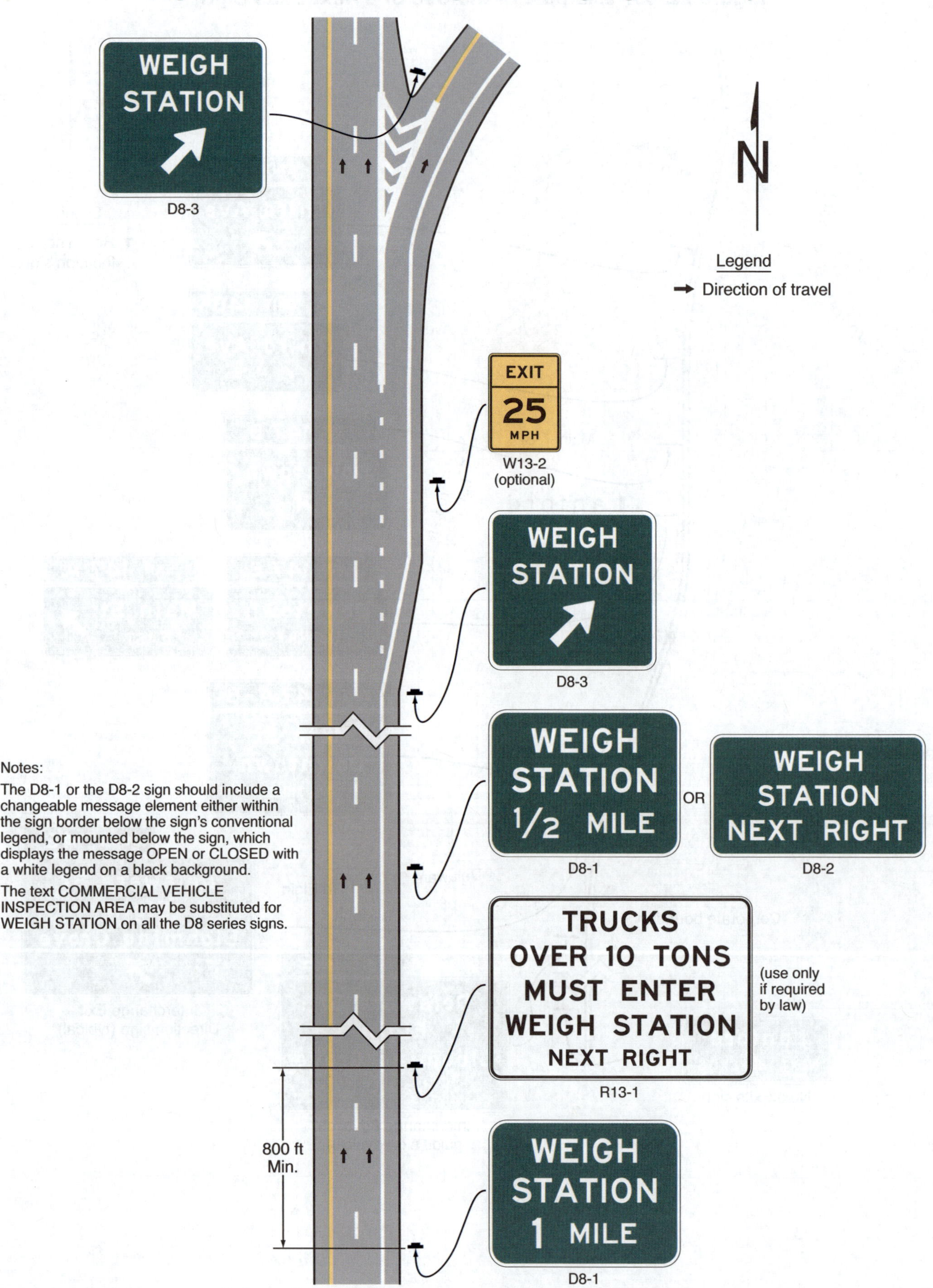

Figure 2E-60. Interstate, Off-Interstate, and U.S. Route Signs

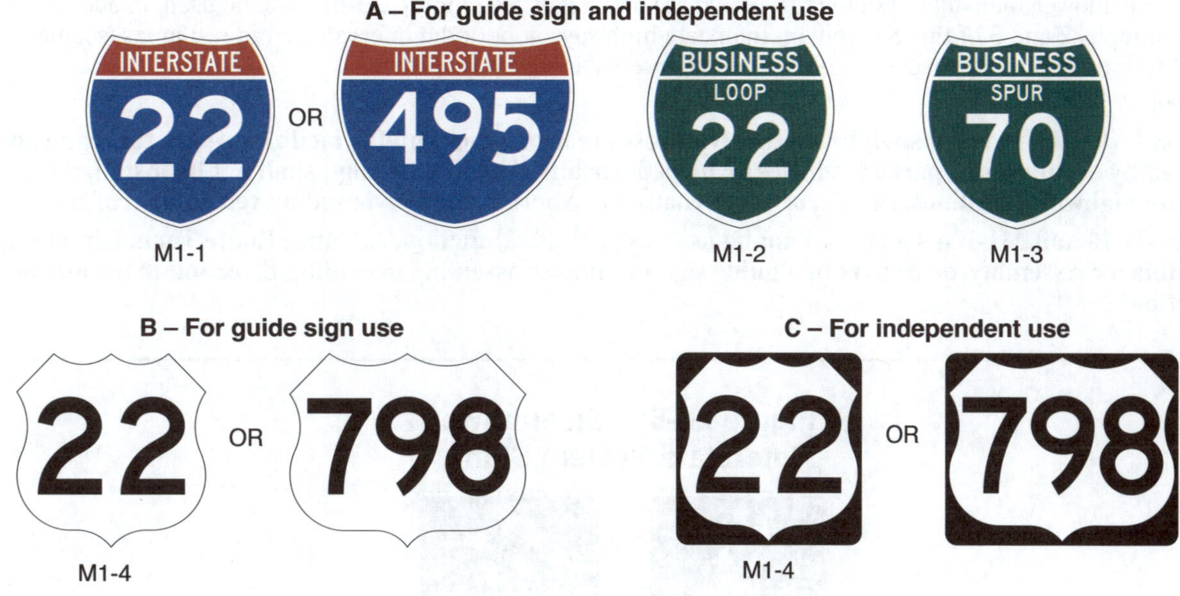

02   *Route signs and auxiliary plaques showing junctions and turns should be used for guidance on approach roads, for route confirmation just beyond entrances and exits, and for reassurance along the freeway or expressway. When used along the freeway or expressway, the Route signs should be enlarged to a 36 x 36-inch minimum size for route numbers with one or two digits and to a 45 x 36-inch minimum size for route numbers with three digits as shown in the "Standard Highway Signs" publication (see Section 1A.05). When independently-mounted Route signs are used in place of Pull-Through signs (see Section 2E.27), they should be located just beyond the exit.*

Option:

03   The standard Trailblazer Assembly (see Section 2D.34) may be used on roads leading to the freeway or expressway. Component messages of the Trailblazer Assembly may be incorporated into a single sign in accordance with the provisions of Section 2D.12. Independently-mounted Route signs may be used instead of Pull-Through signs as confirmation information.

Support:

04   Section 2D.58 contains information regarding the design of signs for Auto Tour Routes.

Option:

05   The commonly-used name or trailblazer route sign for a toll highway (see Chapter 2F) may be displayed on non-toll sections of the Interstate Highway System at:

    A.   The last exit before entering a toll section of the Interstate Highway System;
    B.   The interchange or connection with a toll highway, whether or not the toll highway is a part of the Interstate Highway System; and
    C.   Other locations within a reasonable approach distance of toll highways when the name or trailblazer symbol for the toll highway would provide better guidance to road users unfamiliar with the area than would place names and route numbers.

06   The toll highway name or route sign may be included as a part of the guide sign installations on intersecting highways and approach roads to indicate the interchange with a toll section of an Interstate route. Where needed for the proper direction of traffic, a trailblazer for a toll highway that is part of the Interstate Highway System may be displayed with the Interstate Trailblazer Assembly.

Support:

07   Chapter 2F contains additional information regarding signing for toll highways.

Section 2E.56 Eisenhower Interstate System Signs (M1-10 and M1-10a)

Option:

01   The Eisenhower Interstate System (M1-10 and M1-10a) signs (see Figure 2E-61) may be used, in accordance with Paragraphs 2 and 3 of this Section, on Interstate highways at periodic intervals and in rest areas, scenic overlooks, or other similar roadside facilities on the Interstate Highway System.

Standard:

02   **If used, the M1-10a sign shall be used only in rest areas or other similar facilities where the sign can be viewed by occupants of parked vehicles or by pedestrians. The M1-10a sign shall not be installed on Interstate highway mainlines, ramps, or other roadways where it can be viewed by vehicular traffic.**

03   **The M1-10 and M1-10a signs shall not be used as part of a Junction, Advance Route Turn, Directional, or Trailblazer Assembly, or as part of a guide sign or similar assembly providing direction to a route or destination.**

Figure 2E-61. Eisenhower Interstate System Signs

M1-10

M1-10a

SIGNS FOR ROUTE DIVERSION BY VEHICLE CLASS

Section 2E.57 <u>Signs for Route Diversion by Vehicle Class</u>

Support:

01   On some highways, a physical condition or highway feature might limit certain types or classes of vehicles from proceeding along that route through the site of that condition beyond which those vehicles are otherwise allowed. Examples include, but are not limited to, a restriction on taller legal-height vehicles through a tunnel with a low clearance; a restriction of hazardous materials through a tunnel or over a bridge; and a restriction on wider vehicles, such as large trucks, over a viaduct with narrow lanes. In such cases, the restricted vehicles might be diverted along another route to reach a destination beyond the location of the limiting condition.

Guidance:

02   *Where certain vehicles are prohibited at a downstream location along a route and those vehicles must divert to reach a through destination beyond that location, regulatory, warning, and/or guide signs advising those vehicle operators of the diversion should be installed in advance of the decision point to leave the through route for the diversion route.*

Option:

03   The interchange and pull-through guide signs for the last point at which restricted vehicles must exit may be modified to incorporate regulatory and/or warning panels with word legends to display the regulations and/or warning messages relative to the vehicle class restriction.

04   Standard post-mounted regulatory and warning signs, such as the No Hazardous Materials (R14-3) or Advance Low Clearance (W12-2) signs, may be used as provided elsewhere in this Manual at independent locations to supplement the regulatory and warning signs and panels referenced in Paragraphs 2 and 3 of this Section.

Support:

05   An example of signing for a route diversion by vehicle class is shown in Figure 2E-62.

Figure 2E-62. Example of Signing for Route Diversion by Vehicle Class

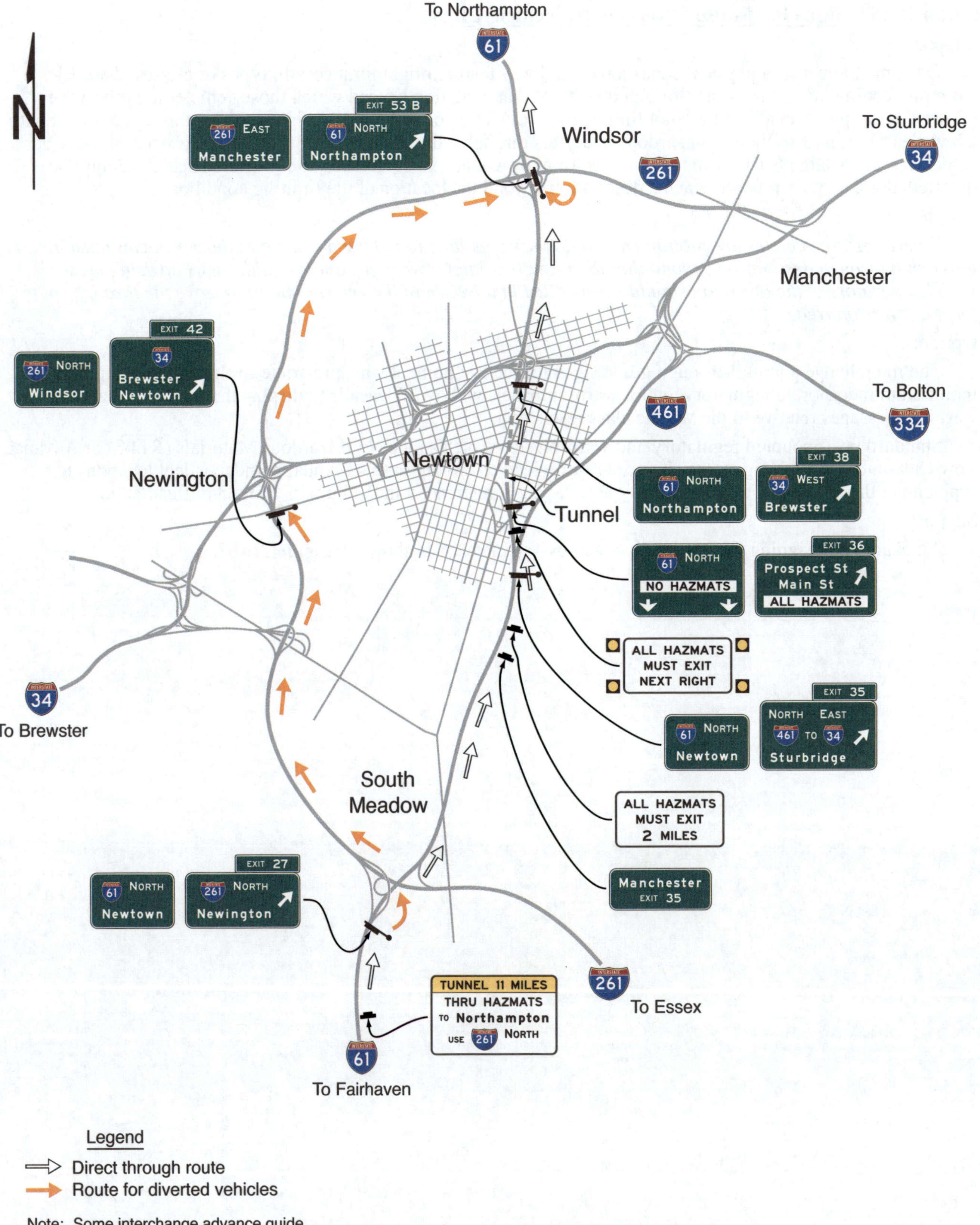

INTERFACE WITH CONVENTIONAL ROADWAYS

Section 2E.58 Signing on Conventional Road Approaches and Connecting Roadways

Support:

01   Section 2D.49 contains information regarding the signing on conventional roads on the approaches to interchanges and the signing on connecting roadways.

Section 2E.59 Wrong-Way Traffic Control at Interchange Ramps

Support:

01   Section 2B.48 contains information regarding the use of regulatory signs to deter wrong-way movements at intersections of freeway or expressway ramps with conventional roads, and in the area where entrance ramps intersect with the mainline lanes.

02   Section 2D.50 contains information regarding the use of a Directional assembly or a guide sign to mark the entrance to a freeway or expressway from a conventional road.

(This page intentionally left blank)

CHAPTER 2F. TOLL ROAD SIGNS

Chapter 2F Subchapter and Section Organization

GENERAL

Section 2F.01 Scope

Support:

01   Toll highways are typically limited-access freeway or expressway facilities. A portion of or an entire route might be a toll highway, or a bridge, tunnel, or other crossing point might be the only toll portion of a highway at which a toll is collected. A toll highway might be a conventional road. The general signing requirements for toll roads will depend on the type of facility and access (freeway, expressway, or conventional road). The provisions of Chapters 2D and 2E will generally apply for guide signs along the toll facility that direct road users within and off the facility where exit points and geometric configurations are not dependent specifically on the collection of tolls. The aspect of tolling and the presence of toll plazas or collection points necessitate additional considerations in the typical signing needs. The notification of the collection of tolls in advance of and at entry points to the toll highway also necessitates additional modifications to the typical signing.

02   The scope of this Section applies to a route or facility on which all lanes are tolled. Chapter 2G contains provisions for the signing of managed lanes within an otherwise non-toll facility that employ tolling or pricing as an operational strategy to manage congestion levels.

Standard:

03   **Except where specifically provided in this Chapter, the provisions of other Chapters in Part 2 shall apply to toll roads.**

Section 2F.02 Sizes of Toll Road Signs and Electronic Toll Collection (ETC) System Pictographs

Standard:

01   **Except as provided in Section 2A.07, the minimum sizes of toll road signs that have standardized designs shall be as shown in Table 2F-1.**

Support:

02   Section 2A.07 contains information regarding the applicability of the various columns in Table 2F-1.

Option:

03   Signs larger than those shown in Table 2F-1 may be used (see Section 2A.07).

Standard:

04   **The ETC system pictograph (see Section 2A.04) shall be of a size that makes it a prominent feature of the sign legend as necessary for conspicuity for those road users with registered ETC accounts seeking such direction, as well as for those road users who do not have ETC accounts so that it is clear to them to avoid such direction when applicable.**

Guidance:

05   *Except as provided in Paragraph 6 of this Section, an ETC pictograph that is in the shape of a horizontally-oriented rectangle should have a minimum height of 1.5 times the upper-case letter height of the principal legend on the sign. The width of an ETC pictograph in the shape of a horizontal rectangle should be between approximately 2 and 3 times the height of the pictograph.*

06   *When the pictograph is the principal legend on the sign, such as for advance guide signs for open-road tolling lanes (see Section 2F.15), the minimum height of a horizontally-oriented rectangular ETC pictograph should be consistent with that of a route shield prescribed for the particular application and type of sign.*

07   *For ETC pictographs whose shape is square, circular, or otherwise similar in height and width, or is a vertically-oriented rectangle, the same basic principles for conspicuity and placement should be followed. ETC pictographs whose shape is not in that of a horizontally-oriented rectangle should be suitably sized to facilitate conspicuity as described in Paragraph 4 of this Section and should be of a similar approximate area as the horizontally-oriented rectangular pictographs designed in accordance with the height and width as provided in Paragraph 5 of this Section.*

Section 2F.03 Use of Color on Toll Signs

Standard:

01   **Use of the color purple on any sign shall comply with the provisions of Sections 1D.05 and 2A.06. Except as provided in Sections 2F.05 and 2F.16, purple as a background color shall be used only when the information associated with the appropriate ETC account is displayed on that portion of the sign. The background color of the remaining portion of such signs shall comply with the provisions of Sections 1D.05 and 2A.06 as appropriate for a regulatory, warning, or guide sign. Purple shall not be used as a background color to display a destination, action message, or other legend that is not a display of the requirement for all vehicles to have a registered ETC account.**

Table 2F-1. Toll Road Sign and Plaque Minimum Sizes

Sign or Plaque	Sign Designation	Section	Conventional Road		Expressway	Freeway	Minimum	Oversized
			Single Lane	Multi-Lane				
Toll Rate	R3-28	2F.04	—	—	114 x 48	114 x 48	—	—
Pay Toll (plaque)	R3-29P	2F.04	—	—	24 x 18	24 x 18	—	—
Take Ticket (plaque)	R3-30P	2F.04	—	—	24 x 18	24 x 18	—	—
ETC Account-Only	R3-31	2F.05	24 x 24	24 x 24	36 x 36	36 x 36	24 x 24	36 x 36
No Cash (plaque)	R3-32P	2F.05	24 x 12	24 x 12	36 x 18	36 x 18	24 x 12	36 x 18
Pay Toll XX Miles Cars (price)	W9-6	2F.06	96 x 66	96 x 66	96 x 66	96 x 66	—	—
Stop Ahead Pay Toll Cars (price)	W9-6a	2F.08	114 x 66	114 x 66	114 x 66	114 x 66	—	—
Pay Toll XX Miles Cars (price) (plaque)	W9-6bP	2F.07	312* x 30	312* x 30	312* x 30	312* x 30	—	—
Stop Ahead Pay Toll (plaque)	W9-6cP	2F.09	264* x 30	264* x 30	264* x 30	264* x 30	—	—
Stop Ahead Pay Toll (plaque)	W9-6dP	2F.09	114 x 48	114 x 48	114 x 48	114 x 48	—	—
Take Ticket XX Miles	W9-6e	2F.06	114 x 48	114 x 48	114 x 48	114 x 48	—	—
Stop Ahead Take Ticket	W9-6f	2F.08	114 x 48	114 x 48	114 x 48	114 x 48	—	—
Take Ticket XX Miles (plaque)	W9-6gP	2F.07	216 x 30	216 x 30	216 x 30	216 x 30	—	—
Stop Ahead Take Ticket (plaque)	W9-6hP	2F.09	282 x 30	282 x 30	282 x 30	282 x 30	—	—
Last Exit Before Toll (1-line plaque)	W16-16P	2F.10	—	—	252* x 36	252* x 36	—	—
Last Exit Before Toll (2-line plaque)	W16-16aP	2F.10	—	—	144 x 54	144 x 54	—	—
Toll (plaque)	W16-17P	2F.11	24 x 12	24 x 12	36 x 18	36 x 18	24 x 12	36 x 18
Toll Collector Symbol Panel	M4-17	2F.12	—	—	48 x 48	48 x 48	—	—
Exact Change Symbol Panel	M4-18	2F.12	—	—	48 x 48	48 x 48	—	—

* The width shown represents the minimum dimension. The width shall be increased as appropriate to match the width of the guide sign.

Notes: 1. Larger signs may be used when appropriate

2. Dimensions are shown as width x height, in inches

---

02   **If only vehicles with registered ETC accounts are allowed to use a highway lane, a toll plaza lane, an open-road tolling lane, or all lanes of a toll highway or connection, the guide signs for such lanes or highways shall incorporate the pictograph (see Section 2A.04) adopted by the toll facility's ETC payment system and the regulatory message ONLY. Except for ETC pictographs whose predominant background color is purple, if incorporated within the green background of a guide sign, the ETC pictograph shall be on a white rectangular or square panel set on a purple underlay panel with a white border. For rectangular ETC pictographs whose predominant background color is purple, a white border shall be used at the outer edges of the purple rectangle to provide contrast between the pictograph and the sign background color.**

03   **If an ETC pictograph is used on a separate plaque in a route sign assembly (see Section 2F.05) or on a header panel within a guide sign, the plaque or the header panel shall have a purple background with a white border and the ETC pictograph shall have a white border to provide contrast between the pictograph and the background of the plaque or header panel.**

04   **Purple underlay panels for ETC pictographs or purple backgrounds for plaques and header panels shall only be used in the manner described in Paragraphs 1 through 3 of this Section to convey the requirement of a registered ETC account on signs for lanes reserved exclusively for vehicles with such an account and on directional signs to an ETC Account-Only facility from a non-toll facility or from a toll facility that accepts multiple payment forms.**

Support:

05   Figure 2F-1 shows examples of ETC account pictographs, their use with various background colors, and modifications involving underlay panels.

06   Section 2F.02 contains provisions regarding the size of pictographs for ETC accounts.

Figure 2F-1. Examples of ETC Account Pictographs
and Use of Purple Backgrounds and Underlay Panels

A – Pictograph design with a purple background and a white contrasting border

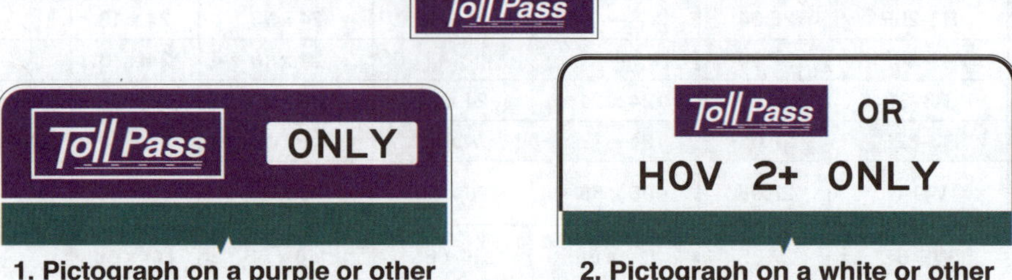

1. Pictograph on a purple or other non-contrasting background

2. Pictograph on a white or other contrasting background

B – Pictograph design with a background color other than purple, shown in a purple underlay panel with a white contrasting border

1. Pictograph on a purple background

2. Pictograph with a purple underlay on a non-contrasting background

3. Pictograph with a purple underlay panel on a white or other contrasting background

REGULATORY SIGNS

Section 2F.04 Regulatory Signs for Toll Plazas

Support:

01   Toll plaza operations often include lane-specific restrictions on vehicle type, forms of payment accepted, and speed limits or required stops. Vehicles are typically required to come to a stop to pay the toll or receive a toll ticket in the attended and exact change or automatic lanes. Electronic toll collection (ETC) lanes with favorable geometrics typically allow vehicles to move through the toll plaza without stopping, but usually within a set regulatory speed limit or advisory speed. In some ETC lanes and in most lanes that accommodate non-ETC vehicles, a stop might be required while the ETC payment is processed because of geometric or other conditions.

Guidance:

02   *Regulatory signs applicable only to a particular lane or lanes should be located in a position that makes their lane applicability clear to road users approaching the toll plaza.*

03   *Regulatory signs, or regulatory panels within guide signs, indicating restrictions on vehicle type and forms of toll payment accepted at a specific toll plaza lane should be installed over the applicable lane either on the toll plaza canopy or on a separate structure immediately in advance of the canopy located in a manner such that each sign is clearly related to an individual toll lane.*

Support:

04   Section 2F.12 contains information regarding the incorporation of regulatory messages into guide signs for toll plazas.

05   Section 2F.16 contains information regarding the design and use of toll plaza canopy signs.

Guidance:

06   *One or more Speed Limit (R2-1) signs (see Section 2B.21) should be installed in the locations provided in Paragraph 8 of this Section for an ETC-Only lane at a toll plaza in which an enforceable regulatory speed limit is established for a lane in which it is intended that vehicles move through the toll plaza without stopping while toll payments requiring stops occur in other lanes at the toll plaza. The speed limit displayed on the signs should be based on an engineering study taking into account the geometry of the toll plaza and the lanes, as well as other appropriate safety and operational factors.*

07   *A Speed Limit (R2-1) sign should not be installed for a toll plaza lane that is controlled by a STOP (R1-1) sign or where a stop is required.*

08   *Where speed limit signs are installed over a toll plaza lane on the toll plaza canopy, on the approach end of the toll booth island, on the toll booth itself, or on a vertical element of the canopy structure, then down arrows or diagonally downward-pointing directional arrows should be used to supplement the speed limit signs if there is a need to clarify the applicability of a sign to a specific lane or to improve compliance.*

Standard:

09   **A STOP (R1-1) sign shall not be installed for a toll plaza lane that is operated as an ETC-Only lane and that is designed for tolls to be collected while vehicles continue moving.**

Option:

10   A STOP (R1-1) sign may be installed to require all vehicles to come to a complete stop to pay a toll in an attended or exact change lane, even if that lane is also available for optional use by vehicles with registered ETC accounts. A PAY TOLL (R3-29P) or TAKE TICKET (R3-30P) plaque (see Figure 2F-2), as appropriate to the operation, may be installed directly under the STOP (R1-1) sign for a toll plaza lane, if needed.

11   The mounting height of the STOP sign and any supplemental plaque may be less than the normal mounting height requirements if constrained by the physical features of the toll island or toll plaza.

Figure 2F-2. Toll Plaza Regulatory Signs and Plaques

12   The lateral offset of a STOP or other regulatory sign located within a toll plaza island may be reduced to a minimum of 1 foot from the face of the toll island or raised barrier to the nearest edge of the sign.

Guidance:

13   *If used, a STOP (R1-1) sign for a toll plaza cash payment lane should be located in a longitudinal position as near as practical to the point where a vehicle is expected to stop to pay the toll or take a ticket.*

Option:

14   A Toll Rate (R3-28) sign (see Figure 2F-2) may be installed in advance of the toll plaza to indicate the toll applicable to the various vehicle types.

Guidance:

15   *If used, the Toll Rate (R3-28) sign should be located between the toll plaza and the first advance sign informing road users of the toll plaza.*

16   *The R3-28 sign should not contain more than three lines of legend. Each line that shows a toll amount should display only a single toll amount.*

Option:

17   Additional toll rate information exceeding three lines of legend may be displayed on the toll booth adjacent to the payment window of an attended lane or the payment receptacle of an exact change or automatic lane where it is visible to a road user who has stopped to pay the toll, but is not visible to approaching road users who have not yet entered the toll lane.

Section 2F.05 Electronic Toll Collection (ETC) Account-Only Regulatory Sign and Plaque (R3-31 and R3-32P)

Standard:

01   **In any route sign assembly providing directions to a toll facility, or to a tolled segment of a highway, where electronic toll collection (ETC) is the only payment method accepted and all vehicles are required to have a registered ETC account, the ETC Account-Only (R3-31) sign (see Figure 2F-3) shall be mounted directly below the route sign of the numbered or named toll facility. The R3-31 sign shall have a white border and purple background and incorporate the pictograph adopted by the toll facility's ETC payment system and the word ONLY in black letters on a white panel set on the purple background of the sign.**

Option:

02   The NO CASH (R3-32P) plaque (see Figure 2F-3) with a black legend and border on a white background may be mounted directly below the R3-31 sign in a Directional or other sign assembly.

Figure 2F-3. ETC Account-Only Auxiliary Signs for Use in Route Sign Assemblies

R3-31

R3-32P

Note: The ETC pictograph shown is an example only. The pictograph for the toll facility's adopted ETC system shall be used.

W16-17P

M3-2P

M1-4

R3-31

M5-1P

Example route sign assembly

WARNING SIGNS

Section 2F.06 Pay Toll and Take Ticket Advance Warning Signs (W9-6 and W9-6e)

Standard:

01   **The Pay Toll (W9-6) and Take Ticket (W9-6e) Advance Warning signs shall display the distance to the toll plaza and, except for toll-ticket facilities, the toll for passenger or 2-axle vehicles (see Figure 2F-4). Where the toll for passenger or 2-axle vehicles is variable by time of day, a changeable message element shall be incorporated into the W9-6 sign to display the toll in effect.**

Guidance:

02   *The Pay Toll Advance Warning (W9-6) sign should be installed at approximately 1 mile and ½ mile in advance of mainline toll plazas at which some or all lanes are required to come to a stop to pay a toll (see Sections 2F.14 and 2F.15).*

03   *The Take Ticket Advance Warning (W9-6e) sign should be installed overhead at approximately 1 mile and ½ mile in advance of mainline toll plazas at which some or all lanes are required to come to a stop to take a toll ticket (see Sections 2F.14 and 2F.15).*

04   *The Pay Toll and Take Ticket Advance Warning signs should be mounted overhead.*

Option:

05   If there is insufficient space for the W9-6 or W9-6e sign at the 1-mile or ½-mile advance location, the Pay Toll or Take Ticket Advance Warning (W9-6bP or W9-6gP) plaque (see Section 2F.07) may be installed at those advance locations above the appropriate guide sign(s) that relate to toll payment types.

06   An additional W9-6 or W9-6e sign may be installed approximately 2 miles in advance of a mainline toll plaza. This sign may be either mounted overhead or post-mounted.

07   If the visibility of a ramp toll plaza at which some or all lanes are required to come to a stop to pay a toll or take a ticket is limited, the W9-6 or W9-6e sign may also be installed in advance of the ramp toll plaza.

Section 2F.07 Pay Toll and Take Ticket Advance Warning Plaques (W9-6bP, W9-6dP, and W9-6gP)

Option:

01   The Pay Toll or Take Ticket Advance Warning (W9-6bP and W9-6gP) plaques (see Figure 2F-4) may be installed above the appropriate guide sign(s) relating to toll payment types at the 1-mile and/or ½-mile advance locations on the approach to a toll plaza if there is insufficient space for the W9-6 or W9-6e sign (see Section 2F.06) at those advance locations.

Standard:

02   **The W9-6bP and W9-6gP plaques shall display the distance to the toll plaza and, except for toll-ticket facilities, the toll for passenger or 2-axle vehicles. Where the toll for passenger or 2-axle vehicles is variable by time of day, a changeable message element shall be incorporated into the W9-6bP plaque to display the toll in effect.**

Option:

03   The distance to the toll plaza may be omitted from the W9-6bP and W9-6gP plaques if the distance is displayed on the guide sign that the plaque accompanies.

04   The Pay Toll (W9-6dP) plaque may be used if the toll information is displayed on the guide sign that the plaque accompanies.

05   The toll for passenger or 2-axle vehicles may be omitted from the W9-6bP plaque if the toll information is displayed on the guide sign that the plaque accompanies.

Section 2F.08 Stop Ahead Pay Toll and Take Ticket Warning Signs (W9-6a and W9-6f)

Standard:

01   **The Stop Ahead Pay Toll (W9-6a) warning sign (see Figure 2F-4) shall display the toll for passenger or 2-axle vehicles. Where the toll for passenger or 2-axle vehicles is variable by time of day, a changeable message element shall be incorporated into the W9-6a sign to display the toll in effect.**

Guidance:

02   *The Stop Ahead Pay Toll (W9-6a) sign should be installed downstream from the W9-6 sign that is ½ mile in advance of a mainline toll plaza where some or all of the lanes are required to come to a stop to pay a toll (see Sections 2F.14 and 2F.15).*

Figure 2F-4. Toll Plaza Warning Signs and Plaques

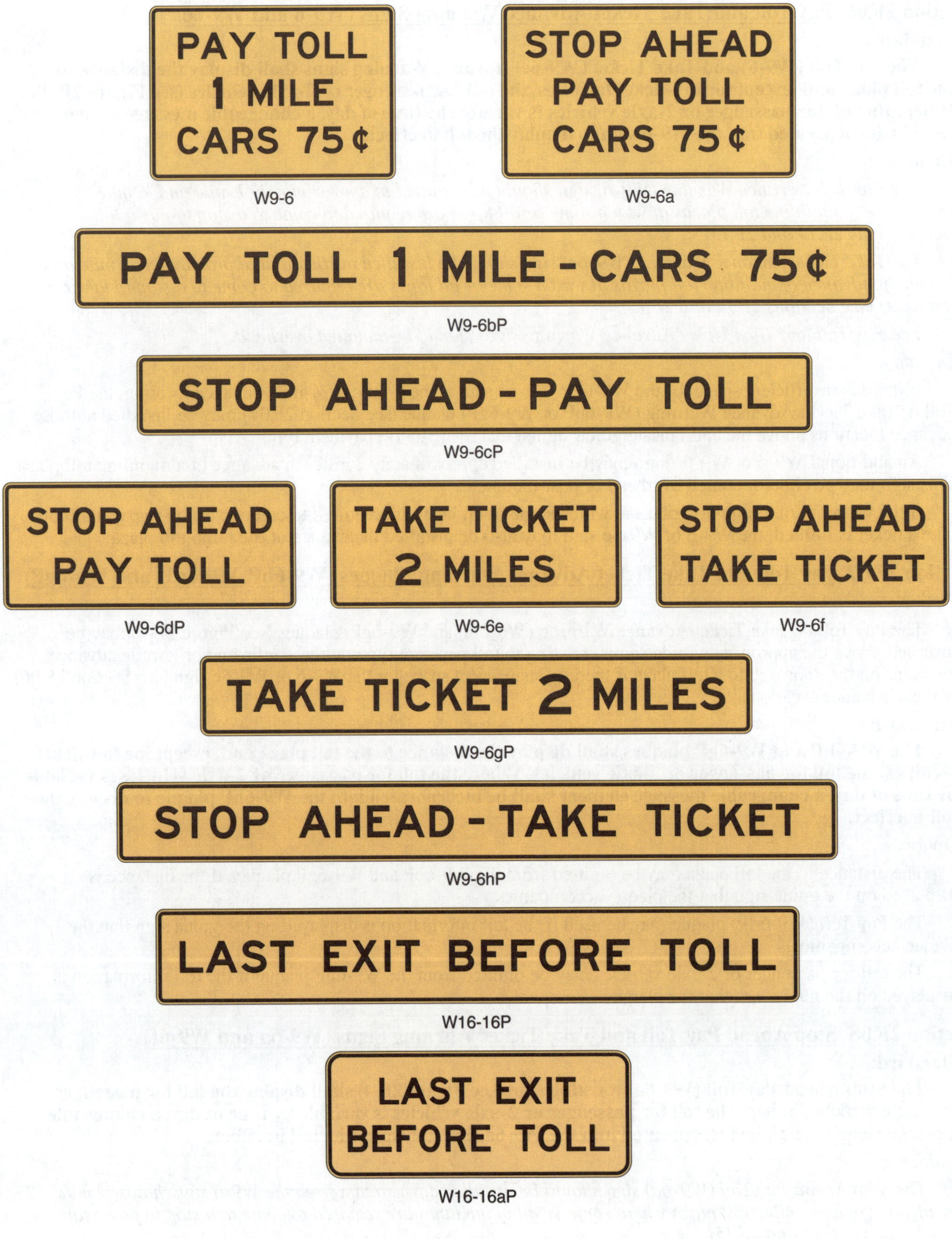

W9-6

W9-6a

W9-6bP

W9-6cP

W9-6dP

W9-6e

W9-6f

W9-6gP

W9-6hP

W16-16P

W16-16aP

03   *The Take Ticket (W9-6f) warning sign (see Figure 2F-4) should be installed downstream from the W9-6e sign that is ½ mile in advance of a mainline toll plaza where some or all of the lanes are required to come to a stop to take a toll ticket (see Sections 2F.14 and 2F.15).*

04   *The W9-6a and W9-6f signs should be mounted overhead. The location of the overhead sign should coincide with the approximate location where the mainline lanes begin to widen on the approach to the toll plaza lanes.*

05   *Where open-road tolling is used in addition to a toll plaza at a particular location, the W9-6a or W9-6f sign should be located such that the message is clearly related to the lanes that access the toll plaza and not to the open-road tolling lanes.*

Option:

06   If there is insufficient space for the W9-6a or W9-6f sign at the recommended location, the Stop Ahead Pay Toll (W9-6cP) or the Stop Ahead Take Ticket (W9-6hP) plaque (see Section 2F.09) may be installed at that location above the appropriate guide sign that relates to toll payment types.

07   If the visibility of a ramp toll plaza at which some or all lanes are required to come to a stop to pay a toll or take a ticket is limited, the W9-6a or W9-6f sign may also be installed in advance of the ramp toll plaza.

Section 2F.09 Stop Ahead Pay Toll and Take Ticket Warning Plaques (W9- 6cP and W9-6hP)

Option:

01   The Stop Ahead Pay Toll (W9-6cP) warning plaque (see Figure 2F-4) may be installed above the appropriate guide sign at the location specified for the Stop Ahead Pay Toll (W9-6a) sign (see Section 2F.08) if there is insufficient space for the W9-6a sign at that location and the toll information is displayed on the guide sign that the plaque accompanies.

02   The Take Ticket (W9-6hP) warning plaque (see Figure 2F-4) may be installed above the appropriate guide sign at the location specified for the Take Ticket (W9-6f) sign (see Section 2F.08) if there is insufficient space for the W9-6f sign at that location.

Section 2F.10 LAST EXIT BEFORE TOLL Warning Plaques (W16-16P and W16-16aP)

Guidance:

01   *The LAST EXIT BEFORE TOLL (W16-16P or W16-16aP) warning plaque (see Figure 2F-4) should be used to notify road users of the last exit from a highway before it becomes a facility on which toll payments are required. The plaque should be installed above the appropriate guide signs for the exit (see Sections 2E.23 and 2E.25), but below the Exit Number or LEFT plaque if used.*

Section 2F.11 TOLL Warning Plaque (W16-17P)

Standard:

01   **The TOLL (W16-17P) warning plaque (see Figure 2F-3) shall have a black legend and border on a yellow background and shall be mounted directly above the route sign of a numbered toll highway or, if used, above the cardinal direction and alternative route auxiliary signs, in any route sign assembly providing direction to a toll highway or to a segment of a highway on which the payment of a toll is required.**

GUIDE SIGNS

Section 2F.12 Toll Facility and Toll Plaza Guide Signs – General

Support:

01   Toll plazas are used on many toll highways, bridges, and tunnels for collection of tolls from road users. Electronic toll collection and/or open-road tolling might also be used on such facilities, either in addition to or in place of collecting toll payments at toll plazas.

02   Chapter 2G contains information regarding signs for preferential and managed lanes that are applicable to toll roads.

03   Chapter 3E contains information regarding pavement markings for certain toll plaza applications.

Standard:

04   **Directional assemblies for entrances to a toll highway, or to a road leading directly to a toll highway with no opportunity to exit before paying or being charged a toll, shall clearly indicate that the facility is a toll facility. Except where the State Toll Route sign (see Paragraph 8 of this Section) is used, the TOLL (W16-17P) warning plaque (see Section 2F.11) shall be used above the route sign of a numbered toll facility in any route sign assembly that provides directions to the toll route from another highway (see Figure 2F-5).**

05   **Except where the State Toll Route sign (see Paragraph 8 of this Section) is used and on Exit Gore signs or destination guide (D1 series) signs, a rectangular panel with the black legend TOLL on a yellow background shall be incorporated into the guide signs leading road users to a tolled highway (see Figures 2F-6 through 2F-8).**

06   **Guide signs for toll highways, toll plazas, and tolled or priced managed lanes (see Chapter 2G) shall have white legends and borders on green backgrounds, except as specifically provided by Sections 2F.12 through 2F.16.**

Option:

07   A State Toll Route sign (see Paragraph 8 of this Section) may be used in lieu of the State Route (M1-5) sign in combination with the TOLL (W16-17P) warning plaque or the TOLL panel (see Paragraphs 10 and 11 of this Section).

Figure 2F-5. Example of Minor Interchange Crossroad Signing for an Approach to a Toll Highway

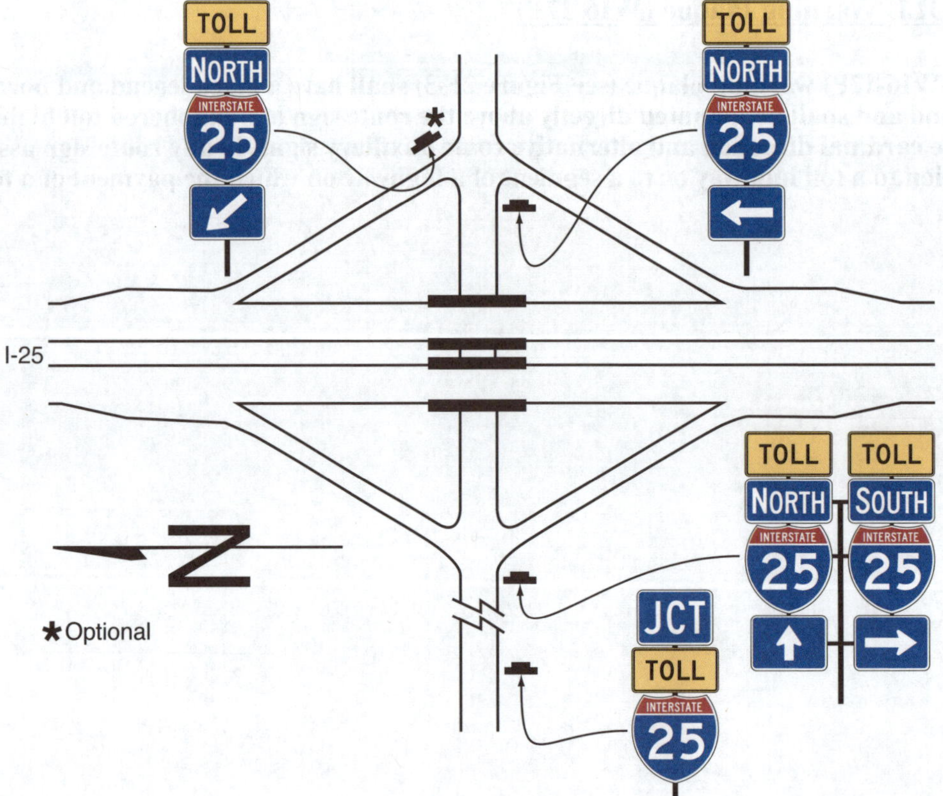

Figure 2F-6. Example of Interchange Crossroad Signing for a One-Lane Approach to a Toll Highway

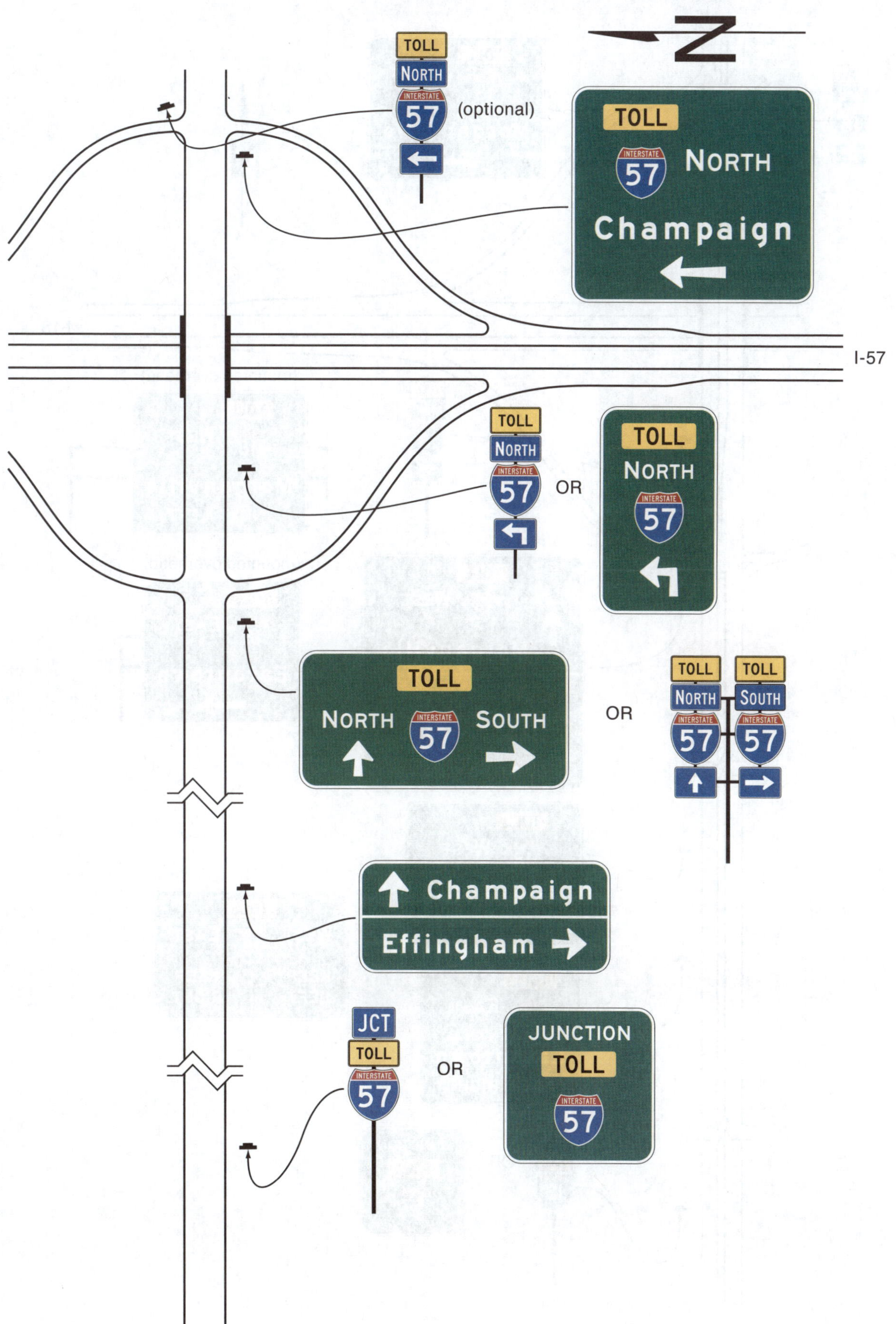

Figure 2F-7. Examples of Multi-Lane Crossroad Signing for a Diamond Interchange

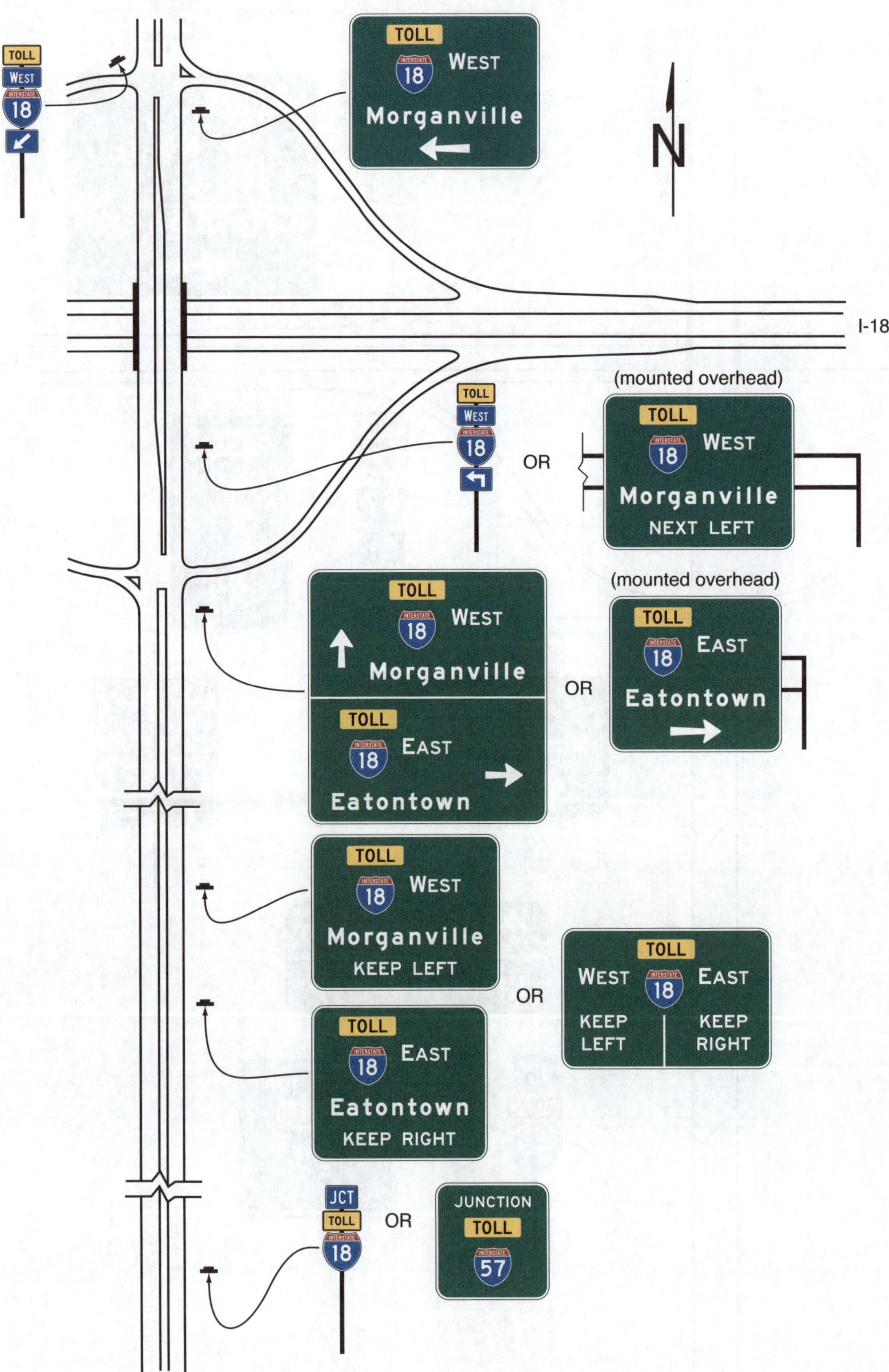

Figure 2F-8. Examples of Guide Signs for Entrances to Toll Highways or Ramps

**A – Entrance to a toll highway on which
registration in a toll account program is not required**

**B – Entrance to an ETC Account-Only toll highway
or entrance to a toll highway via an ETC Account-Only ramp**

**C – Entrance to a non-toll highway via
an ETC Account-Only toll entrance ramp**

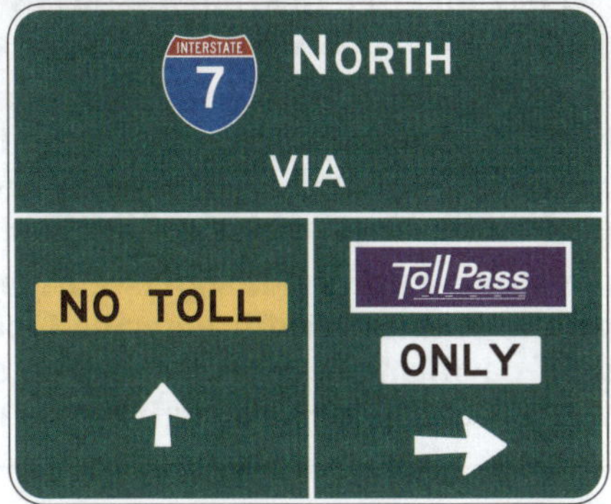

(the toll entrance is the only
connection provided in the vicinity)

(an alternate non-toll entrance is
provided in the vicinity)

Note: The ETC pictographs shown are examples only.
The pictograph for the toll facility's adopted ETC
system shall be used.

Standard:

08   **A State Toll Route sign shall incorporate into its design the word TOLL using the same letter height,
legend, background colors, and overall plaque dimensions specified for the W16-17P warning plaque.**

09   **The Interstate, Off-Interstate, and U.S. Route signs shall not be modified for tolled facilities.**

Option:

10   Where conditions do not accommodate separate signs, or where it is important to associate a particular
regulatory or warning message with specific guidance information, regulatory and/or warning messages may be
combined with guide signs for toll plazas using plaques, header panels, or rectangular regulatory or warning panels
incorporated within the guide signs, as long as the proper legend and background colors are preserved.

Standard:

11   **When regulatory messages are incorporated within a guide sign, they shall be on a rectangular panel with a black legend on a white background. When warning messages are incorporated within a guide sign, they shall be on a rectangular panel with a black legend on a yellow background.**

Guidance:

12   *Guide signs for toll plazas should be designed in accordance with the general principles of guide signs and the specific provisions of Chapter 2E.*

13   *Signs for toll plazas should systematically provide road users with advance and toll plaza lane-specific information regarding:*

    A. The amount of the toll, the types of payment accepted, and the type(s) of registered ETC accounts accepted for payment;

    B. Which lane or lanes are required or allowed to be used for each available payment type; and

    C. Restrictions on the use of a toll plaza lane or lanes by certain types of vehicles (such as cars only or no trucks).

Standard:

14   **Signs for attended lanes at toll plazas shall incorporate the Toll Collector (M4-17) symbol panel (see Figure 2F-9).**

Option:

15   Signs for attended lanes at toll plazas may also display word legends, such as FULL SERVICE, CASH, CHANGE, or RECEIPTS (see Figure 2F-9), to supplement the required symbol panel when lanes have different services available through them.

Standard:

16   **Signs for Exact Change lanes at toll plazas shall incorporate the Exact Change (M4-18) symbol panel and, except for ticketed systems, display the amount of the toll for passenger vehicles (see Figure 2F-9).**

Option:

17   Signs for Exact Change lanes at toll plazas may include an appropriate word legend, such as EXACT CHANGE (see Figure 2F-9), to supplement the required symbol panel.

Standard:

18   **When used, the M4-17 and M4-18 symbol panels shall be used only as panels within guide signs. The M4-17 and M4-18 symbols or panels shall not be used as an independent sign or within a sign assembly.**

19   **If only vehicles with registered ETC accounts are allowed to use a toll plaza lane, the signs for such lanes shall incorporate the pictograph adopted by the toll facility's ETC payment system and the regulatory message ONLY (see Figures 2F-1 and 2F-8 through 2F-11). The use, size, and placement of the ETC pictograph shall comply with the provisions of Sections 2F.02 and 2F.03.**

20   **An Overhead Arrow-per-Lane guide sign (see Figure 2F-10) shall be used in advance of a location where the mainline lanes split to separate traffic entering Open-Road ETC lanes from lanes entering a toll plaza where other methods of payment are accepted and an option lane is provided at the split (see Figure 2F-11). An Overhead Arrow-per-Lane guide sign shall not be used if there is no option lane at the split.**

Figure 2F-9. Examples of Conventional Toll Plaza Advance Signs

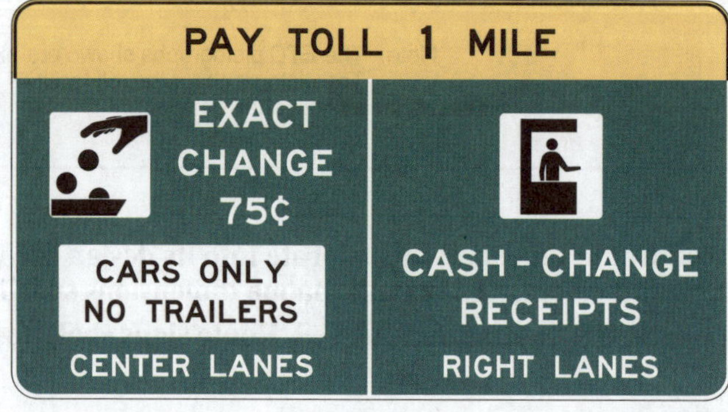

Note: The ETC pictograph that is shown is only an example. The pictograph for the toll facility's adopted ETC system shall be used.

**Figure 2F-10. Example of an Overhead Arrow-per-Lane Guide
Sign and Arrow for a Split with an Option Lane**

Example of an Overhead
Arrow-per-Lane guide sign for a split
with an option lane

Arrow for the option lane
(see the "Standard Highway
Signs" publication for details)

Note: The ETC pictograph that is shown is only an example.
The pictograph for the toll facility's adopted ETC system
shall be used.

Option:

21   The ETC payment system's pictograph, without a purple underlay or purple header panel, may be used on signs for Exact Change or attended lanes at toll plazas to indicate that vehicles with registered ETC accounts may also use those lanes.

Section 2F.13 Electronic Toll Collection (ETC) Signs – General

Support:

01   Figure 2F-8 shows examples of guide signs for entrances to various types of toll highways and for ETC Account-Only entrances to non-toll highways.

Standard:

02   **Signing for entrances to toll highways where ETC is employed only through license plate character recognition such that road users are not required to establish a toll account or register their vehicle equipment shall comply with the provisions of Paragraphs 4 and 5 of Section 2F.12.**

Support:

03   Figure 2F-12 shows examples of guide signs for the entrance to a toll highway on which tolls are collected electronically only and registration in a toll-account program is not required.

Standard:

04   **If only vehicles with registered ETC accounts are allowed to use a toll highway, the guide signs for entrances to such facilities shall incorporate the pictograph adopted by the toll facility's ETC payment system and the regulatory message ONLY (see Figures 2F-1 and 2F-8 through 2F-12). The use, size, and placement of the ETC pictograph and the use and color of the background and underlay panel shall comply with the provisions of Sections 2F.02 and 2F.03.**

Support:

05   Sections 2F.05, 2F.11, and 2F.17 contain additional provisions regarding signs for toll highways that only accept ETC payments.

06   Sections 2G.16 through 2G.19 contain additional provisions regarding signs for priced managed lanes that only accept ETC payments.

07   Figure 2F-13 shows an example of guide signs for alternative toll and non-toll ramp connections to a non-toll highway (see Section 2F.18).

08   Many different ETC payment systems are used by the various toll facility operators. Some of these systems accept payment from other systems' accounts.

Figure 2F-11. Examples of Guide Signs for a Split with an Option Lane for a Mainline Toll Plaza on a Diverging Alignment from Open-Road ETC Lanes

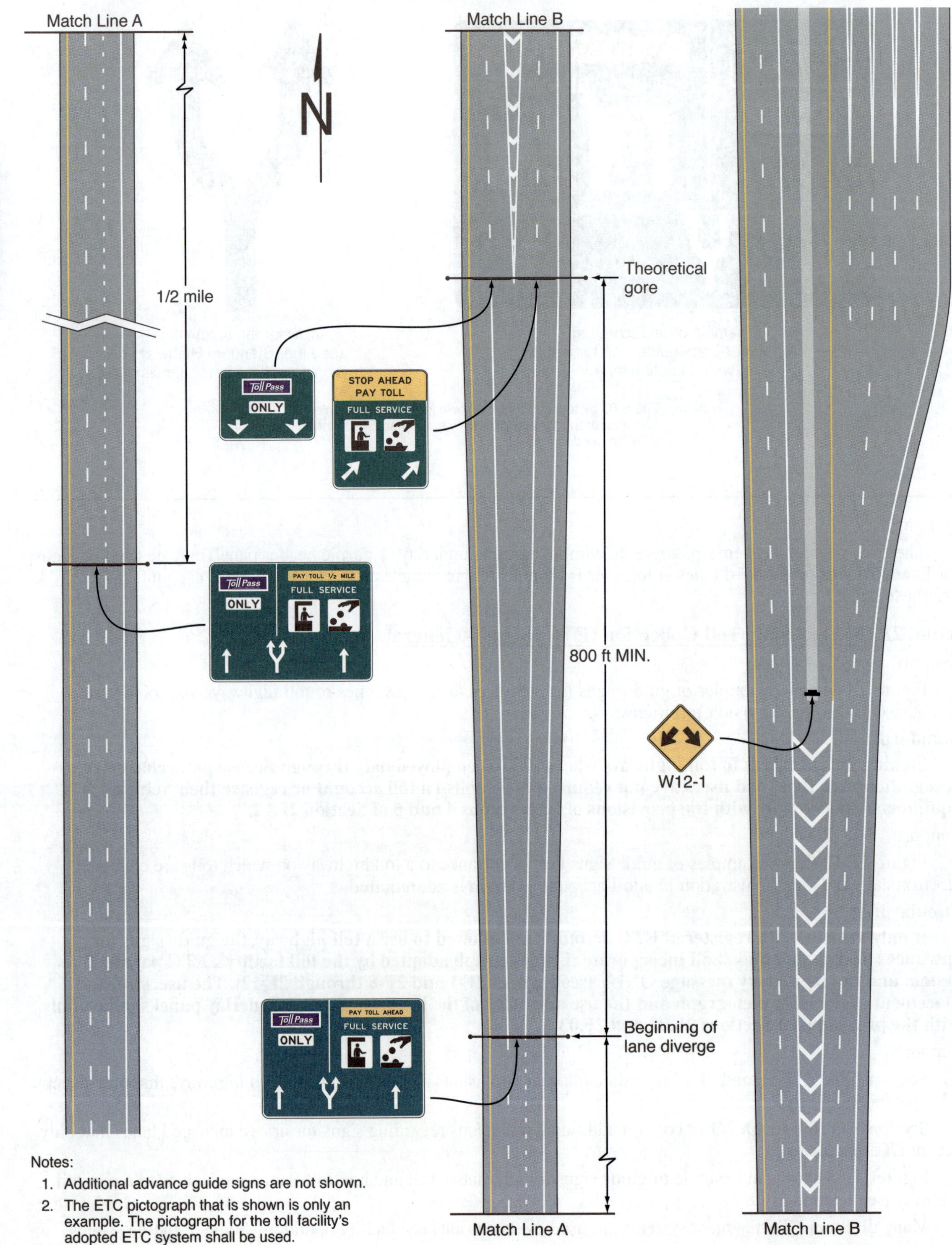

Notes:

1. Additional advance guide signs are not shown.

2. The ETC pictograph that is shown is only an example. The pictograph for the toll facility's adopted ETC system shall be used.

Figure 2F-12. Examples of Guide Signs for the Entrance to a Toll Highway on which Tolls are Collected Electronically Only (Sheet 1 of 3)

A – All tolls are billed through a license plate character recognition system

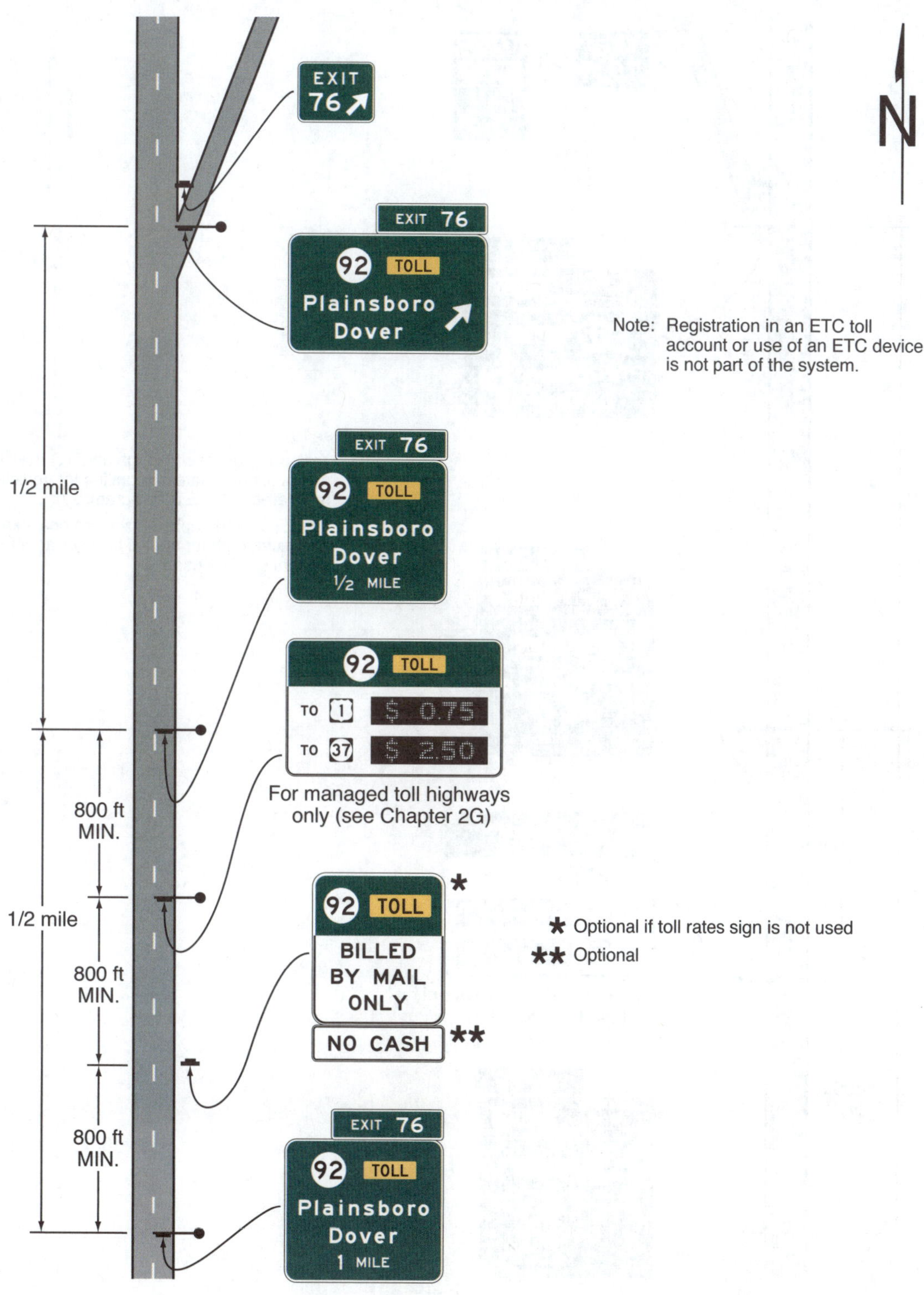

Option:

09   Where a facility will accept payments from other systems' accounts in addition to its primary ETC-account payment system, such information may be displayed on a separate information sign near the entrances to such a facility or in advance of a toll plaza or open-road tolling lanes, as space allows between primary signs.

Figure 2F-12. Examples of Guide Signs for the Entrance to a Toll Highway on which Tolls are Collected Electronically Only (Sheet 2 of 3)

B – Registration in an ETC account program is required for using the tollway

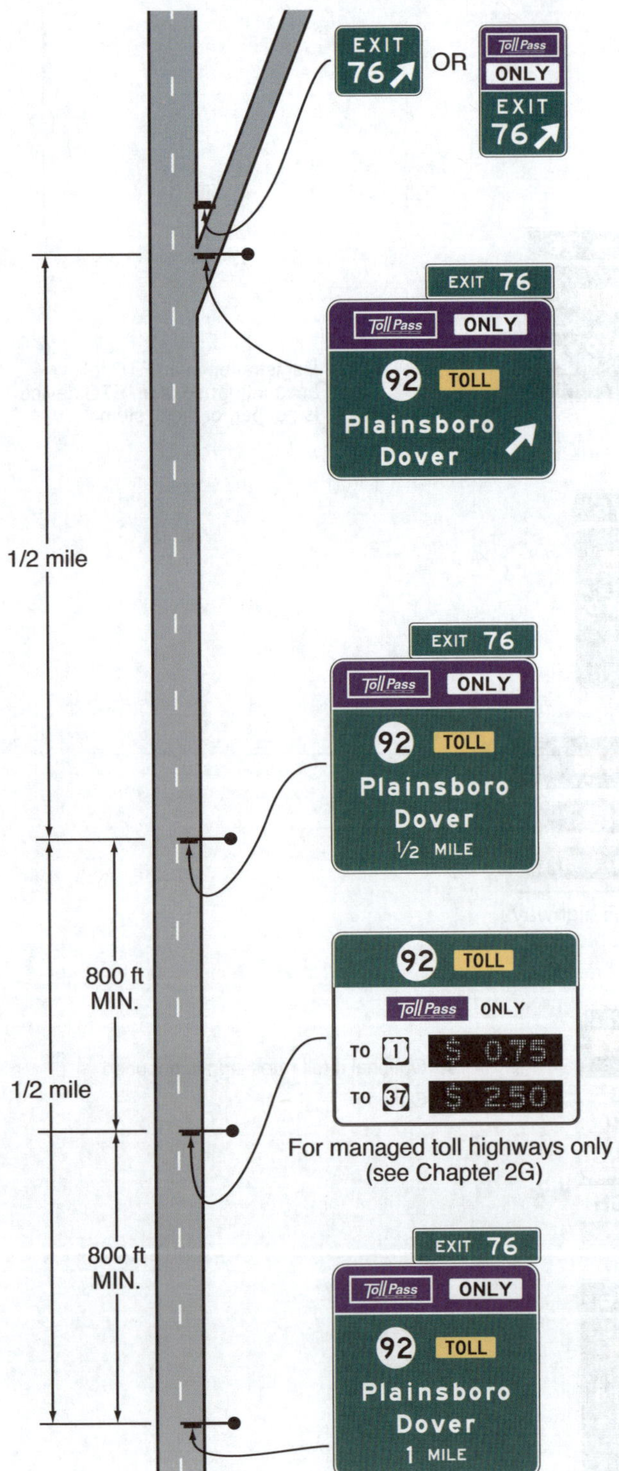

Notes:

1. Tolls are processed though an ETC device and/or license plate recognition for vehicles registered in the ETC program.

2. The ETC pictographs shown are only examples. The pictograph for the toll facility's adopted ETC system shall be used.

Figure 2F-12. Examples of Guide Signs for the Entrance to a Toll Highway on which Tolls are Collected Electronically Only (Sheet 3 of 3)

C – Tolls are billed through either a license plate character recognition system or an ETC account device

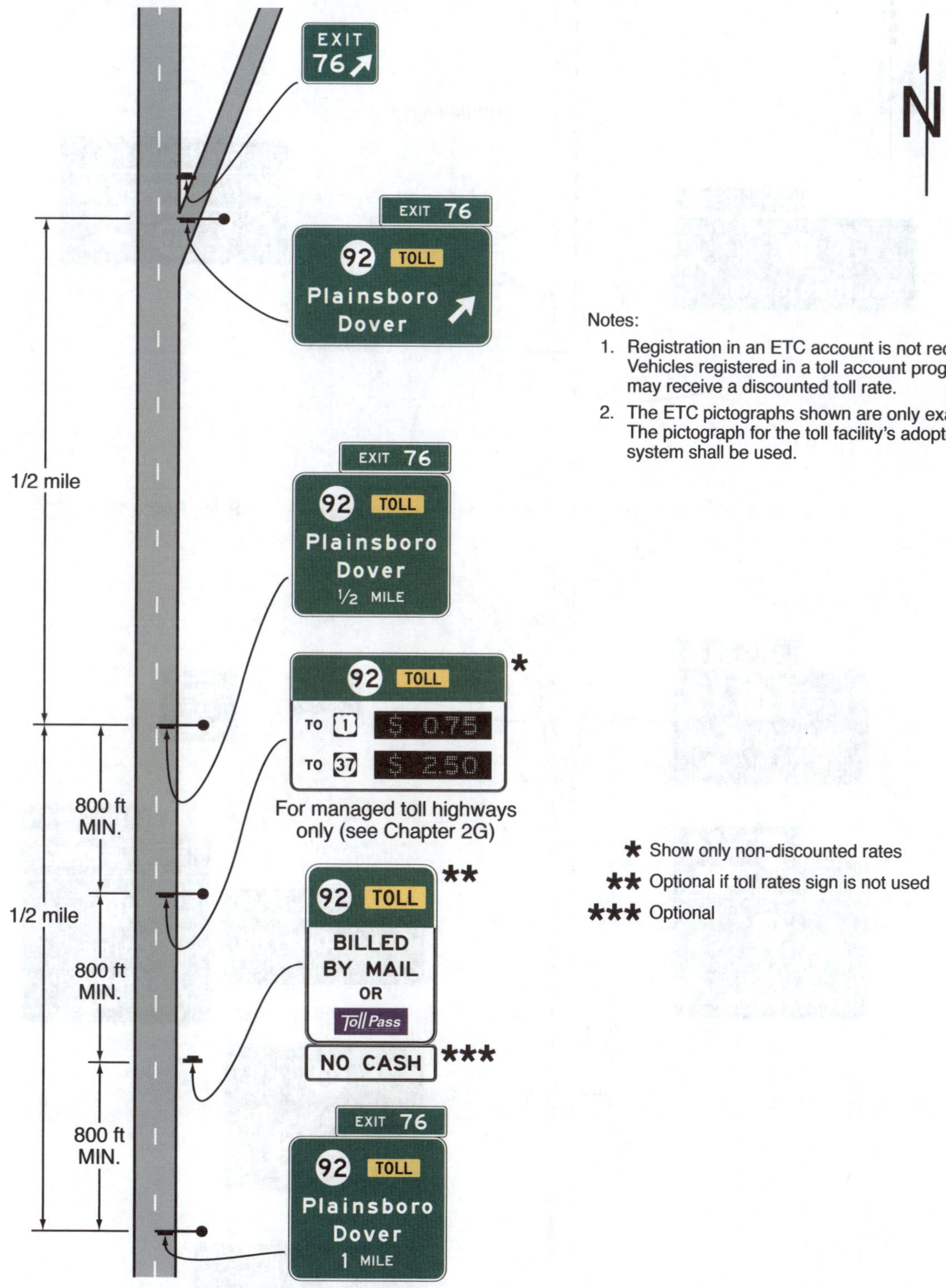

Notes:

1. Registration in an ETC account is not required. Vehicles registered in a toll account program may receive a discounted toll rate.

2. The ETC pictographs shown are only examples. The pictograph for the toll facility's adopted ETC system shall be used.

* Show only non-discounted rates

** Optional if toll rates sign is not used

*** Optional

Figure 2F-13. Examples of Guide Signs for Alternative Toll and Non-Toll Ramp Connections to a Non-Toll Highway

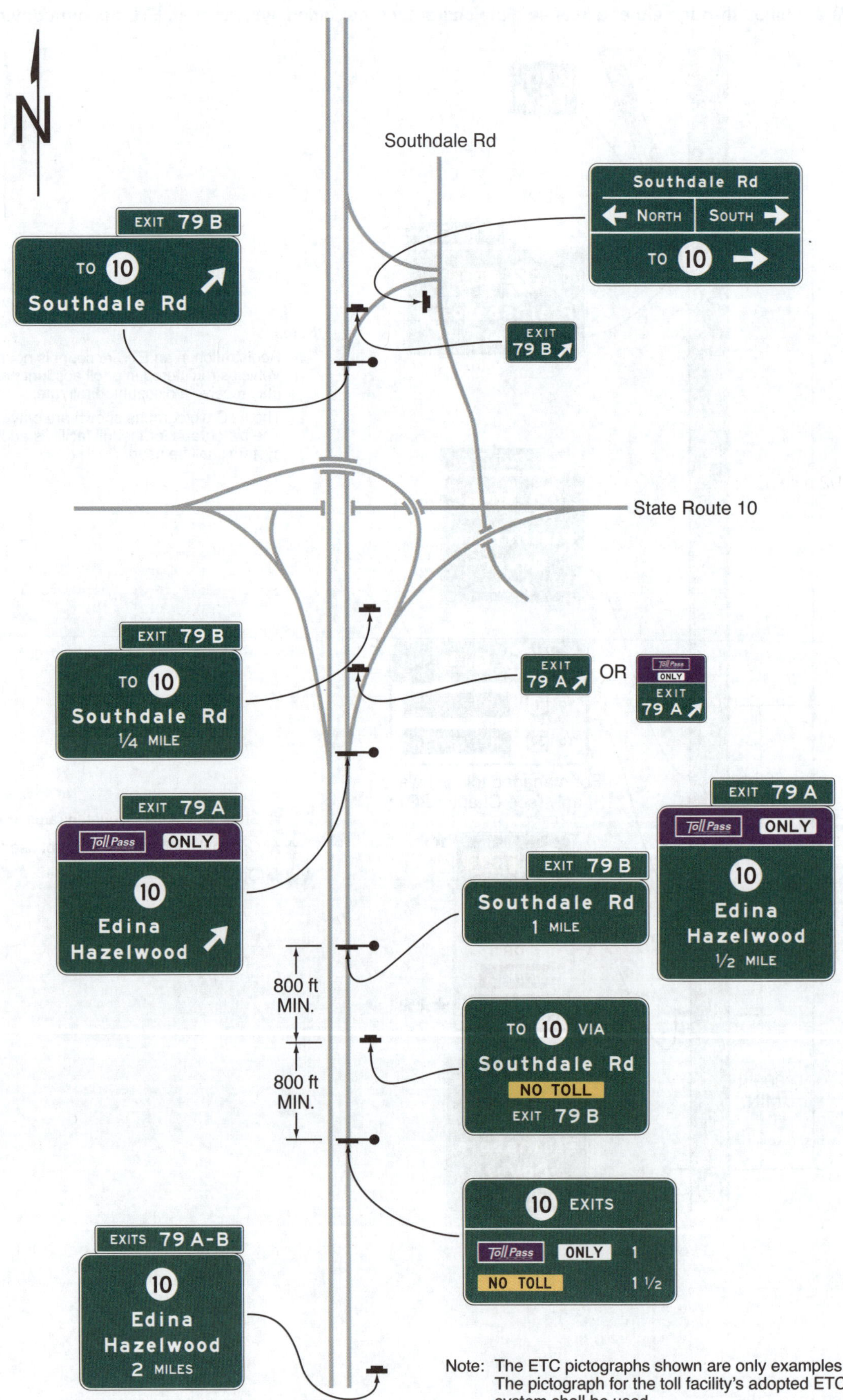

Note: The ETC pictographs shown are only examples. The pictograph for the toll facility's adopted ETC system shall be used.

Section 2F.14 Advance Signs for Conventional Toll Plazas

Guidance:

01   *For conventional toll plazas (those without a divergence onto a separate alignment from mainline-aligned open-road tolling or ETC-Only lanes), one or more sets of overhead advance guide signs complying with the provisions of this Section should be provided. The advance guide signs for multi-lane toll plazas should provide information regarding which lanes to use for all of the toll payment methods accepted at the toll plaza. These signs should include toll plaza lane numbers (if used), or action messages or lane-use information such as LEFT LANE(S), CENTER LANE(S), RIGHT LANE(S), or down arrows over the approximate center of each applicable lane. These signs should also incorporate regulatory messages indicating any restrictions or prohibitions on the use of the lanes associated with the various types of payment methods by certain types of vehicles. For mainline toll plazas, these signs should be at least ½ mile in advance of the toll plaza, and farther if practical.*

02   *Additional guide signs with lane information for the toll payment types should be provided between approximately ¼ mile and 800 feet in advance of the toll plaza at a location that avoids or minimizes obstruction of toll plaza canopy signs (see Section 2F.16) and lane-use control signals.*

03   *The number, mounting, and/or spacing of sets of advance signs for approaches to toll plazas on ramps, toll bridges, or tunnels, to accommodate a limited distance to the plaza from an intersection or from the start of the approach road to the bridge or tunnel, should be based on an engineering study or engineering judgment.*

Support:

04   Figure 2F-14 shows examples of advance signs for a conventional toll plaza.

Section 2F.15 Advance Signs for Toll Plazas on Diverging Alignments from Open-Road ETC Account-Only Lanes

Support:

01   Open-Road ETC lanes are sometimes located on the normal mainline alignment while the lanes for other toll payment methods are located at a toll plaza on a separate alignment (see Figure 2F-15). Since road users paying cash tolls must diverge from the mainline alignment, similar to a movement for an exit, it is important that the guide signs in advance of and at the point of divergence clearly indicate the required lane use and/or movements.

Guidance:

02   *For toll plazas located on a separate alignment that diverges from mainline-aligned Open-Road ETC lanes where vehicles are required to have a registered ETC account to use the Open-Road Tolling lanes, overhead advance signs should be provided at approximately 1 mile and ½ mile in advance of the divergence point. Both the 1-mile and ½-mile advance signs should include:*

    A. *The ETC (pictograph) Account-Only guide sign (see Figures 2F-9 and 2F-15) with a down arrow over the approximate center of each lane that will become an Open-Road ETC lane;*

    B. *For the lane or lanes which will diverge to a toll plaza, guide signs conforming to the provisions of Section 2F.12, indicating which lane or lanes will diverge to the toll plaza for the various cash toll payment methods; and*

    C. *Regulatory signs, plaques, or panels within the guide signs, indicating any restrictions or prohibitions of certain types of vehicles from toll plaza lanes associated with the various types of payment methods.*

03   *At or near the theoretical gore of the divergence point, an additional set of overhead guide signs should be provided and should include:*

    A. *The ETC (pictograph) Account-Only guide sign (see Figures 2F-9 and 2F-15) with a down arrow over the approximate center of each Open-Road ETC lane;*

    B. *Guide signs conforming to the provisions of Section 2F.12 and 2F.13, with diagonally upward-pointing directional arrow(s) over the approximate center of each lane indicating the direction of the divergence, and providing lane information for all types of payment methods accepted at the toll plaza; and*

    C. *Regulatory signs, plaques, or panels within the guide signs, indicating any restrictions or prohibitions on the use of the toll plaza lanes associated with the various types of payment methods by certain types of vehicles.*

04   *Approximately 800 feet in advance of the toll plaza at a location that avoids or minimizes any obstruction of the toll plaza canopy signs (see Section 2F.16) and lane-use control signals, an additional set of overhead advance signs with lane information for the toll payment types should be provided.*

Standard:

05   **The use of down and directional arrows on the signs at the locations described in Paragraphs 2 through 4 of this Section shall comply with the provisions of Section 2D.08.**

Support:

06   Figure 2F-15 shows an example of advance signs for toll plazas on a diverging alignment from Open-Road ETC Account-Only Lanes.

07   Section 4R.02 contains information regarding the use of lane-use control signals for Open-Road ETC lanes for temporary lane closure purposes.

Figure 2F-14. Examples of Mainline Toll Plaza Approach and Canopy Signing

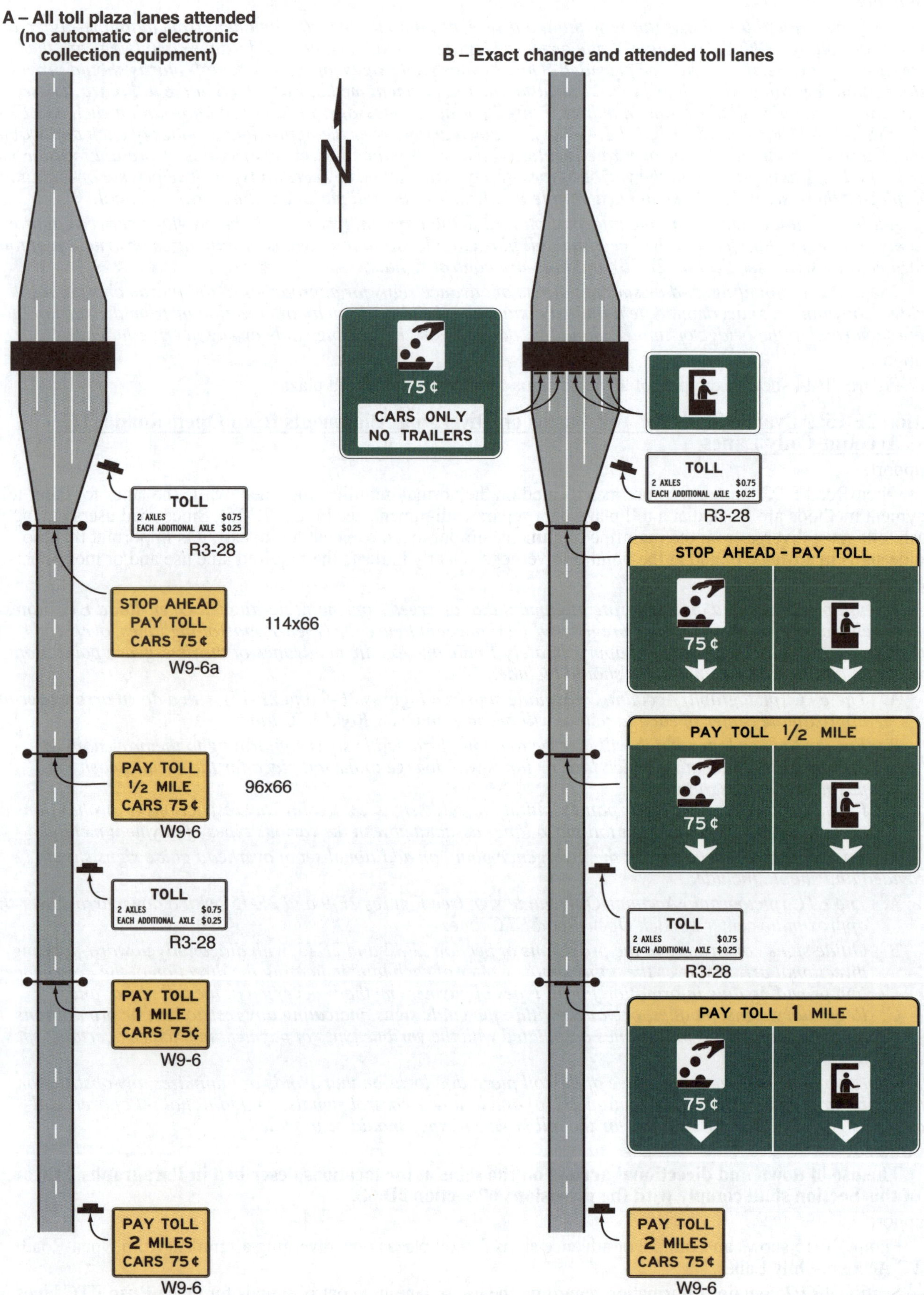

Figure 2F-15. Examples of Guide Signs for a Mainline Toll Plaza on a Diverging Alignment from Open-Road ETC Lanes

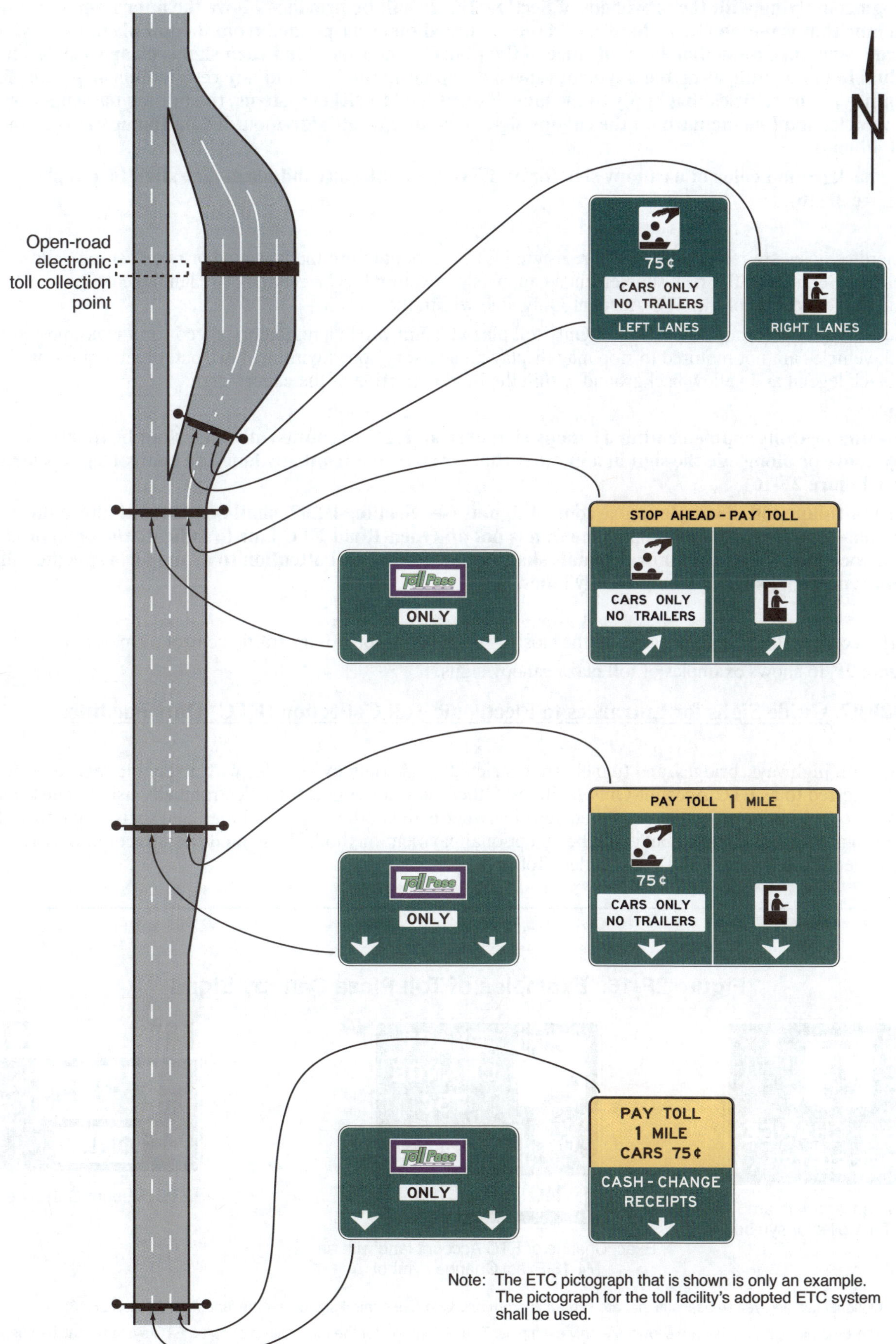

Note: The ETC pictograph that is shown is only an example. The pictograph for the toll facility's adopted ETC system shall be used.

Section 2F.16 Toll Plaza Canopy Signs

Standard:

01   **A sign complying with the provisions of Section 2F.12 shall be provided above the approximate center of each lane that is not an Open-Road ETC lane, mounted on or suspended from the toll plaza canopy, or on a separate structure immediately in advance of the plaza located such that each sign is clearly related to an individual toll lane, indicating the payment type(s) accepted in the lane and any restrictions or prohibitions of certain types of vehicles that apply to the lane. Except for toll-ticket systems, the toll for passenger or 2-axle vehicles shall be included on the canopy sign or on a separate sign mounted on the upstream side of the tollbooth.**

02   **The background color of a canopy sign for an ETC Account-Only toll plaza lane shall be purple (see Figure 2F-16).**

Option:

03   Where vehicles are required to have a registered ETC account to use the lane, one or two flashing yellow beacons (see Section 4R.03) may supplement a canopy sign over an ETC Account-Only lane to call special attention to the location of the ETC Account-Only lane within the plaza.

04   The canopy sign for an ETC Account-Only toll plaza lane in which a regulatory speed limit is not posted and in which vehicles are not required to stop may display an advisory speed within a horizontal rectangular panel with a black legend and yellow background within the bottom portion of the canopy sign.

Standard:

05   **Flashing beacons supplementing a canopy sign over an ETC Account-Only lane shall be mounted directly above or alongside the sign in a manner that is separated from any lane-use control signals for that lane (see Figure 2F-16).**

06   **For multi-lane toll plazas, lane-use control signals (see Section 4R.02) shall be provided above the approximate center of each toll plaza lane that is not an Open-Road ETC lane to indicate the open or closed status of each lane. Lane-use control signals shall not be used to call attention to a lane for a specific toll payment type such as ETC Account-Only lanes.**

Support:

07   Part 6 contains information regarding the closing of a lane for temporary traffic control purposes.

08   Figure 2F-16 shows examples of toll plaza canopy signs.

Section 2F.17 Guide Signs for Entrances to Electronic Toll Collection (ETC) Only Facilities

Support:

01   Some toll highways, bridges, and tunnels are restricted to use only by vehicles with a specific registered ETC account, referred to as ETC Account-Only facilities. Other facilities collect tolls electronically using license plate character recognition in which the registered vehicle owner is then billed by postal mail and registration in an ETC account program is not required but could be an optional payment method. These facilities are commonly referred to as All-Electronic Tolling (AET) or Cashless Tolling.

Figure 2F-16. Examples of Toll Plaza Canopy Signs

Attended lane with an
M4-17 Toll Collector symbol

Exact Change or ETC Account lane with an
M4-18 Exact Change symbol

ETC Account-Only lane

★ Optional flashing yellow beacons that are separated from any lane-use control signals for the lane (see Section 2F.16)

★★ The ETC pictographs that are shown are only examples. The pictograph for the toll facility's adopted ETC system shall be used.

Standard:

02   **Guide signs for facilities that only collect tolls electronically shall comply with the applicable provisions of Chapter 2E and Section 2F.13.**

03   **Guide signs for the entrance ramps to ETC Account-Only facilities shall incorporate the pictograph of the toll facility's primary ETC payment system and the word ONLY in a header panel or plaque designed in accordance with the provisions of Section 2F.13 (see Figure 2F-8).**

Option:

04   A separate information sign displaying the route number, the TOLL warning panel (see Section 2F.11), and the legend NO CASH may be located within the sequence of the Advance guide signs on the approach to the entrance to an ETC Account-Only facility (see Drawing B in Figure 2F-12).

05   Exit Gore signs for entrance ramps to such ETC Account-Only facilities may incorporate the pictograph of the toll facility's ETC payment system and the word ONLY in a header panel or plaque designed in accordance with the provisions of Section 2F.13 (see Figure 2F-13 and Drawing B in Figure 2F-12).

06   If more than one ETC account program is accepted for toll payments on an ETC Account-Only facility, the additional accepted ETC account program pictographs may be displayed on a separate informational guide sign with the legend, ALSO ACCEPTED, within the sequence of advance guide signs for the entrance to the facility.

Support:

07   Section 2F.05 contains information regarding ETC Account-Only signs and plaques for use with route signs in route sign assemblies.

Standard:

08   **Where vehicles are not required to have a registered ETC account to use an ETC-Only facility, guide signs for the facility shall comply with the applicable provisions of Chapter 2E and specifically with the applicable provisions of Section 2F.13.**

09   **Advance and Exit Direction guide signs for the entrances to facilities that do not require registration in an ETC toll account program shall not display a pictograph of an accepted ETC payment system or use purple as a background color on any portion of the signs.**

10   **Information on accepted toll payment methods for a facility that does not require registration in an ETC toll account program shall only be provided on a separate informational guide sign, if used, that displays one of the following legends (see Drawing C in Figure 2F-12):**

    A. TOLL BILLED BY MAIL ONLY, if there is no alternative payment method; or

    B. TOLL BILLED BY MAIL OR [ETC Account Pictograph], if the facility also accepts payments from registered users of an ETC account program.

Option:

11   If there is more than one ETC toll account program accepted, all ETC account program pictographs of the accepted ETC accounts may be displayed on the separate informational guide sign below the TOLL BILLED BY MAIL OR legend. A plaque with the legend NO CASH may be added below the signs described in Paragraph 10 of this Section.

Guidance:

12   *The signs described in Paragraph 10 of this Section should be located within the sequence of Advance Guide signs for the entrance to the facility and/or at a location along the facility itself (see Drawing C in Figure 2F-12).*

Option:

13   If the ETC-Only facility also accepts payments from multiple ETC account programs, but does not require registration in the primary ETC account program associated with the facility in order to use the facility, then the pictographs of the other accepted ETC account programs may be displayed on the separate information sign beneath the legend TOLL BILLED BY MAIL and the word OR.

14   If, in addition to a toll, a nominal surcharge (not a fine, penalty, or violation) is assessed road users not registered in the toll account program, or registered toll account users are assessed a discounted toll, such information may be displayed on a separate information sign on the approach to the entrance to the facility.

Section 2F.18 Guide Signs for ETC-Only Entrance Ramps to Non-Toll Highways

Support:

01   In some cases, access to or from a non-toll route might be provided by a ramp on which a toll is charged in order to manage congestion, limit access, or for other reasons. The toll ramp might be provided as an alternative to or in lieu of a ramp providing similar access without charging a toll. Figures 2F-8 and 2F-13 show examples of guide signs for a ramp on which a toll is charged to enter a non-toll route.

Standard:

02   **Guide signs for ETC-Only Entrance Ramps to non-toll highways shall comply with the provisions of Section 2F.17.**

Option:

03   A NO TOLL panel with a black legend and a yellow background may be included on the top section of the Exit Gore sign for an exit that provides access to the facility without charging a toll (see Figure 2F-8).

Section 2F.19 <u>ETC Account Program Information Signs</u>

Standard:

01   **Except as provided in Paragraph 2 of this Section, signs that inform road users of telephone numbers, Internet addresses, including domain names and uniform resource locators (URLs), or e-mail addresses for enrolling in an ETC account program of a toll facility or managed lane, obtaining an ETC transponder, and/or obtaining ETC account program information shall only be installed in rest areas, parking areas, or similar roadside facilities where the signs are viewed only by pedestrians or occupants of parked vehicles.**

Option:

02   ETC account program information signs displaying telephone numbers that have no more than four characters may be installed on roadways in locations where they will not obscure the road user's view of higher priority traffic control devices and that are removed from key decision points where the road user's view is more appropriately focused on other traffic control devices, roadway geometry, or traffic conditions, including exit and entrance ramps, intersections, toll plazas, temporary traffic control zones, and areas of limited sight distance.

CHAPTER 2G. PREFERENTIAL AND MANAGED LANE SIGNS

Chapter 2G Subchapter and Section Organization

GENERAL

Section 2G.01 Scope
Support:

01   Preferential lanes are lanes designated for special traffic uses such as high-occupancy vehicles (HOVs), light rail, buses, or taxis. Preferential lane treatments might be as simple as restricting a turning lane to a certain class of vehicles during peak periods, or as sophisticated as providing a separate roadway system within a highway corridor for certain vehicles.

02   Preferential lanes might be barrier-separated (on a separate alignment or physically separated from the other travel lanes by a barrier or median), buffer-separated (separated from the adjacent general-purpose lanes only by a narrow buffer area created with longitudinal pavement markings), or contiguous (separated from the adjacent general-purpose lanes only by a lane line). Preferential lanes might allow continuous access with the adjacent general-purpose lanes or restrict access only to designated locations. Preferential lanes might be operated in a constant direction or operated as reversible lanes. Some reversible preferential lanes on a divided highway might be operated counter-flow to the direction of traffic on the immediately adjacent general-purpose lanes.

03   Preferential lanes might be operated on a 24-hour basis, for extended periods of the day, during peak travel periods only, during special events, or during other activities.

04   Open-road tolling lanes and toll plaza lanes that segregate traffic based on payment method are not considered preferential lanes. Chapter 2F contains information regarding signing of open-road tolling lanes and toll plaza lanes.

05   Managed lanes typically restrict access with the adjacent general-purpose lanes to designated locations only.

06   Under certain operational strategies, such as the occupancy requirement of an HOV lane changing in response to actual congestion levels, a managed lane is a special type of preferential lane (see Sections 2G.03 through 2G.07).

07   A managed lane operated on a real-time basis in response to changing conditions might be operated as an HOV lane for a period of time as needed to manage congestion levels.

08   Sections 2G.17 through 2G.19 contain additional information regarding signs for managed lanes that use tolling or pricing as a management strategy.

09   Section 9B.04 contains information regarding Preferential Lane signs for bicycle lanes.

Standard:

10   **Unless otherwise provided, the provisions of this Chapter shall not apply to bicycle lanes.**

Section 2G.02 Sizes of Preferential and Managed Lane Signs
Standard:

01   **Except as provided in Section 2A.07, the minimum sizes of preferential and managed lane signs that have standardized designs shall be as shown in Table 2G-1.**

Support:

02   Section 2A.07 contains information regarding the applicability of the various columns in Table 2G-1.

Option:

03   Signs larger than those shown in Table 2G-1 may be used (see Section 2A.07).

Table 2G-1. Managed and Preferential Lanes Sign and Plaque Minimum Sizes (Sheet 1 of 2)

Sign or Plaque	Sign Designation	Section	Conventional Road		Expressway	Freeway	Oversized
			Single Lane	Multi-Lane			
Preferential Lane Vehicle Occupancy Definition (post-mounted)	R3-10,10a	2G.04	30 x 42	30 x 42	36 x 60	78 x 96	78 x 96
Preferential Lane Operation - High-Occupancy Vehicles (post-mounted)	R3-11 series	2G.05	30 x 42	30 x 42	36 x 60	78 x 96	78 x 96
Motorcycles Allowed (plaque)	R3-11hP	2G.05	30 x 15	30 x 15	36 x 18	78 x 36	78 x 36
Preferential Lane Ahead or Ends - High-Occupancy Vehicles (post-mounted)	R3-12, 12a, 12b, 12c, 12d, 12e	2G.06 2G.07	30 x 42	30 x 42	36 x 60	48 x 84	48 x 84
Preferential Lane Ahead or Ends (post-mounted)	R3-12f, 12g	2G.06 2G.07	30 x 36	30 x 36	36 x 48	48 x 60	48 x 60
Preferential Lane Ends (post-mounted)	R3-12h	2G.07	36 x 48	36 x 48	48 x 66	60 x 84	60 x 84
Preferential Lane Vehicle Occupancy Definition (overhead)	R3-13,13a	2G.04	66 x 36	66 x 36	84 x 48	144 x 78	144 x 78
HOV Lane Operation (overhead)	R3-14,14a	2G.05	72 x 60	72 x 60	96 x 72	144 x 108	144 x 108
HOV Lane Operation (overhead)	R3-14b	2G.05	72 x 60	72 x 60	96 x 72	120 x 96	120 x 96
Preferential Lane Operation (overhead)	R3-14c	2G.05	90 x 60	90 x 60	108 x 72	156 x 102	156 x 102
HOV Lane Ahead (overhead)	R3-15	2G.06	66 x 36	66 x 36	84 x 48	102 x 60	102 x 60
HOV Lane Begins XX Miles (overhead)	R3-15a	2G.06	78 x 48	114 x 72	144 x 84	150 x 108	150 x 108
HOV Lane Ends (overhead)	R3-15b,15c	2G.07	66 x 36	66 x 36	84 x 48	102 x 60	102 x 60
Preferential Lane Ahead or Ends (overhead)	R3-15d,15e	2G.06 2G.07	42 x 36	42 x 36	54 x 48	72 x 60	72 x 60
Priced Managed Lane Vehicle Occupancy Definition (post-mounted)	R3-40	2G.18	—	—	54 x 66	54 x 66	66 x 78
Priced Managed Lane Ends (post-mounted)	R3-42	2G.18	—	—	48 x 60	48 x 60	60 x 78
Priced Managed Lane Ends Advance (post-mounted)	R3-42a	2G.18	—	—	48 x 66	48 x 66	60 x 84
Priced Managed Lane Restriction Ends (post-mounted)	R3-42b	2G.18	—	—	48 x 60	48 x 60	60 x 78
Priced Managed Lane Restriction Ends Advance (post-mounted)	R3-42c	2G.18	—	—	48 x 66	48 x 66	60 x 84
Priced Managed Lane Vehicle Occupancy Definition	R3-43	2G.18	—	—	138 x 66	138 x 66	—
Priced Managed Lane Operation (overhead)	R3-44	2G.18	—	—	90 x 84	90 x 84	—
Priced Managed Lane Operation (overhead)	R3-44a	2G.18	—	—	132 x 84	132 x 84	—
Priced Managed Lane Operation (overhead)	R3-44b	2G.18	—	—	90 x 72	90 x 72	—
Priced Managed Lane Ends (overhead)	R3-45	2G.18	—	—	90 x 66	90 x 66	—
Priced Managed Lane Restriction Ends (overhead)	R3-45a	2G.18	—	—	114 x 66	114 x 66	—
Priced Managed Lane Toll Rate	R3-48	2G.18	—	—	Varies	Varies	—
Priced Managed Lane Toll Rate	R3-48a	2G.18	—	—	Varies	Varies	—
Part-Time Travel on Shoulder Operation	R3-51	2G.21	—	—	66 x 78	66 x 78	—
No Trucks (plaque)	R3-51aP	2G.21	—	—	66 x 12	66 x 12	—
No Trucks or Buses (plaque)	R3-51bP	2G.21	—	—	66 x 24	66 x 24	—
Emergency Stopping Only Other Times (plaque)	R3-51cP	2G.21	—	—	54 x 42	54 x 42	—

Table 2G-1. Managed and Preferential Lanes Sign and Plaque Minimum Sizes (Sheet 2 of 2)

Sign or Plaque	Sign Designation	Section	Conventional Road		Expressway	Freeway	Oversized
			Single Lane	Multi-Lane			
Part-Time Travel on Shoulder Variable Operation	R3-51d	2G.21	—	—	66 x 84	66 x 84	—
Part-Time Travel on Shoulder on Green Arrow	R3-51e	2G.24	—	—	66 x 84	66 x 84	—
Part-Time Travel on Shoulder Ends	R3-52	2G.21	—	—	66 x 72	66 x 72	—
Part-Time Travel on Shoulder Ends Advance	R3-52a	2G.21	—	—	66 x 72	66 x 72	—
Shoulder Must Exit Advance	R3-52b	2G.21	—	—	66 x 72	66 x 72	—
Part-Time Travel on Shoulder Begins Advance	R3-52c	2G.21	—	—	66 x 72	66 x 72	—
Begin Exit Lane	R3-56	2G.21	—	—	48 x 72	48 x 72	—
To Traffic on Shoulder (plaque)	R3-57P	2G.21	—	—	36 x 18	36 x 18	—
Traffic Using Shoulder	W3-9	2G.22	—	—	36 x 48	48 x 60	—
High-Occupancy Vehicles (plaque)	W16-11P	2G.09	24 x 12	24 x 12	30 x 18	30 x 18	30 x 18
Preferential Lane Entrance Gore	E8-1	2G.10	—	—	48 x 96	48 x 96	—
Preferential Lane Intermediate Entrance Gore	E8-1a	2G.10	—	—	48 x 84	48 x 84	—
Preferential Lane Entrance Direction (overhead)	E8-2	2G.10	—	—	228 x 72	228 x 72	—
Preferential Lane Entrance Direction (post-mounted)	E8-2a	2G.10	—	—	186 x 108	186 x 108	—
Preferential Lane Entrance Advance	E8-3	2G.10	—	—	186 x 96	186 x 96	—
Preferential Lane Direct Exit Gore	E8-4	2G.15	—	—	60 x 78	60 x 78	—
Preferential Lane Intermediate Egress Direction	E8-5	2G.13	—	—	Varies x 90	Varies x 90	—
Preferential Lane Intermediate Egress Advance	E8-6	2G.13	—	—	Varies x 84	Varies x 84	—

Notes: 1. Larger signs may be used when appropriate

2. Dimensions are shown as width x height, in inches

REGULATORY SIGNS

Section 2G.03 Regulatory Signs for Preferential Lanes – General

Standard:

01   **When a preferential lane is established, the Preferential Lane regulatory signs (see Figure 2G-1) and pavement markings (see Chapter 3E) for these lanes shall be used to advise road users.**

Support:

02   Preferential Lane (R3-10 series through R3-15 series) regulatory signs (see Figure 2G-1) consist of several different general types of regulatory signs as follows:

    A. Vehicle Occupancy Definition signs define the vehicle occupancy requirements applicable to an HOV lane (such as "2 OR MORE PERSONS PER VEHICLE") or types of vehicles not meeting the minimum occupancy requirement (such as motorcycles or Inherently Low Emission Vehicles (ILEVs)) that are allowed to use an HOV lane (see Section 2G.04).

    B. Preferential Lane Operation signs notify road users of the days and hours during which the preferential restrictions are in effect (see Section 2G.05).

    C. Preferential Lane Advance signs notify road users that a preferential lane restriction begins ahead (see Section 2G.06).

    D. Preferential Lane Ends signs notify users of the termination point of the preferential lane restrictions (see Section 2G.07).

Standard:

03   **Regulatory signs applicable only to a preferential lane shall be distinguished from regulatory signs applicable to general-purpose lanes by the inclusion of the applicable symbol(s) and/or word(s) (see Figure 2G-1).**

Support:

04   The symbol and word message displayed on a particular Preferential Lane regulatory sign will vary based on the specific type of allowed traffic and on other related operational constraints that have been established for a particular lane, such as an HOV lane, a bus lane, or a taxi lane.

Option:

05   Changeable message signs may supplement, substitute for, or be incorporated into static Preferential Lane regulatory signs where travel conditions change or where multiple types of operational strategies (such as variable occupancy requirements or vehicle types) are used and varied throughout the day or week, or on a real-time basis, to manage the use of, control of, or access to preferential lanes.

Support:

06   Figure 2G-1 illustrates examples of changeable messages incorporated into static Preferential Lane regulatory signs displaying open and closed status using lane-use control signal indications (see Chapter 4T).

Standard:

07   **When changeable message signs (see Chapter 2L) are used as regulatory signs for preferential lanes, they shall be the required sign size and shall display the required letter height and legend format that corresponds to the type of roadway facility and design speed.**

Guidance:

08   *When Preferential Lane regulatory signs are used on conventional roads, the decision regarding whether to use a post-mounted or overhead version of a particular type of sign should be based on an engineering study that considers the available space, the existing signs for the adjacent general-purpose traffic lanes, roadway and traffic characteristics, the proximity to existing overhead signs, the ability to install overhead signs, and any other unique local factors.*

09   *If overhead regulatory signs applicable only to a preferential lane are located in approximately the same longitudinal position along the highway as overhead signs applicable only to the general-purpose lanes, the signs for the preferential lane should be separated laterally from the signs for the general-purpose lanes to the maximum extent practicable to minimize conflicting information, while maintaining their visual relationship to the lanes below necessitated by specific legend or arrows indicating lane assignment.*

Standard:

10   **If used, overhead Preferential Lane (R3-13 series, R3-14 series, and R3-15 series) regulatory signs shall be installed on the side of the roadway where the entrance to the preferential lane is located and any appropriate adjustments shall be made to the sign message.**

Option:

11   Where a median of sufficient width is available, the R3-13 series and R3-15 series signs may be post-mounted.

Figure 2G-1. Preferential Lane Regulatory Signs and Plaque (Sheet 1 of 2)

A – Post-mounted Preferential Lane regulatory signs and plaque

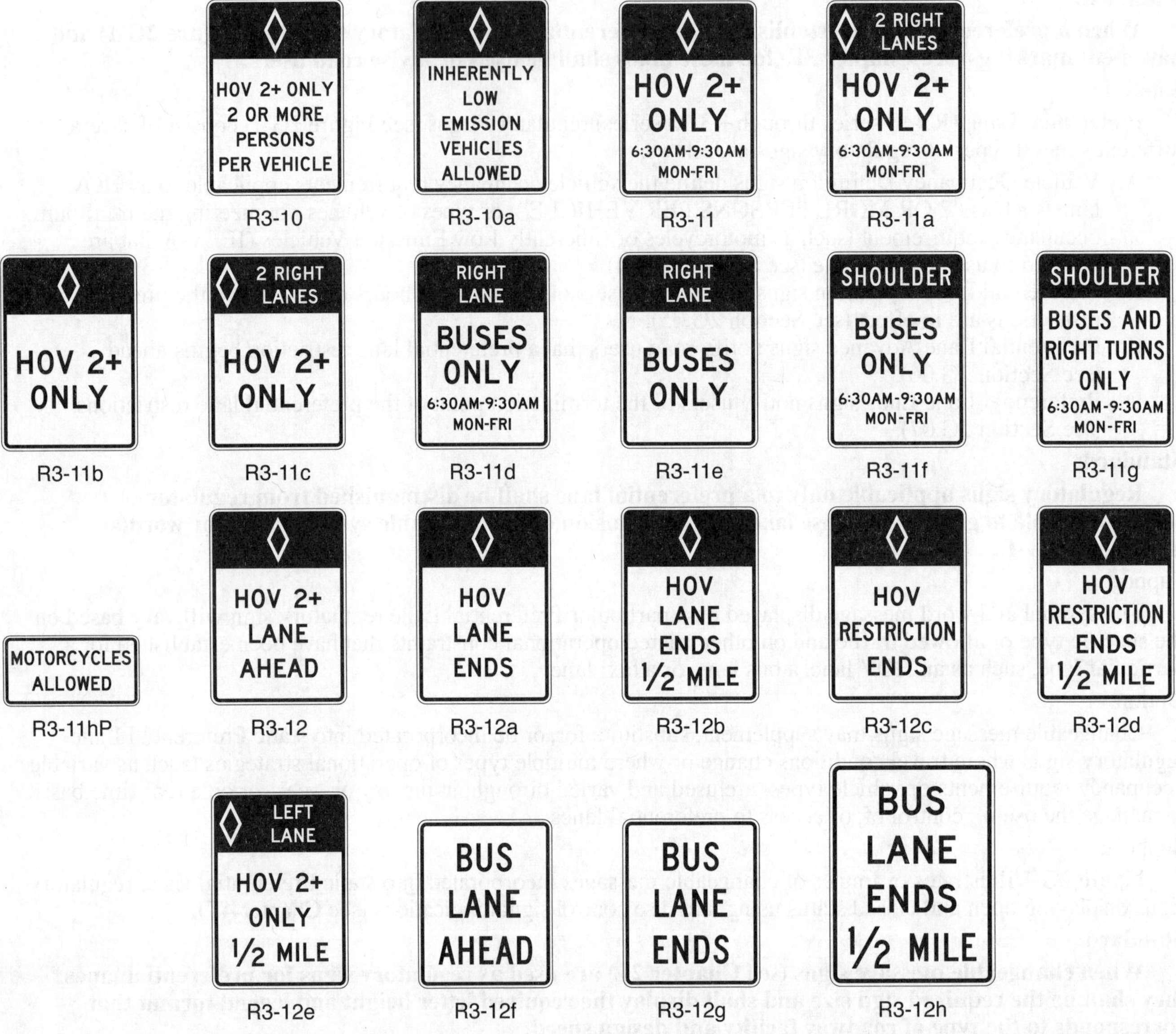

Notes:
1. The minimum vehicle occupancy requirement may vary for each facility (such as 2+, 3+, 4+).
2. The occupancy requirement may be added to the first line of the R3-12a, R3-12b, R3-12c, and R3-12d signs.
3. Some of the legends shown on these signs are for example purposes only. The specific legend for a particular application should be based upon local conditions, ordinances, and State statutes.

Support:

12   The sizes for Preferential Lane regulatory signs will differ to reflect the design speeds for each type of roadway facility. Table 2G-1 provides sizes for each type of roadway facility.

Guidance:

13   *The edges of Preferential Lane regulatory signs that are post-mounted on a median barrier should not project beyond the outer edges of the barrier, including in areas where lateral clearance is limited.*

Option:

14   Where lateral clearance is limited, Preferential Lane regulatory signs that are post-mounted on a median barrier and that are 72 inches or less in width may be skewed up to 45 degrees in order to fit within the barrier width or may be mounted higher, such that the vertical clearance to the bottom of the sign, light fixture, or structural support, whichever is lowest, is not less than 17 feet above any portion of the pavement and shoulders.

Figure 2G-1. Preferential Lane Regulatory Signs and Plaque (Sheet 2 of 2)

B – Overhead Preferential Lane regulatory signs

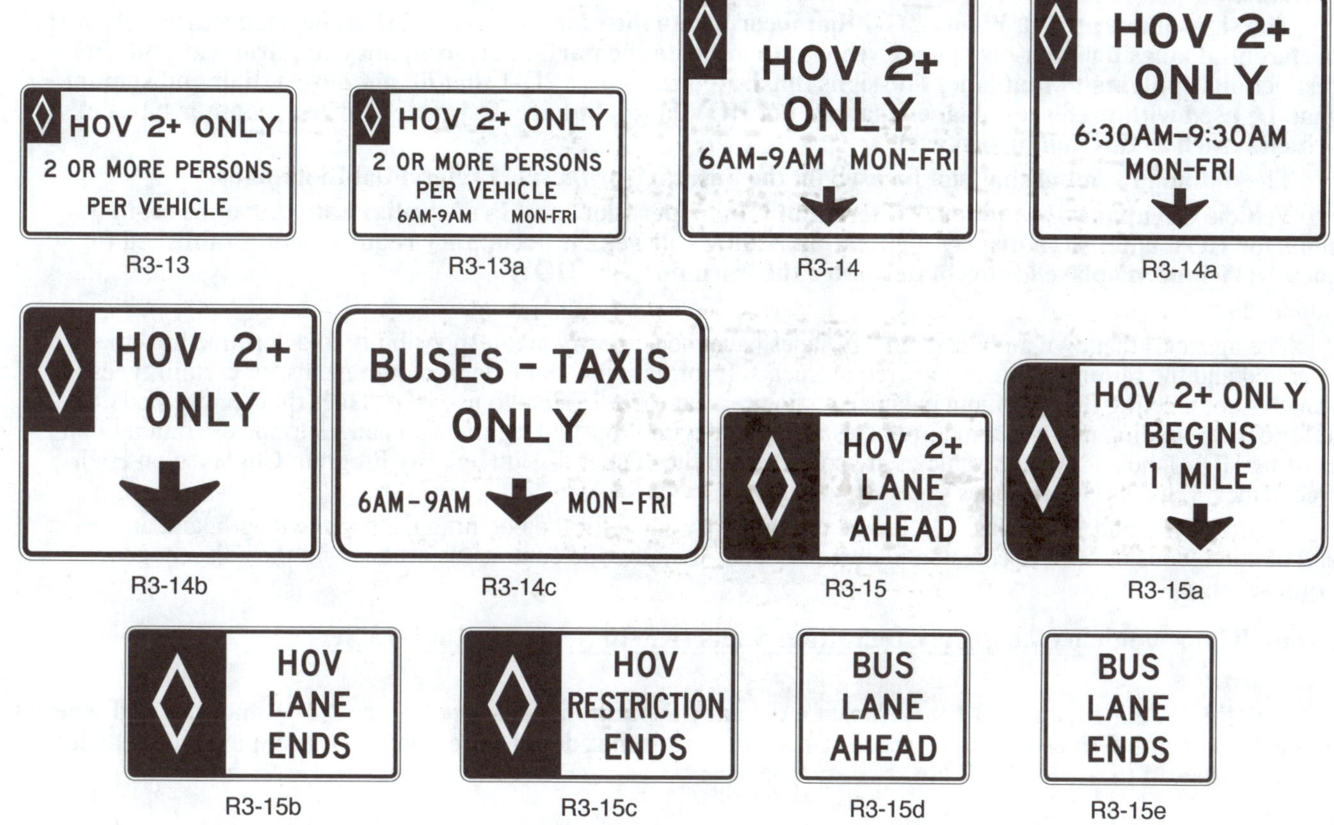

A lane-use control signal may be incorporated into an overhead preferential lane regulatory sign to indicate the status of a reversible operation as shown in the following example:

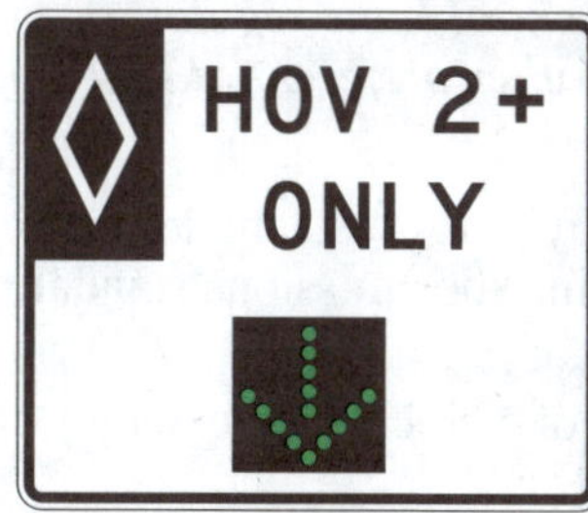

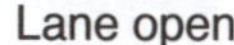

Lane open   Lane closed

Notes:

1. The minimum vehicle occupancy requirement may vary for each facility (such as 2+, 3+, 4+).

2. The occupancy requirement may be added to the first line of the R3-15b and R3-15c signs.

3. Some of the legends shown on these signs are for example purposes only. The specific legend for a particular application should be based upon local conditions, ordinances, and State statutes.

4. Where sufficient median width is available, the R3-13 series and R3-15 series signs may be post-mounted.

Standard:

15   **Where lateral clearance is limited, Preferential Lane regulatory signs that are post-mounted on a median barrier and that are wider than 72 inches shall be mounted with a vertical clearance that complies with the provisions of Section 2A.14 for overhead mounting if any portion of the sign extends over the roadway.**

Guidance:

16   *On conventional roadways, Preferential Lane regulatory sign spacing should be determined by engineering judgment based on speed, block length, distances from adjacent intersections, and other site-specific considerations.*

Support:

17   Sections 2G.04 and 2G.05 contain provisions regarding the placement of Preferential Lane regulatory signs on freeways and expressways.

Standard:

18   **The signs illustrated in Figure 2G-1 that incorporate the diamond symbol shall be used exclusively with preferential lanes for high-occupancy vehicles to indicate the particular occupancy requirement and time restrictions applying to that lane. The signs illustrated in Figure 2G-1 that do not have a diamond symbol shall be used with preferential lanes that are not HOV lanes, but are designated for use by other types of vehicles (such as bus and/or taxi use).**

19   **The diamond symbol shall not be used on the bus, taxi, or bicycle Preferential Lane signs.**

20   **Vehicle Occupancy Definition, Preferential Lane Operation, and Preferential Lane Advance regulatory signs for HOV lanes shall display the minimum allowable vehicle occupancy requirement established for each HOV lane, displayed immediately after the word message HOV.**

Support:

21   The agencies that own and operate HOV lanes have the authority and responsibility to determine how they are operated and the minimum occupancy requirements. Information about federal requirements for certain types of vehicles not meeting the minimum occupancy requirement to be eligible to use HOV lanes that receive Federal-aid program funding and about requirements associated with proposed significant changes to the operation of an existing HOV lane and certain vehicles are contained in the "Federal-Aid Highway Program Guidance on High Occupancy Vehicle (HOV) Lanes."

22   Figures 2G-2 and 2G-3 illustrate the use of regulatory signs for the beginning, along the length, and at the end of contiguous or buffer-separated preferential lanes that provide continuous access with the adjacent general-purpose lanes.

Section 2G.04 Vehicle Occupancy Definition Signs (R3-10 Series and R3-13 Series)

Standard:

01   **The R3-10, R3-13, and R3-13a Vehicle Occupancy Definition signs (see Figure 2G-1) shall be used where agencies determine that it is appropriate to provide a sign that defines the minimum occupancy of vehicles that are allowed to use an HOV lane.**

Guidance:

02   *The Inherently Low Emission Vehicle (ILEV) (R3-10a) sign (see Figure 2G-1) should be used when it is permissible for a properly labeled and certified ILEV, regardless of the number of occupants, to use an HOV lane. When used, the ILEV signs should be post-mounted in advance of and at intervals along the HOV lane based upon engineering judgment and the placement of other Preferential Lane regulatory signs. The R3-10a sign is only applicable to HOV lanes and should not to be used with other preferential lane applications.*

Support:

03   ILEVs are defined by the Environmental Protection Agency (EPA) as vehicles having no fuel vapor (hydrocarbon) emissions and are certified by the EPA as meeting the emissions standards and requirements specified in 40 CFR §88.311-93 and 40 CFR §88.312-93(c).

04   Section 2G.18 contains information regarding the legends of Vehicle Occupancy Definition signs for a priced managed lane that has an occupancy requirement for non-toll travel.

Standard:

05   **For barrier-separated, buffer-separated, or contiguous preferential lanes where access between the preferential and general-purpose lanes is restricted to designated locations on freeways and expressways, an overhead Vehicle Occupancy Definition (R3-13 or R3-13a) sign shall be installed at least ½ mile in advance of the beginning of or initial entry point to an HOV lane. These signs shall only be displayed in advance of the beginning of or initial or intermediate entry point to HOV lanes.**

06   **For buffer-separated or contiguous HOV lanes where access is restricted to designated locations on freeways and expressways, the sequence of a post-mounted Preferential Lane Operation (R3-11a) sign (see Section 2G.05) followed by a post-mounted Vehicle Occupancy Definition (R3-10) sign shall be located at intervals not greater than ½ mile along the length of designated gaps where vehicles are allowed to legally access the HOV lane, and within designated enforcement areas as defined by the operating agency.**

Option:

07   For buffer-separated or contiguous HOV lanes where access is restricted to designated locations on freeways and expressways, the sequence of a post-mounted Preferential Lane Operation (R3-11a) sign (see Section 2G.05) followed by a post-mounted Vehicle Occupancy Definition (R3-10) sign may be located at intervals of approximately ½ mile along the length of the HOV lane.

Figure 2G-2. Example of Signing for an Added Continuous-Access Contiguous or Buffer-Separated HOV Lane

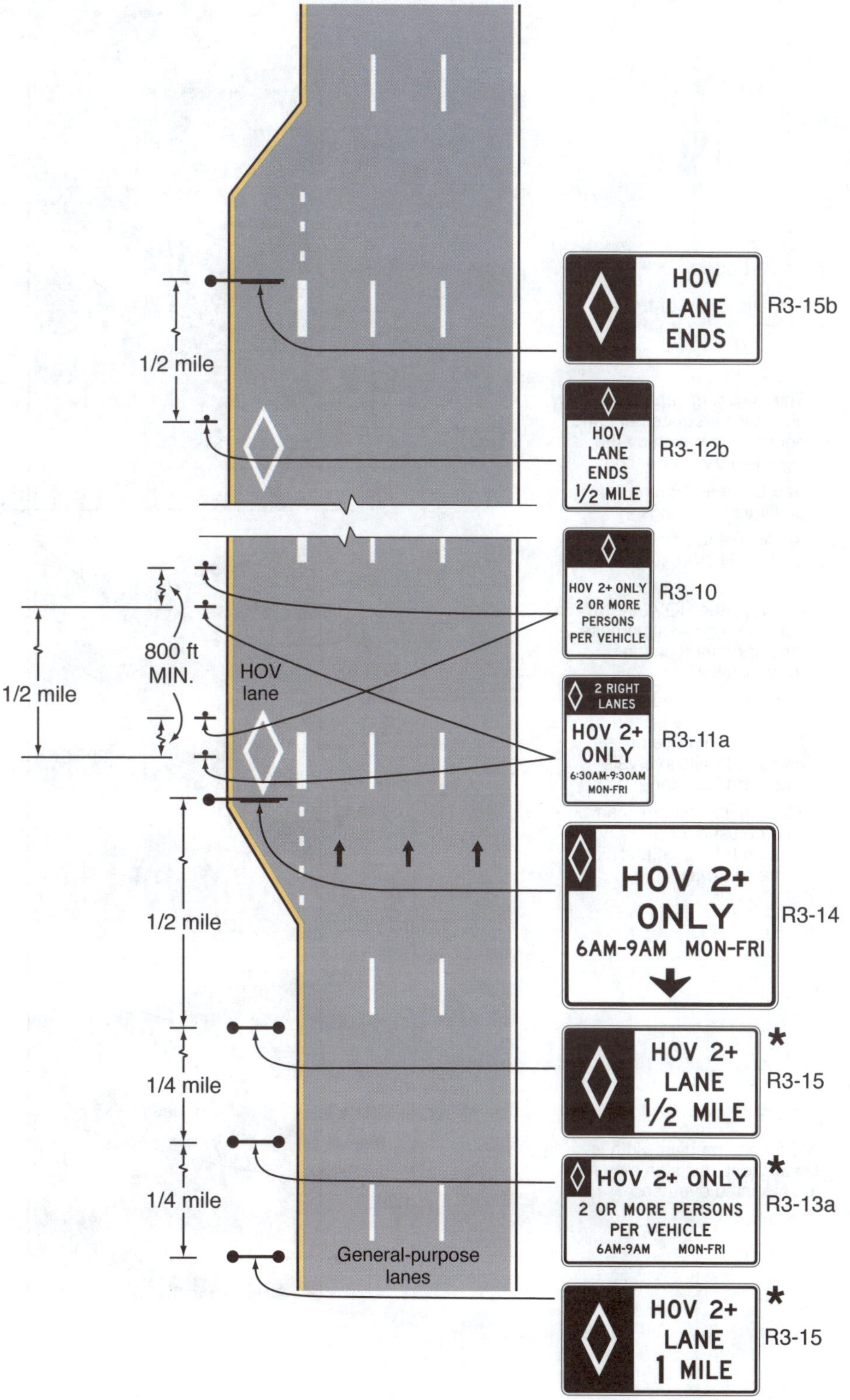

Figure 2G-3. Example of Signing for a General-Purpose Lane that Becomes a Continuous-Access Contiguous or Buffer-Separated HOV Lane

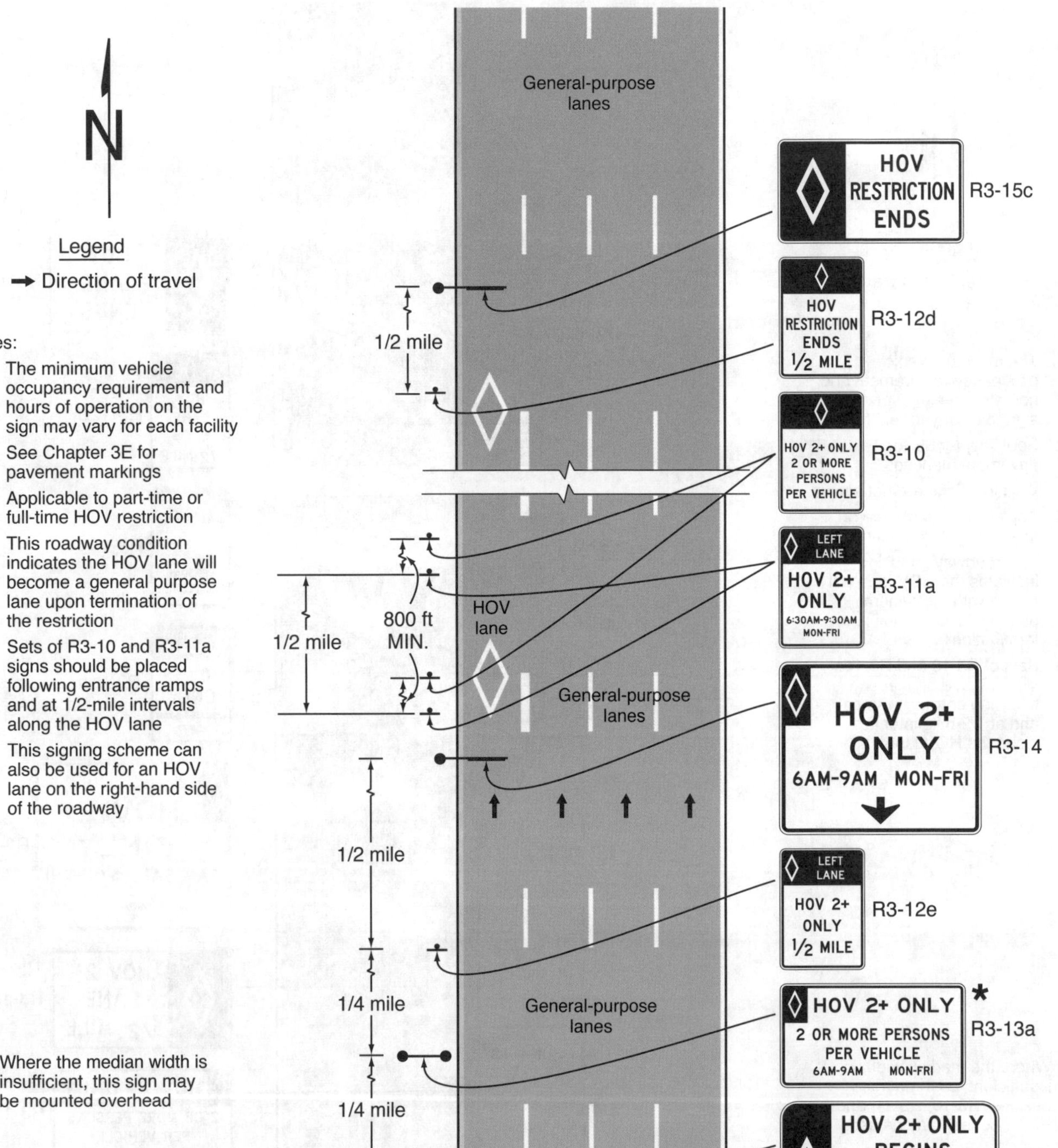

Notes:

1. The minimum vehicle occupancy requirement and hours of operation on the sign may vary for each facility

2. See Chapter 3E for pavement markings

3. Applicable to part-time or full-time HOV restriction

4. This roadway condition indicates the HOV lane will become a general purpose lane upon termination of the restriction

5. Sets of R3-10 and R3-11a signs should be placed following entrance ramps and at 1/2-mile intervals along the HOV lane

6. This signing scheme can also be used for an HOV lane on the right-hand side of the roadway

∗ Where the median width is insufficient, this sign may be mounted overhead

08   For barrier-separated HOV lanes on freeways and expressways, the sequence of a post-mounted Preferential Lane Operation (R3-11a) sign (see Section 2G.05) followed by a post-mounted Vehicle Occupancy Definition (R3-10) sign may be located at intervals of approximately ½ mile along the length of the HOV lane, at intermediate entry points, and at designated enforcement areas as defined by the operating agency.

Standard:

09   **For buffer-separated or contiguous HOV lanes where continuous access with the adjacent general-purpose lanes is provided on freeways and expressways, the sequence of a post-mounted Preferential Lane Operation (R3-11a) sign (see Section 2G.05) followed by a post-mounted Vehicle Occupancy Definition (R3-10) sign, and ILEV (R3-10a) signs, if appropriate, shall be located at intervals not greater than ½ mile along the length of the HOV lane.**

Guidance:

10   *On freeways and expressways, the signs within each Preferential Lane regulatory sign sequence should be separated by a minimum distance of 800 feet and a maximum distance of 1,000 feet.*

11   *On conventional roads, the distance between Preferential Lane regulatory signs within each sequence should be determined by engineering judgment based on speed, block length, distances from adjacent intersections, and other site-specific considerations.*

Standard:

12   **For all types of direct access ramps that provide access to or lead to HOV lanes, a post-mounted Vehicle Occupancy Definition (R3-10) sign, and an ILEV (R3-10a) sign, if appropriate, shall be used at the beginning or initial entry point for the direct access ramp.**

Section 2G.05 Preferential Lane Operation Signs (R3-11 Series and R3-14 Series)

Support:

01   The standardized sizes of post-mounted Preferential Lane Operation (R3-11 series) signs are consistent to accommodate any future addition or removal of a single line of legend for each sign. Each size accommodates two lines of legend for the times of day and days of week that the regulation is in effect. Consistent sign sizes are beneficial for agencies when ordering sign materials, as well as when making legend changes to existing signs if changes occur to operating times or occupancy restrictions in the future.

Guidance:

02   *Where the regulation is in effect during more than one time period of the day, such as during the morning and afternoon peak periods, the height of the R3-11 and R3-14 series signs should be suitably increased to accommodate the additional line(s) of legend.*

Standard:

03   **When used, the post-mounted Preferential Lane Operation (R3-11 series) signs shall be located adjacent to the preferential lane, and the overhead Preferential Lane Operation (R3-14 series) signs shall be mounted directly over the lane.**

04   **The legend format of the post-mounted Preferential Lane Operation (R3-11 series) signs shall have the following sequence:**

    A. Top Lines: Lanes applicable, such as "RIGHT LANE" or "2 RIGHT LANES"
    B. Middle Lines: Eligible uses, such as "HOV 2+ ONLY" (or 3+ or 4+, if appropriate) or "BUSES ONLY" or other applicable uses or eligible turning movements
    C. Bottom Lines: Applicable times and days, such as "7 AM – 9 AM" or "6:30 AM – 9:30 AM, MON-FRI"

05   **The legend format of the overhead Preferential Lane Operation (R3-14 series) signs shall have the following sequence:**

    A. Top Lines: Eligible uses, such as "HOV 2+ ONLY" (or 3+ or 4+, if appropriate) or "BUSES ONLY" or other applicable uses or eligible turning movements
    B. Bottom Lines: Applicable times and days, with the time and day placed above the down arrow, such as "7 AM – 9 AM" or "6:30 AM – 9:30 AM, MON-FRI." When the operating periods exceed the available line width, the hours and days of the week shall be stacked as shown for the R3-14a sign in Figure 2G-1.

06   **For preferential lane restrictions that are in effect on a full-time (24 hours per day, seven days per week) basis, the Preferential Lane Operation (R3-11b, R3-11c, R3-11e, R3-11g, R3-14b, R3-14e, and R3-14g) signs that include a period of operation legend shall be modified to display no legend relative to the period of operation except as provided in Paragraph 7 of this Section.**

Option:

07   In lieu of removing the period of operation text from the Preferential Lane Operation signs for full-time operation, the legend 24 HOURS may be substituted for the times and days of the week on the Preferential Lane Operation (R3-11, R3-11a, R3-11d, R3-11f, R3-14, R3-14d, and R3-14f) signs for preferential lane restrictions that are in effect on a full-time basis.

Support:

08   The 24 HOURS legend displayed on the R3-11c sign reinforces the full-time operation where several facilities in the same area have different hours of operation—some part time, others full time, or where the same lane changes from part-time to full-time operation somewhere along its length.

Standard:

09   **The full-time Preferential Lane Operation (R3-11b, R3-11c, R3-11e, R3-11g, R3-14b, R3-14e, and R3-14g) signs that do not display a period of operation legend, as described in Paragraph 6 of this Section, shall not be used where the preferential lane restriction is in effect only on a part-time basis.**

Option:

10   Where additional movements are allowed from a preferential lane on an approach to an intersection by vehicles not meeting the preferential lane regulation, the format and words used in the legend in the middle lines on the post-mounted Preferential Lane Operation (R3-11 series) signs and on the top line of the overhead Preferential Lane Operation (R3-14 series) signs may be modified to accommodate the allowable movements (such as "HOV 2+ AND RIGHT TURNS ONLY").

11   The MOTORCYCLES ALLOWED (R3-11hP) plaque (see Figure 2G-1) may be used where motorcycles, regardless of the number of occupants, are allowed to use an HOV lane.

Standard:

12   **If used, the MOTORCYCLES ALLOWED plaque shall be mounted below a post-mounted Preferential Lane Operation (R3-11, R3-11a, or R3-11b) sign.**

13   **For all barrier-separated or buffer-separated or contiguous preferential lanes where access is restricted to designated locations, an overhead Preferential Lane Operation (R3-14 series) sign shall be used at the beginning or initial entry point on freeways, expressways, and at locations on conventional roadways where the preferential lane is not the outermost (far right or far left) lane of the roadway, and at any intermediate entry points or gaps in the barrier or buffer where vehicles are allowed to legally enter the access-restricted preferential lanes. For all barrier-separated and buffer-separated preferential lanes, post-mounted Preferential Lane Operation (R3-11 series) signs shall be used only as a supplement to the overhead signs on freeways, expressways, and at locations on conventional roadways where the preferential lane is not the outermost lane of the roadway at the beginning or initial entry point, or at any intermediate entry points or gaps in the barrier or buffer.**

14   **For buffer-separated or contiguous preferential lanes where continuous access with the adjacent general-purpose lanes is provided, including those where a preferential lane is added to the roadway (see Figure 2G-2 for HOV lanes) and those where a general-purpose lane transitions into a preferential lane (see Figure 2G-3 for HOV lanes), an overhead Preferential Lane Operation (R3-14 series) sign shall be used at the beginning or initial entry point of the preferential lane on freeways and expressways.**

Option:

15   On conventional roads where preferential lane operations exist, R3-11 series post-mounted signs may be used in lieu of or in addition to overhead R3-14 series signs, except where overhead signs are required as provided in Paragraph 14 of this Section.

16   Additional overhead (R3-14 series) or post-mounted (R3-11 series) Preferential Lane Operation signs may be provided along the length of any type of preferential lane.

Standard:

17   **For all types of direct access ramps that provide access to or lead to preferential lanes, a post-mounted Preferential Lane Operation (R3-11 series) sign shall be used at the beginning or initial entry point of the direct access ramp.**

Option:

18   For direct access ramps to preferential lanes, an overhead Preferential Lane Operation (R3-14 series) sign may be used at the beginning or initial entry point to supplement the required post-mounted signs.

19   Lane-use control signals (see Chapter 4T) may be used at access points to preferential lanes to indicate that a ramp or access roadway leading to the preferential lane or facility, or one or more specific lanes of the facility, are open or closed (see Figure 2G-1).

Section 2G.06 Preferential Lane Advance Signs (R3-12, R3-12e, R3-12f, R3-15, R3-15a, and R3-15d)

Guidance:

01   *The Preferential Lane Advance (R3-12, R3-12f, R3-15, and R3-15d) signs (see Figure 2G-1) should be used for advance notification of a barrier-separated, buffer-separated, or contiguous preferential lane that is added to the general-purpose lanes (see Figure 2G-2).*

02   *The Preferential Lane Advance (R3-12e and R3-15a) signs (see Figure 2G-1) should be used for advance notification of a general-purpose lane that becomes a preferential lane (see Figure 2G-3).*

Option:

03   The legends on the R3-12f and R3-15d signs may be modified to suit the type of preferential lane.

Guidance:

04   *On conventional roads, for general-purpose lanes that become preferential lanes, a post-mounted (R3-12e) or overhead (R3-15a) Preferential Lane Advance sign should be installed in advance of the beginning of or initial entry point to the preferential lane at a distance determined by engineering judgment based on speed, traffic characteristics, and other site-specific considerations. The distance selected should provide adequate opportunity for ineligible vehicles to vacate the lane prior to the beginning of the restriction.*

05   *On freeways and expressways, for general-purpose lanes that become preferential lanes, an overhead Preferential Lane Advance (R3-15a) sign should be installed at least 1 mile in advance of the beginning of the preferential lane restriction.*

Option:

06   Additional post-mounted or overhead Preferential Lane Advance signs may be placed farther in advance of or closer to the beginning or initial entry points to a preferential lane.

Section 2G.07 Preferential Lane Ends Signs (R3-12a, R3-12b, R3-12c, R3-12d, R3-12g, R3-12h, R3-15b, R3-15c, and R3-15e)

Standard:

01   **A post-mounted Preferential Lane Ends (R3-12b or R3-12h) sign (see Figure 2G-1) shall be installed at least ½ mile in advance of the termination of a preferential lane on freeways and expressways.**

02   **Except as provided in Paragraph 7 of this Section, a post-mounted Preferential Lane Ends (R3-12a or R3-12g) sign shall be installed at the point where a preferential lane and restriction end and traffic must merge into the general-purpose lanes.**

03   **A post-mounted Preferential Lane Ends (R3-12d) sign (see Figure 2G-1) shall be installed at least ½ mile in advance of the point where a preferential lane restriction ends and the lane becomes a general-purpose lane on freeways and expressways.**

04   **Except as provided in Paragraph 8 of this Section, a post-mounted Preferential Lane Ends (R3-12c) sign shall be installed at the point where a preferential lane restriction ends and the lane becomes a general-purpose lane.**

Guidance:

05   *On conventional roads, the distance at which Preferential Lane Ends signs are installed in advance of the termination of a preferential lane and/or restriction should be determined by engineering judgment.*

Option:

06   The legends on the R3-12g and R3-15e signs may be modified to suit the type of preferential lane.

07   An overhead Preferential Lane Ends (R3-15b or R3-15e) sign may be installed instead of or in addition to a post-mounted R3-12a or R3-12g sign at the point where a preferential lane and restriction ends and traffic must merge into the general-purpose lanes.

08   An overhead Preferential Lane Ends (R3-15c) sign may be installed instead of or in addition to a post-mounted R3-12c sign at the point where the preferential lane restriction ends and the lane becomes a general-purpose lane.

WARNING SIGNS AND PLAQUE

Section 2G.08 Warning Signs on Median Barriers for Preferential Lanes

Option:

01   When a warning sign applicable only to a preferential lane is installed on a median barrier with limited lateral clearance to the adjacent travel lanes or shoulders, the warning sign may have a vertically-oriented rectangular shape. For a high-occupancy vehicle lane, such signs may be used instead of using the HOV (W16-11P) plaque (see Section 2G.09) with a standard diamond-shaped warning sign.

Standard:

02   **When a vertically-oriented rectangular-shaped warning sign applicable only to a preferential lane is installed on a median barrier, the top portion of the sign shall be comprised of a white symbol or legend denoting the type of preferential lane (such as the diamond symbol for HOV or the legend BUS LANE) on a black background with a white border, and the bottom portion of the sign shall be comprised of the standard word message or symbol of the standard warning sign as a black legend on a yellow background with a black border (see Figure 2G-4).**

Guidance:

03   *Where lateral clearance is limited, such as when a post-mounted warning sign applicable only to a preferential lane is installed on a median barrier, the edges of the sign should not project beyond the outer edges of the barrier.*

Figure 2G-4. Examples of Warning Signs and Plaques Applicable Only to Preferential Lanes

A – Barrier-mounted rectangular warning signs

W4-1L (modified)

W4-2L (modified)

W13-2 (modified)

B – Warning plaque for use above standard diamond-shaped warning signs

W16-11P

Note: An HOV lane example (diamond symbol) is illustrated. For other types of preferential lanes, the appropriate symbol or word message (see Section 2G.03) shall be displayed in white on the black background of the top portion of these signs.

Option:

04   Where lateral clearance is limited, warning signs applicable only to a preferential lane that are post-mounted on a median barrier and that are 72 inches or less in width may be skewed up to 45 degrees in order to fit within the barrier width or may be mounted higher, such that the vertical clearance to bottom of the sign, light fixture, or its structural support, whichever is lowest, is not less than 17 feet above any portion of the pavement and shoulders.

Standard:

05   **Where lateral clearance is limited, Preferential Lane warning signs that are post-mounted on a median barrier and that are wider than 72 inches shall be mounted with a vertical clearance that complies with the provisions of Section 2A.15 for overhead mounting.**

Section 2G.09 High-Occupancy Vehicle (HOV) Plaque (W16-11P)

Option:

01   In situations where there is a need to warn drivers in an HOV lane of a specific condition, the HOV (W16-11P) plaque (see Figure 2G-4) may be used above a warning sign. The HOV plaque may be used to differentiate a warning sign applicable to the HOV lanes when the sign is also visible to traffic on the adjacent general-purpose roadway. Among the warning signs that may be possible applications of the HOV plaque are the Advisory Exit Speed, Added Lane, and Merge signs.

02   The diamond symbol may be used instead of the word message HOV on the W16-11P plaque. When appropriate, the words LANE or ONLY may be used on this plaque.

Support:

03   Section 2G.08 contains information regarding warning signs that can be mounted on barriers for HOV or other types of preferential lanes.

GUIDE SIGNS

Section 2G.10 Preferential Lane Guide Signs – General

Support:

01   Preferential lanes are used on freeways, expressways, and conventional roads. Except as otherwise provided, Sections 2G.10 through 2G.15 apply only to guide signs for preferential lanes on freeways and expressways.

Guidance:

02   *On conventional roads, guide signs applicable only to preferential lanes are ordinarily not needed, but if used, they should comply with the provisions for guide signs in Chapter 2D and any principles for Preferential Lane guide signs in Sections 2G.10 through 2G.15 that engineering judgment finds to be appropriate for the conditions.*

Support:

03   Additional guidance and standards related to the designation, operational considerations, signs, pavement markings, and other considerations for preferential lanes are provided in Sections 2G.03 through 2G.07, 2G.09, and Chapter 3E.

Guidance:

04   *The appropriate combinations of pavement markings and standard overhead and post-mounted regulatory, warning, and guide signs for a specific preferential lane application should be selected based on an engineering study.*

05   *If overhead signs applicable only to a preferential lane are located in approximately the same longitudinal position along the highway as overhead signs applicable only to the general-purpose lanes, the signs for the preferential lane should be separated laterally from the signs for the general-purpose lanes to the maximum extent practicable to minimize conflicting information.*

06   *The Preferential Lane signs should be designed and located to avoid overloading the road user. The order of priority of guide signs should be Advance Guide, Preferential Lane Entrance Direction, and finally Preferential Lane Exit Destination Supplemental guide signs.*

Standard:

07   **Signs applicable only to a preferential lane shall be distinguished from signs applicable to general-purpose lanes by the inclusion of the applicable symbol(s) and/or word(s).**

Support:

08   The symbol and/or word message that appears on a particular guide sign applicable only to a preferential lane will vary based on the specific type of traffic allowed and on other related operational constraints that have been established for a particular lane, such as an HOV lane, a bus lane, or a taxi lane.

Standard:

09   **For HOV lanes, the diamond symbol shall appear on each Advance Guide (E8-3) sign (see Figure 2G-5), Preferential Lane Entrance Direction (E8-2 or E8-2a) sign (see Figure 2G-6), and Preferential Lane Entrance Gore (E8-1 or E8-1a) sign (see Figure 2G-7) for the designated entry and exit points for barrier-separated and buffer-separated geometric configurations and direct access ramps to or from such lanes. The diamond symbol shall not be used with preferential lanes for other types of traffic, such as bus lanes or taxi lanes.**

10   **Signing for an HOV lane that is managed by means of varying the occupancy requirement in response to changing conditions shall also comply with these provisions.**

Figure 2G-5. Example of an Overhead Advance Guide Sign for a Preferential Lane Entrance

Note: An example of an HOV lane (diamond symbol) sign is illustrated. For other types of preferential lanes, the appropriate symbol or word message (see Section 2G.03) is displayed in white on the black background of the left-hand portion of this sign.

Figure 2G-6. Examples of Overhead or Post-Mounted Preferential Lane Entrance Direction Signs

E8-2
(overhead only)

E8-2a
(post-mounted only)

A changeable message sign may be incorporated into an overhead Preferential Lane guide sign to indicate the status of a reversible operation as shown in the following example:

Lane open   Lane closed

Note: Examples of HOV lane (diamond symbol) signs are illustrated. For other types of preferential lanes, the appropriate symbol or word message (see Section 2G.03) is displayed in white on the black background of the top left-hand portion of these signs.

Figure 2G-7. Entrance Gore Signs for Barrier-Separated Preferential Lanes

Note: Examples of HOV lane (diamond symbol) signs are illustrated. For other types of preferential lanes, the appropriate symbol or word message (see Section 2G.03) is displayed in white on the black background of the top portion of these signs.

E8-1   E8-1a

11   **The diamond symbol shall be displayed in the legend of each Preferential Lane guide sign at the designated entry and exit points for all types of HOV lanes (including barrier-separated, buffer-separated, contiguous, and direct access ramps) in order to alert motorists that there is a minimum allowable vehicle occupancy requirement for vehicles to use the HOV lanes. Guide signs shall not display the occupancy requirement for the preferential lane.**

12   **A combination of guide and regulatory signs shall be used in advance of and at the initial entry point and all intermediate entry points from general-purpose lanes or facilities to contiguous, barrier-separated, and buffer-separated preferential lanes where access between the preferential and general-purpose lanes is restricted to designated locations. The regulatory signs shall comply with the provisions of Sections 2G.03 through 2G.07.**

13 **Regulatory signs alone shall be used in advance of, at the beginning of, and at periodic intervals along contiguous or buffer-separated preferential lanes that provide continuous access between the adjacent general-purpose lanes and the preferential lane (see Figures 2G-2 and 2G-3). The design and placement of the regulatory signs shall comply with the provisions of Sections 2G.03 through 2G.07.**

14 **Except as otherwise provided in Sections 2G.10 through 2G.13, guide signs applicable to a preferential lane with a vehicle occupancy requirement shall be distinguished from those applicable to general-purpose lanes by displaying the white diamond symbol on a black background at the left-hand edge of these signs.**

Option:

15 When post-mounted guide signs applicable only to a preferential lane are installed on a median barrier with limited lateral clearance to the adjacent travel lanes or shoulders, the guide signs may have a vertically-oriented rectangular shape.

Standard:

16 **When vertically-oriented rectangular-shaped guide signs applicable only to a preferential lane are installed on a median barrier, the top portion of the signs shall be comprised of the applicable white symbol or white word message that identifies the type of preferential lane (such as the diamond symbol for an HOV lane) on a black background with a white border, and the bottom portion of the sign shall be comprised of the appropriate guide sign legend on a green background with a white border (see Figures 2G-7 through 2G-9).**

Guidance:

17 *Where lateral clearance is limited, such as when a post-mounted Preferential Lane guide sign is installed on a median barrier, the edges of the sign should not project beyond the outer edges of the barrier.*

Option:

18 Where lateral clearance is limited, Preferential Lane guide signs that are 72 inches or less in width may be skewed up to 45 degrees in order to fit within the barrier width or may be mounted higher, such that the vertical clearance to the bottom of the sign, light fixture, or its structural support, whichever is lowest, is not less than 17 feet above any portion of the pavement and shoulders.

Standard:

19 **Where lateral clearance is limited, Preferential Lane guide signs that are post-mounted on a median barrier and that are wider than 72 inches shall be mounted with a vertical clearance that complies with the provisions of Section 2A.15 for overhead mounting.**

Option:

20 Lane-use control signals (see Chapter 4T) may be used at access points to preferential lanes to indicate that a ramp or access roadway leading to or from the preferential lane or facility, or one or more specific lanes of the facility, are open or closed.

21 Changeable message signs may supplement, substitute for, or be incorporated into static guide signs (see Figure 2G-6) where travel conditions change or where multiple types of operational strategies (such as variable occupancy requirements, vehicle types, or pricing policies) are used and varied throughout the day or week to manage the use of, control of, or access to preferential lanes.

Standard:

22 **When changeable message signs (see Chapter 2L) are used as guide signs for preferential lanes, they shall be the required sign size and shall display the required letter height and legend format that correspond to the type of roadway facility and design speed.**

23 **Advance Guide signs, Preferential Lane Entrance Direction signs, and Preferential Lane Entrance Gore signs for the initial entry point and intermediate entry points into a preferential lane from the general-purpose lanes on the same designated route shall not identify the entry point as an exit by using the word "EXIT" on the sign or on a plaque.**

Guidance:

24 *Advance Guide signs and Preferential Lane Entrance Direction signs for initial and intermediate entry points into a preferential lane should use the word "ENTRANCE," such as "HOV LANE ENTRANCE" (see Figures 2G-5 and 2G-6), to convey the fact that vehicles are not leaving the designated route.*

25 *Preferential Lane Entrance Gore signs (see Figure 2G-7) at the initial entry point to a preferential lane should use the word "ENTRANCE." Preferential Lane Entrance Gore signs at intermediate entry points to a barrier-separated preferential lane where the sign would be located immediately adjacent to and directly viewed by traffic in the preferential lane should not use the word "ENTRANCE."*

Standard:

26 **When the entry point is on the left-hand side of the general-purpose lanes, a LEFT (E1-5aP) plaque (see Figure 2E-9) shall be added to the top left edge of the Advance Guide and Preferential Lane Entrance Direction signs. The LEFT plaque shall not be used on a preferential lane regulatory sign.**

Section 2G.11 <u>Signing for Initial Entry Points to Preferential Lanes</u>

Standard:

01   **Except where a buffer-separated or contiguous preferential lane is added or where a general-purpose lane becomes a buffer-separated or contiguous preferential lane, and provides continuous access with the adjacent general-purpose lanes as illustrated in Figures 2G-2 and 2G-3, an Advance Guide sign shall be provided at least ½ mile prior to the initial entry point to all types of preferential lanes in any type of geometric configuration on freeways and expressways. A Preferential Lane Entrance Direction sign shall also be provided at the initial entry point. Advance Guide and Preferential Lane Entrance Direction signs for such entry points shall not include the word "EXIT" (see Section 2G.10).**

02   **Where a general-purpose lane becomes a preferential lane that does not provide continuous access with the adjacent general-purpose lanes, an Advance Guide sign shall also be provided at approximately 1 mile in advance of the initial entry point. The Advance Guide and Entrance Direction signs in this sequence shall include a panel at the bottom of the sign with a black legend and border on a yellow background displaying a down arrow and the word ONLY as illustrated in Figure 2G-8.**

Guidance:

03   *Unless an Advanced Guide Sign is already required in Paragraph 2 of this Section, an Advance Guide sign should also be installed and located approximately 1 mile in advance of the initial entry point to a preferential lane on freeways and expressways that restricts access to the adjacent general-purpose lanes.*

Option:

04   An Advance Guide sign may also be installed and located approximately 2 miles in advance of the initial entry point to a preferential lane that restricts access with the adjacent general-purpose lanes to designated locations.

Standard:

05   **For barrier-separated, buffer-separated, or contiguous preferential lanes where entry is restricted to only designated points on freeways and expressways, the Advance Guide and Preferential Lane Entrance Direction signs shall be mounted overhead.**

Guidance:

06   *Preferential Lane Exit Destination guide signs, identifying final destination and downstream exit locations accessible from the preferential lane (see Figures 2G-8, 2G-9, 2G-14, 2G-16, and 2G-17), should be installed in advance of the initial entry points to access-restricted preferential lanes (such as barrier-separated and buffer-separated). These signs should be located based on the priority of the message, the available space, the existing signs on adjacent general-purpose traffic lanes, roadway and traffic characteristics, the proximity to existing overhead signs, the ability to install overhead signs, and other unique local factors.*

Standard:

07   **Advance destination guide signs for preferential lanes shall include an upper section displaying a black legend that includes the type of preferential lane and the word "EXITS," such as "HOV EXITS," on a white background. For preferential lanes that incorporate a vehicle occupancy requirement, the white diamond symbol on a black background shall be displayed at the left-hand edge of this upper section (see Figure 2G-9).**

Support:

08   Figure 2G-8 shows an example of signing for a general-purpose lane that becomes a preferential lane that does not provide continuous access with the adjacent general-purpose lanes.

09   Figure 2G-9 shows an example of signs for the initial entry point to a preferential lane.

Section 2G.12 <u>Signing for Intermediate Entry Points to Preferential Lanes</u>

Standard:

01   **For barrier-separated, buffer-separated, and contiguous preferential lanes where entry is restricted only to designated points, an overhead Preferential Lane Entrance Direction sign shall be provided at intermediate entry points to the preferential lane from the general-purpose lanes.**

Guidance:

02   *For barrier-separated and buffer-separated preferential lanes where intermediate entry from the general-purpose lanes is provided via a separate lane or ramp (see Figure 2G-10), at least one Advance Guide sign should be provided in addition to the Preferential Lane Entrance Direction sign.*

03   *For access-restricted preferential lanes where intermediate entrance and egress are at the same designated access location, the Preferential Lane Entrance Direction sign should be located between ½ and ¼ of the length of the designated entry area, as measured from the downstream end of the entry area (see Figure 2G-11).*

Figure 2G-8. Example of Advance Guide and Entrance Direction Signs for a General-Purpose Lane that Becomes a Preferential Lane

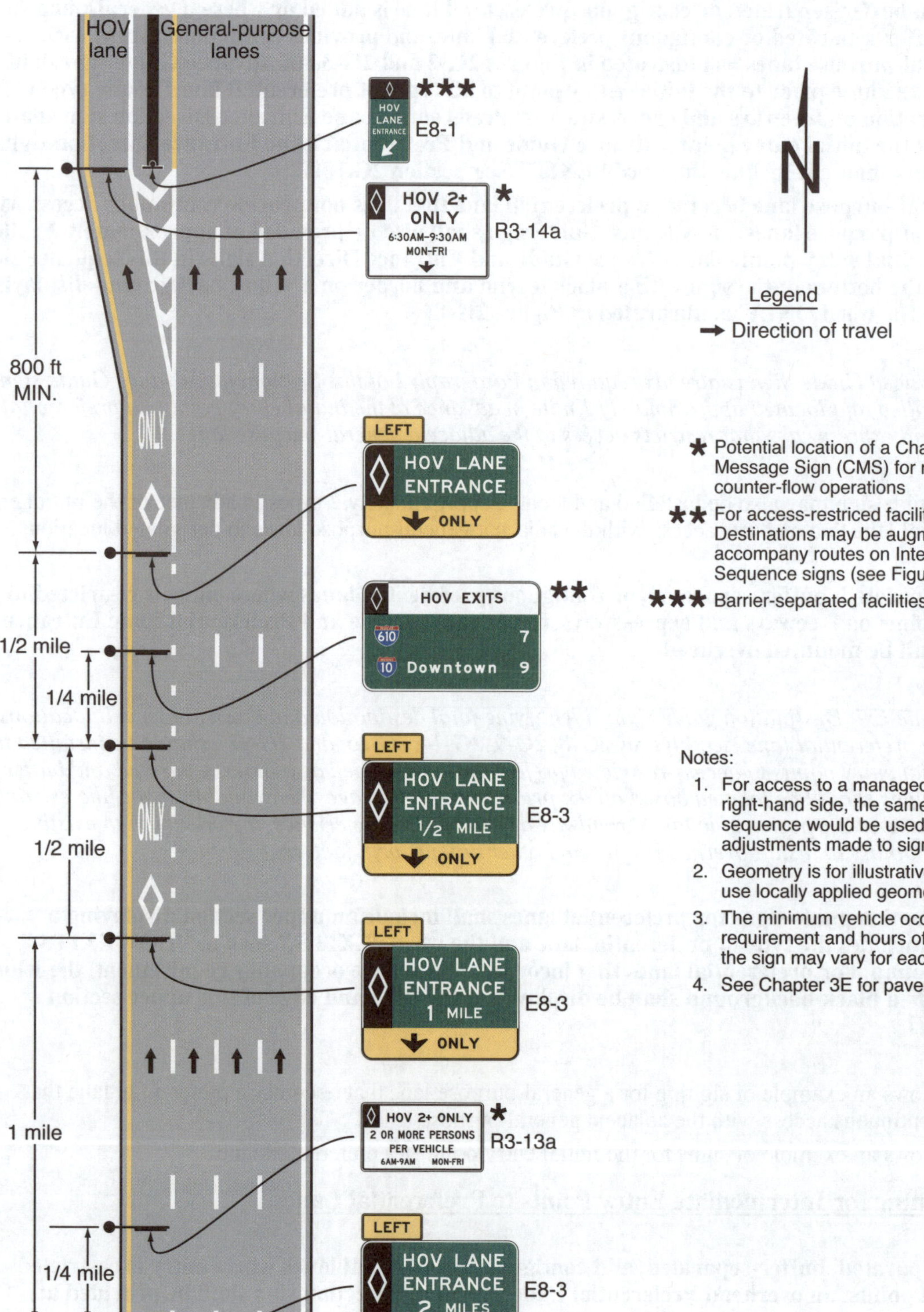

* Potential location of a Changeable Message Sign (CMS) for reversible or counter-flow operations

** For access-restricted facilities. Destinations may be augmented to accompany routes on Interchange Sequence signs (see Figure 2E-10)

*** Barrier-separated facilities only

Notes:

1. For access to a managed lane on the right-hand side, the same signing sequence would be used with adjustments made to sign messages

2. Geometry is for illustrative purposes only; use locally applied geometric criteria

3. The minimum vehicle occupancy requirement and hours of operation on the sign may vary for each facility

4. See Chapter 3E for pavement markings

Figure 2G-9. Example of Signing for an Entrance to Access-Restricted HOV Lanes

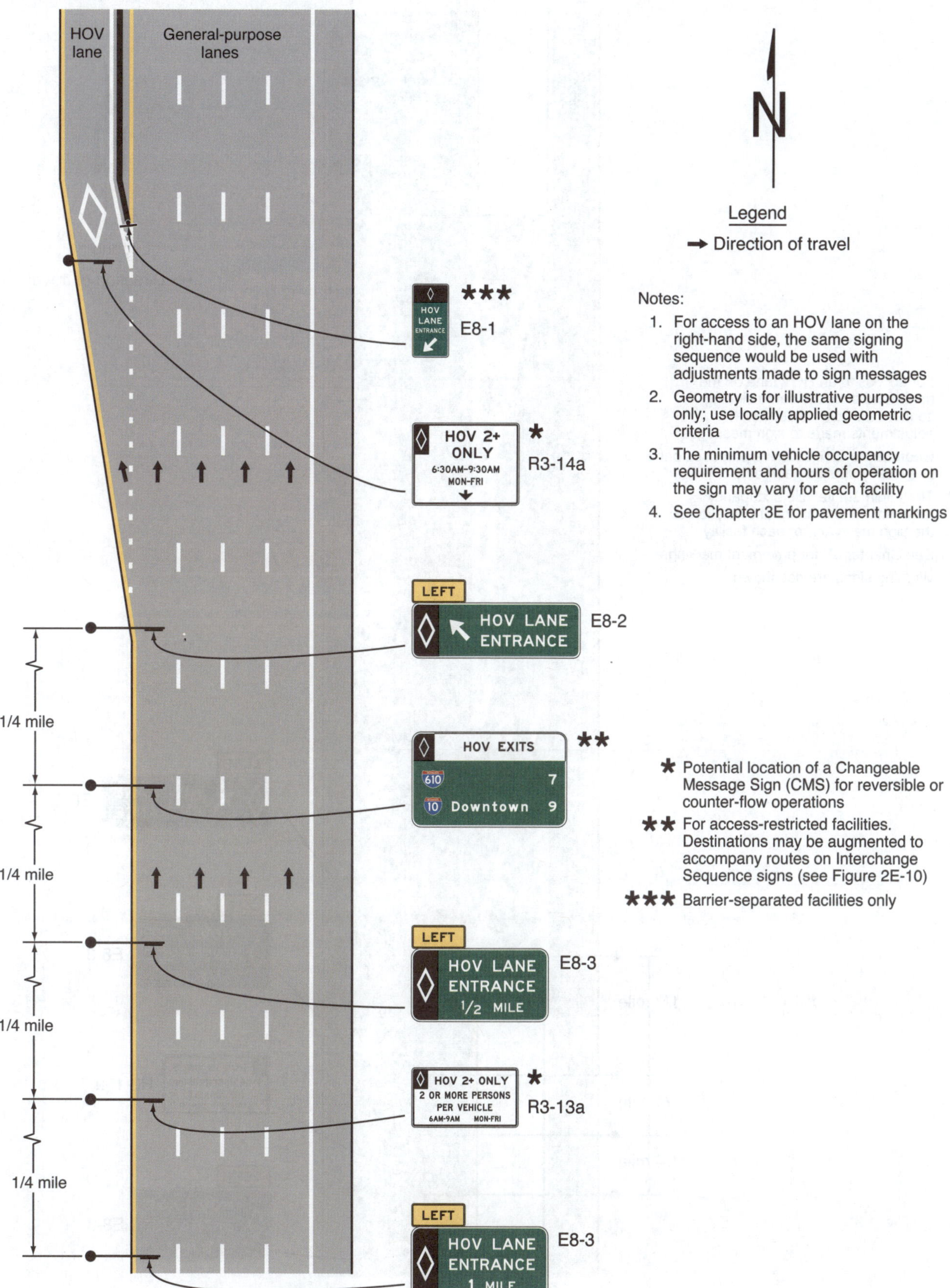

Figure 2G-10. Example of Signing for an Intermediate Entry to a Barrier-Separated or Buffer-Separated HOV Lane

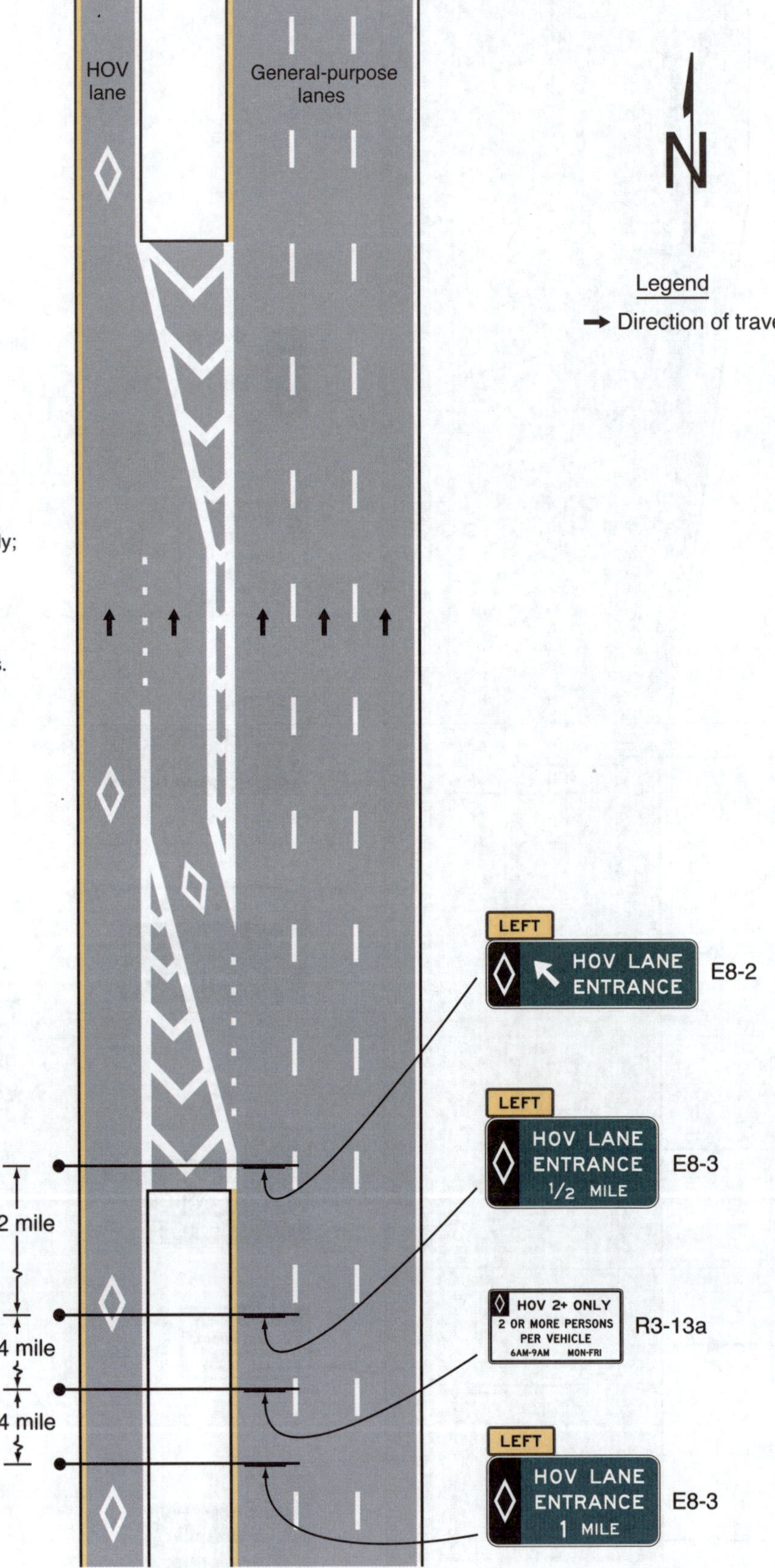

Figure 2G-11. Example of Signing for the Intermediate Entry to, Egress from, and End of Access-Restricted HOV Lanes

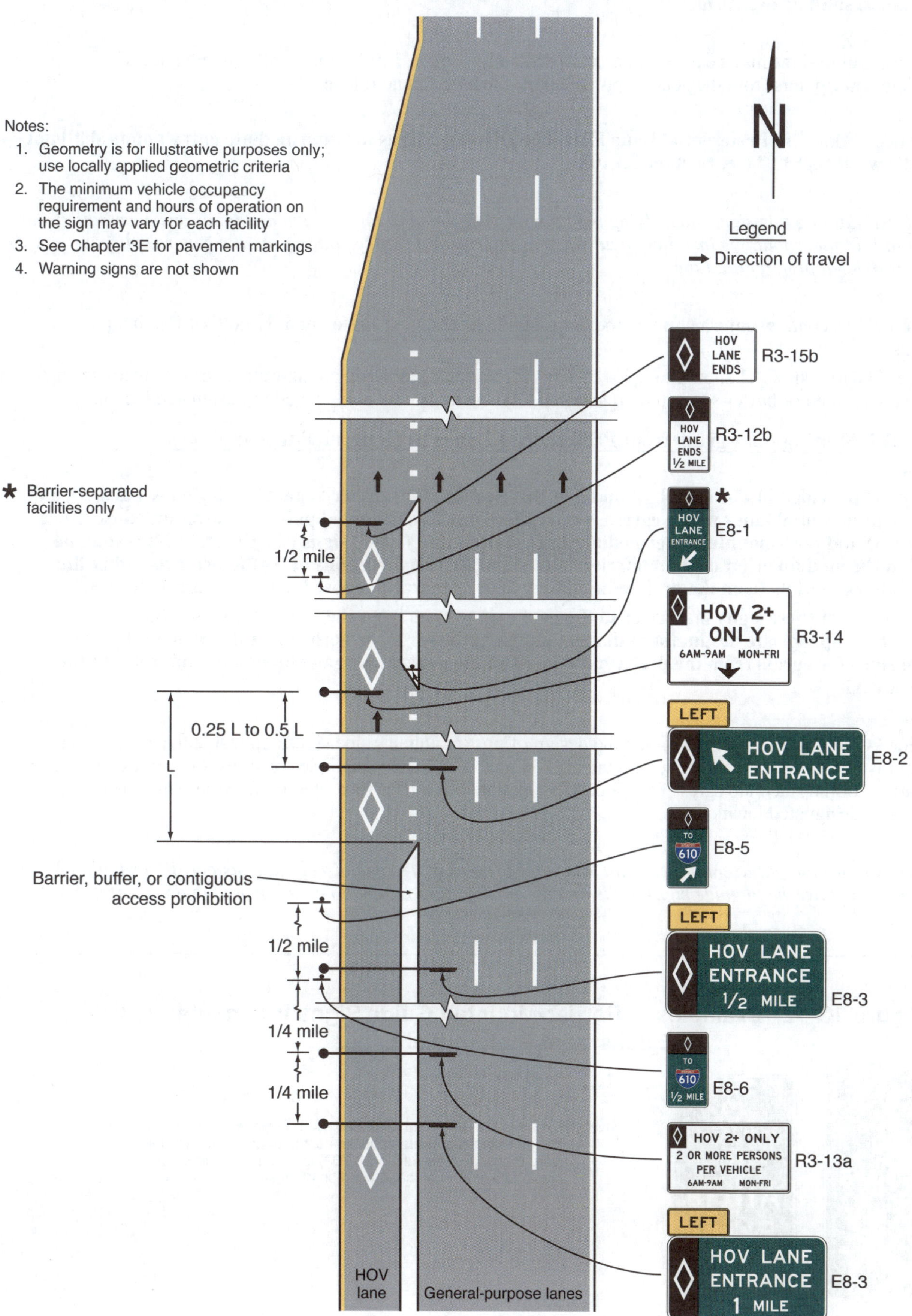

Standard:

04   **The Advance Guide signs, if used for intermediate entry points to a preferential lane from the general-purpose lanes, shall be overhead.**

Option:

05   Advance Guide signs may be provided at approximately ½ mile, 1 mile, and 2 miles in advance of intermediate entry points from the general-purpose lanes to a preferential lane.

Standard:

06   **Advance Guide and Preferential Lane Entrance Direction signs for intermediate entry points shall not include the word "EXIT" (see Section 2G.10).**

Guidance:

07   *Exit Destination guide signs, identifying the final destination and downstream exit locations accessible from the preferential lane, should be installed in advance of intermediate entry points from the general-purpose lanes to access-restricted preferential lanes.*

Support:

08   Section 2G.11 contains information on the design and placement of Preferential Lane Exit Destination guide signs.

09   Figures 2G-10 and 2G-11 show examples of signs for various geometric configurations of intermediate entry to a barrier-separated or buffer-separated preferential lane where access is restricted to designated locations.

Section 2G.13 Signing for Egress from Preferential Lanes to General-Purpose Lanes

Standard:

01   **Except as provided in Paragraphs 4 and 5 of this Section, for barrier-separated, buffer-separated, and contiguous preferential lanes where egress is restricted only to designated points, post-mounted Advance Guide (E8-6) and post-mounted Intermediate Egress Direction (E8-5) signs (see Figure 2G-12) shall be installed in the median or on median barriers that separate two directions of traffic prior to and at the intermediate exit points from the preferential lanes to the general-purpose lanes (see Figure 2G-11).**

02   **The legends of these signs shall refer to the next exit or exits from the general-purpose lanes by displaying the appropriate destination information, exit number(s), or both. The Intermediate Egress Direction signs for egress from the preferential lanes to the general-purpose lanes shall not refer to the egress as an exit.**

Support:

03   Section 2G.10 contains information on the design of post-mounted guide signs applicable to a preferential lane when installed on a median barrier. Figures 2G-11 and 2G-13 show examples of signs for various geometric configurations of intermediate egress from a barrier-separated or buffer-separated preferential lane where access is restricted to designated locations.

Guidance:

04   *Where two or more adjacent preferential lanes are present in a single direction, consideration should be given to the use of overhead guide signs to display the information related to egress from the preferential lanes.*

Figure 2G-12. Examples of Barrier-Mounted Guide Signs for an Intermediate Egress from Preferential Lanes

Note: Examples of HOV lane (diamond symbol) signs are illustrated. For other types of preferential lanes, the appropriate symbol or word message (see Section 2G.03) is displayed in white on the black background of the top portion of these signs.

E8-5   E8-6

05   *For barrier-separated and buffer-separated preferential lanes where egress from a preferential lane to the general-purpose lanes is restricted only to designated points via a separate lane or ramp, the Advance Guide and Intermediate Egress Direction signs for the egress should be mounted overhead and a Pull-Through sign should be mounted with the Intermediate Egress Direction sign (see Figure 2G-13).*

Standard:

06   **For preferential lanes that incorporate a vehicle occupancy requirement, the design of the overhead Advance Guide and Egress Direction signs for intermediate egress from the preferential lanes to the general-purpose lanes shall display a white diamond symbol on a black background at the left-hand edge of the signs.**

07   **The design of Pull-Through signs when used in conjunction with an Egress Direction sign at an intermediate egress from the preferential lanes to the general-purpose lanes shall be distinguished from those applicable to general-purpose lanes by inclusion of an upper section with the applicable black legend on a white background, such as HOV LANE. For preferential lanes that incorporate a vehicle occupancy requirement, the white diamond symbol on a black background shall be displayed at the left-hand edge of this upper section.**

Section 2G.14 Signing for Direct Entrances to Preferential Lanes from Another Highway

Standard:

01   **For direct access ramps to preferential lanes from a transit facility (such as a park-and-ride lot or a transit station or terminal) that is accessible from surface streets, advance guide signs shall be provided along the adjoining surface streets to direct traffic into and through the transit facility to the preferential lane (see Figure 2G-14).**

Support:

02   Figure 2G-14 provides examples of recommended uses and layouts of signs for HOV lanes for direct access ramps, park-and-ride lots, and access from surface streets.

Section 2G.15 Signing for Direct Exits from Preferential Lanes to Another Highway

Standard:

01   **For contiguous preferential lanes on the left-hand side of the roadway, Advance Guide signs, Exit Direction signs, and Exit Gore (E8-4) signs (see Figure 2G-15) specifically applicable to the preferential lanes shall be used for exits to direct access ramps, such as HOV lane ramps (see Figure 2G-16) or ramps to park-and-ride facilities.**

02   **The design of Advance Guide, Exit Direction, and Pull-Through signs for direct exits from preferential lanes shall be distinguished from those applicable to general-purpose lanes by inclusion of an upper section with the applicable black legend on a white background, such as HOV LANE (for Pull-Through signs) or HOV EXIT (for Advance Guide and Exit Direction signs). For preferential lanes that incorporate a vehicle occupancy requirement, the white diamond symbol on a black background shall be displayed at the left-hand edge of this upper section (see Figures 2G-16 and 2G-17).**

Guidance:

03   *Advance Guide and Exit Direction signs for exits to direct access ramps from a preferential lane should be mounted overhead. A Pull-Through sign over the preferential lane should be used with the Exit Direction sign at exits to direct access ramps.*

Standard:

04   **Post-mounted guide signs in a vertically-oriented rectangular shape installed on a median barrier shall not be used for the Advance Guide and Exit Direction signs for exits to direct access ramps.**

05   **Because direct access ramps for preferential lanes at interchanges connecting two freeways are typically left-hand side exits and typically have design speeds similar to the preferential lane, overhead Advance Guide signs and overhead Exit Direction signs shall be provided in advance of and at the entry point to each freeway-to-freeway preferential lane ramp (see Figure 2G-17).**

Guidance:

06   *The use of guide signs for preferential lanes at freeway interchanges should comply with the provisions for guide signs established in Chapter 2E of this Manual.*

Support:

07   Guide signs for direct access ramps for preferential lanes at interchanges connecting two freeways are similar to those for a connecting ramp between two freeway facilities.

Figure 2G-13. Examples of Signs for an Intermediate Egress from a Barrier-Separated or Buffer-Separated HOV Lane

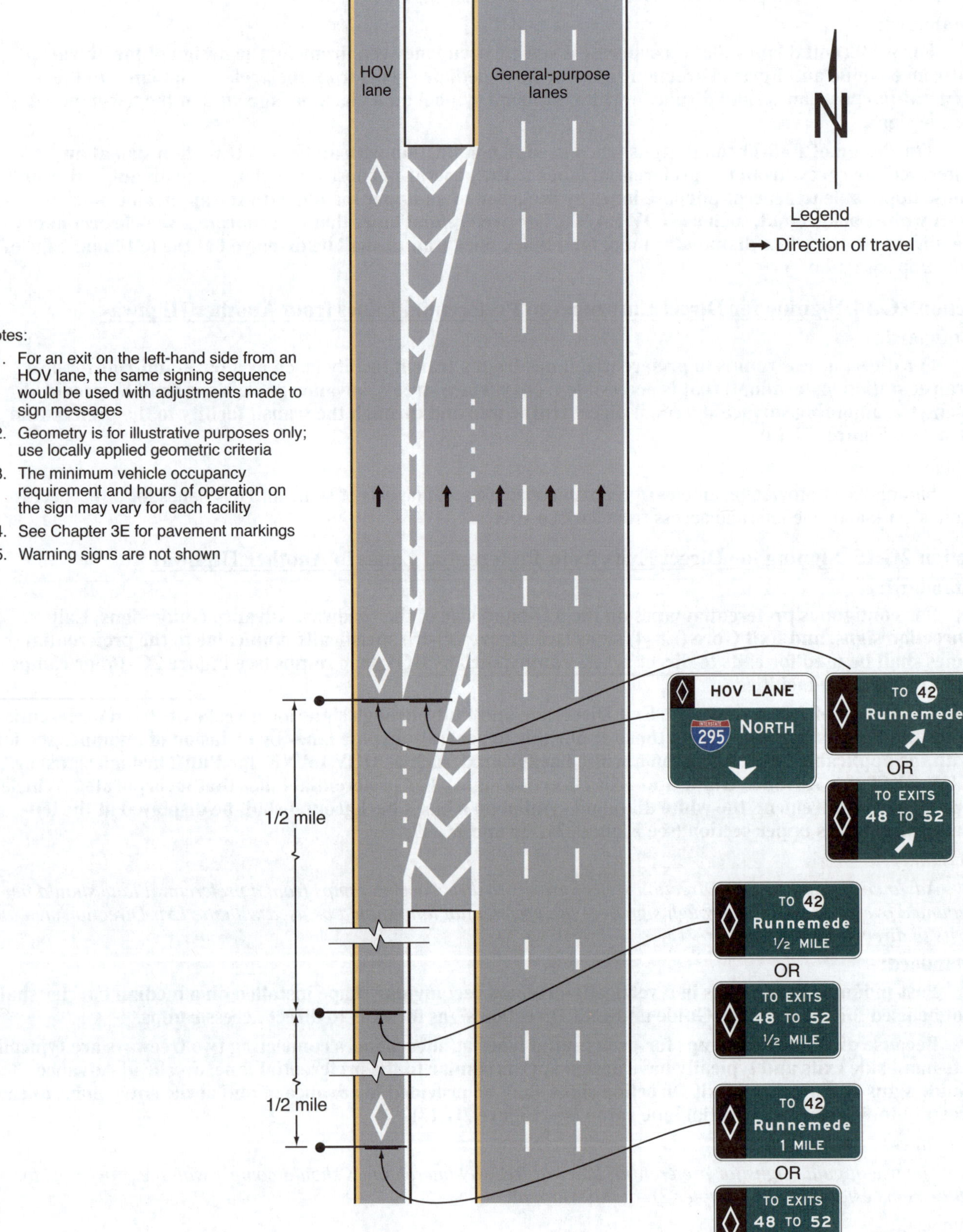

Notes:

1. For an exit on the left-hand side from an HOV lane, the same signing sequence would be used with adjustments made to sign messages

2. Geometry is for illustrative purposes only; use locally applied geometric criteria

3. The minimum vehicle occupancy requirement and hours of operation on the sign may vary for each facility

4. See Chapter 3E for pavement markings

5. Warning signs are not shown

Figure 2G-14. Example of Signing for a Direct Entrance Ramp to an HOV Lane from a Park-and-Ride Facility and a Local Street

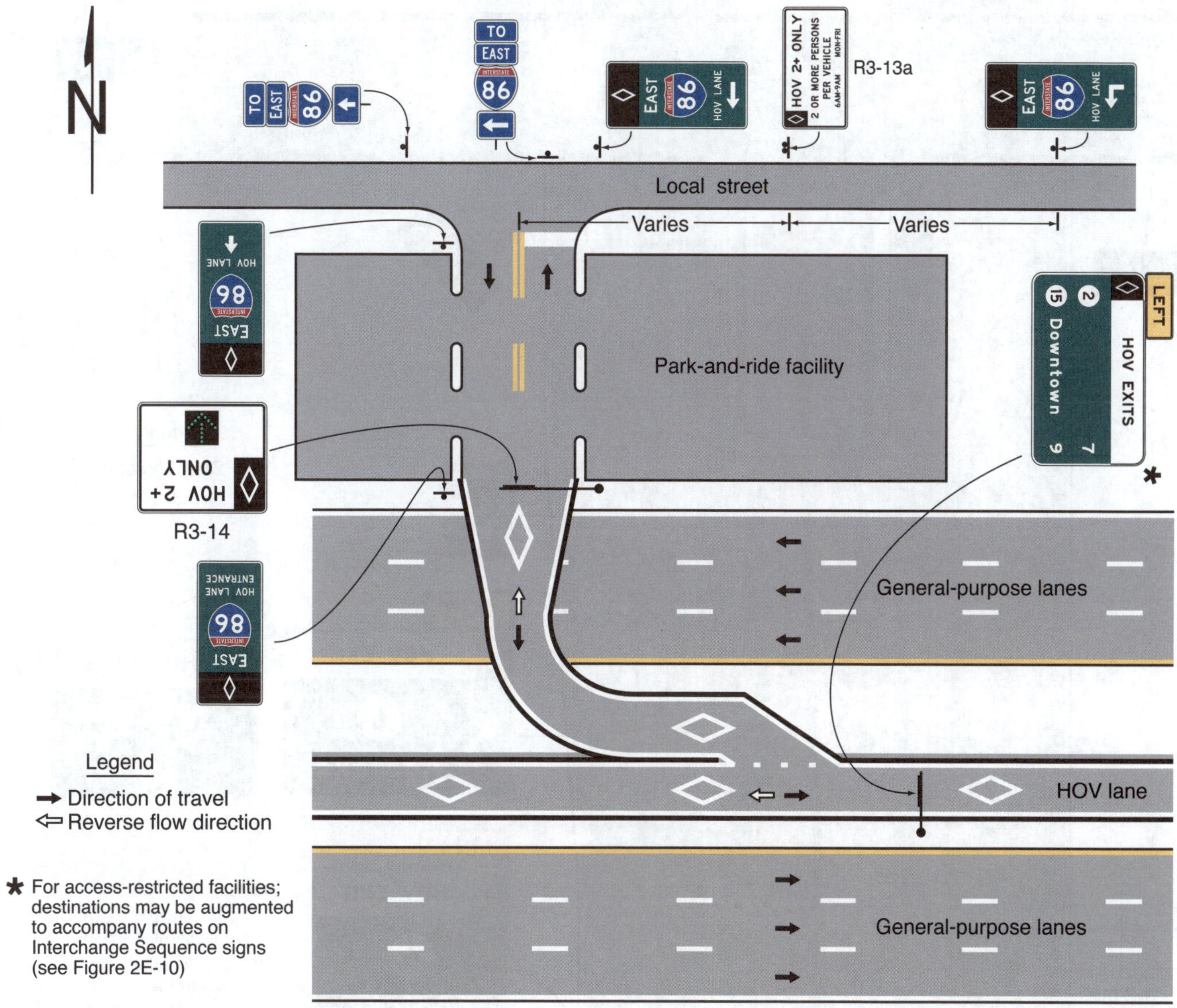

Notes:

1. The minimum vehicle occupancy requirement on the sign may vary for each facility
2. See Chapter 3E for pavement markings
3. Warning signs are not shown
4. Sign locations are approximate
5. Additional signs may be required to direct drivers from the surrounding streets into the park-and-ride lot and the HOV lane
6. Additional signs are required on the adjoining surface streets to inform non-HOVs that they should not enter the HOV facility
7. This figure illustrates a reversible HOV lane with a direct access ramp
8. The guide signs directing local street traffic to the HOV lane should include the word ENTRANCE when the direct access ramp does not traverse a park-and-ride facility

Figure 2G-15. Exit Gore Sign for a Direct Exit from a Preferential Lane

E8-4

Note: An example of an HOV lane (diamond symbol) sign is illustrated. For other types of preferential lanes, the appropriate symbol or word message (see Section 2G.03) is displayed in white on the black background of the top portion of this sign.

Figure 2G-16. Examples of Guide Signs for Direct HOV Lane Entrance and Exit Ramps

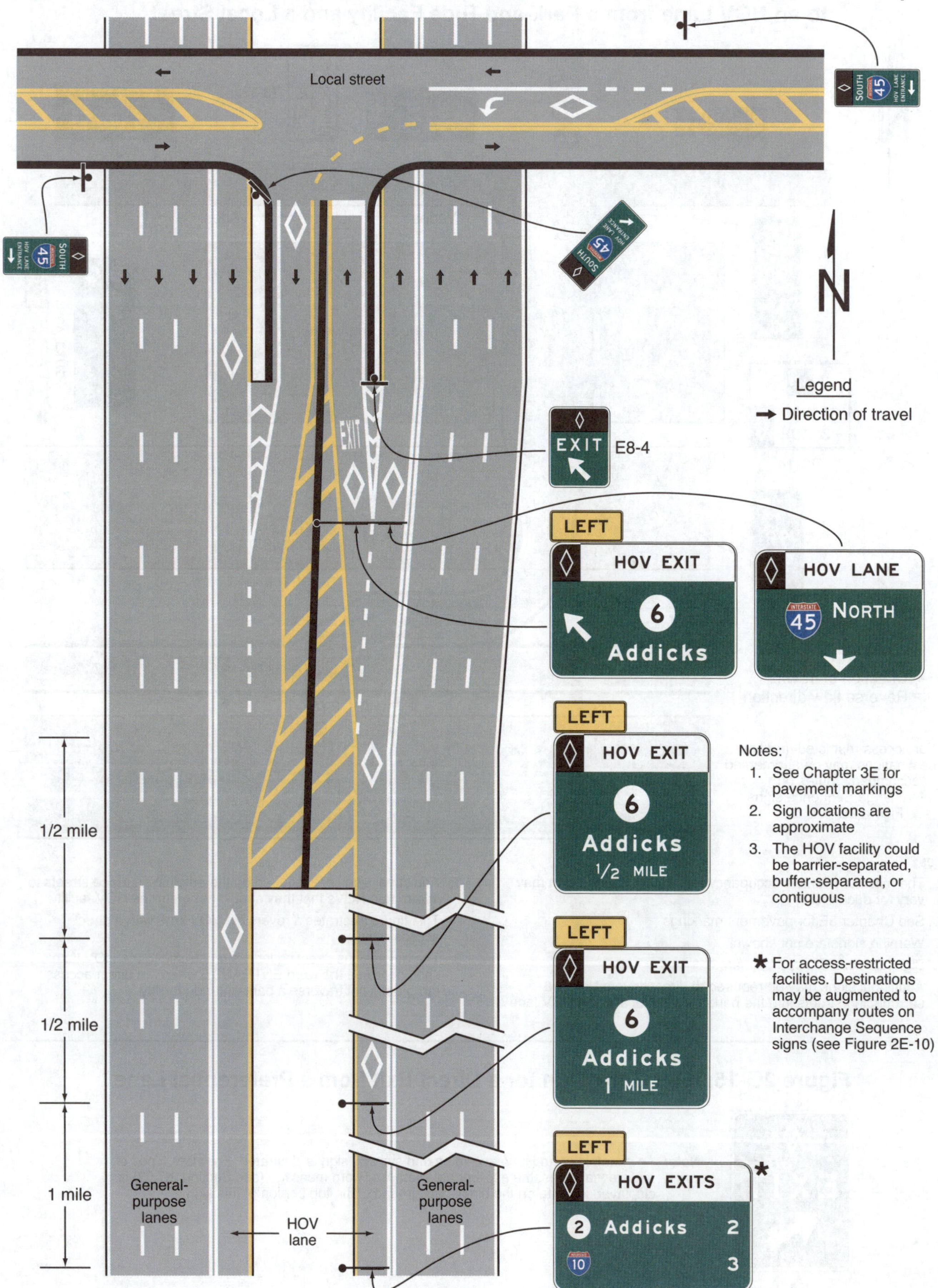

Figure 2G-17. Examples of Guide Signs for a Direct Access Ramp between HOV Lanes on Separate Freeways

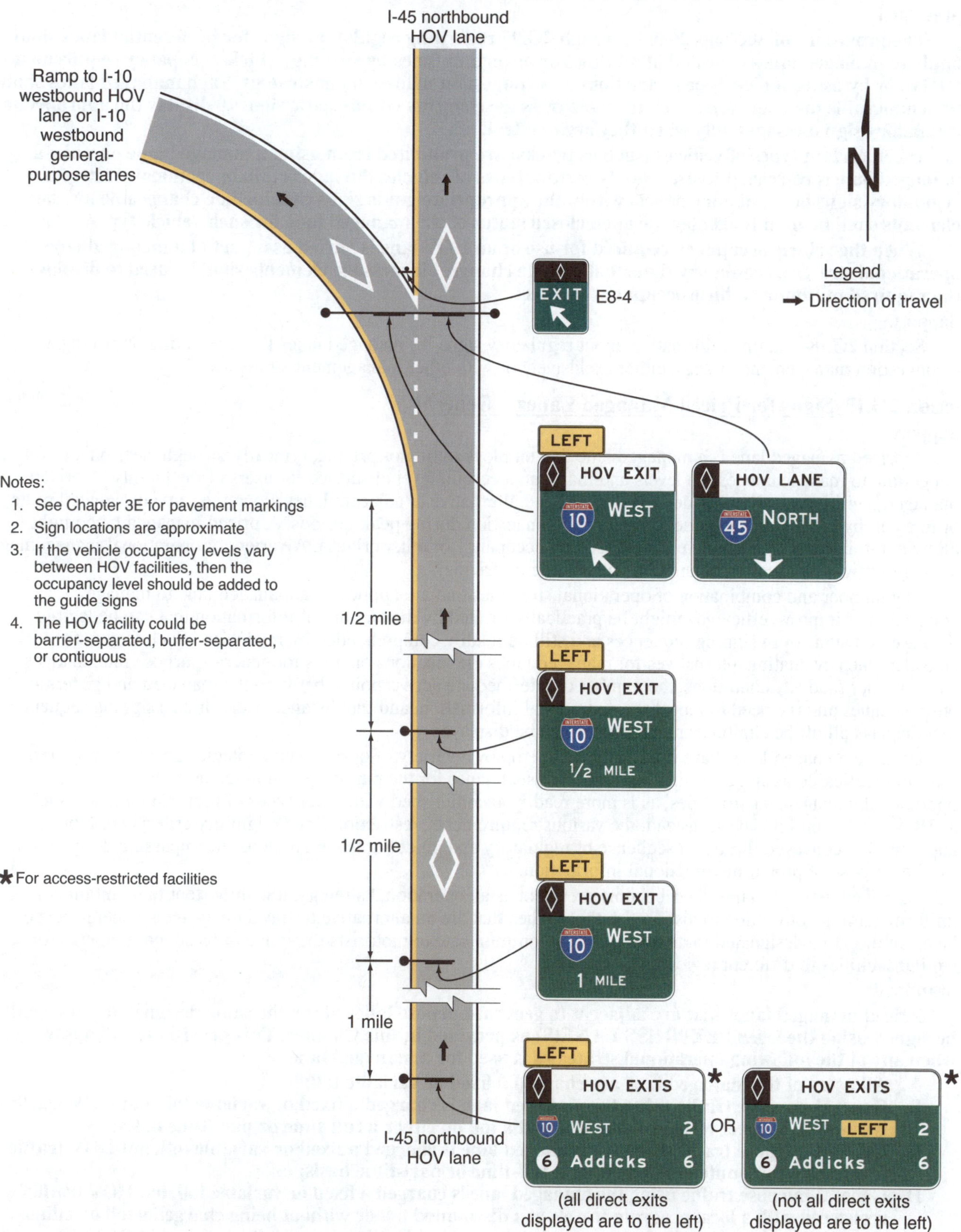

MANAGED LANE SIGNS, PLAQUES, AND LANE-USE CONTROL SIGNALS

Section 2G.16 Signs for Managed Lanes – General

Standard:

01   The provisions of Sections 2G.03 through 2G.07 regarding regulatory signs for preferential lanes shall apply to managed lanes operated at all times or at certain times by varying vehicle occupancy requirements (HOV) or by using vehicle type restrictions as a congestion management strategy. Such managed lanes shall use changeable message signs or changeable message elements within static signs to display the appropriate regulatory sign messages only when they are in effect.

02   When certain types of vehicles (such as trucks) are prohibited from using a managed lane or when a managed lane is restricted to use by only certain types of vehicles during certain operational strategies, regulatory signs or regulatory panels within the appropriate guide signs that include changeable message elements shall be used to display the open/closed status of the managed lane for such vehicle types.

03   When the vehicle occupancy required for use of an HOV lane is varied as a part of a managed lane operational strategy, regulatory signs that include changeable message elements shall be used to display the required minimum vehicle occupancy in effect.

Support:

04   Section 2G.18 contains information about regulatory signs for managed lanes that use tolling or pricing as a congestion management strategy, either exclusively or with other management strategies.

Section 2G.17 Signs for Priced Managed Lanes – General

Support:

01   A priced managed lane is a managed lane that employs tolling or pricing, typically through electronic toll collection, to manage congestion levels and maintain a certain level of service for users of the facility. A priced managed facility typically provides a less-congested alternative to adjacent lanes along the same designated route, or to a nearby facility, that experience recurring congestion during peak periods. A priced managed lane might allow non-toll travel by certain vehicles based on occupancy or other criteria. A variety of operational management strategies might be used in conjunction with tolling or pricing.

02   The number and combination of operational strategies that are applied to a managed lane to manage congestion or improve efficiency might be practically limited by the amount of information that can be legibly displayed on signs or in signing sequences and still be readily comprehended by road users. Such factors to consider when evaluating alternatives for managed lanes are locations of signs for general-purpose interchanges and for other roadway conditions, the number of intermediate access points between the managed and general-purpose lanes and the need to repeat the operational information, and the distance over which a signing sequence that displays all of the eligibility requirements can be displayed.

03   Because managed lanes have the capability to employ a variety of operational strategies on a changing basis, it is not practical to assign a naming convention to such lanes for the purpose of signing based on the specific operational management strategies, as is more readily accomplished with other types of preferential lanes, such as HOV, bus, or bicycle lanes. Instead, the various requirements, restrictions, and eligibility criteria are more appropriately conveyed through a sequence of regulatory and guide signs with a more encompassing designation for the purpose of providing directional information.

04   As priced managed lanes have become prevalent as an operational strategy, it is important to maintain a uniform naming convention to distinguish those lanes that are an alternative to travel on adjacent general-purpose lanes on the same designated route to effectively communicate to motorists the range of basic requirements for similar facilities in different regions.

Standard:

05   Priced managed lanes that are adjacent to general-purpose lanes along the same designated route shall be signed using the legend EXPRESS LANE(S) as provided in this Chapter. This provision shall apply when any of the following operational strategies is used for a managed lane:

    A. All users of the managed lane are charged a fixed or variable toll;

    B. General-purpose traffic using the managed lane is charged a fixed or variable toll, but HOV traffic is allowed to travel without being charged a toll on either a full-time or part-time basis;

    C. General-purpose traffic using the managed lane is charged a fixed or variable toll, but HOV traffic is offered a discounted toll on either a full-time or part-time basis; or

    D. General-purpose traffic using the managed lane is charged a fixed or variable toll, but HOV traffic registered with a local program travels at a discounted toll or without being charged a toll on either a full-time or part-time basis (a transponder or other identifier is typically required of HOVs to indicate registration in conjunction with electronic or visual enforcement and verification of vehicle occupancy).

06   The legends EXPRESS and EXPRESS LANE(S) shall not be used on signs for entrances to highways on which all lanes are managed and there are no adjacent general-purpose lanes on the same designated route. The legends EXPRESS and EXPRESS LANE(S) shall not be used on signs for a managed ramp connection that provides an alternative to a general-purpose ramp connection (see Figure 2F-13), except where the ramp leads directly to a managed lane as described in Section 2G.14. The legends EXPRESS and EXPRESS LANE(S) shall not be used on signs for open-road tolling lanes that bypass a conventional toll plaza (see Chapter 2F).

Section 2G.18 Regulatory Signs for Priced Managed Lanes

Standard:

01   Except as otherwise provided in this Section, the provisions of Sections 2G.03 through 2G.07 regarding regulatory signs for Preferential lanes shall apply to priced managed lanes operated at all times or at certain times with a toll payment requirement of some or all vehicles to use the lane(s). Such managed lanes shall use changeable message signs or changeable message elements within static signs to display the appropriate regulatory sign messages only when they are in effect.

02   Regulatory signs for preferential lanes shall be appropriately modified for adaptation to a priced managed lane, where applicable, as shown in Figure 2G-18.

03   Regulatory signs shall be used to indicate the toll charged. If the toll varies the R3-48 and R3-48a signs that are shown in Figure 2G-18 shall be used to display the actual toll amount in effect at any given time.

04   When only vehicles with a registered ETC account are allowed to use a managed lane where some or all vehicles are charged a toll, regulatory signs to indicate such a restriction shall be provided and shall incorporate the pictograph adopted by the toll facility's ETC payment system and the word ONLY (see Section 2G.19 for the incorporation of such regulatory legends into the guide signs for the entrances to such facilities). The display of the ETC system pictograph shall comply with the provisions of Sections 2F.02 and 2F.03 as shown in Figures 2G-18 and 2G-19.

05   When HOV traffic is allowed to use a priced managed lane without paying a toll and registration in a local program is not required to receive the toll exemption, the Vehicle Occupancy Definition (R3-10 or R3-13) signs (see Section 2G.04) shall be modified to delete the diamond symbol to create priced managed lane Vehicle Occupancy Definition (R3-40 and R3-43) signs to indicate the minimum occupancy related to the management strategy (see Figure 2G-19).

06   A priced managed lane HOV Lane Operation (R3-44 or R3-44a) sign (see Figure 2G-18) shall be installed at the beginning or initial entry point, and at any intermediate entry points where vehicles are allowed to legally enter an access-restricted priced managed lane.

07   When the vehicle occupancy required for non-toll use of a managed lane is varied as a part of a priced managed lane operational strategy, regulatory signs that include changeable message elements shall be used to display the required vehicle occupancy in effect for non-toll travel.

Option:

08   Where registration in a local program or ETC account is required for HOV traffic to travel in a priced managed lane without being charged a toll or by being charged a discounted toll, such information may be displayed on a separate sign within the sequence of the required regulatory and guide signs.

Standard:

09   R3-42 series and R3-45 series signs (see Figure 2G-18) shall be installed in accordance with the provisions of Section 2G.07 to indicate the termination of a priced managed lane or restriction. The R3-42, R3-42a, and R3-45 signs shall be used only where the managed lane and restriction end and traffic must merge into the general-purpose lanes. The R3-42b, R3-42c, and R3-45a signs shall be used only where the managed lane restriction ends and the lane becomes a general-purpose lane.

Section 2G.19 Guide Signs for Priced Managed Lanes

Standard:

01   Except as otherwise provided in this Section, guide signs for barrier-separated, buffer-separated, and contiguous managed lanes shall follow the specific provisions for Preferential Lane guide signs contained in Sections 2G.10 through 2G.15. Except as otherwise provided in this Section, guide signs for highways on which all lanes are managed shall follow the general provisions for freeway and expressway guide signs as contained in Chapter 2E as a whole. Guide signs for highways on which all lanes are managed and tolling or pricing is used as a management strategy shall follow the applicable provisions for toll road guide signs as contained in Chapter 2F, in addition to the general provisions of Chapter 2E.

02   If fixed or variable tolls are used as an operational strategy for a managed lane, the guide signs shall comply with the provisions of Sections 2F.02, 2F.03, and 2F.17 regarding the use, size, and placement of ETC-account pictographs.

Figure 2G-18. Regulatory Signs for Managed Lanes

Notes:
1. The ETC pictograph shown is an example only. The pictograph for the toll facility's adopted ETC system shall be used.
2. Changeable message sign elements shall be used for the numerals displayed for the variable tolls.

Example of a regulatory sign with changeable message elements

03   **Guide signs at the initial and intermediate entry points to a priced managed lane in which all general-purpose passenger vehicles are allowed shall include the legend EXPRESS LANE(S). Except as provided in Paragraph 5 of this Section, the guide signs shall incorporate the pictograph of the ETC account system into a header panel within the guide sign in accordance with Sections 2F.02, 2F.03, and 2F.17. For a priced managed lane that allows non-toll travel by HOV traffic without registration in a local program, the header panel shall be modified to a regulatory format to display both the pictograph of the ETC account system and the minimum occupancy requirement for non-toll travel with a black legend on a white background (see Figure 2G-19).**

04   **Guide signs at the initial and intermediate entry points to a managed lane that allows only HOV traffic with either a fixed or variable occupancy requirement shall follow the provisions of Sections 2G.10 through 2G.12 and 2G.14.**

05   **If registration in a toll-account program is not required for travel in a managed lane in which tolls are charged, then the ETC-account pictographs shall not be displayed on primary guide signs directing traffic to the managed lane. In such cases, the purple header panel shall be replaced with a warning header panel with a black legend and border on a yellow background displaying the word TOLL as illustrated in Figure 2G-20.**

Option:

06   If the managed lane does not accept toll payments from an ETC account system and collects tolls only by post-travel billing of registered vehicle owners, then the legend TOLL BILLED BY MAIL ONLY may be displayed on a separate information sign within the sequence of primary guide signs in advance of the entrance to the managed lane.

07   If the managed lane accepts payments from registered ETC accounts, but does not require registration to use the lane, then the pictographs of the accepted ETC account programs may be displayed on a separate information sign within the sequence of primary guide signs in advance of the entrance to the managed lane. The information sign may also display the legend TOLL BILLED BY MAIL OR in addition to the pictograph of the accepted ETC account program.

Support:

08   Figure 2G-19 shows examples of Guide signs for entrances to priced managed lanes and other ETC account-only toll facilities that incorporate header panels with ETC account pictographs and regulatory legends.

09   Figures 2G-21 through 2G-24 show examples of guide signs for various configurations of initial and intermediate entrances to a priced managed lane.

Figure 2G-19. Examples of Guide Signs for Entrances to Priced Managed Lanes

A – Entrance to a priced managed lane from a general-purpose lane

B – Direct entrance to a priced managed lane from a crossroad

Notes:

1. The ETC pictographs shown are examples only. The pictograph for the toll facility's adopted ETC system shall be used.

2. The examples shown are for facilities on which registration in a toll account program is required for toll payments.

Figure 2G-20. Example of Signing for the Entrance to an Access-Restricted Priced Managed Lane on which Registration in a Toll Account Program Is Not Required

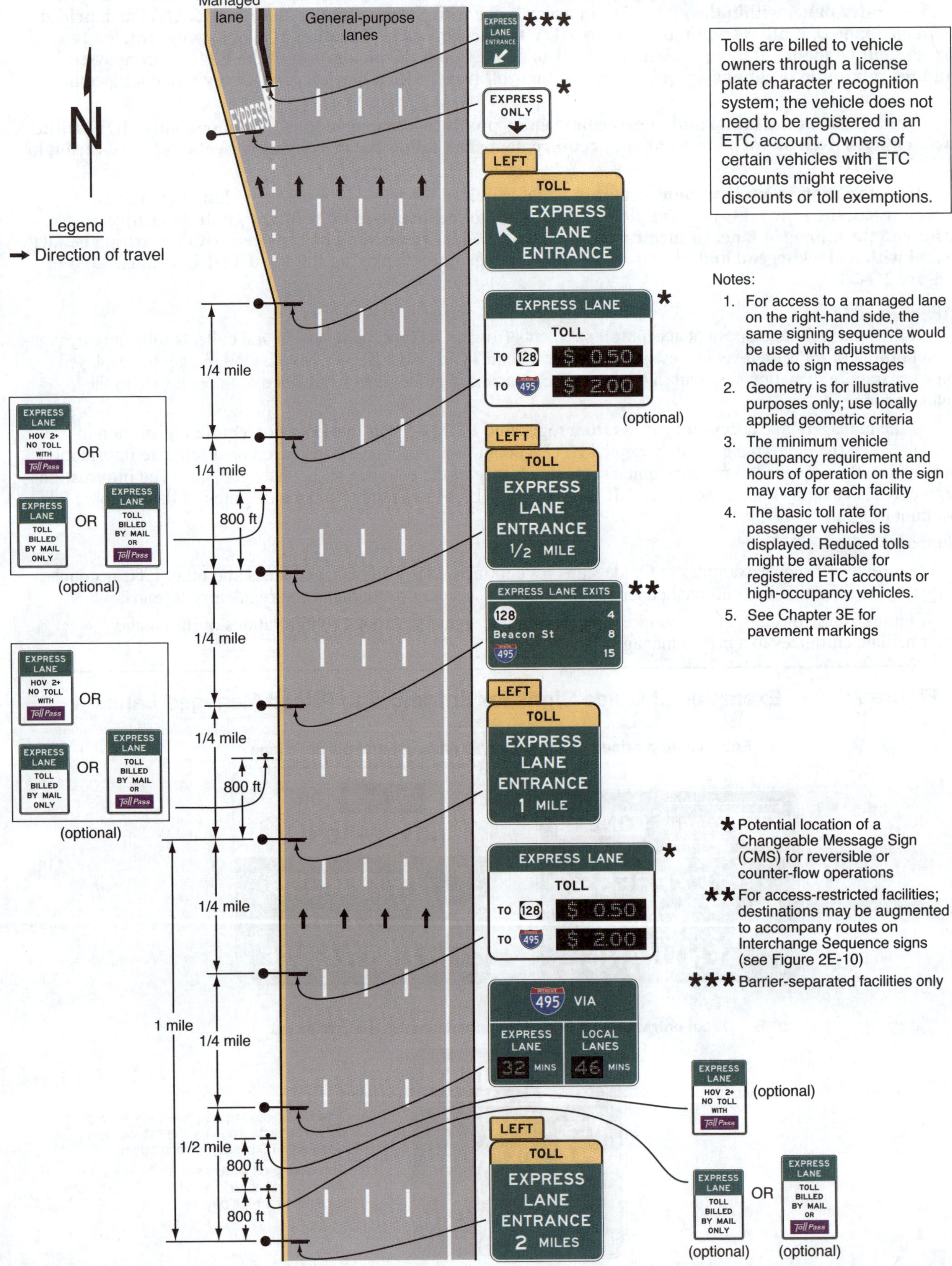

Figure 2G-21. Example of Signing for the Entrance to an Access-Restricted Priced Managed Lane – ETC Account Required

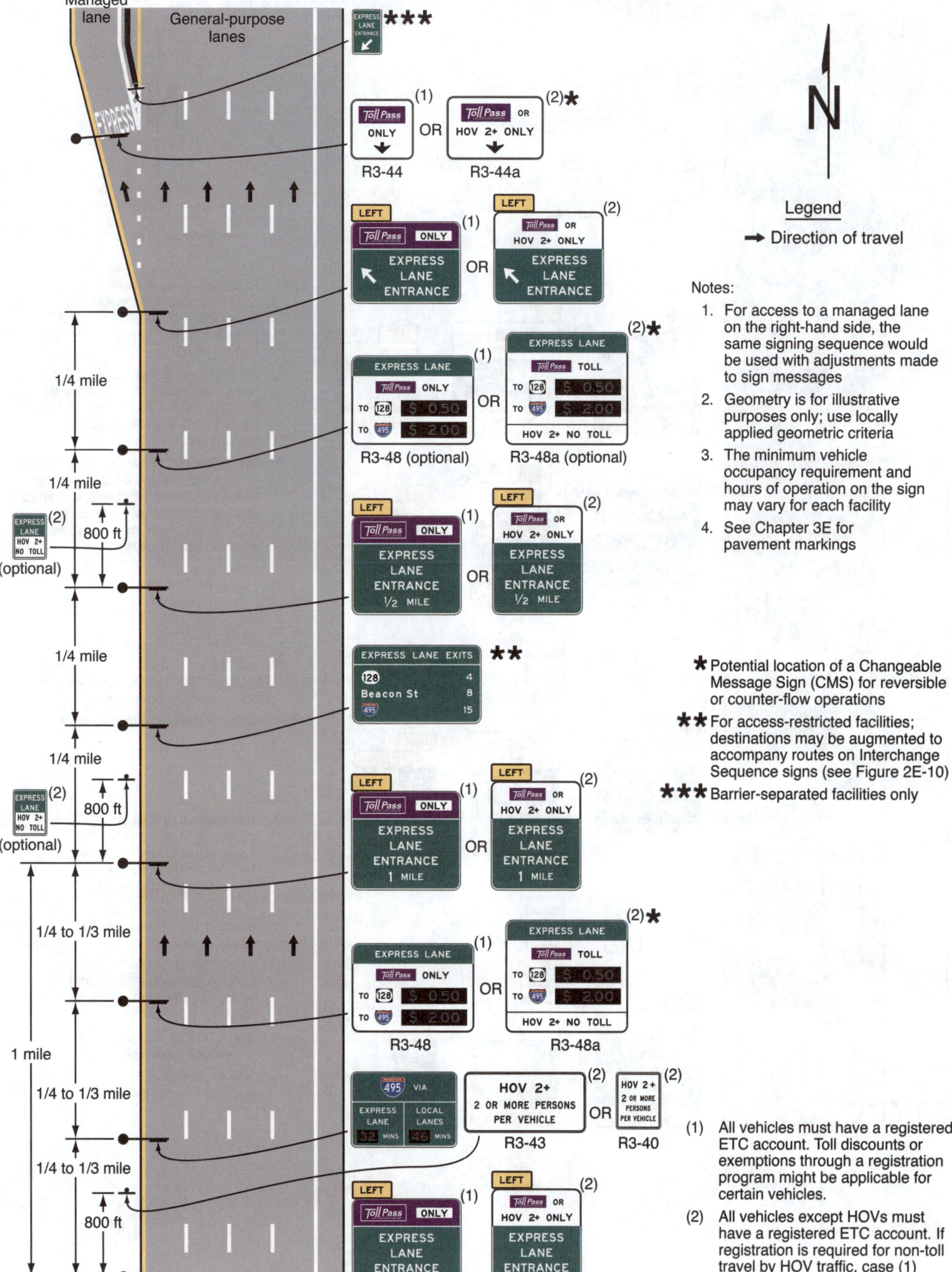

Notes:

1. For access to a managed lane on the right-hand side, the same signing sequence would be used with adjustments made to sign messages

2. Geometry is for illustrative purposes only; use locally applied geometric criteria

3. The minimum vehicle occupancy requirement and hours of operation on the sign may vary for each facility

4. See Chapter 3E for pavement markings

* Potential location of a Changeable Message Sign (CMS) for reversible or counter-flow operations

** For access-restricted facilities; destinations may be augmented to accompany routes on Interchange Sequence signs (see Figure 2E-10)

*** Barrier-separated facilities only

(1) All vehicles must have a registered ETC account. Toll discounts or exemptions through a registration program might be applicable for certain vehicles.

(2) All vehicles except HOVs must have a registered ETC account. If registration is required for non-toll travel by HOV traffic, case (1) signing shall be used.

Figure 2G-22. Example of Signing for the Entrance to an Access-Restricted Priced Managed Lane Where a General-Purpose Lane Becomes the Managed Lane

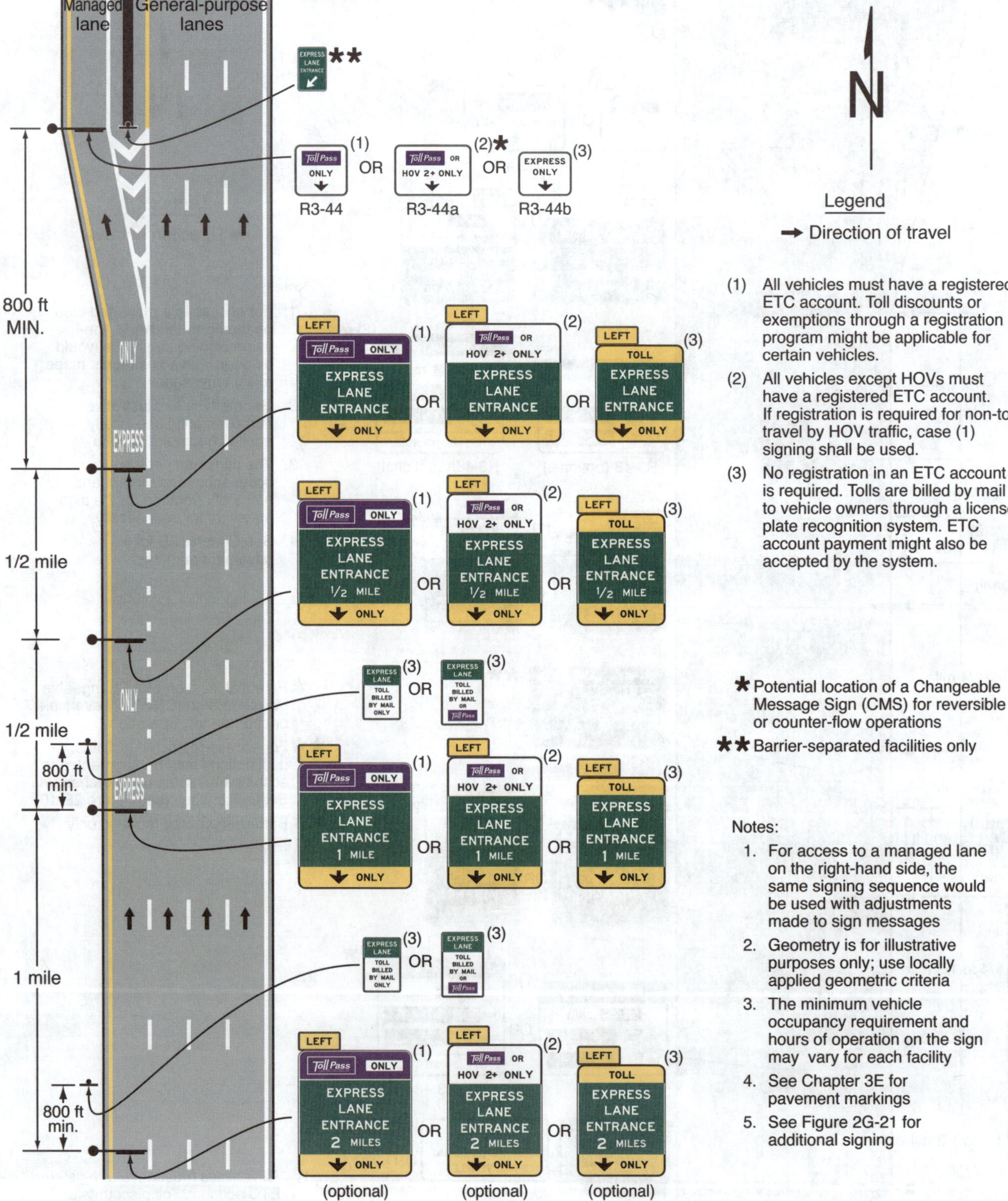

Figure 2G-23. Example of Signing for an Intermediate Entry to a Barrier-Separated or Buffer-Separated Priced Managed Lane

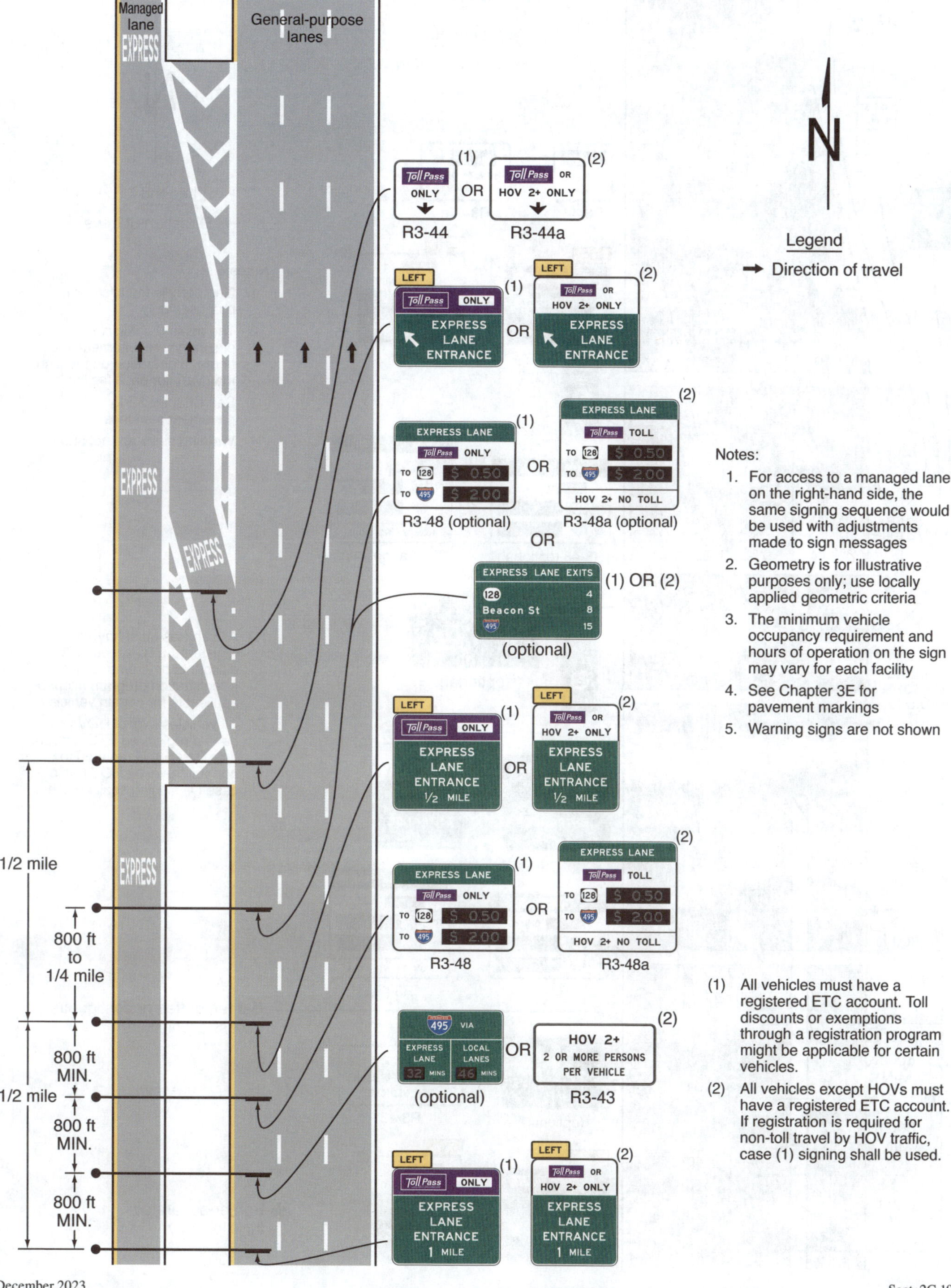

Notes:

1. For access to a managed lane on the right-hand side, the same signing sequence would be used with adjustments made to sign messages

2. Geometry is for illustrative purposes only; use locally applied geometric criteria

3. The minimum vehicle occupancy requirement and hours of operation on the sign may vary for each facility

4. See Chapter 3E for pavement markings

5. Warning signs are not shown

(1) All vehicles must have a registered ETC account. Toll discounts or exemptions through a registration program might be applicable for certain vehicles.

(2) All vehicles except HOVs must have a registered ETC account. If registration is required for non-toll travel by HOV traffic, case (1) signing shall be used.

Figure 2G-24. Example of Signing for the Intermediate Entry to, Egress from, and End of Access-Restricted Priced Managed Lanes

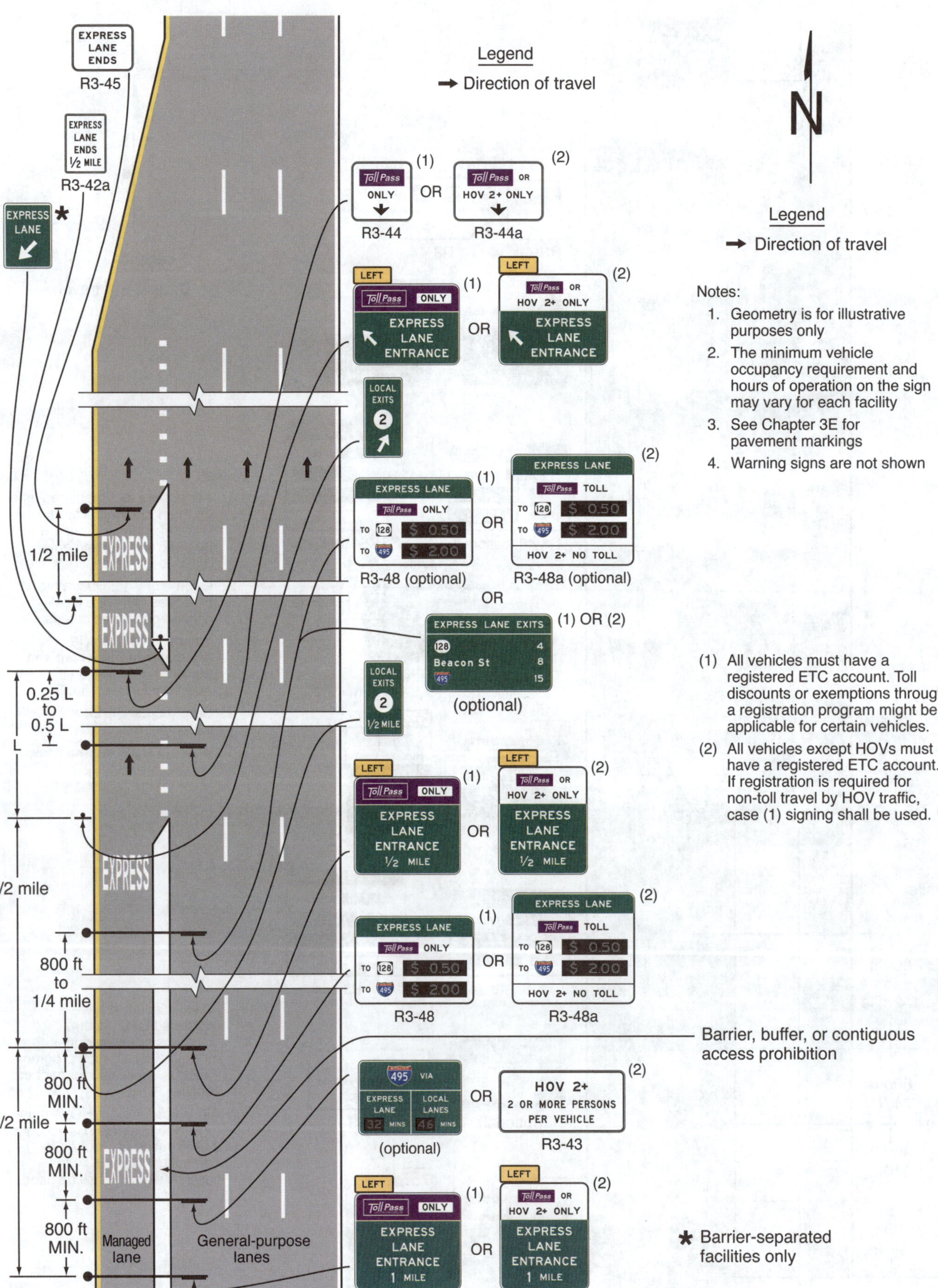

Guidance:

10   *Exit Destination supplemental guide signs, identifying final destination and downstream exit locations accessible from the managed lane (see Figure 2G-25), should be installed in advance of the initial entry points to priced managed lanes. These signs should be located in accordance with the provisions of Paragraph 6 of Section 2G.11.*

11   *For managed lanes that are available as an alternative to travel on adjacent general-purpose lanes on the same designated route, changeable message signs indicating the comparative travel times or congestion levels using the managed lanes versus the general-purpose lanes (see Figure 2G-26) should be installed in advance of the initial and intermediate entry points to the managed lanes.*

Option:

12   Changeable message signs may also be used on non-managed highways to display comparative travel times or congestion levels for a nearby managed highway.

Standard:

13   **The use and locations of guide signs for intermediate egress locations and direct exits from a priced managed lane (see Figures 2G-24 and 2G-27 through 2G-29) shall comply with the provisions of Sections 2G.13 and 2G.15. The signs shall be suitably modified to display header messages of white legend on a green background that relate the guide sign legends to the managed lane(s) as appropriate in accordance with the following:**

   A. Post-mounted or overhead-mounted Advance Guide signs for intermediate egress to the general-purpose lanes shall include the legend LOCAL EXITS in a header panel within the guide signs, destination information or the exit number(s) for the next exit(s) accessible from the general-purpose lanes, and the appropriate distance information to the location of the egress (see Figure 2G-27).

**Figure 2G-25. Example of an Exit Destinations Sign
for a Managed Lane**

**Figure 2G-26. Example of a Comparative Travel Time
Information Sign for Preferential or Managed Lanes**

Note: CMS elements shall be used for the numerals displayed for the estimated travel times.

Figure 2G-27. Examples of Guide Signs for an Intermediate Egress from a Barrier-Separated or Buffer-Separated Managed Lane

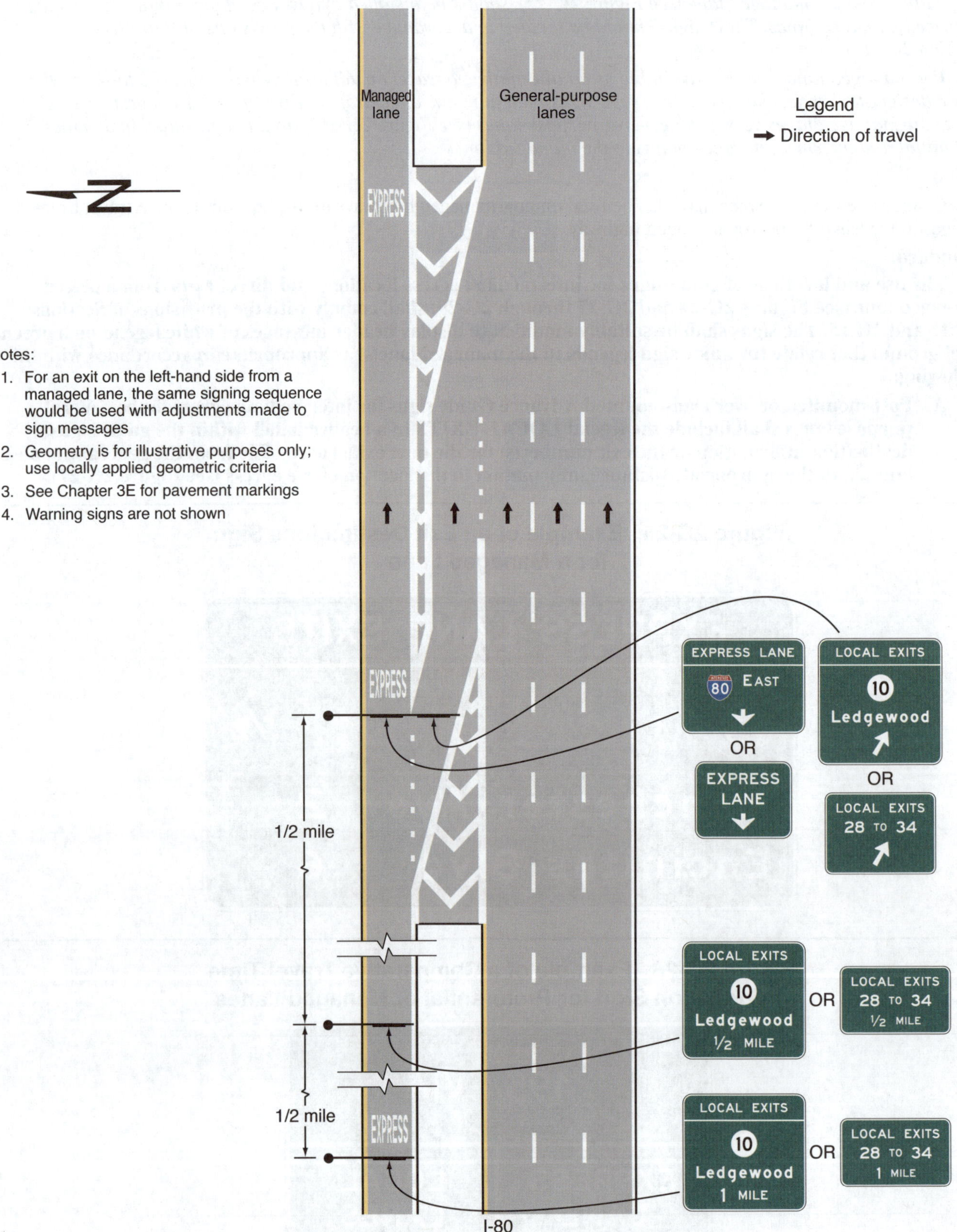

Figure 2G-28. Examples of Guide Signs for Direct Managed Lane Entrance and Exit Ramps

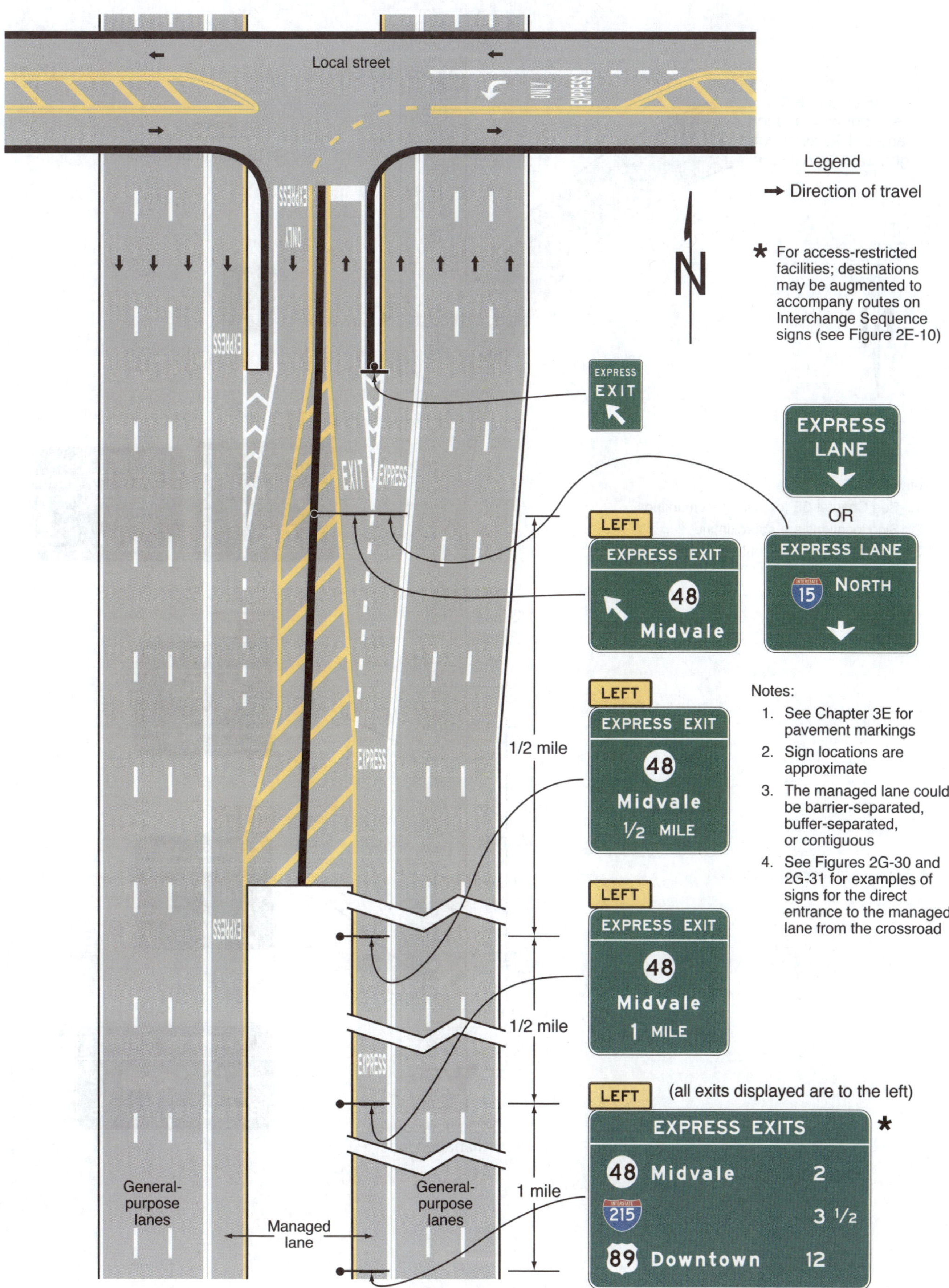

Figure 2G-29. Examples of Guide Signs for a Direct Access Ramp between Managed Lanes on Separate Freeways

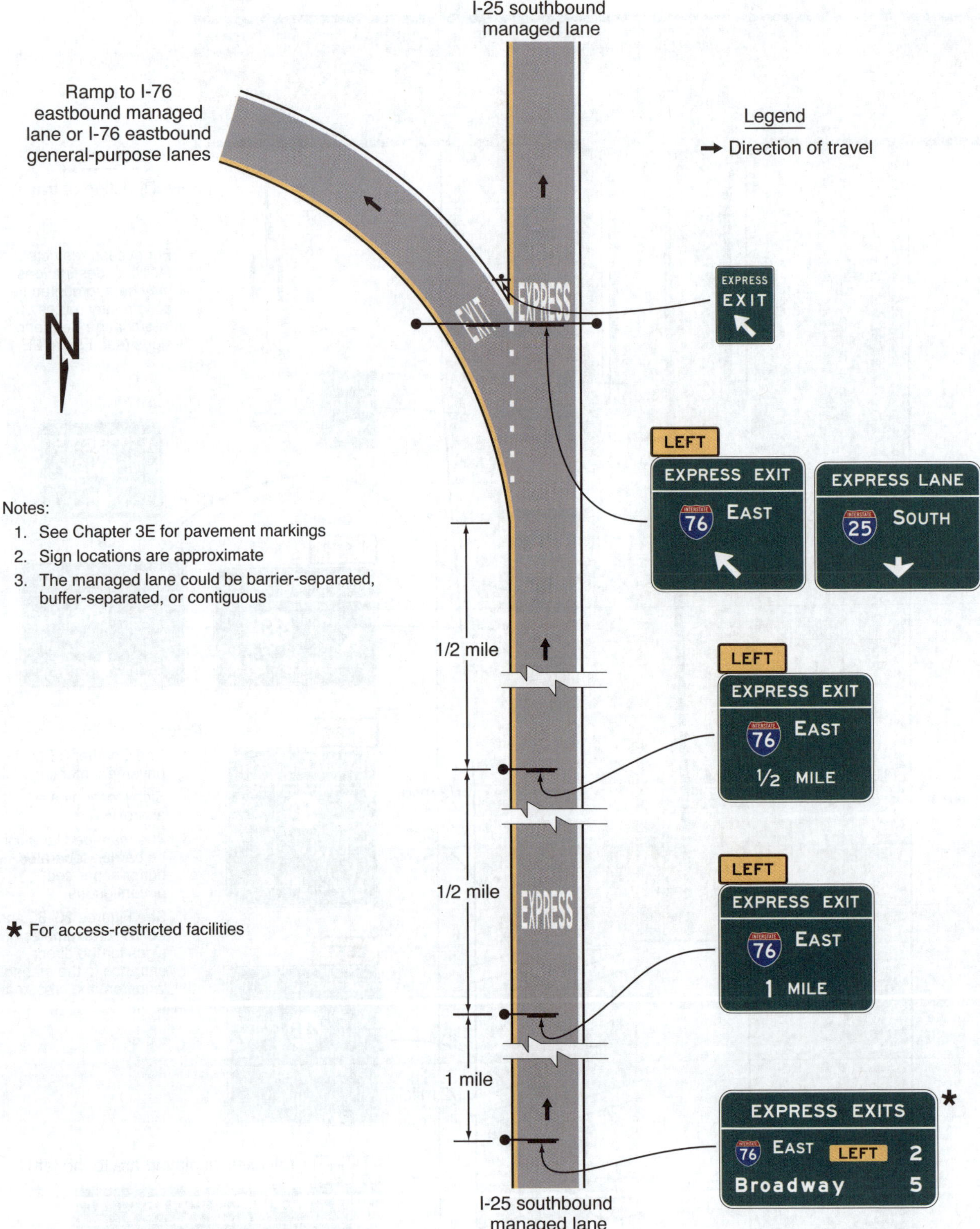

   B. Post-mounted or overhead-mounted Intermediate Egress Direction signs shall include the legend
LOCAL EXITS in a header panel within the signs, the destination information or the exit number(s)
of the next exit(s) accessible from the general-purpose lanes, and a diagonally upward-pointing
directional arrow (see Figure 2G-27).
   C. For direct exits to another roadway, the legend EXPRESS EXIT shall be used on the Advance
Guide and Exit Direction signs (see Figure 2G-28).
   D. For Pull-Through signs, the legend EXPRESS LANE(S) shall be used, either as a header panel
within the Pull-Through sign or as the principal legend of the sign without a header panel
(see Figures 2G-27 through 2G-29).

Support:

14   Section 2G.13 contains information on the use of overhead-mounted guide signs for intermediate egress to the general-purpose lanes.

15   Figures 2G-30 and 2G-31 show examples of guide signing for direct entrances to a priced managed lane from a crossroad or surface street.

Section 2G.20 <u>Signs for Part-Time Travel on a Shoulder – General</u>

Support:

01   In some cases, paved shoulders are allowed to be used for driving use during peak periods to manage congestion. Configurations might be on freeways and expressways, as well as on conventional roads. Travel on the shoulder during these periods might be restricted to certain classes of vehicles, such as buses or HOV, or might be open to general traffic. When the part-time travel on a shoulder is limited to certain classes of vehicles, the signing is similar to that for preferential lanes. Additional signing is typically used to advise road users that the shoulder is not available for emergency use during these periods. Part-time travel on a shoulder might also employ lane-use control signals and/or blank-out signs to inform traffic of the allowable use of the shoulder. Depending on the design of exit ramp terminals and auxiliary lanes, guide signs must account for exit maneuvers during both shoulder use conditions and might necessitate changeable legend elements. However, additional guide signs are not normally necessary specifically for the condition when the shoulder is used for travel. The pavement markings might also be modified where travel allowed on the shoulder begins and ends.

02   Figure 2G-32 shows an example of signing for part-time travel on a shoulder.

Standard:

03   **A shoulder that has been opened to travel on a permanent, full-time basis shall be considered a travel lane and shall be signed and marked in accordance with other provisions of this Manual.**

Support:

04   Section 3E.04 contains provisions regarding the placement of markings on paved shoulders that are open for part-time travel.

Section 2G.21 <u>Regulatory Signs and Plaques for Part-Time Travel on a Shoulder</u>

Standard:

01   **Regulatory signs shall be used to notify road users of the periods of operation that travel is allowed on a paved shoulder. The Part-Time Travel on Shoulder Operation (R3-51) sign (see Figure 2G-32) shall be used where traffic is allowed to travel on the shoulder during certain fixed periods of operation. The Part-Time Travel on Shoulder Variable Operation (R3-51d) sign (see Figure 2G-32) with two flashing beacons (see Chapter 4S) mounted above it shall be used when the period of operation is variable.**

02   **If certain classes of vehicles are not allowed to use the shoulder during these periods, then a Selective Exclusion (R3-51aP or R3-51bP) plaque shall be mounted below the R3-51 or R3-51d sign. If the travel on the shoulder is restricted to certain classes of vehicles, then the regulatory signs shall display that information.**

Option:

03   The EMERGENCY STOPPING ONLY OTHER TIMES (R3-51cP) plaque may be mounted below the R3-51 sign if the R3-51aP or R3-51bP plaque is not used.

Guidance:

04   *The TRAVEL ON SHOULDER BEGINS ½ MILE (R3-52c) sign should be used in advance of the location where part-time travel on shoulder first begins and followed by the DO NOT DRIVE ON SHOULDER (R4-17) sign appropriately spaced downstream.*

Figure 2G-30. Examples of Guide Signs for a Direct Entrance Ramp to a Priced Managed Lane and Trailblazing to a Nearby Entrance to the General-Purpose Lanes

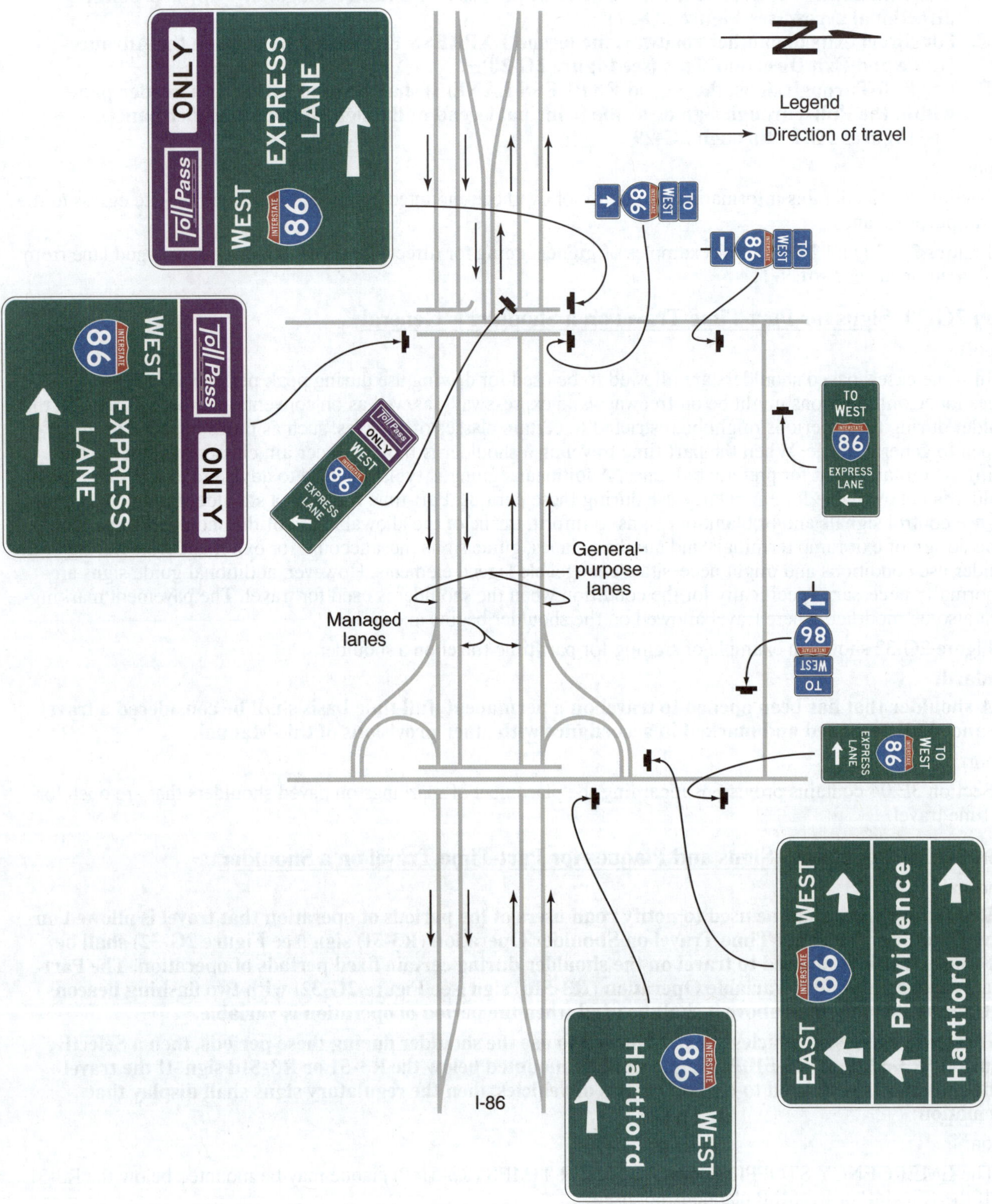

Figure 2G-31. Examples of Guide Signs for Separate Entrance Ramps to General-Purpose and Priced Managed Lanes from the Same Crossroad

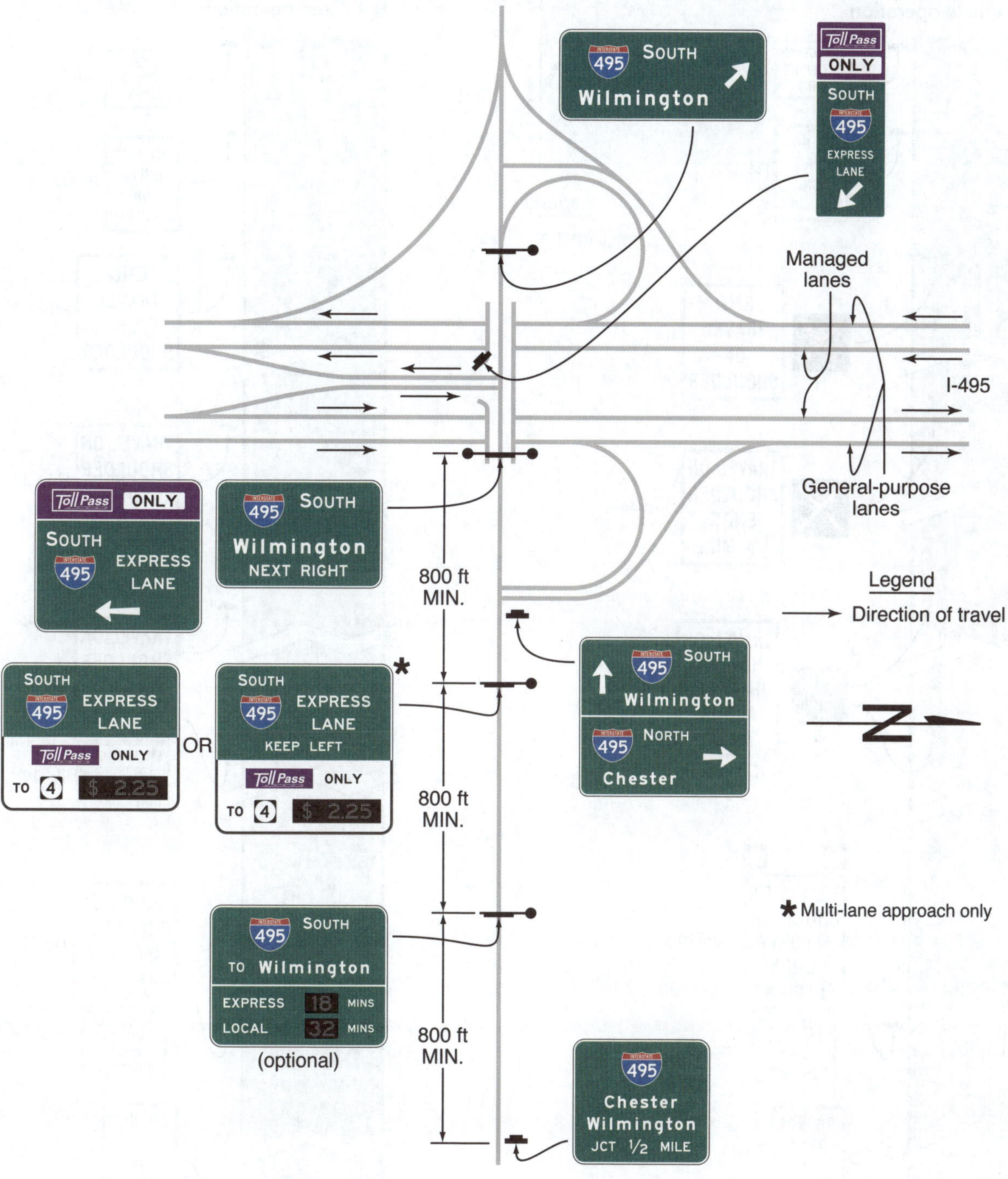

Figure 2G-32. Examples of Signing for Part-Time Travel on a Shoulder (Sheet 1 of 4)

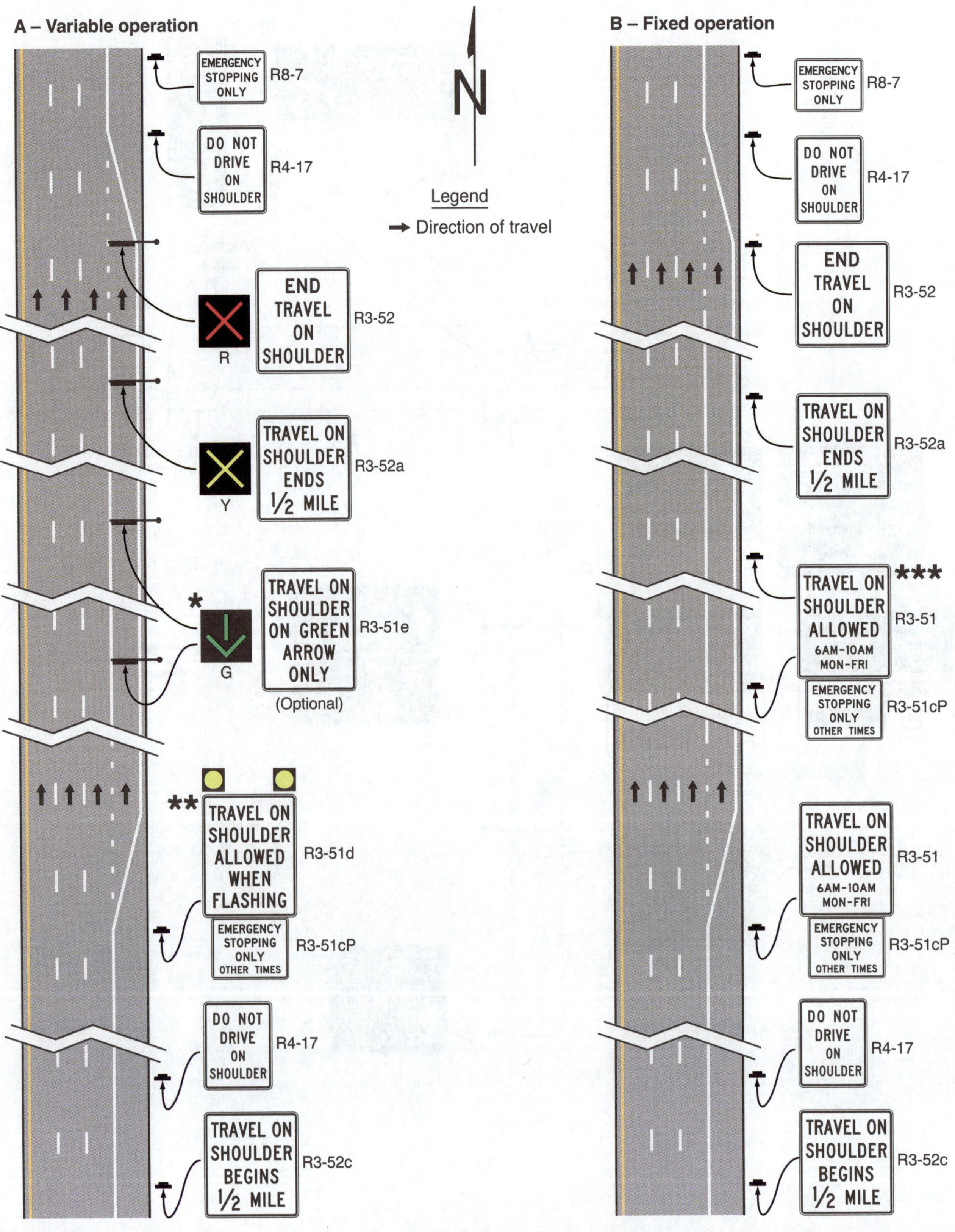

Note: Consideration should be given to using blank-out signs in place of the R3-51 and R3-52 series conventional signs for part-time operations.

*   Lane-use control signals (see Chapter 4T) located every 1/2 mile or less to reaffirm shoulder travel allowed (DOWNWARD GREEN ARROW) or prohibited (RED X).

** A post-mounted TRAVEL ON SHOULDER ALLOWED WHEN FLASHING sign (R3-51d) with beacons may be used lieu of lane-use control signals at the same intervals

*** The R3-51 sign is located every 1/2 mile or less to affirm when travel on shoulder is allowed for fixed period applications.

Figure 2G-32. Examples of Signing for Part-Time Travel on a Shoulder (Sheet 2 of 4)

C – Signing at interchange ramps

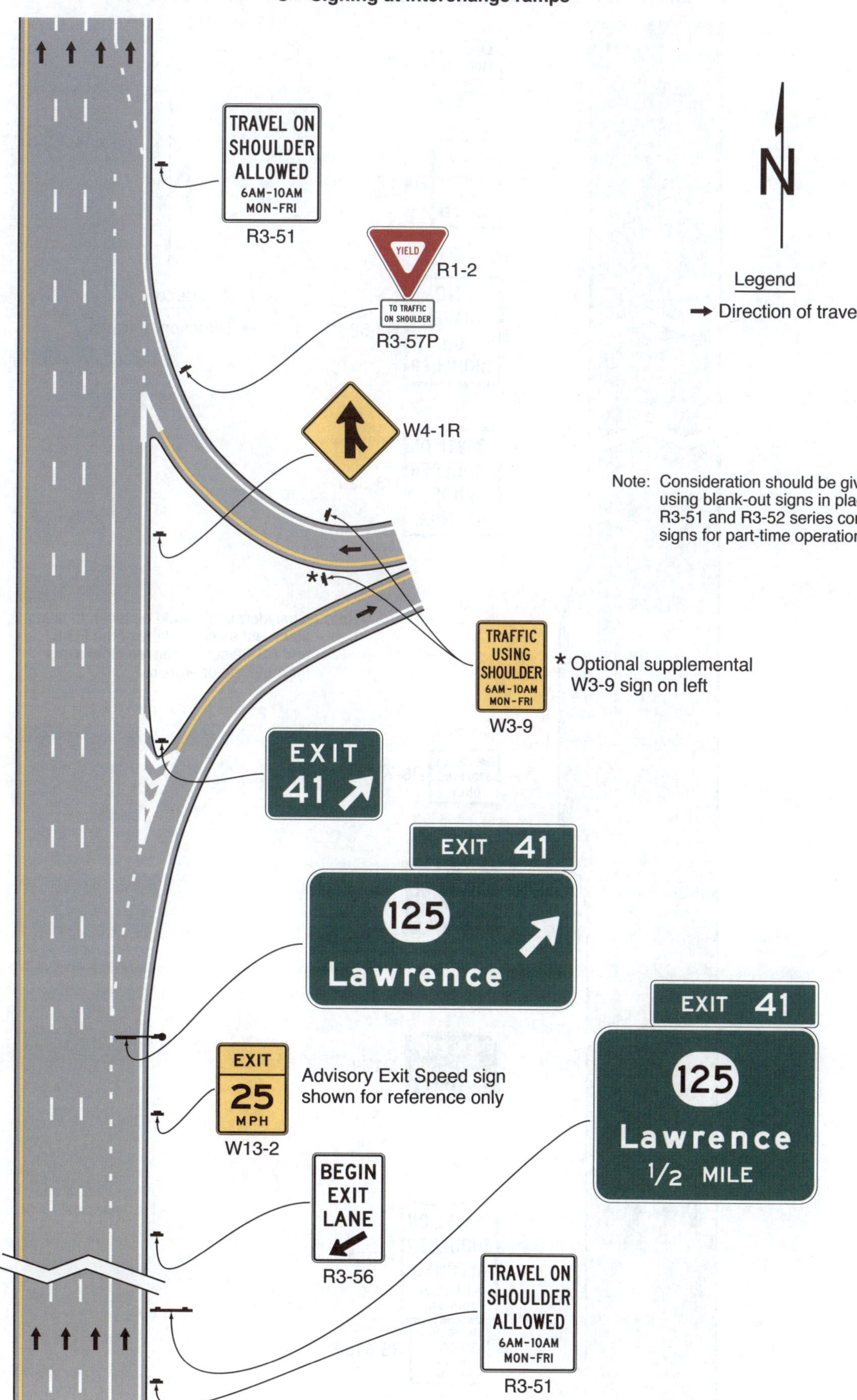

Figure 2G-32. Examples of Signing for Part-Time Travel on a Shoulder (Sheet 3 of 4)

D – Emergency turnout signing for travel on shoulder applications

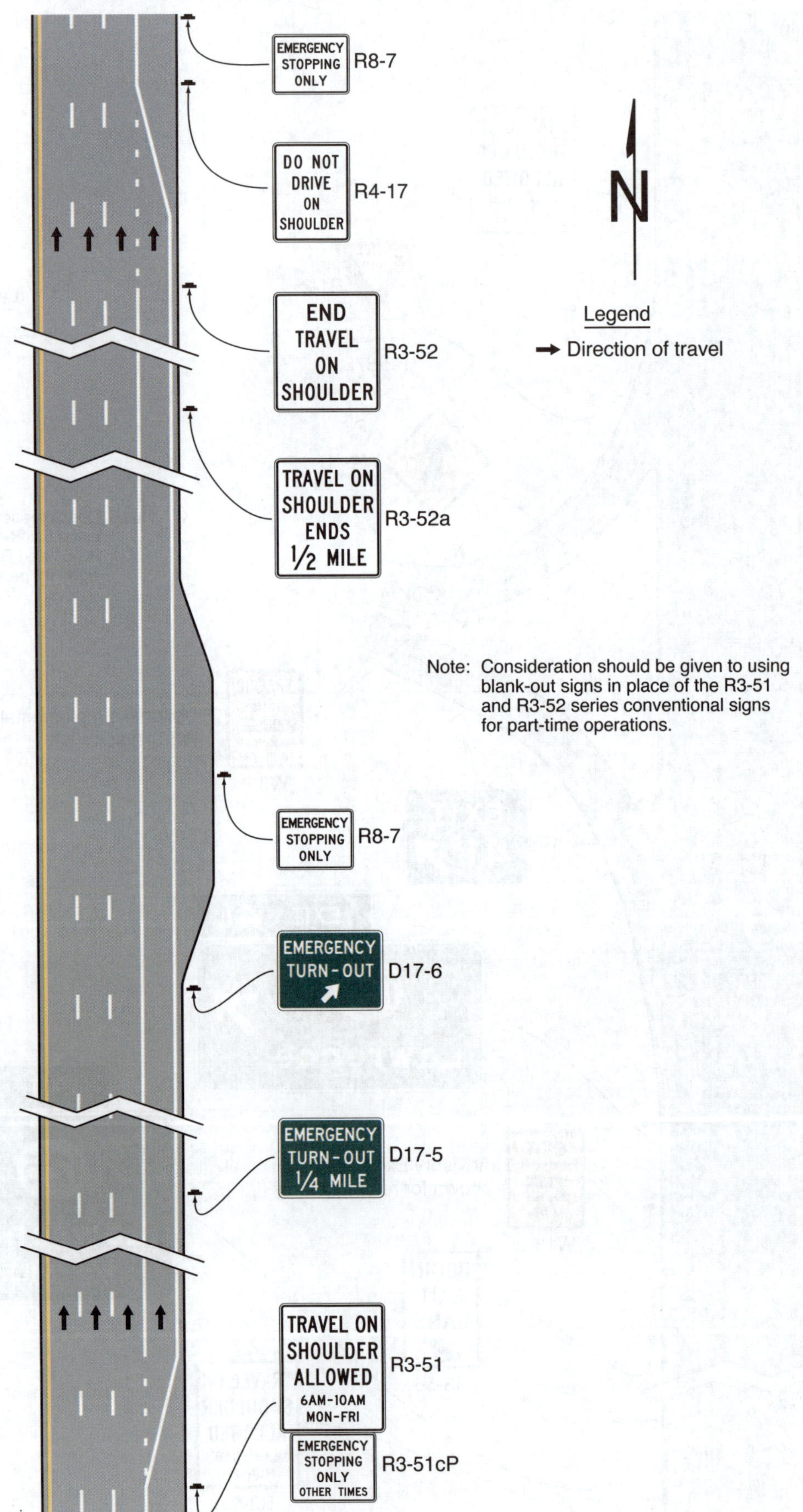

Note: Consideration should be given to using blank-out signs in place of the R3-51 and R3-52 series conventional signs for part-time operations.

Figure 2G-32. Examples of Signing for Part-Time Travel on a Shoulder (Sheet 4 of 4)

E – Travel on shoulder ends at an interchange exit ramp

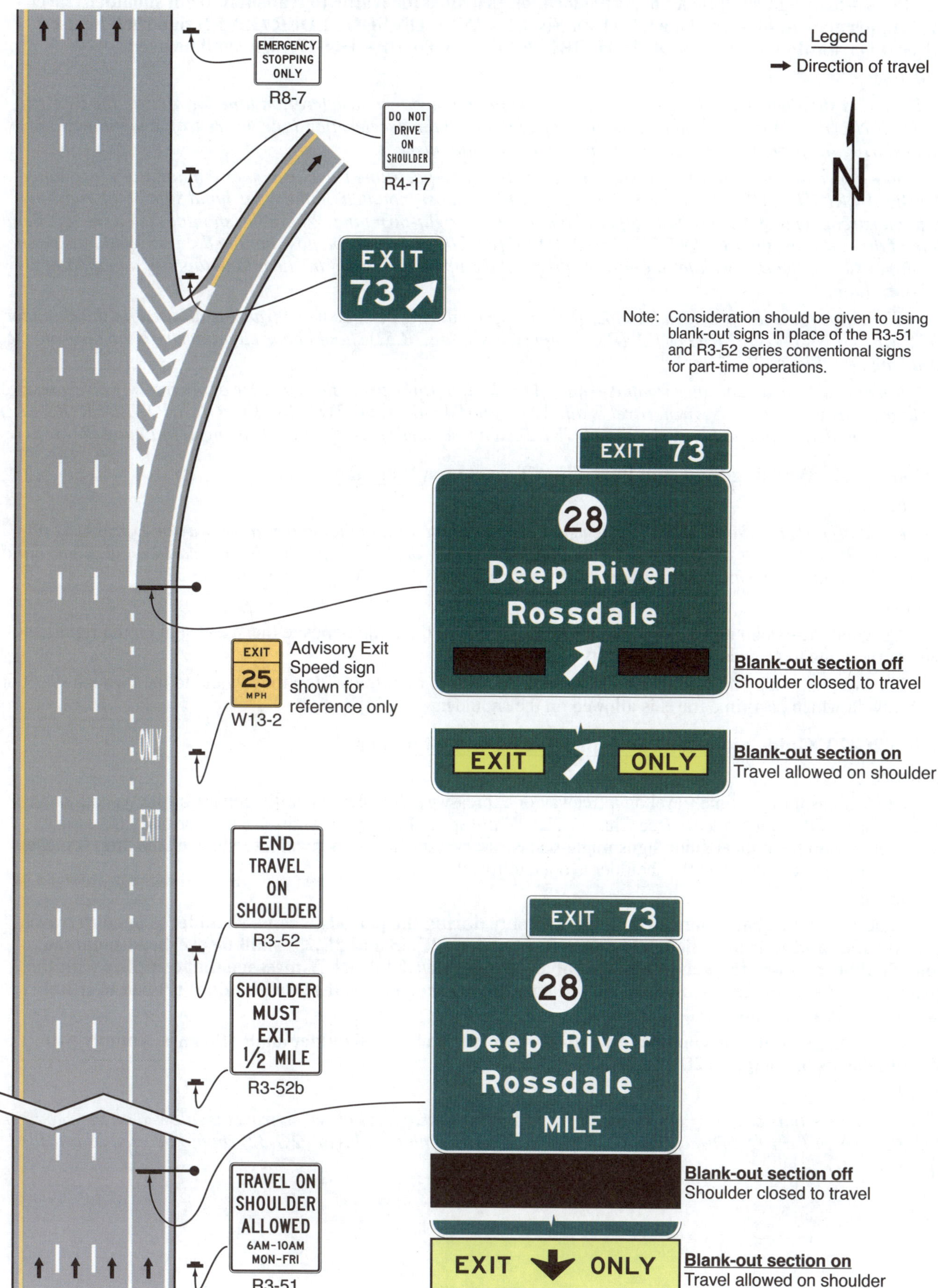

Standard:

05   **Approximately ½ mile from where part-time travel on shoulder ends, the TRAVEL ON SHOULDER ENDS (R3-52a) sign shall be used. At the location provided for traffic to transition from shoulder travel back to permanent highway lane travel, an END TRAVEL ON SHOULDER (R3-52) sign shall be used. After this transition location a DO NOT DRIVE ON SHOULDER (R4-17) sign shall be used.**

Guidance:

06   *Where a shoulder that allows part-time travel is interrupted by a deceleration lane for an exit, the BEGIN EXIT LANE (R3-56) sign should be used at the beginning of the deceleration lane where traffic is allowed to enter during the periods that travel is prohibited on the shoulder.*

07   *On a conventional road where a shoulder that is open to part-time travel becomes a mandatory turn lane, the BEGIN RIGHT TURN LANE (R3-20R) sign should be post-mounted on the right-hand side of the roadway at the upstream end of the turn lane taper of a mandatory right-turn lane. Where the shoulder is on the left-hand side of the roadway, the BEGIN LEFT TURN LANE (R3-20L) sign may be post-mounted on a median (or on the left-hand side of the roadway for a one-way street) at the upstream end of the turn lane taper of a mandatory left-turn lane.*

08   *Where turn-outs are provided for emergency stopping during periods when travel is allowed on the shoulder, the EMERGENCY STOPPING ONLY (R8-7) sign (see Section 2B.52) should be used adjacent to the turn-out (see Drawing D in Figure 2G-32).*

09   *Where traffic on an entrance ramp is required to yield to traffic using the shoulder of the freeway or expressway mainline during the periods when travel is allowed on the shoulder, the TO TRAFFIC ON SHOULDER (R3-57P) plaque should be mounted below the YIELD (R1-2) sign (see Section 2B.05 and Drawing C in Figure 2G-32).*

Section 2G.22 Warning Signs for Part-Time Travel on a Shoulder

Guidance:

01   *The Traffic Using Shoulder (W3-9) sign should be used on a ramp that enters a freeway or expressway on which part-time travel is allowed on the shoulder. When used, the W3-9 sign should be located on the right-hand side of the ramp from which the shoulder traffic approaches (see Drawing C in Figure 2G-32).*

Option:

02   A second W3-9 sign may be used on the left-hand side of the ramp opposite the W3-9 sign on the right-hand side of the ramp to provide greater visibility to oncoming traffic.

03   The W3-9 sign may be used on a conventional road that is required to stop for or yield to the through street or highway on which part-time travel is allowed on the shoulder.

Section 2G.23 Guide Signs for Part-Time Travel on a Shoulder

Support:

01   Guide signs for part-time travel on a freeway or expressway shoulder generally consist of the typical interchange guide sign sequence (see Chapter 2E). While specialized guide signs are not normally necessary, modifications to the typical guide signs might be necessary, especially where an interchange lane drop is created only during the periods when the shoulder is open to travel.

Standard:

02   **Where an interchange lane drop is created only during the periods when a shoulder is open to travel, the Advance and Exit Direction guide signs (see Sections 2E.23 and 2E.25) shall be overhead-mounted and shall be modified to include a blank-out or changeable EXIT ONLY message that complies with the provisions of Section 2E.28 and is displayed only during the periods that the shoulder is open to travel (see Drawing E in Figure 2G-32).**

03   **Guide signs located in conjunction with part-time travel on a shoulder shall otherwise comply with the provisions of Chapters 2D and 2E.**

Guidance:

04   *Where turn-outs are provided for emergency stopping during periods when travel is allowed on the shoulder, the Emergency Turn-Out Directional (D17-6) sign (see Drawing D in Figure 2G-32) should be used as provided in Section 2D.54.*

Section 2G.24 Lane-Use Control Signals for Part-Time Travel on a Shoulder

Support:

01   Lane-use control signals (see Chapter 4T) are sometimes used for part-time travel on a paved shoulder (see Drawing A in Figure 2G-32), in addition to signs, to indicate the allowable use of the shoulder.

Option:

02   Overhead lane-use control signals may be used above a shoulder on which part-time travel is allowed.

Standard:

03   **Except as otherwise provided in this Section, lane-use control signals that are used for part-time travel on a shoulder shall comply with the provisions of Chapter 4T. When used for part-time travel on a shoulder, lane-use control signals shall not be required above the lanes adjacent to the shoulder. When used for part-time travel on a shoulder, a steady RED X signal indication shall be displayed when the shoulder is available for emergency stopping only and travel on the shoulder is otherwise prohibited.**

04   **When part-time travel on shoulder is allowed for variable periods of operation, lane-use control signals (see Chapter 4T) shall be used and evenly spaced approximately every ½ mile or less and centered over the shoulder to indicate when the shoulder is open or closed to vehicle travel. The lane-use control signals shall display a steady DOWNWARD GREEN ARROW signal indication during times when travel is allowed on the shoulder, followed by a steady YELLOW X signal indication just before the shoulder is to be closed to travel, and a steady RED X signal indication when shoulder travel is discontinued. Additionally, during the period when travel is allowed on the shoulder, a lane-use control signal that continuously displays a steady YELLOW X signal indication shall be used approximately ½ mile in advance of the location where part-time travel on the shoulder ends, and then displays a steady RED X signal indication when the travel on shoulder ends. A lane-use control signal with a steady RED X signal indication shall be displayed at all times at the location where part-time travel on the shoulder ends.**

Option:

05   For part-time travel on shoulder with variable periods of operation, post-mounted TRAVEL ON SHOULDER ALLOWED WHEN FLASHING (R3-51d) signs (see Drawing A in Figure 2G-32) with flashing beacons may be used in lieu of the lane-use control signals at the same intervals.

06   The TRAVEL ON SHOULDER ON GREEN ARROW ONLY (R3-51e) sign (see Drawing A in Figure 2G-32) may be used with a lane-use control signal and may be mounted adjacent to the signal head, elsewhere on the signal support, or post-mounted next to, or in advance of, the signal.

Section 2G.25 Lane-Use Control Signals for Active Lane Management on Freeways and Expressways

Support:

01   Active lane management is a component of active traffic management in which the use of travel lanes and speed limits might be varied in real time in response to traffic conditions to manage congestion. Active lane management might employ lane-use control signals (see Chapter 4T) and/or changeable message signs (see Chapter 2L). Figure 2G-33 shows an example of lane-use control signals and variable Speed Limit signs for active lane management during an incident

Standard:

02   **Except as otherwise provided in this Section, lane-use control signals that are used for active lane management shall comply with the provisions of Chapter 4T. When used for active lane management on a freeway or expressway, a steady YELLOW X signal indication shall be displayed to warn road users to vacate the lane when the next downstream lane-use control signal over the same lane is displaying a steady RED X signal indication.**

Option:

03   A steady YELLOW X signal indication may be displayed on one or more lane-use control signals in advance of the steady YELLOW X signal indication required by Paragraph 2 of this Section as conditions warrant to warn road users to vacate the lane.

Support:

04   Using too many steady YELLOW X signal indications could diminish the effectiveness of the steady YELLOW X signal indication in conveying the lane is closed a short distance ahead and the road user needs to vacate the lane soon.

Standard:

05   **When operated in conjunction with a temporary planned lane closure, lane-use control signals shall only supplement the temporary traffic control devices as provided in Part 6 of this Manual.**

Figure 2G-33. Example of Lane-Use Control Signals and Variable Speed Limit Signs for Active Lane Management During an Incident

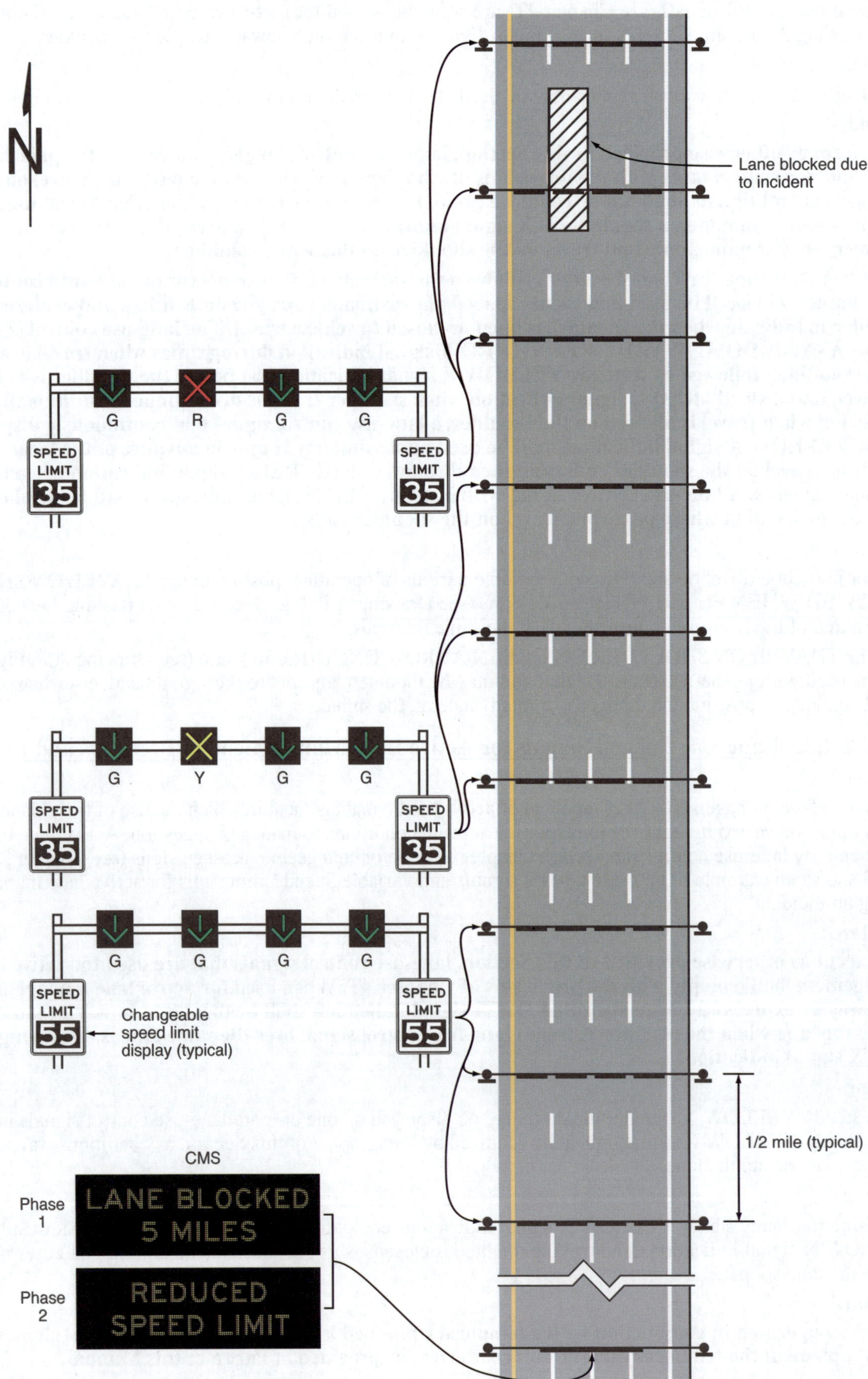

Guidance:

06   *Spacing of lane-use control signals for active lane management on freeways and expressways should be at ½-mile intervals. Closer spacing should be used where the viewing distance is limited by the roadway geometry, overcrossings or other sight obstructions, or where traffic entering from intervening interchange ramps is not adequately served by the ½-mile spacing.*

07   *Combining lane-use control signals with overhead sign support structures should be minimized to avoid overloading road users with too much information or conflicting or incorrect messages, such as exclusive lane use or lane drop implied by the display of a steady DOWNWARD GREEN ARROW signal indication below a guide sign.*

Section 2G.26 <u>Variable Speed Limits for Active Traffic Management on Freeways and Expressways</u>

Support:

01   Active traffic management on freeways and expressways might employ variable speed limits as an element of an overall congestion management plan using variable Speed Limit (R2-1) sign (see Section 2B.21).

02   Careful consideration is needed in locating variable Speed Limit signs along the roadway and potential positioning adjacent to guide signs or lane-use control signals so that the speed displayed is clearly associated with the lane or lanes intended to be regulated and not other adjacent lanes, ramps, or roadways. This might result in the need to place variable Speed Limit signs on separate supports away from guide and other signs or lane-use control signals.

Standard:

03   **The regulatory speed displayed on a variable Speed Limit sign shall comply with Paragraph 13 of Section 2B.21 and the "Standard Highway Signs" publication (see Section 1C.05).**

Guidance:

04   *The location and positioning of variable Speed Limit signs should clearly associate the speed displayed to the lane or lanes intended to be regulated such that it would not present a conflict or confusion with other posted speed limits or advisory speeds for adjacent lanes, ramps, or roadways. Variable Speed Limit signs should not be located on overhead guide sign installations (see Section 2E.43).*

05   *In addition to the post-interchange Speed Limit sign (see Paragraph 17 of Section 2E.47), the spacing of variable Speed Limit signs on freeways and expressways should be based on an engineering study that considers such factors as recurring congestion, high-volume interchanges, weaving sections, and other location-specific factors that are known to affect travel speeds. The variable Speed Limit signs should be placed far enough in advance of known congestion points to adequately adjust the operating speed to minimize the extent of vehicle queuing.*

(This page intentionally left blank)

CHAPTER 2H. GENERAL INFORMATION SIGNS

Section 2H.01 Scope

Support:

01   General Information signs provide road users with navigational or orientation, geographic, or other information useful for traffic operational purposes. They include such items as State lines, city limits, time zones, stream names, elevations, landmarks, and similar geographic features. Chapter 2M contains recreational and cultural interest area symbol signs that are sometimes used in combination with General Information Signs. Section 1D.09 contains information on unnecessary traffic control devices. Section 2A.20 contains information on the excessive use of signs and sign clutter.

Option:

02   A General Information (I3-5 through I4-2) symbol sign (see Figure 2H-1) may be used to provide direction to a transportation (I3 series signs) or other (I4 series signs) facility. The symbol sign may be supplemented by an educational plaque where necessary. The name of the facility may be used, if needed, to distinguish between similar facilities in the same area.

03   The Advance Turn (M5 series) or Directional Arrow (M6 series) auxiliary plaques (see Figure 2H-1) with white arrows on green backgrounds may be used with General Information symbol signs to create a General Information Directional Assembly.

04   The Recycling Center (I4-2) symbol sign may be used to direct road users to recycling centers.

Guidance:

05   *The Recycling Center symbol sign should not be used on freeways and expressways.*

Option:

06   The Passengers Only Ferry Terminal (I3-10) symbol sign may be used with the FERRY (I3-10P) plaque (see Figure 2H-1) mounted below it in a directional assembly to direct road users to passenger-only ferry terminals.

Figure 2H-1. General Information and Miscellaneous Information Signs and Plaques

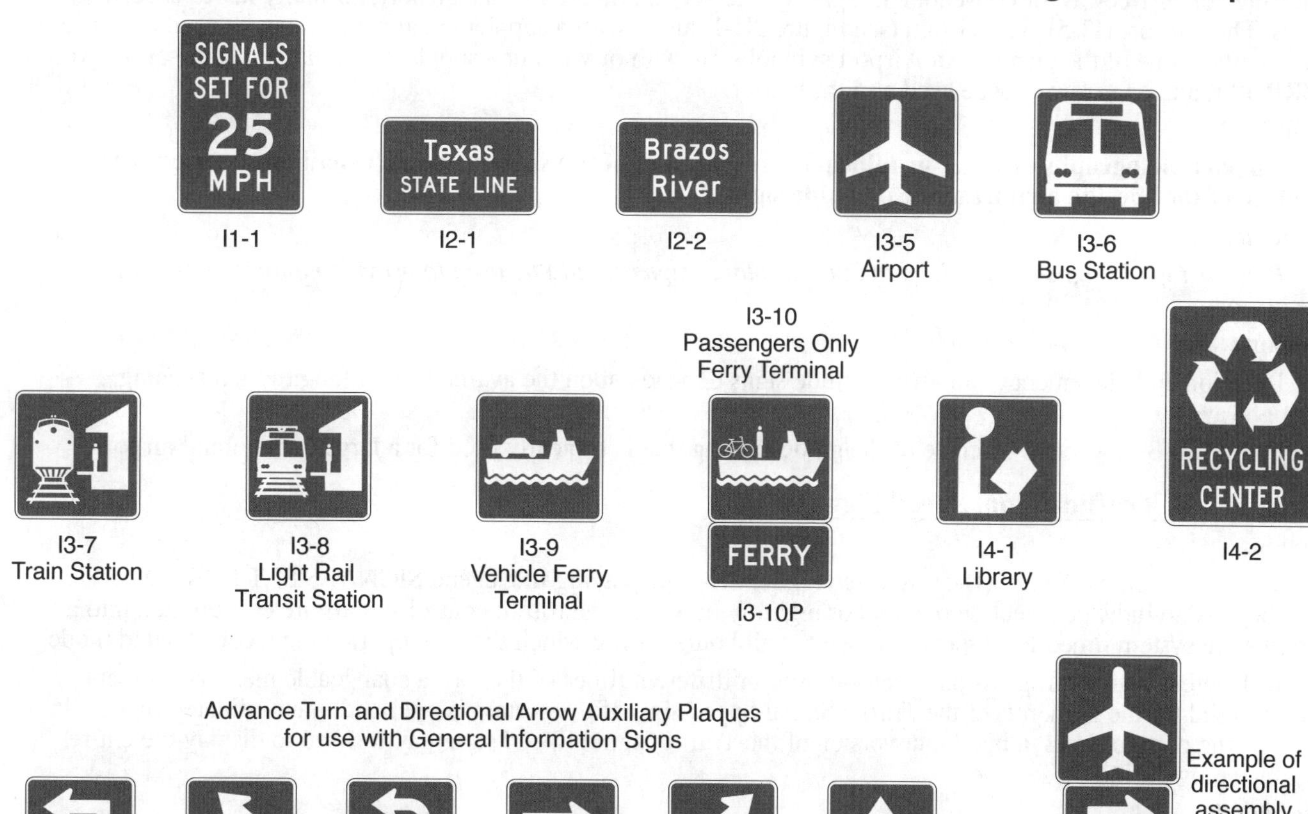

I1-1

I2-1

I2-2

I3-5
Airport

I3-6
Bus Station

I3-10
Passengers Only
Ferry Terminal

I3-7
Train Station

I3-8
Light Rail
Transit Station

I3-9
Vehicle Ferry
Terminal

I3-10P

I4-1
Library

I4-2

Advance Turn and Directional Arrow Auxiliary Plaques
for use with General Information Signs

Example of
directional
assembly

M5-1P

M5-2P

M5-3P

M6-1P

M6-2P

M6-3P

Guidance:

07 *General Information signs should not be installed within a series of guide signs, or at other equally critical locations, unless there are specific reasons for orienting the road user or identifying control points for activities that are clearly in the public interest. On all such signs, the designs should be simple and dignified, devoid of any tendency toward advertising, such as complex graphics or unnecessary messages, and in general compliance with other guide signing.*

Standard:

08 **Promotional descriptive messages that are not relevant to navigation and orientation, such as "Scenic" or "Historic," shall not be included in the legends of General Information signs, except as otherwise provided in this Chapter or in cases in which these terms are part of an official name, such as for a Scenic Byway or Historic District.**

09 **Except for State Welcome signs (see Section 2H.07), Acknowledgment signs (see Section 2H.13), and Alternative Fuels Corridor signs (see Section 2H.14), General Information signs shall have white legends and borders on green rectangular-shaped backgrounds.**

Section 2H.02 Sizes of General Information Signs

Standard:

01 **Except as provided in Section 2A.07, the sizes of General Information signs that have a standardized design shall be as shown in Table 2H-1.**

Support:

02 Section 2A.07 contains information regarding the applicability of the various columns in Table 2H-1.

Option:

03 Signs larger than those shown in Table 2H-1 may be used (see Section 2A.07), except where a maximum allowable size is specified.

Section 2H.03 Airport Signs

Support:

01 Guide signs for commercial service airports and general aviation airports may be provided from the nearest Interstate, other freeway, or conventional highway intersection directly to the airport, normally not to exceed 15 miles. The Airport (I3-5) symbol sign (see Figure 2H-1) along with a supplemental plaque may be used to indicate the specific name of the airport. An Airport symbol sign, with or without a supplemental name plaque or the word AIRPORT, and an arrow may be used as a trailblazer.

Standard:

02 **Airport pictographs or other graphical representation of the specific airport shall not be used with or in place of the specific airport name on guide signs.**

Guidance:

03 *If airport guide signs are used, adequate trailblazer signs should be used to provide motorist direction to the airport.*

Support:

04 Location and placement of all airport guide signs depends upon the availability of longitudinal spacing on highways.

05 Figure 2D-39 shows an example of the guide signing that is typically used for a large commercial airport.

Section 2H.04 Traffic Signal Speed Sign (I1-1)

Option:

01 The Traffic Signal Speed (I1-1) sign (see Figure 2H-1) displaying the legend SIGNALS SET FOR XX MPH may be used to indicate a section of street or highway on which the traffic control signals are coordinated into a progressive system timed for a specified speed at all hours during which they are operated in a coordinated mode.

02 If different system progression speeds are set for different times of the day, a changeable message element may be used for the numerals of the Traffic Signal Speed sign. If the system is operated in coordinated mode only during certain times, a blank-out version of the Traffic Signal Speed sign may be used to display the entire message only during those times.

Standard:

03 **An electronic-display changeable section of the Traffic Signal Speed sign shall be a white legend on a black opaque or green background.**

Table 2H-1. General Information Sign and Plaque Sizes

Sign	Sign Designation	Section	Conventional Road	Freeway or Expressway
Alternative Fuels Corridor	D9-19	2H.14	24 x 24	36 x 36
Alternative Fuels Corridor (1 line) (plaque)	D9-19aP	2H.14	30 x 9	42 x 12
Alternative Fuels Corridor (2 lines) (plaque)	D9-19bP	2H.14	30 x 12	42 x 18
Reference Location (1 digit)	D10-1	2H.11	10 x 18	12 x 24
Intermediate Reference Location (2 digits)	D10-1a	2H.11	10 x 27	12 x 36
Reference Location (2 digits)	D10-2	2H.11	10 x 27	12 x 36
Intermediate Reference Location (3 digits)	D10-2a	2H.11	10 x 36	12 x 48
Reference Location (3 digits)	D10-3	2H.11	10 x 36	12 x 48
Intermediate Reference Location (4 digits)	D10-3a	2H.11	10 x 48	12 x 60
Enhanced Reference Location	D10-4	2H.12	12 x 30 18 x 54 (O)	18 x 54
Intermediate Enhanced Reference Location	D10-5	2H.12	12 x 36 18 x 60 (O)	18 x 60
Traffic Signal Speed	I1-1	2H.04	24 x 36	–
Jurisdictional Boundary	I2-1	2H.05	Varies x 18** Varies x 24 (O)	Varies x 36** Varies x 42 (O)
Geographical Feature	I2-2	2H.06	Varies x 18** Varies x 24 (O)	Varies x 36**
Grade Separation Identification	I2-3	2H.10	–	Varies x 18
Grade Separation Identification (2 lines)	I2-3a	2H.10	–	Varies x 24
Future Interstate Corridor	I2-4	2H.08	54 x 36	72 x 48
Future I-XX Corridor	I2-4a	2H.08	48 x 36	66 x 48
Project Information	I2-5	2H.09	96 x 48	156 x 72
Airport	I3-5	2H.01	24 x 24	30 x 30
Bus Station	I3-6	2H.01	24 x 24	30 x 30
Train Station	I3-7	2H.01	24 x 24	30 x 30
Light Rail Transit Station	I3-8	2H.01	24 x 24	–
Vehicle Ferry Terminal	I3-9	2H.01	24 x 24	30 x 30
Passenger Only Ferry Terminal	I3-10	2H.01	24 x 24	30 x 30
Ferry (plaque)	I3-10P	2H.01	24 x 12	30 x 18
Library	I4-1	2H.01	24 x 24	–
Recycling Center	I4-2	2H.01	30 x 36	–
Acknowledgment	I20-1	2H.13	36 x 30*	72 x 48*
Acknowledgment	I20-2	2H.13	36 x 30*	72 x 48*
Acknowledgment	I20-3	2H.13	42 x 24*	96 x 36*
Acknowledgment - Rest Area	I20-4	2H.13	56 x 36*	72 x 48*
Acknowledgment - Welcome Center	I20-4a	2H.13	56 x 36*	72 x 48*
Acknowledgment (plaque)	I20-5P	2H.13	Varies x Varies***	Varies x Varies***
Last In Corridor (plaque)	W16-19P	2H.14	24 x 18	24 x 18

* The size shown is the maximum size for the corresponding roadway classification. The size of the sign and acknowledgment logo should be appropriately reduced where shorter legends are used.

** The size shown is for the typical sign illustrated in the figure. The size should be determined based on the number of lines of legend on the sign.

*** Limitations on the size of Acknowledgment plaques are provided in Section 2H.13.

Notes: 1. Larger signs may be used when appropriate, except for the I20 series signs and plaque

2. (O) denotes Oversized

3. Dimensions are in inches shown as width x height

Guidance:

04   *If used, the Traffic Signal Speed sign should be mounted as near as practical to each intersection where the timed speed changes, and at intervals of several blocks throughout any section where the timed speed remains constant.*

Section 2H.05 Jurisdictional Boundary Signs (I2-1)

Option:

01   The Jurisdictional Boundary (I2-1) sign may be used to mark the location of the jurisdictional boundary of a State, county, or municipality or the limits of an unincorporated municipal-level community, Tribal Nation, or governmental district where legal jurisdiction, road maintenance responsibility, or emergency response obligation changes.

Guidance:

02   *If used, the Jurisdictional Boundary sign should be located at or as near as practicable to the jurisdictional boundary without interfering with higher-priority traffic control devices. Notices of statutes or local ordinances should be located separately using regulatory signs (see Chapter 2B).*

03   *If used for an unincorporated community, the community should be one that is readily identifiable on official maps and be consistent with postal mailing addresses.*

Standard:

04   **In accordance with Section 2H.01, the Jurisdictional Boundary sign shall be rectangular in shape and shall have a white legend on a green background. The sign shall display only the name of the State, county, municipality, Tribal Nation, or other identifiable community, and an appropriate legend such as ENTERING, STATE LINE, County, or the municipal classification.**

05   **Names of elected officials or promotional messages, such as notable accomplishments or claims, shall not be displayed on a Jurisdictional Boundary sign or added as a supplemental sign or plaque.**

Option:

06   A pictograph representing the jurisdiction may be displayed on the Jurisdictional Boundary sign.

Standard:

07   **If a pictograph is displayed on the Jurisdictional Boundary sign, it shall be the official seal of the jurisdiction and shall comply with the provisions of Section 2A.04. The pictograph shall be placed to the left of the legend. The height of the pictograph shall not exceed 2 times the height of the initial upper-case letter of the principal legend.**

Guidance:

08   *Signs should not be used to identify the boundaries of special-purpose governmental districts, such as school districts, sanitary districts, or improvement districts, as such signs are generally promotional in nature and do not provide navigational or orientation assistance in conjunction with official maps that are available to the general public.*

Support:

09   Section 2H.07 contains information on State Welcome signs.

Section 2H.06 Geographical Feature Signs (I2-2)

Option:

01   The Geographical Feature (I2-2) sign may be used to mark the locations of land features such as river or stream crossings, and summits, that are identifiable on maps or serve as landmarks in providing navigational orientation or reference to the road user.

Guidance:

02   *If used, the Geographical Feature sign should display only the name of the geographical feature. Additional information that is unnecessary for navigational or orientation purposes, such as watershed or tributary names, should not be displayed on the sign.*

Section 2H.07 State Welcome Signs

Support:

01   The design, placement, and function of State Welcome signs that are used to identify State lines differ from Jurisdictional Boundary (I2-1) signs (see Section 2H.05). Because of these differences, it is necessary to distinguish State Welcome signs from State line Jurisdictional Boundary signs.

Option:

02   A State Welcome sign may be located at or in the vicinity of the State boundary except as prohibited in Paragraph 4 of this Section.

03   State Welcome signs may display the State seal or the State flag, the officially-adopted State motto or slogan, and the name of the Governor, in addition to the State name. State Welcome signs may use legend and background colors that provide adequate visual contrast rather than the standard sign colors.

Standard:

04   **State Welcome signs shall be located separate from other signs where they will not interfere with or detract from other traffic control devices.**

05   **State Welcome signs shall not display changeable or other electronic-display messages (see Chapter 2L). State Welcome signs shall not display messages that emulate promotional advertising of any type. State Welcome signs shall not incorporate Acknowledgment signs or messages (see Section 2H.13), or business identification sign panels or logos (see Section 2J.03) into their legends or assemblies. In accordance with Section 2A.04 of this Manual, telephone numbers, Internet addresses, and e-mail addresses, including domain names and uniform resource locators (URLs), and scanning graphics for the purpose of obtaining information shall not be displayed in the legends of State Welcome signs or on their supports.**

Guidance:

06   *State Welcome signs should be located farther from the edge of the roadway than other traffic control devices.*

07   *The maximum size of a State Welcome sign should be consistent with the prevailing size of other guide signs based on the roadway type.*

Section 2H.08 Future Interstate Corridor Signs (I2-4 and I2-4a)

Option:

01   The Future Interstate Corridor (I2-4 and I2-4a) signs (see Figure 2H-2) may be used sparingly along an existing route that will be reconstructed as an Interstate route or along an existing route adjacent to a corridor through which an Interstate route will be constructed, in accordance with the Policy and Conditions stated in 23 CFR 470, Appendix C.

02   Where the route number has been approved by the FHWA, either the I2-4 or I2-4a sign may be used.

Standard:

03   **The I2-4a sign shall not be used where the route number has not been approved by the FHWA.**

04   **Future Interstate Corridor signs shall not be located where they could interfere with or detract from other traffic control devices. If used, Future Interstate Corridor signs shall be installed as independent, post-mounted sign assemblies.**

05   **Future Interstate Corridor signs shall not imply that an existing route has already been designated and marked as an Interstate route. Signs indicating that an existing route is designated as a future Interstate route or corridor shall not provide directional or distance information. Route Sign assemblies (see Section 2D.29) of any type shall not be used to sign a route as a future Interstate or other route. The Interstate route marker or likeness thereof shall not be displayed on the Future Interstate Corridor signs.**

Figure 2H-2. Future Interstate Signs (I2-4, I2-4a)

I2-4   I2-4a

Guidance:

06 *Future Interstate Corridor signs should be limited to strategic locations, such as at the beginning of the designated route or corridor, or beyond interchanges connecting from existing Interstate highways.*

Section 2H.09 Project Information Sign (I2-5)

Support:

01 The Project Information (I2-5) sign (see Figure 2H-3) provides limited information to road users about a highway construction project on which work is imminently forthcoming or ongoing.

Standard:

02 **The Project Information sign legend shall be limited to the following project information:**

    A. The roadway name or route number,
    B. A brief description or title of the project,
    C. The completion date expressed in either a month or season (Spring, Summer, Fall, or Winter), and
    D. The agency name.

Option:

03 Project Information signs installed more than one week prior to commencement of work may include a start date.

Standard:

04 **Project Information signs shall not be installed more than one month prior to the commencement of work. When installing Project Information signs prior to the commencement of work, the jurisdiction shall have a policy on when the Project Information signs are to be installed. Project Information signs shall be removed at the conclusion of work on the project, even if the final inspection or project closeout has not yet occurred.**

05 **The number of Project Information signs shall be limited to one per direction of travel on the roadway on which the project is based. The location of the Project Information sign shall not interfere with the temporary traffic control zone devices.**

06 **The Project Information sign shall have a white legend on a green background and shall not display Internet addresses, e-mail addresses, or telephone numbers (see Section 2A.04).**

Section 2H.10 Grade-Separated Roadway Identification Signs (I2-3 and I2-3a)

Option:

01 The Grade-Separated Roadway Identification (I2-3 and I2-3a) signs (see Figure 2H-4) may be used to identify a grade separation of another highway or other transportation facility such as a railway, bikeway, or pathway.

Guidance:

02 *Except as provided in Paragraph 4 of this Section, when used to identify an overcrossing structure, the I2-3 sign should be mounted above the travel lanes or shoulder of the highway below.*

03 *When used to identify an undercrossing structure, the I2-3 or I2-3a sign should be post-mounted in advance of the structure as near to it as practicable.*

Option:

04 When used to identify an overcrossing structure, the I2-3 or I2-3a sign may be post-mounted in front of an overcrossing or may be mounted to the abutment of the overcrossing facing approaching traffic.

Figure 2H-3. Examples of Project Information Signs

* The actual name of the State where the improvements are being implemented should be used instead of "State".

Figure 2H-4. Examples and Typical Placement of Grade Separation Identification Signs

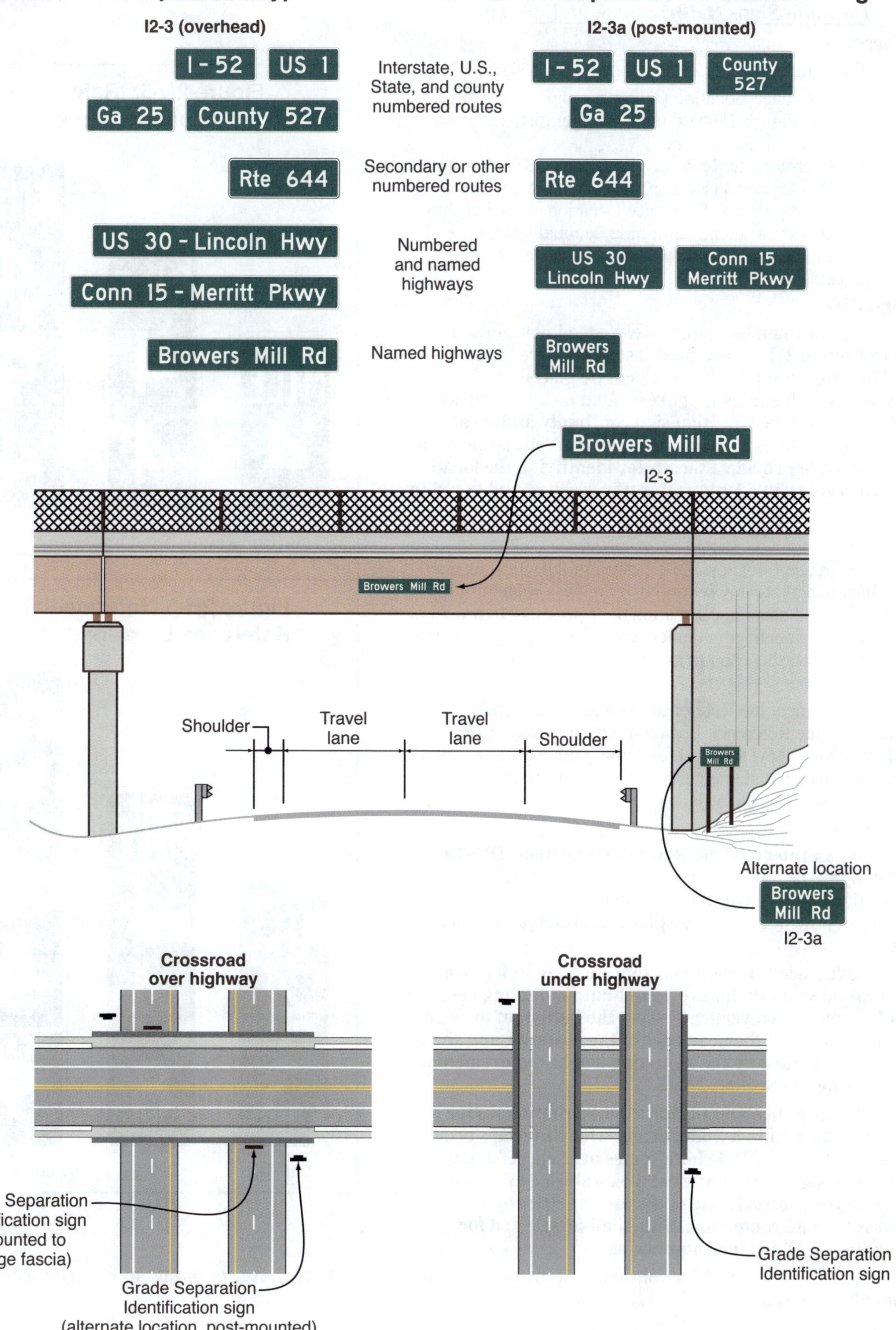

Section 2H.11 Reference Location Signs (D10-1 through D10-3) and Intermediate Reference Location Signs (D10-1a through D10-3a)

Support:

01 There are two types of reference location signs:

A. Reference Location (D10-1 through D10-3) signs (see Figure 2H-5) show an integer distance point along a highway, and

B. Intermediate Reference Location (D10-1a through D10-3a) signs (see Figure 2H-6) show the same information as Reference Location signs, but they also show a tenth-of-a-mile decimal so that they can be installed between integer distance points along a highway.

Standard:

02 **Except when Enhanced Reference Location signs (see Section 2H.12) are used instead, Reference Location (D10-1 through D10-3) signs shall be placed on all expressway facilities that are located on a route where there is reference location sign continuity and on all freeway facilities to assist road users in estimating their progress, to provide a means for identifying the location of emergency incidents and traffic crashes, and to aid in highway maintenance and servicing.**

Option:

03 Reference Location (D10-1 through D10-3) signs may be installed along any section of a highway route or ramp to assist road users in estimating their progress, to provide a means for identifying the location of emergency incidents and traffic crashes, and to aid in highway maintenance and servicing.

04 To augment the Reference Location sign system, Intermediate Reference Location (D10-1a through D10-3a) signs, which show the tenth of a mile with a decimal point, may be installed at one tenth of a mile, two tenths of a mile, or one-half mile intervals.

Standard:

05 **When Intermediate Reference Location (D10-1a through D10-3a) signs are used to augment the reference location sign system, the reference location sign at the integer mile point shall display a decimal point and a zero numeral.**

06 **Reference Location and Intermediate Reference Location signs shall have a minimum mounting height of 4 feet, measured vertically from the bottom of the sign to the elevation of the near edge of the roadway, and shall not be governed by the mounting height requirements prescribed in Section 2A.15.**

07 **The distance numbering shall be continuous for each route within a State, except where overlaps occur (see Section 2E.22). Where routes overlap, reference location sign continuity shall be established for only one of the routes. If one of the overlapping routes is an Interstate route, that route shall be selected for continuity of distance numbering.**

08 **The route selected for continuity of distance numbering shall also have continuity in interchange exit numbering (see Section 2E.22).**

Figure 2H-5. Reference Location Signs

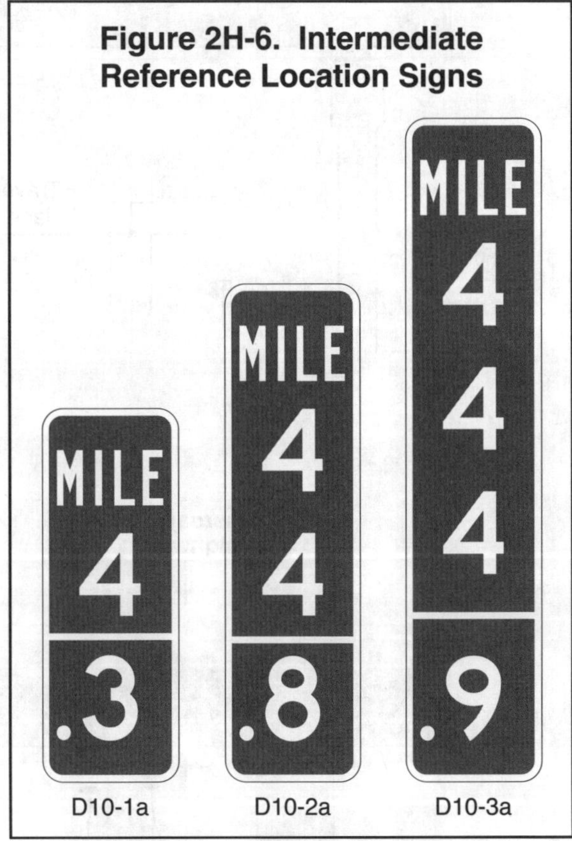

Figure 2H-6. Intermediate Reference Location Signs

Guidance:

09   *On a route without continuity of distance numbering, the first reference location sign beyond the overlap should indicate the total distance traveled on the route (including on the portion that did not have continuity of distance numbering) so that road users will have a means of correlating their travel distance between reference location signs with that shown on their odometer.*

Standard:

10   **For divided highways, the distance measurement shall be made on the northbound and eastbound roadways. The reference location signs for southbound or westbound roadways shall be set at locations directly opposite the reference location signs for the northbound or eastbound roadways.**

11   **Zero distance shall begin at the south and west State lines, or at the south and west terminus points where routes begin within a State.**

12   **Except as provided in Paragraph 13 of this Section, reference location signs shall be installed on the right-hand side of the roadway.**

Option:

13   Where conditions limit or restrict the use of reference location signs on the right-hand side of the roadway, they may be installed in the median. On two-lane conventional roadways, reference location signs may be installed on one side of the roadway only and may be installed back-to-back. Reference location signs may be placed up to 30 feet from the edge of the pavement.

14   If a reference location sign cannot be installed in the correct location, it may be moved in either direction as much as 50 feet.

Guidance:

15   *If a reference location sign cannot be placed within 50 feet of the correct location, it should be omitted.*

Section 2H.12 Enhanced Reference Location Signs (D10-4) and Intermediate Enhanced Reference Location Signs (D10-5)

Support:

01   There are two types of enhanced reference location signs:

    A. Enhanced Reference Location (D10-4) signs (see Figure 2H-7), and
    B. Intermediate Enhanced Reference Location (D10-5) signs (see Figure 2H-7).

Option:

02   An Enhanced Reference Location (D10-4) sign, which enhances the reference location sign system by identifying the route, may be placed on freeways or expressways (instead of reference location signs) or on conventional roads.

03   To augment an enhanced reference location sign system, an Intermediate Enhanced Reference Location (D10-5) sign, which shows the tenth of a mile with a decimal point, may be installed along any section of a highway route or ramp at one tenth of a mile, two tenths of a mile, or one-half mile intervals.

Standard:

04   **When an Intermediate Enhanced Reference Location (D10-5) sign is used to augment the reference location sign system, the Enhanced Reference Location sign at the integer mile point shall display a decimal point and a zero numeral.**

05   **Except as provided in Paragraph 6 of this Section, if enhanced reference location signs are used, they shall be vertical signs having a green background with a white legend and border, except for the route shield, which shall be the standard color and shape. The top line shall display the cardinal direction for**

the roadway. The second line shall display the applicable route shield for the roadway. The third line shall identify the mile reference for the location and the bottom line of the Intermediate Enhanced Reference Location sign shall give the tenth of a mile reference for the location preceded by a decimal point.

Support:

06   The provisions in Section 2H.11 regarding mounting height, distance numbering and measurements, sign continuity, and placement with respect to the right-hand shoulder and/or median for reference location signs also apply to enhanced reference location signs.

Section 2H.13 Acknowledgment Signs and Plaques (I20 Series)

Support:

01   Acknowledgment signs and plaques (see Figure 2H-8) are a way of recognizing a company, business, or volunteer group that provides or sponsors a highway-related service. Acknowledgment signs include sponsorship signs for adopt-a-highway litter removal programs, maintenance of a parkway or interchange, and other highway maintenance or beautification sponsorship programs.

Guidance:

02   *A State or local highway agency that elects to have a sponsorship acknowledgement program should develop a policy on Acknowledgment signs and plaques. The policy should require that eligible sponsoring organizations comply with State laws prohibiting discrimination based on race, religion, color, age, sex, national origin, and other applicable laws.*

Figure 2H-8. Examples of Acknowledgment Sign Designs

Standard:

03   **The State or local acknowledgment sign policy shall include all of the provisions regarding placement and design of Acknowledgment signs and plaques that are contained in this Section.**

04   **Because regulatory, warning, and guide signs have a higher priority, Acknowledgment signs shall only be installed where adequate spacing is available between the Acknowledgment sign and other higher priority signs. Acknowledgment signs shall not be installed in a position where they would obscure the road users' view of other traffic control devices.**

05   **Acknowledgment signs shall not be installed at any of the following locations:**

    A. **On the front or back of, adjacent to, or around any other traffic control device, including traffic signs, highway traffic signals, and changeable message signs;**

    B. **On the front or back of, adjacent to, or around the supports or structures of other traffic control devices, or bridge piers; or**

    C. **At key decision points where a road user's attention is more appropriately focused on other traffic control devices, roadway geometry, or traffic conditions, including exit and entrance ramps, merging or weaving areas, lane terminations, intersections, grade crossings, toll plazas, temporary traffic control zones, and areas of limited sight distance.**

06   **Acknowledgment signs and plaques shall have a white legend and border on a blue background. Acknowledgment signs shall be independent post-mounted roadside installations only and shall not be mounted overhead.**

Option:

07   An Acknowledgment sign may be used to acknowledge the sponsor of a rest area or welcome center.

Standard:

08   **Acknowledgment signs for a rest area, when located on the highway mainline, shall be limited to one sign per direction of travel from which the rest area is accessible, shall be located at least 500 feet from other traffic control devices, and shall not display names or representations of specific products or services provided by the sponsor within the rest area. Acknowledgment signs for rest areas shall display the legend REST AREA as the program activity, such as REST AREA SPONSORED BY. In accordance with Paragraph 5 of this Section, the Rest Area and Welcome Center Acknowledgment (I20-4 and I20-4a) signs shall not be combined in the same sign assembly with or substitute for the Rest Area General Service guide signs (see Section 2I.05).**

Option:

09   An additional Acknowledgment sign may be used within the rest area provided that it is not visible from the highway mainline or ramps to and from the rest area.

10   If a State has officially adopted and is actively promoting a program to encourage the use of safety rest areas through the use of a program name, then that program name may be displayed in smaller lettering below the legend REST AREA on the Rest Area Acknowledgment sign.

Standard:

11   **Program names or slogans, as described in Paragraph 14 of this Section, shall not be displayed on the Rest Area General Service guide signs or other types of traffic signs.**

Guidance:

12   *The minimum spacing between Acknowledgment signs and any other traffic control signs, except parking regulation signs, should be:*

    A. *150 feet on roadways with speed limits of less than 30 mph,*

    B. *200 feet on roadways with speed limits of 30 to 45 mph, and*

    C. *500 feet on roadways with speed limits greater than 45 mph.*

13   *If the placement of a newly-installed higher-priority traffic control device, such as a higher-priority sign, a highway traffic signal, or a temporary traffic control device, conflicts with an existing Acknowledgment sign, the Acknowledgment sign should be relocated, covered, or removed.*

Option:

14   State or local highway agencies may use their own pictograph (see definition in Section 1C.02) and/or a brief jurisdiction-wide program name, such as "Adopt-A-Highway" or "Litter Removal," as part of any portion of the Acknowledgment sign, provided that the signs comply with the provisions for shape, sign and legend size, color, and lettering style in this Chapter and in Chapter 2A.

Guidance:

15   *Acknowledgment signs should clearly indicate the type of highway services provided by the sponsor.*

Standard:

16   **In addition to the general provisions for signs described in Chapter 2A and the sign design principles covered in the "Standard Highway Signs" publication (see Section 1A.05), Acknowledgment sign and plaque designs developed by State or local highway agencies shall comply with the following provisions:**

A.   **Neither the sign or plaque design nor the sponsor acknowledgment name or logo shall contain any contact information, directions, slogans (other than a brief jurisdiction-wide program name, if used), telephone numbers, e-mail or Internet addresses, including domain names and uniform resource locators (URLs), metadata tags ("hash-tags"), or quick-response (QR) codes, bar codes, or similar scanning graphics (see Section 2A.04);**

B.   **Except for the sponsor acknowledgment logo, all of the lettering shall be in upper-case letters of the Standard Alphabets as provided in the "Standard Highway Signs" publication (see Section 1A.05);**

C.   **If a logo, instead of a word legend, is used to represent the sponsor, the logo shall be the primary logo that identifies the sponsoring entity. Secondary or alternate logos, slogans, products, mascots, spokespersons, or other items associated with the sponsoring entity's commercial advertising or marketing shall not be displayed on Acknowledgment signs or plaques;**

D.   **In order to keep the main focus on the highway-related service and not on the sponsor acknowledgment name or logo, the area reserved for the sponsor acknowledgment name or logo shall not be located at the top of the sign or plaque, shall be a maximum of 8 square feet in area, and shall not exceed ⅓ of the total area of the sign;**

E.   **The entire sign display area of an Acknowledgment sign assembly shall not exceed 24 square feet;**

F.   **The sign or plaque shall not contain any messages, lights, symbols, or logos that resemble any official traffic control devices;**

G.   **The sign or plaque shall not contain any external or internal illumination, light-emitting diodes, luminous tubing, fiber optics, luminescent panels, or other flashing, moving, or animated features;**

H.   **The sign or plaque shall not distract from official traffic control messages such as regulatory, warning, or guidance messages;**

I.   **The area of the plaque shall not exceed the lesser of ⅓ the area of the General Service sign below which it is mounted or 24 square feet;**

J.   **The plaque size shall be based on the standard sizes as specified in Table 2H-1. If the size of the General Service sign is oversized for its application (greater than the size specified for the corresponding roadway application in Table 2H-1), or if the size of the General Service sign increases due to modification of the sign legend, a corresponding increase in the size of the plaque shall not be allowed; and**

K.   **The sign or plaque shall not display promotional or contact information about the agency's sponsorship program, including if the sign or plaque does not currently display a sponsor.**

Option:

17   If a specific outlet of a business with multiple locations in the same area is the sponsoring entity, such as a franchisee, the area reserved for the sponsor acknowledgment name or logo may include the name of the municipality or neighborhood in which the sponsoring entity is located.

18   An Acknowledgment plaque may be mounted below the following General Service signs to acknowledge the sponsor of a corridor-based or region-based highway-related service:

A.   Radio-Weather Information (D12-1) sign (see Section 2I.09);

B.   Radio-Traffic Information (D12-1a) sign (see Section 2I.09);

C.   TRAVEL INFO CALL 511 (D12-5 and D12-5a) signs (see Section 2I.12); and

D.   Roadside Assistance (D12-6) sign (see Section 2I.13).

Standard:

19   **An Acknowledgment plaque shall not be mounted in conjunction with any other sign or traffic control device. An Acknowledgment plaque shall not be used alone or without one of the General Service signs specified in Paragraph 18 of this Section.**

20   **The general restrictions on the type of content allowed for display on Acknowledgment signs (see Paragraph 16 of this Section) shall apply to the legends of Acknowledgment plaques.**

Section 2H.14 Alternative Fuels Corridor Sign (D9-19)

Option:

01   The Alternative Fuels Corridor (D9-19) sign (see Figure 2H-9) may be used to inform motorists of an alternative fuels corridor highway segment that has been designated by the Secretary of Transportation as "Corridor Ready."

Standard:

02   **Alternative Fuels Corridor signs shall only be used to designate alternative fuels corridor highway segments that have been designated by the Federal Highway Administration as "Corridor Ready." The appropriate General Service signs or plaques identifying the alternative fuels available in the corridor shall be included with the Alternative Fuels Corridor sign in a sign assembly. The alternative fuel services for an alternative fuels corridor shall be limited to electric vehicle charging, compressed natural gas, liquid natural gas, liquified petroleum, and hydrogen.**

Support:

03   The General Service (D9-11a, D9-11b, D9-11d, D9-11e, and D9-11f) symbol signs for use with an Alternative Fuels Corridor sign are shown in Figure 2I-1.

Standard:

04   **Alternative Fuels Corridor signs shall only be post-mounted on the side of the road and shall not be mounted overhead.**

05   **State or agency variations of the Alternative Fuels Corridor sign shall not be allowed. Acknowledgments of sponsors shall not be allowed in Alternative Fuels Corridor sign assemblies.**

06   **Except as provided in Paragraph 7 of this Section, Alternative Fuels Corridor signs shall be limited to one sign at or near the beginning of the alternative fuels corridor in each direction of travel.**

Figure 2H-9. Examples of Signs for Alternative Fuels Corridors

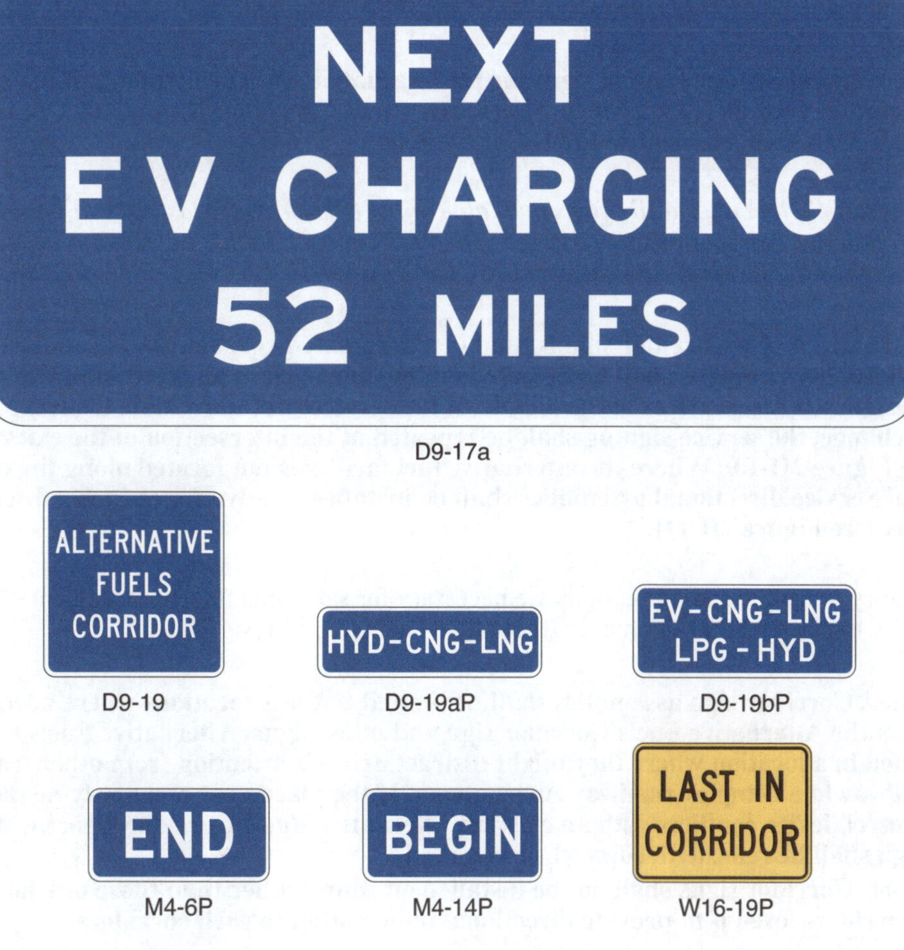

Option:

07   For long corridors, such as segments connecting control cities or major urban areas, additional signs may be located beyond major intersections or major interchanges following the typical post-interchange sign sequence.

08   The beginning of an alternative fuels corridor may be indicated with a BEGIN (M4-14P) plaque (see Figure 2H-9) with a white legend and border on a blue background mounted above the alternative fuels corridor sign in the sign assembly.

09   The end of an alternative fuels corridor may be indicated with an END (M4-6P) plaque (see Figure 2H-9) with a white legend and border on a blue background mounted above the Alternative Fuels Corridor sign in the sign assembly.

Standard:

10   **The General Service signs shall not be used in the sign assembly indicating the end of a corridor.**

11   **When the availability of one or more of the alternative fuel facilities discontinues in an alternative fuels corridor, the LAST IN CORRIDOR (W16-19P) plaque (see Figure 2H-9) shall be included on the last General Service directional assembly on the approach to the interchange or intersection.**

Option:

12   When the availability of one or more of the alternative fuel facilities discontinues in an alternative fuels corridor, an Alternative Fuels Corridor sign with accompanying General Service signs indicating the types of fuels still available in the corridor may be provided beyond the intersection or interchange where the last discontinued fuel facilities were available.

13   When the distance between electric vehicle (EV) charging services in an alternative fuels corridor is greater than 50 miles, the Next EV Charging (D9-17a) sign (see Figure 2H-9) may be located after the EV charging directional assembly, but before the EV charging service exit or turn, to inform road users of the extended distance to the next EV charging service.

Standard:

14   **The Alternative Fuels Corridor (D9-19) sign shall not be used as a directional sign in a directional assembly, or be combined with other signs, except as provided in this Section.**

Option:

15   Up to three General Service symbol signs arranged horizontally displaying the alternative fuels available in the designated corridor may be installed below the Alternative Fuels Corridor sign (see Figure 2H-10).

Standard:

16   **The size of the General Service symbol signs for the alternative fuels available shall not exceed 18 x 18 inches when mounted with the 24 x 24-inch Alternative Fuels Corridor sign and 24 x 24 inches when mounted with the 36 x 36-inch Alternative Fuels Corridor sign.**

Guidance:

17   *When the number of eligible alternative fuels available in the corridor exceeds three, a separate plaque with the two-letter or three-letter designations (D9-19aP or D9-19bP) of each of the fuels available (see Figure 2H-9) should be used in place of the General Service symbol signs.*

Standard:

18   **When the Alternative Fuels Corridor sign is used in a designated corridor on a freeway or expressway, the applicable General Service signs shall be installed on the approach to an interchange in the corridor from which the designated fuel services are available. If the services are not visible from the ramp of a single-exit interchange, the service signing shall be repeated at the intersection of the exit ramp and the crossroad (see Figure 2H-10). Where the alternative fuel facility is not located along the crossroad, additional General Service directional assemblies shall be installed in advance of each subsequent turn to reach the facility (see Figure 2H-11).**

Support:

19   Because regulatory, warning, and guide signs are necessary for safe and efficient movement of traffic, they have a higher priority in placement location over Alternative Fuels Corridor signs.

Standard:

20   **Alternative Fuels Corridor sign assemblies shall be limited to those locations where adequate spacing is available between the Alternative Fuels Corridor sign and other signs. Alternative Fuels Corridor signs shall not be installed in a location where they might distract driver's attention from other traffic control devices or the roadway in a complex roadway environment. If the placement of a newly-installed, higher-priority traffic control device conflicts with an existing Alternative Fuels Corridor sign, the Alternative Fuels Corridor sign shall be relocated, covered, or removed.**

21   **Alternative Fuels Corridor signs shall not be installed on routes other than those officially designated as alternative fuels corridors, even if to provide directional information to such corridors.**

Figure 2H-10. Example of Signing for an Alternative Fuels Corridor

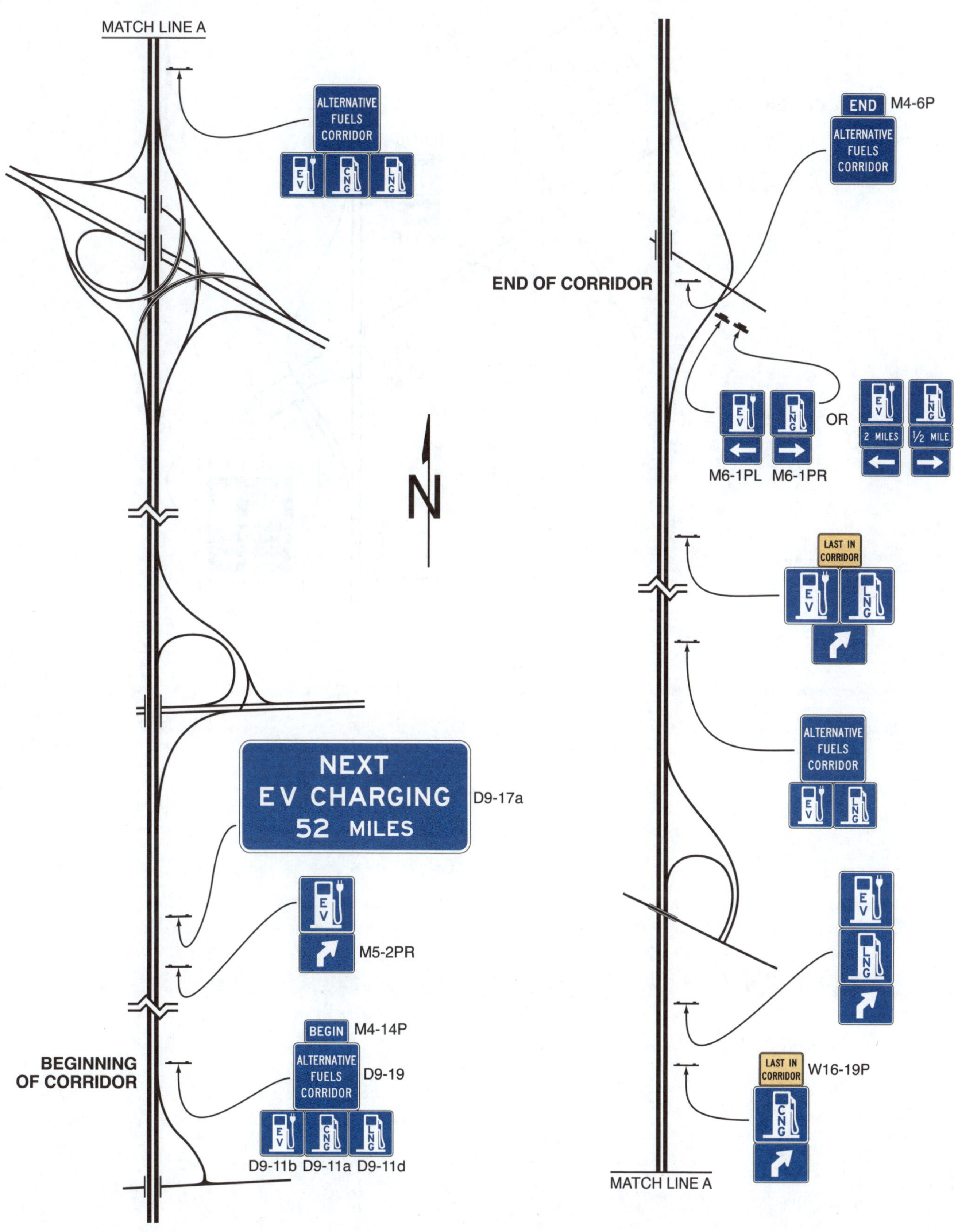

Note: Exit numbering may be used in place of directional arrows on the highway mainline.

Figure 2H-11. Typical Signing from a Freeway Exit Ramp to a Service Facility

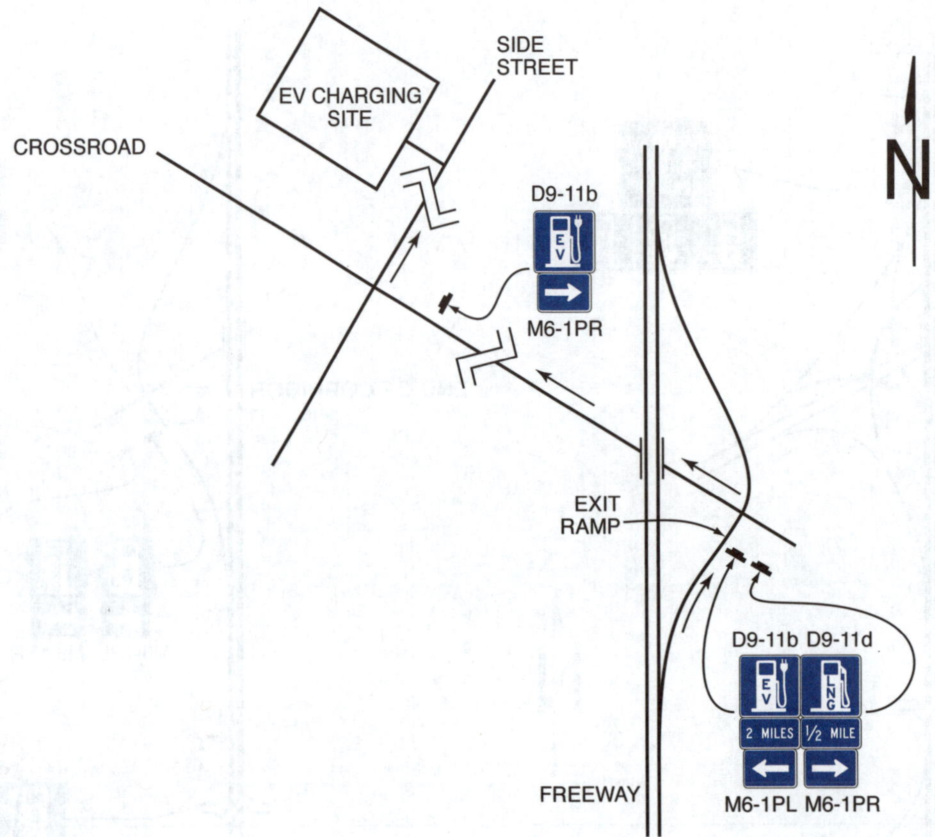

CHAPTER 2I. GENERAL SERVICE SIGNS

Section 2I.01 Sizes of General Service Signs

Standard:

01 **Except as provided in Section 2A.07, the sizes of General Service signs that have a standardized design shall be as shown in Table 2I-1.**

Support:

02 Section 2A.07 contains information regarding the applicability of the various columns in Table 2I-1.

Option:

03 Signs larger than those shown in Table 2I-1 may be used (see Section 2A.07).

Section 2I.02 General Service Signs for Conventional Roads

Support:

01 On conventional roads, commercial services such as gas, food, and lodging generally are within sight and are available to the road user at reasonably frequent intervals along the route. Consequently, on this class of road there usually is no need for special signs calling attention to these services. Moreover, General Service signing is usually not needed in urban areas except for hospitals, law enforcement assistance, tourist information centers, and camping.

Option:

02 General Service signs (see Figure 2I-1) may be used on conventional roads where such services are infrequent and are found only on an intersecting highway or crossroad.

Standard:

03 **All General Service signs and supplemental sign panels shall have a white legend and border on a blue background.**

Guidance:

04 *General Service signs should be installed at a suitable distance in advance of the turn-off point or intersecting highway.*

05 *States that elect to provide General Service signing should establish a statewide policy or warrant for its use, and criteria for the availability of services. Local jurisdictions electing to use such signing should follow State policy for the sake of uniformity.*

Option:

06 Individual States may sign for whatever alternative fuels are available at appropriate locations.

Standard:

07 **To be eligible for an EV Charging General Service sign on a conventional road, the EV chargers provided shall meet the criteria for Direct Current Fast Chargers provided in 23 CFR 680.106 and be in continuous operation at least 16 hours per day, 7 days per week.**

08 **General Service signs, if used at intersections, shall be accompanied by a directional message.**

Option:

09 The Advance Turn (M5 series) or Directional Arrow (M6 series) auxiliary plaques (see Figure 2I-1) with white arrows on blue backgrounds may be used with General Service symbol signs to create a General Service directional assembly.

10 The General Service sign legends may be either symbols or word messages.

Standard:

11 **Symbols and word message General Service legends shall not be intermixed on the same sign.**

12 **The Pharmacy (D9-20) sign shall only be used to indicate the availability of a pharmacy that is open, with a State-licensed pharmacist present and on duty, 24 hours per day, 7 days per week, and that is located within 3 miles of an interchange on the Federal-aid system. The D9-20 sign shall have a 24 HR (D9-20aP) plaque mounted below it.**

13 **Use of the Hospital (D9-2) sign or the HOSPITAL (D9-13aP) plaque (see Figure 2I-1) shall be limited to facilities that operate 24 hours per day, 7 days per week.**

Option:

14 The Emergency Medical Services (D9-13) sign (see Figure 2I-1 and Paragraph 20 of this Section) may be used for facilities that provide emergency medical care but do not operate on a full-time basis.

Support:

15 Formats for displaying different combinations of these services are described in Section 2I.03.

Table 2I-1. General Service Sign and Plaque Sizes (Sheet 1 of 2)

Sign or Plaque	Sign Designation	Section	Conventional Road	Freeway or Expressway
Rest Area Advance	D5-1	2I.05	78 x 36*	132 x 60* (F) 114x 48* (E)
Rest Area Advance Direction	D5-1a	2I.05	78 x 36*	132 x 60* (F) 114 x 48* (E)
Rest Area Entrance Direction	D5-2	2I.05	78 x 36*	132 x 66* (F) 114 x 60 (E)
Rest Area Gore	D5-2a	2I.05	42 x 48*	78 x 78* (F) 66 x 72* (E)
Rest Area Directional	D5-5	2I.05	42 x 48*	—
Next Rest Area	D5-6	2I.05	78 x 54*	132 x 78* (F) 108 x 66* (E)
Rest Area Tourist Info Center Advance	D5-7	2I.08	90 x 72*	156 x 108* (F) 132 x 96* (E)
Rest Area Tourist Info Center Advance Direction	D5-7a	2I.08	90 x 72*	156 x 108* (F) 132 x 96* (E)
Rest Area Tourist Info Center Entrance Direction	D5-8	2I.08	84 x 72*	138 x 108* (F) 120 x 96* (E)
Parking Area Advance	D5-9	2I.05	96 x 36*	162 x 60* (F) 138 x 48* (E)
Parking Area Entrance Direction	D5-9a	2I.05	96 x 36*	162 x 60* (F) 138 x 54* (E)
Parking Area Gore	D5-9b	2I.05	62 x 48*	108 x 78* (F) 88 x 66* (E)
Picnic Area (Roadside Table, Roadside Park) Advance	D5-10	2I.05	84 x 36*	144 x 60* (F) 120 x 48* (E)
Picnic Area (Roadside Table, Roadside Park) Entrance Direction	D5-10a	2I.05	84 x 36*	144 x 60* (F) 120 x 54* (E)
Picnic Area (Roadside Table, Roadside Park) Gore	D5-10b	2I.05	54 x 48*	92 x 78* (F) 76 x 66* (E)
Scenic Area (Scenic View, Scenic Overlook) Advance	D5-11	2I.05	84 x 36*	144 x 60* (F) 120 x 48* (E)
Scenic Area (Scenic View, Scenic Overlook) Entrance Direction	D5-11a	2I.05	84 x 36*	144 x 60* (F) 120 x 54* (E)
Scenic Area (Scenic View, Scenic Overlook) Gore	D5-11b	2I.05	54 x 48*	92 x 78* (F) 76 x 66* (E)
Interstate Oasis	D5-12	2I.04	—	198 x 60 (F) 162 x 48 (E)
Interstate Oasis (plaque)	D5-12aP	2I.04	—	114 x 48
Interstate Oasis Directional	D5-12b	2I.04	—	48 x 36
Brake Check Area Advance	D5-13	2I.06	96 x 54	132 x 66
Brake Check Area Entrance Direction	D5-14	2I.06	96 x 54	132 x 78
Chain-Up Area Advance	D5-15	2I.07	72 x 54	102 x 66
Chain-Up Area Entrance Direction	D5-16	2I.07	72 x 54	102 x 78
Telephone	D9-1	2I.02	24 x 24	30 x 30
Hospital	D9-2	2I.02	24 x 24	30 x 30
Camping	D9-3	2I.02	24 x 24	30 x 30
Litter Container	D9-4	2I.02	24 x 30	36 x 48
International Symbol of Accessibility	D9-6	2I.02	24 x 24	30 x 30
Van Accessible (plaque)	D9-6P	2I.02	18 x 9	—
Gas	D9-7	2I.02	24 x 24	30 x 30
Food	D9-8	2I.02	24 x 24	30 x 30
Lodging	D9-9	2I.02	24 x 24	30 x 30
Tourist Information	D9-10	2I.02	24 x 24	30 x 30
Diesel Fuel	D9-11	2I.02	24 x 24	30 x 30
Alternative Fuel - Compressed Natural Gas	D9-11a	2I.02	24 x 24***	30 x 30***
Electric Vehicle Charging	D9-11b	2I.02	24 x 24***	30 x 30***
Electric Vehicle Charging (plaque)	D9-11bP	2I.02	24 x 18	30 x 24
Alternative Fuel - Ethanol	D9-11c	2I.02	24 x 24	30 x 30

Table 2I-1. General Service Sign and Plaque Sizes (Sheet 2 of 2)

Sign or Plaque	Sign Designation	Section	Conventional Road	Freeway or Expressway
Alternative Fuel - Liquefied Natural Gas	D9-11d	2I.02	24 x 24***	30 x 30***
Alternative Fuel - Liquefied Petroleum Gas	D9-11e	2I.02	24 x 24***	30 x 30***
Alternative Fuel - Hydrogen	D9-11f	2I.02	24 x 24***	30 x 30***
Alternative Fuel - Biofuel	D9-11g	2I.02	24 x 24	30 x 30
RV Sanitary Station	D9-12	2I.02	24 x 24	30 x 30
Emergency Medical Services	D9-13	2I.02	24 x 24	30 x 30
Hospital (plaque)	D9-13aP	2I.02	24 x 12	30 x 12
Ambulance Station (plaque)	D9-13bP	2I.02	24 x 12	30 x 15
Emergency Medical Care (plaque)	D9-13cP	2I.02	24 x 18	30 x 24
Trauma Center (plaque)	D9-13dP	2I.02	24 x 12	30 x 15
Police	D9-14	2I.02	24 x 24	30 x 30
Truck Parking	D9-16	2I.03	24 x 24	30 x 30
Truck External Power (plaque)	D9-16aP	2I.03	24 x 24	30 x 30
Truck Parking Availability - Exit Number	D9-16b	2I.15	Varies x 144	Varies x 144
Truck Parking Availability - Distance	D9-16c	2I.15	Varies x 144	Varies x 144
Truck Parking Availability - Rest Area	D9-16d	2I.15	Varies x Varies	Varies x Varies
Truck Parking Availability - Combined	D9-16e	2I.15	Varies x Varies	Varies x Varies
Next Services Advance (plaque)	D9-17P	2I.02	72 x 24	114 x 30
Next EV Charging	D9-17a	2H, 2J	—	126 x 60
General Services (up to 6 symbols) with Exit Number	D9-18	2I.03	108 x 84	132 x 114 (F) 132 x 108 (E)
General Services with Exit Number	D9-18a	2I.03	72 x 60	132 x 108** (F) 102 x 84** (E)
General Services (up to 6 symbols) with Action Message	D9-18b	2I.03	108 x 84	132 x 114 (F) 132 x 108 (E)
General Services with Action Message	D9-18c	2I.03	72 x 60**	132 x 108** (F) 102 x 84** (E)
Rural Interchange General Services (up to 3 symbols) (plaque)	D9-18dP	2I.03	—	120 x 36
Rural Interchange General Services (1 line) (plaque)	D9-18eP	2I.03	—	Varies x 24
Rural Interchange General Services (2 line) (plaque)	D9-18fP	2I.03	—	Varies x 42
Pharmacy	D9-20	2I.02	24 x 24	30 x 30
24-Hour (plaque)	D9-20aP	2I.02	24 x 12	30 x 12
Telecommunications Device for the Deaf	D9-21	2I.02	24 x 24	30 x 30
Wireless Internet	D9-22	2I.02	24 x 24	30 x 30
Radio - Weather Information	D12-1	2I.09	84 x 48	132 x 84
Radio - Traffic Information	D12-1a	2I.09	96 x 48	120 x 60
Urgent Message When Flashing (plaque)	D12-1bP	2I.09	84 x 30	108 x 36
Carpool Information	D12-2	2I.14	60 x 42	96 x 66
Channel 9 Monitored	D12-3	2I.10	84 x 48	132 x 84
Emergency Call 911	D12-4	2I.11	66 x 30	96 x 48
Travel Info Call 511 (pictograph)	D12-5	2I.12	48 x 60	66 x 72
Travel Info Call 511	D12-5a	2I.12	48 x 36	66 x 48
Roadside Assistance	D12-6	2I.13	60 x 42	78 x 54

* The size shown is for a sign with a REST AREA, PARKING AREA, PICNIC AREA, SCENIC AREA, and/or TOURIST INFO CENTER legend. The size should be appropriately adjusted if an alternate legend is used.

** The size shown is for a sign with four lines of services. The size should be appropriately adjusted depending on the amount of legend displayed.

*** The Standard Highway Signs publication contains layouts for the 18 x 18-inch and 24 x 24-inch alternative fuels symbol signs mounted with the Alternative Fuels Corridor sign in accordance with Section 2H.14.

Notes: 1. Larger signs may be used when appropriate.

2. Dimensions in inches are shown as width x height

3. Where two sizes are shown, the larger size is for freeways (F) and the smaller size is for expressways (E)

Figure 2I-1. General Service Signs and Plaques

D9-1
Telephone

D9-2
Hospital

D9-3
Camping

D9-4
Litter Container

D9-6
International Symbol
of Accessibility

D9-6P

D9-7
Gas

D9-8
Food

D9-9
Lodging

D9-10
Tourist Information

D9-11
Diesel Fuel

D9-11a
Alternative Fuel-
Compressed
Natural Gas

D9-11b
Electric Vehicle
Charging

D9-11bP
Electric Vehicle
Charging

D9-11c
Alternative Fuel-
Ethanol

D9-11d
Liquefied Natural
Gas (LNG)

D9-11e
Liquefied
Petroleum Gas
(LPG)

D9-11f
Hydrogen Fuel
(HYD)

D9-11g
Biofuel (BIO)

D9-12
RV Sanitary
Station

D9-13
Emergency
Medical Services

D9-13aP
Hospital

D9-13bP
Ambulance
Station

D9-13cP
Emergency
Medical Care

D9-13dP
Trauma Center

D9-14
Police

D9-16
Truck Parking

D9-16aP
Truck External
Power

D9-20
Pharmacy

D9-20aP
24-Hour

D9-21
Telecommunications
Device for the Deaf

D9-22
Wireless
Internet

Advance Turn and Directional Arrow auxiliary plaques for use with General Service signs

M5-1P

M5-2P

M5-3P

M6-1P

M6-2P

M6-3P

Example of
directional assembly

Option:

16   If the distance to the next point at which services are available is 10 miles or more, a Next Services Advance (D9-17P) plaque (see Figure 2I-2) may be installed below the General Service sign.

17   The International Symbol of Accessibility (D9-6) sign (see Figure 2I-1) may be used beneath General Service signs where paved ramps and rest room facilities accessible to, and usable by, persons with disabilities are provided.

Figure 2I-2. Example of Next Services Plaque

Guidance:

18   *When the D9-6 sign is used in accordance with Paragraph 16 of this Section, and van-accessible parking is available at the facility, a VAN ACCESSIBLE (D9-6P) plaque (see Figure 2I-1) should be mounted below the D9-6 sign.*

Option:

19   The Recreational Vehicle Sanitary Station (D9-12) sign (see Figure 2I-1) may be used as needed to indicate the availability of facilities designed for the use of dumping wastes from recreational vehicle holding tanks.

20   The Litter Container (D9-4) sign (see Figure 2I-1) may be placed in advance of roadside turn-outs or rest areas, unless it distracts the driver's attention from other more important regulatory, warning, or directional signs.

21   The Emergency Medical Services (D9-13) symbol sign (see Figure 2I-1) may be used to identify medical service facilities that have been included in the Emergency Medical Services system under a signing policy developed by the State and/or local highway agency.

Standard:

22   **The Emergency Medical Services symbol sign shall not be used to identify services other than qualified hospitals, ambulance stations, and qualified free-standing emergency medical treatment centers. If used, the Emergency Medical Services symbol sign shall be supplemented by a sign or plaque, as provided in Paragraph 22 of this Section, identifying the type of service provided.**

Option:

23   The Emergency Medical Services symbol sign may be used above the HOSPITAL (D9-13aP) plaque or above a plaque with the legend AMBULANCE STATION (D9-13bP), EMERGENCY MEDICAL CARE (D9-13cP), or TRAUMA CENTER (D9-13dP). The Emergency Medical Services symbol sign may also be used to supplement Telephone (D9-1), Channel 9 Monitored (D12-3) (see Figure 2I-8), or POLICE (D9-14) signs.

Standard:

24   **The legend EMERGENCY MEDICAL CARE shall not be used for services other than qualified free-standing emergency medical treatment centers.**

Guidance:

25   *Each State should develop a policy for the implementation of the Emergency Medical Services symbol sign.*

26   *The State should consider the following guidelines in the preparation of its policy:*

A. *AMBULANCE*
   1. *24-hour service, 7 days per week.*
   2. *Staffed by two State-certified persons trained at least to the basic level.*
   3. *Vehicular communications with a hospital emergency department.*
   4. *Operator should have successfully completed an emergency-vehicle operator training course.*

B. *HOSPITAL*
   1. *24-hour service, 7 days per week.*
   2. *Emergency department facilities with a physician (or emergency care nurse on duty within the emergency department with a physician on call) trained in emergency medical procedures on duty.*
   3. *Licensed or approved for definitive medical care by an appropriate State authority.*
   4. *Equipped for radio voice communications with ambulances and other hospitals.*

C. *Channel 9 Monitored*
   1. *Provided by either professional or volunteer monitors.*
   2. *Available 24 hours per day, 7 days per week.*
   3. *The service should be endorsed, sponsored, or controlled by an appropriate government authority to guarantee the level of monitoring.*

Section 2I.03 <u>General Service Signs for Freeways and Expressways</u>

Support:

01   General Service (D9-18 series) signs (see Figure 2I-3) are generally not appropriate at major interchanges (see definition in Section 2E.11) and in urban areas.

Standard:

02   **General Service signs shall have a white legend and border on a blue background. Letter and numeral sizes shall comply with the minimum requirements of Tables 2E-2 through 2E-5. All approved symbols shall be permitted as alternatives to word messages, but symbols and word service messages shall not be intermixed on the same sign. If the services are not visible from the ramp of a single-exit interchange, the service signing shall be repeated in smaller size at the intersection of the exit ramp and the crossroad. Such service signs shall use arrows to indicate the direction to the services.**

Figure 2I-3. Examples of General Service Signs with and without Exit Numbering

Guidance:

03   *Where General Service signs are used along routes with exit numbering, the General Service sign should include the exit number within the sign face as shown in Figure 2I-3.*

04   *Distance to services should be displayed on General Service signs along the exit ramp where distances are more than 1 mile from the ramp intersection with the crossroad.*

05   *General Service signing should only be provided at locations where the road user can return to the freeway or expressway and continue in the same direction of travel.*

Guidance:

06   *Only services that fulfill the needs of the road user should be displayed on General Service signs. If State or local agencies elect to provide General Service signing, there should be a statewide policy for such signing and criteria for the eligibility and availability of the various types of services. The criteria should consider the following:*

    A. *Gas, diesel, and/or alternative fuels, except for electric vehicle (EV) charging, if all of the following are available:*
        1. *Vehicle services such as gas, oil, and water;*
        2. *Modern sanitary facilities and drinking water; and*
        3. *Continuous operations at least 16 hours per day, 7 days per week.*
    B. *Food if all of the following are available:*
        1. *Licensing or approval, where required;*
        2. *Continuous operation to serve at least two meals per day, at least 6 days per week; and*
        3. *Modern sanitary facilities.*
    C. *Lodging if all of the following are available:*
        1. *Licensing or approval, where required;*
        2. *Adequate sleeping accommodations; and*
        3. *Modern sanitary facilities.*
    D. *Public telephone if continuous operation, 7 days per week is available.*
    E. *Hospital if continuous emergency care capability, with a physician on duty 24 hours per day, 7 days per week is available. A physician on duty would include the following criteria and should be signed in accordance with the priority as follows:*
        1. *Physician on duty within the emergency department;*
        2. *Registered nurse on duty within the emergency department, with a physician in the hospital on call; or*
        3. *Registered nurse on duty within the emergency department, with a physician on call from office or home.*
    F. *24-Hour Pharmacy if a pharmacy is open, with a State-licensed pharmacist present and on duty, 24 hours per day, 7 days per week and is located within 3 miles of an interchange on the Federal-aid system.*
    G. *Camping if all of the following are available:*
        1. *Licensing or approval, where required;*
        2. *Adequate parking accommodations; and*
        3. *Modern sanitary facilities and drinking water.*

Standard:

07   **To be eligible for an EV Charging General Service sign on freeways and expressways, the EV chargers provided shall meet the criteria for Direct Current Fast Chargers provided in 23 CFR 680.106 and be in continuous operation at least 16 hours per day, 7 days per week.**

Support:

08   Motorist expectations for facilities providing alternative fuels, such as EV Charging, compressed natural gas, liquefied natural gas, liquefied petroleum gas, and hydrogen, vary considerably and alternative fuel vehicles might have different needs than conventional fuel vehicles.

Guidance:

09   *The policy criteria for alternative fuel vehicles should take into account the needs, convenience, and safety of alternative fuel vehicle users (see Section 2H.14).*

Standard:

10   **For any service that is operated on a seasonal basis only, the General Service signs shall be removed or covered during periods when the service is not available.**

11   **The General Service signs shall be mounted in an effective location, between the Advance Guide sign and the Exit Direction sign, in advance of the exit leading to the available services.**

Option:

12   If the distance to the next point where services are available is greater than 10 miles, a Next Services Advance (D9-17P) plaque (see Figure 2I-2) may be installed below the Exit Direction sign.

Standard:

13   **Signs for services shall comply with the format for General Service signs (see Section 2I.02) and as provided in this Manual. No more than six general road user services shall be displayed on one sign, which includes any appended supplemental signs or plaques. General Service signs shall display the legends for one or more of the following services: Food, Gas, EV Charging, Lodging, Camping, Phone, Hospital, 24-Hour Pharmacy, or Tourist Information.**

14   **The qualified services available shall be displayed at specific locations on the sign.**

Guidance:

15   *To provide for future services that might become available, the sign space normally reserved for a given service symbol or word should be left blank when that service is not present.*

16   *The standard display of word messages should be FOOD and PHONE in that order on the top line, and GAS and LODGING on the second line. If used, HOSPITAL, 24-HOUR PHARMACY, and CAMPING should be on separate lines (see Figure 2I-3).*

Option:

17   Signing for EV Charging, DIESEL, LP-GAS, or other alternative fuel services may be substituted for any of the general services or appended to such signs. The International Symbol of Accessibility (D9-6) sign (see Figure 2I-1) may be used for facilities that qualify.

Guidance:

18   *When symbols are used for the road user services, they should be displayed as follows:*

    A. Six services:

        1. Top row—GAS, FOOD, and LODGING

        2. Bottom row—PHONE, HOSPITAL, and CAMPING

    B. Four services:

        1. Top row—GAS and FOOD

        2. Bottom row—LODGING and PHONE

    C. Three services:

        1. Top row—GAS, FOOD, and LODGING

Option:

19   Substitutions of other services for any of the services described in Paragraph 18 of this Section may be made by placing the substitution in the lower right (four or six services) or extreme right (three services) portion of the sign. An action message or an interchange number may be used for symbol signs in the same manner as they are used for word message signs. The Diesel Fuel (D9-11) symbol or the LP-GAS (D9-11e) symbol may be substituted for the symbol representing fuel or appended to such assemblies. The Tourist Information (D9-10) or the 24-Hour Pharmacy (D9-20 and D9-20aP) symbol may be substituted on any of the configurations provided in Paragraph 18 of this Section.

20   At rural interchange areas where limited road user services are available and where it is unlikely that additional services will be provided within the near future, a Rural Interchange General Services (D9-18dP, D9-18eP, or D9-18fP) plaque displaying one to three services (words or symbols) may be mounted below a post-mounted Interchange Advance guide sign.

Standard:

21   **If more than three services become available at rural interchange areas where limited road user services were anticipated, the appended supplemental plaque described in Paragraph 20 of this Section shall be removed and replaced with an independently-mounted General Service sign as described in this Section.**

Option:

22   A separate Telephone Service (D9-1) sign (see Figure 2I-1) may be installed if telephone facilities are located adjacent to the route at places where public telephones would not normally be expected.

23   The Recreational Vehicle Sanitary Station (D9-12) sign (see Figure 2I-1) may be used as needed to indicate the availability of facilities designed for dumping wastes from recreational vehicle holding tanks.

24   In some locations, signs may be used to indicate that services are not available.

25   A separate Truck Parking (D9-16) sign (see Figure 2I-1) may be mounted below the other general road user services to direct truck drivers to designated parking areas.

26   **A TRUCK EXTERNAL POWER (D9-16aP) plaque (see Figure 2I-1) may be mounted below the Truck Parking (D9-16) sign to indicate the availability of receptacles providing power for electrical devices within the truck.**

Section 2I.04 Interstate Oasis Signing (D5-12 Series)

Support:

01 An Interstate Oasis is a facility near an Interstate highway that provides products and services to the public, 24-hour access to public restrooms, and parking for automobiles and heavy trucks. Interstate Oasis guide signs inform road users on Interstate highways as to the presence of an Interstate Oasis at an interchange and which businesses have been designated by the State within which they are traveling as having met the eligibility criteria of the Federal Highway Administration's Interstate Oasis policy. The FHWA's policy, which is dated October 18, 2006, and which can be viewed on the MUTCD Web site at http://mutcd.fhwa.dot.gov/res-policy.htm, provides a more detailed definition of an Interstate Oasis and specifies the eligibility criteria for an Interstate Oasis designation in compliance with the requirements of laws enacted by Congress.

Guidance:

02 *If a State elects to provide or allow Interstate Oasis signing (see Figure 2I-4), there should be a statewide policy, program, procedures, and criteria for the designation and signing of a facility as an Interstate Oasis that complies with the FHWA's policy and with the provisions of this Section.*

03 *States electing to provide or allow Interstate Oasis signing should use the following signing practices on the freeway for any given exit to identify the availability of a designated Interstate Oasis:*

   A. *If adequate sign spacing allows, a separate Interstate Oasis (D5-12) sign should be installed in an effective location with spacing of at least 800 feet from other adjacent guide signs, including any Specific Service signs. This Interstate Oasis sign should be located upstream from the Advance Guide sign or between the Advance Guide sign and the Exit Direction sign for the exit leading to the Interstate Oasis. The Interstate Oasis sign should display the words INTERSTATE OASIS and the exit number or, for an unnumbered interchange, an action message such as NEXT RIGHT.*

   B. *If the spacing of the other guide signs precludes the use of a separate sign as described in Item A of this Paragraph, an INTERSTATE OASIS (D5-12aP) supplemental plaque should be mounted below an existing D9-18 series General Service sign for the interchange.*

Option:

04 If Specific Service signing is provided at the interchange, a business designated as an Interstate Oasis and having a business identification sign panel on the Food and/or Gas Specific Service signs may use the bottom portion of the business identification sign panel to display the word OASIS.

05 If Specific Service signing is not provided at the interchange, the name of the business designated as an Interstate Oasis may be displayed on a business identification sign panel, in compliance with the provisions of Sections 2J.03 through 2J.05, below the INTERSTATE OASIS legend on the D5-12 sign.

Figure 2I-4. Examples of Interstate Oasis Signs and Plaques

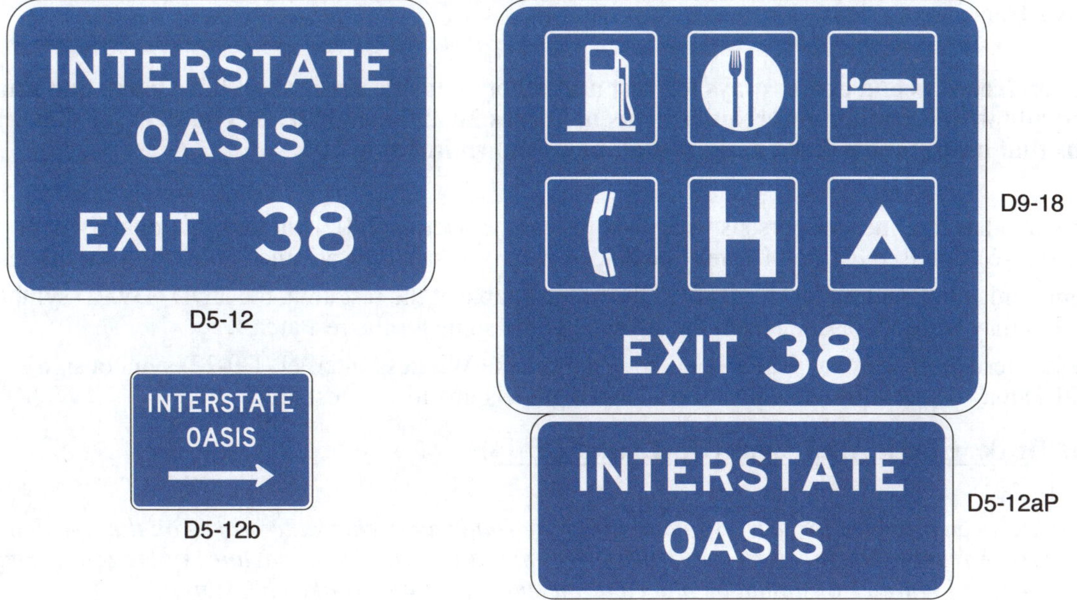

Standard:

06   **If Specific Service signs containing the OASIS legend as a part of the business identification sign panel(s) are not used on the ramp and if the Interstate Oasis is not clearly visible and identifiable from the exit ramp, an Interstate Oasis Directional (D5-12b) sign shall be provided on the exit ramp to indicate the direction and distance to the Interstate Oasis.**

07   **If needed, additional trailblazer guide signs shall be used along the crossroad to guide road users to an Interstate Oasis.**

Section 2I.05 <u>Rest Area and Other Roadside Area Signs (D5-1 through D5-11 Series)</u>

Standard:

01   **Rest Area signs (see Figure 2I-5) shall have a retroreflective white legend and border on a blue background.**

02   **Signs that include the legend REST AREA shall be used only where parking and restroom facilities are available.**

Guidance:

03   *A roadside area that does not contain restroom facilities should be signed to indicate the major road user service that is provided. For example, the sign legends for an area with only parking should use the words PARKING AREA (D5-9 series) instead of REST AREA. The sign legends for an area with only picnic tables and parking should use words such as PICNIC AREA, ROADSIDE TABLE, or ROADSIDE PARK (D5-10 series) instead of REST AREA.*

04   *Rest areas that have tourist information and welcome centers should be signed as provided in Section 2I.08.*

05   *Scenic area signing should be consistent with that provided for rest areas, except that the legends should use words such as SCENIC AREA, SCENIC VIEW, or SCENIC OVERLOOK (D5-11 series) instead of REST AREA.*

06   *If a rest area or other roadside area is provided on a conventional road, a D5-1 and/or D5-1a sign should be installed in advance of the rest area or other roadside area to permit the driver to reduce speed in preparation for leaving the highway. A D5-5 sign (or a D5-2 sign if an exit ramp is provided) should be installed at the turn-off point where the driver needs to leave the highway to access the rest area or other roadside area.*

07   *If a rest area or other roadside area is provided on a freeway or expressway, a D5-1 sign should be placed 1 mile and/or 2 miles in advance of the rest area.*

Standard:

08   **A D5-2a sign shall be placed at the rest area or other roadside area exit gore.**

Option:

09   A D5-1a sign may be placed between the D5-1 sign and the exit gore on a freeway or expressway. A second D5-1 sign may be used in place of the D5-1a sign with a distance to the nearest ½ or ¼ mile displayed as a fraction rather than a decimal for distances of less than 1 mile.

10   To provide the road user with information on the location of succeeding rest areas, a Next Rest Area (D5-6) sign (see Figure 2I-5) may be installed independently or as a supplemental sign mounted below one of the REST AREA advance guide signs.

Standard:

11   **All signs on freeways and expressways for rest and other roadside areas shall have letter and numeral sizes that comply with the minimum requirements of Tables 2E-2 through 2E-5. The sizes for General Service signs that have standardized designs shall be as shown in Table 2I-1.**

Option:

12   If the rest area has facilities for persons with disabilities (see Section 2I.02), the International Symbol of Accessibility (D9-6) sign (see Figure 2I-1) may be placed with or beneath an advance guide sign for the rest area.

13   If telecommunication devices for the deaf (TDD) are available at the rest area, the TDD (D9-21) symbol sign (see Figure 2I-1) may be used to supplement the advance guide signs for the rest area.

14   If wireless Internet services are available at the rest area, the Wireless Internet (D9-22) symbol sign (see Figure 2I-1) may be used to supplement the advance guide signs for the rest area.

Section 2I.06 <u>Brake Check Area Signs (D5-13 and D5-14)</u>

Guidance:

01   *If an area has been provided for drivers to pull off of the roadway to check the brakes on their vehicle, a Brake Check Area Advance (D5-13) sign (see Figure 2I-6) should be installed in advance of the brake check area, and a D5-14 sign (see Figure 2I-6) should be placed at the entrance to the brake check area.*

Figure 2I-5. Rest Area and Other Roadside Area Signs

Figure 2I-6. Brake Check Area and Chain-Up Area Signs

Section 2I.07 Chain-Up Area Signs (D5-15 and D5-16)

Guidance:

01   *If an area has been provided for drivers to pull off of the roadway to install chains on their tires, a Chain-Up Area Advance (D5-15) sign (see Figure 2I-6) should be installed in advance of the chain-up area, and a D5-16 sign (see Figure 2I-6) should be placed at the entrance to the chain-up area.*

Section 2I.08 Tourist Information and Welcome Center Signs (D5-7 Series and D5-8)

Support:

01   Tourist information and welcome centers have been constructed within rest areas on freeways and expressways and are operated by either a State or a private organization. Others have been located within close proximity to these facilities and operated by civic clubs, chambers of commerce, or private enterprise.

Guidance:

02   *The number of supplemental sign panels installed with Tourist Information or Welcome Center signs should be limited to three so as not to impose an undue informational load on the road user.*

Standard:

03   **Tourist Information or Welcome Center signs (see Figure 2I-7) shall have a white legend and border on a blue background. Continuously staffed or unstaffed operation at least 8 hours per day, 7 days per week, shall be required.**

04   **If operated only on a seasonal basis, the Tourist Information or Welcome Center signs shall be removed or covered during the off seasons.**

Guidance:

05   *For freeway or expressway rest area locations that also serve as tourist information or welcome centers, the following signing criteria should be used:*

A. *The locations for tourist information and welcome center Advance Guide, Exit Direction, and Exit Gore signs should meet the General Service signing requirements described in Section 2I.03.*

B. *If the signing for the tourist information or welcome center is to be accomplished in conjunction with the initial signing for the rest areas, the message on the Rest Area Tourist Info Center Advance (D5-7) sign should be REST AREA, TOURIST INFO CENTER, XX MILES or REST AREA, STATE NAME (optional), WELCOME CENTER XX MILES. On the Rest Area Tourist Info Center Entrance Direction (D5-8) sign the message should be REST AREA, TOURIST INFO CENTER with a diagonally upward-pointing directional arrow (or NEXT RIGHT), or REST AREA, STATE NAME (optional), WELCOME CENTER with a diagonally upward-pointing directional arrow (or NEXT RIGHT).*

C. *If the initial rest area Advance Guide and Exit Direction signing is in place, these signs should include, on supplemental signs, the legend TOURIST INFO CENTER or STATE NAME (optional), WELCOME CENTER.*

D. *The Exit Gore sign should contain only the legend REST AREA with the arrow and should not be supplemented with any legend pertaining to the tourist information center or welcome center.*

Option:

06   As an alternative to the supplemental TOURIST INFO CENTER legend, the Tourist Information (D9-10) sign (see Figure 2I-1) may be appended beneath the REST AREA advance guide sign.

Figure 2I-7. Examples of Tourist Information and Welcome Center Signs

D5-7

D5-7a

D5-8

Note: Alternate legends may be substituted for the TOURIST INFO CENTER legend, such as WELCOME CENTER and (State Name) WELCOME CENTER.

07   The name of the State or local jurisdiction may appear on the Advance Guide and Exit Direction tourist information/welcome center signs if the jurisdiction controls the operation of the tourist information or welcome center and the center meets the operating criteria set forth in this Manual and is consistent with State policies.

Guidance:

08   *For tourist information centers that are located off the freeway or expressway facility, additional signing criteria should be as follows:*

    A. *Each State should adopt a policy establishing the maximum distance that a tourist information center can be located from the interchange in order to be included on official signs.*

    B. *The location of signing should be in accordance with requirements pertaining to General Service signing (see Section 2I.03).*

    C. *Signing along the crossroad should be installed to guide the road user from the interchange to the tourist information center and back to the interchange.*

Option:

09   As an alternative, the Tourist Information (D9-10) sign (see Figure 2I-1) may be appended to the guide signs for the exit that provides access to the tourist information center. As a second alternative, the Tourist Information sign may be combined with General Service signing.

Section 2I.09 Radio Information Signing (D12-1 Series)

Option:

01   A Radio-Weather Information (D12-1) sign (see Figure 2I-8) may be used in areas where difficult driving conditions commonly result from weather systems. Radio-Traffic Information (D12-1a) signs may be used in conjunction with traffic management systems.

Standard:

02   **Radio-Weather and Radio-Traffic Information signs shall have a white legend and border on a blue background. Only the numerical indication of the radio frequency shall be used to identify a station broadcasting travel-related weather or traffic information. No more than three frequencies shall be displayed on each sign. Only radio stations whose signal will be of value to the road user and who agree to broadcast either of the following two items shall be identified on Radio-Weather and Radio-Traffic Information signs:**

    A. **Periodic weather warnings at a rate of at least once every 15 minutes during periods of adverse weather; or**

    B. **Driving condition information (affecting the roadway being traveled) at a rate of at least once every 15 minutes, or when required, during periods of adverse traffic conditions, and when supplied by an official agency having jurisdiction.**

03   **If a station to be considered operates only on a seasonal basis, its signs shall be removed or covered during the off season.**

Guidance:

04   *The radio station should have a signal strength to adequately broadcast at least 70 miles along the route. Signs should be spaced as needed for each direction of travel at distances determined by an engineering study. The stations to be included on the signs should be selected in cooperation with the association(s) representing major broadcasting stations in the area to provide: (1) maximum coverage to all road users on both AM and FM frequencies; and (2) consideration of 24 hours per day, 7 days per week broadcast capability.*

Option:

05   The URGENT MESSAGE WHEN FLASHING (D12-1bP) plaque may be mounted below the D12-1 or D12-1a sign if supplemented by Warning Beacons (see Section 4S.03) that flash only when a message related to adverse travel conditions is being broadcast.

06   In roadway rest area locations, a smaller sign using a greater number of radio frequencies, but of the same general design, may be used.

Standard:

07   **Radio-Weather and Radio-Traffic Information signs installed in rest areas shall be positioned such that they are not visible from the main roadway.**

Figure 2I-8. Radio, Telephone, and Carpool Information Signs

WEATHER INFO
TUNE RADIO TO
750 AM 1230 AM
96.3 FM

D12-1

CAR POOL
INFO
CALL *CAR

D12-2

TRAVEL INFO
TUNE RADIO TO
1610 AM

D12-1a

URGENT MESSAGE
WHEN FLASHING

D12-1bP

MICHIGAN
STATE POLICE
MONITORS
CB CHANNEL 9

D12-3

EMERGENCY
CALL 911

D12-4

D12-5*

TRAVEL
INFO
CALL 511

D12-5a

ROADSIDE
ASSISTANCE
CALL #95

D12-6

*The pictograph of the transportation agency or the travel information service or program may be used in place of the 511 pictograph (see Section 2I.12)

Section 2I.10 Channel 9 Monitored Sign (D12-3)

Option:

01   A Channel 9 Monitored (D12-3) sign (see Figure 2I-8) may be installed as needed. Official public agencies or their designees may be displayed as the monitoring agency on the sign.

Standard:

02   **Only official public agencies or their designee shall be displayed as the monitoring agency on the Channel 9 Monitored sign.**

Section 2I.11 EMERGENCY CALL 911 Sign (D12-4)

Option:

01   An Emergency CALL 911 (D12-4) sign (see Figure 2I-8) may be used for cellular telephone communications.

Section 2I.12 TRAVEL INFO CALL 511 Signs (D12-5 and D12-5a)

Option:

01   A TRAVEL INFO CALL 511 (D12-5 or D12-5a) sign (see Figure 2I-8) may be installed if a 511 travel information services telephone number is available to road users for obtaining traffic, public transportation, weather, construction, or road condition information.

02   The pictograph of the transportation agency or the travel information service or program that is providing the travel information may be displayed in place of the 511 pictograph on the D12-5 sign above the TRAVEL INFO CALL 511 legend.

Standard:

03   **The logo of a commercial entity shall not be incorporated within the TRAVEL INFO CALL 511 signs.**

04   **If the pictograph of the transportation agency or the travel information service or program is used in place of the 511 pictograph on the D12-5 sign (see Paragraph 2 of this Section), the maximum height of the pictograph shall not exceed the height of the 511 pictograph on the standard sign size specified for the roadway classification in Table 2H-1.**

05   **The TRAVEL INFO CALL 511 signs shall have a white legend and border on a blue background.**

Section 2I.13 Roadside Assistance Sign (D12-6)

Option:

01   A Roadside Assistance (D12-6) sign (see Figure 2I-8) displaying the Highway Assistance cellular telephone code designated for that roadway or jurisdiction may be used along a highway that is served by an authorized roadside assistance program with authorized service vehicles and personnel that provide roadside vehicle repair assistance to road users free of charge.

Section 2I.14 Carpool and Ridesharing Signing (D12-2)

Option:

01   In areas having carpool matching services, a Carpool Information (D12-2) sign (see Figure 2I-8) may be provided adjacent to highways with preferential lanes or along any other highway.

02   Carpool Information signs may include an Internet domain name or telephone number of more than four characters within the legend.

Standard:

03   **If a local transit pictograph or carpool symbol is incorporated into the Carpool Information sign, the maximum vertical dimension of the pictograph or symbol shall not exceed 18 inches and the maximum horizontal dimension shall not exceed 30 inches.**

Section 2I.15 Signing for Truck Parking Availability (D9-16b through D9-16e)

Option:

01   General Service signs may be used to display the number of available truck parking spaces at roadside areas such as rest areas, welcome centers, and weigh stations, and at facilities off a highway that are open to the public and provide parking for commercial vehicles 24 hours per day, 7 days per week.

Standard:

02   **The Truck Parking Availability General Service (D9-16b through D9-16e) signs (see Figure 2I-9) shall include a changeable message element with a white changeable legend on a black opaque background that displays only the number of parking spaces currently available at each location or the legend FULL. The upper section of the sign shall display the Truck Parking (D9-16) symbol sign and the legend SPACES OPEN. The sign shall display the number of available truck parking spaces for no more than three parking facilities. Where two lines of legend, such as the location and a distance, are displayed for a parking facility, not more than two parking facilities shall be displayed on the sign.**

03   **Where the truck parking facility is located off the main highway and is accessed from the crossroad, directional assemblies with the Truck Parking (D9-16) sign shall be installed along the ramp and along crossroads where the route to the facility requires a turn, where it is unclear as to which roadway to follow, or where additional guidance is needed.**

Support:

04   Displaying the number of parking spaces available at a facility when the number is low could result in truckers choosing to continue to a distant facility that no longer has available spaces by the time they arrive.

Figure 2I-9. Examples of Truck Parking Availability Signs

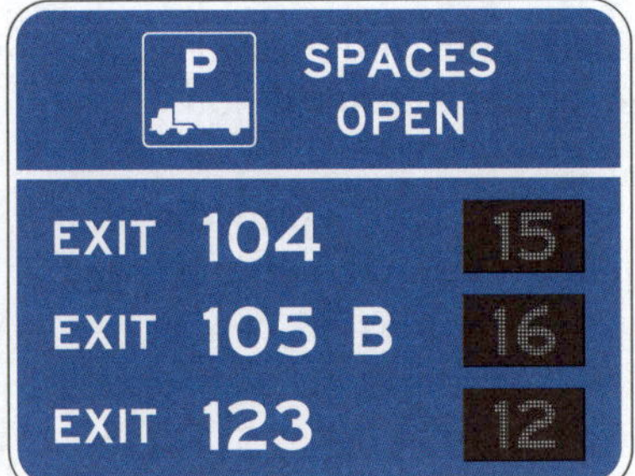

D9-16b
Reference Location Sign
Exit Numbering Method

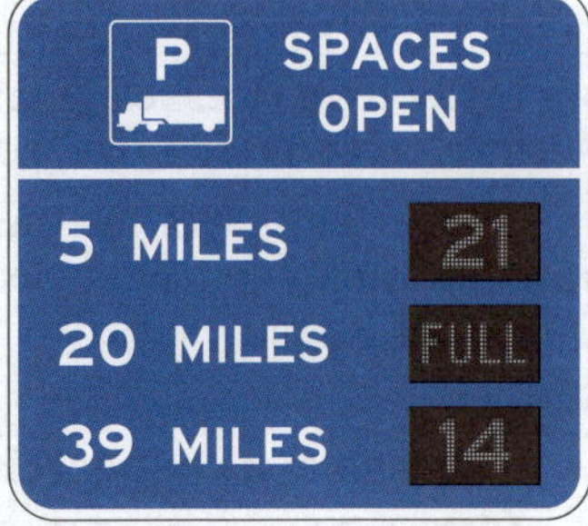

D9-16c
Unnumbered Exits

D9-16d
Rest Areas Only

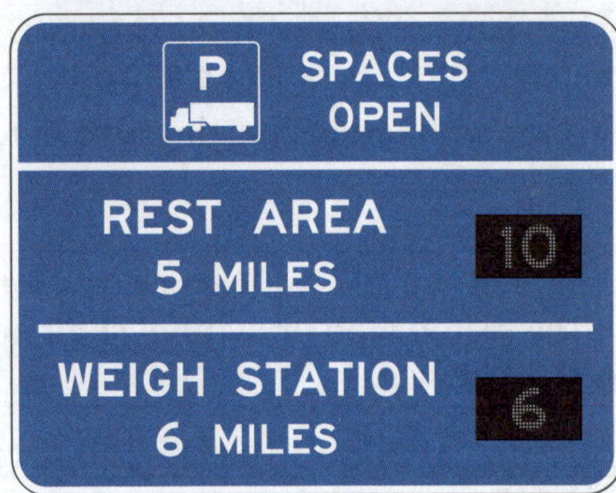

D9-16e
Combined Signing for Roadside Areas
and Off-System Sites

Option:

05   The word FULL in a white legend may be displayed on changeable message elements of a Truck Parking Availability General Service sign when the number of truck parking spaces available at the associated facility reaches a predetermined lower threshold.

Guidance:

06   *Truck Parking Availability signs should be located 3 to 5 miles in advance of the nearest parking facility. The parking facilities displayed on the sign should be no more than 60 miles from the sign location.*

Support:

07   Examples of uses of Truck Parking Availability signs are shown in Figure 2I-10.

Figure 2I-10. Examples of Use of Truck Parking Availability Signs

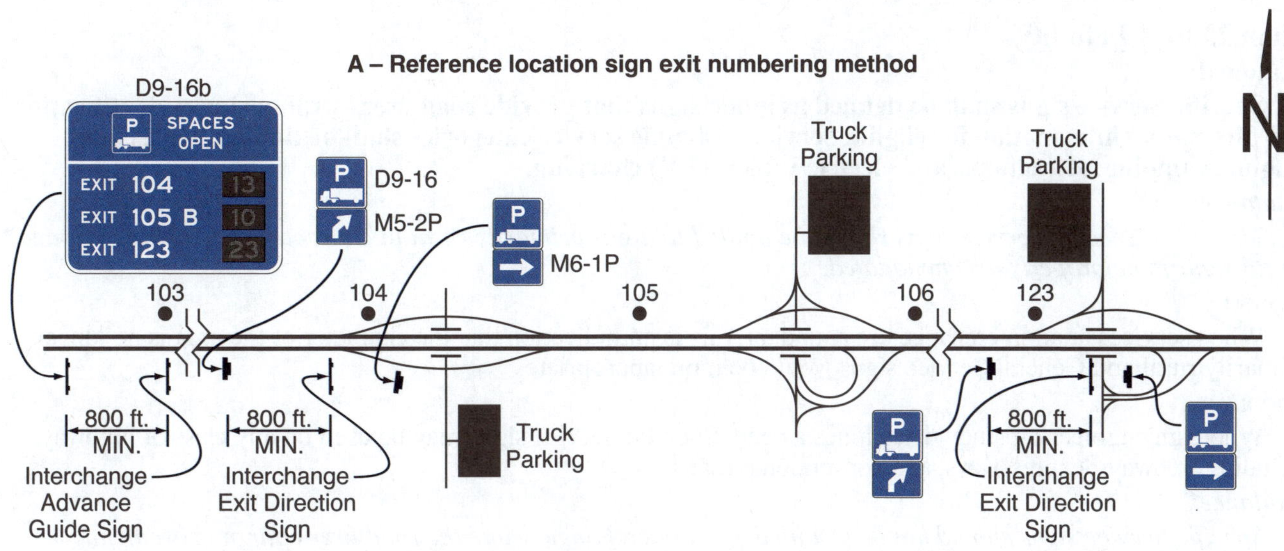

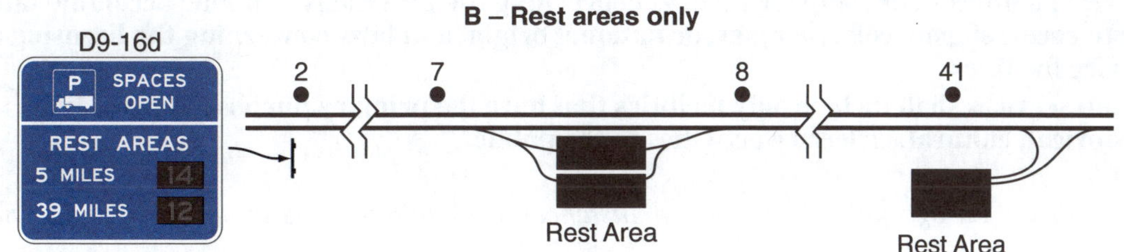

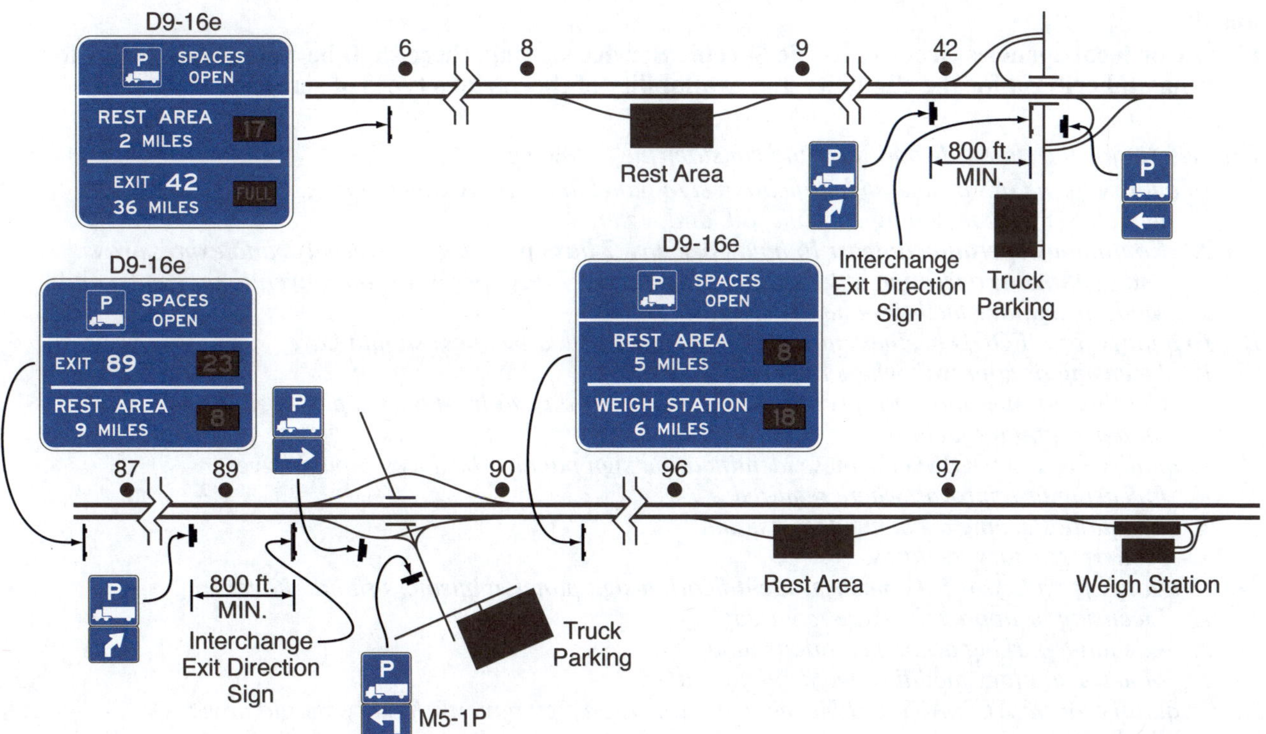

Note: Exit numbering may be used in place of directional arrows on the highway mainline.

CHAPTER 2J. SPECIFIC SERVICE SIGNS

Section 2J.01 Eligibility

Standard:

01   **Specific Service signs shall be defined as guide signs that provide road users with business identification and directional information for eligible services. Eligible service categories shall be limited to gas, food, lodging, camping, attractions, and electric vehicle (EV) charging.**

Guidance:

02   *The use of Specific Service signs should be limited to areas primarily rural in character with adequate space for all signs to be properly accommodated.*

Support:

03   When services at an interchange are abundant, this is an indication that the character of the area is no longer primarily rural and General Service signs would be more appropriate.

Option:

04   Where an engineering study determines a need, Specific Service signs may be used on any class of highway, including freeways, expressways, and conventional roads.

Guidance:

05   *Specific Service signs should not be installed at an interchange where the road user cannot conveniently reenter the freeway or expressway and continue in the same direction of travel.*

Standard:

06   **Eligible service facilities shall comply with laws concerning the provisions of public accommodations without regard to race, religion, color, age, sex, or national origin, and laws concerning the licensing and approval of service facilities.**

07   **The attraction services shall include only facilities that have the primary purpose of providing amusement, historical, cultural, or leisure activities to the public.**

Guidance:

08   *Except as provided in Paragraph 9 of this Section, distances to eligible services should not exceed 3 miles in any direction.*

Option:

09   If, within the 3-mile limit, facilities for the services being considered are not available or choose not to participate in the program, the limit of eligibility may be extended in 3-mile increments until one or more facilities for the services being considered chooses to participate, or until 15 miles is reached, whichever comes first.

Standard:

10   **If State or local agencies elect to provide Specific Service signing, there shall be a statewide policy for such signing and criteria for the eligibility and availability of the various types of services.**

Guidance:

11   *The criteria for the statewide policy should consider the following:*

    A. To qualify for a GAS business identification sign panel, a business should have:
        1. Vehicle services including gasoline, oil, and water;
        2. Continuous operation at least 16 hours per day, 7 days per week for freeways and expressways, and continuous operation at least 12 hours per day, 7 days per week for conventional roads; and
        3. Modern sanitary facilities and drinking water.

    B. To qualify for a FOOD business identification sign panel, a business should have:
        1. Licensing or approval, where required;
        2. Continuous operations to serve at least 2 meals per day, at least 6 days per week; and
        3. Modern sanitary facilities.

    C. To qualify for a LODGING business identification sign panel, a business should have:
        1. Licensing or approval, where required;
        2. Adequate sleeping accommodations; and
        3. Modern sanitary facilities.

    D. To qualify for a CAMPING business identification sign panel, a business should have:
        1. Licensing or approval, where required;
        2. Adequate parking accommodations; and
        3. Modern sanitary facilities and drinking water.

    E. To qualify for an ATTRACTION business identification sign panel, a facility should have:
        1. Regional significance, in compliance with the provisions of Paragraph 7 of this Section; and
        2. Adequate parking accommodations.

Standard:

12   **To be eligible for an Electric Vehicle (EV) CHARGING business identification sign panel, the EV chargers provided shall meet the criteria for Direct Current Fast Chargers provided in 23 CFR 680.106 and be in continuous operation at least 16 hours per day, 7 days per week.**

Option:

13   Business identification sign panels for a proprietary electric vehicle charging service may be included on an EV Charging Specific Service sign if it meets the eligibility criteria in Paragraph 12 of this Section.

Support:

14   Section 2J.12 contains additional information on criteria for the statewide policy regarding signing.

15   Section 2I.04 contains information regarding the Interstate Oasis program.

Section 2J.02 Application

Support:

01   Examples of Specific Service signs are shown in Figure 2J-1.

02   Examples of sign locations are shown in Figure 2J-2.

Standard:

03   **The number of Specific Service signs along an approach to an interchange or intersection, regardless of the number of service types displayed, shall be limited to a maximum of four. Except as provided in Paragraph 4 of this Section, in the direction of traffic flow, successive Specific Service signs shall be for attraction, camping, lodging, food, EV charging, and gas services, in that order.**

Option:

04   When spacing does not allow EV Charging Specific Service signs to be located as described in Paragraph 3 of this Section, then the EV Charging Specific Service signs may be located anywhere within the successive Specific Service sign order where adequate spacing between signs allows.

Guidance:

05   *The Specific Service signs should be located to take advantage of natural terrain, to have the least impact on the scenic environment, and to avoid visual conflict with other signs within the highway right-of-way.*

06   *Where a service type is displayed on two signs, the signs for that service should follow one another in succession.*

Standard:

07   **A Specific Service sign shall display the word message GAS, EV CHARGING, FOOD, LODGING, CAMPING, or ATTRACTION, an appropriate directional legend such as the word message EXIT XX, NEXT RIGHT, SECOND RIGHT, or directional arrows, and the related business identification sign panels. Distances to eligible facilities shall not be displayed on the Specific Service signs on the approach to an interchange.**

08   **A business that does not offer gasoline, but offers alternative fuels, shall not be signed using GAS Specific Service signs.**

Option:

09   A business that does not offer gasoline but offers alternative fuels may be signed using General Service signs for the alternative fuel provided.

Support:

10   General Service signs for facilities providing alternative fuels, including EV charging, compressed natural gas, liquefied natural gas, liquefied petroleum gas, and hydrogen, are provided in Chapter 2I.

Guidance:

11   *Due to the unique and widely varying characteristics of the services that qualify as attractions, and lesser recognition of their business identification sign panels (see Paragraph 12 of this Section), ATTRACTION Specific Service signs should have no more than four business identification sign panels.*

Support:

12   The types of services that meet the definition of attraction, such as those providing amusement, historical, cultural, or leisure activities to the public, vary considerably. In most cases, attractions do not include well-known services or easily recognizable logos, making it more difficult and requiring more time to distinguish between types of attractions shown on an ATTRACTION sign than for other categories of Specific Service signs.

Figure 2J-1. Examples of Business Identification Panel Arrangements on Specific Service Signs

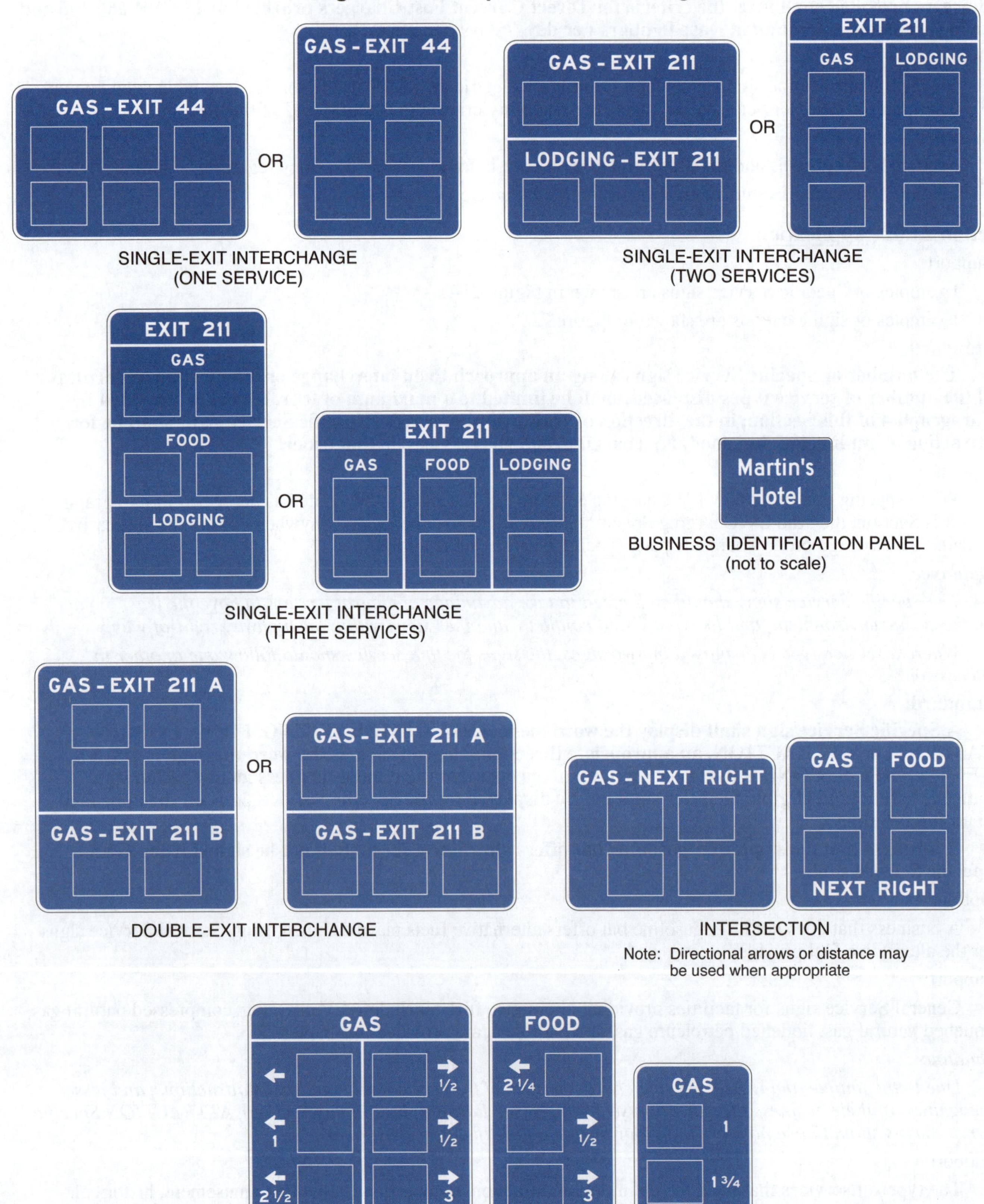

Figure 2J-2. Examples of Specific Service Sign Locations

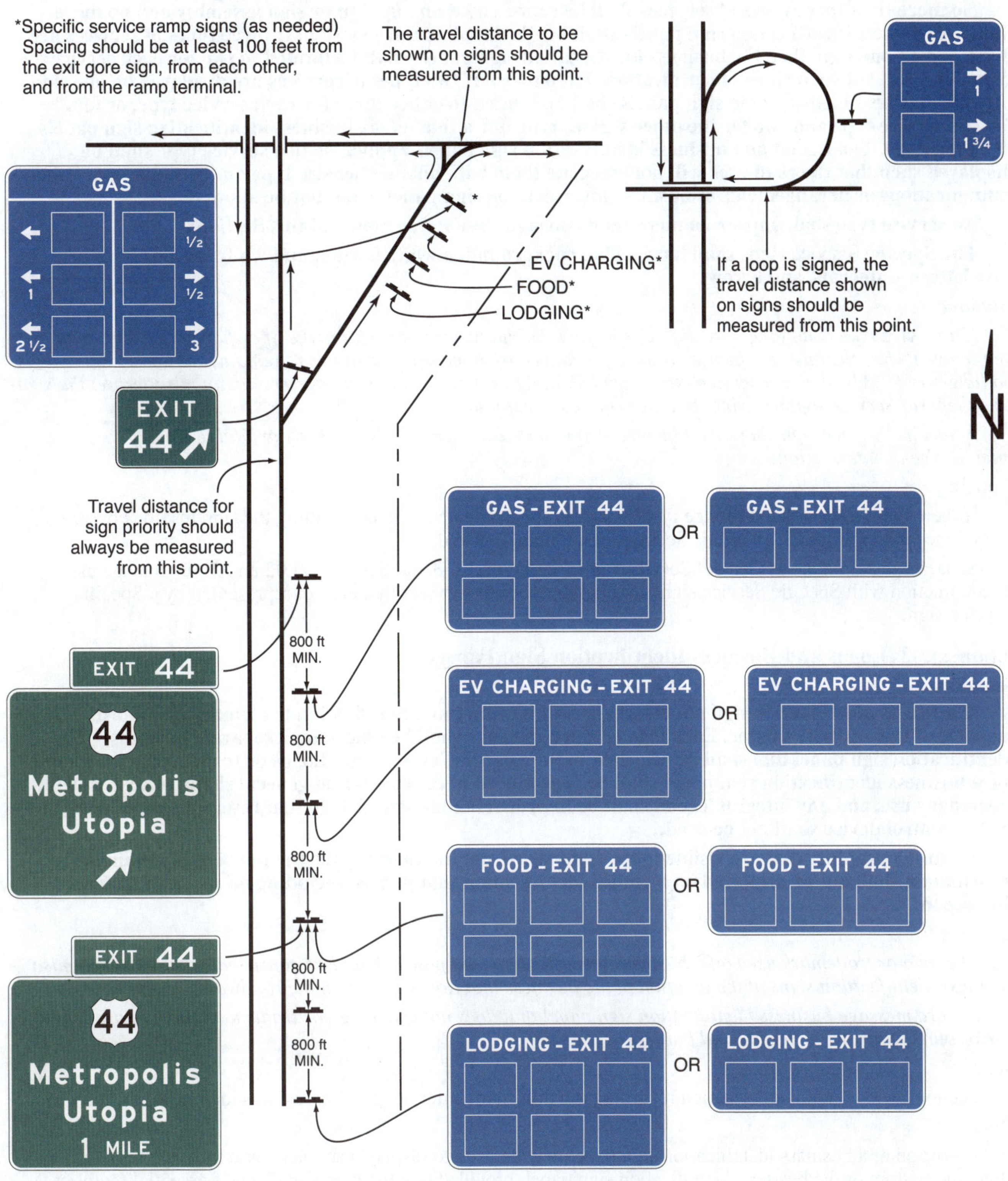

Standard:

13   **No more than three types of services shall be represented on any sign or sign assembly and no more than six business identification sign panels shall be displayed on any one sign. If three types of services are displayed on one sign, then the business identification sign panels shall be limited to two for each service type (for a total of six business identification sign panels). If two types of services are displayed on one sign, then the business identification sign panels shall be limited to either three for each service type, or four for one service type and two for the other service type (for a total of six business identification sign panels in either case). The legend and business identification sign panels applicable to a service type shall be displayed such that the road user will not associate them with another service type on the same sign. Other configurations or arrangements of business identification sign panels shall not be allowed.**

14   **No service type shall appear on more than two signs (see Paragraph 6 of this Section).**

15   **The Specific Service signs shall have a blue background, a white border, and white legends of upper-case letters, numerals, and arrows.**

Guidance:

16   *If a service type is no longer available from an interchange or intersection, the Specific Service sign should be removed when the business identification sign panels are removed. If a sign is to remain, but the service type is no longer available, then the service type legend should be covered so that road users do not misinterpret the sign as a General Service sign implying that the service is available.*

17   *A Specific Service sign should not be installed unless a service type is currently available from an interchange or intersection.*

Option:

18   If there is indication that a service type will again be available in the near future, the sign may be covered, in accordance with Paragraph 16 of this Section, rather than removed.

19   Separate installations of General Service signs (see Figure 2J-3 and Sections 2I.02 and 2I.03) may be used in conjunction with Specific Service signs for eligible types of services that are not represented by a Specific Service sign.

Section 2J.03 <u>Logos and Business Identification Sign Panels</u>

Standard:

01   **A business identification sign panel legend shall be either an identification trademark or a word message of the business's name. Each logo or word message shall be placed on a separate business identification sign panel that shall be attached to the Specific Service sign. Logos or trademarks used alone for a business identification sign panel shall be reproduced in the colors and general shape consistent with customary use, and any integral legend shall be in proportionate size. A logo that resembles an official traffic control device shall not be used.**

02   **Scanning graphics that are visible to the road user from the roadway for the purpose of obtaining information shall not be displayed on business identification sign panels, including on any logo displayed thereupon.**

Guidance:

03   *The logo or trademark used on a business identification sign panel should be consistent with the on-premise business identification signs at the location of the business that are visible from the roadway.*

04   *A word message business identification sign panel that does not use a logo or trademark should have a blue background with a white legend and border.*

Support:

05   Section 2J.05 contains information regarding the minimum letter heights for business identification sign panels.

Option:

06   A portion of a business identification sign panel may be used to display a supplemental message horizontally along the bottom of the business identification sign panel, provided that the message displays essential motorist information consistent with the service category type and related to the operation of the business (see Figure 2J-4).

Standard:

07   **All supplemental messages shall be displayed within the business identification sign panel and shall have letters and numerals that comply with the minimum height requirements shown in Table 2J-1. Supplemental messages promoting the availability of products, amenities, or services that are not directly related to the service category and/or those not available to non-patrons of the primary service provided for the service category, such as car wash, automated teller machines, Internet, lottery, or swimming pool, shall not be displayed on business identification sign panels.**

Figure 2J-3. Example of General Service Signs Used in Conjunction with Specific Service Signs

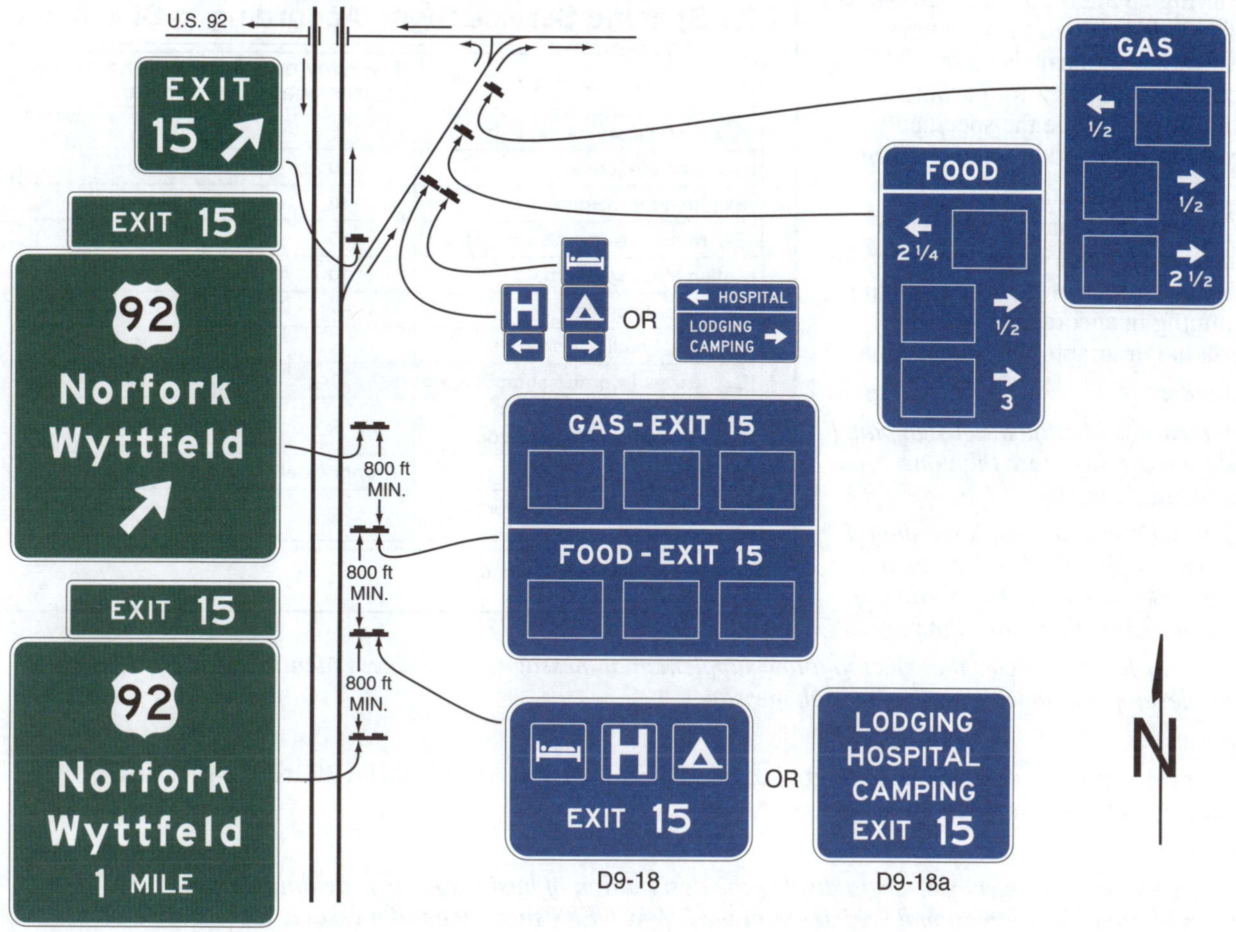

Note: If specific service ramp signs are used, their spacing should be at least 100 feet from the exit gore sign, from each other, and from the ramp terminal.

Figure 2J-4. Examples of Supplemental Messages on Business Identification Sign Panels

08   **Messages related to the promotion or availability of business identification sign panel space shall not be displayed on Specific Service signs.**

09   **To be eligible for an EV CHARGING supplemental message on a business identification sign panel, the business shall:**

    A. Offer electric vehicle charging to the general public without purchasing the primary service (gas, food, lodging, camping, or attraction, as appropriate); and

    B. For the service categories of gas, food, and attraction, provide EV chargers meeting the criteria for Direct Current Fast Chargers (DCFC) provided in 23 CFR 680.106; or

    C. For the service categories of camping and lodging, provide EV chargers meeting the criteria for DCFCs provided in 23 CFR 680.106 and/or AC Level 2 Charging.

Option:

10   A Supplemental message identifying an alternative fuel available may be added only to the business identification sign panels on the GAS Specific Services sign for gasoline facilities that provide the specified alternative fuel in addition to gasoline.

11   The Supplemental message EV CHARGING may be added to a business identification sign panel for the service categories of gas, food, lodging, or camping in accordance with the criteria in Paragraph 9 of this Section.

Guidance:

12   *A business identification sign panel should not display more than one supplemental message.*

13   *The supplemental message should be displayed in a black legend on a yellow background for that portion of the business identification sign panel.*

Table 2J-1. Minimum Letter and Numeral Sizes for Specific Service Signs According to Sign Type

Type of Sign	Freeway or Expressway	Conventional Road or Ramp
A. Specific Service Signs		
Service Categories	10	6
Exit Number Words	10	—
Exit Number Numerals and Letters	10	—
Action Message Words	10	6
Distance Numerals	—	6
Distance Fraction Numerals	—	4
B. Business Identification Sign Panels		
Words and Numerals (Non-Trademark/Graphic Logo)	8	4
Trademark/Graphic Logo	Proportional	Proportional
Supplemental Message Words and Numerals	5	2.5

Note: Sizes are shown in inches

14   *State or local agencies that elect to allow supplemental messages on business identification sign panels should develop a statewide policy for such messages.*

Support:

15   Typical supplemental messages might include DIESEL, LP-GAS, EV CHARGING, 24 HOURS, CLOSED SUNDAY, and RV ACCESS.

Guidance:

16   *If a State or local agency elects to display the designation of businesses as providing on-premise accommodations for recreational vehicles with the RV ACCESS supplemental message, there should be a statewide policy for such designation and criteria for qualifying businesses. The criteria should include such site conditions as access between the public roadway and the site, on-premise geometry, and parking.*

Option:

17   If a business designated as an Interstate Oasis (see Section 2I.04) has a business identification sign panel on the Food and/or Gas Specific Service signs, the word OASIS may be displayed on the bottom portion of the business identification sign panel for that business.

Standard:

18   **A business identification sign panel shall not display the identification logo/trademark or name of more than one business. A business identification sign panel shall not display more than one name or identification logo/trademark for the same business. Slogans, such as marketing slogans associated with the business, shall not be displayed on business identification sign panels or the Specific Service sign itself.**

Section 2J.04 Number and Size of Signs and Business Identification Sign Panels

Guidance:

01   *Sign sizes should be determined by the amount and height of legend and the number and size of business identification sign panels attached to the sign. All business identification sign panels on a sign should be the same size.*

Standard:

02   **Each Specific Service sign or sign assembly shall be limited to no more than six business identification sign panels.**

Option:

03   Where more than six businesses of a specific service type are eligible for business identification sign panels at the same interchange, additional business identification sign panels of that same specific service type may also be displayed in accordance with the provisions of Paragraph 4 of this Section. The additional business identification sign panels may be displayed either by placing more than one specific service type on the same sign (see Paragraph 13 of Section 2J.02) or by using a second Specific Service sign of that specific service type if the additional sign can be added without exceeding the limit of four Specific Service signs at an interchange or intersection approach (see Paragraph 3 of Section 2J.02).

Standard:

04   **Where business identification sign panels for more than six businesses of a specific service type are displayed at the same interchange or intersection approach, the following provisions shall apply:**

   A. No more than 12 business identification sign panels of a specific service type shall be displayed on no more than two Specific Service signs or sign assemblies;

   B. No more than six business identification sign panels shall be displayed on a single Specific Service sign; and

   C. No more than four Specific Service signs shall be displayed on the approach.

Support:

05   Section 2J.08 contains information regarding Specific Service signs for double-exit interchanges.

06   Section 2J.09 contains information regarding Specific Service signs for multiple interchanges that are accessed from collector-distributor roadways rather than from the highway mainline.

Standard:

07   **Each business identification sign panel attached to a Specific Service sign shall be a horizontally-oriented rectangle with a width longer than the height. A business identification sign panel on signs for freeways and expressways shall not exceed 60 inches in width and 36 inches in height (see Table 2J-2). A business identification sign panel on signs for conventional roads and freeway and expressway ramps shall not exceed 30 inches in width and 18 inches in height (see Table 2J-2). The vertical and horizontal spacing between business identification sign panels shall not exceed 8 inches and 12 inches, respectively.**

Table 2J-2. Maximum Business Identification Sign Panel Sizes by Roadway Classification

Roadway Classification	Sign Panel Size
Freeway or Expressway	60 x 36
Conventional Road or Ramp	30 x 18

Note: Sizes are shown in inches as width x height

Support:

08   Sections 2A.10, 2E.13, and 2E.14 contain information regarding borders, interline spacing, and edge spacing.

Section 2J.05 Size of Lettering

Standard:

01   **All Specific Service signs and business identification sign panels shall have letter and numeral sizes that comply with the minimum requirements of Table 2J-1.**

Guidance:

02   *Any legend on a business identification graphic/trademark should be proportional to the size of the graphic trademark.*

Section 2J.06 Signs at Interchanges

Standard:

01   **The Specific Service signs shall be installed between the preceding interchange and at least 800 feet in advance of the Exit Direction sign at the interchange from which the services are available (see Figure 2J-2).**

02   **Specific Service signs shall not be used at freeway-to-freeway interchanges (see Section 2E.37), except where the exit ramp also provides direct access to a conventional road within that interchange (see Figure 2J-5).**

Guidance:

03   *There should be at least an 800-foot spacing between the Specific Service signs, except for Specific Service ramp signs. Excessive spacing should not be used between Specific Service signs, as this is not desirable either.*

Figure 2J-5. Example of Specific Services Signing for a Conventional Road Accessed within a Freeway-to-Freeway Interchange

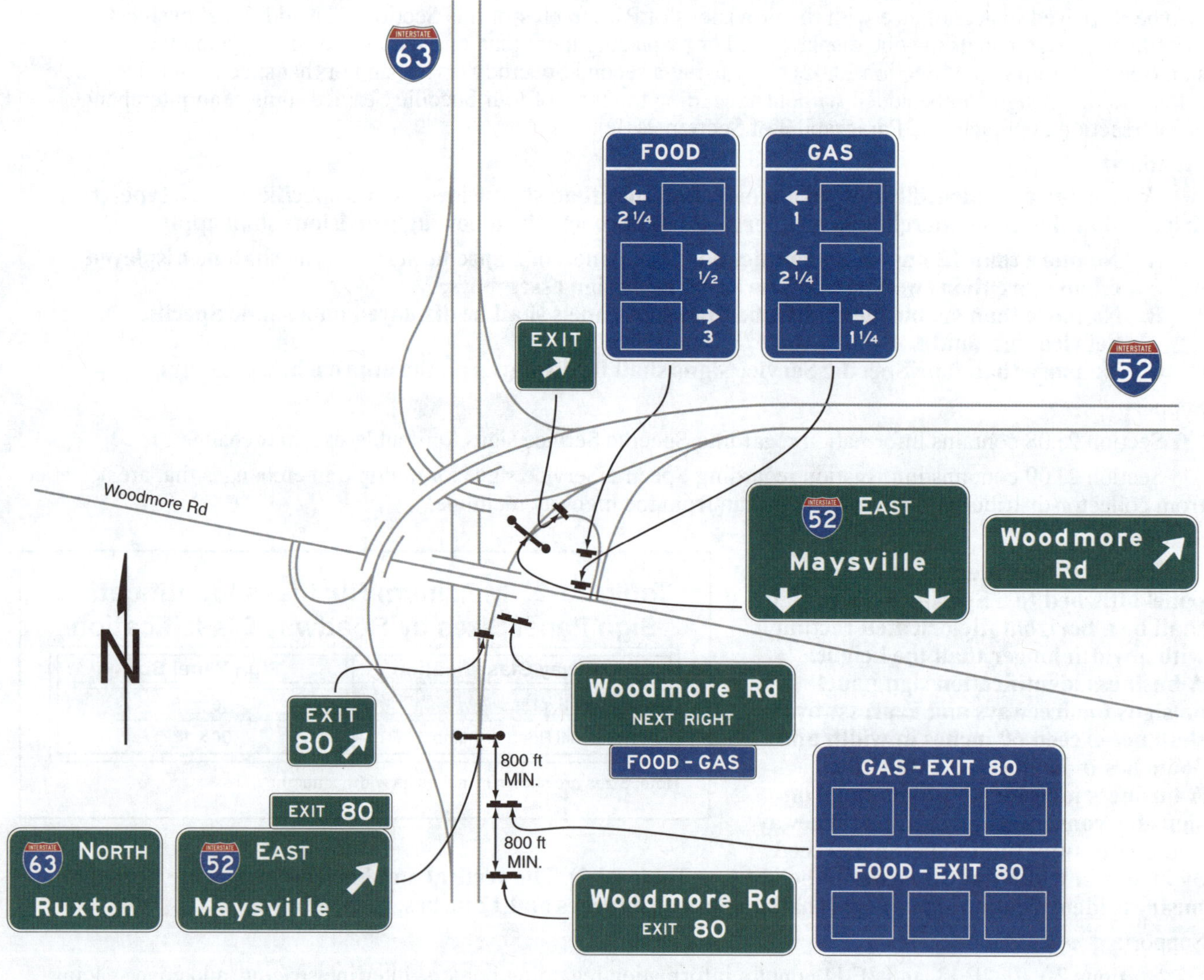

04	*Specific Service ramp signs should be spaced at least 100 feet longitudinally beyond the Exit Gore sign, from each other, and from the ramp terminal. Specific Service ramp signs should be spaced at least 200 feet longitudinally from any Destination guide signs along the ramp. Longer longitudinal spacing should be provided between Specific Service ramp signs and any warning or regulatory signs along the ramp, and any intersection traffic control devices at the ramp terminal.*

05	*When the distance to the next exit providing access to EV charging service is 50 miles or greater, the Next EV Charging (D9-17a) sign should be used (see Figure 2H-9). When used, the Next EV Charging sign should be located directly after the General Service sign for the fuel type displayed in the signing sequence for the exit (see Figure 2H-10).*

Section 2J.07 Single-Exit Interchanges
Standard:

01	**At numbered single-exit interchanges, the name of the service type followed by the exit number shall be displayed on one line above the business identification sign panels. At unnumbered interchanges, the directional legend NEXT RIGHT (LEFT) shall be used in place of the exit number.**

02	**At single-exit interchanges where traffic is allowed to turn onto the crossroad in either direction from the ramp, Specific Service ramp signs shall be installed along the ramp or opposite the ramp terminal for facilities that have business identification sign panels displayed along the main roadway if the facilities are not readily visible from the ramp terminal. Directions to the service facilities shall be indicated by arrows**

on the ramp signs. Business identification sign panels on Specific Service ramp signs shall be duplicates of those displayed on the Specific Service signs located in advance of the interchange, but shall be reduced in size (see Paragraph 7 of Section 2J.04).

Option:

03   Specific Service ramp signs may display distances (see Paragraphs 14 and 15 of Section 2A.08) to a service facility when the facility is not visible from ramp intersection with the crossroad.

Guidance:

04   *Distances of less than ¼ mile, when displayed, should be displayed to the nearest ¹⁄₁₀ mile.*

Section 2J.08 Double-Exit Interchanges

Guidance:

01   *At double-exit interchanges, the Specific Service signs should consist of two sections, one for each exit (see Figure 2J-1).*

Standard:

02   **At a double-exit interchange, the top section shall display the business identification sign panels for the first exit and the bottom section shall display the business identification sign panels for the second exit. At numbered interchanges, the name of the service type and the exit number shall be displayed above the business identification sign panels in each section. At unnumbered interchanges, the word message NEXT RIGHT (LEFT) and SECOND RIGHT (LEFT) shall be used in place of the exit number. The number of business identification sign panels on the sign (total of both sections) or the sign assembly shall be limited to six.**

Guidance:

03   *At a double-exit interchange, where a service type is displayed on two Specific Service signs in accordance with the provisions of Section 2J.04, one of the signs should display the business identification sign panels for that service type for the businesses that are accessible from one of the two exits and the other sign should display the business identification sign panels for that service type for the businesses that are accessible from the other exit.*

Option:

04   At a double-exit interchange where there are four business identification sign panels to be displayed for one of the exits and one or two business identification sign panels to be displayed for the other exit, the business identification sign panels may be arranged in three rows with two business identification sign panels per row.

05   At a double-exit interchange, where a service is to be signed for only one exit, one section of the Specific Service sign may be omitted, or a single exit interchange sign may be used.

06   Signs on ramps and crossroads as described in Section 2J.07 may be used at a double-exit interchange.

Section 2J.09 Collector-Distributor Roadways for Successive Interchanges

Support:

01   Examples of Specific Service signs used in advance of interchanges for collector-distributor roadways that provide access to multiple interchanges are shown in Figure 2J-6.

Option:

02   If services are available from more than one of the interchanges along the collector-distributor roadway and those services are signed with Specific Service signs as described in Paragraph 4 of this Section, then Specific Service signs may be used on the mainline in conformance with the provisions of this Chapter.

Standard:

03   **No more than four Specific Service signs shall be displayed on a highway mainline approach to a collector-distributor roadway.**

04   **If Specific Service signs are located on the highway mainline for services accessed from the collector-distributor roadway, then the business identification sign panels displayed on the collector-distributor roadway shall be only duplicates of those displayed on the highway mainline.**

05   **If more than four Specific Services signs would be required on the mainline in advance of the collector-distributor roadway in order to display all the business identification sign panels used on Specific Service signs in advance of the collector-distributor roadway exits, then General Service signs shall be used on the mainline to identify the types of services displayed on Specific Service signs on the collector-distributor roadway.**

Figure 2J-6. Example of Signing for Services Accessed from a Collector-Distributor Road Adjacent to a Freeway

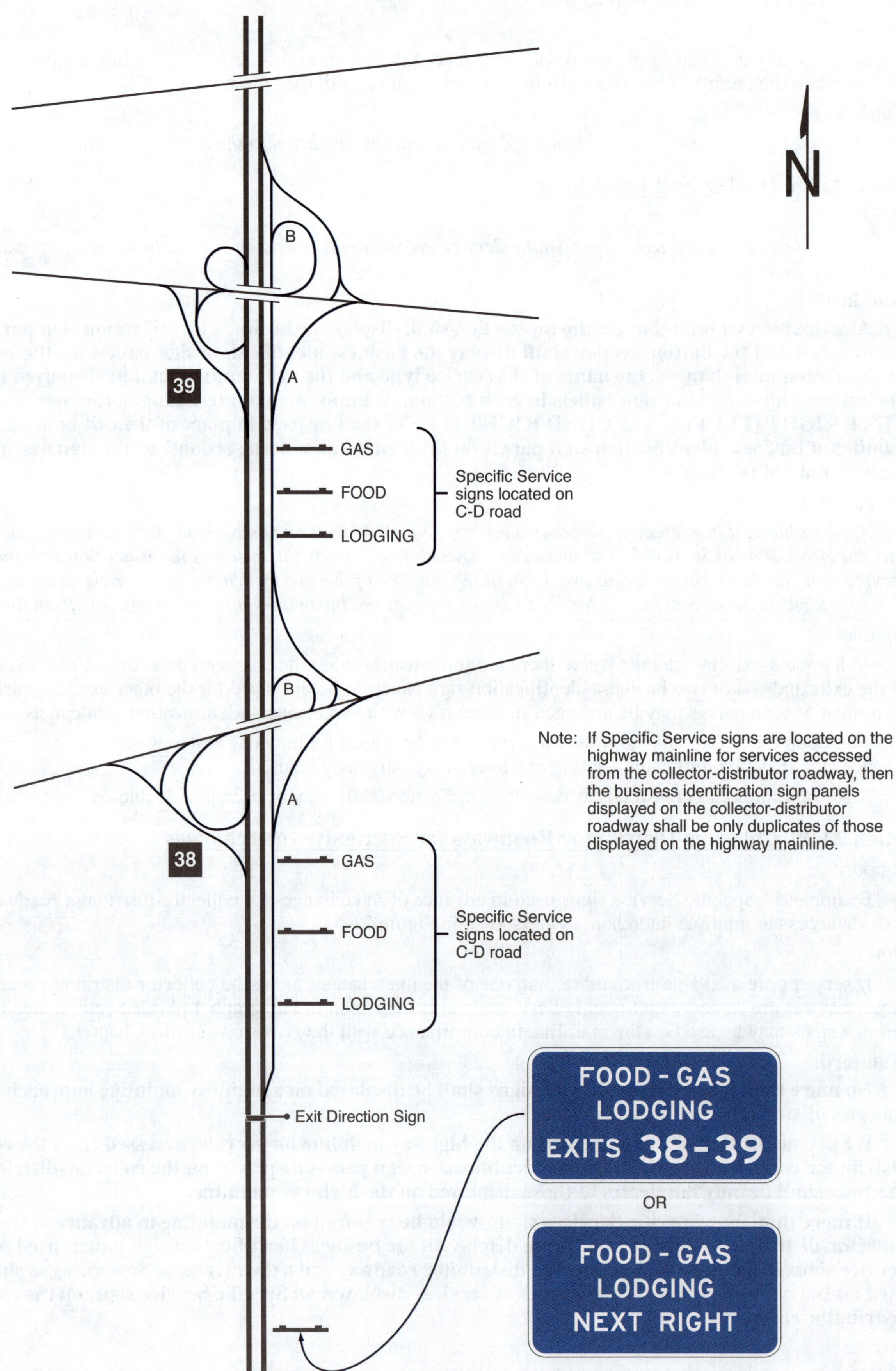

Section 2J.10 Specific Service Trailblazer Signs

Support:

01 Specific Service trailblazer signs (see Figure 2J-7) are guide signs with one to four business identification sign panels that display business identification and directional information for services and eligible attractions. Specific Service trailblazer signs are installed along crossroads for facilities that have business identification sign panels displayed along the main roadway and ramp, and that require additional vehicle maneuvers or are a long distance from the ramp along the crossroad.

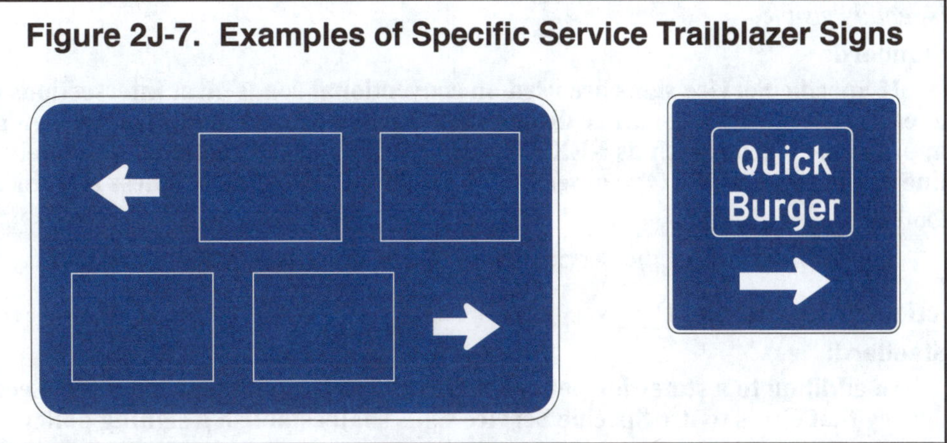

Standard:

02 **Specific Service trailblazer signs shall be installed along crossroads where the route to the business requires a direction change, where it is questionable as to which roadway to follow, or where additional guidance is needed. Where it is not feasible or practical to install Specific Service trailblazer signs to such businesses, those businesses shall not be considered eligible for signing from the ramp and main roadway. A Specific Service trailblazer sign shall not be required at the point where the business is visible from the roadway and its access is readily apparent.**

Guidance:

03 *If used, a Specific Service trailblazer sign should be located a maximum of 500 feet in advance of any required turn.*

Standard:

04 **The location of other traffic control devices shall take precedence over the location of a Specific Service trailblazer sign.**

05 **When used, each Specific Service trailblazer sign or sign assembly shall be limited to no more than four business identification sign panels. The business identification sign panels on Specific Service trailblazer signs shall be duplicates of those displayed on the Specific Service ramp signs.**

06 **Appropriate legends, such as directional arrows or the action message NEXT RIGHT or SECOND RIGHT, shall be displayed with the business identification sign panel to provide proper guidance. The directional legend and border shall be white and shall be displayed on a blue background.**

Option:

07 Specific Service trailblazer signs may contain various types of services on a single sign or on a sign assembly.

08 Specific Service trailblazer signs may be placed farther from the edge of the road than other traffic control signs.

Section 2J.11 Signs at Intersections

Guidance:

01 *If both tourist-oriented information (see Chapter 2K) and specific service information are proposed to be used at the same intersection, the tourist-oriented directional and Specific Service signs should be spaced sufficiently apart from one another, as well as apart from other guide, warning, and regulatory signs, to avoid confusion and allow sufficient time for road users to read and react to the information.*

Standard:

02 **If sufficient space to provide appropriate reading and reaction to all proposed signs is not available, higher priority shall be given to guide, warning, and regulatory signs and either the tourist-oriented directional signs or the Specific Service signs, or both, shall not be used.**

Guidance:

03   *If Specific Service signs are used on conventional roads or at intersections on expressways, they should be installed between the previous interchange or intersection and at least 300 feet in advance of the intersection from which the services are available.*

04   *Business identification sign panels should not be displayed for a type of service for which a qualified facility is readily visible.*

Standard:

05   **If Specific Service signs are used on conventional roads or at intersections on expressways, the name of each type of service shall be displayed above its business identification sign panel(s), together with an appropriate legend, such as NEXT RIGHT (LEFT) or a directional arrow, either displayed on the same line as the name of the type of service or displayed below the business identification sign panel(s).**

Option:

06   Signs similar to Specific Service ramp signs as described in Section 2J.07 may be provided on the crossroad.

Section 2J.12 Signing Policy

Standard:

01   **In addition to a statewide policy for eligibility of service providers (see Section 2J.01), each highway agency that elects to use Specific Service signs shall establish a signing policy.**

Guidance:

02   *The signing policy should include, at a minimum, the provisions of Section 2J.01 and at least the following criteria:*

    A. *Selection of eligible businesses;*
    B. *Distances to eligible services;*
    C. *The use of business identification sign panels, legends, and signs complying with the provisions of this Manual and State design requirements;*
    D. *Removal or covering of business identification sign panels during off seasons for businesses that operate on a seasonal basis;*
    E. *The circumstances, if any, under which Specific Service signs are permitted to be used in non-rural areas; and*
    F. *Determination of the costs to businesses for initial permits, installations, annual maintenance, and removal of business identification sign panels.*

CHAPTER 2K. TOURIST-ORIENTED DIRECTIONAL SIGNS

Section 2K.01 Purpose and Application

Support:

01   Tourist-oriented directional signs are post-mounted guide sign assemblies with one or more signs that display the business identification of and directional information for eligible business, service, and activity facilities.

Standard:

02   **A facility shall be eligible for tourist-oriented directional signs only if it derives its major portion of income or visitors during the normal business season from road users not residing in the area of the facility.**

Option:

03   Tourist-oriented directional signs may include businesses involved with seasonal agricultural products.

Standard:

04   **The use of tourist-oriented directional signs shall be limited to rural highways (see definition in Section 1C.02). Tourist-oriented directional signs shall not be installed on conventional roads in urban or urbanized areas or on freeway or expressway main roadways or ramps.**

Option:

05   Tourist-oriented directional signs may be used in conjunction with General Service signs (see Section 2I.02).

Support:

06   Section 2K.07 contains information on the adoption of a State policy for States that elect to use tourist-oriented directional signs.

Section 2K.02 Design

Standard:

01   **Tourist-oriented directional sign assemblies shall have one or more signs (see Figure 2K-1) for the purpose of displaying the business identification of and directional information for eligible facilities. Except as provided in Paragraph 7 of this Section, each sign shall be rectangular in shape and shall have a white legend and border on a blue background.**

02   **The content of the legend on each sign shall be limited to the identification and directional information for no more than one eligible business, service, or activity facility. The legends shall not include promotional advertising.**

Guidance:

03   *Each sign should have a maximum of two lines of legend including no more than one symbol (see Paragraph 4 of this Section), a separate directional arrow, and the distance to the facility displayed beneath the arrow. Arrows pointing to the left or up should be at the extreme left of the sign panel. Arrows pointing to the right should be at the extreme right of the sign panel. Symbols, when used, should be to the left of the word legend or business identification sign panel (see Paragraphs 6 and 9 of this Section).*

Option:

04   The General Service sign symbols (see Section 2I.02) and the symbols for recreational and cultural interest area signs (see Chapter 2M) may be used on tourist-oriented directional signs.

05   Based on engineering judgment, the hours of operation may be displayed on the sign.

06   Business identification sign panels (see Section 2J.03) for specific businesses, services, and activities may be used in place of word legends on tourist-oriented direction signs.

Standard:

07   **When used, recreational and cultural interest area symbols shall be white on a brown background.**

08   **When used, symbols shall be an appropriate size (see Section 2K.04).**

09   **When used, business identification sign panels shall not exceed 24 inches in width and 15 inches in height. Logos resembling official traffic control devices shall not be permitted.**

Option:

10   The word message TOURIST ACTIVITIES may be displayed at the top of the tourist-oriented directional sign assembly.

Standard:

11   **The TOURIST ACTIVITIES word message shall have a white legend in all upper-case letters and a white border on a blue background. If used, it shall be placed above and in addition to the directional signs.**

Support:

12   Examples of tourist-oriented directional signs are shown in Figures 2K-1 and 2K-2.

Figure 2K-1. Examples of Tourist-Oriented Directional Signs

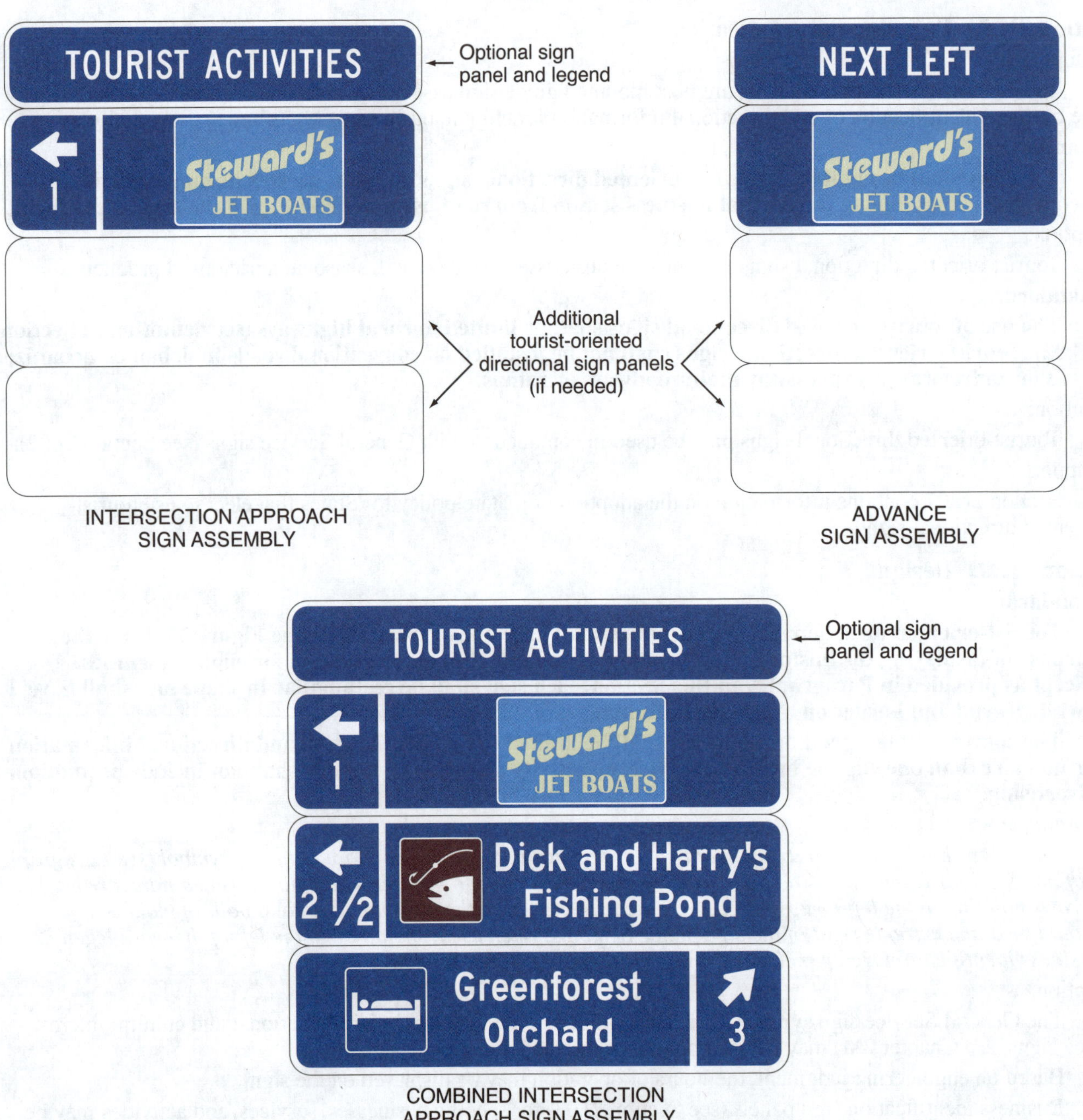

Section 2K.03 Style and Size of Lettering

Guidance:

01   *All letters and numbers on tourist-oriented directional signs, except on the business identification sign panels, should be upper-case and at least 6 inches in height. Any legend on a business identification sign panel should be proportional to the size of the business identification sign panel.*

Standard:

02   **Design standards for letters, numerals, and spacing shall be as provided in the "Standard Highway Signs" publication (see Section 1A.05).**

Figure 2K-2. Examples of Tourist-Oriented Directional Signs on an Intersection Approach

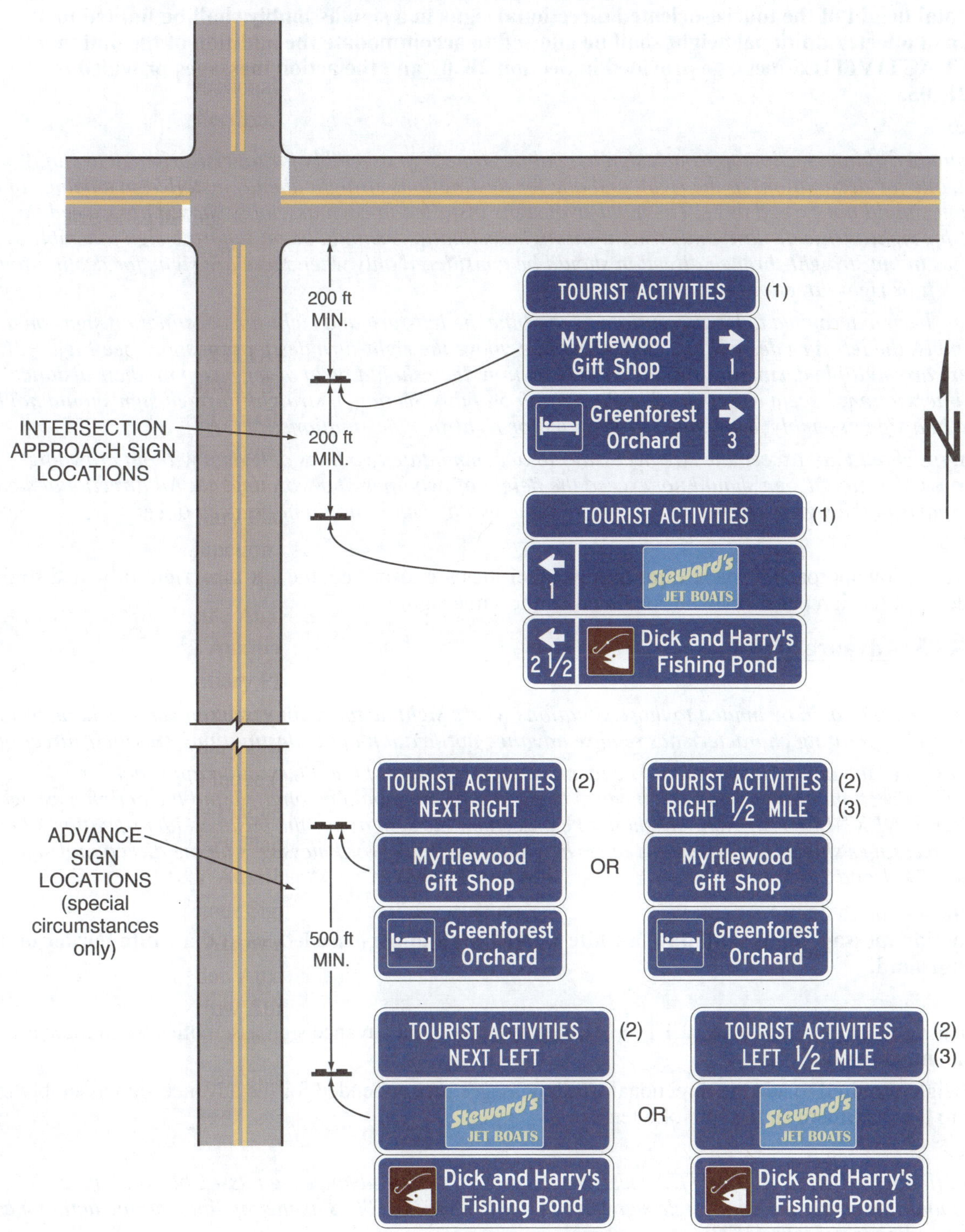

(1) Optional sign panel and legend
(2) Optional legend
(3) Use if there is an intervening intersection

Section 2K.04 Arrangement and Size of Signs

Standard:

01   **The total height of the tourist-oriented directional signs in a sign assembly shall be limited to a maximum of 6 feet. Additional height shall be allowed to accommodate the addition of the optional TOURIST ACTIVITIES message provided in Section 2K.02 and the action messages provided in Section 2K.05.**

Guidance:

02   *The number of intersection approach sign assemblies (one sign assembly for tourist-oriented destinations to the left, one for destinations to the right, and one for destinations straight ahead) installed in advance of an intersection should not exceed three. The number of signs installed in each assembly should not exceed three. The signs for right-turn, left-turn, and straight-through destinations should be on separate sign assemblies. Signs for facilities in the straight-through direction should be considered only when there are signs for destinations in either the left or right direction.*

03   *If it has been determined to be appropriate to combine the left-turn and right-turn destination signs on a single sign assembly, the left-turn destination signs should be above the right-turn destination signs (see Figure 2K-1). When there are multiple destinations in the same direction, they should be in order based on their distance from the intersection. Except as provided in Paragraph 5 of this Section, a straight-through sign should not be combined in a sign assembly displaying left-turn and/or right-turn destinations.*

04   *The signs should not exceed the size necessary to accommodate two lines of legend without crowding. Symbols on a directional sign should not exceed the height of two lines of a word legend. All directional signs and other parts of the sign assembly should be the same width, which should not exceed 6 feet.*

Option:

05   At intersection approaches where three or fewer facilities are displayed, the left-turn, right-turn, and straight-through destination sign panels may be combined on the same sign.

Section 2K.05 Advance Signs

Guidance:

01   *Advance signs should be limited to those situations where sight distance, intersection vehicle maneuvers, or other vehicle operating characteristics require advance notification of the destinations and their directions.*

02   *The design of the advance sign should be identical to the design of the intersection approach sign. However, the directional arrows and distances to the destinations should be omitted and the action messages NEXT RIGHT, NEXT LEFT, or AHEAD should be placed on the sign above the business identification signs. The action messages should have the same letter height as the other word messages on the directional signs (see Figures 2K-1 and 2K-2).*

Standard:

03   **The action message signs shall have a white legend in all upper-case letters and a white border on a blue background.**

Option:

04   The legend RIGHT ½ MILE or LEFT ½ MILE may be used on advance sign assemblies when there are intervening minor roads.

05   The height required to add the directional word messages recommended for the advance sign assembly may be added to the maximum sign height of 6 feet.

Guidance:

06   *The optional TOURIST ACTIVITIES message, when used on an advance sign assembly, and the action message should be combined on a single sign with TOURIST ACTIVITIES as the top line and the action message as the bottom line (see Figure 2K-2).*

Section 2K.06 Sign Locations

Guidance:

01   *If used, the intersection approach signs should be located at least 200 feet in advance of the intersection. Sign assemblies should be spaced at least 200 feet apart and at least 200 feet from other traffic control devices.*

02   *If used, advance signs should be located approximately ½ mile from the intersection with 500 feet between these sign assemblies. In the direction of travel, the order of advance sign placement should be to show the destinations to the left first, then destinations to the right, and last, the destinations straight ahead (see Figure 2K-2).*

03   *Position, height, and lateral offset of sign assemblies should be governed by Chapter 2A except as permitted in this Section.*

Option:

04   Tourist-oriented directional signs may be placed farther from the edge of the road than other traffic control signs.

Standard:

05   **The location of other traffic control devices shall take precedence over the location of tourist-oriented directional signs.**

Section 2K.07 State Policy

Standard:

01   **To be eligible for tourist-oriented directional signing, facilities shall comply with applicable State and Federal laws concerning the provisions of public accommodations without regard to race, religion, color, age, sex, or national origin, and with laws concerning the licensing and approval of service facilities. Each State that elects to use tourist-oriented directional signs shall adopt a policy that complies with these provisions.**

Guidance:

02   *The State policy should include:*

    A. *A definition of tourist-oriented business, service, and activity facilities.*
    B. *Eligibility criteria for signs for facilities.*
    C. *Provision for covering signs during off seasons for facilities operated on a seasonal basis.*
    D. *Provisions for signs to facilities that are not located on the crossroad when such facilities are eligible for signs.*
    E. *A definition of the immediate area in compliance with the provisions of Paragraph 2 of Section 2K.01.*
    F. *Maximum distances to eligible facilities. The maximum distance should be 5 miles.*
    G. *Provision for information centers (plazas) when the number of eligible sign applicants exceeds the maximum permissible number of sign panel installations.*
    H. *Provision for limiting the number of signs when there are more applicants than the maximum number of signs permitted.*
    I. *Criteria for use at intersections on expressways.*
    J. *Provisions for controlling or excluding those businesses which have illegal signs as defined by the Highway Beautification Act of 1965 (23 U.S.C. 131).*
    K. *Provisions for States to charge fees to cover the cost of signs through a permit system.*
    L. *A definition of the conditions under which the time of operation is displayed.*
    M. *Provisions for determining if advance signs will be permitted, and the circumstances under which they will be installed.*

CHAPTER 2L. CHANGEABLE MESSAGE SIGNS

Section 2L.01 Description of Changeable Message Signs
Support:

01   A changeable message sign (CMS) is a traffic control device that is capable of displaying one or more alternative messages. Some CMS have a blank mode when no message is displayed, while others display multiple messages with only one of the messages displayed at a time (such as OPEN/CLOSED signs at weigh stations).

02   The provisions in this Chapter apply to both permanent and portable changeable message signs with electronic displays or the electronic display portion of an otherwise conventional static sign. Additional provisions that only apply to portable changeable message signs (PCMS) can be found in Section 6L.05. The provisions in this Chapter generally do not apply to CMS with non-electronic displays that are changed either manually or electromechanically, such as a hinged-panel, rotating-drum, or back-lit curtain or scroll CMS.

03   The CMS is a traffic control device at all times regardless of the type of message being displayed. Accordingly, the limitations on design, format, and manner of display of a message conveyed on a conventional sign apply to CMS regardless of the type of message being displayed at any given time. Some of the general provisions regarding traffic control devices are reiterated in this Chapter. However, this Chapter is not an independent or stand-alone reference for CMS. Users of CMS are expected to consult the other chapters in this Manual for criteria on how to develop effective messages that comply with this Manual and that meet the expectancy and limitations of the road user. In this regard, the engineering processes applied to decisions about whether to use a particular sign, for example, are no different for the decisions about the type and content of the message under consideration for display on a CMS. The other limited-use messages allowed on CMS as provided for in this Chapter likewise fall under the same MUTCD provisions as the primary-use traffic operation regulatory, warning, and guidance messages except as stated otherwise in this Chapter.

04   CMS messaging can be subject to habituation, a phenomenon by which repeated exposure to a stimulus results in diminished response. CMS habituation can occur through repeated exposure to messages, especially those messages that might not be perceived as having relevance to the road user, resulting in diminished responsiveness of the road user to that message. Because messages can be changed or extinguished, the effectiveness of CMS is tied more to the messages displayed thereon, the frequency of displayed messages, and the relevance to the road user, rather than to the installation of the signs themselves.

Guidance:

05   *Changeable message signs should be used judiciously to avoid habituation and preserve their effectiveness during the display of real-time messages about traffic conditions or traffic advisories.*

Standard:

06   **The design of legends for non-electronic display CMS shall comply with the provisions of Chapters 2A through 2K, 2M, and 2N of this Manual. Other CMS shall comply with the design and application principles established in this Chapter, Chapter 2A, and provisions elsewhere in this Manual for specific signs.**

07   **No items other than inventory or maintenance-related information (see Section 2A.04) shall be displayed on the front or back of a CMS or portable CMS. Names or logos of the manufacturer, brand, or model shall not be displayed on a CMS or portable CMS, either in the message display itself or on the exterior housing.**

Guidance:

08   *Blank-out signs that display only single-phase, predetermined electronic-display legends that are limited by their composition and arrangement of pixels or other illuminated forms in a fixed arrangement (such as a blank-out sign indicating a part-time turn prohibition, a blank-out or changeable lane-use sign, or a changeable OPEN/CLOSED sign for a weigh station) should comply with the provisions of the applicable Section for the specific type of sign, provided that the letter forms, symbols, and other legend elements are duplicates of the conventional messages as detailed in the "Standard Highway Signs" publication (see Section 1A.05). Because such a sign is effectively an illuminated version of a conventional sign, the size of its legend elements, the overall size of the sign, and the placement of the sign should comply with the applicable provisions for the conventional version of the sign.*

Section 2L.02 Applications of Changeable Message Signs
Standard:

01   **CMS shall display only traffic operational, regulatory, warning, and guidance information except as otherwise provided in this Chapter. Advertising or other messages not related to traffic control shall not be displayed on a CMS or on its supports or other equipment.**

Option:

02   CMS may display traffic safety campaign messages (see Section 2L.07), transportation-related messages, emergency homeland security messages, and America's Missing: Broadcast Emergency Response (AMBER) alert messages, all as provided for in this Chapter.

03   Transportation-related messages for the purpose of improving traffic conditions, such as those providing information on alternative means of transportation, electronic toll collection, or carpooling may be displayed to remind or inform drivers of relevant options or opportunities for transportation.

Support:

04   Messages regarding broader transportation items not related to improving traffic conditions, such as reminders of driver's license or vehicle registration renewal, vehicle recall information, and vehicle maintenance, do not meet the purpose of a transportation-related message.

05   Examples of transportation-related messages include "STADIUM EVENT SUNDAY, DELAYS NOON TO 4 PM" and "OZONE ALERT—USE TRANSIT."

Guidance:

06   *A CMS should not be used to display a transportation-related message if doing so could adversely affect respect for the sign. "CONGESTION AHEAD" or other overly simplistic or vague messages should not be displayed alone. These messages should be supplemented with a message on the location or distance to the congestion or incident, delay and travel time, alternative route, or other similar messages.*

07   *CMS should not be used in place of conventional signs for conditions that do not change, except for blankout type signs used to display regulatory, warning, and guidance information that routinely reoccurs, but only on a part-time basis. Similarly, when only certain elements of a message on a non-changeable sign are subject to change, only those elements of the sign should be in an electronic display, for example the prices shown on the R3-48 and R3-48a signs (see Figure 2G-18).*

Support:

08   The purpose of CMS is to provide real-time traffic regulatory, warning, or guidance messages as follows:

    A. Incident management and route diversion;
    B. Warning of adverse roadway travel conditions due to weather;
    C. Special event applications associated with traffic control or conditions;
    D. Lane, ramp, and roadway control;
    E. Priced or other types of managed lanes;
    F. Travel times;
    G. Warning situations;
    H. Traffic regulations;
    I. Speed control or warning;
    J. Variable destination guidance;
    K. Supporting temporary traffic control; or
    L. Active Traffic Management.

09   CMS provide significant flexibility and capability in communicating many types of real-time traffic control messages to road users. While their intended purpose is the display of traffic regulatory, warning, or guidance information, other limited uses are also allowed under certain conditions, as provided in this Chapter. Their integrity as an official traffic control device rests significantly on their judicious use and proper messaging format and content, regardless of the message type being displayed.

Standard:

10   **State and local highway agencies that have permanently-installed or positioned CMS shall issue and maintain a policy regarding the use and display of all types of messages to be used on their CMS. The policy shall define the types of messages that will be allowed, the priority of messages, the proper syntax of messages, the timing of messages, and other important messaging elements to ensure messages displayed meet the basic principles that govern the design and use of traffic control devices in general (see Section 1D.01) and traffic signs in particular as provided for in this Manual.**

Guidance:

11   *State and local agencies that use CMS, but do not have permanently-installed or positioned signs, should develop and establish a policy as discussed in Paragraph 10 of this Section.*

12   *When CMS are used at multiple locations to address a specific situation, the message displays should be consistent along the roadway corridor and adjacent corridors, which might necessitate coordination among different operating agencies.*

13   *AMBER alerts (see Paragraph 2 of this Section), when displayed, should not preempt messages related to traffic or travel conditions. AMBER alert messages should be kept as brief as possible and, when possible, direct road users to another source, such as broadcast or highway advisory radio, for detailed information about the alert.*

Standard:

14   **Types of "alert" messages other than AMBER alerts that are unrelated to traffic or travel conditions shall not be displayed on CMS.**

15   **The format of CMS displays shall not be of a type that could be considered similar to advertising or promotional displays.**

Support:

16   In times of a declared state of emergency, it might be appropriate to display messages related to evacuation, homeland security, or emergency information. Traffic patterns, movement, or other situations might be atypical due to the emergency, necessitating unique messaging not specifically related to traffic conditions.

Standard:

17   **Homeland security and emergency messages shall only be displayed in declared states of emergency when there is an imminent threat to the general population. Generic security or personal safety messages shall not be displayed when there is no context of a declared state of emergency or known imminent national security threat. Homeland security and emergency messages shall not be promotional or advisory in nature, including the message design, layout, or manner of display.**

Guidance:

18   *Homeland Security and emergency messages should undergo significant levels of scrutiny prior to being approved for broadcast to ensure accuracy and consistency with emergency conditions. These messages should be designed to convey a clear and simple meaning in a similar format to traffic control messages.*

Support:

19   Section 2B.21 contains information regarding the design of CMS that are used to display variable speed limits that change based on ambient or operational conditions on the variable Speed Limit (R2-1) sign.

20   Section 2C.13 contains information regarding the design of CMS that are used to display the speed at which approaching vehicles are traveling on the Vehicle Speed Feedback (W13-20 and W13-20aP) sign and plaque.

21   Section 2H.04 contains information regarding the design of CMS that are used to display variable speeds for traffic signal progression on the Traffic Signal Speed (I1-1) sign.

22   Section 5B.01 contains provisions for LEDs used in electronic-display signs to accommodate driving automation systems.

Section 2L.03 Legibility and Visibility of Changeable Message Signs

Support:

01   The maximum distance at which a driver can first correctly identify letters and words on a sign is called the legibility distance of the sign. Legibility distance is affected by the characteristics of the sign design and the visual capabilities of drivers. Visual capabilities, and thus legibility distances, vary among drivers.

02   For the more common types of CMS, the longest measured legibility distances on sunny days occur during mid-day when the sun is overhead. Legibility distances are much shorter when the sun is behind the sign face, when the sun is on the horizon and shining on the sign face, or at night.

03   Visibility is the characteristic that enables a CMS to be seen. Visibility is associated with the point where the CMS is first detected, whereas legibility is the point where the message on the CMS can be read. Environmental conditions such as rain, fog, and snow impact the visibility of CMS and can reduce the available legibility distances. During these conditions, there might not be enough viewing time for drivers to read the message.

Guidance:

04   *CMS used on roadways with speed limits of 55 mph or higher should be visible from ½ mile under both day and night conditions. The message should be designed to be legible from a minimum distance of 600 feet for nighttime conditions and 800 feet for normal daylight conditions. When environmental conditions that reduce visibility and legibility are present, or when the legibility distances stated in the previous sentences in this paragraph cannot be practically achieved, messages composed of fewer units of information should be used and consideration should be given to limiting the message to a single phase (see Section 2L.05 for information regarding the lengths of messages displayed on CMS).*

05   *The electronic display of standardized regulatory and warning signs used individually or as part of the legend for a larger sign should meet the size and legend requirements for those specific signs in Chapters 2B and 2C.*

Section 2L.04 Design Characteristics of Messages

Standard:

01   **Except as provided in Paragraph 2 of this Section, messages shall not include animation, flashing, dissolving, exploding, scrolling, or other dynamic display elements.**

02   **When a portable CMS is used as an arrow board that uses a flashing or sequential display for a lane or shoulder closure, the display and operation shall be considered that of an arrow board and shall comply with the provisions of Sections 6L.05 and 6L.06.**

Guidance:

03   *In developing messages for display on CMS, the provisions of Section 1D.01 should be consulted for the principles of an effective traffic control device.*

Standard:

04   **All message displays on CMS, whether for traffic operational, regulatory, warning, or guidance information, or for the other allowable message types as defined in this Chapter, shall follow the same design and display principles found in this Manual used for other traffic control signs, except as provided elsewhere in this Chapter.**

Guidance:

05   *Except in the case of a limited-legend CMS (such as a blank-out or a part-time regulatory sign display) that is used in place of a conventional regulatory sign or an activated blank-out warning sign that supplements a conventional warning sign at a separate location, the signs should be used as a supplement to and not as a substitute for conventional signs and markings unless otherwise provided for in this Manual.*

Support:

06   When CMS are overused for messages not directly associated with real-time driving conditions, road users might pay less attention to the sign, thereby limiting their effectiveness as traffic control devices.

Guidance:

07   *Warning Beacons (see Section 4S.03) should not be installed on CMS, rather CMS should be used predominately to display messages that are critical to real-time travel conditions. CMS word messages should be limited to no more than three lines, with no more than 20 characters per line.*

08   *The spacing between characters in a word should be between 25 and 40 percent of the letter height. The spacing between words in a message should be between 75 and 100 percent of the letter height. Spacing between the message lines should be between 50 and 75 percent of the letter height. Table 2L-1 contains information for spacing between characters, words, and lines of text.*

Table 2L-1. Spacing between Message Characters, Words, and Lines of Text

Height of Letters Used on CMS	Spacing between Characters in Words	Horizontal Spacing between Words	Vertical Spacing between Lines of Text
12	3 - 5	9 - 12	6 - 9
18	4 ½ - 7	13 ½ - 18	9 - 13 ½

Note: All units are in inches

09   *Except as otherwise provided in this Manual, word messages on CMS should be composed of all upper-case letters. The minimum letter height should be 18 inches for CMS on roadways with speed limits of 45 mph or higher. The minimum letter height should be 12 inches for CMS on roadways with speed limits of less than 45 mph. When a message is composed of two phases and higher informational load (see Section 2L.05), the letter height should be 18 inches, regardless of the speed limit, to optimize legibility distance and available viewing time.*

Option:

10   CMS used to replicate a conventional sign may use the character size of the conventional sign being replicated.

Support:

11   Using letter heights of more than 18 inches will not result in proportional increases in legibility distance.

Guidance:

12   *The width-to-height ratio of the sign characters should be between 0.7 and 1.0. The stroke width-to-height ratio should be 0.2.*

Support:

13   The width-to-height ratio is commonly accomplished using a minimum font matrix density of five pixels wide by seven pixels high.

Standard:

14   **CMS shall automatically adjust their brightness under varying light conditions to maintain legibility.**

Guidance:

15   *The luminance design of a CMS should meet industry criteria for daytime and nighttime conditions. Luminance contrast design should be between 8 and 12 for all conditions.*

Support:

16   CMS maintenance and replacement practices might need to account for the reduction of LED luminance and luminance contrast that occurs naturally over time and might substantially impact legibility.

Guidance:

17   *Contrast orientation of CMS should always be positive, that is, with luminous characters on a dark or less-luminous background.*

Support:

18   Legibility distances for negative-contrast CMS are likely to be at least 25 percent shorter than those of positive-contrast messages. In addition, the increased light emitted by negative-contrast CMS has not been shown to improve detection distances and might visually overwhelm the darker characters of the sign legend.

Standard:

19   **The colors used for the legends and backgrounds on CMS shall be as provided in Table 2A-2.**

20   **Except as provided for in Paragraph 21 of this Section, if a black background is used, the color used for the legend on a CMS shall match the background color that would be used on a standard sign for that type of legend as specified in Table 2A-2.**

Option:

21   CMS that use only yellow or amber LEDs may display a yellow or amber legend that does not match the background color used on a standard sign for that type of legend as specified in Table 2A-2.

Standard:

22   **If a green background is used for a guide message on a CMS or if a blue background is used for a motorist services message on a CMS, the background color shall be provided by green or blue lighted pixels such that the entire CMS would be lighted, not just the white legend.**

Support:

23   Some CMS that employ newer technologies have the capability to display a near duplicate of a standard sign or other sign legend using standard symbols, the Standard Alphabets and letter forms, route shields, and other typical sign legend elements with no apparent loss of resolution or recognition to the road user when compared with a conventional version of the same sign legend. Such signs are of the full-matrix type and can typically display full-color legends. Figure 2L-1 shows comparative examples of the effects of varying pixel densities on legend form.

Guidance:

24   *If used, the CMS described in Paragraph 23 of this Section should not display symbols or route shields unless they can do so in the appropriate legend and background color combinations. Where an LED matrix is used to form the changeable legend, signs with pixel spacing greater than 20 millimeters should display only word legends and no symbols or route shields.*

25   *For a single-phase message where the Standard Alphabets and other legend elements of standard designs are used, the lettering style, size, and line spacing should comply with the applicable provisions for the type of message displayed as provided elsewhere in this Manual. For two-phase messages, larger legend heights should be used as described previously in this Section because of the need for such messages to be legible at a greater distance. Regardless of the number of phases, the CMS should comply with the legibility and visibility provisions of Section 2L.03.*

Section 2L.05 Message Length and Units of Information

Guidance:

01   *The maximum length of a message should be dictated by the number of units of information contained in the message, in addition to the size of the CMS. A unit of information, which is a single answer to a single question that a driver can use to make a decision, should not be more than four words.*

Support:

02   In order to illustrate the concept of units of information, Table 2L-2 shows an example message that is comprised of four units of information.

Figure 2L-1. Example of CMS Capability to Display Sign Legends Based on Pixel Pitch

Color full-matrix CMS with pixel pitches of 20 millimeters or less are typically capable of displaying legends nearly identical to conventional sign legends, including route shields and symbols as provided in the MUTCD.

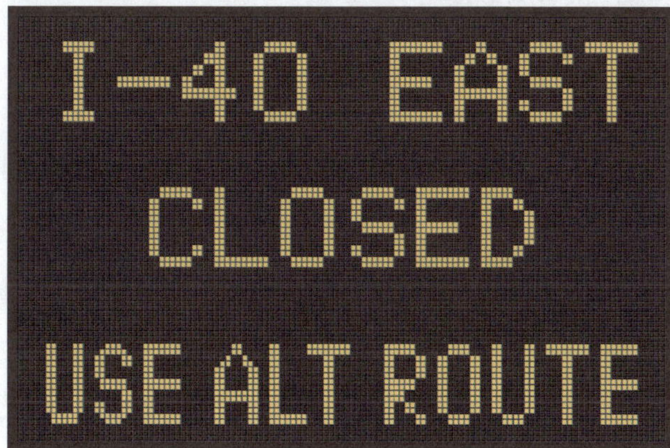

CMS with insufficient pixel density, typically with pixel pitches greater than 20 millimeters – whether full color or monochrome – are generally not capable of adequately displaying conventional sign legends with sufficient clarity and should only display monochrome word messages.

Notes:
1. Pixel pitch is the distance from the center of a pixel to the center of an adjacent pixel.
2. The pixel pitch is described in Metric units because sign manufacturers only use Metric units.

---

03   The maximum allowable number of units of information in a CMS message is based on the principles described in this Section, the current highway operating speed, the legibility characteristics of the CMS, and the lighting conditions.

Standard:

04   **Each message shall consist of no more than two phases. A phase shall consist of no more than three lines of text. Each phase shall be understood by itself, and the meaning of the entire message shall be the same, regardless of the sequence in which the phases are read. Each line of legend shall be centered on the sign. Except for signs located on toll plaza structures or other facilities with a similar booth-lane arrangement, if more than one CMS is visible to road users, then only one sign shall display a sequential message at any given time.**

Table 2L-2. Example of Units of Information

Question	Answer	Number of Information Units
What happened?	MAJOR CRASH	1
Where?	AT EXIT 12	1
Who is the advisory for?	Drivers heading TO NEW YORK	1
What is advised?	USE ROUTE 46	1

Note: The following is an example of a two-phase message that could be developed from the four information units shown in this table:

MAJOR CRASH **AT EXIT 12**	**USE ROUTE 46** **TO NEW YORK**
Phase 1	Phase 2

Option:

05   A legend on a CMS that replicates a legend on a conventional sign that would not normally be center justified may be left justified or right justified as appropriate, such as a travel time or a variable rate toll display.

Standard:

06   **Abbreviations displayed on CMS shall comply with the provisions of Section 1D.08.**

Guidance:

07   *When designing and displaying messages on CMS, the following principles should be used:*

    A. The minimum time that an individual phase is displayed should be based on 1 second per word or 2 seconds per unit of information, whichever produces a lesser value. The display time for a phase should never be less than 2 seconds.

    B. The maximum cycle time of a two-phase message should be 8 seconds.

    C. The duration between the display of two phases should not exceed 0.3 seconds.

    D. No more than three units of information should be displayed in a message phase.

    E. No more than four units of information should be in a message when the traffic operating speeds are 35 mph or more.

    F. No more than five units of information should be in a message when the traffic operating speeds are less than 35 mph.

    G. Only one unit of information should appear on each line of the CMS.

Support:

08   Table 2L-2 provides an example of the number of units of information in a message.

Option:

09   A unit of information consisting of more than one word may be displayed on more than one line. An additional CMS at a downstream location may be used for the purpose of allowing the entire message to be read twice.

10   If more than two phases would be needed to display the necessary information, additional CMS may be used to display this information as a series of two distinct, independent messages with a maximum of two phases at each location, in accordance with the provisions of Paragraph 4 of this Section.

Support:

11   Tables 2L-3 and 2L-4 provide examples of message construction for CMS. Each example shows the message content, layout, and phasing for a potential message and an improved message. The improved message for each example has been optimized for recognition, comprehension, and effectiveness.

Section 2L.06 Travel Time Messages

Support:

01   Travel times provide road users useful information about the level of congestion on segments of highways where motorists experience frequent incidents that slow traffic. Travel times are only helpful to the road user if they have a general understanding of the length of the road segment the travel time is related to so that they can compare that to the time it takes them to travel a similar distance on a highway without congestion. However, travel time messages require road users to read and process a significant amount of information and careful consideration is needed to ensure the overall message is not overloading the motorist.

Guidance:

02   *Travel times should be tied to the distance to a particular destination or junction so that road users can estimate the level of congestion based on the time to travel that distance. When travel times are displayed on CMS, such as during peak traffic conditions, the message should comply with the provisions of Sections 2E.49 and 2E.50. If both a travel time and a distance are displayed, the sign should display only one destination. A distance displayed as part of a travel time message should be rounded to the nearest whole mile.*

Option:

03   When comparative travel time displays are used providing travel times on different routes to one destination, distances to that destination may be eliminated.

04   A reference-location-based exit number (see Section 2E.22) may be displayed in lieu of a destination name or junction thereby providing the necessary distance information to the road user. If reference-location-based exit numbers are displayed, then up to two travel times may be displayed provided that the distance to the exit is not also displayed.

Table 2L-3. Examples of Message Construction for CMS

Example	Phase	Potential Message	Improved Message	Comments
1	1	EXIT 10	EXIT 10 / CLOSED	**Diversionary message:** Each message phase should convey a complete thought independent of the other message phase. The entire message should also make sense regardless of which phase is read first.
	2	CLOSED / USE / EXIT 12	USE / EXIT 12	
2	1	ROADWORK / AHEAD	ROAD WORK / AHEAD	**Advance warning message:** Condensing ROAD and WORK into single word is not necessary since sign width will accommodate the conventional 2-word message. A general CAUTION message is not specific enough to be actionable by the road user. Message should not be repeated to fill sign. Phase 2 of the improved message can be eliminated without any loss of meaning to Phase 1.
	2	CAUTION / CAUTION / CAUTION	FINES / DOUBLE	
3	1	RIGHT / LANE / CLOSED	RIGHT LANE / CLOSED / 1 MILE	**Advance warning message:** Use of single phase message reduces time necessary to read and glances away from the road. Second phase does not provide a complete message.
	2	1 MILE	N/A / Single-Phase Message	
4	1	RT LN CLSD / 1 MI	RIGHT LANE / CLOSED / 1 MILE	**Advance warning message:** Less common abbreviations (see Table 1D-2) are not warranted when the sign can accommodate the full message. Abbreviations in Table 1D-2 should be limited only to portable CMS where the number of characters per line is limited.
	2	N/A / Single-Phase Message	N/A / Single-Phase Message	
5	1	9TH AVENUE / SOUTHWEST / KEEP RIGHT	9TH AVE SW / KEEP RIGHT	**Directional message:** Conventional abbreviations for street name descriptors (see Table 2D-3) are used for consistency with standard signs to improve recognition and reduce the apparent amount of legend.
	2	N/A / Single-Phase Message	N/A / Single-Phase Message	
6	1	EXPWY CONGESTED / USE 101 / FOR AIRPORT	US 19 / CONGESTED	**Diversionary message:** Lack of Expressway route number is vague to unfamiliar road user. Adding exit number for diversion route simplifies message. Diversion message is stated in reverse order and requires more words as a result.
	2	N/A / Single-Phase Message	AIRPORT / USE EXIT 101	
7	1	TRAVEL TIME TO / I-89 / 13 MINUTES	I-89 JCT / 12 MILES / 13 MINS	**Travel time information:** TRAVEL TIME legend is extraneous and out of context for the distance message. Changing only one line of legend between phases compromises recognition of the message.
	2	TRAVEL TIME TO / I-89 / 12 MILES	N/A / Single-Phase Message	
8	1	SEAT BELTS / SAVE LIVES	STATE LAW / FASTEN / SEAT BELTS	**Safety campaign regulatory message:** Slogan-type message does not convey the legal requirement. As an alternative, the STATE LAW legend could be eliminated and the fine for violations displayed on a second phase to convey the regulatory nature of the message.
	2	N/A / Single-Phase Message	N/A / Single-Phase Message	
9	1	DONT TEXT / JUST DRIVE	NO HAND-HELD / PHONE / BY DRIVER	**Regulatory message.** Slogan-type message does not convey the legal requirement. Phase 2 of the improved message can be eliminated without any loss of meaning to Phase 1.
	2	IT CAN WAIT	$250 FINE / AND POINTS	

Note: Examples shown are for single-color CMS with pixel spacing greater than 20 mm and use all upper-case lettering. Multi-color, full-matrix CMS with pixel spacing of 20 mm or less should use upper- and lower-case lettering where appropriate and proper legend and background colors.

Table 2L-4. Examples of Message Construction for Portable CMS*

Example	Phase	Potential Message	Improved Message	Comments
1	1	EXIT 10	EXIT 10 CLOSED	**Diversionary message:** Each phase conveys a complete thought.
	2	CLOSED USE EXIT 12	USE EXIT 12	
2	1	ROADWORK AHEAD	ROAD WORK AHEAD	**Advance warning message:** Condensing ROAD and WORK into a single word is unnecessary because the sign width will accommodate the conventional 2-word phrase. A general CAUTION message is not specific enough to be useful to the road user. Message should not be repeated to fill the sign. Phase 2 of the improved message can be eliminated without any loss of meaning to Phase 1.
	2	CAUTION CAUTION CAUTION	FINES DOUBLE	
3	1	RIGHT LANE CLOSED	RIGHT LN CLOSED 1 MILE	**Advance warning message:** Separation of the message into 2 phases is unnecessary. Second phase does not provide a complete message.
	2	1 MILE	N/A Single-Phase Message	
4	1	RT LN CLSD 1 MI	RIGHT LN CLOSED 1 MILE	**Advance warning message:** Less common abbreviations (see Table 1D-2) are not warranted when the sign can accommodate the full message.
	2	N/A Single-Phase Message	N/A Single-Phase Message	
5	1	9TH AVENUE SW	9 AVE SW KEEP RIGHT	**Directional message:** Conventional abbreviations for street name descriptors (see Table 2D-3) are used for consistency with standard signs to improve recognition and reduce the apparent amount of legend.
	2	KEEP RIGHT	N/A Single-Phase Message	
6	1	ROAD WORK	ROADWORK NEXT 3 MILE	**Advance warning message:** Condensing ROAD and WORK into single word (see Table 1D-2) accommodates a single-phase message.
	2	NEXT 3 MILES	N/A Single-Phase Message	
7	1	SEAT BELTS	FASTEN SEAT BELTS	**Safety campaign regulatory message:** Slogan-type message does not convey the legal requirement. As an alternative, the STATE LAW legend could be eliminated and the fine for violations displayed on a second phase to convey the regulatory nature of the message. Phase 2 of the improved message can be eliminated without any loss of meaning to Phase 1.
	2	SAVE LIVES	STATE LAW	
8	1	DONT TEXT	NO HAND-HELD PHONE	**Regulatory message:** Slogan-type message does not convey the legal requirement.
	2	JUST DRIVE	BY DRIVER	

* Examples shown are for a portable CMS where the display width is generally limited to 8 characters per line of legend.

Section 2L.07 <u>Traffic Safety Campaign Messages</u>

Support:

01   An allowable ancillary use of CMS is the display of traffic safety messages in conjunction with a traffic safety campaign that includes other forms of media as the primary communication and education mechanism.

Standard:

02   **Traffic control messages shall have priority over traffic safety campaign messages.**

Guidance:

03   *When a CMS is used to display a traffic safety campaign, the message should be simple, direct, brief, legible, and clear (see Section 1D.01). Traffic safety campaign messages should be relevant to the road user on the roadway on which the message is displayed. For example, messages regarding school bus stop safety should not be displayed on freeways where school bus stops are not found.*

04   *A CMS should not be used to display a traffic safety campaign message if doing so could adversely affect respect for the sign. Messages with obscure or secondary meanings, such as those with popular culture references, unconventional sign legend syntax, or that are intended to be humorous, should not be used as they might be misunderstood or understood only by a limited segment of road users and require greater time to process and understand. Similarly, slogan-type messages and the display of statistical information should not be used.*

05   *The broad traffic safety campaign marketing message should be appropriately shortened or otherwise modified to comply with the provisions of Section 2L.05 when a traffic safety campaign message is displayed on a CMS.*

06   *Traffic safety campaign messages should emphasize the applicable regulation or warning and should reference any penalties associated with violations of the regulation. Traffic safety campaigns using CMS should include coordinated enforcement efforts where penalties or enforcement type warnings are part of the message displayed on the CMS.*

07   *Traffic safety campaign messages should not be displayed on CMS unless they are part of an active, coordinated safety campaign that uses other media forms as the primary means of outreach. For consistency on a national level, traffic safety campaigns should be coordinated with those on the National Highway Transportation Safety Administration's annual communications calendar.*

Support:

08   Examples of traffic safety campaign messages include "UNBUCKLED SEAT BELTS FINE + POINTS" and "IMPAIRED DRIVERS LOSE LICENSE + JAIL."

Section 2L.08 <u>Permanently-Located Changeable Message Signs</u>

Support:

01   Careful consideration of CMS installation location is important to having a safe and effective message, taking into account several factors. CMS message length and complexity will vary and often include two-phase displays, all of which might require longer glance times by motorists than would be required for conventional sign messages.

02   Permanently-located CMS are generally used on higher-speed, multi-lane facilities with high traffic volumes where more time might be required to properly respond to a message, such as by changing lanes or reducing speed. It also is common for other signs to be in the same vicinity of the desired location for a permanently-located CMS raising the concern of overloading road users with information.

Guidance:

03   *A CMS that is used in place of a conventional sign (such as a blank-out or variable-legend regulatory sign) should be located in accordance with the provisions of Chapter 2A and the provisions for the conventional sign it replaces.*

04   *Permanently-located CMS should:*

    A. *Be located sufficiently upstream of known bottlenecks and high crash locations to enable road users to select an alternate route or take other appropriate action in response to a recurring condition.*

    B. *Be located sufficiently upstream of major diversion decision points, such as interchanges, to provide adequate distance over which road users can change lanes to reach one destination or the other.*

    C. *Not be located within an interchange except for toll plazas or managed lanes.*

    D. *Not be positioned at locations where the information load on drivers is already high because of guide signs and other types of information.*

    E. *Not be located in areas where drivers frequently perform lane-changing maneuvers in response to guide sign information, or because of merging or weaving conditions.*

Support:

05   Many of the factors in locating permanently-located CMS apply to PCMS. Information regarding the design and application of PCMS in temporary traffic control zones is contained in Section 6L.05.

CHAPTER 2M. RECREATIONAL AND CULTURAL INTEREST AREA SIGNS

Section 2M.01 Scope

Support:

01   Recreational or cultural interest areas are attractions or traffic generators that are open to the general public for the purpose of play, amusement, or relaxation. Recreational attractions include such facilities as parks, campgrounds, game-hunting facilities, and ski areas, while examples of cultural attractions include museums, art galleries, and historical buildings or sites.

02   The purpose of recreation and cultural interest area signs is to guide road users to a general area and then to specific facilities or activities within the area.

Option:

03   Recreational and cultural interest area guide signs directing road users to significant traffic generators may be used on freeways and expressways where there is direct access to these areas as provided in Section 2M.09.

04   Recreational and cultural interest area signs may be used off the road network, as appropriate.

Section 2M.02 Application of Recreational and Cultural Interest Area Signs

Support:

01   Provisions for signing recreational or cultural interest areas are subdivided into two different types of signs: (1) symbol signs and (2) destination guide signs.

Guidance:

02   *Highway agencies providing recreational and cultural interest area signing should establish a policy with signing criteria for the eligibility of the various types of services, accommodations, and facilities. These signs should not be used where they might be confused with other traffic control signs.*

Option:

03   Recreational and cultural interest area guide signs may be used in recreational or cultural interest areas for signing non-vehicular events and amenities such as trails, structures, and facilities.

Support:

04   Symbols for use only within recreational and cultural interest area facilities are noted in Table 2M-1.

05   Section 2A.09 contains information regarding the use of recreational and cultural interest area symbols on other types of signs.

Section 2M.03 Regulatory and Warning Signs

Standard:

01   **All regulatory and warning signs installed on roads and streets open to public travel within recreational and cultural interest areas shall comply with the requirements elsewhere in this Manual.**

Section 2M.04 General Design Requirements for Recreational and Cultural Interest Area Symbol Guide Signs

Standard:

01   **When a General Information symbol contained in Chapter 2H (see Figure 2H-1) is used in conjunction with recreational and cultural interest area signing on roadways outside a recreational and cultural interest area facility, the legend and background color of the General Information symbol sign shall be as prescribed in Chapter 2H.**

02   **When a General Service symbol contained in Chapter 2I (see Figure 2I-1) is used in conjunction with recreational and cultural interest area signing on roadways outside a recreational and cultural interest area facility, the legend and background color of the General Service symbol sign shall be as prescribed in Chapter 2I.**

Option:

03   For roadways inside a recreational and cultural interest area, General Information symbol signs and General Service symbol signs may have a white legend on a brown background (see Figures 2H-1 and 2I-1).

Table 2M-1. Category Chart for Recreational and Cultural Interest Area Symbols

General	
Bear Viewing Area	RS-012
Bus Stop *	RS-031
Campfires *	RS-042
Deer Viewing Area	RS-011
Fire Extinguisher *	RS-090
Lighthouse	RS-007
Lookout Tower	RS-006
Nature Study Area	RS-141
Pick-Up Trucks *	RS-140
Recycling *	RS-200
Sea Plane	RS-115
Tunnel *	RS-005
Viewing Area	RS-036

Accommodations	
Men's Restroom *	RS-021
Parking	RS-034
Recreational Vehicle Site *	RS-104
Restrooms *	RS-022
Sleeping Shelter	RS-037
Smoking *	RS-002
Trailer Site *	RS-040
Women's Restroom *	RS-023

Services	
Electrical Hook-Up *	RS-150
First Aid *	RS-024
Laundromat *	RS-085
Picnic Shelter	RS-039
Picnic Site	RS-044
Post Office *	RS-026
Showers *	RS-035
Tramway	RS-071
Trash Dumpster *	RS-091

Land Recreation	
All-Terrain Trail	RS-095
Archery	RS-116
Baseball	RS-096
Climbing	RS-082
Golfing	RS-128
Hiking Trail	RS-068
Horse Trail	RS-064
In-Line Skating *	RS-125
Skateboarding *	RS-098
Spelunking/Caves *	RS-084
Technical Rock Climbing *	RS-081
Tennis	RS-129
Wildlife Viewing	RS-076

Water Recreation	
Beach	RS-145
Boat Ramp	RS-054
Canoeing	RS-079
Fishing Area	RS-063
Hand Launch/Small Boat Launch *	RS-117
Jet Ski/Personal Watercraft	RS-121
Marina *	RS-053
Motorboating	RS-055
Scuba Diving	RS-060
Seal Viewing	RS-106
Swimming	RS-061
Waterskiing	RS-058
Whale Viewing	RS-107

Winter Recreation	
Chair Lift/Ski Lift	RS-105
Cross Country Skiing	RS-046
Dog Sledding	RS-143
Sledding	RS-049
Snow Tubing	RS-144
Snowshoeing	RS-078
Winter Recreational Area	RS-077

*For use only within recreational and cultural interest areas where speed limits are 25 mph or less.

Standard:

04   **Except as provided in Section 2M.09, recreational and cultural interest area symbol guide signs shall be square or rectangular in shape and shall have a white symbol or message and white border on a brown background. The symbols shall be grouped into the following usage and series categories:**

    A. General Applications,
    B. Accommodations,
    C. Services,
    D. Land Recreation,
    E. Water Recreation, and
    F. Winter Recreation.

Support:

05   Table 2M-1 contains a listing of the symbols within each series category.

Option:

06   Mirror images of symbols may be used where the reverse image will better convey the message (see Section 2A.09).

Section 2M.05 Symbol Sign Sizes

Guidance:

01   *Recreational and cultural interest area symbol signs should be 24 x 24 inches. Where greater visibility or emphasis is needed, larger sizes should be used. Symbol sign enlargements should be in 6-inch increments.*

02   *Recreational and cultural interest area symbol signs should be 30 x 30 inches when used on guide signs on freeways or expressways.*

Option:

03   A smaller size of 18 x 18 inches may be used on low-speed, low-volume roadways and on non-road applications.

Section 2M.06 Use of Educational Plaques

Guidance:

01   *Educational plaques should accompany all initial installations of recreational and cultural interest area symbol signs. If used, the educational plaque should be the same width as the symbol sign.*

Option:

02   Symbol signs that are readily recognizable by the public may be installed without educational plaques.

Support:

03   Figure 2M-1 illustrates some examples of the use of educational plaques.

Section 2M.07 Use of Prohibitive Circle and Diagonal for Non-Road Applications

Standard:

01   **Where it is necessary to indicate a prohibition of an activity or an item within a recreational or cultural interest area for non-road use and a standard regulatory sign for such a prohibition is not provided in Chapter 2B, the appropriate recreational and cultural interest area symbol shall be used in combination with a red prohibitive circle and diagonal. The recreational and cultural interest area symbol and the sign border shall be black and the sign background shall be white. The symbol shall be scaled proportionally to fit completely within the circle. The diagonal shall be oriented from the upper left to the lower right portions of the circle as shown in Figure 2M-1 and as detailed in the "Standard Highway Signs" publication.**

02   **Requirements for retroreflection of the red circle and diagonal shall be the same as those requirements for backgrounds, legends, symbols, arrows, and borders.**

Figure 2M-1. Examples of Use of Arrows, Educational Plaques, and Prohibitive Circles and Diagonals

A – Directional signs

B – Directional assemblies

C – Directional assembly with educational plaque

D – Prohibited activities and educational plaque for non-road use*

*Standard regulatory signs shall be used where provided elsewhere in this Manual

Section 2M.08 Placement of Recreational and Cultural Interest Area Symbol Signs

Standard:

01   **If used, recreational and cultural interest area symbol signs shall be placed in accordance with the general requirements contained in Chapter 2A. The symbol(s) shall be placed as sign panels in the uppermost part of the sign and the directional information shall be placed below the symbol(s).**

02   **If the name of the recreational or cultural interest area facility or activity is displayed on a destination guide sign (see Section 2M.09) and a symbol is used, the symbol shall be placed below the name (see Figure 2M-2).**

Option:

03   The symbols displayed with the facility or activity name may be placed below the destination guide sign as illustrated in Figure 2M-2 instead of as sign panels placed with the destination guide sign.

04   Secondary symbols of a smaller size (18 x 18 inches) may be placed beneath the primary symbols (see Drawing A in Figure 2M-1), where needed.

Standard:

05   **Recreational and cultural interest area symbols installed for non-road use shall be placed in accordance with the general sign position requirements of the authority having jurisdiction.**

Support:

06   Figure 2M-3 illustrates typical height and lateral mounting positions. Figure 2M-4 illustrates some examples of the placement of symbol signs within a recreational or cultural interest area. Figures 2M-5 through 2M-10 illustrate some of the symbols that can be used.

Guidance:

07   *The number of symbols used in a single sign assembly should not exceed four.*

Option:

08   The Advance Turn (M5 series) or Directional Arrow (M6 series) auxiliary signs (see Figure 2D-6) with white arrows on brown backgrounds may be used with recreational and cultural interest area symbol guide signs to create a recreational and cultural interest area directional assembly. The symbols may be used singularly, or in groups of two, three, or four on a single sign assembly (see Figures 2M-1, 2M-3, and 2M-4).

Section 2M.09 Destination Guide Signs

Standard:

01   **When recreational or cultural interest area destinations are displayed on a Supplemental guide sign (see Section 2E.51), the sign shall be rectangular in shape with a white legend on a green or brown background.**

Option:

02   Trapezoidal-shaped signs (see Figure 2M-2) may be used to display recreational and cultural interest area destinations on conventional roads.

Standard:

03   **Whenever the trapezoidal shape is used, the color combination shall be a white legend and border on a brown background. When the trapezoidal shape is used for a sign with a directional arrow, a right-angled trapezoid with the wider dimension of the bases (parallel sides) at the top of the sign shall be used. The diagonal leg of the trapezoid shall be oriented in the same direction as the directional arrow. When the trapezoidal shape is used for an advance sign legend, such as with a distance or action message, an isosceles trapezoid with the wider dimension of the bases at the top of the sign shall be used.**

Option:

04   Destination guide signs with a white legend and border on a brown background may be posted at the first point where an access or crossroad intersects a highway where recreational or cultural interest areas are a significant destination along conventional roads, expressways, or freeways. Supplemental guide signs with a white legend and border on a brown background may be used along conventional roads, expressways, or freeways to direct road users to recreational or cultural interest areas. Where access or crossroads lead exclusively to the recreational or cultural interest area, the Advance guide sign (see Section 2E.23) and the Exit Direction sign (see Section 2E.25) may have a white legend and border on a brown background.

Standard:

05   **All Exit Gore (E5-1 series) signs (see Section 2E.26) shall have a white legend and border on a green background. The background color of the interchange Exit Number (E1-5P or E1-5bP) plaque (see Section 2E.22) shall match the background color of the guide sign above which it is mounted. Design characteristics of conventional road, expressway, or freeway guide signs shall comply with Chapter 2D or 2E except as provided in this Section for color combination.**

Figure 2M-2. Examples of Recreational and Cultural Interest Area Guide Signs

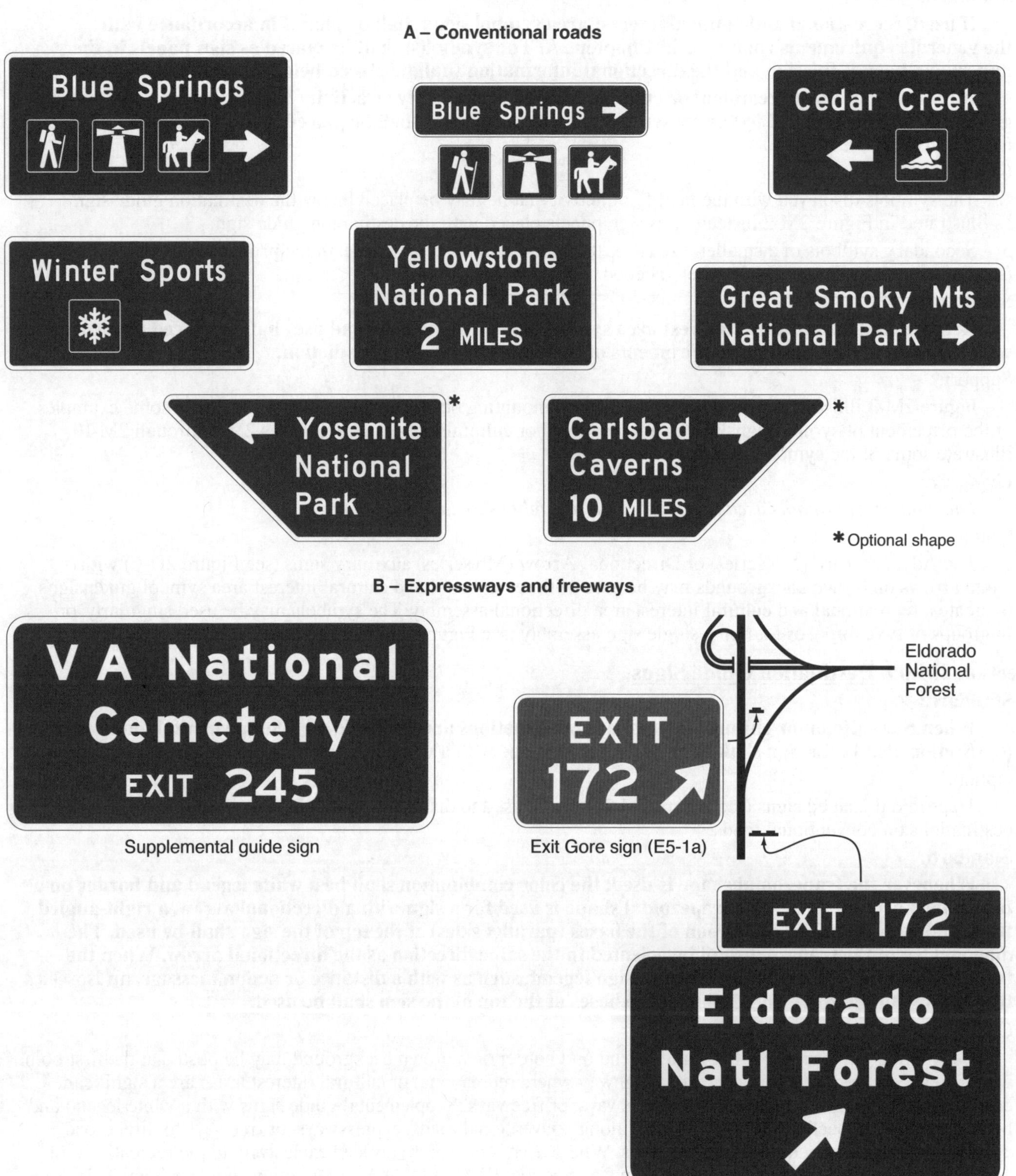

Figure 2M-3. Arrangement, Height, and Lateral Position of Signs Located within Recreational and Cultural Interest Areas

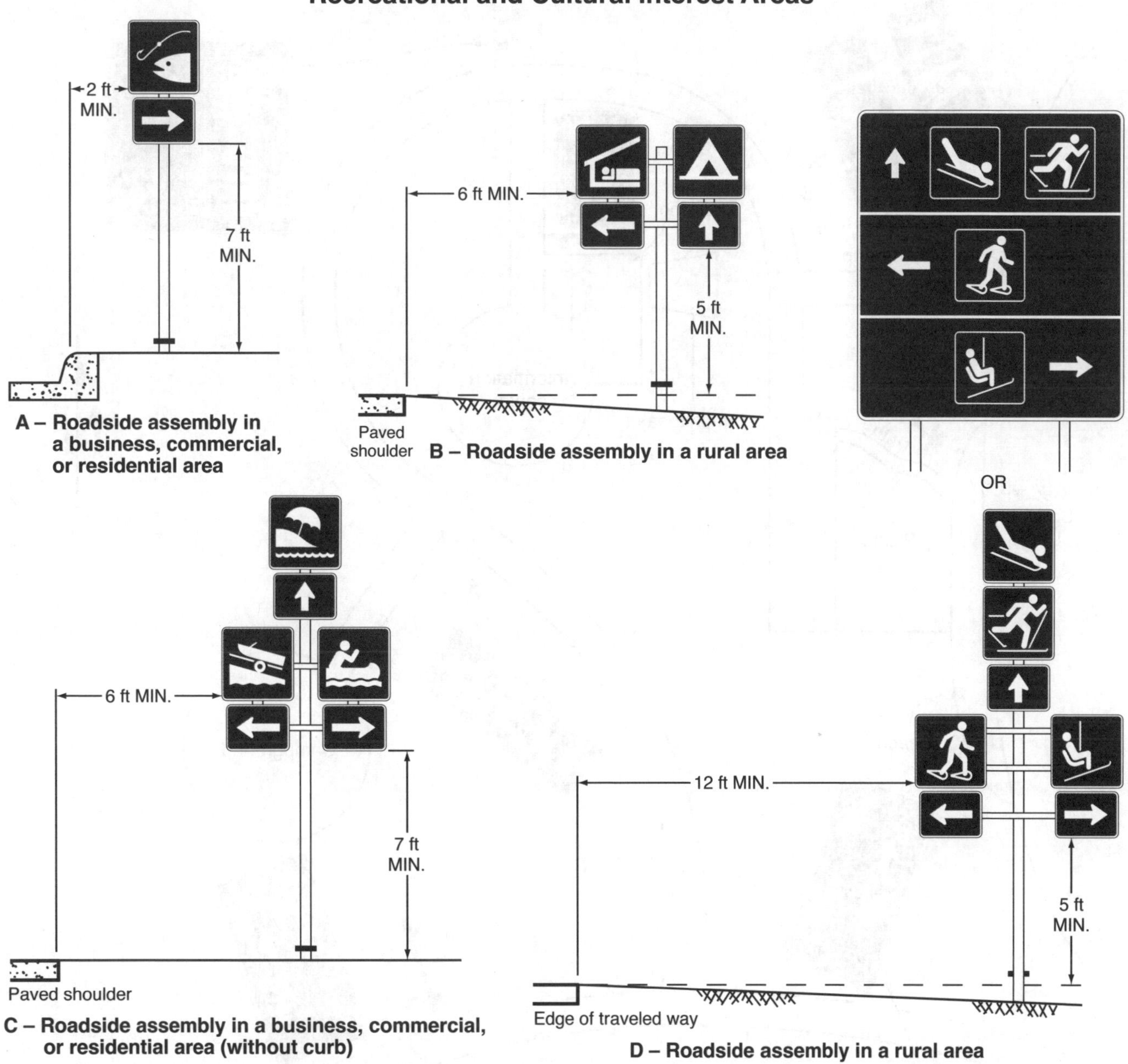

Note: See Section 2A.16 for reduced lateral offset distances that may be used in areas where lateral offsets are limited, and in urban areas where sidewalk width is limited or where existing poles are close to the curb

06   **The Advance guide sign and the Exit Direction sign shall retain the white-on-green color combination where the crossroad also leads to a destination other than a recreational or cultural interest area.**

Support:

07   Figure 2M-2 illustrates destination guide signs commonly used for identifying recreational or cultural interest areas or facilities.

Figure 2M-4. Examples of Symbol and Destination Guide Signing Layout

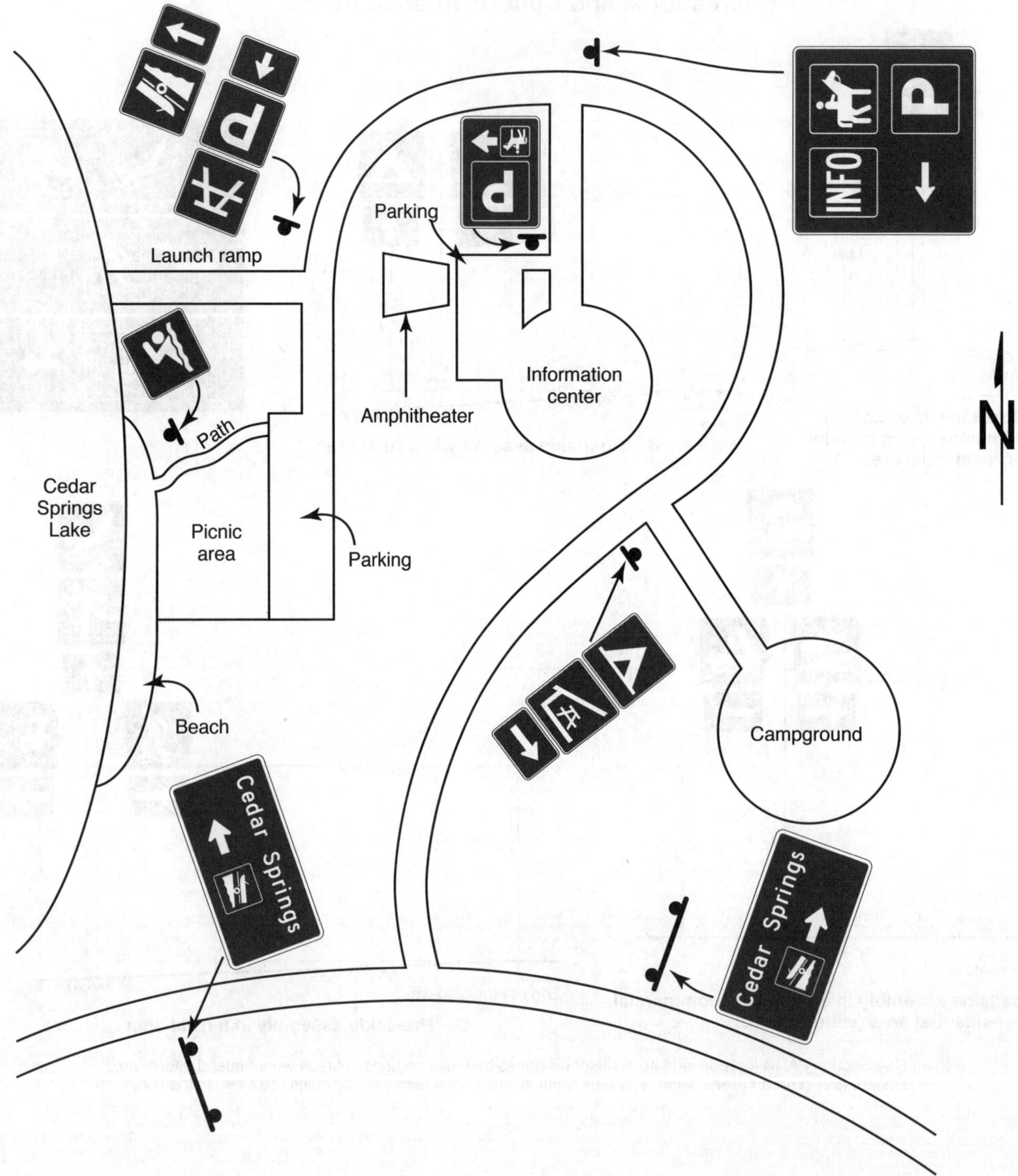

Figure 2M-5. Recreational and Cultural Interest Area Symbol Signs for General Applications

RS-002*
Smoking

RS-005*
Tunnel

RS-006
Lookout Tower

RS-007
Lighthouse

RS-011
Deer Viewing Area

RS-012
Bear Viewing Area

RS-031*
Bus Stop

RS-036
Viewing Area

RS-042*
Campfires

RS-090*
Fire Extinguisher

RS-115
Sea Plane

RS-140*
Pick-Up Trucks

RS-141
Nature Study Area

RS-200*
Recycling

*For use only within recreational and cultural interest areas where speed limits are 25 mph or less.

Figure 2M-6. Recreational and Cultural Interest Area Symbol Signs for Accommodations

RS-021*
Men's Restroom

RS-022*
Restrooms

RS-023*
Women's Restroom

RS-034
Parking

RS-037
Sleeping Shelter

RS-040*
Trailer Site

RS-104*
Recreational Vehicle Site

*For use only within recreational and cultural interest areas where speed limits are 25 mph or less.

Figure 2M-7. Recreational and Cultural Interest Area Symbol Signs for Services

RS-024*
First Aid

RS-026*
Post Office

RS-035*
Showers

RS-039
Picnic Shelter

RS-044
Picnic Site

RS-071
Tramway

RS-085*
Laundromat

RS-091*
Trash Dumpster

RS-150*
Electrical Hook-Up

*For use only within
recreational and cultural
interest areas where speed
limits are 25 mph or less.

Figure 2M-8. Recreational and Cultural Interest Area Symbol Signs for Land Recreation

RS-064
Horse Trail

RS-068
Hiking Trail

RS-076
Wildlife Viewing

RS-081*
Technical Rock Climbing

RS-082
Climbing

RS-084*
Spelunking/Caves

RS-095
All-Terrain Trail

RS-096
Baseball

RS-098*
Skateboarding

RS-116
Archery

RS-125*
In-Line Skating

RS-128
Golfing

RS-129
Tennis

*For use only within recreational and
cultural interest areas where speed
limits are 25 mph or less.

Figure 2M-9. Recreational and Cultural Interest Area Symbol Signs for Water Recreation

RS-053*
Marina

RS-054
Boat Ramp

RS-055
Motorboating

RS-058
Waterskiing

RS-060
Scuba Diving

RS-061
Swimming

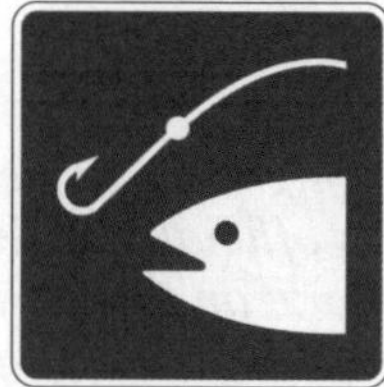

RS-063
Fishing Area

RS-079
Canoeing

RS-106
Seal Viewing

RS-107
Whale Viewing

RS-117*
Hand Launch/
Small Boat Launch

RS-121
Jet Ski/Personal
Watercraft

RS-145
Beach

*For use only within recreational and
cultural interest areas where speed
limits are 25 mph or less.

Figure 2M-10. Recreational and Cultural Interest Area Symbol Signs for Winter Recreation

RS-046
Cross Country Skiing

RS-049
Sledding

RS-077
Winter Recreational Area

RS-078
Snowshoeing

RS-105
Chair Lift/Ski Lift

RS-143
Dog Sledding

RS-144
Snow Tubing

Section 2M.10 <u>Memorial or Dedication Signing</u>

Support:

01   Legislative bodies will occasionally adopt an act or resolution memorializing or dedicating a highway, bridge, or other component of the highway.

02   Named highways (see Section 2D.56) are officially designated and shown on official maps and serve the purpose of providing route guidance, primarily on unnumbered highways, and property addresses. A highway designated as a memorial or dedication is not considered to be a named highway for the purposes of highway signing or road user navigation and orientation.

03   Section 2A.20 contains information regarding excessive use of signs. Because memorial or dedication names are not official highway names, memorial and dedication signing is not essential to providing navigational guidance.

Guidance:

04   *Such memorial or dedication names should not appear on or along a highway, or be placed on bridges or other highway components. If a route, bridge, or highway component is officially designated as a memorial or dedication, and if notification of the memorial or dedication is to be made on the highway right-of-way, such notification should consist of installing a memorial or dedication marker in a rest area, scenic overlook, recreational area, or other appropriate location where parking is provided with the signing inconspicuously located relative to vehicle operations along the highway.*

05   *Memorial or dedication signs should have a white legend and border on a brown background. On all such signs, the design should be simple and dignified, devoid of any appearance of advertising, and in general compliance with other signing.*

06   *The lettering for the name of the person or entity being recognized should be composed of a combination of lower-case letters with initial upper-case letters.*

Standard:

07   **Where such memorial or dedication signs are installed on the highway mainline because the provisions of Paragraph 4 of this Section cannot be met, (1) memorial or dedication names shall not appear on directional guide signs, (2) memorial or dedication signs shall not interfere with the placement of any other traffic control devices, and (3) memorial or dedication signs shall not compromise the safety or efficiency of traffic flow. The memorial or dedication signing shall be limited to one sign at an appropriate location in each route direction, each as an independent post-mounted sign installation.**

08   **Memorial or dedication signs shall be rectangular in shape. The legend displayed on memorial or dedication signs shall be limited to the name of the person or entity being recognized and a simple message preceding the name, such as "Dedicated to." Additional legend, such as biographical information, shall not be displayed on memorial or dedication signs. Decorative or graphical elements, pictographs, logos, or symbols shall not be displayed on memorial or dedication signs. All letters and numerals displayed on memorial or dedication signs shall be as provided in the "Standard Highway Signs" publication (see Section 1A.05). The route number or officially mapped name of the highway shall not be displayed on the memorial or dedication sign.**

09   **Memorial or dedication signs shall not imply that a highway has been officially renamed.**

10   **Memorial or dedication names shall not appear on supplemental signs or on any other information sign on or along the highway or its intersecting routes.**

Guidance:

11   *Freeways and expressways should not be signed as memorial or dedicated highways.*

12   *When used, memorial or dedication signs should be located in accordance with the provisions for excessive use of signs (see Section 2A.20).*

Support:

13   Paragraph 36 of Section 2D.45 contains provisions regarding the use of memorial or dedication signing in conjunction with Street Name signs.

CHAPTER 2N. EMERGENCY MANAGEMENT SIGNS

Section 2N.01 Emergency Management

Guidance:

01   *Contingency planning for an emergency evacuation should be considered by all State and local jurisdictions and should consider the use of all applicable roadways.*

02   *In the event of a disaster where highways that cannot be used will be closed, a successful contingency plan should account for the following elements: a controlled operation of certain designated highways, the establishment of traffic operations for the expediting of essential traffic, and the provision of emergency centers for civilian aid.*

Section 2N.02 Design and Use of Emergency Management Signs

Standard:

01   **Emergency Management signs (see Figure 2N-1) shall be used to guide and control highway traffic during an emergency.**

Guidance:

02   *During an emergency, permanently-installed regulatory and warning signs that conflict with Emergency Management signs should be removed or covered until such time as the Emergency Management signs are no longer necessary.*

03   *Except for Evacuation Route signs, Emergency Management signs that are no longer necessitated by the emergency should be promptly removed and signs that normally provide regulation, warning, or guidance that were removed or covered during the emergency should be promptly displayed again.*

Standard:

04   **Advance planning for transportation operations emergencies shall be the responsibility of State and local authorities.**

Figure 2N-1. Emergency Management Signs

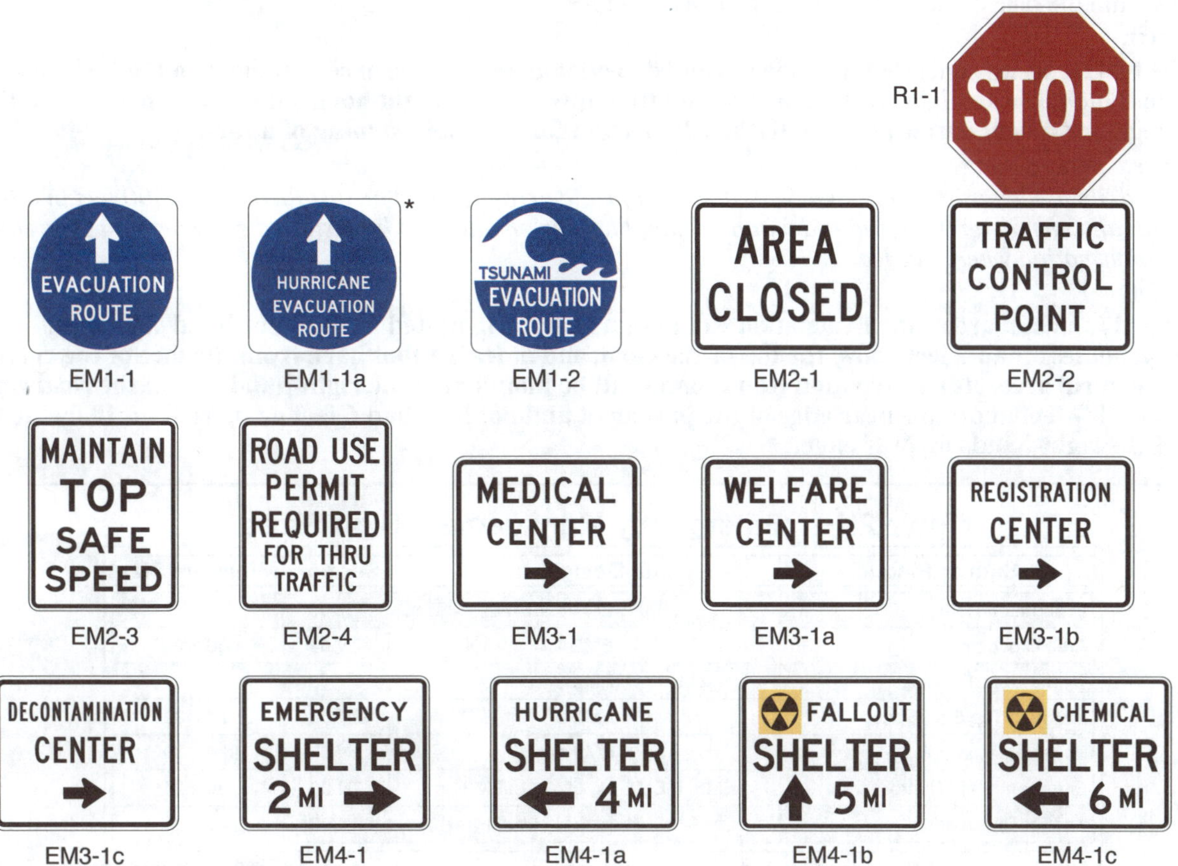

*HURRICANE is an example of one type of evacuation route. Legends for other types may also be used.

Support:

05   The Federal Government provides guidance to the States as necessitated by changing circumstances.

Standard:

06   **Except as provided in Section 2A.07, the sizes for Emergency Management signs shall be as shown in Table 2N-1.**

Support:

07   Section 2A.07 contains information regarding the applicability of the various columns in Table 2N-1.

Option:

08   Signs larger than those shown in Table 2N-1 may be used (see Section 2A.07).

Guidance:

09   *As conditions permit, the Emergency Management signs should be replaced or augmented by standard signs.*

10   *Except where specifically required elsewhere in this Chapter, the background of Emergency Management signs should be retroreflective.*

11   *Because Emergency Management signs might be needed in large numbers for temporary use during an emergency, consideration should be given to their fabrication from any light and economical material that can serve through the emergency period.*

Option:

12   Any Emergency Management sign that is used to mark an area that is contaminated by biological or chemical warfare agents or radioactive fallout may be accompanied by the standard symbol that is illustrated in the upper left corner of the EM4-1b and EM4-1c signs in Figure 2N-1.

Section 2N.03 Evacuation Route Signs (EM1 Series)

Standard:

01   **An Advance Turn Arrow (M5 series) or Directional Arrow (M6 series) auxiliary plaque (see Figure 2D-6) shall be installed below the EM1-2 sign. The Advance Turn Arrow or Directional Arrow auxiliary plaque shall have a white arrow and border on a blue background when used with an EM1-2 sign.**

Option:

02   Where different evacuation conditions use different evacuation routes in the same area, the word HURRICANE, or a word that describes some other type of evacuation route, may be added above the EVACUATION ROUTE legend within the blue circular symbol on the EM1-1a sign.

Standard:

03   **The EM1-1 series signs shall include a white directional arrow. The arrow designs on the EM1-1 series signs shall include a straight, vertical arrow pointing upward, a straight horizontal arrow pointing to the left or right, or a bent arrow pointing to the left or right for advance warning of a turn.**

Guidance:

04   *If used, the Evacuation Route sign, with the appropriate arrow, should be installed 150 to 300 feet in advance of, and at, any turn in an approved evacuation route. The sign should also be installed elsewhere for straight-ahead confirmation where needed.*

Standard

05   **If used in urban areas, the Evacuation Route sign shall be mounted at the right-hand side of the roadway, not less than 7 feet above the top of the curb, and at least 1 foot back from the face of the curb. If used in rural areas, the Evacuation Route sign shall be mounted at the right-hand side of the roadway, not less than 7 feet above the near edge of the pavement and not less than 6 feet or more than 10 feet to the right of the right-hand roadway edge.**

Table 2N-1. Emergency Management Sign Sizes

Sign or Plaque	Sign Designation	Section	Minimum Size
Evacuation Route	EM1-1, EM-1a, EM-1-2	2N.03	24 x 24*
Area Closed	EM2-1	2N.04	30 x 24
Traffic Control Point	EM2-2	2N.05	30 x 24
Maintain Top Safe Speed	EM2-3	2N.06	24 x 30
Permit Required	EM2-4	2N.07	24 x 30
Emergency Aid Center	EM3-1, EM3-1a, EM3-1b, EM3-1c	2N.08	30 x 24
Shelter Directional	EM4-1, EM4-1a, EM4-1b, EM4-1c	2N.09	30 x 24

* A minimum size of 18 x 18 may be used on low-volume roadways or roadways with speeds of 25 mph or less

Notes: 1. Larger signs may be used when appropriate

2. Dimensions are shown as width x height, in inches

06	**Evacuation Route signs shall not be placed where they will conflict with other signs. Where a conflict in placement would occur between the Evacuation Route sign and a standard regulatory sign, the regulatory sign shall take precedence.**

Option:

07	In case of a conflict with guide or warning signs, the Evacuation Route sign may take precedence.

Guidance:

08	*Placement of Evacuation Route signs should be made under the supervision of the officials having jurisdiction over the placement of other traffic signs. Coordination with Emergency Management authorities and agreement between contiguous political entities should occur to assure continuity of routes.*

09	*Use of the specific Evacuation Route (EM1-1a and EM1-2) signs should be limited to areas where different evacuation conditions use different evacuation routes.*

Section 2N.04 AREA CLOSED Sign (EM2-1)

Guidance:

01	*The AREA CLOSED (EM2-1) sign (see Figure 2N-1) should be used to close a roadway in order to prohibit traffic from entering the area. It should be installed on the shoulder as near as practical to the right-hand edge of the roadway, or preferably, on a portable mounting or barricade partly or entirely in the roadway.*

02	*For best visibility, particularly at night, the sign height should not exceed 4 feet measured vertically from the pavement to the bottom of the sign. Unless adequate advance warning signs are used, it should not be placed to create a complete and unavoidable blocked route. Where feasible, the sign should be located at an intersection that provides a detour route.*

Section 2N.05 TRAFFIC CONTROL POINT Sign (EM2-2)

Guidance:

01	*The TRAFFIC CONTROL POINT (EM2-2) sign (see Figure 2N-1) should be used to designate a location where an official traffic control point has been set up to impose such controls as are necessary to limit congestion, expedite emergency traffic, exclude unauthorized vehicles, or protect the public.*

02	*The sign should be installed in the same manner as the AREA CLOSED sign (see Section 2N.04), and at the point where traffic must stop to be checked.*

03	*A STOP (R1-1) sign (see Section 2B.04) should be used in conjunction with the TRAFFIC CONTROL POINT sign.*

04	*The TRAFFIC CONTROL POINT sign should be mounted directly below the STOP sign.*

Section 2N.06 MAINTAIN TOP SAFE SPEED Sign (EM2-3)

Option:

01	The MAINTAIN TOP SAFE SPEED (EM2-3) sign (see Figure 2N-1) may be used on highways where conditions are such that it is prudent to evacuate or traverse an area as quickly as possible.

02	Where an existing Speed Limit (R2-1) sign is in a suitable location, the MAINTAIN TOP SAFE SPEED sign may be mounted directly over the face of the speed limit sign that it supersedes.

Support:

03	Since any speed zoning would be impractical under such emergency conditions, no minimum speed limit can be prescribed by the MAINTAIN TOP SAFE SPEED sign in numerical terms. Where traffic is supervised by a traffic control point, official instructions will usually be given verbally, and the sign will serve as an occasional reminder of the urgent need for maintaining the proper speed.

Guidance:

04	*The sign should be installed as needed, in the same manner as other standard speed signs.*

Standard:

05	**If used in rural areas, the MAINTAIN TOP SAFE SPEED sign shall be mounted on the right-hand side of the road at a horizontal distance of not less than 6 feet or more than 10 feet from the roadway edge, and at a minimum height, measured vertically from the bottom of the sign to the elevation of the near edge of the traveled way, of 5 feet. If used in urban areas, the minimum height, measured vertically from the bottom of the sign to the top of the curb, or in the absence of curb, measured vertically from the bottom of the sign to the elevation of the near edge of the traveled way, shall be 7 feet, and the nearest edge of the sign shall be not less than 1 foot back from the face of the curb.**

Section 2N.07 Permit Required Sign (EM2-4)

Support:

01   The intent of the Permit Required (EM2-4) sign (see Figure 2N-1) is to notify road users of the presence of the traffic control point so that those who do not have priority permits issued by designated authorities can take another route, or turn back, without making a needless trip and without adding to the screening load at the post. Local traffic, without permits, can proceed as far as the traffic control post.

Standard:

02   **If used, the Permit Required (EM2-4) sign shall be used at an intersection that is an entrance to a route on which a traffic control point is located.**

03   **If used, the EM2-4 sign shall be installed in a manner similar to that of the MAINTAIN TOP SAFE SPEED sign (see Section 2N.06).**

Section 2N.08 Emergency Aid Center Signs (EM3-1 Series)

Standard:

01   **In the event of emergency, State and local authorities shall establish various centers for civilian relief, communication, medical service, and similar purposes. To guide the public to such centers a series of directional signs shall be used.**

02   **Emergency Aid Center (EM3-1 series) signs (see Figure 2N-1) shall display the designation of the center and an arrow indicating the direction to the center. They shall be installed as needed, at intersections and elsewhere, on the right-hand side of the roadway, in urban areas at a minimum height, measured vertically from the bottom of the sign to the top of the curb, or in the absence of curb, measured vertically from the bottom of the sign to the elevation of the near edge of the traveled way, of 7 feet, and not less than 1 foot back from the face of the curb, and in rural areas at a minimum height, measured vertically from the bottom of the sign to the elevation of the near edge of the traveled way, of 5 feet, and at a horizontal distance of not less than 6 feet or more than 10 feet from the roadway edge.**

03   **Emergency Aid Center signs shall display one of the following legends, as appropriate, or others designating similar emergency facilities:**

    A. MEDICAL CENTER (EM3-1),
    B. WELFARE CENTER (EM3-1a),
    C. REGISTRATION CENTER (EM3-1b), or
    D. DECONTAMINATION CENTER (EM3-1c).

04   **The Emergency Aid Center sign shall be a horizontally-oriented rectangle. Except as provided in Paragraph 5 of this Section, the Emergency Aid Center signs shall have a black legend and border on a white background.**

Option:

05   When Emergency Aid Center signs are used in an incident situation, such as during the aftermath of a nuclear or biological attack, the background color may be fluorescent pink (see Chapter 6O).

Section 2N.09 Shelter Directional Signs (EM4-1 Series)

Standard:

01   **Shelter Directional (EM4-1 series) signs (see Figure 2N-1) shall be used to direct the public to selected shelters that have been licensed and marked for emergency use.**

02   **The installation of Shelter Directional signs shall comply with established signing standards. Where used, the signs shall not be installed in competition with other necessary highway regulatory, guide, and warning signs.**

03   **The Shelter Directional sign shall be a horizontally-oriented rectangle. Except as provided in Paragraph 4 of this Section, the Shelter Directional signs shall have a black legend and border on a white background.**

Option:

04   When Shelter Directional signs are used in an incident situation, such as during the aftermath of a nuclear or biological attack, the background color may be fluorescent pink (see Chapter 6O).

05   The distance to the shelter may be omitted from the sign when appropriate.

06   Shelter Directional signs may display one of the following legends, or others designating similar emergency facilities:

    A. EMERGENCY (EM4-1),
    B. HURRICANE (EM4-1a),
    C. FALLOUT (EM4-1b), or
    D. CHEMICAL (EM4-1c).

07   If appropriate, the name of the facility may be used.

08   The Shelter Directional signs may be installed on the Interstate Highway System or any other major highway system when it has been determined that a need exists for such signs as part of a State or local shelter plan.

09   The Shelter Directional signs may be used to identify different routes to a shelter to provide for rapid movement of large numbers of persons.

Guidance:

10   *The Shelter Directional sign should be used sparingly and only in conjunction with approved plans of State and local authorities.*

11   *The Shelter Directional sign should not be posted more than 5 miles from a shelter.*

(This page intentionally left blank)

PART 3
MARKINGS

CHAPTER 3A. GENERAL

Section 3A.01 Standardization of Application

Support:

01   Markings are used to supplement other traffic control devices such as signs, signals, and other markings. In other instances, markings are used alone to effectively convey regulations, warnings, or guidance in ways not obtainable by the use of other devices.

02   Markings can take many forms including road surface markings, curb markings, delineators, colored pavements, and channelizing devices.

Standard:

03   **Each standard marking shall be used only to convey the meaning prescribed for that marking in this Manual, including when used for applications not described in this Manual.**

04   **Except as provided in Chapter 3H, markings that must be visible at night shall be retroreflective unless the markings are adequately visible under street or highway lighting. All markings on Interstate highways shall be retroreflective.**

05   **Markings that are no longer applicable for roadway conditions or restrictions and that might cause confusion for the road user shall be removed or obliterated to be unidentifiable as a marking as soon as practicable.**

Option:

06   Until they can be removed or obliterated, markings that are no longer applicable for roadway conditions or restrictions may be temporarily masked with non-reflective, preformed tape that is approximately the same color as the pavement surface.

Section 3A.02 Materials

Guidance:

01   *The materials used for markings should provide the specified color throughout their useful life.*

02   *Consideration should be given to selecting pavement marking materials that will minimize tripping or loss of traction for road users, including pedestrians, bicyclists, and motorcyclists.*

Option:

03   Marking systems that consist of clumps or droplets of material with visible open spaces of bare pavement between the material droplets, which can function in a manner that is similar to the marking systems that completely cover the pavement surface, may be used as pavement markings if they meet the other pavement marking requirements of the highway agency.

Section 3A.03 Colors

Standard:

01   **Markings shall be yellow, white, red, blue, or purple. The colors for markings shall conform to the standard highway colors.**

Option:

02   Black markings may be used in combination with the colors mentioned in Paragraph 1 of this Section to enhance the contrast with a light-colored pavement.

Standard:

03   **When used, yellow markings for longitudinal lines shall delineate:**

    A. The separation of traffic traveling in opposite directions,
    B. The left-hand edge of the roadways of divided highways and one-way streets or ramps, or
    C. The separation of two-way left-turn lanes and reversible lanes from other lanes.

04   **When used, white markings for longitudinal lines shall delineate:**

    A. The separation of traffic flows in the same direction,
    B. The right-hand edge of the roadway, or
    C. Both the right-hand edge and left-hand edge of a reversible roadway.

05 **When used, red raised pavement markers or delineators shall delineate:**

    A. Truck escape ramps, or

    B. One-way roadways, ramps, or travel lanes that shall not be entered or used in the direction from which the markers are visible.

06 **When used, blue markings shall supplement white markings for parking spaces for persons with disabilities.**

07 **When used, purple markings shall be in accordance with the provisions of Chapter 3F to identify toll plaza approach lanes restricted to use only by vehicles with registered electronic toll collection accounts.**

08 **When pavement markings that simulate route signs are used (see Section 3B.22), the colors shall be the same as those that are used for the route signs (see Section 2D.11).**

Support:

09 Provisions regarding colored pavements are contained in Chapter 3H.

Section 3A.04 Functions, Widths, and Patterns of Longitudinal Pavement Markings

Standard:

01 **The general functions of longitudinal lines shall be as follows:**

    A. A double line indicates maximum or special restrictions.

    B. A solid line discourages or prohibits crossing (depending on the specific application).

    C. A broken line indicates a permissive condition.

    D. A dotted lane line provides warning of a downstream change in lane function.

    E. A dotted line used as a lane line or edge line extension guides vehicles through an intersection, a taper area, or an interchange ramp area.

02 **The widths and patterns of longitudinal lines shall be as follows:**

    A. Normal line—4 to 6 inches wide.

    B. Wide line—at least twice the width of a normal line.

    C. Double line—two parallel lines separated by a discernible space. The pavement surface shall be visible between the lines in the same way that it is visible outside the lines, except where contrast markings are used in combination with the double line (see Section 3A.03).

    D. Broken line—normal width line segments separated by gaps.

    E. Dotted line—noticeably shorter line segments separated by shorter gaps than used for a broken line. The width of a dotted line extension shall be at least the same as the width of the line it extends.

Guidance:

03 *To be recognized as a double line rather than two separate, disassociated single lines, the discernible space separating the parallel lines of a double line should not exceed two times the line width of a single line.*

Support:

04 The width of the line indicates the degree of emphasis.

05 Increasing edge line width from 4 inches to 6 inches has been shown to be a beneficial countermeasure to enhance safety at locations with a history of run-off-the-road crashes (see Section 3B.09). Wider normal lines with a 6-inch width instead of the minimum 4-inch width can be beneficial to both human drivers and driving automation systems (see Section 5B.02).

Guidance:

06 *Broken lines should consist of 10-foot line segments and 30-foot gaps, or dimensions in a similar ratio of line segments to gaps as appropriate for traffic speeds and the need for delineation.*

07 *A dotted line used as a lane line (see Section 3B.07) should consist of 3-foot line segments and 9-foot gaps. A dotted line for line extensions within an intersection, taper area, or interchange ramp area (see Section 3B.11) should consist of 2-foot line segments and 2-foot to 6-foot gaps.*

Support:

08 Section 5B.02 contains information on pavement marking considerations for driving automation systems.

Section 3A.05 Maintaining Minimum Pavement Marking Retroreflectivity

Standard:

01 **Except as provided in Paragraph 5 of this Section, a method designed to maintain retroreflectivity at or above 50 mcd/m^2/lx under dry conditions shall be used for longitudinal markings on roadways with speed limits of 35 mph or greater.**

Guidance:

02 *Except as provided in Paragraph 5 of this Section, a method designed to maintain retroreflectivity at or above 100 mcd/m²/lx under dry conditions should be used for longitudinal markings on roadways with speed limits of 70 mph or greater.*

03 *The method used to maintain retroreflectivity should be one or more of those described in "Methods for Maintaining Pavement Marking Retroreflectivity" (FHWA-SA-22-028), 2022 Edition, FHWA or developed from an engineering study based on the values in Paragraphs 1 and 2 of this Section.*

Support:

04 Retroreflectivity levels for pavement markings are measured with an entrance angle of 88.76 degrees and an observation angle of 1.05 degrees. This geometry is also referred to as 30-meter geometry. The units of pavement marking retroreflectivity are reported in mcd/m²/lx, which means millicandelas per square meter per lux.

Option:

05 The following markings may be excluded from the provisions established in Paragraphs 1 and 2 of this Section:

    A. Markings where ambient illumination assures that the markings are adequately visible;
    B. Markings on streets or highways that have an ADT of less than 6,000 vehicles per day;
    C. Dotted extension lines that extend a longitudinal line through an intersection, major driveway, or interchange area (see Section 3B.11);
    D. Curb markings;
    E. Parking space markings; and
    F. Shared-use path markings.

Support:

06 The provisions of this Section do not apply to non-longitudinal pavement markings including, but not limited to, the following:

    A. Transverse markings;
    B. Word, symbol, and arrow markings;
    C. Crosswalk markings; and
    D. Chevron, diagonal, and crosshatch markings.

07 Special circumstances will periodically cause pavement marking retroreflectivity to be below the minimum levels. These circumstances include, but are not limited to, the following:

    A. Isolated locations of abnormal degradation;
    B. Periods preceding imminent resurfacing or reconstruction;
    C. Unanticipated events such as equipment breakdowns, material shortages, and contracting problems; and
    D. Loss of retroreflectivity resulting from snow maintenance operations.

08 When such circumstances occur, compliance with Paragraphs 1 and 2 of this Section is still considered to be achieved if a reasonable course of action is taken to resume maintenance of minimum retroreflectivity in a timely manner according to the maintaining agency's method(s), policies, and procedures.

CHAPTER 3B. PAVEMENT AND CURB MARKINGS

Section 3B.01 Yellow Center Line Pavement Markings

Standard:

01 **Center line pavement markings, when used, shall be the pavement markings used to delineate the separation of traffic lanes that have opposite directions of travel on a roadway and shall be yellow.**

Option:

02 Center line pavement markings may be placed at a location that is not the geometric center of the roadway.

03 On roadways without continuous center line pavement markings, short sections may be marked with center line pavement markings to control the position of traffic at specific locations, such as around curves, over hills, on approaches to grade crossings, at grade crossings, and at bridges.

Standard:

04 **The center line markings on two-lane, two-way roadways shall be one of the following as shown in Figure 3B-1:**

> A. **Two-direction passing zone markings consisting of a normal width broken yellow line where crossing the center line markings for passing with care is permitted for traffic traveling in either direction;**
>
> B. **One-direction no-passing zone markings consisting of a double yellow line, one of which is a normal width broken yellow line and the other is a normal width solid yellow line, where crossing the center line markings for passing with care is permitted for the traffic traveling adjacent to the broken line, but is prohibited for traffic traveling adjacent to the solid line; or**
>
> C. **Two-direction no-passing zone markings consisting of two normal width solid yellow lines where crossing the center line markings for passing is prohibited for traffic traveling in either direction.**

05 **A single solid yellow line shall not be used as a center line marking on a two-way roadway.**

06 **Except where a reversible lane (see Section 3B.04) or a two-way left-turn lane (see Section 3B.05) is present, the center line markings on undivided two-way roadways with four or more lanes for moving motor vehicle traffic always available shall be the two-direction no-passing zone markings consisting of normal width double solid yellow lines as shown in Figure 3B-2.**

Guidance:

07 *Section 3B.11 contains information for application of pavement markings through intersections or interchanges.*

08 *On two-way roadways with three through lanes for moving motor vehicle traffic, two lanes should be designated for traffic in one direction by using one-direction or two-direction no-passing zone markings as shown in Figure 3B-3.*

Figure 3B-1. Yellow Center Lines for Two-Lane, Two-Way Applications

A – Two-lane, two-way marking with passing permitted in both directions

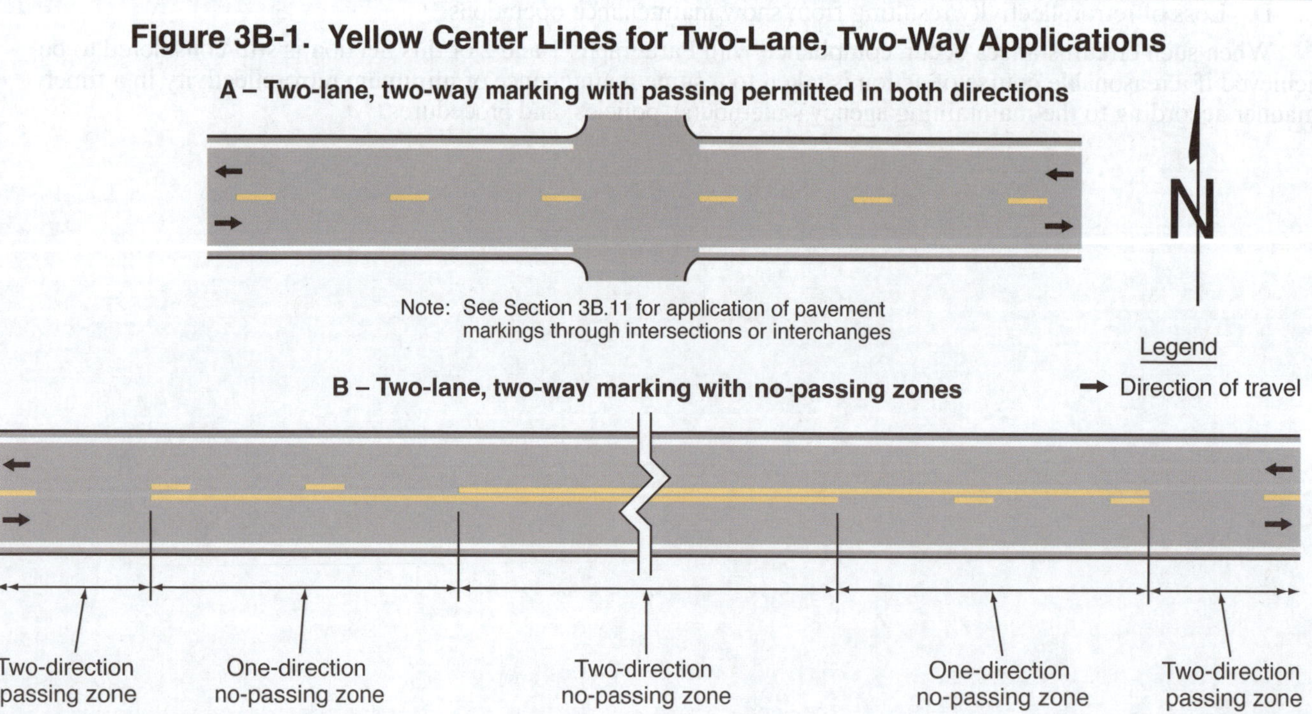

Figure 3B-2. Yellow Center Lines for Four-or-More Lane, Two-Way Applications

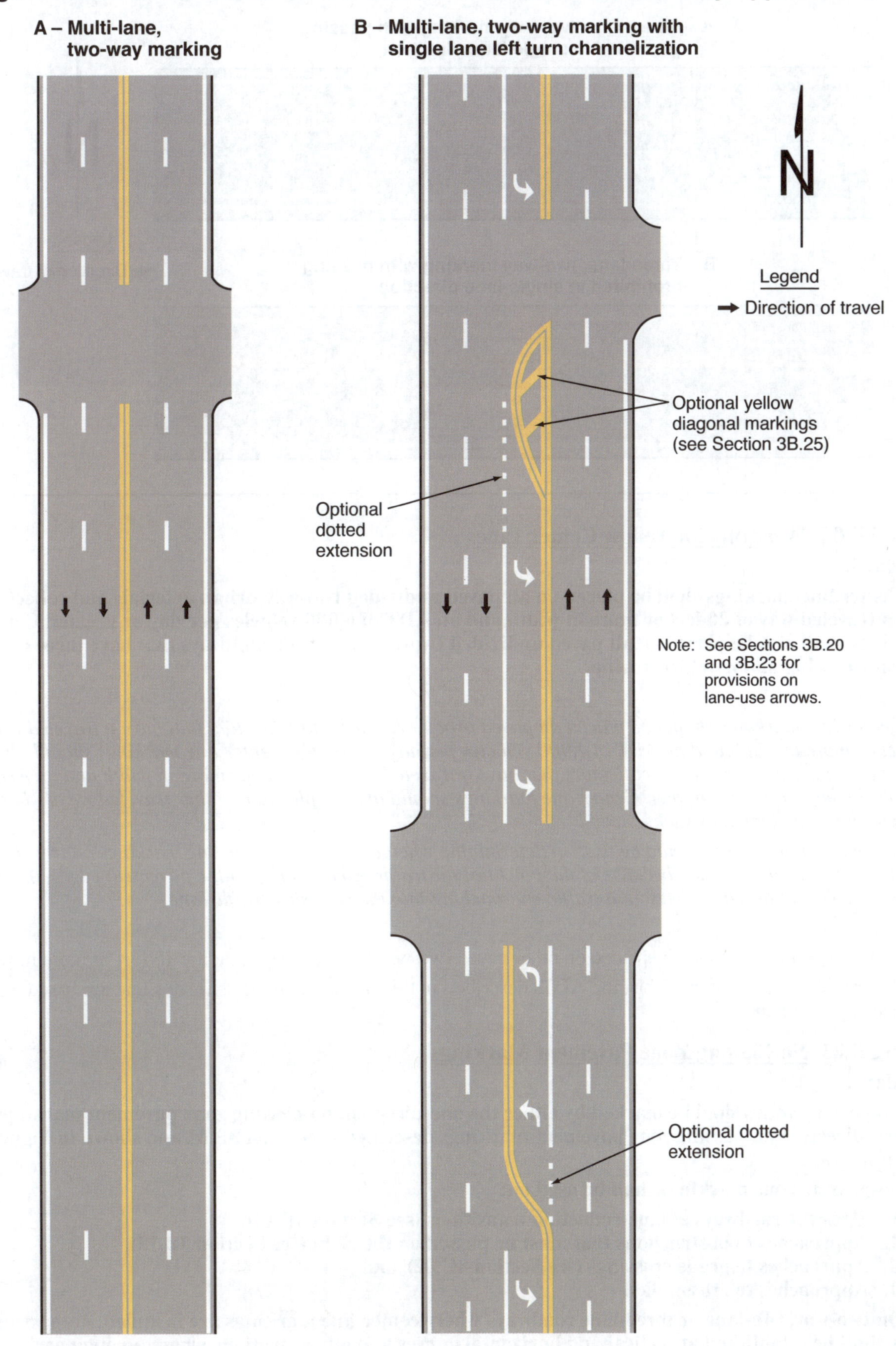

Figure 3B-3. Yellow Center Lines for Three-Lane, Two-Way Applications

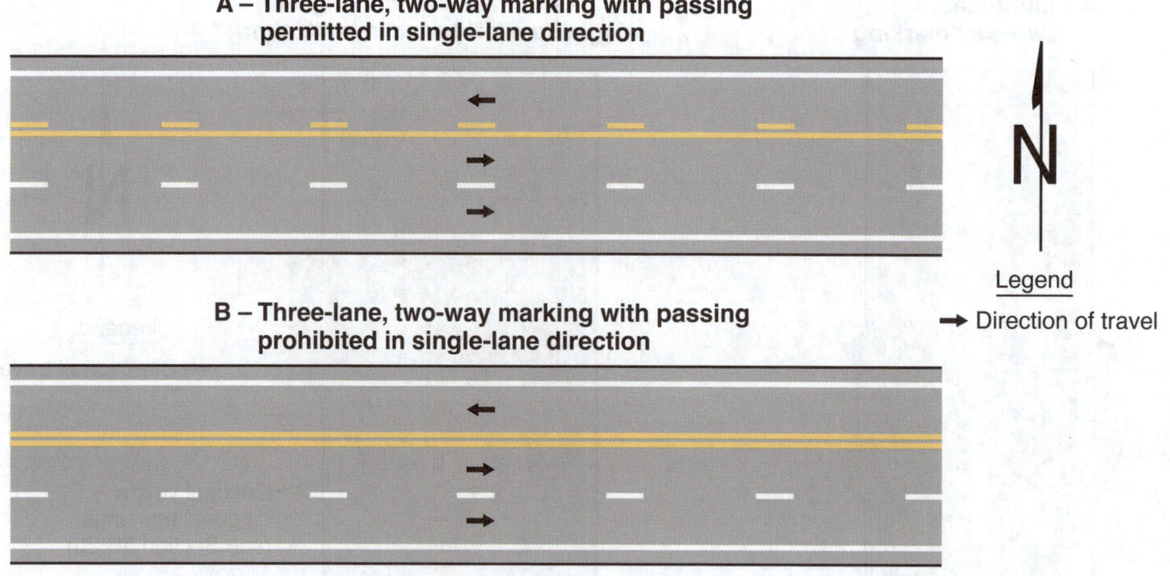

Section 3B.02 Warrants for Yellow Center Lines

Standard:

01   **Center line markings shall be placed on all paved undivided two-way urban arterials and collectors that have a traveled way of 20 feet or more in width and an ADT of 6,000 vehicles per day or greater. Center line markings shall also be placed on all paved undivided two-way streets or highways that have three or more lanes for moving motor vehicle traffic.**

Guidance:

02   *Center line markings should be placed on paved urban arterials and collectors that have a traveled way of 20 feet or more in width and an ADT of 4,000 vehicles per day or greater. Center line markings should also be placed on all rural arterials and collectors that have a traveled way of 18 feet or more in width and an ADT of 3,000 vehicles per day or greater. Center line markings should also be placed on other traveled ways where an engineering study indicates such a need.*

03   *Engineering judgment should be used in determining whether to place center line markings on traveled ways that are less than 16 feet wide because of the potential for traffic encroaching on the pavement edges, traffic being affected by parked vehicles, and traffic encroaching into the opposing traffic lane.*

Option:

04   Center line markings may be placed on other paved two-way traveled ways that are 16 feet or more in width.

05   If a traffic count is not available, the ADTs described in this Section may be estimates that are based on engineering judgment.

Section 3B.03 No-Passing Zone Pavement Markings

Standard:

01   **No-passing zones shall be marked by either the one-direction no-passing zone pavement markings or the two-direction no-passing zone pavement markings described in Section 3B.01 and shown in Figures 3B-1 and 3B-3.**

02   **No-passing zone markings shall be used on:**

   A. Two-way roadways at lane-reduction transitions (see Section 3B.12),
   B. Approaches to obstructions that must be passed on the right (see Section 3B.13),
   C. Approaches to grade crossings (see Section 8C.02), and
   D. Approaches to crosswalks.

03   **On two-way, two-lane or three-lane roadways where center line markings are installed, no-passing zones shall be established at vertical and horizontal curves and other locations where an engineering study indicates that passing must be prohibited because of inadequate sight distances or other special conditions.**

04   **On roadways with center line markings, no-passing zone markings shall be used at horizontal or vertical curves where the passing sight distance is less than the minimum shown in Table 3B-1 for the 85th-percentile speed or the speed limit.**

Support:

05   The passing sight distance on a vertical curve is the distance at which an object 3.5 feet above the pavement surface can be seen from a point 3.5 feet above the pavement (see Figure 3B-4). Similarly, the passing sight distance on a horizontal curve is the distance measured along the center line (or right-hand lane line of a three-lane roadway) between two points 3.5 feet above the pavement on a line tangent to the embankment or other obstruction that cuts off the view on the inside of the curve (see Figure 3B-4).

Table 3B-1. Minimum Passing Sight Distances for No-Passing Zone Markings	
85th-Percentile or Speed Limit	Minimum Passing Sight Distance
25 mph	450 feet
30 mph	500 feet
35 mph	550 feet
40 mph	600 feet
45 mph	700 feet
50 mph	800 feet
55 mph	900 feet
60 mph	1,000 feet
65 mph	1,100 feet
70 mph	1,200 feet

06   The upstream end of a no-passing zone at point "a" in Figure 3B-4 is that point where the sight distance first becomes less than that specified in Table 3B-1. The downstream end of the no-passing zone at point "b" in Figure 3B-4 is that point at which the sight distance again becomes greater than the minimum specified.

Guidance:

07   *Where the distance between successive no-passing zones is less than 400 feet, no-passing zone markings should connect the zones.*

Support:

08   No-passing zone signs (see Sections 2B.36, 2B.37, and 2C.53) are sometimes used to emphasize the existence and extent of a no-passing zone.

Standard:

09   **On three-lane roadways where the direction of travel in the center lane transitions from one direction to the other, a no-passing buffer zone, consisting of a flush median island (see Section 3J.03) at least 50 feet in length, shall be provided in the center lane as shown in Figure 3B-5. A lane-reduction transition (see Section 3B.12) shall be provided approaching each end of the buffer zone.**

Section 3B.04 Yellow Pavement Markings for Reversible Lanes

Standard:

01   **If reversible lanes are used, the lane line pavement markings on each side of reversible lanes shall consist of a normal width broken double yellow line to delineate the edge of a lane in which the direction of travel is reversed from time to time, such that each of these markings serve as the center line markings of the roadway during some period (see Figure 3B-6).**

02   **Signs (see Section 2B.34), lane-use control signals (see Chapter 4T), or both shall be used to supplement reversible lane pavement markings.**

Support:

03   Section 3E.02 contains additional applications of pavement markings for counter-flow preferential lanes that also operate as reversible lanes.

Section 3B.05 Pavement Markings for Two-Way Left-Turn Lanes

Standard:

01   **If a two-way left-turn lane that is never operated as a reversible lane is used, the lane line pavement markings on each side of the two-way left-turn lane shall consist of a normal width broken yellow line and a normal width solid yellow line to delineate the edges of a lane that can be used by traffic in either direction as part of a left-turn maneuver. These markings shall be placed with the broken line toward the two-way left-turn lane and the solid line toward the adjacent traffic lane as shown in Figure 3B-7.**

Guidance:

02   *White two-way left-turn lane-use arrows should be used at or just downstream from the beginning of a two-way left-turn lane.*

Figure 3B-4. Method of Locating and Determining the Limits of No-Passing Zones at Curves

A – No-passing zone at VERTICAL CURVE

Profile View

Note: No-passing zones in opposite directions might or might not overlap, depending on alignment

B – No-passing zone at HORIZONTAL CURVE

Plan View

Note: No-passing zones in opposite directions might or might not overlap, depending on alignment

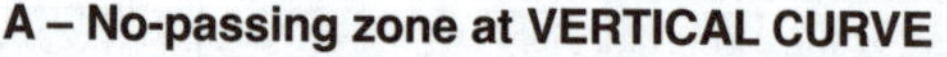

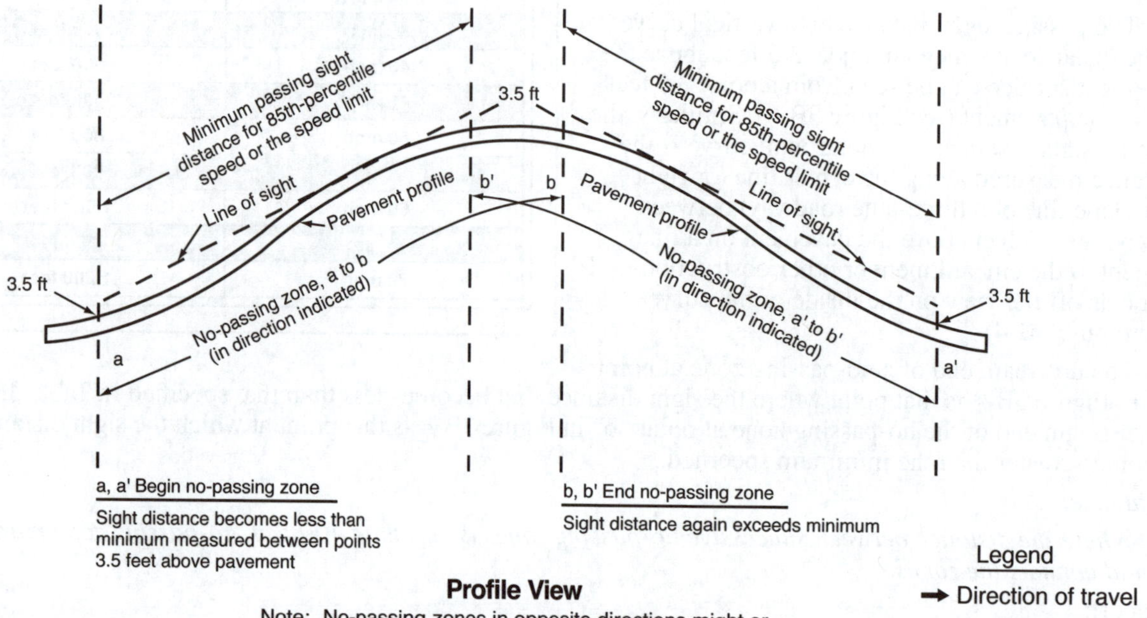

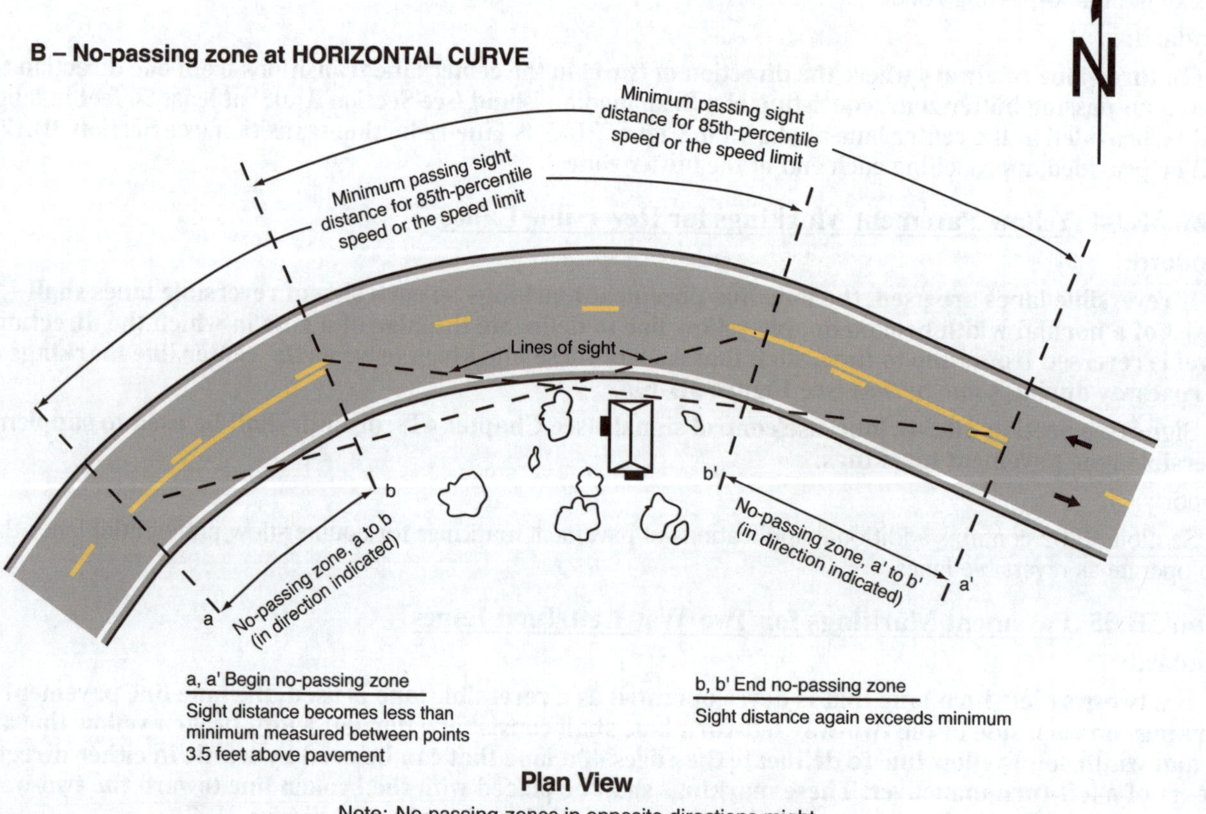

Figure 3B-5. Application of Three-Lane, Two-Way Markings for Changing the Direction of the Center Lane

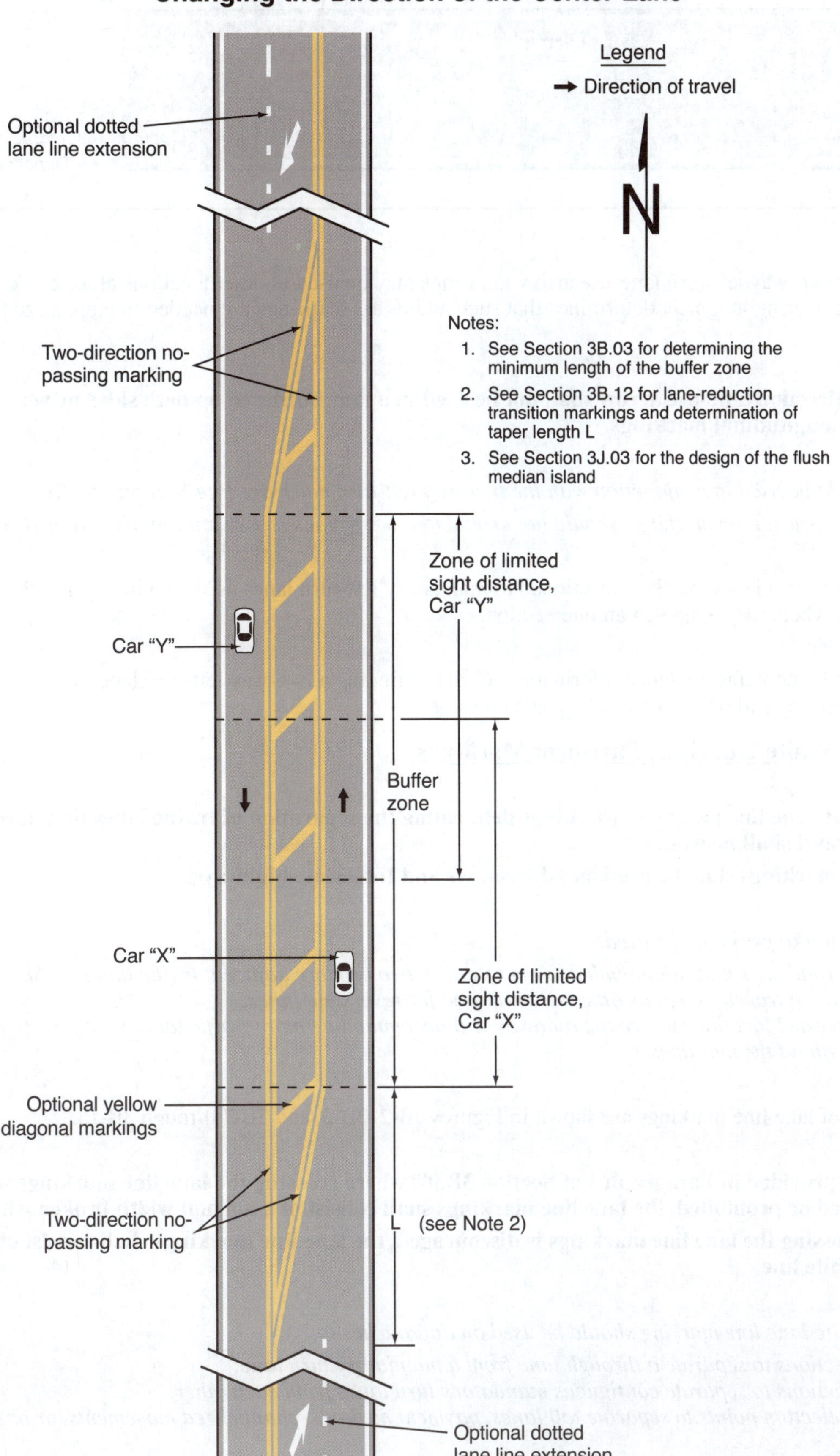

Figure 3B-6. Example of Yellow Pavement Markings for Reversible Lanes

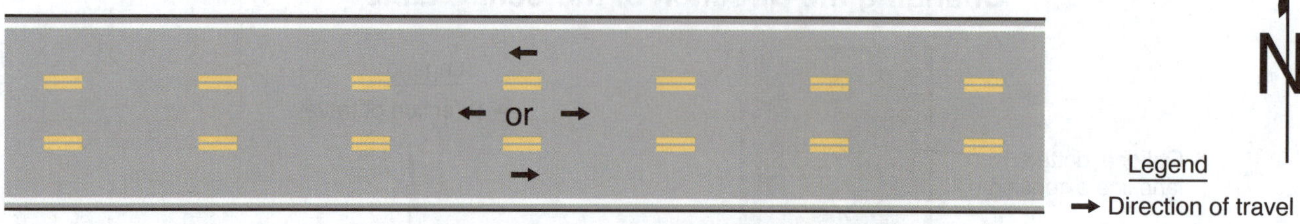

Option:

03   Additional two-way left-turn lane-use arrow markings may be used at other locations along a two-way left-turn lane where engineering judgment determines that such additional markings are needed to emphasize the proper use of the lane.

Standard:

04   **A single-direction lane-use arrow shall not be used in a lane bordered on both sides by yellow two-way left-turn lane longitudinal markings.**

Guidance:

05   *Signs should be used in conjunction with the two-way left-turn markings (see Section 2B.32).*

06   *Two-way left-turn lane markings should not extend to intersections (see definition in Section 1C.02).*

Option:

07   Two-way left-turn lanes may be transitioned to mandatory left-turn lanes as shown in Figure 3B-7 or painted median islands where they approach an intersection.

Support:

08   Section 8A.06 contains guidance information for discontinuing a two-way left-turn lane in the immediate vicinity of a highway-rail or highway-LRT grade crossing.

Section 3B.06 White Lane Line Pavement Markings

Standard:

01   **When used, lane line pavement markings delineating the separation of traffic lanes that have the same direction of travel shall be white.**

02   **Lane line markings shall be used on all freeways and Interstate highways.**

Guidance:

03   *Lane line markings should be used:*

   A. *On all roadways that are intended to operate with two or more adjacent traffic lanes in the same direction of travel, except as otherwise required for reversible lanes.*
   B. *At congested locations where the roadway will accommodate more traffic lanes with lane line markings than without the markings.*

Support:

04   Examples of lane line markings are shown in Figures 3B-2, 3B-3, and 3B-7 through 3B-13.

Standard:

05   **Except as provided in Paragraph 1 of Section 3B.07, where crossing the lane line markings with care is not discouraged or prohibited, the lane line markings shall consist of a normal width broken white line.**

06   **Where crossing the lane line markings is discouraged, the lane line markings shall consist of a normal width solid white line.**

Guidance:

07   *A solid white lane line marking should be used on approaches to:*

   A. *Intersections to separate a through lane from a mandatory turn lane.*
   B. *Intersections to separate contiguous mandatory turn lanes from each other.*
   C. *Toll collection points to separate toll lanes, payment methods, channelized movements, or obstructions.*

Figure 3B-7. Examples of Two-Way Left-Turn Lane Marking Applications

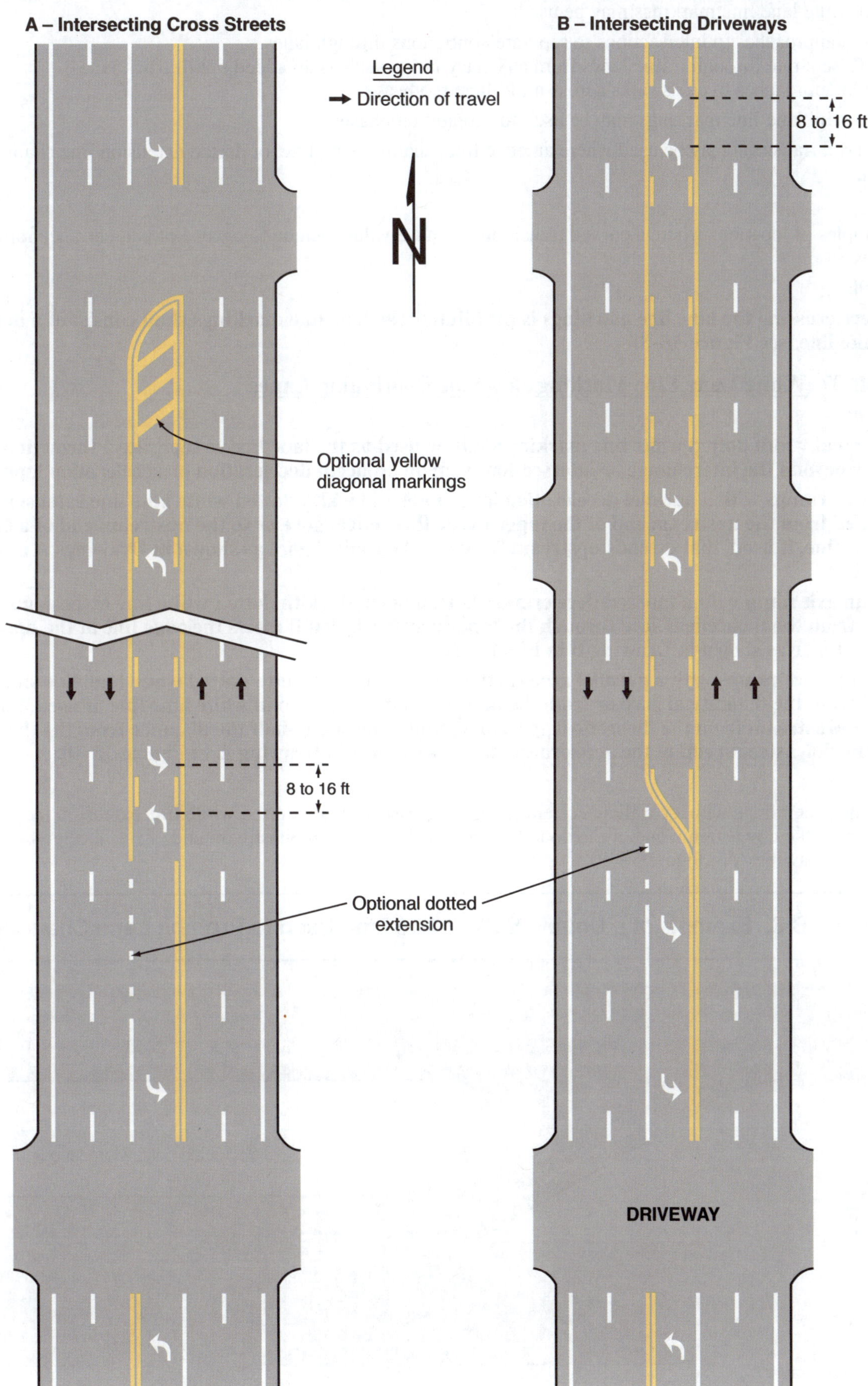

Option:

08   Solid white lane line markings may be used:

    A. On approaches to intersections to separate contiguous through lanes.
    B. To separate through traffic lanes from auxiliary lanes, such as an added uphill truck lane.
    C. On approaches to crosswalks across multi-lane roadways.

09   Wide solid lane line markings may be used for greater emphasis.

10   A curved transition may be used where an edge line, channelizing line, or dotted extension line changes direction.

Support:

11   Examples of locations where a curved transition can have value include freeway exit and entrance ramps, and turn lanes.

Standard:

12   **Where crossing the lane line markings is prohibited, the lane line markings shall consist of a double solid white line (see Figure 3B-8).**

Section 3B.07 White Lane Line Markings for Non-Continuing Lanes

Standard:

01   **A normal width dotted white line marking shall be used as the lane line to separate a through lane that continues beyond the interchange or intersection from an adjacent deceleration or acceleration lane.**

02   **For exit ramps with a parallel deceleration lane, a normal width dotted white lane line extension shall be installed from the upstream end of the taper to the theoretical gore or to the upstream end of a solid white lane line, if used, that extends upstream from the theoretical gore as shown in Drawings A and C in Figure 3B-9.**

03   **For an exit ramp with a tapered deceleration lane, a normal width dotted white line extension shall be installed from the theoretical gore through the taper area such that it meets the edge line at the upstream end of the taper as shown in Drawing B in Figure 3B-9.**

04   **For entrance ramps with a parallel acceleration lane, a normal width dotted white lane line shall be installed from the theoretical gore or from the downstream end of a solid white lane line, if used, that extends downstream from the theoretical gore, to a point at least one-half the distance from the theoretical gore to the downstream end of the acceleration taper, as shown in Drawing A in Figure 3B-10.**

Option:

05   For entrance ramps with a parallel acceleration lane, a normal width dotted white line extension may be installed from the downstream end of the dotted white lane line to the downstream end of the acceleration taper, as shown in Drawing A in Figure 3B-10.

Figure 3B-8. Example of a Double Solid White Line Used to Prohibit Lane Changing

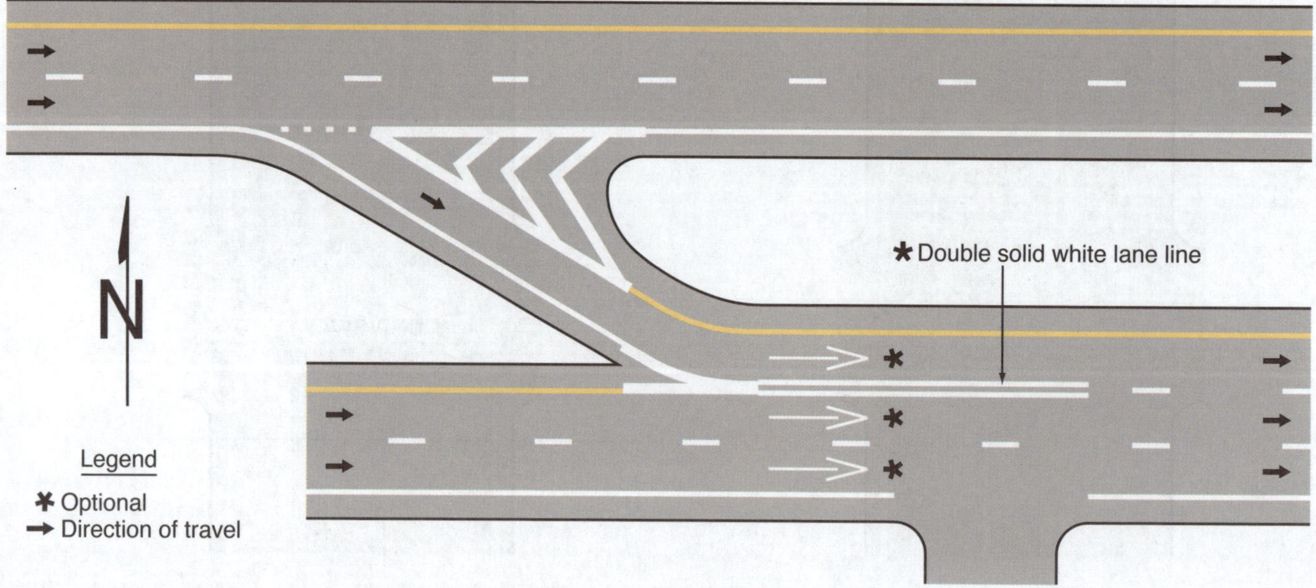

Figure 3B-9. Examples of Dotted Line and Channelizing Line Applications for Exit Ramp Markings (Sheet 1 of 2)

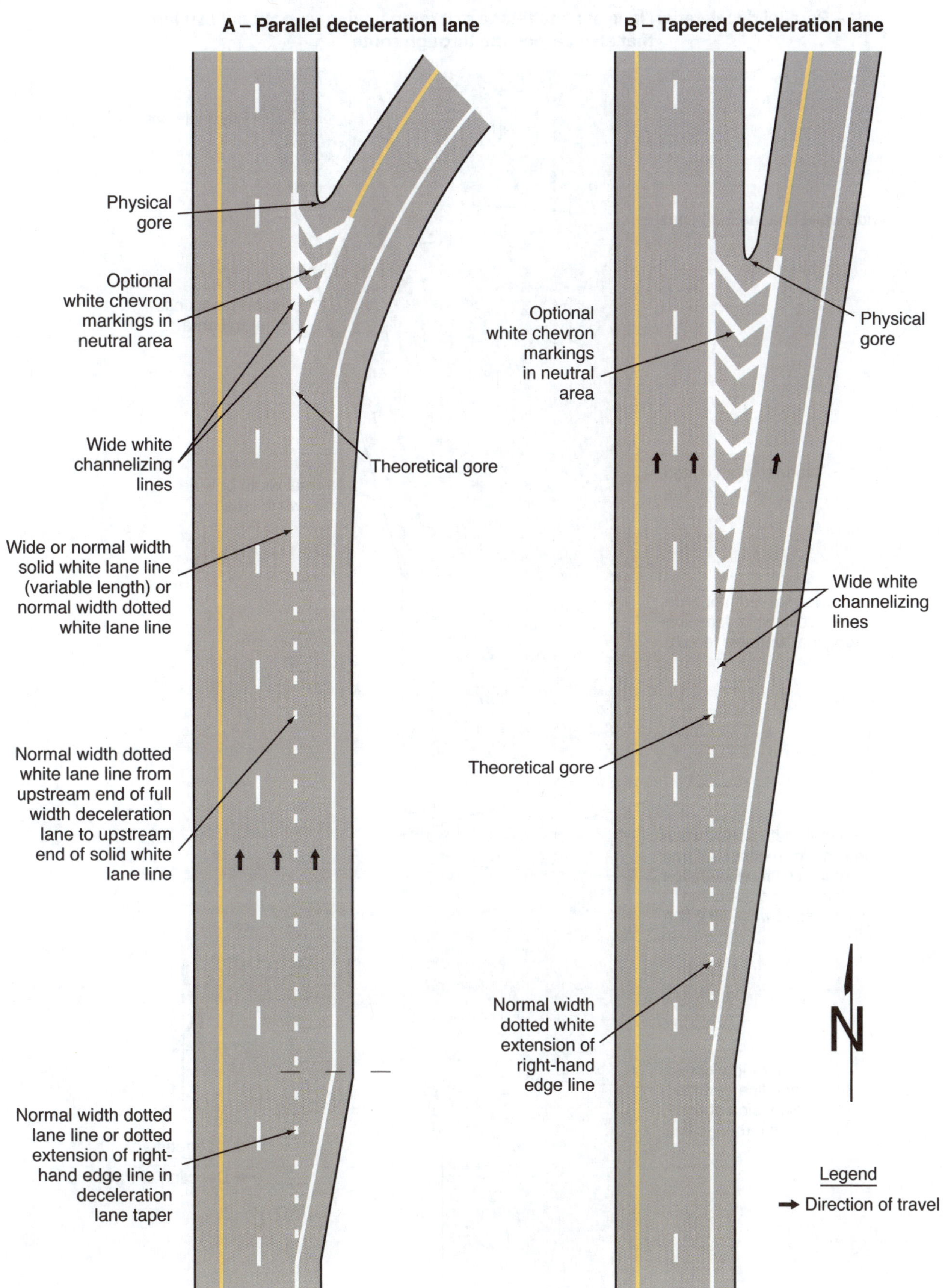

Figure 3B-9. Examples of Dotted Line and Channelizing Line Applications for Exit Ramp Markings (Sheet 2 of 2)

C – Parallel deceleration lane at a multi-lane exit ramp having an optional exit lane that also carries the through route

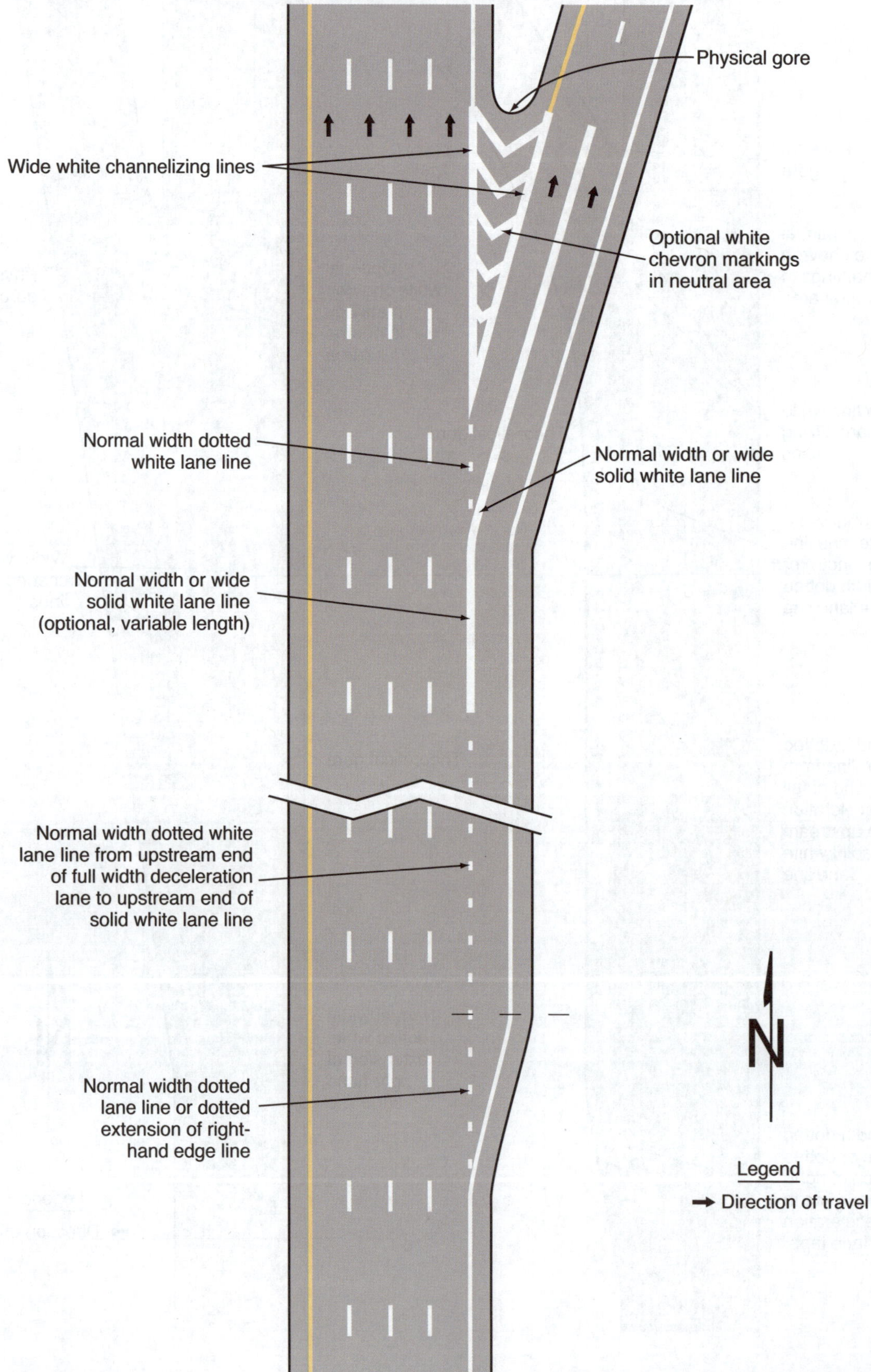

Figure 3B-10. Examples of Dotted Line and Channelizing Line Applications for Entrance Ramp Markings (Sheet 1 of 2)

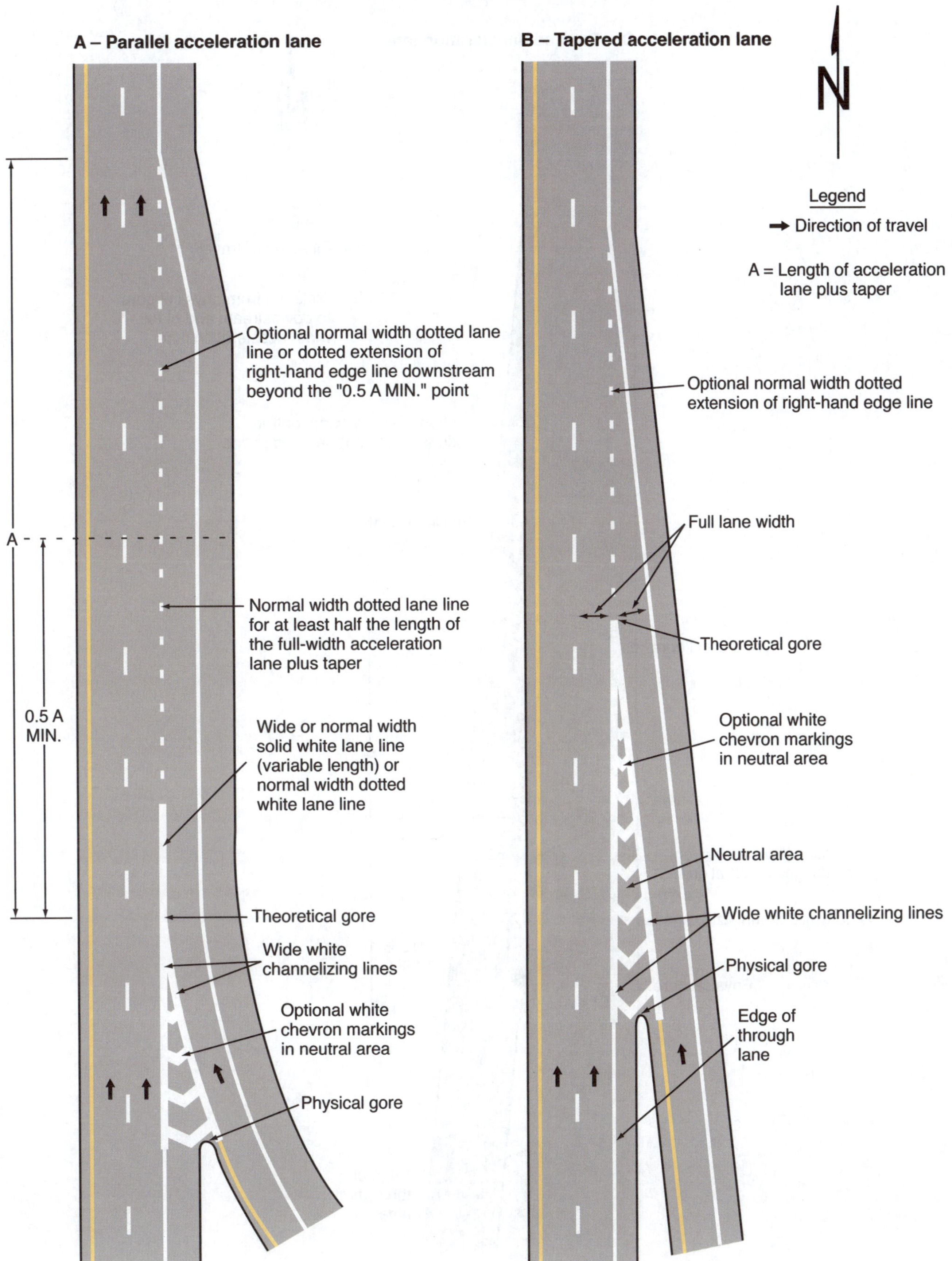

Figure 3B-10. Examples of Dotted Line and Channelizing Line Applications for Entrance Ramp Markings (Sheet 2 of 2)

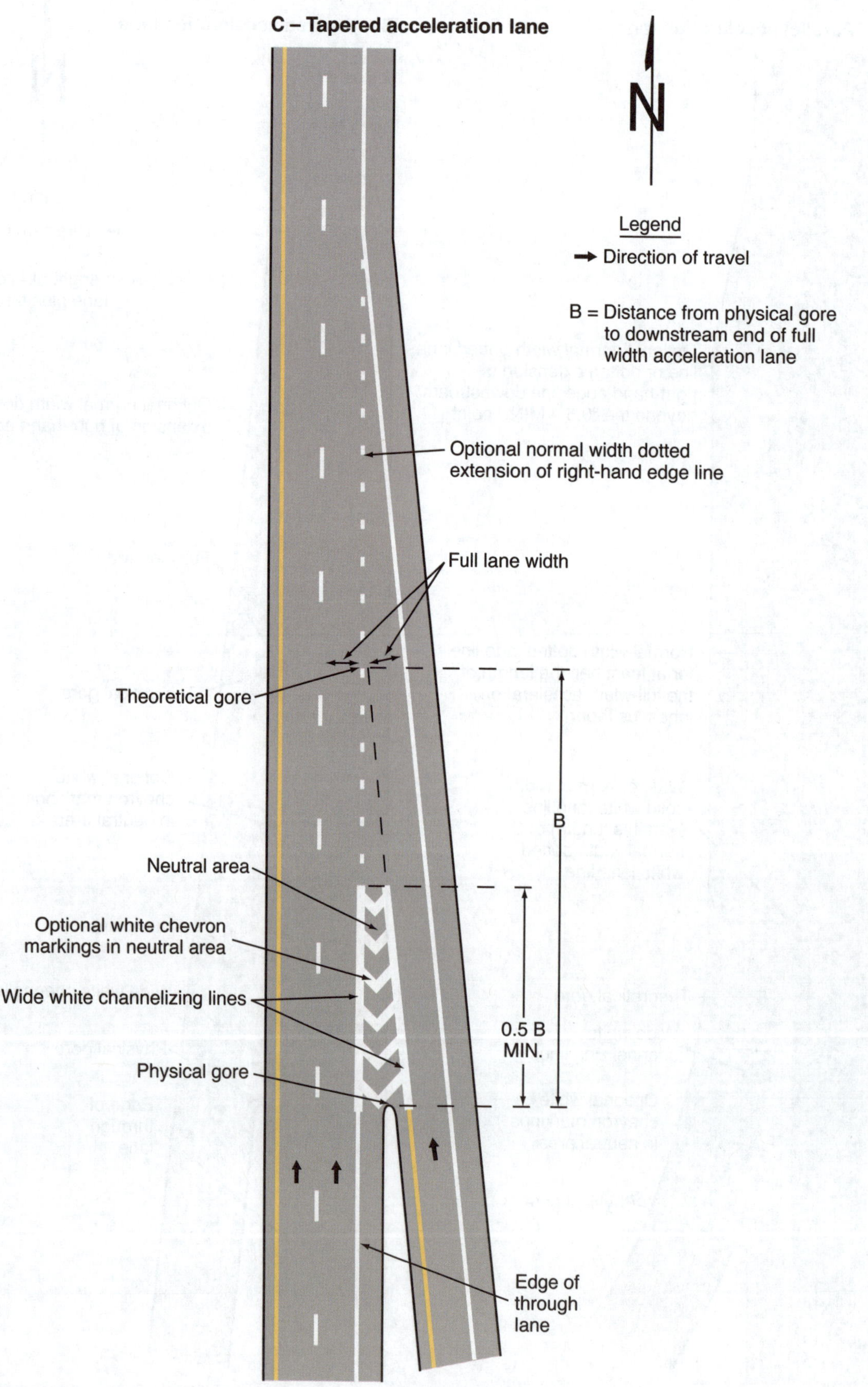

06   For entrance ramps with a tapered acceleration lane, a normal width dotted white line extension may be installed from the downstream end of the channelizing line adjacent to the through lane to the downstream end of the acceleration taper, as shown in Drawings B and C in Figure 3B-10.

Standard:

07   **A wide dotted white lane line shall be used:**

    A. As a lane drop marking in advance of lane drops at exit ramps to distinguish a lane drop from a normal exit ramp (see Drawings A, B, and C in Figure 3B-11),

    B. In advance of freeway route splits with dedicated lanes (see Drawing D in Figure 3B-11),

    C. In advance of freeway route splits with an option lane (see Drawing E in Figure 3B-11),

    D. To separate a through lane that continues beyond an interchange from an adjacent continuous auxiliary lane between an entrance ramp and an exit ramp (see Drawing F in Figure 3B-11),

    E. As a lane drop marking in advance of lane drops at intersections to distinguish a lane drop from an intersection through lane (see Drawing A in Figure 3B-12), and

    F. To separate a through lane that continues beyond an intersection from an adjacent auxiliary lane between two intersections (see Drawing B in Figure 3B-12).

Guidance:

08   *Lane drop markings used in advance of lane drops at freeway and expressway exit ramps should begin at least ½ mile in advance of the theoretical gore.*

09   *On the approach to a multi-lane exit ramp having an optional exit lane that also carries through traffic, lane line markings should be used as illustrated in Drawing B in Figure 3B-11.*

10   *Lane drop markings used in advance of lane drops at intersections should begin a distance in advance of the intersection that is determined by engineering judgment as suitable to enable drivers who do not desire to make the mandatory turn to move out of the lane being dropped prior to reaching the queue of vehicles that are waiting to make the turn. The lane drop markings should begin no closer to the intersection than the most upstream regulatory or warning sign associated with the lane drop.*

11   *The dotted white lane lines that are used for lane drop markings and that are used as a lane line separating through lanes from auxiliary lanes should consist of line segments that are 3 feet in length separated by 9-foot gaps.*

Support:

12   Sections 3B.21 and 3B.23 contain information regarding other markings that are associated with lane drops, such as ONLY word pavement markings and lane-use arrows.

13   Section 3B.12 contains information about the lane line markings that are to be used for transition areas where the number of through lanes is reduced at a location that is not at an interchange or intersection.

Option:

14   In the case of a lane drop at an exit ramp or intersection, a solid white line may replace a portion, but not all of the length, of the wide dotted white lane line.

Section 3B.08 Channelizing Lines

Support:

01   Channelizing lines are used to form neutral areas where traffic traveling in the same general direction is permitted on both sides including entrance and exit ramps, access and egress points to and from managed lanes, toll-plaza bypasses, and left-turn lanes separated from through lanes. Channelizing lines are also sometimes used to alter travel paths for speed management or other purposes.

02   Chapter 3J contains information for the application of channelizing lines used in conjunction with islands.

Standard:

03   **Except as provided in Section 3E.04 and Paragraph 6 of Section 3J.05, a channelizing line shall be a solid wide or double solid white line.**

Support:

04   Examples of channelizing line applications are shown in Figures 3B-9, 3B-10, 3B-11, Drawing C in Figure 3B-15, Figures 3J-1 through 3J-5, and Drawing B in Figure 3J-6.

Standard:

05   **For all exit ramps and for entrance ramps with parallel acceleration lanes, channelizing lines shall be placed on both sides of the neutral area (see Figures 3B-9 and 3B-11 and Drawing A in Figure 3B-10).**

Figure 3B-11. Examples of Applications of Freeway and Expressway Lane-Drop Markings (Sheet 1 of 6)

A – Lane drop at a single lane exit ramp

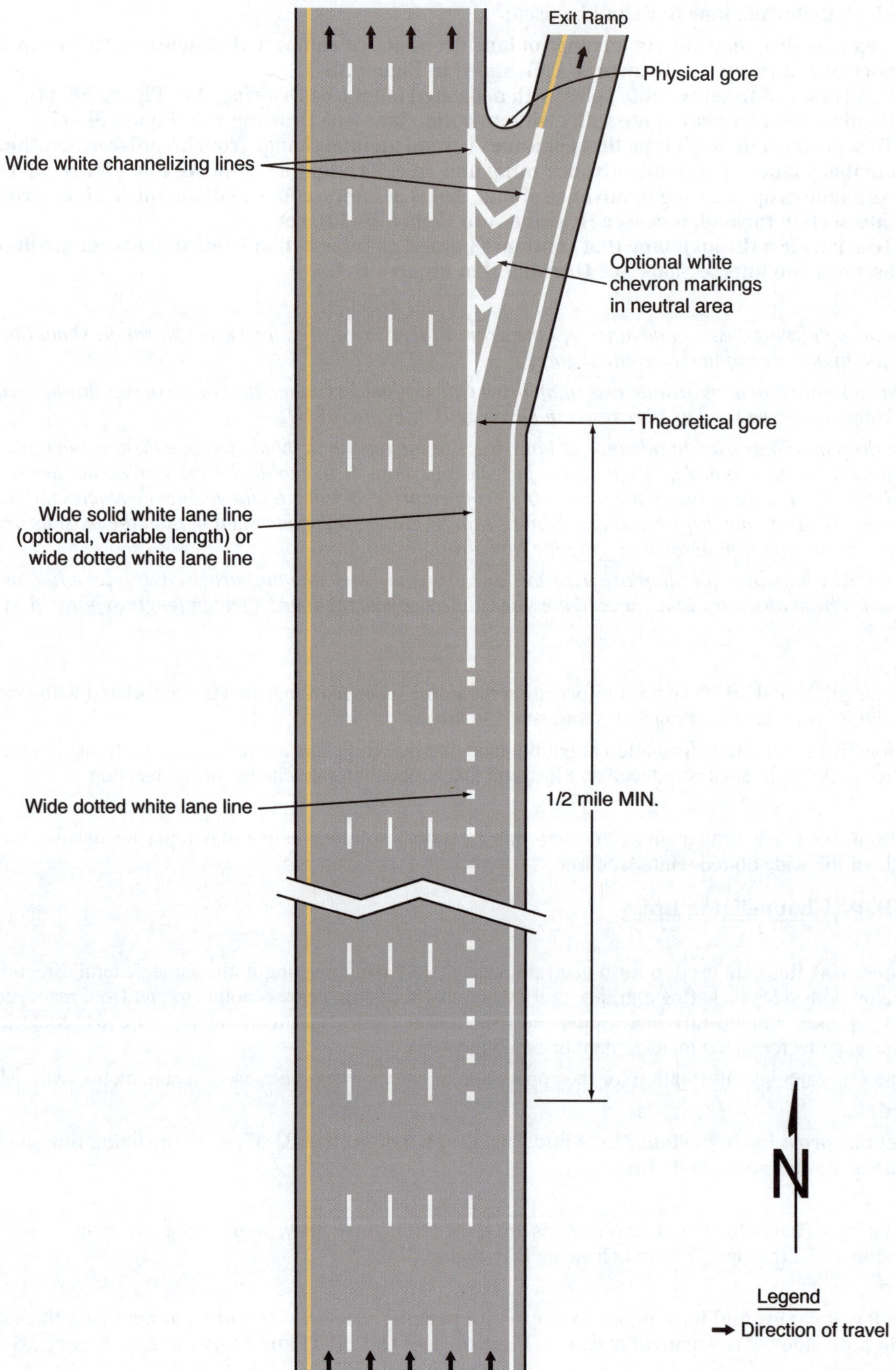

Figure 3B-11. Examples of Applications of Freeway and Expressway Lane-Drop Markings (Sheet 2 of 6)

B – Lane drop at a multi-lane exit ramp having an optional exit lane that also carries the through route

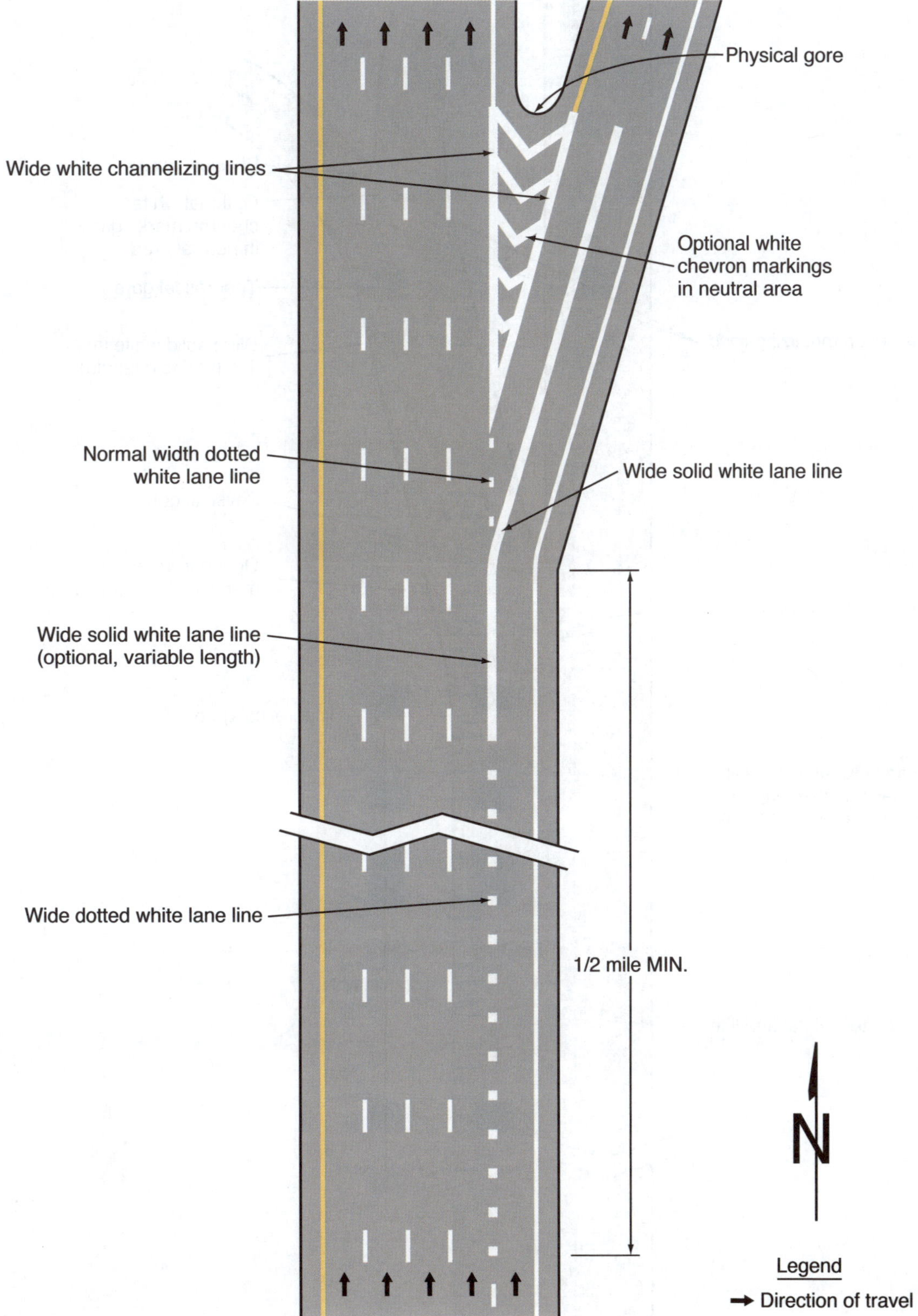

Figure 3B-11. Examples of Applications of Freeway and Expressway Lane-Drop Markings (Sheet 3 of 6)

C – Two-lane lane drop at an exit ramp

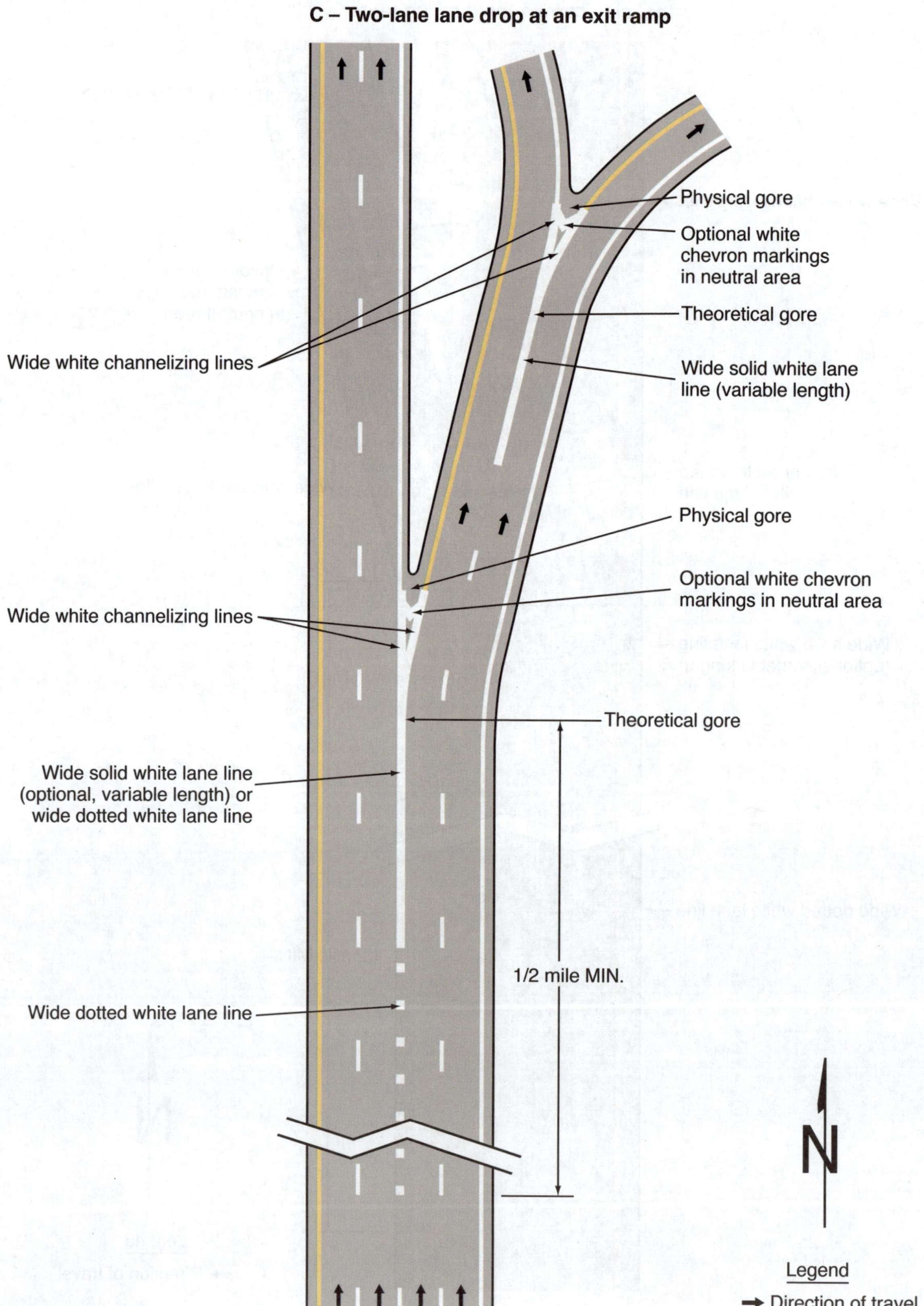

Figure 3B-11. Examples of Applications of Freeway and Expressway Lane-Drop Markings (Sheet 4 of 6)

D – Route split with dedicated lanes

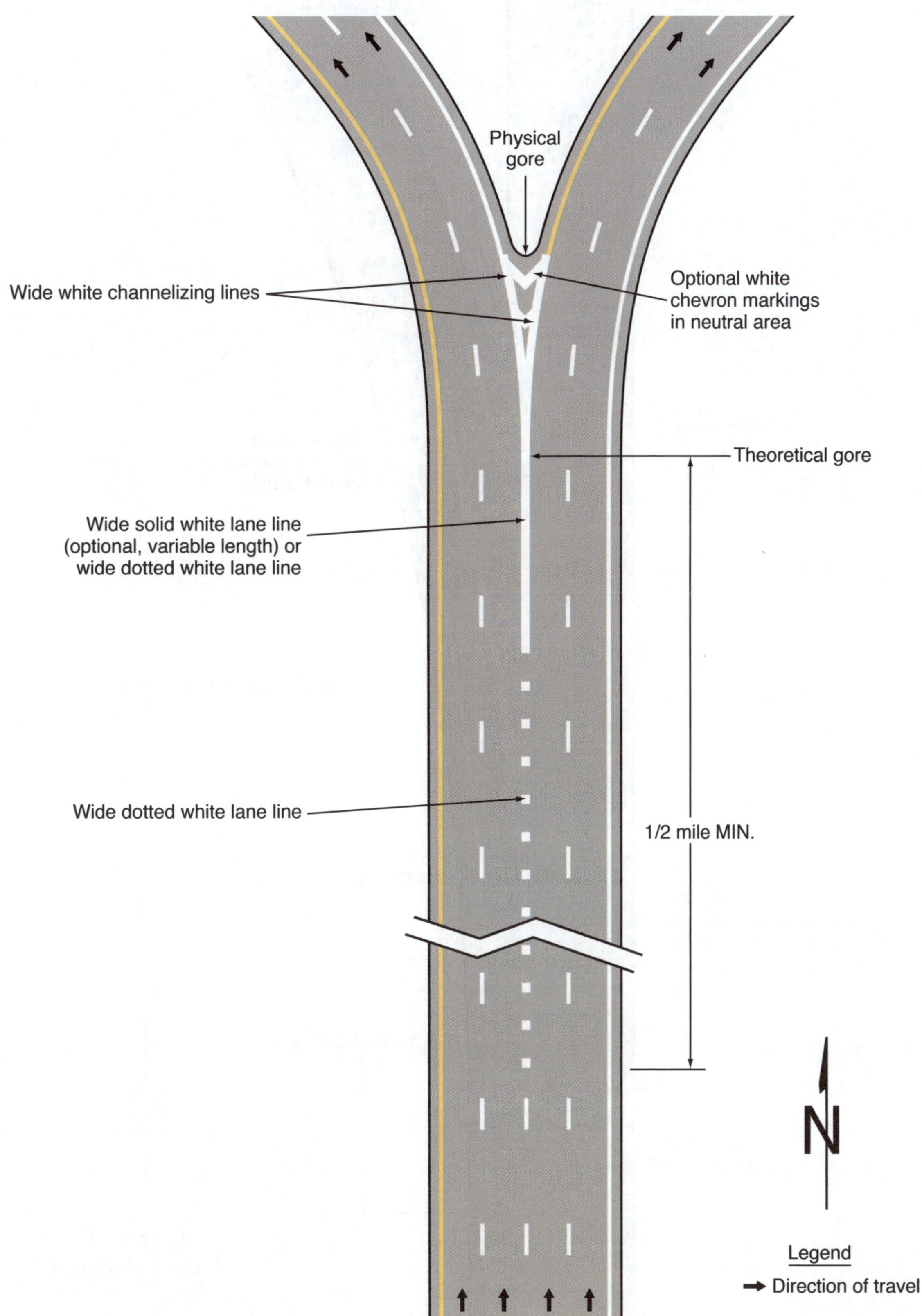

Figure 3B-11. Examples of Applications of Freeway and Expressway Lane-Drop Markings (Sheet 5 of 6)

E – Route split with option lane

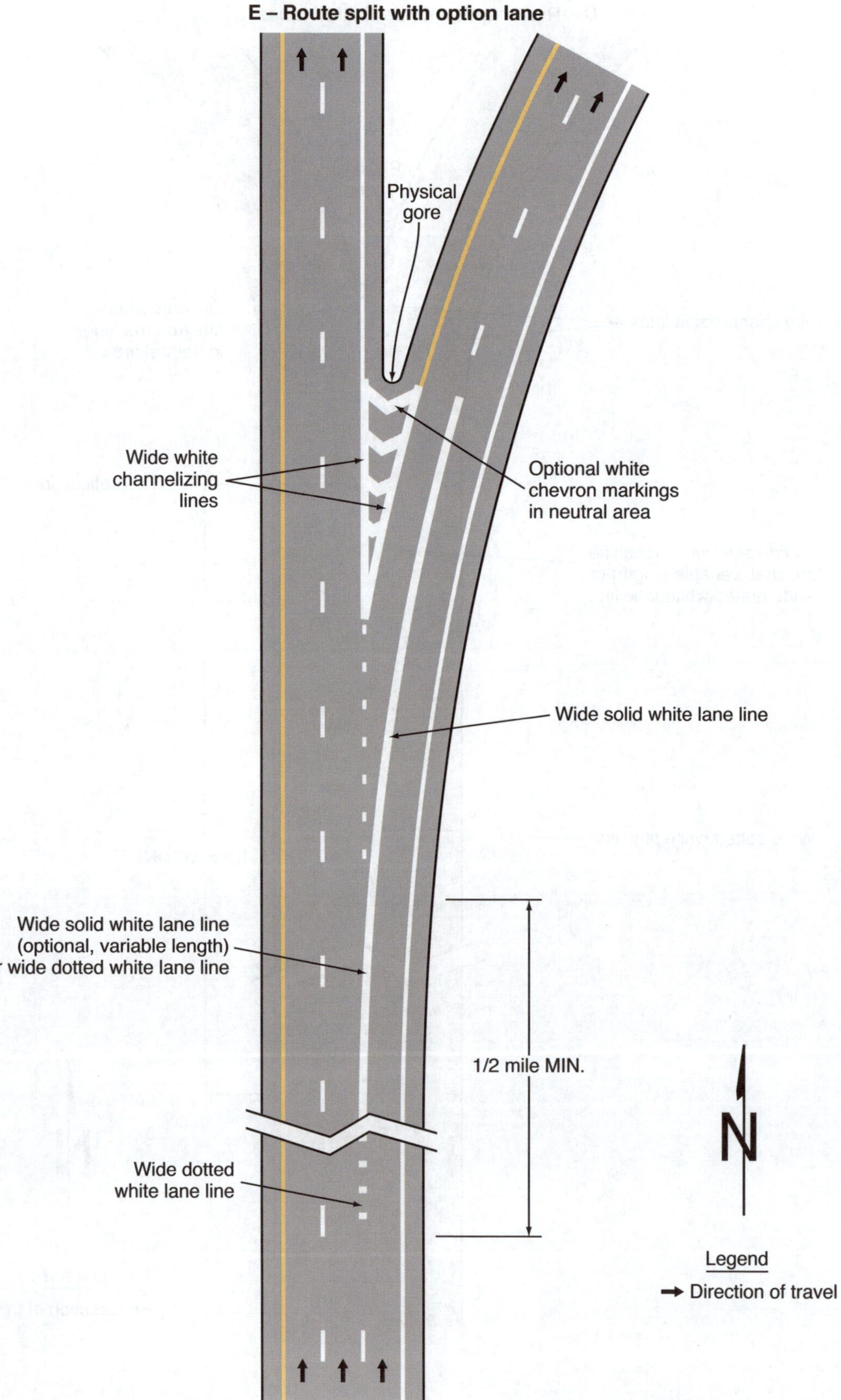

Figure 3B-11. Examples of Applications of Freeway and Expressway Lane-Drop Markings (Sheet 6 of 6)

F – Continuous auxiliary lane, such as at a cloverleaf interchange

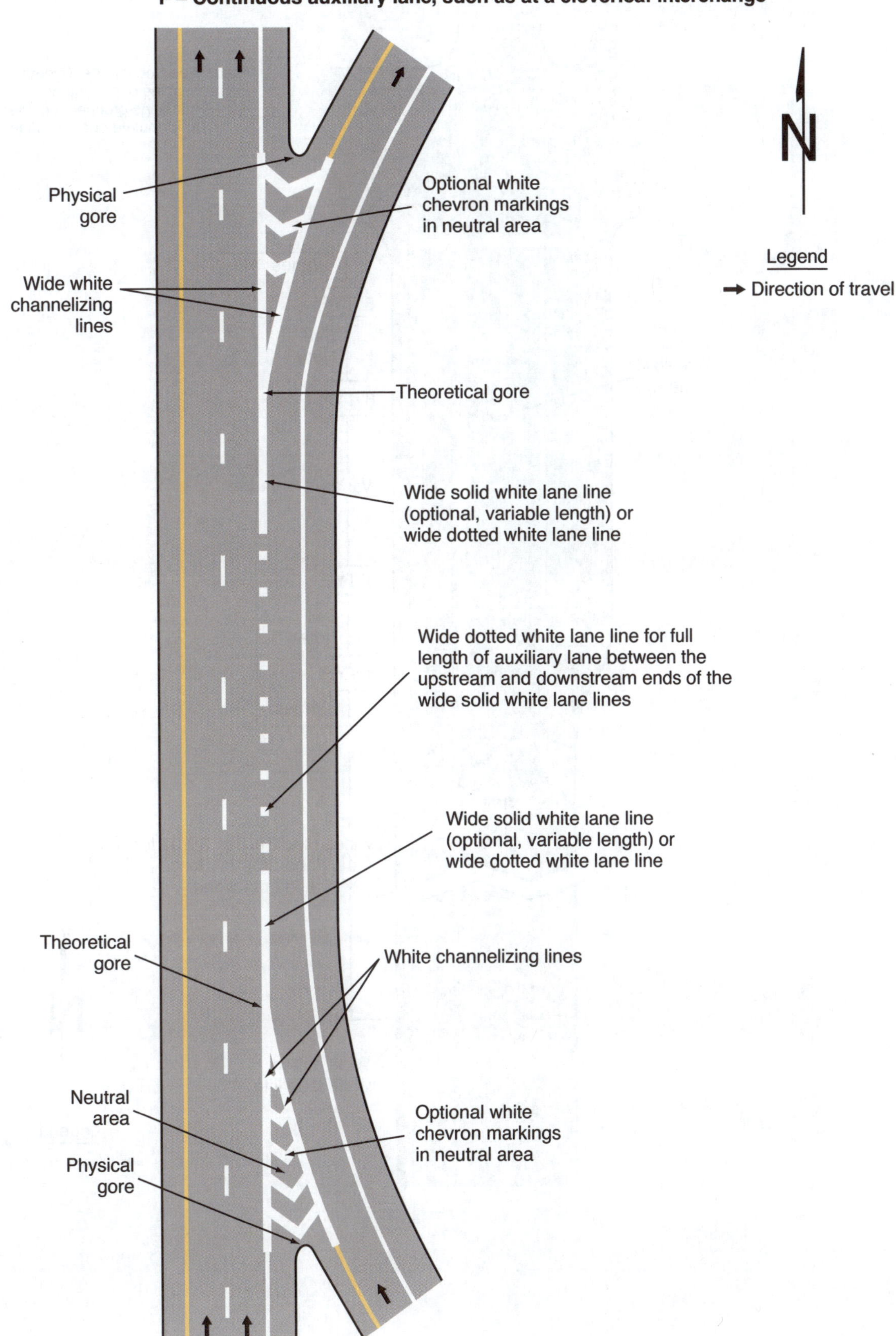

Figure 3B-12. Examples of Applications of Conventional Road Lane-Drop Markings
(Sheet 1 of 2)

A – Lane drop at an intersection

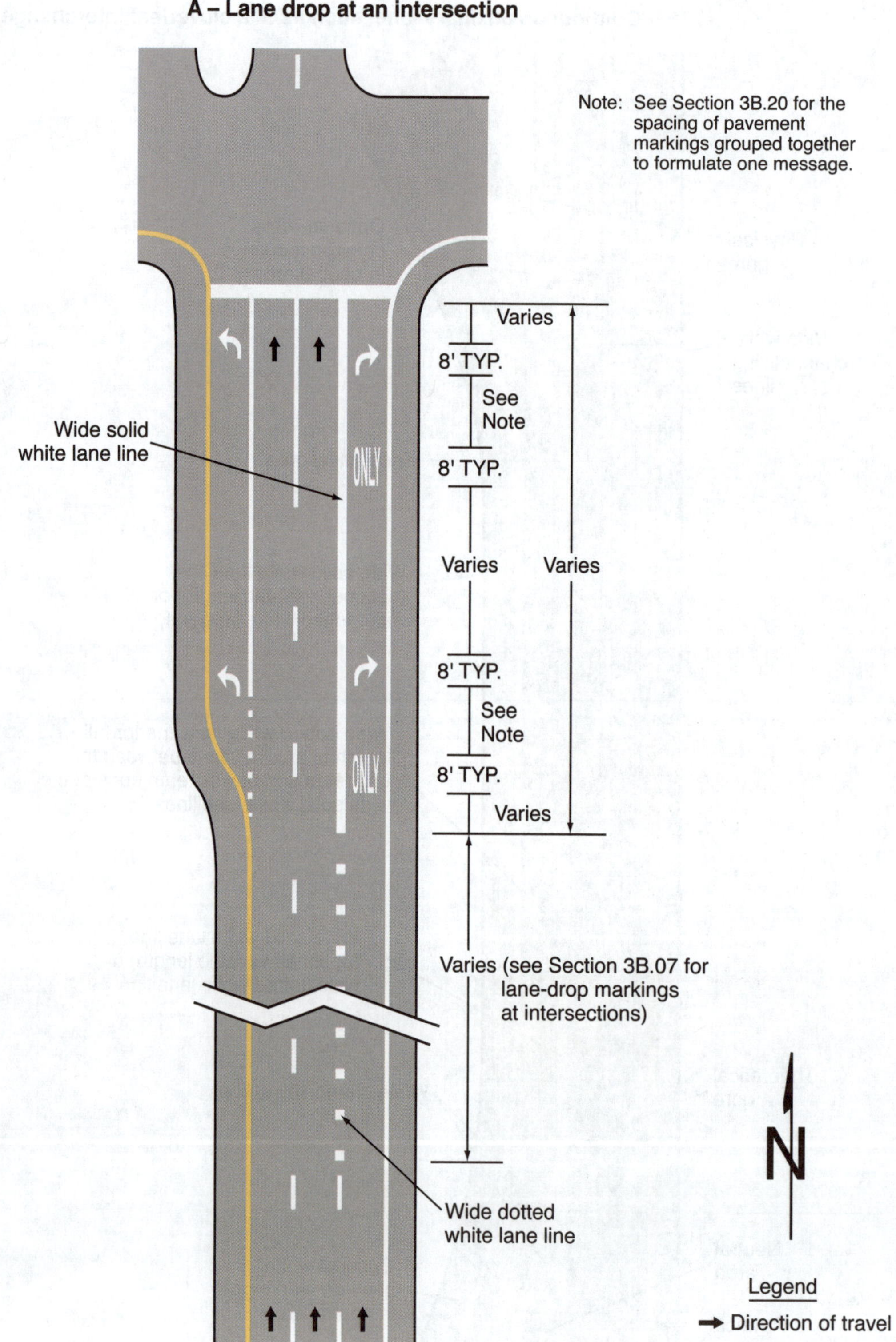

Figure 3B-12. Examples of Applications of Conventional Road Lane-Drop Markings
(Sheet 2 of 2)

B – Auxiliary lane between intersections

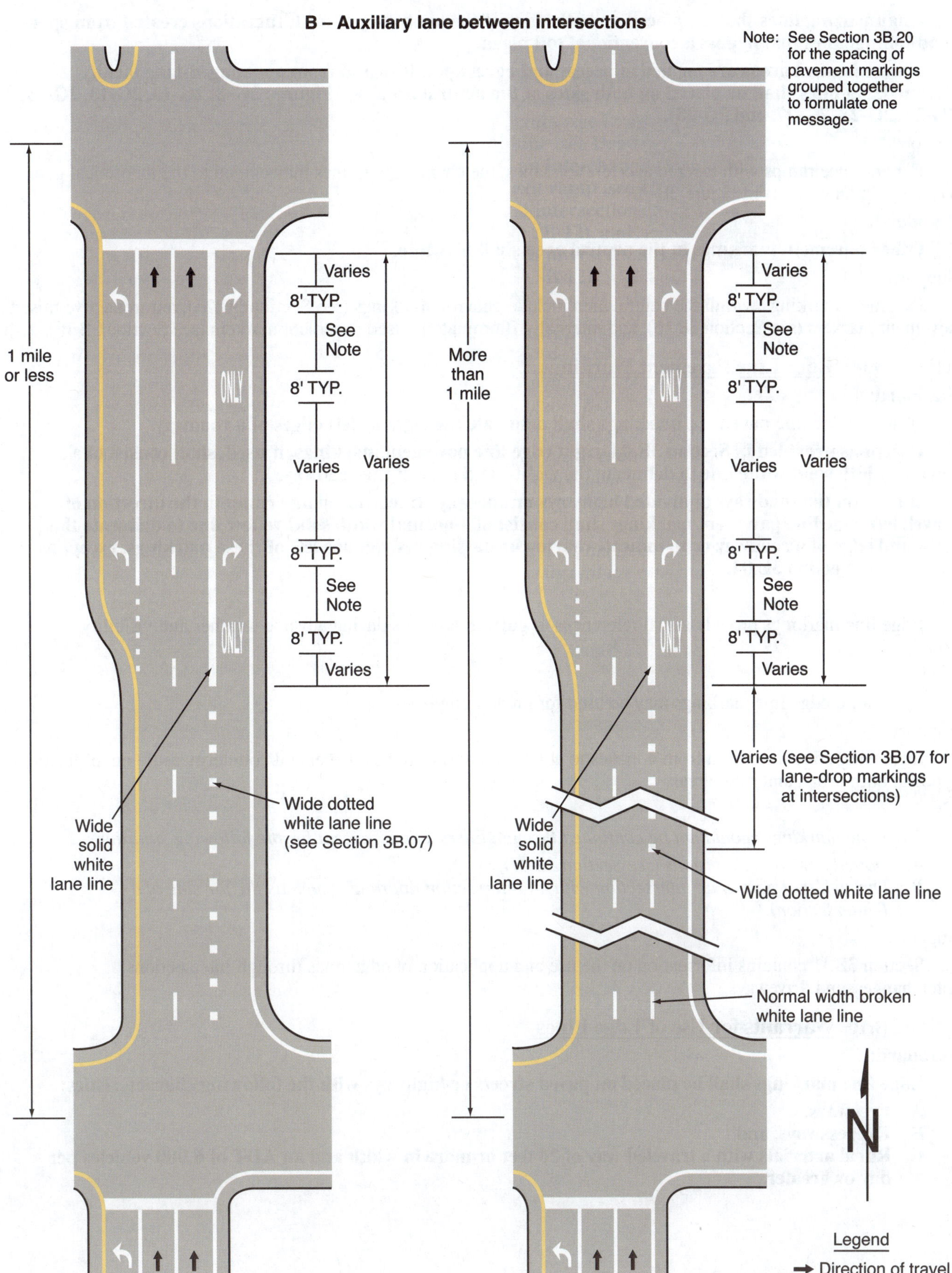

06   **For entrance ramps with tapered acceleration lanes, channelizing lines shall be placed along both sides of the neutral area to a point at least one-half of the distance to the theoretical gore (see Drawing C in Figure 3B-10).**

07   **Channelizing lines shall be placed on both sides of the neutral area for bifurcations created from open-road tolling lanes that bypass a conventional toll plaza.**

08   **Where neutral areas are formed at access and egress points to and from a managed-lane facility, channelizing lines shall be placed on both sides of the neutral area (see Figures 2G-8, 2G-10, 2G-13, 2G-16, 2G-22, 2G-23, 2G-27, and 2G-28).**

Option:

09   For entrance ramps with tapered acceleration lanes, the channelizing lines may extend to the theoretical gore as shown in Drawing B in Figure 3B-10.

Standard:

10   **Other pavement markings in the neutral area shall be white.**

Support:

11   Pavement markings within the neutral area include chevron markings (see Section 3B.25), retroreflective raised pavement markers (see Section 3B.16), and internally illuminated raised pavement markers (see Section 3B.16).

Section 3B.09 <u>Edge Line Pavement Markings</u>

Standard:

01   **If used, edge line pavement markings shall delineate the right or left edges of a roadway.**

02   **Except as provided in Section 3E.04, right edge line pavement markings, if used, shall consist of a normal width solid white line to delineate the right-hand edge of the roadway.**

03   **If used on the roadways of divided highways or one-way streets, or on any ramp in the direction of travel, left edge line pavement markings shall consist of a normal width solid yellow line to delineate the left-hand edge of a roadway or to indicate driving or passing restrictions left of these markings, except as provided in Section 3E.04.**

Support:

04   Edge line markings provide visual references to guide road users during adverse weather and visibility conditions.

Option:

05   Wide solid edge line markings may be used for greater emphasis.

Support:

06   Increasing edge line width from 4 inches to at least 6 inches can be a beneficial countermeasure on all facility types in both urban and rural areas.

Guidance:

07   *Edge line markings should not be continued through intersections, except for the following situations:*
   A. *Dotted edge line extensions (see Section 3B.11), or*
   B. *Through that part of an intersection with no intersection approach (such as the far side of a T-intersection).*

Support:

08   Section 3B.11 contains information on the use and application of edge lines through intersections, interchanges, and driveways.

Section 3B.10 <u>Warrants for Use of Edge Lines</u>

Standard:

01   **Edge line markings shall be placed on paved streets or highways with the following characteristics:**
   A. **Freeways,**
   B. **Expressways, and**
   C. **Rural arterials with a traveled way of 20 feet or more in width and an ADT of 6,000 vehicles per day or greater.**

Guidance:

02   *Edge line markings should be placed on paved streets or highways with the following characteristics:*

   A. *Rural arterials and collectors with a traveled way of 20 feet or more in width and an ADT of 3,000 vehicles per day or greater.*

   B. *On other paved streets and highways where an engineering study indicates a need for edge line markings.*

03   *Edge line markings should not be placed where an engineering study or engineering judgment indicates that providing them is likely to decrease safety for all road users.*

Option:

04   Edge line markings may be placed on streets and highways with or without center line markings.

05   Edge line markings may be excluded, based on engineering judgment, for reasons such as if the traveled way edges are delineated by curbs, parking, or other markings.

06   If a bicycle lane is marked on the outside portion of the traveled way, the edge line that would mark the outside edge of the bicycle lane may be omitted.

07   Edge line markings may be used where edge delineation is desirable to minimize unnecessary driving on paved shoulders or on refuge areas that have lesser structural pavement strength than the adjacent roadway.

Section 3B.11 Application of Pavement Markings through Intersections or Interchanges

Standard:

01   **Pavement markings extended into or continued through an intersection or interchange area shall be the same color as the line markings they extend (see Figure 3B-13).**

Guidance:

02   *Pavement markings extended into or continued through an intersection or interchange area should be at least the same width as the line markings they extend.*

03   *Where highway design or reduced visibility conditions make it desirable to provide control or to guide vehicles through an intersection or interchange, such as at offset, skewed, complex, or multi-leg intersections, on curved roadways, where multiple turn lanes are used, or where offset left-turn lanes might cause driver confusion, dotted lane line extension markings consisting of 2-foot line segments and 2-foot to 6-foot gaps should be used to extend longitudinal line markings through an intersection or interchange area.*

04   *Where greater restriction is preferred, solid lane lines or channelizing lines should be extended into or continued through intersections.*

Standard:

05   **Extensions of center lines through intersections shall be dotted lines.**

Guidance:

06   *Where a double line is extended through an intersection, a single line of equal width to one of the lines of the double line should be used.*

Standard:

07   **Solid lines shall not be used to extend edge lines into or through intersections except through that part of an intersection with no intersecting approach (such as at the far side of a T-intersection).**

Guidance:

08   *Edge line markings should be discontinued across intersecting approaches at intersections or interchanges.*

09   *Driveways that do not meet the definition of an intersection (see Section 1C.02) should have edge line markings maintained across the intersecting approach of the driveway.*

Option:

10   Dotted edge line extensions may be placed through intersections.

Support:

11   Section 3B.31 contains information about edge lines through diverging diamond interchanges with a transposed alignment crossroad.

12   Section 3D.03 provides information for edge lines through roundabouts.

13   Section 5B.02 contains information on edge line extensions for driving automation system considerations.

14   Section 8C.05 contains information about the extension of edge lines through grade crossing areas.

15   Section 9E.03 contains information for the extensions of bicycle lanes through intersections.

Figure 3B-13. Examples of Line Extensions through Intersections (Sheet 1 of 2)

A – Typical pavement markings with offset lane lines continued through the intersection

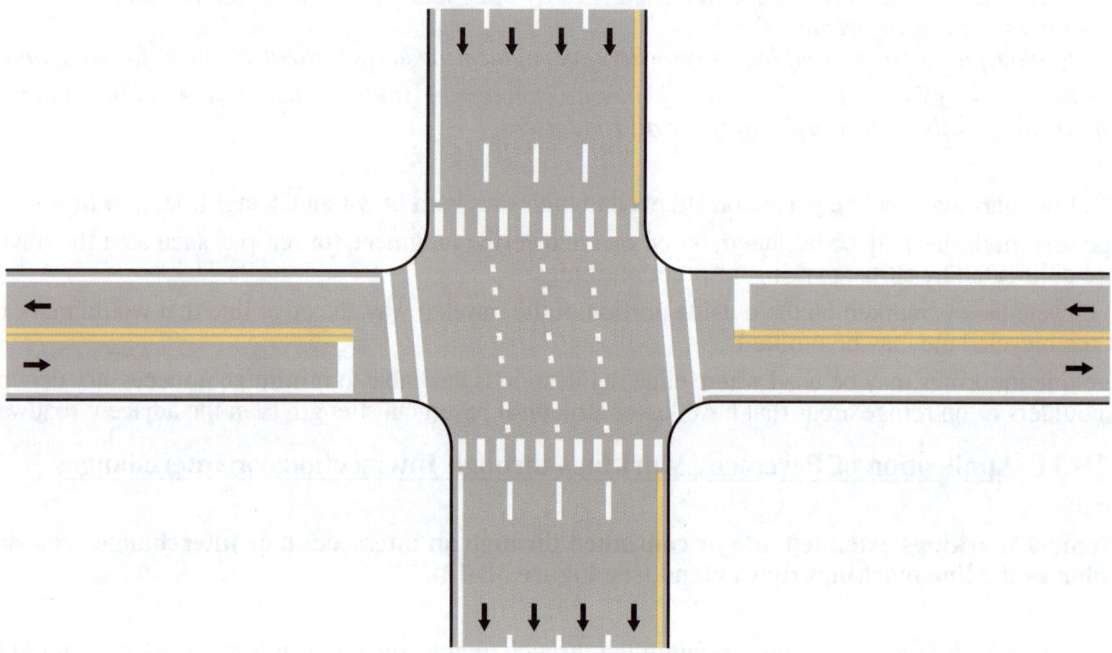

B – Typical pavement markings with line extensions into intersection for double turns

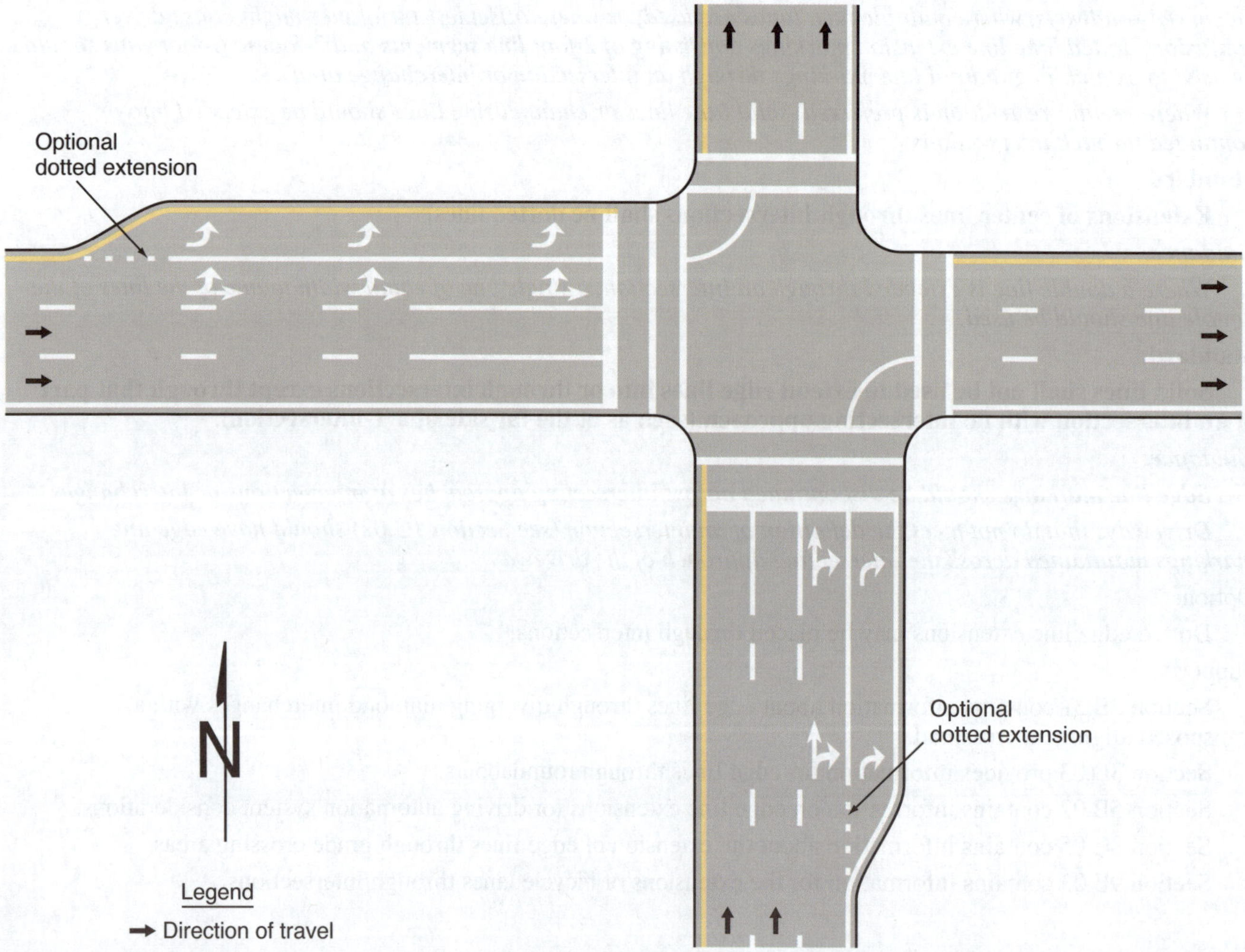

Figure 3B-13. Examples of Line Extensions through Intersections (Sheet 2 of 2)

C – Typical dotted line markings to extend lane line markings into the intersection

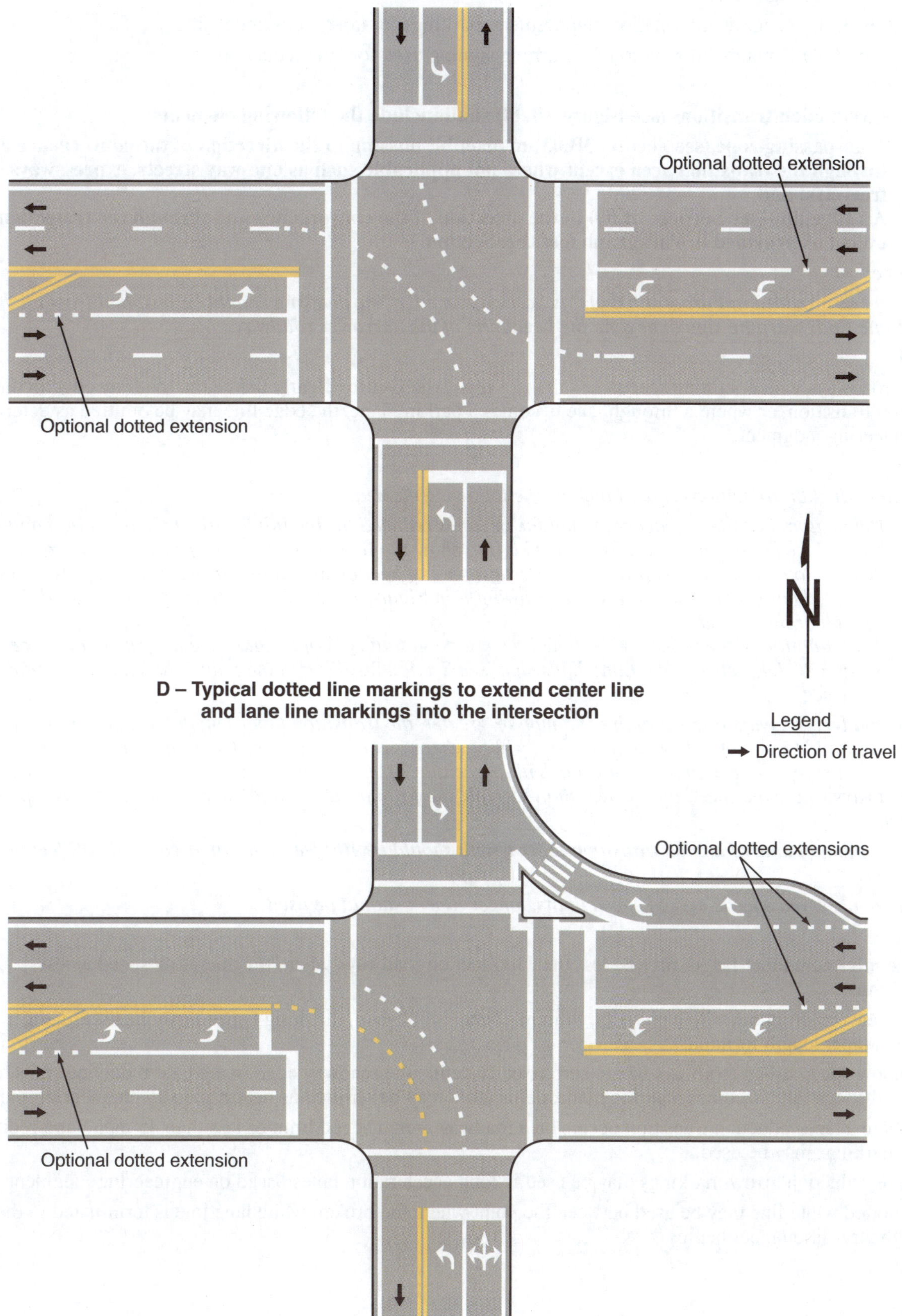

D – Typical dotted line markings to extend center line and lane line markings into the intersection

Section 3B.12 <u>Lane-Reduction Transitions</u>

Support:

01   A lane-reduction is where the number of through lanes is reduced at a location that is not at an interchange or intersection because of narrowing of the roadway or because of a section of on-street parking in what would otherwise be a through lane.

02   Section 3B.07 contains information on pavement markings for lane drops and splits.

03   Section 2C.47 contains information for warning signing used for lane reductions.

Standard:

04   **Lane-reduction transitions (see Figure 3B-14) shall include the following elements:**

   A. A no-passing zone (see Section 3B.03) to prohibit passing in the direction of the convergence and through the transition area except where not applicable such as one-way streets, expressways, and freeways; and

   B. An edge line (see Section 3B.09) in the direction of the convergence and through the transition area, except as provided in Paragraph 6 of this Section.

Guidance:

05   *Except as provided in Paragraph 6 of this Section, the edge line marking should be installed from the location of the Lane Ends warning sign to beyond the beginning of the narrower roadway.*

Option:

06   On roadways with operating speeds less than 25 mph where curbs clearly define the roadway edge in the lane-reduction transition, or where a through lane becomes a parking lane, the edge line may be omitted as determined by engineering judgment.

Guidance:

07   *Lane-reduction transitions should include the following elements:*

   A. *Delineators installed adjacent to the lane or lanes reduced for the full length of the transition and should be so placed and spaced (see Section 3G.04) to show the reduction except as provided in Paragraph 13 of this Section and except as provided in Paragraph 2 of Section 3G.03 for freeways and expressways,*

   B. *Lane-reduction arrow markings (see Drawing F in Figure 3B-21) on the roadway with a speed limit of 45 mph or more, and*

   C. *A termination of the broken white lane line at a point that is ¼ of the advance placement distance (see Section 2C.04) between the Lane Ends sign (see Section 2C.47) and the point where the transition taper begins.*

08   *For roadways having a speed limit of 45 mph or greater, the transition taper length for a lane-reduction transition should be computed by the formula $L = WS$, where L equals the taper length in feet, W equals the width of the offset distance in feet, and S equals the 85th-percentile speed or the speed limit in mph, whichever is higher. For roadways where the speed limit is less than 45 mph, the formula $L = WS^2/60$ should be used to compute the taper length.*

09   *The minimum lane reduction transition taper length should be 100 feet in urban areas and 200 feet in rural areas.*

10   *Where observed speeds exceed speed limits, longer tapers should be used.*

Option:

11   The minimum taper length may be less than 100 feet on roadways where the operating speed is less than 25 mph.

12   On new construction, where no speed limit has been established, the design speed may be used in the transition taper length formula.

13   On low-speed urban roadways where curbs clearly define the roadway edge in the lane-reduction transition, or where a through lane becomes a parking lane, delineators may be omitted as determined by engineering judgment.

14   Where a lane-reduction transition occurs on a roadway with a speed limit of less than 45 mph, lane-reduction arrow markings may be used.

15   Lane-reduction arrow markings may be used in long acceleration lanes based on engineering judgment.

16   A dotted white line may be used between the point where the broken white lane line is terminated to the point where the transition taper begins.

Figure 3B-14. Examples of Applications of Lane-Reduction Transition Markings

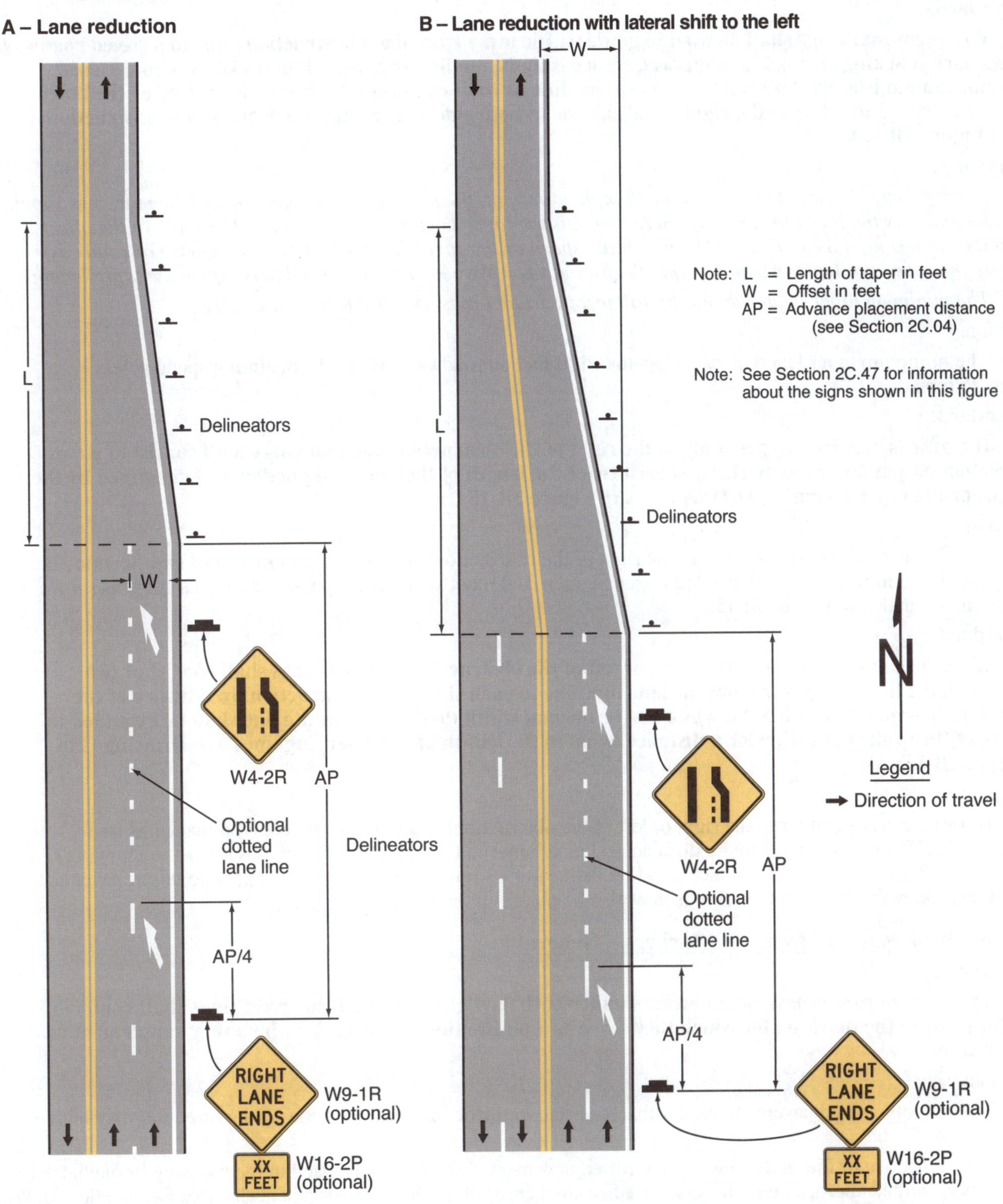

Section 3B.13 Approach Markings for Obstructions

Standard:

01   **Pavement markings shall be used to guide traffic away from fixed obstructions within a paved roadway. Approach markings for bridge supports, refuge islands, median islands, toll plaza islands, and raised channelization islands shall consist of a tapered line or lines extending from the center line or the lane line to a point 1 to 2 feet to the right-hand side, or to both sides, of the approach end of the obstruction (see Figure 3B-15).**

Guidance:

02   *For roadways having a speed limit of 45 mph or greater, the taper length of the tapered line markings should be computed by the formula $L = WS$, where L equals the taper length in feet, W equals the width of the offset distance in feet, and S equals the 85th-percentile speed or the speed limit, whichever is higher. For roadways where the speed limit is less than 45 mph, the formula $L = WS^2/60$ should be used to compute the taper length.*

03   *The minimum taper length should be 100 feet in urban areas and 200 feet in rural areas.*

Option:

04   The minimum taper length may be less than 100 feet on roadways where the operating speed is less than 25 mph.

Standard:

05   **If traffic is required to pass only to the right of the obstruction, the markings shall consist of a two-direction no-passing zone marking at least twice the length of the diagonal portion as determined by the appropriate taper formula (see Drawing A in Figure 3B-15).**

Option:

06   If traffic is required to pass only to the right of the obstruction, yellow diagonal markings (see Section 3B.25) may be placed in the flush median islands (see Section 3J.03) between the no-passing zone markings as shown in Drawings A and B in Figure 3B-15.

Standard:

07   **If traffic can pass either to the right or left of the obstruction, the markings shall consist of two channelizing lines diverging from the lane line, one to each side of the obstruction. In advance of the point of divergence, a wide solid white line or normal width double solid white line shall be extended in place of the broken lane line for a distance equal to the length of the diverging lines (see Drawing C in Figure 3B-15).**

Option:

08   If traffic can pass either to the right or left of the obstruction, additional white chevron markings (see Section 3B.25) may be placed in the flush neutral area between the channelizing lines as shown in Drawing C in Figure 3B-15. Other markings, such as white delineators, white channelizing devices, white raised pavement markers, and white crosswalk markings may also be placed in the flush neutral area.

Section 3B.14 Raised Pavement Markers – General

Standard:

01   **The color of raised pavement markers under both daylight and nighttime conditions shall conform to the color of the marking for which they serve as a positioning guide, or for which they supplement or substitute.**

Option:

02   The side of a raised pavement marker that is visible to traffic proceeding in the wrong direction may be red (see Section 3A.03).

03   Retroreflective or internally illuminated raised pavement markers may be used in the roadway immediately adjacent to curbed approach ends of raised medians and curbs of islands, or on top of such curbs (see Section 3J.06).

Standard:

04   **When used, internally illuminated raised pavement markers shall be steadily illuminated and shall not be flashed.**

Support:

05   Flashing raised pavement markers are considered to be In-Roadway Lights (see Chapter 4U).

Guidance:

06   *The spacing of raised pavement markers used to supplement or substitute for other types of longitudinal markings should correspond with the pattern of broken lines for which the markers supplement or substitute.*

Figure 3B-15. Examples of Applications of Markings for Obstructions in the Roadway
(Sheet 1 of 2)

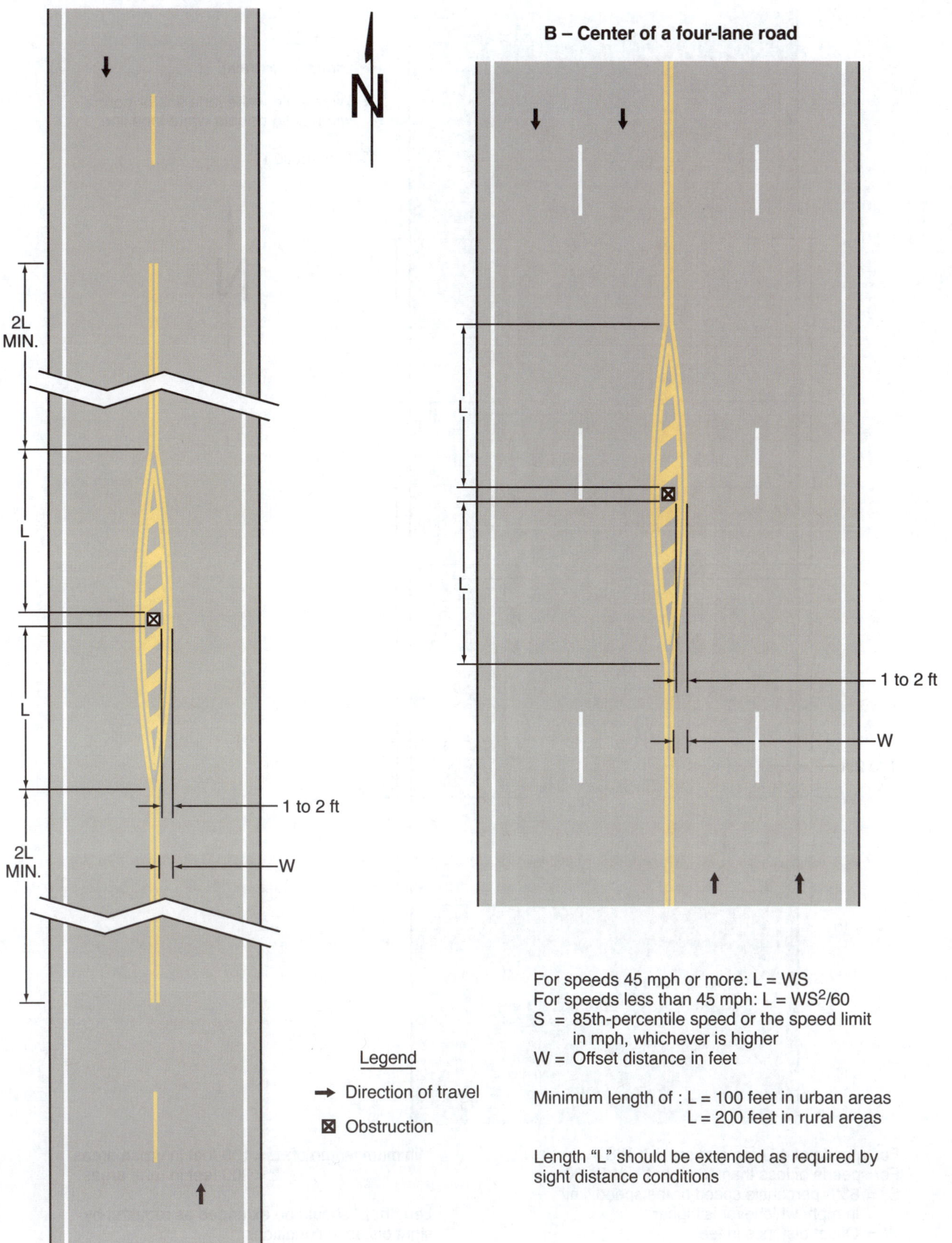

Figure 3B-15. Examples of Applications of Markings for Obstructions in the Roadway
(Sheet 2 of 2)

C – Traffic passing in the same direction on both sides of an obstruction

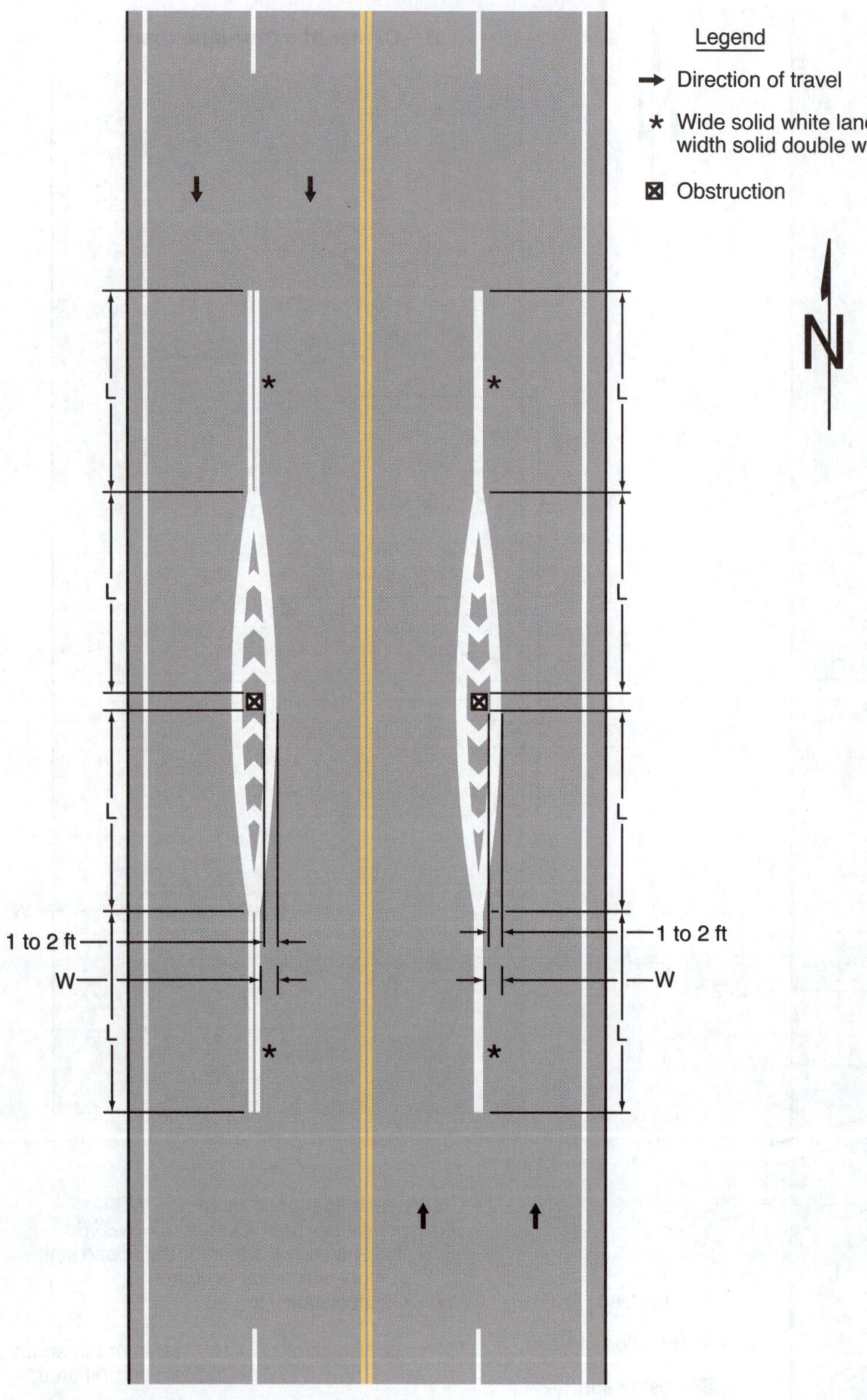

For speeds of 45 mph or more: L = WS
For speeds of less than 45 mph: L = WS^2/60
S = 85th-percentile speed or the speed limit in mph, whichever is higher
W = Offset distance in feet

Minimum length of: L = 100 feet in urban areas
L = 200 feet in rural areas

Length "L" should be extended as required by sight distance conditions

Standard:

07   **The value of N cited in Sections 3B.15 through 3B.17 for the spacing of raised pavement markers shall equal the length of one line segment plus one gap of the broken lines used on the highway.**

Option:

08   For additional emphasis, retroreflective raised pavement markers may be spaced closer than described in Sections 3B.15 through 3B.17, as determined by engineering judgment or engineering study.

Support:

09   Section 9A.03 contains information for the application of raised pavement markers to bicycle facilities.

Section 3B.15 Raised Pavement Markers as Vehicle Positioning Guides with Other Longitudinal Markings

Option:

01   Retroreflective or internally illuminated raised pavement markers may be used as positioning guides with longitudinal line markings without necessarily conveying information to the road user about passing or lane-use restrictions. In such applications, markers may be positioned in line with or immediately adjacent to a single line marking, or positioned between the two lines of a double center line or double lane line marking.

Guidance:

02   *Except as otherwise provided in Paragraphs 3 and 4 of this Section, the spacing for such applications should be 2N (see Section 3B.14).*

Option:

03   Where it is desired to alert the road user to changes in the travel path, such as on sharp curves or on transitions that reduce the number of lanes or that shift traffic laterally, the spacing may be reduced to N or less.

04   On freeways and expressways, the spacing may be increased to 3N for relatively straight and level roadway segments where engineering judgment indicates that such spacing will provide adequate delineation under wet night conditions.

Section 3B.16 Raised Pavement Markers Supplementing Other Markings

Guidance:

01   *The use of retroreflective or internally illuminated raised pavement markers for supplementing longitudinal line markings should comply with the following:*

    A. *Lateral Positioning*

        1. *When supplementing double line markings, pairs of raised pavement markers placed laterally in line with or immediately outside of the two lines should be used.*

        2. *When supplementing wide line markings, pairs of raised pavement markers placed laterally adjacent to each other should be used.*

    B. *Longitudinal Spacing*

        1. *When supplementing solid line markings, raised pavement markers at a spacing no greater than N (see Section 3B.14) should be used, except that when supplementing channelizing lines or edge line markings, a spacing of no greater than N/2 should be used.*

        2. *When supplementing broken line markings, a spacing no greater than 3N should be used. However, when supplementing broken line markings identifying reversible lanes, a spacing of no greater than N should be used.*

        3. *When supplementing dotted lane line markings, a spacing appropriate for the application should be used.*

        4. *When supplementing longitudinal line extension markings through at-grade intersections, one raised pavement marker for each short line segment should be used.*

        5. *When supplementing line extensions through freeway interchanges, a spacing of no greater than N should be used.*

02   *Raised pavement markers should not supplement right-hand edge lines unless an engineering study or engineering judgment indicates the benefits of enhanced delineation of a curve or other location would outweigh possible impacts on bicyclists using the shoulder, and the spacing of raised pavement markers on the right-hand edge does not simulate a broken line during wet night conditions.*

Option:

03   Raised pavement markers also may be used to supplement other markings such as channelizing islands, gore areas, approaches to obstructions, or wrong-way arrows.

04   To improve the visibility of horizontal curves, center lines may be supplemented with retroreflective or internally illuminated raised pavement markers for the entire curved section as well as for a distance in advance of the curve that approximates 5 seconds of travel time.

Section 3B.17 Raised Pavement Markers Substituting for Pavement Markings

Option:

01   Retroreflective or internally illuminated raised pavement markers, or non-retroreflective raised pavement markers supplemented by retroreflective or internally illuminated markers, may be substituted for markings of other types.

Guidance:

02   *If used, the pattern of the raised pavement markers should simulate the pattern of the markings for which they substitute.*

Standard:

03   **Non-retroreflective raised pavement markers shall not be used alone, without supplemental retroreflective or internally illuminated markers, as a substitute for other types of pavement markings.**

Support:

04   Section 6J.02 contains information for flexible temporary pavement markers used during surface treatment paving operations.

Standard:

05   **If raised pavement markers are used to substitute for broken line markings, a group of three to five markers equally spaced at a distance no greater than N/8 (see Section 3B.14) shall be used. If N is other than 40 feet, the markers shall be equally spaced over the line segment length (at ½ points for three markers, at ⅓ points for four markers, and at ¼ points for five markers). At least one retroreflective or internally illuminated marker per group shall be used or a retroreflective or internally illuminated marker shall be installed midway in each gap between successive groups of non-retroreflective markers.**

06   **When raised pavement markers substitute for solid line markings, the markers shall be equally spaced at no greater than N/4, with retroreflective or internally illuminated units at a spacing no greater than N/2.**

Guidance:

07   *Raised pavement markers should not substitute for right-hand edge line markings unless an engineering study or engineering judgment indicates the benefits of enhanced delineation of a curve or other location would outweigh possible impacts on bicyclists using the shoulder, and the spacing of raised pavement markers on the right-hand edge line does not simulate a broken line during wet night conditions.*

Standard:

08   **When raised pavement markers substitute for dotted lines, they shall be spaced at no greater than N/4, with not less than one raised pavement marker per dotted line segment. At least one raised marker every N shall be retroreflective or internally illuminated.**

Option:

09   When substituting for wide lines, raised pavement markers may be placed laterally adjacent to each other to simulate the width of the line.

Support:

10   Section 5B.02 contains information on raised pavement marker considerations for driving automation systems.

Section 3B.18 Curb Markings for Parking Regulations

Guidance:

01   *Except as provided in Paragraph 4 of this Section, since yellow and white curb markings are frequently used for curb delineation and visibility, parking regulations should be established through the installation of standard signs (see Sections 2B.53 and 2B.54).*

02   *Where curbs are marked to convey parking regulations in areas where curb markings are frequently obscured by snow and ice accumulation, signs should be used with the curb markings except as provided in Paragraph 4 of this Section.*

03   *Except as provided in Paragraph 4 of this Section, when curb markings are used without signs to convey parking regulations, a legible word marking regarding the regulation (such as "No Parking" or "No Standing") should be placed on the curb.*

Option:

04   Curb markings without word markings or signs may be used to convey a general prohibition by statute of parking within a specified distance of a STOP sign, YIELD sign, driveway, fire hydrant, or crosswalk.

05   Local highway agencies may prescribe special colors for curb markings to supplement standard signs for parking regulation.

Section 3B.19 <u>Stop and Yield Lines</u>

Option:

01   Stop lines may be used to indicate the point behind which vehicles are required to stop in compliance with a STOP (R1-1) sign, a Stop Here for Pedestrians (R1-5b) sign, a Stop Here for School Crossing (R1-5c) sign, a Stop Here for Trail Crossing (R-5e) sign, or some other traffic control device that requires vehicles to stop, except YIELD signs that are not associated with passive grade crossings.

Standard:

02   **Stop lines shall consist of solid white lines extending across approach lanes to indicate the point at which the stop is intended or required to be made.**

03   **Except as provided in Section 8C.03, stop lines shall not be used at locations where drivers are required to yield in compliance with a YIELD (R1-2) sign, a Yield Here to Pedestrians (R1-5) sign, a Yield Here to School Crossings (R1-5a) sign, a Yield Here to Trail Crossings (R1-5d) sign, or at locations on uncontrolled approaches where drivers or bicyclists are required by State law to yield to pedestrians.**

Guidance:

04   *Stop lines should be used to indicate the point behind which vehicles are required to stop in compliance with a traffic control signal (see Section 4D.08).*

05   *Stop lines should be 12 to 24 inches wide.*

Option:

06   Stop lines may be omitted at ramp control signals.

Support:

07   Section 4J.02 contains information regarding the use and application of stop lines in conjunction with a pedestrian hybrid beacon.

Standard:

08   **If used, a yield line pavement marking shall not be installed without a Yield (R1-2) sign, a Yield Here to Pedestrians (R1-5) sign, a Yield Here to School Crossings (R1-5a) sign, a Yield Here to Trail Crossings (R1-5d) sign, or some other traffic control device that requires vehicles to yield (see Figure 3B-16).**

09   **Yield lines shall not be used at locations where drivers are required to stop in compliance with a STOP (R1-1) sign, a Stop Here for Pedestrians (R1-5b) sign, a Stop Here for School Crossing (R1-5c) sign, a Stop Here for Trail Crossing (R1-5e) sign, a traffic control signal, or some other traffic control device.**

10   **Yield lines shall consist of a row of solid white isosceles triangles pointing toward approaching vehicles extending across approach lanes to indicate the point at which the yield is intended or required to be made.**

Option:

11   If a yield line marking is used on a bicycle facility, a Bicycles Yield to Pedestrians (R9-6) sign (see Section 9B-12) may be used.

Guidance:

12   *The individual triangles comprising the yield line should have a base of 12 to 24 inches wide and a height equal to 1.5 times the base. The space between the triangles should be 3 to 12 inches.*

13   *If used, stop and yield lines should be placed a minimum of 4 feet in advance of the nearest crosswalk line at controlled intersections, except for yield lines at roundabouts as provided for in Section 3D.04 and at midblock crosswalks. In the absence of a marked crosswalk, the stop line or yield line should be placed at the desired stopping or yielding point, but should not be placed more than 30 feet or less than 4 feet from the nearest edge of the intersecting traveled way.*

Standard:

14   **If yield (stop) lines are used at a crosswalk that crosses an uncontrolled multi-lane approach, Yield Here to (Stop Here for) Pedestrians (R1-5 series) signs (see Section 2B.19) shall be used.**

Guidance:

15   *If yield (stop) lines are used at a crosswalk that crosses an uncontrolled multi-lane approach, the yield (stop) line should be placed 20 to 50 feet in advance of the nearest crosswalk line (see Drawing B in Figure 3B-16).*

16   *If yield or stop lines are used in advance of a crosswalk that crosses an uncontrolled multi-lane approach, parking should be prohibited in the area between the yield or stop line and the crosswalk.*

Figure 3B-16. Examples of Yield Line Applications

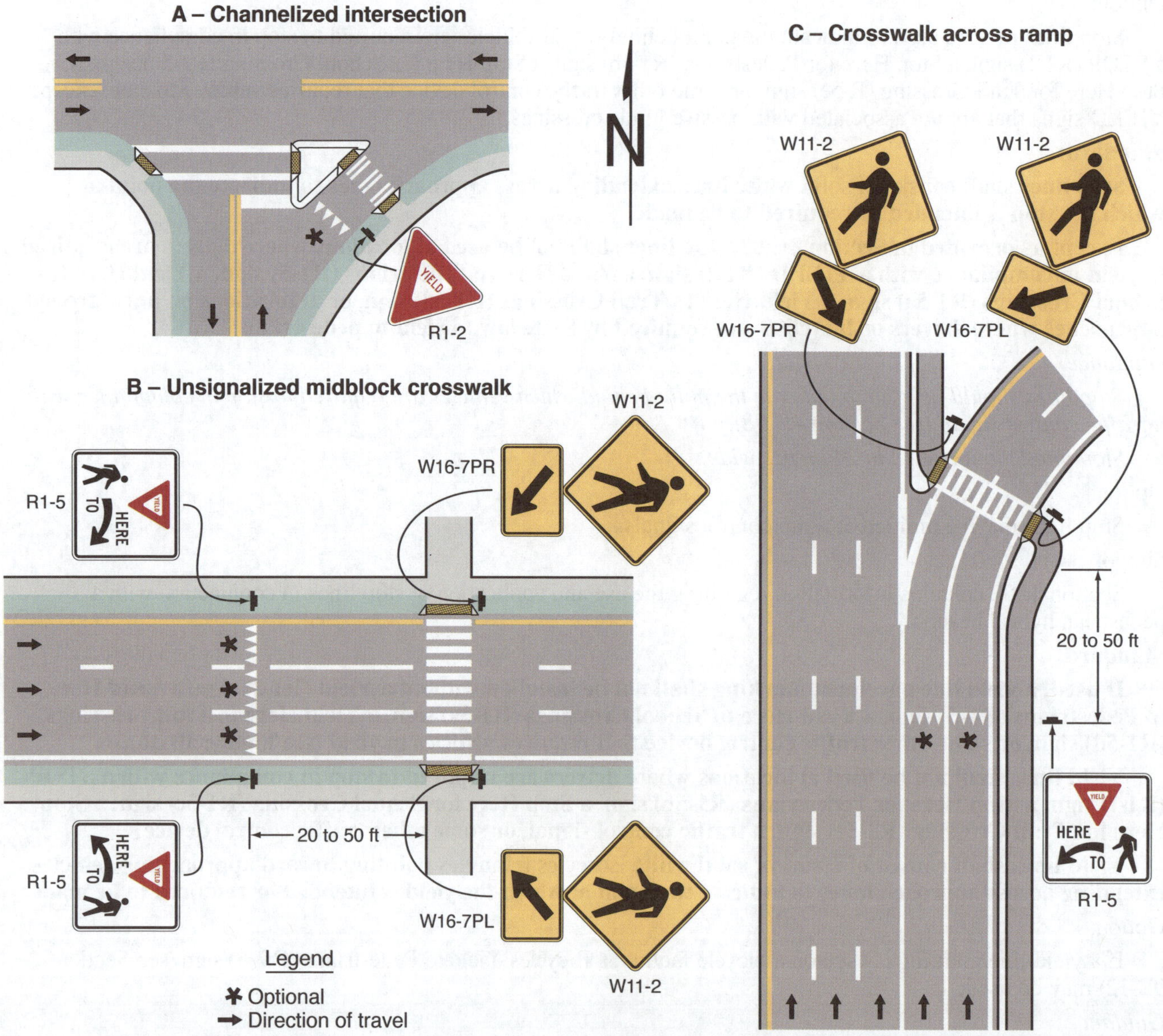

Support:

17   Section 9B.12 contains information for providing signing applicable to bicyclists also subject to a yielding requirement at a crosswalk that crosses an uncontrolled approach.

Guidance:

18   *Yield (stop) lines and Yield Here to (Stop Here for) Pedestrians signs should not be used in advance of crosswalks that cross an approach to or departure from a circular intersection.*

Support:

19   Section 8C.03 contains information regarding the use of stop lines and yield lines at grade crossings.

Option:

20   Stop and yield lines may be staggered longitudinally on a lane-by-lane basis (see Drawing D in Figure 3B-13).

Support:

21   Staggered stop lines and staggered yield lines can improve the driver's view of pedestrians, provide better sight distance for turning vehicles, and increase the turning radius for left-turning vehicles.

Section 3B.20 Word, Symbol, and Arrow Pavement Markings – General

Support:

01   Word, symbol, and arrow markings on the pavement are used for the purpose of regulating, warning, or guiding traffic. These pavement markings can be helpful to road users in some locations by supplementing signs and providing additional emphasis for important regulatory, warning, or guidance messages, because the markings do not require diversion of the road user's attention from the roadway surface. Symbol messages are preferable to word messages. Examples of standard word and arrow pavement markings are shown in Figures 3B-17 and 3B-21, respectively.

Option:

02   Word, symbol, and arrow pavement markings may be used as determined by engineering judgment to supplement signs and/or to provide additional emphasis for regulatory, warning, or guidance messages provided by other devices.

Support:

03   Section 8C.04 contains information for arrow pavement markings in the vicinity of grade crossings.

Standard:

04   **Word, symbol, and arrow markings shall be white, except as otherwise provided in this Section.**

05   **Pavement marking letters, numerals, symbols, and arrows shall be installed in accordance with the design details in the Pavement Markings chapter of the "Standard Highway Signs" publication (see Section 1A.05).**

Guidance:

06   *Word, symbol, and/or arrow markings that are grouped together to formulate one interrelated message should not exceed three lines of information.*

07   *Except for the two opposing white arrows of a two-way left-turn lane marking (see Figure 3B-7) and the pavement word marking messages described in Items B and D of Paragraph 2 of Section 3B.26, the longitudinal space between word, symbol, and/or arrow markings that are used together to formulate one interrelated message should be at least four times the height of the characters for low-speed roads, but not more than ten times the height of the characters under any conditions.*

08   *Except for the SCHOOL word marking (see Section 7C.02), pavement word, symbol, and arrow markings should be no more than one lane in width.*

09   *Pavement word, symbol, and arrow markings should be proportionally scaled to fit within the width of the facility upon which they are applied.*

Option:

10   On narrow, low-speed shared-use paths, the pavement words, symbols, and arrows may be smaller than suggested, but to the relative scale.

11   On roadways where the operating speed is less than 25 mph, word, symbol, and arrow markings may be proportionally reduced by 25 percent.

Section 3B.21 Word Pavement Markings

Guidance:

01   *Letters and numerals should be 6 feet or more in height, except as provided in Section 9E.01 for the BIKE LANE word pavement marking and in Section 9E.15 for a bicycle detector symbol and WAIT HERE FOR GREEN word pavement marking.*

02   *If a pavement marking word message consists of more than one line of information, it should read in the direction of travel. The first word of the message should be nearest to the road user.*

Standard:

03   **The word STOP shall not be placed on the pavement in advance of a stop line, unless every vehicle is required to stop at all times.**

Guidance:

04   *Where through lanes approaching an intersection become mandatory turn lanes, ONLY word pavement markings (see Figure 3B-17) should be used in addition to signs (see Sections 2B.27 and 2B.28) and the required lane-use arrow markings (see Section 3B.23).*

Option:

05   The ONLY word marking may be used to supplement the lane-use arrow markings in lanes that are designated for the exclusive use of a single movement such as turn bays.

06   The ONLY word marking may be used to supplement a preferential lane word or symbol marking (see Section 3E.03).

07   On roadways where the operating speed is less than 25 mph, word markings may be proportionally reduced by 25 percent.

Standard:

08   **The ONLY word marking shall not be used in a lane that is shared by more than one movement.**

Section 3B.22   Symbol Pavement Markings

Support:

01   Section 3E.03 contains information on the diamond-shaped symbol for high-occupancy vehicle (HOV) lanes.

02   Chapter 9E contains information on symbol markings that can be used for bicycle lanes.

Option:

03   Pavement markings simulating Interstate, U.S., State, and County route signs (see Figure 2D-4) with appropriate route numbers, but elongated for proper proportioning when viewed as a marking, may be used to guide road users to their destinations (see Figure 3B-18).

Guidance:

04   *If route sign markings are provided to guide road users, those route sign markings should be provided in option lanes if markings are provided in any lanes.*

05   *If two route sign markings are provided in an option lane, they should be placed in sequence and not divided around an optional lane arrow.*

Figure 3B-17. Example of Elongated Letters for Word Pavement Markings

Figure 3B-18. Examples of Elongated Route Shields and Markers Applied as Pavement Markings

A – Interstate Shield on dark or light pavement

B – U.S. Route Shield on dark pavement

C – U.S. Route Shield on light pavement

D – State Route Marker on dark pavement

E – State Route Marker on light pavement

Note: See the "Standard Highway Signs" publication for sizes and details

Support:

06   Section 3A.03 provides information on route sign colors.

07   Section 9E.14 contains information on route markers for designated bicycle routes that can be used on shared-use paths.

Guidance:

08   *The International Symbol of Accessibility parking space marking (see Figure 3B-19) should be placed in each parking space designated for use by persons with disabilities.*

Option:

09   A blue background with white border may supplement the wheelchair symbol as shown in Figure 3B-19.

10   A yield-ahead triangle symbol or YIELD AHEAD word pavement marking may be used on approaches to intersections where the approaching traffic will encounter a YIELD sign at the intersection.

Standard:

11   **The yield-ahead triangle symbol or YIELD AHEAD word pavement marking shall not be used unless a YIELD sign (see Section 2B.05) is in place at the intersection. The yield-ahead triangle symbol marking shall be as shown in Figure 3B-20.**

Option:

12   A pedestrian symbol pavement marking may be used on portions of facilities that are reserved exclusively for pedestrian use, such as where a shared-use path transitions to become separate facilities for different types of users.

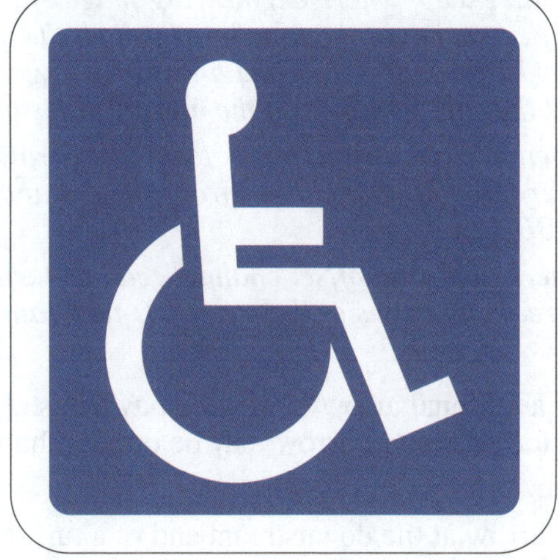

Figure 3B-19. International Symbol of Accessibility Parking Space Marking

Notes:
1. See the "Standard Highway Signs" publication for sizes and details
2. The blue-colored background with white border is optional

Figure 3B-20. Yield Ahead Triangle Symbols

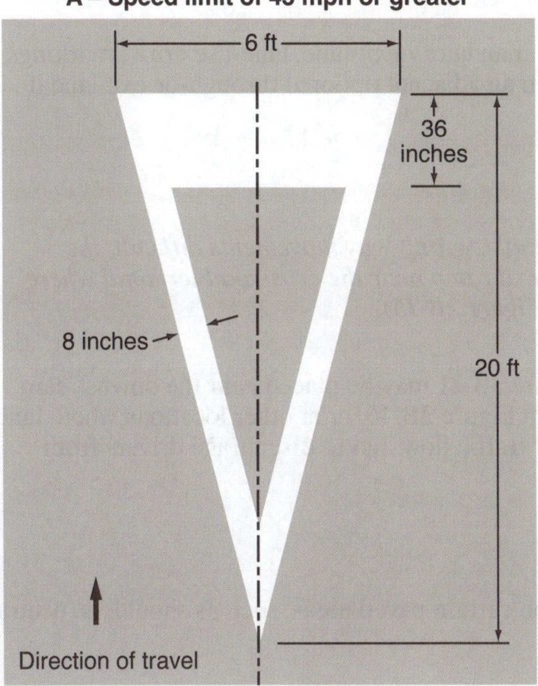

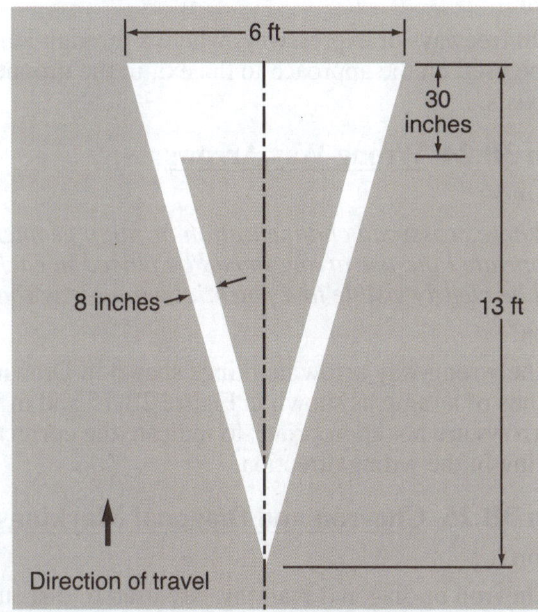

Section 3B.23 Lane-Use Arrows

Support:

01   Lane-use arrow markings (see Figure 3B-21) are used to indicate the mandatory or permissible movements in certain lanes (see Figure 3B-22) and in two-way left-turn lanes (see Figure 3B-7).

02   Section 8C.04 contains information about the placement of lane-use arrow markings in the vicinity of grade crossings.

Guidance:

03   *Lane-use arrow markings should be used in lanes and turn bays designated for the exclusive use of a turning movement, except where engineering judgment determines that physical conditions or other markings (such as a dotted extension of the lane line through the taper into the turn bay) clearly discourage unintentional use of a turn bay by through vehicles. Lane-use arrow markings should also be used in lanes from which movements are allowed that are contrary to the normal rules of the road (see Drawing B in Figure 3B-13).*

04   *When used in turn lanes, at least two arrows should be used, one at or near the upstream end of the full-width turn lane and one an appropriate distance upstream from the stop line or intersection (see Drawing A in Figure 3B-12).*

05   *Where opposing offset channelized left-turn lanes exist, lane-use arrow markings should be placed near the downstream terminus of the offset left-turn lanes to reduce wrong-way movements (see Figure 2B-20).*

Option:

06   An additional arrow or arrows may be used in a turn lane. When arrows are used for a short turn lane, the second (downstream) arrow may be omitted based on engineering judgment.

Support:

07   An arrow at the downstream end of a turn lane can help to prevent wrong-way movements.

Standard:

08   **Where through lanes approaching an intersection become mandatory turn lanes, turn lane-use arrow markings (see Drawing A in Figure 3B-12 and Figure 3B-21) shall be used and shall be accompanied by standard signs (see Section 2B.28).**

Guidance:

09   *Where through lanes approaching an intersection become mandatory turn lanes, ONLY word markings (see Figure 3B-17) should be used in addition to signs (see Sections 2B.27 and 2B.28) and the required turn lane-use arrow markings. These signs and markings should be placed well in advance of the turn and should be repeated as necessary to provide the through motorist advance notification to vacate the lane prior to reaching a point where roadway geometrics or a queue of waiting vehicles forces the motorist to make an unintended turn.*

Option:

10   On freeways or expressways where a through lane becomes a mandatory exit lane, lane-use arrow markings may be used on the approach to the exit in the dropped lane and in an adjacent optional through-or-exit lane if one exists.

Section 3B.24 Wrong-Way Arrows

Guidance:

01   *Where crossroad channelization or ramp geometrics do not make wrong-way movements difficult, the appropriate lane-use arrow should be placed in each lane of an exit ramp near the crossroad terminal where it will be clearly visible to a potential wrong-way road user (see Figure 2B-15).*

Option:

02   The wrong-way arrow markings shown in Drawing G in Figure 3B-21 may be placed near the downstream terminus of a ramp as shown in Figure 2B-15 and in Drawing A in Figure 2B-16, or at other locations where lane-use arrows are not appropriate, to indicate the correct direction of traffic flow and to discourage drivers from traveling in the wrong direction.

Section 3B.25 Chevron and Diagonal Markings

Support:

01   Chevron or diagonal markings are used to discourage travel on certain paved areas, such as shoulders, neutral areas, and flush median islands.

Figure 3B-21. Examples of Standard Arrows for Pavement Markings

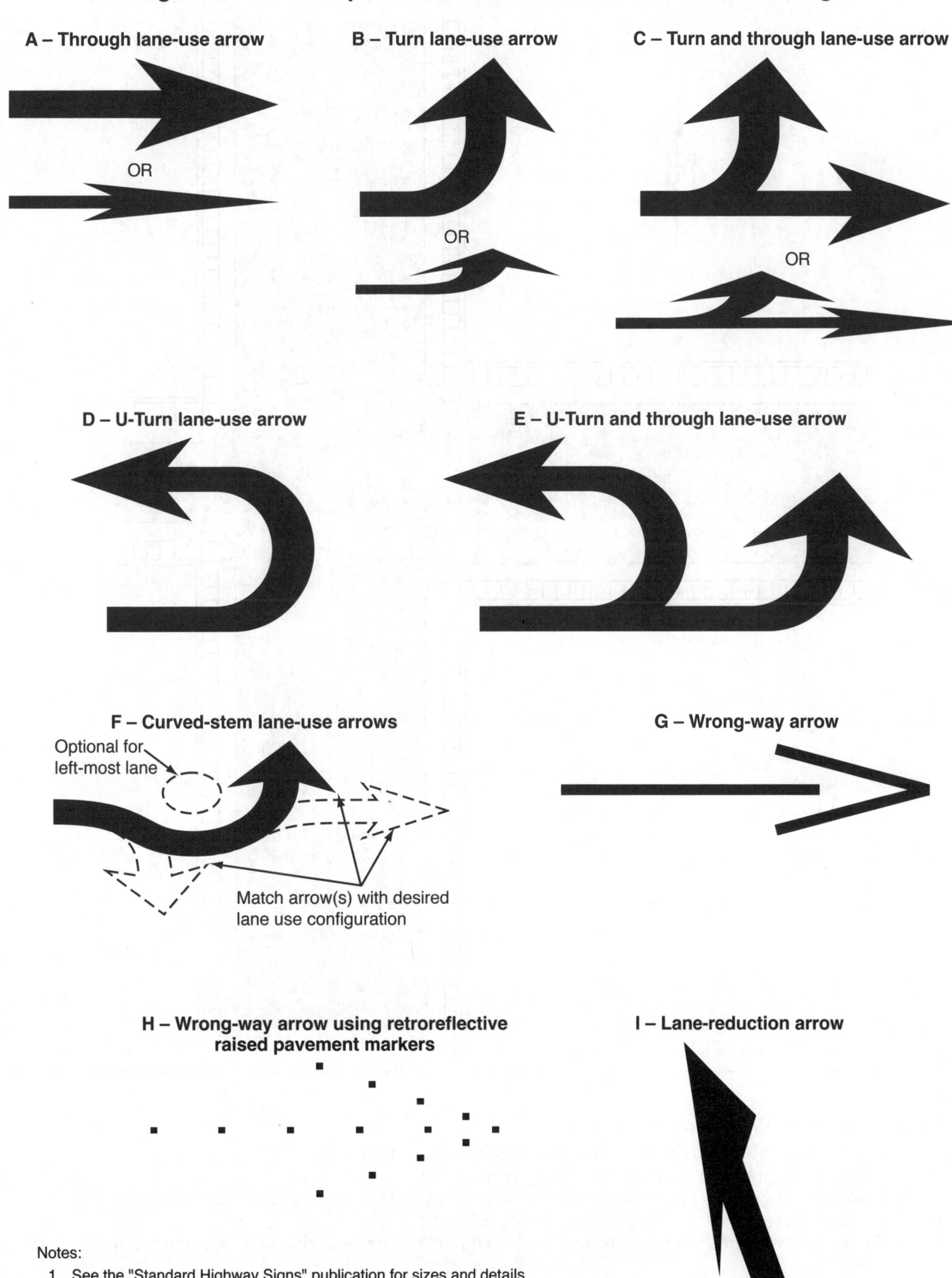

Notes:
1. See the "Standard Highway Signs" publication for sizes and details
2. Refer to Section 3B.20 for guidance and options for proportionally scaling the arrows

Figure 3B-22. Examples of Lane-Use Control Word and Arrow Pavement Markings

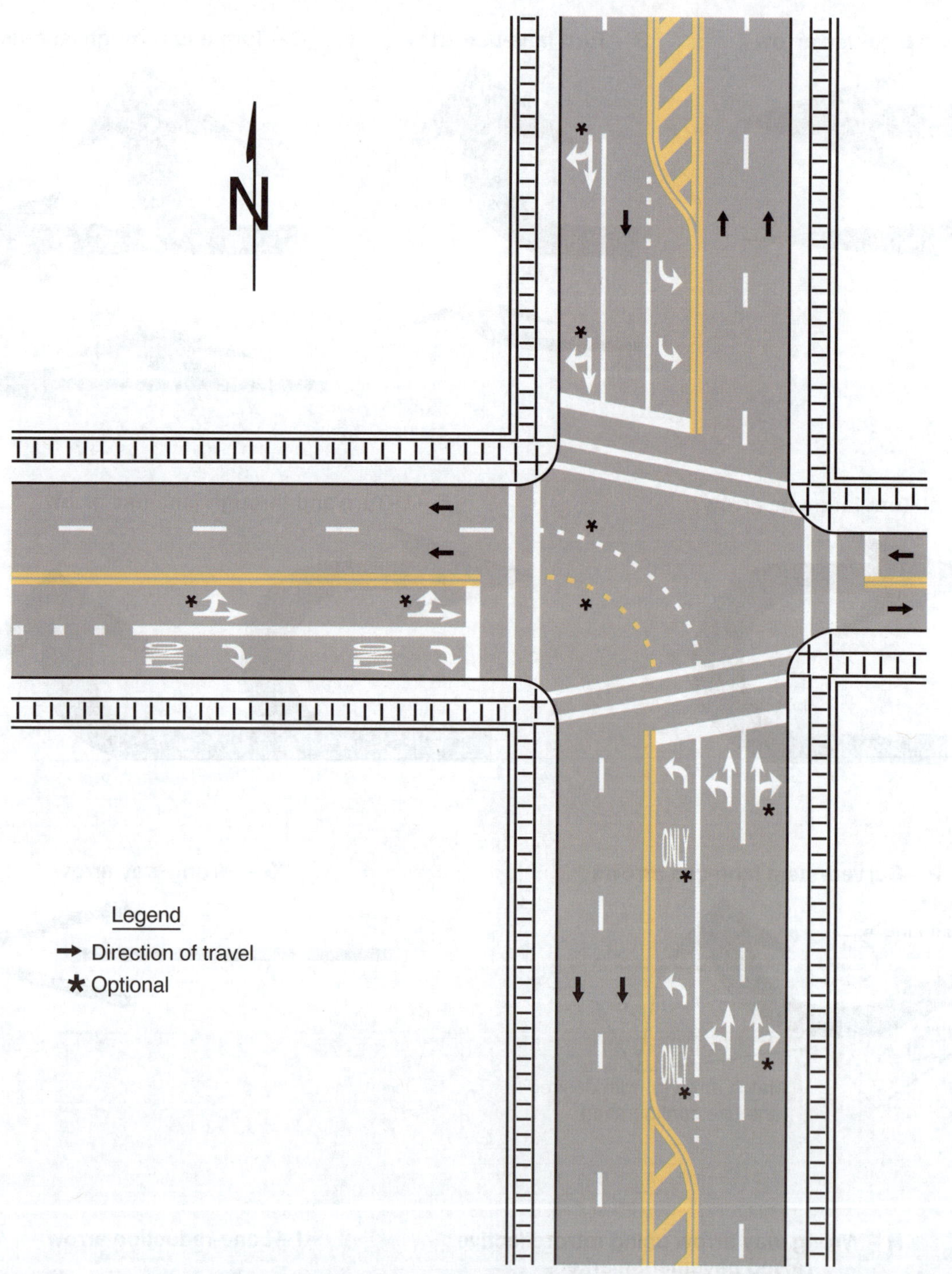

Option:

02   Chevron and diagonal markings may be used:

    A. On approaches to obstructions in the roadway (see Sheet 2 of Figure 3B-15),

    B. For channelized travel paths on approaches to intersections,

    C. In buffer spaces between preferential lanes and general-purpose lanes (see Drawing A in Figure 3E-2),

    D. In the neutral area gores (see Figures 3B-9 through 3B-11),

    E. In the neutral area of bifurcations created from open-road tolling lanes that bypass a conventional toll plaza,

    F. In the neutral areas at access and egress points to and from a managed-lane facility (see Figures 2G-8, 2G-10, 2G-22, and 2G-23), and

    G. In the neutral areas of islands.

03   Chevron markings may be supplemented with white retroreflective or internally illuminated raised pavement markers (see Section 3B.16) for enhanced nighttime visibility.

Support:

04   Section 5B.02 contains information on chevron markings for driving automation system considerations.

Standard:

05   **Chevron markings shall be white, with the point of each chevron facing toward approaching traffic, as shown in Figures 3B-9 through 3B-11, and Drawing C in Figure 3B-15.**

Option:

06   Diagonal markings for opposing directions of traffic may be used:
    A. On approaches to obstructions in the roadway (see Drawings A and B in Figure 3B-15),
    B. In flush median islands between double solid yellow center line markings (see Figure 3B-5), and
    C. In buffer spaces between preferential lanes and general-purpose lanes (see Drawing D in Figure 3E-4).

07   Diagonal markings may be used on paved shoulders or in no-parking zones, or other locations for special emphasis.

Standard:

08   **When diagonal markings are used between opposing directions of traffic or on the left shoulder of a one-way or divided roadway, they shall be yellow and slant away from traffic in the adjacent travel lanes, as shown in Figures 3B-2 and 3B-5, and Drawings A and B in Figure 3B-15.**

09   **When diagonal markings are used on the right shoulder or in no-parking zones (see Figure 3B-23), they shall be white and slant away from traffic in the adjacent travel lane.**

Guidance:

10   *Except as provided in Paragraph 11 of this Section, chevrons and diagonal markings should be at least 12 inches wide for roadways having a speed limit of 45 mph or greater, and at least 8 inches wide for roadways having a speed limit of less than 45 mph. The longitudinal spacing of the chevrons or diagonal lines should be determined by engineering judgment considering factors such as speeds and desired visual impacts. The chevrons and diagonal lines should form an angle of approximately 30 to 45 degrees with the longitudinal lines that they intersect.*

Option:

11   Diagonal markings used in no-parking zones or on roadways with operating speeds of less than 25 mph may be 4 inches wide (see Figure 3B-23).

Section 3B.26 Do Not Block Intersection Markings

Option:

01   Do Not Block Intersection markings may be used to mark the edges of an intersection area that is in close proximity to a signalized intersection, railroad crossing, or other nearby traffic control that might cause vehicles to stop within the intersection and impede other traffic entering the intersection. If authorized by law, Do Not Block Intersection markings with appropriate signs may also be used at other locations.

Standard:

02   **If used, Do Not Block Intersection markings (see Figure 3B-24) shall consist of one of the following alternatives:**
    A. **Wide solid white lines that outline the intersection area that vehicles must not block;**
    B. **Wide solid white lines that outline the intersection area that vehicles must not block and a white word message such as DO NOT BLOCK or KEEP CLEAR;**
    C. **Wide solid white lines that outline the intersection area that vehicles must not block and white cross-hatching within the intersection area; or**
    D. **A white word message, such as DO NOT BLOCK or KEEP CLEAR, within the intersection area that vehicles must not block.**

03   **DO NOT BLOCK INTERSECTION markings shall be accompanied by one or more DO NOT BLOCK INTERSECTION (DRIVEWAY) (CROSSING) (R10-7) signs (see Section 2B.59), one or more DO NOT STOP ON TRACKS (R8-8) signs (see Section 8B.07), or one or more similar signs.**

Section 3B.27 Parking Space Markings

Standard:

01   **On-street parking space markings shall be white.**

Support:

02   Examples of on-street parking space markings are shown in Figure 3B-23.

Figure 3B-23. Examples of Parking Space Markings

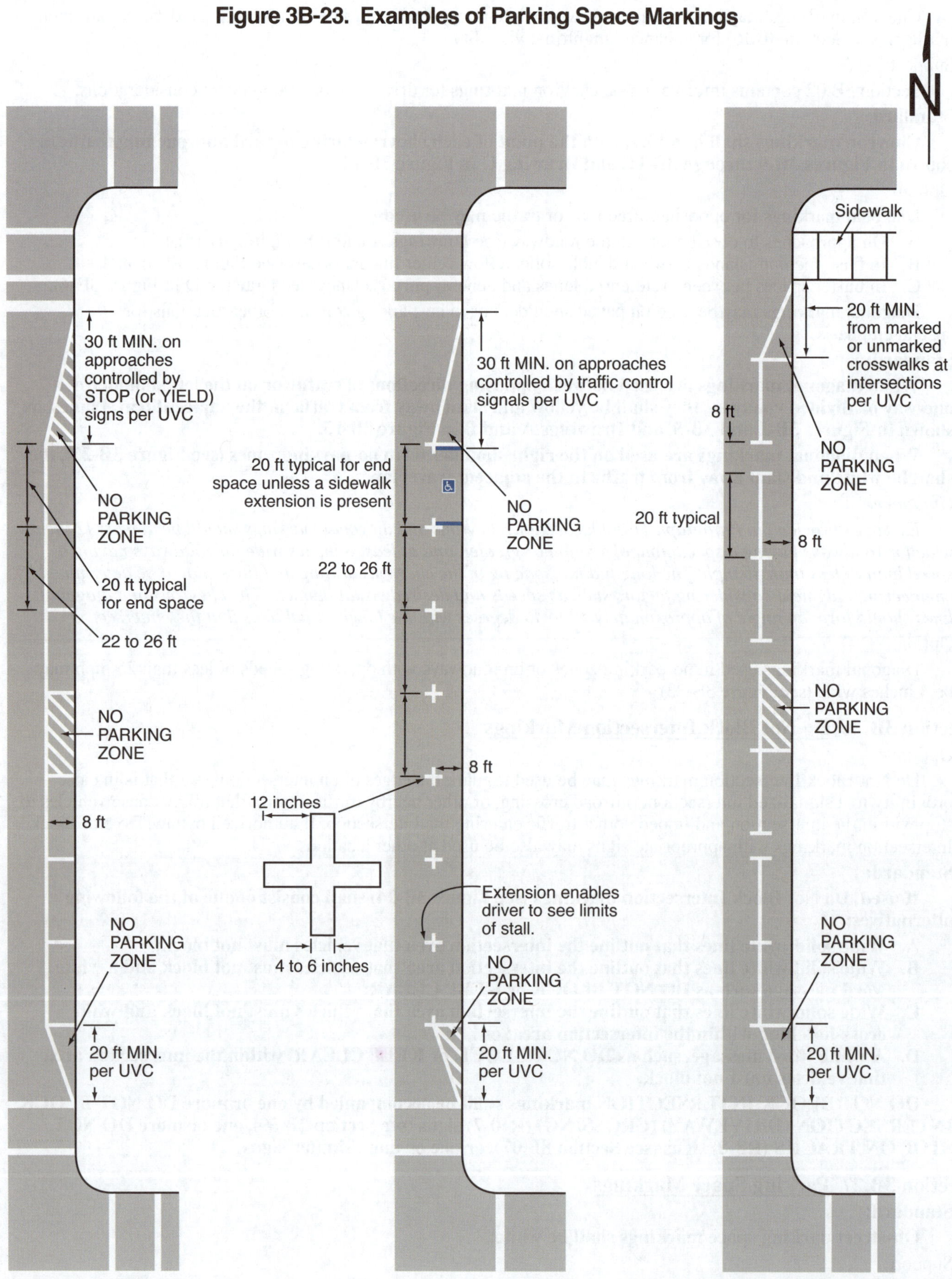

Figure 3B-24. Do Not Block Intersection Markings

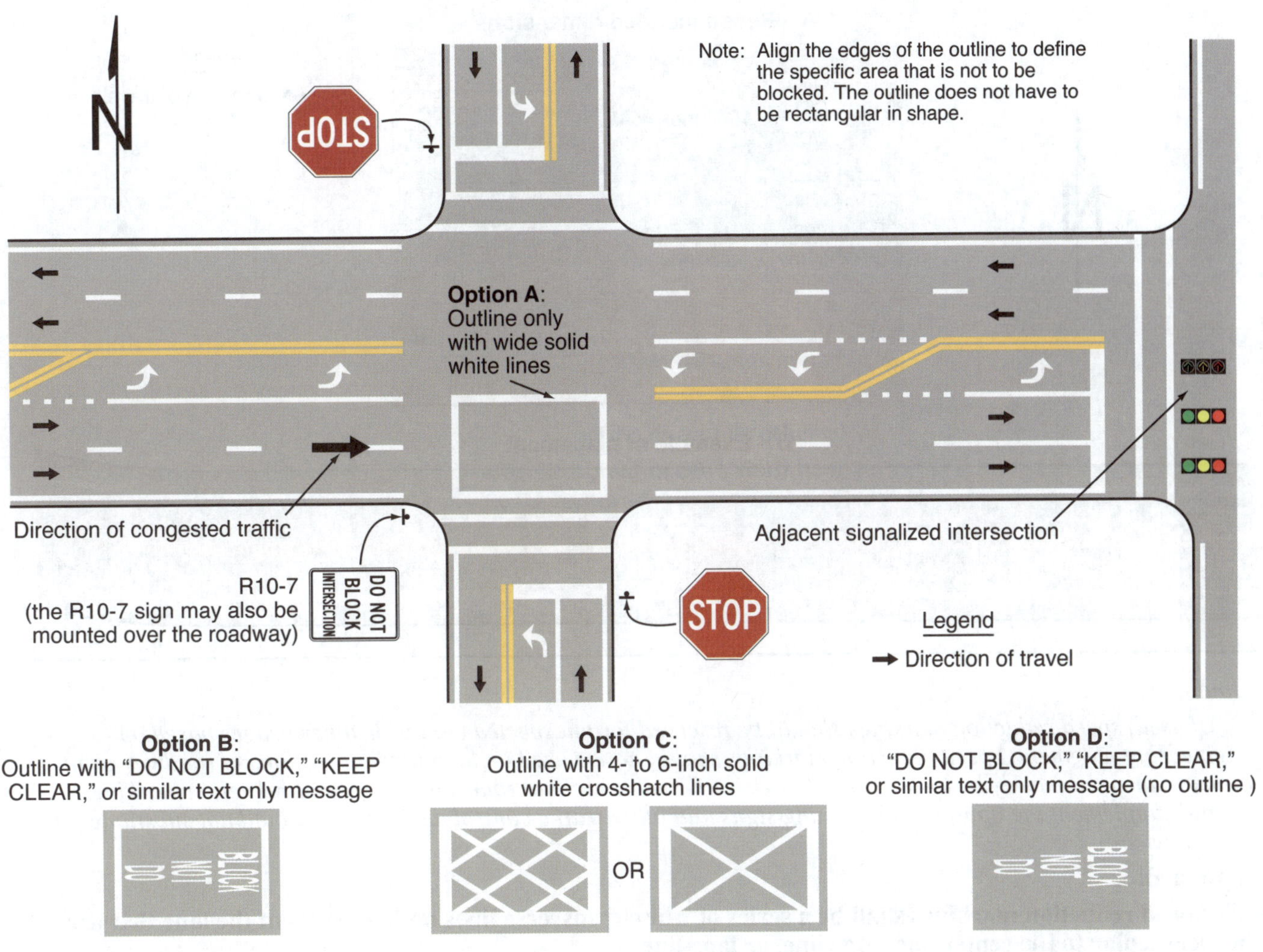

Option:

03   Blue lines may supplement white parking space markings of each parking space designated for use only by persons with disabilities (see Figure 3B-23).

Support:

04   Additional parking space markings for the purpose of designating spaces for use only by persons with disabilities are discussed in Section 3B.22 and illustrated in Figure 3B-19.

Section 3B.28 Speed Reduction Markings

Support:

01   Speed reduction markings (see Figure 3B-25) are transverse markings that are placed on the roadway within a lane (along both edges of the lane) in a pattern of progressively reduced spacing to give drivers the impression that their speed is increasing.

02   Speed reduction markings have been shown to enhance safety around curves and locations with a history of run-off-the-road crashes when applied in combination with horizontal alignment warning signs (see Section 2C.05).

Option:

03   Speed reduction markings may be placed in advance of an unexpectedly severe horizontal or vertical curve or other roadway feature where drivers need to decelerate prior to reaching the feature and where the desired reduction in speeds has not been achieved by the installation of warning signs and/or other traffic control devices.

Figure 3B-25. Example of the Application of Speed Reduction Markings

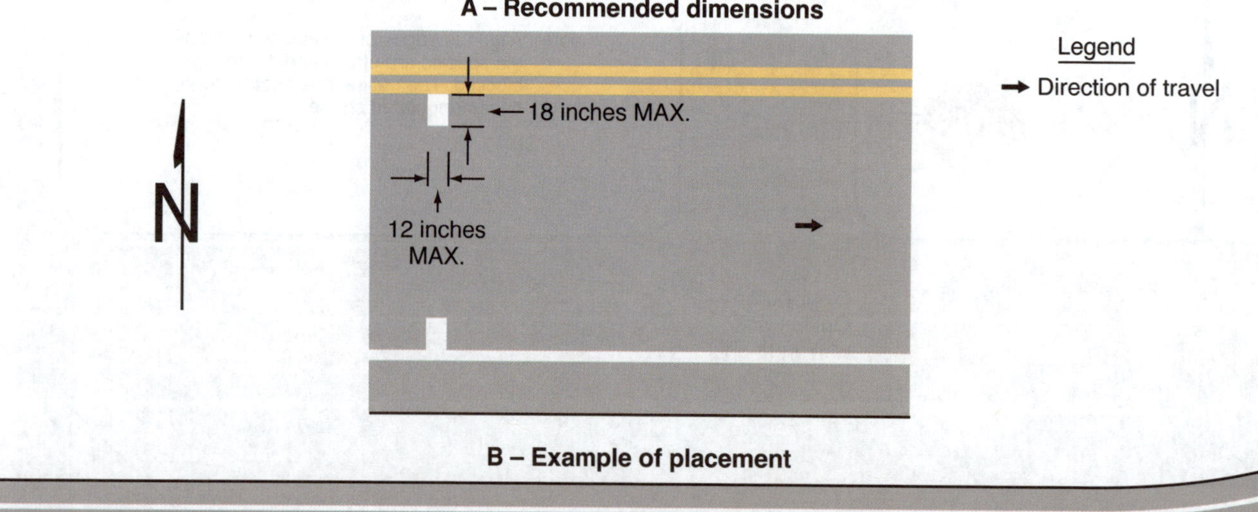

Guidance:

04   *If used, speed reduction markings should be reserved for unexpected curves or other usages based on engineering judgment. Speed reduction markings should not be used on long tangent sections of roadway or in areas frequented mainly by local or familiar drivers, such as school zones. If used, speed reduction markings should supplement the appropriate warning signs and other traffic control devices and should not substitute for these devices.*

Standard:

05   **Speed reduction markings shall be a series of white transverse lines on both sides of the lane that are perpendicular to the center line, edge line, or lane line.**

Guidance:

06   *The longitudinal spacing between the markings should be progressively reduced from the upstream to the downstream end of the marked portion of the lane.*

07   *Speed reduction markings should not be greater than 12 inches in width, and should not extend more than 18 inches into the lane.*

Standard:

08   **Speed reduction markings shall be used only in lanes that have a longitudinal line (center line, edge line, or lane line) on both sides of the lane.**

Section 3B.29 Speed Hump and Speed Table Markings

Standard:

01   **If speed hump markings are used, they shall be a series of white markings placed on a speed hump to identify its location. If markings are used for a speed hump that does not also function as a crosswalk or speed table, the markings shall comply with Option A, B, or C shown in Figure 3B-26. If markings are used for a speed hump that also functions as a crosswalk or speed table, the markings shall comply with Option A or B shown in Figure 3B-27.**

Option:

02   Where used, center line markings, lane line markings, and edge lines may be discontinued on the profile of the speed hump.

Standard:

03   **Where a speed hump or a speed table specifically incorporates a crossing movement for pedestrians, bicyclists, or equestrians, and functions as a raised crosswalk, crosswalk markings (see Chapter 3C) shall be provided.**

Figure 3B-26. Pavement Markings for Speed Humps without Crosswalks

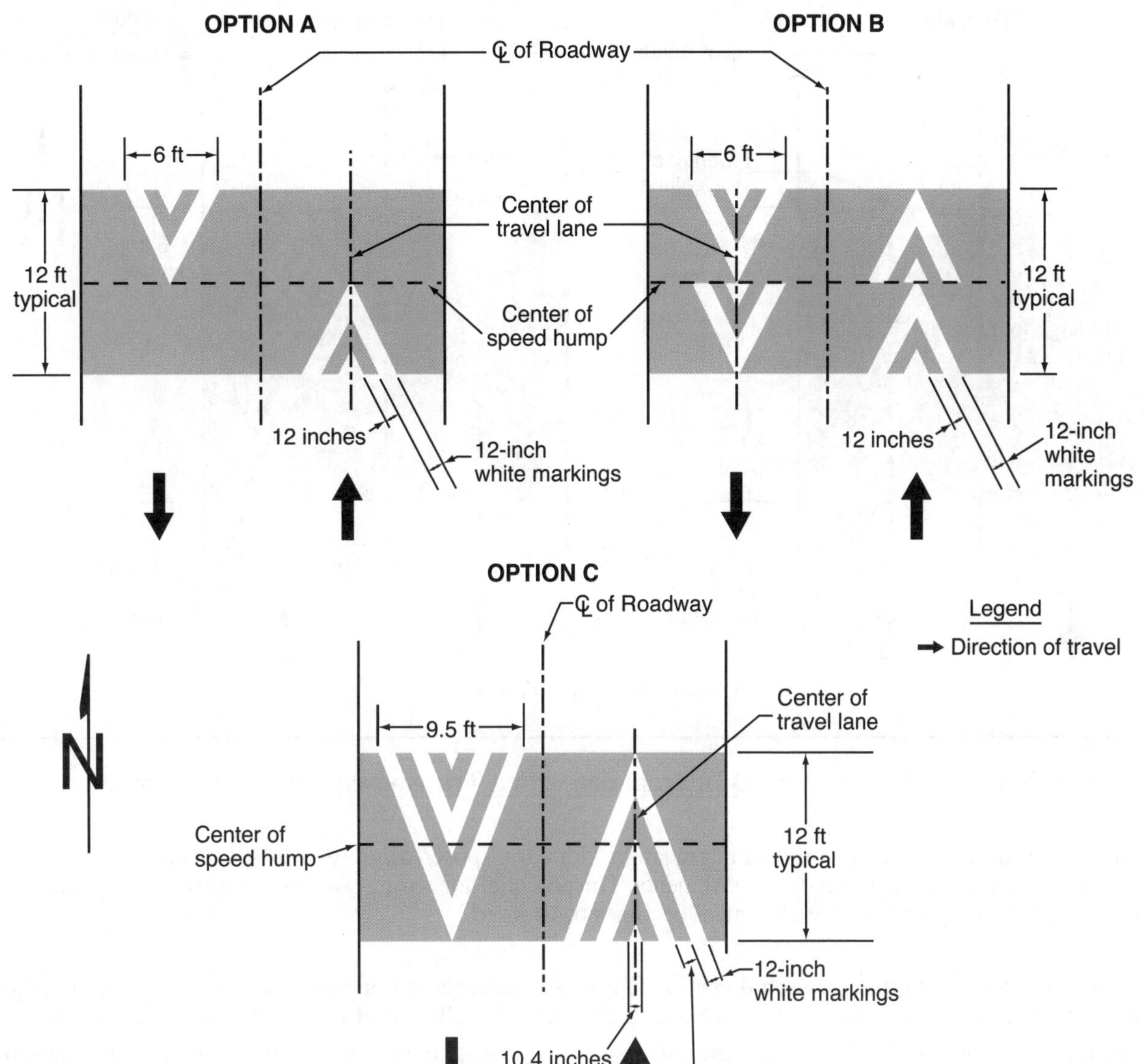

Section 3B.30 Advance Speed Hump and Speed Table Markings

Option:

01   Advance speed hump markings (see Figure 3B-28) may be used in advance of speed humps or other engineered vertical roadway deflections such as dips where added visibility is desired or where such deflection is not expected.

02   Advance word pavement markings such as BUMP or HUMP (see Section 3B.20) may be used on the approach to a speed hump either alone or in conjunction with advance speed hump markings. Appropriate advance warning signs may be used in compliance with Section 2C.27.

Standard:

03   **If advance speed hump or speed table markings are used, they shall be a series of eight white 12-inch transverse lines that become longer and are spaced closer together as the vehicle approaches the speed hump or other deflection. If advance markings are used, they shall comply with the detailed design shown in Figure 3B-28.**

Guidance:

04   *If used, advance speed hump markings should be installed in each approach lane.*

Figure 3B-27. Pavement Markings for Speed Tables or Speed Humps with Crosswalks

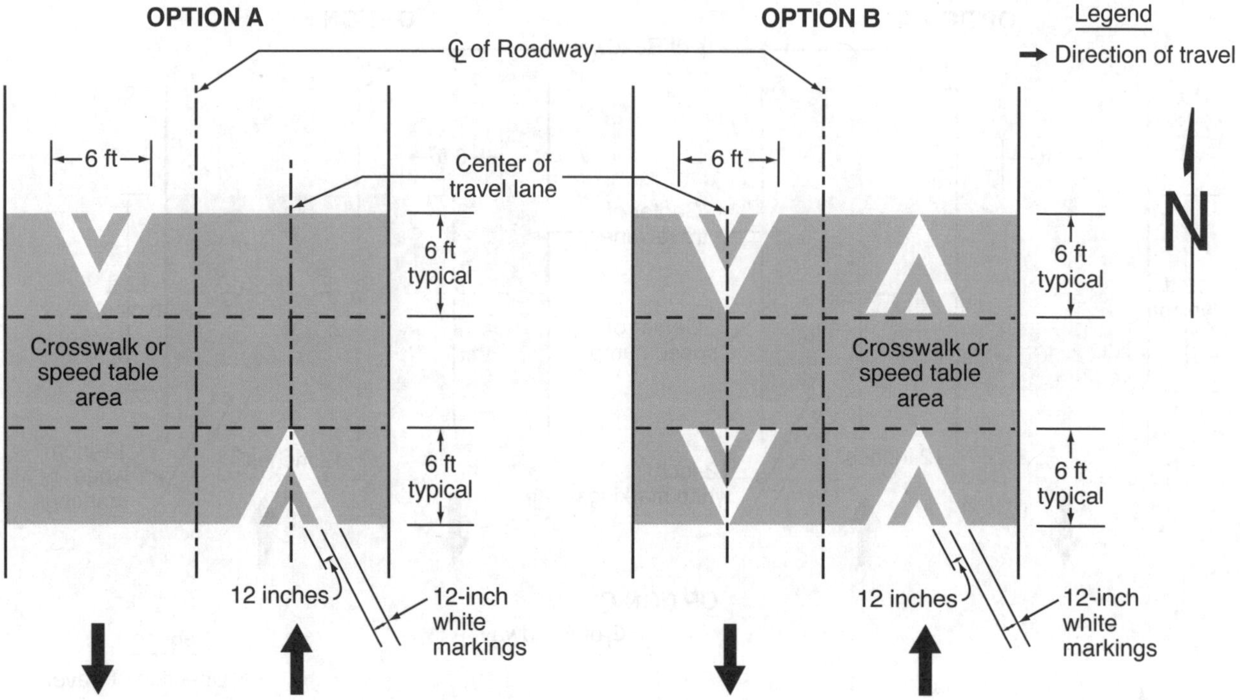

OPTION A

OPTION B

Legend

→ Direction of travel

℄ of Roadway

Center of travel lane

6 ft

6 ft typical

Crosswalk or speed table area

6 ft typical

12 inches

12-inch white markings

6 ft

6 ft typical

Crosswalk or speed table area

6 ft typical

12 inches

12-inch white markings

N

Note: Crosswalk lines are not shown

Section 3B.31 Markings for a Diamond Interchange with a Transposed Alignment Crossroad

Support:

01   Markings used in a diverging diamond interchange with a transposed alignment crossroad can be advantageous for minimizing wrong-way movements. The potential for wrong-way movements is greatest at the crossover intersections where the alignment becomes transposed.

Standard:

02   **On the transposed alignment, each direction shall be considered a one-way roadway whereas the edge line convention shall be in accordance with Section 3B.09. Both yellow and white edge lines shall be used.**

03   **A lane-use arrow (see Section 3B.23) shall be used in each approach lane at the crossover intersection.**

Support:

04   Section 3C.11 contains information on crosswalks and pedestrian movements for diverging diamond interchanges with a transposed alignment crossroad.

Standard:

05   **Flush median islands (see Section 3J.03) shall not be used to divide the inverted flow of traffic.**

Guidance:

06   *Edge line and lane line extensions (see Section 3B.11) should be provided through the crossing points.*

Support:

07   Figure 3B-29 illustrates an example of pavement markings for a diverging diamond interchange with a transposed alignment crossroad.

Figure 3B-28. Advance Warning Markings for Speed Humps or Speed Tables

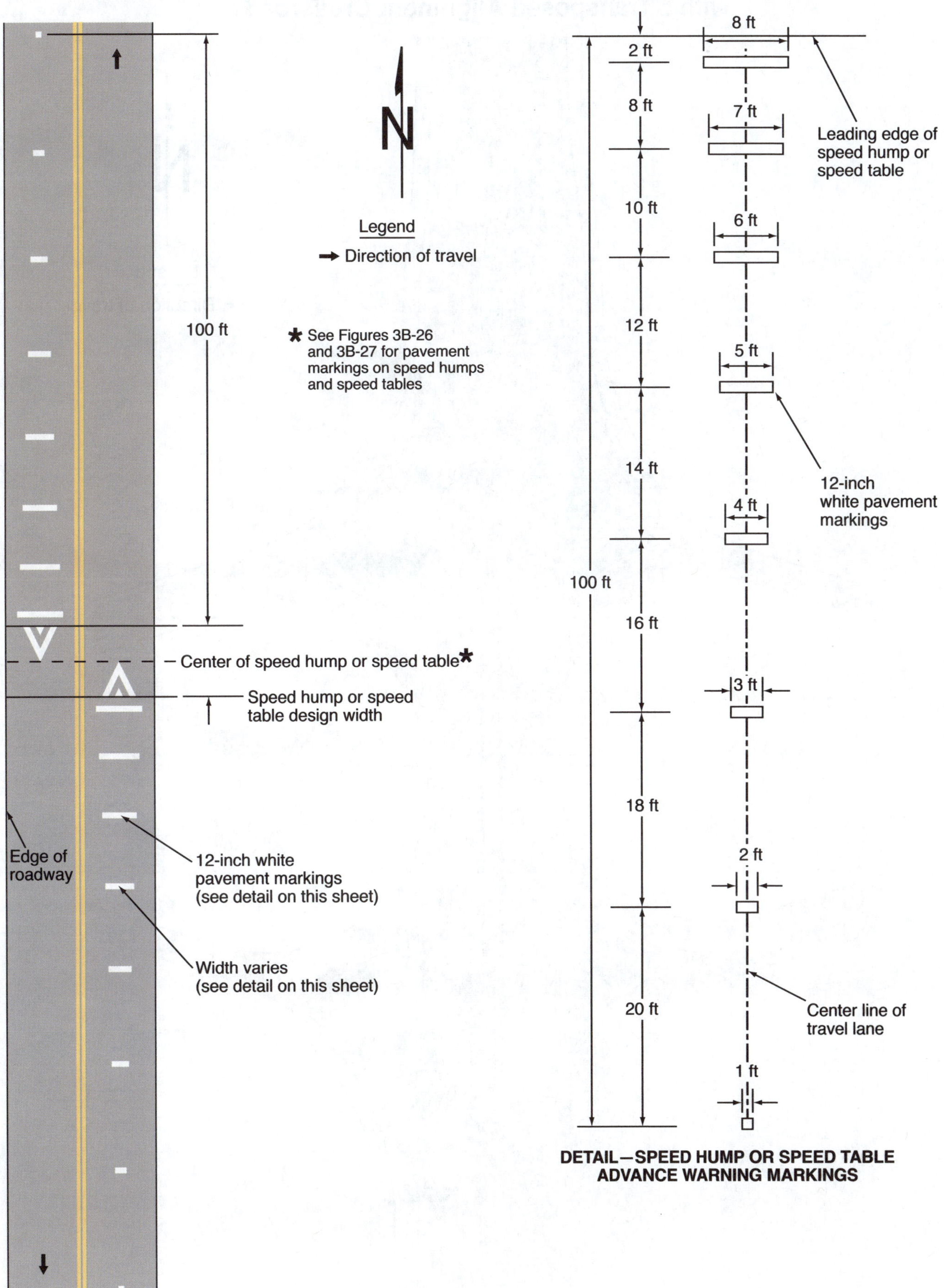

Figure 3B-29. Example of Pavement Markings for a Diamond Interchange with a Transposed Alignment Crossroad

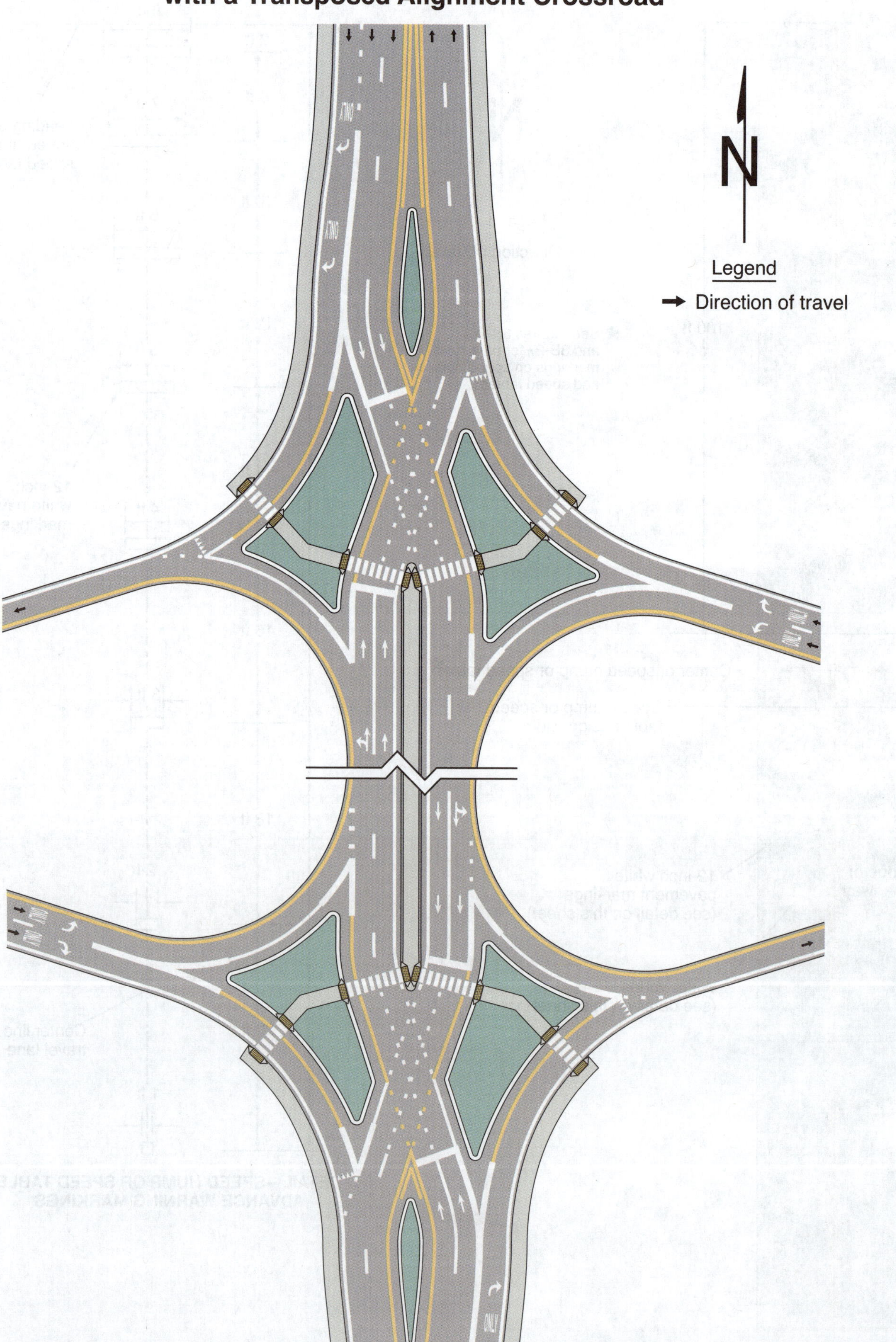

CHAPTER 3C. CROSSWALK MARKINGS

Section 3C.01 General

Support:

01   Crosswalk markings provide guidance for pedestrians who are crossing roadways by defining and delineating paths on approaches to and within signalized intersections, and on approaches to other intersections where traffic stops.

02   In conjunction with signs and other measures, crosswalk markings help to alert road users of a designated pedestrian crossing point across roadways at locations that are not controlled by traffic control signals or STOP or YIELD signs.

03   At non-intersection locations, crosswalk markings legally establish the crosswalk.

04   Detectable warning surfaces mark boundaries between pedestrian and vehicular ways where there is no raised curb. Detectable warning surfaces are typically installed where curb ramps are constructed at the junction of sidewalks and the roadway or shoulder, for marked and unmarked crosswalks. Detectable warning surfaces contrast visually with adjacent walking surfaces, either light-on-dark, or dark-on-light. The U.S. Department of Justice 2010 ADA Standards for Accessible Design, September 15, 2010, 28 CFR 35 and 36, Americans with Disabilities Act of 1990 contains specifications for the design of detectable warning surfaces.

Section 3C.02 Application of Crosswalk Markings

Guidance:

01   *At locations controlled by traffic control signals, crosswalk markings should be installed.*

Option:

02   Crosswalk markings may be omitted if engineering judgment indicates they are not needed to direct pedestrians to the proper crossing path(s).

Guidance:

03   *On approaches controlled by STOP or YIELD signs, crosswalk markings should be installed where engineering judgment indicates they are needed to direct pedestrians to the proper crossing path(s).*

04   *At uncontrolled approaches, an engineering study should be performed before a marked crosswalk is installed. The following criteria should be considered:*

    A. Total number of approach lanes,
    B. The presence of a median,
    C. The distance from adjacent signalized intersections or other controlled crossings,
    D. Projected pedestrian and bicyclist volumes,
    E. Pedestrian and bicyclist paths of travel,
    F. Pedestrian ages and abilities,
    G. Pedestrian and bicyclist delays,
    H. Location or frequency of public transit stops,
    I. Average daily traffic (ADT),
    J. Speed limit or the 85th-percentile speed,
    K. The horizontal and vertical geometry of the crossing location,
    L. The possible consolidation of multiple crossing points,
    M. The availability of street lighting, and
    N. Other appropriate factors.

Standard:

05   **Crosswalk markings shall be provided at legally established crosswalks at non-intersection locations.**

Guidance:

06   *The installation of other traffic control devices and other measures designed to reduce traffic speeds, shorten crossing distances, enhance driver awareness of the crossing, and/or provide active warning of pedestrian presence, should be considered in addition to a new marked crosswalk and signs across an uncontrolled roadway where any of the following conditions exist:*

    A. The roadway has four or more lanes of travel without a raised median or pedestrian refuge island and an ADT of 12,000 vehicles per day or greater; or
    B. The roadway has four or more lanes of travel with a raised median or pedestrian refuge island and an ADT of 15,000 vehicles per day or greater, or
    C. The posted speed limit is 40 mph or greater, or
    D. A crash study reveals that multiple-threat crashes are the predominant crash type on a multi-lane approach, or
    E. When adequate visibility cannot be provided by parking prohibitions.

Support:

07   Chapter 4J contains information on pedestrian hybrid beacons.

08   Chapter 4L contains information on rectangular rapid flashing beacons.

09   Section 4S.03 contains information regarding Warning Beacons to provide active warning of a pedestrian's presence.

10   Section 4U.02 contains information regarding In-Roadway Warning Lights at crosswalks.

11   Chapter 7C contains information on school crosswalks.

12   Chapter 7D contains information regarding school crossing supervision.

13   Section 9E.13 contains information on crosswalk markings for shared-use path crossings.

Section 3C.03 Design of Crosswalk Markings

Support:

01   Section 3B.19 contains information regarding placement of stop line markings and yield line markings near crosswalk markings.

02   Crosswalk markings are classified as either transverse line or high-visibility. Transverse crosswalk markings consist of two transverse lines. High-visibility markings consist of longitudinal lines parallel to traffic flow with or without transverse lines. Figure 3C-1 presents crosswalk marking designs.

Standard:

03   **Crosswalk markings shall be white. When used, transverse lines shall not be less than 6 inches or greater than 24 inches in width.**

Support:

04   The allowable upper limit approaching 24 inches for the width of the transverse lines is normally applied where no stop or yield line is used in advance of the crosswalk or when approach speeds exceed 35 miles per hour.

Standard:

05   **Except as provided in Paragraph 6 of this Section, the minimum width of a marked crosswalk shall be 6 feet.**

06   **At a non-intersection crosswalk where the posted speed limit is 40 mph or greater, the minimum width of the crosswalk shall be 8 feet.**

Figure 3C-1. Crosswalk Markings

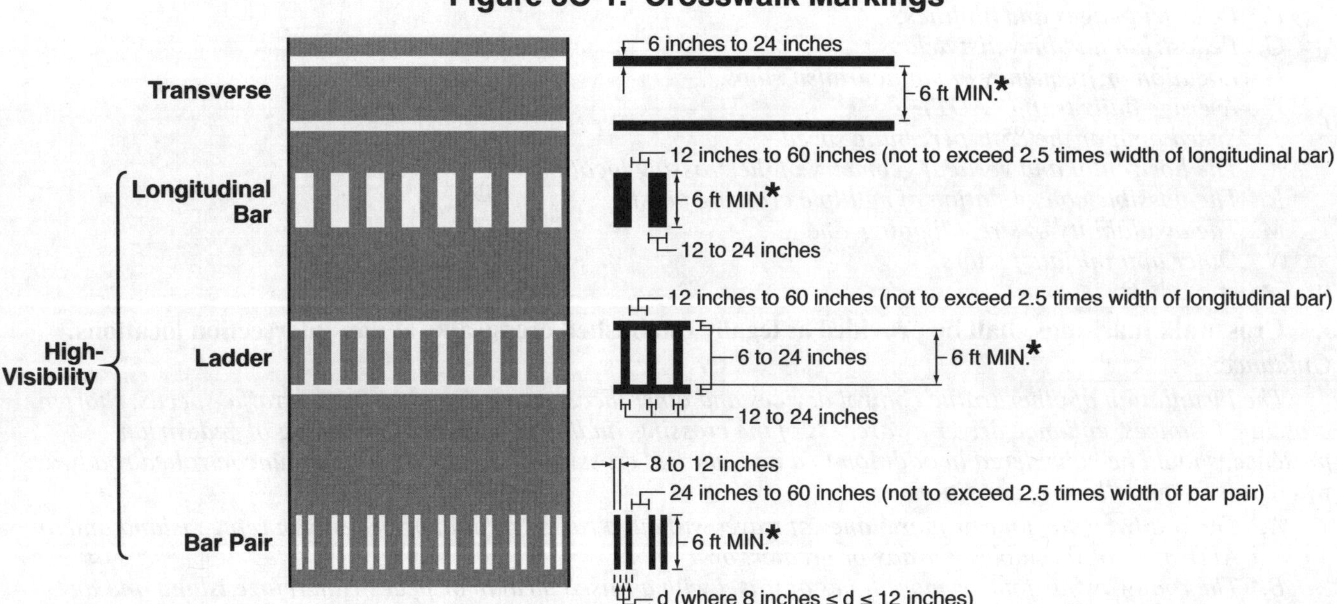

Guidance:

07   *High-visibility crosswalk markings (such as shown in Figure 3C-1) and warning signs (see Section 2C.55) should be installed for all crosswalks at non-intersection locations.*

08   *Added visibility should be provided by parking prohibitions on the approach to marked crosswalks at non-intersection locations.*

Standard:

09   **Where curb ramps are provided, crosswalk markings shall be located so that the curb ramps are within the extension of the crosswalk markings.**

Guidance:

10   *Transverse line crosswalk markings should extend across the full width of pavement or to the edge of the intersecting crosswalk to discourage diagonal walking between crosswalks.*

Support:

11   Provisions for aesthetic treatments for the interior portion of a legally-established crosswalk are contained in Section 3H.03.

Standard:

12   **If paving materials are used to function as the white transverse lines to establish a marked crosswalk, white additives shall be part of the mixture to produce a white surface. The white paving materials shall be retroreflective.**

Section 3C.04 <u>Transverse Line Crosswalks</u>

Guidance:

01   *Transverse line crosswalk markings should be limited to locations controlled by traffic control signals or on approaches controlled by STOP or YIELD signs.*

Support:

02   Transverse line crosswalk marking design consists of two parallel transverse lines (see Figure 3C-1).

03   Transverse line crosswalk markings can provide benefits to crosswalk operations including:

    A. Define where the channelization of pedestrians or other non-motorized users is necessary to facilitate crossing the roadway.
    B. Alert motorists to the location of where pedestrians and other non-motorized users might be expected when crossing the roadway.
    C. Emphasize a crosswalk at a controlled intersection.
    D. Fulfill a legal need to mark the crosswalk.

Section 3C.05 <u>High-Visibility Crosswalks</u>

Option:

01   High-visibility crosswalk markings may be used where additional conspicuity is desired for a crosswalk over transverse line crosswalk markings.

Support:

02   High-visibility crosswalk markings include the longitudinal bar, ladder, and bar pair designs (see Figure 3C-1).

03   High-visibility crosswalk markings can provide benefits to crosswalk operations including:

    A. Providing greater detection distances for the approaching motorist.
    B. Emphasizing a crosswalk where substantial numbers of pedestrians cross without any other traffic control device.
    C. Emphasizing a crosswalk at an uncontrolled approach.
    D. Emphasizing the location where a high number of conflicts between turning motorists and users of the crosswalk are expected.
    E. Improving visibility of the crosswalk location for otherwise difficult-to-detect pedestrians or other non-motorized users of the crosswalk.
    F. Emphasizing a school crossing.

Standard:

04   **The minimum number of individual longitudinal elements to establish a high-visibility crosswalk shall be three. For the bar pair crosswalk design (see Section 3C.08), a coupling set of two longitudinal bars shall be considered to be one individual longitudinal element.**

Guidance:

05   *The dimensions of the individual longitudinal element and the lateral spacing between subsequent individual longitudinal elements for a high-visibility crosswalk should be uniform when establishing the crosswalk.*

06   *The dimensions of the individual longitudinal element and the lateral spacing between subsequent individual longitudinal elements for a high-visibility crosswalk should be uniform when establishing separate crosswalks on multiple approaches to the same intersection and on both sides of a median refuge if one is present.*

07   *The individual longitudinal elements of a high-visibility crosswalk should be angled such that they are parallel to the travel path of approaching traffic.*

Option:

08   The lateral spacing between longitudinal elements may be staggered to avoid wheel paths, center lines, and lane lines.

Section 3C.06 Longitudinal Bar Crosswalks

Support:

01   The longitudinal bar crosswalk marking design (see Figure 3C-1) provides for improved detection and recognition over the transverse line crosswalk for people with low vision and cognitive impairments.

Standard:

02   **The width of an individual longitudinal bar shall not be less than 12 inches or greater than 24 inches.**

03   **The lateral spacing between subsequent longitudinal bars shall not be less than 12 inches or greater than 60 inches. The lateral spacing of the longitudinal bars shall not exceed 2.5 times the width of a longitudinal bar.**

Section 3C.07 Ladder Crosswalks

Support:

01   Ladder crosswalks (see Figure 3C-1) implement a pattern where interior longitudinal bars and transverse lines are used to define the limits of the crosswalk.

02   The ladder crosswalk marking design provides for improved detection and recognition over the transverse crosswalk for people with low vision and cognitive impairments.

03   Since the longitudinal component of the ladder crosswalk marking design is similar to the benefits provided by the longitudinal bar crosswalk design, the ladder crosswalk design is normally used to discourage or prohibit diagonal walking between crosswalks.

Standard:

04   **The transverse lines used to establish the limits of the ladder crosswalk shall not be less than 6 inches or greater than 24 inches in width.**

05   **The width of an individual interior longitudinal bar shall not be less than 12 inches or greater than 24 inches.**

06   **The lateral spacing between subsequent interior longitudinal bars shall not be less than 12 inches or greater than 60 inches. The lateral spacing of the interior longitudinal bars shall not exceed 2.5 times the width of an interior longitudinal bar.**

Option:

07   Where it might be necessary to alleviate a parallax phenomenon due to approaching roadway geometry that curves or to accommodate low approach angles of the approaching motorist, the interior longitudinal bars may be rotated up to 45 degrees to the transverse lines to remain parallel to approaching traffic.

Section 3C.08 Bar Pair Crosswalks

Support:

01   Bar pair crosswalks (see Figure 3C-1) can provide the same benefits as other high-visibility crosswalk designs with the opportunity for less maintenance.

02   Bar pair crosswalks can be useful in locations that are susceptible to slip and fall incidents exacerbated by extreme or inclement weather, or in locations where high motorcycle or bicycle use is expected, in order to maximize wheel traction with the road surface.

Standard:

03   **The width of an individual longitudinal bar that establishes one-half of the bar pair shall not be less than 8 inches or greater than 12 inches. The lateral spacing between successive individual longitudinal bars within the same bar pair shall be equal to the width of one longitudinal bar.**

04   **The lateral spacing between subsequent longitudinal bar pairs shall not be less than 24 inches or greater than 60 inches, or 2.5 times the width of the total width of a bar pair.**

05   **Longitudinal bar pair crosswalks shall not be installed with accompanying transverse lines.**

Section 3C.09 Crosswalk Markings at Circular Intersections

Standard:

01   **Crosswalk markings shall not be provided to or from the central island of a roundabout.**

Guidance:

02   *If pedestrian facilities are provided, crosswalks should be marked across roundabout entrances and exits to indicate where pedestrians are intended to cross.*

03   *On an approach to a circular intersection controlled by a YIELD sign and at uncontrolled exits, crosswalks should be a minimum of 20 feet from the edge of the circulatory roadway.*

Support:

04   Chapter 3D provides figures that illustrate examples of crosswalk markings for roundabouts.

Section 3C.10 Crosswalks for Exclusive Pedestrian Phases that Permit Diagonal Crossings

Option:

01   When an exclusive pedestrian phase that permits diagonal crossing of an intersection is provided at a traffic control signal, a marking as shown in Figure 3C-2 may be used for the crosswalk.

Guidance:

02   *The segments of the crosswalk marking that facilitate the diagonal crossing should not use high-visibility crosswalk markings.*

Section 3C.11 Crosswalks at Diamond Interchanges with a Transposed Alignment Crossroad

Support:

01   Diverging diamond interchanges, also known as double-crossover diamond interchanges, include directional crossovers on either side of the interchange that transpose the crossroad which results in vehicles traveling on the left-hand side of the street or highway between the crossover intersections. The potential for altered travel paths for pedestrians and the associated, unique, operational aspects such as traffic approaching from unexpected directions and unfamiliar signal phasing schemes are important considerations.

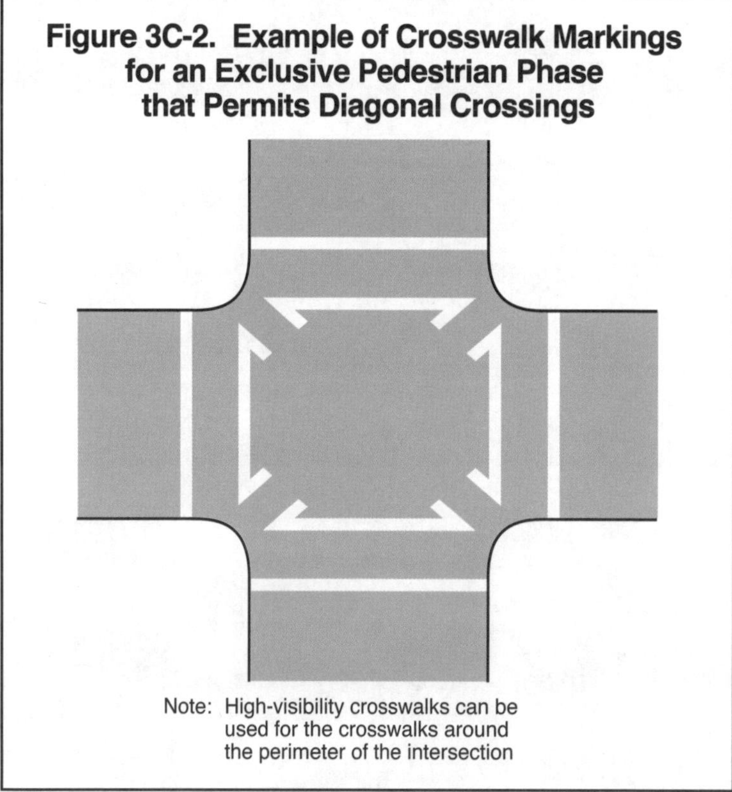

Note: High-visibility crosswalks can be used for the crosswalks around the perimeter of the intersection

Guidance:

02   *If pedestrian facilities are provided, pedestrian crossing movements of the crossroads at a diverging diamond interchange should be marked at the crossover intersections where motor vehicle traffic becomes transposed.*

03   *If pedestrian facilities are provided, crosswalks should be marked across ramp terminals at diverging diamond interchanges to indicate where pedestrians are intended to cross.*

04   *Crosswalks across diverging diamond interchange ramps with yield-controlled vehicle movements should be located a minimum of 20 feet from the edge of an intersecting ramp.*

Support:

05   Section 3B.31 contains information on markings, such as edge lines, lane lines, and lane-use arrows, for diverging diamond interchanges.

06   Figure 3B-29 shows an example of pedestrian crossing locations at a diverging diamond interchange.

Section 3C.12 <u>Pedestrian Islands and Medians</u>

Support:

01   Raised islands or raised medians of sufficient width that are placed in the center area of a street or highway can serve as a place of refuge for pedestrians who are attempting to cross at a midblock or intersection location. Center islands or medians allow pedestrians to find an adequate gap in one direction of traffic at a time, as the pedestrians are able to stop, if necessary, in the center island or median area and wait for an adequate gap in the other direction of traffic before crossing the second half of the street or highway. The U.S. Department of Justice 2010 ADA Standards for Accessible Design, September 15, 2010, 28 CFR 35 and 36, Americans with Disabilities Act of 1990 contains specifications for the design of detectable warning surfaces and provides technical requirements that can be used to determine the minimum width for accessible refuge islands.

CHAPTER 3D. CIRCULAR INTERSECTION MARKINGS

Section 3D.01 General

Guidance:

01   *Pavement markings and signing for a circular intersection should be integrally designed to correspond to the geometric design and intended lane use of a circular intersection.*

02   *Markings on the approaches to a circular intersection and on the circulatory roadway should be compatible with each other to provide a consistent message to road users. The markings should supplement the signing, both conveying the optional and mandatory movements such that road users will know to choose the proper lane in the approach to the circular intersection and remain in that lane throughout departure from the circulatory roadway.*

Support:

03   Common circular intersection types include roundabouts, rotaries, and traffic circles (see definitions in Section 1C.02). Traffic circles and rotaries are often much larger than roundabouts. Modern roundabouts feature channelized, curved approaches that reduce vehicle speed. Traffic calming circles are smaller and are typically used on urban or suburban neighborhood streets.

04   Figure 3D-1 provides an example of the pavement markings for approach and circulatory roadways at a roundabout. Figures 3D-2 through 3D-8 illustrate examples of markings for roundabouts of various geometric and lane-use configurations.

05   Actuated LED pedestrian warning signs (see Section 2A.12), traffic control signals, pedestrian hybrid beacons, and rectangular rapid flashing beacons (see Part 4) are sometimes used at roundabouts to facilitate the crossing of pedestrians or to meter traffic.

06   Section 8A.12 provides information about circular intersections that contain or are in close proximity to grade crossings.

07   Section 9E.05 contains information regarding bicycle lane markings at circular intersections.

08   Section 3C.09 contains information regarding crosswalks at circular intersections.

Section 3D.02 White Lane Line Pavement Markings for Roundabouts

Standard:

01   **Multi-lane approaches to roundabouts shall have lane lines.**

02   **A through lane on a roadway that becomes a dropped lane (mandatory left-turn or right-turn lane) at a roundabout shall be marked with a dotted white lane line in accordance with Section 3B.07.**

Guidance:

03   *Multi-lane roundabouts should have lane line markings within the circulatory roadway to continuously channelize traffic in the circulatory roadway and through the departure movement.*

Standard:

04   **Continuous concentric lane lines shall not be used within the circulatory roadway of a roundabout.**

Option:

05   Channelizing lines (see Section 3B.08) and chevron and diagonal markings (see Section 3B.25) may be used on the approaches to and within the circulatory roadway of multi-lane roundabouts to separate traffic lanes, discourage lane changing, and/or compensate for off-tracking of larger trucks and vehicles.

Support:

06   Reducing the spacing between lines of a broken lane line allows better delineation of the lower-radius curves typically found in circular intersections.

Section 3D.03 Edge Line Pavement Markings for Roundabout Circulatory Roadways

Guidance:

01   *A white edge line should be used on the outer (right-hand) edge of the circulatory roadway.*

02   *Where a white edge line is used for the circulatory roadway, it should be as follows (see Figure 3D-1):*
    A. A solid line adjacent to the splitter island, and
    B. A wide dotted line across the lane(s) entering the roundabout.

Standard:

03   **Edge lines and edge line extensions shall not be placed across the exits from the circulatory roadway at roundabouts.**

Option:

04   A yellow edge line may be placed around the inner (left-hand) edge of the circulatory roadway (see Figure 3D-1) and may be used to channelize traffic (see Drawing B in Figure 3D-3).

Figure 3D-1. Example of Markings for Approach and Circulatory Roadways at a Roundabout

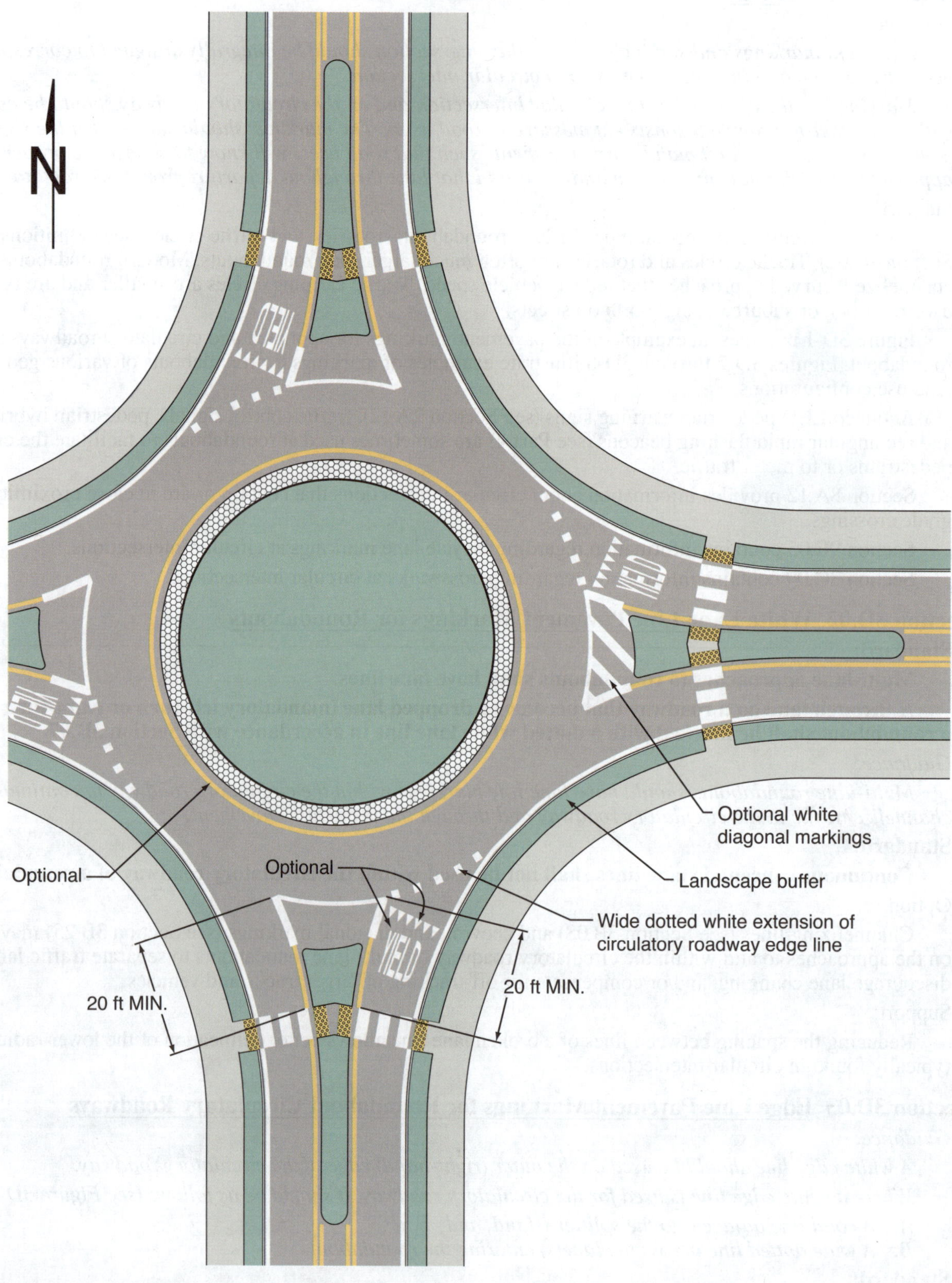

Figure 3D-2. Example of Markings for a One-Lane Roundabout

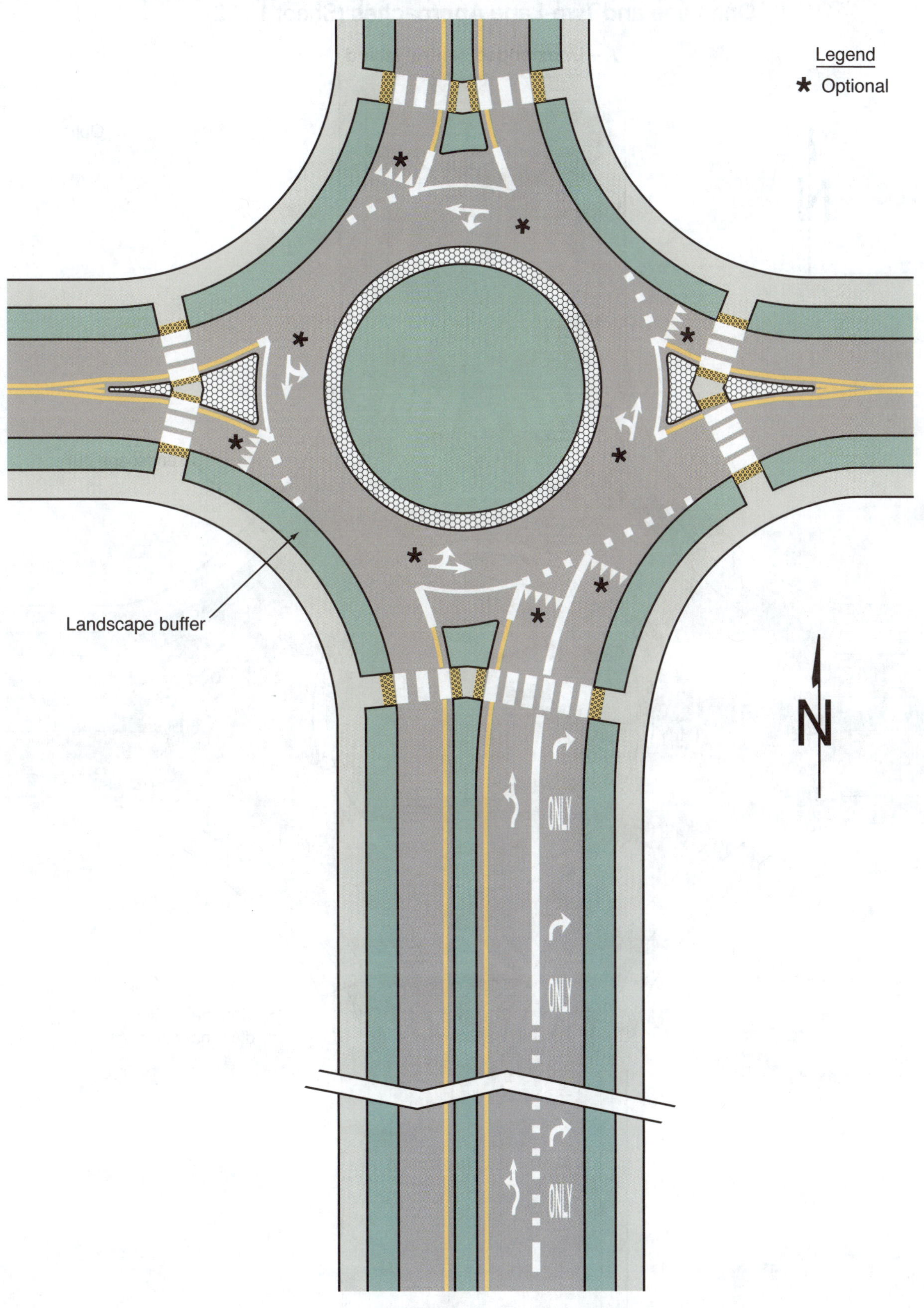

Figure 3D-3. Example of Markings for a Two-Lane Roundabout with One-Lane and Two-Lane Approaches (Sheet 1 of 2)

A – Unextended central island

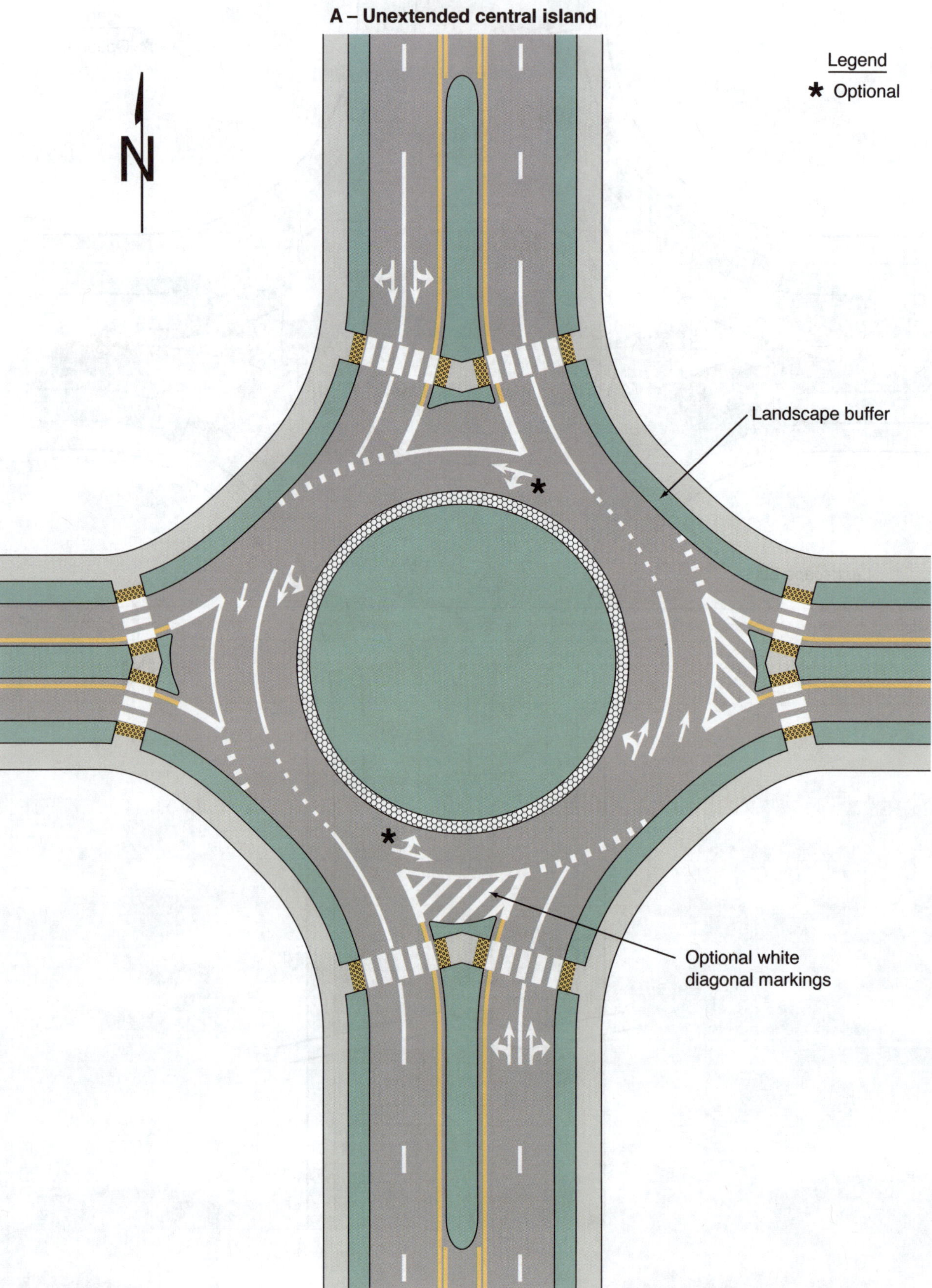

Figure 3D-3. Example of Markings for a Two-Lane Roundabout with One-Lane and Two-Lane Approaches (Sheet 2 of 2)

B – Central island extended by pavement markings

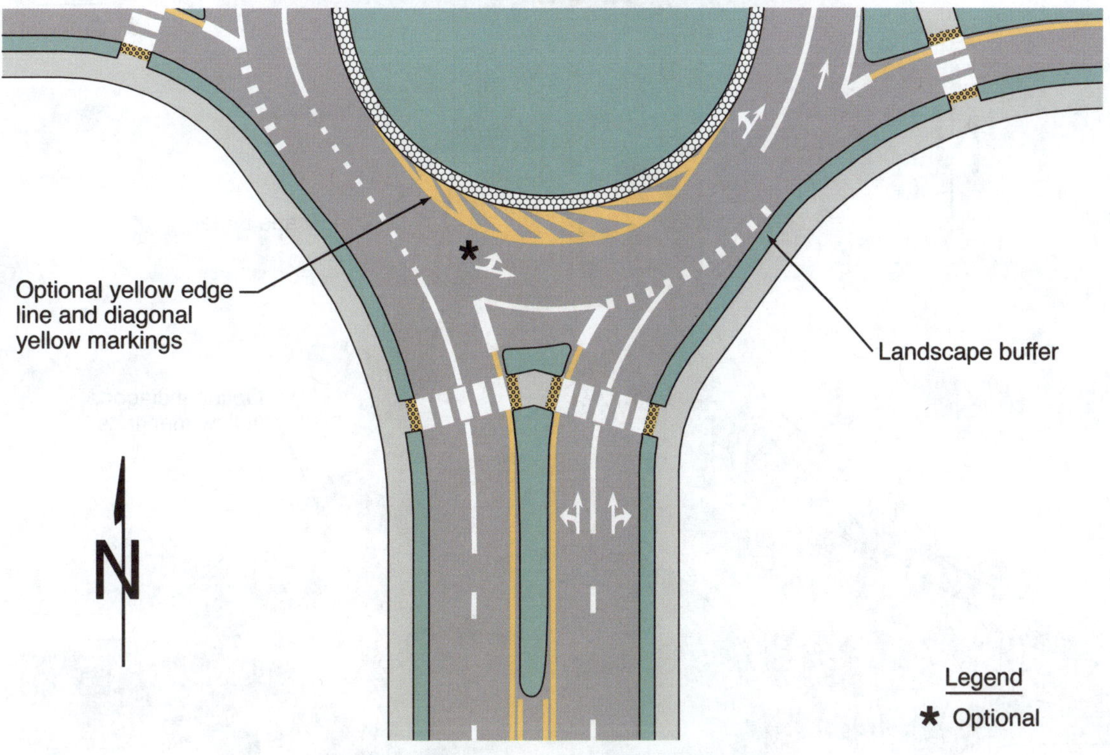

C – Central island extended by a truck apron

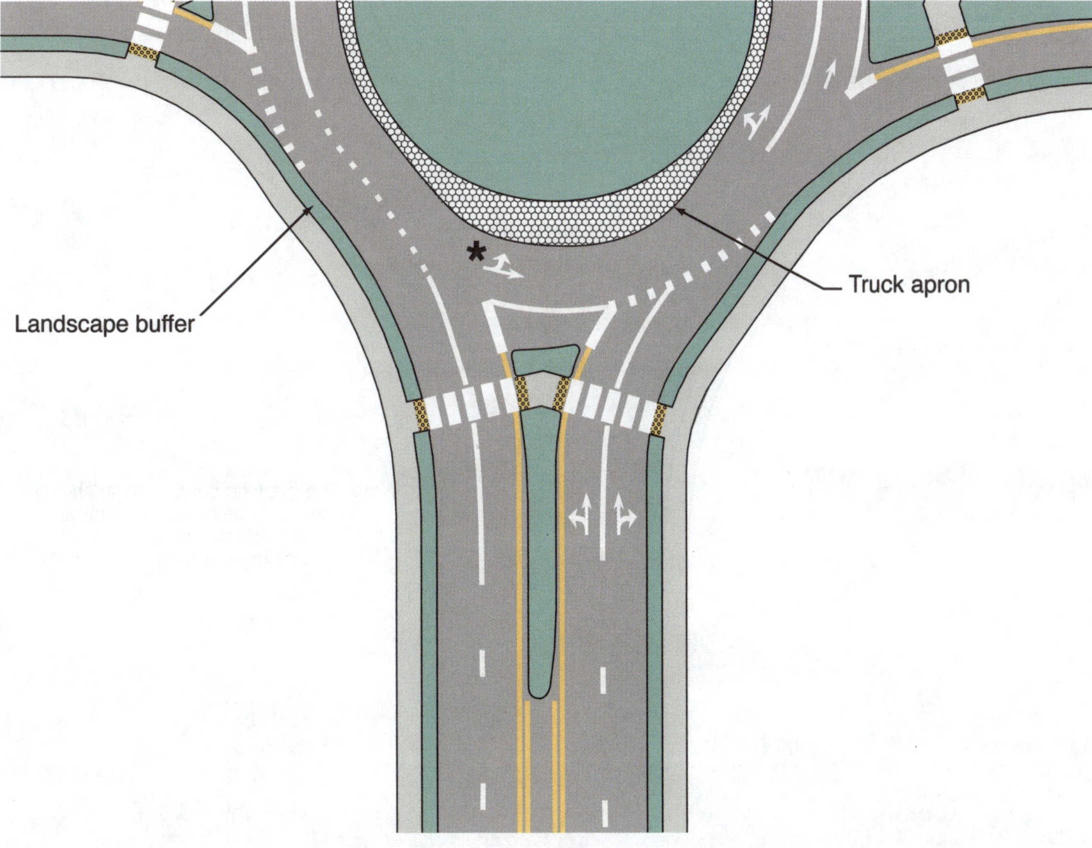

Figure 3D-4. Example of Markings for a Two-Lane Roundabout with One-Lane Exits

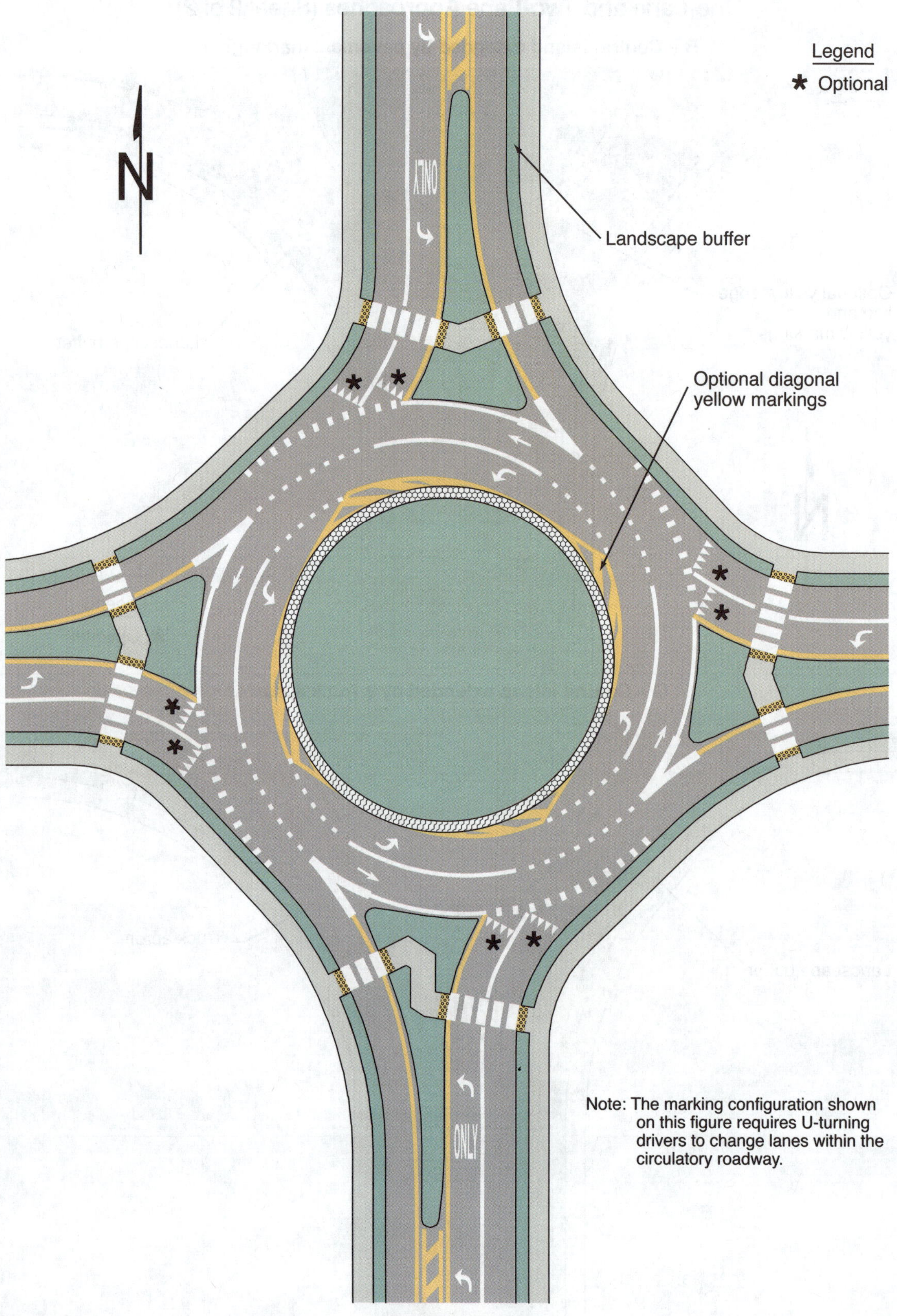

Figure 3D-5. Example of Markings for a Two-Lane Roundabout with Two-Lane Exits

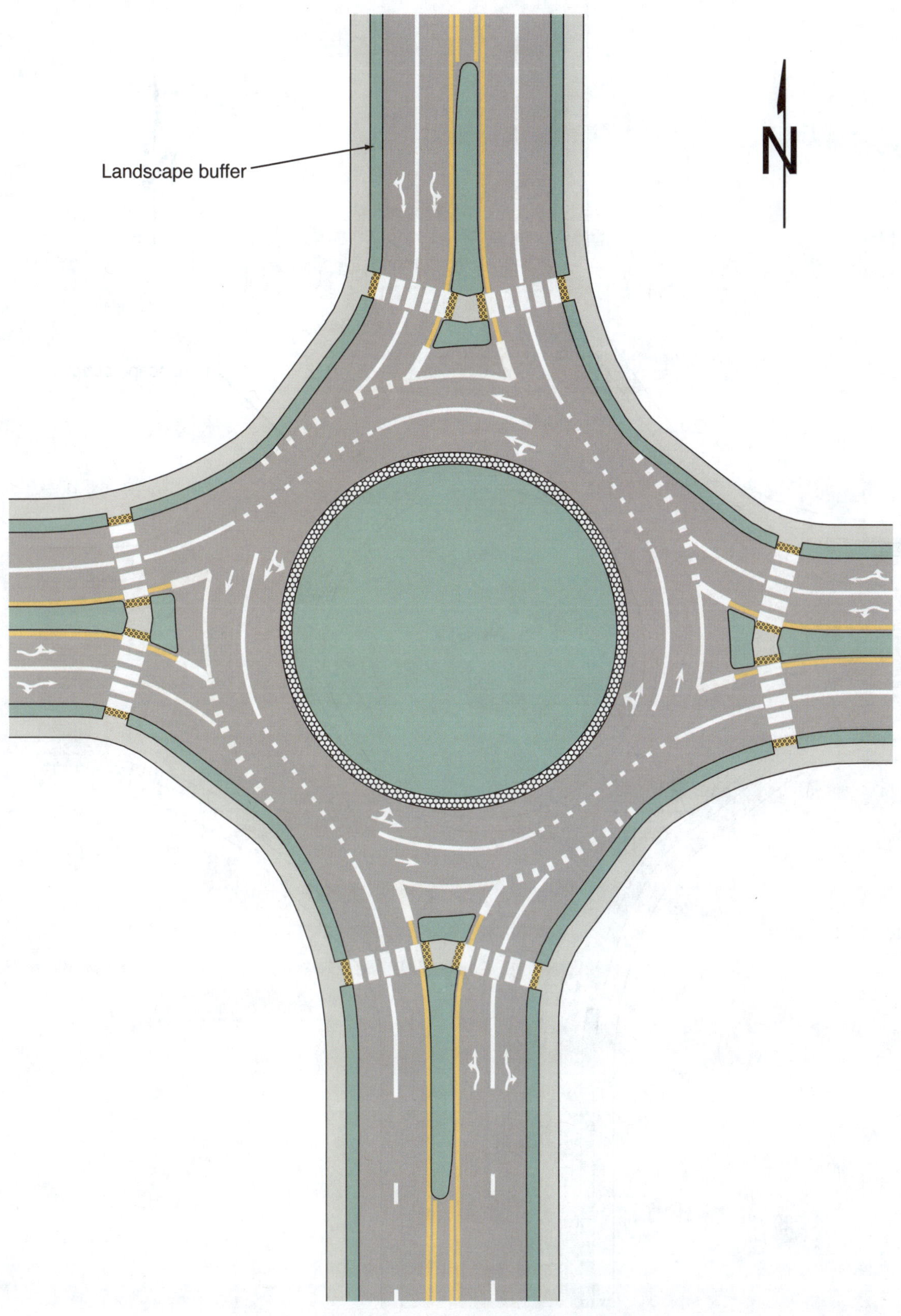

Figure 3D-6. Example of Markings for a Two-Lane Roundabout with a Double Left Turn

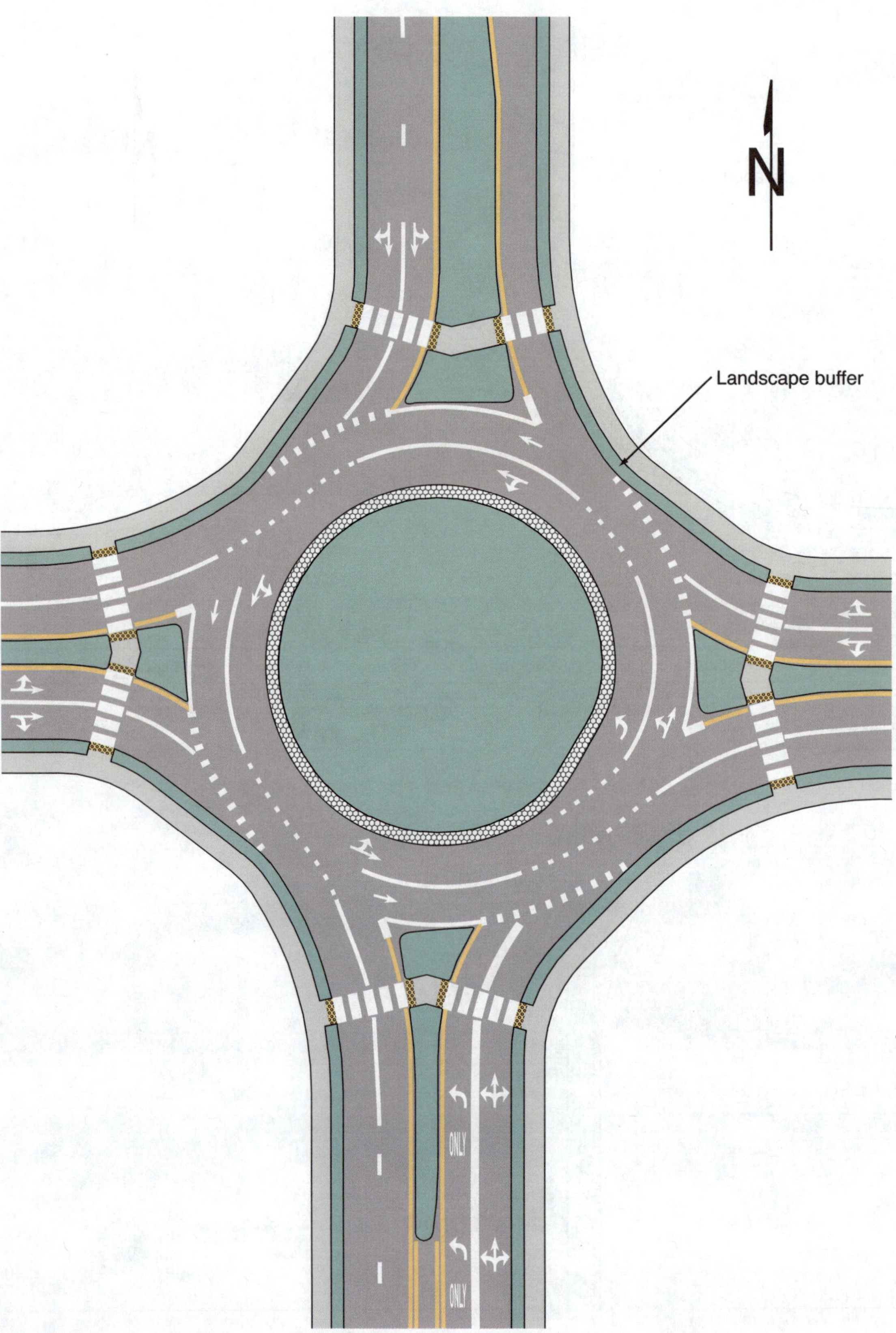

Figure 3D-7. Example of Markings for a Two-Lane Roundabout with a Double Right Turn

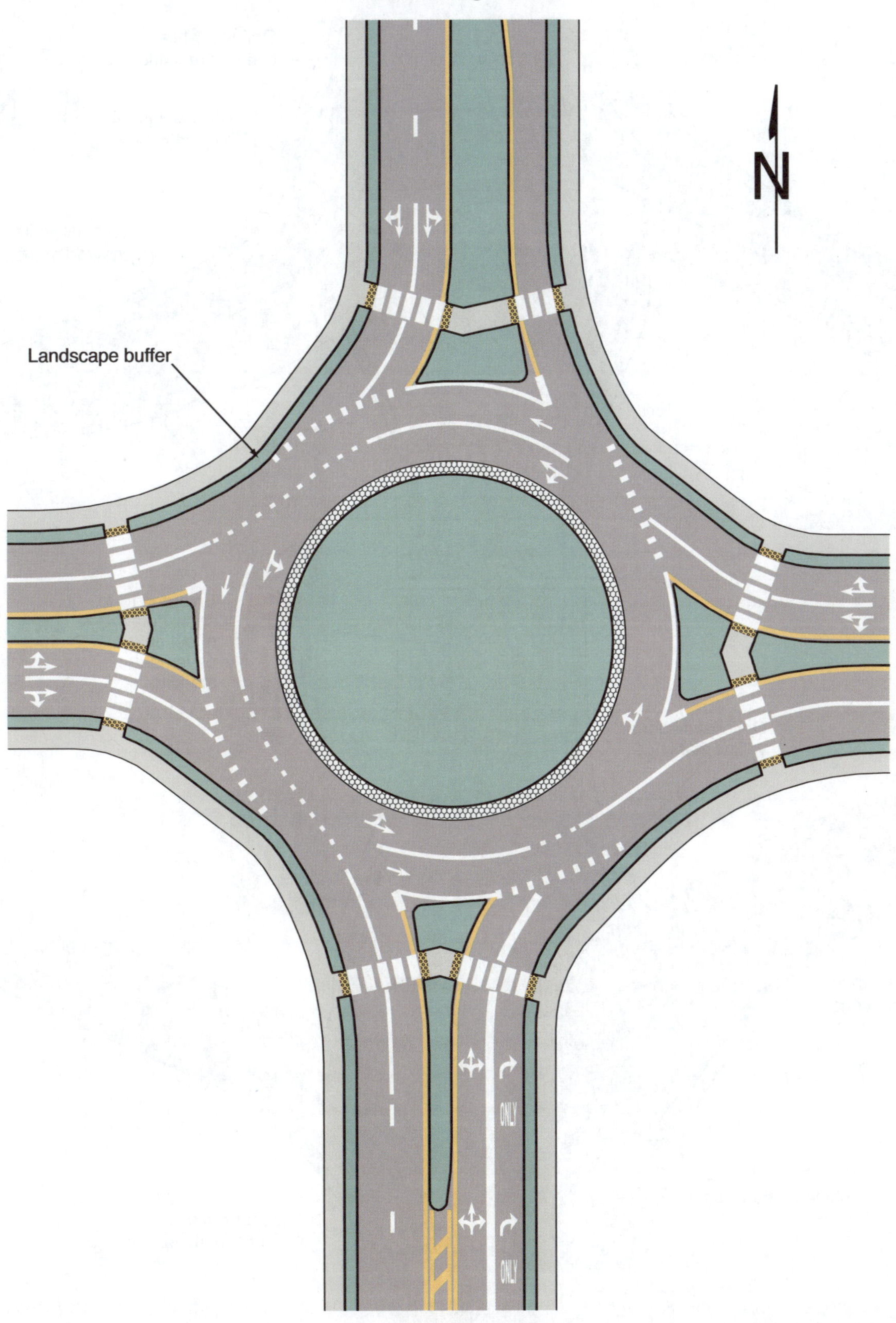

Figure 3D-8. Example of Markings for a Diamond Interchange with Two Circular-Shaped Roundabout Ramp Terminals

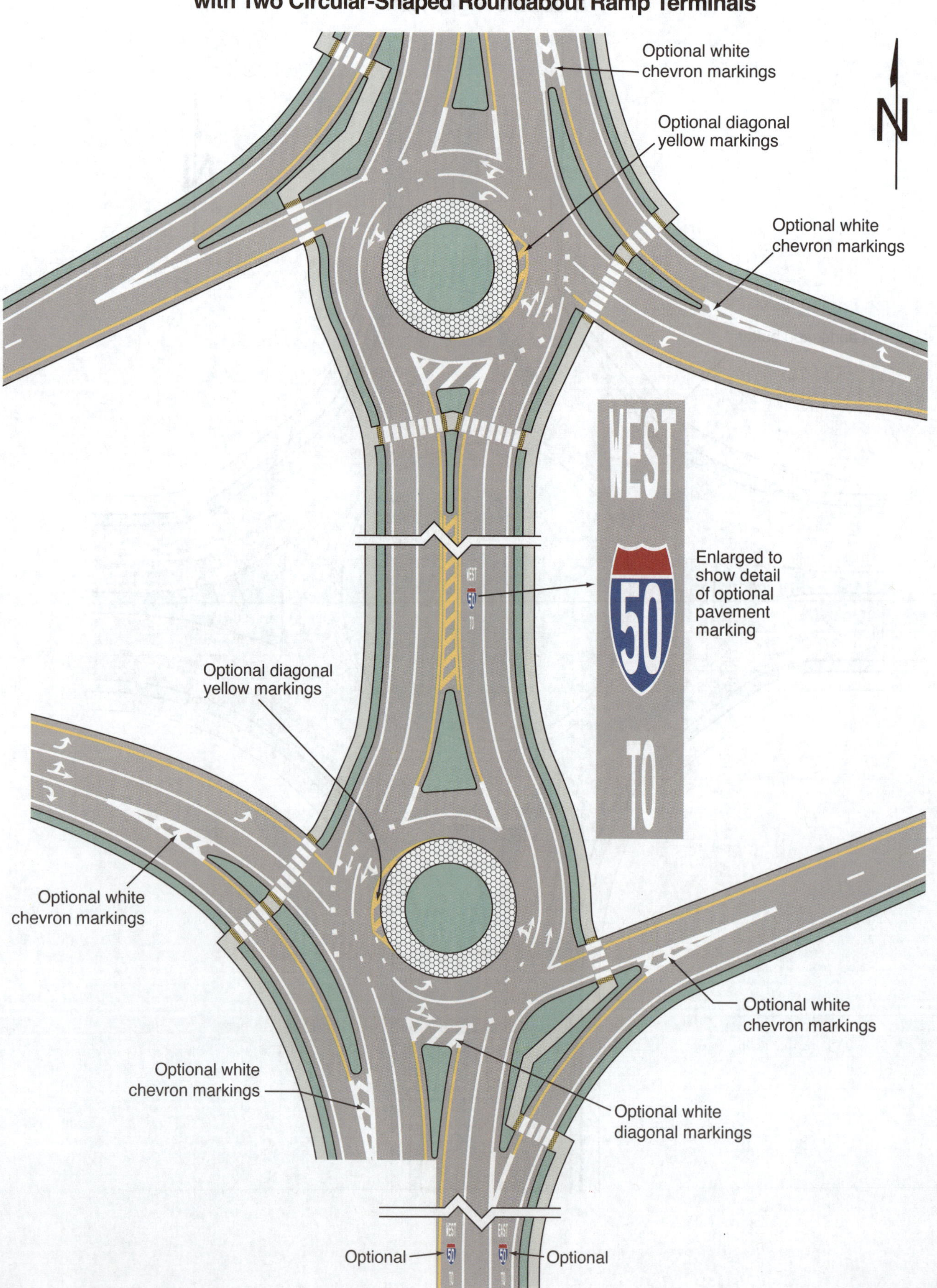

Section 3D.04 <u>Yield Lines for Roundabouts</u>

Support:

01   Section 2B.18 contains information regarding the TO ALL LANES (R1-2cP) plaque that can be used beneath the YIELD sign.

Option:

02   A yield line (see Section 3B.19) may be used to indicate the point behind which vehicles are required to yield at the entrance to a roundabout (see Figures 3D-1 and 3D-2).

Section 3D.05 <u>Word and Symbol Pavement Markings for Roundabouts</u>

Option:

01   YIELD (word) (see Figure 3D-1) and YIELD AHEAD (symbol or word) pavement markings may be used on approaches to roundabouts.

02   Word and/or route shield pavement markings may be used on an approach to or within the circulatory roadway of a roundabout to provide route and/or destination guidance information to road users (see Figure 3D-8).

Section 3D.06 <u>Arrow Pavement Markings for Roundabouts</u>

Guidance:

01   *Lane-use arrow pavement markings should not be used on single-lane approaches to circular intersections.*

02   *Lane-use arrows should be used on approaches to circular intersections with double left or double right turns.*

Standard:

03   **Lane-use arrow pavement markings shall not be provided between a crosswalk and a wide dotted line across the lane(s) entering the circular roadway.**

Option:

04   Where lane-use arrows are used on the approaches to a roundabout, they may be either normal or curved-stem (see Drawing F in Figure 3B-21).

05   An oval or circle may be used with the lane-use arrows to symbolize the central island (see Drawing F in Figure 3B-21).

Guidance:

06   *If lane-use arrows are used on the approaches to a roundabout, the style used should match the style of the lane-use arrows (normal or curved-stem) used on the regulatory lane-use signs on the approach.*

07   *If lane-use arrow pavement markings are used within the circulatory roadway of multi-lane roundabouts, normal lane-use arrows (see Section 3B.23 and Figure 3B-21) should be used.*

Support:

08   Details and sizes of the standard and curved-stem arrows that can be used for circular intersections are contained in the "Standard Highway Signs" publication (see Section 1A.05).

Section 3D.07 <u>Markings for Other Circular Intersections</u>

Option:

01   The markings shown in this Chapter may be used at other circular intersections if engineering judgment indicates that their presence will benefit drivers, pedestrians, or other road users. Figure 2B-21 provides an example of markings at a mini-roundabout.

CHAPTER 3E. PREFERENTIAL LANE MARKINGS FOR MOTOR VEHICLES

Section 3E.01 General

Support:

01   Preferential lanes are established for one or more of a wide variety of special uses, including, but not limited to, high-occupancy vehicle (HOV) lanes, electronic toll collection (ETC) lanes, price-managed lanes, bus only lanes, taxi only lanes, and light rail transit only lanes.

02   This Chapter contains the pavement marking provisions for preferential lanes used by motor vehicles and light rail transit. Part 9 contains information for pavement markings for bicycle lanes.

03   Chapter 3H contains information for the use and application of colored pavement that can be used in preferential lanes to supplement the pavement markings described in this Chapter.

Section 3E.02 Longitudinal Markings

Support:

01   Preferential lanes can take many forms depending on the level of usage and the design of the facility. They might be barrier-separated or buffer-separated from the adjacent general-purpose lanes, or they might be contiguous with the adjacent general-purpose lanes. Barrier-separated preferential lanes might be operated in a constant direction or be operated as reversible lanes. Some reversible preferential lanes on a divided highway might be operated counter-flow to the direction of traffic on the immediately adjacent general-purpose lanes. Section 1C.02 contains definitions of these terms.

02   Preferential lanes might be operated full-time (24 hours per day on all days), for extended periods of the day, part-time (restricted usage during specific hours on specified days), or on a variable basis (such as a strategy for a managed lane).

Standard:

03   **The left-hand and right-hand edge lines and lane lines used for preferential lanes that are adjacent to general-purpose lanes where traffic is flowing in the same direction shall be in accordance with Table 3E-1.**

04   **If there are two or more preferential lanes for traffic moving in the same direction, the lane lines between the preferential lanes shall be normal width broken white lines.**

05   **Preferential lanes for motor vehicles shall have appropriate regulatory signs in accordance with Sections 2G.03 through 2G.07.**

Support:

06   Figure 3E-1 illustrates pavement markings used for barrier-separated preferential lanes. Figure 3E-2 illustrates pavement markings used for buffer-separated preferential lanes. Figure 3E-3 illustrates pavement markings used for contiguous preferential lanes.

Guidance:

07   *Engineering judgment should determine the need for supplemental devices such as tubular markers, traffic cones, or other channelizing devices (see Chapter 3I).*

08   *Where preferential lanes and other travel lanes are separated by a buffer space wider than 4 feet and crossing the buffer space is prohibited, chevron markings (see Section 3B.25) should be placed in the buffer area (see Drawing A in Figure 3E-2).*

09   *The buffer space for a conventional road should be designed so that it is not misinterpreted as on-street parking, a bicycle lane, or any other type of lane.*

Option:

10   If a full-time or part-time contiguous preferential lane is separated from the other travel lanes by a wide broken single white line (see Drawing C in Figure 3E-3), the spacing or skip pattern of the line may be reduced and the width of the line may be increased.

Support:

Standard:

11   **At direct exits from a preferential lane, dotted white line markings shall be used to separate the tapered or parallel deceleration lane for the direct exit (including the taper) from the adjacent continuing preferential through lane, to reduce the chance of unintended exit maneuvers.**

12   **Signs (see Section 2B.34), lane-use control signals (see Chapter 4T), or both shall be used to supplement the reversible lane markings on a divided highway where a part-time counter-flow preferential lane is present.**

Table 3E-1. Standard Edge Line and Lane Line Markings for Preferential Lanes

Type of Preferential Lane	Left-Hand Line	Right-Hand Line
Barrier-Separated, Non-Reversible	A normal solid single yellow edge line	A normal solid single white edge line (see Drawing A in Figure 3E-1)
Barrier-Separated, Reversible	A normal solid single white edge line	A normal solid single white edge line (see Drawing B in Figure 3E-1)
Buffer-Separated, Left-Hand Side	A normal solid single yellow edge line	A wide solid double white line along both edges of the buffer space where crossing is prohibited (see Drawing A in Figure 3E-2) A wide solid single white line along both edges of the buffer space where crossing is discouraged (see Drawing B in Figure 3E-2) A wide broken single white line along both edges of the buffer space, or a wide broken single white line within the buffer space (resulting in wider lanes), where crossing is permitted (see Drawing C in Figure 3E-2)
Buffer-Separated, Right-Hand Side	A wide solid double white line along both edges of the buffer space where crossing is prohibited, or a wide solid single white line along both edges of the buffer space where crossing is discouraged (see Drawing D in Figure 3E-2) A wide broken single white line along both edges of the buffer space, or a wide broken single white line within the buffer space (resulting in wider lanes), where crossing is permitted (see Drawing D in Figure 3E-2) A wide dotted single white line within the buffer space (resulting in wider lanes) where crossing is permitted for any vehicle to perform a right-turn maneuver (see Drawing D in Figure 3E-2)	A normal solid single white edge line (if warranted)
Contiguous, Left-Hand Side	A normal solid single yellow edge line	A wide solid double white lane line where crossing is prohibited (see Drawing A in Figure 3E-3) A wide solid single white lane line where crossing is discouraged (see Drawing B in Figure 3E-3) A wide broken single white lane line where crossing is permitted (see Drawing C in Figure 3E-3)
Contiguous, Right-Hand Side	A wide solid double white lane line where crossing is prohibited (see Drawing D in Figure 3E-3) A wide solid single white lane line where crossing is discouraged (see Drawing D in Figure 3E-3) A wide broken single white lane line where crossing is permitted (see Drawing D in Figure 3E-3) A wide dotted single white lane line where crossing is permitted for any vehicle to perform a right-turn maneuver (see Drawing D in Figure 3E-3)	A normal solid single white lane line (if warranted)

Notes: 1. If there are two or more preferential lanes, the lane lines between the preferential lanes shall be normal broken white lines.

    2. The standard lane markings listed in this table are provided in a tabular format for reference.

Figure 3E-1. Markings for Barrier-Separated Preferential Lanes

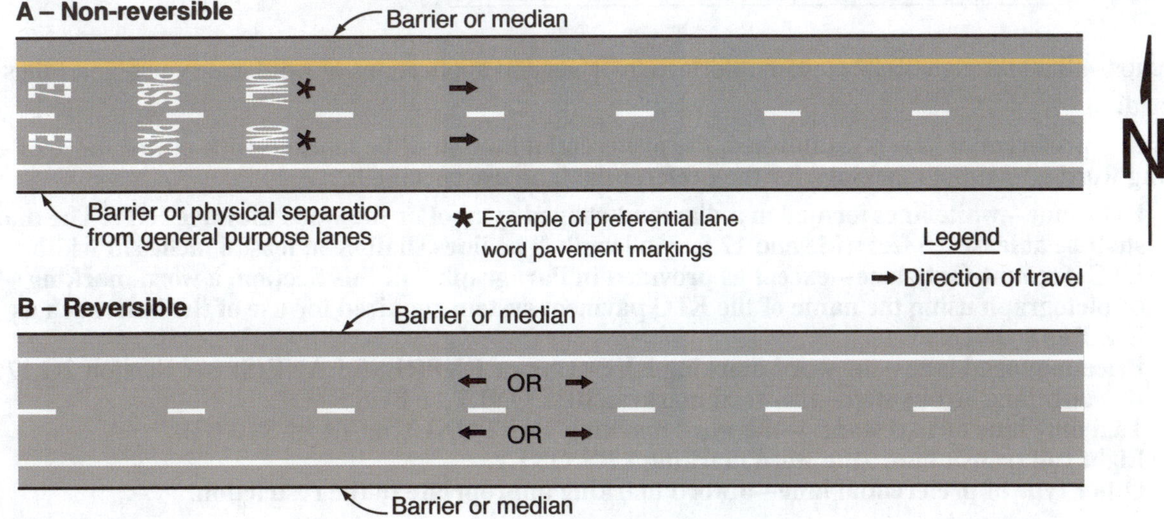

Figure 3E-2. Markings for Buffer-Separated Preferential Lanes (Sheet 1 of 2)

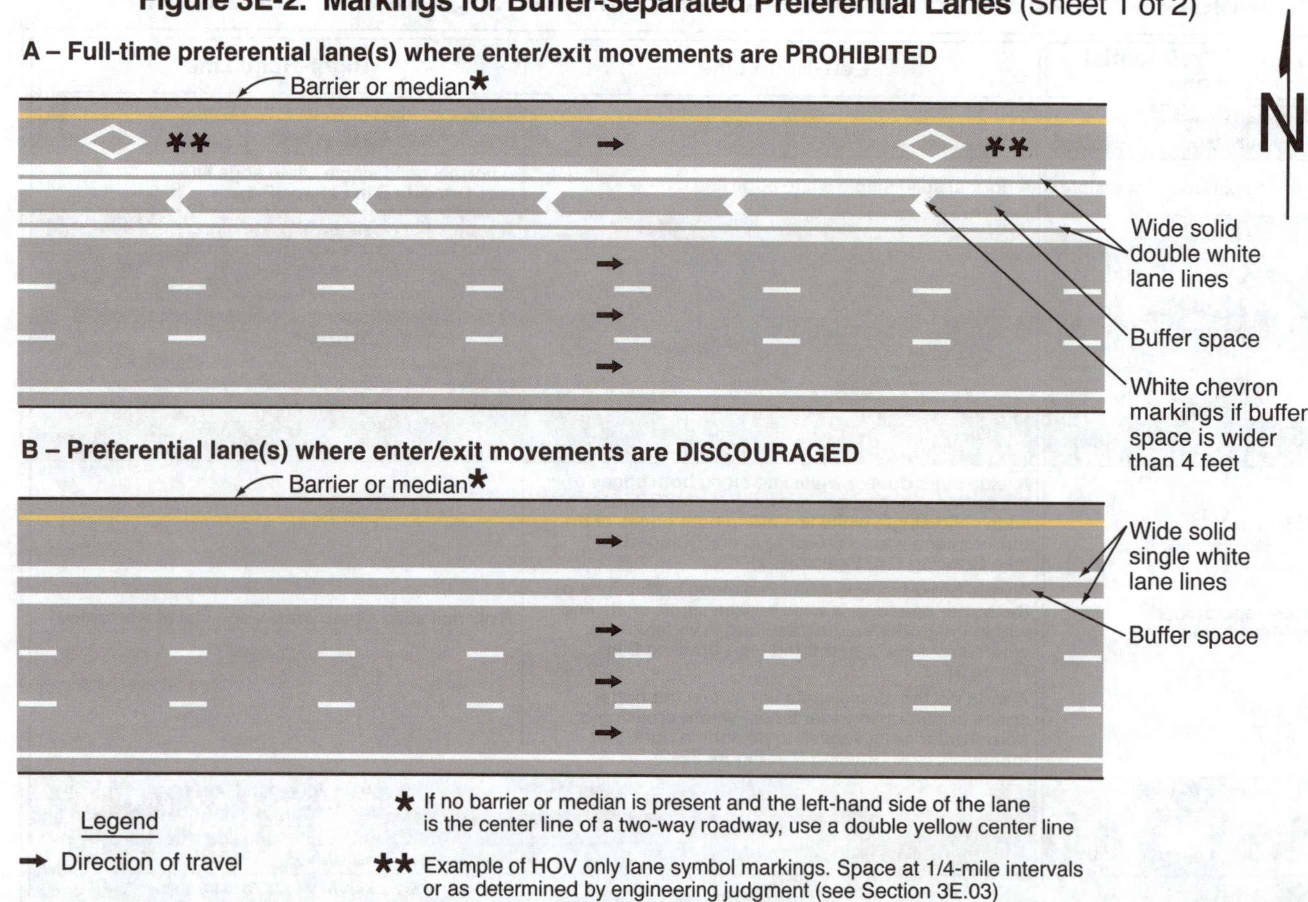

13   The longitudinal pavement markings used for preferential lanes that are adjacent to general-purpose lanes where traffic is flowing in the opposite direction (see Figure 3E-4) shall be in accordance with Table 3E-2.

Support:

14   Figure 3E-4 illustrates pavement markings used for counter-flow preferential lanes on divided highways or on transitions to and from other divided highways such as bridges and crossovers.

Option:

15   Cones, tubular markers, or other channelizing devices (see Chapter 3I) may also be used in addition to longitudinal markings to separate the opposing lanes when a counter-flow preferential lane operation is in effect.

Section 3E.03 Preferential Lane Word and Symbol Markings

Support:

01   Sections 3B.20 through 3B.22 contain information on general applications of word and symbol markings.

Standard:

02   When a preferential lane is established, the preferential lane shall be marked with one or more of the following word or symbol markings for the preferential lane use specified:

A. HOV lane—white lines formed in a diamond-shaped symbol or the word message HOV. The diamond shall be at least 2.5 feet wide and 12 feet in length. The lines shall be at least 6 inches in width.

B. ETC Account-Only lane—except as provided in Paragraph 8 of this Section, a word marking or pictograph using the name of the ETC payment system required for use of the lane, such as E-Z PASS ONLY.

C. Price-managed lane—the word marking EXPRESS or EXPRESS LANE(S) (see Section 2G.17).

D. Bus only lane or bus stop—the word marking BUS ONLY or BUS STOP.

E. Taxi only lane or taxi stand —the word marking TAXI ONLY or TAXI STAND.

F. Light rail transit lane—the word marking LRT ONLY.

G. Other type of preferential lane—a word marking appropriate to the restriction.

Figure 3E-2. Markings for Buffer-Separated Preferential Lanes (Sheet 2 of 2)

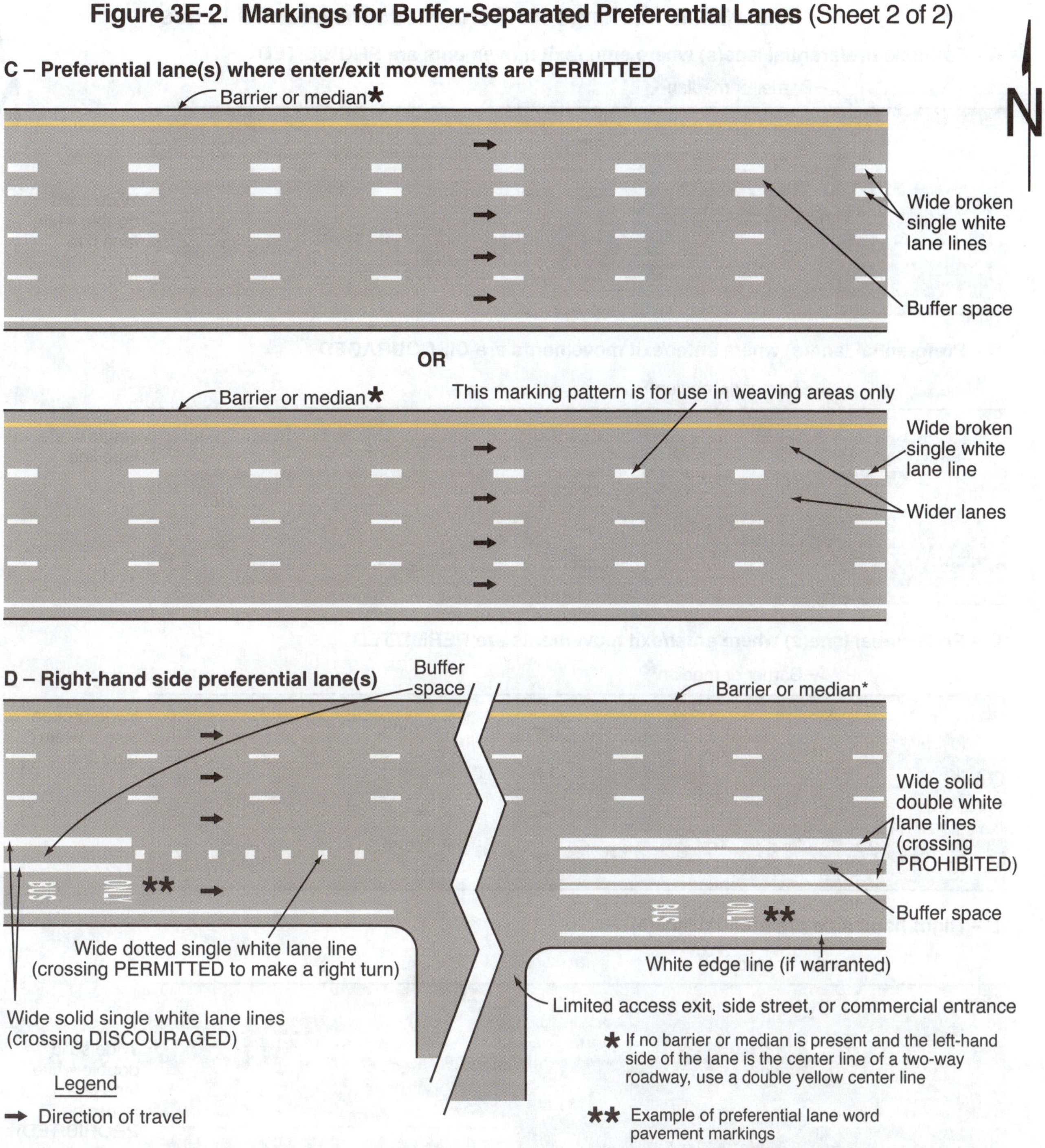

Table 3E-2. Standard Center Line and Edge Line Markings for Counter-Flow Preferential Lanes on Divided Highways

Type of Preferential Lane	Center Line on Left-Hand Side	Edge Line on Left-Hand Side
Part-Time Contiguous	A normal width broken double yellow line	A normal solid single white line (if warranted)
Part-Time Buffer-Separated	A normal width broken double yellow line along both edges of the buffer space	A normal solid single white line (if warranted)
Full-Time Contiguous	A normal width solid double yellow line	A normal solid single white line (if warranted)
Full-Time Buffer-Separated	A normal width solid double yellow line along both edges of the buffer space	A normal solid single white line (if warranted)

Figure 3E-3. Markings for Contiguous Preferential Lanes

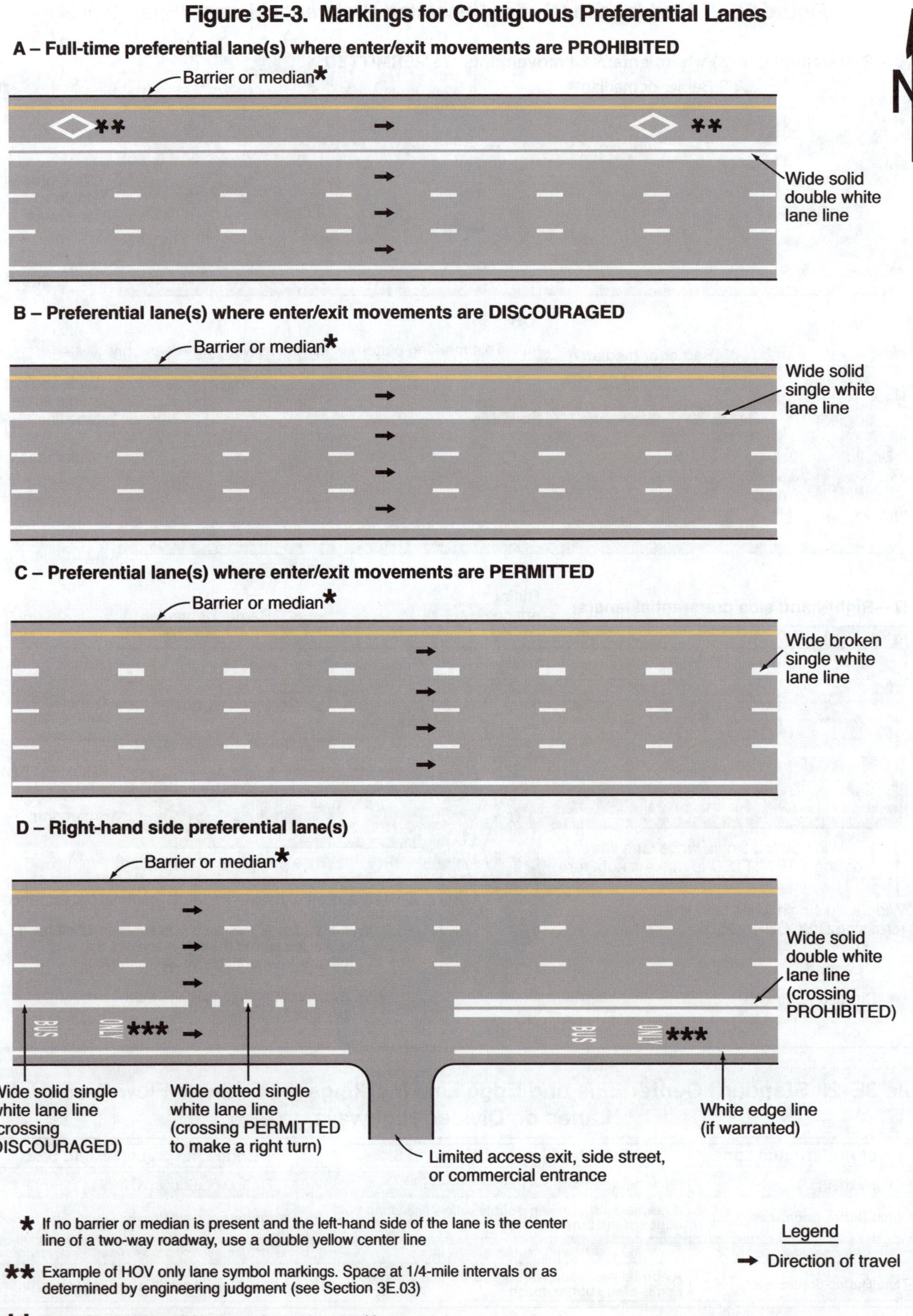

Figure 3E-4. Markings for Contra-Flow Preferential Lanes on Divided Highways

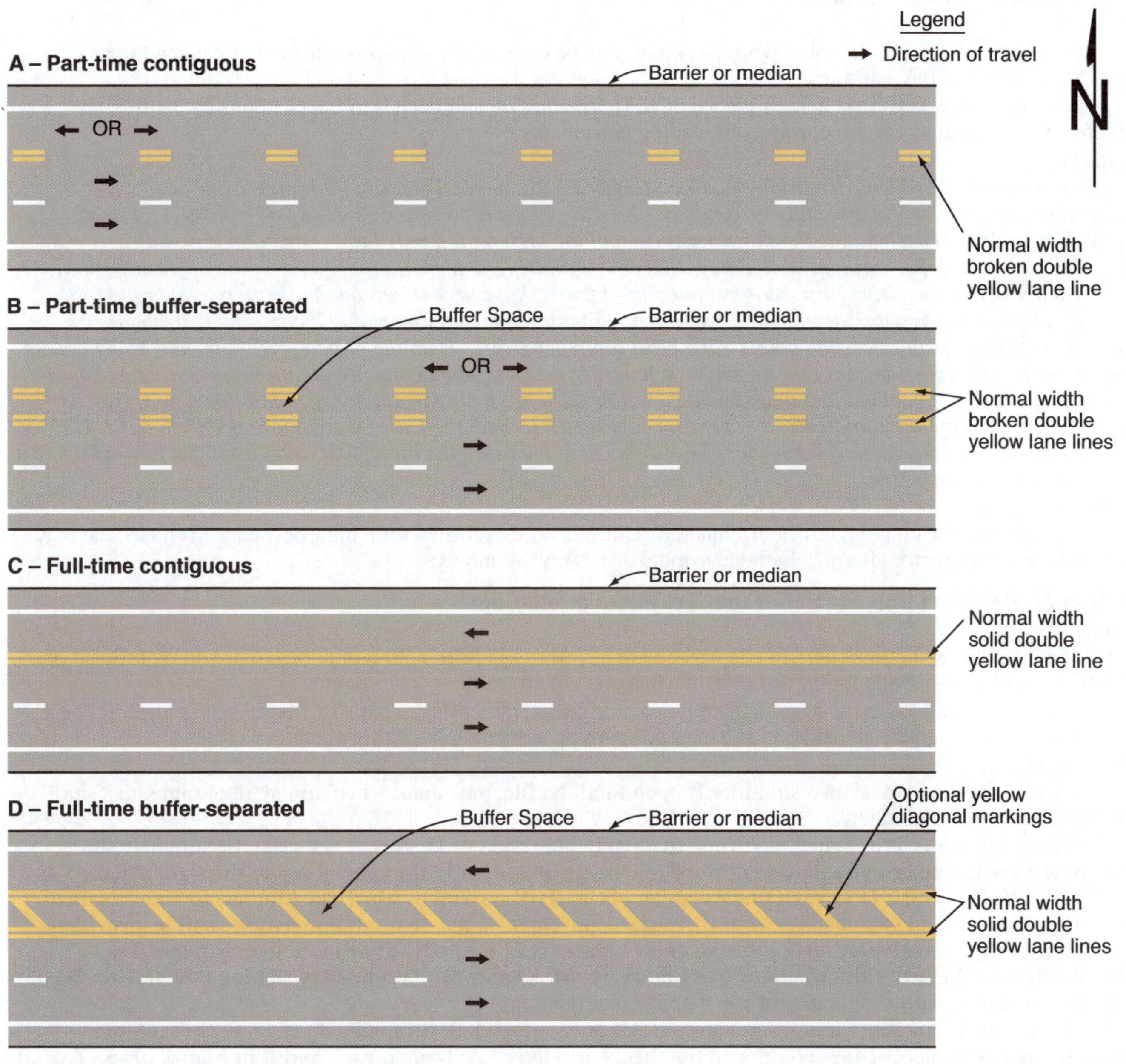

Guidance:

03   *If multiple preferential lane uses are allowed in a single lane, the word or symbol marking for each preferential lane should be used.*

Standard:

04   **Pavement word or symbol markings for motorcycles and Inherently Low Emission Vehicles (ILEV) shall not be used to mark the preferential lane if motorcycles and ILEVs are allowed to use the preferential lane.**

Support:

05   Motorcycles and Inherently Low Emission Vehicles (ILEV) that are allowed to use a preferential lane are granted an exception such as through an established High-Occupancy Vehicle (HOV) regulation. Communicating that motorcycles and ILEVs are allowed to use the preferential lane is accomplished through regulatory signing (see Sections 2G.03 and 2G.04) that complements HOV signing.

Standard:

06   **Static or changeable message regulatory signs (see Sections 2G.03 through 2G.07) shall be used with preferential lane word or symbol markings.**

07 **All preferential lane word and symbol markings shall be white and shall be positioned laterally in the approximate center of the preferential lane.**
Option:

08 Preferential lane-use symbol or word markings may be omitted at toll plazas where physical conditions preclude the use of the markings.

09 Lane-use arrow markings may be placed on the curb lanes on approaches to intersections to signify non-preferential road users can use the lane for turning movements.
Guidance:

10 *All longitudinal pavement markings, as well as word and symbol pavement markings, associated with a preferential lane should end at approximately where the Preferential Lane Ends (R3-12a or R3-12c) sign (see Section 2G.07) designating the downstream end of the preferential only lane restriction is installed.*

11 *The spacing of the markings should be based on engineering judgment that considers the operating speed, block lengths, distance from intersections, and other factors that affect clear communication to the road user.*

12 *In addition to a regular spacing interval, the preferential lane marking should be placed at strategic locations such as major decision points, direct exit ramp departures from the preferential lane, and along access openings to and from adjacent general-purpose lanes. At decision points, the preferential lane marking should be placed on all applicable lanes and should be visible to approaching traffic for all available departures. At direct exits from preferential lanes where extra emphasis is needed, the use of word markings (such as "EXIT" or "EXIT ONLY") in the deceleration lane for the direct exit and/or on the direct exit ramp itself just beyond the exit gore should be considered.*
Option:

13 A numeral indicating the vehicle occupancy requirements established for a high-occupancy vehicle lane may be included in sequence after the diamond symbol or HOV word message.

Section 3E.04 <u>Markings for Part-Time Travel on a Shoulder</u>
Support:

01 Shoulders are sometimes used to add capacity to a roadway in peak hour conditions to provide for transit or HOV priority or to provide higher throughput when open to all traffic.

02 A shoulder that has been opened to travel on a permanent, rather than a part-time basis is considered to be a travel lane and is signed and marked in accordance with other provisions of this Manual.
Standard:

03 **When part-time travel on a shoulder is open to all traffic, pavement word and symbol markings shall not be used in the shoulder.**

04 **When a shoulder is assigned part-time to a particular class or classes of vehicles, the shoulder shall be marked with one or more pavement word markings that identify the special use of the shoulder such as BUS ONLY, TRANSIT ONLY, HOV, or instead of the HOV pavement word marking, white lines formed in a diamond-shaped symbol (see Section 3E.03). A pavement word or symbol marking shall be provided in the shoulder immediately after exit and entry ramps (see Figure 3E-5) or immediately departing an intersection at the full-width shoulder (see Figure 3E-6). Appropriate regulatory signing (see Section 2G.03) shall be installed with the pavement word or symbol markings.**

05 **The channelizing line emanating from the entrance ramp shall be a wide dotted line through the intersecting alignment of the shoulder to the theoretical gore (see Drawings A and B in Figure 3E-5). At exit ramps, the channelizing line proceeding from the theoretical gore across the intersecting alignment of the shoulder shall be a wide dotted line (see Figure 3E-5).**

06 **If used, the extension of the channelizing line at entrance ramps proceeding from the theoretical gore across the opening of the on-ramp alignment shall be a wide dotted line (see Drawing C in Figure 3E-5) where it is demonstrated that traffic entering from an on-ramp stops or yields to traffic on the shoulder of the highway mainline.**

07 **An additional outside solid edge line shall be provided on the shoulder in accordance with Sections 3B.09 and 3B.10.**
Guidance:

08 *Changes in edge line pattern or direction should occur at appropriate regulatory signs.*
Option:

09 At locations where traffic is allowed to enter, exit, or merge with the shoulder, a dotted edge line may be used either in a continuous manner or angled to the pavement edge (see Figure 3E-6).

10 Red-colored pavement (see Section 3H.07) may be used on shoulders that allow only transit vehicles.
Standard:

11 **If used, red-colored pavement shall be discontinued on the shoulder through the influence area of the ramp (see Figure 3E-5).**

Figure 3E-5. Markings for Part-Time Travel on Shoulder and Application of Pavement Word Markings (Sheet 1 of 3)

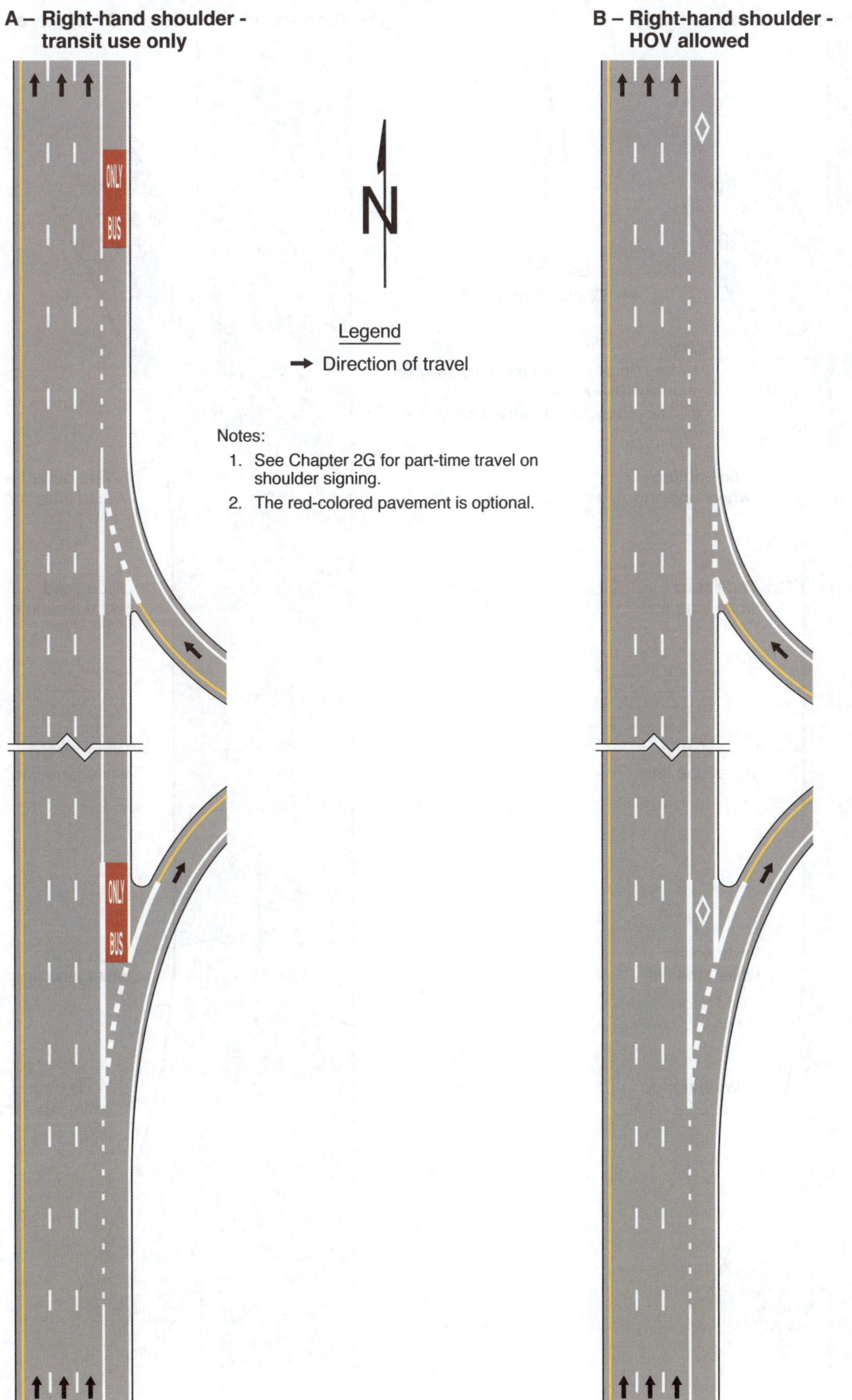

Figure 3E-5. Markings for Part-Time Travel on Shoulder and Application of Pavement Word Markings (Sheet 2 of 3)

C – Right-hand shoulder - transit use only

D – Right-hand shoulder - HOV allowed

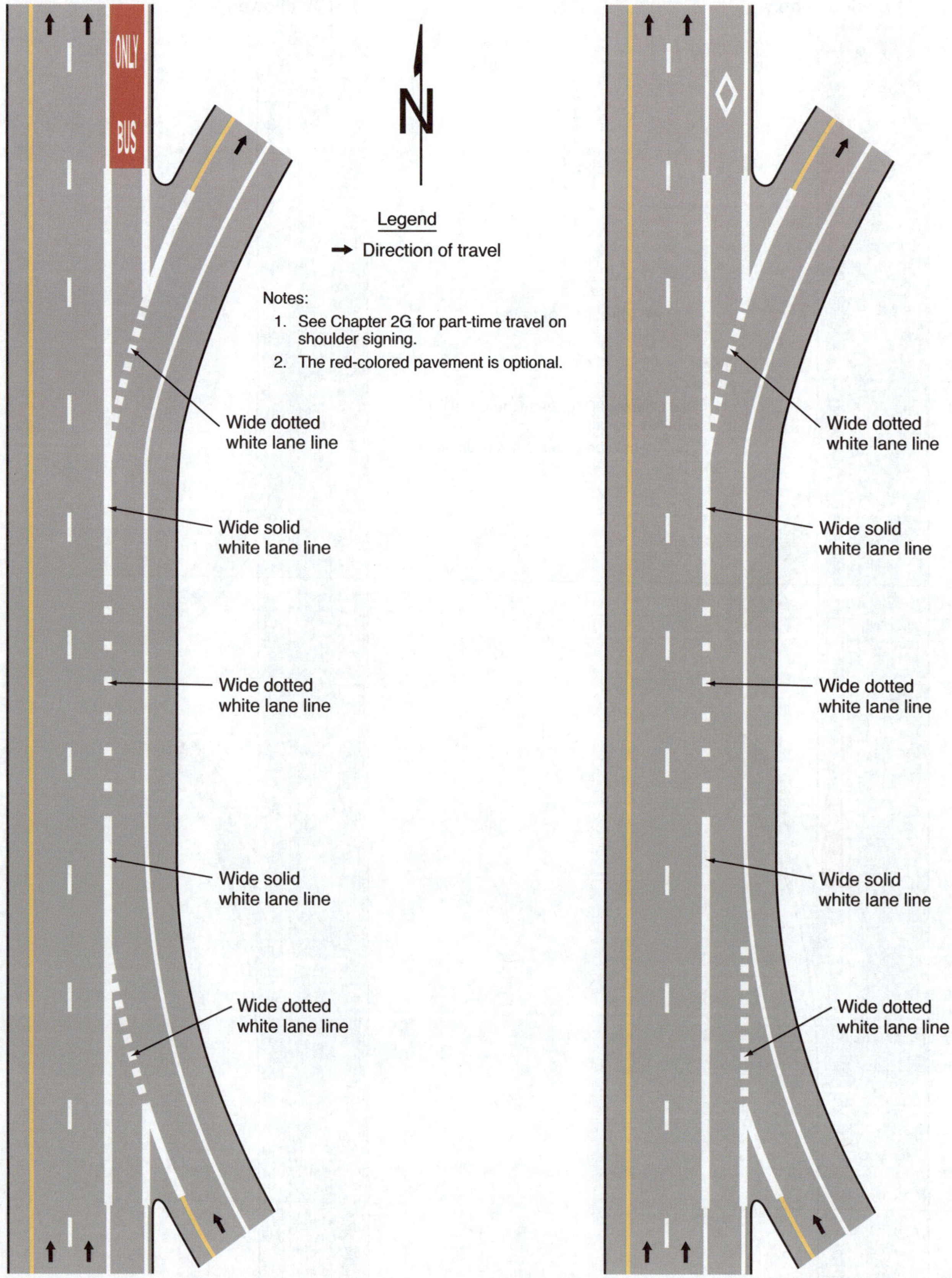

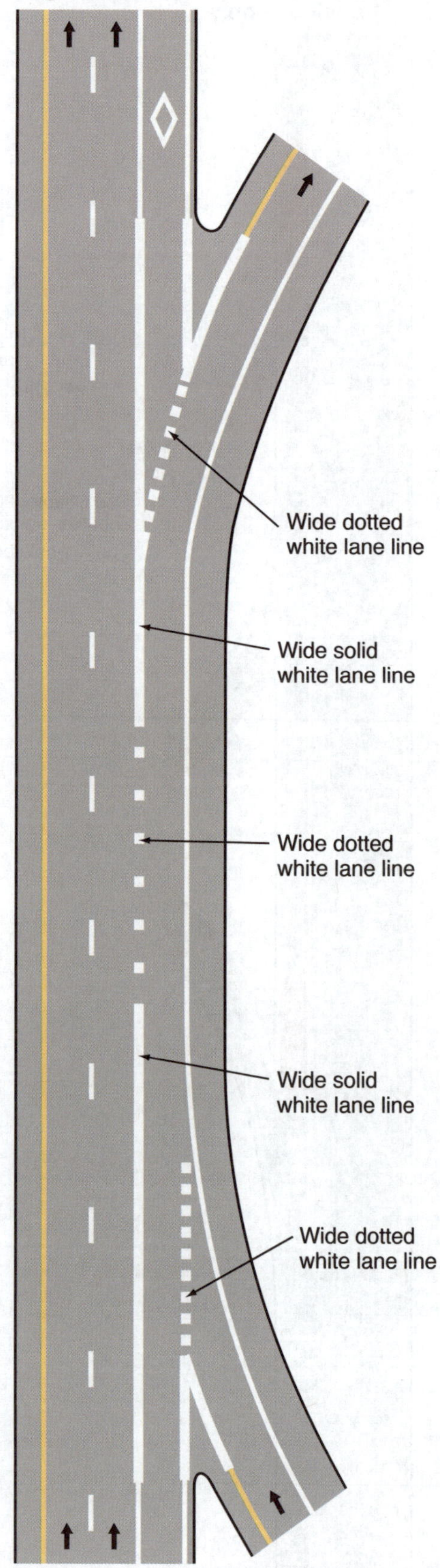

Figure 3E-5. Markings for Part-Time Travel on Shoulder and Application of Pavement Word Markings (Sheet 3 of 3)

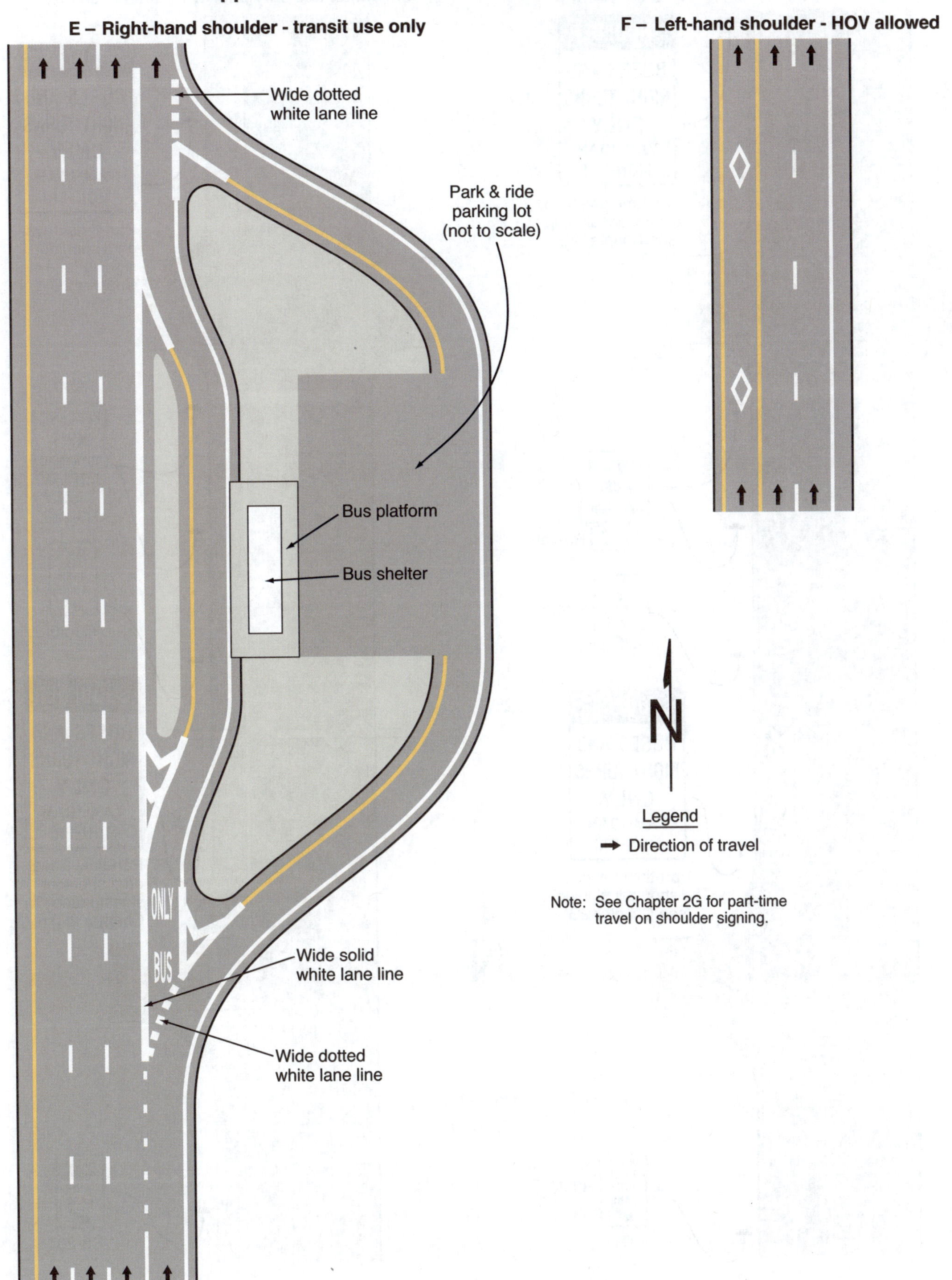

Figure 3E-6. Markings for Part-Time Travel on a Shoulder through an Intersection

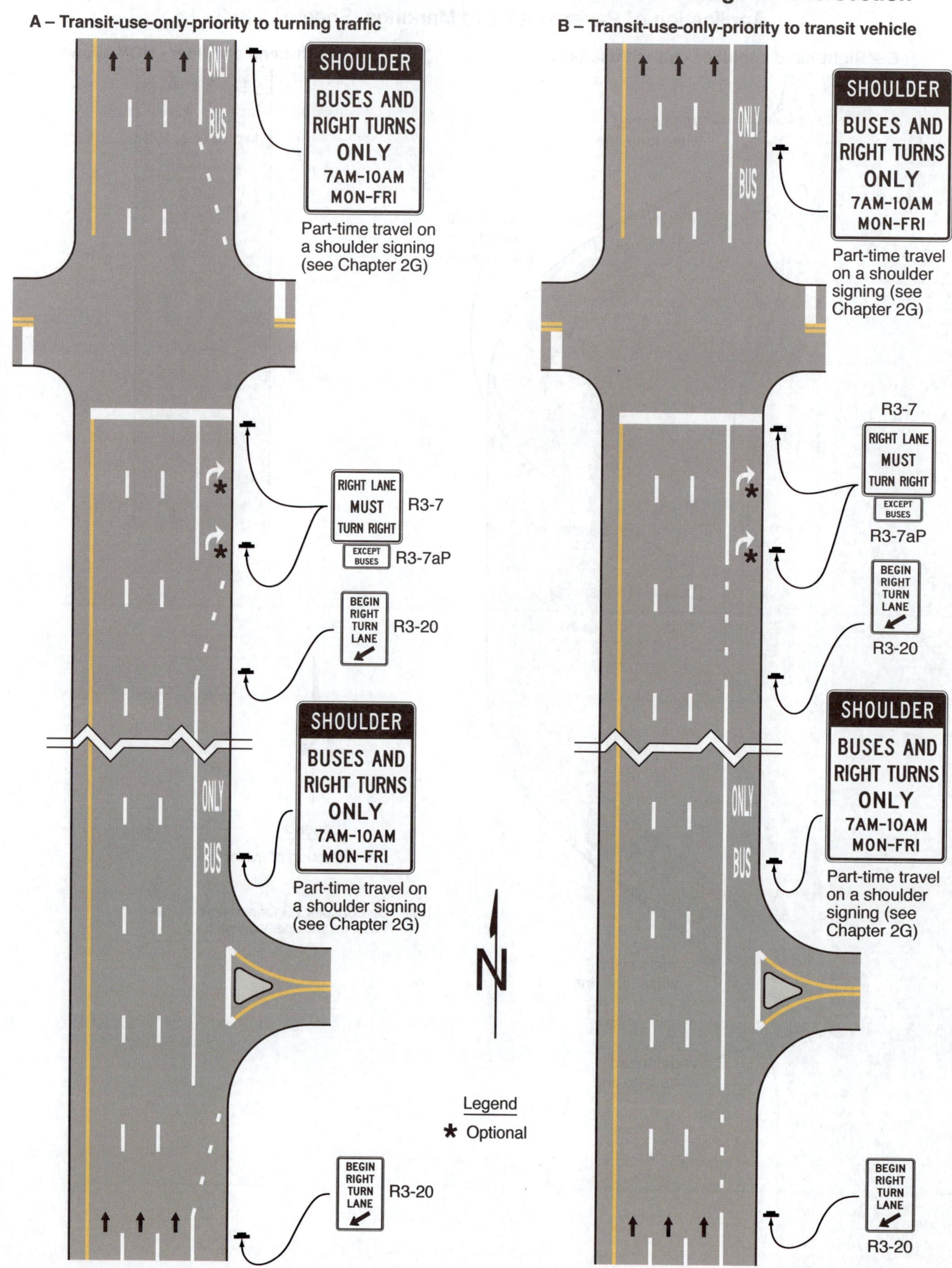

CHAPTER 3F. MARKINGS FOR TOLL PLAZAS

Section 3F.01 General

Support:

01   At toll plazas, pavement markings help road users identify the proper lane(s) to use for the type of toll payment they plan to use, to channelize movements into the various lanes, and to delineate obstructions in the roadway.

02   Section 3H.08 contains information on the use and application of purple-colored pavement at toll plazas for vehicles with registered electronic toll collection (ETC) accounts.

Section 3F.02 Longitudinal Markings

Guidance:

01   *Solid white lane line markings should be used to separate toll lanes, payment methods, or to channelize movements at toll plazas.*

02   *Solid white lane line markings should begin at the upstream end of the full-width toll lane and be continued to the toll plaza.*

Option:

03   For a toll plaza approach lane that is restricted to use only by vehicles with registered ETC accounts, the solid white lane line or edge line on the right-hand side of the ETC Account-Only lane and the solid white lane line or solid yellow edge line on the left-hand side of the ETC Account-Only lane may be supplemented with purple solid longitudinal markings placed contiguous to the inside edges of the lines defining the lane.

Standard:

04   **If the purple solid longitudinal markings described in Paragraph 3 of this Section are used, the purple markings shall be at least 3 inches wide.**

Guidance:

05   *If the purple solid longitudinal markings described in Paragraph 3 of this Section are used, the purple markings should not be wider than the line they supplement.*

Standard:

06   **Toll booths and the islands on which they are located are considered to be obstructions in the roadway and they shall be provided with markings that comply with the provisions of Section 3B.13 and Chapter 3J.**

Option:

07   Longitudinal pavement markings may be omitted alongside toll booth islands between the approach markings and any departure markings.

Section 3F.03 Pavement Word and Symbol Markings

Support:

01   Section 3E.03 contains information on the use of pavement word and symbol markings for ETC Account-Only lanes not specific to toll plazas.

Standard:

02   **Except as provided in Paragraph 4 of this Section, when a lane on the approach to a toll plaza is restricted to use only by vehicles with registered ETC accounts, the ETC Account-Only lane word markings or pictograph described in Section 3E.03 shall be used (see Drawing A in Figure 3H-6).**

03   **When one or more open-road tolling (ORT) lanes that are restricted to use only by vehicles with registered ETC accounts bypass a mainline toll plaza on a separate alignment, pavement word markings or pictographs shall be used on the approach to the point where the ORT lanes diverge from the lanes destined for the mainline toll plaza (see Drawings B and C in Figure 3H-6).**

Option:

04   Preferential lane-use markings may be omitted at toll plazas where physical conditions preclude the use of the markings.

Guidance:

05   *If an ORT lane that is immediately adjacent to a mainline toll plaza is not separated from adjacent cash payment toll plaza lanes by a curb or barrier, then channelizing devices (see Section 3I.01), and/or longitudinal pavement markings that discourage or prohibit lane changing should be used to separate the ORT lane from the adjacent cash payment lane. This separation should begin on the approach to the mainline toll plaza at approximately the point where the vehicle speeds in the adjacent cash lanes drop below 30 mph during off-peak periods and should extend downstream beyond the toll plaza approximately to the point where the vehicles departing the toll plaza in the adjacent cash lanes have accelerated to 30 mph.*

CHAPTER 3G. DELINEATORS

Section 3G.01 General

Support:

01   Delineators are particularly beneficial at locations where the alignment might be confusing or unexpected, such as at lane-reduction transitions and curves. Delineators are effective guidance devices at night and during adverse weather. An important advantage of delineators in certain locations is that they remain visible when the roadway is wet or covered by snow.

02   Delineators are considered guidance devices to help road users navigate the roadway alignment, rather than warning devices.

Option:

03   Delineators may be used on long continuous sections of highway or through short stretches where there are changes in horizontal alignment.

Section 3G.02 Design

Standard:

01   **Delineators shall consist of retroreflective devices that are capable of clearly retroreflecting light under normal atmospheric conditions from a distance of 1,000 feet when illuminated by the high beams of standard automobile lights. They shall be mounted on crashworthy (see definition in Section 1C.02) supports.**

02   **Retroreflective elements for delineators shall have a minimum vertical and horizontal dimension of 3 inches, or a minimum diameter dimension of 3 inches when circular.**

Support:

03   Within a series of delineators along a roadway, delineators for a given direction of travel at a specific location are referred to as single delineators if they have one retroreflective element for that direction, double delineators if they have two identical retroreflective elements for that direction mounted together, or vertically-elongated delineators if they have a single retroreflective element with an elongated vertical dimension to approximate the vertical dimension of two separate single delineators.

Option:

04   A vertically-elongated delineator of appropriate size may be used in place of a double delineator.

Section 3G.03 Application

Standard:

01   **The color of delineators shall comply with the color of edge lines stipulated in Sections 3A.03 and 3B.09.**

02   **A series of single delineators shall be provided on the right-hand side of freeways and expressways and on at least one side of interchange ramps, except when either Condition A or Condition B is met, as follows:**

    A. On tangent sections of freeways and expressways when both of the following conditions are met:

        1. Raised pavement markers are used continuously on lane lines throughout all curves and on all tangents to supplement pavement markings, and

        2. Roadside delineators are used to lead into all curves, or

    B. On sections of roadways where continuous lighting is in operation between interchanges.

Option:

03   Delineators may be provided on other classes of roads.

04   A series of single delineators may be provided on the left-hand side of roadways.

05   Chevron Alignment (W1-8) signs may be used instead of or in addition to standard delineators, as provided in Section 2C.08.

Standard:

06   **Delineators on the left-hand side of a two-way roadway shall be white (see Figure 3G-1).**

Guidance:

07   *A series of single delineators should be provided on the outside of curves on interchange ramps.*

08   *Where median crossovers are provided for official or emergency use on divided highways and where these crossovers are to be marked with pavement markings, a double yellow delineator should be placed on the left-hand side of the through roadway on the far side of the crossover for each roadway.*

Figure 3G-1. Examples of Delineator Placement

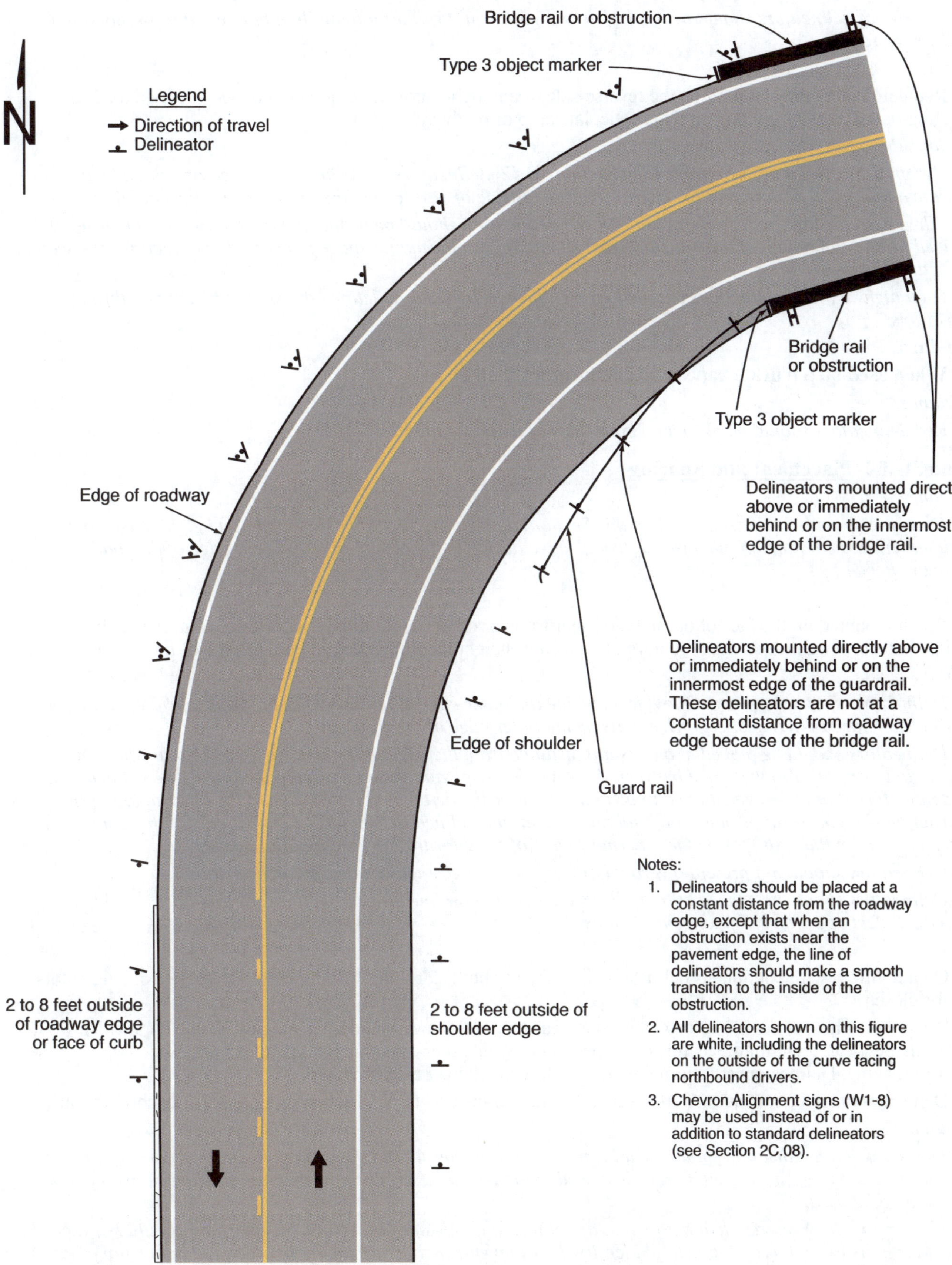

09   *Double or vertically-elongated delineators should be installed at approximately 100-foot intervals along acceleration and deceleration lanes.*

10   *A series of delineators should be used wherever guardrail or other longitudinal barriers are present along a roadway or ramp.*

Option:

11   Red delineators may be used on the reverse side of any delineator where it would be viewed by a road user traveling in the wrong direction on that particular ramp or roadway.

Guidance:

12   *Except as provided in Paragraph 13 of Section 3B.12, delineators of the appropriate color should be used to indicate a lane-reduction transition where either an outside or inside lane merges into an adjacent lane.*

13   *When used for lane-reduction transitions, the delineators should be installed adjacent to the lane or lanes reduced for the full length of the transition and should be so placed and spaced to show the reduction (see Section 3B.12 and Figure 3B-14).*

14   *On a highway with continuous delineation on either or both sides, delineators should be carried through transitions.*

Standard:

15   **When used on a truck escape ramp, delineators shall be red.**

Guidance:

16   *Red delineators should be placed on both sides of truck escape ramps.*

Section 3G.04 Placement and Spacing

Guidance:

01   *Except as provided in Paragraph 2 of this Section, delineators should be mounted at a height, measured vertically from the bottom of the lowest retroreflective device to the elevation of the near edge of the roadway, of approximately 4 feet.*

Option:

02   When mounted on the face of or on top of guardrails or other longitudinal barriers, delineators may be mounted at a lower elevation than the normal delineator height recommended in Paragraph 1 of this Section.

Guidance:

03   *Delineators should be placed 2 to 8 feet outside the outer edge of the shoulder, or if appropriate, in line with the roadside barrier that is 8 feet or less outside the outer edge of the shoulder.*

04   *Delineators should be placed at a constant distance from the edge of the roadway, except that where an obstruction intrudes into the space between the pavement edge and the extension of the line of the delineators, the delineators should be transitioned to be in line with or inside the innermost edge of the obstruction. If the obstruction is a guardrail or other longitudinal barrier, the delineators should be transitioned to be just behind, directly above (in line with), or on the innermost edge of the guardrail or longitudinal barrier.*

05   *Delineators should not present a vertical or horizontal clearance obstacle for pedestrians.*

06   *Delineators should be spaced 200 to 530 feet apart on mainline tangent sections. Delineators should be spaced 100 feet apart on ramp tangent sections.*

Option:

07   On a highway with continuous delineation on either or both sides, the spacing between a series of delineators may be closer.

08   When uniform spacing is interrupted by such features as driveways and intersections, delineators which would ordinarily be located within the features may be relocated in either direction for a distance not exceeding ¼ of the uniform spacing. Delineators still falling within such features may be eliminated.

09   Delineators may be transitioned in advance of a lane transition or obstruction as a guide for oncoming traffic.

Guidance:

10   *The spacing of delineators should be adjusted on approaches to and throughout horizontal curves so that several delineators are always simultaneously visible to the road user. The approximate spacing shown in Table 3G-1 should be used.*

11   *The spacing between red delineators that are placed on both sides of a truck escape ramp should not exceed 50 feet for a distance that is sufficient to identify the ramp entrance. The spacing between red delineators that are placed beyond the ramp entrance should be such that adequate guidance is provided based on the length and design of the escape ramp.*

Option:

12   When needed for special conditions, delineators of the appropriate color may be mounted in a closely-spaced manner on the face of or on top of guardrails or other longitudinal barriers to form a continuous or nearly-continuous "ribbon" of delineation.

Support:

13   Examples of delineator installations are shown in Figure 3G-1.

Table 3G-1. Approximate Spacing for Delineators on Horizontal Curves

Radius (R) of Curve	Approximate Spacing (S) on Curve
50 feet	20 feet
115 feet	25 feet
180 feet	35 feet
250 feet	40 feet
300 feet	50 feet
400 feet	55 feet
500 feet	65 feet
600 feet	70 feet
700 feet	75 feet
800 feet	80 feet
900 feet	85 feet
1,000 feet	90 feet

Notes: 1. Spacing for specific radii may be interpolated from table.

2. The minimum spacing should be 20 feet.

3. The spacing on curves should not exceed 300 feet.

4. In advance of or beyond a curve, and proceeding away from the end of the curve, the spacing of the first delineator is 2S, the second 3S, and the third 6S, but not to exceed 300 feet.

5. S refers to the delineator spacing for specific radii computed from the formula $S = 3\sqrt{R-50}$.

6. The distances for S shown in the table above were rounded to the nearest 5 feet.

CHAPTER 3H. COLORED PAVEMENT

Section 3H.01 Standardization of Application

Support:

01   Colored pavements consist of differently-colored road paving materials, such as colored asphalt or concrete. Other surface treatments can be applied to the surface of a road, island, or area outside the traveled way to simulate a colored pavement.

02   If non-retroreflective colored pavement is used as a purely aesthetic surface treatment (see Section 3H.03) within the provisions of this Chapter and are not intended to communicate regulations, warnings, guidance, or other information to road users, the colored pavement is not considered to be a traffic control device, even if it is located between the lines of a crosswalk.

Standard:

03   **If colored pavement is used within the traveled way, on flush or raised islands, or on shoulders to communicate regulations, warnings, guidance, or other information to road users, or if retroreflectivity is used, the colored pavement shall be considered a traffic control device and shall be limited to the colors and applications specified in this Chapter.**

04   **Except as provided in Paragraph 5 of Section 3H.07, colored pavement shall only be used if the corresponding regulations, warnings, or guidance are applicable at all times.**

Guidance:

05   *Colored pavements used as traffic control devices should only be used where the color pavement contrasts significantly with adjoining paved areas.*

Support:

06   The chromaticity coordinates that define the ranges of acceptable colors for traffic control devices are found in the Appendix to Subpart F of 23 CFR 655.

Standard:

07   **If used, colored pavement shall only be used to supplement other markings as provided in this Manual.**

Support:

08   Longitudinal pavement markings, crosswalks, pavement marking symbols, and elongated route markers are not considered colored pavements.

Section 3H.02 Materials

Option:

01   Colored pavements may be either retroreflective or non-retroreflective, in accordance with the provisions of this Chapter for specific applications.

Guidance:

02   *If surface treatments are applied to the surface of a road, island, or other area outside the traveled way to simulate a colored pavement, pavement marking materials should be selected that will minimize loss of traction for road users (see Paragraph 2 of Section 3A.02).*

Support:

03   Providing for retroreflectivity, such as incorporating glass beads, can affect the skid resistance of pavement markings.

04   Installation of colored pavement to one lane or an area or portion of a multi-lane traveled way can create differentials in skid resistance values between the areas of colored pavement and non-colored pavement that might be unexpected by the road user.

Section 3H.03 Aesthetic Surface Treatments

Support:

01   Aesthetic surface treatments are sometimes used between the transverse lines within a crosswalk, in islands, in medians, in shoulders, within sidewalk extensions designated by pavement markings, or in other areas outside of the traveled way.

02   Common examples of materials used as aesthetic surface treatments include brick, paving bricks, paving stones, or other materials designed to simulate such paving. Some examples of geometries for aesthetic surface treatments include honeycomb, lattice, mesh, grid, and regular polygon patterns.

03   Surfaces with individual units laid out of plane and those that are heavily-textured, rough, or chamfered, could increase rolling resistance and subject pedestrians who use wheelchairs to the effects of vibration; it is desirable to minimize surface discontinuities.

04   Common examples of colors for aesthetic surface treatments incorporated into the material or geometry are brick red, rust, brown, burgundy, clay, tan, or similar earth-tone equivalents (see Figure 3H-1).

Standard:

05   **Aesthetic surface treatments shall not interfere with traffic control devices.**

06   **Aesthetic surface treatments shall not be of a surface that can confuse pedestrians with vision disabilities that rely on tactile treatments or cues for navigation.**

07   **Colors used for aesthetic surface treatments shall be outside the chromaticity coordinates that define the ranges of acceptable colors for traffic control devices.**

08   **Patterns that constitute a purely aesthetic surface treatment shall be devoid of advertising and shall not contain elements of retroreflectivity.**

09   **Patterns that constitute a purely aesthetic surface treatment for the interior area of a crosswalk shall not be designed to encourage road users to remain in the crosswalk, engage or interact with the pattern, or otherwise inhibit users from crossing the street in a safe and efficient manner.**

Guidance:

10   *Aesthetic surface treatments should not use colors or patterns that degrade the contrast of markings used to delineate an area, or that might be mistaken by road users as a traffic control application.*

11   *To provide contrast, a gap of at least one-half the width of the white transverse line used to establish the crosswalk, but not less than 6 inches, should be used between the white crosswalk lines and the aesthetic surface treatment, such as unmarked pavement or a black contrast line (see Section 3A.03).*

12   *To provide contrast, a gap of at least the width of the longitudinal line used to establish the area should be used between the longitudinal line and the aesthetic surface treatment, such as unmarked pavement or a black contrast line (see Section 3A.03). If the longitudinal line is a double line, the gap should be at least the width of one of the lines that makes up the double line.*

13   *Aesthetic surface treatments should not contain pictographs, illustrations, or symbols.*

Section 3H.04 Yellow-Colored Pavement

Support:

01   Yellow-colored pavement is used to enhance the conspicuity of areas separating traffic traveling in opposite directions of travel and the left-hand edge of the roadway.

Standard:

02   **If used, yellow-colored pavement shall be limited to:**

    A. Flush or raised median islands separating traffic flows in opposite directions,
    B. Left-hand shoulders of divided highways, and
    C. Left-hand shoulders of one-way streets or ramps.

03   **Yellow-colored pavement shall not be incorporated into elements of the roadway that function as reversible lanes or two-way left-turn lanes.**

04   **Yellow-colored pavement shall not be used on channelizing islands where traffic travels in the same general direction on both sides.**

Figure 3H-1. Aesthetic Treatments for Transverse Crosswalks

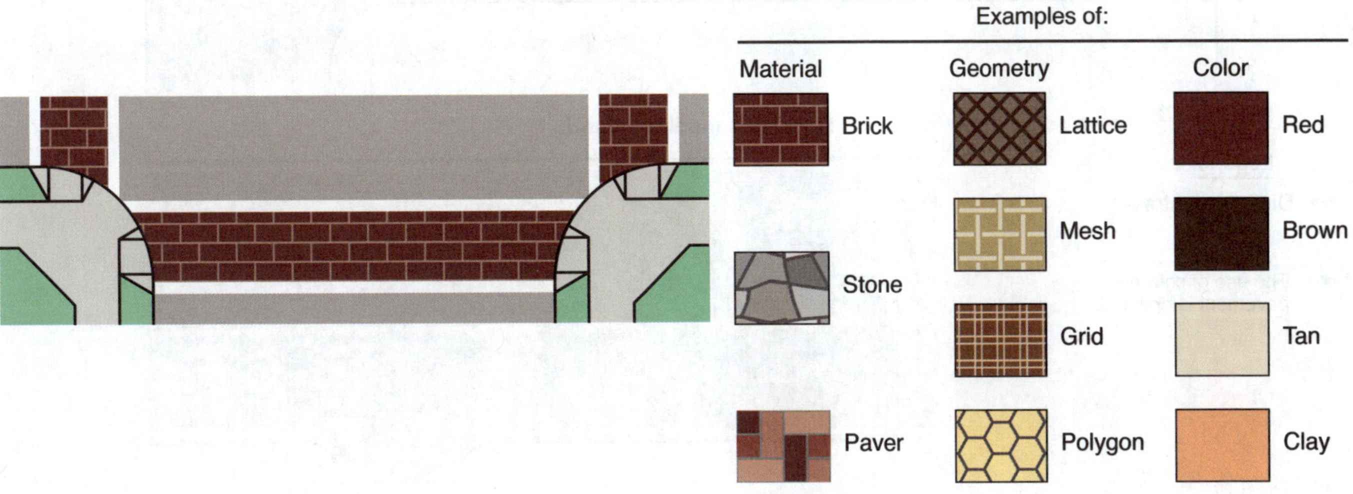

Option:

05   Yellow-colored pavement may be installed for the entire length of the roadway, island, or shoulder, or for only a portion or portions of the roadway, island, or shoulder.

Support:

06   An example of an application of yellow-colored pavement is shown in Figure 3H-2.

Section 3H.05 <u>White-Colored Pavement</u>

Support:

01   White-colored pavement is used to enhance the conspicuity of areas separating traffic traveling in the same direction of travel and the right-hand edge of the roadway.

Standard:

02   **If used, white-colored pavement shall be limited to:**

    A. Flush or raised channelizing islands where traffic passes on both sides in the same general direction,
    B. Right-hand shoulders,
    C. Exit gore areas, and
    D. Entrance gore areas.

Guidance:

03   *When used on right-hand shoulders, white-colored pavement should be limited to areas not intended for use by motor vehicle traffic except those shoulders designated for emergency use.*

Option:

04   White-colored pavement may be installed for the entire length of the roadway, island, or shoulder, or for only a portion or portions of the roadway, island, or shoulder.

05   White-colored pavement may be used instead of chevron markings (see Sections 3B.13 and 3B.25) in neutral areas.

Support:

06   An example of an application of white-colored pavement is shown in Figure 3H-3.

Figure 3H-2. Examples of Yellow-Colored Pavement Applications

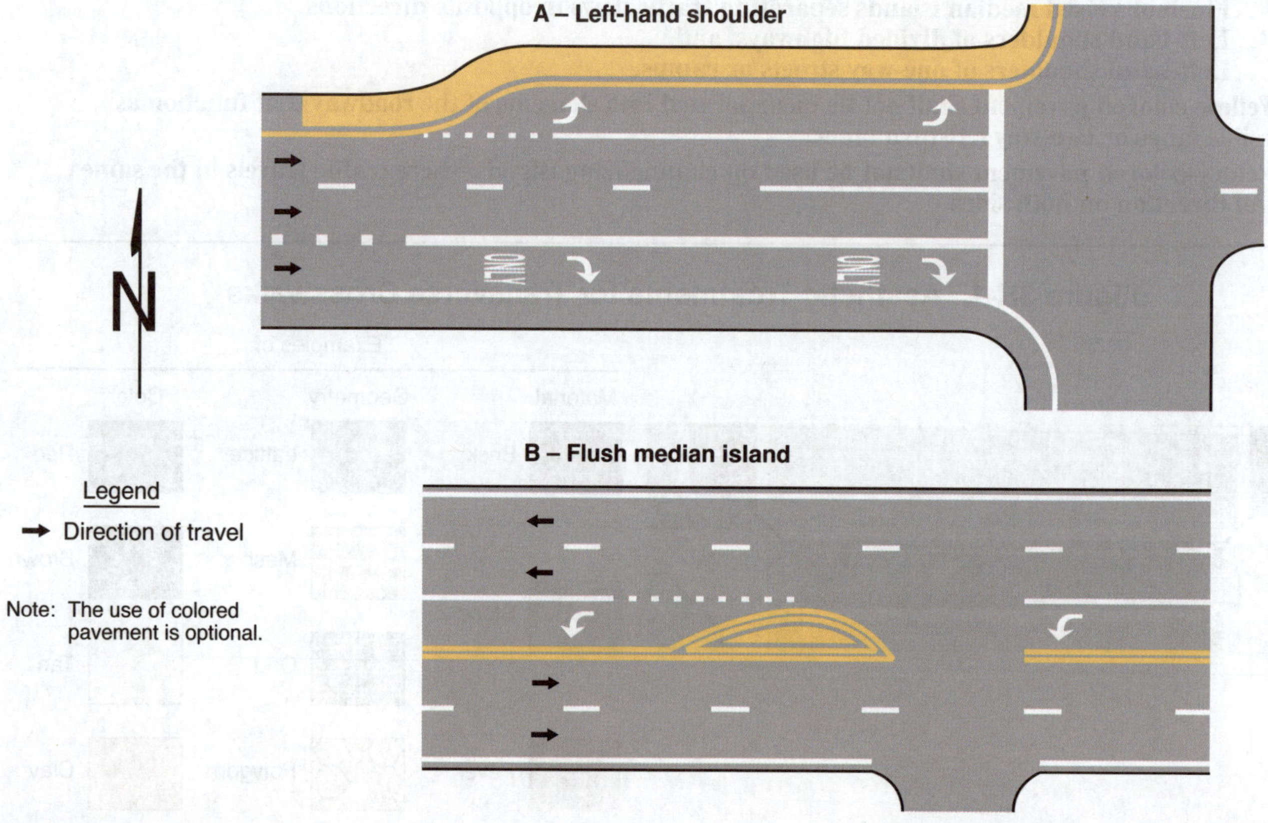

Figure 3H-3. Examples of White-Colored Pavement Applications

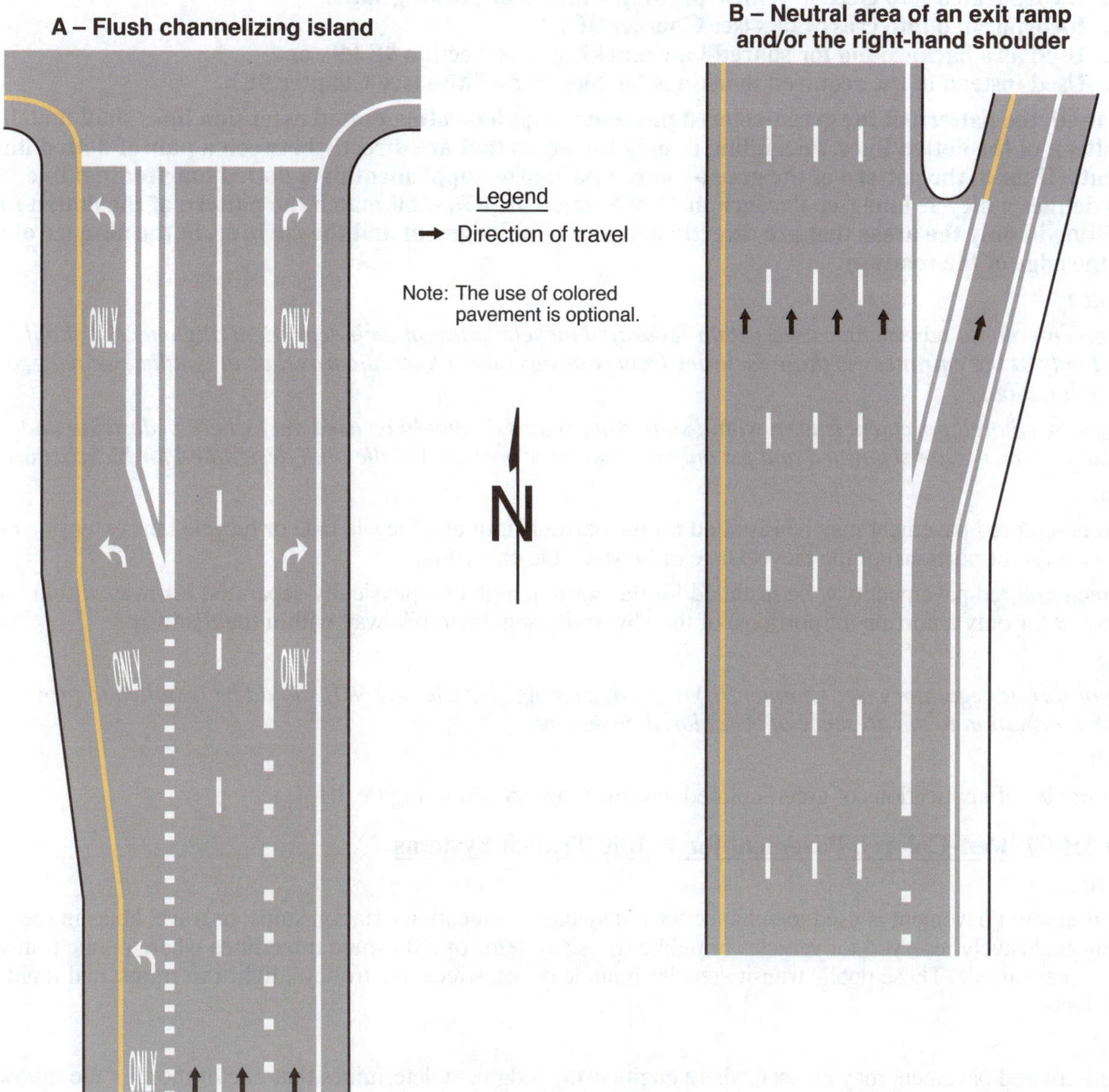

Section 3H.06 Green-Colored Pavement for Bicycle Facilities

Support:

01   Green-colored pavement is used to enhance the conspicuity of locations where bicyclists are expected to operate, and areas where bicyclists and other traffic might have potentially conflicting, weaving, or crossing movements. Green-colored pavement is also used to enhance the conspicuity of word, symbol, and/or arrow pavement markings when these markings are used in certain bicycle facilities.

Standard:

02   If used, green-colored pavement shall be limited to:

A. Bicycle lanes (see Sections 9E.01, 9E.06, 9E.07, and 9E.08),
B. Extensions of bicycle lanes through intersections (see Section 9E.03),
C. Extensions of bicycle lanes through areas where motor vehicles enter a mandatory turn lane in which motor vehicles must weave across bicyclists in bicycle lanes (see Section 9E.02),
D. Two-stage bicycle turn boxes (see Section 9E.11),
E. Bicycle Boxes (see Section 9E.12), and
F. As a background for bicycle detector symbols (see Section 9E.15).

03   **Green-colored pavement shall not be:**

    A. **Incorporated into electric-vehicle parking stations or parking stalls,**
    B. **Incorporated into crosswalks (see Chapter 3C),**
    C. **Used as a background for shared-lane markings (see Section 9E.09), or**
    D. **Used instead of the required markings for bicycle facilities (see Chapter 9E).**

04   **If used, the pattern of the green-colored pavement supplementing dotted extension lines shall match the pattern of the dotted lines, thus filling in only the areas that are directly between a pair of dotted line segments. If used, the pattern of the green-colored pavement supplementing a dotted longitudinal line, which defines a bicycle lane (see Paragraph 11 of Section 9E.02), shall match the pattern of the dotted line, thus filling in only the areas that are directly between a line segment and the curb, or, in the absence of a curb, the edge of the roadway.**

Guidance:

05   *If green-colored pavement is used within separated bicycle lanes on an independent alignment, it should be used only at the entrances to those facilities from roadways open to public travel or at conflict, weaving, or crossing locations.*

06   *If green-colored pavement is used within shared-use paths, it should be used only where pedestrian and bicyclist movements are separated and for only a portion (or portions) of the path designated for bicyclist use.*

Option:

07   Green-colored pavement may be installed for the entire length of a bicycle lane or bicycle lane extension or for only a portion (or portions) of the bicycle lane or bicycle lane extension.

08   Green-colored pavement may be installed for the entire length of a physically-separated bikeway within the roadway or for only a portion (or portions) of the physically-separated bikeway within the roadway.

Guidance:

09   *Appropriate regulatory (see Chapter 9B) or guide signing (see Chapter 9D) should be installed to provide related information to the presence of the colored pavement.*

Support:

10   Examples of applications of green-colored pavement are shown in Figure 3H-4.

Section 3H.07 <u>Red-Colored Pavement for Public Transit Systems</u>

Support:

01   Red-colored pavement is used to enhance the conspicuity of locations, station stops, or travel lanes in the roadway exclusively reserved for vehicles of public transit systems or multi-modal facilities where public transit is the primary mode. These public transit vehicles include buses, streetcars, trolleys, light-rail trains, and rapid transit fleets.

Option:

02   Red-colored pavement may be used where engineering judgment determines that one or more of the following conditions are expected to result from its application:

    A. Increased travel speeds will be expected by the public transport vehicle after an exclusive lane or facility is provided,
    B. Reduced overall service time through the corridor will be expected by the public transport vehicle,
    C. Decreased rates of illegal parking or occupation of the transit or multi-mode lane or facility will be expected.

Standard:

03   **If used, red-colored pavement shall be applied only in lanes, areas, or locations where general-purpose traffic is not allowed to use, queue, wait, idle, or otherwise occupy the lane, area, or location where red-colored pavement is used.**

04   **Red-colored pavement shall be installed for the full width of the lane.**

Option:

05   Red-colored pavement may be used for full-time or part-time operations.

06   Red-colored pavement may be installed for the entire length of a restricted lane or for only a portion (or portions) of the restricted lane.

07   Red-colored pavement may be installed in a broken pattern where entrance into the transit lane is permitted by general traffic, for example where general traffic is allowed in a transit lane in advance of a turn.

Figure 3H-4. Examples of Green-Colored Pavement Applications

A – Applied to the entire corridor

B – Limited to the bicycle symbol and arrow

C – Applied approaching and departing an intersection

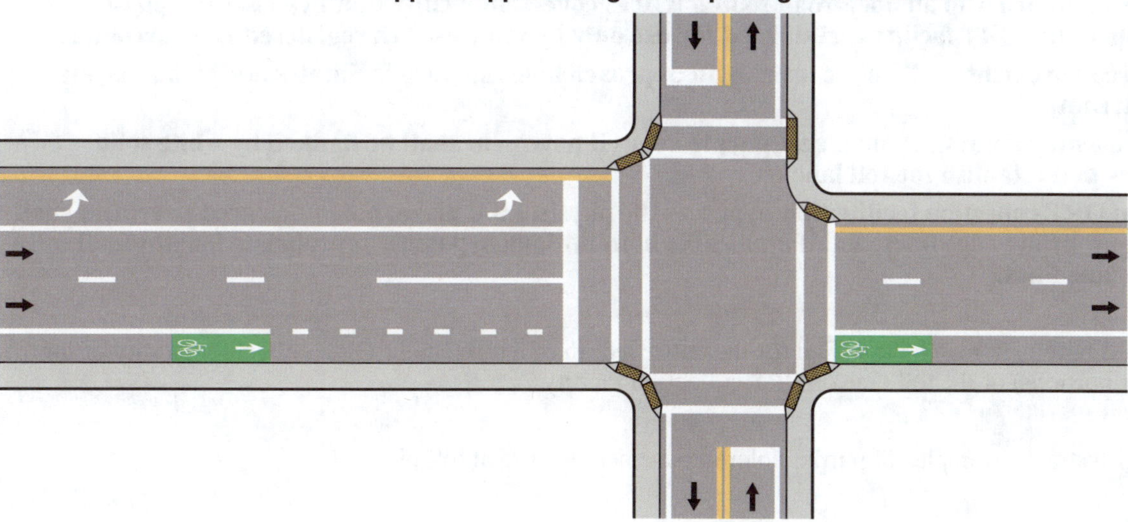

D – Applied supplementing the dotted line approaching intersections and/or the dotted extensions of bicycle lanes through intersections

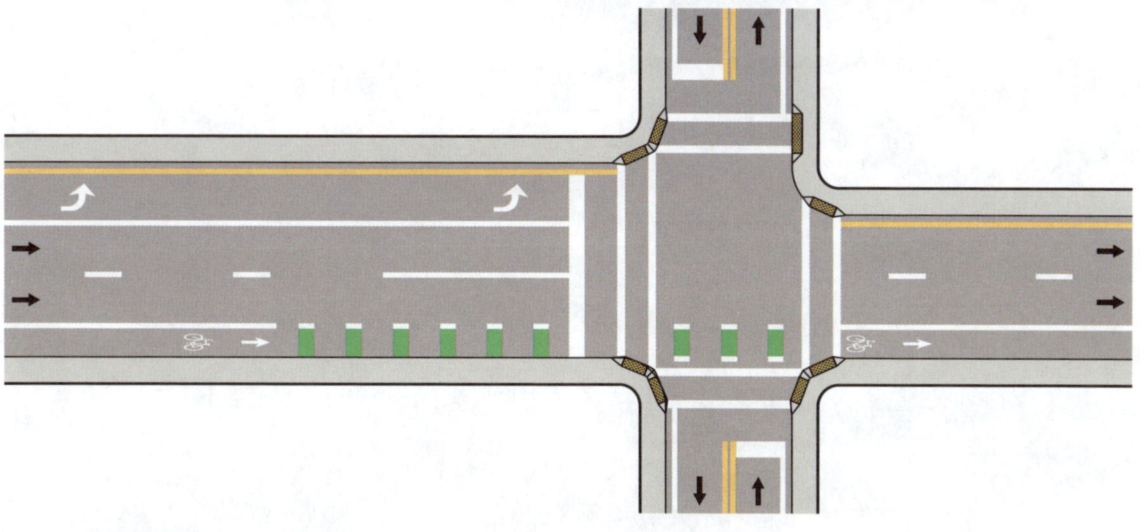

Notes:
1. The use of colored pavement is optional.
2. See Chapter 9E for bicycle facility markings.

Legend

→ Direction of travel

Standard:

08   **Regulatory signs (see Sections 2B.02 and 2G.03) shall be used to establish the allowable use of the lane, area, or location. Regulatory signs shall also be used when it is determined that other vehicles will be allowed to enter the lane to turn or bypass queues.**

Guidance:

09   *If red-colored pavement is used on public transit facilities separated from the roadway or on facilities on an independent alignment, it should be used only at the entrances to those facilities from roadways open to public travel.*

Support:

10   Examples of applications of red-colored pavement are shown in Figure 3H-5.

Section 3H.08 Purple-Colored Pavement for Electronic Toll Collection (ETC) Account-Only Preferential Lanes

Standard:

01   **Purple-colored pavement shall be limited to:**

   A. **Lanes on the approach to a toll plaza where the lane is restricted to use only with a registered ETC account; and**

   B. **Lanes or approaches to an open-road tolling (ORT) collection facility that bypasses the physical toll plaza, where the ORT facility is restricted for use only by vehicles with registered ETC accounts.**

02   **Purple-colored pavement shall not be used in an approach lane that also facilitates additional payment methods downstream.**

03   **If used approaching a physical toll plaza, purple-colored pavement shall be flanked by white solid longitudinal lines that establish the toll lane.**

04   **If used on an ORT collection facility that bypasses the physical toll plaza, purple-colored pavement shall be flanked by appropriate edge lines, and if applicable in multi-lane bypasses, appropriate longitudinal solid or broken white lane lines.**

Option:

05   Purple-colored pavement may be installed for the entire length of a toll lane or ORT collection facility or for only a portion (or portions) of the toll lane or ORT collection facility.

Support:

06   Figure 3H-6 illustrates examples of purple-colored pavement for use at toll plazas.

Figure 3H-5. Examples of Red-Colored Pavement Applications

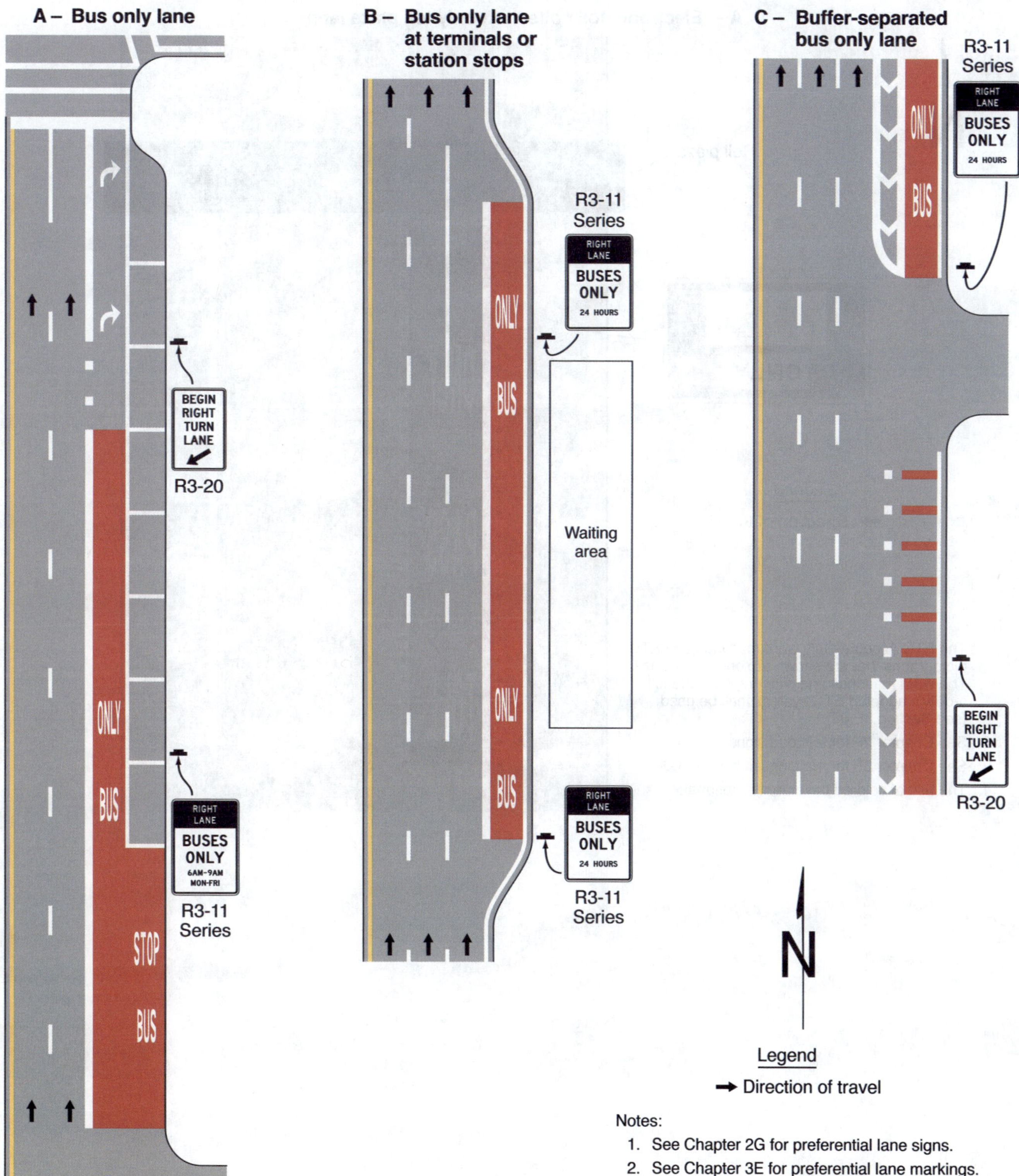

Notes:
1. See Chapter 2G for preferential lane signs.
2. See Chapter 3E for preferential lane markings.
3. The use of colored pavement is optional.

Figure 3H-6. Examples of Purple-Colored Pavement Applications (Sheet 1 of 2)

A – Electronic toll collection only toll plaza lane

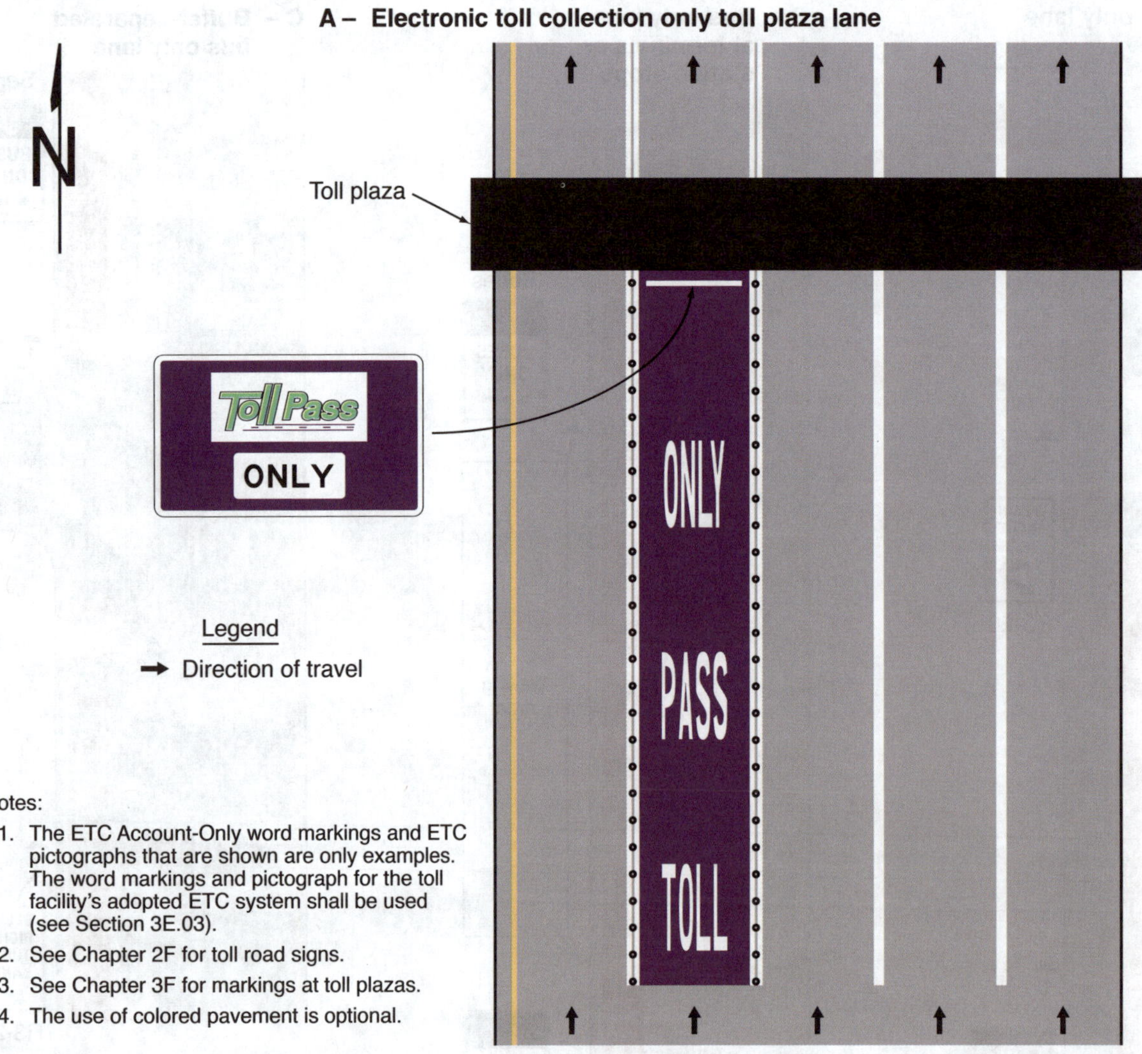

Notes:

1. The ETC Account-Only word markings and ETC pictographs that are shown are only examples. The word markings and pictograph for the toll facility's adopted ETC system shall be used (see Section 3E.03).

2. See Chapter 2F for toll road signs.

3. See Chapter 3F for markings at toll plazas.

4. The use of colored pavement is optional.

Figure 3H-6. Examples of Purple-Colored Pavement Applications (Sheet 2 of 2)

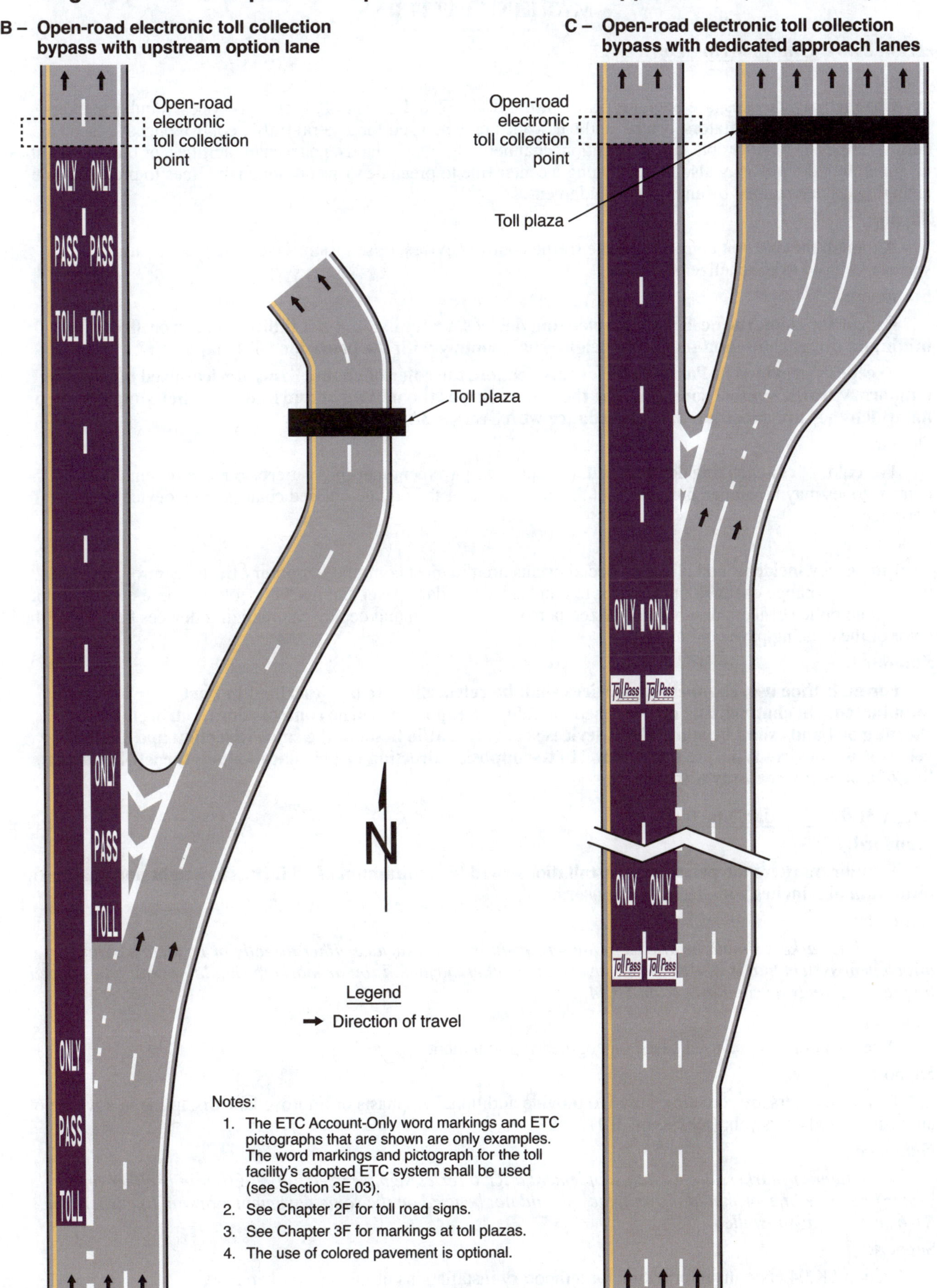

Notes:

1. The ETC Account-Only word markings and ETC pictographs that are shown are only examples. The word markings and pictograph for the toll facility's adopted ETC system shall be used (see Section 3E.03).

2. See Chapter 2F for toll road signs.

3. See Chapter 3F for markings at toll plazas.

4. The use of colored pavement is optional.

CHAPTER 3I. CHANNELIZING DEVICES USED FOR EMPHASIS OF PAVEMENT MARKING PATTERNS

Section 3I.01 Channelizing Devices

Option:

01   Channelizing devices (see Sections 6K.01 through 6K.07 and Figure 6K-1) such as cones, tubular markers, vertical panels, lane separators, drums, and barricades may be used for general traffic control purposes such as adding emphasis to reversible lane delineation, channelizing lines, islands, pedestrian facilities, or bicycle facilities. Channelizing devices may also be used along a center line to preclude turns or along lane lines to preclude lane changing, as determined by engineering judgment.

Support:

02   Although they are not considered to be traffic control devices, raised islands (see Chapter 3J) are also sometimes used to channelize traffic.

Standard:

03   **Except for color, the design of channelizing devices, including, but not limited to, retroreflectivity, minimum dimensions, and mounting height, shall comply with the provisions of Chapter 6K.**

04   **Except as provided in Paragraph 5 of this Section, the color of channelizing devices used outside of temporary traffic control zones shall be the same color as the pavement marking that they supplement, or for which they are substituted, in accordance with Section 3A.03.**

Option:

05   The color of channelizing devices used to emphasize pavement marking patterns outside of temporary traffic control zones may be orange provided that the application of the orange-colored channelizing device is not permanent.

Support:

06   Emergency incidents and planned special events are the most common temporary traffic control zones that would justify orange channelizing devices to emphasize standard pavement marking colors. These events do not necessitate police officers or other authorized personnel to obtain and deploy channelizing devices that match the color of the existing pavement marking.

Standard:

07   **For nighttime use, channelizing devices shall be retroreflective (as described in Part 6) or internally illuminated. On channelizing devices used outside of temporary traffic control zones, retroreflective sheeting or bands shall be white if the devices separate traffic flows in the same direction and shall be yellow if the devices separate traffic flows in the opposite direction or are placed along the left-hand edge line of a one-way roadway or ramp.**

Section 3I.02 Tubular Markers

Standard:

01   **Tubular markers for permanent installations shall be a minimum of 28 inches in height and shall be a minimum of 2 inches wide facing road users.**

Guidance:

02   *Tubular markers should be affixed to the pavement or other surface either directly or by means of an attachment system that is affixed to the pavement or other surface. Tubular markers should normally be spaced no greater than N as cited in Section 3B.14.*

Option:

03   Other spacing may be used based on engineering judgment.

Support:

04   Tubular markers are sometimes used to provide additional emphasis or improve lane discipline in advance of an unsignalized crosswalk (see Figure 3I-1).

Guidance:

05   *When tubular markers are used to supplement a R1-6 series sign (see Section 2B.20) that is either on the center line, lane line, or median island, they should not be used on the same pavement marking line where the R1-6 series sign is installed.*

Support:

06   Section 6K.04 contains information for temporary installations of tubular markers.

Figure 3I-1. Examples of Tubular Markers Supplementing Pavement Markings in Advance of an Unsignalized Crosswalk (Sheet 1 of 2)

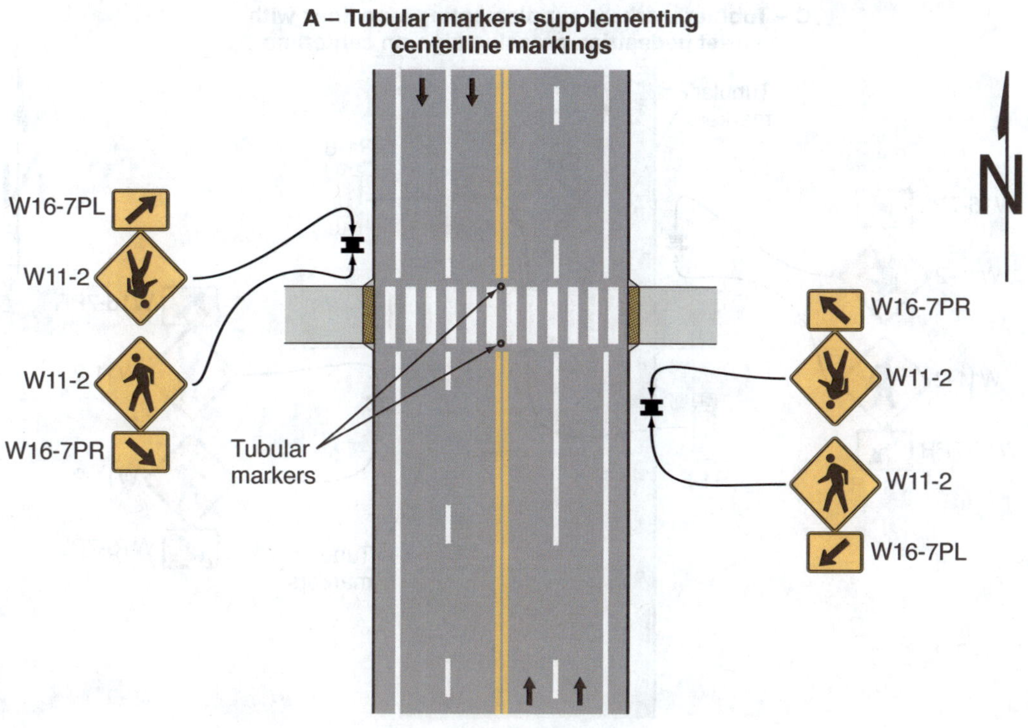

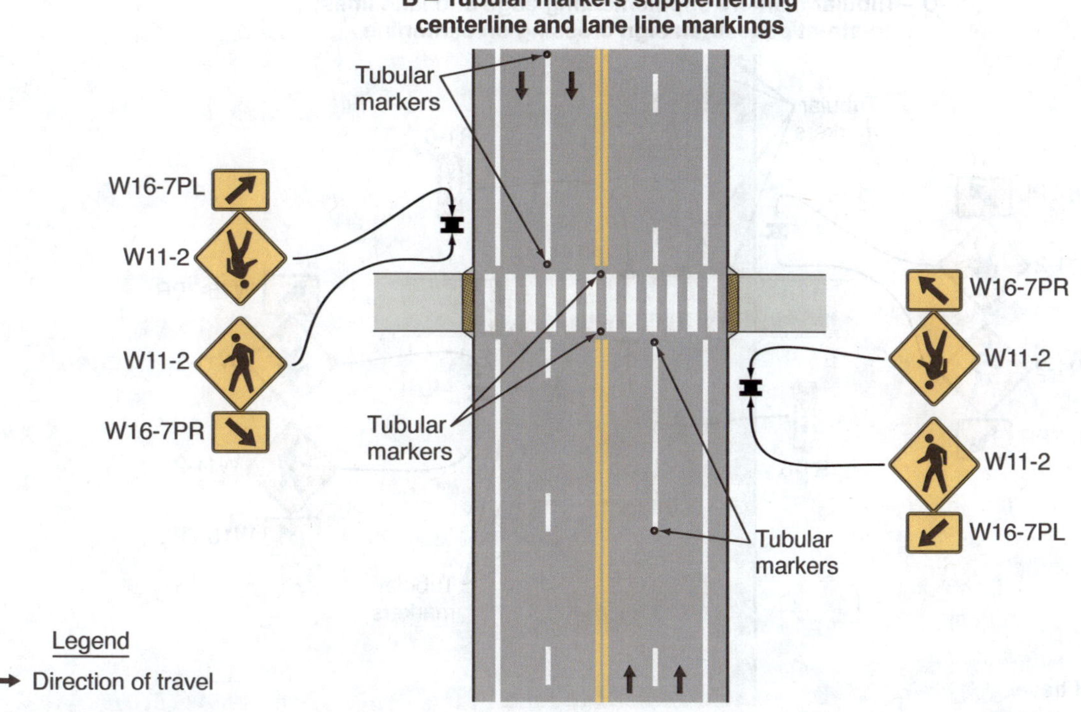

Figure 3I-1. Examples of Tubular Markers Supplementing Pavement Markings in Advance of an Unsignalized Crosswalk (Sheet 2 of 2)

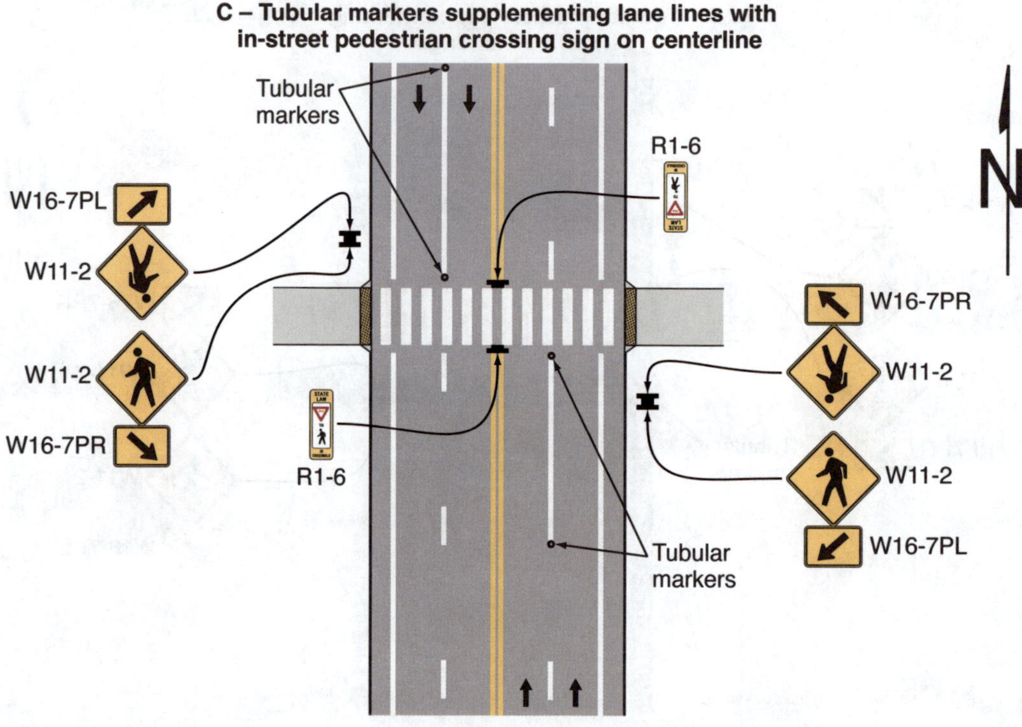

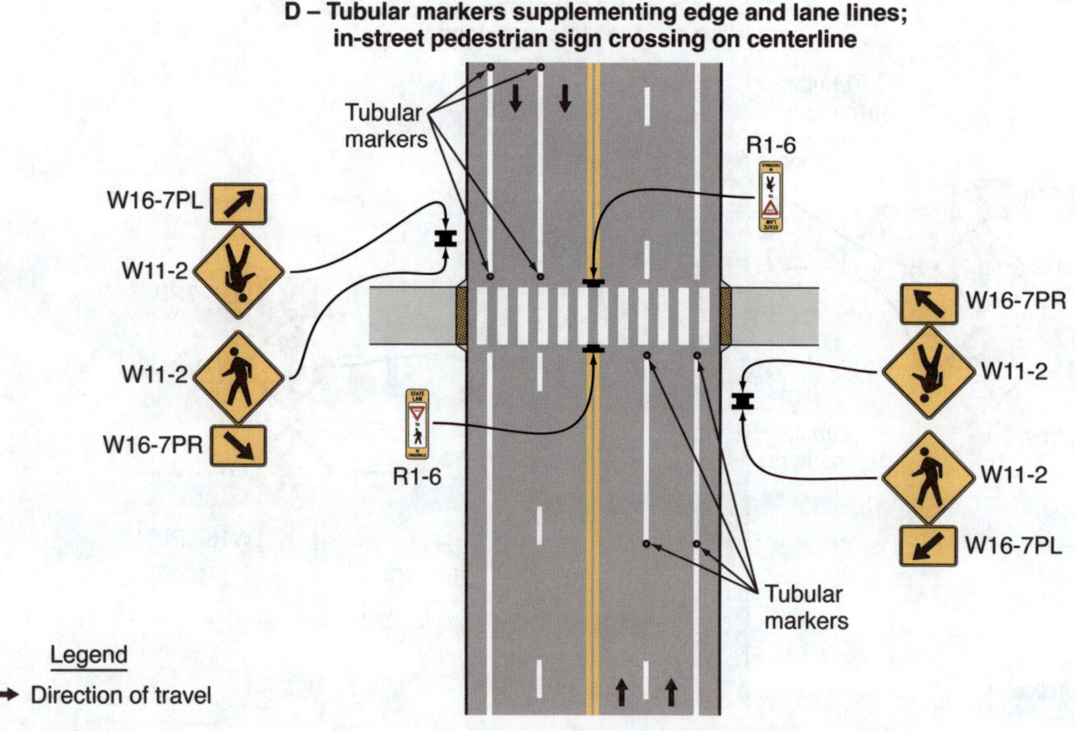

CHAPTER 3J. MARKING AND DELINEATION OF ISLANDS AND SIDEWALK EXTENSIONS

Section 3J.01 General

Support:

01   This Chapter addresses the marking and delineation of islands (see definition in Section 1C.02) and sidewalk extensions designated by pavement markings. Definitions, types, sizes, and other criteria for the design of islands are set forth in "A Policy on Geometric Design of Highways and Streets," 2018 Edition, AASHTO."

02   Section 3C.12 contains information on pedestrian islands and medians.

03   Sections 3H.04 and 3H.05 contain information on colored pavement that can be used within islands.

Option:

04   An island may be designated by curbs, pavement edges, pavement markings, channelizing devices, or other devices.

Section 3J.02 Approach-End Treatment

Support:

01   An approach-end treatment to an island consists of longitudinal pavement markings and/or channelizing devices upstream of the island followed by a divergence of those pavement markings and/or channelizing devices concluding with a transition to other pavement markings that demarcate or outline the island (see Figure 3J-1).

02   Section 3B.13 contains information on pavement markings that function as approach-end treatments for obstructions.

Guidance:

03   *The ends of islands first approached by traffic should be marked with an approach-end treatment, curb markings (see Section 3J.04), or both to guide vehicles into desired paths of travel along the island edge.*

04   *When raised bars or buttons that project more than 1 inch above the pavement surface are used to create a rumble section in the neutral area, the raised bars or buttons should be marked with white or yellow retroreflective materials, as determined by the direction or directions of travel they separate.*

Section 3J.03 Islands Designated by Pavement Markings

Standard:

01   **Except as provided in Paragraph 2 of this Section, islands formed by pavement markings only shall be established using channelizing lines, and shall be white when separating traffic flows in the same general direction or yellow when separating opposing directions of traffic.**

02   **If a continuous flush median island separating travel in opposite directions is used, two sets of double solid yellow lines shall be used to form the island (see Figure 3B-5). Other markings in the median island area, such as diagonal lines (see Section 3B.25), shall also be yellow, except crosswalk markings which shall be white (see Chapter 3C).**

03   **If used, chevron or diagonal markings (see Section 3B.25) within the island shall be the same color as the channelizing line.**

Figure 3J-1. Example of Markings for an Approach-End Treatment to an Island

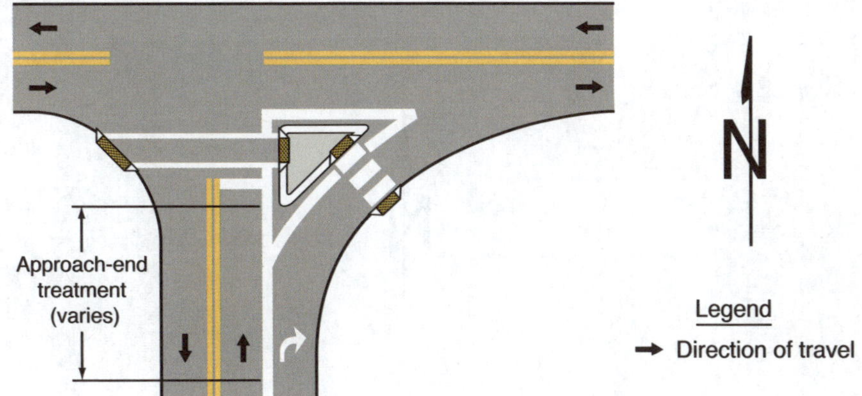

Option:

04   Both chevron and diagonal markings of the same color may be used within the same island based on engineering judgment.

05   The area within the flush island delineated by pavement markings may use colored pavement in accordance with the provisions of Chapter 3H.

Support:

06   Figure 3J-2 illustrates examples of islands designated by pavement markings.

Section 3J.04 Curb Markings for Raised Islands

Standard:

01   **Where curbs are marked for delineation or visibility purposes, the colors shall comply with the general principles of markings (see Section 3A.03).**

Guidance:

02   *Retroreflective solid yellow curb markings should be placed on the approach ends of raised medians and curbs of islands that are located in the line of traffic flow where the curb serves to channel traffic to the right of the obstruction (see Figure 3J-3).*

03   *Retroreflective solid white curb markings should be used when traffic is permitted to pass on either side of the island (see Figure 3J-3).*

04   *The retroreflective area should be of sufficient length to denote the general alignment of the edge of the island along which vehicles travel, including the approach end, when viewed from the approach to the island.*

Option:

05   Where the curbs of the islands become parallel to the direction of traffic flow or where the island is illuminated or marked with delineators, curb markings may be discontinued based on engineering judgment or study.

06   Curb markings at openings in a continuous median island may be omitted based on engineering judgment or study.

Section 3J.05 Pavement Markings for Raised Islands

Support:

01   Pavement markings for raised islands include the approach-end treatment (see Section 3J.02), channelizing lines, edge lines, and chevron or diagonal markings.

Option:

02   Solid yellow edge lines (see Sections 3B.09 and 3B.10) may be used adjacent to raised islands separating travel in opposite directions (see Drawing A in Figure 3J-3).

Figure 3J-2. Examples of Islands Designated by Pavement Markings

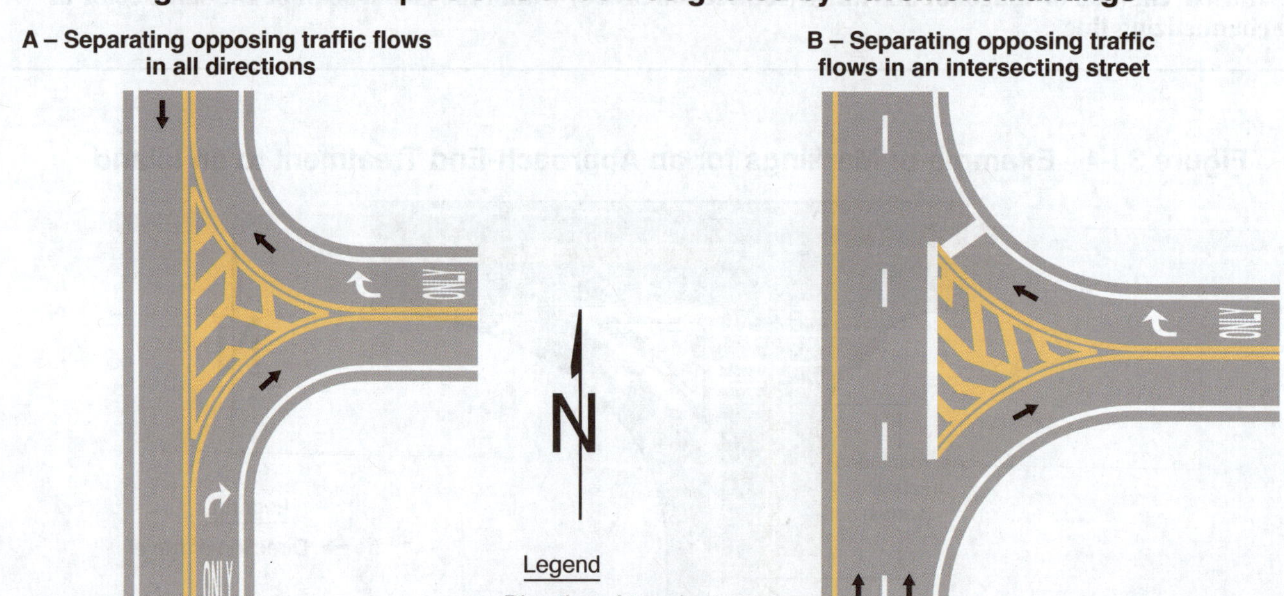

Figure 3J-3. Examples of Curb Markings for Raised Islands

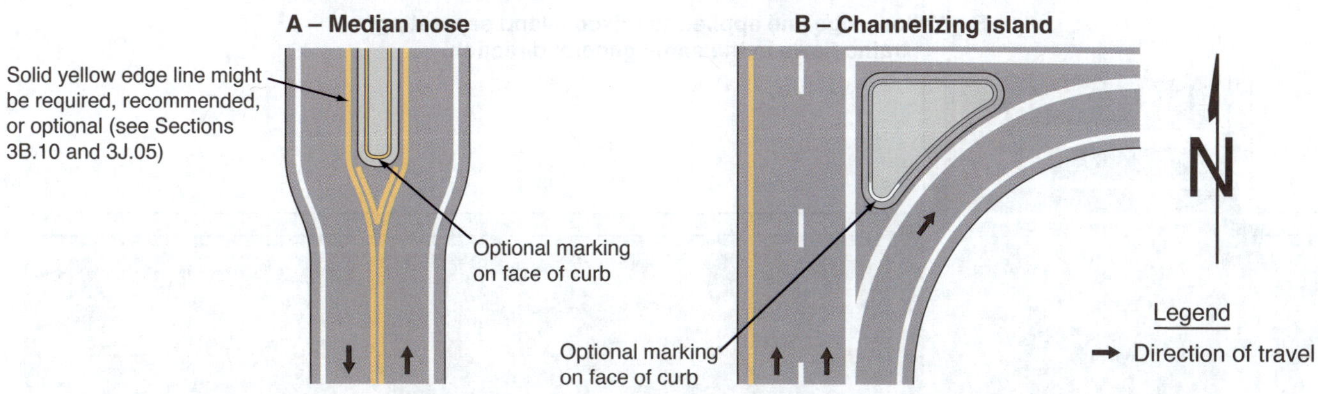

Standard:

03   **Except as provided in Paragraphs 4 and 6 of this Section, raised islands separating traffic flows in the same general direction shall be outlined with white channelizing lines (see Drawing A in Figure 3J-4).**

Option:

04   Pavement markings for smaller raised islands may be omitted based on engineering judgment.

Guidance:

05   *Smaller raised islands without marked channelizing lines, edge lines, or chevron or diagonal markings should use curb markings (see Section 3J.04).*

06   *Where traffic passes on the right of a raised island separating traffic flows in the same general direction, a yellow edge line should be used adjacent to raised islands of discernible size or length instead of continuing the white channelizing line from the approach-end treatment (see Drawing B in Figure 3J-4).*

Figure 3J-4. Examples of Pavement Markings for Raised Islands (Sheet 1 of 2)

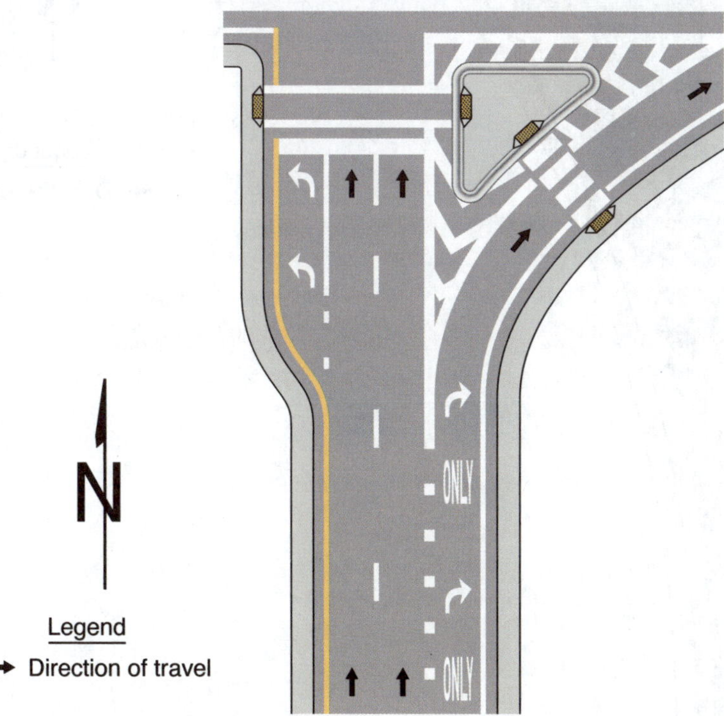

Figure 3J-4. Examples of Pavement Markings for Raised Islands (Sheet 2 of 2)

B – Yellow edge line applied to raised island separating traffic flows in the same general direction

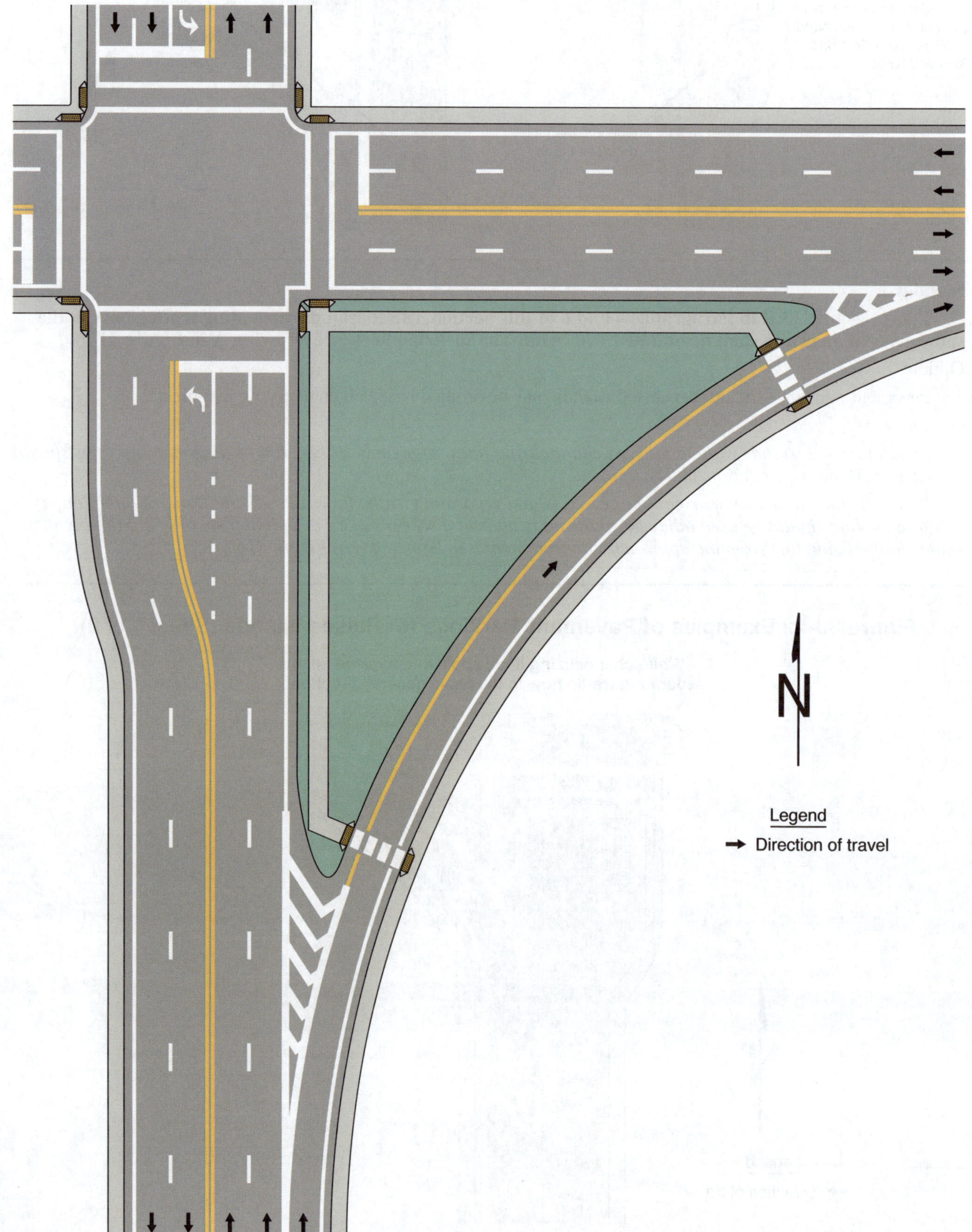

Support:

07   Yellow edge lines adjacent to raised islands that separate traffic flows in the same general direction can be advantageous as a countermeasure for wrong-way entry or travel if the yellow edge line is of discernible length.

Option:

08   Chevron markings may be used in neutral areas formed by diverging channelizing lines at raised islands separating traffic flows in the same general direction.

09   Diagonal markings of an appropriate color may be used in buffer areas between the channelizing line and the raised island (see Figure 3J-5).

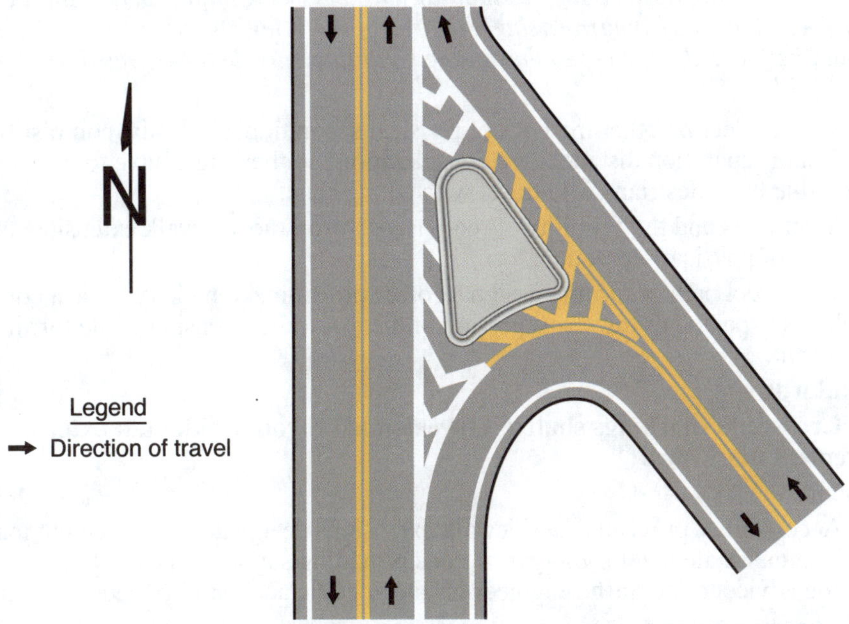

Section 3J.06 Island Delineation

Standard:

01   **Delineators installed on islands shall be the same colors as the related channelizing or edge lines except that, when only facing wrong-way traffic, they shall be red (see Section 3G.03).**

02   **Each roadway through an intersection shall be considered separately in positioning delineators to assure maximum effectiveness.**

Option:

03   Retroreflective or internally illuminated raised pavement markers of the appropriate color may be placed on the pavement in front of the curb and/or on the top of curbed approach ends of raised medians and curbs of islands, as a supplement to or as a substitute for retroreflective curb markings.

Figure 3J-5. Example of Optional Chevron and Diagonal Pavement Markings at a Raised Island

Section 3J.07 Sidewalk Extensions Designated by Pavement Markings

Support:

01   Sidewalk extensions reclaim a portion of the roadway, sometimes including a portion of parking lanes, shoulders, and/or the traveled way, and repurpose that area for non-vehicular uses. They extend the sidewalk or other pedestrian space, shorten pedestrian crossing distances, alter the roadway geometry for speed management or channelizing, or serve other purposes.

02   Sidewalk extensions, sometimes referred to as curb extensions, neckdowns, or bulb-outs, typically are created by physical infrastructure including concrete or asphalt to form a physical narrowing of the roadway with the finished surface at the same level as the adjoining sidewalk.

03   Sidewalk extensions can also be designated by pavement markings for temporary or semi-permanent applications in which the finished surface is at the same level as the vehicular travel pavement. Where an adjoining curb and raised sidewalk are present, this type of application results in a multi-level sidewalk due to the difference in elevation between the adjoining pedestrian surfaces.

04   Sidewalk extensions designated by pavement markings differ from other paved areas designated by pavement markings that are intended to be traversable by a vehicle for authorized or emergency purposes.

Standard:

05   **Sidewalk extensions designated by pavement markings shall be established using double solid lines connecting to the outside physical curb or, in the absence of a curb, to the edge of the roadway. The color of the double solid line shall comply with the provisions of Section 3A.03.**

Support:

06   The paved area between the double solid line forming the sidewalk extension designated by pavement markings and the sidewalk or other roadside area is not part of the roadway. Sidewalk extensions designated by

pavement markings formed by double solid lines are distinct from areas such as shoulders or gore areas where travel is discouraged by the presence of a single line, or flush medians where travel is prohibited by a double solid line. Sidewalk extensions designated by pavement markings with double solid lines designate areas outside the roadway where vehicle traversal is prohibited.

07   Areas formed by a single wide line are sometimes used to alter the roadway geometry for speed management or channelizing, or to serve other purposes, where pedestrians are not expected (see Drawing B in Figure 3J-6). These areas are not considered a sidewalk extension, and provisions to delineate areas where vehicle traversal is discouraged include channelizing lines (see Section 3B.08), edge lines (see Section 3B.09), and diagonal markings (see Section 3B.25).

Guidance:

08   *Channelizing devices such as tubular markers (see Chapter 3I) should be used to provide conspicuity for, and to prevent vehicles from traversing, the area of the sidewalk extension designated by pavement markings. They should be located adjacent to the double solid line outside the traveled way.*

Support:

09   When selecting other methods of physical separation, the visual contrast from adjoining pavement and maximum separation distances are considerations so they are visible to pedestrians having limited vision and detectable by pedestrians who travel with a long cane.

10   Sight lines and the visibility of road users within the sidewalk extension area are considerations when selecting methods of physical separation.

11   The swept path of turning design or other prevailing vehicle types is a consideration, especially if a larger vehicle is expected to traverse a portion of the sidewalk extension while turning where pedestrians might be present.

Standard:

12   **Crosswalk markings shall not be extended through sidewalk extensions designated by pavement markings.**

Support:

13   Accessibility provisions at sidewalk extensions designated by pavement markings are outside the scope of this Manual. State and local organizations providing support services to pedestrians with vision disabilities can provide advice to the traffic engineer on site-specific accessibility decisions. In addition, orientation and mobility

---

Figure 3J-6. Examples of Sidewalk Extensions Designated by Pavement Markings and Channelization

A – Sidewalk extension to reduce the pedestrian crossing distance

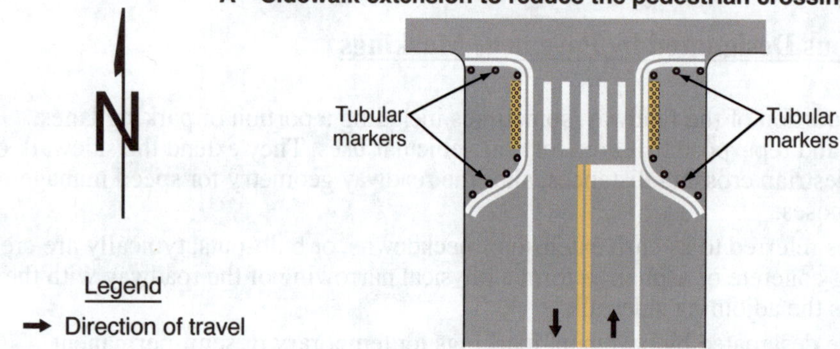

B – Channelizing for speed control and altered travel paths

specialists or similar staff can provide advice to inform such decisions. The U.S. Access Board (www.access-board.gov) provides technical assistance for making pedestrian facilities accessible to persons with disabilities.

Guidance:

14   *Traffic control devices that are critical to the specific conditions at the sidewalk extension, such as STOP or YIELD signs or Pedestrian Crossing signs, should be located within the sidewalk extension designated by pavement markings. Their lateral offset (see Section 2A.16) should be measured from the center of the double solid line designating a sidewalk extension rather than from the physical curb line behind the sidewalk extension area so that the signs are more visible to approaching traffic and not occluded by any physical features placed within the sidewalk extension area.*

Support:

15   The location of accessible pedestrian signals (see Section 4K.02) is a consideration when providing a sidewalk extension designated by pavement markings.

16   Aesthetic surface treatments (see Chapter 3H) are sometimes used in sidewalk extensions designated by pavement markings to emphasize that the area is outside of the traveled way.

Standard:

17   **In accordance with the provisions of Section 3H.03, aesthetic surface treatments, if used within a sidewalk extension designated by pavement markings, shall be non-retroreflective.**

Support:

18   Figure 3J-6 illustrates an example of a sidewalk extension designated by pavement markings and an example of channelizing.

CHAPTER 3K. RUMBLE STRIP MARKINGS

Section 3K.01 Longitudinal Rumble Strip Markings

Support:

01   Longitudinal rumble strips consist of a series of rough-textured or slightly raised or depressed road surfaces intended to alert inattentive drivers through vibration and sound that their vehicle has left the travel lane. Shoulder rumble strips are typically installed along the shoulder near the travel lane. On divided highways, rumble strips are sometimes installed on the median side (left-hand side) shoulder as well as on the outside (right-hand side) shoulder. On two-way roadways, rumble strips are sometimes installed along the center line.

02   This Manual contains no provisions regarding the design and placement of longitudinal rumble strips. The provisions in this Manual address the use of markings in combination with a longitudinal rumble strip. Figure 3K-1 illustrates markings used with or near longitudinal rumble strips.

03   Section 6M.06 contains information related to longitudinal rumble strips.

Option:

04   An edge line or center line may be located over a longitudinal rumble strip to create a rumble stripe.

Standard:

05   **The color of an edge line or center line associated with a longitudinal rumble stripe shall be in accordance with Section 3A.03.**

06   **An edge line shall not be used in addition to a rumble stripe that is located along a shoulder.**

Section 3K.02 Transverse Rumble Strip Markings

Support:

01   Transverse rumble strips consist of intermittent narrow, transverse areas of rough-textured or slightly raised or depressed road surface that extend across the travel lanes to alert drivers to unusual vehicular traffic conditions. Through noise and vibration, they attract the attention of road users to features such as unexpected changes in alignment and conditions requiring a reduction in speed or a stop.

02   This Manual contains no provisions regarding the design and placement of transverse rumble strips that approximate the color of the pavement. The provisions in this Manual address the use of markings in combination with a transverse rumble strip.

03   Section 6M.06 contains information related to transverse rumble strips.

Standard:

04   **Except as otherwise provided in Section 6M.06 for TTC zones, if the color of a transverse rumble strip used within a travel lane is not the color of the pavement, the color of the transverse rumble strip shall be either black or white.**

Guidance:

05   *White transverse rumble strip markings used in a travel lane should not be placed in locations where they could be confused with other transverse markings such as stop lines or crosswalks.*

Figure 3K-1. Examples of Longitudinal Rumble Strip Markings

A – Edge line not on rumble strip

B – Edge line on rumble strip

C – Center line on rumble strip

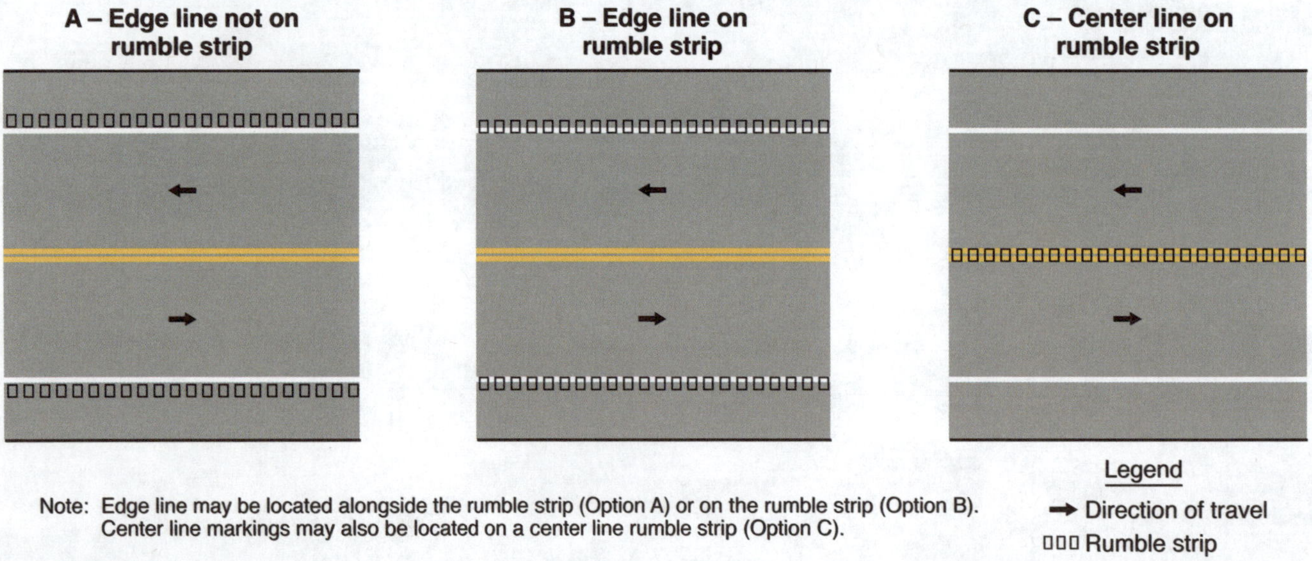

Note: Edge line may be located alongside the rumble strip (Option A) or on the rumble strip (Option B). Center line markings may also be located on a center line rumble strip (Option C).

Legend

➜ Direction of travel

▯▯▯ Rumble strip

PART 4
HIGHWAY TRAFFIC SIGNALS

CHAPTER 4A. GENERAL

Section 4A.01 Types

Support:

01   The following types and uses of highway traffic signals are discussed in Part 4: traffic control signals; bicycle signal faces; pedestrian signal heads; hybrid beacons; rectangular rapid flashing beacons; emergency-vehicle traffic control signals; traffic control signals for one-lane, two-way facilities; traffic control signals for freeway entrance ramps; movable bridge traffic signals; toll plaza traffic signals; flashing beacons; lane-use control signals; and in-roadway warning lights.

Section 4A.02 Meanings of Signal Indications

Support:

01   The "Uniform Vehicle Code" (see Section 1A.06) is the primary source for the standards for the meanings of vehicular signal indications to both vehicle operators and pedestrians as provided in Sections 4A.03 and 4A.04, the standards for the meanings of separate bicycle signal face indications as provided in Section 4A.05, and the standards for the meanings of separate pedestrian signal head indications as provided in Section 4A.06.

02   The physical area that is defined as being "within the intersection" is dependent upon the conditions that are described in the definition of an intersection in Section 1C.02.

Section 4A.03 Meanings of Steady Vehicular Signal Indications

Standard:

01   **The following meanings shall be given to steady highway traffic signal indications for vehicles and pedestrians:**

    A. Steady green signal indications shall have the following meanings:

        1. Vehicular traffic facing a CIRCULAR GREEN signal indication is permitted to proceed straight through or turn right or left or make a U-turn movement except as such movement is modified by lane-use signs, turn prohibition signs, lane markings, roadway design, separate turn signal indications, or other traffic control devices.

        Such vehicular traffic, including vehicles turning right or left or making a U-turn movement, shall yield the right-of-way to:

        (a) Pedestrians lawfully within an associated crosswalk, and
        (b) Other vehicles lawfully within the intersection.

        In addition, vehicular traffic turning left or making a U-turn movement to the left shall yield the right-of-way to other vehicles approaching from the opposite direction so closely as to constitute an immediate hazard during the time when such turning vehicle is moving across or within the intersection.

        2. Vehicular traffic facing a GREEN ARROW signal indication, displayed alone or in combination with another signal indication, is permitted to cautiously enter the intersection only to make the movement indicated by such arrow, or such other movement as is permitted by other signal indications displayed at the same time.

        Such vehicular traffic, including vehicles turning right or left or making a U-turn movement, shall yield the right-of-way to:

        (a) Pedestrians lawfully within an associated crosswalk, and
        (b) Other vehicles lawfully within the intersection.

        3. Pedestrians facing a CIRCULAR GREEN signal indication, unless otherwise directed by a pedestrian signal indication or other traffic control device, are permitted to proceed across the roadway within any marked or unmarked associated crosswalk. The pedestrian shall yield the right-of-way to vehicles lawfully within the intersection or so close as to create an immediate hazard at the time that the green signal indication is first displayed.

        4. Pedestrians facing a GREEN ARROW signal indication, unless otherwise directed by a pedestrian signal indication or other traffic control device, shall not cross the roadway.

B. Steady yellow signal indications shall have the following meanings:

1. Vehicular traffic facing a steady CIRCULAR YELLOW signal indication is thereby warned that the related green movement or the related flashing arrow movement is being terminated or that a steady red signal indication will be displayed immediately thereafter when vehicular traffic shall not enter the intersection. The rules set forth concerning vehicular operation under the movement(s) being terminated shall continue to apply while the steady CIRCULAR YELLOW signal indication is displayed.

2. Vehicular traffic facing a steady YELLOW ARROW signal indication is thereby warned that the related GREEN ARROW movement or the related flashing arrow movement is being terminated. The rules set forth concerning vehicular operation under the movement(s) being terminated shall continue to apply while the steady YELLOW ARROW signal indication is displayed.

3. Pedestrians facing a steady CIRCULAR YELLOW or YELLOW ARROW signal indication, unless otherwise directed by a pedestrian signal indication or other traffic control device shall not start to cross the roadway.

C. Steady red signal indications shall have the following meanings:

1. Vehicular traffic facing a steady CIRCULAR RED signal indication, unless entering the intersection to make another movement permitted by another signal indication, shall stop at a clearly marked stop line; but if there is no stop line, traffic shall stop before entering the crosswalk on the near side of the intersection; or if there is no crosswalk, then before entering the intersection; and shall remain stopped until a signal indication to proceed is displayed, or as provided below.

    Except when a traffic control device is in place prohibiting a turn on red or a steady RED ARROW signal indication is displayed, vehicular traffic facing a steady CIRCULAR RED signal indication is permitted to enter the intersection to turn right, or to turn left from a one-way street into a one-way street, after stopping. The right to proceed with the turn shall be subject to the rules applicable after making a stop at a STOP sign.

2. Vehicular traffic facing a steady RED ARROW signal indication shall not enter the intersection to make the movement indicated by the arrow and, unless entering the intersection to make another movement permitted by another signal indication, shall stop at a clearly marked stop line; but if there is no stop line, before entering the crosswalk on the near side of the intersection; or if there is no crosswalk, then before entering the intersection; and shall remain stopped until a signal indication or other traffic control device permitting the movement indicated by such RED ARROW is displayed.

    When a traffic control device is in place permitting a turn on a steady RED ARROW signal indication, vehicular traffic facing a steady RED ARROW signal indication is permitted to enter the intersection to make the movement indicated by the arrow signal indication, after stopping. The right to proceed with the turn shall be limited to the direction indicated by the arrow and shall be subject to the rules applicable after making a stop at a STOP sign.

3. Unless otherwise directed by a pedestrian signal indication or other traffic control device, pedestrians facing a steady CIRCULAR RED or steady RED ARROW signal indication shall not enter the roadway.

Section 4A.04 Meanings of Flashing Vehicular Signal Indications

Standard:

01 The following meanings shall be given to flashing highway traffic signal indications for vehicles and pedestrians:

A. A flashing green signal indication has no meaning and shall not be used.

B. Flashing yellow signal indications shall have the following meanings:

1. Vehicular traffic, on an approach to an intersection, facing a flashing CIRCULAR YELLOW signal indication is permitted to cautiously enter the intersection to proceed straight through or turn right or left or make a U-turn except as such movement is modified by laneuse signs, turn prohibition signs, lane markings, roadway design, separate turn signal indications, or other traffic control devices.

    Such vehicular traffic, including vehicles turning right or left or making a U-turn, shall yield the right-of-way to:

(a) Pedestrians lawfully within an associated crosswalk, and

(b) Other vehicles lawfully within the intersection.

In addition, vehicular traffic turning left or making a U-turn to the left shall yield the right-of-way to other vehicles approaching from the opposite direction so closely as to constitute an immediate hazard during the time when such turning vehicle is moving across or within the intersection.

2. Vehicular traffic, on an approach to an intersection, facing a flashing YELLOW ARROW signal indication, displayed alone or in combination with another signal indication, is permitted to cautiously enter the intersection only to make the movement indicated by such arrow, or other such movement as is permitted by other signal indications displayed at the same time.

Such vehicular traffic, including vehicles turning right or left or making a U-turn, shall yield the right-of-way to:

(a) Pedestrians lawfully within an associated crosswalk, and
(b) Other vehicles lawfully within the intersection.

In addition, vehicular traffic turning left or making a U-turn to the left shall yield the right-of-way to other vehicles approaching from the opposite direction so closely as to constitute an immediate hazard during the time when such turning vehicle is moving across or within the intersection.

3. Pedestrians facing any flashing yellow signal indication at an intersection, unless otherwise directed by a pedestrian signal indication or other traffic control device, are permitted to proceed across the roadway within any marked or unmarked associated crosswalk. Pedestrians shall yield the right-of-way to vehicles lawfully within the intersection at the time that the flashing yellow signal indication is first displayed.

4. When a flashing CIRCULAR YELLOW signal indication(s) is displayed as a beacon (see Chapter 4S) to supplement another traffic control device, road users are notified that there is a need to pay extra attention to the message contained thereon or that the regulatory or warning requirements of the other traffic control device, which might not be applicable at all times, are currently applicable.

C. Flashing red signal indications shall have the following meanings:

1. Vehicular traffic, on an approach to an intersection, facing a flashing CIRCULAR RED signal indication shall stop at a clearly marked stop line; but if there is no stop line, before entering the crosswalk on the near side of the intersection; or if there is no crosswalk, at the point nearest the intersecting roadway where the driver has a view of approaching traffic on the intersecting roadway before entering the intersection. The right to proceed shall be subject to the rules applicable after making a stop at a STOP sign.

2. Vehicular traffic, on an approach to an intersection, facing a flashing RED ARROW signal indication if intending to turn in the direction indicated by the arrow shall stop at a clearly marked stop line; but if there is no stop line, before entering the crosswalk on the near side of the intersection; or if there is no crosswalk, at the point nearest the intersecting roadway where the driver has a view of approaching traffic on the intersecting roadway before entering the intersection. The right to proceed with the turn shall be limited to the direction indicated by the arrow and shall be subject to the rules applicable after making a stop at a STOP sign.

3. Pedestrians facing any flashing red signal indication at an intersection, unless otherwise directed by a pedestrian signal indication or other traffic control device, are permitted to proceed across the roadway within any marked or unmarked associated crosswalk. Pedestrians shall yield the right-of-way to vehicles lawfully within the intersection at the time that the flashing red signal indication is first displayed.

4. When a flashing CIRCULAR RED signal indication(s) is displayed as a beacon (see Chapter 4S) to supplement another traffic control device, road users are notified that there is a need to pay extra attention to the message contained thereon or that the regulatory requirements of the other traffic control device, which might not be applicable at all times, are currently applicable. Use of this signal indication shall be limited to supplementing STOP (R1-1), DO NOT ENTER (R5-1), or WRONG WAY (R5-1a) signs, and to applications where compliance with the supplemented traffic control device requires a stop at a designated point.

Section 4A.05 Meanings of Bicycle Symbol Signal Indications

Standard:

01 The following meanings shall be given to bicycle symbol signal indications for bicyclists:

A. Bicyclists facing a steady GREEN BICYCLE signal indication are permitted to enter the intersection only to make the movement indicated by the lane-use arrow(s) displayed on the Bicycle Signal sign (see Section 9B.22) that is located immediately adjacent to the signal face, Bicyclists proceeding into the intersection during the display of the steady GREEN BICYCLE signal indication shall yield the right-of-way to:

1. Pedestrians lawfully within an associated crosswalk, and
2. Other vehicles lawfully within the intersection.

B. Bicyclists facing a steady YELLOW BICYCLE signal indication are thereby warned that the related green movement is being terminated and that a steady RED BICYCLE signal indication will be displayed immediately thereafter when bicyclists shall not enter the intersection. The rules set forth concerning bicycle operation under the movement being terminated shall continue to apply while the steady YELLOW BICYCLE signal indication is displayed.

C. Bicyclists facing a steady RED BICYCLE signal indication shall not enter the intersection to make the movement indicated by the lane-use arrow(s) displayed on the Bicycle Signal sign (see Section 9B.22) that is located immediately adjacent to the signal face and, unless entering the intersection to make another movement permitted by another bicycle symbol signal indication, shall stop at a clearly marked stop line; but if there is no stop line, before entering the crosswalk on the near side of the intersection; or if there is no crosswalk, then before entering the intersection; and shall remain stopped until a GREEN BICYCLE signal indication permitting the movement indicated by such RED BICYCLE signal indication is displayed.

Except when a traffic control device is in place prohibiting a turn on red, bicyclists facing a steady RED BICYCLE signal indication are permitted to enter the intersection to turn right if there are no approach lanes for motor vehicle traffic to their right. The right to proceed with the turn shall be subject to the rules applicable after making a stop at a STOP sign.

D. A flashing GREEN BICYCLE signal indication has no meaning and shall not be used.
E. A flashing YELLOW BICYCLE signal indication has no meaning and shall not be used.
F. Bicyclists facing a flashing RED BICYCLE signal indication shall stop at a clearly marked stop line; but if there is no stop line, before entering the crosswalk on the near side of the intersection; or if there is no crosswalk, at the point nearest the intersecting roadway where the bicyclist has a view of approaching traffic on the intersecting roadway before entering the intersection. The right to proceed in the direction indicated by the lane-use arrow(s) displayed on the Bicycle Signal sign (see Section 9B.22) that is located immediately adjacent to the signal face shall be subject to the rules applicable after making a stop at a STOP sign.

Section 4A.06 Meanings of Pedestrian Signal Indications

Standard:

01 Pedestrian signal indications shall have the following meanings:

A. A flashing WALKING PERSON (symbolizing WALK) signal indication has no meaning and shall not be used.
B. Pedestrians facing a steady WALKING PERSON (symbolizing WALK) signal indication shall be permitted to start to cross the roadway in the direction of the signal indication, possibly in conflict with turning vehicles. Pedestrians shall yield the right-of-way to vehicles lawfully within the intersection at the time that the WALKING PERSON (symbolizing WALK) signal indication is first shown.
C. Pedestrians facing a flashing UPRAISED HAND (symbolizing DONT WALK) signal indication shall not start to cross the roadway in the direction of the signal indication. Any pedestrian who has already started to cross the roadway on a steady WALKING PERSON (symbolizing WALK) signal indication shall continue to proceed to the far side of the traveled way of the street or highway, unless otherwise directed by a traffic control device to proceed only to the median of a divided highway or only to some other island or pedestrian refuge area (see Section 3C.12).
D. Pedestrians facing a steady UPRAISED HAND (symbolizing DONT WALK) signal indication shall not enter the roadway in the direction of the signal indication.

Section 4A.07 Lateral Offset of Signal Supports and Cabinets

Guidance:

01 *The following items should be considered when placing signal supports and cabinets:*

    A. *Reference should be made to the "Roadside Design Guide," 4th Edition, 2011, AASHTO, and to the U.S. Department of Justice 2010 ADA Standards for Accessible Design, September 15, 2010, 28 CFR 35 and 36, Americans with Disabilities Act of 1990.*

    B. *Signal supports should be placed as far as practicable from the edge of the traveled way without adversely affecting the visibility of the signal indications.*

    C. *Where supports cannot be located based on the recommended AASHTO clearances, consideration should be given to the use of appropriate safety devices.*

    D. *No part of a concrete foundation for a signal support should extend more than 4 inches above the ground level at any point. This limitation does not apply to the concrete foundation for a rigid support.*

    E. *In order to minimize hindrance to the passage of persons with physical disabilities, a signal support or controller cabinet should not obstruct the sidewalk, or access from the sidewalk to the crosswalk.*

    F. *Controller cabinets should be located as far as practicable from the edge of the roadway.*

    G. *On medians, the minimum clearances provided in Items A through E for signal supports should be obtained, if practicable.*

Section 4A.08 Use of Signs at Signalized Locations

Support:

01 Traffic signal signs are sometimes used at highway traffic signal locations to instruct or guide pedestrians, bicyclists, or motorists. Among the signs typically used at or on the approaches to signalized locations are movement prohibition signs (see Section 2B.26), lane-control signs (see Sections 2B.27 through 2B.29), pedestrian crossing signs (see Section 2B.57), pedestrian and bicycle actuation signs (see Section 2B.58), traffic signal signs (see Sections 2B.59 and 2C.44), No Turn on Red signs (see Section 2B.60), Signal Ahead warning signs (see Section 2C.35), Street Name signs (see Section 2D.45), and Advance Street Name signs (see Section 2D.46).

Guidance:

02 *Regulatory, warning, and guide signs should be used at highway traffic signal locations as provided in Part 2 and as specifically provided elsewhere in Part 4.*

Support:

03 Section 2B.27 contains information regarding the use of overhead lane-control signs on signalized approaches where lane drops, multiple-lane turns involving combined through-and-turn lanes, or other lane-use regulations that would be unexpected by unfamiliar road users are present.

Guidance:

04 *If used, illuminated traffic signal signs should be designed and mounted in such a manner as to avoid glare and reflections that seriously detract from the signal indications. Highway traffic signal faces should be given dominant position and brightness to maximize their priority in the overall display.*

Standard:

05 **The minimum vertical clearance and horizontal offset of the total assembly of traffic signal signs (see Section 2B.59) shall comply with the provisions of Sections 4D.09 and 4D.10.**

06 **STOP signs shall not be used in conjunction with any highway traffic signal operation, except in either of the following cases:**

    A. **If the signal indication for an approach is a flashing red at all times, or**

    B. **If a minor street or driveway is located within or adjacent to an area controlled by a traffic control signal, but does not require separate traffic signal control because an extremely low potential for conflict exists.**

Section 4A.09 Use of Pavement Markings at Signalized Locations

Support:

01 Pavement markings that clearly communicate the operational plan of an intersection to road users play an important role in the effective operation of highway traffic signals. By designating the number of lanes, the use of each lane, the length of additional lanes on the approach to an intersection, crosswalks, and the proper stopping points, the engineer can design the signal phasing and timing to best match the goals of the operational plan.

Guidance:

02 *Pavement markings should be used at highway traffic signal locations as provided in Part 3. If the road surface will not retain pavement markings, signs should be installed to provide the needed road user information.*

Section 4A.10 Responsibility for Operation and Maintenance

Guidance:

01 *Prior to installing any highway traffic signal, the responsibility for the maintenance of the signal and all of the appurtenances, hardware, software, and the timing plan(s) should be clearly established by the responsible agency.*

02 *To this end the agency should:*

A. *Keep every controller assembly in effective operation in accordance with its predetermined timing schedule, check the operation of the controller assembly frequently enough to verify that it is operating in accordance with the predetermined timing schedule, and establish a policy to maintain a record of all timing changes and that only authorized persons are permitted to make timing changes;*

B. *Clean the optical system of the signal sections and replace the light sources as frequently as experience proves necessary;*

C. *Clean and service equipment and other appurtenances as frequently as experience proves necessary;*

D. *Provide for alternate operation of the traffic control signal during a period of failure, using flashing mode or manual control, or manual traffic direction by proper authorities as might be required by traffic volumes or congestion, or by erecting other traffic control devices;*

E. *Have properly-skilled maintenance personnel available without undue delay for all signal malfunctions and signal indication failures;*

F. *Provide spare equipment to minimize the interruption of highway traffic signal operation as a result of equipment failure;*

G. *Provide for the availability of properly-skilled maintenance personnel for the repair of all components; and*

H. *Maintain the appearance of the signal displays and equipment.*

CHAPTER 4B. TRAFFIC CONTROL SIGNALS—GENERAL

Section 4B.01 General

Support:

01   Words such as pedestrians and bicyclists are used redundantly in selected Sections of Part 4 to encourage sensitivity to these elements of "traffic."

02   Standards for traffic control signals are important because traffic control signals need to attract the attention of a variety of road users, including those who are older, those with vision disabilities, as well as those who are fatigued or distracted, or who are not expecting to encounter a signal at a particular location.

Section 4B.02 Advantages and Disadvantages of Traffic Control Signals

Support:

01   When properly used, traffic control signals are valuable devices for safety and the control of vehicular and vulnerable road user traffic. They control the various traffic movements by alternating between directing them to stop and permitting them to proceed and thereby profoundly influence traffic flow. This accomplishes the need to safely separate road users in time in order to prevent crashes.

02   Traffic control signals that are properly designed, located, operated, and maintained will have one or more of the following advantages:

    A. They reduce the frequency and severity of certain types of crashes, especially right-angle collisions and those involving vulnerable road users.

    B. They provide for the orderly movement of traffic.

    C. They increase the traffic-handling capacity of the intersection if:

        1. Proper physical layouts and control measures are used, and

        2. The signal operational parameters are reviewed and updated (if needed) on a regular basis (as engineering judgment determines that significant traffic flow and/or land use changes have occurred) to maximize the ability of the traffic control signal to satisfy current traffic demands.

    D. They are coordinated to provide for continuous or nearly-continuous movement of traffic at a definite speed along a given route under favorable conditions.

    E. They are used to interrupt heavy traffic at intervals to permit other traffic, vehicular or pedestrian, to cross.

03   Traffic control signals are often considered a panacea for all traffic problems at intersections. This belief has led to traffic control signals being installed at many locations where they are not needed, adversely affecting the safety and efficiency of motor vehicle, bicycle, and pedestrian traffic.

04   Traffic control signals, even when justified by traffic and roadway conditions, can be ill-designed, ineffectively placed, improperly operated, or poorly maintained. Improper or unjustified traffic control signals can result in one or more of the following disadvantages:

    A. Excessive delay,

    B. Excessive disobedience of the signal indications,

    C. Increased use of less-adequate routes as road users attempt to avoid the traffic control signals, and

    D. Significant increases in the frequency of collisions (especially rear-end collisions).

Section 4B.03 Alternatives to Traffic Control Signals

Guidance:

01   *Since road user delay and the frequency of some types of crashes are sometimes higher under traffic signal control than under STOP sign control, consideration should be given to providing alternatives to traffic control signals even if one or more of the signal warrants (see Chapter 4C) has been satisfied.*

Option:

02   These alternatives may include, but are not limited to, the following:

    A. Installing signs along the major street to warn road users approaching the intersection;

    B. Installing a roundabout to reduce fatal and serious injury crashes and vehicular conflicts that result in fatal and serious injury crashes (see Section 8A.12 if the location is in close proximity to a grade crossing);

    C. Installing a pedestrian hybrid beacon (see Chapter 4J), rectangular rapid flashing beacons (see Chapter 4L), pedestrian-actuated Warning Beacons (see Chapter 4S), or In-Roadway Warning Lights (see Chapter 4U) if pedestrian safety is the major concern;

    D. Relocating the stop line(s) and making other changes to improve the sight distance at the intersection;

    E. Installing measures designed to reduce speeds on the approaches;

    F. Installing a flashing beacon at the intersection to supplement STOP sign control;

G. Installing flashing beacons on warning signs in advance of a stop-controlled intersection on the major-street and/or minor-street approaches;

H. Adding one or more lanes on a minor-street approach to reduce the number of vehicles per lane on the approach;

I. Revising the geometrics at the intersection to channelize vehicular movements and reduce the time required for a vehicle to complete a movement, which could also assist pedestrians;

J. Revising the geometrics at the intersection to add pedestrian median refuge islands and/or curb extensions;

K. Installing roadway lighting if a disproportionate number of crashes occur at night;

L. Restricting one or more turning movements, perhaps on a time-of-day basis, if alternate routes are available;

M. If the warrant is satisfied, installing multi-way stop control;

N. Employing other alternatives, depending on conditions at the intersection.

Section 4B.04 <u>Basis of Installation of Traffic Control Signals</u>

Support:

01   A careful analysis of traffic operations, pedestrian and bicyclist needs, and other factors at a large number of signalized and unsignalized locations, coupled with engineering judgment, has provided a series of signal warrants, described in Chapter 4C, that define the minimum conditions under which installing traffic control signals might be justified.

Guidance:

02   *The design (including the phasing, operation, and timing) of new traffic control signals should be based on an engineering study of roadway, traffic, and other conditions.*

Section 4B.05 <u>Basis of Removal of Traffic Control Signals</u>

Guidance:

01   *Engineering judgment should be applied in the review of operating traffic control signals to determine whether the type of installation and the timing program meet the current requirements of all forms of traffic.*

02   *If changes in traffic patterns eliminate the need for a traffic control signal, consideration should be given to removing it and replacing it with appropriate alternative traffic control devices, if any are needed.*

03   *If the engineering study indicates that the traffic control signal is no longer justified, and a decision is made to remove the signal, the removal should be accomplished using the following steps:*

A. *Determine the appropriate traffic control to be used after the removal of the signal.*

B. *Remove any sight-distance restrictions as necessary.*

C. *Inform the public of the removal study.*

D. *Flash or cover the signal heads for a minimum of 90 days, and install the appropriate STOP sign control or other traffic control devices.*

E. *Remove the signal if the engineering data collected during the removal study period confirms that the signal is no longer needed.*

Option:

04   Because Items C, D, and E in Paragraph 3 of this Section are not relevant when a temporary traffic control signal (see Section 4D.11) is removed, a temporary traffic control signal may be removed immediately after Items A and B are completed.

05   Instead of total removal of a traffic control signal, the poles, controller cabinet, and cables may remain in place after removal of the signal heads for continued analysis.

CHAPTER 4C. TRAFFIC CONTROL SIGNAL NEEDS STUDIES

Section 4C.01 Studies and Factors for Justifying Traffic Control Signals

Standard:

01   **Except for a temporary traffic control signal (see Section 4D.11) installed in a temporary traffic control zone, before a traffic control signal is installed at a particular location, an engineering study of traffic conditions, pedestrian characteristics, and physical characteristics of the location shall be performed to determine whether installation of a traffic control signal is justified at that location.**

02   **The investigation of the need for a traffic control signal shall include an analysis of factors related to the existing operation and safety at the study location and the potential to improve these conditions, and the applicable factors contained in the following traffic signal warrants:**

Warrant 1, Eight-Hour Vehicular Volume
Warrant 2, Four-Hour Vehicular Volume
Warrant 3, Peak Hour
Warrant 4, Pedestrian Volume
Warrant 5, School Crossing
Warrant 6, Coordinated Signal System
Warrant 7, Crash Experience
Warrant 8, Roadway Network
Warrant 9, Intersection Near a Grade Crossing

03   **The satisfaction of a traffic signal warrant or warrants shall not in itself require the installation of a traffic control signal.**

Support:

04   Sections 8D.08 and 8D.14 contain information regarding the use of traffic control signals instead of gates and/ or flashing-light signals at grade crossings.

Guidance:

05   *When considering the installation of a traffic control signal, alternatives to traffic control signals, including those listed in Section 4B.03, should also be considered.*

06   *A traffic control signal should not be installed unless one or more of the factors described in this Chapter are met.*

07   *A traffic control signal should not be installed unless an engineering study indicates that installing a traffic control signal will improve the overall safety and/or operation of the intersection.*

08   *The study should consider the effects of the right-turning vehicles from the minor-street approaches. Engineering judgment should be used to determine what, if any, portion of the right-turning traffic is subtracted from the minor-street traffic count when evaluating the count against the signal warrants listed in Paragraph 2 of this Section.*

09   *Engineering judgment should also be used in applying various traffic signal warrants to cases where major-street approaches consist of one lane plus one left-turn or right-turn lane. The site-specific traffic characteristics should dictate whether a major-street approach is considered as one lane or two lanes. For example, for a major-street approach with one lane for through and right-turning traffic plus a left-turn lane, if engineering judgment indicates that it should be considered a one-lane approach because the traffic using the left-turn lane is minor, the total traffic volume approaching the intersection should be applied against the signal warrants as a one-lane approach. The major-street approach should be considered two lanes if approximately half of the traffic on the approach turns left and the left-turn lane is of sufficient length to accommodate all left-turning vehicles.*

10   *Similar engineering judgment and rationale should be applied to a minor-street approach with one through/ left-turn lane plus a right-turn lane. In this case, the degree of conflict of minor-street right-turning traffic with traffic on the major street should be considered. Thus, right-turning traffic should not be included in the minor-street volume if the movement enters the major street with minimal conflict. The minor-street approach should be evaluated as a one-lane approach with only the traffic volume in the through/left-turn lane considered.*

11   *If a minor-street approach has one combined through/right-turn lane plus a left-turn lane, the approach should either be analyzed as a two-lane approach based on the sum of the traffic volumes using both lanes or as a one-lane approach based on only the traffic volume in the approach lane with the higher volume.*

12   *At a location that is under development or construction or at a location where it is not possible to obtain a traffic count that would represent future traffic conditions, hourly volumes should be estimated as part of an engineering study for comparison with traffic signal warrants. Except for locations where the engineering study uses the satisfaction of Warrant 8 to justify a signal, a traffic control signal installed under projected conditions should have an engineering study done within 1 year of putting the signal into steady (stop-and-go) operation to determine if the signal is justified. If not justified, the signal should be taken out of steady (stop-and-go) operation or removed.*

Option:

13   For signal warrant analysis, a location with a wide median may be analyzed as one intersection or as two intersections (see Section 2A.23) based on engineering judgment.

14   At an intersection with a high volume of left-turning traffic from the major street, the signal warrant analysis may be performed in a manner that considers the higher of the major-street left-turn volumes as the "minor-street" volume and the corresponding single direction of opposing traffic on the major street as the "major-street" volume.

15   For signal warrants requiring conditions to be present for a certain number of hours in order to be satisfied, any four consecutive 15-minute periods may be considered as 1 hour if the separate 1-hour periods used in the warrant analysis do not overlap each other and both the major-street volume and the minor-street volume are for the same specific 1-hour periods.

16   For signal warrant analysis, bicyclists may be counted as either vehicles or pedestrians.

Support:

17   When performing a signal warrant analysis, bicyclists riding in the street with other vehicular traffic are usually counted as vehicles and bicyclists who are clearly using pedestrian facilities are usually counted as pedestrians.

Option:

18   Engineering study data may include the following:

A. The number of vehicles entering the intersection in each hour from each approach during 12 hours of an average day. It is desirable that the hours selected contain the greatest percentage of the 24-hour traffic volume.

B. Vehicular volumes for each traffic movement from each approach, classified by vehicle type (heavy trucks, passenger cars and light trucks, public-transit vehicles, and, in some locations, bicycles), during each 15-minute period of the 2 hours in the morning and 2 hours in the afternoon during which the total traffic entering the intersection is the greatest.

C. Pedestrian volume counts on each crosswalk during the same periods as the vehicular counts in Item B and during the hours of highest pedestrian volume. Where young, elderly, and/or persons with physical or vision disabilities need special consideration, the pedestrians and their crossing times may be classified by general observation.

D. Information about nearby facilities and activity centers that serve the young, elderly, and/or persons with disabilities, including requests from persons with disabilities for accessible crossing improvements at the location under study. These persons might not be adequately reflected in the pedestrian volume count if the absence of a signal restrains their mobility.

E. The posted or statutory speed limit or the 85th-percentile speed on the uncontrolled approaches to the location.

F. A condition diagram showing details of the physical layout, including such features as intersection geometrics, channelization, grades, sight-distance restrictions, transit stops and routes, parking conditions, pavement markings, roadway lighting, driveways, nearby railroad crossings, distance to the nearest traffic control signals, utility poles and fixtures, and adjacent land use.

G. A collision diagram showing crash experience by type, location, direction of movement, severity, weather, time of day, date, and day of week for at least 1 year.

19   The following data, which are desirable for a more precise understanding of the operation of the intersection, may be obtained during the periods described in Item B of Paragraph 18 of this Section:

A. Vehicle-hours of stopped-time delay determined separately for each approach.

B. The number and distribution of acceptable gaps in vehicular traffic on the major street for entrance from the minor street.

C. The posted or statutory speed limit or the 85th-percentile speed on controlled approaches at a point near to the intersection but unaffected by the control.

D. Pedestrian delay time for at least two 30-minute peak pedestrian delay periods of an average weekday or like periods of a Saturday or Sunday.

E. Queue length on stop-controlled approaches.

Support:

20   The safe and efficient movement of all road users is the primary consideration in the engineering study to determine whether to install a traffic control signal or to install some other type of control or roadway configuration. Installation of a traffic control signal does not necessarily result in improved safety in every case. In some cases, the installation of a traffic control signal at an inappropriate location could adversely impact safety for one or more types of road users. The purpose of the engineering study is to evaluate all of the factors that are relevant to a specific location. The satisfaction of a warrant (or warrants) is one of the relevant factors in the

engineering study, but it is not intended to be the only factor or even the overriding consideration. Agencies can install a traffic control signal at a location where no warrants are met, but only after conducting an engineering study that documents the rationale for deciding that the installation of a traffic control signal is the best solution for improving the overall safety and/or operation at the location.

Section 4C.02 Warrant 1, Eight-Hour Vehicular Volume

Support:

01   The Minimum Vehicular Volume, Condition A (see Table 4C-1), is intended for application at locations where a large volume of intersecting traffic is the principal reason to consider installing a traffic control signal.

02   The Interruption of Continuous Traffic, Condition B (see Table 4C-1), is intended for application at locations where Condition A is not satisfied and where the traffic volume on a major street is so heavy that traffic on a minor intersecting street suffers excessive delay or conflict in entering or crossing the major street.

03   It is intended that Warrant 1 be treated as a single warrant. If Condition A is satisfied, then Warrant 1 is satisfied and analyses of Condition B and the combination of Conditions A and B are not needed. Similarly, if Condition B is satisfied, then Warrant 1 is satisfied and an analysis of the combination of Conditions A and B is not needed.

Guidance:

04   *The need for a traffic control signal should be considered if an engineering study finds that one of the following conditions exist for each of any 8 hours of an average day:*

    A. *The vehicles per hour given in both of the 100 percent columns of Condition A in Table 4C-1 exist on the major street and the more critical minor-street approach, respectively, to the intersection; or*

    B. *The vehicles per hour given in both of the 100 percent columns of Condition B in Table 4C-1 exist on the major street and the more critical minor-street approach, respectively, to the intersection.*

Standard:

05   **These major-street and minor-street volumes shall be for the same 8 hours for each condition; however, the 8 hours that are selected for the Condition A analysis shall not be required to be the same 8 hours that are selected for the Condition B analysis.**

Table 4C-1. Warrant 1, Eight-Hour Vehicular Volume

Condition A—Minimum Vehicular Volume

Number of lanes for moving traffic on each approach		Vehicles per hour on major street (total of both approaches)				Vehicles per hour on more critical minor-street approach (one direction only)			
Major Street	Minor Street	100%[a]	80%[b]	70%[c]	56%[d]	100%[a]	80%[b]	70%[c]	56%[d]
1	1	500	400	350	280	150	120	105	84
2 or more	1	600	480	420	336	150	120	105	84
2 or more	2 or more	600	480	420	336	200	160	140	112
1	2 or more	500	400	350	280	200	160	140	112

Number of lanes for moving traffic on each approach		Vehicles per hour on major street (total of both approaches)				Vehicles per hour on more critical minor-street approach (one direction only)			
Major Street	Minor Street	100%[a]	80%[b]	70%[c]	56%[d]	100%[a]	80%[b]	70%[c]	56%[d]
1	1	750	600	525	420	75	60	53	42
2 or more	1	900	720	630	504	75	60	53	42
2 or more	2 or more	900	720	630	504	100	80	70	56
1	2 or more	750	600	525	420	100	80	70	56

[a] Basic minimum hourly volume

[b] Used for combination of Conditions A and B after adequate trial of other remedial measures

[c] May be used when the major-street speed exceeds 40 mph or in an isolated community with a population of less than 10,000

[d] May be used for combination of Conditions A and B after adequate trial of other remedial measures when the major-street speed exceeds 40 mph or in an isolated community with a population of less than 10,000

Support:

06   On the minor street, the more critical volume is not required to be on the same approach during each of these 8 hours. The more critical minor-street volume is the one that meets the warranting criteria for that approach, and in the case of a one-lane minor-street approach that is opposite from a multi-lane minor-street approach might not have the higher volume.

Option:

07   If the posted or statutory speed limit or the 85th-percentile speed on the major street exceeds 40 mph, or if the intersection lies within the built-up area of an isolated community having a population of less than 10,000, the traffic volumes in the 70 percent columns in Table 4C-1 may be used in place of the 100 percent columns.

Guidance:

08   *The combination of Conditions A and B is intended for application at locations where Condition A is not satisfied and Condition B is not satisfied and should be applied only after an adequate trial of other alternatives that could cause less delay and inconvenience to traffic has failed to solve the traffic problems.*

09   *The need for a traffic control signal should be considered if an engineering study finds that both of the following conditions exist for each of any 8 hours of an average day:*

   A. *The vehicles per hour given in both of the 80 percent columns of Condition A in Table 4C-1 exist on the major street and the more critical minor-street approach, respectively, to the intersection; and*
   B. *The vehicles per hour given in both of the 80 percent columns of Condition B in Table 4C-1 exist on the major street and the more critical minor-street approach, respectively, to the intersection.*

Standard:

10   **These major-street and minor-street volumes shall be for the same 8 hours for each condition; however, the 8 hours satisfied in Condition A shall not be required to be the same 8 hours satisfied in Condition B.**

Support:

11   On the minor street, the more critical volume is not required to be on the same approach during each of the 8 hours. The more critical minor-street volume is the one that meets the warranting criteria for that approach, and in the case of a one-lane minor-street approach that is opposite from a multi-lane minor-street approach might not have the higher volume.

Option:

12   If the posted or statutory speed limit or the 85th-percentile speed on the major street exceeds 40 mph, or if the intersection lies within the built-up area of an isolated community having a population of less than 10,000, the traffic volumes in the 56 percent columns in Table 4C-1 may be used in place of the 80 percent columns.

Section 4C.03 Warrant 2, Four-Hour Vehicular Volume

Support:

01   The Four-Hour Vehicular Volume signal warrant conditions are intended to be applied where the volume of intersecting traffic is the principal reason to consider installing a traffic control signal.

Guidance:

02   *The need for a traffic control signal should be considered if an engineering study finds that, for each of any 4 hours of an average day, the plotted points representing the vehicles per hour on the major street (total of both approaches) and the corresponding vehicles per hour on the more critical minor-street approach (one direction only) all fall above the applicable curve in Figure 4C-1 for the existing combination of approach lanes.*

Support:

03   On the minor street, the more critical volume is not required to be on the same approach during each of these 4 hours. The more critical minor-street volume is the one that meets the warranting criteria for that approach, and in the case of a one-lane minor-street approach that is opposite from a multi-lane minor-street approach might not have the higher volume.

Option:

04   If the posted or statutory speed limit or the 85th-percentile speed on the major street exceeds 40 mph, or if the intersection lies within the built-up area of an isolated community having a population of less than 10,000, Figure 4C-2 may be used in place of Figure 4C-1.

Figure 4C-1. Warrant 2, Four-Hour Vehicular Volume

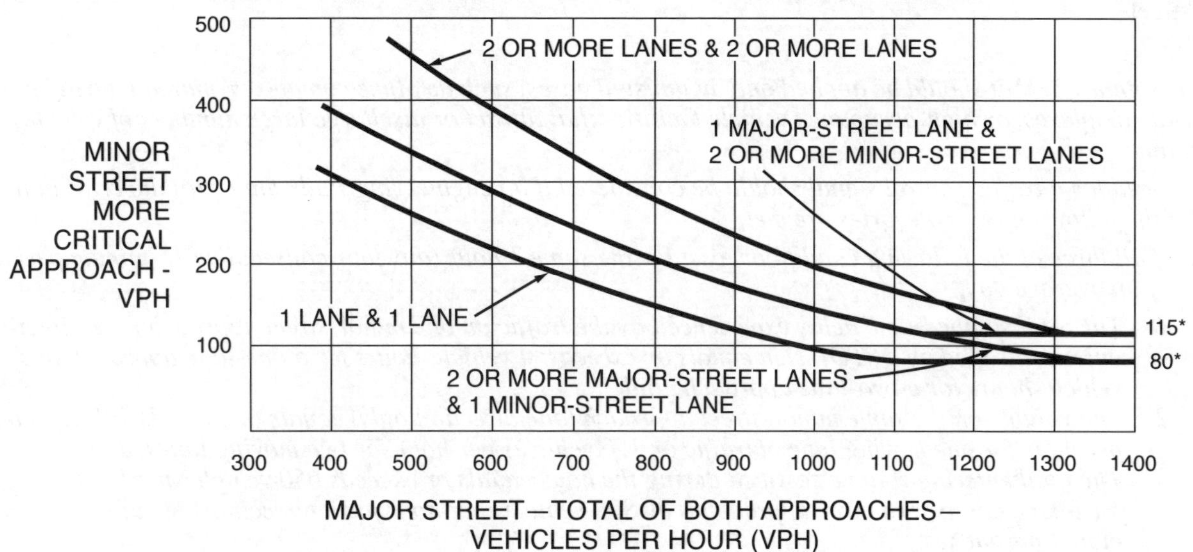

*Note: 115 vph applies as the lower threshold volume for a minor-street approach with two or more lanes and 80 vph applies as the lower threshold volume for a minor-street approach with one lane

Figure 4C-2. Warrant 2, Four-Hour Vehicular Volume (70% Factor)

(COMMUNITY LESS THAN 10,000 POPULATION OR ABOVE 40 MPH ON MAJOR STREET)

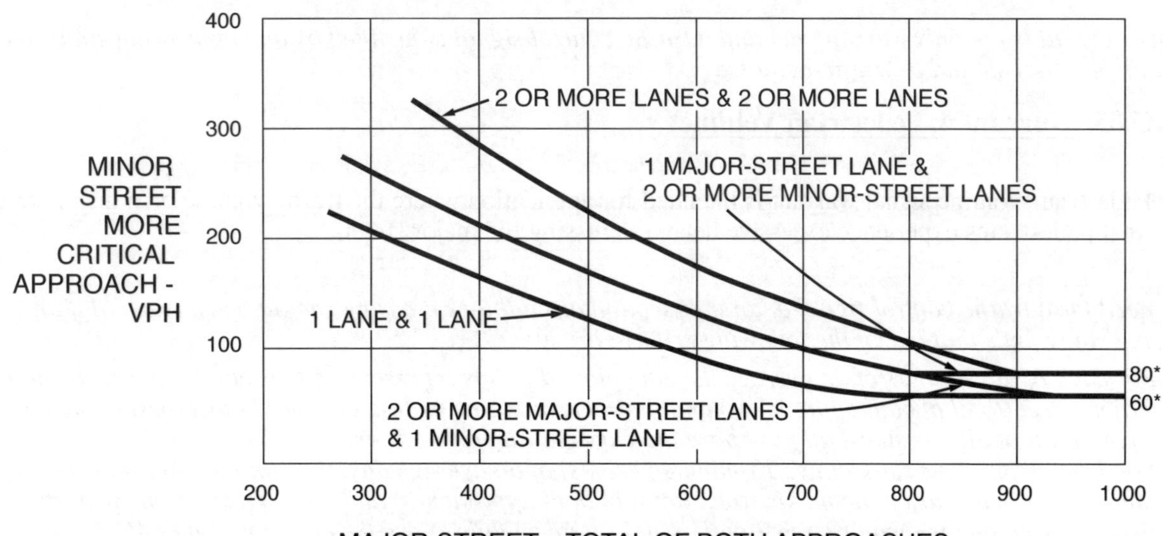

*Note: 80 vph applies as the lower threshold volume for a minor-street approach with two or more lanes and 60 vph applies as the lower threshold volume for a minor-street approach with one lane

Section 4C.04 Warrant 3, Peak Hour

Support:

01   The Peak Hour signal warrant is intended for use at a location where traffic conditions are such that for a minimum of 1 hour of an average day, the minor-street traffic suffers undue delay when entering or crossing the major street.

Guidance:

02   *This signal warrant should be applied only in unusual cases, such as office complexes, manufacturing plants, industrial complexes, or high-occupancy vehicle facilities that attract or discharge large numbers of vehicles over a short time.*

03   *The need for a traffic control signal should be considered if an engineering study finds that the criteria in either of the following two categories are met:*

    A. *If all three of the following conditions exist for the same 1 hour (any four consecutive 15-minute periods) of an average day:*

        1. *The total stopped-time delay experienced by the traffic on one minor-street approach (one direction only) controlled by a STOP sign equals or exceeds: 4 vehicle-hours for a one-lane approach or 5 vehicle-hours for a two-lane approach, and*

        2. *The volume on the same minor-street approach (one direction only) equals or exceeds 100 vehicles per hour for one moving lane of traffic or 150 vehicles per hour for two moving lanes, and*

        3. *The total entering volume serviced during the hour equals or exceeds 650 vehicles per hour for intersections with three approaches or 800 vehicles per hour for intersections with four or more approaches.*

    B. *The plotted point representing the vehicles per hour on the major street (total of both approaches) and the corresponding vehicles per hour on the more critical minor-street approach (one direction only) for 1 hour (any four consecutive 15-minute periods) of an average day falls above the applicable curve in Figure 4C-3 for the existing combination of approach lanes.*

Option:

04   If the posted or statutory speed limit or the 85th-percentile speed on the major street exceeds 40 mph, or if the intersection lies within the built-up area of an isolated community having a population of less than 10,000, Figure 4C-4 may be used in place of Figure 4C-3 to evaluate the criteria in Item B of Paragraph 3 in this Section.

05   If this warrant is the only warrant met and a traffic control signal is justified by an engineering study, the traffic control signal may be operated in the flashing mode during the hours that the volume criteria of this warrant are not met.

Guidance:

06   *If this warrant is the only warrant met and a traffic control signal is justified by an engineering study, the traffic control signal should be traffic-actuated.*

Section 4C.05 Warrant 4, Pedestrian Volume

Support:

01   The Pedestrian Volume signal warrant is intended for application where the traffic volume on a major street is so heavy that pedestrians experience excessive delay in crossing the major street.

Guidance:

02   *The need for a traffic control signal at an intersection or midblock crossing should be considered if an engineering study finds that one of the following criteria is met:*

    A. *For each of any 4 hours of an average day, the plotted points representing the vehicles per hour on the major street (total of both approaches) and the corresponding pedestrians per hour crossing the major street (total of all crossings) all fall above the curve in Figure 4C-5; or*

    B. *For 1 hour (any four consecutive 15-minute periods) of an average day, the plotted point representing the vehicles per hour on the major street (total of both approaches) and the corresponding pedestrians per hour crossing the major street (total of all crossings) falls above the curve in Figure 4C-6.*

Option:

03   If the posted or statutory speed limit or the 85th-percentile speed on the major street exceeds 35 mph, or if the intersection lies within the built-up area of an isolated community having a population of less than 10,000, Figure 4C-7 may be used in place of Figure 4C-5 to evaluate Item A in Paragraph 2 of this Section, and Figure 4C-8 may be used in place of Figure 4C-6 to evaluate Item B in Paragraph 2 of this Section.

Figure 4C-3. Warrant 3, Peak Hour

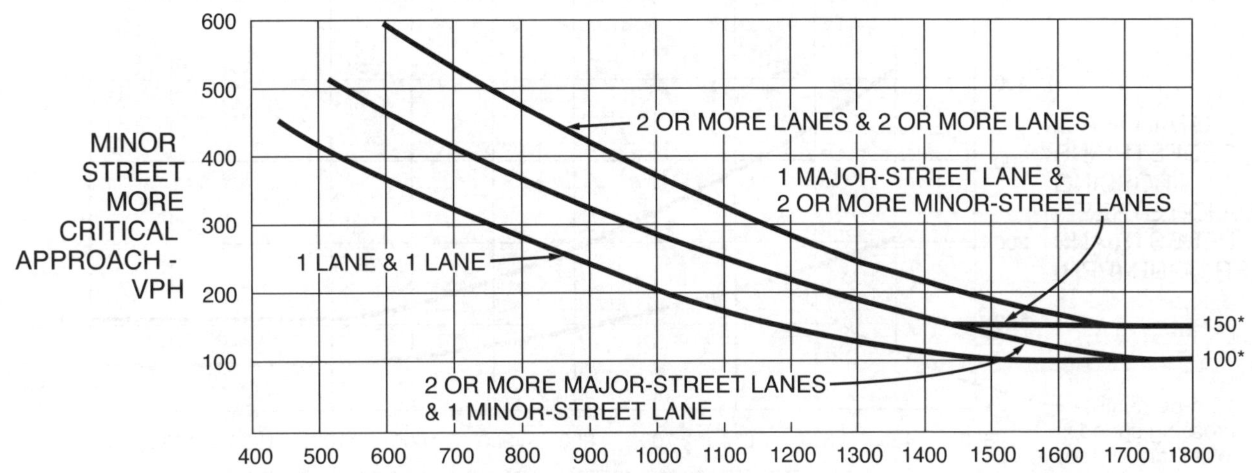

*Note: 150 vph applies as the lower threshold volume for a minor-street approach with two or more lanes and 100 vph applies as the lower threshold volume for a minor-street approach with one lane

Figure 4C-4. Warrant 3, Peak Hour (70% Factor)

(COMMUNITY LESS THAN 10,000 POPULATION OR ABOVE 40 MPH ON MAJOR STREET)

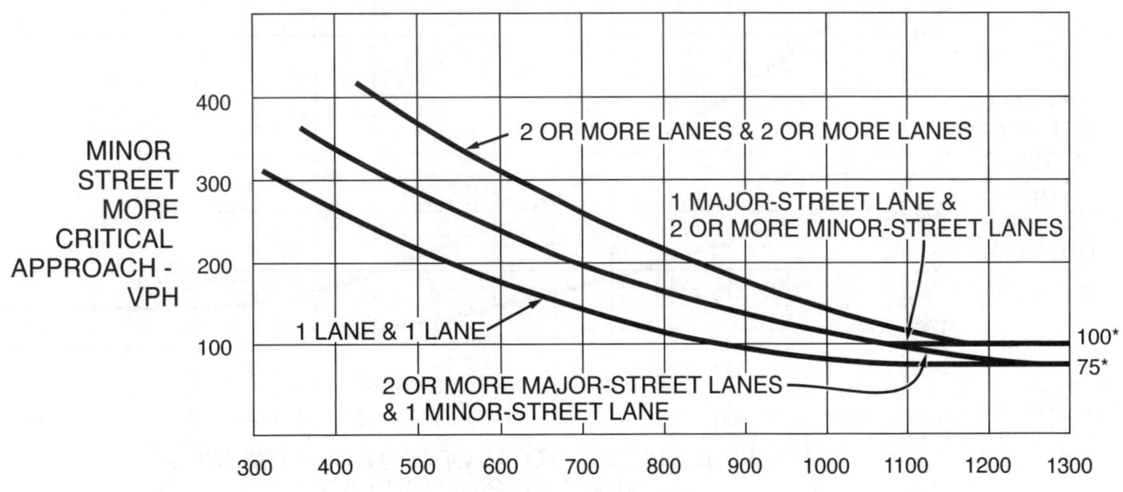

*Note: 100 vph applies as the lower threshold volume for a minor-street approach with two or more lanes and 75 vph applies as the lower threshold volume for a minor-street approach with one lane

Figure 4C-5. Warrant 4, Pedestrian Four-Hour Volume

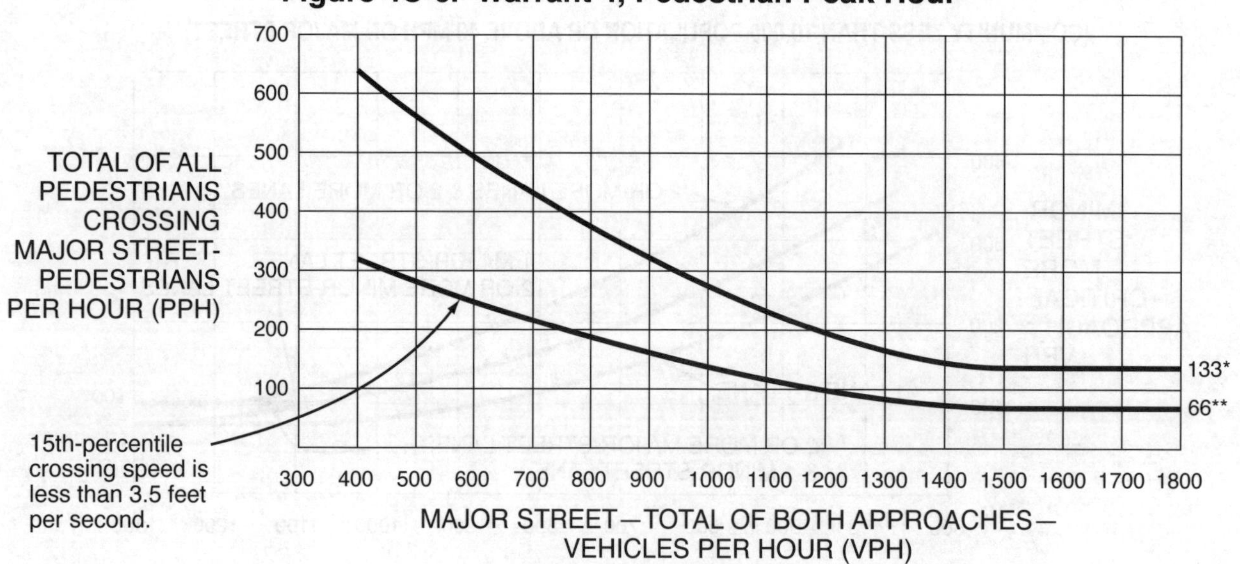

Figure 4C-6. Warrant 4, Pedestrian Peak Hour

Figure 4C-7. Warrant 4, Pedestrian Four-Hour Volume (70% Factor)

(COMMUNITY LESS THAN 10,000 POPULATION OR ABOVE 40 MPH ON MAJOR STREET)

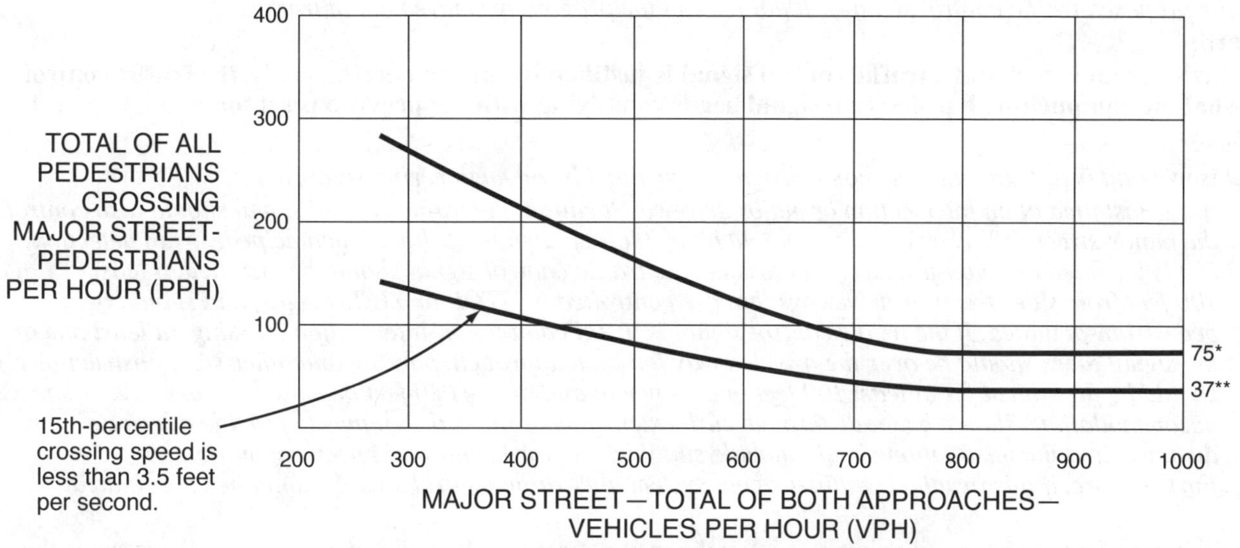

* 75 pph applies as the lower threshold volume

** 37 pph applies as the lower threshold volume if the 15th-percentile crossing speed is less than 3.5 feet per second

Figure 4C-8. Warrant 4, Pedestrian Peak Hour (70% Factor)

(COMMUNITY LESS THAN 10,000 POPULATION OR ABOVE 40 MPH ON MAJOR STREET)

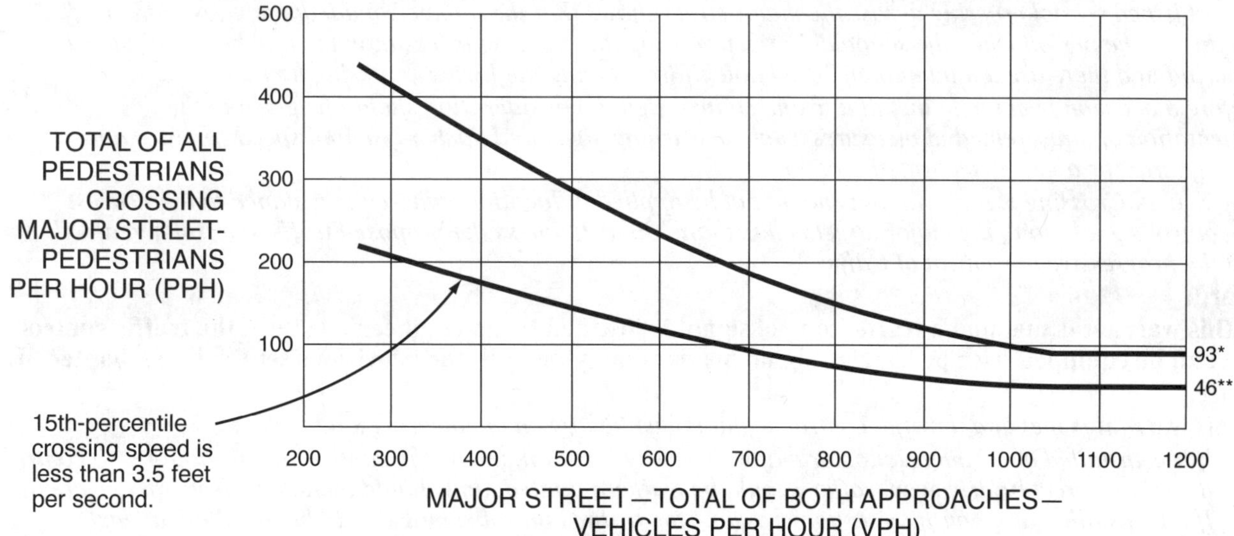

* 93 pph applies as the lower threshold volume

** 46 pph applies as the lower threshold volume if the 15th-percentile crossing speed is less than 3.5 feet per second

04   Where there is a divided street having a median of sufficient width for pedestrians to wait, the criteria in Items A and B of Paragraph 2 of this Section may be applied separately to each direction of vehicular traffic.

Guidance:

05   *The Pedestrian Volume signal warrant should not be applied at locations where the distance to the nearest traffic control signal or STOP sign controlling the street that pedestrians desire to cross is less than 300 feet, unless the proposed traffic control signal will not restrict the progressive movement of traffic.*

Standard:

06   **If this warrant is met and a traffic control signal is justified by an engineering study, the traffic control signal shall be equipped with pedestrian signal heads complying with the provisions set forth in Chapter 4I.**

Guidance:

07   *If this warrant is met and a traffic control signal is justified by an engineering study, then:*

    A. *If it is installed at an intersection or major driveway location, the traffic control signal should also control the minor-street or driveway traffic, should be traffic-actuated, and should include pedestrian detection.*

    B. *If it is installed at a non-intersection crossing, the traffic control signal should be installed at least 100 feet from side streets or driveways that are controlled by STOP or YIELD signs, and should be pedestrian-actuated. If the traffic control signal is installed at a non-intersection crossing, at least one of the signal faces should be over the traveled way for each approach, parking and other sight obstructions should be prohibited for at least 100 feet in advance of and at least 20 feet beyond the crosswalk or site accommodations should be made through curb extensions or other techniques to provide adequate sight distance, and the installation should include suitable standard signs and pavement markings.*

    C. *Furthermore, if it is installed within a signal system, the traffic control signal should be coordinated.*

Option:

08   The criterion for the pedestrian volume crossing the major street may be reduced as much as 50 percent if the 15th-percentile crossing speed of pedestrians is less than 3.5 feet per second (see Figures 4C-5 through 4C-8).

09   A traffic control signal may not be needed at the study location if adjacent coordinated traffic control signals consistently provide gaps of adequate length for pedestrians to cross the street.

Section 4C.06 Warrant 5, School Crossing

Support:

01   The School Crossing signal warrant is intended for application where the fact that schoolchildren cross the major street is the principal reason to consider installing a traffic control signal. For the purposes of this warrant, the word "schoolchildren" includes elementary through high school students.

Guidance:

02   *The need for a traffic control signal should be considered when an engineering study of the frequency and adequacy of gaps in the vehicular traffic stream as related to the number and size of groups of schoolchildren at an established school crossing across the major street shows that the number of adequate gaps in the traffic stream during the period when the schoolchildren are using the crossing is less than the number of minutes in the same period and there are a minimum of 20 schoolchildren during the highest crossing hour.*

03   *Before a decision is made to install a traffic control signal, consideration should be given to the implementation of other remedial measures, such as warning signs and flashers, school speed zones, school crossing guards, or a grade-separated crossing.*

04   *The School Crossing signal warrant should not be applied at locations where the distance to the nearest traffic control signal along the major street is less than 300 feet, unless the proposed traffic control signal will not restrict the progressive movement of traffic.*

Standard:

05   **If this warrant is met and a traffic control signal is justified by an engineering study, the traffic control signal shall be equipped with pedestrian signal heads complying with the provisions set forth in Chapter 4I.**

Guidance:

06   *If this warrant is met and a traffic control signal is justified by an engineering study, then:*

    A. *If it is installed at an intersection or major driveway location, the traffic control signal should also control the minor-street or driveway traffic, should be traffic-actuated, and should include pedestrian detection.*

    B. *If it is installed at a non-intersection crossing, the traffic control signal should be installed at least 100 feet from side streets or driveways that are controlled by STOP or YIELD signs, and should be pedestrian-actuated. If the traffic control signal is installed at a non-intersection crossing, at least one of the signal faces should be over the traveled way for each approach, parking and other sight obstructions should be prohibited for at least 100 feet in advance of and at least 20 feet beyond the crosswalk or site accommodations should be made through curb extensions or other techniques to provide adequate sight distance, and the installation should include suitable standard signs and pavement markings.*

    C. *Furthermore, if it is installed within a signal system, the traffic control signal should be coordinated.*

Section 4C.07 <u>Warrant 6, Coordinated Signal System</u>

Support:

01 Progressive movement in a coordinated signal system sometimes necessitates installing traffic control signals at intersections where they would not otherwise be needed in order to maintain proper platooning of vehicles.

Guidance:

02 *The need for a traffic control signal should be considered if an engineering study finds that one of the following criteria is met:*

    A. *On a one-way street or a street that has traffic predominantly in one direction, the adjacent traffic control signals are so far apart that they do not provide the necessary degree of vehicular platooning.*

    B. *On a two-way street, adjacent traffic control signals do not provide the necessary degree of platooning and the proposed and adjacent traffic control signals will collectively provide a progressive operation.*

03 *The Coordinated Signal System signal warrant should not be applied where the resultant spacing of traffic control signals would be less than 1,000 feet.*

Section 4C.08 <u>Warrant 7, Crash Experience</u>

Support:

01 The Crash Experience signal warrant conditions are intended for application where the severity and frequency of crashes are the principal reasons to consider installing a traffic control signal.

Guidance:

02 *The need for a traffic control signal should be considered if an engineering study finds that all of the following criteria are met:*

    A. *Adequate trial of alternatives with satisfactory observance and enforcement has failed to reduce the crash frequency; and*

    B. *At least one of the following conditions applies to the reported crash history (where each reported crash considered is related to the intersection and apparently exceeds the applicable requirements for a reportable crash):*

        1. *The number of reported angle crashes and pedestrian crashes within a 1-year period equals or exceeds the threshold number in Table 4C-2 for total angle crashes and pedestrian crashes (all severities); or*

        2. *The number of reported fatal-and-injury angle crashes and pedestrian crashes within a 1-year period equals or exceeds the threshold number in Table 4C-2 for total fatal-and-injury angle crashes and pedestrian crashes; or*

        3. *The number of reported angle crashes and pedestrian crashes within a 3-year period equals or exceeds the threshold number in Table 4C-3 for total angle crashes and pedestrian crashes (all severities); or*

        4. *The number of reported fatal-and-injury angle crashes and pedestrian crashes within a 3-year period equals or exceeds the threshold number in Table 4C-3 for total fatal-and-injury angle crashes and pedestrian crashes; and*

    C. *For each of any 8 hours of an average day, the vehicles per hour (vph) given in both of the 80 percent columns of Condition A in Table 4C-1 (see Section 4C.02), or the vph in both of the 80 percent columns of Condition B in Table 4C-1 exists on the major street and the more critical minor-street approach, respectively, to the intersection, or the volume of pedestrian traffic is not less than 80 percent of the requirements specified in the Pedestrian Volume warrant (see Section 4C.05).*

Table 4C-2. Minimum Number of Reported Crashes in a One-Year Period

Number of through lanes on each approach		Total of angle and pedestrian crashes (all severities)[a]		Total of fatal-and-injury angle and pedestrian crashes[a]	
Major Street	Minor Street	Four Legs	Three Legs	Four Legs	Three Legs
1	1	5	4	3	3
2 or more	1	5	4	3	3
2 or more	2 or more	5	4	3	3
1	2 or more	5	4	3	3

[a] Angle crashes include all crashes that occur at an angle and involve one or more vehicles on the major street and one or more vehicles on the minor street

Table 4C-3. Minimum Number of Reported Crashes in a Three-Year Period

Number of through lanes on each approach		Total of angle and pedestrian crashes (all severities)[a]		Total of fatal-and-injury angle and pedestrian crashes[a]	
Major Street	Minor Street	Four Legs	Three Legs	Four Legs	Three Legs
1	1	6	5	4	4
2 or more	1	6	5	4	4
2 or more	2 or more	6	5	4	4
1	2 or more	6	5	4	4

[a] Angle crashes include all crashes that occur at an angle and involve one or more vehicles on the major street and one or more vehicles on the minor street

Standard:

03   **These major-street and minor-street volumes shall be for the same 8 hours.**

Support:

04   On the minor street, the more critical volume is not required to be on the same approach during each of these 8 hours. The more critical minor-street volume is the one that meets the warranting criteria for that approach, and in the case of a one-lane minor-street approach that is opposite from a multi-lane minor-street approach might not have the higher volume.

Option:

05   If the posted or statutory speed limit or the 85th-percentile speed on the major street exceeds 40 mph, or if the intersection lies within the built-up area of an isolated community having a population of less than 10,000:

A. The traffic volumes in the 56 percent columns in Table 4C-1 may be used in place of the 80 percent columns.
B. Tables 4C-4 and 4C-5 may be used in place of Tables 4C-2 and 4C-3, respectively.

Table 4C-4. Minimum Number of Reported Crashes in a One-Year Period

Community less than 10,000 population or above 40 mph on major street					
Number of through lanes on each approach		Total of angle and pedestrian crashes (all severities)[a]		Total of fatal-and-injury angle and pedestrian crashes[a]	
Major Street	Minor Street	Four Legs	Three Legs	Four Legs	Three Legs
1	1	4	3	3	3
2 or more	1	10	9	6	6
2 or more	2 or more	10	9	6	6
1	2 or more	4	3	3	3

[a] Angle crashes include all crashes that occur at an angle and involve one or more vehicles on the major street and one or more vehicles on the minor street

Table 4C-5. Minimum Number of Reported Crashes in a Three-Year Period

Community less than 10,000 population or above 40 mph on major street					
Number of through lanes on each approach		Total of angle and pedestrian crashes (all severities)[a]		Total of fatal-and-injury angle and pedestrian crashes[a]	
Major Street	Minor Street	Four Legs	Three Legs	Four Legs	Three Legs
1	1	6	5	4	4
2 or more	1	16	13	9	9
2 or more	2 or more	16	13	9	9
1	2 or more	6	5	4	4

[a] Angle crashes include all crashes that occur at an angle and involve one or more vehicles on the major street and one or more vehicles on the minor street

Option:

06   Agencies may calibrate Highway Safety Manual (HSM) (AASHTO, 2010) safety performance functions (SPFs) to their own crash data or develop their own SPFs to produce agency specific average crash frequency values. When documented as part of the engineering study, these agency specific crash frequency values may be used instead of the values shown in Tables 4C-2 through 4C-5 when applying the Crash Experience signal warrant.

Support:

07   The values in Tables 4C-2 through 4C-5 for Minimum Number of Reported Crashes that correspond to the Crash Experience signal warrant were derived using the safety performance functions (SPFs) in the Highway Safety Manual (HSM) (AASHTO, 2010) for stop-controlled and signalized intersections with characteristics that are considered typical. The values in Tables 4C-2 through 4C-5 are representative of average crash frequency for the given intersection condition. The values correspond to the threshold at which the signalized intersection safety performance outperforms the stop-controlled intersection, for otherwise identical conditions and equivalent traffic.

Section 4C.09 Warrant 8, Roadway Network

Support:

01   Installing a traffic control signal at some intersections might be justified to encourage concentration and organization of traffic flow on a roadway network.

Guidance:

02   *The need for a traffic control signal should be considered if an engineering study finds that the common intersection of two or more major routes meets one or both of the following criteria:*

    A. The intersection has a total existing, or immediately projected, entering volume of at least 1,000 vehicles per hour during the peak hour of a typical weekday and has 5-year projected traffic volumes, based on an engineering study, that meet one or more of Warrants 1, 2, and 3 during an average weekday; or

    B. The intersection has a total existing or immediately projected entering volume of at least 1,000 vehicles per hour for each of any 5 hours of a non-normal business day (Saturday or Sunday).

03   *A major route as used in this signal warrant should have at least one of the following characteristics:*

    A. It is part of the street or highway system that serves as the principal roadway network for through traffic flow;

    B. It includes rural or suburban highways outside, entering, or traversing a city; or

    C. It appears as a major route on an official plan, such as a major street plan in an urban area traffic and transportation study.

Section 4C.10 Warrant 9, Intersection Near a Grade Crossing

Support:

01   The Intersection Near a Grade Crossing signal warrant is intended for use at a location where none of the conditions described in the other eight traffic signal warrants are met, but the proximity of a grade crossing on an approach controlled by a STOP or YIELD sign at a highway-highway intersection is the principal reason to consider installing a traffic control signal.

Guidance:

02   *This signal warrant should be applied only after adequate consideration has been given to other alternatives or after a trial of an alternative has failed to alleviate the safety concerns associated with the grade crossing. Among the alternatives that should be considered or tried are:*

    A. Providing additional pavement that would enable vehicles to clear the track or that would provide space for an evasive maneuver, or

    B. Reassigning the stop controls at the highway-highway intersection to make the approach across the track a non-stopping approach.

03   *The need for a traffic control signal should be considered if an engineering study finds that both of the following criteria are met:*

    A. A grade crossing exists on an approach controlled by a STOP or YIELD sign at a highway-highway intersection and the center of the track nearest to the intersection is within 140 feet of the stop line or yield line on the approach; and

    B. During the highest traffic volume hour during which rail traffic uses the crossing, the plotted point representing the vehicles per hour on the major street (total of both approaches) of the highway-highway intersection and the corresponding vehicles per hour on the minor-street approach that crosses the track (one direction only, approaching the intersection) falls above the applicable curve in Figure 4C-9 or 4C-10 for the existing combination of approach lanes over the track and the distance D, which is the clear storage distance as defined in Section 1C.02.

04   *The following considerations apply when plotting the traffic volume data on Figure 4C-9 or 4C-10:*

    A. *Figure 4C-9 should be used if there is only one lane approaching the highway-highway intersection at the track crossing location and Figure 4C-10 should be used if there are two or more lanes approaching the highway-highway intersection at the track crossing location.*

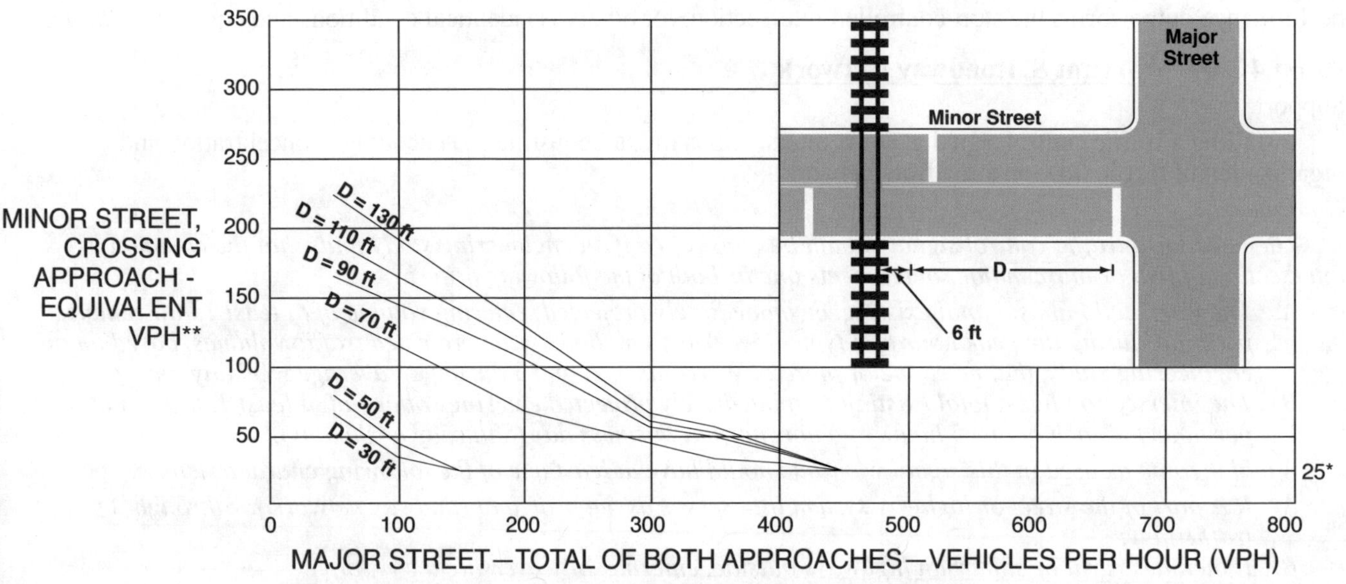

Figure 4C-9. Warrant 9, Intersection Near a Grade Crossing (One Approach Lane at the Track Crossing)

    * 25 vph applies as the lower threshold volume

    ** VPH after applying the adjustment factors in Tables 4C-6, 4C-7, and/or 4C-8, if appropriate

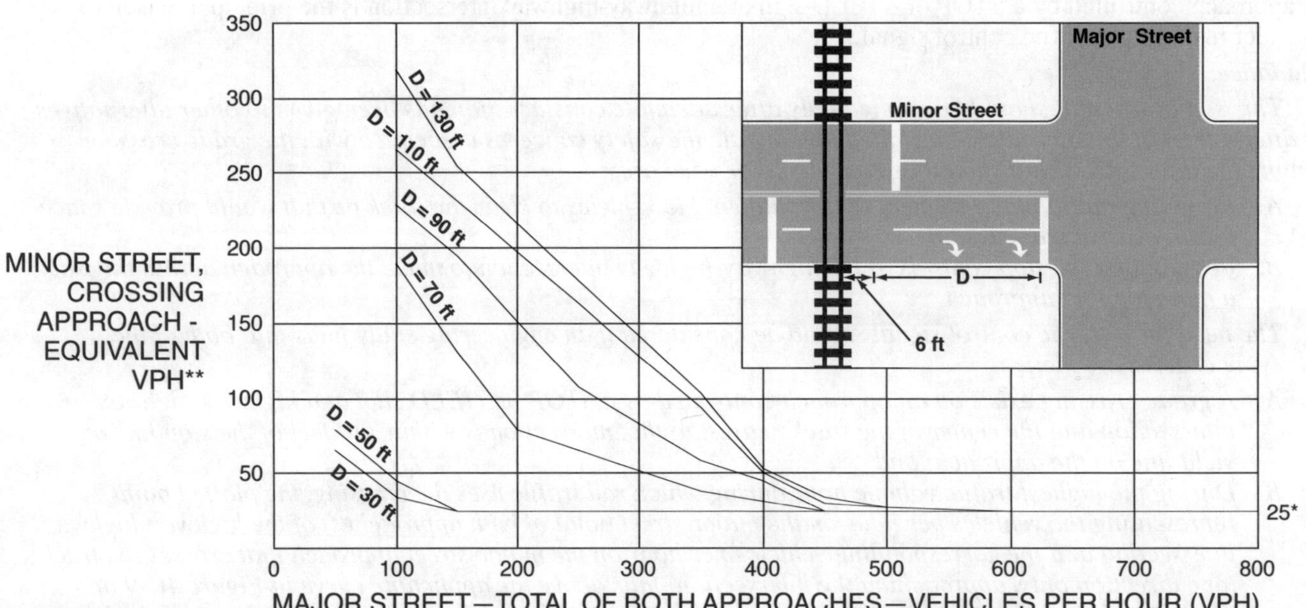

Figure 4C-10. Warrant 9, Intersection Near a Grade Crossing (Two or More Approach Lanes at the Track Crossing)

    * 25 vph applies as the lower threshold volume

    ** VPH after applying the adjustment factors in Tables 4C-6, 4C-7, and/or 4C-8, if appropriate

B. After determining the actual distance D, the curve for the distance D that is nearest to the actual distance D should be used. For example, if the actual distance D is 95 feet, the plotted point should be compared to the curve for D=90 feet.

C. If the rail traffic arrival times are unknown, the highest traffic volume hour of the day should be used.

Option:

05   The traffic volume on the minor-street approach to the highway-highway intersection may be multiplied by up to three adjustment factors as provided in Paragraphs 6 through 8 of this Section.

06   Because the curves are based on an average of four occurrences of rail traffic per day, the vehicles per hour on the minor-street approach may be multiplied by the adjustment factor shown in Table 4C-6 for the appropriate number of occurrences of rail traffic per day.

07   Because the curves are based on typical vehicle occupancy, if at least 2% of the vehicles crossing the track are buses carrying at least 20 people, the vehicles per hour on the minor-street approach may be multiplied by the adjustment factor shown in Table 4C-7 for the appropriate percentage of high-occupancy buses.

08   Because the curves are based on tractor-trailer trucks comprising 10% of the vehicles crossing the track, the vehicles per hour on the minor-street approach may be multiplied by the adjustment factor shown in Table 4C-8 for the appropriate distance and percentage of tractor-trailer trucks.

Standard:

09   **If this warrant is met and a traffic control signal at the highway-highway intersection is justified by an engineering study, then:**

    A. The traffic control signal shall have actuation on the minor street,

    B. Preemption control shall be provided in accordance with Sections 4F.19 and 8D.09, and

    C. The grade crossing shall have flashing-light signals (see Section 8D.02).

Guidance:

10   *If this warrant is met and a traffic control signal at the highway-highway intersection is justified by an engineering study, the grade crossing should have automatic gates (see Section 8D.03).*

Table 4C-6. Warrant 9, Adjustment Factor for Daily Frequency of Rail Traffic

Rail traffic per day	Adjustment factor
1	0.67
2	0.91
3 to 5	1.00
6 to 8	1.18
9 to 11	1.25
12 or more	1.33

Table 4C-7. Warrant 9, Adjustment Factor for Percentage of High-Occupancy Buses

% of high-occupancy buses* on minor-street approach	Adjustment factor
0%	1.00
2%	1.09
4%	1.19
6% or more	1.32

* A high-occupancy bus is defined as a bus occupied by at least 20 people.

Table 4C-8. Warrant 9, Adjustment Factor for Percentage of Tractor-Trailer Trucks

% of tractor-trailer trucks on minor-street approach	Adjustment factor	
	D less than 70 feet	D of 70 feet or more
0% to 2.5%	0.50	0.50
2.6% to 7.5%	0.75	0.75
7.6% to 12.5%	1.00	1.00
12.6% to 17.5%	2.30	1.15
17.6% to 22.5%	2.70	1.35
22.6% to 27.5%	3.28	1.64
More than 27.5%	4.18	2.09

CHAPTER 4D. DESIGN FEATURES OF TRAFFIC CONTROL SIGNALS

Section 4D.01 General

Support:

01   The features of traffic control signals of interest to road users are the location, design, and meaning of the signal indications. Uniformity in the design features that affect the traffic to be controlled, as set forth in this Manual, is especially important for the safety and efficiency of operations.

02   Traffic control signals can be operated in pretimed, semi-actuated, or full-actuated modes. For isolated (non-interconnected) signalized locations on rural high-speed highways, full-actuated mode with advance vehicle detection on the high-speed approaches is typically used. These features are designed to reduce the frequency with which the onset of the yellow change interval is displayed when high-speed approaching vehicles are in the "dilemma zone" such that the drivers of these high-speed vehicles find it difficult to decide whether to stop or proceed.

Standard:

03   **The design and operation of traffic control signals shall take into consideration the needs of all modes of traffic including access and safety.**

04   **When a traffic control signal is not in operation, such as before it is placed in service, during seasonal shutdowns, or when it is not desirable to operate the traffic control signal, the signal faces shall be covered, turned, or taken down to clearly indicate that the traffic control signal is not in operation.**

Guidance:

05   *If a cover is placed over a traffic control signal face that is not in operation and that has a yellow retroreflective strip along the perimeter of its signal backplate (see Paragraph 21 in Section 4D.06), the entire signal face, including the backplate, should be covered. If a traffic control signal face that is not in operation and that has a yellow retroreflective strip along the perimeter of its signal backplate is turned, the turned signal face should be oriented such that the yellow backplate border will not reflect light back to road users on any of the approaches to the intersection.*

Support:

06   Seasonal shutdown is a condition in which a permanent traffic control signal is turned off or otherwise made non-operational during a particular season when its operation is not justified. This might be applied in a community where tourist traffic during most of the year justifies the permanent signalization, but a seasonal shutdown of the signal during an annual period of lower tourist traffic would reduce delays; or where a major traffic generator, such as a large factory, justifies the permanent signalization, but the large factory is shut down for an annual factory vacation for a few weeks in the summer.

Standard:

07   **A traffic control signal shall control traffic only at the intersection or midblock location where the signal faces are placed.**

Guidance:

08   *Midblock crosswalks should not be signalized if they are located within 300 feet from the nearest traffic control signal, unless supported by an engineering study or engineering judgment that indicates safe and efficient operation of the closely-spaced traffic control signals can be achieved.*

09   *Midblock crosswalks should not be signalized if they are located within 100 feet from side streets or driveways that are controlled by STOP signs or YIELD signs, unless supported by an engineering study or engineering judgment that considers restricting turning movements from the side street or driveway to eliminate conflicts with pedestrian and bicyclist movements.*

10   *Engineering judgment should be used to determine the proper phasing and timing for a traffic control signal. Since traffic flows and patterns change, phasing and timing should be reevaluated regularly and updated if needed.*

11   *Traffic control signals within ½ mile of one another along a major route or in a network of intersecting major routes should be coordinated, preferably with interconnected controller units. Where traffic control signals that are within ½ mile of one another along a major route have a jurisdictional boundary or a boundary between different signal systems between them, coordination across the boundary should be considered.*

Support:

12   Signal coordination need not be maintained between control sections that operate on different cycle lengths.

13   Sections 4F.19, 4Q.03, and 8D.09 contain information about coordination of traffic control signals with grade crossing signals and movable bridge signals.

Section 4D.02 Provisions for Pedestrians

Support:

01   Chapter 4I contains additional information regarding pedestrian control features, Chapter 4J contains additional information regarding pedestrian hybrid beacons, and Chapter 4K contains additional information regarding accessible pedestrian signals and detectors.

Standard:

02   **Pedestrian signal heads shall be used in conjunction with vehicular traffic control signals under any of the following conditions, unless the pedestrian crossing is prohibited:**

    A. If the basis for traffic signal installation was justified by an engineering study and meeting either Warrant 4, Pedestrian Volume or Warrant 5, School Crossing (see Chapter 4C);

    B. If an exclusive pedestrian signal phase or a leading pedestrian interval (LPI) is provided with all conflicting vehicular movements being stopped;

    C. At an established signalized school crossing; or

    D. Where there are existing pedestrian accommodations and engineering judgment determines that multi-phase signal indications (such as split-phase timing) would tend to confuse or cause conflicts with pedestrians using a crosswalk guided only by vehicular signal indications.

Guidance:

03   *Pedestrian signal heads should be installed for each marked crosswalk at a location controlled by a traffic control signal.*

04   *Where pedestrian movements regularly occur, pedestrians should be provided with sufficient time to cross the roadway by adjusting the traffic control signal operation and timing to provide sufficient crossing time every cycle or by providing pedestrian detectors.*

05   *Where certain pedestrian movements are prohibited at a traffic control signal location, a No Pedestrian Crossing (R9-3) sign (see Section 2B.57) should be used if it is impracticable to provide a barrier or other physical feature to physically discourage the pedestrian movements.*

Support:

06   Accessible pedestrian signals (see Chapter 4K) that provide information in non-visual formats (such as audible tones and/or speech messages, and vibrating surfaces) enhance safety and accessibility at signalized crossings for pedestrians with vision disabilities.

Option:

07   Pedestrian signal heads may be used under other conditions based on engineering judgment.

Section 4D.03 Provisions for Bicyclists

Standard:

01   **At installations where visibility-limited signal faces are used, signal faces shall be adjusted so bicyclists for whom the indications are intended can see the signal indications. If the visibility-limited signal faces cannot be aimed to serve the bicyclist, then separate signal faces (see Chapter 4H) shall be provided for the bicyclist.**

02   **On bikeways, signal timing and actuation shall be reviewed and adjusted to consider the needs of bicyclists.**

Option:

03   Where it is desired to provide separate signal indications to control bicyclist movements at a traffic control signal, bicycle signal faces may be used (see Chapter 4H).

Support:

04   Sections 9B.02, 9B.11, 9B.20, 9B.22, 9E.02, 9E.06, 9E.07, 9E.08, 9E.11, 9E.12, and 9E.15 contain additional provisions regarding bicyclist movements and actuation at traffic control signals.

Section 4D.04 Provisions for Transit Vehicles

Option:

01   Where it is desired to provide separate signal indications to control transit vehicles at a traffic control signal, LRT signal indications may be used at intersections where special signal phases are used for transit vehicles (see Section 8D.15).

Section 4D.05 <u>Number of Signal Faces on an Approach</u>

Standard:

01   **The signal faces for each approach to an intersection or a midblock location shall be provided as follows:**

   A. **If a signalized motor vehicle through movement exists on an approach, a minimum of two primary signal faces shall be provided for the through movement. Except for single lane approaches, if a signalized motor vehicle through movement does not exist on an approach, a minimum of two primary signal faces shall be provided for the signalized motor vehicle turning movement that is considered to be the major movement from the approach (also see Section 4F.16).**
   B. **See Sections 4F.02 through 4F.08 for left-turn (and U-turn to the left) signal faces.**
   C. **See Sections 4F.09 through 4F.15 for right-turn (and U-turn to the right) signal faces.**

Option:

02   Where a movement (or a certain lane or lanes) at the intersection never conflicts with any other signalized vehicular or pedestrian movement, a continuously-displayed single-section GREEN ARROW signal indication may be used to inform road users that the movement is free-flow and does not need to stop.

Support:

03   In some circumstances where the through movement never conflicts with any other signalized vehicular or pedestrian movement at the intersection, such as at T-intersections with appropriate geometrics and/or pavement markings and signing, an engineering study might determine that the through movement (or certain lanes of the through movement) can be free-flow and not signalized.

Guidance:

04   *If two or more left-turn lanes are provided for a separately-controlled left-turn movement, or if a left-turn movement represents the major movement from an approach, two or more primary left-turn signal faces should be provided.*

05   *If two or more right-turn lanes are provided for a separately-controlled right-turn movement, or if a right-turn movement represents the major movement from an approach, two or more primary right-turn signal faces should be provided.*

Support:

06   Locating primary signal faces overhead on the far side of the intersection has been shown to provide safer operation by reducing intersection entries late in the yellow interval and by reducing red signal violations, as compared to post-mounting signal faces at the roadside or locating signal faces overhead within the intersection on a diagonally-oriented mast arm or span wire. On approaches with two or more lanes for the through movement, one signal face per through lane, centered over each through lane, has also been shown to provide safer operation.

Guidance:

07   *If the posted or statutory speed limit or the 85th-percentile speed on an approach to a signalized location is 45 mph or higher, signal faces should be provided as follows for all new or reconstructed signal installations (see Figure 4D-1):*

   A. *The minimum number and location of primary (non-supplemental) signal faces for through traffic should be provided in accordance with Table 4D-1.*
   B. *If the number of overhead primary signal faces for through traffic is equal to the number of through lanes on an approach, one overhead signal face should be located approximately over the center of each through lane.*
   C. *Except for shared left-turn and right-turn signal faces, any primary signal face required by Sections 4F.02 through 4F.16 for a mandatory turn lane should be located overhead approximately over the center of each mandatory turn lane.*
   D. *All primary signal faces should be located on the far side of the intersection.*
   E. *In addition to the primary signal faces, one or more supplemental pole-mounted or overhead signal faces should be considered to provide added visibility for approaching traffic that is traveling behind large vehicles.*
   F. *All signal faces should have backplates.*

08   *This layout of signal faces should also be considered for any major urban or suburban arterial street with four or more lanes and for other approaches with speeds of less than 45 mph.*

Figure 4D-1. Recommended Vehicular Signal Faces for Approaches with Posted, Statutory, or 85th-Percentile Speed of 45 mph or Higher

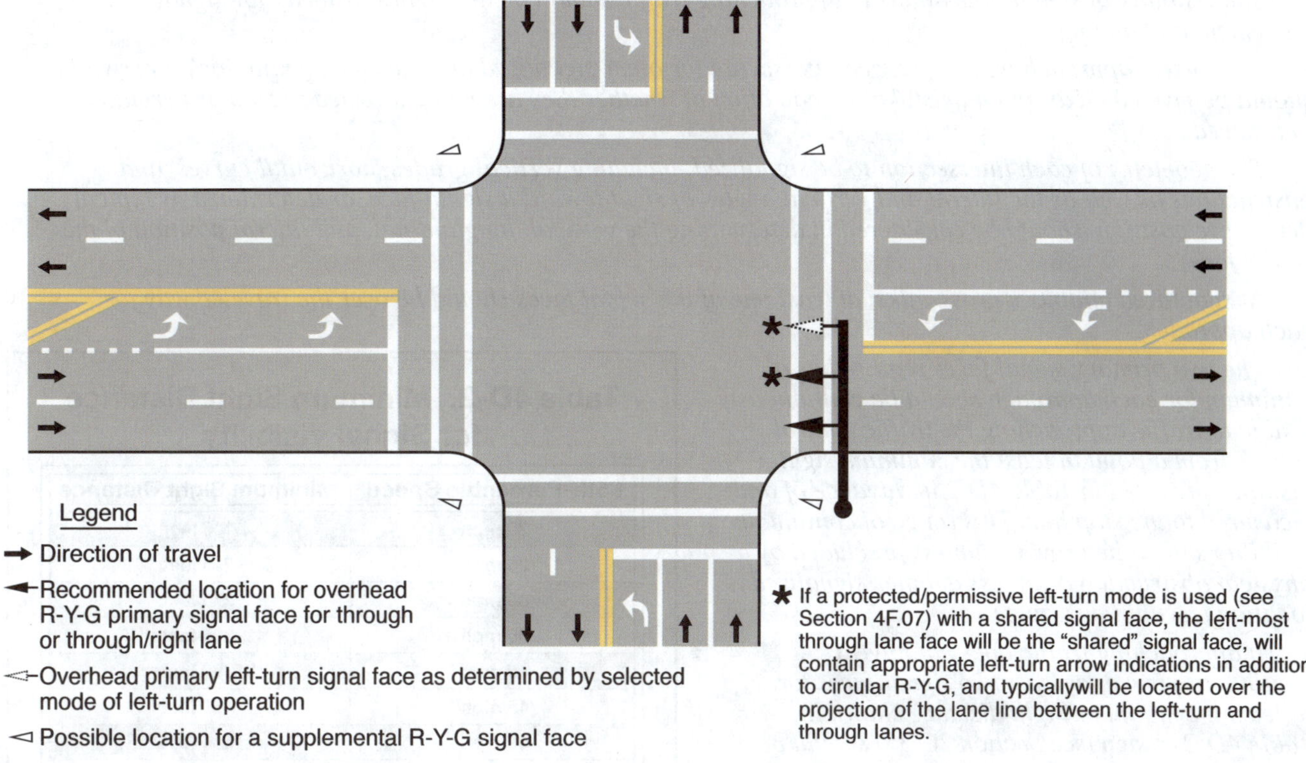

Notes:

1. Signal faces for only one direction and only one possible set of geometrics (number of lanes, etc.) are illustrated. If there are fewer or more than two through lanes on the approach, see Table 4D-1.

2. Any primary left-turn and/or right-turn signal faces, as determined by Sections 4F.02 through 4F.15, should be overhead for each mandatory turn lane.

3. One or more pole-mounted or overhead supplemental faces should be considered, based on the geometrics of the approach, to maximize visibility for approaching traffic.

4. All signal faces should have backplates.

Table 4D-1. Recommended Minimum Number of Primary Signal Faces for Through Traffic on Approaches with Posted, Statutory, or 85th-Percentile Speed of 45 mph or Higher

Number of Through Lanes on the Approach	Total Number of Primary Through Signal Faces for the Approach*	Minimum Number of Overhead-Mounted Primary Through Signal Faces for the Approach
1	2	1
2	2	1
3	3	2**
4 or more	4 or more	3**

Notes:

* A minimum of two through signal faces is always required (see Section 4D.05). These recommended numbers of through signal faces may be exceeded. Also, see cone of vision requirements otherwise indicated in Section 4D.07.

** If practical, all of the recommended number of primary through signal faces should be located overhead.

Section 4D.06 Visibility, Aiming, and Shielding of Signal Faces

Guidance:

01   *The visibility of signal indications to approaching traffic should be the highest priority for signal face placement and aiming.*

02   *Road users approaching a signalized intersection or other signalized area, such as a midblock crosswalk, should be given a clear and unmistakable indication of whether they are being directed to stop or permitted to proceed.*

03   *The geometry of each intersection to be signalized, including vertical grades, horizontal curves, and obstructions as well as the lateral and vertical angles of sight toward a signal face, as determined by typical driver-eye position, should be considered in determining the vertical, longitudinal, and lateral position of the signal face.*

04   *At signalized midblock crosswalks, at least one of the signal faces should be over the traveled way for each approach.*

05   *The two primary signal faces required as a minimum for each approach should be continuously visible to traffic approaching the traffic control signal, from a point at least the minimum sight distance provided in Table 4D-2 in advance of and measured to the stop line. This range of continuous visibility should be provided unless precluded by a physical obstruction or unless another signalized location is within this range.*

06   *If approaching traffic does not have a continuous view of at least two signal faces for at least the minimum sight distance shown in Table 4D-2, a sign (see Section 2C.35) should be installed to warn approaching traffic of the traffic control signal.*

Option:

07   If a sign is installed to warn approaching road users of the traffic control signal, the sign may be supplemented by a Warning Beacon (see Section 4S.03).

Table 4D-2. Minimum Sight Distance for Signal Visibility

85th-Percentile Speed	Minimum Sight Distance
20 mph	175 feet
25 mph	215 feet
30 mph	270 feet
35 mph	325 feet
40 mph	390 feet
45 mph	460 feet
50 mph	540 feet
55 mph	625 feet
60 mph	715 feet

Note: Distances in this table are derived from stopping sight distance plus an assumed queue length for shorter cycle lengths (60 to 75 seconds).

08   A Warning Beacon used in this manner may be interconnected with the traffic signal controller assembly in such a manner as to flash yellow during the period when road users passing this beacon at the legal speed for the roadway might encounter a red signal indication (or a queue resulting from the display of the red signal indication) upon arrival at the signalized location.

09   If the sight distance to the signal faces for an approach is limited by horizontal or vertical alignment, supplemental signal faces aimed at a point on the approach at which the signal indications first become visible may be used.

Guidance:

10   *Supplemental signal faces should be used if engineering judgment has shown that they are needed to achieve intersection visibility both in advance and immediately before the signalized location.*

11   *If supplemental signal faces are used, they should be located to provide optimum visibility for the movement to be controlled.*

12   *In cases where irregular street design necessitates placing signal faces for different street approaches with a comparatively small angle between their respective signal indications, each signal indication should, to the extent practical, be visibility-limited by signal visors, signal louvers, or other means so that an approaching road user's view of the signal indication(s) controlling movements on other approaches is minimized.*

Standard:

13   **Signal visors exceeding 12 inches in length shall not be used on free-swinging signal faces.**

Guidance:

14   *Signal visors should be used on signal faces to aid in directing the signal indication specifically to approaching traffic, as well as to reduce "sun phantom," which can result when external light enters the lens.*

15   *The use of signal visors, or the use of signal faces or devices that direct the light without a reduction in intensity, should be considered as an alternative to signal louvers because of the reduction in light output caused by signal louvers.*

Option:

16   Special signal faces, such as visibility-limited signal faces, may be used such that the road user does not see signal indications intended for other approaches before seeing the signal indications for their own approach, especially if simultaneous viewing of both signal indications could cause the road user to be misdirected.

Guidance:

17   *If the posted or statutory speed limit or the 85th-percentile speed on an approach to a signalized location is 45 mph or higher, signal backplates should be used on all of the signal faces that face the approach. Signal backplates should also be considered for use on signal faces on approaches with posted or statutory speed limits or 85th-percentile speeds of less than 45 mph where sun glare, bright sky, and/or complex or confusing backgrounds indicate a need for enhanced signal face target value.*

Support:

18   The use of backplates enhances the contrast between the traffic signal indications and their surroundings for both day and night conditions, which is also helpful to older drivers.

Standard:

19   **If backplates are used, ancillary legends of any kind that identify the purpose or operation of the signal face shall not be placed on the backplate.**

20   **The inside of signal visors (hoods), the entire surface of louvers and fins, and the front surface of backplates shall have a dull black finish to minimize light reflection and to increase contrast between the signal indication and its background.**

Option:

21   A yellow retroreflective strip with a minimum width of 1 inch and a maximum width of 3 inches may be placed along the perimeter of the face of a signal backplate to project a rectangular appearance at night.

Section 4D.07 <u>Lateral Positioning of Signal Faces</u>

Standard:

01   **At least one and preferably both of the minimum of two primary signal faces required for the through movement (or the major turning movement if there is no through movement) on the approach shall be located between two lines intersecting with the center of the approach at a point 10 feet behind the stop line, one making an angle of approximately 20 degrees to the right of the center of the approach extended, and the other making an angle of approximately 20 degrees to the left of the center of the approach extended. The signal face that satisfies this requirement shall simultaneously satisfy the longitudinal placement requirement described in Section 4D.08 (see Figure 4D-2).**

02   **If both of the minimum of two primary signal faces required for the through movement (or the major turning movement if there is no through movement) on the approach are post-mounted, they shall both be on the far side of the intersection, one on the right and one on the left of the approach lane(s).**

03   **The required signal faces for through traffic on an approach shall be located not less than 8 feet apart measured horizontally perpendicular to the approach between the centers of the signal faces.**

04   **If more than one separate turn signal face is provided for a turning movement and if one or both of the separate turn signal faces are located over the roadway, the signal faces shall be located not less than 8 feet apart measured horizontally perpendicular to the approach between the centers of the signal faces.**

Guidance:

05   *If horizontally-arranged or clustered signal faces are used, the minimum 8-foot horizontal separation between the two signal faces should be measured from the center of the right-most signal indication in the signal face on the left to the center of the left-most signal indication in the signal face on the right.*

06   *Except as provided in Paragraph 7 of this Section, for signal faces located over the roadway, separate turn signal faces should be located at least 8 feet from the nearest traffic signal face for a different movement on the same approach measured horizontally perpendicular to the approach between the centers of the signal faces.*

Option:

07   For modifications to existing traffic signals, the minimum horizontal separation between a separate turn signal face and the nearest traffic signal face for a different movement may be reduced to 3 feet.

Figure 4D-2. Lateral and Longitudinal Location of Primary Signal Faces

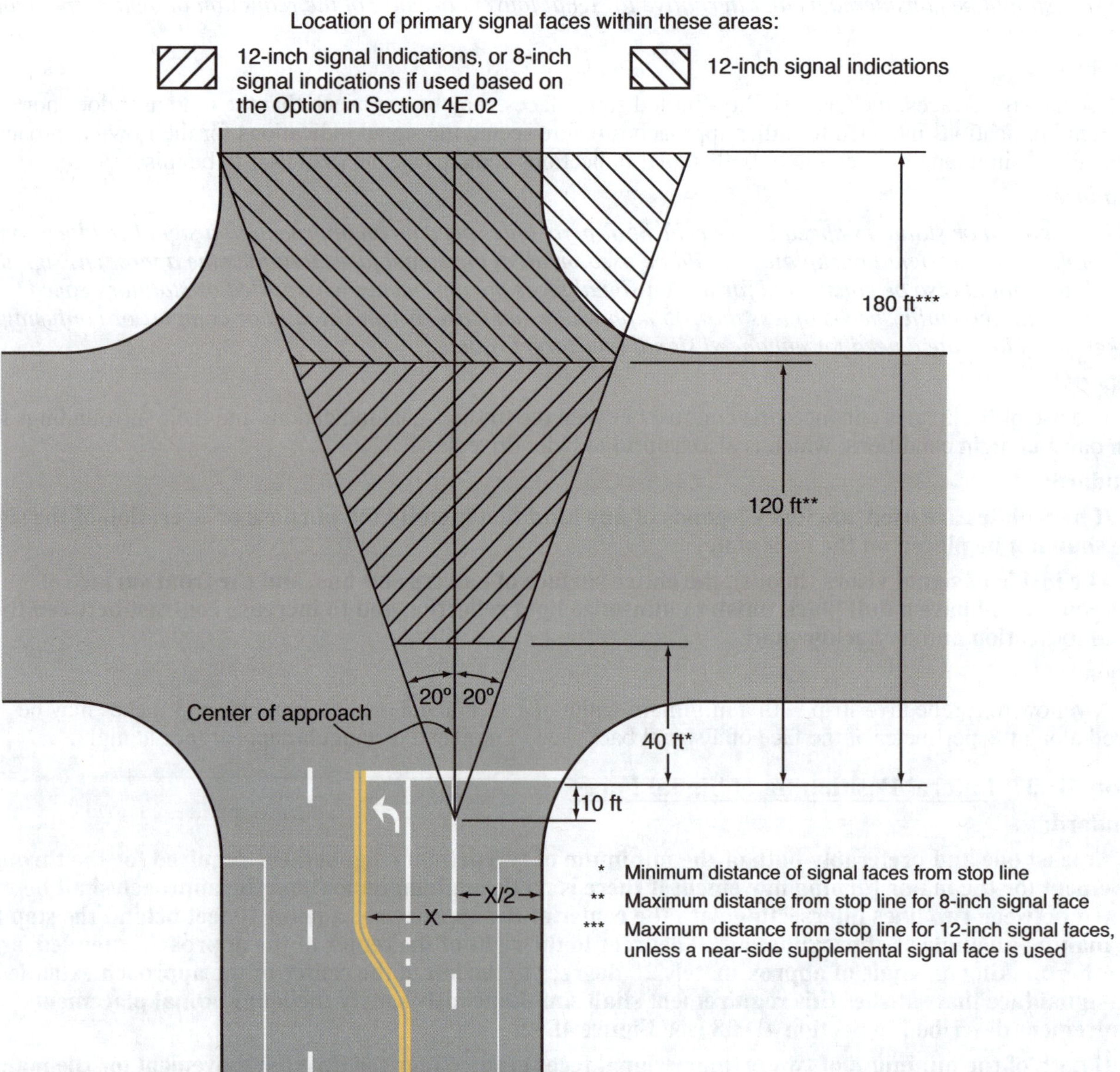

Guidance:

08   *If a signal face controls a specific lane or lanes of an approach, its position should make it readily visible to road users making that movement.*

Support:

09   Section 4D.05 contains additional provisions regarding lateral positioning of signal faces for approaches having a posted or statutory speed limit or an 85th-percentile speed of 45 mph or higher.

Guidance:

10   *If a mandatory left-turn, right-turn, or U-turn lane is present on an approach and if a primary separate turn signal face controlling that lane is mounted over the roadway, the primary separate turn signal face should not be positioned any farther to the right than the extension of the right-hand edge of the mandatory turn lane or any farther to the left than the extension of the left-hand edge of the mandatory turn lane.*

Support:

11   Supplemental turn signal faces mounted over the roadway are not subject to the positioning recommendations in Paragraph 10 of this Section.

Guidance:

12   *For new or reconstructed signal installations, on an approach with a mandatory turn lane(s) for a permissive left-turn (or U-turn to the left) movement, signal faces that display a CIRCULAR GREEN signal indication should not be post-mounted on the far-side median or mounted overhead above the mandatory turn lane(s) or the extension of the lane(s).*

Standard:

13   **If supplemental post-mounted signal faces are used, the following limitations shall apply:**

    A. Left-turn arrows and U-turn arrows to the left shall not be used in near right signal faces that are located to the right of the through and/or right-turn lanes.

    B. Right-turn arrows and U-turn arrows to the right shall not be used in far left signal faces that are located to the left of the through and/or left-turn lanes. A far-side median-mounted signal face shall be considered a far left signal face for this application.

Section 4D.08 <u>Longitudinal Positioning of Signal Faces</u>

Standard:

01   **Except where the width of an intersecting roadway or other conditions make it physically impractical, the signal faces for each approach to an intersection or a midblock location shall be provided as follows:**

    A. A signal face installed to satisfy the requirements for primary left-turn signal faces (see Sections 4F.02 through 4F.08) and primary right-turn signal faces (see Sections 4F.09 through 4F.15), and at least one and preferably both of the minimum of two primary signal faces required for the through movement (or the major turning movement if there is no through movement) on the approach shall be located:

        1. No less than 40 feet beyond the stop line, and

        2. No more than 180 feet beyond the stop line unless a supplemental near-side signal face is provided.

    B. The primary signal faces that are used to satisfy the requirements of Item A shall simultaneously satisfy the lateral placement requirement described in Section 4D.07 (see Figure 4D-2).

Guidance:

02   *Where the nearest signal face is located between 150 and 180 feet beyond the stop line, engineering judgment of the conditions, including the worst-case visibility conditions, should be used to determine if the provision of a supplemental near-side signal face would be beneficial.*

03   *Supplemental near-side signal faces should be located as near as practicable to the stop line.*

Support:

04   Section 4D.05 contains additional provisions regarding longitudinal positioning of signal faces for approaches having a posted or 85th-percentile speed of 45 mph or higher.

Section 4D.09 <u>Mounting Height of Signal Faces</u>

Standard:

01   **The bottom of the signal housing and any related attachments to a vehicular signal face located over any portion of a highway that can be used by motor vehicles shall be at least 15 feet above the pavement.**

02   **The bottom of the signal housing (including brackets) of a vehicular signal face that is vertically arranged or horizontally arranged and not located over a roadway:**

    A. Shall be a minimum of 8 feet above the sidewalk or, if there is no sidewalk, above the pavement grade at the center of the roadway.

    B. Shall be a minimum of 4.5 feet above the median island grade of a center median island if located on the near side of the intersection.

Guidance:

03   *The top of the signal housing of a vehicular signal face located over any portion of a highway that can be used by motor vehicles should not be more than 25.6 feet above the pavement.*

04   *For viewing distances between 40 and 53 feet from the stop line, the maximum mounting height to the top of the signal housing of a vehicular signal face located over any portion of a highway that can be used by motor vehicles should be as shown in Figure 4D-3.*

Figure 4D-3. Maximum Mounting Height of Signal Faces that Are Located between 40 Feet and 53 Feet from the Stop Line

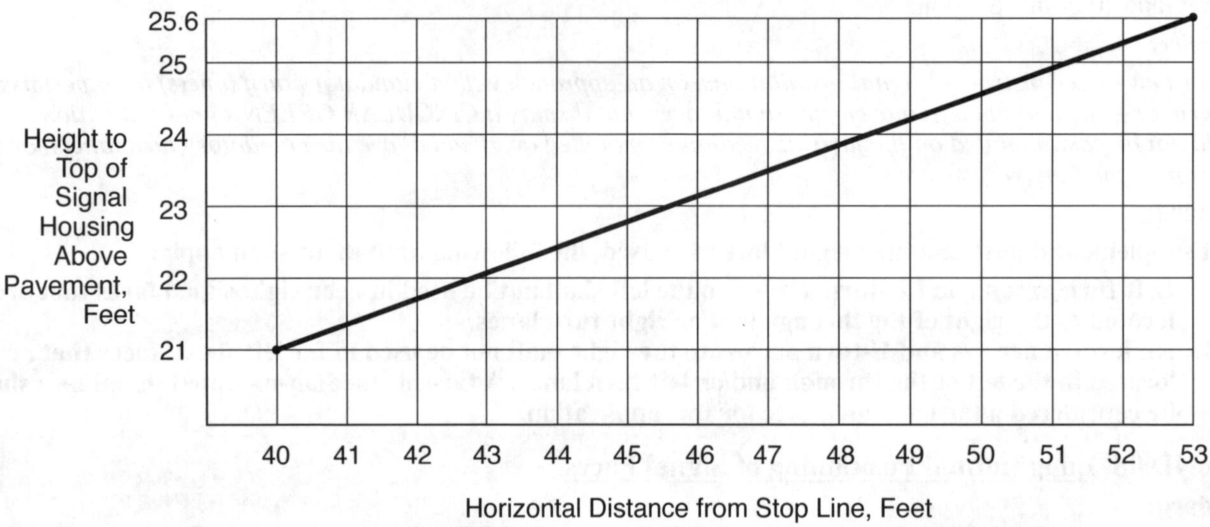

05   *The bottom of the signal housing (including brackets) of a vehicular signal face that is vertically arranged and not located over a roadway or shoulder:*

    A. *Should be a maximum of 19 feet above the sidewalk or, if there is no sidewalk, above the pavement grade at the center of the roadway.*

    B. *Should be a maximum of 19 feet above the median island grade of a center median island if located on the near side of the intersection.*

06   *The bottom of the signal housing (including brackets) of a vehicular signal face that is horizontally arranged and not located over a roadway or shoulder:*

    A. *Should be a maximum of 22 feet above the sidewalk or, if there is no sidewalk, above the pavement grade at the center of the roadway.*

    B. *Should be a maximum of 22 feet above the median island grade of a center median island if located on the near side of the intersection.*

Section 4D.10 Lateral Offset (Clearance) of Signal Faces

Guidance:

01   *Signal faces mounted at the side of a roadway at less than 15 feet from the bottom of the housing and any related attachments should have a horizontal offset of not less than 2 feet from the face of a vertical curb, or if there is no curb, not less than 2 feet from the edge of a shoulder.*

Section 4D.11 Temporary and Portable Traffic Control Signals

Support:

01   A temporary traffic control signal is generally installed using methods that minimize the costs of installation, relocation, and/or removal. Typical temporary traffic control signals are for specific purposes, such as for one-lane, two-way facilities in temporary traffic control zones (see Chapter 4O), for a haul-road intersection, or for access to a site that will have a permanent access point developed at another location in the near future. Portable traffic signals are temporary traffic signals.

02   Because a portable traffic control signals is considered to be a type of temporary traffic control signal, the provisions for temporary traffic control signals are also applicable to portable traffic control signals.

Standard:

03   **Advance signing shall be used when employing a temporary traffic control signal.**

04   **A temporary traffic control signal shall:**

    A. Meet the physical display and operational requirements of a conventional traffic control signal;

    B. Be removed when no longer needed; and

    C. Except as provided in Paragraph 5 of this Section, be placed in the flashing mode during periods when it is not desirable to operate the signal in the steady mode, or the signal heads shall be covered, turned, or taken down to indicate that the signal is not in operation.

Option:

05   If the temporary traffic control signal is capable of being operated in a semi-actuated mode, such that green signal indications are continually shown to major-street traffic except when responding to a minor-street approach vehicle call, it may be operated in a semi-actuated mode instead of being placed in a flashing mode.

Guidance:

06   *A temporary traffic control signal should be used only if engineering judgment indicates that installing the signal will improve the overall safety and/or operation of the location.*

07   *The use of temporary traffic control signals by a work crew on a regular basis in their work area should be subject to the approval of the jurisdiction having authority over the roadway.*

08   *A temporary traffic control signal should not operate longer than 30 days unless associated with a longer-term temporary traffic control zone project.*

09   *Section 6L.01 contains information about the use of temporary traffic control signals in temporary traffic control zones.*

CHAPTER 4E. TRAFFIC CONTROL SIGNAL INDICATIONS

Section 4E.01 Signal Indications – Design, Illumination, Color, and Shape

Standard:

01   The illuminated part of each signal indication shall be circular or arrow, except those used for bicycle symbol signal indications, pedestrian signal heads, light rail transit signal indications, and lane-use control signals.

02   Letters or numbers (including those associated with countdown displays) shall not be displayed as part of a vehicular signal indication.

03   Strobes shall not be used within or adjacent to any signal indication.

04   Except for the flashing vehicular and pedestrian signal indications and the distinctive indications for emergency-vehicle preemption (see Section 4F.19) that are expressly allowed by the provisions of this Part, flashing displays shall not be used within or adjacent to any signal indications.

05   Each circular signal indication shall emit a single color: red, yellow, or green.

06   Except as provided in Paragraph 7 of this Section, each arrow signal indication shall emit a single color: red, yellow, or green.

Option:

07   A bimodal signal section that is capable of alternating between the display of a GREEN ARROW signal indication and the display of a YELLOW ARROW signal indication, both pointing in the same direction, may be used provided that both colors are never displayed simultaneously.

Standard:

08   The arrow, which shall show only one direction, shall be the only illuminated part of an arrow signal indication.

09   Arrows shall be pointed:

A. Vertically upward to indicate a straight-through movement,
B. Horizontally in the direction of the turn to indicate a turn at approximately or greater than a right angle,
C. Upward with a slope at an angle approximately equal to that of the turn if the angle of the turn is substantially less than a right angle, or
D. In a manner that directs the driver through the turn if a U-turn arrow is used (see Figure 4E-1).

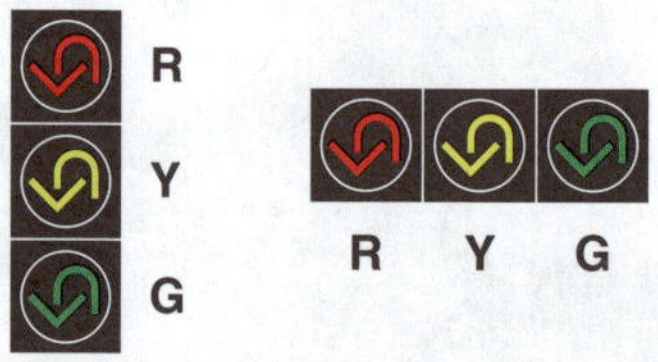

10   Except as provided in Paragraph 11 of this Section, the requirements of Chapters 1 and 2 of the publication entitled "Equipment and Materials Standards of the Institute of Transportation Engineers" that pertain to the aspects of the signal head design that affect the display of the signal indications shall be met for signal optical units that use incandescent lamps within optical assemblies that include lenses. Except as provided in Paragraph 11 of this Section, the requirements of the Institute of Transportation Engineers' publications entitled "Vehicle Traffic Control Signal Heads: Light Emitting Diode (LED) Circular Signal Supplement," 2005, ITE, and "Vehicle Traffic Control Signal Heads: Light Emitting Diode (LED) Vehicle Arrow Traffic Signal Supplement," 2008, ITE, that pertain to the aspects of the signal head design that affect the display of the signal indications shall be met for light-emitting diode (LED) traffic signal modules.

Guidance:

11   *The intensity and distribution of light from each illuminated signal lens or LED signal module should comply with the publications specified in Paragraph 10 of this Section, as appropriate.*

Support:

12   References to signal lenses in this section are not intended to limit signal optical units to incandescent lamps within optical assemblies that include lenses. Research has resulted in signal optical units that are not lenses, such as, but not limited to, light-emitting diode (LED) traffic signal modules. Some units are practical for all signal indications, and some are practical for specific types such as visibility-limited signal indications.

Guidance:

13   *If a signal indication is so bright that it causes excessive glare during nighttime conditions, some form of automatic dimming should be used to reduce the brilliance of the signal indication.*

Section 4E.02 Size of Vehicular Signal Indications

Standard:

01   **There shall be three nominal diameter sizes for vehicular signal indications: 4 inches, 8 inches, and 12 inches.**

02   **Four-inch signal indications shall only be used for bicycle signal faces per Section 4H.07.**

03   **Twelve-inch signal indications shall be used for all arrow signal indications.**

04   **Except as provided in Paragraph 5 of this Section, 12-inch signal indications shall be used for all circular signal indications in all new signal faces.**

Option:

05   Eight-inch circular signal indications may be used in new signal faces only for:

   A. The green or flashing yellow signal indications in an emergency-vehicle traffic control signal (see Section 4M.02);

   B. The circular indications in signal faces controlling the approach to the downstream location where two adjacent signalized locations are close to each other and it is impractical because of factors such as high approach speeds, horizontal or vertical curves, or other geometric factors to install visibility-limited signal faces for the downstream approach;

   C. The circular indications in a signal face that is located less than 120 feet from the stop line on a roadway with a posted or statutory speed limit or operating speed of 30 mph or less;

   D. The circular indications in a supplemental near-side signal face;

   E. The circular indications in a supplemental signal face installed for the sole purpose of controlling pedestrian movements rather than vehicular movements; and

   F. The circular indications in a flashing beacon (see Chapter 4S).

06   Different sizes of signal indications may be used in the same signal face or signal head, provided that the signal face or signal head complies with the requirements contained in Paragraphs 3 through 5 of this Section.

Section 4E.03 Positions of Signal Indications within a Signal Face – General

Support:

01   Standardization of the number and arrangements of signal sections in vehicular traffic control signal faces enables road users who are color vision deficient to identify the illuminated color by its position relative to other signal sections.

Standard:

02   **Unless otherwise provided in this Manual for a particular application, each signal face at a signalized location shall have three, four, or five signal sections. Unless otherwise provided in this Manual for a particular application, if a vertical signal face includes a cluster (see Section 4E.04), the signal face shall have at least three vertical positions.**

03   **A single-section signal face shall be permitted at a traffic control signal if it consists of a continuously-displayed GREEN ARROW signal indication that is being used to indicate a continuous movement.**

04   **The signal sections in a signal face shall be arranged in a vertical or horizontal straight line, except as otherwise provided in Section 4E.04.**

05   **The arrangement of adjacent signal sections in a signal face shall follow the relative positions listed in Sections 4E.04 or 4E.05, as applicable.**

06   **If a signal section that displays a CIRCULAR YELLOW signal indication is used, it shall be located between the signal section that displays the red signal indication and all other signal sections.**

07   **If a U-turn arrow signal section is used in a signal face for a U-turn to the left, its position in the signal face shall be the same as stated in Sections 4E.04 and 4E.05 for a left-turn arrow signal section of the same color. If a U-turn arrow signal section is used in a signal face for a U-turn to the right, its position in the signal face shall be the same as stated in Sections 4E.04 and 4E.05 for a right-turn arrow signal section of the same color.**

08   **A U-turn arrow signal indication pointing to the left shall not be used in a signal face that also contains a left-turn arrow signal indication. A U-turn arrow signal indication pointing to the right shall not be used in a signal face that also contains a right-turn arrow signal indication.**

Option:

09   Within a signal face, two identical CIRCULAR RED or RED ARROW signal indications may be displayed immediately horizontally adjacent or immediately vertically adjacent to each other in a vertical signal face (see Drawing A in Figure 4E-2) or immediately horizontally adjacent to each other in a horizontal signal face (see Drawing B in Figure 4E-2) for emphasis.

10   Horizontally-arranged and vertically-arranged signal faces may be used on the same approach provided they are separated to meet the lateral separation spacing required in Section 4D.07.

Support:

11   Figure 4E-2 illustrates some of the typical arrangements of signal sections in signal faces that do not control separate turning movements. Figures 4F-1 through 4F-7 illustrate the typical arrangements of signal sections in left-turn signal faces. Figures 4F-8 through 4F-14 illustrate the typical arrangements of signal sections in right-turn signal faces.

Section 4E.04 Positions of Signal Indications within a Vertical Signal Face

Standard:

01   **In each vertically-arranged signal face, all signal sections that display red signal indications shall be located above all signal sections that display yellow and green signal indications.**

02   **In vertically-arranged signal faces, each signal section that displays a YELLOW ARROW signal indication shall be located above the signal section that displays the GREEN ARROW signal indication to which it applies.**

03   **The relative positions of signal sections in a vertically-arranged signal face, from top to bottom, shall be as follows:**

CIRCULAR RED
Steady and/or flashing left-turn RED ARROW
Steady and/or flashing right-turn RED ARROW
CIRCULAR YELLOW
CIRCULAR GREEN
Straight-through GREEN ARROW
Steady left-turn YELLOW ARROW
Flashing left-turn YELLOW ARROW
Left-turn GREEN ARROW
Steady right-turn YELLOW ARROW
Flashing right-turn YELLOW ARROW
Right-turn GREEN ARROW

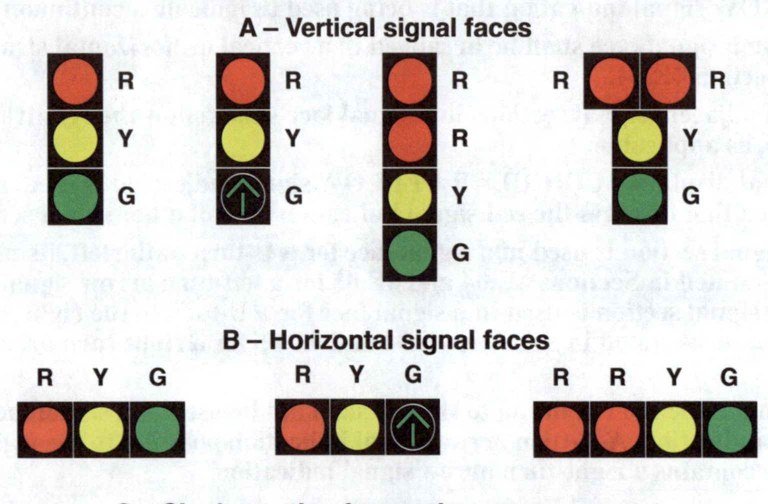

Figure 4E-2. Typical Arrangements of Signal Sections in Signal Faces That Do Not Control Turning Movements

04   **If a bimodal signal section (see Section 4E.01) is used in a vertically-arranged signal face, the bimodal signal section shall occupy the same position relative to the other sections as the signal section that displays the GREEN ARROW signal indication in a vertically-arranged signal face would occupy.**

Option:

05   In a vertically-arranged signal face, signal sections that display signal indications of the same color may be arranged horizontally adjacent to each other at right angles to the basic straight line arrangement to form a clustered signal face (see Figures 4E-2, 4F-4, 4F-6, 4F-10, 4F-11, 4F-13, and 4F-15).

Standard:

06   **Such clusters shall be limited to the following:**

   A. Two identical signal sections,
   B. Two or three different signal sections that display signal indications of the same color, or
   C. For only the specific case described in Section 4F.16 (see Drawing B in Figure 4F-15), two signal sections, one of which displays a GREEN ARROW signal indication and the other of which displays a flashing YELLOW ARROW signal indication.

07   **Except as otherwise provided in Sections 4F.04, 4F.08, 4F.11, and 4F.15 for a three-section separate turn signal face with a bimodal signal section that displays a flashing YELLOW ARROW signal indication, the signal section that displays a flashing yellow signal indication during steady mode operation:**

   A. Shall not be placed in the same vertical position as the signal section that displays a steady yellow signal indication, and
   B. Shall be placed below the signal section that displays a steady yellow signal indication.

Support:

08   Sections 4J.02 and 4N.02 contain exceptions to the provisions of this Section that are applicable to hybrid beacons.

Section 4E.05 <u>Positions of Signal Indications within a Horizontal Signal Face</u>

Standard:

01   **In each horizontally-arranged signal face, all signal sections that display red signal indications shall be located to the left of all signal sections that display yellow and green signal indications.**

02   **In horizontally-arranged signal faces, each signal section that displays a YELLOW ARROW signal indication shall be located to the left of the signal section that displays the GREEN ARROW signal indication to which it applies.**

03   **The relative positions of signal sections in a horizontally-arranged signal face, from left to right, shall be as follows:**

   CIRCULAR RED
   Steady and/or flashing left-turn RED ARROW
   Steady and/or flashing right-turn RED ARROW
   CIRCULAR YELLOW
   Steady left-turn YELLOW ARROW
   Flashing left-turn YELLOW ARROW
   Left-turn GREEN ARROW
   CIRCULAR GREEN
   Straight-through GREEN ARROW
   Steady right-turn YELLOW ARROW
   Flashing right-turn YELLOW ARROW
   Right-turn GREEN ARROW

04   **If a bimodal signal section (see Section 4E.01) is used in a horizontally-arranged signal face, the signal section that displays the dual left-turn arrow signal indication shall be located immediately to the right of the signal section that displays the CIRCULAR YELLOW signal indication, the signal section that displays the straight-through GREEN ARROW signal indication shall be located immediately to the right of the signal section that displays the CIRCULAR GREEN signal indication, and the signal section that displays the dual right-turn arrow signal indication shall be located to the right of all other signal sections.**

05   **Except as otherwise provided in Sections 4F.04, 4F.08, 4F.11, and 4F.15 for a three-section separate turn signal face with a flashing YELLOW ARROW signal indication, the signal section that displays a flashing yellow signal indication during steady mode operation:**

   A. Shall not be placed in the same horizontal position as the signal section that displays a steady yellow signal indication, and
   B. Shall be placed to the right of the signal section that displays a steady yellow signal indication.

CHAPTER 4F. STEADY (STOP-AND-GO) OPERATION OF TRAFFIC CONTROL SIGNALS

Section 4F.01 Application of Steady and Flashing Signal Indications during Steady (Stop-and-Go) Operation

Standard:

01   When a traffic control signal is being operated in a steady (stop-and-go) mode, at least one indication in each signal face shall be displayed at any given time.

02   A signal face(s) that controls a particular vehicular movement during any interval of a cycle shall control that same movement during all intervals of the cycle.

03   Steady and flashing signal indications shall be applied as follows:

A. A steady CIRCULAR RED signal indication:

1. Shall be displayed when it is intended to prohibit traffic, except pedestrians directed by a pedestrian signal head, from entering the intersection or other controlled area. Turning after stopping is permitted as stated in Item C.1 in Paragraph 1 of Section 4A.03.

2. Shall be displayed with the appropriate GREEN ARROW signal indications when it is intended to permit traffic to make a specified turn or turns, and to prohibit traffic from proceeding straight ahead through the intersection or other controlled area, except in protected only mode operation (see Sections 4F.06 and 4F.13), or in protected/permissive mode operation with separate turn signal faces (see Sections 4F.08 and 4F.15).

B. A steady CIRCULAR YELLOW signal indication:

1. Shall be displayed following a CIRCULAR GREEN or straight-through GREEN ARROW signal indication in the same signal face.

2. Shall not be displayed in conjunction with the change from the CIRCULAR RED signal indication to the CIRCULAR GREEN signal indication.

3. Shall be followed by a CIRCULAR RED signal indication except that, when entering preemption operation, the return to the previous CIRCULAR GREEN signal indication shall be permitted following a steady CIRCULAR YELLOW signal indication (see Section 4F.19).

4. Shall not be displayed to an approach from which drivers are turning left permissively using a shared signal face or making a U-turn to the left permissively using a shared signal face unless one of the following conditions exists:

(a) A steady CIRCULAR YELLOW signal indication is also simultaneously being displayed to the opposing approach;

(b) An engineering study has determined that, because of unique intersection conditions, the condition described in Item (a) cannot reasonably be implemented without causing significant operational or safety problems and that the volume of impacted left-turning or U-turning traffic is relatively low, and those left-turning or U-turning drivers are advised that a steady CIRCULAR YELLOW signal indication is not simultaneously being displayed to the opposing traffic if this operation occurs continuously by the installation of a W25-1 sign (see Section 2C.44) with the legend ONCOMING TRAFFIC HAS EXTENDED GREEN; or

(c) Drivers are advised of the operation if it occurs only occasionally, such as during a preemption sequence, by the installation of a W25-2 sign (see Section 2C.44) with the legend ONCOMING TRAFFIC MAY HAVE EXTENDED GREEN.

C. A steady CIRCULAR GREEN signal indication shall be displayed only when it is intended to permit traffic to proceed in any direction that is lawful and practical.

D. A steady RED ARROW signal indication shall be displayed when it is intended to prohibit traffic, except pedestrians directed by a pedestrian signal head, from entering the intersection or other controlled area to make the indicated turn. Except as described in Item C.2 in Paragraph 1 of Section 4A.03, turning on a steady RED ARROW signal indication shall not be permitted.

E. A flashing RED ARROW signal indication shall be displayed as part of a steady (stop-and-go) mode of operation only when it is intended to permit traffic, after coming to a full stop, to cautiously enter the intersection to make a turn in the direction indicated by the arrow after yielding to pedestrians, if any, and/or to opposing traffic, if any.

F. A steady YELLOW ARROW signal indication:

1. Shall be displayed in the same direction as a GREEN ARROW signal indication following a GREEN ARROW signal indication in the same signal face, unless:

(a) The GREEN ARROW signal indication and a CIRCULAR GREEN (or straight-through GREEN ARROW) signal indication terminate simultaneously in the same signal face, or

(b) The green arrow is a straight-through GREEN ARROW (see Item B.1 in this Paragraph).

2. Shall be displayed in the same direction as a flashing YELLOW ARROW signal indication or flashing RED ARROW signal indication following a flashing YELLOW ARROW signal indication or flashing RED ARROW signal indication in the same signal face, when the flashing arrow indication is displayed as part of a steady mode operation, if the signal face will subsequently display a steady red signal indication.

3. Shall not be displayed in conjunction with the change from a steady RED ARROW, flashing RED ARROW, or flashing YELLOW ARROW signal indication to a GREEN ARROW signal indication, except when entering preemption operation as provided in Item F.5(a) of this Paragraph.

4. Shall not be displayed when any conflicting vehicular movement has a green or yellow signal indication (except for the situation regarding U-turns to the left provided in Paragraph 4 of this Section) or any conflicting pedestrian movement has a WALKING PERSON (symbolizing WALK) or flashing UPRAISED HAND (symbolizing DONT WALK) signal indication, except that a steady left-turn (or U-turn to the left) YELLOW ARROW signal indication used to terminate a flashing left-turn (or U-turn to the left) YELLOW ARROW or a flashing left-turn (or U-turn to the left) RED ARROW signal indication in a signal face controlling a permissive left-turn (or U-turn to the left) movement as described in Sections 4F.04 and 4F.08 shall be permitted to be displayed when a CIRCULAR YELLOW signal indication is displayed for the opposing through movement. Vehicles departing in the same direction shall not be considered in conflict if, for each turn lane with moving traffic, there is a separate departure lane, and pavement markings or raised channelization clearly indicate which departure lane to use.

5. Shall not be displayed to terminate a flashing arrow signal indication on an approach from which drivers are turning left permissively or making a U-turn to the left permissively unless one of the following conditions exists:

   (a) A steady CIRCULAR YELLOW signal indication is also simultaneously being displayed to the opposing approach;

   (b) An engineering study has determined that, because of unique intersection conditions, the condition described in Item (a) cannot reasonably be implemented without causing significant operational or safety problems and that the volume of impacted left-turning or U-turning traffic is relatively low, and those left-turning or U-turning drivers are advised that a steady CIRCULAR YELLOW signal indication is not simultaneously being displayed to the opposing traffic if this operation occurs continuously by the installation of a W25-1 sign (see Section 2C.44) with the legend ONCOMING TRAFFIC HAS EXTENDED GREEN; or

   (c) Drivers are advised of the operation if it occurs only occasionally, such as during a preemption sequence, by the installation of a W25-2 sign (see Section 2C.44) with the legend ONCOMING TRAFFIC MAY HAVE EXTENDED GREEN.

6. Shall be terminated by a RED ARROW signal indication for the same direction or a CIRCULAR RED signal indication except:

   (a) When entering preemption operation, the display of a GREEN ARROW signal indication or a flashing arrow signal indication shall be permitted following a steady YELLOW ARROW signal indication.

   (b) When the movement controlled by the arrow is to continue on a permissive mode basis during an immediately following signal phase, the display of a CIRCULAR GREEN signal indication or flashing YELLOW ARROW signal indication shall be permitted following a steady YELLOW ARROW signal indication. To provide a red clearance interval, it shall be permitted to display a steady left-turn RED ARROW signal indication immediately following the steady left-turn YELLOW ARROW signal indication.

G. A flashing YELLOW ARROW signal indication shall be displayed as part of a steady (stop-and-go) mode of operation only when it is intended to permit traffic to cautiously enter the intersection to make a turn in the direction indicated by the arrow after yielding to pedestrians, if any, and/or to opposing traffic, if any.

H. A steady GREEN ARROW signal indication:

1. Shall be displayed only to allow vehicular movements, in the direction indicated, that are not in conflict with other vehicles moving on a green or yellow signal indication (except for the situation regarding U-turns provided in Paragraph 4 of this Section and straight-through GREEN ARROWs provided in Paragraph 5 of this Section), even if the other vehicles are required to yield the right-of-way to the traffic moving on the GREEN ARROW signal

    indication, and are not in conflict with pedestrians crossing in compliance with a WALKING PERSON (symbolizing WALK) or flashing UPRAISED HAND (symbolizing DONT WALK) signal indication. Vehicles departing in the same direction shall not be considered in conflict if, for each turn lane with moving traffic, there is a separate departure lane, and pavement markings or raised channelization clearly indicate which departure lane to use.

    2. Shall be displayed on a signal face that controls a left-turn movement when said movement is not in conflict with other vehicles moving on a green or yellow signal indication (except for the situation regarding U-turns provided in Paragraph 4 of this Section) and is not in conflict with pedestrians crossing in compliance with a WALKING PERSON (symbolizing WALK) or flashing UPRAISED HAND (symbolizing DONT WALK) signal indication. Vehicles departing in the same direction shall not be considered in conflict if, for each turn lane with moving traffic, there is a separate departure lane, and pavement markings or raised channelization clearly indicate which departure lane to use.

    3. Shall not be required on the stem of a T-intersection or for turns from a one-way street.

Option:

04   If U-turns are permitted from the approach and a right-turn GREEN ARROW signal indication is simultaneously being displayed to road users making a right turn from the conflicting approach to the left, road users making a U-turn may be advised of the operation by the installation of a U-TURN YIELD TO RIGHT TURN (R10-16) sign (see Section 2B.59).

05   A steady straight-through GREEN ARROW signal indication may be used instead of a CIRCULAR GREEN signal indication in a signal face to discourage wrong-way turns under the following conditions, even if opposed by a simultaneous permissive left-turn movement:

    A. On an approach intersecting a one-way street;

    B. On an approach intersecting an interchange exit ramp;

    C. On an approach with unique geometric design that prohibits turns; or

    D. On an approach with pre-signals and the adjacent lanes are controlled separately (see Sections 8D.11 and 8D.12).

06   If not otherwise prohibited, steady red, yellow, and green turn arrow signal indications may be used instead of steady circular red, yellow, and green signal indications in a signal face on an approach where all traffic is required to turn or where the straight-through movement is not physically possible.

Support:

07   Section 4F.16 contains information regarding the signalization of approaches that have a combined left-turn/ right-turn lane and no through movement.

08   Section 4D.07 contains information regarding limitations on left-turn arrows, right-turn arrows, and U-turn arrows in supplemental signal faces.

Standard:

09   **A straight-through RED ARROW signal indication or a straight-through YELLOW ARROW signal indication shall not be displayed on any signal face, either alone or in combination with any other signal indication.**

10   **The following combinations of signal indications shall not be simultaneously displayed on any one signal face:**

    A. CIRCULAR YELLOW with CIRCULAR RED,

    B. CIRCULAR GREEN with CIRCULAR RED, or

    C. Straight-through GREEN ARROW with CIRCULAR RED.

11   **Except as provided in Paragraph 13 of this Section, the above combinations shall not be simultaneously displayed on an approach as a result of the combination of displays from multiple signal faces unless the display is created by a signal face(s) devoted exclusively to the control of a right-turn movement and:**

    A. The signal face(s) controlling the right-turn movement is visibility limited from the adjacent through movement or positioned to minimize potential confusion to approaching road users, or

    B. A RIGHT TURN SIGNAL (R10-10) sign (see Sections 4F.09, 4F.11, 4F.13, and 4F.15) is mounted adjacent to the signal face(s) controlling the right-turn movement.

12   **Except as provided in Paragraph 13 of this Section, the following combinations of signal indications shall not be simultaneously displayed on any one signal face or as a result of the combination of displays from multiple signal faces on an approach:**

    A. CIRCULAR GREEN with CIRCULAR YELLOW,

    B. Straight-through GREEN ARROW with CIRCULAR YELLOW,

C. **GREEN ARROW with YELLOW ARROW pointing in the same direction,**
D. **YELLOW ARROW with RED ARROW pointing in the same direction, or**
E. **GREEN ARROW with RED ARROW pointing in the same direction.**

13   **If a separate signal face is provided at a pre-signal (see Section 8D.11) or at a queue cutter signal (see Section 8D.12) for a left-turn and/or right-turn lane that extends from the downstream signalized intersection back to and across a grade crossing, the following combinations of signal indications shall be permitted to be simultaneously displayed as a result of the combination of displays from multiple signal faces at the pre-signal or queue cutter signal:**

A. **Straight-through GREEN ARROW with CIRCULAR RED,**
B. **Straight-through GREEN ARROW with CIRCULAR YELLOW, and**
C. **CIRCULAR YELLOW with CIRCULAR RED.**

14   **Except as otherwise provided in Sections 4F.08, 4F.15, 4J.03, and 4N.03, the same signal section shall not be used to display both a flashing yellow and a steady yellow indication during steady mode operation. Except as otherwise provided in Sections 4F.04, 4F.08, 4F.11, and 4F.13, the same signal section shall not be used to display both a flashing red and a steady red indication during steady mode operation.**

Guidance:

15   *No movement that creates an unexpected crossing of pathways of moving vehicles or pedestrians should be allowed during any green or yellow interval, except when all three of the following conditions are met:*

A. *The movement involves only slight conflict, and*
B. *Serious traffic delays are substantially reduced by permitting the conflicting movement, and*
C. *Drivers and pedestrians subjected to the unexpected conflict are effectively warned thereof by a sign.*

Section 4F.02 Signal Indications for Left-Turn Movements – General

Support:

01   In Sections 4F.03 through 4F.08, provisions applicable to left-turn movements and left-turn lanes are also applicable to signal indications for U-turns to the left that are provided at locations where left turns are prohibited or not geometrically possible.

02   Left-turning traffic is controlled by one of four modes as follows:

A. Permissive Only Mode—turns made on a CIRCULAR GREEN signal indication, a flashing left-turn YELLOW ARROW signal indication, or a flashing left-turn RED ARROW signal indication after yielding to pedestrians, if any, and/or opposing traffic, if any.
B. Protected Only Mode—turns made only when a left-turn GREEN ARROW signal indication is displayed.
C. Protected/Permissive Mode—both modes can occur on an approach during the same cycle.
D. Variable Left-Turn Mode—the operating mode changes among the protected only mode and/or the protected/permissive mode and/or the permissive only mode during different periods of the day or as traffic conditions change.

Option:

03   In areas having a high percentage of older drivers, special consideration may be given to the use of protected only mode left-turn phasing, when appropriate.

Standard:

04   **During a permissive left-turn movement, the signal faces for through traffic on the opposing approach shall simultaneously display green or steady yellow signal indications. If pedestrians crossing the lane or lanes used by the permissive left-turn movement to depart the intersection are controlled by pedestrian signal heads, the signal indications displayed by those pedestrian signal heads shall not be limited to any particular display during the permissive left-turn movement.**

05   **During a protected left-turn movement, the signal faces for through traffic on the opposing approach shall simultaneously display steady CIRCULAR RED signal indications. During a protected left-turn movement, a GREEN ARROW or a YELLOW ARROW signal indication shall not simultaneously be displayed to right-turning traffic on the opposing approach, except where a separate departure lane is available for each left-turn and right-turn lane with moving traffic and pavement markings or raised channelization clearly indicate which departure lane to use (see Item H.1 in Paragraph 3 in Section 4F.01). If pedestrians crossing the lane or lanes used by the protected left-turn movement to depart the intersection are controlled by pedestrian signal heads, the pedestrian signal heads shall display a steady UPRAISED HAND (symbolizing DONT WALK) signal indication during the protected left-turn movement.**

06   **If a combined left-turn/through lane exists on an approach, a left-turn GREEN ARROW or left-turn YELLOW ARROW signal indication or a flashing left-turn RED ARROW signal indication shall not be displayed to the approach simultaneously with a CIRCULAR RED signal indication for the through movement, and a left-turn RED ARROW signal indication shall not be displayed to the approach simultaneously with a CIRCULAR GREEN or CIRCULAR YELLOW signal indication for the through movement.**

07   **A yellow change interval for the left-turn movement shall not be displayed when the status of the left-turn operation is changing from permissive to protected within any given signal sequence.**

08   **If the operating mode changes among the protected only mode and/or the protected/permissive mode and/or the permissive only mode during different periods of the day or as traffic conditions change, the requirements in Sections 4F.03 through 4F.08 that are appropriate to that mode of operation shall be met, subject to the following:**

   A. **The CIRCULAR GREEN and CIRCULAR YELLOW signal indications shall not be displayed when operating in the protected only mode.**

   B. **The left-turn GREEN ARROW and left-turn YELLOW ARROW signal indications shall not be displayed when operating in the permissive only mode.**

Option:

09   When variable left-turn mode phasing is used for an approach that has a combined left-turn/straight-through lane and a flashing yellow arrow is used as the permissive turn display, a five-section shared left-turn signal face containing both circular and arrow indications may be used in combination with one or more separate left-turn signal faces for the mandatory left-turn lane(s), if any are present, on the same approach. The steady left-turn YELLOW ARROW signal indication and the flashing left-turn YELLOW ARROW signal indication may be displayed in the same section of the five-section shared left-turn signal face.

10   Additional static signs or changeable message signs may be used to meet the requirements for the variable left-turn mode or to inform drivers that left-turn green arrows will not be available during certain times of the day.

Support:

11   Sections 4F.03 through 4F.08 describe the use of the following two types of signal faces for controlling left-turn movements:

   A. Shared signal face – This type of signal face controls both the left-turn movement and the adjacent movement (usually the through movement) and can serve as one of the two required primary signal faces for the adjacent movement. A shared signal face always displays the same color of circular indication that is displayed by the signal face or faces for the adjacent movement. If a shared signal face that provides protected/permissive mode left turns is mounted overhead for an approach that includes a mandatory left-turn lane, it is usually positioned over or slightly to the right of the extension of the lane line separating the left-turn lane from the adjacent lane. If a shared signal face that provides protected/permissive mode left turns is mounted overhead for an approach that does not include a mandatory left-turn lane, it is usually positioned over the center of the combined left-turn/straight-through lane.

   B. Separate left-turn signal face – This type of signal face controls only the left-turn movement and cannot serve as one of the two required primary signal faces for the adjacent movement (usually the through movement) because it displays signal indications that are applicable only to the left-turn movement. This type of signal face is used only for an approach that has a mandatory left-turn lane(s). If a separate left-turn signal face is mounted overhead at the intersection, it is positioned over the extension of the mandatory left-turn lane. In a separate left-turn signal face, a flashing left-turn YELLOW ARROW signal indication or a flashing left-turn RED ARROW signal indication is used to control permissive left-turn movements.

12   Section 4D.07 contains provisions regarding the lateral positioning of signal faces that control left-turn movements.

13   It is not necessary that the same mode of left-turn operation or same type of left-turn signal face be used on every approach to a signalized location. Selecting different modes and types of left-turn signal faces for the various approaches to the same signalized location is acceptable.

Option:

14   A signal face that is shared by left-turning and right-turning traffic may be provided for a combined left-turn/right-turn lane on an approach that has no through traffic (see Section 4F.16).

Section 4F.03 Signal Indications for Permissive Only Mode Left-Turn Movements in a Shared Signal Face

Standard:

01 **If a shared signal face is provided for a permissive only mode left turn, it shall meet the following requirements (see Figure 4F-1):**

> **A. It shall be capable of displaying the following signal indications: steady CIRCULAR RED, steady CIRCULAR YELLOW, and CIRCULAR GREEN. Only one of the three indications shall be displayed at any given time.**
>
> **B. During the permissive left-turn movement, a CIRCULAR GREEN signal indication shall be displayed.**
>
> **C. A permissive only shared signal face, regardless of where it is positioned and regardless of how many adjacent through signal faces are provided, shall always simultaneously display the same color of circular indication that the adjacent through signal face or faces display.**
>
> **D. If the permissive only mode is not the only left-turn mode used for the approach, the signal face shall be the same shared signal face that is used for the protected/permissive mode (see Section 4F.07) except that the left-turn GREEN ARROW and left-turn YELLOW ARROW signal indications shall not be displayed when operating in the permissive only mode.**

Section 4F.04 Signal Indications for Permissive Only Mode Left-Turn Movements in a Separate Signal Face

Standard:

01 **A separate left-turn signal face shall not be used for an approach that does not include a mandatory left-turn lane.**

02 **If a separate left-turn signal face is being operated in a permissive only left-turn mode, a CIRCULAR GREEN signal indication shall not be used in that face.**

03 **If a separate left-turn signal face is being operated in a permissive only left-turn mode and a flashing left-turn YELLOW ARROW signal indication is provided, it shall meet the following requirements (see Figure 4F-2):**

> **A. It shall be capable of displaying the following signal indications: steady left-turn RED ARROW, steady left-turn YELLOW ARROW, and flashing left-turn YELLOW ARROW. Only one of the three indications shall be displayed at any given time.**
>
> **B. During the permissive left-turn movement, a flashing left-turn YELLOW ARROW signal indication shall be displayed.**

Figure 4F-1. Typical Position and Arrangements of Shared Signal Faces for Permissive Only Mode Left Turns

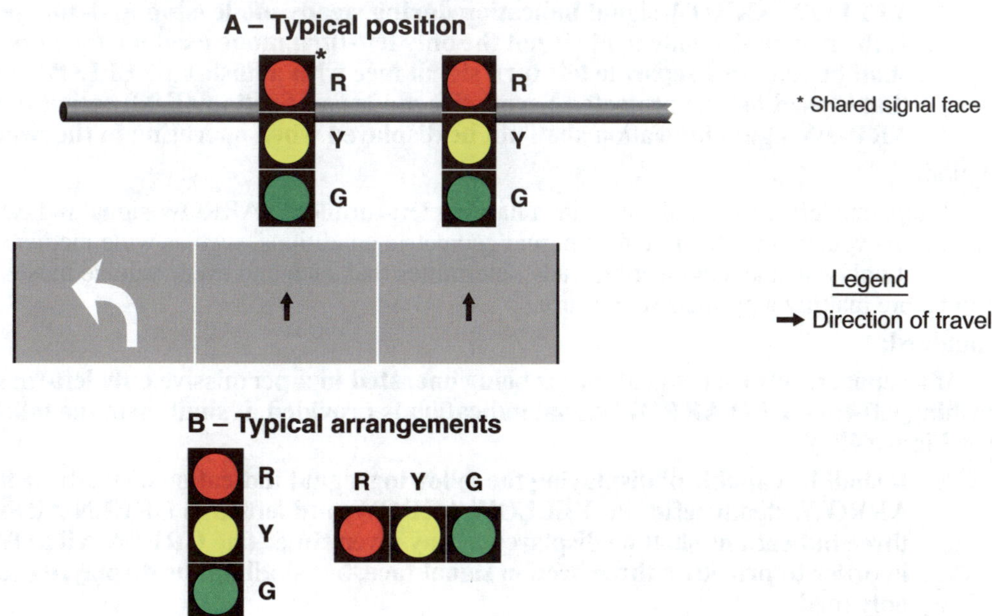

Figure 4F-2. Typical Position and Arrangements of Separate Signal Faces with Flashing Yellow Arrow for Permissive Only Mode Left Turns

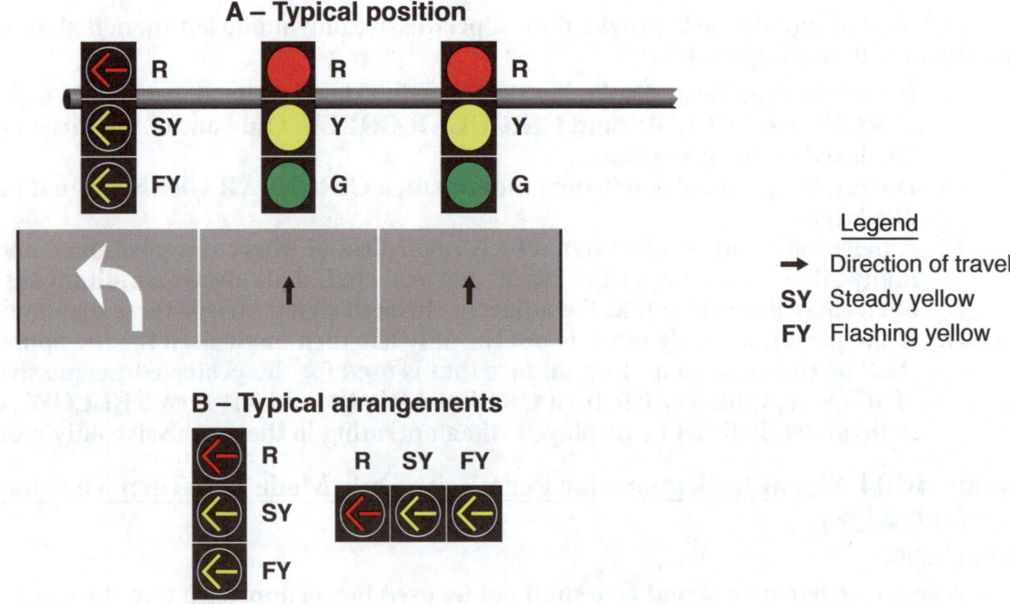

C. A steady left-turn YELLOW ARROW signal indication shall be displayed following the flashing left-turn YELLOW ARROW signal indication.

D. It shall be permitted to display a flashing left-turn YELLOW ARROW signal indication for a permissive left-turn movement while the signal faces for the adjacent through movement display steady CIRCULAR RED signal indications and the opposing left-turn signal faces display left-turn GREEN ARROW signal indications for a protected left-turn movement.

E. During steady mode (stop-and-go) operation, the signal section that displays the steady left-turn YELLOW ARROW signal indication during change intervals shall not be used to display the flashing left-turn YELLOW ARROW signal indication for permissive left turns unless a bimodal signal section capable of alternating between the display of a steady YELLOW ARROW and a flashing YELLOW ARROW signal indication is used to operate variable left-turn mode phasing.

F. During flashing mode operation (see Section 4G.01), the display of a flashing left-turn YELLOW ARROW signal indication shall be only from the signal section that displays a steady left-turn YELLOW ARROW signal indication during steady mode (stop-and-go) operation.

G. If the permissive only mode is not the only left-turn mode used for the approach, the signal face shall be the same separate left-turn signal face with a flashing YELLOW ARROW signal indication that is used for the protected/permissive mode (see Section 4F.08) except that the left-turn GREEN ARROW signal indication shall not be displayed when operating in the permissive only mode.

Option:

04   A separate left-turn signal face with a flashing left-turn RED ARROW signal indication during the permissive left-turn movement may be used for unusual geometric conditions, such as wide medians with offset left-turn lanes, but only when an engineering study determines that each and every vehicle must successively come to a full stop before making a permissive left turn.

Standard:

05   **If a separate left-turn signal face is being operated in a permissive only left-turn mode and a flashing left-turn RED ARROW signal indication is provided, it shall meet the following requirements (see Figure 4F-3):**

A. **It shall be capable of displaying the following signal indications: steady or flashing left-turn RED ARROW, steady left-turn YELLOW ARROW, and left-turn GREEN ARROW. Only one of the three indications shall be displayed at any given time. The GREEN ARROW indication is required in order to provide a three-section signal face, but shall not be displayed during the permissive only mode.**

Figure 4F-3. Typical Position and Arrangements of Separate Signal Faces with Flashing Red Arrow for Permissive Only Mode and Protected/Permissive Mode Left Turns

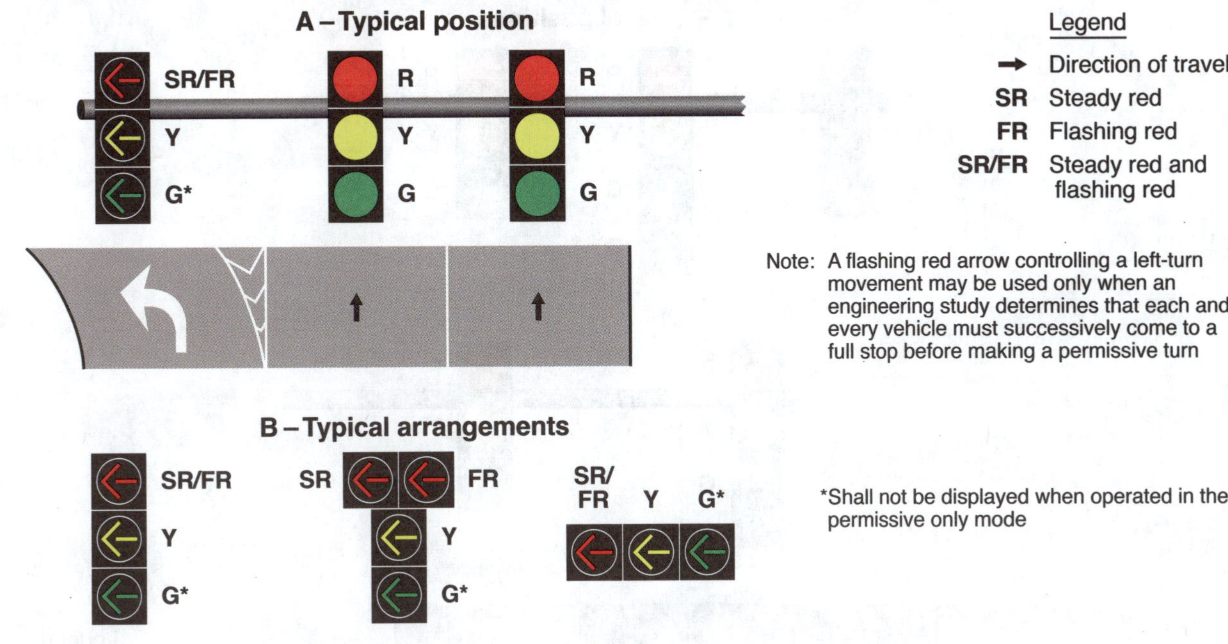

B. During the permissive left-turn movement, a flashing left-turn RED ARROW signal indication shall be displayed, thus indicating that each and every vehicle must successively come to a full stop before making a permissive left turn.

C. A steady left-turn YELLOW ARROW signal indication shall be displayed following the flashing left-turn RED ARROW signal indication.

D. It shall be permitted to display a flashing left-turn RED ARROW signal indication for a permissive left-turn movement while the signal faces for the adjacent through movement display steady CIRCULAR RED signal indications and the opposing left-turn signal faces display left-turn GREEN ARROW signal indications for a protected left-turn movement.

E. A supplementary sign shall not be required. If used, it shall be a LEFT TURN YIELD ON FLASHING RED ARROW AFTER STOP (R10-27) sign (see Section 2B.59).

Option:

06   The requirements of Item A in Paragraph 5 of this Section may be met by a vertically-arranged signal face with a horizontal cluster of two left-turn RED ARROW signal indications, the left-most of which displays a steady indication and the right-most of which displays a flashing indication (see Figure 4F-3).

Section 4F.05 Signal Indications for Protected Only Mode Left-Turn Movements in a Shared Signal Face

Standard:

01   A shared signal face shall not be used for protected only mode left turns unless the CIRCULAR GREEN and left-turn GREEN ARROW signal indications always begin and terminate together. If a shared signal face is provided for a protected only mode left turn, it shall meet the following requirements (see Figure 4F-4):

A. It shall be capable of displaying the following signal indications: steady CIRCULAR RED, steady CIRCULAR YELLOW, CIRCULAR GREEN, and left-turn GREEN ARROW. Only one of the three colors shall be displayed at any given time.

B. During the protected left-turn movement, the shared signal face shall simultaneously display both a CIRCULAR GREEN signal indication and a left-turn GREEN ARROW signal indication.

C. The shared signal face shall always simultaneously display the same color of circular indication that the adjacent through signal face or faces display.

D. If the protected only mode is not the only left-turn mode used for the approach, the signal face shall be the same shared signal face that is used for the protected/permissive mode (see Section 4F.07).

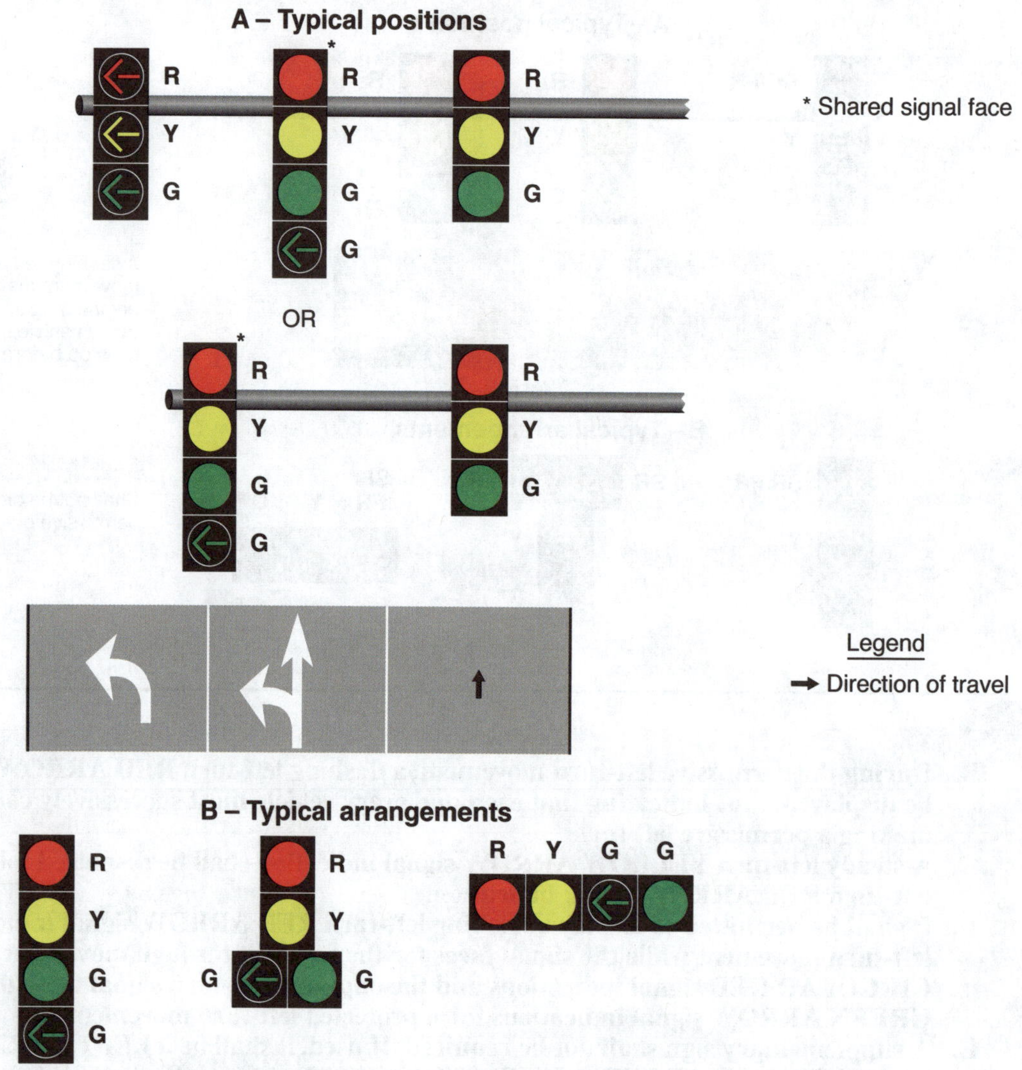

Option:

02   A straight-through GREEN ARROW signal indication may be used instead of the CIRCULAR GREEN signal indication in Items A and B in Paragraph 1 of this Section on an approach where a straight-through GREEN ARROW signal indication is also used instead of a CIRCULAR GREEN signal indication in the other signal face(s) for through traffic.

Section 4F.06 Signal Indications for Protected Only Mode Left-Turn Movements in a Separate Signal Face

Standard:

01   A separate left-turn signal face shall not be used for an approach that does not include a mandatory left-turn lane.

02   If a separate left-turn signal face is provided for a protected only mode left turn, it shall meet the following requirements (see Figure 4F-5):

A. It shall be capable of displaying, the following signal indications: steady left-turn RED ARROW, steady left-turn YELLOW ARROW, and left-turn GREEN ARROW. Only one of the three indications shall be displayed at any given time.

**Figure 4F-5. Typical Position and Arrangements of Separate Signal Faces
for Protected Only Mode Left Turns**

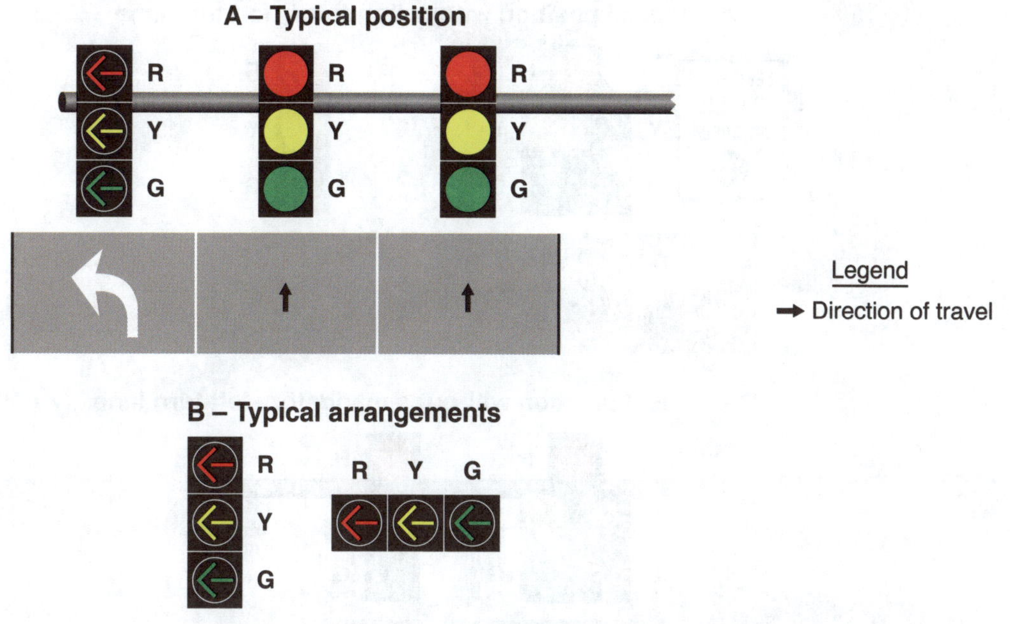

B. During the protected left-turn movement, a left-turn GREEN ARROW signal indication shall be
displayed.

C. A steady left-turn YELLOW ARROW signal indication shall be displayed following the left-turn
GREEN ARROW signal indication.

D. If the protected only mode is not the only left-turn mode used for the approach, the signal face shall
be the same separate left-turn signal face that is used for the protected/permissive mode (see Section
4F.08 and Figures 4F-3 and 4F-7) except that the flashing left-turn YELLOW ARROW or flashing
left-turn RED ARROW signal indication shall not be displayed when operating in the protected
only mode.

Section 4F.07 Signal Indications for Protected/Permissive Mode Left-Turn Movements in a Shared Signal Face

Standard:

01   If a shared signal face is provided for a protected/permissive mode left turn, it shall meet the following
requirements (see Figure 4F-6):

A. It shall be capable of displaying the following signal indications: steady CIRCULAR RED, steady
CIRCULAR YELLOW, CIRCULAR green, steady left-turn YELLOW ARROW, and left-turn
GREEN ARROW. Only one of the three circular indications shall be displayed at any given time.
Only one of the two arrow indications shall be displayed at any given time. If the left-turn GREEN
ARROW signal indication and the CIRCULAR GREEN signal indication(s) for the adjacent
through movement are always terminated together, the steady left-turn YELLOW ARROW signal
indication shall not be required.

B. During the protected left-turn movement, the shared signal face shall simultaneously display a left-
turn GREEN ARROW signal indication and a circular signal indication that is the same color as
the signal indication for the adjacent through lane on the same approach as the protected left turn.

C. A steady left-turn YELLOW ARROW signal indication shall be displayed following the left-turn
GREEN ARROW signal indication, unless the left-turn GREEN ARROW signal indication and the
CIRCULAR GREEN signal indication(s) for the adjacent through movement are being terminated
together. When the left-turn GREEN ARROW and CIRCULAR GREEN signal indications are
being terminated together, the required display following the left-turn GREEN ARROW signal
indication shall be either the display of a CIRCULAR YELLOW signal indication alone or the
simultaneous display of the CIRCULAR YELLOW and left-turn YELLOW ARROW signal
indications.

D. During the permissive left-turn movement, the shared signal face shall display only a CIRCULAR
GREEN signal indication.

Figure 4F-6. Typical Position and Arrangements of Shared Signal Faces for Protected/Permissive Mode Left Turns

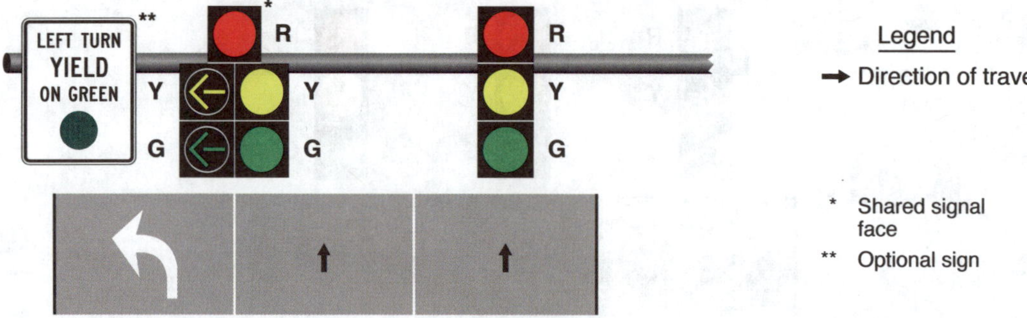

A – Typical position with a mandatory left turn lane

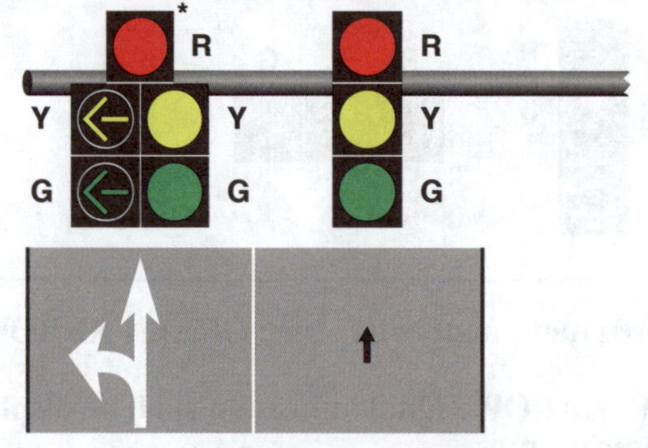

B – Typical position without a mandatory left turn lane

C – Typical arrangements

E. A protected/permissive shared signal face, regardless of where it is positioned and regardless of how many adjacent through signal faces are provided, shall always simultaneously display the same color of circular indication that the adjacent through signal face or faces display.

F. A supplementary sign shall not be required. If used, it shall be a LEFT TURN YIELD ON GREEN (symbolic circular green) (R10-12) sign (see Section 2B.59).

Section 4F.08 Signal Indications for Protected/Permissive Mode Left-Turn Movements in a Separate Signal Face

Standard:

01   A separate left-turn signal face shall not be used for an approach that does not include a mandatory left-turn lane.

02   If a separate left-turn signal face is being operated in a protected/permissive left-turn mode, a CIRCULAR GREEN signal indication shall not be used in that face.

03   If a separate left-turn signal face is being operated in a protected/permissive left-turn mode and a flashing left-turn yellow arrow signal indication is provided, it shall meet the following requirements (see Figure 4F-7):

A. It shall be capable of displaying the following signal indications: steady left-turn RED ARROW, steady left-turn YELLOW ARROW, flashing left-turn YELLOW ARROW, and left-turn GREEN ARROW. Only one of the four indications shall be displayed at any given time.

B. During the protected left-turn movement, a left-turn GREEN ARROW signal indication shall be displayed.

C. A steady left-turn YELLOW ARROW signal indication shall be displayed following the left-turn GREEN ARROW signal indication. It shall be permitted to display a steady left-turn RED ARROW signal indication immediately following the steady left-turn YELLOW ARROW signal indication to provide a red clearance interval.

D. During the permissive left-turn movement, a flashing left-turn YELLOW ARROW signal indication shall be displayed.

E. A steady left-turn YELLOW ARROW signal indication shall be displayed following the flashing left-turn YELLOW ARROW signal indication if the permissive left-turn movement is being terminated and the separate left-turn signal face will subsequently display a steady left-turn RED ARROW indication.

F. It shall be permitted to display a flashing left-turn YELLOW ARROW signal indication for a permissive left-turn movement while the signal faces for the adjacent through movement display steady CIRCULAR RED signal indications and the opposing left-turn signal faces display left-turn GREEN ARROW signal indications for a protected left-turn movement.

G. When a permissive left-turn movement is changing to a protected left-turn movement, a left-turn GREEN ARROW signal indication shall be displayed immediately upon the termination of the flashing left-turn YELLOW ARROW signal indication. A steady left-turn YELLOW ARROW signal indication shall not be displayed between the display of the flashing left-turn YELLOW ARROW signal indication and the display of the steady left-turn GREEN ARROW signal indication.

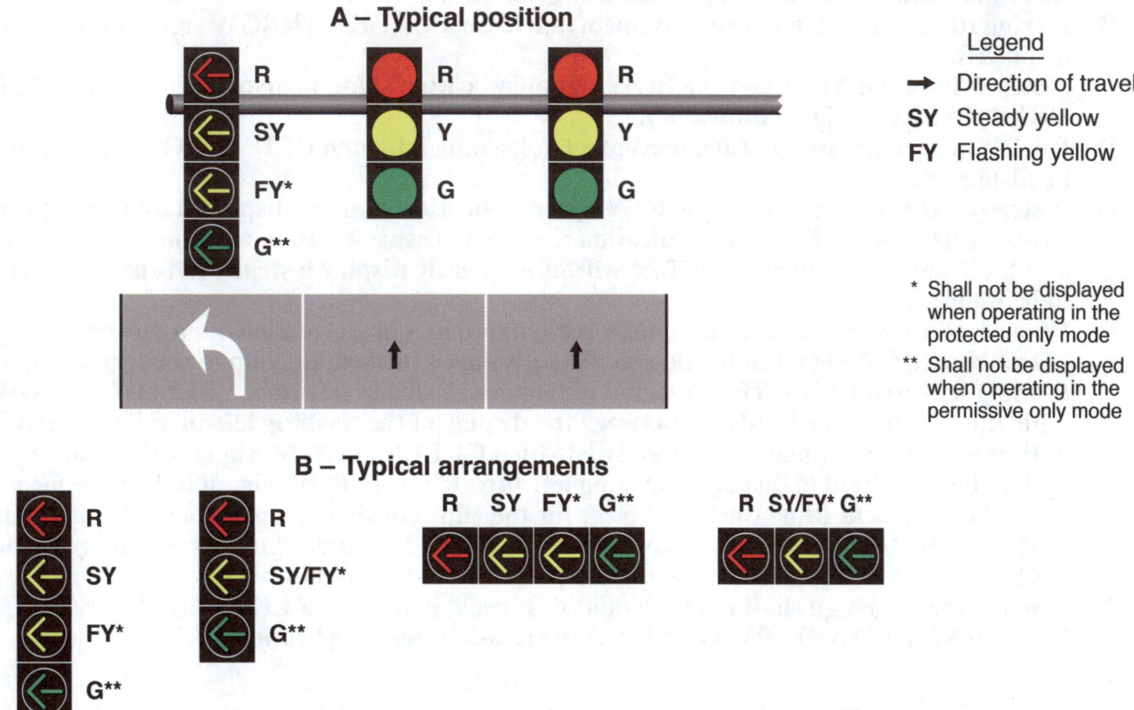

Figure 4F-7. Typical Position and Arrangements of Separate Signal Faces with Flashing Yellow Arrow for Protected/Permissive Mode and Variable Mode Left Turns

    H. **The display shall be either:**

      1. **A four-section signal face with the steady left-turn YELLOW ARROW signal indication being displayed in a different section than the flashing left-turn YELLOW ARROW signal indication, or**

      2. **A three-section signal face with the steady left-turn YELLOW ARROW signal indication and the flashing left-turn YELLOW ARROW signal indication being displayed in the same bimodal signal section.**

    I. **During steady mode (stop-and-go) operation where a four-section signal face is used, the signal section that displays the steady left-turn YELLOW ARROW signal indication during change intervals shall not be used to display the flashing left-turn YELLOW ARROW signal indication for permissive left turns.**

    J. **During flashing mode operation (see Chapter 4G) where a four-section signal face is used, the display of a flashing left-turn YELLOW ARROW signal indication shall be only from the signal section that displays a steady left-turn YELLOW ARROW signal indication during steady mode (stop-and-go) operation.**

Option:

04   A bimodal signal section (capable of displaying a GREEN ARROW for the protected left-turn movement and a flashing YELLOW ARROW for the permissive left-turn movement) along with a steady left-turn YELLOW ARROW signal indication and a steady left-turn RED ARROW signal indication may be used for a separate left-turn signal face and may be considered to be a four-section signal face that is compliant with Item H.1 of Paragraph 3 of this Section.

05   A separate left-turn signal face with a flashing left-turn RED ARROW signal indication during the permissive left-turn movement may be used for unusual geometric conditions, such as wide medians with offset left-turn lanes, but only when an engineering study determines that each and every vehicle must successively come to a full stop before making a permissive left turn.

Standard:

06   **If a separate left-turn signal face is being operated in a protected/permissive left-turn mode and a flashing left-turn RED arrow signal indication is provided, it shall meet the following requirements (see Figure 4F-3):**

    A. **It shall be capable of displaying the following signal indications: steady or flashing left-turn RED ARROW, steady left-turn YELLOW ARROW, and left-turn GREEN ARROW. Only one of the three indications shall be displayed at any given time.**

    B. **During the protected left-turn movement, a left-turn GREEN ARROW signal indication shall be displayed.**

    C. **A steady left-turn YELLOW ARROW signal indication shall be displayed following the left-turn GREEN ARROW signal indication.**

    D. **During the permissive left-turn movement, a flashing left-turn RED ARROW signal indication shall be displayed.**

    E. **A steady left-turn YELLOW ARROW signal indication shall be displayed following the flashing left-turn RED ARROW signal indication if the permissive left-turn movement is being terminated and the separate left-turn signal face will subsequently display a steady left-turn RED ARROW indication.**

    F. **When a permissive left-turn movement is changing to a protected left-turn movement, a left-turn GREEN ARROW signal indication shall be displayed immediately upon the termination of the flashing left-turn RED ARROW signal indication. A steady left-turn YELLOW ARROW signal indication shall not be displayed between the display of the flashing left-turn RED ARROW signal indication and the display of the steady left-turn GREEN ARROW signal indication.**

    G. **It shall be permitted to display a flashing left-turn RED ARROW signal indication for a permissive left-turn movement while the signal faces for the adjacent through movement display steady CIRCULAR RED signal indications and the opposing left-turn signal faces display left-turn GREEN ARROW signal indications for a protected left-turn movement.**

    H. **A supplementary sign shall not be required. If used, it shall be a LEFT TURN YIELD ON FLASHING RED ARROW AFTER STOP (R10-27) sign (see Section 2B.59).**

Option:

07   The requirements of Item A in Paragraph 6 of this Section may be met by a vertically-arranged signal face with a horizontal cluster of two left-turn RED ARROW signal indications, the left-most of which displays a steady indication and the right-most of which displays a flashing indication (see Figure 4F-3).

Section 4F.09 Signal Indications for Right-Turn Movements – General

Support:

01   In Sections 4F.10 through 4F.15, provisions applicable to right-turn movements and right-turn lanes are also applicable to signal indications for U-turns to the right that are provided at locations where right turns are prohibited or not geometrically possible.

02   Right-turning traffic is controlled by one of four modes as follows:

A. Permissive Only Mode—turns made on a CIRCULAR GREEN signal indication, a flashing right-turn YELLOW ARROW signal indication, or a flashing right-turn RED ARROW signal indication after yielding to pedestrians, if any.

B. Protected Only Mode—turns made only when a right-turn GREEN ARROW signal indication is displayed.

C. Protected/Permissive Mode—both modes occur on an approach during the same cycle.

D. Variable Right-Turn Mode—the operating mode changes among the protected only mode and/or the protected/permissive mode and/or the permissive only mode during different periods of the day or as traffic conditions change.

Standard:

03   **During a permissive right-turn movement, the signal faces, if any, that exclusively control U-turn traffic that conflicts with the permissive right-turn movement (see Item H.1 in Paragraph 3 in Section 4F.01) shall simultaneously display steady U-turn RED ARROW signal indications. If pedestrians crossing the lane or lanes used by the permissive right-turn movement to depart the intersection are controlled by pedestrian signal heads, the signal indications displayed by those pedestrian signal heads shall not be limited to any particular display during the permissive right-turn movement.**

04   **During a protected right-turn movement, a GREEN ARROW or a YELLOW ARROW signal indication shall not simultaneously be displayed to left-turning traffic on the opposing approach, except where a separate departure lane is available for each left-turn and right-turn lane with moving traffic and pavement markings or raised channelization clearly indicate which departure lane to use (see Item H.1 in Paragraph 3 in Section 4F.01). Signal faces, if any, that exclusively control U-turn traffic that conflicts with the protected right-turn movement shall simultaneously display steady RED ARROW signal indications. If pedestrians crossing the lane or lanes used by the protected right-turn movement to depart the intersection are controlled by pedestrian signal heads, the pedestrian signal heads shall display a steady UPRAISED HAND (symbolizing DONT WALK) signal indication during the protected right-turn movement.**

05   **If a combined right-turn/through lane exists on an approach, a right-turn GREEN ARROW or right-turn YELLOW ARROW signal indication or a flashing right-turn RED ARROW signal indication shall not be displayed to the approach simultaneously with a CIRCULAR RED signal indication for the through movement, and a right-turn RED ARROW signal indication shall not be displayed to the approach simultaneously with a CIRCULAR GREEN or CIRCULAR YELLOW signal indication for the through movement.**

06   **If the operating mode changes among the protected only mode and/or the protected/permissive mode and/or the permissive only mode during different periods of the day or as traffic conditions change, the requirements in Sections 4F.10 through 4F.15 that are appropriate to that mode of operation shall be met, subject to the following:**

A. **The CIRCULAR GREEN and CIRCULAR YELLOW signal indications shall not be displayed when operating in the protected only mode.**

B. **The right-turn GREEN ARROW and right-turn YELLOW ARROW signal indications shall not be displayed when operating in the permissive only mode.**

Option:

07   When variable right-turn mode phasing is used for an approach that has a combined right-turn/straight-through lane and a flashing yellow arrow is used as the permissive turn display, a five-section shared right-turn signal face containing both circular and arrow indications may be used in combination with one or more separate right-turn signal faces for the mandatory right-turn lane(s), if any are present, on the same approach. The steady right-turn YELLOW ARROW signal indication and the flashing right-turn YELLOW ARROW signal indication may be displayed in the same section of the five-section shared right-turn signal face.

08   Additional static signs or changeable message signs may be used to meet the requirements for the variable right-turn mode or to inform drivers that right-turn green arrows will not be available during certain times of the day.

Support:

09   Sections 4F.10 through 4F.15 describe the use of the following two types of signal faces for controlling right-turn movements:

    A. Shared signal face – This type of signal face controls both the right-turn movement and the adjacent movement (usually the through movement) and can serve as one of the two required primary signal faces for the adjacent movement. A shared signal face always displays the same color of circular indication that is displayed by the signal face or faces for the adjacent movement.

    B. Separate right-turn signal face – This type of signal face controls only the right-turn movement and cannot serve as one of the two required primary signal faces for the adjacent movement (usually the through movement) because it displays signal indications that are applicable only to the right-turn movement. If a separate right-turn signal face is mounted overhead at the intersection, it is positioned over the extension of the mandatory right-turn lane. In a separate right-turn signal face, a flashing right-turn YELLOW ARROW signal indication or a flashing right-turn RED ARROW signal indication is used to control permissive right-turn movements.

10   Section 4D.07 contains provisions regarding the lateral positioning of signal faces that control right-turn movements.

11   It is not necessary that the same mode of right-turn operation or same type of right-turn signal face be used on every approach to a signalized location. Selecting different modes and types of right-turn signal faces for the various approaches to the same signalized location is acceptable.

Option:

12   A signal face that is shared by left-turning and right-turning traffic may be provided for a combined left-turn/right-turn lane on an approach that has no through traffic (see Section 4F.16).

Section 4F.10 Signal Indications for Permissive Only Mode Right-Turn Movements in a Shared Signal Face

Standard:

01   If a shared signal face is provided for a permissive only mode right turn, it shall meet the following requirements (see Figure 4F-8):

    A. It shall be capable of displaying the following signal indications: steady CIRCULAR RED, steady CIRCULAR YELLOW, and CIRCULAR GREEN. Only one of the three indications shall be displayed at any given time.

    B. During the permissive right-turn movement, a CIRCULAR GREEN signal indication shall be displayed.

    C. A permissive only shared signal face, regardless of where it is positioned and regardless of how many adjacent through signal faces are provided, shall always simultaneously display the same color of circular indication that the adjacent through signal face or faces display.

    D. If the permissive only mode is not the only right-turn mode used for the approach, the signal face shall be the same shared signal face that is used for the protected/permissive mode (see Section 4F.14) except that the right-turn GREEN ARROW and right-turn YELLOW ARROW signal indications shall not be displayed when operating in the permissive only mode.

Section 4F.11 Signal Indications for Permissive Only Mode Right-Turn Movements in a Separate Signal Face

Standard:

01   A separate right-turn signal face shall not be used for an approach that does not include a mandatory right-turn lane.

02   If a separate right-turn signal face is being operated in a permissive only right-turn mode, a CIRCULAR GREEN signal indication shall not be used in that face.

03   If a separate right-turn signal face is being operated in a permissive only right-turn mode and a flashing right-turn yellow arrow signal indication is provided, it shall meet the following requirements (see Figure 4F-9):

    A. It shall be capable of displaying one of the following sets of signal indications:

        1. Steady right-turn RED ARROW, steady right-turn YELLOW ARROW, and flashing right-turn YELLOW ARROW. Only one of the three indications shall be displayed at any given time.

        2. Steady CIRCULAR RED, steady right-turn YELLOW ARROW, and flashing right-turn YELLOW ARROW. Only one of the three indications shall be displayed at any given time. If the CIRCULAR RED signal indication is sometimes displayed when the signal faces for the adjacent through lane(s) are not displaying a CIRCULAR RED signal indication, a RIGHT TURN SIGNAL (R10-10R) sign (see Section 2B.59) shall be used unless the CIRCULAR RED signal indication in the separate right-turn signal face is shielded, hooded, louvered, positioned, or designed such that it is not readily visible to drivers in the through lane(s).

Figure 4F-8. Typical Positions and Arrangements of Shared Signal Faces for Permissive Only Mode Right Turns

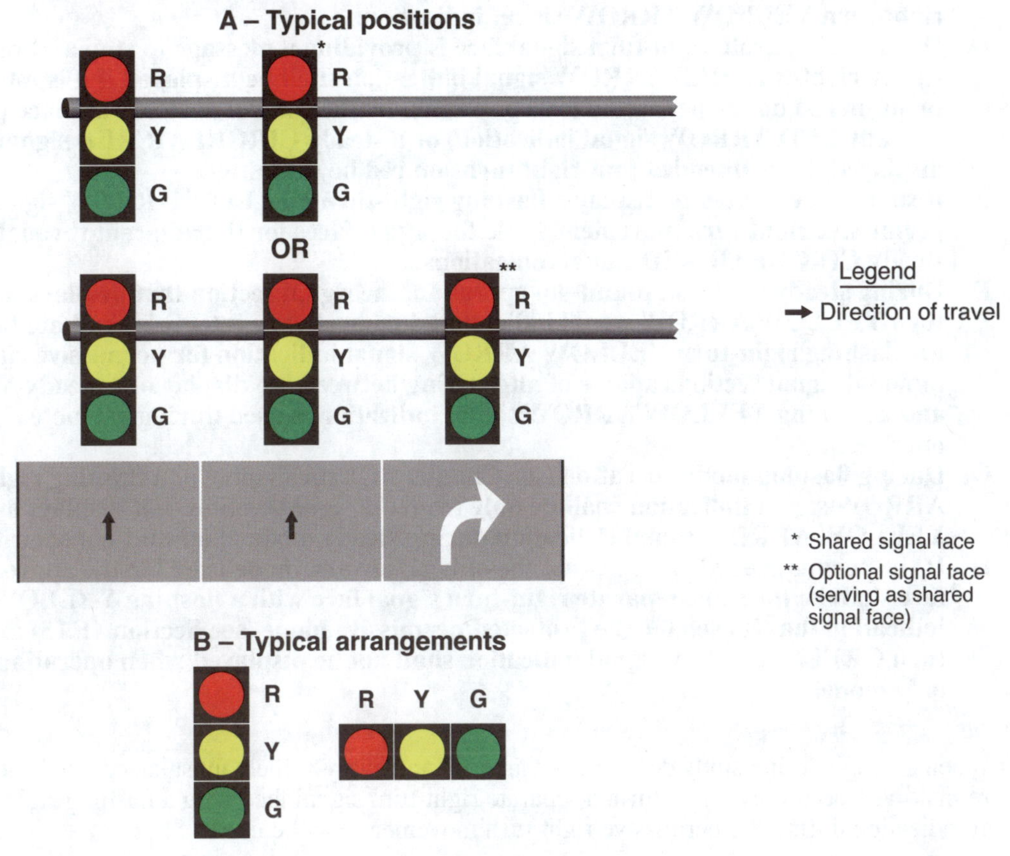

Figure 4F-9. Typical Position and Arrangements of Separate Signal Faces with Flashing Yellow Arrow for Permissive Only Mode Right Turns

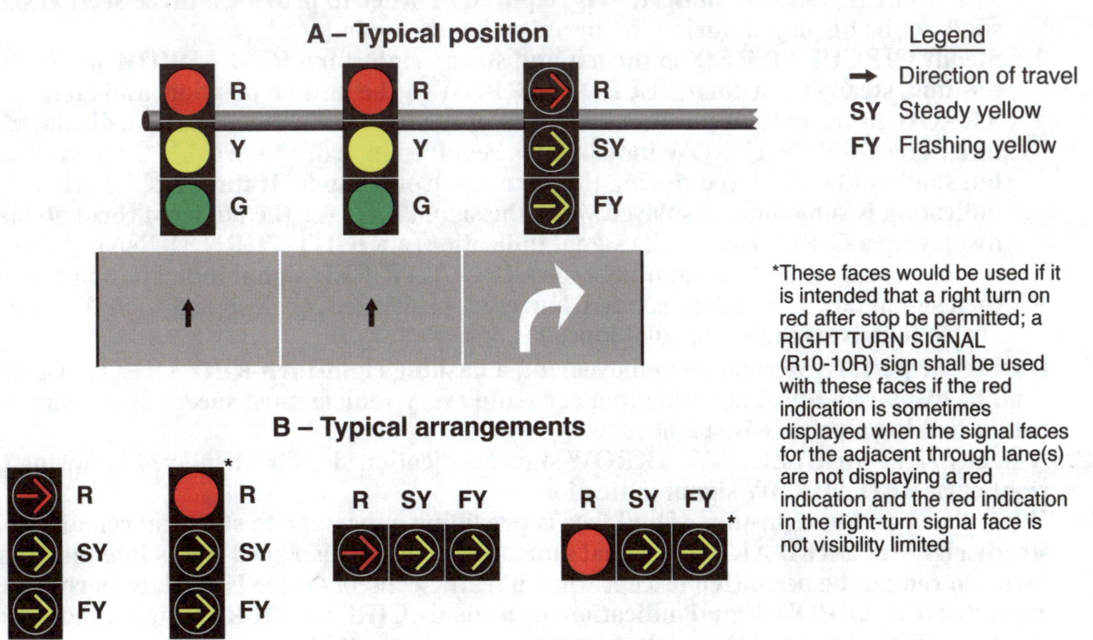

B. During the permissive right-turn movement, a flashing right-turn YELLOW ARROW signal indication shall be displayed.

C. A steady right-turn YELLOW ARROW signal indication shall be displayed following the flashing right-turn YELLOW ARROW signal indication.

D. When the separate right-turn signal face is providing a message to stop and remain stopped, a steady right-turn RED ARROW signal indication shall be displayed if it is intended that right turns on red not be permitted (except when a traffic control device is in place permitting a turn on a steady RED ARROW signal indication) or a steady CIRCULAR RED signal indication shall be displayed if it is intended that right turns on red be permitted.

E. It shall be permitted to display a flashing right-turn YELLOW ARROW signal indication for a permissive right-turn movement while the signal faces for the adjacent through movement display steady CIRCULAR RED signal indications.

F. During steady mode (stop-and-go) operation, the signal section that displays the steady right-turn YELLOW ARROW signal indication during change intervals shall not be used to display the flashing right-turn YELLOW ARROW signal indication for permissive right turns unless a bimodal signal section capable of alternating between the display of a steady YELLOW ARROW and a flashing YELLOW ARROW signal indication is used during variable right-turn mode operation.

G. During flashing mode operation (see Chapter 4G), the display of a flashing right-turn YELLOW ARROW signal indication shall be only from the signal section that displays a steady right-turn YELLOW ARROW signal indication during steady mode (stop-and-go) operation.

H. If the permissive only mode is not the only right-turn mode used for the approach, the signal face shall be the same separate right-turn signal face with a flashing YELLOW ARROW signal indication that is used for the protected/permissive mode (see Section 4F.15) except that the right-turn GREEN ARROW signal indication shall not be displayed when operating in the permissive only mode.

Option:

04   When an engineering study determines that each and every vehicle must successively come to a full stop before making a permissive right turn, a separate right-turn signal face with a flashing right-turn RED ARROW signal indication during the permissive right-turn movement may be used.

Standard:

05   If a separate right-turn signal face is being operated in a permissive only right-turn mode and a flashing right-turn RED arrow signal indication is provided, it shall meet the following requirements (see Figure 4F-10):

A. It shall be capable of displaying one of the following sets of signal indications:

1. Steady or flashing right-turn RED ARROW, steady right-turn YELLOW ARROW, and right-turn GREEN ARROW. Only one of the three indications shall be displayed at any given time. The GREEN ARROW indication is required in order to provide a three-section signal face, but shall not be displayed during the permissive only mode.

2. Steady CIRCULAR RED on the left and steady right-turn RED ARROW on the right of the top position, steady right-turn YELLOW ARROW in the middle position, and right-turn GREEN ARROW in the bottom position. Only one of the four indications shall be displayed at any given time. The GREEN ARROW indication is required in order to provide three vertical positions, but shall not be displayed during the permissive only mode. If the CIRCULAR RED signal indication is sometimes displayed when the signal faces for the adjacent through lane(s) are not displaying a CIRCULAR RED signal indication, a RIGHT TURN SIGNAL (R10-10R) sign (see Section 2B.59) shall be used unless the CIRCULAR RED signal indication in the separate right-turn signal face is shielded, hooded, louvered, positioned, or designed such that it is not readily visible to drivers in the through lane(s).

B. During the permissive right-turn movement, a flashing right-turn RED ARROW signal indication shall be displayed, thus indicating that each and every vehicle must successively come to a full stop before making a permissive right turn.

C. A steady right-turn YELLOW ARROW signal indication shall be displayed following the flashing right-turn RED ARROW signal indication.

D. When the separate right-turn signal face is providing a message to stop and remain stopped, a steady right-turn RED ARROW signal indication shall be displayed if it is intended that right turns on red not be permitted (except when a traffic control device is in place permitting a turn on a steady RED ARROW signal indication) or a steady CIRCULAR RED signal indication shall be displayed if it is intended that right turns on red be permitted.

Figure 4F-10. Typical Position and Arrangements of Separate Signal Faces with Flashing Red Arrow for Permissive Only Mode and Protected/Permissive Mode Right Turns

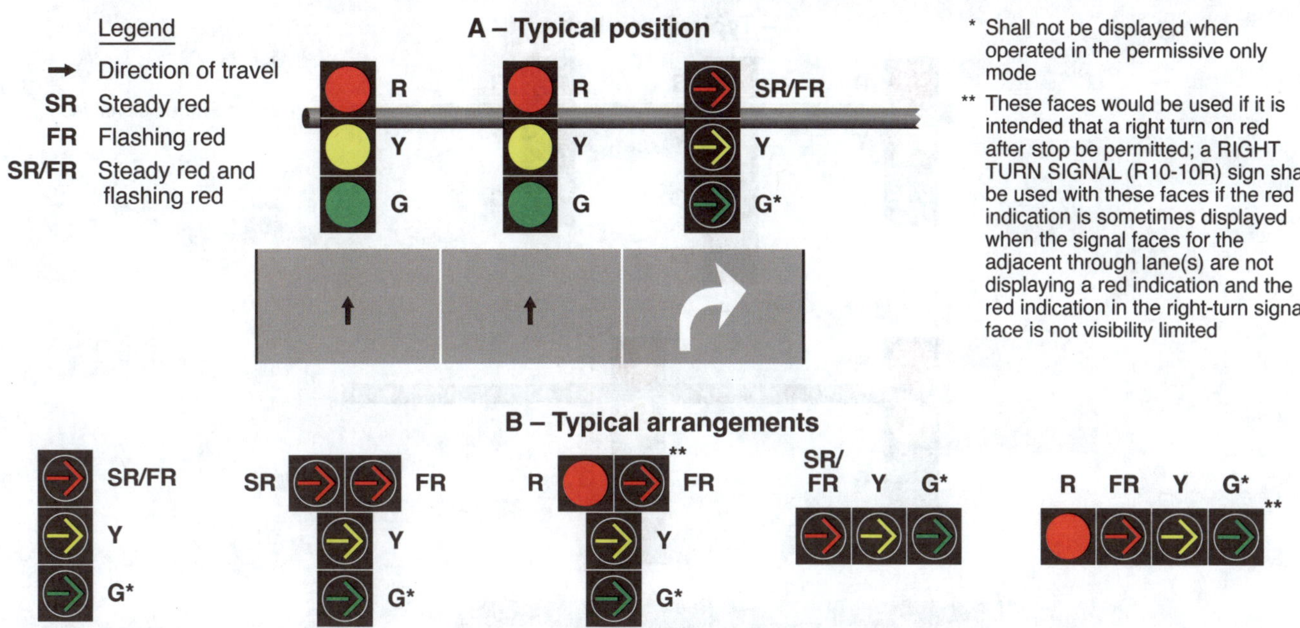

E. **The display of a flashing right-turn RED ARROW signal indication for a permissive right-turn movement while the signal faces for the adjacent through movement display steady CIRCULAR RED signal indications and the opposing left-turn signal faces display left-turn GREEN ARROW signal indications for a protected left-turn movement shall be permitted.**

F. **A supplementary sign shall not be required. If used, it shall be a RIGHT TURN YIELD ON FLASHING RED ARROW AFTER STOP (R10-27) sign (see Section 2B.59).**

Option:

06   The requirements of Item A.1 in Paragraph 5 of this Section may be met by a vertically-arranged signal face with a horizontal cluster of two right-turn RED ARROW signal indications, the left-most of which displays a steady indication and the right-most of which displays a flashing indication (see Figure 4F-10).

Section 4F.12 Signal Indications for Protected Only Mode Right-Turn Movements in a Shared Signal Face

Standard:

01   **A shared signal face shall not be used for protected only mode right turns unless the CIRCULAR GREEN and right-turn GREEN ARROW signal indications always begin and terminate together. If a shared signal face is provided for a protected only right turn, it shall meet the following requirements (see Figure 4F-11):**

A. **It shall be capable of displaying the following signal indications: steady CIRCULAR RED, steady CIRCULAR YELLOW, CIRCULAR GREEN, and right-turn GREEN ARROW. Only one of the three colors shall be displayed at any given time.**

B. **During the protected right-turn movement, the shared signal face shall simultaneously display both a CIRCULAR GREEN signal indication and a right-turn GREEN ARROW signal indication.**

C. **The shared signal face shall always simultaneously display the same color of circular indication that the adjacent through signal face or faces display.**

D. **If the protected only mode is not the only right-turn mode used for the approach, the signal face shall be the same shared signal face that is used for the protected/permissive mode (see Section 4F.15).**

Option:

02   A straight-through GREEN ARROW signal indication may be used instead of the CIRCULAR GREEN signal indication in Items A and B in Paragraph 1 of this Section on an approach where a straight-through GREEN ARROW signal indication is also used instead of a CIRCULAR GREEN signal indication in the other signal face(s) for through traffic.

Figure 4F-11. Typical Positions and Arrangements of Shared Signal Faces for Protected Only Mode Right Turns

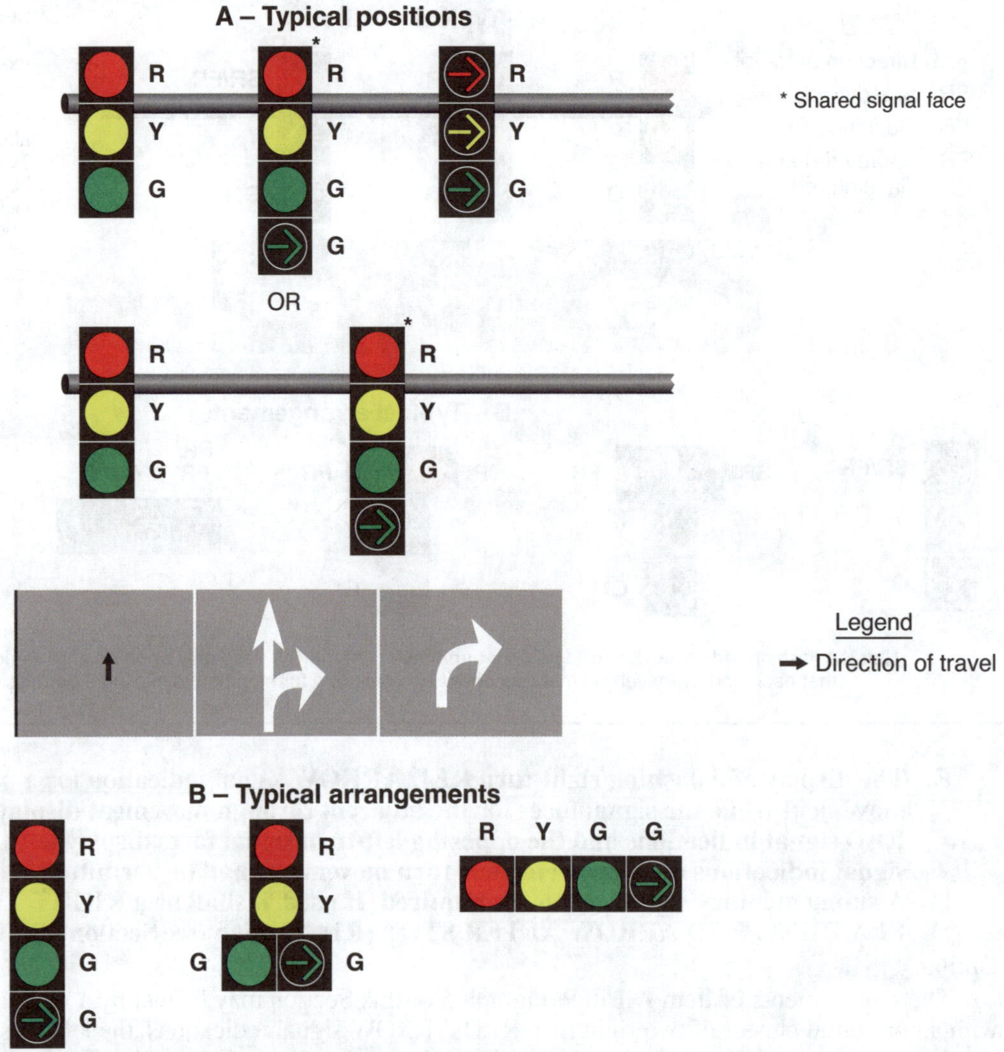

Section 4F.13 Signal Indications for Protected Only Mode Right-Turn Movements in a Separate Signal Face

Standard:

01   **A separate right-turn signal face shall not be used for an approach that does not include a mandatory right-turn lane.**

02   **If a separate right-turn signal face is provided for a protected only mode right turn, it shall meet the following requirements (see Figure 4F-12):**

    A. It shall be capable of displaying one of the following sets of signal indications:

        1. Steady right-turn RED ARROW, steady right-turn YELLOW ARROW, and right-turn GREEN ARROW. Only one of the three indications shall be displayed at any given time.

        2. Steady CIRCULAR RED, steady right-turn YELLOW ARROW, and right-turn GREEN ARROW. Only one of three indications shall be displayed at any given time. If the CIRCULAR RED signal indication is sometimes displayed when the signal faces for the adjacent through lane(s) are not displaying a CIRCULAR RED signal indication, a RIGHT TURN SIGNAL (R10-10R) sign (see Section 2B.59) shall be used unless the CIRCULAR RED signal indication is shielded, hooded, louvered, positioned, or designed such that it is not readily visible to drivers in the through lane(s).

Figure 4F-12. Typical Position and Arrangements of Separate Signal Faces for Protected Only Mode Right Turns

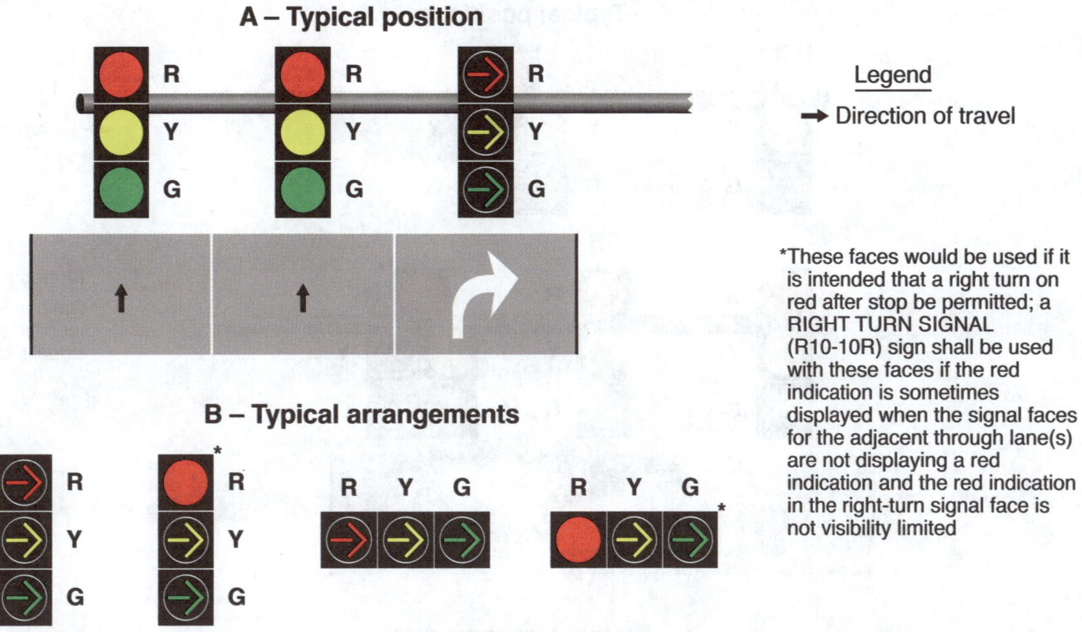

B. During the protected right-turn movement, a right-turn GREEN ARROW signal indication shall be displayed.

C. A steady right-turn YELLOW ARROW signal indication shall be displayed following the right-turn GREEN ARROW signal indication.

D. When the separate signal face is providing a message to stop and remain stopped, a steady right-turn RED ARROW signal indication shall be displayed if it is intended that right turns on red not be permitted (except when a traffic control device is in place permitting a turn on a steady RED ARROW signal indication) or a steady CIRCULAR RED signal indication shall be displayed if it is intended that right turns on red be permitted.

E. If the protected only mode is not the only right-turn mode used for the approach, the signal face shall be the same separate right-turn signal face that is used for the protected/permissive mode (see Section 4F.15) except that a flashing right-turn YELLOW ARROW or flashing right-turn RED ARROW signal indication shall not be displayed when operating in the protected only mode.

Section 4F.14 Signal Indications for Protected/Permissive Mode Right-Turn Movements in a Shared Signal Face

Standard:

01 If a shared signal face is provided for a protected/permissive mode right turn, it shall meet the following requirements (see Figure 4F-13):

A. It shall be capable of displaying the following signal indications: steady CIRCULAR RED, steady CIRCULAR YELLOW, CIRCULAR green, steady right-turn YELLOW ARROW, and right-turn GREEN ARROW. Only one of the three circular indications shall be displayed at any given time. Only one of the two arrow indications shall be displayed at any given time. If the right-turn GREEN ARROW signal indication and the CIRCULAR GREEN signal indication(s) for the adjacent through movement are always terminated together, the steady right-turn YELLOW ARROW signal indication shall not be required.

B. During the protected right-turn movement, the shared signal face shall simultaneously display a right-turn GREEN ARROW signal indication and a circular signal indication that is the same color as the signal indication for the adjacent through lane on the same approach as the protected right turn.

Figure 4F-13. Typical Positions and Arrangements of Shared Signal Faces for Protected/Permissive Mode Right Turns

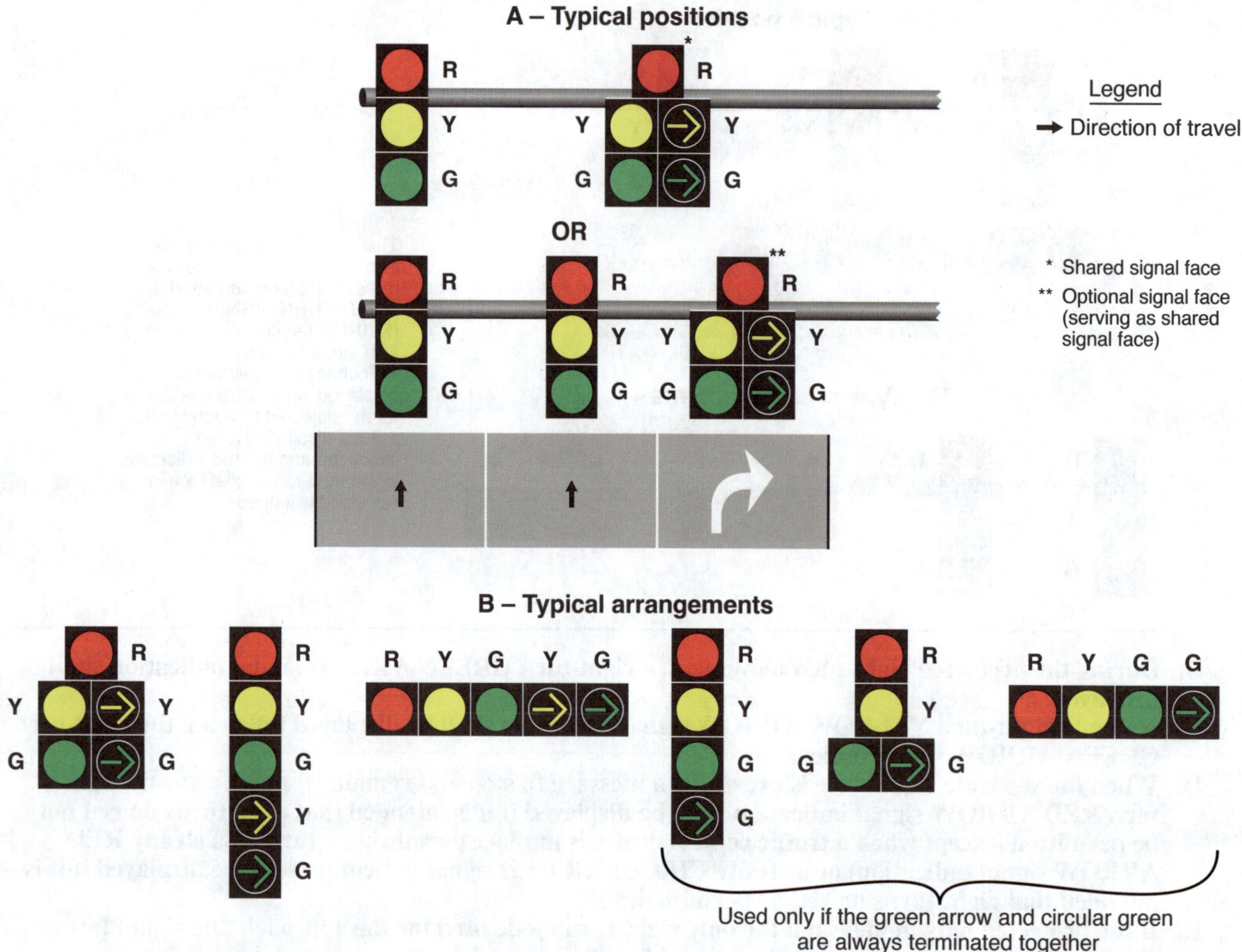

C. A steady right-turn YELLOW ARROW signal indication shall be displayed following the right-turn GREEN ARROW signal indication, unless the right-turn GREEN ARROW signal indication and the CIRCULAR GREEN signal indication(s) for the adjacent through movement are being terminated together. When the right-turn GREEN ARROW and CIRCULAR GREEN signal indications are being terminated together, the required display following the right-turn GREEN ARROW signal indication shall be either the display of a CIRCULAR YELLOW signal indication alone or the simultaneous display of the CIRCULAR YELLOW and right-turn YELLOW ARROW signal indications.

D. During the permissive right-turn movement, the shared signal face shall display only a CIRCULAR GREEN signal indication.

E. A protected/permissive shared signal face, regardless of where it is positioned and regardless of how many adjacent through signal faces are provided, shall always simultaneously display the same color of circular indication that the adjacent through signal face or faces display.

Section 4F.15 Signal Indications for Protected/Permissive Mode Right-Turn Movements in a Separate Signal Face

Standard:

01   A separate right-turn signal face shall not be used for an approach that does not include a mandatory right-turn lane.

02   If a separate right-turn signal face is being operated in a protected/permissive right-turn mode, a CIRCULAR GREEN signal indication shall not be used in that face.

03 **If a separate right-turn signal face is being operated in a protected/permissive right-turn mode and a flashing right-turn yellow arrow signal indication is provided, it shall meet the following requirements (see Figure 4F-14):**

A. **It shall be capable of displaying one of the following sets of signal indications:**

1. **Steady right-turn RED ARROW, steady right-turn YELLOW ARROW, flashing right-turn YELLOW ARROW, and right-turn GREEN ARROW. Only one of the four indications shall be displayed at any given time.**

2. **Steady CIRCULAR RED, steady right-turn YELLOW ARROW, flashing right-turn YELLOW ARROW, and right-turn GREEN ARROW. Only one of the four indications shall be displayed at any given time. If the CIRCULAR RED signal indication is sometimes displayed when the signal faces for the adjacent through lane(s) are not displaying a CIRCULAR RED signal indication, a RIGHT TURN SIGNAL (R10-10R) sign (see Section 2B.59) shall be used unless the CIRCULAR RED signal indication in the separate right-turn signal face is shielded, hooded, louvered, positioned, or designed such that it is not readily visible to drivers in the through lane(s).**

B. **During the protected right-turn movement, a right-turn GREEN ARROW signal indication shall be displayed.**

C. **A steady right-turn YELLOW ARROW signal indication shall be displayed following the right-turn GREEN ARROW signal indication. It shall be permitted to display a steady right-turn RED ARROW signal indication immediately following the steady right-turn YELLOW ARROW signal indication to provide a red clearance interval.**

D. **During the permissive right-turn movement, a flashing right-turn YELLOW ARROW signal indication shall be displayed.**

Figure 4F-14. Typical Position and Arrangements of Separate Signal Faces with Flashing Yellow Arrow for Protected/Permissive Mode and Variable Mode Right Turns

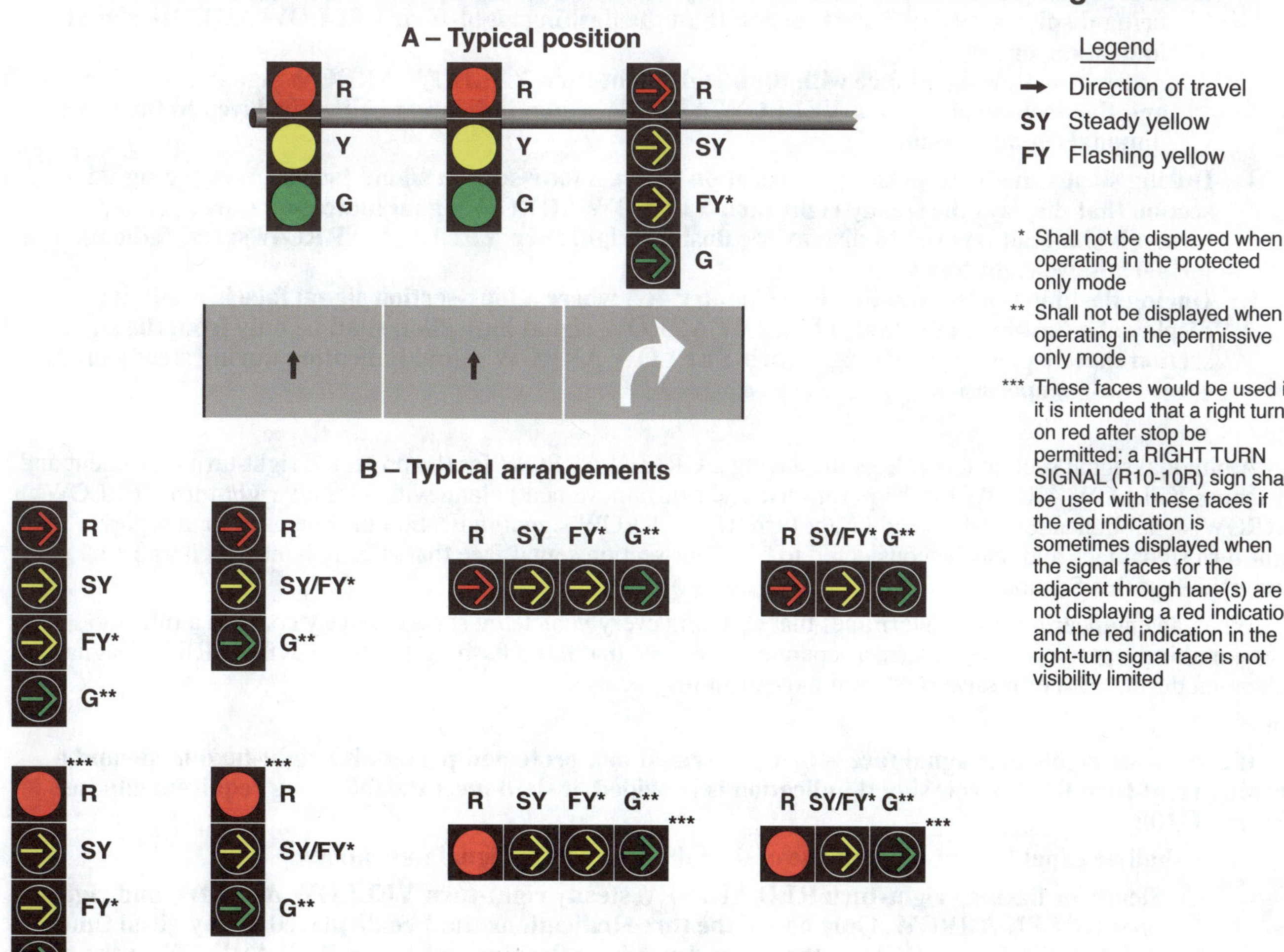

E. A steady right-turn YELLOW ARROW signal indication shall be displayed following the flashing right-turn YELLOW ARROW signal indication if the permissive right-turn movement is being terminated and the separate right-turn signal face will subsequently display a steady red indication.

F. When a permissive right-turn movement is changing to a protected right-turn movement:

1. If a permissive left-turn movement from the opposing approach is being terminated simultaneously with the termination of the permissive right-turn movement, a steady right-turn YELLOW ARROW signal indication shall be displayed following the flashing right-turn YELLOW ARROW signal indication. To provide a red clearance interval, it shall be permitted to display a steady right-turn RED ARROW signal indication immediately following the steady right-turn YELLOW ARROW signal indication.

2. If a permissive left-turn movement from the opposing approach that is being terminated simultaneously with the termination of the permissive right-turn movement is not present, a right-turn GREEN ARROW signal indication shall be displayed immediately upon the termination of the flashing right-turn YELLOW ARROW signal indication. In this situation, a steady right-turn YELLOW ARROW signal indication shall not be displayed between the display of the flashing right-turn YELLOW ARROW signal indication and the display of the steady right-turn GREEN ARROW signal indication.

G. When the separate right-turn signal face is providing a message to stop and remain stopped, a steady right-turn RED ARROW signal indication shall be displayed if it is intended that right turns on red not be permitted (except when a traffic control device is in place permitting a turn on a steady RED ARROW signal indication) or a steady CIRCULAR RED signal indication shall be displayed if it is intended that right turns on red be permitted.

H. It shall be permitted to display a flashing right-turn YELLOW ARROW signal indication for a permissive right-turn movement while the signal faces for the adjacent through movement display steady CIRCULAR RED signal indications.

I. The display shall be either:

1. A four-section signal face with the steady right-turn YELLOW ARROW signal indication being displayed in a different section than the flashing right-turn YELLOW ARROW signal indication, or

2. A three-section signal face with the steady right-turn YELLOW ARROW signal indication and the flashing right-turn YELLOW ARROW signal indication being displayed in the same bimodal signal section.

J. During steady mode (stop-and-go) operation where a four-section signal face is used, the signal section that displays the steady right-turn YELLOW ARROW signal indication during change intervals shall not be used to display the flashing right-turn YELLOW ARROW signal indication for permissive right turns.

K. During flashing mode operation (see Chapter 4G) where a four-section signal face is used, the display of a flashing right-turn YELLOW ARROW signal indication shall be only from the signal section that displays a steady right-turn YELLOW ARROW signal indication during steady mode (stop-and-go) operation.

Option:

04   A bimodal signal section (capable of displaying a GREEN ARROW for the protected right-turn movement and a flashing YELLOW ARROW for the permissive right-turn movement) along with a steady right-turn YELLOW ARROW signal indication and a steady right-turn RED ARROW signal indication may be used for a separate right-turn signal face and may be considered to be a four-section signal face that is compliant with Item I.1 of Paragraph 3 of this Section.

05   When an engineering study determines that each and every vehicle must successively come to a full stop before making a permissive right turn, a separate signal face that has a flashing right-turn RED ARROW signal indication during the permissive right-turn movement may be used.

Standard:

06   If a separate right-turn signal face is being operated in a protected/permissive right-turn mode and a flashing right-turn RED arrow signal indication is provided, it shall meet the following requirements (see Figure 4F-10):

A. It shall be capable of displaying one of the following sets of signal indications:

1. Steady or flashing right-turn RED ARROW, steady right-turn YELLOW ARROW, and right-turn GREEN ARROW. Only one of the three indications shall be displayed at any given time.

2. Steady CIRCULAR RED on the left and steady or flashing right-turn RED ARROW on the right of the top position, steady right-turn YELLOW ARROW in the middle position, and right-turn

> **GREEN ARROW in the bottom position. Only one of the four indications shall be displayed at any given time. If the CIRCULAR RED signal indication is sometimes displayed when the signal faces for the adjacent through lane(s) are not displaying a CIRCULAR RED signal indication, a RIGHT TURN SIGNAL (R10-10R) sign (see Section 2B.59) shall be used unless the CIRCULAR RED signal indication in the separate right-turn signal face is shielded, hooded, louvered, positioned, or designed such that it is not readily visible to drivers in the through lane(s).**

B. During the protected right-turn movement, a right-turn GREEN ARROW signal indication shall be displayed.

C. A steady right-turn YELLOW ARROW signal indication shall be displayed following the right-turn GREEN ARROW signal indication.

D. During the permissive right-turn movement, the separate right-turn signal face shall display a flashing right-turn RED ARROW signal indication.

E. A steady right-turn YELLOW ARROW signal indication shall be displayed following the flashing right-turn RED ARROW signal indication if the permissive right-turn movement is being terminated and the separate right-turn signal face will subsequently display a steady red indication.

F. When a permissive right-turn movement is changing to a protected right-turn movement, a right-turn GREEN ARROW signal indication shall be displayed immediately upon the termination of the flashing right-turn RED ARROW signal indication. A steady right-turn YELLOW ARROW signal indication shall not be displayed between the display of the flashing right-turn RED ARROW signal indication and the display of the steady right-turn GREEN ARROW signal indication.

G. When the separate right-turn signal face is providing a message to stop and remain stopped, a steady right-turn RED ARROW signal indication shall be displayed if it is intended that right turns on red not be permitted (except when a traffic control device is in place permitting a turn on a steady RED ARROW signal indication) or a steady CIRCULAR RED signal indication shall be displayed if it is intended that right turns on red be permitted.

H. It shall be permitted to display a flashing right-turn RED ARROW signal indication for a permissive right-turn movement while the signal faces for the adjacent through movement display steady CIRCULAR RED signal indications and the opposing left-turn signal faces display left-turn GREEN ARROW signal indications for a protected left-turn movement.

I. A supplementary sign shall not be required. If used, it shall be a RIGHT TURN YIELD ON FLASHING RED ARROW AFTER STOP (R10-27) sign (see Section 2B.59).

Option:

07 The requirements of Item A.1 in Paragraph 6 of this Section may be met by a vertically-arranged signal face with a horizontal cluster of two right-turn RED ARROW signal indications, the left-most of which displays a steady indication and the right-most of which displays a flashing indication (see Figure 4F-10).

Section 4F.16 Signal Indications for Approaches with No Through Movement

Support:

01 The provisions of this section apply only to approaches where no through movement exists, such as the stem of a T-intersection or where the opposite approach is a one-way roadway in the opposing direction.

Standard:

02 **Except for single-lane approaches, a minimum of two primary signal faces shall be provided for the signalized turning movement that is considered to be the major movement from the approach (see Section 4D.05).**

Option:

03 The required two primary signal faces and any supplemental primary signal faces may continuously display a steady CIRCULAR RED signal indication while steady or flashing YELLOW and steady GREEN ARROW signal indications are displayed during times when the traffic control signal is being operated in the steady (stop-and-go) mode. The continuous display of steady CIRCULAR RED is intended to reinforce that there is no through movement for safety-critical locations.

Standard:

04 **CIRCULAR GREEN and CIRCULAR YELLOW signal indications shall not be displayed to an approach with no through movement if:**

A. The posted or statutory speed limit on the approach is 35 mph or higher,

B. The one-way roadway that opposes the approach is an exit ramp from a freeway or expressway, or

C. The one-way roadway that opposes the approach has a posted or statutory speed limit of 35 mph or higher.

Support:

05   A lane from which left-turn and right-turn movements can both be made is sometimes provided on an approach that has no through movement, either as the only approach lane or as one of several approach lanes.

Option:

06   If all of the lanes on the approach are designated as mandatory turn lanes and no lane is designated as a combined left-turn/right-turn lane, the left-turn and right-turn movements may start and terminate independently, and the left-turn and right-turn movements each may be operated in one or more of the modes of operation as described in Sections 4F.02 through 4F.15.

Standard:

07   **When a combined left-turn/right-turn lane exists on an approach, the left-turn and right-turn movements shall start and terminate simultaneously and the red signal indication used in each of the signal faces on the approach shall be a CIRCULAR RED.**

Support:

08   This requirement for the use of CIRCULAR RED signal indications in signal faces for approaches having a combined lane for left-turn and right-turn movements is a specific exception to other provisions in this Chapter that would otherwise require the use of RED ARROW signal indications.

Standard:

09   **The signal faces provided for an approach with a combined left-turn/right-turn lane and no through movement shall be one of the following:**

   A. **Except as provided in Paragraph 6 of Section 4F.01 and Paragraph 4 of this Section, two or more signal faces, each capable of displaying CIRCULAR RED, CIRCULAR YELLOW, and CIRCULAR GREEN signal indications, shall be provided for the approach. This display shall be permissible regardless of the number of mandatory left-turn and/or right-turn lanes that exist on the approach in addition to the combined left-turn/right-turn lane and regardless of whether or not there are pedestrian or opposing vehicular movements that conflict with the left-turn or right-turn movements. However, if there is an opposing approach and the signal phasing protects the left-turn movement on the approach with the combined left-turn/right-turn lane from conflicts with the opposing vehicular movements and any signalized pedestrian movements, a left-turn GREEN ARROW signal indication shall also be included in the left-most signal face and shall be displayed simultaneously with the CIRCULAR GREEN signal indication.**

   B. **If the approach has one or more mandatory turn lanes in addition to the combined left-turn/right-turn lane and there is no conflict with a signalized vehicular or pedestrian movement, and GREEN ARROW signal indications are used in place of CIRCULAR GREEN signal indications on the approach, the signal faces for the approach shall be:**

   1. **A signal face(s) capable of displaying CIRCULAR RED, YELLOW ARROW, and GREEN ARROW signal indications for the mandatory turn lane(s), with the arrows pointing in the direction of the turn, and**
   2. **A shared left-turn/right-turn signal face capable of displaying CIRCULAR RED, left-turn YELLOW ARROW, left-turn GREEN ARROW, right-turn YELLOW ARROW, and right-turn GREEN ARROW signal indications, in an arrangement of signal sections that complies with the provisions of Section 4E.04 or 4E.05.**

   C. **If the approach has one or more mandatory turn lanes in addition to the combined left-turn/right-turn lane and there is a conflict with a signalized vehicular or pedestrian movement, and flashing YELLOW ARROW signal indications are used in place of CIRCULAR GREEN signal indications on the approach, the signal faces for the approach shall be as described in Items B.1 and B.2 of this Paragraph, except that flashing YELLOW ARROW signal indications shall be used in place of the GREEN ARROW signal indications for the turning movement(s) that conflicts with the signalized vehicular or pedestrian movement.**

Support:

10   Figure 4F-15 illustrates application of these Standards on approaches that have only a combined left-turn/right-turn lane, and on approaches that have one or more mandatory turn lanes in addition to the combined left-turn/right-turn lane.

Figure 4F-15. Signal Indications for Approaches with a Combined Left-Turn/Right-Turn Lane and No Through Movement (Sheet 1 of 3)

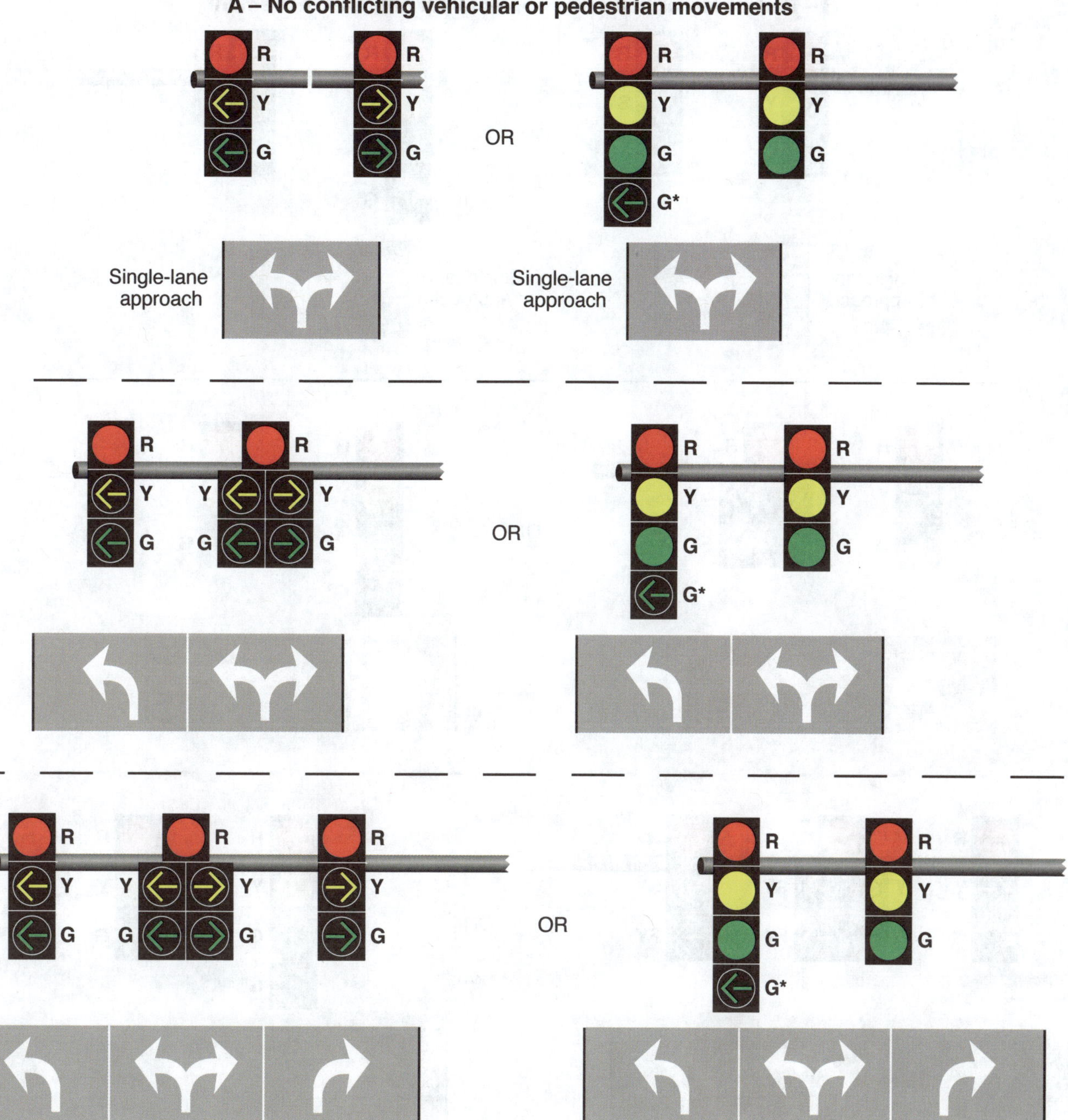

Notes:

1. Horizontally-aligned signal faces may also be used.
2. Shared signal faces may also be 5 sections in a vertical straight line instead of a cluster.

*Left-turn GREEN ARROW section shall be included if there is an opposing one-way approach and the signal phasing eliminates conflicts.

Figure 4F-15. Signal Indications for Approaches with a Combined Left-Turn/Right-Turn Lane and No Through Movement (Sheet 2 of 3)

B – Pedestrian or vehicular conflict with one turn movement

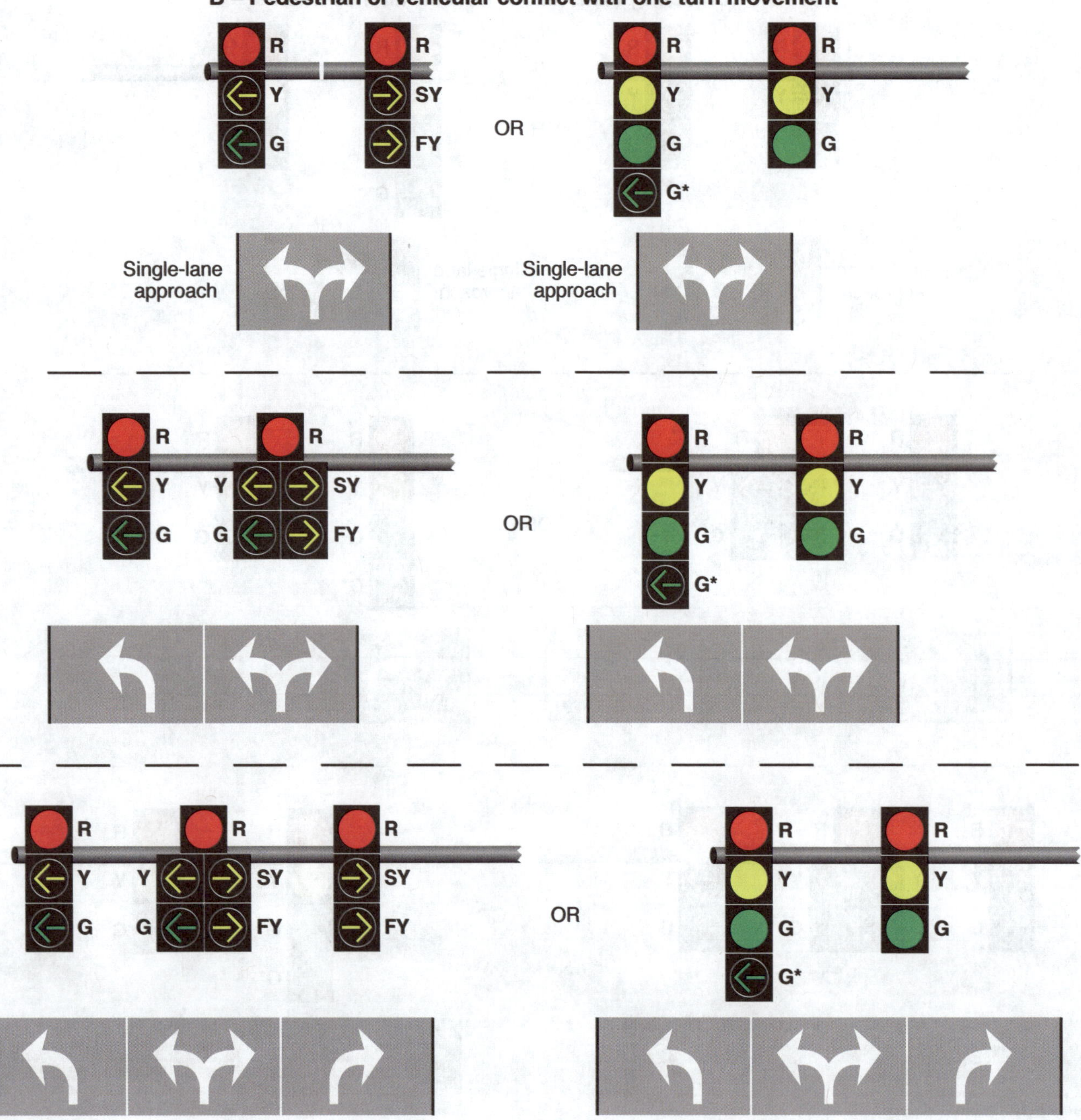

Notes:

1. A conflict with the right-turn movement is illustrated.
2. Horizontally-aligned signal faces may also be used.
3. Shared signal faces may also be 5 sections in a vertical straight line instead of a cluster.

*Left-turn GREEN ARROW section shall be included if there is an opposing one-way approach and the signal phasing eliminates conflicts.

Figure 4F-15. Signal Indications for Approaches with a Combined Left-Turn/Right-Turn Lane and No Through Movement (Sheet 3 of 3)

C – Pedestrian or vehicular conflicts with both turn movements

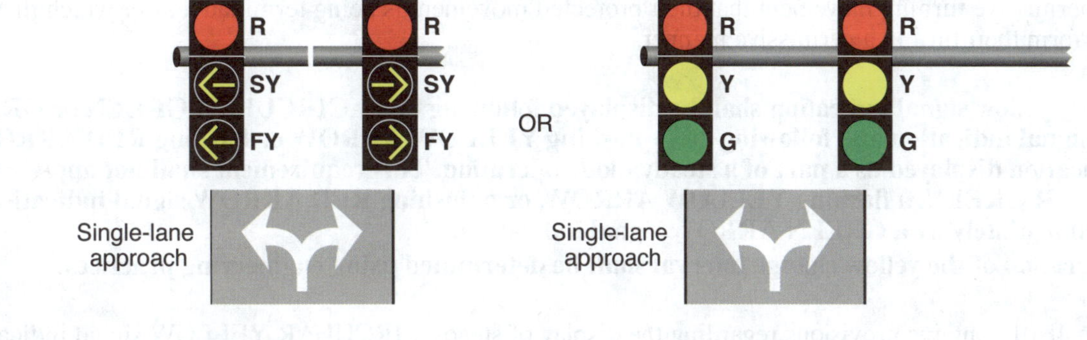

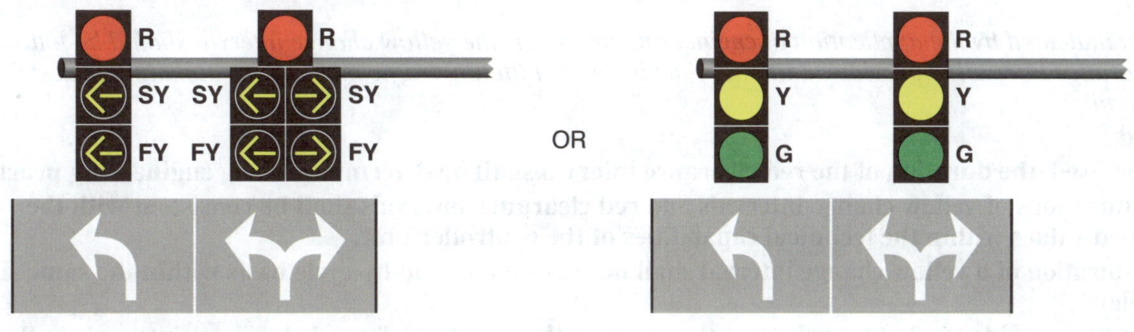

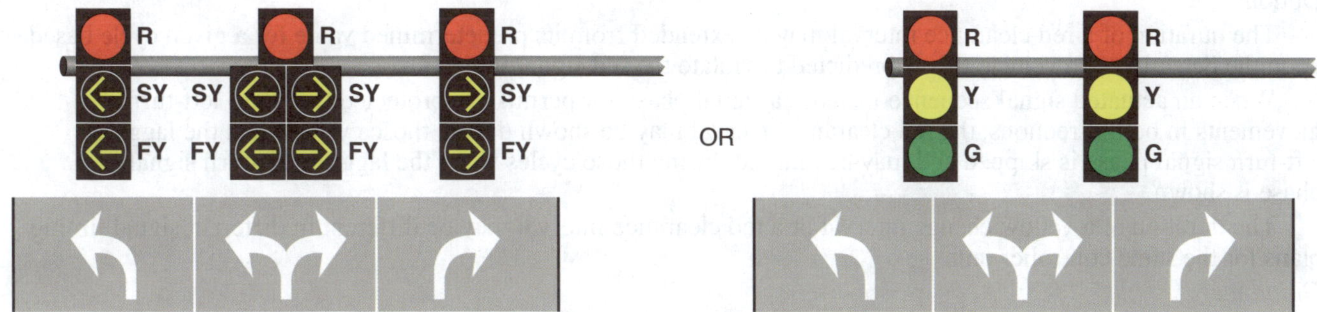

Notes:
1. Horizontally-aligned signal faces may also be used.
2. Shared signal faces may also be 5 sections in a vertical straight line instead of a cluster.

Option:

11   If the lane-use regulations on an approach are variable such that at certain times all of the lanes on the approach are designated as mandatory turn lanes and no lane is designated as a combined left-turn/right-turn lane:

    A. During the times that no lane is designated as a combined left-turn/right-turn lane, the left-turn and right-turn movements may start and terminate independently, and the left-turn and right-turn movements may be operated in one or more of the modes of operation as described in Sections 4F.02 through 4F.15; and

    B. If a protected/permissive mode is used, the operation of the shared left-turn/right-turn signal face provided in Paragraph 9 may be modified to display the steady left-turn (right-turn) YELLOW ARROW signal indication and the flashing left-turn (right-turn) YELLOW ARROW signal indication in the same section in order to not exceed the maximum of five sections per signal face provided in Section 4E.03.

Section 4F.17 <u>Yellow Change and Red Clearance Intervals</u>

Support:

01   The exclusive function of the yellow change interval is to warn traffic approaching a signalized location that their permission to proceed is being terminated after which they will be directed to stop, or in the case of a protected/permissive turning movement that their protected movement is being terminated after which they will need to perform their turn in a permissive manner.

Standard:

02   **A steady yellow signal indication shall be displayed following every CIRCULAR GREEN or GREEN ARROW signal indication and following every flashing YELLOW ARROW or flashing RED ARROW signal indication displayed as a part of a steady mode operation. This requirement shall not apply when a CIRCULAR GREEN, a flashing YELLOW ARROW, or a flashing RED ARROW signal indication is followed immediately by a GREEN ARROW signal indication.**

03   **The duration of the yellow change interval shall be determined using engineering practices.**

Support:

04   Section 4F.01 contains provisions regarding the display of steady CIRCULAR YELLOW signal indications to approaches from which drivers are allowed to make permissive left turns.

Guidance:

05   *When indicated by the application of engineering practices, the yellow change interval should be followed by a red clearance interval to provide additional time before conflicting traffic movements, including pedestrians, are released.*

Standard:

06   **When used, the duration of the red clearance interval shall be determined using engineering practices.**

07   **The durations of yellow change intervals and red clearance intervals shall be consistent with the determined values within the technical capabilities of the controller unit.**

08   **The duration of a yellow change interval shall not vary on a cycle-by-cycle basis within the same signal timing plan.**

09   **Except as provided in Paragraph 10 of this Section, the duration of a red clearance interval shall not be decreased or omitted on a cycle-by-cycle basis within the same signal timing plan.**

Option:

10   The duration of a red clearance interval may be extended from its predetermined value for a given cycle based upon the detection of a vehicle that is predicted to violate the red signal indication.

11   When an actuated signal sequence includes a signal phase for permissive/protected (lagging) left-turn movements in both directions, the red clearance interval may be shown during those cycles when the lagging left-turn signal phase is skipped and may be omitted during those cycles when the lagging left-turn signal phase is shown.

12   The duration of a yellow change interval or a red clearance interval may be different in different signal timing plans for the same controller unit.

Guidance:

13   *A yellow change interval should have a minimum duration of 3 seconds, and a maximum duration of 6 seconds. The longer intervals should be reserved for use on approaches with higher speeds. Except when clearing a one-lane, two-way facility (see Section 4O.02) or when clearing an exceptionally wide intersection, a red clearance interval should have a duration not exceeding 6 seconds.*

Standard:

14   **Except for Warning Beacons mounted on advance warning signs on the approach to a signalized location (see Section 2C.35), signal displays that are intended to provide a "pre-yellow warning" interval, such as flashing green signal indications, vehicular countdown displays, or other similar displays, shall not be used at a signalized location.**

Support:

15   The use of signal displays (other than Warning Beacons mounted on advance warning signs) that convey a "pre-yellow warning" have been found by research to increase the frequency of crashes.

Section 4F.18 Preemption and Priority Control of Traffic Control Signals – General

Option:

01   Traffic control signals may be designed and operated to respond to certain classes of approaching vehicles by altering the normal signal timing and phasing plan(s) during the approach and passage of those vehicles. The alternative plan(s) may be as simple as extending a currently displayed green interval or as complex as replacing the entire set of signal phases and timing.

Support:

02   Some types or classes of vehicles supersede others when a traffic control signal responds to more than one type or class. In general, a vehicle that is more difficult to control supersedes a vehicle that is easier to control.

Option:

03   Preemption or priority control of traffic control signals may also be a means of indicating to specified classes of vehicles at certain non-intersection locations, such as on approaches to one-lane bridges and tunnels, movable bridges, highway maintenance and construction activities, metered freeway entrance ramps, and transit operations, that they are permitted to proceed.

Guidance:

04   *When a traffic control signal that is returning to a steady mode from a dark mode (typically upon restoration from a power failure) receives a preemption or priority request, care should be exercised to minimize the possibility of vehicles or pedestrians being misdirected into a conflict with the vehicle making the request.*

Option:

05   During the change from a dark mode to a steady mode under a preemption or priority request, the display of signal indications that could misdirect road users may be prevented by one or more of the following methods:

    A. Having the traffic control signal remain in the dark mode,
    B. Having the traffic control signal remain in the flashing mode,
    C. Altering the flashing mode,
    D. Executing the normal start-up routine before responding, or
    E. Responding directly to initial or dwell period.

Guidance:

06   *Traffic control signals operating under preemption control or under priority control should be operated in a manner designed to keep traffic moving.*

07   *Traffic control signals that are designed to respond under preemption or priority control to more than one type or class of vehicle should be designed to respond in the relative order of importance or difficulty in stopping the type or class of vehicle. The order of priority should be: train, boat, heavy vehicle (fire vehicle, emergency medical service), light vehicle (law enforcement), light rail transit, rubber-tired transit.*

Option:

08   If engineering judgment indicates that light rail transit signal indications would reduce road user confusion that might otherwise occur if standard traffic signal indications were used to control these movements, light rail transit signal indications complying with Section 8D.15 and as illustrated in Figure 8D-3 may be used for preemption or priority control of the following exclusive movements at signalized intersections:

    A. Public transit buses in "queue jumper" lanes, and
    B. Public transit buses in semi-exclusive or mixed-use alignments.

Section 4F.19 Preemption Control of Traffic Control Signals

Support:

01   Preemption control (see definition in Section 1C.02) is typically given to trains, boats, emergency vehicles, and light rail transit.

02   Examples of preemption control include the following:

    A. The prompt displaying of green signal indications at signalized locations ahead of fire vehicles, law enforcement vehicles, ambulances, and other official emergency vehicles;
    B. A special sequence of signal phases and timing to expedite and/or provide additional clearance time for vehicles to clear the tracks prior to the arrival of rail traffic; and
    C. A special sequence of signal phases to display a steady red indication to prohibit turning movements toward the tracks during the approach or passage of rail traffic.

Standard:

03   **During the transition into preemption control, the yellow change interval, and any red clearance interval that follows, shall not be shortened or omitted.**

Option:

04   During the transition into preemption control:

A. A. Any pedestrian walk interval and/or pedestrian change interval may be shortened or omitted.
B. The red clearance interval, if any, may be omitted so that the return to the previous green signal indication follows a steady yellow signal indication in the same signal face.

Standard:

05   **During preemption control and during the transition out of preemption control:**

A. **Any yellow change interval, and any red clearance interval that follows, shall not be shortened or omitted.**
B. **A signal indication sequence from a steady yellow signal indication to a green signal indication shall not be permitted.**

Option:

06   A distinctive indication may be provided at the intersection to inform law enforcement personnel who are escorting traffic (such as a parade or funeral procession) that the traffic control signal has changed to a red indication not because of normal cycling, but because it has been preempted by rail traffic approaching an adjacent grade crossing or by boat traffic approaching an adjacent movable bridge.

07   A distinctive indication may be provided at the intersection to show that an emergency vehicle has been given control of the traffic control signal (see Section 11-106 of the "Uniform Vehicle Code"). In order to assist in the understanding of the control of the traffic control signal, a common distinctive indication may be used where drivers from different agencies travel through the same intersection when responding to emergencies.

Guidance:

08   *Except for traffic control signals interconnected with light rail transit systems, traffic control signals with railroad preemption or coordinated with flashing-light signal systems should be provided with a back-up power supply.*

09   *If a traffic control signal or hybrid beacon is installed near or within a grade crossing or if a grade crossing with active traffic control devices is within or near a signalized highway intersection, Chapter 8D should be consulted.*

Support:

10   Section 8D.09 contains additional information regarding preemption for grade crossings. Section 8D.10 contains information regarding prohibiting movements toward the grade crossing during preemption. Sections 8D.11 and 8D.12 contain additional information regarding pre-signals and queue cutter signals, respectively, for grade crossings.

Section 4F.20 Priority Control of Traffic Control Signals

Support:

01   Priority control (see definition in Section 1C.02) is typically given to certain non-emergency vehicles such as light-rail transit vehicles operating in a mixed-use alignment and buses.

02   Examples of priority control include the following:

A. The displaying of early or extended green signal indications at an intersection to assist public transit vehicles improve operations, and
B. Special phasing to assist public transit vehicles in entering the travel stream ahead of other waiting traffic.

Standard:

03   **During priority control and during the transition into or out of priority control:**

A. **The shortening or omission of any yellow change interval, and of any red clearance interval that follows, shall not be permitted.**
B. **The shortening of any pedestrian walk interval below that time described in Section 4I.06 shall not be permitted.**
C. **The omission of a pedestrian walk interval and its associated change interval shall not be permitted unless the associated vehicular phase is also omitted or the pedestrian phase is exclusive.**
D. **The shortening or omission of any pedestrian change interval shall not be permitted.**
E. **A signal indication sequence from a steady yellow signal indication to a green signal indication shall not be permitted.**

CHAPTER 4G. FLASHING OPERATION OF TRAFFIC CONTROL SIGNALS

Section 4G.01 Flashing Operation of Traffic Control Signals – General

Standard:

01   The light source of a flashing signal indication shall be flashed continuously at a rate of not less than 50 or more than 60 times per minute.

02   The displayed period of each flash shall be a minimum of ½ and a maximum of ⅔ of the total flash cycle.

03   Flashing signal indications shall comply with the requirements of other Sections of this Manual regarding visibility limiting or positioning of conflicting signal indications, except that flashing yellow signal indications for through traffic shall not be required to be visibility limited or positioned to minimize visual conflict for road users in separately-controlled turn lanes.

04   Each traffic control signal shall be provided with an independent flasher mechanism that operates in compliance with this Chapter.

05   A manual switch shall be provided to initiate the flashing mode. A conflict monitor (malfunction management unit) circuit and, if appropriate, an automatic means shall also be provided to initiate the flashing mode.

06   The flashing operation shall not be terminated by removing or turning off the controller unit or the conflict monitor (malfunction management unit) or both.

Option:

07   Based on engineering study or engineering judgment, traffic control signals may be operated in the flashing mode on a scheduled basis during one or more periods of the day rather than operated continuously in the steady (stop-and-go) mode.

Support:

08   Sections 4I.06 and 4K.04 contain information regarding the operation of pedestrian signal heads and accessible pedestrian signal detector push button locator tones, respectively, during flashing operation.

Section 4G.02 Flashing Operation – Transition into Flashing Mode

Option:

01   The transition from steady (stop-and-go) mode to flashing mode, if initiated by a conflict monitor (malfunction management unit) or by a manual switch, may be made at any time.

Standard:

02   Programmed changes from steady (stop-and-go) mode to flashing mode shall be made under either of the following circumstances:

    A. At the end of the common major-street red interval (such as just prior to the start of the green in both directions on the major street), or

    B. Directly from a CIRCULAR GREEN signal indication to a flashing CIRCULAR YELLOW signal indication, or from a GREEN ARROW signal indication to a flashing YELLOW ARROW signal indication, or from a flashing YELLOW ARROW signal indication (see Sections 4F.02, 4F.04, 4F.08, 4F.09, 4F.11, and 4F.15) to a flashing YELLOW ARROW signal indication (in a different signal section if the signal face displays the steady YELLOW ARROW signal indication in a different section than the flashing YELLOW ARROW signal indication).

03   During programmed changes into flashing mode, no green signal indication or flashing yellow signal indication shall be terminated and immediately followed by a steady red or flashing red signal indication without first displaying the steady yellow signal indication.

Section 4G.03 Flashing Operation – Signal Indications during Flashing Mode

Guidance:

01   *When a traffic control signal is operated in the flashing mode, a flashing yellow signal indication should be used for the major street and a flashing red signal indication should be used for the other approaches unless flashing red signal indications are used on all approaches.*

Standard:

02   When a traffic control signal is operated in the flashing mode, all of the green signal indications at the signalized location shall be dark (non-illuminated) and shall not be displayed in either a steady or flashing manner, except for single-section GREEN ARROW signal indications as provided in Paragraph 6 of this Section.

03   **Flashing yellow signal indications shall be used on more than one approach to a signalized location only if those approaches do not conflict with each other.**

04   **Except as provided in Paragraph 5 of this Section, when a traffic control signal is operated in the flashing mode, one and only one signal indication in every signal face at the signalized location shall be flashed.**

Option:

05   If a signal face has two identical CIRCULAR RED or RED ARROW signal indications (see Section 4E.04), both of those identical signal indications may be flashed simultaneously.

Standard:

06   **No steady indications, other than a single-section signal face consisting of a continuously-displayed GREEN ARROW signal indication that is used alone to indicate a continuous movement in the steady (stop-and-go) mode, shall be displayed at the signalized location during the flashing mode. A single-section GREEN ARROW signal indication shall remain continuously displayed when the traffic control signal is operated in the flashing mode.**

07   **If a signal face includes both circular and arrow signal indications of the color that is to be flashed, only the circular signal indication shall be flashed.**

08   **All signal faces that are flashed on an approach shall flash the same color, either yellow or red, except that separate turn signal faces (see Sections 4F.04, 4F.06, 4F.08, 4F.11, 4F.13, and 4F.15) shall be permitted to flash a RED ARROW signal indication when the adjacent through movement signal indications are flashed yellow. Shared signal faces (see Sections 4F.03, 4F.05, 4F.07, 4F.10, 4F.12, and 4F.14) for turn movements shall not be permitted to flash a CIRCULAR RED signal indication when the adjacent through movement signal indications are flashed yellow.**

09   **The appropriate RED ARROW or YELLOW ARROW signal indication shall be flashed when a signal face consists entirely of arrow indications. A signal face that consists entirely of arrow indications and that provides a protected only turn movement during the steady (stop-and-go) mode or that provides a flashing yellow arrow or flashing red arrow signal indication for a permissive turn movement during the steady (stop-and-go) mode shall be permitted to flash the YELLOW ARROW signal indication during the flashing mode if the adjacent through movement signal indications are flashed yellow and if it is intended that a permissive turn movement not requiring a full stop by each turning vehicle be provided during the flashing mode.**

Section 4G.04 <u>Flashing Operation – Transition Out of Flashing Mode</u>

Standard:

01   **All changes from flashing mode to steady (stop-and-go) mode shall be made under one of the following procedures:**
   A. **Yellow-red flashing mode: Changes from flashing mode to steady (stop-and-go) mode shall be made at the beginning of the major-street green interval (when a green signal indication is displayed to through traffic in both directions on the major street), or if there is no common major-street green interval, at the beginning of the green interval for the major traffic movement on the major street.**
   B. **Red-red flashing mode: Changes from flashing mode to steady (stop-and-go) mode shall be made by changing the flashing red indications to steady red indications followed by appropriate green indications to begin the steady mode cycle. These green indications shall be the beginning of the major-street green interval (when a green signal indication is displayed to through traffic in both directions on the major street) or if there is no common major-street green interval, at the beginning of the green interval for the major traffic movement on the major street.**

Guidance:

02   *The steady red clearance interval provided during the change from red-red flashing mode to steady (stop-and-go) mode should have a minimum duration of 6 seconds.*

03   *When changing from the yellow-red flashing mode to steady (stop-and-go) mode at a location where there is a common major-street green interval, the flashing red signal indications for the minor street should immediately change to steady red signal indications, and the flashing yellow signal indications for the through movements on the major street should change to green signal indications in both directions (after the minor-street signal indications have been steady red for a short time, if desired), or the flashing yellow signal indications for the through movements on the major street should change to steady yellow signal indications followed by a steady red clearance interval before changing to green signal indications in both directions.*

04   *When changing from the yellow-red flashing mode to steady (stop-and-go) mode at a location where there is no common major-street green interval, the flashing red signal indications for the minor street should immediately change to steady red signal indications, and the flashing yellow signal indications for the through movements on the major street should change to steady yellow signal indications followed by a steady red clearance interval before changing to green signal indications for the major traffic movement on the major street.*

Standard:

05   **During programmed changes out of flashing mode, no flashing yellow signal indication shall be terminated and immediately followed by a steady red or flashing red signal indication without first displaying a steady yellow signal indication.**

Option:

06   Because special midblock signals that rest in flashing circular yellow in the position normally occupied by the green signal indication do not have a green signal indication in the signal face, these signals may go directly from flashing circular yellow (in the position normally occupied by the green signal indication) to steady yellow without going first to a green signal indication.

CHAPTER 4H. BICYCLE SIGNALS

Section 4H.01 Use of Bicycle Signal Faces

Option:

01   A bicycle signal face may be used to provide separate control of a bicyclist movement for various situations, including the following:

   A. To provide a protected bicycle signal phase or a leading or lagging bicycle interval;
   B. To continue a through bicycle lane on the right-hand side of a mandatory right-turn lane (or on the left-hand side of a mandatory left-turn lane) that would otherwise be in non-compliance with Paragraph 1 of Section 9E.02 or Paragraph 7 of Section 9E.06;
   C. To provide a bicycle interval for a counter-flow bicycle facility; or
   D. To provide for unusual or unexpected arrangements of the bicyclist movement through complex intersections, conflict areas, or signal control.

02   A bicycle signal face may be used at a mid-block traffic control signal where there are no motor vehicle movements parallel to the bicycle crossing.

Support:

03   Chapter 4C contains information on warrants for the installation of a new traffic control signal.

Guidance:

04   *The decision as to whether to incorporate a bicycle signal face(s) into a new traffic control signal design should be made during the engineering study performed in accordance with Paragraph 1 of Section 4C.01.*

05   *Engineering judgment should be exercised in determining whether or not it would be advantageous or beneficial to install a bicycle signal face(s) at an existing traffic control signal.*

Support:

06   Retrofitting existing circular traffic signals that are operated as bicycle signal faces with bicycle symbol signal faces is analogous to retrofitting existing traffic signals with pedestrian signals where such a determination is not required through an engineering study.

07   For the purpose of warrant analyses, provisions for classifying bicycles are provided in Paragraph 16 of Section 4C.01 and Paragraph 2 of Section 9F.01.

Standard:

08   **If used, a bicycle signal face shall only be used to control bicyclist movements from a designated bicycle lane or from a separate facility, such as a shared-use path.**

09   **If used, a bicycle signal face shall only be used to control bicyclist movements where bicyclists moving on a GREEN BICYCLE or YELLOW BICYCLE signal indication are not in conflict with any simultaneous motor vehicle movement at the signalized location, including right (or left) turns on red.**

Guidance:

10   *If used where motor vehicle traffic can make the same movements as bicyclists, a bicycle signal face should only be used if the bicyclist movement controlled by the bicycle signal face is sometimes allowed to proceed or sometimes required to stop at times when motor vehicle traffic, making the same movement and controlled by other vehicular signal faces, is required to stop or allowed to proceed, respectively.*

Section 4H.02 Prohibited Uses of Bicycle Signal Faces

Standard:

01   **Bicycle signal faces shall not be used to control conflicting bicyclist movements from perpendicular or nearly perpendicular directions.**

02   **Bicycle signal faces shall not be used for controlling any bicyclist movement that is sharing an approach lane with motor vehicle traffic.**

03   **Bicycle signal faces shall not be used in any manner with respect to the design and operation of a hybrid beacon.**

Section 4H.03 Bicycle Signal Signs

Support:

01   The primary purposes of the Bicycle Signal (R10-40, R10-40a, R10-41, R10-41a, R10-41b) sign (see Section 9B.22) are to inform road users that the signal indications in the bicycle signal face are intended only for bicyclists, and to inform bicyclists which specific bicyclist movements are controlled by the bicycle signal face.

Standard:

02   **Except as provided in Paragraph 3 of this Section, a Bicycle Signal (R10-40, R10-40a, R10-41, R10-41a, or R10-41b) sign shall be installed immediately adjacent to (including above or below) every bicycle signal face. The Bicycle Signal sign shall have a minimum size of 24 inches x 36 inches if it is placed next to an overhead-mounted bicycle signal face and shall have a minimum size of 12 inches x 21 inches if it is placed next to a post-mounted bicycle signal face.**

Option:

03   The Bicycle Signal sign may be omitted adjacent to a supplemental near-side bicycle signal face containing 4-inch indications.

Section 4H.04 Application of Bicycle Symbol Signal Indications during Steady (Stop-and-Go) Operation

Standard:

01   **Steady bicycle symbol signal indications shall be applied as follows:**

    A. A steady RED BICYCLE signal indication shall be displayed when it is intended to prohibit bicyclists in a designated bicycle lane or from a separate facility such as a shared-use path from entering the intersection or other controlled area. Turning after stopping shall be permitted as stated in Item C in Paragraph 1 of Section 4A.05.

    B. A steady YELLOW BICYCLE signal indication shall be displayed following a GREEN BICYCLE signal indication in the same bicycle signal face. A YELLOW BICYCLE signal indication shall not be displayed in conjunction with the change from the RED BICYCLE signal indication to a GREEN BICYCLE signal indication. The YELLOW BICYCLE signal indication shall be followed by a RED BICYCLE signal indication.

    C. A steady GREEN BICYCLE signal indication shall be displayed only when it is intended to permit bicyclists in a designated bicycle lane or from a separate facility such as a shared-use path to enter the intersection as discussed in Section 4A.05.

Section 4H.05 Application of Bicycle Symbol Signal Indications during Flashing Operation

Standard:

01   **The mode of operation of the bicycle signal faces at a traffic control signal shall be the same as the mode of operation of the other traffic signal faces at the same signalized location. Bicycle signal faces shall operate in the steady (stop-and-go) mode when the other traffic signal faces are operating in the steady (stop-and-go) mode. Bicycle signal faces shall operate in the flashing mode when the other signal faces are operating in the flashing mode. Bicycle signal faces shall not be placed in a dark mode when other vehicular traffic signal faces are operating in the flashing mode.**

Guidance:

02   *When a traffic control signal is operated in the flashing mode, bicycle signal faces should display a flashing RED BICYCLE signal indication if the other vehicular signal faces on the same approach are displaying flashing red signal indications or if there are no other vehicular signal faces on the same approach.*

03   *When a traffic control signal is operated in the flashing mode, bicycle signal faces should display a flashing YELLOW BICYCLE signal indication if the other vehicular signal faces for the through lanes on the same approach are displaying flashing yellow signal indications unless it is determined by engineering judgment that a flashing RED BICYCLE signal indication would provide a safer operation.*

Section 4H.06 Layout of Bicycle Signal Faces

Standard:

01   **Bicycle signal faces shall consist of all bicycle symbol signal indications (see Figure 4H-1). Circular or arrow signal indications shall not be used in a bicycle signal face.**

Option:

02   Bicycle signal faces may be oriented vertically or horizontally.

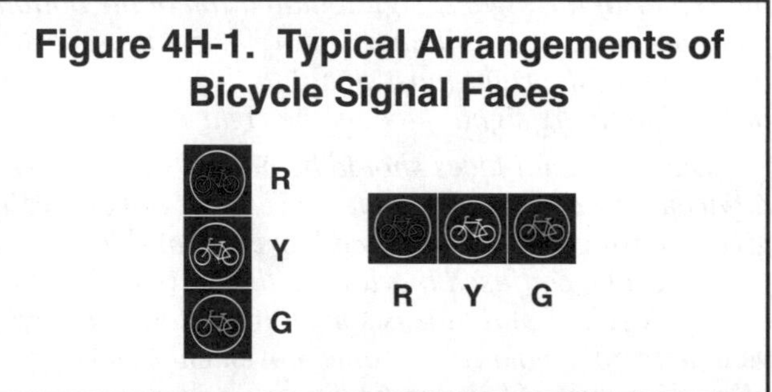

Figure 4H-1. Typical Arrangements of Bicycle Signal Faces

Standard:

03   The layouts and arrangements of the bicycle signal face shall be in accordance with the following provisions:

    A. Only the bicycle symbol shown on Page 6-7 in the 2004 Standard Highway Signs publication (see Section 1A.05) shall be used for bicycle symbol signal indications and shall be proportioned to fit within the signal lens. The bicycle symbol shall only be positioned horizontally and shall face to the left.

    B. The RED BICYCLE, YELLOW BICYCLE, and GREEN BICYCLE signal indications shall be in the same relative position to each other as specified for the CIRCULAR RED, CIRCULAR YELLOW, and CIRCULAR GREEN signal indications, respectively, in Sections 4E.04 and 4E.05.

    C. As a specific exception to Paragraph 5 of Section 4E.04, two YELLOW BICYCLE signal indications or two GREEN BICYCLE signal indications shall not be arranged horizontally adjacent to each other at right angles to the basic straight line arrangement to form a clustered signal face.

Option:

04   Backplates (see Paragraphs 18 and 19 in Section 4D.06) may be used with bicycle signal faces.

05   If a bicycle signal face having 4-inch signal indications is used, the accompanying visors may be omitted.

Section 4H.07 Size of Bicycle Symbol Signal Indications

Standard:

01   There shall be three nominal diameter sizes for bicycle signal indications: 4 inches, 8 inches, and 12 inches.

02   All signal indications in a bicycle signal face shall be of the same size.

03   Four-inch signal indications shall not be used for any bicycle signal face other than a supplemental, post-mounted, near-side bicycle signal face.

Section 4H.08 Placement of Bicycle Signal Faces

Standard:

01   The provisions of Sections 4D.05 through 4D.08 shall apply to the placement of the bicycle signal faces except as follows:

    A. As a specific exception to Item A in Paragraph 1 of Section 4D.05, a minimum of one primary bicycle signal face shall be provided to control traffic for the bicyclist movement, even if a bicyclist through movement exists.

    B. The primary bicycle signal face shall have either 8-inch or 12-inch signal indications, even if it is located at the near side of the signal-controlled location.

    C. When the primary bicycle signal face is located more than 120 feet beyond the stop line, a supplemental near-side bicycle signal face shall be provided.

Guidance:

02   *When the primary bicycle signal face is located more than 80 feet and up to 120 feet beyond the stop line, a supplemental near-side bicycle signal face should be provided.*

03   *A bicycle signal face should be separated horizontally or vertically from the nearest vehicular traffic signal face for the same approach by at least 3 feet measured either horizontally perpendicular to the approach between the centers of the signal faces or vertically from the center of the lowest signal indication of the top signal face to the center of the highest signal indication of the bottom signal face. If horizontally-arranged or clustered signal faces are used, the minimum 3-foot horizontal separation between the two signal faces should be measured from the center of the right-most signal indication in the signal face on the left to the center of the left-most signal indication in the signal face on the right.*

04   *Bicycle signal faces should be placed such that visibility is maximized for bicyclists and minimized for adjacent or conflicting vehicle movements not controlled by the bicycle signal face. Consideration should be given to using visibility-limited bicycle signal faces in situations where drivers not controlled by the bicycle signal face might be confused by viewing the bicycle signal indications, such as when the bicyclist movement controlled by the bicycle signal face is sometimes allowed to proceed or sometimes required to stop at times when motor vehicle traffic, making the same movement and controlled by other vehicular signal faces, is required to stop or allowed to proceed, respectively.*

Section 4H.09 Mounting Height of Bicycle Signal Faces

Standard:

01   **The provisions of Section 4D.09 shall apply to the mounting height of bicycle signal faces except as follows:**

    A. The bottom of the signal housing (including brackets) of a bicycle signal face that is not located over a roadway or shoulder shall be a minimum of 7 feet above the sidewalk or ground, and

    B. If 4-inch signal indications are used in a supplemental, post-mounted, near-side bicycle signal face, the bottom of the signal housing (including brackets) shall be a minimum of 4 feet and a maximum of 8 feet above the sidewalk or ground. Bicycle signal faces with 4-inch signal indications installed above a pedestrian sidewalk or pathway shall not project more than 4 inches into the pedestrian facility.

Section 4H.10 Intensity and Light Distribution of Bicycle Signal Faces

Guidance:

01   *Except for the 4-inch nominal size of the lens diameter, the intensity and distribution of light from each illuminated bicycle signal face should be similar to that recommended for vehicular traffic signal faces in accordance with Paragraph 11 of Section 4E.01 to the extent practical.*

Section 4H.11 Yellow Change and Red Clearance Intervals for Bicycle Signal Faces

Standard:

01   **The provisions of Section 4F.17 shall apply to the duration of the yellow change and the red clearance intervals of a bicycle signal phase.**

Guidance:

02   *The minimum duration of the yellow change interval of a bicycle signal phase should be 3 seconds.*

Support:

03   The function of the yellow change interval is to warn bicyclists approaching a signalized location that their permission to proceed is being terminated after which they will be directed to stop. Providing clearance time for a bicyclist to travel through the intersection or conflict area is the purpose of the red clearance interval rather than the yellow change interval.

Section 4H.12 Bicycle Push Buttons

Option:

01   Bicycle push buttons may be used for bicycle detection.

Support:

02   The location of bicycle push buttons intended only for use by bicyclists and not pedestrians are determined by engineering judgment considering a reasonable reach without requiring most bicyclists to dismount.

Standard:

03   **Where used, push buttons intended to be used by both pedestrians and bicyclists shall be located and operated to meet all accessibility requirements (see Section 4I.05).**

04   **Bicycle push buttons shall be accompanied by an appropriate regulatory sign (R10-4, R10-24, or R10-26) explaining the purpose and operation of the push button (see Sections 2B.58 and 9B.20).**

CHAPTER 4I. PEDESTRIAN CONTROL FEATURES

Section 4I.01 Pedestrian Signal Heads

Support:

01   Pedestrian signal heads provide special types of traffic signal indications exclusively intended for controlling pedestrians. These signal indications consist of the illuminated symbols of a WALKING PERSON (symbolizing WALK) and an UPRAISED HAND (symbolizing DONT WALK).

02   Section 4D.02 contains information on when to use pedestrian signal heads.

03   Accessible pedestrians signals (see Chapter 4K) where pedestrian signal heads are used provide information in non-visual formats (such as audible tones and/or speech messages, and vibrating surfaces) so that a pedestrian with vision disabilities can know when to cross the street.

04   Chapter 4J contains information regarding the use of pedestrian hybrid beacons and Chapter 4U contains information regarding the use of In-Roadway Warning Lights at unsignalized marked crosswalks.

Section 4I.02 Size, Design, and Illumination of Pedestrian Signal Head Indications

Standard:

01   All new pedestrian signal head indications shall be displayed within a rectangular background and shall consist of symbolized messages (see Figure 4I-1), except that existing pedestrian signal head indications with lettered or outline style symbol messages shall be permitted to be retained for the remainder of their useful service life. The symbol designs that are set forth in the "Standard Highway Signs" publication (see Section 1A.05) shall be used. Each pedestrian signal head indication shall be independently displayed and emit a single color.

02   If a two-section pedestrian signal head is used, the UPRAISED HAND (symbolizing DONT WALK) signal section shall be mounted directly above the WALKING PERSON (symbolizing WALK) signal section. If a one-section pedestrian signal head is used, the symbols shall be either overlaid upon each other or arranged side-by-side with the UPRAISED HAND symbol to the left of the WALKING PERSON symbol, and a light source that can display each symbol independently shall be used.

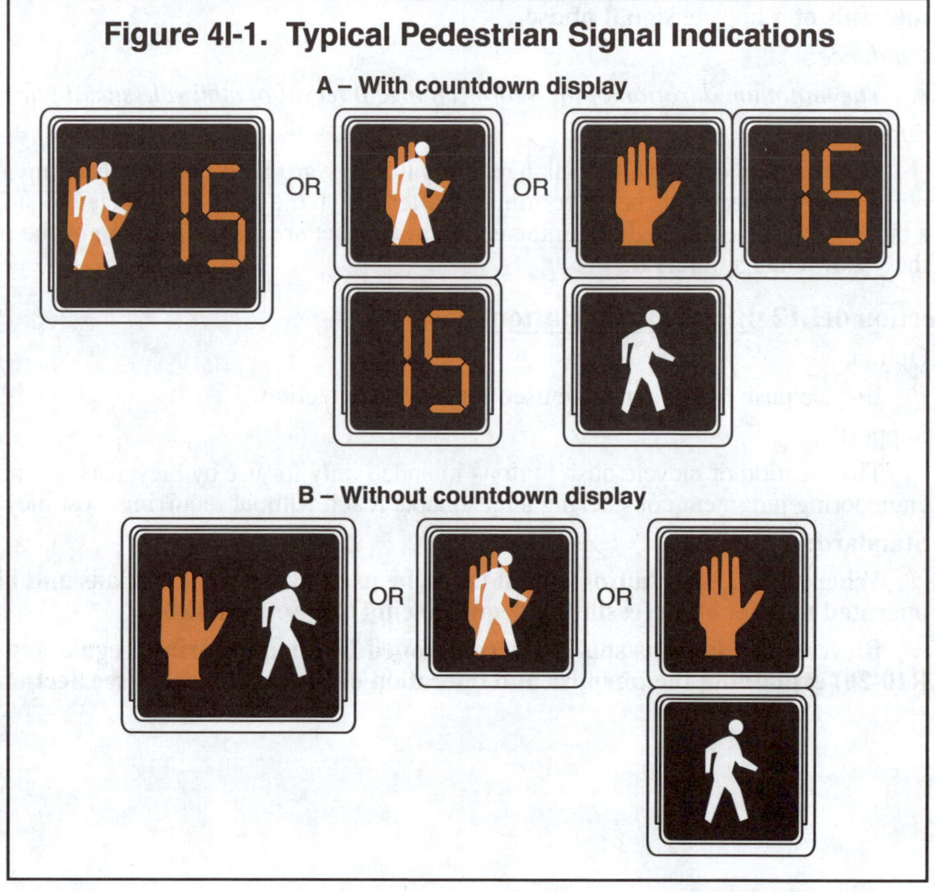

03   The WALKING PERSON (symbolizing WALK) signal indication shall be white, with all except the symbol obscured by an opaque material for signal optical units that use incandescent lamps within optical assemblies that include lenses. The UPRAISED HAND (symbolizing DONT WALK) signal indication shall be Portland orange, with all except the symbol obscured by an opaque material for signal optical units that use incandescent lamps within optical assemblies that include lenses.

04   **Except as provided in Paragraph 5 of this Section, the requirements of Chapter 3 of the publication entitled "Equipment and Materials Standards of the Institute of Transportation Engineers," 2008, ITE, that pertain to the aspects of the pedestrian signal head design that affect the display of the signal indications shall be met for signal optical units that use incandescent lamps within optical assemblies that include lenses. Except as provided in Paragraph 5 of this Section, the requirements of the publication entitled "Pedestrian Traffic Control Signal Indicators – Light Emitting Diode (LED) Signal Modules," 2011, ITE, that pertain to the aspects of the signal head design that affect the display of the signal indications shall be met for light-emitting diode (LED) pedestrian signal head modules.**

Guidance:

05   *The intensity and distribution of light from each illuminated pedestrian signal lens or LED pedestrian signal head module should comply with the publications specified in Paragraph 4 of this Section, as appropriate.*

06   *When not illuminated, the WALKING PERSON (symbolizing WALK) and UPRAISED HAND (symbolizing DONT WALK) symbols should not be visible to pedestrians at the far end of the crosswalk that the pedestrian signal head indications control.*

Standard:

07   **For pedestrian signal head indications, the symbols shall be at least 6 inches high.**

08   **The light source of a flashing UPRAISED HAND (symbolizing DONT WALK) signal indication shall be flashed continuously at a rate of not less than 50 or more than 60 times per minute. The displayed period of each flash shall be a minimum of ½ and a maximum of ⅔ of the total flash cycle.**

Guidance:

09   *Pedestrian signal head indications should be conspicuous and recognizable to pedestrians at all distances from the beginning of the controlled crosswalk to a point 10 feet from the end of the controlled crosswalk during both day and night.*

10   *For crosswalks where the pedestrian enters the crosswalk more than 100 feet from the pedestrian signal head indications, the symbols should be at least 9 inches high.*

11   *If the pedestrian signal indication is so bright that it causes excessive glare in nighttime conditions, some form of automatic dimming should be used to reduce the brilliance of the signal indication.*

Option:

12   An animated eyes symbol may be added to a pedestrian signal head in order to prompt pedestrians to look for vehicles in the intersection during the time that the WALKING PERSON (symbolizing WALK) signal indication is displayed.

Standard:

13   **If used, the animated eyes symbol shall consist of an outline of a pair of white steadily illuminated eyes with white eyeballs that scan from side to side at a rate of approximately once per second. The animated eyes symbol shall be at least 12 inches wide with each eye having a width of at least 5 inches and a height of at least 2.5 inches. The animated eyes symbol shall be illuminated at the start of the walk interval and shall terminate at the end of the walk interval.**

Section 4I.03 Location and Height of Pedestrian Signal Heads

Standard:

01   **Pedestrian signal heads shall be mounted with the bottom of the signal housing including brackets not less than 7 feet or more than 10 feet above sidewalk level, and shall be positioned and adjusted to provide maximum visibility at the beginning of the controlled sidewalk.**

Guidance:

02   *If pedestrian signal heads are mounted on the same support as vehicular signal heads, there should be a physical separation between them.*

Section 4I.04 Countdown Pedestrian Signals

Standard:

01   **All pedestrian signal heads used at crosswalks where the pedestrian change interval is more than 7 seconds shall include a pedestrian change interval countdown display in order to inform pedestrians of the number of seconds remaining in the pedestrian change interval.**

Option:

02   Pedestrian signal heads used at crosswalks where the pedestrian change interval is 7 seconds or less may include a pedestrian change interval countdown display in order to inform pedestrians of the number of seconds remaining in the pedestrian change interval.

Standard:

03   **Where countdown pedestrian signals are used, the countdown shall always be displayed simultaneously with the flashing UPRAISED HAND (symbolizing DONT WALK) signal indication displayed for that crosswalk.**

04   **Countdown pedestrian signals shall consist of Portland orange numbers that are at least 6 inches in height on a black opaque background. The countdown pedestrian signal shall be located immediately adjacent to the associated UPRAISED HAND (symbolizing DONT WALK) pedestrian signal head indication (see Figure 4I-1).**

05   **The display of the number of remaining seconds shall begin only at the beginning of the pedestrian change interval (flashing UPRAISED HAND). After the countdown displays zero, the display shall remain dark until the beginning of the next countdown.**

06   **The countdown pedestrian signal shall display the number of seconds remaining until the termination of the pedestrian change interval (flashing UPRAISED HAND). Countdown displays shall not be used during the walk interval. Countdown displays shall not be used during the red clearance interval of a concurrent vehicular phase that is ending simultaneously with or after the end of the pedestrian phase.**

Guidance:

07   *If used with a pedestrian signal head that does not have a concurrent vehicular phase, the pedestrian change interval (flashing UPRAISED HAND) should be set to be approximately 4 seconds less than the required pedestrian clearance time (see Section 4I.06) and an additional clearance interval (during which a steady UPRAISED HAND is displayed) should be provided prior to the start of the conflicting vehicular phase.*

08   *For crosswalks where the pedestrian enters the crosswalk more than 100 feet from the countdown pedestrian signal display, the numbers should be at least 9 inches in height.*

09   *Because some technology includes the countdown pedestrian signal logic in a separate timing device that is independent of the timing in the traffic signal controller, care should be exercised by the engineer when timing changes are made to pedestrian change intervals.*

10   *If the pedestrian change interval is interrupted or shortened as a part of a transition into a preemption sequence (see Section 4F.19), the countdown pedestrian signal display should be discontinued and go dark immediately upon activation of the preemption transition.*

Section 4I.05 Pedestrian Detectors

Option:

01   Pedestrian detectors may be push buttons or passive detection devices.

Support:

02   Passive detection devices register the presence of a pedestrian in a position indicative of a desire to cross, without requiring the pedestrian to push a button. Some passive detection devices are capable of tracking the progress of a pedestrian as the pedestrian crosses the roadway for the purpose of extending or shortening the duration of certain pedestrian timing intervals.

03   The provisions in this Section place pedestrian push buttons within easy reach of pedestrians who are intending to cross each crosswalk and make it obvious which push button is associated with each crosswalk. These provisions also position push button poles in optimal locations for installation of accessible pedestrian signals (see Chapter 4K). Information regarding reach ranges can be found in the U.S. Department of Justice 2010 ADA Standards for Accessible Design, September 15, 2010, 28 CFR 35 and 36, Americans with Disabilities Act of 1990.

Guidance:

04   *If pedestrian push buttons are used, they should be capable of easy activation requiring no more than 5 pounds of force, should not require tight grasping, pinching, or twisting of the wrist, and should be conveniently located near each end of the crosswalks. Except as provided in Paragraphs 5 and 6 of this Section, pedestrian push buttons should be located to meet all of the following criteria (see Figure 4I-2):*

    A.   *Unobstructed and accessible within one or more of the reach ranges specified in Section 308, and from a clear ground space as specified in Section 305, of the 2010 ADA Standards for Accessible Design;*
    B.   *To provide a wheelchair accessible route from the push button to the ramp;*
    C.   *On the side of the curb ramp which is farthest from the center of the intersection;*

Figure 4I-2. Preferred Push Button Location Area

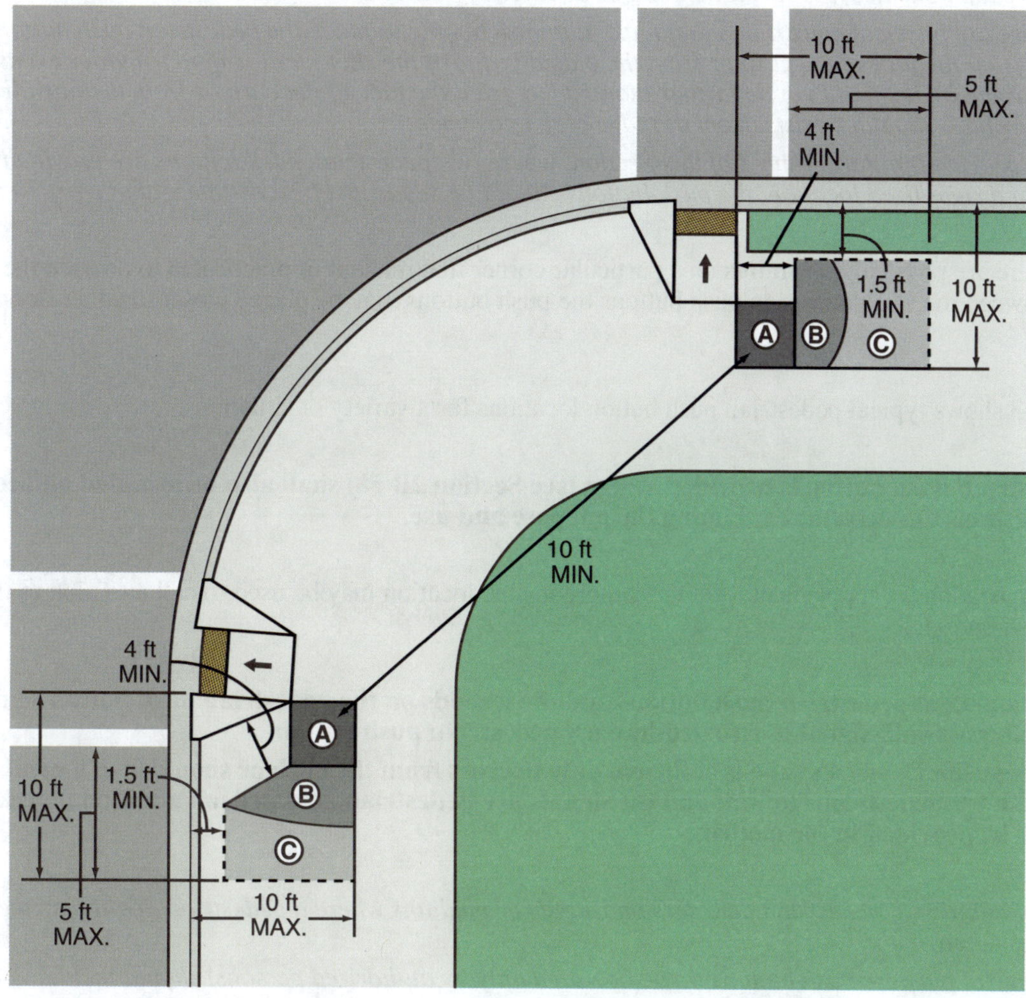

Notes:

1. The push button detector should be located 5 feet or less from the outside edge of the marked crosswalk farthest from the intersection.

2. The push button detector should be located no farther from the crosswalk than the stop line, if one is present.

3. A 4-foot minimum unobstructed pedestrian access route should be maintained.

4. The maximum (MAX.) and minimum (MIN.) dimensions shown in this figure are recommendations.

5. Two pedestrian push buttons on the same corner should be separated by at least 10 feet. The 10-foot dimension shown in this figure is in reference to the placement of the push buttons within their respective areas.

6. Figure 4I-3 shows typical push button locations.

7. This figure is not drawn to scale.

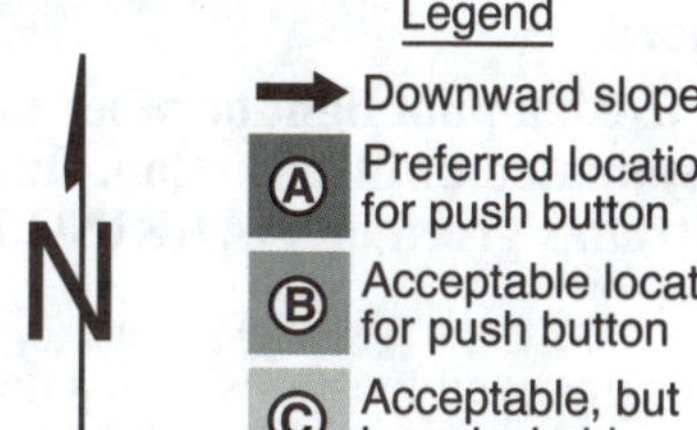

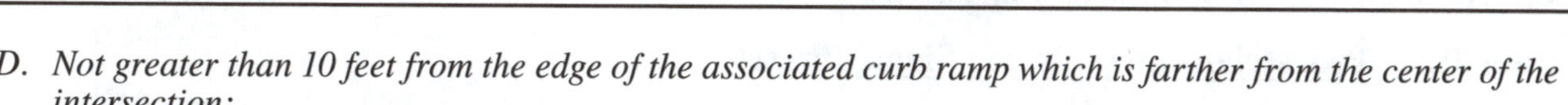

D. Not greater than 10 feet from the edge of the associated curb ramp which is farther from the center of the intersection;

E. Not greater than 5 feet from the outside edge of the marked crosswalk farthest from the center of the intersection;

F. Not farther from the crosswalk than the stop line is, if present;

G. Between 1.5 and 6 feet from the face of the curb or from the outside edge of the shoulder (or if no shoulder exists, from the edge of the pavement);

H. With the face of the push button parallel to the crosswalk to be used;

I. At a mounting height of approximately 3.5 feet, but no more than 4 feet, above the sidewalk;

J. Allowing a minimum 4-foot continuous clear width for a pedestrian access route; and

K. Outside the flared side of the curb ramp, if present.

05   *Where there are physical constraints that make it impracticable to place the pedestrian push button adjacent to a level all-weather surface, the surface should be as level as feasible.*

06   *Where there are physical constraints that make it impracticable to place the pedestrian push button between 1.5 and 6 feet from the face of the curb or from the outside edge of the shoulder (or if no shoulder exists, from the edge of the pavement), it should not be farther than 10 feet from the face of the curb or from the outside edge of the shoulder (or if no shoulder exists, from the edge of the pavement).*

07   *Except as provided in Paragraph 8 of this Section, where two pedestrian push buttons are provided on the same corner of a signalized location, the push buttons should be separated by a distance of at least 10 feet.*

Option:

08   Where there are physical constraints on a particular corner that make it impracticable to provide the 10-foot separation between the two pedestrian push buttons the push buttons may be placed closer together or on the same pole.

Support:

09   Figure 4I-3 shows typical pedestrian push button locations for a variety of situations.

Standard:

10   **If a pedestrian push button is provided, a sign (see Section 2B.58) shall also be installed adjacent to the pedestrian push button detector explaining the purpose and use.**

Option:

11   At certain locations, a supplemental sign in a more visible location may be used to call attention to the pedestrian push button.

Standard:

12   **The positioning of pedestrian push buttons and the legends on the pedestrian push button signs shall indicate which crosswalk signal is actuated by each pedestrian push button.**

13   **If the pedestrian clearance time is sufficient only to cross from the curb or shoulder to a median of sufficient width for pedestrians to wait and the signals are pedestrian actuated, an additional pedestrian detector shall be provided in the median.**

Guidance:

14   *The use of additional pedestrian detectors on islands or medians where a pedestrian might become stranded should be considered.*

15   *If used, special purpose push buttons (to be operated only by authorized persons) should include a housing capable of being locked to prevent access by the general public and do not need an instructional sign.*

Standard:

16   **If used, a pilot light or other means of indication installed with a pedestrian push button shall not be illuminated until actuation. Once it is actuated, the pilot light shall remain illuminated until the pedestrian's green or WALKING PERSON (symbolizing WALK) signal indication is displayed.**

Option:

17   At signalized locations with a demonstrated need and subject to equipment capabilities, pedestrians with special needs may be provided with additional crossing time by means of an extended push button press.

Standard:

18   **If additional crossing time is provided by means of an extended push button press, a PUSH BUTTON FOR 2 SECONDS FOR EXTRA CROSSING TIME (R10-32P) plaque (see Figure 2B-27) shall be installed adjacent to the pedestrian push button detector.**

Section 4I.06 Pedestrian Intervals and Signal Phases

Standard:

01   **At intersections equipped with pedestrian signal heads, the pedestrian signal indications shall be displayed except when the vehicular traffic control signal is being operated in the flashing mode. At those times, the pedestrian signal indications shall not be displayed.**

02   **Except as provided in Paragraph 3 of Section 4J.03, when the pedestrian signal heads associated with a crosswalk are displaying either a steady WALKING PERSON (symbolizing WALK) or a flashing UPRAISED HAND (symbolizing DONT WALK) signal indication, a steady red signal indication shall be shown to any conflicting vehicular movement that is approaching the intersection or midblock location perpendicular or nearly perpendicular to the crosswalk.**

Figure 4I-3. Typical Push Button Locations

A – Perpendicular curb ramps with crosswalks close together

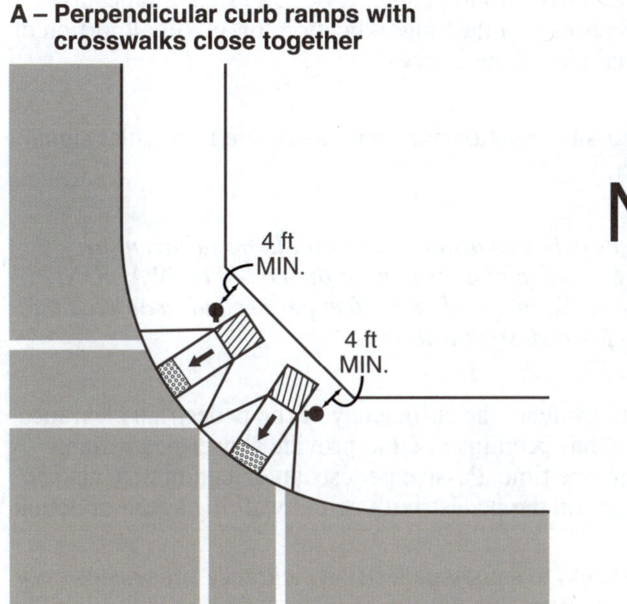

B – Parallel curb ramps

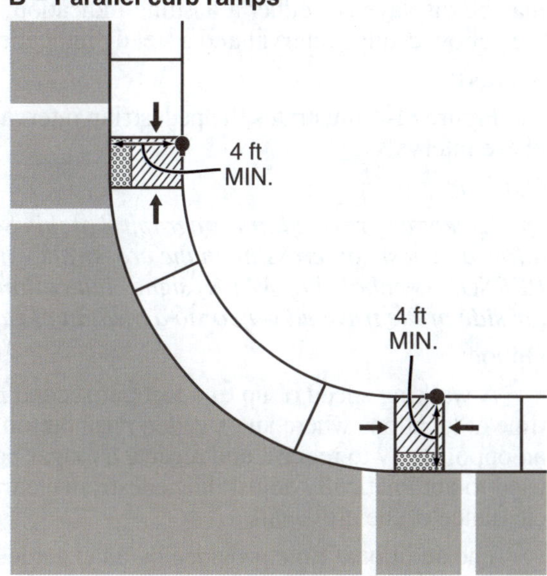

C – Perpendicular curb ramps with shared landing area

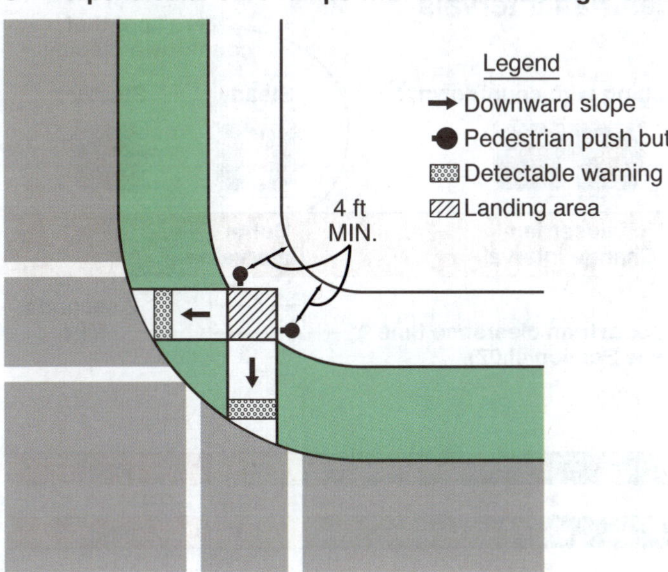

D – Perpendicular curb ramps with continuous surface between ramps

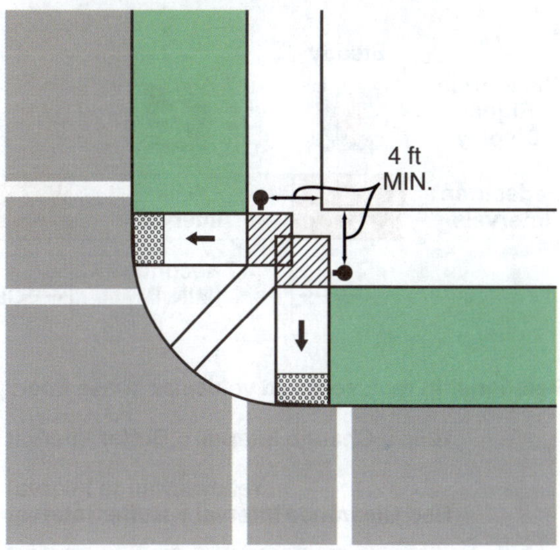

Notes:

1. This figure is not drawn to scale.

2. These drawings are intended to describe the typical locations for pedestrian push button installations. They are not intended to be a guide for the design of curb ramps.

03   **When pedestrian signal heads are used, a WALKING PERSON (symbolizing WALK) signal indication shall be displayed only when pedestrians are permitted to leave the curb or shoulder.**

04   **A pedestrian change interval consisting of a flashing UPRAISED HAND (symbolizing DONT WALK) signal indication shall begin immediately following the WALKING PERSON (symbolizing WALK) signal indication. Following the pedestrian change interval, a buffer interval consisting of a steady UPRAISED HAND (symbolizing DONT WALK) signal indication shall be displayed for at least 2 seconds prior to the release of any conflicting vehicular movement. The sum of the time of the pedestrian change interval and the buffer interval shall not be less than the calculated pedestrian clearance time (see Paragraphs 7 through 16 of this Section). The buffer interval shall not begin later than the beginning of the red clearance interval, if used.**

Option:

05   During the yellow change interval, the UPRAISED HAND (symbolizing DON'T WALK) signal indication may be displayed as either a flashing indication, a steady indication, or a flashing indication for an initial portion of the yellow change interval and a steady indication for the remainder of the interval.

Support:

06   Figure 4I-4 illustrates the pedestrian intervals and their possible relationships with associated vehicular signal phase intervals.

Guidance:

07   *Except as provided in Paragraph 8 of this Section, the pedestrian clearance time should be sufficient to allow a pedestrian crossing in the crosswalk who left the curb or edge of pavement at the end of the WALKING PERSON (symbolizing WALK) signal indication to travel at a walking speed of 3.5 feet per second to at least the far side of the traveled way or to a median of sufficient width for pedestrians to wait.*

Option:

08   A walking speed of up to 4 feet per second may be used to evaluate the sufficiency of the pedestrian clearance time at locations where an extended push button press function has been installed to provide slower pedestrians an opportunity to request and receive a longer pedestrian clearance time. Passive pedestrian detection may also be used to automatically adjust the pedestrian clearance time based on the pedestrian's actual walking speed or actual clearance of the crosswalk.

09   The additional time provided by an extended push button press to satisfy pedestrian clearance time needs may be added to either the walk interval or the pedestrian change interval.

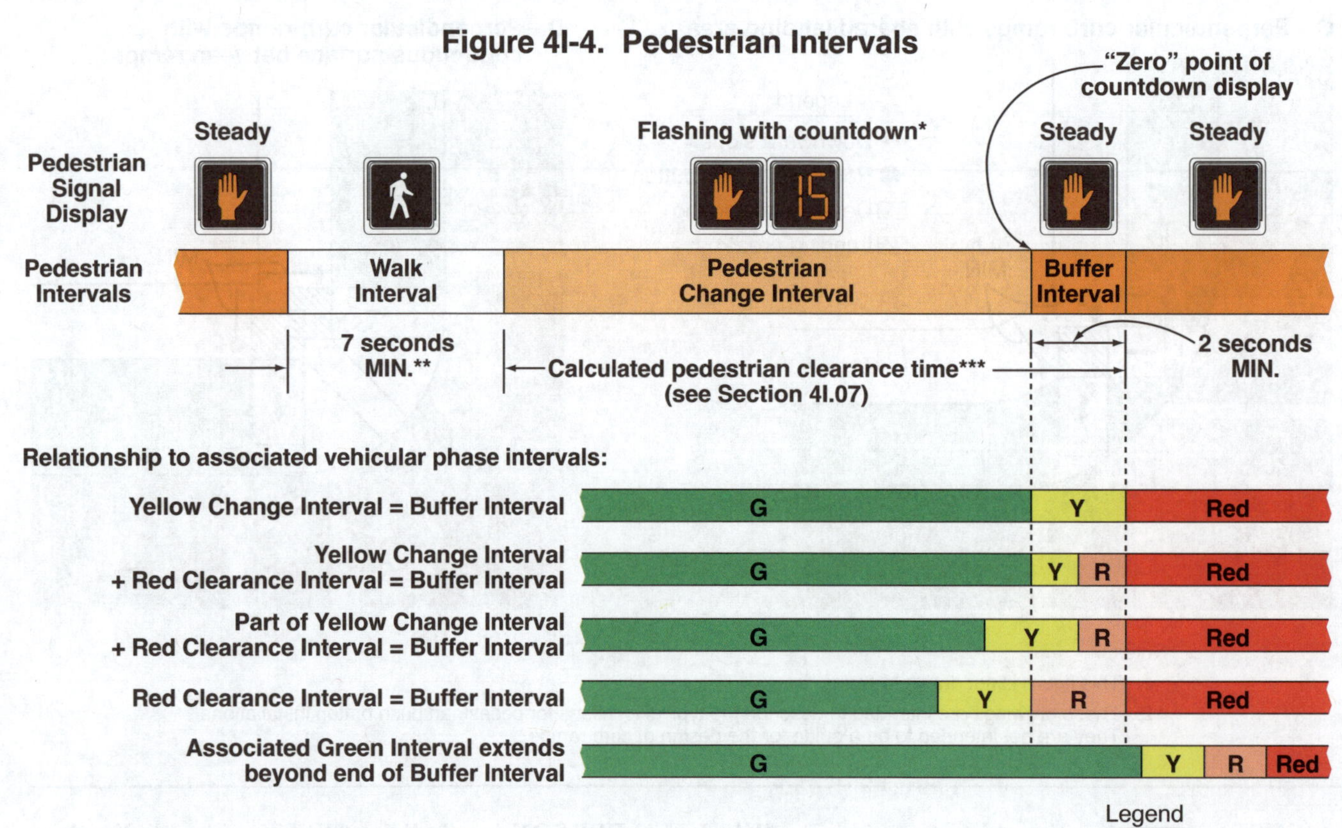

Figure 4I-4. Pedestrian Intervals

Guidance:

10   *Where pedestrians who walk slower than 3.5 feet per second, or pedestrians who use wheelchairs, routinely use the crosswalk, a walking speed of less than 3.5 feet per second should be considered in determining the pedestrian clearance time.*

11   *Except as provided in Paragraph 12 of this Section, the walk interval should be at least 7 seconds in length so that pedestrians will have adequate opportunity to leave the curb or shoulder before the pedestrian clearance time begins.*

Option:

12   If pedestrian volumes and characteristics do not require a 7-second walk interval, walk intervals as short as 4 seconds may be used.

Support:

13   The walk interval is intended for pedestrians to start their crossing. The pedestrian clearance time is intended to allow pedestrians who started crossing during the walk interval to complete their crossing. Longer walk intervals are often used when the duration of the vehicular green phase associated with the pedestrian crossing is long enough to allow it.

Guidance:

14   *The total of the walk interval and pedestrian clearance time should be sufficient to allow a pedestrian crossing in the crosswalk who left the pedestrian detector (or, if no pedestrian detector is present, a location 6 feet behind the face of the curb or 6 feet behind the edge of the pavement) at the beginning of the WALKING PERSON (symbolizing WALK) signal indication to travel at a walking speed of 3 feet per second to the far side of the traveled way being crossed or to the median if a two-stage pedestrian crossing sequence is used. Any additional time that is required to satisfy the conditions of this paragraph should be added to the walk interval.*

Option:

15   On a street with a median of sufficient width for pedestrians to wait, a pedestrian clearance time that allows the pedestrian to cross only from the curb or shoulder to the median may be provided.

Standard:

16   **Where the pedestrian clearance time is sufficient only for crossing from the curb or shoulder to a median of sufficient width for pedestrians to wait, median-mounted pedestrian signals, with pedestrian detectors (see Sections 4I.05 and 4K.01) if actuated operation is used, shall be provided and signing such as the R10-3d sign (see Section 2B.58) shall be provided to notify pedestrians to cross only to the median to await the next WALKING PERSON (symbolizing WALK) signal indication.**

Support:

17   Accessible pedestrian signals (see Chapter 4K) where median-mounted pedestrian signals and detectors are used provide information in non-visual formats (such as audible tones and/or speech messages, and vibrating surfaces) so that a pedestrian with vision disabilities can know when to resume crossing the street after crossing to the median.

Option:

18   During the transition into preemption, the walk interval and the pedestrian change interval may be shortened or omitted as described in Section 4F.19.

19   At intersections with high pedestrian volumes and high conflicting turning vehicle volumes, a brief leading pedestrian interval, during which an advance WALKING PERSON (symbolizing WALK) indication is displayed for the crosswalk while red indications continue to be displayed to parallel through and/or turning traffic, may be used to reduce conflicts between pedestrians and turning vehicles.

Support:

20   Accessible pedestrian signals (see Chapter 4K) where leading pedestrian intervals are used provide information in non-visual formats (such as audible tones and/or speech messages, and vibrating surfaces) so that a pedestrian with vision disabilities can know when to cross the street in the absence of the audible cues normally provided when the onset of the vehicular and pedestrian movements coincide.

21   If a leading pedestrian interval is used without accessible features, pedestrians with vision disabilities might begin crossing at the onset of the vehicular movement when vehicle operators are not expecting them to begin crossing.

Guidance:

22   *If a leading pedestrian interval is used, it should be at least 3 seconds in duration and should be timed to allow pedestrians to cross at least one lane of traffic or, in the case of a large corner radius, to travel far enough for pedestrians to establish their position ahead of the turning traffic before the turning traffic is released.*

23   *If a leading pedestrian interval is used, consideration should be given to prohibiting turns across the crosswalk during the leading pedestrian interval.*

24   *At locations where a leading pedestrian interval is used, the minimum time for the WALKING PERSON (symbolizing WALK) indication should be the time provided for the leading pedestrian interval plus 7 seconds.*

Support:

25   At intersections with pedestrian volumes that are so high that drivers have difficulty finding an opportunity to turn across the crosswalk, the duration of the green interval for a parallel concurrent vehicular movement is sometimes intentionally set to extend beyond the pedestrian clearance time to provide turning drivers additional green time to make their turns while the pedestrian signal head is displaying a steady UPRAISED HAND (symbolizing DONT WALK) signal indication after pedestrians have had time to complete their crossings.

CHAPTER 4J. PEDESTRIAN HYBRID BEACONS

Section 4J.01 <u>Application of Pedestrian Hybrid Beacons</u>

Support:

01   A pedestrian hybrid beacon is a special type of hybrid beacon used to warn and control traffic at an unsignalized location to assist pedestrians in crossing a street or highway at a marked crosswalk.

Option:

02   A pedestrian hybrid beacon may be considered for installation to facilitate pedestrian crossings at a location that does not meet traffic signal warrants (see Chapter 4C), or at a location that meets traffic signal warrants under Sections 4C.05 and/or 4C.06 but a decision is made to not install a traffic control signal.

Standard:

03   **If used, pedestrian hybrid beacons shall be used in conjunction with signs and pavement markings (see Section 4J.02) to warn and control traffic at locations where pedestrians enter or cross a street or highway. A pedestrian hybrid beacon shall only be installed at a marked crosswalk.**

Guidance:

04   *If one of the signal warrants of Chapter 4C is met and a traffic control signal is justified by an engineering study, and if a decision is made to install a traffic control signal, it should be installed based upon the provisions of Chapters 4D through 4I and 4K.*

05   *If a traffic control signal is not justified under the signal warrants of Chapter 4C and if gaps in traffic are not adequate to permit pedestrians to cross, or if the speed for vehicles approaching on the major street is too high to permit pedestrians to cross, or if pedestrian delay is excessive, the need for a pedestrian hybrid beacon should be considered on the basis of an engineering study that considers major-street volumes, speeds, widths, and gaps in conjunction with pedestrian volumes, walking speeds, and delay.*

06   *For a major street where the posted or statutory speed limit or the 85th-percentile speed is 35 mph or less, the need for a pedestrian hybrid beacon should be considered if the engineering study finds that the plotted point representing the vehicles per hour on the major street (total of both approaches) and the corresponding total of all pedestrians crossing the major street for 1 hour (any four consecutive 15-minute periods) of an average day falls above the applicable curve in Figure 4J-1 for the length of the crosswalk.*

07   *For a major street where the posted or statutory speed limit or the 85th-percentile speed exceeds 35 mph, the need for a pedestrian hybrid beacon should be considered if the engineering study finds that the plotted point representing the vehicles per hour on the major street (total of both approaches) and the corresponding total of all pedestrians crossing the major street for 1 hour (any four consecutive 15-minute periods) of an average day falls above the applicable curve in Figure 4J-2 for the length of the crosswalk.*

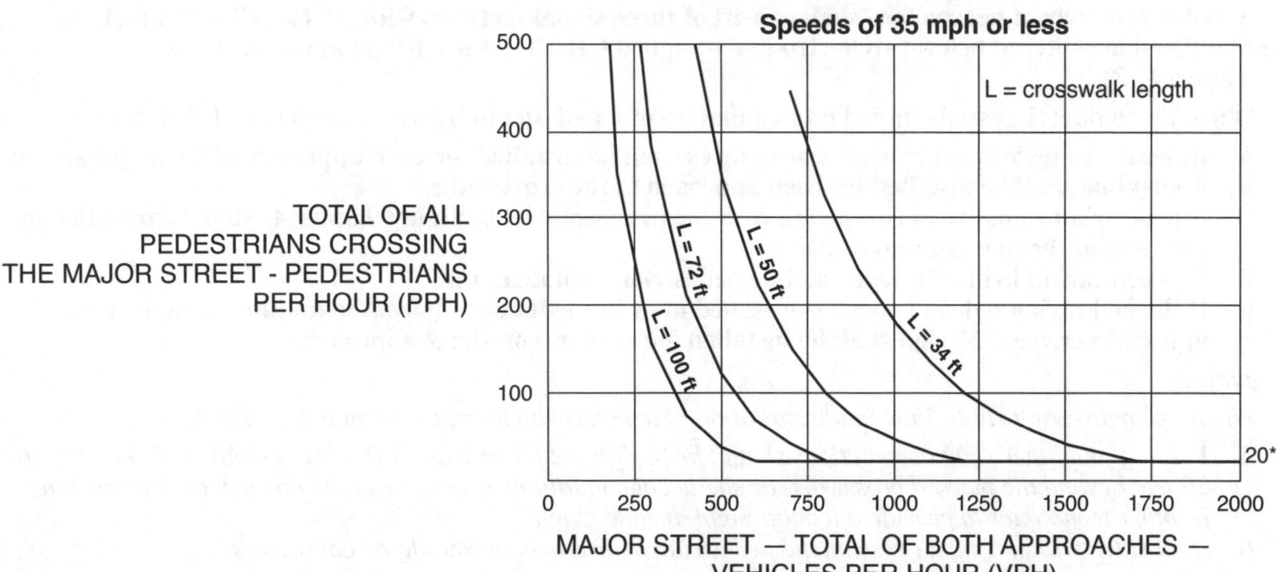

Figure 4J-1. Guidelines for the Installation of Pedestrian Hybrid Beacons on Low-Speed Roadways

* Note: 20 pph applies as the lower threshold volume

Figure 4J-2. Guidelines for the Installation of Pedestrian Hybrid Beacons on High-Speed Roadways

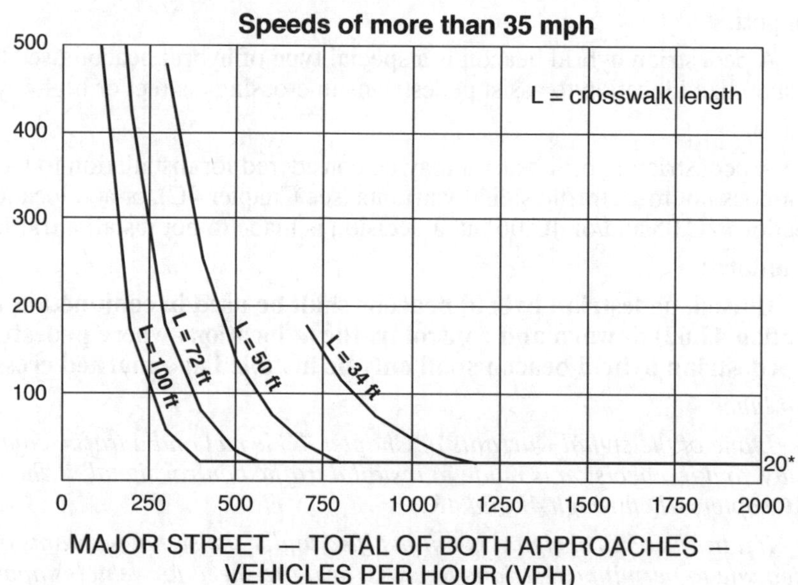

08 *For crosswalks that have lengths other than the four that are specifically shown in Figures 4J-1 and 4J-2, the values should be interpolated between the curves.*

Option:

09 The criteria for the pedestrian volume crossing the major street shown in Figures 4J-1 and 4J-2 may be reduced as much as 50 percent if the 15th-percentile crossing speed of pedestrians is less than 3.5 feet per second.

10 Where there is a divided street having a median of sufficient width for pedestrians to wait, the criteria for the major-street traffic volume shown in Figures 4J-1 and 4J-2 may be applied separately to each direction of vehicular traffic.

Section 4J.02 Design of Pedestrian Hybrid Beacons

Standard:

01 **Except as otherwise provided in this Section, a pedestrian hybrid beacon shall meet the provisions of Chapters 4D through 4G, 4I, and 4J.**

02 **A pedestrian hybrid beacon face shall consist of three signal sections, with a CIRCULAR YELLOW signal indication centered below two horizontally-aligned CIRCULAR RED signal indications (see Figure 4J-3).**

03 **When an engineering study finds that installation of a pedestrian hybrid beacon is justified, then:**

  A. **At least two pedestrian hybrid beacon faces shall be installed for each approach of the major street;**
  B. **A stop line shall be installed for each approach to the crosswalk;**
  C. **A pedestrian signal head complying with the provisions set forth in Chapter 4I shall be installed at each end of the marked crosswalk;**
  D. **The pedestrian hybrid beacon shall be pedestrian actuated; and**
  E. **If the pedestrian hybrid beacon is installed at or immediately adjacent to an intersection with a minor street, a STOP sign shall be installed for each minor-street approach.**

Guidance:

04 *When an engineering study finds that installation of a pedestrian hybrid beacon is justified, then:*

  A. *Parking and other sight obstructions should be prohibited for at least 100 feet in advance of and at least 20 feet beyond the marked crosswalk, or site accommodations should be made through curb extensions or other techniques to provide adequate sight distance; and*
  B. *If installed within a signal system, the pedestrian hybrid beacon should be coordinated.*

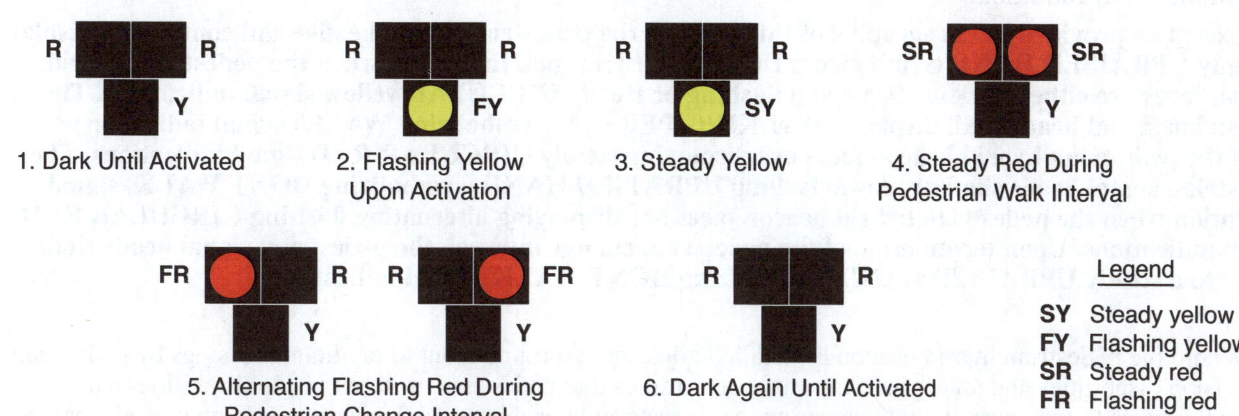

Figure 4J-3. Sequence for a Pedestrian Hybrid Beacon

Note: An optional steady red clearance interval may be added after Interval 3, and an optional short buffer interval (alternating flashing red while the pedestrian signal heads are displaying a steady UPRAISED HAND) may be added after Interval 5.

---

05   *On approaches having posted or statutory speed limits or 85th-percentile speeds in excess of 35 mph and on approaches having traffic or operating conditions that would tend to obscure visibility of roadside hybrid beacon face locations, both of the minimum of two pedestrian hybrid beacon faces should be installed over the roadway.*

06   *On multi-lane approaches having posted or statutory speed limits or 85th-percentile speeds of 35 mph or less, either a pedestrian hybrid beacon face should be installed on each side of the approach (if a median of sufficient width exists) or at least one of the pedestrian hybrid beacon faces should be installed over the roadway.*

07   *A pedestrian hybrid beacon should comply with the signal face location provisions described in Sections 4D.05 through 4D.10.*

Option:

08   A CROSSWALK—STOP ON RED (symbolic circular red) (R10-23) or a STOP ON STEADY RED—YIELD ON FLASHING RED AFTER STOP (R10-23a) sign (see Section 2B.59) may be installed facing each major street approach.

09   A W11-2 (Pedestrian), S1-1 (School), or W11-15 (Trail) crossing warning sign with an AHEAD (W16-9P) supplemental plaque may be placed in advance of a pedestrian hybrid beacon. A Warning Beacon may be installed to supplement the W11-2, S1-1, or W11-15 sign.

10   Backplates (see Section 4D.06) may be used with pedestrian hybrid beacons.

Support:

11   Accessible pedestrian signals (see Chapter 4K) where a pedestrian hybrid beacon is used provide information in non-visual formats (such as audible tones and/or speech messages, and vibrating surfaces) so that a pedestrian with vision disabilities can know when to cross the street.

Guidance:

12   *If a Warning Beacon supplements a W11-2 sign in advance of a pedestrian hybrid beacon, it should be programmed to flash only when the pedestrian hybrid beacon is not in the dark mode.*

Standard:

13   **If a Warning Beacon is installed to supplement the W11-2 sign, the design and location of the Warning Beacon shall comply with the provisions of Sections 4S.01 and 4S.03.**

14   **Bicycle signal faces (see Chapter 4H) shall not be used at a pedestrian hybrid beacon.**

Section 4J.03 Operation of Pedestrian Hybrid Beacons

Standard:

01   **Pedestrian hybrid beacon indications shall be dark (not illuminated) during periods between actuations.**

02   **Following an actuation by a pedestrian, a pedestrian hybrid beacon face shall display a flashing CIRCULAR yellow signal indication, followed by a steady CIRCULAR yellow signal indication, followed by both steady CIRCULAR RED signal indications during the pedestrian walk interval, followed by alternating flashing CIRCULAR RED signal indications during the pedestrian change interval (see Figure 4J-3). Upon**

termination of the pedestrian change interval, the pedestrian hybrid beacon faces shall revert to a dark (not illuminated) condition.

03   Except as provided in Paragraph 4 of this Section, the pedestrian signal heads shall continue to display a steady UPRAISED HAND (symbolizing DONT WALK) signal indication when the pedestrian hybrid beacon faces are either dark or displaying flashing or steady CIRCULAR yellow signal indications. The pedestrian signal heads shall display a WALKING PERSON (symbolizing WALK) signal indication when the pedestrian hybrid beacon faces are displaying steady CIRCULAR RED signal indications. The pedestrian signal heads shall display a flashing UPRAISED HAND (symbolizing DONT WALK) signal indication when the pedestrian hybrid beacon faces are displaying alternating flashing CIRCULAR RED signal indications. Upon termination of the pedestrian change interval, the pedestrian signal heads shall revert to a steady UPRAISED HAND (symbolizing DONT WALK) signal indication.

Option:

04   Where the pedestrian hybrid beacon is installed adjacent to a roundabout to facilitate crossings by pedestrians with vision disabilities and an engineering study determines that pedestrians without vision disabilities can be allowed to cross the roadway without actuating the pedestrian hybrid beacon, the pedestrian signal heads may be dark (not illuminated) when the pedestrian hybrid beacon faces are dark.

Guidance:

05   *The duration of the flashing yellow interval should be determined by engineering judgment.*

06   *The duration of the flashing yellow interval should not vary on a cycle-by-cycle basis.*

07   *If the pedestrian hybrid beacon is coordinated as a part of a signal system, it should remain in the dark condition after a pedestrian actuation has been received until the point in the background cycle when the predetermined duration of the flashing yellow interval needs to be initiated in order to achieve the appropriate coordinated offset.*

Option:

08   If a minimum dark time between activations of the pedestrian hybrid beacon has been set on the controller, the pedestrian hybrid beacon may remain in the dark condition after a pedestrian actuation has been received until the minimum dark time has been provided.

Support:

09   The minimum dark time is a preprogrammed time set in the controller that provides time between the pedestrian actuation and beginning of the flashing yellow interval. At locations in coordinated signal systems, the dark time can be variable based on when the pedestrian actuation occurs in the coordinated signal timing sequence.

Standard:

10   **The duration of the steady yellow change interval shall be determined using engineering practices in accordance with the provisions in Section 4F.17.**

Guidance:

11   *A steady yellow change interval should have a minimum duration of 3 seconds and a maximum duration of 6 seconds (see Section 4F.17). The longer intervals should be reserved for use on approaches with higher speeds.*

Option:

12   A steady red clearance interval may be used after the steady yellow change interval.

13   The alternating flashing CIRCULAR RED signal indications may continue to flash for a short period after the pedestrian change interval has terminated to provide a buffer interval for pedestrians.

Guidance:

14   *A pedestrian hybrid beacon that is located 200 feet or less from an active grade crossing should be preempted in accordance with the applicable provisions in Sections 4F.19 and 8D.09.*

Standard:

15   **If a pedestrian hybrid beacon is placed into a flashing mode by a conflict monitor (malfunction management unit) or by a manual switch, the pedestrian hybrid beacon faces shall display flashing CIRCULAR YELLOW signal indications to each approach of the major street and the pedestrian signal heads shall revert to a dark (not illuminated) condition.**

CHAPTER 4K. ACCESSIBLE PEDESTRIAN SIGNALS AND DETECTORS

Section 4K.01 General

Support:

01   Accessible pedestrian signals and detectors provide information in non-visual formats (such as audible tones and/or speech messages, and vibrating surfaces). The decision of when to use accessible pedestrian signals is subject to requirements of the Americans with Disabilities Act and Section 504 of the Rehabilitation Act of 1973.

02   The primary technique that pedestrians with vision disabilities use to cross streets at signalized locations is to initiate their crossing when they hear the traffic in front of them stop and the traffic alongside them begin to move, which often corresponds to the onset of the green interval. The existing environment is often not sufficient to provide the information that pedestrians with vision disabilities need to cross a roadway at a signalized location.

03   The following factors are relevant in determining whether a particular signalized location presents difficulties for pedestrians with vision disabilities to cross the roadway:

    A. Potential demand for accessible pedestrian signals;
    B. A request for accessible pedestrian signals;
    C. Traffic volumes during times when pedestrians might be present, including periods of low traffic volumes or high turn-on-red volumes;
    D. The complexity of the traffic signal phasing (such as split phases, protected turn phases, leading pedestrian intervals, and exclusive pedestrian phases); and
    E. The complexity of the intersection geometry.

04   The factors that make crossing at a signalized location difficult for pedestrians with vision disabilities include: increasingly quiet vehicles, turns on red (which mask the beginning of the through phase), continuous turning movements, complex signal operations, circular intersections, and wide streets. In addition, low traffic volumes might make it difficult for pedestrians with vision disabilities to discern signal phase changes.

05   State and local organizations providing support services to pedestrians with vision and/or hearing disabilities can provide advice to the traffic engineer on site-specific accessibility decisions. In addition, orientation and mobility specialists or similar staff can provide advice to inform such decisions. The U.S. Access Board (www.access-board.gov) provides technical assistance for making pedestrian signal information accessible to persons with vision disabilities.

Standard:

06   **When used, accessible pedestrian signals shall be used in combination with pedestrian signal timing.**

07   **The information provided by an accessible pedestrian signal shall indicate which pedestrian crossing is served by each device.**

08   **Under steady (stop-and-go) operation, accessible pedestrian signals shall not be limited in operation by the time of day or day of week.**

Option:

09   Accessible pedestrian signal detectors may be push buttons or passive detection devices.

10   At locations with pretimed traffic control signals or non-actuated approaches, pedestrian push buttons may be used to activate the accessible pedestrian signals.

Support:

11   Accessible pedestrian signals are typically integrated into the pedestrian detector (push button), so the audible tones and/or messages come from the push button housing. They have a push button locator tone and a vibrotactile arrow, and can include audible beaconing and other special features.

Option:

12   The name of the street to be crossed may also be provided in accessible format, such as Braille or raised characters. Tactile maps of crosswalks may also be provided.

Support:

13   Specifications regarding Braille or raised characters can be found in the U.S. Department of Justice 2010 ADA Standards for Accessible Design, September 15, 2010, 28 CFR 35 and 36, Americans with Disabilities Act of 1990.

Standard:

14   **At accessible pedestrian signal locations where pressing the pedestrian push button is necessary to activate the walk interval, pressing the pedestrian push button shall activate both the walk interval and the accessible pedestrian signals.**

Section 4K.02 Location

Support:

01   Accessible pedestrian signals that are located as close as possible to pedestrians waiting to cross the street provide the clearest and least ambiguous indication of which pedestrian crossing is served by a device.

Guidance:

02   *Push buttons for accessible pedestrian signals should be located in accordance with the provisions of Section 4I.05 and should be located as close as possible to the crosswalk line furthest from the center of the intersection and as close as possible to the curb ramp.*

Standard:

03   **Except for the situation regarding simultaneous walk indications for all crosswalks, if two accessible pedestrian push buttons are placed less than 10 feet apart or on the same pole (see Paragraphs 7 and 8 in Section 4I.05), each accessible pedestrian push button shall be provided with the following features:**

   A. A push button locator tone,
   B. A vibrotactile walk indication,
   C. A speech walk message for the WALKING PERSON (symbolizing WALK) indication (see Section 4K.03), and
   D. A speech push button information message (see Section 4K.05).

04   **If the pedestrian clearance time is sufficient only to cross from the curb or edge of pavement to a median of sufficient width for pedestrians to wait and accessible pedestrian signal detectors are used, an additional accessible pedestrian signal detector shall be provided in the median.**

Section 4K.03 Walk Indications

Support:

01   Technology that provides different sounds for each non-concurrent signal phase has frequently been found to provide ambiguous information. Research indicates that a rapid percussive tone for each crossing coming from accessible pedestrian signal devices on separated poles located close to each crosswalk provides unambiguous information to pedestrians with vision disabilities. Vibrotactile indications provide information to pedestrians who are blind and deaf and are also used by pedestrians who are blind or who have low vision to confirm the walk signal in noisy situations.

Standard:

02   **Accessible pedestrian signals shall have both audible and vibrotactile walk indications.**

03   **Vibrotactile walk indications shall be provided by a vibrotactile arrow that is located on the push button (see Paragraph 1 in Section 4K.04). The vibrotactile arrow shall vibrate during the walk interval.**

04   **Accessible pedestrian signals shall have an audible walk indication during the walk interval only.**

05   **The audible walk indication shall be audible at the beginning of the associated crosswalk. The accessible walk indication shall have the same duration as the pedestrian walk signal except when the pedestrian signal rests in walk.**

Guidance:

06   *If the pedestrian signal rests in walk, the accessible walk indication should be limited to the first 7 seconds of the walk interval. The accessible walk indication should be recalled by a button press during the walk interval provided that the crossing time remaining is longer than the pedestrian change interval.*

Standard:

07   **Where two accessible pedestrian signals on one corner, or in a median, that are associated with different phases are placed less than 10 feet apart, the audible walk indication shall be a speech walk message (see Paragraph 3 in Section 4K.02). In all other cases, including at midblock crossings, on corners where only one accessible pedestrian signal is present, in a median, and on corners where two accessible pedestrian signals are separated by a distance of at least 10 feet, the audible walk indication shall be a percussive tone.**

08   **Audible tone walk indications shall repeat at eight to ten ticks per second. Audible tones used as walk indications shall consist of multiple frequencies with a dominant component at 880 Hz.**

Guidance:

09   *The volume of audible walk indications and push button locator tones (see Section 4K.04) should be set to be a maximum of 5 dBA louder than ambient sound, except when audible beaconing is provided in response to an extended push button press.*

Standard:

10   **Automatic volume adjustment up to a maximum volume of 100 dBA in response to ambient traffic sound level shall be provided.**

Guidance:

11   *The sound level of audible walk indications and push button locator tones should be adjusted to be low enough to avoid misleading pedestrians with vision disabilities when the following conditions exist:*

    A. Where there is an island that allows unsignalized right turns across a crosswalk between the island and the sidewalk.

    B. Where multi-leg approaches or complex signal phasing require more than two pedestrian phases, such that it might be unclear which crosswalk is served by each audible tone.

    C. At intersections where a diagonal pedestrian crossing is allowed, or where one street receives a WALKING PERSON (symbolizing WALK) signal indication simultaneously with another street.

Option:

12   An alert tone, which is a very brief burst of high-frequency sound at the beginning of the audible walk indication that rapidly decays to the frequency of the walk tone, may be used to alert pedestrians to the beginning of the walk interval.

Support:

13   An alert tone can be particularly useful if the walk tone is not easily audible in some traffic conditions.

14   Speech walk messages communicate to pedestrians which street has the walk interval. To be a useful system, the words and their meaning need to be correctly understood by all users in the context of the street environment where they are used. Because of this, tones are the preferred means of providing audible walk indications except where two accessible pedestrian signals on one corner are not separated by a distance of at least 10 feet.

15   If speech walk messages are used, pedestrians have to know the names of the streets that they are crossing in order for the speech walk messages to be unambiguous. In getting directions to travel to a new location, pedestrians with vision disabilities do not always get the name of each street to be crossed. Therefore, it is desirable to give users of accessible pedestrian signals the name of the street controlled by the push button. This can be done by means of a speech push button information message (see Section 4K.05) during the flashing or steady UPRAISED HAND intervals, or by raised print and Braille labels on the push button housing.

16   By combining the information from the push button message or Braille label, the vibrotactile arrow aligned in the direction of travel on the relevant crosswalk, and the speech walk message, pedestrians with vision disabilities are able to correctly respond to speech walk messages even if there are two push buttons on the same pole.

Standard:

17   **If speech walk messages are used to communicate the walk interval, they shall provide a clear message that the walk interval is in effect, as well as to which crossing it applies.**

Guidance:

18   *Speech walk messages that are used at intersections having pedestrian phasing that is concurrent with vehicular phasing should be patterned after the model: "Broadway. Walk sign is on to cross Broadway."*

19   *Speech walk messages that are used at intersections having exclusive pedestrian phasing should be patterned after the model: "Walk sign is on for all crossings."*

20   *Speech walk messages should not contain any additional information, except they should include designations such as "Street" or "Avenue" where this information is necessary to avoid ambiguity at a particular location.*

21   *Speech walk messages should not state or imply a command to the pedestrian, such as "Cross Broadway now." Speech walk messages should not tell pedestrians that it is "safe to cross," because it is always the pedestrian's responsibility to check actual traffic conditions.*

Standard:

22   **A speech walk message is not required at times when the walk interval is not timing, but, if provided:**

    A. It shall begin with the term "wait."

    B. It need not be repeated for the entire time that the walk interval is not timing.

23   **If a pilot light (see Section 4I.05) is used at an accessible pedestrian signal location, each actuation shall be accompanied by the speech message "wait."**

Option:

24   Accessible pedestrian signals that provide speech walk messages may provide similar messages in languages other than English, if needed.

Standard:

25   **If used, speech walk messages in a language other than English shall be stated first in English, and then repeated in the second language, alternating back and forth while the walk interval is timing.**

Section 4K.04 Vibrotactile Arrows and Locator Tones

Standard:

01   **To enable pedestrians with vision disabilities to distinguish and locate the appropriate push button at an accessible pedestrian signal location, and to help them align with the crosswalk, each push button shall clearly indicate by means of a vibrotactile arrow which crosswalk signal is actuated by the push button. Vibrotactile arrows shall be located on the button of the push button assembly, shall have high visual contrast (light on dark or dark on light), and shall be aligned parallel to the direction of travel on the associated crosswalk.**

02   **A locator tone shall be incorporated into the accessible pedestrian signal equipment to help pedestrians with vision disabilities locate the vibrotactile arrow, and the associated push button if a push button is provided.**

Support:

03   A push button locator tone is a repeating sound that informs approaching pedestrians that a push button to actuate pedestrian timing or receive additional information exists, and that enables pedestrians with vision disabilities to locate the push button.

Standard:

04   **Push button locator tones shall have a duration of 0.15 seconds or less, and except as provided in Paragraph 5 of this Section, push button locator tones shall repeat at 1-second intervals at all times that the audible walk indication is not active, including during the pedestrian change interval and during the time that the pedestrian signal is resting in walk (see Paragraph 6 in Section 4K.03).**

Option:

05   The push button locator tone may default to a deactivated mode during periods when the steady UPRAISED HAND (symbolizing DONT WALK) signal indication is being displayed for the associated crosswalk if a passive pedestrian detection system is implemented that activates the locator tone at all times (other than when the audible walk indication is active) that a pedestrian is present within a 12-foot radius from the push button location. Where pedestrian facilities (such as sidewalks) are present, the passive detection requirement may be reduced such that it only applies to pedestrians who are on the pedestrian facilities within the 12-foot radius from the push button location.

Standard:

06   **Push button locator tones shall be deactivated when the traffic control signal or pedestrian hybrid beacon is operating in a flashing mode. This requirement shall not apply to traffic control signals or pedestrian hybrid beacons that are activated from a flashing or dark mode to a steady (stop-and-go) mode by pedestrian actuations.**

07   **Push button locator tones shall be intensity responsive to ambient sound.**

Guidance:

08   *Push button locator tones should be audible 6 to 12 feet from the push button, or to the building line, whichever is less.*

Support:

09   Section 4K.03 contains additional provisions regarding the volume and sound level of push button locator tones.

Section 4K.05 Extended Push Button Press Features

Option:

01   Pedestrians may be provided with additional features such as increased crossing time, audible beaconing, or a speech push button information message as a result of an extended push button press.

Standard:

02   **If an extended push button press (see Paragraph 18 in Section 4I.05) is used to provide any additional feature(s), a push button press of less than one second shall actuate only the pedestrian timing and any**

associated accessible walk indication, and a push button press of one second or more shall actuate the pedestrian timing, any associated accessible walk indication, and any additional feature(s).

03   **If additional crossing time is provided by means of an extended push button press, a PUSH BUTTON FOR 2 SECONDS FOR EXTRA CROSSING TIME (R10-32P) plaque (see Section 4I.05) shall be installed adjacent to the pedestrian push button detector.**

Support:

04   Audible beaconing is the use of an audible signal in such a way that pedestrians with vision disabilities can home in on the signal that is located on the far end of the crosswalk as they cross the street.

05   Not all crosswalks at an intersection need audible beaconing. Audible beaconing is not appropriate at locations with channelized turns or split phasing, because of the possibility of confusion.

Guidance:

06   *Audible beaconing should be considered following an engineering study at:*

    A. Crosswalks longer than 70 feet, unless those crosswalks are divided by a median that has another accessible pedestrian signal with a locator tone;
    B. Crosswalks that are skewed;
    C. Intersections with irregular geometry, such as more than four legs;
    D. Crosswalks where audible beaconing is requested by a person with vision disabilities; or
    E. Other locations where a study indicates audible beaconing would be beneficial.

07   *If audible beaconing is used, it should be initiated by an extended push button press.*

Standard:

08   **If audible beaconing is used, the volume of the push button locator tone during the pedestrian change interval of the called pedestrian phase shall be increased up to a maximum of 100 dBA, and shall come from a loudspeaker that is mounted at the far end of the crosswalk at a height of 7 to 10 feet above the pavement.**

Guidance:

09   *The audible beaconing loudspeaker mounted at the far end of the crosswalk should be within the width of the crosswalk.*

Support:

10   When the locator tone is active during the pedestrian change interval at a traffic control signal or pedestrian hybrid beacon where audible beaconing is used, the locator tone from the audible beaconing loudspeaker is at an elevated volume, while the locator tone from the accessible pedestrian signal is at its normal, quiet setting.

Option:

11   The sound level of the accessible pedestrian signal walk indication and subsequent push button locator tone may be increased by an extended push button press.

12   Speech push button information messages may provide intersection identification, as well as information about unusual intersection signalization and geometry, such as notification regarding exclusive pedestrian phasing, leading pedestrian intervals, split phasing, diagonal crosswalks, and medians or islands.

Standard:

13   **If speech push button information messages are made available by actuating the accessible pedestrian signal detector, they shall only be actuated when the walk interval is not timing. They shall begin with the term "Wait," followed by intersection identification information modeled after: "Wait to cross Broadway at Grand." If information on intersection signalization or geometry is also given, it shall follow the intersection identification information.**

Guidance:

14   *Speech push button information messages should not be used to provide landmark information or to inform pedestrians with vision disabilities about detours or temporary traffic control situations.*

Support:

15   Additional information on the structure and wording of speech push button information messages is included in the Institute of Transportation Engineers' "Electronic Toolbox for Making Intersections More Accessible for Pedestrians Who Are Blind or Visually Impaired."

CHAPTER 4L. RECTANGULAR RAPID FLASHING BEACONS

Section 4L.01 Application of Rectangular Rapid Flashing Beacons

Option:

01   A pedestrian-activated and/or bicyclist-activated rectangular rapid flashing beacon (RRFB) may be used to provide supplemental emphasis to pedestrian, school, and trail warning signs at marked crosswalks across uncontrolled approaches.

Standard:

02   **An RRFB shall only be installed to function as a Warning Beacon (see Section 4S.03). Except as otherwise provided in this Chapter, all other provisions of the MUTCD applicable to Warning Beacons shall apply to RRFBs.**

03   **An RRFB shall only be used to supplement a post-mounted W11-2 (Pedestrian), S1-1 (School), or W11-15 (Trail) crossing warning sign with a diagonal downward arrow (W16-7P) plaque, or an overhead-mounted W11-2 , S1-1, or W11-15 crossing warning sign, located at or immediately adjacent to a marked crosswalk.**

04   **Except for crosswalks across the approach to or egress from a roundabout, or crosswalks across free-flow turn lanes separated by a channelizing island, an RRFB shall not be used for crosswalks across approaches controlled by YIELD signs, STOP signs, traffic control signals, or pedestrian hybrid beacons.**

Option:

05   An additional RRFB may be installed on that approach in advance of the crosswalk, as a Warning Beacon to supplement a W11-2 (Pedestrian), S1-1 (School), or W11-15 (Trail) crossing warning sign with an AHEAD (W16-9P) or distance (W16-2P or W16-2aP) plaque.

Standard:

06   **If an additional RRFB is installed on the approach in advance of the crosswalk, it shall be supplemental to and not a replacement for the RRFB at the crosswalk itself.**

Section 4L.02 Design of Rectangular Rapid Flashing Beacons

Standard:

01   **Each RRFB unit shall consist of two rapidly-flashed rectangular-shaped yellow indications, each with an LED-array based pulsing light source. The size of each RRFB indication shall be at least 5 inches wide by at least 2 inches high.**

02   **The two RRFB indications for each RRFB unit shall be aligned horizontally, with the longer dimension horizontal and with a minimum space between the two indications of at least 7 inches, measured from nearest edge of one indication to the nearest edge of the other indication. The outside edges of the RRFB indications, including any housings, shall not project beyond the outside edges of the W11-2, S1-1, or W11-15 sign that it supplements.**

03   **An RRFB unit shall not be installed independent of the crossing warning signs for the approach that the RRFB faces. If the RRFB unit is supplementing a post-mounted sign, the RRFB unit shall be installed on the same support as the associated W11-2, S1-1, or W11-15 crossing warning sign and plaque. If the RRFB unit is supplementing an overhead-mounted sign, the RRFB unit shall be mounted directly above the top of sign or below the bottom of the sign.**

Option:

04   As a specific exception to Paragraph 6 of Section 4S.01, the RRFB unit associated with a post-mounted sign and plaque may be located between and immediately adjacent to the bottom of the crossing warning sign and the top of the supplemental downward diagonal arrow plaque (or, in the case of a supplemental advance sign, the AHEAD or distance plaque) or within 12 inches above the crossing warning sign, rather than the recommended minimum of 12 inches above or below the sign assembly.

05   Signal visors and backplates, with or without a yellow retroreflective strip, may be used with RRFB units based on provisions in Section 4D.06.

Standard:

06   **For any approach on which RRFBs are used to supplement post-mounted signs, at least two W11-2 , S1-1, or W11-15 crossing warning signs (each with an RRFB unit and a W16-7P plaque) shall be installed at the crosswalk, one on the right-hand side of the roadway and one on the left-hand side of the roadway.**

Guidance:

07   *On a divided highway, the left-hand side RRFB assembly should be installed on the median, if practicable, rather than on the far left side of the highway.*

Standard:

08   **For any approach on which RRFBs are used to supplement an overhead-mounted sign, at least one W11-2 , S1-1, or W11-15 crossing warning sign (without a W16-7P plaque) located approximately over the center of the lanes of the approach (or where optimum visibility can be achieved) shall be installed at the crosswalk.**

Standard:

09   **If used at intersections, the design of the RRFBs shall conform to the requirements for post-mounted or overhead placement described in Paragraph 3 of this Section.**

Option:

10   RRFBs may be installed at intersections with more than one crosswalk on the same uncontrolled approach (see Figure 4L-1).

11   If used at intersections with two crosswalks on an uncontrolled approach, post-mounted RRFBs may be installed to face only one direction of travel at the first crosswalk that traffic encounters (see Figure 4L-1).

Figure 4L-1. Example of RRFBs at Uncontrolled, Marked Crosswalks at an Intersection

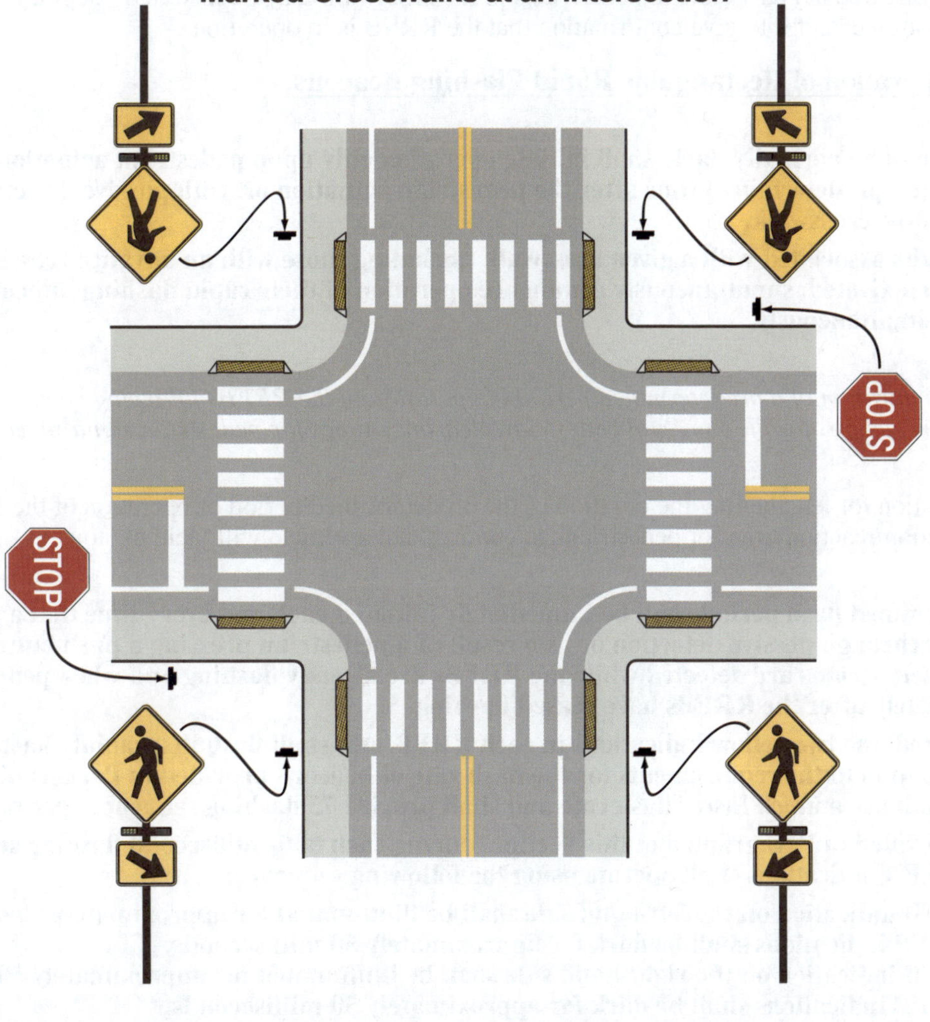

Notes:

1. When activated, the RRFBs on both approaches shall simultaneously commence operation of their rapid flashing indications and shall cease operation simultaneously.

2. If placed overhead, follow the requirements of Paragraph 8 of Section 4L.02, except that the signs may be placed approximately over the center of the intersection.

Standard:

12   **The light intensity of the yellow indications during daytime conditions shall meet the minimum specifications for Class 1 yellow peak luminous intensity in the publication "Directional Flashing Optical Warning Devices for Authorized Emergency, Maintenance, and Service Vehicles J595," 2005, Society of Automotive Engineers (SAE).**

Option:

13   If the RRFB indications are so bright that they cause excessive glare during nighttime conditions, an automatic signal dimming device may be used to reduce the brilliance of the RRFB indications during nighttime conditions.

Standard:

14   **If pedestrian push button detectors (rather than passive detection) are used to actuate the RRFB indications, a PUSH BUTTON TO TURN ON WARNING LIGHTS/AWAIT GAP IN TRAFFIC (R10-25) sign (see Section 2B.58) shall be installed explaining the purpose and use of the pedestrian push button detector.**

Support:

15   Section 4I.05 contains further information about pedestrian push button detector location criteria.

16   Section 4H.12 contains information about bicyclist push buttons.

Guidance:

17   *An audible information device should be used with RRFBs to assist pedestrians with vision disabilities.*

Option:

18   A small light directed at and visible to pedestrians in the crosswalk may be installed integral to the RRFB or pedestrian push button detector to give confirmation that the RRFB is in operation.

Section 4L.03 Operation of Rectangular Rapid Flashing Beacons

Standard:

01   **The RRFB shall be normally dark, shall initiate operation only upon pedestrian actuation, and shall cease operation at a predetermined time after the pedestrian actuation or, with passive detection, after the pedestrian clears the crosswalk.**

02   **All RRFB units associated with a given crosswalk (including those with an advance crossing sign, if used) shall, when activated, simultaneously commence operation of their rapid flashing indications and shall cease operation simultaneously.**

Guidance:

03   *The minimum duration of a predetermined period of operation of the RRFBs following each actuation should be based on the procedures for the timing of pedestrian clearance times for pedestrian signals (see Section 4I.06).*

Support:

04   One consideration for lengthening the duration of the predetermined period of operation of the RRFBs is adding the perception/reaction time for pedestrians to confirm that a vehicle will yield or stop.

Standard:

05   **The predetermined flash period shall be immediately initiated each and every time that a pedestrian is detected either through passive detection or as a result of a pedestrian pressing a push button detector, including when pedestrians are detected while the RRFBs are already flashing and when pedestrians are detected immediately after the RRFBs have ceased flashing.**

06   **When activated, the two yellow indications in each RRFB unit shall flash in a rapidly flashing sequence. As a specific exception to the requirements for the flash rate of beacons provided in Paragraph 3 of Section 4S.01, RRFBs shall use a much faster flash rate and shall provide 75 flashing sequences per minute.**

07   **Except as provided in Paragraph 8 of this Section, during each 800-millisecond flashing sequence, the left and right RRFB indications shall operate using the following sequence:**

    A. **The RRFB indication on the left-hand side shall be illuminated for approximately 50 milliseconds.**
    B. **Both RRFB indications shall be dark for approximately 50 milliseconds.**
    C. **The RRFB indication on the right-hand side shall be illuminated for approximately 50 milliseconds.**
    D. **Both RRFB indications shall be dark for approximately 50 milliseconds.**
    E. **The RRFB indication on the left-hand side shall be illuminated for approximately 50 milliseconds.**
    F. **Both RRFB indications shall be dark for approximately 50 milliseconds.**
    G. **The RRFB indication on the right-hand side shall be illuminated for approximately 50 milliseconds.**

    H. Both RRFB indications shall be dark for approximately 50 milliseconds.
    I. Both RRFB indications shall be illuminated for approximately 50 milliseconds.
    J. Both RRFB indications shall be dark for approximately 50 milliseconds.
    K. Both RRFB indications shall be illuminated for approximately 50 milliseconds.
    L. Both RRFB indications shall be dark for approximately 250 milliseconds.

08   **The flash rate of each individual RRFB indication, as applied over the full flashing sequence, shall not be more than 5 flashes per second, to avoid frequencies that might cause seizures.**

09   **If an audible information device is used in conjunction with an RRFB, the audible information device shall not use vibrotactile indications or percussive indications.**

Guidance:

10   *If an audible information device is used in conjunction with an RRFB, the audible message should be a speech message that says, "Warning lights are flashing." The audible message should be spoken twice.*

CHAPTER 4M. TRAFFIC CONTROL SIGNALS FOR EMERGENCY-VEHICLE ACCESS

Section 4M.01 Application of Emergency-Vehicle Traffic Control Signals

Support:

01   An emergency-vehicle traffic control signal is a special traffic control signal that directs all conflicting traffic to stop in order to permit the driver of an authorized emergency vehicle to proceed into the roadway or intersection.

Option:

02   An emergency-vehicle traffic control signal may be installed at a location that does not meet other traffic signal warrants such as at an intersection or other location to permit direct access from a building housing the emergency vehicle.

03   An emergency-vehicle hybrid beacon may be installed instead of an emergency-vehicle traffic control signal under the conditions described in Section 4N.01.

Guidance:

04   *If a traffic control signal is not justified under the signal warrants of Chapter 4C and if gaps in traffic are not adequate to permit the timely entrance of emergency vehicles, or the stopping sight distance for vehicles approaching on the major street is insufficient for emergency vehicles, installing an emergency-vehicle traffic control signal should be considered. If one of the signal warrants of Chapter 4C is met and a traffic control signal is justified by an engineering study, and if a decision is made to install a traffic control signal, it should be installed based upon the provisions of Chapters 4D through 4I and 4K.*

05   *The sight distance determination should be based on the location of the visibility obstruction for the critical approach lane for each street or drive and the posted or statutory speed limit or 85th-percentile speed on the major street, whichever is higher.*

Section 4M.02 Design of Emergency-Vehicle Traffic Control Signals

Standard:

01   **Except as otherwise provided in this Section, an emergency-vehicle traffic control signal shall meet the requirements of this Manual.**

02   **An Emergency Vehicle (W11-8) sign (see Section 2C.54) with an EMERGENCY SIGNAL AHEAD (W11-12P) supplemental plaque shall be placed in advance of all emergency-vehicle traffic control signals. If a Warning Beacon is installed to supplement the W11-8 sign, the design and location of the beacon shall comply with the Standards of Sections 4S.01 and 4S.03.**

Guidance:

03   *At least one of the two required signal faces for each approach on the major street should be located over the roadway.*

04   *The following size signal indications should be used for emergency-vehicle traffic control signals: 12-inch diameter for steady red and steady yellow circular signal indications and any arrow indications, and 8-inch diameter for green or flashing yellow circular signal indications.*

Standard:

05   **An EMERGENCY SIGNAL (R10-13) sign (see Section 2B.59) shall be installed facing each major-street approach.**

06   **If an overhead signal face is provided, the EMERGENCY SIGNAL sign shall be mounted adjacent to the overhead signal face.**

Option:

07   An approach that only serves emergency vehicles may be provided with only one signal face consisting of one or more signal sections.

08   Besides using an 8-inch diameter signal indication, other appropriate means to reduce the flashing yellow light output may be used.

Section 4M.03 Operation of Emergency-Vehicle Traffic Control Signals

Standard:

01   **Green signal indications for emergency vehicles at signalized locations operating in the steady (stop-and-go) mode shall be obtained as provided in Section 4F.19.**

02   **As a minimum, the signal indications, sequence, and manner of operation of an emergency-vehicle traffic control signal installed at a midblock location shall be as follows:**

    A. **The signal indication, between emergency-vehicle actuations, shall be either green or flashing yellow. If the flashing yellow signal indication is used instead of the green signal indication, it shall be displayed in the normal position of the green signal indication, while the steady red and steady yellow signal indications shall be displayed in their normal positions.**

    B. **When an emergency-vehicle actuation occurs, a steady yellow change interval followed by a steady red interval shall be displayed to traffic on the major street.**

    C. **A yellow change interval is not required following the green interval for the emergency-vehicle driveway.**

Guidance:

03   *Emergency-vehicle traffic control signals located at intersections should either be operated in the flashing mode (see Sections 4G.01 and 4G.03) between emergency-vehicle actuations or be full-actuated or semi-actuated to accommodate normal vehicular and pedestrian traffic on the streets.*

04   *Warning Beacons, if used with an emergency-vehicle traffic control signal, should be flashed only:*

    A. For an appropriate time in advance of and during the steady yellow change interval for the major street, and

    B. During the steady red interval for the major street.

05   *The duration of the steady red interval for traffic on the major street should be determined by on-site test-run time studies, but should not exceed 1.5 times the time required for the emergency vehicle to clear the path of conflicting vehicles.*

Option:

06   An emergency-vehicle traffic control signal sequence may be initiated manually from a local control point such as a fire station or law enforcement headquarters or from an emergency vehicle equipped for remote operation of the signal.

CHAPTER 4N. HYBRID BEACONS FOR EMERGENCY-VEHICLE ACCESS

Section 4N.01 Application of Emergency-Vehicle Hybrid Beacons

Standard:

01 **Emergency-vehicle hybrid beacons shall be used only in conjunction with signs to warn and control traffic at an unsignalized location where emergency vehicles enter or cross a street or highway. Emergency-vehicle hybrid beacons shall be actuated only by authorized emergency or maintenance personnel.**

Guidance:

02 *Emergency-vehicle hybrid beacons should only be used when all of the following criteria are satisfied:*

   A. *The conditions justifying an emergency-vehicle traffic control signal (see Section 4M.01) are met;*
   B. *An engineering study, considering the road width, approach speeds, and other pertinent factors, determines that emergency-vehicle hybrid beacons can be designed and located in compliance with the requirements contained in this Chapter and in Section 4S.01, such that they effectively warn and control traffic at the location; and*
   C. *The location is not at or within 100 feet from an intersection or driveway where the side road or driveway is controlled by a STOP or YIELD sign.*

Section 4N.02 Design of Emergency-Vehicle Hybrid Beacons

Standard:

01 **Except as otherwise provided in this Section, an emergency-vehicle hybrid beacon shall meet the requirements of this Manual.**

02 **An emergency-vehicle hybrid beacon face shall consist of three signal sections, with a CIRCULAR YELLOW signal indication centered below two horizontally-aligned CIRCULAR RED signal indications (see Figure 4N-1).**

03 **At least two emergency-vehicle hybrid beacon faces shall be installed for each approach of the major street.**

Guidance:

04 *On approaches having posted or statutory speed limits or 85th-percentile speeds in excess of 40 mph, and on approaches having traffic or operating conditions that would tend to obscure visibility of roadside beacon faces, both of the minimum of two emergency-vehicle hybrid beacon faces should be installed over the roadway.*

05 *On multi-lane approaches having posted or statutory speed limits or 85th-percentile speeds of 40 mph or less, either an emergency-vehicle hybrid beacon face should be installed on each side of the approach (if a median of sufficient width exists) or at least one of the emergency-vehicle hybrid beacon faces should be installed over the roadway.*

06 *An emergency-vehicle hybrid beacon should comply with the signal face location provisions described in Sections 4D.05 through 4D.10.*

Standard:

07 **Stop lines and EMERGENCY SIGNAL—STOP ON FLASHING RED (R10-14 or R10-14a) signs (see Section 2B.59) shall be used with emergency-vehicle hybrid beacons for each approach of the major street.**

Figure 4N-1. Sequence for an Emergency-Vehicle Hybrid Beacon

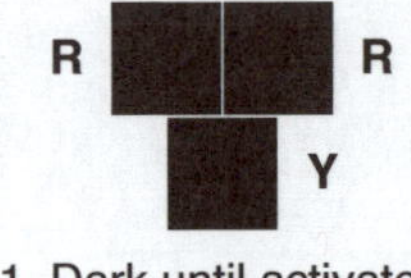

1. Dark until activated

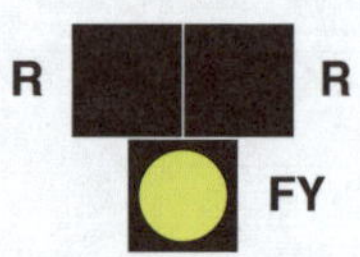

2. Flashing yellow upon activation

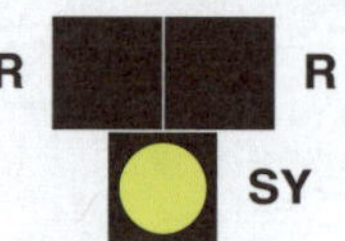

3. Steady yellow

4. Alternating flashing red during egress of the emergency vehicle(s)

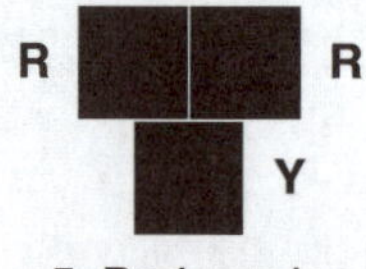

5. Dark again until activated

Note: An optional steady red clearance interval may be used after Interval 3 and before Interval 4.

Option:

08   If needed for extra emphasis, a STOP HERE ON FLASHING RED (R10-14b) sign (see Section 2B.59) may be installed with an emergency-vehicle hybrid beacon.

09   Emergency-vehicle hybrid beacons may be equipped with a light or other display visible to the operator of the egressing emergency vehicle to provide confirmation that the beacons are operating.

10   Emergency-vehicle hybrid beacons may be supplemented with an advance warning sign, which may also be supplemented with a Warning Beacon (see Section 4S.03).

Guidance:

11   *If a Warning Beacon is used to supplement the advance warning sign, it should be programmed to flash only when the emergency-vehicle hybrid beacon is not in the dark mode.*

Section 4N.03 <u>Operation of Emergency-Vehicle Hybrid Beacons</u>

Standard:

01   **Emergency-vehicle hybrid beacons shall be placed in a dark mode (no indications displayed) during periods between actuations.**

02   **Upon actuation by authorized emergency personnel, the emergency-vehicle hybrid beacon faces shall each display a flashing yellow signal indication, followed by a steady yellow change interval, prior to displaying two CIRCULAR RED signal indications in an alternating flashing array for a duration of time adequate for egress of the emergency vehicles (see Figure 4N-1). The alternating flashing red signal indications shall only be displayed when it is required that drivers on the major street stop and then proceed subject to the rules applicable after making a stop at a STOP sign. Upon termination of the flashing red signal indications, the emergency-vehicle hybrid beacons shall revert to a dark mode (no indications displayed) condition.**

Guidance:

03   *The duration of the flashing yellow interval should be determined by engineering judgment.*

Standard:

04   **The duration of the steady yellow change interval shall be determined using engineering practices in accordance with the provisions in Section 4F.17.**

Guidance:

05   *A yellow change interval should have a minimum duration of 3 seconds and a maximum duration of 6 seconds (see Section 4F.17). The longer intervals should be reserved for use on approaches with higher speeds.*

Option:

06   A steady red clearance interval may be used after the steady yellow change interval.

Guidance:

07   *An emergency-vehicle hybrid beacon that is located 200 feet or less from an active grade crossing should be preempted in accordance with the applicable provisions in Sections 4F.19 and 8D.09.*

Standard:

08   **If an emergency-vehicle hybrid beacon is placed into a flashing mode by a conflict monitor (malfunction management unit) or by a manual switch, the emergency-vehicle hybrid beacon faces shall display flashing yellow signal indications to each approach of the major street.**

CHAPTER 4O. TRAFFIC CONTROL SIGNALS FOR ONE-LANE, TWO-WAY FACILITIES

Section 4O.01 Application of Traffic Control Signals for One-Lane, Two-Way Facilities

Support:

01   A traffic control signal at a narrow bridge, tunnel, or roadway section that is not of sufficient width for two opposing vehicles to pass is a special signal that alternates which direction of travel is permitted to proceed.

02   Temporary traffic control signals (see Sections 4D.11 and 6L.01) are the most frequent application of traffic control signals for one-lane, two-way facilities.

Guidance:

03   *Sight distance across or through the one-lane, two-way facility should be considered as well as the approach speed and sight distance approaching the facility when determining whether traffic control signals should be installed.*

Option:

04   At a narrow bridge, tunnel, or roadway section where a traffic control signal is not justified under the conditions of Chapter 4C, a traffic control signal may be used if gaps in opposing traffic do not permit the flow of traffic through the one-lane section of roadway.

Section 4O.02 Design of Traffic Control Signals for One-Lane, Two-Way Facilities

Standard:

01   **The provisions of Chapters 4D through 4G shall apply to traffic control signals for one-lane, two-way facilities, except that:**

    A. The durations of the red clearance intervals shall be adequate to clear the one-lane section of conflicting vehicles; and

    B. Adequate means, such as interconnection, shall be provided to prevent conflicting signal indications, such as green and green, at opposite ends of the section.

Section 4O.03 Operation of Traffic Control Signals for One-Lane, Two-Way Facilities

Guidance:

01   *Traffic control signals at one-lane, two-way facilities should operate in a manner consistent with traffic requirements.*

02   *Adequate time should be provided to allow traffic to clear the narrow facility before opposing traffic is allowed to move. Engineering judgment should be used to determine the proper timing for the signal.*

Standard:

03   **When in the flashing mode, the signal indications shall flash red.**

CHAPTER 4P. TRAFFIC CONTROL SIGNALS FOR FREEWAY ENTRANCE RAMPS

Section 4P.01 Application of Freeway Entrance Ramp Control Signals

Support:

01   Ramp control signals are traffic control signals that control the flow of traffic entering the freeway facility. This is often referred to as "ramp metering."

02   Freeway entrance ramp control signals are sometimes used if controlling traffic entering the freeway could reduce the total expected delay to traffic in the freeway corridor, including freeway ramps and local streets.

Guidance:

03   *The installation of ramp control signals should be preceded by an engineering study of the physical and traffic conditions on the highway facilities likely to be affected. The study should include the ramps and ramp connections and the surface streets that would be affected by the ramp control, as well as the freeway section concerned.*

Support:

04   Information on conditions that might justify freeway entrance ramp control signals, factors to be evaluated in traffic engineering studies for ramp control signals, design of ramp control signals, and operation of ramp control signals can be found in the FHWA's "Ramp Management and Control Handbook."

Section 4P.02 Design of Freeway Entrance Ramp Control Signals

Standard:

01   **Ramp control signals shall meet all of the standard design specifications for traffic control signals, except as otherwise provided in this Section.**

02   **The signal face for freeway entrance ramp control signals shall be either a two-section signal face containing red and green signal indications or a three-section signal face containing red, yellow, and green signal indications.**

Option:

03   Ramp control signals may be placed in the dark mode (no indications displayed) when not in use.

04   Ramp control signals may be used to control some, but not all, lanes on a ramp, such as when non-metered HOV bypass lanes are provided on a ramp.

Standard:

05   **If only one controlled lane is present on an entrance ramp, or if more than one controlled lane is present on an entrance ramp and the ramp control signals are operated such that green signal indications are always displayed simultaneously to all of the controlled lanes on the ramp, then a minimum of two signal faces per ramp shall face entering traffic.**

06   **If two controlled lanes are present on an entrance ramp and the ramp control signals are operated such that green signal indications are not always displayed simultaneously to both of the controlled lanes on the ramp, then one signal face shall be provided over the approximate center of each separately-controlled lane.**

07   **If three or more controlled lanes are present on an entrance ramp and the ramp control signals are operated such that green signal indications are not always displayed simultaneously to all of the controlled lanes on the ramp, then one signal face shall be provided over the approximate center of each separately-controlled lane.**

Guidance:

08   *Additional side-mounted signal faces should be considered for ramps with three or more separately-controlled lanes.*

Option:

09   For entrance ramps with only one controlled lane, the two required signal faces may both be mounted at the side of the roadway on a single pole (as a specific exception to the normal 8-foot minimum lateral separation of signal faces required by Section 4D.07), with the lower signal face installed at a minimum mounting height of 4.5 feet.

10   For entrance ramps with two or more controlled lanes, if two signal faces are installed for the right-hand lane or for the left-hand lane, the two signal faces for that lane may both be mounted at the closest side of the roadway on a single pole (as a specific exception to the normal 8-foot minimum lateral separation of signal faces required by Section 4D.07), with the lower signal face installed at a minimum mounting height of 4.5 feet.

Guidance:

11   *Ramp control signals should be located and designed to minimize their viewing by mainline freeway traffic.*

12   *Regulatory signs with legends appropriate to the control, such as XX VEHICLE(S) PER GREEN or XX VEHICLE(S) PER GREEN EACH LANE (see Section 2B.61), should be installed.*

13   *When ramp control signals are installed on a freeway-to-freeway ramp, special consideration should be given to assuring adequate visibility of the ramp control signals, and multiple advance warning signs with flashing Warning Beacons should be installed to warn road users of the metered operation.*

Section 4P.03 <u>Operation of Freeway Entrance Ramp Control Signals</u>

Guidance:

01   *Operational strategies for ramp control signals, such as periods of operation, metering rates and algorithms, and queue management, should be determined by the operating agency prior to the installation of the ramp control signals and should be closely monitored and adjusted as needed thereafter.*

02   *When the ramp control signals are operated only during certain periods of the day, a RAMP METERED WHEN FLASHING (W3-8) sign (see Section 2C.37) should be installed in advance of the ramp control signal near the entrance to the ramp, or on the arterial on the approach to the ramp, to alert road users to the presence and operation of ramp meters.*

Standard:

03   **The RAMP METERED WHEN FLASHING sign shall be supplemented with a Warning Beacon (see Section 4S.03) that flashes when the ramp control signal is in operation (controlling the flow of traffic entering the freeway). Flashing light-emitting diode (LED) units shall not be used within the legend or border of the sign.**

CHAPTER 4Q. TRAFFIC CONTROL FOR MOVABLE BRIDGES

Section 4Q.01 Application of Traffic Control for Movable Bridges

Support:

01   Traffic signals for movable bridges are a special type of highway traffic signal installed at movable bridges to notify road users to stop because of a road closure rather than alternately controlling the flow of conflicting traffic movements. The signals are operated in coordination with the opening and closing of the movable bridge, and with the operation of movable bridge warning and resistance gates, or other devices and features used to warn, control, and stop traffic.

02   Movable bridge warning gates installed at movable bridges decrease the likelihood of vehicles and pedestrians passing the stop line and entering an area where potential hazards exist because of bridge operations.

03   A movable bridge resistance gate is sometimes used at movable bridges and located downstream of the movable bridge warning gate. A movable bridge resistance gate provides a physical deterrent to road users when placed in the appropriate position. The movable bridge resistance gates are considered a design feature and not a traffic control device; requirements for them are contained in AASHTO's "Standard Specifications for Movable Highway Bridges."

Standard:

04   **Traffic control at movable bridges shall include both signals and gates, except in the following cases:**

   A. Neither is required if other traffic control devices or measures considered appropriate are used under either of the following conditions:

   1. On low-volume roads (roads of less than 400 vehicles average daily traffic), or
   2. At manually-operated bridges if electric power is not available.

   B. Only signals are required in urban areas if intersecting streets or driveways make gates ineffective.
   C. Only movable bridge warning gates are required if a traffic control signal that is controlled as part of the bridge operations exists within 500 feet of the movable bridge resistance gates and no intervening traffic entrances exist.

Section 4Q.02 Design and Location of Movable Bridge Signals and Gates

Standard:

01   **The signal faces and mountings of movable bridge signals shall comply with the provisions of Chapters 4D through 4G except as provided in this Section.**

02   **Signal faces with 12-inch diameter signal indications shall be used for all new movable bridge signals.**

Option:

03   Existing signal faces with 8-inch diameter lenses may be retained for the remainder of their useful service life.

Standard:

04   **Since movable bridge operations cover a variable range of time periods between openings, the signal faces shall be one of the following types:**

   A. Three-section signal faces with red, yellow, and green signal indications; or
   B. Two one-section signal faces with red signal indications in a vertical array separated by a STOP HERE ON RED (R10-6) sign (see Section 2B.59).

05   **Regardless of which signal type is selected, at least two signal faces shall be provided for each approach to the movable span and a stop line (see Section 3B.19) shall be installed to indicate the point behind which vehicles are required to stop.**

Guidance:

06   *If movable bridge operation is frequent, the use of three-section signal faces should be considered.*

07   *Insofar as practicable, the height and lateral placement of signal faces should comply with the requirements for other traffic control signals in accordance with Chapter 4D. They should be located no more than 50 feet in advance of the movable bridge warning gate.*

Option:

08   Movable bridge signals may be supplemented with audible warning devices to provide additional warning to drivers and pedestrians.

Guidance:

09   *A DRAW BRIDGE (W3-6) sign (see Section 2C.36) should be used in advance of movable bridge signals and gates to give warning to road users, except in urban conditions where such signing would not be practical.*

Standard:

10   **If physical conditions prevent a road user from having a continuous view of at least two signal indications for the distance specified in Table 4D-2, an auxiliary device (either a supplemental signal face or the DRAW BRIDGE (W3-6) sign to which has been added a Warning Beacon that is interconnected with the movable bridge controller unit) shall be provided in advance of movable bridge signals and gates.**

Option:

11   The DRAW BRIDGE (W3-6) sign may be supplemented by a Warning Beacon (see Section 4S.03).

Support:

12   If two sets of gates (both a warning and a resistance gate) are used for a single direction, highway traffic signals are not required to accompany the resistance gate nearest the span opening.

Standard:

13   **Movable bridge warning gates, if used, shall be at least standard railroad size, striped with 16-inch alternate vertical, fully-reflectorized red and white stripes. Flashing red lights in accordance with the Standards for those on railroad gates (see Section 8D.03) shall be included on the gate arm and they shall only be operated if the gate is closed or in the process of being opened or closed.**

Guidance:

14   *In the horizontal position, the top of the gate should be approximately 4 feet above the pavement.*

15   *Movable bridge warning gates should be of lightweight construction. In its normal upright position, the gate arm should provide adequate lateral clearance.*

Option:

16   The movable bridge resistance gates may be delineated, if practical, in a manner similar to the movable bridge warning gate.

Guidance:

17   *Movable bridge warning gates, if used, should extend at least across the full width of the approach lanes if movable bridge resistance gates are used. On divided highways in which the roadways are separated by a barrier median, movable bridge warning gates, if used, should extend across all roadway lanes approaching the span openings.*

18   *If movable bridge resistance gates are not used on undivided highways, movable bridge warning gates, if used, should extend across the full width of the roadway.*

Option:

19   A single full-width gate or two half-width gates may be used.

Support:

20   The locations of movable bridge signals and gates are determined by the location of the movable bridge resistance gate (if used) rather than by the location of the movable spans. The movable bridge resistance gates for high-speed highways are preferably located 50 feet or more from the span opening except for bascule and lift bridges, where they are often attached to, or are a part of, the structure.

Guidance:

21   *Except where physical conditions make it impracticable, movable bridge warning gates should be located 100 feet or more from the movable bridge resistance gates or, if movable bridge resistance gates are not used, 100 feet or more from the movable span.*

22   *On bridges or causeways that cross a long reach of water and that might be hit by large marine vessels, within the limits of practicality, traffic should not be halted on a section of the bridge or causeway that is subject to impact.*

23   *In cases where it is impractical to halt traffic on a span that is not subject to impact, traffic should be halted at least one span from the opening. If traffic is halted by signals and gates more than 330 feet from the movable bridge warning gates (or from the span opening if movable bridge warning gates are not used), a second set of gates should be installed approximately 100 feet from the gate or span opening.*

24   *If the movable bridge is close to a grade crossing and traffic might possibly be stopped on the crossing as a result of the bridge opening, a traffic control device should notify the road users to not stop on the railroad tracks.*

Section 4Q.03 <u>Operation of Movable Bridge Signals and Gates</u>

Standard:

01   **Traffic control devices at movable bridges shall be coordinated with the movable spans, so that the signals, gates, and movable spans are controlled by the bridge tender through an interlocked control.**

02   **If the three-section type of signal face is used, the green signal indication shall be displayed at all times between bridge openings, except that if the bridge is not expected to open during continuous periods in excess of 5 hours, a flashing yellow signal indication shall be permitted to be used. The signal shall display a steady red signal indication when traffic is required to stop. The duration of the yellow change interval between the display of the green and steady red signal indications, or flashing yellow and steady red signal indications, shall be determined using engineering practices (see Section 4F.17).**

03   **If the vertical array of red signal indications is the type of signal face selected, the red signal indications shall flash alternately only when traffic is required to stop.**

Guidance:

04   *Traffic control signals on adjacent streets and highways should be interconnected with the movable bridge control if indicated by engineering judgment. When such interconnection is provided, the traffic control signals at adjacent intersections should be preempted by the operation of the movable bridge in the manner described in Section 4F.19.*

CHAPTER 4R. HIGHWAY TRAFFIC SIGNALS AT TOLL PLAZAS

Section 4R.01 Traffic Signals at Toll Plazas
Standard:

01   **Traffic control signals or devices that closely resemble traffic control signals that use red or green circular indications shall not be used at toll plazas to indicate the open or closed status of the toll plaza lanes.**

Guidance:

02   *Traffic control signals or devices that closely resemble traffic control signals that use red or green circular indications should not be used for new or reconstructed installations at toll plazas to indicate the success or failure of electronic toll payments or to alternately direct drivers making cash toll payments to stop and then proceed.*

Section 4R.02 Lane-Use Control Signals at or Near Toll Plazas
Standard:

01   **Lane-use control signals used at toll plazas shall comply with the provisions of Chapter 4T except as otherwise provided in this Section.**

02   **At toll plazas with multiple lanes where one or more lanes is sometimes closed to traffic, a lane-use control signal shall be installed above the center of each toll plaza lane to indicate the open or closed status of the controlled lane.**

Option:

03   The bottom of the signal housing of a lane-use control signal above a toll plaza lane having a canopy may be mounted lower than 15 feet above the pavement, but not lower than the vertical clearance of the canopy structure.

04   Lane-use control signals may also be used to indicate the open or closed status of an Open-Road ETC lane as a supplement to other devices used for the temporary closure of a lane (see Part 6).

Section 4R.03 Warning Beacons at Toll Plazas
Standard:

01   **Warning Beacons used at toll plazas shall comply with the provisions of Chapter 4S except as otherwise provided in this Section.**

Guidance:

02   *Warning Beacons, if used with a toll plaza canopy sign (see Section 2F.16) to assist drivers of such vehicles in locating the dedicated ETC Account-Only lane(s), should be installed in a manner such that the beacons are distinctly separate from the lane-use control signals (see Section 4T.01) for the toll plaza lane.*

Option:

03   Warning Beacons that are mounted on toll plaza islands, behind impact attenuators in front of toll plaza islands, and/or on toll booth pylons (ramparts) to identify them as objects in the roadway may be mounted at a height that is appropriate for viewing in a toll plaza context, even if that height is lower than the normal minimum of 8 feet above the pavement.

CHAPTER 4S. FLASHING BEACONS

Section 4S.01 General Design and Operation of Flashing Beacons

Support:

01   A flashing beacon is a highway traffic signal with one or more signal sections that operates in a flashing mode. It can provide traffic control when used as an intersection control beacon (see Section 4S.02) or it can provide warning when used in other applications (see Sections 4S.03, 4S.04, and 4S.05).

Standard:

02   **Flashing beacon units, their mountings, signal visors, and backplates shall comply with the provisions of Chapters 4D and 4E, except as otherwise provided in this Chapter.**

03   **Beacons shall be flashed at a rate of not less than 50 or more than 60 times per minute. The illuminated period of each flash shall be a minimum of ½ and a maximum of ⅔ of the total cycle.**

04   **A beacon shall not be included within the border of a sign except for Interchange Exit Direction signs with advisory speed panels (see Section 2E.25).**

05   **There shall be two nominal diameter sizes for flashing beacon signal indications: 8 inches and 12 inches.**

Guidance:

06   *If used to supplement a warning or regulatory sign, the edge of the beacon signal housing should normally be located no closer than 12 inches outside of the nearest edge of the sign or from the nearest edge of any of the signs and plaques in a sign assembly.*

Option:

07   An automatic dimming device may be used to reduce the brilliance of flashing yellow signal indications during night operation.

08   Backplates (see Section 4D.06) may be used with flashing beacons.

Section 4S.02 Intersection Control Beacon

Standard:

01   **An Intersection Control Beacon shall consist of one or more signal faces directed toward each approach to an intersection. Each signal face shall consist of one or more signal sections of a standard traffic signal face, with flashing CIRCULAR YELLOW or CIRCULAR RED signal indications in each signal face. They shall be installed and used only at an intersection to control two or more directions of travel.**

02   **Application of Intersection Control Beacon signal indications shall be limited to the following:**
    A. Yellow on one route (normally the major street) and red for the remaining approaches that are controlled by STOP signs, or
    B. Red for all approaches (if all of the intersection approaches are controlled by STOP signs).

03   **Flashing yellow signal indications shall not face conflicting vehicular approaches.**

04   **A STOP sign (see Section 2B.04) shall be used on approaches to which a flashing red signal indication is displayed on an Intersection Control Beacon.**

05   **If two horizontally-aligned red signal indications are used on an approach for an Intersection Control Beacon, they shall be flashed simultaneously to avoid being confused with grade crossing flashing-light signals. If two vertically-aligned red signal indications that have a physical separation between them are used on an approach for an Intersection Control Beacon, they shall be flashed alternately.**

06   **Twelve-inch signal indications shall be used for Intersection Control Beacons facing approaches where:**
    A. Road users view both Intersection Control Beacon and lane-use control signal indications simultaneously; or
    B. The nearest Intersection Control Beacon signal face is more than 120 feet beyond the stop line, unless a supplemental near-side Intersection Control Beacon signal face is provided.

Guidance:

07   *Twelve-inch signal indications should be used for Intersection Control Beacons facing approaches where:*
    A. The posted or statutory speed limit or the 85th-percentile approach speed is higher than 40 mph, or
    B. Where only post-mounted flashing beacon signal faces are used.

08   *An Intersection Control Beacon should not be mounted on a pedestal in the roadway unless the pedestal is within the confines of a traffic or pedestrian island.*

Option:

09   Supplemental signal indications may be used on one or more approaches in order to provide adequate visibility to approaching road users.

10   Intersection Control Beacons may be used at intersections where traffic or physical conditions do not justify conventional traffic control signals but crash rates indicate the possibility of a special need.

11   An Intersection Control Beacon is generally located over the center of an intersection; however, it may be used at other suitable locations.

Section 4S.03 <u>Warning Beacon</u>

Support:

01   Typical applications of Warning Beacons include the following:

    A. As supplemental emphasis to signs or object markers on or in front of obstructions that are in or immediately adjacent to the roadway;
    B. As supplemental emphasis to warning signs;
    C. As emphasis for midblock crosswalks;
    D. As supplemental emphasis to regulatory signs, except STOP, DO NOT ENTER, WRONG WAY, and SPEED LIMIT signs; and
    E. In conjunction with a regulatory or warning sign that includes the phrase WHEN FLASHING in its legend or on a supplemental plaque to indicate that the regulation is in effect or that the condition is present only at certain times. Section 2A.12 prohibits the use flashing light-emitting diode (LED) units within the legend or border of the sign in conjunction with the phrase WHEN FLASHING in its legend or on a supplemental plaque.

Standard:

02   **A Warning Beacon shall consist of one or more signal sections of a standard traffic signal face with a flashing CIRCULAR YELLOW signal indication in each signal section.**

03   **A Warning Beacon shall be used only to supplement an appropriate warning or regulatory sign or marker.**

04   **Warning Beacons, if used at intersections, shall not face conflicting vehicular approaches.**

Guidance:

05   *The condition or regulation justifying Warning Beacons should largely govern their location with respect to the roadway.*

06   *If an obstruction is in or adjacent to the roadway, illumination of the lower portion or the beginning of the obstruction or illumination of the sign on or in front of the obstruction, in addition to the beacon, should be considered.*

07   *Warning Beacons should be operated only during those periods or times when the condition or regulation exists.*

Option:

08   If Warning Beacons have more than one signal section, they may be flashed either alternately or simultaneously.

09   A Warning Beacon interconnected with a traffic signal controller assembly may be used with a BE PREPARED TO STOP (W3-4) sign and a WHEN FLASHING (W16-13P) plaque (see Section 2C.35).

10   Warning Beacons that are actuated by pedestrians, bicyclists, or other road users may be used as appropriate to provide additional warning to vehicles approaching a crossing or other location.

Guidance:

11   *An audible information device should be used with pedestrian-actuated Warning Beacons to assist pedestrians with vision disabilities.*

Standard:

12   **If an audible information device is used in conjunction with a pedestrian-actuated Warning Beacon at a pedestrian crossing, the audible information device shall not use vibrotactile indications or percussive indications.**

Guidance:

13   *If an audible information device is used in conjunction with a pedestrian-actuated Warning Beacon at a pedestrian crossing, the audible message should be a speech message that says, "Warning lights are flashing." The audible message should be spoken twice.*

Section 4S.04 Speed Limit Sign Beacon

Standard:

01 **A Speed Limit Sign Beacon shall be used only to supplement a Speed Limit sign.**

02 **A Speed Limit Sign Beacon shall consist of one or more signal sections of a standard traffic control signal face, with a flashing CIRCULAR YELLOW signal indication in each signal section. If two or more signal indications are used, they shall be alternately flashed.**

Option:

03 A Speed Limit Sign Beacon may be used with a fixed or variable Speed Limit sign. If applicable, a flashing Speed Limit Sign Beacon (with an appropriate accompanying sign) may be used to indicate that the displayed speed limit is in effect.

Section 4S.05 Stop Beacon

Standard:

01 **A Stop Beacon shall be used only to supplement a STOP sign, a DO NOT ENTER sign, or a WRONG WAY sign.**

02 **A Stop Beacon shall consist of one or more signal sections of a standard traffic signal face with a flashing CIRCULAR RED signal indication in each signal section. If two horizontally-aligned signal indications are used for a Stop Beacon, they shall be flashed simultaneously to avoid being confused with grade crossing flashing-light signals. If two vertically-aligned signal indications are used for a Stop Beacon, they shall be flashed alternately.**

Guidance:

03 *The edge of the signal housing of a Stop Beacon should be not less than 12 inches or more than 24 inches from the nearest edge of the STOP sign, DO NOT ENTER sign, or WRONG WAY sign that it supplements.*

CHAPTER 4T. LANE-USE CONTROL SIGNALS

Section 4T.01 Application of Lane-Use Control Signals

Support:

01   Lane-use control signals are special overhead signals that permit or prohibit the use of specific lanes of a street or highway or that indicate the impending prohibition of their use. Lane-use control signals are distinguished by placement of special signal faces over a certain lane or lanes of the roadway, over a shoulder where driving is permitted at certain times, and by their distinctive shapes and symbols. Supplementary signs are sometimes used to explain their meaning and intent.

02   Lane-use control signals are most commonly used for reversible lane control, but are also used in certain non-reversible-lane applications and for toll plaza lanes (see Section 4R.02).

Guidance:

03   *An engineering study should be conducted to determine whether a reversible lane operation can be controlled satisfactorily by static signs (see Section 2B.34) or whether lane-use control signals are necessary. Lane-use control signals should be used to control reversible lane operations if any of the following conditions are present:*

    A. *More than one lane is reversed in direction;*
    B. *Two-way or one-way left turns are allowed during peak-period reversible operations, but those turns are from a different lane than used during off-peak periods;*
    C. *Other unusual or complex operations are included in the reversible lane pattern;*
    D. *Demonstrated crash experience occurring with reversible lane operation controlled by static signs that can be corrected by using lane-use control signals at the times of transition between peak and off-peak patterns; and/or*
    E. *An engineering study indicates that the safety and efficiency of the traffic operations of a reversible lane system would be improved by lane-use control signals.*

Standard:

04   **Pavement markings (see Section 3B.04) shall be used in conjunction with reversible lane control signals.**

Option:

05   Lane-use control signals may also be used if there is no intent or need to reverse lanes, but there is a need to indicate the open or closed status of one or more lanes, such as:

    A. On a freeway, if it is desired to close certain lanes at certain hours to facilitate the merging of traffic from a ramp or other freeway;
    B. On a freeway, near its terminus, to indicate a lane that ends;
    C. On a freeway or long bridge, to indicate that a lane may be temporarily blocked by a crash, breakdown, construction or maintenance activities, or similar temporary conditions; and
    D. On a conventional road or driveway, at access or egress points to or from a facility, such as a parking garage, where one or more lanes of the access or egress are opened or closed at various times.

06   A USE LANE(S) WITH GREEN ARROW (R10-8) sign (see Section 2B.59) may be used in conjunction with lane-use control signals.

Section 4T.02 Meaning of Lane-Use Control Signal Indications

Standard:

01   **The meanings of lane-use control signal indications (see Figure 4T-1) shall be as follows:**

    A. **A steady DOWNWARD GREEN ARROW signal indication shall mean that the lane which the arrow signal indication is located over is open to vehicle travel in that direction.**
    B. **A steady YELLOW X signal indication shall mean that the lane which the Yellow X signal indication is located over is about to be closed to vehicle traffic in that direction and shall be followed by a steady RED X signal indication (either within the same signal face or in a downstream signal face).**
    C. **A steady RED X signal indication shall mean that the lane which the Red X signal indication is located over is closed to vehicle traffic in the direction viewed by the road user.**
    D. **A steady WHITE TWO-WAY LEFT-TURN ARROW signal indication shall mean that the lane which the turning arrows indication is located over is open to traffic making a left turn from either direction of travel, but not for through travel.**
    E. **A steady WHITE ONE-WAY LEFT-TURN ARROW signal indication shall mean that the lane which the turning arrow indication is located over is open to traffic making a left turn in that direction (without opposing turns in the same lane), but not for through travel.**

Figure 4T-1. Lane-Use Control Signal Indications

Downward
Green Arrow   Yellow X   Red X   White Two-Way
Left-Turn Arrows   White One-Way
Left-Turn Arrow

Section 4T.03 Design of Lane-Use Control Signals

Standard:

01   **All lane-use control signal indications shall be in units with rectangular signal faces and shall have opaque backgrounds. Except as provided in Paragraph 13 of this Section, the nominal minimum height and width of each DOWNWARD GREEN ARROW, YELLOW X, and RED X signal face shall be 18 inches for typical applications. Except as provided in Paragraph 13 of this Section, the WHITE TWO-WAY LEFT-TURN ARROW and WHITE ONE-WAY LEFT-TURN ARROW signal faces shall have a nominal minimum height and width of 30 inches.**

02   **Each lane to be reversed or closed shall have signal faces with at least a DOWNWARD GREEN ARROW and a RED X symbol.**

03   **Each reversible-lane that also operates as a two-way or one-way left-turn lane during certain periods shall have signal faces that also include the applicable WHITE TWO-WAY LEFT-TURN ARROW or WHITE ONE-WAY LEFT-TURN ARROW symbol.**

04   **Each non-reversible-lane immediately adjacent to a reversible-lane shall have signal indications that display a DOWNWARD GREEN ARROW to traffic traveling in the permitted direction and a RED X to traffic traveling in the opposite direction.**

05   **If in separate signal sections, the relative positions, from left to right, of the signal indications shall be RED X, YELLOW X, DOWNWARD GREEN ARROW, WHITE TWO-WAY LEFT-TURN ARROW, WHITE ONE-WAY LEFT-TURN ARROW.**

Guidance:

06   *The color of lane-use control signal indications should be clearly visible for at least 2,300 feet at all times under normal atmospheric conditions, unless otherwise physically obstructed.*

07   *Lane-use control signal faces should be located approximately over the center of the controlled lane.*

08   *If the area to be controlled is more than 2,300 feet in length, or if the vertical or horizontal alignment is curved, intermediate lane-use control signal faces should be located over each controlled lane at frequent intervals. This location should be such that road users will at all times be able to see at least one signal indication and preferably two along the roadway, and will have a definite indication of the lanes specifically reserved for their use.*

09   *All lane-use control signal faces should be located in a straight line across the roadway approximately at right angles to the roadway alignment.*

10   *On roadways having intersections controlled by traffic control signals, the lane-use control signal face should be located sufficiently far in advance of or beyond such traffic control signals to prevent them from being misconstrued as traffic control signals.*

Standard:

11   **Except as provided in Paragraph 12 of this Section, the bottom of the signal housing of any lane-use control signal face shall be a minimum of 15 feet and a maximum of 19 feet above the pavement grade.**

Option:

12   The bottom of a lane-use control signal housing may be lower than 15 feet above the pavement if it is mounted on a canopy or other structure over the pavement, but not lower than the vertical clearance of the structure.

13   Except for lane-use control signals at toll plazas (see Section 4R.02), lane-use control signal faces with nominal height and width of 12 inches for the DOWNWARD GREEN ARROW, YELLOW X, and RED X signal faces, and lane-use control signal faces with nominal height and width of 18 inches for the WHITE TWO-WAY LEFT-TURN ARROW and WHITE ONE-WAY LEFT-TURN ARROW signal faces may be used in areas with minimal visual clutter and with speeds of less than 40 mph.

14   Other sizes of lane-use control signal faces larger than 18 inches with proportional dimensions and with message recognition distances appropriate to signal spacing may be used for the DOWNWARD GREEN ARROW, YELLOW X, and RED X signal faces.

15   Non-reversible lanes not immediately adjacent to a reversible-lane on any street so controlled may also be provided with signal indications that display a DOWNWARD GREEN ARROW to traffic traveling in the permitted direction and a RED X to traffic traveling in the opposite direction.

16   The signal indications provided for each lane may be in separate signal sections or may be superimposed in the same signal section.

Section 4T.04 Operation of Lane-Use Control Signals

Standard:

01   **All lane-use control signals shall be coordinated so that all the signal indications along the controlled section of roadway are operated uniformly and consistently. The lane-use control signal system shall be designed to reliably guard against showing any prohibited combination of signal indications to any traffic at any point in the controlled lanes.**

02   **For reversible lane control signals, the following combination of signal indications shall not be simultaneously displayed over the same lane to both directions of travel:**
   A. **DOWNWARD GREEN ARROW in both directions,**
   B. **YELLOW X in both directions,**
   C. **WHITE ONE-WAY LEFT-TURN ARROW in both directions,**
   D. **DOWNWARD GREEN ARROW in one direction and YELLOW X in the other direction,**
   E. **WHITE TWO-WAY LEFT-TURN ARROW or WHITE ONE-WAY LEFT-TURN ARROW in one direction and DOWNWARD GREEN ARROW in the other direction,**
   F. **WHITE TWO-WAY LEFT-TURN ARROW in one direction and WHITE ONE-WAY LEFT-TURN ARROW in the other direction, and**
   G. **WHITE ONE-WAY LEFT-TURN ARROW in one direction and YELLOW X in the other direction.**

03   **A moving condition in one direction shall be terminated either by the immediate display of a RED X signal indication or by a YELLOW X signal indication followed by a RED X signal indication.**

04   **In either case, the duration of the RED X signal indication shall be an appropriate duration to allow traffic time to vacate the lane before any moving condition is allowed in the opposing direction.**

05   **Whenever a DOWNWARD GREEN ARROW signal indication is changed to a WHITE TWO-WAY LEFT-TURN ARROW signal indication, the RED X signal indication shall continue to be displayed to the opposite direction of travel for an appropriate duration to allow traffic time to vacate the lane being converted to a two-way left-turn lane.**

06   **If an automatic control system is used, a manual control to override the automatic control shall be provided.**

Guidance:

07   *The type of control provided for reversible lane operation should be such as to permit either automatic or manual operation of the lane-use control signals.*

Standard:

08   **If used, lane-use control signals shall be operated continuously, except that lane-use control signals that are used only for special events or other infrequent occurrences and lane-use control signals on non-reversible freeway lanes are permitted to be darkened when not in operation. The change from normal operation to non-operation shall occur only when the lane-use control signals display signal indications that are appropriate for the lane use that applies when the signals are not operated. The lane-use control signals shall display signal indications that are appropriate for the existing lane use when changed from non-operation to normal operations. Also, traffic control devices shall clearly indicate the proper lane use when the lane-use control signals are not in operation.**

Support:

09   Section 2B.34 contains additional information concerning considerations involving left-turn prohibitions in conjunction with reversible lane operations. Section 2G.24 contains additional information concerning lane-use control signals used for part-time travel on a shoulder. Section 2G.25 contains additional information concerning lane-use control signals used for active lane management on freeways and expressways.

CHAPTER 4U. IN-ROADWAY WARNING LIGHTS

Section 4U.01 Application of In-Roadway Warning Lights

Support:

01   In-Roadway Warning Lights are special types of highway traffic signals installed in the roadway surface to warn road users that they are approaching a condition on or adjacent to the roadway that might not be readily apparent and might require the road users to reduce their speed and/or come to a stop. This includes situations warning of marked school crosswalks, marked midblock crosswalks, marked crosswalks on uncontrolled approaches, marked crosswalks in advance of roundabouts as described in Chapter 3D, and other roadway situations involving pedestrian crossings.

Standard:

02   **In-Roadway Warning Lights shall not be used for any application that is not described in this Chapter.**

03   **When used, In-Roadway Warning Lights shall be flashed and shall not be steadily illuminated.**

Support:

04   Steadily illuminated lights installed in the roadway surface are considered to be internally illuminated raised pavement markers (see Section 3B.14).

Option:

05   In-Roadway Warning Lights may be flashed in a manner that includes a continuous flash of varying intensity and time duration that is repeated to provide a flickering effect (see Section 4U.02).

Guidance:

06   *If used, In-Roadway Warning Lights should not exceed a height of ¾ inch above the roadway surface.*

Section 4U.02 In-Roadway Warning Lights at Crosswalks

Option:

01   In-Roadway Warning Lights may be installed at certain marked crosswalks, based on an engineering study or engineering judgment, to provide additional warning to road users.

Standard:

02   **If used, In-Roadway Warning Lights at crosswalks shall be installed only at marked crosswalks with applicable warning signs. They shall not be used at crosswalks controlled by YIELD signs, STOP signs, traffic control signals, or pedestrian hybrid beacons.**

If In-Roadway Warning Lights are used at a crosswalk, the following requirements shall apply:

A. Except as provided in Paragraphs 7 and 8 of this Section, they shall be installed along both sides of the crosswalk and shall span its entire length.

B. They shall initiate operation based on pedestrian actuation and shall cease operation at a predetermined time after the pedestrian actuation or, with passive detection, after the pedestrian clears the crosswalk.

C. They shall display a flashing yellow light when actuated. The flash rate shall be at least 50, but not more than 60, flash periods per minute. If they are flashed in a manner that includes a continuous flash of varying intensity and time duration that is repeated to provide a flickering effect, the flickers or pulses shall not repeat at a rate that is between 5 and 30 per second to avoid frequencies that might cause seizures.

D. They shall be installed in the area between the outside edge of the crosswalk line and 10 feet from the outside edge of the crosswalk.

E. They shall face away from the crosswalk if unidirectional, or shall face away from and across the crosswalk if bidirectional.

03   **If used on one-lane, one-way roadways, a minimum of two In-Roadway Warning Lights shall be installed on the approach side of the crosswalk. If used on two-lane roadways, a minimum of three In-Roadway Warning Lights shall be installed along both sides of the crosswalk. If used on roadways with more than two lanes, a minimum of one In-Roadway Warning Light per lane shall be installed along both sides of the crosswalk.**

Guidance:

04   *If used, In-Roadway Warning Lights should be installed in the center of each travel lane, at the center line of the roadway, at each edge of the roadway or parking lanes, or at other suitable locations away from the normal tire track paths.*

05   *The location of the In-Roadway Warning Lights within the lanes should be based on engineering judgment.*

Option:

06   On one-way streets, In-Roadway Warning Lights may be omitted on the departure side of the crosswalk.

07   Based on engineering judgment, the In-Roadway Warning Lights on the departure side of the crosswalk on the left-hand side of a median may be omitted.

08   Unidirectional In-Roadway Warning Lights installed at crosswalk locations may have an optional, additional yellow light indication in each unit that is visible to pedestrians in the crosswalk to indicate to pedestrians in the crosswalk that the In-Roadway Warning Lights are in fact flashing as they cross the street. These yellow lights may flash with and at the same flash rate as the light module in which each is installed.

Guidance:

09   *If used, the period of operation of the In-Roadway Warning Lights following each actuation should be sufficient to allow a pedestrian crossing in the crosswalk to leave the curb or shoulder and travel at a walking speed of 3.5 feet per second to at least the far side of the traveled way or to a median of sufficient width for pedestrians to wait. Where pedestrians who walk slower than 3.5 feet per second, or pedestrians who use wheelchairs, routinely use the crosswalk, a walking speed of less than 3.5 feet per second should be considered in determining the period of operation.*

10   *An audible information device should be used with In-Roadway Warning Lights to provide assistance for pedestrians with vision disabilities.*

Standard:

11   **If pedestrian push buttons (rather than passive detection) are used to actuate the In-Roadway Warning Lights, a Push Button To Turn On Warning Lights/WAIT FOR GAP IN TRAFFIC (R10-25) sign (see Section 2B.58) shall be installed explaining the purpose and use of the pedestrian push button detector.**

12   **Where the period of operation is sufficient only for crossing from a curb or shoulder to a median of sufficient width for pedestrians to wait, median-mounted pedestrian actuators shall be provided.**

13   **If an audible information device is used in conjunction with In-Roadway Warning Lights, the audible information device shall not use vibrotactile indications or percussive indications.**

Guidance:

14   *If an audible information device is used in conjunction with In-Roadway Warning Lights, the audible message during the time that the lights are flashing should be a speech message that says, "Warning lights are flashing." The audible message should be spoken twice.*

PART 5
TRAFFIC CONTROL DEVICE CONSIDERATIONS FOR AUTOMATED VEHICLES
CHAPTER 5A. GENERAL

Section 5A.01 Scope and Purpose

Support:

01 The scope of the provisions in this Part are intended for consideration of traffic control devices that are specifically being designed to accommodate automated vehicles capable of performing partial or full real-time operational functions in general traffic on a sustained basis. This Part does not require agencies to use these provisions in their accommodation of automated vehicles on their roadways. Rather, the purpose of these provisions is to provide agencies with general considerations and guidance for traffic control devices that can be more helpful in the accommodation of such vehicles, while at the same time being more beneficial to road users.

02 It is important for early implementers of automated vehicles to understand the ramifications of traffic control devices in a mixed fleet environment and to consider the needs of both human and machine-led road users. Partial automation technologies are already commercially available in the vehicle fleet and are operating under current infrastructure conditions. The overall effectiveness of the automation is impacted by the uniformity and consistent application of the highway infrastructure, including traffic control devices.

03 This Chapter provides an overview of foundational driving automation system (see definition in Section 5A.03) technology terminology, key principles, considerations for traffic control device selection, and topics for agencies to consider. The MUTCD does not address standardization of digital infrastructure, geometric road design, traffic control device maintenance levels, minimum pavement conditions, or other items that might be important for safe and effective operation of driving automation system technologies.

Section 5A.02 Overview of Automated Vehicles and Connected Vehicles

Support:

01 Driving automation system technology automates some or all aspects of the driving tasks to assist or replace the human driver and can include driver assistance technology generally known as advanced driver assistance systems (ADAS). Automated vehicles (AVs) are vehicles in which at least one element of vehicle control (such as steering, speed control, or braking) occurs without direct human driver input. AVs function by gathering information from a suite of sensors that can include, but are not limited to:

 A. Cameras,
 B. Radar,
 C. Light detection and ranging (LiDAR),
 D. Ultrasonic, and
 E. Infrared.

02 AVs can combine sensor data with other inputs including detailed map data and information from other connected vehicles or infrastructure. AVs might be able to detect and classify objects in their surroundings and might predict how they are likely to behave.

03 Connected vehicle technology enables cars, buses, trucks, trains, roads, and roadside infrastructure, as well as other devices such as cellular telephones, to communicate with one another. Connected vehicle technology enabling vehicles to communicate with each other is known as vehicle-to-vehicle (V2V). Connected vehicle technology enabling vehicles to communicate with infrastructure is known as vehicle-to-infrastructure (V2I). Connected vehicle technology enables equipped vehicles on the road to be aware of the location and status of other nearby equipped vehicles or devices. Road users could receive notifications and alerts of dangerous situations, such as a vehicle that is about to run a red traffic signal as it nears an intersection, or an oncoming car, that is out of sight beyond a curve swerving into the opposing lane to avoid an object on the road.

Section 5A.03 Definitions and Terms

Support:

01 The definitions and terms shown in Items A through G below, which are found in the Society of Automotive Engineers standard SAE J3016 and other sources, are used extensively in automated vehicle technology. Their definitions, which are summarized below for reference and for use with the provisions of this Manual, are as follows:

 A. Advanced Driver Assistance Systems (ADAS) – Electronic systems that aid a vehicle driver with one or more driving tasks while driving. They are intended to increase the safe operation of a vehicle and include applications such as automatic braking, lane keeping assistance, adaptive cruise control, and others.

B. Automated Driving System (ADS) – The hardware and software that are collectively capable of performing the entire Dynamic Driving Task (DDT) on a sustained basis, regardless of whether it is limited to a specific Operational Design Domain (ODD); this term is used specifically to describe a Level 3, 4, or 5 driving automation system.

C. Automation Levels – The levels of automation that are described in Table 5A-1.

D. Cooperative Automation – Technology that enables communication with other vehicles and the infrastructure to coordinate automated vehicle operation.

E. Driving Automation System – The hardware and software that are collectively capable of performing part or all of the DDT on a sustained basis; this term is used generically to describe any system capable of Levels 1 through 5 driving automation.

F. Dynamic Driving Task (DDT) – All of the real-time operational and tactical functions required to operate a vehicle in on-road traffic, excluding the strategic functions such as trip scheduling and selection of destinations and waypoints.

G. Operational Design Domain (ODD) – Operating conditions under which a given driving automation system or feature thereof is specifically designed to function, including, but not limited to, environmental, geographical, and time-of-day restrictions, and/or the requisite presence or absence of certain traffic or roadway characteristics.

Table 5A-1. Automation Levels

Automation Level	Description	Automation Category	Automation Type
Level 0	The full-time performance by the human driver of all aspects of the Dynamic Driving Task, even when enhanced by warning or momentary intervention systems.	None*	None
Level 1	The driving mode specific execution by a sustained driver assistance system of either steering or acceleration/deceleration using information about the driving environment and with the expectation that the human driver performs all remaining aspects of the Dynamic Driving Task.	Advanced Driver Assistance Systems (ADAS)	Driving Automation System
Level 2	The driving mode specific execution by one or more sustained driver assistance systems of both steering and acceleration/deceleration using information about the driving environment and with the expectation that the human driver performs all remaining aspects of the Dynamic Driving Task.		
Level 3	The driving mode specific sustained performance by an ADS of all aspects of the Dynamic Driving Task within a given ODD with the expectation that the human driver will respond appropriately to a request to intervene.	Automated Driving System (ADS)	
Level 4	The driving mode specific sustained performance by an ADS of all aspects of the Dynamic Driving Task, even if a human driver does not respond appropriately to a request to intervene.		
Level 5	The full-time sustained performance by an ADS of all aspects of the Dynamic Driving Task under all roadway and environmental conditions that can be managed by a human driver.		

*NOTE: Level 0 might include some ADAS features, but they are considered to be warning or momentary intervention systems at this level.

Section 5A.04 Traffic Control Device Design and Use Considerations

Support:

01   The interaction of traffic control devices with driving automation systems can create many challenges for agencies in determining traffic control device selection and application. The lack of tolerance of driving automation systems for non-uniformity in traffic control device design and application is a limiting factor of current driving automation system sophistication. This is because driving automation systems have a limited ability to interpolate across gaps in traffic control device cues to the vehicle in the following types of situations:

A. The driving automation system technology's ability to adapt to existing traffic control device design and typical quality, such as the refresh rates of electronic changeable message sign displays or the overall quality of a device that has been in service on the roadway for many years;

B. The color perception of signs;

C. The electronically perceptible conspicuity and contrast of markings in different environments and lighting conditions;

D. The driving automation system camera technology and device photometric characteristics in interpreting various types of traffic signals

E. The ability to discern and comprehend temporary traffic control devices and their varying applications, such as active electronic display devices or flaggers; and

    F. The ability to decipher traffic control at highway-rail or highway-LRT grade crossings, especially at passive grade crossings.

02   These and other challenges might limit the functionality of driving automation systems, thus making them less effective or functional, with potential implications for safety and traffic operations. The uniform design and consistent application of standardized traffic control devices supports the functionality of driving automation system technology in many situations. Similarly, good traffic control device maintenance practices and programs will help improve the potential for driving automation systems to operate properly in many roadway environments.

Guidance:

03   *Agencies should adopt traffic control device maintenance policies and or practices with consideration to both the human driver and driving automation system technology needs (see Sections 1D.10, 2A.19, 3A.05 and 4A.10).*

04   *Engineering judgment (see Section 1D.03) used to determine traffic control device selection and placement should consider uniformity in application and location needed to support both the human driver and driving automation system technology.*

Support:

05   A systematic approach to traffic control device selection, application, and maintenance taking into consideration certain fundamental principles, will help agencies considering the inclusion of AVs on their roadways. Generally, improvements to traffic control device uniformity and improved maintenance policies and practices that keep traffic control devices in good working order with high levels of conspicuity that are beneficial to the human driver will be beneficial to AVs as well.

Guidance:

06   *Agencies should apply the following fundamental principles and considerations as they evaluate traffic control devices and other maintenance practices to support driving automation system technologies during maintenance and infrastructure improvements:*

    A. *Applying uniform and consistent traffic control devices on each type of roadway, and applying a similar approach to traffic control at similar locations in similar situations.*
    B. *Establishing maintenance policies that incorporate effective practices to identify and then fix or replace in a timely manner any traffic control device that is reaching the end of its useful life, or that is damaged or otherwise no longer serviceable.*
    C. *Making sure that temporary or emergency traffic control, to the extent practicable, is planned in advance using devices that comply with the provisions of this Manual and that follow policies designed to provide uniformity throughout the site and across jurisdictions.*
    D. *Removing extraneous devices that are no longer necessary or that provide limited benefit to vehicle operation or navigation.*

CHAPTER 5B. PROVISIONS FOR TRAFFIC CONTROL DEVICES

Section 5B.01 <u>Signs</u>

Support:

01   Driving automation systems use sensors, algorithms, and processing to locate, read, and comprehend traffic signs and assist the human driver or AV in appropriately making vehicle operational decisions. Location, condition, uniformity, design characteristics, and consistent application all affect the ability of driving automation systems to perform these functions.

Standard:

02   **When scanning graphics (see Section 2A.04) of any type are used on a sign for support of driving automation systems, the scanning graphics shall not be visible to the human eye and the sign shall have no apparent loss of resolution or recognition for the road user.**

Guidance:

03   *Agencies seeking to better accommodate driving automation systems to support AVs, while also potentially benefitting human drivers, should consider:*

    A. Clearly associating the sign location and application with the displayed message to the specific lane or road to which it applies, such as in the case of parallel roads or lanes with different speed limits or restrictions.

    B. The practice of sign and information spreading (see Section 2A.20) to limit the amount of information displayed in one location or on one sign to minimize sign clutter.

    C. Signs with designs that are otherwise not provided for in this Manual or the "Standard Highway Signs" publication (see Section 1A.05) are designed based on the standardized sign design practices and features as provided for in this Manual for the type of sign, the location, and the characteristics of the roadway on which it is used.

    D. The refresh rate of LEDs in the illuminated portion of electronic-display signs to provide for greater consistency in driving automation system detection.

Section 5B.02 <u>Markings</u>

Support:

01   Driving automation systems use sensors, algorithms, and processing to locate, read, and comprehend pavement markings. Location, condition, uniformity, design characteristics, and consistent application all have some effect on the ability of driving automation systems to perform this function. Certain pavement marking applications and practices have been shown through research to better support driving automation system technology, while also benefitting, or at least not detracting from, the performance of the human operator.

Guidance:

02   *Agencies seeking to better accommodate driving automation system to support AVs, while also potentially benefitting human drivers, should consider:*

    A. Normal width longitudinal lines of at least 6 inches in width (see Section 3A.04).

    B. Edge lines of at least 6 inches in width (see Sections 3A.04 and 3B.09).

    C. Dotted edge line extensions along all entrance and exit ramps, all auxiliary lanes, and all tapers where a deceleration or auxiliary lane is added (see Section 3B.11).

    D. Chevron markings in the neutral areas of exit gores to distinguish them from travel lanes (see Section 3B.25).

    E. Raised pavement markers only as a supplement to, rather than as a substitute for, pavement markings (see Sections 3B.16 and 3B.17).

    F. Uniform contrast markings on light-colored pavements to create greater contrast.

    G. Broken lines with uniform marking and gap length (see Section 3A.04).

Section 5B.03 <u>Highway Traffic Signals</u>

Guidance:

01   *Agencies seeking to better accommodate driving automation systems to support AVs, while also potentially benefitting human drivers, should consider:*

    A. Consistent signal face placement along corridors with respect to overhead mounting versus post mounting on the side of the roadway.

    B. Consistent number of signal faces for approach lanes and the selection of signal indications and signal clusters along a corridor to promote uniform displays for identical or similar situations.

    C. The refresh rate of LED traffic signals to provide for greater consistency in driving automation system detection.

    D. *Providing signal faces with backplates (see Section 4D.06) having retroreflective borders to enhance signal face conspicuity and detection by driving automation system sensors.*

    E. *Using FLASHING YELLOW ARROW signal indications for permissive turns.*

Support:

02   Signal faces that display a CIRCULAR GREEN indication and that are located over or directly in line with a mandatory turn lane can be less effective for driving automation systems to recognize as a traffic signal face controlling permissive turning movements.

03   Achieving uniformity along a corridor is desirable for driving automation systems, but can be challenging. Multiple options are available for traffic signal displays to allow design variations based on specific intersection variables such as available overhead clearance, utility conflicts, signal support design constraints, and other factors. V2I capabilities can complement driving automation system recognition of traffic signals to provide redundancy, and to improve reliability and accuracy.

Section 5B.04 Temporary Traffic Control

Guidance:

01   *Agencies seeking to better accommodate driving automation systems to support, while also potentially benefitting human drivers, in and through temporary traffic control (TTC) zones should consider:*

    A. *Consistent type, spacing, and mounting height of signs (see Sections 6B.04 and 6F.02).*

    B. *Use of the END ROAD WORK (G20-2) sign to establish the end of the TTC zone (see Section 6H.36).*

    C. *Wider retroreflective material on, or reduced spacing of, channelizing devices to better accommodate driving automation system sensors in nighttime and adverse weather conditions (see Chapter 6K).*

    D. *Continuous markings at the beginning of TTC zones and in lane transitions.*

    E. *Temporary raised pavement markers only as a supplement to, rather than as a substitute for, pavement markings.*

    F. *Removal or obliteration of pavement markings that are no longer applicable as soon as practicable, for long-term stationary operations in the temporary traveled way (see Section 6J.01).*

Support:

02   Pavement markings that are not fully removed and pavement scarring are of particular concern as there can be misinterpretation by driving automation systems that can result in erroneous vehicle positioning in TTC zones.

03   V2I communications can complement driving automation systems recognition in TTC zones by communicating the presence of a TTC zone to vehicles.

04   Section 6J.01 describes the use of pavement markings in TTC zones and the removal or obliteration of existing pavement markings.

05   Section 6J.02 describes the use of temporary pavement markings in TTC zones.

Section 5B.05 Traffic Control for Highway-Rail and Highway-Light Rail Transit Grade Crossings

Guidance:

01   *Agencies seeking to better accommodate driving automation systems to support AVs, while also potentially benefitting human drivers, at grade crossings should consider:*

    A. *Consistent placement of signs and markings for passive and active grade crossings along a corridor to promote uniformity and to improve the ability of driving automation system technology to recognize grade crossings.*

    B. *Removal of signs and pavement markings associated with grade crossings that are out of service (see Section 8A.09).*

Support:

02   V2I communications can complement driving automation system recognition of grade crossings to improve reliability and accuracy, and to relay information on the arrival or presence of a train or LRT vehicle at a grade crossing.

Section 5B.06 Traffic Control for Bicycle Facilities

Guidance:

01   *Agencies seeking to better accommodate driving automation systems to support AVs, while also potentially benefitting human road users, should consider:*

    A. *Use of an END (R3-9dP) plaque with a BIKE LANE (R3-17) sign to indicate the end of a bicycle lane that is merging with other traffic (see Sections 2B.33 and 9B.04).*

    B. *Use of Bicycle Lane Ends (W9-5) and Bicycle Merging (W9-5a) warning signs in advance of the end of a bicycle lane and where a merging maneuver might occur (see Section 9C.07).*

Support:

02   Bicycle facilities that are physically separated from motor vehicle traffic using vertical objects or vertical separation can facilitate detection from driving automation system sensors (see Section 9E.07).

(This page intentionally left blank)

PART 6
TEMPORARY TRAFFIC CONTROL

CHAPTER 6A. GENERAL

Section 6A.01 General

Support:

01   Whenever the acronym "TTC" is used in Part 6, it refers to "temporary traffic control."

Standard:

02   **The needs and control of all road users (motorists, bicyclists, and pedestrians within the highway, or on a site roadway open to public travel (see definition in Section 1C.02), including persons with disabilities) through a TTC zone shall be an essential part of highway construction, utility work, maintenance operations, and the management of traffic incidents.**

Support:

03   When the normal function of the roadway, or a site roadway open to public travel, is suspended, TTC planning provides for continuity of the movement of motor vehicle, bicycle, and pedestrian traffic (including accessible passage); transit operations; and access (and accessibility) to property and utilities.

04   The primary function of TTC is to facilitate movement of road users through or around TTC zones while protecting road users, workers, responders to traffic incidents, and equipment.

05   Of equal importance to the public traveling through the TTC zone is the safety of workers performing the many varied tasks within the work space. TTC zones present constantly changing conditions that are unexpected by the road user. This creates an even higher degree of vulnerability for the workers and incident management responders on or near the roadway (see Section 6C.04). At the same time, the TTC zone provides for the efficient completion of whatever activity interrupted the normal use of the roadway.

06   Consideration for road user safety, worker and responder safety, and the efficiency of road user flow is an integral element of every TTC zone, from planning through completion. A concurrent objective of the TTC is the efficient construction and maintenance of the highway and the efficient resolution of traffic incidents.

07   No one set of TTC devices can satisfy all conditions for a given project or incident. At the same time, defining details that would be adequate to cover all applications is impractical. Instead, Part 6 displays typical applications that depict common applications of TTC devices. The TTC selected for each situation depends on the type of highway, road user conditions, the duration of operation, physical constraints, and the nearness of the work space or incident management activity to road users.

08   The TTC needs on low-volume and special purpose roads will sometimes be minimal, especially for shorter-term durations and for lower-speed roads. The use of maintenance vehicle warning flashers, a limited number of signs, or a single flagger could be adequate for these situations.

09   Improved road user performance might be realized through a well-prepared public relations effort that covers the nature of the work, the time and duration of its execution, the anticipated effects upon road users, and possible alternate routes and modes of travel. Such programs have been found to result in a significant reduction in the number of road users traveling through the TTC zone, which reduces the possible number of conflicts.

10   Operational improvements might be realized by using intelligent transportation systems (ITS) in work zones. The use in work zones of ITS technology, such as portable camera systems, highway advisory radio, variable speed limits, ramp metering, traveler information, merge guidance, warning systems for vehicles exiting the work space, and queue detection information, is aimed at increasing safety for both workers and road users and helping to ensure a more efficient traffic flow. The use in work zones of ITS technologies has been found to be effective in providing traffic monitoring and management, data collection, and traveler information.

Standard:

11   **TTC plans and devices shall be the responsibility of the public body or official or the owners of site roadways open to public travel having jurisdiction for guiding road users.**

Guidance:

12   *There should be adequate statutory authority for the implementation and enforcement of needed road user regulations, parking controls, speed zoning, and the management of traffic incidents. Such statutes should provide sufficient flexibility in the application of TTC to meet the needs of changing conditions in the TTC zone.*

Support:

13   The provisions of Part 6 apply to both rural and urban areas. A rural highway is normally characterized by lower volumes, higher speeds, fewer turning conflicts, and less conflict with pedestrians or other vulnerable road users. An urban street is typically characterized by relatively low speeds, wide ranges of road user volumes, narrower roadway lanes, frequent intersections and driveways, significant vulnerable road user activity, and more businesses and houses.

14   The determination as to whether a particular facility at a particular time of day can be considered to be a high-volume roadway or can be considered to be a low-volume roadway is made by the public agency or official having jurisdiction.

15   Special plans preparation and coordination with transit, other highway agencies, law enforcement and other emergency units, utilities, schools, trucking associations, and railroad companies might be needed to reduce unexpected and unusual road user operation situations.

Section 6A.02 <u>Fundamental Principles of Temporary Traffic Control</u>

Guidance:

01   *Road user and worker safety and accessibility in TTC zones should be an integral and high-priority element of every project from planning through design and construction. Similarly, maintenance and utility work should be planned and conducted with the safety and accessibility of all motorists, bicyclists, pedestrians (including those with disabilities), and workers being considered at all times. If the TTC zone includes a grade crossing, early coordination with the railroad company or light rail transit agency should take place.*

02   *The following are the seven fundamental principles of TTC:*

    A. *General plans or guidelines should be developed to provide safety for motorists, bicyclists, pedestrians, workers, enforcement/emergency officials, and equipment, with the following factors being considered:*

        1. *The basic safety principles governing the design of permanent roadways and roadsides should also govern the design of TTC zones. The goal should be to route road users through such zones using roadway geometrics, roadside features, and TTC devices as nearly as possible comparable to those for normal highway situations.*

        2. *A TTC plan, in detail appropriate to the complexity of the work project or incident, should be prepared and understood by all responsible parties before the site is occupied. Any changes in the TTC plan should be approved by an official who is knowledgeable (for example, trained and/or certified) in proper TTC practices.*

    B. *Road user movement should be inhibited as little as practical, based on the following considerations:*

        1. *TTC at work and incident sites should be designed on the assumption that drivers will only reduce their speeds if they clearly perceive a need to do so (see Section 6B.01).*

        2. *Frequent and abrupt changes in geometrics such as lane narrowing, dropped lanes, or main roadway transitions that require rapid maneuvers, should be avoided.*

        3. *Work should be scheduled in a manner that minimizes the need for lane closures or alternate routes, while still getting the work completed quickly and the lanes or roadway open to traffic as soon as possible.*

        4. *Attempts should be made to reduce the volume of traffic using the roadway or freeway to match the restricted capacity conditions. Road users should be encouraged to use alternative routes. When the roadway capacity is reduced because of lane closures, the demand could exceed the available capacity, which might result in either a lengthy stopped or slow moving queue of vehicles that might extend past the normal location of the signs shown in the typical advance warning area. An assessment of the expected queue length, which should be a part of the TTC plan design process, might result in adjustments to the sign spacing and number of signs as well as the use of more conspicuous devices to increase the distance and conspicuity of the advance warning area. For high-volume roadways and freeways, the closure of selected entrance ramps or other access points and the use of signed diversion routes should be evaluated.*

        5. *Bicyclists and pedestrians, including those with disabilities, should be provided with access and passage through the TTC zone.*

        6. *If work operations permit, lane closures on high-volume streets and highways should be scheduled during off-peak hours. Night work should be considered if the work can be accomplished with a series of short-term operations.*

        7. *Early coordination with officials having jurisdiction over the affected cross streets and providing emergency services should occur if significant impacts to roadway operations are anticipated.*

C. *Motorists, bicyclists, and pedestrians should be guided in a clear and positive manner while approaching and traversing TTC zones and incident sites. The following principles should be applied:*
   1. *Adequate warning, delineation, and channelization should be provided to assist in guiding road users in advance of and through the TTC zone or incident site by using proper pavement marking, signing, or other devices that are effective under varying conditions. Information should be provided in usable formats for pedestrians with vision disabilities.*
   2. *TTC devices inconsistent with intended travel paths through TTC zones should be removed or covered. However, in intermediate-term stationary, short-term, and mobile operations, where visible permanent devices are inconsistent with intended travel paths, devices that highlight or emphasize the appropriate path should be used. Traffic control devices should provide information in usable formats for pedestrians with vision disabilities.*
   3. *Flagging procedures, when used, should provide positive guidance to road users traversing the TTC zone.*

D. *To provide acceptable levels of operations, routine day and night inspections of TTC elements should be performed as follows:*
   1. *Individuals who are knowledgeable (for example, trained and/or certified) in the principles of proper TTC should be assigned responsibility for safety in TTC zones. The most important duty of these individuals is to check that TTC devices on the project are consistent with the TTC plan and are effective for motorists, bicyclists, pedestrians, and workers.*
   2. *As the work progresses, temporary traffic controls and/or working conditions should be modified, as needed, to facilitate road user movement and provide worker safety. The individual responsible for TTC should have the authority to halt work until applicable or remedial safety measures are taken.*
   3. *TTC zones should be carefully monitored under varying conditions of road user volumes, light, and weather to check that applicable TTC devices are effective, clearly visible, clean, and in compliance with the TTC plan.*
   4. *When warranted, an engineering study should be made (in cooperation with law enforcement officials) of reported crashes occurring within the TTC zone. Crash records in TTC zones should be monitored to identify the need for changes in the TTC zone.*

E. *Attention should be given to the maintenance of roadside safety during the life of the TTC zone by applying the following principles:*
   1. *To accommodate run-off-the-road incidents, disabled vehicles, or emergency situations, unencumbered roadside recovery areas or clear zones should be provided where practical.*
   2. *Channelization of road users should be accomplished by the use of pavement markings, signing, and crashworthy, detectable channelizing devices.*
   3. *Work equipment, workers' private vehicles, materials, and debris should be stored in such a manner to reduce the probability of being impacted by run-off-the-road vehicles.*

F. *Each person whose actions affect TTC zone safety, from the upper-level management through the field workers, should receive training appropriate to the job decisions each individual is required to make. Only those individuals who are trained in proper TTC practices and have a basic understanding of the principles (established by applicable standards and guidelines, including those of this Manual) should supervise the selection, placement, and maintenance of TTC devices used for TTC zones and for incident management.*

G. *Good public relations should be maintained by applying the following principles:*
   1. *The needs of all road users should be assessed such that appropriate advance notice is given and clearly defined alternative paths are provided.*
   2. *The cooperation of the various news media should be sought in publicizing the existence of and reasons for TTC zones because news releases can assist in keeping the road users well informed.*
   3. *The needs of abutting property owners, residents, and businesses should be assessed and appropriate accommodations made.*
   4. *The needs of emergency service providers (law enforcement, fire, and medical) should be assessed and appropriate coordination and accommodations made.*
   5. *The needs of railroads and transit should be assessed and appropriate coordination and accommodations made.*
   6. *The needs of operators of commercial vehicles such as buses and large trucks should be assessed and appropriate accommodations made.*
   7. *Early coordination should occur with school officials to discuss potential impacts on picking up and dropping off schoolchildren, on school bus routing, and on safe routes to school patterns.*

Section 6A.03 TTC Devices

Guidance:

01 *The design and application of TTC devices used in TTC zones should consider the needs of all road users (motorists, bicyclists, and pedestrians), including those with disabilities.*

Standard:

02 **Traffic control devices shall be defined as all signs, signals, markings, channelizing devices, or other devices that use colors, shapes, symbols, words, sounds, or tactile information for the primary purpose of communicating a regulatory, warning, or guidance message to road users on a street, highway, pedestrian facility, bikeway, pathway, or site roadway open to public travel.**

03 **All traffic control devices used for construction, maintenance, utility, or incident management operations on a street, highway, pedestrian facility, bikeway, pathway, or site roadway open to public travel shall comply with the applicable provisions of this Manual.**

04 **All TTC devices shall be removed as soon as practical when they are no longer needed. When work is suspended for short periods of time, TTC devices that are no longer appropriate shall be removed or covered.**

Section 6A.04 Crashworthiness of TTC Devices

Support:

01 Various Sections of the MUTCD require certain traffic control devices, their supports, and/or related appurtenances to be crashworthy (see definition in Section 1C.02). Such MUTCD crashworthiness provisions apply to all streets, highways, and site roadways open to public travel.

Section 6A.05 Night Work

Support:

01 Conducting highway construction and maintenance activities during night hours could provide an advantage when traditional daytime traffic control strategies cannot achieve an acceptable balance between worker and public safety, traffic and community impact, and constructability. The two basic advantages of working at night are reduced traffic congestion and less involvement with business activities. However, the two basic conditions that must normally be met for night work to offer any advantage are reduced traffic volumes and easy set up and removal of the traffic control patterns on a nightly basis.

02 Shifting work activities to night hours, when traffic volumes are lower and normal business is less active, might offer an advantage in some cases, as long as the necessary work can be completed and the worksite restored to essentially normal operating conditions to carry the higher traffic volume during non-construction hours.

03 Although working at night might offer advantages, it also includes safety issues. Reduced visibility inherent in night work impacts the performance of both drivers and workers. Because traffic volumes are lower and congestion is minimized, speeds are often higher at night necessitating greater visibility at a time when visibility is reduced. Finally, the incidence of impaired (alcohol or drugs), fatigued, or drowsy drivers might be higher at night.

04 Working at night also involves other factors, including construction productivity and quality, social impacts, economics, and environmental issues. A decision to perform construction or maintenance activities at night normally involves some consideration of the advantages to be gained compared to the safety and other issues that might be impacted.

05 Section 6N.18 contains specific provisions on TTC for work during nighttime hours.

CHAPTER 6B. TEMPORARY TRAFFIC CONTROL ELEMENTS

Section 6B.01 <u>Temporary Traffic Control Plans</u>

Support:

01   Each TTC zone is different. Many variables, such as location of work, highway type, geometrics, vertical and horizontal alignment, intersections, interchanges, road user volumes, road user mix (motorists, bicyclists, and pedestrians), road vehicle mix (buses, trucks, and cars), and road user speeds affect the needs of each zone. The goal of TTC in work zones is safety with minimum disruption to road users. The key factor in promoting TTC zone safety is proper judgment.

02   A TTC plan describes TTC measures to be used for facilitating road users through a work zone or an incident area. TTC plans play a vital role in facilitating road user flow when a work zone, incident, or other event temporarily disrupts normal road user flow. Important auxiliary provisions that cannot conveniently be specified on project plans can easily be incorporated into Special Provisions within the TTC plan.

03   TTC plans range in scope from being very detailed to simply referencing typical drawings contained in this Manual, standard approved highway agency drawings and manuals, or specific drawings contained in the contract documents. The degree of detail in the TTC plan depends entirely on the nature and complexity of the situation.

04   During TTC activities, commercial vehicles might need to follow a different route from passenger vehicles because of bridge, weight, clearance, or geometric restrictions. Also, vehicles carrying hazardous materials might need to follow a different route from other vehicles. The Hazardous Materials and National Network signs are included in Sections 2B.67 and 2B.68, respectively.

Guidance:

05   *A TTC plan should be developed for planned activities that will affect road users. A TTC plan should be developed for unplanned and emergency situations where practicable.*

06   *The TTC plan should start in the planning phase and continue through the design, construction, and restoration phases. The TTC plans and devices should follow the principles set forth in Part 6. The management of traffic incidents should follow the principles set forth in Chapter 6O.*

07   *TTC plans should be prepared by persons knowledgeable (for example, trained and/or certified) about the fundamental principles of TTC and work activities to be performed. The design, selection, and placement of TTC devices for a TTC plan should be based on engineering judgment.*

08   *Coordination should be made between adjacent or overlapping projects to check that duplicate signing is not used and to check compatibility of traffic control between adjacent or overlapping projects.*

09   *Traffic control planning should be completed for all highway construction, utility work, maintenance operations, and incident management including minor maintenance and utility projects prior to occupying the TTC zone. Planning for all road users should be included in the process.*

10   *For any planned special event that will have an impact on the traffic on any street or highway, a TTC plan should be developed in conjunction with and be approved by the agency or agencies that have jurisdiction over the affected roadways.*

11   *Provisions for effective continuity of accessible circulation paths for pedestrians should be incorporated into the TTC plan.*

Option:

12   Provisions may be incorporated into the project bid documents that enable contractors to develop an alternate TTC plan.

13   Modifications of TTC plans may be necessary because of changed conditions or a determination of better methods of safely and efficiently handling road users.

Guidance:

14   *This alternate or modified plan should have the approval of the responsible highway agency or owner of site roadways open to public travel prior to implementation.*

15   *Provisions for effective continuity of transit service should be incorporated into the TTC planning process because often public transit buses cannot efficiently be detoured in the same manner as other vehicles (particularly for short-term maintenance projects). Where applicable, the TTC plan should provide for features such as accessible temporary bus stops, pull-outs, and satisfactory waiting areas for transit patrons, including persons with disabilities (see Section 8A.13 for additional light rail transit issues to consider for TTC).*

16   *Provisions for effective continuity of railroad service and acceptable access to abutting property owners and businesses should also be incorporated into the TTC planning process.*

17   *Reduced speed zoning (lowering the regulatory speed limit) should be avoided as much as practical because drivers will reduce their speeds only if they clearly perceive a need to do so.*

18   *If reduced speed limits are used, they should be used only in the specific portion of the TTC zone where conditions or restrictive features are present. However, frequent changes in the speed limit should be avoided. A TTC plan should be designed so that vehicles can travel through the TTC zone with a speed limit reduction of no more than 10 mph.*

19   *A reduction of more than 10 mph in the speed limit should be used only when required by restrictive features in the TTC zone. Where restrictive features justify a speed reduction of more than 10 mph, additional driver notification should be provided. The speed limit should be stepped down in advance of the location requiring the lowest speed, and additional TTC warning devices should be used.*

Support:

20   Research has demonstrated that large reductions in the speed limit, such as a 30-mph reduction, increase speed variance and the potential for crashes. Smaller reductions in the speed limit of up to 10 mph cause smaller changes in speed variance and lessen the potential for increased crashes. A reduction in the regulatory speed limit of only up to 10 mph from the normal speed limit has been shown to be more effective.

21   Chapter 6P contains typical applications (TAs) of TTC zones that are organized according to duration, location, type of work, and highway type. Table 6P-1 is an index of these typical applications. These typical applications include the use of various TTC methods, but do not include a layout for every conceivable work situation.

22   Decisions regarding the selection of the most appropriate typical application to use as a guide for a specific TTC zone require an understanding of each situation. Although there are many ways of categorizing TTC zone applications, the typical applications illustrated in Chapter 6P are characterized by work duration, work location, work type, and highway type.

Guidance:

23   *Typical applications should be altered, when necessary, to fit the conditions of a particular TTC zone.*

Option:

24   Other devices may be added to supplement the devices shown in the typical applications. The sign spacings and taper lengths may be increased to provide additional time or space for driver response.

25   Devices labeled as optional in the typical applications may be deleted.

Support:

26   Formulating specific plans for TTC at traffic incidents is difficult because of the variety of situations that can arise.

27   Well-designed TTC plans for planned special events will likely be developed from a combination of treatments from several of the typical applications.

Section 6B.02 Temporary Traffic Control Zones

Support:

01   A TTC zone is an area of a highway where road user conditions are changed because of a work zone, an incident zone, or a planned special event through the use of TTC devices, uniformed law enforcement officers, or other authorized personnel.

02   A work zone is an area of a highway with construction, maintenance, or utility work activities. A work zone is typically marked by signs, channelizing devices, barriers, pavement markings, and/or work vehicles. It extends from the first warning sign or high-intensity rotating, flashing, oscillating, or strobe lights on a vehicle to the END ROAD WORK sign or the last TTC device.

03   An incident zone is an area of a highway where temporary traffic controls are imposed by authorized officials in response to a traffic incident (see Section 6O.01). It extends from the first warning device (such as a sign, light, or cone) to the last TTC device or to a point where road users return to the original lane alignment and are clear of the incident.

04   A planned special event often creates the need to establish altered traffic patterns to handle the increased traffic volumes generated by the event. The size of the TTC zone associated with a planned special event can be small, such as closing a street for a festival, or can extend throughout a municipality for larger events. The duration of the TTC zone is determined by the duration of the planned special event.

Section 6B.03 Components of Temporary Traffic Control Zones

Support:

01   A TTC zone is often divided into four areas as needed, based on engineering judgment: the advance warning area, the transition area, the activity area, and the termination area. Figure 6B-1 illustrates the four areas typically included in a TTC zone. These four areas are described in Sections 6B.04 through 6B.07.

Section 6B.04 Advance Warning Area

Support:

01   The advance warning area is the section of highway where road users are informed about the upcoming transition and activity areas or incident area.

Option:

02   The advance warning area may vary from a single sign or high-intensity rotating, flashing, oscillating, or strobe lights on a vehicle to a series of signs in advance of the TTC zone activity area.

Guidance:

03   *Typical distances for placement of advance warning signs on freeways and expressways should be longer because drivers are conditioned to uninterrupted flow. Therefore, the advance warning sign placement should extend on these facilities as far as ½ mile or more.*

04   *On urban streets, the effective placement of the nearest warning sign to the TTC zone, in feet, should range from 4 to 8 times the speed limit in mph, with the high end of the range being used when speeds are relatively high. When two or more advance warning signs are used on higher-speed streets, such as major arterials, the advance warning area should extend a greater distance (see Table 6B-1).*

Option:

05   When a single advance warning sign is used (in cases such as low-speed residential streets), the advance warning area may be as short as 100 feet.

Guidance:

06   *Since rural highways are normally characterized by higher speeds, the effective placement of the first warning sign in feet should be substantially longer—from 8 to 12 times the speed limit in mph. Since two or more advance warning signs are normally used for these conditions, the advance warning area should extend 1,500 feet or more for open highway conditions (see Table 6B-1).*

07   *The distances contained in Table 6B-1 are approximate, are intended for guidance purposes only, and should be applied with engineering judgment. These distances should be adjusted for field conditions, if necessary, by increasing or decreasing the recommended distances.*

Support:

08   The need to provide additional reaction time for a condition is one example of justification for increasing the sign spacing. Conversely, decreasing the sign spacing might be justified in order to place a sign immediately downstream of an intersection or major driveway such that traffic turning onto the roadway in the direction of the TTC zone will be warned of the upcoming condition.

Option:

09   Advance warning may be eliminated when the activity area is sufficiently removed from the road users' path so that it does not interfere with the normal flow.

Section 6B.05 Transition Area

Support:

01   The transition area is that section of highway where road users are redirected out of their normal path. Transition areas usually involve strategic use of tapers, which because of their importance are discussed separately in detail.

Standard:

02   **Except for mobile operations, when redirection of the road users' normal path is required, road users shall be directed from the normal path to a new path with appropriate channelizing devices, traffic control devices, and/or TTC methods.**

Option:

03   Because it is impracticable in mobile operations to redirect the road users' normal path with stationary channelization, more dominant vehicle-mounted traffic control devices, such as arrow boards, portable changeable message signs, and high-intensity rotating, flashing, oscillating, or strobe lights, may be used instead of channelizing devices to establish a transition area.

Figure 6B-1. Component Parts of a Temporary Traffic Control Zone

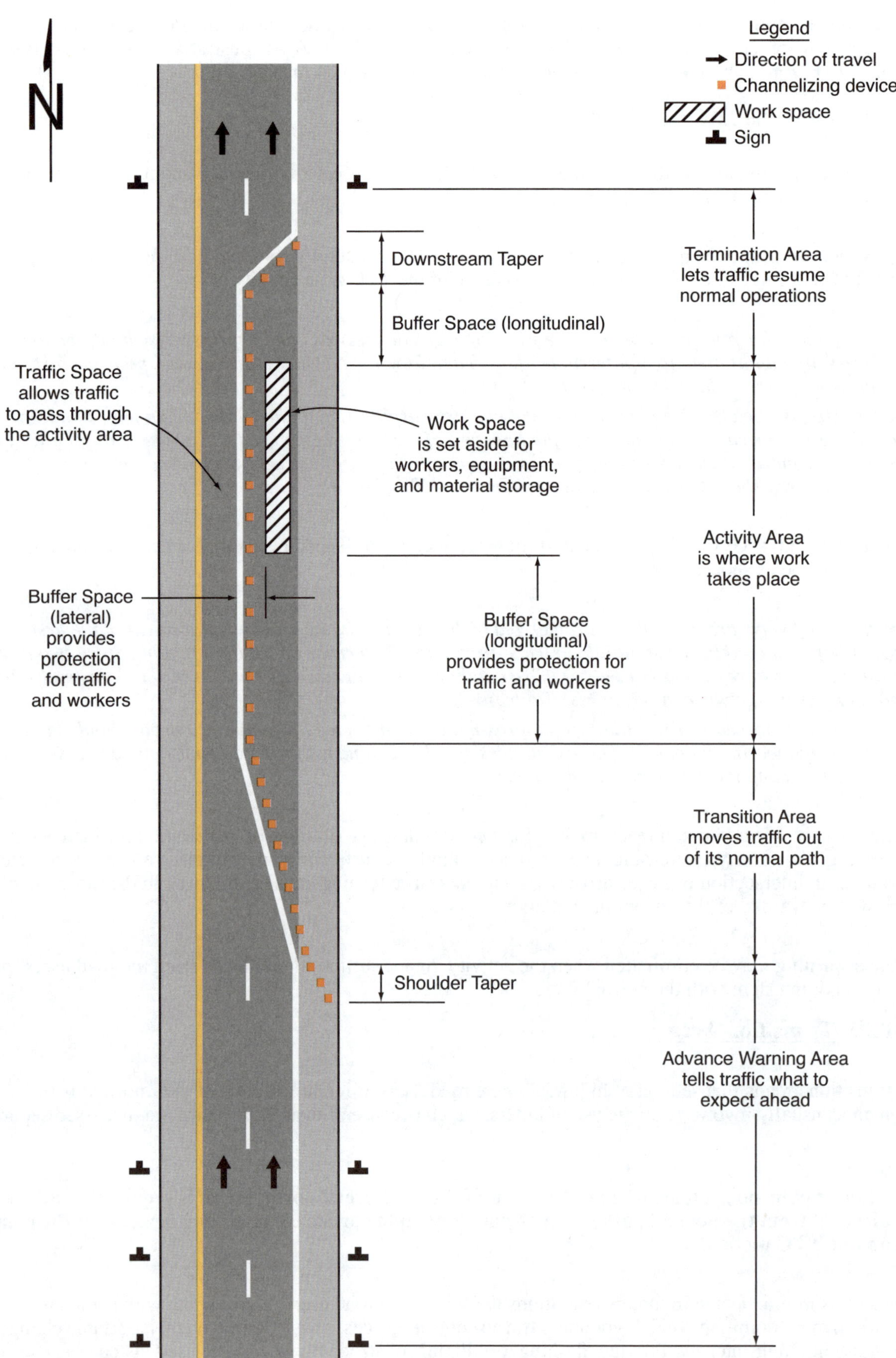

Table 6B-1. Recommended Advance Warning Sign Minimum Spacing

Road Type	Distance between Signs**		
	A	B	C
Urban (low speed)*	100 feet	100 feet	100 feet
Urban (high speed)*	350 feet	350 feet	350 feet
Rural	500 feet	500 feet	500 feet
Expressway / Freeway	1,000 feet	1,500 feet	2,640 feet

* Speed category to be determined by the highway agency or owner of site roadways open to public travel.

** The column headings A, B, and C are the dimensions shown in Figures 6P-1 through 6P-54 The A dimension is the distance from the transition or point of restriction to the first sign. The B dimension is the distance between the first and second signs. The C dimension is the distance between the second and third signs. (The "first sign" is the sign in a three-sign series that is closest to the TTC zone. The "third sign" is the sign that is furthest upstream from the TTC zone.)

Section 6B.06 Activity Area

Support:

01   The activity area is the section of the highway where the work activity takes place. It is comprised of the work space, the traffic space, and the buffer space.

02   The work space is that portion of the highway closed to road users and set aside for workers, equipment, and material, and a shadow vehicle if one is used upstream. Work spaces are usually delineated for road users by channelizing devices or, to exclude vehicles and pedestrians, by temporary barriers.

Option:

03   The work space may be stationary or may move as work progresses.

Guidance:

04   *Since there might be several work spaces (some even separated by several miles) within the project limits, each work space should be adequately signed to inform road users and reduce confusion.*

Support:

05   The traffic space is the portion of the highway in which road users are routed through the activity area.

06   The buffer space is a lateral and/or longitudinal area that separates road user flow from the work space or an unsafe area, and might provide some recovery space for an errant vehicle.

Guidance:

07   *Neither work activity nor storage of equipment, vehicles, or material should occur within a buffer space.*

Option:

08   Buffer spaces may be positioned either longitudinally or laterally with respect to the direction of road user flow. The activity area may contain one or more lateral or longitudinal buffer spaces.

09   A longitudinal buffer space may be placed in advance of a work space.

10   The longitudinal buffer space may also be used to separate opposing road user flows that use portions of the same traffic lane, as shown in Figure 6B-2.

11   If a longitudinal buffer space is used, the values shown in Table 6B-2 may be used to determine the length of the longitudinal buffer space.

Support:

12   Typically, the buffer space is formed as a traffic island and defined by channelizing devices.

13   When a shadow vehicle, arrow board, or changeable message sign is placed in a closed lane in advance of a work space, only the area upstream of the vehicle, arrow board, or changeable message sign constitutes the buffer space.

Option:

14   The lateral buffer space may be used to separate the traffic space from the work space, as shown in Figures 6B-1 and 6B-2, or such areas as excavations or pavement-edge drop-offs. A lateral buffer space also may be used between two travel lanes, especially those carrying opposing flows.

Figure 6B-2. Types of Tapers and Buffer Spaces

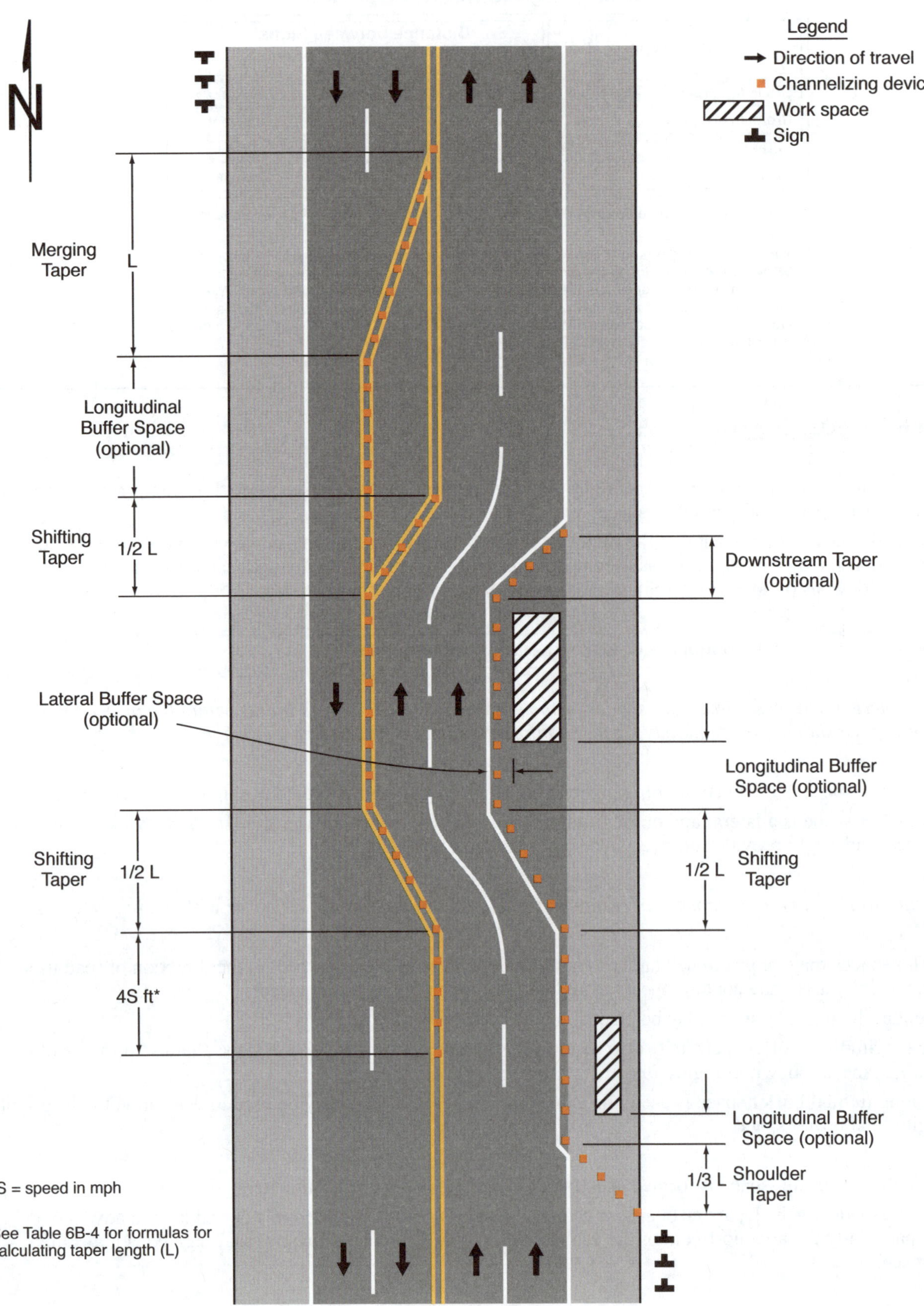

Guidance:

15   *The width of a lateral buffer space should be determined by engineering judgment.*

Option:

16   When work occurs on a high-volume, highly-congested facility, a vehicle storage or staging space may be provided for incident response and emergency vehicles (for example, tow trucks and fire apparatus) so that these vehicles can respond quickly to road user incidents.

Section 6B.07 Termination Area

Support:

01   The termination area is the section of the highway where road users are returned to their normal driving path. The termination area extends from the downstream end of the work area to the last TTC device such as END ROAD WORK signs, if posted.

Option:

02   An END ROAD WORK sign, a Speed Limit sign, or other signs may be used to inform road users that they can resume normal operations.

03   A longitudinal buffer space may be used between the work space and the beginning of the downstream taper.

Section 6B.08 Tapers

Option:

01   Tapers may be used in both the transition and termination areas. Whenever tapers are to be used in close proximity to an interchange ramp, crossroads, curves, or other influencing factors, the length of the tapers may be adjusted.

Support:

02   Tapers are created by using a series of channelizing devices and/or pavement markings to move traffic out of or into the normal path. Types of tapers are shown in Figure 6B-2.

03   Longer tapers are not necessarily better than shorter tapers (particularly in urban areas with characteristics such as short block lengths or driveways) because extended tapers tend to encourage sluggish operation and to encourage drivers to delay lane changes unnecessarily. The test concerning adequate lengths of tapers involves observation of driver performance after TTC plans are put into effect.

Guidance:

04   *The appropriate taper length (L) should be determined using the criteria shown in Tables 6B-3 and 6B-4.*

Support:

05   A merging taper requires the longest distance because drivers are required to merge into common road space.

Table 6B-2. Stopping Sight Distance as a Function of Speed

Speed*	Distance
20 mph	115 feet
25 mph	155 feet
30 mph	200 feet
35 mph	250 feet
40 mph	305 feet
45 mph	360 feet
50 mph	425 feet
55 mph	495 feet
60 mph	570 feet
65 mph	645 feet
70 mph	730 feet
75 mph	820 feet

* Posted speed, off-peak 85th-percentile speed prior to work starting, or the anticipated operating speed

Table 6B-3. Taper Length Criteria for Temporary Traffic Control Zones

Type of Taper	Taper Length
Merging Taper	at least L
Shifting Taper	at least 0.5 L
Shoulder Taper	at least 0.33 L
One-Lane, Two-Way Traffic Taper	50 feet minimum, 100 feet maximum
Downstream Taper	50 feet minimum, 100 feet maximum

Note: Use Table 6B-4 to calculate L

Table 6B-4. Formulas for Determining Taper Length

Speed (S)	Taper Length (L) in feet
40 mph or less	$L = \dfrac{WS^2}{60}$
45 mph or more	$L = WS$

Where: L = taper length in feet
W = width of offset in feet
S = posted speed limit, or off-peak 85th-percentile speed prior to work starting, or the anticipated operating speed in mph

Guidance:

06   *A merging taper should be long enough to enable merging drivers to have adequate advance warning and sufficient length to adjust their speeds and merge into an adjacent lane before the downstream end of the transition.*

Support:

07   A shifting taper is used when a lateral shift is needed. When more space is available, a longer than minimum taper distance can be beneficial. Changes in alignment can also be accomplished by using horizontal curves designed for normal highway speeds.

Guidance:

08   *A shifting taper should have a length of approximately ½ L (see Tables 6B-3 and 6B-4).*

Support:

09   A shoulder taper might be beneficial on a high-speed roadway where shoulders are part of the activity area and are closed, or when improved shoulders might be mistaken as a driving lane. In these instances, the same type, but abbreviated, closure procedures used on a normal portion of the roadway can be used.

Guidance:

10   *If used, shoulder tapers should have a length of approximately ⅓ L (see Tables 6B-3 and 6B-4). If a shoulder is used as a travel lane, either through practice or during a TTC activity, a normal merging or shifting taper should be used.*

Support:

11   A downstream taper might be useful in termination areas to provide a visual cue to the driver that access is available back into the original lane or path that was closed.

Guidance:

12   *If used, a downstream taper should have a minimum length of 50 feet and a maximum length of 100 feet with devices placed at a spacing of approximately 20 feet.*

Support:

13   The one-lane, two-way taper is used in advance of an activity area that occupies part of a two-way roadway in such a manner that a portion of the road is used alternately by traffic in each direction.

Guidance:

14   *A taper having a minimum length of 50 feet and a maximum length of 100 feet with channelizing devices at approximately 20-foot spacing should be used to guide traffic into the one-lane section, and a downstream taper should be used to guide traffic back into their original lane.*

Support:

15   An example of a one-lane, two-way traffic taper is shown in Figure 6B-3.

Section 6B.09 Detours and Diversions

Support:

01   A detour is a temporary rerouting of road users onto an existing highway in order to avoid a TTC zone.

Guidance:

02   *Detours should be clearly signed over their entire length so that road users can easily use existing highways to return to the original highway.*

Support:

03   A diversion is a temporary rerouting of road users onto a temporary highway or alignment placed around the work area.

Figure 6B-3. Example of a One-Lane, Two-Way Traffic Taper

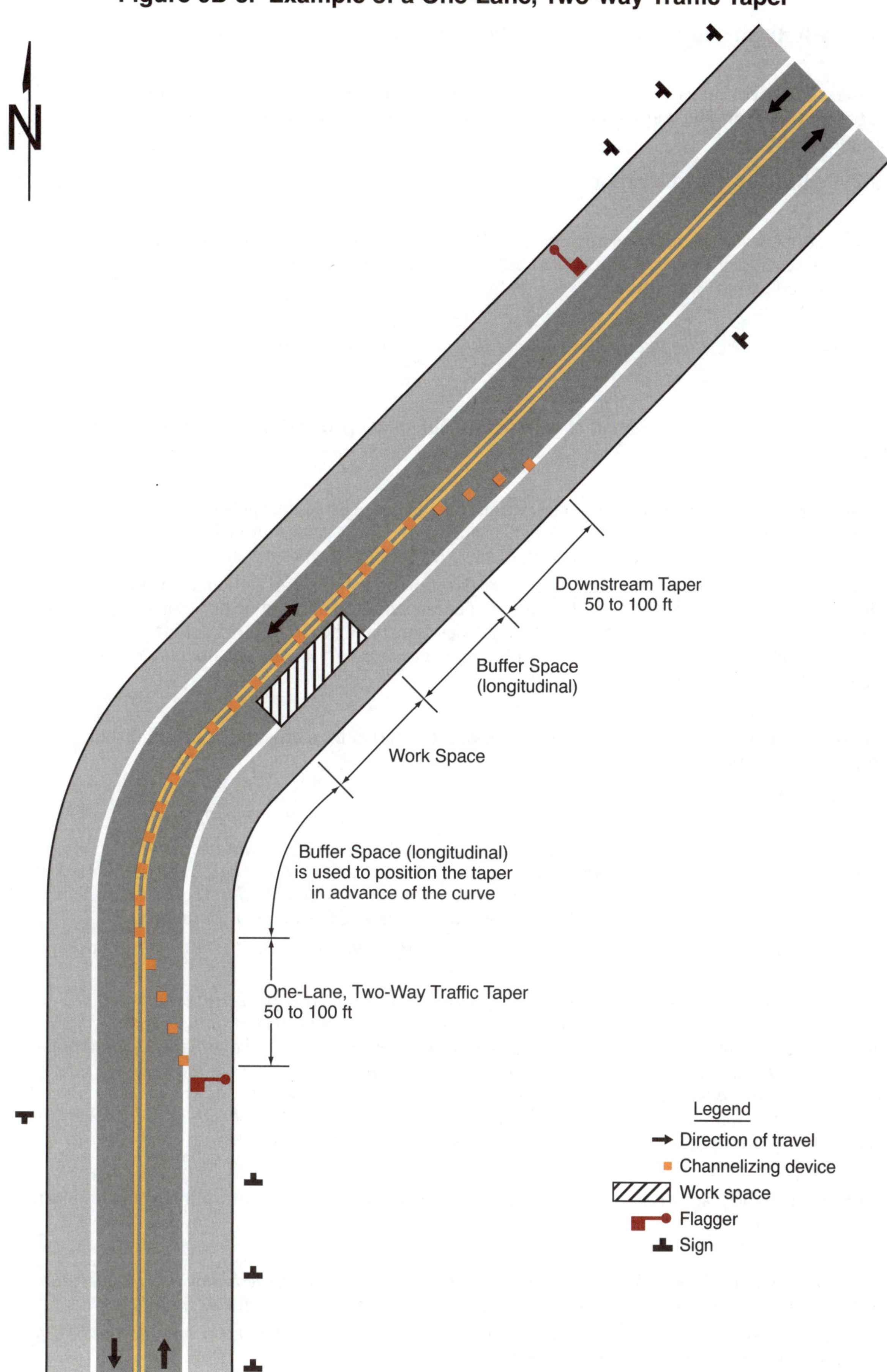

CHAPTER 6C. PEDESTRIAN AND WORKER SAFETY

Section 6C.01 Pedestrian and Worker Safety – General

Standard:

01   **The various TTC provisions for pedestrian and worker safety set forth in Part 6 shall be applied by knowledgeable (for example, trained and/or certified) persons after appropriate evaluation and engineering judgment.**

Section 6C.02 Pedestrian Considerations

Support:

01   A wide range of pedestrians might be affected by TTC zones, including the young, elderly, and people with disabilities such as hearing, vision, or mobility. Pedestrians need a clearly delineated and usable travel path. Considerations for pedestrians with disabilities are addressed in Section 6C.03.

Guidance:

02   *Prior to closing a sidewalk or other pedestrian facility, the maintaining agency should advise users of the future closure.*

Standard:

03   **If the TTC zone affects the movement of pedestrians, adequate pedestrian access and walkways shall be provided.**

Option:

04   If establishing or maintaining an alternate pedestrian route is not feasible during the project, an alternate means of providing for pedestrians may be used, such as adding free bus service around the project or assigning someone the responsibility to assist pedestrians with disabilities through the project limits.

05   If an existing pedestrian route is impacted by a short-duration or a short-term stationary work zone that is attended with project personnel, establishing an alternate pedestrian route may not be necessary if the work can be stopped and pedestrians can navigate the work zone. Pedestrians may be delayed for a short period of time for project personnel to move equipment and material to facilitate passage. Work zone personnel may also provide assistance to pedestrians as necessary.

Support:

06   Pedestrians are reluctant to retrace their steps to a prior intersection for a crossing or to add distance or out-of-the-way travel to a destination.

Guidance:

07   *The following three items should be considered when planning for pedestrians in TTC zones:*

   A. Pedestrians should not be led into conflicts with vehicles, equipment, and operations.
   B. Pedestrians should not be led into conflicts with vehicles moving through or around the worksite.
   C. Pedestrians should be provided with a convenient and accessible path that replicates as nearly as practical the most desirable characteristics of the existing sidewalk(s) or footpath(s).

08   *A pedestrian route should not be severed and/or moved for non-construction activities such as parking for vehicles and equipment.*

09   *TTC zones should be designed to minimize conflicts between vehicular and pedestrian movements. Consideration should be made to separate pedestrian movements from both worksite activity and vehicular traffic. Unless an acceptable route that does not involve crossing the roadway can be provided, pedestrians should be appropriately directed with advance signing that encourages them to cross to the opposite side of the roadway. In urban and suburban areas with high vehicular traffic volumes, these signs should be placed at intersections (rather than midblock locations) so that pedestrians are not confronted with midblock worksites that will induce them to attempt skirting the worksite or making a midblock crossing.*

Support:

10   Figures 6P-28 and 6P-29 show typical TTC device usage and techniques for pedestrian movement through work zones.

Guidance:

11   *To accommodate the needs of pedestrians, including those with disabilities, the following considerations should be addressed when temporary pedestrian pathways in TTC zones are designed or modified:*

   A. Provisions for continuity of accessible paths for pedestrians should be incorporated into the TTC plan.
   B. Access to transit stops should be maintained.

    C. *A smooth, continuous hard surface should be provided throughout the entire length of the temporary pedestrian facility. There should be no curbs or abrupt changes in grade or terrain that could cause tripping or be a barrier to pedestrians with disabilities. The geometry and alignment of the facility should meet the applicable requirements of the "U.S. Department of Justice 2010 ADA Standards for Accessible Design, September 15, 2010, 28 CFR 35 and 36, Americans with Disabilities Act of 1990."*

    D. *The width of the existing pedestrian facility should be provided for the temporary facility if practical. Traffic control devices and other construction materials and features should not intrude into the usable width of the sidewalk, temporary pathway, or other pedestrian facility. When it is not possible to maintain a minimum width of 60 inches throughout the entire length of the pedestrian pathway, a 60 x 60-inch passing space should be provided at least every 200 feet to allow individuals in wheelchairs to pass.*

    E. *Blocked routes, alternate crossings, and sign and signal information should be communicated to pedestrians with vision disabilities by providing devices such as audible information devices or barriers and channelizing devices that are detectable to the pedestrians traveling with the aid of a long cane or who have vision disabilities.*

    F. *When channelization is used to delineate a pedestrian pathway, a continuous detectable edging should be provided throughout the length of the facility such that pedestrians using a long cane can follow it. These detectable edgings should comply with the provisions of Section 6M.04.*

    G. *Signs and other devices mounted lower than 7 feet above the temporary pedestrian pathway should not project more than 4 inches into accessible pedestrian facilities.*

Support:

12   Where pedestrians in TTC zones are routed on temporary pedestrian pathways, providing information in non-visual formats (such as accessible pedestrian signals with audible tones and/or speech messages, and vibrotactile surfaces) aids pedestrians with vision disabilities so they can navigate the temporary pathway. Section 6C.03 contains additional information on accessibility considerations in TTC zones. Section 4K.01 contains information on accessible pedestrian signals.

Option:

13   Whenever it is feasible, the worksite may be closed off from pedestrian intrusion if doing so is determined to be preferable to channelizing pedestrians along the site with TTC devices.

Guidance:

14   *Fencing should not create sight distance restrictions for road users. Fences should not be constructed of materials that would be hazardous if impacted by vehicles. Wooden railing, fencing, and similar systems placed immediately adjacent to motor vehicle traffic should not be used as substitutes for crashworthy temporary traffic barriers.*

15   *Ballast for TTC devices should be kept to the minimum amount needed and should be mounted low to prevent penetration of the vehicle windshield.*

16   *Movement by work vehicles and equipment across designated pedestrian paths should be minimized and, when necessary, should be controlled by flaggers or other TTC. Staging or stopping of work vehicles or equipment along the side of pedestrian paths should be avoided, since it encourages movement of workers, equipment, and materials across the pedestrian path.*

17   *Access to the work space by workers and equipment across pedestrian walkways should be minimized because the access often creates unacceptable changes in grade, and rough or muddy terrain, and pedestrians will tend to avoid these areas by attempting non-intersection crossings where no curb ramps are available.*

Option:

18   A canopied walkway may be used to protect pedestrians from falling debris, and to provide a covered passage for pedestrians.

Guidance:

19   *Covered walkways should be sturdily constructed and adequately lighted for nighttime use.*

20   *When pedestrian and vehicle paths are rerouted to a closer proximity to each other, consideration should be given to separating them by a temporary traffic barrier.*

21   *If a temporary traffic barrier is used to shield pedestrians, it should be designed to accommodate site conditions.*

Support:

22   Depending on the possible vehicular speed and angle of impact, temporary traffic barriers might deflect upon impact by an errant vehicle. Guidance for locating and designing temporary traffic barriers can be found in Chapter 9 of the "Roadside Design Guide," 4th Edition, 2011, AASHTO.

Standard:

23   **Normal vertical curbing shall not be used as a substitute for temporary traffic barriers when temporary traffic barriers are needed.**

Option:

24   Temporary traffic barriers or longitudinal channelizing devices may be used to discourage pedestrians from unauthorized movements into the work space. They may also be used to inhibit conflicts with vehicular traffic by minimizing the possibility of midblock crossings.

Support:

25   A major concern for pedestrians is building construction encroaching onto the contiguous sidewalks, which forces pedestrians off the curb into direct conflict with moving vehicles.

Guidance:

26   *If a significant potential exists for vehicle incursions into the pedestrian path, pedestrians should be rerouted or temporary traffic barriers should be installed.*

Support:

27   TTC devices, temporary traffic barriers, and wood or chain link fencing with a continuous detectable edging can satisfactorily delineate a pedestrian path.

Guidance:

28   *Tape, rope, or plastic chain strung between devices should not be used as a control for pedestrian movements because they are not detectable and are therefore not accessible to and usable by individuals with disabilities.*

29   *In general, pedestrian routes should be preserved in urban and commercial suburban areas. Alternative routing should be discouraged.*

30   *The highway agency in charge of the TTC zone should regularly inspect the activity area so that effective pedestrian TTC is maintained.*

Section 6C.03 Accessibility Considerations

Support:

01   Additional information on the design and construction of accessible temporary facilities is found in the "Guidelines for Accessible Pedestrian Signals (NCHRP Web-Only Document 117B)," 2008 Edition (TRB) and the U.S. Department of Justice 2010 ADA Standards for Accessible Design, September 15, 2010, 28 CFR 35 and 36, Americans with Disabilities Act of 1990.

02   Where pedestrians are detoured to a temporary traffic control signal, an accessible pedestrian signal (see Chapter 4K) provides information in non-visual formats (such as audible tones and/or speech messages, and vibrating surfaces) so that a pedestrian with vision disabilities can know when to cross the street along the alternate route.

Guidance:

03   *Adequate provisions should be made for pedestrians with disabilities. The extent of needs for such provisions should be determined through engineering judgment or by the individual responsible for each TTC zone situation.*

Standard:

04   **When existing pedestrian facilities are disrupted, closed, or relocated in a TTC zone, the temporary facilities shall be detectable and include accessibility features consistent with the features present in the existing pedestrian facility. A barrier that is detectable by a person with a vision disability traveling with the aid of a long cane shall be placed across the full width of the closed pedestrian facility.**

Support:

05   Maintaining a detectable, channelized pedestrian route is much more useful to pedestrians with vision disabilities than closing a walkway and providing audible directions to an alternate route involving additional crossings and a return to the original route. Braille is not useful in conveying such information because it is difficult to find. Audible instructions might be provided, but the extra distance and additional street crossings might add complexity to a trip.

Guidance:

06   *Because printed signs and surface delineation are not usable by pedestrians with vision disabilities, blocked routes, alternate crossings, and sign and signal information should be communicated to pedestrians with vision disabilities by providing audible information devices, tactile and/or vibrating surface devices, and barriers and channelizing devices that are detectable to pedestrians traveling with the aid of a long cane or who have vision disabilities.*

Support:

07   The most desirable way to provide information to pedestrians with vision disabilities that is equivalent to visual signing for notification of sidewalk closures is a speech message provided by an audible information device. Devices that provide speech messages in response to passive pedestrian actuation are the most desirable. Other devices that continuously emit a message, or that emit a message in response to use of a pushbutton, are also acceptable. Audible information devices might not be needed if detectable channelizing devices make an alternate route of travel evident to pedestrians with vision disabilities.

Guidance:

08   *If a pushbutton is used to provide equivalent TTC information to pedestrians with vision disabilities, the pushbutton should be equipped with a locator tone to notify pedestrians with vision disabilities that a special accommodation is available, and to help them locate the pushbutton.*

Section 6C.04 <u>Worker Safety Considerations</u>

Support:

01   Equally as important as the safety of road users traveling through the TTC zone is the safety of workers. TTC zones present temporary and constantly changing conditions that are unexpected by road users. This creates an even higher degree of vulnerability for workers on or near the roadway.

02   Maintaining TTC zones with road user flow inhibited as little as possible, and using TTC devices that get the road users' attention and provide positive direction are of particular importance. Likewise, equipment and vehicles moving within the activity area create a risk to workers on foot. When possible, the separation of moving equipment and construction vehicles from workers on foot provides the operators of these vehicles with a greater separation clearance and improved sight lines to minimize exposure to the hazards of moving vehicles and equipment.

Guidance:

03   *The following are the key elements of worker safety and TTC management that should be considered to improve worker safety:*

  A. *Training—all workers should be trained on how to work next to motor vehicle traffic in ways that minimize their vulnerability. Workers having specific TTC responsibilities should be trained in TTC techniques, device usage, and placement.*

  B. *Temporary Traffic Barriers—temporary traffic barriers should be placed along the work space depending on factors such as lateral clearance of workers from adjacent traffic, speed of traffic, duration and type of operations, time of day, and volume of traffic.*

  C. *Speed Management—reducing the speed of vehicular traffic, mainly through regulatory speed zoning, funneling, lane reduction, and/or the use of speed safety cameras, uniformed law enforcement officers, or flaggers should be considered.*

  D. *Activity Area—operations entering and departing the work space, and within the work space, should be planned to minimize backing maneuvers by construction vehicles and equipment to minimize the risk of run-over and back-over crashes.*

  E. *Worker Safety Planning—a trained person designated by the employer should conduct a basic hazard assessment for the worksite and job classifications required in the activity area. This safety professional should determine whether engineering, administrative, or personal protection measures should be implemented. This plan should be in accordance with the Occupational Safety and Health Act of 1970, as amended, "General Duty Clause" Section 5(a)(1) - Public Law 91-596, 84 Stat. 1590, December 29, 1970, as amended, and with the requirement to assess worker risk exposures for each job site and job classification, as per 29 CFR 1926.20 (b)(2) of "Occupational Safety and Health Administration Regulations, General Safety and Health Provisions."*

Option:

04   The following are additional elements of TTC management that may be considered to improve worker safety:

  A. Shadow Vehicle—in the case of mobile and constantly moving operations, such as pothole patching and striping operations, a shadow vehicle, equipped with appropriate lights and warning signs, may be used to protect the workers from impacts by errant vehicles. The shadow vehicle may be equipped with a rear-mounted impact attenuator.

  B. Road Closure—if alternate routes are available to handle road users, the road may be closed temporarily to facilitate project completion and thus further reduce worker vulnerability.

  C. Law Enforcement Use—in highly vulnerable work situations, particularly those of relatively short-duration, law enforcement units may be stationed to heighten the awareness of passing vehicular traffic and to improve safety through the TTC zone.

  D. Lighting—for nighttime work, the TTC zone and approaches may be lighted.

E.   Special Devices—these include rumble strips, changeable message signs, hazard identification beacons, flags, and warning lights. Intrusion warning devices may be used to alert workers to the approach of errant vehicles.

Support:

05   Judicious use of the special devices described in Item E in Paragraph 4 of this Section might be helpful for certain difficult TTC situations, but misuse or overuse of special devices or techniques might lessen their effectiveness.

Section 6C.05 High-Visibility Safety Apparel

Standard:

01   **For daytime and nighttime activity, all workers, including emergency responders, within the right-of-way who are within the TTC zone shall wear high-visibility safety apparel that meets the Performance Class 2 or 3 requirements of the ANSI/ISEA 107–2015 publication entitled "American National Standard for High-Visibility Safety Apparel and Headwear," or equivalent revisions, except as provided in Paragraph 4 of this Section. A person designated by the employer to be responsible for worker safety shall make the selection of the appropriate class of garment.**

02   **The apparel background (outer) material color shall be fluorescent orange-red, fluorescent yellow-green, or a combination of the two as defined in the ANSI standard. The retroreflective material shall be orange, yellow, white, silver, yellow-green, or a fluorescent version of these colors.**

03   **When uniformed law enforcement personnel are used to direct traffic, to investigate crashes, or to handle lane closures, obstructed roadways, and disasters, high-visibility safety apparel as described in this Section shall be worn by the law enforcement personnel.**

Option:

04   Emergency and incident responders and law enforcement personnel within the TTC zone may wear high-visibility safety apparel that meets the performance requirements of the ANSI/ISEA 207-2006 publication entitled "American National Standard for High-Visibility Public Safety Vests," or equivalent revisions, and labeled as ANSI 207-2006, in lieu of ANSI/ISEA 107-2015 apparel.

Standard:

05   **Except as provided in Paragraph 6 of this Section, firefighters or other emergency responders working within the right-of-way shall wear high-visibility safety apparel as described in this Section.**

Option:

06   Firefighters or other emergency responders working within the right-of-way and engaged in emergency operations that directly expose them to flame, fire, heat, and/or hazardous materials may wear retroreflective turn-out gear that is specified and regulated by other organizations, such as the National Fire Protection Association.

Guidance:

07   *For flagger wear during nighttime activity, high-visibility safety apparel that meets the Performance Class 3 requirements of the ANSI/ISEA 107–2015 publication entitled "American National Standard for High-Visibility Apparel and Headwear," or equivalent revision, and labeled as meeting the ANSI 107-2015 standard performance for Class 3 risk exposure should be worn.*

CHAPTER 6D. FLAGGER CONTROL

Section 6D.01 Qualifications for Flaggers

Guidance:

01   *Because flaggers are responsible for public safety and make the greatest number of contacts with the public of all highway workers, they should be trained in proper traffic control practices and public contact techniques. Flaggers should be able to satisfactorily demonstrate the following abilities:*

    A. Ability to receive and communicate specific instructions clearly, firmly, and courteously;
    B. Ability to move and maneuver quickly in order to avoid danger from errant vehicles;
    C. Ability to control signaling devices (such as paddles and flags) in order to provide clear and positive guidance to drivers approaching a TTC zone in frequently changing situations;
    D. Ability to understand and apply proper traffic control practices, sometimes in stressful or emergency situations; and
    E. Ability to recognize dangerous traffic situations and warn workers in sufficient time to avoid injury.

Section 6D.02 STOP/SLOW Paddle for Hand-Signaling

Guidance:

01   *The STOP/SLOW paddle (see Figure 6D-1 and Table 6G-1) should be the primary and preferred hand-signaling device because the STOP/SLOW paddle gives road users more positive guidance than red flags.*

Standard:

02   **The STOP/SLOW paddle (R1-1 and W20-8) shall have an octagonal shape on a rigid handle. When used at night, the STOP/SLOW paddle shall be retroreflectorized.**

Option

03   A STOP/STOP or a SLOW/SLOW paddle may be used in certain situations (see Section 6D.05), provided the device meets the size and shape requirements for the STOP/SLOW paddle.

Guidance:

04   *The STOP/SLOW paddle should be fabricated from light semi-rigid material.*

Support:

05   The optimum method of displaying a STOP or SLOW message is to place the STOP/SLOW paddle on a rigid staff that is tall enough that when the end of the staff is resting on the ground, the message is high enough to be seen by approaching or stopped traffic.

Option:

06   The STOP/SLOW paddle may be modified to improve conspicuity by incorporating either white or red flashing lights on the STOP face, and either white or yellow flashing lights on the SLOW face. The flashing lights may be arranged in any of the following patterns:

    A. Two white or red lights, one centered vertically above and one centered vertically below the STOP legend; and/or two white or yellow lights, one centered vertically above and one centered vertically below the SLOW legend;
    B. Two white or red lights, one centered horizontally on each side of the STOP legend; and/or two white or yellow lights, one centered horizontally on each side of the SLOW legend;
    C. One white or red light centered below the STOP legend; and/or one white or yellow light centered below the SLOW legend;
    D. A series of eight or more small white or red lights no larger than ¼ inch in diameter along the outer edge of the paddle, arranged in an octagonal pattern at the eight corners of the border of the STOP face; and/or a series of eight or more small white or yellow lights no larger than ¼ inch in diameter along the outer edge of the paddle, arranged in a diamond pattern along the border of the SLOW face; or
    E. A series of white lights forming the shapes of the letters in the legend.

Standard:

07   **If flashing lights are used on the STOP face of the paddle, their colors shall be all white or all red. If flashing lights are used on the SLOW face of the paddle, their colors shall be all white or all yellow.**

08   **If more than eight flashing lights are used, the lights shall be arranged such that they clearly convey the octagonal shape of the STOP face of the paddle and/or the diamond shape of the SLOW face of the paddle.**

09   **If flashing lights are used on the STOP/SLOW paddle, the flash rate shall be at least 50, but not more than 60, flashes per minute.**

Figure 6D-1. Use of Hand-Signaling Devices by Flaggers

PREFERRED METHOD
STOP/SLOW Paddle

EMERGENCY SITUATIONS ONLY
Red Flag

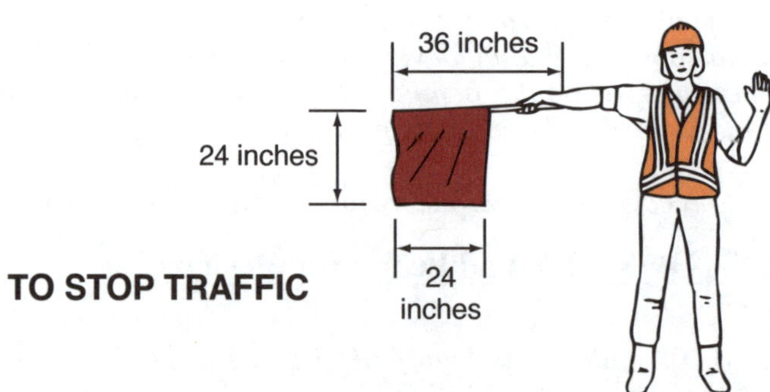

TO STOP TRAFFIC

TO LET
TRAFFIC PROCEED

TO ALERT AND
SLOW TRAFFIC

Section 6D.03 Flag for Hand-Signaling

Guidance:

01   *Use of flags should be limited to emergency situations.*

Standard:

02   **Flags, when used, shall be red or fluorescent orange-red in color, shall be a minimum of 24 inches square, and shall be securely fastened to a staff that is approximately 36 inches in length.**

Guidance:

03   *The free edge of a flag should be weighted so the flag will hang vertically, even in heavy winds.*

Standard:

04   **When used at nighttime, flags shall be retroreflectorized.**

Section 6D.04 Flashlight for Hand-Signaling

Option:

01   When flagging in an emergency situation at night in a non-illuminated flagger station, a flagger may use a flashlight with a red glow cone to supplement the STOP/SLOW paddle or flag.

Standard:

02   **When a flashlight is used for flagging in an emergency situation at night in a non-illuminated flagger station, the flagger shall hold the flashlight in the left hand, shall hold the paddle or flag in the right hand as shown in Figure 6D-1, and shall use the flashlight in the following manner to control approaching road users:**

    A. To inform road users to stop, the flagger shall hold the flashlight with the left arm extended and pointed down toward the ground, and then shall slowly wave the flashlight in front of the body in a slow arc from left to right such that the arc reaches no farther than 45 degrees from vertical.

    B. To inform road users to proceed, the flagger shall point the flashlight at the vehicle's bumper, slowly aim the flashlight toward the open lane, then hold the flashlight in that position. The flagger shall not wave the flashlight.

    C. To alert or slow traffic, the flagger shall point the flashlight toward oncoming traffic and quickly wave the flashlight in a Figure eight motion.

Section 6D.05 Flagger Procedures

Support:

01   The use of paddles and flags by flaggers is illustrated in Figure 6D-1.

Standard:

02   **Flaggers shall use a STOP/SLOW paddle, a flag, or an Automated Flagger Assistance Device (AFAD) (see Sections 6L.02 through 6L.04) to control road users approaching a TTC zone. The use of hand movements alone without a paddle, flag, or AFAD to control road users shall be prohibited when controlling traffic in a one-lane two-way operation except when the control is provided by emergency responders at incident scenes as described in Section 6O.01 or provided by uniformed law enforcement officers.**

03   **The following methods of signaling with a paddle shall be used:**

    A. To stop road users, the flagger shall face road users and aim the STOP paddle face toward road users in a stationary position with the arm extended horizontally away from the body. The free arm shall be held with the palm of the hand above shoulder level toward approaching traffic.

    B. To direct stopped road users to proceed, the flagger shall face road users with the SLOW paddle face aimed toward road users in a stationary position with the arm extended horizontally away from the body. The flagger shall motion with the free hand for road users to proceed.

    C. To alert or slow traffic, the flagger shall face road users with the SLOW paddle face aimed toward road users in a stationary position with the arm extended horizontally away from the body.

Option:

04   To further alert or slow traffic, the flagger holding the SLOW paddle face toward road users may motion up and down with the free hand, palm down.

Standard:

05   **The following methods of signaling with a flag shall be used:**

    A. To stop road users, the flagger shall face road users and extend the flag staff horizontally across the road users' lane in a stationary position so that the full area of the flag is visibly hanging below the staff. The free arm shall be held with the palm of the hand above shoulder level toward approaching traffic.

    B. **To direct stopped road users to proceed, the flagger shall face road users with the flag and arm lowered from the view of the road users, and shall motion with the free hand for road users to proceed. Flags shall not be used to signal road users to proceed.**

    C. **To alert or slow traffic, the flagger shall face road users and slowly wave the flag in a sweeping motion of the extended arm from shoulder level to straight down without raising the arm above a horizontal position. The flagger shall keep the free hand down.**

Guidance:

06   *The flagger should stand either on the shoulder adjacent to the road user being controlled or in the closed lane prior to stopping road users. A flagger should only stand in the lane being used by moving road users after road users have stopped. The flagger should be clearly visible to the first approaching road user at all times. The flagger also should be visible to other road users. The flagger should be stationed sufficiently in advance of the workers to warn them (for example, with audible warning devices such as horns or whistles) of approaching danger by out-of-control vehicles. The flagger should stand alone, away from other workers, work vehicles, or equipment.*

Option:

07   In certain conditions, it may be more appropriate for a flagger to use a STOP/STOP or a SLOW/SLOW paddle to convey the appropriate message to approaching road users and avoid confusing those that are approaching the operation from the opposing direction.

Section 6D.06 <u>Flagger Stations</u>

Standard:

01   **Except as provided in Paragraph 2 of this Section, flagger stations shall be located such that approaching road users will have sufficient distance to stop at an intended stopping point.**

Option:

02   If sufficient stopping sight distance is not achievable, the location of the flagger station may be modified based on engineering judgment.

03   The distances shown in Table 6B-2, which provides information regarding the stopping sight distance as a function of speed, may be used for the location of a flagger station. These distances may be increased for downgrades and other conditions that affect stopping distance.

Guidance:

04   *Flagger stations should be located such that an errant vehicle has additional space to stop without entering the work space. The flagger should identify an escape route that can be used to avoid being struck by an errant vehicle.*

Standard:

05   **Except in emergency situations, flagger stations shall be preceded by an advance warning sign or signs. Except in emergency situations, flagger stations shall be illuminated when flagging is used at night.**

CHAPTER 6E. ONE-LANE, TWO-WAY TRAFFIC CONTROL

Section 6E.01 One-Lane, Two-Way Traffic Control – General

Standard:

01   **Except as provided in Paragraph 4 of this Section, when traffic in both directions must use a single lane for a limited distance, movements from each end shall be coordinated.**

Guidance:

02   *Provisions should be made for alternate one-way movement through the constricted section via methods such as flagger control, a flag transfer, a pilot car, traffic control signals, or stop or yield control.*

03   *Control points at each end should be chosen to permit easy passing of opposing lanes of vehicles.*

Option:

04   If the work space on a low-volume street or road is short and road users from both directions are able to see the traffic approaching from the opposite direction through and beyond the worksite, the movement of traffic through a one-lane, two-way constriction may be self-regulating.

Section 6E.02 Flagger Method

Guidance:

01   *Except as provided in Paragraph 2 of this Section, traffic should be controlled by a flagger at each end of a constricted section of roadway. One of the flaggers should be designated as the coordinator. To provide coordination of the control of the traffic, the flaggers should be able to communicate with each other orally, electronically, or with manual signals. These manual signals should not be mistaken for flagging signals.*

Option:

02   When a one-lane, two-way TTC zone is short enough to allow a flagger to see from one end of the zone to the other, traffic may be controlled by either a single flagger or by a flagger at each end of the section.

Guidance:

03   *When a single flagger is used, the flagger should be stationed on the shoulder opposite the constriction or work space, or in a position where good visibility and traffic control can be maintained at all times. When good visibility and traffic control cannot be maintained by one flagger station, traffic should be controlled by a flagger at each end of the section.*

Section 6E.03 Flag Transfer Method

Support:

01   The driver of the last vehicle proceeding into the one-lane section is given a red flag (or other token) and instructed to deliver it to the flagger at the other end. The opposite flagger, upon receipt of the flag, then knows that traffic can be permitted to move in the other direction. A variation of this method is to replace the use of a flag with an official pilot car that follows the last road user vehicle proceeding through the section.

Guidance:

02   *The flag transfer method should be employed only where the one-way traffic is confined to a relatively short length of a road, usually no more than 1 mile in length.*

Section 6E.04 Pilot Car Method

Option:

01   A pilot car may be used to guide a queue of vehicles through the TTC zone or detour.

Guidance:

02   *The pilot car should have the name of the contractor or contracting authority prominently displayed.*

Standard:

03   **The PILOT CAR FOLLOW ME (G20-4) sign (see Figure 6H-1) shall be mounted on the top or on the rear of the pilot vehicle (see Section 6H.37).**

04   **The pilot car operation shall be coordinated with flagging operations or other methods of control at each end of the one lane section of the work zone.**

05   **If an Automated Flagger Assistance Device (AFAD) (see Section 6L.02) is used in pilot car operations, the AFAD shall be operated by a flagger positioned near and within the line of sight of the AFAD. The AFAD shall not be left unattended at any time that the AFAD is being used.**

Guidance:

06   *If temporary traffic control signals are used in pilot car operations and long wait times will be encountered by road users, consideration should be given to using signs to notify drivers of the wait time and/or pilot car operation, based on engineering judgment.*

Section 6E.05 <u>Temporary Traffic Control Signal Method</u>

Option:

01   Traffic control signals may be used to control vehicular traffic movements in one-lane, two-way TTC zones (see Figure 6P-12 and Chapter 4O).

Section 6E.06 <u>Stop or Yield Control Method</u>

Option:

01   STOP or YIELD signs may be used to control traffic on low-volume roads at a one-lane, two-way TTC zone when drivers are able to see the other end of the one-lane, two-way operation and have sufficient visibility of approaching vehicles.

Guidance:

02   *If the STOP or YIELD sign is installed for only one direction, then the STOP or YIELD sign should face road users who are driving on the side of the roadway that is closed for the work activity area.*

CHAPTER 6F. TEMPORARY TRAFFIC CONTROL ZONE SIGNS – GENERAL

Section 6F.01 General Characteristics of TTC Zone Signs

Support:

01   TTC zone signs convey both general and specific messages by means of words, symbols, and/or arrows and have the same three categories as all road user signs: regulatory, warning, and guide.

Option:

02   Where the color orange is required, the fluorescent orange color may also be used.

Support:

03   The fluorescent version of orange provides higher conspicuity than standard orange, especially during twilight.

Option:

04   Standard orange flags, flashing beacons, and/or flashing warning lights may be used in conjunction with signs.

Standard:

05   **When standard orange flags, flashing beacons, and/or flashing warning lights are used in conjunction with a sign, they shall not block the sign face.**

06   **Except as provided in Section 2A.07, the sizes for TTC signs and plaques shall be as shown in Tables 6G-1, 6H-1, and 6I-1. The sizes in the minimum column shall only be used on low-volume rural roads, local streets, or roadways where the operating speed is 30 mph or less.**

Option:

07   The dimensions of signs and plaques shown in Tables 6G-1, 6H-1, and 6I-1 may be increased wherever necessary for greater legibility or emphasis.

Guidance:

08   *Deviations from standard sizes as prescribed in this Manual should be in 6-inch increments.*

Support:

09   Sign design details are contained in the "Standard Highway Signs" publication (see Section 1A.05).

10   Section 2A.04 contains additional information regarding the design of signs, including an Option allowing the development of special word message signs if a standard word message or symbol sign is not available to convey the necessary regulatory, warning, or guidance information.

Standard:

11   **All signs used at night shall be either retroreflective or illuminated to show the same shape and similar color both day and night.**

12   **The requirement for sign illumination shall not be considered to be satisfied by street, highway, or strobe lighting.**

Option:

13   Sign illumination may be either internal or external.

14   Signs may be made of rigid or flexible material.

Section 6F.02 Sign Placement

Guidance:

01   *Signs should be located on the right-hand side of the roadway unless otherwise provided in this Manual.*

Option:

02   Where special emphasis is needed, signs may be placed on both the left-hand and right-hand sides of the roadway. Signs mounted on portable supports may be placed within the roadway itself. Signs may also be mounted on or above barricades.

Support:

03   The provisions of this Section regarding mounting height apply unless otherwise provided for a particular sign elsewhere in this Manual.

Standard:

04   **The minimum height, measured vertically from the bottom of the sign to the elevation of the near edge of the pavement, of signs installed at the side of the road in rural areas shall be 5 feet (see Figure 6F-1).**

05   **The minimum height, measured vertically from the bottom of the sign to the top of the curb, or in the absence of curb, measured vertically from the bottom of the sign to the elevation of the near edge of the traveled way, of signs installed at the side of the road in business, commercial, or residential areas where parking or pedestrian movements are likely to occur, or where the view of the sign might be obstructed, shall be 7 feet (see Figure 6F-1).**

06   **The minimum height, measured vertically from the bottom of the sign to the sidewalk, of signs installed above sidewalks shall be 7 feet.**

07   **The bottom of a sign mounted on a barricade, or other portable support, shall be at least 1 foot above the traveled way.**

Option:

08   The height to the bottom of a secondary sign mounted below another sign may be 1 foot less than the height provided in Paragraphs 4 through 6 of this Section.

Guidance:

09   *Neither portable nor permanent sign supports should be located on sidewalks, bicycle facilities, or areas designated for pedestrians or bicyclists.*

Standard:

10   **Signs shall be mounted and placed in accordance with Section 307 of the U.S. Department of Justice 2010 ADA Standards for Accessible Design, September 15, 2010, 28 CFR 35 and 36, Americans with Disabilities Act of 1990.**

Guidance:

11   *Except as provided in Paragraph 12 of this Section, signs mounted on portable sign supports that do not meet the minimum mounting heights provided in Part 2 should not be used for a duration of more than 3 days.*

Figure 6F-1. Height and Lateral Location of Signs—Typical Installations

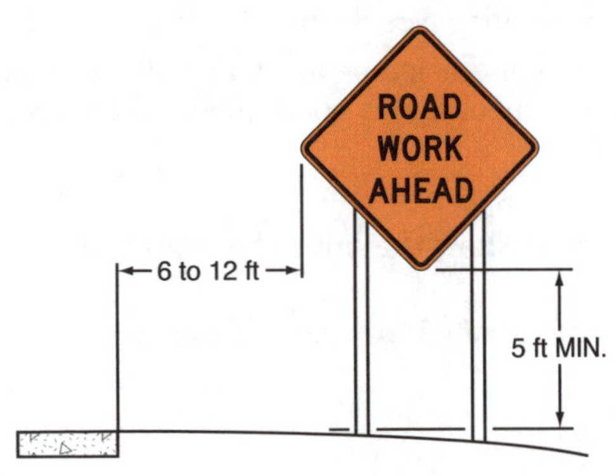

A – Rural area

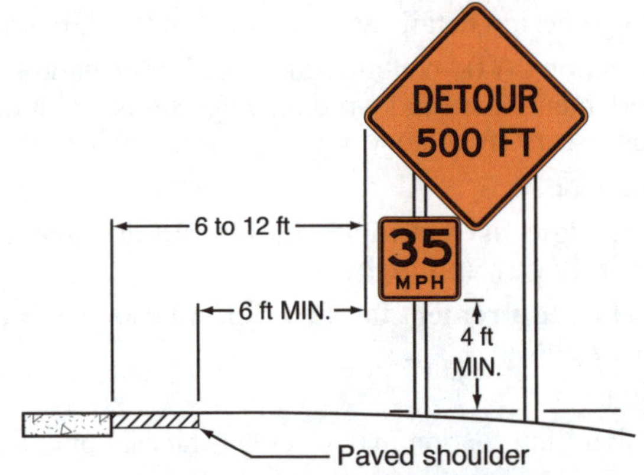

B – Rural area with advisory speed plaque

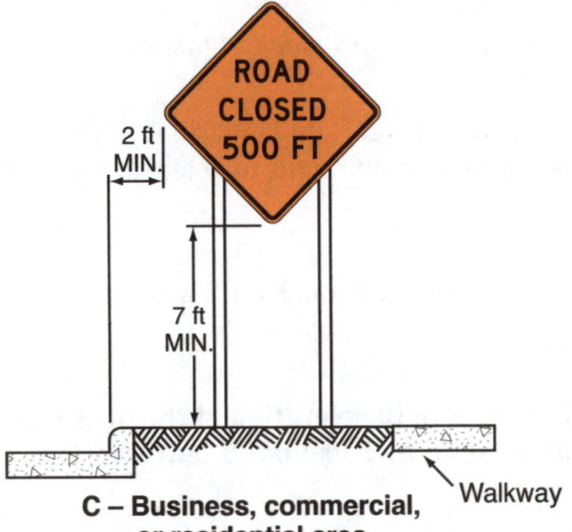

C – Business, commercial, or residential area

D – Business, commercial, or residential area (without curb)

Option:

12   The R9-8 through R9-11a series, R11 series, W1-6 through W1-8 series, M4-10, E5-1, or other similar type signs (see Figures 6G-1, 6H-1, and 6I-1) may be used on portable sign supports that do not meet the minimum mounting heights provided in Part 2 for longer than 3 days.

Support:

13   Methods of mounting signs other than on posts are illustrated in Figure 6F-2.

Guidance:

14   *Signs mounted on Type 3 Barricades should not cover more than 50 percent of the top two rails or 33 percent of the total area of the three rails.*

Standard:

15   **Signs and sign supports used together shall be crashworthy (see Section 6A.04). Where large signs having an area exceeding 50 square feet are installed on multiple breakaway posts, the clearance from the ground to the bottom of the sign shall be at least 7 feet.**

Option:

16   For mobile operations, a sign may be mounted on a work vehicle, a shadow vehicle, or a trailer stationed in advance of the TTC zone or moving along with it.

Section 6F.03 Sign Maintenance

Guidance:

01   *Signs should be properly maintained for cleanliness, visibility, retroreflectivity, and correct positioning.*

02   *Signs that have lost significant legibility should be promptly replaced.*

Support:

03   Section 2A.21 contains information regarding the retroreflectivity of signs, including the signs that are used in TTC zones.

Figure 6F-2. Methods of Mounting Signs Other Than on Posts

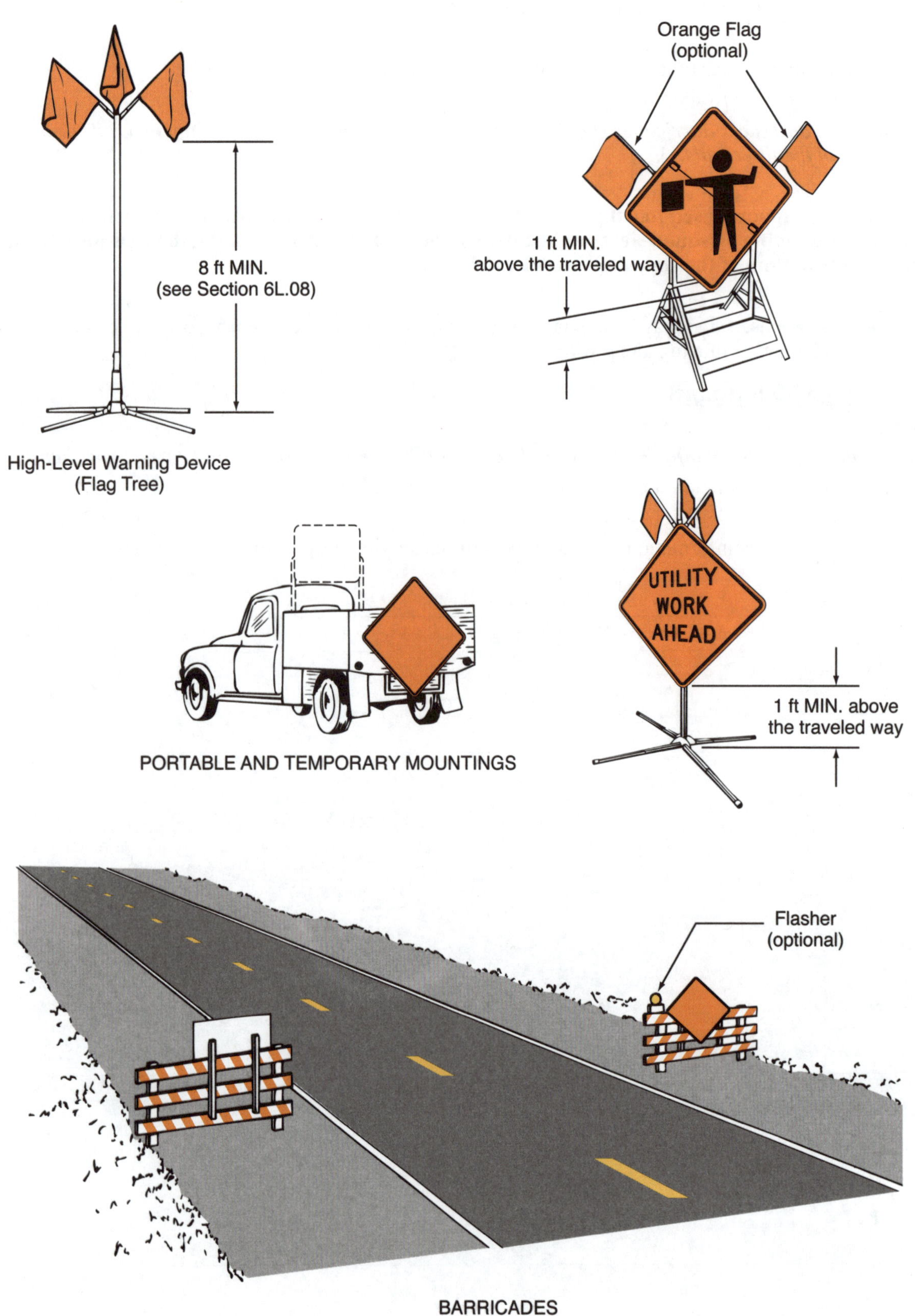

CHAPTER 6G. TTC ZONE REGULATORY SIGNS

Section 6G.01 Regulatory Sign Authority

Support:

01   Regulatory signs such as those shown in Figure 6G-1 inform road users of traffic laws or regulations and indicate the applicability of legal requirements that would not otherwise be apparent.

Standard:

02   **Regulatory signs shall be authorized by the public agency or official having jurisdiction and shall conform with Chapter 2B.**

Section 6G.02 Regulatory Sign Design and Size

Standard:

01   **TTC regulatory signs shall comply with the Standards for regulatory signs presented in Part 2 and in the FHWA's "Standard Highway Signs" publication (see Section 1A.05).**

02   **The sizes for TTC regulatory signs shall be as shown in Table 6G-1.**

Section 6G.03 Regulatory Sign Applications

Standard:

01   **If a TTC zone requires regulatory measures different from those existing, the existing permanent regulatory devices shall be removed or covered and superseded by the appropriate temporary regulatory signs. This change shall be made in compliance with applicable ordinances or statutes of the jurisdiction.**

Section 6G.04 Road Closed Signs (R11-2 Series)

Guidance:

01   *The ROAD CLOSED (R11-2) sign (see Figure 6G-1) should be used when the roadway is closed to all road users except contractors' equipment or officially authorized vehicles. The R11-2 sign should be accompanied by appropriate warning and detour signing.*

Option:

02   STREET CLOSED (R11-2a), BRIDGE OUT (R11-2b), or PATH CLOSED (R11-2c) signs may be substituted for Road Closed signs where applicable.

Guidance:

03   *Road Closed signs should be installed at or near the center of the roadway on or above a Type 3 Barricade that closes the roadway (see Section 6K.07).*

Standard:

04   **Road Closed signs shall not be used where road user flow is maintained through the TTC zone with a reduced number of lanes on the existing roadway or where the actual closure is some distance beyond the sign.**

Section 6G.05 Local Traffic Only Signs (R11-3 Series and R11-4)

Guidance:

01   *The Local Traffic Only signs (see Figure 6G-1) should be used where road user flow detours to avoid a closure some distance beyond the sign, but where local road users can use the roadway to the point of closure. These signs should be accompanied by appropriate warning and detour signing.*

02   *In rural applications, the Local Traffic Only sign should have the legend ROAD CLOSED XX MILES AHEAD, LOCAL TRAFFIC ONLY (R11-3).*

Option:

03   In urban areas, a ROAD (STREET) CLOSED TO THRU TRAFFIC (R11-4) sign or the legend ROAD CLOSED, LOCAL TRAFFIC ONLY may be used.

04   In urban areas, a word message that includes the name of an intersecting street name or well-known destination may be substituted for the words XX MILES AHEAD on the R11-3 sign where applicable.

05   A STREET CLOSED (R11-3a) or BRIDGE OUT (R11-3b) sign may be substituted for an R11-3 sign, where applicable.

06   The words BRIDGE OUT, BRIDGE CLOSED, or STREET CLOSED may be substituted for the words ROAD CLOSED on the R11-4 sign where applicable.

Figure 6G-1. Regulatory Signs and Plaques
in Temporary Traffic Control Zones (Sheet 1 of 2)

R1-1

R1-2

R1-2aP

R1-7

R1-7a

R1-8

G20-5aP

R2-1

R2-6P

R2-6aP

R2-6bP

R2-10

R2-11

R2-12

R3-1

R3-2

R3-3

R3-4

R3-5

R3-6

R3-7

R3-8

R3-18

R3-27

R4-1

R4-2

R4-7

R4-7c

R4-9

R4-9a

R5-1

R5-1a

R6-1

R6-2

R8-3

Note: See Chapter 2B for information on the application of these signs.

Figure 6G-1. Regulatory Signs and Plaques
in Temporary Traffic Control Zones (Sheet 2 of 2)

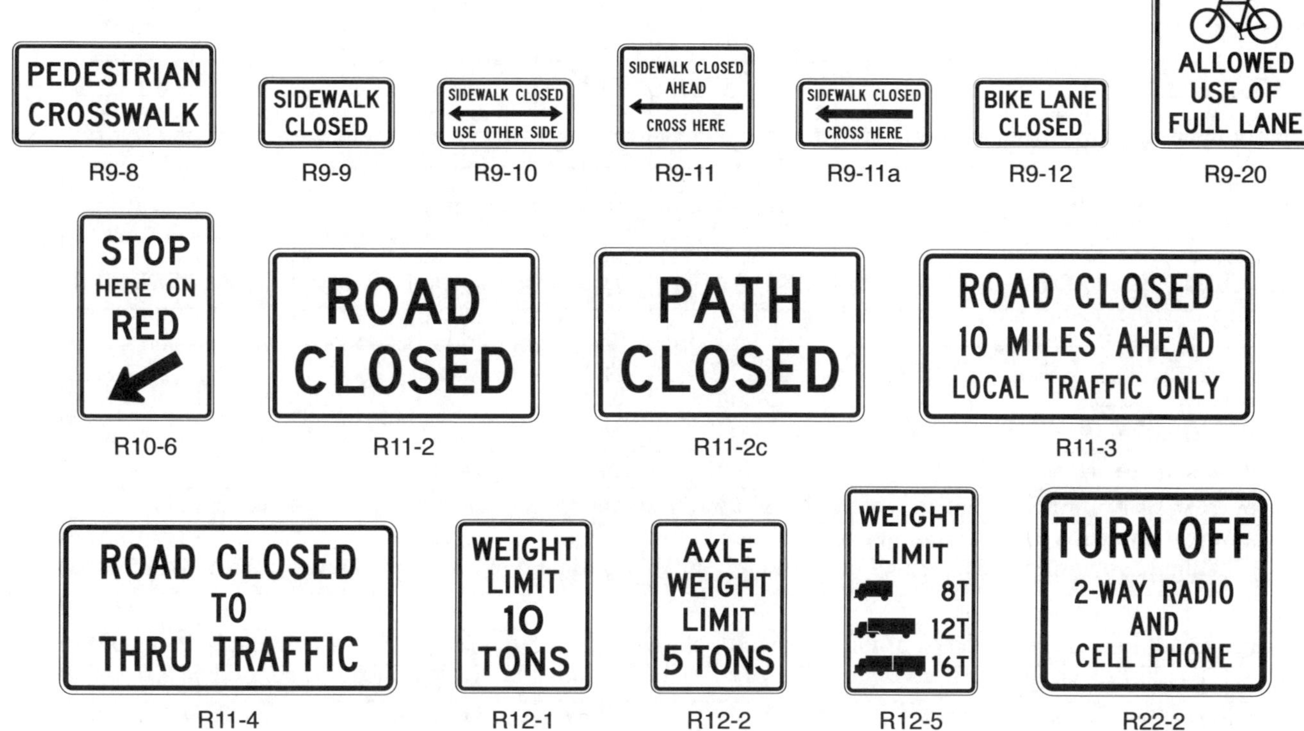

Note: See Chapter 2B for information on the application of these signs.

Section 6G.06 Weight Limit Signs (R12-1, R12-2, and R12-5)

Standard:

01 **A Weight Limit sign (see Figure 6G-1), which shows the gross weight or axle weight that is permitted on the roadway or bridge, shall be consistent with State or local regulations and shall not be installed without the approval of the authority having jurisdiction over the highway.**

02 **When weight restrictions are imposed because of the activity in a TTC zone, a marked detour shall be provided for vehicles weighing more than the posted limit.**

Section 6G.07 STAY IN LANE Signs (R4-9 and R4-9a)

Option:

01 A STAY IN LANE (R4-9) sign (see Figure 6G-1) may be used where a multi-lane shift has been incorporated as part of the TTC on a highway to direct road users around road work that occupies part of the roadway on a multi-lane highway.

Guidance:

02 *A STAY IN LANE TO MERGE POINT (R4-9a) sign (see Figure 6G-1) should be used during late merge operations (see Section 6N.19) to direct traffic to use all available lanes until the merge point is reached.*

Section 6G.08 Work Zone and Higher Fines Signs and Plaques

Option:

01 A WORK ZONE (G20-5aP) plaque (see Figure 6G-1) may be mounted above a Speed Limit sign to emphasize that a reduced speed limit is in effect within a TTC zone. An END WORK ZONE SPEED LIMIT (R2-12) sign (see Figure 6G-1) may be installed at the downstream end of the reduced speed limit zone.

Guidance:

02 *A BEGIN HIGHER FINES ZONE (R2-10) sign (see Figure 6G-1) should be installed at or near the beginning of a TTC zone where increased fines are imposed for traffic violations, and an END HIGHER FINES ZONE (R2-11) sign (see Figure 6G-1) should be installed at or near the downstream end of the TTC zone.*

Table 6G-1. Temporary Traffic Control Zone Regulatory Sign and Plaque Sizes

Sign or Plaque	Sign Designation	Section	Conventional Road	Freeway or Expressway	Minimum
Stop	R1-1	6G.02	30 x 30*	—	—
Stop (on Stop/Slow Paddle)	R1-1	6D.02	18 x 18	—	—
Yield	R1-2	6G.02	36 x 36 x 36*	—	30 x 30 x 30
To Oncoming Traffic (plaque)	R1-2aP	6G.02	36 x 30	48 x 36	24 x 18
Wait on Stop	R1-7	6L.03	24 x 30	24 x 30	—
Wait on Stop - Go on Slow	R1-7a	6G.03	30 x 36	30 x 36	—
Go on Slow	R1-8	6L.03	24 x 30	24 x 30	—
Speed Limit	R2-1	6G.08	24 x 30*	36 x 48	—
Fines Higher (plaque)	R2-6P	6G.08	24 x 18	36 x 24	—
Fines Double (plaque)	R2-6aP	6G.08	24 x 18	36 x 24	—
$XX Fine (plaque)	R2-6bP	6G.08	24 x 18	36 x 24	—
Begin Higher Fines Zone	R2-10	6G.08	24 x 30	36 x 48	—
End Higher Fines Zone	R2-11	6G.08	24 x 30	36 x 48	—
End Work Zone Speed Limit	R2-12	6G.08	24 x 36	36 x 54	—
Movement Prohibition	R3-1,2,3,4	6G.02	24 x 24*	36 x 36	—
Mandatory Movement Lane Control - Turn Only	R3-5	6G.02	30 x 36	—	—
Optional Movement Lane Control - Thru and Turn	R3-6	6G.02	30 x 36	—	—
Right (Left) Lane Must Turn Right (Left)	R3-7	6G.02	30 x 30*	—	—
Advance Intersection Lane Control (2 lanes)	R3-8	6G.02	30 x 30	—	—
Movement Prohibition - No U or Left Turn	R3-18	6G.02	24 x 24*	36 x 36	—
Movement Prohibition - No Straight Through	R3-27	6G.02	24 x 24*	36 x 36	—
Do Not Pass	R4-1	6G.02	24 x 30	36 x 48	—
Pass With Care	R4-2	6G.02	24 x 30	36 x 48	—
Keep Right	R4-7	6G.02	24 x 30	36 x 48	—
Narrow Keep Right	R4-7c	6G.02	18 x 30	—	—
Stay in Lane	R4-9	6G.07	24 x 30	36 x 48	—
Stay In Lane To Merge Point	R4-9a	6G.07	36 x 48	36 x 48	—
Do Not Enter	R5-1	6G.02	30 x 30*	36 x 36	—
Wrong Way	R5-1a	6G.02	36 x 24*	42 x 30	—
One Way	R6-1	6G.02	36 x 12*	48 x 18	—
One Way	R6-2	6G.02	24 x 30*	36 x 48	—
No Parking (symbol)	R8-3	6G.02	24 x 24*	36 x 36	—
Pedestrian Crosswalk	R9-8	6G.09	36 x 18	—	—
Sidewalk Closed	R9-9	6G.10	24 x 12	—	—
Sidewalk Closed, Use Other Side	R9-10	6G.10	24 x 12	—	—
Sidewalk Closed Ahead, Cross Here	R9-11	6G.10	24 x 18	—	—
Sidewalk Closed, Cross Here	R9-11a	6G.10	24 x 12	—	—
Bike Lane Closed	R9-12	6P.01	24 x 12	—	—
Stop Here on Red	R10-6	6L.04	24 x 36	—	—
Road Closed	R11-2, 2a, 2b, 2c	6G.04	48 x 30	—	—
Road Closed - Local Traffic Only	R11-3, 3a, 3b, 4	6G.05	60 x 30	—	—
Weight Limit	R12-1, 2	6G.06	24 x 30	36 x 48	—
Weight Limit	R12-5	6G.06	24 x 36	36 x 48	—
Turn Off 2-Way Radio and Cell Phone	R22-2	6G.11	42 x 36	42 x 36	—
Work Zone (plaque)	G20-5aP	6G.08	24 x 18	30 x 24	—

* See Table 2B-1 for minimum size required for signs facing traffic on multi-lane conventional roads

Notes:

1. Larger signs may be used wherever necessary for greater legibility or emphasis

2. Dimensions are shown in inches and are shown as width x height

Option:

03   Alternate legends such as BEGIN (or END) DOUBLE FINES ZONE may also be used for the R2-10 and R2-11 signs.

04   A FINES HIGHER, FINES DOUBLE, or $XX FINE plaque (see Section 2B.25 and Figure 6G-1) may be mounted below the Speed Limit sign if increased fines are imposed for traffic violations within the TTC zone.

05   Individual signs and plaques for work zone speed limits and higher fines may be combined into a single sign or may be displayed as an assembly of signs and plaques.

Section 6G.09 PEDESTRIAN CROSSWALK Sign (R9-8)

Option:

01   The PEDESTRIAN CROSSWALK (R9-8) sign (see Figure 6G-1) may be used to indicate where a temporary crosswalk has been established.

Standard:

02   **If a temporary crosswalk is established, it shall be accessible to pedestrians with disabilities in accordance with Section 6C.03.**

Section 6G.10 SIDEWALK CLOSED Signs (R9-9, R9-10, R9-11, and R9-11a)

Guidance:

01   *SIDEWALK CLOSED signs (see Figure 6G-1) should be used where pedestrian flow is restricted. Bicyclist/ Pedestrian Detour (M4-9a) signs or Pedestrian Detour (M4-9b) signs should be used where pedestrian flow is rerouted (see Section 6I.02).*

02   *The SIDEWALK CLOSED (R9-9) sign should be installed at the beginning of the closed sidewalk, at the intersections preceding the closed sidewalk, and elsewhere along the closed sidewalk as needed.*

03   *The SIDEWALK CLOSED, (ARROW) USE OTHER SIDE (R9-10) sign should be installed at the beginning of the restricted sidewalk when a parallel sidewalk exists on the other side of the roadway.*

04   *The SIDEWALK CLOSED AHEAD, (ARROW) CROSS HERE (R9-11) sign should be used to indicate to pedestrians that sidewalks beyond the sign are closed and to direct them to open crosswalks, sidewalks, or other travel paths.*

05   *The SIDEWALK CLOSED, (ARROW) CROSS HERE (R9-11a) sign should be installed just beyond the point to which pedestrians are being redirected.*

Support:

06   These signs are typically mounted on a detectable barricade to encourage compliance and to communicate with pedestrians that the sidewalk is closed. Printed signs are not useful to many pedestrians with vision disabilities. A barrier or barricade detectable by a person with a vision disability is sufficient to indicate that a sidewalk is closed. If the barrier is continuous with detectable channelizing devices for an alternate route, accessible signing might not be necessary.

Section 6G.11 TURN OFF 2-WAY RADIO AND CELL PHONE Sign (R22-2)

Standard:

01   **The TURN OFF 2-WAY RADIO AND CELL PHONE (R22-2) sign (see Figure 6G-1) shall be used to require road users to turn off mobile radio transmitters and cellular telephones where blasting operations occur.**

Support:

02   Section 6H.25 contains information about the full sequence of signs for blasting zones and the specific requirements for location of this regulatory sign.

Section 6G.12 Other Regulatory Signs

Option:

01   Regulatory word message signs other than those classified and specified in this Manual and the "Standard Highways Signs" publication (see Section 1A.05) may be developed and used based on engineering judgment to aid the enforcement of other laws or regulations in TTC zones.

Guidance:

02   *Special regulatory signs should comply with the general requirements of color, shape, and alphabet size and series. The sign message should be brief, legible, and clear.*

CHAPTER 6H. TTC ZONE WARNING SIGNS

Section 6H.01 Warning Sign Function, Design, and Application

Support:

01   TTC zone warning signs (see Figure 6H-1) notify road users of specific situations or conditions on or adjacent to a roadway that might not otherwise be apparent.

Standard:

02   **TTC warning signs shall comply with the Standards for warning signs presented in Part 2 and in the FHWA's "Standard Highway Signs" publication (see Section 1A.05).**

03   **The sizes for TTC warning signs shall be as shown in Table 6H-1.**

04   **Except as provided in Paragraph 5 of this Section, TTC warning signs shall be diamond-shaped with a black legend and border on an orange background, except for the Grade Crossing Advance Warning (W10-1) sign, which shall have a black legend and border on a yellow background.**

Option:

05   Warning signs that are required or recommended in Parts 2 or 7 to have a fluorescent yellow-green background may have that color background in TTC zones.

06   Existing warning signs with a yellow background that are still applicable may remain in place.

07   Warning signs used for TTC incident management situations may have a black legend and border on a fluorescent pink background.

08   Mounting or space considerations may justify a change from the standard diamond shape to a rectangular shape.

09   In emergencies, available warning signs having yellow backgrounds may be used if signs with orange or fluorescent pink backgrounds are not at hand.

Guidance:

10   *Where roadway or road user conditions require greater emphasis, larger than standard size warning signs should be used, with the symbol or legend enlarged approximately in proportion to the outside dimensions.*

11   *Where any part of the roadway is obstructed or closed by work activities or incidents, advance warning signs should be installed to alert road users well in advance of these obstructions or restrictions.*

12   *Where road users include pedestrians, the provision of supplemental audible information or detectable barriers or barricades should be provided for people with vision disabilities.*

Support:

13   Detectable barriers or barricades communicate very clearly to pedestrians who have vision disabilities that they can no longer proceed in the direction that they are traveling.

Option:

14   Advance warning signs may be used singly or in combination.

15   Where distances are not displayed on warning signs as part of the message, a supplemental plaque with the distance legend may be mounted immediately below the sign on the same support.

Section 6H.02 Position of Advance Warning Signs

Guidance:

01   *Where highway conditions permit, warning signs should be placed in advance of the transition and activity areas at varying distances depending on roadway type, condition, and posted speed. Table 6B-1 contains information regarding the spacing of advance warning signs. Where a series of two or more advance warning signs is used, the closest sign to the transition and activity areas should be placed approximately 100 feet for low-speed urban streets to 1,000 feet or more for freeways and expressways.*

02   *Where multiple advance warning signs are needed on the approach to a transition and activity area, the ROAD WORK AHEAD (W20-1) sign should be the first advance warning sign encountered by road users.*

Support:

03   Various conditions, such as limited sight distance or obstructions that might require a driver to reduce speed or stop, might require additional advance warning signs.

Option:

04   As an alternative to a specific distance on advance warning signs, the word AHEAD may be used.

Figure 6H-1. Warning Signs and Plaques in Temporary Traffic Control Zones (Sheet 1 of 4)

Note: See Chapter 2C for information on the application of these signs.

Figure 6H-1. Warning Signs and Plaques in Temporary Traffic Control Zones (Sheet 2 of 4)

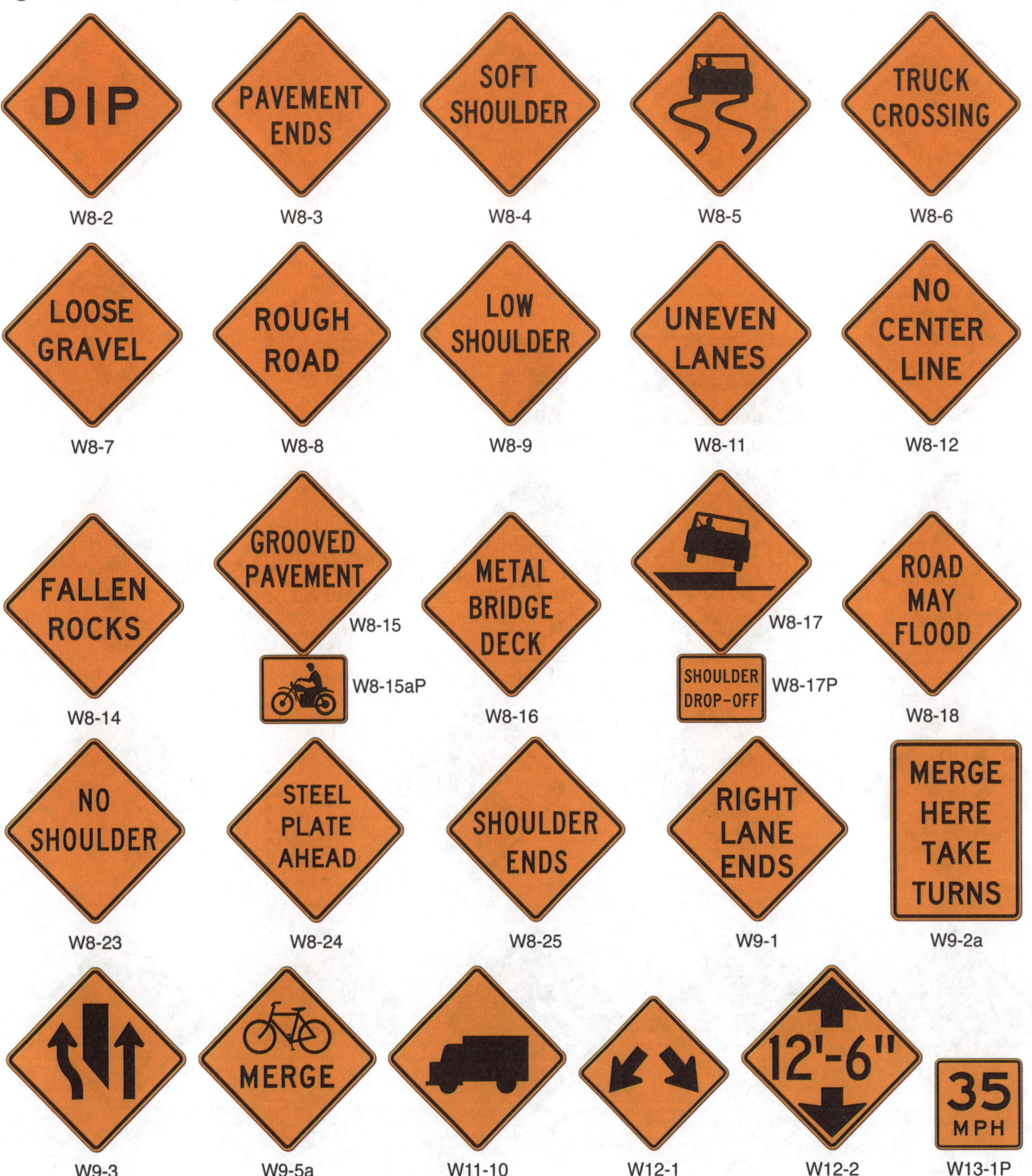

Note: See Chapter 2C for information on the application of these signs.

Figure 6H-1. Warning Signs and Plaques in Temporary Traffic Control Zones (Sheet 3 of 4)

Note: See Chapter 2C for information on the application of these signs.

* An optional STREET WORK word message sign is shown in the "Standard Highway Signs" publication.

** An optional STREET CLOSED word message sign is shown in the "Standard Highway Signs" publication.

*** An optional FLAGGER (W20-7a) word message sign is shown in the "Standard Highway Signs" publication.

**** An optional FRESH TAR word message sign is shown in the "Standard Highway Signs" publication.

Figure 6H-1. Warning Signs and Plaques in Temporary Traffic Control Zones (Sheet 4 of 4)

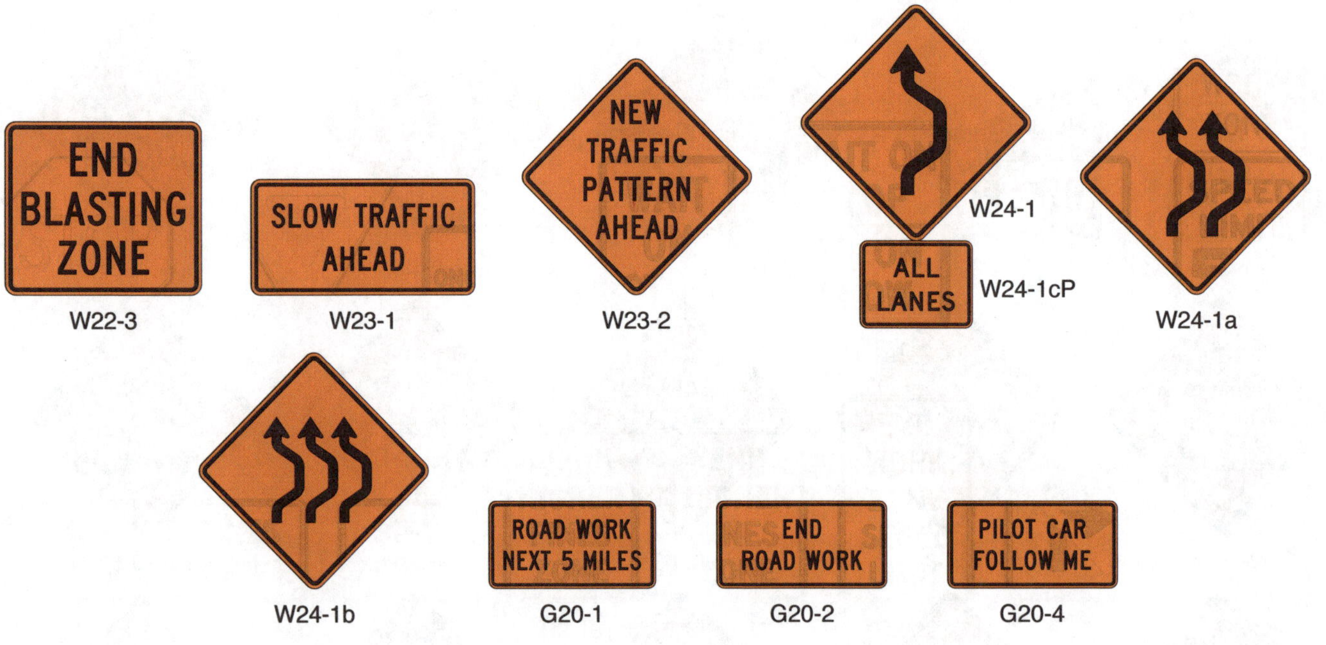

Note: See Chapter 2C for information on the application of these signs.

Support:

05   At TTC zones on lightly-traveled roads, all of the advance warning signs prescribed for major construction might not be needed.

Option:

06   Utility work, maintenance, or minor construction can occur within the TTC zone limits of a major construction project, and additional warning signs may be needed.

Guidance:

07   *Utility, maintenance, and minor construction signing and TTC should be coordinated with appropriate authorities so that road users are not confused or misled by the additional TTC devices.*

Section 6H.03 ROAD (STREET) WORK Sign (W20-1)

Guidance:

01   *The ROAD (STREET) WORK (W20-1) sign (see Figure 6H-1), which serves as a general warning of obstructions or restrictions, should be located in advance of the work space or any detour, on the road where the work is taking place.*

02   *Where traffic can enter a TTC zone from a crossroad or a major (high-volume) driveway, an advance warning sign should be used on the crossroad or major driveway.*

Option:

03   The legend STREET may be substituted for ROAD and the distance legend may be either XX FEET, XX MILES, or AHEAD.

Section 6H.04 DETOUR Sign (W20-2)

Guidance:

01   *The DETOUR (W20-2) sign (see Figure 6H-1) should be used in advance of a road user detour over a different roadway or route.*

Option:

02   The distance legend may be either XX FEET, XX MILES, or AHEAD.

Table 6H-1. Temporary Traffic Control Zone Warning Sign and Plaque Sizes (Sheet 1 of 2)

Sign or Plaque	Sign Designation	Section	Conventional Road	Freeway or Expressway	Minimum
Turn and Curve Signs	W1-1,2,3,4	6H.01	36 x 36	48 x 48	30 x 30
Reverse Curve (2 or more lanes)	W1-4b,4c	6H.30	36 x 36	48 x 48	30 x 30
Large Arrow (1-direction)	W1-6	6H.01	48 x 24	60 x 30	—
Chevron Alignment	W1-8	6H.01	18 x 24	30 x 36	—
Stop Ahead	W3-1	6H.01	36 x 36	48 x 48	30 x 30
Yield Ahead	W3-2	6H.01	36 x 36	48 x 48	30 x 30
Signal Ahead	W3-3	6H.01	36 x 36	48 x 48	30 x 30
Be Prepared to Stop	W3-4	6H.01	36 x 36	48 x 48	30 x 30
Reduced Speed Limit Ahead	W3-5	6H.01	36 x 36	48 x 48	30 x 30
XX MPH Speed Zone Ahead	W3-5a	6H.01	36 x 36	48 x 48	30 x 30
Merging Traffic	W4-1,5	6H.01	36 x 36	48 x 48	36 x 36
Lane Ends	W4-2	6H.08	36 x 36	48 x 48	30 x 30
Added Lane	W4-3,6	6H.01	36 x 36	48 x 48	30 x 30
No Merge Area (plaque)	W4-5aP	6H.01	18 x 24	24 x 30	—
Road Narrows	W5-1	6H.01	36 x 36	48 x 48	30 x 30
Narrow Bridge	W5-2	6H.01	36 x 36	48 x 48	30 x 30
One Lane Bridge	W5-3	6H.01	36 x 36	48 x 48	30 x 30
Ramp Narrows	W5-4	6H.10	36 x 36	48 x 48	30 x 30
Divided Highway	W6-1	6H.01	36 x 36	48 x 48	30 x 30
Divided Highway Ends	W6-2	6H.01	36 x 36	48 x 48	30 x 30
Two-Way Traffic	W6-3	6H.16	36 x 36	48 x 48	30 x 30
Narrow Two-Way Traffic	W6-4	6H.17	12 x 18	12 x 18	—
Hill	W7-1	6H.01	36 x 36	48 x 48	30 x 30
Next XX Miles (plaque)	W7-3aP	6H.33	24 x 18	36 x 30	—
Bump	W8-1	6H.01	36 x 36	48 x 48	24 x 24
Dip	W8-2	6H.01	36 x 36	48 x 48	24 x 24
Pavement Ends	W8-3	6H.01	36 x 36	48 x 48	30 x 30
Soft Shoulder	W8-4	6H.26	36 x 36	48 x 48	30 x 30
Slippery When Wet	W8-5	6H.01	36 x 36	48 x 48	30 x 30
Truck Crossing	W8-6	6H.21	36 x 36	48 x 48	30 x 30
Loose Gravel	W8-7	6H.01	36 x 36	48 x 48	30 x 30
Rough Road	W8-8	6H.01	36 x 36	48 x 48	24 x 24
Low Shoulder	W8-9	6H.26	36 x 36	48 x 48	24 x 24
Uneven Lanes	W8-11	6H.27	36 x 36	48 x 48	30 x 30
No Center Line	W8-12	6H.29	36 x 36	48 x 48	30 x 30
Fallen Rocks	W8-14	6H.01	36 x 36	48 x 48	30 x 30
Grooved Pavement	W8-15	6H.01	36 x 36	48 x 48	30 x 30
Motorcycle (plaque)	W8-15aP	6H.34	24 x 18	30 x 24	—
Metal Bridge Deck	W8-16	6H.34	36 x 36	48 x 48	30 x 30
Shoulder Drop Off (symbol)	W8-17	6H.26	36 x 36	48 x 48	30 x 30
Shoulder Drop-Off (plaque)	W8-17P	6H.26	24 x 18	30 x 24	—
Road May Flood	W8-18	6H.01	36 x 36	48 x 48	24 x 24
No Shoulder	W8-23	6H.01	36 x 36	48 x 48	30 x 30
Steel Plate Ahead	W8-24	6H.28	36 x 36	48 x 48	30 x 30
Shoulder Ends	W8-25	6H.01	36 x 36	48 x 48	30 x 30
Lane Ends	W9-1,2	6H.01	36 x 36	48 x 48	30 x 30
Merge Here Take Turns	W9-2a	6N.19	36 x 48	36 x 48	—
Interior Lane Shift Ahead	W9-3	6H.07	36 x 36	48 x 48	30 x 30

Table 6H-1. Temporary Traffic Control Zone Warning Sign and Plaque Sizes (Sheet 2 of 2)

Sign or Plaque	Sign Designation	Section	Conventional Road	Freeway or Expressway	Minimum
Bicycles Merging	W9-5a	6P.01	30 x 30	—	18 x 18
Grade Crossing Advance Warning	W10-1	6H.01	36 dia.	48 Dia.	—
Truck	W11-10	6H.21	36 x 36	48 x 48	24 x 24
Double Arrow	W12-1	6H.01	30 x 30	36 x 36	
Low Clearance	W12-2	6H.01	36 x 36	48 x 48	30 x 30
Advisory Speed (plaque)	W13-1P	6H.32	18 x 18	24 x 24	18 x 18
On Ramp (plaque)	W13-4P	6H.09	36 x 36	36 x 36	—
No Passing Zone (pennant)	W14-3	6H.01	48 x 48 x 36	64 x 64 x 48	40 x 40 x 30
XX Feet (2-line plaque)	W16-2P	6H.01	24 x 18	30 x 24	—
Road Work (with distance)	W20-1	6H.03	36 x 36	48 x 48	30 x 30
Path Work (with distance)	W20-1b	6P.01	36 x 36	—	30 x 30
Detour (with distance)	W20-2	6H.04	36 x 36	48 x 48	30 x 30
Bike Detour (with distance)	W20-2a	6P.01	36 x 36	—	30 x 30
Bike Diversion (with distance)	W20-2b	6P.01	36 x 36	—	30 x 30
Road Closed (with distance)	W20-3	6H.05	36 x 36	48 x 48	30 x 30
Path Closed (with distance)	W20-3a	6P.01	36 x 36	—	30 x 30
One Lane Road (with distance)	W20-4	6H.06	36 x 36	48 x 48	30 x 30
Lane(s) Closed (with distance)	W20-5,5a	6H.07	36 x 36	48 x 48	30 x 30
Bike Lane Closed (with distance)	W20-5b	6P.01	36 x 36	—	30 x 30
Flagger (symbol)	W20-7	6H.15	36 x 36	48 x 48	30 x 30
Flagger	W20-7a	6H.15	36 x 36	48 x 48	30 x 30
Slow (on Stop/Slow Paddle)	W20-8	6D.02	18 x 18	—	—
Workers	W21-1,1a	6H.18	36 x 36	48 x 48	30 x 30
Fresh Oil	W21-2	6H.19	36 x 36	48 x 48	30 x 30
Road Machinery Ahead	W21-3	6H.20	36 x 36	48 x 48	30 x 30
Slow Moving Vehicle	W21-4	6N.05	36 x 18	—	—
Shoulder Work	W21-5	6H.22	36 x 36	48 x 48	30 x 30
Shoulder Closed	W21-5a	6H.22	36 x 36	48 x 48	30 x 30
Shoulder Closed (with distance)	W21-5b	6H.22	36 x 36	48 x 48	30 x 30
Survey Crew	W21-6	6H.23	36 x 36	48 x 48	30 x 30
Utility Work (with distance)	W21-7	6H.24	36 x 36	48 x 48	30 x 30
Mowing Ahead	W21-8	6N.05	36 x 36	48 x 48	30 x 30
Blasting Zone Ahead	W22-1	6H.25	36 x 36	48 x 48	30 x 30
End Blasting Zone	W22-3	6H.25	42 x 36	42 x 36	36 x 30
Slow Traffic Ahead	W23-1	6H.11	48 x 24	48 x 24	—
New Traffic Pattern Ahead	W23-2	6H.14	36 x 36	48 x 48	30 x 30
Double Reverse Curve (1 lane)	W24-1	6H.31	36 x 36	48 x 48	30 x 30
Double Reverse Curve (2 lanes)	W24-1a	6H.31	36 x 36	48 x 48	30 x 30
Double Reverse Curve (3 lanes)	W24-1b	6H.31	36 x 36	48 x 48	30 x 30
All Lanes (plaque)	W24-1cP	6H.31	24 x 18	30 x 24	—
Road Work Next XX Miles	G20-1	6H.35	36 x 18	48 x 24	—
End Road Work	G20-2	6H.36	36 x 18	48 x 24	—
Pilot Car Follow Me	G20-4	6H.37	36 x 18	—	—

* See Table 2C-1 for minimum size required for signs facing traffic on multi-lane conventional roads

Notes:

1. Larger signs may be used wherever necessary for greater legibility or emphasis
2. Dimensions are shown in inches and are shown as width x height

Section 6H.05 ROAD (STREET) CLOSED Sign (W20-3)

Guidance:

01   *The ROAD (STREET) CLOSED (W20-3) sign (see Figure 6H-1) should be used in advance of the point where a highway is closed to all road users, or to all but local road users.*

Option:

02   The legend STREET may be substituted for ROAD and the distance legend may be either XX FEET, XX MILES, or AHEAD.

Section 6H.06 ONE LANE ROAD Sign (W20-4)

Standard:

01   **The ONE LANE ROAD (W20-4) sign (see Figure 6H-1) shall be used only in advance of that point where motor vehicle traffic in both directions must use a common single lane (see Section 6E.01).**

Option:

02   The distance legend may be either XX FEET, XX MILES, or AHEAD.

Section 6H.07 Lane(s) Closed Signs (W20-5, W20-5a, and W9-3)

Standard:

01   **The Lane(s) Closed sign (see Figure 6H-1) shall be used in advance of that point where one or more through lanes of a multi-lane roadway are closed.**

02   **For a single lane closure, the Lane Closed (W20-5) sign (see Figure 6H-1) shall use the legend RIGHT (LEFT) LANE CLOSED. Where two or more adjacent lanes are closed, the W20-5a sign (see Figure 6H-1) shall use the legend XX RIGHT (LEFT) LANES CLOSED.**

Option:

03   The distance legend may be either XX FEET, XX MILES, or AHEAD.

Guidance:

04   *The Interior Lane Shift (W9-3) sign (see Figure 6H-1) should be used in advance of that point where work occupies an interior lane(s) and approaching motor vehicle traffic is directed to the right or left of the work zone in the lane(s) by using a shifting taper to route traffic around the closed interior lane(s).*

Section 6H.08 Lane Ends Signs (W4-2 and W9-2a)

Option:

01   The Lane Ends (W4-2) sign (see Figure 6H-1) may be used to warn drivers of the reduction in the number of lanes for moving motor vehicle traffic in the direction of travel on a multi-lane roadway.

Guidance:

02   *The MERGE HERE TAKE TURNS (W9-2a) sign (see Figure 6H-1) should be used to identify the merge point at which vehicles from alternate lanes take turns merging during Late Merge applications (see Section 6N.19).*

Section 6H.09 ON RAMP Plaque (W13-4P)

Guidance:

01   *When work is being done on a ramp, but the ramp remains open, the ON RAMP (W13-4P) plaque (see Figure 6H-1) should be used to supplement the advance ROAD WORK sign.*

Section 6H.10 RAMP NARROWS Sign (W5-4)

Guidance:

01   *The RAMP NARROWS (W5-4) sign (see Figure 6H-1) should be used in advance of the point where work on a ramp reduces the normal width of the ramp along a part or all of the ramp.*

Section 6H.11 SLOW TRAFFIC AHEAD Sign (W23-1)

Option:

01   The SLOW TRAFFIC AHEAD (W23-1) sign (see Figure 6H-1) may be used on a shadow vehicle, usually mounted on the rear of the most upstream shadow vehicle, along with other appropriate signs for mobile operations to warn of slow moving work vehicles. A ROAD WORK (W20-1) sign may also be used with the SLOW TRAFFIC AHEAD sign.

Section 6H.12 <u>EXIT OPEN and EXIT CLOSED Signs (E5-2 and E5-2a)</u>

Option:

01   An EXIT OPEN (E5-2) or EXIT CLOSED (E5-2a) sign (see Figure 6H-1) may be used to supplement other warning signs where work is being conducted in the vicinity of an exit ramp and where the exit maneuver for vehicular traffic using the ramp is different from the normal condition.

Section 6H.13 <u>EXIT ONLY Sign (E5-3)</u>

Option:

01   An EXIT ONLY (E5-3) sign (see Figure 6H-1) may be used to supplement other warning signs where work is being conducted in the vicinity of an exit ramp and where the exit maneuver for vehicular traffic using the ramp is different from the normal condition.

Section 6H.14 <u>NEW TRAFFIC PATTERN AHEAD Sign (W23-2)</u>

Option:

01   A NEW TRAFFIC PATTERN AHEAD (W23-2) sign (see Figure 6H-1) may be used on the approach to an intersection or along a section of roadway to provide advance warning of a change in traffic patterns, such as revised lane usage, roadway geometry, or signal phasing.

Guidance:

02   *To retain its effectiveness, the W23-2 sign should be displayed for up to 2 weeks, and then it should be covered or removed until it is needed again.*

Section 6H.15 <u>Flagger Signs (W20-7 and W20-7a)</u>

Guidance:

01   *The Flagger (W20-7) sign (see Figure 6H-1) should be used in advance of any point where a flagger is stationed to control road users.*

Option:

02   A distance legend may be displayed on a supplemental plaque below the Flagger sign. The sign may be used with appropriate legends or in conjunction with other warning signs, such as the BE PREPARED TO STOP (W3-4) sign (see Figure 6H-1).

03   The FLAGGER (W20-7a) word message sign with a distance legend may be substituted for the Flagger (W20-7) sign.

Section 6H.16 <u>Two-Way Traffic Sign (W6-3)</u>

Guidance:

01   *When one roadway of a normally-divided highway is closed, with two-way vehicular traffic maintained on the other roadway, the Two-Way Traffic (W6-3) sign (see Figure 6H-1) should be used at the beginning of the two-way vehicular traffic section and at intervals to remind road users of opposing vehicular traffic.*

Section 6H.17 <u>Narrow Two-Way Traffic Sign (W6-4)</u>

Standard:

01   **The Narrow Two-Way Traffic (W6-4) sign (see Figure 6H-1) shall be an upright, retroreflective orange-colored sign placed on a flexible support and sized at least 12 inches wide by 18 inches high.**

Support:

02   The Narrow Two-Way Traffic (W6-4) sign is intended for mounting only on a flexible support in a series along the center line to separate opposing vehicular traffic on a two-lane, two-way operation.

Standard:

03   **Narrow Two-Way Traffic signs shall not be placed within pedestrian crossings.**

Section 6H.18 <u>Workers Signs (W21-1 and W21-1a)</u>

Option:

01   A Workers (W21-1) sign (see Figure 6H-1) may be used to alert road users of workers in or near the roadway.

Guidance:

02   *In the absence of other warning devices, a Workers sign should be used when workers are in the roadway.*

Option:

03   The WORKERS (W21-1a) word message sign may be used as an alternate to the Workers (W21-1) symbol sign.

Section 6H.19 FRESH OIL (TAR) Sign (W21-2)

Guidance:

01   *The FRESH OIL (TAR) (W21-2) sign (see Figure 6H-1) should be used to warn road users of the surface treatment.*

Section 6H.20 ROAD MACHINERY AHEAD Sign (W21-3)

Option:

01   The ROAD MACHINERY AHEAD (W21-3) sign (see Figure 6H-1) may be used to warn of machinery operating in or adjacent to the roadway.

Section 6H.21 Motorized Traffic Signs (W8-6 and W11-10)

Option:

01   Motorized Traffic (W8-6 and W11-10) signs may be used to alert road users to locations where unexpected travel on the roadway or entries into or departures from the roadway by construction vehicles might occur. The TRUCK CROSSING (W8-6) word message sign may be used as an alternate to the Truck (W11-10) symbol sign (see Figure 6H-1) where there is an established construction vehicle crossing of the roadway.

Support:

02   These locations might be relatively confined or might occur randomly over a segment of roadway.

Section 6H.22 Shoulder Work Signs (W21-5, W21-5a, and W21-5b)

Support:

01   Shoulder Work signs (see Figure 6H-1) warn of maintenance, reconstruction, or utility operations on the highway shoulder where the roadway is unobstructed.

Standard:

02   **The Shoulder Work sign shall have the legend SHOULDER WORK (W21-5), RIGHT (LEFT) SHOULDER CLOSED (W21-5a), or RIGHT (LEFT) SHOULDER CLOSED XX FT or AHEAD (W21-5b).**

Option:

03   The Shoulder Work sign may be used in advance of the point on a non-limited access highway where there is shoulder work. It may be used singly or in combination with a ROAD WORK NEXT XX MILES or ROAD WORK AHEAD sign.

Guidance:

04   *On freeways and expressways, the RIGHT (LEFT) SHOULDER CLOSED XX FT or AHEAD (W21-5b) sign followed by RIGHT (LEFT) SHOULDER CLOSED (W21-5a) sign should be used in advance of the point where the shoulder work occurs and should be preceded by a ROAD WORK AHEAD sign.*

Section 6H.23 SURVEY CREW Sign (W21-6)

Guidance:

01   *The SURVEY CREW (W21-6) sign (see Figure 6H-1) should be used to warn of surveying crews working in or adjacent to the roadway.*

Section 6H.24 UTILITY WORK Sign (W21-7)

Option:

01   The UTILITY WORK (W21-7) sign (see Figure 6H-1) may be used as an alternate to the ROAD (STREET) WORK (W20-1) sign for utility operations on or adjacent to a highway.

Support:

02   Typical examples of where the UTILITY WORK sign is used appear in Figures 6P-4, 6P-6, 6P-10, 6P-15, 6P-18, 6P-21, 6P-22, 6P-26, and 6P-33.

Option:

03   The distance legend may be either XX FEET, XX MILES, or AHEAD.

Section 6H.25 Signs for Blasting Areas

Support:

01   Radio-Frequency (RF) energy can cause the premature firing of electric detonators (blasting caps) used in TTC zones.

Standard:

02   **Road users shall be warned where blasting operations occur. A sequence of signs shall be prominently displayed to warn all road users of blasting operations and to direct operators of mobile radio equipment, including cellular telephones, to turn off transmitters in a blasting area. These signs shall be covered or removed when there are no explosives in the area or the area is otherwise secured.**

03   **The BLASTING ZONE AHEAD (W22-1) sign (see Figure 6H-1) shall be used in advance of any TTC zone where explosives are being used. The TURN OFF 2-WAY RADIO AND CELL PHONE (R22-2) and END BLASTING ZONE (W22-3) signs shall be used in sequence with this sign.**

04   **The TURN OFF 2-WAY RADIO AND CELL PHONE (R22-2) sign (see Section 6G.11 and Figure 6G-1) shall follow the BLASTING ZONE AHEAD (W22-1) sign and shall be placed at least 1,000 feet before the beginning of the blasting zone.**

05   **The END BLASTING ZONE (W22-3) sign (see Figure 6H-1) shall be placed a minimum of 1,000 feet past the blasting zone.**

Option:

06   The END BLASTING ZONE sign may be placed either with or preceding the END ROAD WORK sign.

Section 6H.26 Shoulder Signs and Plaque (W8-4, W8-9, W8-17, and W8-17P)

Option:

01   The SOFT SHOULDER (W8-4) sign (see Figure 6H-1) may be used to warn of a soft shoulder condition.

02   The LOW SHOULDER (W8-9) sign (see Figure 6H-1) may be used to warn of a shoulder condition where there is an elevation difference of 3 inches or less between the shoulder and the travel lane.

Guidance:

03   *The Shoulder Drop Off (W8-17) sign (see Figure 6H-1) should be used when an unprotected shoulder drop-off, adjacent to the travel lane, exceeds 3 inches in depth for a continuous length along the roadway, based on engineering judgment.*

Option:

04   A SHOULDER DROP-OFF (W8-17P) supplemental plaque (see Figure 6H-1) may be mounted below the W8-17 sign.

Section 6H.27 UNEVEN LANES Sign (W8-11)

Guidance:

01   *The UNEVEN LANES (W8-11) sign (see Figure 6H-1) should be used during operations that create a difference in elevation between adjacent lanes that are open to travel.*

Section 6H.28 STEEL PLATE AHEAD Sign (W8-24)

Option:

01   A STEEL PLATE AHEAD (W8-24) sign (see Figure 6H-1) may be used to warn road users that the presence of a temporary steel plate(s) might make the road surface uneven and might create slippery conditions during wet weather.

Section 6H.29 NO CENTER LINE Sign (W8-12)

Guidance:

01   *The NO CENTER LINE (W8-12) sign (see Figure 6H-1) should be used when the work obliterates the center line pavement markings. This sign should be placed at the beginning of the TTC zone and repeated at 2-mile intervals in long TTC zones.*

Support:

02   Section 6J.02 contains information regarding temporary markings.

Section 6H.30 Reverse Curve Signs (W1-4 Series)

Guidance:

01   *In order to give road users advance notice of a lane shift, a Reverse Curve (W1-4, W1-4b, or W1-4c) sign (see Figure 6H-1) should be used when a lane (or lanes) is being shifted to the left or right. If the design speed of the curves is 30 mph or less, a Reverse Turn (W1-3) sign should be used.*

Standard:

02   **If a Reverse Curve (or Turn) sign is used, the direction of the reverse curve (or turn) shall be appropriately illustrated. Except as provided in Paragraph 3 of this Section, the number of lanes illustrated on the sign shall be the same as the number of through lanes available to road users.**

Option:

03   Where two or more lanes are being shifted, a W1-4 (or W1-3) sign with an ALL LANES (W24-1cP) plaque (see Figure 6H-1) may be used instead of a sign that illustrates the number of lanes.

04   Where more than three lanes are being shifted, the Reverse Curve (or Turn) sign may be rectangular.

Section 6H.31 Double Reverse Curve Signs (W24-1 Series)

Option:

01   The Double Reverse Curve (W24-1, W24-1a, or W24-1b) sign (see Figure 6H-1) may be used where the tangent distance between two reverse curves is less than 600 feet, thus making it difficult for a second Reverse Curve (W1-4 series) sign to be placed between the curves. If the design speed of the curves is 30 mph or less, Double Reverse Turn signs may be used.

Standard:

02   **If a Double Reverse Curve (or Turn) sign is used, the direction of the double reverse curve (or turn) shall be appropriately illustrated. Except as provided in Paragraph 3 of this Section, the number of lanes illustrated on the sign shall be the same as the number of through lanes available to road users.**

Option:

03   Where two or more lanes are being shifted, a W24-1 (or Double Reverse Turn sign showing one lane) sign with an ALL LANES (W24-1cP) plaque (see Figure 6H-1) may be used instead of a sign that illustrates the number of lanes.

04   Where more than three lanes are being shifted, the Double Reverse Curve (or Turn) sign may be rectangular.

Section 6H.32 Advisory Speed Plaque (W13-1P)

Option:

01   In combination with a warning sign, an Advisory Speed (W13-1P) plaque (see Figure 6H-1) may be used to indicate a recommended speed through the TTC zone.

Standard:

02   **The Advisory Speed plaque shall not be used in conjunction with any sign other than a warning sign, nor shall it be used alone. When used with orange TTC zone signs, this plaque shall have a black legend and border on an orange background. The plaque shall be at least 24 x 24 inches in size when used with a sign that is 36 x 36 inches or larger. Except in emergencies, an Advisory Speed plaque shall not be mounted until the recommended speed is determined by the highway agency.**

Support:

03   Warning signs with advisory speed plaques (see Section 2C.59) inform drivers of the recommended operating speed based on temporary conditions within a TTC zone. Examples include narrow lanes, temporary diversion (reverse curves), lane shifts, sight distance restrictions, rough road surface, bumps, low/no shoulder, workers on foot, work vehicles or equipment close to the open travel lane, or other conditions that indicate the need for reduced speed.

04   AASHTO and ITE design documents contain established engineering practices for the determination of the recommended advisory speeds for horizontal curves or locations with limited sight distance.

Section 6H.33 Supplementary Distance Plaque (W7-3aP)

Option:

01   In combination with a warning sign, a Supplementary Distance (W7-3aP) plaque (see Figure 6H-1) with the legend NEXT XX MILES may be used to indicate the length of highway over which a work activity is being conducted, or over which a condition exists in the TTC zone.

02   In long TTC zones, Supplementary Distance plaques with the legend NEXT XX MILES may be placed in combination with warning signs at regular intervals within the zone to indicate the remaining length of highway over which the TTC work activity or condition exists.

Standard:

03   **The Supplementary Distance plaque with the legend NEXT XX MILES shall not be used in conjunction with any sign other than a warning sign, nor shall it be used alone. When used with orange TTC zone signs, this plaque shall have a black legend and border on an orange background. The plaque shall be at least 30 x 24 inches in size when used with a sign that is 36 x 36 inches or larger.**

Guidance:

04   *When used in TTC zones, the Supplementary Distance plaque with the legend NEXT XX MILES should be placed below the initial warning sign designating that, within the approaching zone, a temporary work activity or condition exists.*

Section 6H.34 Motorcycle Plaque (W8-15P)

Option:

01   A Motorcycle (W8-15P) plaque (see Figure 6H-1) may be mounted below a LOOSE GRAVEL (W8-7) sign, a GROOVED PAVEMENT (W8-15) sign, a METAL BRIDGE DECK (W8-16) sign, or a STEEL PLATE AHEAD (W8-24) sign if the warning is intended to be directed primarily to motorcyclists.

Section 6H.35 ROAD WORK NEXT XX MILES Sign (G20-1)

Guidance:

01   *The ROAD WORK NEXT XX MILES (G20-1) sign (see Figure 6H-1) should be installed in advance of TTC zones that are more than 2 miles in length.*

Option:

02   The ROAD WORK NEXT XX MILES sign may be mounted on a Type 3 Barricade. The sign may also be used for TTC zones of shorter length.

Standard:

03   **The distance displayed on the ROAD WORK NEXT XX MILES sign shall be stated to the nearest whole mile.**

Section 6H.36 END ROAD WORK Sign (G20-2)

Guidance:

01   *When used, the END ROAD WORK (G20-2) sign (see Figure 6H-1) should be placed near the downstream end of the termination area, as determined by engineering judgment.*

Option:

02   The END ROAD WORK sign may be installed on the back of a warning sign facing the opposite direction of road users or on the back of a Type 3 Barricade.

Section 6H.37 PILOT CAR FOLLOW ME Sign (G20-4)

Standard:

01   **The PILOT CAR FOLLOW ME (G20-4) sign (see Figure 6H-1) shall be mounted in a conspicuous position on the top or on the rear of a vehicle used for guiding one-way vehicular traffic through or around a TTC zone (see Section 6E.04).**

Section 6H.38 Other Warning Signs

Option:

01   Advance warning signs may be used by themselves or with other advance warning signs.

02   Besides the warning signs specifically related to TTC zones, several other warning signs in Part 2 may apply in TTC zones.

03   Word message warning signs other than those classified and specified in this Manual and the "Standard Highway Signs" publication (see Section 1A.05) may be developed and used based on engineering judgment to warn of special conditions in TTC zones.

Standard:

04   **Except as provided in Sections 6F.01 and 6H.01, other warning signs that are used in TTC zones shall have black legends and borders on an orange background.**

Guidance:

05   *Other warning signs should comply with the general requirements of color, shape, and alphabet size and series. The sign message should be brief, legible, and clear.*

CHAPTER 6I. TTC ZONE GUIDE SIGNS

Section 6I.01 Guide Signs – General

Support:

01   Guide signs along highways provide road users with information to help them along their way through the TTC zone. The design of guide signs is presented in Part 2.

Guidance:

02   *The following guide signs should be used in TTC zones as needed:*

    A. Standard route markings where temporary route changes are necessary,
    B. Directional signs and street name signs, and
    C. Special guide signs relating to the condition or work being done.

Standard:

03   **If additional temporary guide signs are used in TTC zones, they shall have a black legend and border on an orange background.**

Option:

04   Guide signs used in TTC incident management situations may have a black legend and border on a fluorescent pink background.

05   When temporary directional signs and temporary street name signs are used in conjunction with detour routing, these signs may have a black legend and border on an orange background.

06   When permanent directional signs or permanent street name signs are used in conjunction with detour signing, they may have a white legend on a green background (see Sections 2D.35 and 2D.45).

07   The sizes for TTC guide signs shall be as shown in Table 6I-1.

Section 6I.02 Detour Signs and Plaques (M4-8P, M4-8a, M4-8bP, M4-9, M4-9a, M4-9b, M4-9c, and M4-10)

Standard:

01   **Each detour shall be adequately marked with standard temporary route signs and destination signs.**

Option:

02   Detour signs in TTC incident management situations may have a black legend and border on a fluorescent pink background.

03   The Detour Arrow (M4-10) sign (see Figure 6I-1) may be used where a detour route has been established.

04   The DETOUR (M4-8P) plaque (see Figure 6I-1) may be mounted at the top of a route sign assembly to mark a temporary route that detours from a highway, bypasses a section closed by a TTC zone, and rejoins the highway beyond the TTC zone.

Table 6I-1. Temporary Traffic Control Zone Guide Sign and Plaque Sizes

Sign or Plaque	Sign Designation	Section	Conventional Road	Freeway or Expressway	Minimum
Exit Open	E5-2	6H.12	48 x 36	48 x 36	—
Exit Closed	E5-2a	6H.12	48 x 36	48 x 36	—
Exit Only	E5-3	6H.13	48 x 36	48 x 36	—
Detour	M4-8P	6I.02	24 x 12	30 x 15	—
End Detour	M4-8a	6I.02	24 x 18	24 x 18	—
End (plaque)	M4-8bP	6I.02	24 x 12	24 x 12	—
Detour	M4-9	6I.02	30 x 24	48 x 36	—
Bike/Pedestrian Detour	M4-9a	6I.02	30 x 24	—	—
Pedestrian Detour	M4-9b	6I.02	30 x 24	—	—
Bike Detour (with arrow)	M4-9c	6I.02	30 x 24	—	—
Detour	M4-10	6I.02	48 x 18	—	—

Notes:

    1. Larger signs may be used wherever necessary for greater legibility or emphasis

    2. Dimensions are shown in inches and are shown as width x height

Figure 6I-1. Exit Open and Closed and Detour Signs and Plaques

Guidance:

05   *The Detour Arrow (M4-10) sign should normally be mounted just below the ROAD CLOSED (R11-2, R11-3a, or R11-4) sign. The Detour Arrow sign should include a horizontal arrow pointed to the right or left as required.*

06   *The DETOUR (M4-9) sign (see Figure 6I-1) should be used for unnumbered highways, for emergency situations, for periods of short durations, or where, over relatively short distances, road users are guided along the detour and back to the desired highway without route signs.*

07   *A Street Name sign should be placed above, or the street name should be incorporated into, a DETOUR (M4-9) sign to indicate the name of the street being detoured.*

Option:

08   The END DETOUR (M4-8a) sign or the END (M4-8bP) plaque (see Figure 6I-1) may be used to indicate that the detour has ended.

Guidance:

09   *When the END DETOUR sign is used on a numbered highway, the sign should be mounted above a route sign after the downstream end of the detour.*

10   *The Pedestrian/Bicyclist Detour (M4-9a) sign (see Figure 6I-1) should be used where a pedestrian/bicyclist detour route has been established because of the closing of a pedestrian/bicycle facility to through traffic.*

Standard:

11   **If used, the Pedestrian/Bicyclist Detour sign shall have an arrow pointing in the appropriate direction.**

Option:

12   The arrow on a Pedestrian/Bicyclist Detour sign may be on the sign face or on a supplemental plaque.

13   The Pedestrian Detour (M4-9b) sign or Bicyclist Detour (M4-9c) sign (see Figure 6I-1) may be used where a pedestrian or a bicyclist detour route (not both) has been established because of the closing of the pedestrian or bicycle facility to through traffic.

Section 6I.03 EXIT CLOSED Panel

Guidance:

01   *When an exit ramp is closed, an EXIT CLOSED sign panel with a black legend and border on an orange background should be placed diagonally across the interchange/intersection guide signs.*

CHAPTER 6J. TTC ZONE PAVEMENT MARKINGS

Section 6J.01 Pavement Markings in TTC Zones

Support:

01   Pavement markings are installed or existing markings are maintained or enhanced in TTC zones to provide road users with a clearly defined path for travel through the TTC zone in day, night, and twilight periods under both wet and dry pavement conditions.

Guidance:

02   *The work should be planned and staged to provide for the placement and removal of the pavement markings in a way that minimizes the disruption to traffic flow approaching and through the TTC zone during the placement and removal process.*

Standard:

03   **Existing pavement markings shall be maintained in all long-term stationary (see Section 6N.01) TTC zones in accordance with Chapters 3A and 3B, except as otherwise provided for temporary pavement markings in Section 6J.02. Pavement markings shall match the alignment of the markings in place at both ends of the TTC zone. Pavement markings shall be placed along the entire length of any paved detour or temporary roadway prior to the detour or roadway being opened to road users.**

Guidance:

04   *For long-term stationary operations, pavement markings in the temporary traveled way that are no longer applicable should be removed or obliterated as soon as practical. Pavement marking obliteration should remove the non-applicable pavement marking material, and the obliteration method should minimize pavement scarring.*

Standard:

05   **Painting over existing pavement markings with black paint or spraying with asphalt shall not be accepted as a substitute for removal or obliteration.**

Option:

06   Removable, non-reflective, preformed tape that is approximately the same color as the pavement surface may be used where markings need to be covered temporarily.

Section 6J.02 Temporary Markings

Support:

01   Temporary markings are those pavement markings or devices that are placed within TTC zones to provide road users with a clearly defined path of travel through the TTC zone when the permanent markings are either removed or obliterated during the work activities. Temporary markings are typically needed during the reconstruction of a road while it is open to traffic, such as overlays or surface treatments or where lanes are temporarily shifted on pavement that is to remain in place.

Guidance:

02   *Unless justified based on engineering judgment, temporary pavement markings should not remain in place for more than 14 days after the application of the pavement surface treatment or the construction of the final pavement surface on new roadways or over existing pavements.*

03   *The temporary use of edge lines, channelizing lines, lane-reduction transitions, gore markings, and other longitudinal markings, and the various non-longitudinal markings (such as stop lines, railroad crossings, crosswalks, words, symbols, or arrows) should be in accordance with the State's or highway agency's policy.*

Standard:

04   **Warning signs, channelizing devices, and delineation shall be used to indicate required road user paths in TTC zones where it is not possible to provide a clear path by pavement markings.**

05   **Except as otherwise provided in this Section, all temporary pavement markings for no-passing zones shall comply with the requirements of Chapters 3A and 3B. All temporary broken line pavement markings shall use the same cycle length as permanent markings and shall have line segments that are at least 2 feet long.**

Guidance:

06   *All pavement markings and devices used to delineate road user paths should be reviewed during daytime and nighttime periods.*

Option:

07   Half-cycle lengths with a minimum of 2-foot stripes may be used on roadways with severe curvature (see Section 3A.04) for broken line center lines in passing zones and for lane lines.

08   For temporary situations of 14 days or less, for a two- or three-lane road, no-passing zones may be identified by using DO NOT PASS (R4-1), PASS WITH CARE (R4-2), and NO PASSING ZONE (W14-3) signs (see Sections 2B.36, 2B.37, and 2C.53) rather than pavement markings. Also, DO NOT PASS, PASS WITH CARE, and NO PASSING ZONE signs may be used instead of pavement markings on roads with low volumes for longer periods in accordance with the State's or highway agency's policy.

Guidance:

09   *If used, the DO NOT PASS, PASS WITH CARE, and NO PASSING ZONE signs should be placed in accordance with Sections 2B.36, 2B.37, and 2C.53.*

10   *If used, the NO CENTER LINE sign should be placed in accordance with Section 6H.29.*

Section 6J.03 Temporary Raised Pavement Markers

Option:

01   Retroreflective or internally illuminated raised pavement markers, or non-retroreflective raised pavement markers supplemented by retroreflective or internally illuminated markers, may be substituted for markings of other types in TTC zones.

Standard:

02   **If used, the color and pattern of the raised pavement markers shall simulate the color and pattern of the markings for which they substitute.**

03   **If temporary raised pavement markers are used to substitute for broken line segments, a group of at least three retroreflective markers equally spaced at no greater than 5 feet shall be installed every 40 feet.**

04   **If temporary raised pavement markers are used to substitute for solid lines, the markers shall be equally spaced at no greater than 10 feet, with retroreflective or internally illuminated units at a spacing no greater than 20 feet.**

Option:

05   Temporary raised pavement markers may be used to substitute for broken line segments by using at least two retroreflective markers placed at each end of a segment of 2 to 5 feet in length, using the same cycle length as permanent markings.

Guidance:

06   *Raised pavement markers should be considered for use along surfaced detours or temporary roadways, and other changed or new travel-lane alignments.*

Option:

07   Retroreflective or internally illuminated raised pavement markers, or non-retroreflective raised pavement markers supplemented by retroreflective or internally illuminated markers, may also be used in TTC zones to supplement markings as prescribed in Chapters 3A and 3B.

Section 6J.04 Delineators

Option:

01   Delineators may be used in TTC zones to indicate the alignment of the roadway and to outline the required vehicle path through the TTC zone.

Standard:

02   **When used, delineators shall combine with or supplement other TTC devices. They shall be mounted on crashworthy supports and shall be in accordance with Chapter 3G.**

Guidance:

03   *Spacing along roadway curves should be as set forth in Section 3G.04 and should be such that several delineators are visible to an approaching driver.*

CHAPTER 6K. TTC ZONE CHANNELIZING DEVICES

Section 6K.01 Channelizing Devices – General

Standard:

01 **Designs of various channelizing devices shall be as shown in Figure 6K-1. All channelizing devices shall be crashworthy (see definition in Section 1C.02).**

Support:

02 The function of channelizing devices is to warn road users of conditions created by work activities in or near the roadway and to guide road users. Channelizing devices include cones, tubular markers, vertical panels, drums, barricades, and longitudinal channelizing devices.

03 Channelizing devices provide for smooth and gradual vehicular traffic flow from one lane to another, onto a bypass or detour, or into a narrower traveled way. They are also used to channelize traffic away from the work space, pavement drop-offs, pedestrian or shared-use paths, bicycle facilities, or opposing directions of vehicular traffic.

Guidance:

04 *The spacing between cones, tubular markers, vertical panels, drums, and barricades should not exceed a distance in feet equal to 1 times the speed limit in mph when used for taper channelization, and should not exceed a distance in feet equal to 2 times the speed limit in mph when used for tangent channelization.*

05 *When channelizing devices have the potential of leading vehicular traffic out of the intended vehicular traffic space as shown in Figure 6P-39, the channelizing devices should be extended a distance in feet of 2 times the speed limit in mph beyond the downstream end of the transition area.*

Option:

06 A gap not exceeding 2 inches between the bottom rail and the ground surface may be used to facilitate drainage.

07 Warning lights (see Section 6L.07) may be added to channelizing devices in areas with frequent fog, snow, or severe roadway curvature, or where visual distractions are present.

08 A series of sequential flashing warning lights may be placed on channelizing devices that form a merging taper in order to increase driver detection and recognition of the merging taper.

Support:

09 The flashing rates and patterns for warning lights used on channelizing devices are specified in Section 6L.07.

Standard:

10 **The retroreflective material used on channelizing devices shall display a similar color day or night.**

11 **Except as provided in Paragraph 12 of this Section, information identifying the owner or manufacturer of the channelizing device shall not be displayed on any portion of the device that can be seen by road users approaching the device.**

Option:

12 The name and telephone number of the highway agency, contractor, or supplier may be displayed on the non-retroreflective surface of all types of channelizing devices.

Standard:

13 **The area containing the name and telephone number shall be non-retroreflective and not over 2 inches in height.**

Guidance:

14 *Particular attention should be given to maintaining the channelizing devices to keep them clean, visible, and properly positioned at all times.*

Standard:

15 **Channelizing devices that are no longer serviceable (see definition in Section 1C.02) shall be replaced.**

Section 6K.02 Pedestrian Channelizing Devices

Support:

01 Pedestrian channelizing devices indicate a suitable path of pedestrian travel around or through the work zone.

Figure 6K-1. Examples of Channelizing Devices

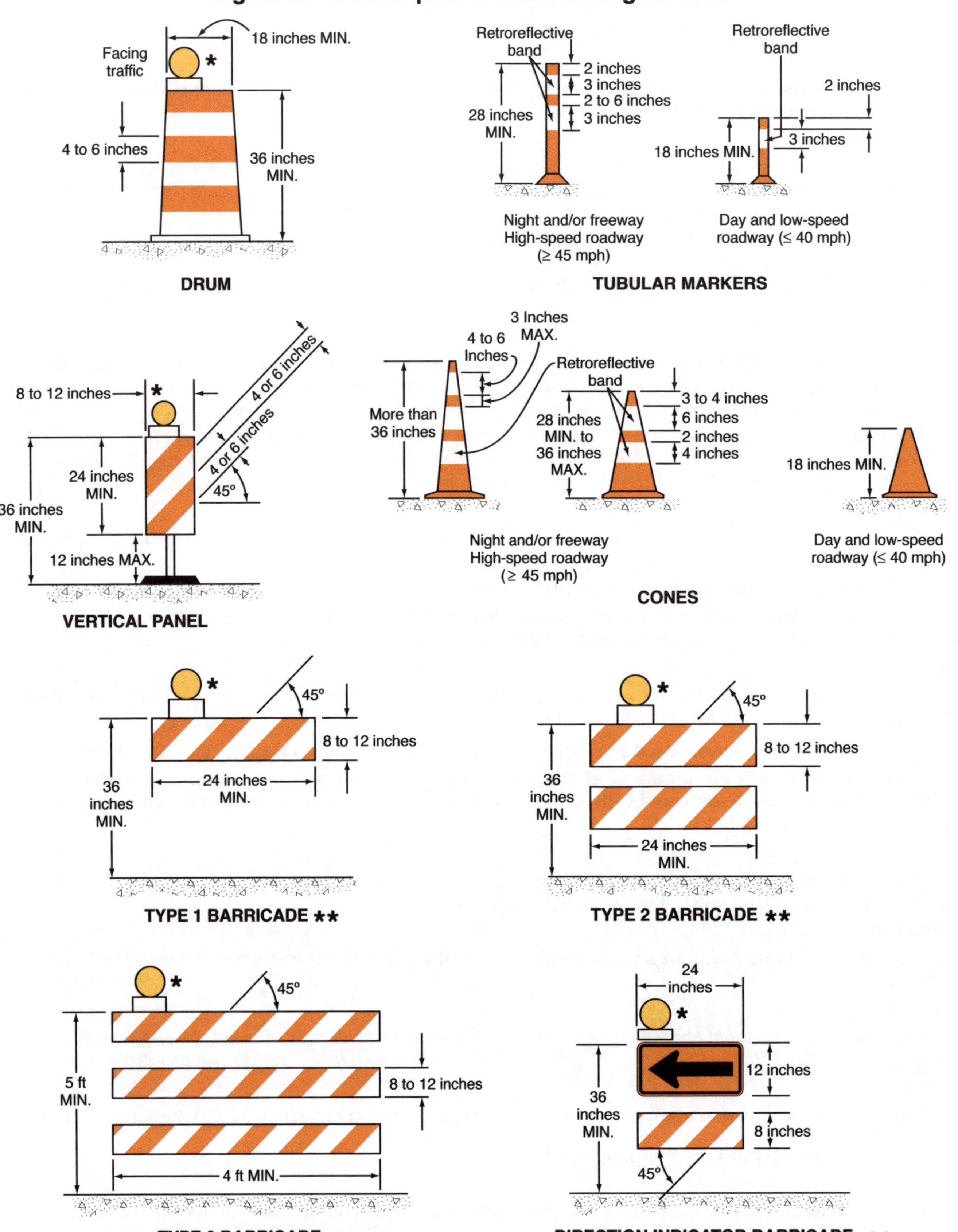

Guidance:

02   *Pedestrian channelizing devices should be provided when work activities impact sidewalks or other pedestrian facilities or when the design of the temporary pedestrian facility does not otherwise include accessibility features consistent with the features in the existing pedestrian facility.*

03   *The pedestrian channelizing devices should be used both to close sidewalks and to delineate an alternate route.*

Support:

04   An example of a pedestrian channelizing device is depicted in Figure 6K-2.

Standard:

05   **Pedestrian channelizing devices shall be crashworthy (see definition in Section 1C.02) when exposed to vehicular traffic.**

06   **Devices used to channelize pedestrians shall be detectable to users of long canes and visible to pedestrians with vision disabilities.**

07   **When used as a sidewalk closure, the device shall cover the entire width of the sidewalk.**

Figure 6K-2. Pedestrian Channelizing Device

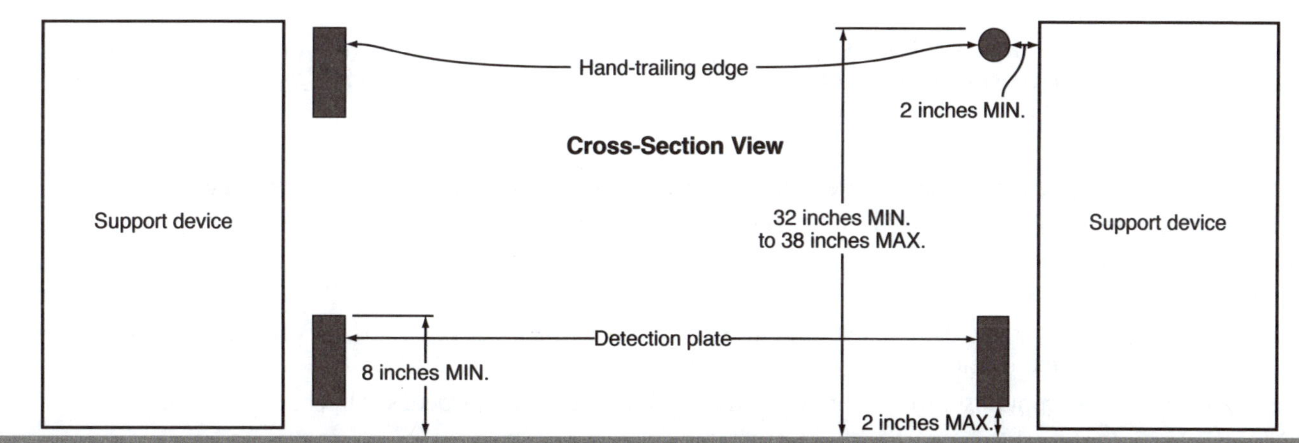

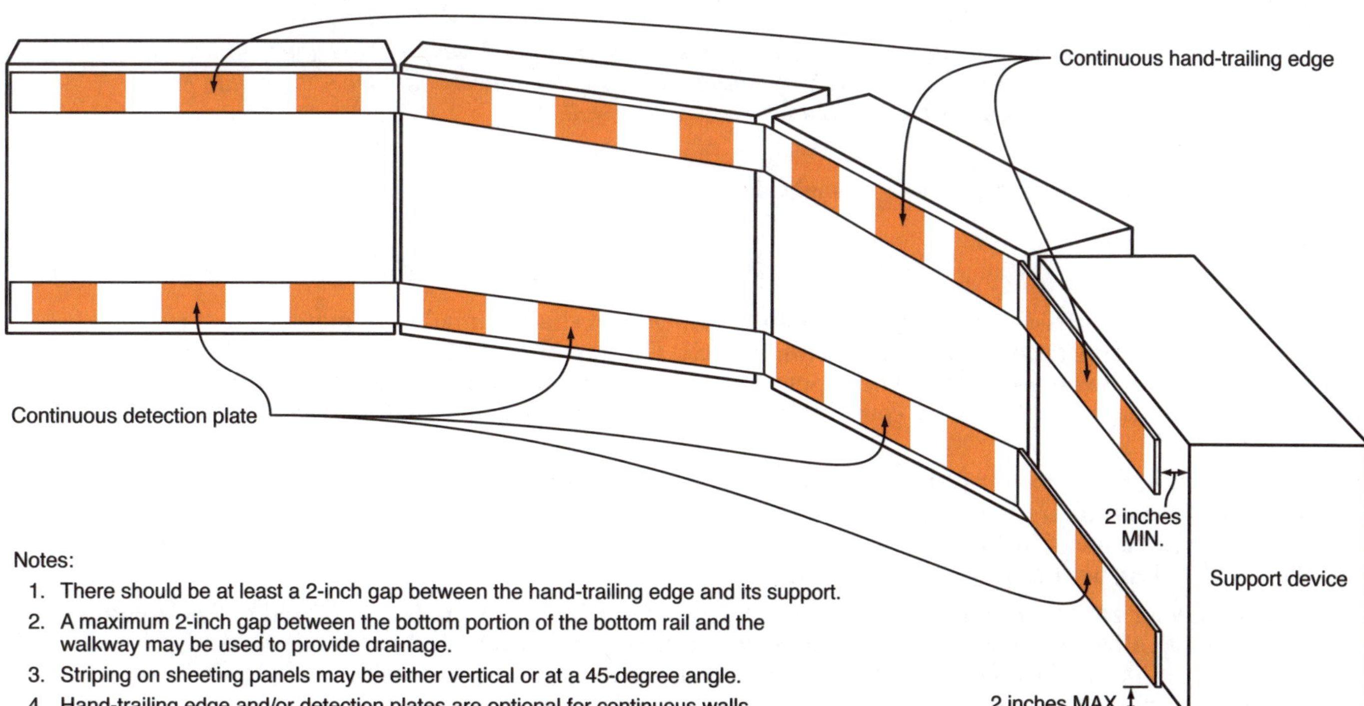

Notes:
1. There should be at least a 2-inch gap between the hand-trailing edge and its support.
2. A maximum 2-inch gap between the bottom portion of the bottom rail and the walkway may be used to provide drainage.
3. Striping on sheeting panels may be either vertical or at a 45-degree angle.
4. Hand-trailing edge and/or detection plates are optional for continuous walls.

08   **Pedestrian channelizing devices shall have continuous detection plates and hand-trailing edges. The bottom of the detection plate shall be no higher than 2 inches above the walkway. The top edge of the detection plate shall be at least 8 inches above the walkway. The top of the hand-trailing edge shall be no lower than 32 inches and no higher than 38 inches above the walkway. The top surface of the hand-trailing edge shall be smooth to optimize hand trailing. Both the detection plate and the hand-trailing edge shall share a common vertical plane.**

Guidance:

09   *When pedestrian channelizing devices are combined in a series, the gap between devices should not exceed 1 inch.*

Support:

10   The hand-trailing edge is the upper rail on a pedestrian channelizing device, as shown in Figure 6K-2. It is provided to allow pedestrians with vision disabilities to follow the pedestrian channelizing device with their hand. The hand-trailing edge is not a weight-bearing railing.

Guidance:

11   *There should be at least a 2-inch gap between the hand-trailing edge and its support.*

Standard:

12   **When visible to vehicular traffic the detection plate and the hand-trailing edge of the pedestrian channelizing device shall have retroreflective sheeting complying with Paragraph 10 of Section 6K.01.**

Guidance:

13   *When not visible to vehicular traffic, the pedestrian channelizing device should have a contrasting pattern in alternating light and dark colors to provide visual contrast on the upper surface consisting of a minimum of 6 inches of sheeting or other contrasting materials.*

Option:

14   Non-retroreflective materials may be used on the pedestrian side of the pedestrian channelizing device.

15   The sheeting on the pedestrian side of the pedestrian channelizing device may have stripes that are oriented either vertically or at a 45-degree angle.

Support:

16   The contrast of the light and dark stripes on the barricade sheeting assists pedestrians with vision disabilities in following the designated detour.

17   Section 6M.04 also contains information regarding detectable edging for pedestrian channelization.

Option:

18   A continuous wall may be used as a pedestrian channelizing device.

Guidance:

19   *When used, a continuous wall should have a lower edge no more than 2 inches above the walkway, should extend a minimum of 32 inches above the walkway, should have a common vertical face, and should have alternating, contrasting sheeting positioned 32 inches above the walkway.*

Option:

20   The continuous wall may extend to any height above the 32-inch minimum.

Section 6K.03 Cones

Standard:

01   **Cones (see Figure 6K-1) shall be predominantly orange and shall be made of a material that can be struck without causing damage to the impacting vehicle. For daytime and low-speed roadways, cones shall be not less than 18 inches in height. When cones are used on freeways and other high-speed highways or at night on all highways, or when more conspicuous guidance is needed, cones shall be a minimum of 28 inches in height.**

02   **For nighttime use, cones shall be retroreflectorized or equipped with lighting devices for maximum visibility. Retroreflectorization of cones that are 28 to 36 inches in height shall be provided by a 6-inch wide white band located 3 to 4 inches from the top of the cone and an additional 4-inch wide white band located approximately 2 inches below the 6-inch band.**

03   **Retroreflectorization of cones that are more than 36 inches in height shall be provided by horizontal, circumferential, alternating orange and white retroreflective stripes that are 4 to 6 inches wide. Each cone shall have a minimum of two orange and two white stripes with the top stripe being orange. Any non-retroreflective spaces between the retroreflective stripes shall not exceed 3 inches in width.**

Option:

04   Traffic cones may be used to channelize road users, divide opposing vehicular traffic lanes, divide lanes when two or more lanes are kept open in the same direction, and delineate short-duration maintenance and utility work.

Guidance:

05   *Steps should be taken to minimize the possibility of cones being blown over or displaced by wind or moving vehicular traffic.*

Option:

06   Cones may be doubled up to increase their weight.

Support:

07   Some cones are constructed with bases that can be filled with ballast. Others have specially weighted bases, or weight such as sandbag rings, that can be dropped over the cones and onto the base to provide added stability.

Guidance:

08   *Ballast should be kept to the minimum amount needed.*

Section 6K.04 Tubular Markers

Standard:

01   **Tubular markers (see Figure 6K-1) shall be predominantly orange for TTC zone applications and shall be not less than 18 inches high and 2 inches wide facing road users. They shall be made of a material that can be struck without causing damage to the impacting vehicle.**

02   **Tubular markers shall be a minimum of 28 inches in height when they are used on freeways and other high-speed highways, on all highways during nighttime, or whenever more conspicuous guidance is needed.**

03   **For nighttime use, tubular markers shall be retroreflectorized. Retroreflectorization of tubular markers that have a height of less than 42 inches shall be provided by two 3-inch wide white bands placed a maximum of 2 inches from the top with a maximum of 6 inches between the bands. Retroreflectorization of tubular markers that have a height of 42 inches or more shall be provided by four 4-inch to 6-inch wide alternating orange and white stripes with the top stripe being orange.**

Guidance:

04   *Tubular markers have less visible area than other devices and should be used only where space restrictions do not allow for the use of other more visible devices.*

05   *Tubular markers should be stabilized by affixing them to the pavement, by using weighted bases, or by using weights such as sandbag rings that can be dropped over the tubular markers and onto the base to provide added stability. Ballast should be kept to the minimum amount needed.*

Option:

06   Tubular markers may be used effectively to divide opposing lanes of road users, divide vehicular traffic lanes when two or more lanes of moving vehicular traffic are kept open in the same direction, and to delineate the edge of a pavement drop off where space limitations do not allow the use of larger devices.

Standard:

07   **A tubular marker shall be attached to the pavement to display the minimum 2-inch width to the approaching road users.**

Section 6K.05 Vertical Panels

Standard:

01   **Vertical panels (see Figure 6K-1) shall have retroreflective striped material that is 8 to 12 inches in width and at least 24 inches in height. They shall have alternating diagonal orange and white retroreflective stripes sloping downward at an angle of 45 degrees in the direction vehicular traffic is to pass.**

02   **Where the height of the retroreflective material on the vertical panel is 36 inches or more, a stripe width of 6 inches shall be used.**

Option:

03   Where the height of the retroreflective material on the vertical panel is less than 36 inches, a stripe width of 4 inches may be used.

04   Where space is limited, vertical panels may be used to channelize vehicular traffic, divide opposing lanes, or replace barricades.

Section 6K.06 Drums

Standard:

01 **Drums (see Figure 6K-1) used for road user warning or channelization shall be constructed of lightweight, deformable materials. They shall be a minimum of 36 inches in height and have at least an 18-inch minimum width regardless of orientation. Metal drums shall not be used. The markings on drums shall be horizontal, circumferential, alternating orange and white retroreflective stripes 4 to 6 inches wide. Each drum shall have a minimum of two orange and two white stripes with the top stripe being orange. Any non-retroreflectorized spaces between the horizontal orange and white stripes shall not exceed 3 inches wide. Drums shall have closed tops that will not allow collection of construction debris or other debris.**

Support:

02 Drums are highly visible, have good target value, give the appearance of being formidable obstacles and, therefore, command the respect of road users. They are portable enough to be shifted from place to place within a TTC zone in order to accommodate changing conditions, but are generally used in situations where they will remain in place for a prolonged period of time.

Option:

03 Although drums are most commonly used to channelize or delineate road user flow, they may also be used alone or in groups to mark specific locations.

Guidance:

04 *Drums should not be weighted with sand, water, or any material to the extent that would make them hazardous to road users or workers when struck. Drums used in regions susceptible to freezing should have drain holes in the bottom so that water will not accumulate and freeze causing a hazard if struck by a road user.*

Standard:

05 **Ballast shall not be placed on the top of a drum.**

Section 6K.07 Type 1, 2, or 3 Barricades

Support:

01 A barricade is a portable or fixed device having from one to three rails with appropriate markings and is used to control road users by closing, restricting, or delineating all or a portion of the right-of-way.

02 As shown in Figure 6K-1, barricades are classified as Type 1, Type 2, or Type 3.

Standard:

03 **Stripes on barricade rails shall be alternating orange and white retroreflective stripes sloping downward at an angle of 45 degrees in the direction road users are to pass. Except as provided in Paragraph 4 of this Section, the stripes shall be 6 inches wide.**

Option:

04 When rail lengths are less than 36 inches, 4-inch wide stripes may be used.

Standard:

05 **The minimum length for Type 1 and Type 2 Barricades shall be 24 inches, and the minimum length for Type 3 Barricades shall be 48 inches. Each barricade rail shall be 8 to 12 inches wide. Barricades used on freeways, expressways, and other high-speed roadways shall have a minimum of 270 square inches of retroreflective area facing road users.**

Guidance:

06 *Where barricades extend entirely across a roadway, the stripes should slope downward in the direction toward which road users must turn.*

07 *Where both right and left turns are provided, the barricade stripes should slope downward in both directions from the center of the barricade or barricades.*

08 *Where no turns are intended, the stripes should be positioned to slope downward toward the center of the barricade or barricades.*

09 *Barricade rails should be supported in a manner that will allow them to be seen by the road user, and in a manner that provides a stable support that is not easily blown over or displaced.*

10 *The width of the existing pedestrian facility should be provided for the temporary facility if practical. Traffic control devices and other construction materials and features should not intrude into the usable width of the sidewalk, temporary pathway, or other pedestrian facility. When it is not possible to maintain a minimum width of 60 inches throughout the entire length of the pedestrian pathway, a 60 x 60-inch passing space should be provided at least every 200 feet to allow individuals in wheelchairs to pass.*

11 *Barricade rail supports should not project into pedestrian circulation routes more than 4 inches from the support between 27 and 80 inches from the surface as described in Section 307 of the U.S. Department of Justice 2010 ADA Standards for Accessible Design, September 15, 2010, 28 CFR 35 and 36, Americans with Disabilities Act of 1990.*

Option:

12 For Type 1 Barricades, the support may include other unstriped horizontal rails necessary to provide stability.

Guidance:

13 *On high-speed expressways or in other situations where barricades might be susceptible to overturning in the wind, ballasting should be used.*

Option:

14 Sandbags may be placed on the lower parts of the frame or the stays of barricades to provide the required ballast.

Support:

15 Type 1 or Type 2 Barricades are intended for use in situations where road user flow is maintained through the TTC zone.

Option:

16 Barricades may be used alone or in groups to mark a specific condition or they may be used in a series for channelizing road users.

17 Type 1 Barricades may be used on conventional roads or urban streets.

Guidance:

18 *Type 2 or Type 3 Barricades should be used on freeways and expressways or other high-speed roadways. Type 3 Barricades should be used to close or partially close a road.*

Option:

19 Type 3 Barricades used at a road closure may be placed completely across a roadway or from curb to curb.

Guidance:

20 *Where provision is made for access of authorized equipment and vehicles, the responsibility for Type 3 Barricades should be assigned to a person who will provide proper closure at the end of each work day.*

Support:

21 When a highway is legally closed but access must still be allowed for local road users, barricades usually are not extended completely across the roadway.

Standard:

22 **A sign shall be installed with the appropriate legend concerning permissible use by local road users (see Section 6G.05).**

Guidance:

23 *Adequate visibility of the barricades from both directions should be provided.*

Option:

24 Signs may be installed on barricades (see Section 6F.02).

Section 6K.08 Direction Indicator Barricades

Standard:

01 **The Direction Indicator Barricade (see Figure 6K-1) shall consist of a One-Direction Large Arrow (W1-6) sign mounted above a diagonal striped, horizontally-aligned, retroreflective rail.**

02 **The One-Direction Large Arrow (W1-6) sign shall have a black legend and border on an orange background. The stripes on the bottom rail shall be alternating orange and white retroreflective stripes sloping downward at an angle of 45 degrees in the direction road users are to pass. The stripes shall be 4 inches wide. The One-Direction Large Arrow (W1-6) sign shall be 24 x 12 inches. The bottom rail shall have a length of 24 inches and a height of 8 inches.**

Option:

03 The Direction Indicator Barricade may be used in tapers, transitions, and other areas where specific directional guidance to drivers is necessary.

Guidance:

04 *If used, Direction Indicator Barricades should be used in a series to direct the driver through the transition and into the intended travel lane.*

Section 6K.09 Temporary Traffic Barriers as Channelizing Devices

Support:

01   Temporary traffic barriers (see Section 6M.02) are not TTC devices in themselves; however, when placed in a position identical to a line of channelizing devices and marked and/or equipped with appropriate channelization features to provide guidance and warning both day and night, they serve as TTC devices.

Standard:

02   **Temporary traffic barriers serving as TTC devices shall comply with requirements for such devices as set forth throughout Part 6.**

03   **Temporary traffic barriers (see Section 6M.02) shall not be used solely to channelize road users, but also to protect the work space. If used to channelize vehicular traffic, the temporary traffic barrier shall be supplemented with delineation, pavement markings, or channelizing devices for improved daytime and nighttime visibility.**

Guidance:

04   *Temporary traffic barriers should not be used for a merging taper except in low-speed urban areas.*

05   *When it is necessary to use a temporary traffic barrier for a merging taper in low-speed urban areas or for a constricted/restricted TTC zone, the taper length should be designed to optimize road user operations considering the available geometric conditions.*

Standard:

06   **When it is necessary to use a temporary traffic barrier for a merging taper in low-speed urban areas or for a constricted/restricted TTC zone, the taper shall be delineated using channelizing devices, and/or an edge line, and/or delineators on the barrier.**

Guidance:

07   *When used for channelization, temporary traffic barriers should be of a light color for increased visibility.*

Section 6K.10 Longitudinal Channelizing Devices

Support:

01   Longitudinal channelizing devices are lightweight, deformable devices that are highly visible, have good target value, and can be connected together.

Standard:

02   **If used singly as Type 1, 2, or 3 barricades, longitudinal channelizing devices shall comply with the general size, color, stripe pattern, retroreflectivity, and placement characteristics established for the devices described in this Chapter.**

Guidance:

03   *If used to channelize vehicular traffic at night, longitudinal channelizing devices should be supplemented with retroreflective material or delineation for improved nighttime visibility.*

Option:

04   Longitudinal channelizing devices may be used instead of a line of cones, drums, or barricades.

05   Longitudinal channelizing devices may be hollow and filled with water as a ballast.

06   Longitudinal channelizing devices may be used for pedestrian traffic control.

Standard:

07   **If used for pedestrian traffic control, longitudinal channelizing devices shall be interlocked to delineate or channelize flow. The interlocking devices shall not have gaps that allow pedestrians to stray from the channelizing path.**

Guidance:

08   *Longitudinal channelizing devices have not met the crashworthy requirements for temporary traffic barriers and should not be used to shield obstacles or provide positive protection for pedestrians or workers.*

Section 6K.11 Temporary Lane Separators

Option:

01   Temporary lane separators may be used to channelize road users, to divide opposing vehicular traffic lanes, and to divide lanes when two or more lanes are open in the same direction.

Standard:

02   **Temporary lane separators shall consist of a longitudinal base component with a maximum height of 4 inches and a maximum width of 1 foot. The longitudinal base shall have sloping sides in order to facilitate crossover by emergency vehicles. One or more types of channelizing devices, such as tubular markers, vertical panels, or a Narrow Two-Way Traffic (W6-4) sign (see Section 6H.17) mounted on flexible supports, shall be affixed to the longitudinal base.**

03   **Channelizing devices affixed to the longitudinal base of a temporary lane separator shall be retroreflectorized to provide nighttime visibility.**

Guidance:

04   *A temporary lane separator should be stabilized by affixing it to the pavement in a manner suitable to its design, while allowing the unit to be intentionally moved from place to place within the TTC zone in order to accommodate changing conditions.*

05   *Temporary Lane Separators should not be used to shield obstacles or provide positive protection for pedestrians or workers because these devices have not met the crashworthy requirements for temporary traffic barriers.*

Standard:

06   **At pedestrian crossing locations, temporary lane separators shall have an opening or be shortened to provide a pathway that is at least 60 inches wide for crossing pedestrians.**

Section 6K.12 Other Channelizing Devices

Option:

01   Channelizing devices other than those described in this Chapter may be used in special situations based on an engineering study.

Guidance:

02   *Other channelizing devices should comply with the general size, color, stripe pattern, retroreflection, and placement characteristics established for the devices described in this Chapter.*

CHAPTER 6L. OTHER TTC ZONE TRAFFIC CONTROL DEVICES

Section 6L.01 Temporary Traffic Control Signals

Standard:

01   **Temporary traffic control signals (see Section 4D.11) used to control road user movements through TTC zones and in other TTC situations shall comply with the applicable provisions of Part 4.**

Support:

02   Temporary traffic control signals are typically used in TTC zones such as temporary haul road crossings; temporary one-way operations along a one-lane, two-way highway; temporary one-way operations on bridges, reversible lanes, and intersections.

Standard:

03   **A temporary traffic control signal that is used to control traffic through a one-lane, two-way section of roadway shall comply with the provisions of Section 4O.02.**

Guidance:

04   *When temporary traffic control signals are used, conflict monitors typical of traditional traffic control signal operations should be used.*

Support:

05   Where pedestrians are detoured to a temporary traffic control signal, an accessible pedestrian signal (see Chapter 4K) provides information in non-visual formats (such as audible tones and/or speech messages, and vibrating surfaces) so that a pedestrian with vision disabilities can know when to cross the street along the alternate route.

Option:

06   Temporary traffic control signals may be portable or temporarily mounted on fixed supports.

Guidance:

07   *Temporary traffic control signals should only be used in situations where temporary traffic control signals are preferable to other means of traffic control, such as changing the work staging or work zone size to eliminate one-way vehicular traffic movements, using flaggers to control one-way or crossing movements, using STOP or YIELD signs, and using warning devices alone.*

Support:

08   Factors related to the design and application of temporary traffic control signals include the following:
   A. Safety and road user needs;
   B. Work staging and operations;
   C. The feasibility of using other TTC strategies (for example, flaggers, providing space for two lanes, or detouring road users, including bicyclists and pedestrians);
   D. Sight distance restrictions;
   E. Human factors considerations (for example, lack of driver familiarity with temporary traffic control signals);
   F. Road-user volumes including roadway and intersection capacity;
   G. Affected side streets and driveways;
   H. Vehicle speeds;
   I. The placement of other TTC devices;
   J. Parking;
   K. Turning restrictions;
   L. Pedestrians;
   M. The nature of adjacent land uses (such as residential or commercial);
   N. Legal authority;
   O. Signal phasing and timing requirements;
   P. Full-time or part-time operation;
   Q. Actuated, fixed-time, or manual operation;
   R. Power failures or other emergencies;
   S. Inspection and maintenance needs;
   T. Need for detailed placement, timing, and operation records; and
   U. Operation by contractors or by others.

09   Although temporary traffic control signals can be mounted on trailers or lightweight portable supports, fixed supports offer superior resistance to displacement or damage by severe weather, vehicle impact, and vandalism.

Guidance:

10   *Other TTC devices should be used to supplement temporary traffic control signals, including warning and regulatory signs, pavement markings, and channelizing devices.*

11   *Temporary traffic control signals not in use should be covered or removed.*

12   *If a temporary traffic control signal is located within ½ mile of an adjacent traffic control signal, consideration should be given to interconnected operation.*

Standard:

13   **Temporary traffic control signals shall not be located within 200 feet of a grade crossing unless the temporary traffic control signal is provided with preemption in accordance with Sections 4F.18, 4F.19, and 8D.09, or unless a uniformed officer or flagger is provided at the crossing to prevent vehicles from stopping within the crossing.**

Section 6L.02 Automated Flagger Assistance Devices – General

Support:

01   Automated Flagger Assistance Devices (AFADs) enable a flagger(s) to be positioned out of the lane of traffic and are used to control road users through TTC zones. These devices are designed to be remotely operated either by a single flagger at one end of the TTC zone or at a central location, or by separate flaggers near each device's location.

02   There are two types of AFADs:

A.   An AFAD (see Section 6L.03) that uses a remotely controlled STOP/SLOW sign on either a trailer or a movable cart system to alternately control right-of-way.
B.   An AFAD (see Section 6L.04) that uses remotely controlled red and yellow lenses and a gate arm to alternately control right-of-way.

03   AFADs might be appropriate for short-term and intermediate-term activities (see Section 6N.01). Typical applications include TTC activities such as, but not limited to:

A.   Bridge maintenance,
B.   Haul road crossings, and
C.   Pavement patching.

Standard:

04   **AFADs shall only be used in situations where there is only one lane of approaching traffic in the direction to be controlled.**

05   **When used at night, the AFAD location shall be illuminated in accordance with Section 6D.06.**

Guidance:

06   *AFADs should not be used for long-term stationary work (see Section 6N.01).*

Standard:

07   **Because AFADs are not traffic control signals, they shall not be used as a substitute for or a replacement for a continuously operating temporary traffic control signal as described in Section 6L.01.**

08   **AFADs shall meet the crashworthy (see definition in Section 1C.02) performance criteria contained in Section 6A.04.**

Guidance:

09   *If used, AFADs should be located in advance of one-lane, two-way tapers and downstream from the point where approaching traffic is to stop in response to the device.*

Standard:

10   **If used, AFADs shall be placed so that all of the signs and other items controlling traffic movement are readily visible to the driver of the initial approaching vehicle with advance warning signs alerting other approaching traffic to be prepared to stop.**

11   **If used, an AFAD shall be operated only by a flagger (see Section 6D.01) who has been trained on the operation of the AFAD. The flagger(s) operating the AFAD(s) shall not leave the AFAD(s) unattended at any time while the AFAD(s) is being used.**

12   **The use of AFADs shall conform to one of the following methods:**

A.   **An AFAD at each end of the TTC zone (Method 1), or**
B.   **An AFAD at one end of the TTC zone and a flagger at the opposite end (Method 2).**

13   **Except as provided in Paragraph 14 of this Section, two flaggers shall be used when using either Method 1 or Method 2.**

Option:

14   A single flagger may simultaneously operate two AFADs (Method 1) or may operate a single AFAD on one end of the TTC zone while being the flagger at the opposite end of the TTC zone (Method 2) if both of the following conditions are present:

   A. The flagger has an unobstructed view of the AFAD(s), and
   B. The flagger has an unobstructed view of approaching traffic in both directions.

Guidance:

15   *When an AFAD is used, the advance warning signing should include a ROAD WORK AHEAD (W20-1) sign, a ONE LANE ROAD (W20-4) sign, and a BE PREPARED TO STOP (W3-4) sign.*

Standard:

16   **When the AFAD is not in use, the signs associated with the AFAD, both at the AFAD location and in advance, shall be removed or covered.**

Guidance:

17   *A State or local agency that elects to use AFADs should adopt a policy, based on engineering judgment, governing AFAD applications. The policy should also consider more detailed and/or more restrictive requirements for AFAD use, such as the following:*

   A. Conditions applicable for the use of Method 1 and Method 2 AFAD operation,
   B. Volume criteria,
   C. Maximum distance between AFADs,
   D. Conflicting lenses/indications monitoring requirements,
   E. Fail-safe procedures,
   F. Additional signing and pavement markings,
   G. Application consistency,
   H. Larger signs or lenses to increase visibility, and
   I. Use of backplates.

Section 6L.03 <u>STOP/SLOW Automated Flagger Assistance Devices</u>

Standard:

01   **A STOP/SLOW Automated Flagger Assistance Device (AFAD) shall include a STOP/SLOW sign that alternately displays the STOP (R1-1) face and the SLOW (W20-8) face of a STOP/SLOW paddle (see Figure 6L-1).**

02   **The AFAD's STOP/SLOW sign shall have an octagonal shape, shall be fabricated of rigid material, and shall be mounted with the bottom of the sign a minimum of 6 feet above the pavement on an appropriate support. The size of the STOP/SLOW sign shall be at least 24 x 24 inches with letters at least 8 inches high. The background of the STOP face shall be red with white letters and border. The background of the SLOW face shall be diamond-shaped and orange with black letters and border. Both faces of the STOP/SLOW sign shall be retroreflectorized.**

03   **The AFAD's STOP/SLOW sign shall have a means to positively lock, engage, or otherwise maintain the sign assembly in a stable condition when set in the STOP or SLOW position.**

04   **The AFAD's STOP/SLOW sign shall be supplemented with active conspicuity devices by incorporating either:**

   A. **White or red flashing lights within the STOP face and white or yellow flashing lights within the SLOW face meeting the provisions contained in Section 6D.02; or**
   B. **A Stop Beacon (see Section 4S.05) mounted a maximum of 24 inches above the STOP face and a Warning Beacon (see Section 4S.03) mounted a maximum of 24 inches above, below, or to the side of the SLOW face. The Stop Beacon shall not be flashed or illuminated when the SLOW face is displayed, and the Warning Beacon shall not be flashed or illuminated when the STOP face is displayed. Except for the mounting locations, the beacons shall comply with the provisions of Chapter 4S.**

Option:

05   Type B warning light(s) (see Section 6L.07) or strobe lights may be used in lieu of the Warning Beacon during the display of the SLOW face of the AFAD's STOP/SLOW sign.

Standard:

06   **If Type B warning lights or strobe lights are used in lieu of a Warning Beacon, they shall flash continuously when the SLOW face is displayed and shall not be flashed or illuminated when the STOP face is displayed.**

Figure 6L-1. Example of the Use of a STOP/SLOW Automated Flagger Assistance Device (AFAD)

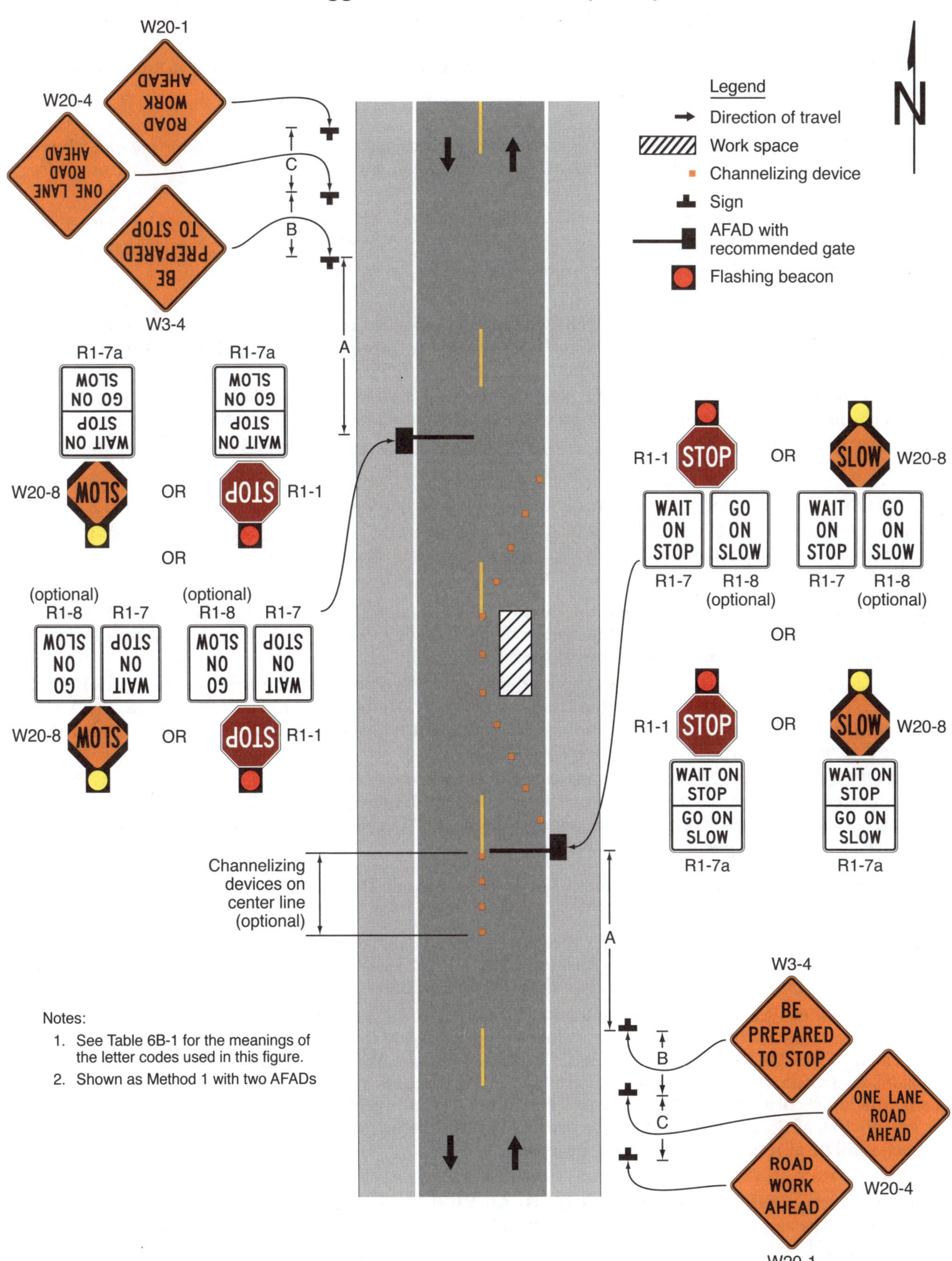

Notes:
1. See Table 6B-1 for the meanings of the letter codes used in this figure.
2. Shown as Method 1 with two AFADs

Option:

07 The faces of the AFAD's STOP/SLOW sign may include louvers to improve the stability of the device in windy or other adverse environmental conditions.

Standard:

08 **If louvers are used, the louvers shall be designed such that the full sign face is visible to approaching traffic at a distance of 50 feet or greater.**

Guidance:

09 *The STOP/SLOW AFAD should include a gate arm that descends to a down position across the approach lane of traffic when the STOP face is displayed and then ascends to an upright position when the SLOW face is displayed.*

Option:

10 In lieu of a stationary STOP/SLOW sign with a separate gate arm, the STOP/SLOW sign may be attached to a mast arm that physically blocks the approach lane of traffic when the STOP face is displayed and then moves to a position that does not block the approach lane when the SLOW face is displayed.

Standard:

11 **Gate arms, if used, shall be fully retroreflectorized on both sides, and shall have vertical alternating red and white stripes at 16-inch intervals measured horizontally as shown in Figure 8D-1. When the arm is in the down position blocking the approach lane:**
   A. The minimum vertical aspect of the arm and sheeting shall be 2 inches, and
   B. The end of the arm shall reach at least to the center of the lane being controlled.

12 **A WAIT ON STOP (R1-7) sign (see Figure 6L-1) shall be displayed to road users approaching the AFAD.**

Option:

13 A GO ON SLOW (R1-8) sign (see Figure 6L-1) may also be displayed to road users approaching the AFAD.

14 The WAIT ON STOP/ GO ON SLOW (R1-7a) sign (see Figure 6L-1) may also be used to display both messages to approaching road users.

Standard:

15 **The GO ON SLOW sign, if used, and the WAIT ON STOP sign shall be positioned on the same support structure as the AFAD or immediately adjacent to the AFAD such that they are in the same direct line of view of approaching traffic as the sign faces of the AFAD.**

16 **To inform road users to stop, the AFAD shall display the STOP face and the red or white lights, if used, within the STOP face shall flash or the Stop Beacon shall flash. To inform road users to proceed, the AFAD shall display the SLOW face and the yellow or white lights, if used, within the SLOW face shall flash or the Warning Beacon or the Type B warning lights shall flash.**

17 **If STOP/SLOW AFADs are used to control traffic in a one-lane, two-way TTC zone, safeguards shall be incorporated to prevent the flagger(s) from simultaneously displaying the SLOW face at each end of the TTC zone. Additionally, the flagger(s) shall not display the AFAD's SLOW face until all oncoming vehicles have cleared the one-lane portion of the TTC zone.**

Section 6L.04 Red/Yellow Lens Automated Flagger Assistance Devices

Standard:

01 **A Red/Yellow Lens Automated Flagger Assistance Device (AFAD) shall alternately display a steadily illuminated CIRCULAR RED lens and a flashing CIRCULAR YELLOW lens to control traffic without the need for a flagger in the immediate vicinity of the AFAD or on the roadway (see Figure 6L-2).**

02 **Red/Yellow Lens AFADs shall have at least one set of CIRCULAR RED and CIRCULAR YELLOW lenses that are 12 inches in diameter. Unless otherwise provided in this Section, the lenses and their arrangement, CIRCULAR RED on top and CIRCULAR YELLOW below, shall comply with the applicable provisions for traffic signal indications in Part 4. If the set of lenses is post-mounted, the bottom of the housing (including brackets) shall be at least 7 feet above the pavement. If the set of lenses is located over any portion of the highway that can be used by motor vehicles, the bottom of the housing (including brackets) shall be at least 15 feet above the pavement.**

Option:

03 Additional sets of CIRCULAR RED and CIRCULAR YELLOW lenses, located over the roadway or on the left-hand side of the approach and operated in unison with the primary set, may be used to improve visibility and/ or conspicuity of the AFAD.

Figure 6L-2. Example of the Use of a Red/Yellow Lens Automated Flagger Assistance Device (AFAD)

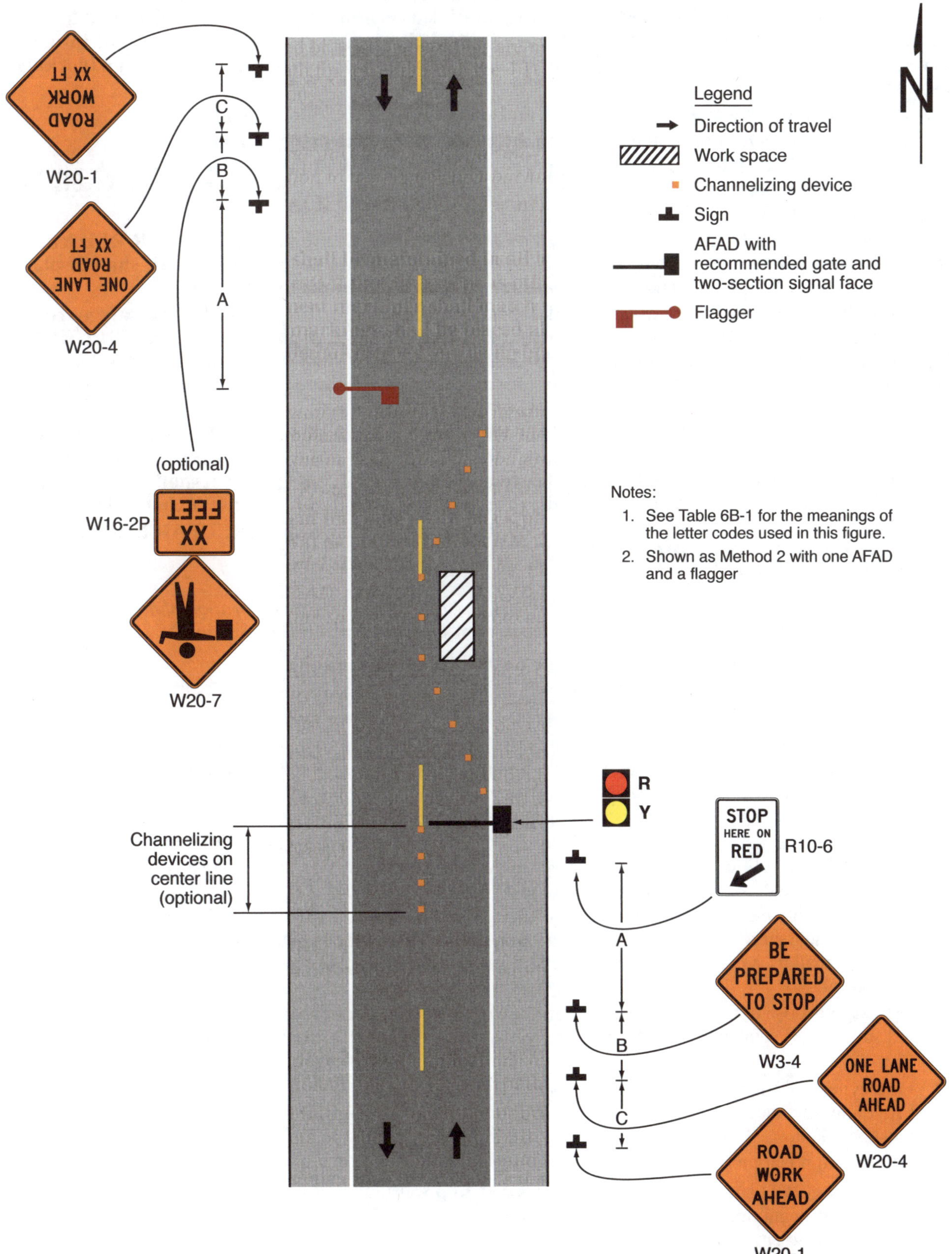

Standard:

04   **A Red/Yellow Lens AFAD shall include a gate arm that descends to a down position across the approach lane of traffic when the steady CIRCULAR RED lens is illuminated and then ascends to an upright position when the flashing CIRCULAR YELLOW lens is illuminated. The gate arm shall be fully retroreflectorized on both sides, and shall have vertical alternating red and white stripes at 16-inch intervals measured horizontally as shown in Figure 8D-1. When the arm is in the down position blocking the approach lane:**

    A. **The minimum vertical aspect of the arm and sheeting shall be 2 inches, and**

    B. **The end of the arm shall reach at least to the center of the lane being controlled.**

05   **A Stop Here On Red (R10-6 or R10-6a) sign (see Section 2B.59) shall be installed on the right-hand side of the approach at the point at which drivers are expected to stop when the steady CIRCULAR RED lens is illuminated (see Figure 6L-2).**

06   **To inform road users to stop, the AFAD shall display a steadily illuminated CIRCULAR RED lens and the gate arm shall be in the down position. To inform road users to proceed, the AFAD shall display a flashing CIRCULAR YELLOW lens and the gate arm shall be in the upright position.**

07   **If Red/Yellow Lens AFADs are used to control traffic in a one-lane, two-way TTC zone, safeguards shall be incorporated to prevent the flagger(s) from actuating a simultaneous display of a flashing CIRCULAR YELLOW lens at each end of the TTC zone. Additionally, the flagger shall not actuate the AFAD's display of the flashing CIRCULAR YELLOW lens until all oncoming vehicles have cleared the one-lane portion of the TTC zone.**

08   **A change interval shall be provided as the transition between the display of the flashing CIRCULAR YELLOW indication and the display of the steady CIRCULAR RED indication. During the change interval, the CIRCULAR YELLOW lens shall be steadily illuminated. The gate arm shall remain in the upright position during the display of the steadily illuminated CIRCULAR YELLOW change interval.**

09   **A change interval shall not be provided between the display of the steady CIRCULAR RED indication and the display of the flashing CIRCULAR YELLOW indication.**

Guidance:

10   *The steadily illuminated CIRCULAR YELLOW change interval should have a duration of at least 5 seconds, unless a different duration, within the range of durations recommended by Section 4F.17, is justified by engineering judgment.*

Section 6L.05 <u>Portable Changeable Message Signs</u>

Support:

01   Portable changeable message signs (PCMS) are TTC devices installed for temporary use with the flexibility to display a variety of messages. In most cases, portable changeable message signs follow the same provisions for design and application as those given for changeable message signs in Chapter 2L. The information in this Section describes situations where the provisions for portable changeable message signs differ from those given in Chapter 2L.

02   Portable changeable message signs are used most frequently on high-density urban freeways, but have applications on all types of highways where highway alignment, road user routing problems, or other pertinent conditions require advance warning and information.

03   Portable changeable message signs have a wide variety of applications in TTC zones including: roadway, lane, or ramp closures; incident management; width restriction information; speed control or reductions; advisories on work scheduling; road user management and diversion; warning of adverse conditions or special events; and other operational control.

04   The primary purpose of portable changeable message signs in TTC zones is to advise the road user of unexpected situations. Portable changeable message signs are particularly useful as they are capable of:

    A. Conveying complex messages,

    B. Displaying real time information about conditions ahead, and

    C. Providing information to assist road users in making decisions prior to the point where actions must be taken.

05   Some typical applications include the following:

    A. Where the speed of vehicular traffic is expected to drop substantially;

    B. Where significant queuing and delays are expected;

    C. Where adverse environmental conditions are present;

    D. Where there are changes in alignment or surface conditions;

    E. Where advance notice of ramp, lane, or roadway closures is needed;

    F. Where crash or incident management is needed; and/or

    G. Where changes in the road user pattern occur.

Guidance:

06   *The components of a portable changeable message sign should include: a message sign, control systems, a power source, and mounting and transporting equipment. The front face of the sign should be covered with a protective material.*

Standard:

07   **Portable changeable message signs shall comply with the applicable design and application principles established in Chapter 2A. Portable changeable message signs shall display only traffic operational, regulatory, warning, and guidance information, and shall not be used for advertising messages.**

Support:

08   Section 2L.02 contains information regarding overly simplistic or vague messages that is also applicable to portable changeable message signs.

Standard:

09   **The colors used for legends on portable changeable message signs shall comply with those shown in Table 2A-5.**

Support:

10   Section 2L.04 contains information regarding the luminance, luminance contrast, and contrast orientation that is also applicable to portable changeable message signs.

Guidance:

11   *Portable changeable message signs should be visible from ½ mile under both day and night conditions.*

Support:

12   Section 2B.21 contains information regarding the design of portable changeable message signs that are used to display speed limits that change based on operational conditions, or are used to display the speed at which approaching drivers are traveling.

Guidance:

13   *A portable changeable message sign should be limited to three lines of eight characters per line or should consist of a full matrix display.*

14   *Except as provided in Paragraph 15 of this Section, the letter height used for portable changeable message sign messages should be a minimum of 18 inches.*

Option:

15   For portable changeable message signs mounted on service patrol trucks or other incident response vehicles, a letter height as short as 10 inches may be used. Shorter letter sizes may also be used on a portable changeable message sign used on low speed facilities provided that the message is legible from at least 650 feet.

16   The portable changeable message sign may vary in size.

Guidance:

17   *Messages on a portable changeable message sign should consist of no more than two phases, and a phase should consist of no more than three lines of text. Each phase should be capable of being understood by itself, regardless of the order in which it is read. Messages should be centered within each line of legend. If more than one portable changeable message sign is simultaneously legible to road users, then only one of the signs should display a sequential message at any given time.*

Support:

18   Road users have difficulties in reading messages displayed in more than two phases on a typical three-line portable changeable message sign.

Standard:

19   **Except when being used to simulate an Arrow Board display (see Section 6L.06), techniques of message display such as animation, rapid flashing, dissolving, exploding, scrolling, traveling horizontally or vertically across the face of the sign, or other dynamic elements shall not be used.**

Guidance:

20   *When a message is divided into two phases, the display time for each phase should be at least 2 seconds, and the sum of the display times for both of the phases should be a maximum of 8 seconds.*

21   *All messages should be designed with consideration given to the principles provided in this Section and also taking into account the following:*

    A. *The message should be as brief as possible and should contain three thoughts (with each thought preferably shown on its own line) that convey:*
       1. *The problem or situation that the road user will encounter ahead,*
       2. *The location of or distance to the problem or situation, and*
       3. *The recommended driver action.*

    B. *If more than two phases are needed to display a message, additional portable changeable message signs should be used. When multiple portable changeable message signs are needed, they should be placed on the same side of the roadway and they should be separated from each other by a distance of at least 1,000 feet on freeways and expressways, and by a distance of at least 500 feet on other types of highways.*

Standard:

22   **When the word messages shown in Tables 1D-1 or 1D-2 need to be abbreviated on a portable changeable message sign, the provisions described in Section 1D.08 shall be followed.**

23   **In order to maintain legibility, portable changeable message signs shall automatically adjust their brightness under varying light conditions.**

24   **The control system shall include a display screen upon which messages can be reviewed before being displayed on the message sign. The control system shall be capable of maintaining memory when power is unavailable.**

25   **Portable changeable message signs shall be equipped with a power source and a battery back-up to provide continuous operation when failure of the primary power source occurs.**

26   **The mounting of portable changeable message signs on a trailer, a large truck, or a service patrol truck shall be such that the bottom of the message sign shall be a minimum of 7 feet above the roadway in urban areas and 5 feet above the roadway in rural areas when it is in the operating mode.**

Guidance:

27   *Portable changeable message signs should be used as a supplement to and not as a substitute for conventional signs and pavement markings.*

28   *When portable changeable message signs are used for route diversion, they should be placed far enough in advance of the diversion to allow road users ample opportunity to perform necessary lane changes, to adjust their speed, or to exit the affected highway.*

29   *Portable changeable message signs should be sited and aligned to provide maximum legibility and to allow time for road users to respond appropriately to the portable changeable message sign message.*

30   *Portable changeable message signs should be placed off the shoulder of the roadway and behind a traffic barrier, if practicable. Where a traffic barrier is not available to shield the portable changeable message sign, it should be placed off the shoulder and outside of the clear zone. If a portable changeable message sign has to be placed on the shoulder of the roadway or within the clear zone, it should be delineated with retroreflective TTC devices.*

31   *When portable changeable message signs are used in TTC zones, they should display only TTC messages.*

32   *When portable changeable message signs are not being used to display TTC messages, they should be relocated such that they are outside of the clear zone or shielded behind a traffic barrier and turned away from traffic. If relocation or shielding is impracticable, they should be delineated with retroreflective TTC devices.*

33   *Portable changeable message sign trailers should be delineated on a permanent basis by affixing retroreflective material, known as conspicuity material, in a continuous line on the face of the trailer as seen by oncoming road users.*

Section 6L.06 Arrow Boards

Standard:

01   **An arrow board shall be a sign with a matrix of elements capable of either flashing or sequential displays. This sign shall provide additional warning and directional information to assist in merging and controlling road users through or around a TTC zone.**

Guidance:

02   *An arrow board in the arrow or chevron mode should be used to advise approaching traffic of a lane closure along major multi-lane roadways in situations involving heavy traffic volumes, high speeds, and/or limited sight distances, or at other locations and under other conditions where road users are less likely to expect such lane closures.*

03 *If used, an arrow board should be used in combination with appropriate signs, channelizing devices, or other TTC devices.*

04 *An arrow board should be placed on the shoulder of the roadway or, if practicable, farther from the traveled lane. It should be delineated with retroreflective TTC devices. When an arrow board is not being used, it should be removed; if not removed, it should be shielded; or if the previous two options are not feasible, it should be delineated with retroreflective TTC devices.*

Standard:

05 **Arrow boards shall meet the minimum size, legibility distance, number of elements, and other specifications shown in Figure 6L-3.**

Support:

06 Type A arrow boards are appropriate for use on low-speed urban streets. Type B arrow boards are appropriate for intermediate-speed facilities and for maintenance or mobile operations on high-speed roadways. Type C arrow boards are intended to be used on high-speed, high-volume motor vehicle traffic control projects. Type D arrow boards are intended for use on vehicles authorized by the State or local agency.

Standard:

07 **Type A, B, and C arrow boards shall have solid rectangular appearances. A Type D arrow board shall conform to the shape of the arrow.**

08 **All arrow boards shall be finished in non-reflective black. The arrow board shall be mounted on a vehicle, a trailer, or other suitable support.**

Guidance:

09 *The minimum mounting height, measured vertically from the bottom of the board to the roadway below it or to the elevation of the near edge of the roadway, of an arrow board should be 7 feet, except on vehicle-mounted arrow boards, which should be as high as practicable.*

10 *A vehicle-mounted arrow board should be provided with remote controls.*

Standard:

11 **Arrow board elements shall be capable of at least a 50 percent dimming from full brilliance. The dimmed mode shall be used for nighttime operation of arrow boards.**

Guidance:

12 *Full brilliance should be used for daytime operation of arrow boards.*

Standard:

13 **The arrow board shall have suitable elements capable of the various operating modes. The color presented by the elements shall be yellow.**

Guidance:

14 *If an arrow board consisting of a bulb matrix is used, the elements should be recess-mounted or equipped with an upper hood of not less than 180 degrees.*

Standard:

15 **The minimum element on-time shall be 50 percent for the flashing mode, with equal intervals of 25 percent for each sequential phase. The flashing rate shall be not less than 25 or more than 40 flashes per minute.**

16 **An arrow board shall have the following three mode selections:**

 A. A Flashing Arrow, Sequential Arrow, or Sequential Chevron mode;
 B. A flashing Double Arrow mode; and
 C. A flashing Caution or Alternating Diamond mode.

17 **An arrow board in the arrow or chevron mode shall be used only for stationary or moving lane closures on multi-lane roadways.**

18 **For shoulder work, for blocking the shoulder, for roadside work near the shoulder, or for temporarily closing one lane on a two-lane, two-way roadway, an arrow board shall be used only in the caution mode.**

Guidance:

19 *For a stationary lane closure, the arrow board should be located on the shoulder at the beginning of the merging taper.*

20 *Where the shoulder is narrow, the arrow board should be located in the closed lane.*

Figure 6L-3. Advance Warning Arrow Board Display Specifications

Operating Mode	Display (Type C arrow board illustrated)

1. At least one of the three following modes shall be provided:

   Flashing Arrow

   Merge Right

   (right arrow shown; left is similar)

   Sequential Arrow

   Merge Right

   Sequential Chevron

   Merge Right

2. The following mode shall be provided: Flashing Double Arrow

   Merge Right or Left

3. At least one of the following modes shall be provided: Flashing Caution or Alternating Diamond Caution

   Flashing Caution   Flashing Caution   Alternating Diamond Caution

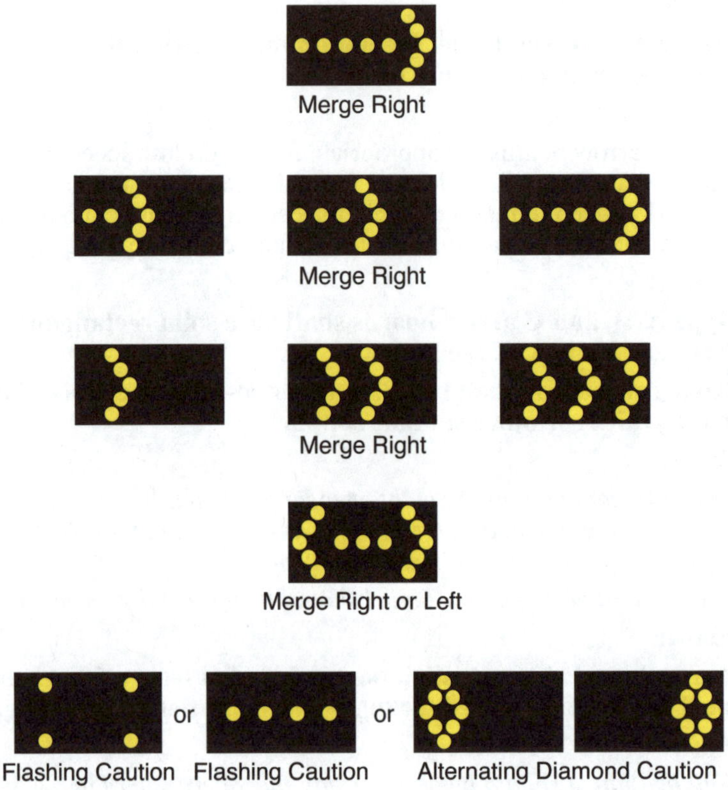

Arrow Board Type	Minimum Size	Minimum Legibility Distance	Minimum Number of Elements
A	48 x 24 inches	1/2 mile	12
B	60 x 30 inches	3/4 mile	13
C	96 x 48 inches	1 mile	15
D	None*	1/2 mile	12

*Length of arrow equals 48 inches, width of arrowhead equals 24 inches

Standard:

21   **When arrow boards are used to close multiple lanes, a separate arrow board shall be used for each closed lane.**

Guidance:

22   *When arrow boards are used to close multiple lanes, if the first arrow board is placed on the shoulder, the second arrow board should be placed in the first closed lane at the upstream end of the second merging taper (see Figure 6P-37). When the first arrow board is placed in the first closed lane, the second arrow board should be placed in the second closed lane at the downstream end of the second merging taper.*

23 *For mobile operations where a lane is closed, the arrow board should be located to provide adequate separation from the work operation to allow for appropriate reaction by approaching drivers.*

Standard:

24 **A vehicle displaying an arrow board shall be equipped with high-intensity rotating, flashing, oscillating, or strobe lights.**

25 **Arrow boards shall only be used to indicate a lane closure. Arrow boards shall not be used to indicate a lane shift.**

Option:

26 A portable changeable message sign may be used to simulate an arrow board display.

Section 6L.07 <u>Flashing Beacons and Warning Lights</u>

Guidance:

01 *Lighting devices should be provided in TTC zones based on engineering judgment.*

Option:

02 Flashing beacons (see Chapter 4S) and/or warning lights may be used to supplement retroreflectorized signs, barriers, and channelizing devices.

Support:

03 Type A, Type B, Type C, and Type D 360-degree warning lights are portable, powered, yellow, lens-directed, enclosed lights.

Standard:

04 **Warning lights shall comply with the provisions in Chapter 13 of the publication entitled, "Equipment and Materials Standards of the Institute of Transportation Engineers," 1998, Institute of Transportation Engineers.**

05 **When warning lights are used, they shall be mounted on signs or channelizing devices in a manner that, if hit by an errant vehicle, they will not be likely to penetrate the windshield.**

Guidance:

06 *The maximum spacing for warning lights should be identical to the channelizing device spacing requirements.*

Support:

07 The light weight and portability of warning lights are advantages that make these devices useful as supplements to the retroreflectorization on signs and channelizing devices. The flashing lights are effective in attracting road users' attention.

Option:

08 Warning lights may be used in either a steady-burn or flashing mode.

Standard:

09 **Warning lights shall flash when placed on channelizing devices used alone or in a cluster to warn of a condition.**

10 **Except for the sequential flashing warning lights discussed in Paragraph 12 of this Section, warning lights placed on channelizing devices used in a series to channelize road users shall be steady-burn.**

11 **Except for the sequential flashing warning lights that are described in Paragraph 12 of this Section, flashing warning lights shall not be used for delineation, as a series of flashers fails to identify the desired vehicle path.**

12 **If a series of sequential flashing warning lights is used on channelizing devices that form a merging taper, the successive flashing of the lights shall occur from the upstream end of the merging taper to the downstream end of the merging taper in order to identify the desired vehicle path. Each flashing warning light in the sequence shall be flashed at a rate of not less than 55 or more than 75 times per minute.**

13 **Type A Low-Intensity Flashing warning lights, Type C Steady-Burn warning lights, and Type D 360-degree Steady-Burn warning lights shall be maintained so as to be capable of being visible on a clear night from a distance of 3,000 feet. Type B High-Intensity Flashing warning lights shall be maintained so as to be capable of being visible on a sunny day when viewed without the sun directly on or behind the device from a distance of 1,000 feet.**

14 **Warning lights shall have a minimum mounting height of 30 inches to the bottom of the lens.**

Support:

15   Type A Low-Intensity Flashing warning lights are used to warn road users during nighttime hours that they are approaching or proceeding in a potentially hazardous area.

Option:

16   Type A warning lights may be mounted on channelizing devices.

Support:

17   Type B High-Intensity Flashing warning lights are used to warn road users during both daylight and nighttime hours that they are approaching a potentially hazardous area.

Option:

18   Type B warning lights are designed to operate 24 hours per day and may be mounted on advance warning signs or on independent supports.

19   Type C Steady-Burn warning lights and Type D 360-degree Steady-Burn warning lights may be used during nighttime hours to delineate the edge of the traveled way.

Guidance:

20   *When used to delineate a curve, Type C and Type D 360-degree warning lights should only be used on devices on the outside of the curve, and not on the inside of the curve.*

Section 6L.08 <u>High-Level Warning Devices (Flag Trees)</u>

Option:

01   A high-level warning device (flag tree) may supplement other TTC devices in TTC zones.

Support:

02   A high-level warning device is designed to be seen over the top of typical passenger cars. A typical high-level warning device is shown in Figure 6F-1.

Standard:

03   **A high-level warning device shall consist of a minimum of two flags with or without a Type B high-intensity flashing warning light. The distance from the roadway to the bottom of the lens of the light and to the lowest point of the flag material shall be not less than 8 feet. The flag shall be 16 inches square or larger and shall be orange or fluorescent red-orange in color.**

Option:

04   An appropriate warning sign may be mounted below the flags.

Support:

05   High-level warning devices are most commonly used in high-density road user situations to warn road users of short-term operations.

CHAPTER 6M. OTHER TTC ZONE DESIGN FEATURES AND SAFETY DEVICES

Section 6M.01 General

Support:

01   Although certain devices and design features, such as lighting, barriers, dividers, crash cushions, and screens, are sometimes used in TTC zones to supplement traffic control devices or enhance traffic operations or safety for road users, they are not considered to be traffic control devices. The following Sections describe the most commonly used devices and design features. Section 1D.04 contains additional information about these devices and design features.

Section 6M.02 Positive Protection and Temporary Traffic Barriers

Support:

01   Temporary traffic barriers, including portable or movable barriers, are devices designed to help prevent penetration by vehicles while minimizing injuries to vehicle occupants, and to protect workers, bicyclists, and pedestrians.

Guidance:

02   *Except as otherwise required, at a minimum, longitudinal traffic barriers and/or other positive protection devices should be considered in work zone situations that place workers at increased risk from motorized traffic, and where positive protection devices offer the highest potential for improved safety for workers and road users.*

Support:

03   Considerations for positive protection include, but are not limited to, the following circumstances:

    A. Work zones that provide workers no means of escape from motorized traffic such as tunnels or bridges;
    B. Long-term stationary work zones of two weeks or more resulting in substantial worker exposure to motorized traffic;
    C. Projects with anticipated operating speeds of 45 mph or greater, especially when combined with high traffic volumes;
    D. Work operations that place workers, pedestrians, or bicyclists close to travel lanes open to traffic; and
    E. Roadside hazards, such as drop-offs or unfinished bridge decks, that will remain in place overnight or longer.

04   Work zone setups vary depending on the nature of the positive protection used.

05   23 CFR Part 630.1108(a) contains additional requirements for certain projects.

Option:

06   Temporary traffic barriers may be used to separate two-way vehicular traffic.

Standard:

07   **Temporary traffic barriers shall be supplemented with standard delineation, pavement markings, or channelizing devices for improved daytime and nighttime visibility if they are used to channelize vehicular traffic. The delineation color shall match the applicable pavement marking color.**

08   **Temporary traffic barriers, including their end treatments, shall be crashworthy (see definition in Section 1C.02).**

09   **Short intermittent segments of temporary traffic barrier shall not be used because they nullify the containment and redirective capabilities of the temporary traffic barrier, increase the potential for serious injury both to vehicle occupants and pedestrians, and encourage the presence of blunt leading ends. Adjacent temporary traffic barrier segments shall be properly connected in order to provide the overall strength required for the temporary traffic barrier to perform properly.**

Option:

10   Steady-burn warning lights (see Section 6L.07) may be mounted on temporary traffic barrier installations.

Support:

11   Temporary traffic barrier includes portable concrete, portable steel, or movable barrier which can all be moved laterally and/or longitudinally when needed and/or from site to site. More specific information on the use of temporary traffic barriers is contained in Chapters 8 and 9 of "Roadside Design Guide," 4th Edition, 2011, AASHTO.

Section 6M.03 Temporary Raised Islands

Standard:

01 **Temporary raised islands shall be used only in combination with pavement striping and other suitable channelizing devices.**

Option:

02 A temporary raised island may be used to separate vehicular traffic flows in two-lane, two-way operations on roadways having a vehicular traffic volume range of 4,000 to 15,000 average daily traffic (ADT) and on freeways having a vehicular traffic volume range of 22,000 ADT to 60,000 ADT.

03 Temporary raised islands also may be used in other than two-lane, two-way operations where physical separation of vehicular traffic from the TTC zone is not required.

Guidance:

04 *Temporary raised islands should have the basic dimensions of 4 inches high by at least 12 inches wide and have rounded or chamfered corners.*

05 *The temporary raised islands should not be designed in such a manner that they would cause a motorist to lose control of the vehicle if the vehicle inadvertently strikes the temporary raised island. If struck, pieces of the island should not be dislodged to the extent that they could penetrate the occupant compartment or involve other vehicles.*

Standard:

06 **At pedestrian crossing locations, temporary raised islands shall have an opening or be shortened to provide at least a 60-inch wide passageway for the crossing pedestrian.**

Section 6M.04 Detectable Edging for Pedestrians

Support:

01 Individual channelizing devices, tape or rope used to connect individual devices, other discontinuous barriers and devices, and pavement markings are not detectable by persons with vision disabilities and are incapable of providing detectable path guidance on temporary or realigned sidewalks or other pedestrian facilities.

Guidance:

02 *A continuously-detectable edging should be provided throughout the length of a temporary pedestrian facility such that it can be followed by pedestrians using long canes for guidance. This edging should extend at least 8 inches above the surface of the sidewalk or pathway, with the bottom of the edging a maximum of 2 inches above the surface. This edging should be continuous throughout the length of the facility except for gaps at locations where pedestrians or vehicles will be turning or crossing. This edging should consist of a prefabricated or formed-in-place curbing or other continuous device that is placed along the edge of the sidewalk or walkway. This edging should be firmly attached to the ground or to other devices. Adjacent sections of this edging should be interconnected such that the edging is not displaced by pedestrian or vehicular traffic or work operations, and such that it does not constitute a hazard to pedestrians, workers, or other road users.*

Support:

03 Examples of detectable edging for pedestrians include:

   A. Prefabricated lightweight sections of plastic, metal, or other suitable materials that are interconnected and fixed in place to form a continuous edge.
   B. Prefabricated lightweight sections of plastic, metal, or other suitable materials that are interconnected, fixed in place, and placed at ground level to provide a continuous connection between channelizing devices located at intervals along the edge of the sidewalk or walkway.
   C. Sections of lumber interconnected and fixed in place to form a continuous edge.
   D. Formed-in-place asphalt or concrete curb.
   E. Prefabricated concrete curb sections that are interconnected and fixed in place to form a continuous edge.
   F. Continuous temporary traffic barrier or longitudinal channelizing barricades placed along the edge of the sidewalk or walkway that provides a pedestrian edging at ground level.
   G. Chain link or other fencing equipped with a continuous bottom rail.

Guidance:

04 *Detectable pedestrian edging should be orange, white, or yellow and should match the color of the adjacent channelizing devices or traffic control devices, if any are present.*

Section 6M.05 Crash Cushions

Support:

01   Crash cushions are systems that mitigate the effects of errant vehicles that strike obstacles, either by smoothly decelerating the vehicle to a stop when hit head-on, or by redirecting the errant vehicle. The two types of crash cushions that are used in TTC zones are stationary crash cushions and truck-mounted attenuators. Crash cushions in TTC zones help protect the drivers from the exposed ends of barriers, fixed objects, shadow vehicles, and other obstacles. Specific information on the use of crash cushions can be found in "Roadside Design Guide," 4th Edition, 2011, AASHTO.

Standard:

02   **Crash cushions shall be crashworthy (see definition in Section 1C.02). They shall also be designed for each application to stop or redirect errant vehicles under prescribed conditions. Crash cushions shall be periodically inspected to verify that they have not been hit or damaged. Damaged crash cushions shall be promptly repaired or replaced to maintain their crashworthiness.**

Support:

03   Stationary crash cushions are used in the same manner as permanent highway installations to protect drivers from the exposed ends of barriers, fixed objects, and other obstacles.

Standard:

04   **Stationary crash cushions shall be designed for the specific application intended.**

05   **Truck-mounted attenuators shall be energy-absorbing devices attached to the rear of shadow trailers or trucks and shall be used in accordance with the manufacturer's specifications. If used, the shadow vehicle with the attenuator shall be located in advance of the work area, workers, or equipment to reduce the severity of rear-end crashes from errant vehicles.**

Support:

06   Trucks or trailers are often used as shadow vehicles to protect workers or work equipment from errant vehicles. These shadow vehicles are normally equipped with flashing arrows, changeable message signs, and/or high-intensity rotating, flashing, oscillating, or strobe lights and are located properly in advance of the workers and/or equipment that they are protecting. However, these shadow vehicles might themselves cause injuries to occupants of the errant vehicles if they are not equipped with truck-mounted attenuators.

Guidance:

07   *The shadow truck should be positioned a sufficient distance in advance of the workers or equipment being protected so that there will be sufficient distance, but not so much so that errant vehicles will travel around the shadow truck and strike the protected workers and/or equipment.*

Support:

08   Chapter 9 of "Roadside Design Guide," 4th Edition, 2011, AASHTO contains additional information regarding the use of shadow vehicles.

Section 6M.06 Rumble Strips

Support:

01   Transverse rumble strips consist of intermittent, narrow, transverse areas of rough-textured or slightly-raised or depressed road surface that extend across the travel lanes to alert drivers to unusual vehicular traffic conditions. Through noise and vibration they attract the driver's attention to such features as unexpected changes in alignment and to conditions requiring a stop.

02   Longitudinal rumble strips consist of a series of rough-textured or slightly-raised or depressed road surfaces located along the shoulder to alert road users that they are leaving the travel lanes.

Standard:

03   **If it is desirable to use a color other than the color of the pavement for a longitudinal rumble strip, the color of the rumble strip shall be the same color as the longitudinal line the rumble strip supplements.**

04   **If the color of a transverse rumble strip used within a travel lane is not the color of the pavement, the color of the rumble strip shall be white, black, or orange.**

Option:

05   Intervals between transverse rumble strips may be reduced as the distance to the approached conditions is diminished in order to convey an impression that a closure speed is too fast and/or that an action is imminent. A sign warning drivers of the onset of rumble strips may be placed in advance of any transverse rumble strip installation.

Guidance:

06   *Transverse rumble strips should be placed transverse to vehicular traffic movement. They should not adversely affect overall pavement skid resistance under wet or dry conditions.*

07   *In urban areas, even though a closer spacing might be warranted, transverse rumble strips should be designed in a manner that does not promote unnecessary braking or erratic steering maneuvers by road users.*

08   *Transverse rumble strips should not be placed on sharp horizontal or vertical curves.*

09   *Rumble strips should not be placed through pedestrian crossings or on bicycle routes.*

10   *Transverse rumble strips should not be placed on roadways used by bicyclists unless a minimum clear path of 4 feet is provided at each edge of the roadway or on each paved shoulder.*

11   *Longitudinal rumble strips should not be placed on the shoulder of a roadway that is used by bicyclists unless a minimum clear path of 4 feet is also provided on the shoulder.*

Section 6M.07 Screens

Support:

01   Screens are used to block the road users' view of activities that can be distracting. Screens might improve safety and motor vehicle traffic flow where volumes approach the roadway capacity because they discourage gawking and reduce headlight glare from oncoming motor vehicle traffic.

Guidance:

02   *Screens should not be mounted where they could adversely restrict road user visibility and sight distance and adversely affect the operation of vehicles.*

Option:

03   Screens may be mounted on the top of temporary traffic barriers that separate two-way motor vehicle traffic.

Guidance:

04   *Design of screens should be in accordance with Chapter 9 of "Roadside Design Guide," 4th Edition, 2011, AASHTO.*

Section 6M.08 Lighting for Night Work

Support:

01   Utility, maintenance, or construction activities on highways are frequently conducted during nighttime periods when vehicular traffic volumes are lower. Large construction projects are sometimes operated on a double-shift basis requiring night work (see Section 6N.18).

Guidance:

02   *When nighttime work is being performed, floodlights should be used to illuminate the work area, equipment crossings, and other areas.*

03   *When used, floodlighting should be installed in a manner that minimizes glare to approaching road users, flaggers, or workers.*

04   *The adequacy of the floodlight placement and elimination of potential glare should be determined by driving through and observing the floodlighted area from each direction on all approaching roadways after the initial floodlight setup, at night, and periodically. Lighting should be sufficient so as to give road users the capability to identify a worker as a person. Care should be taken to minimize the potential for shadows to conceal workers within the work area.*

Support:

05   Desired illumination levels vary depending upon the nature of the task involved. An average horizontal luminance of 5 foot candles can be adequate for general activities. Tasks requiring high levels of precision and extreme care can require an average horizontal luminance of 20 foot candles.

Standard:

06   **Except in emergency situations, flagger stations shall be illuminated at night.**

CHAPTER 6N. TYPES OF TEMPORARY TRAFFIC CONTROL ZONE ACTIVITIES

Section 6N.01 Work Duration

Support:

01   Work duration is a major factor in determining the number and types of devices used in TTC zones.
The duration of a TTC zone is defined relative to the length of time a work operation occupies a spot location.

Standard:

02   **The five categories of work duration and their time at a location shall be defined as follows:**

    A. Long-term stationary is work that occupies a location more than 3 days.

    B. Intermediate-term stationary is work that occupies a location more than one daylight period up to 3 days, or nighttime work lasting more than 1 hour.

    C. Short-term stationary is daytime work that occupies a location for more than 1 hour within a single daylight period.

    D. Short duration is work that occupies a location up to 1 hour.

    E. Mobile is work that moves intermittently or continuously.

Support:

03   At long-term stationary TTC zones, there is ample time to install and realize benefits from the full range of TTC procedures and devices that are available for use. Larger channelizing devices, temporary roadways, and temporary traffic barriers are frequently used.

Standard:

04   **Since long-term operations extend into nighttime, retroreflective and/or illuminated devices shall be used in long-term stationary TTC zones.**

Support:

05   In intermediate-term stationary TTC zones, it might not be feasible or practical to use procedures or devices that would be desirable for long-term stationary TTC zones, such as altered pavement markings, temporary traffic barriers, and temporary roadways. The increased time to place and remove these devices in some cases could significantly lengthen the project, thus increasing exposure time.

Standard:

06   **Since intermediate-term operations extend into nighttime, retroreflective and/or illuminated devices shall be used in intermediate-term stationary TTC zones.**

Support:

07   Most maintenance and utility operations are short-term stationary work.

08   As compared to stationary operations, mobile and short-duration operations are activities that might involve different treatments. Devices having greater mobility might be necessary such as signs mounted on trucks. Devices that are larger, more imposing, or more visible can be used effectively and economically. The mobility of the TTC zone is important.

Guidance:

09   *Safety in short-duration or mobile operations should not be compromised by using fewer devices simply because the operation will frequently change its location.*

Support:

10   During short-duration work, it often takes longer to set up and remove the TTC zone than to perform the work. Workers face hazards in setting up and taking down the TTC zone. Also, since the work time is short, delays affecting road users are significantly increased when additional devices are installed and removed.

Option:

11   Considering these factors, simplified control procedures may be warranted for short-duration work. A reduction in the number of devices may be offset by the use of other more dominant devices such as high-intensity rotating, flashing, oscillating, or strobe lights on work vehicles.

Support:

12   Mobile operations often involve frequent short stops for activities such as litter cleanup, pothole patching, or utility operations, and are similar to short-duration operations.

Option:

13   Flags and/or channelizing devices may additionally be used and moved periodically to keep them near the mobile work area.

14   Flaggers may be used for mobile operations that often involve frequent short stops.

Support:

15   Mobile operations also include work activities where workers and equipment move along the road without stopping, usually at slow speeds. The advance warning area moves with the work area.

Guidance:

16   *When mobile operations are being performed, a shadow vehicle equipped with an arrow board or a sign should follow the work vehicle, especially when vehicular traffic speeds or volumes are high. Where feasible, warning signs should be placed along the roadway and moved periodically as work progresses.*

17   *To avoid high-volume conditions, consideration should be given to scheduling mobile operations work during off-peak hours.*

18   *If there are mobile operations on a high-speed travel lane of a multi-lane divided highway, arrow boards should be used.*

Standard:

19   **Mobile operations shall have appropriate devices on the equipment (that is, high-intensity rotating, flashing, oscillating, or strobe lights, signs, or special lighting), or shall use a separate vehicle with appropriate warning devices. Although vehicle hazard warning lights are permitted to be used to supplement high-intensity rotating, flashing, oscillating, or strobe lights, they shall not be used instead of these devices.**

Option:

20   For mobile operations that move at speeds of less than 3 mph, mobile signs or stationary signing that is periodically retrieved and repositioned in the advance warning area may be used.

Support:

21   A rolling roadblock is a method of TTC used to slow or stop traffic as a means of temporarily removing traffic from a roadway segment downstream of the road block. The rolling roadblock closes all lanes of traffic by using pacing vehicles to create a gap so that construction activities can be performed. Rolling roadblocks are used where long-term road closures using TTC devices are not needed. A rolling roadblock consists of one blocking/pacing vehicle per lane of traffic, a clearing vehicle, and an advance warning vehicle. The rolling roadblock is normally performed by law enforcement officers during off-peak hours.

Section 6N.02 <u>Location of Work</u>

Support:

01   Chapter 6C and Sections 6M.04 and 6N.04 contain additional information regarding the steps to follow when pedestrian or bicycle facilities are affected by the worksite.

02   The choice of TTC needed for a TTC zone depends upon where the work is located. As a general rule, the closer the work is to road users (including bicyclists and pedestrians), the greater the number of TTC devices that are needed. Procedures are described later in this Chapter for establishing TTC zones in the following locations:

    A. Outside the shoulder,
    B. On the shoulder with no encroachment,
    C. On the shoulder with minor encroachment,
    D. Within the median, and
    E. Within the traveled way.

Standard:

03   **When the work space is within the traveled way, except for short-duration and mobile operations, advance warning shall provide a general message that work is taking place and shall supply information about highway conditions. TTC devices shall clearly delineate the path roadway users are to follow through the TTC zone.**

Section 6N.03 <u>Modifications to Fulfill Special Needs</u>

Support:

01   The typical applications in Chapter 6P illustrate commonly encountered situations in which TTC devices are employed.

Option:

02   Other devices may be added to supplement the devices provided in the typical applications, and device spacing may be adjusted to provide additional reaction time. When conditions are less complex than those depicted in the typical applications, fewer devices may be needed.

Guidance:

03   *When conditions are more complex, typical applications should be modified by giving particular attention to the provisions set forth in Chapter 6A and by incorporating appropriate devices and practices from the following list:*

    A. Additional devices:
       1. Signs
       2. Arrow boards
       3. More channelizing devices at closer spacing (see Section 6M.04 for information regarding detectable edging for pedestrians)
       4. Temporary raised pavement markers
       5. High-level warning devices
       6. Portable changeable message signs
       7. Temporary traffic control signals (including accessible pedestrian signals where not otherwise required)
       8. Temporary traffic barriers
       9. Crash cushions
       10. Screens
       11. Rumble strips
       12. More delineation

    B. Upgrading of devices:
       1. A full complement of standard pavement markings
       2. Brighter and/or wider pavement markings
       3. Larger and/or brighter signs
       4. Channelizing devices with greater conspicuity
       5. Temporary traffic barriers in place of channelizing devices

    C. Improved geometrics at detours or crossovers

    D. Increased distances:
       1. Longer advance warning area
       2. Longer tapers

    E. Lighting:
       1. Temporary roadway lighting
       2. Steady-burn lights used with channelizing devices
       3. Flashing lights for isolated hazards
       4. Illuminated signs
       5. Floodlights

    F. Pedestrian routes and temporary facilities

    G. Bicycle diversions and temporary facilities

Section 6N.04 <u>Work Affecting Pedestrian and Bicycle Facilities</u>

Support:

01   It is not uncommon, particularly in urban areas, that road work and the associated TTC will affect existing pedestrian or bicycle facilities. It is essential that the needs of all road users, including pedestrians with disabilities, are considered in TTC zones.

02   In addition to specific provisions identified in Sections 6N.05 through 6N.13, there are a number of provisions that might be applicable for all of the types of activities identified in this Chapter.

Guidance:

03   *Where pedestrian or bicyclist usage is high, the typical applications should be modified by giving particular attention to the provisions set forth in Chapter 6C, this Chapter, Sections 6K.02 and 6M.04, and in other Sections of Part 6 related to accessibility and detectability provisions in TTC zones.*

04   *Pedestrians should be separated from the worksite by appropriate devices that maintain the accessibility and detectability for pedestrians with disabilities.*

05   *Bicyclists and pedestrians should not be exposed to unprotected excavations, open utility access, overhanging equipment, or other such conditions.*

06   *Except for - and mobile operations, when a highway shoulder is occupied, a SHOULDER WORK (W21-5) sign should be placed in advance of the activity area. When work is performed on a paved shoulder 8 feet or more in width, channelizing devices should be placed on a taper having a length that conforms to the requirements of a shoulder taper. Signs should be placed such that they do not narrow any existing pedestrian passages to less than 48 inches.*

07   *Pedestrian detours should be avoided since pedestrians rarely observe them and the cost of providing accessibility and detectability might outweigh the cost of maintaining a continuous route. Whenever possible, work should be done in a manner that does not create a need to detour pedestrians from existing routes or crossings.*

Standard:

08   **Where pedestrian routes are closed, alternate pedestrian routes shall be provided.**

09   **When existing pedestrian facilities are disrupted, closed, or relocated in a TTC zone, the temporary facilities shall be detectable and shall include accessibility features consistent with the features present in the existing pedestrian facility.**

Guidance:

10   *The continuity of a bikeway should be maintained through the TTC zone if practical.*

Support:

11   The continuity of a bikeway through the TTC zone is particularly important where bicyclists have been traveling on a shoulder, bicycle lane, or shared-use path adjacent to a general-purpose lane (having a speed limit greater than or equal to 35 miles per hour) and there would be a significant safety concern if bicyclists were to share that general-purpose lane through the TTC zone.

12   On roadways which are not bikeways but where bicyclists (when present) typically share lanes with motor vehicle traffic, the TTC plan and Typical Applications for general traffic will usually be adequate for bicyclists as well.

13   In order to maintain room for bicycle lanes through the TTC zone on a multi-lane roadway, one or more travel lanes could be closed.

Guidance:

14   *If a bikeway detour is unavoidable, it should be as short and direct as practical.*

15   *On-road bicyclists should not be directed onto a path or sidewalk intended for pedestrian use except where such a path or sidewalk is a shared-use path, or where no practical alternative is available (such as might be the case on a bridge in the course of a rehabilitation project).*

16   *If a portion of a bikeway is to be closed due to construction activities and the detoured bikeway follows a complex path not in the original bikeway corridor, then a full detour plan should be developed and implemented. The TTC for the detour of the bikeway should include all necessary advance warning (W21 series) signs, detour (W4-9 series) signs, and any other TTC devices necessary to guide bicyclists along the detour route.*

Support:

17   Figures 6P-47 through 6P-51 provide examples and contain additional information for accommodating bicycles through or around typical TTC zones.

Option:

18   If an on-street bikeway had a wide travel lane or lanes in which bicyclists traveled side by side with motor vehicles prior to construction, and construction activities reduce the lane width(s) to less than 14 feet through the TTC zone, then the BICYCLISTS ALLOWED USE OF FULL LANE (R9-20) sign may be used.

Standard:

19   **The minimum TTC sign and plaque sizes for shared-use paths shall conform to those shown in Table 9A-1. The minimum TTC sign and plaque sizes for on-street bikeways shall conform to Chapters 6G, 6H, and 6I.**

Section 6N.05 Work Outside of the Shoulder

Support:

01   When work is being performed beyond the shoulders, but within the right-of-way, little or no TTC might be needed. TTC generally is not needed where work is confined to an area 15 feet or more from the edge of the traveled way. However, TTC is appropriate where distracting situations exist, such as vehicles parked on the shoulder, vehicles accessing the worksite via the highway, and equipment traveling on or crossing the roadway to perform the work operations (for example, mowing). A typical application for work beyond the shoulder is shown in Figure 6P-1.

Guidance:

02   *Where the situations described in Paragraph 1 of this Section exist, a single warning sign, such as ROAD WORK AHEAD (W20-1), should be used. If the equipment travels on the roadway, the equipment should be equipped with appropriate flags, high-intensity rotating, flashing, oscillating, or strobe lights, and/or a SLOW MOVING VEHICLE (W21-4) sign.*

03   *If work vehicles are on the shoulder, a SHOULDER WORK (W21-5) sign should be used.*

04   *A general warning sign like ROAD MACHINERY AHEAD (W21-3) should be used if workers and equipment must occasionally move onto the shoulder.*

Option:

05   For mowing operations, the sign MOWING AHEAD (W21-8) may be used.

06   Where the activity is spread out over a distance of more than 2 miles, the SHOULDER WORK (W21-5) sign may be repeated every 1 mile.

07   A supplementary plaque with the message NEXT XX MILES (W7-3aP) may be used.

Section 6N.06 <u>Work on the Shoulder with No Encroachment</u>

Support:

01   The provisions of this Section apply to short-term through long-term stationary operations.

Standard:

02   **When paved shoulders having a width of 8 feet or more are closed, at least one advance warning sign shall be used. In addition, channelizing devices shall be used to close the shoulder in advance to delineate the beginning of the work space and direct motor vehicle traffic to remain within the traveled way.**

Guidance:

03   *When paved shoulders having a width of 8 feet or more are closed on freeways and expressways, road users should be warned about potential disabled vehicles that cannot get off the traveled way. An initial general warning sign, such as ROAD WORK AHEAD (W20-1), should be used, followed by a RIGHT or LEFT SHOULDER CLOSED (W21-5a) sign. Where the downstream end of the shoulder closure extends beyond the distance that can be perceived by road users, a supplementary plaque bearing the message NEXT XX FEET (W16-4P) or MILES (W7-3aP) should be placed below the SHOULDER CLOSED (W21-5a) sign. On multi-lane, divided highways, signs advising of shoulder work or the condition of the shoulder should be placed only on the side of the affected shoulder.*

04   *When an improved shoulder is closed on a high-speed roadway, it should be treated as a closure of a portion of the road system because road users expect to be able to use it in emergencies. Road users should be given ample advance warning that shoulders are closed for use as refuge areas throughout a specified length of the approaching TTC zone. The sign(s) should read SHOULDER CLOSED (W21-5a) with distances indicated. The work space on the shoulder should be closed off by a taper or channelizing devices with a length of ⅓ L using the formulas in Tables 6B-3 and 6B-4.*

05   *When the shoulder is not occupied but work has adversely affected its condition, the LOW SHOULDER (W8-9) or SOFT SHOULDER (W8-4) sign should be used, as appropriate.*

06   *Where the condition extends over a distance in excess of 1 mile, the sign should be repeated at 1-mile intervals.*

Option:

07   In addition, a supplementary plaque bearing the message NEXT XX MILES (W7-3aP) may be used.

Support:

08   Temporary traffic barriers might be needed to inhibit encroachment of errant vehicles into the work space and to protect workers.

Standard:

09   **When used for shoulder work, arrow boards shall operate only in the caution mode.**

Support:

10   A typical application for stationary work operations on shoulders is shown in Figure 6P-3. A typical application for short-duration or mobile work on shoulders is shown in Figure 6P-4. A typical application for work on freeway shoulders is shown in Figure 6P-5.

Section 6N.07 <u>Work on the Shoulder with Minor Encroachment</u>

Support:

01   Chapter 6C and Sections 6M.04 and 6N.04 contain additional information regarding the steps to follow when pedestrian or bicycle facilities are affected by the worksite.

Guidance:

02   *When work takes up part of a lane, vehicular traffic volumes, vehicle mix (buses, trucks, cars, and bicycles), speed, and capacity should be analyzed to determine whether the affected lane should be closed. Unless the lane encroachment permits a remaining lane width of 10 feet, the lane should be closed.*

03   *Truck off-tracking should be considered when determining whether the minimum lane width of 10 feet is adequate.*

Option:

04   A lane width of 9 feet may be used for short-term stationary work on low-volume, low-speed roadways when vehicular traffic does not include longer and wider heavy commercial vehicles.

Support:

05   Figure 6P-6 illustrates a method for handling vehicular traffic where the stationary or short-duration work space encroaches slightly into the traveled way.

Section 6N.08 <u>Work within the Median</u>

Support:

01   Chapter 6C and Sections 6M.04 and 6N.04 contain additional information regarding the steps to follow when pedestrian or bicycle facilities are affected by the worksite.

Guidance:

02   *If work in the median of a divided highway is within 15 feet from the edge of the traveled way for either direction of travel, TTC should be used through the use of advance warning signs and channelizing devices.*

Section 6N.09 <u>Work within the Traveled Way of a Two-Lane Highway</u>

Support:

01   Chapter 6C and Sections 6M.04 and 6N.04 contain additional information regarding the steps to follow when pedestrian or bicycle facilities are affected by the worksite.

02   Detour signs are used to direct road users onto another roadway. At diversions, road users are directed onto a temporary roadway or alignment placed within or adjacent to the right-of-way. Typical applications for detouring or diverting road users on two-lane highways are shown in Figures 6P-7, 6P-8, and 6P-9. Figure 6P-7 illustrates the controls around an area where a section of roadway has been closed and a diversion has been constructed. Channelizing devices and pavement markings are used to indicate the transition to the temporary roadway.

Guidance:

03   *When a detour is long, Detour (M4-8, M4-9) signs should be installed to remind and reassure road users periodically that they are still successfully following the detour.*

04   *When an entire roadway is closed, as illustrated in Figure 6P-8, a detour should be provided and road users should be warned in advance of the closure, which in this example is a closure 10 miles from the intersection. If local road users are allowed to use the roadway up to the closure, the ROAD CLOSED AHEAD, LOCAL TRAFFIC ONLY (R11-3a) sign should be used. The portion of the road open to local road users should have adequate signing, marking, and delineation.*

05   *Detours should be signed so that road users will be able to traverse the entire detour route and back to the original roadway as shown in Figure 6P-9.*

Support:

06   Techniques for controlling vehicular traffic under one-lane, two-way conditions are described in Section 6E.01.

Option:

07   Flaggers may be used as shown in Figure 6P-10.

08   STOP/YIELD sign control may be used on roads with low traffic volumes as shown in Figure 6P-11.

09   A temporary traffic control signal may be used as shown in Figure 6P-12.

Section 6N.10 <u>Work within the Traveled Way of an Urban Street</u>

Support:

01   Chapter 6C and Sections 6M.04 and 6N.04 contain additional information regarding the steps to follow when pedestrian or bicycle facilities are affected by the worksite.

02   In urban TTC zones, decisions are needed on how to control vehicular traffic, such as how many lanes are required, whether any turns need to be prohibited at intersections, and how to maintain access to business, industrial, and residential areas.

03   Pedestrian traffic needs separate attention. Chapter 6C contains information regarding pedestrian movements near TTC zones.

Standard:

04   **If the TTC zone affects the movement of bicyclists, adequate access to the roadway or shared-use paths shall be provided (see Part 9).**

05   **Where transit stops are affected or relocated because of work activity, both pedestrian and vehicular access to the affected or relocated transit stops shall be provided.**

Guidance:

06   *If a designated bicycle route is closed because of the work being done, a signed alternate route should be provided. Bicyclists should not be directed onto the path used by pedestrians.*

07   *Worksites within the intersection should be protected against inadvertent pedestrian incursion by providing detectable channelizing devices.*

Support:

08   Utility work takes place both within and outside the roadway to construct and maintain services such as power, gas, light, water, or telecommunications. Operations often involve intersections, since that is where many of the network junctions occur. The work force is usually small, only a few vehicles are involved, and the number and types of TTC devices placed in the TTC zone is usually minimal.

Guidance:

09   *As discussed under short-duration projects, however, the reduced number of devices in utility TTC zones should be offset by the use of high-visibility devices, such as high-intensity rotating, flashing, oscillating, or strobe lights on work vehicles or high-level warning devices.*

Support:

10   Figures 6P-6, 6P-10, 6P-15, 6P-18, 6P-21, 6P-22, 6P-23, 6P-26, and 6P-33 are examples of typical applications for utility operations. Other typical applications might apply as well.

Section 6N.11 <u>Work within the Traveled Way of a Multi-Lane, Non-Access Controlled Highway</u>

Support:

01   Chapter 6C and Sections 6M.04 and 6N.04 contain additional information regarding the steps to follow when pedestrian or bicycle facilities are affected by the worksite.

02   Work on multi-lane (two or more lanes of moving motor vehicle traffic in one direction) highways is divided into right-lane closures, left-lane closures, interior-lane closures, multiple-lane closures, and closures on five-lane roadways.

Standard:

03   **When a lane is closed on a multi-lane road for other than a mobile operation, a transition area containing a merging taper shall be used.**

Guidance:

04   *When justified by an engineering study, temporary traffic barriers (see Section 6K.09) should be used to prevent incursions of errant vehicles into hazardous areas or work space.*

Support:

05   Figure 6P-34 illustrates a lane closure in which temporary traffic barriers are used.

Option:

06   When the right-hand lane is closed, TTC similar to that shown in Figure 6P-33 may be used for undivided or divided four-lane roads.

Guidance:

07   *If morning and evening peak hour vehicular traffic volumes in the two directions are uneven and the greater volume is on the side where the work is being done in the right-hand lane, consideration should be given to closing the inside lane for opposing vehicular traffic and making the lane available to the side with heavier vehicular traffic, as shown in Figure 6P-31.*

08   *If the larger vehicular traffic volume changes to the opposite direction at a different time of the day, the TTC should be changed to allow two lanes for opposing vehicular traffic by moving the devices from the opposing lane to the center line. When it is necessary to create a temporary center line that is not consistent with the pavement markings, channelizing devices should be used and closely spaced.*

Option:

09   When closing a left-hand lane on a multi-lane undivided road, as vehicular traffic flow permits, the two interior lanes may be closed, as shown in Figure 6P-30, to provide drivers and workers additional lateral clearance and to provide access to the work space.

Standard:

10   **When only the left-hand lane is closed on undivided roads, channelizing devices shall be placed along the center line as well as along the adjacent lane.**

Guidance:

11   *When an interior lane is closed, an adjacent lane should also be considered for closure to provide additional space for vehicles and materials and to facilitate the movement of equipment within the work space.*

12   *When multiple lanes in one direction are closed, a capacity analysis should be made to determine the number of lanes needed to accommodate motor vehicle traffic needs. Vehicular traffic should be moved over one lane at a time. As shown in Figure 6P-37, the tapers should be separated by a distance of 2L, with L being determined by the formulas in Tables 6B-3 and 6B-4.*

Option:

13   If operating speeds are 40 mph or less and the space approaching the work area does not permit moving traffic over one lane at a time, a single continuous taper may be used.

Standard:

14   **When a directional roadway is closed, inapplicable WRONG WAY signs and markings, and other existing traffic control devices at intersections within the temporary two-lane, two-way operations section shall be covered, removed, or obliterated.**

Option:

15   When half the road is closed on an undivided highway, both directions of vehicular traffic may be accommodated as shown in Figure 6P-32. When both interior lanes are closed, temporary traffic controls may be used as provided in Figure 6P-30. When a roadway must be closed on a divided highway, a median crossover may be used (see Section 6N.15).

Support:

16   TTC for lane closures on five-lane roads is similar to other multi-lane undivided roads. Figure 6P-32 can be adapted for use on five-lane roads. Figure 6P-35 can be used on a five-lane road for short-duration and mobile operations.

Section 6N.12 <u>Work within the Traveled Way at an Intersection</u>

Support:

01   Chapter 6C and Sections 6M.04 and 6N.04 contain additional information regarding the steps to follow when pedestrian or bicycle facilities are affected by the worksite.

02   The typical applications for intersections are classified according to the location of the work space with respect to the intersection area (as defined by the extension of the curb or edge lines). The three classifications are near side, far side, and in-the-intersection. Work spaces often extend into more than one portion of the intersection. For example, work in one quadrant often creates a near-side work space on one street and a far-side work space on the cross street. In such instances, an appropriate TTC plan is obtained by combining features shown in two or more of the intersection and pedestrian typical applications.

03   TTC zones in the vicinity of intersections might block movements and interfere with normal road user flows. Such conflicts frequently occur at more complex signalized intersections having such features as traffic signal heads over particular lanes, lanes allocated to specific movements, multiple signal phases, signal detectors for actuated control, and accessible pedestrian signals and detectors.

Guidance:

04   *The effect of the work upon signal operation should be considered, and temporary corrective actions should be taken, if necessary, such as revising signal phasing and/or timing to provide adequate capacity, maintaining or adjusting signal detectors, and relocating signal heads to provide adequate visibility as described in Part 4.*

Standard:

05   **When work will occur near an intersection where operational, capacity, or pedestrian accessibility problems are anticipated, the highway agency having jurisdiction shall be contacted.**

Guidance:

06   *For work at an intersection, advance warning signs, devices, and markings should be used on all cross streets, as appropriate. The typical applications depict urban intersections on arterial streets. Where the posted speed limit, the off-peak 85th-percentile speed prior to the work starting, or the anticipated speed exceeds 40 mph, additional warning signs should be used in the advance warning area.*

07   *Pedestrian crossings near TTC sites should be separated from the worksite by appropriate barriers that maintain the accessibility and detectability for pedestrians with disabilities.*

Support:

08   Near-side work spaces, as depicted in Figure 6P-21, are simply handled as a midblock lane closure. A problem that might occur with near-side lane closure is a reduction in capacity, which during certain hours of operation could result in congestion and back-ups.

Option:

09   When near-side work spaces are used, a mandatory turn lane may be used for through vehicular traffic.

10   Where space is restricted in advance of near-side work spaces, as with short block spacings, two warning signs may be used in the advance warning area, and a third action-type warning or a regulatory sign (such as Keep Left) may be placed within the transition area.

Support:

11   Far-side work spaces, as depicted in Figures 6P-22 through 6P-25, involve additional treatment because road users typically enter the activity area by straight-through and left-turn or right-turn movements.

Guidance:

12   *When a lane through an intersection must be closed on the far side, it should also be closed on the near-side approach to preclude merging movements within the intersection.*

Option:

13   If there are a significant number of vehicles turning from a near-side lane that is closed on the far side, the near-side lane may be converted to a mandatory turn lane.

Support:

14   Figures 6P-26 and 6P-27 provide guidance on applicable procedures for work performed within the intersection.

Option:

15   If the work is within the intersection, any of the following strategies may be used:

    A.   A small work space so that road users can move around it, as shown in Figure 6P-26;
    B.   Flaggers or uniformed law enforcement officers to direct road users, as shown in Figure 6P-27;
    C.   Work in stages so the work space is kept to a minimum; and
    D.   Road closures or upstream diversions to reduce road user volumes.

Guidance:

16   *Depending on road user conditions, a flagger(s) and/or a uniformed law enforcement officer(s) should be used to control road users.*

Support:

17   Figures 6P-52 through 6P-54 provide guidance on applicable procedures for work performed within a circular intersection.

Section 6N.13 <u>Work within the Traveled Way of a Freeway or Expressway</u>

Support:

01   Special conditions encountered where vehicular traffic must be moved through or around TTC zones on high-speed, high-volume roadways can pose challenges to the TTC. Although the general principles outlined in other Sections of this Manual are applicable to all types of highways, high-speed, access-controlled highways need special planning and attention in order to accommodate vehicular traffic while also protecting road users and workers. The traffic volumes, vehicle mix (buses, trucks, cars, and bicycles, if permitted), and speed of vehicles on these facilities require that careful TTC procedures be implemented, for example, to induce critical merging maneuvers well in advance of work spaces and in a manner that creates minimum turbulence and delay in the vehicular traffic stream.

02   When the roadway capacity is reduced as a result of lane closures, the demand might exceed the available capacity and result in either a lengthy stopped or slow moving queue of vehicles that might extend past the normal signs used in the typical advance warning area.

Guidance:

03   *An assessment of the expected queue length should be a part of the TTC plan design process and adjustments to the sign spacing and number of signs as well as the possibility of using more conspicuous devices should be considered to increase the distance and conspicuity of the advance warning area.*

Support:

04   One strategy often employed to mitigate the extended queue issue is to work during off peak hours or at night. When the work is limited to night hours, increased use of warning lights, illumination of work spaces, and intelligent advance warning systems might be necessary.

05   TTC for a typical lane closure where a queue is not anticipated to accumulate on a divided highway is shown in Figures 6P-33 and 6P-34. Temporary traffic controls for short-duration and mobile operations on freeways are shown in Figure 6P-35. A typical application for shifting vehicular traffic lanes around a work space is shown in Figure 6P-36. TTC for multiple and interior lane closures on a freeway is shown in Figures 6P-37 and 6P-38.

Guidance:

06   *The method for closing an interior lane when the open lanes have the capacity to carry vehicular traffic should be as shown in Figure 6P-37. When the capacity of the other lanes is needed, the method shown in Figure 6P-38 should be used.*

Section 6N.14 <u>Two-Lane, Two-Way Traffic on One Roadway of a Normally-Divided Highway</u>

Support:

01   Two-lane, two-way operation on one roadway of a normally-divided highway is a typical procedure that requires special consideration in the planning, design, and work phases, because unique operational problems (for example, increasing the risk of head-on crashes) can arise with the two-lane, two-way operation.

Standard:

02   **When two-lane, two-way traffic control must be maintained on one roadway of a normally-divided highway, opposing vehicular traffic shall be separated with either temporary traffic barriers (concrete safety-shape or approved alternate), channelizing devices, Narrow Two-Way Traffic (W6-4) signs on flexible supports (see Section 6H.17), or a temporary raised island throughout the length of the two-way operation. The use of markings and complementary signing, by themselves, shall not be used.**

Support:

03   Figure 6P-39 shows the procedure for two-lane, two-way operation. Treatments for entrance and exit ramps within the two-way roadway segment of this type of work are shown in Figures 6P-40 and 6P-41.

Section 6N.15 <u>Crossovers</u>

Guidance:

01   *The following are considered good guiding principles for the design of crossovers:*

    A. *Tapers for lane drops should be separated from the crossovers, as shown in Figure 6P-39.*

    B. *Crossovers should be designed for speeds no lower than 10 mph below the posted speed, the off-peak 85th-percentile speed prior to the work starting, or the anticipated operating speed of the roadway, unless unusual site conditions require that a lower design speed be used.*

    C. *A good array of channelizing devices, delineators, and full-length, properly placed pavement markings should be used to provide drivers with a clearly defined travel path.*

    D. *The design of the crossover should accommodate all vehicular traffic, including trucks and buses.*

Support:

02   Temporary traffic barriers and the excessive use of TTC devices cannot compensate for poor geometric and roadway cross-section design of crossovers.

Section 6N.16 Interchanges

Guidance:

01   *Access to interchange ramps on limited-access highways should be maintained even if the work space is in the lane adjacent to the ramps. Access to exit ramps should be clearly marked and delineated with channelizing devices. For long-term projects, conflicting pavement markings should be removed and new ones placed. Early coordination with officials having jurisdiction over the affected cross streets and providing emergency services should occur before ramp closings.*

Option:

02   If access is not possible, ramps may be closed by using signs and Type 3 Barricades. As the work space changes, the access area may be changed, as shown in Figure 6P-42. A TTC zone in the exit ramp may be handled as shown in Figure 6P-43.

03   When a work space interferes with an entrance ramp, a lane may need to be closed on the freeway (see Figure 6P-44). A TTC zone in the entrance ramp may require shifting ramp vehicular traffic (see Figure 6P-44).

Section 6N.17 Work in the Vicinity of a Grade Crossing

Standard:

01   **When grade crossings exist either within or in the vicinity of a TTC zone, lane restrictions, flagging, or other operations shall not create conditions where vehicles can be queued across the tracks. If the queuing of vehicles across the tracks cannot be avoided, a uniformed law enforcement officer or flagger shall be provided at the crossing to prevent vehicles from stopping on the tracks, even if automatic warning devices are in place.**

Support:

02   Figure 6P-46 shows work in the vicinity of a grade crossing.

03   Section 8A.13 contains additional information regarding TTC zones in the vicinity of grade crossings.

Guidance:

04   *Early coordination with the railroad company or transit agency should occur before work starts.*

Section 6N.18 Work during Nighttime Hours

Support:

01   Section 6A.05 contains additional information regarding considerations for conducting work operations during nighttime hours.

Guidance:

02   *Considering the safety issues inherent to night work, consideration should be given to enhancing traffic controls (see Section 6N.03) to provide added visibility and driver guidance, and increased protection for workers.*

03   *In addition to the enhancements listed in Section 6N.03, consideration should be given to providing additional lights and retroreflective markings to workers, work vehicles, and equipment.*

Option:

04   Where reduced traffic volumes at night make it feasible, the entire roadway may be closed by detouring traffic to alternate facilities, thus removing the traffic risk from the activity area.

Guidance:

05   *Consideration should be given to stationing uniformed law enforcement officers and lighted patrol cars at night work locations where there is a concern that high speeds or impaired drivers might result in undue risks for workers or other drivers.*

Standard:

06   **Except in emergencies, temporary lighting shall be provided at all flagger stations used during nighttime work.**

Support:

07   Desired illumination levels vary depending upon the nature of the task involved. An average horizontal luminance of 5 foot candles can be adequate for general activities. An average horizontal luminance of 10 foot candles can be adequate for activities around equipment. Tasks requiring high levels of precision and extreme care can require an average horizontal luminance of 20 foot candles.

Section 6N.19 <u>Late Merge</u>

Support:

01   The Late Merge is designed to use all available lanes until the merge point is reached at the lane closure taper rather than merging as soon as possible into the open lane. The Late Merge addresses many of the challenges that are associated with traffic operations in advance of lane closures at TTC zones such as queue length, capacity, and driver satisfaction.

Option:

02   Late Merge systems may consist of static or portable changeable message signs.

Guidance:

03   *Static Late Merge signing should consist of the STAY IN LANE TO MERGE POINT (R9-4a) sign and the MERGE HERE TAKE TURNS (W9-2a) sign (see Figure 6N-1).*

Option:

04   The following messages may be used on changeable message signs at an upstream location during the Late Merge application:

    A. "STAY IN YOUR LANE/MERGE AHEAD"
    B. "STAY IN YOUR LANE/MERGE AHEAD XX MILES"
    C. "USE BOTH LANES/TO MERGE POINT"
    D. "USE BOTH LANES/STOPPED TRAFFIC AHEAD"
    E. "SLOW TRAFFIC AHEAD/USE BOTH LANES"

05   The following messages are typically used on changeable message signs at the merge point during the Late Merge application:

    A. "TAKE YOUR TURN/MERGE HERE"
    B. "MERGE HERE/TAKE TURNS"

Figure 6N-1. Late Merge

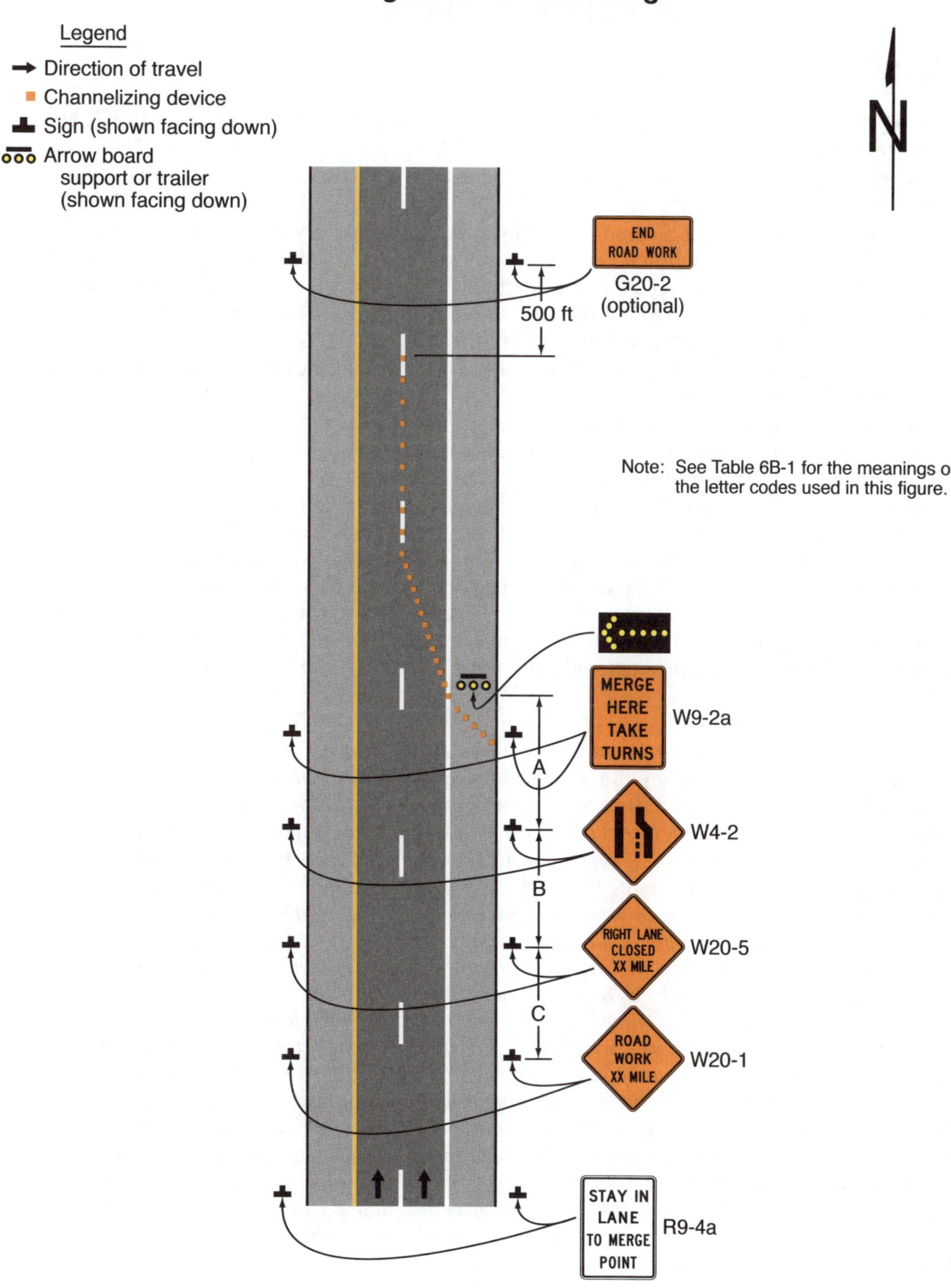

Late Merge Signing

CHAPTER 6O. CONTROL OF TRAFFIC THROUGH TRAFFIC INCIDENT MANAGEMENT AREAS

Section 6O.01 General

Support:

01   The National Incident Management System (NIMS) requires the use of the Incident Command System (ICS) at traffic incident management scenes.

02   A traffic incident is an emergency road user occurrence, a natural disaster, or other unplanned event that affects or impedes the normal flow of traffic.

03   A traffic incident management area is an area of a highway where temporary traffic controls are installed, as authorized by a public authority or the official having jurisdiction of the roadway, in response to a road user incident, natural disaster, hazardous material spill, or other unplanned incident. It is a type of TTC zone and extends from the first warning device (such as a sign, light, or cone) to the last TTC device or to a point where vehicles return to the original lane alignment and are clear of the incident.

04   Traffic incidents can be divided into three general classes of duration, each of which has unique traffic control characteristics and needs. These classes are:

    A. Major—expected duration of more than 2 hours,
    B. Intermediate—expected duration of 30 minutes to 2 hours, and
    C. Minor—expected duration under 30 minutes.

05   The primary functions of TTC at a traffic incident management area are to inform road users of the incident and to provide guidance information on the path to follow through the incident area. Alerting road users and establishing a well-defined path to guide road users through the incident area will serve to protect the incident responders and those involved in working at the incident scene and will aid in moving road users expeditiously past or around the traffic incident, will reduce the likelihood of secondary traffic crashes, and will preclude unnecessary use of the surrounding local road system. Examples include a stalled vehicle blocking a lane, a traffic crash blocking the traveled way, a hazardous material spill along a highway, and natural disasters such as floods and severe storm damage.

Guidance:

06   *In order to reduce response time for traffic incidents, highway agencies, appropriate public safety agencies (law enforcement, fire and rescue, emergency communications, emergency medical, and other emergency management), and private sector responders (towing and recovery and hazardous materials contractors) should mutually plan for occurrences of traffic incidents along the major and heavily traveled highway and street system.*

07   *On-scene responder organizations should train their personnel in TTC practices for accomplishing their tasks in and near traffic and in the requirements for traffic incident management contained in this Manual. On-scene responders should take measures to move the incident off the traveled roadway or to provide for appropriate warning. All on-scene responders and news media personnel should constantly be aware of their visibility to oncoming traffic and wear high-visibility apparel. Planning and training should include incorporation of estimated time durations to clear the event as part of their initial incident estimate. When events are deemed as probable Major Traffic Incidents that could generate prolonged lane or road closures, notification of all affected agencies should be initiated as part of the initial incident report that is provided to the emergency communications center who would then be responsible for making notifications to appropriate state, regional, and local agencies and resources for the purpose of ramping up and responding as quickly as possible thus facilitating a more rapid transition from emergency TTC to an MUTCD-compliant TTC zone when warranted.*

08   *Emergency vehicles arriving at an incident should be positioned in a manner that attempts to protect both the responders performing their duties and road users traveling through the incident scene, while minimizing, to the extent practical, disruption of the adjacent traffic flow. Emergency vehicle positions should optimize traffic flow through the incident scene. All emergency vehicles that subsequently arrive should be positioned in a manner that does not interfere with the established temporary traffic flow.*

09   *Responders arriving at a traffic incident should estimate the magnitude of the traffic incident, the expected time duration of the traffic incident, and the expected vehicle queue length, and then should set up the appropriate temporary traffic controls for these estimates.*

Option:

10   Warning and guide signs used for TTC traffic incident management situations may have a black legend and border on a fluorescent pink background (see Figure 6O-1).

Support:

11   While some traffic incidents might be anticipated and planned for, emergencies and disasters might pose more severe and unpredictable problems. The ability to quickly install proper temporary traffic controls might greatly reduce the effects of an incident, such as secondary crashes or excessive traffic delays. An essential part of fire, rescue, spill clean-up, highway agency, and enforcement activities is the proper control of road users through the

Figure 6O-1. Examples of Traffic Incident Management Area Signs

traffic incident management area in order to protect responders, victims, and other personnel at the site. These operations might need corroborating legislative authority for the implementation and enforcement of appropriate road user regulations, parking controls, and speed zoning. It is desirable for these statutes to provide sufficient flexibility in the authority for, and implementation of, TTC to respond to the needs of changing conditions found in traffic incident management areas.

Option:

12   For traffic incidents, particularly those of an emergency nature, TTC devices on hand may be used for the initial response as long as they do not themselves create unnecessary additional hazards.

Support:

13   The establishment, maintenance, and prompt removal of lane diversions can be effectively managed by interagency planning that includes representatives of highway and public safety agencies.

Guidance:

14   *All traffic control devices needed to set up the TTC at a traffic incident should be available so that they can be readily deployed for all major traffic incidents. The TTC should include the proper traffic diversions, tapered lane closures, and upstream warning devices to alert traffic approaching the queue and to encourage early diversion to an appropriate alternative route.*

15   *Attention should be paid to the upstream end of the traffic queue such that warning is given to road users approaching the back of the queue.*

16   *If manual traffic control is needed, it should be provided by qualified flaggers or uniformed law enforcement officers.*

Option:

17   If flaggers are used to provide traffic control for an incident management situation, the flaggers may use appropriate traffic control devices that are readily available or that can be brought to the traffic incident scene on short notice.

Guidance:

18   *When light sticks or flares are used to establish the initial traffic control at incident scenes, channelizing devices (see Section 6K.01) should be installed as soon thereafter as practical.*

Option:

19   The light sticks or flares may remain in place if they are being used to supplement the channelizing devices.

Guidance:

20   *The light sticks, flares, and channelizing devices should be removed after the incident is terminated.*

Section 6O.02 Major Traffic Incidents

Support:

01   Major traffic incidents are typically traffic incidents involving hazardous materials, fatal traffic crashes involving numerous vehicles, and other natural or man-made disasters. These traffic incidents typically involve closing all or part of a roadway facility for a period exceeding 2 hours.

Guidance:

02   *If the traffic incident is anticipated to last more than 24 hours, applicable procedures and devices set forth in other Chapters of Part 6 should be used.*

Support:

03   A road closure can be caused by a traffic incident such as a road user crash that blocks the traveled way. Road users are usually diverted through lane shifts or detoured around the traffic incident and back to the original roadway. A combination of traffic engineering and enforcement preparations is needed to determine the detour route, and to install, maintain, or operate, and then to remove the necessary traffic control devices when the detour is terminated. Large trucks are a significant concern in such a detour, especially when detouring them from a controlled-access roadway onto local or arterial streets.

04   During traffic incidents, large trucks might need to follow a route separate from that of automobiles because of bridge, weight, clearance, or geometric restrictions. Also, vehicles carrying hazardous material might need to follow a different route from other vehicles.

05   Some traffic incidents such as hazardous material spills might require closure of an entire highway. Through road users must have adequate guidance around the traffic incident. Maintaining good public relations is desirable. The cooperation of the news media in publicizing the existence of, and reasons for, traffic incident management areas and their TTC can be of great assistance in keeping road users and the general public well informed.

Section 6O.03 Intermediate Traffic Incidents

Support:

01   Intermediate traffic incidents typically affect travel lanes for a time period of 30 minutes to 2 hours, and usually require traffic control on the scene to divert road users past the blockage. Full roadway closures might be needed for short periods during traffic incident clearance to allow traffic incident responders to accomplish their tasks.

Section 6O.04 Minor Traffic Incidents

Support:

01   Minor traffic incidents are typically disabled vehicles and minor crashes that result in lane closures of less than 30 minutes. On-scene responders are typically law enforcement and towing companies, and occasionally highway agency service patrol vehicles.

02   Diversion of traffic into other lanes is often not needed or is needed only briefly. It is not generally possible or practical to set up a lane closure with traffic control devices for a minor traffic incident. Traffic control is the responsibility of on-scene responders.

Guidance:

03   *When a minor traffic incident blocks a travel lane, the vehicles involved in the incident should be moved from the blocked lane to the shoulder as quickly as possible.*

Section 6O.05 Use of Emergency-Vehicle Lighting

Support:

01   The use of emergency-vehicle lighting (such as high-intensity rotating, flashing, oscillating, or strobe lights) is essential, especially in the initial stages of a traffic incident, for the safety of emergency responders and persons involved in the traffic incident, as well as road users approaching the traffic incident. Emergency-vehicle lighting, however, provides warning only and provides no effective traffic control. The use of too many lights at an incident scene can be distracting and can create confusion for approaching road users, especially at night. Road users approaching the traffic incident from the opposite direction on a divided facility are often distracted by emergency-vehicle lighting and slow their vehicles to look at the traffic incident posing a hazard to themselves and others traveling in their direction.

02   The use of emergency-vehicle lighting can be reduced if good traffic control has been established at a traffic incident scene. This is especially true for major traffic incidents that might involve a number of emergency vehicles. If good traffic control is established through placement of advance warning signs and traffic control devices to divert or detour traffic, then public safety agencies can perform their tasks on scene with minimal emergency-vehicle lighting.

Guidance:

03   *Public safety agencies should examine their policies on the use of emergency-vehicle lighting, especially after a traffic incident scene is secured, with the intent of reducing the use of this lighting as much as possible while not endangering those at the scene. Special consideration should be given to reducing or extinguishing forward facing emergency-vehicle lighting, especially on divided roadways, to reduce distractions to oncoming road users.*

04   *Because the glare from floodlights or vehicle headlights can impair the nighttime vision of approaching road users, any floodlights or vehicle headlights that are not needed for illumination, or to provide notice to other road users of an incident response vehicle being in an unexpected location, should be turned off at night.*

CHAPTER 6P. TYPICAL APPLICATIONS

Section 6P.01 Typical Applications

Support:

01 Chapter 6N contains discussions of typical TTC activities. Section 6A.02 contains discussions on development of TTC plans for the various activities. This Chapter presents typical applications for a variety of situations commonly encountered. While not every situation is addressed, the information illustrated can generally be adapted to a broad range of conditions. In many instances, an appropriate TTC plan is achieved by combining features from various typical applications. For example, work at an intersection might present a near-side TTC zone for one street and a far-side TTC zone for the other street. These treatments are found in two different typical applications, while a third typical application shows how to handle pedestrian crosswalk closures.

02 In general, the procedures illustrated represent minimum solutions for the situations depicted. Except for the notes (which are clearly classified using headings as being Standard, Guidance, Option, or Support), the information presented in the typical applications can generally be regarded as Guidance.

Option:

03 TTC plans may deviate from the typical applications described in this Chapter to allow for conditions and requirements of a particular site or jurisdiction.

04 Other devices may be added to supplement the devices and device spacing may be adjusted to provide additional reaction time or delineation. Fewer devices may be used based on field conditions.

Support:

05 Figures and tables found throughout Part 6 provide information for the development of TTC plans.

06 Table 6P-1 is an index of the 54 typical applications. In the printed version, the typical applications are shown on the right-hand page with notes on the facing page to the left. In the electronic version, the notes are shown on the page preceding the figure. The legend for the symbols used in the typical applications is provided in Table 6P-2. In many of the typical applications, sign spacings and other dimensions are indicated by letters using the criteria provided in Table 6B-1. The formulas for determining taper lengths are provided in Table 6B-4.

07 Most of the typical applications show TTC devices for only one direction.

Table 6P-1. Index to Typical Applications (Sheet 1 of 2)

Typical Application Description	Typical Application Number
Work Outside of the Shoulder (see Section 6N.05)	
Work Beyond the Shoulder	TA-1
Blasting Zone	TA-2
Work on the Shoulder (see Sections 6N.06 and 6N.07)	
Work on the Shoulders	TA-3
Short-Duration or Mobile Operation on a Shoulder	TA-4
Shoulder Closure on a Freeway	TA-5
Shoulder Work with Minor Encroachment	TA-6
Work within the Traveled Way of a Two-Lane Highway (see Section 6N.09)	
Road Closed with a Diversion	TA-7
Roads Closed with an Off-Site Detour	TA-8
Overlapping Routes with a Detour	TA-9
Lane Closure on a Two-Lane Road Using Flaggers	TA-10
Lane Closure on a Two-Lane Road with Low Traffic Volumes	TA-11
Lane Closure on a Two-Lane Road Using Traffic Control Signals	TA-12
Temporary Road Closure	TA-13
Haul Road Crossing	TA-14
Work in the Center of a Road with Low Traffic Volumes	TA-15
Surveying Along the Center Line of a Road with Low Traffic Volumes	TA-16
Mobile Operations on a Two-Lane Road	TA-17
Work within the Traveled Way of an Urban Street (see Section 6N.10)	
Lane Closure on a Minor Street	TA-18
Detour for One Travel Direction	TA-19
Detour for a Closed Street	TA-20
Work within the Traveled Way at an Intersection and on Sidewalks (see Section 6N.12)	
Lane Closure on the Near Side of an Intersection	TA-21
Right-Hand Lane Closure on the Far Side of an Intersection	TA-22
Left-Hand Lane Closure on the Far Side of an Intersection	TA-23
Half Road Closure on the Far Side of an Intersection	TA-24
Multiple Lane Closures at an Intersection	TA-25
Closure in the Center of an Intersection	TA-26
Closure at the Side of an Intersection	TA-27
Sidewalk Detour or Diversion	TA-28
Crosswalk Closures and Pedestrian Detours	TA-29
Work within the Traveled Way of a Multi-Lane, Non-Access Controlled Highway (see Section 6N.11)	
Interior Lane Closure on a Multi-Lane Street	TA-30
Lane Closure on a Street with Uneven Directional Volumes	TA-31
Half Road Closure on a Multi-Lane, High-Speed Highway	TA-32
Stationary Lane Closure on a Divided Highway	TA-33
Lane Closure with a Temporary Traffic Barrier	TA-34
Mobile Operation on a Multi-Lane Road	TA-35

Table 6P-1. Index to Typical Applications (Sheet 2 of 2)

Typical Application Description	Typical Application Number
Work within the Traveled Way of a Freeway or Expressway (see Section 6N.13)	
Lane Shift on a Freeway	TA-36
Double Lane Closure on a Freeway	TA-37
Interior Lane Closure on a Freeway	TA-38
Median Crossover on a Freeway	TA-39
Median Crossover for an Entrance Ramp	TA-40
Median Crossover for an Exit Ramp	TA-41
Work in the Vicinity of an Exit Ramp	TA-42
Partial Exit Ramp Closure	TA-43
Work in the Vicinity of an Entrance Ramp	TA-44
Temporary Reversible Lane Using Movable Barriers	TA-45
Work in the Vicinity of a Grade Crossing (see Section 6N.17)	
Work in the Vicinity of a Grade Crossing	TA-46
Work in the Vicinity of Bicycle Lanes and Shared Use Paths (see Section 6N.04)	
Bicycle Lane Closure without a Detour	TA-47
Bicycle Lane Closure with an On-Road Detour	TA-48
Shared-Use Path Closure with a Diversion	TA-49
On-Road Detour for a Shared-Use Path	TA-50
Paved Shoulder Closure with a Bicycle Diversion onto a Temporary Path	TA-51
Work in the Traveled Way of Roundabouts	
Short-Term or Short-Duration Work in a Circular Intersection	TA-52
Flagging Operation on a Single-Lane Circular Intersection	TA-53
Inside Lane Closure on a Multi-Lane Circular Intersection	TA-54

Table 6P-2. Meaning of Symbols on Typical Application Diagrams

Symbol	Meaning	Symbol	Meaning
	Arrow board		Shadow vehicle
	Arrow board support or trailer (shown facing down)		Sign (shown facing left)
	Changeable message sign or support trailer		Surveyor
	Channelizing device		Temporary barrier
	Crash cushion		Temporary barrier with warning light
	Direction of temporary traffic detour		Traffic or pedestrian signal
	Direction of travel		Truck-mounted attenuator
	Flagger		Type 3 barricade
	High-level warning device (Flag tree)		Warning light
	Longitudinal channelizing device		Work space
	Luminaire		Work vehicle
	Pavement markings that should be removed for a long-term project		

Notes for Figure 6P-1—Typical Application 1
Work Beyond the Shoulder

Guidance:

1. *If the work space is in the median of a divided highway, an advance warning sign should also be placed on the left-hand side of the directional roadway.*

Option:

2. The ROAD WORK AHEAD sign may be replaced with other appropriate signs such as the SHOULDER WORK sign. The SHOULDER WORK sign may be used for work adjacent to the shoulder.
3. The ROAD WORK AHEAD sign may be omitted where the work space is behind a barrier, more than 24 inches behind the curb, or 15 feet or more from the edge of any roadway.
4. For short-term, short-duration or mobile operation, all signs and channelizing devices may be eliminated if a vehicle with activated high-intensity rotating, flashing, oscillating, or strobe lights is used.
5. Vehicle hazard warning signals may be used to supplement high-intensity rotating, flashing, oscillating, or strobe lights.

Standard:

6. **Vehicle hazard warning signals shall not be used instead of the vehicle's high-intensity rotating, flashing, oscillating, or strobe lights.**

Figure 6P-1. Work Beyond the Shoulder (TA-1)

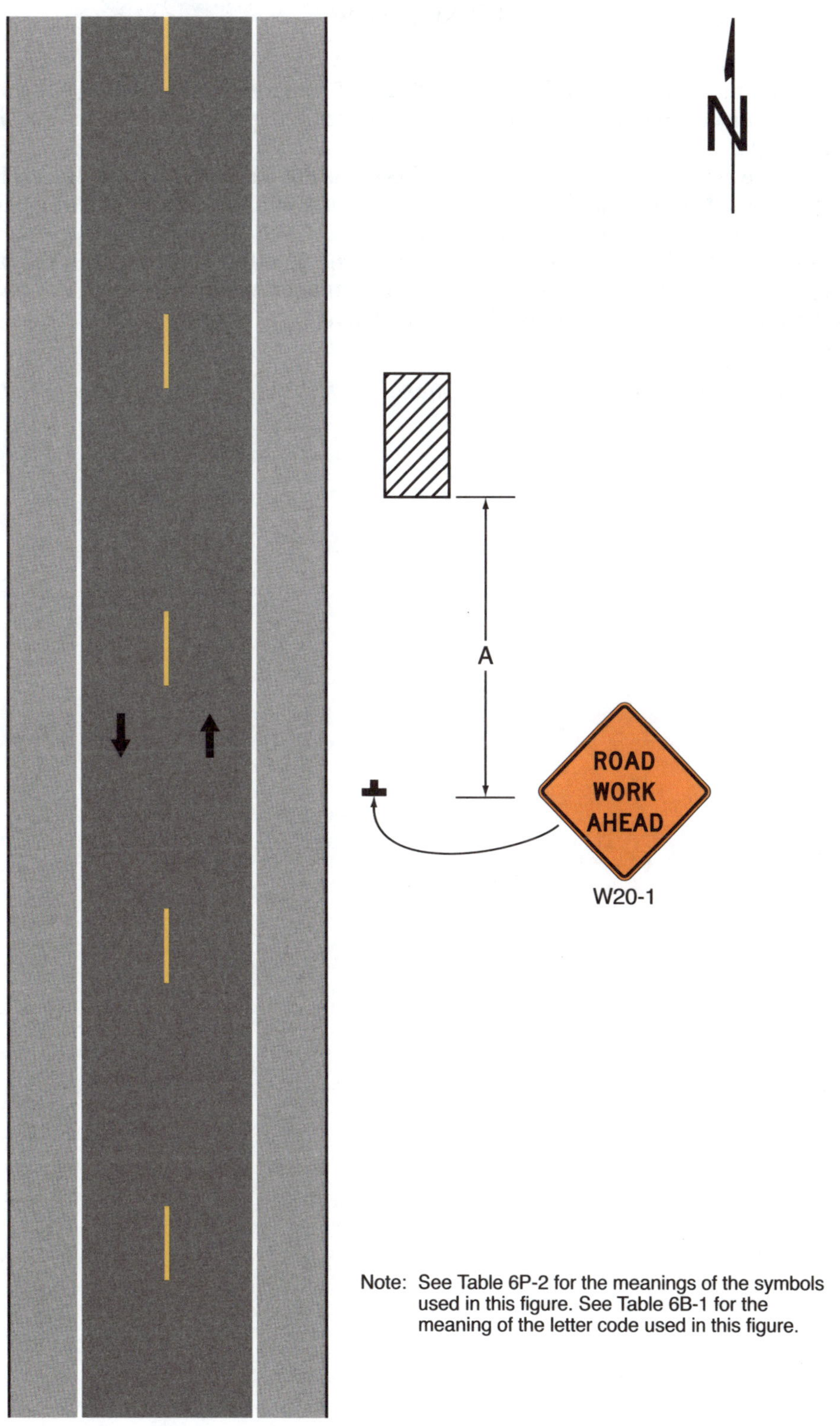

Note: See Table 6P-2 for the meanings of the symbols
used in this figure. See Table 6B-1 for the
meaning of the letter code used in this figure.

Typical Application 1

Notes for Figure 6P-2—Typical Application 2
Blasting Zone

Standard:

1. **Whenever blasting caps are used within 1,000 feet of a roadway, the signing shown shall be used.**
2. **The signs shall be covered or removed when there are no explosives in the area or the area is otherwise secure.**
3. **Whenever a side road intersects the roadway between the BLASTING ZONE AHEAD sign and the END BLASTING ZONE sign, or a side road is within 1,000 feet of any blasting cap, similar signing, as on the mainline, shall be installed on the side road.**
4. **Prior to blasting, the blaster in charge shall determine whether road users in the blasting zone will be endangered by the blasting operation. If there is danger, road users shall not be permitted to pass through the blasting zone during blasting operations.**

Guidance:

5. *On a divided highway, the signs should be mounted on both sides of the directional roadways.*

Figure 6P-2. Blasting Zone (TA-2)

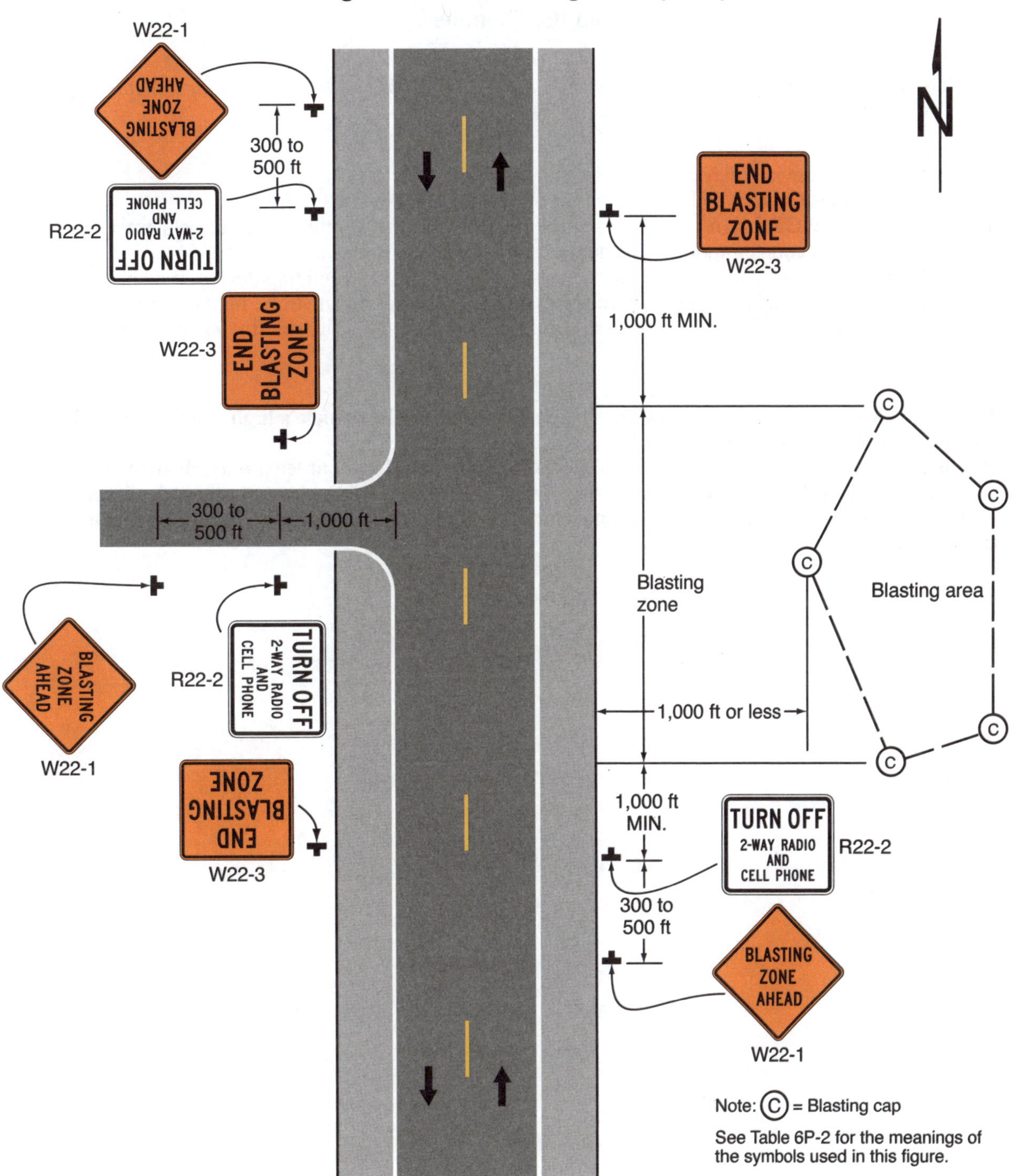

Typical Application 2

Notes for Figure 6P-3—Typical Application 3
Work on the Shoulders

Guidance:

1. *A SHOULDER WORK sign should be placed on the left-hand side of the roadway for a divided or one-way street only if the left-hand shoulder is affected.*

Option:

2. Positive protection devices may be used per Section 6M.02.
3. The Workers symbol signs may be used instead of SHOULDER WORK signs.
4. The SHOULDER WORK sign on an intersecting roadway may be omitted where drivers emerging from that roadway will encounter another advance warning sign prior to this activity area.
5. For short-duration operations of 60 minutes or less, all signs and channelizing devices may be eliminated if a vehicle with activated high-intensity rotating, flashing, oscillating, or strobe lights is used.
6. Vehicle hazard warning signals may be used to supplement high-intensity rotating, flashing, oscillating, or strobe lights.

Standard:

7. **Vehicle hazard warning signals shall not be used instead of the vehicle's high-intensity rotating, flashing, oscillating, or strobe lights.**
8. **When paved shoulders having a width of 8 feet or more are closed, at least one advance warning sign shall be used. In addition, channelizing devices shall be used to close the shoulder in advance to delineate the beginning of the work space and to direct vehicular traffic to remain within the traveled way.**

Figure 6P-3. Work on the Shoulders (TA-3)

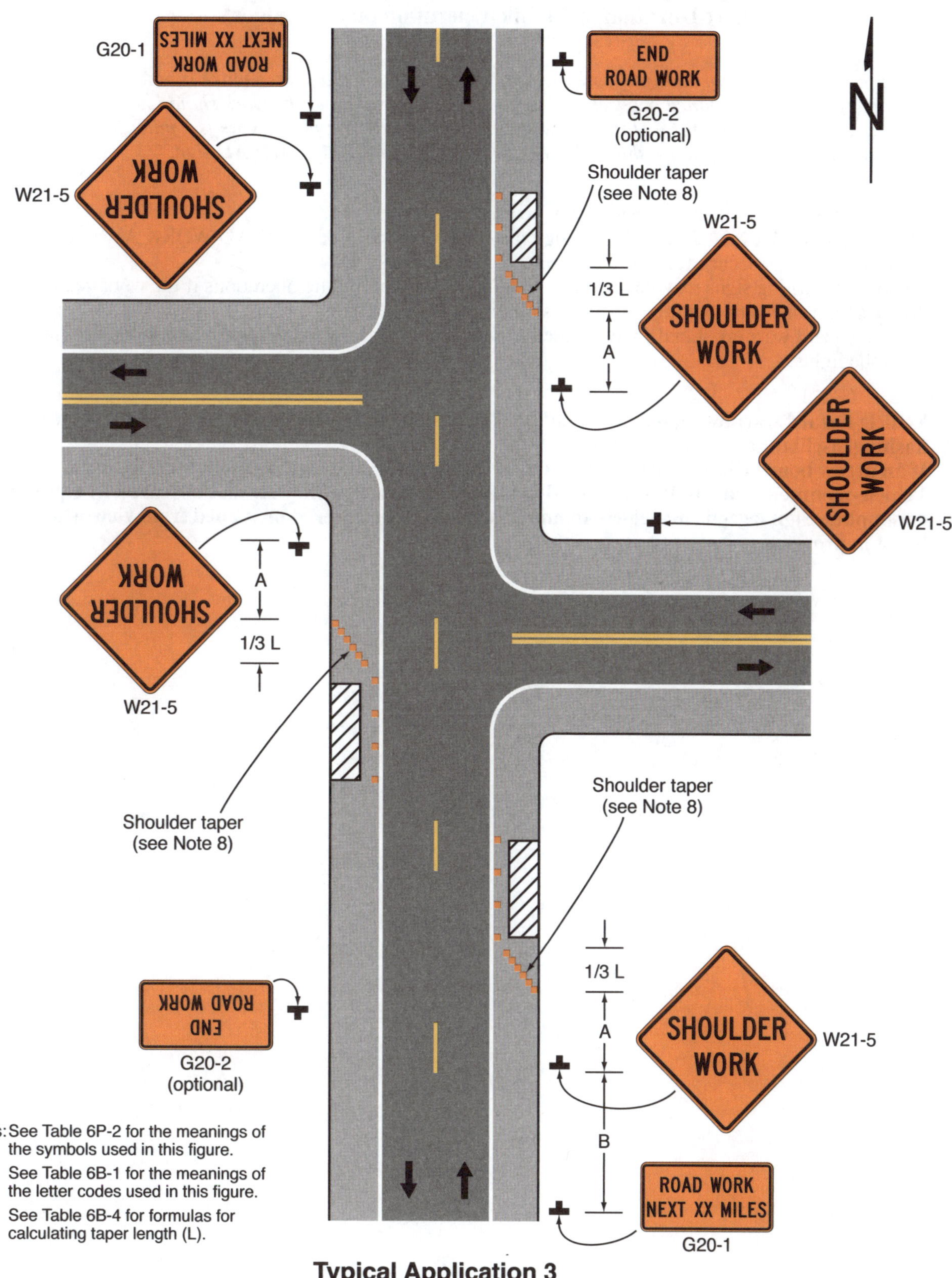

Notes: See Table 6P-2 for the meanings of the symbols used in this figure.

See Table 6B-1 for the meanings of the letter codes used in this figure.

See Table 6B-4 for formulas for calculating taper length (L).

Typical Application 3

Notes for Figure 6P-4—Typical Application 4
Short-Duration or Mobile Operation on a Shoulder

Guidance:

1. *In those situations where multiple work locations within a limited distance make it practicable to place stationary signs, the distance between the advance warning sign and the work should not exceed 5 miles.*
2. *In those situations where the distance between the advance signs and the work is 2 miles to 5 miles, a Supplemental Distance plaque should be used with the ROAD WORK AHEAD sign.*

Option:

3. Additional positive protection devices may be used per Section 6M.02.
4. The ROAD WORK NEXT XX MILES sign may be used instead of the ROAD WORK AHEAD sign if the work locations occur over a distance of more than 2 miles.
5. Stationary warning signs may be omitted for short-duration or mobile operations if the work vehicle displays high-intensity rotating, flashing, oscillating, or strobe lights.
6. Vehicle hazard warning signals may be used to supplement high-intensity rotating, flashing, oscillating, or strobe lights.

Standard:

7. **Vehicle hazard warning signals shall not be used instead of the vehicle's high-intensity rotating, flashing, oscillating, or strobe lights.**
8. **If an arrow board is used for an operation on the shoulder, the caution mode shall be used.**
9. **Vehicle-mounted signs shall be mounted in a manner such that they are not obscured by equipment or supplies. Sign legends on vehicle-mounted signs shall be covered or turned from view when work is not in progress.**

Figure 6P-4. Short-Duration or Mobile Operation on a Shoulder (TA-4)

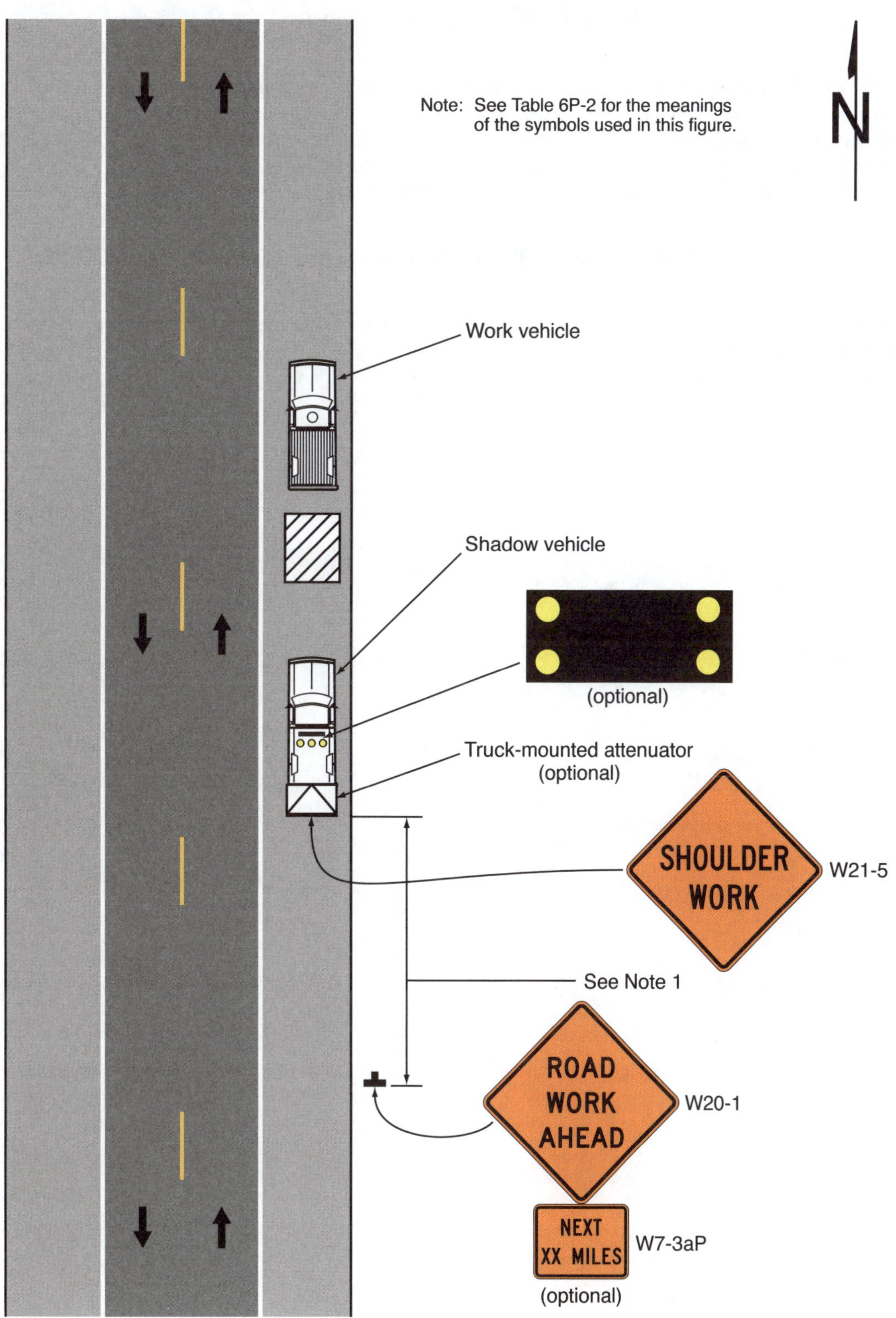

Typical Application 4

Notes for Figure 6P-5—Typical Application 5
Shoulder Closure on a Freeway

Guidance:

  1. *RIGHT (LEFT) SHOULDER CLOSED signs should be used on limited-access highways where there is no opportunity for disabled vehicles to pull off the roadway.*
  2. *If drivers cannot see a pull-off area beyond the closed shoulder, information regarding the length of the shoulder closure should be provided in feet or miles, as appropriate.*
  3. *The use of a temporary traffic barrier should be based on engineering judgment.*

Standard:

  4. Temporary traffic barriers, if used, shall comply with the provisions of Section 6M.02.

Option:

  5. The barrier shown in this typical application is an example of one method that may be used to close a shoulder of a long-term project.
  6. The warning lights shown on the barrier may be used.

Figure 6P-5. Shoulder Closure on a Freeway (TA-5)

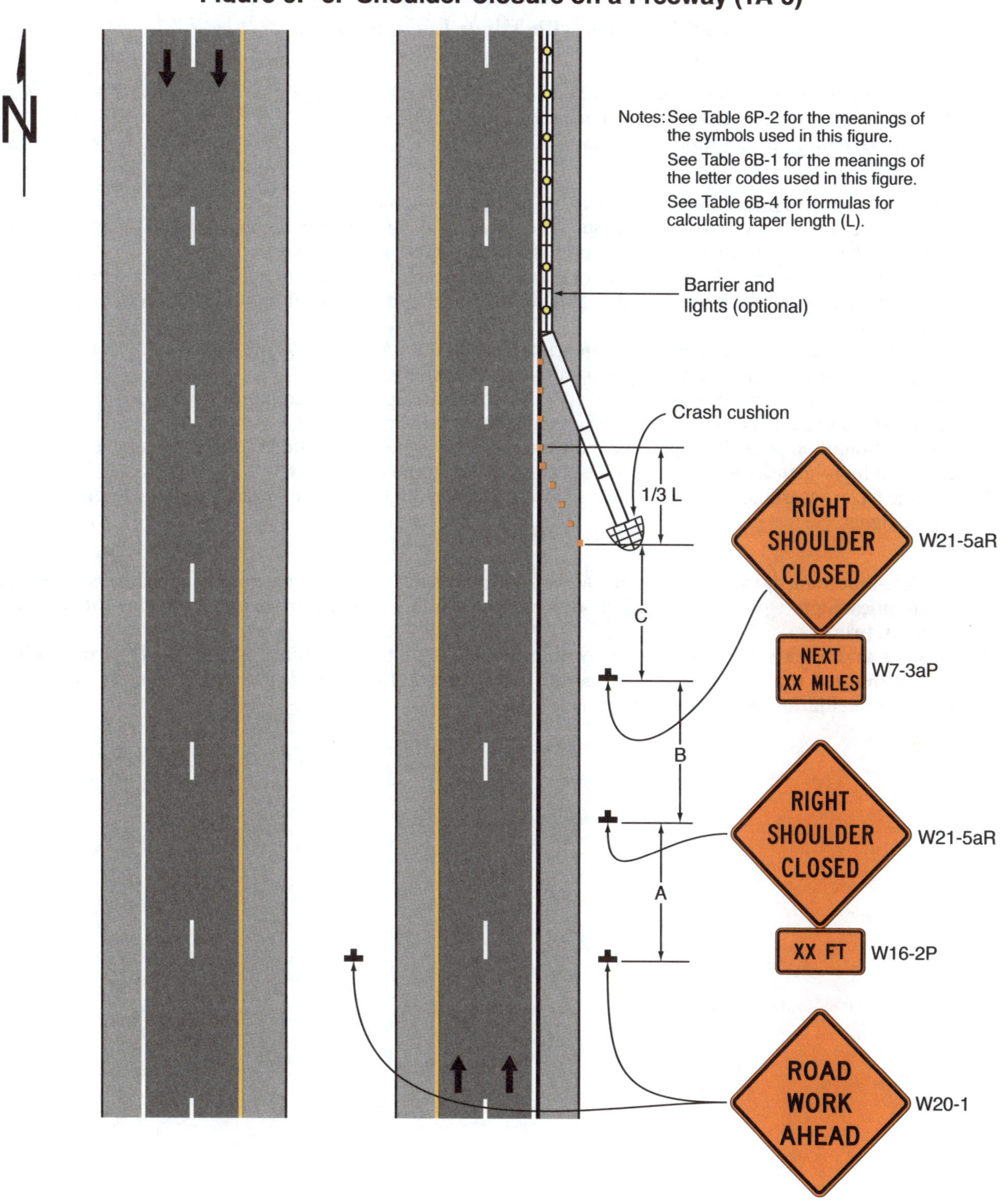

Typical Application 5

Notes for Figure 6P-6—Typical Application 6
Shoulder Work with Minor Encroachment

Guidance:
1. *All lanes should be a minimum of 10 feet in width as measured to the near face of the channelizing devices.*
2. *The treatment shown should be used on a minor road having low speeds. For higher-speed traffic conditions, a lane closure should be used.*

Option:
3. Additional positive protection devices may be used per Section 6M.02.
4. For short-term use on low-volume, low-speed roadways with vehicular traffic that does not include longer and wider heavy commercial vehicles, a minimum lane width of 9 feet may be used.
5. Where the opposite shoulder is suitable for carrying vehicular traffic and of adequate width, lanes may be shifted by use of closely-spaced channelizing devices, provided that the minimum lane width of 10 feet is maintained.
6. Additional advance warning may be appropriate, such as a ROAD NARROWS sign.
7. Temporary traffic barriers may be used along the work space.
8. The shadow vehicle may be omitted if a taper and channelizing devices are used.
9. A truck-mounted attenuator may be used on the shadow vehicle.
10. For short-duration work, the taper and channelizing devices may be omitted if a shadow vehicle with activated high-intensity rotating, flashing, oscillating, or strobe lights is used.
11. Vehicle hazard warning signals may be used to supplement high-intensity rotating, flashing, oscillating, or strobe lights.

Standard:
12. **Vehicle-mounted signs shall be mounted in a manner such that they are not obscured by equipment or supplies. Sign legends on vehicle-mounted signs shall be covered or turned from view when work is not in progress.**
13. **Shadow and work vehicles shall display high-intensity rotating, flashing, oscillating, or strobe lights.**
14. **Vehicle hazard warning signals shall not be used instead of the vehicle's high-intensity rotating, flashing, oscillating, or strobe lights.**

Figure 6P-6. Shoulder Work with Minor Encroachment (TA-6)

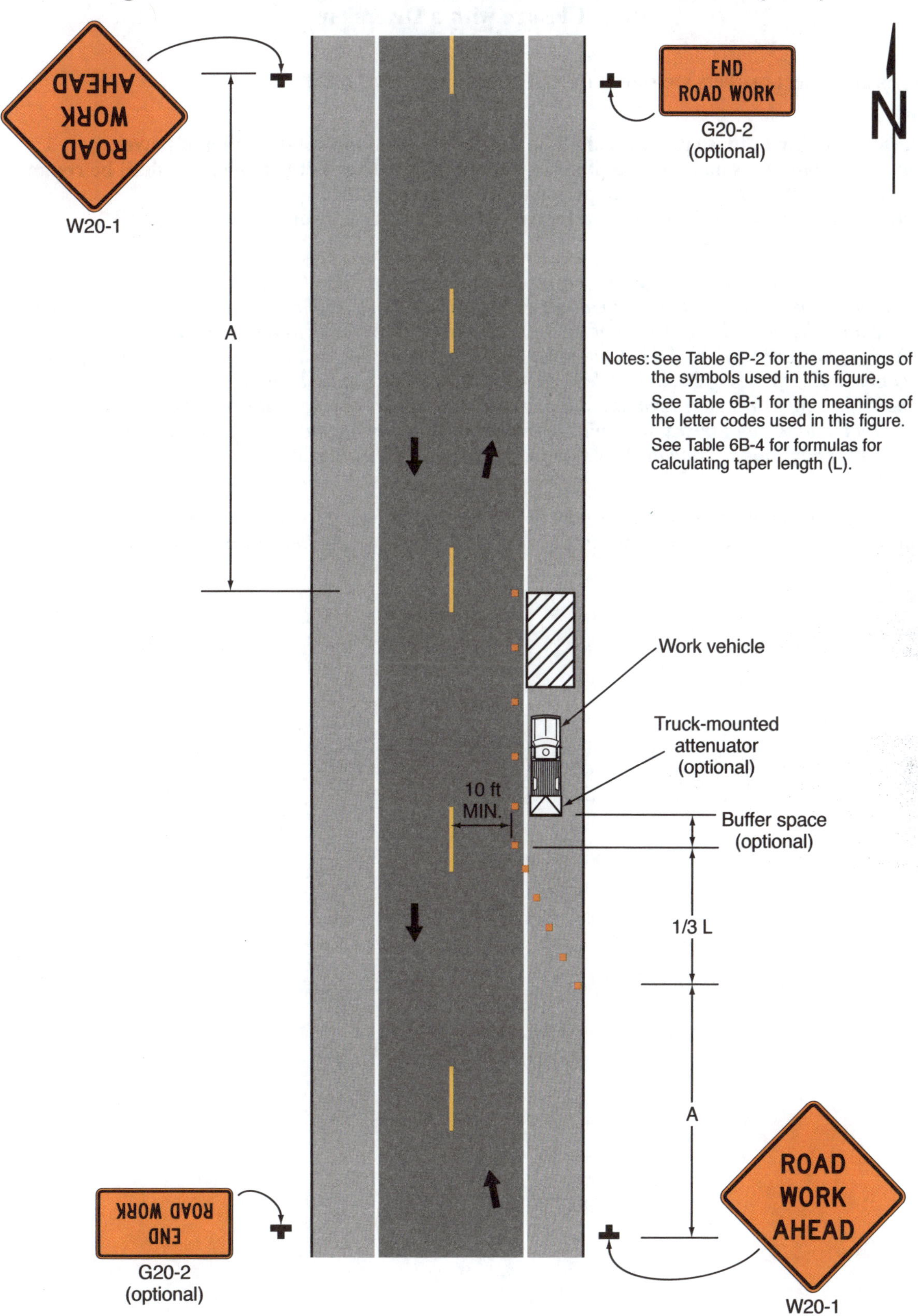

Typical Application 6

Notes for Figure 6P-7—Typical Application 7
Road Closure with a Diversion

Support:

1. Signs and object markers are shown for one direction of travel only.

Standard:

2. **Devices similar to those depicted shall be placed for the opposite direction of travel.**
3. **Pavement markings no longer applicable to the traffic pattern of the roadway shall be removed or obliterated before any new traffic patterns are open to traffic.**
4. **Temporary traffic barriers and end treatments shall be crashworthy.**

Guidance:

5. *If the tangent distance along the temporary diversion is more than 600 feet, a Reverse Curve sign, left first, should be used instead of the Double Reverse Curve sign, and a second Reverse Curve sign, right first, should be placed in advance of the second reverse curve back to the original alignment.*
6. *When the tangent section of the diversion is more than 600 feet, and the diversion has sharp curves with recommended speeds of 30 mph or less, Reverse Turn signs should be used.*
7. *Where the temporary pavement and old pavement are different colors, the temporary pavement should start on the tangent of the existing pavement and end on the tangent of the existing pavement.*
8. *Delineators or channelizing devices should be used along the diversion.*

Option:

9. Flashing warning lights and/or flags may be used to call attention to the warning signs.
10. On sharp curves, large arrow signs may be used in addition to other advance warning signs.

Figure 6P-7. Road Closure with a Diversion (TA-7)

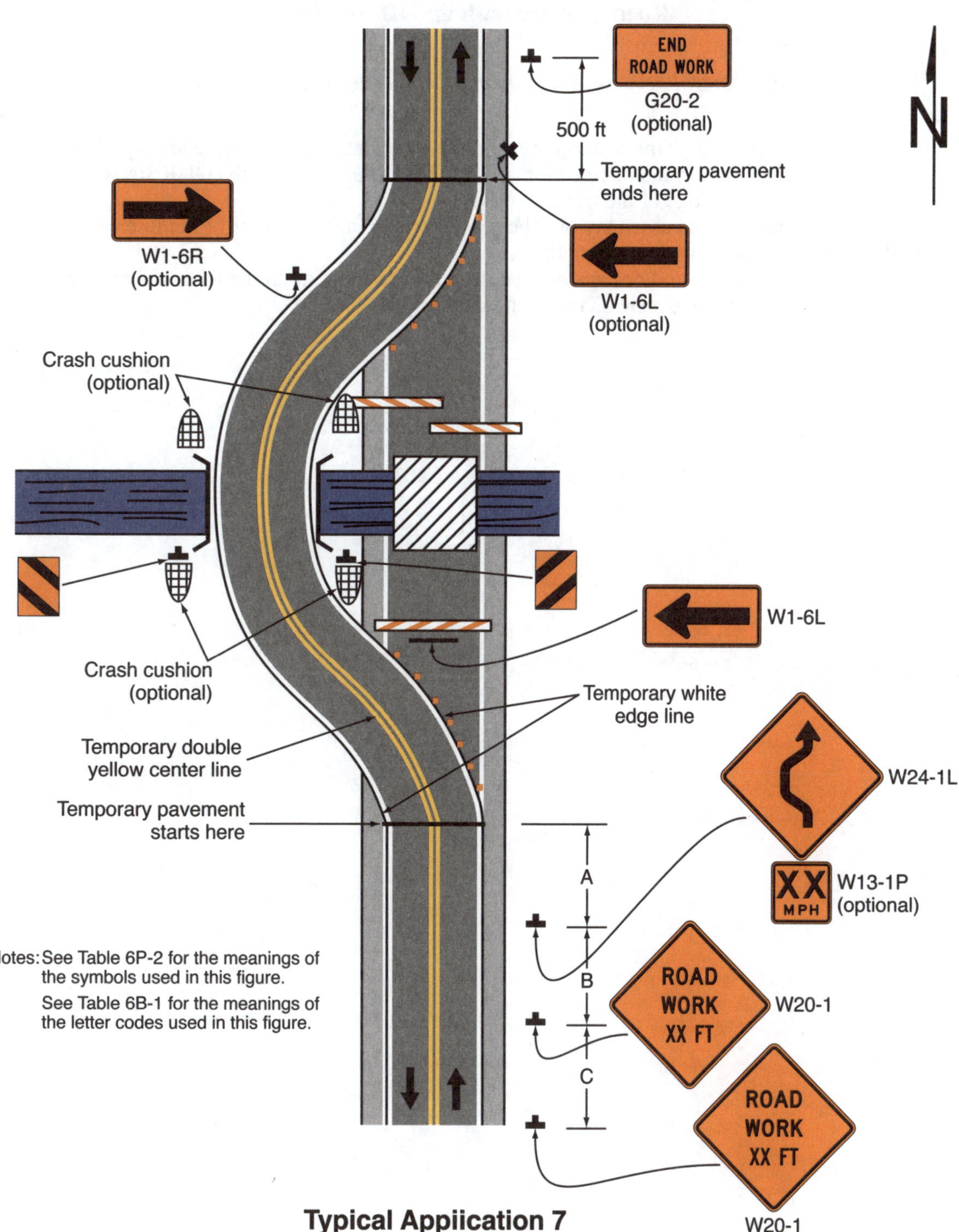

Typical Appiication 7

Notes for Figure 6P-8—Typical Application 8
Road Closure with an Off-Site Detour

Guidance:

1. *Regulatory traffic control devices should be modified as needed for the duration of the detour.*

Option:

2. If the road is opened for some distance beyond the intersection and/or there are significant origin/destination points beyond the intersection, the ROAD CLOSED and DETOUR signs on Type 3 Barricades may be located at the edge of the traveled way.
3. A Route Sign Directional assembly may be placed on the far left corner of the intersection to augment or replace the one shown on the near right corner.
4. Flashing warning lights and/or flags may be used to call attention to the advance warning signs.
5. Cardinal direction plaques may be used with route signs.

Figure 6P-8. Road Closure with an Off-Site Detour (TA-8)

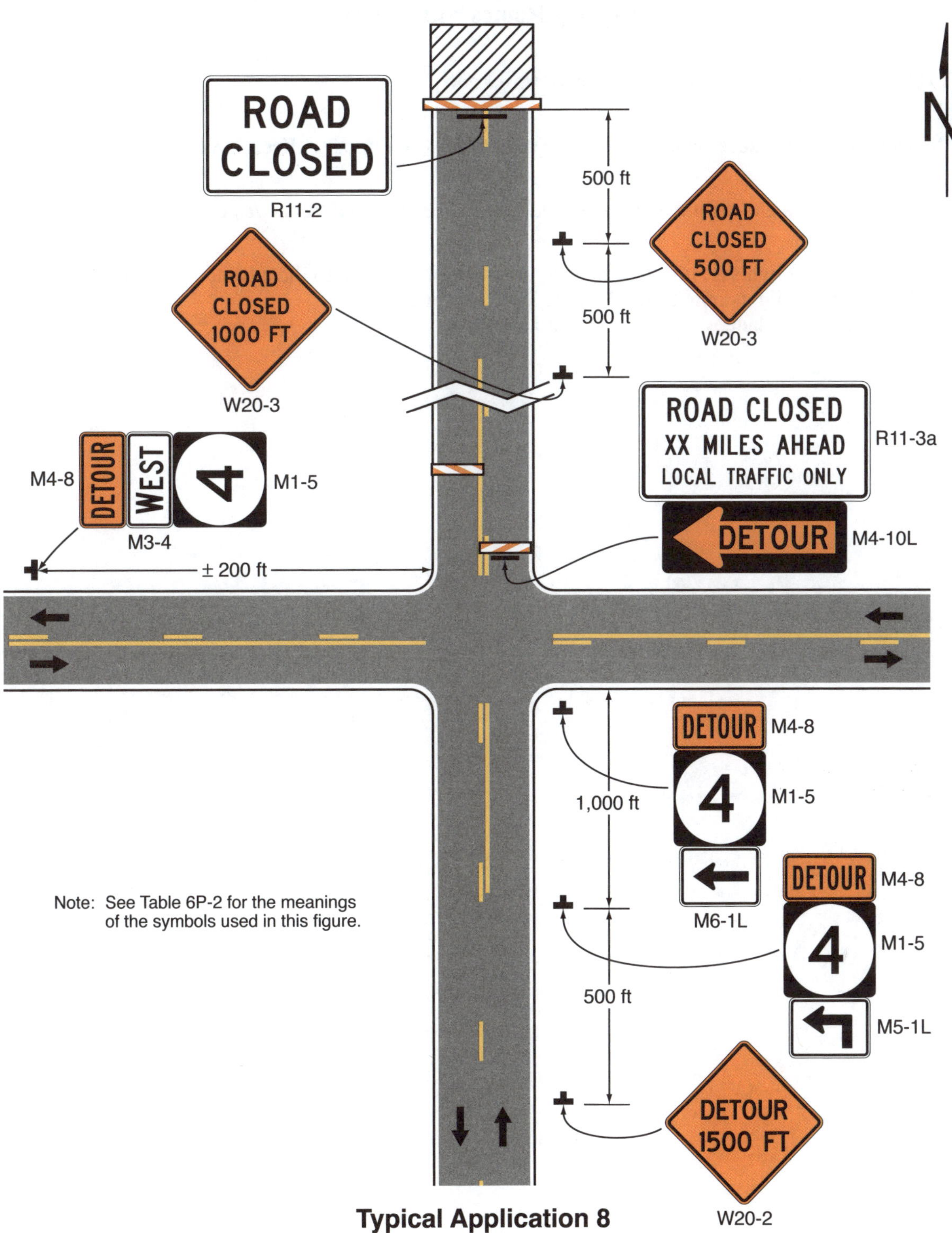

Typical Application 8

Notes for Figure 6P-9—Typical Application 9
Overlapping Routes with a Detour

Support:

1. TTC devices are shown for one direction of travel only.

Standard:

2. Devices similar to those depicted shall be placed for the opposite direction of travel.

Guidance:

3. *STOP or YIELD signs displayed to side roads should be installed as needed along the temporary route.*

Option:

4. Flashing warning lights and/or flags may be used to call attention to the advance warning signs.
5. Flashing warning lights may be used on the Type 3 Barricades.
6. Cardinal direction plaques may be used with route signs.

Figure 6P-9. Overlapping Routes with a Detour (TA-9)

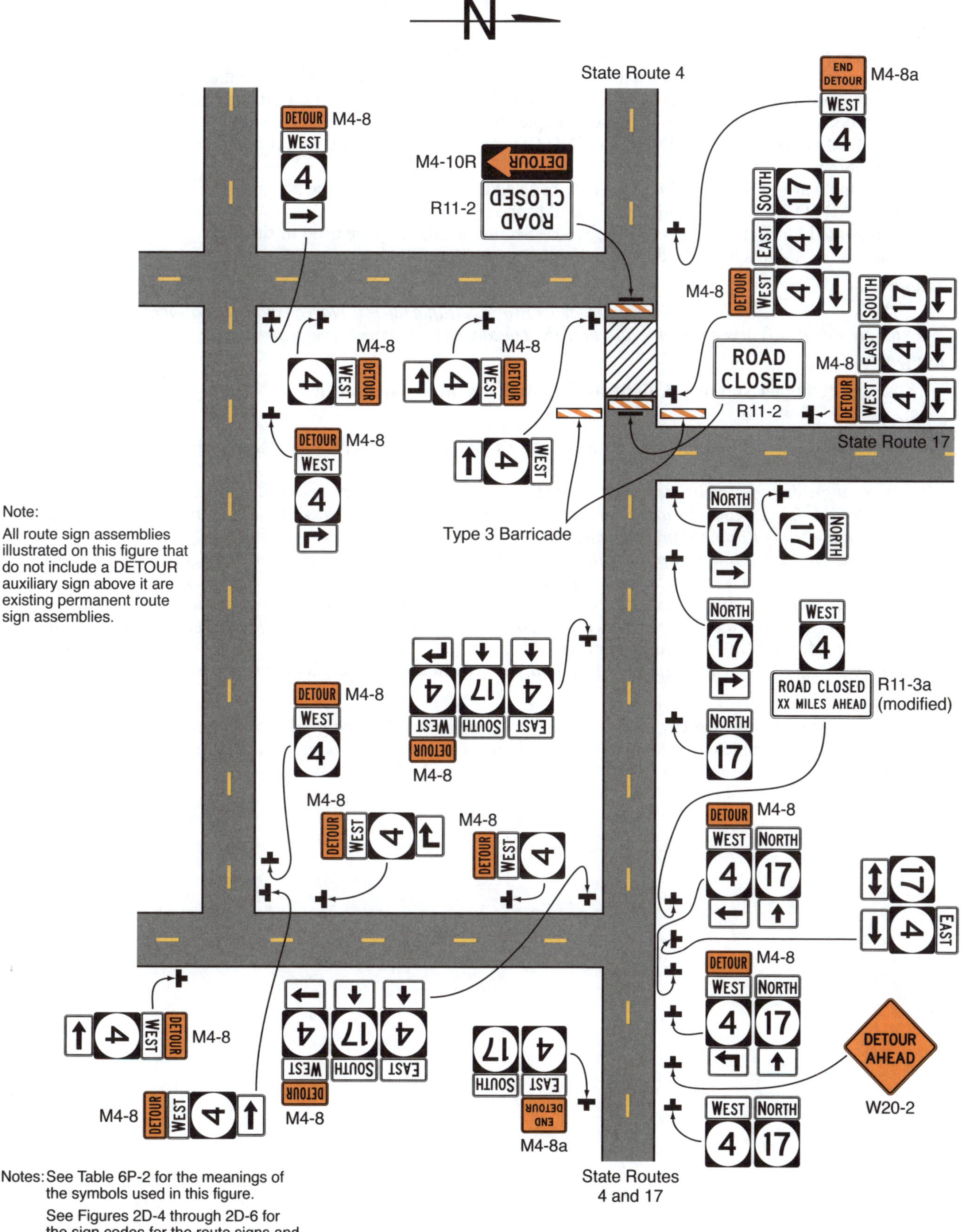

Notes: See Table 6P-2 for the meanings of the symbols used in this figure.

See Figures 2D-4 through 2D-6 for the sign codes for the route signs and the directional and arrow auxiliary signs associated with them.

Typical Application 9

Notes for Figure 6P-10—Typical Application 10
Lane Closure on a Two-Lane Road Using Flaggers

Option:

1. Positive protection devices may be used per Section 6M.02.
2. For low-volume situations with short TTC zones on straight roadways where the flagger is visible to road users approaching from both directions, a single flagger, positioned to be visible to road users approaching from both directions, may be used (see Chapter 6D).
3. The ROAD WORK AHEAD and the END ROAD WORK signs may be omitted for short-duration operations.
4. Flashing warning lights and/or flags may be used to call attention to the advance warning signs. A BE PREPARED TO STOP sign may be added to the sign series.
5. Automated Flagger Assistance Devices (see Section 6L.02) may be used in situations where there is only one lane of approaching traffic in the direction to be controlled.

Guidance:

6. *The buffer space should be extended so that the two-way traffic taper is placed before a horizontal (or crest vertical) curve to provide adequate sight distance for the flagger and a queue of stopped vehicles.*

Standard:

7. **At night, flagger stations shall be illuminated, except in emergencies.**

Guidance:

8. *When used, the BE PREPARED TO STOP sign should be located between the Flagger sign and the ONE LANE ROAD sign.*
9. *When a grade crossing exists within or upstream of the transition area and it is anticipated that queues resulting from the lane closure might extend through the grade crossing, the TTC zone should be extended so that the transition area precedes the grade crossing.*
10. *When a grade crossing equipped with active warning devices exists within the activity area, provisions should be made for keeping flaggers informed as to the activation status of these warning devices.*
11. *When a grade crossing exists within the activity area, drivers operating on the left-hand side of the normal center line should be provided with comparable warning devices as for drivers operating on the right-hand side of the normal center line.*
12. *Early coordination with the railroad company or transit agency should occur before work starts.*

Option:

13. A flagger or a uniformed law enforcement officer may be used at the grade crossing to minimize the probability that vehicles are stopped within 15 feet of the grade crossing, measured from both sides of the outside rails.

Figure 6P-10. Lane Closure on a Two-Lane Road Using Flaggers (TA-10)

Notes: See Table 6P-2 for the meanings of the symbols used in this figure.

See Table 6B-1 for the meanings of the letter codes used in this figure.

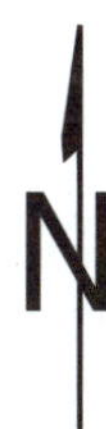

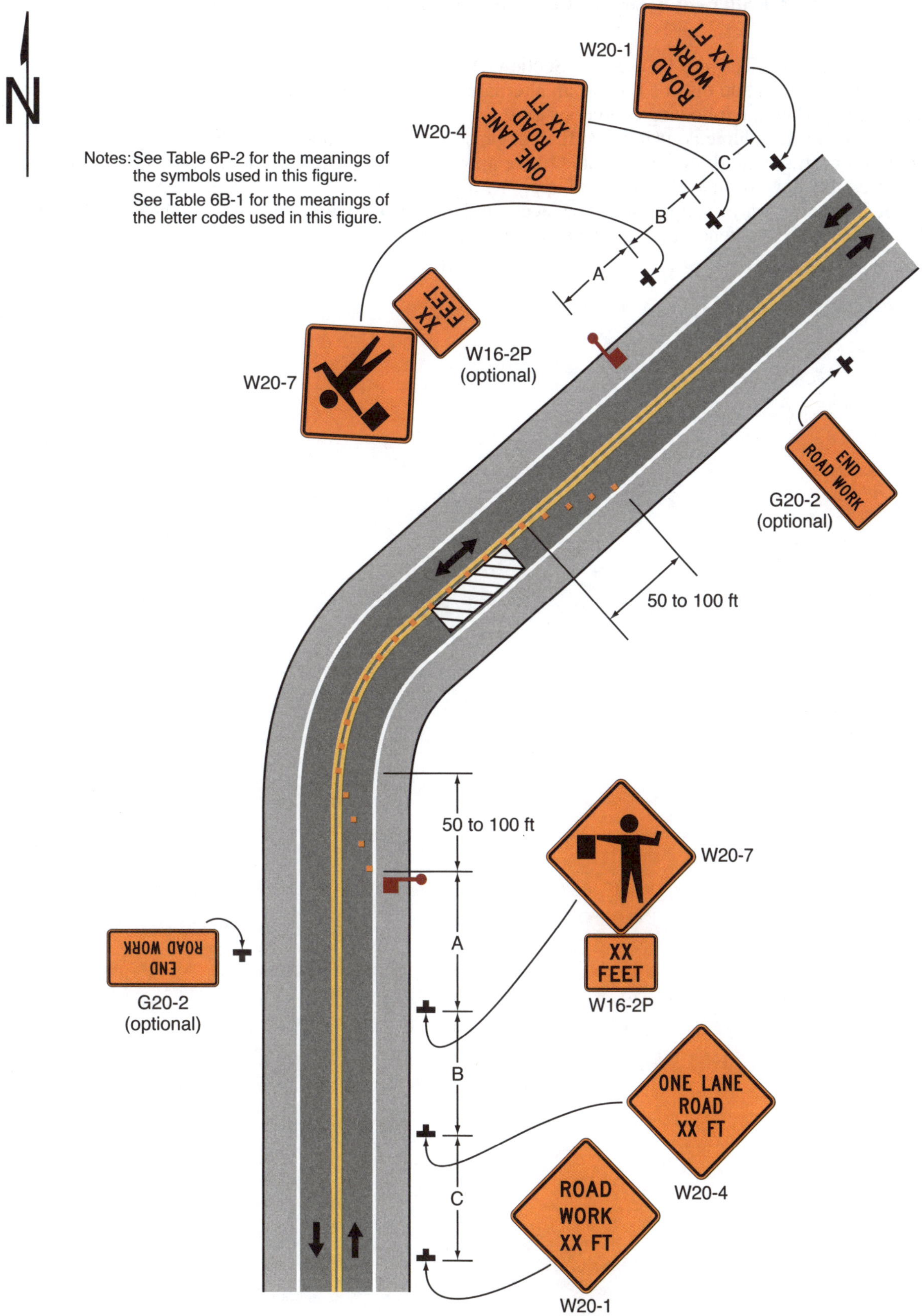

Typical Application 10

Notes for Figure 6P-11—Typical Application 11
Lane Closure on a Two-Lane Road with Low Traffic Volumes

Option:

1. Positive protection devices may be used per Section 6M.02.
2. This TTC zone application may be used as an alternate to the TTC application shown in Figure 6P-10 (using flaggers) when the following conditions exist:
   a. Vehicular traffic volume is such that sufficient gaps exist for vehicular traffic that must yield.
   b. Road users from both directions are able to see approaching vehicular traffic through and beyond the worksite and have sufficient visibility of approaching vehicles.
3. The Type B flashing warning lights may be placed on the ROAD WORK AHEAD and the ONE LANE ROAD AHEAD signs whenever a night lane closure is necessary.

Figure 6P-11. Lane Closure on a Two-Lane Road with Low Traffic Volumes (TA-11)

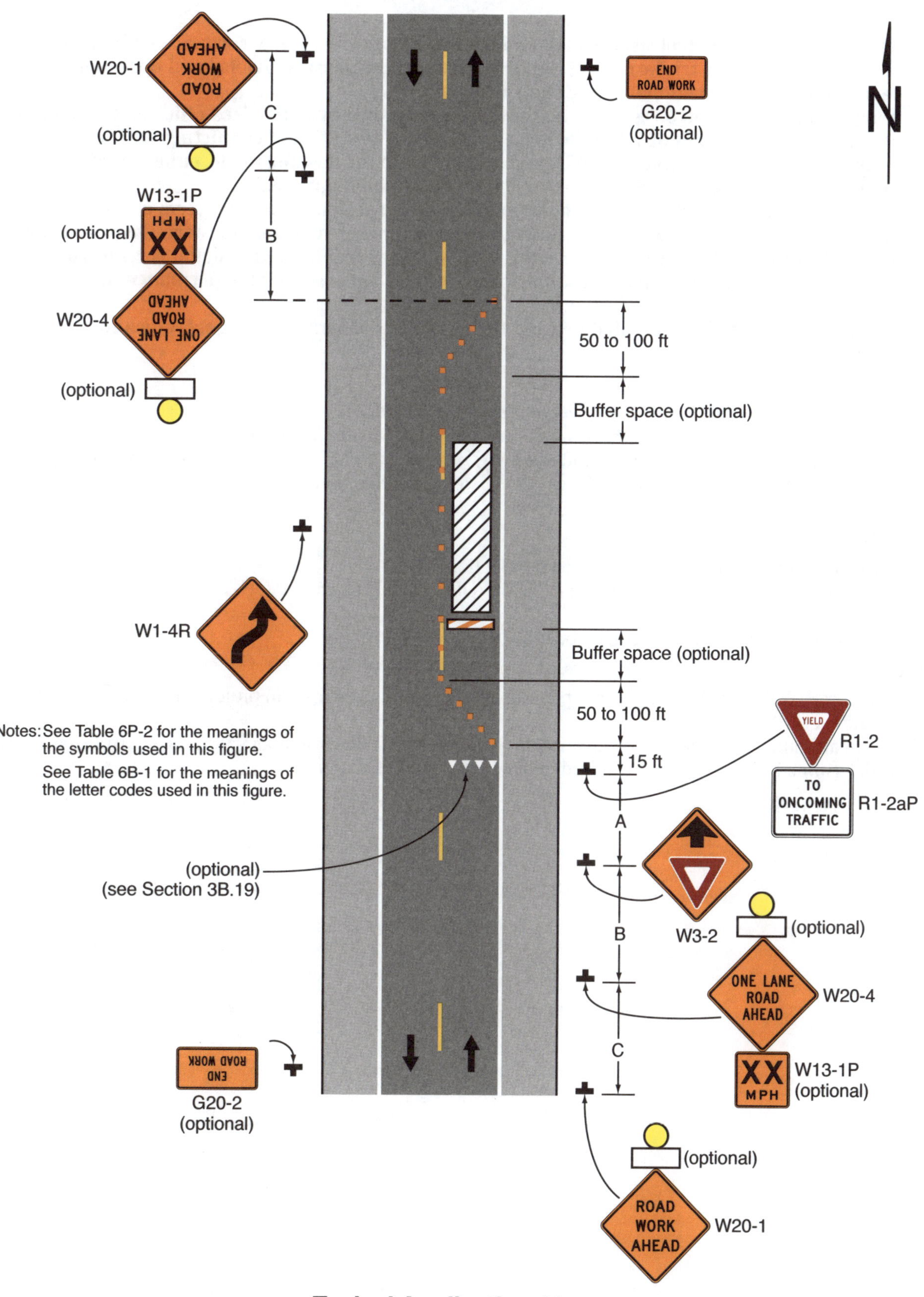

Typical Application 11

Notes for Figure 6P-12—Typical Application 12
Lane Closure on a Two-Lane Road Using Temporary Traffic Control Signals

Standard:

1. **Temporary traffic control signals shall be installed and operated in accordance with the provisions of Part 4. Temporary traffic control signals shall meet the physical display and operational requirements of conventional traffic control signals.**
2. **Temporary traffic control signal timing shall be established by authorized officials. Durations of red clearance intervals shall be adequate to clear the one-lane section of conflicting vehicles.**
3. **When the temporary traffic control signal is changed to the flashing mode, either manually or automatically, red signal indications shall be flashed to both approaches.**
4. **Stop lines shall be installed with temporary traffic control signals for long-term closures. Existing conflicting pavement markings and raised pavement marker reflectors between the activity area and each stop line shall be removed. After the temporary traffic control signal is removed, the stop lines and other temporary pavement markings shall be removed and the permanent pavement markings restored.**
5. **Safeguards shall be incorporated to avoid the possibility of conflicting signal indications at each end of the TTC zone.**

Guidance:

6. *Where no-passing lines are not already in place, they should be added.*
7. *Adjustments in the location of the advance warning signs should be made as needed to accommodate the horizontal or vertical alignment of the roadway, recognizing that the distances shown for sign spacings are minimums. Adjustments in the height of the signal heads should be made as needed to conform to the vertical alignment.*

Option:

8. Positive protection devices may be used per Section 6M.02.
9. Flashing warning lights shown on the ROAD WORK AHEAD and the ONE LANE ROAD AHEAD signs may be used.
10. Removable pavement markings may be used.

Support:

11. Temporary traffic control signals are preferable to flaggers for long-term projects and other activities that would require flagging at night.
12. The maximum length of activity area for one-way operation under temporary traffic control signal control is determined by the capacity required to handle the peak demand.

Figure 6P-12. Lane Closure on a Two-Lane Road Using Temporary Traffic Control Signals (TA-12)

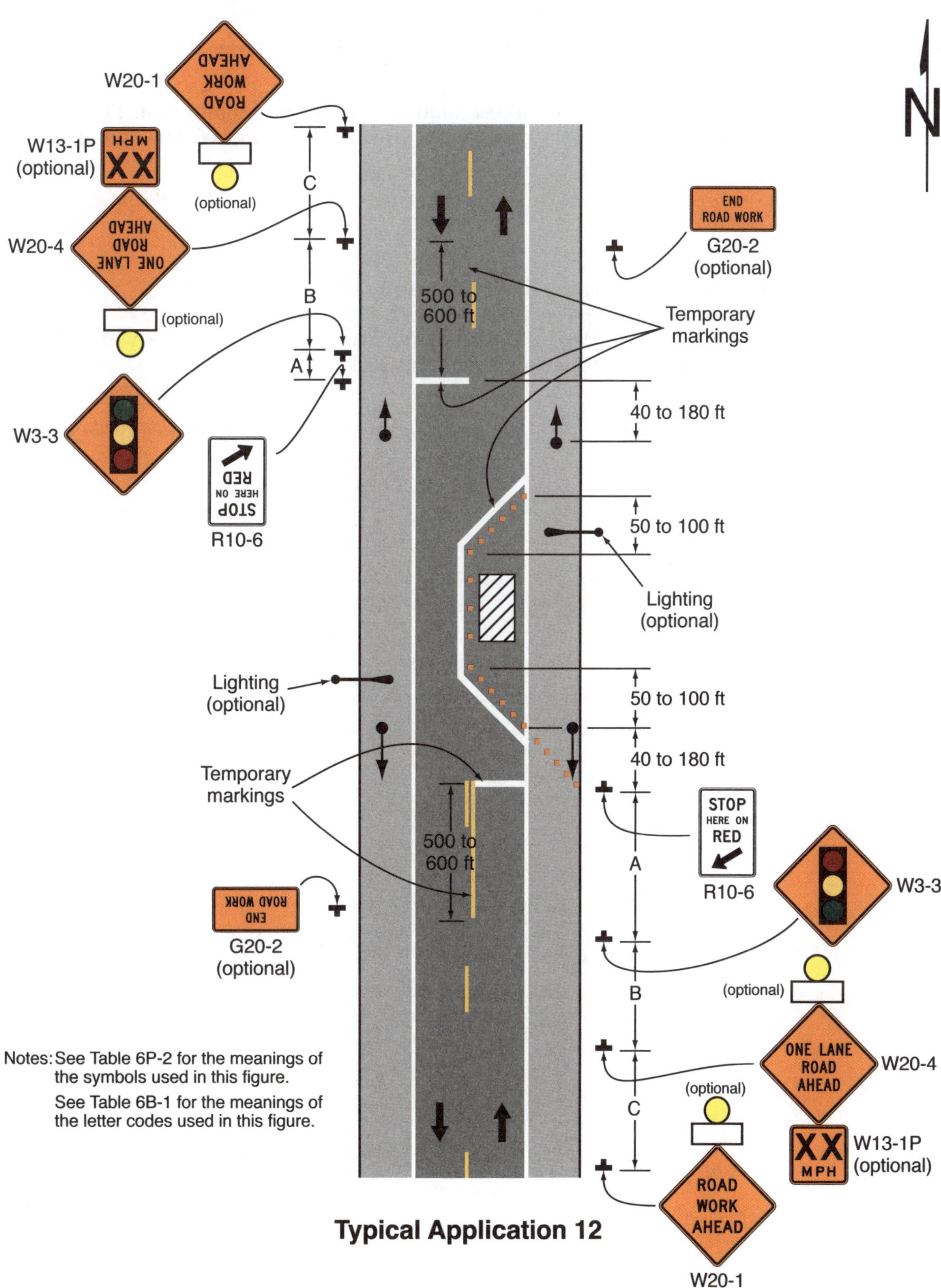

Typical Application 12

Notes for Figure 6P-13—Typical Application 13
Temporary Road Closure

Support:

1. Conditions represented are a planned closure not exceeding 20 minutes during the daytime.

Standard:

2. **A flagger or uniformed law enforcement officer shall be used for this application. The flagger, if used for this application, shall follow the procedures provided in Sections 6D.05 and 6D.06.**

Guidance:

3. *The uniformed law enforcement officer, if used for this application, should follow the procedures provided in Sections 6D.05 and 6D.06.*

Option:

4. A BE PREPARED TO STOP sign may be added to the sign series.
5. Positive protection devices may be used per Section 6M.02.
6. Automated Flagger Assistance Devices (see Section 6L.02) may be used in situations where there is only one lane of approaching traffic in the direction to be controlled.

Guidance:

7. *When used, the BE PREPARED TO STOP sign should be located before the Flagger symbol sign.*

Figure 6P-13. Temporary Road Closure (TA-13)

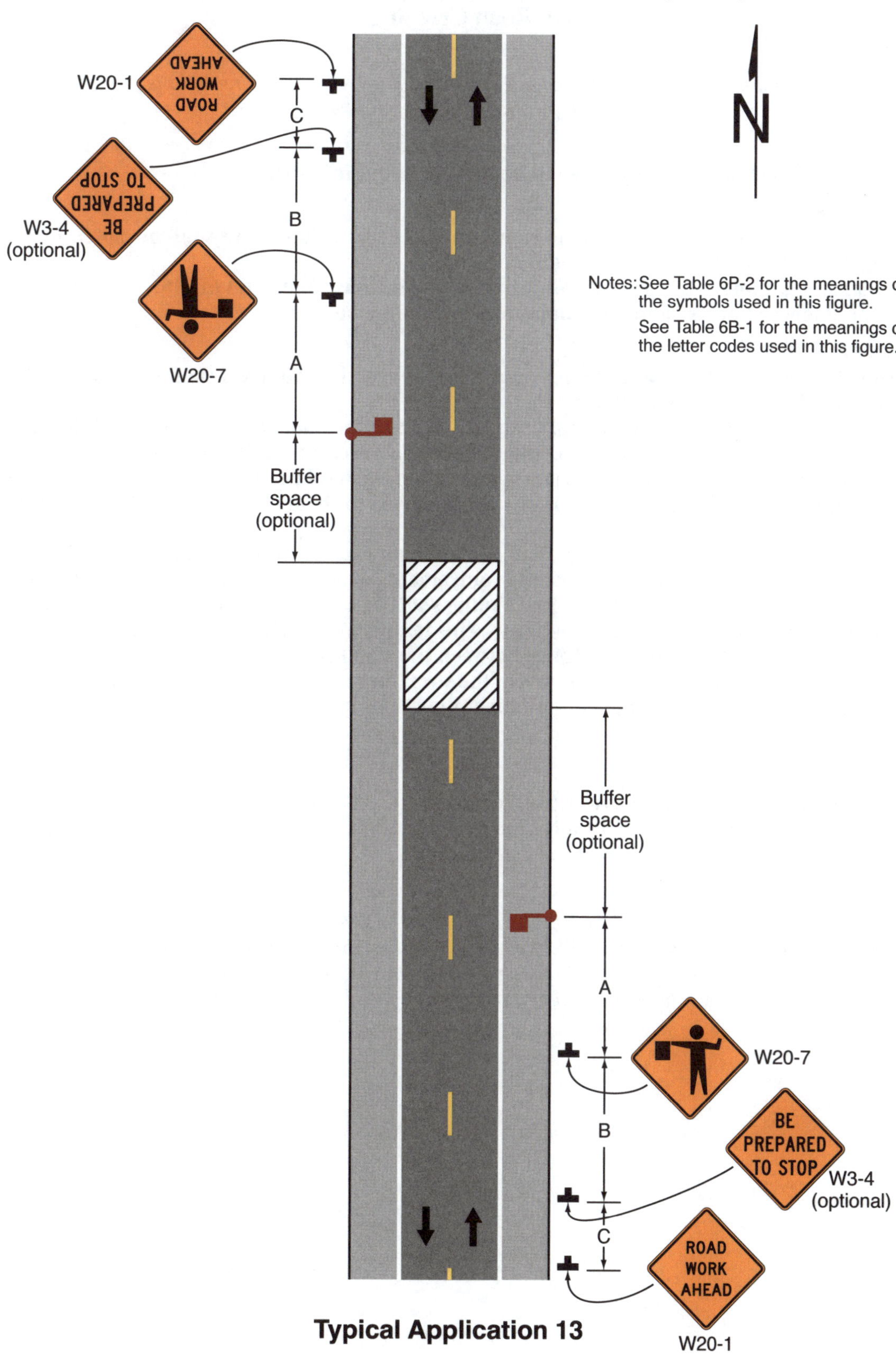

Typical Application 13

Notes for Figure 6P-14—Typical Application 14
Haul Road Crossing

Guidance:

1. *Floodlights should be used to illuminate haul road crossings where existing light is inadequate.*
2. *Where no-passing lines are not already in place, they should be added.*

Standard:

3. **The traffic control method selected shall be used in both directions.**

Flagging Method

4. **When a road used exclusively as a haul road is not in use, the haul road shall be closed with Type 3 Barricades and the Flagger symbol signs covered.**
5. **The flagger shall follow the procedures provided in Sections 6D.05 and 6D.06.**
6. **At night, flagger stations shall be illuminated, except in emergencies.**

Signalized Method

7. **When a road used exclusively as a haul road is not in use, the haul road shall be closed with Type 3 Barricades. The signals shall either:**
   a. **Flash yellow on the main road and flash red on the haul road or be covered, and the Signal Ahead and STOP HERE ON RED signs shall be covered or hidden from view; or**
   b. **Display green on the main road and steady red on the haul road, but only if actuated signal operation is used such that green is always displayed to the main road except when a vehicle is detected on the haul road.**
8. **The temporary traffic control signals shall control both the highway and the haul road and shall meet the physical display and operational requirements of conventional traffic control signals as described in Part 4. Traffic control signal timing shall be established by authorized officials.**
9. **Stop lines shall be used on existing highways with temporary traffic control signals.**
10. **Existing conflicting pavements markings between the stop lines shall be removed. After the temporary traffic control signal is removed, the stop lines and other temporary pavement markings shall be removed and the permanent pavement markings restored.**

Option:

Flagging Method

11. Automated Flagger Assistance Devices (see Section 6L.02) may be used in situations where there is only one lane of approaching traffic in the direction to be controlled.

Guidance:

Signalized Method

12. *If actuated signal operation is used (see Item b in Note 7 above) and pedestrian facilities, such as sidewalks, are present in the area of the haul road crossing, then consideration should be given to providing pedestrian actuation capability at the temporary traffic control signal to accommodate any pedestrians who might be depending upon a pedestrian phase to cross the main road.*

Figure 6P-14. Haul Road Crossing (TA-14)

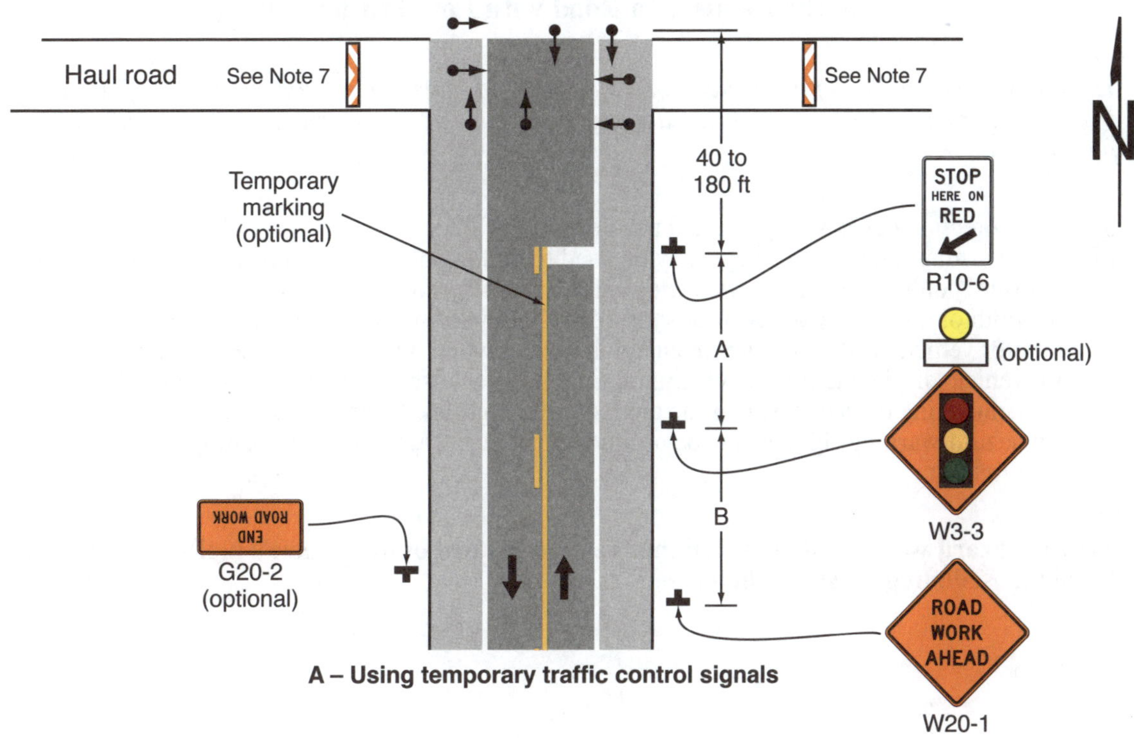

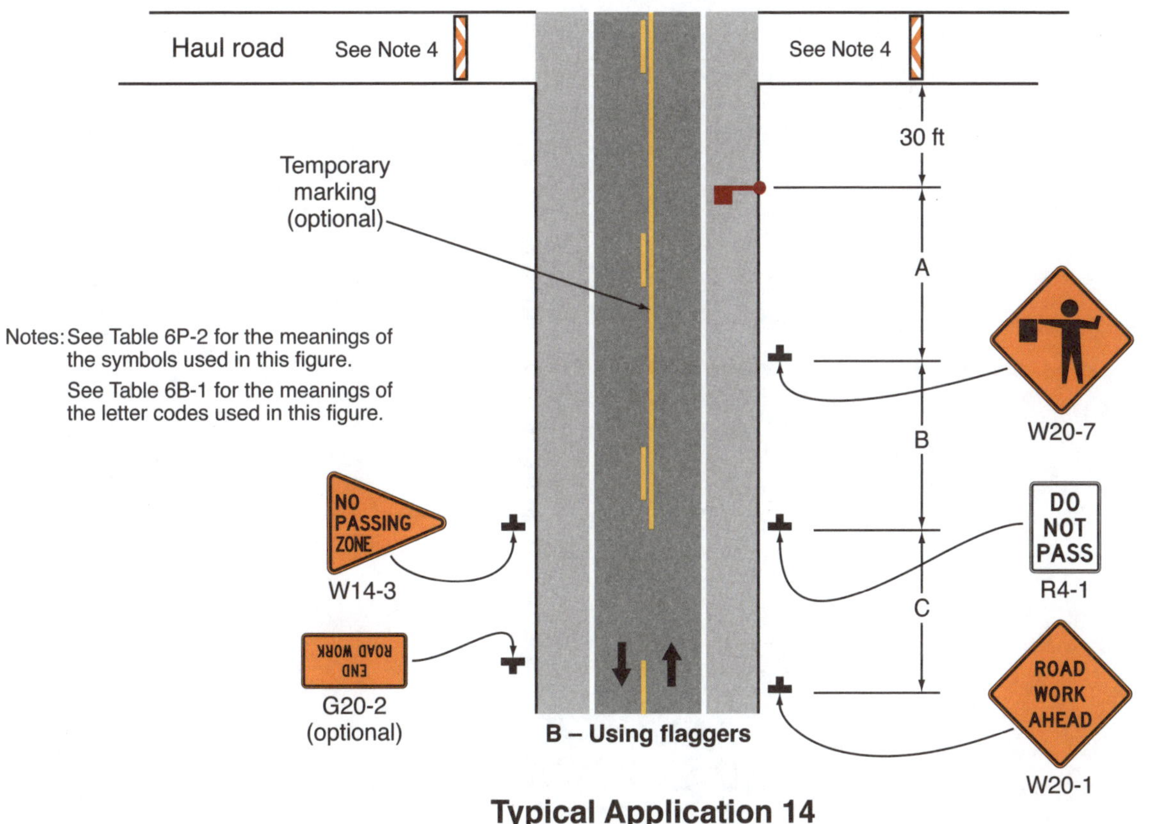

Typical Application 14

Notes for Figure 6P-15—Typical Application 15
Work in the Center of a Road with Low Traffic Volumes

Guidance:

1. *The lanes on either side of the center work space should have a minimum width of 10 feet as measured from the near edge of the channelizing devices to the edge of the pavement or the outside edge of the paved shoulder.*

Option:

2. Positive protection devices may be used per Section 6M.02.
3. Flashing warning lights and/or flags may be used to call attention to the advance warning signs.
4. If the closure continues overnight, warning lights may be used on the channelizing devices.
5. A lane width of 9 feet may be used for short-term stationary work on low-volume, low-speed roadways when motor vehicle traffic does not include longer and wider heavy commercial vehicles.
6. A work vehicle displaying high-intensity rotating, flashing, oscillating, or strobe lights may be used instead of the channelizing devices forming the tapers or the high-level warning devices.
7. Vehicle hazard warning signals may be used to supplement high-intensity rotating, flashing, oscillating, or strobe lights.

Standard:

8. **Vehicle hazard warning signals shall not be used instead of the vehicle's high-intensity rotating, flashing, oscillating, or strobe lights.**

Figure 6P-15. Work in the Center of a Road with Low Traffic Volumes (TA-15)

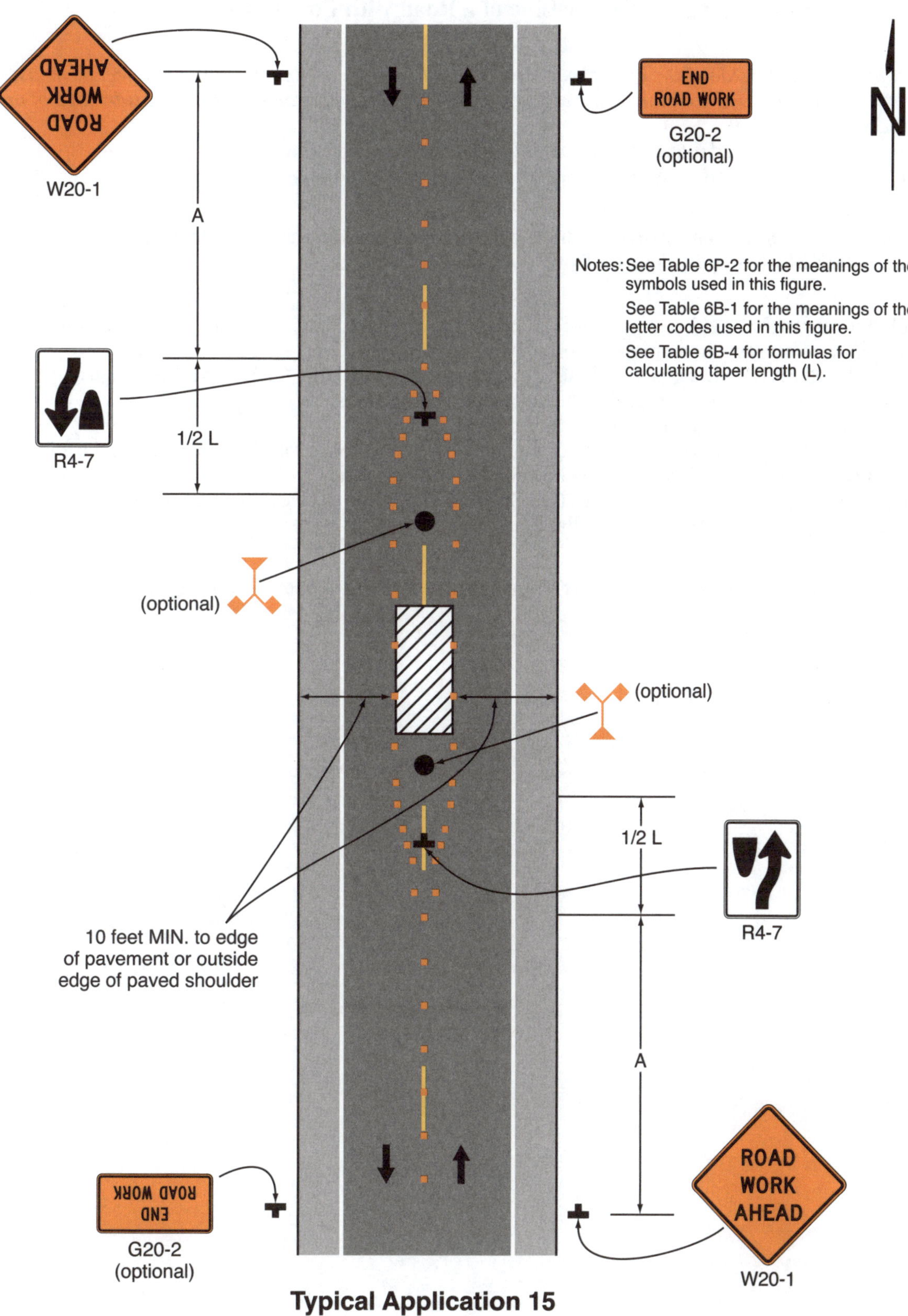

Typical Application 15

Notes for Figure 6P-16—Typical Application 16
Surveying Along the Center Line of a Road with Low Traffic Volumes

Guidance:

1. *The lanes on either side of the center work space should have a minimum width of 10 feet as measured from the near edge of the channelizing devices to the edge of the pavement or the outside edge of the paved shoulder.*
2. *Cones should be placed 6 to 12 inches on either side of the center line.*
3. *A flagger should be used to warn workers who cannot watch road users.*

Standard:

4. **For surveying on the center line of a high-volume road, one lane shall be closed using the information illustrated in Figure 6P-10.**

Option:

5. A high-level warning device may be used to protect a surveying device, such as a target on a tripod.
6. Cones may be omitted for a cross-section survey.
7. ROAD WORK AHEAD signs may be used in place of the SURVEY CREW AHEAD signs.
8. Flags may be used to call attention to the advance warning signs.
9. If the work is along the shoulder, the flagger may be omitted.
10. For a survey along the edge of the road or along the shoulder, cones may be placed along the edge line.
11. A BE PREPARED TO STOP sign may be added to the sign series.
12. Automated Flagger Assistance Devices (see Section 6L.02) may be used in situations where there is only one lane of approaching traffic in the direction to be controlled.

Guidance:

13. *When used, the BE PREPARED TO STOP sign should be located before the Flagger symbol sign.*

Figure 6P-16. Surveying Along the Center Line of a Road with Low Traffic Volumes (TA-16)

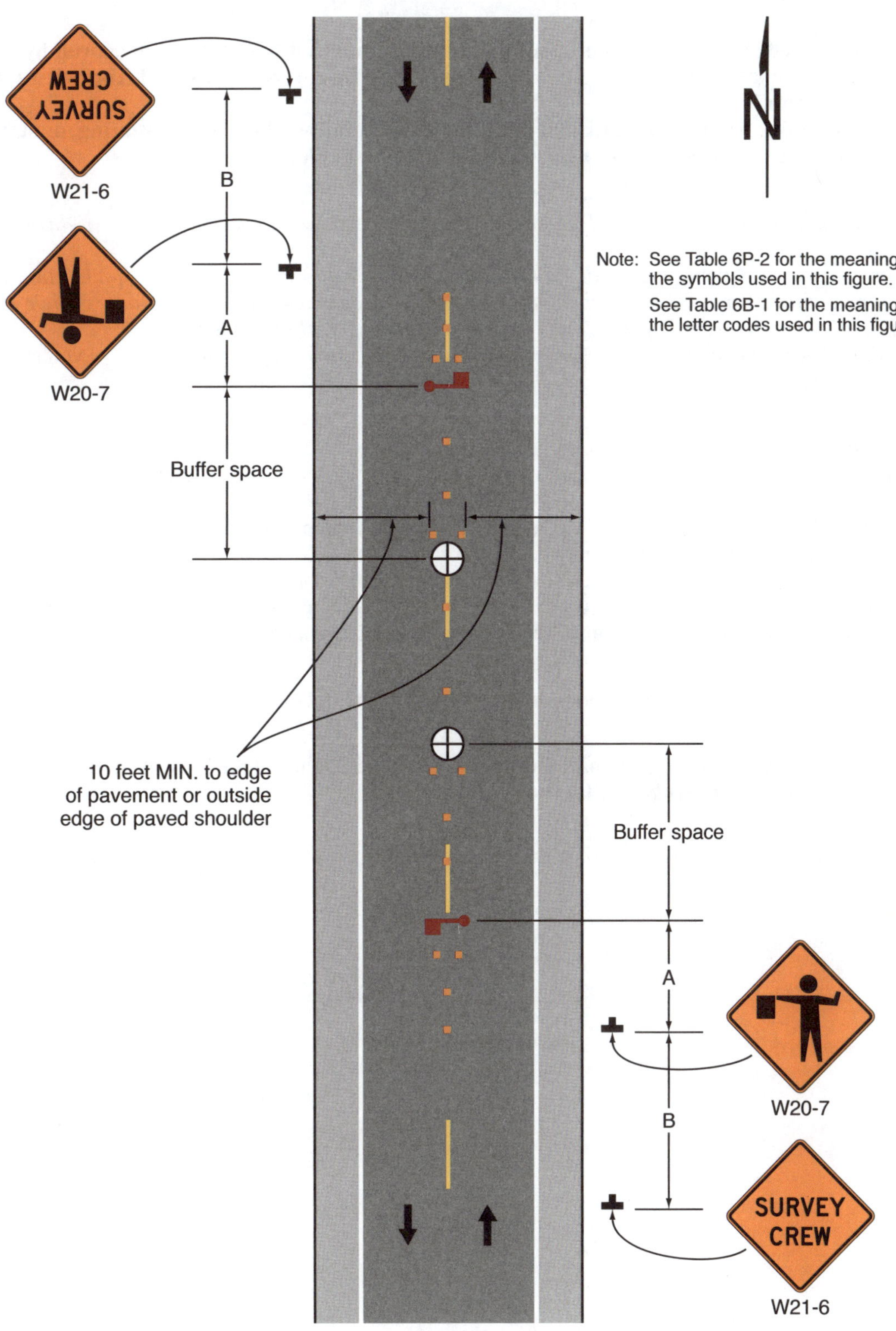

Typical Application 16

Notes for Figure 6P-17—Typical Application 17
Mobile Operations on a Two-Lane Road

Standard:

1. **Vehicle-mounted signs shall be mounted in a manner such that they are not obscured by equipment or supplies. Sign legends on vehicle-mounted signs shall be covered or turned from view when work is not in progress.**
2. **Shadow and work vehicles shall display high-intensity rotating, flashing, oscillating, or strobe lights.**
3. **If an arrow board is used, it shall be used in the caution mode.**

Guidance:

4. *Where practical and when needed, the work and shadow vehicles should pull over periodically to allow vehicular traffic to pass.*
5. *Whenever adequate stopping sight distance exists to the rear, the shadow vehicle should maintain the minimum distance from the work vehicle and proceed at the same speed. The shadow vehicle should slow down in advance of vertical or horizontal curves that restrict sight distance.*
6. *The shadow vehicles should also be equipped with two high-intensity flashing lights mounted on the rear, adjacent to the sign.*

Option:

7. Positive protection devices may be used per Section 6M.02.
8. The distance between the work and shadow vehicles may vary according to terrain, paint drying time, and other factors.
9. Additional shadow vehicles to warn and reduce the speed of oncoming or opposing vehicular traffic may be used. Law enforcement vehicles may be used for this purpose.
10. A truck-mounted attenuator may be used on the shadow vehicle or on the work vehicle.
11. If the work and shadow vehicles cannot pull over to allow vehicular traffic to pass frequently, a DO NOT PASS sign may be placed on the rear of the vehicle blocking the lane.

Support:

12. Shadow vehicles are used to warn motor vehicle traffic of the operation ahead.

Standard:

13. **Vehicle hazard warning signals shall not be used instead of the vehicle's high-intensity rotating, flashing, oscillating, or strobe lights.**

Figure 6P-17. Mobile Operations on a Two-Lane Road (TA-17)

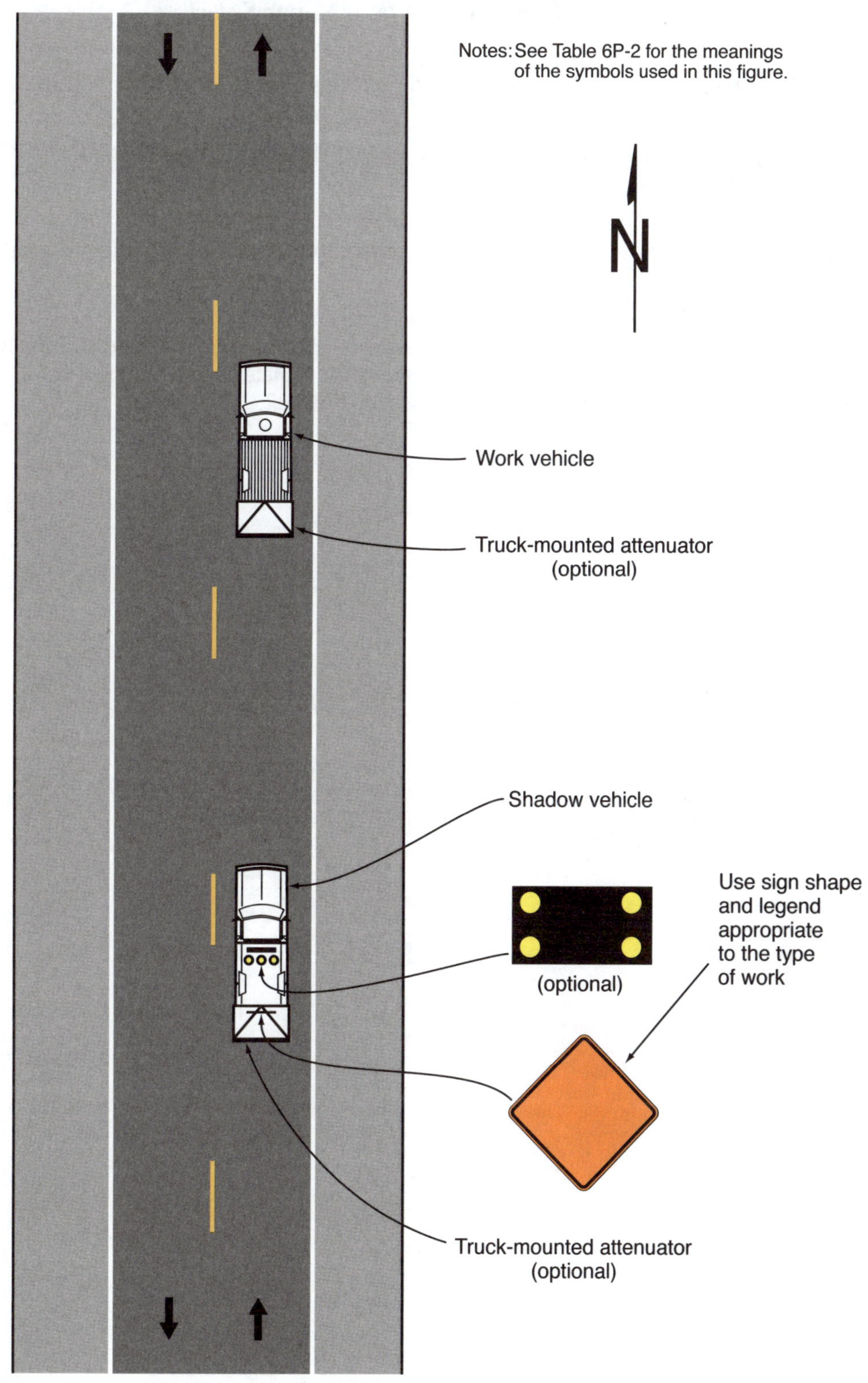

Typical Application 17

Notes for Figure 6P-18—Typical Application 18
Lane Closure on a Minor Street

Standard:

  1. **This TTC shall be used only for low-speed facilities having low traffic volumes.**

Option:

  2. Where the work space is short, where road users can see the roadway beyond, and where volume is low, vehicular traffic may be self-regulating.

Standard:

  3. **Where vehicular traffic cannot effectively self-regulate, one or two flaggers shall be used as illustrated in Figure 6P-10.**

Option:

  4. Flashing warning lights and/or flags may be used to call attention to the advance warning signs.
  5. A truck-mounted attenuator may be used on the work vehicle and the shadow vehicle.
  6. Positive protection devices may be used per Section 6M.02.

Figure 6P-18. Lane Closure on a Minor Street (TA-18)

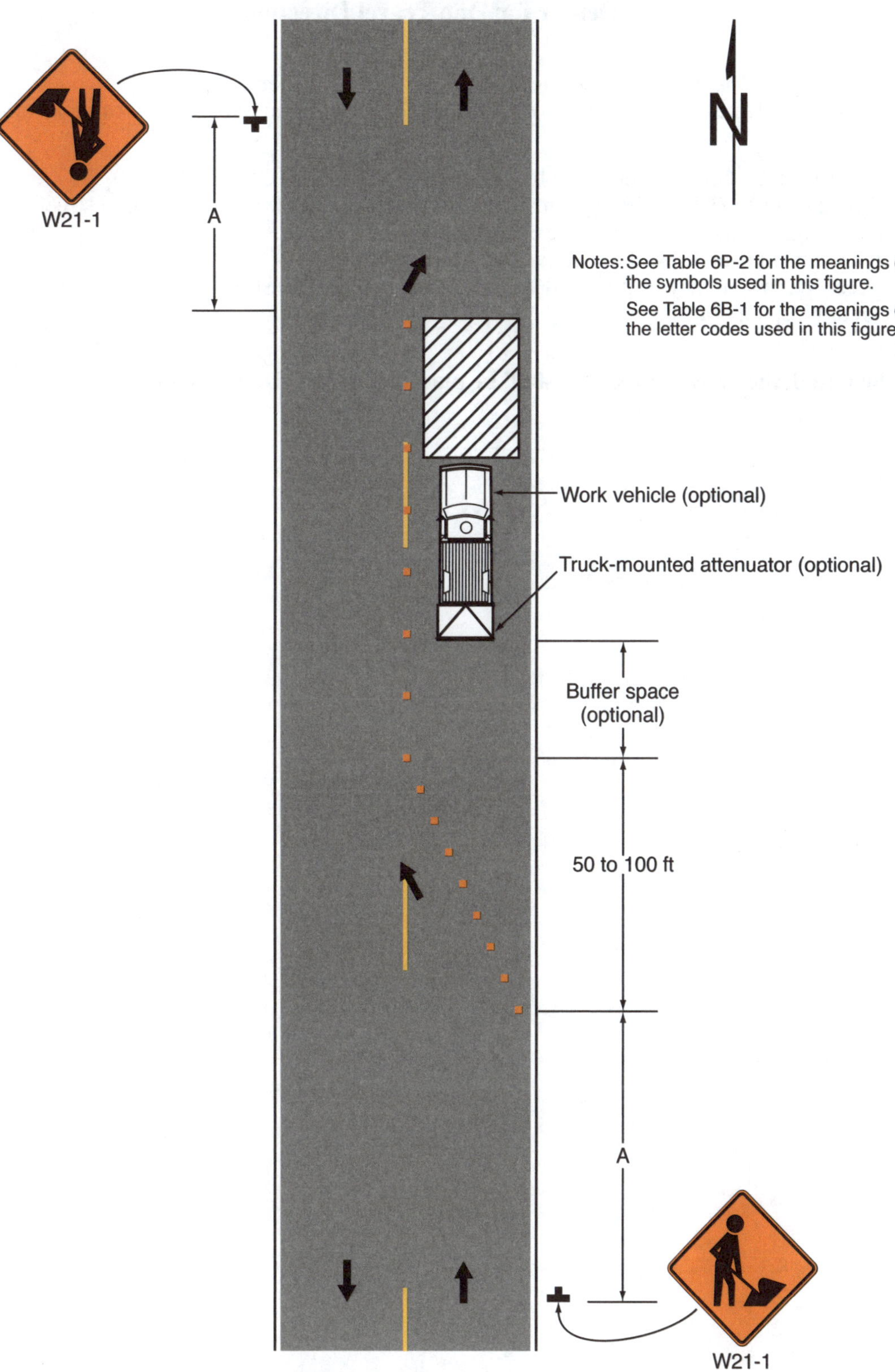

Typical Application 18

Notes for Figure 6P-19—Typical Application 19
Detour for One Travel Direction

Guidance:

1. *This plan should be used for streets without posted route numbers.*
2. *On multi-lane streets, Detour signs with an Advance Turn Arrow should be used in advance of a turn.*

Option:

3. The STREET CLOSED legend may be used in place of ROAD CLOSED.
4. Additional DO NOT ENTER signs may be used at intersections with intervening streets.
5. Warning lights may be used on Type 3 Barricades.
6. Detour signs may be located on the far side of intersections.
7. A Street Name sign may be mounted with the Detour sign. The Street Name sign may be either white on green or black on orange.

Standard:

8. **When used, the Street Name sign shall be placed above the Detour sign.**

Figure 6P-19. Detour for One Travel Direction (TA-19)

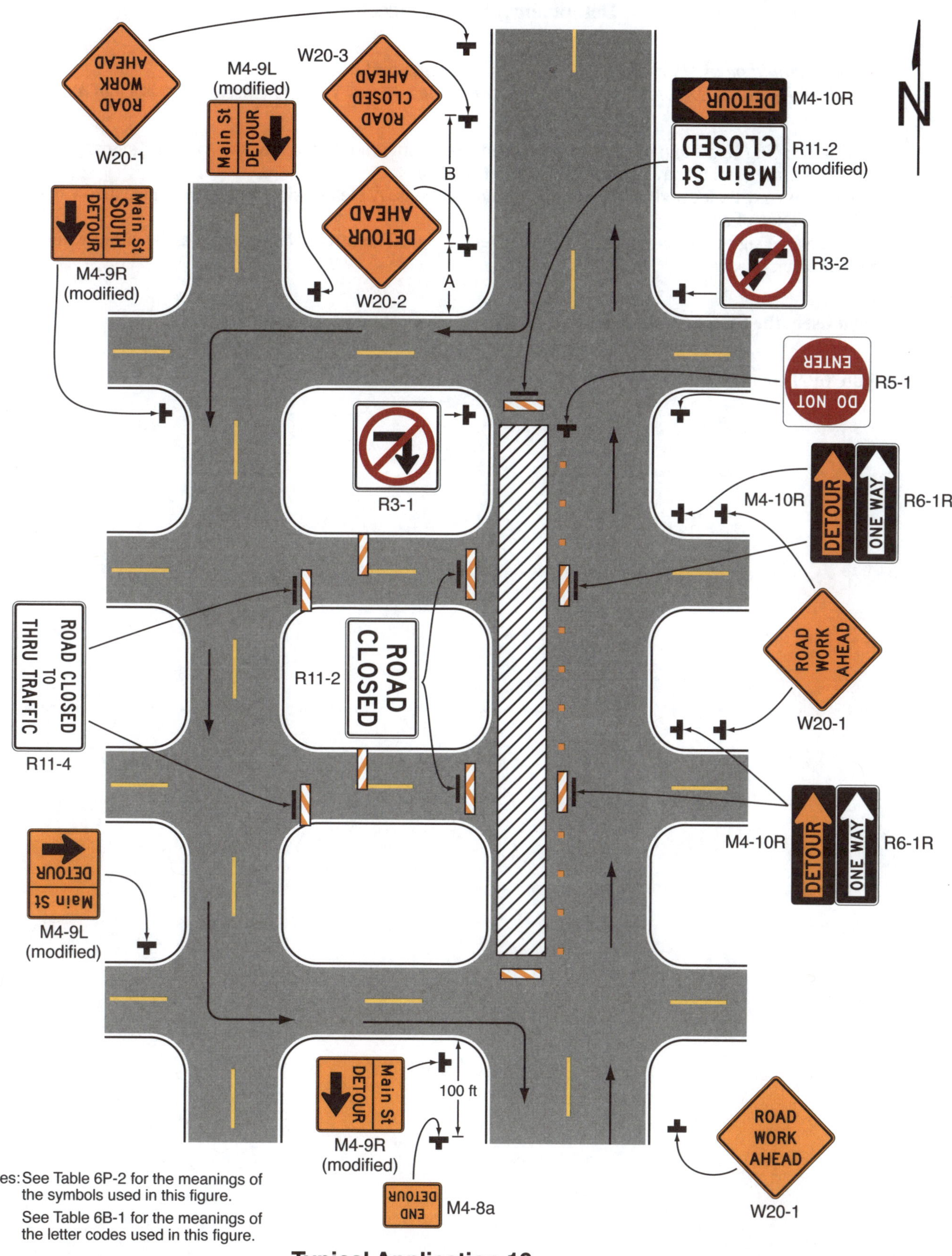

Notes: See Table 6P-2 for the meanings of the symbols used in this figure.

See Table 6B-1 for the meanings of the letter codes used in this figure.

Typical Application 19

Notes for Figure 6P-20—Typical Application 20
Detour for a Closed Street

Guidance:
1. *This plan should be used for streets without posted route numbers.*
2. *On multi-lane streets, Detour signs with an Advance Turn Arrow should be used in advance of a turn.*

Option:
3. Flashing warning lights and/or flags may be used to call attention to the advance warning signs.
4. Flashing warning lights may be used on Type 3 Barricades.
5. Detour signs may be located on the far side of intersections. A Detour sign with an advance arrow may be used in advance of a turn.
6. A Street Name sign may be mounted with the Detour sign. The Street Name sign may be either white on green or black on orange.

Standard:
7. **When used, the Street Name sign shall be placed above the Detour sign.**

Support:
8. Figure 6P-9 contains the information for detouring a numbered highway.

Figure 6P-20. Detour for a Closed Street (TA-20)

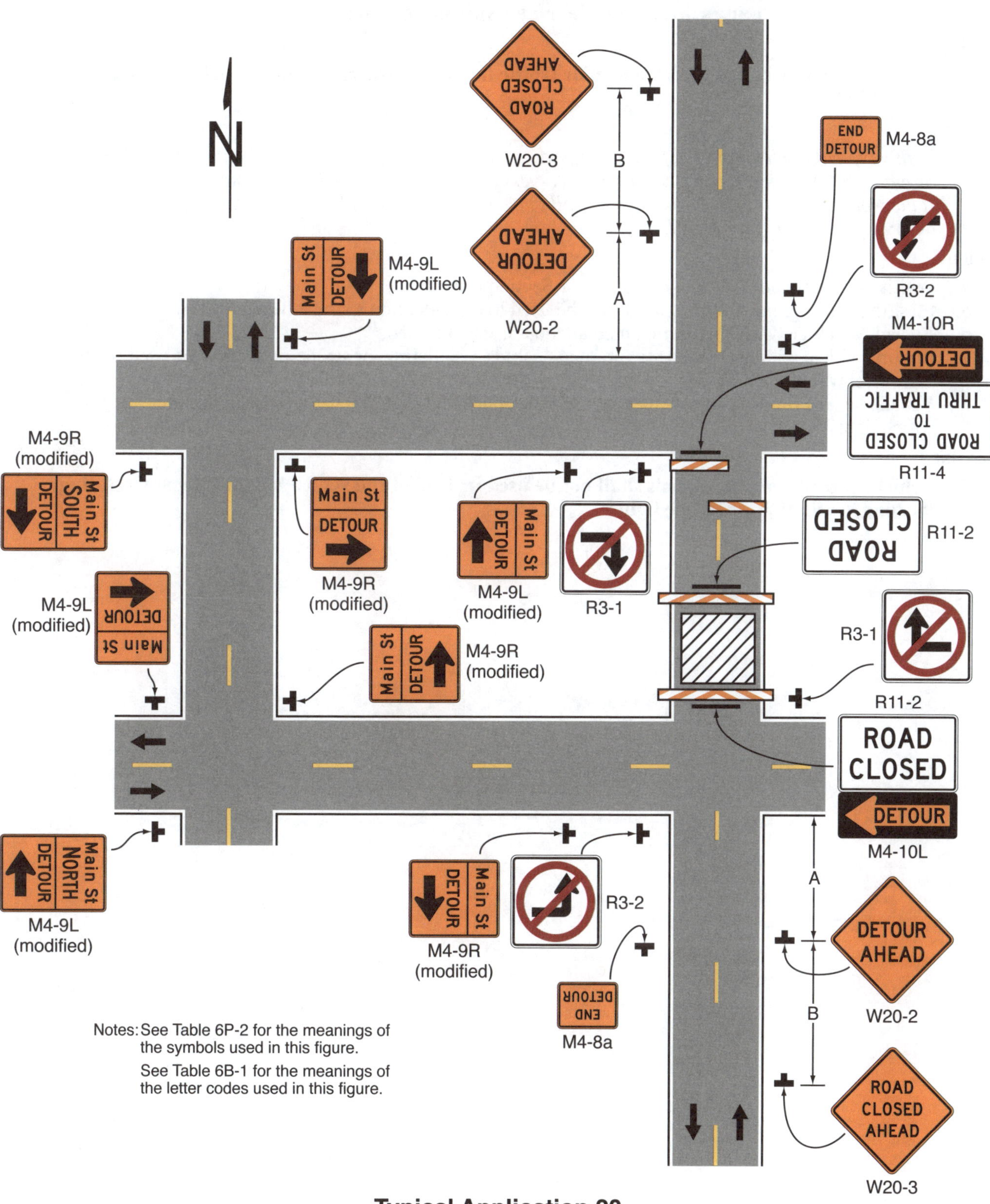

Notes: See Table 6P-2 for the meanings of the symbols used in this figure.

See Table 6B-1 for the meanings of the letter codes used in this figure.

Typical Application 20

Notes for Figure 6P-21—Typical Application 21
Lane Closure on the Near Side of an Intersection

Standard:

1. **The merging taper shall direct vehicular traffic into either the right-hand or left-hand lane, but not both.**

Guidance:

2. *In this typical application, a left taper should be used so that right-turn movements will not impede through motor vehicle traffic. However, the reverse should be true for left-turn movements.*
3. *If the work space extends across a crosswalk, the crosswalk should be closed using the information and devices shown in Figure 6P-29.*

Option:

4. Positive protection devices may be used per Section 6M.02.
5. Flashing warning lights and/or flags may be used to call attention to the advance warning signs.
6. A shadow vehicle with a truck-mounted attenuator may be used.
7. A work vehicle with high-intensity rotating, flashing, oscillating, or strobe lights may be used with the high-level warning device.
8. Vehicle hazard warning signals may be used to supplement high-intensity rotating, flashing, oscillating, or strobe lights.

Standard:

9. **Vehicle hazard warning signals shall not be used instead of the vehicle's high-intensity rotating, flashing, oscillating, or strobe lights.**

Figure 6P-21. Lane Closure on the Near Side of an Intersection (TA-21)

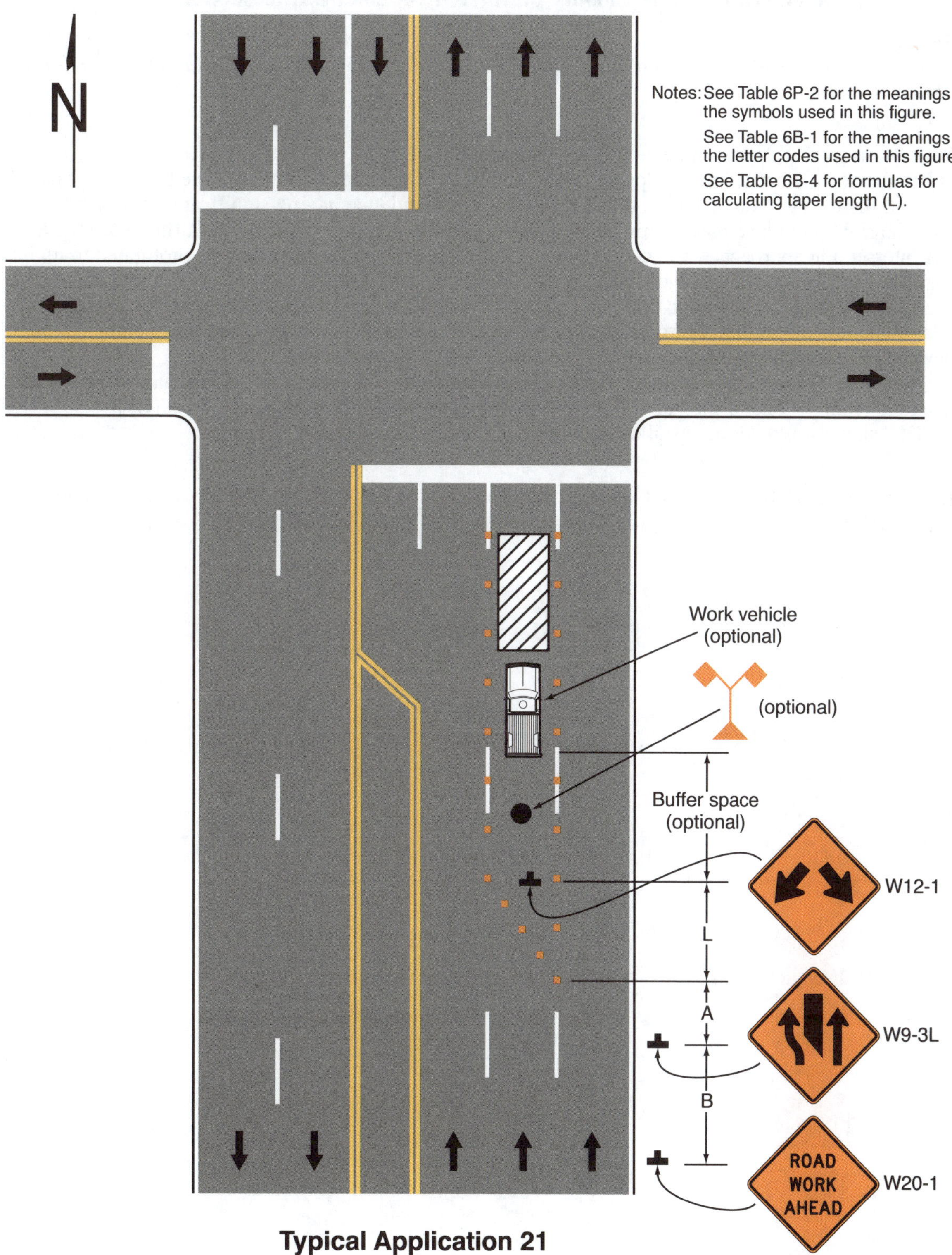

Typical Application 21

Notes for Figure 6P-22—Typical Application 22
Right-Hand Lane Closure on the Far Side of an Intersection

Guidance:

1. *If the work space extends across a crosswalk, the crosswalk should be closed using the information and devices shown in Figure 6P-29.*

Option:

2. Positive protection devices may be used per Section 6M.02.
3. When the normal procedure of closing on the near side of the intersection any lane that is not carried through the intersection results in the closure of a right-hand lane having significant right-turn movements, then the right-hand lane may be restricted to right turns only, requiring through traffic to use the left lane.
4. For intersection approaches reduced to a single lane, left-turn movements may be prohibited to maintain capacity for through vehicular traffic.
5. Flashing warning lights and/or flags may be used to call attention to the advance warning signs.
6. Where the turning radius is large, it may be possible to create a right-turn island using channelizing devices or pavement markings.
7. If dimension "A" is not available to create a temporary right-turn lane, continuous channelizers may be installed from the end of the taper to the intersection and, as a result, the RIGHT LANE MUST TURN RIGHT signs would not be installed.

Support:

8. By first closing off the right-hand lane and then reopening it as a turn bay, the capacity of the through lane is preserved by separating the right-turning vehicles from the through vehicles.

Figure 6P-22. Right-Hand Lane Closure on the Far Side of an Intersection (TA-22)

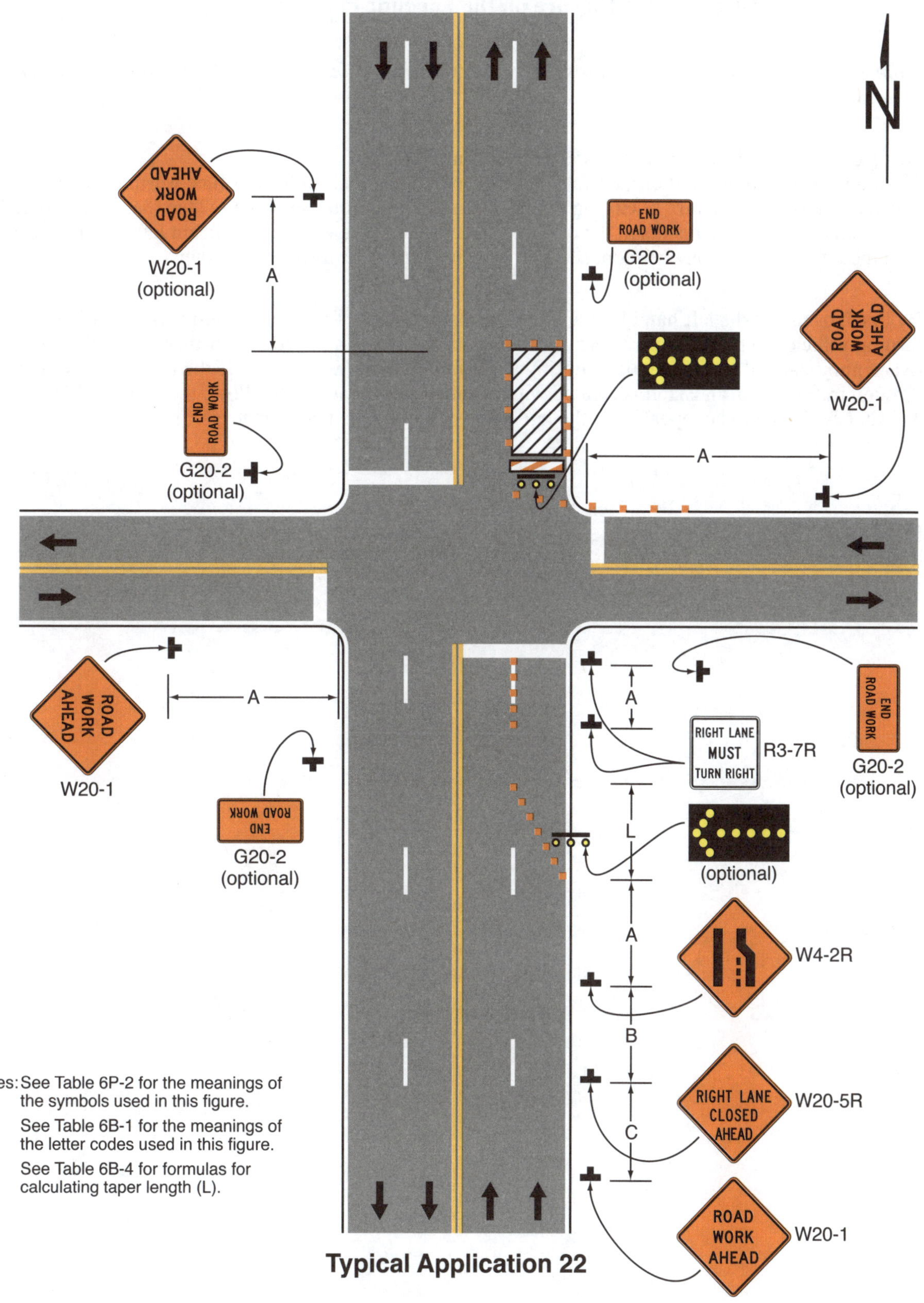

Notes: See Table 6P-2 for the meanings of the symbols used in this figure.

See Table 6B-1 for the meanings of the letter codes used in this figure.

See Table 6B-4 for formulas for calculating taper length (L).

Typical Application 22

Notes for Figure 6P-23—Typical Application 23
Left-Hand Lane Closure on the Far Side of an Intersection

Guidance:

1. *If the work space extends across a crosswalk, the crosswalk should be closed using the information and devices shown in Figure 6P-29.*

Option:

2. Positive protection devices may be used per Section 6M.02.
3. Flashing warning lights and/or flags may be used to call attention to the advance warning signs.
4. When the normal procedure of closing on the near side of the intersection any lane that is not carried through the intersection results in the closure of a left-hand lane having significant left-turn movements, then the left-hand lane may be reopened as a turn bay for left turns only, as shown.

Support:

5. By first closing off the left-hand lane and then reopening it as a turn bay, the left-turn bay allows storage of turning vehicles so that the movement of through traffic is not impeded. A left-turn bay that is long enough to accommodate all turning vehicles during a traffic signal cycle will provide the maximum benefit for through traffic. Also, an island is created with channelizing devices that allows the LEFT LANE MUST TURN LEFT sign to be repeated on the left adjacent to the lane that it controls.

Figure 6P-23. Left-Hand Lane Closure on the Far Side of an Intersection (TA-23)

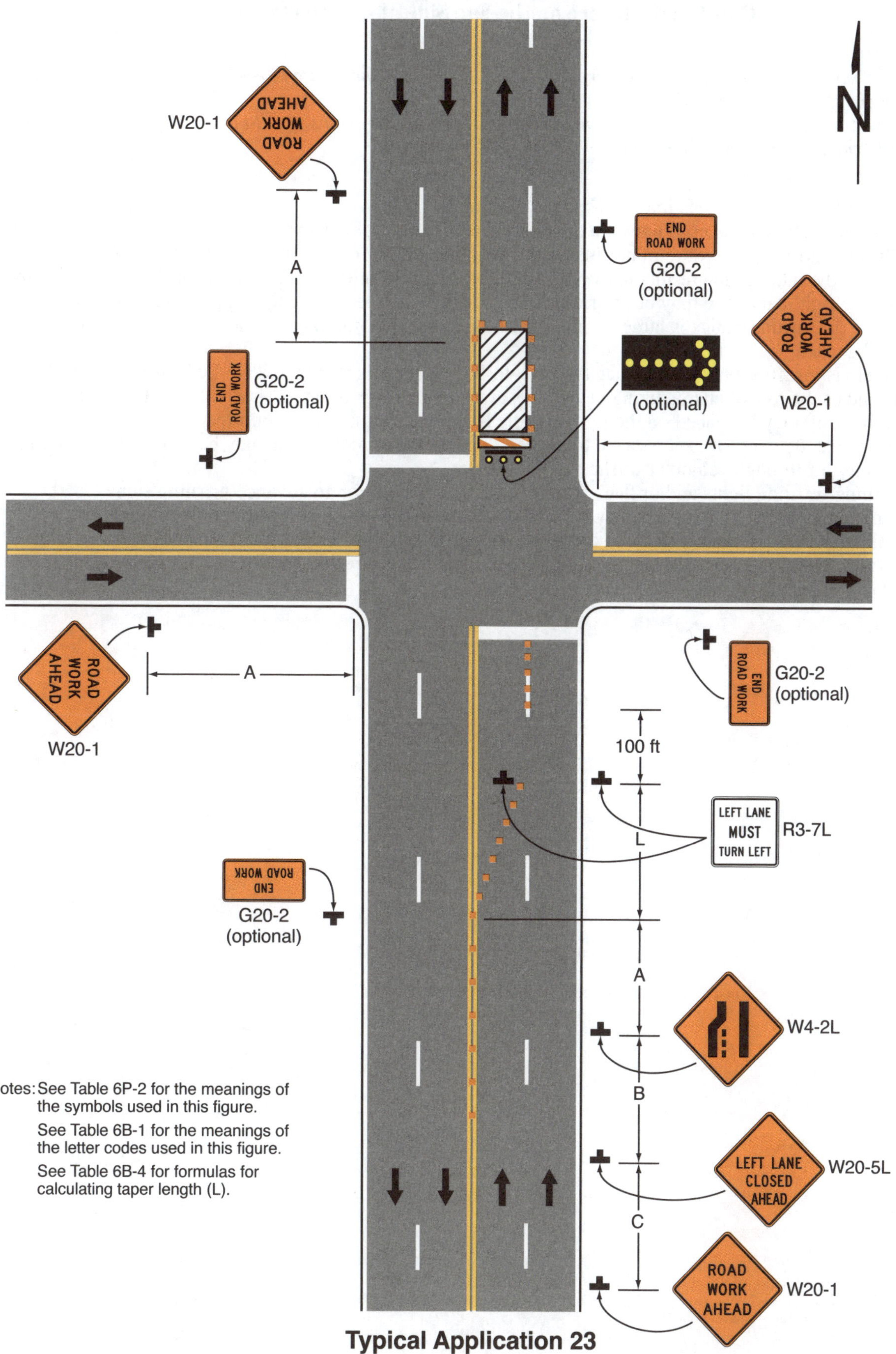

Notes: See Table 6P-2 for the meanings of the symbols used in this figure.

See Table 6B-1 for the meanings of the letter codes used in this figure.

See Table 6B-4 for formulas for calculating taper length (L).

Typical Application 23

Notes for Figure 6P-24—Typical Application 24
Half Road Closure on the Far Side of an Intersection

Guidance:

1. *If the work space extends across a crosswalk, the crosswalk should be closed using the information and devices shown in Figure 6P-29.*
2. *When turn prohibitions are implemented, two turn prohibition signs should be used, one on the near side and, space permitting, one on the far side of the intersection.*

Option:

3. Positive protection devices may be used per Section 6M.02.
4. A buffer space may be used between opposing directions of vehicular traffic as shown in this application.
5. When the normal procedure of closing on the near side of the intersection any lane that is not carried through the intersection results in the closure of a right-hand lane having significant right-turn movements, then the right-hand lane may be restricted to right turns only, requiring through traffic to use the left lane.
6. Where the turning radius is large, a right-turn island using channelizing devices or pavement markings may be used.
7. If there is insufficient space to place the back-to-back Keep Right sign and No Left Turn symbol signs at the end of the row of channelizing devices separating opposing vehicular traffic flows, the No Left Turn symbol sign may be placed on the right and the Keep Right sign may be omitted.
8. For intersection approaches reduced to a single lane, left-turn movements may be prohibited to maintain capacity for through vehicular traffic.
9. Flashing warning lights and/or flags may be used to call attention to advance warning signs.
10. Temporary pavement markings may be used to delineate the travel path through the intersection.
11. If dimension "A" is not available to create a temporary right turn lane, continuous channelizers may be installed from the end of the taper to the intersection and, as a result, the RIGHT LANE MUST TURN RIGHT signs would not be installed.

Figure 6P-24. Half Road Closure on the Far Side of an Intersection (TA-24)

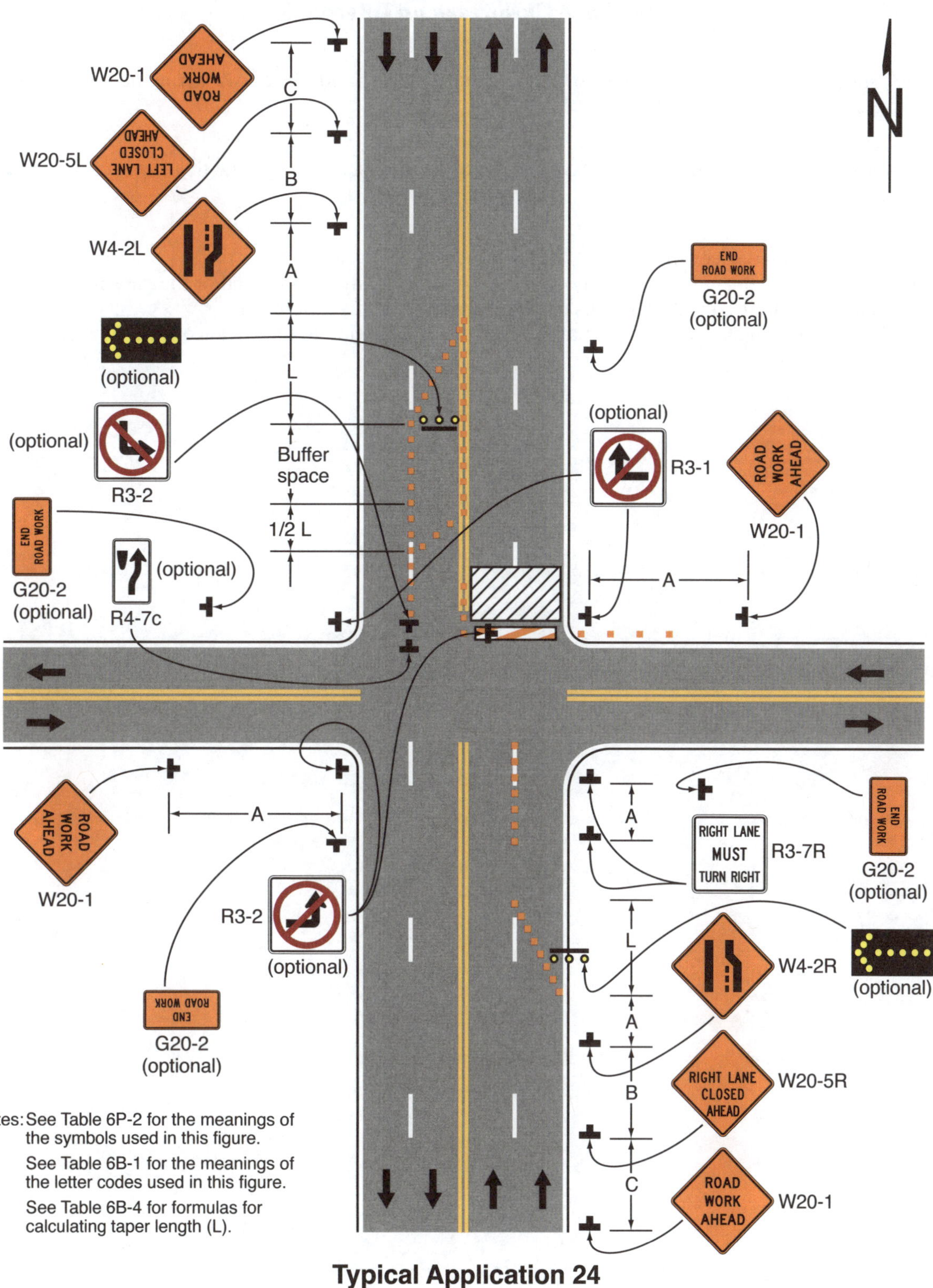

Notes: See Table 6P-2 for the meanings of the symbols used in this figure.

See Table 6B-1 for the meanings of the letter codes used in this figure.

See Table 6B-4 for formulas for calculating taper length (L).

Typical Application 24

Notes for Figure 6P-25—Typical Application 25
Multiple Lane Closures at an Intersection

Guidance:

1. *If the work space extends across a crosswalk, the crosswalk should be closed using the information and devices shown in Figure 6P-29.*

Support:

2. The normal procedure is to close on the near side of the intersection any lane that is not carried through the intersection, as shown.

Option:

3. Positive protection devices may be used per Section 6M.02.
4. If the left-turn movement that normally uses the closed turn bay is small and/or the gaps in opposing vehicular traffic are frequent, left turns may be permitted on that approach.
5. Flashing warning lights and/or flags may be used to call attention to the advance warning signs.

Figure 6P-25. Multiple Lane Closures at an Intersection (TA-25)

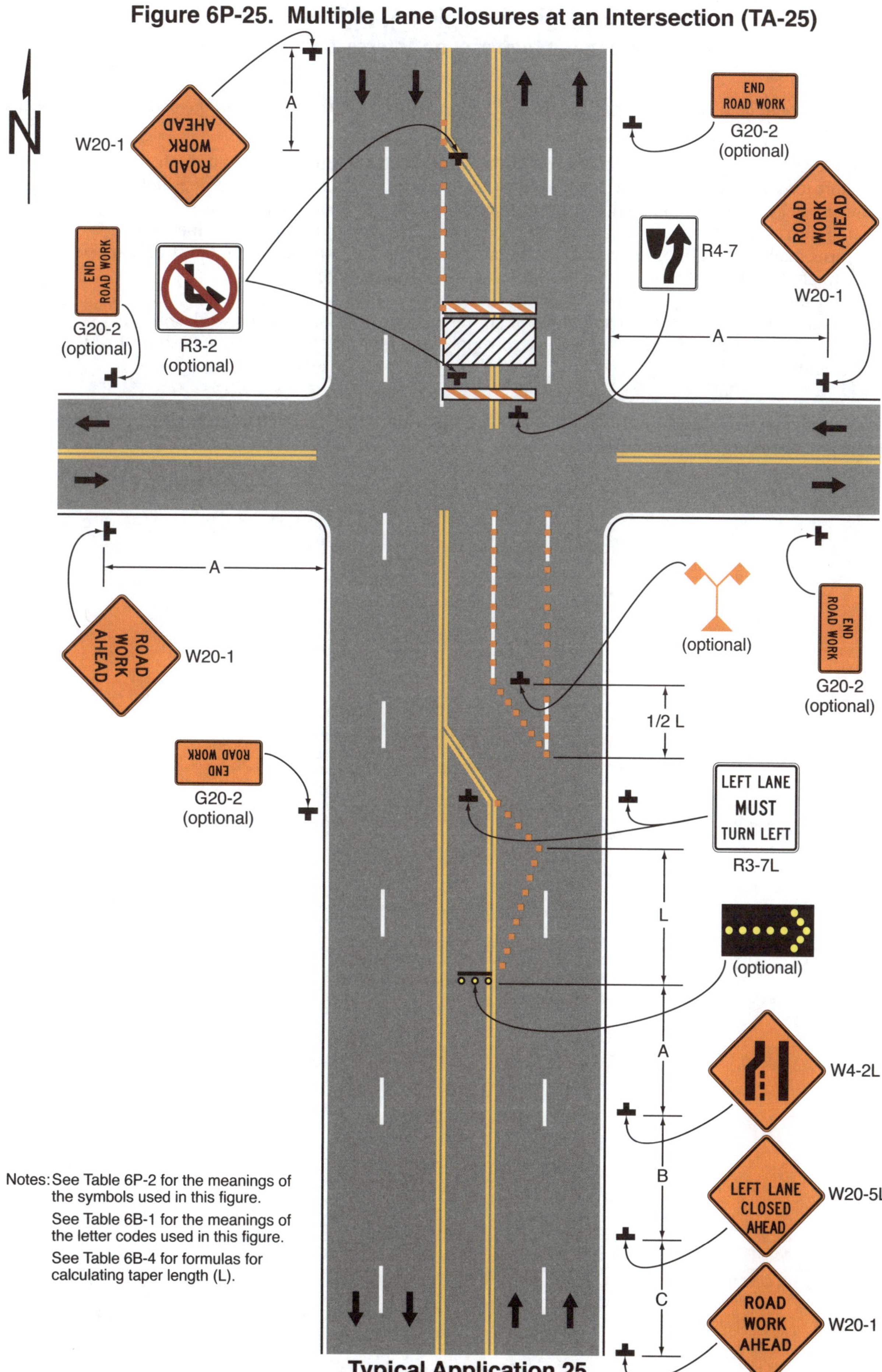

Notes: See Table 6P-2 for the meanings of the symbols used in this figure.

See Table 6B-1 for the meanings of the letter codes used in this figure.

See Table 6B-4 for formulas for calculating taper length (L).

Typical Application 25

Notes for Figure 6P-26—Typical Application 26
Closure in the Center of an Intersection

Guidance:

1. *All lanes should be a minimum of 10 feet in width as measured to the near face of the channelizing devices.*

Option:

2. A high-level warning device may be placed in the work space, if there is sufficient room.
3. For short-term use on low-volume, low-speed roadways with vehicular traffic that does not include longer and wider heavy commercial vehicles, a minimum lane width of 9 feet may be used.
4. Flashing warning lights and/or flags may be used to call attention to advance warning signs.
5. Left turns may be prohibited as required by geometric conditions, such as where the streets are so narrow that it might be physically impossible to turn left, especially for large vehicles.
6. For short-duration work operations, the channelizing devices may be eliminated if a vehicle displaying high-intensity rotating, flashing, oscillating, or strobe lights is positioned in the work space.
7. Vehicle hazard warning signals may be used to supplement high-intensity rotating, flashing, oscillating, or strobe lights.

Standard:

8. **Vehicle hazard warning signals shall not be used instead of the vehicle's high-intensity rotating, flashing, oscillating, or strobe lights.**

Figure 6P-26. Closure in the Center of an Intersection (TA-26)

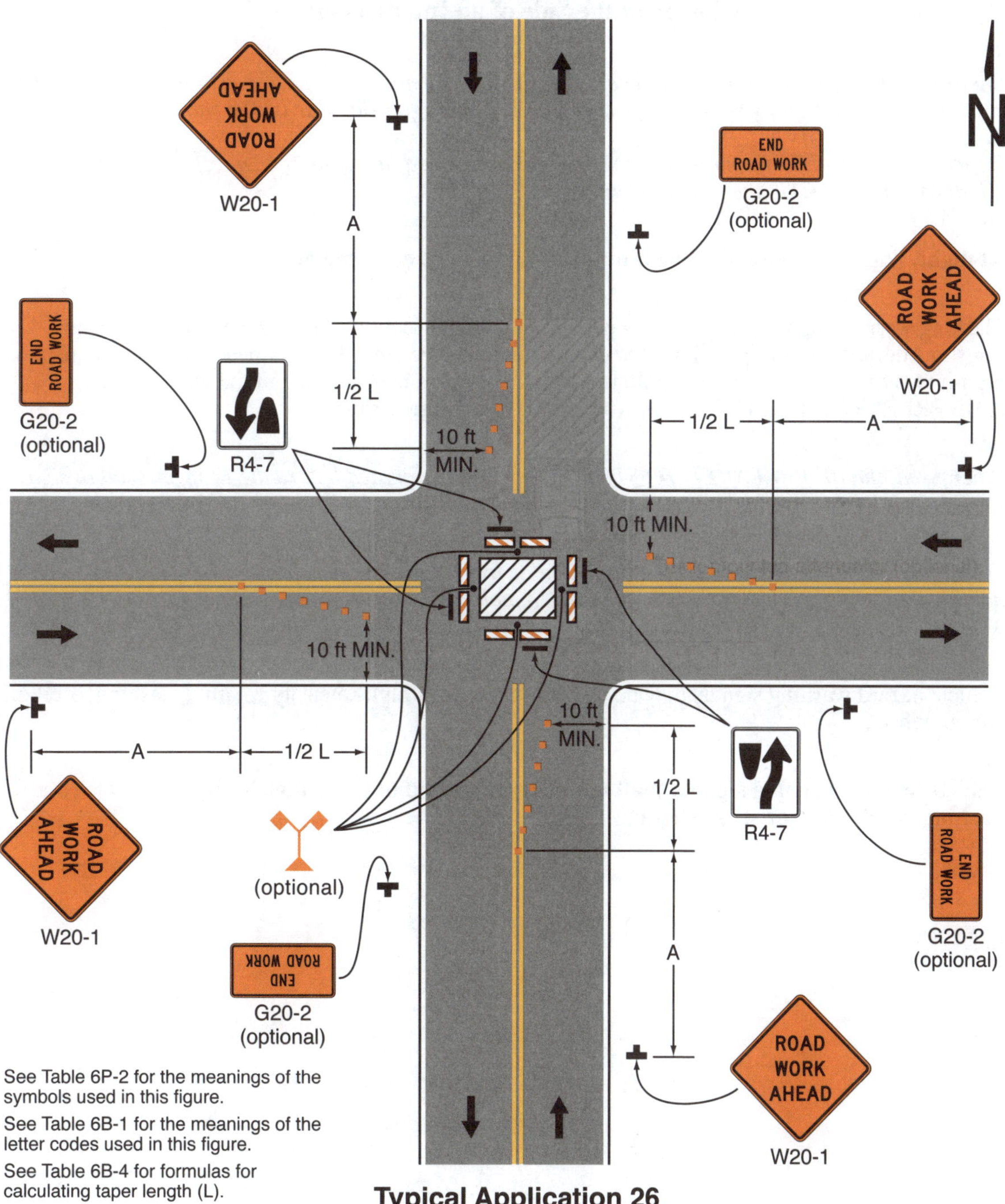

Note: See Table 6P-2 for the meanings of the symbols used in this figure.

See Table 6B-1 for the meanings of the letter codes used in this figure.

See Table 6B-4 for formulas for calculating taper length (L).

Typical Application 26

Notes for Figure 6P-27—Typical Application 27
Closure at the Side of an Intersection

Guidance:

1. *The situation depicted can be simplified by closing one or more of the intersection approaches. If this cannot be done, and/or when capacity is a problem, through vehicular traffic should be directed to other roads or streets.*
2. *Depending on road user conditions, flagger(s) or uniformed law enforcement officer(s) should be used to direct road users within the intersection.*

Standard:

3. **At night, flagger stations shall be illuminated, except in emergencies.**

Option:

4. Flashing warning lights and/or flags may be used to call attention to the advance warning signs.
5. For short-duration work operations, the channelizing devices may be eliminated if a vehicle displaying high-intensity rotating, flashing, oscillating, or strobe lights is positioned in the work space.
6. A BE PREPARED TO STOP sign may be added to the sign series.

Guidance:

7. *When used, the BE PREPARED TO STOP sign should be located before the Flagger symbol sign.*
8. *ONE LANE ROAD AHEAD signs should also be used to provide adequate advance warning.*

Support:

9. *Turns may be prohibited as required by vehicular traffic conditions, such as where the streets are so narrow that it might be physically impossible to make certain turns, especially for large vehicles.*

Option:

10. Positive protection devices may be used per Section 6M.02.
11. Vehicle hazard warning signals may be used to supplement high-intensity rotating, flashing, oscillating, or strobe lights.

Standard:

12. **Vehicle hazard warning signals shall not be used instead of the vehicle's high-intensity rotating, flashing, oscillating, or strobe lights.**

Figure 6P-27. Closure at the Side of an Intersection (TA-27)

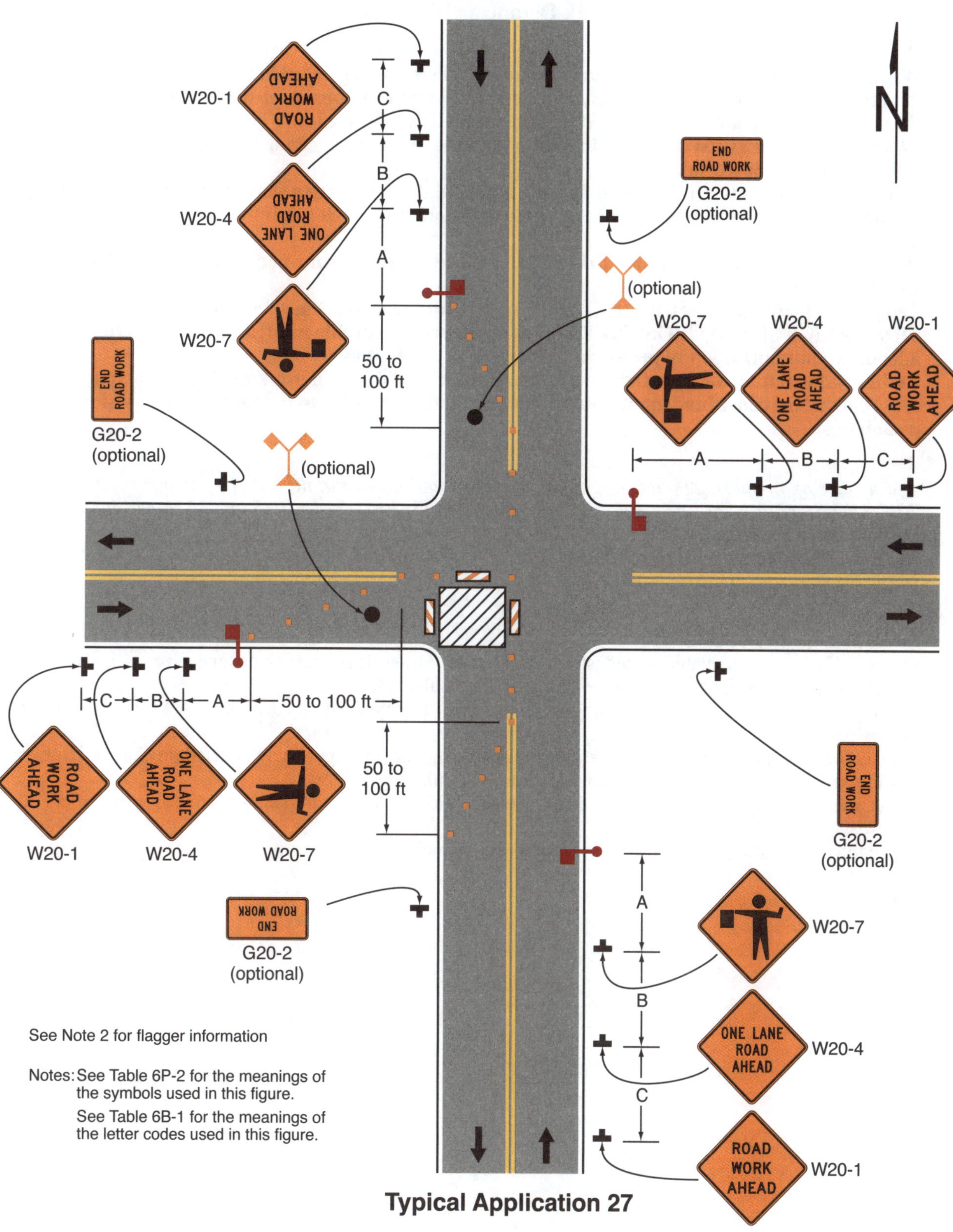

See Note 2 for flagger information

Notes: See Table 6P-2 for the meanings of the symbols used in this figure.

See Table 6B-1 for the meanings of the letter codes used in this figure.

Typical Application 27

Notes for Figure 6P-28—Typical Application 28
Sidewalk Detour or Diversion

Standard:

1. **When existing pedestrian facilities are disrupted, closed, or relocated in a TTC zone, the temporary facilities shall be detectable and include accessibility features consistent with the features present in the existing pedestrian facility. A pedestrian channelizing device (see Figure 6K-2) that is detectable by a person with a vision disability traveling with the aid of a long cane shall be placed across the full width of the closed sidewalk.**
2. **When used, temporary ramps shall provide a 12:1 (8.33%) or flatter slope, with a slip-resistant surface. The ramp landing area shall provide a 48-inch x 48-inch minimum area with a 2% or flatter cross-slope.**
3. **When used, Longitudinal Channelizing Devices used for temporary pedestrian routes shall comply with Section 6K.02.**
4. **Temporary traffic barriers, if used, shall comply with the provisions of Section 6M.02.**
5. **SIDEWALK CLOSED CROSS HERE signs shall include audible information devices to provide adequate communication to pedestrians with vision disabilities.**
6. **Audible information devices shall be provided where midblock sidewalk closings and changed crosswalk areas cause inadequate communication to be provided to pedestrians with vision disabilities.**

Guidance:

7. *The surface of an alternate pathway should meet the requirements of the U.S. Department of Justice 2010 ADA Standards for Accessible Design, September 15, 2010, 28 CFR 35 and 36, Americans with Disabilities Act of 1990.*
8. *The protective requirements of a TTC situation have priority in determining the need for temporary traffic barriers and their use in this situation should be based on engineering judgment.*

Option:

9. Street lighting may be considered.
10. Only the TTC devices related to pedestrians are shown. Other devices, such as lane closure signing or ROAD NARROWS signs, may be used to control vehicular traffic.
11. For nighttime closures, Type A Flashing warning lights may be used on barricades that support signs and close sidewalks.
12. Type C Steady-Burn or Type D 360-degree Steady-Burn warning lights may be used on channelizing devices separating the temporary sidewalks from vehicular traffic flow.
13. Signs, such as KEEP RIGHT (LEFT), may be placed along a temporary sidewalk to guide or direct pedestrians.
14. The width of the alternate pedestrian route may be 48 inches with a passing area of 60 inches every 200 feet.

Figure 6P-28. Sidewalk Detour or Diversion (TA-28)

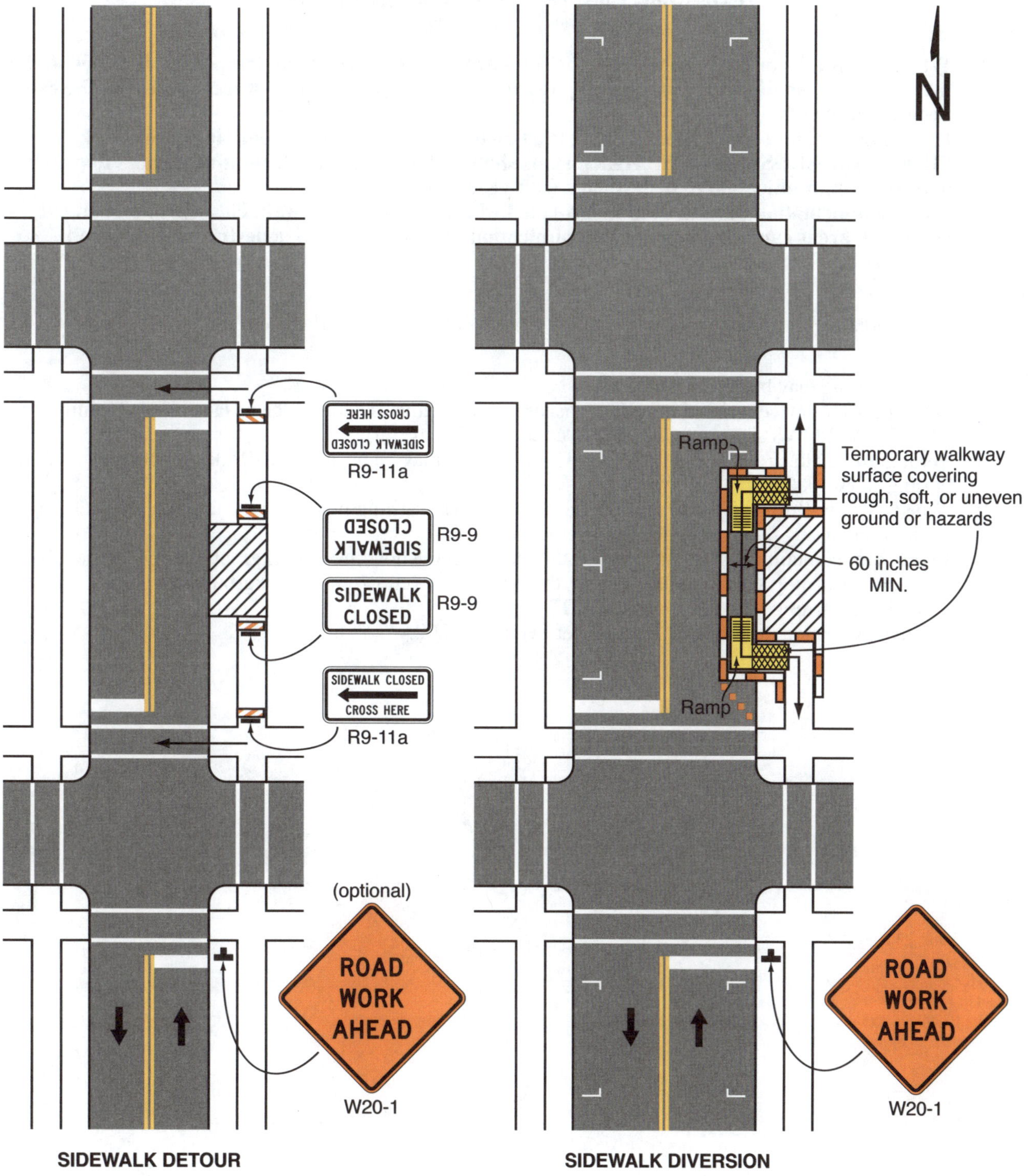

Typical Application 28

Note: See Table 6P-2 for the meanings
of the symbols used in this figure.

Notes for Figure 6P-29—Typical Application 29
Crosswalk Closures and Pedestrian Detours

Standard:

1. **When crosswalks or other pedestrian facilities are closed or relocated, temporary facilities shall be detectable and shall include accessibility features consistent with the features present in the existing pedestrian facility.**
2. **Curb parking shall be prohibited for at least 50 feet in advance of the midblock crosswalk.**
3. **SIDEWALK CLOSED CROSS HERE signs shall include audible information devices to provide adequate communication to pedestrians with vision disabilities.**
4. **Audible information devices shall be provided where midblock sidewalk closings and changed crosswalk areas cause inadequate communication to be provided to pedestrians with vision disabilities.**

Guidance:

5. *Pedestrian traffic signal displays controlling closed crosswalks should be covered or deactivated.*

Option:

6. Street lighting may be considered.
7. Only the TTC devices related to pedestrians are shown. Other devices, such as lane closure signing or ROAD NARROWS signs, may be used to control vehicular traffic.
8. For nighttime closures, Type A Flashing warning lights may be used on barricades supporting signs and closing sidewalks.
9. Type C Steady-Burn or Type D 360-degree Steady-Burn warning lights may be used on channelizing devices separating the work space from vehicular traffic.
10. In order to maintain the systematic use of the fluorescent yellow-green background for pedestrian, bicycle, and school warning signs in a jurisdiction, the fluorescent yellow-green background for pedestrian, bicycle, and school warning signs may be used in TTC zones.
11. Positive protection devices may be used per Section 6M.02.

Figure 6P-29. Crosswalk Closures and Pedestrian Detours (TA-29)

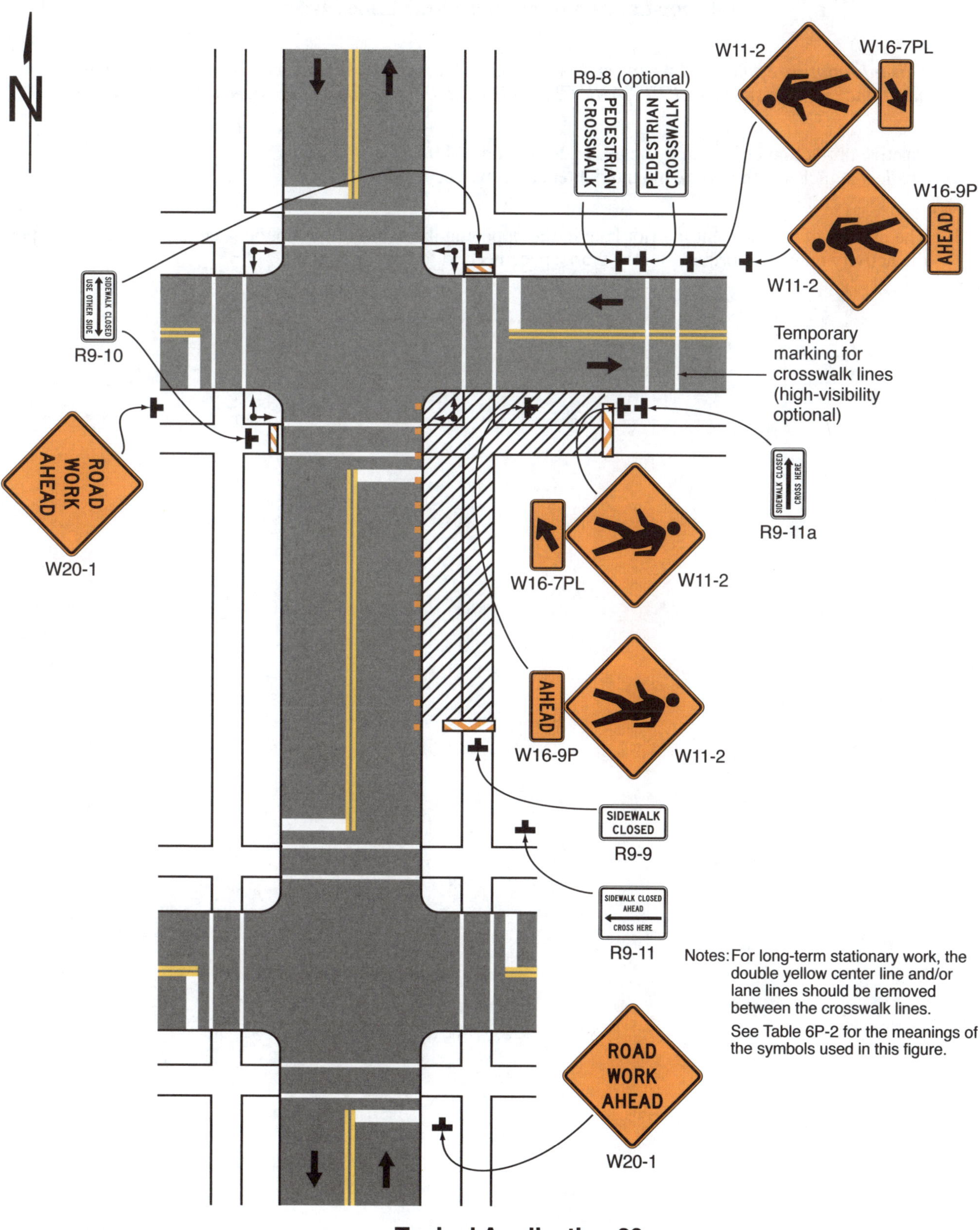

Typical Application 29

Notes for Figure 6P-30—Typical Application 30
Interior Lane Closure on a Multi-Lane Street

Guidance:

  1. *This information applies to low-speed, low-volume urban streets. Where speed or volume is higher, additional signing such as LEFT LANE CLOSED XX FT should be used between the signs shown.*

Option:

  2. Positive protection devices may be used per Section 6M.02.
  3. Shadow vehicles with a truck-mounted attenuator may be used.

Support:

  4. The closure of the adjacent interior lane in the opposing direction might not be necessary, depending upon the activity being performed and the work space needed for the operation.

Figure 6P-30. Interior Lane Closure on a Multi-Lane Street (TA-30)

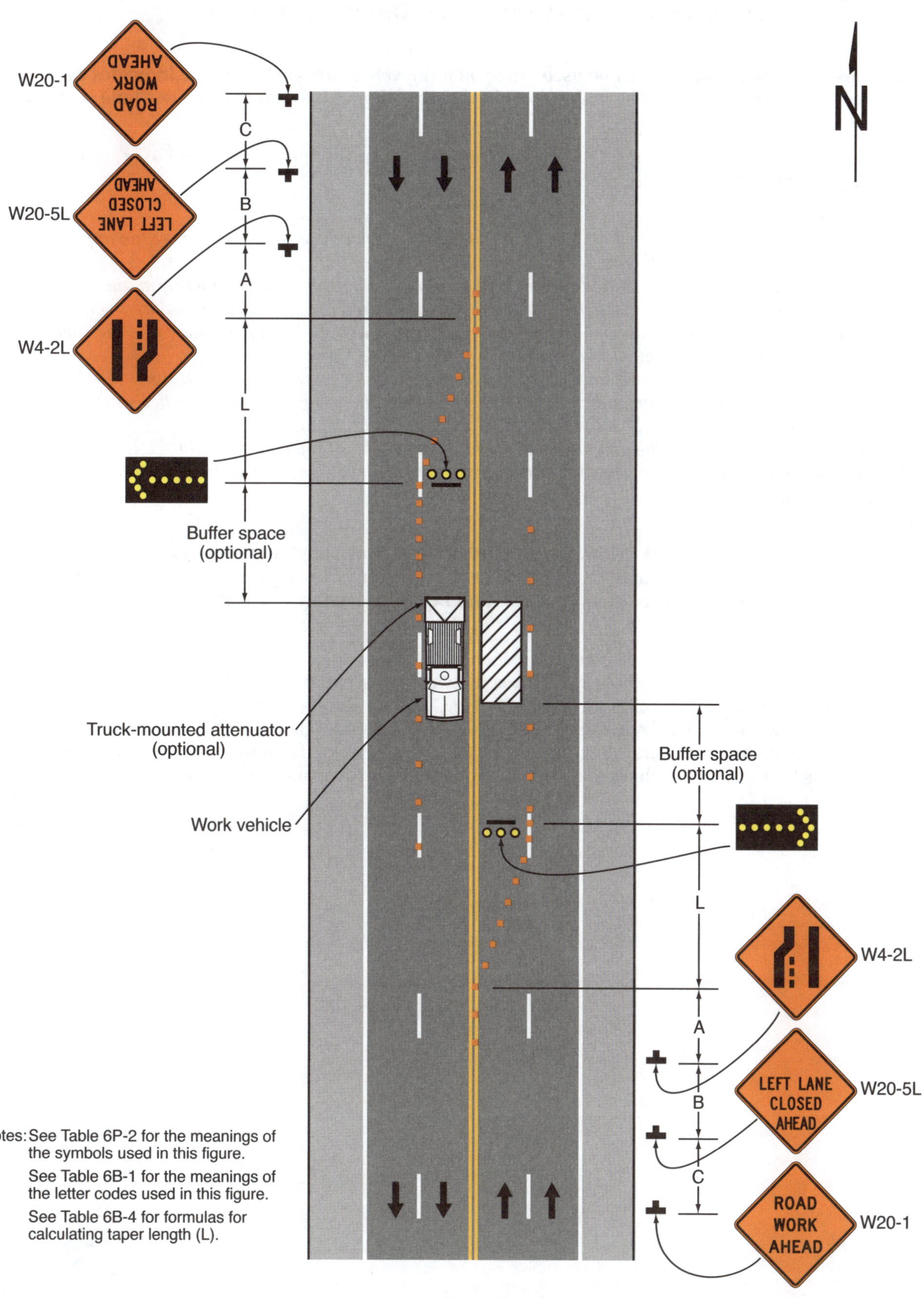

Typical Application 30

Notes for Figure 6P-31—Typical Application 31
Lane Closure on a Street with Uneven Directional Volumes

Standard:

1. **The illustrated information shall be used only when the vehicular traffic volume indicates that two lanes of vehicular traffic shall be maintained in the direction of travel for which one lane is closed.**

Option:

2. The procedure may be used during a peak period of vehicular traffic and then changed to provide two lanes in the other direction for the other peak.

Guidance:

3. *For high speeds, a LEFT LANE CLOSED XX FT sign should be added for vehicular traffic approaching the lane closure, as shown in Figure 6P-32.*
4. *Conflicting pavement markings should be removed for long-term projects. For short-term and intermediate-term projects where this is impracticable, the channelizing devices in the area where the pavement markings conflict should be placed at a maximum spacing of ½ S feet where S is the speed in mph. Temporary markings should be installed where needed.*
5. *If the lane shift has curves with recommended speeds of 30 mph or less, Reverse Turn signs should be used.*
6. *Where the shifted section is long, a Reverse Curve sign should be used to show the initial shift and a second sign should be used to show the return to the normal alignment.*
7. *If the tangent distance along the temporary diversion is less than 600 feet, the Double Reverse Curve sign should be used at the location of the first Two Lane Reverse Curve sign. The second Two Lane Reverse Curve sign should be omitted.*

Standard:

8. **Except as provided in Note 11 below, the number of lanes illustrated on the Reverse Curve or Double Reverse Curve signs shall be the same as the number of through lanes available to road users, and the direction of the reverse curves shall be appropriately illustrated.**

Option:

9. Positive protection devices may be used per Section 6M.02.
10. A longitudinal buffer space may be used in the activity area to separate opposing vehicular traffic.
11. Where two or more lanes are being shifted, a Reverse Curve (or Reverse Turn) sign with an ALL LANES plaque (see Figure 6H-1) may be used instead of a sign that illustrates the number of lanes.
12. Where more than three lanes are being shifted, the Reverse Curve (or Turn) sign may be rectangular.
13. A work vehicle or a shadow vehicle may be equipped with a truck-mounted attenuator.

Figure 6P-31. Lane Closure on a Street with Uneven Directional Volumes (TA-31)

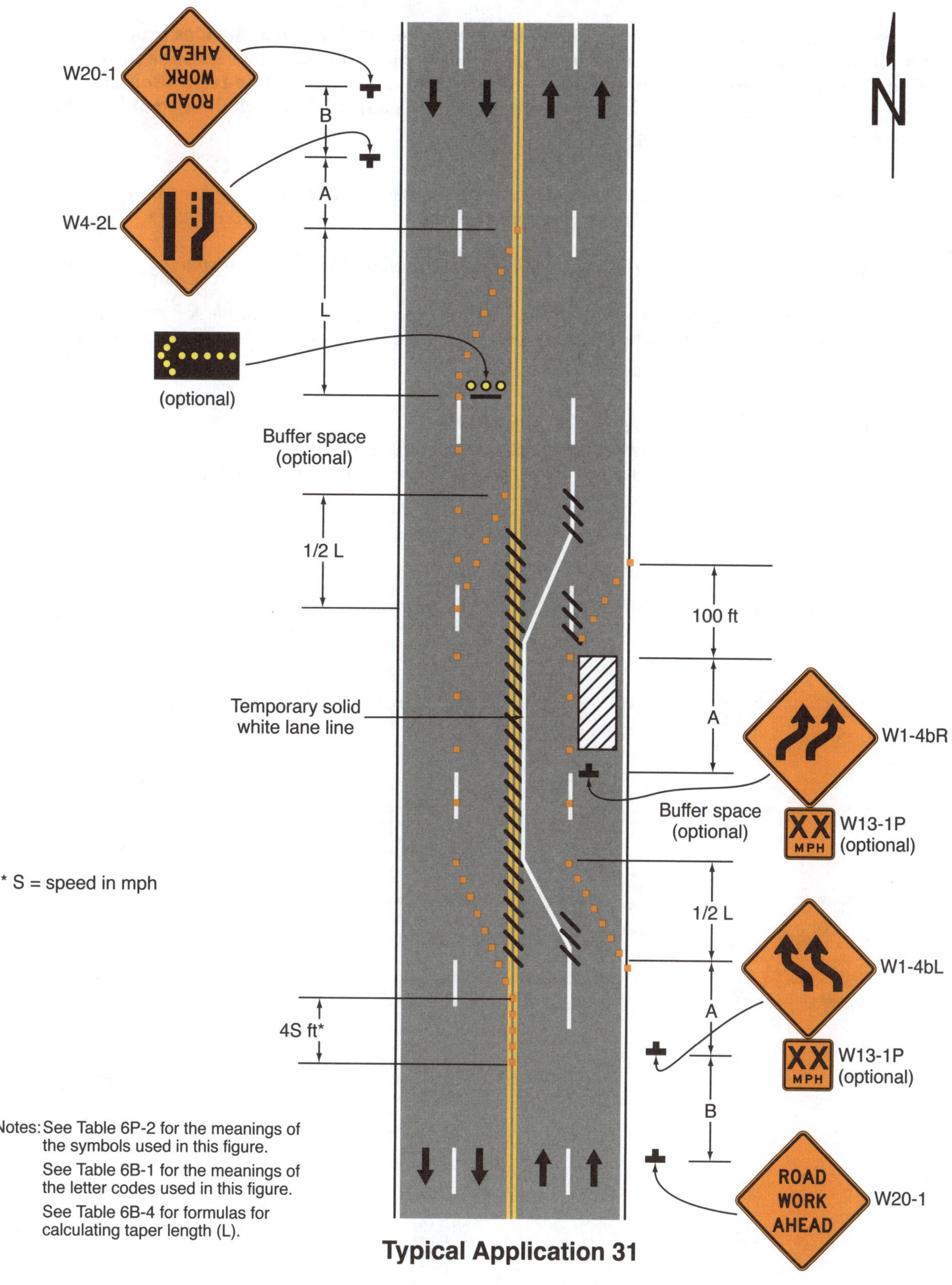

Typical Application 31

Notes for Figure 6P-32—Typical Application 32
Half Road Closure on a Multi-Lane, High-Speed Highway

Standard:

1. **Pavement markings no longer applicable shall be removed or obliterated as soon as practical. Except for intermediate-term and short-term situations, temporary markings shall be provided to clearly delineate the temporary travel path. For short-term and intermediate-term situations where it is not feasible to remove and restore pavement markings, channelization shall be made dominant by using a very close device spacing.**

Guidance:

2. *When paved shoulders having a width of 8 feet or more are closed, channelizing devices should be used to close the shoulder in advance of the merging taper to direct vehicular traffic to remain within the traveled way.*
3. *Where channelizing devices are used instead of pavement markings, the maximum spacing should be ½ S feet where S is the speed in mph.*
4. *If the tangent distance along the temporary diversion is less than 600 feet, a Double Reverse Curve sign should be used instead of the first Reverse Curve sign, and the second Reverse Curve sign should be omitted.*

Option:

5. Positive protection devices may be used per Section 6M.02.
6. Warning lights may be used to supplement channelizing devices at night.
7. A truck-mounted attenuator may be used on the work vehicle and/or the shadow vehicle.

Figure 6P-32. Half Road Closure on a Multi-Lane, High-Speed Highway (TA-32)

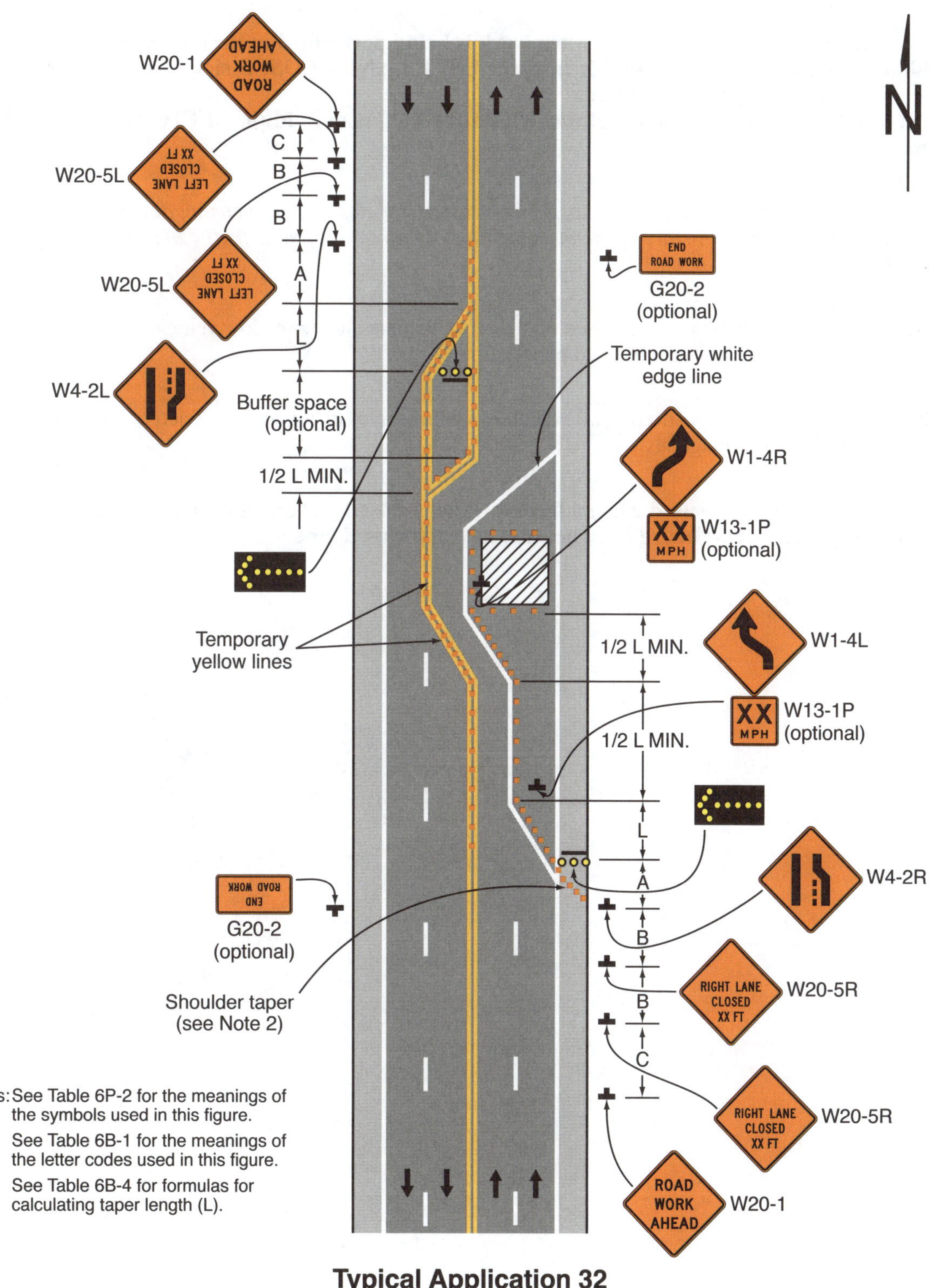

Typical Application 32

Notes for Figure 6P-33—Typical Application 33
Stationary Lane Closure on a Divided Highway

Standard:

1. **This information also shall be used when work is being performed in the lane adjacent to the median on a divided highway. In this case, the LEFT LANE CLOSED signs and the corresponding Lane Ends signs shall be substituted.**
2. **When a side road intersects the highway within the TTC zone, additional TTC devices shall be placed as needed.**

Guidance:

3. *When paved shoulders having a width of 8 feet or more are closed, channelizing devices should be used to close the shoulder in advance of the merging taper to direct vehicular traffic to remain within the traveled way.*

Option:

4. A truck-mounted attenuator may be used on the work vehicle and/or shadow vehicle.
5. Positive protection devices may be used per Section 6M.02.

Support:

6. Where conditions permit, restricting all vehicles, equipment, workers, and their activities to one side of the roadway might be advantageous.

Standard:

7. **An arrow board shall be used when a freeway lane is closed. When more than one freeway lane is closed, a separate arrow board shall be used for each closed lane.**

Figure 6P-33. Stationary Lane Closure on a Divided Highway (TA-33)

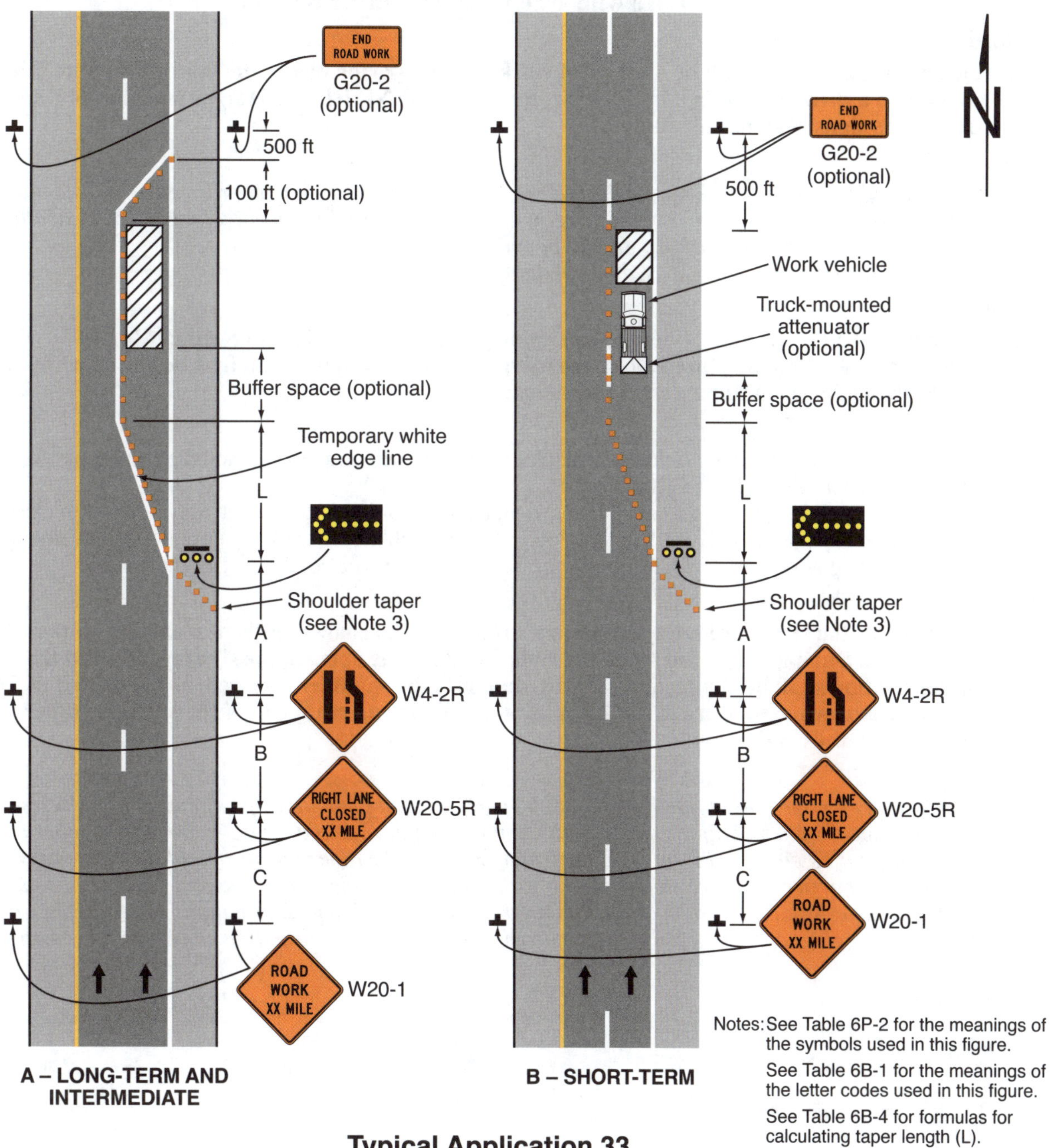

Typical Application 33

Notes for Figure 6P-34—Typical Application 34
Lane Closure with a Temporary Traffic Barrier

Standard:

1. **This information also shall be used when work is being performed in the lane adjacent to the median on a divided highway. In this case, the LEFT LANE CLOSED signs and the corresponding Lane Ends signs shall be substituted.**

Guidance:

2. *For long-term lane closures on facilities with permanent edge lines, a temporary edge line should be installed from the upstream end of the merging taper to the downstream end of the downstream taper, and conflicting pavement markings should be removed.*
3. *The use of a barrier should be based on engineering judgment.*

Standard:

4. **Temporary traffic barriers, if used, shall comply with the provisions of Section 6M.02.**
5. **The barrier shall not be placed along the merging taper. The lane shall first be closed using channelizing devices and pavement markings.**

Option:

6. Type C Steady-Burn warning lights may be placed on channelizing devices and the barrier parallel to the edge of pavement for nighttime lane closures.
7. The barrier shown in this typical application is an example of one method that may be used to close a lane for a long-term project. If the work activity permits, a movable barrier may be used and relocated to the shoulder during non-work periods or peak-period vehicular traffic conditions, as appropriate.

Standard:

8. **If a movable barrier is used, the temporary white edge line shown in the typical application shall not be used. During the period when the right-hand lane is opened, the sign legends and the channelization shall be changed to indicate that only the shoulder is closed, as illustrated in Figure 6P-5. The arrow board, if used, shall be placed at the downstream end of the shoulder taper and shall display the caution mode.**

Guidance:

9. *If a movable barrier is used, the shift should be performed in the following manner. When closing the lane, the lane should be initially closed with channelizing devices placed along a merging taper using the same information employed for a stationary lane closure. The lane closure should then be extended with the movable-barrier transfer vehicle moving with vehicular traffic. When opening the lane, the movable-barrier transfer vehicle should travel against vehicular traffic from the termination area to the transition area. The merging taper should then be removed using the same information employed for a stationary lane closure.*

Figure 6P-34. Lane Closure with a Temporary Traffic Barrier (TA-34)

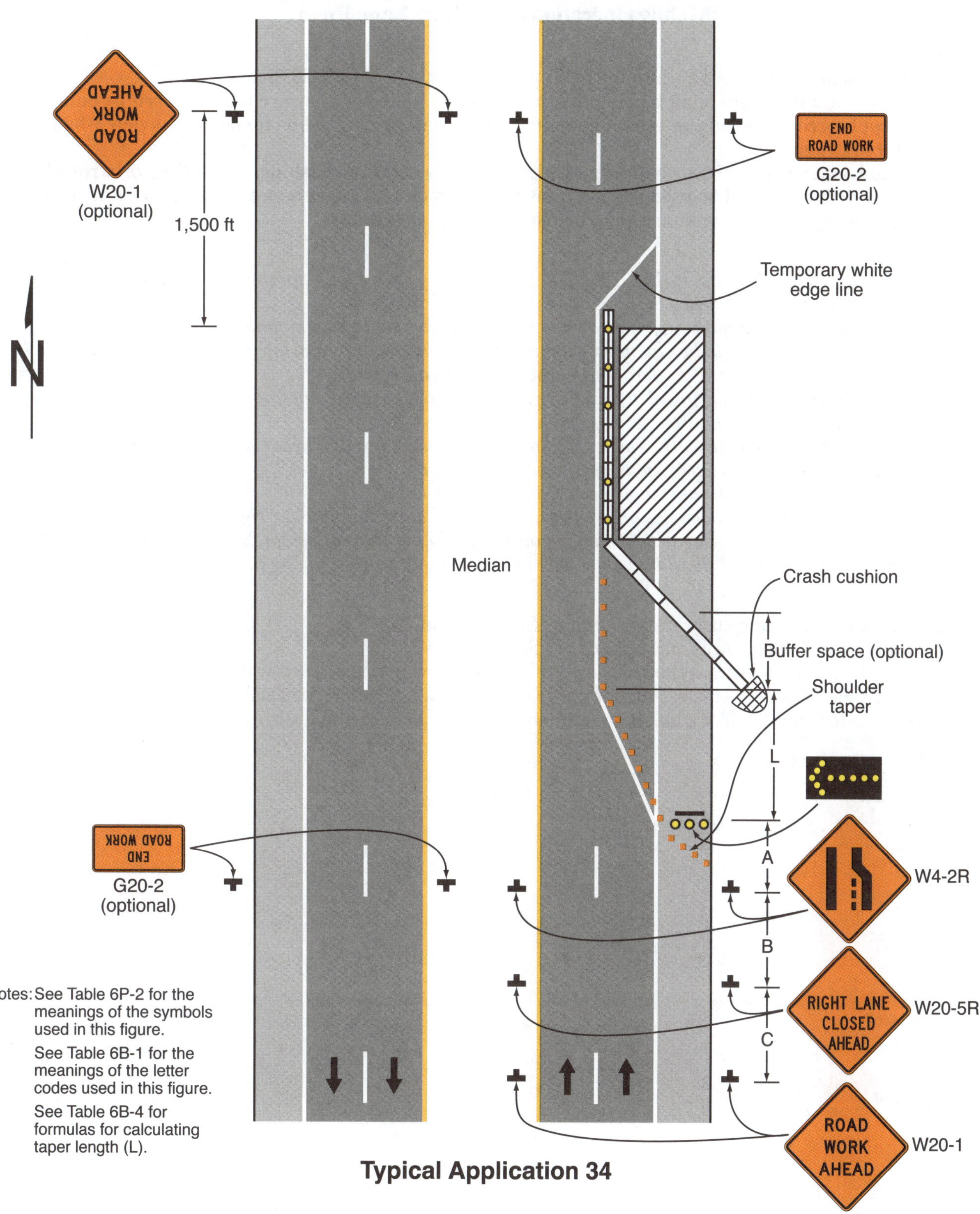

Typical Application 34

Notes for Figure 6P-35—Typical Application 35
Mobile Operation on a Multi-Lane Road

Standard:
1. **Arrow boards shall, as a minimum, be Type B, with a size of 60 x 30 inches.**
2. **Vehicle-mounted signs shall be mounted in a manner such that they are not obscured by equipment or supplies. Sign legends on vehicle-mounted signs shall be covered or turned from view when work is not in progress.**
3. **Shadow and work vehicles shall display high-intensity rotating, flashing, oscillating, or strobe lights.**
4. **An arrow board shall be used when a freeway lane is closed. When more than one freeway lane is closed, a separate arrow board shall be used for each closed lane.**

Guidance:
5. *Vehicles used for these operations should be made highly visible with appropriate equipment, such as flags, signs, or arrow boards.*
6. *Shadow Vehicle 1 should be equipped with an arrow board and truck-mounted attenuator.*
7. *Shadow Vehicle 2 should be equipped with an arrow board. An appropriate lane closure sign should be placed on Shadow Vehicle 2 so as not to obscure the arrow board.*
8. *Shadow Vehicle 2 should travel at a varying distance from the work operation so as to provide adequate sight distance for vehicular traffic approaching from the rear.*
9. *The spacing between the work vehicles and the shadow vehicles, and between each shadow vehicle, should be minimized to deter road users from driving in between.*
10. *Work should normally be accomplished during off-peak hours.*
11. *When the work vehicle occupies an interior lane (a lane other than the far right or far left) of a directional roadway having a right-hand shoulder 10 feet or more in width, Shadow Vehicle 2 should drive on the right-hand shoulder with a sign indicating that work is taking place in the interior lane.*

Option:
12. A truck-mounted attenuator may be used on Shadow Vehicle 2.
13. Positive protection devices may be used per Section 6M.02.
14. On high-speed roadways, a third shadow vehicle (not shown) may be used with Shadow Vehicle 1 in the closed lane, Shadow Vehicle 2 straddling the edge line, and Shadow Vehicle 3 on the shoulder.
15. Where adequate shoulder width is not available, Shadow Vehicle 3 may also straddle the edge line.

Figure 6P-35. Mobile Operation on a Multi-Lane Road (TA-35)

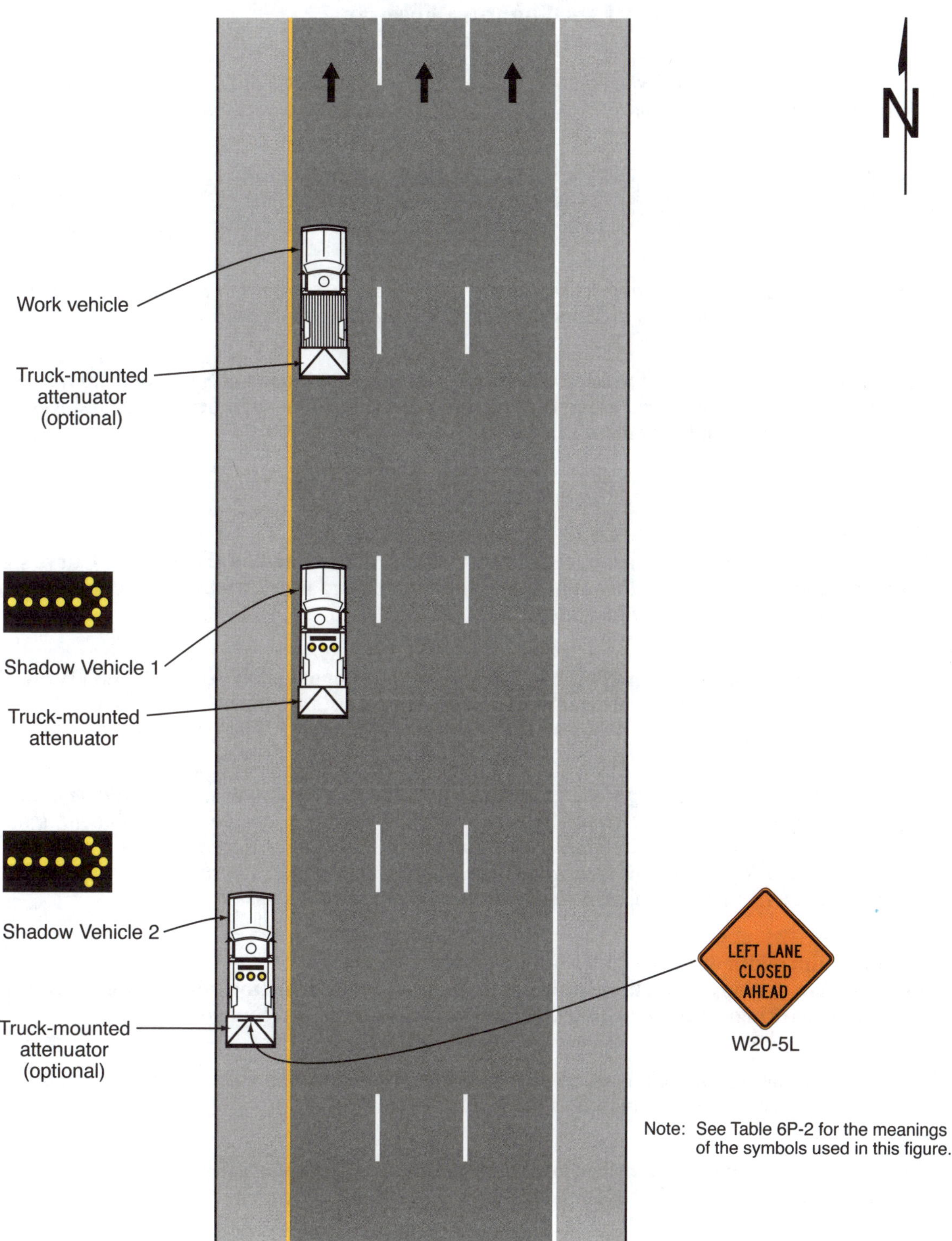

Typical Application 35

Notes for Figure 6P-36—Typical Application 36
Lane Shift on a Freeway

Guidance:

1. *The lane shift should be used when the work space extends into either the right-hand or left-hand lane of a divided highway and it is impracticable, for capacity reasons, to reduce the number of available lanes.*

Support:

2. When a lane shift is accomplished by using (1) geometry that meets the design speed at which the permanent highway was designed, (2) full normal cross-section (full lane width and full shoulders), and (3) complete pavement markings, then only the initial general work-zone warning sign is required.

Guidance:

3. *When the conditions in Note 2 above are not met, the information shown in the typical application should be employed and the provisions in Notes 4 through 17 below are applicable.*

Standard:

4. **Temporary traffic barriers, if used, shall comply with the provisions of Section 6M.02.**
5. **The barrier shall not be placed along the shifting taper. The lane shall first be shifted using channelizing devices and pavement markings.**

Guidance:

6. *A warning sign should be used to show the changed alignment.*

Standard:

7. **Except as provided in Note 8 below, the number of lanes illustrated on the Reverse Curve signs shall be the same as the number of through lanes available to road users, and the direction of the reverse curves shall be appropriately illustrated.**

Option:

8. Where two or more lanes are being shifted, a W1-4 (or W1-3) sign with an ALL LANES (W24-1cP) plaque (see Figure 6H-1) may be used instead of a sign that illustrates the number of lanes.
9. Where more than three lanes are being shifted, the Reverse Curve (or Turn) sign may be rectangular.

Guidance:

10. *Where the shifted section is longer than 600 feet, one set of Reverse Curve signs should be used to show the initial shift and a second set should be used to show the return to the normal alignment. If the tangent distance along the temporary diversion is less than 600 feet, a Double Reverse Curve sign should be used instead of the first Reverse Curve sign, and the second Reverse Curve sign should be omitted.*
11. *If a STAY IN LANE sign is used, then solid white lane lines should be used.*

Standard:

12. **The minimum width of the shoulder lane shall be 10 feet.**
13. **For long-term stationary work, existing conflicting pavement markings shall be removed and temporary markings shall be installed before traffic patterns are changed.**

Option:

14. For short-term stationary work, lanes may be delineated by channelizing devices or removable pavement markings instead of temporary markings.

Guidance:

15. *If the shoulder cannot adequately accommodate trucks, trucks should be directed to use the travel lanes.*
16. *The use of a barrier should be based on engineering judgment.*

Option:

17. Type C Steady-Burn warning lights may be placed on channelizing devices and the barrier parallel to the edge of the pavement for nighttime lane closures.

Figure 6P-36. Lane Shift on a Freeway (TA-36)

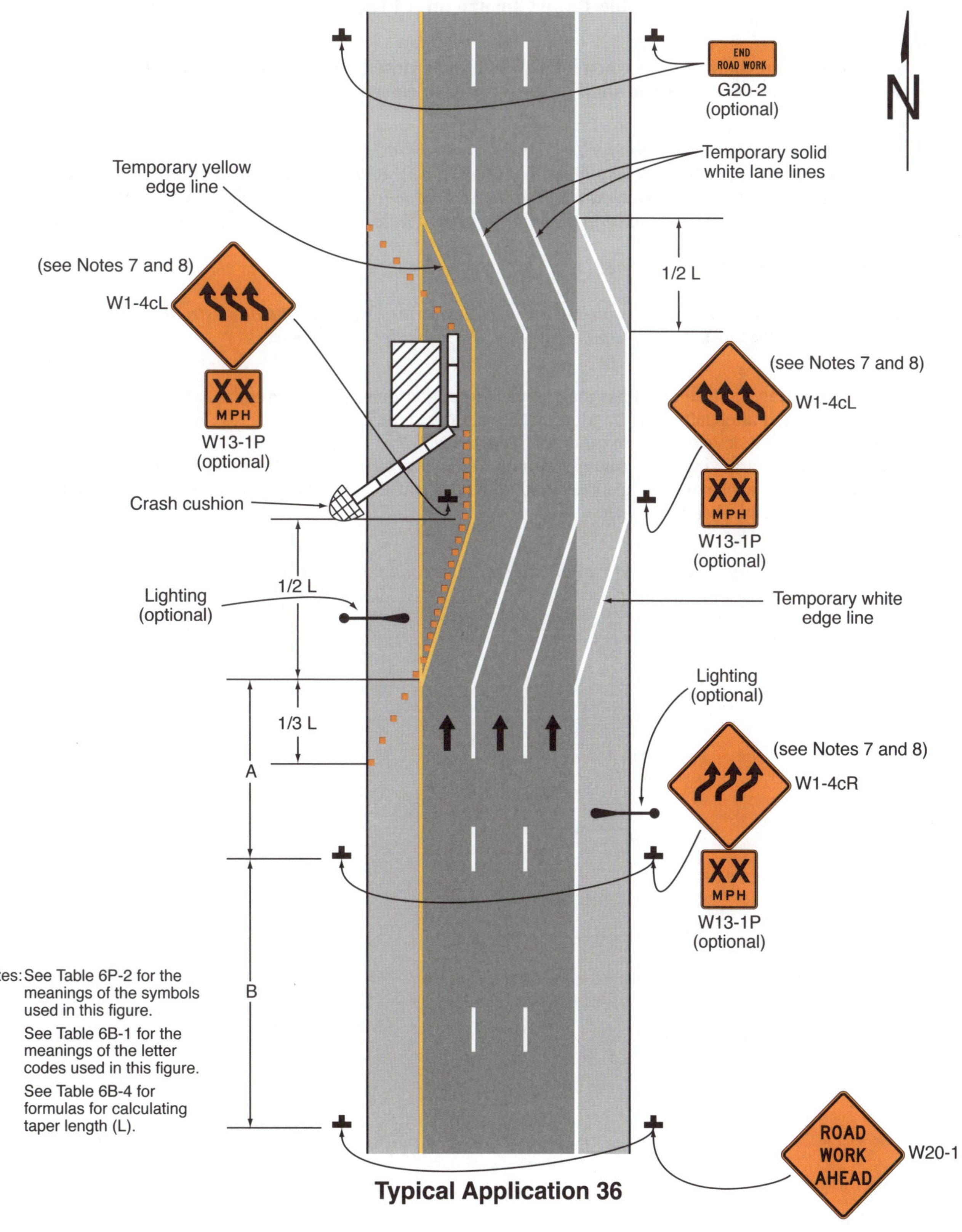

Typical Application 36

Notes for Figure 6P-37—Typical Application 37
Double Lane Closure on a Freeway

Standard:

    1. An arrow board shall be used when a freeway lane is closed. When more than one freeway lane is closed, a separate arrow board shall be used for each closed lane.

Guidance:

    2. Ordinarily, the preferred position for the second arrow board is in the closed exterior lane at the upstream end of the second merging taper. However, the second arrow board should be placed in the closed interior lane at the downstream end of the second merging taper in the following situations:
        a. When a shadow vehicle is used in the interior closed lane, and the second arrow board is mounted on the shadow vehicle;
        b. If alignment or other conditions create any confusion as to which lane is closed by the second arrow board; and
        c. When the first arrow board is placed in the closed exterior lane at the downstream end of the first merging taper (the alternative position when the shoulder is narrow).

Option:

    3. Flashing warning lights and/or flags may be used to call attention to the initial warning signs.
    4. A truck-mounted attenuator may be used on the shadow vehicle.
    5. Positive protection devices may be used per Section 6M.02.
    6. If a paved shoulder having a minimum width of 10 feet and sufficient strength is available, the left-hand and adjacent interior lanes may be closed and vehicular traffic carried around the work space on the right-hand lane and a right-hand shoulder.

Guidance:

    7. When a shoulder lane is used that cannot adequately accommodate trucks, trucks should be directed to use the normal travel lanes.

Figure 6P-37. Double Lane Closure on a Freeway (TA-37)

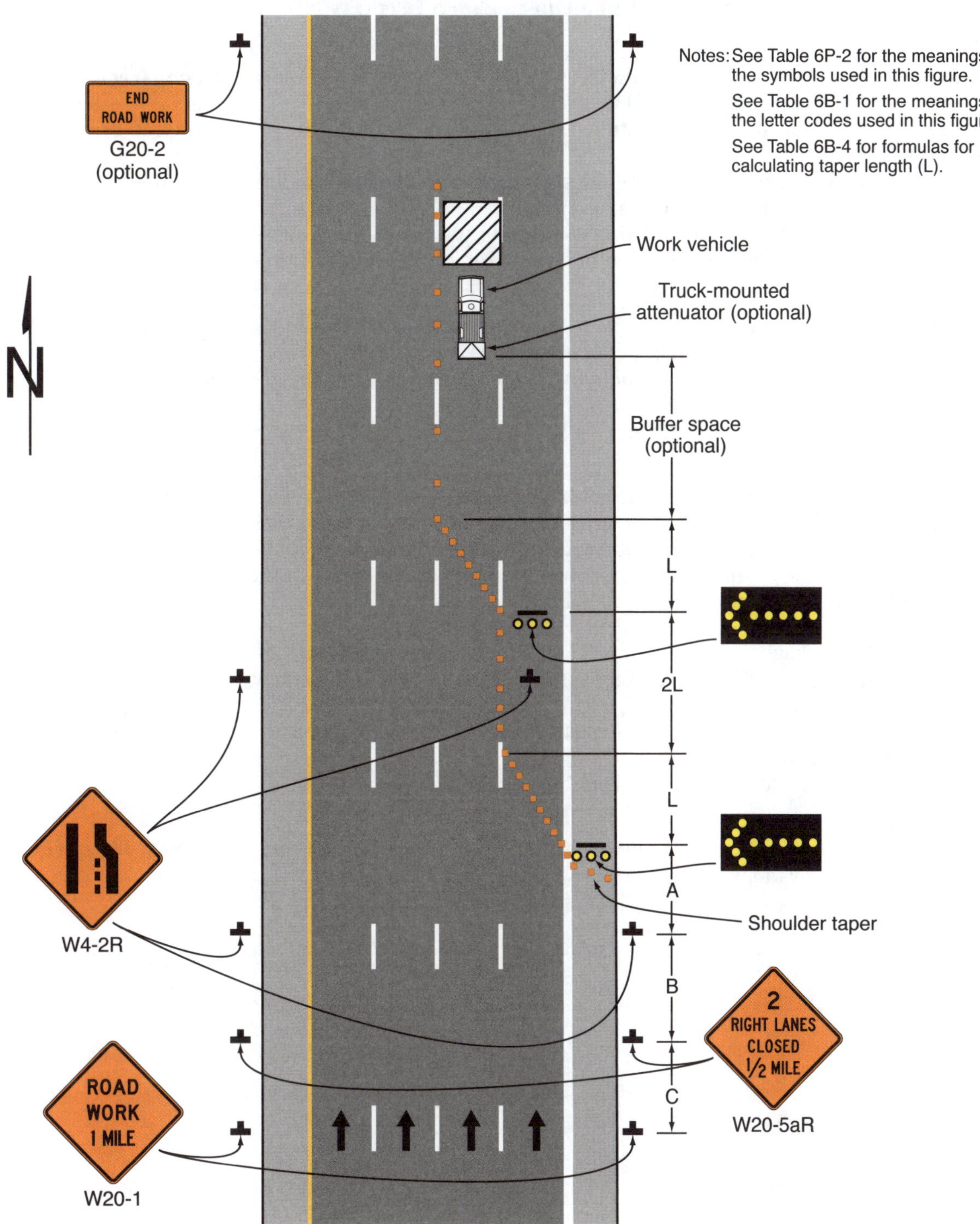

Typical Application 37

Notes for Figure 6P-38—Typical Application 38
Interior Lane Closure on a Freeway

Standard:

1. **An arrow board shall be used when a freeway lane is closed. When more than one freeway lane is closed, a separate arrow board shall be used for each closed lane.**
2. **If temporary traffic barriers are installed, they shall comply with the provisions and requirements in Section 6M.02.**
3. **The barrier shall not be placed along the shifting taper. The lane shall first be shifted using channelizing devices and pavement markings.**
4. **For long-term stationary work, existing conflicting pavement markings shall be removed and temporary markings shall be installed before traffic patterns are changed.**

Guidance:

5. *For a long-term closure, a barrier should be used to provide additional safety to the operation in the closed interior lane. A buffer space should be used at the upstream end of the closed interior lane.*
6. *An arrow board displaying an arrow pointing to the right should be placed on the left-hand shoulder at the beginning of the taper.*
7. *For long-term use, the broken lane lines should be made solid white in the two-lane section.*

Option:

8. As an alternative to initially closing the left-hand lane, as shown in the typical application, the right-hand lane may be closed in advance of the interior lane closure with appropriate channelization and signs. The Interior Lane Shift Ahead symbol sign may be mirrored to indicate a right lane shift.
9. A short, single row of channelizing devices in advance of the vehicular traffic split to restrict vehicular traffic to their respective lanes may be added.
10. DO NOT PASS signs may be used.
11. If a paved shoulder having a minimum width of 10 feet and sufficient strength is available, the left-hand and center lanes may be closed and motor vehicle traffic carried around the work space on the right-hand lane and a right-hand shoulder.
12. A work vehicle with a truck-mounted attenuator may be used within the closed interior lane between the buffer space and the work area.
13. Positive protection devices may be used per Section 6M.02.

Guidance:

14. *When a shoulder lane is used that cannot adequately accommodate trucks, trucks should be directed to use the normal travel lanes.*

Figure 6P-38. Interior Lane Closure on a Freeway (TA-38)

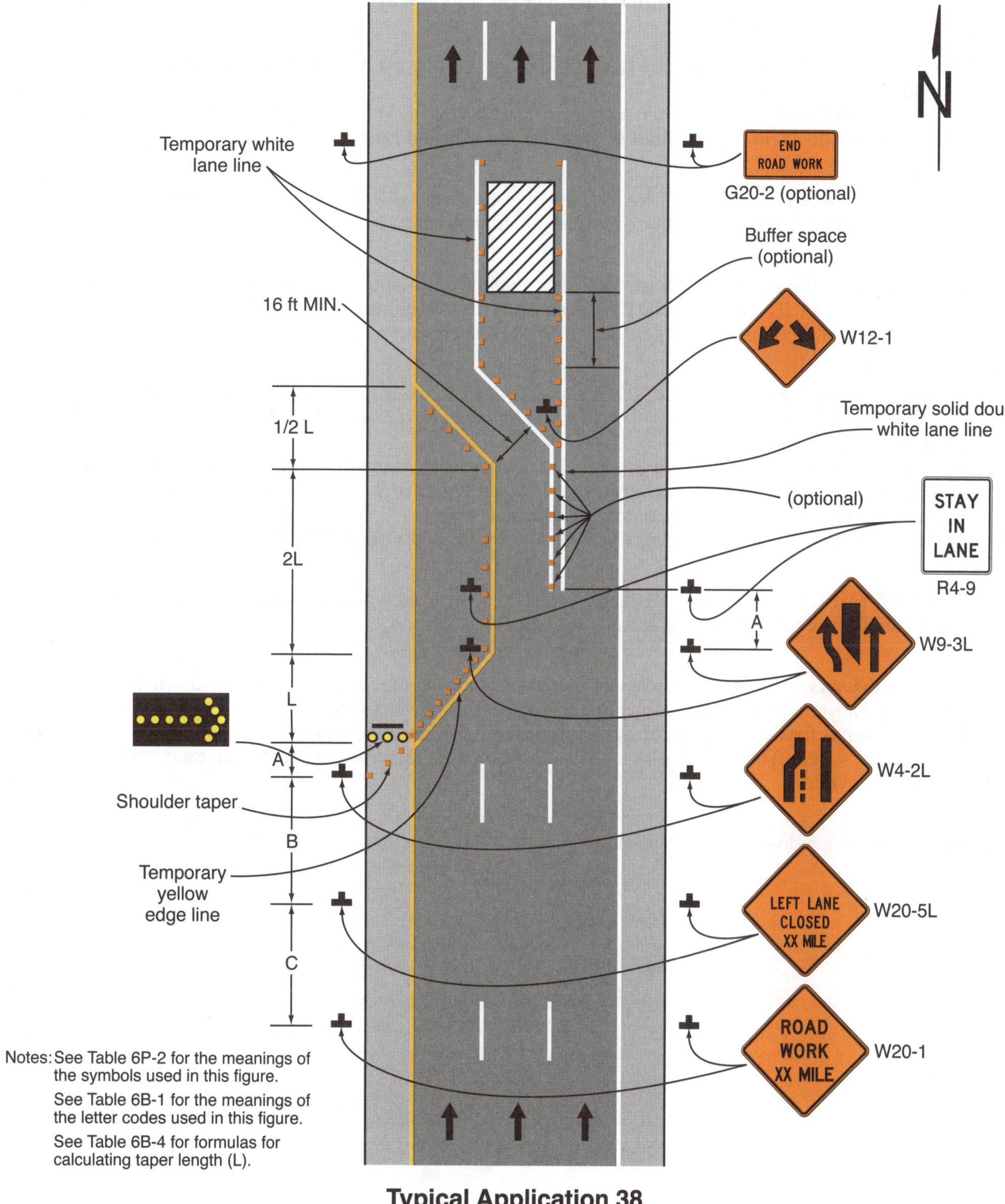

Typical Application 38

Notes for Figure 6P-39—Typical Application 39
Median Crossover on a Freeway

Standard:

1. **Channelizing devices or temporary traffic barriers shall be used to separate opposing vehicular traffic.**
2. **An arrow board shall be used when a freeway lane is closed. When more than one freeway lane is closed, a separate arrow board shall be used for each closed lane.**

Guidance:

3. *For long-term work on high-speed, high-volume highways, consideration should be given to using a temporary traffic barrier to separate opposing vehicular traffic.*

Option:

4. When a temporary traffic barrier is used to separate opposing vehicular traffic, the Two-Way Traffic, Do Not Pass, KEEP RIGHT, and DO NOT ENTER signs may be eliminated.
5. The alignment of the crossover may be designed as a reverse curve.

Guidance:

6. *When the crossover follows a curved alignment, the design criteria contained in the "AASHTO Green Book – A Policy On Geometric Design Of Highways And Streets," 7th Edition, 2018, AASHTO should be used.*
7. *When channelizing devices have the potential of leading vehicular traffic out of the intended traffic space, the channelizing devices should be extended a distance in feet of 2 times the speed limit in mph beyond the downstream end of the transition area as depicted.*
8. *Where channelizing devices are used, the Two-Way Traffic signs should be repeated every 1 mile.*

Option:

9. NEXT XX MILES Supplemental Distance plaques may be used with the Two-Way Traffic signs, where XX is the distance to the downstream end of the two-way section.

Support:

10. When the distance is sufficiently short that road users entering the section can see the downstream end of the section, they are less likely to forget that there is opposing vehicular traffic.
11. The sign legends for the four pairs of signs approaching the lane closure for the non-crossover direction of travel are not shown. They are similar to the series shown for the crossover direction, except that the left-hand lane is closed.

Option:

12. Positive protection devices may be used per Section 6M.02.

Figure 6P-39. Median Crossover on a Freeway (TA-39)

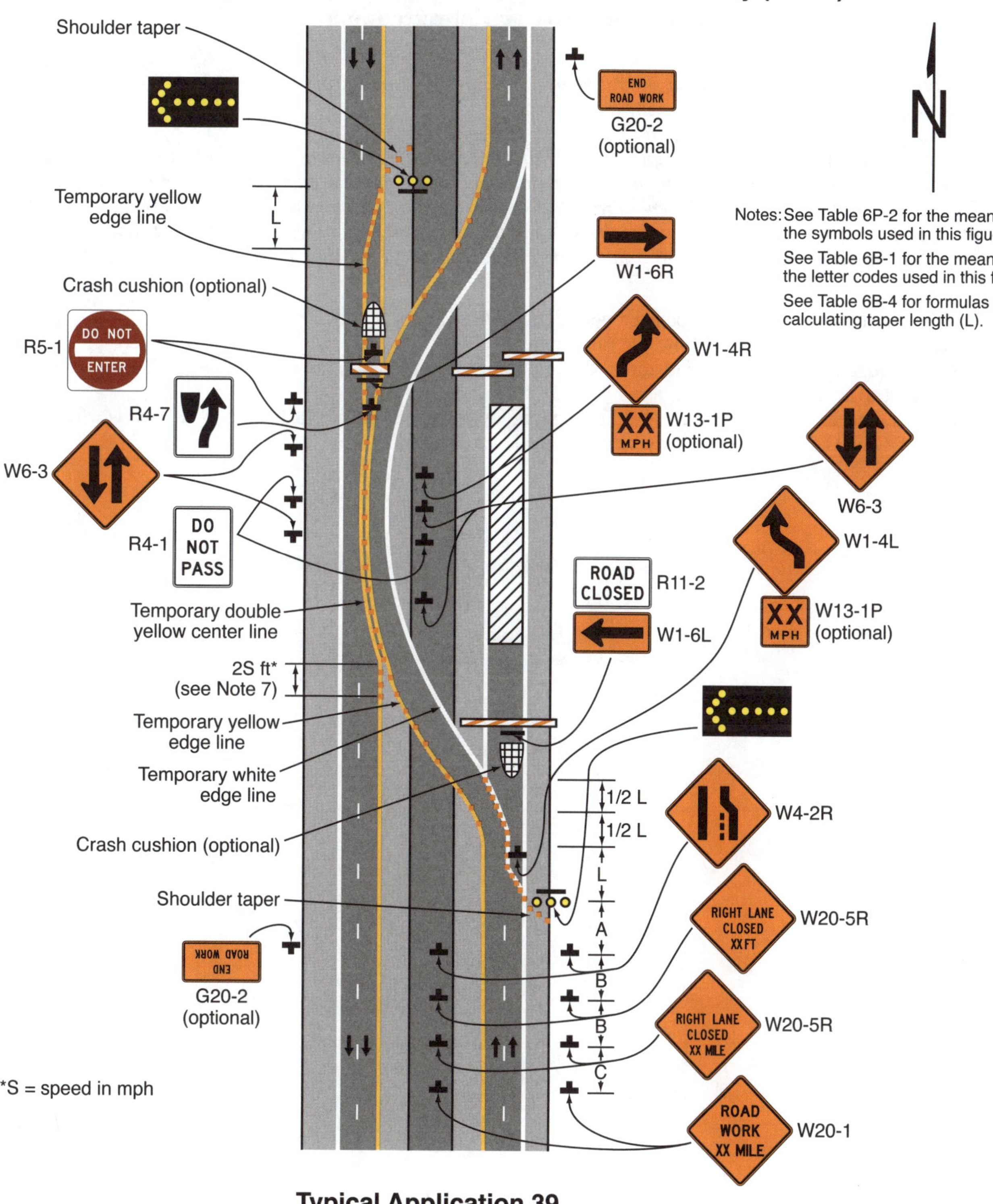

Typical Application 39

Notes for Figure 6P-40—Typical Application 40
Median Crossover for an Entrance Ramp

Guidance:
1. *The typical application illustrated should be used for carrying an entrance ramp across a closed directional roadway of a divided highway.*
2. *A temporary acceleration lane should be used to facilitate merging.*
3. *When used, the YIELD or STOP sign should be located far enough forward to provide adequate sight distance of oncoming mainline vehicular traffic to select an acceptable gap, but should not be located so far forward that motorists will be encouraged to stop in the path of the mainline traffic. If needed, yield or stop lines should be installed across the ramp to indicate the point at which road users should yield or stop. Also, a longer acceleration lane should be provided beyond the sign to reduce the gap size needed.*

Option:
4. Positive protection devices may be used per Section 6M.02.
5. If vehicular traffic conditions allow, the ramp may be closed.
6. A broken edge line may be carried across the temporary entrance ramp to assist in defining the through vehicular traffic lane.
7. When a temporary traffic barrier is used to separate opposing vehicular traffic, the Two-Way Traffic signs and the DO NOT ENTER signs may be eliminated.

Figure 6P-40. Median Crossover for an Entrance Ramp (TA-40)

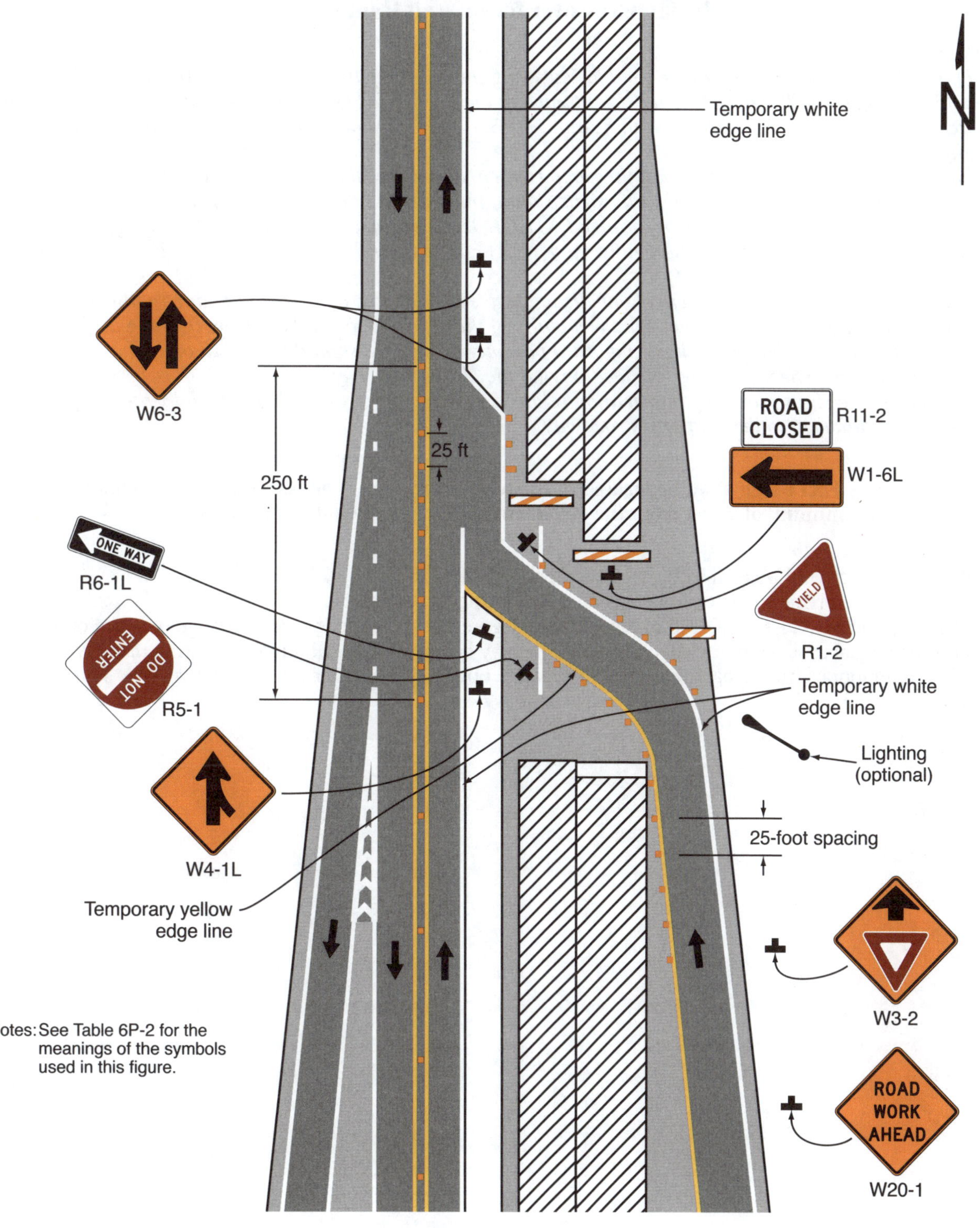

Typical Application 40

Notes for Figure 6P-41—Typical Application 41
Median Crossover for an Exit Ramp

Guidance:

1. *This typical application should be used for carrying an exit ramp across a closed directional roadway of a divided highway. The design criteria contained in the "AASHTO Green Book – A Policy On Geometric Design Of Highways And Streets," 7th Edition, 2018, AASHTO should be used for determining the curved alignment.*
2. *The guide signs should indicate that the ramp is open, and where the temporary ramp is located. Conversely, if the ramp is closed, guide signs should indicate that the ramp is closed.*
3. *When the exit is closed, a black-on-orange EXIT CLOSED sign panel should be placed diagonally across the interchange/intersection guide signs and channelizing devices should be placed to physically close the ramp.*
4. *In the situation (not shown) where channelizing devices are placed along the mainline roadway, the devices' spacing should be reduced in the vicinity of the off ramp to emphasize the opening at the ramp itself. Channelizing devices and/or temporary pavement markings should be placed on both sides of the temporary ramp where it crosses the median and the closed roadway.*
5. *Advance guide signs providing information related to the temporary exit should be relocated or duplicated adjacent to the temporary roadway.*

Standard:

6. **A temporary EXIT sign shall be located in the temporary gore. For better visibility, it shall be mounted a minimum of 7 feet from the pavement surface to the bottom of the sign.**

Option:

7. Positive protection devices may be used per Section 6M.02.
8. Guide signs referring to the exit may need to be relocated to the median.
9. The temporary EXIT sign placed in the temporary gore may be either black on orange or white on green.
10. In some instances, a temporary deceleration lane may be useful in facilitating the exiting maneuver.
11. When a temporary traffic barrier is used to separate opposing vehicular traffic, the Two-Way Traffic signs may be omitted.

Figure 6P-41. Median Crossover for an Exit Ramp (TA-41)

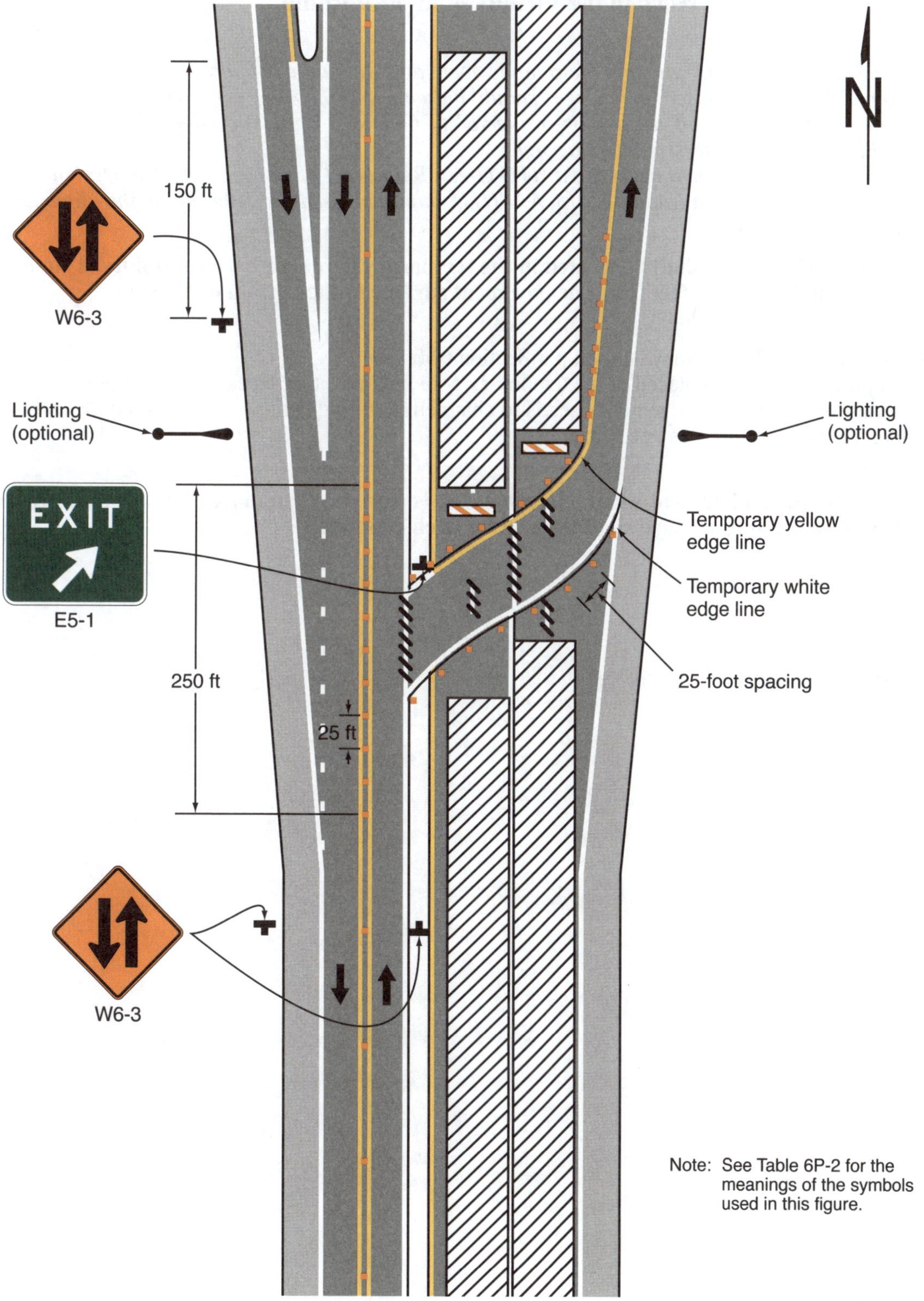

Typical Application 41

Notes for Figure 6P-42—Typical Application 42
Work in the Vicinity of an Exit Ramp

Guidance:

1. *The guide signs should indicate that the ramp is open, and where the temporary ramp is located. However, if the ramp is closed, guide signs should indicate that the ramp is closed.*
2. *When the exit ramp is closed, a black-on-orange EXIT CLOSED sign panel should be placed diagonally across the interchange/intersection guide signs.*
3. *The design criteria contained in the "AASHTO Green Book – A Policy On Geometric Design Of Highways and Streets," 7th Edition, 2018, AASHTO should be used for determining the alignment.*

Standard:

4. **A temporary EXIT sign shall be located in the temporary gore. For better visibility, it shall be mounted a minimum of 7 feet from the pavement surface to the bottom of the sign.**

Option:

5. Positive protection devices may be used per Section 6M.02.
6. The temporary EXIT sign placed in the temporary gore may be either black on orange or white on green.
7. An alternative procedure that may be used is to channelize exiting vehicular traffic onto the right-hand shoulder and close the lane as necessary.

Standard:

8. **An arrow board shall be used when a freeway lane is closed. When more than one freeway lane is closed, a separate arrow board shall be used for each closed lane.**

Figure 6P-42. Work in the Vicinity of an Exit Ramp (TA-42)

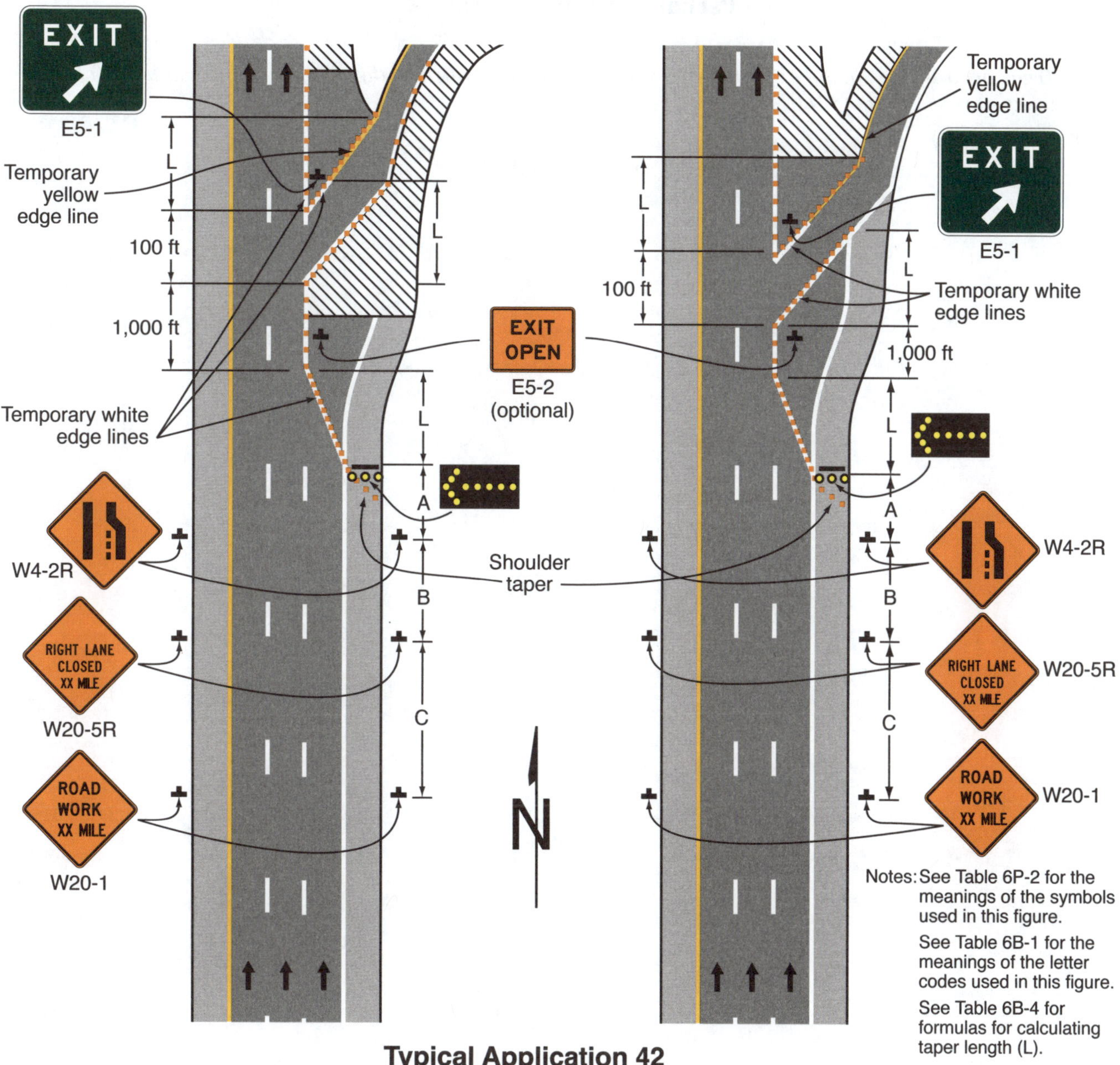

Typical Application 42

Notes for Figure 6P-43—Typical Application 43
Partial Exit Ramp Closure

Guidance:

1. *Truck off-tracking should be considered when determining whether the minimum lane width of 10 feet is adequate (see Section 6N.07).*

Option:

2. Positive protection devices may be used per Section 6M.02.

Figure 6P-43. Partial Exit Ramp Closure (TA-43)

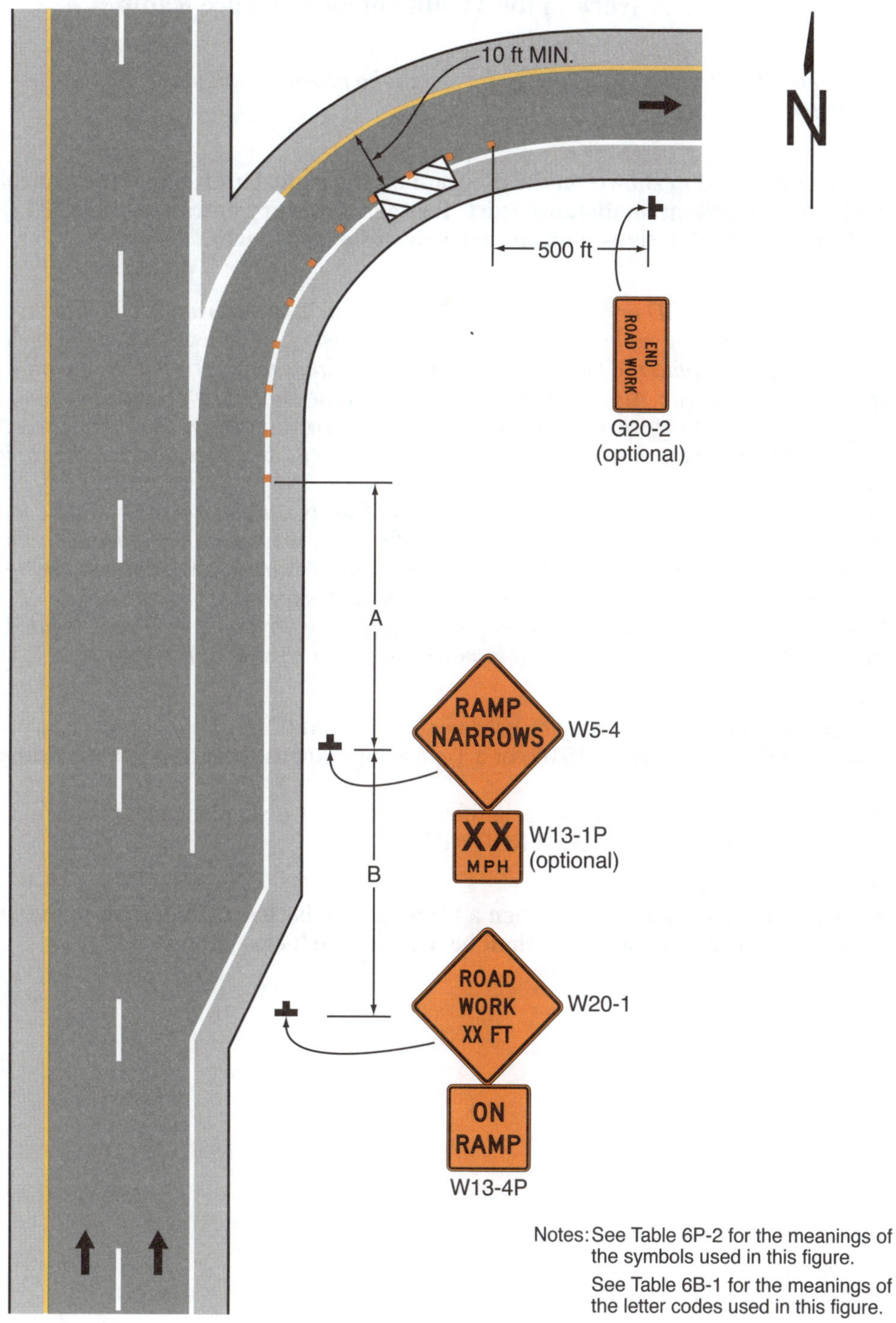

Typical Application 43

Notes for Figure 6P-44—Typical Application 44
Work in the Vicinity of an Entrance Ramp

Guidance:

1. *An acceleration lane of sufficient length should be provided whenever possible as shown on the diagram on the left.*

Standard:

2. **For the information shown on the diagram on the right-hand side of the typical application, where inadequate acceleration distance exists for the temporary entrance, the YIELD sign shall be replaced with STOP signs (one on each side of the approach).**

Guidance:

3. *When used, the YIELD or STOP sign should be located so that ramp vehicular traffic has adequate sight distance of oncoming mainline vehicular traffic to select an acceptable gap in the mainline vehicular traffic flow, but should not be located so far forward that motorists will be encouraged to stop in the path of the mainline traffic. Also, a longer acceleration lane should be provided beyond the sign to reduce the gap size needed. If sufficient gaps are not available, consideration should be given to closing the ramp.*
4. *Where a STOP sign is used, a temporary stop line should be placed across the ramp at the desired stop location.*
5. *The mainline merging taper with the arrow board at its starting point should be located sufficiently in advance so that the arrow board is not confusing to drivers on the entrance ramp, and so that the mainline merging vehicular traffic from the lane closure has the opportunity to stabilize before encountering the vehicular traffic merging from the ramp.*
6. *If the ramp curves sharply to the right, warning signs with advisory speeds located in advance of the entrance terminal should be placed in pairs (one on each side of the ramp).*

Option:

7. Positive protection devices may be used per Section 6M.02.
8. A Stop Beacon (see Section 4S.05) or a Type B high-intensity warning flasher with a red lens may be placed above the STOP sign.
9. Where the acceleration distance is significantly reduced, a supplemental plaque may be placed below the Yield Ahead sign reading NO MERGE AREA.

Standard:

10. **An arrow board shall be used when a freeway lane is closed. When more than one freeway lane is closed, a separate arrow board shall be used for each closed lane.**

Figure 6P-44. Work in the Vicinity of an Entrance Ramp (TA-44)

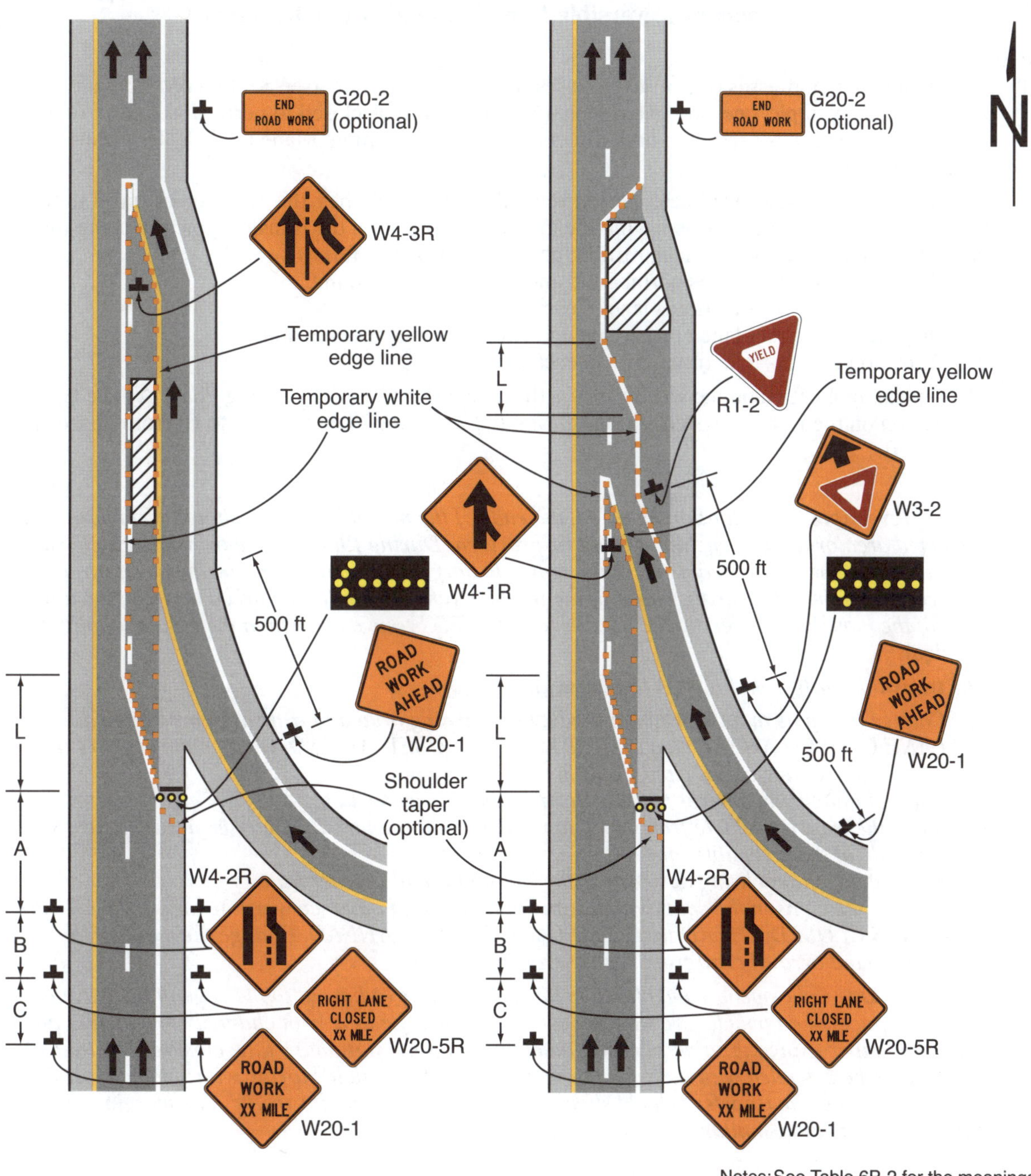

A – Added lane

B – Merge required

Notes: See Table 6P-2 for the meanings of the symbols used in this figure.

See Table 6B-1 for the meanings of the letter codes used in this figure.

See Table 6B-4 for formulas for calculating taper length (L).

Typical Application 44

Notes for Figure 6P-45—Typical Application 45
Temporary Reversible Lane Using Movable Barriers

Support:

1. This application addresses one of several uses for movable barriers (see Section 6M.02) in highway TTC zones. In this example, one side of a 6-lane divided highway is closed to perform the work operation, and vehicular traffic is carried in both directions on the remaining 3-lane roadway by means of a median crossover.

   To accommodate unbalanced peak-period vehicular traffic volumes, the direction of travel in the center lane is switched to the direction having the greater volume, with the transfer typically being made twice daily. Thus, there are four vehicular traffic phases described as follows:
   a. Phase A—two travel lanes northbound and one lane southbound;
   b. Transition A to B—one travel lane in each direction;
   c. Phase B—one travel lane northbound and two lanes southbound; and
   d. Transition B to A—one travel lane in each direction.

   The typical application on the left illustrates the placement of devices during Phase A. The typical application on the right shows conditions during the transition (Transition A to B) from Phase A to Phase B.

Guidance:

2. *For the reversible lane situation depicted, the ends of the movable barrier should terminate in a protected area or a crash cushion should be provided. During Phase A, the transfer vehicle should be parked behind the downstream end of the movable barrier for southbound traffic as shown in the typical application on the left. During Phase B, the transfer vehicle should be parked between the downstream ends of the movable barriers at the north end of the TTC zone as shown in the typical application on the right.*

   The transition shift from Phase A to B should be as follows:

   a. *Change the signs in the northbound advance warning area and transition area from a LEFT LANE CLOSED AHEAD to a 2 LEFT LANES CLOSED AHEAD. Change the mode of the second northbound arrow board from Caution to Right Arrow.*
   b. *Place channelizing devices to close the northbound center lane.*
   c. *Move the transfer vehicle from south to north to shift the movable barrier from the west side to the east side of the reversible lane.*
   d. *Remove the channelizing devices closing the southbound center lane.*
   e. *Change the signs in the southbound transition area and advance warning area from a 2 LEFT LANES CLOSED AHEAD to a LEFT LANE CLOSED AHEAD. Change the mode of the second southbound arrow board from Right Arrow to Caution.*

3. *Where the lane to be opened and closed is an exterior lane (adjacent to the edge of the traveled way or the work space), the lane closure should begin by closing the lane with channelizing devices placed along a merging taper using the same information employed for a stationary lane closure. The lane closure should then be extended with the movable-barrier transfer vehicle moving with vehicular traffic. When opening the lane, the transfer vehicle should travel against vehicular traffic. The merging taper should be removed in a method similar to a stationary lane closure.*

Option:

4. The procedure may be used during a peak period of vehicular traffic and then changed to provide two lanes in the other direction for the other peak.
5. A longitudinal buffer space may be used in the activity area to separate opposing vehicular traffic.
6. A work vehicle or a shadow vehicle may be equipped with a truck-mounted attenuator.

Standard:

7. **An arrow board shall be used when a freeway lane is closed. When more than one freeway lane is closed, a separate arrow board shall be used for each closed lane.**

Figure 6P-45. Temporary Reversible Lane Using Movable Barriers (TA-45)

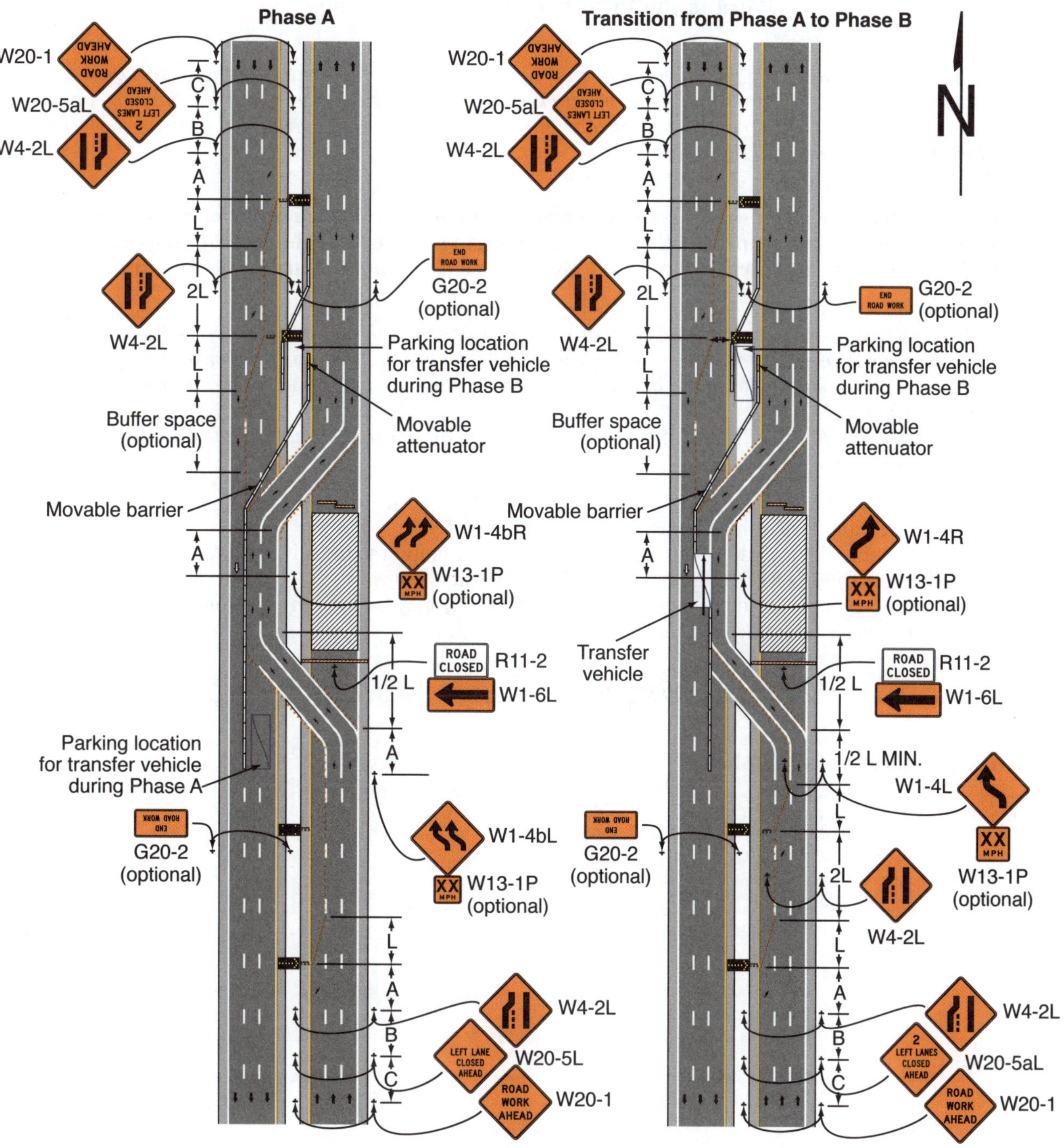

Typical Application 45

Notes: See Table 6P-2 for the meanings of the symbols used in this figure.
See Table 6B-1 for the meanings of the letter codes used in this figure.
See Table 6B-4 for formulas for calculating taper length (L).

Notes for Figure 6P-46—Typical Application 46
Work in the Vicinity of a Grade Crossing

Guidance:

1. *When grade crossings exist either within or in the vicinity of roadway work activities, extra care should be taken to minimize the probability of conditions being created, by lane restrictions, flagging, or other operations, where vehicles might be stopped within the grade crossing, considered as being 15 feet on either side of the closest and farthest rail.*

Standard:

2. **If the queuing of vehicles across active rail tracks cannot be avoided, a uniformed law enforcement officer or flagger shall be provided at the grade crossing to prevent vehicles from stopping within the grade crossing (as described in Note 1 above), even if automatic warning devices are in place.**

Guidance:

3. *Early coordination with the railroad company or transit agency should occur before work starts.*
4. *In the example depicted, the buffer space of the activity area should be extended upstream of the grade crossing (as shown) so that a queue created by the flagging operation will not extend across the grade crossing.*
5. *The DO NOT STOP ON TRACKS sign should be used on all approaches to a grade crossing within the limits of a TTC zone.*

Option:

6. Positive protection devices may be used per Section 6M.02.
7. Flashing warning lights and/or flags may be used to call attention to the advance warning signs.
8. A BE PREPARED TO STOP sign may be added to the sign series.
9. Automated Flagger Assistance Devices (see Section 6L.02) may be used in situations where there is only one lane of approaching traffic in the direction to be controlled.

Guidance:

10. *When used, the BE PREPARED TO STOP sign should be located before the Flagger symbol sign.*

Standard:

11. **At night, flagger stations shall be illuminated, except in emergencies.**

Figure 6P-46. Work in the Vicinity of a Grade Crossing (TA-46)

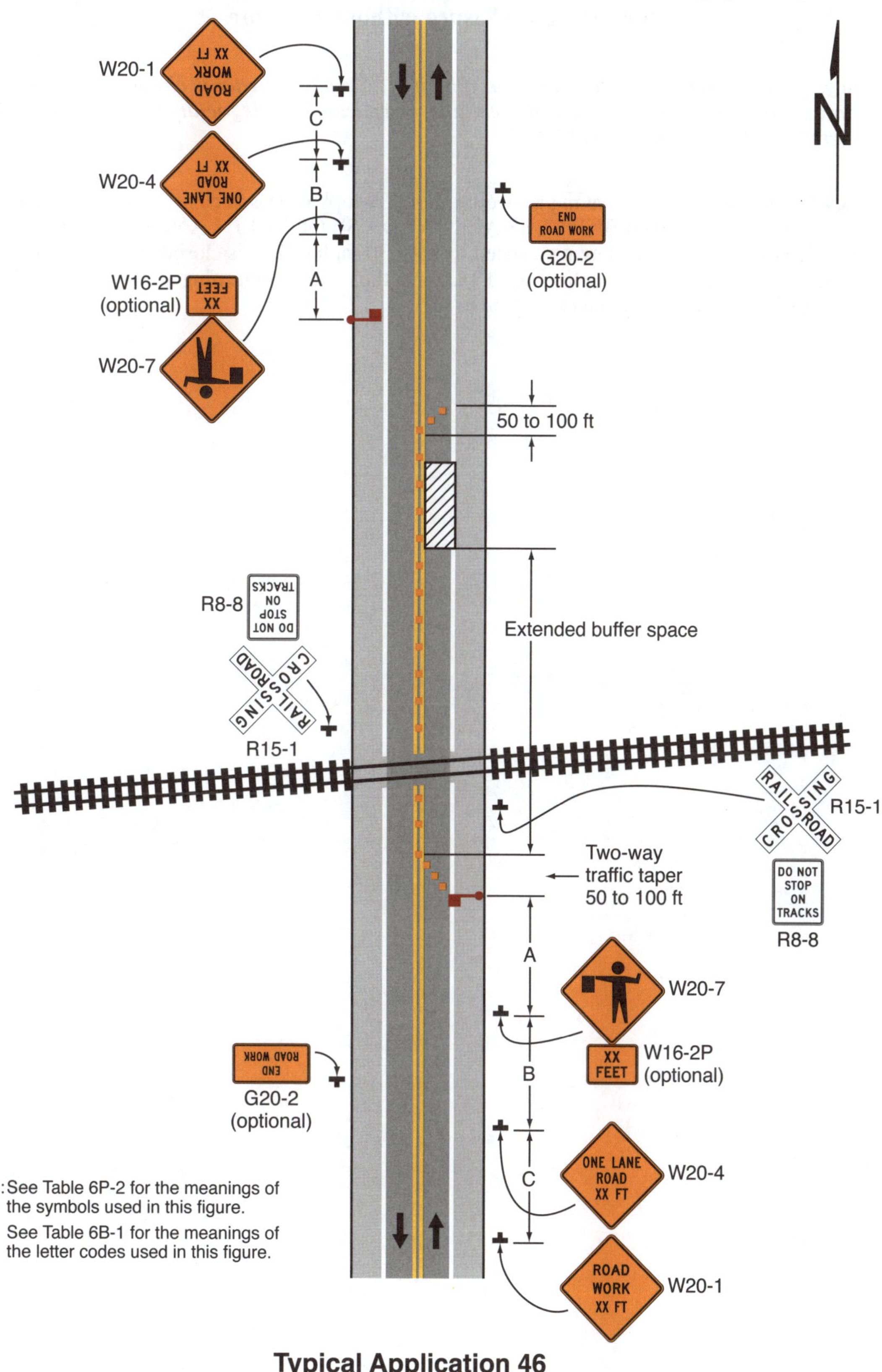

Typical Application 46

Notes for Figure 6P-47—Typical Application 47
Bicycle Lane Closure without a Detour

Guidance:

1. *If a bicycle lane on a roadway having a speed limit of 35 mph or higher is closed and conditions are not appropriate to direct bicyclists into a shared lane, a separate bicycle facility or detour route should be considered (see Figures 6P-48 and 6P-51).*

Option:

2. If a bicycle lane on a roadway having a speed limit of 30 mph or less is closed, and the adjacent travel lane is less than 14 feet wide, then BICYCLES ALLOWED USE OF FULL LANE signs may be used.

3. If a bicycle lane on a roadway having a speed limit of 30 mph or less is closed, and the adjacent travel lane is at least 14 feet wide throughout the TTC zone, then Bicycle Warning signs in association with IN STREET or IN ROADWAY plaques may be used.

Figure 6P-47. Bicycle Lane Closure without a Detour (TA-47)

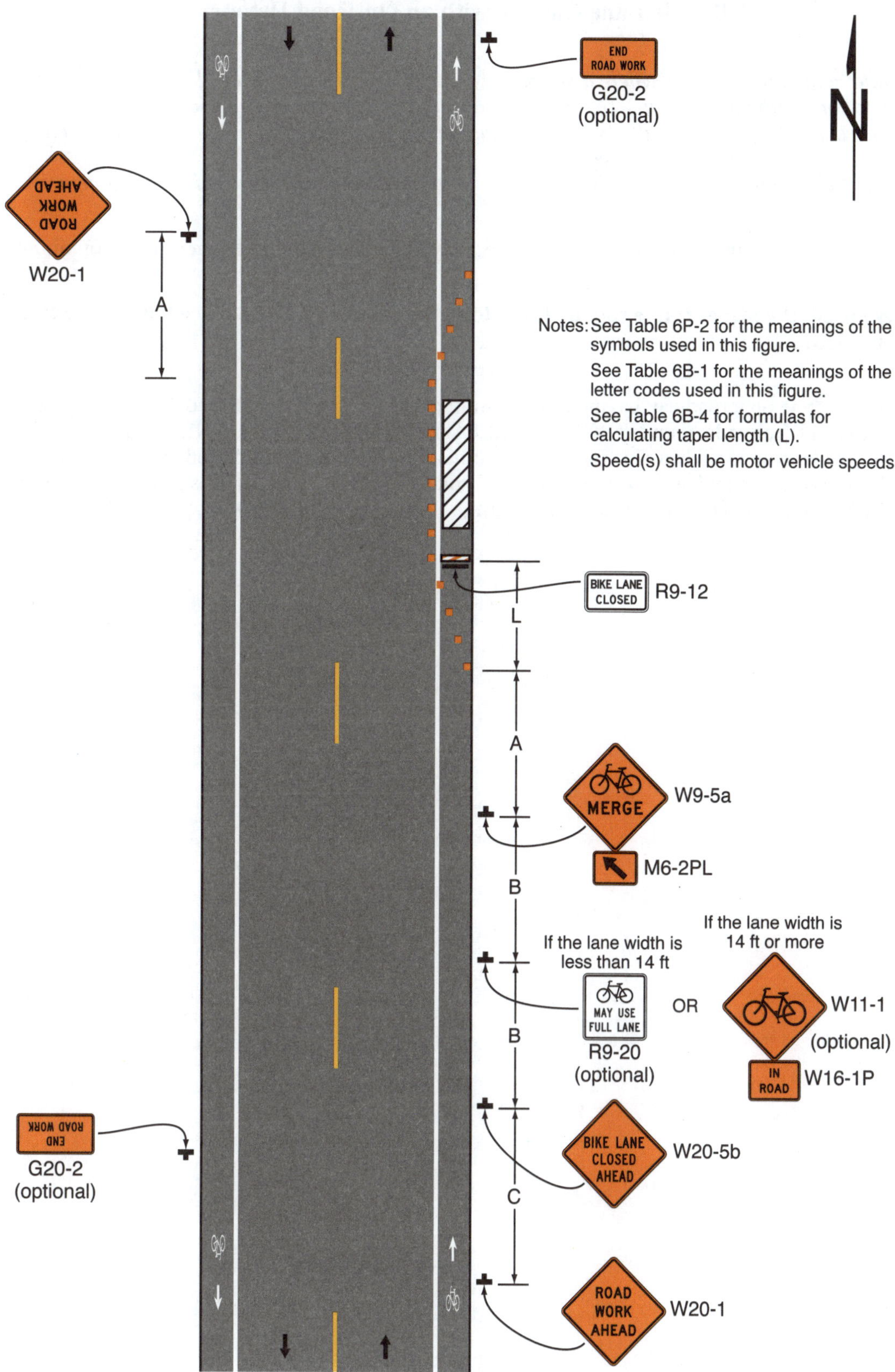

Typical Application 47

Notes for Figure 6P-48—Typical Application 48
Bicycle Lane Closure with an On-Road Detour

Guidance:

1. *A detour route for bicyclists where a section of bicycle lane is closed should use the most direct route practical on roadways or shoulders where conditions are appropriate for bicycling.*
2. *Bicycle related regulatory and/or warning signs should be considered along the bicycle detour based on engineering judgment and traffic conditions.*
3. *A Street Name sign or Bike Route Name sign should be mounted with the Bike Detour sign.*

Option:

4. The Street Name sign or Bike Route Name sign may be either white on green or black on orange.

Standard:

5. **Where used, the Street Name sign or Bike Route Name sign shall be placed above the Bike Detour sign.**

Option:

6. If a bicycle lane on a roadway having a speed limit of 30 mph or less is closed, and the adjacent travel lane is less than 14 feet wide, then BICYCLES ALLOWED USE OF FULL LANE signs may be used.
7. If a bicycle lane on a roadway having a speed limit of 30 mph or less is closed, and the adjacent travel lane is at least 14 feet wide throughout the TTC zone, then Bicycle Warning signs in association with IN STREET or IN ROADWAY plaques may be used.

Figure 6P-48. Bicycle Lane Closure with an On-Road Detour (TA-48)

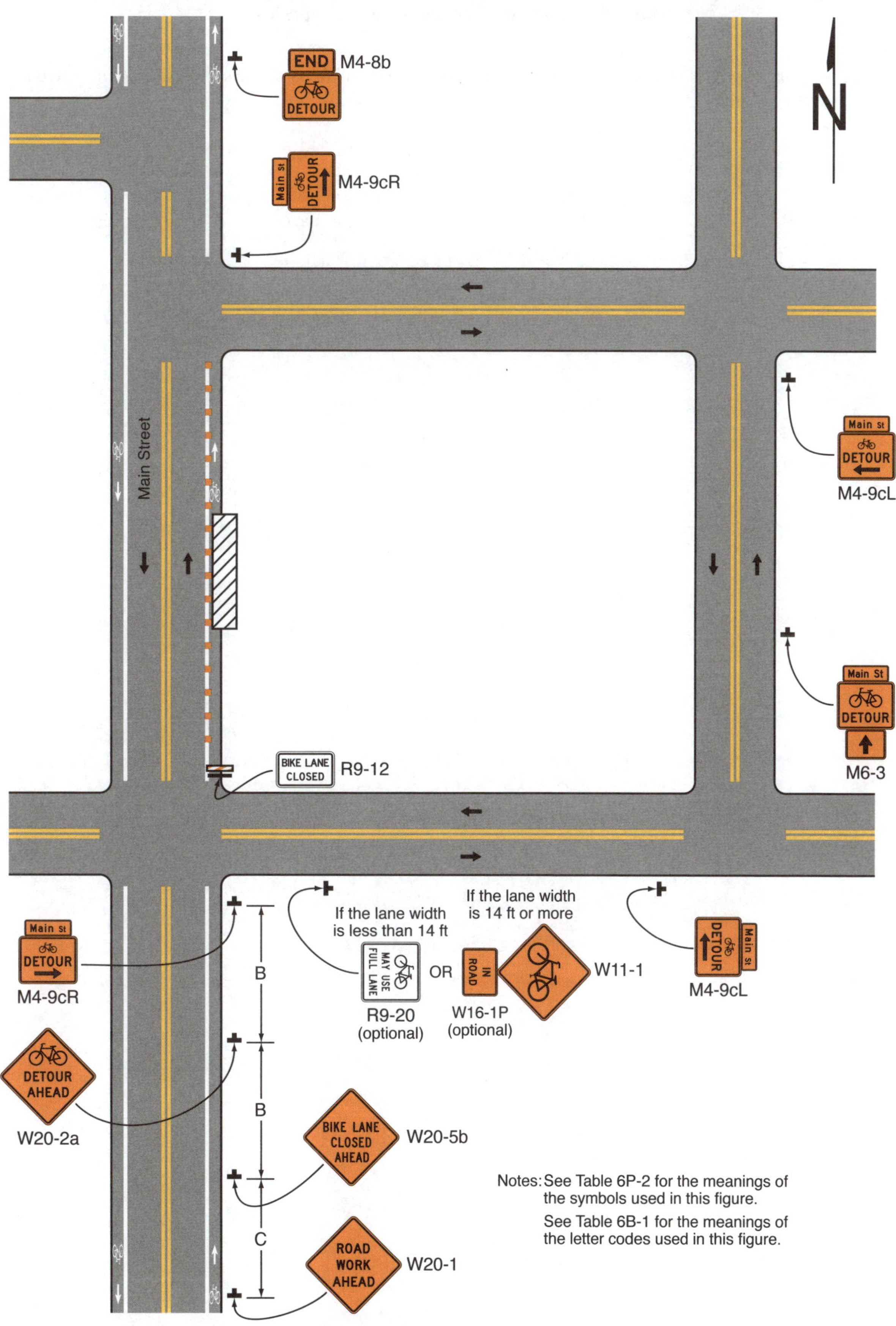

Typical Application 48

Notes for Figure 6P-49—Typical Application 49
Shared-Use Path Closure with a Diversion

Guidance:

1. *The temporary paved shared-use path should be at least as wide as the shared-use path that was temporarily closed.*

Figure 6P-49. Shared-Use Path Closure with a Diversion (TA-49)

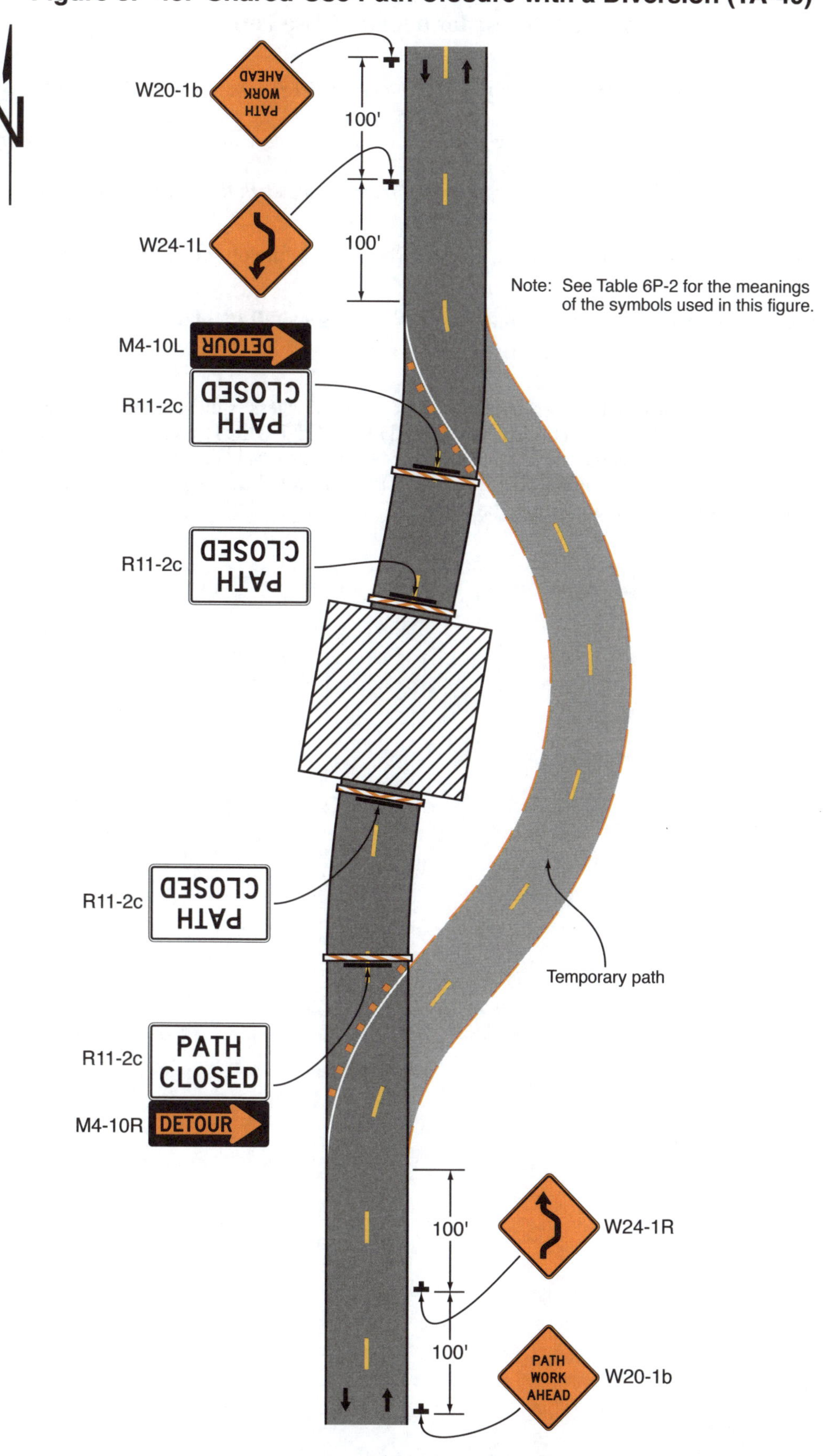

Typical Application 49

Notes for Figure 6P-50—Typical Application 50
On-Road Detour for a Shared-Use Path

Guidance:

1. *The on-road detour route for bicyclists should use the most direct route practical on roadways or shoulders where conditions are appropriate for bicycling.*
2. *Bicycle related regulatory and/or warning signs should be considered along the bicycle detour based on engineering judgment and traffic conditions.*
3. *A Street Name sign or Bike Route Name sign should be mounted with the Bike Detour sign.*

Option:

4. The Street Name sign or Bike Route Name sign may be either white on green or black on orange.

Standard:

5. **Where used the Street Name sign or Bike Route Name sign shall be placed above the Bike Detour sign.**

Option:

6. If a bicycle lane on a roadway having a speed limit of 30 mph or less is closed, and the adjacent travel lane is less than 14 feet wide, then BICYCLES ALLOWED USE OF FULL LANE signs may be used.
7. If a bicycle lane on a roadway having a speed limit of 30 mph or less is closed, and the adjacent travel lane is at least 14 feet wide throughout the TTC zone, then Bicycle Warning signs in association with IN STREET or IN ROADWAY plaques may be used.

Figure 6P-50. On-Road Detour for a Shared-Use Path (TA-50)

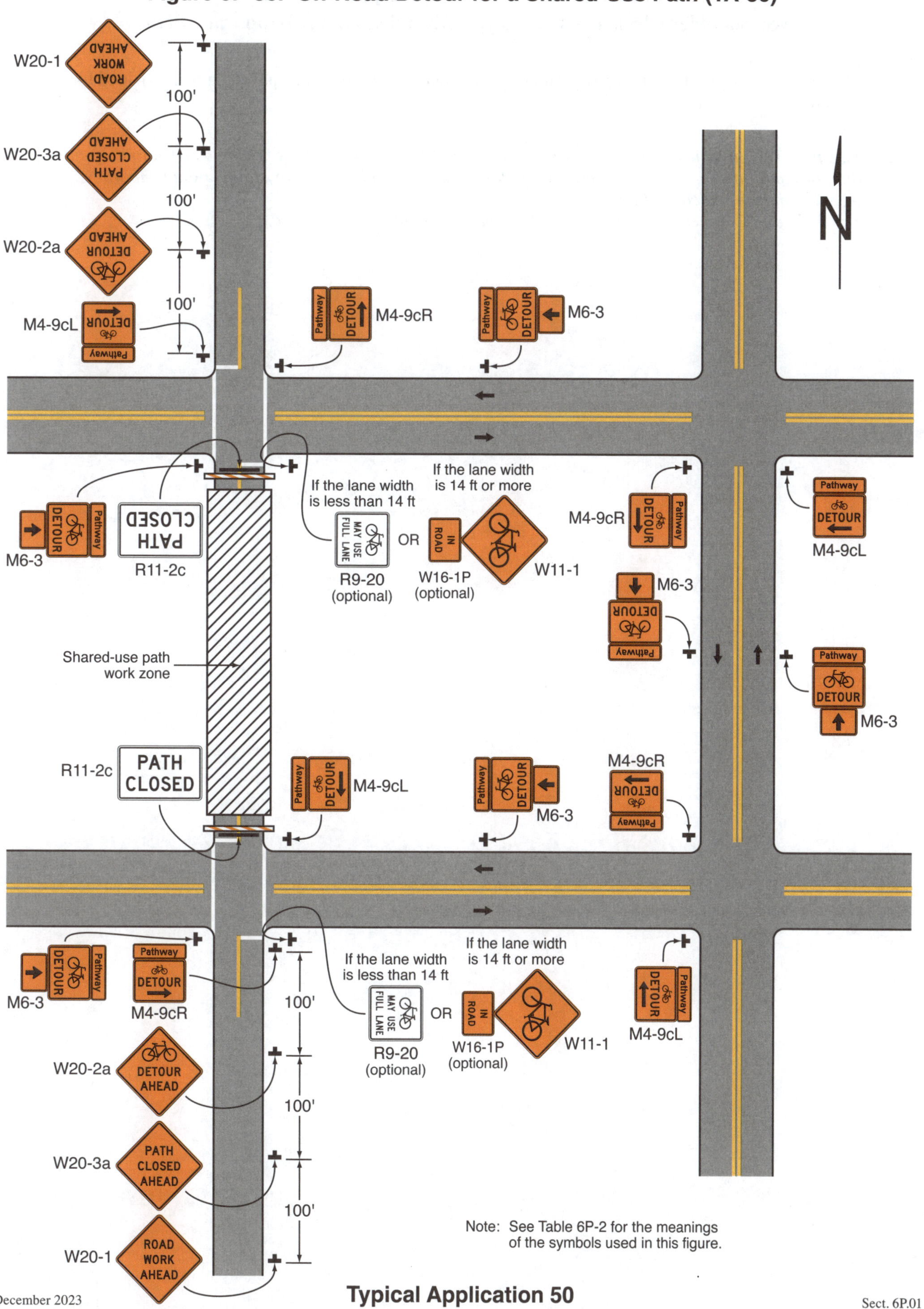

Note: See Table 6P-2 for the meanings of the symbols used in this figure.

Typical Application 50

Notes for Figure 6P-51—Typical Application 51
Paved Shoulder Closure with a Bicycle Diversion onto a Temporary Path

Option:

1. This plan may be used where a paved shoulder is closed and a temporary paved path is provided for bicyclists.

Guidance:

2. *This plan should be used where a paved shoulder is closed on a roadway having a speed limit greater than or equal to 45 mph that is part of a bikeway system (local, county or state) and a temporary paved path is provided for bicyclists.*
3. *The A, B, and C dimensions should be based on anticipated bicycle speeds.*

Figure 6P-51. Paved Shoulder Closure with a Bicycle Diversion onto a Temporary Path (TA-51)

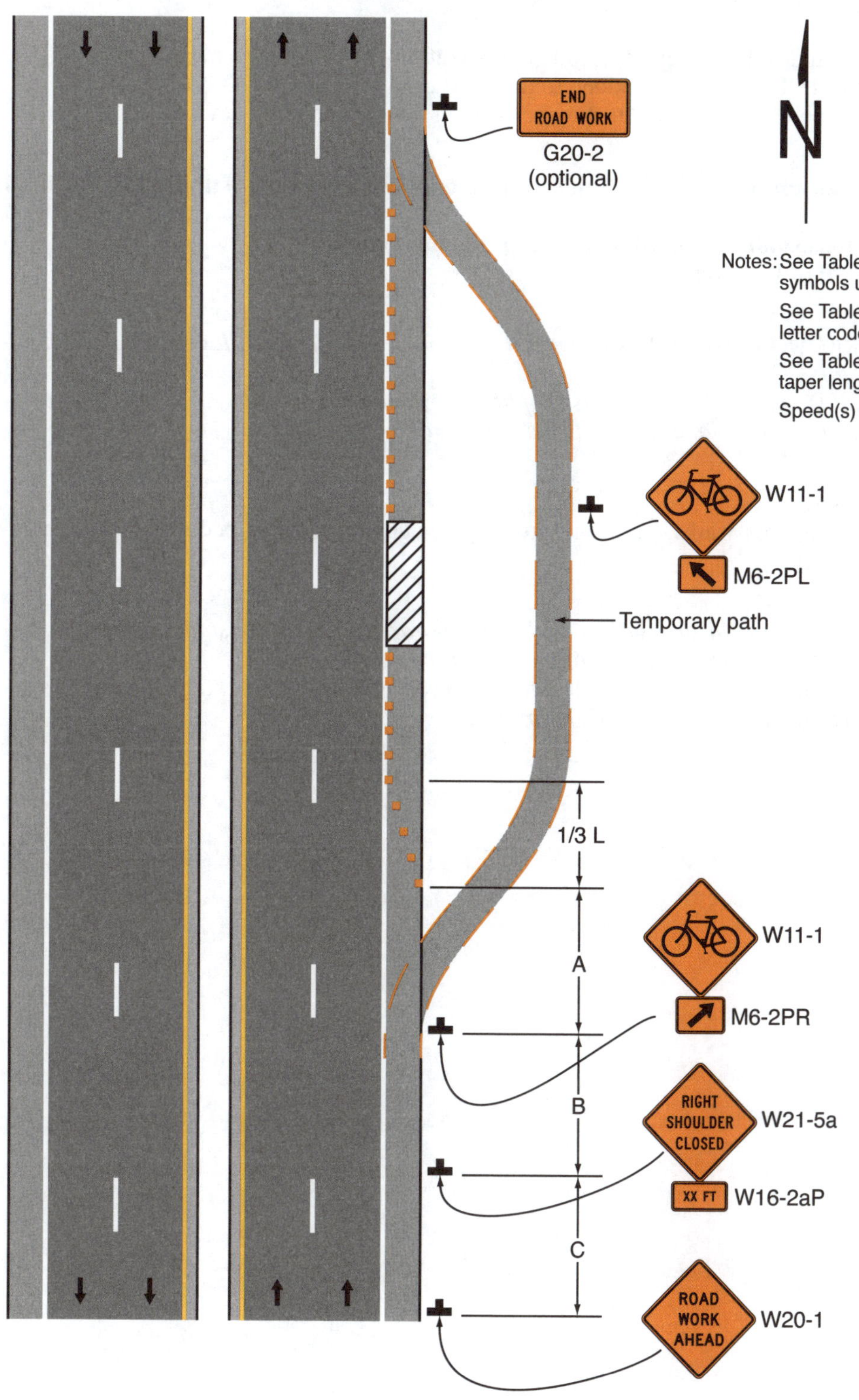

Typical Application 51

Notes for Figure 6P-52—Typical Application 52
Short-Term or Short-Duration Work in a Circular Intersection

Option:

1. Flashing warning lights and/or flags may be used to call attention to the advance warning signs. A BE PREPARED TO STOP sign may be added to the sign series.
2. If closure continues overnight, warning lights may be used on the channelizing devices.

Standard:

3. **Where a quadrant of the circular intersection is closed, only one direction of approach traffic shall be released at a time.**
4. **At night, flagger stations shall be illuminated, except in emergencies.**
5. **WRONG WAY signs shall be covered.**

Guidance:

6. *When used, the BE PREPARED TO STOP sign should be located between the Flagger sign and the ONE LANE ROAD sign.*
7. *YIELD, ONE WAY, and Directional arrow signs should be covered or removed.*
8. *Confusing or misleading guide or lane-use control signs should be covered.*

Option:

9. Crosswalks may be closed.
10. As an alternative to closing crosswalks, warning signs may be added informing pedestrians that there is traffic coming from the left.

Guidance:

11. *Since the geometrics of the circular intersection will be temporarily altered, consideration should be given to establishing a truck detour for the duration of the project.*
12. *For intermediate or long-term work, the circular intersection should be closed and traffic detoured, with appropriate detour signing (see Figure 6P-8) provided.*

Figure 6P-52. Short-Term or Short-Duration Work in a Circular Intersection (TA-52)

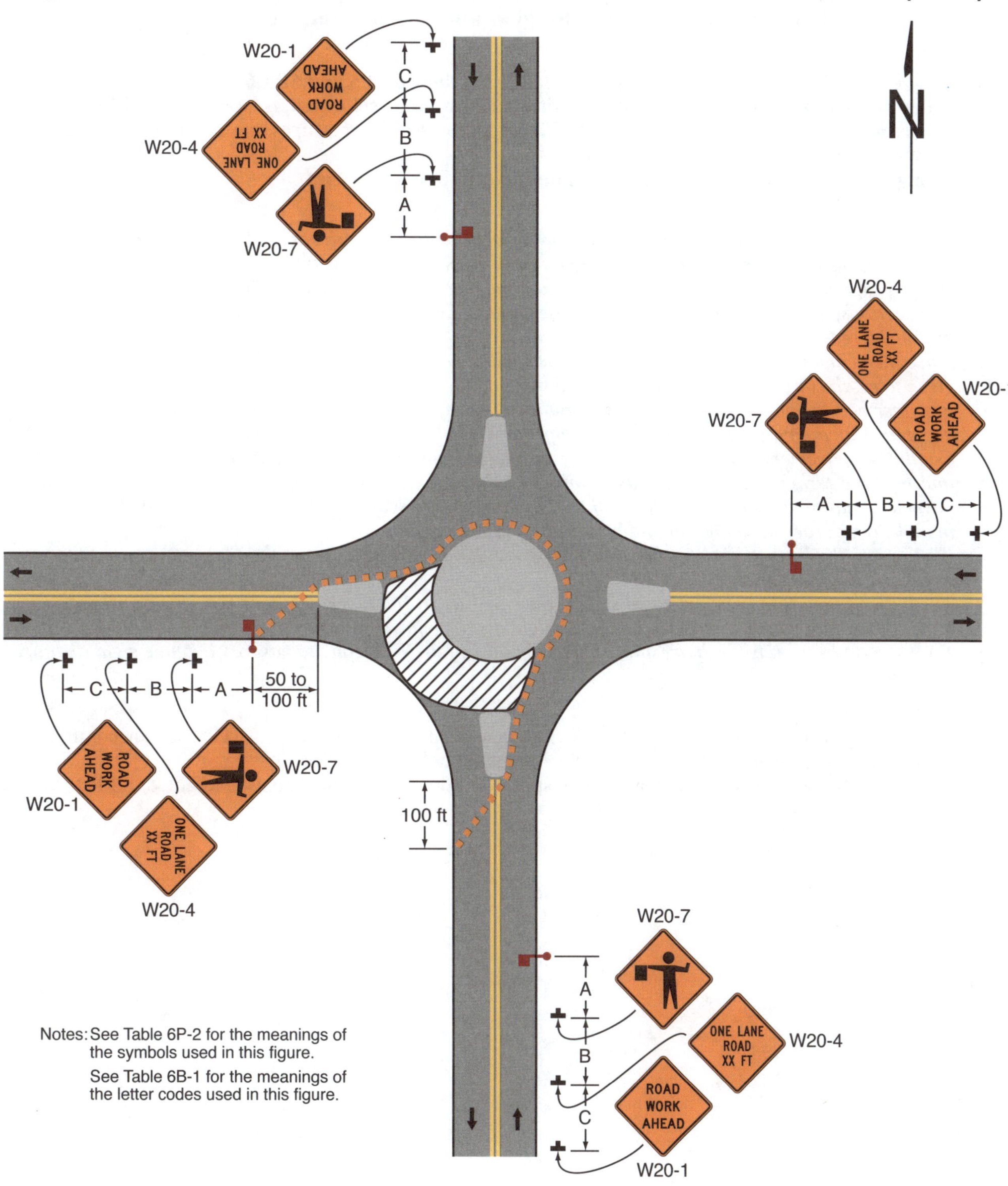

Typical Application 52

Notes for Figure 6P-53—Typical Application 53
Flagging Operation on a Single-Lane Circular Intersection

Standard:
1. **Flaggers shall follow the procedures provided in Sections 6D.05 and 6D.06.**
2. **When crosswalks or other pedestrian facilities are closed or relocated, temporary facilities (see Figure 6P-29) shall be detectable and shall include accessibility features consistent with the features present in the existing pedestrian facility.**
3. **At night, flagger stations shall be illuminated, except in emergencies.**

Guidance:
4. *Flaggers on each approach to the intersection should coordinate with each other so that traffic proceeds through the circular intersection from only one entry point at any one time.*
5. *When designing the TTC and installing the channelizing devices for work activities at circular intersections, accommodations for the turning radius of wider heavy commercial vehicles should be considered.*
6. *Since the geometrics of the circular intersection will temporarily be altered, consideration should be given to establishing a truck detour for the duration of the project.*
7. *For intermediate or long-term work, the circular intersection should be closed if traffic cannot be accommodated, and traffic detoured with appropriate detour signing (see Figure 6P-8) provided.*
8. *Conflicting pavement markings should be removed for long-term projects. For short-term and intermediate-term projects where this is impracticable, the channelizing devices in the area where the pavement markings conflict should be placed at a maximum spacing of ½ S feet where S is the speed in mph. Temporary markings should be installed where needed.*
9. *When used, the BE PREPARED TO STOP sign should be located between the Flagger sign and the ONE LANE ROAD sign.*
10. *The buffer space should be extended so that the two-way traffic taper is placed before a horizontal (or crest vertical) curve to provide adequate sight distance for the flagger and a queue of stopped vehicles.*
11. *Care should be exercised when establishing the limits of the TTC zone to ensure adequate sight distance in advance of the transition.*

Option:
12. Periodic adjustments to the channelizing devices may be allowed in an active TTC zone to accommodate the turning movements of tractor trailer vehicles and other large vehicles.
13. On the approaches where traffic flow will be split, two pilot vehicles may be used to guide traffic through the circular intersection.

Figure 6P-53. Flagging Operation on a Single-Lane Circular Intersection (TA-53)

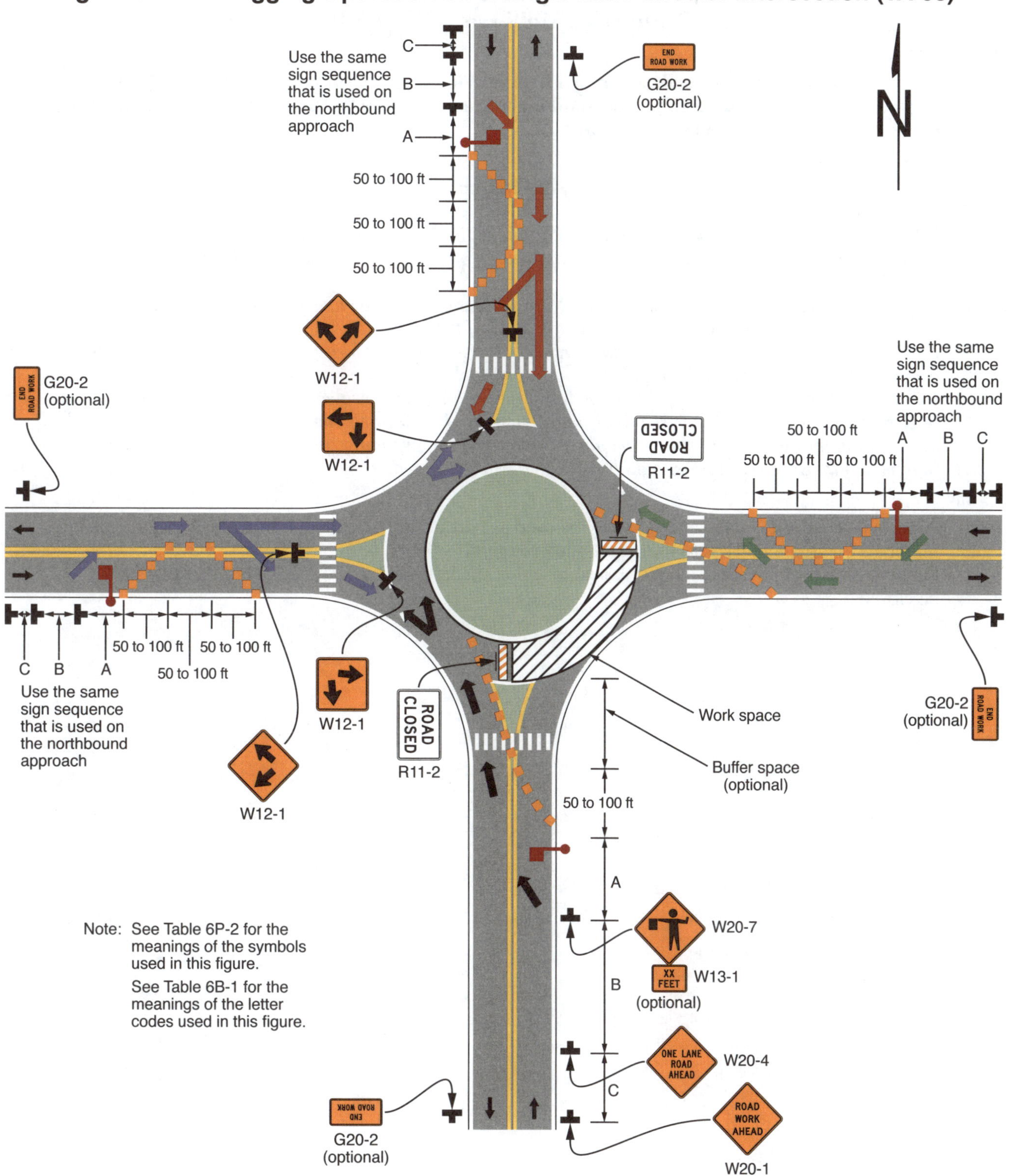

Typical Application 53

Notes for Figure 6P-54—Typical Application 54
Inside Lane Closure on a Multi-Lane Circular Intersection

Standard:

1. **When crosswalks or other pedestrian facilities are closed or relocated, temporary facilities (see Figure 6P-29) shall be detectable and shall include accessibility features consistent with the features present in the existing pedestrian facility.**

Guidance:

2. *Care should be exercised when establishing the limits of the TTC zone to ensure adequate sight distance in advance of the transition.*
3. *When designing the TTC and installing the channelizing devices for work activities at circular intersections, accommodations for the turning radius of wider heavy commercial vehicles should be considered.*
4. *Since the geometrics of the circular intersection will temporarily be altered, consideration should be given to establishing a truck detour for the duration of the project.*
5. *For intermediate or long-term work, the circular intersection should be closed if traffic cannot be accommodated, and traffic detoured with appropriate detour signing provided (see Figure 6P-8).*
6. *Conflicting pavement markings should be removed for long-term projects. For short-term and intermediate-term projects where this is impracticable, the channelizing devices in the area where the pavement markings conflict should be placed at a maximum spacing of ½ S feet where S is the speed in mph. Temporary markings should be installed where needed.*

Option:

7. A portable changeable message sign may be used as part of the TTC plan to provide clear guidance to motorists on all approaches to the circular intersection.
8. On a multi-lane approach, a lane (or lanes) on either the left-hand side or the right-hand side may be closed.

Figure 6P-54. Inside Lane Closure on a Multi-Lane Circular Intersection (TA-54)

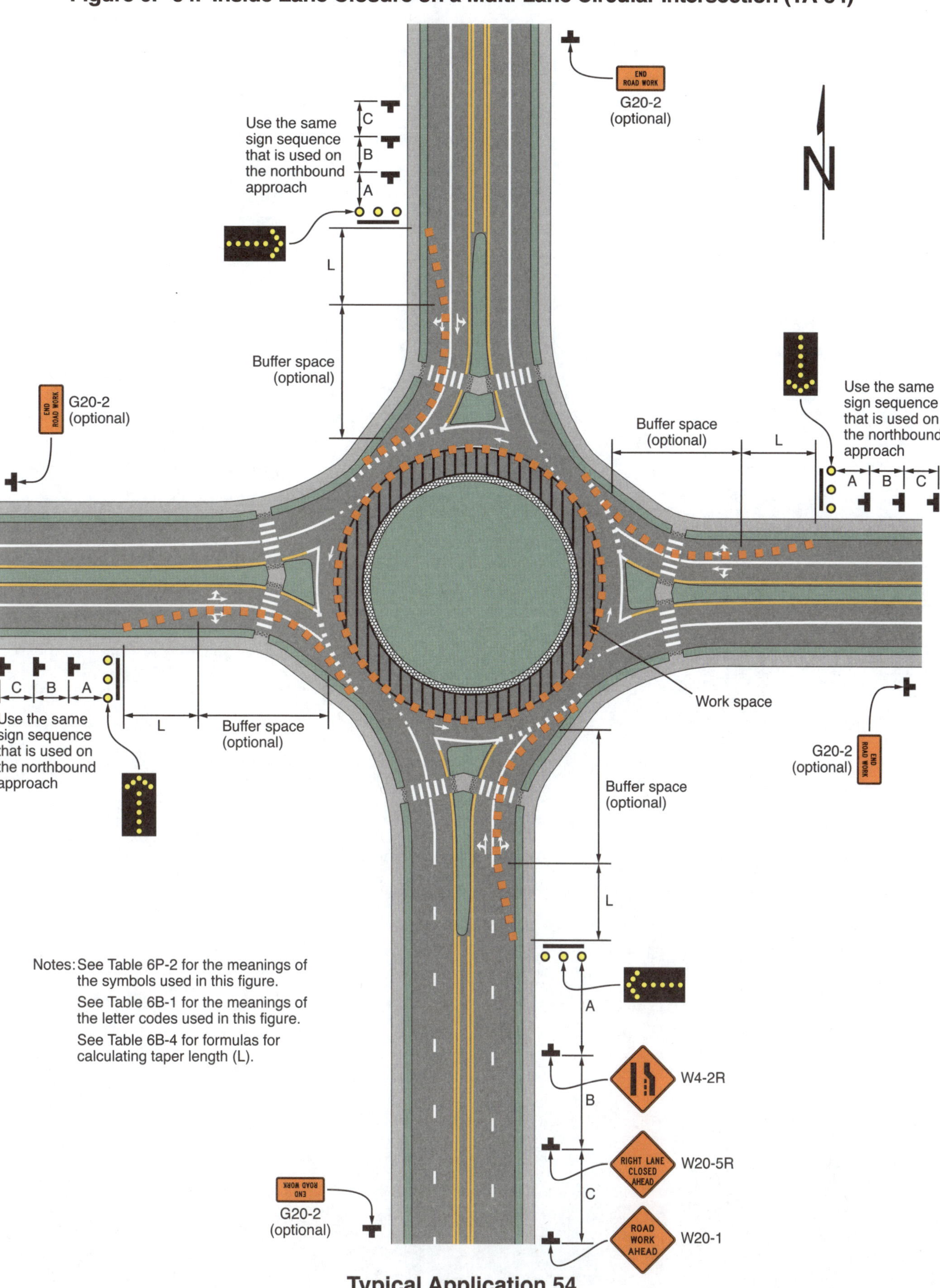

Typical Application 54

(This page intentionally left blank)

PART 7
TRAFFIC CONTROL FOR SCHOOL AREAS
CHAPTER 7A. GENERAL

Section 7A.01 Introduction

Support:

01   Part 7 sets forth basic principles and prescribes standards for the design, application, installation, and maintenance of all traffic control devices (including signs, signals, and markings) and other controls (including adult crossing guards) for the special pedestrian conditions in school areas.

Section 7A.02 School Route Plans and School Crossings

Guidance:

01   *A school route plan for each school serving elementary to high school students should be prepared in order to develop uniformity in the use of school area traffic controls and to serve as the basis for a school traffic control plan for each school.*

02   *The school route plan, developed in a systematic manner by the school, law enforcement, and traffic officials responsible for school pedestrian safety, should consist of a map (see Figure 7A-1) showing streets, the school, existing traffic controls, established school walk routes, and established school crossings.*

Figure 7A-1. Example of School Route Plan Map

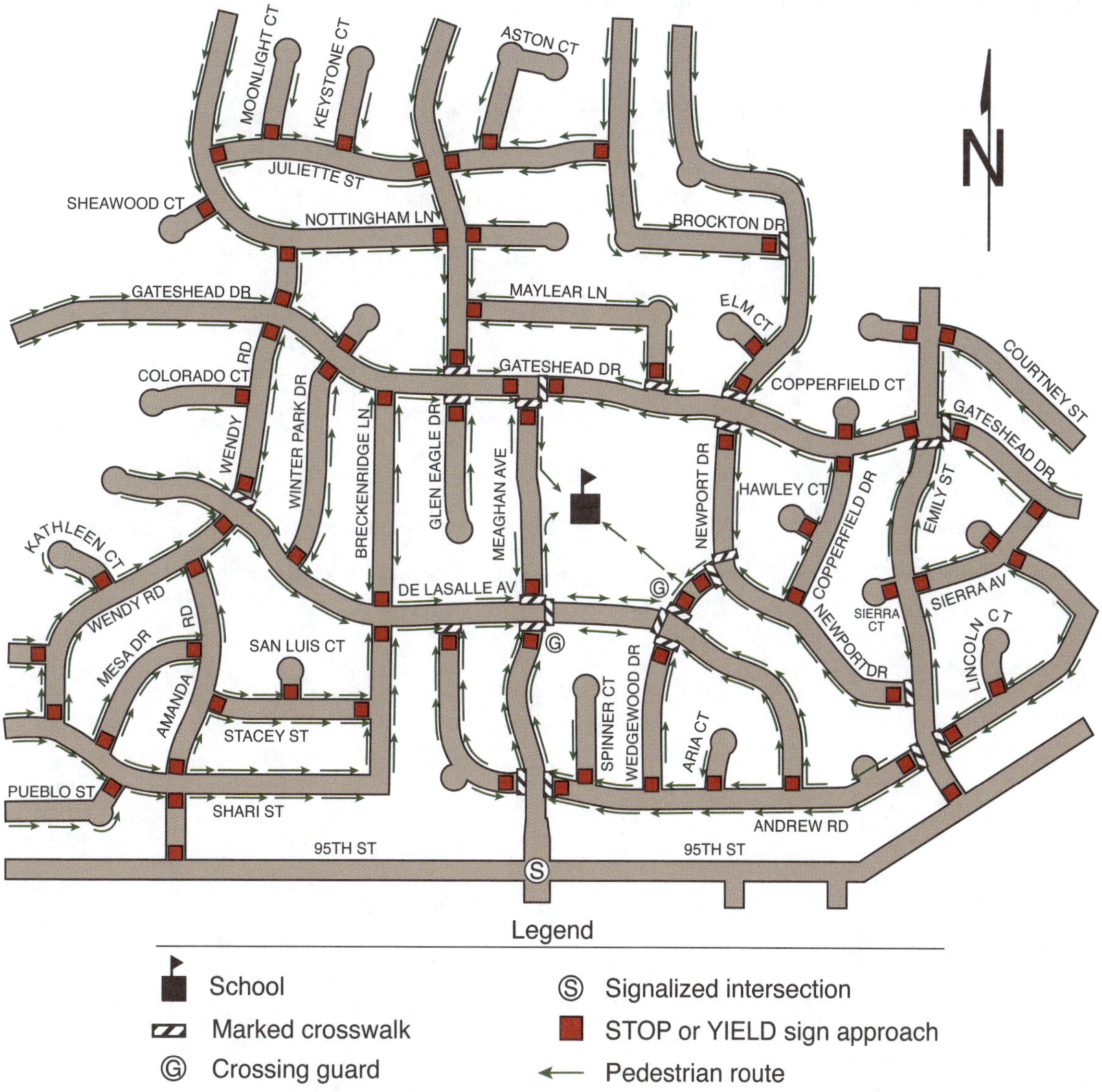

Legend

03   *Bicycle use as a mode of transportation, as applicable, should also be considered if students biking to and from school are not allowed to use the sidewalks along the pedestrian route.*

04   *The type(s) of school area traffic control devices used, either warning or regulatory, should be related to the volume and speed of vehicular traffic, street width, and the number and age of the students using the crossing.*

05   *School area traffic control devices should be included in a school traffic control plan.*

Support:

06   To establish a safer route to and from school for schoolchildren, the application of planning criterion for school walk routes might make it necessary for children to walk an indirect route to an established school crossing located where there is existing traffic control and to avoid the use of a direct crossing where there is no existing traffic control.

07   The frequency of gaps in the traffic stream that are sufficient for student crossing is different at each crossing location. When the delay between the occurrences of adequate gaps becomes excessive, students might become impatient and endanger themselves by attempting to cross the street during an inadequate gap. In these instances, the creation of sufficient gaps needs to be considered to accommodate the crossing demand.

Guidance:

08   *School walk routes should be planned to take advantage of existing traffic controls.*

09   *The following factors should be considered when determining the feasibility of requiring children to walk a longer distance to a crossing with existing traffic control:*

    A. *The availability of adequate sidewalks or other pedestrian walkways to and from the location with existing control,*
    B. *The number of students using the crossing,*
    C. *The age levels of the students using the crossing, and*
    D. *The total extra walking distance.*

Support:

10   A School Crossing signal warrant is provided in Section 4C.06.

CHAPTER 7B. SIGNS

Section 7B.01 Design of School Signs

Standard:

01 **Except as provided in Section 2A.07, the sizes of signs and plaques to be used on conventional roadways in school areas shall be as shown in Table 7B-1.**

02 **The sizes in the Oversized column in Table 7B-1 shall be used on expressways in school areas.**

Guidance:

03 *The sizes in the Oversized column should be used on roadways that have four or more lanes with posted speed limits of 40 mph or higher.*

Table 7B-1. School Area Sign and Plaque Sizes

Sign	Sign Designation	Section	Conventional Road	Minimum	Oversized
School	S1-1	7B.02	36 x 36	30 x 30	48 x 48
School Bus Stop Ahead	S3-1	7B.04	36 x 36	30 x 30	48 x 48
School Bus Turn Ahead	S3-2	7B.04	36 x 36	30 x 30	48 x 48
Reduced School Speed Limit Ahead	S4-5, S4-5a	7B.05	36 x 36	30 x 30	48 x 48
School Speed Limit XX When Flashing	S5-1	7B.05	24 x 48	—	36 x 72
End School Zone	S5-2	7B.02	24 x 30	—	36 x 48
End School Speed Limit	S5-3	7B.05	24 x 30	—	36 x 48
Yield (Stop) Here for School Crossing	R1-5a, R1-5c	7B.03	36 x 36	—	—
In-Street Ped or School Crossing	R1-6, R1-6a, R1-6b, R1-6c	7B.03	12 x 36	—	—
Overhead School Crossing	R1-9b, R1-9c	7B.03	90 x 24	—	—
Speed Limit (School Use)	R2-1	7B.05	24 x 30	—	36 x 48
Begin Higher Fines Zone	R2-10	7B.06	24 x 30	—	36 x 48
End Higher Fines Zone	R2-11	7B.06	24 x 30	—	36 x 48

Plaque	Sign Designation	Section	Conventional Road	Minimum	Oversized
Time of Day X:XX to X:XX AM X:XX to X:XX PM	S4-1P	7B.06	24 x 12	—	36 x 18
When Children Are Present	S4-2P	7B.06	24 x 12	—	36 x 18
School	S4-3P	7B.02, 7B.05	24 x 9	—	36 x 12
When Flashing	S4-4P	7B.05, 7B.06	24 x 12	—	36 x 18
Days of Week Mon-Fri	S4-6P	7B.05	24 x 12	—	36 x 18
All Year	S4-7P	7B.02	24 x 12	—	30 x 18
Fines Higher	R2-6P	7B.06	24 x 18	—	36 x 24
Fines Double	R2-6aP	7B.06	24 x 18	—	36 x 24
$XX Fine	R2-6bP	7B.06	24 x 18	—	36 x 24
XX Feet (2-line)	W16-2P	7B.02, 7B.03	24 x 18	—	30 x 24
XX Ft (1-line)	W16-2aP	7B.02, 7B.03	24 x 12	—	30 x 18
Directional Arrow	W16-5P	7B.02, 7B.03	21 x 15	—	30 x 21
Advance Turn Arrow	W16-6P	7B.02, 7B.03	21 x 15	—	30 x 21
Downward Diagonal Arrow	W16-7P	7B.02, 7B.03	21 x 15	—	30 x 21
Ahead	W16-9P	7B.02, 7B.03	24 x 12	—	30 x 18

Notes: 1. Larger sizes may be used when appropriate

2. Dimensions are shown in inches and are shown as width x height

3. Minimum sign sizes for multi-lane conventional roads shall be as shown in the Conventional Road column

Option:

04   Signs and plaques larger than those shown in Table 7B-1 may be used (see Section 2A.07).

Standard:

05   **School warning signs, including the "SCHOOL" portion of the School Speed Limit (S5-1) sign and
including any supplemental plaques used in association with these warning signs, shall have a fluorescent
yellow-green background with a black legend and border unless otherwise provided in this Manual for a
specific sign.**

06   **The signs used for school area traffic control shall be retroreflective or illuminated.**

Support:

07   Sections 2A.13 and 2A.14 contain provisions regarding the installation, placement, and location of signs.

08   Section 2A.15 contains provisions regarding the mounting heights of signs.

09   Section 2A.16 contains provisions regarding the lateral offsets of signs.

10   The "Standard Highway Signs" publication (see Section 1A.05) contains information regarding sign lettering.

Option:

11   In-roadway signs for school traffic control areas may be used consistent with the requirements of Sections
2B.20 and 7B.03.

Section 7B.02 <u>School Area Signs and Plaques</u>

Support:

01   Many state and local jurisdictions find it beneficial to advise road users that they are approaching a school that
is adjacent to a highway, where additional care is needed, even though no school crossing is involved and the speed
limit remains unchanged. Additionally, some jurisdictions designate school zones that have a unique legal standing
in that fines for speeding or other traffic violations within designated school zones are increased or special
enforcement techniques such as photo radar systems are used. It is important and sometimes legally necessary to
mark the beginning and end points of these designated school zones so that the road user is given proper notice.

02   The School (S1-1) sign (see Figure 7B-1) has the following four applications:

  A.   School Area – the S1-1 sign can be used to warn road users that they are approaching a school area
       that might include school buildings or grounds, a school crossing, or school related activity adjacent to
       the highway.
  B.   School Zone – the S1-1 sign can be used to identify the location of the beginning of a designated
       school zone.
  C.   School Advance Crossing – if combined with an AHEAD (W16-9P) plaque or an XX FEET (W16-2P or
       W16-2aP) plaque to comprise the School Advance Crossing assembly (see Figure 7B-1), the S1-1 sign can
       be used to warn road users that they are approaching a crossing where schoolchildren cross the roadway
       (see Section 7B.03).
  D.   School Crossing – if combined with a diagonal downward-pointing arrow (W16-7P) plaque to comprise
       the School Crossing assembly (see Figure 7B-1), the S1-1 sign can be used to warn approaching road users
       of the location of a crossing where schoolchildren cross the roadway (see Section 7B.03).

Option:

03   If a school area or school zone is located on a cross street in close proximity to the intersection, a School (S1-1)
sign with a supplemental arrow (W16-5P or W16-6P) plaque (see Figure 7B-1) may be installed on each approach
of the street or highway to warn road users making a turn onto the cross street that they will encounter a school
area soon after making the turn.

Standard:

04   **If a school zone has been designated under State or local statute, a School (S1-1) sign (see Figure 7B-1)
shall be installed to identify the beginning point(s) of the designated school zone (see Figure 7B-2).**

Option:

05   A School Zone (S1-1) sign may be supplemented with a SCHOOL (S4-3P) plaque (see Figure 7B-1).

06   A School Zone (S1-1) sign may be supplemented with an ALL YEAR (S4-7P) plaque (see Figure 7B-1) if the
school operates on a 12-month schedule.

07   The downstream end of a designated school zone may be identified with an END SCHOOL ZONE (S5-2) sign
(see Figures 7B-1 and 7B-2).

Figure 7B-1. Signs in School Areas and at School Crossings (Sheet 1 of 2)
A – School area signs

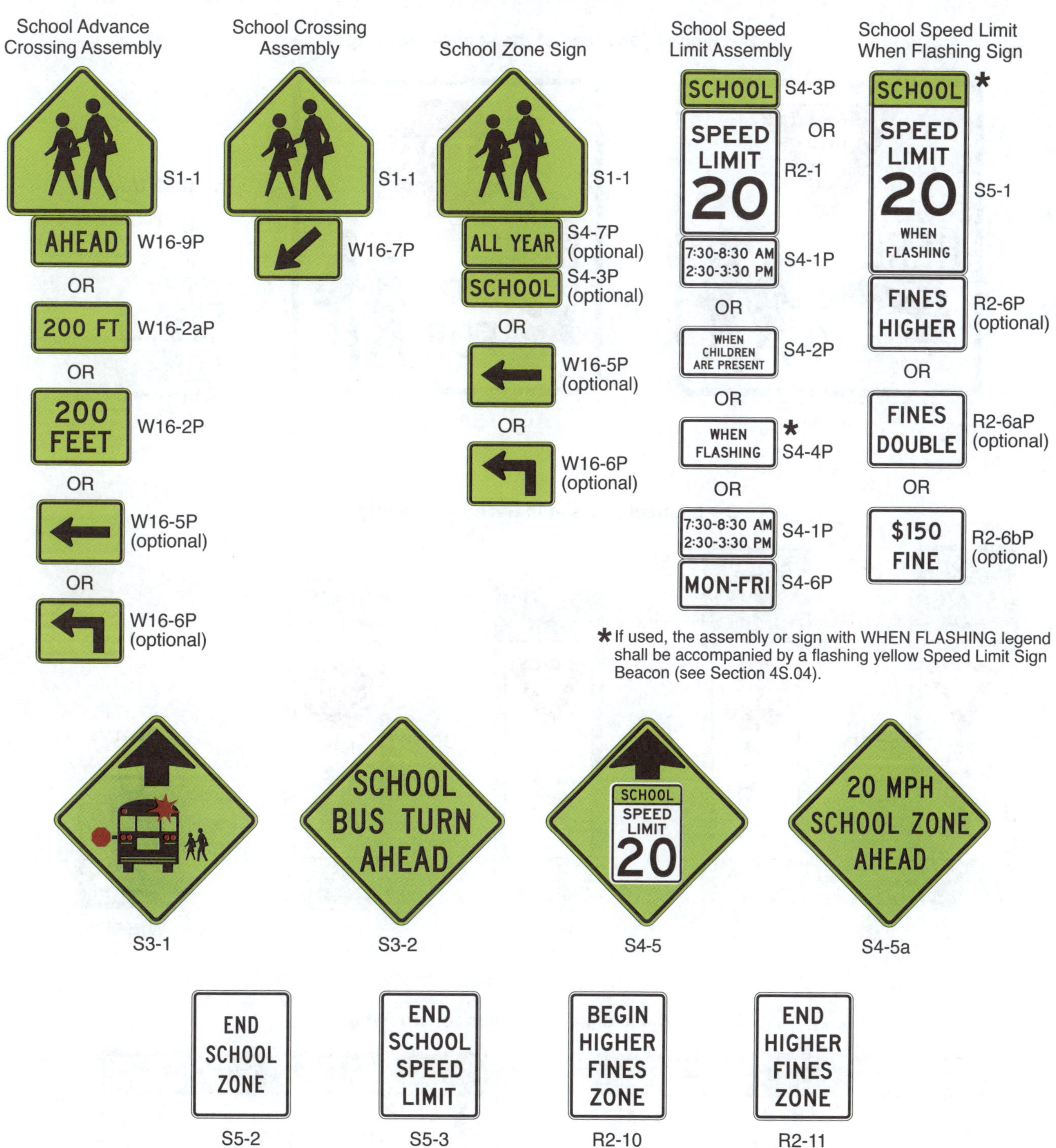

Figure 7B-1. Signs in School Areas and at School Crossings (Sheet 2 of 2)

B – Signs in advance of the school crossing

R1-5a

R1-5c

In-Street Use

S1-1*

W16-9P*

C – In-street signs at the school crossing

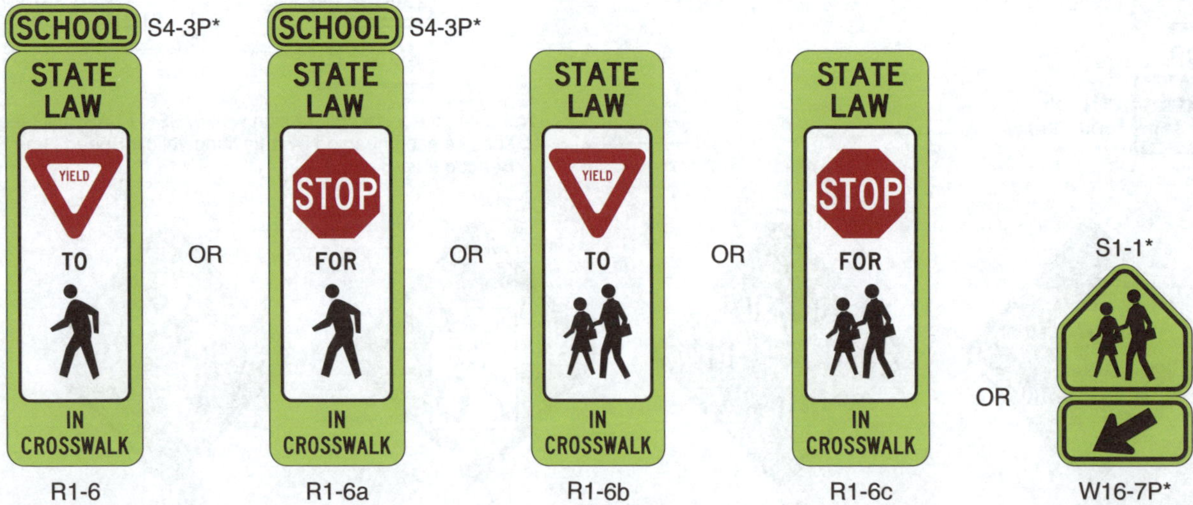

D – Overhead signs at the school crossing

R1-9b (shown reduced 50%
proportional to other signs on page)

R1-9c (shown reduced 50%
proportional to other signs on page)

Notes:

1. The use of the STATE LAW legend is optional on the R1-6 series and R1-9 series signs (see Section 7B.03).

2. The use of the SCHOOL plaque above the R1-6 and R1-6a signs is optional.

3. Signs are shown in proportion to their designated sizes unless otherwise noted.

*Reduced size signs for in-street use (see Section 7B.03):

S1-1	12 x 12 inches
S4-3P	12 x 4 inches
W16-7P	12 x 6 inches
W16-9P	12 x 6 inches

Figure 7B-2. Example of Signing for a School Zone with a School Speed Limit and a School Crossing

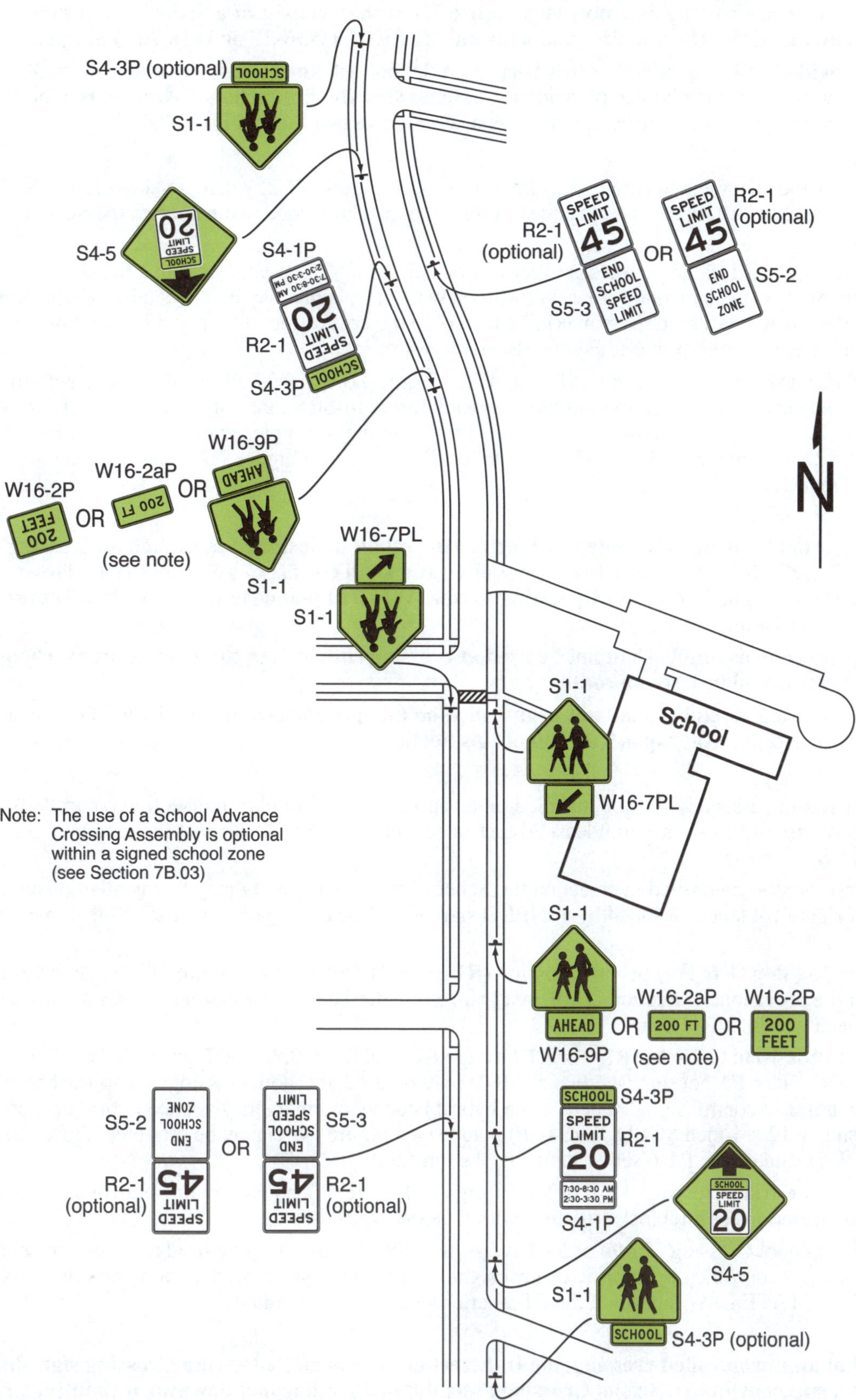

Section 7B.03 School Crossing Signs

Standard:

01   **The School Advance Crossing assembly (see Figure 7B-1) shall consist of a School (S1-1) sign supplemented with an AHEAD (W16-9P) plaque or an XX FEET (W16-2P or W16-2aP) plaque.**

02   **Except as provided in Paragraph 3 of this Section, a School Advance Crossing assembly shall be used in advance (see Table 2C-3 for advance placement guidelines) of the first School Crossing assembly that is encountered in each direction as traffic approaches a school crosswalk (see Figure 7B-3).**

Option:

03   The School Advance Crossing assembly may be omitted (see Figure 7B-2) where a School Zone (S1-1) sign (see Section 7B.02) is installed to identify the beginning of a school zone in advance of the School Crossing assembly.

04   If a school crosswalk is located on a cross street in close proximity to an intersection, a School Advance Crossing assembly with a supplemental arrow (W16-5P or W16-6P) plaque may be installed on each approach of the street or highway to warn road users making a turn onto the cross street that they will encounter a school crosswalk soon after making the turn (see Figure 7B-3).

05   A 12-inch reduced size in-street School (S1-1) sign (see Figure 7B-1), installed in compliance with the mounting height and special mounting support requirements for an In-Street Pedestrian Crossing (R1-6 or R1-6a) sign (see Section 2B.20), may be used in advance of a school crossing to supplement the post-mounted school warning signs. A 12 x 6-inch reduced size AHEAD (W16-9P) plaque (see Figure 7B-1) may be mounted below the reduced size in-street School (S1-1) sign.

Standard:

06   **If used, the School Crossing assembly (see Figure 7B-1) shall be installed at the school crossing (see Figures 7B-2 and 7B-3), or as close to it as possible, and shall consist of a School (S1-1) sign supplemented with a diagonal downward-pointing arrow (W16-7P) plaque (see Section 2C.63) to show the location of the crossing.**

07   **The School Crossing assembly shall not be used at crossings other than those adjacent to schools and those on established school pedestrian routes.**

08   **The School Crossing assembly shall not be installed on an approach controlled by a STOP or a YIELD sign except as provided in Paragraphs 9 and 10 of this Section.**

Option:

09   The School Crossing assembly may be installed on an approach to a circular intersection controlled by a YIELD sign where the crosswalk is at least 20 feet in advance of the yield point at the entrance to a circulatory roadway.

10   At a signalized or stop-controlled intersection the School Crossing assembly may be installed on an approach to a channelized right turn lane controlled by a YIELD sign where the crosswalk is at least 20 feet in advance of the yield point.

11   A Yield Here To (Stop Here For) School Crossing (R1-5a or R1-5c) sign (see Figure 7B-4) may be used, in accordance with the provisions of Section 2B.19, in advance of a marked crosswalk that crosses an uncontrolled multi-lane approach within school zones.

12   The In-Street Pedestrian Crossing (R1-6 or R1-6a) sign (see Section 2B.20 and Figure 7B-1) or the In-Street School Crossing (R1-6b or R1-6c) sign (see Figure 7B-1) may be used at school crossings on approaches that are not controlled by a traffic control signal, a pedestrian hybrid beacon, or emergency-vehicle hybrid beacon. If used at a school crossing, a 12 x 4-inch SCHOOL (S4-3P) plaque (see Figure 7B-1) may be mounted above the sign. The STATE LAW legend on the R1-6 series signs may be omitted.

13   The In-Street Pedestrian Crossing (R1-6 or R1-6a) sign or In-Street School Crossing (R1-6b or R1-6c) sign may be used at intersections or midblock crossings with flashing beacons.

14   The Overhead School Crossing (R1-9b or R1-9c) sign (see Figure 7B-1) may be used at school crossings on approaches that are not controlled by a traffic control signal, pedestrian hybrid beacon, or an emergency-vehicle hybrid beacon. The STATE LAW legend on the R1-9 series signs may be omitted.

Standard:

15   **When used at an uncontrolled crossing, the In-Street or Overhead Pedestrian Crossing sign shall be used only as a supplement to a School Crossing assembly with a diagonal downward-pointing arrow (W16-7P) plaque at the crosswalk location.**

Figure 7B-3. Example of Signing for a School Crossing Outside of a School Zone

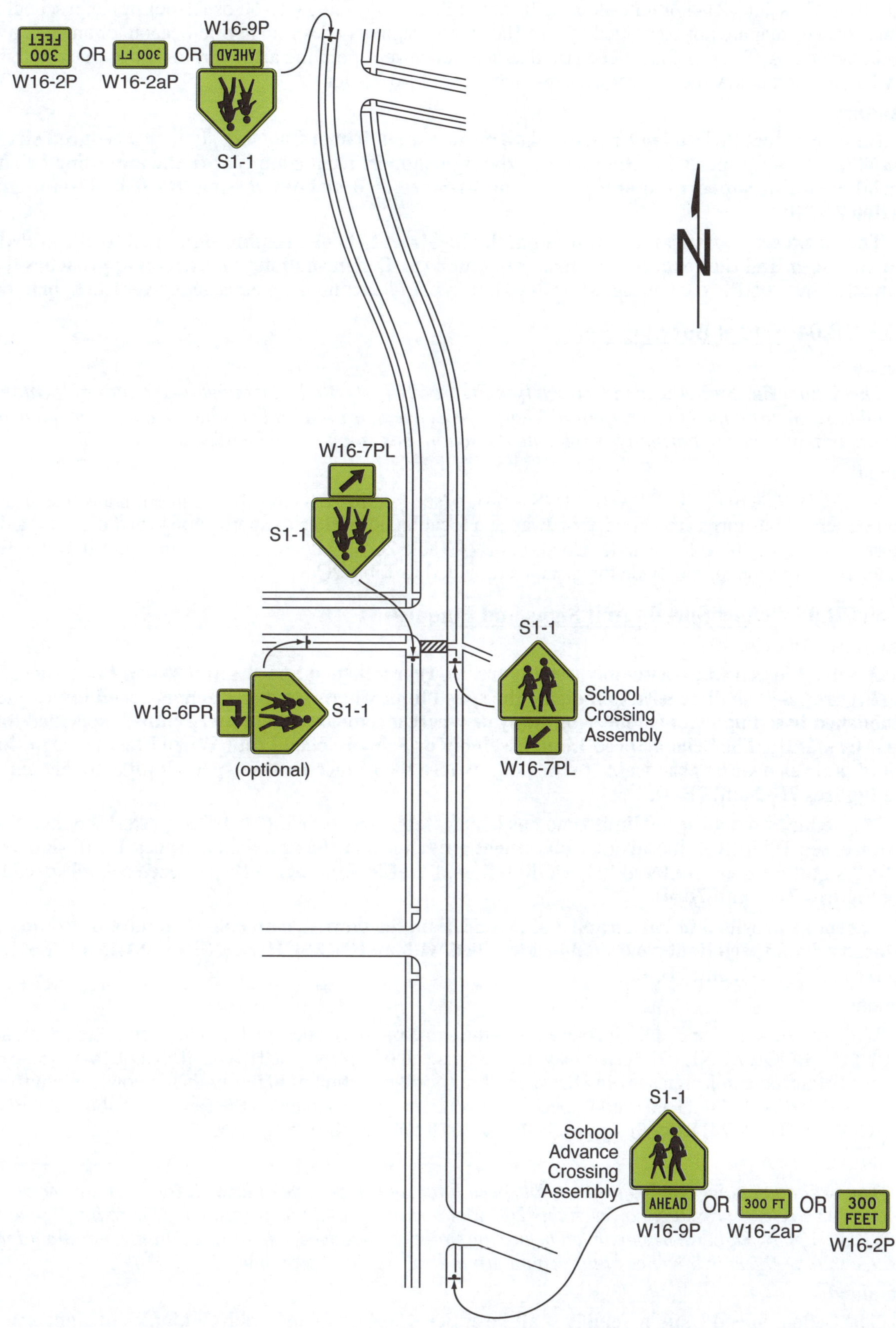

Option:

16   A 12-inch reduced size in-street School (S1-1) sign (see Figure 7B-1) may be used instead of the In-Street Pedestrian Crossing (R1-6 or R1-6a) or the In-Street School Crossing (R1-6b or R1-6c) sign at a school crossing on approaches that are not controlled by a traffic control signal, pedestrian hybrid beacon, or an emergency-vehicle hybrid beacon. A 12 x 6-inch reduced size diagonal downward-pointing arrow (W16-7P) plaque (see Figure 7B-1) may be mounted below the reduced size in-street School (S1-1) sign.

Standard:

17   **If an In-Street Pedestrian Crossing sign, an In-Street School Crossing sign, or a reduced size in-street School (S1-1) sign is placed in the roadway, the sign support shall comply with the mounting height and special mounting support requirements for an In-Street Pedestrian Crossing (R1-6 or R1-6a) sign (see Section 2B.20).**

18   **The In-Street Pedestrian Crossing sign, the In-Street School Crossing sign, the Overhead Pedestrian Crossing sign, and the reduced size in-street School (S1-1) sign shall not be used on approaches that are controlled by a traffic control signal, pedestrian hybrid beacon, or an emergency-vehicle hybrid beacon.**

Section 7B.04 School Bus Stop Signs

Guidance:

01   *The School Bus Stop Ahead (S3-1) sign (see Figure 7B-1) should be installed in advance of locations where a school bus, when stopped to pick up or discharge passengers, is not visible to road users for an adequate distance and where there is no opportunity to relocate the school bus stop to provide adequate sight distance.*

Option:

02   The SCHOOL BUS TURN AHEAD (S3-2) sign (see Figure 7B-1) may be installed in advance of locations where a school bus turns around on a roadway at a location not visible to approaching road users for a distance as determined by the "0" column under Condition B of Table 2C-3, and where there is no opportunity to relocate the school bus turn around to provide the distance provided in Table 2C-3.

Section 7B.05 School Speed Limit Signs and Plaques

Standard:

01   **A School Speed Limit assembly (see Figure 7B-1) or a School Speed Limit When Flashing (S5-1) sign (see Figure 7B-1) shall be used to indicate the speed limit where a reduced school speed limit zone has been established based upon an engineering study or where a reduced school speed limit is specified for such areas by statute. The School Speed Limit assembly or School Speed Limit When Flashing sign shall be placed at or as near as practicable to the point where the reduced school speed limit zone begins (see Figures 7B-2 and 7B-4).**

02   **If a reduced school speed limit zone has been established, a School (S1-1) sign shall be installed in advance (see Table 2C-3 for advance placement guidelines) of the first School Speed Limit sign assembly or S5-1 sign that is encountered in each direction as traffic approaches the reduced school speed limit zone (see Figures 7B-2 and 7B-4).**

03   **Except as provided in Paragraph 4 of this Section, the downstream end of an authorized and posted reduced school speed limit zone shall be identified with an END SCHOOL SPEED LIMIT (S5-3) sign (see Figures 7B-1, 7B-2, and 7B-4).**

Option:

04   If a reduced school speed limit zone ends at the same point as a designated school zone (see Section 7B.02), an END SCHOOL ZONE (S5-2) sign may be used instead of an END SCHOOL SPEED LIMIT (S5-3) sign. A standard Speed Limit sign showing the speed limit for the section of highway that is downstream from the authorized and posted reduced school speed limit zone may be mounted on the same post above the END SCHOOL SPEED LIMIT (S5-3) sign or the END SCHOOL ZONE (S5-2) sign.

Guidance:

05   *The beginning point of a reduced school speed limit zone should be at least 200 feet in advance of the school grounds or a school crossing; however, this 200-foot distance should be increased if the reduced school speed limit is 30 mph or higher. The maximum beginning point of a reduced school speed limit zone should not be greater than 500 feet in advance of the school grounds or a school crossing.*

Standard:

06   **The School Speed Limit assembly shall be either a static sign assembly, a blank-out sign, or a changeable message sign (see Chapter 2L).**

Figure 7B-4. Example of Signing for a School Zone with a School Speed Limit and Higher Fines Only for Speeding

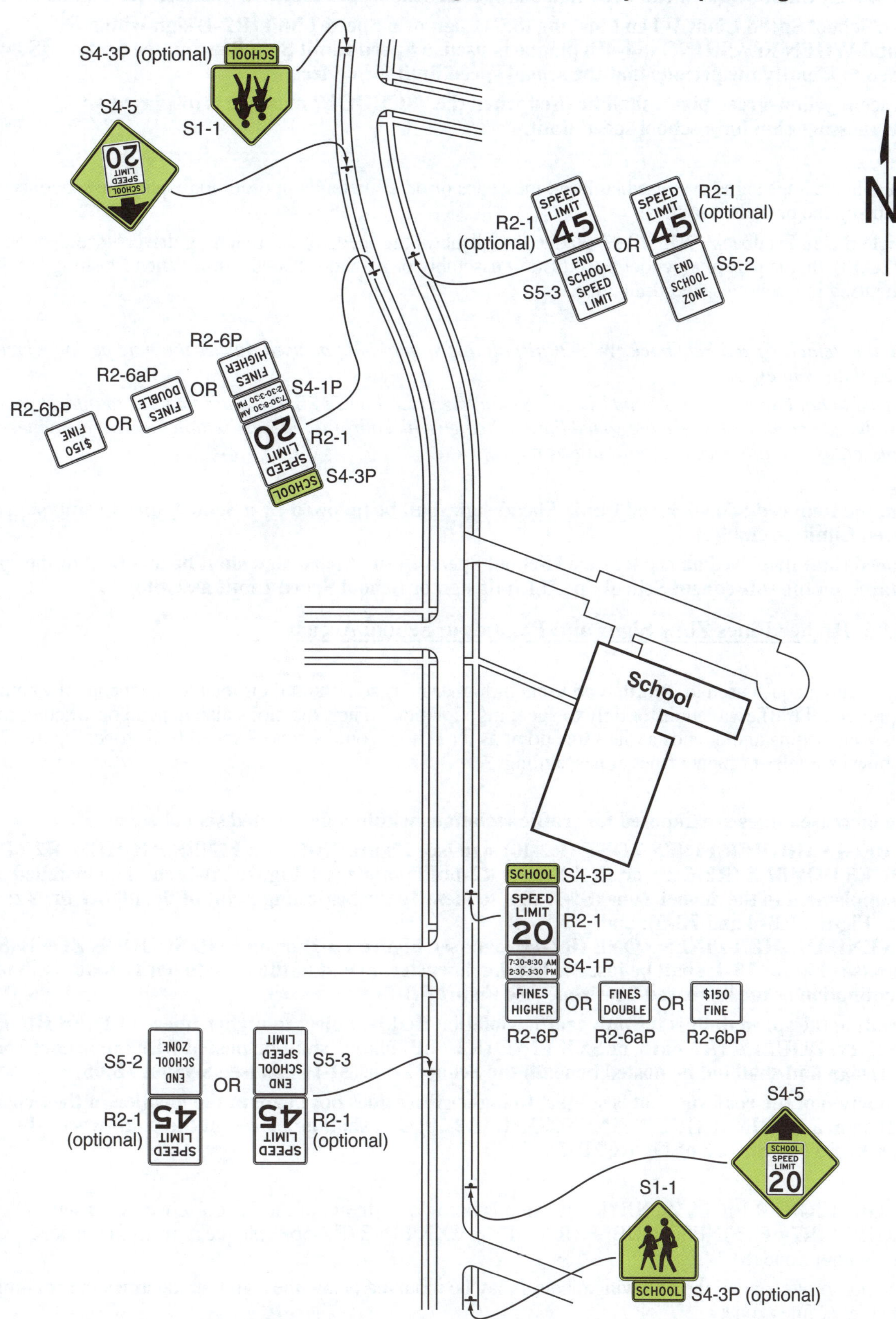

07   The static School Speed Limit assembly shall consist of a top plaque (S4-3P) with the legend SCHOOL, a Speed Limit (R2-1) sign, and a bottom plaque (S4-1P, S4-2P, S4-4P, or S4-6P) indicating the specific periods of the day and/or days of the week that the special school speed limit is in effect (see Figure 7B-1).

08   When a School Speed Limit When Flashing (S5-1) sign or a Speed Limit (R2-1) sign with a supplemental WHEN FLASHING (S4-4P) plaque is used, a Speed Limit Sign Beacon (see Section 4S.04) shall be used to identify the periods that the school speed limit is in effect.

09   Fluorescent yellow-green pixels shall be used when the "SCHOOL" message is displayed on a changeable message sign for a school speed limit.

Option:

10   Changeable message signs may use blank-out messages or other methods in order to display the school speed limit only during the periods it applies.

11   A Vehicle Speed Feedback (W13-20aP) plaque that displays the speed of approaching drivers (see Sections 2B.21 and 2C.13), that is part of a School Speed Limit assembly or a School Speed Limit When Flashing (S5-1) sign, may be used in a school speed limit zone.

Guidance:

12   *If used, the Vehicle Speed Feedback (W13-20aP) plaque should only be used during the time period when the school speed limit is in effect.*

13   *A Reduced School Speed Limit Ahead (S4-5or S4-5a) sign (see Figure 7B-1) should be used to inform road users of a reduced speed zone where the speed limit is being reduced by more than 10 mph, or where engineering judgment indicates that advance notice would be appropriate.*

Standard:

14   **If used, the Reduced School Speed Limit Ahead sign shall be followed by a School Speed Limit sign or a School Speed Limit assembly.**

15   **The speed limit displayed on the Reduced School Speed Limit Ahead sign shall be identical to the speed limit displayed on the subsequent School Speed Limit sign or School Speed Limit assembly.**

Section 7B.06 Higher Fines Zone Signs and Plaques in School Areas

Support:

01   The signs and plaques used to inform road users of higher fines zones and their locations depend on whether the fines apply to all traffic violations or only to speeding violations. Their locations also depend on whether the higher fines zone begins and/or ends at the same point as the school zone or school speed limit zone. Figures 7B-4 and 7B-5 show examples of higher fines zones signing.

Standard:

02   **Where increased fines are imposed for traffic violations within a designated school zone:**

    A. A BEGIN HIGHER FINES ZONE (R2-10) sign (see Figure 7B-1) or a FINES HIGHER (R2-6P), FINES DOUBLE (R2-6aP), or $XX FINE (R2-6bP) plaque (see Figure 7B-1) shall be installed as a supplement to the School Zone (S1-1) sign to identify the beginning point of the higher fines zone (see Figures 7B-4 and 7B-5); and

    B. An END HIGHER FINES ZONE (R2-11) sign (see Figure 7B-1) or an END SCHOOL ZONE (S5-2) sign (see Figure 7B-1) shall be installed at the downstream end of the zone to notify road users of the termination of the increased fines zone (see Figure 7B-5).

03   **If exceeding the speed limit is the only traffic violation that is subject to higher fines, a FINES HIGHER (R2-6P), FINES DOUBLE (R2-6aP), or $XX FINE (R2-6bP) plaque shall be posted with the School Speed Limit (S5-1) sign and shall not be posted beneath the School Zone (S1-1) sign (see Section 7B.05).**

04   **If the portion of the roadway that is subject to higher fines does not begin at the location of the School Zone (S1-1) sign, a BEGIN HIGHER FINES ZONE (R2-10) sign shall be placed at the point where the higher fines begin (see Sheet 2 of Figure 7B-5).**

Option:

05   If a BEGIN HIGHER FINES ZONE (R2-10) sign is used downstream of the School Zone (S1-1) sign, a FINES HIGHER (R2-6P), FINES DOUBLE (R2-6aP), or $XX FINE (R2-6bP) plaque may also be placed beneath the School Zone (S1-1) sign.

06   Where appropriate, one of the following plaques may be mounted below the sign that identifies the beginning point of the higher fines zone:

    A. A S4-1P plaque (see Figure 7B-1) specifying the times that the higher fines are in effect,

    B. A WHEN CHILDREN ARE PRESENT (S4-2P) plaque (see Figure 7B-1), or

    C. A WHEN FLASHING (S4-4P) plaque (see Figure 7B-1) if used in conjunction with a yellow flashing beacon.

Figure 7B-5. Example of Signing for a School Zone with Higher Fines for All Traffic Violations (Sheet 1 of 2)

A – Higher fines zone begins at the beginning of the school zone

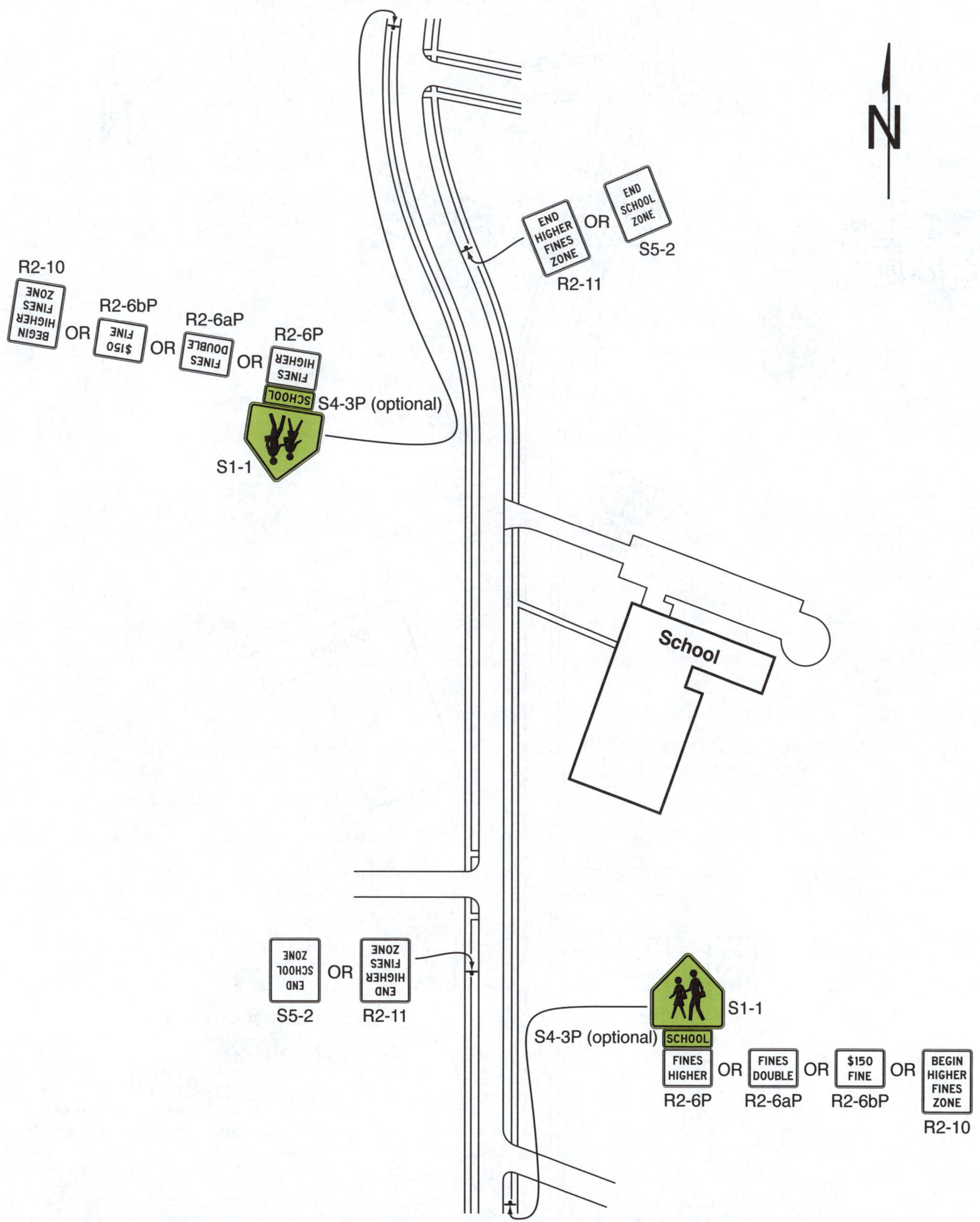

Figure 7B-5. Example of Signing for a School Zone with Higher Fines
for All Traffic Violations (Sheet 2 of 2)

B – Higher fines zone begins after the beginning of the school zone

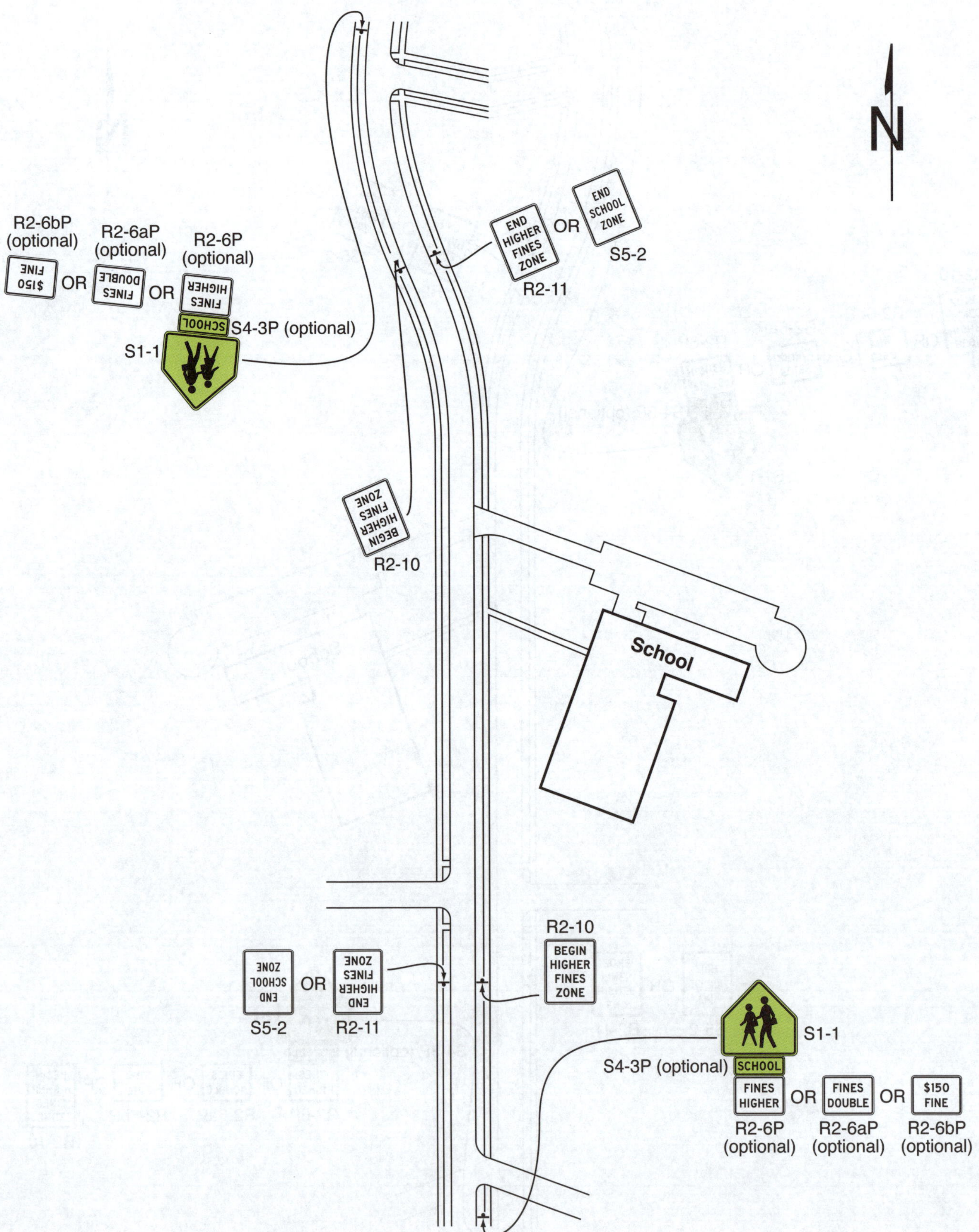

Guidance:

07   *If other traffic violations in addition to exceeding the speed limit are subject to higher fines, then the duplicate FINES HIGHER (R2-6P), FINES DOUBLE (R2-6aP), or $XX FINE (R2-6bP) plaque should be omitted from the School Speed Limit When Flashing (S5-1) sign (see Section 7B.05).*

Option:

08   If a higher fines zone ends at the same point as a reduced school speed limit zone, an END SCHOOL ZONE (S5-2) sign may be used instead of a combination of an END HIGHER FINES ZONE (R2-11) sign and an END SCHOOL SPEED LIMIT (S5-3) sign (see Figure 7B-5).

09   Where the higher fines zone is established by statute, the BEGIN HIGHER FINES ZONE (R2-10) sign, FINES HIGHER (R2-6P), FINES DOUBLE (R2-6aP), and $XX FINE (R2-6bP) plaques may be omitted.

Section 7B.07 <u>Parking and Stopping (R7 and R8 Series) Signs</u>

Option:

01   Parking and stopping regulatory signs may be used to prevent parked or waiting vehicles from blocking pedestrians' views, and drivers' views of pedestrians, and to control vehicles as a part of the school traffic plan.

Support:

02   Parking signs and other signs governing the stopping and standing of vehicles in school areas cover a wide variety of regulations. Typical examples of regulations are as follows:

    A. NO PARKING X:XX AM to X:XX PM SCHOOL DAYS ONLY
    B. NO STOPPING X:XX AM to X:XX PM SCHOOL DAYS ONLY,
    C. XX MIN LOADING X:XX AM to X:XX PM SCHOOL DAYS ONLY, and
    D. NO STANDING X:XX AM to X:XX PM SCHOOL DAYS ONLY.

03   Sections 2B.53 through 2B.55 contain information regarding the signing of parking regulations in school zone areas.

CHAPTER 7C. MARKINGS

Section 7C.01 Crosswalk Markings

Guidance:

01   *Crosswalks should be marked at all intersections on established routes to a school where there is substantial conflict between motorists, bicyclists, and student movements; where students are encouraged to cross between intersections; where students would not otherwise recognize the proper place to cross; or where motorists or bicyclists might not expect students to cross (see Figure 7A-1).*

02   *An engineering study considering the factors described in Section 3C.02 should be performed before a marked crosswalk is installed at a location away from a traffic control signal or an approach controlled by a STOP or YIELD sign.*

03   *Because non-intersection school crossings are generally unexpected by the road user, warning signs (see Section 7B.03) should be installed for all marked school crossings at non-intersection locations. Adequate visibility of students by approaching motorists and of approaching motorists by students should be provided by parking prohibitions or other appropriate measures.*

Support:

04   Section 3C.03 contains provisions regarding the placement and design of crosswalks, and Section 3B.19 contains provisions regarding the placement and design of the stop lines and yield lines that are associated with them. Provisions regarding the curb markings that can be used to establish parking regulations on the approaches to crosswalks are contained in Section 3B.18.

Section 7C.02 Pavement Word, Symbol, and Arrow Markings

Option:

01   If used, the SCHOOL word marking may extend to the width of two approach lanes (see Figure 7C-1).

Guidance:

02   *If the two-lane SCHOOL word marking is used, the letters should be 10 feet or more in height.*

Support:

03   Section 3B.20 contains provisions regarding other word, symbol, and arrow pavement markings that can be used to guide, warn, or regulate traffic.

Figure 7C-1. Two-Lane Pavement Marking of "SCHOOL"

CHAPTER 7D. CROSSING SUPERVISION

Section 7D.01 Adult Crossing Guards

Option:

01   Adult crossing guards may be used to provide gaps in traffic at school crossings where an engineering study has shown that adequate gaps need to be created, and where authorized by law.

Support:

02   Adult crossing guards can also add conspicuity at the crossing where children, who are typically smaller in stature, might not be as visible.

03   High standards for selection of adult crossing guards are essential because they are responsible for the safety of and the efficient crossing of the street by schoolchildren within and in the immediate vicinity of school crosswalks.

Guidance:

04   *Jurisdictions should have policies and procedures for the qualifications, selection, and training of adult crossing guards.*

Section 7D.02 Operating Procedures for Adult Crossing Guards

Standard:

01   **Law enforcement officers performing school crossing supervision and adult crossing guards shall wear high-visibility retroreflective safety apparel labeled as ANSI 107-2020 standard performance for Class 2, Type R, as described in Section 6C.05.**

02   **Adult crossing guards shall not direct traffic in the usual law enforcement regulatory sense. In the control of traffic, they shall pick opportune times to create a sufficient gap in the traffic flow. At these times, they shall stand in the roadway to indicate that pedestrians are about to use or are using the crosswalk, and that all vehicular traffic must stop.**

03   **Adult crossing guards shall use a STOP paddle. The STOP paddle shall be the primary hand-signaling device.**

04   **The STOP paddle shall comply with the provisions for a STOP/SLOW paddle (see Section 6D.02) except both sides shall be a STOP face.**

05   **The paddle shall be retroreflective or illuminated when used during hours of darkness.**

(This page intentionally left blank)

PART 8
TRAFFIC CONTROL FOR RAILROAD AND LIGHT RAIL TRANSIT GRADE CROSSINGS

CHAPTER 8A. GENERAL

Section 8A.01 Introduction

Support:

01   Where the acronym "LRT" is used in Part 8, it refers to "light rail transit."

02   Chapters 8A, 8B, 8C, and 8D describe the traffic control devices that are used at highway-rail and highway-LRT grade crossings. Unless otherwise provided in the text or on a figure or table, the provisions of Part 8 are applicable to both highway-rail and highway-LRT grade crossings. Where the phrase "grade crossing" is used by itself without the prefix "highway-rail" or "highway-LRT," it refers to both highway-rail and highway-LRT grade crossings.

03   Chapter 8E describes the traffic control devices that are used at pathway and sidewalk grade crossings.

04   Traffic control for grade crossings includes all signs, signals, markings, other warning devices, and their supports along highways approaching and at grade crossings. The function of this traffic control is to promote safety and provide effective operation of rail and/or LRT and highway traffic at grade crossings.

05   For purposes of design, installation, operation, and maintenance of traffic control devices at grade crossings, it is recognized that the crossing of the highway and rail tracks is situated on a right-of-way available for the joint use of both highway traffic and railroad or LRT traffic.

06   Grade crossings and the traffic control devices that are associated with them are unique in that in many cases, the highway agency or authority with jurisdiction, the regulatory agency with statutory authority (if applicable), and the railroad company or transit agency are jointly involved in the development of engineering judgment or the performance of an engineering study. This joint process is accomplished through the efforts of a Diagnostic Team made up of the highway agency with jurisdiction, the regulatory agency with statutory authority (if applicable), and the railroad company and/or transit agency (if applicable).

07   In Part 8, the combination of traffic control devices selected or installed at a specific grade crossing is referred to as a "traffic control system."

08   The combination of railroad or LRT active traffic control devices used to inform road users at a grade crossing of the approach or presence of rail traffic and the necessary control equipment for the devices are referred to as a "grade crossing warning system." The "2023 AREMA Communications and Signals Manual" published by the American Railway Engineering and Maintenance-of-Way Association (AREMA) contains further information about grade crossing warning systems.

Standard:

09   **Except at grade crossings of privately-owned roadways, pathways, and sidewalks, the traffic control devices, systems, and practices described in this Manual shall be used at all grade crossings open to public travel, consistent with Federal, State, and local laws and regulations.**

Support:

10   23 CFR 655.603 contains information on the applicability of this Manual at private grade crossings.

Section 8A.02 Highway-LRT Grade Crossings

Support:

01   Part 8 also describes the traffic control devices that are used in locations where light rail transit (LRT) vehicles are operating along streets and highways in mixed traffic with road users.

02   LRT is a mode of public transportation that employs LRT vehicles (commonly known as light rail vehicles, streetcars, or trolleys) that operate on rails in streets in mixed traffic, and LRT traffic that operates in semi-exclusive rights-of-way, or in exclusive rights-of-way. Where the phrase "LRT" is used in Part 8, it refers to light rail vehicles, streetcars, and trolleys. Grade crossings with LRT can occur at intersections or at midblock locations, including public and private driveways.

03   An initial educational campaign along with an ongoing program to continue to educate new drivers is beneficial when introducing LRT operations to an area and, hence, new traffic control devices.

04   LRT alignments can be grouped into one of the following three types (see definitions in Section 1C.02):

A.   Exclusive: An LRT right-of-way that is grade-separated or protected by a fence or traffic barrier. Motor vehicles, pedestrians, and bicycles are prohibited within the right-of-way. This type of alignment does not have grade crossings and is not further addressed in Part 8.

   B. Semi-exclusive: An LRT alignment that is in a separate right-of-way or along a street or railroad right-of-way where motor vehicles, pedestrians, and bicyclists have limited access and cross at designated locations only, such as at grade crossings where road users must yield the right-of-way to the light rail transit traffic.

   C. Mixed-use: An alignment where LRT operates in mixed traffic with all types of road users. In a mixed-use alignment, the light rail transit traffic does not have the right-of-way over other road users at grade crossings and intersections. If the LRT traffic is controlled by traffic control signals or LRT signal faces at an intersection with a roadway, the alignment is considered to be mixed-use even if some of the approaches to the intersection are used exclusively by LRT traffic.

Guidance:

05   *If a highway-LRT grade crossing is equipped with flashing-light signals and is located 200 feet or less from an intersection or midblock location controlled by a traffic control signal, a pedestrian hybrid beacon, or an emergency-vehicle hybrid beacon, the intersection should be provided with rail preemption in accordance with Sections 4F.19 and 8D.09 unless otherwise determined by the Diagnostic Team.*

Option:

06   Where LRT vehicles are operating in a mixed-use alignment, traffic signal priority or preemption may be used as determined by a Diagnostic Team.

Standard:

07   **Where LRT and railroads use the same tracks or adjacent tracks, the traffic control devices, systems, and practices for highway-rail grade crossings shall be used.**

Section 8A.03 <u>Traffic Control Systems and Practices at Grade Crossings</u>

Support:

01   Because of the large number of significant variables to be considered, no single standard system of traffic control devices is universally applicable for all grade crossings.

Standard:

02   **Before any new grade crossing traffic control system is installed or before modifications are made to an existing system, approval shall be obtained from the highway agency with jurisdiction, the regulatory agency with statutory authority (if applicable), and the railroad company and/or transit agency.**

03   **The Diagnostic Team members shall make a recommendation, documented in an engineering study (see Section 8A.05), on new grade crossing traffic control systems and on proposed changes to an existing grade crossing traffic control system. The Diagnostic Team recommendation shall be made based on the Diagnostic Team's site visits, meetings, conference calls, or a combination of some or all of these methods.**

04   **Except as provided in Paragraph 7 of this Section, operational changes made to a grade crossing traffic control system shall be evaluated by a Diagnostic Team.**

05   **Among the types of changes at a grade crossing for which a Diagnostic Team shall conduct an engineering study are: additions, removals, or modifications of the lanes approaching or traversing the grade crossing; addition or removal of tracks; significant changes in the number or speed of trains; significant changes in the number or speed of vehicles; addition of vehicle access near the grade crossing; additions or modifications to sidewalks; additions or modifications to bicycle lanes, especially if a counter-flow bicycle lane is added on a one-way street; changes to roadway use, including conversion to or from one-way operation or reversible lanes; and the installation of or significant operational changes to traffic control signals that might affect the grade crossing.**

Option:

06   A Diagnostic Team may conduct an engineering study and make recommendations as part of the Quiet Zone establishment process (see Section 8A.11).

07   Where determined by the responsible public agency, the railroad company, and/or the transit agency, general maintenance activities or minor operational changes to the grade crossing traffic control system that do not have a negative impact on the overall operation of the traffic control system may be made without a review and determination by a Diagnostic Team.

Support:

08   Many other details of grade crossing traffic control systems that are not set forth in Part 8 are contained in publications such as the "2023 AREMA Communications and Signals Manual" published by the American Railway Engineering and Maintenance-of-Way Association (AREMA), the Third Edition of "Highway-Rail Crossing Handbook" published by the FHWA and the FRA, and the 2nd Edition of "Preemption of Traffic Signals Near Railroad Crossings" published by the Institute of Transportation Engineers (ITE).

Section 8A.04 Traffic Control Systems at Highway-LRT Grade Crossings

Support:

01   The combination of devices selected or installed at a specific highway-LRT grade crossing is referred to as a "Light Rail Transit Traffic Control System."

02   The normal rules of the road and traffic control priority identified in the "Uniform Vehicle Code" (see Section 1A.06) govern the order assigned to the movement of vehicles at an intersection unless the local agency determines that it is appropriate to assign a higher priority to LRT vehicles. Examples of different types of LRT priority control include separate traffic control signal phases for LRT movements, restriction of movement of roadway vehicles in favor of LRT operations, and preemption of highway traffic signal control to accommodate LRT movements.

Standard:

03   **Highway-LRT grade crossings in semi-exclusive alignments outside of a roadway shall be equipped with flashing-light signals, with or without automatic gates, unless a Diagnostic Team determines that the use of Crossbuck Assemblies, STOP signs, or YIELD signs alone would be adequate.**

Section 8A.05 Engineering Studies at Grade Crossings

Standard:

01   **The appropriate traffic control system to be used at a grade crossing shall be determined based on an engineering study conducted by a Diagnostic Team involving the highway agency with jurisdiction, the regulatory agency with statutory authority (if applicable), and the railroad company and/or transit agency (as applicable).**

Option:

02   The regulatory agency with statutory authority (if applicable) may approve the grade crossing traffic control system.

Guidance:

03   *Among the factors that should be considered in the determination by a Diagnostic Team of which traffic control devices would be appropriate to install at a grade crossing are road geometrics, stopping sight distance, clearing sight distance, the proximity of nearby roadway intersections (including the traffic control devices at the intersections), adjacent driveways, traffic volume across the grade crossing, extent of queuing upstream or downstream from the grade crossing, train volume, pedestrian and bicycle volumes, operation of passenger trains, presence of nearby passenger station stops, maximum allowable train speeds, variable train speeds, accelerating and decelerating trains, multiple tracks, high-speed train operation, number of school buses or hazardous material haul vehicles, and the crash history at or near the location.*

Option:

04   The engineering study may include the Highway-Rail Intersection (HRI) components of the National Intelligent Transportation Systems (ITS) architecture, which is a USDOT accepted method for linking the highway, vehicles, and traffic management systems with rail operations and wayside equipment.

Support:

05   More detail on Highway-Rail Intersection components is available from the USDOT's Federal Railroad Administration, 1200 New Jersey Avenue, SE, Washington, DC 20590, or www.fra.dot.gov.

Section 8A.06 Uniform Provisions

Standard:

01   **All signs used in grade crossing traffic control systems shall be retroreflective or illuminated as described in Section 2A.21 to show the same shape and similar color to an approaching road user during both day and night.**

02   **No sign or signal shall be located in the center of an undivided highway, unless it is crashworthy (breakaway, yielding, or shielded with a longitudinal barrier or crash cushion) or unless it is placed on a raised island.**

Guidance:

03   *Any signs or signals placed on a raised island in the center of an undivided highway should be installed with a clearance of at least 2 feet from the outer edge of the raised island to the nearest edge of the sign or signal, except as permitted in Section 2A.16.*

04   *Where a raised median island is installed supplemental to an automatic gate to discourage road users from driving around a lowered gate, the Diagnostic Team should consider the length of the vehicle queues that typically form on the approach to the grade crossing when determining how far in advance of the grade crossing to extend the island.*

05   *If the roadway at a grade crossing includes a two-way left-turn lane (see Section 3B.05), the two-way left-turn lane should be discontinued in the immediate vicinity of the grade crossing by installing median islands, by designating the lane for left turns in one direction only, or by installing yellow diagonal markings in the lane (see Figure 3B-5). If yellow diagonal markings are used, the use of channelizing devices (see Section 3I.01), such as supplemental tubular markers, should also be considered.*

Option:

06   If yellow diagonal markings are used, extending the automatic gate across the lane may be considered.

Section 8A.07 Minimum Track Clearance Distance and Clear Storage Distance

Support:

01   The upstream point of the minimum track clearance distance is determined in the following manner:

    A. If an automatic gate is present on the approach, the upstream point is the portion of the automatic gate arm that is farthest from the nearest rail.

    B. If an automatic gate is not present on the approach, the upstream point is the portion of the stop line that is farthest from the nearest rail.

    C. If the roadway is not paved, the upstream point is the point that is farthest from the nearest rail that is 10 feet measured perpendicular from the nearest rail.

02   The downstream point of the minimum track clearance distance is 6 feet beyond the track(s) or the edge of the downstream highway-highway intersection, whichever is closer, and is measured perpendicular to the farthest rail, along the center line or edge line of the highway, as appropriate, to obtain the longer distance. Where an Exit Gate system (see Section 8D.05) is present, the downstream point is the point where the rear of the vehicle would be clear of the exit gate arm. In cases where the exit gate arm is not perpendicular to the highway, the distance is measured either along the center line or edge line of the highway, as appropriate, to obtain the longer distance.

03   Where two adjacent grade crossings (see Section 8A.08) are located within 200 feet of each other as measured along the highway, the minimum track clearance distance is measured from a point that is upstream of the first grade crossing to a point that is downstream from the second grade crossing.

04   Where a highway-highway intersection is located beyond a grade crossing, the clear storage distance defines on a lane-by-lane basis the area of the roadway between the downstream point of the minimum track clearance distance and the intersection stop line, yield line, or normal stopping point on the highway.

05   The Highway-Rail Crossing Handbook contains an illustration of the minimum track clearance distance and the clear storage distance.

06   The minimum track clearance distance and the clear storage distance are used by the Diagnostic Team to determine the appropriate traffic control devices and/or roadway treatments to be used at the grade crossing, and to determine the queue start-up and queue clearance time necessary where a traffic signal or hybrid beacon is interconnected with a grade crossing active warning system.

Section 8A.08 Adjacent Grade Crossings

Support:

01   Adjacent grade crossings sometimes exist within 200 feet of each other as measured along the highway between the inside rails. These closely-spaced grade crossings sometimes result from separate railroads or from a railroad and an LRT alignment operating in parallel corridors.

Guidance:

02   *Where adjacent grade crossings are located within 200 feet of each other along the highway as measured along the highway between the inside rails, the Diagnostic Team should consider the possibility that rail traffic might arrive at a grade crossing when rail traffic is already occupying the adjacent grade crossing.*

03   *Where the shortest distance between the tracks at adjacent grade crossings, measured along the highway between the inside rails, is 100 feet or less, the grade crossings should be treated as one individual grade crossing.*

04   *Where the shortest distance between the tracks at adjacent grade crossings, measured along the highway between the inside rails, is more than 100 feet and less than 200 feet, additional signs or other appropriate traffic control devices should be used to inform approaching road users of the long distance to cross the tracks.*

05   *Where active traffic control devices are installed between adjacent grade crossings that are more than 100 feet apart and less than 200 feet apart as measured along the highway between the inside rails, the operation of the devices should provide additional time for vehicles to clear the extended minimum track clearance distance (see Section 8A.07) that results from the closely-spaced grade crossings.*

06   *Where the shortest distance between the tracks at adjacent grade crossings, measured along the highway between the inside rails, is more than 200 feet, the grade crossings should be treated as individual grade crossings and traffic control devices should be installed between the grade crossings.*

Support:

07   The "2023 AREMA Communications and Signals Manual" published by the American Railway Engineering and Maintenance-of-Way Association (AREMA) contains further information and recommendations about the location and operation of active traffic control devices at adjacent grade crossings that are located within 200 feet of each other.

Section 8A.09 Grade Crossing Elimination

Option:

01   If a particular grade crossing appears to be redundant or unnecessary for motor vehicle, pedestrian, and bicycle traffic, an engineering study may be conducted to determine the costs and benefits of eliminating the crossing.

Guidance:

02   *If an engineering study is conducted, any necessary improvements to adjacent grade crossings and the surrounding roadway network to accommodate diverted traffic should also be included in the analysis.*

03   *If the conclusion of the engineering study is that the grade crossing should be eliminated, a Diagnostic Team should use the engineering study to determine the appropriate steps that need to be taken to accomplish the grade crossing elimination.*

Standard:

04   **Where a grade crossing is eliminated, the traffic control devices for the crossing shall be removed, and shall be covered or turned from view in the interim period prior to removal.**

Guidance:

05   *If the existing traffic control devices at a multiple-track grade crossing become improperly placed or are no longer applicable because of the removal of some of the tracks, the existing devices should be relocated and/ or modified.*

06   *Where a roadway is removed from a grade crossing, the roadway approaches in the railroad or LRT right-of-way should also be removed and appropriate signs and object markers should be placed at the roadway end in accordance with Section 2C.73.*

07   *Where a railroad or LRT is eliminated at a grade crossing, the tracks should be removed or paved over.*

Option:

08   Based on engineering judgment, the TRACKS OUT OF SERVICE (R8-9) sign (see Figure 8B-1) may be temporarily installed until the tracks are removed or covered. The length of time before the tracks will be removed or covered may be considered in making the decision as to whether to install the sign.

Section 8A.10 Illumination at Grade Crossings

Support:

01   Illumination is sometimes installed at or adjacent to a grade crossing in order to provide better nighttime visibility of trains or LRT equipment and the grade crossing (for example, where a substantial amount of railroad or LRT operations are conducted at night, where grade crossings are blocked for extended periods of time, or where crash history indicates that road users experience difficulty in seeing trains or LRT equipment or traffic control devices during hours of darkness).

02   Recommended types and locations of luminaires for illuminating grade crossings are contained in the American National Standards Institute's "ANSI/IES RP-8-22, Recommended Practice: Lighting Roadway and Parking Facilities," which is available from the Illuminating Engineering Society.

Section 8A.11 Quiet Zone Treatments at Highway-Rail Grade Crossings

Support:

01   49 CFR Part 222 (Use of Locomotive Horns at Highway-Rail Grade Crossings; Final Rule) prescribes Quiet Zone requirements and treatments.

Standard:

02   **Any traffic control device and its application where used as part of a Quiet Zone shall comply with all applicable provisions of the MUTCD.**

Section 8A.12 Grade Crossings within or in Close Proximity to Circular Intersections

Support:

01   At circular intersections, such as roundabouts and traffic circles, that include or are within close proximity to a grade crossing, a queue of vehicular traffic could cause motor vehicles to stop on the grade crossing.

Standard:

02   **Where circular intersections include or are within 200 feet of a grade crossing, an engineering study shall be made to determine if queuing could impact the grade crossing.**

Guidance:

03   *The Diagnostic Team should review the findings of the engineering study and determine the appropriate measures to clear highway traffic from the grade crossing prior to the arrival of rail traffic.*

Support:

04   Among the actions that can be taken to keep the grade crossing clear of traffic or to clear traffic from the grade crossing prior to the arrival of rail traffic are the following:

    A. Grade crossing regulatory and warning devices;
    B. Highway traffic signals;
    C. Traffic metering devices;
    D. Activated signs;
    E. Geometric design revisions, including reconstruction or elimination of the circular intersection; or
    F. A combination of these or other actions.

Section 8A.13 Temporary Traffic Control Zones

Support:

01   Temporary traffic control planning provides for continuity of operations (such as movement of motor vehicle traffic, pedestrians, and bicyclists, transit operations, and access to property/utilities) when the normal function of a roadway at a grade crossing is suspended because of temporary traffic control operations. Temporary traffic control planning is also needed when traffic is detoured over an existing grade crossing.

Standard:

02   **Traffic controls for temporary traffic control zones that include grade crossings shall be as provided in Part 6.**

Guidance:

03   *Where a temporary traffic control zone extends over an active grade crossing (see Section 6N.17), and where the direction of traffic in any lane is reversed over the grade crossing, the railroad company or transit agency should be part of the temporary traffic control planning process. Where a grade crossing warning system is not modified to support the temporary traffic control operation, at least one uniformed law enforcement officer should be in place at all times that rail traffic might approach or occupy the grade crossing.*

04   *Where traffic is detoured over an existing passive grade crossing, a temporary traffic control plan (see Section 6B.01) should be prepared.*

05   *Public and private agencies, emergency services, businesses, and railroad companies or transit agencies should meet to plan appropriate traffic detours and the necessary signing, marking, signalization, and flagging requirements for operations during temporary traffic control zone activities or during the period when traffic is being detoured over an existing passive grade crossing. Consideration should be given to the length of time that the grade crossing is to be closed, the length of time that a detour is to be in place, the type of rail or LRT and highway traffic affected, the time of day, and the materials and techniques of repair.*

06   *The agencies responsible for the operation of the LRT and highway should be contacted when the initial planning begins for any temporary traffic control zone that might directly or indirectly influence the flow of traffic on facilities where LRT vehicles operate on a mixed-use alignment.*

07   *Temporary traffic control operations should minimize the inconvenience, delay, and crash potential to affected traffic. Prior notice should be given to affected public or private agencies, emergency services, businesses, railroad companies or transit agencies, and road users before the free movement of road users or rail traffic is infringed upon or blocked.*

Support:

08   Section 6N.17 contains additional information regarding temporary traffic control zones in the vicinity of grade crossings, and Figure 6P-46 shows an example of a typical situation that might be encountered.

CHAPTER 8B. SIGNS

Section 8B.01 Purpose and Application

Support:

01   Passive traffic control systems, consisting of signs and pavement markings only, identify and direct attention to the location of a grade crossing and advise road users to reduce their speed or stop at the grade crossing as necessary in order to yield to any rail traffic occupying, or approaching and in proximity to, the grade crossing.

02   Signs and markings regulate, warn, and guide the road users so that they, as well as LRT vehicle operators on mixed-use alignments, can take appropriate action when approaching a grade crossing.

03   Unless otherwise provided in this Chapter, the provisions of Part 2 are applicable to the design and location of signs at grade crossings.

Section 8B.02 Sizes of Grade Crossing Signs

Standard:

01   **The minimum sizes of grade crossing signs shall be as shown in Table 8B-1.**

Option:

02   Signs larger than those shown in Table 8B-1 may be used (see Section 2A.07).

Section 8B.03 Grade Crossing (Crossbuck) Sign (R15-1) and Number of Tracks Plaque (R15-2P) at Active and Passive Grade Crossings

Standard:

01   **The Grade Crossing (R15-1) sign (see Figure 8B-1), commonly identified as the Crossbuck sign, shall be retroreflective white with the words RAILROAD CROSSING in black lettering, mounted as shown in Figure 8B-2.**

Support:

02   In most States, the Crossbuck sign requires road users to yield the right-of-way to rail traffic at a grade crossing.

Standard:

03   **As a minimum, one Crossbuck sign shall be used on each highway approach to every highway-rail grade crossing, alone or in combination with other traffic control devices.**

04   **As a minimum, one Crossbuck sign shall be used on each highway approach to every highway-LRT grade crossing where flashing-light signals or automatic gates are used, alone or in combination with other traffic control devices.**

Option:

05   A Crossbuck sign may be used on a highway approach to a highway-LRT grade crossing where flashing-light signals or automatic gates are not used, alone or in combination with other traffic control devices.

Standard:

06   **If there are two or more tracks at a grade crossing, the number of tracks shall be indicated on a supplemental Number of Tracks (R15-2P) plaque (see Figure 8B-1) of inverted T shape mounted below the Crossbuck sign in the manner shown in Figure 8B-2.**

07   **On each approach to a highway-rail grade crossing and, if used, on each approach to a highway-LRT grade crossing, the Crossbuck sign shall be installed on the right-hand side of the highway on each approach to the grade crossing. Where restricted sight distance or unfavorable highway geometry exists on an approach to a grade crossing, or where there is a one-way multi-lane approach, an additional Crossbuck sign shall be installed on the left-hand side of the highway, possibly placed back-to-back with the Crossbuck sign for the opposite approach, or otherwise located so that two Crossbuck signs are displayed for that approach.**

08   **A strip of retroreflective white material not less than 2 inches in width shall be used on the back of each blade of each Crossbuck sign for the length of each blade at all passive grade crossings, except those where Crossbuck signs have been installed back-to-back or where double-faced Crossbuck signs have been installed.**

Option:

09   A strip of retroreflective white material not less than 2 inches in width may be used on the back of each blade of each Crossbuck sign for the length of each blade at active grade crossings.

Table 8B-1. Grade Crossing Sign and Plaque Minimum Sizes (Sheet 1 of 2)

Sign or Plaque	Sign Designation	Section	Conventional Road		Expressway	Minimum	Oversized
			Single Lane	Multi-Lane			
Stop	R1-1	8B.04, 8B.05	30 x 30	36 x 36	36 x 36	—	48 x 48
Yield	R1-2	8B.04, 8B.05	30 x 30 x 30	36 x 36 x 36	36 x 36 x 36	—	48 x 48 x 48
No Right Turn - Train (symbol)	R3-1a	8D.10	24 x 30	30 x 36	—	—	—
No Left Turn - Train (symbol)	R3-2a	8D.10	24 x 30	30 x 36	—	—	—
Do Not Stop on Tracks	R8-8	8B.07	24 x 30	24 x 30	36 x 48	—	36 x 48
Tracks Out of Service	R8-9	8B.08	24 x 24	24 x 24	36 x 36	—	36 x 36
Stop Here When Flashing	R8-10	8B.09	24 x 36	24 x 36	—	—	36 x 48
Stop Here When Flashing	R8-10a	8B.09	24 x 30	24 x 30	—	—	36 x 42
Stop Here on Red	R10-6	8B.10	24 x 36	24 x 36	—	—	36 x 48
Stop Here on Red	R10-6a	8B.10	24 x 30	24 x 30	—	—	36 x 42
Left (Right) Lane Signal	R10-10b	8D.11	30 x 36	30 x 36	—	24 x 30	—
Left (Right) Turn Lane Signal	R10-10c	8D.11	30 x 42	30 x 42	—	24 x 36	—
Grade Crossing (Crossbuck)	R15-1	8B.03	48 x 9	48 x 9	—	—	—
Number of Tracks (plaque)	R15-2P	8B.03	27 x 18	27 x 18	—	—	—
Exempt (plaque)	R15-3P	8B.11	24 x 12	24 x 12	—	—	—
Light Rail Only Right Lane	R15-4a	8B.12	24 x 30	24 x 30	—	—	—
Light Rail Only Left Lane	R15-4b	8B.12	24 x 30	24 x 30	—	—	—
Light Rail Only Center Lane	R15-4c	8B.12	24 x 30	24 x 30	—	—	—
Light Rail Do Not Pass (symbol)	R15-5	8B.13	24 x 30	24 x 30	—	—	—
Do Not Pass Stopped Train	R15-5a	8B.13	24 x 30	24 x 30	—	—	—
No Motor Vehicles On Tracks (symbol)	R15-6	8B.14	24 x 24	24 x 24	—	—	—
Do Not Drive On Tracks	R15-6a	8B.14	24 x 30	24 x 30	—	—	—
Light Rail Divided Highway (symbol)	R15-7	8B.15	24 x 24	24 x 24	—	—	—
Light Rail Divided Highway (T-Intersection) (symbol)	R15-7a	8B.15	24 x 24	24 x 24	—	—	—
Look	R15-8	8E.03	—	—	—	18 x 9	—
Grade Crossing Advance Warning	W10-1	8B.06	36 Dia.	36 Dia.	48 Dia.	—	48 Dia.
Exempt (plaque)	W10-1aP	8B.11	24 x 12	24 x 12	—	—	—
Grade Crossing and Intersection Advance Warning (symbol)	W10-2,3,4	8B.06	36 x 36	36 x 36	48 x 48	—	48 x 48
Low Ground Clearance (symbol)	W10-5	8B.16	36 x 36	36 x 36	48 x 48	—	48 x 48
Low Ground Clearance (plaque)	W10-5P	8B.16	30 x 24	30 x 24	—	—	—
Light Rail Activated Blank-Out (symbol)	W10-7	8B.17	24 x 24	24 x 24	—	—	—
Trains May Exceed 80 MPH	W10-8	8B.19	36 x 36	36 x 36	48 x 48	—	48 x 48
No Train Horn	W10-9	8B.20	36 x 36	36 x 36	48 x 48	—	48 x 48
No Train Horn (plaque)	W10-9P	8B.20	30 x 24	30 x 24	—	—	—
Storage Space (symbol)	W10-11	8B.21	36 x 36	36 x 36	48 x 48	—	48 x 48
Storage Space XX Feet between Tracks & Highway	W10-11a	8B.21	30 x 36	30 x 36	—	—	—
Storage Space XX Feet between Highway & Tracks Behind You	W10-11b	8B.21	30 x 36	30 x 36	—	—	—
Skewed Crossing (symbol)	W10-12	8B.22	36 x 36	36 x 36	48 x 48	—	48 x 48
No Gates or Lights (plaque)	W10-13P	8B.23	30 x 24	30 x 24	—	—	—
Next Crossing (plaque)	W10-14P	8B.24	30 x 24	30 x 24	—	—	—
Use Next Crossing (plaque)	W10-14aP	8B.24	30 x 24	30 x 24	—	—	—

Table 8B-1. Grade Crossing Sign and Plaque Minimum Sizes (Sheet 2 of 2)

Sign or Plaque	Sign Designation	Section	Conventional Road		Expressway	Minimum	Oversized
			Single Lane	Multi-Lane			
Rough Crossing (plaque)	W10-15P	8B.25	30 x 24	30 x 24	—	—	36 x 30
Another Train Coming Activated Blank-Out	W10-16	8B.18	30 x 30	30 x 30	—	—	—
Busway Crossing	W10-21	8B.06	36x36	36x36	48x48	—	48x48
Signal Ahead (plaque)	W10-21aP	8B.06	30x24	30x24	—	—	—
Emergency Notification	I13-1	8B.27	—	—	—	12 x 9	—
Push to Exit	I13-2	8E.06	—	—	—	18 x 9	—

Notes: 1. Larger signs may be used when appropriate

2. Dimensions in inches are shown as width x height

3. Table 9A-1 shows the minimum sizes that may be used for grade crossing signs and plaques that face shared-use paths and pedestrian facilities

---

Guidance:

10 *Minimum clearance dimensions for crossbuck signs relative to the proximity to the nearest rail should conform to the requirements of the railroad company and/or transit agency, and the regulatory agency with statutory authority (if applicable).*

11 *Except as provided in Paragraph 12 of this Section, the mounting height of Crossbuck signs, measured vertically from the center of the sign to the elevation of the near edge of the pavement, should be approximately 9 feet (see Figure 8B-2).*

Option:

12 The 9-foot mounting height for the Crossbuck sign may be varied as required by local conditions and may be increased to accommodate signs mounted below the Crossbuck sign.

Section 8B.04 Crossbuck Assemblies with YIELD or STOP Signs at Passive Grade Crossings

Standard:

01 **A Crossbuck Assembly shall consist of a Crossbuck (R15-1) sign, and a Number of Tracks (R15-2P) plaque if two or more tracks are present, that complies with the provisions of Section 8B.03, and either a YIELD (R1-2) or STOP (R1-1) sign installed on the same support, except as provided in Paragraph 10 of this Section. YIELD or STOP signs used at passive grade crossings shall be installed in compliance with the provisions of Section 2B.18, and Figures 8B-2 and 8B-3.**

02 **At all public highway-rail grade crossings that are not equipped with the active traffic control systems that are described in Chapter 8D, except crossings where road users are directed by an authorized person on the ground to not enter the crossing at all times that an approaching train is about to occupy the crossing, a Crossbuck Assembly shall be installed on the right-hand side of the highway on each approach to the highway-rail grade crossing.**

03 **If a Crossbuck sign is used on a highway approach to a public highway-LRT grade crossing that is not equipped with the active traffic control systems that are described in Chapter 8D, a Crossbuck Assembly shall be installed on the right-hand side of the highway on each approach to the highway-LRT grade crossing.**

04 **Where restricted sight distance or unfavorable highway geometry exists on an approach to a grade crossing that has a Crossbuck Assembly, or where there is a one-way multi-lane approach, an additional Crossbuck Assembly shall be installed on the left-hand side of the highway.**

05 **A YIELD sign shall be the default traffic control device for Crossbuck Assemblies on all highway approaches to passive grade crossings unless an engineering study performed by the regulatory agency or highway authority having jurisdiction over the roadway approach determines that a STOP sign is appropriate.**

Guidance:

06 *The use of STOP signs at passive grade crossings should be limited to unusual conditions where requiring all motor vehicles to make a full stop is determined to be necessary by a Diagnostic Team. Among the factors that should be considered by the Diagnostic Team are the line of sight to approaching rail traffic (giving due consideration to seasonal crops or vegetation beyond both the highway and railroad or LRT rights-of-ways), the number of tracks, the speeds of trains or LRT equipment and motor vehicles, and the crash history at the grade crossing.*

Figure 8B-1. Regulatory Signs and Plaques for Grade Crossings

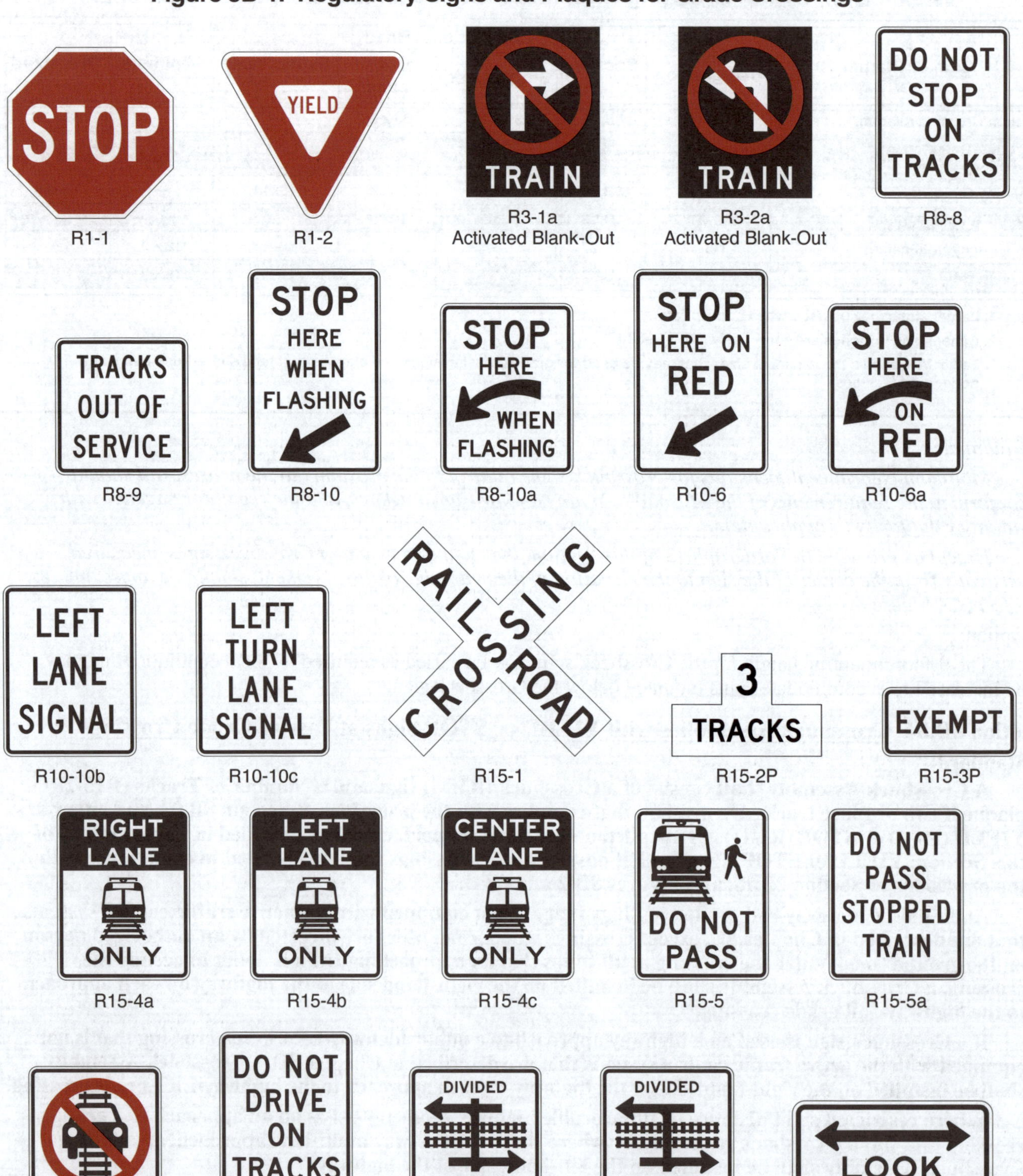

Figure 8B-2. Crossbuck Assembly with a YIELD or STOP Sign
on the Crossbuck Sign Support

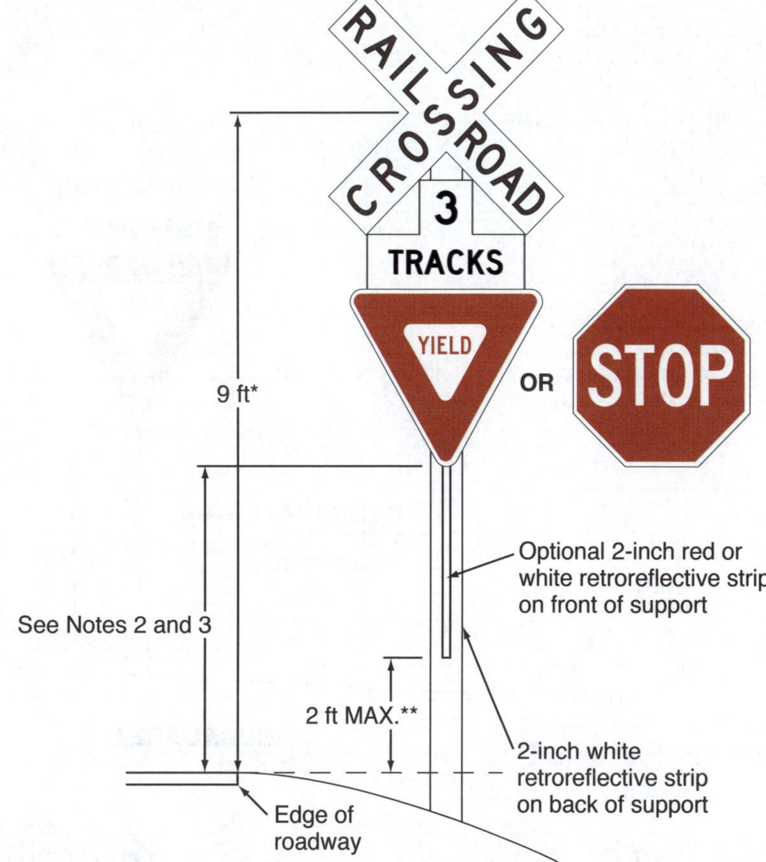

Notes:

1. YIELD or STOP signs are used only at passive crossings. A STOP sign is used only if an engineering study determines that it is appropriate for that particular approach.
2. Mounting height shall be at least 4 feet for installations of YIELD or STOP signs on existing Crossbuck sign supports.
3. Mounting height shall be at least 5 feet for new installations in rural areas and at least 7 feet for new installations in areas where parking or pedestrian movements are likely to occur.

07 *Where a passive grade crossing is located on a stop-controlled approach and the clear storage distance is less than the length of the design vehicle, and where adequate sight distance to oncoming traffic on the parallel roadway is available to road users stopped on the approach to the grade crossing, consideration should be given to installing a STOP sign at the Crossbuck Assembly instead of at the highway-highway intersection. If the STOP sign is installed at the Crossbuck Assembly instead of at the highway-highway intersection, the Diagnostic Team should consider installing some other intersection traffic control device at the highway-highway intersection.*

Standard:

08 **If a Crossbuck Assembly is installed on the approach to a passive grade crossing located at a highway-highway intersection controlled by a traffic control signal that is not interconnected with the grade crossing and not preempted by the approach of rail traffic, a Diagnostic Team shall be convened to determine the appropriate traffic control devices. A STOP sign shall not be installed on a Crossbuck Assembly in this situation.**

Support:

09 Sections 8A.01 through 8A.05 contain information regarding the responsibilities of the Diagnostic Team, highway agency, regulatory agency with statutory authority (if applicable), and the railroad company or transit agency regarding the selection, design, and operation of traffic control devices placed at grade crossings.

Option:

10 If a YIELD or STOP sign is installed for a Crossbuck Assembly at a grade crossing, it may be installed on the same support as the Crossbuck sign or it may be installed on a separate support at a point where the motor vehicle is to stop, or as near to that point as practicable, but in either case, the YIELD or STOP sign is considered to be a part of the Crossbuck Assembly.

Figure 8B-3. Crossbuck Assembly with a YIELD or STOP Sign on a Separate Sign Support (Sheet 1 of 2)

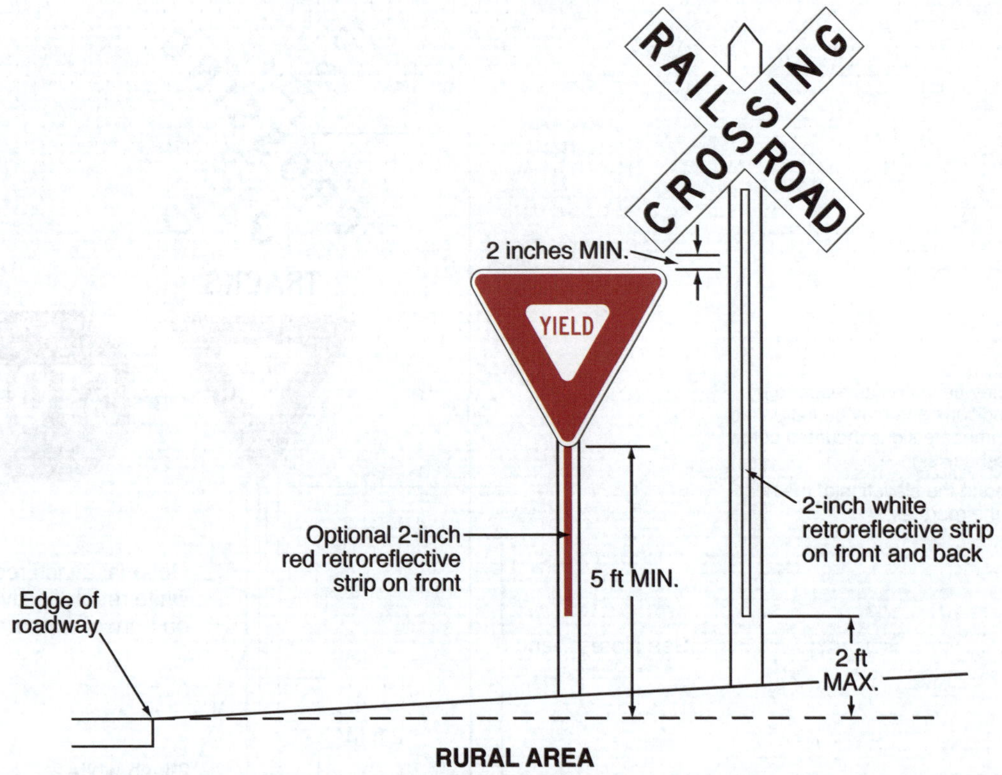

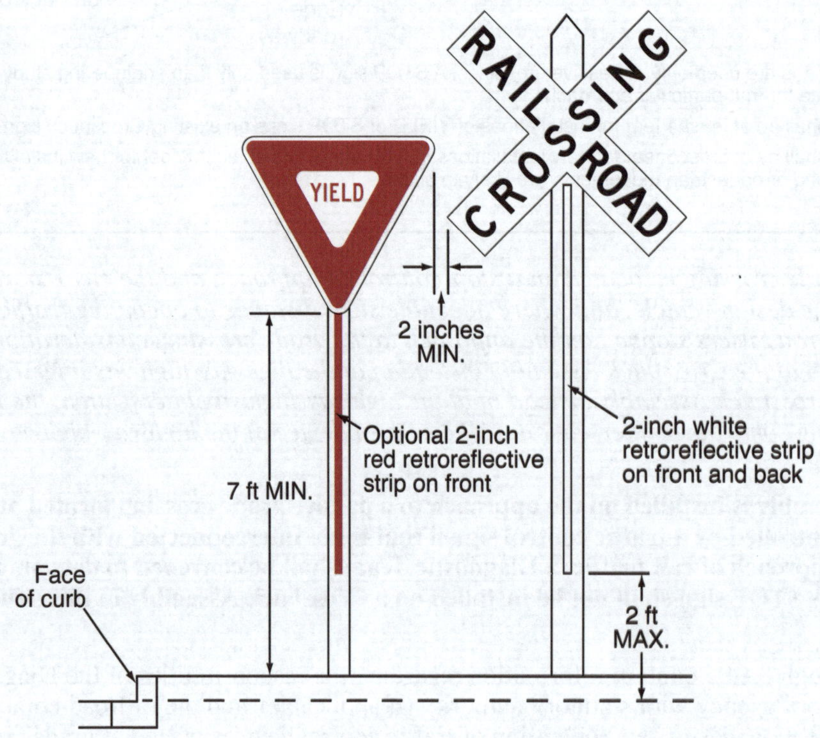

Notes:

1. YIELD signs are used only at passive crossings.

2. Place the face of the signs in the same plane and place the YIELD sign closer to the traveled way. Provide a 2-inch minimum separation between the edge of the Crossbuck sign and the edge of the YIELD sign.

Figure 8B-3. Crossbuck Assembly with a YIELD or STOP Sign on a Separate Sign Support (Sheet 2 of 2)

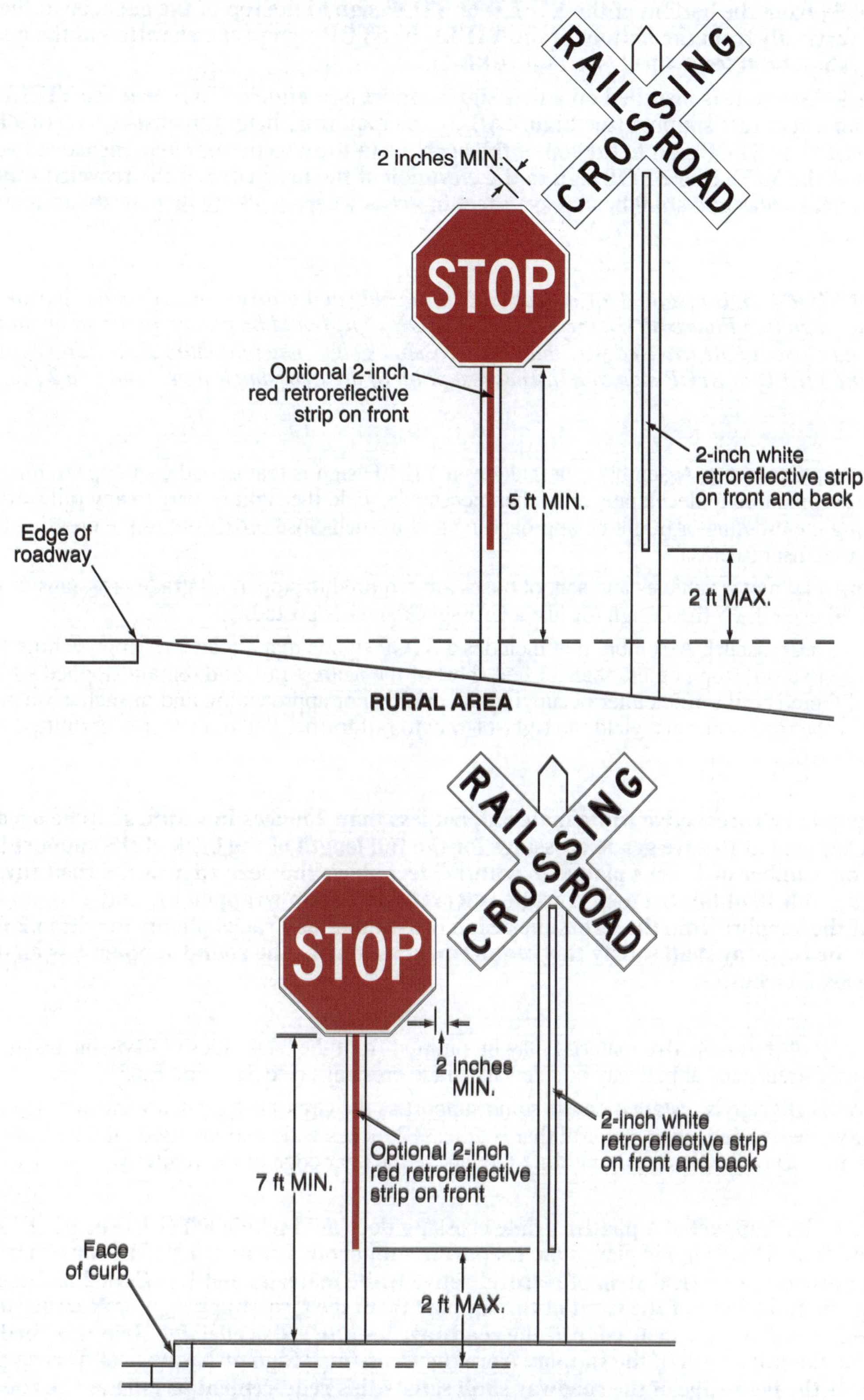

Notes:

1. STOP signs are used only at passive crossings and only if an engineering study determines that it is appropriate for that particular approach.

2. Place the face of the signs in the same plane and place the STOP sign closer to the traveled way. Provide a 2-inch minimum separation between the edge of the Crossbuck sign and the edge of the STOP sign.

Standard:

11   If a YIELD or STOP sign is installed on an existing Crossbuck sign support, the mounting height, measured vertically from the bottom of the YIELD or STOP sign to the top of the curb, or in the absence of curb, measured vertically from the bottom of the YIELD or STOP sign to the elevation of the near edge of the traveled way, shall be at least 4 feet (see Figure 8B-2).

12   If a Crossbuck Assembly is installed on a new sign support (see Figure 8B-2) or if the YIELD or STOP sign is installed on a separate support (see Figure 8B-3), the mounting height, measured vertically from the bottom of the YIELD or STOP sign to the top of the curb, or in the absence of curb, measured vertically from the bottom of the YIELD or STOP sign to the elevation of the near edge of the traveled way, shall be at least 5 feet in rural areas and shall be at least 7 feet in areas where parking or pedestrian movements are likely to occur.

Guidance:

13   *If a YIELD or STOP sign is installed for a Crossbuck Assembly at a grade crossing on a separate support than the Crossbuck sign (see Figure 8B-3), the YIELD or STOP sign should be placed in the same plane as the Crossbuck sign and closer to the traveled way than the Crossbuck sign. The minimum separation between the nearest point of the YIELD or STOP sign and the nearest point of the Crossbuck sign should be 2 inches as shown in Figure 8B-3.*

Support:

14   The meaning of a Crossbuck Assembly that includes a YIELD sign is that a road user approaching the grade crossing needs to be prepared to decelerate, and when necessary, yield the right-of-way to any rail traffic that might be occupying the crossing or might be approaching and in such close proximity to the crossing that it would be unsafe for the road user to cross.

15   Certain commercial motor vehicles and school buses are required to stop at all grade crossings in accordance with 49 CFR 392.10 even if a YIELD sign (or just a Crossbuck sign) is posted.

16   The meaning of a Crossbuck Assembly that includes a STOP sign is that a road user approaching the grade crossing must come to a full stop not less than 15 feet short of the nearest rail, and remain stopped while the road user determines if there is rail traffic either occupying the crossing or approaching and in such close proximity to the crossing that the road user must yield the right-of-way to rail traffic. The road user is permitted to proceed when it is safe to cross.

Standard:

17   **A vertical strip of retroreflective white material, not less than 2 inches in width, shall be used on each Crossbuck support at passive grade crossings for the full length of the back of the support from the Crossbuck sign or Number of Tracks plaque to within 2 feet above the near edge of the roadway, except as provided in Paragraph 18 of this Section. A white retroreflective strip wrapped around a round support for the full length of the support from the Crossbuck Sign or Number of Tracks plaque to within 2 feet above the near edge of the roadway shall satisfy this requirement as long as the round support has an outside diameter of at least 2 inches.**

Option:

18   The vertical strip of retroreflective material may be omitted from the back sides of Crossbuck sign supports installed on one-way streets and at pathway or sidewalk grade crossings (see Section 8E.05).

19   If a YIELD or STOP sign is installed on the same support as the Crossbuck sign, a vertical strip of red (see Section 2A.11) or white retroreflective material that is at least 2 inches wide may be used on the front of the support from the YIELD or STOP sign to within 2 feet above the near edge of the roadway.

Standard:

20   **If a Crossbuck sign support at a passive grade crossing does not include a YIELD or STOP sign (either because the YIELD or STOP sign is placed on a separate support or because a YIELD or STOP sign is not present on the approach), a vertical strip of retroreflective white material, not less than 2 inches in width, shall be used for the full length of the front of the support from the Crossbuck sign or Number of Tracks plaque to within 2 feet above the near edge of the roadway. A white retroreflective strip wrapped around a round support for the full length of the support from the Crossbuck Sign or Number of Tracks plaque to within 2 feet above the near edge of the roadway shall satisfy this requirement as long as the round support has an outside diameter of at least 2 inches.**

21   **At all grade crossings where YIELD or STOP signs are installed, Yield Ahead (W3-2) or Stop Ahead (W3-1) signs shall also be installed if the criteria for their installation in Section 2C.35 is met.**

Support:

22   Section 8C.03 contains provisions regarding the use of stop lines or yield lines at grade crossings.

Section 8B.05 Use of STOP (R1-1) or YIELD (R1-2) Signs without Crossbuck Signs at Highway-LRT Grade Crossings

Guidance:

01   *The use of only STOP or YIELD signs for road users at highway-LRT grade crossings should be limited to those crossings where the need and feasibility is determined by the Diagnostic Team. Such crossings should have all of the following characteristics:*

   A.   *The crossing roadways are secondary in character (such as a minor street with one lane in each direction, an alley, or a driveway) with low traffic volumes and low speed limits. The specific thresholds of traffic volumes and speed limits should be determined by the local agencies.*
   B.   *The line of sight for an approaching LRT operator is adequate from a sufficient distance such that the operator can sound an audible signal and bring the LRT equipment to a stop before arriving at the crossing.*
   C.   *The road user has sufficient sight distance at the stop line to permit the vehicle to cross the tracks before the arrival of the LRT equipment.*
   D.   *If at an intersection of two roadways, the intersection does not meet the warrants for a traffic control signal as provided in Chapter 4C.*
   E.   *The LRT tracks are located such that motor vehicles are not likely to stop on the tracks while waiting to enter a crossroad or highway.*

Standard:

02   **For all highway-LRT grade crossings where only STOP (R1-1) or YIELD (R1-2) signs are installed, the placement shall comply with the requirements of Section 2B.18. Stop Ahead (W3-1) or Yield Ahead (W3-2) Advance Warning signs shall also be installed if the criteria for their installation given in Section 2C.35 is met.**

Section 8B.06 Grade Crossing Advance Warning Signs (W10-1 through W10-4)

Standard:

01   **A Grade Crossing Advance Warning (W10-1) sign (see Figure 8B-4) shall be used on each highway in advance of every grade crossing, except in the following circumstances:**

   A.   **On an approach to a grade crossing from an intersection with a parallel highway if the distance from the nearest rail of the tracks to the edge of the parallel roadway is less than 100 feet and W10-2, W10-3, or W10-4 signs are used on the approaches of the parallel highway (see Paragraph 5 of this Section);**
   B.   **On low-volume, low-speed highways crossing minor spurs or other tracks that are infrequently used and road users are directed by an authorized person on the ground to not enter the crossing at all times that approaching rail traffic is about to occupy the crossing;**
   C.   **In business or commercial areas where active grade crossing traffic control systems are in use;**
   D.   **Where physical conditions do not permit even a partially effective display of the sign; or**
   E.   **At highway-LRT grade crossings where Crossbuck signs are not used (see Section 8B.03).**

02   **The placement of the Grade Crossing Advance Warning sign shall be in accordance with Section 2C.04 and Table 2C-3.**

03   **If a YIELD or STOP sign is present at a passive grade crossing, a Yield Ahead (W3-2) or Stop Ahead (W3-1) Advance Warning sign shall also be installed if the criteria for their installation given in Section 2C.35 is met. If a Yield Ahead or Stop Ahead sign is installed on the approach to the crossing, the W10-1 sign shall be installed upstream from the Yield Ahead or Stop Ahead sign. The Yield Ahead or Stop Ahead sign shall be located in accordance with Table 2C-3. The minimum distance between the signs shall be in accordance with Section 2C.04 and Table 2C-3.**

Option:

04   On divided highways and one-way streets, an additional W10-1 sign may be installed on the left-hand side of the roadway.

Standard:

05   **If the distance between the tracks and a parallel highway, from the nearest rail of the tracks to the edge of the parallel roadway, is less than 100 feet, a W10-2, W10-3, or W10-4 sign (see Figure 8B-4) shall be installed on each approach of the parallel highway to warn road users making a turn that they will encounter a grade crossing soon after making a turn, and a W10-1 sign for the approach to the tracks shall not be required to be between the tracks and the parallel highway.**

06   **If the W10-2, W10-3, or W10-4 sign is used, sign placement in accordance with the guidelines for Intersection Warning signs in Table 2C-3 using the speed of through traffic shall be measured from the highway intersection.**

Figure 8B-4. Warning Signs and Plaques for Grade Crossings

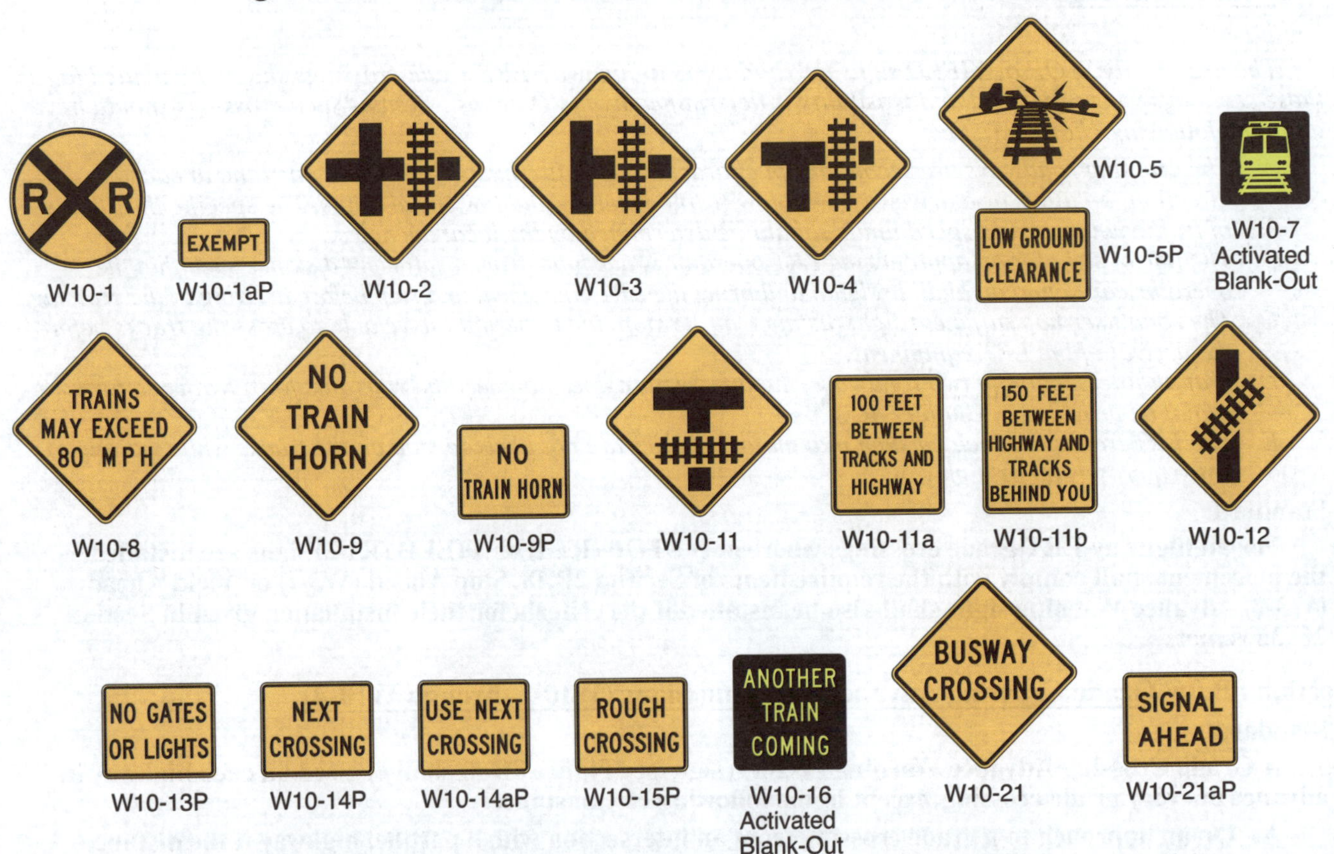

Note: The W10-11 sign is a W10-3 sign modified for geometrics. Other signs can be oriented
or revised as needed to better portray the geometrics of the roadways and the tracks.

Guidance:

07   *If the distance between the tracks and the parallel highway, from the nearest rail of the tracks to the edge of the parallel roadway, is 100 feet or more, a W10-1 sign should be installed in advance of the grade crossing, and the W10-2, W10-3, or W10-4 sign should not be used on the parallel highway.*

Section 8B.07 DO NOT STOP ON TRACKS Sign (R8-8)

Guidance:

01   *If motor vehicle queues are likely to extend onto the tracks, a DO NOT STOP ON TRACKS (R8-8) sign (see Figure 8B-1) should be used.*

Support:

02   Locations where motor vehicles could queue onto the grade crossing include intersections where a STOP or YIELD sign is installed downstream of the grade crossing, where there is a downstream circular intersection, or where there is a pre-signal installed at the grade crossing.

Guidance:

03   *The R8-8 sign, if used, should be located on the right-hand side of the highway on either the near or far side of the grade crossing, depending upon which position provides better visibility to approaching drivers.*

Option:

04   DO NOT STOP ON TRACKS signs may be placed on both sides of the track.

05   On divided highways and one-way streets, a second DO NOT STOP ON TRACKS sign may be placed on the near or far left-hand side of the highway at the grade crossing to further improve the visibility of the sign.

Section 8B.08 <u>TRACKS OUT OF SERVICE Sign (R8-9)</u>

Option:

01   The TRACKS OUT OF SERVICE (R8-9) sign (see Figure 8B-1) may be used at a grade crossing instead of a Crossbuck (R15-1) sign and a Number of Tracks (R15-2P) plaque or instead of a Crossbuck Assembly where railroad or LRT tracks have been temporarily or permanently abandoned, but only until such time that the tracks are removed or covered.

Standard:

02   **Where tracks are out of service, except as provided in Paragraphs 3 and 4 of this Section, traffic control devices and gate arms shall be removed and the signal heads shall be removed or hooded or turned from view to clearly indicate that they are not in operation.**

03   **Where tracks are out of service, even if a TRACKS OUT OF SERVICE (R8-9) sign has been installed, an Emergency Notification System (I13-1) sign (see Section 8B.27) shall be retained at the grade crossing and shall be visible to road users.**

Guidance:

04   *Warning signs, such as the Low Ground Clearance Grade Crossing (W10-5) sign and the Skewed Crossing (W10-12) sign, that warn road users about physical roadway conditions at the grade crossing should be left in place after the tracks are taken out of service, until the physical condition is no longer present..*

Standard:

05   **The R8-9 sign shall be removed when the tracks have been removed or paved over or when the grade crossing is returned to service. The Emergency Notification System (I13-1) sign shall be removed when the tracks have been removed or paved over.**

Section 8B.09 <u>STOP HERE WHEN FLASHING Signs (R8-10 and R8-10a)</u>

Option:

01   The STOP HERE WHEN FLASHING (R8-10 and R8-10a) signs (see Figure 8B-1) may be used at a grade crossing to inform drivers of the location of the stop line or the point at which to stop when the flashing-light signals (see Section 8D.02) are activated.

Section 8B.10 <u>STOP HERE ON RED Signs (R10-6 and R10-6a)</u>

Support:

01   The STOP HERE ON RED (R10-6 or R10-6a) sign (see Figure 8B-1) defines and facilitates observance of stop lines at traffic control signals.

Option:

02   STOP HERE ON RED signs may be used at locations where motor vehicles frequently violate the stop line or where it is not obvious to road users where to stop.

Guidance:

03   *If possible, stop lines should be placed at a point where the motor vehicle driver has adequate sight distance along the track.*

Section 8B.11 <u>EXEMPT Grade Crossing Plaques (R15-3P and W10-1aP)</u>

Option:

01   Where authorized by law or regulation, an EXEMPT (R15-3P) plaque (see Figure 8B-1) with a white background may be used below the Crossbuck sign or Number of Tracks plaque, if present, at the grade crossing, and an EXEMPT (W10-1aP) plaque (see Figure 8B-4) with a yellow background may be used below the Grade Crossing Advance Warning (W10-1 through W10-4) sign.

02   Where neither the Crossbuck sign nor the advance warning signs exist for a particular highway-LRT grade crossing, an EXEMPT (R15-3P) plaque with a white background may be placed on its own post on the near right-hand side of the approach to the crossing.

Support:

03   These plaques inform drivers of motor vehicles carrying passengers for hire, school buses carrying students, or motor vehicles carrying hazardous materials that a stop is not required at certain designated grade crossings, except when rail traffic is approaching or occupying the grade crossing, or the driver's view is blocked.

Section 8B.12 <u>Light Rail Transit Only Lane Signs (R15-4 Series)</u>

Support:

01   The Light Rail Transit Only Lane (R15-4 series) signs (see Figure 8B-1) are used for multi-lane operations, where road users might need additional guidance on lane use and/or restrictions.

Option:

02   Light Rail Transit Only Lane signs may be used on a roadway lane limited to only LRT use to indicate the restricted use of a lane in semi-exclusive and mixed alignments.

Guidance:

03   *If used, the R15-4a, R15-4b, and R15-4c signs should be installed on posts adjacent to the roadway containing the LRT tracks or overhead above the LRT only lane.*

Option:

04   If the trackway is paved, preferential lane markings (see Chapter 3E) may be installed, but only in combination with Light Rail Transit Only Lane signs.

Support:

05   The trackway is the continuous way designated for LRT, including the entire dynamic envelope. Section 8C.06 contains more information regarding the dynamic envelope.

Section 8B.13 <u>Do Not Pass Light Rail Transit Signs (R15-5 and R15-5a)</u>

Support:

01   A Do Not Pass Light Rail Transit (R15-5) sign (see Figure 8B-1) is used to indicate that motor vehicles are not allowed to pass LRT vehicles that are loading or unloading passengers where there is no raised platform or physical separation from the lanes upon which other motor vehicles are operating.

Option:

02   The R15-5 sign may be used in mixed-use alignments and may be mounted overhead where there are multiple lanes.

03   Instead of the R15-5 symbol sign, a regulatory sign with the word message DO NOT PASS STOPPED TRAIN (R15-5a) may be used (see Figure 8B-1).

Guidance:

04   *If used, the R15-5 or R15-5a sign should be located immediately before the LRT boarding area.*

Section 8B.14 <u>No Motor Vehicles on Tracks Signs (R15-6 and R15-6a)</u>

Support:

01   The No Motor Vehicles On Tracks (R15-6) sign (see Figure 8B-1) is used where there are adjacent traffic lanes separated from the LRT lane by a curb or pavement markings.

Guidance:

02   *The DO NOT ENTER (R5-1) sign should be used where a road user could wrongly enter an LRT only street.*

Option:

03   A No Motor Vehicles On Tracks sign may be used to deter motor vehicles from driving on the trackway. It may be installed on a 3-foot flexible post between double tracks, on a post alongside the tracks, or overhead.

04   Instead of the R15-6 symbol sign, a regulatory sign with the word message DO NOT DRIVE ON TRACKS (R15-6a) may be used (see Figure 8B-1).

05   A reduced size of 12 x 12 inches may be used if the R15-6 sign is installed between double tracks.

Standard:

06   **The smallest size for the R15-6 sign shall be 12 x 12 inches.**

Section 8B.15 <u>Divided Highway with Light Rail Transit Crossing Signs (R15-7 Series)</u>

Option:

01   The Divided Highway with Light Rail Transit Crossing (R15-7 or R15-7a) sign (see Figure 8B-1) may be used as a supplemental sign on the approach legs of a roadway that intersects with a divided highway where LRT equipment operates in the median. The sign may be placed beneath a STOP sign or mounted separately.

Guidance:

02   *The number of tracks displayed on the R15-7 or R15-7a sign should be the same as the actual number of tracks.*

Standard:

03   **When the Divided Highway with Light Rail Transit Crossing sign is used at a four-leg intersection, the R15-7 sign shall be used. When used at a T-intersection, the R15-7a sign shall be used.**

Section 8B.16 Low Ground Clearance Grade Crossing Sign (W10-5)

Guidance:

01   *If the highway profile conditions are sufficiently abrupt to create a hang-up situation for long wheelbase vehicles or for trailers with low ground clearance, the Low Ground Clearance Grade Crossing (W10-5) sign (see Figure 8B-4) should be installed in advance of the grade crossing.*

Standard:

02   **Because this symbol might not be readily recognizable by the public, the Low Ground Clearance Grade Crossing (W10-5) warning sign shall be accompanied by a LOW GROUND CLEARANCE (W10-5P) educational plaque. The LOW GROUND CLEARANCE educational plaque shall remain in place for at least 3 years after the initial installation of the W10-5 sign (see Section 2A.09).**

Guidance:

03   *Because other vehicle types and combinations also face the potential risk of hanging up at a grade crossing, word message warning signs and selective exclusion regulatory signs (see Section 2B.45) for specific vehicle types and combinations should be used in addition to, or in place of, the Low Ground Clearance Grade Crossing (W10-5) sign.*

Support:

04   While not all inclusive, some potential low ground clearance vehicles and combinations include single-unit trucks, buses, motor coaches, low-boy trailers, car carriers, and recreational vehicles.

Guidance:

05   *Auxiliary plaques such as AHEAD, NEXT CROSSING, or USE NEXT CROSSING (with appropriate arrows), or a supplemental distance plaque should be placed below the W10-5 sign at the nearest intersecting highway where a vehicle can detour or at a point on the highway wide enough to permit a U-turn.*

06   *If engineering judgment of roadway geometric and operating conditions confirms that motor vehicle speeds across the tracks should be below the posted speed limit, a W13-1P advisory speed plaque should be posted.*

07   *A signed detour should be installed to guide potential hang-up vehicles to alternate nearby crossings to avoid the potential hang-up condition.*

Support:

08   Information on ground clearance requirements at grade crossings is available in the 2019 edition of the "American Railway Engineering and Maintenance-of-Way Association's Engineering Manual," or in "A Policy on Geometric Design of Highways and Streets," 2018 Edition, AASHTO.

09   An inventory of crossings with low ground clearance concerns, including a list of potential vehicle types that could hang up on the crossing, can be useful in tracking locations of low ground clearance crossings. Specific geometric conditions, known incidents, or anecdotal evidence of vehicle hang-ups can also be used to identify crossings with low ground clearance concerns.

Section 8B.17 Light Rail Transit Approaching-Activated Blank-Out Warning Sign (W10-7)

Support:

01   The Light Rail Transit Approaching-Activated Blank-Out (W10-7) warning sign (see Figure 8B-4) supplements the traffic control devices to warn road users crossing the tracks of approaching LRT equipment.

Option:

02   A Light Rail Transit Approaching-Activated Blank-Out warning sign may be used at signalized intersections near highway-LRT grade crossings or at crossings controlled by STOP signs or automatic gates.

Support:

03   The provisions contained in Chapter 2L for blank-out signs are applicable to the W10-7 sign.

Section 8B.18 Another Train Coming Sign (W10-16)

Support:

01   Conflicts between vehicles or vulnerable road users and multiple trains can occur at multi-track crossings on sidewalks, pathways, and at crossings in station areas where grade crossing users might not consider the arrival of another train on a different track.

Guidance:

02   *The decision to provide notification of another train should be made by a Diagnostic Team. In making this determination, the Diagnostic Team should consider the pedestrian usage, pedestrian collision history, train speeds and volumes, operating plans and/or schedules, and the presence of a nearby station or transit center.*

Option:

03   An ANOTHER TRAIN COMING (W10-16) train-activated blank-out sign (see Figure 8B-4) may be used to provide notification of another train coming. For added sign conspicuity, a Warning Beacon may be used in accordance with the requirements of Section 4S.03.

Section 8B.19 TRAINS MAY EXCEED 80 MPH Sign (W10-8)

Guidance:

01   *Where trains are permitted to travel at speeds exceeding 80 mph, a TRAINS MAY EXCEED 80 MPH (W10-8) sign (see Figure 8B-4) should be installed facing road users approaching the highway-rail grade crossing.*

02   *If used, the TRAINS MAY EXCEED 80 MPH signs should be installed between the Grade Crossing Advance Warning (W10-1 through W10-4) sign (see Figure 8B-4) and the highway-rail grade crossing on all approaches to the highway-rail grade crossing. The locations should be determined based on specific site conditions.*

Section 8B.20 NO TRAIN HORN Sign or Plaque (W10-9 and W10-9P)

Standard:

01   **Either a NO TRAIN HORN (W10-9) sign (see Figure 8B-4) or a NO TRAIN HORN (W10-9P) plaque (see Figure 8B-4) shall be installed in each direction at each highway-rail grade crossing where a Quiet Zone has been established in compliance with 49 CFR Part 222. If a W10-9P plaque is used, it shall supplement and be mounted directly below the Grade Crossing Advance Warning (W10-1 through W10-4) sign (see Figure 8B-4).**

Section 8B.21 Storage Space Signs (W10-11, W10-11a, and W10-11b)

Guidance:

01   *A Storage Space (W10-11) sign supplemented by a word message Storage Space Ahead (W10-11a) sign (see Figure 8B-4) should be used where there is a highway intersection in close proximity to the grade crossing and the Diagnostic Team determines that adequate space is not available to store a design vehicle(s) between the highway intersection and the train or LRT equipment dynamic envelope.*

02   *The Storage Space (W10-11 and W10-11a) signs should be mounted in advance of the grade crossing at an appropriate location to advise drivers of the space available for motor vehicle storage between the highway intersection and the grade crossing.*

Option:

03   A word message Storage Space Behind (W10-11b) sign (see Figure 8B-4) may be mounted beyond the grade crossing at the highway intersection under the STOP or YIELD sign or just prior to the signalized intersection to remind drivers of the storage space between the tracks and the highway intersection.

Standard:

04   **A Storage Space (W10-11) sign shall not be used as a replacement for the required Advance Warning (W10-1) sign. If used, the Storage Space sign shall be used in addition to the W10-1 sign and shall be mounted on a separate post.**

Section 8B.22 Skewed Crossing Sign (W10-12)

Option:

01   The Skewed Crossing (W10-12) sign (see Figure 8B-4) may be used at a skewed grade crossing to warn road users that the tracks are not perpendicular to the highway.

Guidance:

02   *If the Skewed Crossing sign is used, the symbol should show the direction of the crossing (near left to far right as shown on the sign image in Figure 8B-4, or the mirror image if the track goes from far left to near right).*

Standard:

03   **The Skewed Crossing sign shall not be used as a replacement for the required Advance Warning (W10-1) sign. If used, the Skewed Crossing sign shall be used in addition to the W10-1 sign and shall be mounted on a separate post.**

Section 8B.23 NO GATES OR LIGHTS Plaque (W10-13P)

Option:

01   The NO GATES OR LIGHTS (W10-13P) plaque (see Figure 8B-4) may be mounted below the Grade Crossing Advance Warning (W10-1 through W10-4) sign at grade crossings that are not equipped with automatic gates or automated signals.

Section 8B.24 Next Crossing Plaques (W10-14P and W10-14aP)

Option:

01   The NEXT CROSSING (W10-14P) plaque (see Figure 8B-4) may be mounted below the Low Ground Clearance (W10-5) sign (see Section 8B.16) or Skewed Crossing (W10-12) sign to indicate to a road user that the warning is associated with the next grade crossing. This plaque may be used where multiple grade crossings exist in close proximity to one another.

02   Where recommended by a Diagnostic Team, the USE NEXT CROSSING (W10-14aP) plaque (see Figure 8B-4) may be mounted below the Low Ground Clearance (W10-5) sign (see Section 8B.16) to advise a road user with a low clearance load to use the crossing after the upcoming crossing to avoid encountering a low ground clearance situation.

Section 8B.25 ROUGH CROSSING Plaque (W10-15P)

Option:

01   The ROUGH CROSSING (W10-15P) plaque (see Figure 8B-4) may be mounted below the Grade Crossing Advance Warning (W10-1 through W10-4) sign on the approach to a grade crossing to provide supplemental information that the surface or condition of the grade crossing might require a reduced speed or some other appropriate action by the road user.

02   If the grade crossing is rough, word message signs such as BUMP, DIP, or ROUGH CROSSING may be installed. A W13-1P advisory speed plaque may be installed below the word message sign in advance of rough crossings.

Section 8B.26 Light Rail Transit Station Sign (I3-8)

Option:

01   The Light Rail Transit Station (I3-8) sign (see Section 2H.01) may be used to direct road users to an LRT station or boarding location. It may be supplemented by the name of the transit system and by arrows as provided in Section 2D.08.

Section 8B.27 Emergency Notification System Sign (I13-1)

Standard:

01   **The Emergency Notification System (I13-1) sign (see Figure 8B-5) shall be installed on each approach at all highway-rail grade crossings, and at all highway-LRT grade crossings with automatic gates or flashing light-signals, to provide information to road users so that they can notify the railroad company or transit agency about emergencies or malfunctioning traffic control devices.**

02   **At a highway-rail grade crossing, the Emergency Notification System sign shall, at a minimum, include the USDOT grade crossing inventory number and the emergency contact telephone number.**

03   **Where Emergency Notification System signs are used at a highway-LRT grade crossing, they shall, at a minimum, include a unique crossing identifier and the emergency contact telephone number.**

04   **The minimum width of the Emergency Notification System sign shall be 12 inches and the minimum height shall be 9 inches. The lettering on Emergency Notification System signs for the telephone number, the grade crossing inventory number, and the explanation of the purpose of the sign shall be composed of numerals and upper-case letters that are at least 1 inch in height.**

05   **Emergency Notification System signs shall be retroreflective.**

06   **Except as provided in Paragraph 7 of this Section, Emergency Notification System signs shall have a white legend and border on a blue background.**

Figure 8B-5. Example of an Emergency Notification System Sign

I13-1

Option:

07   The seven-character grade crossing inventory number may be shown on the sign as a black legend on a white rectangular background.

Guidance:

08   *Except as provided in Paragraph 12 of this Section, Emergency Notification System signs should be attached to the Crossbuck Assemblies or grade crossing signal masts on the right-hand side of each roadway approach to the grade crossing rather than on the railroad or LRT signal control equipment housings. Emergency Notification System signs should be oriented so the face of the sign is approximately parallel or approximately perpendicular to the edge of the roadway or pathway and is visible to road users or pathway users. The visibility of the Emergency Notification System sign should not be obstructed by automatic gates in either the vertical or horizontal position.*

09   *The Emergency Notification System signs should be positioned so as to not obstruct any traffic control devices or limit the view of rail traffic approaching the grade crossing.*

10   *Emergency Notification System signs mounted on Crossbuck Assemblies or signal masts should only be large enough to provide the necessary contact information. Use of larger signs on Crossbuck Assemblies or signal masts that might obstruct the view of rail traffic or other motor vehicles should be avoided.*

11   *At station crossings, Emergency Notification System signs or information should be posted in a conspicuous location.*

Option:

12   Emergency Notification System signs may be located on a separate post. Where located on a separate post, the size of the Emergency Notification System sign may be increased for improved visibility.

13   Where the improvement of the conspicuity of an Emergency Notification System sign is desired, a solid yellow rectangular header panel with a legend of "NOTICE" in black letters may be used (see Section 2A.11).

14   Additional Emergency Notification System signs may be installed at a grade crossing.

CHAPTER 8C. MARKINGS

Section 8C.01 Purpose and Application

Support:

01   Passive traffic control systems, consisting of signs and pavement markings only, identify and direct attention to the location of a grade crossing and advise road users to reduce their speed or stop at the grade crossing as necessary in order to yield to any rail traffic occupying, or approaching and in proximity to, the grade crossing.

02   Signs and pavement markings regulate, warn, and guide the road users so that they, as well as LRT vehicle operators on mixed-use alignments, can take appropriate action when approaching a grade crossing.

03   Unless otherwise provided in this Chapter, the provisions of Part 3 are applicable to the design and location of pavement markings at grade crossings.

Section 8C.02 Grade Crossing Pavement Markings

Standard:

01   **On paved roadways, grade crossing pavement markings shall consist of an X, the letters RR, a no-passing zone marking (on two-lane, two-way highways with center line markings in compliance with Section 3B.01), and certain transverse lines as shown with detailed dimensions in Figures 8C-1 and 8C-2.**

02   **Except as provided in Paragraphs 3 and 4 of this Section, grade crossing pavement markings shall be placed in each approach lane on all paved approaches to highway-rail grade crossings where signals or automatic gates are located, and at all other grade crossings where the posted or statutory highway speed is 40 mph or higher.**

03   **Grade crossing pavement markings shall not be required at highway-rail grade crossings where the posted or statutory highway speed is less than 40 mph if the Diagnostic Team determines that other installed devices provide suitable warning and control.**

04   **Grade crossing pavement markings shall not be required at highway-rail grade crossings in urban areas if the Diagnostic Team determines that other installed devices provide suitable warning and control.**

05   **Grade crossing pavement markings shall be placed in each approach lane on all paved approaches to highway-LRT grade crossings where a Crossbuck sign is placed at the grade crossing.**

06   **If grade crossing pavement markings are used on a multi-lane approach to a grade crossing, identical markings shall be placed in each approach lane that crosses the tracks.**

07   **All grade crossing pavement markings shall be retroreflective white. All other markings shall be in accordance with Part 3.**

Guidance:

08   *Where grade crossing pavement markings are used, a portion of the X symbol should be directly opposite the Grade Crossing Advance Warning sign.*

Option:

09   Where determined by the Diagnostic Team, supplemental pavement marking symbol(s) may be placed between the Grade Crossing Advance Warning sign and the grade crossing.

Guidance:

10   *If supplemental pavement marking symbol(s) are placed between the Grade Crossing Advance Warning sign and the grade crossing, the downstream transverse line should be at least 50 feet upstream from the stop or yield line at the grade crossing.*

Section 8C.03 Stop and Yield Lines

Guidance:

01   *On paved roadway approaches to passive grade crossings where a STOP sign is installed in conjunction with the Crossbuck sign, a stop line should be installed to indicate the point behind which motor vehicles are required to stop or as near to that point as practicable.*

Option:

02   On paved roadway approaches to passive grade crossings where a YIELD sign is installed in conjunction with the Crossbuck sign, a yield line (see Section 3B.19) or a stop line may be installed to indicate the point behind which motor vehicles are required to yield or stop or as near to that point as practicable.

Figure 8C-1. Example of Placement of Warning Signs and Pavement Markings at Grade Crossings

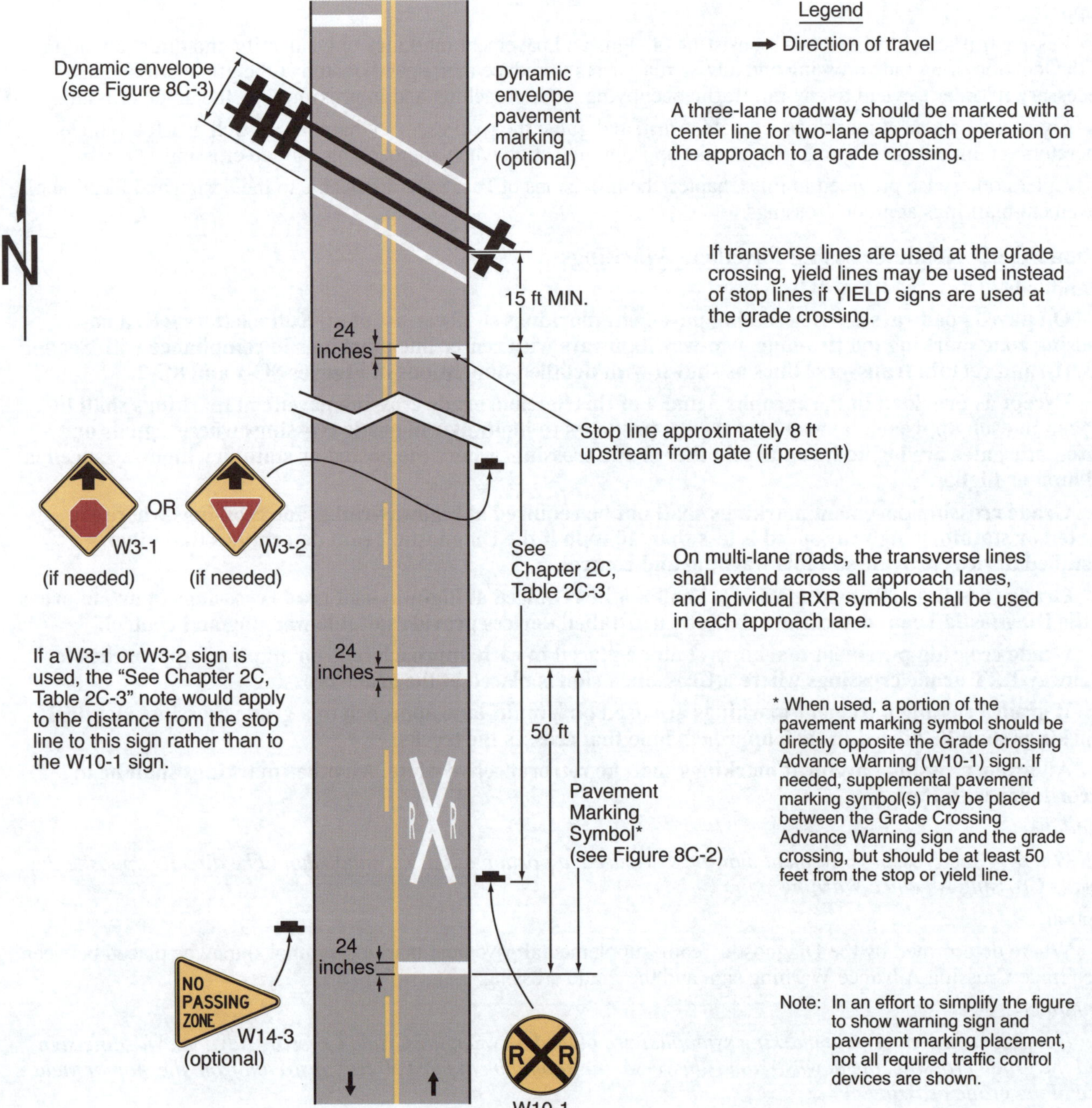

Figure 8C-2. Grade Crossing Pavement Markings

A – Grade crossing pavement marking symbol

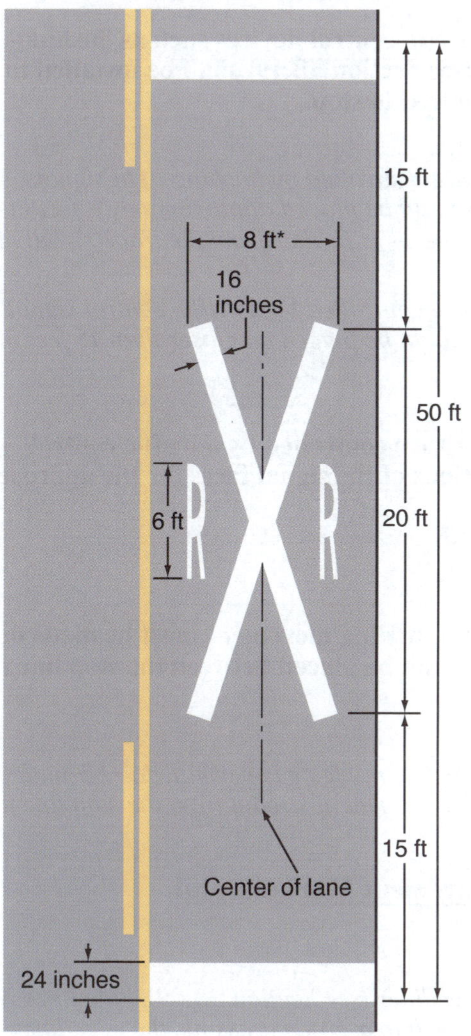

*Width may vary according to lane width

B – Grade crossing alternative (narrow) pavement marking symbol

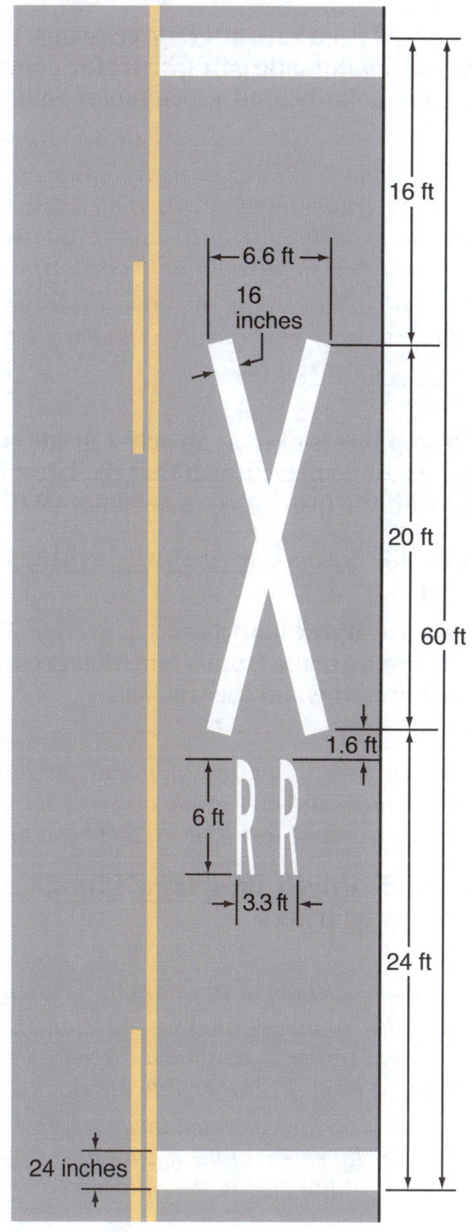

Note: Refer to Figure 8C-1 for placement

Guidance:

03   *If a yield line (see Figure 3B-16) or stop line is used at a passive grade crossing, it should be a transverse line at a right angle to the traveled way and should be placed no closer than 15 feet in advance of the nearest rail.*

Standard:

04   **On paved roadways at grade crossings that are equipped with active control devices such as flashing-light signals, automatic gates, or traffic control signals, a stop line (see Section 3B.19) shall be installed to indicate the point behind which motor vehicles are or might be required to stop.**

Guidance:

05   *If a stop line is used at an active grade crossing where road users are controlled by flashing-light signals, it should be a transverse line at a right angle to the traveled way and should be placed approximately 8 feet in advance of the flashing-light signals or automatic gate (if present), whichever is farther from the track(s), but no closer than 15 feet in advance of the nearest rail (see Figure 8C-1).*

06   *If a stop line is used at an active grade crossing where road users are controlled by a traffic control signal, it should be a transverse line at a right angle to the traveled way and should be placed no closer than 15 feet in advance of the nearest rail.*

Standard:

07   **If a stop line is used at an active grade crossing where road users are controlled by a traffic control signal, it shall be placed such that the lateral and longitudinal positions of the signal faces for the approach comply with the provisions of Sections 4D.07 and 4D.08.**

Section 8C.04 Lane-Use Arrow Markings

Standard:

01   **Lane-use arrow markings (see Section 3B.23) that indicate that a turning movement must be made or is permitted to be made from a lane that crosses a grade crossing shall not be placed between the stop line for the grade crossing and the track(s).**

Guidance:

02   *Lane-use arrow markings that indicate that a turning movement must be made or is permitted to be made from a lane that crosses a grade crossing should not be placed less than 100 feet upstream from the stop line for the grade crossing or less than 20 feet beyond the farthest rail.*

Section 8C.05 Edge Lines, Lane Lines, Center Lines, Raised Pavement Markers, and Tubular Markers

Guidance:

01   *Except as provided in Paragraphs 3 through 5 of this Section, if edge lines (see Section 3B.09), lane lines (see Section 3B.06), or center lines (see Section 3B.01) are used on an approach to a grade crossing, the edge lines, lane lines, and center lines should extend up to and across the grade crossing to reduce the likelihood that road users might inadvertently turn into the track area.*

02   *If crossing surface maintenance or highway approach maintenance is performed that alters the markings, the removal or replacement of the markings, raised pavement markers, and/or tubular markers should be coordinated with the road authority and the railroad company or transit agency.*

Option:

03   Edge lines, lane lines, and center lines may be omitted on or between the rails to conform to the requirements of the railroad company and/or transit agency.

04   Edge lines, lane lines, and center lines may be omitted on or between the rails where the highway profile is sufficiently abrupt to create a hang-up situation for pavement marking equipment with low ground clearance.

05   The edge lines, lane lines, and center lines may be omitted from the highway surface at a grade crossing if the surface cannot retain the application of the edge line, lane line, or center line marking.

06   If recommended by a Diagnostic Team, raised pavement markers (see Section 3B.16) may be used to supplement the edge lines, lane lines, or center lines that extend up to and across the grade crossing.

07   If recommended by a Diagnostic Team, tubular markers (see Section 3I.02) may be used to supplement the edge lines that extend up to and across the grade crossing.

Guidance:

08   *Tubular markers should be installed in accordance with the clearance requirements of the railroad company and/or transit agency.*

Standard:

09 **The color under both daytime and nighttime conditions of raised pavement markers or tubular markers that are used at a grade crossing shall be the same color as the edge line, lane line, or center line that they supplement.**

Section 8C.06 Dynamic Envelope and Do Not Block Pavement Markings

Option:

01 Dynamic envelope markings may be installed at a grade crossing to mark the edges of the train dynamic envelope.

Standard:

02 **If used, pavement markings for indicating the dynamic envelope shall comply with the provisions of Part 3 and shall be a solid white line not less than 4 inches or greater than 24 inches in width.**

Option:

03 Contrasting pavement color (see Section 3A.03 and Chapter 3H) and/or contrasting pavement texture may be used alone or in combination with pavement markings to indicate the dynamic envelope.

Guidance:

04 *If a solid white line is used to convey the dynamic envelope, the line should be placed completely outside of the dynamic envelope. If used, dynamic envelope pavement markings should be placed parallel to the nearest rail in accordance with the railroad company or transit agency requirements. If used, dynamic envelope pavement markings should extend across the roadway as shown in Figure 8C-3. Dynamic envelope pavement markings should not be placed perpendicular to the roadway at skewed grade crossings.*

Figure 8C-3. Example of Dynamic Envelope Pavement Markings at Grade Crossings

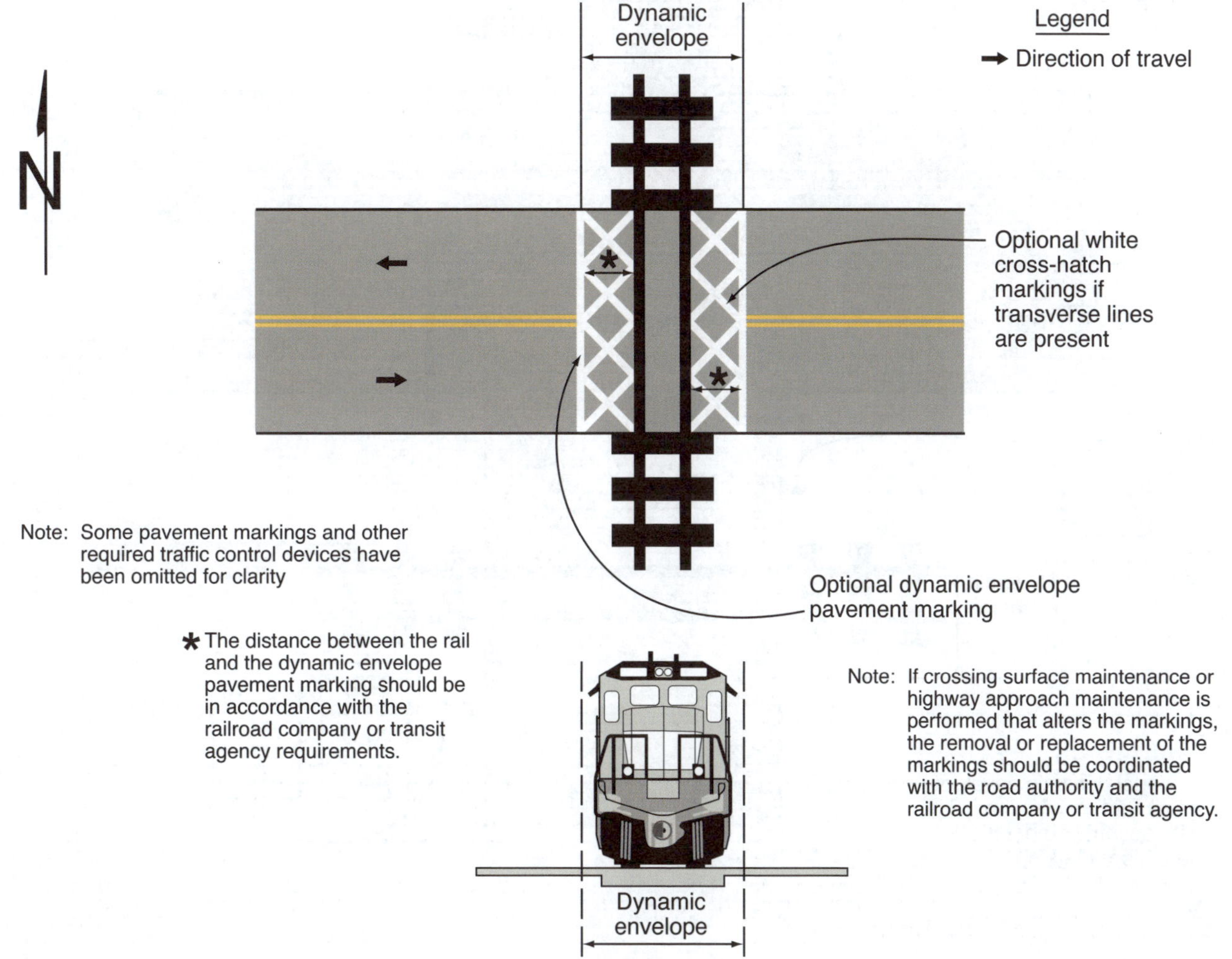

Option:

05   If solid white lines are used to indicate the dynamic envelope, white cross-hatching lines (see Figure 8C-3) may also be placed on the highway pavement within the dynamic envelope as a supplement to, but not as a substitute for, the solid white lines. White cross-hatching lines (see Section 3B.26) may also be placed on the pavement to mark areas adjacent to the dynamic envelope where vehicles are not intended to stop or stand as shown in Figure 8C-4.

06   In semi-exclusive LRT alignments, the dynamic envelope markings may be along the LRT trackway between intersections where the trackway is immediately adjacent to travel lanes and no physical barrier is present.

07   In mixed-use LRT alignments, the dynamic envelope markings may be continuous between intersections (see Figure 8C-5).

08   In mixed-use LRT alignments, pavement markings for adjacent travel or parking lanes may be used instead of dynamic envelope markings if the lines are outside the dynamic envelope.

Figure 8C-4. Example of Dynamic Envelope and Do Not Block Pavement Markings at Grade Crossings

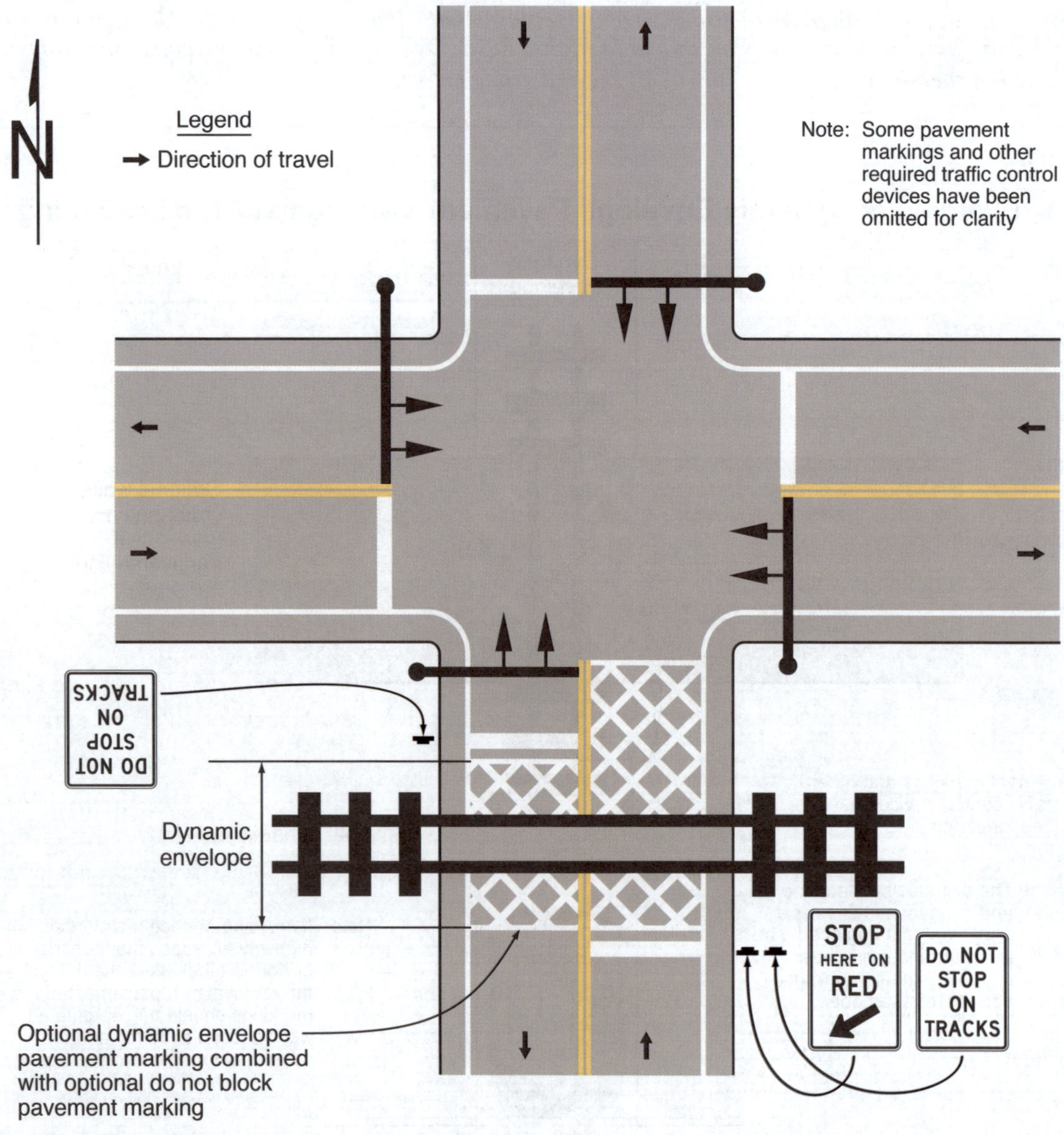

Figure 8C-5. Examples of Light Rail Transit Vehicle Dynamic Envelope Markings for Mixed-Use Alignments

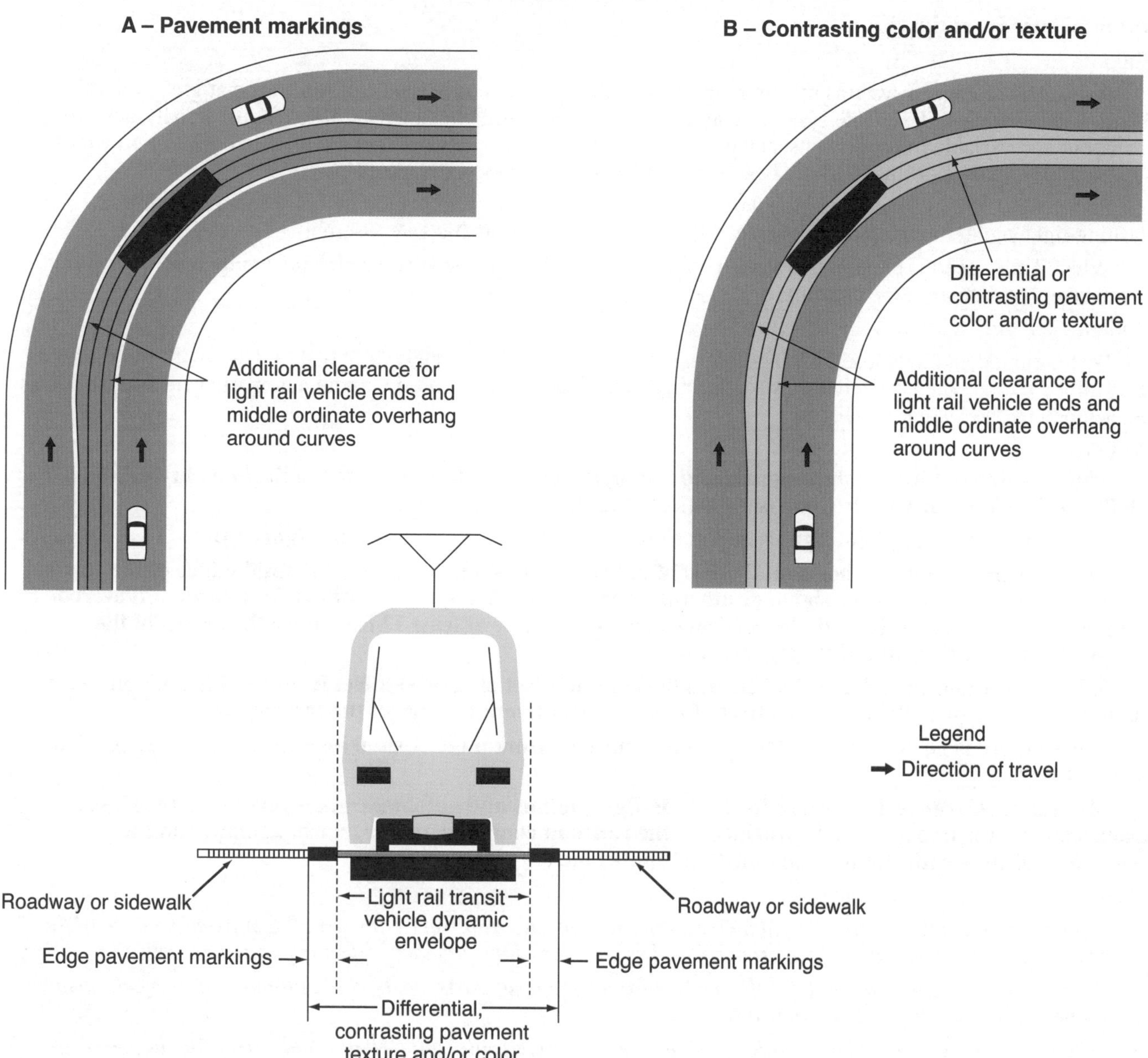

CHAPTER 8D. FLASHING-LIGHT SIGNALS, AUTOMATIC GATES, AND TRAFFIC CONTROL SIGNALS

Section 8D.01 Introduction

Support:

01   Active traffic control systems inform road users of the approach or presence of rail traffic at grade crossings. These systems include Exit Gate systems, automatic gates, flashing-light signals, traffic control signals, actuated blank-out and variable message signs, and other active traffic control devices that are used in conjunction with the signs and pavement markings that are described in Chapters 8B and 8C, respectively.

02   Figure 8D-1 shows a post-mounted flashing-light signal (two light units mounted in a horizontal line), a flashing-light signal mounted on an overhead structure, and an automatic gate assembly.

03   Where LRT speed is cited in this Part, it refers to the maximum speed at which LRT equipment is permitted to traverse a particular grade crossing.

Option:

04   Post-mounted and overhead flashing-light signals may be used separately or in combination with each other as determined by the Diagnostic Team. Also, flashing-light signals may be used without automatic gate assemblies, as determined by the Diagnostic Team.

Standard:

05   **The meaning of flashing-light signals and automatic gates shall be as stated in Sections 11-701 and 11-703 of the Uniform Vehicle Code (see Section 1A.06).**

06   **Location for flashing-light signals and automatic gates shall be as shown in Figure 8D-1.**

07   **Where there is a curb, a horizontal offset of at least 2 feet shall be provided from the face of the vertical curb to the nearest part of the signal or automatic gate arm in its upright position. Where a cantilevered-arm flashing-light signal is used, the vertical clearance shall be at least 17 feet above the crown of the highway to the lowest point of the signal unit.**

08   **Where there is a shoulder, but no curb, a horizontal offset of at least 2 feet from the edge of a paved shoulder shall be provided, with an offset of at least 6 feet from the edge of the traveled way.**

09   **Where there is no curb or shoulder, the minimum horizontal offset shall be 6 feet from the edge of the traveled way.**

10   **Minimum clearance dimensions for flashing-light signals and automatic gates relative to the closest track shall conform to standards provided by the railroad company and/or transit agency, and the regulatory agency with statutory authority (if applicable).**

Guidance:

11   *Equipment housings (controller cabinets) should have a lateral offset of at least 30 feet from the edge of the highway, and where railroad or LRT property and conditions allow, at least 25 feet from the nearest rail.*

12   *If a pedestrian route is provided, sufficient clearance from supports, posts, and automatic gate mechanisms should be maintained for pedestrian travel.*

13   *Where determined by the Diagnostic Team, a lateral escape route to the right-hand side of the highway in advance of the grade crossing traffic control devices should be kept free of guardrail or other ground obstructions. Where guardrail is not deemed necessary or appropriate, barriers should not be used for protecting signal supports.*

14   *The same lateral offset and roadside safety features should apply to flashing-light signal and automatic gate locations on both the right-hand and left-hand sides of the roadway.*

Option:

15   In industrial or other areas involving only low-speed highway traffic or where signals are vulnerable to damage by turning truck traffic, guardrail may be installed to provide protection for the signal assembly.

Guidance:

16   *Where both traffic control signals and flashing-light signals (with or without automatic gates) are in operation at the same highway-LRT grade crossing, the operation of the devices should be coordinated to avoid any display of conflicting signal indications.*

Option:

17   If highway traffic signals must be located within close proximity to the flashing-light signals, the highway traffic signals may be mounted on the same overhead structure as the flashing-light signals.

Figure 8D-1. Composite Drawing of Active Traffic Control Devices for Grade Crossings Showing Clearances

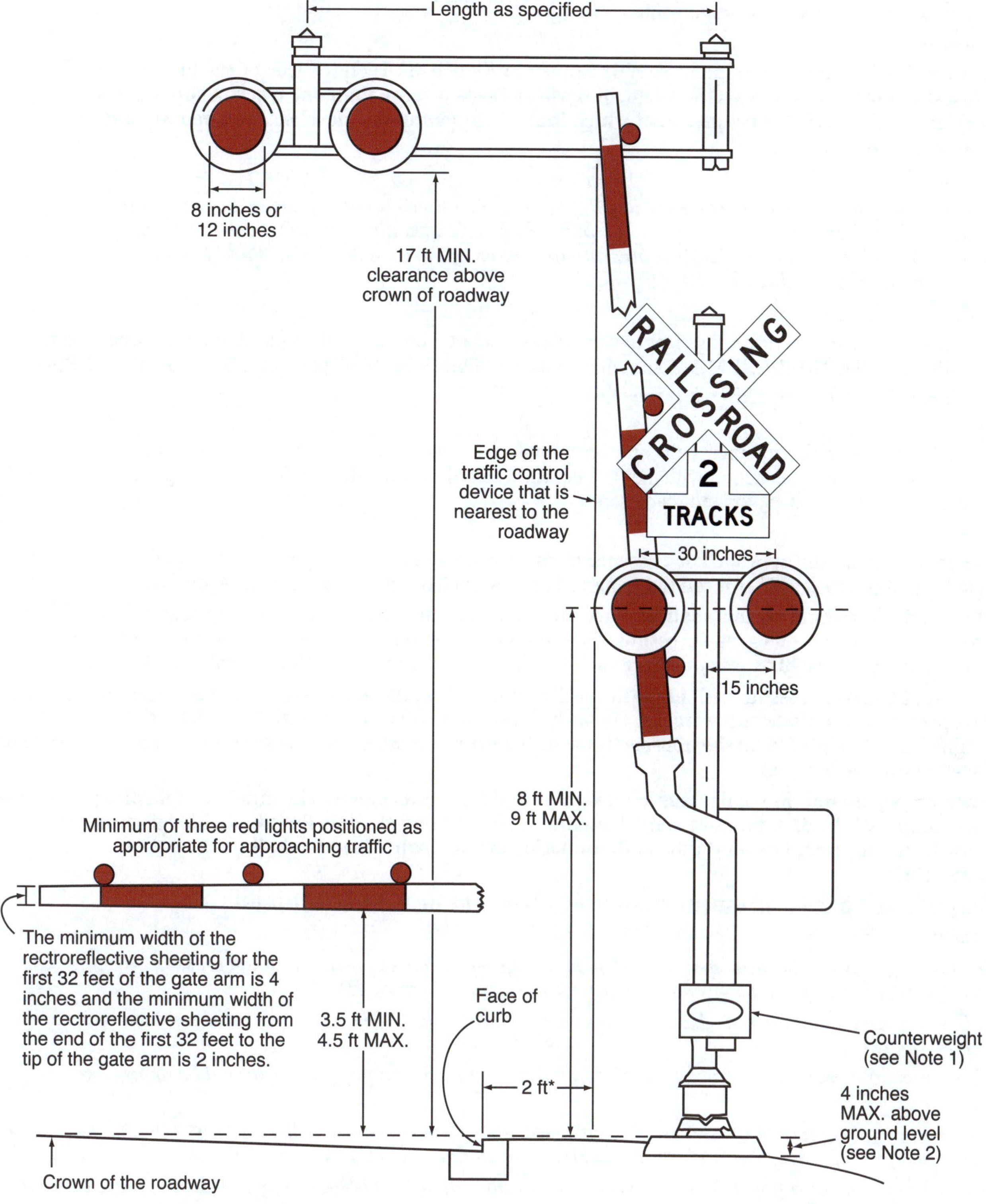

*For locating this reference line on an approach that does not have a curb, see Section 8D.01.

Notes:

1. Where gates are located in the median, additional median width may be required to provide the minimum clearance for the counterweight supports.

2. The top of the signal foundation should be no more than 4 inches above the surface of the ground and should be at the same elevation as the crown of the roadway. Where site conditions would not allow this to be achieved, the shoulder side slope should be re-graded or the height of the signal post should be adjusted to meet the 17-foot vertical clearance requirement.

Section 8D.02 Flashing-Light Signals

Support:

01   Section 8D.04 contains additional information regarding flashing-light signals at highway-LRT grade crossings in semi-exclusive and mixed-use alignments.

Standard:

02   **If used, the flashing-light signal assembly (shown in Figure 8D-1) on the side of the highway shall include a standard Crossbuck (R15-1) sign, and where there is more than one track, a supplemental Number of Tracks (R15-2P) plaque, all of which indicate to motorists, bicyclists, and pedestrians the location of a grade crossing.**

Guidance:

03   *The bottom of the Number of Tracks (R15-2P) plaque (when used) should be located as low as practicable above the flashing-light backgrounds. The Crossbuck (R15-1) sign should be located just above the Number of Tracks (R15-2P) plaque or, if no plaque is present, the bottom of the Crossbuck sign should be located as low as practicable above the flashing-light backgrounds.*

Support:

04   Additional information regarding sizes and clearances of components used on flashing-light signals can be found in Part 3 of the "2023 AREMA Communications and Signals Manual" published by the American Railway Engineering and Maintenance-of-Way Association (AREMA).

Option:

05   At highway-rail grade crossings, bells or other audible warning devices may be included in the assembly and may be operated in conjunction with the flashing-light signals to provide additional warning for pedestrians, bicyclists, and/or other non-motorized road users.

Standard:

06   **When indicating the approach or presence of rail traffic, the flashing-light signal shall display toward approaching highway traffic two red lights mounted in a horizontal line flashing alternately.**

07   **If used, flashing-light signals shall be placed to the right-hand side of approaching highway traffic on all highway approaches to a grade crossing. They shall be located laterally with respect to the highway in compliance with Figure 8D-1 except where such location would adversely affect signal visibility.**

08   **If used at a grade crossing with highway traffic in both directions, back-to-back flashing-light signals shall be placed on each side of the tracks. On multi-lane one-way streets and divided highways, flashing-light signals shall be placed on the approach side of the grade crossing on both sides of the roadway or shall be placed above the highway.**

09   **Each red signal unit in the flashing-light signal shall flash alternately. The number of flashes per minute for each lamp shall be 35 minimum and 65 maximum. Each lamp shall be illuminated for approximately the same length of time. The total time of illumination of each pair of lamps shall be the entire operating time.**

10   **Flashing-light units shall use either 8-inch or 12-inch nominal diameter lenses.**

Guidance:

11   *In choosing between the 8-inch or 12-inch nominal diameter lenses for use in grade crossing flashing-light signals, consideration should be given to the principles stated in Section 4E.02.*

12   *If flashing-light signals are used, at least one pair of flashing lights should be provided for each approach lane of the roadway.*

13   *The center-to-center distance between the two red lights in a flashing-light unit should be approximately 30 inches.*

14   *The mounting height of the flashing-light units, measured from the center of the flashing-light unit housing to the elevation of the crown of the roadway, should be between 8 feet and 9 feet .*

15   *The top of the support pole foundation should be no more than 4 inches above the surface of the ground and should be at the same elevation as the crown of the roadway.*

Standard:

16   **Grade crossing flashing-light signals shall operate at a low voltage using storage batteries either as a primary or stand-by source of electrical energy. Provision shall be made to provide a source of energy for charging batteries.**

Option:

17   Additional flashing-light signals may be mounted on the same supporting post and directed toward vehicular traffic approaching the grade crossing from other than the principal highway route, such as where there are approaching routes on highways closely adjacent to and parallel to the track(s).

Guidance:

18   *Where the storage distance for vehicles approaching a grade crossing is less than a design vehicle length, the Diagnostic Team should consider providing additional flashing-light signals aligned toward the movement turning toward the grade crossing.*

19   *The Diagnostic Team should consider the use of additional flashing-light signals to provide supplemental warning to pedestrians, especially on one-way streets and divided highways.*

Standard:

20   **References to lenses in this Section shall not be used to limit flashing-light signal optical units to incandescent lamps within optical assemblies that include lenses.**

Support:

21   Research has resulted in flashing-light signal optical units that are not lenses, such as, but not limited to, light-emitting diode (LED) flashing-light signal modules.

Option:

22   If a Diagnostic Team determines that it is appropriate, the flashing-light signals may be installed on overhead structures or cantilevered supports as shown in Figure 8D-1 where needed for additional emphasis, or for better visibility to approaching traffic, particularly on multi-lane approaches or highways with profile restrictions.

23   If it is determined by a Diagnostic Team that one flashing-light signal on the cantilever arm is not sufficiently visible to road users, one or more additional flashing-light signals may be mounted on the supporting post and/or on the cantilever arm.

Standard:

24   **Breakaway or frangible bases shall not be used on the supporting posts for overhead structures or cantilevered arms that support overhead flashing-light signals.**

Section 8D.03 Automatic Gates

Support:

01   An automatic gate is a traffic control device used in conjunction with flashing-light signals.

Standard:

02   **The automatic gate (see Figure 8D-1) shall consist of a drive mechanism and a fully retroreflective red-and-white-striped gate arm with lights. When in the down position, the gate arm shall extend across the approaching lanes of highway traffic.**

03   **In the normal sequence of operation, unless constant warning time detection or other advanced system requires otherwise, the flashing-light signals and the lights on the gate arm (in its normal upright position) shall be activated immediately upon detection of approaching rail traffic. The gate arm shall start its downward motion not less than 3 seconds after the flashing-light signals start to operate, shall reach its horizontal position at least 5 seconds before the arrival of the rail traffic, and shall remain in the down position until the rail traffic completely clears the grade crossing.**

04   **When the rail traffic clears the grade crossing, and if no other rail traffic is detected, the gate arm shall ascend to its upright position, following which the flashing-light signals and the lights on the gate arm shall cease operation.**

05   **Gate arms shall be fully retroreflective on both sides and shall have vertical stripes alternately red and white at 16-inch intervals measured horizontally. The width (which becomes the height of the retroreflective sheeting when the automatic gate is in the down position) of the retroreflective sheeting on the front of the gate arm shall be at least 4 inches for the first 32 feet of gate arm length measured from the center of the gate mast. The front of the gate arm beyond 32 feet to the tip of the gate shall have retroreflective sheeting at least 2 inches in width.**

Support:

06   It is acceptable to replace a damaged gate arm with a gate arm having vertical stripes even if the other existing gate arms at the same grade crossing have diagonal stripes; however, it is also acceptable to replace a damaged gate arm with a gate arm having diagonal stripes if the other existing gate arms at the same grade crossing have diagonal stripes in order to maintain consistency per the provisions of Paragraph 13 of Section 1B.03.

Standard:

07	**Gate arms shall have at least three red lights as shown in Figure 8D-1.**

08	**When activated, the gate arm light nearest the tip shall be illuminated continuously and the other lights shall flash alternately in unison with the flashing-light signals such that the left-most flashing gate arm light(s) flashes simultaneously with the left-hand light of the flashing-light signals and the right-most flashing gate arm light(s) flashes simultaneously with the right-hand light of the flashing-light signals.**

Support:

09	Typical gate arm lights are approximately 4 inches in diameter if they are circular. Rectangular gate arm lights with approximately the same illuminated surface area are sometimes used on gate arms instead of circular lights.

Standard:

10	**The entrance gate arm mechanism shall be designed to fail safe in the down position.**

Guidance:

11	*The gate arm should ascend to its upright position in 12 seconds or less.*

12	*In its normal upright position, when no rail traffic is approaching or occupying the grade crossing, the gate arm should be approximately vertical (see Figure 8D-1).*

13	*In the design of individual installations, consideration should be given to timing the operation of the gate arm to accommodate large and/or slow-moving motor vehicles.*

14	*The gate arms should cover the approaching highway to block all motor vehicles from being driven around the gate arms without crossing the center line.*

15	*The height of the gate arm when it is in the down position should be between 3.5 feet and 4.5 feet above the crown of the roadway. When the gate arm is in the down position, no portion of the counterweight should extend into the traveled way, sidewalk, or pathway.*

Option:

16	Channelizing devices and/or raised median islands may be used to discourage driving around lowered automatic gates.

Guidance:

17	*Where sufficient space is available, median islands should be at least 60 feet in length.*

Option:

18	Where automatic gates are located in the median, additional median width may be required to provide the minimum clearance for the counterweight supports.

19	Automatic gates may be supplemented by cantilevered flashing-light signals (see Figure 8D-1) where there is a need for additional emphasis or better visibility.

Section 8D.04 Use of Active Traffic Control Systems at LRT Grade Crossings

Standard:

01	**At highway-LRT grade crossings where LRT speeds exceed 25 mph, active traffic control systems (see Section 8D.01) shall be used.**

02	**At highway-LRT grade crossings where LRT speeds exceed 40 mph, the active traffic control system shall include automatic gates.**

Option:

03	The Diagnostic Team may recommend an active traffic control system with automatic gates at highway-LRT grade crossings where LRT speeds do not exceed 40 mph.

Guidance:

04	*At highway-LRT grade crossings where LRT speeds are 25 mph or less, active traffic control systems should be used unless the Diagnostic Team determines that the use of Crossbuck Assemblies, STOP signs alone, or YIELD signs alone would be adequate.*

05	*Where the highway-LRT grade crossing is at a location other than an intersection and LRT speeds exceed 20 mph, traffic control signals should not be used in lieu of flashing-light signals.*

Support:

06	Sections 8D.02 and 8D.03 contain additional provisions regarding the design and operation of flashing-light signals and automatic gates, respectively.

Standard:

07	**If flashing-light signals are in operation at a highway-LRT crossing that is used by pedestrians, bicyclists, and/or other non-motorized road users, an audible device such as a bell shall also be provided and shall be operated in conjunction with the flashing-light signals.**

Section 8D.05 Exit Gate and Four-Quadrant Gate Systems

Option:

01   Exit Gate systems may be installed to improve safety at grade crossings where a Diagnostic Team determines that less restrictive measures, such as automatic gates and median islands, are not effective.

Support:

02   A grade crossing that includes exit gates on some, but not all, of the exiting lanes is an Exit Gate system, but is not considered to be a Four-Quadrant Gate system.

03   The term Four-Quadrant Gate system is used in a generic sense in that it refers to the fact that all entrances and exits from a grade crossing are controlled by automatic gates in order to provide a full closure to all entering and exiting lanes. The term Four-Quadrant Gate system does not refer to the number of gates installed, but rather the fact that a full closure is provided.

Standard:

04   **The Exit Gate system shall use a series of automatic gates with fully retroreflective red-and-white-striped gate arms with lights, and when in the down position the gate arms extend individually across the entrance and exit lanes of the roadway as shown in Figure 8D-2. The provisions contained in Section 8D.02 for flashing-light signals shall be followed for signal specifications, location, and clearance distances.**

05   **Gate arm design, colors, and lighting requirements shall be in accordance with the provisions contained in Section 8D.03.**

Support:

06   The provisions contained in Section 8D.03 for automatic gates are applicable to exit gates.

Standard:

07   **In the normal sequence of operation, unless constant warning time detection or other advanced system requires otherwise, the flashing-light signals and the lights on the gate arms (in their normal upright positions) shall be activated immediately upon the detection of approaching rail traffic. The entrance gate arms shall start their downward motion not less than 3 seconds after the flashing-light signals start to operate and shall reach their horizontal position at least 5 seconds before the arrival of the rail traffic. Exit gate arm activation and downward motion shall be based on detection or timing requirements established by a Diagnostic Team. If an Exit Gate system is present, the queue exit gate clearance time (see AREMA Manual) shall be long enough to permit the exit gate arm to lower after a design vehicle of maximum length is clear of the minimum track clearance distance (see Section 8A.07). The gate arms shall remain in the down position as long as the rail traffic occupies the grade crossing.**

08   **When the rail traffic clears the grade crossing, and if no other rail traffic is detected, the gate arms shall ascend to their upright positions, following which the flashing-light signals and the lights on the gate arms shall cease operation.**

09   **Except as provided in Paragraph 21 of this Section, the exit gate arm mechanism shall be designed to fail-safe in the up position.**

10   **At locations where gate arms are offset a sufficient distance for motor vehicles to drive between the entrance and exit gate arms, median islands (see Figure 8D-2) shall be installed in accordance with the needs determined by the Diagnostic Team.**

Guidance:

11   *The gate arm should ascend to its upright position in 12 seconds or less.*

12   *Constant warning time detection circuits should be used with Exit Gate systems where practical.*

13   *The operating mode of the exit gates should be determined by a Diagnostic Team.*

14   *If the Timed Exit Gate Operating Mode is used, the Diagnostic Team should also determine the Exit Gate Clearance Time (see definition in Section 1C.02).*

15   *If the Dynamic Exit Gate Operating Mode is used, highway vehicle intrusion detection devices that are part of a system that incorporates processing logic to detect the presence of motor vehicles within the minimum track clearance distance (see Section 8A.07) should be installed to control exit gate operation. Exit gates should be independently controlled for each direction of roadway traffic.*

16   *Regardless of which exit gate operating mode is used, the Exit Gate Clearance Time should be considered when determining additional time requirements for the Minimum Warning Time.*

Support:

17   The minimum warning time is the least amount of time that active warning devices operate prior to the arrival of rail traffic at a grade crossing.

Figure 8D-2. Examples of Location Plan for Flashing-Light Signals and Four-Quadrant Gates

The locations of the gates and the placement of medians or islands
between the gates are to be determined by the Diagnostic Team.

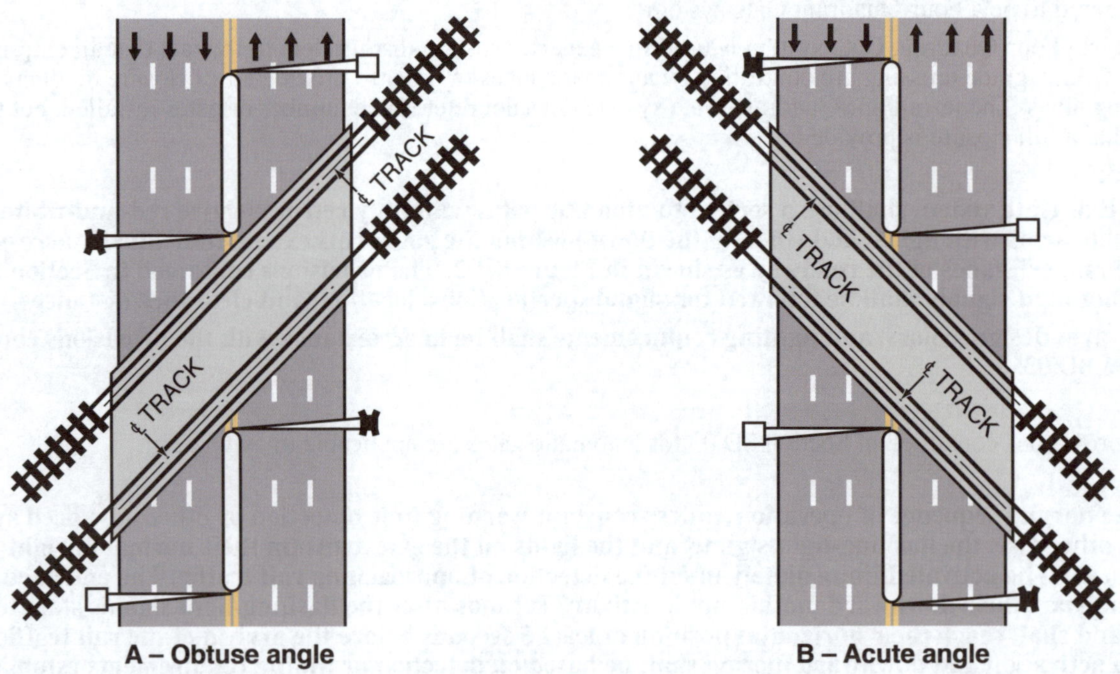

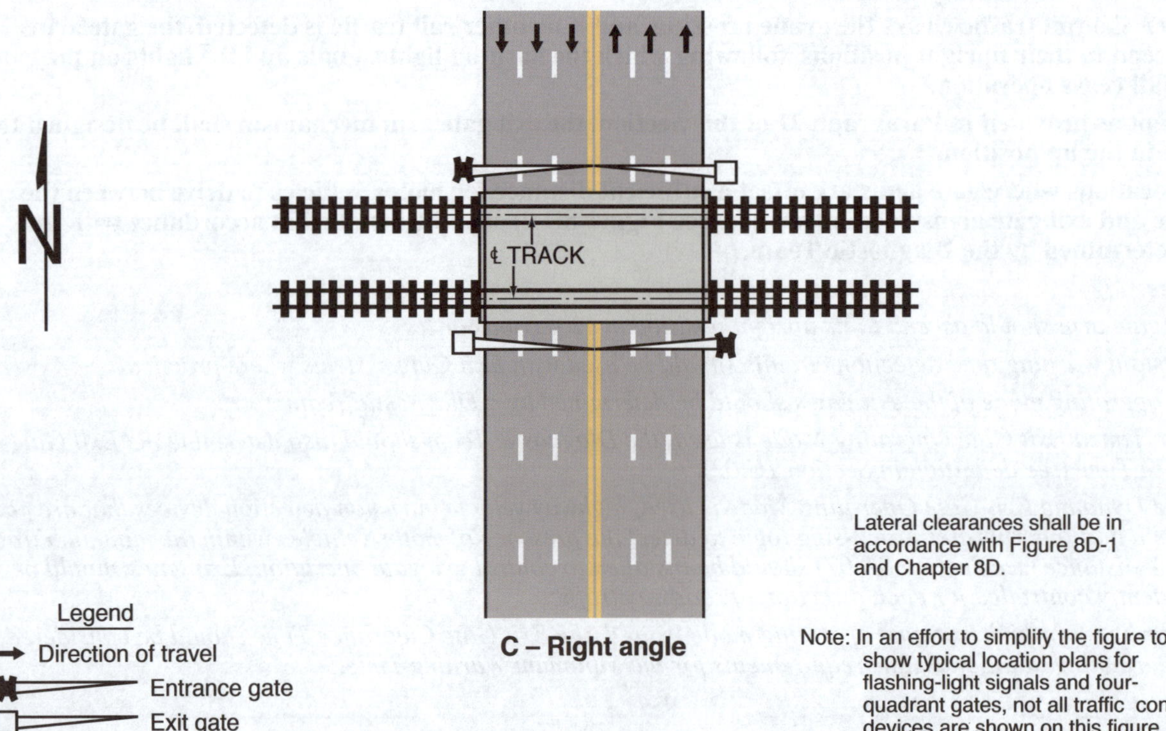

Guidance:

18   *If an Exit Gate system is used at a location that is adjacent to an intersection that could cause motor vehicles to queue within the minimum track clearance distance (see Section 8A.07), the Dynamic Exit Gate Operating Mode should be used unless the Diagnostic Team determines otherwise.*

19   *If an Exit Gate system is interconnected with a highway traffic signal (see Section 8D.09), back-up or standby power should be considered for the highway traffic signal. Also, circuitry should be installed to prevent the highway traffic signal from leaving the track clearance green interval until all of the gates are lowered.*

20   *Exit Gate systems should include remote health (status) monitoring capable of automatically notifying railroad or LRT signal maintenance personnel when anomalies have occurred within the system.*

Option:

21   Exit gate arms may fail in the down position if the grade crossing is equipped with remote health (status) monitoring.

22   Exit Gate system installations may include median islands between opposing lanes on an approach to a grade crossing.

Guidance:

23   *Where sufficient space is available, median islands should be at least 60 feet in length.*

Section 8D.06 <u>Wayside Horn Systems</u>

Option:

01   A wayside horn system (see definition in Section 1C.02) may be installed in compliance with 49 CFR Part 222 to provide audible warning directed toward the road users at a highway-rail grade crossing or at a pathway grade crossing.

Standard:

02   **Wayside horn systems used at grade crossings where the locomotive horn is not sounded shall be equipped and shall operate in compliance with the requirements of Appendix E to 49 CFR Part 222.**

Guidance:

03   *The same lateral clearance and roadside safety features should apply to wayside horn systems as described in the provisions contained in Section 8D.01. Wayside horn systems, when mounted on a separate pole assembly, should be installed no closer than 15 feet from the center of the nearest track and should be positioned to not obstruct the motorists' line of sight of the flashing-light signals.*

Section 8D.07 <u>Rail Traffic Detection</u>

Standard:

01   **The devices employed in active traffic control systems shall be actuated by some form of rail traffic detection.**

02   **Rail traffic detection circuits, insofar as practical, shall be designed on the fail-safe principle.**

03   **Flashing-light signals shall operate for at least 20 seconds before the arrival of any rail traffic, except as provided in Paragraph 4 of this Section.**

Option:

04   On tracks where all rail traffic operates at less than 20 mph and where road users are directed by an authorized person on the ground to not enter the crossing at all times that approaching rail traffic is about to occupy the crossing, a shorter signal operating time for the flashing-light signals may be used.

05   Additional warning time may be provided when determined by an engineering study.

Guidance:

06   *Where the speeds of different rail traffic on a given track vary considerably under normal operation, special devices or circuits should be installed to provide reasonably uniform notice in advance of all rail traffic movements over the grade crossing. Special control features should be used to eliminate the effects of station stops and switching operations within approach control circuits to prevent excessive activation of the traffic control devices while rail traffic is stopped on or switching upon the approach track control circuits.*

Section 8D.08 Use of Traffic Control Signals at Grade Crossings

Standard:

01   **Except as provided in Paragraph 2 of this Section, traffic control signals shall not be used instead of flashing-light signals to control road users at a highway-rail grade crossing.**

Option:

02   Traffic control signals may be used instead of flashing-light signals to control road users at industrial highway-rail grade crossings and other places where the maximum speed of trains is 10 mph or less.

Support:

03   Sections 8D.04 and 8D.14 contain information regarding the use of traffic control signals at highway-LRT grade crossings.

Standard:

04   **The appropriate provisions of Part 4 relating to traffic control signal design, installation, and operation shall be applicable where traffic control signals are used instead of flashing-light signals to control road users at grade crossings.**

Section 8D.09 Preemption of Highway Traffic Signals at or Near Grade Crossings

Support:

01   Traffic signal preemption for grade crossings is a complex topic that requires a specific understanding of grade crossing warning systems and highway traffic signal operations. While most traffic signal operations are governed only by the traffic signal controller unit and the associated traffic signal equipment, preemption for grade crossings is also governed by the grade crossing warning system. Active grade crossing warning systems include flashing-light signals and possibly automatic gates, as well as various types of train detection equipment. Where the traffic signal controller unit is interconnected with the grade crossing warning system for the purpose of preemption, a combined system is created. It is the combined system that requires a thorough understanding of the design and operating parameters in order to provide proper operation of the preemption system.

02   The Federal Railroad Administration (FRA) has issued two documents that provide additional information relating to preemption of highway traffic signals at or near grade crossings. The first document is "Technical Bulletin S-12-01, Guidance Regarding the Appropriate Process for the Inspection of Highway-Rail Grade Crossing Warning System Pre-emption Interconnections with Highway Traffic Signals" and the second document is "Safety Advisory 2010-02, Signal Recording Devices for Highway-Rail Grade Crossing Active Warning Systems that are Interconnected with Highway Traffic Signal Systems."

Guidance:

03   *If a grade crossing is equipped with flashing-light signals and is located 200 feet or less from an intersection or midblock location controlled by a traffic control signal, a pedestrian hybrid beacon, or an emergency-vehicle hybrid beacon, the intersection should be provided with rail preemption in accordance with Section 4F.19 unless otherwise determined by the Diagnostic Team.*

04   *Coordination with the flashing-light signals, such as using queue detection and queue cutter signals, blank-out signs, or other alternatives, should be considered where a traffic control signal, a pedestrian hybrid beacon, or an emergency-vehicle hybrid beacon is located more than 200 feet from the grade crossing. Factors to be considered should include traffic volumes, highway vehicle mix, highway vehicle and train approach speeds, frequency of trains, presence of midblock driveways or unsignalized intersections, and the potential for vehicular queues resulting from an adjacent downstream grade crossing or highway traffic signal to extend into the minimum track clearance distance (see Section 8A.07).*

05   *The highway agency or authority with jurisdiction and the regulatory agency with statutory authority, if applicable, should jointly determine the preemption operation and the timing of highway traffic signals interconnected with grade crossings adjacent to signalized locations.*

06   *If a highway traffic signal is installed 200 feet or less from a passive grade crossing, unless otherwise determined by the Diagnostic Team, an active grade crossing warning system should be installed at the grade crossing to provide a means to preempt the highway traffic signal in order to clear vehicles from the minimum track clearance distance (see Section 8A.07) upon approach of rail traffic.*

07   *If a highway traffic signal is interconnected with flashing-light signals, the flashing-light signals should be provided with automatic gates to prevent additional vehicles from being drawn into the minimum track clearance distance (see Section 8A.07) during the track clearance interval prior to the arrival of rail traffic unless a Diagnostic Team determines otherwise.*

Support:

08   Regular joint inspections by the highway agency or authority with jurisdiction, the regulatory agency with statutory authority, if applicable, and the railroad company or transit agency are a best practice and typically include verification of the preemption operation, the amount of warning time and/or preemption time being provided by the grade crossing warning system, and the timing of highway traffic signals interconnected and/or coordinated with the flashing-light signals.

09   Section 4F.19 includes a recommendation that traffic control signals that are adjacent to highway-rail grade crossings and that are coordinated with the flashing-light signals at the grade crossing or that include railroad preemption features be provided with a back-up power supply.

Standard:

10   **Information regarding the type of preemption and any related timing parameters shall be provided to the railroad company or transit agency so that the railroad company or transit agency can design the appropriate train detection circuitry.**

11   **If preemption is provided, unless otherwise determined by a Diagnostic Team, the normal sequence of highway traffic signal indications shall be preempted upon the approach of a train to provide a track clearance interval to provide an opportunity for motor vehicles at the grade crossing to clear the minimum track clearance distance (see Section 8A.07) prior to the arrival of rail traffic.**

Option:

12   Where train switching or train restarts occur close to a grade crossing, the Diagnostic Team may determine that the preemption time can be reduced in accordance with the operating requirements of the railroad company and/or transit agency.

Standard:

13   **Where flashing-light signals are in place at a grade crossing, any highway traffic signal faces installed within 50 feet of any rail shall be preempted upon the approach of rail traffic. The Diagnostic Team shall determine the signal indications displayed by the highway traffic signal faces that control movements across the grade crossing in accordance with Section 4F.19 in order to avoid the display of signal indications that conflict with the flashing-light signals.**

Guidance:

14   *Where the flashing-light signals are in place at a grade crossing, the operation of any flashing yellow beacon installed within 50 feet of any rail should be considered by a Diagnostic Team to determine whether the operation of the beacon should be terminated during the approach and passage of rail traffic.*

Standard:

15   **The preemption special control mode shall be activated by a supervised preemption interconnection using fail-safe design principles between the control circuits of the grade crossing warning system and the traffic signal controller unit. The approach of rail traffic to a grade crossing shall de-energize the interconnection or send a message via a fail-safe data communication protocol (such as the "IEEE Standard for the Interface Between the Rail Subsystem and the Highway Subsystem at a Highway Rail Intersection," 1570-2002 (R2008), Institute of Electrical and Electronics Engineers), which in turn shall activate the traffic signal controller preemption sequence. This shall establish and maintain the preemption condition during the time the grade crossing warning system is activated, except that when automatic gates exist, the preemption condition shall not be terminated until the automatic gates are energized to start their upward movement.**

Support:

16   Advance preemption is the notification of approaching rail traffic that is forwarded to the highway traffic signal controller unit or assembly by the railroad or light rail transit equipment in advance of the activation of the grade crossing warning system.

17   The maximum preemption time is the maximum amount of time needed following initiation of the preemption sequence for the highway traffic signals to complete the timing of the right-of-way transfer time, queue clearance time, and separation time.

18   The separation time is the component of maximum preemption time during which the minimum track clearance distance is clear of vehicular traffic prior to the arrival of rail traffic.

19   Simultaneous preemption is the notification of approaching rail traffic that is forwarded to the highway traffic signal controller unit or assembly and grade crossing warning system at the same time.

20   The right-of-way transfer time is the amount of time needed prior to display of the track clearance interval. This includes any time needed by the railroad, light rail transit, or highway traffic signal control equipment to react to a preemption call, and any traffic control signal green, pedestrian walk and clearance if used (see Section 4F.19), yellow change, and red clearance intervals for conflicting traffic.

21   A supervised preemption interconnection is one that incorporates both a normally-open and a normally-closed circuit from the grade crossing warning system to verify the proper operation of the interconnection.

Option:

22   Instead of supervision, a double-break preemption interconnection circuit that uses two normally-closed circuits that open both the source and return energy circuits may be used.

23   A preemption interconnection may incorporate both supervision and double-break circuits.

Guidance:

24   *Where train detection circuits are present at a passive grade crossing, the operation of the preemption interconnection should be treated as if active traffic control devices exist at the crossing and the preemption operation should be determined by a Diagnostic Team.*

25   *Where left turns are permitted at a downstream highway-highway traffic control signal from the roadway approach that crosses the track and a delayed or impeded left-turn movement could prevent vehicles from clearing the track, a protected left-turn movement should be provided during the track clearance interval if green signal indications are displayed to the approach for track clearance.*

26   *The decision to implement simultaneous or advance preemption should include consideration of the right-of-way transfer time, the queue clearance time, and the separation time in order to determine the maximum preemption time. These time periods should be compared to and verified with the operation of the grade crossing traffic control devices in order to evaluate the operation of the highway traffic signal and the preemption operation. These factors should be considered regardless of whether simultaneous or advance preemption operation is implemented as they are based on traffic signal minimum timing, vehicle acceleration characteristics, and physical distances along the roadway.*

Support:

27   Preemption time variability occurs when the traffic signal controller enters the preemption clearance interval with less than the maximum design right-of-way transfer time or when the speed of a train approaching the grade crossing varies.

28   The time interval between the initiation of advance preemption and the operation of the grade crossing warning system for rail traffic will decrease in situations when rail traffic is accelerating or increase in situations when rail traffic is decelerating.

Guidance:

29   *Where preemption is used and automatic gates are present, the possibility that an automatic gate might descend upon a vehicle should be analyzed.*

30   *If simultaneous preemption is used, an analysis of extended grade crossing warning time requirements should be conducted.*

31   *If advance preemption is used, an analysis of preemption operation, traffic signal sequencing, and traffic signal phasing should be conducted to identify preemption time variability. The analysis should include both the condition requiring the longest amount of time to enter the track clearance interval and the condition requiring the shortest amount of time to enter the track clearance interval.*

Standard:

32   **Where automatic gates are present and green signal indications are displayed at the downstream traffic control signal during the track clearance interval, the preemption sequence shall be designed such that the green signal indications are not terminated until the automatic gate(s) that controls access over the grade crossing toward the downstream intersection is fully lowered.**

Support:

33   The following are two examples of mutually-exclusive methods to resolve preemption time variability:

   A. Gate-down circuitry provides a means to hold the traffic signal controller sequence in the track clearance interval until the automatic gate(s) that controls access over the grade crossing toward the downstream intersection is fully lowered.

   B. Timing correction resolves preemption time variability by adding the right-of-way transfer time to the track clearance interval in the traffic signal controller unit and setting a fixed maximum period of time between the start of advance preemption and the operation of the flashing-light signals.

34   The Third Edition of the "Railroad-Highway Grade Crossing Handbook" and the "2023 AREMA Communications and Signals Manual" published by the American Railway Engineering and Maintenance-of-Way Association (AREMA) provide additional information about preemption time variability.

Standard:

35   **Where gate-down circuitry is used to resolve preemption time variability and an automatic gate is broken or is not fully lowered, the crossing control circuits shall not terminate the track clearance interval before the rail traffic has entered the grade crossing.**

36   **Where timing correction is used to resolve preemption time variability, a timing circuit shall be used to maintain a maximum time interval between the initiation of advance preemption and the operation of the grade crossing warning system when the approaching rail traffic is decelerating.**

Guidance:

37   *Where a highway-highway intersection controlled by traffic control signals is interconnected with a grade crossing equipped with exit gates, advance preemption should be used because of the additional operating time that is required for the exit gates.*

38   *Where rail traffic routinely stops and re-starts within or just outside of the approaches to a grade crossing that is interconnected with highway traffic signals, the effects of rail traffic operations on the preemption operation should be analyzed.*

39   *Highway traffic signal control equipment should be capable of providing immediate re-service of successive requests for preemption from the railroad warning devices, even if the initial preemption sequence has not been completed. As appropriate, the highway traffic signal control equipment should be able to promptly return to the start of the track clearance interval at any time that the demand for preemption is cancelled and then reactivated. The highway traffic signal control equipment should have the ability to provide this immediate re-service at any point in the preemption sequence.*

Standard:

40   **Where traffic control signals are programmed to operate in a flashing mode during the preemption dwell interval (the period following the track clearance interval that lasts for the duration of the preemption interconnection activation), the beginning of the preemption dwell flashing mode shall not occur until the grade crossing equipment indicates that the rail traffic has entered the grade crossing.**

41   **At locations where conflicting preemption calls might be received to serve boats and trains, the Diagnostic Team shall determine the relative priority when conflicting preemption calls occur (see Section 4F.19). Where the boat and the train do not conflict with each other, the Diagnostic Team shall determine the preemption sequence when both preemption calls are occurring simultaneously. The United States Coast Guard or other appropriate authority that regulates the operation of the waterway shall be invited to participate on the Diagnostic Team and/or to provide input to the Diagnostic Team.**

Support:

42   Section 4C.10 describes the Intersection Near a Grade Crossing signal warrant that is intended for use at a location where the proximity to the intersection of a grade crossing on an intersection approach controlled by a STOP or YIELD sign is the principal reason to consider installing a traffic control signal.

43   Section 4F.19 describes additional considerations regarding preemption of traffic control signals at or near grade crossings.

Section 8D.10 <u>Movements Prohibited During Preemption</u>

Guidance:

01   *At a signalized intersection that is located within 100 feet of a grade crossing and the intersection traffic control signals are preempted by the approach of rail traffic, all existing permissive-only turning movements toward the grade crossing should be prohibited, steady red arrow signal indications should be shown to all existing protected/permissive and protected-only turning movements toward the grade crossing, and red signal indications should be shown to the straight-through movement toward the grade crossing during the signal preemption sequences. The prohibition of a permissive-only turning movement toward the grade crossing during preemption should be accomplished through the installation of a blank-out turn prohibition (R3-1a or R3-2a) sign (see Figure 8B-1).*

Option:

02   All movements toward the track may be prohibited at a signalized intersection that is preempted by the approach of rail traffic, even if the clear storage distance is more than 100 feet.

Support:

03   Including the word "TRAIN" as part of the blank-out turn prohibition sign informs road users that the turn prohibition being displayed by the sign is in effect because rail traffic is approaching or occupying a nearby rail grade crossing, and that the turn prohibition will be terminated after the rail traffic has cleared the grade crossing.

04   Rail operations can include the use of activated blank-out turn prohibition (R3-1a or R3-2a) signs at unsignalized highway-highway intersections in the vicinity of grade crossings, such as where a semi-exclusive or mixed-use alignment is within or parallel to the roadway where road users are normally permitted to turn across the tracks.

Guidance:

05   *An LRT-activated blank-out turn prohibition (R3-1a or R3-2a) sign should be used during preemption where all three of the following conditions are present:*

    A. *There is no active warning system for the LRT grade crossing,*

    B. *Vehicles traveling along a parallel roadway would normally be permitted to turn left or right to travel across tracks that are located within 100 feet of the highway-highway intersection or within the median of the intersection, and*

    C. *The drivers turning at the highway-highway intersection are not controlled by a traffic control signal.*

Standard:

06   **Blank-out turn prohibition signs that are associated with preemption shall display their message only when a preemption signal is being received from the railroad or LRT equipment or while the automatic gate is activated.**

Support:

07   The provisions contained in Chapter 2L for blank-out signs are applicable to R3-1a and R3-2a signs.

Section 8D.11 Pre-Signals at or Near Grade Crossings

Guidance:

01   *If a grade crossing is located in close proximity to an intersection controlled by a traffic control signal and the clear storage distance is less than the design vehicle length, the use of pre-signals to control traffic approaching the grade crossing in the direction toward the intersection should be considered.*

02   *If a grade crossing equipped with flashing-light signals, but without automatic gates, is located within 200 feet of an intersection controlled by a traffic control signal, a pre-signal should be provided.*

Support:

03   A pre-signal is generally used where the grade crossing is located less than 200 feet from a downstream signalized intersection. Section 8D.12 contains information for grade crossings located 200 feet or more from a downstream signalized intersection.

04   Other measures that could be considered instead of or in addition to a pre-signal to minimize the possibility of vehicles queuing across the grade crossing include providing additional lanes, reducing the cycle length, using split phasing, using protected turn phasing, and/or providing an extended green interval for the approach.

Standard:

05   **Pre-signal faces shall display a steady red signal indication during the track clearance interval of the signal preemption sequence to prohibit additional motor vehicles from entering the minimum track clearance distance (see Section 8A.07).**

06   **Pre-signal faces shall not display green signal indications when the grade crossing flashing-light signals are displaying flashing red indications.**

Guidance:

07   *Consideration should be given to using visibility-limited signal faces (see definition in Section 1C.02) at the intersection for the downstream signal faces that control the approach that is equipped with pre-signals.*

Option:

08   The duration of the extended green interval may be adjusted by vehicle detection located between the pre-signal and the downstream signalized intersection.

09   The pre-signal phase sequencing may be timed with an offset from the downstream signalized intersection such that the pre-signal's green signal indication terminates prior to the downstream intersection's green signal indication to minimize the possibility of stopping motor vehicles within the minimum track clearance distance (see Section 8A.07) and the clear storage distance .

Guidance:

10   *If pre-signals are used, the queue clearance time (see Section 8D.09) should be long enough to allow a design vehicle of maximum length stopped just inside the minimum track clearance distance (see Section 8A.07) to start up and move through the downstream intersection, or to clear the minimum track clearance distance if there is sufficient clear storage distance.*

Support:

11   The storage area for mandatory left-turn and right-turn lanes at signalized intersections that are downstream from grade crossings sometimes extends from the signalized intersection back to and across the grade crossing. In such cases, drivers that are in the turn lane are required to make a straight-through movement when they cross the track(s) and then are required to make a turning movement when they reach the downstream signalized intersection.

Guidance:

12 *A separate pre-signal face for the mandatory left-turn lane and/or right-turn lane should be provided in addition to the pre-signal signal faces provided for the through movement where both of the following conditions are met:*

    A. *The storage area for the turn lane extends from the downstream signalized intersection back to and across the grade crossing, and*
    B. *The green interval for the turning movement at the downstream intersection does not always begin and end simultaneously with the green interval for the adjacent through movement at the downstream intersection.*

Standard:

13 **Where adjacent lanes at a pre-signal are controlled separately, all of the signal faces shall be capable of displaying the following signal indications: CIRCULAR RED, CIRCULAR YELLOW, and straight-through GREEN ARROW. Left-turn GREEN ARROW and right-turn GREEN ARROW signal indications shall not be used in pre-signal faces. CIRCULAR GREEN signal indications shall not be used in pre-signal faces where adjacent lanes are controlled separately.**

Option:

14 Where all adjacent lanes at a pre-signal are controlled together, CIRCULAR GREEN signal indications may be used in pre-signal faces.

Standard:

15 **If a separate signal face is provided at a pre-signal for separate control of a mandatory left-turn and/or right-turn lane that extends from the downstream signalized intersection back to and across the grade crossing, the separate signal face shall be devoted exclusively to controlling traffic in the turn lane separately from adjacent lanes, and:**

    A. **Shall be visibility-limited from the adjacent through movement, or**
    B. **A LEFT (RIGHT) LANE SIGNAL (R10-10b) sign (see Figure 8B-1) shall be mounted adjacent to the separate signal face controlling traffic in a single turn lane or in the turn lane that is farthest from the adjacent through lane(s) if multiple turn lanes are present for a particular turning movement, and a LEFT (RIGHT) TURN LANE SIGNAL (R10-10c) sign (see Figure 8B-1) shall be mounted adjacent to the separate signal face controlling traffic in the other turn lanes if multiple turn lanes are present for a particular turning movement.**

Support:

16 Because the signal faces at a pre-signal do not always display the same signal indications as the downstream signalized intersection, the approach to the pre-signal is considered to be a separate approach from the approach to the downstream signalized intersection. This means that the provisions in Sections 4D.05 through 4D.08 regarding the number of signal faces, the visibility and aiming of the signal faces, and the lateral and longitudinal positioning of the signal faces apply separately to the approach to the pre-signal.

17 The provisions in Section 4D.07 regarding the lateral positioning of separate turn signal faces are applicable to the separate signal faces that are provided at pre-signals for a mandatory turn lane that extends from the downstream signalized intersection back to and across the grade crossing.

Guidance:

18 *A STOP HERE ON RED (R10-6 or R10-6a) sign should be installed at the pre-signal's stop line.*

Standard:

19 **If a pre-signal is installed upstream from a signalized intersection, a No Turn on Red (R10-11, R10-11a, or R10-11b) sign (see Section 2B.60) shall be installed at the pre-signal for the approach that crosses the track if turns on red would otherwise be permitted at the downstream intersection.**

Option:

20 DO NOT STOP ON TRACKS (R8-8) signs may be installed in conjunction with a pre-signal.

21 Pre-signal faces may be located either upstream or downstream from the grade crossing in order to provide the most effective display to road users approaching the grade crossing.

22 If pre-signal faces must be located within close proximity to the flashing-light signals, the pre-signal faces may be mounted on the same overhead structure as the flashing-light signals.

Section 8D.12 Queue Cutter Signals at or Near Grade Crossings

Support:

01   A queue cutter signal is a traffic control signal that controls one direction of traffic at a grade crossing to minimize the possibility of vehicles stopping within the minimum track clearance distance (see Section 8A.07). Although a queue cutter signal has a similar purpose as a pre-signal (see Section 8D.11), the difference is that a queue cutter signal is independent from the downstream signalized intersection, whereas a pre-signal is part of the downstream signal.

Option:

02   At grade crossing locations where the queue from a bottleneck (usually a signalized intersection) that is downstream from the grade crossing frequently extends back to and across the grade crossing, a queue cutter signal may be installed.

03   A queue cutter signal may be operated in one of the following modes:

    A. Actuated mode – the queue cutter signal operation is dependent on downstream detection of a growing queue.
    B. Non-actuated mode – the queue cutter signal operates on a time-of-day plan based on anticipated downstream queues. This mode could be similar to the functional operation of a pre-signal.
    C. Variable mode – the queue cutter signal operation varies between the actuated mode and the non-actuated mode based on the time of day, on queue detection, or both.

Support:

04   A non-actuated queue cutter signal is generally used where the grade crossing is located between 200 feet and 400 feet from a downstream bottleneck. An actuated queue cutter signal is generally used where the grade crossing is located more than 400 feet from a downstream bottleneck. Section 8D.11 contains information for grade crossings located less than 200 feet from a downstream signalized intersection.

Standard:

05   **Where adjacent lanes at a queue cutter signal are controlled separately, all of the signal faces shall be capable of displaying the following signal indications: CIRCULAR RED, CIRCULAR YELLOW, and straight-through GREEN ARROW. Left-turn GREEN ARROW and right-turn GREEN ARROW signal indications shall not be used in queue cutter signal faces. CIRCULAR GREEN signal indications shall not be used in queue cutter signal faces where adjacent lanes are controlled separately.**

Option:

06   Where all adjacent lanes at a pre-signal are controlled together, CIRCULAR GREEN signal indications may be used in queue cutter signal faces.

07   Queue cutter signal faces may be located either upstream or downstream from the grade crossing in order to provide the most effective display to road users approaching the grade crossing.

08   If queue cutter signal faces must be located within close proximity to the flashing-light signals, the queue cutter signal faces may be mounted on the same overhead structure as the flashing-light signals.

Guidance:

09   *A STOP HERE ON RED (R10-6 or R10-6a) sign should be installed at the queue cutter signal's stop line.*

Option:

10   DO NOT STOP ON TRACKS (R8-8) signs may be installed in conjunction with a queue cutter signal.

Guidance:

11   *Where a queue cutter signal operates in an actuated mode based on vehicle presence detection, the queue detector should be located to provide adequate distance to detect a growing queue, permit the queue cutter signal to complete any programmed minimum green or yellow change interval time, and then allow a design vehicle that lawfully crosses the queue cutter signal's stop line during the yellow change interval to clear the minimum track clearance distance (see Section 8A.07) before the growing queue extends to the grade crossing.*

12   *A queue cutter signal that is operating in an actuated mode and that is displaying CIRCULAR RED signal indications should continue to display CIRCULAR RED signal indications as long as the downstream detection system continues to detect the presence of a vehicular queue at the detection point on the departure side of the grade crossing.*

13   *Where a queue cutter signal operates in actuated mode based on vehicle presence detection, consideration should be given to the potential for turning movements between the grade crossing and the downstream bottleneck that could create an intermediate queue of vehicles. Supplemental queue detectors should be considered to detect the formation of these intermediate queues to activate the queue cutter signal.*

14   *Where a queue cutter signal is operated in a non-actuated mode, the queue cutter signal should be coordinated with adjacent signals to provide for the progressive movement of traffic.*

Option:

15 Where a queue cutter signal is always operated in a non-actuated mode based on anticipated queues, the queue cutter signal may be operated in a flashing mode at times when the downstream queues are not expected to extend back to and across the grade crossing.

16 When a variable-mode queue cutter signal is operating in the non-actuated mode, the queue detector may be used to extend the display of the CIRCULAR RED signal indication as long as the downstream detection system continues to detect the presence of a vehicular queue at the detection point on the departure side of the grade crossing.

Standard:

17 **A queue cutter signal shall be interconnected with the flashing-light signals at the grade crossing.**

18 **Queue cutter signal faces shall not display green signal indications when the grade crossing flashing-light signals are displaying flashing red indications.**

19 **When a queue cutter signal that is displaying straight-through GREEN ARROW signal indications (when operating in a steady, stop-and-go mode) or flashing CIRCULAR YELLOW signal indications (when operating in a programmed flashing mode) is preempted by the approach of rail traffic, it shall immediately display steady CIRCULAR YELLOW signal indications during the yellow change interval (see Section 4F.17) followed by steady CIRCULAR RED signal indications. The queue cutter signal shall continue to display the steady CIRCULAR RED signal indications until the rail traffic clears the grade crossing and no other rail traffic is detected.**

20 **A queue cutter signal operating in an actuated mode shall display straight-through GREEN ARROW signal indications except when it receives an actuation from the downstream vehicle presence detection system or is preempted by the approach of rail traffic. When it receives an actuation from the vehicle presence detection system, the queue cutter signal shall finish timing any active minimum green interval, if used, and then display steady CIRCULAR YELLOW signal indications during the yellow change interval (see Section 4F.17) followed by steady CIRCULAR RED signal indications. When no preemption call is present and the queue length is such that no vehicles are detected in the detection zone of the downstream vehicle presence detection system, the queue cutter signal shall finish timing any active minimum red interval, if used, and then return to the display of straight-through GREEN ARROW signal indications.**

21 **The failure modes of the queue cutter signal control system and vehicle presence detection circuitry shall be evaluated and accounted for in the design of any such system. Fail-safe design techniques shall be used in the system design. If a queue detector fails, the queue cutter signal shall display flashing CIRCULAR RED signal indications until the normal functioning of the detection system is restored.**

Support:

22 The storage area for mandatory left-turn and right-turn lanes at signalized intersections that are downstream from grade crossings sometimes extends from the signalized intersection back to and across the grade crossing. In such cases, drivers that are in the turn lane are required to make a straight-through movement when they cross the track(s) and then are required to make a turning movement when they reach the downstream signalized intersection.

Guidance:

23 *A separate queue cutter signal face for the mandatory left-turn lane and/or right-turn lane should be provided in addition to the queue cutter signal faces provided for the through movement where both of the following conditions are met:*

  A. *The storage area for the turn lane extends from the downstream signalized intersection back to and across the grade crossing, and*
  B. *The green interval for the turning movement at the downstream intersection does not always begin and end simultaneously with the green interval for the adjacent through movement at the downstream intersection.*

Standard:

24 **If a separate signal face is provided at a queue cutter signal for separate control of a mandatory left-turn and/or right-turn lane that extends from the downstream signalized intersection back to and across the grade crossing, the separate signal face shall be devoted exclusively to controlling traffic in the turn lane separately from adjacent lanes, and:**

  A. **Shall be visibility-limited from the adjacent through movement, or**
  B. **A LEFT (RIGHT) LANE SIGNAL (R10-10b) sign (see Figure 8B-1) shall be mounted adjacent to the separate signal face controlling traffic in a single turn lane or in the turn lane that is farthest from the adjacent through lane(s) if multiple turn lanes are present for a particular turning movement, and a LEFT (RIGHT) TURN LANE SIGNAL (R10-10c) sign (see Figure 8B-1) shall be mounted adjacent to the separate signal face controlling traffic in the other turn lanes if multiple turn lanes are present for a particular turning movement.**

Support:

25   Because the signal faces at a queue cutter signal do not always display the same signal indications as the downstream signalized intersection, the approach to the queue cutter signal is considered to be a separate approach from the approach to the downstream signalized intersection. This means that the provisions in Sections 4D.05 through 4D.08 regarding the number of signal faces, the visibility and aiming of the signal faces, and the lateral and longitudinal positioning of the signal faces apply separately to the approach to the queue cutter signal.

26   The provisions in Section 4D.07 regarding the lateral positioning of separate turn signal faces are applicable to the separate signal faces that are provided at queue cutter signals for a turn lane that extends from the downstream signalized intersection back to and across the grade crossing.

27   While queue cutter signals and queue jumping signals have similar names, their purpose, design, and operation are quite different. Care must be taken to avoid confusion between queue cutter signals used in conjunction with a grade crossing and queue jumping signals used with transit operations.

Section 8D.13 Warning Beacons or LED-Enhanced Warning Signs at Grade Crossings

Option:

01   Warning Beacons (see Section 4S.03) or LEDs within the legend, symbol, or border of the sign (see Section 2A.12) may be used to supplement warning signs installed at or on an approach to a grade crossing if additional emphasis is desired for the warning sign. The Warning Beacon or LED-enhanced sign may operate continuously or be activated upon the approach or presence of rail traffic.

Support:

02   Most of the warning signs that are used at or on an approach to a grade crossing warn of physical conditions that exist at the grade crossing regardless of whether rail traffic is approaching or occupying the grade crossing. In these cases, a Warning Beacon or LED-enhanced sign would typically be operated continuously to enhance the conspicuity of the sign.

03   Some warning signs, such as a BE PREPARED TO STOP (W3-4) sign (see Section 2C.35), if used in advance of a grade crossing and supplemented with a WHEN FLASHING (W16-13P) plaque, provide information that is typically not applicable except when rail traffic is approaching or occupying the grade crossing. Likewise, a special word message sign (see Section 2A.04) with a legend such as TRAIN WHEN FLASHING provides notice of a condition that only exists when rail traffic is approaching or occupying the grade crossing. These signs would not typically be operated continuously, but instead only when the condition is present.

Standard:

04   **If a Warning Beacon or LEDs within the legend, symbol, or border of the sign is activated by the approach or presence of rail traffic in conjunction with a warning sign that includes the legend WHEN FLASHING either on the sign itself or on a supplemental plaque, the activation of the Warning Beacon or LEDs shall be accomplished by a supervised preemption interconnection using fail-safe design principles (see Section 8D.09) between the control circuits of the grade crossing warning system and the Warning Beacon or LED-enhanced sign.**

Support:

05   In the event of a system failure, the normal fault state using a fail-safe interconnection for a Warning Beacon or LED-enhanced sign that is activated by the approach or presence of rail traffic at the grade crossing would be for the Warning Beacon or LEDs to operate when no rail traffic is present.

Option:

06   A Warning Beacon or LED-enhanced sign that is activated by the approach or presence of rail traffic at the grade crossing may continue to operate for a period of time following the passage of the rail traffic to permit the standing queue to dissipate.

Guidance:

07   *If a Warning Beacon or LED-enhanced sign is activated by the approach or presence of rail traffic at the grade crossing, the Warning Beacon or LED-enhanced sign should begin operating prior to the activation of the flashing-light signals at the grade crossing based upon the typical travel time from the location of the Warning Beacon or LED-enhanced sign to the stop line for the grade crossing.*

08   *If a Warning Beacon or LED-enhanced sign that is activated by the approach or presence of rail traffic at the grade crossing is operated by commercial AC power, a back-up power system should be provided.*

Section 8D.14 <u>Traffic Control Signals at or Near Highway-LRT Grade Crossings</u>

Support:

01   There are two types of traffic control signals for controlling vehicular and LRT movements at interfaces of the two modes. The first is the standard traffic control signal described in Part 4, which is the focus of this Section. The other type of signal is referred to as an LRT signal and is discussed in Section 8D.15.

Standard:

02   **The provisions of Part 4 and Sections 8D.08 through 8D.12 relating to traffic control signal design, installation, and operation, including interconnection with nearby automatic gates or flashing-light signals, shall be applicable as appropriate where traffic control signals are used at highway-LRT grade crossings.**

03   **If traffic control signals are in operation at an LRT grade crossing that is used by pedestrians, bicyclists, and/or other non-motorized road users, an audible device such as a bell shall also be provided and shall be operated in conjunction with the traffic control signals.**

Guidance:

04   *If the highway traffic signal has emergency-vehicle preemption capability, it should be coordinated with LRT operation.*

05   *Where LRT operates in a wide median, motor vehicles crossing the tracks and being controlled by both near and far side traffic signal faces should receive a protected left-turn phase from the far side signal face to clear motor vehicles from the crossing when LRT traffic is approaching the crossing.*

Option:

06   Signal indications that permit the movement of motor vehicles, pedestrians, and bicyclists and do not conflict with LRT movements may be provided during LRT phases.

07   A traffic control signal may be installed in addition to Exit Gate systems and automatic gates at a highway-LRT grade crossing if the crossing occurs within a highway-highway intersection and if the installation of the traffic control signal can be justified based on the warrants described in Chapter 4C.

08   Where a highway-LRT grade crossing is at a location other than an intersection and LRT operating speeds are less than 25 mph, traffic control signals may be used in lieu of flashing-light signals.

Support:

09   Typical circumstances for using traffic control signals might include:

    A. Geometric conditions preclude the installation of highway-LRT grade crossing warning devices,
    B. LRT vehicles share the same roadway with road users, or
    C. Traffic control signals already exist.

10   Section 4F.18 contains information regarding traffic control signals at or near highway-LRT grade crossings that are not equipped with highway-LRT grade crossing warning devices.

11   Section 4C.10 describes the Intersection Near a Grade Crossing signal warrant that is intended for use at a location where the proximity to the intersection of a grade crossing on an intersection approach controlled by a STOP or YIELD sign is the principal reason to consider installing a traffic control signal.

Guidance:

12   *When a highway-LRT grade crossing exists within a signalized intersection, consideration should be given to providing separate turn signal faces (see definition in Section 1C.02) for the movements crossing the tracks.*

Standard:

13   **Separate turn signal faces that are provided for turn movements toward the crossing shall display a steady red indication during the approach and/or passage of LRT traffic.**

Support:

14   Section 8D.10 contains information regarding the prohibition of turning movements toward the crossing during preemption.

Section 8D.15 Use of LRT Signals for Control of LRT Vehicles at Highway-LRT Grade Crossings

Option:

01   LRT signal indications may be used at grade crossings and at intersections in mixed-use alignments in conjunction with standard traffic control signals where special LRT signal phases are used to accommodate turning LRT vehicles or where additional LRT clearance time is desirable.

02   LRT signal indications may be used at intersections where special signal phases are used for bus movements.

Standard:

03   **If the LRT crossing control is separate from the intersection control, the two shall be interconnected. The LRT signal phase shall not be terminated until after the LRT vehicle has cleared the crossing or intersection.**

04   **If a separate set of standard traffic control signal indications (red, yellow, and green circular and arrow indications) is used to control LRT movements, the indications shall be positioned so they are not visible to motorists, pedestrians, and bicyclists (see Section 4D.06).**

Guidance:

05   *If a signal face used to control LRT movements cannot be positioned where the indications are not visible to road users, the LRT signal indications shown in Figure 8D-3 should be used.*

Standard:

06   **If special LRT signal indications such as those shown in Figure 8D-3 are used, the color of the signal indications shall be white.**

Option:

07   If used, individual LRT signal sections may be displayed to form clustered signal faces or multiple LRT signal indications may be displayed in an individual housing.

Guidance:

08   *LRT signal faces should be located at least 3 feet from the nearest highway traffic signal face for the same approach measured either horizontally perpendicular to the approach between the centers of the signal faces or vertically from the center of the lowest signal indication of the top signal face to the center of the highest signal indication of the bottom signal face.*

Support:

09   Section 4F.18 contains information about the use of the LRT signal indications shown in Figure 8D-3 for the control of exclusive bus movements at "queue jumper lanes" and for the control of exclusive bus rapid transit movements on mixed-use alignments.

CHAPTER 8E. PATHWAY AND SIDEWALK GRADE CROSSINGS

Section 8E.01 Purpose

Support:

01   Traffic control for pathway and sidewalk grade crossings includes all signs, signals, markings, other warning devices, and their supports at pathway and sidewalk grade crossings and along pathway and sidewalk approaches to grade crossings. The function of this traffic control is to promote safety and provide effective operation of both rail and pathway or sidewalk traffic at pathway or sidewalk grade crossings.

02   Other physical treatments that are described in this Chapter that are also applicable to pathways and sidewalks at grade crossings, such as detectable warnings, swing gates, and fencing, provide increased safety for pathway and sidewalk users.

03   Crosswalk markings at intersections where pedestrians cross LRT tracks in mixed-use alignments are covered by the provisions of Chapter 3C rather than by the provisions of this Chapter.

04   Figure 8E-1 illustrates the difference between a pathway grade crossing and a sidewalk grade crossing. A pathway is frequently placed in its own right-of-way on an alignment that is independent of any roadway. If a pathway is built parallel to a roadway, it is physically separated from the roadway by an open space or barrier such that the traffic control devices for the roadway grade crossing do not exert an influence over or provide adequate warning to pathway users. A sidewalk runs parallel to a roadway within the highway right-of-way and is close enough to the edge of the roadway's traveled way that the traffic control devices for the roadway grade crossing can frequently exert an influence over or provide adequate warning to sidewalk users. Pathways are typically used by both pedestrians and bicyclists, whereas sidewalks are typically used only by pedestrians.

Section 8E.02 Use of Standard Devices, Systems, and Practices

Guidance:

01   *The pathway or sidewalk user's ability to detect the presence of approaching rail traffic should be considered in determining the type and placement of traffic control devices at pathway or sidewalk grade crossings.*

02   *The traffic control devices, including the appropriate traffic control system to be used, and other physical treatments at a pathway or sidewalk grade crossing should be determined by a Diagnostic Team that includes the agency with jurisdiction over the pathway or sidewalk.*

03   *At skewed grade crossings, the adjustment, re-alignment, or relocation of existing sidewalk grade crossings should be considered when determining the placement of traffic control devices for roadway users.*

Support:

04   The safety of pathway and sidewalk users is enhanced when pathways and sidewalks are designed such that they do not cross the tracks at a narrow angle. The casters of wheelchairs and the wheels of bicycles could fall into and might be constrained in the flangeway gap at a skewed crossing. The flangeway gap is typically 2.5 inches wide at LRT grade crossings and 3 inches wide at railroad grade crossings.

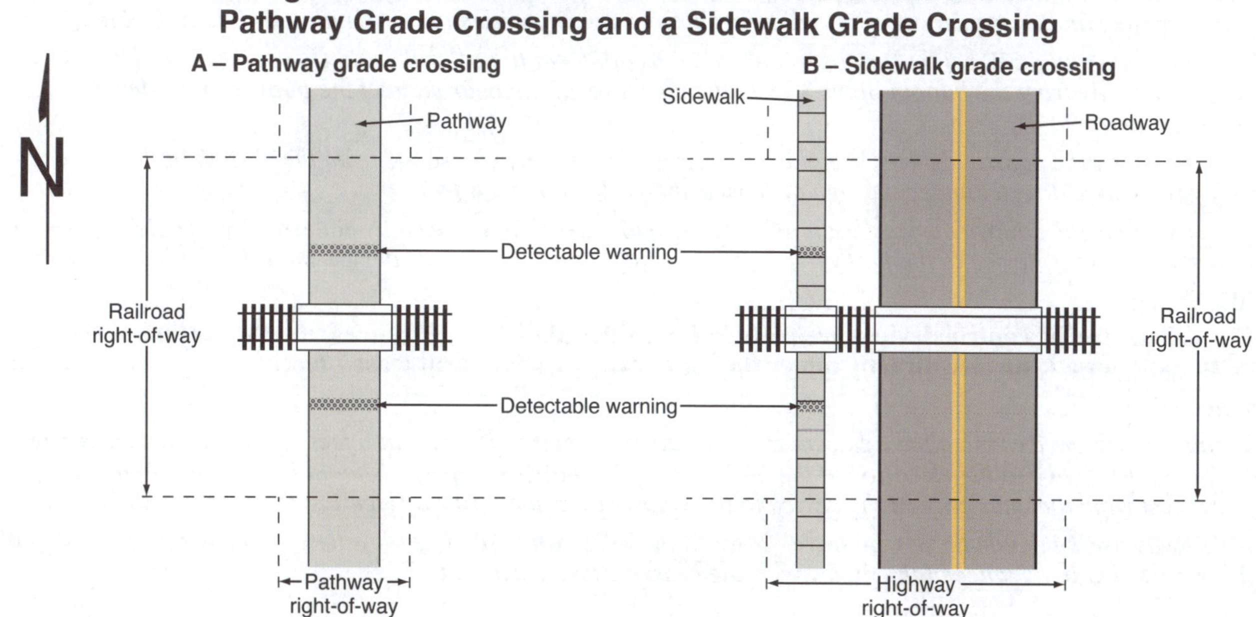

**Figure 8E-1. Illustration of the Difference between a
Pathway Grade Crossing and a Sidewalk Grade Crossing**

05   It is desirable that pathways and sidewalks be designed such that they maintain a relatively consistent horizontal alignment and profile from the nearest rail to the detectable warning (if present) or from the nearest rail to the stop line (if present) on each approach to the crossing. Providing a pedestrian refuge area in advance of the stop line or the detectable warning surface so that pedestrians have a place to wait while rail traffic approaches and occupies the crossing can be beneficial to pedestrian safety.

06   When designing new sidewalk grade crossings, placing the sidewalk outside of the area occupied by grade crossing traffic control devices for vehicular traffic is desirable (see Figure 8E-2). This includes making sure that the counterweights and support arms for the automatic gates for vehicular traffic do not obstruct the sidewalk when the gate is fully lowered.

07   Additional information regarding the design of pathways and sidewalks is contained in the U.S. Department of Justice 2010 ADA Standards for Accessible Design, September 15, 2010, 28 CFR 35 and 36, Americans with Disabilities Act of 1990.

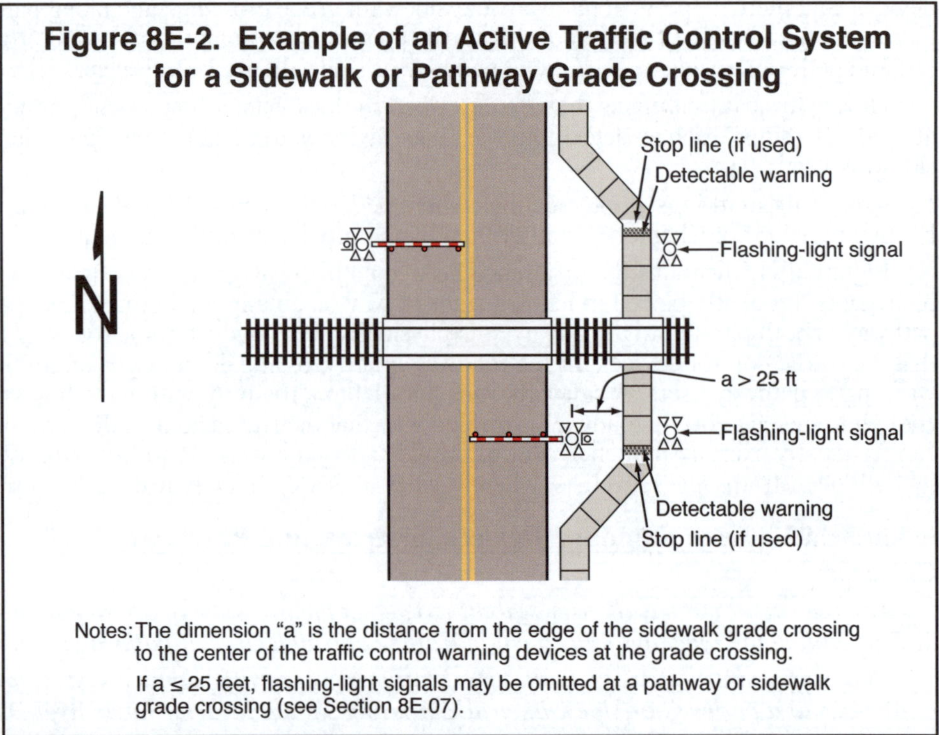

Notes: The dimension "a" is the distance from the edge of the sidewalk grade crossing to the center of the traffic control warning devices at the grade crossing.

If a ≤ 25 feet, flashing-light signals may be omitted at a pathway or sidewalk grade crossing (see Section 8E.07).

Section 8E.03 Pathway and Sidewalk Grade Crossing Signs and Markings

Standard:

01   **Pathway and sidewalk grade crossing signs shall be standard in shape, legend, and color.**

02   **The minimum sizes of sidewalk grade crossing signs that are intended to be viewed only by sidewalk users and of pathway grade crossing signs shall be as shown in the shared-use path column in Table 9A-1.**

Guidance:

03   *No portion of a traffic control device or its support should protrude into the pathway or sidewalk grade crossing. Sidewalk and pathway grade crossing traffic control devices should be located such that all physical features of the device, including the support hardware, conform to clearance requirements provided by the railroad company and/or transit agency, and the regulatory agency with statutory authority (if applicable).*

04   *The minimum mounting height for post-mounted signs adjacent to pathways and sidewalks should be 4 feet, measured vertically from the bottom of the sign to the elevation of the near edge of the pathway or sidewalk surface (see Figure 9A-1).*

05   *If overhead traffic control devices are placed above pathways, the clearance from the bottom of the device to the pathway surface directly under the sign or device should be at least 8 feet.*

06   *If overhead traffic control devices are placed above pathways that are used by equestrians, the clearance from the bottom of the device to the pathway surface directly under the sign or device should be at least 10 feet.*

Standard:

07   **If overhead traffic control devices are placed above sidewalks, the clearance from the bottom of the device to the sidewalk surface directly under the sign or device shall be at least 7 feet.**

Guidance:

08   *Traffic control devices mounted adjacent to pathways at a height of less than 8 feet measured vertically from the bottom of the device to the elevation of the near edge of the pathway surface should have a minimum lateral offset of 2 feet from the near edge of the device to the near edge of the pathway (see Figure 9A-1).*

09   *If pathway users include those who travel faster than pedestrians, such as bicyclists or skaters, warning signs should be installed in advance of the pathway grade crossing (see Figure 8E-3).*

**Figure 8E-3. Example of Signing and Markings
at a Pathway Grade Crossing**

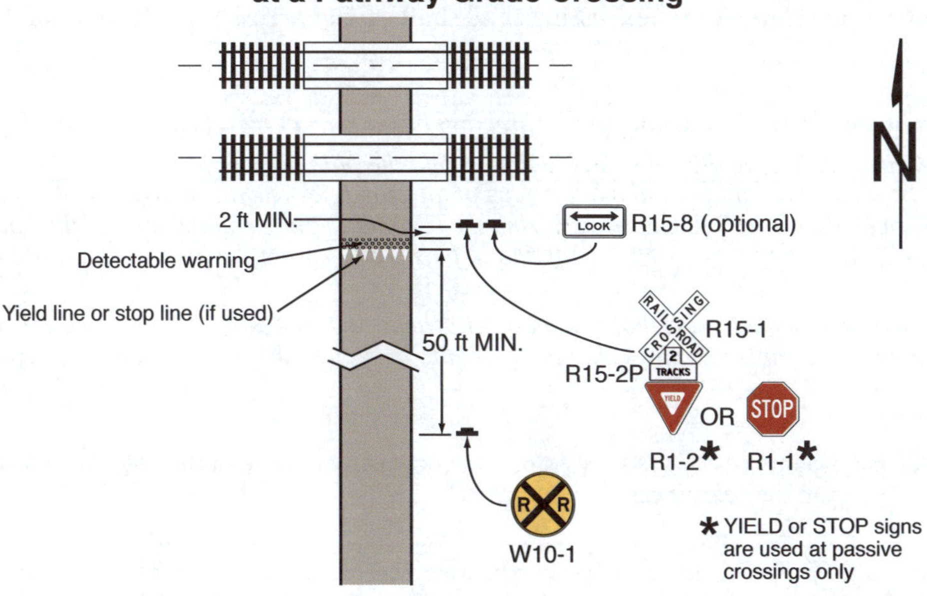

Option:

10   The Skewed Crossing (W10-12) sign (see Section 8B.22) may be used at a skewed pathway or sidewalk grade crossing to warn pathway or sidewalk users that the tracks are not perpendicular to the pathway or sidewalk.

11   The LOOK (R15-8) sign (see Figure 8B-1) may be used at a pathway or sidewalk grade crossing to inform pathway or sidewalk users to look in both directions prior to crossing the track(s).

Guidance:

12   *If a LOOK (R15-8) sign is used at a pathway or sidewalk grade crossing, it should be mounted on a separate post that is farther from the pathway or sidewalk than the Crossbuck sign or Crossbuck Assembly.*

Section 8E.04 Stop Lines, Edge Lines, and Detectable Warnings

Guidance:

01   *A stop line should be provided at a pathway grade crossing if the surface where the marking is to be applied is capable of retaining the application of the marking.*

Option:

02   A stop line may be provided at a sidewalk grade crossing if the surface where the marking is to be applied is capable of retaining the application of the marking.

Guidance:

03   *If used at pathway or sidewalk grade crossings, the stop line should be a transverse line that extends across the full width of the pathway or sidewalk at the point where a pathway or sidewalk user is to stop. If no detectable warning is provided, the stop line should be placed at least 2 feet in advance of the automatic gate, counterweight, flashing-light signals, or Crossbuck Assembly (if any of these are present), and at least 12 feet from the nearest rail.*

Option:

04   Edge lines (see Section 3B.09) to delineate the designated user route may be used on the approach to and across the tracks at a pathway grade crossing, a sidewalk grade crossing, or a station crossing if the surface where the marking is to be applied is capable of retaining the application of the marking.

Support:

05   Edge line delineation can be beneficial where the distance across the tracks is long, commonly because of a skewed grade crossing or because of multiple tracks, or where the pathway or sidewalk surface is immediately adjacent to a traveled way.

06   Information regarding the design of detectable warning surfaces is contained in the U.S. Department of Justice 2010 ADA Standards for Accessible Design, September 15, 2010, 28 CFR 35 and 36, Americans with Disabilities Act of 1990.

Standard:

07   **Detectable warnings (see Chapter 3C) shall be used at pathway grade crossings where pedestrian travel is permitted and at sidewalk grade crossings and shall extend across the full width of the pathway or sidewalk.**

Guidance:

08   *The dimension of the detectable warning in the direction of pedestrian travel should be at least 2 feet.*

09   *Detectable warnings should be placed immediately beyond the pathway or sidewalk stop line (if a stop line is present) or should be incorporated into and made a part of the stop line. The downstream edge of the detectable warning should be located at least 2 feet upstream from the automatic gate, counterweight, flashing-light signals, or Crossbuck Assembly (if any of these are present) and at least 12 feet from the nearest rail (see Figures 8E-2 and 8E-3).*

10   *If the distance between the nearest rail of two adjacent tracks at a sidewalk or pathway grade crossing is 30 feet or more, additional detectable warnings should be used to designate the limits of the pedestrian refuge area (see Figure 8E-4).*

Option:

11   At pathway-LRT and sidewalk-LRT grade crossings, the downstream edge of the detectable warning may be located less than 12 feet from the nearest rail.

Guidance:

12   *The downstream edge of the detectable warning at pathway-LRT and sidewalk-LRT grade crossings should be located at least 2 feet upstream from the automatic gate, counterweight, flashing-light signals, or Crossbuck Assembly (if any of these are present), at least 6 feet from the nearest rail, and in accordance with the requirements of the railroad company and/or transit agency, and regulatory agency with statutory authority (if applicable).*

Figure 8E-4. Example of Detectable Warning and Stop Lines for a Refuge Area at a Sidewalk Grade Crossing

* If A is 30 feet or more, detectable warnings should be used at the pedestrian refuge area (see Section 8E.04)

** Minimum widths for accessible refuge islands are provided in the U.S. Department of Justice 2010 ADA Standards for Accessible Design, September 15, 2010; Code of Federal Regulations Title 28, Parts 35 and 36; and the Americans with Disabilities Act of 1990.

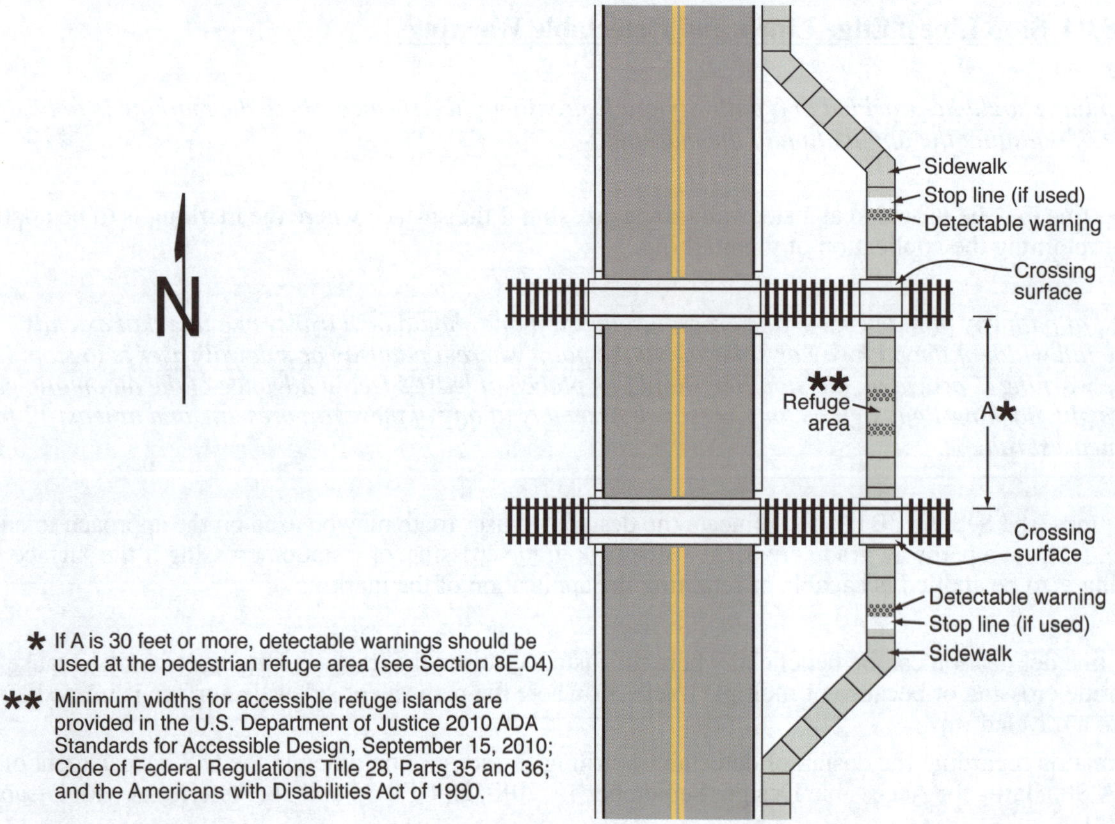

Section 8E.05 Passive Traffic Control Devices – Crossbuck Assemblies

Standard:

01 **Where the nearest edge of a passive pathway or sidewalk grade crossing is located more than 25 feet from the center of the nearest traffic control warning device at the grade crossing, a Crossbuck Assembly (see Figure 8E-5) shall be installed on each approach to the pathway or sidewalk grade crossing. The distance shall be measured perpendicular to the traveled way from the center of the support post of a Crossbuck Assembly at a passive grade crossing or from the center of the mast of an active traffic control warning device at an active grade crossing to the nearest edge of the pathway or sidewalk surface where it crosses the track(s) (see Figure 8E-2).**

Option:

02 A Crossbuck Assembly may be installed on the approaches to a pathway or sidewalk grade crossing where the nearest edge of the pathway or sidewalk is located 25 feet or less from the center of the nearest traffic control warning device at a grade crossing.

03 The Crossbuck Assembly may be omitted at station crossings.

04 The retroreflective strip on the back of the support may be omitted on the Crossbuck support at a pathway or sidewalk grade crossing.

Standard:

05 **The minimum height, measured vertically from the bottom of the YIELD or STOP sign to the elevation of the near edge of the pathway or sidewalk, of Crossbuck Assemblies installed on pathways or sidewalks shall be 4 feet where the lateral offset to the nearest edge of the sign is 2 feet or more and shall be 7 feet where the lateral offset to the nearest edge of the sign is less than 2 feet (see Figure 8E-5).**

06 **The minimum lateral offset, measured horizontally from the nearest edge of the pathway or sidewalk to the nearest edge of the Crossbuck Assembly signs, shall be 0 feet for sidewalks and 2 feet for pathways.**

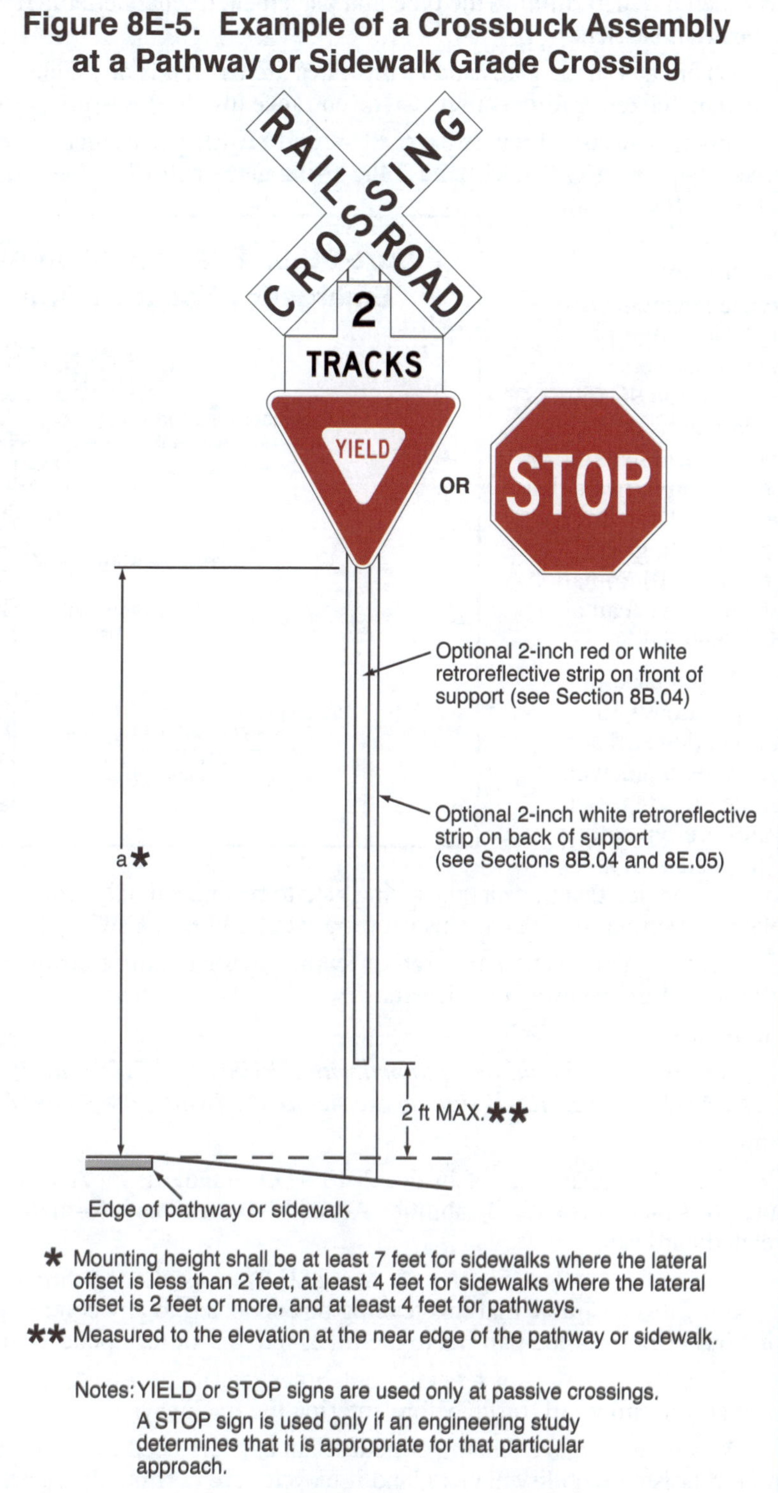

Figure 8E-5. Example of a Crossbuck Assembly at a Pathway or Sidewalk Grade Crossing

Section 8E.06 Channelizing Devices Used with Sidewalk and Pathway Traffic Control Devices

Support:

01   The pathway or sidewalk user's ability to detect the presence of approaching rail traffic needs to be considered in determining the type and placement of channelizing devices such as swing gates, fencing, and pedestrian barriers.

02   Where automatic gates and swing gates are used, it is desirable to design the pathway or sidewalk in a manner that channelizes or directs users to the entrance to and exit from the pathway or sidewalk grade crossing.

03   Swing gates (see Figures 8E-6, 8E-9, and 8E-10) are designed to open away from the track(s) so that pathway or sidewalk users can quickly push the swing gate open when moving away from the track(s), and to automatically return to the closed position after each use.

04   It is important to use retroreflective material, appropriate object markers (see Section 9C.09), and/or signs on swing gates, maze fencing, or pedestrian barriers that are placed at pathway or sidewalk grade crossings. Illumination of such areas can also be beneficial.

05   When used in conjunction with automatic gates at a pathway or sidewalk grade crossing, swing gates are typically equipped with a latching device that permits the swing gate to be opened only from the track side of the swing gate. Push bars, kick plates, or similar devices are also appropriate for use on a swing gate.

Figure 8E-6. Example of an Automatic Pedestrian Gate and an Emergency Escape Route at a Pathway Grade Crossing

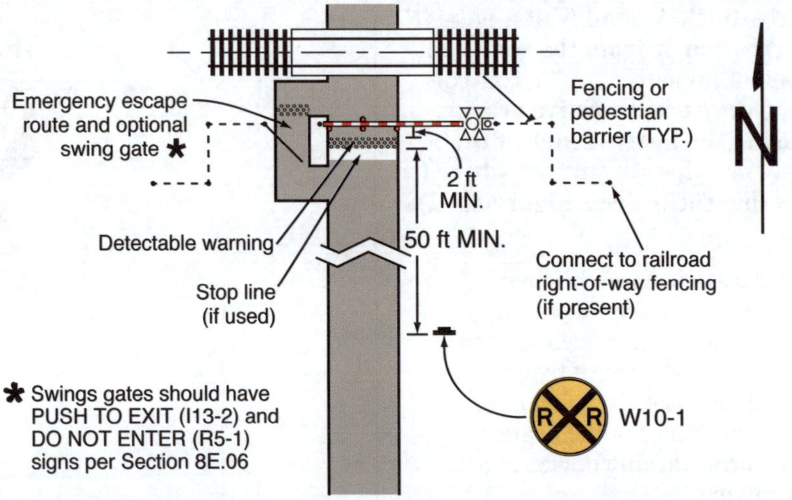

06   Latching devices that are used on swing gates need to be designed in a manner such that they are operable by all users of the pathway or sidewalk.

Guidance:

07   *A swing gate should be equipped with a PUSH TO EXIT (I13-2) sign on the track side of the swing gate, and a DO NOT ENTER (R5-1) sign on the side of the swing gate facing away from the tracks (see Figure 8E-10).*

Support:

08   The U.S. Department of Justice 2010 ADA Standards for Accessible Design, September 15, 2010, 28 CFR 35 and 36, Americans with Disabilities Act of 1990 contains information regarding the design of swing gates and related hardware.

09   Where fencing (see Figures 8E-6 and 8E-9) is installed to direct pathway or sidewalk users to the grade crossing, it is desirable that this fencing be connected to any continuous existing or new fencing or channelization that has been installed parallel to the track(s) to discourage pedestrians from circumventing the grade crossing.

10   Pedestrian barriers or fencing, sometimes referred to as a "maze fencing," direct pathway or sidewalk users to face approaching rail traffic before entering the trackway.

11   Where used, maze fencing or pedestrian barriers need to be designed to permit the passage of wheelchairs and power-assisted mobility devices, and if bicycles are permitted, to permit the passage of dismounted bicyclists with tandem bicycles, cargo bicycles, or bicycles with trailers.

Section 8E.07 Active Traffic Control Systems

Standard:

01   **Except as provided in Paragraph 5 of this Section, at pathway-LRT and sidewalk-LRT grade crossings where LRT operating speeds on a semi-exclusive alignment exceed 25 mph, active traffic control systems shall be used.**

02   **Except as provided in Paragraph 5 of this Section, at pathway-LRT and sidewalk-LRT grade crossings where LRT operating speeds on a semi-exclusive alignment exceed 40 mph, active traffic control systems, including automatic gates, shall be used.**

03   **If used at a pathway or sidewalk grade crossing, an active traffic control system (see Section 8D.01) shall include flashing-light signals (see Figure 8E-7) on each approach to the crossing.**

Guidance:

04   *If used at a pathway or sidewalk grade crossing, an active traffic control system (see Section 8D.01) should include an audible device such as a bell that is operated in conjunction with the flashing-light signals.*

Option:

05   Flashing-light signals, bells, and other audible warning devices may be omitted at pathway or sidewalk grade crossings that are located within 25 feet of an active warning device at a grade crossing that is equipped with those devices.

06   Additional pairs of flashing-light signals, bells, or other audible warning devices may be installed on the active traffic control devices at a grade crossing for pathway or sidewalk users approaching the grade crossing from the back side of those devices.

Guidance:

07   *Where railroad or LRT tracks in a semi-exclusive alignment are parallel and immediately adjacent to a roadway and if adequate space exists, a pedestrian refuge area or island should be provided between the tracks and the roadway to permit pedestrians to stand clear of the tracks while waiting to cross the roadway and to stand clear of the roadway while waiting to cross the tracks. If a pedestrian refuge area or island is provided at a signalized crossing of the roadway, additional pedestrian features (see Chapter 4I), such as signal heads, signing, and detectors, should be installed in the refuge area or on the island.*

Figure 8E-7. Example of a Flashing-Light Signal Assembly at a Pathway or Sidewalk Grade Crossing

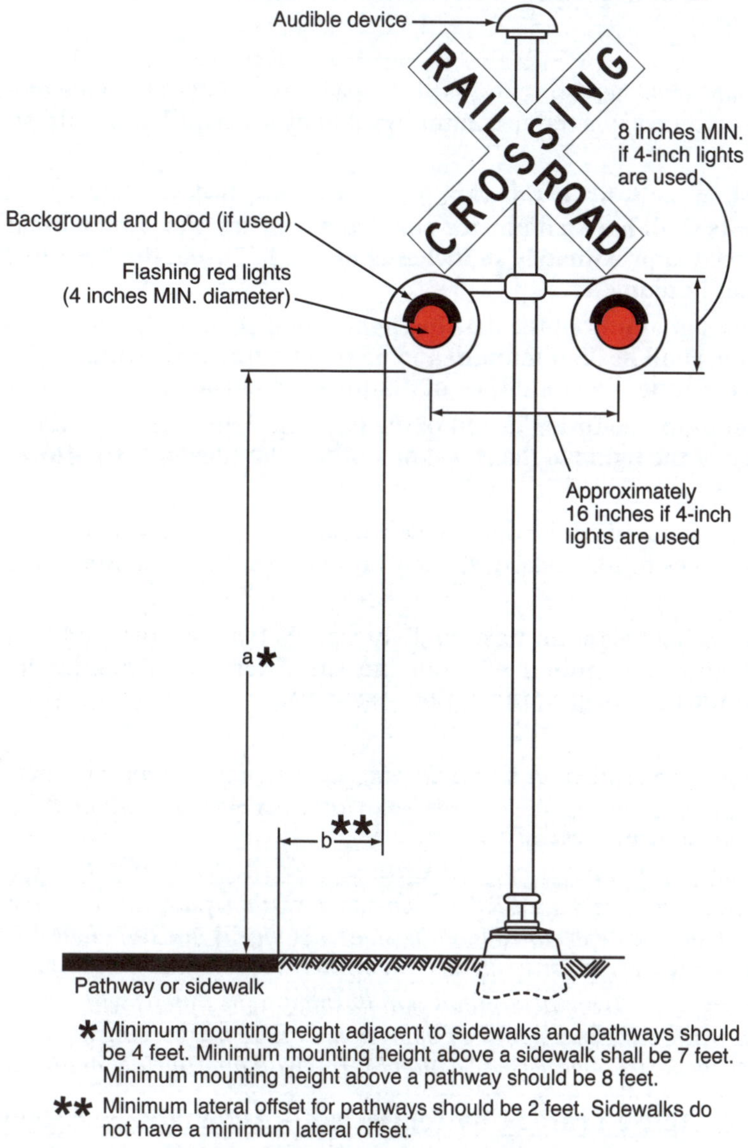

Section 8E.08 <u>Active Traffic Control Devices – Signals</u>

Support:

01   Pedestrian signal heads are typically used at highway-highway intersections where pedestrians have an expectation that other roadway users will sometimes be legally required to yield the right-of-way to them. At grade crossings where rail traffic does not stop, pedestrians will not have the right-of-way yielded to them. Therefore, pedestrian signal heads are not an appropriate traffic control device to use at a pathway or sidewalk grade crossing where rail traffic does not stop. Instead, the universal application of horizontally-aligned, alternately-flashing red lights is the uniform active traffic control device for all grade crossings where rail traffic does not stop including pathway and sidewalk grade crossings.

Standard:

02   **Except as provided in Paragraph 3 of this Section, pedestrian signal heads as described in Chapter 4I comprised of Upraised Hand and Walking Person symbols shall not be used at a pathway or sidewalk grade crossing.**

Option:

03   Pedestrian signal heads may be used at a pathway or sidewalk grade crossing where the movement of LRT vehicles is controlled by a traffic control signal or by special LRT signals (see Section 8D.15).

Standard:

04   **If used at a pathway or sidewalk grade crossing, flashing-light signals shall be aligned horizontally and the light units shall have a diameter of at least 4 inches. For 4-inch diameter light units, the light centers shall be spaced approximately 16 inches apart and, if used, the flashing light unit backgrounds shall be at least 8 inches in diameter.**

05   **Each red signal unit in the flashing-light signal shall flash alternately. The number of flashes per minute for each lamp shall be 35 minimum and 65 maximum. Each lamp shall be illuminated for approximately the same length of time. The total time of illumination of each pair of lamps shall be the entire operating time.**

06   **The minimum mounting height of the flashing-light signals shall be 4 feet, measured vertically from the bottom edge of the lights to the elevation of the near edge of the pathway or sidewalk surface.**

Option:

07   At station, pathway, or sidewalk grade crossings with multiple tracks, traffic control devices may be installed between the tracks in compliance with any railroad clearance requirements.

Standard:

08   **The mounting height for flashing-light signals that are installed between the tracks at multiple-track crossings shall be a minimum of 1 foot, measured vertically from the bottom edge of the lights to the elevation of the near edge of the pathway surface.**

Guidance:

09   *If a Diagnostic Team finds that a flashing-light signal with a Crossbuck sign and an audible device is still not resulting in appropriate pedestrian behavior, consideration should be given to also installing an automatic pedestrian gate (see Section 8E.09).*

10   *Flashing-light signals (see Figure 8E-7) with a Crossbuck (R15-1) sign and an audible device should be installed along semi-exclusive LRT alignments at station, pathway, or sidewalk grade crossings where the Diagnostic Team has determined that the sight distance is not sufficient for pathway or sidewalk users to complete their crossing prior to the arrival of LRT traffic at the crossing.*

11   *If the Diagnostic Team determines that flashing-light signals with a Crossbuck sign and an audible device would not provide sufficient notice of approaching LRT traffic, consideration should be given to also installing an automatic pedestrian gate (see Section 8E.09) with appropriate channelization or fencing.*

Section 8E.09 <u>Active Traffic Control Devices – Automatic Pedestrian Gates</u>

Option:

01   Automatic pedestrian gates (see Figures 8E-6, 8E-8, 8E-9, 8E-11,and 8E-12) may be used at pathway or sidewalk grade crossings.

Standard:

02   **A pathway or sidewalk grade crossing across tracks where trains are permitted to travel at speeds of 80 mph or higher shall be equipped with a system of automatic pedestrian gates and an escape area with swing gates and fencing installed in the vicinity of the crossing to direct users to the pathway or sidewalk grade crossing (see Figure 8E-6) unless the Diagnostic Team determines that other safety treatments for the crossing would be more appropriate.**

Guidance:

03   *Where automatic pedestrian gates are installed across a pathway or sidewalk at a grade crossing, or where a sidewalk is located between the edge of a roadway and the support for an automatic gate arm that extends across the sidewalk and into the roadway, an emergency escape route (see Figures 8E-9 and 8E-10) should be provided to allow pedestrians to egress away from the track area when the automatic pedestrian gates are activated.*

Standard:

04   **Except as provided in Paragraph 6 of this Section, automatic pedestrian gate arms shall be provided with at least one red light as shown in Figures 8E-6, 8E-8, 8E-9, 8E-11, and 8E-12. This light shall be continuously illuminated whenever the warning system is active.**

05   **If any red lights in addition to the continuously-illuminated red light that is required in Paragraph 4 of this Section are provided on the automatic pedestrian gate arm, they shall be installed in pairs and shall be flashed alternately in unison with the other flashing-light units at the crossing.**

Option:

06   The red light on an automatic pedestrian gate arm may be omitted if the pathway or sidewalk grade crossing is located within 25 feet of the traveled way at a highway-rail or highway-LRT grade crossing that is equipped with active warning devices (see Figure 8E-11).

Figure 8E-8. Example of an Automatic Pedestrian Gate at a Pathway or Sidewalk Grade Crossing

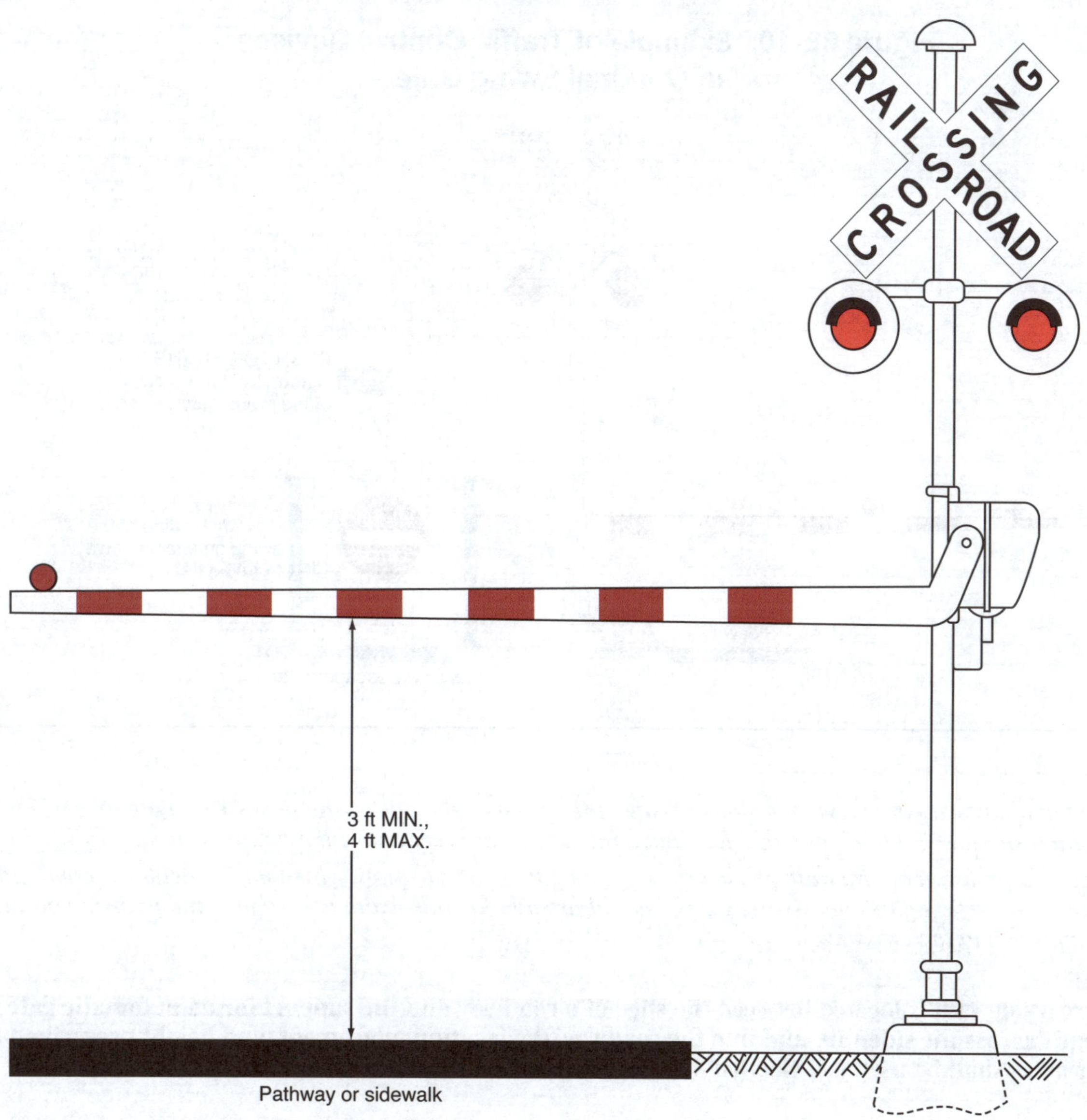

Figure 8E-9. Example of Active Traffic Control Systems with Automatic Pedestrian Gates and Swing Gates at a Sidewalk Grade Crossing

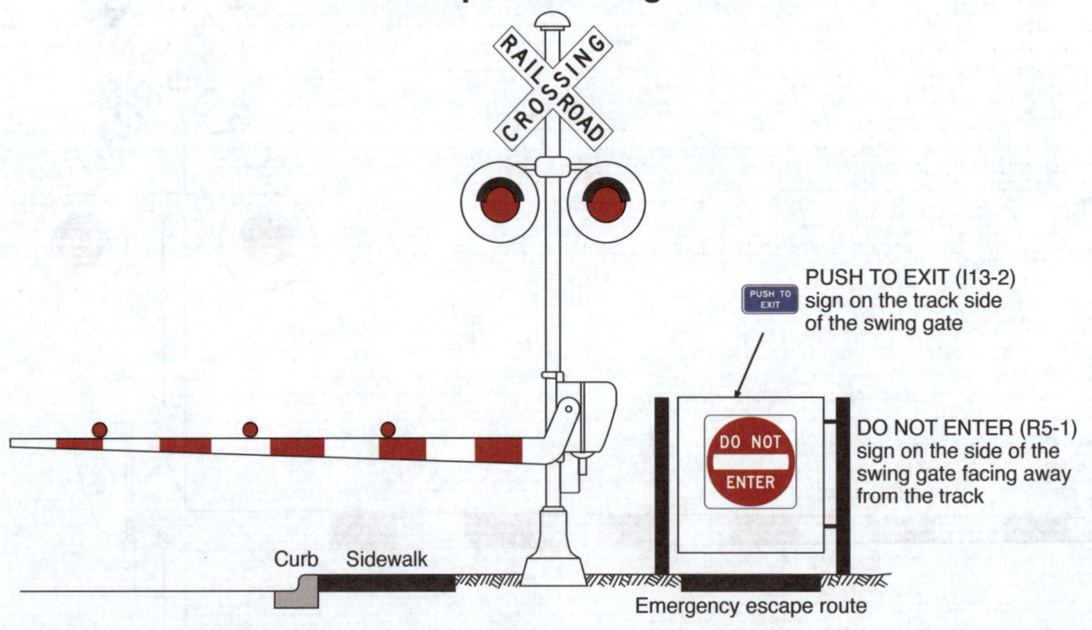

Figure 8E-10. Example of Traffic Control Devices for an Optional Swing Gate

Guidance:

07   *If used at a pathway or sidewalk grade crossing, the height of the automatic pedestrian gate arm when in the down position should be a minimum of 3 feet and a maximum of 4 feet above the pathway or sidewalk.*

08   *If used at a pathway or sidewalk grade crossing, the gate configuration, which might include a combination of automatic pedestrian gates and swing gates, should provide for full-width coverage of the pathway or sidewalk on each approach to the crossing.*

Standard:

09   **Where a sidewalk is located between the edge of a roadway and the support for an automatic gate arm that extends across the sidewalk and into the roadway, the location, placement, and height prescribed for vehicular gates shall be used (see Section 8D.03).**

Figure 8E-11. Example of a Separate Automatic Pedestrian Gate

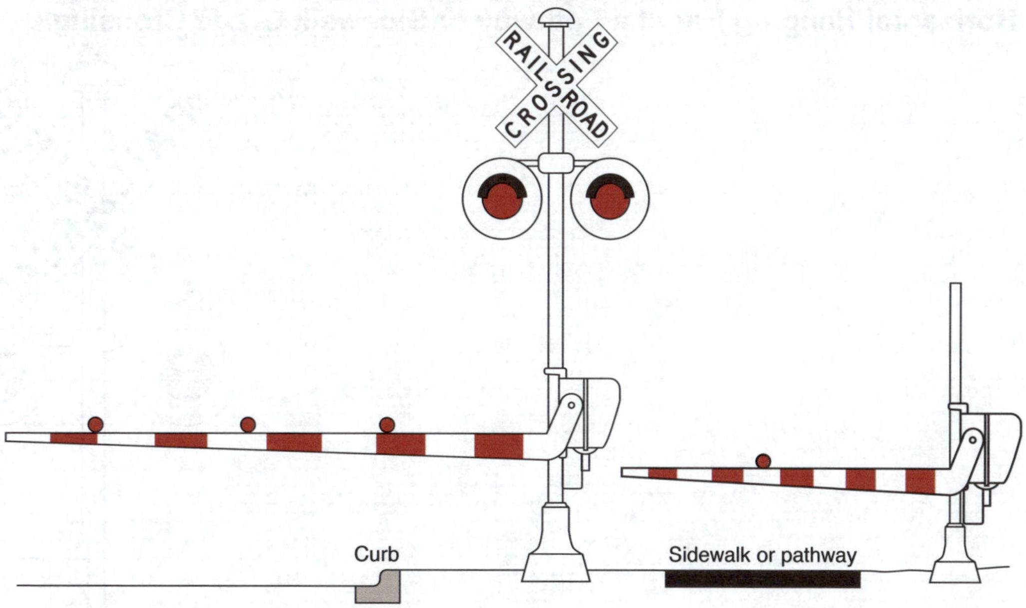

Guidance:

10 *Except as provided in Paragraph 11 of this Section, if a separate automatic pedestrian gate is used for a sidewalk at a highway-rail or highway-LRT grade crossing, instead of a supplemental or auxiliary gate arm installed as a part of the same mechanism as the vehicular gate, a separate mechanism (see Figure 8E-11) should be provided for the separate automatic pedestrian gate so that if a pedestrian manually raises the pedestrian gate arm, it will have no effect on the vehicular gate.*

Option:

11 A supplemental or auxiliary pedestrian gate arm installed as a part of the same mechanism as the vehicular gate may be used if the operating mechanism is designed to prevent the vehicular gate from being raised as a result of a pedestrian manually raising the pedestrian gate arm.

12 A horizontal hanging bar (see Figure 8E-12) may be attached to an automatic pedestrian gate at a pathway or sidewalk grade crossing to inform pedestrians with vision disabilities that the automatic pedestrian gate is in the down position and to reduce the likelihood that pedestrians will violate a lowered crossing gate.

Guidance:

13 *If a horizontal hanging bar is attached to an automatic pedestrian gate, the height of the horizontal hanging bar when in the down position should be a maximum of 26 inches above the pathway or sidewalk.*

Section 8E.10 Active Traffic Control Devices – Multiple-Track Pathway or Sidewalk Grade Crossings

Guidance:

01 *Where railroad or LRT tracks are immediately adjacent to other tracks, the traffic control devices that control pedestrian movements should be designed to avoid having pedestrians wait between sets of tracks.*

Figure 8E-12. Example of an Automatic Pedestrian Gate with a Horizontal Hanging Bar at a Pathway or Sidewalk Grade Crossing

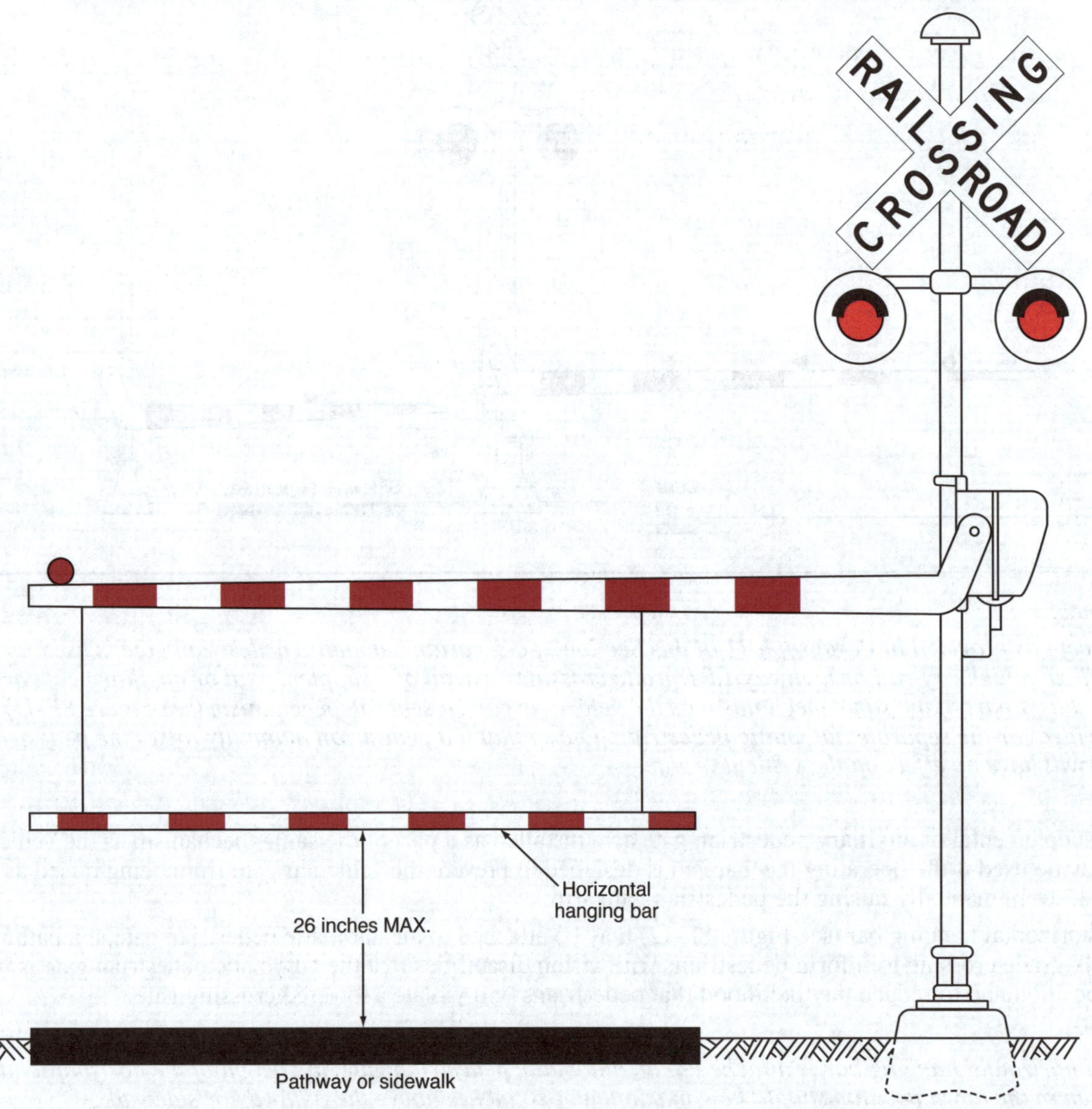

PART 9
TRAFFIC CONTROL FOR BICYCLE FACILITIES

CHAPTER 9A. GENERAL

Section 9A.01 General

Support:

01   Part 9 covers signs and pavement markings specifically related to bicycle operation on roadways, separated bikeways, and shared-use paths. In jurisdictions where small, low-speed, human or electric-powered transportation devices (often referred to as a micromobility devices) are allowed to use bicycle facilities, they can be regulated by signs, pavement markings, and other traffic control devices related to bicycle operations. Part 4 contains information on highway traffic signals and bicycle signal faces. Part 6 contains information on work zones for bicycle facilities and the mitigation of impacts to bicycle travel through work zones.

02   Definitions and acronyms pertaining to Part 9 are provided in Sections 1C.02 and 1C.03.

03   When operating on a roadway, bicycles are typically defined as vehicles and the operator of a bicycle is given the same rights and duties as an operator of a motor vehicle. Bicyclists are also vulnerable road users who have little to no protection from crash forces.

04   Designing bicycle facilities and the traffic control devices on those facilities in a manner that encourages predictable behavior and compliance with traffic laws from all roadway users can improve safety and increase public acceptance of bicyclists from other road users. The misuse of traffic control devices for improperly designed bicycle facilities or non-uniform applications can produce ineffective or counterproductive results. Section 1D.01 provides more information on the importance of uniformity of traffic control devices.

05   The "Bikeway Selection Guide" (FHWA-SA-18-077), FHWA, provides information on the designs and configuration of bicycle facilities.

Support:

06   The operation of bicycles is generally allowed on rights-of-way open to motor vehicles, even if the bicycle-specific traffic control devices outlined in Part 9 are not present.

Guidance:

07   *All signs, signals, and markings, including those on bicycle facilities, should be properly maintained to command respect from all road users. When installing signs and markings on bicycle facilities, an agency should be designated to maintain these devices.*

Section 9A.02 Standardization of Application for Signing

Support:

01   The installation of nonstandard signing on bikeways or modifying standard signing in a manner inconsistent with Chapter 2A of this Manual to draw special attention, educate users or the community, or brand a bicycle facility can contribute to problems with public acceptance and enforcement.

Standard:

02   **Bicycle signs shall comply with the provisions of this Manual for standard shape, legend, and color.**

03   **All signs installed on bikeways shall be retroreflective, including those on shared-use paths and bicycle lane facilities.**

04   **Where signs serve both bicyclists and other road users, vertical mounting height and lateral placement shall be as provided in Sections 2A.15 and 2A.16 of this manual.**

Guidance:

05   *Where used on a shared-use path, no portion of a sign or its support should be placed less than 2 feet laterally from the near edge of the path, or less than 8 feet vertically over the entire width of the shared-use path (see Figure 9A-1).*

06   *Mounting height for post-mounted signs on shared-use paths should be a minimum of 4 feet, measured vertically from the bottom of the sign to the elevation of the near edge of the path surface (see Figure 9A-1).*

07   *Signs for the exclusive use of bicyclists should be located so that other road users are not confused by them.*

08   *The clearance for overhead signs on shared-use paths should be adjusted when appropriate to accommodate path users requiring more clearance, such as equestrians, or typical maintenance or emergency vehicles.*

Figure 9A-1. Sign Placement on Shared-Use Paths

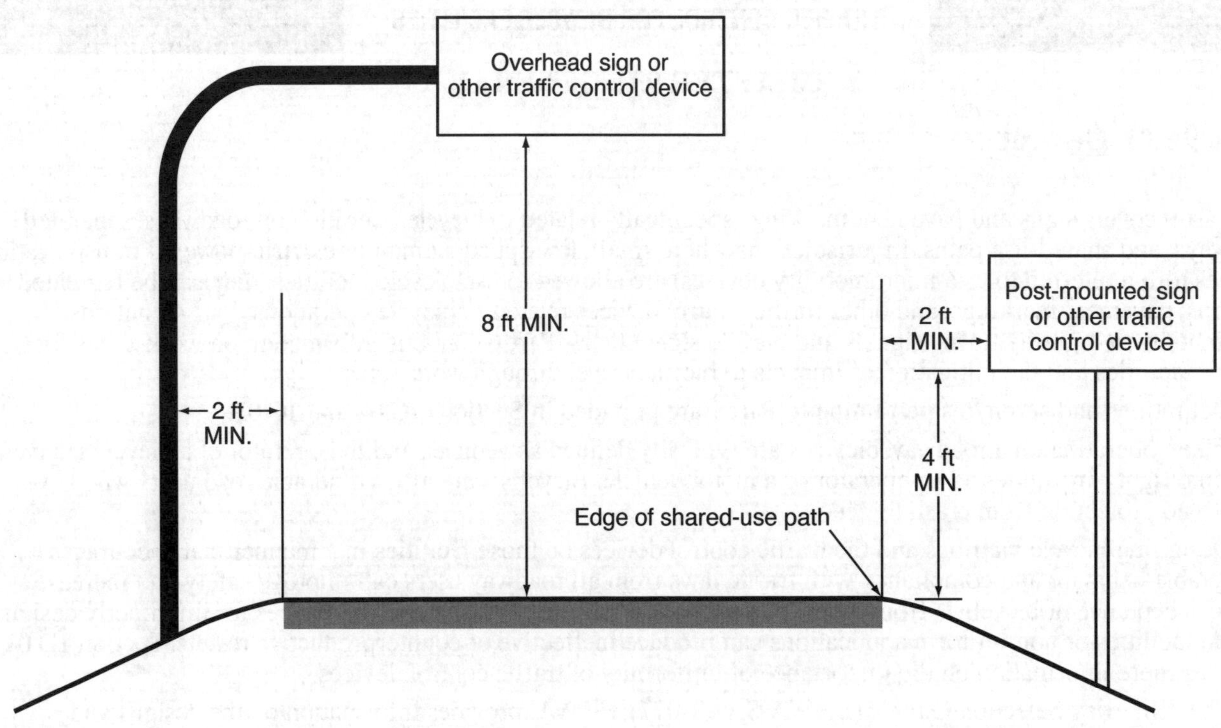

Standard:

09   **If the sign or plaque applies to motorists and bicyclists, then the size shall be as shown for conventional roads in Tables 2B-1, 2C-1, 2D-1, and 8B-1, as applicable.**

10   **The minimum sign and plaque sizes for signs specific to bicycle-only facilities and shared-use paths shall be those shown in Table 9A-1. These sizes shall be used only for signs and plaques installed specifically for bicyclist applications.**

Option:

11   Larger sizes of signs and plaques may be used on bicycle facilities when appropriate (see Section 2A.07).

12   Any diamond-shaped warning sign that is placed such that it is applicable only to bicyclists or pedestrians on shared-use paths or separated bicycle lanes may be 18" x 18".

Guidance:

13   *Except for size, the design of signs and plaques for bicycle facilities should be identical to that provided in this Manual for signs and plaques for streets and highways.*

Support:

14   Uniformity in design of bicycle signs and plaques includes shape, color, symbols, arrows, wording, lettering, and illumination or retroreflectivity.

Section 9A.03 Standardization of Application for Markings

Support:

01   Markings indicate the separation of the lanes for road users, assist the bicyclist by indicating assigned travel paths, indicate correct position for traffic control signal actuation, and provide advance information for turning and crossing maneuvers.

Guidance:

02   *Pavement marking word messages, symbols, and/or arrows should be used in bikeways where appropriate.*

03   *Consideration should be given to selecting pavement marking materials that will minimize loss of traction for bicycles under wet conditions.*

Table 9A-1. Bicycle Facility Sign and Plaque Minimum Sizes (Sheet 1 of 3)

Sign or Plaque	Sign Designation	Section	Off Roadway[1,5]	Roadway[2,5]
Stop	R1-1	9B.01	18 x 18	—
Yield	R1-2	9B.01	18 x 18 x 18	—
Bike Lane (plaque)	R3-5hP	9B.04	—	30 x 12
Except Bicycles (plaque)	R3-7bP	9B.02	—	24 x 12
Advance Intersection Lane Control with Bike Lane	R3-8x Series	9B.03	—	Varies x 36
Bike Lane	R3-17	9B.04	—	24 x 18
Ahead, Ends (plaques)	R3-17aP, R3-17bP	9B.04	—	24 x 9
Movement Restriction	R4-1,2,3,7,16	9B.24	12 x 18	—
Begin Right Turn Lane Yield to Bikes	R4-4	9B.05	—	36 x 30
Bicycle Passing Clearance	R4-19	9B.15	—	30 x 30
Bicycle Wrong Way	R5-1b	9B.06	12 x 18	12 x 18
No Motor Vehicles	R5-3	9B.07	24 x 24	24 x 24
No Bicycles	R5-6	9B.08	18 x 18	—
On Freeway (plaque)	R5-10dP	9B.17	—	24 x 6
No Parking Bike Lane	R7-9,9a	9B.09	—	12 x 18
Back-In Parking	R7-10	9B.10	—	12 x 18
No Pedestrian Crossing	R9-3	9B.08	18 x 18	—
Ride With Traffic (plaque)	R9-3cP	9B.06	12 x 12	12 x 12
Bicycles Use Pedestrian Signal	R9-5	9B.11	12 x 18	12 x 18
Bicycles Yield to Pedestrians	R9-6	9B.12	12 x 18	12 x 18
Shared-Use Path Restriction	R9-7	9B.13	12 x 18*	—
No Skaters	R9-13	9B.08	18 x 18	18 x 18
No Equestrians	R9-14	9B.08	18 x 18	18 x 18
No Snowmobiles	R9-15	9B.08	18 x 18	18 x 18
No All-Terrain Vehicles	R9-16	9B.08	18 x 18	18 x 18
Bicycles Allowed Use of Full Lane	R9-20	9B.14	—	30 x 30
Bicycles Use Shoulder Only	R9-21	9B.16	—	24 x 30
Bicycles Must Exit	R9-22	9B.17	—	24 x 30
Bicycle All Turns from Bike Lane	R9-23	9B.18	—	12 x 18
Bicycle Left Turn from Bike Lane	R9-23a	9B.18	—	12 x 18
Bicycle Left Turn Must Use Turn Box	R9-23b, R9-23c	9B.18	—	12 x 18
Bicycle All Turns	R9-24,24a	9B.19	—	24 x 6
Bicycle U and Left Turns	R9-25,25a,25b	9B.19	—	24 x 9
Bicycle U Turn	R9-26,26a,26b	9B.19	—	24 x 6
Bicycle Left Turn	R9-27,27a,27b	9B.19	—	24 x 6
Push Button for Green	R10-4	9B.20	9 x 12	9 x 12
Left Turn Yield to Bicycle	R10-12b	9B.21	—	30 x 36
Bicycle Detector	R10-22	9B.20	12 x 18	12 x 18
Bike Push Button for Green Light	R10-24	9B.20	9 x 15	9 x 15
Push Button for Warning Lights - Wait for Gap	R10-25	9B.20	9 x 12	9 x 12
Bike Push Button for Green Light (arrow)	R10-26	9B.20	9 x 15	9 x 15
Bicycle Signal	R10-40, R10-40a, R10-41, R10-41a, R10-41b, R10-41c	9B.22	12 x 21	12 x 21
Grade Crossing (Crossbuck)	R15-1	9B.23	24 x 4.5	48 x 9
Number of Tracks (plaque)	R15-2P	9B.23	13.5 x 9	27 x 18
Look	R15-8	9B.23	18 x 9	36 x 18
Horizontal Alignment	W1-1,2,3,4,5	9C.01	18 x 18	—
Large Arrow	W1-6,7	9C.01	24 x 12	—
Intersection Warning	W2-1,2,3,4,5	9C.02	18 x 18	—

Table 9A-1. Bicycle Facility Sign and Plaque Minimum Sizes (Sheet 2 of 3)

Sign or Plaque	Sign Designation	Section	Off Roadway[1,5]	Roadway[2,5]
Stop Ahead, Yield Ahead, Signal Ahead	W3-1,2,3	9C.08	18 x 18	—
Narrow Bridge	W5-2	9C.08	18 x 18	—
Path Narrows	W5-4a	9C.08	18 x 18	—
Hill	W7-5	9C.08	18 x 18	30 x 30
Bump, Dip	W8-1,2	9C.03	18 x 18	—
Pavement Ends	W8-3	9C.03	18 x 18	—
Bicycle Surface Condition	W8-10	9C.03	18 x 18	30 x 30
Slippery When Wet (plaque)	W8-10P	9C.03	12 x 9	12 x 9
Bicycle Lane Ends	W9-5	9C.07	—	30 x 30
Bicycles Merging	W9-5a	9C.07	18 x 18	30 x 30
Grade Crossing Advance Warning	W10-1	9C.08	24 Dia.	36 Dia.
No Train Horn (plaque)	W10-9P	9C.08	18 x 12	30 x 24
Skewed Crossing	W10-12	9C.08	18 x 18	36 x 36
Bicycle	W11-1	9C.04, 9C.08	18 x 18	—
Pedestrian	W11-2	9C.08	18 x 18	—
Trail Crossing	W11-15	9C.04	18 x 18	—
Trail Crossing (plaque)	W11-15P	9C.04	18 x 12	—
Low Clearance	W12-2	9C.08	18 x 18	—
Playground	W15-1	9C.08	18 x 18	—
In Road (plaque)	W16-1P	9C.08	—	18 x 12
In Street (plaque)	W16-1aP	9C.08	—	18 x 12
XX Feet (2-line plaque)	W16-2P	9C.04	18 x 12	—
XX Ft (1-line plaque)	W16-2aP	9C.04	12 x 9	—
Downward Diagonal Arrow (plaque)	W16-7P	9C.04	12 x 9	—
Ahead (plaque)	W16-9P	9C.04	—	24 x 12
Except Bicycles (plaque)	W16-20P	9C.05	—	24 x 12
2-Way Bicycle Cross Traffic (plaque)	W16-21P	9C.06	—	24 x 12
Object Marker Type 3	OM3-L,C,R	9C.09	6 x 18	12 x 36
Destination (1 line)	D1-1, D1-1a	9D.01	Varies x 6	—
Bicycle Destination (1 line)	D1-1b, D1-1c	9D.01	Varies x 6	Varies x 6
Destination (2 lines)	D1-2, D1-2a	9D.01	Varies x 12	—
Bicycle Destination (2 lines)	D1-2b, D1-2c	9D.01	Varies x 12	Varies x 12
Destination (3 lines)	D1-3, D1-3a	9D.01	Varies x 18	—
Bicycle Destination (3 lines)	D1-3b, D1-3c	9D.01	Varies x 18	Varies x 18
Distance (1 line)	D2-1	9D.01	Varies x 6	—
Bike Route Destination and Distance (1 Line)	D2-1a	9D.01	Varies x 12	Varies x 12
Distance (2 line)	D2-2	9D.01	Varies x 9	—
Bike Route Destination and Distance (2 Lines)	D2-2a	9D.01	Varies x 15	Varies x 15
Distance (3 line)	D2-3	9D.01	Varies x 12	—
Bike Route Destination and Distance (3 Lines)	D2-3a	9D.01	Varies x 18	Varies x 18
Street Name (1 line)	D3-1	9D.01	Varies x 6	Varies x 6
Street Name (2 lines)	D3-1	9D.01	Varies x 12	Varies x 12
Bicycle Parking Area Directional	D4-3	9D.09	18 x 12	18 x 12
Bicycle-Sharing Station Directional	D4-4	9D.09	12 x 18	12 x 18
Bicycle Lockers Directional	D4-4a	9D.09	12 x 18	12 x 18
Reference Location (1-digit)	D10-1	9D.10	6 x 9	—
Intermediate Reference Location (2-digits)	D10-1a	9D.10	6 x 15	—
Reference Location (2-digits)	D10-2	9D.10	6 x 15	—
Intermediate Reference Location (3-digits)	D10-2a	9D.10	6 x 21	—

Table 9A-1. Bicycle Facility Sign and Plaque Minimum Sizes (Sheet 3 of 3)

Sign or Plaque	Sign Designation	Section	Off Roadway[1,5]	Roadway[2,5]
Reference Location (3-digits)	D10-3	9D.10	6 x 21	—
Intermediate Reference Location (4-digits)	D10-3a	9D.10	6 x 24	—
Bike Route	D11-1	9D.02	24 x 18	24 x 18
Bike Route (plaque)	D11-1bP	9D.03	18 x 6	18 x 6
Bike Route with Destination	D11-1c	9D.02	24 x 18	24 x 18
Shared-Use Path Destination (1 line)	D11-10a	9D.12	Varies x 6*	—
Shared-Use Path Destination (2 lines)	D11-10b	9D.12	Varies x 12*	—
Shared-Use Path Destination (3 lines)	D11-10c	9D.12	Varies x 18*	—
Shared-Use Path Destination and Distance (1 line)	D11-10d	9D.12	Varies x 6*	—
Shared-Use Path Destination and Distance (2 lines)	D11-10e	9D.12	Varies x 12*	—
Shared-Use Path Destination and Distance (3 lines)	D11-10f	9D.12	Varies x 18*	—
Bicycles Directional	D11-11	9D.11	18 x 18	—
Pedestrians Directional	D11-12	9D.11	18 x 18	—
Skaters Directional	D11-13	9D.11	18 x 18	—
Equestrians Directional	D11-14	9D.11	18 x 18	—
Bicycle Turn Box Guide Signs	D11-20, D11-20a	9D.13	—	12 x 18
State or Local Bicycle Route	M1-8, M1-8a	9D.05	12 x 18	18 x 24
Non-Numbered Bicycle Route	M1-8b, M1-8c	9D.06	12 x 12	18 x 18
U.S. Bicycle Route	M1-9	9D.07	12 x 18	18 x 24
Bicycle Route Auxiliary Signs (plaque)	M2-1P, M3-1P,2P,3P,4P, M4-1P,1aP,2P,3P,5P,6P,7P, 7aP,8P,14P	9D.08	12 x 6	12 x 6
Bicycle Route Arrow Signs (plaque)	M5-1P, 2P, M6-1P, 2P, 3P, 4P, 5P, 6P, 7P	9D.08	12 x 9	12 x 9

* For use on shared-use paths only.

Notes: 1. Includes shared-use paths and bicycle-only facilities outside of the roadway.

2. If the sign or plaque applies to motorists and bicyclists, then the size shall be as shown for conventional roads in Tables 2B-1, 2C-1, 2D-1, or 8B-1.

3. Larger signs may be used when appropriate.

4. Dimensions are shown in inches and are shown as width x height.

5. Separated bicycle lanes (see definition in Section 1C.02) can be located within the roadway or outside the roadway, and the minimum sign sizes for these facilities are shown in the off roadway and roadway columns respectively.

Standard:

04   **Pavement markings on bicycle facilities that must be visible at night or in low-light conditions shall be retroreflective unless the markings are adequately visible under provided lighting.**

05   **The colors, width of lines, patterns of lines, symbols, and arrows used for marking bicycle facilities shall be as defined in Part 3.**

Support:

06   Section 3H.06 contains information on green-colored pavement for use with certain traffic control devices for bicycles and bicycle facilities.

07   Section 9E.17 contains information on the use of channelizing devices to emphasize the pavement markings for bicycle facilities.

Guidance:

08   *Raised pavement markers should not be used on bicycle lanes or shared-use paths.*

09   *If used around bicycle facilities, raised pavement markers should not be placed immediately adjacent to the travel path of bicyclists in a bicycle lane or on a shared-use path.*

Support:

10   Using raised pavement markers creates a collision potential for bicyclists by placing fixed objects immediately adjacent to the travel path of the bicyclist. Raised pavement markers can cause a bicyclist to lose balance and fall, and might not be visible to a bicyclist who is following another bicyclist.

CHAPTER 9B. REGULATORY SIGNS

Section 9B.01 STOP and YIELD Signs (R1-1 and R1-2)

Standard:

01   STOP (R1-1) signs (see Figure 9B-1) shall be installed on bicycle facilities at points where bicyclists are required to stop.

02   YIELD (R1-2) signs (see Figure 9B-1) shall be installed on bicycle facilities at points where bicyclists have an adequate view of conflicting traffic as they approach the sign, and where bicyclists are required to yield the right-of-way to that conflicting traffic.

03   A STOP sign or a YIELD sign shall not be installed in conjunction with a bicycle signal face (see Chapter 4H).

Figure 9B-1. Regulatory Signs and Plaques for Bicycle Facilities (Sheet 1 of 2)

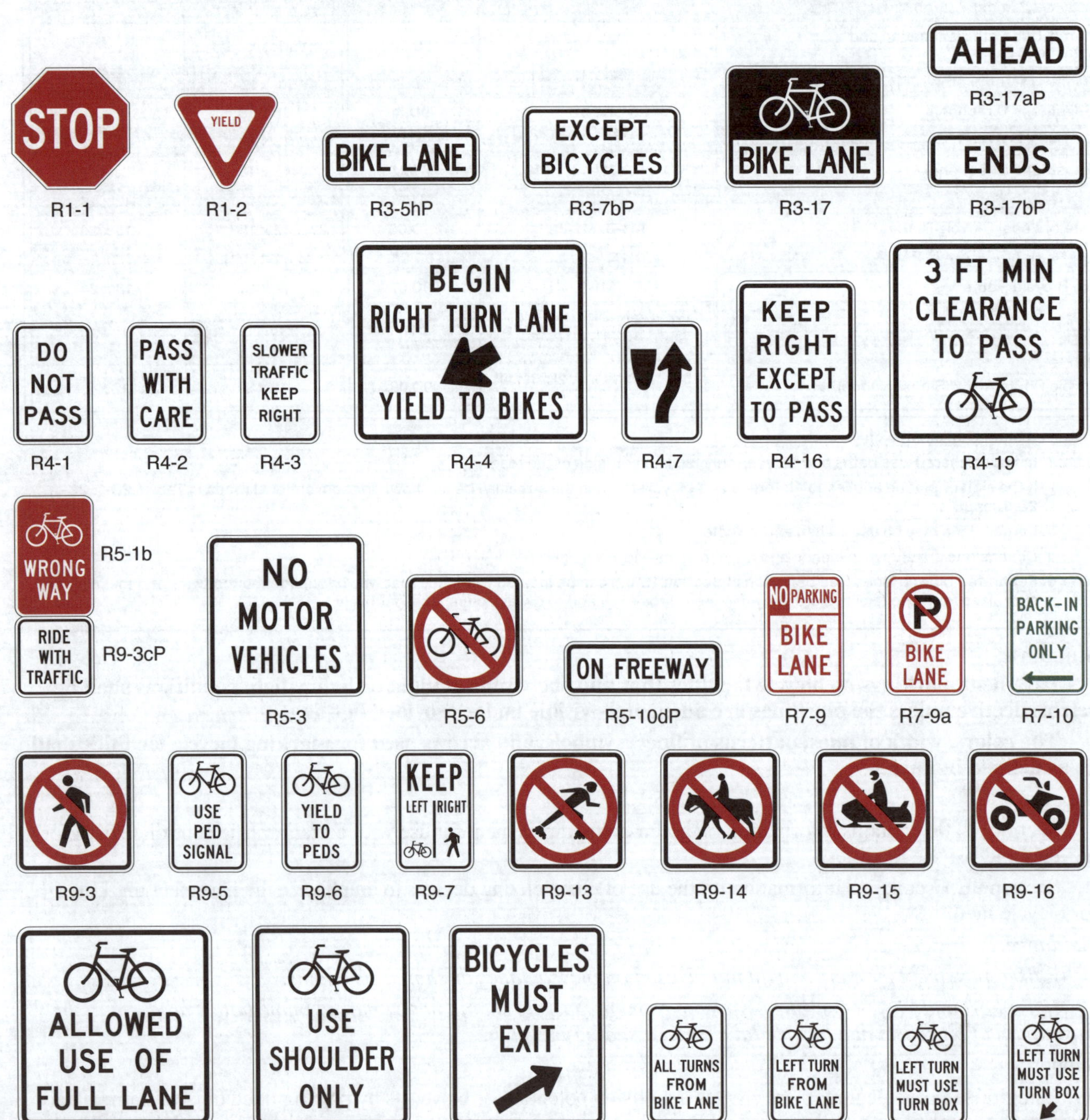

Figure 9B-1. Regulatory Signs and Plaques for Bicycle Facilities (Sheet 2 of 2)

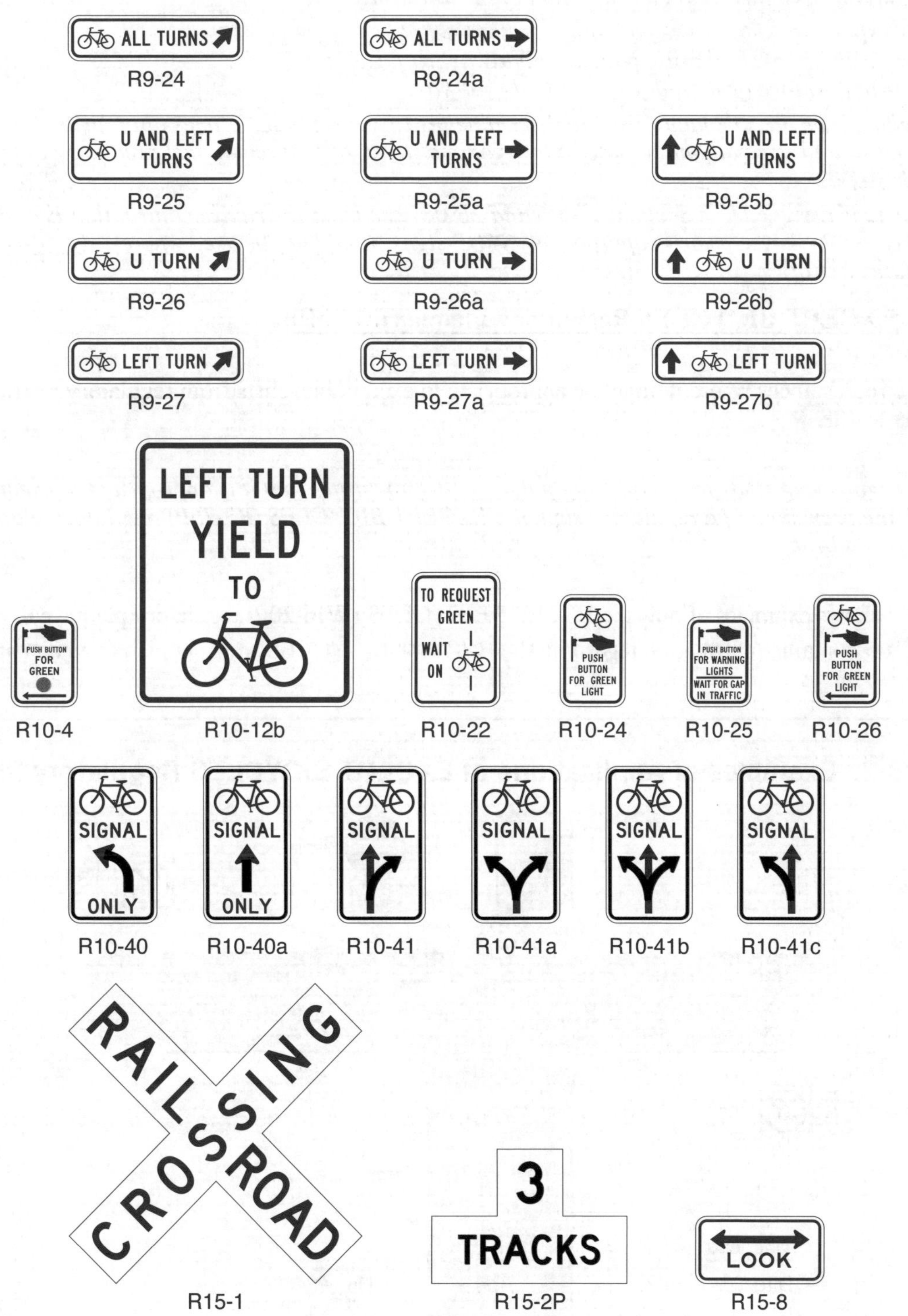

Option:

04   Larger signs may be used on shared-use paths and separated bikeways for added emphasis.

Guidance:

05   *Where conditions require shared-use path users or bicyclists on separated bikeways, but not roadway users, to stop or yield, the STOP or YIELD sign should be placed or shielded so that it is not readily visible to roadway users.*

06   *When the placement of STOP or YIELD signs is being considered, the priority at a shared-use path/roadway intersection should be assigned with consideration of the following:*

   A.   *Relative speeds of shared-use path and roadway users,*
   B.   *Relative volumes of shared-use path and roadway traffic, and*
   C.   *Relative importance of shared-use path and roadway.*

07   *Speed should not be the sole factor used to determine priority, as it is sometimes appropriate to give priority to a high-volume shared-use path that crosses a low-volume street, or to a regional shared-use path that crosses a minor collector street.*

08   *When priority is assigned (see Sections 2B.06 and 2B.08), the least-restrictive control that is appropriate should be placed on the lower-priority approaches. STOP signs should not be used where YIELD signs would provide adequate control.*

Section 9B.02 EXCEPT BICYCLES Regulatory Plaque (R3-7bP)

Support:

01   There are circumstances where it might be appropriate to exempt bicyclists from regulatory restrictions applied to other traffic.

Guidance:

02   *Where an engineering study or engineering judgment demonstrates that it is appropriate to exempt bicyclists from the provisions of a regulatory sign, the EXCEPT BICYCLES (R3-7bP) regulatory plaque (see Figure 9B-1) should be used.*

Support:

03   Figure 9B-2 shows examples of how the EXCEPT BICYCLES (W16-20P) regulatory plaque can be applied.

04   Section 9C.05 contains information regarding the EXCEPT BICYCLES warning plaque when applicable to a warning sign.

Figure 9B-2. Examples of Applications of EXCEPT BICYCLES Regulatory Plaques

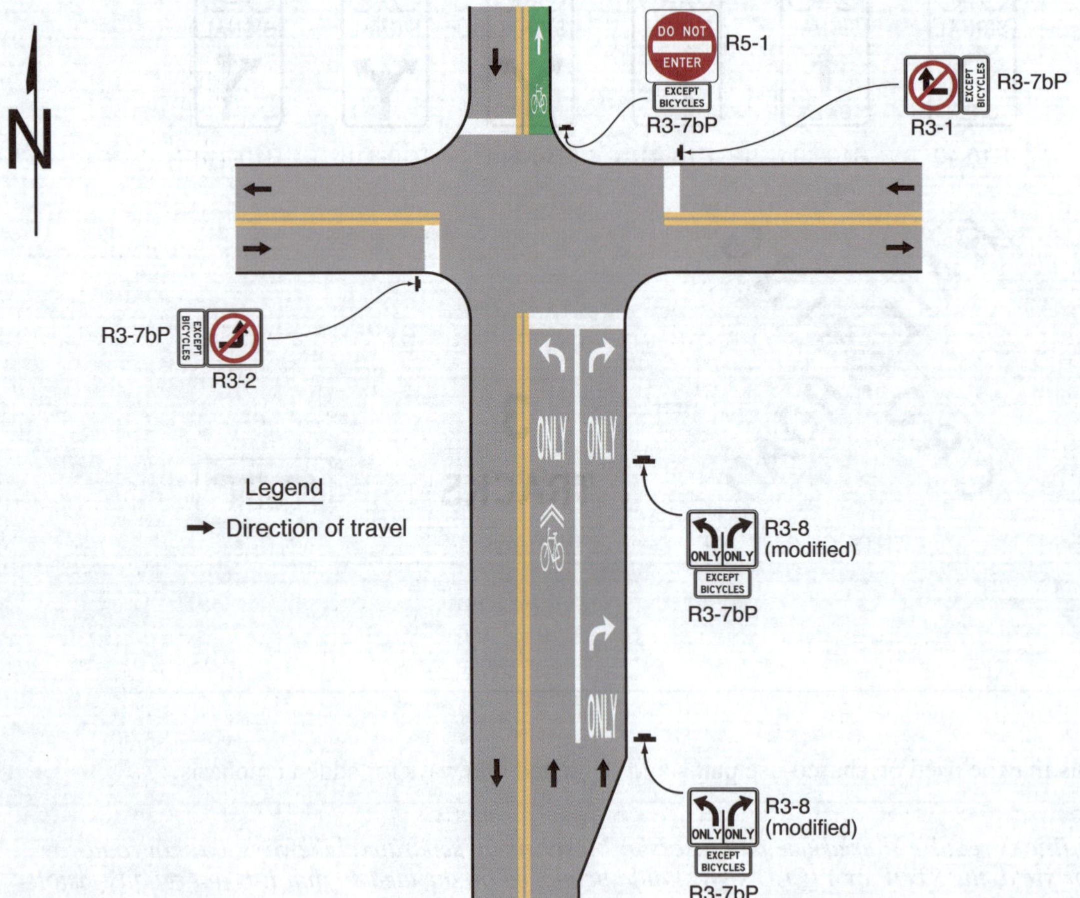

Standard:

05   **The EXCEPT BICYCLES regulatory plaque shall not be used to exempt bicyclists from the legal requirement of a STOP or YIELD sign, Yield Here to Pedestrians Signs, Stop Here for Pedestrians Signs, or a traffic signal indication.**

06   **Where a regulatory sign, such as the No Left Turn (R3-2) sign (see Section 2B.26), is installed on the same post or mounting as a STOP sign or YIELD sign, the EXCEPT BICYCLES regulatory plaque shall not be installed in conjunction with the regulatory sign on that post or mounting that includes the STOP sign or YIELD sign.**

07   **The EXCEPT BICYCLES regulatory plaque shall be placed below the regulatory sign that it supplements.**

Section 9B.03   Advance Intersection Lane Control Signs (R3-8 Series) for Bicycle Lanes

Option:

01   Advance Intersection Lane Control (R3-8 series) signs (see Section 2B.30) may display the arrangement of a conventional or buffer-separated bicycle lane in relation to other lanes in the same direction that are present on a roadway approach to an intersection.

Support:

02   The number and combination of permissible movements by both the motor vehicle and the bicycle on the same approach to an intersection might be practically limited by the amount of information that can be legibly displayed on signs or in signing sequences and still be readily comprehended by road users. The excessive display of all movements by more than one mode can result in unwieldy signs that are difficult to locate and install.

Guidance:

03   *On an approach to an intersection with complex geometry that can include multiple through lanes and multiple turn lanes and also includes a bicycle lane, consideration should be given to displaying all allowable movements on separate signs, such as using Mandatory Movement Lane Control (R3-5) signs (see Section 2B.28) for the through lanes and Mandatory Movement Lane Control (R3-7) signs (see Section 2B.28) for the turn lanes, and guide signs for bicycle routes (see Section 9D.02 through 9D.07) and Bicycle Route Sign auxiliary plaques (see Section 9D.08) for the bicycle movement.*

Standard:

04   **The portion of the sign face for the bicycle lane shall be limited to the relationship of the bicycle lane to the other lanes on the roadway approach to the intersection. The portion of the sign face for the bicycle lane shall not be modified to display specific, supplementary information about the bicycle lane such as bicycle lane extensions, contiguous buffer spaces, or other ancillary bicycle operations such as two-stage turn boxes or bicycle boxes.**

05   **Counter-flow bicycle lanes shall not be displayed on Advance Intersection Lane Control signs.**

06   **The shared-lane marking symbol shall not be displayed on Advance Intersection Lane Control signs.**

07   **Shared-use paths shall not be displayed on Advance Intersection Lane Control signs.**

08   **Advance Intersection Lane Control signs that display the bicycle lane shall use a contrasting white legend on a black background for the bicycle lane (see Figure 2B-4). The portion of the display for the bicycle lane shall not use the color green on the sign face in an attempt to be consistent with the green-colored pavement that might be present on the intersection approach.**

Section 9B.04   Bicycle Lane Signs and Plaques (R3-17, R3-5hP, R3-17aP, and R3-17bP)

Standard:

01   **The Bike Lane (R3-17) sign and the BIKE LANE (R3-5hP), AHEAD (R3-17aP), and ENDS (R3-17bP) plaques (see Figure 9B-1) shall be used only in conjunction with marked bicycle lanes as described in Sections 9E.01, 9E.06, and 9E.07.**

Guidance:

02   *If used, Bicycle Lane signs and plaques should be located at the beginning of the bicycle lane and in advance of the downstream end of the bicycle lane.*

Option:

03   Additional Bicycle Lane signs and plaques may be used at periodic intervals along the bicycle lane as determined by engineering judgment based on the operating speed of bicycle and other traffic, block length, distances from adjacent intersections, and other considerations.

Support:

04   Section 2B.33 contains information for the application of BEGIN and END plaques.

05   Section 9B.03 contains information on displaying the bicycle lane on Advance Intersection Lane Control signs.

Option:

06   Where two or more movements from a bicycle lane are allowed, or where the emphasis of allowed bicycle movements is needed, an Optional Movement Lane Control sign (see Section 2B.29) may be supplemented with a BIKE LANE (R3-5hP) plaque above the Optional Movement Lane Control sign.

07   Where bicycle lanes are located between travel lanes on intersection approaches or where only a single bicycle movement is allowed from a certain bicycle lane, a Mandatory Movement Lane Control sign (see Section 2B.28) may be supplemented with a BIKE LANE plaque to require a bicyclist in a particular bicycle lane at an intersection to stay in the same lane and proceed straight through the intersection, or to indicate a required turn from a particular bicycle lane.

Section 9B.05 BEGIN RIGHT TURN LANE YIELD TO BIKES Sign (R4-4)

Option:

01   Where motor vehicles entering a mandatory right-turn lane must weave across bicyclists in bicycle lanes, the BEGIN RIGHT TURN LANE YIELD TO BIKES (R4-4) sign (see Figure 9B-1) may be used to inform both the motorist and the bicyclist of this weaving maneuver (see Figures 9E-3 and 9E-4).

Guidance:

02   *The R4-4 sign should not be used when bicyclists need to move left because of a right-turn lane drop situation.*

Section 9B.06 Bicycle Wrong Way Sign and RIDE WITH TRAFFIC Plaque (R5-1b and R9-3cP)

Option:

01   The Bicycle WRONG WAY (R5-1b) sign and RIDE WITH TRAFFIC (R9-3cP) plaque (see Figure 9B-1) may be placed facing wrong-way bicyclists, such as on the left-hand side of a roadway.

02   This sign and plaque may be mounted back-to-back with other signs to minimize visibility to other traffic.

Guidance:

03   *The RIDE WITH TRAFFIC plaque should be used only in conjunction with the Bicycle WRONG WAY sign, and should be mounted directly below the Bicycle WRONG WAY sign.*

Section 9B.07 NO MOTOR VEHICLES Sign (R5-3)

Option:

01   The NO MOTOR VEHICLES (R5-3) sign (see Figure 9B-1) may be installed at the entrance to a shared-use path.

Section 9B.08 Selective Exclusion Signs

Option:

01   Selective Exclusion signs (see Figure 9B-1) may be installed at the entrance to a roadway or facility to notify road or facility users that designated types of traffic are excluded from using the roadway or facility.

Support:

02   Typical exclusion messages include:
   A. No Bicycles (R5-6);
   B. No Pedestrians (R9-3);
   C. No Skaters (R9-13);
   D. No Equestrians (R9-14);
   E. No Snowmobiles (R9-15); and
   F. No All-Terrain Vehicles (R9-16).

Option:

03   Where bicyclists, pedestrians, and motor-driven cycles are all prohibited, the R5-10a word message sign (see Section 2B.45) may be used.

Section 9B.09 No Parking Bike Lane Signs (R7-9 and R7-9a)

Standard:

01   **If the installation of signs is necessary to restrict parking, standing, or stopping in a bicycle lane, appropriate signs as described in Sections 2B.53 through 2B.55, or the No Parking Bike Lane (R7-9 or R7-9a) signs (see Figure 9B-1) shall be installed.**

Section 9B.10 Back-In Parking Sign (R7-10)

Option:

01   The Back-In Parking (R7-10) sign (see Section 2B.52 and Figure 9B-1) may be used where back-in parking is required by motor vehicles in the presence of a bicycle lane or movement.

Support:

02   Angled back-in curb parking is commonly applied on streets where a bicycle lane is present so that the scanning behavior of a motorist associated with the back-in angle parking task, both entering and exiting the parking space, would place a bicyclist in a bicycle lane in a more direct view of the motor vehicle operator.

03   Figure 9B-3 shows an example of the use of back-in parking signs in conjunction with bicycle lanes.

Section 9B.11 Bicycles Use Ped Signal Sign (R9-5)

Option:

01   The Bicycles Use Ped Signal (R9-5) sign (see Figure 9B-1) may be used where the crossing of a street by bicyclists is controlled by pedestrian signal indications.

02   In order to remind drivers who are making turns to yield to or stop for pedestrians or bicyclists, a Turning Vehicles Yield to Pedestrians (R10-15) sign, Turning Vehicles Stop for Pedestrians (R10-15a) sign (see Section 2B.59), or Left Turn Yield to Bicycles (R10-12b) sign (see Section 9B.21) may be used.

Guidance:

03   *If used, the R9-5 sign should be installed in the vicinity of where bicyclists will be crossing the street.*

Figure 9B-3. Examples of Bicycle Facilities Adjacent to Back-In Parking

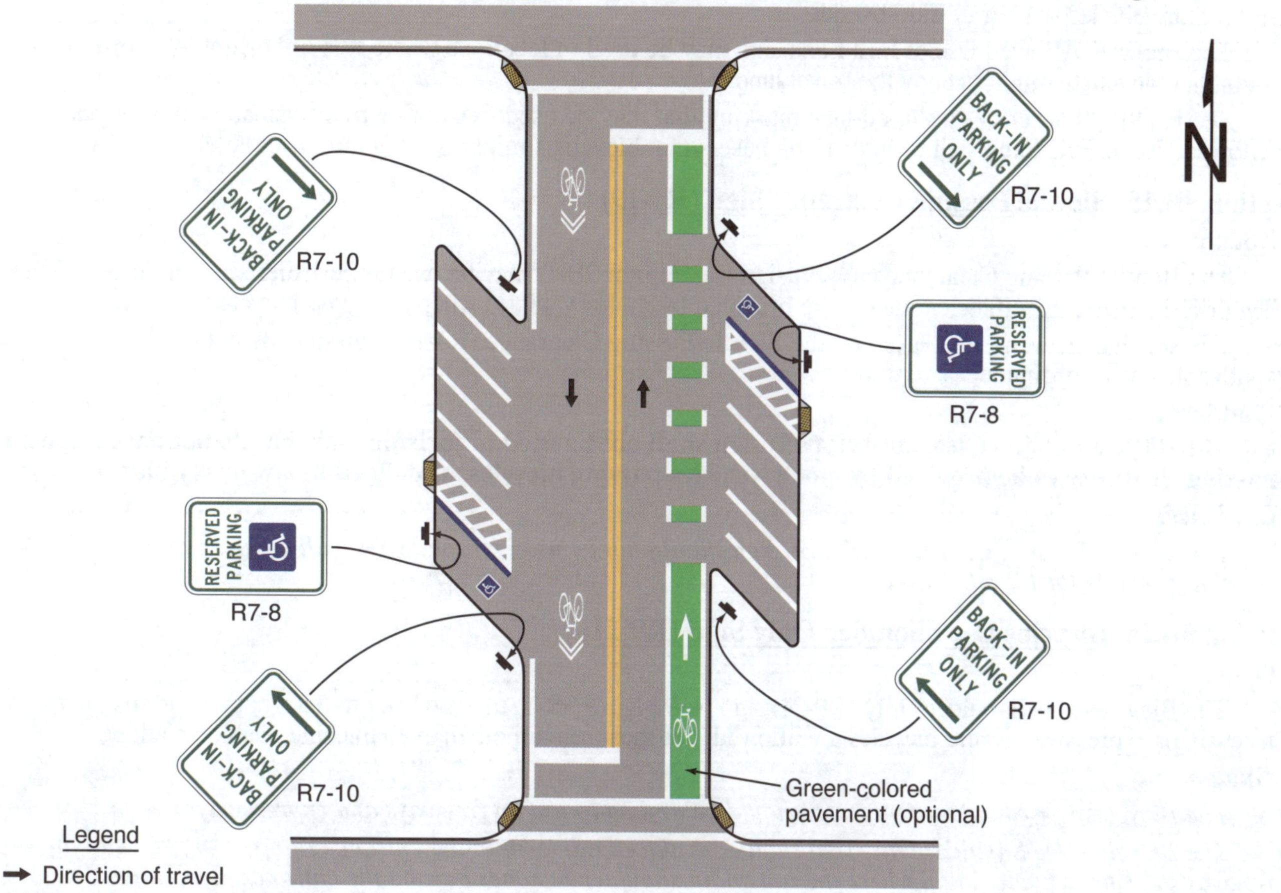

Section 9B.12 Bicycles Yield to Peds Sign (R9-6)

Option:

01   The Bicycles Yield to Peds (R9-6) sign (see Figure 9B-1) may be used at locations where a bicyclist is required to cross or share a facility used by pedestrians and is required to yield to the pedestrians.

Standard:

02   **Where the Bicycles Yield to Peds sign is supported by a yield line pavement marking (see Section 3B.19) to establish the yielding point, the sign and the pavement marking shall be installed adjacent to each other.**

03   **The Bicycles Yield to Peds sign shall not be used in bicycle corridors to establish a programmatic regulation where no yielding point exists.**

04   **The Bicycles Yield to Peds sign shall not be used in conjunction with a STOP or YIELD sign, Yield Here to Pedestrians Sign, or a Stop Here for Pedestrians Sign.**

Section 9B.13 Shared-Use Path Restriction Sign (R9-7)

Option:

01   The Shared-Use Path Restriction (R9-7) sign (see Figure 9B-1) may be installed to supplement a solid white pavement marking line (see Section 9E.13) on facilities that are to be shared by pedestrians and bicyclists in order to provide a separate designated pavement area for each mode of travel. The symbols may be transposed as appropriate.

Guidance:

02   *If two-way operation is allowed on the facility for pedestrians and/or bicyclists, the designated pavement area that is provided for each two-way mode of travel should be wide enough to accommodate both directions of travel for that mode.*

Section 9B.14 Bicycles Allowed Use of Full Lane Sign (R9-20)

Support:

01   The Uniform Vehicle Code (UVC) defines a "substandard width lane" as a "lane that is too narrow for a bicycle and a vehicle to travel safely side by side within the same lane."

Option:

02   The Bicycles Allowed Use of Full Lane (R9-20) sign (see Figure 9B-1) may be used on roadways where no bicycle lanes or adjacent shoulders usable by bicycles are present and where travel lanes are too narrow for bicycles and motor vehicles to operate side-by-side.

03   The Bicycles Allowed Use of Full Lane sign may be used in locations where it is important to inform road users that bicyclists might occupy the travel lane.

04   Section 9E.09 describes a shared-lane marking that may be used in addition to or instead of the Bicycles Allowed Use of Full Lane sign to inform road users that bicyclists might occupy the travel lane.

Section 9B.15 Bicycle Passing Clearance Sign (R4-19)

Option:

01   The Bicycle Passing Clearance (R4-19) sign (see Figure 9B-1) may be used in jurisdictions that have defined in law or ordinance a specific clearance to be provided by motor vehicles when they pass bicycles.

02   The specific clearance displayed on the Bicycle Passing Clearance (R4-19) sign may be adjusted to reflect the applicable law or ordinance.

Standard:

03   **The Bicycle Passing Clearance (R4-19) sign shall not be used in jurisdictions that do not have a specific passing clearance to be provided by motor vehicles passing bicycles, as defined in law or ordinance.**

Guidance:

04   *The Bicycle Passing Clearance (R4-19) sign should not be used on roadways with bicycle lanes or with shoulders usable for bicycle travel.*

Section 9B.16 Bicycles Use Shoulder Only Sign (R9-21)

Option:

01   The Bicycles Use Shoulder Only (R9-21) sign (see Figure 9B-1) may be used to designate locations on a freeway or expressway where bicycles are allowed, but must remain on an available and usable shoulder.

Guidance:

02   *The Bicycles Use Shoulder Only sign should be limited to use on freeways and expressways.*

03   *The Bicycles Use Shoulder Only sign should be placed adjacent to the entrance ramp or entrance to the freeway at or near the location where the full-width shoulder resumes beyond the entrance ramp taper.*

Section 9B.17 <u>Signing for Bicycles on Freeways and Expressways</u>

Standard:

01   **The Bicycles Must Exit (R9-22) sign (see Figure 9B-1) shall be used in advance of a location where a freeway or expressway becomes prohibited to bicycle travel, and shall be placed in advance of the intersection or exit ramp prior to the prohibited segment of roadway (see Figure 9B-4).**

Option:

02   The Bicycles Must Exit sign may be used below a post-mounted Exit Direction sign.

Standard:

03   **If the Bicycles Must Exit sign is used, a No Bicycles (R5-6) sign (see Figure 9B-1) shall be placed downstream from the intersection or exit ramp departure point where the prohibited segment of freeway or expressway begins. The No Bicycles sign shall not be placed below the Exit Gore sign.**

Option:

04   The ON FREEWAY (R5-10dP) plaque (see Figure 9B-1) may be used with an appropriate Selective Exclusion sign to indicate a prohibition along ramps leading to an adjacent or parallel freeway.

Support:

05   Section 2B.45 contains information on regulatory signing for prohibiting bicycles from using particular roadways or facilities.

Section 9B.18 <u>Two-Stage Bicycle Turn Box Regulatory Signing (R9-23 Series)</u>

Support:

01   Where two-stage bicycle turn boxes are provided in an intersection, the design of an approach to that intersection will determine whether the use of a two-stage bicycle turn box is required by bicycles to facilitate a turn.

02   Situations in which a two-stage bicycle turn box might be necessary to facilitate turns include, but are not limited to, those in which:

    A. A separated bicycle facility is provided where upstream access to a lane used to facilitate turns by motor vehicle traffic is physically inaccessible to bicycles;

    B. Left turns are prohibited from the left-most lane, or right turns are prohibited from the right-most lane, at an intersection; or

    C. Locations where physical or operational conditions make it impracticable or unsafe for a bicyclist to merge and make the appropriate turn as would any other vehicle.

Standard:

03   **Where bicycles are required to use a two-stage bicycle turn box (see Figure 9B-5), the Two-Stage Bicycle Turn Box regulatory sign series (see Figure 9B-5) shall be used.**

04   **Where bicycles are required to use a two-stage bicycle turn box, the Bicycles All Turns from Bike Lane (R9-23) or Bicycle Left Turn from Bike Lane (R9-23a) advance regulatory sign shall be mounted in advance of the intersection, and at least one Bicycle Turn Must Use Turn Box (R9-23b or R9-23c) sign shall be used at the intersection.**

05   **Where used, the Bicycle Turn Must Use Turn Box (R9-23b) sign shall be mounted at the near side of the intersection.**

06   **Where used, the Bicycle Turn Must Use Turn Box location (R9-23c) sign shall be mounted at the far side of the intersection.**

Option:

07   Where use of a two-stage bicycle turn box is optional, the Two-Stage Bicycle Turn Box guide sign series (see Section 9D.13) may be used to provide directional information.

08   If used, an appropriately sized Street Name (D3-1) sign (see Section 2D.45) may be installed below the All Turns from Bike Lane sign or Left Turn from Bike Lane sign to identify the crossroad where the turn box will be available.

Support:

09   Section 9E.11 contains information regarding pavement markings and turning restrictions for two-stage turn boxes.

Figure 9B-4. Signing for Termination of Bicycle Access on Freeways and Expressways

Figure 9B-5. Example of
Two-Stage Bicycle Turn Box when Use is Mandatory

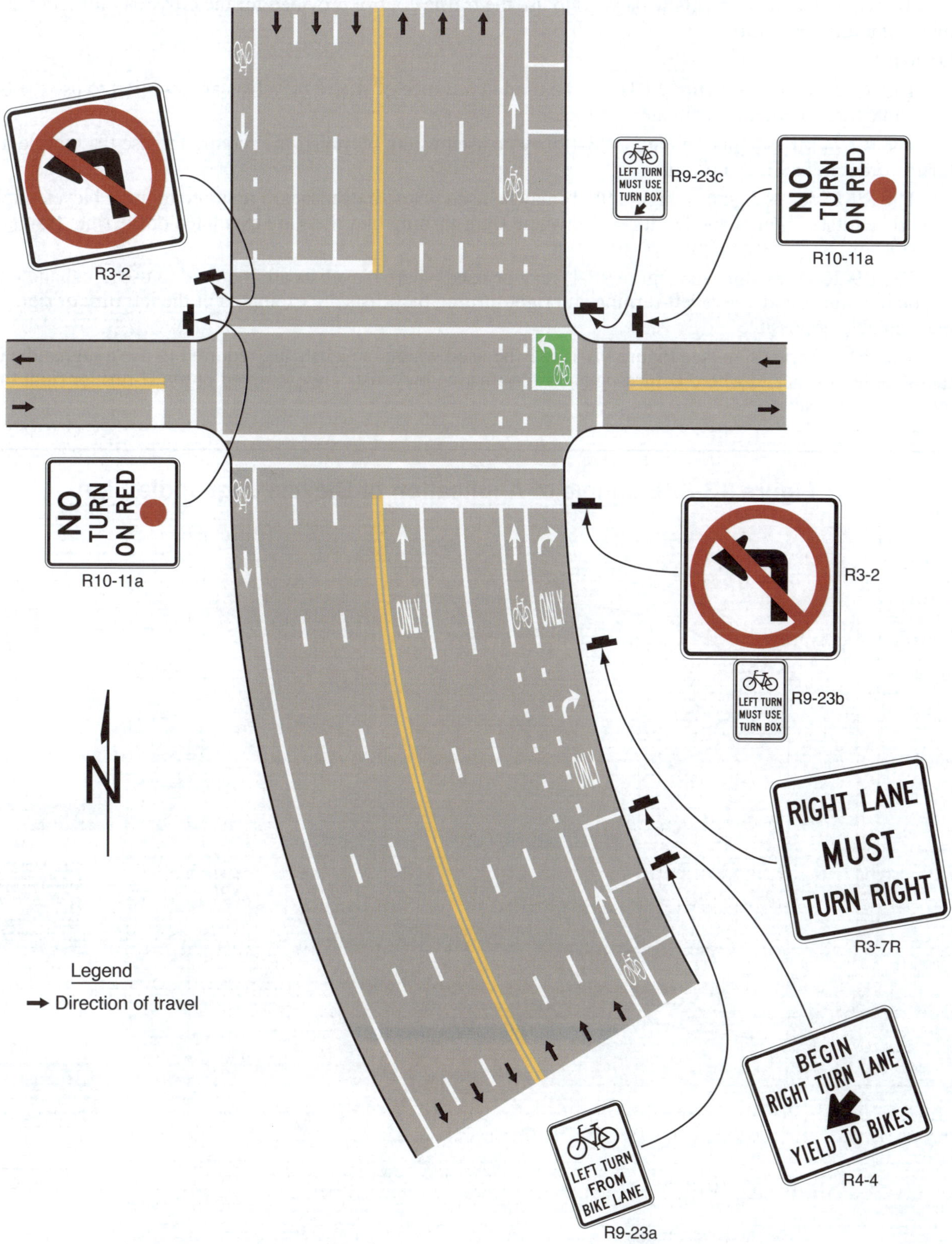

Section 9B.19 Bicycle Jughandle Signs (R9-24, R9-25, R9-26, and R9-27 Series)

Support:

01   Bicycle jughandle turns allow bicycles to use the traffic control provided for the crossroad for facilitating a left turn, right turn, or U-turn.

Option:

02   The R9-23 sign (see Figure 9B-1) may be used in advance of where bicyclists are required to use the bicycle jughandle turn in order to facilitate all turns.

03   The R9-24 series sign (see Figure 9B-1) may be used where bicyclists are required to use the bicycle jughandle turn in order to facilitate all turns.

04   The R9-25 series sign (see Figure 9B-1) may be used where bicyclists are required to use a bicycle jughandle turn to facilitate U-turns and left turns and where right-turning bicyclists are exempted or the right turn is not available or possible (see Figure 9B-6).

05   The R9-26 series sign (see Figure 9B-1) may be used where bicyclists are required to use a jughandle to facilitate U-turns and where left-turning and right-turning bicyclists are exempted or the left turn or right turn is not available or possible.

06   The R9-27 series sign (see Figure 9B-1) may be used where bicyclists are required to use a jughandle to facilitate left turns and where U-turning and right-turning bicyclists are exempted or the U-turn or right turn is not available or possible.

Figure 9B-6. Example of Application of Bicycle Jughandle Sign

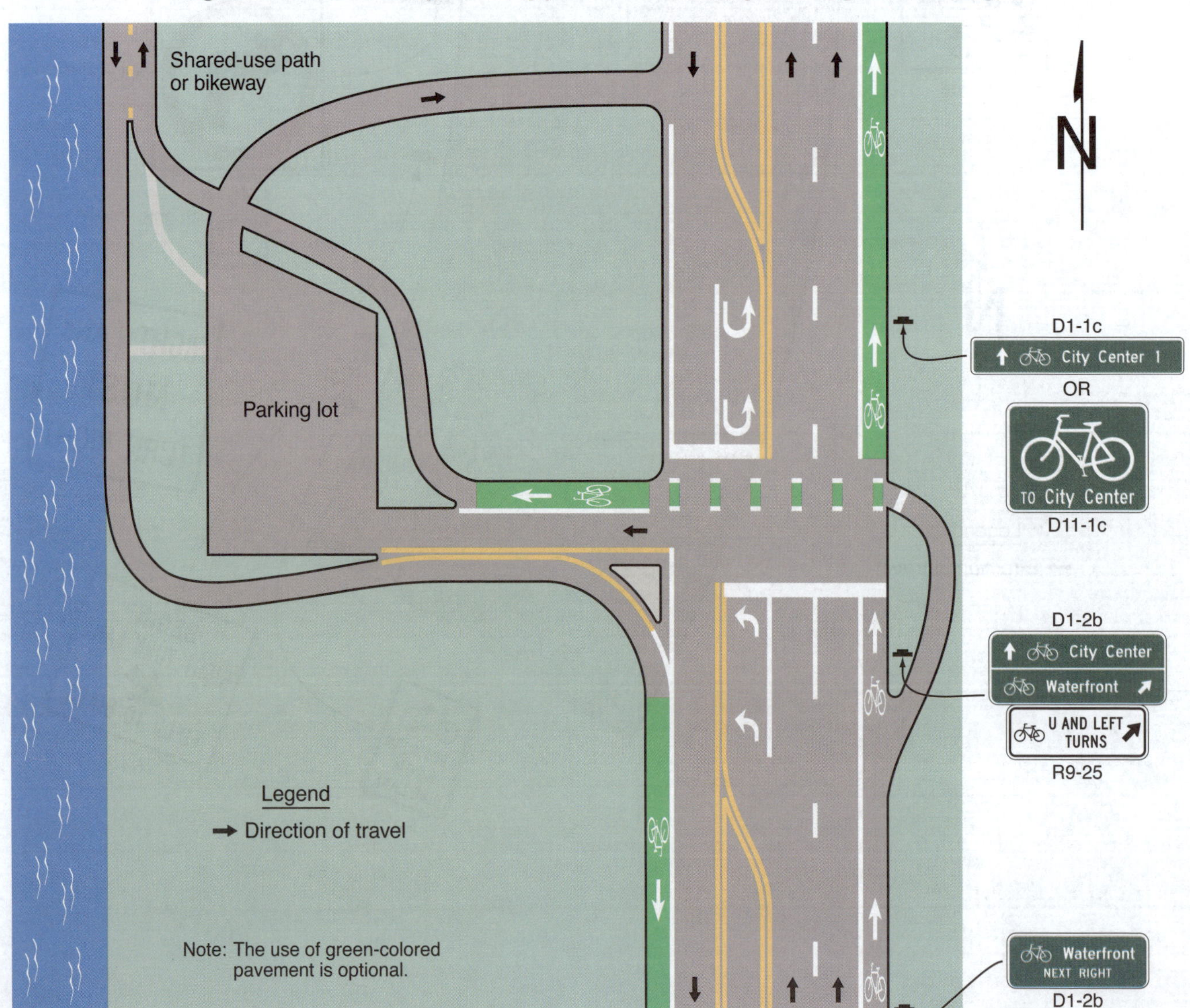

07   A Bicycle Jughandle sign may be used to indicate a jughandle turn initially made by a left turn for a bicycle lane on the left-hand side of a one-way street or for a counter-flow bicycle lane. The legend RIGHT may be substituted for the legend LEFT on Bicycle Jughandle signs to represent bicycle facilities on the left-hand side of the roadway where facilitating a right turn would be applicable.

Guidance:

08   *Applications of Bicycle Jughandle signs should be limited to brief independent alignments either through physical separation or islands formed by pavement markings. Bicycle Jughandle signs should not be used for a turning movement facilitated by a two-stage turn box (see Section 9B.18).*

Support:

09   Bicycle Jughandle signs are designed to be mounted below guide signs.

10   Section 9D.01 contains information regarding the use of Bicycle Destination signs that can be used for jughandles.

Section 9B.20 Bicycle Actuation Signs (R10-4, R10-22, R10-24, R10-25, and R10-26)

Option:

01   Where bicycles are not controlled by pedestrian signal indications, the R10-4, R10-24, or R10-26 sign (see Section 2B.58) may be used.

Guidance:

02   *If used, the R10-4, R10-24, or R10-26 signs (see Figure 9B-1) should be installed in the vicinity of where bicycles will be crossing the street.*

Option:

03   If bicycles are crossing a roadway where In-Roadway Warning Lights (see Section 4U.02) or other warning lights or beacons have been provided, the R10-25 sign may be used.

04   The Bicycle Detector (R10-22) sign (see Figure 9B-1) may be installed at signalized intersections where pavement markings are used to indicate the location where a bicycle is to be positioned to actuate the signal (see Section 9E.15).

Guidance:

05   *If the Bicycle Detector sign is installed, it should be placed at the roadside adjacent to the marking to emphasize the location of the marking.*

Section 9B.21 Left Turn Yield to Bicycles Sign (R10-12b)

Option:

01   The Left Turn Yield to Bicycles (R10-12b) sign (see Figure 9B-1) may be used to emphasize the requirement for motorists to yield to bicyclists in situations where the motorist is turning across a bicycle movement that may be unexpected in direction, location, or some other quality that would be inconsistent with the typical bicycle lane.

Support:

02   Section 2B.59 contains provisions on the placement and use of regulatory Traffic Signal signs.

Section 9B.22 Bicycle Signal Signs (R10-40, R10-40a, R10-41, R10-41a, R10-41b, and R10-41c)

Support:

01   The purposes of the Bicycle Signal signs (see Figure 9B-1) are to inform road users that the signal indications in the bicycle signal face are intended only for bicyclists, and to inform bicyclists which specific bicycle movements are controlled by the bicycle signal face.

02   Section 4H.03 contains information on signs that are used in conjunction with bicycle signal faces.

Standard:

03   **The Bicycle Signal – Mandatory Movement (R10-40 or R10-40a) sign or the Bicycle Signal – Optional Movement (R10-41. R10-41a, R10-41b, or R10-41c) sign shall require bicycles to turn, shall permit turns where such turns would otherwise not be allowed, shall require a bicycle to stay in the same lane and proceed straight through an intersection, or shall indicate allowed movements when a GREEN BICYCLE signal indication is displayed on a bicycle signal face.**

Section 9B.23 LOOK Sign (R15-8)

Option:

01   At railroad or LRT grade crossings with shared-use paths or separated bikeways, the LOOK (R15-8) sign (see Figure 9B-1) may be mounted on the Crossbuck support below the Crossbuck (R15-1) sign or any other signs, or on a separate post in the immediate vicinity of the grade crossing on the railroad or LRT right-of-way.

Guidance:

02   *A LOOK sign should not be mounted on a Crossbuck Assembly that has a YIELD or STOP sign mounted on the same support as the Crossbuck.*

Section 9B.24 Other Regulatory Signs

Option:

01   Other regulatory signs described in Chapters 2B and 8B may be installed on bicycle facilities as appropriate.

CHAPTER 9C. WARNING SIGNS AND OBJECT MARKERS

Section 9C.01 Turn or Curve Warning Signs (W1 Series)

Guidance:

01 *To warn bicyclists of unexpected changes in shared-use path direction, appropriate Turn, Curve, or Large Arrow (W1-1 through W1-7) signs (see Figure 9C-1) should be used.*

02 *The W1-1 through W1-5 signs should be installed at least 50 feet in advance of the beginning of the change of alignment.*

Figure 9C-1. Warning Signs and Plaques and Object Markers for Bicycle Facilities (Sheet 1 of 2)

★ A fluorescent yellow-green background color may be used for this sign or plaque. The background color of the plaque should match the color of the warning sign that it supplements.

Figure 9C-1. Warning Signs and Plaques and Object Markers for Bicycle Facilities (Sheet 2 of 2)

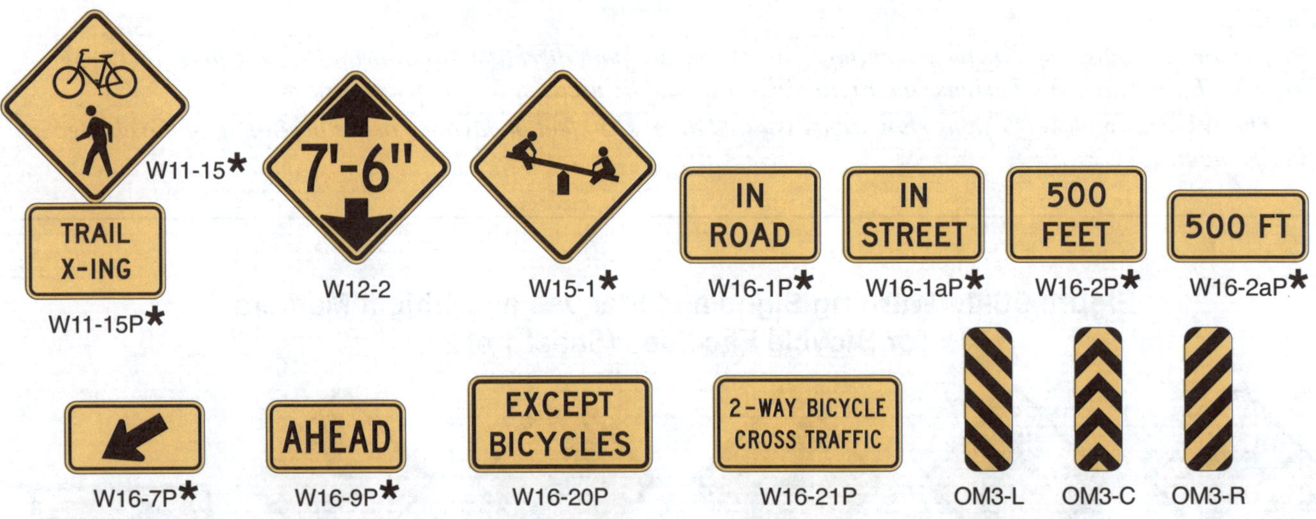

* A fluorescent yellow-green background color may be used for this sign or plaque. The background color of the plaque should match the color of the warning sign that it supplements.

Section 9C.02 Intersection Warning Signs (W2 Series)

Option:

01   Intersection Warning (W2-1 through W2-5) signs (see Figure 9C-1) may be used on a roadway, street, or shared-use path in advance of an intersection to indicate the presence of an intersection and the possibility of turning or entering traffic.

Guidance:

02   *When engineering judgment determines that the visibility of the intersection is limited on the shared-use path approach, Intersection Warning signs should be used.*

03   *Intersection Warning signs should not be used where the shared-use path approach to the intersection is controlled by a STOP sign, a YIELD sign, or a traffic control signal.*

Section 9C.03 Bicycle Surface Condition Warning Sign (W8-10)

Option:

01   The Bicycle Surface Condition Warning (W8-10) sign (see Figure 9C-1) may be installed where roadway or shared-use path conditions could cause a bicyclist to lose control of the bicycle.

02   Signs warning of other conditions that might be of concern to bicyclists, including BUMP (W8-1), DIP (W8-2), PAVEMENT ENDS (W8-3), and any other word message that describes conditions that are of concern to bicyclists, may also be used (see Figure 9C-1).

03   A supplemental plaque may be used to clarify the specific type of surface condition.

Section 9C.04 Bicycle Warning and Trail Crossing Signs (W11-1 and W11-15)

Support:

01   The Bicycle Warning (W11-1) sign (see Figure 9C-1) alerts the road user to unexpected entries into the roadway by bicyclists, and other crossing activities that might cause conflicts. These conflicts might be relatively confined, or might occur randomly over a segment of roadway.

02   Section 9C.06 contains information for Bicycle Cross Traffic Warning plaques that can be used below STOP signs on crossroads or driveways that intersect with bicycle facilities.

Option:

03   The Trail Crossing (W11-15) sign (see Figure 9C-1) may be used where both bicyclists and pedestrians might be crossing the roadway, such as at an intersection with a shared-use path. A TRAIL X-ING (W11-15P) supplemental plaque may be mounted below the W11-15 sign.

04   If used in advance of a trail crossing, a W11-15 or W11-15a sign should be supplemented with an AHEAD (W16-9P) or XX FEET (W16-2P or W16-2aP) plaque to inform road users that they are approaching a point where crossing activity might occur.

Guidance:

05   *If used in advance of a specific crossing point, the Bicycle Warning or Trail Crossing sign should be placed at a distance in advance of the crossing location that complies with Table 2C-3.*

Standard:

06   **Bicycle Warning and Trail Crossing signs, when used at the location of the crossing, shall be supplemented with a diagonal downward-pointing arrow (W16-7P) plaque to show the location of the crossing.**

Option:

07   A fluorescent yellow-green background color with a black legend and border may be used for Bicycle Warning and Trail Crossing signs and supplemental plaques.

Guidance:

08   *When the fluorescent yellow-green background color is used, a systematic approach featuring one background color within a zone or area should be used. The mixing of standard yellow and fluorescent yellow-green backgrounds within a zone or area should be avoided.*

Section 9C.05 EXCEPT BICYCLES Warning Plaque (W16-20P)

Option:

01   Where it might be advantageous to notify bicyclists that the conditions or hazards depicted by a warning sign are not applicable to bicycles, the EXCEPT BICYCLES (W16-20P) warning plaque (see Figure 9C-1) may be used.

Support:

02   Examples of warning signs where an EXCEPT BICYCLES warning plaque can be mounted include DEAD END (W14-1) and NO OUTLET (W14-2) signs (see Section 2C.24).

03   Sections 2C.57 and 2C.58 contain information on the design of supplemental warning plaques.

Section 9C.06 Two-Way Bicycle Cross Traffic Warning Plaque (W16-21P)

Standard:

01   **When used, the Two-Way Bicycle Cross Traffic (W16-21P) warning plaque (see Figure 9C-1) shall be installed below a STOP or YIELD sign.**

Option:

02   The Two-Way Bicycle Cross Traffic warning plaque may be used below STOP or YIELD signs on crossroads and driveways to alert road users of an unexpected bicycle movement.

Support:

03   The Two-Way Bicycle Cross Traffic warning plaque can help minimize overuse or misapplication of other warning signs such as the Bicycle Warning (W11-1) sign.

Guidance:

04   *The Two-Way Bicycle Cross Traffic warning plaque should be used in combination with a STOP or YIELD sign when a counter-flow or two-way bicycle facility has an approach that is counter to the customary scanning behavior of a motorist at that location.*

Section 9C.07 Bicycle Lane Ends Warning Sign (W9-5) and Bicycles Merging Sign (W9-5a)

Support:

01   *Where a warning sign is appropriate, the Bicycle Lane Ends (W9-5) warning sign (see Figure 9C-1) is intended to alert road users that a bicycle lane is ending and that bicycles will share or occupy the travel lane after merging.*

Option:

02   The Bicycle Lane Ends warning sign may be used in advance of the end of a bicycle lane to warn that a bicycle lane will be ending.

03   The Bicycles Merging (W9-5a) sign (see Figure 9C-1) may be used where a bicycle merging maneuver might occur. The Bicycles Merging sign may be used in addition to the Bicycle Lane Ends (W9-5) warning sign.

Guidance:

04   *To avoid excessive use of signs, the Bicycle Lane Ends warning sign should not be used where a bicycle lane is dropped on the approach to an intersection and resumes immediately after the intersection.*

Option:

05   A Bicycles Allowed Use of Full Lane (R9-20) sign (see Section 9B.14) and/or shared-lane markings (see Section 9E.09) may be installed downstream of the merge area.

06   A W16-2aP supplemental warning plaque may be used to inform road users of the distance to the end of the bicycle lane and/or to the bicycle merge.

Section 9C.08 Other Bicycle Warning Signs

Option:

01   Other bicycle warning signs (see Figure 9C-1) such as PATH NARROWS (W5-4a) and Hill (W7-5) may be installed on shared-use paths to warn bicyclists of conditions not readily apparent.

02   In situations where there is a need to warn road users to watch for bicycles traveling along the highway, the Bicycle Warning (W11-1) sign may be used with the IN ROAD (W16-1P) plaque or the IN STREET (W16-1aP) plaque (see Figure 9C-1).

Guidance:

03   *If used, other advance bicycle warning signs should be installed at least 50 feet in advance of the beginning of the condition.*

04   *Where temporary traffic control zones are present on bikeways, appropriate signs from Part 6 should be used.*

Option:

05   Other warning signs described in Chapters 2C and 8C may be installed on bicycle facilities as appropriate.

Section 9C.09 Object Markers

Standard:

01   **Obstructions in a shared-use path shall be marked with retroreflective material or appropriate object markers as described in Section 2C.70.**

Option:

02   Fixed objects adjacent to shared-use paths may be marked with Type 1, Type 2, or Type 3 object markers. If the object marker is not also intended to be seen by motorists, a smaller version of the Type 3 object marker may be used (see Table 9A-1).

CHAPTER 9D. GUIDE AND SERVICE SIGNS

Section 9D.01 Bicycle Destination Signs (D1-1b, D1-1c, D1-2b, D1-2c, D1-3b, D1-3c, D2-1a, D2-2a, and D2-3a)

Support:

01   The purpose of Bicycle Destination (D1-1b, D1-1c, D1-2b, D1-2c, D1-3b, and D1-3c) signs (see Figure 9D-1) and Bicycle Distance (D2-1a, D2-2a, and D2-3a) signs (see Figure 9D-1) is to provide guidance to bicyclists traveling along a bikeway network directing them to typical bicycle destinations or points of interest. The smaller size of Bicycle Destination and Distance signs can deemphasize the messages to motorists, especially when the direction(s) or destination(s) displayed provides access to routes or pathways where the use of motor vehicles is prohibited or discouraged. Examples include, but are not limited to:

    A. Bicycles can go in a direction counter to conventional traffic,
    B. Access to a separated bikeway or shared-use path from a street,
    C. Access to a bicycle route,
    D. Bicycles are directed to another roadway or bikeway that facilitates a parallel or alternative route to the same destination, or
    E. Access to a sidewalk that provides connectivity between bicycle facilities.

02   Section 2D.36 contains information on Destination signs used for when the destinations listed would apply to both motorists and bicyclists.

03   Section 2D.43 contains information on Distance signs used for when the destinations listed would apply to both motorists and bicyclists.

Standard:

04   **Because of their smaller size, Bicycle Destination and Distance signs shall not be used as a substitute for vehicular destination signs when the message is also intended to be applicable to motorists.**

Option:

05   Bicycle Destination and Distance (D1-1b, D1-1c, D1-2b, D1-2c, D1-3b, D1-3c, D2-1a, D2-2a, and D2-3a) signs may be installed to provide direction, destination, and distance information as needed for bicycle travel. If several destinations are to be shown at a single location, they may be placed on a single sign with an arrow (and the distance, if desired) for each name. If more than one destination lies in the same direction, a single arrow may be used for the destinations.

06   Destination (D1-1 and D1-1a) signs (see Section 2D.36) and Street Name (D3-1) signs (see Section 2D.45) may be installed instead of or in addition to Bicycle Destination signs as needed if the Destination or Street Name sign applies to motorists and bicyclists.

07   Distance (D2-1 through D2-3) signs (see Section 2D.43) may be installed instead of, or in addition to, Bicycle Distance (D2-1a through D2-3a) signs, as needed, if the destination and distance information applies to motorists and bicyclists.

Guidance:

08   *Adequate separation should be made between any destination or group of destinations in one direction and those in other directions by suitable design of the arrow, spacing of lines of legend, heavy lines entirely across the sign, or separate signs.*

09   *Where a Bicycle Destination sign with distance information is located less than ½ mile from the destination, the distance displayed should be to the nearest ¼ mile. Where the distance to be displayed on a Bicycle Destination sign is less than ¼ mile, the distance should be displayed in feet, rather than miles, to the nearest 50 feet.*

Option:

10   Distances may be displayed in fractions of a mile to the nearest $\frac{1}{10}$ mile to communicate distance information on Bicycle Destination signs where the distance to a destination is desired to be more precise than ¼-mile increments.

Support:

11   Section 2A.08 contains provisions on the display of fractions on guide signs.

Standard:

12   **An arrow pointing to the right, if used, shall be at the extreme right-hand side of the sign. An arrow pointing left or up, if used, shall be at the extreme left-hand side of the sign. The distance numerals, if used, shall be placed to the right of the destination names.**

13   **Except as provided in Paragraph 14 of this Section, a bicycle symbol shall be placed next to each destination or group of destinations.**

Figure 9D-1. Guide Signs and Plaques for Bicycle Facilities (Sheet 1 of 3)

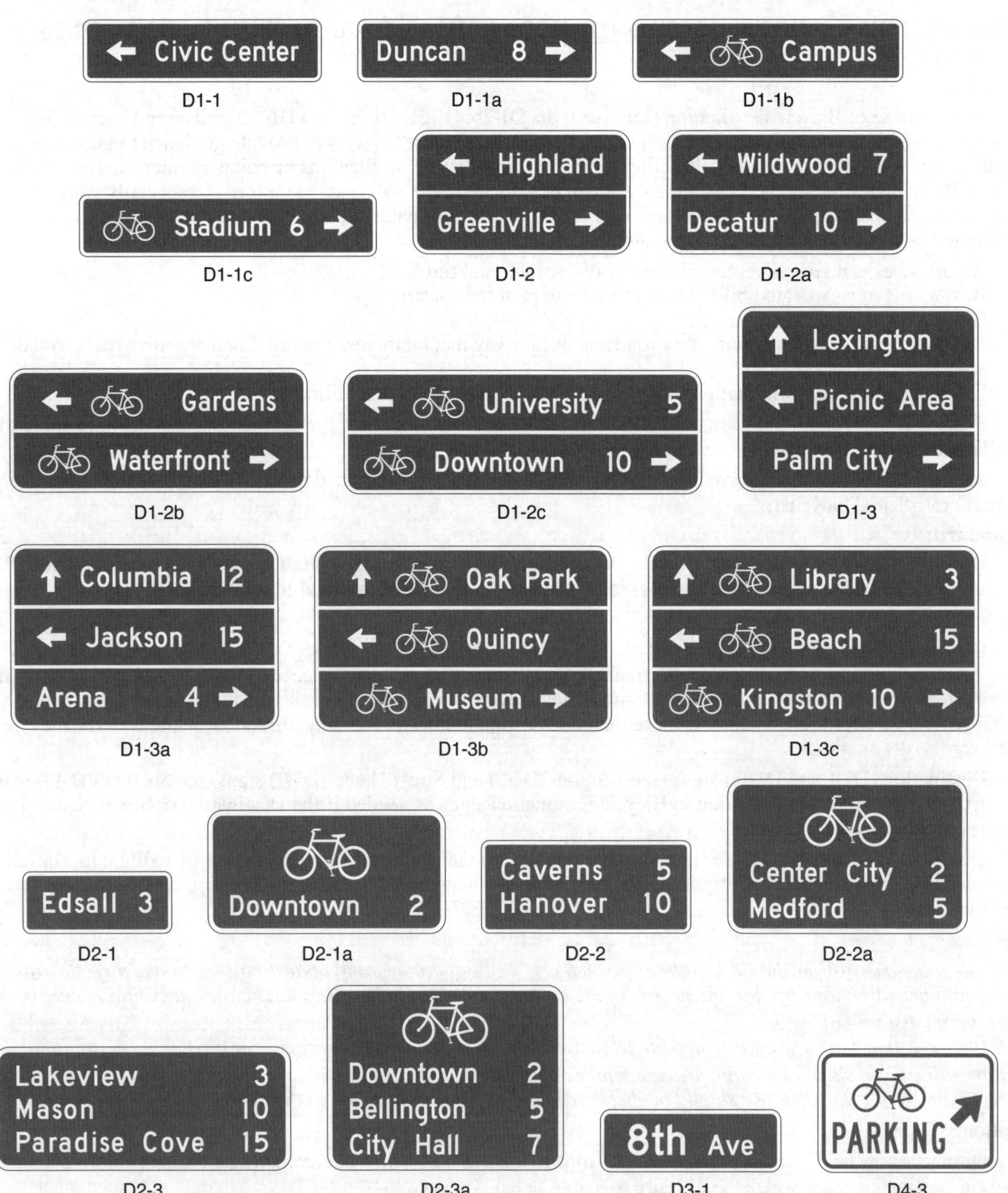

Option:

14   An oversized bicycle symbol may be displayed as the top line of a Bicycle Destination sign instead of individual bicycle symbols for each of the destination/distance lines.

Standard:

15   **If an arrow is at the extreme left, the bicycle symbol shall be placed to the right of the respective arrow.**

Figure 9D-1. Guide Signs and Plaques for Bicycle Facilities (Sheet 2 of 3)

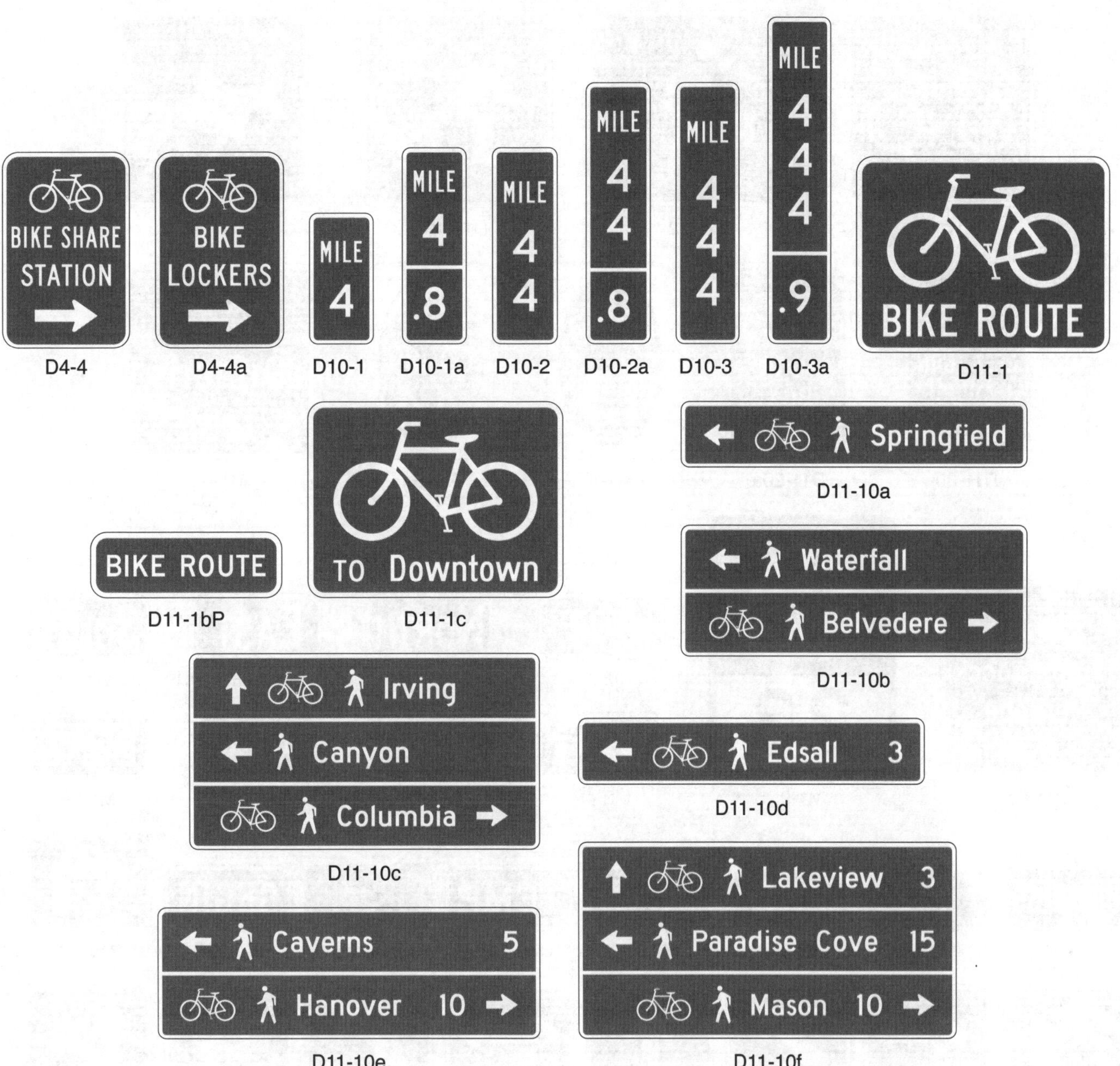

Guidance:

16   *Where the arrow is at the extreme right, the bicycle symbol should be to the left of the destination legend.*

17   *Unless a sloping arrow will convey a clearer indication of the direction to be followed, the directional arrows should be either horizontal or vertical.*

18   *If several individual name signs are assembled into a group, all of the signs in the assembly should have the same horizontal width.*

19   *Travel times should not be used on Bicycle Destination signs.*

Support:

20   Travel times can vary greatly for bicyclists based on a variety of factors including individual speed, bicycle type, and type of facility.

Figure 9D-1. Guide Signs and Plaques for Bicycle Facilities (Sheet 3 of 3)

Section 9D.02 Bike Route Guide Signs (D11-1 and D11-1c)

Support:

01   The Bike Route Guide (D11-1 or D11-1c) sign (see Figure 9D-1) is used where no unique designation of routes is desired. Sections 9D.04 through 9D.07 contain information for Bicycle Route signs where the bicycle route is designated by number, name, or both.

Option:

02   Bike Route Guide signs may be provided along designated unnumbered, unnamed bicycle routes to inform bicyclists of bicycle route direction changes and to confirm route direction and destination.

03   If used, Bike Route Guide signs may be repeated at regular intervals so that bicycles entering from side streets will have an opportunity to know that they are on a bicycle route. Similar guide signing may be used for shared roadways with intermediate signs placed for bicycle guidance.

04   The Alternative Bike Route Guide (D11-1c) sign may be used to display a word legend that provides information on route direction, destination, and/or route name in place of the "BIKE ROUTE" word legend on the D11-1 sign (see Figure 9D-1).

05   Other plaques such as BEGIN (M4-14P) and END (M4-6P) may be used with Bike Route Guide signs.
Guidance:

06   *Travel times should not be used on Bike Route Guide signs.*
Support:

07   Travel times can vary greatly for bicyclists based on a variety of factors including individual speed, bicycle type, and type of facility.

08   Figure 9D-2 shows examples of guide sign applications for bicycle travel.

Section 9D.03 BIKE ROUTE Plaque (D11-1bP)

Option:

01   The BIKE ROUTE (D11-1bP) plaque (see Figure 9D-1) may be installed to supplement:
   A.   The Alternative Bike Route Guide (D11-1c) sign (see Section 9D.02);
   B.   The Bicycle Directional (D11-11) sign (see Section 9D.11) for use on a shared-use path; or
   C.   A Street Name (D3-1) sign (see Section 2D.45).

Figure 9D-2. Examples of Bicycle Guide Signing (Sheet 1 of 2)

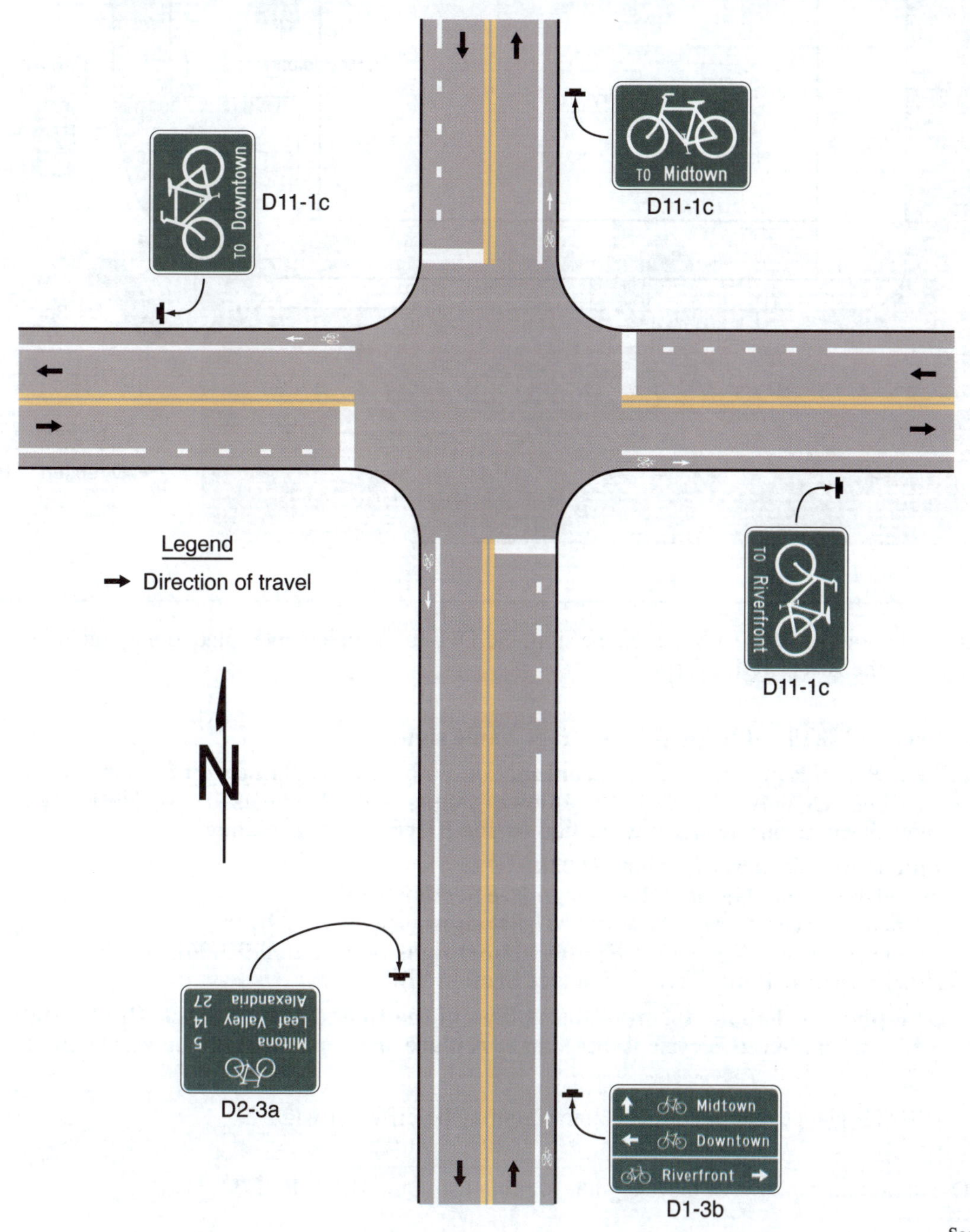

Figure 9D-2. Examples of Bicycle Guide Signing (Sheet 2 of 2)

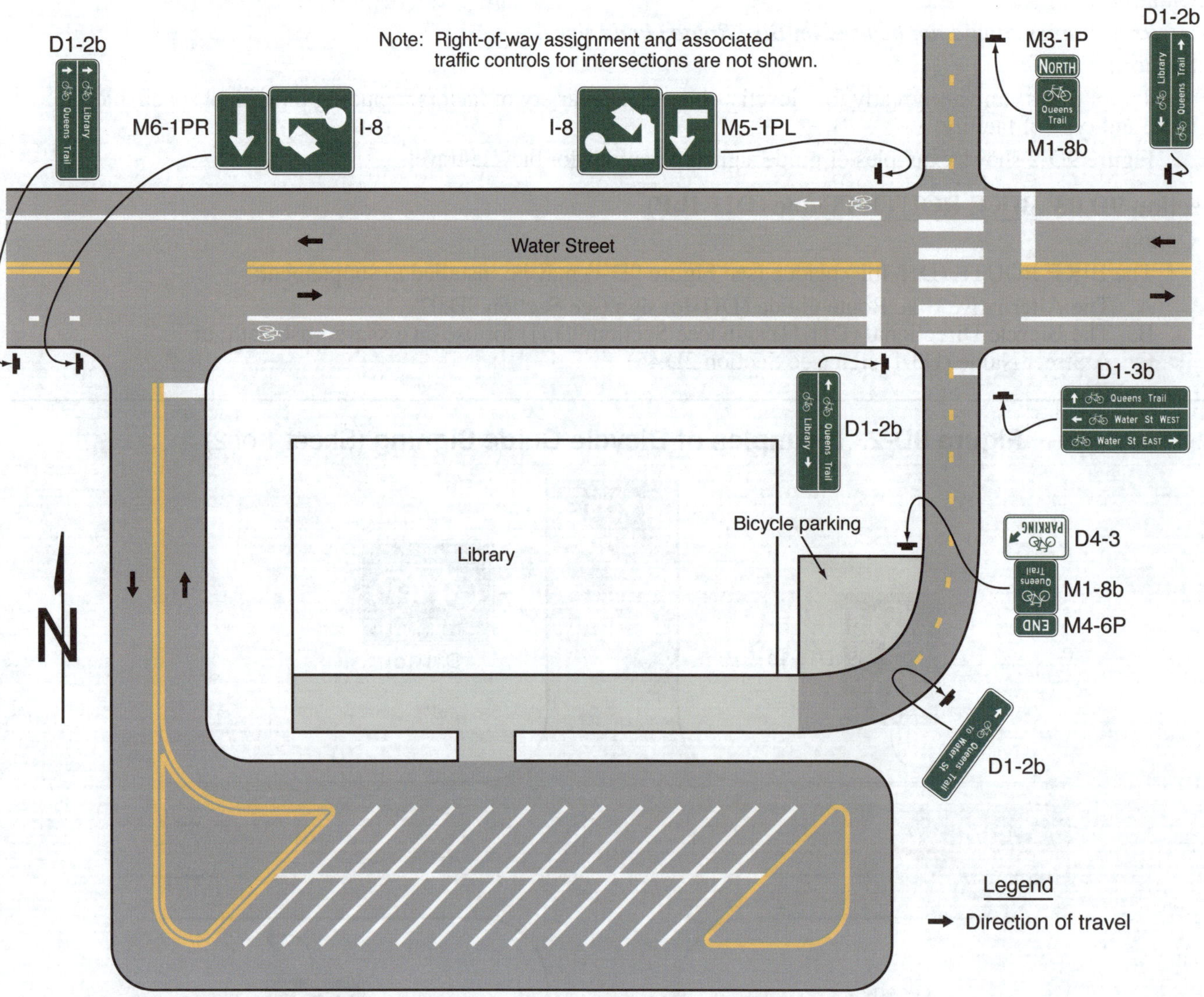

02   When installed above or below a Street Name sign, the D11-1bP supplemental plaque may include a bicycle symbol to the left of the BIKE ROUTE legend.

Standard:

03   **The bicycle symbol shall not be used on a Street Name sign.**

04   **Where a BIKE ROUTE plaque is used in conjunction with a Street Name sign to identify a street that is part of an overall bicycle network, one of the following signs shall also be used systematically to establish the designated bicycle route on the street identified by the BIKE ROUTE plaque:**

   A. **Bike Route Guide signs (see Section 9D.02),**
   B. **Alternative Bike Route Guide (D11-1c) sign (see Section 9D.02),**
   C. **State or Local Bicycle Route (M1-8 and M1-8a) signs (see Section 9D.05),**
   D. **Non-Numbered Bicycle Route (M1-8b and M1-8c) signs (see Section 9D.06), or**
   E. **United States Bicycle Route (M1-9) sign (see Section 9D.07).**

05   **BIKE ROUTE plaques shall not incorporate replicas of the United States Bicycle Route, State or Local Bicycle Route, or Non-Numbered Bicycle Route sign to replace or supplement the bicycle symbol.**

Option:

06   The BIKE ROUTE plaque and the Street Name sign may be different widths.

Support:

07   Figure 9D-3 shows an example of bicycle guide signing using the BIKE ROUTE plaque.

Figure 9D-3. Example of Bicycle Route Guide Signing

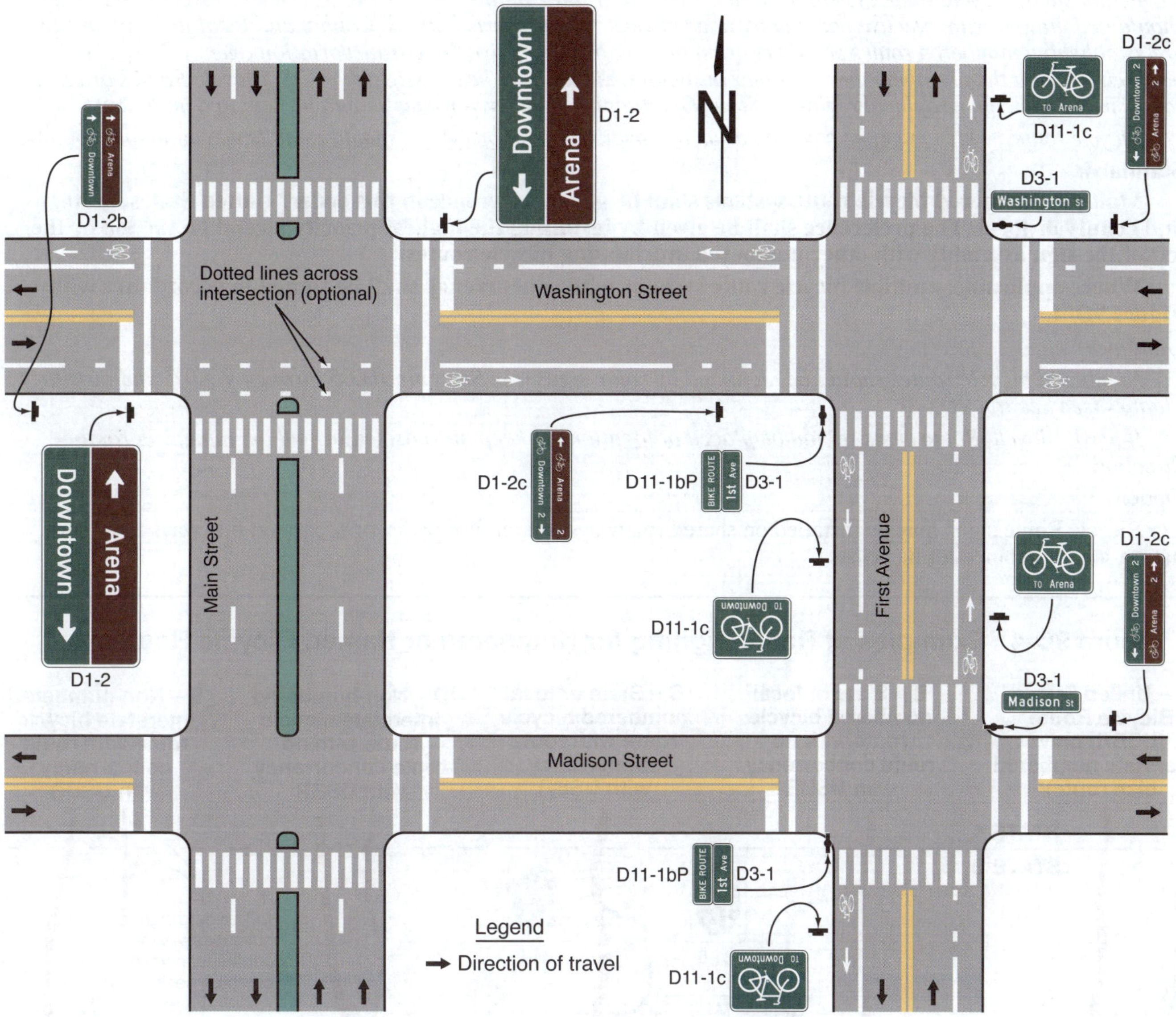

Note: The D1-2 signs are larger than the other guide signs on this figure
because the legends on the D1-2 signs need to be legible to motorists.

Section 9D.04 Numbered Bikeway Systems

Support:

01   The purpose of numbering and signing bikeways and bicycle routes is to identify routes and facilitate travel.

02   The United States Bicycle Routes are numbered by the American Association of State Highway Transportation Officials (AASHTO) upon recommendations of State highway organizations. County and local bikeways and bicycle routes are numbered by the appropriate authorities.

03   Bicycle route sign systems can be used to distinguish junctions, turns, the beginning of routes, and route termination points. Extensive use of reassurance markers is typically not needed.

04   An agency or jurisdiction can use several methods for bicycle route guidance including maps, information guides, or signing.

Guidance:

05   *Establishing bicycle route systems described in Paragraph 2 of this Section and any other bicycle route system should be followed with effective communication between affected jurisdictions. County and local jurisdictions that are establishing numbered routes should coordinate with the respective State transportation agency. Care should be taken to avoid the use of numbers or other designations that have been assigned to U.S. Bicycle Routes or other routes in the same geographical region or State. Overlapping numbered routes should be kept to a minimum.*

06   *Bicycle routes, which might be a combination of various types of bikeways, should establish a continuous routing.*

Standard:

07   **Multiple numbered bicycle route systems shall be given preference in this order: United States, State, and county or local. The preference shall be given by installing the highest-priority legend on the top or the left of the sign assembly with other numbered overlapping bicycle routes.**

08   **Where applicable, multiple bicycle route systems with concurrency shall be signed in accordance with Figure 9D-4.**

Guidance:

09   *Numbered bicycle routes should be identified by route signs (see Sections 9D.05 through 9D.07) and auxiliary plaques (see Section 9D.08).*

10   *If used, Bicycle Route signs should be placed at locations to keep bicyclists informed of changes in route direction.*

Option:

11   Bicycle Route signs may be installed on shared roadways, shared-use paths, or separated bikeways to provide navigational guidance for bicyclists.

Figure 9D-4. Examples of Route Signing for Numbered or Named Bicycle Routes

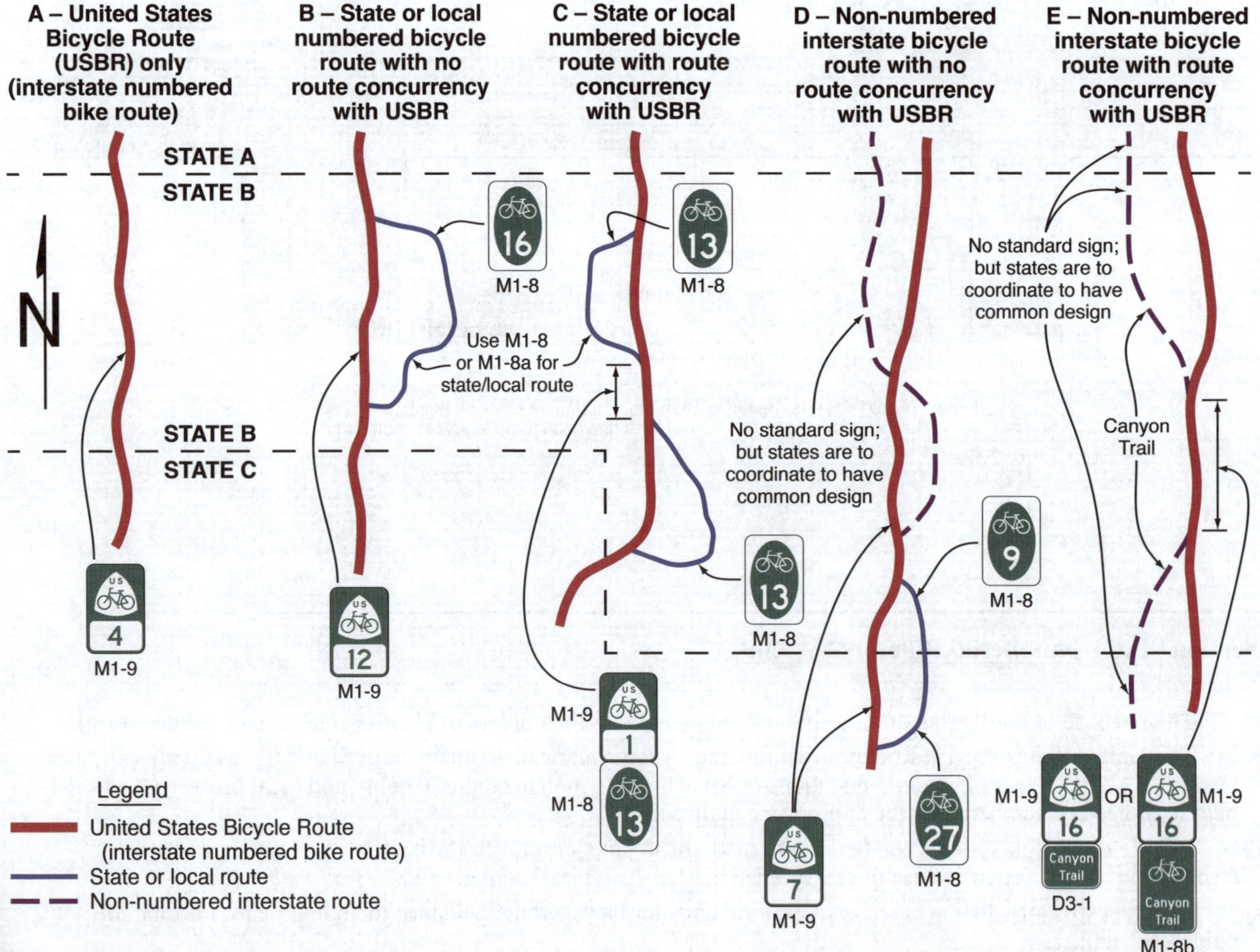

Section 9D.05 <u>Numbered Bicycle Route Signs (M1-8 and M1-8a)</u>

Option:

01   To establish a unique identification (route designation) for a State or local bicycle route, the Bicycle Route (M1-8 or M1-8a) sign (see Figure 9D-1) may be used.

Standard:

02   **The Numbered Bicycle Route (M1-8) sign shall display a route designation and shall have a green background with a white legend and border.**

03   **The Numbered Bicycle Route (M1-8a) sign shall display the same information as the M1-8 sign and in addition shall display a pictograph or words on the upper portion of the sign panel that are associated with the route or with the agency that has jurisdiction over the route.**

04   **If a Numbered Bicycle Route (M1-8 or M1-8a) sign is used on a roadway, it shall include a bicycle symbol.**

Guidance:

05   *If a pictograph is used on the M1-8a sign the maximum dimension (height or width) of the pictograph should not exceed 2 times the height of the route numeral, and should be contained within a green border. The minimum width of the graphic on the M1-8a sign should be 2/3 of the sign width, and the maximum width should be 9/10 of the sign width.*

06   *If a bicycle symbol is used on the M1-8a sign, it should have a minimum height of 1/4 of the M1-8a sign panel height.*

Section 9D.06 <u>Non-Numbered Bicycle Route Signs (M1-8b and M1-8c)</u>

Standard:

01   **Except as provided in Paragraph 2 of this Section, Non-Numbered Bicycle Route (M1-8b and M1-8c) signs (see Figure 9D-1) used on roadways shall have a green background with a white border, and shall include words identifying the bicycle route or a legend consisting of words identifying the bicycle route and a pictograph or bicycle symbol.**

Option:

02   Words identifying the bicycle route may be omitted on Non-Numbered Bicycle Route (M1-8b and M1-8c) signs where a pictograph includes the likeness of a bicycle that clearly identifies the route as a bicycle route.

Support:

03   Bicycle routes are sometimes designated specifically by name or established using a distinctive route identity, but are not numbered or are intentionally excluded from an overall numbered bicycle route system.

04   Section 9D.02 contains information for Bicycle Route signs where no unique designation route is beneficial or desired.

Option:

05   Where a bicycle route is named instead of numbered, the Non-Numbered Bicycle Route sign may be used.

06   A green background or white border may be omitted on Non-Numbered Bicycle Route (M1-8b or M1-8c) signs used on shared-use paths.

Support:

07   Certain uninterrupted, long-distance interstate bicycle routes can largely be on shared-use paths, or other off-roadway facilities. In order to achieve continuity, these bicycle systems might have to share alignments with urban streets, rural highways, or water crossings.

08   Long-distance interstate bicycle routes can be administered by independent organizations serving other non-transportation objectives.

Guidance:

09   *In order to provide signing on a facility managed by a transportation agency, a statewide policy for encouraging independent organizations to adopt the Non-Numbered Bicycle Route sign should be established.*

Section 9D.07 U.S. Bicycle Route Sign (M1-9)

Guidance:

01   *Where a designated bicycle route extends through two or more States, a coordinated submittal by the affected States for an assignment of a U.S. Bicycle Route number designation should be sent to the American Association of State Highway and Transportation Officials.*

Standard:

02   **The U.S. Bicycle Route (M1-9) sign (see Figure 9D-1) shall have a green legend and border with a white background and shall display the route designation as assigned by AASHTO.**

Section 9D.08 Bicycle Route Sign and Auxiliary Plaques

Support:

01   Section 2D.12 contains additional provisions for the design of route sign auxiliary plaques. Sections 2D.29 through 2D.34 contain additional provisions for the general application of route signs.

Guidance:

02   *If a designated or numbered bicycle route is concurrent with a numbered highway, the route sign and auxiliary plaques for the bikeway should be installed as independent assemblies and should not be installed with other Route signs or confirmation assemblies for the numbered or named highway on the same assembly.*

Standard:

03   **Route signs for bikeways shall not be installed on guide signs or overhead.**

Option:

04   Route assemblies for a designated or numbered bicycle route may be installed at locations or distances other than those prescribed in Sections 2D.29 through 2D.34 if engineering judgment indicates that the operation or speed of the bicycle justifies alternate locations or distances.

05   Auxiliary plaques (see Figure 9D-1) may be used in conjunction with Bicycle Route signs as needed.

Guidance:

06   *If used, Junction (M2-1P), Cardinal Direction (M3 series), and Alternative Route (M4 series) auxiliary plaques should be mounted above the appropriate Bicycle Route signs.*

07   *If used, Advance Turn Arrow (M5 series) and Directional Arrow (M6 series) auxiliary plaques should be mounted below the appropriate Bicycle Route signs.*

08   *Except for the M4-8P plaque, all route sign auxiliary plaques should match the color combination of the route sign that they supplement.*

09   *Route sign auxiliary plaques carrying word legends that are used on bicycle routes should have a minimum size of 12 x 6 inches. Route sign auxiliary plaques carrying arrow symbols that are used on bicycle routes should have a minimum size of 12 x 9 inches.*

Standard:

10   **If both the Junction (M2-1P), Cardinal Direction (M3 series), or Alternative Route (M4 series) auxiliary plaque and the Advance Turn Arrow (M5 series) or Directional Arrow (M6 series) auxiliary plaques are used on the same sign assembly as a Bicycle Route sign, the Junction, Cardinal Direction, or Alternative Route auxiliary plaque shall be installed above the Bicycle Route sign, and the Advance Turn Arrow or Directional Arrow auxiliary plaque shall be installed below the Bicycle Route sign.**

Option:

11   With route signs of larger sizes, auxiliary plaques may be suitably enlarged, but not such that they exceed the width of the route sign.

12   A route sign and any auxiliary plaques used with it may be combined on a single sign as a guide sign.

13   Figure 9D-3 shows typical placements of signs for bicycle routes.

Standard:

14   **If used, a Bicycle Route sign assembly shall consist of a route sign and auxiliary plaques that identify the route and indicate the direction.**

Guidance:

15   *If the bicycle route is signed, Bicycle Route sign assemblies should be installed on all approaches where that route intersects with other numbered bicycle routes.*

Standard:

16   **Within groups of assemblies, information for bicycle routes intersecting from the left shall be mounted at the left in horizontal arrangements and at the top or center of vertical arrangements. Similarly, information for bicycle routes intersecting from the right shall be at the right or bottom, and for straight-through bicycle routes at the center in horizontal arrangements or top in vertical arrangements.**

17   **A Junction assembly shall consist of a Junction auxiliary plaque and a Bicycle Route sign. The Bicycle Route sign shall carry the number of the intersected or joined bicycle route.**

Option:

18   The Junction assembly may be installed in advance of intersections where a numbered bicycle route is intersected or joined by another numbered bicycle route.

Standard:

19   **An Advance Bicycle Route Turn assembly shall consist of a Bicycle Route sign, an Advance Turn Arrow or word message auxiliary plaque, and a Cardinal Direction auxiliary plaque, if needed. If used, it shall be installed in advance of an intersection where a turn must be made to remain on the indicated route.**

Option:

20   The Advance Bicycle Route Turn assembly may be used in advance of intersecting routes. On the approach to an intersection with a numbered bicycle route, the Advance Bicycle Route Turn assembly may be used to pre-position turning bicyclists in the correct lane position from which to make their turn.

Standard:

21   **A Directional assembly shall consist of a Cardinal Direction auxiliary plaque, if needed, a route sign, and a Directional Arrow auxiliary plaque.**

Guidance:

22   *The various uses of Directional assemblies should be as follows:*

    A. *Turning movements should be marked by a Directional assembly with a route sign displaying the number of the turning route and a single-headed arrow pointing in the direction of the turn.*

    B. *The beginning of a route should be marked by a Directional assembly with a route sign displaying the number of that route and a single-headed arrow pointing in the direction of the route.*

    C. *An intersected route on a crossroad where the route is designated on both legs should be designated by:*

        1. *Two Directional assemblies, each with a route sign displaying the number of the intersected route, a Cardinal Direction auxiliary plaque, and a single-headed arrow pointing in the direction of movement on that route; or*

        2. *A Directional assembly with a route sign displaying the number of the intersected route and a double-headed arrow, pointing at appropriate angles to the left, right, or ahead.*

    D. *An intersected route on a side road or on a crossroad where the route is designated only on one of the legs should be designated by a Directional assembly with a route sign displaying the number of the intersected route, a Cardinal Direction auxiliary plaque, and a single-headed arrow pointing in the direction of movement on that route.*

Option:

23   Straight-through movements may be indicated by a Directional assembly with a route sign displaying the number of the continuing route and a M6-3P Directional Arrow – Through auxiliary plaque.

Guidance:

24   *A Directional assembly should not be used for a straight-through movement in the absence of other assemblies indicating right or left turns, as the Confirming assembly sign beyond the intersection normally provides adequate guidance.*

25   *Directional assemblies should be located on the near right corner of the intersection. Where unusual conditions exist, the location of a Directional assembly should be determined by engineering judgment.*

Support:

26   It is more important that guide signs be readable, and that the information and direction displayed thereon be readily understood, at the appropriate time and place than to be located with absolute uniformity.

Guidance:

27   *If used, Confirming or Reassurance assemblies should consist of a Cardinal Direction auxiliary plaque and a route sign. Where the Confirming or Reassurance assembly is for an alternative route, the appropriate auxiliary plaque for an alternative route should also be included in the assembly.*

28   *If used, a Confirming assembly should be installed just beyond intersections of numbered routes.*

29   *If used, Reassurance assemblies should be installed between intersections in urban areas as needed, and beyond the built-up area of any incorporated city or town.*

30   *If used, Bicycle Route signs for either confirming or reassurance purposes should be spaced at such intervals as necessary to keep bicyclists informed of their routes.*

Section 9D.09 Bicycle Parking Area, Sharing Station, and Lockers Guide Signs (D4-3, D4-4, and D4-4a)

Support:

01   Bicycle parking areas include bicycle racks or stands, parking stations or structures, sharing systems, or lockers. These facilities can be either regulated or unregulated.

Option:

02   The Bicycle Parking Area (D4-3) guide sign (see Figure 9D-1) may be installed where it is desirable to show the direction to a designated bicycle parking area. The arrow may be reversed as appropriate.

03   The Bicycle-Sharing Station (D4-4) guide sign (see Figure 9D-1) may be installed to provide directional information to a designated bicycle-sharing system. The arrow may be reversed as appropriate.

04   The Bicycle-Sharing Station guide sign may be modified with two lines to accommodate installation in constrained areas.

05   The Bicycle Lockers (D4-4a) guide sign (see Figure 9D-1) may be installed where it is desirable to show the direction to designated bicycle lockers. The arrow may be reversed as appropriate.

Guidance:

06   *If used, the Bicycle-Sharing Station guide sign should be used in conjunction with a regulated bicycle-sharing system such as one that requires the user to pre-register or provide a deposit in order to use a bicycle.*

07   *Where it is determined that unregulated bicycle-sharing parking facilities necessitate a bicycle parking sign, the Bicycle Parking Area guide sign should be used.*

Standard:

08   **In accordance with Section 1D.07, Bicycle Parking Area, Sharing Station, and Lockers guide signs shall not include promotional advertising, business logos, or other identification that would convey the involvement of a public-private partnership for operating the bicycle parking facility or sharing system.**

Section 9D.10 Reference Location Signs (D10-1 through D10-3) and Intermediate Reference Location Signs (D10-1a through D10-3a)

Support:

01   There are two types of reference location signs:
   A. Reference Location (D10-1, D10-2, and D10-3) signs (see Figure 9D-1) show an integer distance point along a shared-use path; and
   B. Intermediate Reference Location (D10-1a, D10-2a, and D10-3a) signs (see Figure 9D-1) show the same information as Reference Locations signs, but they also show a tenth-of-a-mile decimal so that they can be installed between integer distance points along a shared-use path.

Option:

02   Reference Location (D10-1 through D10-3) signs may be installed along any section of a shared-use path to assist users in estimating their progress, to provide a means for identifying the location of emergency incidents and crashes, and to aid in maintenance and servicing.

03   To augment the reference location sign system, Intermediate Reference Location (D10-1a to D10-3a) signs, which show the tenth of a mile with a decimal point, may be installed at one-tenth-of-a-mile intervals, or at some other regular spacing.

Guidance:

04   *If Intermediate Reference Location (D10-1a through D10-3a) signs are used to augment the reference location sign system, the Reference Location sign at the integer mile point should display a decimal point and a zero numeral.*

05   *Reference location signs for shared-use paths should have a minimum mounting height of 2 feet, measured vertically from the bottom of the sign to the elevation of the near edge of the shared-use path, and should not be governed by the mounting height requirements prescribed in Section 9A.02.*

Option:

06   Reference location signs may be installed on one side of the shared-use path only and may be installed back-to-back.

07   If a reference location sign cannot be installed in the correct location, it may be moved in either direction as much as 50 feet.

Guidance:

08   *If a reference location sign cannot be placed within 50 feet of the correct location, it should be omitted.*

09   *Zero distance should begin at the south and west terminus points of shared-use paths.*

Support:

10   Section 2H.11 contains additional information regarding reference location signs.

Section 9D.11 Mode-Specific Directional Guide Signs for Shared-Use Paths (D11-11, D11-12, D11-13, and D11-14)

Option:

01   Where separate pathways are provided for different types of users, mode-specific Directional Guide (D11-11, D11-12, D11-13, and D11-14) signs (see Figure 9D-1) may be used to guide different types of users to the pathway that is intended for their respective modes.

02   Mode-specific Directional Guide signs may be installed at the entrance to shared-use paths where the signed mode(s) are permitted or encouraged, and periodically along these facilities as needed.

03   The Bicycle Directional (D11-11) sign, when combined with the BIKE ROUTE (D11-1bP) supplemental plaque, may be substituted for the D11-1 Bike Route Guide sign on shared-use paths.

04   When some, but not all, non-motorized user types are encouraged or permitted on a shared-use path, mode-specific Directional Guide signs may be placed in combination with each other, and in combination with signs (see Section 9B.08) that prohibit travel by particular modes.

Support:

05   Figure 9D-5 shows an example of signing where separate pathways are provided for different non-motorized user types.

Section 9D.12 Destination Guide Signs for Shared-Use Paths (D11-10a, D11-10b, D11-10c, D11-10d, D11-10e, and D11-10f)

Support:

01   This Section contains information on the application of Destination Guide signs for shared-use paths.

Standard:

02   **Where bicycle traffic is allowed on the shared-use path, Destination Guide signs for shared-use paths and any identification markers shall be retroreflective.**

Guidance:

03   *Destination Guide signs for shared-use paths should be installed on independent assemblies and should not be combined with regulatory and warning signs.*

Option:

04   Destination Guide signs for shared-use paths may use symbols detailed in the "Standard Highway Signs" publication (see Section 1A.05) in addition to the bicycle symbol to display other modes permitted to use the shared-use path.

Standard:

05   **If used, symbols on Destination Guide signs for shared-use paths shall be limited to those where the symbol displayed is an allowable mode on the path or pathway alignment, and where the symbol is supported by other regulatory signs to convey the operation. Symbols unrelated to the allowable modes that would otherwise display directional navigation to a facility, activity, or point of interest shall not be used.**

Support:

06   Chapter 2M contains information for symbol signs used for facilities, activities, and points of interest.

Guidance:

07   *Destination Guide signs for shared-use paths, exclusive of any identification marker used, should be rectangular in shape. Simplicity and uniformity in design, position, and application as described in Section 2A.04 are important and should be incorporated into the sign design.*

08   *Destination Guide signs for shared-use paths should be limited to three destinations per sign (see Section 2D.06).*

09   *Abbreviations (see Section 1D.08) should be kept to a minimum, and should include only those that are commonly recognized and understood.*

**Figure 9D-5. Examples of Mode-Specific
Guide Signing on a Shared-Use Path**

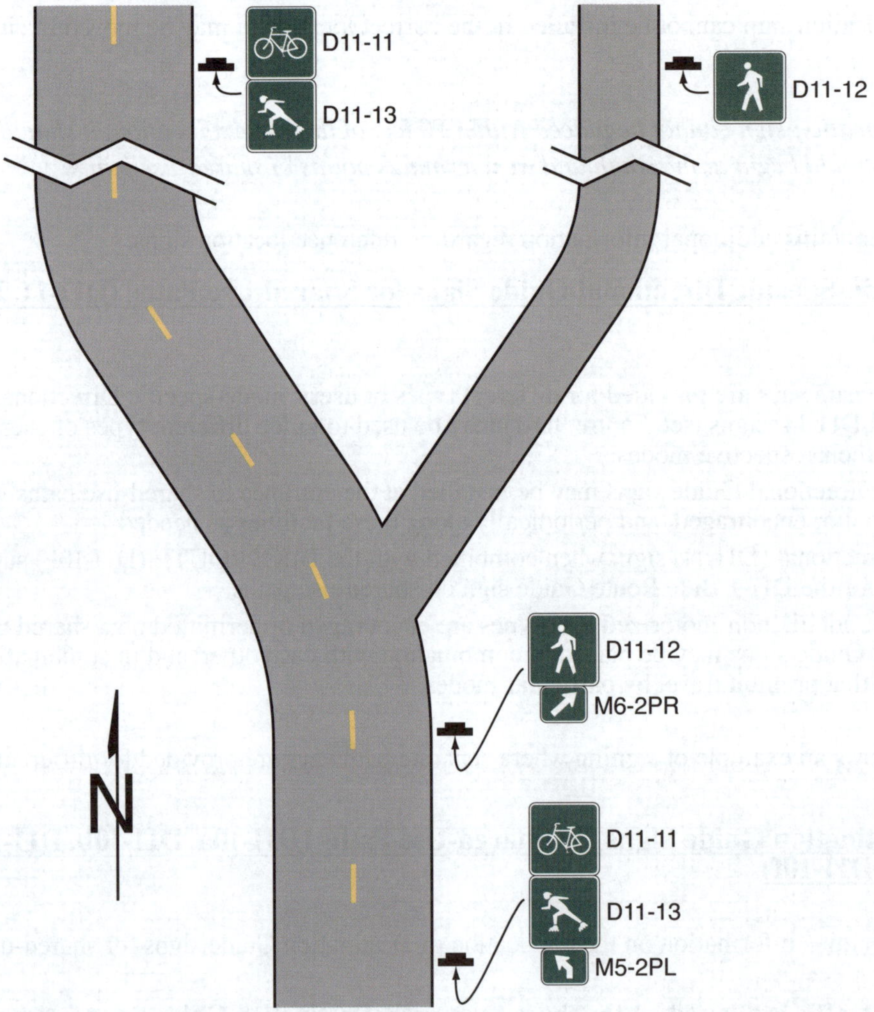

Support:

10   Figure 9D-6 shows an example of a signing system of Destination Guide signs used on shared-use paths.

Standard:

11   **The arrow location and priority order of destinations shall follow the provisions described in Sections 2D.08 and 2D.36. Arrows shall be of the designs provided in Section 2D.08.**

12   **The lettering for destinations on Destination Guide signs for shared-use paths shall be a combination of lower-case letters with initial upper-case letters (see Section 2D.04). All other word messages on Destination Guide signs for shared-use paths shall be in all upper-case letters.**

13   **Except as provided in Paragraph 15 of this Section, the lettering style used for destination and directional legends on Destination Guide signs for shared-use paths shall comply with the provisions of Section 2D.04.**

Option:

14   The distance to the place named may be displayed on the Destination Guide sign. If several destinations are to be displayed at a single point, the several names may be placed on a single sign with an arrow (and the distance, if desired) for each name. If more than one destination lies in the same direction, a single arrow may be used for such a group of destinations.

15   A lettering style other than the Standard Alphabets provided in the "Standard Highway Signs " publication (see Section 1A.05) may be used on Destination Guide signs for shared-use paths if an engineering study determines that the legibility and recognition values for the chosen lettering style at minimum letter heights meet or exceed the values for the Standard Alphabets for the same legend height and stroke width.

Figure 9D-6. Examples of Destination Guide Signs for Shared-Use Paths

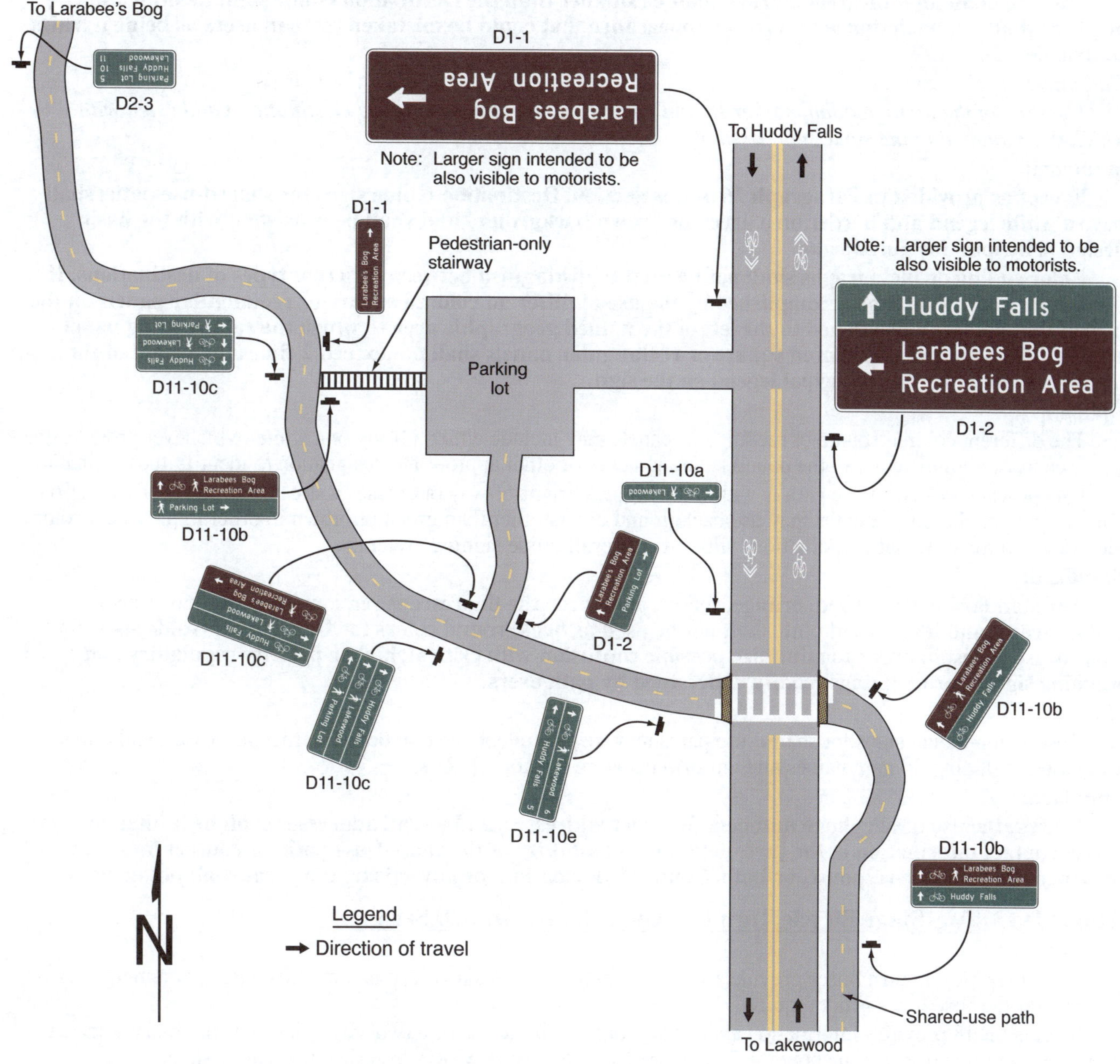

Standard:

16   **Where a shared-use path is within the highway right-of-way or crosses a street or highway, an alternative lettering style shall not be used.**

Option:

17   Pictographs (see definition in Section 1C.02) may be used on Destination Guide signs for shared-use paths.

Standard:

18   **If a pictograph is used, its height shall not exceed 2 times the height of the upper-case letters of the principal legend on the sign.**

19   **Business logos, commercial graphics, or other forms of advertising (see Section 1D.07) shall not be used on Destination Guide signs for shared-use paths or sign assemblies.**

Option:

20   An identification marker may be used in an assembly for Destination Guide signs applied to shared-use paths, or may be incorporated into the overall design of Destination Guide sign, as a means of visually identifying the sign as part of an overall system of signs.

Standard:

21   **The size of an identification marker shall be smaller than the Destination Guide sign. Identification markers shall not be designed to have an appearance that could be mistaken by road users as being a traffic control device.**

Guidance:

22   *The area of the identification marker should not exceed ⅕ of the area of the Destination Guide sign with which it is mounted in the same sign assembly.*

Standard:

23   **Except as provided in Paragraph 26 of this Section, Destination Guide signs for shared-use paths shall have a white legend and border on a green or brown background and shall be consistent with the basic design principles for guide signs.**

24   **Color coding or pictographs shall not be used to distinguish between different types of destinations. If used, color coding shall be accomplished by the use of different colored square or rectangular panels on the face of the sign, each positioned to the left of the named geographic area to which the color-coding panel applies. The height of the colored square or rectangular panels shall not exceed 2 times the height of the upper-case letters of the principal legend on the sign.**

Option:

25   The different colored square or rectangular panels may include either a black or a white (whichever provides the better contrast with the color of the panel) letter, numeral, or other appropriate designation to identify the destination.

26   Except where a shared-use path is within the highway right-of-way or crosses a street or highway, Destination Guide signs for shared-use paths may use background colors other than green or brown in order to provide a color identification for systematic destinations within the overall guide signing system.

Standard:

27   **The standard colors of red, orange, yellow, purple, or the fluorescent versions thereof, fluorescent yellow-green, and fluorescent pink shall not be used as background colors for Destination Guide signs for shared-use paths, in order to minimize possible confusion with critical, higher-priority regulatory and warning sign color meanings readily understood by path users.**

Option:

28   Destination Guide signs for shared-use paths may display telephone numbers, Internet addresses, and e-mail addresses, including domain names and uniform resource locators (URLs).

Standard:

29   **If used, the use of telephone numbers, Internet addresses, and e-mail addresses shall be limited to direct contact information of the jurisdiction with authority of the shared-use path, or contact information for emergency service response, or both. Contact information for advertising purposes shall not be used.**

Section 9D.13 Two-Stage Bicycle Turn Box Guide Signs (D11-20 Series)

Support:

01   Two-stage bicycle turn boxes provide a way for a bicyclist to make a turn in a manner such that a merge across the general-purpose lanes is not required.

02   Section 9B.18 provides information about situations when the use of a two-stage bicycle turn box is required and also contains information about the Two-Stage Bicycle Turn Box (R9-23 series) regulatory signs.

03   Section 9E.11 contains information regarding pavement markings for two-stage bicycle turn boxes.

Option:

04   Where a two-stage bicycle turn box is provided, the Two-Stage Bicycle Turn Box guide sign series (see Figure 9D-1) may be used.

Standard:

05   **Where used, the Two-Stage Bicycle Turn Box Advance (D11-20) guide sign shall be mounted in advance of the intersection where the turn box is located.**

06   **Where used, the Two-Stage Bicycle Turn Box (D11-20a) guide sign shall be mounted on the far side of the intersection.**

Option:

07   Where the Two-Stage Bicycle Turn Box Advance (D11-20) guide sign is used, an additional Two-Stage Bicycle Turn Box Advance guide sign may be mounted on the near side of the intersection where the turn box is located.

08   If used, an appropriately-sized Street Name (D3-1) sign (see Section 2D.45) may be installed below the Two-Stage Bicycle Turn Box Advance guide sign to identify the crossroad where the turn box will be available.

09   Figure 9D-7 shows an example of Two-Stage Bicycle Turn Box guide signs at a location where the use of the turn box is optional.

Section 9D.14 <u>General Service Signing for Bikeways</u>

Option:

01   General Service signs (see Chapter 2I) may be used on bikeways.

Standard:

02   **The sizes of General Service signs intended for viewing by both bicyclists and other road users shall comply with the sizes in Table 2I-1.**

Option:

03   General Service signs intended for the exclusive use of bicyclists may be of reduced size.

Figure 9D-7. Example of Two-Stage Bicycle Turn Box when Use is Optional

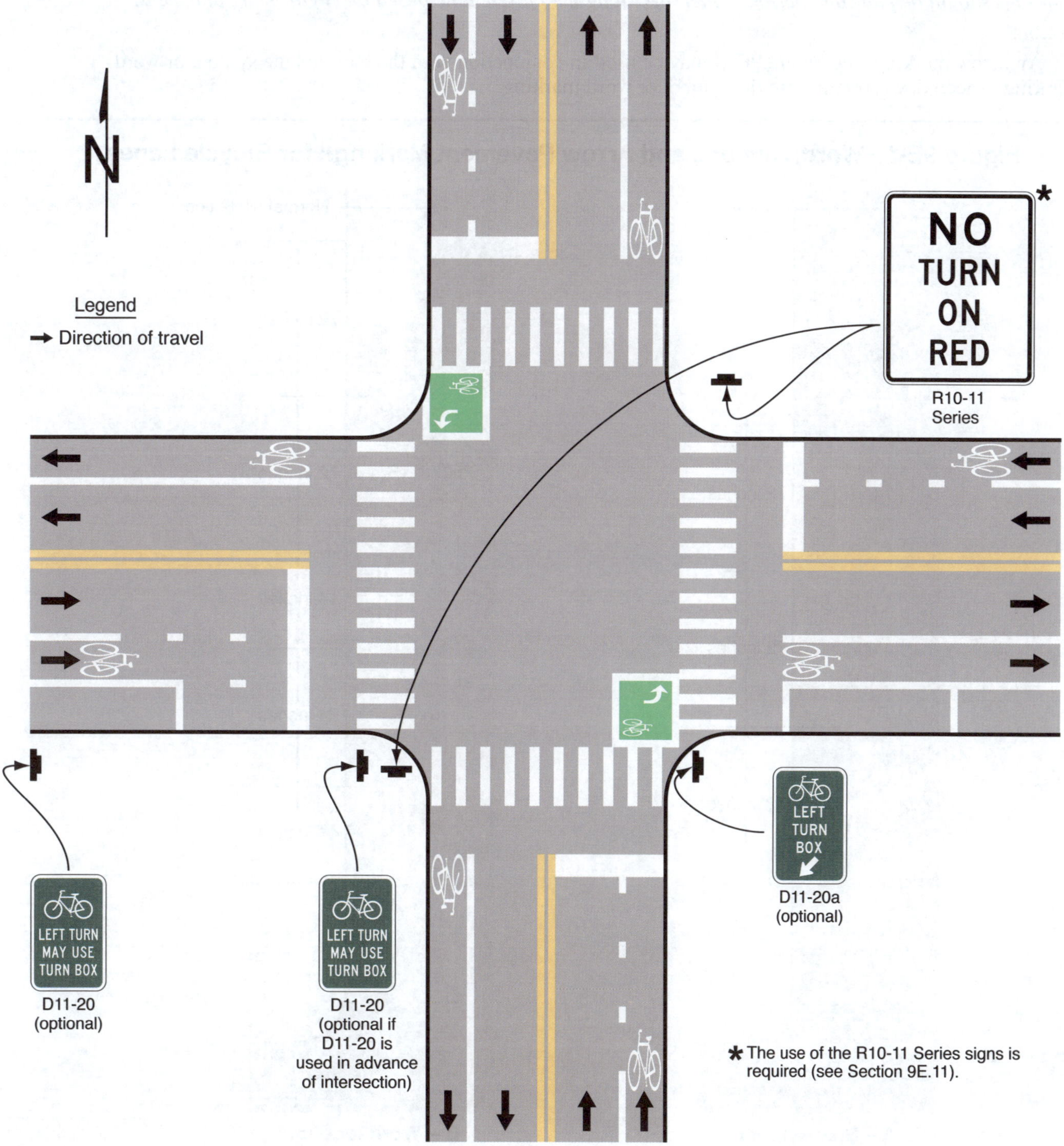

CHAPTER 9E. MARKINGS

Section 9E.01 Bicycle Lanes

Support:

01 Pavement markings designate that portion of the roadway for preferential use by bicyclists. Markings inform all road users of the restricted nature of the bicycle lane.

Standard:

02 **Longitudinal pavement markings and bicycle lane symbol or word markings (see Figure 9E-1) shall be used to define bicycle lanes.**

Guidance:

03 *The first symbol or word marking in a bicycle lane should be placed at the beginning of the bicycle lane and downstream symbol or word markings should be placed after major intersections. Additional symbol or word markings should be placed at periodic intervals along the bicycle lane based on engineering judgment.*

Option:

04 An arrow marking (see Figure 9E-1) may be used in conjunction with the bicycle lane symbol or word marking, placed downstream from the symbol or word marking.

Figure 9E-1. Word, Symbol, and Arrow Pavement Markings for Bicycle Lanes

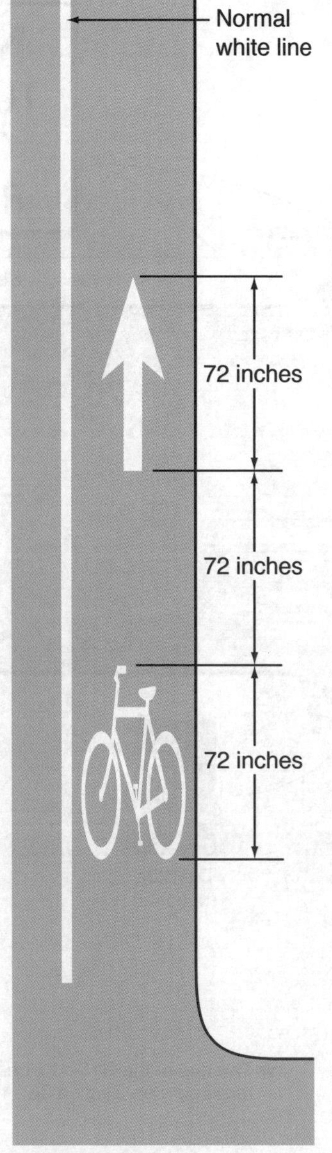

A – Bike symbol

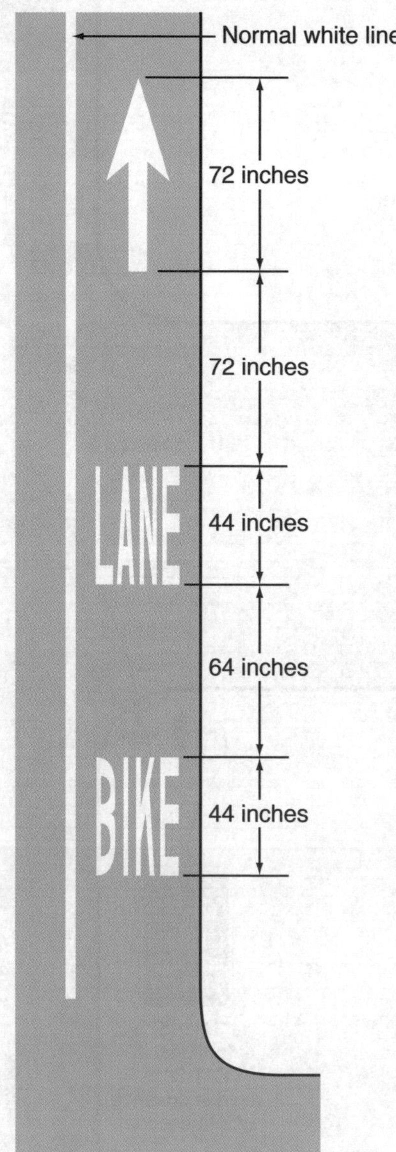

B – Word legends

05   Where the bicycle lane symbols or word markings are used, Bicycle Lane signs (see Section 9B.04) may also be used, but not necessarily adjacent to every set of pavement markings in order to avoid overuse of the signs.

Support:

06   Section 3H.06 contains information on green-colored pavement for use in bicycle lanes.

Standard:

07   **The bicycle symbol or BIKE LANE pavement word marking and the pavement marking arrow shall not be used in a shoulder.**

08   **A portion of the roadway shall not be established as both a shoulder and a bicycle lane.**

Support:

09   Where a shoulder is provided or is of sufficient width to meet the expectation of a highway user in that it can function as a space for emergency, enforcement, or maintenance activities, or avoidance or recovery maneuvers, Section 9B.16 contains information regarding the Bicycles Use Shoulder Only sign that can be used to denote locations on a freeway or expressway where bicycles are permitted on an available and usable shoulder.

10   Examples of pavement markings for bicycle lanes on a two-way street are shown in Figure 9E-2.

Section 9E.02 Bicycle Lanes at Intersection Approaches

Standard:

01   **Except as provided in Paragraph 2 of this Section, a through bicycle lane shall not be positioned to the right of a right turn only lane or to the left of a left turn only lane.**

Option:

02   A through bicycle lane may be positioned to the right of a right turn only lane or to the left of a left turn only lane provided that the bicycle lane is controlled by a traffic signal that displays bicycle signal indications (see Chapter 4H).

Support:

03   Unless controlled by a bicycle signal indication, a bicyclist continuing straight through an intersection from the right of a right turn only lane or from the left of a left turn only lane would be inconsistent with normal traffic behavior and would violate the expectations of right-turning or left-turning motorists.

Guidance:

04   *When the right (left) through lane is dropped to become a mandatory right-turn (left-turn) lane, the bicycle lane markings should stop at least 100 feet before the beginning of the right-turn (left-turn) lane. Through bicycle lane markings should resume to the left (right) of the mandatory right-turn (left-turn) lane.*

05   *Except as provided in Paragraph 2 of this Section, an optional through-right (through-left) turn lane next to a mandatory right-turn (left-turn) lane should not be used where there is a through bicycle lane.*

Standard:

06   **A bicycle lane located on an intersection approach between general-purpose lanes for motor vehicle movements shall be marked with at least one bicycle symbol and at least one arrow pavement marking as provided in Paragraph 4 of Section 9E.01.**

07   **A bicycle lane shall not be marked within a general-purpose lane, either with dotted or any other line markings.**

Option:

08   Where there is insufficient width in the roadway to include both a bicycle lane and a general-purpose turn lane, bicycle travel may be accommodated within the turn lane or general-purpose lane using shared-lane markings.

Standard:

09   **Where a general-purpose turn lane is controlled by a traffic control signal, through bicycle movements shall not be accommodated in the turn lane unless the turning movement is always permitted to proceed simultaneously with the adjacent through movement.**

Support:

10   Examples of bicycle lane markings on approaches to intersections are shown in Figures 9E-3, 9E-4, and 9E-9.

Guidance:

11   *The longitudinal line defining a bicycle lane should be dotted on approaches to intersections where turning vehicles are permitted to cross the path of through-moving bicycles (see Figure 9D-7).*

Figure 9E-2. Example of Pavement Markings for Bicycle Lanes on a Two-Way Street

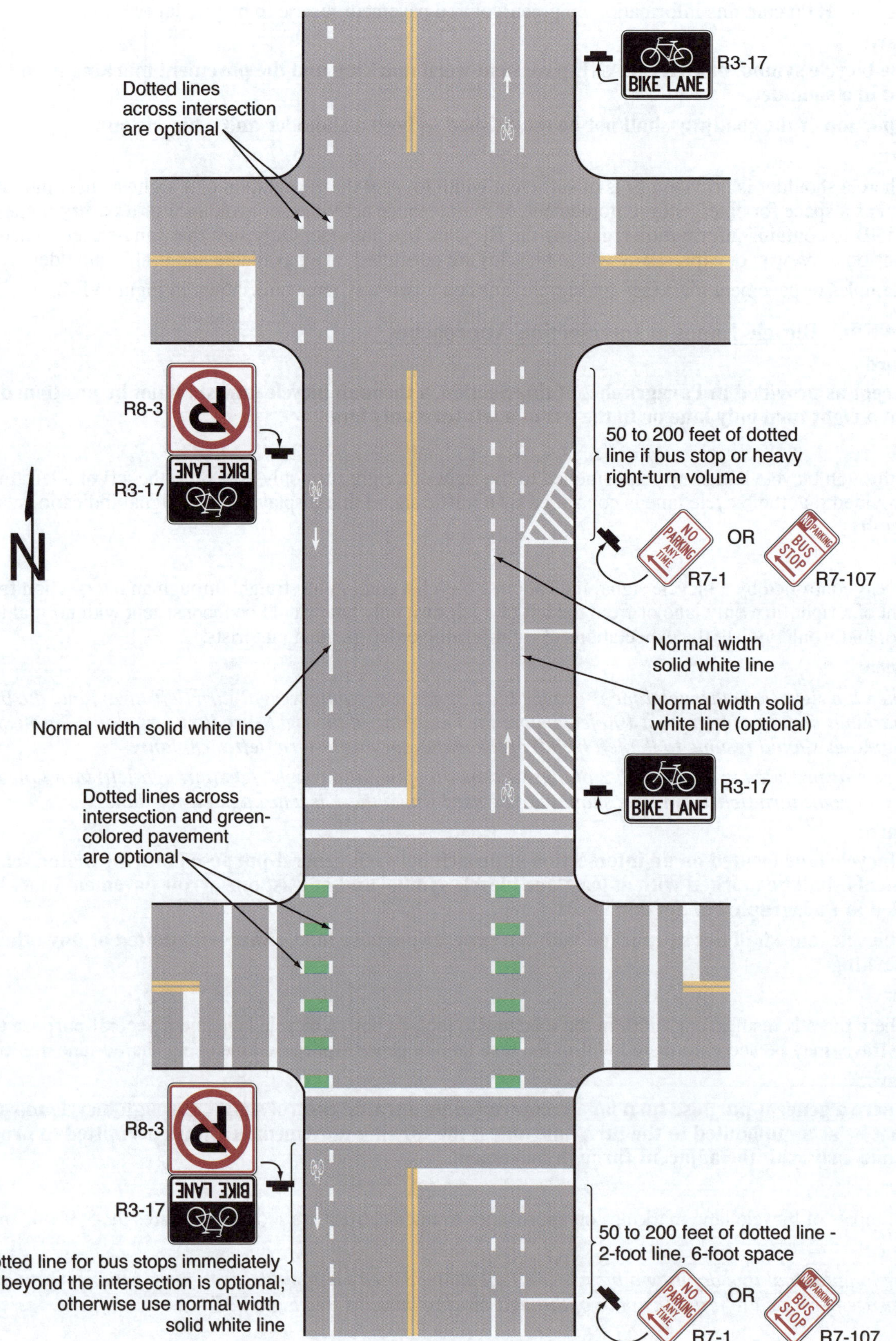

Figure 9E-3. Examples of Bicycle Lane Markings on an Approach to an Intersection
(Sheet 1 of 3)

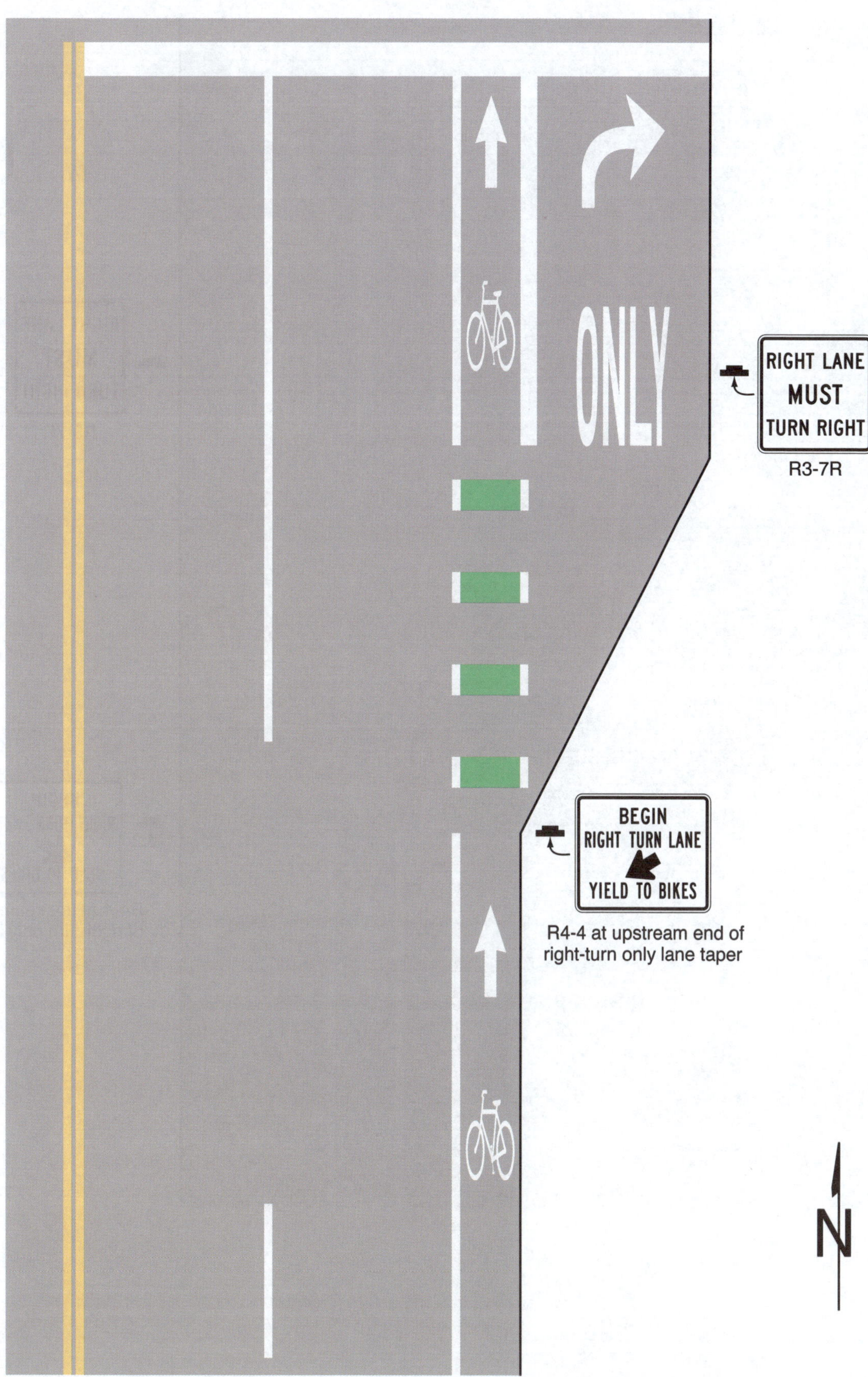

Figure 9E-3. Examples of Bicycle Lane Markings on an Approach to an Intersection
(Sheet 2 of 3)

Figure 9E-3. Examples of Bicycle Lane Markings on an Approach to an Intersection
(Sheet 3 of 3)

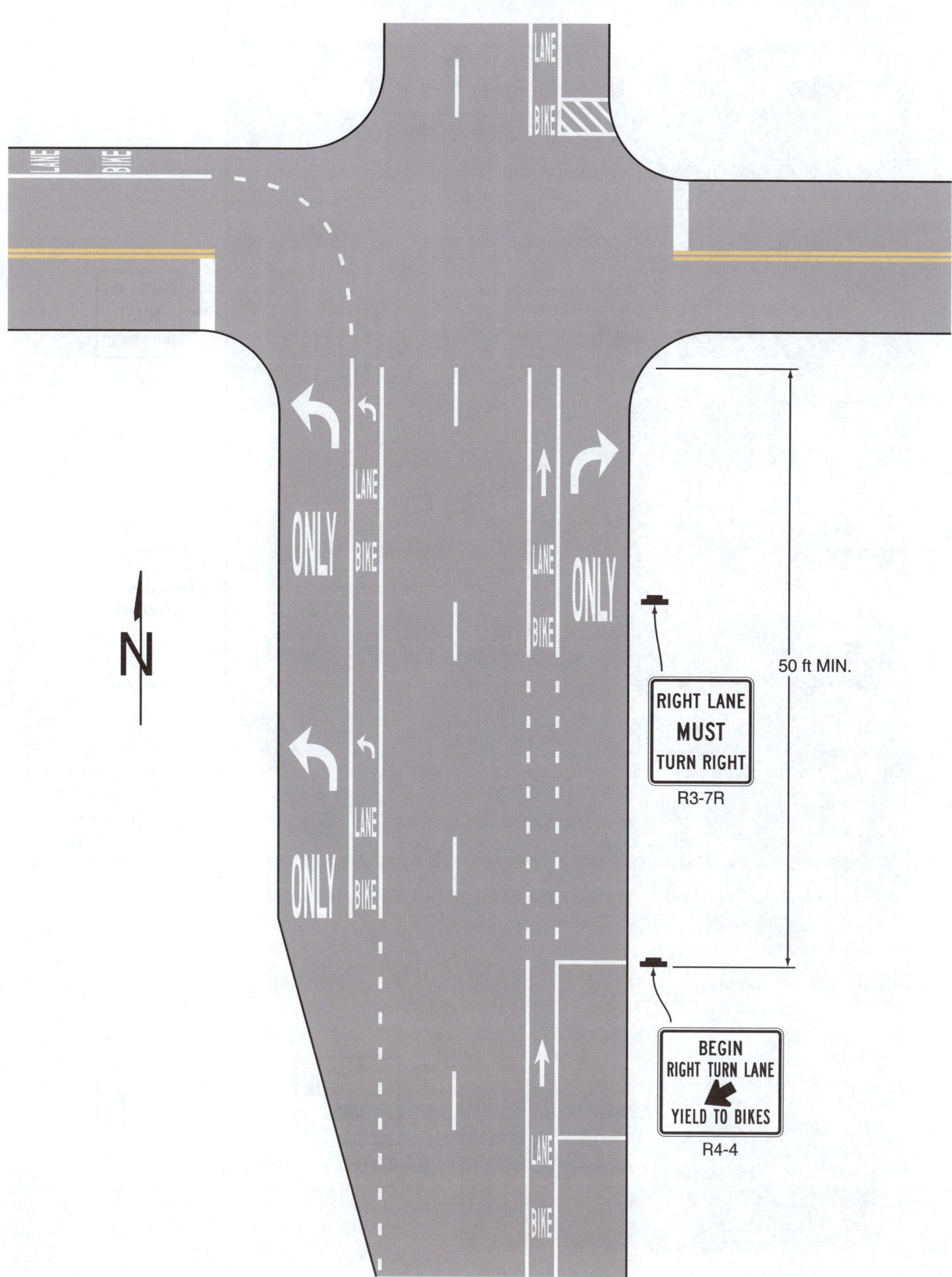

**Figure 9E-4. Example of Bicycle Lane Markings on an Approach
to an Intersection that Transitions from a Shared Lane**

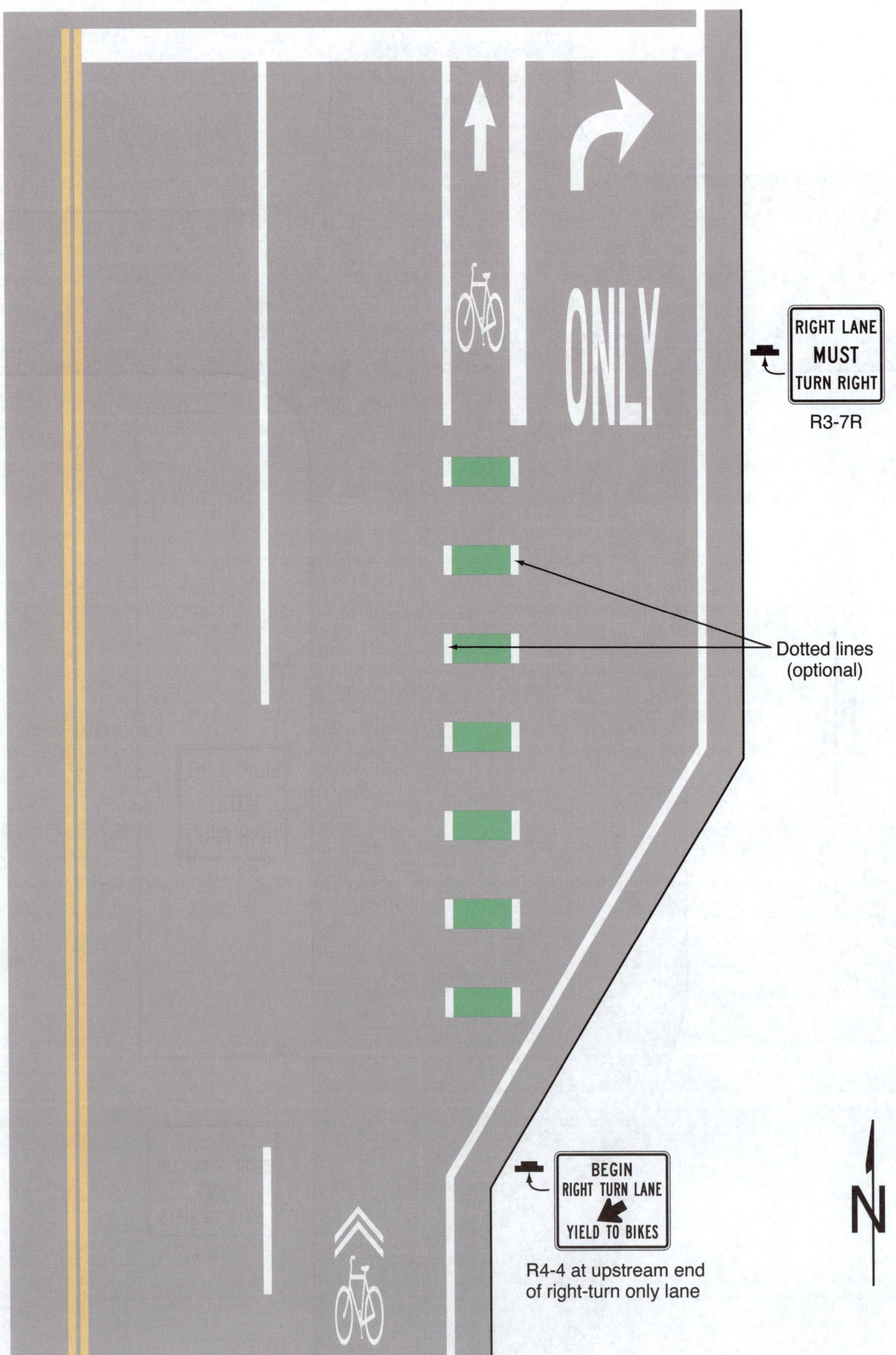

Support:

12   Buffer-separated and separated bicycle lanes require additional considerations at intersections, including sight distances for bicycles and other road users, user expectations, and intersection geometry.

Option:

13   A buffer-separated or separated bicycle lane may be shifted closer to, or farther away from the adjacent general-purpose lane depending upon site-specific conditions (see Drawings D and E in Figure 9E-7).

Support:

14   A buffer-separated or separated bicycle lane shifted away from the adjacent general-purpose lane at an intersection can create space for a motor vehicle to queue between the general-purpose lane and the extension of the bicycle lane. This design can also improve the safety and comfort of bicyclists by reducing the speed of turning motor vehicles, improving sightlines, and creating additional buffer space prior to the conflict point with turning motor vehicles.

15   The purpose of shifting a buffer-separated or separated bicycle lane away from the adjacent general-purpose lane is to allow the driver of a turning vehicle to undertake the tasks of turning and scanning for bicycle cross traffic in isolation versus simultaneously. Sufficient sight distance for both drivers and bicyclists is important in this design (see Drawing E in Figure 9E-7).

16   The purpose of shifting a buffer-separated or separated bicycle lane toward the adjacent general-purpose lane is to improve the visibility of bicyclists to the adjacent traffic and avoid conflicts between turning motor vehicles and bicyclists (see Drawing D in Figure 9E-7).

17   Staggering stop lines (see Section 3B.19) so that general-purpose lanes stop further in advance from the intersection than the bicycle lane can improve the visibility of bicyclists for drivers of turning vehicles (see Drawing D in Figure 9E-7).

Option:

18   Where a general-purpose mandatory turn lane is provided at an intersection and the approach also includes a separated or buffer-separated bicycle lane, a mixing zone may be established to allow general-purpose turning traffic to share the roadway space with bicyclists (see Figure 9E-5).

Standard:

19   **Mixing zones shall be used only where the bicycle lane is one-way in the same direction of travel as the adjacent general-purpose lane.**

20   **Mixing zones with a yielding area shall have yield markings indicating where general-purpose traffic entering the shared space shall yield to bicyclists.**

21   **Where a mixing zone continues to the intersection itself sharing space between bicyclists and general-purpose turning traffic, shared-lane markings and turn arrows shall be provided in the lane.**

Support:

22   Mixing zones require bicycles and general traffic to share space, interrupting a buffer-separated or separated bicycle lane where bicycle traffic is otherwise separated from general traffic. The preference is to provide a dedicated bicycle facility for the intersection approach, If that is not possible, the mixing zone needs to indicate that bicyclists and motorists are entering a shared condition.

Guidance:

23   *Where a mixing zone provides for the re-establishment of a bicycle lane after bicycles and general-purpose lanes cross paths, a buffered or physically-separated space should be provided between the bicycle lane and the adjacent general-purpose lane (see Drawing C in Figure 9E-5).*

Section 9E.03 <u>Extensions of Bicycle Lanes through Intersections</u>

Support:

01   Extensions of bicycle lanes through intersections can help identify the paths of bicyclists and guide them on movements that could be difficult to discern. Extensions of bicycle lanes through intersections also assist other road users of the intersection to identify where bicyclists are expected to operate and to recognize potentially unexpected conflict points.

02   The design, placement, and maintenance of bicycle lane extensions through intersections are important considerations, especially when contiguous to a crosswalk, to avoid potential confusion to pedestrians with vision disabilities.

03   The width and color of lane extension markings are discussed in Section 3B.11.

Figure 9E-5. Examples of Pavement Markings for Mixing Zones

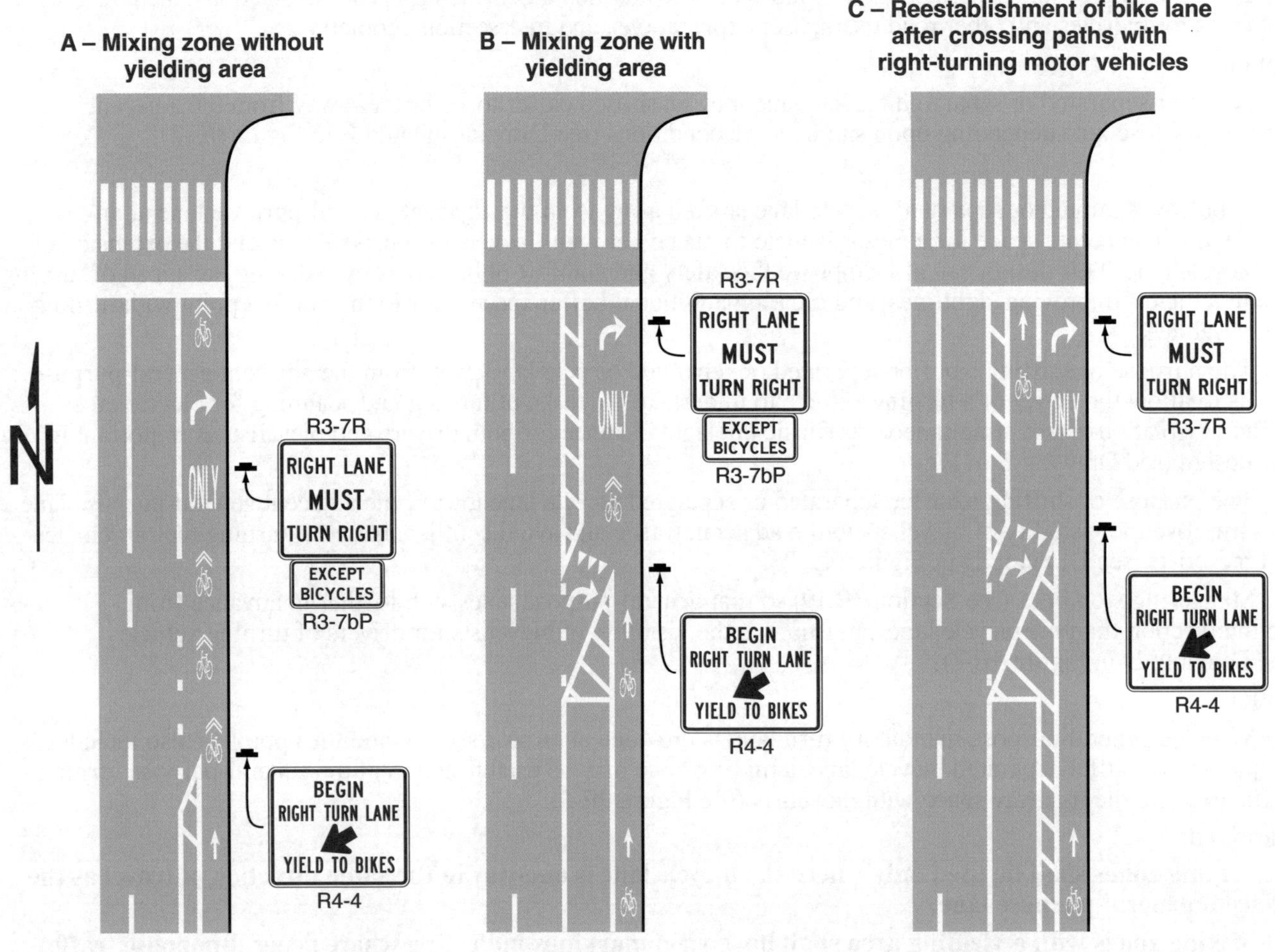

Option:

04   The bicycle symbol, the arrow marking, pavement word markings, or a combination thereof may be used in bicycle lane extensions through intersections.

05   Green-colored pavement may be used in a bicycle lane extension in accordance with the provisions of Section 3H.06.

Standard:

06   **Shared-lane markings or chevron markings shall not be used in bicycle lanes or bicycle lane extensions (see Section 9E.09).**

07   **Extensions of bicycle lanes through intersections shall use dotted line patterns.**

Support:

08   Separated and buffer-separated bicycle lanes may have alignments that are not as obvious within an intersection as a standard bicycle lane, therefore additional conspicuity is important where these types of bicycle lanes cross intersections.

Guidance:

09   *Lane extension markings should be used to extend a buffer-separated or separated bicycle lane through intersections and driveways.*

10   *The extension of a bicycle lane through an intersection should use two lines defining both lateral limits of the bicycle lane.*

Standard:

11   **Where the path of the bicycle lane through the intersection is contiguous to a crosswalk, two longitudinal dotted lines shall be provided to establish the lateral limits of the bicycle lane extension. The transverse line establishing one side of the crosswalk, or the limit of a high-visibility crosswalk pattern (see Section 3C.05) that does not employ a transverse line, shall not be used to demarcate one side of the bicycle lane extension.**

Section 9E.04 Bicycle Lanes at Driveways

Support:

01   The definition of an "Intersection" in Section 1C.02 contains information to determine if a driveway can be considered an intersection.

Option:

02   Bicycle lanes may be continued through a driveway using solid or dotted longitudinal lines.

03   The bicycle symbol, the arrow marking, pavement word markings, or a combination thereof may be used in bicycle lane extensions through driveways.

04   Green-colored pavement (see Section 3H.06) may be used as a background to enhance the conspicuity of the bicycle symbol at driveways.

Section 9E.05 Bicycle Lanes at Circular Intersections

Standard:

01   **Bicycle lanes shall not be provided in the circulatory roadway of an unsignalized circular intersection that includes conflicts at entry or exit points (see Chapter 3D) except as provided in Paragraph 4 of this Section.**

Guidance:

02   *Bicycle lane markings should stop at least 100 feet before the crosswalk, or if no crosswalk is provided, at least 100 feet before the yield line, or if no yield line is provided, then at least 100 feet before the edge of the circulatory roadway.*

03   *If used, bicycle crossings should be a minimum of 20 feet from the edge of the circular roadway.*

Option:

04   Separated bicycle lanes may be used in circular intersections.

Support:

05   Separated bicycle lanes allow bicycles to navigate a circular intersection and its crossing points without merging into traffic and without dismounting and using a crosswalk at the intersection crossing point. This is beneficial at multi-lane and higher-speed circular intersections.

06   Section 9E.10 contains information on using shared-lane markings to facilitate the bicycle movement through a circular intersection.

07   The "Guide for the Development of Bicycle Facilities," 2012 Fourth Edition, American Association of State Highway and Transportation Officials, contains information on designing for bicycles on the sidewalk in lieu of, or in addition to, using shared-lane markings in the circulatory roadway of the intersection.

08   The FHWA's informational guide "Improving Intersections for Pedestrians and Bicyclists" contains information on incorporating separated bicycle lanes and other bicycle facilities into circular intersections.

Section 9E.06 Buffer-Separated Bicycle Lanes

Support:

01   Buffer-separated bicycle lanes provide additional lateral separation between a bicycle lane and a general-purpose lane by a pattern of pavement markings without the presence of vertical elements. Providing a buffer space between a bicycle lane and a general-purpose lane creates more separation between motor vehicles and bicycles, can reduce vehicle encroachment into the bicycle lane, and can increase the comfort of bicyclists.

02   Providing a buffer space between a bicycle lane and a parking lane can reduce crashes involving bicycles and the opening of vehicle doors from the parking lane.

Standard:

03   **If used, and except as provided in Paragraph 5 of this Section, a buffer space shall be marked with a solid white line along both edges of the buffer space where crossing is discouraged.**

Guidance:

04   *Engineering judgment should be used to establish intermittent breaks or interruptions in the buffer space, such as for driveways, transit stops, or on-street parallel parking lanes, in order to convey access points or an otherwise general legal movement to cross the buffer space (see Figure 9E-6).*

Option:

05   Buffer spaces may be established without specific longitudinal lines if contiguous facilities have longitudinal lines or other pavement markings themselves that, when installed, automatically demarcate the buffer space (see Drawing D in Figure 9E-6).

Figure 9E-6. Examples of Markings for Buffer-Separated Bicycle Lanes

A – Buffer-separated bicycle lanes on a two-way street with no parking

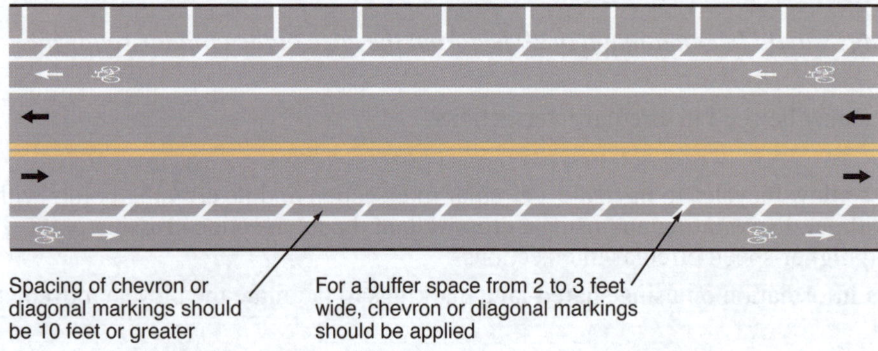

B – Buffer-separated bicycle lanes on a two-way street with on-street parking

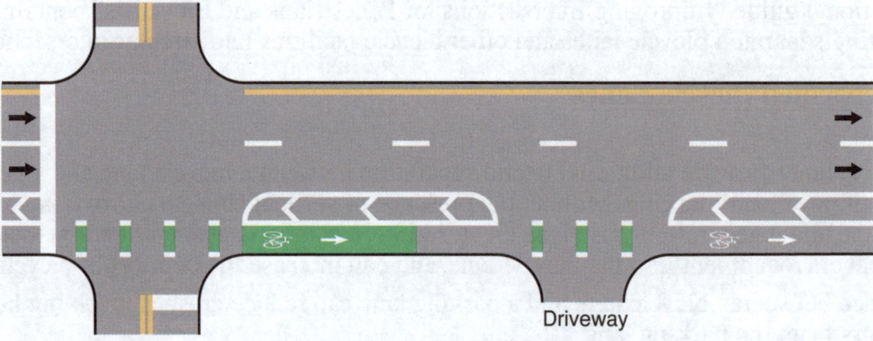

C – Intersection and driveway with no on-street parking

D – Contiguous buffer on both sides of the bicycle lane

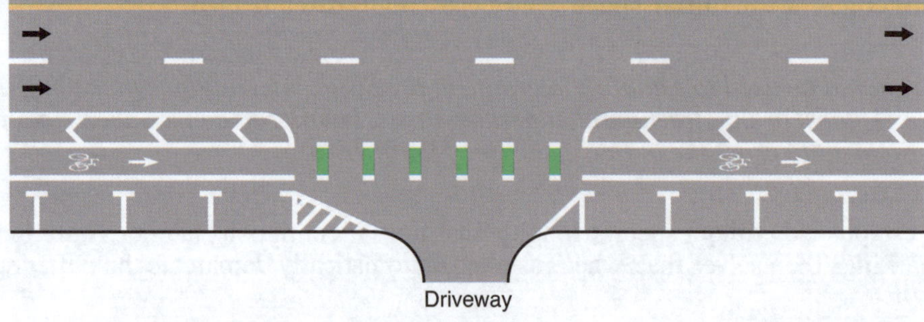

Standard:

06 **Except as provided in Paragraph 7 of this Section, a through buffer-separated bicycle lane shall not be positioned to the right of a mandatory right-turn lane or to the left of a mandatory left-turn lane.**

Option:

07 A buffer-separated bicycle lane may be placed to the right of a mandatory right-turn lane (or to the left of a mandatory left-turn lane) only if a bicycle signal face (see Section 4H.01) is used and the signal phasing and signing eliminates any potential conflicts between the bicycle movement and the turning movement.

Guidance:

08 *The width of the buffer space should be at least 3 times the width of the normal or wide longitudinal line used to mark the buffer space.*

09 *Where a buffer space is 2 to 3 feet wide, chevron or diagonal markings (see Section 3B.25) should be applied within the buffer space.*

Option:

10 Where a buffer space is less than 2 feet wide, diagonal markings or no markings at all in the buffer space may be applied within the buffer space.

Standard:

11 **If used, diagonal markings shall slant away from traffic in the adjacent travel lane for motor-vehicle traffic.**

Guidance:

12 *Where used, the spacing of chevrons or diagonal markings should be 10 feet or greater.*

Support:

13 Chevron and diagonal markings convey that the buffer space is not an additional bicycle lane or other travel lane open to traffic.

Standard:

14 **Where a buffer space is more than 3 feet wide, chevron or diagonal markings shall be applied within the buffer space.**

Guidance:

15 *Lane extension markings should be used to extend a buffer-separated bicycle lane across intersections and driveways.*

Section 9E.07 <u>Separated Bicycle Lanes</u>

Support:

01 Separated bicycle lanes provide a physical separation between a general-purpose lane and a bicycle lane through the use of vertical objects or vertical separation between the general-purpose lane and bicycle lane. Providing a physical separation between a bicycle lane and a general-purpose lane can reduce vehicle encroachment into the bicycle lane beyond a marked buffer alone and can in some cases prevent that encroachment altogether.

02 Physical separation between general-purpose lanes and bicycle lanes introduces additional design considerations over buffer-separated bicycle lanes, including the awareness of a potentially unexpected conflict point for turning motor vehicles and the provision of adequate sight distance for all users at intersections and driveway crossings.

Option:

03 Vertical elements used to provide physical separation between general-purpose lanes and bicycle lanes may include, but are not limited to, tubular markers, raised islands, or parked vehicles.

Support:

04 Where on-street parking is provided adjacent to the buffer area of a separated bicycle lane, pedestrians will need to access those vehicles.

Guidance:

05 *BIKE LANE (R3-17) signs (see Figure 9B-1) should be used to distinguish a separated bicycle lane from a general-purpose lane.*

06 *Where an on-street parking lane serves as the separation between a general-purpose lane and a separated bicycle lane, a buffer space should be provided between the parking lane and the bicycle lane to allow for opening doors of parked vehicles.*

Support:

07   Separated bicycle lanes may be designed for one-way or two-way bicycle travel. Providing one-way separated bicycle lanes in the same direction as and on the right-hand side of the general-purpose lane, whether on a one-way or two-way roadway, accommodates the expectations of road users and might result in fewer conflict points at intersections or driveway crossings.

Option:

08   Separated bicycle lanes may be provided on one or both sides of a roadway or in a center median.

Support:

09   The presence of two-way separated bicycle lanes on one side of a roadway or in a center median can introduce additional challenges and conflict points, which can warrant additional design considerations when selecting the design for a separated bicycle lane. These considerations include design requirements for pedestrians who would interact with the separated bicycle lane.

Standard:

10   **The edge line and lane line colors used for separated bicycle lanes shall conform to the requirements in Chapter 3A (see Figure 9E-7).**

11   **Directional arrows shall be used in conjunction with the bicycle lane symbol or word marking in separated bicycle lanes, placed downstream from the symbol or word marking.**

12   **Turns on red shall be prohibited across separated bicycle lanes while bicyclists are allowed to proceed through the intersection.**

Support:

13   Additional information on signals for bicycle facilities is found in Chapter 4H.

Standard:

14   **The buffer space for a separated bicycle lane shall be marked with solid longitudinal lines.**

15   **A marked buffer space that is 2 feet or wider for a separated bicycle lane, including those buffer spaces where tubular markers are provided, shall use chevron or diagonal markings within the buffer, unless physical separation is provided that occupies the majority of the buffer space, such as raised islands or other physical dividers, or such as where an on-street parking lane occupies the majority of the buffer space.**

Guidance:

16   *Where used in the buffer area of a separated bicycle lane, the spacing of chevrons or diagonal markings should be 10 feet or greater.*

17   *Crosswalks that cross a separated bicycle lane should be marked consistent with the style of crosswalk marking provided across the adjacent general-purpose lane.*

Support:

18   Where on-street parking is provided as the physical separation adjacent to the buffer area of a separated bicycle lane, the chevron or diagonal marking provisions in Section 9E.06 apply to the area outside of the marked parking area within the buffer (see Figure 9E-7).

19   Intersection treatments for separated bicycle lanes can vary depending on the geometric and operational conditions at the intersection (see Section 9E.02).

Section 9E.08 Counter-Flow Bicycle Lanes

Support:

01   Counter-flow bicycle lanes are one-directional and provide a lawful path of travel for bicycles in the opposite direction from general traffic on a roadway that allows general traffic to travel in only one direction.

02   Counter-flow bicycle lanes establish two-way traffic on a roadway. Section 9B.21 contains information on the Left Turn Yield to Bicycles (R10-12b) sign used with traffic signals and counter-flow bicycle lanes.

Guidance:

03   *Where used, a counter-flow bicycle lane should be marked such that bicycles in the counter-flow lane travel on their right-hand side of the road in accordance with normal rules of the road, with opposing traffic on the left.*

Standard:

04   **Counter-flow bicycle lanes located at the edge of the roadway shall use double yellow center line pavement markings (see Section 3B.01), a painted median island, a raised median island (see Chapter 3J), or some form of physical separation where the speed limit is 30 mph or less.**

05   **For speed limits 35 mph or greater, a buffer per Section 3B.25, a painted or raised median island, or some form of physical separation shall be used to separate a counter-flow bicycle lane from the adjacent travel lane.**

Figure 9E-7. Examples of Lane Markings for Separated Bicycle Lanes (Sheet 1 of 2)

A – One-way bicycle lanes on a two-way street

Note: Diagonal or chevron markings shall be used if buffer width is 2 feet or greater for separated bicycle lanes.

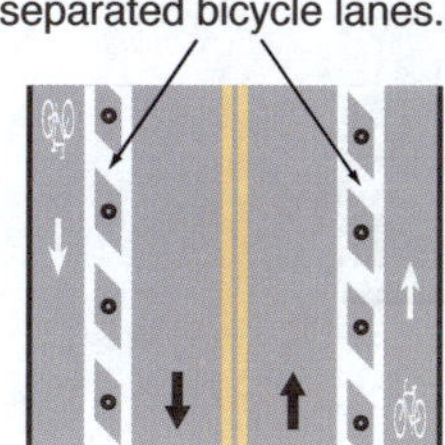

B – One-way bicycle lane on a one-way street behind on-street parking

Note: Parking provides the vertical element of this separated bicycle lane.

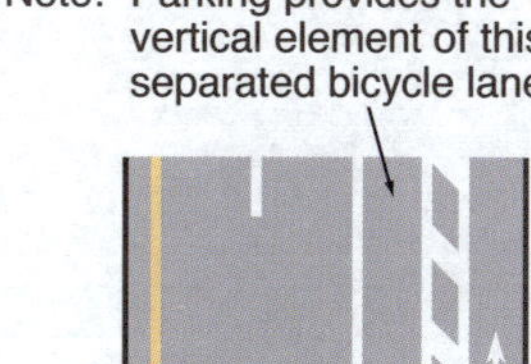

C – Two-way bicycle lane on a one-way street

Note: Diagonal or chevron markings shall be used if buffer width is 2 feet or greater for separated bicycle lanes.

D – Separated bicycle lane shifted toward the adjacent general-purpose lane

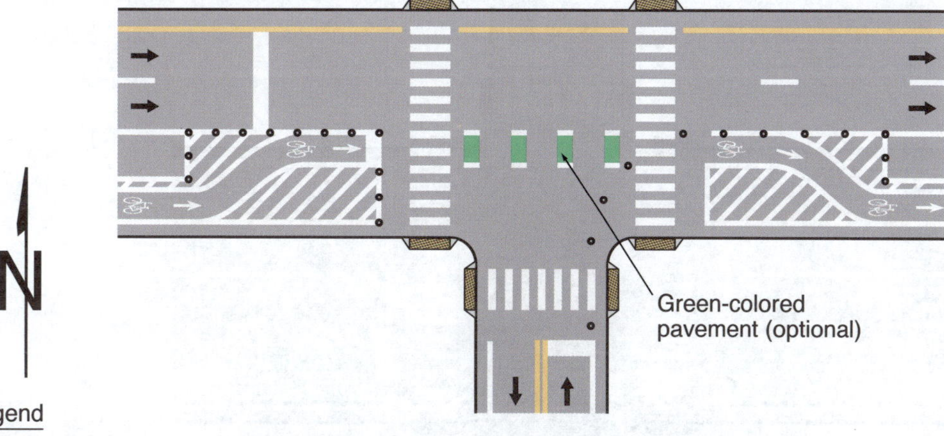

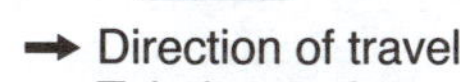

Legend

→ Direction of travel
∘ Tubular marker

E – Separated bicycle lane shifted away from traffic

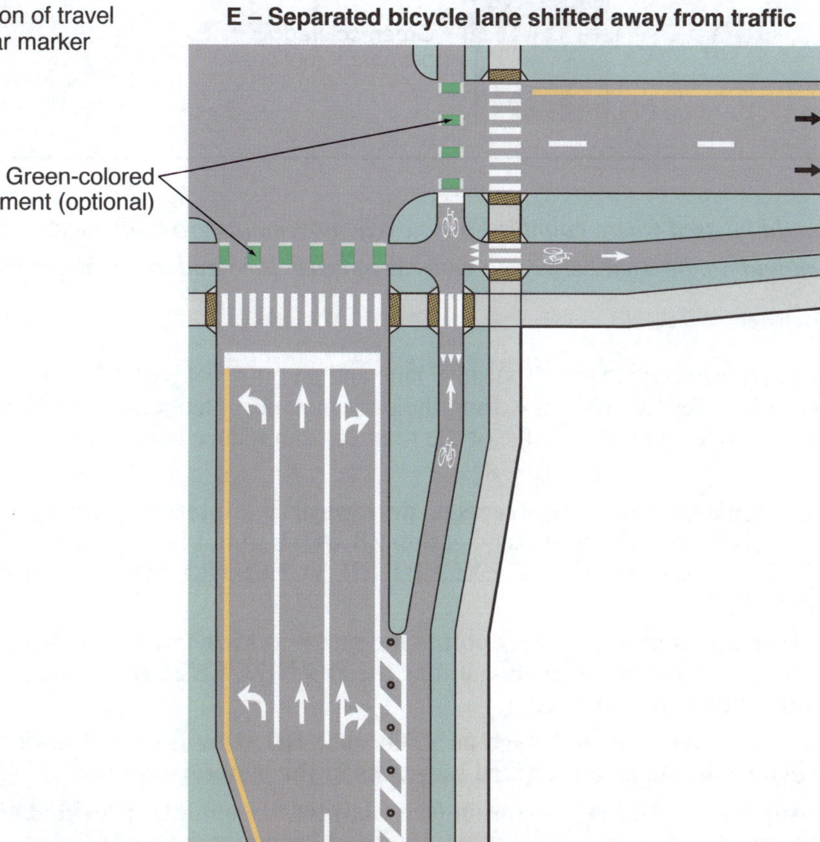

Figure 9E-7. Examples of Lane Markings for Separated Bicycle Lanes (Sheet 2 of 2)

F – Separated bicycle lane at an intersection

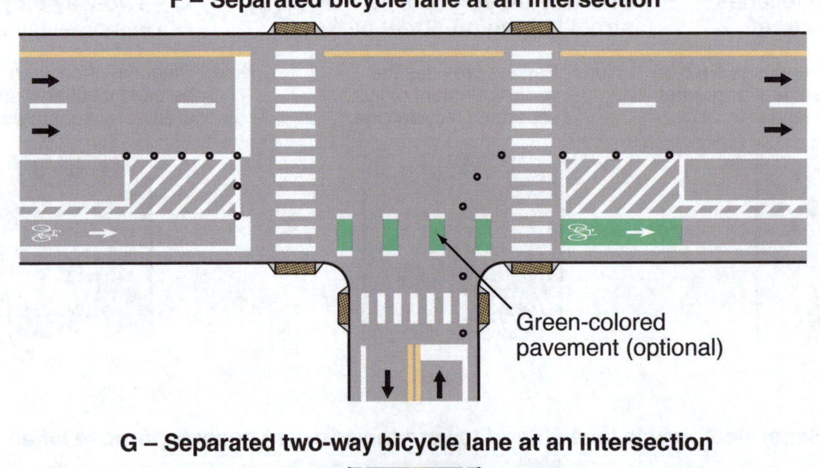

G – Separated two-way bicycle lane at an intersection

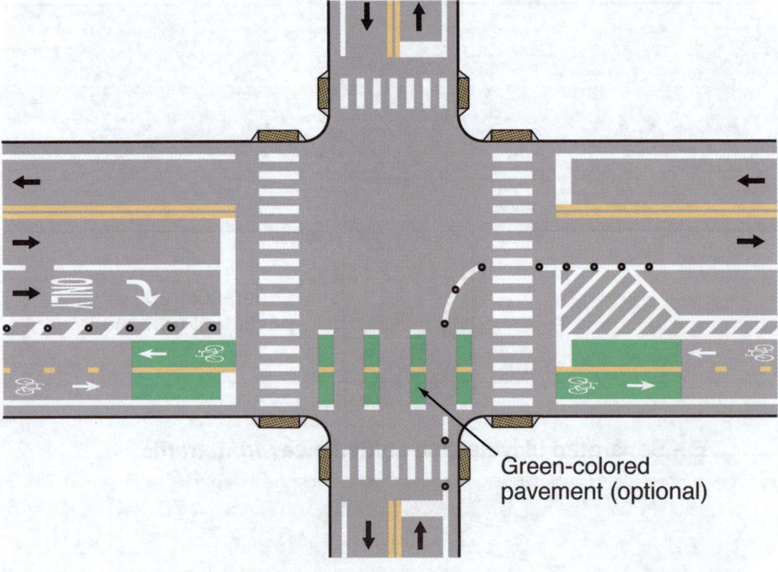

Guidance:

06   *Lane extension markings should be used where counter-flow bicycle movements cross intersections.*

07   *Counter-flow bicycle lanes should not be used between a general-purpose lane and an on-street parallel parking lane for motor vehicles.*

Support:

08   Counter-flow bicycle lanes located between a general-purpose lane and an on-street parallel parking lane for motor vehicles can limit visibility of bicycles for vehicles exiting the parking lane, potentially impacting the safety of bicyclists. Locating counter-flow bicycle lanes at the edge of the roadway can reduce conflicts for bicycles.

Standard:

09   **Where signs are provided to regulate turns from streets or driveways that intersect with a roadway that has a counter-flow bicycle lane, ONE WAY signs (see Section 2B.49) shall not be used. Movement Prohibition signs (see Section 2B.26) with supplemental EXCEPT BICYCLES (R3-7bP) regulatory plaque(s) shall be used (see Figure 9E-8).**

10   **If a DO NOT ENTER (R5-1) sign(s) is used at egress points for motor vehicle traffic, the EXCEPT BICYCLES (R3-7bP) regulatory plaque(s) shall be placed under the DO NOT ENTER sign (see Figure 9E-8) where a counter-flow bicycle lane is used.**

11   **Where intersection traffic controls are provided (such as STOP or YIELD signs or traffic signals), appropriate devices shall be provided and oriented toward bicyclists in the counter-flow lane.**

12   **At signalized locations, appropriate bicycle signalization (see Chapter 9F) shall be provided and oriented toward bicyclists in the counter-flow lane, including a method for counter-flow bicycles to actuate the green phase for the counter-flow movement.**

Support:

13   Higher levels of traffic control or additional signalization, signing, and/or pavement marking treatments can be helpful for intersecting traffic where the counter-flow bicycle movement is unexpected.

Guidance:

14   *A Bicycle Cross Traffic warning plaque (see Section 9C.06) should be used below a STOP sign on the crossroad at intersections where a counter-flow bicycle lane is provided on the primary street.*

Section 9E.09 Shared-Lane Marking

Support:

01   The "Standard Highway Signs" publication (see Section 1A.05) contains details on the shared-lane marking symbol.

Option:

02   The shared-lane marking shown in Figure 9E-9 may be used to:

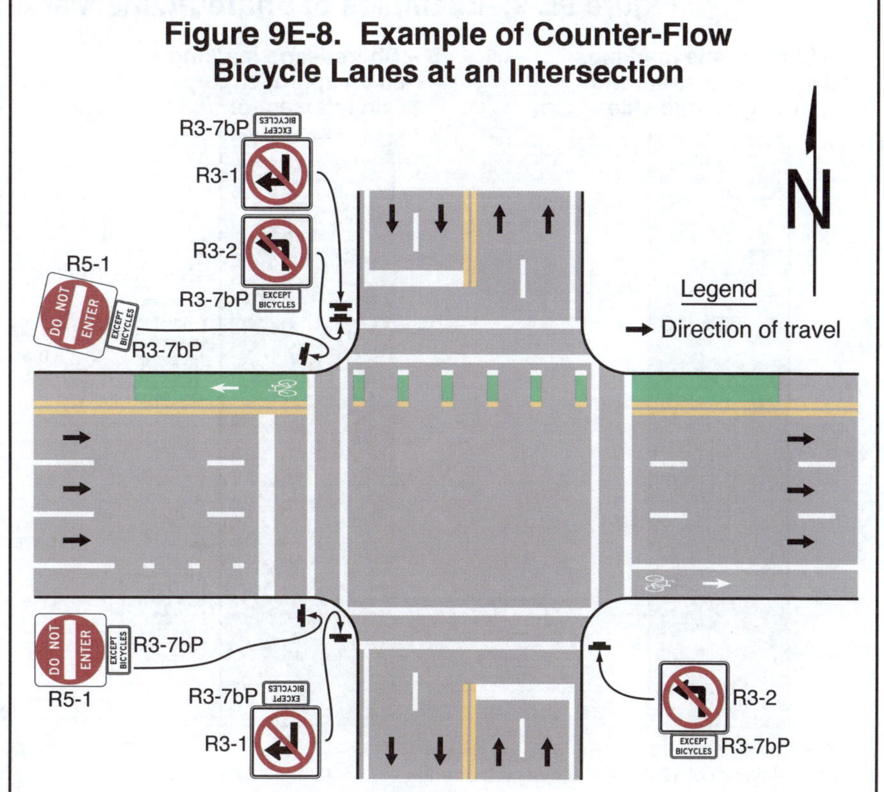

    A. Assist bicyclists with lateral positioning in a shared lane with on-street parallel parking in order to reduce the chance of a bicyclist impacting the open door of a parked vehicle,

    B. Assist bicyclists with lateral positioning in lanes that are too narrow for a motor vehicle and a bicycle to travel side-by-side within the same traffic lane,

    C. Alert road users of the lateral location bicycles are likely to occupy within the traveled way,

    D. Encourage safe passing of bicycles by motor vehicles,

    E. Reduce the incidence of wrong-way bicycling in the roadway, and

    F. Assist bicyclists with lateral positioning in mixing zones.

Guidance:

03   *The shared-lane marking should not be placed on roadways that have a speed limit of 40 mph or greater.*

Standard:

04   **Shared-lane markings shall not be used in:**

    A. Shoulders;

    B. Bicycle lanes or in designated extensions of bicycle lanes through intersections or driveways,

    C. A travel lane in which light-rail transit vehicles also travel;

    D. The transition area where a motor vehicle entering a mandatory turn lane must weave across bicyclists in bicycle lanes;

    E. Two-stage turn boxes;

    F. Bicycle boxes;

    G. Shared-use paths or shared-use path crossings; or

    H. Physically-separated bikeways, either in the roadway or on an independent right-of-way.

05   **Green-colored pavement shall not be applied as a background to shared-lane markings (see Section 3H.06).**

Option:

06   Black background markings (see Section 3A.03) may be used in combination with shared-lane markings to enhance contrast.

Figure 9E-9. Examples of Shared-Lane Marking Applications

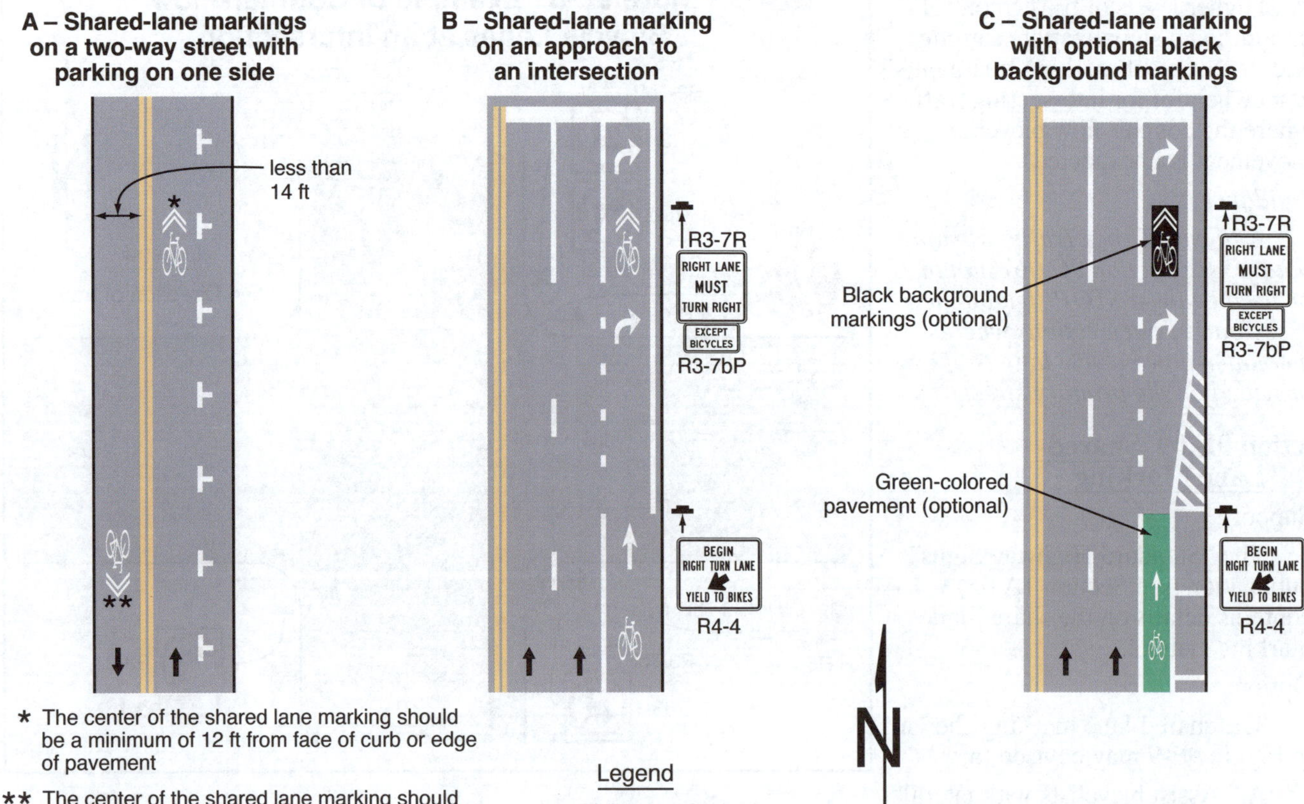

* The center of the shared lane marking should be a minimum of 12 ft from face of curb or edge of pavement

** The center of the shared lane marking should be a minimum of 4 ft from face of curb or edge of pavement

Note: For shared-lane marking design details, see the Pavement Markings chapter of the Standard Highway Signs publication

Guidance:

07   *If used in a shared lane with on-street parallel parking, shared-lane markings should be placed so that the centers of the markings are a minimum of 12 feet from the face of the curb, or from the edge of the pavement where there is no curb.*

08   *If used on a street without on-street parking that has an outside travel lane that is less than 14 feet wide, shared-lane markings should be placed so that the centers of the markings are a minimum of 4 feet from the face of the curb, or from the edge of the pavement where there is no curb.*

09   *At non-intersection locations, the shared-lane marking should be spaced at intervals of not less than 50 feet or greater than 250 feet.*

10   *The first shared-lane marking downstream from an intersection should be placed no more than 50 feet from the intersection.*

Option:

11   Section 9B.14 describes a Bicycles Allowed Use of Full Lane sign that may be used in addition to or instead of the shared-lane marking to inform road users that bicyclists might occupy the travel lane.

Guidance:

12   *If the Bicycles Allowed Use of Full Lane (R9-20) sign is used as an addition to shared-lane marking, the shared-lane marking should be placed so that the center of the marking is in the approximate center of the travel lane.*

Option:

13   The shared-lane marking may be used (see Figure 9E-9) where the width of the roadway is insufficient to continue a bicycle lane or separated bikeway on the approach to the intersection, or it is advantageous to terminate the bicycle lane or separated bikeway in order to provide for a shared lane.

14   The shared-lane marking may be used on an approach to an intersection (see Figure 9E-5) in a mandatory turn lane to indicate a shared space for bicyclists and motorists where there is insufficient width in the roadway for both the bicycle lane and turn lane.

Section 9E.10 Shared-Lane Markings for Circular Intersections

Option:

01   Shared-lane markings may be used in the circulatory roadway of circular intersections.

Guidance:

02   *If used, shared-lane markings should be placed in the center of the lane when used inside of circulatory roadways.*

Support:

03   The "Guide for Development of Bicycle Facilities," 2012 Fourth Edition, American Association of State Highway and Transportation Officials, contains information on designing for bicycles on shared-used paths in lieu of, or in addition to, using shared-lane markings in the circulatory roadway of the intersection.

Section 9E.11 Two-Stage Bicycle Turn Boxes

Support:

01   Two-stage bicycle turn boxes allow bicyclists the opportunity to make turns at an intersection or crossing point instead of requiring them to merge into traffic upstream or to dismount and use a crosswalk at the intersection or crossing point.

02   Section 9B.18 contains information on regulatory signing that shall be used in conjunction with a two-stage bicycle turn box pavement marking where bicyclists are required to use the turn box.

03   Section 9D.13 contains information on guide signing that can be used in conjunction with a two-stage bicycle turn box pavement marking where bicyclists are not required to use the turn box.

Standard:

04   **If used, two-stage bicycle turn boxes shall be located:**

    A. In an area between the closest through bicycle or motor vehicle movement and the parallel crosswalk (see Drawing A in Figure 9E-10),

    B. In an area between the through bicycle movement and the parallel pedestrian crossing movement if no crosswalk is established (see Drawing B in Figure 9E-10),

    C. On the innermost side of the bicycle facility provided that the two-stage turn box is located in a portion of the intersection where parallel or motor vehicle traffic does not travel, such as projections of islands or parking lanes (see Drawing C in Figure 9E-10), or

    D. In an area between the through bicycle movement and a pedestrian facility for T-intersections (see Drawing D in Figure 9E-10).

05   **A two-stage bicycle turn box shall consist of at least one bicycle symbol pavement marking and at least one pavement marking arrow.**

06   **A turn arrow in the appropriate direction shall be used if a two-stage turn box is used with a one-way bicycle lane, and a through arrow in the appropriate direction shall be used if a two-stage turn box is used with a two-way bikeway (see Figure 9E-11).**

07   **A two-stage bicycle turn box shall be bounded on all sides by a solid white line.**

08   **For two-stage bicycle turn boxes that facilitate turns from a one-way bikeway, the bicycle symbol shall precede the pavement marking turn arrow in the direction of bicycle travel (see Figure 9E-10).**

09   **Passive detection of bicycles in the two-stage bicycle turn box shall be provided if the signal phase that permits bicycles to enter the intersection during the second stage of their turn is actuated.**

Guidance:

10   *Engineering judgment should be used to develop the size of the two-stage bicycle turn box. Factors considered should include intersection geometry and keeping queued bicycles away from moving traffic, as well as peak hour bicycle volumes to avoid overflow of the two-stage turn box that subjects any bicyclist to conflicting movements.*

Option:

11   The two-stage turn box may use green-colored pavement.

Standard:

12   **If used, green-colored pavement shall encompass all of the two-stage turn box.**

13   **Where the path of vehicles lawfully turning on red would pass through a two-stage bicycle turn box, a full-time no-turn-on-red prohibition (see Section 2B.60) shall be provided for the crossroad approach.**

Figure 9E-10. Examples of Two-Stage Turn Box Locations at Intersections

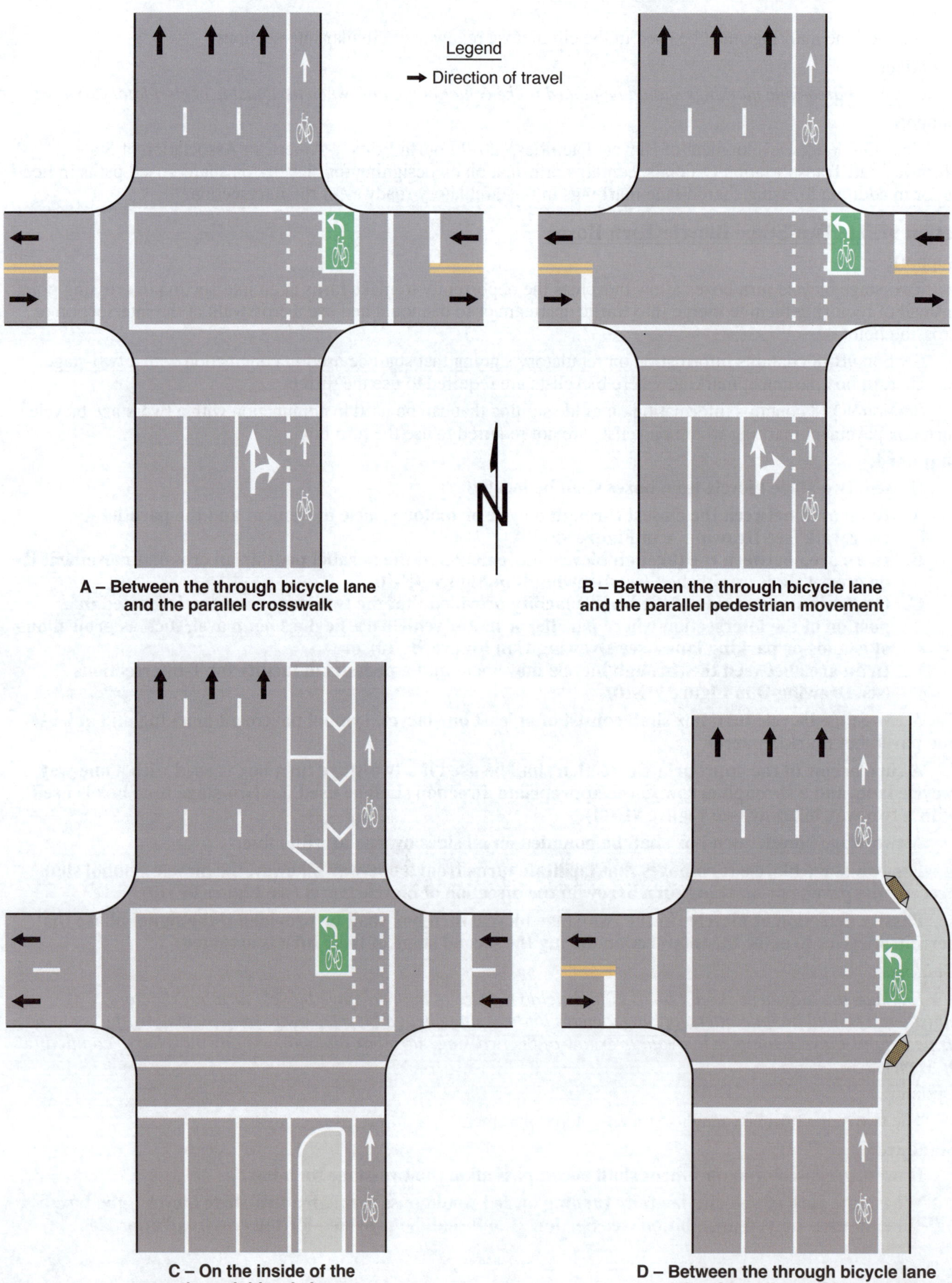

A – Between the through bicycle lane and the parallel crosswalk

B – Between the through bicycle lane and the parallel pedestrian movement

C – On the inside of the through bicycle lane

D – Between the through bicycle lane and the sidewalk at a T-intersection

Section 9E.12 Bicycle Box

Option:

01   A bicycle box (see Figure 9E-12) may be used to increase the visibility of stopped bicycles on the approach to a signalized intersection during the portion of the signal cycle when a red signal indication is being displayed to motor vehicles in the approach lane(s) that is behind the box.

Guidance:

02   *Providing a bicycle box on a signalized intersection approach where a discernible number of conflicts between vehicles turning across through bicycles in a bicycle lane has been demonstrated during the green interval of a signal should be evaluated based on engineering judgment or study.*

03   *Other treatments should be considered for conflicts between turning vehicles and through bicycles such as using leading or exclusive signal phases, or separating turning traffic from through traffic through mandatory turn lanes.*

04   *A bicycle lane should be used on the approach to a bicycle box.*

05   *A bicycle box should not be contiguous with a crosswalk. A stop line on the downstream end of the bicycle box should be used to mark the location where bicycles are required to stop.*

Standard:

06   **If used, the distance from the upstream edge of the bicycle box that is nearest to the stop line for motor vehicles to the downstream edge of the bicycle box that is nearest the crosswalk or intersection shall be at least 10 feet. At least one bicycle symbol marking (see Figure 9E-12) shall be used in the bicycle box.**

07   **Where an existing stop line for motor vehicles is relocated upstream to install a new bicycle box, the yellow change and red clearance intervals (see Section 4F.17) shall be recalculated and if necessary, reprogrammed to accommodate the length of the bicycle box.**

08   **Countdown pedestrian signals (see Section 4I.04) for the crosswalk or pedestrian crossing movement that crosses the approach shall accompany bicycle boxes that extend across more than one approach lane for motor vehicles. Countdown pedestrian signals used with bicycle boxes shall display the pedestrian change interval countdown without the need for actuation.**

09   **Turns on red shall be prohibited from the lane where a bicycle box is placed.**

Support:

10   Countdown pedestrian signals can inform bicyclists whether there is adequate time remaining to an adjacent lane before the onset of the green signal phase for that approach.

Guidance:

11   *Countdown pedestrian signals for the crosswalk or pedestrian crossing movement that crosses the approach should accompany single-lane bicycle boxes where it is demonstrated that bicycles arrive at the intersection at or near the end of the red signal indication being displayed to traffic in the approach lane(s) that is behind the box.*

Option:

12   Green-colored pavement may be used in a bicycle box.

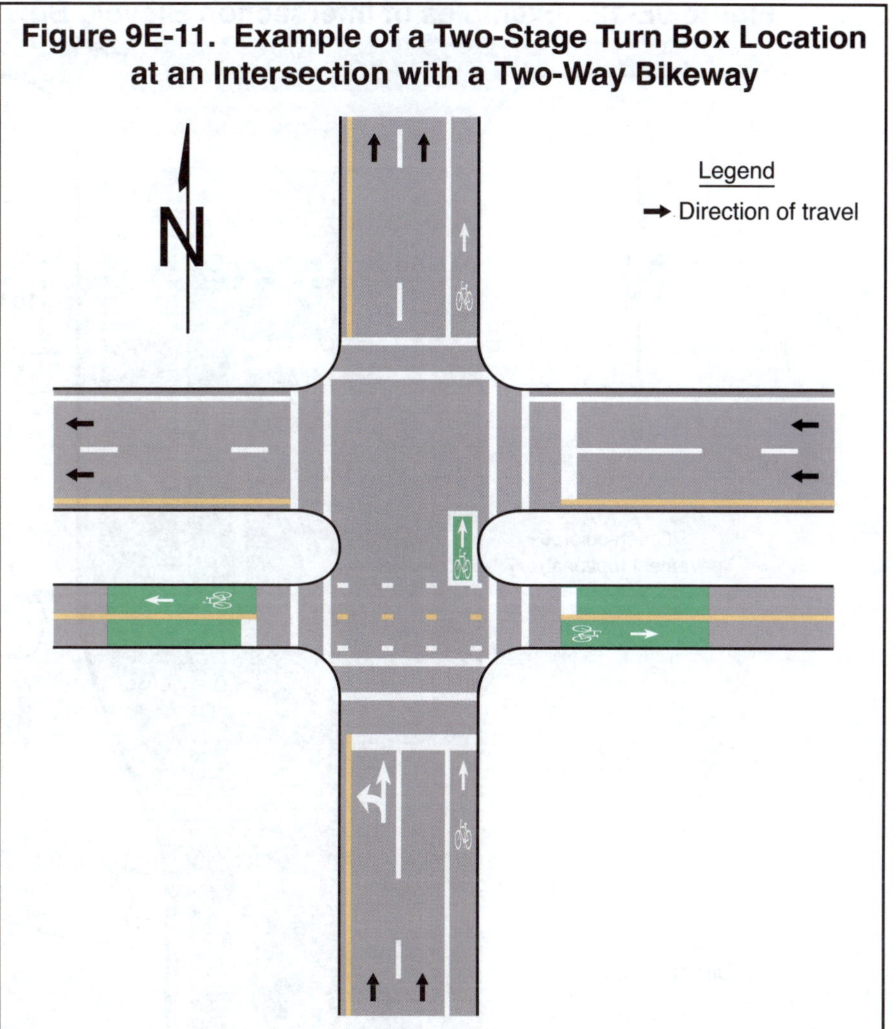

Figure 9E-11. Example of a Two-Stage Turn Box Location at an Intersection with a Two-Way Bikeway

Figure 9E-12. Examples of Intersection Bicycle Boxes (Sheet 1 of 2)

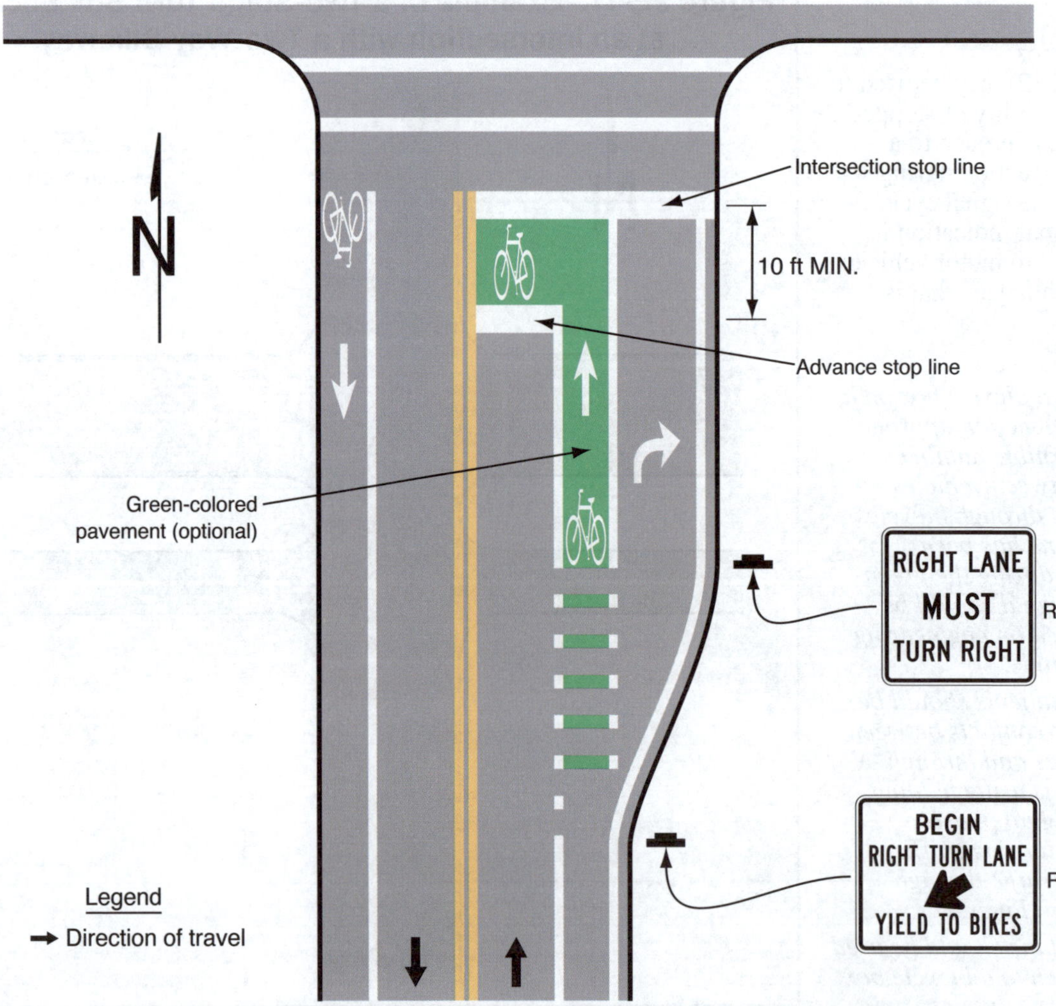

A – Bicycle box across one lane

Standard:

13   **If used, green-colored pavement shall be used in the full limits of the bicycle box.**

Support:

14   Section 9B.02 contains information on the EXCEPT BICYCLES (R3-7bP) regulatory plaque that can be used below the STOP HERE ON RED (R10-6 or R10-6a) sign (see Section 2B.59) to exempt bicyclists from the requirement of the advance stop line.

Section 9E.13 Shared-Use Paths

Option:

01   Where shared-use paths are of sufficient width to designate two minimum width lanes, a solid yellow center line may be used to separate the two directions of travel where passing or traveling to the left of the line is not permitted. A broken yellow center line may be used where passing is permitted (see Figure 9E-13).

Guidance:

02   *Broken lines used on shared-use paths should have a nominal 3-foot segment with a 9-foot gap.*

Option:

03   A solid white line may be used on shared-use paths to separate different types of users in the same direction. The R9-7 sign (see Section 9B.13) may be used to supplement the solid white line.

04   Smaller size pavement word markings and symbols may be used on shared-use paths. Where arrows are needed on shared-use paths, half-size layouts of the arrows may be used (see Section 3B.20).

Standard:

05   **Where a shared-use path crosses a roadway, crosswalk markings shall be used (see Chapter 3C).**

Figure 9E-12. Examples of Intersection Bicycle Boxes (Sheet 2 of 2)

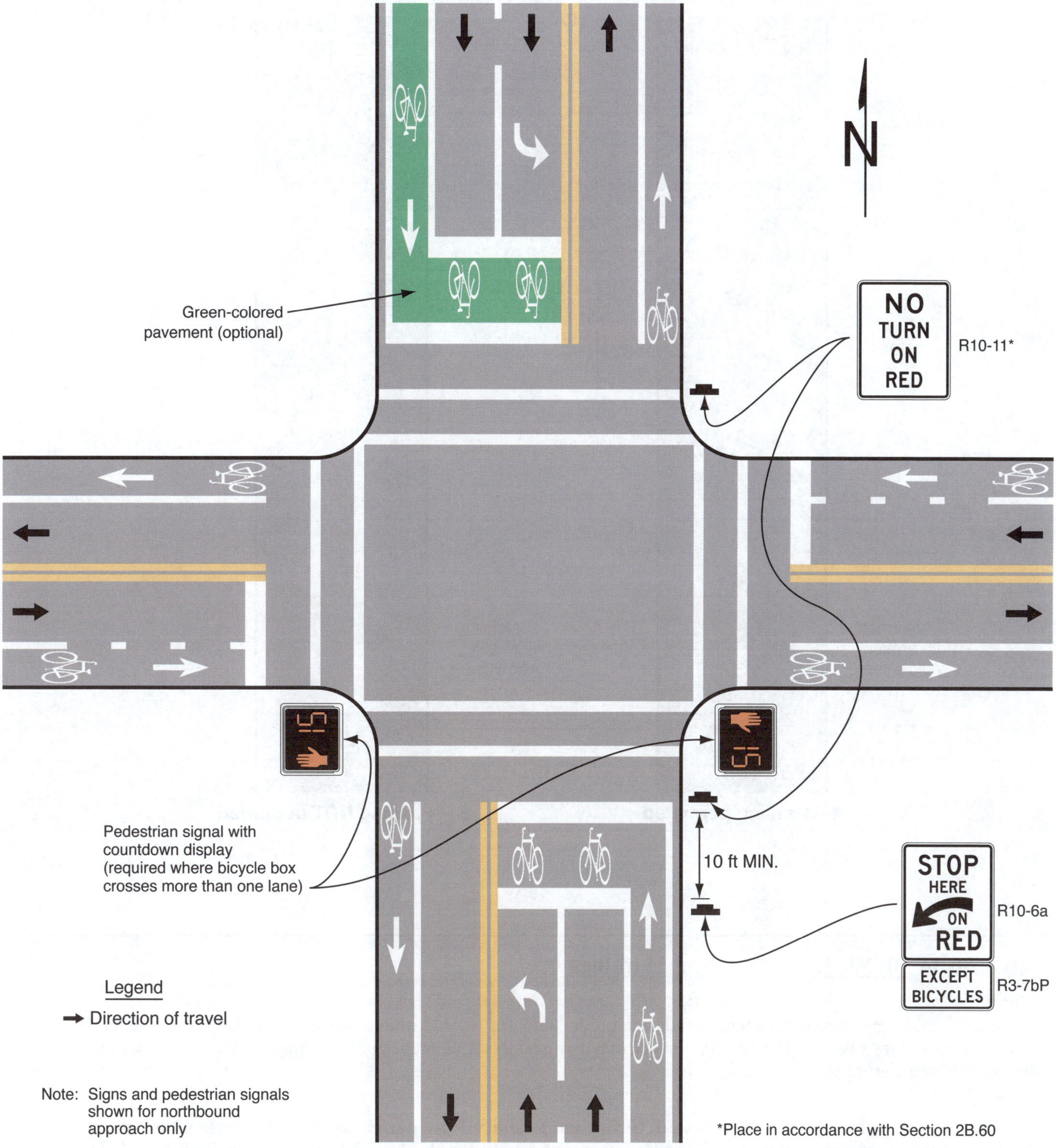

B – Bicycle box across multiple lanes

Option:

06 Where pedestrian and bicycle movements on a shared-use path are separated on the approach to a roadway crossing, parallel bicycle and pedestrian crossing markings may be used as shown in Figure 9E-14.

Guidance:

07 *If parallel bicycle and pedestrian crossing markings are used where a shared-use path crosses a roadway, crossing areas for bicycles should use green-colored pavement if the shared-use path crossing has a high volume of either mode.*

Figure 9E-13. Examples of Center Line Markings for Shared-Use Paths

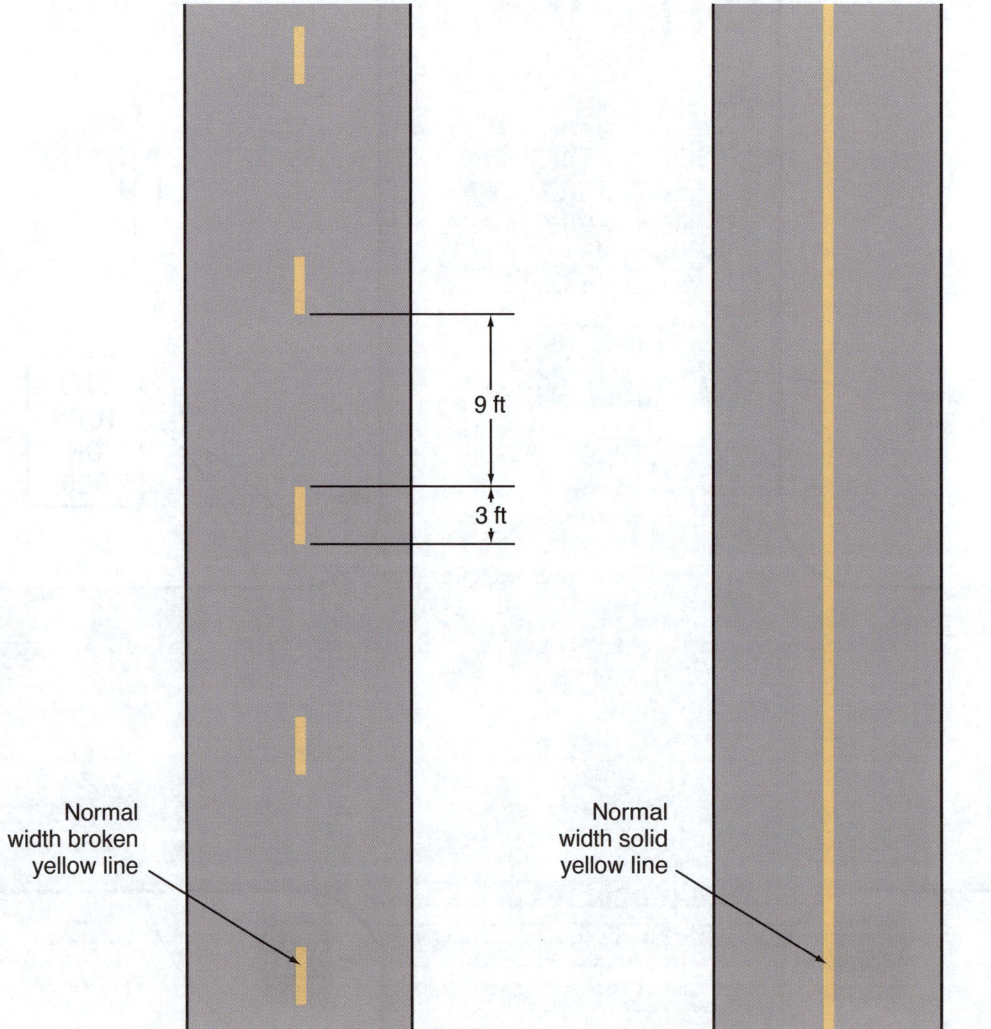

Section 9E.14 Bicycle Route Pavement Markings

Option:

01   Bicycle route pavement markings simulating guide signs for bicycle routes (see Section 9D.02 through 9D.07) and route auxiliary plaques (see Section 9D.08) may be used to supplement guide signing to help bicyclists in navigation (see Figure 9E-15).

Standard:

02   **Bicycle route pavement markings shall be limited to shared-use paths, separated bicycle lanes, or buffer-separated bicycle lanes. Bicycle route pavement markings shall not be used in standard bicycle lanes or in shared lanes.**

Guidance:

03   *A systematic methodology of locating guide signs for bicycle routes adjacent to the bicycle route pavement marking should be used that includes locations where either the sign or the pavement marking can exist alone to avoid overuse of the guide sign or the pavement marking.*

04   *The route marker pavement marking should be elongated.*

05   *The location, size, and materials of the route marker pavement marking should be designed in a manner that will minimize the loss of traction for bicyclists under wet conditions.*

**Figure 9E-14. Example of Pavement Markings for a
Shared-Use Path with Mode Separation**

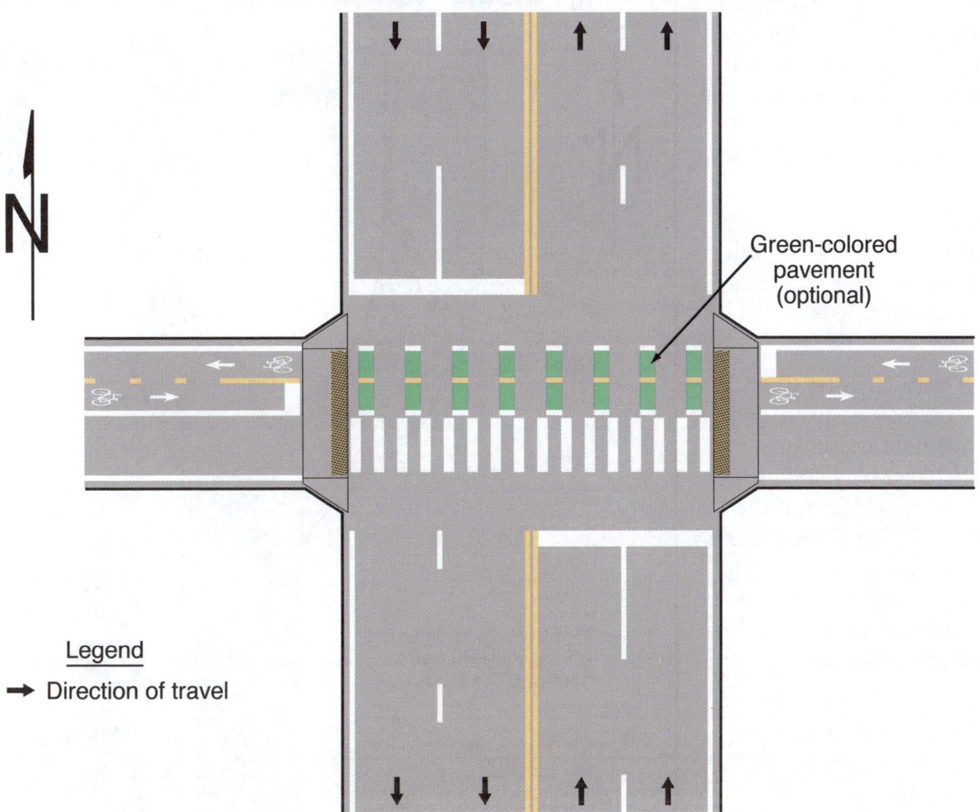

Section 9E.15 Bicycle Detector Symbol

Option:

01   The bicycle detector symbol (see Figure 9E-16) may be placed on the pavement indicating the optimum
position for a bicycle to actuate the signal.

02   Appropriately-sized WAIT HERE FOR GREEN word markings may be placed on the pavement immediately
below the bicycle detector symbol.

03   A R10-22 sign (see Section 9B.20) may be installed to supplement the bicycle detector symbol
pavement marking.

Support:

04   The "Standard Highway Signs" publication (see Section 1A.05) contains details on the bicycle
detector symbol.

05   Section 3H.06 contains information on incorporating green-colored pavement as a background enhancement
to the bicycle detector symbol.

Section 9E.16 Pavement Markings for Obstructions

Guidance:

01   *Markings as shown in Figure 9E-17 should be used at the location of obstructions in the center of a
shared-use path or a physically-separated bikeway, including vertical elements intended to physically prevent
unauthorized motor vehicles from entering the path.*

02   *For roadway situations where it is impracticable to eliminate a drain grate or other roadway obstruction
that is inappropriate for bicycle travel, white markings applied as shown in Figure 9E-17 should be used to guide
bicyclists around the condition.*

Section 9E.17 Raised Devices

Support:

01   Chapter 3I contains information on using channelizing devices to emphasize pavement marking patterns associated with certain bicycle facilities. A common application is the use of flexible raised devices to create separated bicycle lanes (see Section 9E.07).

02   Using inflexible raised devices immediately adjacent to the travel path of a bicyclist without a buffer creates a collision potential for bicyclists.

Option:

03   In accordance with Chapter 3I, channelizing devices may be used to emphasize a pavement marking pattern that establishes a bicycle lane or other bicycle facility provided that the installation of channelizing devices does not prevent motor vehicles from turning when the turn requires the motor vehicle to merge with the bicycle lane or facility as required by law or ordinance.

Guidance:

04   *If used, channelizing devices for bicycle facilities should be tubular markers (see Section 3I.02).*

05   *The selection of a raised device for use with bicycle facilities should consider the collision potential of both the post and the base since the base might still be present in the event the post is struck and missing.*

Support:

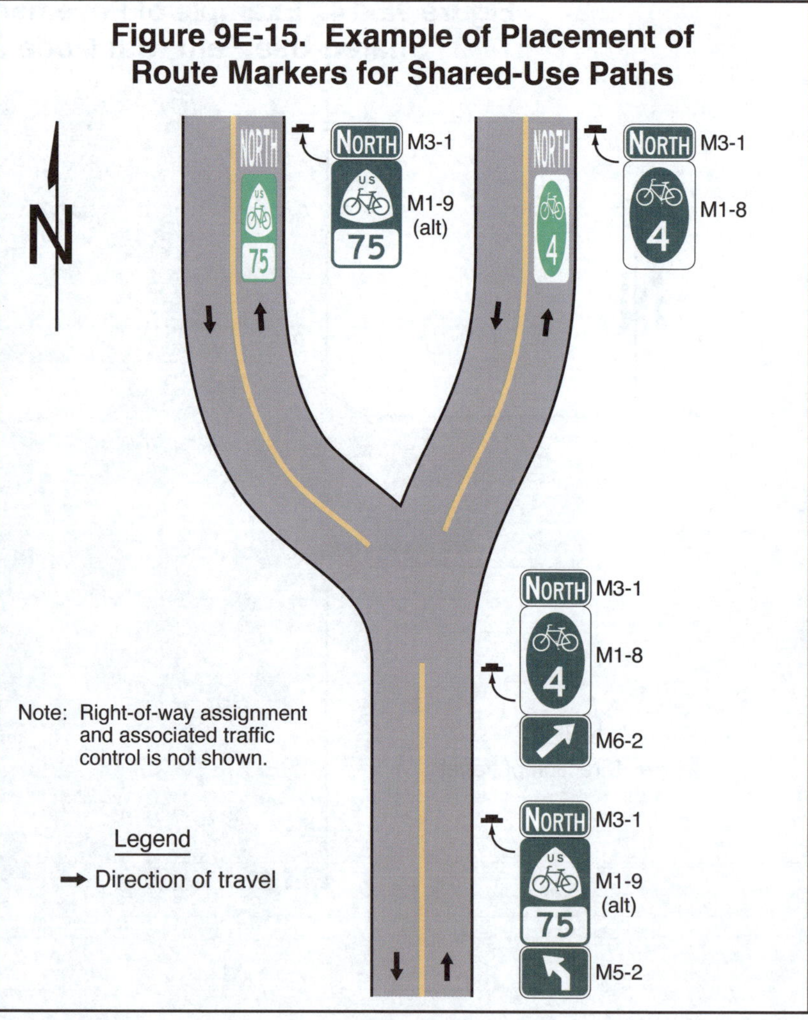

Figure 9E-15. Example of Placement of Route Markers for Shared-Use Paths

06   Measures to reduce the likelihood of a road user striking a channelizing device include marking a buffer space, improving lighting, improving retroreflectivity, or the periodic addition of taller vertical elements within runs of shorter elements.

Standard:

07   **Channelizing devices that are used to emphasize the pavement marking patterns of bicycle facilities shall not incorporate the color green into either the device or its retroreflective element to supplement the presence of green-colored pavement.**

Guidance:

08   *If used in buffer-separated bicycle lanes, channelizing devices should be placed in the buffer space and at least 1 foot from the longitudinal bicycle lane pavement marking.*

Figure 9E-16. Bicycle Detector Pavement Marking

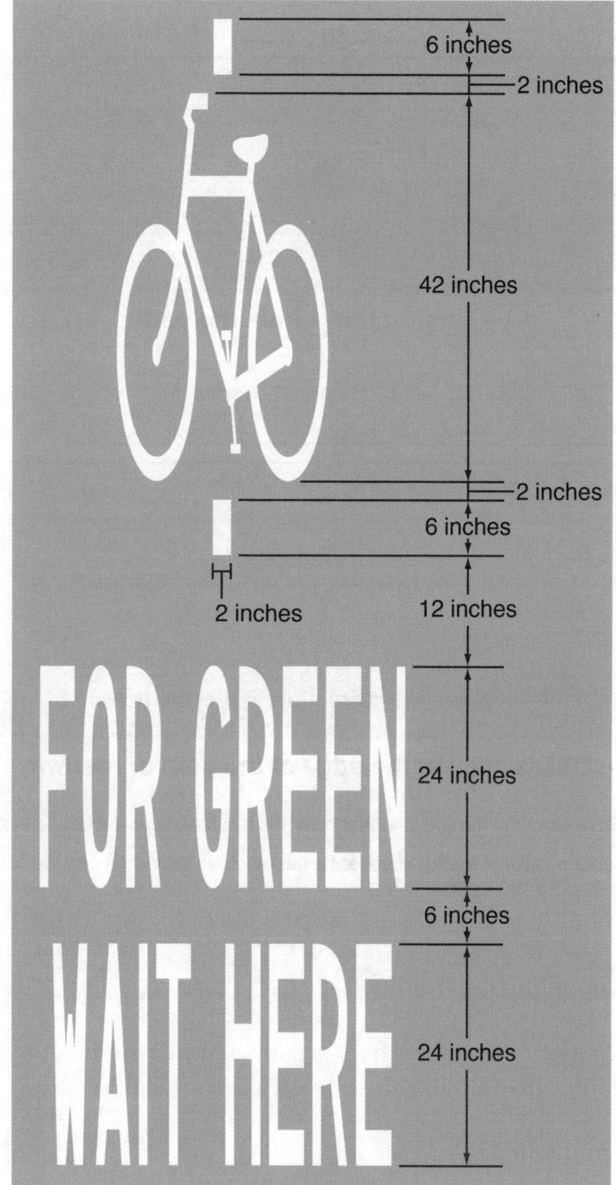

Note: The word pavement markings are optional.

Figure 9E-17. Examples of Obstruction Pavement Markings

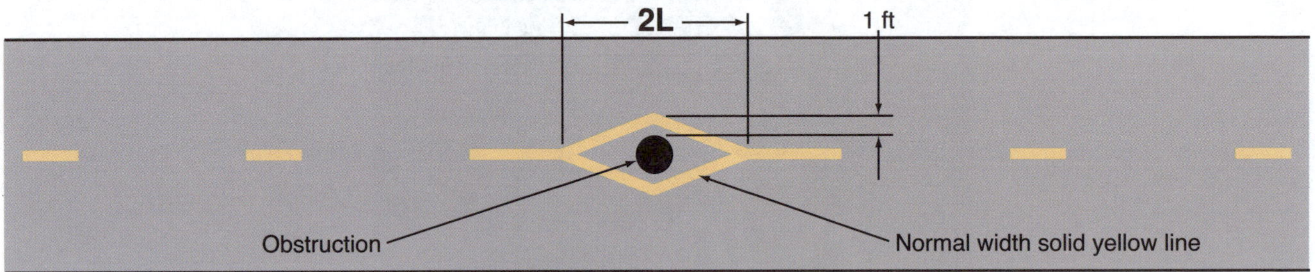

A – Obstruction within the path

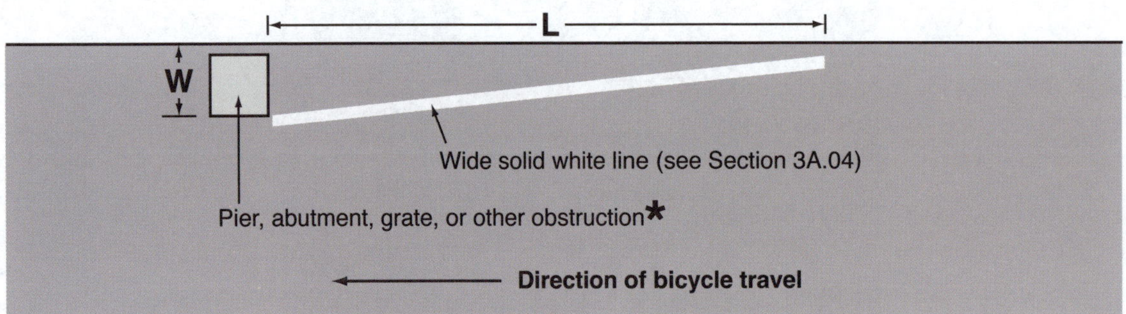

B – Obstruction at the edge of the path or roadway

L = WS, where W is the offset in feet and S is bicycle approach speed in mph

*Provide an additional foot of offset for a raised obstruction and use the formula L = (W+1) S for the taper length

CHAPTER 9F. SIGNALS

Section 9F.01 Application

Support:

01   Part 4 contains information regarding signal warrants and other requirements relating to signal installations.

Option:

02   For purposes of signal warrant evaluation, bicycles may be counted as either vehicles or pedestrians.

Section 9F.02 Bicycle Signal Faces

Support:

01   Chapter 4H contains information on the design and application of bicycle signal faces.

02   Section 9B.22 contains information for the Bicycle Signal sign that is required to be installed with a bicycle signal face.

Section 9F.03 Signal Operations for Bicycles

Standard:

01   **At installations where visibility-limited signal faces are used, signal faces shall be adjusted so bicyclists for whom the indications are intended can see the signal indications. If the visibility-limited signal faces cannot be aimed to serve the bicyclist, then separate signal faces shall be provided for the bicyclist.**

02   **On bikeways, signal timing and actuation shall be reviewed and adjusted to consider the needs of bicyclists.**

(This page intentionally left blank)

Appendices

CONGRESSIONAL ACTIONS

PUBLIC LAW 102-240-DEC. 18, 1991 (INTERMODAL SURFACE TRANSPORTATION EFFICIENCY ACT OF 1991)

Section 1077. REVISION OF MANUAL — Not later than 90 days after the date of the enactment of this Act, the Secretary shall revise the Manual of Uniform Traffic Control Devices and such other regulations and agreements of the Federal Highway Administration as may be necessary to authorize States and local governments, at their discretion, to install stop or yield signs at any rail-highway grade crossing without automatic traffic control devices with 2 or more trains operating across the rail-highway grade crossing per day.

PUBLIC LAW 102-388-OCT. 6, 1992 (DEPARTMENT OF TRANSPORTATION AND RELATED AGENCIES APPROPRIATIONS ACT, 1993)

Section 406 — The Secretary of Transportation shall revise the Manual of Uniform Traffic Control Devices to include —

(a) a standard for a minimum level of retroreflectivity that must be maintained for pavement markings and signs, which shall apply to all roads open to public travel; and

(b) a standard to define the roads that must have a centerline or edge lines or both, provided that in setting such standard the Secretary shall consider the functional classification of roads, traffic volumes, and the number and width of lanes.

PUBLIC LAW 104-59-NOV. 28, 1995 (NATIONAL HIGHWAY SYSTEM DESIGNATION ACT OF 1995)

Section 205. RELIEF FROM MANDATES —

(c) METRIC REQUIREMENTS —

(1) PLACEMENT AND MODIFICATION OF SIGNS — The Secretary shall not require the States to expend any Federal or State funds to construct, erect, or otherwise place or to modify any sign relating to a speed limit, distance, or other measurement on a highway for the purpose of having such sign establish such speed limit, distance, or other measurement using the metric system.

(2) OTHER ACTIONS — Before September 30, 2000, the Secretary shall not require that any State use or plan to use the metric system with respect to designing or advertising, or preparing plans, specifications, estimates, or other documents, for a Federal-aid highway project eligible for assistance under title 23, United States Code.

(3) DEFINITIONS — In this subsection, the following definitions apply:

(A) HIGHWAY — The term 'highway' has the meaning such term has under section 101 of title 23, United States Code.

(B) METRIC SYSTEM — the term 'metric system' has the meaning the term 'metric system of measurement' has under section 4 of the Metric Conversion Act of 1975 (15 U.S.C. 205c).

Section 306. MOTORIST CALL BOXES — Section 111 of title 23, United States Code, is amended by adding at the end the following:

(c) MOTORIST CALL BOXES —

(1) IN GENERAL — Notwithstanding subsection (a), a State may permit the placement of motorist call boxes on rights-of-way of the National Highway System. Such motorist call boxes may include the identification and sponsorship logos of such call boxes.

(2) SPONSORSHIP LOGOS —

(A) APPROVAL BY STATE AND LOCAL AGENCIES — All call box installations displaying sponsorship logos under this subsection shall be approved by the highway agencies having jurisdiction of the highway on which they are located.

(B) SIZE ON BOX — A sponsorship logo may be placed on the call box in a dimension not to exceed the size of the call box or a total dimension in excess of 12 inches by 18 inches.

(C) SIZE ON IDENTIFICATION SIGN — Sponsorship logos in a dimension not to exceed 12 inches by 30 inches may be displayed on a call box identification sign affixed to the call box post.

(D) SPACING OF SIGNS — Sponsorship logos affixed to an identification sign on a call box post may be located on the rights-of-way at intervals not more frequently than 1 per every 5 miles.

(E) DISTRIBUTION THROUGHOUT STATE — Within a State, at least 20 percent of the call boxes displaying sponsorship logos shall be located on highways outside of urbanized areas with a population greater than 50,000.

(3) NONSAFETY HAZARDS — The call boxes and their location, posts, foundations, and mountings shall be consistent with requirements of the Manual on Uniform Traffic Control Devices or any requirements deemed necessary by the Secretary to assure that the call boxes shall not be a safety hazard to motorists.

Section 353(a) SIGNS — Traffic control signs referred to in the experimental project conducted in the State of Oregon in December 1991 shall be deemed to comply with the requirements of Section 2B-4 of the Manual on Uniform Traffic Control Devices of the Department of Transportation.

Section 353(b) STRIPES — Notwithstanding any other provision of law, a red, white, and blue center line in the Main Street of Bristol, Rhode Island, shall be deemed to comply with the requirements of Section 3B-1 of the Manual on Uniform Traffic Control Devices of the Department of Transportation.

PUBLIC LAW 115-141-MAR. 23, 2018 (CONSOLIDATED APPROPRIATIONS ACT, 2018) DIVISION L, TITLE I

Section 125 — For this fiscal year, the Federal Highway Administration shall reinstate Interim Approval IA-5, relating to the provisional use of an alternative lettering style on certain highway guide signs, as it existed before its termination, as announced in the Federal Register on January 25, 2016 (81 Fed. Reg. 4083).

Option:

Series E(modified)-Alternate may be used in place of Series E(modified) for the names of places, streets, and highways on freeway and expressway guide signs in accordance with the provisions of the following paragraph.

Standard:

The use of Series E(modified)-Alternate shall be limited to the display of names of places, streets, and highways on freeway and expressway guide signs. Words shall be composed of lower-case letters with initial upper-case letters. The design and spacing of the letters shall be as provided in the "Standard Highway Signs" publication (see Section 1A.05 of this Manual). The nominal loop height of the lower-case letters shall be 84 percent of the height of the initial upper-case letter. Interline spacing, measured from the baseline of the upper line of legend to the upper limit of the initial upper-case letter of the lower line of legend, shall be at least 96 percent of the initial upper-case letters (equivalent to 84 percent of the initial upper-case letter when measured from the baseline of the upper line of legend to the upper limit of the rising stems of the lower-case letters of the lower line of legend). Edge spacing shall be as provided in Section 2E.13 of this Manual. The size of the sign shall be suitably enlarged to accommodate the larger lower-case letters and interline spacing. When the name of a place, street, or highway contains numerals, the numerals shall be composed of the FHWA Standard Alphabet Series E(modified). Other lettering on the sign, such as for cardinal directions and distance or action messages, and all numerals or special characters, shall be composed of Series B, C, D, E, E(modified), or F of the FHWA Standard Alphabets as provided in this Manual.

Series E(modified)-Alternate shall not be used for any application other than as provided in the two preceding paragraphs.

METRIC CONVERSIONS

Throughout this Manual all dimensions and distances are provided in English units. Tables A2-1 through A2-4 show the equivalent Metric (International System of Units) value for each of the English unit numerical values that are used in this Manual.

Table A2-1. Conversion of Inches to Millimeters

Inches	Millimeters	Inches	Millimeters	Inches	Millimeters	Inches	Millimeters
0.25	6	3.5	87	12	300	36	900
0.4	10	4	100	15	375	42	1050
0.5	13	4.5	113	16	400	48	1200
0.75	19	5	125	18	450	54	1350
1	25	6	150	21	525	60	1500
1.25	31	8	200	24	600	72	1800
2	50	9	225	27	675	84	2100
2.25	56	10	250	28	700	120	3000
2.5	62	10.4	260	30	750		
3	75	10.6	265	32	800		

Note: 1 inch = 25.4 millimeters; 1 millimeter = 0.039 inches

Table A2-2. Conversion of Feet to Meters

Feet	Meters	Feet	Meters	Feet	Meters	Feet	Meters
1	0.3	11	3.4	40	12	200	60
2	0.6	12	3.7	50	15	250	75
2.5	0.75	12.75	3.9	53	16	300	90
3	0.9	14	4.3	60	18	330	100
3.25	1	15	4.6	70	21	400	120
3.5	1.1	16	4.9	72	22	500	150
4	1.2	17	5.2	75	23	530	160
4.5	1.4	18	5.5	80	24	600	180
4.75	1.45	19	5.8	90	27	650	200
5	1.5	20	6.1	95	29	700	210
5.67	1.7	22	6.7	100	30	750	230
6	1.8	23.5	7.2	110	34	800	245
7	2.1	25	7.6	120	37	1,000	300
8	2.4	25.6	7.8	125	38	1,500	450
9	2.7	30	9	130	675	2,000	600
9.25	2.8	32	9.8	140	700	2,300	700
9.5	2.9	33	10	150	750	3,000	900
10	3	36	11	180	800		

Note: 1 foot = 0.3048 meters; 1 meter = 3.28 feet

Table A2-3. Conversion of Miles to Kilometers

Miles	Kilometers	Miles	Kilometers	Miles	Kilometers	Miles	Kilometers
0.25	0.4	1	1.6	5	8	70	110
0.5	0.8	2	3.2	10	16		
0.6	1	3	4.8	15	25		

Note: 1 mile = 1.609 kilometers; 1 kilometer = 0.621 miles

Table A2-4. Conversion of Miles per Hour to Kilometers/Hour

mph	km/h	mph	km/h	mph	km/h	mph	km/h
3	5	25	40	45	70	65	105
10	16	30	50	50	80	65	110
15	20	35	60	55	90	80	130
20	30	40	60	60	100		

Note: 1 mile per hour = 1.609 kilometers/hour; 1 kilometer/hour = 0.621 miles per hour